The Embryologic Basis of Craniofacial Structure

Michael H. Carstens

Editor

The Embryologic Basis of Craniofacial Structure

Developmental Anatomy, Evolutionary Design, and Clinical Applications

Volume I

 Springer

Editor
Michael H. Carstens
Plastic and Reconstructive Surgery
Saint Louis University Plastic and Reconstructive Surgery
Falls Church, VA, USA

ISBN 978-3-031-15635-9   ISBN 978-3-031-15636-6   (eBook)
https://doi.org/10.1007/978-3-031-15636-6

This Springer imprint is published by the registered company Springer Nature Switzerland AG
The registered company address is: Gewerbestrasse 11, 6330 Cham, Switzerland

Paper in this product is recyclable.

This work is dedicated with love to my wife, Socorro Gross Galiano, and to our family: Momo, Lily, Alana, Xavi and Josh.

Foreword

It is with both pride and respect that I express appreciation to Dr. Michael Carstens for the privilege of "story telling" about him personally and providing an admittedly biased commentary on *The Embryologic Basis of Craniofacial Structure: Developmental Anatomy, Evolutionary Design, and Clinical Applications*. Simply said, this book is like none I have ever encountered. What makes this work unique?

That the ***content*** is 100% innovation-embedded will be obvious to anyone. Author, surgeon, "different thinker," Dr. Michael Carstens is a person possessed with an endless, open-minded curiosity and a commitment to basic science principles who, over the past three plus decades, has observed, studied, and recorded his observations and clinical management notes on multiple facial cleft patients. Having been such an individual himself, one who has benefited from excellent surgical care both in infancy and with later revisions, he knows at a gut level what plastic surgery can do for patients. Dr. Carstens has provided a "give back" account of a life committed to advancing our knowledge of how exactly the structures of the head and neck are based on an unorthodox mélange of scientific disciplines, from basic molecular embryology to advances in epigenetics. This treatise is clearly personal and even a "love story" about a passion for understanding coupled with relentless persistence.

But, from my perspective, it is the ***context*** of this work, its relationship to its author, that I find most striking. Behind the scientific and intellectual offerings of every piece of literature there is always a "back story" which can make reading or owning a book or manuscript even more poignant and either highly enjoyable or thoroughly detested. I would like to share with you my take on what this "back story" looks like.

My storytelling about the author begins when Michael first came to us from general surgery residency in Boston to begin training in Plastic and Maxillofacial Surgery at the University of Pittsburgh Medical Center (UPMC). He was initially a Resident in plastic surgery and subsequently our first Fellow in cranio-maxillofacial surgery. Why did we choose him to come to Pitt? As a training program committed to developing dedicated future leaders in both research and clinical "doctoring," we learned the value of selectively recruiting individuals for enthusiasm and work ethic and providing an ambience for trainees to "be themselves". In particular, we were always looking for those with ambition to make a difference, possibly while being different themselves. As a Stanford undergraduate Michael had his own very special mentor, Dr. Donald Laub, chief of Stanford's program in plastic and reconstructive surgery. Don Laub's legacy to our specialty is unorthodox, as his Interplast Foundation (now Resurg) has demonstrated global reach for providing surgical care and teaching for underserved countries, plastic surgery as a way of "giving back" to the world. He is an acknowledged role model for count-less residents and colleagues. After phone calls and personal discussions, Laub assured me that Michael was just the kind of person we needed. His endorsement, in pure "Laubese," was exceptionally strong, "this kid has a *stickity-stick-to-itiveness* that just won't quit." From the very start of his residency, Michael did not disappoint. During his time at Pitt, besides intelli-gence and work ethic, I found myself driven to uncover just what might be locked up in his character. I can recall repeatedly challenging Michael to be "really remarkable." That single phrase encompasses what this book is all about. Despite the overwhelming volume of scientific detail, it has a remarkably personal tone, as Michael adds color to each chapter by his own "get it done" approach to research and problem solving.

This book has had a long gestation but there is no doubt in my mind that the seeds of its author's "out of the box" thinking were sewn during his time at Pitt. The origins began with early discoveries with anatomic dissections, and the author gives worthy acknowledgment to his professors, especially George Sotereanos, DMD, for his support and early work done together on vascular anatomy of the head and buccal mucosa, and the pivotal role periosteum has played as a "biosynthetic envelope" encasing stromal and regenerative stem cells. This work gave rise to the publication of two new flaps, the subgaleal fascia, and with co-resident Guy Stofman, the buccinator, which have contributed significantly to our reconstructive arma-mentarium. These projects also proved in the long run to be a source of important clinical information regarding the developing fetus and subsequently in cleft surgery. I have subse-quently followed Michael's trajectory for over 30 years, with great interest, and even awe. So, years later, it comes as no surprise to me that he should have produced this work. It was all there at the start.

As stated previously, this is a work that cannot be separated from the life experiences that produced it. Perhaps motivated by the early work by Ralph Millard during the Korean war, perhaps under the spell of Don Laub to go back to Latin America after residency and fellow-ship, Michael's career took a wide detour to conflict-ridden Central America, as a surgical consultant for the Pan American Health Organization in Nicaragua. I thought he was nuts at the time, but something also told me that working under those conditions might give him lessons in both surgery and life that would mold his character and define him as a plastic surgeon. Much of Michael's life has been subsequently spent in less developed countries where cleft reconstruction surgery was, and is even today, less available, particularly for the poor. Those patients, who have contributed by their own fate of having facial cleft pathologies, and who have benefitted from Michael's skill and expertise, are "his people." I am sure that much of what he has been able to accomplish is due to these nameless contributors … from whom he has learned and to whom he feels indebted.

Those who have volunteered overseas know that the gain routinely is far more than the contribution. Michael's service taught him invaluable lessons, learned under duress. In areas of the world where the infrastructure is rudimentary and where ancillary components for proper cleft care do not exist, one must create the "system" on his or her own, right on the spot. This led to the establishment of a residency in plastic surgery at the National University of Nicaragua, an innovative program of high quality, supported by the Interplast Foundation, the University of Wisconsin, and the A.O. Foundation from Switzerland. Burn care was a challenging prob-lem as well. In this regard, among mentors who have been especially meaningful in Michael's life are Carlos and Vivian Pellas, significant entrepreneurs and burn survivors after an airplane crash. Their suffering and recovery led them as philanthropists to create—with Michael's input—a plastic surgery and burn foundation APROQUEN which remains self-sufficient to this day, three decades later! Supporting such a national center for burn care has taken capital and resources. Accordingly, the Pellas wisdom and generosity has created a physical infra-structure in the form of the Hospital Vivian Pellas (Joint International Commission-certified!) in Managua. Working in collaboration with multiple others, the work of these innovators cre-ated a "system" for care and learning from scratch. This book by Dr. Carstens might be con-sidered a physical representation of those experiences.

So, let's return to the work at hand. Our specialty has benefited from constant evolution in terms of techniques and innovation. Yet in terms of understanding basic developmental mechanisms the field of congenital anomalies has lain rather fallow. Plastic surgery for these patients has been one of evolving surgical designs, each one hotly contested, yet without significant change in our underlying understanding of the problem itself. Around the turn of the twentieth century, the prominent French anatomist and plastic surgeon Victor Veau was critical of this approach, stating presciently that "cleft surgery is applied embryology." Carstens takes Veau's challenge into the context of molecular biology and to the next level by laying out the role for regenerative medicine and cell therapy as the future of craniofacial surgery in the twenty-first century.

As one approaches this book, perhaps the advice from Henry David Thoreau, written in the 1850s, should be that "the price of anything is the amount of life you exchange for it." For those who commit to these two volumes, the price will be an investment of your own time in "deep learning," to appreciate the principles and mechanisms of development that creates final configuration of our own adult anatomy, as described by Gray. The book is not an easy read, nor is it intended to be. The benefit today for readers from multiple disciplines is new learning and likely a better understanding of the threads woven and completed by nature. The effort made by the author to "connect the dots," to make understandable the complex, is one of devotion and, yes, even love for the hidden story behind our structure. "Developmental Field Theory," born from simple discoveries at Pittsburgh, has morphed into a complex narrative that the author always strives to make comprehensible. As he states, "Anatomy is a knowable truth."

For me, the most riveting and provocative part is Chap. 20 on "Biologics." Expected topics such as the physiology of osteoconduction and osteoinduction are modified by the impact of morphogens but there is also a much wider array of concepts. Chapter 20 is itself a crystal ball which envisions cell therapies as the building blocks of regenerative medicine. Here, the work of Professor Arnold Caplan, PhD, from Case Western Reserve University, is appropriately credited for the identification in 1991 of a construct of cells within human bone marrow that can be harvested, and culture-expanded in vitro. Such cells were found to have the capacity to form multiple mesodermal phenotypes such as cartilage, bone, nerve, etc. leading Caplan to coin the term mesenchymal stem cell (MSC). Thus, an avalanche of research on the properties of "stem cells" was launched. Thirty years later we now know their properties much better. Although these cells do not themselves regenerate missing tissue, they do "home" to sites of injury where they produce a host of molecules resulting in healing process involving reprogramming in situ endogenous stem cells. Caplan has re-conceived MSCs as "mesenchymal signaling cells." And from work in Pittsburgh in the 1990s and beyond it is recognized that these cells also exist in fat, arising as pericytes from the microvasculature.

Over the last decade, Michael's work has focused on the clinical applications of these stromal vascular fraction cells (SVF) derived from adipose tissue. From his faculty position at the Wake Forest University Institute of Regenerative Medicine and as Professor of Surgery at the National University of Nicaragua, he has done pioneering work in proof-of-concept studies with SVF cells in diabetic wound healing and neuropathy, pathologies involving scarring and fibrosis as in post-COVID lung patients, inflammatory conditions such as the aggressive and routine fatal kidney disease known as Mesoamerican nephropathy, and even an application for brain pathology, in particular Parkinson's disease. These studies have been performed under Ministry of Health and national IRB approval. Michael credits this work as being a final lynchpin for him to understand the mechanism of tissue failure underlying craniofacial clefts. This clinical research involving continuing data collection and practice follow-up has proven to be a productive "detour" in a long-range thought process resulting in a model of tissue formation based on normal versus pathologic stem cell function, a "squaring of the circle," as it were. For me, that is what Chap. 20 is all about.

In sum: I believe Michael's work to be of profound and of lasting value. What its final contribution will be … only time will tell. But if I could gaze into a crystal ball to converse with a book reviewer in the future, perhaps writing a century from now, say in 2123, I could envision the following words: "This is clearly a unique offering, now recognized even as a true master-piece Magnum Opus, written 100 years ago by Professor Michael Carstens, MD. The passage of time and collateral supporting information allow one to say now that the author was extremely prescient in his dissertation about how to 'connect the dots of life' and the need for 'silos of medical disciplines' to interrelate through craniofacial embryology, and how specifically the biology of regenerative medicine and cell therapies has evolved." The commentator might continue: "My assessment is that in 2023, when the treatise was first published, it was likely received with skepticism and even suspicion by some. It is now known and accepted that Dr. Carstens, by rejecting much of the dogma of his time and through empirical, open-minded observations, together with an emphasis of principles and process in biologic research, demonstrated that to think creatively and with a vision of the future, a great work can be created, and our scientific-healthcare world has benefitted from his efforts."

In closing, I would like to add a very personal comment, one which Michael might find embarrassing, given his unassuming nature. But this is my chance, and I am going to take it. As one who knows him well, I believe only Michael Carstens could have written this book. My assessment of life-long evidence suggests that he is a true innovator, perhaps at the McArthur genius level, as he has demonstrated exceptional creativity, a track record of significant accomplishments, and promise of a collaborative spirit to foster continued learning and teaching. The AMA (American Medical Association) has recently defined certain characteristics that are most desirable in "doctors of character," and healthcare "system-citizens." These individuals possess critical thinking skills to treat without dogma individual patients and entire populations, while seeing patterns and synthesizing observations as a "system thinker." Culled from both clinical practice and research, they are humble, yet determined, while constantly striving to exceed expectations. They relate to others as trusted team members who are not simply analytical on-lookers but problem-solvers and mentors. In sum, this book, the result of a 30-year pilgrimage, exemplifies both the McArthur Foundation "characteristics" and the AMA Standards of Excellence for "doctors of character" who are also "characters" in their own right. And Dr. Michael Carstens is, indeed, one of these.

So, with my back story in mind, as you read this book—the front story—best of luck and here's a promise: although the onslaught of new thinking awaiting you may seem daunting, take heart! In the end, your persistence will be richly rewarded. Anatomy is, indeed, knowable. You will come away seeing and understanding the human head and neck anatomy in a deeper and more satisfying way than you ever thought possible.

Respectfully,

Professor of Plastic Surgery, Emeritus   J. William Futrell, MD
The University of Pittsburgh
Pittsburgh, Pennsylvania

Foreword

The Open Sesame for Craniofacial Biology

Dr. Michael Carstens' book *The Embryologic Basis of Craniofacial Structure: Developmental Anatomy, Evolutionary Design, and Clinical Applications* presents the students of craniofacial biology with an "open sesame" to reveal the treasures that the field possesses in numberless profusion. Here, the reader finds the vertebrate biology of embryology merged with what is known of the molecular biology accompanying these cellular migrations. It is richly illustrated. The bibliographies are exhaustive. At first glance, one might think it is an entire life's work. In fact, it is the product of many, many lives whose efforts have been diligently searched out and integrated into this comprehensive display of the vastness of the modern craniofacial biology.

The Enlightenment was significant for many things, among them Denis Diderot and his effort to record all useful information in the *Encyclopedia*. Dr. Michael Carstens is clearly our century's Diderot in craniofacial biology! Hopefully like Diderot, he and colleagues will update this magnum opus to periodically keep us readers up to date. The effort will be of lasting value.

There are several reasons to study this massive production. The first is to absorb as much knowledge as you can, or you need. There is no other single resource like it in the world. It represents a carefully curated synthesis of knowledge from many academic fields and their experts. It also references the history of our attempts to correct congenital defects. The collected reference lists are of great value in themselves.

Second, for anyone, whether student or teacher, the illustrations of this text will prove priceless for the preparation of didactic lectures and as background to illustrate the research of the day. These illustrations will also be available online. Generations of students and scholars will be in debt to Dr. Carstens. The amount of professional time saved by lecturers preparing to teach this subject in the future will simply be immense.

Third, anyone working in the field of craniofacial biology—and especially those striving to grow bone where needed—will find this text to be the starting point for his or her lab. Students, residents, and postdocs joining labs will find this to be their Bible for acquiring the knowledge base to build the work ahead of them.

This is not a text that has been dumbed down for easy consumption by casual readers or beginning students. The information seems limitless, but the reader may have to work to get it at times. A diligent reader will be richly rewarded. Given the way the teaching of anatomy and embryology has been steadily reduced in medical schools over recent years, medical personnel of all levels will find this presentation to be far beyond anything they have previously encountered. Not only will this book inform clinical and research work going forward, but it will also provide illustrations and information that may help better inform patients and families of the biology of the defect at hand.

Seldom in one's professional life does one see such a major production of such importance to so many. Michael's effort is truly epic in scope and depth. Many will be in his debt for many years to come. For those of us reading this effort, nothing would please its author more than for his learning that one of us has used this information to advance the field, in a basic science or clinical effort of benefit to patients. For that is where it all began!

Steve Jobs gave the memorable Stanford University graduation address in 2005 famously quoting the parting words of editor Stewart Brand on the back of the last edition of the Whole Earth Catalog: "Stay Hungry; Stay Foolish." Anyone studying *The Embryological Basis of Craniofacial Structure* will be impressed at the level of commitment, time, and work required to prepare and share this opus. He or she will ask, "How could anyone have possibly produced this work while practicing surgery and doing research?" The answer is simply, "Stay Hungry, Stay Foolish." Thank you, Dr. Carstens, for sharing a large part of your life and passion.

Professor of Plastic Surgery Emeritus, The University of Pittsburgh   Ernest K Manders
Pittsburgh, Pennsylvania

Adjunct Clinical Professor of Plastic Surgery
The Ohio State University
Columbus, Ohio

Foreword

The year was 1997, and the meeting involved our hospital's Medical Executive Committee. Of concern was a proposal to initiate an innovative, markedly different approach to congenital facial anomalies in children. In this case, treatment of cleft deformities which had never been previously described. The surgical procedure was intended to maximize cosmetic, but most important, functional outcome. As Surgeon-in-Chief at that time, the concern of some non-surgical leadership of the medical staff was voiced with little understanding of the anatomy, embryology, and science involved. Navigating the "arbitration" of that process was arduous. As surgeons, we are in a dynamic, ever-changing field. Innovation with new procedures and augmenting the established ones is a most important component of surgical advancement. It was early in Dr. Carstens' career at our institution. He had based his theory and proposed procedure on detailed review and inclusion of his work in human fetal embryology and facial development. Ultimately, the new repair was instituted at our hospital; it went on to become the basis for developmental field theory and ultimately gave rise to this book.

In 1999–2000, Michael and our maxillofacial surgeon, Dr. Martin Chin, were the first to implement rhBMP-2 in craniofacial reconstruction under a special FDA-approved compassionate use protocol. This case was the beginning of a fruitful collaboration between Carstens and Chin as they subsequently pioneered in the implementation of morphogen-based bone reconstruction, a concept that is now achieving a place in the clinical armamentarium of craniofacial surgery.

Over the past 3 decades, he has refined long accepted procedures, as well as bringing new ideas and techniques, sharing them with colleagues through juried journal publications. These three volumes reflect his dedication to treating these children and are a memorialization of his life's work. All who practice in this noble calling of pediatric craniofacial surgery will appreciate this important contribution to the discipline. It is a privilege to introduce my colleague, mentor, and friend's seminal work.

Professor and Chief of Surgery for the UCSF   James Betts, MD
Benioff Children's Hospital Oakland
San Francisco, CA, USA

Foreword

Understanding how daily natural processes work is an essential mission of science. This is a special challenge for the clinician-scientist treating children with developmental craniofacial disorders. Explanations of human developmental anatomy taught in medical education failed to provide a framework upon which improved surgical procedures could be engineered. The results of conventional surgical repairs were disappointing, particularly as the child grew into an adult, and it became apparent that the early intervention impaired future growth.

Establishing a complete and accurate model of how an embryo assembles itself has been the objective of scientists and philosophers for centuries. The evolution of these models is gradual. Progress reflects thoughtful observers' extension of science and intuition. The accepted premise is that abnormalities in development result in anatomic deficiencies related to deviations of the normal process of development. It is therefore essential to understand how the process of normal development works. Crafting effective treatments for craniofacial disorders demands the clinician-scientist engineer surgery that complements the underlying biology that gave rise to the problem in the first place. Conceptualizing developmental anatomy is challenged by rapidly changing three-dimensional relationships, changes in tissue volume, and differentiation of cells for specialized function. Simultaneous events occur at the macroscopic, cellular, and molecular levels.

Simple descriptions of how unexpected biologic development results in clinical craniofacial disorders are suitable for basic biology students. When surgeons assume trust of an anxious parent of an infant who has a facial deficiency, it is essential to base treatment on the best model science can deliver.

Michael Carstens spent decades conceptualizing an improved model of developmental anatomy. The mission of this work is to document for students, scientists, and clinicians the complex synthesis of anatomy, emerging biologic science, and intuition required to understand how the embryo assembles itself. This description provides an opportunity for clinicians and scientists to engineer procedures and treatments that are more effective and less invasive. Carstens' dual role of scientist and clinician affords a unique perspective on both immediate surgical outcomes and the result after years of growth. Thoughtful observation of how variations in surgical design can produce more-or-less desirable outcomes as children grow takes time, patience, and will.

It has been an honor to work with Michael for over 30 years. Medical science evolved during that time, allowing us to advance treatments in ways we could not have fully appreciated as students. The route to success must incorporate the difficult and sometimes controversial adoption of new ideas. In the end, we all want to do the best thing for the child. It is our calling as doctors.

Attending Surgeon, California Pacific Medical Center   Martin Chin, DDS
San Francisco, California

Founder, Beyond Faces Foundation

Foreword

Michael Carstens, MD moved to Los Angeles in 2003 to be the craniofacial fellow at Children's Hospital Los Angeles. Dr. Carstens was unique as he didn't fit the mold of more typical post-resident fellows. He had been in solo private practice pediatric plastic surgery for many years at the Children's Hospital Northern California in Oakland. His intense intellectual and surgical curiosity were stifled. Eventually, he decided to follow his drive for a more academic environment by applying for the craniofacial fellowship at CHLA. He left his family in Oakland and rented a small apartment near Children's Hospital. Thus began his academic journey. His time in Los Angeles with me was crucial, as it allowed him to formulate and develop a profound understanding of craniofacial embryogenesis, one that led to many innovations in surgical technique.

This book is the culmination of almost two decades of relentless exploration and mapping of craniofacial embryological development. Dr. Carstens' intention is to offer surgeons a rationale and philosophy for how to think about surgical repair for a wide variety of congenital defects. He presents in complex detail a model to help surgeons in their reconstructive efforts to move closer to nature's intended mechanics.

Dr. Carstens' contributions in this book are embryologically sophisticated and original. They are invaluable to anyone with a deep interest in cleft formation and repair.

I have enjoyed knowing Michael for almost 20 years. He has an optimistic and enthusiastic approach to life. He engages deeply with whatever captures his attention. For example, although Mike was born in Iowa, he developed a deep interest in Latin American culture. Having spent time in Central America, he spoke Spanish like a native speaker. He was so immersed in linguistic fluency, that I would see him warmly trading stories in Spanish with the janitorial crew at the hospital. Over the course of the year, Michael became an integral part of CHLA. He was well-liked by all.

Michael is also unique in that he himself was born with a cleft lip and palate. Perhaps his personal experience is the nidus of his passionate interest in craniofacial development and treatment.

I am honored to witness the fruition of his work.

Los Angeles, CA, USA   John F. Reinisch
October 13, 2022

Foreword

A Unified Theory of Developmental Field Repair

It is an honor and a privilege to be asked to write a foreword to the book written by my friend and colleague Prof. Michael Carstens *The Embryologic Basis of Craniofacial Structure: Developmental Anatomy, Evolutionary Design, and Clinical Applications.*

Let me begin with a short story. I discovered Michael on a sleepy afternoon in the comfort of a couch in the library of Bombay Hospital and Institute of Medical Sciences. The year was 2005, I was browsing through "Clinics in Plastic Surgery" and stumbled upon his article on clefts. Within minutes I was wide awake and fascinated by his completely new approach. It was a delightful and scholarly change from the endless discussions about "Rotation advancement" *vs.* "Triangular flap," etc. for Michael dived deep and tried to bring into focus where and how it all starts and how we have got it wrong all these years. I was then the President-elect of the Indian Society of Cleft lip Palate and Craniofacial Anomalies. I felt a great urge to get this man to come to India and address our members. He came to the annual meeting in Guwahati, Assam State of India and the audience was held spellbound by his analysis.

Michael came often to address various meetings. During the last two decades, Michael has kept working on his theories and worked across disciplines to try and get a coherent whole. We kept corresponding. It was through this association that he has honored me thus and I am grateful.

The book consists of 20 chapters, and I must admit it was a task to read it. Mainly due to my vast ignorance of the various areas of scholarship involved. Once I got the hang of it, the fascination was rekindled, and the pieces of the jigsaw started falling in place.

I want to quote a few snippets from Chap. 1, they are not continuous in the text, but they create a whole as far as the philosophy behind the book is concerned and give you, the reader, the essence of the approach.

1. "Understanding the sequence of field deficiency has bi-directional consequences: it leads us further inward to work out the genetic sequences that produce the field; in the first place; and it leads us outward towards a more developmentally-based and innovative therapeutics."
2. "The well-known dictum of DeMeyer in 1964, 'the face predicts the brain,' can be inverted to 'the brain predicts the face,' as mechanisms of induction are now better understood."
3. "Despite advances in presurgical orthodontics and operative techniques we continue to be faced with results that deteriorate over time; most of our patients requiring multiple secondary interventions. Such a model cannot be correct."

Read in sequence they reflect the essence of the book and the novelty in its approach.

This is a book for the anatomist, vertebrate biologist, and also for surgeons involved in treating craniofacial clefts. For the latter, it ought to be required reading before they start independent surgery as Fellows/Senior residents. Whether they and their mentors adopt the thinking or not, they need to be aware of it to make informed choices.

There are at least four disciplines involved in understanding embryology: paleontology (going back in phylogeny), genetics, molecular biology (gene signaling) and, of course, anatomy, which is influenced by the preceding disciplines in a predictable manner as Michael shows so cogently.

As he freely admits, it is by standing on the shoulders of previous giants that Michael has been able to look farther, and his references give ample and due credit to scores of people whose work he has relied upon. The difference is that he has digested this enormous amount of scholarly material and distilled a holistic narrative from it.

It is nice to note the credit given to Tessier, Talmant, and Delaire among others. The fact that Tessier accurately described the pathology 50 years ago without the knowledge of a lot of later material in gene signaling and regulators like BMP4 is uncanny to say the least and testament to the meticulous observation and surgical mastery of the great man. Prof. Talmant told me he was trained in philosophy by Delaire and in Surgery by Tessier, which made him the best equipped to deal with clefts. It is important to synthesize multiple sources and come to your final technique.

The first 13 chapters lay the ground for the last 7 which are largely clinical with recommendations for rational incisions and rational repair. They are logical, and I hope, despite the enormous influence Prof. Millard still has over the cleft world, people synthesize his ideas with the rationale of Michael so that a newer type of repair evolves.

Finally, this book is for the intrepid reader. It is very heavy going for those only interested in technique. However, without the background in newer sciences and their findings; I think the clinical case for a technical shift cannot be made.

In the end, I want to quote one more nugget from Michael:

"But the iron fact of the matter is this: without a detailed understanding of the developmental anatomy of the face based on modern developmental biology, genetics, comparative anatomy and neuroembryology, paediatric plastic surgery is a collection of techniques in search of a science."

In my view, this is a very clear call for a rethink and reappraisal.

The great physicist Albert Einstein noted the fundamental forces of gravitation, electromagnetism, and the strong and weak nuclear forces, and during the last decades of his life in Princeton tried very hard to bring them all into a single theory which he called the "unified field theory." Unfortunately, he did not succeed and that struggle in theoretical physics goes on.

Dare I say Michael has succeeded in creating a unified field theory of cleft origins, pathology, and anatomy and that the "Developmental Field Reassignment repair' will fill the lacuna long present in our understanding and treatment of clefts. I hope it does. I wish him the best and commend this book to you the reader.

Mumbai, India   Mukund R. Thatte
August 28th, 2022

Foreword

All surgeons involved with cleft care know fully well the frustration of seeing well-executed repairs in infancy transform into a predictable sequence of secondary deformities requiring further correction. Even in the best of hands, re-operation rates may reach as high as 85%. What exists here is not failure of technique but, an inadequate biologic model of the problem in the first place.

"If the pathologic anatomy of the cleft site pivots on a deficiency state in a specific developmental field, and if the surgical correction of the cleft does not include reconstitution of that defective field such that it will grow normally over time, and such that it will cease to perturb the growth of its neighboring fields, then all forms of cleft surgery are condemned over time to varying degrees of relapse." Michael Carstens

Most facial anomalies represent defects in specific *developmental fields*. The success or failure of surgical manipulations permits a more accurate understanding of just exactly where these fields exist and how they act.

When a deficient developmental field is released from normal surrounding fields, subsequent facial growth can be anticipated to be more normal.

We have long wanted a publication on the subject. Michael Carstens needs no introduction to those who have followed this literature. We have recognized the outstanding qualities which made him the logical choice to write this complete research ever since we watched his brilliant and devoted work.

He is introducing a new and clinically relevant model of craniofacial pathogenesis based on concepts of developmental biology and neuroembryology that are yet little known in medicine.

Without a detailed understanding of the developmental anatomy of the face based on modern developmental biology, genetics, comparative anatomy, and neuroembryology, pediatric plastic surgery is a collection of techniques in search of a science.

The clinical significance of the neuromeric organization is that it enables us to map out the anatomic site of origin for all zones of ectoderm and mesoderm supplied by a given zone of the nervous system.

Evolution of new concepts in flap surgery based on expanding knowledge of skin circulation increased understanding of the biology of wound healing and tissue transplantation.

The craniofacial/plastic surgeon can learn from his research in basic sciences, thus enlarging the body of knowledge of the specialty.

Dr. Carstens has left his imprint and injected his own philosophy of diagnosis and treatment throughout this book.

He has offered talent, time, and energy to make possible the completion of this extensive three volumes edition making an original scientific contribution.

We can confidently predict the progressive growth of this area, a specialty with unlimited possibilities engendered by the fertile imagination of this plastic surgeon and hope he is deservedly recognized for his titanic effort.

Post graduate professor, School of Medicine, Nat. Univ. of Buenos Aires   Ricardo Bennun
Director Cleft Lip/palate and Craniofacial Program
Asociación PIEL, Buenos Aires, Argentina

Foreword

The work of Dr. Jean-Claude Talmant, pediatric plastic surgeon in Nantes, France, will perhaps not be fully appreciated for another generation. He has single-handedly explored the relationship between fetal breathing mechanics facial anatomy, using facial clefts as his model. Pressures exerted normally within the nasopharynx and nasal cavity during development play an undeniable role in shaping the morphology of the lip-nose complex at birth. Dr. Talmant documented these effects of fetal ultrasound. But he did not stop with merely proving an anatomic point; he carefully and methodically developed techniques to address these effects during primary surgical repair with a resulting restoration of not only breathing mechanics but of their cerebral control. Jean-Claude's life work also demonstrates a principle that is very much in the tradition of French anatomy: the careful observation and recording of natural events over time. In this regard, his contributions are a surgical version of Marcel Proust's revolution 7-volume masterpiece, *À la Recherche du Temps Perdus* (Remembrance of Things Past). Like Proust, he has lovingly captured the passage of time as reflected in the visage of each one of his patients. Although his case series is small, it has been recorded with the utmost precision and care. No one has done it better and his attention to detail remains a model for surgeons to emulate, both now and in the future.

Commentary—Jean-Claude Talmant

In March 2006, Mukund Thatte, president of the Indian Society of Cleft Lip & Palate had invited us, before the Guwahati congress, to visit the white rhinoceros sanctuary of Kaziranga Reserve. Our convoy crossed the green hills of Assam, covered with groves and tea plantations. I was in the back of the land rover with my wife, Odile. To the left of the driver, an American passenger had dozed off during the trip. From time to time he awoke and seemed to emerge from a bubbling and uninterrupted reflection: It was Michael Carstens. The thirst for understanding and the urgency to transmit from his enthusiastic nature which is his character, even when resting. Both of us, during this congress, felt the convergence of our approaches to cleft lip and palate. He sent me his work revealing the connections between the embryological evolution of facial vasculature and the clinical classification of facial clefts by Paul Tessier. He read the CME lecture (later published by Mukund Thatte in the Indian Journal of Plastic Surgery), following up on many of the arguments that I had put forward in my presentations.

A few years passed, interspersed with shipments of chapters from Michael, as he was immersed in the writing of his work. We came from two different worlds. Michael, through investigations, meetings, and research, has gathered an encyclopedic knowledge of craniofacial embryology. His ability to synthesize and his talent as a storyteller make us forget the aridity of the subject. He has the ability to make developmental anatomy come alive and outline its principles without it being necessary to memorize all the details. To think of anatomy as biologic and conceptual, rather than a collection of unrelated facts is truly a paradigm shift. Michael writes clearly and patiently to explain how each structure can be identified by its neuromeric origin and can be tracked along its pathway from the primitive stages of the embryo into its final form, following an algorithm of steps triggered by molecular signals. He deduced from this a fundamental principal for surgical therapeutics—to restore the deformed and displaced elements to their rightful place and to thereby reconstruct the missing embryological field. Defects of embryogenesis leave behind a scrambled map of mismatched tissues incapable of growing harmoniously. By unscrambling these mismatched fields, normal growth can be restored. The principal of Developmental Field Reassignment is virtually identical to that proposed by Victor Veau almost a century before, but this time is substantiated by modern molecular embryology.

For my part, I learned a lot from Paul Tessier, who clarified the anatomy of facial clefts, exploring them through precise and extensive dissections in the only sub-periosteal and sub-perichondrial planes which would respect the structures (fields), until he could envision them as distinct entities, thus making it possible to identify and reposition them. Jean Delaire, by entrusting me with the sequelae of the many cleft lip and palate cases that he was following, positioned me in a unique to observe and understand the anatomy of previously operated clefts. Thus, I became able to detect the poor outcomes arising from different protocols. I paid all my attention to certain common themes in these deformities, in particular, that of the nostril whose alar cartilage had obviously abnormal relations with the myrtiformis muscle. Michael saw the logic of my analysis and found that it could be perfectly explained as an application of reassignment of displaced fields; he thus integrated it into the DFR model.

I am also indebted to my brother, Jacques Talmant, who was the first to show me a video of fetal ventilation captured shortly before the year 2000 by his wife, sonographer Claude Talmant. It is thus better understood that the facial growth deficit in utero is due to a perfectly normal growth potential locked up in an abnormal anatomical context.

Since the beginning of the 2010s, our meetings have multiplied. Michael came to see us in Nantes with David Matthews, one of Paul Tessier's three musketeers. Michael and David know each other well and like each other. They attended operations and multidisciplinary consultations with my consultant orthodontist, Jean Pierre Lumineau who, since the mid-1980s, has performed anterior maxillary expansion by quad helix in our 4-year-old patients to reconstruct the sector missing from the lateral incisor by gingivoperiosteoplasty and iliac bone graft before the age of 5.5 years. Michael returned for a week in 2012 with Jyotsna Murthy and I remember a passionate lesson, where at the end of our surgical intervention, he drew his famous non-Philtral Prolabium flap (NPP) with a surgical marker right on the operating field drapes!

Michael wanted that we continue to make an impression at the congresses with our complementary presentations. He invited me to Santiago, Chile to visit his friend Luis Monasterio in 2014. We presented again with Ricardo Bennun of Buenos Aires for the Congress of the South American Cleft-Craniofacial Society in 2016 in Salvador de Bahia. We met in 2017 in Chennai, at the invitation of Jyotsna Murthy at Sri Ramachandran Medical Center. These meetings gave us the opportunity to discuss our views and our doubts, addressing our differences frankly. After re-reading Michael's Chaps. 18 and 19 where the description of my protocol is presented, identical in uni- and bilateral clefts, I think a few comments are warranted to clarify what I consider the most important components of a DFR-based cleft lip/nose/palate protocol.

The first operation, done at 6 months (without orthopedic preparation), reconstructs the lip and seeks a complete correction of the nose to introduce the child to nasal ventilation. At the same time, the velum is closed with an intravelar veloplasty according to Sommerlad (a point to which we will return). What is the rationale for the velar closure?

Prior to operation, the child ventilates through his palate cleft; thus, he has not had the opportunity to develop in his brain a cortical representation of the nostril. By separating the nasal and oral passages, the child is forced to choose a nasal mode of ventilation for the first time. At 6 months, the hyoid bone is still positioned so high that there is no space to lower the tongue: Nasal ventilation is the only option at rest and in sleep. An immediately patent nostril, one that is neither too small in its dimensions nor readily collapsible due to the mal-positioning of the nasalis muscle, will allow *nasal ventilation* and *bilabial contact*. This is striking in bilateral clefts where the premaxillary deformity that initially prevents labial occlusion corrects spontaneously in 4 months. *No orthopedic treatment can do this*! Forcing the child to depend upon oral ventilation at this early age is against nature. In the absence of adequate nasal ventilation, the initial programming of the breathing process produces an imprint on the cerebral cortex that experience has shown me is difficult to erase.

In the months following the intervelar veloplasty, the cleft of the hard palate shrinks to such an extent that it can be closed in two planes, without a raw surface, almost always between 14 and 18 months. Fistulas are rare; and they are often so small that locating them is more difficult than closing them.

Around age 4, if necessary, orthopedic expansion of the anterior maxilla by the quad helix device restores intercanine width at least 4 mm greater than its mandibular counterpart to reconstruct the alveolar cleft before age 5.5.

The choice of the procedure implemented for the closing of the soft palate deserves discussion because its implications are considerable.

Let us first review the fetal development of the velum. As soon as its fusion is acquired, nasal ventilation begins. With each exhalation, the bony palate is pulled forward by the projection of the tip of the nose, while the muscular structure of the veil is retained by its continuity with the velopharyngeal sphincter. It is even driven back by the reflux of amniotic fluid into the reservoir of the nasal fossae which follows each exhalation. Thus, the aponeurosis of the tensor veli muscle stretches between the hard palate and the pharynx. It is at birth that it is in proportion the longest. It will then be invaded by the muscles which remain at a distance from the posterior edge of the palatal plates. Only the tensor inserts into the proximal 1/3, the zone of the aponeurosis. The remainder of the velar sling is situated in the middle 1/3 of the velum. In the event of a cleft, the palatine aponeurosis, whatever its primary hypoplasia, is not subject to any distension and remains retracted on the pterygoid process. The palato-pharyngeal and levator veli muscles converge anteriorly into a single muscle body in the cleft. They insert aberrantly into the aponeurosis without inserting directly on the bone. Thus, if the repair does not take into account the pathologic antero-posterior orientation of the muscles, they will remain unable to move back and raise the veil due to this abnormal anchoring.

At the beginning of my career, I sutured these muscles without intravelar dissection and lengthened the veil by suturing the posterior pillars, thus creating a flange limiting the ascent of the veil and ventilation. These soft palates remained motionless and short in their functional part in front of the uvula. For this first generation of my patients with non-syndromic clefts, 50% required pharyngoplasty. My evolution since 1999 toward the posterior transposition of the velar muscles after extensive dissection, rotation of nearly 90° and suturing in tension with overlapping has completely transformed the morphologic outcome. If the muscles are of good quality, the velum lengthens under the effect of the correction of the vector of the muscular action. For non-syndromic clefts, the Sommerlad veloplasty is well-suited to this analysis and achieves the desired anatomic and physiologic objectives.

If I happened to revise by the same technique (Sommerlad) some of my previous soft palate cases in association with a successful lipofilling of the pharynx, I found that I no longer had to do pharyngoplasties in these non-syndromic cases.

Things are different when it comes to syndromic forms. *Whatever the operation, the result will be disappointing if the muscles are pathological.* Pharyngoplasties and sphincteroplasties are then resorted to, which are dangerous due to their impact on nasal ventilation. It is then that the *interposition flap of the cheek mucosa carried by the buccinator muscle* becomes a recommendable option for reconstructing the retracted sector of the palatine aponeurosis because it respects nasal ventilation. I read the work of Michael and that of Robert Mann, both of whom, having developed this technique for 30 years, deserves consideration and respect.

I remember a conference where Michael proposed the buccal flap as the primary treatment for velar clefts in 2016. It had been very severely criticized, and that is not surprising given the context at the time. The cheek mucosa has little to do embryologically with the velopharyngeal structure, especially if one is a follower of Victor Veau. This was a real paradigm shift, even more so than my premise of insisting that the nasal deformity must be repaired in the primary surgery.

These surgical practices that go against everything that has been promoted over the past 50 years are beginning to prove themselves! Robert Mann's idea of dressing up his concept in a more respectable way by speaking of it in terms of modern embryology can be understood, insofar as the foreshortening of the palatine aponeurosis settles in utero, but one should not be misled: Retroposition of the velum en masse by interposing a cheek flap with the vascular supply of the buccinator muscle is an opportunistic technique that is more akin to the repair of tumors than that of malformations.

In syndromic clefts, the soft tissue structures present in the sidewalls of the pharynx and soft palate are not normal, and their repositioning is disappointing. The situation is akin to that of a trauma, where the principle to be implemented is to replace the deficient or missing pathological structure—here, the palatine aponeurosis—by the best equivalent available. On the basis of Robert Mann's 30-year experience, the benefit of soft palate and soft palate scar reduction that accompanies his technique, and the mediocrity of most other solutions, ultimately give the buccal flap a prominent status. It will soon be one of the good answers to the velar insufficiency sequelae of many cleft veloplasties, if not the best, especially if there is a minimum of remaining muscle tone which is still functional. Michael, embryology expert, is not dogmatic. He is pragmatic. He knows that the efficiency of a technique and its harmlessness are its first justifications. Common sense takes precedence over blind respect for repairs with embryologically similar tissue. *If a field is deficient or otherwise absent, it must be recognized and replaced.*

For 50 years, more than two generations of practitioners, entrapped by dogma, have explored very opposing options in the treatment of cleft lip and palate. Only Victor Veau's philosophy of restoring displaced structures to their normal anatomic position and function has stood the test of time. The most effective protocol pursues this logic throughout the course of treatment: *Restore without compromise, with ambition and despite the technical demands, the lip, the nostril, the septum, the alveolar arch, the dental occlusion, the palate, the veil by installing nasal ventilation from the first operation and preserving it.*

Correction of the deformation of the alar cartilage and the septum and reconstruction of the missing embryological field of the lateral incisor are therefore aesthetic and functional prerequisites. The nose, forgotten in recent decades, whose subtle functions and interactions are better understood, will finally take its rightful place: that of the great organizer of facial growth. This is a very demanding standard that only a coherent protocol can satisfy at a cost. Rigorous learning of the underlying biology on the part of all specialties involved is required to achieve a shared awareness, a consensus about what the problem and the principles of its correction.

In pediatric plastic surgery, understanding the nature of a deformity and knowing the normal, going beyond mere appearances to perceive all the functional implications of a repair is the goal of the science of morphology, still in its infancy and hardly taught. "Function creates form," a well-known aphorism by Louis Sullivan, architect of the first skyscraper on the eve of the twentieth century, becomes medical evidence when one devotes one's life to repairing cleft lip and palate. And yet, ignorance of the role of nasal ventilation remains preponderant in the medical profession, as does the prohibition of all infant nasal surgery. Paul Tessier, although little suspected of lacking interest and curiosity for craniofacial anatomy, lamented in an essay later found by Tony Wolfe in his archives, his "ignorance at all times of embryology, functions and functional interactions in the child and adult." He regretted not having been able to really remedy this shortcoming due to lack of time, and no doubt also, because of the difficulties of access to this knowledge. Perhaps for this reason, although he and Michael never met one another, he welcomed two binders of writings and drawings that Michael sent to him from California. These were proof positive that his systematization of craniofacial clefts had a solid basis in developmental anatomy.

It is not certain that anatomy and embryology are much better considered and taught today's curricula and especially if there exists a will to recognizing their ongoing importance.

Fortunately, and we discover it in this book, born of a driving initiative, motivated by the sole ambition for a radical advance in the treatment of malformations. Such initiatives come from practitioners whose experience has sharpened their capacity for judgment without taking away their originality, creativity and, above all, the constancy essential to the pursuit of work that requires rigorous and uncompromising evaluation. True innovators take advantage of their independence to escape from the uniformity of worn-out thought and to question knowledge that is stagnant and wanting in depth. The fundamental and practical research pursued by innovators, whatever their favorite fields, while respectful of medical ethics, is most often done with a curiosity and an open-mindedness that leads them to explore new paths without fear of controversy or confrontation. When like minds meet, they take advantage of their convergences as well as their differences to progress and rise together. They blow a breath of fresh air that invigorates a sleepy world. May this breath, symbolized by the work you hold in your hands, inspire you and many other colleagues, to move forward and innovate for yourself, for in this pathway lies the promise of a better life for your patients.

Cleft Lip and Palate Center
Pays de la Loire, Clinique Jules Verne
Nantes, France

Jean-Claude Talmant, MD

Foreword

This is one of the few books that can be truly called a magnum opus. My first connection with it was in the early 2000s, when Michael first contacted me with some embryological questions. Since then, he and I have been in regular contact. My role has been principally listening to and commenting on a stream of ideas that have put together a remarkable edifice that contains an integration of fundamental principles of fields as diverse as paleontology, anatomy, and reconstructive surgery, with embryology serving as a connecting link. This book represents the culmination of at least two decades of original thought that has resulted in a system for understanding the fundamental relationships that underlie the incredibly complex anatomy of the adult human head and neck and putting that understanding into practical use in reconstructive surgery.

The basis for this system is a recognition of the highly segmental organization of the human body. In the trunk and abdomen, a fundamental segmental organization is evident even to the moderately attuned observer, but it is not so apparent in the craniofacial region. Starting with the neuromeric organization of the central nervous system, Dr. Carstens has extrapolated this fundamental pattern to more peripheral tissues and has been able to produce a system that allows one to coherently organize many complex facets of craniofacial anatomy into understandable segments. In some cases, actual laboratory investigations have not been performed to confirm extrapolations from animal models to humans, but the way that the information is presented in this book allows researchers to devise experiments that would test relationships and mechanisms hypothesized in the book. For students of anatomy, there is no better way to understand the complex anatomy of the head and neck than organizing the myriad of structures on the basis of their embryological organization.

Starting with some early personal research on vascular fields in the head, the author has gone into some rarely cited embryological literature on the vasculature of human embryos. As a result, he has laid out a basis for understanding the basis of adult vasculature, especially of the face, that not only sheds considerable light on the anatomical basis of many adult structures, but more importantly mechanisms underlying the genesis of many facial clefts. In this domain, he has produced a new understanding of the developmental basis for the widely recognized Tessier system for classifying craniofacial clefts.

As he was developing his embryological models for craniofacial development, Dr. Carstens began to dive into the increasingly rich paleontological literature on the phylogenetic development of the head. This enriched his way of viewing the vital role of segmentation as an organizing principle in both phylogeny and ontogeny. I am aware of no book that juxtaposes the anatomy of fossil fishes with discussions of congenital malformations of the craniofacial region and their surgical treatment.

Because of the vast amount of detail, much of which will probably be unfamiliar to many readers, this is not an easy book to read. Nevertheless, with the help of the many illustrations, the reader finally has access to what will be for many years the definitive source of information about both the embryonic and phylogenetic development of the cra-

niofacial region. More importantly, for the practicing surgeon, understanding the embryonic field relationships in the developing head can lead to devising reconstructive procedures
for congenital malformations that are notorious for requiring one or many follow-up
surgeries.

Prefessor of Anatomy, Emeritus University of Michigan   Bruce M. Carlson, MD, PhD
Ann Arbor, MI, USA

Preface

What Brings You to Open This Book?

A preface is a contradiction in terms. Although positioned at the beginning of a book, it represents the last act of its creation and a final reflection upon its value. A preface poses a question which can only be answered by you, the reader ...: Why go forward? Of all subjects in biology, anatomy seems supremely impenetrable, a "known world" of facts ... not much to get excited about here. Could there be more to it than meets the eye? Is it worth your time and effort to seek beneath the surface of cut-and-dried facts to find something else? The answer is *yes*. Anatomy is a Rosetta stone that permits you to understand the hidden language of development. To decipher it requires three keys: a willing suspension of preconceptions, patience, and persistence. So, not unlike an archaeologist unearthing past treasures, your search for the story of development will become its own reward.

The Open Sesame of Structure

It all began with a skull. I was a resident in plastic surgery at the University of Pittsburgh; the year was 1987. On my basement workbench was a model skull, a drill, colored pipe cleaners, Moore's Anatomy, and Grant's Atlas. The model lacked many of the small orifices and fissures as shown on the pages of the atlas. Convinced that these all had an explanation, I set out to find them, drilling out those that were not present. The pipe cleaners served as markers for the exit and entry sites of nerves and vessels; obviously, soft tissue development took place prior to that of the bones. It occurred to me that, with growth, the skull bones would advance toward one another like drifting continents until making contact, and the orifices, foramina, and sutures represented boundary zones between these fields. Thus, vessels and nerves did not penetrate bone fields. They would remain as markers at their peripheries, thus identifying a system of primitive embryonic fields. I never forgot that moment. I was convinced that *anatomy had to make biologic sense*. Hidden in plain sight, within each structure of the face and skull was a secret story, written in code, of how it came to be, one which would explain the reason for its ultimate shape, size, location, and function. If only there were a way to decipher it ...!

Curtains and a Surgical Innovation

Fast-forward to 1996 ... residency, fellowship, and tour of duty with the WHO in Central America were behind me. As I was preparing for board exams at home in Berkeley, California, I found myself thinking about secondary cleft patients I had cared for previously in Nicaragua and puzzling about curtains. The glass wall of the second-floor bedroom looked westward to the Golden Gate Bridge. Late-afternoon light flooded in; there were no curtains. I found myself imagining a pleated curtain attached to the wall on the right. As I pulled it leftward, the pleats, of course, became asymmetrical. I remembered the remarkable similarity of the previously operated patients ... and how, on the side of their repairs, the soft tissues of the face always seemed stretched taut, like a curtain pulled medially from a lateral bony anchorage point. Then I thought, "What if the curtain were detached from the wall, and allowed to move freely to the midline?" Would not the pleats stay symmetrical? And what if, at the time of surgery, the soft-tissue envelope were to be detached from the maxilla and allowed to move forward into the midline? Would not a tension-free closure be achieved? If so, the face could regain its natural symmetry .... The key to this insight was to think backward in time, like a videotape in reverse, to the beginnings of facial development. Little did I know how relevant paleontology would become.

The 4Ds

Subsequent surgical cases proved this prediction to be true. Using a developmental model for cleft repair, with a dissection radically different from the status quo, the on-table results were quite striking. Working in a different plane to shift the tissue into the center, I saw the tension melt away. By moving the soft-tissue "curtain" laden with stem cells into the center, further deposition of bone could take place where it was intended. But I could not understand neither how nor the why this should work ... my thoughts kept returning to the image of tectonic plates. It was as if the cleft side of the face had drifted away from the midline.

Instead of using cookie-cutter designs to rearrange the tissues as we saw them on the table, should we not be attempting to put into reverse a pathology in four dimensions stemming backward in time to the cleft event itself? In this model, an unknown site of **d**eficiency, causing a **d**ivision of tissue, would lead to **d**isplacement of the cleft side of the face as it drifted away from the midline, like a tectonic plate. And of course, with the explosive growth of the facial envelope during the embryo-to-fetus transition, neighboring structures (such as the nose) would become **d**istorted. This process theory made sense—but *what was the original problem*? I felt myself on the edge of a paradigm shift, searching for answers.

Sojourn in Basic Science

Reviewing all three volumes of Millard's *Cleft Craft* and reading its myriad of references produced nothing. If I were going to get *anywhere*, I would have to immerse myself in basic science—several to be exact: embryology, developmental biology, genetics, neuroscience, comparative anatomy, and evolutionary biology. Little did I know that this process would take 20 years. Over time, many colleagues helped me along the way. I first learned about homeobox genes and their potential mapping of the pharyngeal arches from the fetal pathologist Geoffrey Machin at Kaiser Oakland (we were doing vascular injection studies). Two years later, at John Rubenstein's neuroscience lab at UCSF, I found that the homeotic gene map had been extended forward to encompass the entire CNS. Moreover, Michael Depew, Rubenstein's fellow, introduced me to the *Dlx* system of genes, one which further mapped out the pharyngeal arches. He also showed me how to relate comparative anatomy to skull evolution. By 2004, at Children's Hospital Los Angeles, with John Reinisch, I worked out how the developmental field map of the skull could explain synostoses. But it was not until 2006, when I stumbled on the work of Dorcas Paget, medical artist and amateur neuroembryologist at the Carnegie Institution in the 1940s and 1950s, that the actual neurovascular basis for these fields became clear. By 2009–2010, at Saint Louis University, I began writing out these ideas as a series of essays and notes to myself as it were ....

Stem Cell Stimulus

The jump start for this book really came about when I transitioned from surgery into stem cell biology and clinical research with adipose-derived stromal vascular fraction (SVF) cells. The intense focus on stem cell biology made possible a coherent assembly of the developmental story of craniofacial anatomy based on the neuromeric model. It also made it possible to consider regional anatomic organizations such as the pharyngeal arches or the orbit, map them out, and describe their functional role in the formation of the head and neck. I believed that this vision could be a great value and felt compelled to bring it to life. And so, it was in 2017 that the concept of this book was presented to, and approved by, Springer-Verlag.

Qualifications

I suppose that this is as good a time as any to introduce myself. The facts of my professional life are available elsewhere (as in the back of the book), which is where they deserve to stay … since they have nothing to do with why you and I are meeting here in these pages.

So how in the world did I ever get to this place, 35 years in the making? The ideas in this book span a wide spectrum of disciplines. To amalgamate these into something relevant required curiosity, persistence, a whole lot of patience to let ideas mature, and a steely conviction that somehow it would all be worth it in the end. In retrospect, the process by which this happened began in 1972 at the office of Dr. Donald Laub at the Division of Plastic Surgery at Stanford. I had returned to campus after doing thesis research for a year in Ecuador. I had with me a copy of *Where There Is No Doctor* by David Werner, a primary care manual in Spanish for rural communities (it later earned a Guggenheim prize). I had met Werner, and we agreed that I would edit the language during my senior year and carry out a field test in Sinaloa, Mexico. I thought I should sign up for an anatomy course, and Laub, being a friend of Werner's, was my contact at the Medical School. While waiting, I was transfixed by the photos of cleft children from Latin America repaired by him through his brainchild, the Interplast Foundation. My heart was pounding. As I handed him a note of introduction from Werner, written on a 3 × 5 card, Laub looked at me intently [disclosure: I have a repaired cleft lip and palate]. Then, pointing a finger at me, he said with utmost assurance, "You, Michael, *you* are going to be a plastic surgeon!" I felt he had seen right through me, it felt like a hot knife slicing down to the bone, to something undiscovered. It was a challenge, that is for sure. I made up my mind right then and there to take him up on it. So that was it for me, for the rest of my life.

My surgical career morphed into a search for a new model for cleft biology, a way to do things better. Perhaps it was the strange admixture of interests with degrees in Latin American affairs and chemistry that enabled me to appreciate the value of merging seemingly disparate disciplines, like interlocking Venn diagrams, into something distinct. Perhaps it was visualization. From childhood, I loved drawing. I saw anatomy as something dynamic, in four dimensions. Somehow, I was able to create a mental videotape of embryogenesis that would let me move around the conceptus in my mind's eye from different angles at different stages to imagine how development takes place. No matter what the factors, I acquired through experience a core belief: that *nature makes sense*. Beneath its awesome complexity lies an unsurpassed beauty and an ultimate simplicity—something worth loving, worth pursuing, and worth sharing.

In summary, the why of this book is that the beauty of anatomy, its hidden story, has *always* been with us; it is a story that needs to be told. I have been acting through the years as its custodian, but it can no longer remain with me … it is not mine. It is yours, to have, to enjoy, to develop further, and to share with others.

Why Should You Read This Book?

Research requires a leap of faith, an investment of your time in the hope of discovering the unexpected. The contemplation of anatomy is a true case in point … will teach you something far beyond what you imagined. It will speak to you in accordance with your interests. For medical students, residents, and clinicians, it offers insights permitting translation of the *why* of structure into the *how* of therapeutic innovation. For graduate students and for scientists, it will add to, and perhaps challenge, the "known world" of structural biology.

This book can be read in whatever sequence you like, as all its parts are interconnected. The figures and accompanying legends are designed to help you visualize the developmental process. Take your time with them. Develop your own videotape. Push yourself to see familiar structures in a new light. Make this knowledge your own and, by thinking creatively, drive yourself past dogma to find innovative clinical solutions for your patients. Above all, may the beauty of development speak to you. *Open sesame* ... let your own search begin.

Falls Church, VA, USA

Michael H. Carstens

Acknowledgments

This book and the ideas it contains have its own evolutionary history; it spans a lifetime. So many, both living and departed, have contributed … these comments are incomplete and inadequate. The only way I can capture this moment is by use of a timeline:

My parents (deceased), Keith and Virginia Carstens, and sisters, Sue and Lynnea.

Dr. Eduardo Cornejo and Rosemarie Doring (my family in Ecuador and a lifelong connection with Latin America).

Robert Kiekel, Oregon State University, and Alicia de Ferraresi, Stanford (my professors of Spanish and Portuguese).

Casa de la Cultura Ecuatoriana, Quito, Ecuador (sponsorship of *La Novela Ecuatoriana: La Generación de 1930*).

Drs. Rodney D. Chamberlain and Kathleen Conyers Chamberlain (my Palo Alto family who instilled in me the ideal of medicine).

David B. Werner, author of *Where There Is No Doctor* (health care for the poor).

Donald R. Laub, Stanford, plastic surgery and founder of Interplast (my inspiration for plastic surgery and lifetime mentor).

Robert A. Chase, Stanford (for my love affair with anatomy).

Mark Gorney and Ed Falces, St. Francis Hospital Plastic Surgery and the Reconstructive Surgery Foundation, San Francisco (you showed me in the most personal way the transformative power of plastic surgery, supported the initial work in Central America, and lived the values of our profession).

David E. Marcello, Erwin F Hirsch, and Gene A Grindlinger, Boston University (my mentor in general surgery).

J. William Futrell, George F Sotereanos, and Wolf Losken, University of Pittsburgh (for my education in plastic surgery and craniomaxillofacial surgery; anatomy and technique of the subgaleal fascia flap and buccinator flap).

Joachim Prein and Klaus Honigmann, Kantonsspital Basel and the AO Foundation (for unleashing creativity and a lifetime of service to Nicaragua).

Carlos Pellas and Vivian Pellas; the Fundación APROQUEN (philanthropy for burns and plastic surgery), Hospital Vivian Pellas (excellence in medical care), and innovators for regenerative medicine.

Enrique Ochoa, Antonio Fuente del Campo, Luis Monasterio and Ricardo Bennun - masters of Latin American cleft/craniofacial surgery, mentors, and friends.

Dr. Rigoberto Sampson (deceased), the Nicaplast Foundation, the National Autonomous University of Nicaragua – León, Fundación Nicaplast (the foundation of modern reconstructive surgery in Nicaragua).

Jim Betts, Children's Hospital Oakland (creation of the CHO pediatric plastic surgery service), with support from Richard Rowe and Michael Austin.

Martin Chin, Children's Hospital Oakland (innovation in cleft surgery and pioneer with rhBMP-2 reconstruction).

Jerold Z. Kaplan, Alta Bates Hospital Berkeley Burn Center (reconstruction with Integra® dermal matrix).

Alexandra Cabri and Jo Wolters, medical artists (for bringing surgical concepts to life).

Robert Hardesty, Loma Linda University plastic surgery (for the first academic support for the work and for your poolhouse).

John F. Reinisch, Children's Hospital Los Angeles (mentoring pediatric and craniofacial surgery, master craftsman, and friend—for saving my career).

Christian E. Paletta and Robert Johnson, Saint Louis University (chairs at SLU plastic surgery and general surgery—you made innovation possible at Cardinal Glennon Children's Hospital).

Debbie Watters, RN, Cardinal Glennon Children's Hospital, Saint Louis University (for leadership of the cleft-craniofacial team at Glennon).

David C. Matthews, Carolinas Medical Center (true mastery of craniofacial surgery, out-of-the-box thinker, and lifelong friend).

Jean-Claude Talmant, Jean Delaire, Paul Tessier (contributors in the tradition of French anatomy).

Harvey B. Sarnat and Laura Flores Sarnat (beginnings of neuroembryology).

Mukund Thatte and Karoon Agrawal (introduction to the plastic surgery tradition of India).

S.M. Balaji, Chennai (you brought morphogen-based surgery to India and driving force for innovation).

Rolf Ewers, University of Vienna (the anatomic evidence for the Tessier cleft system).

Bruce Carlson, University of Michigan, and William Bemis, Cornell University (mentors in developmental anatomy and evolutionary biology).

Ernie Manders, University of Pittsburgh (for your inspiration and support for work and its place at Pitt Plastic Surgery).

Sonia Castro, Carlos Saenz, and Carlos Cruz, and Ministry of Health of Nicaragua (for you support for innovation in regenerative medicine and for your commitment to health care for all).

Socorro Gross Galiano—the beginning and the end of this work—I owe it all to you.

Contents

Volume I

1 Neuromeric Organization of the Head and Neck. 1
Michael H. Carstens

2 Anatomy of Mesenchyme and the Pharyngeal Arches 51
Michael H. Carstens

3 Embryonic Staging: The Carnegie System. 139
Michael H. Carstens

4 Neurovascular Organization and Assembly of the Face 171
Michael H. Carstens

5 The Neuromeric System: Segmentation of the Neural Tube. 241
Michael H. Carstens and Harvey B. Sarnat

6 Development of the Craniofacial Blood Supply: Intracranial System. 311
Michael H. Carstens

7 Development of the Craniofacial Blood Supply: Extracranial System 415
Michael H. Carstens

8 Developmental Anatomy of the Craniofacial Bones 487
Michael H. Carstens

9 Neuromuscular Development: Motor Columns, Cranial Nerves,
and Pharyngeal Arches. 709
Michael H. Carstens

10 The Neck: Development and Evolution 781
Michael H. Carstens

11 Developmental Anatomy of Craniofacial Skin, Fat, and Fascia 941
Michael H. Carstens

12 The Meninges 1037
Michael H. Carstens

Volume II

13 The Orbit. 1087
Michael H. Carstens

14 Pathologic Anatomy of the Hard Palate 1229
Michael H. Carstens

15   **Alveolar Extension Palatoplasty: The Role of Developmental Field
     Reassignment in the Prevention of Sequential Vascular Isolation and Growth
     Arrest** 1307
     Michael H. Carstens

16   **Pathologic Anatomy of the Soft Palate** 1389
     Michael H. Carstens

17   **Buccinator Interposition Palatoplasty: The Role of Developmental Field
     Reassignment in the Management of Velopharyngeal Insufficiency** 1437
     Michael H. Carstens

18   **Pathologic Anatomy of Nasolabial Clefts: Spectrum of the Microform
     Deformity and the Neuromeric Basis of Cleft Surgery** 1485
     Michael H. Carstens

19   **DFR Cheilorhinoplasty: The Role of Developmental Field
     Reassignment in the Management of Facial Asymmetry
     and the Airway in the Complete Cleft Deformity** 1563
     Michael H. Carstens

20   **Biologics in Craniofacial Reconstruction: Morphogens and Stem Cells** 1643
     Michael H. Carstens

Index 1719

About the Contributors

Karoon Agrawal For over four decades, Dr. Karoon Agrawal, Professor of Plastic Surgery at the National Heart Institute of New Delhi, India, has been a thought leader with extensive contributions in both cleft lip/palate and hypospadias. A tireless academician, he has also proven himself to be a glutton for punishment on two accounts: first, he is the editor of a 6-volume treatise published with Thieme covering the entire specialty; and second, he willingly committed to review this book from cover to cover, contributing careful and in-depth comments. For the time and effort this required, I sincerely hope his wife will forgive me. Karoon cuts a fine-featured, slender and scholarly figure, accompanied by a warm smile and restless intellect. He is also a first-class debater, which I discovered to my discomfiture. We were summoned by the Indian Society for Cleft Lip and Palate to debate on the relative merits of alveolar extension palatoplasty versus the traditional concept of cleft palate repair. Right off, I knew it was a dangerous proposition…no mortal could expect to approximate Karoon's surgical experience. But I had not counted on his sense of humor. There, up on the screen, with his very first slide, in front of the entire audience, was a picture of a crouching King Kong in the posture of a sumo wrestler with my face pasted on top! I suppose it was intended as a back-handed compliment about AEP but at the time I wanted to crawl in a hole. Anyway, the discussion was lively and the whole audience loved it. Later on, as the journal editor for the ISCLP, Karoon invited me to contribute to a special issue to present the developmental science of palatoplasty. Although it was like pulling teeth, he was patient with me and the result was two articles that eventually formed the blueprint for Chaps. 14–17. Karoon's gentle but persistent questioning forced me to focus on mechanism and good writing. I can say, in retrospect, that the consent chapters are as much his thinking as my own. As both colleague and friend, Dr. Karoon Agrawal represents the very best of Indian plastic surgery, a gift for which I will always be indebted.

S. M. Balaji From the vantage point of the Balaji Dental and Craniofacial Hospital in Chennai (formerly Madras), Tamil Nadu, India, Prof. Dr. SM Balaji cuts an imposing figure with an impossibly long list of accomplishments. As his full name is also impossibly long, I shall refer to him as Bala; and he is a true force of nature. A relentless perfectionist, academician, optimist, and general over-achiever, his energy (seemingly boundless) is somehow contained within a sizeable and powerful frame. However, behind the booming voice and irrepressible smile is a restless and innovative mind, rather unencumbered by dogma and quick to seek clinical implementation for concepts he considers of value. Given this constellation of attributes, it is not surprising that Bala has been the driving force behind the introduction of rhBMP-2 to reconstructive surgery for the Indian subcontinent and Southeast Asia. We started out doing cleft cases together, but he quickly found a way to innovate with bone grafts with an eye toward regenerative applications in the future. In the operating theater, I found him to be a master technician and a never-ending source of new ideas. Away from the hospital, Bala and his wife, Sachin, were the most gracious of hosts. But our cooperation did not stop there, for Bala is a tireless organizer and connector—his conferences in the Seychelles and Maldives gathered together like minds with the results that were eye-opening. Bala showed me the promise of a much larger world. I am convinced that innovation in medicine for the twenty-first century will come about as clinicians and scientists from India harness the incredible power of their clinical experience and produce studies that will change the direction of our thinking and techniques. As this story unfolds, I am sure that Dr. SM Balaji will be in the forefront; I hope to follow along to see its denouement.

William Bemis   at Cornell likes to go by "Willy" and that it is totally appropriate. Although I have never had the pleasure of walking into his office, the exuberance of his personality is matched only by the length of his beard. He is the quintessential college professor you never forget and with whom you know the office door will always be open for you. But he is also the author of the single book that changed my professional life and made the one in your hands become a reality. *Comparative Anatomy of the Vertebrates: An Evolutionary Perspective* drove home an important lesson: one cannot understand human adult anatomy without knowing both its development and its prehistory. The final form of our structures and systems must be understood in the context of other vertebrate species extent, and of their historical precursors through the passage of time. Consider the humble ossification centers of any bone (the occipital bone, for example), described in small print in Gray's Anatomy. Do these not represent separate, previously autonomous bone fields which have been melded together in the distant past to create the structure before our eyes today? Bemis text was riveting; each chapter brought new insights into structures that I thought I knew…only to find something different more relevant. Although he wears multiple hats, Willy Bemis lives and breathes his subject, be it paleontology, evolutionary biology, genetics, or comparative anatomy. When I discussed this project with him, very much aware of my own ignorance, he straightaway sent me all the slides from his book and threw open his mind (and his time) to help me out. As a scientist he would be better seen in khakis and a butterfly net than with a lab coat. But I am convinced that any medical curriculum with a serious interest in the developmental biology of the twenty-first century and beyond would do well to make Willy Bemis' text required reading for all premedical curricula. For it is this approach that will make the final interface with genetics and molecular biology. Without Willy Bemis, Chap. 8 (indeed all of them) could never have been imagined. My only regret is not to have had Dr. Bemis as my professor or, nowadays, as that eccentric but delightful neighbor just around the corner.

Ricardo Bennun Alveolar extension palatoplasty as a concept is uniquely South American, both its design and in its surgical verification. AEP, in concept the palatal version of subperiosteal tissue transfer for cleft lip repair, was drawn out on a paper napkin during an airplane flight to Ecuador. Although the operation was described in 1999, it was subsequently picked up by Dr. Luis Monasterio, distinguished cleft surgeon in Santiago, Chile. Lucho invited me to do some cases with him at the Fundación Ganz. Since the incisions required to reposition the entire embryonic field of the hard palate were made on the alveolar ridge, the effect of these on dental eruption was an issue of debate. Dr. Monasterio did his own series, following dental development for several years. He determined that eruption was unaffected (which he subsequently reported to the 12th American Cleft Palate Association meeting). He also introduced me to Dr. Ricardo Bennun from the University of Buenos Aires and director of the Fundación Piel. Ricardo subsequently took the operation to the next level beginning a case series which now extends to over a decade. Ricardo is a true biologic surgeon, reflecting his long-time commitment to burn care; he keeps on asking questions and seeking answers. I had the opportunity to contribute two chapters for his 2015 work, *Cleft Lip and Palate Management: A Comprehensive Atlas*; in the process his attention to detail and insistence of quality forced me to think about the issues more deeply than before. It is his influence that really pushed me over the edge, daring me to write this book, a task that seemed overwhelming to me at the time. True to form, his atlas of AEP cases faithfully recorded herein will stand the test of time as surgical proof that developmental biology can win out over dogma for better patient outcomes. For this, all cleft surgeons will be grateful.

James Betts Some people are larger-than-life. Jim Betts, pediatric surgeon and urologist, Professor and Chief of Surgery for UCSF Benioff Children's Hospital Oakland (formerly Children's Hospital Oakland) and volunteer EMT and fire fighter in Big Sur, has been serving the community for four decades. His modesty is the real deal, but it belies a life of devotion to patients, colleagues, and friends, with a long trajectory from his roots in small town Vermont. He is a true mensch who never fails to answer the call of the moment. In 1989, amidst the chaos and destruction of the Loma Prieta earthquake, Jim Betts was at the scene, rescuing the entrapped, even performing an amputation of a leg to save a boy's life. At 73, he still works on the weekends at the fire station in Big Sur (where he has a home). But there is another side of Jim Betts that underscores his character, one which literally made my own life work possible, because as the phrase goes, "justice, justice shall you pursue." Dr. Jim Betts lives by this…he puts his principles first. When he believes in something, he stands up for it. I had started a hospital-based pediatric plastic surgery service at CHO to meet the needs of our largely disadvantaged community and thus saw Jim in the pediatric emergency room both day and night. When brought my ideas to him about embryologic cleft surgery, he because a stanch ally, despite the controversy that ensued. Unbeknownst to me Jim Betts advocated for the possibility for a new form of cleft surgery, one that he thought had merit.... Time and again, in the board room, at the directors' meeting and in front of the CHO IRB, Jim Betts held the line. The work we were trying to push forward could have been so quietly and very adroitly shut down and swept into the dustheap but Jim believed in what we were doing; he never backed down. In the end, with his support, embryologic cleft surgery and morphogen-driven bone reconstruction became realities, Martin Chin and I kept our heads down, published the work, and never looked back. Now, many years later, the very existence of this book is a tribute to Jim Betts, to his vision for achieving better outcomes for children, and, most of all, to his unshakeable integrity.

Bruce M. Carlson   You cannot separate Bruce Carlson, MD, PhD, from his fish and the pond at his cabin in the backwoods of Michigan. As Professor Emeritus at the Department of Anatomy from the University of Michigan, Dr. Carlson is, of course, the author of the do-no-pass-go textbook *Human Embryology and Developmental Biology*. He also holds a MS in ichthyology—a touchstone for understanding vertebrate evolution This single book (which I got my hands on in 1997) proved to be the very beginning for my own work. Of all the contributors to the book, Bruce stands alone as those who thought it should be done from the very beginning. By making me an honorary member of the department Bruce enabled me to work with the University of Michigan library to amass the thousands of references required for this research. Although I am sure his wife Jean must have questioned his sanity, Bruce did actually go through the manuscript time and again, from its crudest interaction. And though I am just an MD without the benefit of doctoral work in anatomy, Bruce kept track of the good ideas, allowing me to get out from my errors gracefully, while supporting the concept of embryologically based surgery. So, as this book takes shape and can one day sit on your shelf, consider it my tribute back to you, Bruce Carlson, the professor I never had but can always count on.

Michael H. Carstens   Dr. Michael Carstens received his AB in Latin American Studies with honors and Phi Beta Kappa from Stanford University. He subsequently completed a BS in chemistry with honors from Colorado State University. He received his MD from Stanford. He completed residency training in general surgery at Boston University and in plastic surgery at the University of Pittsburgh. He holds two fellowships in craniofacial surgery from the University of Pittsburgh and from the University of Southern California. He is currently working with the Wake Forest Institute of Regenerative Medicine at Wake Forest University and teaches in several medical schools around the world. His academic interests are craniofacial embryology, developmental origin of stem cells, and surgical management of facial clefts.

Dr. Carstens speaks six languages; this makes him a well-respected ambassador for craniofacial surgery around the world. His career has been remarkable for social involvement. In 1990–1992, a consultancy for the World Health Organization brought Dr. Carstens to Nicaragua; he became a co-founder of the renowned APROQUEN foundation for burn care in that country. Currently, he works for WFIRM and serves as a scientific consultant in regenerative medicine for the Ministry of Health of Nicaragua where he continues doing research on developmental embryology and clinical applications of stromal vascular fraction (SVF) cells for complex wounds and chronic inflammatory states.

Martin Chin   In 1992, at Children's Hospital Oakland, I met a man who would become my intellectual partner in crime, the inspiration for innovative surgical procedures, and over the years an unforgettable friend. Oral-maxillofacial surgeon Martin Chin, DDS, is the personification of a "thinking man's surgeon," a consummate professional and humanitarian who looks at problems in a unique way, always just a little different; his first concern is to define the problem from first principles, putting goals and concepts first, then adapting technique to suit the case. When it comes to a device or an alteration, nothing gets in his way. A mechanical genius, ensconced in his garage, Martin can literally make anything, thus enabling him to pioneer in alveolar distraction osteogenesis. Martin combines these talents with a gift for biologic thinking. When we say a little girl in the ICU with a complex set of facial clefts, it was if we both could see the same map, but Martin had the inspiration to ask if we could engineer the bone using a morphogen, rhBMP-2, and thus fill zones she was missing in a completely novel way. In his own quiet way, Martin always got around. He made a point of going to Loma Linda University to collaborate with oral surgeon Dr. Phil Boyne in grafting mandibular defects. But Martin's vision was to create a structure that did not exist. So armed with this wild idea, and against the opposition of other colleagues at CHO, with the support of our Chief of Surgery Dr. Jim Betts, we somehow secured permission from the FDA, treated the child and succeeded far beyond expectation. Martin never stopped going forward; our treatment of alveolar defects, considered by many to be beyond the pale, has now been fully integrated into the surgical armamentarium. In the long run though, Martin Chin is a model of a surgical biologist who sees just a little farther than the rest of us. I strain to keep up with him…but, even if he pushes me past my comfort zone, I know his heart is in the right place and, well, it's good for me…sort of like avocados. All this is to say that developmental field reassignment cleft surgery started at Children's Hospital Oakland and with Martin Chin. Anyone who ever benefits from this, be they patients or professionals, owes him a debt of gratitude. Of course, in his quiet way, he would be the last to say so…this book is my chance to set the record straight.

Rolf Ewers   Over the years, Dr. Rolf Ewers has been not only a friend and trusted colleague, but a key contributor to this work, although he himself is unaware of the fact. When I visited his unit in Vienna, Rolf was (and is) an outstanding thought leader in craniomaxillofacial surgery. At the time I was working out the vascular embryology of the face as a means of understanding the Tessier classification of rare facial clefts. Rolf and Hildegund welcomed me to their home and subsequently directed me over to Federal Pathological-Anatomical Museum of the University of Vienna. This collection is housed in the so-called Narrenturm, a former hospital for the mentally ill founded in 1784. In addition to many anatomical oddities collected from all over Europe over the centuries are skeletal clefts of all types and configurations. Here, before my eyes, was proof positive that Tessier's system was based on focal neurovascular abnormalities. I returned from Vienna convinced that Dorcas Padget's observations were correct; Chaps. 6 and 7 describing the vascular embryology of both the intracranial and extracranial arterial systems are the result. I am indebted to Rolf for his seminal contribution; these chapters are but a very inadequate way to express my appreciation, and to acknowledge the historical role of Vienna in the development of science and medicine.

J. William Futrell Pittsburgh can be bitterly cold in the winter. In 1985, as a third-year resident in general surgery, I found myself trudging through the snow up the hill to Scaife Hall to interview with Dr. J William Futrell for a position at the University of Pittsburgh plastic surgery program, wondering what I was doing there in the first place. A convert to California (from Iowa), the years spent in Boston only reminded me of how much I wanted to return. There was a residency spot waiting for me with David Furnas at UC Irvine. But the storied Pitt program, built up by Bill Futrell, could not be overlooked so I went to check it out. Folksy, quirky, and professorial, he walked in with a novel in his lab coat and reading glasses hanging from one ear. On the wall behind him was a photo of Futrell in the Duke backfield, the quintessential football coach. What I remember from that hour, years later, is the sense of excitement and innovation he projected. Pitt was awesome and challenging, overshadowing the competition. For reasons I have yet to understand, he decided to take me. When his letter arrived, I accepted right away, the snowdrifts be damned. Futrell gave me a career and a way of thinking, encapsulated by "good enough, isn't." All Pitt graduates know the essence of Futrellism, as he kept pushing everyone's buttons, searching to find the way to get the best out of us. Futrell's thinking was a Rorshach: "You want to take a job in Central America? You're crazy! But if you do, you will find out something about clefts that you will *never* learn here." "You say you found a new flap with the puny buccinator muscle? It'll *never* work! But you should do it anyway…" And so it was…for all of us. But he and Annie made a home for all of us, each in a very personal way. With the passing years their door always remained open. I think Bill Futrell had his own ideas about what each of us could do, about how we would (and should) contribute to plastic surgery. As such, the choices I made must have driven him nuts, but he was always there, the perennial coach, and an unwavering advocate for commitment and creativity. So a quarter of a century later, this book is a fitting tribute to J William Futrell. I could sum it up thus: "To whom much is given, much is expected." Bill Futrell gave me his very best. This book is intended as payback, in full measure.

Ernest K. Manders   The early life of Ernie Manders was a true hejira, from a small town in British Columbia, to Alabama, back to the logging town of Coos Bay on the rugged Oregon coast, and thence forward into a surgical career of unceasing innovation. In 1997, when Bill Futrell recruited him for Pitt Plastic Surgery from Penn State, he had already made his mark by engineering a three-dimensional tissue expander for breast reconstruction, a real advance for cancer survivors. This was just the beginning, as his interests translated into tissue biology and bone biology. But the most striking feature of Ernie Manders' career is his combination of imagination, humanity, and service intertwined as a gifted professor and mentor for everyone who sought him out. In practice and struggling with ideas that seemed to not fit any known model, I turned to Ernie, knowing that I would find encouragement and new ideas. He read the drafts of all my early cleft papers and made them better. When it came to tissue engineering with rh-BMP2, his support for what was a yet unproven concept helped Martin Chin and me get FDA authorization for our compassionate-use protocol at Children's Hospital Oakland. Ernie Manders has never forgotten where he came from. He gave back to Coos Bay by establishing a plastic surgery residency rotation there to encourage surgeons to serve in rural communities. Perhaps for this reason he always stood behind my commitment to work in Central America. Finally, Ernie's contributions to this book deserve recognition. Over the years, his criticisms were always on-point, but positive…and eloquently delivered. The last chapter, on biologics, is my tribute back to him.

Robert Mann   For cleft surgeons worldwide, the work of Dr. Robert Mann needs no introduction. For over three decades he has relentlessly explored and refined the concepts of interposition palatoplasty using vascularized buccal tissue, all the while patiently and thoroughly documenting his results. Our collaboration represents a convergence of thought and technique, a phenomenon not uncommon in science. In this case, Robert's work dramatically illustrates the impact for patients when embryologic principles are applied to a clinical problem. It is also a testimony to his unflagging persistence in the face of controversy and to his constant willingness to give of himself to share these new ideas with the coming generation of surgeons. His thinking and the cases herein presented stake out a definitive new standard in the treatment of congenital palate defects.

David C. Matthews   Known to but a fortunate few, the lifework of Dr. David Matthews, pediatric plastic and craniofacial surgeon, can be summarized in two words: Semper Fidelis. A survivor of Vietnam, he transformed the scars of that experience into a lifetime of commitment to children and the underserved. David was the youngest of Paul Tessier's fellows; their emotional bond was strong. Possessed of great technical skills, he never forgot his mission, to provide service to all. He moved to Charlotte, North Carolina. When his practice group there demanded that he change his priorities to augment the bottom line, David went to set up a solo practice of reconstructive plastic service to the poorest of the poor. Year after year, he took the hits for the trauma service at Carolinas Medical

Center, supporting his dedicated office staff, and living on a shoe-string in an apartment above his office. The constancy of his service to the community was unmatched. Despite the human toll exacted by these choices, David's commitment never wavered, nor did his creativity. David's insights into cleft care and bone biology are the best in the world. When we met, as Tessier's disciple, he immediately grasped the significance of development field theory. We share common values as well. I no longer felt alone outside the box and we became friends. With David and his wife Anna I discovered friendship in another dimension, one that has remained constant through the years. With him it is Semper Fi, regardless of the circumstances. When the chips are down, David is there. That's just the way he is. But David Matthews is not done. His thinking has pushed forward into wholistic medicine, acupuncture, nutrition, and electrical properties of cells. He remains ahead of the curve. The rest of us can only try to keep up.

John F. Reinisch   is best known for his pioneering work with ear reconstruction. In 2003–2004 I was fortunate to serve as his fellow in pediatric plastic surgery and craniofacial surgery at Children's Hospital Los Angeles. A true master of soft tissue closure, the techniques I learned from him revolutionized my appreciation of wound healing. A true craftsman, John's open-mindedness encouraged me to go further than I had thought possible to define the developmental fields of the face, recognize their abnormal position in clefts, and relocate them to fulfill their physiologic roles. Working with Dr. Mark Urata, this was the beginning of developmental field reassignment. John's tutelage proved to be the resuscitation of my surgical career. For this, and for the many kindnesses John and his wife Nan showed me during my hejira far from home, I will always be grateful. DFR and JFR will always be inseparable.

Harvey B. Sarnat   In 1998, at Children's Hospital Oakland, Dr. Martin Chin and I were working though the concept of in situ osteogenesis, using the morphogen BMP-2 to convert stem cells in the periosteum into bone-forming cells. In this process we were in contact with two pioneers in biologic surgery, maxillofacial surgeon Dr. Phil Boyne at Loma Linda and craniofacial surgeon and bone biologist Dr. Bernard Sarnat at UCLA. I thus met Dr. Harvey Sarnat, a pediatric neurologist and neuropathologist also at UCLA, and the nephew of the latter. Harvey had written a book about developmental neuroanatomy which timed with the appearance of Butler and Hodos work on comparative vertebrate neuroanatomy. The sources together formed a basis to understand developmental field from the standpoint of Hox genes and segmentation. Later on, during my fellowship year at CHLA with John Reinisch, I got to know Harvey and his wife Laura, also a distinguished pediatric neurologist. Our collaboration grew into two chapters on a new model of craniofacial clefts based on neuroembryology. Over time Harvey's insights into the system became the structural ribcage for developmental fields, welding together the origins of craniofacial and CNS tissues. Despite my ignorance of the many critical details Dr. Sarnat gave generously of his time and talent, making Chap. 5 accessible for all readers, regardless of specialty, to have an internal look into how the neuromeric model really works. Without Harvey's contributions, this book would not have been possible.

Jean-Claude Talmant The work of Jean-Claude Talmant will perhaps not be fully appreciated for another generation. He has single-handedly explored the relationship between fetal breathing mechanics and fetal development using facial clefts as his model. Pressures exerted normally within the nasopharynx and nasal cavity during development play an undeniable role in shaping the morphology of the lip-nose complex at birth. Dr. Talmant documented these effects using fetal ultrasound. But he did not stop with merely proving an anatomic point; he carefully and methodically developed techniques to address these effects during primary surgical repair with a resulting restoration of not only breathing mechanics but of their cerebral control. Jean-Claude's life work also demonstrates a principle that is very much in the tradition of French anatomy: the careful observation and recording of natural events over time. In this regard, his contributions represent a surgeon's version of Marcel Proust's revolutionary 7-volume masterpiece, *À la Recherche du Temps Perdus* (Remembrance of Things Past). Like Proust, he has lovingly captured the passage of time as reflected in the visage of each one of his patients. Although his case series is small, it has been recorded with the utmost precision and care. No one has done it better and his attention to detail remains a model for surgeons to emulate, both now and in the future.

Mukund Thatte Finding Mukund, and finding India, was an extraordinary event that almost did not happen. Dr. Thatte had been looking for me (unsuccessfully) for over a year to speak at an upcoming meeting in Guwahati, Assam, in March 23–25, 2006, for the Indian Cleft Lip and Palate Association. I had been in transit between Children's Hospital Oakland and Saint Louis University; somehow the contact was lost. But, to my surprise and good fortune, we eventually connected and, of course, I would attend. India is a natural place to go for innovation in plastic surgery…the specialty was literally born there from the pioneering work of Suśhruta, author of the world's first treatise of reconstructive surgery, the "Suśhruta Samhita" (Suśhruta's Compendium), written in ancient Sanskrit. The journey was arduous. Assam being tucked into India's northeast corner abuts to the south with Bangladesh. At the end of the road awaited the Kaziranga wild animal preserve, home of the rare white rhinoceros and, within the shelter of a tea plantation, the Borgos Hotel, at the doorway of which stood Dr. Thatte…as striking as were his surroundings. Stately, tall, impeccable in both manner and speech, Mukund cut a figure of charm and gravitas, admixed with warmth and a sly sense of humor. As one of the true doyens of Indian plastic surgery I was stunned to find that he had literally picked me out from a 2004 journal. As I have discovered over the years, Mukund is possessed of an unbounded curiosity and an intellect curiously allergic to dogma. I am not sure that he knows what the inside of a box looks like. At any rate, I owe to him the introduction to the collection of world-class surgeons and, with his wife

Urmila, remarkable friendships. Over the years Mukund has been a staunch advocate of the value of my basic science approach to plastic surgery. With his support, developmental field theory assumed an honored place in the specialty among Indian colleagues with open minds. For this India will always be special...for it was here, in the tradition of Suśhruta, that something utterly new could come safely into existence.

Michael H. Carstens

Introduction

Developmental anatomy is the bedrock upon which all treatment methods for cleft and craniofacial anomalies must be based. Traditionally, facial development has been considered independent of brain development, but recent advances in molecular genetics demonstrate a more intimate neuroembryological relation than was previously appreciated. The well-known dictum of DeMeyer in 1964, "the face predicts the brain," can be inverted to "the brain predicts the face," as mechanisms of induction are now better understood.

Many important insights into neuroembryology can be deduced from craniofacial anomalies and the results obtained by surgical intervention for their correction. Why should a closer collaboration between craniofacial surgeons and neurologists be of mutual benefit? Most facial anomalies represent defects in specific *developmental fields*. The success or failure of surgical manipulations permits a more accurate understanding of just exactly where these fields exist and how they behave.

When a deficient developmental field is released from normal surrounding fields, subsequent facial growth can be anticipated to be more normal. By the same token, persistent patterns of relapse after surgery point an accusatory finger at the site of the pathology. Nowhere is this more apparent than in the treatment of routine clefts of the lip and palate. Despite advances in presurgical orthodontics and operative techniques, we continue to be faced with results that deteriorate over time; most of our patients requiring multiple secondary interventions. Such a model cannot be correct. As an anatomist, I have always struck with by the inability of descriptive embryology to answer very fundamental questions about clefts, unaddressed even in the most contemporary of texts. First, if the mechanisms of unilateral and bilateral cleft formation are the same, why are the surgical approaches so dif-

ferent? Second, what explains the spectrum of severity of clefts? Third, what relation exists between cleft palate and cleft lip? Fourth, why are isolated palatal clefts more likely to be associated with additional birth defects? Finally, if isolated genes are to blame for clefts, why don't we find evidence of such mis-expression all over the body; how is it possible for genetic mis-expression to be unilateral?

The purpose of this chapter is to introduce a new and clinically relevant model of craniofacial pathogenesis based on concepts of developmental biology and neuroembryology that are as yet little known in medicine. Much of the relevant literature is less than 20 years old. Many key discoveries have come from technologies as yet in evolution. All clinicians dealing with congenital anomalies are witnesses to nature's variations. These pathologies may well constitute the great "Rosetta stone" of developmental biology; on the faces of our patients are written the hieroglyphics of embryology. Properly translated these can create, for the first time, a new "gross anatomy of the embryo." Craniofacial surgeons and neurologists have an indispensable role to play in the discovery of this knowledge…for those who confront the experimentations of nature are those closest to her secrets.

Our central tenet is this. All craniofacial tisues, be they mesenchyme (neural crest and paraxial mesoderm), ectoderm (skin), or endoderm (mucosa), originate at specific sites of the embryo during gastrulation. These sites correspond to the embryonic segmentation units of the central nervous system from which their innervation proceeds. These developmental units are called *neuromeres*, the anatomic boundaries of which are defined by a series of position-specifying *homeotic genes*. Thus, if we know the innervation of a structure, we can deduce where its component tissues were originally produced. Many important inferences arise from this system. Applications of neuromeric anatomy provide a potential embryonic "map" of all craniofacial structures with important implications for diagnosis and surgery. Let us consider two examples to see how this works.

Exclusive of the cranial base (basisphenoid and posterior) and parietal bone, the craniofacial skeleton is made exclu-

M. H. Carstens (✉)
Wake Forest Institute of Regenerative Medicine, Wake Forest University, Winston-Salem, NC, USA
e-mail: mcarsten@wakehealth.edu

sively from neural crest. Thus, the cell populations producing the ethmoid and presphenoid all originate from the neural folds in genetic register with the first rhombomere (abbreviated r1). While they are still in residence within the neural folds, the identity of these neural crest cells (NCC) is generalized at first. When the NCCs migrate from the fold, they are simply destined for the basal and frontal part of the forebrain. But upon receiving additional signals from the local environment, they become further specified as tissues of anterior cranial base and frontonasal complex.

In similar fashion, the neural crest cells of the neural folds in register with rhombomeres 2 and 3 populate the rostral and caudal aspects, respectively, of the first arch. Here they are further organized by a set of distal-less, *Dlx genes* into regional identifiers. We shall discuss this process further along. Neurovascular supply develops within the pharyngeal arches; this network defines the future developmental fields. When the process of embryonic folding positions the first arch with respect to the future face, each population supplied by a branch of the V2 stapedial artery (an addition onto the internal maxillary artery) will come into contact with an epithelial program and form a soft tissue-bone complex. This is reflected in the spatial order of the neurovascular pedicles, each accompanied by a sensory branch of V2 that fans out from the pterygopalatine fossa to their respective destination. Most distal along of these is the medial nasopalatine axis giving rise to the premaxilla and vomer. Just behind it, the lateral nasopalatine and descending palatine axes supply the inferior turbinate and palatine bones, respectively. Maxilla and zygoma are next being fed from the anterior superior alveolar and zygomatic axes. The squamous temporal, mandible, malleus, and incus are r3 neural crest bones and have their own neurovascular supply. All of these fields develop in accordance with the extracranial stapedial system as it forms a hybrid with internal maxillary of the external carotid. This vascular anatomy will be discussed in detail in Chaps. 6 and 7.

Disturbances at a particular neuromeric level can affect individual or multiple fields, causing to be deficient or absent. Thus, *isolated cleft of the secondary hard palate* (unassociated with cleft lip) represents a deficiency state of the vomer field. This occurs as a spectrum. As the vomer is progressively smaller, it lifts away from the plane of the palatal shelves and the cleft extends forward toward the incisive foramen. This is intrinsic genetic specification of the vascular system which is clinically relevant. In *Treacher-Collins syndrome*, multiple r2 and r3 developmental fields of the midface are affected: the maxilla, zygoma, and the mandible are all small. Each of these has some degree of compromise in its neurovascular axis. Vomer and premaxilla are unaffected as their medial nasopalatine axis remains intact (Figs. 1.1 and 1.2).

Developmental fields form in a specific spatiotemporal sequence. Each one builds upon its predecessors. Making a face is much akin to assembling a house with magical pieces of Lego®, each one of which will grow over time. Imagine a Lego house made from 20 pieces (4 on the floor and 5 stories high). All pieces are growing independently. If a cornerstone piece is removed, the 19 remaining pieces undergo a deformation and the house tilts into the deficiency site. The missing Lego® piece in cleft lip is the premaxilla. Surgical reconstruction of this field is the key to repair of the common cleft.

We shall explore these concepts further during the course of this chapter.

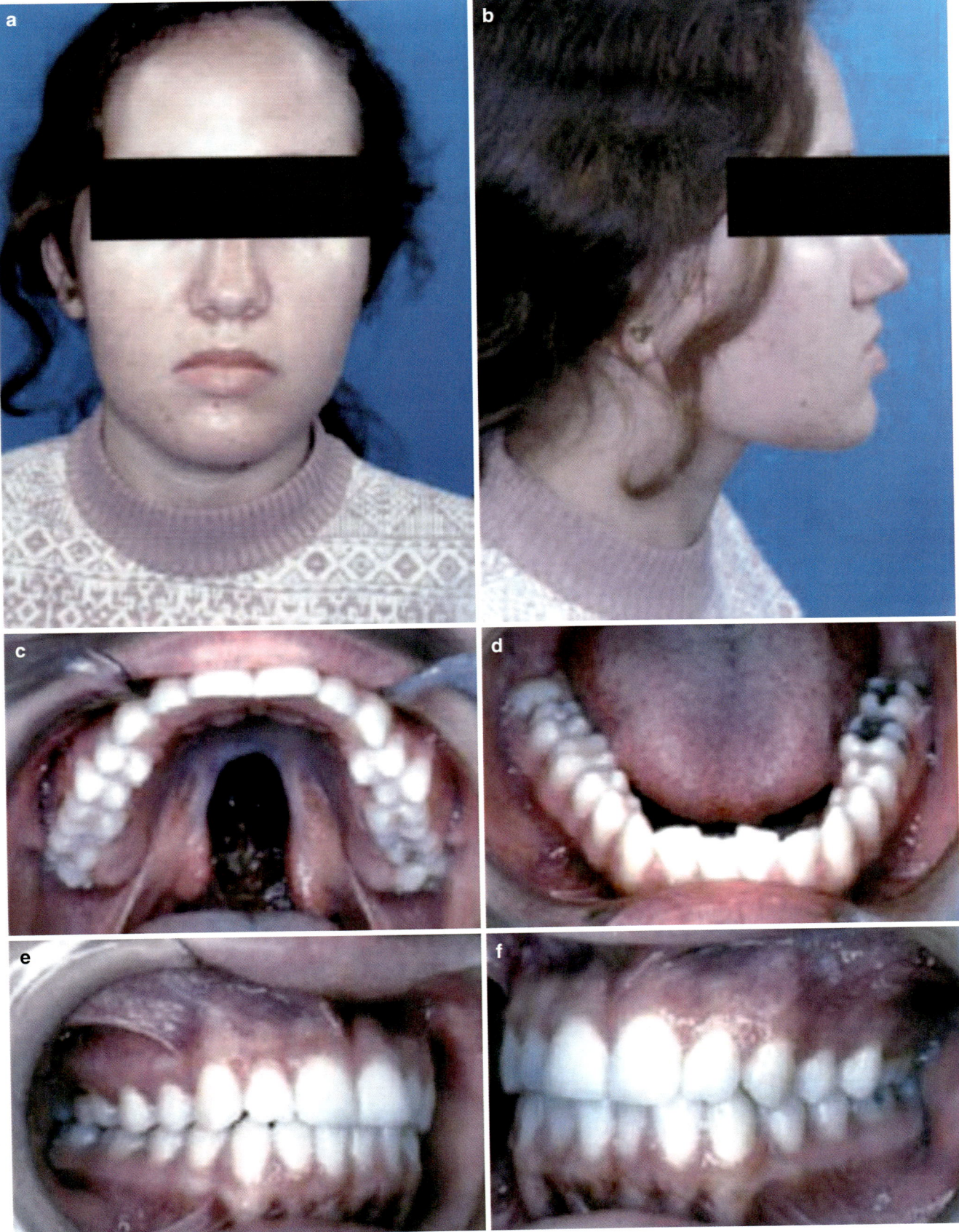

Fig. 1.1 Isolated cleft palate showing the vertically deficient vomer field retracted out of the palatal plane making fusion with the maxillary palatal shelves impossible. (Courtesy of Michael Carstens, MD)

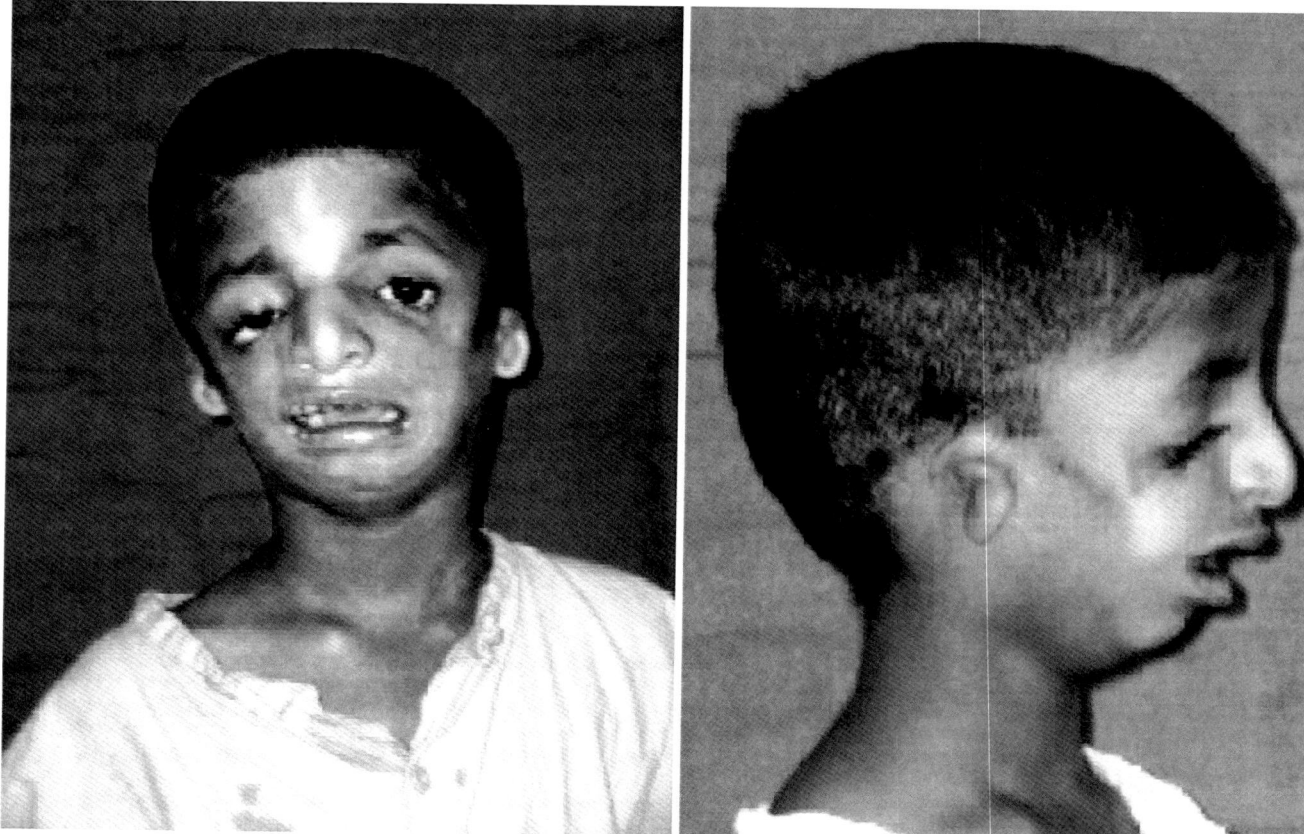

Fig. 1.2 Treacher-Collins Franceshetti syndrome showing deficiencies in malar, maxillary, and mandibular fields. (Reprinted from Goel L, Bennur SK, Jambhale S. "Treacher-collins syndrome-a challenge for anaesthesiologists". *Indian Journal of Anaesthesia.* 2009;**53** (4): 496–500. With permission from Wolters Kluwer Medknow Publications)

Definitions

Anterior visceral endoderm (AVE): Described in mammals; the AVE constitutes the inducer of the forebrain.

Chondral neurocranium: Skull dedicated to protection of the brain and formed in cartilage. (1) Axial cranial base = ethmoid, presphenoid, basisphenoid, basioccipital, exoccipital, and supraoccipital. The first two bones are formed by neural crest. The basisphenoid comes from somitomere 1. The remaining bones are formed from the sclerotomes of somites 1–5 (the occipital somites). (2) Lateral cranial base = temporal bone (petrous and mastoid fields). These bones are non-sclerotomal and originate from somitomeres 6 and 7.

Chondral viscerocranium: Skull dedicated to supporting the pharyngeal arches. Larynx may represent a transformation of previous visceral cartilages.

Frontonasal prominence (obsolete): Neural crest of the caudal prosencephalic neural folds (prosomeric levels p1, p2, p3, and p4) migrates forward to populate the rostral prosencephalic neural folds (prosomeric levels p5 and p6). Note that these rostral folds are "sterile," i.e., they contain neither cells of the CNS nor neural crest. For this reason, they are referred to as **nonneural ectoderm** (NNE). NNE gives rise to the optic, olfactory, and adenohypophyseal placodes, as well as to the epidermis of the forehead, nose, and vestibular lining. The arrival of prosencephalic neural crest beneath the NNE results in creation of a dermis for the forehead (p4 zone), the external nose (p5 zone), and the internal nose (p6 zone). As the frontal lobes grow, the skin envelope is pushed forward and medially on both sides of the midline. The resulting configuration is a p4 forehead, a p5 orbital roof, and a p6 ethmoid-nasal complex. These are **bilateral** structures connected in the midline. There are two p5 optic placodes and two p6 nasal placodes. Initially, both fronto-orbital-nasal masses are widely separated. The initial orbital angle is about 180°. A combination of brain growth, embryonic folding, and the differential growth of the PNC mesenchyme (p6 < p5 < p4) results in a dramatic apoptosis of the midline MNC sphenethmoid complex and causes a narrowing of the orbital angle to its final configuration of 90°.

Intermediate mesoderm (IM): This mesenchyme lies medial to LPM and lateral to PAM; unclear if a separate population or induced from LPM by medial genes; forms urogenital system; sequential formation of urinary organs in craniocaudal order via nephrotomes: pronephros, mesonephros, and metanephros. Origin of pronephros is at fourth cervical neuromere.

Lateral plate mesoderm (LPM): First wave of mesoderm at gastrulation migrates to most distal position of the trilaminar disc; has two potential layers, an inner visceral layer (LPMv) and an outer somatic layer (LPMs). These are split by intraembryonic coelom into two layers. LPMv forms cardiac mesoderm and much of peripheral vasculature and all gut musculature below neck. Unsplit LPM in neck forms cervical esophagus. Neuromeric coding of trunk LPMv may follow the sympathetic nervous system and ventral unpaired branches of the aorta. LPMs forms the fascial envelope of the trunk and all bones of the extremities including the clavicle, the scapula, and the pelvic girdle. LPMs and PAM share common hox genes and segmental innervation.

Membranous neurocranium: Skull dedicated to protection of the brain and formed in membrane. All bones except parietal are exclusively neural crest. Parietal receives mesenchyme from somitomere 4; includes the bones of the nasal and orbital cavities.

Membranous viscerocranium: Skull dedicated to supporting the pharyngeal arches and formed in membrane—includes lisphenoid, inferior turbinate, maxilla, palatine, zygoma, mandible, and ear bones.

Neuromere: Individual developmental zone of the embryonic neural tube. Boundaries defined by a "barcode" of overlapping zones of gene expression. Each neuromere provides spatial segmentation to mesoderm existing outside the neural tube at that level. All neural crest, paraxial mesoderm, and somatic lateral plate mesoderm originating from a given neuromeric level will bear the same *Hox* code.

Paraxial mesoderm (PAM): Final wave of mesoderm to exit the primitive streak; remains closest to neuraxis and is induced by genes from the notochord and neural tube. Because it contains angioblasts, it immediately forms the primitive perineural plexus. Most anterior PAM forms the dorsal aorta, the muscles of the pharyngeal arches, the cranial base caudal to the pituitary, the pharyngeal arch muscles, and the. Derivatives of PAM are discussed in Chap. 2.

Placode: Specialized areas of epithelium seen on the surface of the embryo in close association with the brain. (1) Required by special somatic afferent systems of smell, sight, hearing/balance. Contain neurons with eventual connection to the CNS. (2) Surrounding all placodes are zones of tissue with functional importance for the eventual development of a sensory unit. (3) All placodes remodel and "sink" into the embryonic head; in this manner they internalize functional tissue. (4) Once internalized placodes undergo structural alteration (p6 adenohypophyseal = anterior pituitary, p6 nasal = olfactory bulbs, p5 optic = lens, r2–r3 trigeminal = gasserion ganglion, r5–r6 facio-acoustic = cochlea and vestibular apparatus, r7–r11 statoacoustic placodes (in fishes) produce ventral nuclei of taste ganglia). (5) In all cases, specialized tissue of the sense organ (placode) must make contact with cranial nerve structures in order for perception to occur.

Prechordal plate mesoendoderm (PCM): First cells to exist from Hensen's node form first an endoderm and then a mesoderm. This zone lies directly beneath the future midbrain and forebrain; may form the extraocular muscles innervated by the rostral nucleus of cranial nerve III (inferior rectus, medial rectus, and inferior oblique). In avian model, PCM is a potent inducer of forebrain.

Somite: Organizational unit of paraxial mesoderm, each is derived from a precursor somitomere. The first somite arises from Sm8. The units are completely separated by epithelium and roughly cubic in structure. Each somite has three laminar derivatives: (1) an internal lamina, the *sclerotome*, forms the vertebrae and ribs; (2) an external lamina, the *dermatome*, forms those dermal structures associated with each neuromeres; (3) an intermediate lamina, the *myotome*, is divided into a dorsomedial epaxial zone and a ventrolateral hypaxial zone.

Somitomere: Initial and transient form of segmentation of the paraxial mesoderm corresponding to a given neuromeric level. Somitomeres are incompletely segmented. Each contains level-specific paraxial mesoderm that has exited from the primitive streak during gastrulation. Somitomeres undergo secondary transformation to derivative structures. The walls of the dorsal aortae are formed from the entire length of somitomeres. They provide the mesenchyme for striated muscles and contribute to the cranial base of the middle cranial fossa and petrous complex. In the occipital regions, Sm8–Sm11 contributes for the muscles of the larynx and pharynx prior to formal transition to somites. Beginning with the Sm8 and continuing through Sm11, each somitomere forms first a pharyngeal arch and the remaining mesoderm is transformed into an occipital somite. All somitomeres from 12 onward are transformed exclusively into somites.

Zoologic Abbreviations

Alisphenoid (greater wing of sphenoid): AS
Basisphenoid: BS
Basioccipital: BO
Ear bones: Malleus (M), Incus (I), Stapes (Sp)
Epiotic: EpO
Exoccipital: EO
Frontal: PrF (prefrontal medial), PtF (postfrontal)
Lateral pterygoid plate: LPt (in continuity with AS)
Maxilla: Mx1 (incisor/canine), Mx2 lateral (premolars), Mx3 (molars)
Mandible: Mn1 (incisor/canine), Mn2 (premolars), Mn3 (molars), MnR (ramus)
Medial Pterygoid plate: Mpt (in continuity with the orbitosphenoid)
Palatine: Pl
Parietal: P

Premaxilla: PMx medial (central incisor), PMx lateral (lateral incisor), PMxF (frontal process)

Presphenoid: PS

Prootic: PrO

Pterygoid: Pt

Opisthotic: Op

Orbitosphenoid: OS

Supraoccipital: SOc (chondral), SOm (membranous)

Temporal: Tm (mastoid), Tp (petrous), Ts (squamous), Tt (tympanic)

Zygoma: PO (postorbital), J (jugal)

Developmental Fields: Lessons from the Common Labiomaxillary Cleft

The 4-D Theory of Cleft Formation

The pathologic anatomy of the labiomaxillary cleft is a four-dimensional problem. The principles of its surgical management must be conceptualized in the same manner. How the cleft site appears in the newborn is very different from its anatomy in utero. Indeed, the initiation of the cleft problem may occur as early in embryogenesis as gastrulation (the process by which the germ layers of the embryo are established) at 15–18 days gestation [1, 2]. At the time of initiation of the cleft site, four pathologic processes are unleashed; these processes exert their effects in a strict sequential order [3]. A spectrum of presentations is thereby produced ranging from the "cleft-lip nose" with an apparently normal lip, to a full-blown [4] (Figs. 1.3, 1.4, and 1.5).

The nature of this pathologic sequence has been identified [3–10]. First, a deficiency state exists in the functional matrix responsible for synthesizing the frontal process of the premaxilla that forms the piriform margin. Within this abnormal developmental field, insufficient bone volume results. This causes a stereotypical displacement pattern of the soft tissue envelopment on both sides of the cleft. If the deficiency state is significant enough, it will affect the ability of adjacent developmental fields to perform soft tissue closure of the nostril floor and lip. The resulting division further aggravates tissue displacement. Over time, the effects of deficiency, displacement, and division create a distortion of the soft tissue envelope. This results in an abnormal anatomy of the septum. Ongoing growth of the osteocartilagenous nasal vault,

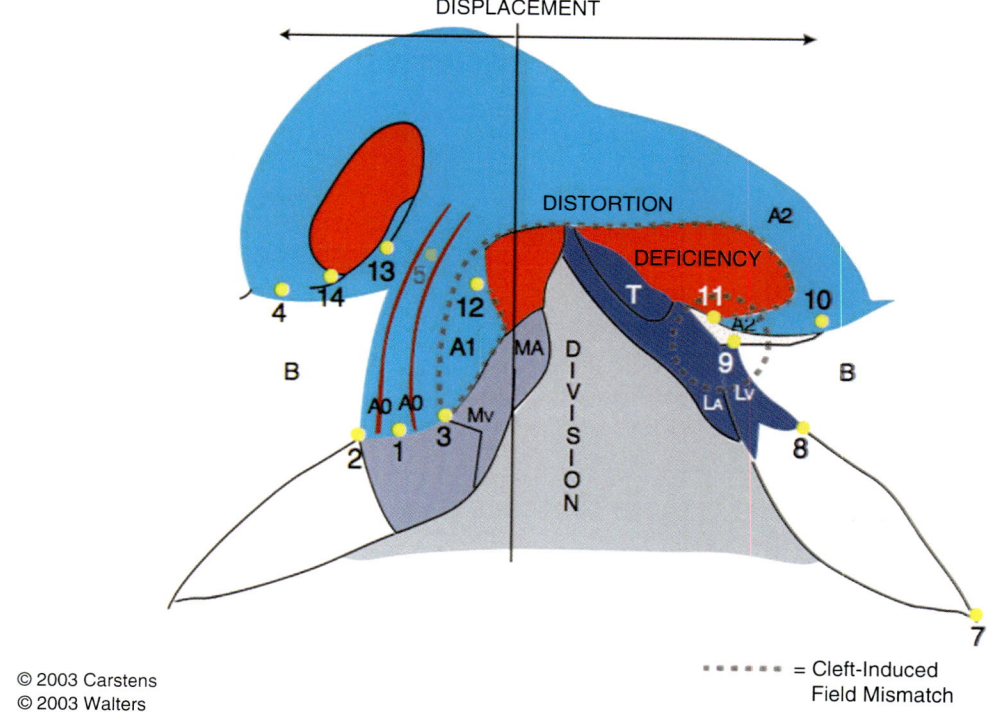

© 2003 Carstens
© 2003 Walters

• • • • • • = Cleft-Induced
Field Mismatch

Fig. 1.3 Four-dimensional model of cleft. A deficiency state in the premaxilla at the lateral piriform fossa results in decrease production of BMP4. This protein diffuses down through the subcutaneous tissues to reach the epithelial edge of A2. Consequently, Sonic hedgehog in the epithelium remains active, epithelial integrity is preserved, and fusion is impossible. This prevents lateral nasal mass A2 from uniting with medial nasal mass A1. The diffusion gradients I rostral-caudal and thus the vermilion is affected first. As the severity of BMP4 deficit increases, the "cleft" ascends. A1 (non-philtral prolabium) represents soft tissues coverage of the missing premaxillary bone. It is supplied by medial sphenopalatine and innervated by V1. It remains attached with the true philtra; prolabium A0. It does not participate in forming the nasal floor. The piriform deficit is accompanied by a deficit in the p6 vestibular lining (red). The alar cartilage is pulled down and the overlying nasalis muscle is misdirected. This explains the appearance of the cleft nose. (Reprinted from Carstens M. Functional matrix cleft repair: principles and techniques. Clinics in Plastic Surgery 2004;31 (2): 159–189. With permission from Elsevier)

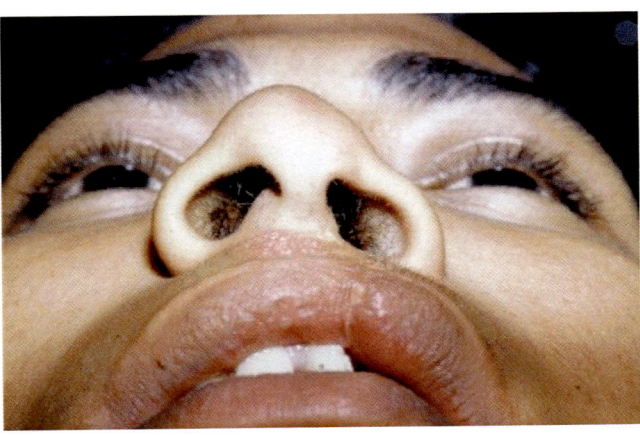

Fig. 1.4 Relapse pattern of left unilateral cleft after apparently successful surgical repair in infancy reveals how unbalanced growth forces act over time to maintain and exacerbate asymmetry of soft tissue and bone. (Courtesy of Michael Carstens, MD)

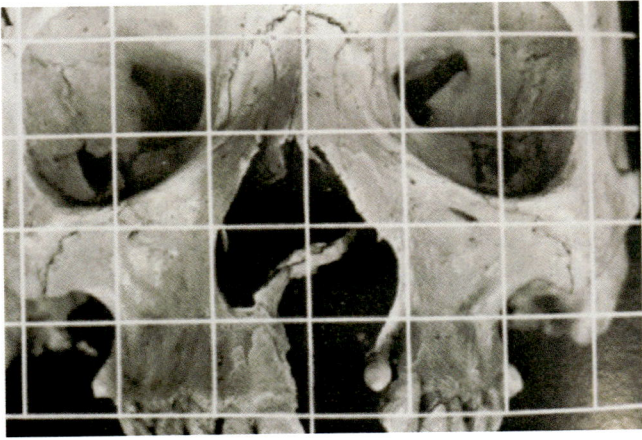

Fig. 1.5 Untreated cleft maxilla grows normally, but remains in a retrocessed position compared to the normal right side. The functional matrix is produced by the embryo long before bone production occurs. In cleft situations, both sides are disconnected and grow asymmetrically. The cleft side functional matrix is thus displaced by the time of osteogenesis. Both soft tissue and bone are therefore normal, but stuck in the wrong position. Without correcting force vectors surgically in infancy, the cleft maxilla will function normally but its overlying periosteal cell layer will continue to produce membranous bone that is malpositioned. (Courtesy of Michael Carstens, MD)

uncoupling of normal relationships between the skeletal elements, and aberrant force vectors exerted by the perioral musculature result in the characteristic "opening-up" of the cleft site so elegantly described by Delaire [11–18].

The above concepts of cleft formation are known as "4 dimensional theory." These can be summarized by the pneumonic of "the 4 D's," each one of which corresponds to a dimension. Deficiency is axial. Displacement is coronal. Division is sagittal. Deficiency is temporal. Interestingly enough, the order of these processes follows the order of axis specification in the embryo: anteroposterior, then mediolateral, and finally right-left.

The Biologic Significance of Relapse

All surgeons involved with cleft care know full well the frustration of seeing well-executed repairs in infancy degenerate into a predictable sequence of secondary deformities requiring further correction. Even in the best of hands, reoperation rates may reach as high as 85% [19]. What exists here is not failure of technique, but an inadequate biologic model of the problem in the first place. If the pathologic anatomy of the cleft site hinges on a deficiency state in a specific developmental field, and if the surgical correction of the cleft does not include reconstitution of that defective field such that it will grow normally over time, and such that it will cease to perturb the growth of its neighboring fields, then all forms of cleft surgery are condemned over time to varying degrees of relapse!

Are there any grounds for optimism? If secondary deformities constitute a type of plastic surgery "lemon," is it not possible to make of them some form of "lemonade?" The answer to this question is overwhelmingly positive because, in pediatric craniofacial surgery, all patterns of relapse point an unequivocal accusatory finger directly at the original pathology. Relapse is nothing more than the manifestation over time of a deficient developmental field. Knowledge of the embryology of the face conceived in terms of specific fields (zones of soft tissue and the bone that they produce) will enable surgeons to conceive new surgical approaches based, not on geometry, but upon developmental anatomy. Cut-and-paste tissue manipulation will be supplanted by biologic principles of field reconstitution and reassignment.

Beyond Descriptive Embryology: Developmental Fields and the Functional Matrix

It may come as a surprise to many readers that the drawings of cleft lip pathogenesis depicted in the most recent of texts reflect an understanding of facial embryogenesis that is nearly century and a half out-of-date. Descriptive embryology as a science began from pioneering observations by Wilhelm His in the 1870s using light microscopy and histologic staining [20, 21]. The approach was morphologic rather than cellular. (To the end of his career, His vigorously opposed the idea that genetic information could be contained in the nucleus.) Terms introduced by His such as "lateral nasal process" can be found in all textbooks. Yet who among us can define just what a "process" is? What are its constituent parts and from where in the embryo do they come? All surgeons are well aware that clefts, for example, occur in a comprehensible spectrum of presentations. Unfortunately, concepts such as "failure of fusion" or "failure of mesoderm penetration" are incapable of providing a rational explanation for the varying degrees of pathology. For most of us,

Fig. 1.6 Wilhelm His (1831–1904), Professor of Anatomy, University of Leipzig. His was the inventor of the microtome for preparation of pathology specimens. He was also the father of descriptive embryology: "… progress in anatomy is most likely to occur when its problems include the study of growth and function, as well as of structure." (Reprinted from Wikimedia. Retrieved from https://commons.wikimedia.org/wiki/File:Wilhelm_His.jpeg)

embryology seems a mere jumble of terms with no clinical relevance. But the iron fact of the matter is this: without a detailed understanding of the developmental anatomy of the face based on modern developmental biology, genetics, comparative anatomy, and neuroembryology, pediatric plastic surgery is a collection of techniques in search of a science (Fig. 1.6 Wilhelm His).

Based upon clinical observations of secondary cleft patterns, this author arrived at the following hypotheses: (1) unidentified developmental fields might constitute the "building blocks" of the face; (2) a mesenchymal deficiency state in such a field would be characterized by an inadequate osteosynthetic capability; (3) an osseous deficiency state

exists in the inferolateral piriform fossa/lateral nasal wall of cleft patients; (4) such a functional matrix deficiency state might account for the relapse pattern observed in cleft patients after primary repair.

Neurovascular Mapping of Developmental Fields: Nasomaxillary Model

To test these ideas, it seemed logical as a first approximation to study the relative contributions of the internal carotid artery (ICA) and external carotid artery (ECA) circulations to the skin and epithelium of the nasal fossa. Contrast injections into isolated internal carotid arteries in a series of aborted fetuses were performed. These results were first reported at the Plastic Surgery Educational Foundation awards program at the 2000 annual meeting in Los Angeles of the American Society of Plastic Surgeons [22] (Figs. 1.7 and 1.8).

The author found, to his surprise, that the upper border of the inferior turbinate combined with the skin/mucosa junction of the inferolateral piriform fossa just anterior to the inferior turbinate constituted a potential field interface zone. At this site, three distinct biologic systems (vascularization, innervation, and genetic programming) functioned in precisely the same manner. (1) The internal carotid supplied the mucosa (but not skin) of the lateral nasal wall, but only as far as the upper border of the inferior turbinate. The mucosa beneath the turbinate and the skin margin along the infracartilagenous nostril were un-perfused. (2) The sensory innervation followed the exact same distribution pattern! The epithelium supplied by the ICA corresponded to sensory supply from the first branch of the trigeminal, while that supplied by the ECA was innervated by V2. (3) The inferolateral piriform fossa also represents an interface zone between three entirely different developmental zones of the embryonic neural crest!

The clinical relevance of this model can be seen in developmental anatomy of the piriform fossa (Figs. 1.7 and 1.8). The lateral wall of the nasal cavity has an *upper* zone populated by prosencephalic neural crest (PNC), innervated by V1, and irrigated by ICA. The *lower* zone of the lateral nasal wall is populated by rhombencephalic neural crest (RNC), innervated by V2, and irrigated by ECA. The "breakpoint" between these two zones is the inferior turbinate. This outer

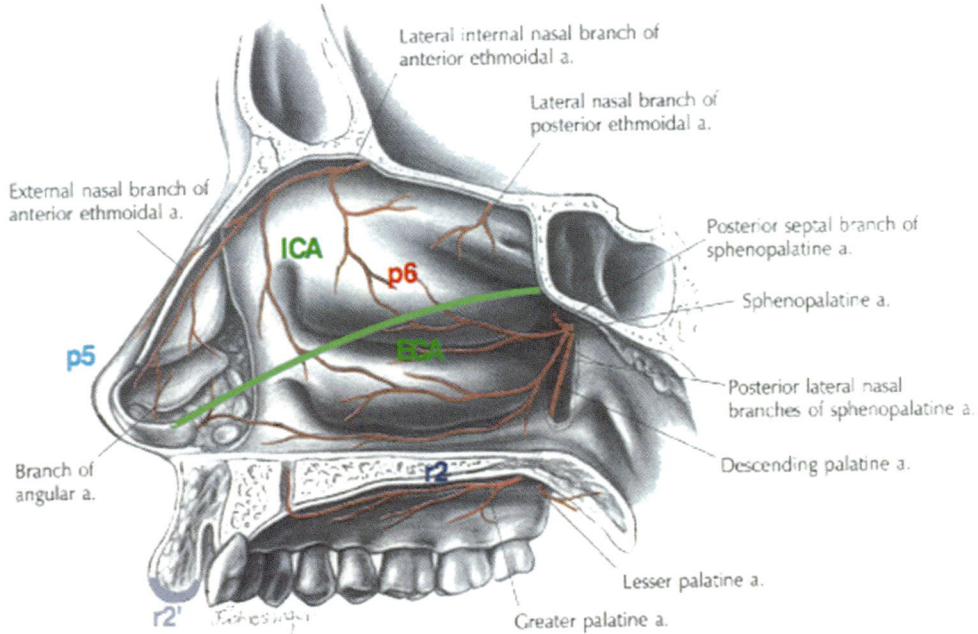

LATERAL NASAL WALL

Fig. 1.7 Lateral nasal wall showing neurovascular "breakpoint." Innervation is V1 above and V2 below. StV1 anterior and posterior ethmoid neuroangiosomes are traditionally considered internal carotid from ophthalmic. They supply r1 MNC mesenchyme: upper and middle turbinates. Vestibular lining posterior to the external nasal valve (lower lateral vs. upper lateral cartilage) is p6 nonneural ectoderm. Nasal vestibular skin anterior to external nasal valve is p5 nonneural ectoderm. StV2 lateral nasopalatine neuroangiosome is traditionally considered external carotid from the internal maxillary. It supplies RNC mesen- chyme: inferior turbinate and assists frontal process of premaxilla and supplies the nasal side of the hard palate. StV2 descending palatine neuroangiosome is also traditionally considered external carotid from the internal maxillary. It supplies the oral side of the hard palate. Oral side mucosa is non-respiratory and has a different color from that of the nasal side. (Reprinted from Carstens M. Functional matrix cleft repair: principles and techniques. Clinics in Plastic Surgery 2004;31 (2): 159– 189. With permission from Elsevier)

lamina of the piriform margin belongs to the *ascending pro- cess* of the maxilla. The medial wall of the nasal cavity has an *upper* zone populated by PNC, innervated by V1, and irri- gated by the ICA. This zone consists of the perpendicular plate of the ethmoid and the septum. The *lower* zone of the medial nasal wall is populated by mesencephalic neural crest (MNC), innervated by V2 and irrigated by the ECA. This zone contains the vomer and the premaxilla.

Details of the experimental data and their many implica- tions appeared in print as a supplement to the Journal of Craniofacial Surgery in February 2002 [9]. The purpose of this publication was to describe a preliminary, but clinically relevant, "map" of developmental fields as they applied to the facial midline. Details of the medial nasal wall fit a model quite analogous to that of the lateral nasal wall. In this model, the septum and perpendicular plate of the ethmoid form a sharp developmental boundary with the vomer and premax- illa characterized by blood supply (ICA vs. ECA), innerva- tion (V1 vs. V2), and neural crest origin (PNC vs. MNC/ RNC). As we shall see, this model will permit us to assemble an accurate picture of how the premaxilla and maxilla develop and interface with one another. Isolated deficits in components of this system explain the pathologic anatomy of all forms of clefts.

The premaxillary developmental field has three subdivi- sions: central incisor, lateral incisor, and a vertical *ascending process*. Formation of an intact alveolar arch involves fusion between the premaxilla and the maxilla. This brings the two

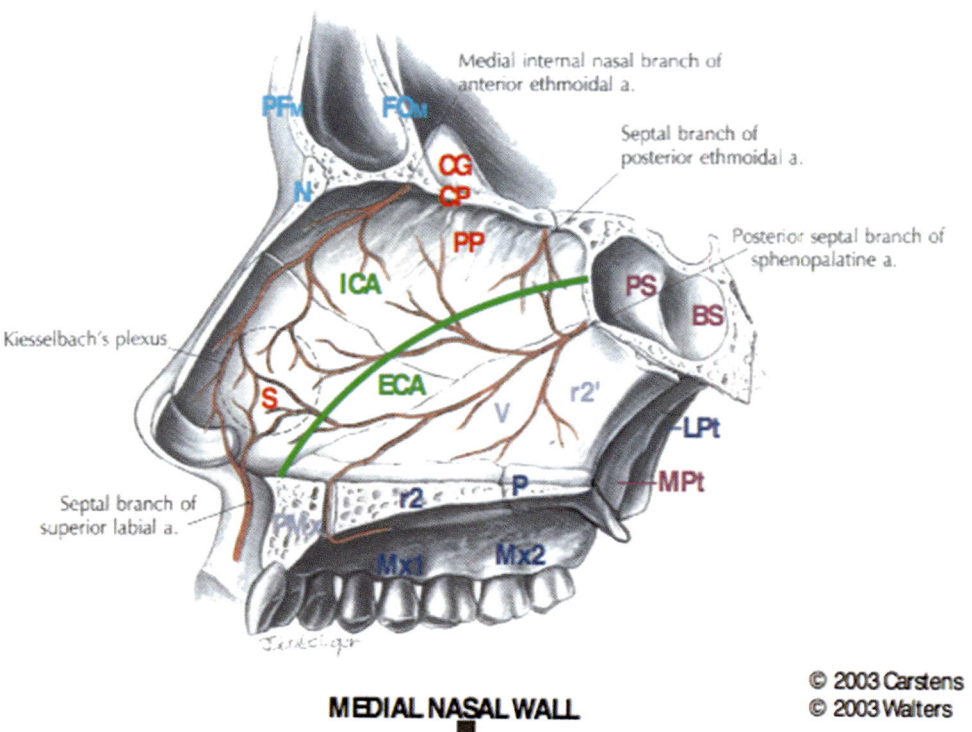

MEDIAL NASAL WALL

© 2003 Carstens
© 2003 Walters

Fig. 1.8 Medial nasal wall showing neurovascular "breakpoint." Innervation is V1 above and V2 below. StV1 anterior and posterior ethmoid neuroangiosomes are traditionally considered internal carotid from ophthalmic. They supply r1 MNC mesenchyme: perpendicular plate of the ethmoid and septum. Septal vestibular lining is p6 nonneural ectoderm. Membranous septal skin and columella are p5 nonneural ectoderm. StV2 medial nasopalatine neuroangiosome is traditionally considered external carotid from the internal maxillary. It supplies RNC mesenchyme: vomer and premaxilla. (Reprinted from Carstens M. Functional matrix cleft repair: principles and techniques. Clinics in Plastic Surgery 2004;31 (2): 159–189. With permission from Elsevier)

ascending processes (premaxilla and maxilla) into apposition and fusion. The piriform margin is therefore bicortical; it therefore serves as a buttress for reconstruction. Loss of the medial (premaxillary) ascending process leads to a scooped-out piriform fossa and is the cause of the cleft nasal deformity (Tessier #2 cleft). Loss of the lateral (maxillary) ascending process creates a cleft extending to the medial inferior orbital rim, but leaving the piriform fossa intact (Tessier #4 cleft). More extensive neural crest deficiency in the premaxillary field leads to loss of alveolar bone in a labiolingual gradient. This creates a cleft of the primary palate and, frequently, loss of the lateral incisor (Figs. 1.9 and 1.10).

Although this model description based on classifying tissues as ICA and ECA derivatives was immediately useful for mapping out the relative roles of premaxilla, vomer, and maxilla, in cleft formation, the vascular embryology behind it proved to be too simplistic. The arterial supply for these fields, in point of fact, represents a new system that arose with the transition between agnathic (jawless) fishes and gnathostomes (fishes with jaws). The process of converting the first and second gill arches to maxilla, mandible, and supporting hyoid bones required reassignment of neural crest from the gills to a new location and shape. This in turn required a disruption in the arterial supply to first and second arches and the creation of a new system.

The stapedial innovation arises from a remnant of the second aortic arch artery, the *hyoid artery*. This original stapedial stem passes through the stapes within the developing tympanic cavity and into intracranial and extracranial divisions. These are guided into position initially by the branches of cranial nerve VII and later by the branches of the trigeminal nerve. The intracranial division ultimately reaches the orbit where it connects with ophthalmic to form all arteries supplying the non-ocular tissues of the orbit and face. The extracranial division joins the distal-most branch of external carotid to form another hybrid system, the maxillomandibular complex. Arteries follow sensory branches of V2 and V3 to supply the neural crest structures of the jaws. We shall discuss this system at length at various junctures through the remainder of this book (Figs. 1.11 and 1.12).

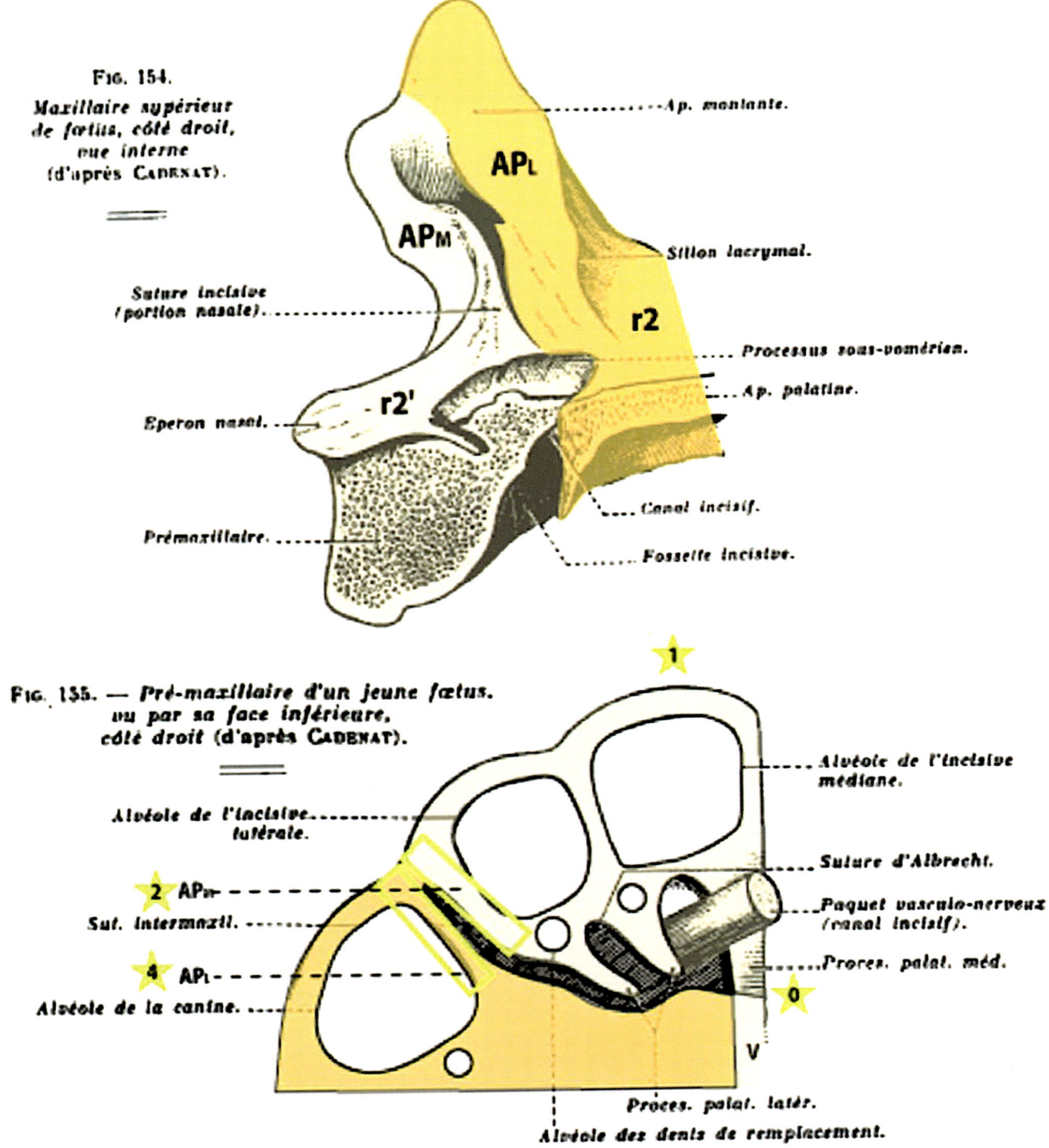

Fig. 1.9 Premaxilla, 1922 drawing by Etienne Cadenat showing intranasal view of frontal processes of premaxilla and maxilla. Lateral nasopalatine field (r2) and medial nasopalatine field (r2′) can each produce a incisor, thereby making the so-called "supranumerary incisor" at site a perfectly normal event. Frontal process of premaxilla is really an vertical extension from the lateral incisor field. (Reprinted from Augier M. Sur le dèveloppement du prémaxillaire humaine. *Compte Rendu l'Assoc Anatomie* 1932; 27:18–28)

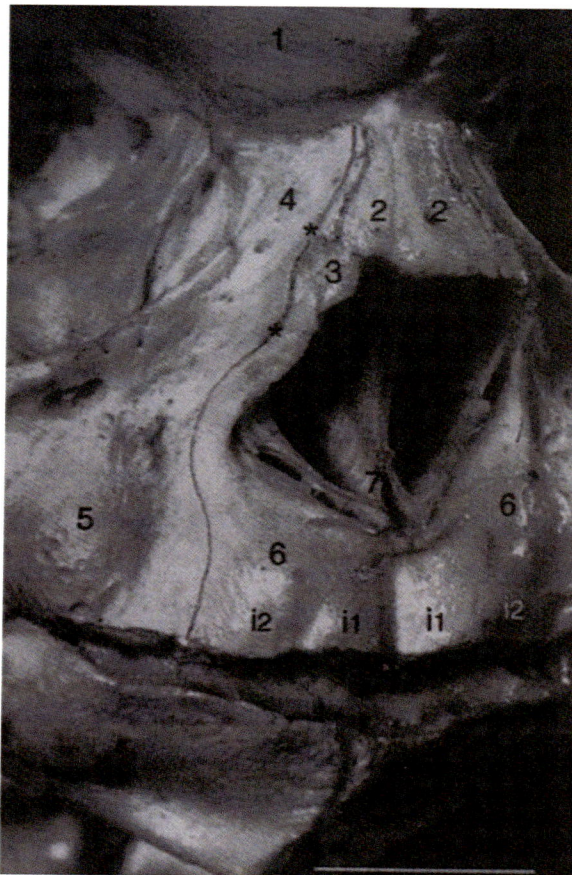

Fig. 1.10 Premaxilla in human fetus. (Reprinted from Barteczko K, Jacob M. A re-evaluation of the premaxillary bone in humans. Anat Embryol (Berlin) 2004; 207 (6):417–437. With permission from Springer Nature)

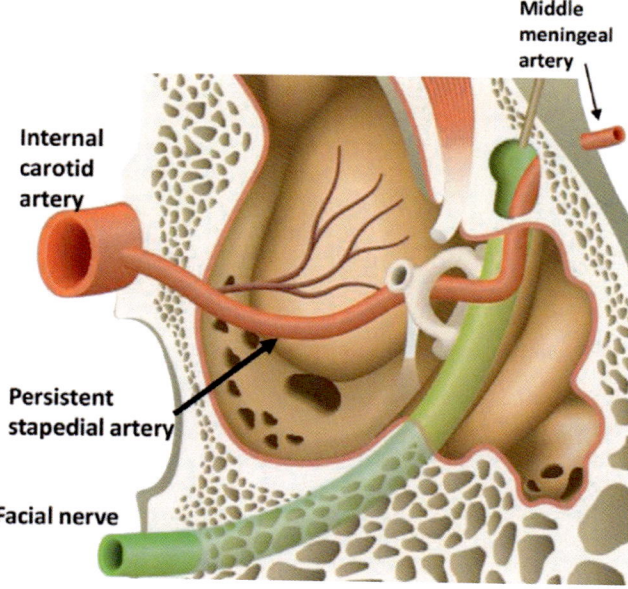

Fig. 1.11 Stapedial artery passing upward through the stapes into the tympanic cavity (diagram). Artery is seen (arrow). (Reprinted from Sugimoto H, Ito M, Hatano M, Yoshizaki T. Persistent stapedial artery with stapes ankylosis. Auris Nasus Larynx 2014; 41 (6): 582–585. With permission from Elsevier)

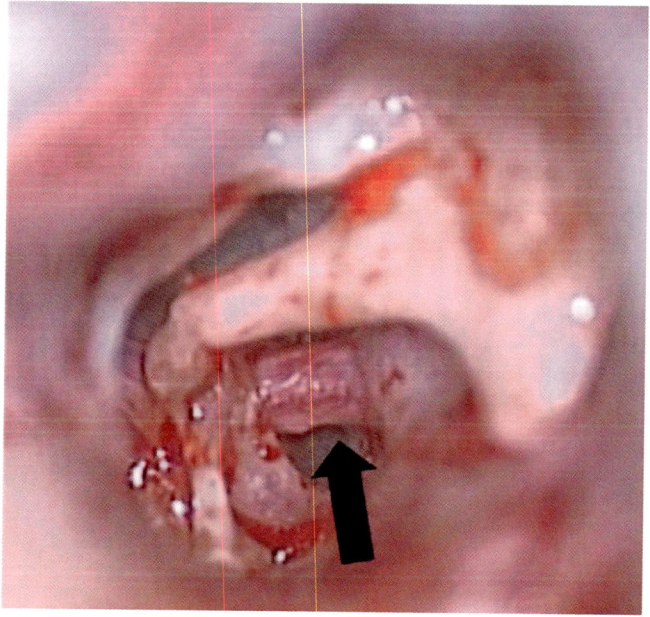

Fig. 1.11 (continued)

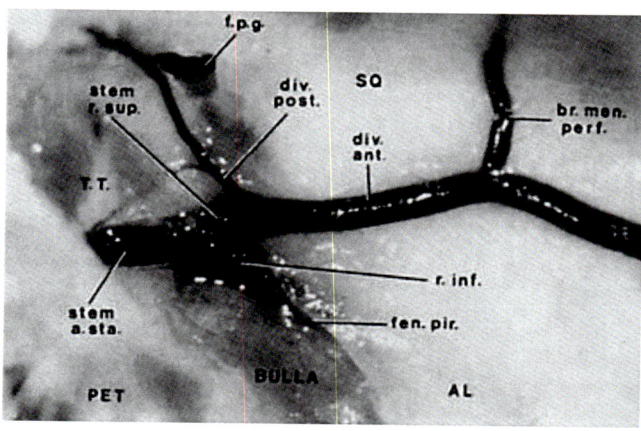

Fig. 1.12 Photograph of stapedial artery traversing the stapes. (Reprinted from Diamond MK. Homologies of the stapedial artery in humans, with a reconstruction of the primitive stapedial artery configuration of euprimates. *Am J Phys Anthro* 1991; 84:433–462. With permission from John Wiley & Sons)

Neural crest cells refer to those cells that arise during embryogenesis from a border zone between the more lateral zone ectoderm responsible for forming skin and the more axial zone of ectoderm responsible for forming the brain and spinal cord. Neural crest cells migrate widely and form many structures such as dermis, bone, and cartilage usually associated with mesoderm. These cells also form components of the nervous system such as Schwann cells and autonomic ganglia. For this reason, neural crest is often referred to as an ectomesenchyme. The great extent of its derivatives has often led authors to refer to it as "the fourth germ layer." It should be born in mind that neural crest cells do not make their appearance until well after gastrulation is complete. The three traditional germ layers all are recognized at the time of gastrulation.

The behavior of neural crest cells in terms of their migration pathways and derivatives stems largely from what part of the neural folds they originate from. The names of these neural crest zones correspond to the original three parts of the developing central nervous system (learned by most of us and promptly forgotten). These are: the *prosencephalon* (forebrain), *mesencephalon* (midbrain), and *rhombencephalon* (hindbrain) (Fig. 1.13 Embryonic brain).

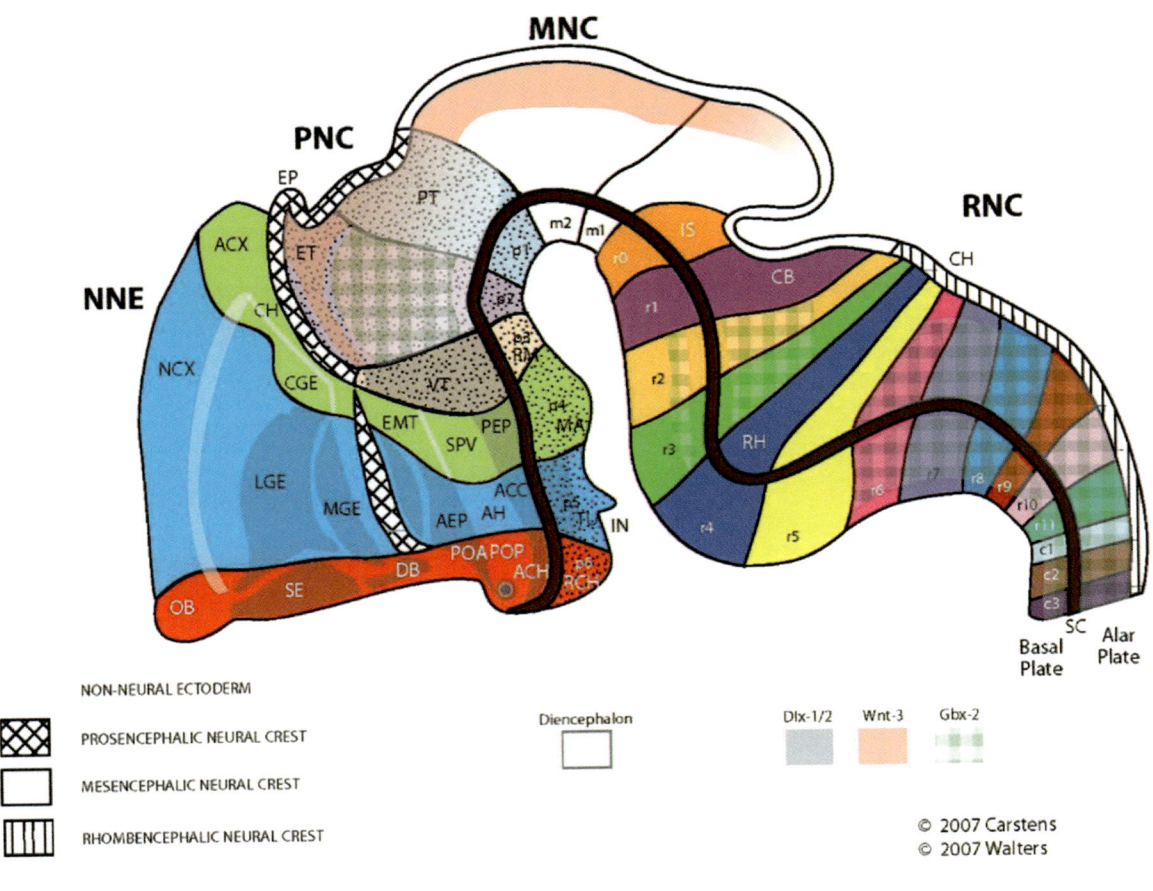

Fig. 1.13 *Neuromeric model of the CNS.* Originally this model postulated 6 prosomeres p1 (aqua), p2 (salmon), p3 (flesh), p4 (light green), p5 (light blue), and p6 (red). More recent mapping is more refined and reflects the basic process of CNS development, summarized as follows: The neural primordium begins as a hollow tube with no subdivisions. Initial differentiation of the rostral part of the neural tube shows the appearance of the forebrain, midbrain, and hindbrain vesicles. The forebrain vesicle then divides into a secondary prosencephalon and the diencephalon. The hindbrain also divides into four regions: the prepontine hindbrain, the pontine hindbrain, the pontomedullary hindbrain, and the medullary hindbrain. Subsequently, more subdivisions appear in the forebrain: caudal secondary prosencephalon (hp1); rostral secondary prosencephalon (hp2); and, within diencephalon, the appearance of three prosomeres (p1–p3). In addition, midbrain subdivides into mesomeres 1 and 2 (m1–m2), and hindbrain forms isthmus (is) and rhombomeres 1–11 (r1–r11). Finally, certain parts of the forebrain become further differentiated: the caudal prosencephalon has formed the main part of the telencephalon; the rostral secondary prosencephalon has formed the preoptic telencephalon (POTel), the terminal hypothalamus (THy), and the peduncular hypothalamus (PedHy); and prosomeres 1–3 form the pretectum (Pt), thalamus (Th), and prethalamus (PTh). Neuromeric development will be discussed in Chap. 5. (Courtesy of Michael Carstens, MD)

Neuromeres and Neuromeric Coding

Neuromeres: The Clinical Significance of the Neuromeric Map

All clinicians are familiar with the concept that certain cells born in the neural plates at the exact boundary between neural ectoderm and nonneural ectoderms are termed *neural crest* and that these cells have a very important role to play in development of the head and neck. Our model here is one in which neural crest can be "mapped" to very precise zones of origin. Each such zone corresponds to a developmental unit of the neural plate. These segmented units, called *neuromeres*, are distributed in a transverse fashion along the entire neural axis of the embryo [23–27]. The nervous system is thus divided into transverse developmental units just like the body of an earthworm. Each such segment has a genetic definition. Certain genes or combinations of genes express products only in a particular zone. The anatomic extent of these protein products constitutes the domain of the neuromere. Each neuromere has a certain neuroanatomic content characterized by nuclei and ascending or descending tracts. The neural crest sitting just outside the neural tube in the domain of a given neuromere will express the same defining set of proteins as those cells within the neural tube. Furthermore, neural crest cells from a given neuromeric level will supply certain zones of ectoderm and mesoderm. This model allows us to see the nervous system as the master integrative agent of development. In our discussion, it will be necessary to introduce a number of terms and concepts from neuroembryology (Fig. 1.14).

Before plunging into further details, let's summarize the highlights of the neuromeric model, so we can stay oriented. The CNS of all vertebrates is divided into three classes of neuromeres [28–32]. The forebrain is formed from six *prosomeres* (formerly considered to be four in number). From caudal to cranial, these are numbered p1 to p6. Prosomeres p1–p3 belong to the diencephalon. They are subdivided into two tiers, dorsal (alar) and ventral (basal). The telencephalon is formed from prosomeres p4 to p6. The basal tier of p6 has much to do with the olfactory system, while basal p5 is associated with the visual apparatus.

Puelles and Rubenstein propose the midbrain to be constructed from two *mesomeres* m1 and m2. These contain, respectively, the superior and inferior colliculi. (An anatomical boundary between the two has not been demonstrated as it is in the borders between the rhombomeres.) In this model, the hindbrain is described as being made up from 12 well-defined *rhombomeres* r0 to r11. *An alternative viewpoint held by Sarnat considers r0 to be the principle neuromere of the midbrain and r1 the neuromere of the isthmic region (metencephalon from which develops the pons and cerebellar cortex* [33].

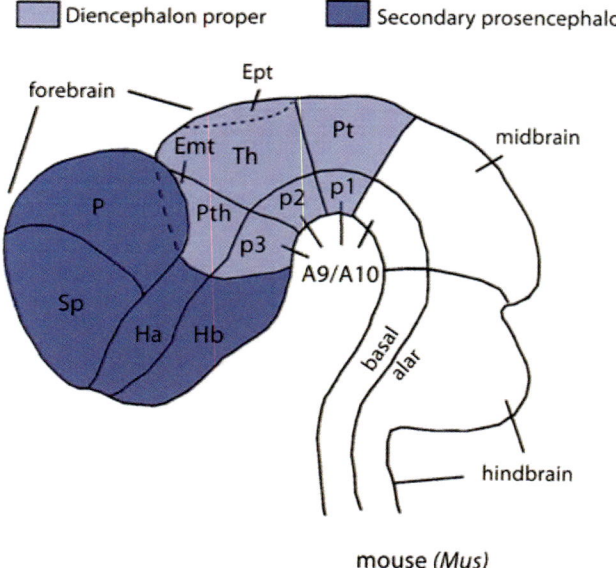

Fig. 1.14 Embryonic brain begins as three parts and then subdivides. Prosencephalon splits into a caudal diencephalon and a rostral secondary prosencephalon, the alar portion of which is telecephalon. Diencephalon contains prosomeres p1–p3. Note the division into basal and alar plates. Schematic diagram of a lateral view of the mouse embryonic brain representing the forebrain and its major divisions and subdivisions. Note that the forebrain shows two major transversal divisions named the diencephalon proper and the secondary prosencephalon (shown in different colors). The secondary prosencephalon produces the telencephalon dorsally (with the pallium and subpallium), the optic vesicle laterally (not represented), and the hypothalamus ventrally (the latter includes alar and basal parts). The amygdala originates in the secondary prosencephalon and includes pallial, subpallial, and alar hypothalamic cell groups. (Reprinted from Medina L, Bupesh M, Abellán A. Contributions of genoarchitecture to understanding forebrain evolution and development, with particular emphasis on amygdala. *Brain, Behav, Evol* 2011; 3: 216–236. With permission from S. Karger AG)

Neural crest from four neuromeres (m1–m2 and r0–r1) behave clinically as a midbrain neural crest. MNC is involved with the construction of the orbit. The midbrain per se has two mesomeres, m1 being rostral and much larger, while m2 (also called the preisthmus) abuts the isthums, a transition zone between midbrain and hindbrain.

Initially, the most rostral rhombomere of the hindbrain is r1, but a small zone emerges at its most anterior aspect that interfaces with m2. This zone corresponds to the isthmus itself and is termed r0. The remaining hindbrain neuromeres r2–r5 are dedicated to the metencephalon (pons) and r7–r11 for the medulla.

Neural crest originating from neural folds associated with rhombomeres r2 to r11 supplies the pharyngeal arch system. When the neural crest cells migrate, they swarm over the surface of the mesoderm lying just outside the neural tube. This mesoderm is called paraxial mesoderm (PAM) and is segmented in direct register with the neuromeric system. Each segment of PAM is called a somitomere (Sm) and is shaped like a ball. The first seven somitomeres (corresponding to

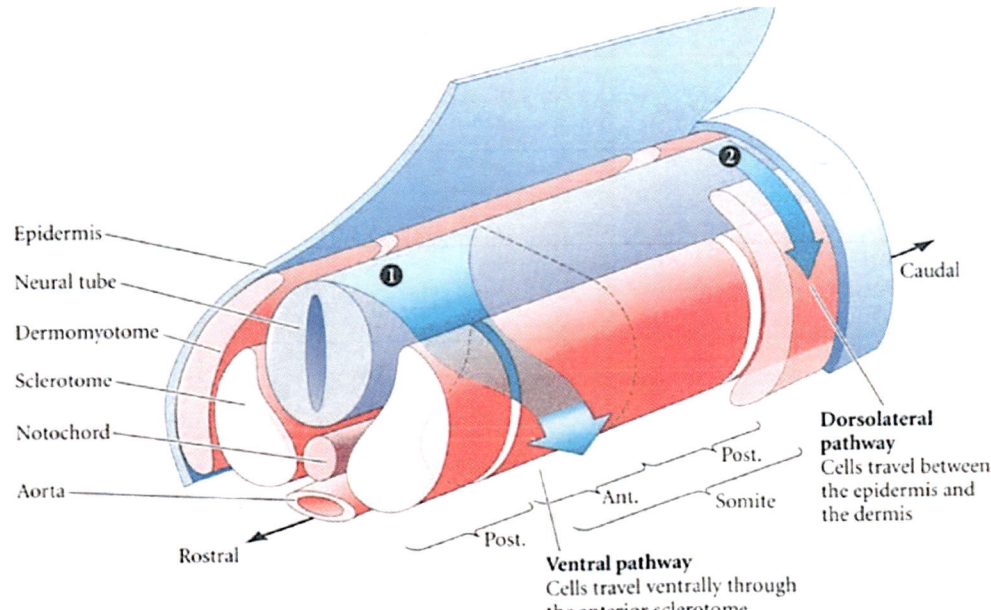

Fig. 1.15 Neural crest migration routes in the trunk. Neural crest of the trunk (from c1 caudal) has two primary fates: neural (ganglia) and neuroectoderma (melanocytes). NCCs migrate in three directions: (1) dorsolaterally to form dorsal root ganglia, (2) ventrally around the neural tube to form the sympathetic ganglia, and (3) rostrocaudally along the axis of the aorta to form the para-aortic ganglia. An additional doso- lateral route beneath the ectoderm populates the skin with melanocytes. (Reprinted from Gilbert SF, Barressi MJF. Developmental Biology, 11th ed. Sunderland, MA: Sinauer; 2016. Copyright © 2016. Oxford Publishing Limited. Reproduced with permission of the Licensor through PLSclear)

r1–r7) are incompletely separated. All somitomeres from Sm8 caudally undergo anatomic rearrangement into somites. Thus, Sm8–Sm11 form the four occipital somites. Sm12 becomes the first cervical somite. Developmental biologists refer to mesoderm from Sm1 to Sm7 as *cephalic mesoderm.*

The mesenchyme of pharyngeal arch consists of neural crest from two rhombomeres plus PAM from a somitomere. The original number of pharyngeal arches in primitive aquatic vertebrates was seven. With the tetrapod transition to a land-based existence, this number was reduced to five, the first two being dedicated to the jaws and oral cavity and the last three to formation of the pharynx. We shall examine how neural crest and mesoderm are mapped out according to the neuromeric system.

Craniofacial Neural Crest

Neural crest migration is a physical process in which the cells move along *pathways of least resistance* (anatomic cleavage planes) or *molecular "guide wires"* such as vimentin. Neural crest cells swarm over the surface of PAM somitomeres like taffy poured over an apple. Neural crest cells migrate following three pathways. (1) They can move laterally outward in the plane between the nonneural ectoderm/ endoderm and the somitomeres. (2) They remain interposed between the neural tube and the somitomeres. (3) They migrate ventral to the neural tube and then travel caudally

stopping along their way to form the sympathetic chain (Fig. 1.15).

Craniofacial neural crest is organized according to the original three-part embryonic brain, each population of which has distinctive migration patterns. Rhombencephalic (hindbrain) neural crest is divided into 12 rhombomeres (r0–r11), the last 10 of which migrate segmentally into the pharyngeal arches and over the face and head. Mesencephalic (midbrain) neural crest populations arise from two mesomeres (m1 and m2) and from the most rostral hindbrain (r0–r1). It migrates as streams into the anterior cranial base, orbit and frontonasal zones. Prosencephalic neural crest is produced by the caudal three (of six) prosomeres (p1–p3). It migrates as a glacier-like sheet forward to populate the frontonasal skin (Fig. 1.16).

Fate mapping experiments show that the r1 neuromere is subdivided into a rostral zone that forms the cerebellar cortex and a caudal zone giving rise to the deep cerebellar nuclei. Neural folds corresponding to the caudal region of r1 neural plate give rise to neural crest with a unique fate, one with direct relevance to the formation of cleft lip and cleft palate. Rapid growth of the head causes the embryo to fold. This *cephalic flexure* forces the future eye to lie ventral to somitomeres 1–5. The eye has already been coated with midbrain neural crest which creates biologic zones of scleral. Thus, the extraocular myoblasts have a direct pathway to access the globe where they are organized by the r1 fascia and attach according to a strict spatiotemporal sequence.

Fig. 1.16 Craniofacial NCC migration routes. Prosencephalic (p1–p3, purple): *sheet* to frontonasal dermis—the epidermis is NNE. Mesencephalic (m1–m2, gold): *streams* to periocular and frontal structures. Mesencephalic (r0–r1, yellow) *streams* to anterior cranial base and sphenethmoid complex. Rhombencephalic (r2–r11, mandibular-red, hyoid-blue, glossopharyngeal-green, vagus1-magenta, vagus2-unlabeled): *segmental* to the pharyngeal arches. (Adapted from Carlson BM. Human Embryology and Developmental Biology, 6th edition. St. Louis, MO: Elsevier; 2019. With permission from Elsevier)

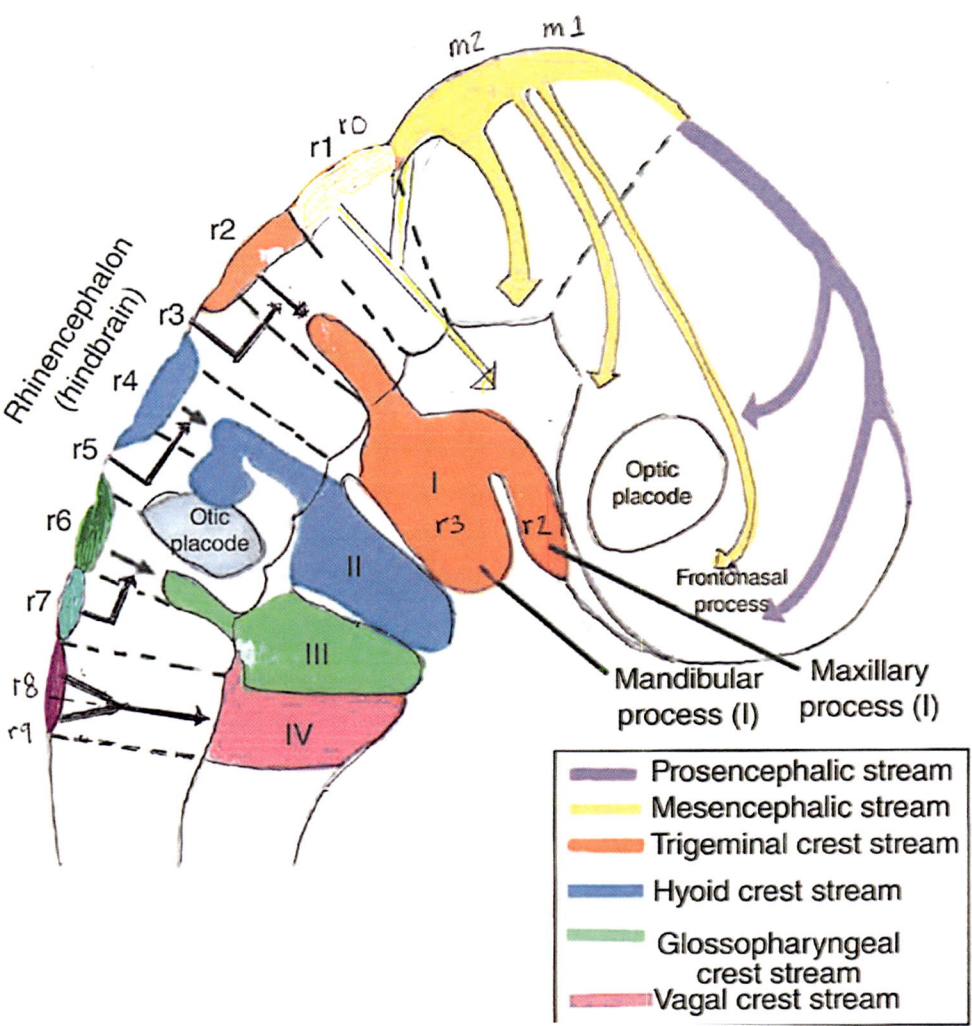

The cephalic fate of midbrain neural crest includes formation of extraocular muscle fascia, all dura innervated by V1, and the presphenoid bone and the frontosphenethmoid complex. All these structures are supplied by the internal carotid artery via the stapedial system, the individual branches of which are programmed by V1. In the first pharyngeal arch r2 neural crest derivatives. These are supplied by the external carotid via branches of V2 stapedial.

Discovery of the rhombomeres came first, dating to the late nineteenth century, but their properties were not understood until the mid 1980s. Furthermore, the exact number of rhombomeres was uncertain. Their anatomic role at the most caudal medulla was unclear. Puelles in 2001 published work in the avian model demonstrating the existence of "pseudo-rhombomeres." Previously, it was thought that the rhombomeric series terminated at r8 that included the entire spinal cord. Puelles' work showed that the final number of rhombomeres was 12 [34, 35]. The existence of neuromeres in the more rostral CNS required advancements in gene mapping. These were not reported until 1993. Over the ensuing decade,

further investigation of the neuromeric system has proceeded at a frenetic pace at neuroscience laboratories around the world. In this paper, we shall be using the Puelles and Rubenstein model as in its latest iteration (2015).

At this juncture, we must take note of a caveat that applies to concepts about prosomeres held widely within the scientific community. Many descriptions of brain anatomy extent in the literature assign four neuromeres to the forebrain, two for the telencephalon and two for the diencephalon. These are known respectively as T2/T1 and D2/D1. This nomenclature is elegantly presented in the brain development section of texts such as that of O'Rahilly and Muller [36–38]. Based on sophisticated gene mapping techniques previously unavailable, the Rubenstein-Puelles model has by no means been universally incorporated into the thinking of neurologists, neuropathologists, and researchers [39, 40]. However, from the standpoint of craniofacial surgery, the R-P model constitutes an extremely sensitive instrument with which to analyze patterns of deformity. Hence, we shall make use of this terminology in our discussion.

The clinical significance of the neuromeric model is that it enables us to map out the anatomic site of origin for all zones of ectoderm and mesoderm supplied by a given zone of the nervous system. The role of neural crest populations in those zones, specifically what structures they make, can be understood as well on the basis of their neuromere of origin. The premaxilla, for example, could develop from a precursor cell population of RNC along the neural fold corresponding to the second rhombomere. When these cells migrate into the first arch, those that populate the anterior and medial sector of the arch will be positioned such that they will come into contact with an spatially specific "program" in the epithelium which will direct them to form the bone fields. This sector will be specifically supplied by the medial nasopalatine branch of the arterial axis. A deficiency state in the population (be it of inadequate cell number, defective migration, abnormal post-migratory rates of mitosis, or cell death) will lead to a small or absent premaxilla. Furthermore, if the premaxillary MNC has several subsets, aligned in craniocaudal order along the neural fold (ascending process, lateral incisor, and central incisor), then the spectrum of deficiency states seen in the premaxilla of cleft patients can be understood as progressively greater degrees of disturbance in the premaxillary MNC precursor population. But let's not get ahead of ourselves…to understand this system, we must first understand the scientific origins of the neuromeric map.

Craniofacial Mesoderm: Extraembryonic Versus Intraembryonic

The Little Appreciated Hypoblast

Formation of the hypoblast and preparation for gastrulation takes place during stage 4 (days 5–6). The process begins when cells of the blastoderm absorb water and then secrete it, causing a separation of blastomerm from the yolk, the subgerminal space. In this process, the deep cells of the blastoderm die creating a monolayer epiblast (light gray). This is known as *zona pellucida*. At the posterior pole of chick blastoderm, the deep cells persist, creating a thickening known as the *zona opaca*. This *posterior marginal zone* (PMZ) is thus bilaminar, having both primitive ectoderm (dark gray) and primitive endoderm (green). The anterior border of PMZ appears like a C-shaped sickle and is termed *Koller's sickle*. Some of the deep cells (green) persist and are physically interposed between the PMZ and the yolk. These cells will give rise to the secondary hypoblast. The C-shaped configuration of the PMZ between the superficial cells and the deep cells is known as *Koller's sickle*. It is from this point, during stage 6, that the primitive will originate (Fig. 1.17).

Later in stage 4, two events take place simultaneously: (1) cells of epiblast become transiently less adherent to each other. This "leaky" state permits epiblast cells of the zone pellucida to delaminate and fall into the subgerminal space, where they will link up again to form the *primary hypoblast* (light gray). (2) Cells below the zona opaca, the posterior marginal zone, proliferate forward beneath the epiblast, forming the *secondary hypoblast* (green) (Fig. 1.18).

By the end of stage 4, primary hypoblast (light gray) and secondary hypoblast (green) unite to form a single layer, thus creating the *bilaminar embryo*. In stage 5, hypoblast (primitive endoderm) will migrate outward to line the entire blastocoele and create the extraembryonic mesoderm that will be forming the placental circulation required to support the metabolic needs of the rapidly growing embryo during gastrulation. Note that at stage 6 primitive streak is initiated at *Koller's sickle* and develops forward as far as the future anterior terminus of the notochord, i.e., at future neuromeres r0–r1. In so doing, the length of primitive streak is likely determined by a nested homeotic code within the epiblast that controls how far forward the primitive streak is permitted to extend (Fig. 1.19).

Accepted dogma about axis formation holds that primitive streak is the responsible event for determination of anterior-posterior, dorsoventral, and mediolateral relationships. In point of fact, axis determination may occur much earlier, at stage 4. Why should cells of the original blastoderm become asymmetrical? Specifically, why should a deeper layer of cells persist at the future posterior pole of the conceptus? What relationship might this have with the development of a homeotic gradient leading to the ultimate induction of the neuromeric segmentation system?

Let's assume for the moment that the entire blastoderm consists of homogeneous cells which are pluripotent, capable of expressing the entire library of homeotic genes. Then let's divide this population into two groups, cells that express a molecular marker, the HNK1 epitope, exhibiting a temporal and spatial distribution that can be related closely to the morphogenesis of tissues involved in the establishment of the craniocaudal axis. Blastoderm cells "choose up sides" with some being HNK1+ and the others being HNK1−. We can represent the positive cells as squares and the negative cells as triangles. A unidentified factor favors the accumulation of HNK1+ at the posterior pole, making the system instantly nonrandom. It could be that cell death of HNK1+ takes place centrally, but that the accumulation simply reflects their enhanced survival posteriorly. Or perhaps, migration of HNK1+ takes place. In any case, a deep layer of cells persists and the overlying single-layer epiblast is now a nonrandom structure, with electrochemical gradients both the AP and mediolateral dimensions. Further research will clarify these issues.

As we move to stage 5, the epiblast cells are pseudostratified columnar with apical microvilli facing the amniotic cavity and a definite basal lamina. Laterally, the epiblast cells of the embryo give way to extraembryonic epiblast cells that

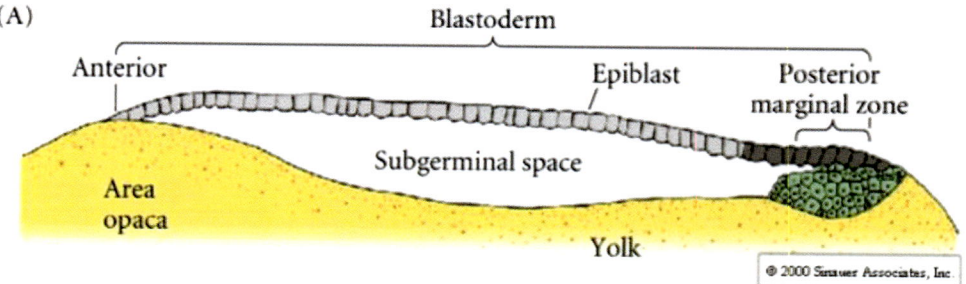

Fig. 1.17 Formation of the hypoblast and preparation for gastrulation. (**a**) In stage 4, cells of the blastoderm absorb water and the secrete it, causing a separation of blastomerm from the yolk, the subgerminal space. In this process, the deep cells of the blastoderm die creating a monolayer epiblast (light gray). This is known as *zona pellucida*. At the posterior pole of chick blastoderm, the deep cells persist, creating a thickening known as the *zona opaca*. This *posterior marginal zone* (PMZ) is thus bilaminar, having both primitive ectoderm (dark gray) and primitive endoderm (green). The anterior border of PMZ appears like a C-shaped sickle and is termed *Koller's sickle*. Some of the deep cells (green) persist and are physically interposed between the PMZ and the yolk. These cells will give rise to the secondary hypoblast. The C-shaped configuration of the PMZ between the superficial cells and the deep cells is known as *Koller's sickle*. It is from this point, during stage 6, that the primitive will originate. (Reprinted from Gilbert SF, Barressi MJF. Developmental Biology, 11th ed. Sunderland, MA: Sinauer; 2016. Copyright © 2016. Oxford Publishing Limited. Reproduced with permission of the Licensor through PLSclear)

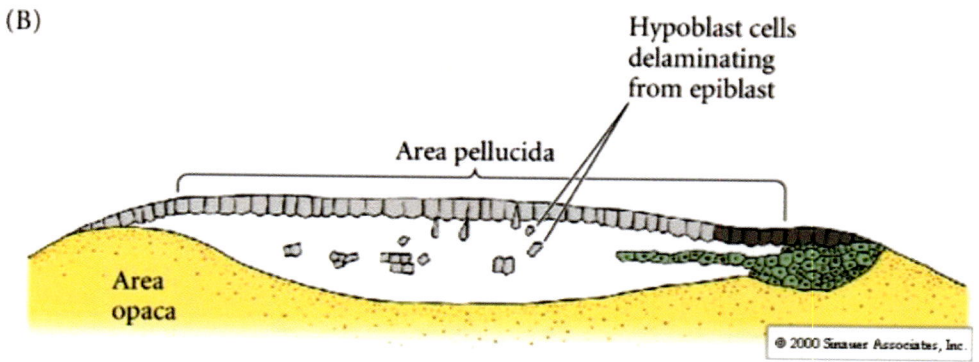

Fig. 1.18 Formation of the hypoblast and preparation for gastrulation. (**b**) Later in stage 4, two events take place simultaneously: (1) cells of epiblast become transiently less adherent to each other. This "leaky" state permits epiblast cells of the zone pellucida delaminate and fall into the subgerminal space, where they will link up again to form the *primary hypoblast* (light gray). (2) Cells below the zona opaca, the poste-rior marginal zone, proliferate forward beneath the epiblast, forming the *secondary hypoblast* (green). (Reprinted from Gilbert SF, Barressi MJF. Developmental Biology, 11th ed. Sunderland, MA: Sinauer; 2016. Copyright © 2016. Oxford Publishing Limited. Reproduced with permission of the Licensor through PLSclear)

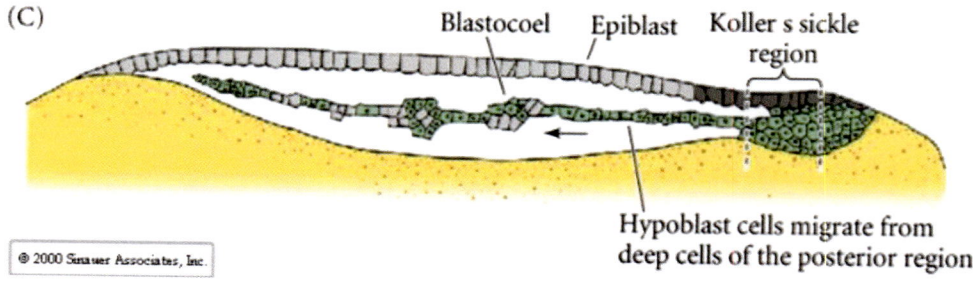

Fig. 1.19 Formation of the hypoblast and preparation for gastrulation. (**c**) By the end of stage 4, primary hypoblast (light gray) and secondary hypoblast (green) unite to form a single layer, thus creating the *bilaminar embryo*. In stage 5, hypoblast (primitive endoderm) will migrate outward to line the entire blastocoele and create the extraembryonic mesoderm that will form the placental circulation required to support the metabolic needs of the rapidly growing embryo during gastrulation. Note that primitive streak in stage 6 is initiated at *Koller's sickle* and develops forward as far as the future anterior terminus of the notochord, i.e., at future neuromeres r0–r1. In so doing, the length of primitive probably follows a nested homeotic code the epiblast that controls how far forward primitive streak can extend. (Reprinted from Gilbert SF, Barressi MJF. Developmental Biology, 11th ed. Sunderland, MA: Sinauer; 2016. Copyright © 2016. Oxford Publishing Limited. Reproduced with permission of the Licensor through PLSclear)

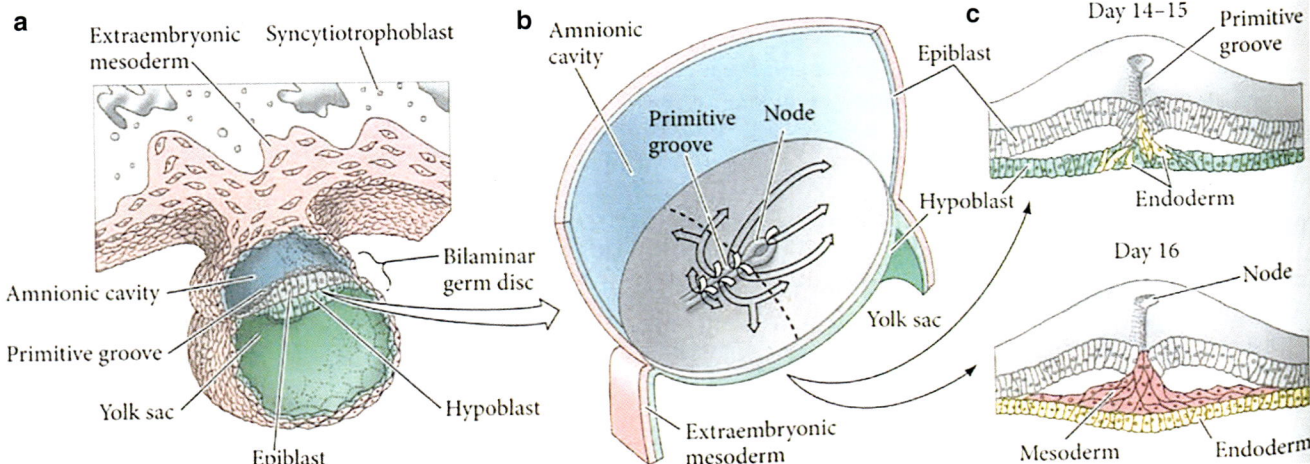

Fig. 1.20 Bilaminar to trilaminar embryo. (**a**) At end of stage 5, implantation is complete. The conceptus, consisting of amniotic cavity, the bilaminar disc, and the yolk sac, is connected to uterus via syncytiotrophoblast (light gray). Note that the external wall of the embryo, cytotrophoblast, not depicted here so that we can see better the extraembryonic mesoderm. Hypoblast (green) has spread outward from the confines of the germ disc to line the entire yolk sac cavity. Hypoblast, extraembryonic endoderm (EEE), gives rise to extraembryonic mesoderm (pink) which is interposed between the inner wall of the yolk sac (green) and the outer wall of the yolk sac, the cytotrophoblast (not shown). Extraembryonic mesoderm can then track all around the embryo, arcing over the roof of the amniotic cavity made from extraembryonic ectoderm (blue). All tissues that are pink are capable of forming blood vessels. Therefore, EEM makes the vascular connection with the trophoblast to form the placenta. (**b**) Gastrulation begins at stage 6 (13–18 days). Extrembryonic mesoderm (pink) can be seen tracking around extraembryonic endoderm (green) lining the yolk sac and extraembryonic ectoderm (blue) lining the amniotic cavity. Primitive streak develops at Koller's sickle and extends forward about 2/3 of the length of the embryo. (**c**) Cell migration occurs in two waves. Epiblast cells most proximal to primitive streak form definite intraembryonic endoderm (yellow), displacing hypoblast completely out from beneath the embryo. 1–2 days later, intraembryonic mesoderm (red) is interposed and spreads outward and forward as shown. Remaining epiblast is now called ectoderm proper. (Reprinted from Schoenwolf GC, Bleyl SB, Brauer PR, Francis-West PH (eds). Larsen's Human Embryology, 5th ed. Philadelphia, PA: Churchill Livingstone; 2014. With permission from Elsevier)

define the amnion. These cells are columnar. The reader is referred to Fig. 1.20 in which all tissues are color-coded (Fig. 1.20).

The stage 5 hypoblast spreads outward following the inner surface of the trophoblast and thus comes to line the entire cavity of the blastocyst. We can therefore distinguish two types of hypoblast, each of which has specific roles in development. The *visceral hypoblast* lies subjacent to the embryo proper. The cells are cuboidal. The surface they present to the blastocyst cavity is uniform with apical microvilli but no basal lamina. It is responsible for the formation of the midline primitive streak in stage 6, the signal event of gastrulation. Removal of hypoblast can result in multiple embryonic axes. In addition, visceral hypoblast is required for induction of the head region and for specification of germline cells. The *parietal hypoblast* forms the inner wall of the blastocyst and is histologically distinct, being composed of squamous cells. It quickly secretes a new layer of cells, the *extraembryonic mesoblast*, between it and the wall of the trophoblast. This results in a two-layer structure, the *primary yolk sac*.

Note that the term *mesoblast* is used until such time as embryonic tissues are in their correct final position. This layer is also called extraembryonic mesoderm.

Extraembryonic mesoblast proliferates wildly tracking upward along the confines of the trophoblast wall and therefore passing over the roof of the amniotic cavity. It is thus interposed beneath the cytotrophoblast of the future placenta. Recall that blood vessels are produced from mesoblast, wherever it is located. All tissues colored pink in Fig. 1.20 and angiogenic. At stage 5, small "fingers" of cytotrophoblast are inserted into the syncytiotrophoblast, but the primary villi do not have a mesodermal core.

Gastrulation begins at stage 6 (13–18 days). Extrembryonic mesoderm can be seen tracking around extraembryonic endoderm lining the yolk sac and extraembryonic ectoderm lining the amniotic cavity. Primitive streak develops at Koller's sickle and extends forward about 2/3 of the length of the embryo. Cell migration occurs in two waves. At day 14, epiblast cells most proximal to primitive streak form definite intraembryonic endoderm, displacing hypoblast completely out from beneath the embryo. At day 16, ingression of intraembryonic mesoderm (red) takes place; it spreads outward and forward. By the end of stage 5, extraembryonic mesoderm has penetrated the overlying cytotrophoblast "fingers" to create a vascular core. This marks the beginning of maternal-embryonic circulation. Having made its contribution of cells to create the remainder of the embryo, the epiblast can rest on its laurels and is now called ectoderm proper.

Gastrulation takes place during embryonic stages 6 and 7 as the bilaminar embryo develops three germ layers.

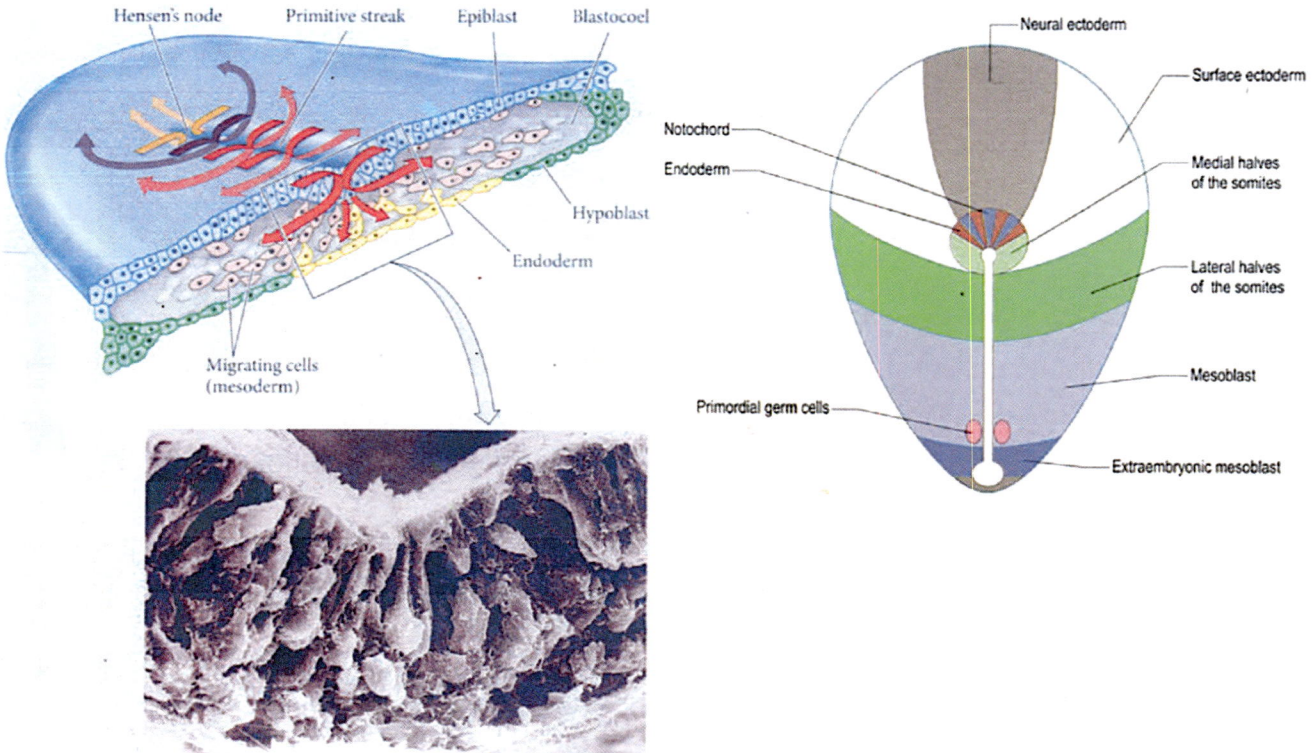

Fig. 1.21 Gastrulation proceeds in a cranial to caudal fashion (here depicted left-to-right). Cells are squeezed like toothpaste into the primitive streak. Cells entering first form the endoderm, here shown pushing the hypoblast out of the way. Cells migrating between the epiblast and the endoderm form the mesoderm. Those cells closest to the streak (1) will be first to enter and, once inside the bilaminar structure, will be pushed most laterally (1'). Cells further away from the streak (2) will enter later in time and, once inside, will remain more medially (2'). When gastrulation is complete, the epiblast becomes the definitive ectoderm. Once in their final position, mesodermal cells are patterned by chemical signals from either the ectoderm or endoderm with which they are in contact. These signals are region-specific, i.e., unique combinations of genes are expressed in specific anatomic sectors of the trilaminar embryo. Gene products causing mesoderm 1' to become lateral plate mesoderm originate at a finite distance from the midline. The organization of these anatomic sectors is neuromeric and organized around the homeobox genes. The hox code at level t1 is different from those of level t2 or t3. (Reprinted from Gilbert SF, Barressi MJF. Developmental Biology, 11th ed. Sunderland, MA: Sinauer; 2016. Copyright © 2016. Oxford Publishing Limited. Reproduced with permission of the Licensor through PLSclear)

Gastrulation proceeds in cranial-caudal order. As soon as mesoderm is produced, it becomes organized in accordance with the neuromere with which it is in genetic register. This anatomy will be the subject of the subsequent chapter. For now, we shall follow the events that create developmental segmentation of paraxial mesoderm (Fig. 1.21 Gastrulation, Fig. 1.22 Erich Blechschmidt, and Fig. 1.23 Blechschmidt hypothesis).

Segmentation as a concept is highly intuitive. Repetitive functional units are observed in the centipede, in the multiple ribs of the snake, and in the dermatomes of the human thorax. All vertebrates possess the fundamental elements of a vertebral body, muscle units attached thereto, sensorimotor nerves arising at that same level to innervate those muscles, and well-defined geographic zones of skin. All the future spinal cord and all vertebrate embryos have primitive organizational blocks of mesoderm called somites [41–58]. Each somite is preceded by an earlier, incompletely segmented structure, called a *somitomere* (vide infra). These flank the neural tube from the cranial base to the tail. Each vertebra at a given neuromeric level is the combination of the caudal half-somite from the neuromeric level above immediately rostral to it and the cranial half-somite at that same neuromeric level. *Human embryos possess 42–44 pairs of somites*: 4 occipital somites contribute to the posterior brain case. These are followed by 8 cervical, 12 thoracic, 5 lumbar, 5 sacral, and 8–10 coccygeal somites. The rostral first cervical somite contributes to the foramen magnum and its caudal half to the atlas. For this reason, there are eight cervical nerves but only seven visible vertebrae.

Given the highly regular organization of the spinal cord, it seemed logical to anatomists of the nineteenth century that a similar pattern might exist in the CNS as well. Darwin's concepts of evolution were bolstered by tremendous progress in paleontology and comparative anatomy. As the anatomic logic of vertebrate structure became defined, it was natural

Fig. 1.22 Eric Blechschmidt (1904–1992). German anatomist founder of the embryology collection at Göttingen University. His theory of development based on *morphogenetic fields* anticipated later developments in molecular biology…and the work you are reading at present. (Courtesy of Dr. Christoph Viebahn Curator of the Blechschmidt Collection, University of Göttingen, Germany)

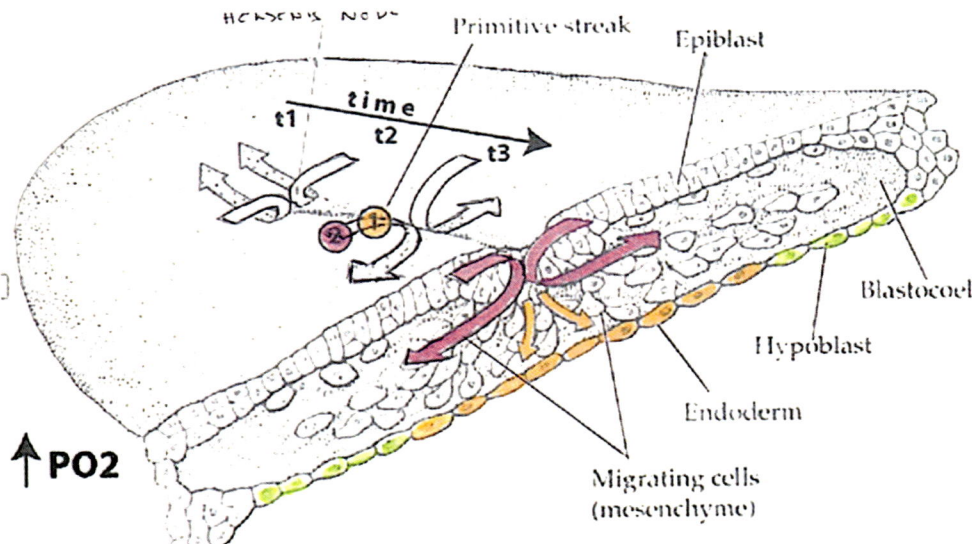

Fig. 1.23 Gastrulation (Blechschmidt hypothesis): The two-layer embryonic disc is attached at its posterior pole (left). The dorsal epiblast will eventually form the entire embryo. The ventral hypoblast is derived from "dropout" of epiblast cells into the yolk sac. These former epiblast cells then coalesce to form a temporary layer (also called the primitive endoderm) designed to permit gastrulation. Blood supply is greatest at the periphery via the vitelline (umbilical) veins. High peripheral p02 leads to high local mitotic rate. (This drives the epiblast cells toward the midline, just like toothpaste squeezed from a tube. The primitive groove allows epiblast cells at its margin to be pushed into it. The first cells to enter push the hypoblast aside to form the definitive endoderm (upper inset)). Cells entering subsequently intercalate themselves between the epiblast and the endoderm. This intermediate layer is the mesoderm (lower inset). Cells of the epiblast that do not participate in gastrulation form the ectoderm. All three primary germ layers are now present. Gastrulation proceeds from cranial to caudal. Homeobox genes expressed in varying combinations along the axis of the embryo define a segmental (neuromeric) organization at the time of gastrulation. Cells ingressing at a particular level will be assigned a hox code specific for that level. Segmentation of cells at gastrulation into genetically distinct anatomic levels provides the basis for the future organization of the embryo. (Reprinted from Gilbert SF, Barressi MJF. Developmental Biology, 11th ed. Sunderland, MA: Sinauer; 2016. Copyright © 2016. Oxford Publishing Limited. Reproduced with permission of the Licensor through PLSclear)

for scientists to look for commonalities in development as well. Neuroanatomists had noted the presence of small bulges on the surface of the embryo at the hindbrain region; these seemed to correlate in a regular way with cranial nerves. They also appeared to relate in some manner to the occipital somites and, rostrally, to the pharyngeal arches. These developmental zones of the rhombencephalon were termed rhombomeres, but their biologic rationale was uncertain. Furthermore, the physical presence of neuromeres appeared to die out at higher levels of the neuraxis.

Elaborate attempts were made by comparative anatomists to understand the organization of the head based on somites. In the 1930s, this culminated in the magnum opus of Goodrich based on fishes emphasizing dorsoventral relationships between mesoderm and cranial nerves [59]. This comparative approach was applied to the bones of the skull by Sir Gavin de Beer and later, in 1980, by Jarvik [60, 61]. Given the limitations of the data, progress in explaining the anatomy of the head and neck remained in a state of gridlock. As so often is the case in science, new technologies would be required before dramatic advances in knowledge could be made, and a new model, based on neuroembryology and genetics, would emerge.

The advent of scanning electron microscopy opened a new window on the morphology of development. Exhaustive work by Hinrichsen described the external development of the face [62]. Within the neural tube, SEM proved the existence of rhombomeres as segmental diverticula of the lateral wall. The boundaries of these neuromeres and their specific anatomic content were confirmed using contrast injections and immunofluorescence (Figs. 1.24 and 1.25) At the same time, Meier and Jacobson observed a somite-like organizational pattern of paraxial mesoderm (that mesoderm closest to the neural tube) in a wide variety of vertebrates [63–70]. In mammalian embryos, seven of these incompletely separated masses termed somitomeres were noted (Figs. 1.26 and 1.27). At the level of the eighth somitomere, a completely separate somite surrounded by an epithelial coat was noted. This is termed "the first occipital somite;" four such somites were documented rostral to the first cervical somite [71–73].

The anatomy of a somitomere consists of a whorl of PAM cells surrounding a central lumen. Though lacking the defined structures of a somite (dermatome, myotome, and sclerotome), the mesoderm of a somitomere has an intrinsic spatial orientation. Distal somitomeric PAM probably represents cells ingressing early at the neuromeric level during gastrulation, while PAM nearer the midline represents later-arriving cells. This spatial organization means that different PAM derivatives may come from distinct "compartments" of the somitomere. Each compartment may be characterized by the expression of different genes or by varying degrees of expression of the same gene. Alternatively, these differences

Fig. 1.24 Klaus Hinrichsen (1927–1997) was chair of Anatomy and Embryology at Ruhr Unversität Bochum. His was the first work to demonstrate the stages of human facial development using scanning electron microscopy. (Courtesy of Ruhr University Bochum)

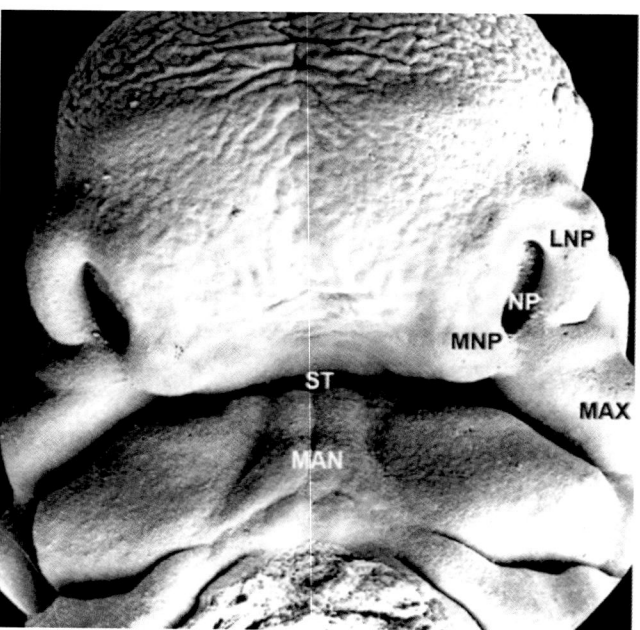

Fig. 1.25 Stage 14 embryo shows with startling detail the stomodeum and first arch. Note the tongue swellings interposed between the mandibular processes. (Courtesy of Ruhr University Bochum)

may be translated into a specific migration sequence of PAM into its pharyngeal arch "target."

Somitomeres provided a revolutionary new construct for the embryogenesis of the head. For the first time, an organization of mesoderm was observed that correlated with the previously described model of pharyngeal arches.

Fig. 1.26 Segments 1–7 are cranial somitomeres; they do not condense into somites, but rather expand as the brain parts with which they are associated expand. Segment 8 is the first somite and appears just caudal to the otic (ear) ectoderm. Segments 8–11 are the occipital somites of the head. (Reprinted from Packard DS, Meier S. An Experimental Study of the Somitomeric Organization of the Avian Segmental plate. *Devel Biol* 1983;97 (1):191–202. With permission from Elsevier)

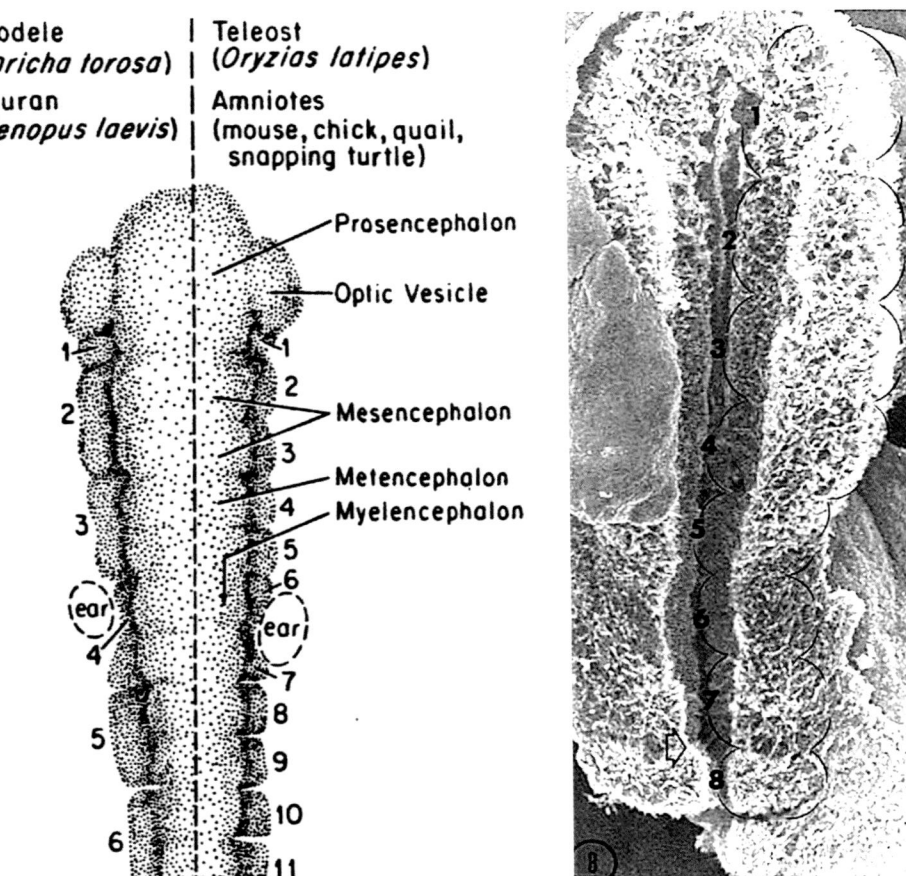

Fig. 1.27 Somitomeres in the segmental plate of *Medaka*, the Japanese rice fish (Orizias latipes) prior to conversion into somites. Recall that the segmental plate refers to the unorganized PAM posterior to the most caudal somite. It first segments into somitomeres and then into somites. The structure of somitomeres is primitive, consisting of a whorl of cells surrounding a central cavity, or somitocoele, seen here. (Reprinted from Martindale MQ, Meier S, Jacobson AG. Mesodermal Metamerism in the teleost: *Oryzias latipes* (the Medaka). *Journal of Morphology* 1987;193:241–252. With permission from John Wiley & Sons)

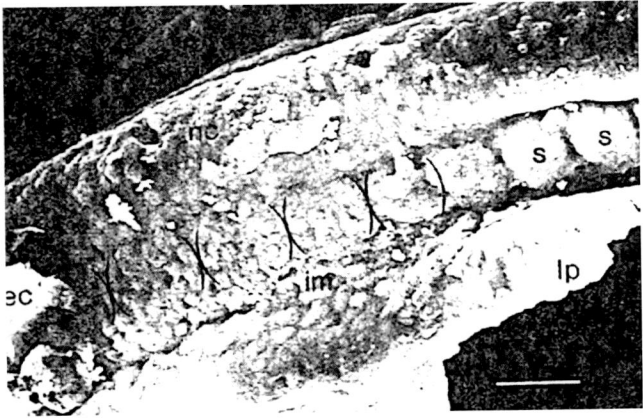

tomic relationships existing among the neuromeres of the medulla, the neural crest cells corresponding to each neuromeric level and the zones of outlying mesoderm innervate by those neuromeres and populated by those neural crest populations. The mesoderm of the human head was conceived as having 11 somitomeres divided into two distinct zones. The first 7 belong to what was termed *head mesoderm*. The next four somitomeres constitute the *segmental plate*. Segmental plate somitomeres go on to condense into somites while head somitomeres do not undergo a condensation. The transitional nature of the eighth somitomere is quite apparent.

The organization of paraxial mesoderm is tightly regulated pointing to the existence of a so-called "cellular clock", with approximately four somitomeres being produced per day until all PAM is organized. Just at the time at which the 19th somitomere makes its appearance, the first somite transformation takes place at the eighth somitomere. A separation distance of 11 somitomeres between the last somite and the newest somitomere is maintained (Figs. 1.28 and 1.29).

A warning: readers who repair their textbooks of descriptive embryology and read about the transformation sequence of somitomeres to somites may be confused by the numbering system used in this paper. Many illustrations are based upon the avian 5-occipital somite model. Mammals have 4 occipital somites. Second, the number of rhombomeres

Somitomeres form in register with the rhombomeres of the hindbrain. As we shall see, the advent of genetic mapping to the neural plate provided a means to understand the ana-

Fig. 1.28 Head mesoderm undergoes two forms of segmentation. From Hensen's node backwards, from levels r0–r1 to r7, paraxial mesoderm segments into 7 pairs of *incompletely* separated somitomeres (circles). Caudal to Sm7, new somitomeres (ovals) continue to appear at the rate of about 4 per day. At the formation of the 19th somitomere, the eighth somitomere undergoes conversion into the first somite (square). beginning with Sm8. Total somites in humans are 4 occipital, 8 cervical, 12 thoracic, 5 lumbar, 5 sacral, and approximately 10 coccygeal—but most of these latter somites degenerate, leaving a total count of 35–38. (Reprinted from Carlson BM. Human Embryology and Developmental Biology, 5th edition. Philadelphia, PA: Elsevier Saunders; 2014. With permission from Elsevier)

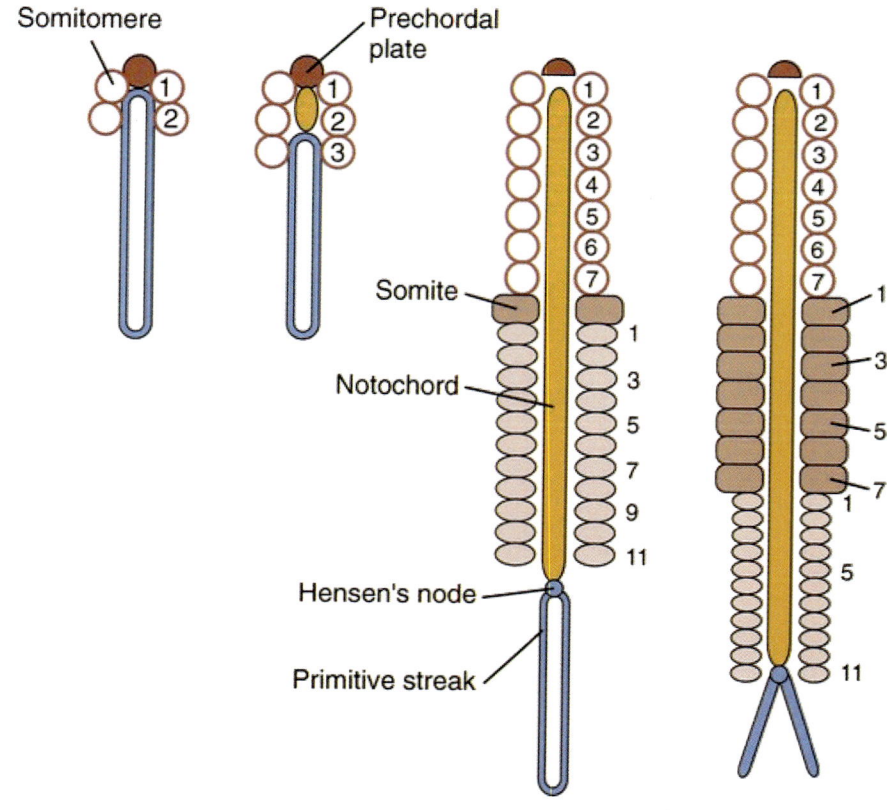

Fig. 1.29 Cranial end of embryo is to the left. On the left are seen three fully separate somites. The fourth somites is beginning to separate itself from the remaining segmental plate. (Reprinted from Gilbert SF, Barressi MJF. Developmental Biology, 11th ed. Sunderland, MA: Sinauer; 2016. Copyright © 2016. Oxford Publishing Limited. Reproduced with permission of the Licensor through PLSclear)

described in most papers is only eight, with the eighth one being depicted as quite large. It remained for Puelles (vide infra) to describe the eighth rhombomere as broken up into four individual segments matching the cranial nerve anatomy and foramina of the skull; these he called "pseudorhombomeres." With the accumulation of further evidence, the "pseudo" prefix has been dropped. Rhombomeres 8–11 are in direct genetic register with the occipital somites. Thus, readers studying over diagrams of somitomere-somite transformation will note the appearance of the first somite at Sm8 matched that of the 12th somitomere because a 4-occipital somite model is being used (Figs. 1.28 and 1.29 Carlson).

What happens in somitogenesis? A somitomere becomes a somite when it displays complete separation from its neighbors, when its outer layer of cells becomes an epithelium, and when it displays the three primary subunits of sclerotome, myotome, and dermatome. The first such transformation (that of the eighth somitomere to the first occipital somite) is externally difficult to recognize. The rostral half of the eighth somitomere is incompletely separated from the seventh somitomere. It behaves just like a somitomere. In contrast, the caudal half of the eighth somitomere undergoes a complete somitic transformation. The back end of the resulting first somite is thus completely separate from the rostral aspect of the second somite (ninth somitomere). Remaining somites are easily distinguished from one another (Fig. 1.30 Somite generation KW Tosney).

What are the unique characteristics of the 4 occipital somites? A brief digression is required here because recent cell labeling studies have conclusively demonstrated that somitogenesis begins at the eighth somitomere but in a subtle manner. The reader will recall that a typical somite is transformed from a cuboidal mass of mesoderm into bone, muscle, and dermis. In development, antecedents of these tissues within the somite occur as transient structures: the sclerotome, myotome, and dermatome. The *sclerotome* produces the chondral bone of the cranial base (basioccipital, exoccipital, and supraoccipital bones) and the vertebrae. The *dorsal dermatome* and *dorsal myotome* produce, respectively, the dermis and paraspinous muscles supplied by the posterior (dorsal) ramus of the spinal nerve. These are known as *epaxial* structures. The dermis of the remainder of the body wall and the extremities and their corresponding muscles are all *hypaxial* structures supplied by the anterior (ventral) ramus of the spinal nerve. The muscles arise from the *ventral myotome*. The dermis arises from lateral plate mesoderm.

In addition, the anatomic organization of occipital somites differs from that of all other somites. Although they have sclerotomes and ventral myotomes, they possess neither dorsal myotomes nor dermatomes. A dorsal myotome appears for the first time in the first cervical somite. This produces the muscles of suboccipital triangle. A true dermatome does not appear until the second cervical somite. For this reason, *there is no skin innervated by C1*. Although C1 is well-described in amphibians, in mammals it appears only transiently, providing neither sensory nor motor innervation in the mature state [74]. C1 is well-defined in amphibians. The dorsal branch of nerve C2 supplies the skin over the retroauricular skin, while the ventral branch of C2 supplies the mastoid and the rostral anterior triangle of the neck as part of the cervical plexus.

For these reasons, documentation of occipital somite anatomy requires a molecular, rather than a strictly morphologic, approach. When these structures are considered in terms of their associated rhombomeres, the anatomic pattern becomes quite clear. It is not surprising that the migration patterns of craniofacial muscles derived from somitomeres and somites are very different (Fig. 1.30).

Somitomeres and occipital somites follow similar trends in muscle formation [75–81]. Pharyngeal arches are hypaxial structures, the muscles of which are organized into two layers, deep and superficial. In this model, internally placed myoblasts of Sm7 produce the musculature of the soft palate, while externally placed myoblasts form stylopharyngeus and superior constrictor. Occipital somites make the muscles of the tongue from the *internal plane* of their myotomes. Sm8 (the first occipital somite) would be expected to produce a tongue muscle associated with the Sm7 palate; this is palatoglossus. Labeling studies show that the sternocleidomastoid and trapezius muscles originate from the *external plane* of the occipital somites. These two muscles logically represent the superficial plane of all four occipital myotomes. Both of these muscles originate from the osseous product of the Sm7, the mastoid process. Neuromeric theory allows for an accurate assignment of muscle origins according to their somitomere/somite of origin (based on the location of the motor nerve within the neuraxis). This approach to origin and insertion is rational and consistent, though in many cases it proves to be the reverse of that described in traditional textbooks of anatomy. We shall cover this subject in Chap. 9.

Comparative anatomy has much to contribute to our understanding of the significance of the r7/r8 transition zone. Fish brains are small. The positions of the roots of cranial nerve XI and the hypoglossal lie *outside* the piscine skull. These nerves can reach their targets without traversing the skull. The expansion of the vertebrate skull associated with tetrapods in the posterior fossa incorporated these nerves into the jugular foramen and four occipital foramina (subsequently reorganized in mammals into a single occipital foramen). This points to the transformation of somites 1–4 from a truncal to a cranial fate.

How Segmentation of the Mesoderm Matches that of the Neural Tube

Scanning EM studies in chicks by Meier in 1980 demonstrated that the segmentation of the somitomeres was also in register with segmentation of the intermediate mesoderm and lateral plate mesoderm (the somatic lamina) [65, 82]. Before continuing on, readers should note the following definitions having to do with varying regions of mesoderm. All forms of mesoderm come into being during gastrulation. The latter term refers to the process by which populations from a single layer of cells, the *epiblast*, rearrange themselves to form a three-layer "sandwich" consisting of ectoderm, mesoderm, and endoderm. A pear-shaped embryonic disc results; in the center of the disc an axially oriented neural plate develops. The ectoderm then becomes divided into two zones, neural ectoderm (the future nervous system) and surface ectoderm, the future epidermis of the body. The edges of the flattened neural plate roll up into cigar-like neural folds. These will ultimately contact each other in the dorsal midline seal up, forming the neural tube. This process is called *neurulation* [83, 84] (Fig. 1.31).

At the time of formation of the trilaminar embryonic disc (and before neurulation), mesoderm spreads out laterally to form various zones [85]. Those mesodermal cells closest to the neural tube and notochord are called paraxial mesoderm (PAM). PAM starts at the rostral tip of the notochord and is organized into somitomeres (and somites) all the way down to the tail. Mesoderm migrating further laterally stays flat; it

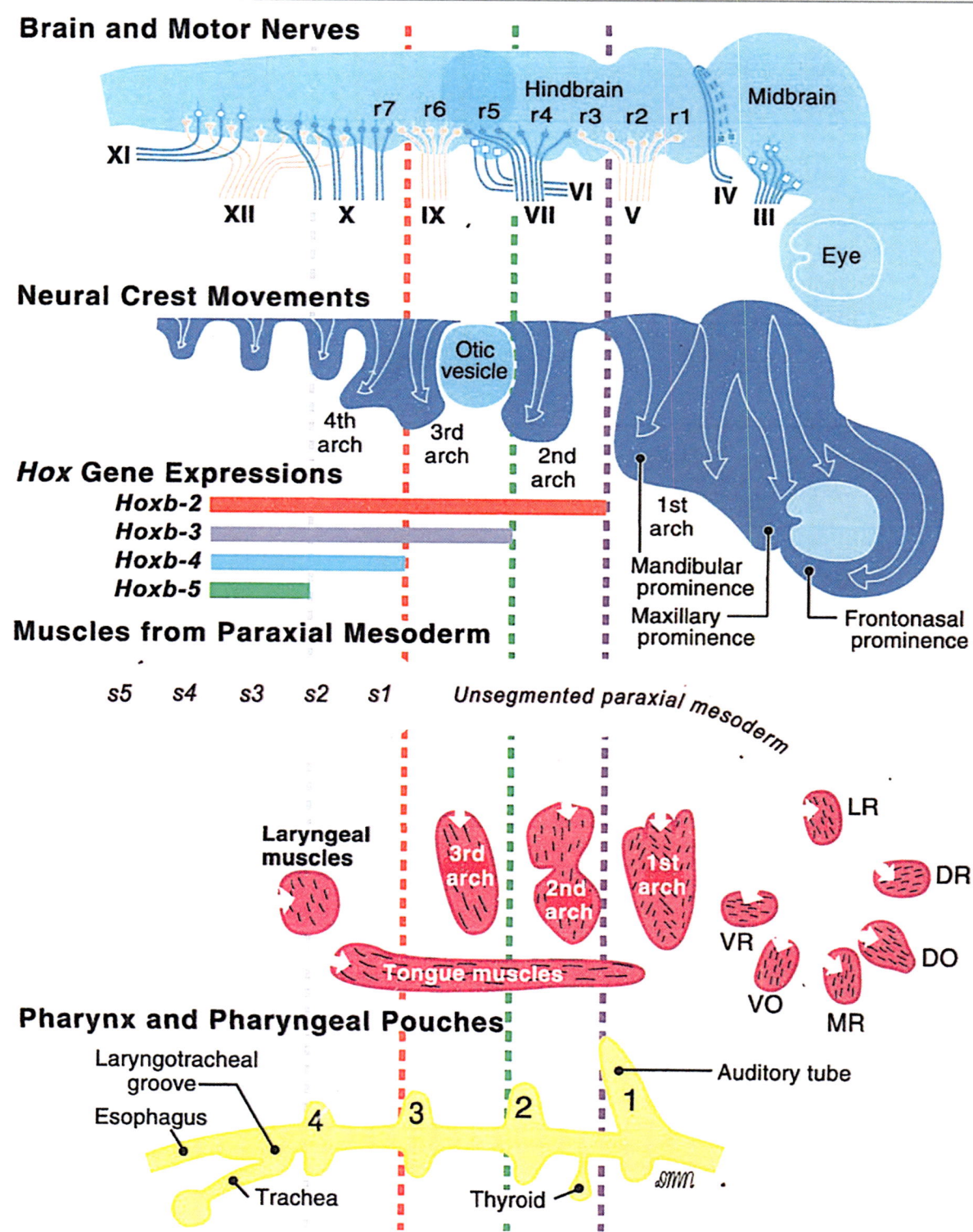

Fig. 1.30 Adapted from Map shows the origins of craniofacial muscles and neuromeric innervation of the pharyngeal arches. Note r6–r7 supply third arch. Fourth arch and fifth arch are supplied, respectively by r8–r9 and by r10–r11. Note the possibility of six head segments, the premandibular segment (not shown) being rostral to the first arch. The sixth segment consists of mesoderm from Sm10 and Sm11, with derivatives formed prior to their conversion to somites 3 and 4. Note that hox genes apply to neuromeres r3 and caudal while neuromeres r2 and cranial are specified by non-hox homeotic genes. Note coding for third arch is r6–r7, whereas fourth arch is r8–r9. Cranial nerve X, although dedicated to the fourth and fifth arches, has representation in r7 and therefore supplies muscles of the third arch. Just a first head segment has three different motor nerves, so fourth head segment is supplied by both IX and X. (Reprinted from Noden D. Relations and interactions between cranial mesoderm and neural crest populations. *J Anat* 2005; 207:575–601. With permission from John Wiley & Sons)

Fig. 1.31 Neurulation takes place in four steps. (**a**) Notochord develops with gastrulation process in early stage 7 (18–19 days). Notochordal signal induces thickening in a midline swatch of surface ectoderm running the length of the embryo. (**b**) The initial shape of the neural plate is that of a key, but this changes in middle stage 7 (20–21 days) due to midline migration of neural plates cells which simultaneously lengthen, a process called *convergent extension*. Note neural crest (green) forming. (**c**) At 22 days, neural folds are thrown up due to *hinge joints* made from specialized cells that become narrow at the apex and broad at the base. As the neural folds approximate each other, neural crest cells separate from the neural tube and begin to migrate. Closure begins at stage 10 at the level of somite 5, the craniocervical junction, and proceeds both cranially and caudally. (**d**) The rostral neuropore closes at stage 11 and the caudal neuropore is closed at stage 12. (Reprinted from Carlson BM. Human Embryology and Developmental Biology, 6th edition. St. Louis, MO: Elsevier; 2019. With permission from Elsevier)

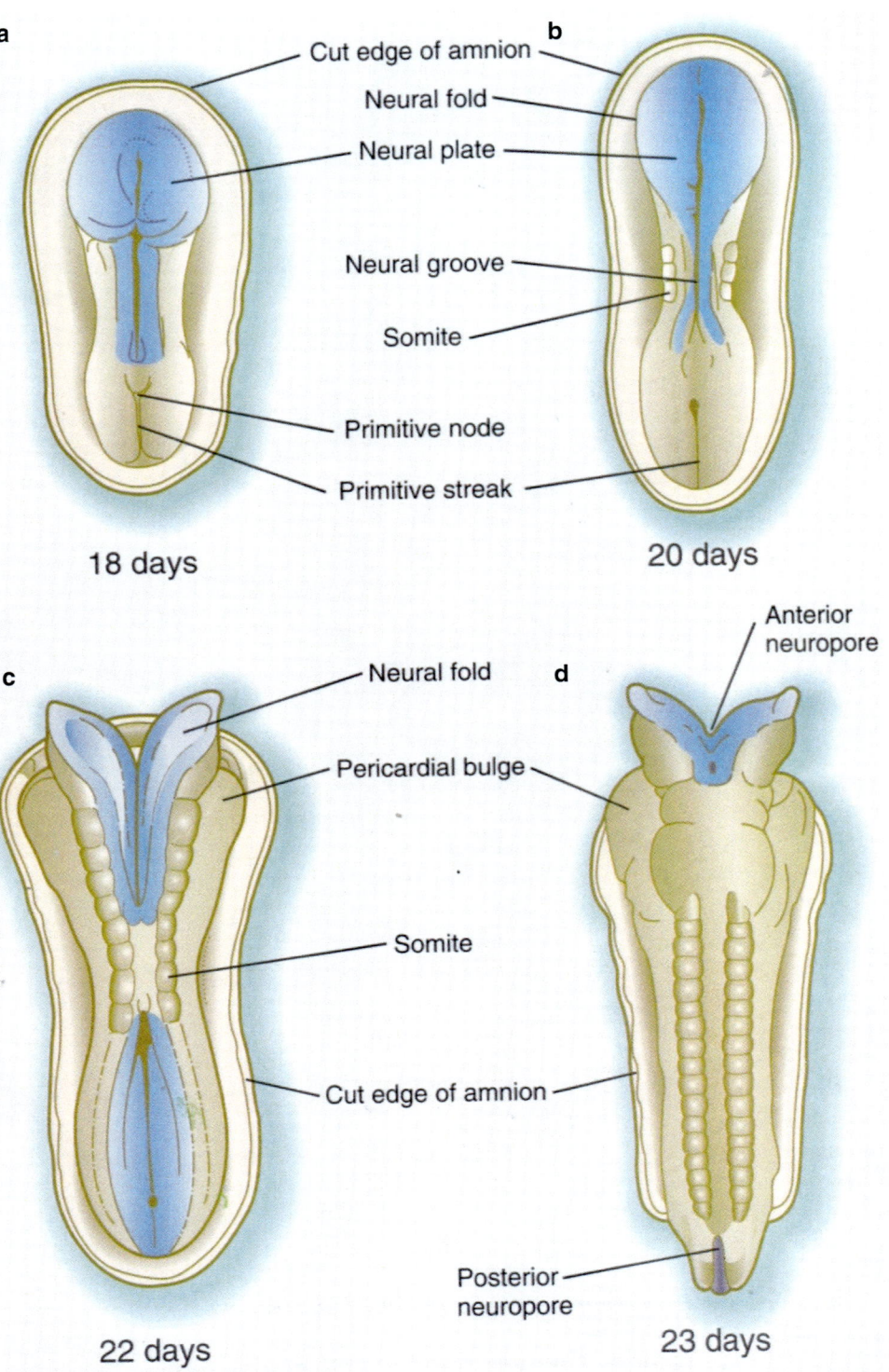

is known as the lateral plate mesoderm (LPM). LPM has a natural separation plane (the embryo will eventually take advantage of this to form the intraembryonic coelom). The outer (dorsal) lamina of the LPM is the called somatic lateral plate mesoderm (LPMs for short). LPMs is responsible for forming all the non-axial bones of the body. From the clavi-

cle on down, every bone (save the vertebrae) develops from LPMs. LPMs is overtly segmented in register with PAM. The inner (ventral) lamina of the LPM is called the visceral lateral plate mesoderm (LPMv for short). LPMv forms the mesoderm of the respiratory system and gut. It also forms the heart and blood vessels. After formation of PAM and LPM,

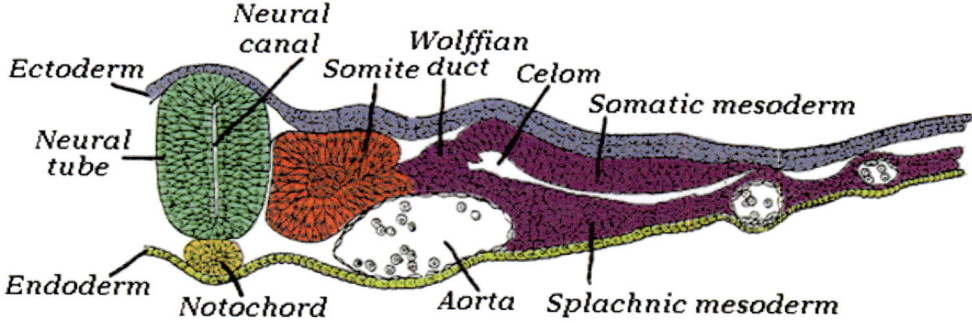

Fig. 1.32 Anatomic sectors of mesoderm. Paraxial mesoderm containing the somites is separated from lateral plate mesoderm by a column of Intermediate mesoderm (labeled here as Wolffian duct). IM is rod-like and runs from level c4 back to lumbosacral junction. It is the source off future genitourinary system. Note the intraembryonic coelom separates lateral plate into a *somatic lamina* (LPMs) dedicated to the body wall, and a *visceral lamina* (LPMv) form which forms the muscle layer of the esophagus and gastrointestinal tract. (Reprinted from Wikimedia. Retrieved from https://commons.wikimedia.org/wiki/File:Gray19_with_color.png)

an intercalated zone of intermediate mesoderm (IM) is formed; this runs down the length of the embryo as a segmented cord. IM is responsible for the production of the genitourinary system (Fig. 1.32).

Neural crest cells constitute a fourth germ layer of the embryo. These cells are found at the interface between the neural ectoderm and the surface ectoderm, i.e., atop the neural fold. In mammals, these neural cells "migrate" well before neural tube closure occurs, but after completion of gastrulation. Neural crest cells are essential to the formation of the aortic arch arteries. Since AA1 is fully formed by stage 9, the initiation of neural crest migration can be assigned to late stage 7 and is in full progress by stage 8. Note: the reader is forewarned that the concept of cellular migration may actually be a misnomer. An alternative viewpoint advocated by Anderson is that neural crest "movements" may represent local forms of cell proliferation that are physically directed by the microanatomy of their environment. In this model, the distinct manner in which neural crest populations reach their targets is a passive consequence of environmental passageways and roadblocks (vide infra).

Readers are warned here of another important revolution in biologic thinking. The germ layer system with which we are all so familiar is probably incorrect. It is certainly useful in terms of describing the organization of the vertebrate body via gastrulation and hence will probably persist in textbooks. But it is now known that a germ layer can produce cells associated with a different germ layer. For example, the hypoblast (the temporary second layer derived from the epiblast) lines the yolk sac which in turn produces the *extraembryonic mesoderm* (EEM). The intraembryonic mesoderm (IEM) produced later in time by gastrulation is in physical continuity with the EEM, but does not produce it! The endodermal lung bud emanating from the esophagus produces its own mesoderm. Thus, the mesenchyme of the lung does not result from the endoderm interacting with another (unknown) source of mesoderm, but actually is produced by the bud itself. Neural crest, of strictly ectodermal lineage, has the ability to produce structures such as fascia and bone normally considered outside the head and neck as strictly mesodermal derivatives (Fig. 1.33).

All physicians are aware that embryonic tissues seem to be organized into morphologic units called pharyngeal arches. With these definitions in mind, we can now explore how the somitomeric system fits this model. We shall then discuss the discovery of homeobox (*Hox*) genes provides the genetic basis for hindbrain segmentation. Thereafter, we shall complete our neuromeric map of the central nervous system by discussing how non-*Hox* genes define a more complex, but still logical, system of neuromeres rostral to the hindbrain. A population of neural crest is associated with each rhombomere, located at its respective sector along the neural fold. In like manner, each rhombomere is matched up with a corresponding somitomere. The exception to this is rhombomere 0. Neural crest from this neuromere interacts with prechordal plate mesoderm (PCM). In this way, the r0 neural crest provides the inferomedial sclera and the fascia for the inferior oblique, inferior rectus, and medial rectus, while the PCM provides the myoblasts (Fig. 1.30).

In humans, the paraxial mesoderm from the first 11 somitomeres and neural crest from r0 to r11 are organized into six head segments [86, 87]. The first segment is much larger than the rest. It receives mesoderm from somitomeres 1–3 and 5. It is aligned with mesomeres m1–m2 and r0–r1. The first segment does not interact with the pharyngeal arch system. It contributes to the orbit by forming all seven extraocular muscles. A new model (vide infra) posits this segment as constituting a new ***premandibular arch*** (herein abbreviated PA0). Each additional head segment is represented by a pharyngeal arch and consists of paired rhombomeres and a single somitomere (Table 1.1).

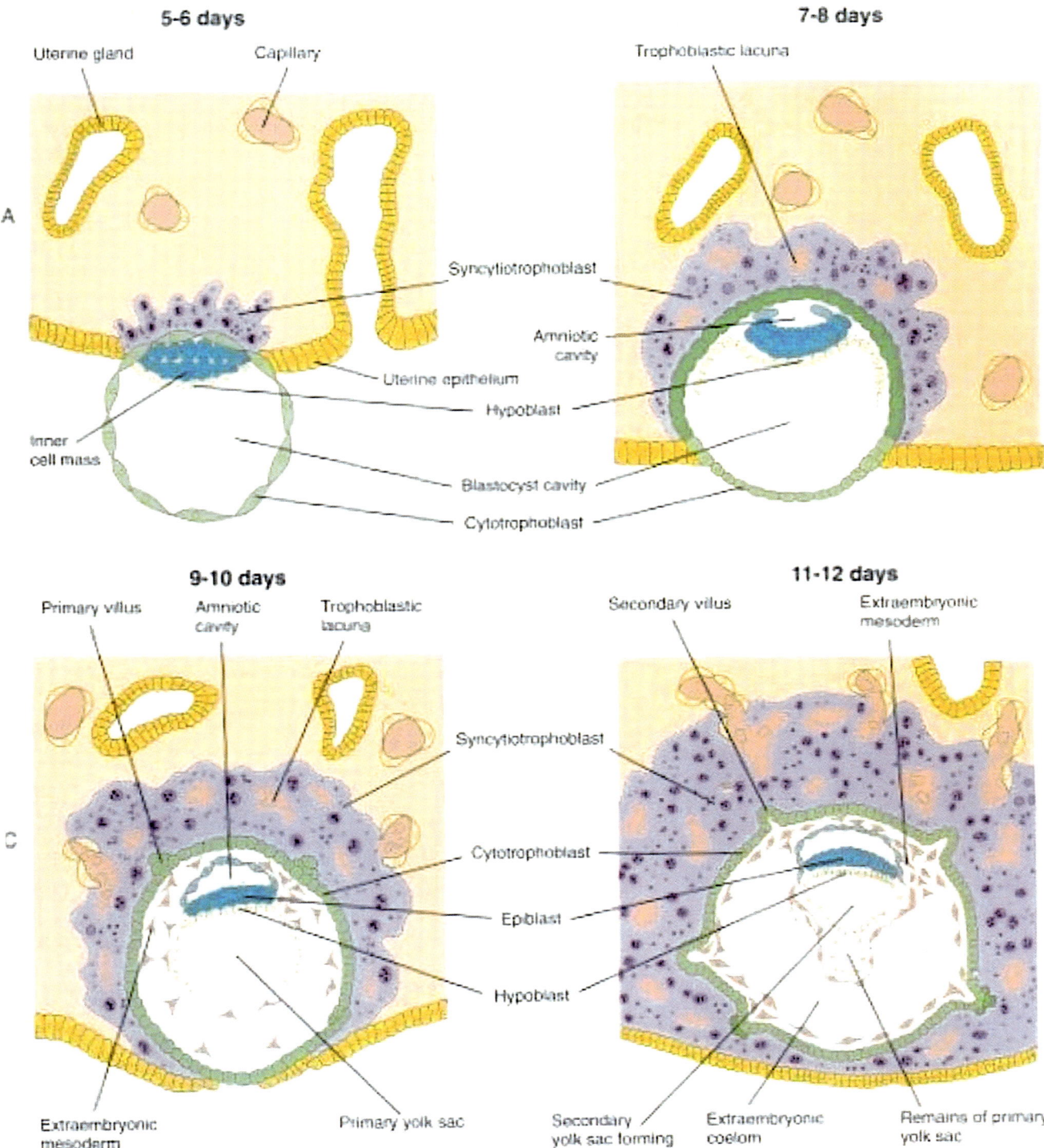

Fig. 1.33 Germ layers are not fixed in stone. One layer can give rise to another. In the stage 5 embryo, the hypoblast (yellow) appears at 5 days and, spreading along the inner surface of cytotrophoblast (green), lines the entire blastocyst cavity. By days 9–10, hypoblast which spreads around the entire embryo secretes a new layer of extraembryonic mesoderm (green). At this time, cytotrophoblast send out solid "fingers," called *primary villi*, into the surrounding synctytiotrophblast. Two days later, extraembryonic mesoderm invades the villi creating a solid core capable of making blood vessels, thus defining *secondary villi*. Another example of germ layer transformation is the productions of lung mesoderm. (Reprinted from Carlson BM. Human Embryology and Developmental Biology, 6th edition. St. Louis, MO: Elsevier; 2019. With permission from Elsevier)

Table 1.1 Anatomic content of head segments

Segment	Rhombo.	Nerve	Somitomere	Muscles
Premandibular	r0–r1	III, IV, VI	Sm1–Sm3, Sm5	Eye muscles
First arch	r2–r3	V	Sm4	Mastication
Second arch	r4–r5	IX	Sm6	Mastication, facial
Third arch	r6–r7	IX, X, XI	Sm7	Palate, sup-constrictor
Fourth arch	r8–r9	X, XI	Sm8–Sm9	Larynx, mid-constrictor
Fifth arch	r10–r11	X, XI	Sm10–Sm11	Inferior constrictor

Fig. 1.34 New scenario for jaw evolution. No fossils are known to have possessed undifferentiated branchial arches, but in all the agnathan species, the mandibular (white with solid line), hyoid (light purple), and postotic arches (purple) were already differentiated to some extent, and the lamprey embryo exhibits the Hox code similar to gnathostome embryos. In the agnathans also, the mandibular arch and premandibular regions appear to be specified by the Hox-code-default state. To obtain the jaw patterning program, Dlx code is required for differentiation of upper and lower components of the jaw, as well as the rearrangement of tissue interactions so that the oral region is specified only from the mandibular arch domain. Note the possibility of six head segments, the premandibular segment (not shown) being rostral to the first arch. The sixth segment consists of mesoderm from Sm10 and Sm11, with derivatives formed prior to their conversion to somites 3 and 4. Note that hox genes apply to neuromeres r3 and caudal, while neuromeres r2 and cranial are specified by non-hox homeotic genes. Note coding for third arch is r6–r7, whereas fourth arch is r8–r9. Cranial nerve X, although dedicated to the fourth and fifth arches, has representation in r7 and therefore supplies muscles of the third arch. Just a first head segment has three different motor nerves, so fourth head segment is supplied by both IX and X. (Reprinted from Kuratani S. Developmental studies of the lamprey and hierarchical evolutionary steps towards the acquisition of the jaw. Journal of Anatomy 2005; 207 (5): 489–499. With permission from John Wiley & Sons)

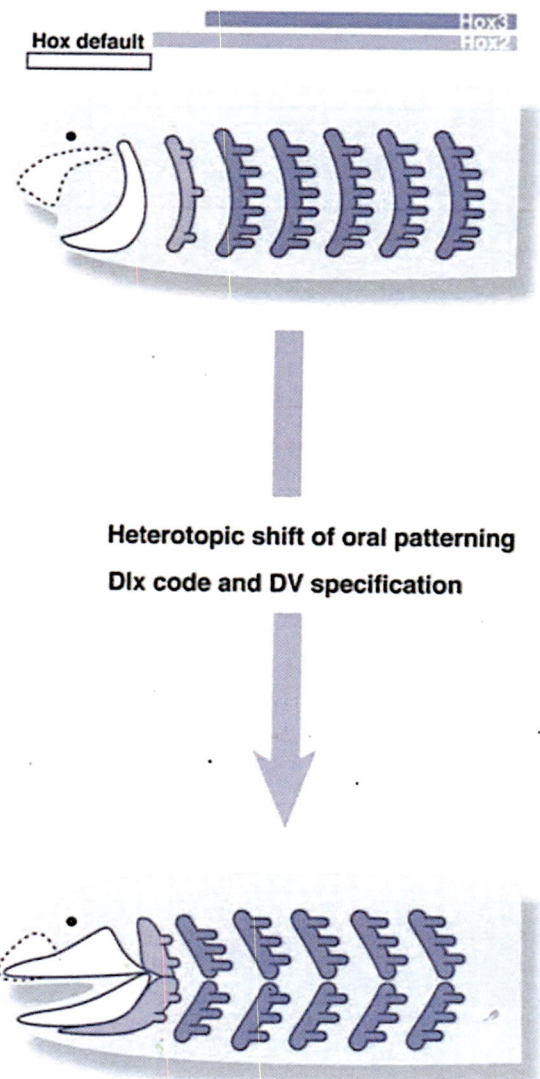

The model as referenced above was based strictly on somitomeres and is inaccurate. However, when corrected for the two rhombomeres per arch, it is plausible. The value of the model is that it is not strictly based on the arches and it takes into account mesenchyme that is dedicated to fronto-orbital structures (Fig. 1.34).

The reader will note that the arch numbering system for the pharyngeal arches differs from that of most texts. Each arch derives from two neuromeres. The ten rhombomeres of tetra-pods (r2–r11) produce five pharyngeal arches. This paper proposes the following principle: *neuromeres are never "lost" in evolution; their anatomic content is merely altered.* Every line of evidence regarding the homeobox genes encoded in the hindbrain suggests that this system is strongly conserved throughout evolution. Thus, a terminology based on a mysterious "loss" of the fifth arch while the sixth is preserved is not logical. The neuromeric approach simplifies discussion of the comparative anatomy of pharyngeal arches [88].

Note that the first two pharyngeal arches are larger and exist as a dimer (pair). The mesenchyme (neural crest + paraxial mesoderm) of the second arch becomes confluent with, and is engulfed by, that of the first arch, like two slices of bread surround a slice of cheese. In this process, the ectoderm of PA2 is reduced to a small sector of skin covering the anterior external auditory canal. The endoderm of PA2 within the oral cavity is found transiently over a central swatch of the dorsum of the tongue, but is quickly absorbed. The last three pharyngeal arches are considerably smaller. The reason that they (PA3–5) are reduced in size is that some of their somitomeric mass must be shared to form somites, whereas the paraxial mesoderm of somitomeres 2–6 is completely dedicated to the formation of arch structures.

Recall that most bone from mesenchyme originating from r0 through r7 is not from PAM but from neural crest. Neural crest cells from each rhombomeric zone migrate downward from the neural fold and spill over the outer surface of their corresponding somitomere much like caramel poured over an apple. The tissue flows into pharyngeal arches. In each arch the neural crest from more cranial member of the rhombomere pair occupies the proximal/dorsal half of the arch, while the more caudal member is located in the distal/ventral half of the arch. An aortic arch artery supplies each pharyngeal arch transiently. The arterial axis enters each arch ventrally and exits dorsally to join with the ipsilateral dorsal aorta. Thus, *the anterior and ventral aspect of each pharyngeal arch is metabolically more "mature."* For this reason, products of the distal arch such as bones and muscles appear before their proximal counterparts. The mandible (r3) is formed prior to the maxilla (r2). The facial muscles of the lower face (r5) appear earlier in time that those supplying the upper face (r4).

Although most craniofacial bones are formed from neural crest, PAM from the somitomeres has important derivatives as well. Sm1–3 do not participate in pharyngeal arch formation. PAM from Sm4 forms the parietal bone and likely basisphenoid. Sm5–Sm6 PAM produce the petrous temporal bone, while the PAM of the non-somitic part of Sm7 produces the mastoid process. Despite their lack of formal myotomes, the first seven somitomeres provide the myoblasts for many important craniofacial muscles.

Beginning with Sm7, all somitomeres change their configuration into fully separate somites having characteristic dermatomes, myotomes, and sclerotomes. The myotome of the first occipital somite produces two large external and ventral muscles, the sternocleidomastoid and the trapezius. Tongue musculature comes from the exclusively internal and ventral myotomes of the first through fourth occipital somites. The fascial envelopes of all muscles of the head and neck are derived from neural crest corresponding the neuromeric level of the myoblasts.

Molecular Basis of Segmentation

Introduction to Homeotic Genes

Vertebrates have a segmental body plan. Molecular biology has, since 1990, literally revolutionized our comprehension of the mechanisms by which segment formation and segment specialization are accomplished. The genes that control this development have an incredible degree of phylogenetic conservatism. Nucleotide sequencing studies show that genes found in primitive organisms such as *Drosophila* and worms exist in mammals as well and that these genes share common functions. A surprisingly limited number of genes are required. A given gene may play multiple roles depending upon the period of development in which it is expressed and the organ in which it is expressed. The molecular products of these genes can be broken down into two main types: (1) intracellular transcription factors, and (2) extracellular signaling molecules.

Transcription factors stay within the cell. They can bind to genes to initiate patterns of expression at key steps in development [89]. One class of transcription factors, the basic helix-loop-helix protein, has a short sequence of amino acids in which two alpha helices are separated by a "loop" of amino acids. Immediately next door to this sequence is a basic region that binds to DNA. The helix-loop-helix causes dimerization. This is typical in muscle development. Another class having an unusual geometry is the zinc finger protein. Symmetrically spaced units of cystine and histidine along the polypeptide chain bind to zinc ions via four ligands. When this happens, the four residues are drawn together and the polypeptide chain puckers up like a finger. This finger can subsequently insinuate itself into specific binding sites of DNA; activation of DNA sequences results.

In understanding embryonic segmentation and, ultimately, formation of the head, the most important class of transcription factors is the homeodomain proteins [24, 90–100]. These all have a helix-loop-helix configuration consisting of the same 61 amino acids. The DNA coding for this region, the homeodomain, is a unique sequence of 183 nucleotides known as the homeobox; every single gene producing this type of protein has the same sequence. Because of this molecular anatomy, these genes are called homeobox genes. Please note that in scientific terminology all genes are written in italics, whereas the protein product is not. Thus, the Sonic hedgehog protein is produced by the gene *Sonic hedgehog* (*Shh* for short). Note that when the gene is human, the entire abbreviation is capitalized, i.e., *SHH*.

As originally described in *Drosophila*, eight homeobox-containing genes exist on a single chromosome. These are divided into two sections, the anterior antennapedia complex and the posterior bithorax complex. Mammals possess 38 Hox genes analogous to those of the fruit fly. These are

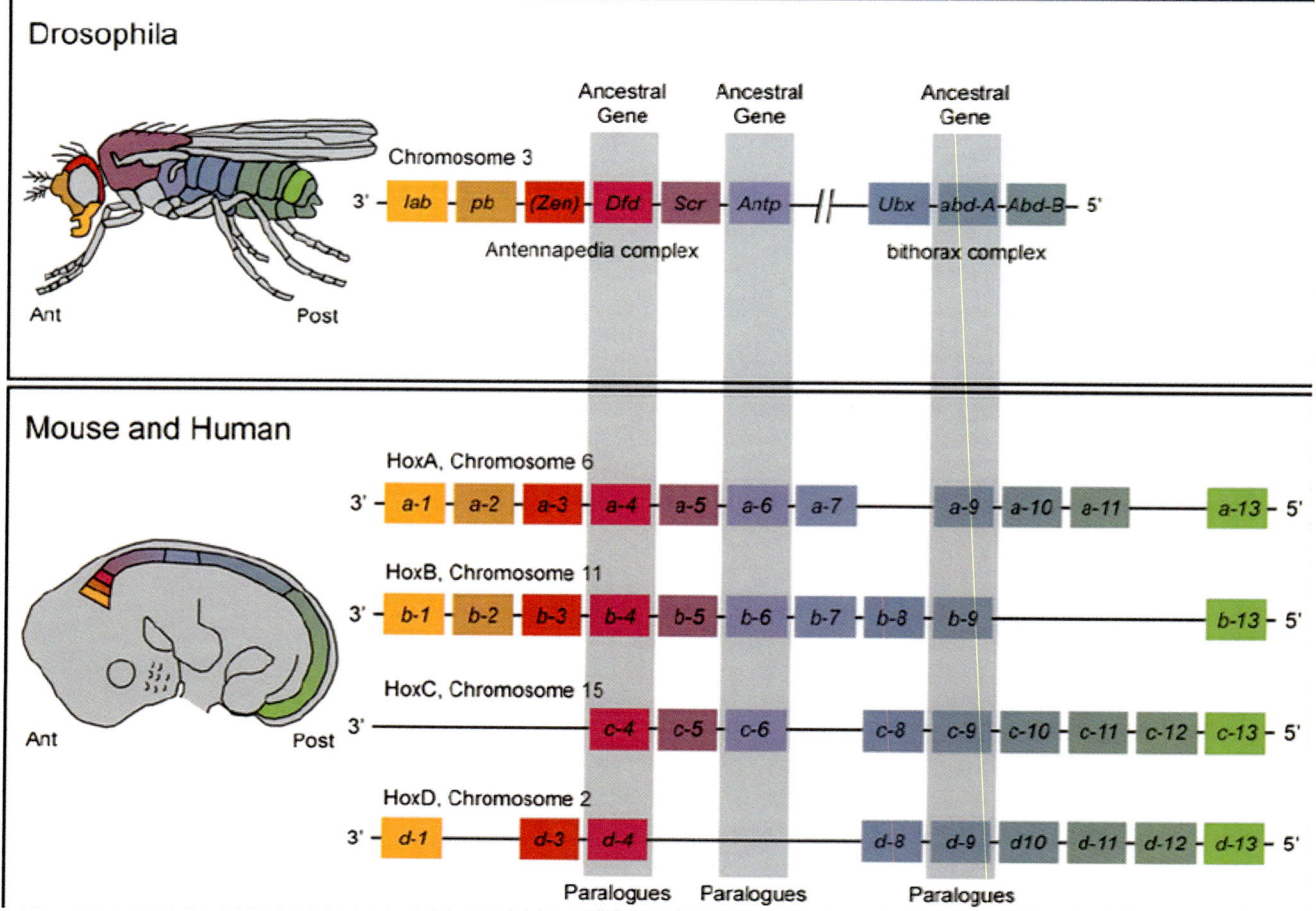

Fig. 1.35 Common to all homeotic genes is a 180 base pair sequence known as the homeobox. This encodes a 60 amino acid sequence that binds to DNA and unlocks it. Homeotic genes are primitive and universal. They specify the embryonic axis and define it in a segmentation system common to both invertebrates and vertebrates. Homeobox genes originally described in the fruitfly are called *hox* genes. Similar *hox* coding is found in mammals from the third rhombomere caudal. Rostral to r3, additional homeotic genes perform the function of specifying the neuromeres. Note the genes are read out in 3′ to 5′ order which corresponds from anterior (head) to posterior (tail). Human hox genes have been duplicated to 38 distributed over 4 chromosomes. (Reprinted from Pang D, Thompson DNP. Embryology and malformations of the craniovertebral junction. *Childs Nerv Sys* 2011; 27 (4):523–564. With permission from Springer Nature)

located on four different chromosomes and are arranged into 13 paralogous groups. What is amazing is that the genes are distributed anatomically along the chromosome. Hox genes for anterior mammalian segments are located at the 3′ end, while more caudal segments are at the 5′ end. Because the genes are activated and expressed in the 3′ to 5′ direction segment, formation in all organisms is a craniocaudal developmental sequence! (Figs. 1.35, 1.36, and 1.37).

Clinical Significance

This order has major experimental and evolutionary significance. Hox genes create characteristic morphology in each segment. Hox specification of neuromeric levels c7 and t1 creates the 12th and 13th somites; the 7th cervical vertebra results. At levels c8 and t1, a different Hox "barcode" results in a morphologically different structure, the first thoracic vertebra. When mutations in *Hox* genes occur, morphological variations are seen in the segmental structures that would normally have been expressed. A posterior-to-anterior transformation occurs when the contents of a give segment resemble those of the next most anterior structure. This is called a *loss of function mutation*. An anterior-to-posterior transformation occurs when the contents of a given segment resemble those of the next most posterior structure. This is called a *gain of function mutation* [101–103] (Fig. 1.37).

Retinoic acid (vitamin A) is a potent posteriorizing agent [104]. Exposure of mouse embryos to retinoic acid at a specific time in development causes an extra cervical vertebra to appear beneath the skull base! This structure, known as the proatlas, articulates with the skull with a single peg [105, 106]. The two condyles disappear. The cervical system, now consisting of 8 vertebrae, is also different. The morphology of the former atlas and axis reverts to that of standard murine cervical vertebrae. By the rules of vertebral formation, as the

(B) Wild-type Misexpression of *Hoxa10* (C) Misexpression of *Hoxb6*

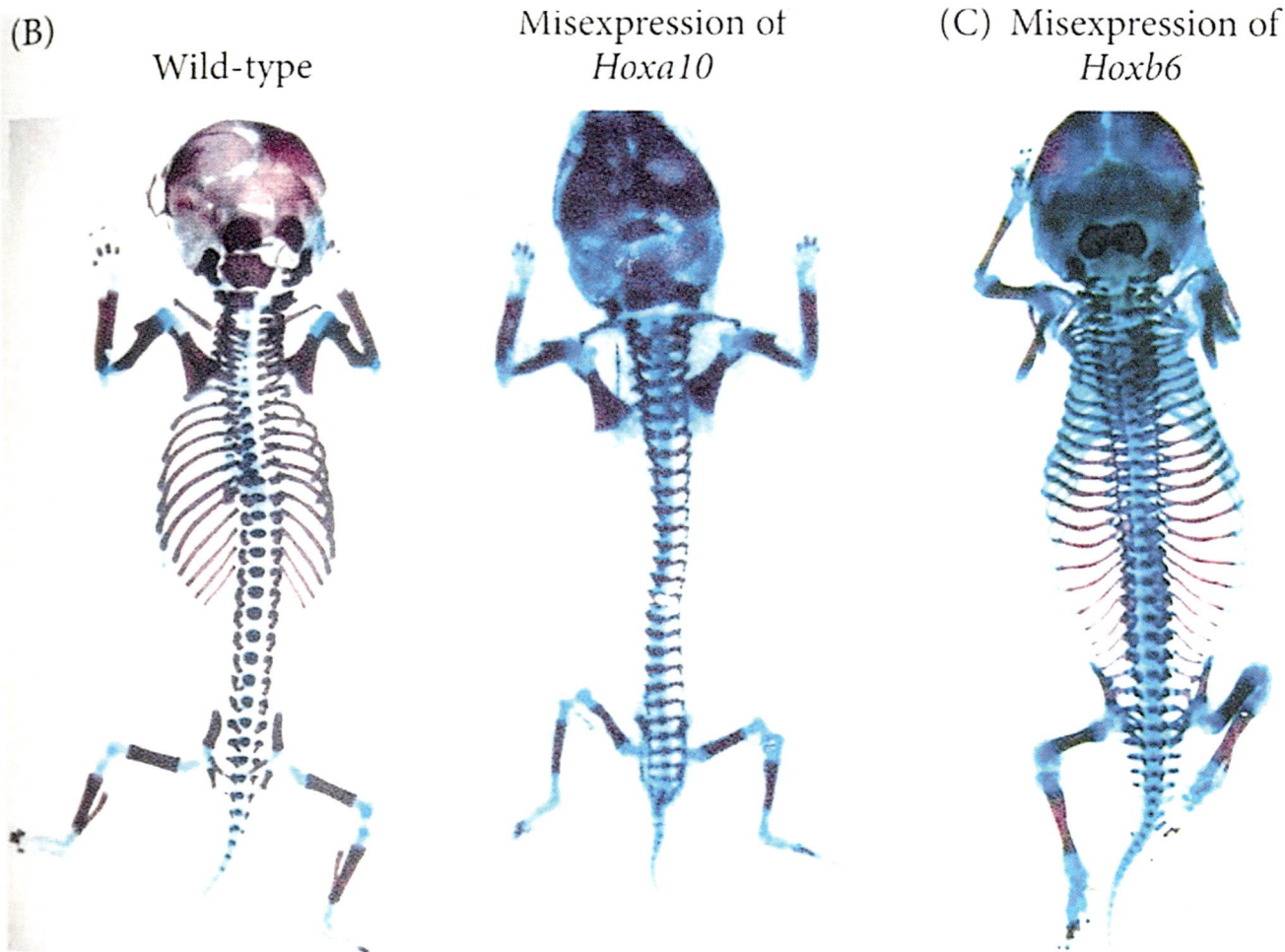

Fig. 1.36 Anterior-posterior relationships in the embryo are neuromere-specific and each neuromeric level has its own specific hox code. Once cells ingress through the primitive streak, their gene activation sequence is determined (at least partially) and they acquire an identity. If presomitic mesoderm from the thoracic region is transplanted into the neck, it differentiates according to its original program and forms ribs. Hox 10a represses rib production. It is normally expressed in all but the thoracic region. If it is expressed throughout the entire PSM, no ribs will form. Hoxb6 promotes rib expression. If it is expressed throughout the PSM, ribs will appear at all vertebral levels. (Reprinted from Gilbert SF, Barressi MJF. Developmental Biology, 11th ed. Sunderland, MA: Sinauer; 2016. Copyright © 2016. Oxford Publishing Limited. Reproduced with permission of the Licensor through PLSclear)

first of eight cervical vertebrae, the proatlas must be produced from the posterior half of the fourth occipital somite and the anterior half of the first cervical somite. The respective neuromeric levels are r11 and c1. Thus, the evolution of the foramen magnum and the double condylar occipito-atlantal joint was quite possible due to a "frameshift" mutation of the Hox code causing a shared "loss of function" between the fifth and sixth cervical somite. The paraxial mesoderm of these somites was "expropriated" from the pro-atlas to (1) expand the posterior braincase, (2) enable a double condyle system formerly between the proatlas and the second cervical vertebra to be "reassigned" to the skull base, and (3) reconfigure geometry of the foramen magnum (Figs. 1.38 and 1.39).

Potential evolutionary consequences of these changes are: (1) expansion of the posterior skull permitted better accommodation for an expanded visual cortex; (2) an increase in rotation, flexion, and extension of the skull was immediately useful for better predation and an augmented repertoire of feeding behaviors, and (3) better biomechanics for the occipitocervical junction adapted to tetrapod behavior on land...in primates this may even have facilitated the transition to an erect posture.

Hox genes are so-named because the homeobox-bearing genes are analogous to the antennapedia/bithorax complex. However, recent work has disclosed other gene families with <u>no</u> relation to *Drosophila,* but <u>with</u> the homeobox and additional conserved sequences. These include groups such as the *Engrailed* or *Lim* genes. The nine paired (Pax) genes all contain a paired domain of 128 amino acids; this is a DNA binding site [107]. *POU* genes (*Pit-1, Oct-1,* and *Unc-86*) all have a common 75 amino acid loop (for DNA binding) as

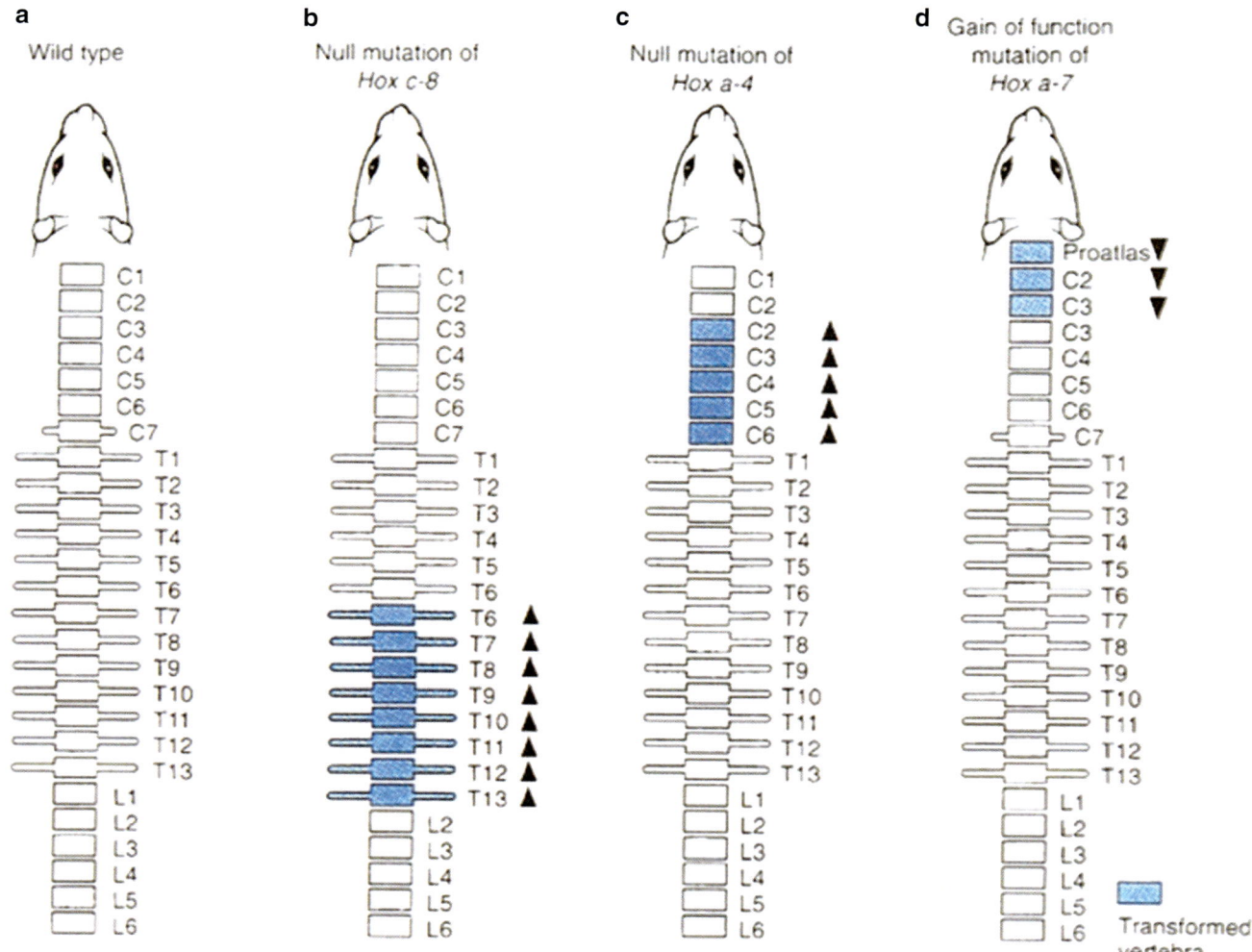

Fig. 1.37 Mammalian segmental organization is based on the *hox* code. Human embryos have 38 somites. Note the 4 occipital somites and the 3 coccygeal somites. There can be up to 10 coccygeal somites, but they involute back to 2–3. Hox coding of the mode and frameshift mutations shown. The mouse has 13 thoracic somites. Application of retinoic acid causes an additional cervical vertebra to appear. This is the primitive *proatlas*. Vertebrae 1 and 2 undergo transformation from atlas and axis to typical cervical vertebrae. The proatlas thus articulates directly with the cervical spine; atlas and axis are eliminated. (Reprinted from Kessel M, Balling, Gruss P. Variations of the cervical vertebrae after expression of the Hox-1.1 transgene in mice. *Cell* 1990; 61 (2):301–308: with permission from Elsevier)

well as a homeobox. As we shall see, many of these non-*Hox* homeobox genes have been used in the mapping of the vertebrate forebrain.

Some molecules produced by cells serve as extracellular "signal carriers." Most inductions in embryology (interactions between tissues such as epithelium and mesenchyme) occur using peptide growth factors. The first molecule, nerve growth factor, was reported with great fanfare in the 1950s. Two large families of these molecules exist and their members are making their way into many scientific papers in our plastic surgery literature. The transforming growth factor-beta family is made up of more than 30 genes that are very active throughout embryogenesis and beyond. Induction of mesoderm, proliferation of myoblasts, and the invasive properties of angiogenic endothelium are all attributable to TGF-B products [108]. Nine genes make up the fibroblast

growth factor family. FGF products perform a plethora of tasks from inducing growth of the limb bud to ensuring neuronal survival (and in their absence, neuronal death) [109]. These molecules are now widely used in and many applications have been made to cranial suture closure. A cautionary note however: the widespread and protean properties of the TGF-B's, coupled with their ability to activate other cascades, make interpretation of laboratory results difficult at best. This is because, once again, the action of the very same gene can vary widely, for it depends heavily on the period of development and the location in which the gene is expressed.

In craniofacial development, one of the most potent signaling families, that of the hedgehog proteins, has been only recently isolated. In mammals, three forms of this protein (called Sonic hedgehog, Indian hedgehog, and Desert hedgehog) come from three different genes with the same name.

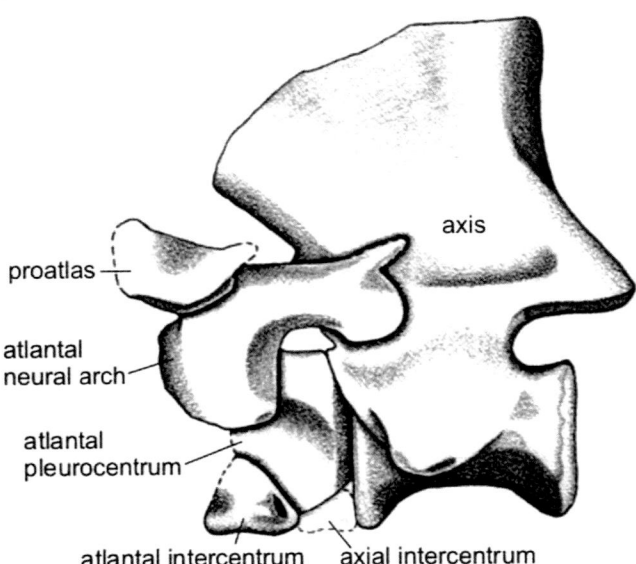

Fig. 1.38 Primordial first cervical vertebra (the proatlas) accounts for discrepancy between number of cervical spinal nerves and vertebral bodies. Proatlas forms by parasegmentation between fifth occipital somite and first cervical somite. Incorporation of proatlas into skull base as condyles, dens, and basioccipital-dental ligament provides dual condyle system for craniovertebral flexion-extension. (Reprinted from Campione NE, Reisz RR. Morphology and evolutionary significance of the atlas-axis complex in varanopod synapsids. *Acta Paleoontologica Polonica* 2011; 56 (4):739–48. With permission from Creative Commons License 4.0: http://creativecommons.org/licenses/by/4.0/)

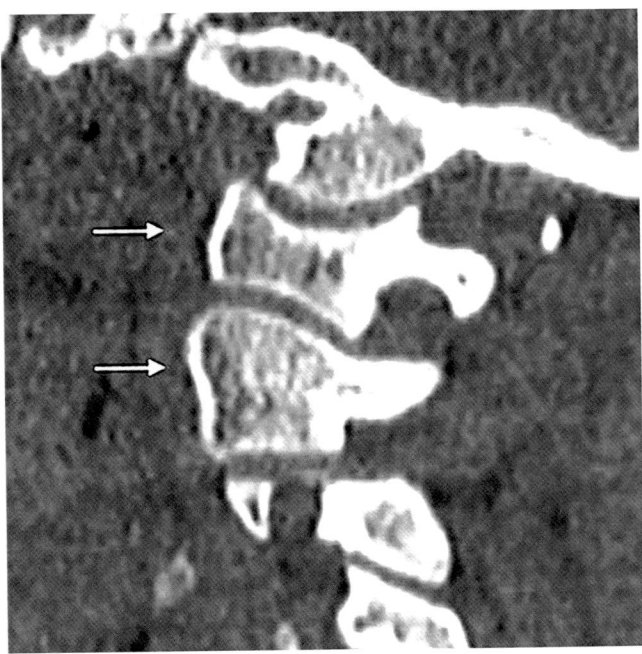

Fig. 1.39 Proatlas see in an 8 year old boy. (Reprinted from Spittank H, Goehmann U, Hage H, Sacher R. Persistent proatlas with additional segmentation of the cranio-vertebral junction: the Tsuang-Goehmann malformation. Radiol Case 2016; 10 (10):15–23. https://www.radiologycases.com/index.php/radiologycases/article/view/2890. With permission from Journal of Radiology Case Reports)

Sites of action of *Shh* genes include the primitive node and notochord, the neural floorplate, the ectoderm of facial "processes" (there's that pesky antiquated term again). Shared *Shh* activity in the apical ectoderm of the second pharyngeal arch and in the epithelial buds of the lungs may explain seemingly unrelated pathologic states [110–116].

Craniocaudal Pattern Formation

Gastrulation involves the creation of a three-layer "sandwich" of ectoderm on top of the future organism that comes from a single sheet of cells, the epiblast, sitting atop a nutrient source, the yolk sac (in the chick) or the gel-filled blastocoele cavity (in mammals). A two-layer sandwich comes first. In the avian model, epiblast cells are observed loosen their intercellular bonds. Cells "drop out," from the epiblast in random fashion, falling down into the blastocoele cavity, where they reconnect to form a second, subjacent sheet [117, 118]. In mammals, the proliferation of epiblast leads to direct formation of an underlying second layer. In either model, the deeper cells are known as the *hypoblast*, or so-called primitive endoderm. A cellular cleavage plane now exists between the layers; this pathway will now permit gastrulation to happen. A groove up the midline axis of the epiblast forms, called the primitive streak. At the rostral end of the streak sits a small, elevated pit called Hensen's node. Just like water draining from the bathtub down the drain, epiblast cells migrate toward the streak and enter the space below the epiblast. They are guided in their migrations by the basal lamina of the epiblast above. In so doing, these migrating cells are like arctic divers plunging through a hole in the ice and then navigating beneath the ice by following its undersurface.

At this juncture, we must raise a critical question, one with great philosophical implications: does cell "migration" exist and, if it does, how do the cells "know" where they should go? If we are conceived of cells following a route to a destination, how was the route set up in the first place? We wind up with the dilemma of attributing to cells a form of "knowledge" they cannot possess. Migration theories border close on the concept of Deus ex machina.

Fortunately, we are rescued from biochemical theism by the pioneering work of German embryologist Erich Blechschmidt (1880–1955) [119, 120]. Blechschmidt conceived embryologic events in very simple terms. Populations of cells in specific anatomic locations would expand their numbers in proportion to their relative access to nutrition. As the populations increase, identifiable anatomic structure within the embryo (ligaments and flexures) would channel the growth and ultimately affect the final three-dimensional form of these populations. Thus, the final shape of a structure is determined by *physical factors*, not a genetic program.

In the case of gastrulation, the Blechschmidt model postulates that the single layer epiblast is attached to an unyielding superstructure, the amnion. Around the periphery of the epiblast/amnion junction run two vascular structures, the vitelline veins. These run from the connecting stalk (or the yolk sac) forward to the primitive heart. This implies that the highest availability of nutrients to the cells of the epiblast occurs at the periphery. Multiplication of this peripheral population creates a physical force directed away from the constraining amnion and toward the midline. This force acts on the cells of the midline, causing them to ingress into the primitive streak. The midline epiblast cells are squeezed like toothpaste and forced out between the epiblast and hypoblast until they reach the periphery again. The situation is not unlike students lined up outside the door of a small theater. Once the doors open, the pressure from the back of the crowd will propel those students directly in front of the doors right into the theater and down to the very first row. Thus, the theater fills up from distal to proximal. We shall see that this same pattern recurs as the embryo "fills up" with mesoderm. The most peripheral mesoderm (extraembryonic or cardiogenic) comes from epiblast cells nearest the primitive streak. These are followed by lateral plate mesoderm cells and finally by the future paraxial mesoderm cells.

Cells ingressing from the epiblasts do so at four general spatial points: (1) Hensen's node, (2) the remainder of the primitive pit, (3) the rostral streak, and finally (4) the caudal streak. The fate of a migrating cell (what layer it will belong to and how far out it shall migrate) is determined by its anatomic site of ingression and the timing of ingression. Another analogy might be that of paratroopers jumping out of a plane from four different doors. Standing in line before their respective doors, each receives instructions on his mission just prior to exiting. So it is that, from the apex of Hensen's node, the very first cells to make their exit will become endoderm and, in the exact midline, the notochord. These are followed by the medial somitic mesoderm, the lateral somitic mesoderm, and finally, the lateral plate mesoderm. When gastrulation is complete at a given level of the embryonic axis, a caudal regression of Hensen's node and of the streak takes place [83, 121–127].

The very first cells to ingress produce the foregut endoderm. Next follows mesoderm tip of the notochord and most anterior mesoderm called the prechordal plate mesoderm (PCM). Gastrulation occurs over time as a craniocaudal process of cellular ingression and reassignment. Sonic hedgehog produced at the PCM and at the notochord does two very important things. First, Shh from the PCM begins to specify the hindbrain, while Shh from the notochord begins to organize the somites, i.e., a decision is made between "brain" and "non-brain." Furthermore, the ingressing cells at a given neuromeric level receive a Hox code "identity" that is identical with those epiblast cells that did not ingress. A unique pattern of Hox gene expression is thus a "barcode" that persists in all the cells that originated at the region, regardless of

where they eventually migrate. This is readily seen in the form of the axial skeleton. Formation of a normal craniocaudal pattern of segment is reflected in the unique combination of Hox genes that specifies all vertebrae [102].

Noden applied these concepts higher in the neuraxis to the rhombencephalon by correlating patterns of gene product expression with cranial nerves and neural crest and rhombomeres [123]. A Hox "barcode" could be found for rhombomeric levels r3 and caudal, similar gene definitions using Krox-20 (the human form of this murine gene is *EGR-2*). *Follistatin, Engrailed,* and *Wnt-1* permitted "mapping out" rhombomeres r0, r1, and r2 [128].

At rhombomere 0 (some texts describe this as the anterior border of r1), *FGF-8* and *Wnt-1* stimulate expression of *engrailed* genes *En-1* and *En-2*. The farther one moves away from the "signaling center," the lower the concentrations of En-1/En-2 proteins become. Indeed, r0 as the principle rhombomere of the mesencephalon expresses many important genes that determine the spatial organization of the brain. These molecules stimulate the rostral region to become mesencephalon (midbrain) and the caudal region at r1 to become metaencephalon (cerebellum). The midbrain subsequently forms two developmental units in which are located the superior and inferior colliculi for visual and auditory connections. Puelles and Rubenstein refer to these as m1 and m2, but for our purposes, since m2 (the preisthmic mesomere) is so small we will just refer to them collectively as m1.

In the 2013 iteration of the prosomeric model, early forebrain is divided into six prosomeres. Just like the rest of the spinal cord, hindbrain, and midbrain the prosencephalon possess a ventral (basal) zone and a dorsal zone (alar) zone. It should be noted that use of the terms basal and alar is a descriptive anlogy based upon the hindbrain. No sulcus limitans exist in the forebrain; therefore true "alar" or "basal" zones do not exist. We make use of this crude analogy simply to better understand the eventual differentiation of the prosencephalon into two major subdivisions. The caudal diencephalon is made from the alar and basal units of p1–p3 and contains the epithalamus, thalamus, and hypothalamus. The rostral diencephalon is formed from the basal units of p4–p6. This contains visual apparatus and olfactory lobe. The telencephalon is made from the three dorsal zones of p4, p5, and p6. It contains the cerebral cortex. Much progress in mapping the derivatives of the prosomeric zones is due to elegant work by Rubenstein and Puelles [32]. Terminology of the prosomeric model is described in Fig. 1.38.

The reader should note that recent changes in neuromeric mapping have been incorporated into the neuromeric model Consider r0 the boundary of midbrain and forebrain, "the organizer," to be located at r0 [129]. The diencephalon is considered to have three prosomeres, whereas the secondary prosencephalon is described in terms of anatomic units.

Fig. 1.40 Prosomeric model of Puelles (2003). Development of the prosomeric system at three stages. Top: Four diencephalic prosomeres identified. Caudal hindbrain not yet differentiated into rhombomers r8–r11. Middle: Beginning with stage 9, flexion at the mesencephalic junction is noted. Anterior prosomeres p4–p6 identified. Bottom: maturing brain stages 17–18 shows cerebellum emerging from r1. Note that the 2013 model does not recognize two mesomeres, nor the isthmic rhombomere r0. (Reprinted from Puelles L. Brain segmentation and forebrain development in amniotes. *Brain Res Bull* 2001; 55 (6):695–710. With permission from Elsevier)

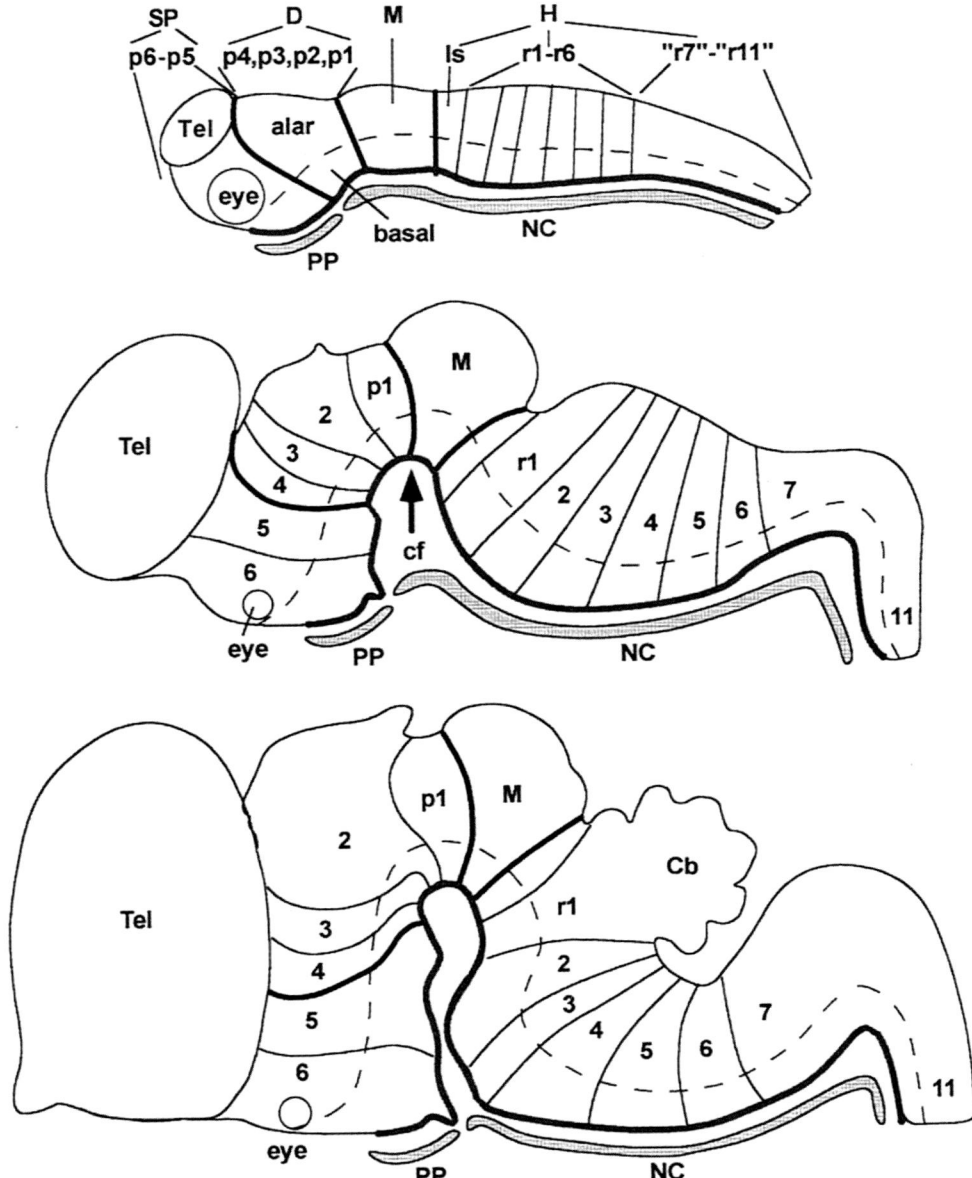

Figures 1.40 and 1.41 demonstrate these differences. For our purposes, since the anterior prosencephalic neural folds do not produce neural crest, and since it comes from above the diencephalon, we shall refer to PNC originating from prosomeres p1 to p3, but we will also describe the zones of the anterior neural folds populated by the PNC in term of p4, p5, and p6 (Fig. 1.42).

Neuromeric Basis of Neural Crest "Migration" and Fate

Now that we have a description of the neuromeric system firmly in mind, we shall need a schematic way to visualize it. Using a mouse embryo in saggital section, Rubenstein

et al. have outlined the boundaries of the neuromeric zones. Key neuroanatomic landmarks are presented in a linear, schematic manner. Basal and alar regions of the entire neuraxis, including the prosencephalon, can be readily appreciated. These relationships are then projected onto an anatomic representation of the embryo. This author has modified these drawing by applying a color code in which each neuromeric zone along the neuraxis has its own specific color. When tracing out all the derivatives of rhombomere 3, all muscles and bones of the mandibular portion of the first pharyngeal arch will be depicted in green. This permits structures such as dermis and dura to be "mapped" backward to their origins at a specific neuromeric level. Let's look at the neural folds in terms of five zones (Figs. 1.42, 1.43, 1.44, and 1.45).

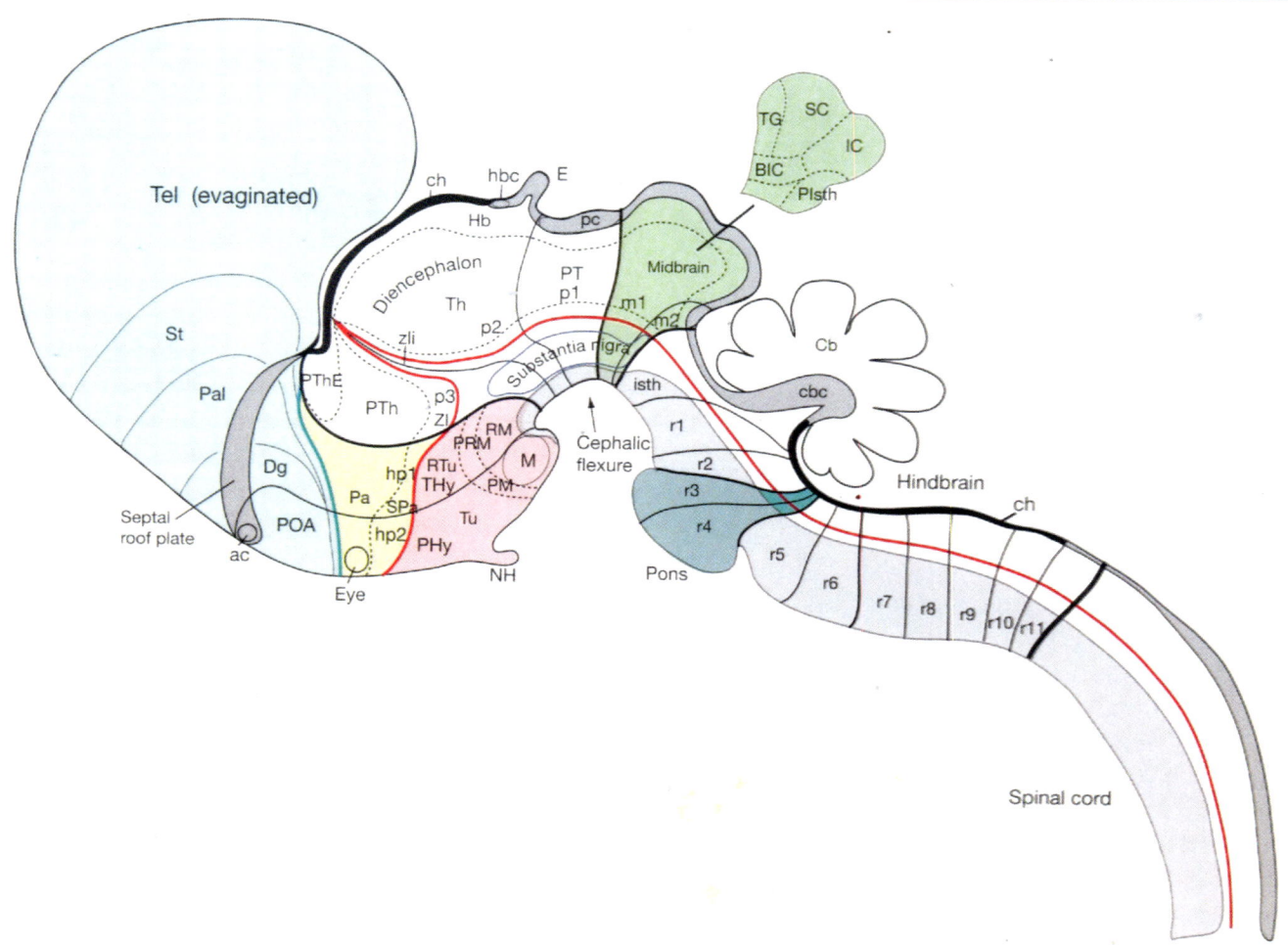

Fig. 1.41 Prosomeric model of Puelles (2013). Division of prosencephalon is now in three parts: a caudal diencephalon and rostrally, telencephalon (blue), and hypothalamus (yellow and pink). Hypothalamus was formerly considered part of diencephalon. Prosomeres p4–p6 no longer used. Anterior neural folds (formerly termed p4–p6) remain populated by neural crest from p1 to p3. Note the red line indicates alar/basal boundary of the brain. Diencephalon consists now of 3 prosomeres, not 4. Midbrain as pre-isthmic mesomere m1. Hindbrain has isthmic rhombomere r0. Telencephalon (blue) is divided into pallium and the subpallial regions (striatum, pallidium, diagonal domain-Dg, and the preoptic area). Within the terminal zone of hypothalamus are found the eye fields yellow, neurophyphysis, and mammillary bodies (pink). The peduncular hypothalamus contains subthalamic nucleus (STh). Roofplate extends from the isthmus forward to anterior commissure. (Reprinted from Puelles L, Harrison M, Paxinos G, Watson C. A developmental ontology for the mammalian brain based on the prosomeric model. Trends Neurosci 2013; 36 (10):570–578. With permission from Elsevier)

Anterior Prosencephalic Neural Fold: Nonneural Ectoderm

Pioneering work by Couly and LeDourain established a detailed fate map of the avian neural plate [130–139]. In particular, they showed that neural crest cells do not occupy the entire length of the neural fold. The neural folds of the anterior prosencephalon (from p6 back to p4–p3) do not have neural crest cells.. The neural folds of the posterior prosencephalon (from p3 to p2 on backward) contain neural crest. All ectoderm lateral to the neural folds is destined to form epidermis. This ectoderm, plus that of the anterior prosencephalic neural folds, is termed *nonneural ectoderm.* We shall describe the anatomic content of the nonneural ectoderm in cranial-to-caudal order, i.e., from p6 to p4. The reader is advised that folding of the embryonic head coupled with growth of the forebrain causes radical changes in the spatial relationships of these three zones to occur, especially between p6 and p5. These changes will be described below.

The most anterior zones of nonneural ectoderm are termed p6. The first important structures of p6 are the *adenohypophyseal placodes* (AP). Described as singular, these fused structures occupy either side of the midline. Pathology in either AP explains the occurrence of epithelial tumor localized to either the right or left pituitary. Posterior to each AH placode, the nonneural ectoderm becomes the *nasal epithelium* (NE). This forms the epithelial lining of the nasopharynx all the way forward from Rathke's pouch and Waldeyer's ring to the true skin of the nasal vestibule. The NE makes a boundary with true skin of the nose and upper lip (in the chick known as upper beak epithelium UBE). Within the nasal epithelium lies the *nasal placode* (NP). This specialized epithelium gives rise to three classes of neurons, all of which migrate into the brain.

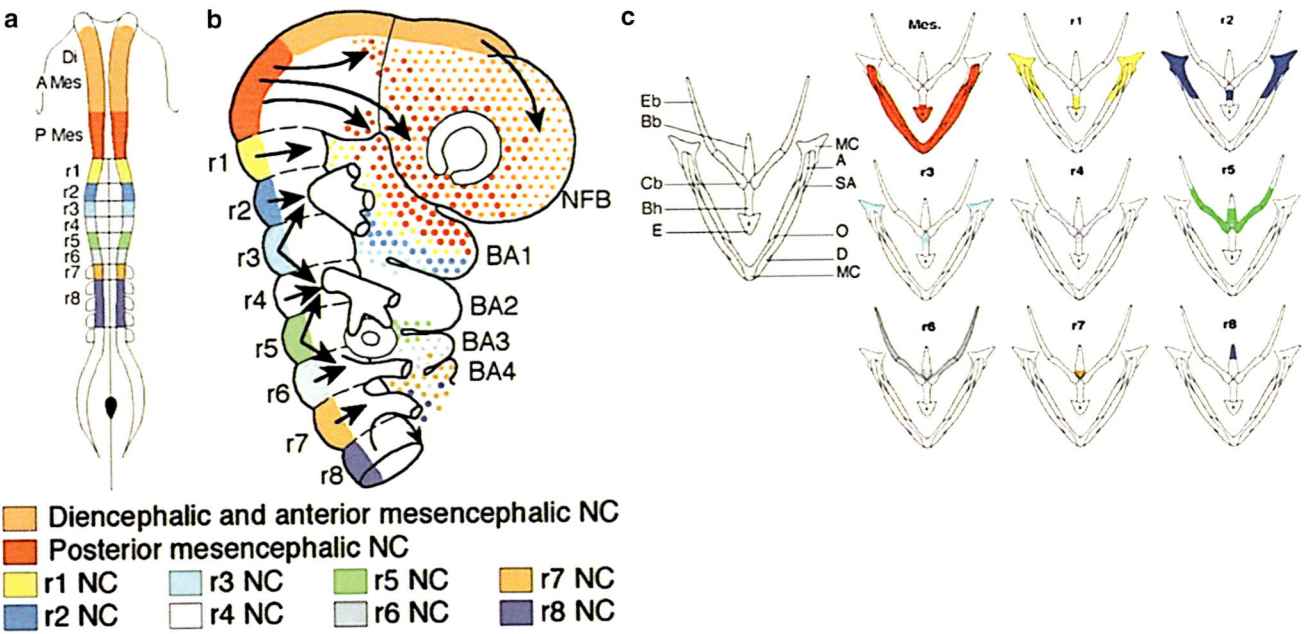

Diencephalic and anterior mesencephalic NC
Posterior mesencephalic NC
☐ **r1 NC** ☐ **r3 NC** ☐ **r5 NC** ☐ **r7 NC**
☐ **r2 NC** ☐ **r4 NC** ☐ **r6 NC** ☐ **r8 NC**

Fig. 1.42 Migration of neural crest. PNC (p3–p2–p1 = orange) migrates as a *sheet* (orange) beneath the more rostral neural folds of nonneural ectoderm. Frontonasal skin consists of epidermis from NNE and dermis from PNC. MNC (m1 = red, r1 = yellow) migrates as individual *streams* from the midbrain forward to form the dura of the anterior cranial fossa, the sphenethmoid structures of the anterior cranial base, the frontonasal bones, and the frontonasaloribal subcutaneous tissues. RNC (r2–r11) migrates as individual *segments* to supply all the pharyngeal arches. (Reprinted from Creuzet S, Couly G, Le Douarin NM. Patterning the neural crest derivatives during development of the vertebrate head: insights from avian studies. J Anat. 2005 Nov; 207 (5): 447–459. With permission from John Wiley & Sons)

The lateral NP contains *olfactory neurons* for conventional odor detection. The medial NP contains accessory *olfactory neurons*, those involved with the detection of complex chemical signals, pheromones being the prime example. Finally, the medial NP contains *neurons associated with gonadotropin hormone releasing hormone* (GnRH). Animal behaviors related to detection of sexual cues from urine and sniffing of genitalia relate to this system. GnRH is related to development of secondary sex characteristics. *Kallman's syndrome* (anosmia and/or hypogonadotropic hypogonadism) results from anatomic defects in the nasal placode [140–144]. Note that the Spanish pathologist, Maestre de San Juan, first described this syndrome [145] (Fig. 1.44 Placodes).

The p5 zone of nonneural ectoderm comprises the skin of the nose and the philtrum of the upper lip. This upper beak epithelium (UBE), the term is derived from the avian model, constitutes the epidermis of the nose and philtrum of the upper lip, but not the forehead. The lateral boundaries of the UBE are marked by sensory innervation of V1 and lie medial to the arcade formed by the facial artery-angular artery. Just caudal (posterior) to the UBE lies the *optic placode* (OP). This forms the lens and is crucial for development of the globe (and ultimately for that of the orbit).

The p4 zone of nonneural ectoderm forms the *calvarial ectoderm*. The epidermis of the forehead and the frontal bone sweep back over the cerebrum within this zone. In birds, the frontal bone is huge, while the parietal bone is diminutive. Formation of the avian calvarium is exclusively from neural crest and is extremely rapid. Sutures are not observed; pathologic craniosynostosis does not occur. In mammals, the parietal bone forms from PAM of r2 and r3 derivation. The varying forms of synostosis all involve boundary between bones of dissimilar developmental derivation. These observations may have important implications for the relative incidence of craniosynostosis observed in humans.

Without a dermis for vascularization and support, epidermis becomes rather worthless. Indeed, all epithelia require a subjacent supporting layer. The production of all dermis and bone in the face is exclusively the responsibility of neural crest. So where does the neural crest come from and how does it "know" what to do? The answer to the first question lies in a clear conception of the three functional types of neural crest and of the manner in which they migrate [146, 147]. The answer to the second question relates to the "programming" function exerted upon the neural crest cells by the epithelial environments they inhabit [148–150]. We shall deal with the first question in the section below and discuss the second at the conclusion of this essay.

Posterior Prosencephalic Neural Fold: Neural Crest (PNC)

The first zone of neural crest extends over the posterior prosencephalon from p3 back to p1. This PNC moves forward in the subectodermal plane as a large vertical ***sheet*** of cells in the midline. The PNC migrates from the dorsal part of the

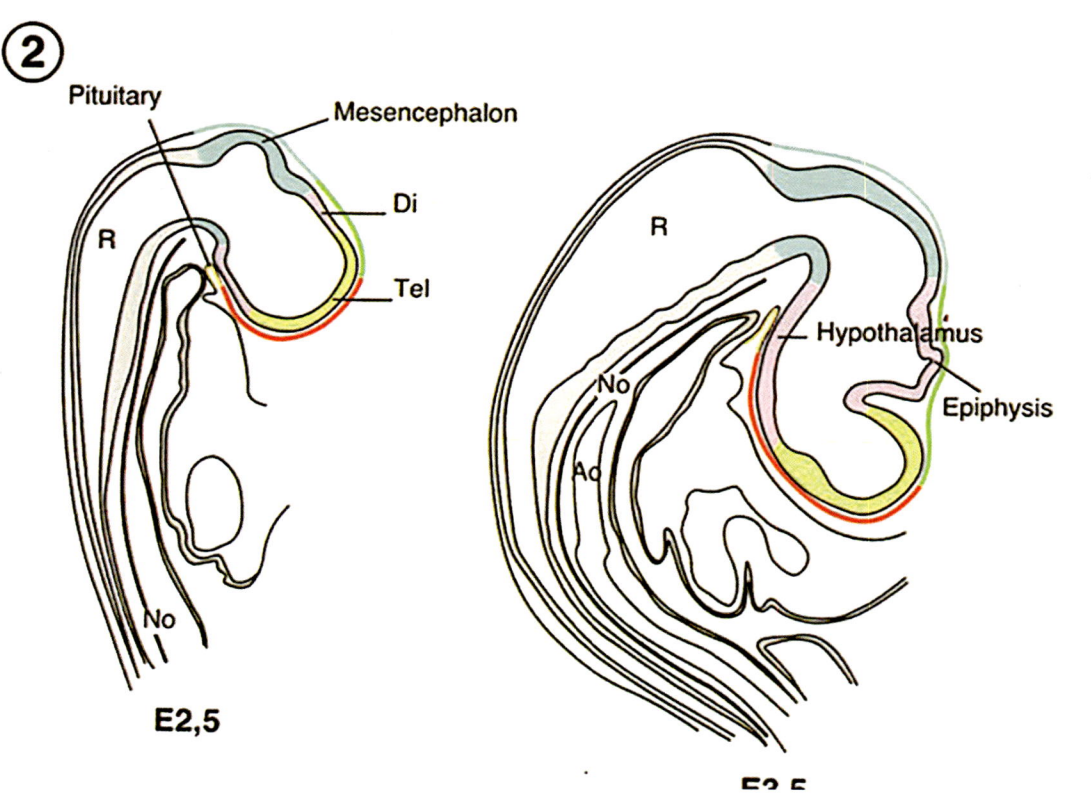

Fig. 1.43 Neural folds showing location of ectodermal placodes. Note how anterior neural folds become tucked beneath the brain to form frontonasal skin. (Reprinted from Le Dourain NM, The avian embryo as a model to study the development of the neural crest: a long and still ongoing story. Mechanisms of Development 2004;121 (9): 1089–1102. With permission from Elsevier)

lamina terminalis, from which also is formed the plate of corpus callosum. For this reason, p6 deficiency associated with holoprosencephaly can result in hypoplasia or absence of the intercanthal ligament; hypertelorism results. This is contrasted to the r1 deficiency state seen in anencephaly. The r1 component of the frontal bone is absent, the absence of r1 dura and neural crest pericytes profoundly affects forebrain development, and the sphenoid bone (an r1 structure) is small and misshapen. (For further details, see Chap. 5 on the forebrain commissure.)

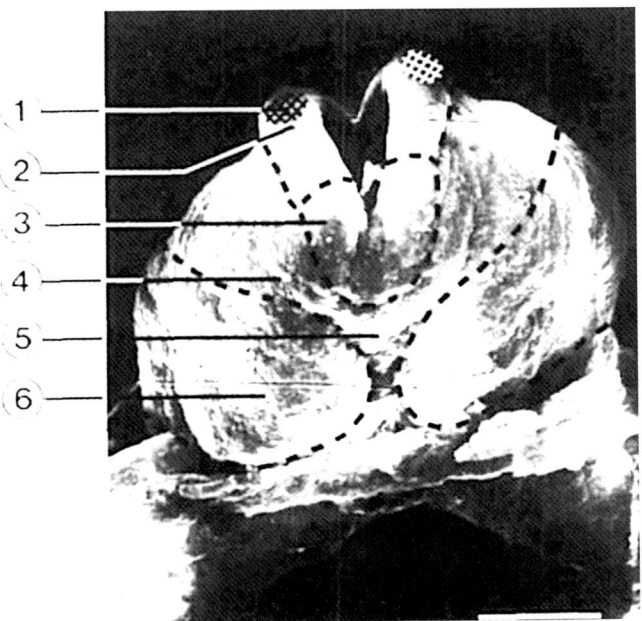

Fig. 1.44 Anterior view of rostral prosencephalic neural folds. (1) Olfactory placode, (2) nasal mucosa, (3) Rathke's pouch, (4) eyelids, (5) decidual ectoderm of the oropharyngeal membrane, (6) first arch ectoderm. (Reprinted from Couly G, Le Dourain NM/Head morphogenesis in embryonic avian chimeras: evidence *for a segmental pattern in the ectoderm corresponding to the neuromeres.* Development 1990; 108 (4):543–558. With permission from the Company of Biologists)

Migration of p3–p1 neural crest beneath the nonneural folds of upper beak ectoderm produces frontonasal skin. We shall describe this skin as p6–p4. Subsequently, subcutaneous tissue are provided by MNC. These cells will form nasal cartilages in accordance with the underlying program of p6–p5 skin.

The most anterior cells of the sheet will reach the p6 skin where they will be "instructed" in situ to make upper lateral cartilages. Into this envelope, MNC neural crest from r1 migrates to form nasal septum, the alar and lateral cartilages, the perpendicular plate of the ethmoid, the ethmoid labyrinth and crista galli and, finally, the upper and middle turbinates. Sixth prosomere neural crest also "activates" the adenophypophyseal and nasal placodes. Without interaction with neural crest, these placodes are nonfunctional.

When MNC arrives in the p5 zone, it forms the nasal bones and the lower lateral cartilages. The *nasal process of the frontal bone* descends deep to the nasal bones to articulate with the *facial process of the premaxilla*. Here, to avoid future confusion, we must make a brief digression. The piriform rim is really a *bilaminar* structure. Its internal aspect is made from the nasal process of the frontal bone (p5) that extends downward just beneath the nasal bone. It abuts the frontal process of the premaxilla, PMxF, a product of r2 NCC supplied by medial/lateral nasopalatine axes. Overlying PMxF is the frontal process of the maxilla, MxF, an r2 derivative supplied by medial branch of the anterior superior alveolar axis. This abuts against the nasal bone itself. The bicortical structure of the piriform rim makes it stronger; it is

capable of holding screws. Surgeons refer to the piriform as a "buttress" and use it for placement of fixation plates. Additional bones programmed by the p6 vestibular lining are the orbital lamina of the ethmoid bone and the lacrimal bone.

The boundary between the neural crest derivatives of zone p5 and those of zone p4 is uncertain at this time. The supraorbital margin of mammals and the roof of the orbit are p5, while the remainder of the forehead would logically be p4. Indeed, the ossification centers of the frontal bone appear in caudal to cranial order. In primitive tetrapods, orbital rim has two distinct fields, prefrontal (PrF) and post frontal (PtF), arranged on either side of the supraorbital neurovascular axis. These bones arise from MNC that arrives prior to that covering the frontal lobe. For this reason, they persist even when frontal bone per se is absent, as in cases of anencephaly.

Mesencephalic Neural Crest (MNC)

The MNC lies over the mesencephalon. Unlike PNC that moves as a large sheet, MNC proliferates as distinct *streams*, described as r0–r1, followed by m1–m2. Recall that the development of the mesencephalon is stimulated by FGF-8/Wnt-1 produced at the isthmus (levels r0 and r1). Thus, migration from r0 to r1 antedates that from m1 to m2. Neural crest associated with each of these neuromeres a remarkable expansion atop the rapidly growing mesencephalon. Although the midbrain is quite small in the mature state, the embryo is enormous. This pre-migratory MNC contributes to large surface areas of dura associated with the sensory distribution of V1.

Mesencephalic neural crest migrates in three successive streams in craniocaudal order. That from r0 and r1 is dedicated to development of extraocular connective tissues and the posterior orbital wall. Neural crest from r0 migrates first and its final destination is the most medial and cranial of all MNC. It makes no bone. It lies internal and caudal to the optic vesicle. It therefore produces the inferomedial sclera and the fascia of the corresponding extraocular muscles innervated from the cranial oculomotor nucleus (inferior oblique, inferior rectus, and medial rectus). A limited amount of dura over the base of the forebrain, the rhinencephalon, would logically come from r0 neural crest. The innervation pattern of the terminal nerve (cranial nerve 0) may well map out the contribution of r0 neural crest.

Next to migrate is MNC from r1. The presence of the optic vesicle forces this stream to assume a superolateral position within the future orbit. It also fills in the space behind the globe and lateral to the future optic nerve. The presphenoid bone is formed from r1 MNC via a chondral intermediate. So too is the lesser wing of the sphenoid, being derived from preexisting orbitosphenoid cartilage. As the superolateral half of the optic vesicle is enveloped by r1 MNC, its scleral coat is acquired as well as the fasciae for the remaining extraocular muscles innervated by the caudal oculomotor nucleus (superior rectus and levator palpebrae superioris). A great deal of r1 neural crest is involved with

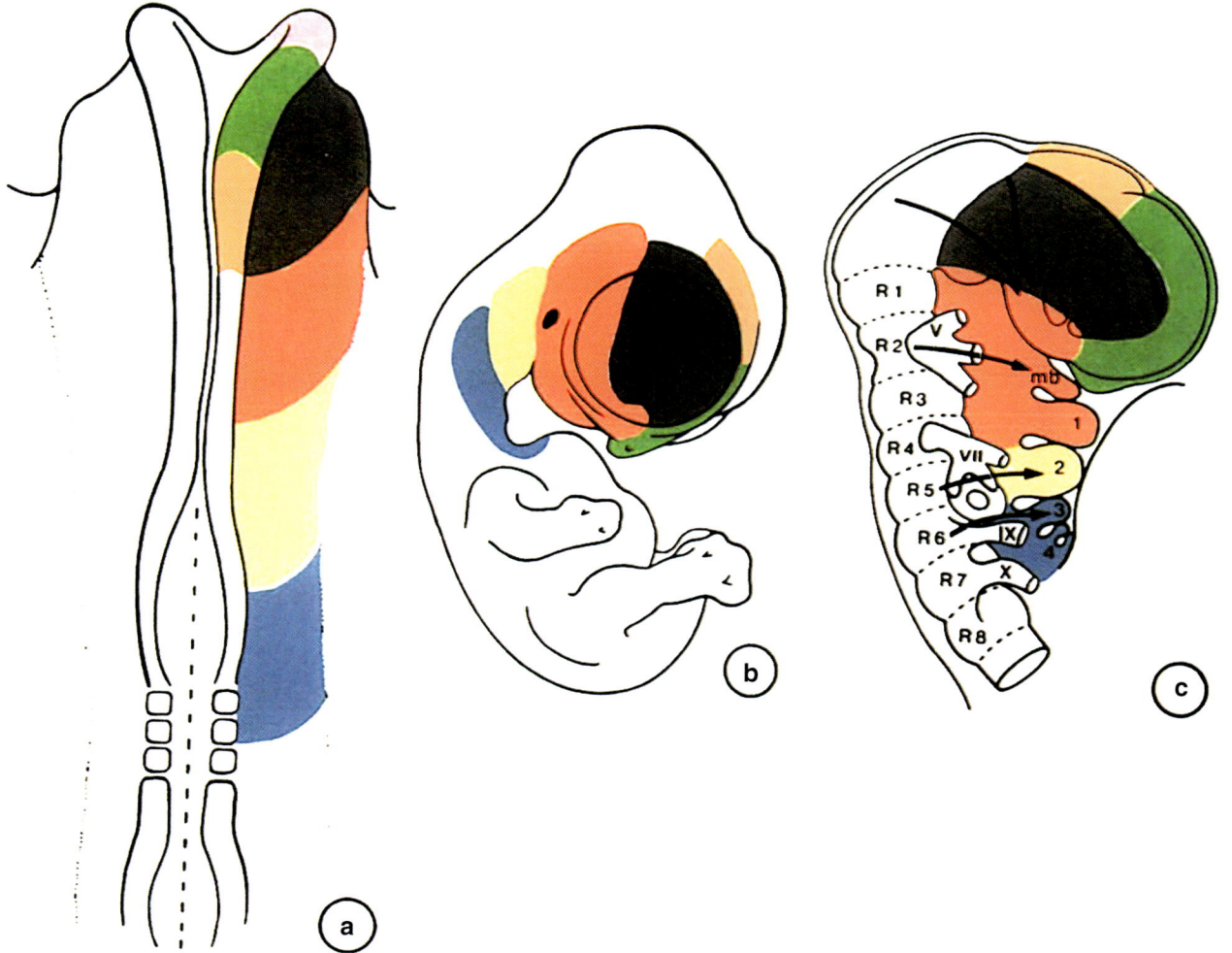

Fig. 1.45 Fate map of ectomeres. Nasal mucosa (pink), nasal skin (green), anterior calvarial ectoderm (orange), eyelid ectoderm (gray), first arch (red), second arch (yellow), third arch (blue). The pink tissue is tucked beneath the midline and flanked by frontonasal skin.

(Reprinted from Couly G, Le Dourain NM/Head morphogenesis in embryonic avian chimeras: evidence *for a segmental pattern in the ectoderm corresponding to the neuromeres.* Development 1990; 108 (4):543–558. With permission from the Company of Biologists)

dura synthesis. It covers the entire cerebral cortex innervated by V1.

In the initial iteration of this model, consideration was given as to whether the neural crest cells had a predetermined identity according to their position along the neural fold. This is unlikely. The neuromeric population migrates into its destination where it becomes organized according to the genetic "street map" of the pharyngeal arch. Each sector of the arch is targeted by a vascular pedicle. In the case of the first arch, these are all branches of the internal maxillomandibular axis which were "added on" as the result of its union with the extracranial stapedial. And the spatial order of these branches may be determined by their position within the arch. Thus, considering the various StV2 branches fanning out from the pterygopalatine fossa, medial nasopalatine may represent the most distal branch; therefore, it peels off away

from the other arteries to maxilla and zygoma, to follow a distinct route into the midline of the oral cavity.

Rhombencephalic Neural Crest (RNC)

Rhombencephalic neural crest does not travel in streams; at each neuromeric level, it proliferates or "migrates" as a *segment* into and around its corresponding somitomere and thence into its designated pharyngeal arch.. Furthermore, cells from one neuromere do not mingle with those of a neighboring neuromere. Under experimental conditions, mixing between even or odd numbered neuromeres is permissible but not in early life. Neural crest from r2 does not mix with that of r3. Because it obeys the "rules," this neural crest from the most rostral rhombomere is named r2 RNC.

RNC can be subdivided into two zones. The *rostral zone* (RNCr) is described neural crest interacting with those somitomeres that do not go on to become full-fledged somites. These include the head mesoderm of Sm1 and the first three pharyngeal arches. As r2–r7 reorganize themselves into arches, the somitomeres seem to "pair up." These dimers reorganize themselves into a single structure. Perhaps because the cellular volume of the first three pharyngeal arches is greater, neural crest migration into them takes longer to complete than that of more caudal zones. RNCr migration is not completed until somite-stage 14.

The *caudal zone* of rhombencephalic neural crest (RNCc) involves neuromeres responsible for populating the second group of pharyngeal arches. PA4–5 are smaller in size. It is therefore reasonable to expect that the neural crest population would take less time to migrate into these arches. Indeed, RNCc migration is complete by somite-stage 11. Later in this series, we shall speculate as to the manner in which the anatomy of a pharyngeal arch comes about and the manner in which this affects the order in which different fields within a pharyngeal arch are expressed (as reflected in the ossification sequence of craniofacial bones).

Spatial Reassignment of Nonneural Ectoderm

Head folding coupled with forebrain growth causes changes in the spatial relationships of nonneural ectoderm p6–p4. Prior to closure of the anterior neural folds, the p6 zone starts from the midline just rostral to the stomodeum, just in front of Rathke's pouch, and projects backward (caudally) along the folds. The apical surface of this future nasal vestibular epithelium (NVE) faces upward (dorsally). With closure of the neural folds, both p6 zones come together in the midline. They acquire a population of PNC from the diencephalon at this time. Head folding forces this zone to roll forward and ventrally much like the tracks of an armored car. The NE that once faced anteriorly is now drawn into the future nasal cavity. At the leading edge of the NVE lies the AH placode and just behind it, the nasal placode (NP). The apical surface of the nasal vestibular epithelium now faces ventrally.

As the p6 NE and NP are pulled into the future nasal cavity, similar changes occur with the p5 zone of nonneural ectoderm. This zone of the upper beak epithelium is slated to become the future epidermis of the nose and upper lip. Being in continuity with the p6 NE, the p5 UBE is pulled forward. The topology of the nasal epithelium resembled the letter U turned 90 degrees onto its side. The upper limb of the U is the keratinized p5 epithelium of the nasal dorsum. Neural crest that has migrated immediately below the p5 epithelium will be programmed by it to form dermis and the nasal bones. The lower limb of the U is the nonkeratinized p6 epithelium of the nasal vestibule. The vomeronasal organ, located in the membranous septum, probably represents the remnant of the nasal placode. Neural crest that has migrated immediately

above the p6 epithelium will be programmed by it to form the alar and triangular cartilages. The future nose consists of two such p6/p5 systems. The medial walls contain p6 neural crest. As these approximate, the perpendicular plate of the ethmoid and the nasal septum result.

Another way to visualize the topology of the nasal chamber is to imagine a condom placed on a flat surface. The tip of the reservoir represents the p6 nasal placode, while the remaining latex of the reservoir is the p6 nasal epithelium. All the rest of the condom is p5 nasal skin. If the tip of the reservoir is glued to the table and the condom inflated, an invagination will occur. This will place the nasal skin on the outside and the nasal epithelium on the inside.

Timing of Neural Crest Migration

Age in embryos is calculated by the presence of anatomic landmarks. Because somites appear at absolutely regular intervals and are readily counted, somite-stage is a reliable means to measure time in a developing embryo. Neural crest migrates from different anatomic sites at different times. To study this pattern, Osumi-Yamashita labeled neural crest populations from four zones: rostral rhomboencephalon RNCr, caudal rhombencephalon RNCc, mesencephalon MNC, and caudal prosencephalon PNC. Neural crest migration from each zone was studied and the timing of its completion (as measured by somite stage) was determined.

At 11-somites, RNCr (neural crest from r2 to r7) migration was complete. This zone populates the first three pharyngeal arches in craniocaudal order. Thus, first arch bones such as the mandible appear before third arch derivatives such as the greater cornu of the hyoid. At the 14-somite stage, two distinct zones complete their migrations. RNCc (neural crest from r7 to r11) completes its migration in cranio-caudal order. MN, on the other hand (from m1 to m2 and r0 to r1), proceeds caudal to cranial. That is r1 migrates then m1. At the 16-somite stage, PNC migration is complete. PNC proliferation appears to proceed in caudal-to-cranial order, like toothpaste being squeezed from a tube. Thus, the p5 zone is populated by PNC prior to p6. Placodes are activated by the presence of underlying neural crest. Experimentally, it is known that the p5 optic placode certainly appears before the p6 olfactory placode.

The take-home message of this work is that cranial neural crest cells depart from the neural folds in strict order according to the neurologic maturation of the corresponding neuromeres. Thus, RNC from r2 to r11 migrates in cranio-caudal order. MNC gets started a bit later, because hindbrain induces midbrain. First, rhombomere, being biologically associated with midbrain, start first and is followed by the mesomere. Whether any significance can be placed upon cells coming from the diminutive neuromeres r0 and m2 is unknown, so we just concentrate on r1 and m1. Finally, since prosencephalon is induced after mesencephalon and develops in caudal to cranial fashion, it makes sense that PNC migration takes place last and follows the order p1 > p2 > p3. Once again, what significance can be attached to the localization of p1 vs. p2 in terms of the derivatives produced is unknown.

Pathologies tend to be more severe the earlier in development they strike. A very late "hit" on neural crest migration to the midline could cause a mild holoprosencephaly due to absence of the ethmoid complex without affecting the orbit. The more severe is this type of neurocrisopathy, the greater the degree of hypotelorism will result.

Fate of the Neural Crest: The Role of Epithelial "Programming"

In the preceding discussion, we have seen how neural crest derivatives receive instructions from the nonneural ectoderm as to what structures they should form. Many important surgical implications flow from these considerations. NVE (nasal vestibular epithelium) from p6 determines the size and shape of the upper lateral and septal cartilages of the nose, whereas the UBE (nasal skin) from p5 will affect the formation of lower lateral nasal cartilages and the nasal bones. Might a similar role be exerted by the foregut endoderm?

Recent experimental work from the laboratories of Couly and LeDouarin is of enormous clinical significance in this regard for it supports the hypothesis that foregut endoderm (FGE) plays the decisive role in the formation of neural crest bones and cartilages of the pharyngeal arches. First, let it be said that neural crest cells if left to themselves in the Petri dish will form cartilage. Next, it is necessary that we recognize that certain membranous bones of neural crest form via cartilaginous intermediates, while others form directly within membrane. By applying techniques of surgical extirpation and transplantation in the previously described quail chick chimera model, these workers mapped out distinct territories of FGE and found that specific zones were responsible for the production of specific bones and cartilages. This yielded a "map" of FGE in which endoderm destined to form Sessel's pouch was found to underlie the frontonasal bud, while the remainder of FGE resulted from the summation of individual outgrowths of endoderm corresponding to the pharyngeal pouches.

Following our previous discussion of gastrulation, it is possible to conceive foregut endoderm as having a neuromeric code as well. *The r1 zone is unclear* because the vestibular lining is all a PNC/MNC complex supplied by V1.

With closure of the palate, this mucosa becomes hidden from view, but it is histologically distinct from that of the p6 nasal epithelium. This is readily apparent on lateral dissection of the septum in which the mucoperiosteum stops at the border of the vomer, the innervation changes from V1 (the medial nasal nerves) to V2 (the sphenopalatine nerve), and the blood supply switches from internal carotid along the upper septum to the external carotid sphenopalatine coursing along the septo-vomerine border. All the remaining endoderm of the pharynx and larynx begins posterior to the buccopharyngeal membrane.

Couly and Le Douarin made an important distinction between neural crest cells emanating from differing parts of the neural fold in terms of the expression of *Hox* genes. They found that rostral neural crest corresponding to PNC and MNC constituted a domain in which *Hox* genes were not expressed (*Hox*-negative) These neural crest cells gave rise to the membranous bones of the neurocranium, the nasal capsule, and the first pharyngeal arch maxilla and mandible. Thus, the *Hox*-negative domain extended to r3. In contrast, the domain corresponding the pharyngeal arches 2–5 was considered *Hox*-positive and yielded the hyoid bone and cartilages of the visceral skeleton. Specific mapping of the hyoid-forming region of FGE revealed that extirpation of specific zones caused failure of formation of corresponding components of the hyoid. When the entire hyoid region was reverse 180 degrees, the orientation of the hyoid bone was reversed as well.

The take-home message of this work and previous studies by the same authors is that pharyngeal endoderm is required from neural crest cells to become cartilage or membranous bones based on cartilage, while ectoderm is required for neural crest to become membranous bone via the classic mechanism (no cartilaginous intermediate). Hox-negative neural crest is exclusively responsible for the generation of the facial skeleton, but does not possess the information required to pattern the skeleton. For this to occur, FGE is required. Defined areas of FGE induce the formation of specific bones and cartilages from cephalic neural crest. Information required to determine the axes of facial bones depend upon the spatial orientation of specific zones of FGE. The ability to respond to patterning cues from FGE is exclusive to non-Hox-expressing cephalic neural crest. Cells form different zones of the *Hox*-negative domain which behave in an equivalent manner. Due to enormous regenerative capability, up to 3/4 of the neural fold responsible for cephalic morphogenesis can be removed with no consequences. This implies that the pathways taken by PNC and MNC are not inherent to the neural cells per se, but are determined by the microenvironment through which the neural crest cells migrate/proliferate. Although the nature of the signaling employed by FGE to instruct neural crest is uncertain, it has been shown that the spatial orientation of branchial pouch endoderm is established prior to the migration of neural crest into the pharyngeal arches. Protein markers such as BMP7, FGF8, and Pax1 are found in different regions of the FGE irrespective of whether neural crest cells are present or not. Indeed, these polarities may well be established prior to formation of the pharyngeal arches, i.e., they may be determined by the spatiotemporal sequence of cell movements at gastrulation itself.

These considerations allow us to think of facial bones as the products of specific biosynthetic units or fields corresponding to the concept of functional matrix first elaborated by Moss years ago. Individual deficits of the facial bones as manifested in craniofacial clefts would thus result from very early insults or aberrations of either epithelial programming units, in the neural crest cells that come to populate them, or in the biochemistry of epithelial-mesenchyme interaction.

Ectomeres and Endomeres: A Final Note

Le Dourain et al. have mapped out developmental fate of craniofacial ectoderm. Recall that gastrulation does not take place anterior to r0–r1. The buccopharyngeal membrane maps to r1 ectoderm and r1 endoderm with no intervening mesoderm. With head flexion and the forward positioning of the pharyngeal arches, BPM is pulled inside the oral cavity. Its anterior surface is ectodermal. This means that *all mucosa of the mouth anterior to BMP is ectoderm from r2 and r3*, not endoderm. Furthermore, the future oral cavity is separated from the nasal cavity by a wall consisting of intraoral r2 ectoderm, a middle layer of MNC, and intranasal p6 vestibular mucosa. When the bottom of the nasal cavity disintegrates, prosencephalic ectoderm directly meets rhombencephalic ectoderm at the border between septum and perpendicular ethmoid (p6) versus vomer (r2). Upper eyelid skin maps to r1. The distribution pattern of V1 scalp demonstrates persistence of PNC skin in the midline all the way back to posterior fontanelle (Fig. 1.46).

Endoderm begins behind the BMP. After it undergoes apoptosis, the oropharynx is mapped with concentric rings first arch, with a small zone of r2 mucosa being dosal. The corresponding to r3 mucosa is much larger and lies rostral to the tonsil. Second arch mucosa is eliminated in the oropharynx such that third arch mucosa abuts first arch behind the tonsillar fossa.

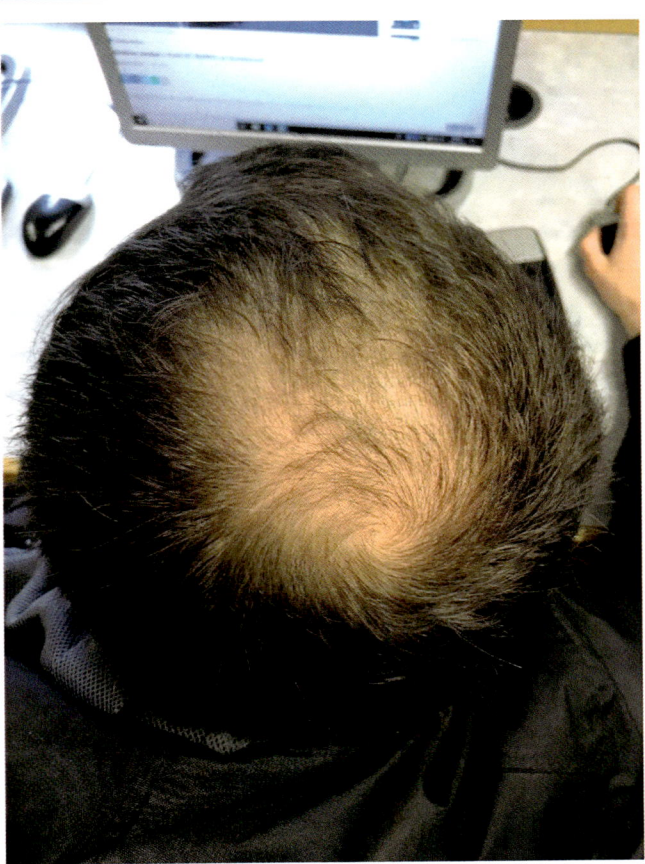

Fig. 1.46 Prosencephalic skin can be mapped on the basis of V1 innervation. StV1 skin remains in the midline backward all the way to the posterior fontanelle. Patterns of balding may reflect deficiencies in specific neural crest populations. (Reprinted from Wikimedia. Retrieved from: https://commons.wikimedia.org/wiki/File:IMG-20190314-WA0000.jpg. With permission from Creative Commons License 4.0: https://creativecommons.org/licenses/by-sa/4.0/deed.en)

Summary

We have now completed a brief synopsis of neuromeric organization. Neuromeres are developmental units of the nervous system with specific neurologic content. Outlying each neuromere are tissues of ectoderm, mesoderm, and endoderm that bear an anatomic relationship to the neuromere in three basic ways. This relationship is physical in that motor and sensory connections exist between a given neuromeric level and its target tissues. The relationship is also developmental because the target cells exit during gastrulation precisely at that same level. Finally, the relationship is chemical because the genetic definition of a neuromere is shared with those tissues with which it interacts. The model developed by

Puelles and Rubenstein is used to describe the neuroanatomy of the neuromeres. Although important details of the model are currently being refined, it has immediate clinical relevance for practicing clinicians.

The physical size and shape of each neuromere are defined by the protein products of genes expressed within the confines of the neuromere. Many of the crucial genes in this system contain a unique sequence of DNA bases, leading to a stereotypical amino acid sequence known as the homeobox. Homeobox genes are master regulators of other genes because the homeobox unit unlocks other DNA sequences. Homeobox genes are divided into two classes: *Hox* genes are homeobox genes analogous to those originally described in *Drosophila*. Non-*Hox* homeobox genes possess the homeobox sequence of bases, but bear no relationship to the *Drosophila* system.

Neural crest developing in the neural fold above a given neuromere bears a similar relationship with that of neuromere. Neural crest cell populations are organized into three main groups depending upon their location along the neuraxis. The physical behavior of neural crest migration is determined by the microanatomy of each of these three environments. Neural crest from the caudal prosencephalon moves forward as a cohesive sheet and populates the nonneural ectoderm of the rostral prosencephalon. PNC is responsible for the mesenchyme producing the bones, cartilages, and connective tissues of the fronto-orbital-nasal mass. The nature, shape, and size of these derivatives is not inherent in the cells of the PNC, but results from the instructions given to PNC by specific "target" zones of ectoderm (termed p6, p5, and p4) and foregut endoderm (r0, r1, r2′) with which the neural crest cells interact.

Mesencephalic neural crest is associated with neuromeres r0, r1, and r2′ and travels as individual cells in anatomically distinct streams. This MNC is responsible for the bulk of the orbit, the sclera, and the sphenoid. Its most caudal portion (r2′) is of great clinical importance because it interacts with r2′ endoderm to produce the vomer and premaxilla. Deficits in this epithelial-mesenchymal system are responsible for the most common forms of clefting involving the primary palate, the lip, and the secondary palate (in association with cleft lip).

Rhombencephalic neural crest begins at neuromeric level r2 and continues to r11. Neural crest migration is segmentally segregated into pairs of outlying somitomeres and somites forming the five pharyngeal arches. This process occurs in cranio-caudal order and is completed over two time periods. Population of the first three arches (the rostral RNC or RNCr) coincides with that of MNC. A second wave (the caudal RNC or RNCc) populates pharyngeal arches 4 and 5.

This paper permits us to think about the head and neck in terms of neuromeric terminology. Relationships between the processes of neurulation and gastrulation have been presented to demonstrate the manner in which neuromeric anatomy is established in the embryo. We are now in a position to describe in detail the static anatomic structures that result from this system. The neuromeric "map" of craniofacial bones, dermis, dura, muscles, and fascia will be the subject of the next part of this series.

References

1. Carlson BM. Human embryology and developmental biology. St. Louis: Mosby; 1999.
2. Gilbert S. Developmental biology. 7th ed. Sunderland, MA: Sinauer Associates; 2003.
3. Carstens MH. The sliding sulcus procedure: simultaneous repair of unilateral clefts of the lip and primary palate—a new technique. J Craniofac Surg. 1999;10:415–34.
4. Carstens MH. The spectrum of minimal clefting: process-oriented cleft management in the presence of an intact alveolus. J Craniofac Surg. 2000;11:270–94.
5. Carstens MH. Correction of the unilateral cleft lip nasal deformity using the sliding sulcus procedure. J Craniofac Surg. 1999;10:346–64.
6. Carstens MH. Correction of the bilateral cleft using the sliding sulcus technique. J Craniofac Surg. 2000;11:137–67.
7. Carstens MH. Sequential cleft management with the sliding sulcus technique and alveolar extension palatoplasty. J Craniofac Surg. 1999;10:503–18.
8. Carstens MH. Functional matrix cleft repair: a common strategy for unilateral and bilateral clefts. J Craniofac Surg. 2000;11:437–69.
9. Carstens MH. Neural tube programming and craniofacial cleft formation. I. The neuromeric organization of the head and neck. Eur J Paediatr Neurol. 2004;8:181–210.
10. Carstens M. Functional matrix cleft repair: principles and techniques. Clin Plast Surg. 2004;31:159–89.
11. Delaire J. The potential role of facial muscles in monitoring maxillary growth and morphogenesis. In: Carlson DS, McNamara Jr JA, editors. Muscle adaptation and craniofacial growth. Craniofacial growth monograph no. 8, Center for Human Growth and Development. Ann Arbor, MI: University of Michigan; 1978. p. 157–80.
12. Delaire J, Precious D, Gordeef A. The advantage of wide subperiosteal exposure in primary surgical correction of labial maxillary clefts. Scan J Plast Reconstr Surg. 1989;22:147–51.
13. La DJ. cheilo-rhinoplastie primaire pour fente labiomaxllarie congenitale unilateral. Revue Stomatolo. 1975;76:193.
14. Delaire J. Theoretical principles and technique of functional closure of the lip and nasal aperture. J Maxillofac Surg. 1975;6:109.
15. Delaire J. The potential role of facial muscles in monitoring maxillary growth and porphogenesis. In: McNamara JA, editor. Monograph 38: "Cranial growth series". Center for Human Growth and Development. Ann Arbor, MI: University of Michigan Press; 1978.
16. Delaire J, Precious S, Gordeef A. The advantage of wide subperiosteal exposure in primary surgical correction of labial maxillary clefts. Scand J Plast Reconstr Surg. 1993;22:710.
17. Markus AF, Delaire J, Smith WP. Facial balance in cleft lip and palate I. Normal development and cleft palate. Br J Oral Surg. 1992;30:287–95.
18. Precious DA, Delaire J. Surgical considerations in patients with cleft deformities. In: Bell WH, editor. Modern practice in orthognathic and reconstructive surgery. Philadelphia: WB Saunders; 1992. p. 390–425.
19. Mulliken JB. Repair of bilateral cleft lip: review, revisions, reflections. J Craniofac Surg. 2003;14:68–76.
20. His W. Beobachtungen zur Geshichte und Gamenbildung beim menschlichen embryo. Kgl Akad Wis. 1901;27.
21. His W. ABh sachs Ges (Akad) Wiss. 1902;27:347.

22. Carstens MH. Developmental anatomy of the facial midline. Plastic Surgery Educational Foundation awards presentation. American Society of Plastic Surgeons annual meeting, Los Angeles, CA, Nov 2002.

23. Berquist H. Studies on the cerebral tube in vertebrates: the neuromeres. Acta Zool. 1952;33:117–23.

24. Graham A, Papalopulu N, Krumlauf R. The murine and Drosophila homeobox gene complexes have common features of organization and expression. Cell. 1989;57:367–78.

25. Landmesser LT, editor. The assembly of the nervous system. New York: Liss; 1989.

26. Vaage S. The segmentation of the primitive neural tube in chick embryos (Gallsu domesticus). Adv Anat Embryol Cell Biol. 1969;41:1–88.

27. Butler AB, Hood M. Comparative vertebrate neuroanatomy. New York: Wiley-Liss; 1997.

28. Puelles L, Rubenstein JLR. Expression patterns of homeobox and other putative regulatory genes in the embryonic mouse forebrain suggest a neuromeric organization. Topics Neurosci. 1993;16:472–80.

29. Rubenstein JLR, Martinez S, Himamura K, Puellas L. The embryonic vertebrate forebrain: the prosomere model. Science. 1994;266:578–80.

30. Rubenstein JLR, Puelles L. Expression patterns of homeobox and other putative regulatory genes in the embryonic mouse forebrain suggest a neuromeric organization. Topics Neurosci. 1993;16:472–80.

31. Rubenstein JLR, Puelles L. Homeobox gene expression during development of the vertebrate brain. Curr Topics Dev Biol. 1994;29:1–63.

32. Puelles L, Rubenstein JLR. Forebrain gene expression domains and the evolving prosomeric model. Trends Neurosci. 2003;26:469–76.

33. Sarnat HB. Personal communication, Nov 2003.

34. Cambronero F, Puelles L. Rostrocaudal nuclear relationships in the avian medulla oblongata: a fate map with quail-chick chimeras. J Comp Neurol. 2000;427:522–45.

35. Wingate RJT, Lumsden A. Persistence of rhombomeric organization in the postsegmental hindbrain. Development. 1996;112:2143–52.

36. Muller F, O'Rahilly R. The timing and sequence of appearance of neuromeres and their derivatives in staged human embryos. Acta Anat. 1997;158:83–99.

37. O'Rahilly R, Gardner E. The timing and sequence of events in the development of the human nervous system during the embryonic period proper. Z Anat Entweck-Gesch. 1974;134:1–12.

38. O'Rahilly R, Muller F. Human embryology and teratology. 4th ed. New York: Springer-Verlag; 2001.

39. Figador MC, Stern CD. Segmental organization of embryonic diencephalons. Nature. 1993;363:630–4.

40. Larsen CW. Boundary formation and compartition in the avian diencephalon. J Neurosci. 2001;21:4699–711.

41. Bonner-Fraser M. Rostrocaudal differences within the somites confer segmental pattern to trunk neural crest migration. In: Ordahl CP, editor. Somitogenesis, part 1. San Diego: Academic Press; 2000. p. 279–96.

42. Brand-Saberi B, Whiting J, Ebesperger C, Christ B. The formation of somite compartments in the avian embryo. Int J Dev Biol. 1995;40:411–20.

43. Burke AC. Hox genes and the global patterning of the somatic mesoderm. In: Ordahl CP, editor. Somitogenesis, part 1. San Diego: Academic Press; 2000. p. 155–81.

44. Christ B, Ordahl P. Early stages of chick somite development. Anat Embryol. 1995;191:381–96.

45. Dietrich S, Schubert FR, Healy C, Sharpe PT, Lumsden A. Specification of hypaxial musculature. Development. 1998;125:2235–49.

46. Fan CM, Tessier-Levigne M. Patterning of mammalian somites by surface ectoderm and notochord: evidence for sclerotome induction by a hedgehog homolog. Cell. 1994;79:1175–86.

47. Hall BK. The embryonic development of bone. Am Sci. 1988;76:174–81.

48. Hall BK, Miyake T. Divide, accumulate, differentiate: cell differentiations in skeletal muscle revisited. Int J Dev Biol. 1995;39:881–93.

49. Kato N, Aoyama H. Dermamyotomal origin of the ribs as revealed by extirpation and transplantation experiments in chick and quail embryos. Development. 1998;125:3437–43.

50. Nowicki JL, Burke AC. Testing Hox genes by surgical manipulation. Dev Biol. 1999;210:238.

51. Ordahl CP. Myogenic lineages within the developing somites. In: Bernfeld M, editor. Molecular basis of morphogenesis. New York: Wiley-Liss; 1993. p. 165–70.

52. Pourquie O, et al. Lateral and axial signals involved in somite patterning: a role for BMP-4. Cell. 1996;84:461.

53. Pourquie O. Segmentation of the paraxial mesoderm and vertebrate somitogenesis. In: Ordahl CP, editor. Somitogenesis, part 1. San Diego: Academic Press; 2000. p. 82–106.

54. Tajbakhsh S, Spurle R. Somite development: constructing the vertebrate body. Cell. 1998;92:127–38.

55. Tam PPL, Behringer RR. Mouse gastrulation: the formation of a mammalian body plan. Mech Dev. 1997;68:3–25.

56. Tam PPL, Trainor PA. Specification and segmentation of the paraxial mesoderm. Anat Embryol. 1994;189:379–90.

57. Tam PPL, Goldman D, Camus A, Schoenwolf GC. Early events of somitogenesis in higher vertebrates: allocation of precursor cells during gastrulation and the organization of a meristic pattern in the paraxial mesoderm. In: Ordahl CP, editor. Somitogenesis, part 1. San Diego: Academic Press; 2000. p. 1–32.

58. Venters SI, Thornsteindottir S, Duxton MI. Early development of the myotome in the mouse. Dev Dyn. 1999;216:219–32.

59. Goodrich ES. Studies on the structure and development of vertebrates. London: Macmillan; 1930.

60. DeBeer GR. The development of the vertebrate skull. Oxford, 1937. Paperback reprint, University of Chicago; 1988.

61. Jarvik E. Basic structure and evolution of vertebrates. New York: Academic Press; 1980.

62. Hinrichsen K. The early development of morphology and patterns of development of the face in the human embryo, Advances in anatomy, embryology and cell biology, vol. 98. New York: Springer Verlag; 1985.

63. Meier SP. Development of the chick mesoblast: pronephros, lateral plate, and early vasculature. J Embrol Exp Morphol. 1980;55:291–306.

64. Meier SP. Morphogenesis of the chick embryo mesoblast: morphogenesis of the prechordal plate and early vasculature. J Embryol Exp Morphol. 1981;83:49–61.

65. Meier SP. The development of segmentation in the cranial region of vertebrate embryos. Scan Electron Microsc. 1982;3:1269–82.

66. Meier SP. The distribution of cranial neural crest cells during ocular morphogenesis. In: Daentl DA, editor. Clinical, structural and biochemical advances in hereditary eye research. New York: Alan R. Liss; 1982. p. 1–15.

67. Meier SP, Tam PPL. Metameric pattern in the embryonic axes of the mouse. I. Differentiation of the cranial region. Differentiation. 1982;21:95–108.

68. Meier SP. Somite formation and its relationship to metameric patterning of the mesoderm. Cell Differ. 1984;14:235–43.

69. Jacobson AG. Somitomeres: the primordial body segments. In: Bellairs R, Ede DA, Lash JW, editors. Somites in developing embryos. New York: Plenum Publ.; 1986.

70. Jacobson AG. Somitomeres: mesodermal segments of the head and neck. In: Hanken J, Hall BK, editors. The skull, Development, vol. I. Chicago: University of Chicago Press; 1993.

71. Huang R, Zhi Q, Ordahl CO, Christ B. The fate of the first avian somite. Anat Embryol. 1997;195:435–49.

72. Huang R, Zhi Q, Patel K, Wilting J, Christ b. Contribution of single somites to the skeleton and muscles of the occipital and cervical regions in avian embryos. Anat Embryol. 2000;202: 375–83.

73. Hunter RM. The development of the anterior occipital somites in the rabbit. J Morphol. 1935;57:501–31.

74. Sarnat HB, Netsky MG. Evolution of the nervous system. 2nd ed. London: Oxford University Press; 1981. ISBN: 978-0195027761.

75. Noden DM. Origins of avian ocular and periocular tissues. Exp Eye Res. 1979;29:27–43.

76. Noden DM. The embryonic origins of avian cephalic and cervical muscles and associated connective tissues. Am J Anat. 1983;168:257–76.

77. Noden DM. The role of neural crest in patterning of avian cranial, skeletal, connective, and muscle tissues. Dev Biol. 1983;96:347–56.

78. Noden DM. Patterning of avian craniofacial muscles. Dev Biol. 1986;116:347–56.

79. Noden DM. Interactions and fates of avian craniofacial mesenchyme. Development (Suppl). 1988;103:121–40.

80. Noden DM. Origins and patterns of craniofacial mesenchymal tissues. J Craniofac Genet Dev Biol. 1991;11:192–213.

81. Bock WJ. The avian skeletomuscular system. Avian Biol. 1974;4:1119–257.

82. Meier SP. Development of the chick mesoblast: morphogenesis of the prechordal plate and cranial segments. Dev Biol. 1981;83:49–61.

83. Schoenwolf G. Cell movements in the epiblast during gastrulation and neurulation in avian embryos. In: Keller R, Clark Jr WH, Griffin F, editors. Gastrulation: movements, patterns, and molecules. New York: Plenum; 1991. p. 1–28.

84. Schoenwolf G. Neurulation: coming to closure. Trends Neurosci. 1997;20:510–7.

85. Larsen WJ. Human embryology. 2nd ed. New York: Churchill Livingstone; 1997. p. 49–71.

86. Christ B, et al. Segmentation of the vertebrate body. Anat Embryol. 1998;197:1–8.

87. DeRobertis EM, Oliver G, Wright CVE. Homeobox genes and the vertebrate body plan. Sci Am. 1990;263:46–52.

88. Kardong KV. Vertebrates: comparative anatomy, function, evolution. 3rd ed. New York: McGraw Hill; 2002.

89. Lobe CG. Transcription factors and mammalian development. Curr Topics Dev Biol. 1992;27:57–63.

90. Duboule D, editor. Guide to the homeobox genes. Oxford: Oxford University Press; 1994.

91. Hunt P, Krumlauf R. Deciphering the Hox code: clues to patterning branchial regions of the head. Cell. 1991;66:1075–8.

92. Hunt P, Krumlauf R. Hox codes and positional specification in vertebrate embryonic axes. Annu Rev Cell Biol. 1992;8:227–56.

93. Keynes R, Lumsden A. Segmentation and the origin of regional diversity in the vertebrate nervous system. Neuron. 1990;2:1–9.

94. Krumlauf P. Hox genes and pattern formation in the branchial region of the vertebrate head. Trends Genet. 1993;9:106–12.

95. Lumsden A, Krumlauf R. Patterning the vertebrate neuraxis. Science. 1996;274:1109–15.

96. Mavilio F. Regulation of vertebrate homeobox-containing genes by morphogens. Eur J Biochem. 1993;212:273–88.

97. Mcginnis W, Krumlauf R. Homeobox genes and axial patterning. Cell. 1992;68:283–302.

98. Puelles L, Rubenstein JLR. In: Rossant J, Tam PPL, editors. Mouse development: patterning, morphogenesis, and organogenesis. San Diego: Academic Press; 2002. p. 37–54.

99. Scott MP. Vertebrate homeobox nomenclature. Cell. 1992;71:551–3.

100. Stein S, Fritsch R, Lenmire L, Kessel M. Checklist: vertebrate homeobox genes. Mech Dev. 1996;55:91–108.

101. Kessel M. Respecification of vertebral identities by retinoic acid. Development. 1992;118:487–501.

102. Kessel M, Balling R, Gruss P. Variations of cervical vertebrae after expression of a Hox 1.1 transgene in mice. Cell. 1990;61: 301–8.

103. Kessel M, Gruss P. Homeotic transformations of muringe vertebrae and concomitant alteration of Hox codes induced by retinoic acid. Cell. 1991;67:89–104.

104. Morris-Kay G. Retinoic acid and development. Pathobiology. 1992;60:264–70.

105. Jenkins FA Jr. The evolution and development of the dens of the mammalian axis. Anat Rec. 1969;164:174–84.

106. Kemp TS. The atlas-axis complex of the mammal-like reptiles. J Zool (London). 1969;159:223–48.

107. Wehr R, Gruss P. Pax and vertebrate development. Int J Dev Biol. 1996;40:369–77.

108. Kingsley DM. The TGF-B superfamily: new members, new receptors, and new genetic tests of function in different organisms. Genes Dev. 1994;8:133–46.

109. Wilkie AOM, et al. Function of FGFs and their receptors. Curr Biol. 1995;5:500–7.

110. Ahlgren SC, Bonner-Fraser M. Inhibition of Sonic hedgehog signaling in vivo results in craniofacial neural crest cell death. Curr Biol. 1999;9:1304–14.

111. Bellon F. Identification of Sonic hedgehog as a candidate gene responsible for holoprosencephaly. Nat Genet. 1996;14: 353–6.

112. Helms JA, Kim CH, Hu D, Minkoff R, Thaller C, Eichele G. Sonic hedgehog participates in craniofacial morphogenesis and is down-regulated by teratogenic doses of retinoic acid. Dev Biol. 1997;187:25–35.

113. Hu D, Helms JA. The role of Sonic hedgehog in normal and abnormal craniofacial morphogenesis. Development. 1999;126:4873–84.

114. Ramalho-Santos M, Melton DA, McMahon AP. Hedgehog signal regulate multiple aspects of gastrointestinal development. Development. 2000;127:2763–72.

115. Roberts DJ, et al. Sonic hedgehog is an andodermal signal inducing BMP-4 and Hox genes during induction and regionalization of the chick hindgut. Development. 1995;121:3163–74.

116. Sukegawa A, Narita T, Kameda T, Saitoh K, Nohna T, Iba H, Yasugi S, Fukuda K. The concentric structure of the developing gut is regulated by Sonic hedgehog derived from endodermal epithelium. Development. 2000;127:1971–80.

117. Azar Y, Eyal-Giladi H. Marginal zone cells: the primitive streak-inducing component of the primary hyoblast in the chick. J Embryol Exp Morphol. 1979;52:79–88.

118. Eyal-Giladi H. The establishment of the axis in chordates: facts and speculations. Development. 1997;124:2285–890.

119. Blechschmidt E. The human embryo. Stuttgart: Schattauer; 1964.

120. Blechschmidt E. The beginnings of life. New York: Springer-Verlag; 1977.

121. Lemire L, Kessel M. Gastrulation and homeobox genes in chick embryos. Mech Dev. 1997;67:3–16.

122. Schoenwolf G, Garcia-Martinez V, Dias MS. Mesoderm movement and fate during avian gastrulation and neurulation. Dev Dyn. 1992;193:235–48.

123. Tam PPL, Beddington RSP. The formation of mesodermal tissues in the mouse embryo during gastrulation and early organogenesis. Development. 1987;99:109–26.

124. Tam PPL, Beddington RSP. Establishment and organization of germ layers in the gastrulating mouse embryo. In: Postimplantation development in the mouse. Ciba Found Symp. 1992;165: 27–49.

125. Tam PPL, Williams EA, Chan WY. Gastrulation in the mouse embryo: ultrastructural and molecular aspects of germ layer morphogenesis. Microsc Res Tech. 1993;26:301–28.

126. Tam PPL, Zhou SX. The allocation of epiblast cells to ectodermal and germ-line lineage is influenced by the position of the cells in the gastrulating mouse embryo. Dev Biol. 1996;178:124–32.

127. Tam PPL, Parameswaran M, Kinder SJ, Weinberger RP. The allocation of epiblast cells to the embryonic heart and other mesodermal lineages: the role of ingression and tissue movement during gastrulation. Development. 1997;124:1631–42.

128. Noden DM. Vertebrate craniofacial development: the relation between ontogenetic process and morphological outcome. Brain Behav Evol. 1991;38:190–225.

129. Rhinn M, Brand M. The midbrain-hindbrain boundary organizer. Curr Opin Neurobiol. 2001;11:34–42.

130. Hall BK. The neural crest in development and evolution. New York: Academic Press; 1997.

131. Le Douarin NM, Kalcheim C. The neural crest. 2nd ed. Cambridge: Cambridge University Press; 1999.

132. Cobos I, et al. Fate map of the avian anterior forebrain at the 4 somite stage, based on the analysis of quail-chick chimeras. Dev Biol. 2001;239:46–67.

133. Couly GF, LeDourain NM. Mapping of the early neural primordium in quail-chick chimeras I. Developmental relationships between placodes, facial ectoderm, and prosencephalon. Dev Biol. 1985;110:422–39.

134. Couly GF, LeDourain NM. Mapping of the early neurala primordium in quail-chick chimeras II. The prosencephalic neural plate and neural folds: implications for the genesis of cephalic human congenital abnormalities. Dev Biol. 1987;120:198–214.

135. Couly GF, Le Douarin NM. Head morphogenesis in embryonic avian chimeras: evidence for a segmental pattern in the ectoderm corresponding to the neuromeres. Development. 1990;108:543–58.

136. Couly GF, Coulty PM, Le Douarin NM. The developmental fate of the cephalic mesoderm in quail-chick chimeras. Development. 1992;114:1–15.

137. Couly GF, Grapin-Botton PM, Coultey P, Le Douarin NM. The regeneration of the cephalic neural crest, a problem revisited: the regenerating cells originate from the contralateral or from the anterior and posterior neural fold. Development. 1996;122:3393–407.

138. Couly G, Grapin-Botton A, Coulty PM, Le Douarin NM. Determination of the identity of the derivatives of the cephalic neural crest: incompatibility between *Hox* gene expression and lower jaw development. Development. 1998;125:3445–59.

139. Fernandez-Garre P, et al. Fate map of the chicken neural plate at stage HH4. Development. 2002;129:2807–22.

140. Le Douarin NM, Catala M, Batini C. Embryonic neural chimeras in the study of vertebrate brain and head development. Int Rev Cytol. 1997;175:241–309.

141. La GG. dysplasie olfacto-genitale: agenesie des lobes olfactifs avec absence de development gonadique a la puberte. Acta Neuroveg. 1960;21:345–94.

142. Rugarli EI, Ballabio A. Kallman syndrome: from genetics to neurobiology. JAMA. 1993;270:2713–6.

143. Lieblich JM, Rogol AD, White BJ, Rosen SW. Syndrome of anosmia with hypogonadic hypogonadism (Kallman Syndrome): clinical and laboratory studies in 23 cases. Am J Med. 1982;73:506–19.

144. Molsted K, Kjaer I, Giwercman A, Vesterhauge S, Shakkbarek NE. Craniofacial morphology in patients with Kallman's syndrome with and without cleft lip. Cleft Palate Craniofac J. 1197;34:417–34.

145. de San Juan AM. Teratologia: falta total de los nervfos olfatorios con anosmia en un individuo en quien existia una atrofia congenita de los testiculos y el miembro viril. El Siglo Med. 1856;3:211.

146. Serbedzija GN, Fraser SE, Bronner-Fraser M. Pathways of neural crest migration in the mouse embryo as revealed by vital dye labeling. Development. 1990;108:605–12.

147. Serbedzija GN, Bonner-Fraser M, Fraser SE. Vital dye analysis of cranial neural crest migration in the mouse embryo. Development. 1992;116:297–307.

148. Tan SS, Morriss-Kay GM. The development and distribution of the cranial neural crest in the rat embryo. Cell Tissue Res. 1985;240:403–16.

149. Tan SS, Morriss-Kay GM. Analysis of cranial neural crest cell migration and early fates in post-implantation rat chimeras. J Embryol Exp Morphol. 1986;98:21–58.

150. Gui T, Osama-Yamashita N, Eto K. Proliferation of nasal epithelia and mesenchymal cells during primary palate formation. J Craniofac Genet Dev Biol. 1993;13:250–8.

Anatomy of Mesenchyme and the Pharyngeal Arches

2

Michael H. Carstens

Introduction

Our previous discussion centered on the anatomic components of the neuromeric system and the identification of their derivatives. The emphasis was on mesenchyme (neural crest and mesoderm) because so many structures (bone, muscles, fascia, and dura) arise from it. We examined how the mesenchyme of any particular structure could be traced back ("assigned") to one or more neuromeric zones in the developing embryo. Excess, deficiency, or outright absence of a developmental field can be attributed to pathologic processes occurring at a neuromeric level very early in development. Craniofacial developmental fields function much like Lego® pieces built one upon another in a tightly regulated sequence. When a field "goes wrong," neighboring fields cannot form correctly. The result is a facial cleft.

We shall now focus on the actual process by which such clefts occur. Our discussion has two parts, static and dynamic. We begin with a detailed analysis of individual fields, where they come from, in what order they form, and what anatomic consequences are seen in the event of field failure. Contributions of various types of mesenchyme will be discussed. We shall then describe the physical manner by which fields are assembled over time. As we shall see, the clinical observations made by Paul Tessier regarding patterns of craniofacial cleft formation (derived on a strictly empiric basis) match closely patterns of neural crest migration [1–3].

Three Caveats

1. *This subject is dense and difficult...but don't be daunted!* Take time to study the definitions section at beginning of the previous chapter and the illustrations section at the back of this chapter. This will give you a visual orientation to the terminology and concepts we are about to discuss. The legends make each illustration self-explanatory. Because one must reference these figures time and again, I have elected to keep them in one section, in a fixed intellectual order that is not in sync in the manuscript. Figure numbers in the text will therefore be out of order. Visualizing gastrulation is key to understand how the neuromeric plan is initiated and carried out. Start with Figs. 2.1, 2.2, 2.3, 2.4, 2.5, 2.6, 2.7, and 2.8
   (a) Figures 1.17–1.19 demonstrates the transformation of a single layer epiblast made of totipotent cells to a bilaminar embryo (stage 4). Note how the hypoblast (primitive endoderm) is derived for the "drop-out" of epiblast cells.
   (b) Figure 2.1 starts with the bilaminar embryo (stage 5). Here you will see the fate map of the epiblast as it gets organized with a primitive streak, but prior to the ingression of mesodermal cells that produces a trilaminar embryo.
   (c) Figure 2.2 is a traditional view of gastrulation (stage 6–7). It is static and therefore deceptive because the

M. H. Carstens (✉)
Wake Forest Institute of Regenerative Medicine, Wake Forest University, Winston-Salem, NC, USA
e-mail: mcarsten@wakehealth.edu

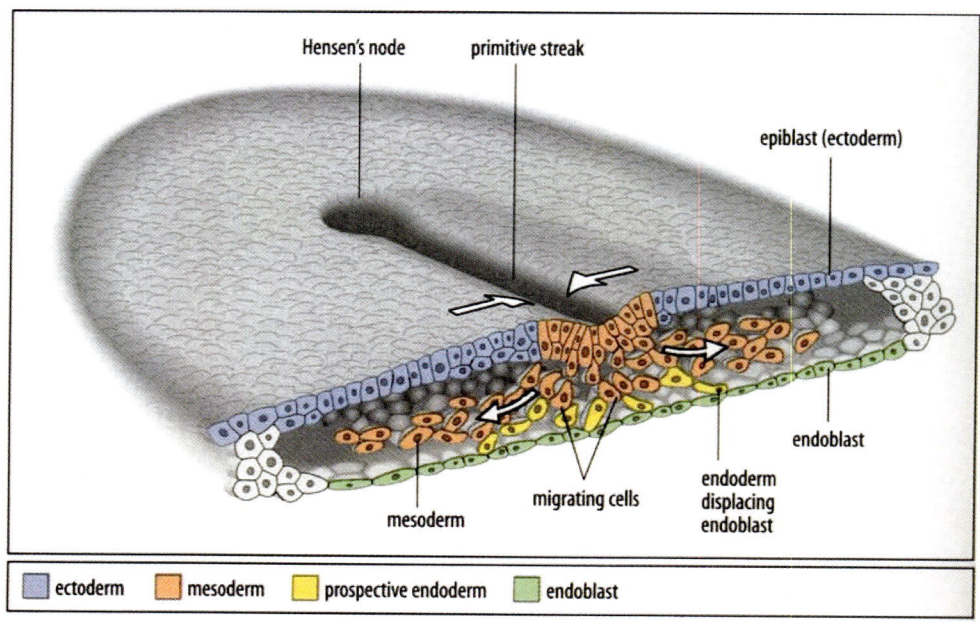

Fig. 2.1 Gastrulation: ingression of mesoderm. This figure shows the classic model of gastrulation. We will expand on this in Figs. 2.2 and 2.3. Note that at each neuromeric level the first wave of cells, future endoderm (yellow) displaces the primitive endoderm (endoblast) outside the boundaries of the embryo proper. The second wave of cells, the future mesoderm (red) moves into the potential space between the epiblast (blue) and the endoderm (yellow). Cells migrating anteriorly through Hensen's node become prechordal plate and anterior-most notochord. Populations just behind them become notochord and somites. Intermediate level of the streak makes intermediate mesoderm (GU system) and lateral plate mesoderm. The most posterior cells of the streak produce the extraembryonic mesoderm. The entire streak ratchets backwards producing these distinct cell populations at every neuromeric level. (Reprinted from Gilbert SF, Barresi M. Developmental Biology, 11th ed. Sinaurer: Sunderland, MA, 2016. Reproduced with permission of the Licensor through PLSclear)

three-dimensional sorting out of germ layers does not happen simultaneously at all levels of the embryo. It is a process that sweeps backward, being directed by the cranio-caudal maturation of neuromeres, each with its unique homeotic signature.

(d) Figure 2.3 goes into greater depth by demonstrating the regional origins of mesoderm and its migrations from initial to mid-streak (stage 6) to definitive streak (7) to neurulation (stage 8). You will see how, with time, lateral plate mesoderm surrounds paraxial mesoderm and is itself surrounded by extraembryonic mesoderm. Physical continuity between LPM and EEM means that blood vessels formed in both of these layers are in continuity, creating the eventual connection between intraembryonic LPM vessels and the placental blood supply.

2. *Consider this a conversation, not a lecture.* Our purpose is to explore an entirely new framework for understanding head and neck anatomy. Get curious! Works by Carlson, O'Rahilly, Gilbert, Liem, Kjaer will greatly enhance your understanding of development [4–8].

3. *This work is interpretive, but testable…*concepts from diverse specialties are woven together to paint a coherent picture of how development might work. The model is my own best guess as to how this system works. You are encouraged to take it further.

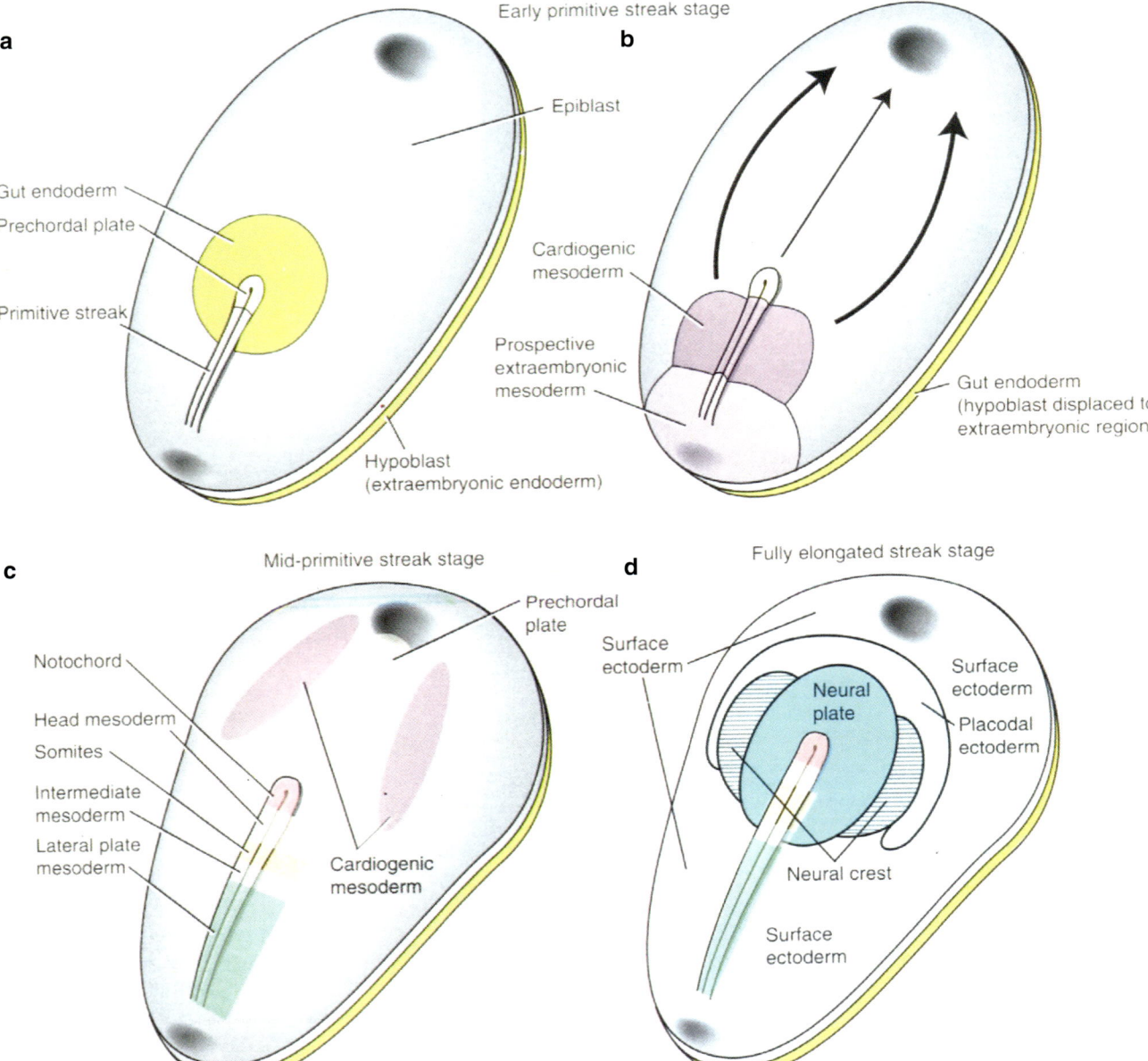

Fig. 2.2 Gastrulation: reorganization of the bilaminar disc. <u>Source</u>: Fate maps of the epiblast (chick + mouse data) showing the regions of epiblast that undergo ingression through and primitive striate and subsequent differentiation. To form the principle components of the trilaminar disc. (**a**) <u>Initial primitive streak stage</u>: Blastoderm consists of upper epiblast(yellow) and a lower hypoblast, (the future extraembryonic endoderm). Note that this layer contains within it the cells that will eventually become mematopoeietic stem cells and somatic stem cells. The prochordal plate is surrounded by prospective true gut endoderm. Oval zones indicated the future buccopharyngeal and cloacal membranes. Even at this stage, the entire epiblast has a primitive set of coordinates. (**b**) <u>Early primitive streak stage</u>: Cardiogenic mesoderm and extra-embryonic mesoderm. Curved arrows indicate the forward flow of cardiogenic mesoderm. Straight arrow shows the forward progression of the prechordal plate. Note that at this stage definitive gut endoderm has displaced the hypoblast to the peripheraly where it becomes the extraembryonic endoderm. (**c**) <u>Mid-primitive streak stage</u>: Demonstrates locations of prospective mesoderm within the epiblast and around the primitive streak. This includes the prospective notochord, head mesoderm (somitomeres) PAM somites, intermediate mesoderm, and lateral plate mesoderm. Note that at this stage the prechordal plate and cardiogenic mesoderm have entered the primitive streak and are now relocated deep to the epiblast. (**d**) Late primitive streak stage shows neural plate, definitive ectoderm, neural crest cells, and placodal ectoderm more peripherally. Cells in the cranial half of the disc have now ingressed. The stage is set for gastrulation. (Reprinted from Schoenwolf GD, Beyl SB, Brauer PB, Phillipa-West PH (eds). Larsen's Human Embryology, 5th ed. Philadelphia, PA: Churchill Livingstone; 2014. With permission from Elsevier)

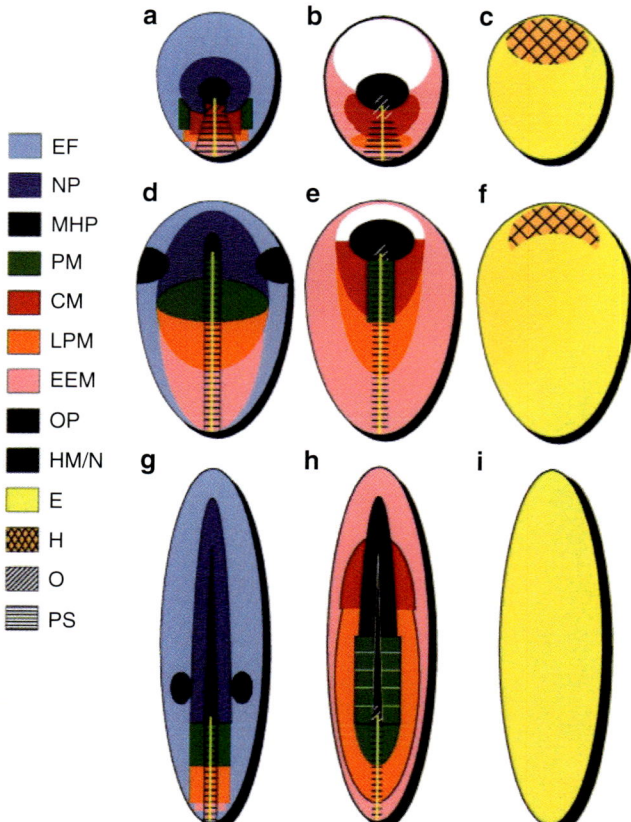

EF
NP
MHP
PM
CM
LPM
EEM
OP
HM/N
E
H
O
PS

Fig. 2.3 Detail of gastrulation. Prospective fate maps of the avian blastoderm at three stages: (**1**) Stage 6: initial-to-mid-streak; (**2**) stage 7: definitive primitive-streak, and (**3**) stage 8: mid-neurula. The epiblast is shown in (**a**), (**d**), and (**g**), the ingressed mesoblast in (**b**), (**e**), and (**h**), and the ingressed endoderm/hypoblast in (**c**), (**f**), and (**i**). At all three ranges of stages, the epiblast contains a peripheral epidermal ectoderm (EE) and a central prospective neural plate (NP), subdivided into prospective median hinge point (MHP or floor plate), neuroepithelial cells and lateral neuroepithelial cells (that is, L neuroepithelial cells and DLHP neuroepithelial cells). The neural plate lies in close proximity to the Organizer (O) at all stages, with the latter located at the rostral end of the primitive streak (PS). At the later two ranges of stages, cells of the prospective otic placodes (OP) are intermixed on each side near the rostrocaudal level of the Organizer with cells of the prospective epidermal ectoderm and neural plate. In more caudal levels of the blastoderm, the central epiblast and primitive streak contain prospective mesodermal cells. At initial-to-mid-streak stages (**a**–**c**), prospective mesoderm is ingressing through the primitive streak in the following rostrocaudal sequence: (1) prospective head mesoderm (HM or prechordal plate; not shown in epiblast); (2) prospective cardiac mesoderm (CM); (3) prospective lateral plate mesoderm (LPM); and (4) prospective extraembryonic mesoderm (EEM); prospective paraxial mesoderm (PM) occupies the epiblast lateral to the primitive streak. At definitive primitive-streak stages (**d**–**f**), prospective mesoderm is ingressing through the primitive streak in the following rostrocaudal sequence: (1) prospective head mesoderm (not shown in epiblast); (2) prospective rostral heart mesoderm (that is, conotruncus; not shown in epiblast); (3) prospective paraxial mesoderm; (4) prospective lateral plate mesoderm; and (5) prospective extraembryonic mesoderm. At mid-neurula stages (**g**–**i**), prospective mesoderm is ingressing through the primitive streak in the following rostrocaudal sequence: (1) prospective notochord (N); (2) prospective paraxial mesoderm; (3) prospective lateral plate mesoderm; and (4) prospective extraembryonic mesoderm. Endoderm (E) is ingressing through the primitive streak at all ranges of stages shown, displacing the hypoblast (H) rostrally to an extraembryonic location. (Reprinted from Schoenwolf G, Smith JL. Neurulation: coming to closure. Trends Neurosci 1997; 20(11): 510–517. With permission from Elsevier)

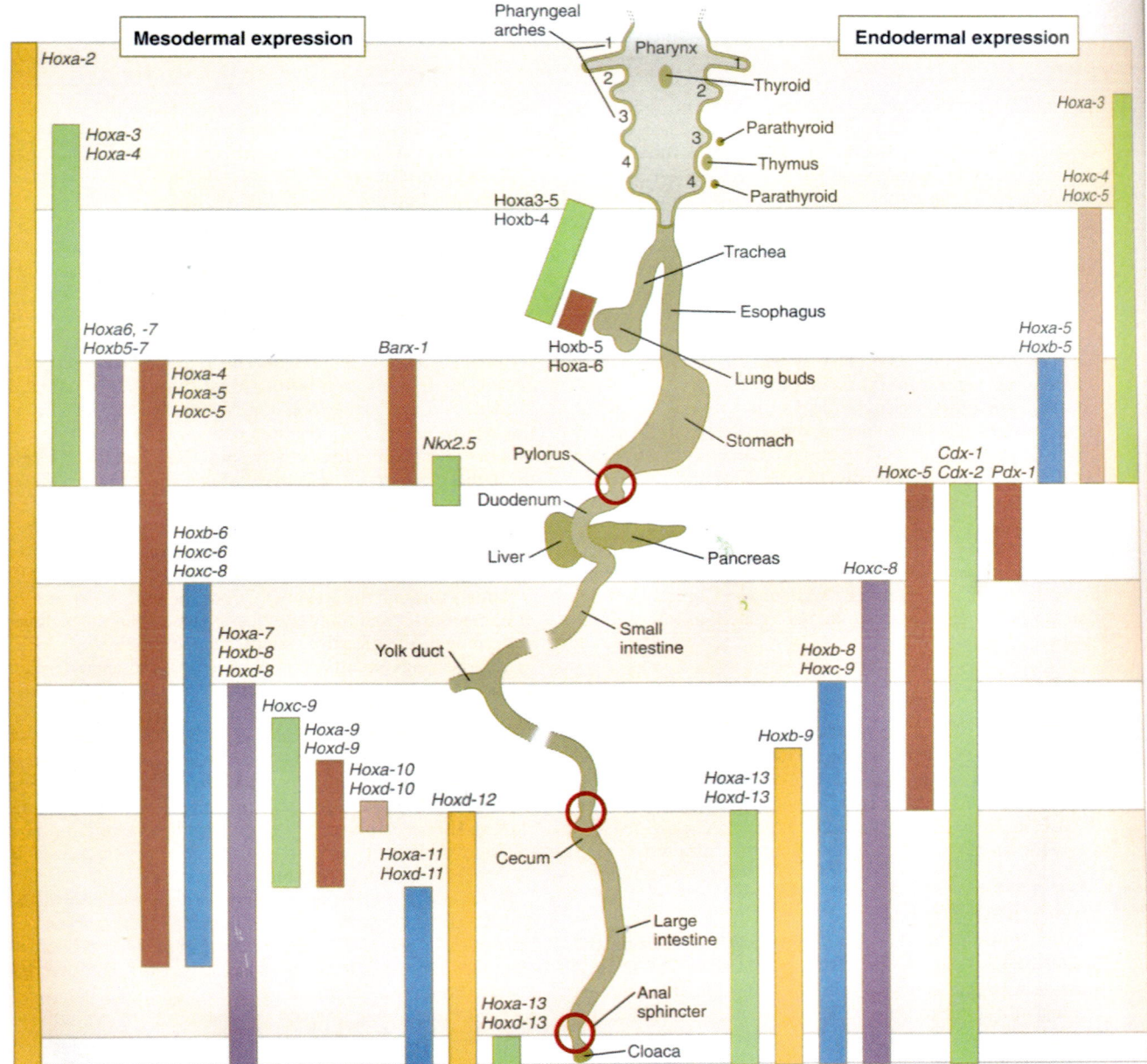

Fig. 2.4 Hox gene expression along the digestive tract. Expression patterns along the mesoderm (left) and the endoderm (right) differ. This shows how, at gastrulation, homeotic coding is imprinted, first, into the endoderm (right), and secondarily into the lateral plate mesoderm (left). For this reason, expression patterns for the same region differ between mucosa and its mesenchymal surround. (Reprinted from Carlson BM. Human Embryology and Developmental Biology, 6th edition. St. Louis, MO: Elsevier; 2019. With permission from Elsevier)

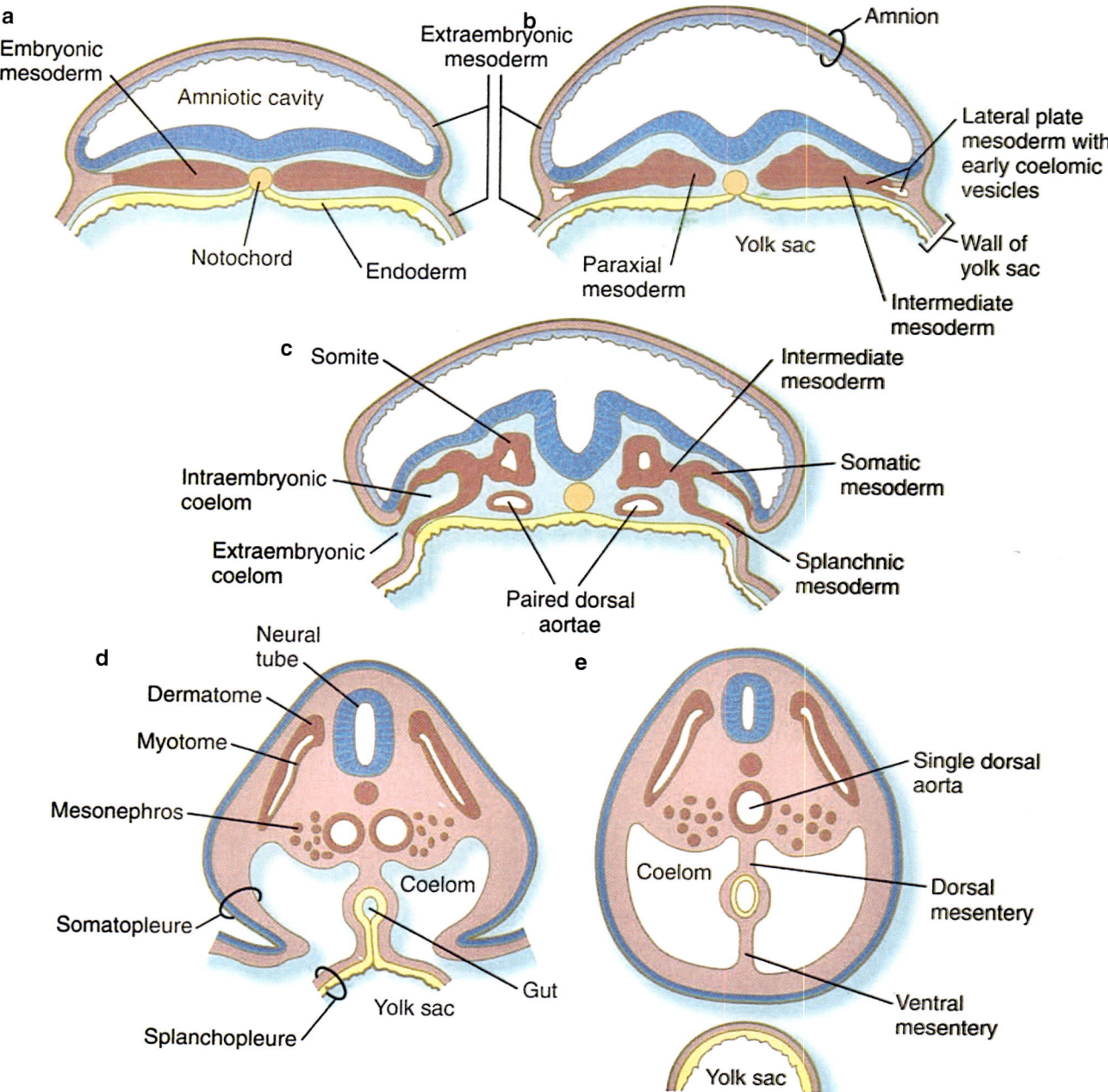

Fig. 2.5 Mesoderm development, both extra and intraembryonic, in staged cross sections. Note: (1) Vesicle formation causing splitting of the lateral plate. (2) Paired dorsal aortae are seen caudal to the heart. In the head, aortic arches ascending through the pharyngeal arches connect to the dorsal aortae (these latter will fuse). (3) Somatic mesoderm **LPMs** makes up body wall, the limb bones, and "programs" the mus-cles of the extremities. (4) Visceral mesoderm **LPMv** remains closely applied to the endoderm, forming the muscle layers surrounding the gut. (Reprinted from Carlson BM. Human Embryology and Developmental Biology, 6th edition. St. Louis, MO: Elsevier; 2019. With permission from Elsevier)

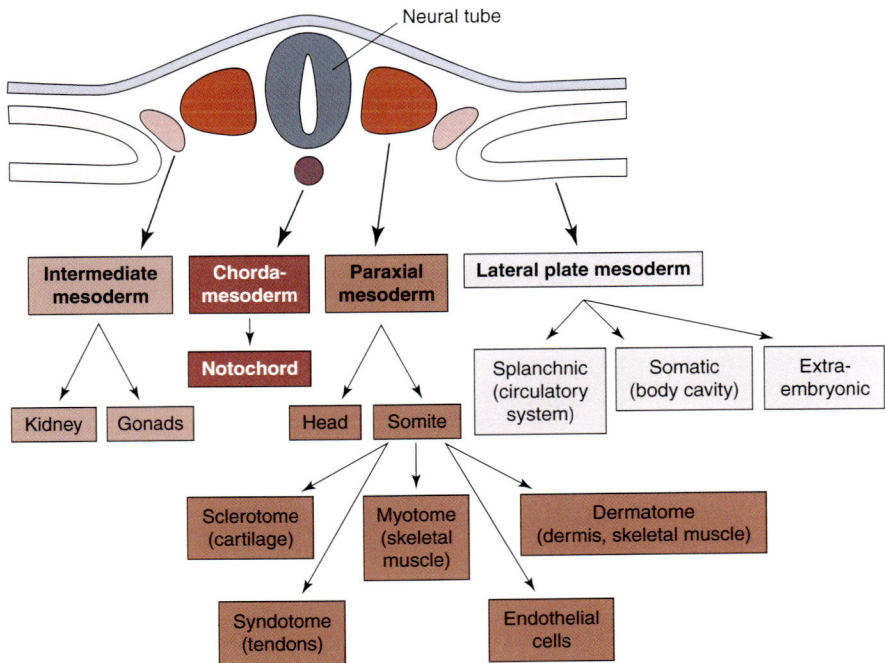

Fig. 2.6 Developmental fate of mesoderm. Blood vessels form in each of the mesodermal compartments. Ectoderm and endoderm are incapable of forming blood vessels. Note that lateral plate is programmed by skin or mucosa. LPM muscles include possibly sternocleidomastoid and trapezius. Also important are diaphragm and lining tissues of body cavities (serosal membranes). Cutaneous contributions are the dermis of the trunk and extremities, and adipose tissue. LPM vascular components populate the fat with stem cells. Thus, stem cells could potentially have neuromeric identities early in their development which separate them into distinct populations. (Reprinted from Gilbert SF, Barresi M. Developmental Biology, 11th ed. Sinaurer: Sunderland, MA, 2016. Reproduced with permission of the Licensor through PLSclear)

Fig. 2.7 Relationship between somitomeres and somites. Mesoderm segments at four somitomeres per day. Cranial somitomeres (open circles) take shape along Hensen's node until seven pairs appear. All somitomeres caudal to Sm7 have a different fate (ovals): they become transformed into somites (rectangles). At formation of the 19th somitomere, Sm8 becomes the 1st somite (S1). Through the duration of the process, the equilibrium between somitogenesis anteriorly and somitomeres synthesis posteriorly keeps the number of caudal somitomeres at 11. (Reprinted from Carlson BM. Human Embryology and Developmental Biology, 6th edition. St. Louis, MO: Elsevier; 2019. With permission from Elsevier)

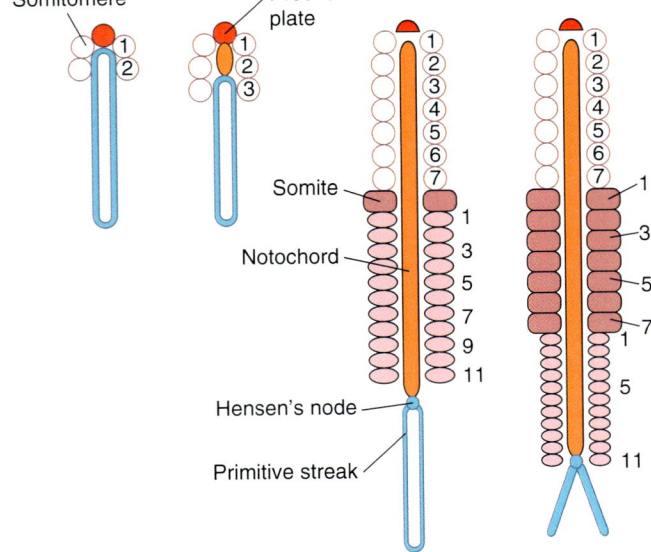

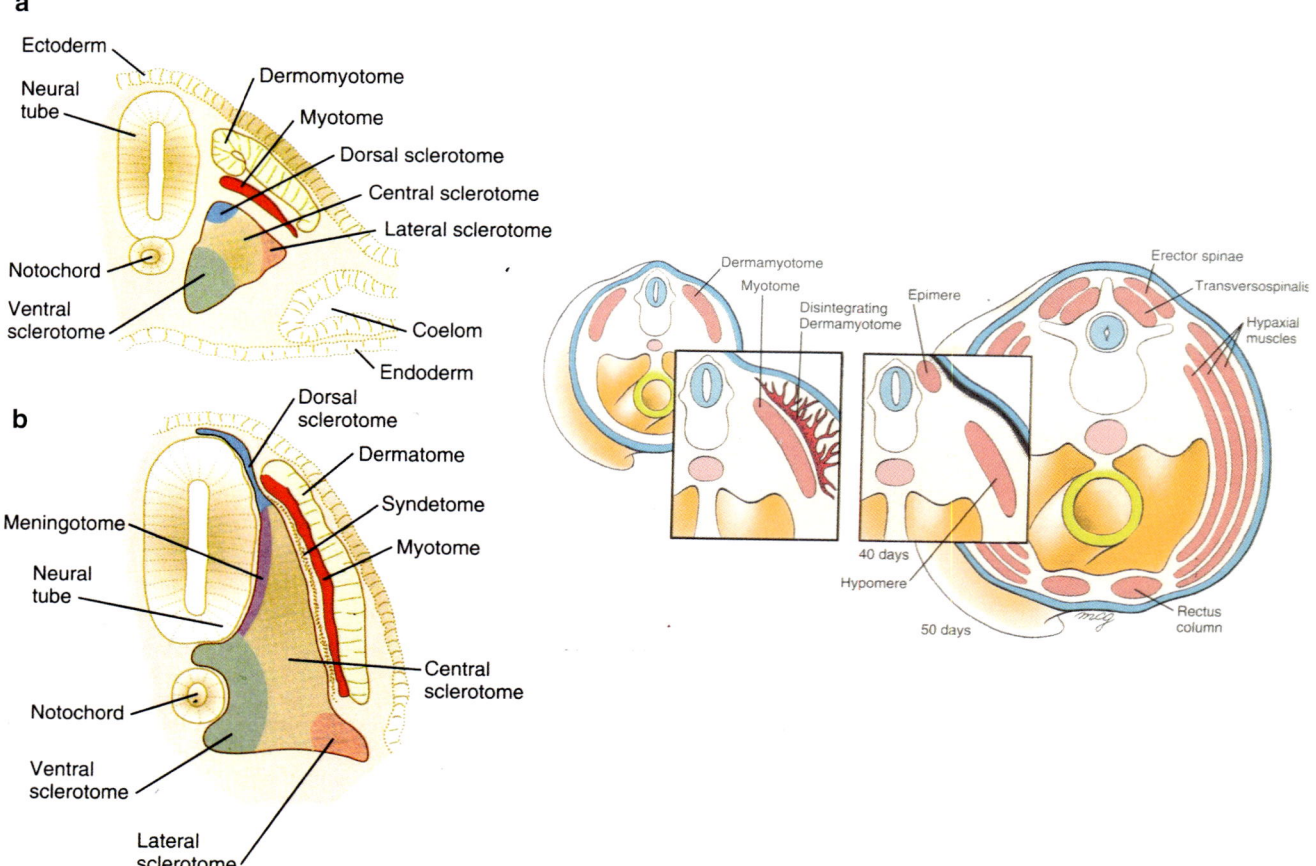

Fig. 2.8 Organization of somites: epaxial vs. hypaxial muscles. Organization of somites at early (**a**) and late (**b**) stages. Fate of the dermomyotome and myotome (cervical neuromere c2 and below). The dermamyotome produced subjacent myotome and dermis. Precursors of dermis (possibly pre-pericytes) migrate up to the surface of their corresponding neuromere. There they interact with lateral plate mesoderm to form the definitive substrate of the future dermis. Meanwhile, the myotome slits into a dorsal epimere and a ventral hypomere. The epimere forms the deep muscles of the back. The hypomere in the thorax forms three layers of the chest wall. In the abdomen, a fourth division ventral segment splits off to form the rectus. This multilevel muscle arises from a vertical fusion from neuromeric levels t12 to 15. (Left: Reprinted from Carlson BM. Human Embryology and Developmental Biology, 6th edition. St. Louis, MO: Elsevier; 2019. With permission from Elsevier. Right: Reprinted from Schoenwolf GC, Beyl SB, Brauer PR, Francis-West PH. Larsen's Human Embryology, 5th ed. Philadelphia, PA: Churchill-Livingstone; 2014. With permission from Elsevier)

Anatomy of Craniofacial Mesoderm

Craniofacial clefts result from hypoplasia or absence of recognizable anatomic structures, such as bone, cartilage, muscle, or dermis. These all originate from mesenchyme; this is to be distinguished from epithelium. Let's get these definitions straight. Epithelial cells are *polar*. One surface faces an external, extracellular environment (such gut lumen or air), while the other surface is joined with a supporting cellular network. Epithelial cells are *interconnected* with tight junctions, gap junctions, and the like. They display *internal polarity*. Organelles such as mitochondria are localized to certain regions within each cell. Mesenchymal cells are *non-polar*; they do not have "sides." Mesenchymal cells are *not attached* to each other; they can migrate within an extracellular environment. Mesenchymal cells are *internally homo-geneous*; organelles are not concentrated in specific zones within the cell.

Mesenchyme of the head and neck come from two main sources: paraxial mesoderm and neural crest. Neural crest mesenchyme will be discussed in a later section of this paper. Paraxial mesoderm results from the physical act of gastrulation, the creation of a trilaminar embryo. During stages 5–6, the embryo is converted from a single layer of cells into a bilaminar structure consisting of two epithelial layers, a dorsal *epiblast* and a ventral *hypoblast* (also known as the *primitive endoderm*). The hypoblast floats on top of the yolk sac. During stage 6, a midline *primitive streak* forms in the epiblast and appears first at its caudal end, the *connecting stalk*; it extends forward about 2/3 of the distance of the epiblast. The primitive streak provides a means by which cells living near its border lose their epithelial characteristics and become individual mesenchymal cells capable of independent migration. The process of gastrulation takes place at stage 7 when these cells pass into the primitive streak. At any given neuromeric level, they form specific new structures *depending upon the timing of their passage into the primitive streak and their anterior-posterior position along the neur-axis* (Figs. 2.1, 2.2, and 2.3).

The first cells to ingress cluster beneath the midline to form the *notochord*; these multiply rapidly in a lateral direction. This causes the hypoblast cells to be pushed out laterally [4]. Eventually, the new layer completely covers the undersurface of the epiblast and lines the yolk sac. It is now called the *definitive endoderm*. Endoderm does not demonstrate overt signs of segmentation. Nevertheless, it is organized into developmental zones, *endomeres*, in perfect register with the neuromeres of the neural plate. Like the notochord, each endomere bears a unique pattern of homeobox gene expression. The gut is organized into homeotic zones [9] (Fig. 2.4).

The next population of epiblast cells to enter the primitive streak now migrates between the epiblast above and the endoderm below to form the mesoderm. They contain a high concentration of glycosoaminoglycans that absorb water, thereby pushing the epiblast and hypoblast apart. Mesoderm fills in this potential space. It consists of three zones. As demonstrated by work of Tam et al., the identities of these zones are determined by timing of epibast cell ingression and this in turn has everything to do with the spatial positioning of these cells prior to ingression [9–14] (Figs. 2.5 and 2.6).

The first zone is produced by the migration of cells closest to the primitive streak. These cells aggregate close to the midline to form the *paraxial mesoderm* (PAM). Genes elaborated in the midline notochord and in the midline neural plate induce the paraxial mesoderm (PAM) in two zones. Anterior to Hensen's node at r0 is *prechordal plate mesoderm*. This specialized mesoderm will play a role in the development of the midbrain and forebrain and of their accompanying vascular supply [15]. Posterior to Hensen's node, PAM rounds up into discrete structures on either side of the notochord called *somitomeres*, as first described by Jacobson [16, 17]. As gastrulation proceeds in an orderly cranio-to-caudal sequence, somitomeres make their appearance at regular time intervals. The first seven somitomeres are incompletely segmented. The PAM cells are oriented around a central cavity, the *somitocoele*. When 11 somitomeres are produced (stage 10), the end of the CNS (i.e., r11) is reached. The function of PAM is the production of the axial skeleton (including the cranial base) and striated muscles. This process begins with the transformation of the eighth somitomere into the first somite and continues throughout the axis of the embryo [18–20].

The second zone is produced by mesodermal cells originating from more peripheral regions of the epiblast. This intermediate mesoderm forms a thin cord of *intermediate mesoderm* (IM) running the length of the embryo. IM gives rise to the genitourinary system [21].

The third zone is called *lateral plate mesoderm* (LPM). Gene products in the lateral aspect of the epiblast will induce the LPM to form a dorsal, *somatic* layer (LPMs) and a ventral, *visceral* layer (LPMv). It makes sense that the LPMs is in register with its overlying ectomere and the LPMv is in register with the underlying endomere (Fig. 2.10). LPMs synthesizes the appendicular skeleton, dermis of the trunk, dura of the spinal cord, and the non-craniofacial fasciae. LPMv produces the cardiovascular system, smooth muscle, and the viscera (internal organs). Note that LPM is unsplit until it reaches level c1. Lateral plate mesoderm is intensively angiogenic. In the head, more than 90% of LPM cells are angioblasts and they invade medially into the pharyngeal arches and the body of the embryo to produce (with assistance from neural crest cells) the craniofacial vascular system [21].

Paraxial Mesoderm: Somitomeres and Somites

The CNS is flanked by a continuous column of PAM. *Head mesoderm* is incompletely segmented up to somitomere 8 and segmented after that. It provides the endothelial components of all blood vessels of the CNS and pharyngeal arches. The first seven somitomeres form the cranial base of the middle cranial fossa. The posterior aspect of the vertebrate braincase is produced, not from somitomeres, but from their conversion products, the *occipital somites*. Fusion of their sclerotomes produces the cranial base: the basioccipital, exoccipital, and supraoccipital bones. Finally, PAM in its somitomeric form produces the myoblasts of all craniofacial striated muscles, save those of the tongue. The myotomes of occipital somites 1–4 are responsible for, going forward, the tongue (pre-hyoid hypobranchial group), and going backward, the sternocleidomastoid and trapezius. Muscles connecting the pharyngeal arches (post-hyoid hypobranchial group) arise from cervical somites 1–4 (Figs. 2.7 and 2.8) (Table 2.1).

The number of occipital somites varies with the organism. Avian embryos possess five, while mammalian embryos possess four (abbreviated O1–O4). Let us first examine the avian model. At the appearance of the 19th somitomere, an anatomic transformation of the 8th somitomere takes place. The caudal end of Sm7 becomes completely separate from that of the eighth somitomere. The cranial end of Sm7 remains incompletely separated from Sm6. One might ask: *is this an anatomic basis for the para segmentation pattern seen in all subsequent somite derivatives?*

In birds, somitomere 7 also develops a sclerotome and a myotome, but *no dermatome*. It becomes the first occipital somite. Production of the 13th somitomere coincides with

Table 2.1 Somitomeres and their derivatives

Somitomere	Motor nerve	Neuromere	Derivative muscles
Sm1	III	m1	Inferior rectus, medial rectus
Sm2	III	m2	Superior rectus, inferior oblique
Sm3	IV	r0	Superior oblique
Sm4	V3	r3	Mastication, quadratus pyramidalis?
Sm5	VI, VII	r4	Lateral rectus, facial?
Sm6	VI, VII	r5	Lateral rectus, mastication, facial
Sm7	IX/X	r6–r7	Palate, pharynx
Sm8//S1	X, XI, XII	r8	Pharynx, larynx//tongue, SCM-trapezius
Sm9//S2	X, XI, XII	r9	Pharynx, larynx//tongue, SCM-trapezius
Sm10//S3	X, XI, XII	r10	Tongue, SCM-trapezius
Sm11//S4	X, XI, XII	r11	Tongue, SCM-trapezius

Sm somitomere, *S* somite

Somitomeres 1–7 constitute the head mesoderm

All somitomeres from and after Sm8 undergo transformation to somites

the transformation of the eighth somitomere into the first occipital somite. This is the first completely epithelialized somite; both its cranial and caudal borders are distinct. Elegant studies by Huang have mapped the avian mastoid process, sternocleidomastoid, and trapezius muscles to the level of the seventh somitomere/first occipital somite. The spatial pattern in which the five avian sclerotomes amalgamate to produce the skull base is well demonstrated as well [22, 23].

In mammals, the pattern is slightly different. The eighth somitomere is transformed into the first occipital somite. This structure is completely epithelialized. All remaining somite formation proceeds in exactly the same manner to produce the vertebral column and peripheral mesodermal structures, all of which are unified in a segmental fashion by the peripheral nervous system. Despite these differences, the topology used by mammalian sclerotomes to produce the cranial base can be considered to follow an analogous pattern. O'Rahilly and Mueller document the contributions of each occipital somite in human material [24].

The notochord and neural tube serve as an axis dividing the embryo into dorsal and ventral sectors. All structures relating to the notochord and nervous system are considered *epaxial*; these are innervated by dorsal motor and sensory nerves. All other structures of the embryo are considered *hypaxial*; these are innervated by ventral motor and sensory nerves. Craniofacial PAM thus has two primary fates.

- Epaxial PAM remains in situ next to the CNS where it forms the endothelial skeleton of blood vessels surrounding the primitive brain (primitive head plexus and hindbrain channels) and later those arising within the brain itself (but always external to the pia mater). Because the eye is considered part of the brain, the extraocular muscles can be considered epaxial as well.
- Hypaxial PAM provides protection for the brain: it forms part of the *chondral neurocranium*. This refers to bone

structures of the skull base posterior to the pituitary (but not posterior fossa) and the petro-mastoid complex. PAM from the fourth somitomere also contributes to the parietal bone. Craniofacial blood vessels such as the dorsal aortae and aortic arches and all remaining muscles are considered hypaxial.

Three Types of Cranium

The craniofacial skeleton consists of three parts. These will be discussed in depth in a subsequent chapter. The oldest, *chondrocranium*, has been mentioned. In evolution, with the advent of gills, *splanchnocranium* arises as the mechanism to support feeding and respiration. The derivation of its mesenchyme is exclusively from neural crest (Fig. 2.19). Careful mapping of derivatives in the chick embryo by Noden demonstrates great homology with mammals [25–28]. This is true across the phylogenetic spectrum (Fig. 2.9).

Note that in the phylogeny (biologic history) of the skull, neural crest cells are almost as primitive as mesoderm. Their forebears appear in the protochordate sea squirts. So, in the production of chordates, with the invention of gills the splanchonocranium and chondrocranium appear simultaneously. In evolution, the transition between chondrichthyans (fishes with an exclusively cartilaginous skeleton, such as sharks) and osteichthyans (bony fishes) demonstrates with the invention of a new covering layer appearing over the skull, the *dermatocranium*. These membranous bones are almost always neural crest with some exceptions, the parietal bone in humans being one of them (Figs. 2.10, 2.11, and 2.12).

Second Iteration of Cranial Mesoderm: The Pharyngeal Arches

The pharyngeal arches of mammals are five in number, the fifth one being vestigial. fifth arch 5 does even not merit its own artery. Pharyngeal arches are sectors of embryo epithelium which, like balloons, are "inflated" with neural crest and project downward from the neuraxis. Each arch is in reg-

Fig. 2.9 Organization of major component of vertebrate skull. Primitive aquatic vertebrate showing chondrocranium (green), viscerocranium (orange), and dermatocranium (brown). (Reprinted from Carlson BM. Human Embryology and Developmental Biology, 6th edition. St. Louis, MO: Elsevier; 2019. With permission from Elsevier)

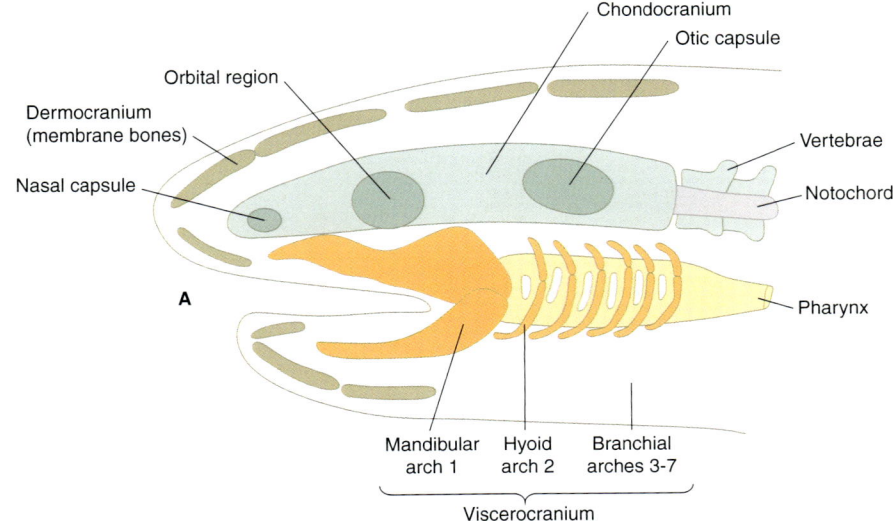

Fig. 2.10 Embryonic skull—avian versus mammal. In mammals, all cartilages anterior to pituitary fossa are neural crest. Schematic (**a**) chick and (**b**) mouse skulls showing the contributions of neural crest, paraxial and lateral mesoderms to the cranial skeleton. The avian map is based on transplantation and retroviral lineage tracings in the chick embryo; hyobranchial structures, all of which are derived from neural crest cells, are not shown. The mouse map is based largely on the location of neural crest cells, as identified by expression of *LacZ* driven by a *Wnt1* promoter in *cre-lox* transgenic embryos [28]. Origins of mouse laryngeal cartilages are by extrapolation from avian data, with the caveat that birds do not have a thyroid cartilage. Blue dots indicate the locations of crest cells present at sites of calvarial sutures. Abbreviations (Figs. 2.5, 2.9 and 2.11): *Ang* angular, *Art* articular, *Bs* basisphenoid, *Den* dentary, *Eth* ethmoid, *Lac* lacrimal, *Ls* laterosphenoid*, *Mc* mandibular cartilage, *Nc* nasal capsule, *Os* orbitosphenoid*, *Pal* palatine, *Pfr* prefrontal, *Po* postorbital, *Ps* presphenoid, *Ptr* pterygoid, *Qd* quadrate, *Qju* quadratojugal, *San* surangular, *Sqm* squamosal, *regions of the pleurosphenoid. (Reprinted from Noden DM, Trainor P. Relations and interactions between cranial mesoderm and neural crest populations. J Anat 2005 207: 575–601. With permission from John Wiley & Sons)

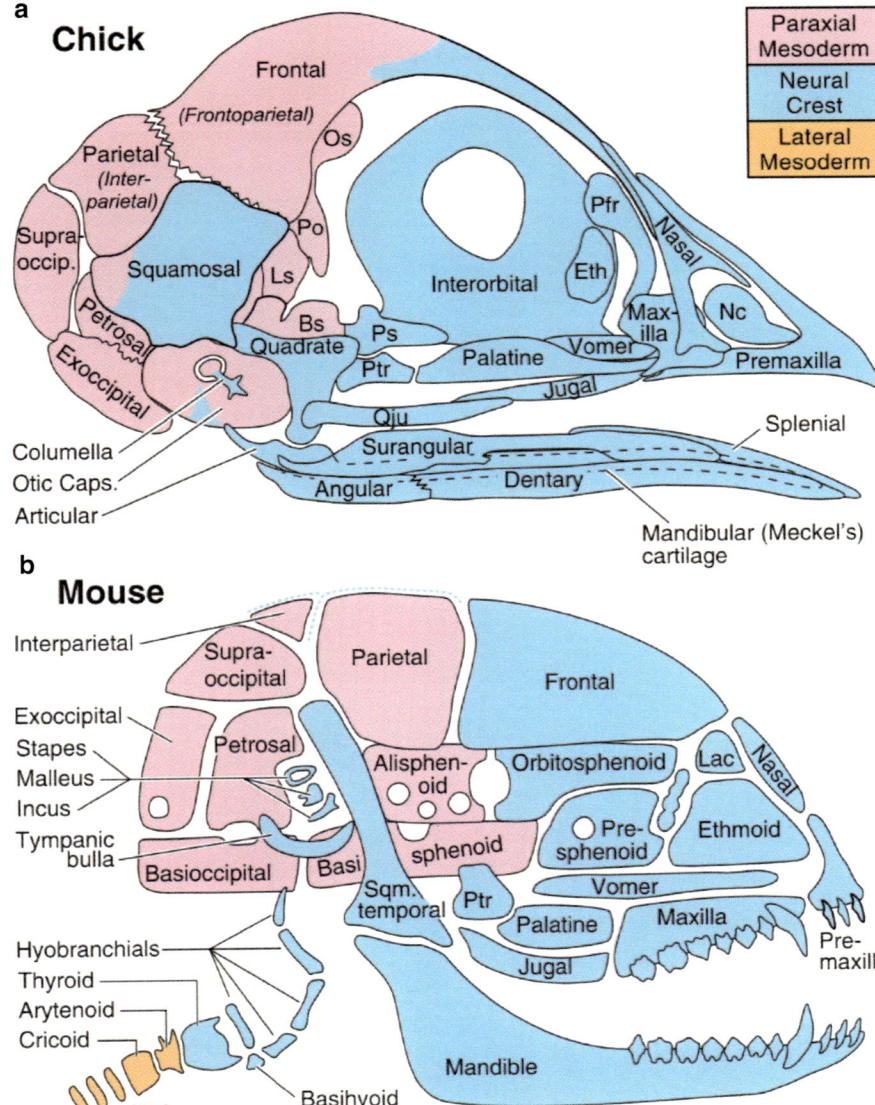

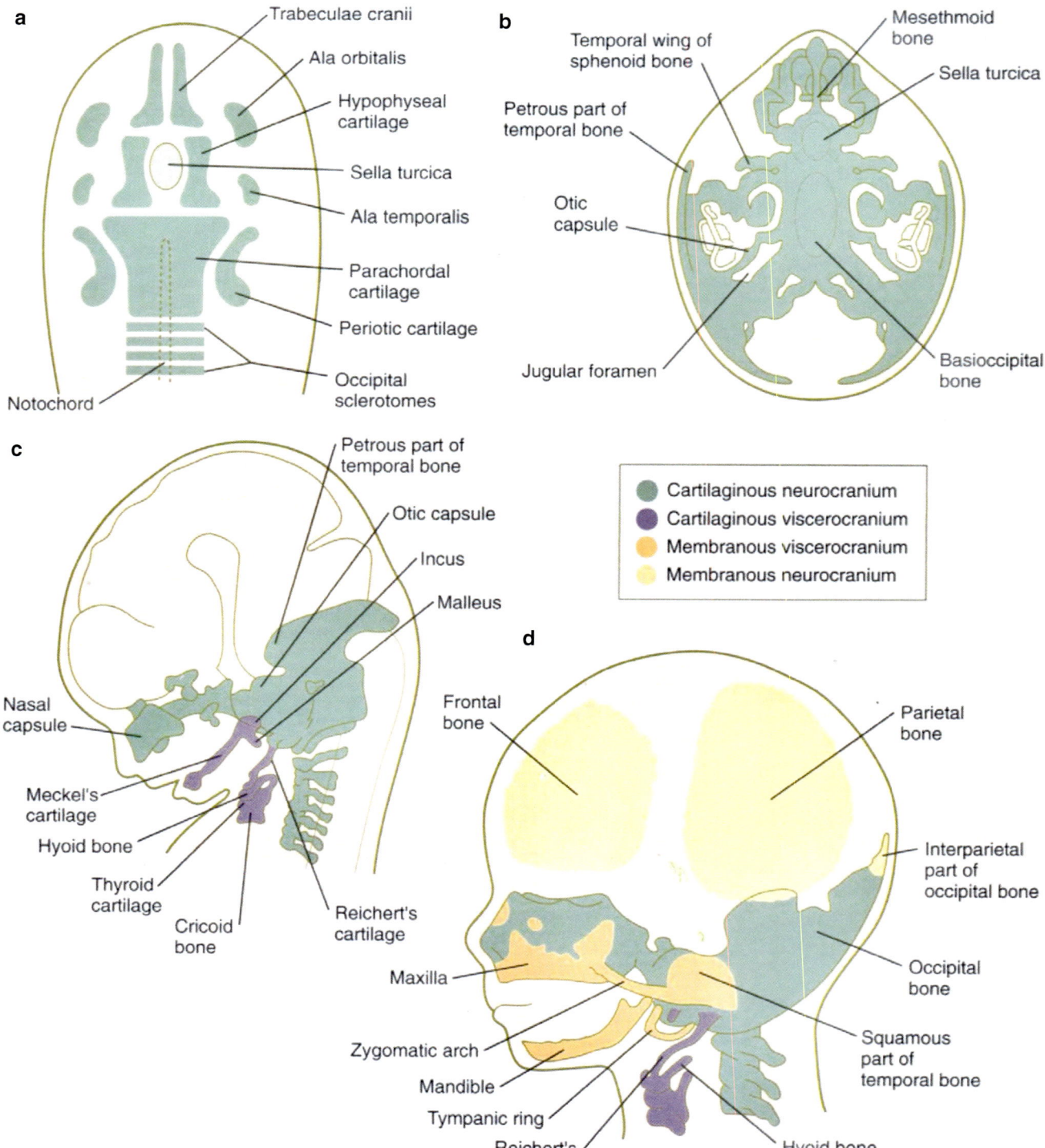

Fig. 2.11 Modern model of the cranium classified by developmental pattern. Note that neural crest can form intermediate cartilages that develop into membranous bone *or* go on to form chondral bone. PAM is exclusively chondral with the exception of scapula where some PAM portions develop in membrane. (Reprinted from Carlson BM. Human Embryology and Developmental Biology, 6th edition. St. Louis, MO: Elsevier; 2019. With permission from Elsevier)

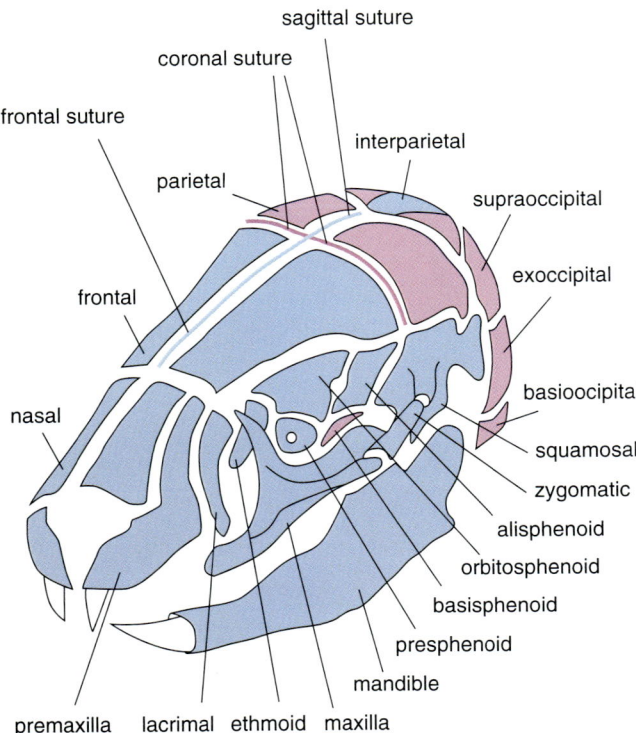

Fig. 2.12 Bones of the skull. Note interparietal has multiple origins—it originally consisted of four distinct bone fields. (Reprinted from Mishima Y, Taylor NS. Neural crest signaling pathways critical to cranial bone development and pathology. *Exp Cell Res* 2014; 325(2):138–147. With permission from Elsevier)

ister with two rhombomeres from which it obtains its neural crest mesenchyme: first arch = r1–r2, second arch = r4–r5, third arch = r6–r7, fourth arch = r8–r9, and fifth arch = r10–r11. Note that the most anterior two rhombomeres, r0 and r1, are not involved in pharyngeal arch formation. Recent evidence supports the existence of a theoretical *premandibular arch* [29, 30]. This structure is better understood as midbrain neural crest (m1–m2, r0–r1) which migrates forward to surround the eye and cover to the anterior forebrain. It is responsible for synthesis of the anterior cranial base (i.e., presphenoid, sphenoid complex, the ethmoid complex, frontal bone, and the superomedial orbit), part of the orbit, and the frontonasal mesenchyme of the upper face and forehead. The importance of the frontonaso-orbital mesenchyme will be stressed time and again in this text.

Head has no bone or cartilage representation in the pharyngeal arches. Its contribution in the arches is very limited: striated muscles. Derivatives of each arch come from spatially distinct sectors of PAM which we have designated as somitomeres. Immediately upon gastrulation, head mesoderm is divided by the otic placode into two distinct zones. *Preotic* PAM refers to the first seven semi-segmented somitomeres. *Postotic* PAM (starting with somitomere 8) is organized into somites, one for each neuromere. These are distinguished by the expression patterns of genes unique to each sector of the arch [31–34].

Somitomeric PAM is assigned to the pharyngeal arches in a pattern that is distinct from that of neural crest. Each arch is in genetic register with two rhombomeres. Neural crest cells from each rhomberomere are distributed to distinct sectors of their "target" arch. For example, r2 forms the cranial half of the first arch (maxillary-zygomatic-palatine complex), while the caudal half of first arch derived from r3 produces mandible and squamous temporal bone. The remainder of the arches are organized along the same principle (Fig. 2.13).

The clinical model of *macrostomia* suggests that an embryologic "fault line" exists along the axis of the first and secondary pharyngeal arches. Embryonic folding turns these arches 90° into contact with the ventral surface of the prosencephalon. The future mesenchyme of the maxilla (r2 neural crest) is now dorsal to that of the mandible (r3 neural crest). We note that second arch fuses into first arch almost as soon as it develops. The lateral facial cleft (#7) running from the oral commissure toward the ear reproduces this genetic boundary zone [22] with r2/r4 being cranial to the cleft and r3/r5 being caudal to it [35]. We shall see (vide infra) that these boundaries follow a common system of Distal-less (Dlx) genes for each arch [36, 37] (Fig. 2.14).

In contrast, the assignment of somitomeric muscle precursors to the arches is *not* symmetric, but *is* related to the location of their motor neurons within the hindbrain. The eye receives Sm1–Sm3 and Sm5 because their motor nerves belong to the *medial motor column* reserved for nonpharyngeal arch muscles. Muscles innervated from the *lateral motor column* are distributed as follows. first arch (Sm4), second arch Sm6, third arch (Sm7), fourth arch (Sm8 and Sm9), and fifth arch (Sm9/10?).

Somitomeres are transient structures. Their derivatives go elsewhere, either directly or becoming somites. In this model, PAM behaves quite differently depending on whether its destination is epaxial or hypaxial. Some epaxial PAM does not transition through an intermediate structure. It retains a physical relationship to the CNS, surrounds the neural tube, and supplies blood vessels. Although vascular PAM is not overtly segmented, the arterial supply it eventually produces is neuromerically organized to serve each developmental unit of the brain and spinal cord (see Chap. 6). Another destiny of PAM is to supply the pharyngeal arches with muscles. The organization of this mesoderm also follows neuromeric principles since each arch is in register with two rhombomeres.

Muscle groups appear to be divided along somitomeric lines. In birds, both Sm5 and Sm6 contain distinct muscles for lateral rectus having motor nuclei in r4 and r5. Upper and lower facial muscles may be supplied by rhombomeres r4 and r5, respectively. Superior constrictor arises from Sm7.

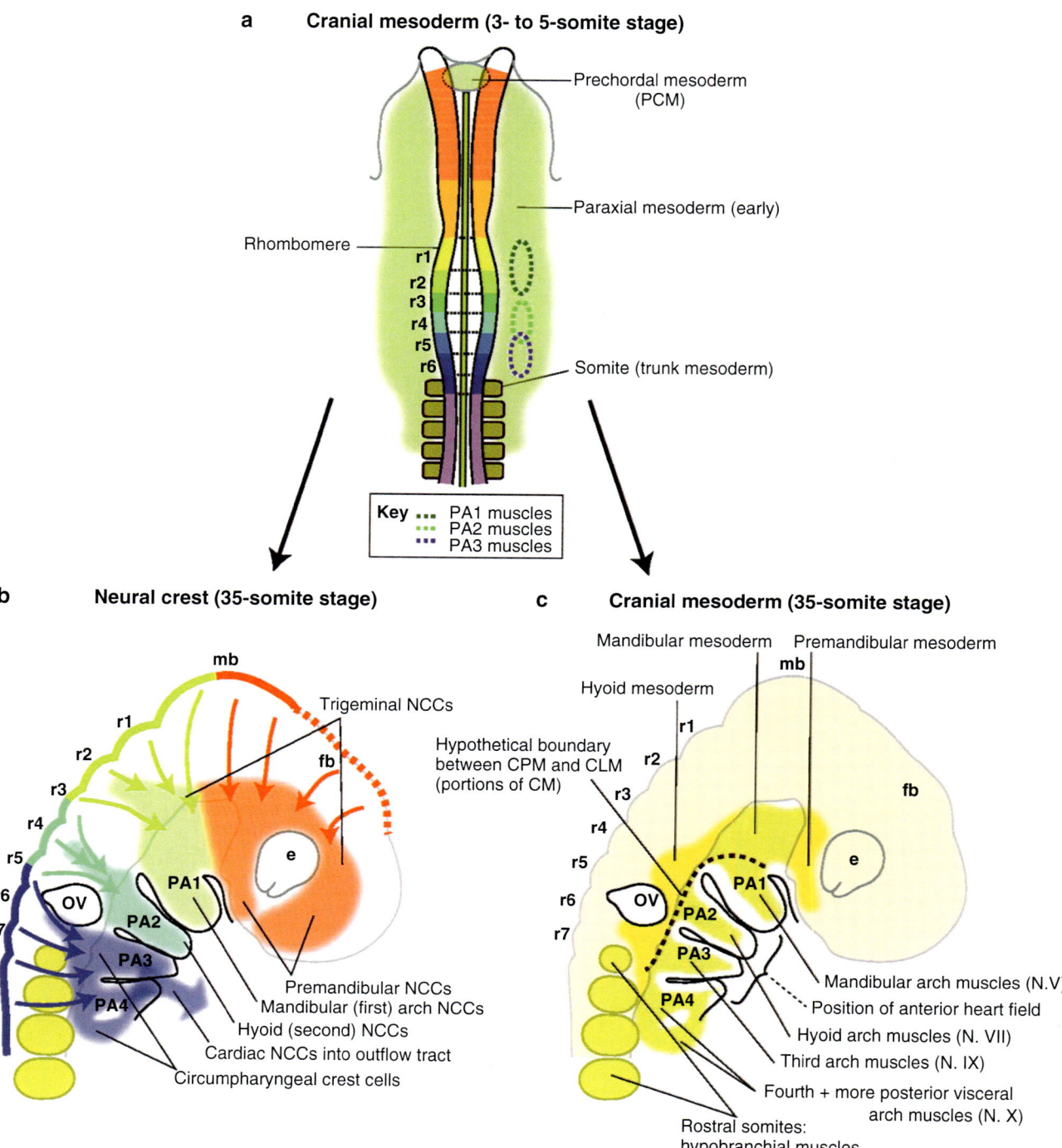

a   Cranial mesoderm (3- to 5-somite stage)

Prechordal mesoderm (PCM)

Paraxial mesoderm (early)

Rhombomere

r1
r2
r3
r4
r5
r6

Somite (trunk mesoderm)

Key   ••• PA1 muscles
      ••• PA2 muscles
      ••• PA3 muscles

b   Neural crest (35-somite stage)

mb

r1
r2
r3
r4
r5
r6
r7

Trigeminal NCCs

fb

e

OV

PA1
PA2
PA3
PA4

Premandibular NCCs
Mandibular (first) arch NCCs
Hyoid (second) NCCs
Cardiac NCCs into outflow tract
Circumpharyngeal crest cells

c   Cranial mesoderm (35-somite stage)

Mandibular mesoderm   Premandibular mesoderm

Hyoid mesoderm   mb

Hypothetical boundary
between CPM and CLM
(portions of CM)

r1
r2
r3
r4
r5
r6
r7

fb

e

OV

PA1
PA2
PA3
PA4

Mandibular arch muscles (N.V)
Position of anterior heart field
Hyoid arch muscles (N. VII)
Third arch muscles (N. IX)
Fourth + more posterior visceral
arch muscles (N. X)

Rostral somites:
hypobranchial muscles
(occipitospinal, cervical nerves)

Fig. 2.13 Migration of PAM and neural crest to pharyngeal arches. Note: avian model includes somitomere 7 as the first somite. In mammals this transition occurs at somitomere 8. (Reprinted from Ramkumar Sambasivan, Shigeru Kuratani, Shahragim Tajbakhsh. An eye on the head: the development and evolution of craniofacial muscles. Development 2011;138: 2401–2415. With permission from The Company of Biologists)

The space above it and the skull base, the *sinus of Morgagni*, transmit ascending palatine artery from second arch facial and palatine branch of third arch ascending pharyngeal, thus demarcating a boundary between second and third arch. The cranial base just above it is probably Sm5–Sm6 in register

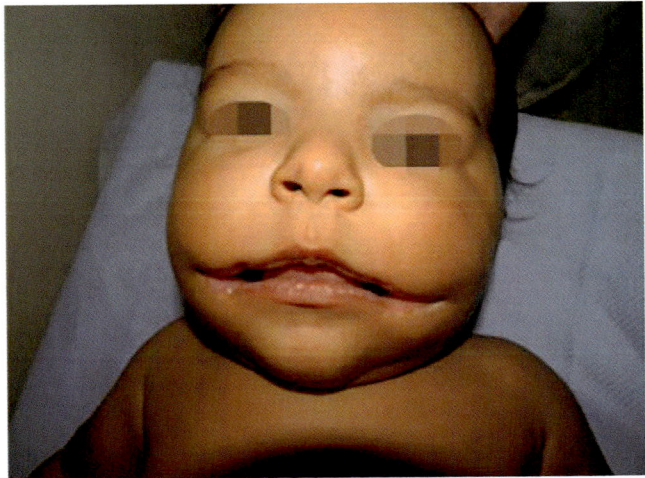

Fig. 2.14 Macrostomia. (Reprinted with permission from Gurgel do Amaral Teles G, Maarques Perfeito D. Surgical correction of unilateral and bilateral macrostoma: case report and review of the literature. *Revista Brasileira de Cirurgia Plástica* 2016; 31(2):273–277)

with second arch, while the superior constrictor per se represents myoblasts in register with third arch (Fig. 2.15).

The third destiny of PAM is to occupy an intermediate state as a somites. Epaxial somite derivatives provide skeletal cover and muscles for the spinal cord. These structures are segmented along rigidly neuromeric lines. Hypaxial somite sclerotomes demonstrate *parasegmentation*. This is most easily recognizable in vertebral column. Each vertebral body is formed from the sclerotome of two adjacent somites. This pattern, in which the cranial half of somite *n* combines with the caudal half of the somite above it, *n − 1*, is known as *parasegmentation*. For example, the third thoracic vertebra forms from the cranial half of somite T3 and the caudal half of somite T2. This segmentation pattern changes in the skull base. Occipital somites 1–3 are fused and incorporate the rostral half of S4 [38–40].

The occipitocervical junction (OCJ) is a critical evolutionary zone. Fishes have but three occipital somites and no joint between the head and body. Parasegmentation in fishes begins between the third occipital somite and the first cervical somite, i.e., level S3–S4. Tetrapods add an additional somite to the posterior fossa and attach it to three cervical somites. Parasegmentation is pushed backward to the S4–S5 level. In so doing, this tetrapod innovation produces a neck and OCJ zone. The atlas is formed from cervical somites 1

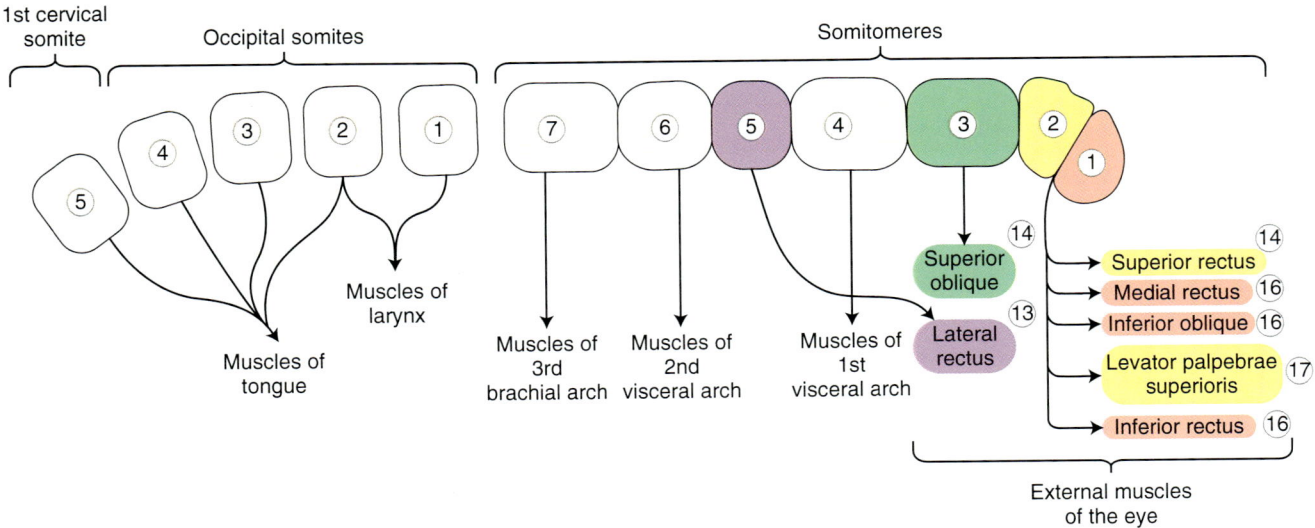

Fig. 2.15 LR/Sm5[13], SO/Sm3[14], SR/Sm2[14], MR/Sm1[16], MR/Sm1[16], IO/Sm1[16], IR/Sm1[16], LPS derived from SR/Sm2[17] First three somitomeres spatially positioned anterior to the notochord and are referred to as *prechordal mesoderm*. Somitomeres 4–5 are positioned at the tip of

the notochord. Thus, all are in position to access the globe. Avian model (depicted) has five occipital somites. In mammalian model, with four occipital somites, tongue muscles arise from S1 to S4

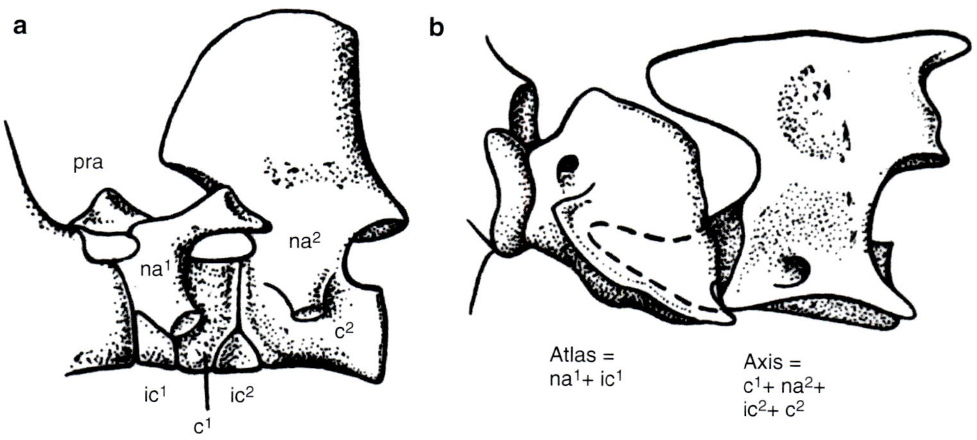

Fig. 2.16 Proatlas (pra) is seen in a primitive reptile (**a**) versus mammal (**b**). Absorption is marked by appearance in mammals of two occipital condyles on the skull base. (Reprinted from Romer AS. Vertebrate Paleontology, 3rd ed. University of Chicago Press; 1956. With permission from University of Chicago Press)

Fig. 2.17 Formation of proatlas. (Courtesy of Michael Carstens, MD)

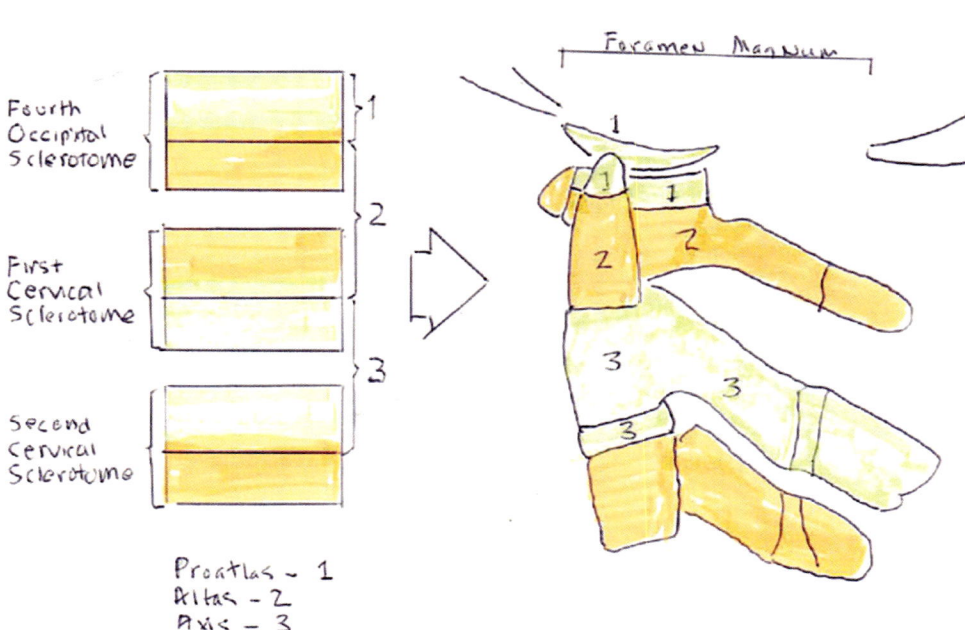

and 2. The ancient *proatlas* (now incorporated into the foramen magnum) is formed from occipital somite 4 and cervical somite 1 [41] (Figs. 2.16, 2.17, and 2.18).

The sutures of the skull represent distinct segmental field boundaries. Those bones (arising from rhombomeric levels 1–7) do not manifest parasegmentation. The mammalian parietal bone is produced by epaxial PAM from somitomeres 4 and r3–r4 neural crest, while the temporal bone is synthesized from r5 to r6 neural crest and somitomeres 6 and 7. In mammals, PAM from somitomeres 5–7 forms petromastoid. Segmentation from occipital somites 1–4 has been mapped out into a series of concentric rings. The basioccipital, exoccipital, and supraoccipital bones each receive contributions from all four occipital somites. These bones are *composite fields*. Remaining bones of the membranous neurocranium are exclusively neural crest.

Let us digress, for the sake of completeness to the behavior of PAM assigned to the neck, trunk, and extremities. We have

seen that, in somites, the lateral dermatome and myotome form the skin and muscles of the body wall and extremities. Because trunk musculature develops in a very straightforward segmental manner, the ventral (hypaxial) motor and sensory nerves that supply it are arranged in a logical, linear, spatial pattern. Muscles such as the external oblique arise from multiple myotomes. Their motor nerves receive contributions from several neuromeres. Sensory nerves, such as the iliohypogastric, reflect multiple dermatomes.

Muscles assigned to the extremities have a different neuroanatomy. They arise from a unique portion of the somitic myotome and undergo complex migratory patterns to seek out their levels of insertion. For this reason, the ventral motor and sensory nerves supplying the limbs are organized into complex "switchyards" called *plexuses*. The topologic arrangement of the roots, trunks, divisions, cords, and branches of each plexus is a *faithful replica of the migratory patterns of the target muscles*.

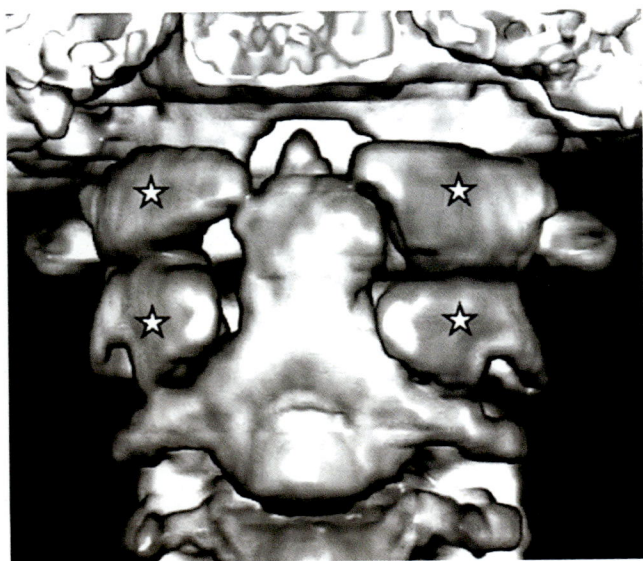

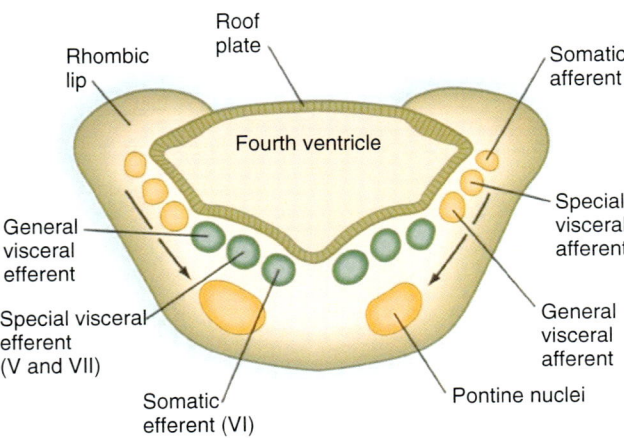

Fig. 2.19 Functional organization of cranial nerves. Medial and lateral motor column. In the brainstem, parasympathetic column, *special visceral efferent*, is interposed between them. (Reprinted from Carlson BM. Human Embryology and Developmental Biology, 6th edition. St. Louis, MO: Elsevier; 2019. With permission from Elsevier)

Fig. 2.18 Clinical case of proatlas in 8-year-old boy, anterior view showing the two lateral masses (upper two asterisks) and the dens. Note upper 1/3 of dens belongs to proatlas, lower 2/3 belongs to atlas and axis (although it appears to arise from the axis alone). (Reprinted from Spittank H, Goehmann U, Hage H, Sacher R. Presistent proatlas with persistent segmentation of the craniovertebral junction—the Tsuang-Goehmann malformation. *J Radiology Case Reports* 2016; 10(10):15–23. https://www.radiologycases.com/index.php/radiologycases/article/view/2890. With permission from Journal of Radiology Case Reports)

The muscles of the neck are a mixture of these two models. Its epaxial muscles are strictly neuromeric. The hypaxial muscles, both the vertebral muscles and the strap muscles, derived from the ancient *coracomandibularis* muscle, are supplied by the cervical plexus. We shall deal with the cervical plexus in a later chapter.

Let us refocus on paraxial mesoderm, its organization, and its relationship to the pharyngeal arches. As previously stated, shortly after gastrulation PAM originating from rhombomeres r0–r7, the so-called *head mesoderm*, has three distinct types of derivatives: (1) the *material prima* of cerebral blood vessels, (2) cartilages of the cranial base posterior to the hypophysis, and (3) craniofacial striated muscles.

- With the process of neurulation, as the neural plate rolls up upon itself to form the neural tube, it drags some PAM upward with it. This mesoderm covers the CNS in its entirety, like the paint over a newly constructed house. It provides the angioblasts which will form the *primitive head plexus* which nourished the embryonic brain between stages 6 and 8 until such time as the internal carotid and longitudinal neural arteries develop at stage 9. It explains the presence of the pial plexus. This model of PAM continues to provide all the way down the spinal cord.
- The second product of somitomeric PAM is the *materia prima* of parachordal cartilages and the otic capsule. This

mesoderm remains sessile and is located beneath the CNS. Cranial base bones anterior to the hypophysis are neural crest.

- Striated muscles are the third product of head mesoderm. These been mapped to seven individual structures within the head mesoderm, somitomeres, which also appear beginning at stage 7 and are produced at the rate of approximately 4 per day. Recall that by day 25 with the appearance of the 19th somitomere, the process of somitogenesis begins at Sm8 when it is transformed into the first occipital somite. The ultimate destiny of myoblasts within a particular somitomere is determined by the neuromeric level from which the PAM originated and the motor nerve associated with that level. Thus, muscles innervated by the *medial motor column* III, IV, and VI are destined for the eye regardless of the spatial position of the somitomere. All remaining somitomeric muscles are innervated by the *lateral motor column* and supply the pharyngeal arches. It should be noted that craniofacial muscles and their motor nerves migrate into position well after the initial formation of the pharyngeal arches themselves (Fig. 2.19).

Let's look a little more closely into the biologic rationale for the timing of this arch/somite sequence. Once again, we must return to gastrulation. Cells exiting the primitive streak at any given neuromeric level do so in a rigid spatiotemporal sequence [42]. This is seen clearly in the distribution of the various muscle blastema within the pharyngeal arches. We can think of an arch as a "pocket" into which mesodermal cells migrate. The earliest cells to enter travel most distally: *they fall to the bottom of the pocket*. They are vascularized first. Subsequent migrations occupy progressively more superficial levels of the pocket. When each pocket is full of cells, a finite amount of time has elapsed and a new pocket begins to form at the next most caudal neuromere.

Maturation and migration of cells also follow a timing sequence. The most distal cells of the pocket are the "oldest;" these are more likely to undergo population expansion than more recently arrived cells. If the distal zone of each arch is vascularized first, this transformation will precede that of the proximal population. Thus, *muscle development in the pharyngeal arches follows the spatial-temporal order of blood supply: caudal-to-cranial, deep-to-superficial, and lateral-to-medial.*

In development, the "decision" as to what zone of mesoderm is fated to become lateral plate and what zone will be transformed into paraxial mesoderm is a chemical one determined by gene products expressed in the overlying ectoderm and neural tube. One could think of these chemical signals as radio waves emitted from broadcasting towers located at specific anatomic zones of the embryo. Signals from the ectoderm induce LPMs, from the endoderm LPMv, and signals from the notochord and neural tube induce PAM. Each chemical signal diffuses outward, its concentration decreasing at greater distances from its source. Mesodermal cells closer to the neuraxis will "listen" more attentively to the stronger signals from the midline than to those from the periphery. They will organize first as somitomeres and then, from Sm8 onward, as somites.

In similar fashion, we can consider each somitomere and somite to be made up of distinctive zones, each one reflecting the presence of specific combinations of gene products unique to that zone. This "genetic map" specifies all future structures (dermis, muscle, cartilage, or bone) that will develop from that somitomere. This process is not spatially obvious in somitomeres. They appear as simple balls of cells around a central cavity. The advent of somitogenesis begins at Sm8 (which we all learned in medical school): sclerotome, myotome, and dermatome. But *the final determination of muscle "identity" is determined by signals received once the myoblasts have arrived at their final destination*, be it the eye, the arches, the pharynx, or the tongue.

Genetic zones exist with each pharyngeal arch using a common "street map" of HOX and DLX genes to establish spatial zones of epithelium. These interact with the incoming neural crest cells to create patterns that determine future structures. For example, on either side of the longitudinal axis of the first arch, the same genes are expressed that will determine tooth-bearing fields. Thus, from distal to proximal, maxillary fields of Mx1 (incisors/canines), Mx2 (premolars), and Mx3 (molars) are set directly opposite Mn1, Mn2, and Mn3. Sensory nerves and arteries will be assigned to each of these fields. This provides a *genetic basis for*

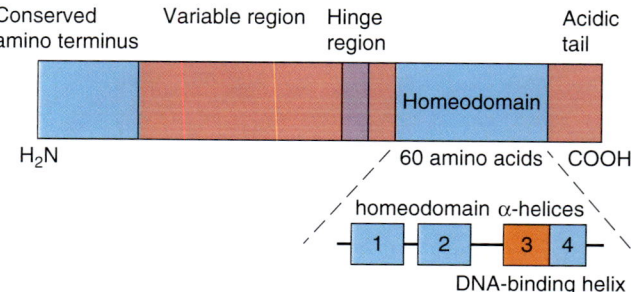

Fig. 2.20 Structure of a homeodomain protein. This highly conserved region contains 60 amino acids that form a DNA-binding helix-loop—helix structure. These are encoded by a 180 nucleotide region called a homeobox. Humans contain 39+ *HOX* genes found in four clusters of four different chromosomes. Neuromeres rostral to r3 are encoded by non-Hox homeotic genes. (Reprinted from Carlson BM. Human Embryology and Developmental Biology, 6th edition. St. Louis, MO: Elsevier; 2019. With permission from Elsevier)

occlusion. We shall see later that this order determines the eruption of teeth in a very logical sequence (Figs. 2.20, 2.21, 2.22, and 2.23).

In this regard, we should note that *the presence of one blastema determines the physical location of a subsequent blastema. One muscle cannot displace its predecessor.* This is well demonstrated by the four extrinsic muscles of the tongue. These muscles arise from occipital somites 1–4 and their motor has four distinct nuclei from r8 to r11. It is reasonable to assume the muscles arise in cranio-caudal order. Thus, styloglossus migrates from S1, assumes a proximal insertion from stylomandibular ligaments, and a distal insertion into the mucosa along the lateral aspect of the tongue. Chondroglossus, hyoglossus, and genioglossue migrate from S2, S3, and S4, respectively, and insert in progressively more medial positions, with genioglossus occupying the midline (Fig. 2.24).

We conclude this important section by stating that all cells of a particular pharyngeal arch share a common *Hox* gene code with their corresponding neuromeric zone in the central nervous system. However, the development specifics of individual structures are based on a very simple system of cellular age and distance from important sources of gene products produced in surrounding structures.

This system is readily seen in the relative amounts of neural crest material "assigned" to form pharyngeal arches versus occipital somites. As one proceeds caudally, the *relative size* of each pharyngeal arch becomes smaller and smaller compared to its corresponding somite. There is also a change in the type of product produced by the pharyngeal arch mesenchyme. Beginning with the fourth pharyngeal arch, neural crest mesenchyme ceases to be transformed into bone. Instead, it forms cartilage.

Fig. 2.21 Lateral view of organization of head and pharynx of a 30 day old human embryo with individual tissue components in register. The identity of germ layer derivatives with respect to the neuromeric level or origin is maintained throughout ontogeny. Unsegmented head mesoderm divided into seven somitomeres (not depicted here). Homeotic genes of Hox series encode from r3 caudal, while neuromeres from r2 rostral are encoded by non-hox homeotic genes. Patterns of gene expression in relation to neuroanatomic landmarks in early mammalian embryo. Bars define craniocaudal limits of expression for gene products. Cranial sensory nerves derived from neural crest and placodal precursor are laid out in proper register. *CRABP* cytoplasmic retinoic acid-binding protein, *RAR* retinoic acid receptor. **Grand generalization** (1) All pharyngeal arches are in register with pairs of rhombomeres. (2) Motor control and general sensory perception for pharyngeal arch structures are in strict neuromeric register with the brainstem via the lateral motor column and the lateral sensory column. Neurons dedicated to a specific arch may arrive at their destination either directly or indirectly, using a different nerve to gain access. (3) *Pharyngeal arch structures should be classified, not by the individual nerves that supply them, but by the neuromeres which supply the neurons in the first place.* (Reprinted from Carlson BM. Human Embryology and Developmental Biology, 6th edition. St. Louis, MO: Elsevier; 2019. With permission from Elsevier)

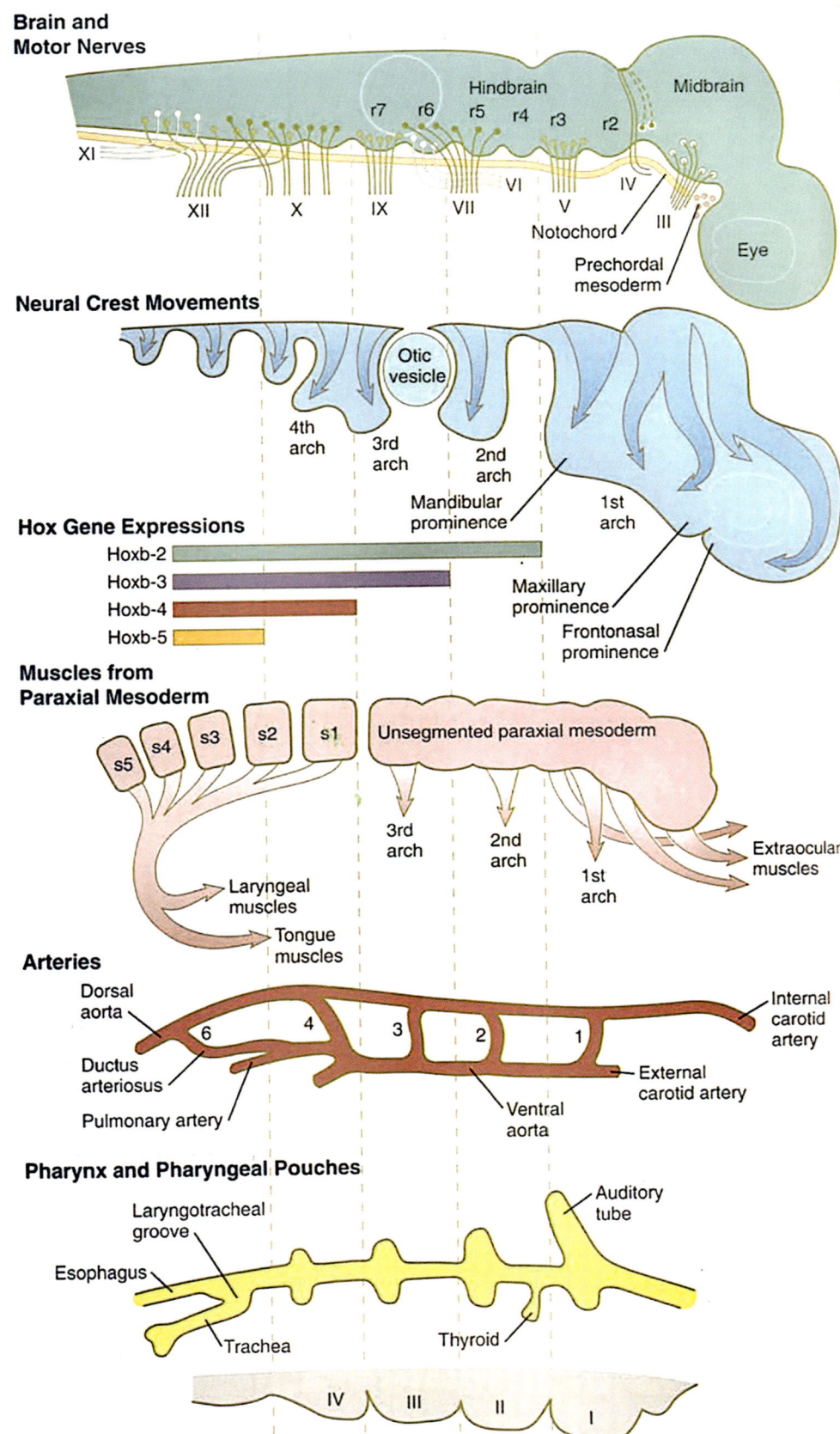

Fig. 2.22 (a) Hox vs. (b) Dlx system. Universal system of Dlx genes defines anterior-posterior zones of pharyngeal arches. (Reprinted from Minoux M, Rijli FM. Molecular mechanisms of cranial neural crest cell migration and patterning in craniofacial development. Development 2010; 137: 2605–2621. With permission from The Company of Biologists)

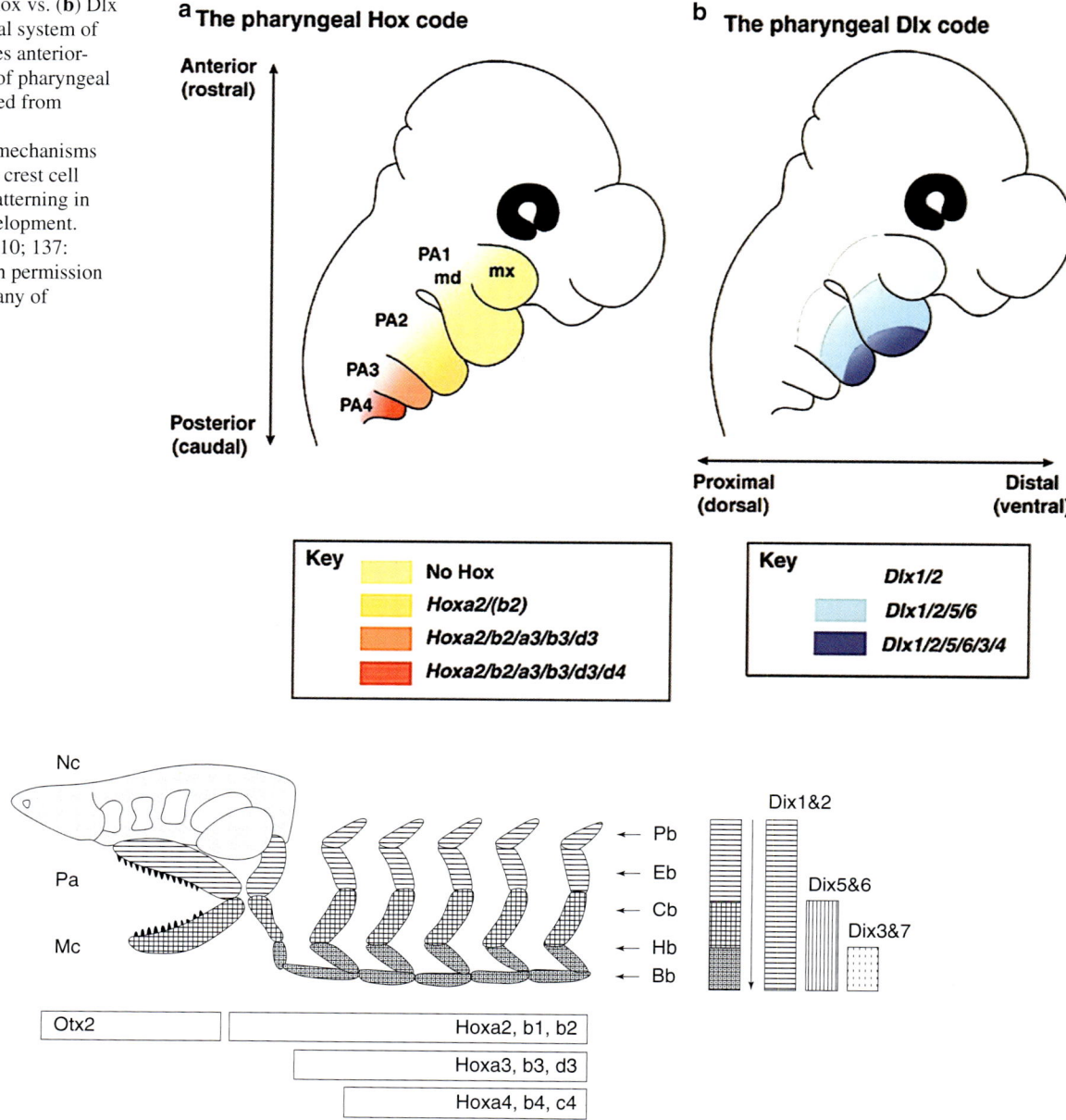

Fig. 2.23 Dlx system Note the neural crest cartilage components of the branchial arches. In humans, these are recapitulated in the formation of bone of second, third, and fourth arches. *PB* prebranchial, *EB* epibranchial, *CB* ceratobranchial, *HB* hypobranchial, *Bb* basibranchial, *NC* nasal cartilage, *Pa* palatoquadrate cartilage, *Mc* Meckel's cartilage.

Note that the homeotic coding of first pharyngeal is based, not on Hox genes, but on Otx genes. (Reprinted from Olsson L, Ericsson R, Cerny R. Vertebrate head development: Segmentation, novelties, and Homology. Theory in Biosciences. 2005 124(2): 143–163. With permission from Springer Nature)

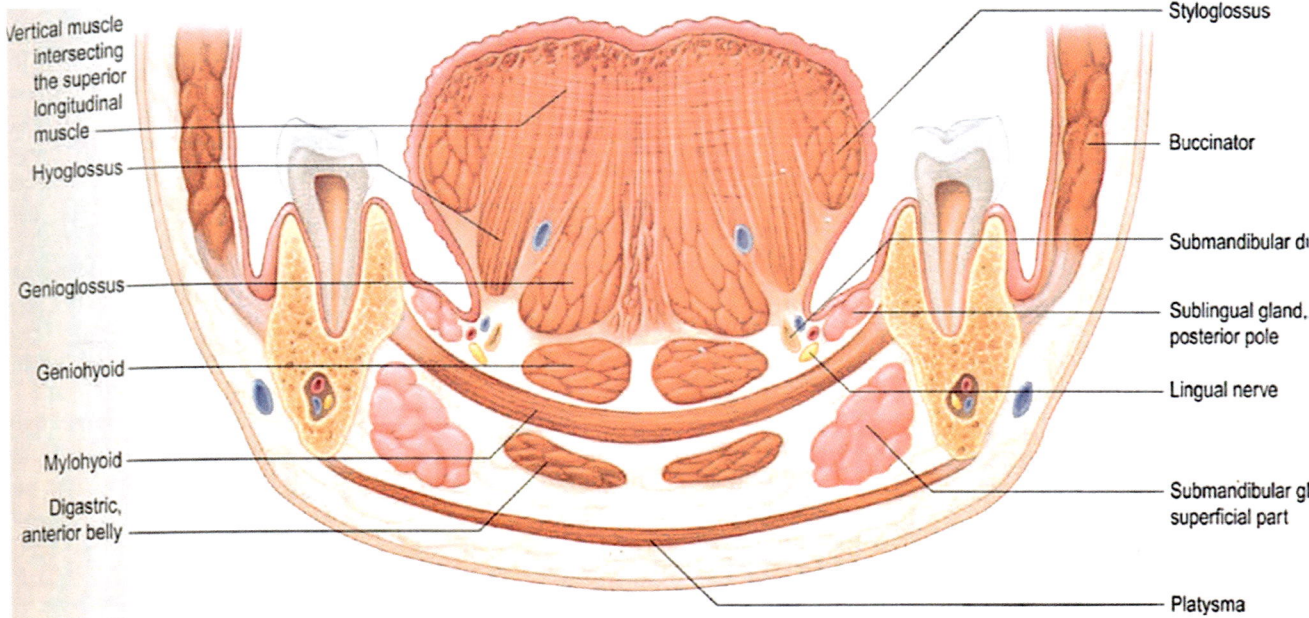

Fig. 2.24 Stacking of the muscles of the tongue. Lateral to medial: styloglossus (S1), hyoglossus (S2), chondroglossus (S3), genioglossus (S4), thus matching the four motor roots of hypoglossal nerve.

(Reprinted from Standring S (ed). Gray's Anatomy: The Anatomic Basis of Clinical Medicine, 40th ed. London, UK: Churchill Livingstone; 2008. With permission from Elsevier)

Table 2.2 Derivatives of pharyngeal arches

Pharyngeal arch	Rhombomere/ nerve	Derivatives
First (mandibular)	r2–r3/V2, V3	Maxilla complex, mandible
First (mandibular)	r2–r3/V3	Non-jaws, mastication
Second (hyoid)	r4–r5/VII	Mastication, animation
Third (glossopharyngeal)	r6–r7/IX, X	Sup constrictor, palate
Fourth (pharyngolaryngeal)	r8–r9/X	Middle constrictor, larynx
Fifth (internal laryngeal):	r10–r11/X	Inferior constrictor

Blood supply: (1) external carotid, (2) stapedial

Fate of Pharyngeal Arch PAM (Table 2.2)

The Mammalian System of Pharyngeal Arches

The pharyngeal arches, numbering five in mammals, appear during embryogenesis between stages 10 and 15 and represent a transient stage in the development of the head and neck. Each arch consists of a mesenchymal core of hindbrain neural crest from paired rhombomeres, surrounded by a discrete epithelium. The arches are situated between the body wall above and the aortic outflow tract that was tucked at stage 9 to lie immediately ventral to the future pharynx. Running the core of each arch is an aortic arch artery spanning from the aortic outflow tract ventral to the future pharynx and the dorsal aorta running along the axis of the embryo. The arteries are constructed from angioblasts that invade from lateral plate mesoderm, from which develops a tube of endothelial cells that subsequently recruit surrounding neural crest to form pericytes that ensheath the vessel.

We have previously pointed out the fallacy of the sixth pharyngeal arch. Its components are in register with r10–r11. Aortic arch 6 is unlike its predecessors. It forms as two distinct plexuses, one from the dorsal aorta and the other from the outflow tract. The ventral plexuses extend into the future pulmonary anlagen. On the left side only, a transient connection exists between outflow tract and dorsal aorta; this forms ductus arteriosus. The components of these vascular structures are in register with r10 and r11, but there are no epithelial structures or neural crest populations associated with these vessels.

Fifth arch does indeed exist but is vestigial. Because fifth aortic arch either disintegrates or fails to form, it is dependent for its survival upon the blood supply to fourth arch, superior thyroid artery. fifth arch neural crest from r10 to r11 is a putative source for cricoid cartilage and the inferior fibers of inferior constrictor (cricothyodeus).

The rhombomeres that are the source of neural crest are even-odd pairs with the even-numbered rhombomere being cranial. Within the arch, neural crest is organized into topographic sectors by a system of *distal-less* (*Dlx*) genes. In knock-out experiments conducted on the first arch, genetic expression in the distal zone affects the mandible, while those causing disturbances in the proximal zone affect the maxilla. Myoblasts for each arch arise from a discrete somitomere, immediately invade the mesenchyme, and are distributed to Dlx-determined compartments as well [36, 43].

Evolutionary Considerations

The evolutionary design of pharyngeal arches follows the neuroanatomic organization of the basal gnathostome. Early agnathic fishes had, in general, 8 branchial arches, all of which were respiratory. However, in some primitive armored fishes (placoderms), up to 15 branchial arches have been

identified. This is an intriguing detail. If we assume two neuromeres per arch, this means that in mammals, the primitive genetic marker for arch production could extend back to somite 24, i.e., the t12–11 junction between thorax and abdomen. A 9-arch system in humans would extend up to and including neuromere c8. More modern jawless craniates are represented today by hagfishes and the slightly more advanced lampey eel, *Petromyzon marinus*. They have a total of seven arches which, in humans, would encompass neuromeric levels r2–c4. As we shall see, descent of the thymus along pretracheal fascia can lead to ectopic locations as far distal as the pericardium (level c4).

The earliest chondricthyan (represented today by sharks) modified the seven-arch system, converting the first two branchial arches into jaws and hyoid structures capable of providing support. This brings the total number of respiratory branchial units to 5. An external carotid artery developed to supply the lower jaw. *The conversion of neural crest from gill structures to jaws was accompanied by the invention of the stapedial system.* Osteicthyces (bony fishes) lost the seventh arch, so assume a 2 + 4 system. Tetrapods further modified this to 2 + 3 configuration (two arches for the jaws and additional visceral arches) (Fig. 2.25).

Examination of the original gnathostome model is very instructive. Arches 1 and 2 have been modified to form jaws and arch 3–7 are respiratory The primitive gill arch (branch)

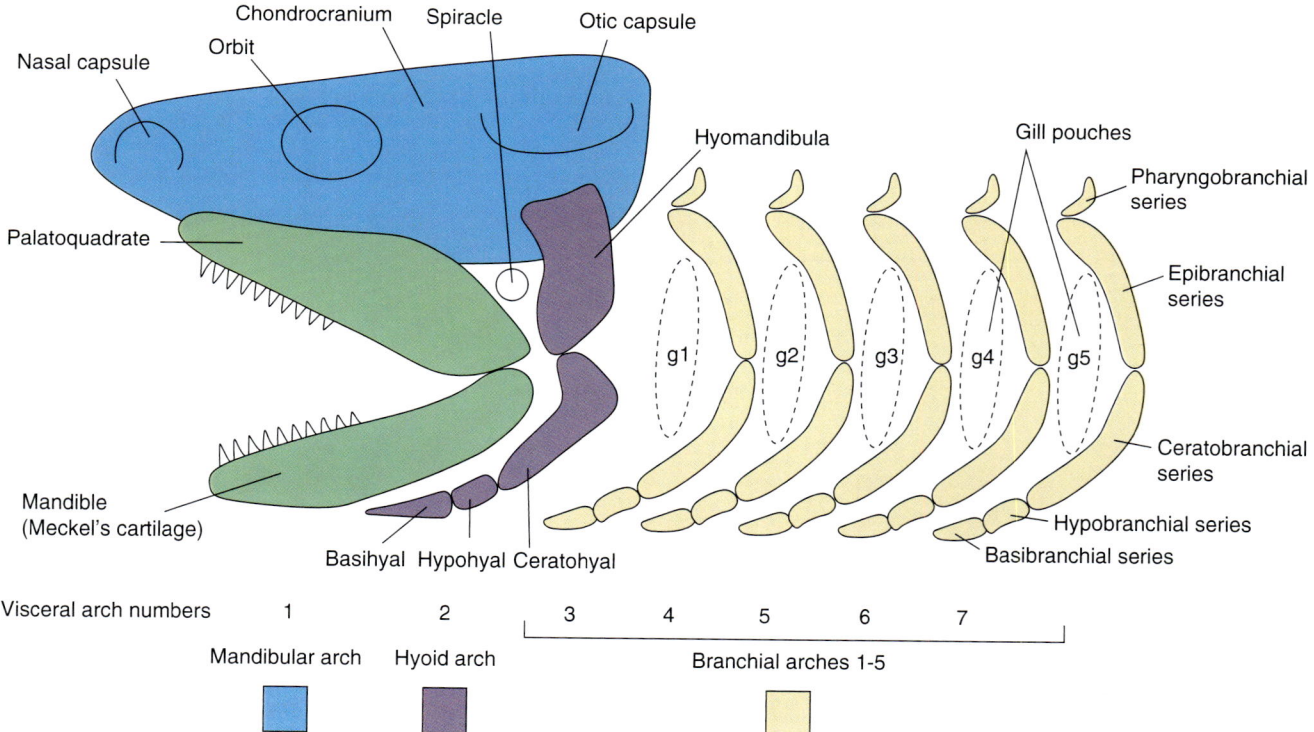

Fig. 2.25 Head segmentation. Jawless fishes: 7 respiratory arches (in primitive forms, up to 15). Cartilagenous fishes: 2 + 5 respiratory arches. Bony fishes: 2 + 4 respiratory arches. Tetrapods: 2 + 3 pharyngeal arches. (Courtesy of William E. Bemis)

Fig. 2.26 Innervation of branchial arches. Sensory nerves are both pretrematic and posttrematic. Motor nerve is posttrematic and accompanies the aortic arch artery in the posteromedial quadrant. (Reprinted from Evans DH, Piermarini PM, Choe KP. The Multifunctional Fish Gill: Dominant Site of Gas Exchange, Osmoregulation, Acid-Base Regulation, and Excretion of Nitrogenous Waste. Physiol Rev. 2005;85(1):97–177. With permission from The American Physiological Society)

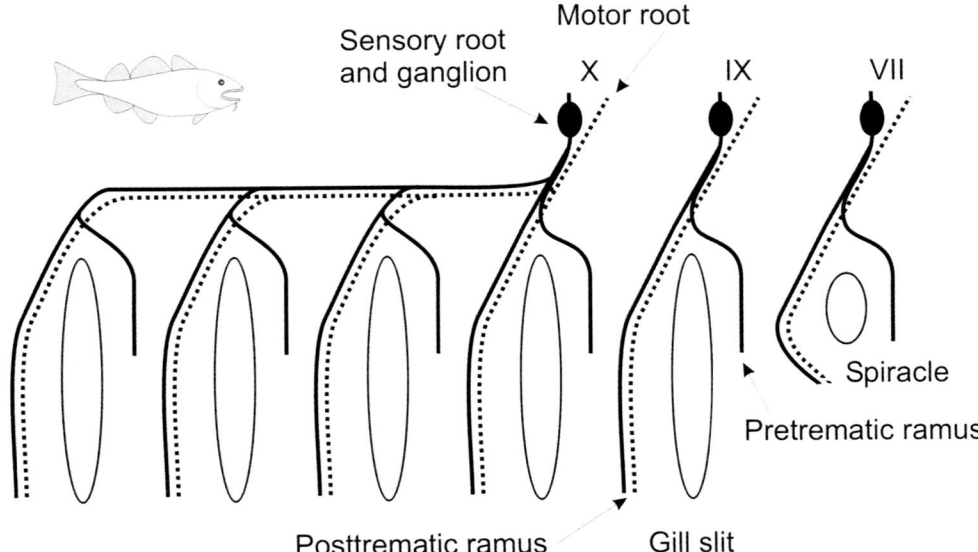

consists of anterior and posterior walls surrounding a gill slit. Each branch is supplied by a mixed motor and sensory nerve that emanates from a ganglion. *The dorsal part of the ganglion is placodal and the ventral part is placodal.* This neuroanatomy exists to this day. As the nerve approaches the gill slit, it branches into three functional parts: (1) a deeply placed *sensory nerve to the pharynx*, (2) a *pretrematic sensory nerve*, and (3) a *postrematic sensory-motor nerve*. The main cranial nerve for each banchiomere is located in the caudal zone of the arch [44, 45].

Reconfiguration of branchial arch structures is seen in the asymmetry of the pharyngeal arch. The gill slit is gone, but is replaced by the pharyngeal pouch. Cross-section of the branchiomere shows four quadrants. The endodermal side has branchiomeric muscle anterior and the skeletal rod posterior. The ectodermal side has the cranial nerve anterior and the aortic arch artery posterior (Figs. 2.26 and 2.27).

In the mammalian example, spatial asymmetry is reflected in the timing with which various anatomic structures make their appearance. This depends upon how quickly they are vascularized. Hypothetically, the aortic arch artery develops from below upwards. Tissues that are distal and caudal are more efficiently supplied. The mandible (r3) forms before the maxilla (r2). Malleus (r3) appears before incus (r2).

Let's take a closer look at the two ear bones. The malleus is a homolog of the *articular*—a dermal bone (neural crest) that ensheaths the proximal end of *Meckel's cartilage* (mandible). Thus, the malleus is an r3 derivative. In fishes, the primitive *palatoquadrate cartilage* of the upper jaw is the analog of Meckel's cartilage because the maxilla will be constructed around it. It has three components seen in the acanthodian skull. Of all jawed fishes (gnathostomes), the acanthodians have the earliest fossil record. The anterior palatoquadrate is known as the *autopalatine cartilage*. This

provides the basis for the maxilla and palate. Immediately behind is the *metapterygoid (epipterygoid) cartilage*, the future template for the alisphenoid bone. Most posterior is the *quadrate cartilage*. This is the origin of the incus. Not surprisingly, it is anchored to the *prootic* temporal bone. Hence, the incus (an r2 derivative) lies internal to the malleus (an r3 derivative). Malleus forms before incus (Fig. 2.28).

The same pattern obtains to derivatives of the second and third arches. Facial muscles develop in an absolutely stereotypical manner: from ventral to dorsal, deep to superficial, and lateral to medial [46]. It is logical to "assign" the muscles of the upper division of the facial nerve in r4 and those of lower division to the nucleus of VII in r5. The source for the upper half and lesser cornu of the hyoid and the styloid process should be r5 neural crest. The hyoid appears well before the styloid, but the latter is attached to the temporal bone. Future work with labelling should tease out these spatiotemporal relationships.

Do somitomeres produce any derivatives prior to being transformed into somites? As previously stated, Sm1–Sm11 are closely applied to the cranial base. They interact with lateral plate angioblasts to produce the *primitive vascular tubes* that will later become the internal carotid and vertebrobasilar arteries. In particular, Sm8–Sm11 produce the caudal longitudinal neural arteries that extend downward from cranial nerve 6 to the foramen magnum. The muscles of the larynx have been traced to somitomeres 8–9. They are not classified as somatic structures. Once Sm8–Sm11 become somites, they make use of somite subdivisions (dermatome, myotome, and sclerotome) to make *non*-pharyngeal arch structures, such as tongue muscles, sternocleidomastoid, and trapezius.

In sum, prior to their conversion into somites 1–4, somitomeres 8–11 produce perineural vasculature and myoblasts

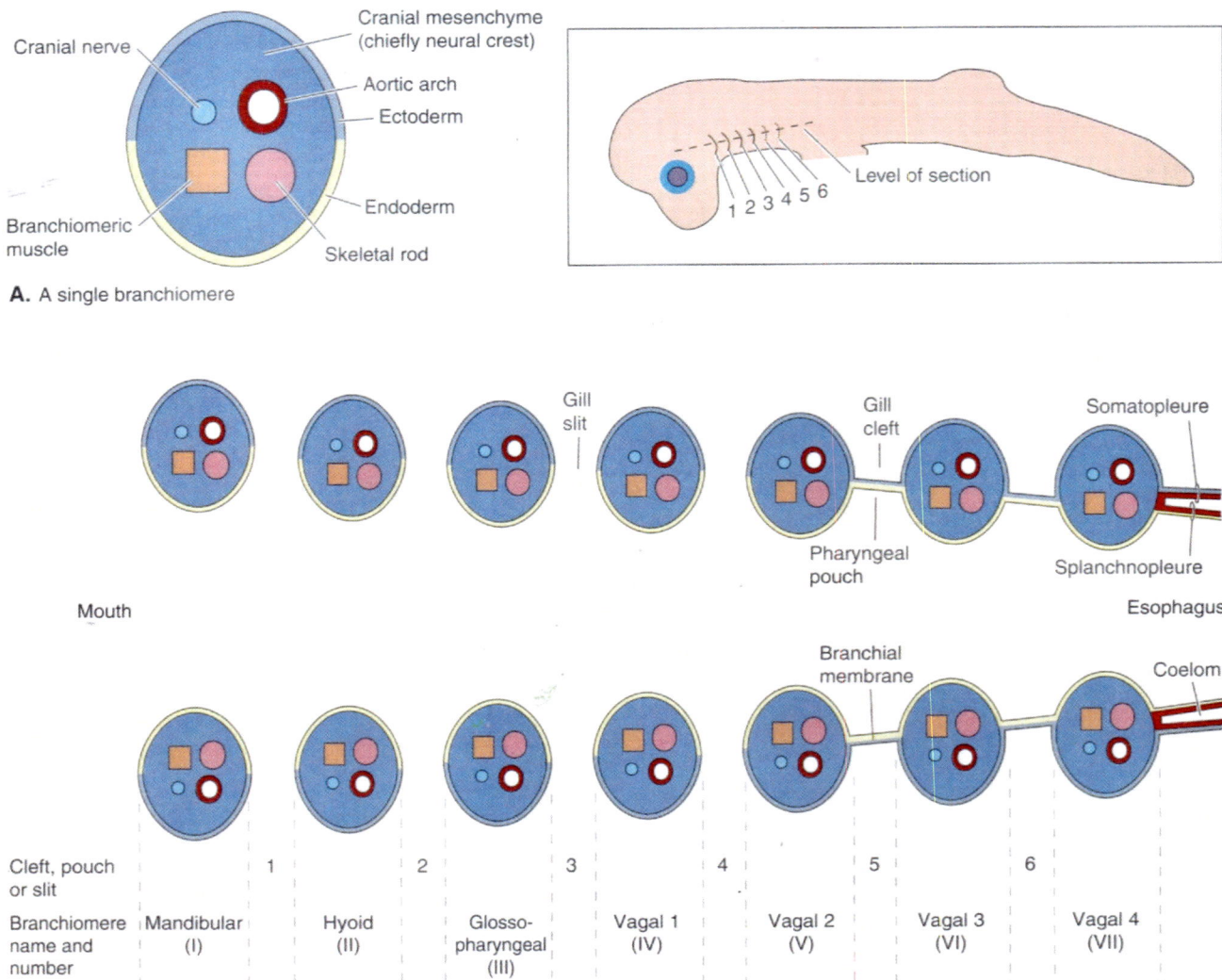

Fig. 2.27 Petromyzon seven-arch system showing the asymmetric position of the artery. Vascularization sequence of the pharyngeal arches likely determines the developmental maturation sequence of arch muscles. (Courtesy of William E. Bemis)

dedicated to pharyngeal arches 4–5. Once out of the head region, paraxial mesoderm in its somitomeric state from Sm12 caudally could potentially provide immediate vascular coverage for the future spinal cord. All further derivatives, including dura, will develop from specific regions of the somite.

Do somitomere and somites have any effect upon neural crest migration routes into the arches? The answer is no; however, there are selective routes by which migration takes place. *Although neural crest cells are produced at each neuromeric level, they appear to migrate out from only even-numbered rhombomeres.* Thus, all neural crest migration into PA1 proceeds via r2, migration into PA2 proceeds via r4, and migration into PA3 proceeds via r6. The pattern is disrupted for the fourth arch as it receives neural crest directly from r8 to r9 (Fig. 2.29).

Some investigators postulate that something "goes wrong" with neural crest cells at odd neuromeric levels. But this flies in the face of evidence regarding programming of neural crest that is neuromere-specific. For example, neural crest cells with the *hox* gene "tattoo" of r3 are never found in the maxillary part of first arch. If one transplants r3 neural crest cells into the proximal portion of PA1, two mandibles are produced. So the selective pathways are "permissive" routes by which neural crest from odd-numbered rhombomeres is forced to migrate with its even-number predecessor.

At this juncture, we have a clear picture of how a pharyngeal arch is constructed. Furthermore, we can "assign" individual bones and muscles to neural crest and PAM from specific neuromeric levels. Kjaer and her coinvestigators painstakingly catalogued the order of appearance of craniofacial bones. When one applied neuromeric mapping to these

Fig. 2.28 *Amia* skull: palatal series. Anterior part of palatoquadrate ensheaths the maxilla followed by entopterygoid, ectoptyerygoid, metapterygoid, then quadrate. (Courtesy of William E. Bemis)

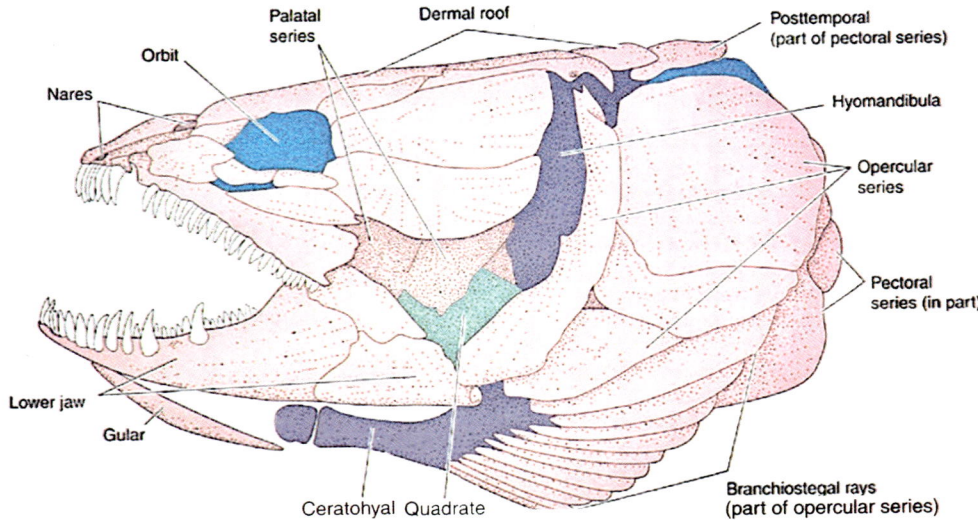

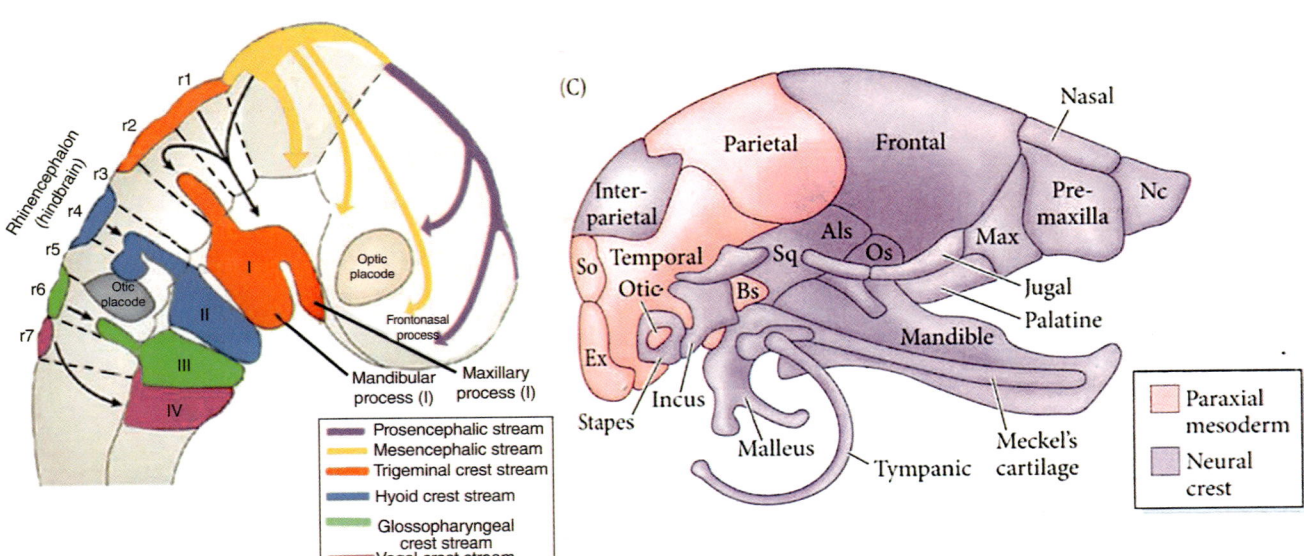

Fig. 2.29 Cranial neural crest migration routes and derivatives. Note that neural crest from r0 to r1 and m1 to m2 acts in concert to populate the frontnasal and periorbital zones. Note that PNC creates the fronto-nasal skin cover. Neural crest bone map shows interparietal as an "island" between paraxial mesoderm bones. (Left: Reprinted from Carlson BM. Human Embryology and Developmental Biology, 6th edition. St. Louis, MO: Elsevier; 2019. With permission from Elsevier. Right: Reprinted from Gilbert SF, Barresi M. Developmental Biology, 11th ed. Sinaurer: Sunderland, MA, 2016. Reproduced with permission of the Licensor through PLSclear)

structures, the initial results often appear contradictory and confusing. If the maxilla comes from level r2 and the mandible from r3, why should mandibular structures appear first? Why should dental development follow a mesial-to-distal plan? What sense can be made of these patterns?

Vascularization of the Pharyngeal Arches: Nutritional Basis of Derivatives

Lateral plate mesoderm is the primary source for angioblasts in the developing embryo, constituting more than 90% of the LPM cell population. They are highly invasive and populate the surrounding tissues with angioblasts in all directions as soon as gastrulation occurs at stage 7. As the LPM spreads out, its anterior-most population occupies a position forward from the future brain, anterior to prechordal plate mesoderm and to the buccopharyngeal membrane. This zone is known as the primitive heart field. At the same time, LPM angioblasts invade paraxial mesoderm to create the major intraembryonic vessels such as dorsal aortae, the head plexus, and primitive hindbrain channels. At stage 9, head folding at the mesencephalic flexures forces the heart field and its accompanying LPM angioblasts into a position just ventral to the developing pharyngeal arches. This puts LPM angioblasts associated with the aortic outflow tract in direct opposition to

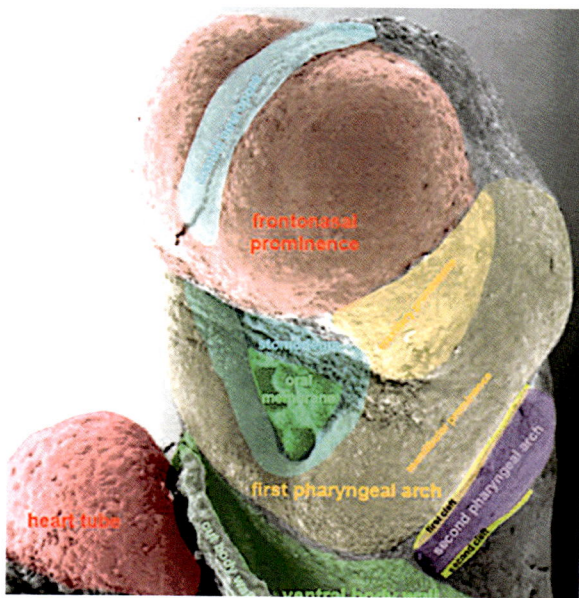

Fig. 2.30 Stage 11–12 Closure of rostral neuropores. First and second pharyngeal arches fully developed with third arch in formation. Note buccopharyngeal membrane (green). Oral ectoderm (blue-green) leading back to BPM is r2–r3 with no expression of second arch ectoderm (purple) in the oral cavity. On the surface, first arch ectoderm will cover over the second arch back to the exernal auditory meatus. (Courtesy of Prof. Kathleen K. Sulik, University of North Carolina)

those found in the vicinity of the dorsal aortae. From these two opposing sources, blood vessels bud outward, migrating toward each other through the intervening neural crest mesenchyme of the pharyngeal arches. The result of this process is the production of a succession of aortic arch arteries beginning with at stage 10 with AA2 (Figs. 2.30 and 2.31).

Successive sprouting of aortic arches from the heart occurs in a strict spatiotemporal sequence [47–51]. Recall that at stage 8, just prior to embryonic folding, the primitive heart is located in front of the forebrain in the anterior most aspect of the embryonic disc. The proximal atrial ends of the primitive heart are connected to the vitelline, umbilical, and cardinal veins, while the distal outflow tract is connected to the dorsal aortae. Pharyngeal arch formation begins at stage 9, with the folding of heart from the anterior aspect of the embryonic disc in front of the forebrain to a new position beneath the future mouth and pharynx. When this takes place, the heart dorsal aortae are dragged downward and backward, creating, passively, paired first aortic arch arteries. During stages 10–12, the ventricular outflow tracts give off an additional aortic arch artery, each one supplying a pharyngeal arch. The fifth aortic arch involutes leaving PA5 dependent upon AA4 for survival. The sixth aortic arches dedicated to the pulmonary circulation appear at stage 14

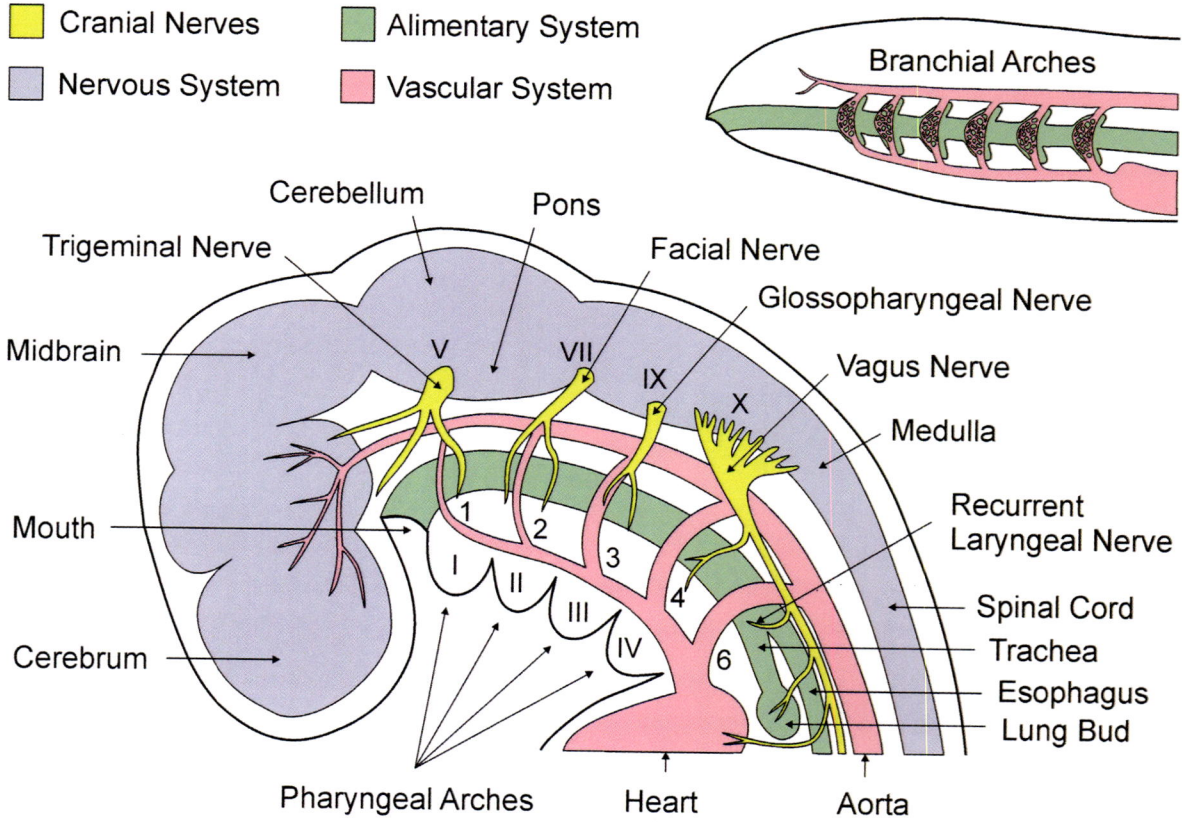

Fig. 2.31 Pharyngeal arches in tetrapods versus fishes. AAs are transient—they unite the ventral aortic outflow tract with dorsal aortae and then undergo extensive remodeling. AA5 involutes—its target, PA5 is supplied by AA4. AA6 assigned to pulmonary circulation There is **no** *sixth pharyngeal arch* in tetrapods. (Reprinted from Carlson BM. Human Embryology and Developmental Biology, 6th edition. St. Louis, MO: Elsevier; 2019. With permission from Elsevier)

Stage 14
32 day
35 somite

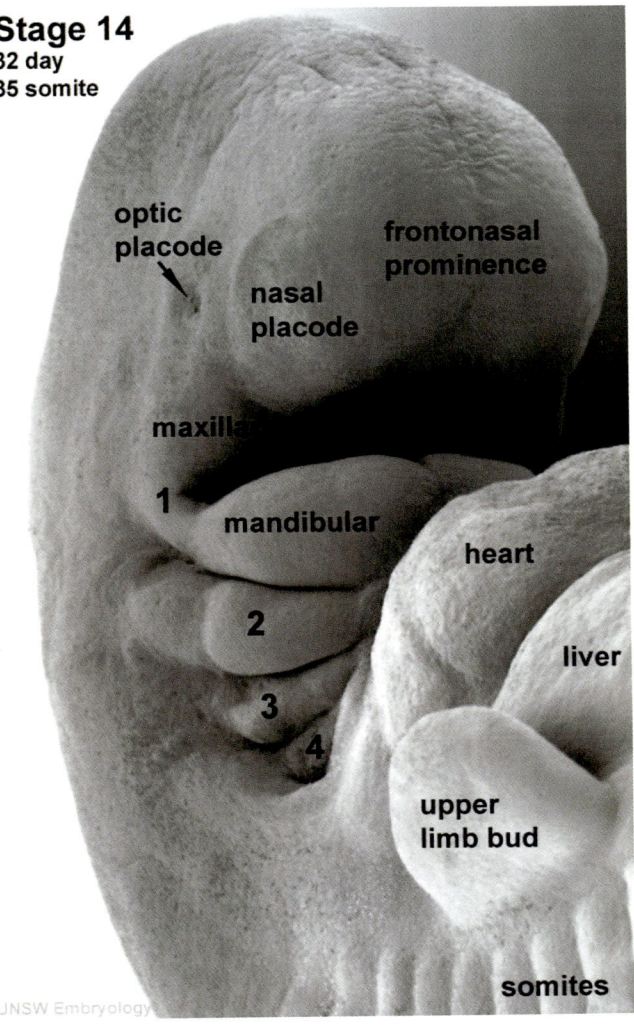

Fig. 2.32 Pharyngeal arches, stage 14: Arch synthesis (stages 9–14) now complete. Arches not melded together; rearrangement begins at stage 15. First cervical somite (S5) barely visible (next to UNSW Embryology). Occipital somites fused and not seen. (Courtesy of Prof. Kathleen K. Sulik, University of North Carolina)

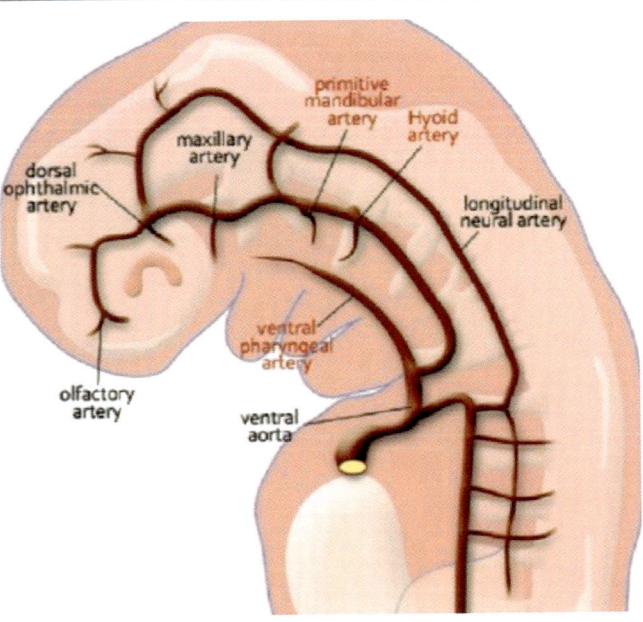

Fig. 2.33 Schematic drawing of embryonic vascular system at stages 14–15 immediately prior to initiation of the stapedial system. Third and fourth arches are present, primitive internal carotid supplies optic vesicle first, via the primitive maxillary and now, at stage 14, by primitive dorsal ophthalmic. (This causes PrMax to be re-directed to the hypophysis.) At stage 12, Internal carotid has joined the longitudinal neural arteries hindbrain channels at the midbrain-hindbrain junction. Segmental arteries such as trigeminal become unnecessary and involute. Obliteration of AA1 and AA2 leaves behind dorsal remnants, mandibular and hyoid arteries, respectively. The first three arches are supplied by ventral pharyngeal artery which originates from the base of the truncus arteriosus in the segment common to both AA3 and AA4. It will morph to external carotid in the next stage. At stage 15–16, hyoid artery gives off the stem of stapedial, sending it upward into tympanic cavity. (Reprinted from Tanoue S, Kayoso H, Mori H. Maxillary Artery: Functional and Imaging Anatomy for Safe and Effective Transcatheter Treatment. Radiographics 2013; 33(7) 209–229. With permission from the Radiological Society of North America)

and the pharyngeal arch period of embryonic development comes to an end by stage 15 (Figs. 2.31 and 2.32).

During the aortic arch period, each successive stage leads to a reconfiguration of the system. In the process of melding first arch and second arch together, their respective aortic arch arteries disintegrate, with blood supply being reestablished from extracranial stapedial and the external carotid system. Paired AA3s create the extracranial segment of internal carotid. The AA4s give rise to the right subclavian artery, but only the left side persists as a segment of aortic arch from which left subclavian arises.

In this process, three remnant arteries deserve mention. Cranial to AA1, the primitive internal carotid gives off the misnamed *primitive maxillary artery*. This vessel constitutes the initial supply for the developing eye, but with the advent of the definitive ophthalmic artery primitive maxillary is subsequently reconfigured as the *inferior hypophyseal artery*. When AA1 breaks down, it leaves behind a *primitive man-*

dibular artery dangling from the dorsal aortae. This vessel involutes. The breakdown of AA2 also gives a dorsal remnant, one that will have an important future. *Hyoid artery* gives off the all-important *stapedial* system which will supply the dura, orbit, and jaws (Fig. 2.33).

The exact anatomic manner in which the outflow tract from the heart supplies the pharyngeal arches is remarkably simple and elegant. Diagrams in books based on the gill arches of fishes are quite misleading. A typical illustration depicts a core artery running up the center of each gill arch in ventral-dorsal fashion to anastomose with the dorsal aorta. How might this happen? One never sees a central vascular core running through the axis of a muscle. Instead, arteries and motor nerves travel in the interstices between muscles, i.e. they follow fascial planes. Each muscle unit is penetrated from without by a nerve. *The arterial supply is derived by induction from the nerve via VEGF* [52]. The arterial supply of the arch arises when neural crest Schwann cells cause the surrounding mesoderm to create a vascular conduit.

In the head and neck, all muscles are surrounded by fascia, derived exclusively from neural crest cells. As neural crest cells spread over each pharyngeal arch, they penetrate the PAM via natural cleavage planes separating genetically myoblast populations. Spatial relationships among the fascial planes reflect the order of development of the muscles. Furthermore, the relative positions of motor neurons in the neural plate (and neural tube) faithfully replicate the spatial location of their muscle "targets". Muscles close to the vertebral axis such as the paraspinous group are classified as *epaxial*. Their motor neurons lie close to the midline in the *medial lamina of the medial motor column* (MMCm). Muscle groups hypaxial to the midline but not assigned to the extremities have neurons in the *lateral lamina of the medial motor column* (MMCl). Muscle groups of the extremities have motor neurons still more laterally located in the neural plate. These form the lateral motor column. Ventral muscles of the limbs are supplied by neurons from the LMCm. Dorsal muscles of the limbs are supplied by neurons from the LMCl. All motor nerves to striated muscles use Schwann cells for axonal insulation. These Schwann cells are of neural crest derivation. *Because neural crest defines the fascial planes between muscles, it is logical that all neurovascular structures make use of these planes in order to access their target muscles.*

The spatiotemporal order of appearance of muscles within the pharyngeal arches has a great deal to do with blood supply. The overall pattern is ventral to dorsal, caudal to cranial, and medial to lateral (in that relative order). Muscle development requires metabolic activity; this in turn requires blood supply. Thus, ***the order in which muscles develop within a pharyngeal arch reflects the pattern of the arterial development to that arch***.

Hypothetical Model of Vascular Development

The heart and outflow tract are rapidly embraced by pharyngeal arches filled with neural crest. From stages 10–14, stems appear simultaneously from both the outflow tract and the dorsal aorta. These follow the core of the pharyngeal arches and unite to create pharyngeal arch arteries. Muscles within the arches are vascularized in a fixed order, depending on their location. This pattern, in turn, determines the temporal and spatial order in which they appear.

Important deductions about the neuromeric basis of the intracranial arterial system can be made from a careful analysis of dural innervation. The meninges consist of *leptomeninx*, the pia mater and arachnoid, and *pachymeninx*, the dura. The leptomeninges are exclusive neural crest structure created by the immediate migration of local neural crest. They are admixed with PAM with forms of the blood vessels found between pia and arachnoid. Pachymeninx covers only the prosencephalon and cerebellum. Intracranial dura is a neural crest tissue. Below the tentorium cerebelli, the midbrain and hindbrain are covered only by pia and arachnoid. The innervation of the entire prosencephalic dura comes from V1 to V3 with small contributions from VII and IX (via X) to the basisphenoid/basioccipital. The periosteum covering the bones of the middle cranial fossa arises from PAM, but shares a common innervation because second and third arches sensation is routed exclusively to the trigeminal nucleus (Fig. 2.34).

Fig. 2.34 Meninges. First wave of local neural crest is admixed mesoderm containing angioblasts to form leptomeninx (pia and archanoid). Second wave of neural crest from hindbrain forms the pachymeninx (dura). (Reprinted from Williams PL (ed). Gray's Anatomy, 38th ed. Philadelphia, PA: Churchill Livingstone; 1997. With permission from Elsevier)

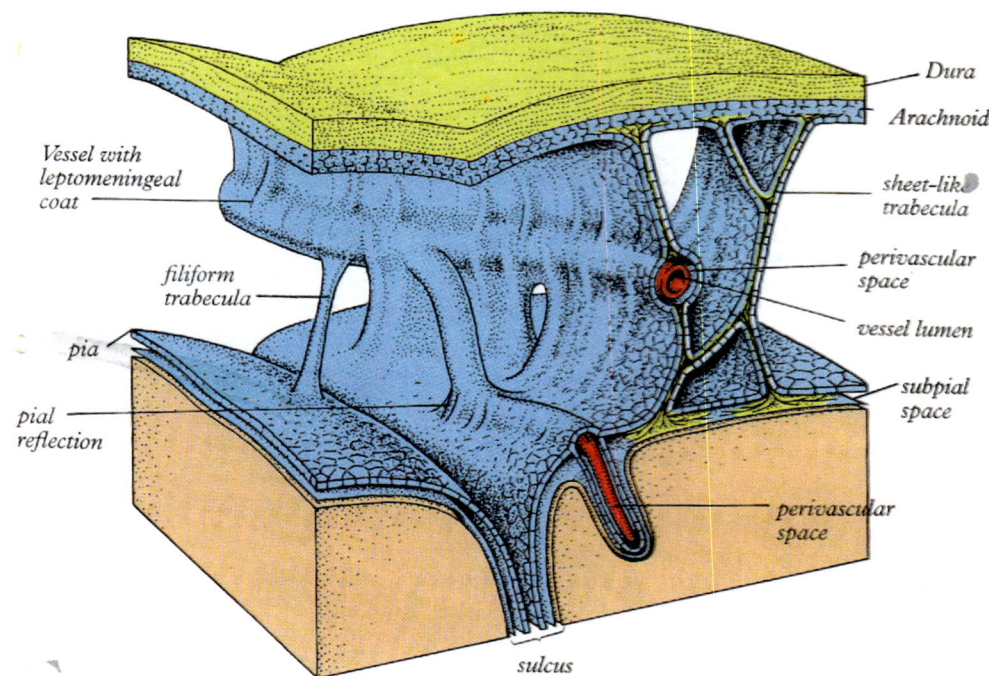

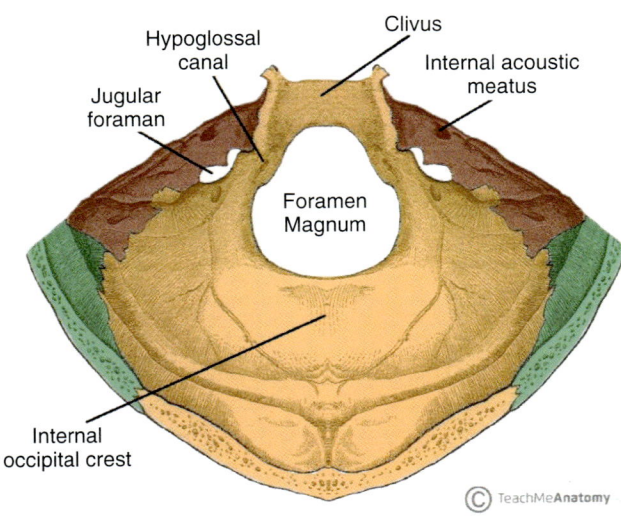

Fig. 2.35 Blood supply of posterior fossa dura: occipital (representing second arch), ascending pharyngeal (representing third arch). Innervation of posterior fossa dura: IX, X via jugular foramen, foramen magnum supplied by branches of C1, C2. (Reprinted from TeachMeAnatomy, courtesy of Dr. Oliver Jones)

As far as the occipital lobes are concerned, they retain V3 innervation. Recall that the cerebral hemispheres arose as balloons from their initial location anterior to pituitary fossa. The parietal and occipital lobes extend backwards to enclose the diencephalon. As such, their dura remains r3 neural crest and is innervated by V3. Periosteal lining (endosteum) of the middle cranial fossa is associated with the neuromeres associated with its bone components, but the innervation remains the same.

Posterior cranial fossa is different. Here, sensory branches of glossopharyngeal and vagal enter the cranium via jugular foramen and are directed anterior to the foramen magnum. The periosteum of this zone refers to IX and X as well as to V3. Thus, the symptom nausea and vomiting are seen with increased intracranial pressure. Posterior to foramen magnum, the periosteum covering the cranial base is mesoderm. Although these bones arise from the occipital somites, their periosteum arises from mesoderm originating from second and third cervical somites that enter the posterior cranial fossa via foraen magnum. For this reason, it is innervated by C2–C3. This explains the stiff neck seen in basilar meningitis. The innervation to the remainder of the calvarium mimics that of the overlying scalp. The dura of the spinal cord from foramen magnum caudal is PAM and is innervated by a strictly neuromeric manner (Fig. 2.35).

Fate of Non-pharyngeal Arch PAM

Not all paraxial mesoderm participates in the formation of pharyngeal arches (strictly hypaxial structures). Somitomeres

1–3 are exclusively dedicated to the extraocular muscles. Sm5 produces lateral rectus and possibly contributes to the petrous temporal complex. Let's look at specific components of the skull formed from paraxial mesoderm [27, 53]. The parietal bone is synthesized from r2 to r3 neural crest and from Sm4 paraxial mesoderm. The squamous temporal bone is formed from r3 neural crest and r4 PAM. Sm6 forms the petrous temporal bone and Sm7 the mastoid temporal bone. The temporal complex likely receives neural crest from r4 to r6. Sm8–Sm11 are subsequently converted into occipital somites with sclerotomes. In the vertebral column proper, the sclerotomes of each somite combine to form vertebral bodies. Signals involved in the induction of the medial somite arise from the notochord and neural tube. Hence, it is not surprising that the vertebral bodies encase these structures. Remnants of the notochord persist as the *nucleus pulposus*.

This situation exists in the cranial base as well. Sm 11 and Sm12 combine to form the proatlas (the first true cervical vertebra, bringing the total to 8). In mammals, a vestigial proatlas (the original first cervical bone) persists as three structures: the rostral tip of the dens, the dento-occipital ligament, and the condyles of the exoccipital bone. The course of the rostral notochord is as follows. Via the dens and the dentooccipital ligament, it gains access to the basioccipital bone (BO). BO is formed as the fusion of sclerotomes from occipital somites 1–4. The notochord then passes through the core of basioccipital bone up to the basisphenoid bone (BS). BS is produced from the PAM of Sm4. The notochord terminates at the junction of the PAM basisphenoid and neural crest presphenoid. Thus, the notochord occupies the center of PAM-derived basicranium, running from r11 forward to r1.

Blood Supply to the Face: An Overview

Summary of Circulations as Determined by Neural Crest

As previously discussed, the rostral rhombencephalon consists of six neuromeres: r2–r7. The first two pharyngeal arches unite as a single functional unit. For evolutionary reasons, both arches produce derivatives very different from their original design. The reassignment of selected first arch derivatives to jaws and dura was so radical that it required an entirely separate system of perfusion. Specifically, *all bone structures of the first arch relating to the jaws and the bones that connect them with the skull are supplied by the extracranial stapedial system.* By this we refer to V2 and V3 neuroangiosomes that depart from the axis of the external carotid maxillary artery. *All non-jaw structures of the first arch (muscles of mastication and their fasciae), glands, and subcutaneous tissue are supplied by the ligual, facial, and*

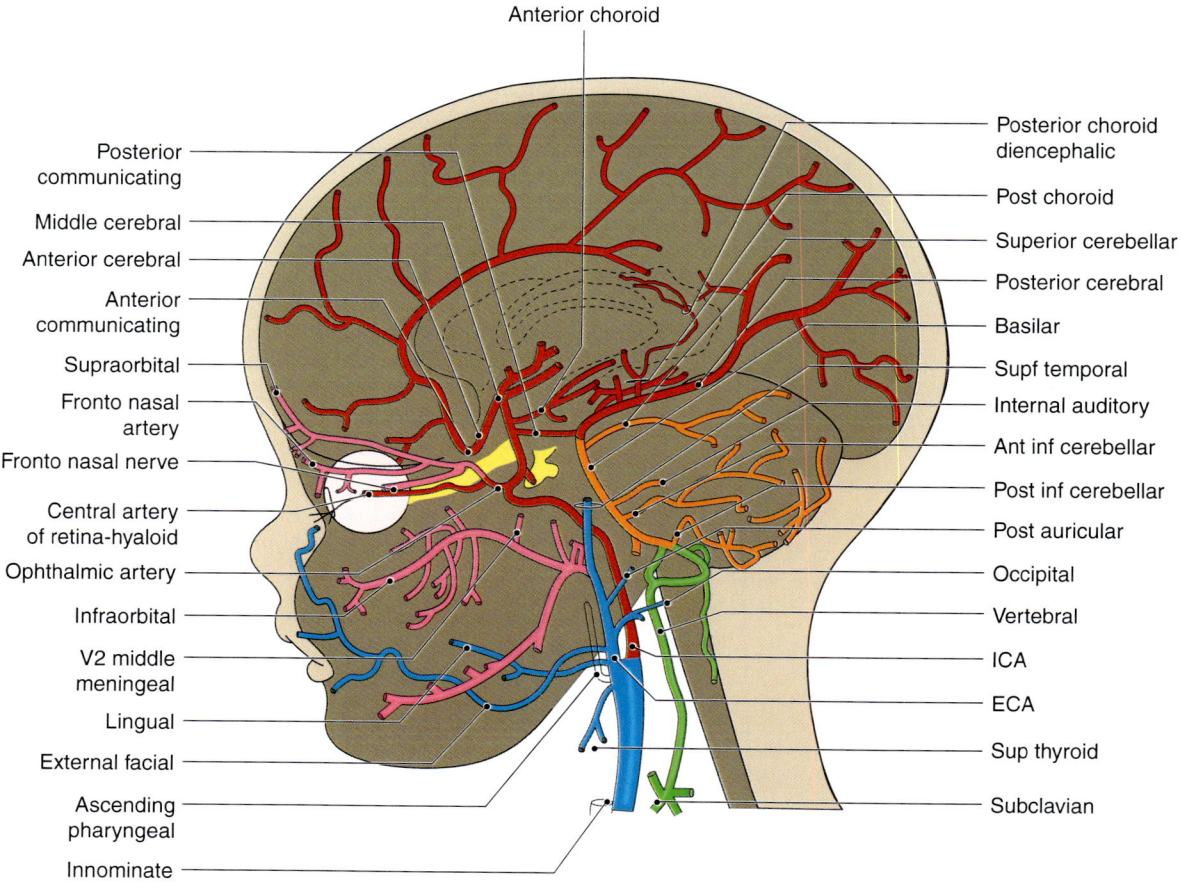

Fig. 2.36 Fetal state. **Stapedial system** (not previously depicted) originates at stage 17 and produces (1) **meningeal aa**, (anast with **external carotid**), (2) **StV1 extraocular ophthalmic aa** (anast with **primitive ophthalmic**), and (3) **StV2 and StV3 sphenopalatine aa** (anast with **external carotid**). (Courtesy of Michael Carstens, MD)

superficial temporal branches of external carotid artery. All remaining arches are supplied by arteries of the external carotid system (Figs. 2.36, 2.37, 2.38, and 2.39).

In evolution, the earliest jawless fishes had eight (or more) gill, or branchial, arches, and a simple circulatory system consisting of parallel ventral and dorsal aortae. Oxygen poor blood pumped by the heart through the ventral aortae passed through gills and the oxygenated blood circulated through the body via the dorsal aortae. This system can still be seen in the agnathic lamprey and hagfish.

In Sum

Frontonasal derivatives containing forebrain neural crest and midbrain neural crest

- Stapedial neuroangiosomes supplied by StV1

Pharyngeal arch derivatives containing hindbrain neural crest

- Stapedial neuroangiosomes supplied by StV2 and StV3
- External carotid neuroangiosomes

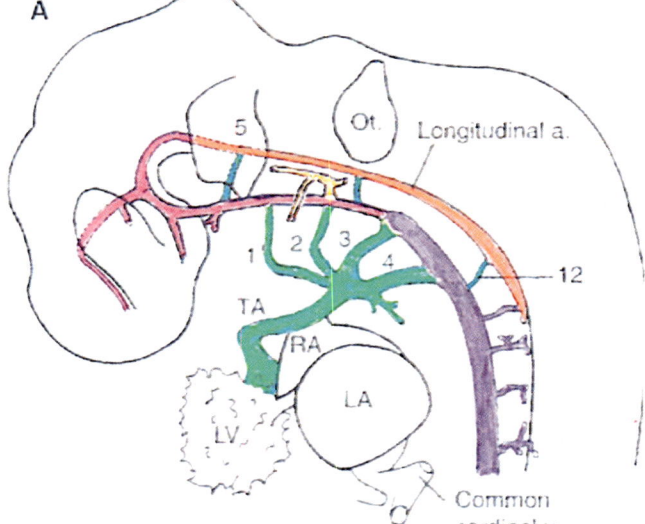

Fig. 2.37 Vascular stages 13–14. **Primitive ICA** to forebrain and midbrain, **Longitudinal neural** (future **basilar**) to hindbrain, connecting **presegmentals** (trigeminal, hyoid, otic, hypoglossa), **aortic arches** (future **external carotid**), **dorsal aorta** giving off **dorsal intersegmentals** (these will later unite longitudinally as the **vertebral artery**). (Courtesy of Michael Carstens, MD)

Fig. 2.38 Vascular stage 17. Ventral pharyngeal artery converting to external carotid. Extracranial division of stapedial (yellow) arrives at V3, meets with ventral pharyngeal artery, and forms distal external carotid to the first arch. (Courtesy of Michael Carstens, MD)

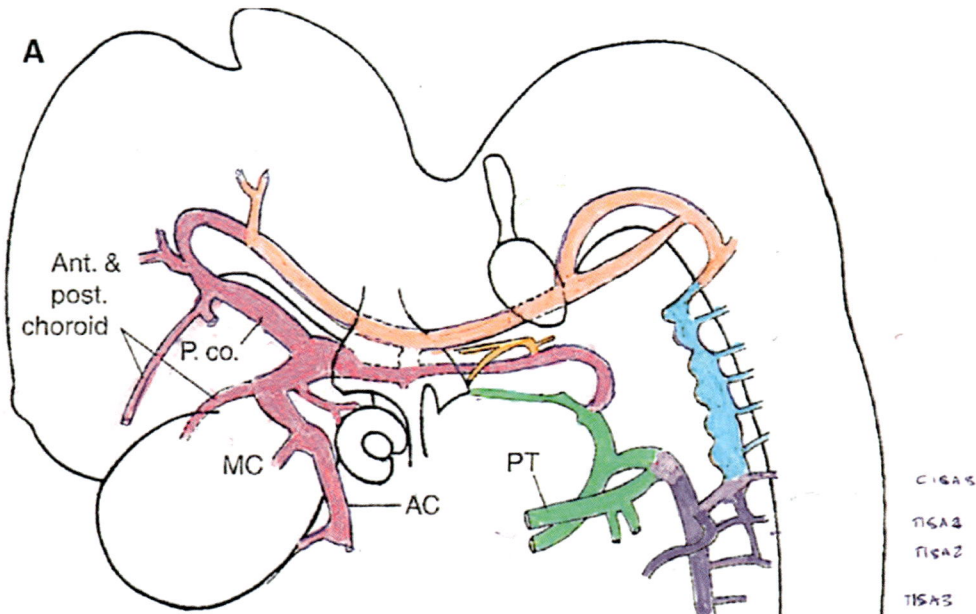

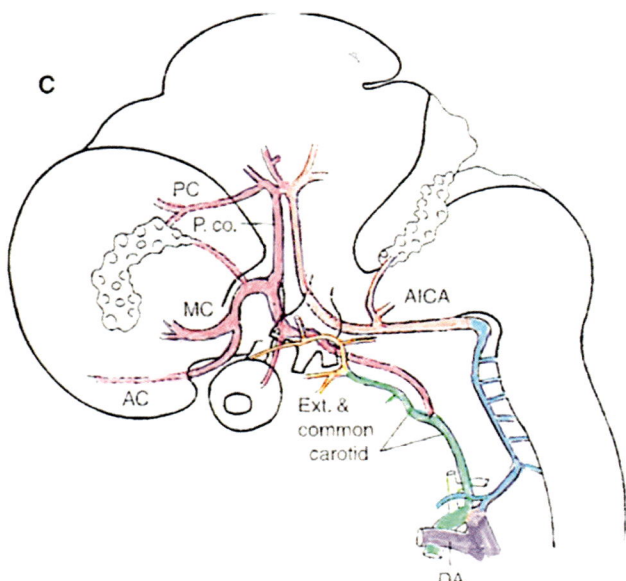

Fig. 2.39 Vascular stage 21. Stapedial connections now complete. Superior division has reached the orbit and united with ophthalmic. (Courtesy of Michael Carstens, MD)

Aortic Arch Precursors

The jawed fishes (gnathostomes) reassigned neural crest from their anterior gill arches to make jaw structures. This change took place in two steps (1) Early cartilagenous fishes (chondrichthyans) such as sharks lose the first branchial arch, the tissues being reassigned from respiration to the creation of jaws. A vascular innovation is required to support tissues reassigned from gill structures to jaws. The upper jaw, eye, and face are supplied by a *stapedial artery* originating from the dorsal aorta at its junction with the second branchial arch. The lower jaw is supplied by an *external carotid* originating from the ventral part of the first collector loop, located junction between second and third arches. Thus, sharks have five gills. (2) Bony fishes (osteoichthyces) reassign *both* first *and* second branchial arches as a composite unit. The hyoid arch is converted into a bony suspension for the jaws. They have four gills. In these species, external carotid now extends to the tissues covering upper jaw as well.

The relationship between the stapedial and external carotid circulations has passed through several iterations during evolution which we shall discuss in a subsequent chapter dealing with vascular embryology. We can summarize the circulations as follows (Fig. 2.40).

Amalgamation of the first and second pharyngeal arches with transformation into jaws requires the dissolution of the original aortic arch arteries AA1 and AA2 in stage 12. In stage 13, assembly of the stapedial and external carotid systems is underway. AA2 leaves behind a dorsal stem, the *hyoid artery* dangling from the dorsal aorta. AA3 and AA4 dissociate. AA3 forms the common carotids and connects an intermediate ventral pharyngeal artery which subsequently morphs into external carotid. The hyoid gives off stapedial, as the first branch of the extracranial internal carotid. Stapedial enters the tympanic cavity and divides, giving off an intracranial *superior division* and an extracranial *inferior division* which exits via the pterotympanic fissure.

Extracranial stapedial follows chorda tympani to the root of V3, at which point it joins with the maxillary branch of external carotid to form the *maxillomandibular artery*. MMA is a hybrid that contains stapedial vessels designated for the

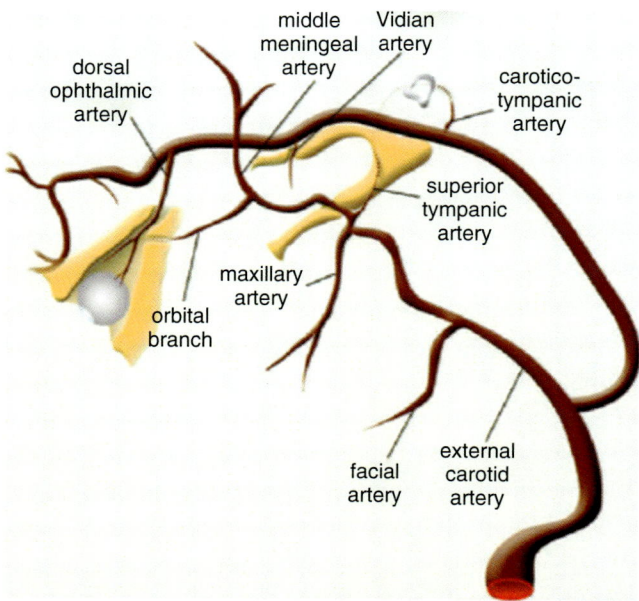

Fig. 2.40 Stapedial contributions to ophthalmic and maxillomandibular systems—embryonic stapedial system has been replaced. **Stapedial stem** originated from the hyoid artery, the remnant of which is *caroticotympanic artery*. It proceeded through stapes and divided. The **inferior maxillomandibular division** followed CN VII chorda tympani nerve to exit the skull and access the external carotid system at the root of V3 in the infratemporal fossa. The adult remnant of this is *superio tympanic artery*. The **superior supraorbital division** (disintegrated) followed CN VII superior petrosalnerve and then trigeminal ganglion V1 and V2 to supply dura branches and the orbital branch. Secondary connection between the two systems was provided by V3-induced middle meningeal which accessed the intracranial division, completing the middle meningeal system. Orbital branch was accessed by ophthalmic making the original connection with middle meningeal optional—thus the meningo-orbital branch of MMA is usually not present in the adult. This topic is discussed in depth in Chaps. 6 and 7. (Reprinted from Tanoue S, Kayoso H, Mori H. Maxillary Artery: Functional and Imaging Anatomy for Safe and Effective Transcatheter Treatment. Radiographics 2013; 33(7):209–229. With permission from the Radiological Society of North America)

jaws and supporting bone fields and external carotid vessels assigned to the soft tissue envelope of the face and mouth (muscles of mastication, muscles of expression, glands, and skin). External carotid to the facial complex consists three branches: maxillary, lingual, and superficial temporal.

During development, the first and second pharyngeal arches merge together as soon as PA2 is formed. PA3, PA4, and the puny PA5 remain distinct. During the merger process, the first and second aortic arch arteries that serve PA1 and PA2 disintegrate, leaving behind dorsal *remnant vessels* dangling from the dorsal aortae. The dorsal remnant of first aortic arch artery, the *mandibular artery*, eventually supplies the pituitary. The dorsal remnant of second aortic arch artery, *the* hyoid artery, gives off the all-important stapedial artery. With growth, the hyoid comes to lie just beneath the future tympanic cavity. It is the only extracranial branch of internal carotid. Hyoid persists as the carotid-tympanic artery.

Birth and Death of the Stapedial Artery System

Hyoid artery, being a second arch structure, is associated with cranial nerve VII. It sends a branch upward (in conjunction with CN VII) into the embryonic middle ear, which is eventually enclosed by tympanic cavity. This is **stapedial stem.** Within the tympanic cavity, stapedial divides. Both divisions of stapedial follow branches of the seventh nerve and exit the tympanic cavity.

Upper division stapedial follows greater petrosal all the way to trigeminal ganglion; it divides to follow V1 and V2. V1 stapedial supplied all dura, bone, and skin and vestibular lining, conjunctiva, and orbital structures innervated by V1. V2 stapedial supplies dura, lacrimal gland, and lateral orbit. Superior division does not have the opportunity to follow V3 because it already has exited the skull.

Lower division stapedial follows chords tympani (VII) to reach V3 — where it anastomoses with internal maxillary from ECA. This pathway is marked by the remnant anterior tympanic artery. From this anastomosis, it immediately sends out a ventral branch, inferior alveolar, following V3 to the mandible and a dorsal branch following V3 to the dura, middle meningeal. It then continues forward to with ECA until it reaches the sphenopalatine fossa where it provides all remaining branches following V2 into the zygomaticomaxillary-palatine complex.

The return of StV3 middle meningeal to the dura has drastic hemodynamic consequences for the previous two intracranial stapedial arteries. Middle menineal ascends and connects at trigeminal ganglion with the original superior division originating at tympanic cavity. Thus, bifurcation of StV1 and StV2, formerly dependent on long distance from a relatively small stapedial stem, is exposed to a high pressure system. Greater flow leads to an involution and disintegration of the proximal upper division associated with greater petrosal leading back to the tympanic cavity. So extracranial SstV3 finds itself in the cranial cavity where it will form the meningeal system and supply the orbit via StV1 branch to superior fissure and StV2 branch to the meningo-orbital foramen. Within the orbit, these two embryologically distinct arteries will unite to form the supraorbital stapedial. They plug into the ophthalmic to supply all not ocular tissue of the orbit.

Note that when StV1 joins primitive ophthalmic, its proximal stem back to the trigeminal ganglion involutes. Thus, middle meningeal does NOT supply StV1 directly. STV1 is fed by primitive ophthalmic from internal carotid. StV1 supplies dura innervated by V1, principally via its ethmoid branches. However, orbit *does* receive StV2 from the meningeal system. It comes into the lateral orbit via a foramen and helps supply the lacrimal gland. This is the axis of Tessier cleft zone 9. This anatomy is explained in greater detail in Chaps. 6 and 7.

Fig. 2.41 Maxillo-mandibular artery
First part
Non-neural crest mesenchyme—ECA
- **Mastication**

Neural crest mesenchyme—stapedial—all V3
- **External ear**
- **Tympanic membrane**
- **Dura**
- **Mandible**

Second part
Nonneural crest mesenchyme—ECA
- **Mastication**

Third part
Neural crest mesenchyme—stapedial—all V2
- **Maxillary fields**

(Reprinted from Lewis, Warren H (ed). Gray's Anatomy of the Human Body, 20th American Edition. Philadelphia, PA: Lea & Febiger, 1918)

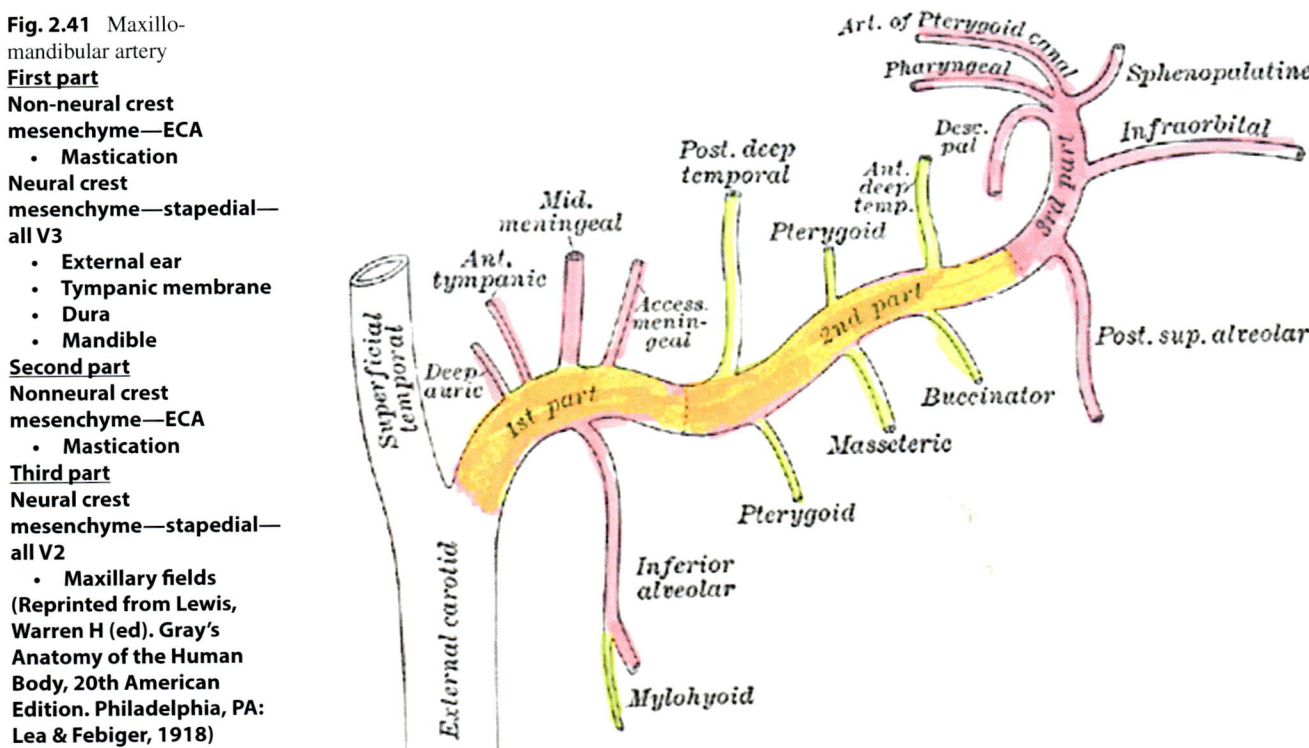

Internal Maxillo-Mandibular Arterial Axis Summarized

Facial circulation is the topic of Chap. 7. But, as our discussion of neuromeric derivatives is based on the blood supply, let's review the IMMA (Fig. 2.41).

ECA neuroangiosomes supply soft tissue fields.

Stapedial neuroangiosomes supply bone fields innervated by V2 and V3.

Segment 1 (Mandibular): Dorsal Branches
- *Deep auricular artery* represents one of two escape routes for the inferior division of stapedial as it exits tympanic cavity. It supplies r3 tympanic membrane and r3 temporomandibular joint.
- *Anterior tympanic* passes through pterotympanic fissure and represents a surviving remnant of stapedial.
- *Middle meningeal* proceeds epaxially with V3 middle meningeal nerve to supply the dura. Entry via foramen spinosum makes an anastomosis with StV2 of the superior division at the takeoff of StV2. The StV2 remnant becomes *anterior branch of middle meningeal* to supply alisphenoid and frontolateral dura. Alisphenoid is innervated by V2 and remains an r2 derivative. The StV3 segment becomes *posterior branch of middle meningeal* to supply the dura of the middle and posterior cranial fossae.
- *Accessory meningeal* goes through foramen ovale to trigeminal ganglion. Represents remnant of the primitive trigeminal artery. Remnant of dorsal aorta.

Segment 1 (Mandibular): Ventral Branches—Mixed
- Inferior alveolar artery: The principal artery to the lower jaw was not necessarily a stapedial derivative in the early gnathostomes Recall that in the primitive state lower jaw is initially supplied by ECA. It is possible that full expression of the maxillo-mandibular artery with stapedial reaching the maxillary stem does not occur until later in evolution. This is evidenced by the evolution of the meninges. Fishes have a single layer of meninx. Reptiles add a second layer, the pachymeninx (dura). Note: the three-layer model of meninges, with an arachnoid and subarachnoid space, is a mammalian invention.
- Mylohyoid artery derived directly from the original stem external carotid: supplies muscles of mastication.

Segment 2 (Infratemporal): Dorsal Branches
- Posterior deep temporal—squamous temporal/muscle
- Pterygoid—mastication
- Anterior deep temporal—squamous temporal/muscle anastomose with lacrimal

Segment 2 (Infratemporal): Ventral Branches
- Pterygoid (ECV3) supplies mandibular ramus
- Masseter (ECV3) supplies ramus
- Buccinator only muscle of second arch supplied by MMA

Segment 3 (Pterygopalatine)
- Zygomatic, temporal: zone 8
- Zygomatic, facial: zone 7
- Posterior superior alveolar: zone 6
- Infraorbital: zones 5 and 4
- Descending palatine zone 3
- Artery of the pterygoid canal: V3 part of auditory canal and tympanic cavity
- Pharyngeal branch: V2 part of auditory canal
- Nasopalatine branch lateral: zone 3—inferior turbinate and maxillary process
- Nasopalatine branch medial: zone 2/zone 1/premaxilla/vomer

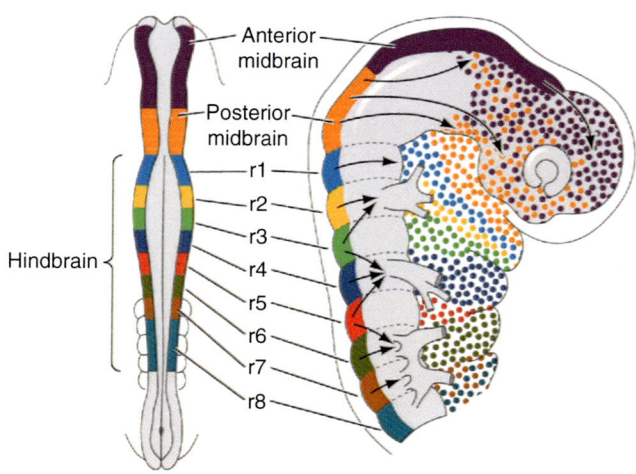

Fig. 2.42 Cranial neural crest map PNC vs. MNC vs. RNC. PNC (p1–p4) = frontonasal dermis (Stapedial V1). MNC (m1, r1) = dura, frontonasal and sphenethmoid mesenchyme (Stapedial V1). RNC (r2–r11) dura, maxilla-zygoma (stapedial V2), dura, mandible (stapedial V3), all remaining structures in first to fourth arches (external carotid system). Defects of midbrain neural crest: (1) Hypertelorism: failure of apoptosis. (2) Encephalocoele: field deficiency defects. (3) Holoprosencephaly: failure of MNC to separate the prechordal mesoderm. (Reprinted from Creuzet S, Couly G, Le Douarin NM. Patterning the neural crest derivatives during development of the vertebrate head: insights from avian studies. J Anat 2005; 207:447–459. With permission from John Wiley & Sons)

Anatomy of Craniofacial Neural Crest

All students of the nervous system are familiar with neural crest cells. The importance of these cells for development is so great that they are often referred to as "the 4th germ layer." The biology of the neural crest is summarized in several authoritative reviews. In this section, we shall detail the neuromeric organization of the neural crest and the manner in which distinct anatomic zone of neural crest migrates in distinctive ways. We shall do so neuromere-by-neuromere. Because we must rely on nomenclature derived from neuroembryology, readers are advised to study because it shows the order in which the various neural crest populations migrate into their final position [54–56].

Building Blocks of the Face

The neural crest entering the face comes from three different sites with three different migration patterns (Fig. 2.42).

Forebrain: Prosencephalic Neural Crest (PNC) > Fronto-Orbito-Nasal Skin

The neural folds above the forebrain have two different zones, anterior and posterior. Tissues from these sites create the skin of the forehead, nose, and upper eyelid. Epidermis arises from *nonneural ectoderm* (NNE) of the anterior zone. Dermis arises from *neural crest* of the posterior zone (PNC). PNC migrates forward underneath NNE as a *single sheet* with distinct genetic zones. These correspond to Tessier cleft zones 10–13. FNO skin is unique; *frontonasal dysplasias* seen in the upper face are a consequence of its development.

Midbrain: Mesencephalic Neural Crest (MNC) > Upper Face

MNC develops from m1 to m2 from midbrain and r0 to r1 from isthmus. We'll refer to them as MNC or r1. These four

populations travel forward in *streams* over the lateral aspect of the forebrain to reach the midline. They produce the entire V1-innervated dura. MNC bone are: the sphenoid complex (except alisphenoid), the ethmoid complex, the lacrimal bone, the frontal bone, and the membranous bones of the orbital series. When the optic cup evaginates from diencephalon, it becomes coated with MNC. The fields of MNC are supplied by V1-induced branches of stapedial ophthalmic axis. These neuroangiosomes are referred to as StV1,

Hindbrain: Rhombencephalic Neural Crest (RNC) > Midface

RNC develops from r2 to r11 in the hindbrain and migrates into the five pharyngeal arches. Each arch is supplied by a pair of rhombomeres. Blood supply for all these derivatives is external carotid. In the first arch, a separate population of RNC from r2 and r3 is responsible for synthesizing the mandible, zygomaticomaxillary complex, and alisphenoid. These *non-pharyngeal arch jaw fields* are supplied by V2 and V3-induced branches of the stapedial maxillo-mandibular axis. These neuroangiosomes are referred to as StV2 and StV3.

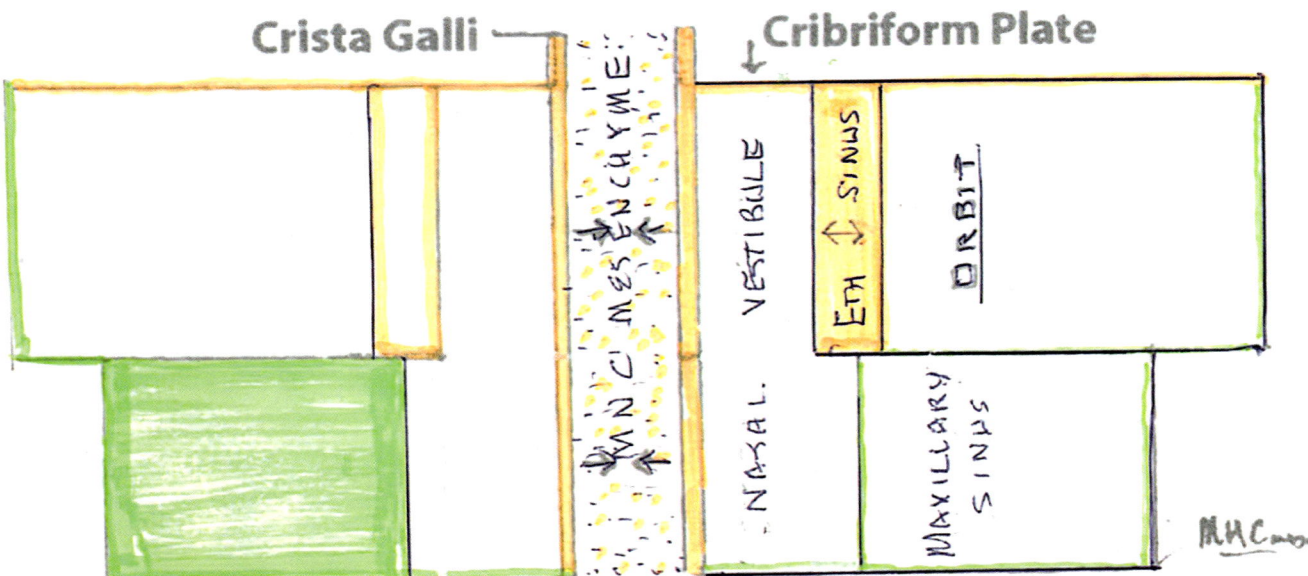

Fig. 2.43 Frontonaso-orbital and maxillary mesenchyme. MNC (orange) is also the source of anterior cranial base frontal bone, ethmoid, and sphenoid. Ethmoid sinuses are separation planes between MNC programmed by r1 sclera (overlying the p5 globe) and p6 vestibular epidermis with r1 submucosa. (Courtesy of Michael Carstens, MD)

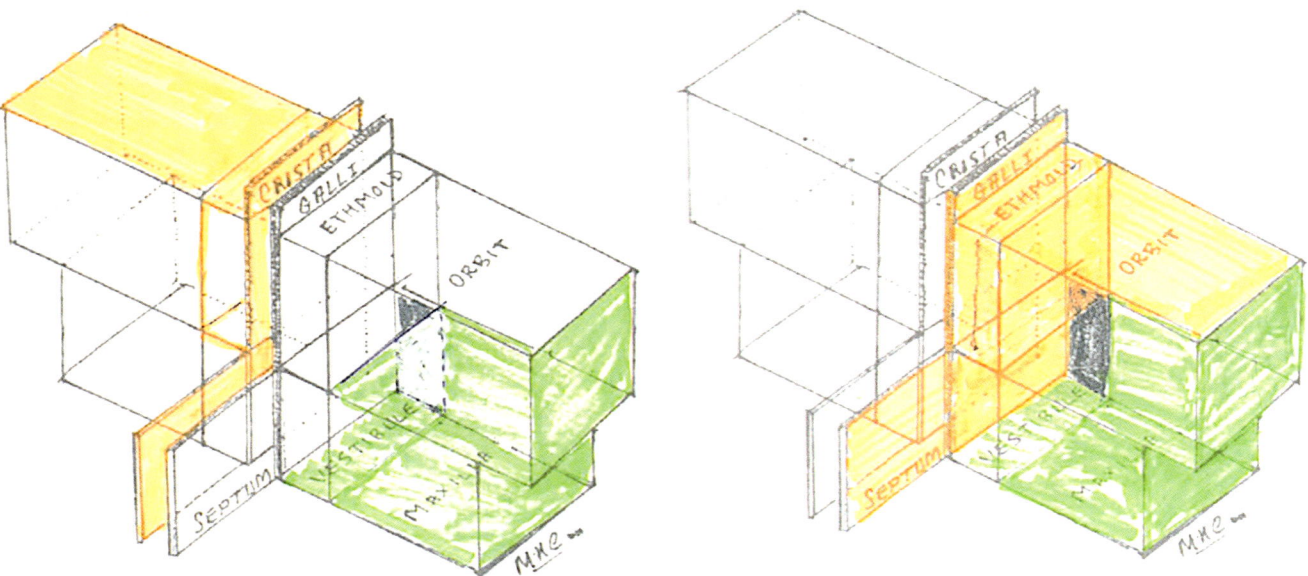

Fig. 2.44 Naso-orbital-maxillary complex: five-sided box. MNC (orange): ethmoid, sphenoid. RNC (green): maxilla, a five-sided box. (Courtesy of Michael Carstens, MD)

Developmental Fields Can Be Lumped into Three Groups: A + B + C

The face is constructed from three blocks of fields on either side of the midline. We can classify them by their blood supply and innervation. This in turn is based on the stapedial arterial system (Figs. 2.43 and 2.44).

A field neuroangiosomes consist of an *FNO skin coverage* and *bone fields of the anterior cranial fossa and medial-*superior orbital walls. The epidermis contains *nasal and optic placodes*, islands of specialized nonneural ectoderm that form the nasal cavity, and complete development of the globe. The MNC bone fields are: sphenoid (except alisphenoid), the ethmoid complex, the frontal bone, and the orbital series (lacrimal, prefrontal, and postfrontal). These are innervated by V1. The blood supply is V1-induced branches of the stapedial ophthalmic axis.

B field neuroangiosomes are **non**-pharyngeal arch bone fields of the jaws and supporting bones: mandible, vomer,

Table 2.3 Neuromeric organization of craniofacial neural crest

Field	NCrest	Neuromeres	Blood supply	Derivatives
A	PNC	p3–p1	StV1	Fronto-naso-orbital dermis
A	MNC	m1–m2, r0–r1	StV1	Fronto-naso-orbital bone fields
B	RNC	r2–r3	StV2/StV3	Jaws and supporting bones to skull
C	RNC	r2–r11	Ext carotid	All other arch structures

premaxilla, maxilla, palatine, zygoma, and alisphenoid. These are innervated by V2. The blood supply is V2-induced branches of the stapedial maxillo-mandibular axis. Mandible, supplied by StV3, is not included in our discussion.

C field neuroangiosomes: *pharyngeal arch tissues* (except the jaws). These are innervated by V2. The blood supply is from the branches of external carotid (Table 2.3).

Migration Patterns of Neural Crest

Neural crest cells from the level of the r0 posteriorly are produced in a cranio-caudal order and from m1 forward in caudal-order. *Three groups of neural crest*, each defined by their level of origin from the embryonic brain, mature in a fixed temporal order, and migrate in unique patterns [57–63].

First to depart from the neural folds are those neural crest cells associated with the mesencephalon. These are, from the midbrain, **m1** and **m2**, and from anterior-most hindbrain, **r0** and **r1**. These MNC cells do not participate in pharyngeal arch formation. Instead, beginning at stage 9, they migrate forward in three distinct *streams* toward the orbit and interorbital midline. These pathways are lengthy; therefore, MNC fronto-nasal migration is not complete until **stage 14** (at the same time after pharyngeal arches are filled).

Next to mature are those cells from the *rostral* rhombencephalon **r2–r7**. These neural crest cells are assigned to pharyngeal arch formation. RNCr cells migrate in a strictly *segmental* fashion. Although these cells begin traveling in time immediately MNC, they have a shorter distance to travel and thus arrive at their destination by **stage 12**. However, these cells have to be positioned in space which takes them until stages 14–15. Cells from the *caudal* rhombencephalon **r8–r11** start later still. These cells also travel in *segmental* fashion over a relatively short distance within their respective segments. It is therefore logical that these RNCc cells arrive at their destination concomitantly with those from the mesencephalon, i.e. at **stage 14**.

Cells from the neural folds above the caudal prosencephalon (i.e. at neuromeric levels **p1**, **p2**, and **p3**) migrate forward as a large *sheet*. They begin their journey after the departure of the MNC. These PNC cells travel a great distance forward; their migration is not complete until **stage 16**. Recall that facial assembly begins at stage 17.

In conclusion, the formation of facial fields occurs in this following sequence: (1) MNC and PNC complete the formation of the frontal fields. (2) RNC$_{ROSTRAL}$ (first and second arches) moves into position. (3) MNC forms the anterior cranial base and orbit. (4) RNC$_{CAUDAL}$ completes formation of the pharynx. (5) PNC produces frontonasal dermis. Let's first take a look at each of these populations. Then we will consider a model of spatial assembly.

Prosencephalic Neural Crest: Anatomic Considerations

The New Prosomeric Model

The model used in this book represents an amalgamation of the neural crest fate mapping studies of Couly and LeDouarin and the neuromeric system of Puelles and Rubenstein. The neuromeric anatomy of the prosencephalon was initially described in 1993 and went through subsequent iterations [63–68]. So far, we have seen a relatively straight-forward relationship between the neuromeres of the hindbrain and midbrain and their derivative tissues. In order to map out structures that relate to the forebrain, here is a brief refresher on the prosomeric system.

First, the developing vertebrate brain is subdivided into transverse building blocks arranged longitudinally down the entire neural tube, much like the watertight compartments of a submarine. These are repetitive (metameric). Although the neuroanatomic content of individual neuromeres can vary, they all share a common dorsoventral organization (roof, alar plate, basal plate, and floor plate). The initial segmentation of the CNS occurs immediately after gastrulation based on *Otx-2* signals from down the anterior visceral endoderm, prechordal plate, and notochord. In the subsequent stage, gradients of Engrailed are laid down on either side of the r0 isthmus (Fig. 2.45).

The anatomic boundaries and neuroanatomy of each neuromere are determined by a unique combination of developmental (homeotic) genes that defines the position of the neuromere along the neuraxis and also what structures it contains. In the mammalian (mouse) model, the activity of homeotic genes within the neuromeres continues to be active into the perinatal period.

The concept of mapping the forebrain into genetically based developmental units, prosomeres, is anatomically sound, but has undergone a number of modifications, The original 1994 model divided the forebrain into six proso-

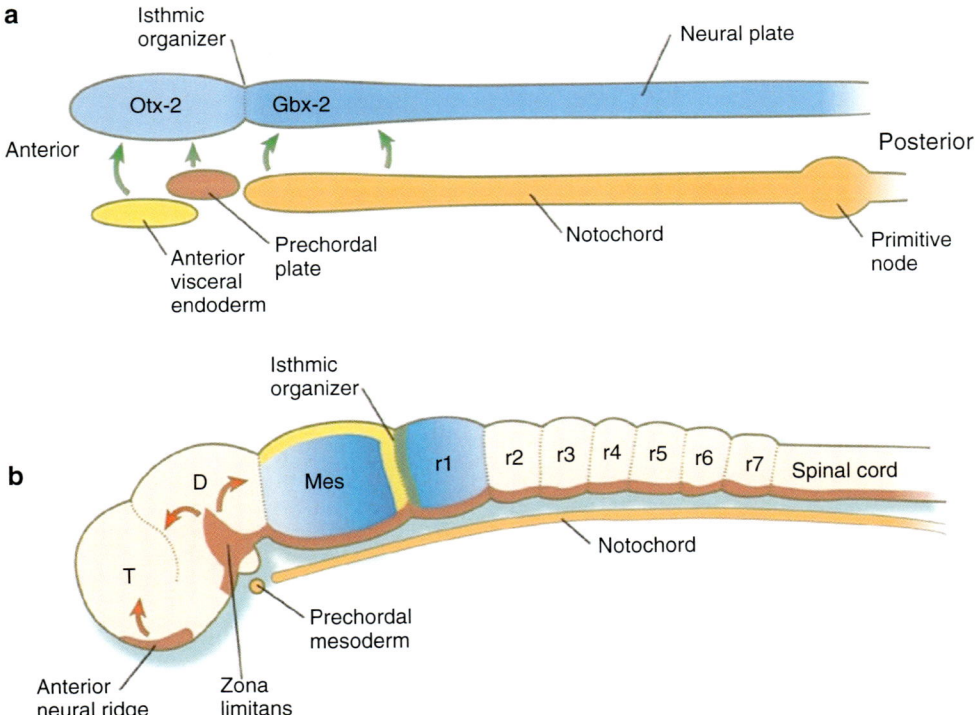

Fig. 2.45 Schematic representation of signaling centers acting on and within the early embryonic brain. (**a**) In response to signals (greens arrows) from the anterior visceral endoderm, the prechordal plate, and the notochord, the neural tube express Oyx-2 in the future forebrain/midbrain and Gbx-2 in the hindbrain/spinal cord. (**b**) Later in development, signals fFGF-8 (green) and Wnt-1 (yellow) from the isthmic organizer induce decreasing gradients of En-1 and En-2 (blue) on either side. Sonic hedgehog (red) is secreted from the other two organizers, the anterior neural ridge and the zona limitans, as well as the ventral part (floor plate) of the neural tube. *D* diencephalon, *Mes* mesencephalon, *r* rhombomere, *T* telencephalon. (Reprinted from Carlson BM. Human Embryology and Developmental Biology, 6th edition. St. Louis, MO: Elsevier; 2019. With permission from Elsevier)

meres, three for the telecephalon and three for the diencephalon. These six prosomeres were discrete developmental units, each with a unique genetic identity. The prosomeres were vertically oriented columns such that the neural folds above the alar zones would share a common homeotic coding with those of the basal zones. Thus, neural crest could be mapped out using the same markers as the base of the brain. This model contradicted well-established studies, done with traditional anatomic methods such as Nissel stains, carried out *without* the use of genetic markers, in which four identifiable units were described [69] (Fig. 2.46a).

Further work showed increasing complexity of the genetic subunits of the telencephalon. The floor plate and basal fields did not run directly upward to the roof, as was previously thought. Time honored dogmas such as the vertical "stacking" of epithalamus, thalamus, and hypothalamus were disproven. As genetic mapping work proceeded, the prosomeric model underwent successive iterations. The 2009 model clarified the anatomy of p1–p3, but made the boundaries of p4–p6 more complex [67]. In 2013, the existence of diencephalon with prosomeres p1–p3 was retained, but the concept of a telencephalon with p6–p4 prosomeres was dropped [68] (Fig. 2.46b).

In Chap. 5, prosomeric system will be discussed in detail but for purposes of this chapter it is important to understand its terminology because this will be used to map out neural crest arising over the prosencephalon.

How can we make use of this model? Caudal forebrain neural folds do *contain neural crest*, so they can be mapped from p1 to p3. This neural crest is the source for the dermis of all frontonasal skin innervated by V1. Rostral forebrain neural folds are utterly *devoid of neural crest* cells. Instead, the rostral folds will form the epidermis of frontonasal skin. They also contain specialized epithelial structures, from proximal to distal: the optic olfactory, and adenophyphyseal placodes. The derivatives of the rostral neural folds have been mapped out by LeDourain and we can extrapolate them to the model of Puelles model. Caudal secondary prosencephalon is zone p4 and contains the *calvarial ectoderm*, the "program" for the upper frontal bone. Rostral secondary prosencephalon has two zones. Zone p5 contains the optic placode, the eventual source of the lens. Zone p6 contains the nasal placode and the adenohypophyseal placode. Although the model proposed by this author is hypothetical, the neural fold fate map continues to be refined in the mouse and correlated with the human prosomeric system.

Fig. 2.46 Previous iterations of the prosomere model. Generalized vertebrate embryo showing neuromeres and distribution of major signaling molecules. Mindbrain/hindbrain signaling regions = arrows just rostral to first rhombomere. Arrows between second and third rhombomeres represent a hypothetical signaling region in the forebrain. Model subsequently revised by 2013 with defined prosomeres restricted to the diencephalon. Prosomeres p6, p5, and p4 are now not assigned to neural folds but map of neural fold derivatives remains valid. (Left: Reprinted from Carlson BM. Human Embryology and Developmental Biology, 6th edition. St. Louis, MO: Elsevier; 2019. With permission from Elsevier. Right: Reprinted from Puelles L, Harrison M, Paxinos G, Watson C. A developmental ontology for the mammalian brain based on the prosomeric model. Trends Neurosci 2013; 36(10):570–578. With permission from Elsevier)

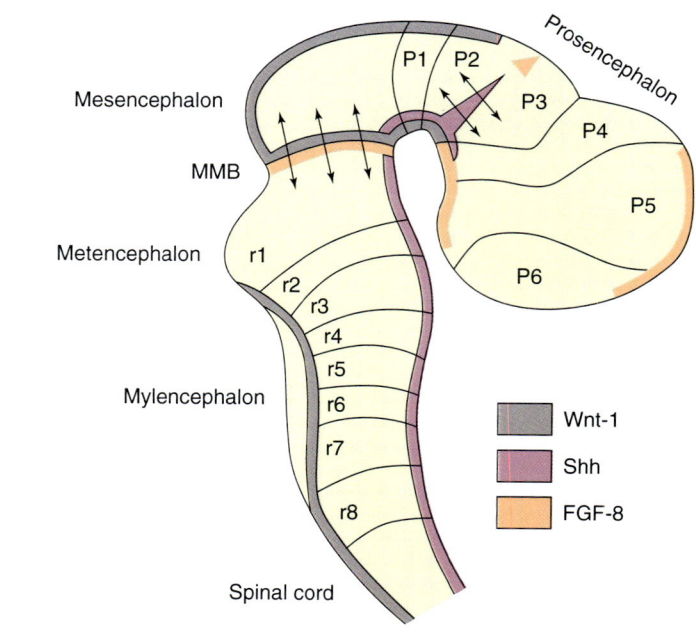

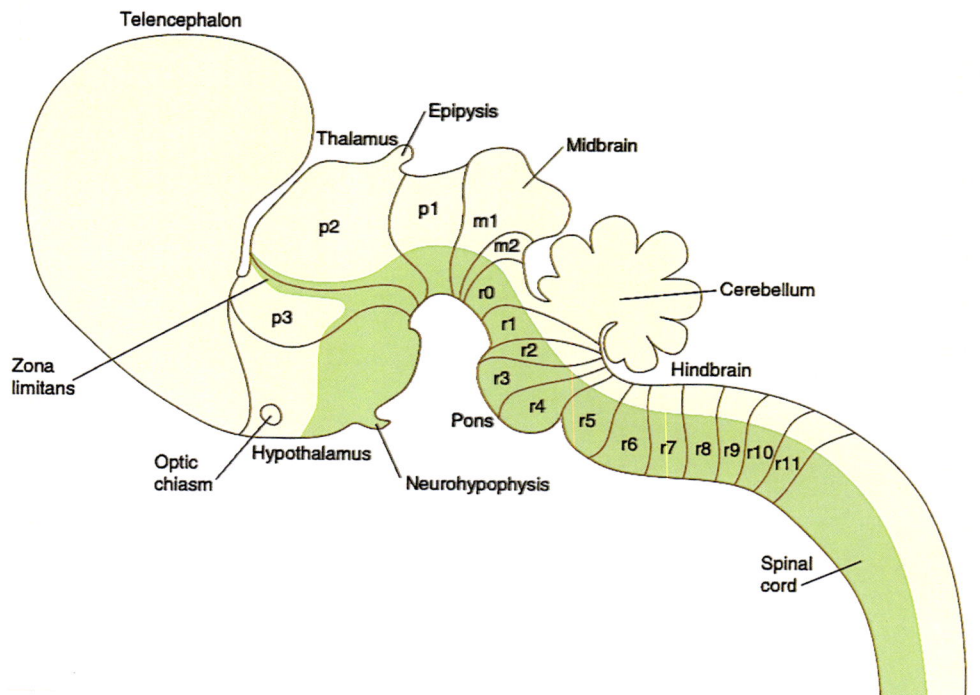

For the purposes of mapping out facial structures, we shall adopt the following conventions:

- Rostral forebrain nonneural ectoderm (NNE) can be divided into zones corresponding to the placodes.
- Placodes are ectodermal structures that maintain a relationship to the CNS.
- Zones p5 and p6 are repositioned basal to the forebrain.
- Zones p4–p6 participate in the development of the face.

- V1 stapedial connects p4–p6 facial fields and the subjacent anterior cerebral artery.
- Field defects in the ACA fields relate to and may predict defect segments of the StV1 system that supply the p4–p6 facial fields.

Prosomeric Mesenchyme and Nonneural Epithelium: General Considerations

Why are these distinctions so important? Recall that frontonasal skin (FNO) is mapped out by the sensory distribution of V1. How this skin develops is utterly unique. Recall that the gastrulation process that produces germ layers at stage 7 does not extend forward from level r0. Thus, *ectoderm and mesoderm are simply not available to make the skin of the eyes, nose, and forehead.* The embryo uses a different strategy using three different tissue sources to accomplish the same thing [68].

- Frontonasal epidermis develops from the neural folds overlying the rostral prosencephalon (telencephalon). This zone is devoid of neural crest and is termed the *non-neural ectoderm* (NNE). The contributions of NNE to craniofacial skin have been mapped by Couly et al. [70].
- Frontonasal dermis arises from the caudal prosencephalon (diencephalon); the folds that produce neural crest are in register with prosomeres p4–p1. We shall refer to the prosencephalic neural crest as PNC.
- Subcutaneous frontonasal tissues are derived from midbrain neural crest (MNC); it covers the entire forebrain and upper face.

Survival of the FNO epidermis requires the acquisition of an underlying support network in order to survive. PNC cells from caudal forebrain migrate forward to populate the sub-epithelial plane of p4–p5–p6 with neural crest-derived dermis. In this way, PNC ensures the viability of the nonneural epithelium.

Cellular movement of PNC into the forward position takes place more in the form of a sheet than as clearly identifiable streams. How do these populations migrate? The order in which the target zones of the rostral folds are populated is from back to front (caudal-to-rostral). This is supported by the known *sequence of placode development,* which is: otic, optic, and then olfactory. *Failure of neural crest to populate the sub-placodal zone leads to placodal dysfunction or outright absence.* We shall discuss the derivatives of placodes that result from these interactions in the next section [71–74].

Prosomeric Placodes

Development of the face is dominated by behavior of three important placodes: pituitary, nasal, and optic. Development of the ear is controlled by the optic placode of the hindbrain. A placodal region is one in which the *specialized zone of ectoderm is in direct contact with the CNS.* Cellular migration of three types takes place from all placodes. (1) Specialized sensory cells are represented by structures such as the otoliths of the labyrinth from the otic placodes of r4–r5 hindbrain and the vomeronasal organs flanking the septum from the medial nasal placodes. (2) Neuroblasts arising from the lateral olfactory placodes provide the apparatus of smell, whereas those from the medial olfactory placodes transmit chemoreceptive data to the brain. (3) Neural crest or PAM cells interact with invading placodal tissue to form into the cartilaginous capsules of the presphenoid, the nasal cavities, the orbits, and the temporal bone (Figs. 2.47, 2.48, 2.49, and 2.50).

The natural behavior of placodes is to penetrate (more accurately to become incorporated) into the embryonic CNS. As these placodes "sink beneath the waves," they interact with the brain to induce additional structures. These simple relationships are readily seen in the neural fold state prior

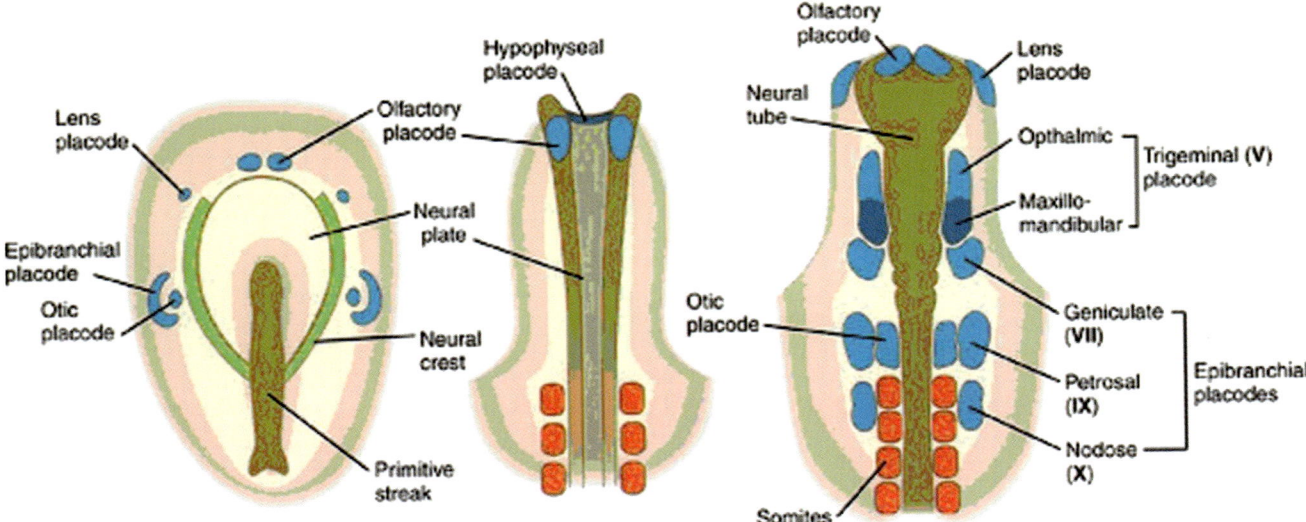

Fig. 2.47 Neural folds and placodes. Early stages in the formation of cranial ectodermal placodes (blue) in the chick embryo, as viewed from the dorsal aspect. These specialized zones of the neural folds interact with neural crest cells to produce the neurosensory-endocrine apparatus of the embryo (the pituitary, olfactory systems, lens, inner ear, and specialized ganglia). (Reprinted from Carlson BM. Human Embryology and Developmental Biology, 6th edition. St. Louis, MO: Elsevier; 2019. With permission from Elsevier)

Fig. 2.48 Nonneural
ectoderm of the anterior folds
and epidermal derivatives.
Adenohyophyseal placode/
Rathke's pouch, p6 (brown),
nasal epithelium, p6 (pink),
upper beak/nose epidermis,
p5 (green), calvarial
epidermis, p4 (orange), eyelid
epidermis/cornea epithelium,
r1 (white), midface oral-
maxilla-zygoma epidermis,
premaxilla mucoperiosteum
r2 (blue), lower face mandible
epidermis, r3 (light blue).
(Reprinted from Standring S
(ed). Gray's Anatomy 40th ed.
New York, NY: Churchill
Livingstone; 2008. With
permission from Elsevier)

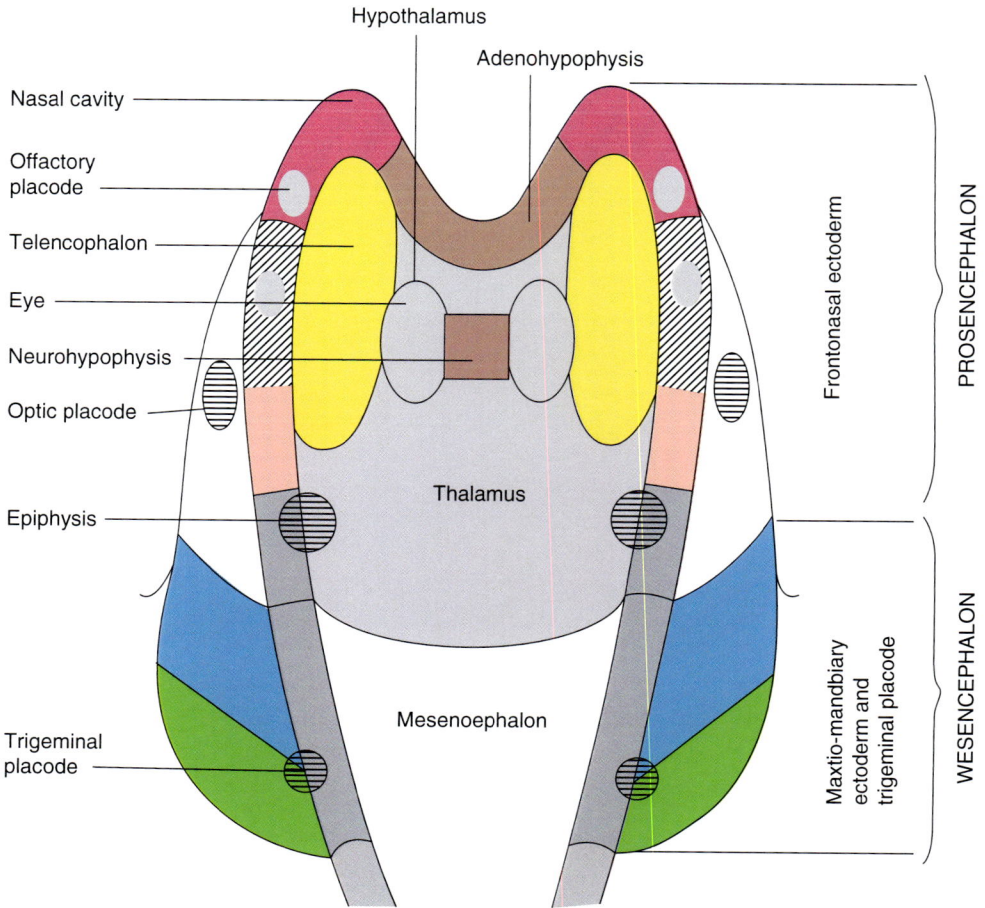

to closure of the neuropore and head folding. Growth of the telencephalon at this point involves the formation of paired hemispheric vesicles that push laterally from the sidewalls of the neural tube. These cerebral vesicles push rostrally and caudally. They expand so greatly that they eventually envelop the entire neuraxis. Tucked within the hemispheres are the midbrain and hindbrain. The cerebellum sprouts forth from r0 and r1; it projects backward beneath the hemispheres. Neuromeric pathologies in r0–r1 can be seen as facial tissue defects and cerebellar hypoplasia or aplasia.

Adenohypophyseal Placode

The most rostral/cranial placodes in p6 neural folds are the underlineadenohypophyseal placodes. These are repositioned by forebrain expansion and folding until they come to lie tucked below the brain into the future nasopharynx at Rathke's pounch. At this point, they fuse to form the epithelial component of the pituitary, the adenophypophysis. Rathke's pouch also marks the epithelial boundary zone between p6 nasal vestibular lining anteriorly and likely third arch endothelium posteriorly (Figs. 2.51 and 2.52) [75, 76].

The most distal zone of each p6 NNE to be populated by PNC is that of the *adenohypophysis (pituitary) placode* (AP). These are bilateral but, at the anterior extreme of the embryo,

the p6 zones are continuous with each other, resembling a handlebar moustache. Textbooks describe the AP as if it were a single entity when, in point of fact, the placodes are *bilateral* and fuse. For this reason, in the presence of forebrain pathology, the ipsilateral adenohypophysis can be smaller. Tumors of a given cell type can also be unilateral.

With formation of the primary head fold, the most anterior nonneural ectoderm is tucked ventrally and caudally below the forebrain until it lies just beneath the presphenoid/basisphenoid junction. This can be visualized by imagining the fingernail of your right index finger as the placode and the metacarpo-phalangeal joint of the same finger as the PS/BS "joint." This is the topology of the neural plate. If you then make a fist with the hands supine, the tip of the index finger (containing the placode) will come to rest beneath your MP (PS/BS) joint.

At any rate, the relocation of the APs into the future nasopharynx beneath the PS/BS junction permits upward "penetration" by these placodes to take place. Once within the "potential space of the sphenoids," the adenohypophysis will ultimately be married up with the neurophypophysis from the caudal diencephalon. Neurohypophysis emerges from basal p1. It descends and is affixed to the posterior aspect of adenohypophysis. At this site, primitive tumors, such as *cra-*

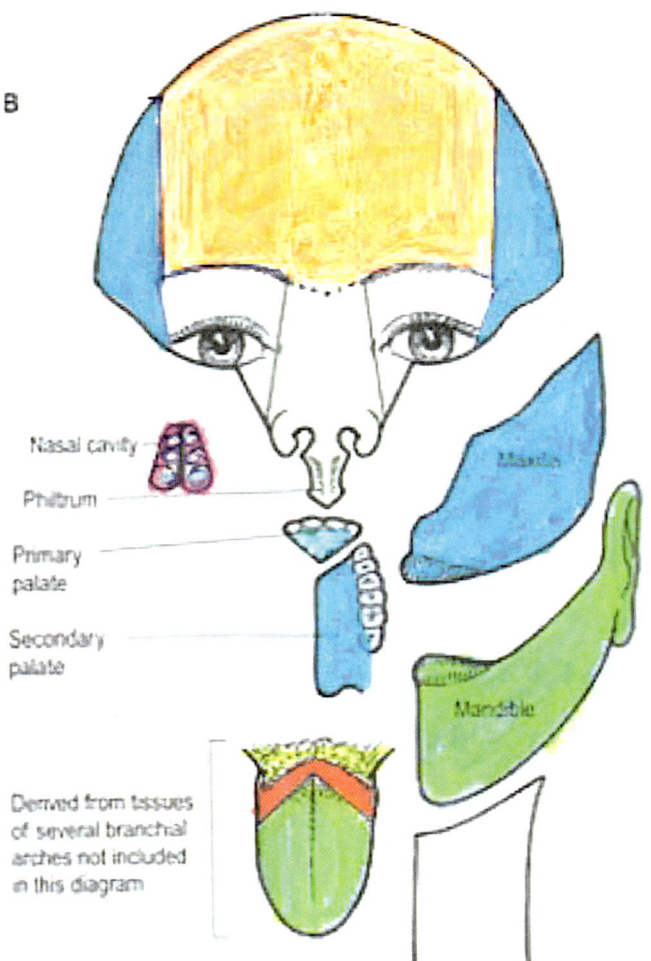

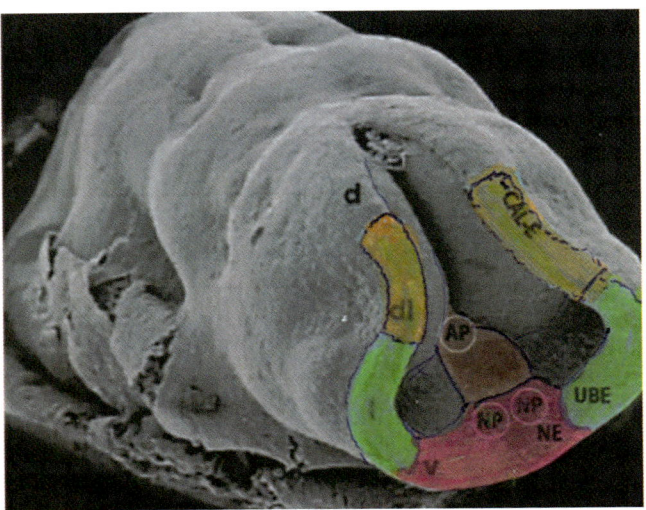

Fig. 2.50 Rostral prosencephalic folds (zones p4–p6). The neural folds of the prosencephalon extend all the way forward from midbrain to neurohypophysis. Caudal folds over diencephalon have neural crest. Rostral folds over secondary prosencephalon have nonneural ectoderm bearing placodes but no neural crest cells. Most anterior folds can be mapped into p6 epithelium (red) surrounding the adenohypophyseal and nasal placodes, p5 epidermis (green) upper beak ectoderm, and p4 calvarial ectoderma (orange). Note the bulge of the optic vesicle just lateral to p5. Ectoderm from r1 is responsible for the epidermis of the upper eyelid and corneal epithelium. Neural crest from the caudal neural folds flows forward beneath the NNE in sheet-like fashion to create dermal substrate for frontonasal skin and the submucosa of the nasal vestibular lining. MNC flowing forward between brain and skin comes in contact with the nasal lining to form the sphenethmoid complex, nasal and lacrimal bones. (Courtesy of Prof. Kathleen K. Sulik, University of North Carolina)

Fig. 2.49 Internal nose: lining (p6), superior and middle turbinates (r1), inferior turbinate (r2). Forehead (p4), external nose (p5). Medial and lateral nasal processes both p5 but induced by their respective half of the nasal placode. Upper eyelid and lacrimal skin (r1) with lower eyelid skin lateral to punctum (r2). Oral cavity floor: anterior 2/3 tongue from first arch (r3) abuts against posterior 1/3 from third arch (r7). Root of tongue fourth arch (r8) abuts against larynx r8–r9. Note: buccopharyngeal membrane separates first and third arches at tongue, anterior palatal pillars, and the proximal 1/3 of the soft palate. (B) Neuromeric derivations of craniofacial soft tissue structure (Reprinted from Standring S (ed). Gray's Anatomy. 40th ed. New York, NY: Churchill Livingstone; 2008. With permission from Elsevier)

niopharyngiomas, can bulge beneath the mucosa into the pharynx. At times, this can even disrupt palatal formation [77–80].

Nasal Placodes

Located more caudally along p6 neural folds, the nasal placodes contribute three separate sets of neurons that ultimately find their way to three different sites in the p6 telencephalon! Without this incorporation, areas of the basal forebrain (the rhinencephalon) are hypoplastic as in *Kallman syndrome* [81–83] (Fig. 2.53).

The anatomic fate of the nasal placode is determined by the furious proliferation of MNC around the placode which throws the surface ectoderm into relief. This produces as an elongation of the snout. It also results in an "invagination" of the nasal placode "backward" into the ethmoid mesenchyme, owing to its physical connection to the primitive basal forebrain, rhinencephalon. As the nasal placode is drawn inward, it pulls along with it the most distal zone of p5 nasal skin into the introitus of the nasal vestibule. In so doing, the p5 skin doubles back on itself. MNC associated with this skin is programmed by the *p5 intranasal dermis* to form the *lower lateral cartilage*. Further inward, MNC associated with *p6 vestibular lining* will form the upper lateral cartilage and septum. Note: Nasal bones should *not* be considered p5 derivatives. They are MNC derivatives programmed by p6 vestibular lining (Fig. 2.54).

Nasal placodes are essential to the formation of the nasal cavity. If a placode is aberrantly placed, *heminose* results with a tubular proboscis marking the ectopic placode. Complete absence of placodes causes arrhinia which can coexist with normal ocular development (Fig. 2.55) [84].

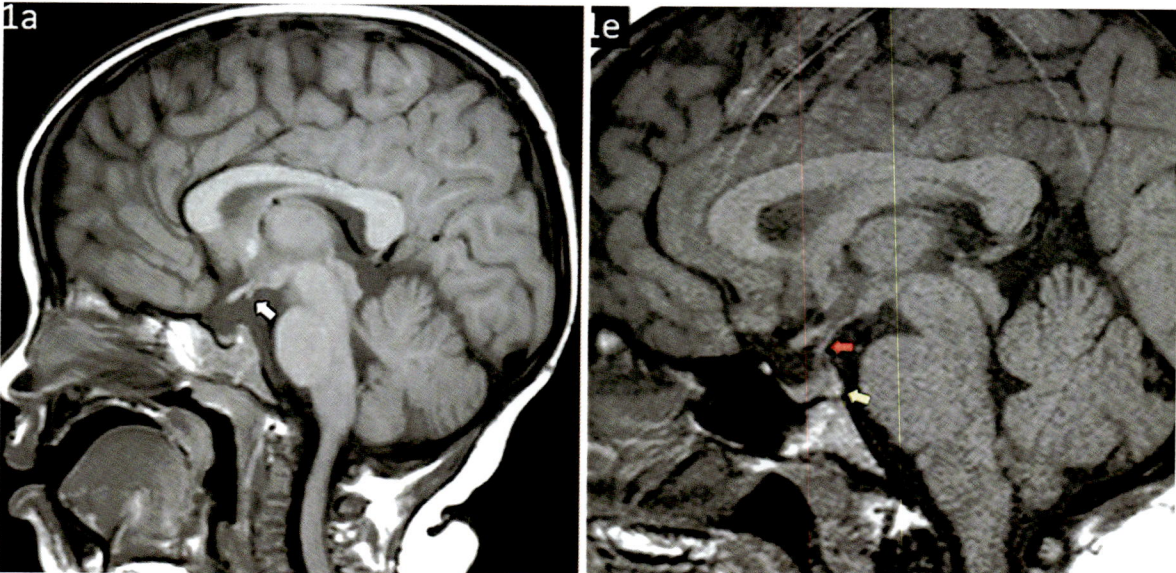

Fig. 2.51 Four-year-old male patient with congenital hypopituitarism presented as global developmental delay and growth retardation. Workup showed hormonal deficiencies and MRI abnormalities. (**a**) T1-weighted MRI shows partially empty sella turcica, pituitary stalk cannot be identified. Anterior pituitary tissue seen along the floor of the sella. The posterior pituitary is ectopic, seen as a 2 mm hyperintense nodule at the median eminence (white arrow). (**b**) T1-weighted MR shows normal 30-year-old patient are provided for comparison. Note the normal infundibulum extends from the hypothalamus to the pitu- itary gland (red arrow). Sella turcica is well-developed and contains both the anterior pituitary lobe and T1 hyperintense posterior pituitary lobe (yellow arrow). (Reprinted from Omer A, Haddad D, Pisinski L, Krauthammer AV. The Missing Link: A case of absent pituitary infun- dibulum and ectopic neurohypophysis in a Pediatric Patient with Heterotaxy Syndrome. J Radiology Case Reports 2017; 11(9): 28–38. DOI: https://doi.org/10.3941/jrcr.v11i9.3046. with permission from Journal of Radiology Reports)

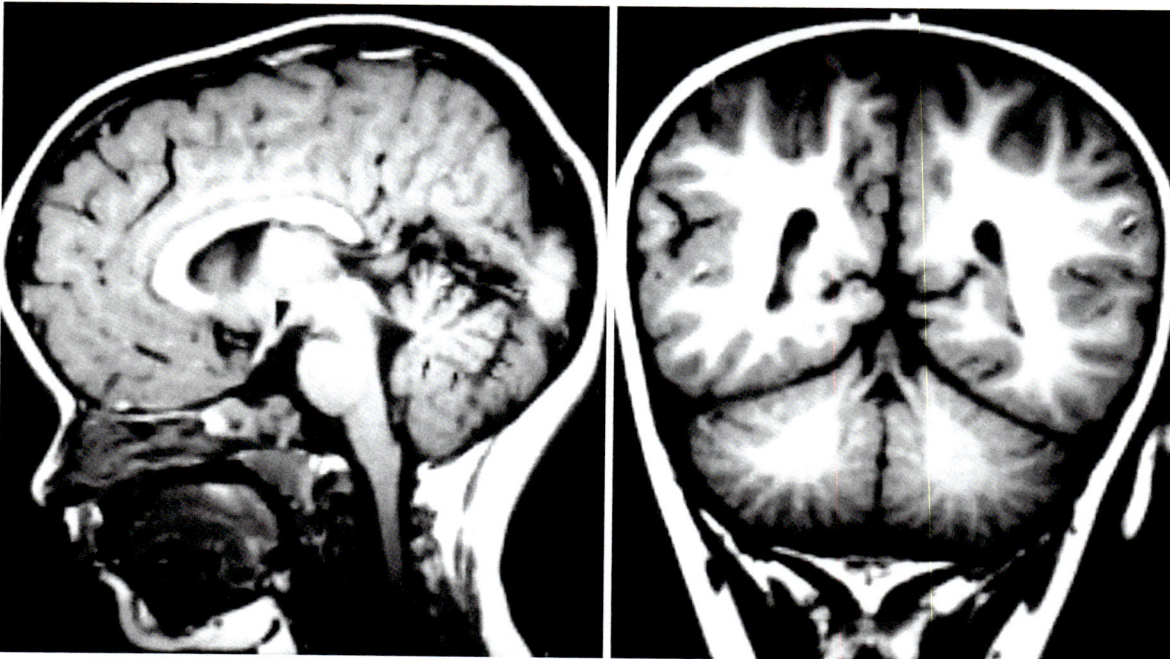

Fig. 2.52 Cerebellum arises from r0 to r1. Sella turcica arises from the same level. MR image shows cerebellar hemispheres are separated by a cleft because the posterior lobe of the vermis is absent. (Reprinted from L I Al-Gazali, L Sztriha, J Punnose, W Shather, M Nork. Absent pitu- itary gland and hypoplasia of the cerebellar vermis associated with par- tial ophthalmoplegia and postaxial polydactyly: a variant of orofaciodigital syndrome VI or a new syndrome? Journal of Medical Genetics 1999; 36(2):161–166. With permission from BMJ Publishing Group Ltd.)

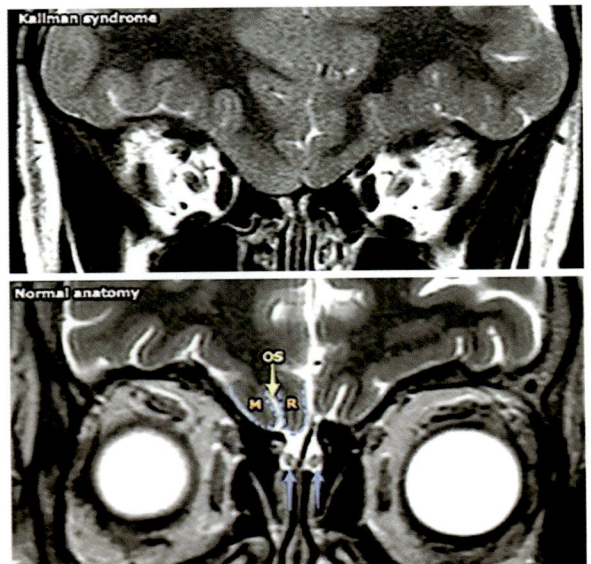

Fig. 2.53 Kallmann syndrome is a rare genetic disorder characterized by hypogonadotropic hypogonadism associated with anosmia or hyposmia. When anosmia is absent, a similar syndrome is referred to as normosmic idiopathic hypogonadotropic hypogonadism. MRI is the modality of choice in assessing for the absence of olfactory bulbs, and coro\nal T2 sequences are most effective. It is a rare disorder with an estimated prevalence of one in 10,000 males and one in 50,000 females. Although patients with Kallmann syndrome are anosmic from birth, this usually is not apparent to either the parents or the child. The diagnosis is only made when puberty does not occur. At that time gonadotropin levels (FSH and LH) and sex hormones (testosterone and estradiol) are low, whereas other pituitary hormones are normal. Associated findings include enlarged paranasal sinuses, small anterior pituitary, septo-optic dysplasia, CL/P, sensorineural deafness, and renal agenesis. The olfactory nerves, bulbs, and sulci are absent (arhinencephaly). Importantly the hypothalamus and pituitary are most often morphologically normal in appearance. However, the anterior pituitary can appear small. (Case courtesy of Assoc. Prof. Frank Gaillard, Radiopaedia.org, rID: 6083. https://radiopaedia.org/cases/6083/)

Optic Placodes

The placodes are located still more caudally along the prosencephalic neural folds, in zone p5. They are required for the proper formation of the eye, specifically the lens apparatus. In *aphakia*, absence of the placode, the globe has no lens. If the Pax-6 gene expressed by the placodes is more globally suppressed, the globe itself will be phthisic (shrunken), a condition known as microphthalmia or absent, resulting in anophthalmia (Fig. 2.56) [85].

Otic Placodes

The otic placodes develop from the neural folds of the hindbrain at neuromeres r4–r5. They share common characteristics with those of the forebrain and are therefore included here for the sake of completion. The otic placodes induce the cochlear and vestibular systems using mechanisms of mechanosensory hairs. Abnormalities of these structures are another form of neurocristopathy (Fig. 2.57) [86].

Blood Supply for Fronto-Orbital-Nasal Skin

Vascular development for FON skin takes place in two phases. Recall that, as a neuroepithelium, PNC has no intrinsic blood vessel-forming capacity. In the initial phase, invasive angioblasts create a primitive head plexus which surrounds the developing forebrain. The origin of these cells is unclear; they may come from the prechordal plate. In any case they provide the mesodermal substrate for the subdermal plexus. The subcutaneous plane is then populated by midbrain neural crest. This combination of mesodermal cells and neural crest cells forms a vascular head plexus. Neuroangiosomes from the StV1 system (lacrimal, supraorbital, and supratrochlear) connect with this plexus and provide stable support for the skin. Later on, as we shall see, the

Fig. 2.54 Nasal bones and upper lateral cartilages are synthesized based on a program embedded in the underlying r1 vestibular lining. The lower lateral cartilages are programmed by the r2 skin. Hypoplastic nasal bones are seen in Down syndrome (43–62%). Absent nasal bones seen in trisomy 21 (60–73%), trisomy 18 (53–57%), trisomy 13 (32–45%), Turner's syndrome (9%). (Left: Reprinted from Lewis, Warren H (ed). Gray's Anatomy of the Human Body, 20th American Edition. Philadelphia, PA: Lea & Febiger, 1918, Gray fig. 582. Right: Case courtesy of Dr. Ayush Goel. Radiopaedia.org, rID: 54486. https://radiopaedia.org/articles/absent-nasal-bone)

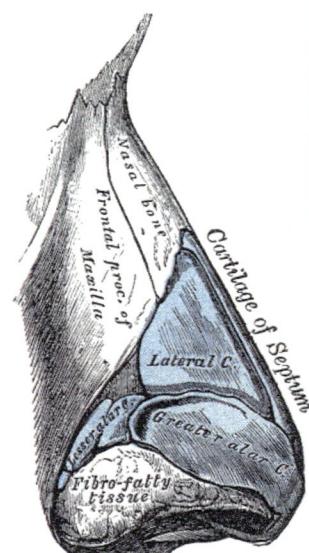

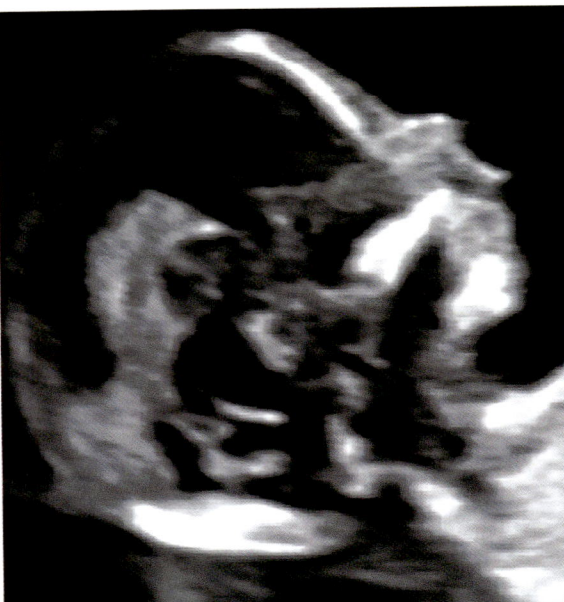

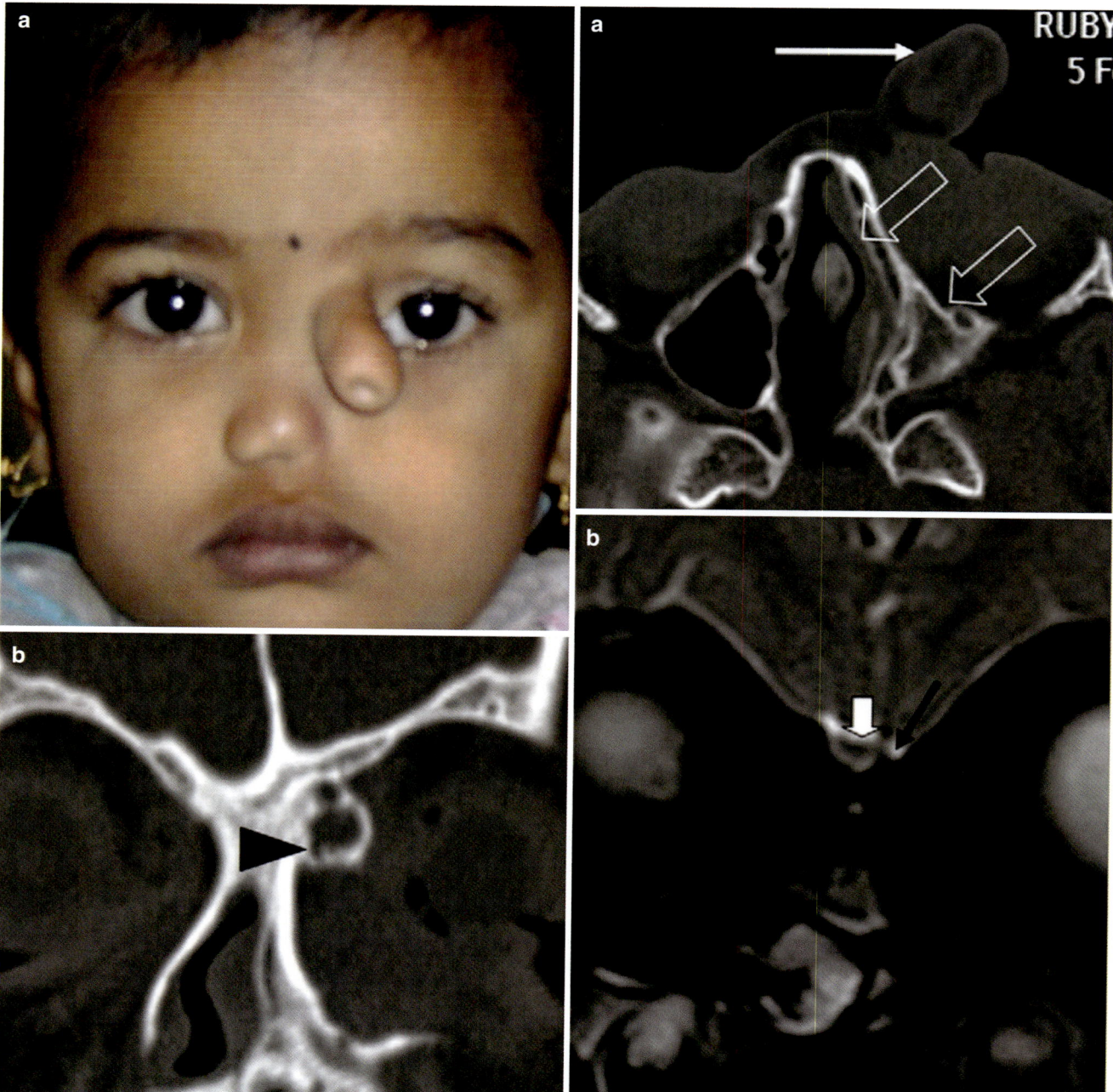

Fig. 2.55 Abortive development of left nasal placode proboscis lateralis in a 2-year-old girl with associated ipsilateral sinonasal aplasia, orbital cyst, absent olfactory bulb and olfactory tract. Ipsilateral olfactory pathway is missing. Upper left: Tubular trunk-like process arising from medial canthus of left orbit. Upper right: HRCT image (**a**) showing a tubular trunk-like process (white arrow) arising from superomedial aspect of the left orbit with aplasia of ipsilateral nasal cavity and maxillary/ethmoid sinuses (open arrows). Note septum pulled over to the left. Lower left: Coronal HRCT image (**b**) shows well-formed bony canal in superomedial compartment of the left orbit (black arrowhead). Lower right: normal right olfactory bulb and groove (white block arrow) and hypoplastic left olfactory groove with absent left olfactory bulb (black arrow). (Reprinted from Vaid S, Shah D, Rawat S, Shukla R. Proboscis lateralis with ipsilateral sinonasal and olfactory pathway aplasia. J Pediatric Surg 2010; 45: 453–456. With permission from Elsevier)

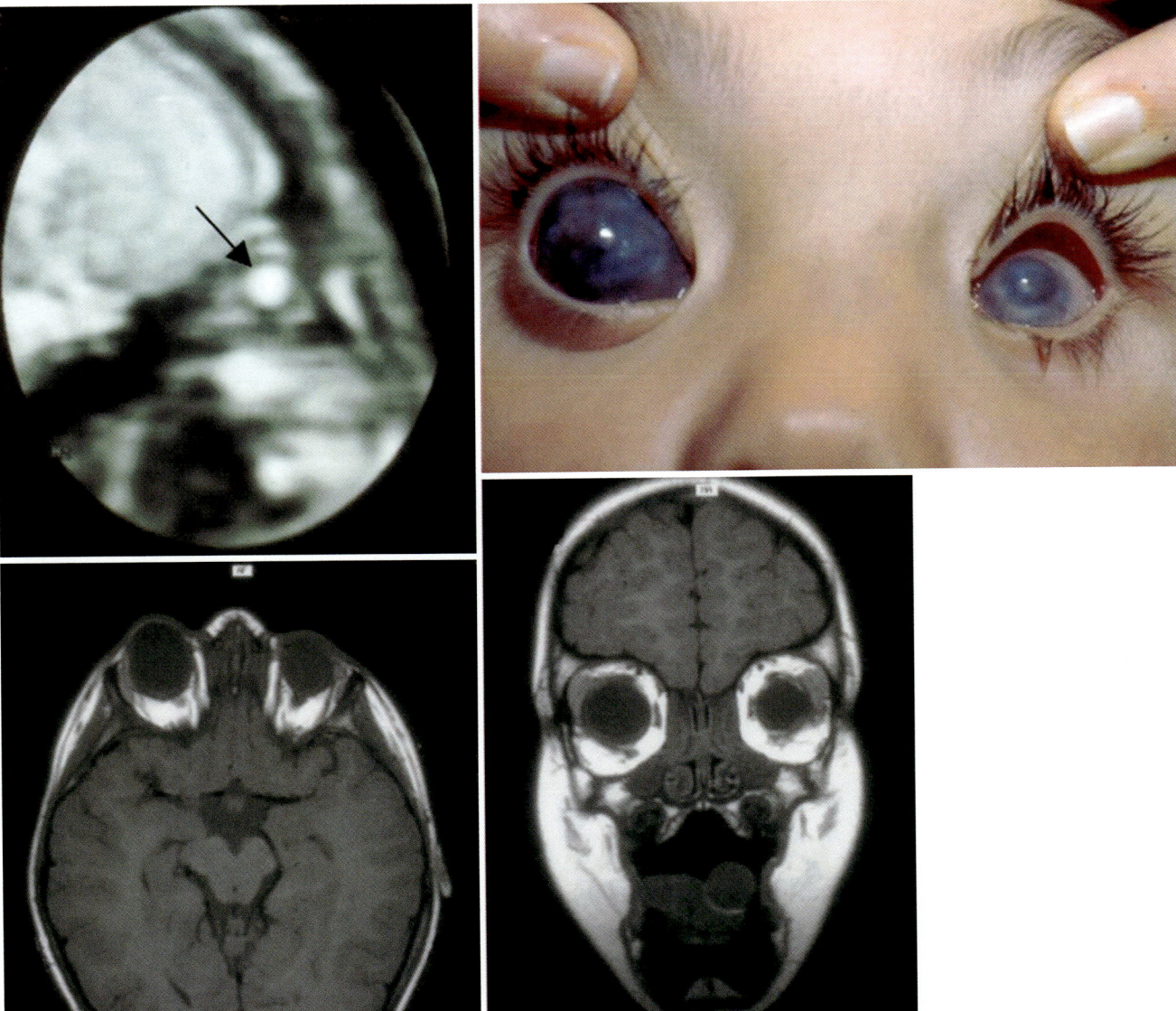

Fig. 2.56 Congenital primary aphakia (CPA) consists of absence of a lens, induction of which takes place in weeks 4–5, causing complete aplasia of the anterior segment (the diagnostic histologic finding in CPA). Primary (congenital) aphakia can occur in isolation or as part of a complex anterior segment abnormality, i.e., microphthalmia, absence of the iris, anterior segment aplasia, and/or sclerocornea. Absent placode at 4–5 weeks prevents the formation of any lens structure in the eye. This leads to complete aplasia of the anterior segment, Secondary aphakia is characterized by absorption of the lens after it is formed. The causation is unknown but may involve a mutation in the *FOXE3* gene. Upper left: MRI of 28.5 week fetus showing eyeball as white sphere without the lens, normally seen inside as a black sphere. Upper right: 1 month baby with left-sided microopthalmia and right-sided CPA. Lower left and right: axial and coronal MRI of 4-year old demonstrating bilateral microopathalmia and right-sided aphakia. The brain tissue is otherwise normal. (Reprinted from Valleix S, Niel F, Nedelec B, et al. Homozygous nonsense mutation inf the FOXE3 gene as a cause of congenital primary aphakia in humans. Am J Hum Genet 2006; 79(2): 358–364. With permission from Elsevier)

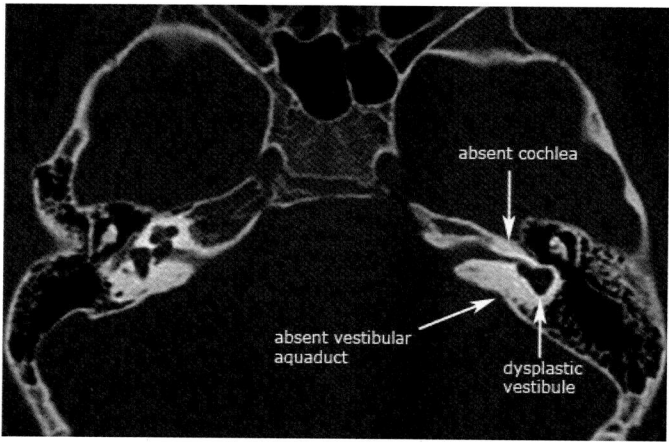 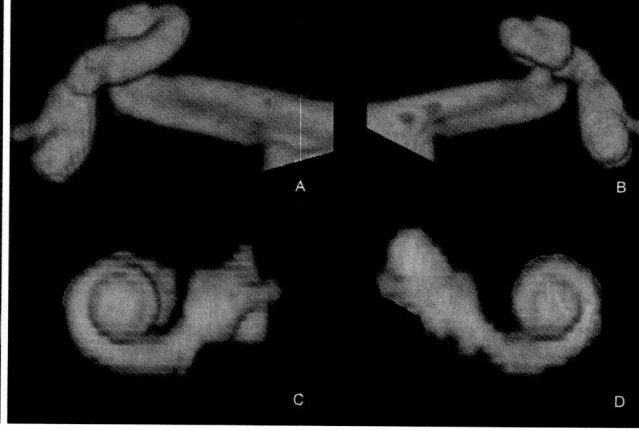

Fig. 2.57 Michel Syndrome. Absent/hypoplastic otic placode follows a sequence: 3–4 weeks: Michel (complete labyrinthine aplasia). 4–5 weeks: cochlea and vestibular system confluent. 5–6 weeks: cochlear aplastia/hypoplasia (vestibular system forms earlier and is spared). Left: Michel syndrome (oculopalatalskeltal). Right: unilateral SCC hypoplastic (left side with intact cochlea and hearing loss). (**a**) Superior left, (**b**) lateral left, (**c**) superior right, (**d**) lateral right. (Left:

Reprinted from Wikimedia. Retrieved from: https://en.wikipedia.org/wiki/Michel_aplasia. With permission from Creative Commons License 1.0: https://creativecommons.org/publicdomain/zero/1.0/deed.en. Right: Reprinted from Breheret R, Brecheteau C, Tanguy J-Y, LacourreyeL. Bilateral semicircular canal hypoplasia European Annals of Otorhinoloaryngology, Head and Neck Diseases 2013; 130:225–228. With permission from Elsevier)

MNC plane is invaded by facial muscles from the second arch that are supplied by laterally based vessels of the facial artery system. This arrangement permits elevation of skin flaps from above the muscle plane based on their vascular pedicles.

MNC Schizophrenia

PNC does not have the capacity to form bone or cartilage. What the skin envelope does have is the "template" of information that instructs underlying substrate to produce bone. The mesenchymal "raw materials" necessary for the synthesis of all cartilage and bone structures of the nose, eyes, anterior cranial fossa, and sphenethmoid complex come from midbrain neural crest. But the instructions for bone formation reside within the epithelial structures of the brain and skin. MNC mesenchyme must obey orders from two different masters and therefore splits into two laminae.

- Epidermis/dermis induces outer ectomeninx to produce frontal bone.
- Brain tissue induces inner endomeninx to make dura.

Recall that membranous ossification requires a "*stimulating mesenchyme*" (neural crest dura or dermis) that contains a genetic program for the bone and a "*responding mesenchyme*" (neural crest or PAM) that forms the bone itself. Recall that neural crest, by virtue of being both epithelial and mesenchymal, can serve an epithelial function. In some parts of the skull, a responding mesenchyme may be sandwiched between two layers of stimulating ectoderm or endoderm. *Bilaminar* membranous bone formation occurs. In the calvarium, the interface between the laminae constitutes a potential space that is later manifested as a *diploic space*

[87]. This model for bilaminar bone fields explains the anatomy of the *sinuses of the skull*. These occur at sites where mucosa from oropharyngeal cavity has anatomic access to that potential space. By exploiting field separation planes, the mucosa will expand into the frontal, ethmoid, sphenoid, and mastoid bones to create their respective sinuses, *but it will never transgress a neuromeric boundary*, i.e., a suture.

All calvarial bones formed by neural crest and PAM cranial to the first motion segment (r11–c1 junction the proatlas) are *segmental*, i.e., the bone boundaries correspond to neuromeres. At the craniocervical junction, vertebrae become, for the first time, *parasegmental*, i.e., each vertebral body results from contributions of the caudal somite above and the rostral somite at that level.

Forebrain Neural Crest Migration

The neural folds of the anterior secondary prosencephalon (zones p4–p6) contain nonneural ectoderm; they do not have neural crest. PNC develops from the neural folds of the posterior secondary prosencephalon (zones p1–p3). It then migrates forward into the FNO in caudal-to-rostral process. That is, it proceeds forward away from the midbrain. PNC cells come in, like guided missiles, seeking "target zones" of nonneural ectoderm: the calvarial epithelium, the optic placode, the upper beak epithelium, the nasal epithelium, the nasal placode, and the adenhypophyseal placode. In every case, the migration of PNC involves a succession of lateral-to-medial cell movements, each zone advancing further forward, building upon its predecessor. Thus, the flow of PNC through prosomeric zone p4 is forward and medial; each zone of p5 is "newer" than its predecessor. Neural crest entering the sixth prosomeric zone is the most recent of all.

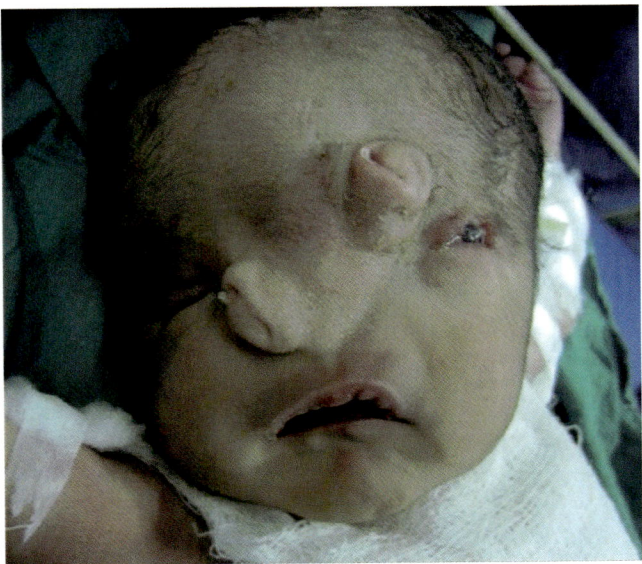

Fig. 2.58 Multiple displaced nasal placodes. Right side with double nostril and left side interorbital position. Note left #10 zone cleft affecting upper eyelid and orbit with concomitant hairline abnormality. (Reprinted from NIH National Library of Medicine Images. Retrieved from: https://openi.nlm.nih.gov/imgs/512/315/3119293/PMC3119293_MEAJO-18-192-g001.png)

Pathologic correlations follow from this anatomy:

1. Placodes can be <u>defective per se</u> due to the dysgenesis of NNE itself.
   - Selective neuroepithelial defects in either the lateral or medial half of the <u>nasal placode</u> can lead either to *anosmia* or to *Kallman's* syndrome.
2. Placodes can also be <u>dysfunctional</u> due to failure to properly interact with PNC.
3. Placode pathology can be unilateral or bilateral.
4. Pathology in one placode may not necessarily affect other placodal zones.
   - Isolated failure of AP, with consequent pituitary insufficiency, can occur in the presence of a normal face.
   - Nasal placode defects resulting in *heminose* or *arhinia* can occur in the presence of a normal eye.
   - Defects in the activity of *Pax-6* may affect the ability of the <u>optic placode</u> to form a lens with the nose remaining normal.
5. Placodes also affect surrounding structures.
   - The nasal placode is required for creation of a nasal passage within the underlying MNC of the ethmoid complex.
6. Finally, multiple placode dysfunctions may occur in the presence of an underlying vascular sequence, such as in complex *craniofacial clefts* (Fig. 2.58).

Development of Frontonasal Bone and Skin: Prosomeric Zones p4 and p5

Development of the frontal bone will be discussed in detail in the osteology chapter. For now, suffice it to say that the frontal bone in an MNC derivative is defined by two distinct sets of fields. The upper (forehead) component is defined by zone p4 *calvarial epithelium*. MNC in this zone is programmed by forebrain dura and forehead skin. The lower (supraorbital) component is defined by zone p5 *perioptic epithelium* surrounding optic placode. MNC in this zone gets its program from the dura of the floor of the anterior cranial fossa, the sclera (an) extension of dura, and caudal forehead skin.

Frontal bone anatomy is thus complex. The upper zone (frontal bone proper) is bilaminar and occurs between two inductive substrates of dermis and rostral forebrain dura. The lower zone (fronto-orbital bone) is also bilaminar, but from different sources: it occurs between the inductive substrates of basal forebrain dura, the periocular extension of dura (i.e., sclera), and p5 skin. Sclera is an MNC tissue. These differences in induction are demonstrated in anencephaly, an r1 deficiency state that always deforms the sphenoid, but has a selective effect on the frontal bone, wiping out the upper zone while sparing the lower zone. The presence of normal orbital and nasal structures seen anencephaly bespeaks of the interactions and reflects of a second and independent source of programming for frontal bone.

Frontal bone development in zones p4 and p5 can be summarized as follows. This bone is a composite structure considered by many comparative anatomists to represent the amalgamation of previous distinct membranous bones in ancestral tetrapods. The first zone of formation of the prefrontal zones occurs from lateral-to-medial to producing *postfrontal bone* (PF) and *prefrontal bone* (PrF). The orbital rim zones may thus represent the previously separate medial and lateral prefrontal bones, separated by the supraorbital neurovascular of V1. Neural crest then stacks up beneath the calvarial epithelium. Ossification of the orbital rims takes place <u>before</u> that of the frontal tubers. This could likely result from two factors: (1) periorbital MNC migration is complete before that of the forehead, or (2) ascending pathway of the arterial axis provides blood supply to the orbit and then to the frontal zone.

MNC in zone p4 occupies a potential space between two osteoinductive substrates: the *r1 dura* over the frontal lobes and *p5 dermis*. Each of these layers will induce membranous osteogenesis to take place in the residual mesenchyme creating a bilaminar frontal bone with sinus cavities. This potential space communicates with the nasal cavity via the frontonasal duct. When the nasal epithelium invades and exploits that potential space, the *frontal zone of the frontal sinus* results.

MNC in zone 5 has a similar fate. It is programmed by r1 dura of the anterior cranial fossa and r1 sclera, so a potential space exists for an orbital zone of the frontal sinus in communication with its predecessor. Let's examine how this process works.

PNC flows around the *optic placode* (OP) and causes it to become "activated." Interaction between the optic placode and the optic vesicle induces the latter to form the *optic cup* into which the placode is incorporated as the lens. At the same time, placodal neural crest mesenchyme flows inward to surround the globe. This neural crest is stimulated to form membranous bone by the dura of the basal forebrain laid down previously by r1 MNC. This can be called the *cranial lamina of the front-orbital bone*. At the same time, r0/r1 dura/sclera associated with the globe serves as an inducing agent causing placodal p5 neural crest to form a transient cartilaginous orbital capsule surrounding the globe. (In birds, this forms a ring of chondral bones arranged much like ball bearings around the orbital rim.) In mammals, the orbital capsule is converted into the *orbital lamina of the fronto-orbital bone*. Exploitation of the potential space between these two layers by nasal epithelium results in the *orbital zone of the frontal sinus*.

Note that dysgenesis/agenesis of the optic placode have manifestations in the development of the globe itself. These can be isolated and total such as *anophthalmia* or partial such as loss of the lens in *congenital aphakia*.

The neural crest making up the two laminae of the fronto-orbital bone is divided into two bone fields by the supraorbital neurovascular pedicle. The lateral *postfrontal bone* field (PF) is the site of the Tessier #10 cleft. Frontal sinus pneumatization usually does not extend into this zone, hence PF tends to be very thin. This may explain why clefts in this zone present as an *encephalocele* herniating through the orbital roof into the lateral orbit. PF of the lateral orbital rim is in continuity with postorbital bone of the lateral orbital rim; therefore, zone 10 clefts tend to affect the lateral eyebrow. Dermoid cysts are common here as well (Fig. 2.59) [88, 89].

The medial *pre-frontal* bone field (PrF) is the site of the Tessier #11 cleft. This zone is always pneumatized. Perhaps this is why encephaloceles are not as common here. PrF of the medial orbital rim is in continuity with the nasal process of frontal bone. Therefore, zone 11 clefts tend to affect the medial eyebrow. Orbital dermoid can occur here as well. Neural crest excess or tumors in either zone of the fronto-orbital rim can result in significant *orbital dystopia*.

The final r1 product within the orbit is the *lacrimal bone*. This bone sits anterior to orbital plate of ethmoid. A medial extension of r2 called the *frontal process of the maxilla* (MxF) forms the most medial aspect of the inferior orbital rim and approaches L, but must straddle around the lacrimal

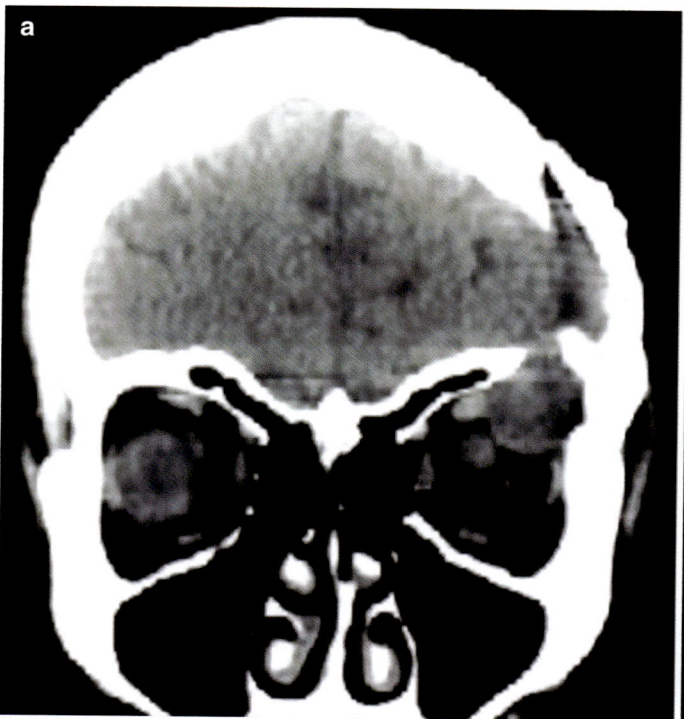

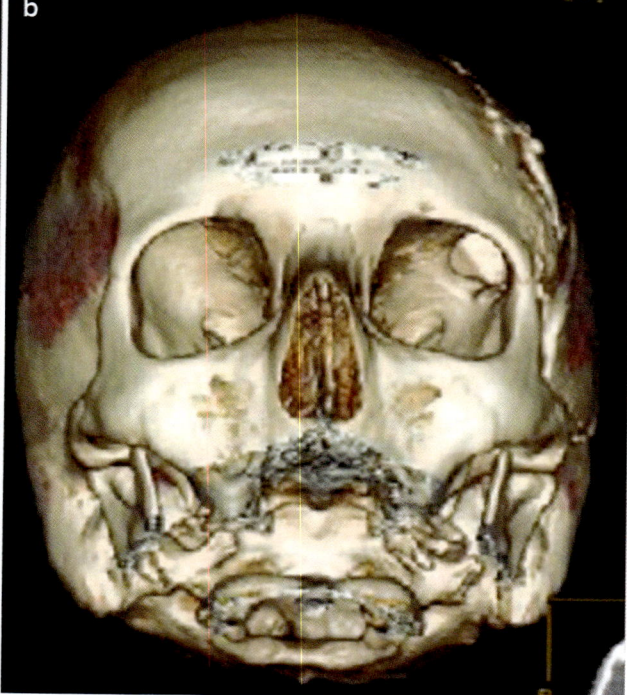

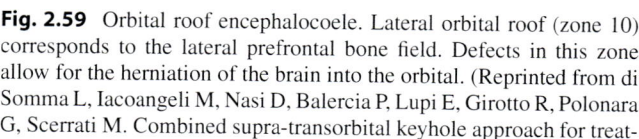

Fig. 2.59 Orbital roof encephalocoele. Lateral orbital roof (zone 10) corresponds to the lateral prefrontal bone field. Defects in this zone allow for the herniation of the brain into the orbital. (Reprinted from di Somma L, Iacoangeli M, Nasi D, Balercia P, Lupi E, Girotto R, Polonara G, Scerrati M. Combined supra-transorbital keyhole approach for treatment of delayed intraorbital encephalocele: A minimally invasive approach for an unusual complication of decompressive craniectomy. Surg Neurol Int Suppl 1, 2016;7(2): S12–S16. With permission from Surgical Neurology International)

sac to reach it. MxF accomplishes this goal by producing two tongues of mesenchyme, anterior and posterior. Where these contact the lacrimal bone, *the anterior and posterior lacrimal crests* are produced. The lacrimal bone extends caudally downward from the orbit as a lamina that terminates just lateral to the inferior turbinate. Lacrimal bone is bilaminar, being programmed externally by p5 and internally by p6 vestibular lining. *The space between these two laminae (r2 medial and r1 lateral) is exploited by mucosa to form the lacrimal sac and duct.* The proper formation of the lacrimal bone is dependent on the physical integrity of the r2 inferior turbinate. If no IT is present, the caudal portion of lacrimal will be deformed or absent. On the other hand, excessive neural crest in the lacrimal field or failure of this field to undergo appropriate apoptosis may result in *lacrimal duct stenosis* or frank *obliteration* of the lacrimal sac.

The skin covering the nasal dorsum and extending down to the columella and philtrum is constructed from p5 nonneural ectoderm populated by PNC and supported by MNC. In birds, this is known as the *upper beak epithelium* (UBE). In mammals, the external nasal skin is continuity with that of the internal nasal skin (nasal vestibular epithelium). Within each nostril, the boundary between p5 skin and p6 vestibular epithelium is readily visible; it lies at the caudal margin of the upper lateral cartilage.

Coda: Additional pathologies of frontal p5 would have to include states of developmental deficiency or excess. *Anencephaly* separates the p4 upper calvarial component of frontal bone from existence of independent p5 prefrontal and postfrontal bone fields of the orbital roof and rim. Clinically, it is manifested as gross absence of the frontal bone in concert with loss of forebrain dura; at the same time, the orbital rims and the remainder of the face are preserved [90–97] (Fig. 2.60). Mesenchymal excess is illustrated by *frontonasal dysplasia* [98–102] (Fig. 2.61). These conditions demonstrate the p5 fields to be shield-shaped with distinct manifestations in the skin. Bone abnormalities in this condition reflect MNC populations that are distinct for that of the sphenethmoid complex. On the other hand, failure of normal process of apoptosis of the entire MNC sphenethmoid complex creates a state of residual excess termed *hypertelorism* with stretched and distorted sphenethmoid fields accompanied by otherwise normal frontal bone fields and frontal lobes [103, 104]. Surgical correction of hypertelorism corrects a physical deformity that is superimposed on an otherwise normal CNS [105, 106] (Figs. 2.62 and 2.63).

- In sum: p4 and p5 nonneural ectoderm is supported by a glacier-like migration of PNC forward from the caudal secondary prosencephalon which reaches the facial midline via successive waves of lateral to medial migration, each more distal, anterior, and midline than its predecessor. The order of formation of p5 components is as follows: orbital rim (postfrontal, prefrontal, nasal process of frontal bone), a separate frontal bone, optic placode flanks by orbital rim of prefrontal and postfrontal bones, labyrinth, and nasal. These are Tessier clefts #10, #11, #12, and #13.

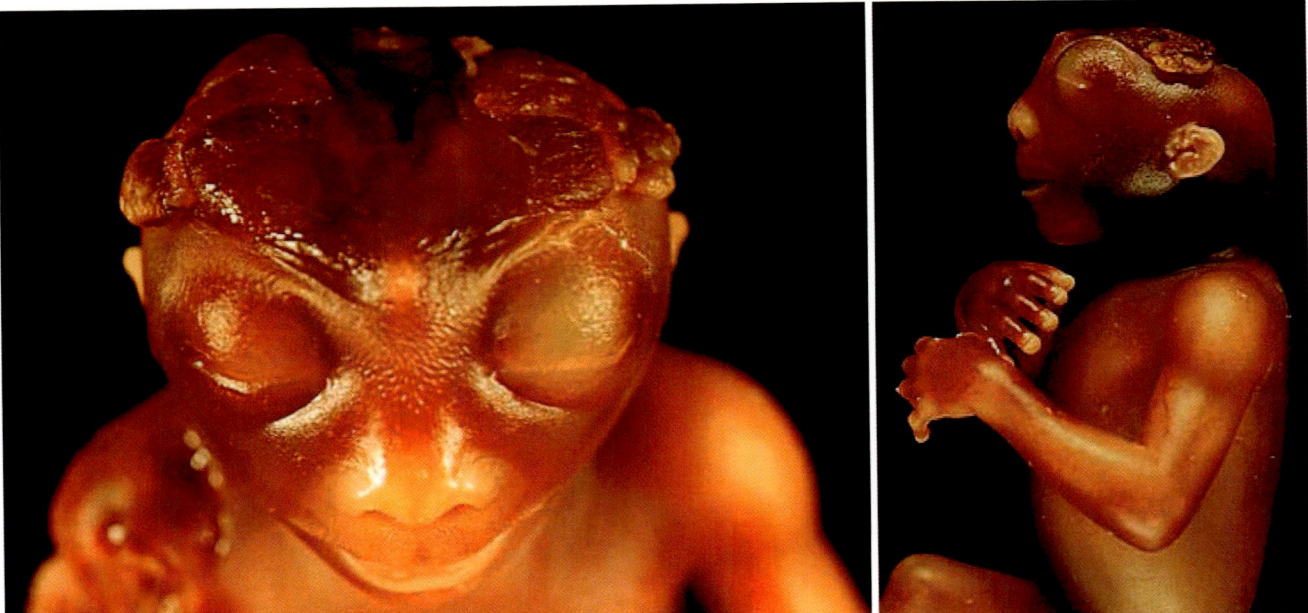

Fig. 2.60 Anencephaly. Note the recessed surface areas of zones 10 and 11. (Left: Reprinted from Wikimedia. Retrieved from: https://commons.wikimedia.org/wiki/File:Anencephaly_front.jpg. Right: Reprinted from Wikimedia. Retrieved from: https://commons.wikimedia.org/wiki/File:Anencephaly_side.jpg)

Fig. 2.61 Frontonasal dysplasia and hypertelorism. Hypertelorism reflects inadequate apoptosis of the central ethmoid fields. It is seful to examine the ethmoid sinuses to see if they are broadened. This condition involves r1 neural crest cranial bone and neural crest midline soft tissues over the nose. (Reprinted from Sin Young Song, Joon Woo Choi, Han Wook Lew, Kyung S Koh. Nasal reconstruction of a fronto-nasal dysplasia deformity using aesthetic rhinoplasty techniques. Arch Plast Surg 2015; 42(5): 637–639. With permission from Archives of Plastic Surgery)

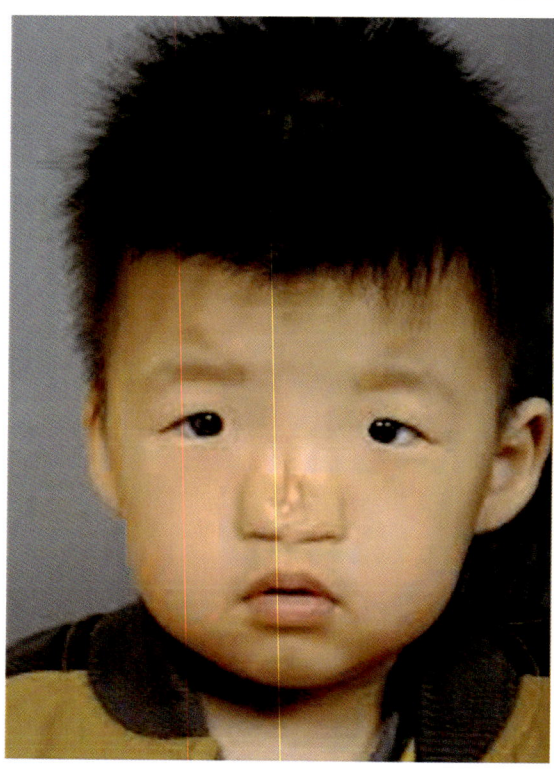

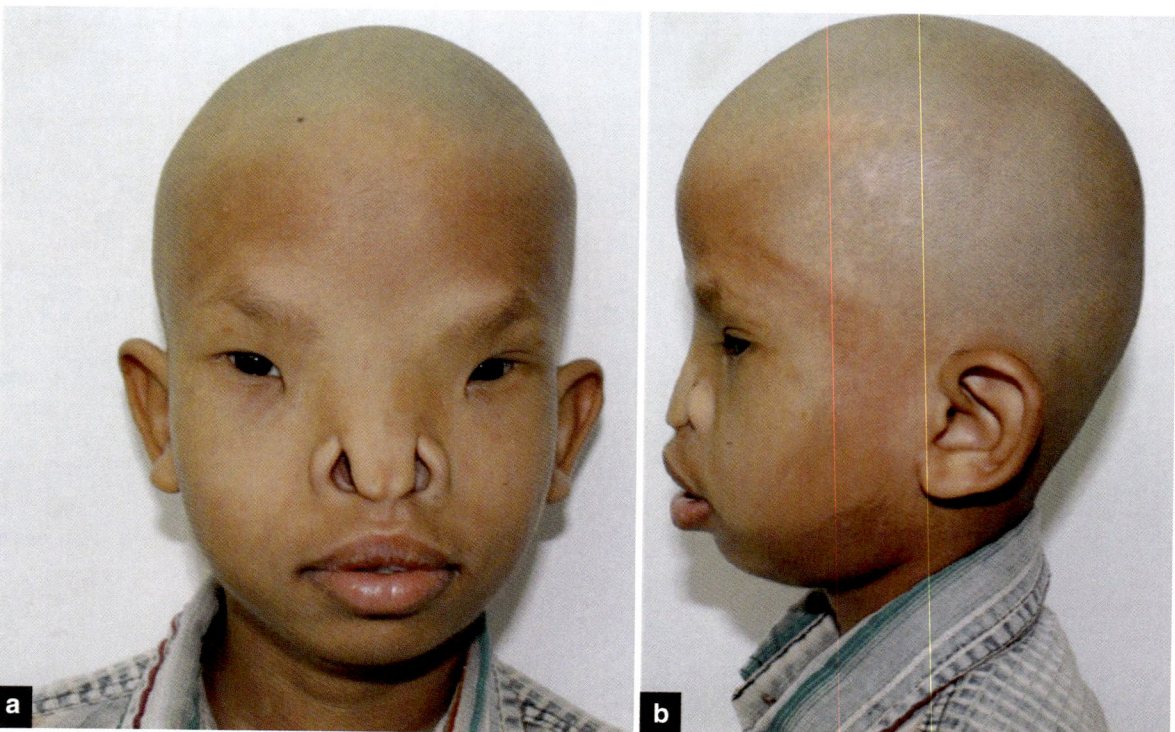

Fig. 2.62 Severe hypertelorism and vertical deficit in zone 11 affecting lateral branch of external nasal field and causing retraction of the alae upward and the medial brow pulled downward. Note notches in the nasal alae. Columella is virtually absent. (Courtesy of Prof. Dr. S. M. Balaji, MDS)

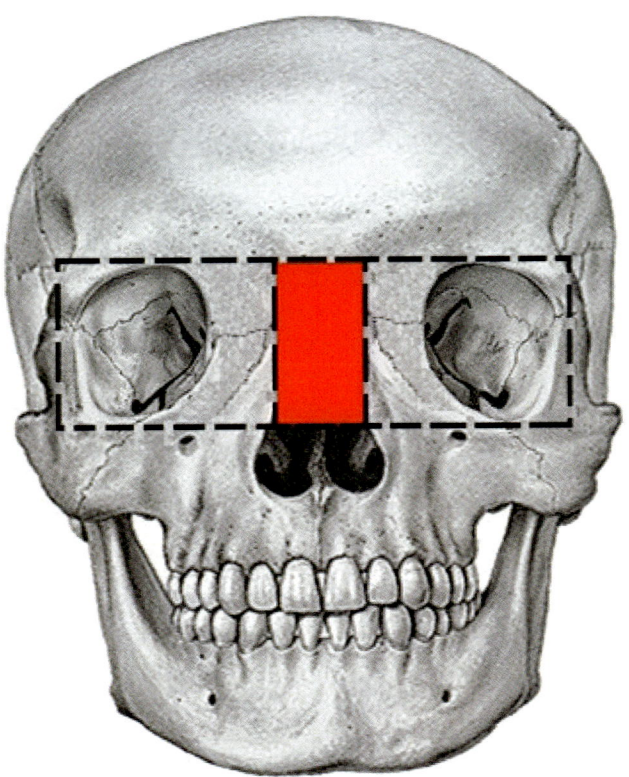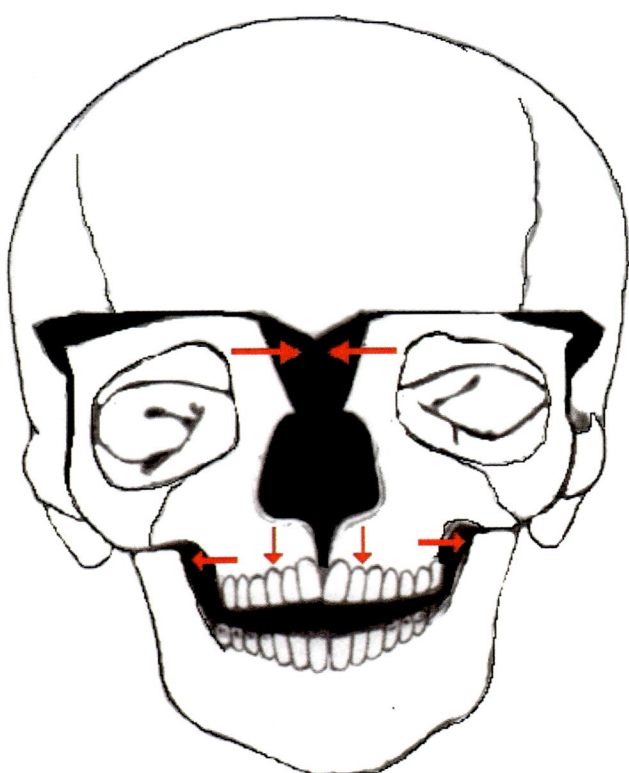

Fig. 2.63 Box osteotomies pioneered by French surgeon Paul Tessier resolved the issue of mesenchymal excess in zones 13 and 12. Hypertelorism affects ethmoid fields in two dimensions: transverse excess and possible vertical deficiency. Surgical correction of hypertelorism recognizes the excess mesenchyme in the ethmoid fields. Choice of procedure depends, in part, on the existence of vertical foreshortening affecting the palate. Left: "box" osteotomy. Right: facial bipartition. (Left: Reprinted from Wikimedia. Retrieved from: https:// commons.wikimedia.org/wiki/File:Picture_box_osteotomy.jpg. With permission from Creative Commons License 3.0: https://creativecommons.org/licenses/by-sa/3.0/deed.en. Right: Reprinted from Wikimedia. Retrieved from: https://commons.wikimedia.org/wiki/File:Picture_median_fasciotomy.jpg. With permission from Creative Commons License 3.0: https://creativecommons.org/licenses/by-sa/3.0/deed.en)

Naso-Oral Lining, Nasal Bones, and Nasal Cartilages: Prosomeric Zones p6–p5

The nonneural ectoderm of the sixth prosomere has three subzones: the *adenohypophyseal placode* (AP), i.e., the pituitary, the *nasal epithelium* (NE), and the *nasal placode* (NP). We shall see how the nature of the placodes, coupled with the proliferation of surrounding MNC mesenchyme, leads to the formation of bilateral nasal chambers. The vestibular epithelium lining the nasal chambers all the way backward from the vibrissae located at the lower border of the upper lateral nasal cartilages is of p6 derivation, whereas the remaining external to the vibrissae is p5 skin.

We shall also see that two factors influence how both sides of the face approximate each other in the midline: (1) the pattern of closure of the rostral neuropore and (2) the natural apoptosis of MNC nasoethmoid mesenchyme. Defects in the former can produce encephalocoels, while defects in the latter cause hypertelorism [107, 108].

Along with proliferation of the brain, mesenchymal structures are also being positioned. The developing eye is tucked directly beneath somitomeres 1–3, providing easy access for the myoblasts of the extraocular muscles to reach their scleral targets. Recall that the eye projects initially 90° from the neuraxis. Hence, the most caudal aspect of the globe is the *lateral sclera*. This is spatially positioned close to the fifth somitomerewhich obligingly sends out lateral rectus. Meanwhile, the fourth somitomere is positioned just posterior to the AP, where it forms the basisphenoid.

At the same time, r1 neural crest invading anterior to AP forms the trabecular and nasal cartilages. Trabecular cartilage gives rise to presphenoid bone and subsequently the remainder of the sphenethmoid complex. The sphenoid sinus occupies the interface between PS and BS These mesenchymal tissues force the adenohypophyseal placode to assume new position facing ventrally within the future pharynx. Work by Kjaer demonstrates that basisphenoid (BS) always forms before presphenoid (PS). This is consistent with the idea that somitomeres are closely apposed to the neuraxis and develop earlier in time compared with neighboring neural crest populations.

A similar process explains how the p6 nasal epithelium gets drawn inside the nasal chamber. The traditional view holds that ingrowth of the nasal placodes, a "burrowing" into the underlying MNC ethmoid complex, is responsible for forming the inner nasal architecture. Persistent placodal neuroanatomic connection to the rhinencephalon of the p6 basal forebrain is the key element here. Explosive expansion of MNC beneath the nasal skin forces the nose to grow *outward* from the face. The adhesion of the p6 nasal placode to the brain is so strong that the nasal epithelium is drawn inwards. At the same time, placodal traction may be responsible neuroectodermal tissue being drawn out from the central nervous system as the olfactory bulbs. Alternatively, placodal structure may simply contain the cell adhesion molecules required for neuronal growth.

Formation of Nasal Cartilages: Prosomeric Zones p6–p5

Zone of p5 nonneural ectoderm is known as *upper beak ectoderm* and becomes the nasal skin. It surrounds nasal placode and proliferates, causing the nasal placode to become internalized. This draws epithelium inside the naris where it become *nasal vestibular epithelium* (NE). When MNC migrates into the nasal tissues, it follows the subcutaneous plane. Later, SMAS fascia lies superficial to it. Inside the naris MNC is contact with skin proximal to the external nasal valve and with vestibular lining distal to the valve.

Every rhinoplasty surgeon takes advantage of this embryology. The upper lateral cartilages are derived exclusively from MNC programmed by p6 vestibular lining. *The actual "program" that determines the size and geometry of the nasal cartilages resides in the p6 NE.* In a similar manner, the lower lateral cartilages are MNC structures programmed by the p5 epithelium. The nasal epithelium dictates to the neural crest exactly where to make the cartilages. For this reason, *variations in vestibular lining may explain differences in size or angulation of the nasal cartilages.* Deficiency states of the p6 nasal fields would explain the clinical picture of *short nasal bones* described by Sheen (Fig. 2.64) [109–111].

Note that septum is an MNC structure programmed by a sandwich between two walls p6 vestibular lining brought together by the programmed death of ethmoid tissue. The floor beneath the septum consists of fused vomerine bone fields of the medial nasopalatine neuroangiosome. The septum develops temporally after the perpendicular plates. In holoprosencephaly MNC, ethmoid tissue is affected: the perpendicular plate and the septum can be completely absent. This is accompanied by absence of the vomer/premaxilla resulting in wide bilateral "cleft." In less severe states, the septum and premaxilla may be attenuated. On the other hand, a widened *bifid septum* results if failure of normal midline apoptosis resulting in persistent MNC mesenchyme prevents approximation of the nasal and optic fields (Fig. 2.65) [112].

Prior to closure of the rostral neuropore (RNP), the two potential nasal cavities are widely separated. The remnant of the RNP is the *foramen cecum* (just above the nasal bones). Recall that closure of the neural folds concludes neurulation. This process is initiated at the fourth occipital somite, i.e., at the 11th rhomobomere. From that direction, closure pro-

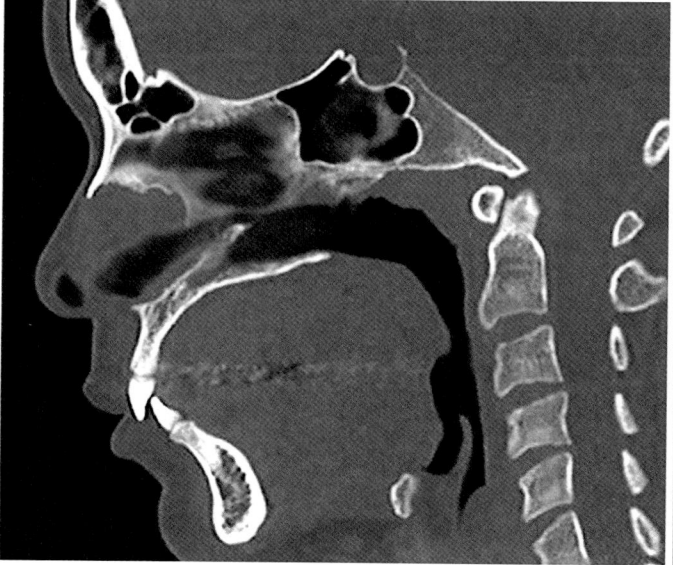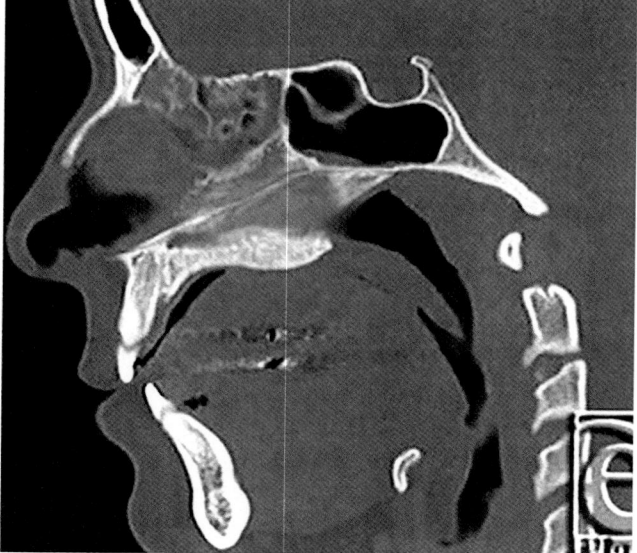

Fig. 2.64 Short nasal bones. (Reprinted from Mowlavi AS, Chamberlain TL, Melgar A, Talle A, Saadat S, Sharifi-Amina S, Willhelmi BJ. Upper Vault Septal Anatomy and Short Nasal Bone Syndrome: Implications for Rhinoplasty Eplasty 2018; 18: e29. With permission from ePlasty)

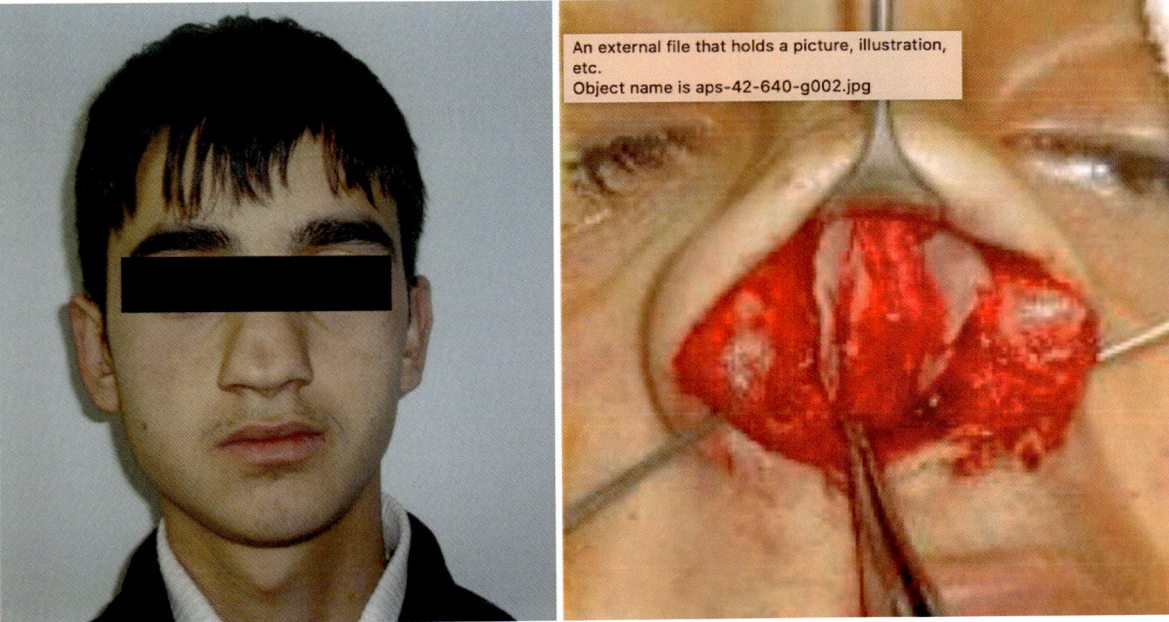

Fig. 2.65 Bifid nasal septum. (Reprinted from Karakor-Altuntas Z, et a. Isolated Congenital Nasal Bifid Septum Separated by a Wide Layer of Soft Tissue. Arch Plast Surg 2015; 42(5): 640–642 With permission from Archives of Plastic Surgery)

ceeds both rostrally and caudally. Closure of the rostral neuropore occurs *one stage before* that of the caudal neuropore. However, closure of the RNP is a *bidirectional* process. Starting from the adenohypophyseal placode, closure is directed *backward*. Rostral to the RNP, the pattern of closure is directed *forward*. These two "zippers" meet at nasofrontal foramen. Defects in this process may lead to a patent foramen cecum and the development of dermoid, encephalocoele or *glioma* (Fig. 2.66).

No nasal chamber can be constructed without a placode. Unilateral or bilateral absence of placodes is the developmental basis for hemi-nose or *arrhinia* [113–119]. The high position of the single nasal placode seen in *ethmocephaly or cyclopia* can be understood in terms of a gross deficiency of p5 mesenchyme [120–122]. In this contracted state, the nasal placode is drawn up to the position of the RNP and therefore appears above the common orbit (Fig. 2.67).

When PNC populates the *nasal placode* (NP), it initiates within the NP three neuronal populations. The lateral zone contains neurons of the <u>olfactory system</u>. These process odors. The medial zone contains neurons of the <u>accessory olfactory system</u>. These process chemicals such as pheromones. Also contained in the medial zone are neurons associated with release of <u>gonadotropin releasing hormone (GnRH)</u>. Each set of neurons is relayed to separate areas of the frontal and temporal lobes. These pathologies are well-illustrated in Kallman syndrome [81].

Deep to the fifth and sixth prosomeric zones lies a vast deposit of midbrain neural crest that forms the sphenethmoid complex, nasal bones, nasal cartilages, and septum.

Mesencephalic Neural Crest Derivatives

Definition of MNC: Mesomeres m1–m2, Rhombomeres r0–r1

Midbrain neural crest has a rather loose definition because, from a functional standpoint, its sources include the most anterior zone of hindbrain. It develops from neuromeres m1, m2, r0, and r1. Midbrain proper has two mesomeres, m1 and m2, with the first mesomere being much larger in size than m2. Known contents of m1 are separate colliculi (hills) for vision and hearing, and two distinct nuclei for cranial nerve III. The *presisthmic tectum*, m2, develops as a subdivision of m1. Its anatomic contents are unclear.

The *isthmus* forms the junction between midbrain and hindbrain. Isthmus is coded r0 and is much smaller than first rhombomere r1, which contains the nucleus of IV. Isthmus is also in direct contact with the midline prechordal mesoderm. Isthmus develops as an anterior subdivision of r1. It contains the decussation of trochlear nerve IV.

For purposes of our discussion, the two populations of greatest interest are m1 and r1 (cf Fig. 2.46).

Although soft tissue and bone derivatives of MNC are known, *the specific contributions of its component neuromeres are not mapped out*. MNC dura covers the entire anterior cerebrum and cerebellum. MNC as broadly defined is responsible for producing the anterior cranial fossa, sphenoid, ethmoid complex, and the anterior cranial fossa, sclera, and orbital contents.

MNC migrates ventrally and anteriorly on its way to the orbit and face. It hugs the axis of the embryo. It moves

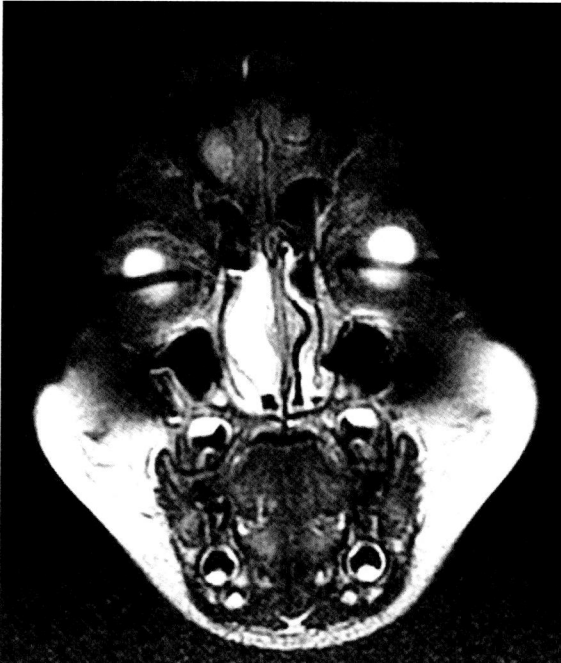
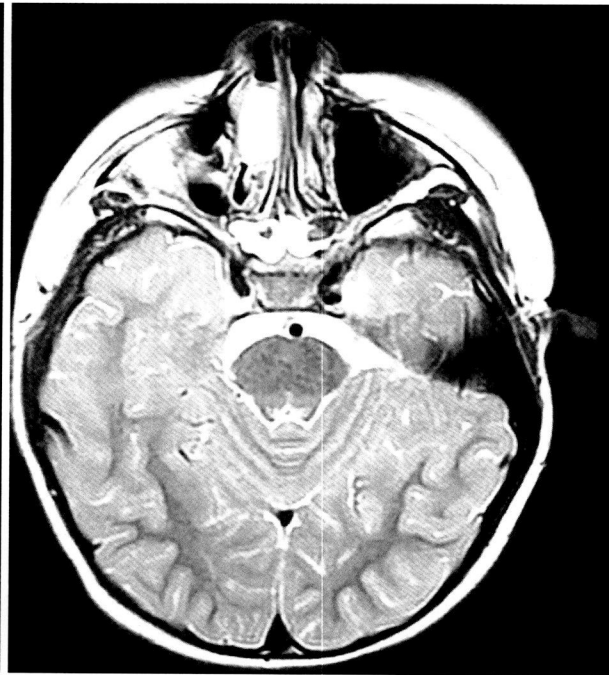

Fig. 2.66 Nasal glioma. Nasal gliomas are composed of dysplastic glial tissue and are congenital non-neoplastic lesions best categorized as heterotopia. They are rarely associated with other congenital malformations [2]. A nasal glioma may be connected to the brain by a stalk of tissue in up to 15% of cases, but the stalk does not contain a direct fluid-filled tract that communicates with the subarachnoid spaces; therefore, a nasal glioma is distinct from a nasal encephalocele. # Types extranasal (60%), intranasal (30%), and mixed (10%). Nasal gliomas occur near the root of the nose (where the cranial portion of the nose joins the forehead). Extra-nasal gliomas are usually seen in a paramedian loca- tion at the bridge of the nose external to the nasal passage, whereas intranasal lesions are usually located within the nasal passage medial to the middle turbinate bone. Ultrasound is useful for determining if the mass is cystic or solid. Doppler flow studies of nasal gliomas reveal a characteristic low arterial flow velocity during the end-diastolic phase Nasal gliomas are often isointense relative to the normal brain at MR imaging, which is the imaging modality of choice. High-resolution surface coil MR imaging is often useful in demonstrating the intracranial stalk. (Reprinted from Radiopaeida. Courtesy of Dr. Francis Deng)

directly in front of the notochord to interact with the prechordal mesoderm located between the rostral notochord and the buccopharyngeal membrane. MNC migration sequence is determined by the spatial order of CNS development. Hindbrain development is rostral-caudal. Thus, r0 migrates first, followed by r1. The two populations probably do not remain distinct. Midbrain development follows that of hindbrain and is caudal-rostral; thus, m2 is followed by m1. Given the small size of isthmus, we can also assume the two populations are combined.

The immediate task of MNC is to protect the brain by providing tissues that will complete the preexisting vascular plexus, lay down the meninges, and support the skin. Synthesis of the anterior cranial base and orbit follows. As the optic primordium emerges from p5 basal diencephalon, it pushes its way through this MNC like a fist through a sock. The future globe thus acquires a coating of neural crest that will provide the future sclera. The globe is genetically divided into two developmental sectors: ventro-nasal and dorso-temporal. Neural crest of the sclera will form binding sites for the extraocular muscles according to this system.

Directly lateral to caudal forebrain are positioned the first three somitomeres Sm1–Sm3. These contain all extraocular muscles except lateral rectus. Sm4 lies directly opposite the tip of the notochord. This mesoderm is sessile and forms the basisphenoid. Sm5 sits just behind it and contains the remaining extraocular muscle, LR. Since m1 and m2 contain the nuclei of III, it makes sense to assign them both to the sclera, although, given the small size of m2, this may not be functionally significant.

In thinking about MNC, keep in mind that it is not that MNC from r1 or m1 carries a particular "message" that will produce a derivative, rather it is the *interaction* of MNC with the local environment it populates that will determine the outcome.

The embryonic axis defined above has many implications for the future developmental sequence of the eye. The initial blood supply for the eye is from primitive maxillary artery. This vessel is ultimately reassigned to supply the hypothalamus. It is replaced first by primitive dorsal ophthalmic and later by primitive ventral ophthalmic. These arteries are held together to form the definitive ophthalmic artery (prior to arrival of stapedial). Via prDOA and prVOA (primitive dor-

Fig. 2.67 Holoprosencephaly with hypotelorism sequence. Upper left: bilateral cleft lip and palate with hypoplastic premaxilla (loss of lateral incisor subfields) and vomer; Upper right: cleft lip and palate with absent premaxilla and/or vomer; Lower left: cebocephaly (single chamber proboscis in correct position); Lower center: ethmocephaly (proboscis in the ethmoid field, bilateral orbits); Lower right: cyclopia (proboscis above the plane of single orbit). (Reprinted from Bianchi D. Fetology: Diagnosis and Management of the Fetal Patient, 2nd ed. McGraw Hill; 2010. With permission from McGraw-Hill Education)

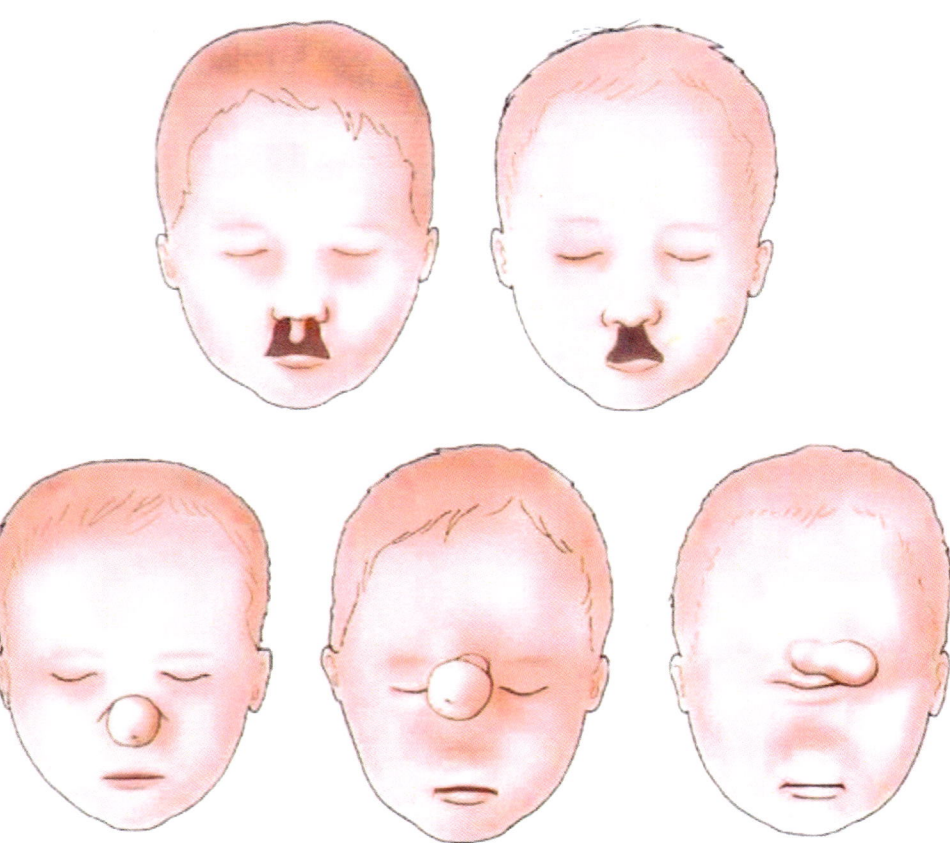

sal ophthalmic and primitive ventral ophthalmic arteries, respectively), neural crest cells arrive at the iris to form the intrinsic striated muscles of the eye.

Given the above, can we deduce which population of MNC supplies what sector of the globe? The temporal-dorsal sector of the optic vesicle (supplied by prDOA) is biologically "older" than the nasal-ventral sector [123–127].

The insertion sequence of the extraocular muscles is as follows: LR[13] (Sm5) > SO[14] (Sm3) > SR[14] (Sm2) > MR[16] (Sm1) > IO[16] (Sm2) > IR[17] (Sm1) > LPS[19–21] (Sm2).

Dura innervated by V1 is continuous with the sclera of that side of the globe as well. For this reason, sensory nerves to the eyes are exclusively V1. Myoblasts from Sm1 to Sm3 and Sm5 migrate into the orbit where they are ensheathed by MNC fascia. This means that the cell source for the muscular arteries will be V1 stapedial.

MNC Bone Fields

Mesencephalic bone fields can be abbreviated as: presphenoid (PS), orbitosphenoid (OS), lateral pterygoid plate (MPt), ethmoid lamina (EL), ethmoid perpendicular plate (PP), and septum (S).

Bone fields of the anterior cranial fossa and orbit develop from neural crest via two mechanisms: (1) by the formation of an intermediate cartilage that becomes a membranous bone; or (2) via direct osteogenesis in membrane. Of particu-

lar value is the construction of the sphenoid complex. The principle cartilages are demonstrated in Figs. 2.68 and 2.69. Development follows a pattern ventral to dorsal, caudal to rostral, and medial to lateral. Bone fields build quickly, one upon another to produce the sphenoid complex (Figs. 2.70 and 2.71).

We arrive therefore at three extremely important generalizations.

- At a given neuromeric level, *bone derived from somitomeric PAM will be laid down* **prior** *to that derived from neural crest.* Basisphenoid precedes presphenoid. Temporal bone precedes squamous occipital bone.
- PAM bones are not found in the viscerocranium (pharyngeal arches).
- PAM bones of the skull are chondral, the sole exception being the parietal bone in which r2–3 neural crest is admixed with Sm4 PAM.

Rhombencephalic Neural Crest Derivatives: r2–r7

The relationships between the homeobox genetic code in the hindbrain, neural crest migration, and the pharyngeal arches were worked out in the late 1980s with contributions by

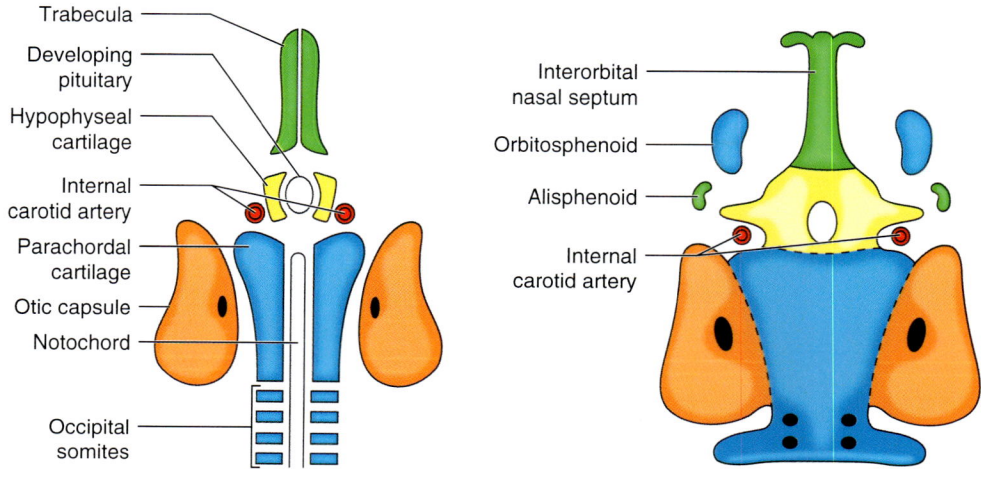

Fig. 2.68 Chondrocranium 1. Occipital somites [4] consolidate to form the basioccipital bone

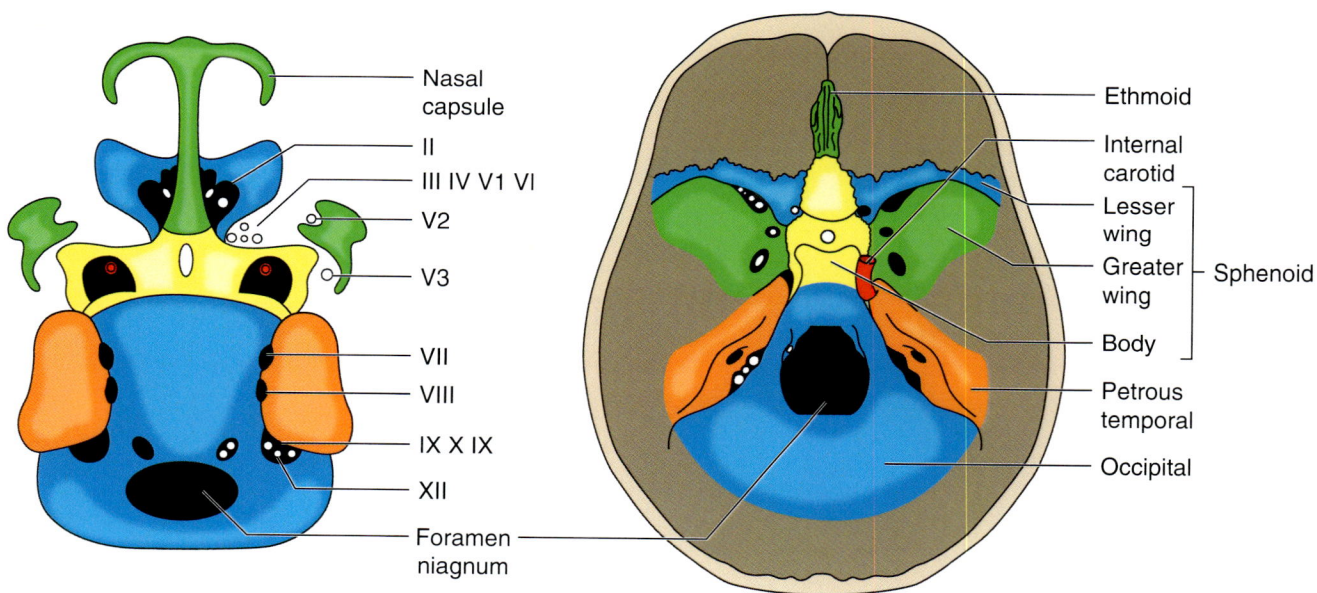

Fig. 2.69 Chondrocranium 3. Note sphenoid mesenchyme condensing around the nerves

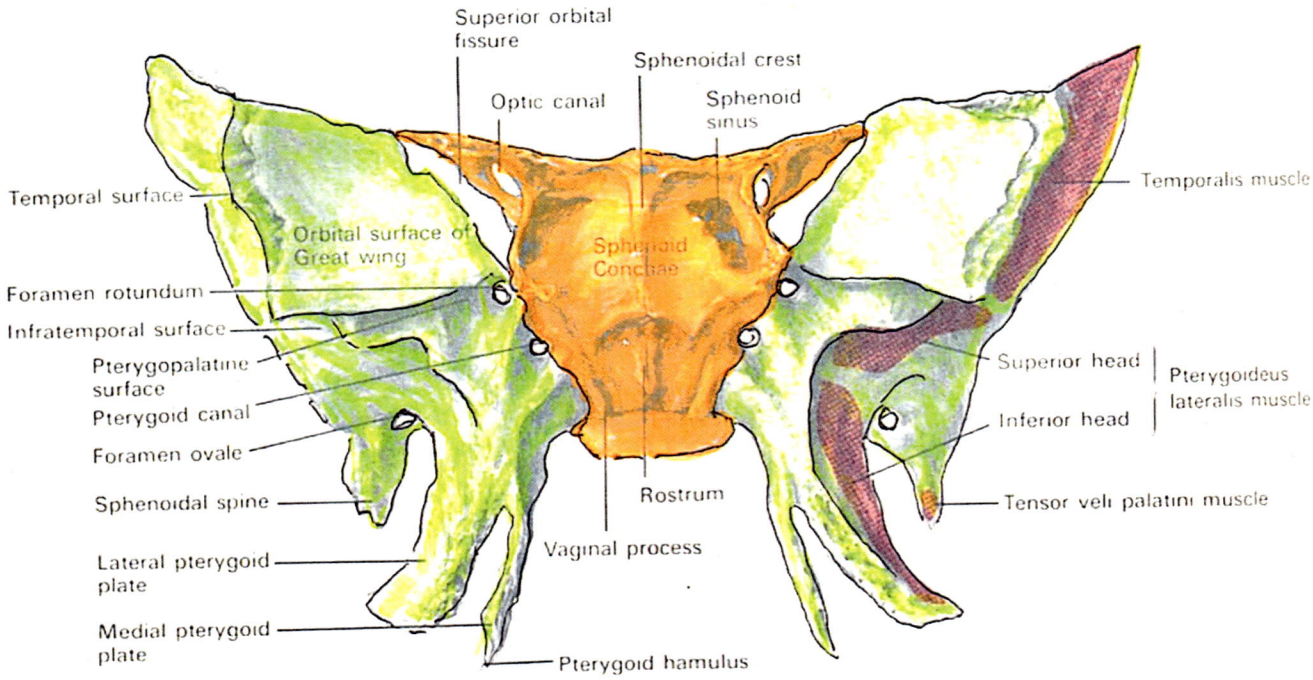

Fig. 2.70 Sphenoid, anterior. (Reprinted from Lewis, Warren H (ed). Gray's Anatomy of the Human Body, 20th American Edition. Philadelphia, PA: Lea & Febiger, 1918)

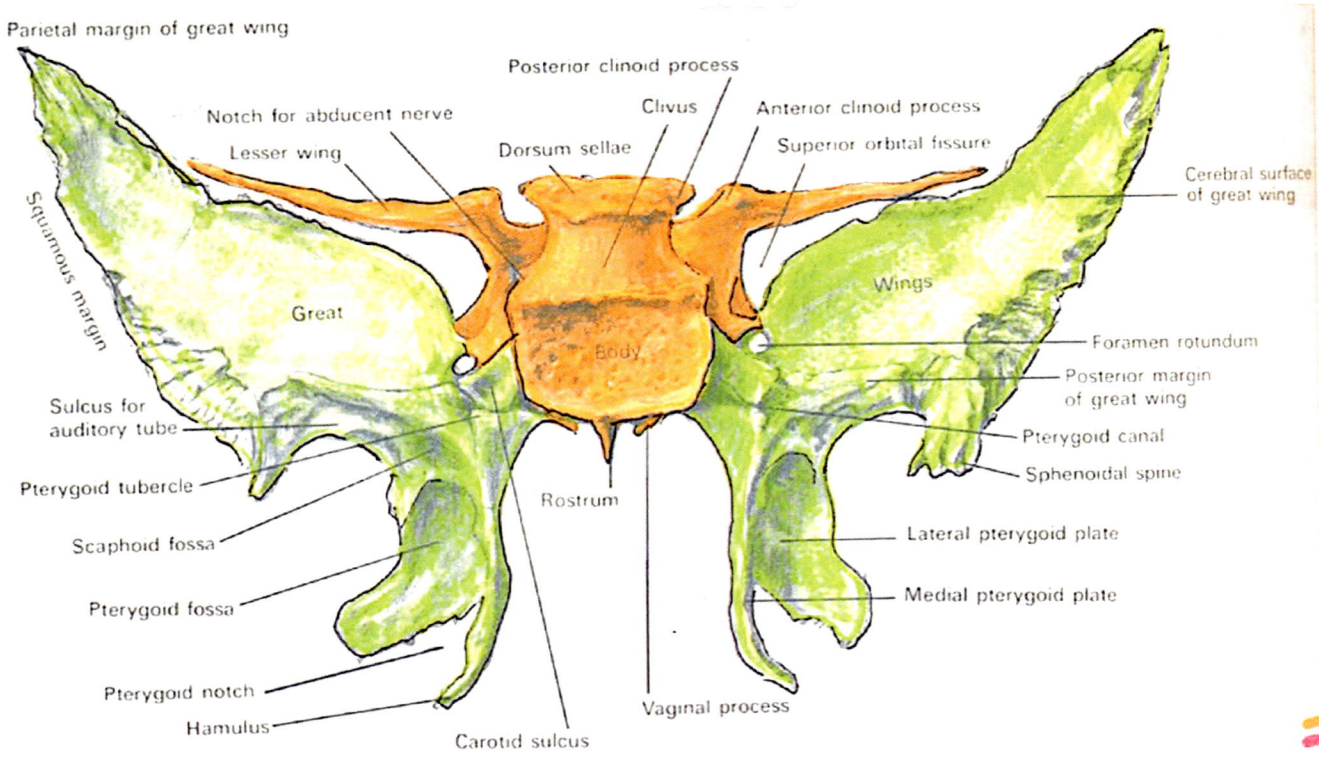

Fig. 2.71 Sphenoid, posterior. (Reprinted from Lewis, Warren H (ed). Gray's Anatomy of the Human Body, 20th American Edition. Philadelphia, PA: Lea & Febiger, 1918)

Lumsden and from Krumlauf [32, 33, 128–130]. Herein we will concern ourselves with the patterning of the first three pharyngeal arches.

The earliest neural crest migrations to be completed are those originating from the rostral rhombencephalon, i.e., from rhombomeres r2 to r7. This type of neural crest (RNCr) is intimately involved with the formation of the first three pharyngeal arches. RNCr pathways are *segmentally restricted* between rhombomeres; RNC cannot participate in craniofacial development independent of the pharyngeal arch to which it is assigned. Subsequently, RNC is physically transported to the face (frontonasal mesenchyme) by the processes of pharyngeal arch translocation and embryonic folding. The very first fold in the neuraxis, called the *mesencephalic flexure*, is hinged at the midbrain. It results in the CNS being tucked back on itself almost 150°. The most rostral zones of the prosencephalon (p6 and p5) are brought ventrally until they are in almost direct contact with the rapidly growing first and second pharyngeal arch amalgam. Epithelial contact between the frontonasal and lateral masses results in fusion. Neural crest and PAM mesenchymal derivatives then move into their appropriate locations.

This type of RNCr migration is very different from that observed in neural crest of prosencephalic or mesencephalic origin. The biologic behavior of PNC and MNC is not constrained by a secondary reorganization into a pharyngeal arch. These neural crest cells move directly under the epithelium to their destinations, set up shop, and go to work.

Timing is everything, especially in craniofacial development. We know that RNCr migration into the first three arches is complete by the 11-somite stage. On the other hand, the mesenchyme to which the RNCr will directly attach to form the facial skeleton comes from MNC. However, MNC migration is not complete until the 14-somite stage. What happens during the intervening time? *CNS folding occurs in this interval.* At this point in time, the first and second pharyngeal arch "sandwich" has swung into position. All potential RNCr fields are now ready to make contact with their MNC counterparts. Note further that, although MNC interacts with PNC via the r1 perpendicular ethmoid plates, *no direct anatomic contact exists from RNCr and those from PNC.*

When the pharyngeal arches arrive on the scene, they are composite entities. Recall that the five arches are not completely separate. The external surface of the arches shows deep grooves in the epithelium. These grooves are not of full thickness; it would be theoretically possible for neural crest and PAM from one arch to meld into its neighbor. For this reason, the first and second pharyngeal arches, once formed, promptly meld into each other to create a "sandwich." In this case, neural crest cells from r2 interact with those of r4, whereas those from r3 interact and meld with those of r5. For this reason, myoblasts from r4 populate the maxillary fields and those from r5 populate the mandibular fields.

Let's turn our attention to the osseous derivatives of the first and second pharyngeal arches. By the time the PA1/PA2 sandwich arrives on the scene, it has two visible components. An upper maxillary mass contains derivatives of r2 and r4. It comes into direct contact with the frontonasal "process" (sic), i.e., the orbito-spheno-ethmoid complex. A lower mandibular mass contains derivatives of r3 and r5. Unlike the maxillary mass, the mandibular mass bears no physical relationship to the orbito-sphenoid complex. It is suspended from the cranial base r6 petrous temporal bone at the temporomandibular joint.

First Pharyngeal Arch: Rhombomere 2

Migration Pathways

The anatomic pathways taken by the r2 neural crest are dictated by the neuroanatomy of V2 and by preexisting MNC fields of the orbit and nose, the presence of which force it to follow along their boundaries or to make detours around them. Perigrinations require a pathway for the pilgrim to follow. *The mechanism of r2 RNC migration is to take advantage of previously established MNC structures.* Just as in gastrulation mesoderm has a "leading edge," the pioneering cells of which leave a trail for later mesoderm to follow, so do the initial cells of a neural crest pathway leave signals behind for their successors. Succeeding waves of angiosomes will follow each other *based upon their ordinal anatomic site of origin* along the vascular axis.

RNC from r2 follows two principal pathways forward. An *intracranial pathway* pursues the course of the meningeal branch of V2 which is given off before foramen rotundum. It proceeds to the backwall of the future orbit where it deadends as alisphenoid. The *extracranial pathway* courses through foramen rotundum as described below.

Playing "follow the leader" means respecting obstacles. The r1 MNC blastema responsible for synthesis of the *lesser sphenoid wing of the sphenoid* and the lateral pterygoid plate "pushes" RNC laterally. It therefore completes the back wall of the orbit by producing greater wing of sphenoid. Once the blastema of *alisphenoid* mesenchyme is already in place, additional r2 RNC pathway is forced to pass beneath the greater wing. By this "underhanded move," it gains entrance to the pterygopalatine fossa. There it breaks up into individual pathways dictated by the branches of V2 that fan out from the fossa to the oro-nasal cavity, maxillary complex, and the zygomatico-orbital complex.

The Premaxillary-Vomerine Complex

We shall follow the pathway of the terminal branch of StV2, the nasopalatine axis (NP), as it turns medially from the pterygopalatine fossa and heads toward the midline. In so doing, it encounters a physical obstacle, the upwardly growing *palatine bone*. NP inserts itself between palatine bone and the cranial base. The palatine bone is derived from STV2 descending palatine, the stem of which is given off prior to that of NP. The palatine bone consists of a *horizontal plate* that forms the posterior hard palate and a *perpendicular plate* projecting up into the orbit from the superior border of which are projected upward a posteriorly directed *sphenoidal process* and an anteriorly directed *orbital process*. These are two separate processes because of the presence of the nasopalatine axis, which they must straddle. Thus, the presence of the sphenopalatine notch represents a "footprint" of the migratory pathway used by NPA on its way to the midline. *All notches and foramina of the skull are tacit recognitions for the prior existence of neurovascular structures.*

After passing over the palatine, NPA bifurcates. A lateral branch (NPm) is dedicated to the nasal wall of the maxillary complex and the medial branch (NPm) is directed toward the nasal midline. Here it encounters the previously synthesized perpendicular plate (PPE) of ethmoid. As it grows forward, the neurovascular axis of NPm is forced to follow its inferior border of perpendicular plate.

We have said before that the derivatives of NPm are the premaxilla and the vomer and that these are *bilateral* field complexes. Beneath the presphenoid is a midline axial structure (probably bilateral in origin), the *sphenoidal rostrum*. Inserted on either side of the rostrum are two flanking lamine from the posterior margins of the vomerine bones. These are known as the *vaginal processes*. As the vomers extend anteriorly from the vaginal processes, they descend beneath the perpendicular plates of the ethmoid (PPE). The ability of NPm to follow along the fused midline laminae of r1 (i.e., the perpendicular ethmoid plates), determines whether or not one or both premaxillary bones will be produced. The r1 PPE is a bilateral structure.

Holoprosencephaly (HPE) involves neurovascular territories of the ventral and anterior forebrain and the r1 sphenethmoid complex upon which the brain rests [131–135]. Because r1 neural crest is involved in the synthesis of the branches of anterior cerebral circulation, an r1 deficiency state may be a possible cause of HPE. HPE can exist in a unilateral form that includes ipsilateral attenuation or absence of the perpendicular plate. In these cases, hypoplasia or frank absence of the ipsilateral premaxilla can be present. If the palatal shelves must fuse to vomer in order to achieve union at the midline, then unilateral HPE, by destroying the pathway for medial nasopalatine migration, will create an ipsilateral palatal cleft.

The lamination sequence of NPm is completely different from that of r0 and r1. Building upon the PAM derivative basisphenoid, r1 neural crest spreads to form the central anterior cranial fossa. One does not see absence or deformation of the sphenoid in the presence of a normal ethmoid. The ethmoid capsule forms a lattice for r2 migration. Earliest in time to migrate is the premaxillary neural crest. At about

the time PMx neural crest arrives at the anterior base of the presphenoid, the r1 perpendicular plates of the ethmoid are forming. These make up the "perpendicular plate pathway" by which PMx passes along the future roof of the mouth and arrives at its final position in the facial midline. As PMx migrates, it leaves behind it a molecular "slime trail" for vomerine neural crest to follow [136].

Once in the midline, PMx forms a common alveolar process that subsequently splits into two independent zones. Neural crest "flow" into zone PMxA creates the central incisor and alveolus. This field is a pie-shaped wedge with the apex pointing toward the incisive foramen. A "split-second" later, a second neural crest population containing the future lateral incisor and ascending process of the premaxilla arrives on the scene. It too has the form of a pie-shaped wedge with its apex at the incisive foramen. Thus, zone 2 does not "spill over" from zone 1, but arises just posterolateral to it. Zones 1 and 2 are in continuity, just as are zones 3, 4, and 5.

From its most distal aspect, PMx produces a cellular outgrowth of vital importance to plastic surgeons. This *frontal process* of the premaxilla (PMxF) is responsible for the internal piriform rim. *Deficiency states in PMxF are the cause of the isolated cleft lip nasal deformity* (vide infra) [137–139].

The vomerine neural crest migrates into position later in time than that of PMx. The premigratory position of V field cells is thus *caudal to PMx along the neural fold*. Isolated deficiency states of the V population can exist. It is therefore possible to have hypoplastic vomer in the presence of a perfectly normal lip and an intact premaxilla. This can lead to a spectrum of problems. Thus, even though primary contact between the premaxilla maxilla has resulted in fusion of the alveolar arch, small vomer bone may be physically unavailable for fusion with the ipsilateral r2 palatal shelf, leading to secondary contact failure and a midline cleft palate.

PMx and V are supplied from a common neurovascular axis with which they can have two different types of relationship. In the first scenario, they follow each other "*Indian file*" down a common pathway. PMx neural crest is "pathfinder" for subsequent vomerine development. In other words, PMx is required but not sufficient for vomer. An absent premaxilla will always be accompanied by an absent vomer. On the other hand, an intact premaxilla could be accompanied by hypoplasia of vomer. A second scenario is on which each field represents an independent stem from the main axis. This model would permit a situation where vomer could be fully developed in the presence of an abnormal premaxilla. Clinical experience supports the first model. A third (unrelated) scenario occurs if there is developmental failure of the ethmoid fields, as in holoprosencephaly. Failure of the r1 "perpendicular plate pathway" makes migration of the medial nasopalatine axis impossible and leads to outright absence both vomer and premaxilla.

Organization of r2 RNC

Some superomedial r2 and r3 neural crest remains sessile alongside the developing brains. In conjunction with r1 neural crest, these populations produce the dura covering the prosencephalon. The neuromeric territories are defined by the sensory distributions of trigeminal to the dura.

The neural crest responsible for synthesizing the maxilla and lateral orbit can be organized into *three distinct osteogenic populations*. The segments arrive at the scene via spatially distinct migratory pathways. These pathways do not conflict with each other. The first population is the *orbital complex*. It follows intracranial V2 and is directed toward the orbit where it produces lateral pterygoid and greater wing of sphenoid. Neural crest cells within each segment produce bone derivatives in a strict time sequence. The second population contains the various components of the *maxillary complex*. The overall shape of these fields is that of a five-sided box, the medial wall of which is partially open into the nose. The third segment contains the two fields of the *zygomatic complex*, which forms secondarily over the malar eminence of the maxilla and the posterolateral orbit. We shall now describe the order in which these fields make their appearance.

Retro-Orbital Complex: Alisphenoid (AS) and Lateral Pterygoid (LPt)

Alisphenoid (AS) and *lateral pterygoid lamina* (LPt) are membranous bone with different blood supplies. These two fields are confluent with each other at the pterygopalatine fossa. From their juncture, each field is joined to the presphenoid bone by an osseous process. Where these processes straddle, the preexisting sensory nerve V2, the *foramen rotundum*, is created. They thus predate the remaining r2 neural crest populations as they pass through the foramen. Within the orbit, isolated deficiencies of the alisphenoid have been described that communicate with the external postorbital region. Alisphenoid develops from StV2 intracranial as the meningo-orbital artery. These may range from the reduced volume of the lateral orbit seen in trigonocephaly due to lacrimal defects to outright loss of the alisphenoid posterior wall.

The alisphenoid would appear to be an extension of the cranial base. It represents, however, an attachment of PA2 to the skull. This is a very ancient arrangement. The primitive palatoquadrate cartilage (from which the maxilla is derived) had three components, *autopalatine*, *metapterygoid*, and quadrate. The most anterior is the precursor of the maxilla and palate, the middle one forms the *alisphenoid*, and the latter forms the incus.

Lateral pterygoid lamina is the remnant of the ancient ectopterygoid bone of the palate series. It provides the primary insertion for lateral pterygoid muscle which spans to r3 mandible. Its arterial supply comes from the artery of the same name that is programmed by the second zone of IMMA and from StV2 lesser palatine artery.

Clefts occurring below the axis of the orbit are classified by Tessier as #3–7. All involve neural crest bones originating from the second rhombomere. The best way to understand the anatomic distribution of these clefts is to consider the neurovascular supply of the maxillary dentition. The anterior superior alveolar artery (ASAA), also known as infraorbital artery, produces medial and lateral branches to distinct groups of dental units. The posterior superior alveolar artery (PSAA) arises in the pterygopalatine fossa from the third part of the internal maxillary artery, proximal to the infraorbital artery. Thus, a developmental relationship exists between the developmental fields of the ASAA and the PSAA.

Maxillary Complex: Dental Zones (Mx1, Mx2, Mx3), Inferior Turbinate (It), and Palatine (Pl)

The Maxilla Is a Five-Sided Box

The maxilla is a five-sided box, the medial wall of which opens into the nose, permitting the development of a large maxillary sinus. Two additional bones, inferior turbinate and palatine, "patch up" the medial wall defect. The major supply for maxilla is the superior alveolar neuroangiosome. It trifurcates into three dental zones with each branch running through the membranous bony wall to reach the teeth creating within the anterior wall of the maxilla three developmental field zones.

The ASAN departs from the infraorbital nerve midway along its course through the canal. It traverses the *canalis sinosus* in the anterior wall of the maxillary sinus. The canalis sinosus swerves laterally away from the infraorbital canal and then hooks downward and medially to pass beneath the infraorbital foramen. It then runs downward to the canine and the incisors. The presence of the canalis is a tacit recognition of the prior existence of the anterior superior neurovascular bundle supplying those structures. The anterior superior alveolar artery arises from the infraorbital artery just prior to its exit from the infraorbital foramen. Just like its companion nerve, its course defines the canalis sinosus. Tessier clefts #3 and #4 occur here.

Mx1 is defined by *medial branch of anterior superior alveolar nerve* (MSA) and comprises those structures located medial to a vertical line dropped from the inferior orbital foramen. Mx1 is composed of: (1) the *mesial alveolus containing the lateral incisor and canine* and (2) the *anterior maxillary wall medial to the infraobital foramen*. The orbital rim and floor medial to the infraorbital fissure are formed by Mx1. This pedicle bifurcates into a medial branch that supplies the lateral incisor and a lateral branch supplying the canine. The dental units develop in a mesio-distal sequence, except that in most the lateral incisor from maxilla fails to develop, being supplanted by the lateral incisor unit from premaxilla. Note that it is *normal* for both premaxilla and maxilla to produce an incisor tooth in the "lateral" position. Note that it is biologically normal to produce a second lateral incisor. That is why ectopic incisors are so common. In addition, in cases of complete cleft lip, where the premaxilla does not produce a lateral incisor, the presence of this tooth in the maxillary segment can be explained by this duality. It bifurcates as well to supply the medial and lateral premolars. Defects in MSA are the source of Tessier #4 cleft zone, while those of LSA are the basis for the Tessier #5 cleft zone.

From the medial edge and from the superomedial edge of Mx1 arise two extremely important structures for plastic surgeons; these are crucial for understanding Tessier clefts #3 and #4. These structures are the *inferior turbinate bone* (IT) and the *frontal process of the maxilla* (MxF). This thin lamina extends upward from the canine region, past the lateral margin of the nasal bone, and terminates by abutting with the prefrontal bone (PrF), i.e., the medial orbital rim. It receives a separate vessel from MSA. Defects of IT cause the #3 cleft, while defects of MxF cause the #4 cleft. More about inferior turbinate below.

Mx2 is defined by the *lateral branch of anterior superior alveolar nerve* (LSA). This arises from the inferior alveolar nerve in its course along the infraorbital groove. The nerve then tracks laterally and forward in the lateral maxillary wall. It bifurcates to supply the premolar teeth. These develop in a mesio-distal sequence. Defects of LSA produce the Tessier #5 cleft.

Mx2 is the field set for LSA (middle superior alveolar nerve) and is composed of the alveolar housing for the premolar teeth (supplied by the middle superior alveolar nerve and artery) and the maxillary wall lateral to the infraorbital foramen. The orbital rim lateral to the intraorbital foramen and the orbital floor lateral to the infraorbital fissure are formed by Mx2. It is thus in contact with the greater wing of the sphenoid. In zoologic term, this is called the alisphenoid (AS). As we shall see, neural crest forming AS arrives after Mx2 and thus potentially must be constructed upon it. An isolated defect of Mx2 causes a *Tessier #5 cleft*. When the alisphenoid is also affected, one sees (rarely) a *Tessier #5, #9 cleft*.

Mx3 is defined by the *posterior superior alveolar nerve* (PSA). The pedicle trifircates to supply the molar in mesial-distal fashion. The PSAN departs from V2 in the pterygopalatine fossa. It pierces the maxilla just along its infratemporal surface and descends beneath the sinus mucosa to supply the molars. This will be the site of Tessier cleft #6.

Mx3 is the field set for posterior superior alveolar nerve and consists of the alveolar housing for the molars and the maxillary wall projecting upward and lateral, the *zygomatic process of the maxilla* (buttress). The *Tessier #6 cleft* is associated with hypoplastic states in this zone. The postorbital field (PO) of the zygoma sits directly above Mx3 buttress.

The *Tessier #8 cleft* localizes to PO. Dual affectation leads to the *Tessier #6, #8 cleft*.

The Sixth Side of the Box: Inferior Turbinate

This is a discrete bone forming 1/3 of the lower half of the lateral nasal wall. This structure is of critical importance in explaining the pathologic anatomy of the Tessier clefts #2, #3, and #4. To understand the clinical presentations of these clefts, we must dig into the role that IT plays in the separation of the nasal cavity from the maxillary sinus. Anterior to IT lies the ascending process of the premaxilla. Posterior to IT is the vertical plate of the palatine bone (Pl). As IT projects into the nasal cavity, it forms a caudally directed scroll. Beneath this scroll one encounters the terminus of the lacrimal duct. Thus, lower half of the lateral nasal wall is formed by three r2 neural crest derivatives (in antero-posterior order): MxF, IT, and Pl.

The upper half of the lateral nasal wall can likewise be divided into three discrete zones. The *posterior zone* is made from the vertical plate of the r2 palatine bone. The <u>central zone</u> contains the middle turbinate. This bone represents the caudal border of the r1 ethmoid complex. MT sits above IT and has the form of a Roman arch. An aperture between MT and IT results. This is hidden behind a scimitar-like projection of MT into the nasal cavity, the infundibulum. It is via this field boundary that the maxillary sinus drains. The <u>anterior zone</u> of the lateral nasal wall is made from the ascending process of the maxilla. The cranial margin of MxF abuts against the r1 bones: nasal bone medially, nasal process of the frontal bone posteriorly, and the r1 lacrimal bone laterally. Thus, the maxillary sinus is a five-sided box, five sides of which are exclusively of r2 derivation. The sixth side, the medial wall (the lateral nasal wall), is the combination of an upper tier of r1 fields with a lower tier of exclusively r2 fields.

As a mucous-producing structure, the maxillary sinus must have an obligatory escape route for its secretions. Fortunately, our six-sided box is not watertight. The field boundary between the anterior and posterior zones of the p6 lower ethmoid provides just the exit point for the maxillary sinus. It thus drains into the inferior aspect of the infundibulum, dripping out from beneath the middle turbinate. This messy situation neatly reinforces our previously described model of sinus development. ***All sinuses result from the expansion of oral mucosa into a potential space between fields***. Thus, the mucosa seizes the opportunity to insert itself into the opening between the anterior and posterior ethmoid zones and insert itself into the potential cavity between p6 and r2.

The structural integrity of the inferior turbinate is the prerequisite for the proper formation of two other structures that appear later in time. First, the r2′ frontal process of the premaxilla (PMxF) is constructed upon the scaffolding of IT. Second, p5 neural crest cells descend along the lateral surface of IT to create the lacrimal bone. Because the inferior turbinate is such a crucial component of the lateral nasal wall, IT deficiency states will create a severe cleft condition that begins at the lateral piriform margin, eliminates the medial wall of the maxillary sinus, and ascends into the lacrimal system. This is the pathology of the *Tessier #3 cleft*.

The Sixth Side of the Box: Frontal Process of Maxilla

The second important element to be produced from Mx1 is the frontal process of the maxilla. As stated before, MxF in all but humans and higher primates is physically distinct from PMxF. Although MxF originates in the MSA zone of maxilla, it also depends on IT for its construction. The physical presence or absence of PMxF is *not* required for MxF synthesis. When MxF is deficient, additional forms of clefting can occur. Recall that the inner aspect of the piriform margin is composed of two laminae; a small upper one coming down from p5 and the lower one ascending from r2′. The superior lamina, the *descending nasal process of the frontal bone* (Fn), buttresses the undersurface of the nasal bones. The inferior lamina, the *frontal process of the premaxilla* (PMxF), arises from the distal (versus mesial) margin of the premaxilla just above region of the lateral incisor region.

The lateral piriform margin and the medial piriform margin are initially separated by the *nasolacrimal groove*. This closes over during fetal development, causing the lateral piriform rims to approach the midline from either side. They eventually overlap the medial piriform margins. *Due to this overlap, no suture is observed in the term fetus.* In the past, this led physical anthropologist Sir Ashley Montague to question the existence of the premaxilla in humans [140]. Definitive proof regarding the premaxilla has been provided by Barteczko and Trevizian [141, 142]. In the *Tessier #4 cleft*, the persistence of this groove spares the medial piriform margin. All the incisors ipsilateral to the cleft are intact. Tessier #4 and #5 clefts occur as gradations of severity within the Mx1 field. Therefore, the #5 is simply a more severe form extending all the way to the infraorbital foramen. Loss of dental units within Mx1 can also occur.

The developmental significance of the bilaminar piriform margin is that deficits of the r2′ PMxF will allow the r2 MxF to sit more laterally and the *vertical height of the lateral piriform margin will be lower*. This describes the pathologic osseous anatomy of the isolated cleft lip nose. In those r2′ deficiency states in which the PMxF is hypoplastic or missing, the piriform margin is unilaminar and extremely thin. This means that the actions of the nasalis and paranasalis muscle complexes acting over time will exert a more pronounced distracting force to displace the piriform margin out laterally.

Mammalian evolution demonstrates the development of a medial projection from Mx1 to Mx3, the hard palate. The palatal shelves extend from each of these fields. Palatal shelves have a bilaminar vascular supply. The nasal side is supported by lateral nasopalatine artery, while the oral size is supplied by greater palatine artery. For this reason, the hard palate is potentially bilaminar and can, on occasion, contain a sinus.

Just above the palatal shelves is located the medial maxillary wall. This is made up of three coplanar structures all lined up in a row. The fused frontal processes (PMxF and MxF) form the anterior third of the wall. The *inferior turbinate* (IT) bone forms the middle third of the wall. It is a derivative of Mx1m. The "sprouting" of IT from Mx1 occurs *prior to* that of MxF. The lateral ascending process is constructed upon an intact inferior turbinate. Thus, *a cleft in zone #3 will always destroy zone #4 but not vice versa.* An isolated defect of MxF causes a *Tessier #4 cleft*.

The Sixth Side of the Box: Palatine Bone

Immediately behind Mx3 lies the palatine bone (P). This structure ossifies after the maxilla. Descending palatine artery supplies the perpendicular plate; its lesser palatine branch is assigned to the horizontal plate.

Laterofacial clefts occur at the interface between the maxillary and mandibular regions of the first arch. They can penetrate deeply, affecting multiple fields, including the palatine bone. Deficiency states have been described with severe hypoplasia or absence of this structure. In such cases, the soft palate musculature will have nothing to insert upon. These muscles will remain as unfulfilled mesenchymal blobs over their sites of origin, i.e., the Eustachian tube (tensor veli palatini) and petrous apex (levator veli palatini). In the literature, this condition is described as an *ipsilateral absence of the soft palate*. This has also been reported in the context of severe Goldenhar's syndrome.

The Zygomatic Complex: Jugal (J) and Postorbital (PO)

This segment of r2 neural crest belongs to the zygoma and is ossified in cartilage after the maxilla is formed. Its arterial axis is the StV2 *zygomatic artery*. The zone is the composite of two previous bones seen in lower vertebrates. The *jugal bone* forms the temporal process of the zygoma and the inferior half of the malar eminence. It is the most distal and is synthesized first. It begins along the axis of the zygomatico-facial neurovascular bundle and spreads downward to contact the Mx buttress. It also spreads forward to contact the Mx2 lateral orbital rim. The *postorbital bone* forms the frontal process of the zygoma and the superior half of the malar eminence. It also begins at ZF axis and spreads upward to contact the p5 zygomatic process of the frontal bone. It also spreads backward to contact the r3 zygomatic process of the squamous temporal bone. The anatomic split between PO and J is indicated by the zygomatico-facial neurovascular axis. *Tessier clefts #8 and 7* correspond to these two fields, respectively. These are commonly affected in Treacher-Collins syndrome, where the zygomatic arch is absent and the malar eminence is reduced or absent.

First Pharyngeal Arch: Rhombomere 3

The original lower jaw of tetrapods consisted of nine membranous bones organized around Meckel's cartilage. Over time, some of these fields disappeared (or were amalgamated). The most proximal bone fields became incorporated as components of the ear. Only the dentary bone remains to form the mammalian mandible. The mandible can be divided into four developmental fields. Three of these belong to the tooth-bearing alveolar bone. The ramus field contains also the coronoid process and the condyle. It develops after the alveolus. For this reason, in cases of craniofacial microsomia or Treacher-Collins, partial or total absence of the ramus can occur with preservation of the dental mandible.

Details of the formation of these fields are well-described by Kjaer. No clearer account of mandibular development can be found in the literature. Our purpose here is simply to identify the individual r3 fields and discuss their temporal order of formation. These concepts serve to rationalize the Kjaer's observations and permit additional clinical correlations (such as the anatomic rationale of the muscles of mastication).

Some superomedial r3 neural crest remains sessile alongside the developing brain. It helps form the dura covering the prosencephalon. This r3 zone is defined by the sensory distribution of V3 to the dura. Remaining r3 neural crest produces the bones of the second arch.

First r3 Segment: Mn1, Mn2, and Mn3

The alveolar zones of the mandible are similar to those of the maxilla. Three distinct sensory nerves supply the dental units of each zone. Mn1 contains both the incisors *and* the canine. Mn2 contains the premolars. Mn3 contains the molars. Each zone is supplied by a distinct sensory nerve. Early in development, each nerve has its own separate canal. These canals are organized in a strict time sequence. Derivatives in zone Mn1 represent the "oldest" mesenchyme; the incisors are the first teeth to erupt. Accordingly, the nerve to Mn1 occupies the most caudal canal. Just cranial to it lies the canal lies for nerve Mn2. Mn3 follows the same pattern. The three canals eventually become *roofed over* by the medial lamina of the mandible growing upward from its lower border. Thus, the inferior alveolar nerve is "three nerves in one!"

Individual variations in the number of dental units (both absence and excess) have been well-studied. These will affect the overall size and shape of the mandible from within their defined zone. Assuming that the number of teeth is constant, the AP length of the mandible is determined by the postdental segment. Under the same conditions, the vertical dimensions of the chin are determined by the fact that it represents a separate bone field from that of the dentary. *The majority of dentofacial skeletal deformities of the mandible result from deficiencies or excesses of mesenchyme in these two zones.*

Second r3 Segment: Ramus, Condyle, and Coronoid

The ramus forms via membranous ossification later in time than the alveolar bone. A portion of this periosteum converts to a cartilage cap that eventually forms the condyle. Pathologies affecting this zone may result in hypoplasia or absence but tend to spare neighboring fields. For example, craniofacial microsomia tends not to affect the tooth-bearing region of the alveolus. The coronoid forms at about the same time as the condyle. The cartilage goes on to form chondral bone. From this same PAM, the temporalis muscle is formed. For this reason, in Treacher Collins syndrome, absence of the coronoid is associated with an absent temporalis muscle.

Third r3 Segment: Derivatives "Assigned" to the Ear

The incorporation of the original angular bone becomes the tympanic bone. The reflected lamina of the angular held the original tympanic membrane; this is incorporated into the modern tympanic bone. The prearticular forms the anterior process of the malleus, the articular forms the malleus proper, and the quadrate forms the incus. In nonmammalian vertebrates, the jaw joint is represented by the quadrate and articular bones.

Second Pharyngeal Arch: Rhombomeres 4–5

The homeotic code defining rhombomere 4 is distinct from that defining rhombomere 5. As these populations sweep into the second arch, they do so at slightly different times. It is likely that they remain spatially distinct. We can hypothesize that r4 occupies the anterior (rostral) half of the arch and that r5 is distributed to the posterior (caudal) half of the arch. Motor branch of VII has upper and lower divisions, distributed to the anterior and posterior zones of the arch. The spatial arrangement of the branches of external carotid follows the geometry of the second arch once it has become repositioned.

Neural crest from r4 may represent the ancient hyoid suspension of the mandible. It likely forms the *styloid process* of the temporal bone. Fascia: all Sm6 muscles innervated by upper division of facial nerve are enveloped by r4 neural crest fascia. Neural crest from r5 forms the stylohyoid ligament and upper half of the hyoid bone. Sm6 muscles innervated by lower division of facial nerve are suspended in r5 neural crest fascia. Stapes is likely an r5 derivative.

The Extensive Distribution of Second Arch Mesenchyme

Second arch produces two layers of fascia which enclose functionally distinct muscles. Muscles of mastication work in concert with those of the first arch are located in the deep plane of the face and are suspended in deep investing fascia (DIF). Thus, rostrally located buccinator has r4 fascia and caudally located posterior belly of digastric has r5 fascia. Muscles of facial animation are located in a superficial plane and are suspended in superficial investing fascia (SIF), otherwise known by surgeons as the superficial musculoaponeurotic system or SMAS. The SMAS is continuous with the epicranius layer of the head and therefore second arch facial muscles are distributed over the entire cranium and anterior neck down to the clavicles. This is because there is an uninterrupted subcutaneous plane which second arch can exploit and therefore distribute the facial muscles over a large surface area. The expansion respects the superficial cervical fascia and stays within the confines of the skull.

Third Pharyngeal Arch: Rhombomeres 6–7

This arch is often given short shrift in the literature with only a single muscle, stylopharyngeus, attributed to glossopharyngeal nerve. In reality, it plays a significant role in the construction of the upper pharynx and contains within its fascia multiple muscles the muscles of the palate and superior constrictor. The sensory distribution of third arch is exceptionally broad, from the skull base to the larynx and forward to the posterior third of the tongue. The muscles of the third arch all take origin in somitomere 7, which also provides mesenchyme for the mastoid component of the petromastoid complex. third arch is in register with r6–r7 and receives neural crest from these levels. Its motor innervation resides in the lateral motor column of medulla, specifically in cranial third of nucleus ambiguus. As we shall see in greater detail later, nucleus ambiguus contains neurons for glossopharyngeal in r6–r7, for vagus to larynx in r8–r9 (superior and inferior laryngeal nerves), for vagus to pharyngeal plexus in r8–r11, and for accessory nerve r8–r11. Additional fibers of vagus to the pharyngeal plexus are also found in r7 admixed with those for cranial nerve IX.

Gray's anatomy considers innervation of all the constrictors to come from accessory, in that XI is branchiomotor and takes over for IX. This is simply a question of neuromeric supply within nucleus ambiguus. In this model, soft palate and superior constrictor have motor nuclei in r6–7, middle constrictor is supplied from r8 to r9, and inferior constrictor is innervated from r10 to r11.

Third arch can be said to enclose the terminus of the nasopharynx with the hypopharynx being the province of the fourth and fifth arches. All palate muscles (with exception of tensor veli palatini) and superior constrictor are supplied by r6–r7, but fibers can arrive either via IX per se or via vagus nerve to pharyngeal plexus. This vagal contribution continues to cause confusion because fourth and fifth arches are associated with vagus and third arch is not. Soft palate is definitely not considered a fourth arch structure—in fact, its definition in most texts is rather vague. In sum,

palate and constrictor are third arch structures—it does not matter what nerve conveys motor fibers to these muscles—the important neuroanatomic fact is that they are supplied from r6 and r7.

Localization of derivatives and fascia within third arch has not been well mapped-out, but we can use some common-sense neuromeric concepts to deduce its boundaries. Because second arch mucosa is eliminated within the oropharynx, third arch abuts with first arch along a line extending from pterygoid Hamulus to pterygomandibular raphe. It extends all the way back to pharyngeal tubercle of the occipital bone mylohyoid line and the posterior 2/3 of the tongue. Soft palate is halfway between the base of the tongue and the posterior pharynx. Laterally, it interfaces with first arch along Waldeyer's ring and the site of the ancient buccopharyngeal membrane. If there is indeed zonation, the r6 zone should be anterior and that of the palate and posterior to the cranial base should represent r7. The inferior boundary of superior constrictor contains stylopharyngeus muscle and cranial nerve IX, so it obviously borders with fourth arch. Motor innervation for the soft palate has not been mapped out, but is likely a mix between r6–r7. In terms of bone derivatives, r7 neural crest (possibly in combination with lateral plate mesoderm) forms the caudal half of the hyoid bone, the greater cornu, and the mastoid temporal bone.

Fourth Pharyngeal Arch: Rhombomeres 8–9

Fourth arch marks the transition from a nasopharynx to the hypopharynx. Its external boundaries are defined by middle constrictor enclosing an endodermal outgrowth for the respiratory apparatus. Fourth arch is in register with rhombomeres r8–r9, each of which contains a motor nucleus supplying the larynx via vagus nerves. Superior laryngeal nerve to cricothyroid originates from r8, while inferior laryngeal nerve to the remaining laryngeal muscles originates from r9. The laryngeal muscles are likely to arise and migrate from somitomeres 8 and 9 prior to their transformation into occipital somites 1–2. For that reason, they are not routinely listed as somite derivatives. Recall that the principle products of occipital somite myotomes S1–S4 are sternocleidomastoid and trapezius.

Middle constrictor is larger and thicker than its superior counterpart. It extends backward from second and third arch components, lesser and greater cornu. It is shaped like a fan with a broad posterior extension from the lower border of superior constrictor down to the lower pharynx. The lower border of middle constrictor with inferior constrictor admits internal laryngeal nerve and superior thyroid artery, laryngeal branch. For these reasons, it is likely that middle constrictor (like its laryngeal counterparts) arises from *both* Sm8 and Sm9. Motor supply originates from r8 to r9 of nucleus ambiguus via the vagus to the pharyngeal plexus (or accessory). Thyroid cartilage is a fourth arch derivative.

Derivatives of the Fifth Pharyngeal Arch: Rhombomeres 10–11

Fifth arch remains poorly defined. Some consider it nonexistent. But since rhombomeres r10 and r11 remain intact and produce neural crest, since somitomeres Sm10 and Sm11 are present, and since blood supply is available from superior thyroid with collaterals from inferior thyroid, it seems reasonable to assign the final derivatives of the pharynx to this arch. Fifth arch is the putative source of cricoid cartilage. This is an obvious boundary. Below it, trachea forms from cervical lateral plate mesoderm, as does the esophagus, both in register with neuromere c1 and below.

Inferior constrictor is the largest of the three constrictor and overlaps its middle counterpart. It has two parts, *thryopharyngeus* and *cricopharyngeus*. Since the latter abuts with esophagus, it must arise from Sm11; the former arises from Sm10 or perhaps Sm9–Sm10. Passing under the lower border of inferior constrictor, one finds the recurrent laryngeal nerve and the laryngeal branch of inferior thyroid artery. Innervation of thryopharyngeus is from nucleus ambiguus r10–r11, while that of cricopharyngeus may also be recurrent laryngeal nerve.

Formation of the Larynx and Trachea

The exact contributions of neural crest and lateral plate mesoderm to the laryngeal, arytenoid, and cricoid cartilages are unclear, but these structures arise from levels r8–r11. The multiple nuclei of the vagus nerves supplying these structures are distributed along the length of this segment of medulla. The tracheal cartilages display a unique structure that is anticipated by the cricoid. These cartilage units are U-shaped, being incomplete posteriorly. Because endoderm provides the pattern for so many other cartilages in the oropharynx, it is tempting to think that lateral plate mesoderm responds in the same way. The endodermal bud grows downward into an unorganized central mass of LPM and forces it to form rings of cartilage. But what might explain the periodicity of the trachea? Let's speculate a bit.

The origins of LPM in the neck that could contribute to the trachea are the very same as that of the cervical esophagus. The cervical esophagus displays a curious vascular pattern of axial segmental vessels, 4–5 in number (one per neuromeric level). At the level of the thorax, the esophageal vasculature comes from the celiac axis. In mammals, this situation suggests that four neuromeric units are involved in producing the esophagus, i.e., from c5 to c8. At the same time, the musculature of the cervical esophagus gradually changes from striated to smooth, the transition being complete at the thoracic inlet.

It would appear that there is something segmental about the cervical esophagus. The reason for this might be that LPMs and LPMv are unsplit in the neck. Readers will recall that the formation of the bony body wall requires that these two layers separate with LPMv forming the internal mesoderm of the body cavity, while LPMs encircles the visceral organs. Indeed, the anatomy of ribs is nothing more than segmental extensions of PAM from each vertebra that bud outward into the mass of LPMs. We know that somatic LPM follows the same *hox* code as the vertebrae. Thus, within the neck, lateral plate mesoderm maintains an occult segmental pattern. Although cervical LPM is an unsplit fusion of visceral and somatic laminae, it is the somatic component that imposes a segmental order to the esophagus and trachea. *As soon as LPMs and LPMv part company (at the thoracic inlet), all segmentation of visceral structures disappears.*

We therefore arrive at the following hypothetical mechanism for tracheal development. The endoderm confers cartilage-forming signals on the unsplit LPM from neuromeric levels c1 to t1–2. The spaces between the tracheal rings are occult manifestations of individual neuromeric units of LPM. As soon as LPM splits apart at the inlet, gene expression within LPMv responds by formation a carina and the appearance of lung buds. The lungs are appropriately enclosed within the ensheathing visceral pleura. Sensory innervation of the visceral pleura is non-segmental. It is organized around the neuromeric units of the sympathetic autonomic nervous system (SANS). The parietal pleura, on the other hand, is formed from LPMs, innervated segmentally and thus is capable of exquisite, localizing pain.

Spatial Relationships of Pharyngeal Arches

Mammalian facial muscle anatomy represents a significant departure from the initial arrangement of pharyngeal arch muscles in other tetrapods. These muscles were designed to be co-planar. Deep layer muscles fulfill a sphincteric function, while superficial muscles act as dilators. The original branchial arch gill structure had anterior (pre-trematic) and posterior (post-trematic) elements. We advance the hypothesis that the neuromeric arrangement of the pharyngeal arches is a holdover of this anatomic plan. Every arch is in register with two rhombomeres. The cranial neuromere produces anterior structures, while the caudal neuromere produces posterior structures. With embryonic folding, these become to rotate 90°. Hence, the r2 maxillary complex is cranial to r3 mandible.

Only a few second arch muscles remain co-planar with those of the first arch. Most second arch mesenchyme is positioned superficial to envelop the first arch with the facial muscles positioned superficially with an entirely separate fascia, the SMAS. The third arch (r6–r7) remains co-planar with the first two, however. It abuts up against the r8/first occipital somite muscle, the sternocleidomastoid. One could surmise that the amalgamation of PA2 with PA1 occurs concomitant with or shortly after the formation of PA3. The fourth and fifth arches are formed in continuity with the caudal aspect of PA3 and thus are "tucked in" deep to the plane of the first three arches. Thus, at the level of PA3, the pharynx becomes a *tube within a tube*.

Formation of the Cranial Base

Posterior Cranial Base

Analysis of the order of ossification of the facial skeleton from a neuromeric perspective reveals several important observations. The development of the neurocranium is in direct proportion to the need for brain coverage. Bone made from somitomeric PAM will at all times form before that formed by neural crest. The chondral neurocranium preceeds the membranous neurocranium. For example, basipostsphenoid (PAM from the first somitomere) forms before the presphenoid (r1 neural crest). Why might this be so?

We can envision a spatial relation between the neuraxis and the somitomeres versus the eventual position of the pharyngeal arches. Although the somitomere may not possess a sclerotome per se, the physical position of the PAM next to the neural tube makes it likely to be converted into paraxial bone. Just like company of firefighters provides the "first response" to an emergency in their neighborhood, so does PAM represent the mesodermal "first response" to brain formation and the need for structural support.

The timing of neural crest migration occurs well after somite formation. Neural crest thus represents a "second response" to brain formation. Just as in the formation of a house, the foundation of the cranial base is first laid down by PAM, because this mesenchyme is physically situated at the ground floor. Subsequently neural crest takes over, making the ensheathing bones of the calvarial roof. This happens as the sidewalls of the neural tube grow upward and approximate. Neural crest from the neural folds follows this process right along. It is thus no accident that the ossification sequence of the frontal and parietal bones is ventral to dorsal.

The layout of the chondrocranium demonstrates these concepts. Somitomeres Sm1–Sm3 lie just lateral to forebrain. They do not contribute bone. Sm4 makes basisphenoid. From the sphenooccipital synchondrosis backwards, basioccipital is laid down by parachordal cartilages from Sm5 to Sm7. Sm6–Sm7 contribute laterally to the otic capsule. Sm8–Sm11 become occipital somites S1–S4. Because the somitomeres are sessile and located immediately next to the notochord, they can be rapidly converted to cartilage. From basisphenoid forward, all derivatives are mesencephalic neural crest from r0/r1. These include the trabecular and ethmoid cartilages that fuse to make presphenoid and ethmoid. Lateral to these cartilages are the nasal and optic capsules. Refer again to Figs. 2.68 and 2.69.

Cranial base anatomy displays both segmentation and parasegmentation. PAM forming the basisphenoid is strictly an r1 derivative. So too, the temporal bone comes from Sm6–7 plus corresponding r6–r7 neural crest. *Parasegmentation begins at the spheno-occipital suture.* Each of the four occipital somites contribute to the basioccipital, exoccipital, and supraoccipital bones in a laminar manner. If the sclerotomes of these somites produced vertebrae, it would be as if the four bodies were fused in succession like Russian dolls forming BO. The lateral elements around the foramina would likewise be laminated, with r8 on the outside and r11 closest to the ring. These laminae would produce the exoccipital bone. Finally, fusion of the neural arches would yield the supraoccipital bone (Figs. 2.72, 2.73, 2.74, and 2.75).

Experimental work provides solid evidence that this is so. The avian model described by Huang has five occipital somites [22, 23]. Using tracing maker methodology, the contributions of each somite to the skull base are depicted in color. O'Rahilly demonstrated the mammalian model to have only four somites [24]. The pattern of these is similar to that of birds, so it is reasonable to suppose that the same topology of assembly holds true as well.

Anterior Cranial Base

We have completed our survey of the mesenchymal derivatives involved in head and neck. We are now in a position to understand facial bone formation in terms of neuromeric derivatives. This is a visual exercise involving migratory pathways by which neural crest and PAM arrive at their final destinations in the face and skull. A description of this entire process is beyond the scope of this paper. Nonetheless, we can use a model of the *fronto-orbito-sphenoid complex* as a convenient model to study how these migrations can be understood on a neuromere-by-neuromere basis. (Assembly of the facial midline will be covered in greater detail in Chap. 3).

This science has direct relevance for the surgical correction of congenital based on the painstaking clinical observations of Paul Tessier regarding the various forms of craniofacial clefting [1–3]. Tessier documented that these anomalies seemed to follow certain occult anatomic patterns that he classified numerically. The Tessier system, derived purely on an empiric basis, has a remarkable fit with neuromeric fields. By understanding the face as an assembly of fields with a neuromeric basis, clefting takes on an entirely new relevance for neuroscientists and surgeons alike.

Analysis of craniofacial clefts provides the means by which we may understand the developmental anatomy of the

Fig. 2.72 Occipital bone complex of basal amniote. (Courtesy of Augustus T. White, Palaeos.com. http:// palaeos.com/vertebrates/ bones/braincase/images/ Opisthotic2.jpg)

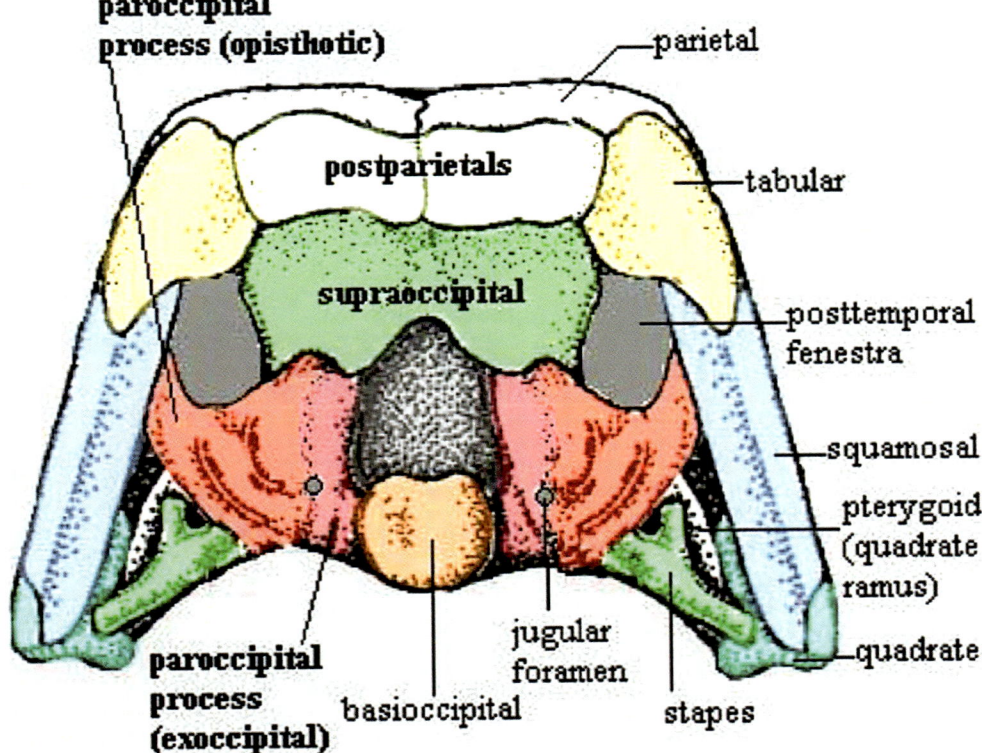

Fig. 2.73 Posterior cranial fossa development 1. Embryogenesis of the posterior cranial base involves amalgamation of the first four occipital somites. These can be considered as primitive vertebrae. The ventral hemal arches combine to make the basioccipital bone(s), the lateral pedicles produce the exoccipital bones, and the neural arches form the supraoccipital bone(s). Purple = r1 presphenoid and basisphenoid. Turquoise = r8 first occipital somite. Red = r9 second occipital somite. Pink = r10 third occipital somite. Blue-green = r11 fourth occipital somite. (Courtesy of Michael Carstens, MD)

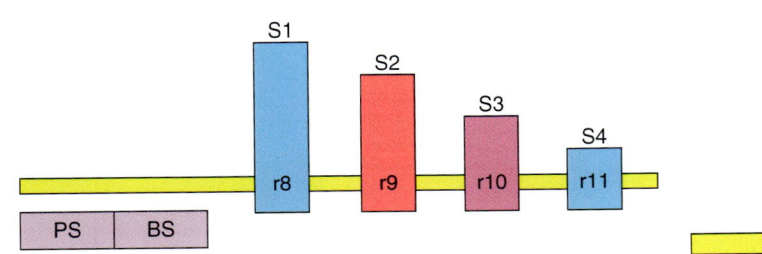

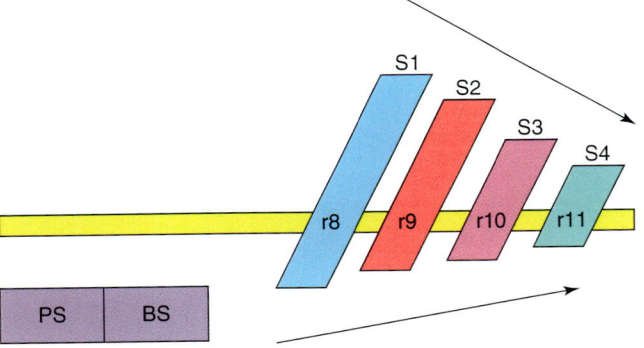

Fig. 2.74 Posterior cranial fossa development 2. Although somites 1–4 initially appear as somite-like masses, they undergo a topologic transformation in which they become inclined and stack up sequentially inside each other like Russian dolls. Under the influence of occipital lobe, development of the fused neural arches undergoes expansion posteriorly and superiorly. Each of the occipital somite myotomes contributes myoblasts to the tongue. The low hairline and large tongue seen in Down's syndrome reflect misallocation of paraxial mesoderm away from the occipital braincase and toward the tongue muscles. (Courtesy of Michael Carstens, MD)

Fig. 2.75 Cranial base
sutures

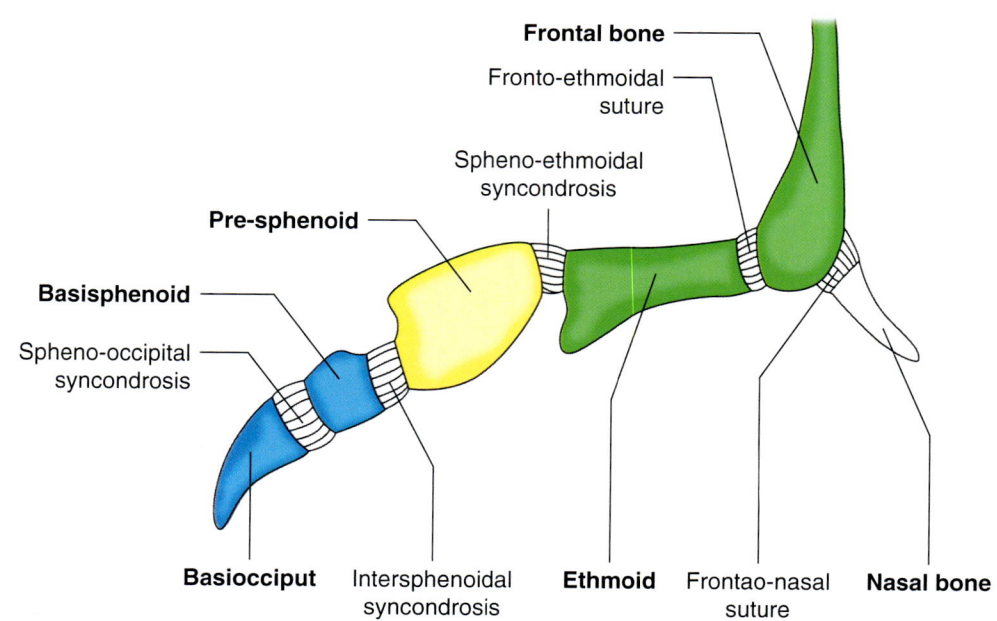

fields themselves. It is a secret window into the spatiotemporal order of neural crest migration. Let us examine a theoretical diagram of the anterior cranial base in which each field is color-coded to match its neuromeric level of origin. The pathways by which neural crest cells arrive at the orbit and the structures created by each neural crest are indicated by color-code arrows. The physical location of all neural crest cells is understood to be from sites above and behind the orbitosphenoid complex. Derivatives listed most distally along each pathway migrate earlier and develop earlier.

The discerning reader will note that most neural crest bones of the face form via membranous ossification. However, when neural crest cells are grown in isolation in vitro, they form cartilage. Craniofacial neural crest, in order to form a membranous bone, requires signaling from epithelium. In the absence of such interaction, cartilage will form. But this does not determine that the cartilage will automatically become chondral bone. Subsequent signals can convert a neural crest cartilage to a membranous bone. Another example of this process is the zygoma, attached first to the maxilla as a neural crest cartilage derivative and subsequently converted into membranous bone.

Orbitosphenoid Derivatives Neural Crest
These bones originate from MNC and are programmed by dura and sclera.

The anterior surface of the neural crest presphenoid (PS) is facing us. Hidden behind it is the basisphenoid (BS) produced from paraxial mesoderm of somitomere 4. Although these two bones have differing mesenchymal sources, both PS and BS belong to the axial cranial base. Both are formed by chondral ossification. *All components of the cranial base, be they from paraxial mesoderm or from neural crest, develop*

as cartilage. Recall that pituitary represents the dividing line for the anterior cranial base. All bone fields anterior to it are neural crest; all those posterior to it are PAM (Fig. 2.75).

Arriving later in time, two neural crest *orbital cartilages* flank the presphenoid. These ossify to make the orbitosphenoid (OS), better known as the lesser wing. OS is joined to PS via two bony pedicles. These form an arch around the preexistent optic nerve, the optic foramen. Projecting like a beak from the anterior-inferior (rostral-ventral) surface of presphenoid is the *pharyngeal tubercle*. Hanging down caudally from the underbelly of PS are two neural crest laminae, the medial pterygoid plates (MPt). These are remnants of ancient pterygoid. A secondary chondral extension later develops: the pterygoid hamulus, around which the sling of tensor veli palatini will be draped.

Fronto-Orbital Derivatives
These bones originate from MNC programmed by p5 dermis and r1 dura (light blue). MNC sweeps into position over the frontal lobes and below FNO skin producing dura and subcutaneous tissue. The programming for frontal bone relates to frontal lobe developmental fields. The ossification center for each frontal bone is centered between two branches of V1 stapedial: frontal artery via supratrochlear supplies Tessier cleft zones 13 and 12, while supraorbital supplies zones 11 and 10.

Below the horizontal axis of the artery, the orbital rim has two fields that appear to be in continuity with frontal but are biologically distinct. Medial is *prefrontal* bone (PrF) and lateral is *postfrontal* bone (PF): they belong to zones 11 and 10, respectively. These bones form by a wave of ossification which starts inside the orbital roof, moves outward and ascends from the orbital rim. Recall that MNC makes the dura of the anterior cranial base and that sclera is an extension of dura as well. Thus, the orbital roof component of these two fields is potentially bilaminar and can make a sinus.

These upper frontal fields are exquisitely sensitive to perturbations in the underlying r1 zone, whereas PrF and PF are resistant. Thus, in *anencephaly*, the orbital rims remain intact. The mediolateral axis of the frontal fields displays a clear-cut time line. The more medial fields form last. Thus, global deficiency states first manifest themselves medially and, with greater severity, progress laterally. Trigonocephaly, frequently accompanied by frontal lobe deficits,

Palatine Derivatives

Attached only the ventrolateral aspect of the presphenoid is the alisphenoid bone (AS), forming the greater wing of the sphenoid. AS is the remnant of ancient epipterygoid. Its axis is meningo-orbital artery which is sole anterior branch from StV2 intracranial to the orbit. Ossification of AS occurs in two ways. The lateral margin of the superior orbital fissure forms in cartilage. The remainder of the AS is a membranous bone. Hugging the external face of the r1 MPt is a second laminar bone; the lateral pterygoid plate (LPt) is the remnant of ectopterygoid and is r2 RNC. Just like MPt, LPt is formed in membrane.

Sitting directly in front of the medial pterygoid plates just like two bookends are the r2 palatine bones. They develop from the descending palatine neuroangiosome. These strictly membranous bones have perpendicular and horizontal laminae joined at a 90° angle. At the anterior and superior corner of the perpendicular lamina, each palatine bone projects a quadrangular, superiorly directed *orbital process*. These orbital processes are not obvious within the eye socket. (Designed by nature expressly for purposes of the plastic surgery oral board examination.) Just posterior to the orbital processes, smaller prominences project upward to make contact with the anterolateral corners of the presphenoid. Between these anterior orbital processes and the posterior sphenoid processes lies the sphenopalatine foramen. *This is an absolutely critical anatomic landmark with enormous significance for the formation of cleft lip and palate.*

Nasopalatine Derivatives

Through the sphenopalatine foramen passes the most medial branch of the StV2, the axis of medial nasopalatine artery (NPm). Neurovascular pedicles are like paleontologic footprints; they serve as evidence of the earliest cellular migrations in the embryo. In this case, the sphenopalatine axis shows us the pathway by which neural crest passes forward toward the orbit, is forced by *preexistent r1 mesenchyme* to take a posterior and inferior route, and gains access to the midline of the future nasal cavity. The nasopalatine field contains within it two neural crest osteosynthetic "packets." The most anterior is that of the premaxilla; this is followed up by the vomer. The presence of r1 neural crest in the roof of the nasal midline guides NPm into the midline of the future face. Without this r1 "superhighway," vomer and premaxilla will be unable to arrive at their destination in zones 1 and 2. Pathologies that affect the ethmoid complex, such as holoprosencephaly, can create varying degrees of deficiency or outright absence of the vomer/premaxilla complex.

Assembly of the Face

Herein we present a preview of a topic which is the subject of Chap. 4. What we have accomplished in this chapter is to catalog one-by-one the mesenchymal components that are used to construct the face based upon their neuromeric level of origin. So far, our discussion has been static. We will now illustrate the dynamics of this assembly process. This topic is so important that it merits in-detail treatment in Chap. 4. Our purpose in this section is to integrate what we have covered so far and show the basics of how this process works.

Facial construction occurs in three related steps: (1) The frontonasal skin envelope is created by epidermis from rostral forebrain nonneural ectoderm and dermis from caudal forebrain neural crest. (2) A massive migration of midbrain neural crest creates paired fronto-naso-orbital units that (with apotosis and the approximation of the nasal placodes) fit together like the letter "T." (3) Rostral hindbrain neural crest + PAM from somitomeres 4 and 6 creates the non-jaw components of first and second arches. The two arches are amalgamated masses of soft tissue. They move medially under the wings of the "T" to fill out the facial midline. The timeline for this process is very tight and precise.

Like an origami puzzle, the face "comes together" when two lateral components (the pharyngeal arch complexes on either side) sprout out from the rhombencephalic zone of the embryo, extend towards each other in the midline like pincers, and make contact with a central pair of units organized around the forebrain. This process is driven by apoptosis of nasoethmoid MNC discussed in greater detail in Chap. 4.

An apology straightaway to the reader is due here. Although experimental work has defined the end point of neural crest migration, the start point on a zone-by-zone basis is less clear. ***Here are our working hypotheses.***

1. Neural crest populations *migrate in a strict time sequence*.
2. A developmental zone contains *several populations*, each producing a specific anatomic structure or field.
3. Fields develop according to their vascularization.
4. Adjoining developmental fields are interdependent. Substances produced in one field may affect the function of another.
5. Developmental zones are autonomous; if a neural crest field N in zone X is deficient, it will not affect the ability of fields to develop appropriate cell mass in adjacent zone Y.

6. The presence of a deficiency in zone X can affect the physical shape of fields in adjacent zone Y. These other normal fields will have normal volume and surface area, but, with growth, they can undergo *secondary deformation* as they collapse into the deficiency site. This is known as *deficiency-induced field mismatch*.

Neuromeric Production of Soft Tissues

In subsequent sections, we shall be discussing the mechanisms by which a spectrum of clefts is produced using a neuromeric field model in which deficits in specific fields lead to specific types of clefts. To do so, we shall need to expand our vocabulary one step further. Neuromeric concepts must be applied to soft tissue structures as well.

To this point, the anatomy of facial fields has been presented in terms of the mesenchymal components that make them up, emphasizing neural crest and paraxial mesoderm. Our discussion of neural crest has focused on: (1) *where* the neural crest components arise in the embryo prior to their migration, (2) the *sequence* in which these neural crest fields migrate to the face, (3) the *pathways* they use to get there, (4) the *spatial arrangement* of the fields once in final position, and (5) the *developmental relationships* that exist among fields that enable them, like Lego® pieces, to fit together in a precise time sequence to build the face.

Our discussion of mesoderm has focused on: (1) distinguishing its various *anatomic types* (paraxial PAM, lateral plate LPM), (2) its *mechanism of formation* during gastrulation, (3) the *neuromeric basis of segmentation* as it applies to the gastrulation process, (4) *somitomeres and somites* as segmental units, (5) *neural crest contributions* with these mesodermal structures, (6) the *reorganization* of facial mesenchyme into head mesoderm, pharyngeal arches, and occipital somites, and (7) the *spatial origami of head folding* by which these units are positioned for final integration into the face.

Because the facial bones are so readily distinguished and because they form in such a strict time sequence, much of our discussion of developmental fields has to date centered on these bony building blocks. But we must now change our emphasis completely to that of the soft tissues. ***The bones of the craniofacial skeleton, like all bones in the body, are merely the products of soft tissue developmental fields*** (functional matrices, if you will). Any defect in a bony is merely the manifestation of a problem in the functional matrix that produced it. A field/functional matrix disturbance can therefore manifest itself in bone, in soft tissue structures, or in both.

Skin

Cutaneous coverage of the head and neck has its own separate chapter. For our purposes, consider the following. The *skin of the forehead, nose, and philtrum* is composed of p5 nonneural ectoderm epidermis and p5 neural crest dermis. The *upper eyelid* epidermis comes from p5 NNE and the dermis is made from r1 MNC. The epidermis of the *lower eyelid* is made of r2 ectoderm, while the dermis is produced by r2 neural crest. For this reason, the sensory supply to the upper lid *and the entire conjunctiva* is from V1, while that of the lower lid is from V2. The remainder of the facial skin is produced in accordance with its innervation patterns. Ectoderm corresponding to r2–r11 is initially present in the five pharyngeal arches, but as these fold up upon one another, r6–r11 ectoderm disappears. The remaining facial skin is produced from r2 and r3 (with a small representation of second arch r4–r5 skin over the external auricle).

The dermatomes of the occiput come from cervical neuromeres c2–4. Neural crest likely produces the *hair of the face and scalp* from r1 to r3 and from c2 to c4. It is intriguing that the hair pattern corresponds to the dermis innervated by the dorsal (epaxial) branches of C2–4, while the dermis supplied by ventral (hypaxial) branches of C2–4 is hairless. In the same manner, a dorsal/ventral pattern might be seen in V2 and V3-derived dermis. The epaxial dermis would produce hair, while the hypaxial dermis would produce beard. Most patterns of human hair formation fit this model. Eyebrows, for example, would occur at the interface between r1 RNC dermis and p5 PNC dermis. They also correspond to the surface anatomy of orbital bone fields: prefrontal and postfrontal.

Mucosa

Nasal vestibular lining develops from the nonneural ectoderm of the nasal placode. *Oral mucosa* develops from ectoderm of pharyngeal arches 1–2. Representation from the second arch is quickly crowded out and disappears due to expansion of the first and third arches. The oropharynx posterior to Waldeyer's ring is produced from endoderm of pharyngeal arches 3 and 4, like a series of Michelin tires stacked one aside the other.

Facial Muscles, Fascia, Fat

Muscles of mastication originate from somitomere 4 (first arch) and somitomere 6 (second arch). They are covered by neural crest deep investing fascia from r2 to r3 and r4 to r5, respectively.

Facial muscles develop from somitomere 6. Those supplied by the upper division of the facial nerve are covered in neural crest fascia likely to originate from the proximal second arch (i.e., from r4), while the fascia corresponding the muscles supplied by the lower division of the facial nerve would originate from the distal second arch (i.e., from r5). *Salivary gland* formation would result from neural crest mesenchyme from r4 to r5 invaded by oral epithelium. The connective tissue within the salivary glands is r3 oral epithelium. This has implications for understanding the derivation of the parotid gland. Because this structure is penetrated by the facial nerve, it may be reasonable to assign it to r4 and r5 (being distributed along the upper and lower divisions, respectively). Facial *fat* is a neural crest derivative.

Biologic Basis for Developmental Fields

At this juncture, the discerning reader must be queried by nettlesome questions such as: Does the functional matrix concept have any provable experimental basis? Why should I bother learning this stuff? The best answers to the first question reside in a plethora of papers stemming from the quail-chick chimera system popularized in developmental biology literature by Couly and LeDourain. These investigators showed that visible differences in neural crest cells existed in quail and chick embryos such that microsurgical extirpation experiments could be carried out. When NC cells from one type of embryo are transplanted into the other, their derivatives can be distinguished under the light microscope. Using the neuromeric map, these authors were able to demonstrate what derivatives came from what levels. For example, the mandible, malleus, and incus are all neural crest products from level r3.

The Couly-LeDourain model also allowed for assessment of programming function. How does r2 neural crest "know" to make zygoma or r5 neural crest to make hyoid bone? The answer is not surprising. Mesenchyme responds to an occult *program in the surrounding epithelium* [143–147]. Ectodermal zone patterns influence the development of dermal bones. More recently, foregut endoderm (FE) has been shown to instruct neural crest what to do. By taking out a certain zone of FE, specific parts of the hyoid bone fail to appear. The spatial layout of the hyoid fields is faithfully reflected in the organization of the FE. Finally, if a given zone of FE is reversed 180°, then the corresponding part of the hyoid bone is reversed as well.

Thus, all neural crest bone and cartilage derivatives associated with the foregut arise as the result of programming embedded in the endoderm. Because the endoderm in each region arose from a spatially dedicated zone of epiblast cells, it can truly be said that the overall organizational plan of the organism is set up prior to gastrulation.

Future research will undoubtedly result in the understanding of the fate of foregut endoderm. Specific zones will correspond to specific structures. Each zone will ultimately be categorized by a unique pattern of gene expression, alterations of which will lead to predictable abnormalities in the neural crest products associated with that zone.

Developmental fields do not occur in isolation. Interaction with other fields is often required. This is particularly well demonstrated in the formation of facial bones. Many of these structures result when specific populations of neural crest migrate from their nascent position in the neural fold to distant locations. Here they interact with local epithelial cells (ectoderm or endoderm) from which they receive signals that determine cellular mitotic rate (volume) and the physical confines in which such population expansion may take place (shape). Moreover, the presence of one field may be required in order for another field to correctly develop. The footplate of the lacrimal bone is positioned just lateral to the inferior turbinate (IT). *The formation of the inferior turbinate occurs earlier in time than does that of the lacrimal bone.* A disturbance in the formation of IT may lead to defective or absent lacrimal system. By the same token, problems within the lacrimal system (stenosis of the duct) can occur without affecting the inferior turbinate.

It should be thus apparent that developmental fields are the results of tightly regulated sequence of field creation, field positioning, and field assembly. Process such as flexion of the embryonic neuraxis, spatial repositioning of pharyngeal arches, and programmed cell death (apoptosis) are required in order for field assembly to take place correctly. All students of the nervous system will recognize that brain growth is critical to the development of the face. It should be understood, up front, that the migration of neural crest cells refers to a point in time before embryonic folding takes place. All the populations have arrived and are positioned with respect to epithelial developmental zones and pharyngeal arches.

Recall that positional genes, such as the *DLX* system, map out the arches into spatial regions, each with its own developmental fate. ***Thus, the spatial position of a neural crest population on the neural fold or in the pharyngeal arch does not confer specificity. Once the cells arrive at their destination, they are "instructed" as to their final anatomic form.*** Thus, r1 neural crest does not "know" to become the sphenethmoid complex. Those cells that become physically positioned in front of the notochord will be instructed by it to form the trabecular cartilages from which develop first the presphenoid and later the ethmoid.

Toward a Neuromeric Theory of Facial Cleft Formation

The purpose of this chapter has been to demonstrate how neuromeric concepts can be used to understand facial development. From this, an integrative theory of cleft formation will be presented. We shall begin with facial bones and then proceed to soft tissue structures. This distinction is completely artificial! Recall that facial bones are mesenchymal responses to an epithelial (soft tissue) environment. *Bone does not "grow itself": it is the product of a developmental field.* Each developmental field is neuromeric in nature and includes ***all soft tissues associated with a given bone.***

Muscle insertion into its "target" bone is a good example. Muscles take their origin from a specific somitomere, somite, or group of somites. This origin corresponds to the neuromeric level of its motor nerve. The muscle then makes a primary insertion into a bone developing from the same neuromeric level. Supraspinatus is innervated by C5 and therefore originates from somite S9. The supraspinous fossa of scapula arises from neuromeric level c5 as well; therefore, it receives the primary insertion of the muscle. The secondary insertion of a muscle is into the nearest available binding site, preferably one arising from the same neuromeric level. Hence, supraspinatus seeks out the c5 developmental zone of humerus. Infraspinatus is larger. It is innervated by both C5 and C6 and therefore takes origin (distal to infraspinatus) from S9 and S10. Its primary insertion into infraspinous fossa occurs because this zone of scapula arises from both levels c5 and c6—and therefore has a larger surface area. It then seeks a secondary insertion more distal on the humerus (Fig. 2.76).

Understanding the pathways by which neural crest cells migrate into position is a crucial first step in our discussion. Facial bone synthesis occurs in a rigid spatiotemporal order There are three general populations of neural crest involved in constructing the face; the behavior of each depends on its anatomic zone of origin.

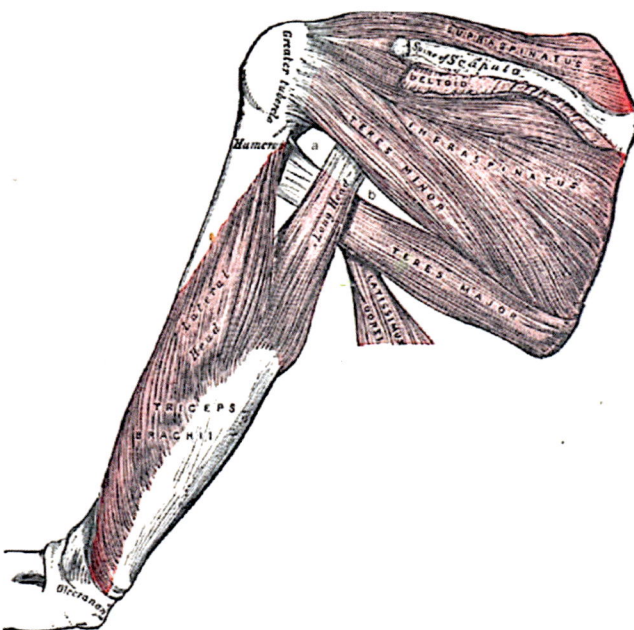

Fig. 2.76 Muscle origin/primary insertion/secondary insertion: the scapula model. Suprapinatus C5 suprascapular nerve, infrasinatus C5–C6 and teres minor C5–C6 axillary nerve—c5 zone of humerus. Teres major C5–C6 lower subscapular nerve—more distal has c6 zone. Note triceps is c6–c7 zone on the humerus. (Courtesy of Michael Carstens, MD)

- Prosencephalic neural crest (PNC) arises from the neural folds above the caudal forebrain. PNC migrates forward like a *glacier-like sheet* to populate the neural folds of the rostral forebrain (which are lacking in neural crest). PNC has a limited but important role. It produces fronto-naso-orbital skin.
- Mesencephalic neural crest (MNC) arises from the neural folds above the midbrain proper (an enormous structure in the embryo) and from rhombomeres r0 and r1. MNC departs and the midbrain develops—in caudal-cranial order—as *individual streams* and migrates forward beneath the nonneural ectoderm outside the brain where it is responsible for the synthesis of forebrain dura, anterior cranial fossa, the sphenethmoid complex, the primary nasal cavity, and the superomedial orbit.
- Rhombencephalic neural crest (RNC) arises from the neural folds above the hindbrain. RNC migrates in strictly defined *segments* into the pharyngeal arches. It produces the remainder of the inferolateral orbit, supporting bones, fascia, the caudal components of the nasal chamber, (e.g., vomer, premaxilla, inferior turbinate, and hard palate), the oropharynx, and larynx.

A detailed description regarding the anatomy of these populations, how they relate to the neuromeric system, and how isolated defects in a component population lead to a craniofacial cleft is the subject of this chapter. Our model will focus on the fronto-naso-orbito-maxillary complex—as this is where facial clefts occur. We shall examine first how the bone fields form and then turn our attention to the soft tissue anatomy of the nose and mouth. The interactions between underlying bone fields and soft tissue fields that characterize the common labiomaxillary cleft will be detailed as these provide a common biologic model for craniofacial cleft formation in general.

Zygomatico-Maxillary Complex

The ZMC consists of the upper jaw and the r2 bones that constitute its medial wall (palatine, inferior turbinate) and connect it with the face.

Melding of individual fields to form the face is *not* a slow process. Although the four types of neural crest that migrate into the head and neck (RNCc, MNC, and RNCc, PNC) complete their migrations at different times of embryonic development, all neural crest are in place within a matter of 15 h or so. These components of the future face are initially located in parts of the embryo that are widely separated. The PNC and MNC are positioned together initially around the brain, while the rostral RNC is found in the pharyngeal arches. Furthermore, by the time PNC has populated the future frontonasal zone of the embryo (Carnegie stage 10), reorganization of the first 11 somitomeres into pharyngeal arches and occipital somites is complete. These quite disparate components must be physically approximated in order for facial assembly to occur. Cephalic folding accomplishes this goal in less than 24 h.

Human embryos at day 22 have cephalic neural folds that are broad and thick. These stick up in the air like the fins of some ancient Cadillac limousine. The reason for this neural fold projection is the tremendous proliferation of head mesenchyme lying just beneath them. This mesenchyme is, of course, a product of the explosive growth of MNC. Somitomeres lie astride the future cranial base like saddlebags. Their individual contributions have been described.

The rapidly growing embryo lies atop a yolk sac that is not growing much at all. As the cephalic part of the embryonic axis expands, its yolk sac "tether" forces the neural plate to bend at specific sites. The first of these flexures occurs at the site of the future mesencephalon. At day 22, the *cranial (mesencephalic) flexure* bends the prosencephalon ventrally toward the pharyngeal arches. In less than 24 h, the angle between the forebrain and the rest of the neuraxis decreases from >150° to <100°. Voila! The first and second pharyngeal arches now have ready access to the frontonasal mesenchyme. Although these mesenchymal masses are covered with epithelium, when contact is made between the frontonasal and lateral masses, epithelial fusion quickly ensues. The underlying mesenchymal fields can now interact.

What does the field geography look like at the time of arrival of the pharyngeal arches? Let's pretend we are standing in front of the embryo, directly opposite the future embryonic mouth. In front of us on either side we see the combined first and second arch complexes containing all the future bone and muscle fields of PA1 and PA2. The mesenchyme of these arches has had ample time to fuse, thus forming the first-second arch "sandwich." The upper (maxillary) half of our sandwich contains all the bone fields produced by r2 and r4 RNC. Although the maxillary mesenchymal mass seems just a shapeless blob, in reality these fields are all lined up in precise spatial order, ready to develop into specific bone and soft tissue structures. Superficial to these bone fields lie the blastema of the future facial muscles produced by Sm5 PAM with the fascia provided by r5 RNC. Laid out in exactly the same manner is the lower (mandibular) half of our sandwich, the bones and muscles of which are made from the same precursors. A hidden "fault line" exists in the pharyngeal arch separating the soft tissue mesenchyme of the maxilla and the mandible. In the Tessier #7 cleft (lateral orofacial cleft), the soft tissues are divided by a fissure extending from the oral commissure back to the ear.

In the neuromeric model, the development of each pharyngeal arch follows certain hypothetical rules. We shall discuss them and see how they are applied to the formation of the face. (1) Tetrapods possess five pharyngeal arches. The fate of the original sixth and seventh arches in tetrapods is to become transformed into the upper neck (c1–c4). (2) Aortic arch arteries are formed by the fusion of a bud from the dorsal aortae and a bud from the ventral aortic outflow tract. (3) In each arch, blood supply becomes preferentially available to the caudal mesenchyme such that the distal end of each pharyngeal arch is first to develop. (4) In each arch, the bone derivatives of the caudal and distal sectors will form earlier in time than those of the cranial and proximal sectors.

Initially, the pharyngeal arches project outward from the embryo like sidearms. Explosive growth of the PA1/PA2 arch complex takes place at the very same time as the embryo undergoes head flexion. This differential growth causes the sidearms to physically become repositioned toward the ventral midline of the embryo. This takes place concomitant with massive apoptosis of the MNC nasoethmoid complex. In so doing, they come into a "docking position" below the frontonasal mesenchymal mass. The PA1/PA2 field complex locks on to, and then interacts with the previously constructed fronto-naso-orbital mass. Thus, preexisting pathology of FNO fields can impact upon the development of the RNC fields.

The concept that a given field is the prerequisite for proper development of subsequent fields has a sound experimental basis. Neural crest cells respond differently according to their epithelial environment. In the presence of pharyngeal endoderm, NC will form cartilage, whereas, when in contact with ectoderm, NC forms membranous bone. Furthermore, neural crest-derived cartilages can serve as precursors of membranous bone. When strips of foregut endoderm were removed in chick embryos, specific cartilaginous bones failed to develop. Adjacent neural crest membranous bones normally destined to ossify later on from PNC or PA1 also failed to develop. Absence of Meckel's cartilage (subsequently the quadrate and articular bones) led to developmental failure of the pterygoid, quadratojugal, angular, supra-angular, opercular, and dentary bones.

We shall now describe how MNC and RNC become physically positioned. The facial midline consists of three complexes of fields.

- A fields: MNC fronto-naso-orbital skin and bone fields are supplied by V1 stapedial branches of ophthalmic artery.
- B fields: $RNC_{ROSTRAL}$ mesenchyme forms the jaws and supporting bone fields; it is supplied by V2 and V3 stapedial branches of internal maxilla-mandibular artery.
- C fields: $RNC_{ROSTRAL}$ mesenchyme forms all remaining structures of the first and second arch complex (except the jaws) and is supplied by branches of external carotid artery.

The <u>maxillary fields</u> are assembled from seven populations (fields) of r2 neural crest, all of which sweep forward toward the developing face in a strict spatiotemporal order according to their site of origin on the neural fold. Their target, the r0/r1 primordium of the sphenethmoid and orbit, is already in place. The blood supply to each field comes from the stapedial V2 branches of internal maxillary. *The physical anatomy of these arteries, the order in which they form the internal maxillary axis, replicates the spatiotemporal order of the fields they serve.* First to arrive is the r2′ premaxillary and vomerine MNC supplied by the terminal branch of the IMA, the *medial sphenopalatine artery.* (All subsequent fields come from r2 proper.) Next on the scene is the r2 inferior turbinate field supplied by the *lateral sphenopalatine artery.* Behind IT comes the palatine bone supplied by the *greater palatine artery.* Note the horizontal plate of palatine bone and the anterior palatine aponeurosis are supplied by *lesser palatine artery.* Mx1, Mx2, and Mx3 follow in succession, each supplied by their respective superior alveolar branches off the *infraorbital artery. Zygomatic artery* arises from infraorbital and supplies the latera wall of the orbit.

Construction of the <u>malar fields</u> occurs around the axis of the zygomatic nerve. In zoologic terms, the zygoma has a cranial field, the *postorbital bone*, and a caudal field, the *jugal bone.* Persistence of this transverse separation is occasionally seen as the *os japonicum* [142]. The post-orbital bone articulates with the zygomatic process of the frontal bone to form the lateral orbital rim. Isolated failure of the field is the Tessier #8 cleft. It also articulates with the posterior maxillary wall Mx3. Hence, the association between the Tessier #6 cleft with the #8 cleft. The more caudal field, the jugal bone, bridges between r3 zygomatic process of the squamous temporal bone and the maxillary buttress above the first molar. The zygoma is an example of a neural crest derivative that begins as a cartilage and then is converted into membranous bone. The ossification process is exactly analogous to that of the coronoid process of the mandible.

Connecting the ZMC with the Cranium

Previously, we have seen how brain growth forces the embryo to flex. This brings the cranial base of the anterior fossa into contact with pharyngeal arch mesenchyme to assemble the face. Interaction between A and B fields at critical contact points results in horizontal (lateral to medial) and vertical (cranial to caudal) approximation. The tissues for the letter "T." These processes are well depicted in the SEM work by Hinrichsen and schematically illustrated by this author in a previous communication [143].

The first relevant A-B contact is between the r1 sphenethmoid mesenchyme and the r2 premaxillary mesenchyme. As the future PMx/V fields travel toward the face, they encounter a physical obstacle, the previously synthesized r1 back wall of the orbit. Unable to advance further, they are forced ventrally. They duck beneath the orbit, seeking the midline. PMx and V pile up beneath the presphenoid. They then "see"

their respective ethmoid lamina and beneath which they "hitch a ride" to their final destination. Apoptosis of the MNC ethmoid mesenchym in the center of the face causes the nasoethmoid-premaxillary-vomerine masses to approximate each other. They eventually fuse, uniting the facial midline. Failure of this to take place is the basis of hypertelorism.

The external appearance of the early embryonic face is dominated by huge disc-like nasal placodes made from p6 epithelium. Placodal adherence to the brain is the key to understanding the formation of nasal cavities. Rapid proliferation of MNC mesenchyme surrounding the p6 placodes forces the surrounding skin to be pushed forward creating a "hemi-nose." The topology of this process can be envisioned by a humble analogy. The tip of an elastic structure shaped like a condom is glued to a flat surface. The peripheral rim is likewise glued. A needle is then placed and the structure is insufflated. The central disc represents the placode, while the periphery is the facial skin. A donut-like chamber results.

In the hemi-nose, the internal skin is p6 while the surrounding outer skin is p5. The heminasal chambers approximate in the midline and fuse. Into the common p6 medial wall mesenchyme of the future nose, the r1 perpendicular ethmoid plate and septum develop. The process of nasal fusion takes place from inside-out/back-to-front. Thus, the vomerine bones approximate from the sphenoid forward. The two premaxillae follow suit. The presence of bifid frenulum and a wide diastema between the central incisors are forem fruste signs of inadequate premaxillary approximation (the Tessier #0 cleft). The process of palatal development is vividly depicted by Kaufman [91].

Various nasal anomalies can occur from defective embryogenesis. Absence of a nasal placode will lead to hemi-nose. Very rarely, complete nasal duplication is seen. This is likely due to additional nasal placode on either side of the midline. A notch in the nasal rim (sometimes with defect between the central and lateral incisors) is the Tessier #1 cleft. This represents a "fault line" in the soft tissues of the nasal roof between the medial nasal process and the lateral nasal process.

(A + B) + C contact between the maxilla and orbit follows a similar closure pattern. At the postero-lateral corner of the nasal cavity, ascending processes of the r2 palatine bone make contact with the p5 orbit and the r1 sphenoid. This represents a "hinge" for what will be a lateral to medial rotation. When this is complete, closure of the palate can take place. This requires two contact points. Proliferation of premaxilla and Mx1 provides the *first contact point* between the maxilla and the midline. The frontal processes of these two fields (PMxF and MxF) ascend to make contact with the r1 nasal and lacrimal bones. Lamination between frontal process of maxilla and the lacrimal provides a potential space by which the lacrimal duct gains access to the nose. Successful contact between PMxF and MxF positions the internal aspect of the respective alveolar processes in space such that fusion of the primary palate can occur. This is initiated just in anterior to the nasaopalatine nerve and takes place *posterior to anterior.*

The second contact point is between the more proximal fields of both ethmoid complex and r2. Just behind the premaxillae, the vomerine bones represent neural crest mesenchyme that migrated a bit later than premaxilla. It is thus biologically "younger," and will ossify later than that of the premaxilla. The palatal shelf develops from Mx1 later in time than either the frontal process or the inferior turbinate. Palatal shelf projection and elevation take place in an anteroposterior sequence. Mx1 is developmentally "older" than Mx2 and Mx3; hence, it produces the shelf first. Successful contact between the vomer and palatal process of the maxilla takes place just posterior to the nasopalatine nerve. The fusion pattern is *anterior to posterior.*

Assembly of the Oronasal Soft Tissues

Time and again, we have emphasized that craniofacial osseous structures are mere by-products of soft tissue function matrices, i.e., of developmental fields that were preexistent in the embryo, are correctly positioned by folding, and interact in a tightly controlled time sequence to produce the recognizable anatomic features of the fetus. Why so much obsession with bones? What about the soft tissues?

The answer to the first question stems (in large part) from a plethora of experimental data pertaining to ossification patterns. Radiologic studies by Kjaer of these patterns constitute a treasure trove of incalculable worth for all those interested in craniofacial development. Most all the bones in question are neural crest derivatives. When the ossification sequence is combined with neuromeric compartments, a neural crest "map" of the embryo can be constructed. Such a map displays all bone fields organized along the neural folds into their respective neuromeric zones. Within each zone, the relative position of each bone field to its confreres is reconstructed.

Bone fields have readily defined sutures and ossification centers. Sutures represent truly separate compartments; neural crest cells have been demonstrated diving into every one. Bone fields allow us to think neuromerically about the overlying soft tissue. For example, we can position the lateral border of the p5 nasal skin envelope precisely over the interface between r2 frontal process of maxilla and the r1 nasal bone. The medial border of the cheek sandwich with r2 dermis lies directly over r2 MxF. In the palate, the tensor veli palatini originates from somitomere 4 and spans from the r3 lateral Eustachian tube to the r2 palatine horizontal lamina. The levator veli palatini originates from Sm6 and the Sm6

petrous apex. Levator forms later in time than tensor. TVP inserts into the lateral margin of the anterior palatine aponeurosis, an r2 derivative supplied by lesser palatine artery. LVP inserts the middle 1/3 of palatine aponeurosis, a derivative of r5–r6 neural crest supplied by ascending palatine branch of facial to the muscle and ascending pharyngeal to the fascia.

Muscle origins and insertions, from a neuromeric perspective, develop mathematically. The *origin* of a muscle corresponds to its somitomere or somite(s). After migration, the myoblasts, "packaged" by fascia, will seek out the nearest site of exposed collagen II. The *primary insertion* is to a bone that develops in the neuromeric zone as its motor nerve. The *secondary insertion* will occur at the first available bone in the surrounding environment displaying an ossification center.

Formation of the Normal Lip and Prolabium

As the nasal chambers move toward midline fusion, the r2 premaxilla is covered over by a layer of p5 skin and MNC mesenchyme. This becomes the columella and philtrum as follows [148]. Note the importance of BMP4 in midline fusion [149] (Fig. 2.77).

Recall that dorsal nasal skin has p5 dermis and MNC subcutaneous tissue. With r1 septal growth, the nasal tip rises and projects. This stretches the skin anterior to the septum to form the columella. The skin remaining atop the premaxilla, the prolabium is completely devoid of muscle. When the prolabium unites with the lateral lip elements, it encounters biplanar orbicularis. Superficial orbicularis will bind to the mesenchyme of prolabium, but cannot penetrate it. It stays bunched up on the sides to create the *philtral columns* of *Cupid's bow.* Deep orbicularis penetrates below the plane of prolabium and above the bone to unite with itself across the midline. How does this take place? How do the outlying maxillary fields gain access to the midline? What causes epithelial breakdown such that these skin-covered fields can fuse with each other?

At Carnegie stage 13, no true mouth-nose distinction can be made. Breakdown of the floor of the primitive nasal cavities occurs at stage 14, thus creating the primary choane. During this stage, the maxillary process fills out the lateral nasal process with r2 mesenchyme, while the medial nasal process is filled by MNC mesenchyme. Beneath these soft tissue structures lies bony support: That of the lateral nasal process is Mx1 (Tessier zone 4), while the support for medial nasal process is premaxilla. Between stages 15 and 16, an epithelial edge emanates from the LNP termed "Simonart's band" and attaches to the MNP. Multiple studies confirm that this process involves a burst of oxygen consumption and RNA synthesis at the alar base. In mice, fusion of lateral to medial nasal process takes just 4 h. Oxygen deprivation or environmental exposure, such as carbon monoxide, can interfere with this critical event [150].

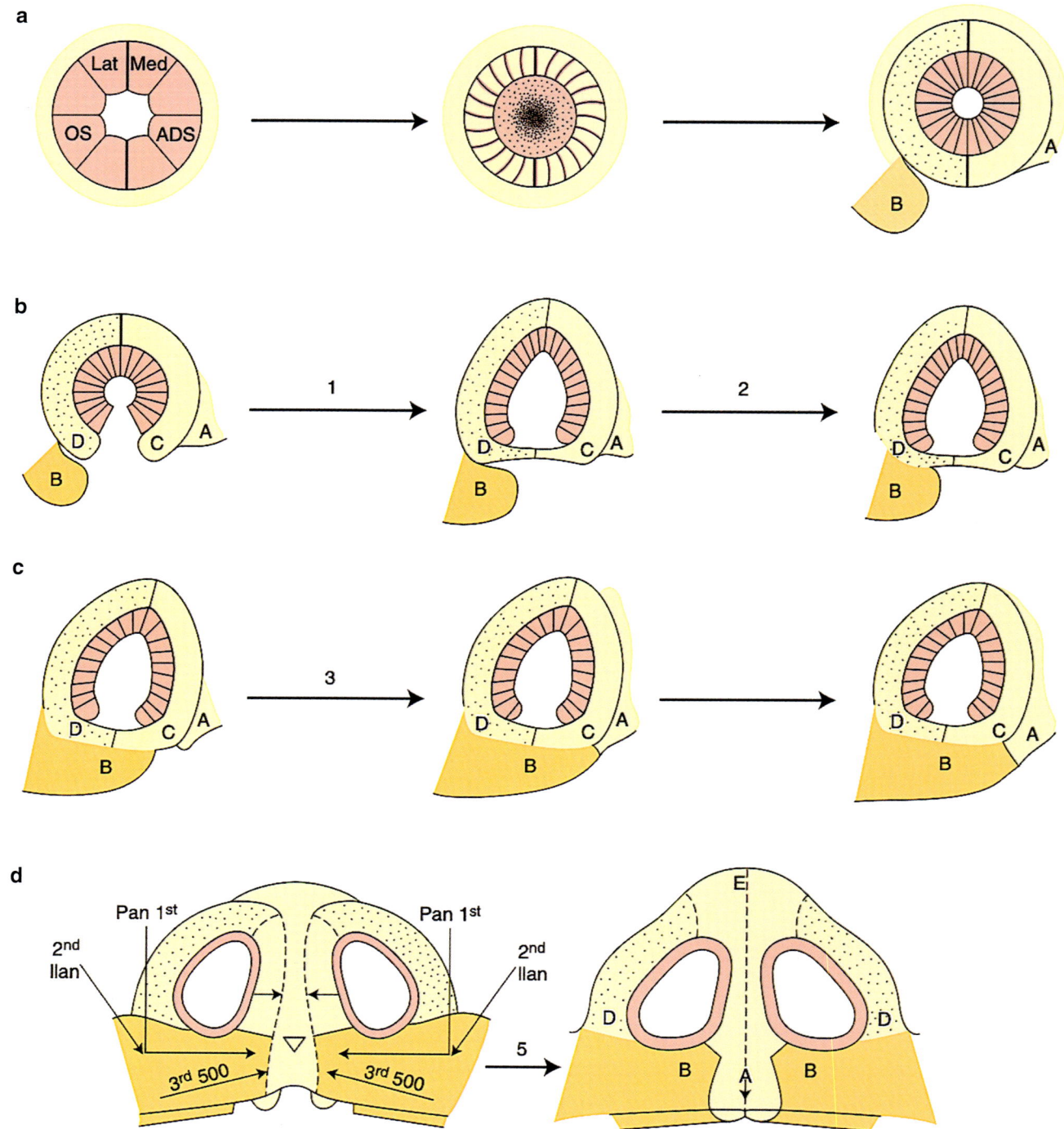

Fig. 2.77 Lip closure sequence. (Courtesy of Michael Carstens, MD)

persistence of epithelium, and failure of mesenchymal fusion. Since the signal emanates from the PMx, it diffuses down from the piriform fossa to the lip. Therefore, as the BMP-4 signal is progressively weaker, lip cleft severity worsens: i.e., from a vermilion notch, to an incomplete form involving half of the lip, and finally to a complete cleft lip. Variations of this mechanism are the likely basis for the final pathologies seen in the entire spectrum of Tessier clefts [151, 152] (Figs. 2.81 and 2.82).

In summation, *the volume of the premaxilla determines whether or not lip closure can occur*. First, small premaxillae make small amounts of BMP-4. The amount of BMP-4 produced is critical to produce the epithelial breakdown necessary to permit mesenchymal merger. Second, when the premaxilla is too small, the physical distance between it and maxilla will exceed a critical dimension. Epithelial bridge formation between the alar base and the prolabium cannot occur. Third, if this critical distance exists at the

Concomitant with the biochemical activity of the LNP, the underlying maxillary bone fields are developing as well. The zone #3 inferior turbinate develops immediately behind zone #4 Mx1. Building on the scaffolding of IT, the maxillary frontal process is synthesized. Construction of neural crest bone can be monitored by BMP-4. The cellular mass of Mx1 and PMx determines the transverse distance between LNP and MNP. Work by Johnston in cleft-forming rodents demonstrated a consistently *abnormal angle* between MNP and LNP. When a critical transverse distance between these fields exists, bridge formation will fail and a soft tissue cleft results. Reduction in physical size of the premaxilla can also cause the *critical contact distance* to be exceeded.

Prolabium is constructed from p5 skin and MNC mesenchyme; it has a vermillion consisting of r2 mucosa covering the premaxilla. The mucosa has an odd "flaky" appearance for the lack of underlying muscle. Epithelial breakdown allows maxillary myoblasts to gain access to the prolabium. In accordance with other pharyngeal arch derivatives, first arch muscle maturation follows a strict sequence: deep-to-superficial, caudal-to-cranial, and lateral-to-medial. The orbicularis muscles, being very medial, are late-forming (compared to platysma). The deep (sphincter) layer of orbicularis (DOO) forms well before the superficial (dilator) layer. DOO shows common characteristics with the buccinator. Both belong to the deep plane and develop in contact with the oral mucosa. They are innervated by VII from above. Because DOO is programmed by the mucosa, it curls around the lip but terminates at the white roll. SOO forms later in time. It makes physical contact with the p5 prolabial mesenchyme with which it fuses. For this reason, when one pares the edge of a unilateral cleft prolabium only a single muscle layer is observed: DOO from the opposite, non-cleft side.

The Pathologic Anatomy of Cleft Formation

The pathologic anatomy of unilateral and bilateral labiomaxillary clefts stems from a tissue deficiency state localized to the lower lateral piriform fossa. The developmental field at fault is the premaxilla. Such clefts always have an osseous component consisting of a scooped-out nasal floor. The extent of bone involvement is variable, up to and including a complete cleft of the primary palate. Soft tissue involvement occurs likewise as a spectrum (Figs. 2.78, 2.79, 2.80, 2.81, 2.82, and 2.83).

Formation of the premaxilla results from interactions between these tissues. The premaxilla has several anatomic subcomponents; these are assembled in a strict sequence. The medial incisor field (PMxA) forms first, followed by the lateral incisor field (PMxB). This can be understood as the "flow" of neural crest mesenchyme forward from the ipsilateral vomer. The time sequence of dental eruption (central incisor A > lateral incisor B) is a manifestation of the relative biologic "maturity" of the mesenchymal field within which

each tooth develops. The frontal process field (PMxF) is a vertical offshoot of PMxB; this subfield is the biologically "newest" tissue. These zones of premaxilla correspond to the vascular axis of medial nasopalatine artery.

Pathology affecting the premaxilla occurs as a spectrum based on this original developmental pattern. Deficits in the neuroangiosome leading to a deficiency state of the premaxilla first occur in the most distal aspect of the frontal process (i.e., at its most cranial extent). As the mesenchymal deficit worsens, frontal process will be reduced in a cranial-caudal gradient. "Scooping out" of the piriform rim results; the nasal lining is pulled down as well. This causes depression of the alar base and a downward-lateral displacement of the lateral crus. Biologic signals from PMxF do not affect lip formation. Therefore, the *forme fruste* manifestation of premaxillary deficiency is a cleft lip nose with a perfectly normal lip.

Once the frontal process is eliminated, the deficiency state shows up in the lateral incisor field. Progressive degrees of premaxillary deficiency in the lateral incisor field cause progressive loss of alveolar bone substance. Alveolar bone development follows a gradient from the incisive foramen forward. Mild deficiency causes notching on the labial surface. As the deficiency worsens, the notch deepens *backward* toward the incisive foramen. A critical lack of alveolar bone mass results in outright failure of lateral incisor development.

The reader should now be quite comfortable with the mechanisms by which bone is affected in a cleft. But a little more detail is required to spell out how a deep osseous field can affect the overlying soft tissue. Tessier's series includes colobomas, eyebrow absence, as well as skin deficits. It is herein proposed that the soft tissue deficits seen in craniofacial clefts represent either failures of fusion or failures of formation. The notch in the alar rim is a boundary zone fusion failure, whereas the absence of eyelashes on the lateral lower eyelid represents the failure of that field to produce a product. Neural crest mesenchymal cells that participated in lash formation in an adjacent field fail to behave correctly in the target field. Failure of formation is a more difficult topic to discuss (and out of our scope here), but the mechanism of fusion may well be universal throughout the head and neck.

Let's take the role of the LNP-MNP nasal bridge as a case in point. Migration of myoblasts containing mesenchyme along the Simonart's band cannot occur without a generalized breakdown of the epithelium covering the lateral lip element, the skin bridge, and the p5 prolabial skin. Stability of the epithelia in facial processes is maintained by repression activity of *Sonic Hedgehog* (*SHH*) within the skin. BMP-4 causes derepression of *SHH*; epithelial breakdown results. Thus, absence or deficiency of an appropriate BMP-4 signal will lead to restricted expression of SHH, abnormal

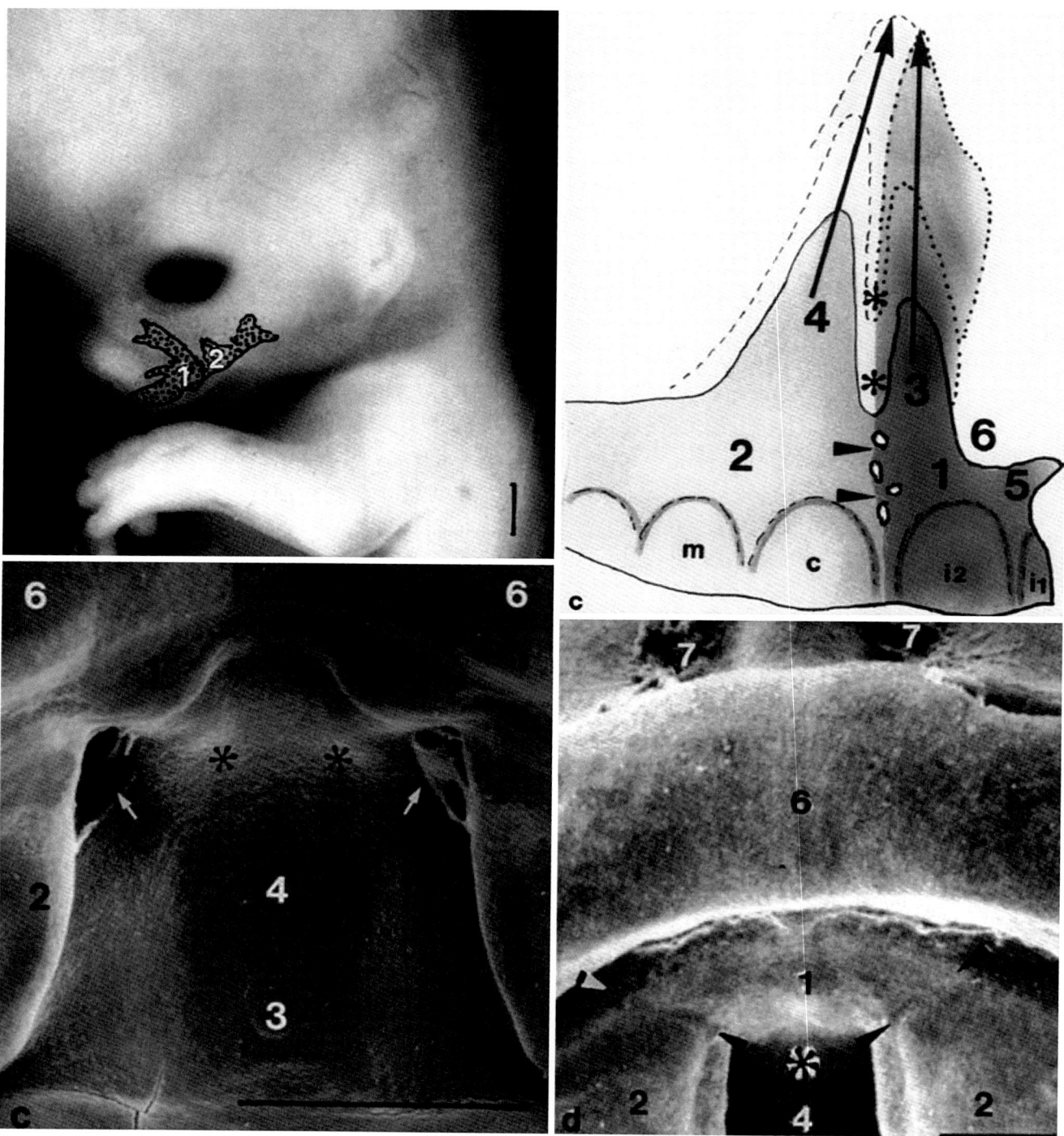

Fig. 2.78 Premaxillary frontal process seen before it becomes overlapped by the frontal process of the maxilla. <u>Upper left</u>: embryo stage 23. <u>Upper right</u>: Reconstruction upper jaw in a human fetus in the fifth month. Schematic drawing to demonstrate the progressive growth of the facial processes from maxilla as well as premaxilla including the fusion line. 1. Premaxilla, 2. maxilla, 3. processus frontalis premaxillaris, 4. processus frontalis maxillaris, 5. processus stenonianus/spina nasalis, 6. apertura piriformisremaxilla, i1 medial incisor, i2 lateral incisor, c canine, m molar, arrowheads former sutura incisiva, *sutura incisiva, long arrow direction of growth towards the os frontale. <u>Lower</u> <u>left</u>: Embryo stage 19. 1. Premaxillary anlage/primary palate, 2. secondary palate, 3. adenohypophysis/remnant of Rathke's pouch, 4. roof of the oral cavity, 5. mandibular arch and floor of the oral cavity removed, 6. upper lip anlage, 7. nasal plug, *intermaxillary bulge at the ventral roof of oral cavity, short arrows choanae, arrowheads sutura incisiva between primary and secondary palate. <u>Lower right</u>: Embryo stage 23. (Reprinted from Bartezco K, Jacob M. A re-evaluation of the premaxillary bone in humans. Anat Embryol (Berl) 2004 Mar;207(6):417–37. With permission from Springer Nature)

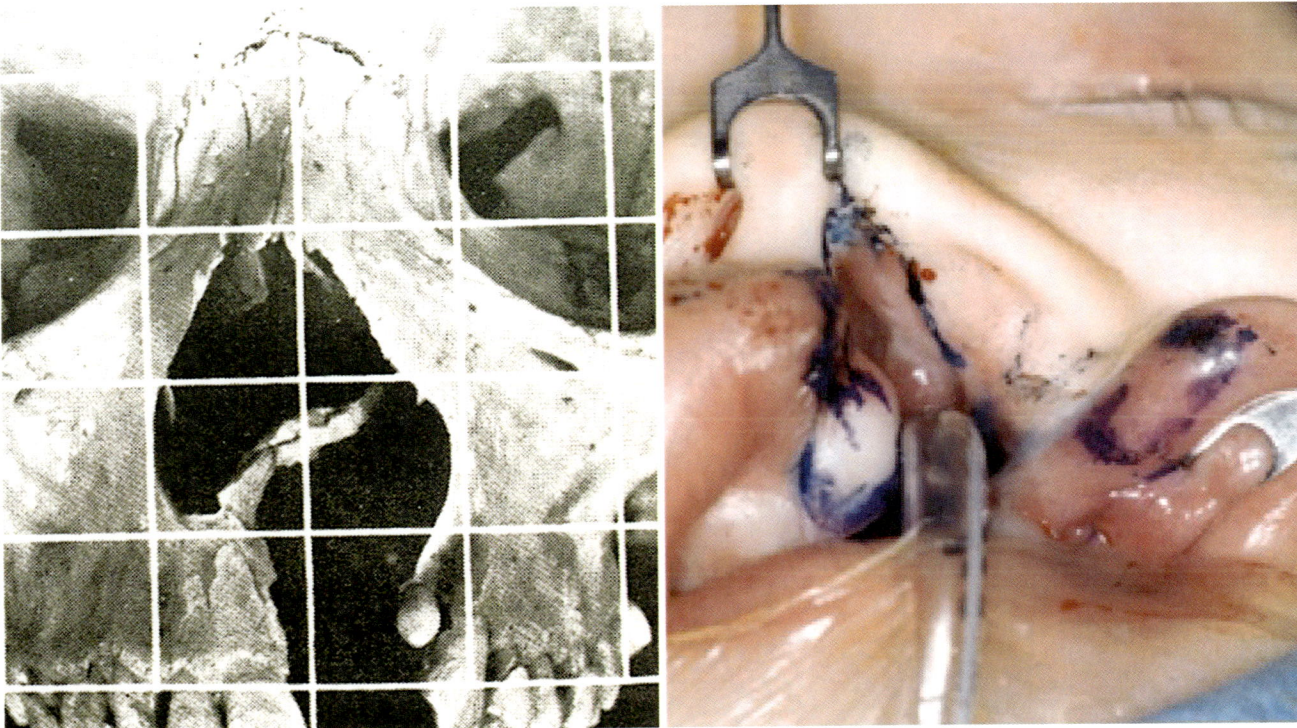

Fig. 2.79 UCL soft tissue and bone defects. Note the scooping out of the piriform fossa on the side of the cleft. Note as well that the maxilla is capable of producing an incisor which is not really "ectopic," but rather constitutes part of three dental units supplied by the medial branch of anterior superior alveolar neuroangiosome. Soft tissues are displaced downward and laterally to fit into the piriform fossa. (Courtesy of Michael Carstens, MD)

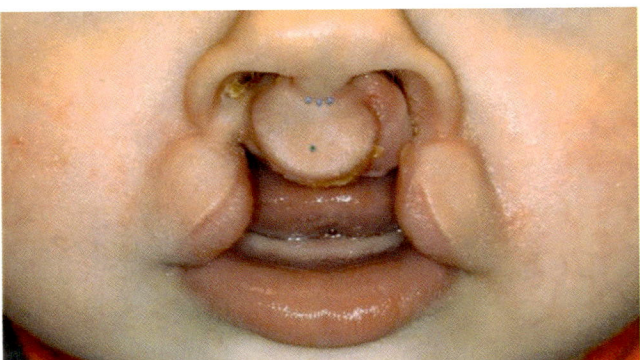

Fig. 2.80 Bilateral cleft lip showing the prolabium to consist of four neurovascular fields. <u>Philtral prolabium (PP)</u>: V1 anterior ethmoid: these two fields are the width of the columella. <u>Non-philtral prolabium (NPP)</u>: V2 medial sphenopalatine: these fields flank the PP. (Courtesy of Michael Carstens, MD)

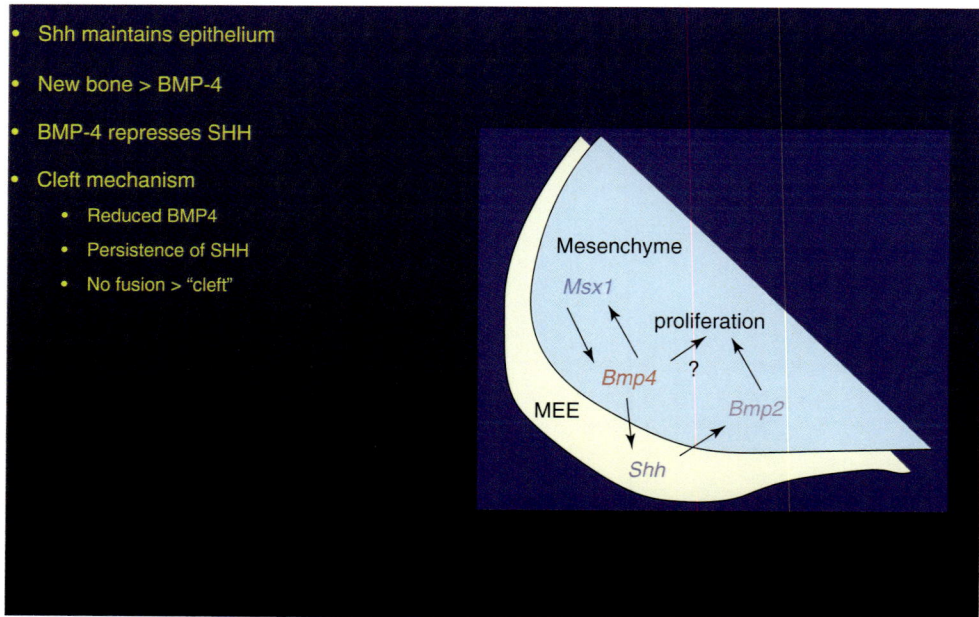

- Shh maintains epithelium
- New bone > BMP-4
- BMP-4 represses SHH
- Cleft mechanism
  - Reduced BMP4
  - Persistence of SHH
  - No fusion > "cleft"

Mesenchyme

Msx1

proliferation

Bmp4

?

Bmp2

MEE

Shh

Fig. 2.81 BMP4/SHH interaction controls lip fusion. Epithelial integrity/cohesion is regulated by signals from the underlying mesenchyme. Epithelial cohesion depends on local expression of Sonic Hedgehog (Shh). Osteogenesis of membranous bone underlying the epithelium involves expression of BMP-4; the protein diffuses outward from the piriform fossa into the soft tissues and thence downward into the lip. The most distal target of BMP-4 in upper lip is the vermillion border. BMP-4 represses expression of Shh and permits epithelial disintegration. Mechanism of cleft lip: (1) Fusion of the lateral lip element to the prolabium involves breakdown of epithelium which requires presence of adequate BMP-4. (2) Reduced volume of underlying membranous bone > reduced (BMP-4). (3) This permits the persistent expression of Shh in epithelium of the lateral lip element and/or prolabium. (4) Intact epithelium resists fusion processes—creating a "cleft" which proceeds from caudal to cranial and deep to superficial. (Reprinted from Zhang Z, Song Y, Zhao X, et al. Rescue of cleft palate in Msx-1 deficient mice by transgenic Bmp4 reveals a network of BMP4 and Shh signaling in the regulation of mammalian palatogenesis. Development 2002; 129:4135–4146. With permission from the Company of Biologists)

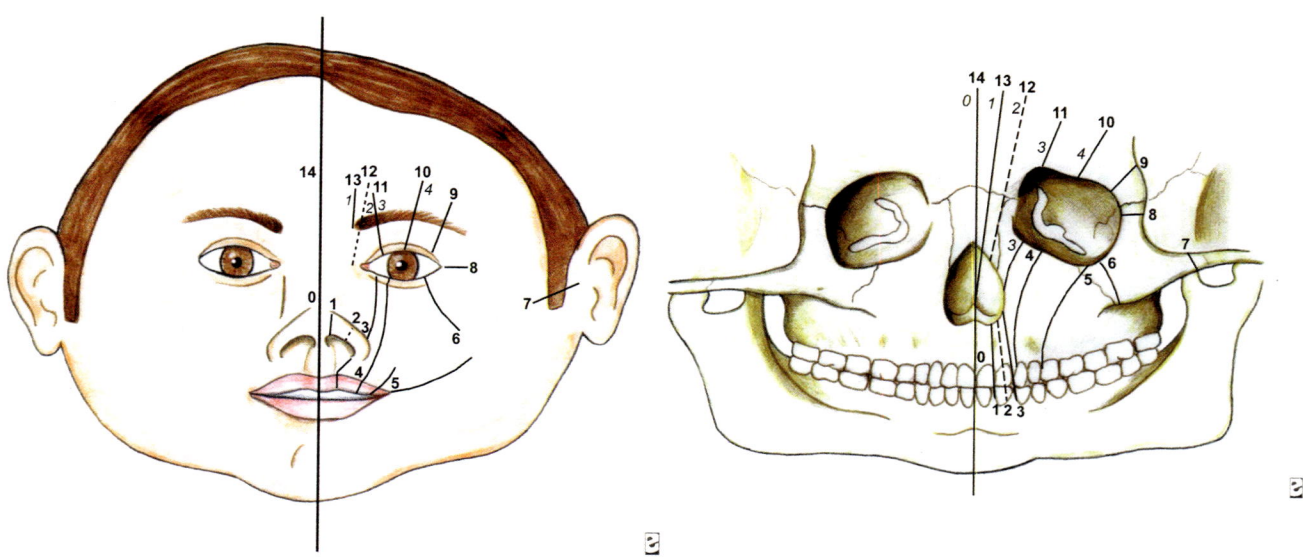

Fig. 2.82 Paul Tessier's numeric classification of craniofacial clefts. This model recognized two tiers, above and below the orbit. Maxillary clefts are numbered 0–7, while orbitofrontal clefts are numbered 8–14. Empirically, the two zones display pairing in which the sum of the lower and upper clefts is 14, the system has several minor flaws. It does not distinguish between states of field deficiency and field excess. Lumping together of developmental fields occurs thus, the common cleft involving PMx belongs to zone #2 rather than the inferior turbinate zone #3. Clefts in zone #3 are much more devastating because they involve an entirely different mechanism. (Reprinted from Sari E. Tessier Number 30 Facial Cleft: A rare maxillofacial anomaly. Turkish J Plast Surg 2018; 26: 12–19. With permission from Turkish Journal of Plastic Surgery)

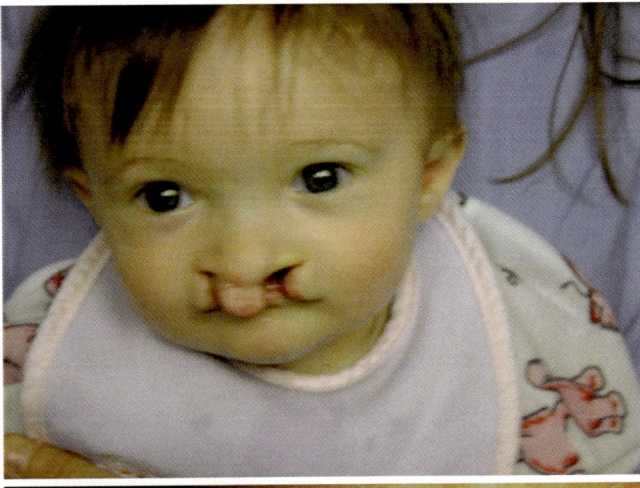
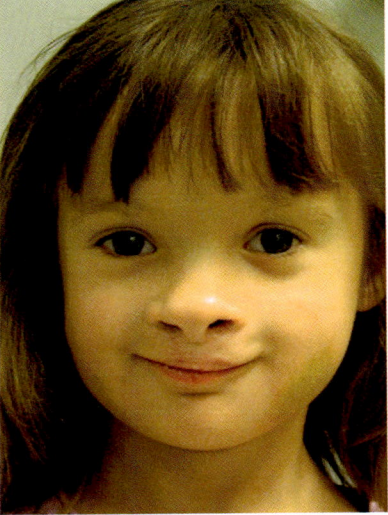

5 years
no NAM

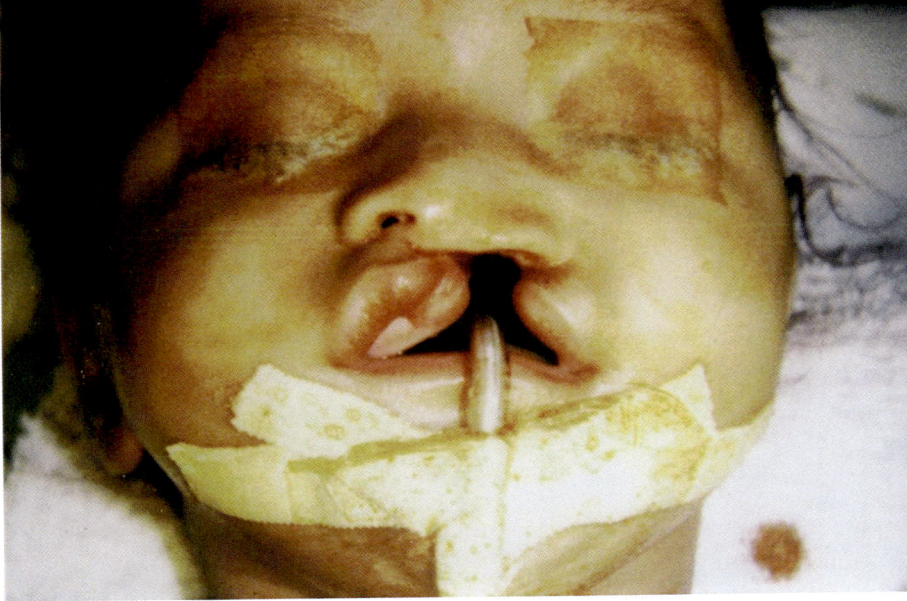
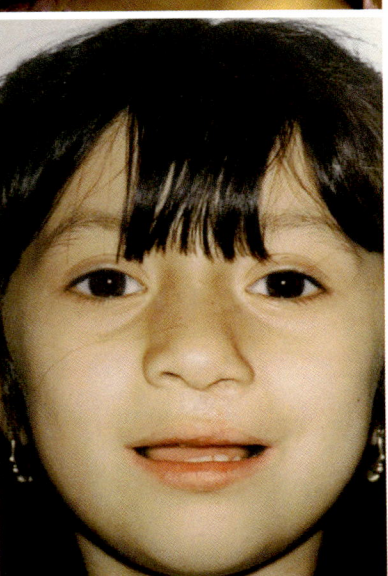

Fig. 2.83 Developmental field reassignment. Repositioning fields into their embryonic relationships facilitate formal symmetrical facial growth over time. Upper left and right: secondary repair. Lower left and right primary repair. (Courtesy of Michael Carstens, MD)

level of the incisive foramen, a cleft of the secondary palate will form. This is because the horizontal repositioning of the palatal shelf from the maxilla must make contact with the vomer just posterior to the incisive foramen. The process is just like a zipper. If initial contact is not made, fusion of the palatal shelf to the vomer cannot take place. Even if initial contact is made, a secondary palatal cleft can still result due to displacement of the vomer away from the midline. The vomer can become warped by the inequality of growth forces on either side of the cleft. Thus, the zipper may get started anteriorly, but as it proceeds posteriorly, when it encounters the deviated vomer, a palatal cleft will ensue.

Clinical Consequences of the Developmental Field Model Applied to Facial Clefts

- The repair of facial clefts involves the identification and repositioning of misplaced fields.
- Fields should be dissected to conserve the entire neuroangiosome basis—virtually no tissue should be discarded.
- Such fields should, when replaced into their proper anatomic relationships, grow symmetrically making use of the entire mesenchymal substrate.
- Reconstructive procedures based on embryologic principles offer the best chance to maintain near-normal growth and preserve their esthetic relationships.

How to Understand the Assembly of the Face: A Method of Study

Understanding of how component fields of paraxial mesoderm and neural are assembled together to form the face and skull is not easy task. It involves visualizing structure arising from one anatomic site and then moving into new positions via migration or folding. This is a *four-dimensional process*. The physical development of certain fields may be dependent upon the correct development of precursor fields. Errors in this developmental sequence lead to varying forms of craniofacial clefts.

Craniofacial development is thus like a complex play in several acts. The challenge for both author and reader alike is to set the stage wherein all the actors are properly introduced; their respective roles and relationships with one another are clearly presented. Then and only then can the action begin and the plot unfold. Shakespeare understood this well. His plays commonly begin with a *dramatis personae*. After the characters are introduced, the temporal logic of the acts must be presented. When and where does each act take place? Now the reader is ready to follow the dramatic action.

Step 1: Summary of Ideas

Major ideas introduced by previous papers in this series include: (1) the segmental organization of the early embryo based upon the neuromeric system, (2) the mechanism by which gastrulation forms a trilaminar embryo, (3) the anatomy of the resulting germ layers, including the neural crest, (4) the segmental reorganization of mesenchyme into somitomeres and somites, (5) formation of pharyngeal arches and neural crest migration patterns, and (6) derivative analysis of final anatomic structures.

Step 2: Key Definitions and Point of Clarification

Craniofacial development is an interdisciplinary study. Interested readers may find themselves entering an uncharted territory between the basic science laboratory and the operating room. Several contemporary texts are worth consulting. The reader should review the definitions and anatomic abbreviations presented at the beginning of part 1.

Step 3: Staging of Embryos (General Principles)

Embryology is the analysis of relationships between order and form. The timing with which anatomic structures make their appearance is all-important to understand the overall process. Several parameters traditionally used to describe development in the first 8 weeks of life make the literature confusing. These parameters are: (1) time/age, (2) crown-rump length, (3) the observable number of somite pairs, and (4) Carnegie stage. Each system has its own degree of (im)precision and overlap.

Lumping the events of embryogenesis into weeks is quite imprecise. Critical events may take place within the space of a single day. Fusion between the lateral nasal prominence and the medial nasal prominence may occur in as little as 6 h. Embryo size (measured as C-R length in millimeters) is a more accurate measurement of maturation. The Carnegie staging system, originally described by Streeter and refined by O'Rahilly, categorizes development into a series of stages defined by the formation and maturation of key structures. The system derives its name from the leadership role played by the Carnegie Institution as a sponsor of embryological research in the early twentieth century. Each stage is tightly linked to C-R length, intervals of which are as small as 2 mm through stage 15 (3–36 days).

Morphologic studies correlating developmental stage with the number of observable somites are useful but are limited to a certain window of embryogenesis. Carnegie stage 9 (days 20–21) is defined by the appearance of the first 1–3 somite pairs. The first occipital somite, previously described by von Baer and others and confirmed by Huang, may indeed be present at Carnegie stage 8. Why the confusion? It turns out that the first occipital somite is incompletely segmented at its rostral end (i.e., from Sm7), but fully epithelialized at its caudal end. Thus, the boundary between somitomeres 7 and somite 1 is indistinct but that between somite 1 and somite 2 is complete. In this scenario, Carnegie stage 9 would include up to four somite pairs. Using this numbering system, formation of the 31st pair (observed between days 29 and 31) would define Carnegie stage 13. The bottom line is the human embryo.

Although the number of somite pairs ceases to be of importance in defining subsequent embryonic stages, these entities play a vital role in the segmentation of the spine and the organization of the trunk. In addition to four occipital somites, human spines are constructed of 8 cervical, 12 thoracic, 5 lumbar, and 5 sacral somites. Fusion of these latter to form the sacrum is evidenced by the presence of four individual foramina. The coccyx is constructed with an additional 2–5 somites bringing the total count in humans to approximately 37–40 pairs. The final somite bearing a myotome is the first coccygeal; therefore the total number of spinal nerves sums to 31 pairs. The time required for somitogenesis is 10 days.

Step 4: Carnegie Staging System: With Special Reference to the Head and Neck

The best way to follow the events of craniofacial formation is by Carnegie stage. Key aspects of this system are summarized below as they apply to specific anatomic sites (e.g., cranial base synthesis, the formation of the nose). Carnegie staging will be discussed and illustrated in depth in Chap. 3. Our purpose here is paint an overview of the developmental timeline so that the reader may refer to in through the course of the test. Measurements of crown-rump length are in millimeters. The beautiful SEM studies of embryos, both human and murine, constitute an excellent way to visualize these processes; the reader is encouraged to obtain a copy of this invaluable works and study them with care [153–156].

Major Themes of Craniofacial Development

First: The formation of mesenchyme is a more primordial event than the development of the brain. Gastrulation sets up the trilaminar embryo in 48 h during stage 7. Neural plate and folds appear in stage 8. From a teleological standpoint, proper protection of the nervous system must be ensured. Brain development must take place in the presence of a preexistent mesenchyme into which it can expand. The covering layers outside the CNS come from either neural crest NC or paraxial mesoderm PAM.

Second: Formation of mesoderm via gastrulation is a craniocaudal process beginning at the level of the tip of the notochord. The anterior extent of the notochord lies at the junction of the presphenoid and basisphenoid bones. This corresponds to neuromeric level r0. Mesodermal segmentation results as a response to gene signals emanating from different levels of the notochord. No definitive endoderm and intraembryonic mesoderm can be produced anterior to r0. All mesodermal and endodermal tissues anterior to the notochord represent forward migrations of tissues produced by gastrulation.

Third: The forebrain sitting in front of the notochord is an evolutionary "after-thought." It arises from induction signals from the tip of the notochord, the anterior visceral endoderm (just in front of the notochord), and from a zone of ectoderm located at the extreme anterior aspect of the trilaminar embryo. No mesoderm is associated with forebrain development. All mesenchyme associated with forebrain coverage comes via two mechanisms: (a) migration of neural crest from the caudal prosencephalic neural folds and (b) migration of midbrain neural crest from m1 to m2, and r0 to r1.

Fourth: The forebrain arises as a vesicular structure that, under normal conditions, becomes divided into two hemispheres. Only the forebrain and r0/r1 cerebellum have dura and it is derived exclusively from neural crest. There is no dura around the hindbrain. Dura coverage of the spinal cord begins at foramen magnum and is exclusively of paraxial mesoderm. Growth of the forebrain occurs in all directions such that the two hemispheres eventually surround the midbrain and hindbrain. *Intracranial dural anatomy can be mapped on the basis of sensory supply to neural crest from rhombomeres 1, 2, and 3.*

Fifth: Epithelium surrounding the brain is not stable without a supporting layer of mesenchyme. Neural crest from diencephalon creates the *dermis* underlying ectoderm above the skull (the future skin of the upper face and scalp). Neural crest creates the *submucosa* underlying the ectoderm of the oropharynx beneath the skull base.

Sixth: The brain case (neurocranium) has two primary units. The *membranous neurocranium* consists of dermal bones formed by signals (and some PAM), differentiates due to signals from the dermis above and meninges below to an intermediate layer of mesenchyme. *The chondral neurocranium* forward from the sella turcia is of neural crest origin. Backward from the sella turcica it all comes from PAM.

Seventh: By Carnegie stage 10 all sources of PAM for the growing brain have been synthesized. Eleven somitomeres have formed. The dorsal aortae and primitive hindbrain channels that constitute circulation of the embryo are derived somitomeres. Transformation of the last four (Sm8–Sm11) creates the occipital somites.

Eighth: By Carnegie stage 11, the physical provision of mesenchyme to the brain and face is complete. This requires four elements: (a) neural crest, (b) paraxial mesoderm for striated muscles, (c) pharyngeal arches; and (d) folding. Organization of paraxial mesoderm into somitomeres begins with notochord development in stage 8. In stage 9, all somitomeres contributing to the head are present. This includes the transformation of somitomeres 8–11 into the four occipital somites. All neural crest migration to the head (begun at stage 9 in mammals) is complete by stage 11. Pharyngeal arches first appear as when populated by neural crest from rostral hindbrain. Embryonic flexion brings all these mesenchymal sources into contact with the forebrain.

Summary

The idea of common neuromeric definition provides us with a new understanding of the anatomic rationale behind the bones, muscles, and fascia of the craniofacial skeleton. These structures, be they derived from neural crest, PAM or both, can be traced back to the level of the embryo from which its cells originated. The neuromeric system offers a unique perspective on craniofacial deformities. Pathologic states involving a particular neuromeric level can affect one or all of its derivatives. Deformities involving seemingly unrelated bones or muscles can be understood in terms of common neuromeric levels of origin. Genes found at multiple levels of the embryo may be mis-expressed within a single neuromeric level or on only one side of the body.

The purpose of this communication was to outline the principles of the neuromeric system and to describe the manner in which the craniofacial skeleton originates. Each bone was assigned to a neuromere(s) of origin using a color code. The time course of assembly of the craniofacial bones was discussed. In subsequent chapters, this information will be related to the clinical patterns of the common craniofacial conditions. The application of these principles to the rare craniofacial clefts described by Tessier enables us to understand that these pathologies are variations in the craniofacial field system.

References

1. Tessier P. Fentes orbito-faciales verticales et obliques (colobomas) completes et frustes. Ann Chir Plast. 1969;19:301–11.
2. Tessier P. Anatomical classifications of facial, craniofacial, and laterofacial clefts. J Maxillofac Surg. 1976;4:69–92.
3. Tessier P. Plastic surgery of the orbit and eyelids (trans: SA Wolfe). Philadelphia: Mosby Inc. (Masson); 1981.
4. Carlson BR. Human embryology and developmental biology. 6th ed. St. Louis: Mosby/Elsevier; 2020.
5. O'Rahilly R, Muller F. Human embryology and teratology. 3rd ed. New York: Wiley-Liss; 2001.
6. Gilbert SF, Barressi MJF. Developmental biology. 12th ed. New York: Sinauer/Churchill Livingstone; 2020.
7. Liem K, Bemis WE, Walker WF, Grande L. Functional anatomy of the vertebrates: an evolutionary perspective. 3rd ed. Belmont, CA: Thompson; 2001.
8. Kjaer I, Fischer-Hansen B. The prenatal human cranium. Copenhagen: Munsksgaard; 1999.
9. Tan PPL, Beddington RSP. The formation of mesodermal tissues in the mouse embryo during gastrulation and early organogenesis. Development. 1987;99:109–26.
10. Tan PPL, Trainor PA. Specification and segmentation of the paraxial mesoderm. Anat Embryol. 1994;189:379–90.
11. Tan PPL, Ziou SX. The allocation of epiblast cells to ectodermal and germ-line lineage is influenced by the position of the cells in the gastrulating mouse embryo. Dev Biol. 1996;1778: 124–32.
12. Tan PPL, Behringer RR. Mouse gastrulation: the formation of a mammalian body plan. Mech Dev. 1997;68:3–25.
13. Tan PPL, Parameswaran M, Kinder SJ, Weinberger RP. The allocation of epiblast cells to the embryonic heart and other mesodermal lineages: the role of ingression and tissue movement during gastrulation. Development. 1997;124:1631–42.
14. Tan PPL, Goldman D, Camus A, Schoenwolf GC. Early events of somitogenesis in higher vertebrates: allocation of precursor cells during gastrulation and the organization of a meristic pattern in the paraxial mesoderm. In: Ordahl CP, editor. Somitogenesis, part I. San Diego: Academic Press; 2000. p. 1–32.
15. Meier SP. Development of the chick mesoblast: morphogenesis of the prechordal plate and cranial segments. Dev Biol. 1981;83:4.
16. Jacobson AG. Somitomeres: the primordial body segments. In: Bellairs R, Ede DA, Lash JW, editors. Somites in developing embryos. New York: Plenum; 1986.
17. Jacobson AG. Somitomeres: mesodermal segments of the head and neck. In: Hanken J, Bk H, editors. The skull, Development, vol. II. Chicago: University of Chicago Press; 1993.
18. Meier SP, Tam PPL. Metameric pattern in the embryonic axes of the mouse. I. Differentiation of the cranial region. Differentiation. 1982;21:95–108.
19. Pourquie O. Segmentation of the paraxial mesoderm and vertebrate somitogenesis. In: Ordhal CP, editor. Somitogenesis, part I. San Diego: Academic Press; 2000. p. 165–70.
20. Bronner Fraser M. Rostrocaudal differences within the somites confer segmental pattern to trunk neural crest migration. In: Ordahl CP, editor. Somitogenesis, part I. San Diego: Academic Press; 2000. p. 279–96.
21. Meier SP. Development of the chick mesoblast: pronephros, lateral plate and early vasculature. Dev Biol. 1980;55:299–306.
22. Huang R, Zhi Q, Ordahl CO, Christ B. The fate of the first avian somite. Anat Embryol. 1997;195:435–49.
23. Huang R, Zhi Q, Patel K, Wilting J, Christ B. Contribution of single somites to the skeleton and muscles of the occipital and cervical regions in avian embryos. Anat Embryol. 2000;202:375–v383.
24. Müller F, O'Rahilly R. Segmentation in staged human embryos: the occipitocervical region revisited. J Anat. 2003;203:297–315.
25. Noden DW. Origins and patterning of craniofacial mesenchymal tissues. J Craniofac Genet Dev Biol Suppl. 1985;2:15–31.
26. Noden DW. Cell movements and control of patterned tissue assembly during craniofacial development. J Craniofac Genet Dev Biol. 1991;11:191–213.
27. Noden DM, Trainor PA. Relations and interactions between cranial mesoderm and neural crest populations. J Anat. 2005;207:575–601.
28. Jiang X, Iseki S, Maxxon RE, et al. Tissue origins and interactions in the mammalian skull vault. Dev Biol. 2002;241:106–16.
29. Kuratani S, Matsuo I, Aizawa S. Developmental patterning and evolution of the mammalian viscerocranium: genetic insights into comparative morphology. Dev Dyn. 1997;209:139–55.
30. Kuratani S. Craniofacial development and the evolution of the vertebrates: the old problems on a new background. Zool Sci. 2003;22:1–19.
31. Lumsden A, Keynes R. Segmental patterns of neuronal development in the chick hindbrain. Nature. 1981;337:424–8.
32. Lumsden A, Sprawson N, Graham A. Segmental origin and migration of neural crest cells in the hindbrain region of the chick embryo. Development. 1991;113:1281–91.
33. Krumlauf R. Hox genes and pattern formation in the branchial region of the vertebrate head. Trends Genet. 1993;9:106–12.
34. Lumsden A, Krumlauf R. Patterning the vertebrate neuraxis. Science. 1996;1996(274):1109–15.
35. Carstens MH, Chin M, Ng T, Tom WK. Reconstruction of #7 facial cleft with distraction assisted in situ osteogenesis (DISO): role of recombinant human bone morphogenetic protein-2 with Helistat activated collagen sponge. J Craniofac Surg. 2005;16: 1023–32.
36. Depew M, Lufkin T, Rubenstein JLR. The specification of jaw subdivisions by *DLX* genes. Science. 2002;298(5592):381–4.
37. Depew MJ, Simpson CA. 21st Century neontology and the comparative development of the vertebrate skull. Dev Dyn. 2005;235:1256–91.
38. Brand-Saberi B, Whiting J, Ebesperger C, Christ B. The formation of somite compartments in the avian embryo. Int J Dev Biol. 1995;40:411–20.
39. Christ B, Schmidt C, Huang R, et al. Segmentation of the vertebrate body. Anat Embryol. 1998;197:1–8.
40. Burke AC. Hox genes and the global patterning of somatic mesoderm. In: Ordahl CP, editor. Somitogenesis part 1. San Diego: Academic Press; 2000. p. 155–81.
41. Kemp TS. The atlas-axis complex of the mammal-like reptiles. J Zool (Lond). 1969;159:223–48.
42. Schoenwolf G. Cell movements in the epiblast during gastrulation and neurulation in chick embryos. In: Kellar R, Clark Jr WH, Griffin F, editors. Gastrulation: movements, patterns, and molecules. New York: Plenum; 1991. p. 1–28.
43. Lemire L, Kessel M. Gastrulation and homeobox genes in chick embryos. Mech Dev. 1997;67:3–16.

44. Graham A, Smith A. Patterning of the pharyngeal arches. BioEssays. 2001;23:54–61.

45. Evans DH, Piermarini PM, Choe KP. The multifunctional fish gill: dominant site of gas exchange, osmoregulation, acid-base regulation, and excretion of nitrogenous waste. Physiol Rev. 2005;85(1):97–177.

46. Gasser RF. Development of the facial muscles in man. Am J Anat. 1966;120:357–75.

47. Padget DH. The development of the cranial arteries in the human embryo. Contrib Embryol. 1938;32:205–61.

48. Noden DM. Development of craniofacial blood vessels. In: Feinberg RN, Silver GK, Auerbach R, editors. The development of the vascular system. Basel: S Karger; 1991. p. 1–24.

49. Ruberte J, Carretero A, Marcucio R, Noden DM. Morphogenesis of blood vessels in the head muscles of the avian embryo. Spatial, temporal and VEGF expression analyses. Dev Dyn. 2000;27:470–83.

50. Hiruma T, Nakajima Y, Nakamura H. Development of pharyngeal arch arteries in the early mouse embryo. J Anat. 2002;201(15):29.

51. Etchevers HC, Couly G, Le Douarin NM. Morphogenesis of the branchial vascular sector. Trends Cardiovasc Med. 2002;12:299–304.

52. Mukouyama Y-S, Shi D, Britsch S, et al. Sensory nerves determine the pattern of arterial differentiation and blood vessel branching in the skin. Cell. 2002;109:693–705.

53. Morris-Kay GM. Derivation of the mammalian skull vault. J Anat. 2001;199:143–51.

54. LeDouarin NM, Kalcheim C. The neural crest. 2nd ed. Cambridge: Cambridge University Press; 2001.

55. Hall BK. The neural crest and neural crest cells in development and evolution. 2nd ed. New York: Springer-Verlag; 2009.

56. Trainor P. Neural crest cells: evoution, development, and disease. San Diego: Academic Press; 2013.

57. Serbedzija GN, Fraser SE, Bronner-Fraser M. Pathways of neural crest migration in the mouse embryo as revealed by vital dye labeling. Development. 1990;108:605–12.

58. Serbedzija GN, Bonner-Fraser M, Fraser SE. Vital dye analysis of cranial neural crest migration in the mouse embryo. Development. 1992;116:297–307.

59. Osumi-Yamashita N, Ninomiya Y, Doi H, Eto K. The contribution of both forebrain and midbrain crest cells to the mesenchyme in the frontonasal mass of mouse embryos. Dev Biol. 1994;164:409–19.

60. Osumi-Yamashita N, Ninomiya Y, Eto K. Mammalian craniofacial embryology in vitro. Int J Dev Biol. 1997;41:187–94.

61. Couly GF, Le Douarin NM. Mapping of the early neural primordium in quail-chick chimeras I. Developmental relationships between placodes, facial ectoderm, and prosencephalon. Dev Biol. 1985;110:422–39.

62. Couly GF, Le Douarin NM. Mapping of the early neurala primordium in quail-chick chimeras II. The prosencephalic neural plate and neural folds: implications for the genesis of cephalic human congenital abnormalities. Dev Biol. 1987;120:198–214.

63. Creuzet S, Couly G, Le Douarin NM. Patterning the neural crest derivatives during development of the vertebrate head: insights from avian studies. J Anat. 2005;207:447–59.

64. Puelles L, Rubenstein JLR. Expression patterns of homeobox and other putative regulatory genes suggest a neuromeric organization. Trends Neurosci. 1993;16(11):472–9.

65. Rubenstein JLR, Shimamura K, Martinez S, Puelles L. Regionalization of the prosencephalic neural plate. Annu Rev Neurosci. 1998;21:445–77.

66. Puelles L, Rubenstein JLR. Forebrain gene expression domains and the evolving prosomeric model. Trends Neurosci. 2003;26:469–76.

67. Puelles L. Forebrain development: prosomere model. In: Squire LR, editor. Encyclopedia of neuroscience, vol. 4. Oxford: Academic Press; 2009. p. 315–9.

68. Puelles L, Harrison M, Paxinos G. A developmental ontology for the mammalian brain based on the prosomeric model. Trends Neurosci. 2013;36(10):570–8.

69. O'Rahilly R, Muller F. The embryonic human brain: an atlas of developmental stages. 3rd ed. New York: Wiley-Liss; 2004.

70. Couly GF, Le Douarin NM. Head morphogenesis in avian chimeras: evidence for a segmental pattern in the ectoderm corresponding to the neuromeres. Development. 1990;108:543–58.

71. Webb JF, Noden DM. Ectodermal placodes: contributions to the development of the vertebrate head. Am Zool. 1993;33:434–47.

72. Streit A, Streit A. Early development of the cranial sensory nervous system: from a common field to individual placodes. Dev Biol. 2004;276:1–15.

73. Singh S, Graves AK. Molecular basis of craniofacial placode development. Wiley Interdiscp Rev Dev Biol. 2016;5(3):363–76.

74. Pathey C, Schlosser G, Schimeld SM. Evolutionary history of vertebrate cranial placodes. Dev Biol. 2014;389(1):82–97.

75. Omer A, Haddad D, Pisinski L, Krauthammer AV. The missing link: a case of absent pituitary infundibulum and ectopic neurohypophysis in a pediatric patient with heterotaxy syndrome. J Radiol Case Rep. 2017;11(9):28–38. https://doi.org/10.3941/jrcr.v11i9.3046.

76. Al-Gazali LI, Sztriha L, Punnose J, Shather W, Nork M. Absent pituitary gland and hypoplasia of the cerebellar vermis associated with partial ophthalmoplegia and postaxial polydactyly: a variant orofaciodigital syndrome VI or a new syndrome? J Med Genet. 1999;36:161–6.

77. May JA, Krieger MD, Bowen I, Geffner ME. Craniopharyngioma in childhood. Adv Pediatr Infect Dis. 2006;53:183–209.

78. Garrè ML, Cama A. Craniopharyngioma: modern concepts in pathogenesis and treatment. Curr Opin Pediatr. 2007;19(4):471–9. https://doi.org/10.1097/MOP.0b013e3282495a22.

79. DiRocco C, Caldarelli M, Tamburrini G, Massimi L. Surgical management of craniopharyngiomas—experience with a pediatric series. J Pediatr Endocrinol Metab. 2006;19:355–66.

80. Koral K, Weprin B. Sphenoid sinus craniopharyngioma simulating mucocele. Acta Radiol. 2006;47:494–6.

81. Dodé C, Hardelin J-P. Kallman syndrome. J Hum Genet. 2008;17:139–46.

82. Junklass J. Atypical presentation of a patient with both Kallman's syndrome and craniopharyngioma: case report and literature review. Case Rep. 2005;11(1):30–6.

83. Boehm U, Bouloux PM, Dattani MT, et al. Expert consensus document: European Consensus Statement on congenital hypogonadotropic hypogonadism—pathogenesis, diagnosis and treatment. Nat Rev Endocrinol. 2015;11(9):547–64. https://doi.org/10.1038/nrendo.2015.112.

84. Vaid S, Shah D, Rawat S, Shukla R. Proboscis lateralis with ipsilateral sinonasal and olfactory pathway aplasia. J Pediatr Surg. 2010;45:453–6.

85. Vulleix S, Niel F, Nedelec B, et al. Homozygous nonsense mutation of the FOXE3 gene as a cause of congenital primary aphakia in humans. Am J Hum Genet. 2006;79:358–64.

86. Breheret R, Brecheteau C, Tanguy J-Y, Lacourreye L. Bilateral semicircular canal hypoplasia. Eur Ann Otorhinoloaryngol Head Neck Dis. 2013;130:225–8. https://doi.org/10.1016/j.anorl.2012.10.005.

87. Hu D, Marcucio RS, Helms JA. A zone of frontonasal ectoderm regulates patterning and growth in the face. Development. 2003;130:1749–58.

88. Tirumandas M, Sharma A, Gbenimacho I, Shoha MM, Tubbs RS, Oakes WJ, Loukas M. Nasal encephalocoeles: a review of the etiology, pathophysiology, clinical presentations, diagnosis, treatment, and complications. Childs Nerv Syst. 2013;29(5):739–44. https://doi.org/10.1007/s00381-012-1998-z.

89. Di Somma L, Iacoangeli M, Nasi D, Balercia P, Lupi E, Girotto R, Polonara G, Scerrati M. Combined supra-transorbital key-

hole approach for treatment of delayed intraorbital encephalocele: a minimally invasive approach for an unusual complication of decompressive craniectomy. Surg Neurol Int. 2016;7(Suppl 1):S12–6.

90. Siebert JR, Kokich VG, Warkany J, Lemire RJ. Atelencephalic microcephaly: craniofacial anatomy and morphologic comparisons with holosprosencephaly and anencephaly. Teratology. 1987;36:279–85.

91. Marin-Padilla M. Study of the sphenoid bone in human cranioschisis and craniorachischisis. Virchows Arch A Pathol Anat Histopathol. 1965;339:245–53.

92. Kjaer I, Keeling JW, Graem N. Midline maxillofacial skeleton in human anencephalic fetuses. Cleft Palate Craniofac J. 1994;31(4):250–6.

93. Medical Taskforce on Anencephaly. The infant with anencephaly. N Engl J Med. 1990;322:699–74. https://doi.org/10.1056/NEJM199003083221006.

94. Dambska M, Schmidt-Sidor B, Maslinska D, et al. Anomalies of cerebral structures in acranial neonates. Clin Neuropathol. 2003;22:291–5.

95. Dias MS, Partington M. Embryology of myelomeningocele and anencephaly. Neurosurg Focus. 2004;16:E1.

96. Lemire RJ, Cohen MM Jr, Beckwith JB, Kokich VG, Siebert JR. The facial features of holoprosencephaly in anencephalic human specimens I. Historical review and associated malformations. Teratology. 1981;23:297–303.

97. Siebert JR, Cohen MM Jr, Sulik KK, Shaw C-E, Lemire RJ. The facial features of holoprosencephaly in anencephalic human specimens II. Craniofac Anat Teratol. 1981;23:305–15.

98. Guion-Almeida ML, Richeiri CA. Fronto-nasal dysplasia, macroblepharon, eyelid colobomas, ear anomalies, macrostomia, mental retardation and CNS structural abnormalities defining the phenotype. Clin Dysmorphol. 2001;10:191–202.

99. Richieri Costa A, Guion-Almeida ML. The syndrome of frontal-nasal dysplasia, callosal agenesis, basal encephalocele, and eye anomalies: phenotypical and etiological considerations. Int J Med Sci. 2004;1:34–42.

100. Kawamoto HK, Keller JB, Mell MW. Craniofacial-nasal dysplasia: Surgical treatment algorithm. Plast Reconstr Surg. 2007;120(7):1943–56.

101. Balci S, Mavili ME, Son YE, et al. A female patient with frontonasal dysplasia sequence and frontonasal encephalocele. Ann Plast Surg. 1999;43:457–9.

102. Song SY, Cho JW, Lew HW, Koh KS. Nasal reconstruction of a frontonasal dysplasia deformity using aesthetic rhinoplasty techniques. Arch Plast Surg. 2015;42(5):637–9. https://doi.org/10.5999/aps.2015.42.5.637.

103. Sharma R. Hypertelorism. Indian J Plast Surg. 2014;47(3):284–92.

104. Lightwood RC, Sheldon WPH. Hypertelorism: a unilateral case. Arch Dis Child. 1928;3(15):168–72. https://doi.org/10.1136/adc.3.15.168.

105. Tessier P, Guiot G, Derome P. Orbital hypertelorism: II. Definite treatment of hypertelorism (OR.H.) by craniofacial or by extra-cranial osteotomies. Scand J Plast Reconstr Surg. 1973;7:39–58.

106. Balaji SM, Modified facial bipartition. Ann Maxillofac Surg. 2012;2(20):170–3.

107. Smith JL, Schoenwolf GC. Neurulation: coming to closure. Trends Neurosci. 1997;20(11):510–7. https://doi.org/10.1016/s0166-2236(97)01121-1.

108. Copp AJ. Neurulation in the cranial region: normal and abnormal. J Anat. 2005;207:623–5.

109. Sheen J, Sheen AP. Aesthetic rhinoplasty. 2nd ed. St. Louis: Quality Medical Publishers; 1998.

110. Gruber RP, Tabbal GN, Sheen J, Toriumi D. Treatment of complex nasal deformities. Aesthet Surg J. 1999;19(6):475–82.

111. Mowlavi AS, Chabelian TL, Melgar A, et al. Upper septal vault and short nasal bone syndrome: implications for rhinoplasty. Eplasty. 2018;18:e29.

112. Karakor-Altuntas Z, et al. Isolated congenital nasal bifid septum separated by a wide layer of soft tissue. Arch Plast Surg (Korean Society of Aesthetic Plastic Surgery). 2015;42(5):640–2.

113. Shino M, Chikamatsu K, Yasuoka Y, et al. Congenital arhinia: a case report and functional evaluation. Laryngoscope. 2005;115:1118–23.

114. Zhang M-M, Hu Y-H, He W, Hu K-K. Congenital arrhinia: a rare case. Am J Case Rep. 2014;15:115–8.

115. Newman MH, Burdi AB. Congenital alar field defects: clinical and embryologic considerations. Cleft Palate J. 1989;19(6):475–82.

116. Abulezzi T. Case of heminasal aplasia: clinical picture, radiologic findings, and follow-up after early surgical treatment. J Craniofac Surg. 2019;30(3):e199–202.

117. Meyer R. Total external and internal reconstruction in arhinia. Plast Reconstr Surg. 1997;99:534–42.

118. Gong A. Proboscis lateralis type IV: a report from the Indian subcontinent. Acta Chir Plast. 1991;33:34–9.

119. Olsen OE, Gjelland K, Reigstad H, Rosendahl K. Congenital absence of the nose: a case report and literature review. Pediatr Radiol. 2001;31:225–32.

120. Toraynski E, Jacobiec FA. Cyclopia and synophthalmia: a model of embryologic interactions. In: Jacobiec FA, Duane TD, Jaeger EA, editors. Ocular anatomy, embryology, and teratology, Biomedical foundations of ophthalmology, vol. I. Philadelphia: JB Lippincott; 1982.

121. England SJ, Blanchard GB, Mehadevan L, Adams RJ. A dynamic fate map of the forebrain shows how vertebrate eyes form and explains two causes of cyclopia. Development. 2006;133:4613–7.

122. Bianchi D. Fetology: diagnosis and management of the fetal patient. 2nd ed. New York: McGraw-Hill; 2010.

123. Suzukui DG, Fukumoto Y, Yoshimura M, Yamazaki Y, Kosaka J, Kuratani S, Wada H. Comparative morphology and development of extraocular muscles in the lamprey and gnathostomes reveal the ancestral state and developmental patterns of the vertebrate head. Zool Lett (Zoological Society of Japan). 2016;2:10.

124. Plock J, Contaldo C, Von Ludinghausen M. Extraocular muscles in human fetuses with craniofacial malformations: anatomical findings and clinical relevance. Clin Anat. 2007;23(3):239–45.

125. Noden DM. Patterning of avian craniofacial muscles. Dev Biol. 1986;116:347–56.

126. Noden DM, Francis-West P. Differentiation and morphogenesis of craniofacial muscles. Dev Dyn. 2006;235:1194–2018.

127. Ziermann JM, Diogo R, Noden DM. Neural crest and the patterning of vertebrate craniofacial muscles. Wiley Genesis. 2018;56(6–7):e23097. https://doi.org/10.1002/dvg.23097.

128. Lumsden A, Keynes R. Segmental patterns of neuronal development in the chick hindbrain. Nature. 1989;337:424–8.

129. Lumsden A, Krumlauf R. Patterning the vertebrate neuraxis. Science. 1996;274:1109–15.

130. Hunt P, Krumlauf R. Hox codes and positional specification in vertebrate embryonic axes. Annu Rev Cell Biol. 1992;8:227–56.

131. Cohen MM, Sulik KK. Perspectives on holoprosencephaly: part I. Epidemiology, genetics and syndromology. Teratology. 1989;40(3):211–35.

132. Cohen MM, Sulik KK. Perspectives on holoprosencephaly: part II. Central nervous system, craniofacial anatomy, syndrome commentary, diagnostic approach, and experimental studies. J Craniofac Genet Dev Biol. 1992;12(4):196–244.

133. Cohen MM, Sulik KK. Perspectives on holoprosencephaly: part III. Spectra, distinctions, continuities and discontinuities. Am J Med Genet. 1989;34(2):271–88.

134. Hendry JM, Nemerofsky R, Stolman C, Granick MS. Plastic surgery considerations for holoprosencephaly patients. J Craniofac Surg. 2004;15:675–7.

135. Nagase T, Nagase M, Osumi N. Craniofacial anomalies of the cultured mouse embryo induced by inhibition of sonic hedgehog signaling: an animal model of holoprosencephaly. J Craniofac Surg. 2005;16:80–8.

136. Gui T, Osama-Yamashita N, Eto K. Proliferation of nasal epithelia and mesenchymal cells during primary palate formation. J Craniofac Genet Dev Biol. 1993;13:250–8.

137. Carstens MH. The spectrum of minimal clefting: process-oriented cleft management in the presence of an intact alveolus. J Craniofac Surg. 2000;11(3):270–94.

138. Carstens MH. Functional matrix cleft repair: principles and techniques. Clin Plast Surg. 2004;31:159–89.

139. Carstens MH. Developmental field reassignment in unilateral cleft lip: reconstruction of the premaxilla. In: Losee J, Kirschner R, editors. Comprehensive cleft care. Boca Raton, FL: CRC Press/ Taylor & Francis; 2007.

140. Montague A. The premaxilla in primates. Q Rev Biol. 1935;10(1):32–59.

141. Barteczko Barteczko K, Jacob M. A re-evaluation of the premaxillary bone in humans. Anat Embryol (Berl). 2004;207:417–37.

142. Trevizan M, Consolaro A. Premaxilla: an independent bone that can base therapeutics for middle third growth. Dental Press J Orthod. 2017;22(2):21–6. https://doi.org/10.1590/2177-6709.22.2.021-026.oin.

143. Burdi AR, Lawton TJ, Grosslight J. Prenatal pattern emergence in early human facial development. Cleft Palate Craniofac J. 1988;25:8–15.

144. Graham A, Smith A. Patterning the pharyngeal arches. BioEssays. 2001;23:54–61.

145. Couly G, Cruezet S, Bennaceur S, et al. Interactions between Hox-negative cephalic neural crest cells and the foregut endoderm in patterning the facial skeleton in the vertebrate head. Development. 2002;129:1061–73.

146. Ruthin B, Creuzet S, Vincent C, et al. Patterning of the hyoid cartilage depends upon signals arising from the ventral foregut endoderm. Dev Dyn. 2003;228:239–46.

147. Tapadia MD, Cordero D, Helms JÁ. It's all in your head: new insights into craniofacial morphogenesis. Development. 2005;132:851–61.

148. Johnston MC, Millicovsky G. Normal and abnormal development of the lip and palate. Clin Plast Surg. 1985;2:521.

149. Gong S-G, Guo C. Bmp4 gene is expressed at the putative site of fusion in the midfacial region. Differentiation. 2003;71:228–36.

150. Carstens M. Development of the facial midline. J Craniofac Surg. 2002;13:129–87.

151. Ashique AM, Fu K, Richman JM. Endogenous bone morphogenetic proteins regulate outgrowth and epithelial survival during avian lip fusion. Development. 2002;129:4647–60.

152. Zhang Z, Song Y, Zhao X, et al. Rescue of cleft palate in Msx-1 deficient mice by transgenic *bmp4* reveals a network of BMP and Shh signaling in the regulation of mammalian palatogenesis. Development. 2002;129:4135–40.

153. Jirasek JE. Atlas of human prenatal morphogenesis. Amsterdam: Martinus Nijhof; 1983.

154. Jirasek JE. An atlas of the human embryo and fetus. CRC Press; 2000.

155. Hinrichsen K. The early development of morphology and patterns of the face in the human embryo. In: Advances in anatomy, embryology, and cell biology, vol. 98. New York: Springer; 1985. p. 1–72.

156. Kaufman MH. The atlas of mouse development. San Diego: Academic Press; 1992. p. 429.

Michael H. Carstens

Historical Background

The science of embryology in the United States was born out of a unique combination of personalities and events involving philanthropy, the importation of a new system, medical education, friendships made abroad, and artistic talent.

Johns Hopkins (1795–1873) was born to Quaker parents of strong convictions. His father, a tobacco planter, in response to the early abolitionist movement, freed his slaves in 1807. Johns went to work to help his family at age 12, beginning the grocery business and quickly demonstrated an acumen for business. Unable to marry his cousin for religious reasons, he devoted his life to his work and became highly successful, investing in the Baltimore & Ohio railroad, of which he eventually became director in 1847. He was active in banking and real estate and was able to retire at age 52 (Fig. 3.1).

Social involvement and charity were core values for Johns Hopkins. A lifelong abolitionist, during the Civil War he worked with the Rev. Henry Ward Beecher and supported President Lincoln's vision for post-war emancipation. Later in life, he founded Johns Hopkins Orphanage for Colored Children. When the city of Baltimore hit upon financial straits, he gave the city financial support.

Cholera and yellow fever epidemics in Baltimore made a deep impression upon Hopkins who saw the city's need for a medical facility, particularly in view of the advances in medicine that took place during the Civil War. He was also impressed by the German university system, with its emphasis on research. Accordingly, in 1870, he made a will of $7 million in B%O stock to establish a research university, a free hospital with university-based schools of medical and nursing. Half of the endowment went toward the creation of Johns Hopkins University as the first formally research-based institution in the United States. Johns Hopkins Hospital was founded in 1876 and the School of Medicine opened its doors in 1893 (Fig. 3.2).

Johns Hopkins occupies a unique place in the history of American medical education. Based upon the Heidelberg model, Hopkins imposed strict entrance requirements and

Fig. 3.1 Johns Hopkins. (1) Whitehall tobacco plantation: (a) Hannah Hopkins $10,000. (2) B&O Railroad. (3) 1870 Johns Hopkins University. (4) 1889 Johns Hopkins Hospital. (5) 1893 Johns Hopkins Medical School. (Reprinted from Wikimedia. Retrieved from: https://commons.wikimedia.org/wiki/File:Hopkinsp.jpg)

M. H. Carstens (✉)
Wake Forest Institute of Regenerative Medicine, Wake Forest University, Winston-Salem, NC, USA
e-mail: mcarsten@wakehealth.edu

Fig. 3.2 Johns Hopkins
Hospital. (Reprinted from
Wikimedia. Retrieved from:
https://commons.wikimedia.
org/wiki/File:Johns_Hopkins_
Hospital_in_
Baltimore_1900s.jpg)

academic structures, thus becoming the birthplace of the modern residency system. The original core faculty members, the "big four", were not only renowned contributors in their fields, but colorful characters as well. William Henry Welch (pathology) was a confirmed bachelor and gourmand who could consume up to five desserts in a single sitting. Howard Kelly (obstetrics and gynecology), an evangelist with an interest in saving souls, kept snakes for pets. William Osler (medicine) was a practical joker. William Stewart Halsted (surgery) was pathologically shy, but an iron taskmaster with his students. His later experiments with local anesthesia led to a debilitation addiction to cocaine. In addition to setting up the first Ph.D. programs in medical science, Hopkins established brand new departments for the history of medicine and for art as applied to medicine under the direction of world famous medical illustrator Max Brödel (Fig. 3.3).

Andrew Carnegie (1835–) was a Sottish-born industrialist and philanthropist who made an indelible contribution to the study of human development. His biography is compelling story and the subject of a PBS documentary. Born in Dunfermline, Scotland, he was influenced by his father, William, who engaged in radical politics and was involved in the creation of a Trademen's Subscription Library. At age 13, the family moved to Allegeney, Pennsylvania. He was a voracious reader using a library. When he was denied access to it, he fought for his own education, writing to the owners

to the effect that no one should be denied access to learning based on their station in life (Figs. 3.4 and 3.5).

He got his start as a telegraph operator with the Pennsylvania Railroad and transitioned into the railroad car business. In 1885, he started a new venture with the Keystone Bridge Company. This morphed into the Carnegie Steel Company. By understanding the new technology of the Bessamer process, he revolutionized the industry. In 1891, Carnegie Steel was sold to J.P. Morgan of US Steel for the unheard sum of $480 million. Carnegie vowed to devote the remainder of his life to philanthropy.

In keeping with its founder's belief in education, the Carnegie Foundation created 2509 public libraries worldwide. The Foundation sponsored and International Endowment for Peace. Most relevant for us, in 1902, in cooperation with President Theodore Roosevelt, the Carnegie Institution for Science was founded with a $22 million endowment in Washington, DC., as a think tank for the advancement of knowledge. A Department of Embryology was created in 1913 in affiliation with the Department of Anatomy at nearby Johns Hopkins Medical School, the first purely university-based medical institution in the United States. This effort was captained by Prof. Franklin Mall (Figs. 3.6 and 3.7).

Franklin P. Mall was born in 1862 in Belle Plaine, Iowa. He studied medicine at the University of Michigan (one of his classmates was William P. Mayo). In 1884, he followed

Fig. 3.3 The Four Doctors by John Singer Sargent, portrait in oil, 1906. (Reprinted from Wikimedia. Retrieved from: https://commons.wikimedia.org/wiki/File:Four_doctors_1907.jpg. With permission from Free Art License 1.3 (FAL 1.3): http://artlibre.org/licence/lal/en/)

his research interests in Leipzig, Germany, where he studied under embryologist Wihelm His. There he met pathologist William Welch, the future head of the Johns Hopkins Medical School In 1886, Mall returned to study pathology at the Johns Hopkins Hospital. 1893 saw the opening of the medical school at Johns Hopkins and Mall was immediately appointed by Welch as Chairman of Anatomy (Fig. 3.8 Mall).

From 1910 to 1912, Mall edited with Franz Keibel the Handbook of Human Embryology and published several monographs in conjunction with the Carnegie Institution. With this impetus, he successfully lobbied for the creation of the Embryology Department at Carnegie in 1913, which he chaired until his death in 1917. Mall was succeeded by George Streeter who studied Mall's embryo collection and began work on a staging system consisting of 23 Horizons. Streeter retired in 1940 and work was continued by the third Chairman, Dr. George Corner, whose beautiful three-dimensional drawings can be accessed through the Carnegie Institution and through the UNSW website.

In 1973, the entire embryo collection was transferred to the University of California Davis Medical School along with Drs. Ronan O'Rahilly and Fabiola Müller. They further refined the Carnegie system and their work, *Developmental Stages in Human Embryos*, was published as Carnegie Institution Publication No. 637 in 1987. Following O'Rahilly and Müllers retirement to Europe, the collection was transferred in 1990 to the National Museum of Health and Medicine in Washington, DC, where it remains as a resource for research. In 2005, the Carnegie Department of

Fig. 3.4 Andrew Carnegie (1835–1919) was born in Dunfermline, Scotland to William Carnegie, a craft weaver who espoused radical politics and created a Tradesmaen's Subscription Library. He emigrated to the United in 1848 and settled in Allentown, Pennsymvania. He subsequently built Carnegie Steel which he sold to J.P. Morgan in 1901 for the sum in today's US dollars of $10 billion. As a champion of philanthropy used his immense wealth to build 2,508 public libraries world-wide, create an International Endowment for Peace, and establish the Carnegie Institute for Science which included a Department of Embryology. (Reprinted from Wikimedia. Retrieved from: https://commons.wikimedia.org/wiki/File:Andrew_Carnegie_by_Francis_Luis_Mora.jpg)

Fig. 3.5 Carnegie Homestead Steel Works, Johnston, PA. (Reprinted from Wikimedia. Retrieved from: https://commons.wikimedia.org/wiki/File:Homestead_Steel_Works,_Homestead,_Pa._(det.4a10138).tif)

Fig. 3.6 Carnegie Institution, cartoon. The Institution was founded with an astounding $22 million (almost half a billion today). (Reprinted from Washington Evening Star; 1911)

Embryology moved to the Maxine Singer building on the Johns Hopkins Homewood campus.

Our foray into the history of embryology at the Carnegie Institution would not be complete without recognizing the unique contributions made by medical artist Dorcas Padget, a self-taught neuroembryologist and researcher whose work is the foundation of our understanding for the vascular development of the head and neck. Her story is inextricably woven with that of pioneer neurosurgeon Walter Dandy at Johns Hopkins and of Max Brödel, world famous medical illustrator and the founder of the Department for Art as Anatomy at Hopkins (Fig. 3.9 left).

Dandy was himself a talented artist. As a student at Hopkins, his work was noted by Franklin Mall, who stimulated Dandy to reconstruct an early human embryo from his collection. Brödel was Dandy's instructor as he worked drawings of the neurovascular supply of the pituitary (Fig. 3.9 right) As chairman of Neurosurgery at Hopkins, Dandy's neurosurgical research progressed to the point where a full-time illustrator was needed for his department. Dorcas Padget was born in Albany, NY, in 1906 and was interested in science throughout her life. She received a full scholarship to Vassar College, where she studied art under Professor Treadwell. Recognizing her talent, in 1926, he referred Padget to Max Brödel to specialize in medical art.

Fig. 3.7 Carnegie Institution for Science. (Reprinted from Wikimedia. Retrieved from: https://commons.wikimedia. org/wiki/File:Administration_ Building_-_Carnegie_ Institution_of_Washington. JPG. With permission from Creative Commons License 3.0: https://creativecommons. org/licenses/by-sa/3.0/deed. en)

Fig. 3.8 (Left) Franklin P. Mall and (right) George Streeter. The first and second chairmen of the Carnegie Institution Department of Embryology. (Reprinted with permission from Johns Hopkins University Department of Medical Art—Gary Lees. Carnegie Institution—Sonya Bajwa)

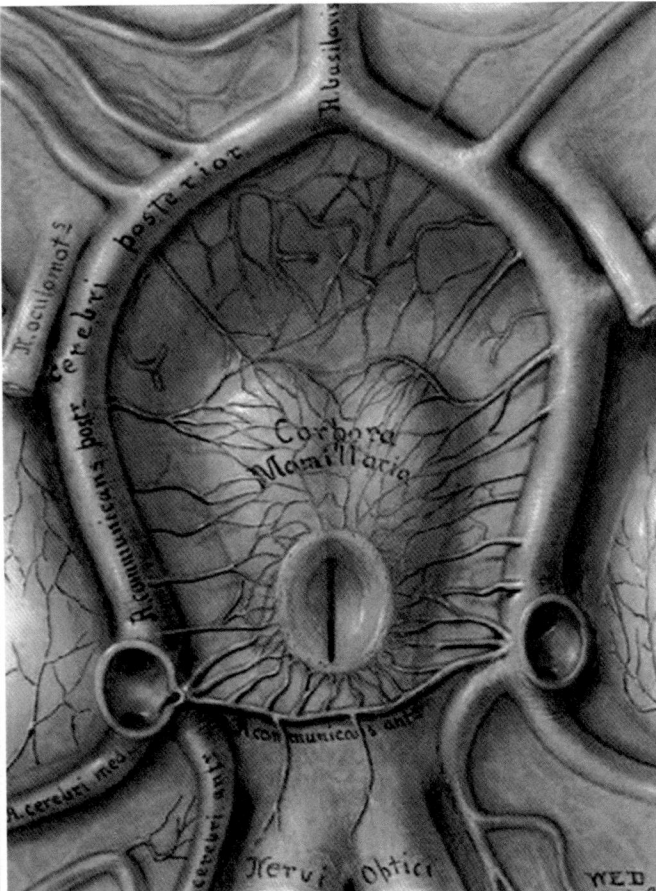

Fig. 3.9 (Left) Walter Dandy, Johns Hopkins neurosurgeon and medical artist. He was influenced by Mall and by Brödel. (Right) Dandy's illustration of the vascular supply to the pituitary. (Reprinted with permission from Johns Hopkins University Department of Medical Art—Gary Lees. Carnegie Institution—Sonya Bajwa)

Brödel offered her a 1-year position with the proviso of permanent employment upon graduation. Padget was exceptional from the start and won a full-time position. But the need to help her sisters get through college combined with a steady income resulted Padget forgoing her final year at Vassar. Though she would carry out world class research, she never got her degree and never achieved her goal to become a physician (Fig. 3.10).

For the next 22 years, Padget's work was the backbone of Hopkins Neurosurgery publication efforts, providing the illustrations for many articles and for Dandy's book, *Surgery of the Brain*. Her work is characterized by careful attention to didactic points, elegant design, and exquisite detail. Her depiction of the extirpation of a large hypoph-

yseal cyst is a masterpiece of three-dimensional design and see-through techniques to show multiple planes (Fig. 3.11 left).

Upon Dandy's death in 1946, Padget began work at the Carnegie Institution. She performed painstaking dissection and tracking for arterial vessel in 22 separate embryos and her monograph, published in 1948, is the foundation of neurovascular developmental anatomy. As the Carnegie Horizons had not yet been worked out and were not available to her, she divided the embryos into seven stages which fit remarkably well into the subsequent system. Over the remainder of her career, Padget authored numerous papers on congenital anomalies of the nervous system. Despite a long struggle with cancer beginning in the 1950s, she remained productive

Fig. 3.10 (Left) Max Brödel and (right) Dorcas Padget. Brödel was brought to Johns Hopkins to found the first program in medical illustration in the U.S. Brodel secured an appointment for Dorcas Hagar with the JHU Neurosurgery Department. Portrait of Dorcas Hagar Padget by Audrey Juliet Artnott (also a student of Brodel). (Reprinted with permission from Johns Hopkins University Department of Medical Art—Gary Lees. Carnegie Institution—Sonya Bajwa)

until her death at Johns Hopkins Hospital in 1973. Her work set the standard for medical art and her contributions to neu-

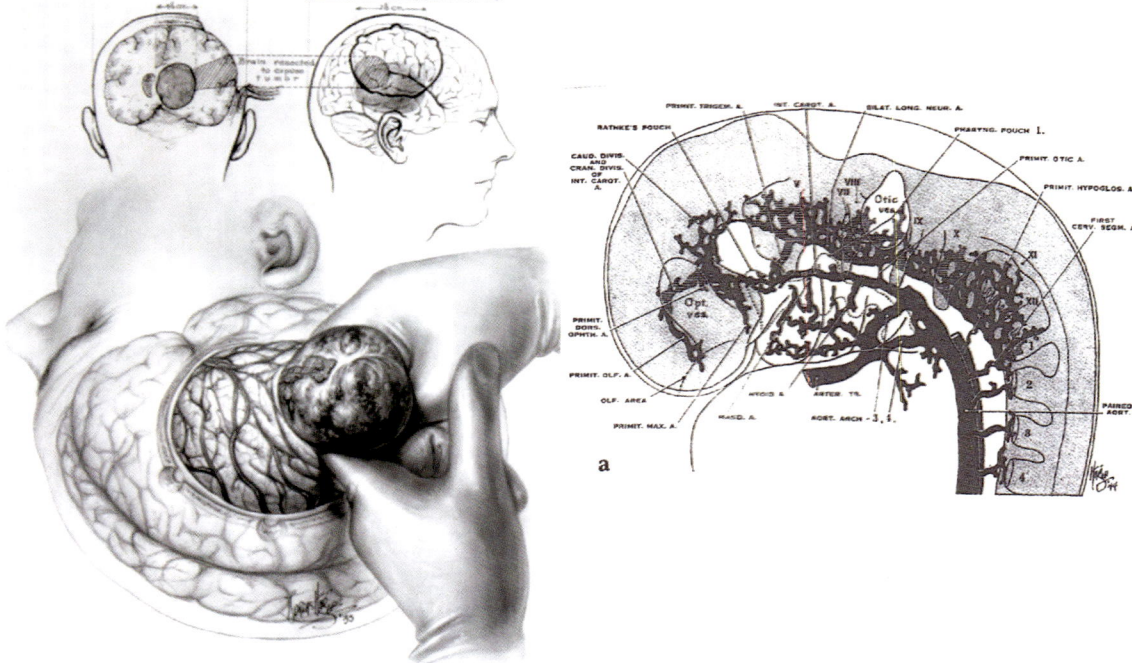

Fig. 3.11 Artwork by Dorcas Padget. (Reprinted with permission from Johns Hopkins University Department of Medical Art—Gary Lees. Carnegie Institution—Sonya Bajwa)

roembryology are indisputable (Fig. 3.11 right).

How to Use This Chapter

Human development is a four-dimentional process in which multiple anatomic structures change over time. To understand it, we must create in our minds a videotape of the embryo on multiple levels, from external features, to assembly of individual parts such as the tongue, to the relationships among structures and system such as the evolution of blood supply to the neck. To accomplish this, a reference system to keep track of these changes is required.

Original descriptions of the human embryo suffered from a lack of standardized parameters. Embryos of the same age could vary in size. Gestational age can be inexact. These problems were addressed by Streeter, Corner, and others at the Carnegie Institution by the creation of a staging for embryos based on reproducible landmarks. The system was first described in a series of monographs published by the Carnegie Institution. It was codified by O'Rahilly R and Muller F, *Developmental stages in human embryos*, Publication 637, The Carnegie Institution of Washington, DC, 1987. The Carnegie system was fully implemented in O'Rahilly R, Müller F: *Human Embryology and Teratology*, 3rd ed. New York, Wiley-Liss, 2001.

How can one go about learning the Carnegie system? Modern embryology references such as Carlson refer to the staging system, but no systematic description is available. O'Rahilly and Müller provide reference to the system throughout the text in a very useful way, but leave a description of the Carnegie system to a separate reference work by the same authors (Fig. 3.12).

A comprehensive outline of the Carnegie system can easily expand into a synopsis of embryonic development in general…one can easily get lost with this approach. Our purpose here is to simply outline some of the major events, characterized by major external features with emphasis on the face. Certain key features of the brain, vascular, digestive, and respiratory systems are also listed, inasmuch as these have bearing on craniofacial development.

Devoting time to study this system will give you a means to think about development in a dynamic and esthetically satisfying way. You are encouraged to plow through the descriptions, referring to accompanying figures. Legends contain additional information on visible structures such as the extremities.

A full visual account has been assembled at University of New South Wales by Dr. Mark Hill: https://embryology.med. unsw.edu.au/embryology/index.php/Main_Page.

Monographs on the Carnegie system and by Dr. Mark Hill and on the Kyoto University Embryo Collection by Dr. Mark

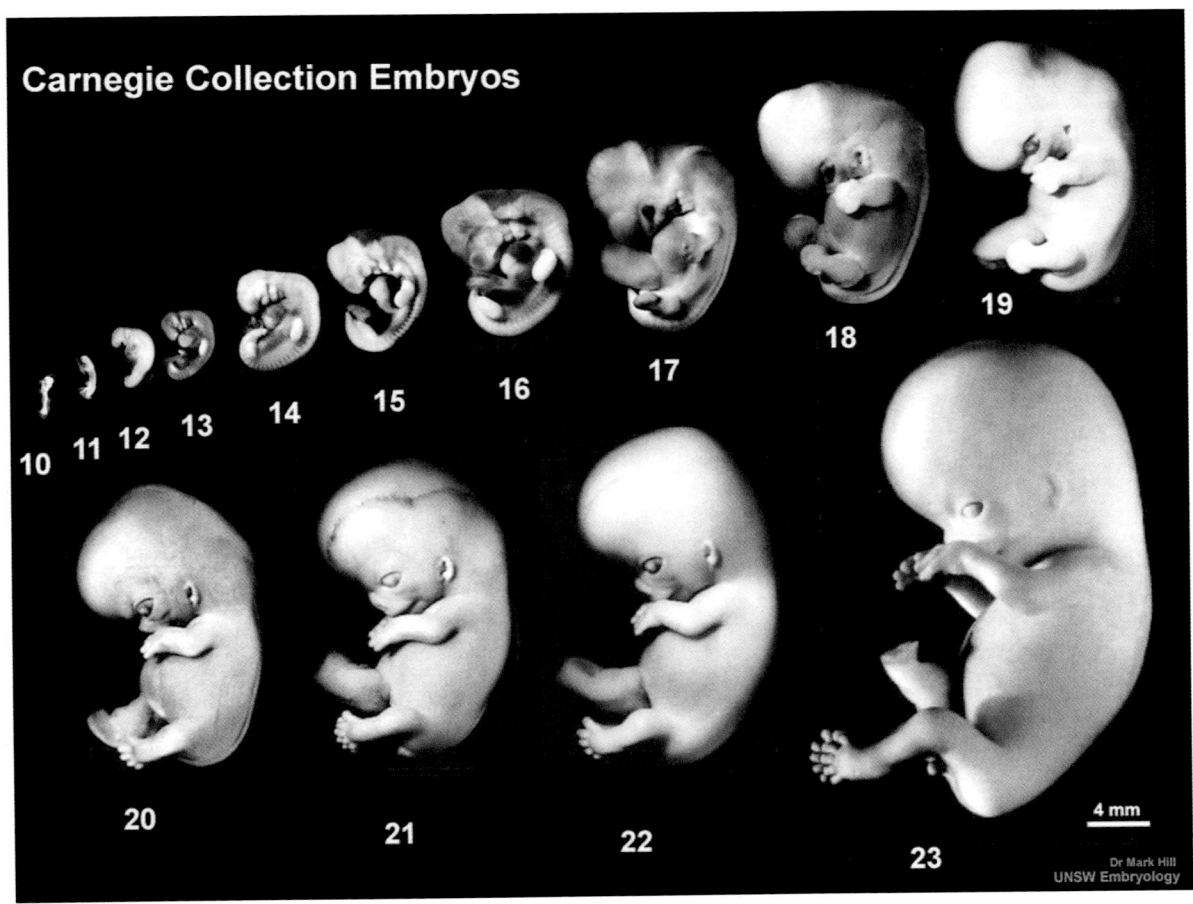

Fig. 3.12 Carnegie stages summary. (Reprinted with permission from Johns Hopkins University Department of Medical Art—Gary Lees. Carnegie Institution—Sonya Bajwa)

Hill, Dr. Shigehito Yamada, and Dr. Cecilia Lo are available online through the iBooks store and are compatible with both iPad and laptop.

Stage-By-Stage Description

Major characteristics of the 23 stages are listed. These emphasize key developmental events and external features. In subsequent chapters, this form of stage-by-stage summary will be useful to chart the events of individual structures. Our purpose here is to gain a general visual orientation to key events in development. Although our focus is on the head and neck, many aspects of staging have to do with anatomic features that are easy to identify. We constantly stress developmental events in the CNS, so these are included in the capsule summaries. Hopefully, with repetition, these structures will become second-nature for the reader. One should range back and forth from the capsule illustrations to the figures. This material is closely correlated with the embryology texts of Carlson and O'Rahilly. Note that there is not a consensus on the exact number of days allotted to some of the stages. The numbering used in this text is as per Carlson. For an alternative, consult the UNSW website.

Stage 1 (24 h, 0.1–0.15 mm): Fertilization (Fig. 3.13)

Fertilization takes place in the ampulla of uterine tube

Stage 1a: The *penetrated oocyte* contains all genetic material plus redundant chromosomes enclosed within a single cell membrane.

Stage 1b: The *pronuclear embryo* characterized by resuming arrested meiosis II, followed by anaphase and telophase, in which redundant chromosomes are packaged as the second polar body and expelled from the cell. Pronuclear embryos have two separate haploid components, male pronucleus and female pronucleus. These move to the midline of the cell and press together.

Stage 1c: The *syngamic embryo* has fused pronuclear cell membranes—parental chromosomes come together and assume a position on the spindle. The genome is now formed.

Carnegie Stage 1

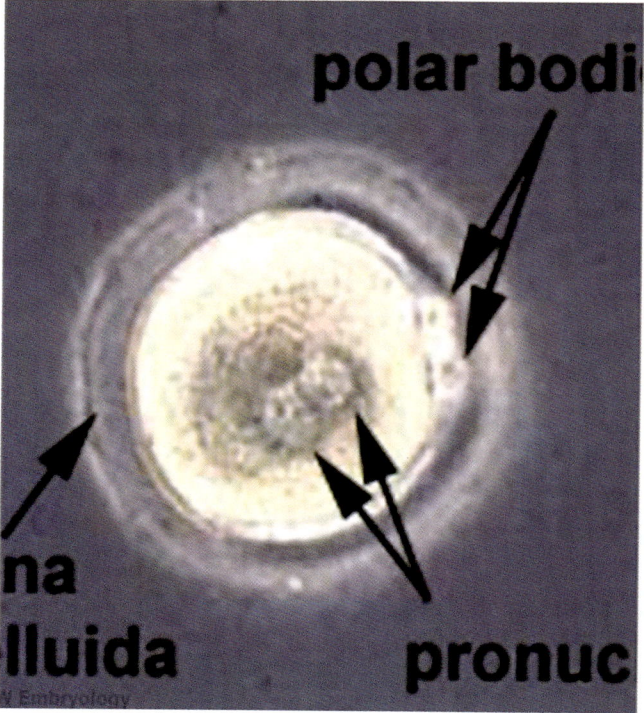

Fig. 3.13 Stage 1: 1 day, 0.1–0.15 mm. (1) Haploid pronuclei have not joined. (2) Polar bodies. (3) Zona pellucida. (Reprinted from Carlson BM. Human Embryology and Developmental Biology, 6th edition. St. Louis, MO: Elsevier; 2019. With permission from Elsevier)

Carnegie Stage 2

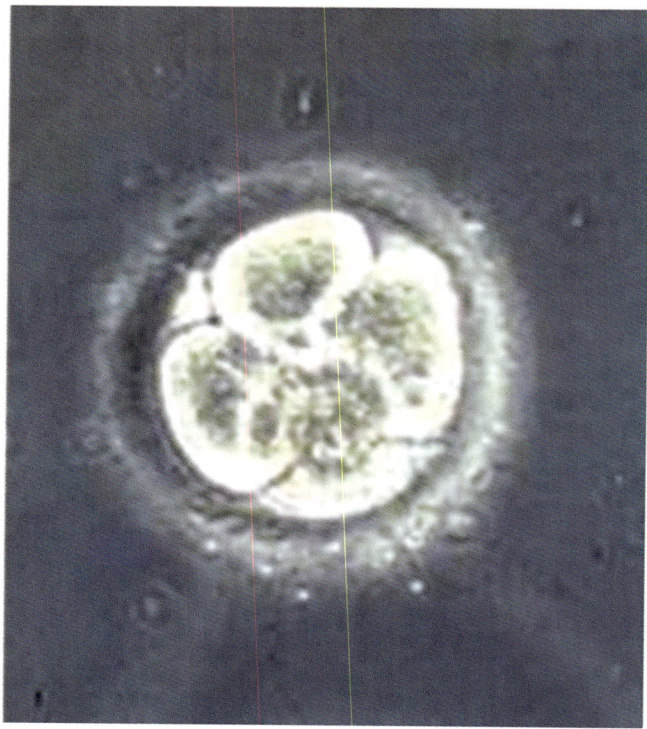

Fig. 3.14 Stage 2: 2–3 days, 0.1–0.2 mm. Early mitosis: (1) No G1 or G2 phase. (2) Reduction in volume. (Reprinted from Carlson BM. Human Embryology and Developmental Biology, 6th edition. St. Louis, MO: Elsevier; 2019. With permission from Elsevier)

Stage 2 (2–3 Days, 0.1–0.2 mm): Cleavage
(Fig. 3.14)

Zygote undergoes cleavage with up to 16blastomeres, no change in size but DNA content rapidly expanding. At morula stage, fate of individual blastomeres not yet determined.

Stage 3 (4–5 Days, 0.1–0.2 mm): Free-Floating Blastocyst (Fig. 3.15)

Embryo passes through four phases.

Cavitation produces a blastocyst. No change in volume, so the fluid is produced by the metabolism of the cells and not be importation from the exterior. *Collapse and expansion* take place prior to escape from zona pellucida. Collapse is a sudden event, happening in less than 5 min. *Hatching* refers to escape at days 7–8 post-insemination via the zona pellucida. Embryo now gets bigger. External trophoblast cells tightly bound together.

Stage 4 (6 Days): Attachment of Blastocyst
(Figs. 3.16, 3.17, and 3.18)

Adplantation presages the onset of implantation. Outer trophoblast fimbriae create multiple adhesion sites. Trophoblast differentiates into inner cytotrophoblast and outer syncytiotrophoblast which dissolves the epithelium of the endometrium. Embryo is now bilaminar with appearance of hypoblast just below the inner cell mass (epiblast).

Stage 5 (7–12 Days, 0.1–0.2 mm): Implantation, Bilaminar Disc, Trophoblast Development, No Villous Development (Figs. 3.18 and 3.19)

Thick outer syncytiotrophoblast does not have cell boundaries. Thin inner cytotrophoblast has cell boundaries. Extraembryonic mesoblast, the chorion, is present but lacks villous structures.

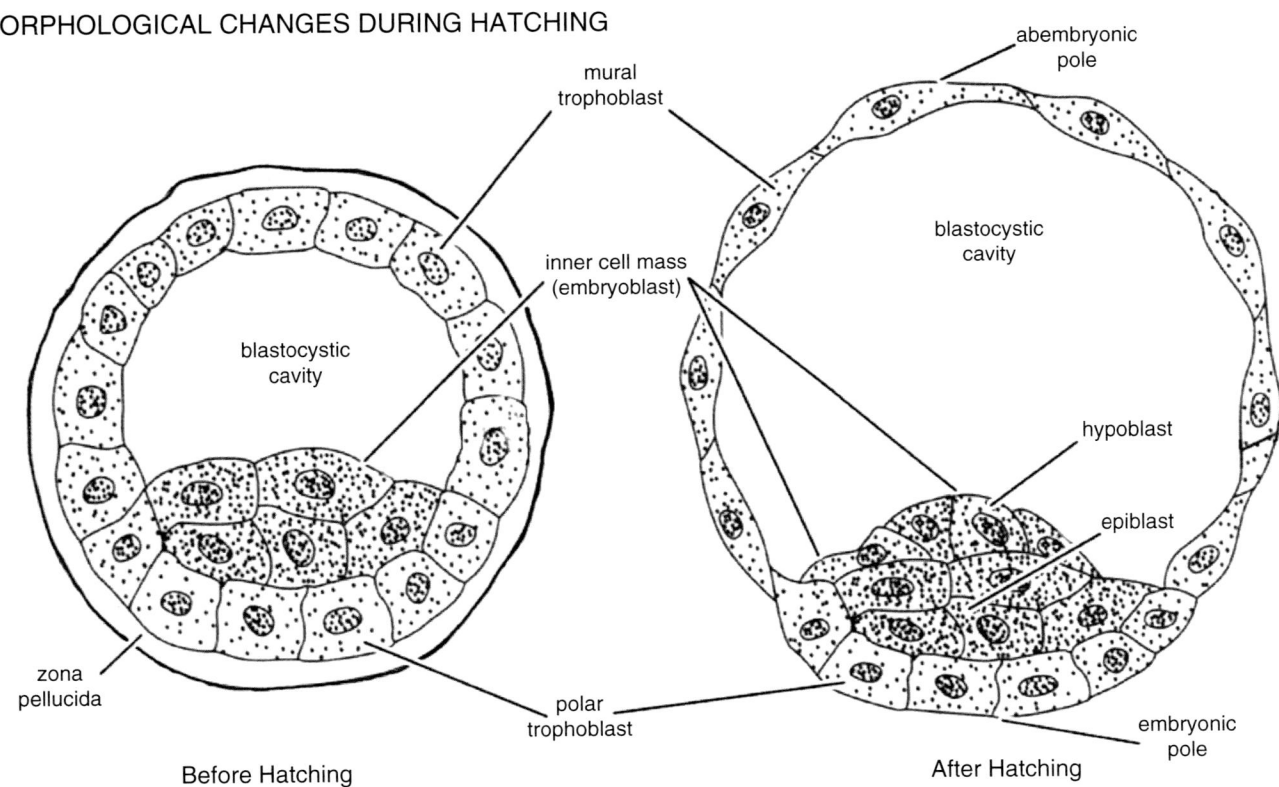

MORPHOLOGICAL CHANGES DURING HATCHING

Before Hatching

After Hatching

Fig. 3.15 Stage 3: 5 days, 0.1–02 mm. (1) Free floating blastocyst. (2) Zona pellucida > "hatch". (3) Inner cell mass. (4) 8 cells = TOTIPOTENT. (5) Hypoblast not present—hypoblast defines the next stage 4. (6) Squamous trophectoderm. (7) Must "hatch" in order to implant.

(Reprinted from Veeck LL, Zaninović N. An atlas of human blastocysts. New York, NY: Parthenon Publishing Group; 2003. With permission from Taylor & Francis)

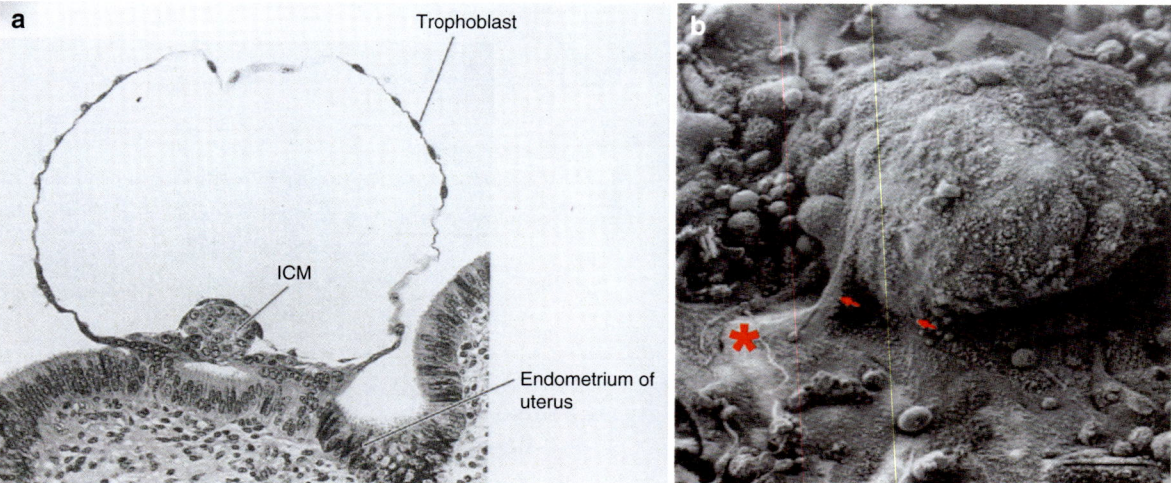

Fig. 3.16 Stage 4 is reserved for the attaching blastocyst that is adhering to the endometrial lining. Attachment is complete at 6 days. Implantation is completed during stage 5. (**a**) Photomicrograph showing beginnings of hypoblast immediately under the epiblast. Cells are now PLURIPOTENT—they can produce all cell types *except* umbilical cord and placenta. (**b**) Attachments of the blastocyst. (**a**: Reprinted from Gilbert SF, Barresi M. Developmental Biology, 11th ed. Sinaurer: Sunderland, MA, 2016. Reproduced with permission of the Licensor through PLSclear. **b**: Reprinted from Bentin-Ley U, Sjogren A, Nilsson L, et al. Presence of uterine pinopodes at the embryo-endometrial interface during human implantation *in vitro*. *Human Reprod.* 1999; 14:515–520. With permission from Oxford University Press)

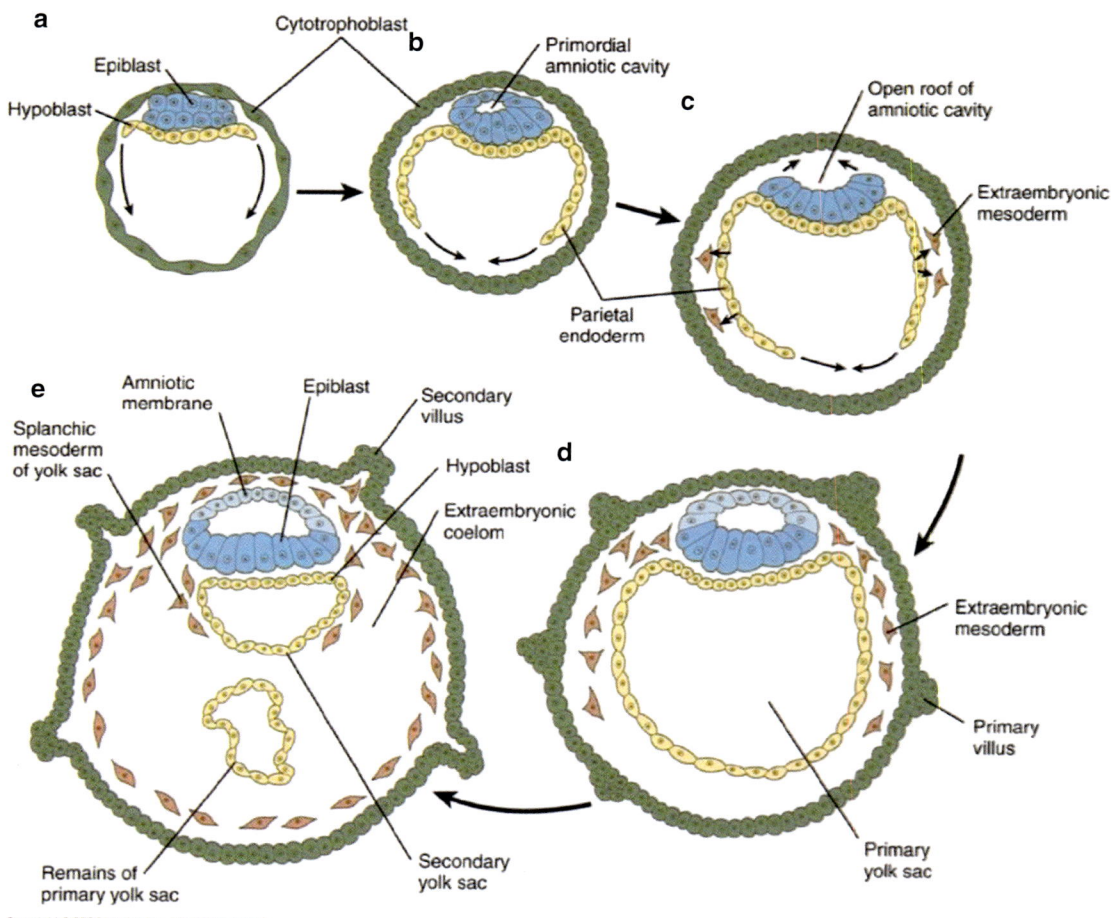

Fig. 3.17 Carnegies stages 4–5 extraembryonic gastrulation. Cytrophoblast (green)—syncytiotrophoblast not shown; epiblast (blue), hypoblast/extraembryonic endoderm (yellow); extraembryonic mesoderm (red). **A**, 6d = implantation starts; **B**, 7.5d = implantation complete, **C**, 8d = extraembryonic mesoderm; **D**, 9d = extrembryonic ectoderm = amnion; **E**, 2 weeks = EEM invades villi; surrounds primitive endoderm remnant (secondary yolk sac) > vitelline vessels. (Reprinted from Carlson BM. Human Embryology and Developmental Biology, 6th edition. St. Louis, MO: Elsevier; 2019. With permission from Elsevier)

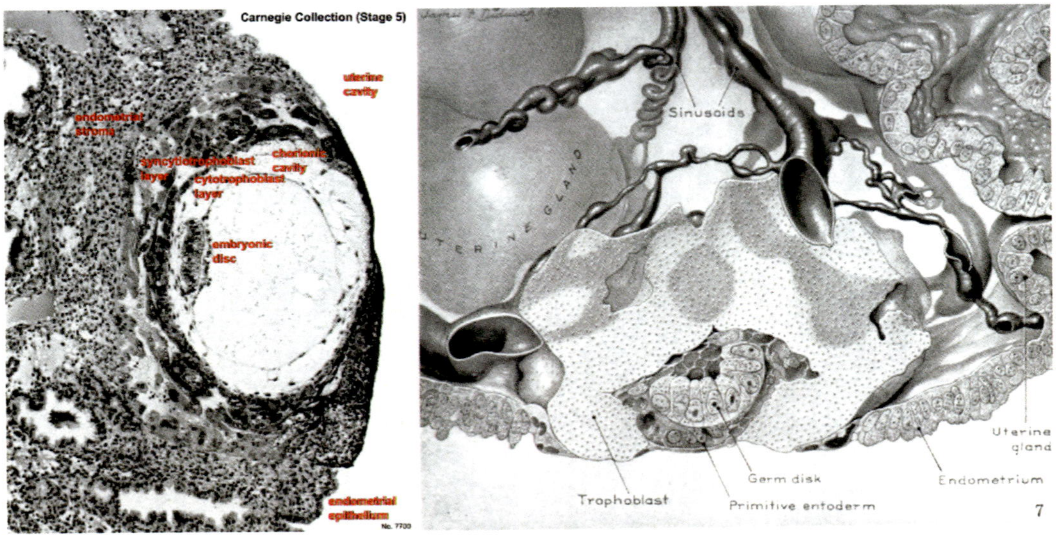

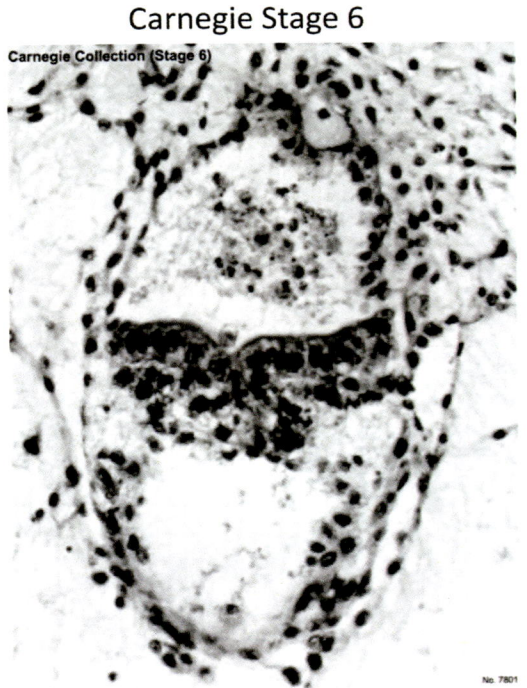

Fig. 3.18 Stage 5 0.1–02 mm, 7–12 days. (1) Implantation complete: trophoblast differentiates, amniotic cavity. (2) Decidual reaction. (3) Bilaminar embryo. (4) Hypoblast > extraembryonic mesoderm (pink) > blood vessels. (5) Primary umbilical vesicle, yolk sac. Note new layer of extraembryonic mesoderm (EEM) produced by the hypoblast. The EEM surrounds the entire embryo and invades into projections of the cytotrophoblast into the syncytiotrophoblast called *primary villi*. When these are filled with EEM, they become *secondary villi*. Left: conceptus is fully implanted with a bilaminar disc and primary yolk sac. Right: Schematic drawing by James F. Dudisch. (Left: Reprinted from Hertig AT, Rock J, Adams EC, Mulligan WJ. On the preimplantation stages of the human ovum: A description of four normal and four abnormal specimens ranging from the second to the fifth day of development. Contrib Embryol. Carnegie Institution of Washington 1954;35:199–220. Right: Reprinted with permission from Johns Hopkins University Department of Medical Art—Gary Lees. Carnegie Institution—Sonya Bajwa)

Carnegie Stage 6

Fig. 3.19 Stage 6a: Bilaminar embryo just before gastrulation. 0.2 mm and 14–15 days stage *VERY FAST*. SEM showing the primitive streak. (1) Chorionic villi (functional placenta). (2) Trophoblasts invade decidua > maternal lakes. (3) Spiral arteries *held open* by trophoblast. (4) Connecting stalk (extraembryonic mesoderm) contains: (a) diverticulum of the allantois, (b) omphaloenteric duct, (c) amniotic somatopleure. (5) Blood islands. (6) Primitive streak: caudal > cranial: (a) interaction with endoderm, (b) tissue movements begin at stage 7. (Reprinted with permission from Johns Hopkins University Department of Medical Art—Gary Lees. Carnegie Institution—Sonya Bajwa)

Stage 5a: 7–8 days with solid trophoblast. Endometrial stroma is very edematous. On the inner surface of the trophoblast, a two-layer embryonic disc has distinct epiblast and hypoblast.

Stage 5b: lacunae form within the syncytiotrophoblast. These communicate with each other and with the maternal sinusoids, but maternal blood is scant. Within the endometrium, "fingers" of syncytiotrophoblast become filled with cytotrophoblast. These are the future villi. Yolk sac appears and is bounded by an external exoceloemic membrane. The embryonic disc now has two forms of cells: the epiblast epithelium being pseudostratified columnar and the hypoblast is simple cuboidal.

Stage 5c: large lacunar spaces interconnect and are blood-filled. At the caudal end 5 of the embryonic disc is a clump of extraembryonic mesoblasts. Hypoblast lines the coelomic cavity. Extraembryonic mesoderm appears between hypoblast and trophoblast.

Stage 6 (13–18 Days, 0.2–0.3 mm): Gastrulation Begins (Figs. 3.20 and 3.21)

This stage is marked by the initiation of gastrulation by formation of the primitive streak and cell movements, as discussed previously. It results in a tri-laminar embryo.

Fig. 3.20 Stage 6b
gastrulation. (Reprinted from
Gilbert SF, Barresi
M. Developmental Biology,
11th ed. Sinaurer: Sunderland,
MA, 2016. Reproduced with
permission of the Licensor
through PLSclear)

Stage 6 primitive streak > gastrulation begins
ENDODERM displaces hypoblast
MESODERM: extraembryonic, lateral plate, paraxial, axial

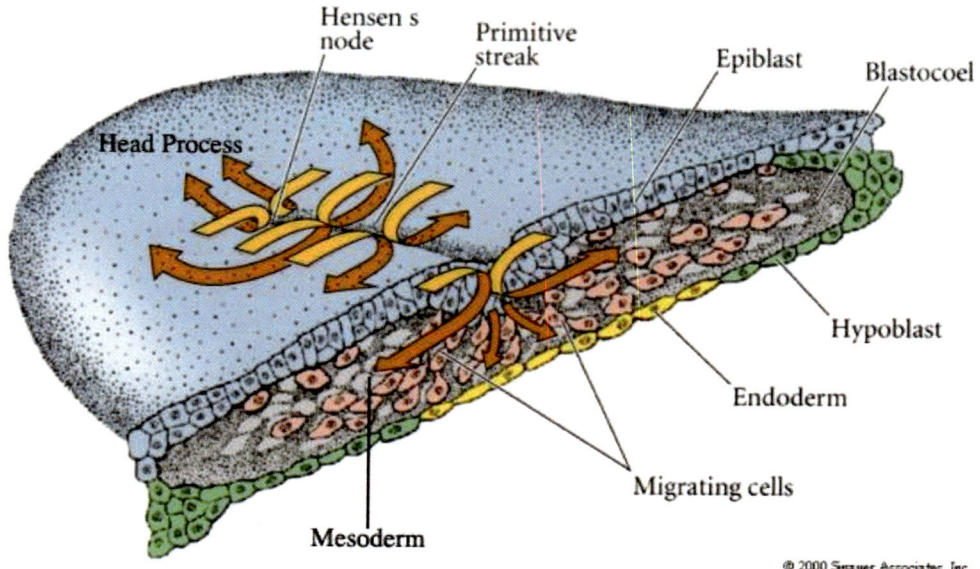

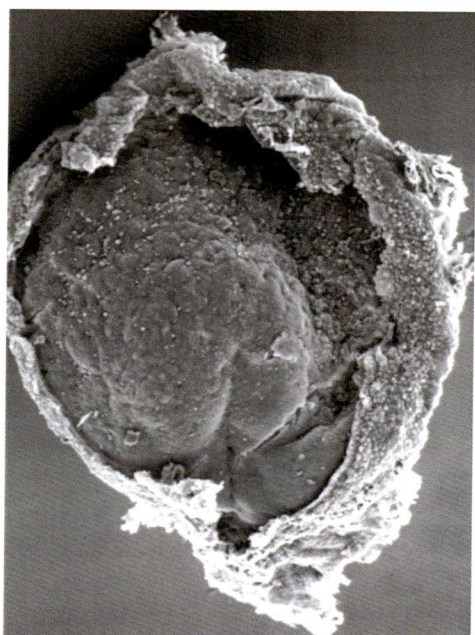

Fig. 3.21 Stage 7: 0.4 mm, 15–17 days. (1) Gastrulation. (2) Blood
vessels: (a) chorionic villi, umbilical vesicle. (3) Rostral to PS, notochordal process: (a) arises from the primitive node, (b) extends to the
prechordal plate, (c) floor breaks down, (d) True notochord present. (4)
Order of ingresssion: (a) intraembryonic endoderm, (b) extraembryonic
mesoderm, (c) intraembryonic mesoderm: PAM then LPM. (5) Epiblast
and endoderm fuse at two sites: (a) bucco-pharyngeal membrane, (b)
cloacal membrane. (Courtesy of Prof. Kathleen K. Sulik, University of
North Carolina)

Stage 7 (19–22 Days, 0.4 mm): Notochord
(Fig. 3.22)

Gastrulation continues with the formation of mesoderm and
the caudal-cranial development of a notochordal process
which develops in the opposite direction of gastrulations.
The establishment of notochord sets the stage for induction
of the neural plate and eventual embryonic folding.

Stage 8 (23–24 Days, 0.5–3 mm): Neurulation
(Fig. 3.23)

The embryo is shaped like a pear. Primitive node and notochord visible.

Characterized by the primitive pit, notochordal canal, and
neurenteric canal. Its mesoderm is presomitic, meaning that
somitomeres have formed (up to 18), the first 7 of which
represent the so-called mesoderm supplying the orbit and
pharyngeal arches 1–3. The initial seven somitomeres do not
undergo transformation to somites.

At this time, uteroplacental circulation was established.
Decidua capsularis covers over the embryo.

Carnegie Stage 8

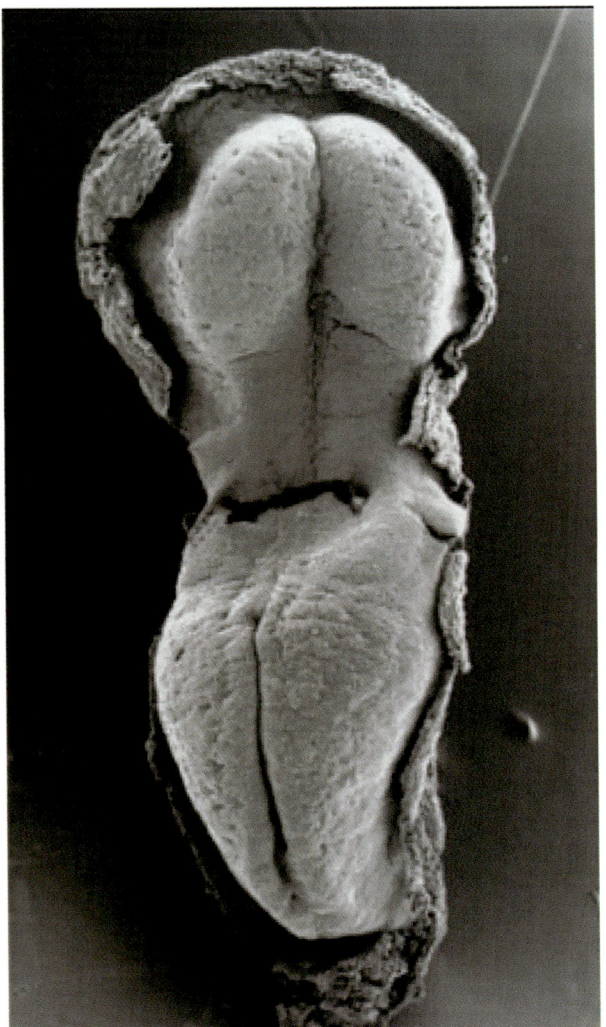

Fig. 3.22 Stage 8: 17–19 days. 1.0–1.5 mm. (1) Gastrulation complete. (2) Notochordal process extends *opposite* to primitive streak. (3) Segmentation: cranial > caudal. (4) Homeotic code for each segment. (5) Hindbrain Otx-2 induces midbrain and forebrain. (6) Headfold. (Courtesy of Prof. Kathleen K. Sulik, University of North Carolina)

Stage 9 (25–27 Days, 1.5–2.5 mm, 1–3 Somites, First Aortic Arch) (Fig. 3.24)

Histologic Features

The "embryo proper" is defined by the neural groove.

Embryonic disc sufficiently developed to body systems.
Primitive streak involves 1/4 to 1/3 of length of embryo.
Mesoderm organized longitudinally. Intermediate mesoderm is present, but no kidney structures present, so not identifiable.
Coelom splits lateral plate mesoderm.

Carnegie Stage 9

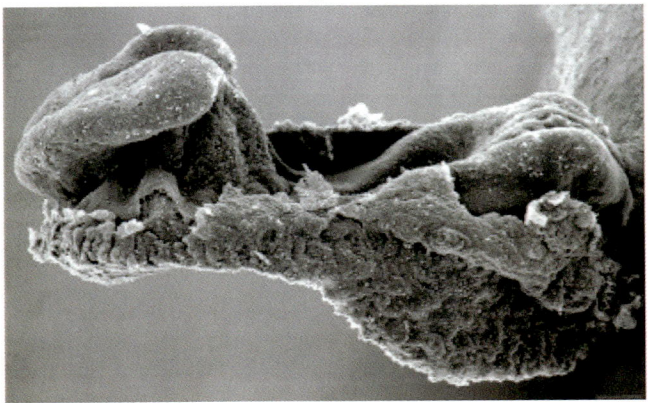

Fig. 3.23 Stage 9: 19–21 days, 1.5–2.5 mm, 1–3 somites. (1) Neural groove, neural folds, neural crest migrates. (2) Lateral plate mesoderm vacuolates. (3) Endoderm opens into yolk sac. (4) Folding of dorsal aortae. (5) Connection of DA with PHC. (6) First pharyngeal arch (first aortic arch—transient). (Courtesy of Prof. Kathleen K. Sulik, University of North Carolina)

Carnegie Stage 10

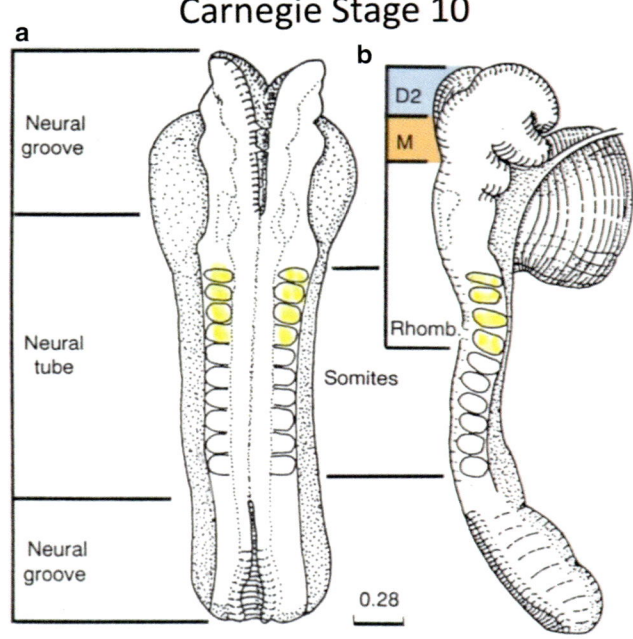

Fig. 3.24 Stage 10: 22–23 days, 2–3.5 mm, 4–12 somites. (1) Neurulation. (2) Heart tubes. (3) Second pharyngeal arch (second aortic arch—transient). (Reprinted from O'Rahilly R, Müller F. The Embryonic Human Brain: An Atlas of Developmental Stages, 3rd ed. Hoboken, NJ: Wiley-Liss; 2006. With permission from John Wiley & Sons)

Pericardial cavity (intraembryonic coelom) is a horseshoe that starts with occipital and develops via fusion of adjacent cavities.
Notochordal cavity
Notochordal plate

Vascular System

Blood vessels arise in diverse sites: chorion, connecting stalk, umbilical vesicle, and the embryo proper. In embryo, the primitive head plexus and primitive hindbrain channels are flanked by dorsal aortae that connect to the heart via the first aortic arch arteries. Few or no blood cells at this stage. Circulation is ebb-and-flow.

Heart: Endocardial plexus has three parts: atrial (connected to vitelline circulation), ventricular, and conal (bulbar outflow tract).

Digestive System

Foregut assembled from folding. Caudal portal not defined. Foregut related to the neural groove. Although a pit is present in the ventral foregut wall, the thyroid does not appear until stage 10.

Nervous System: Major Divisions of the Brain

Neural folds make their debut, but become dominant later in stage 10. Note that neural folds are separated from each other by a rostral notch which leads directly to the oropharyngeal membrane—at r0. Neural folds are closest together at the junction of the hindbrain and spinal cord. It is here, at stage 10, that neural tube closure will commence.

Neural crest: Head ectoderm differentiating—midbrain, trigeminal, facial, otic, and occipital zones have mitotic figures.

Eye: Optic primorida not yet present.

Ear: Otic disc appears.

Stage 10 (28 Days, 2.5–3.5 mm, 4–12 Somites, Second Aortic Arch) (Figs. 3.25, 3.26, and 3.27)

Characteristics of this stage are as follows: fusion of neural folds in progress, optic sulcus often present, pharyngeal arch

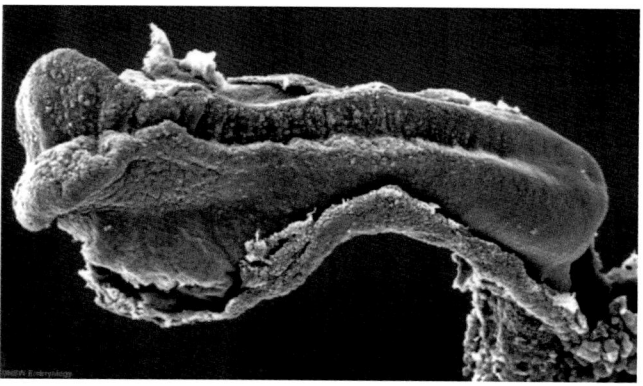

Fig. 3.25 Stage 10 SEM showing 12 somites, posterior PAM still in somitomeric form. (Courtesy of Prof. Kathleen K. Sulik, University of North Carolina)

1 visible, cardiac loop seen, laryngeal sulcus, intermediate mesoderm identifiable (future genitourinary system).

Histologic Features

Primitive streak: Limited to the posterior (caudal) end of the embryo (where the sperm entered). With the rapid growth of the embryonic plate, the length of the primitive streak (PS) in relation to the embryo gets progressively smaller. Caudal to the neurenteric canal are dense axial cells representing PS. Caudal embryo consists of thickened endoderm of the hindgut, the PS, and, alongside of the neurenteric canal, undifferentiated mesenchyme that will later become differentiated mesoderm. Above, is the neural plate.

Somites: 4–12 pairs are present, the first four being the occipital somites. Recall that Sm11 becomes S4 and that, in tetrapods, this is no longer part of the trunk, but is added on to the cranial base. Fishes have a three occipital somite cranial base.

Notochordal plate: This is present senso stricto only where the notochordal cells have become completely separated from the underlying endoderm. Rostral to the neurenteric canal, the notochordal plate remains fused with the endoderm and forms the roof of the foregut. This zone is marked in the future by Rathke's pouch and demarcates the location of the future basisphenoid.

Neurenteric canal: Although this was prominent at stage 9, it is now disappearing.

Prechordal plate: This most-anterior mesoderm lies just beneath the prosencephalon. A basement membrane separates it from the brain. Prochordal cells migrate during stage 9–10. They are interposed between the eye fields and push out transversely at 180 degrees. It was thought that PCM was the source of the extraocular muscles, but Noden has shown them to originate from somitomeres.

Umbilical vesicle: This structure is trilaminar on SEM— mesothelium, mesenchyme, and endoderm.

Cardiovascular System

Second aortic arch artery develops from the conus. Pericardial cavity is present and communicates with intraembryonic and extraembryonic coeloms. These spaces serve through stage 10 as passive conduits for nutritional fluids, while the blood vessels are getting organized to handle cardiac-based circulation.

Heart: Cardiac contractions begin at this stage. The wall of the heart consists of endothelium, cardiac jelly, and a thin covering of myocardium. Three steps in heart development are recognized during stage 10. (1) The endocardial primordium consists of a plexus on either side of the foregut. (2) These endocardial tubes fuse to form a single tube. At the time of 7–10 somites, an S-shaped flexion forms. It has, from anterior to posterior, a conotruncus, that will give off aortic

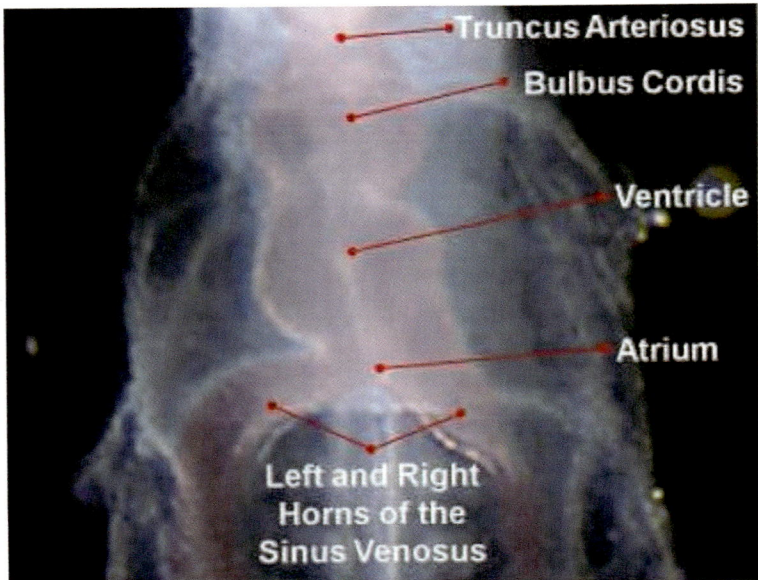

Carnegie Stage 10b

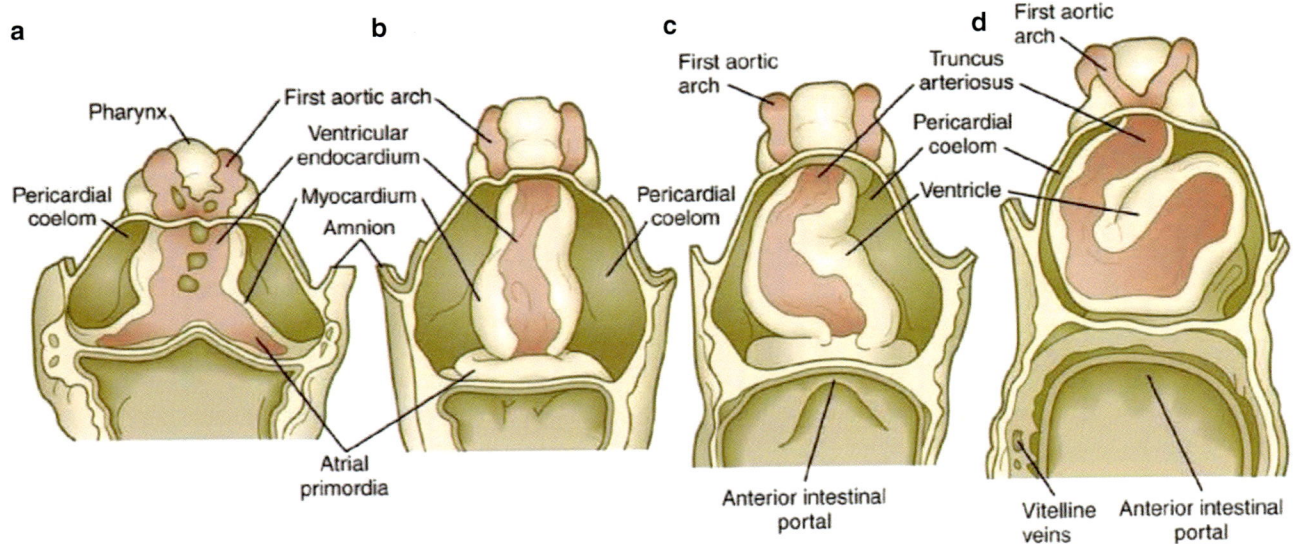

Fig. 3.26 Stage 10 showing folding of the heart and first aortic arches. (Left: Courtesy of Prof. Kathleen K. Sulik, University of North Carolina. Right: Reprinted from Carlson BM. Human Embryology and Developmental Biology, 6th edition. St. Louis, MO: Elsevier; 2019. With permission from Elsevier)

arch arteries, ventricles, and, caudally, the future sinus venosus (not seen at stage 10). (3) S-curve is formed by the ventricular loop, causing asymmetry.

Digestive/Respiratory Systems

The first pharyngeal arch is visible, clumped around the first aortic arch artery. At the apex of the foregut, the oropharyngeal membrane is present. It will eventually demarcate the zone between oral ectoderm infolded from the pharyngeal arches and the true endermal lining of the gut. Nonneural ectoderm of the most anterior neural folds produces paired adenohypophyseal placodes that will be tucking inside the future mouth to form the nonneural pituitary gland; thyroid primordium in the ventral wall of the foregut. Ventral pharynx has a laryngotracheal sulcus at the caudal end of which is the pulmonary primordium.

Carnegie Stage 11

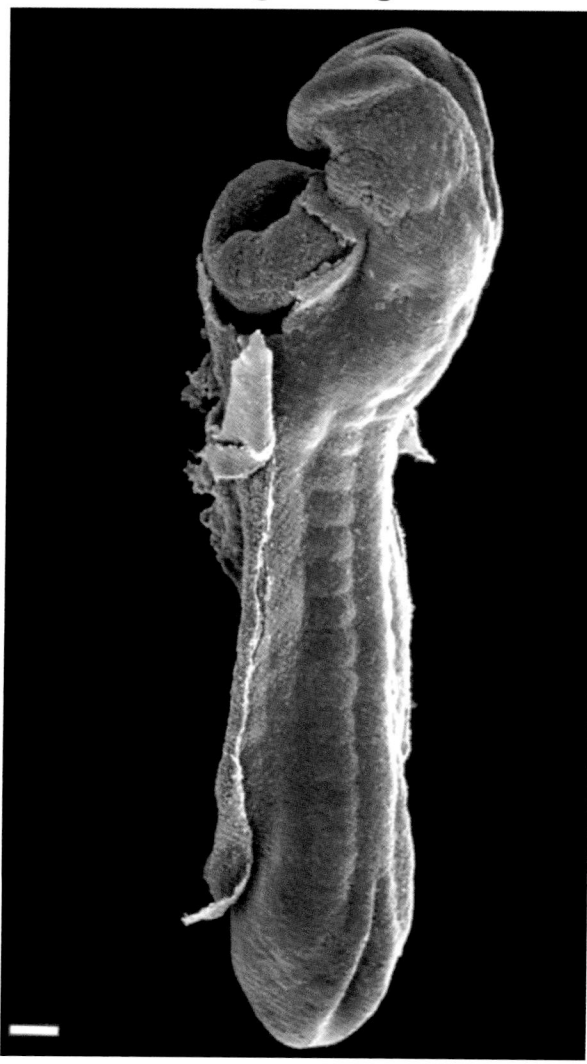

Fig. 3.27 Stage 11: 23–26 days, 2.5–4.5 mm, 13–20 somites. (1) Ectoderm: rostral neuropore closes. (2) Third pharyngeal arch (third aortic arch—persists). (Courtesy of Prof. Kathleen K. Sulik, University of North Carolina)

Nervous System: The Neural Tube and Optic Primordium

Neural folds elevate. The hindbrain subdivides, first into the rhombomeres as described by O'Rahilly with four segments, A–D. A terminal notch seen at stage 10 in the forebrain marks the future telencephalon. The neural folds of diencephalon are divided into an optic part (D1) and a post-optic part (D2). A midline thickening of D1 will become the chiasm. Rhombomere D (r8–r11) and the cervical spine are grown fastest and closure of the neural folds starts here. CNS gets longer proportional to addition of new somites. Closure of the neural groove gets started at time of somite 5 (neuromere c1). It proceeds cranially and caudally with the caudal neuropore always located opposite the latest pair of somites.

Neural crest: NC cells form the cranial ganglia and migrate around the brain and into the future face.

Eye: At the 7–8 somite stage, the neural folds of D1 form the optic primordium which extends medially as the chiasmatic plate. An indented zone of D1 will be the optic sulcus. The ventricular surface of D1 is smooth but that of D2 has a rounded prominence that projects into the ventrical as the future thalamic nuclei.

Ear: Otic plate/placode contributes cells to the vestibule-acoustic (facial) neural crest.

Stage 11 (29 Days, 2.5–4.5 mm 13–20 Somites, Third Aortic Arch) (Figs. 3.28, 3.29, and 3.30)

External: Rostral neuropore closing. second pharyngeal arch. Otic invagination.

Internal: Optic vesicle. Sinus venosus, oropharyngeal membrane breaks down—relates to future division between pharyngeal arches 2 and 3.

Histologic Features

Somites: 13–20 pairs. Note that somite 1 is quite small and lies adjacent to r8 vagal/accessory neural crest. S1 has almost no contact with overlying ectoderm. Somitomeres/somites form every 6 h. Process of forming 44 units involves 11 days (day 20–31).

Coelom: This refers to a walled-off space present in the intraembryonic or extraembryonic mesoblast (primitive undifferentiated mesoderm). Mesoblastic cells pass through the primitive streak and migrate outward, coalescing to form vesicles, the tubes, then cavities. A horseshoe-shaped cavity forms in the rostral end of the embryo and extends backwards on either side to become confluent with the extraembryonic coelom. These passages along extraembryonic fluids with nutrients gain access to the interior of the embryo during stages 9–10 until the cardiovascular system takes over. From the internal surface of the coelom mesoblastic cells the visceral lining layers of the internal organs are produced. Coelomic epithelium can readily transition to mesenchyme or vice versa. At the level somite 1, the coelom is involved with the sinus venosus and liver. It contributes the mesenchymal elements to the liver, whereas the epithelial elements come from the gut.

Fig. 3.28 Stage 11 showing the pharyngeal arches: the second pharyngeal arch is well shown, third arch is in formation. (Courtesy of Prof. Kathleen K. Sulik, University of North Carolina)

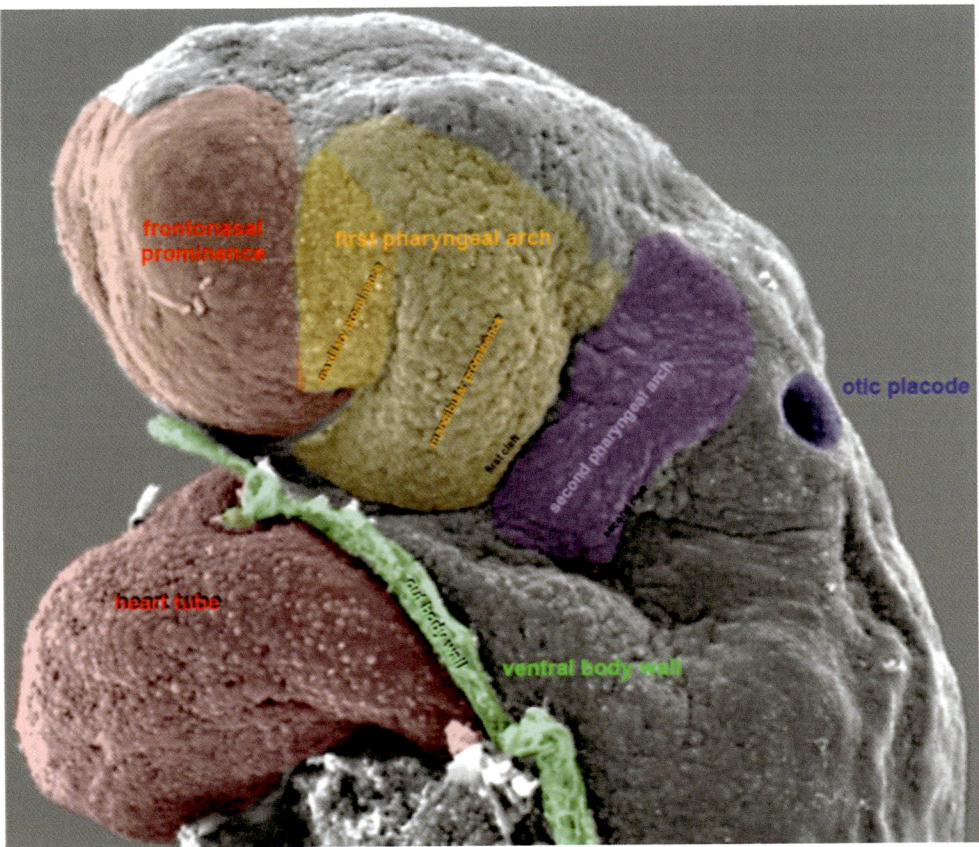

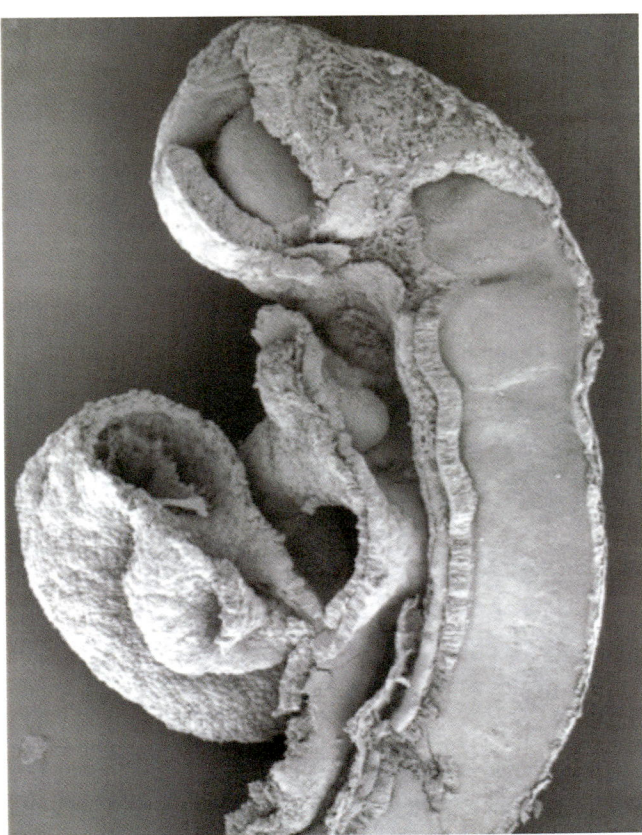

Fig. 3.29 Stage 11: midline section showing rhombomeres, bucco-pharygeal membrane, yolk sac. (Courtesy of Prof. Kathleen K. Sulik, University of North Carolina)

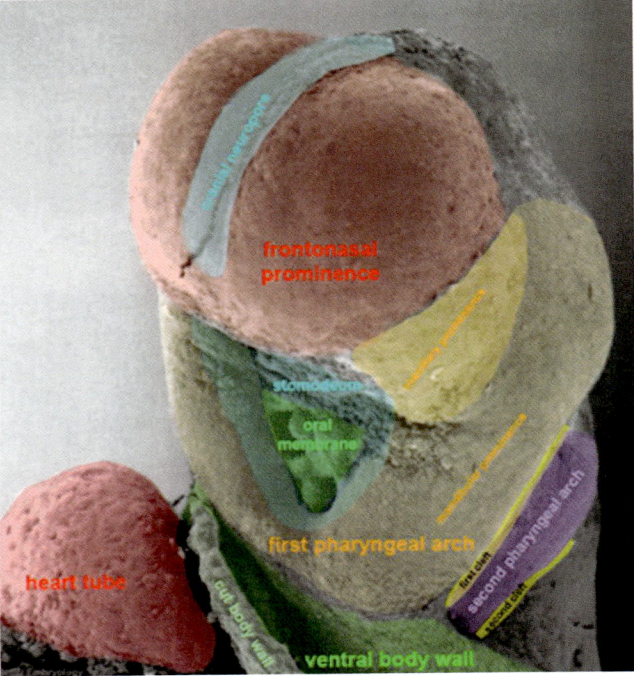

Fig. 3.30 Bucco-pharyngeal membrane constructed from non-neural ectoderm externally and endoderm internally. Lining of oral cavity is first, second, and third arch but second arch is overtaken. BPM marks the boundary in the pharynx between third arch and fourth arch. (Courtesy of Prof. Kathleen K. Sulik, University of North Carolina)

Cardiovascular System

The third aortic arch artery develops from conus. Most embryonic vessels are simple endothelial tubes except in the heart. Cardiac endothelium is surrounded by jelly and enclosed by contractile myocardial tissue, itself derived from coelomic epithelium and bathed in coelomic fluid. Vascular system maturation is cranial-caudal. Capillaries from the rostral umbilical vesicle communicate with the heart. By 16 somites, the sinus venosus is present. Common cardinal veins have not yet reached the atria, so the true circulation is not yet established. Embryonic blood cells are very few, but blood islands are active in the umbilical vesicle and will produce hematopoietic cells in the next stage. CNS angiogenesis is proceeding. In sum, the embryo at stage 11 lives via a system of endothelial channels that transport a virtually cell-free fluid via ebb-and-flow and the primitive pulsations of the myocardium. By the end of stage 11, the sequence of cardiac structures is: right atrium, left atrium, atrioventricular canal, left ventricle, right ventricle, conotruncus, and aortic arch arteries.

Digestive and Respiratory Systems

The second pharyngeal arch is present. Oropharyngeal membrane breaks down in anticipation of third arch at next stage. Esophago-pulmonary groove is present. Liver diverticulum grows into septum transversum.

Nervous System: Closure of the Rostral Neuropore

Closure of rostral neuropore is bidirectional; from hindbrain forward to midbrain and from optic chiasm backward toward roof of diencephalon to form the lamina terminalis with closure complete to this level at somite 14. By somite 20, the entire forebrain is closed. Neural ectoderm becomes isolated from contact with amniotic fluid; compensatory proliferation of capillaries all over the tube takes place. As the walls of the tube thicken, capillaries send in penetrating branches to supply deep-lying cells at ventricular cavities. Forebrain is still quite simple with D1 being the optic primordium and midline chiasmatic plate, the caudal limit of which is the optic recess. The floor of D2 contains the neurohypophysis, opposite of which lies, in the future mouth, the adenohyphysis in the apex of the oropharyngeal membrane. Rostral to D1 is telencephalon created by the fusion of the rostral neuropore and marked by lamina terminalis and the commissural plate. Mesencephalon at this point is large and poorly-defined consisting, at this point, of a single segment (it will eventually have two mesomeres).

Neuromeres: As defined by O'Rahilly, neuromeres are visually identifiable. Recall that this is from morphology, not genetic mapping (as per Puelles and Rubinstein), so it is more simplistic (see Chap. 5 on the neuromeric system).

Rhombomeres are identified by ganglia, with trigeminal defining r2, faciovestibulocochlear defining r4, glossopharyngeal defining r6, and vagal determining r7. Midbrain has in stage 11 a single mesomere.

Neural crest: Cells are produced at all levels, but appear to be "given off" at this stage. MNC spreads toward the optic region.

Eye: Right and left optic primordia are a U-shaped continuum at the optic chiasma. At 14 somites, optic evagination occurs with continuity between the optic ventricle and forebrain. At 14–16 somites, it acquires surrounding midbrain neural crest. At 17–19 somites, the optic vesicle is formed. It is kept separate from overlying ectoderm by the ensheathing neural crest.

Ear: Otic disc is sharply demarcated from surrounding ectoderm. Recall the neural ectoderm arises from stem

Carnegie Stage 12

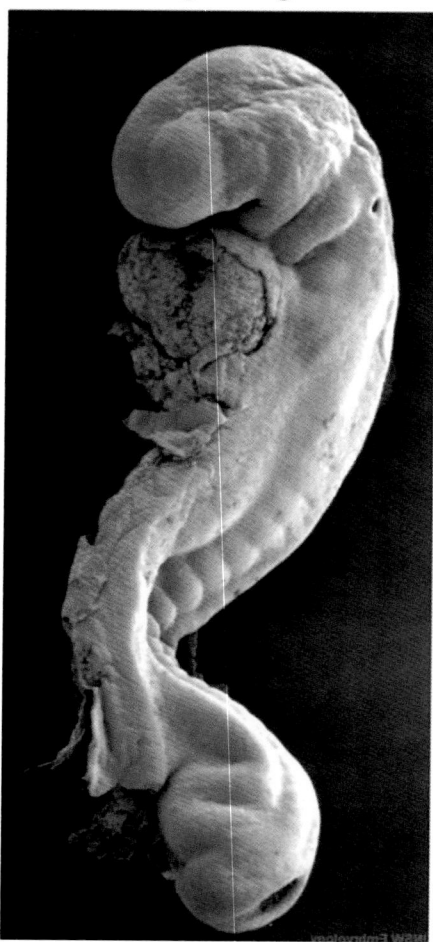

Fig. 3.31 Stage 12: 26–30 days, 3–5 mm, 21–29 somites. (1) Ectoderm: caudal neuropore closes, forebrain developing. (2) Placodes: optic, otic (open to the skin). (3) Mesoderm: stomodeum, heart large, upper limb bud. (4) Fourth pharyngeal arch (fourth aortic arch—persists). (Courtesy of Prof. Kathleen K. Sulik, University of North Carolina)

cells in the epiblast and is segregated from the epibast at stage 8.

Stage 12 (30–31 Days, 3–5 mm, 21–29 Somites, Fourth Aortic Arch) (Fig. 3.31)

External: 21–29 somite pairs present. third pharyngeal arch visible. The first and second arches are fusing and their respective aortic arteries are disintegrating. Otic vesicle almost closed. Caudal neuropore is closing. Upper limb buds appear.

Internal: present are the interventricular septum, lung buds, and the dorsal pancreas.

External Form

Four occipital somites are lined up with the primordia of the roots of hypoglossal nerve. With the three pharyngeal arches, three points of contact (bars) exist between ectoderm and endoderm. Furthermore, each arch is subdivided into dorsal and ventral sectors; these correspond to neuromeres (first arch r2–r3, second arch r4–r5, and third arch r6–r7). Just caudal to third arch is condensed mesoblast of the future larynx. This will be enclosed at stage 13 by the fourth pharyngeal arch. Note the third, fourth, and fifth pharyngeal arches become internalized. Cervical sinus, the boundary between the externalized first/second arch complex and the cervical somatic derivatives, is located here. Otocyst maintains a pore.

Rostral neuropore closed at stage 11 and caudal neuropore closes at stage 12. Shape of embryo changes to a C-shaped curve with the increasing mass of the spinal cord, somites, and surrounding mesoderm.

The extensive aperture between gut and umbilical vesicle (yolk sac) undergoes a pursestring-like contraction with definition of the umbilical stalk. This will include the paired umbilical arteries and veins. Upper limb bud appears opposite somite 8–10 (neuromeres c4–c7).

Cardiovascular System

Fourth aortic arch artery develops from conus.

During stage 11, embryonic nutrition depends upon exocoelomic fluid brought into direct contact with the interior of the embryo by the coelomic cavities. This was an ebb-and-flow circuit, abetted by early contractions of the primitive heart. By 30 somites, the requirements of the embryonic mesoblast exceed this system. The brain and skin ectoderm remain independent of this system, being nourished by amniotic fluid. When the neural plate rolls into the neural tube and sinks beneath the skin, a new system is required. As early as stage 10, a primitive head plexus and primitive hindbrain channels develop alongside the CNS. These are supplied by dorsal segmental arteries from the dorsal aorta,

primitive trigeminal, facial, otic, and hypoglossal arteries. Simultaneously, a separate plexus differentiates in situ in the mesoblast lateral to the neural tube. These channels will connect with the external carotid system to form the vasculature of the future scalp. At 14 somites, umbilical arteries and veins extend to the allantoic diverticulum and thence to the connecting stalk.

Stage 12 presents us with the first connected circulatory circuit. This requires (1) overhaul of the venous end of the heart, (2) creation of dedicated CNS vasculature, (3) cardinal veins, (4) the liver plexus, and (5) modifications in the vitelline plexus into large trunks leading to the atria of the heart. Enlarged parts of vitelline plexus become right and left atria, each side of which enlarges to make common cardinal vein. Note that left atrium becomes cut off from the sinus venosus which now leads exclusively to right atrium. Sinus venosus receives all venous blood flow from the embryo. Placenta does not yet exist.

Digestive/Respiratory Systems

Stage 12 is marked by a dramatic expansion in epithelium of the gut. In the pharynx, the roof proliferates compared to the floor which remains thin. Respiratory tissue expands.

Nervous System: Closure of Caudal Neuroport, Secondary Neurulation Begins

The roof of rhombencephalon becomes thin. The basal plate contains the motor nuclei. Twelve rhombomeres are present. The eighth rhombomere was originally considered to cover the entire medulla, but now it was recognized by Puelles that r8 is broken up into distinct genetic units (r8–r11). Telencephalon enlarges. Area adjacent to the former rostral neuropore is termed the commissural plate. Diencephalon has future thalamic structures. Midbrain now has two segments, m1 and m2. Nasal discs become thickened.

Eye: Optic neural crest at its maximum. Beginning of the *sclera*.

Ear: Otocyst, originally open to the external surface via a pore. Represents a detached island of neurectoderm having simple skin epithelium; has nothing in common with surrounding skin.

Stage 13 (32 Days, 4–6 mm, 30–31 Somites, Fifth Aortic Arch—Abortive) (Fig. 3.32)

Summary External: 30 + somite pairs (somite count no longer useful after this stage). fourth pharyngeal arch forms; otic vesicle now closed; optic disc not indented.

Internal: retinal and lens discs present; trachea begins; right and left lung buds are visible; heart has septum primum.

Carnegie Stage 13

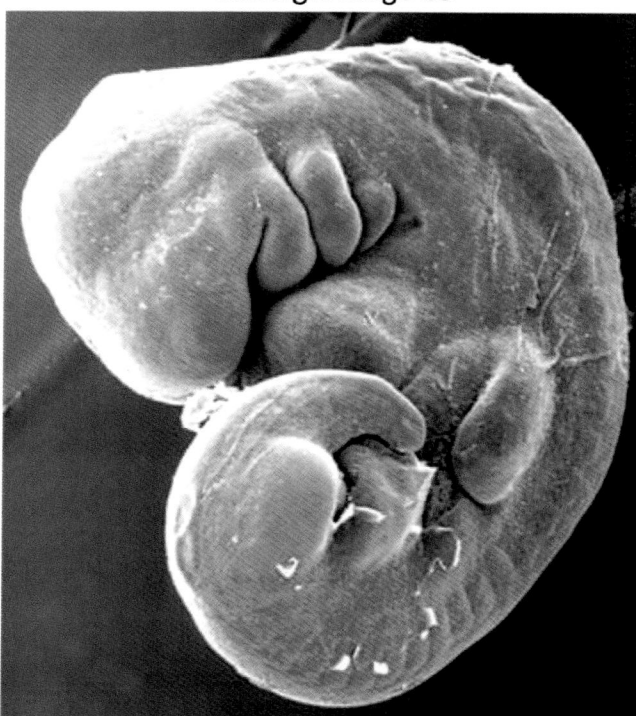

Fig. 3.32 Stage 13: 26–30 days, 26–30 mm, 30–31 somites. (1) Nasal placodes. (2) Lower limb bud. (3) Fifth pharyngeal arch (fifth aortic arch—fails or disintegrates). (Courtesy of Prof. Kathleen K. Sulik, University of North Carolina)

External Form

Three-dimensional form of the embryo determined by the shape of the rapidly growing CNS. Through the glossopharyngeal arch runs the third aortic arch artery and just behind it is a depressed zone where the surface ectoderm is in contact with the pharyngeal endoderm. The floor ceases to be visible in stage 14 due to growth of tissues in front of and behind this depression such that the floor of this triangular area becomes covered over, thus forming the cervical sinus. A new fourth pharyngeal arch is forming in this region. Both upper and lower limb buds are seen.

Cardiovascular System

This stage is marked by further development of embryonic-material circulation. fifth aortic arch artery from conus and dorsal aorta appears and then fails to develop. The enormous CNS is well perfused through the surface capillary network, even though distinct arterial structures are in formation. The lateral plate mesoderm that will form lungs and gut has numerous small arteries in situ. The caudal end of the embryo is rapidly developing and contains a vascular plexus from which arise the umbilical arteries. These progress backward into the connecting stalk via which they unite with the chorionic circulation. Interchange now occurs between the embryonic blood in the chorionic villi and maternal blood circulating in the intervillous spaces. Venous return via a plexus in the connecting stalk dumps into the right and left umbilical veins and thence into the sinus venosus of the heart.

An additional circulation develops between the body and the vitelline plexus consisting of a single vitelline artery from the splanchnic aorta and right and left vitelline veins that connect with the liver sinusoids. This hepatic plexus empties into the sinus venosus. Thus, sinus venosus received venous return from three sources: (1) from the embryo via the cardinal veins, (2) from the placenta via the umbilical veins, and (3) from the yolk sac via the vitelline veins to the liver. The venous system was initially symmetrical but, with the advent of the sinus venosus and the hepatic plexus, a one-sided portal system becomes established. We shall not spend more detail on this here.

The heart begins to undergo functional partitioning in stage 12.

The liver has the following blood vessels. (1) The intrinsic hepatic plexus arises from coelomic mesoderm. (2) This intrinsic system receives flow from the yolk sac via the vitelline veins. (3) The left umbilical vein makes an anastomosis with the hepatic plexus to form a single ductus venosus that empties into the inferior vena cava. Hepatocardiac veins coalesce to drain the hepatic plexus into the sinus venosus.

Digestive/Respiratory Systems

In stage 12, certain zones of endoderm become determined to form lung, dorsal pancreas, and liver. In the floor of the pharynx, the two lobes of the thyroid primordium appear. Pharyngeal pouch 3, interposed between the second and third pharyngeal arch masses, produces thymus.

Trachea appears. The right primary bronchus descends vertically, whereas the left bronchus is directed laterally.

Nervous System: Neural Tube Is Closed, Cerebellus Appears from r1

In form, the CNS is still tubular, with basal and alar plates. Early neurons are emerging as ventral roots; dorsal roots are lagging behind in development. Rhombomeres are clearly distinguished. Motor columns are present in the brain stem and those of the spinal cord are in continuity with hypoglossal. Midbrain has two distinct segments. Diencephalon also has two main components: D1 gives rise to the optic evagination and D2 to the thalamus.

Eye: Optic retina, pigmented retina, and optic stalk are seen. On the outer surface of the retinal disc, a marginal zone of nuclei forms a meshwork which will eventually send fibers backwards into the optic nerve. Pigmented retina is much thinner. The optic vesicle is covered by a basement membrane external to which, in the mesenchyme that originated from MNC, angiogenesis is taking place. Retinal disc moves outward to come into contact with overlying p5 ectoderm, the optic placode. Induction of the lens results.

Ear: Otic vesicle interacts with surrounding mesoderm likely from PAM of somitomeres 5–7 plus facioacoustic neural crest from r4-r5 causing angiogenesis. The vestibular part of vestibulocochlear ganglion appears.

Carnegie Stage 14

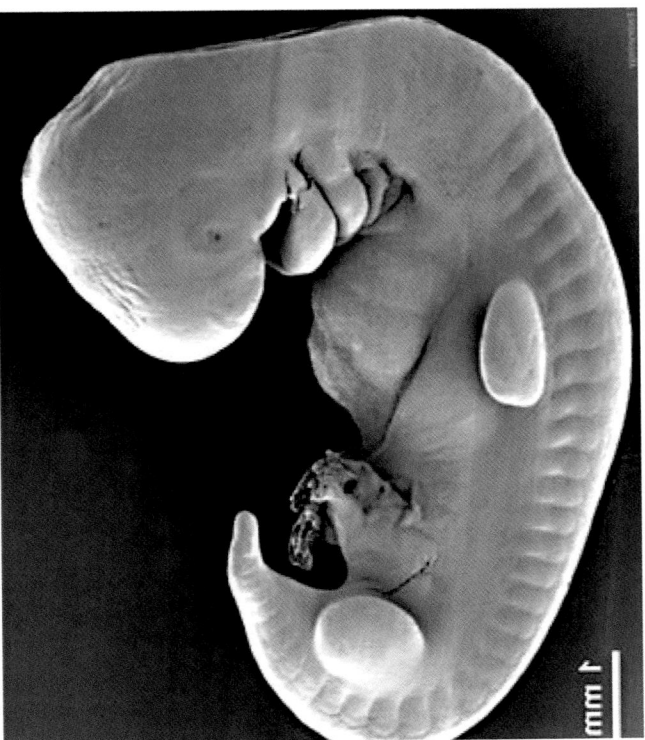

Fig. 3.33 Stage 14: 31–35 days, 5–7 mm, 32+ somites. (1) Endoderm: gut development. (2) End of pharyngeal arch period. (3) No sixth pharyngeal arch (sixth aortic arch is a plexus for pulmonary circulation). (Courtesy of Prof. Kathleen K. Sulik, University of North Carolina)

Stage 14 (33–35 Days, 5–7 mm, Sixth Aortic Arch for Pulmonary Circulation) (Figs. 3.33 and 3.34)

External: Lens disc invaginated and open to the exterior, lengthened upper limb buds.
Internal: Fifth pharyngeal arch, ventral pancreas, dorsal growth of the lung sacs, cerebellar hemispheres.

External Form

Cervical bend at the level of sclerotomes 5–6 (*Nackengrube of His*) anticipates the later flexure sequence of the vertebral column. first and second arches have not yet coalesced; third pharyngeal arch is beginning to become internalized and concealed with respect to them. second arch has notable dorsal (r4) and ventral (r5) zones. Because the head is largely translucent at stage 14, the following can be observed: the optic cup, trigeminal ganglion with V2 and V3 cells flowing into the dorsal/maxillary (r2) and ventral/mandibular (r3) zones of first arch, faciovestibular mass of neural crest enters dorsal (r4) zone of second arch, otic vesicle with well-defined endolymphatic appendage, and both glossopharyngeal and vagus nerves. Nasal plate is flat; it does not have the distinct elevated rim that defines it at later stages.

Vascular System

Sixth aortic arch artery present. Its formation and destiny are unique. Recall that first aortic arch is the passive deformation of the dorsal aortae leading to the heart as it is retropositioned beneath the pharynx. Aortic arches 2–5 were formed from the union of a dorsally directed sprout from the aortic sac and a ventrally directed sprout from the dorsal aortae. The two components of aortic arch 5 never unite and it aborts. Aortic arch 6 is formed but loses contact with the dorsal aorta to become dedicated to the lung. No remnant corresponding to AA6 exists from the dorsal aortae. The four divisions of the cardiac tube are still visible. Atrioventricular cushions are not united. The outflow tract is still a single lumen but beginnings of septation are taking place.

Digestive System

Stage 14 marks the close of the aortic arch period. There is no outcropping of endoderm distal to AA6. Thus, there can be no sixth pharyngeal arch. Furthermore, the fate of AA6 is strictly pulmonary. The fifth pharyngeal arch survives by perfusion from the external carotid supply dedicated to the fourth pharyngeal arch (superior thyroid). Fifth arch neural crest from r10 to r11 produces the arytenoids.

Prior to stage 14, the pharyngeal pouches were simply lateral extension of pharyngeal endoderm interposed between the aortic arch arteries and making contact with the external ectoderm such that barriers to neural crest migration were established. Thus, neural crest flowing down around the aor-

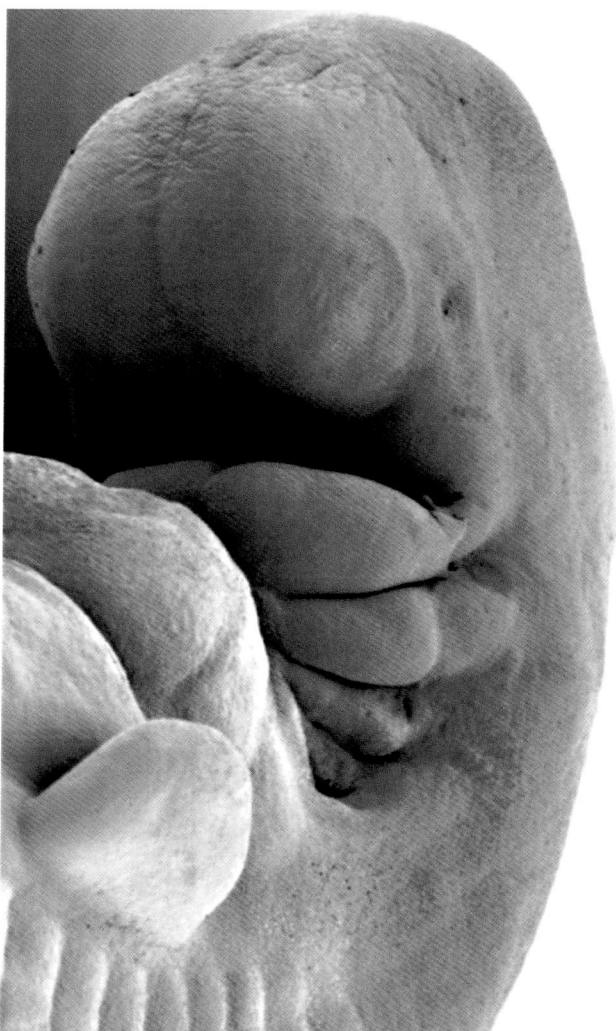

Fig. 3.34 Stage 14, 32 days 35 somites Adjacent to upper limb bud is liver and above that, the heart. Four pharyngeal arches seen, with the fifth arch internalized. Optic placode has a pit to the surface ectoderm. (Courtesy of Prof. Kathleen K. Sulik, University of North Carolina)

tic arch arteries created the mesenchymal expansions known as the pharyngeal arches, a process much like filling up a balloon with water.

The pharyngeal pouches are now pocket-like, each with future derivatives. Thymus and parathyroid tissues are identified. Thyroid is elongating downward, but remains connected to the epithelium of the oropharynx. It has two defined lobes and an intervening isthmus.

Merger of the first and second arches affects the oropharyngeal distribution of second arch. Sensory representation of second arch is eliminated from the epithelium (although it persists in the gustatory tissues of the tongue). Thus, the pharyngeal lining has V2 and V3 apposed to third arch IX. This reduction in representation of second arch to the tubotympanic recess means that the Eustachian tube represents the

interface between first and third arches with the lining of the tube being somatic sensory to V3 and conducts r6-r7 fibers of vagus into the ear canal as the Arnold's nerve.

Intestinal growth proceeds with individual zones defined by homeotic genes. Ventral pancreas arises as an evagination from the bile duct.

Respiratory System

Arytenoid swellings corresponding to the fifth pharyngeal arch appear in the pharyngeal floor. Trachea and esophagus are separate structures. The epithelium structure of the lungs, previously dominant, is now surrounded by mesenchyme from the adjacent coelomic walls—these being ultimately derived from visceral lateral plate mesoderm. The proliferation of the coelomic cells took place at stage 12. As they migrate to surround the pulmonary and alimentary epithelium, they leave behind an external "slime trail" of cells which constitute the lining mesothelium.

Within the coelomic mesenchyme ensheathing the lungs, an extensive capillary network develops by angiogenesis. These channels connect with the sixth arch pulmonary arteries.

Nervous System: Future Cerebral Hemispheres Are Defined

The neural tube corresponding to spinal cord develops from cranial to caudal, from medial (ventral) to lateral (dorsal), and from internal to external. It has three zones: (1) the ventricular zone contains the stem cells from which ependymal cells, glial cells, and neurons will develop; (2) the mantle zone at stage 14 contains the earliest neurons which give rise to ventral roots; and (3) the marginal zone constitutes free space for the neurons to grow into. Hindbrain cranial nerves have formed and individual hypoglossal roots are uniting. In the midbrain, oculomotor is now present. Cerebellum arises from the alar plate of r1. In the telencephalon, neural crest migration is now in contact with the brain at the olfactory area.

Eye: Optic cup is present. In the outer uveal layer, a distinct capillary network appears. Lens placode communicates with the surface with a definite pit.

Ear: The optic vesicle does not change much in size, but elongates due to the growth of the cochlear duct. Endolymphatic duct appears. Within the walls of the vesicle, a thickening develops for the future labyrinth.

Stage 15 (36–37 Days, 7–9 mm) (Fig. 3.35)

External: lens vesicle now closed; nasal placodes develop pits, hand plates.

Carnegie Stage 15

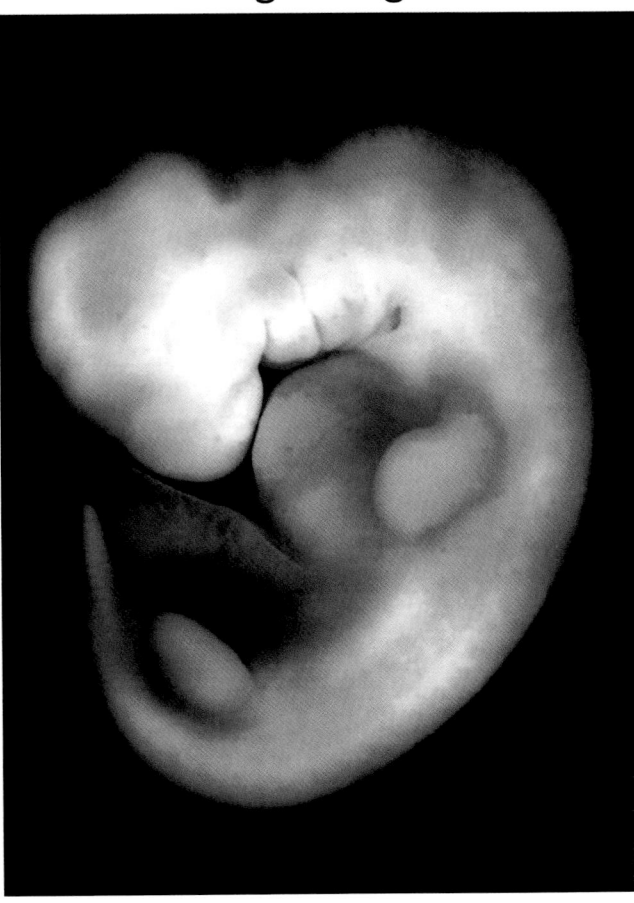

Fig. 3.35 Stage 15: 35–38 days, 7–9 mm. (1) Ectoderm: lens pit, otocyst, nasal pit. (2) Hand plate. (3) Umbilical cord. (4) Head: maxillary/mandibular fields, cervical groove. (5) This is the last stage at which somites can be seen on the surface. (Reprinted from The Kyoto Collection, Kyoto University Graduate School of Medicine. Courtesy of Prof. Kohei Shiota and Shigehito Yamada)

Internal: heart has foramen secundum, intestine for loop with distinct cecum, lobar buds, ureteral pelvis, defined cerebral hemispheres, retinal pigment.

External Form

Lens vesicles are closed and the external pit with which they communicated disappears. Nasal discs appear to sink into the MNC mesenchyme. Proliferation of mesenchyme around the rim of the discs created elevated prominences, the future nostrils. Ventral first arch forms auricular hillocks. In stage 15, the superficial tissues are thin, permitting visualization of somites, muscles, and ganglia along the length of the embryo. In subsequent stages, interposition of proliferating mesenchyme makes them no longer visible.

Cardiovascular System

Arterial part of the heart is distinct from the venous part of the heart which was derived from the vitelline plexus. Mesenchyme is distributed asymmetrically between ventricles and atria. Blood flow through the atrioventricular canal is now divided into left and right streams. Semilunar valves appear. Recall the cardiac neural crest has descended into the heart to contribute to valves and conduction system.

Thyroid primordium now detaches from the pharynx and descends. Primary bronchi are located in front of esophagus which has lengthened.

Respiratory System

Through stage 15, caudal trachea remains in close proximity to anterior wall of esophagus. This presents opportunities for congenital anomalies to occur. Esophagus begins to separate from trachea.

Nervous System: Diencephalon Develops Longitudinal Zones

Brain surrounded by identifiable primary meninx. Mesenchyme of secondary meninx is present but not defined. Rhombomeres remain identifiable. Geniculate ganglio of VII separates from the vestibulocochlear ganglion of VIII. Cochlar part of ganglion VIII now distinct from vestibular part. Abducent nerve seen. Decussation of trochlear nerve constant.

Eye: Lens now has its own capsule and contains lens fibers. Neural crest cells surrounding lens but nothing formed as yet. Lens vesicle and optic cup just below the surface. Surface ectoderm constitutes epithelium of future cornea.

Ear: Mesenchyme condenses to form otic capsule (not yet chondrified). Groove on the membranous labyrinth designates future semicircular ducts. Fibers from vestibular reach the otocyst. From the ventral segment of second arch, a subsegment forms the tragus.

Stage 16 (38–40 Days, 8–11 mm) (Figs. 3.36 and 3.37)

External: *Nasal pits* coming to midline, retinal pigment, arch 3 is disappearing from the surface, auricular hillocks, lower extremity (thigh, leg, and foot).

Internal: Foramen secundum; intestinal mesentery is present and rotation of the gut is starting; bipartite pelvis, no longitudinal cerebral fissure; neurohypophyseal evagination; lens pit D-shaped; semicircular ducts appear as thickenings.

Carnegie Stage 16

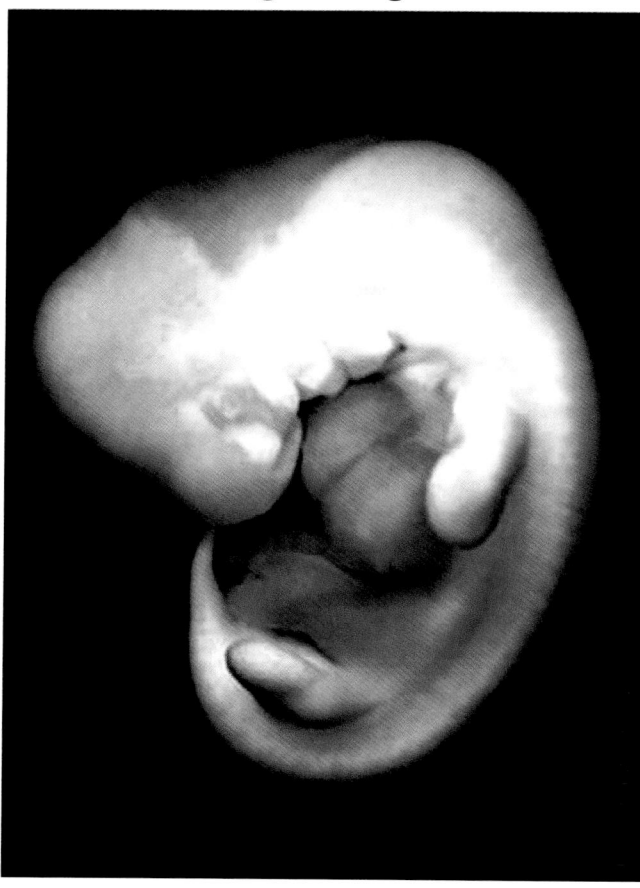

Fig. 3.36 Stage 16: 37–42 days, 8–11 mm. (1) Nasal pits now positioned ventral. (2) Nasolacrimal groove. (3) Eye: retinal pigment. (4) Auricular hillocks. (5) Arm. (Reprinted from The Kyoto Collection, Kyoto University Graduate School of Medicine. Courtesy of Prof. Kohei Shiota and Shigehito Yamada)

External Form

Midbrain and thalamic regions larger than stage 15. Details of special regions emerge.

Nasal discs have elevated borders and appear concave; moving ventrally. Nasofrontal groove marks the beginning of the nose.

Hyoid arch appears massive because of precursors of hillocks. Both first and second arches have grooves marking dorsal and ventral segments.

Cervical sinus was visible at stage 14, its floor is formed by third and fourth arches. At stage 16, these are receding and the sinus is closing.

Hand has a central carpus and digital flanges. Lower extremity has 3 growth centers. Rostrolateral (femoral, obturator nerve of lumbar plexus), caudomedial (peroneal and tibial nerve of sacral plexus), and foot (extension of tibial nerve).

Spinal ganglia present throughout the trunk.

Face

Roof of the mouth slopes ventrally between the nasal pits (Hinrichsen)—no sign of the upper jaw. Skin ectoderm is thickened over areas of activity (lower jaw, hyoid).

Between head ectoderm and brain is thin primary meninx carrying blood vessels to brain wall and cranial nerves. Nasal and otic proliferation.

Maxillary fields are bulging inferolateral to eye and nose, fields are widely separated. Premaxillary condensations.

Cardiovascular separation of aortic and pulmonic circulations into fourth and sixth AAs. Pulmonary develops bronchial tree.

Nervous System: Neurohypophysis Evaginates

Ascending fibers of posterior columns (dorsal funiculus) reach the medulla. Rhombomeres still present. Cells of somatic and visceral efferent nuclei of 5, 7, 9, 11 migrating laterally. Isthmus; Groove between cerebellum and midbrain. Midbrain still has two parts. SANS forms a solid trunk.

Eye forms D-shaped lens cavity. Perilentil vessels present from posterior and anterior chorodoidal. Shallow grooves appear above and below the eye; these mark the future eyelids.

Ear: Auricular hillocks all present long endolymphatic sac Wall thickening in otic vesicle preseages semicircular canals.

Stage 17 (41–43 Days, 11–14 mm) (Fig. 3.38)

External: Head enlarged due to brain. Trunk straighter with slight lumbar curve. Nasal pit now ventral, so nostril not seen on profile; full complement of ear hillocks both first and second arches; hand plate has digital rays; foot now a rounded plate. Surface manifestations of somites now

Fig. 3.37 Stage 16 showing nasolacrimal groove. At the top of the first cleft, adjacent to the otic vesicle, is future external auditory canal, flanked by auricular cartilage hillocks from first and second arches. (Reprinted from The Kyoto Collection, Kyoto University Graduate School of Medicine. Courtesy of Prof. Kohei Shiota and Shigehito Yamada)

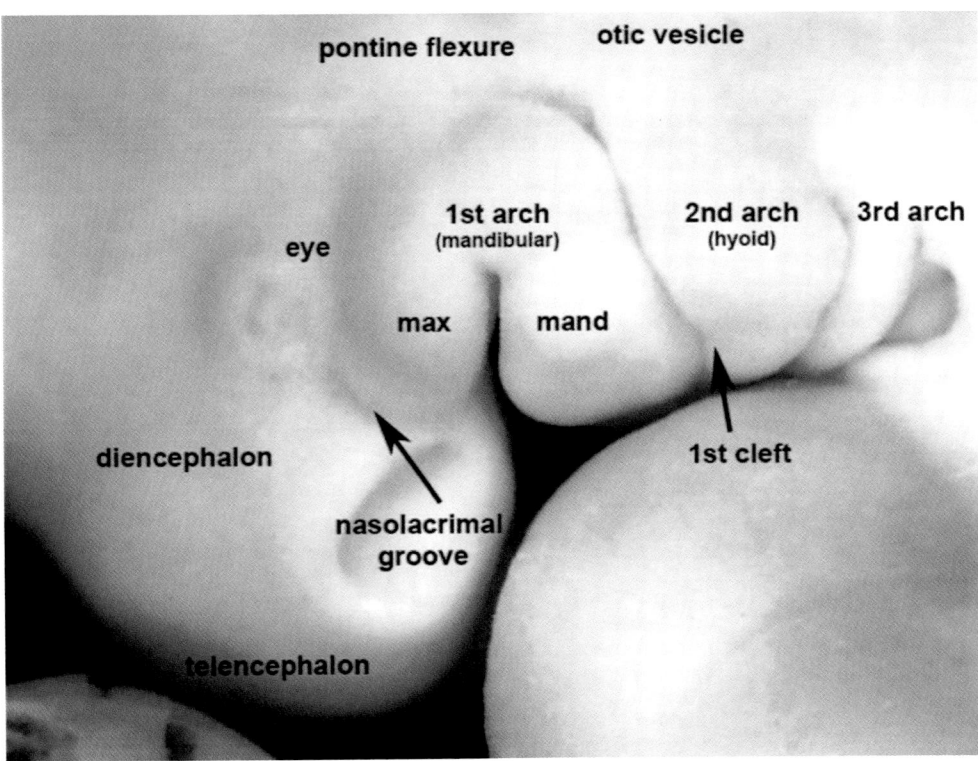

Carnegie Stage 17

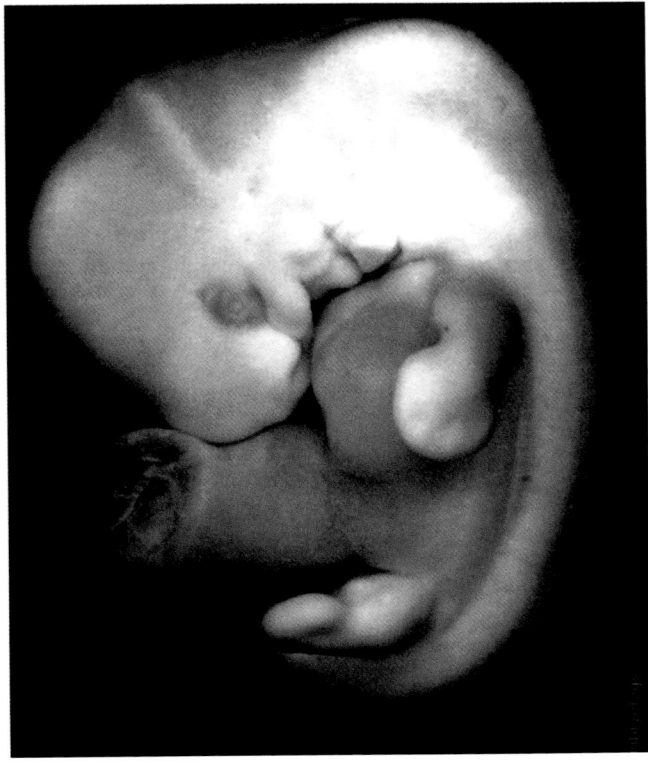

Fig. 3.38 Stage 17: 42–44 days, 11–14 mm. (1) Lens Pit, eye pigment. (2) External acoustic meatus. (3) Nasal pit moved ventrally (well seen here). (4) Mesonephric ridge. (5) Hand digital rays. (6) Thigh, ankle, footplate. (Reprinted from The Kyoto Collection, Kyoto University Graduate School of Medicine. Courtesy of Prof. Kohei Shiota and Shigehito Yamada)

seen only in lumbar region—mesenchymal development of the neck and thorax.

Internal: Foramen primum being eliminated, foramen secundum and semilunar cusps appear, fusion of A-V cushions (stages 16–18); palate begins; fusion pancreatic buds; bronchial tree has segmentations; functioning mesonephros with calices; olfactory bulb; retinal fissure almost closed and lens is crescent-shaped, semicircular canals imminent but not present; ear bones present.

External Form

Head and neck proportions increase—now have same profile area as entire body; fronto-mesencephalic length increased, just in front of precerebellar notch. Nostrils can only be seen on decapitated specimens; auricular hillocks fully developed—those of second arch are larger. First arch ventral hillock is tragus dorsal 2 hillocks crus helicis—hillocks are coalescing.

Upper limb bud has finger rays. Crenation characteristic of fingertips starts, but is prominent in next stage. Lower limb has rounded digital plate (paddle). Evidence of growth of muscles and the pelvic girdle. Surface marking from somites only lumbar and sacral.

Face

Furrows between processes (fields) are smoothing out—coalescence elevates the furrows. Intermaxillary field. Nasal passages end in cul-de-sac, the *hinteren Blindsack* of Peter, terminate in bucconasal membrane.

Pits become respiratory passages—formation of MNC fields. Regional specialization of epithelium. Outlines of nose seen, median rim of nostrils.

Six zones of odontogenic epithelium.

Nervous System: Future Olfactory Bulges, Future Amygdaloid Nuclei

Rhombomeres are beginning to recede; migration from medioventral cell column to pharyngeal arch (visceral) motor nuclei still in progress; Trigeminal is first to form. Geniculate and vestibulocochlear ganglia now separate. Olfactory tubercle and future olfactory bulb are prominent; olfactory fibers can be separated into medial (vomeronasal and terminal) vs. lateral (olfactory).

Eye: Retinal fissure almost closed. Retina differentiates. Lens vesicle is crescentic. Lower eyelid fold.

Ear has large endolymphatic sac, cochlear duct elongating. Walls of vestibular labyrinth thinning. Semicircular ducts not present. Six hillocks present and develop anterior to posterior and ventral to dorsal: first arch tragus (1), crus helicis (2, 3); second arch antitragus (6) helix (5, 4). Concha and external meatus formed from the first pha-

ryngeal cleft. Eustachian tube forms from first pharyngeal pouch. Note anterior wall is first arch and posterior wall is second arch. Mucosal representation of second arch between the first and second pouches is eliminated such that first arch mucosa abuts third arch mucosa.

Stage 18 (44–45 Days, 13–17 mm) (Figs. 3.39 and 3.40)

External form alone is not sufficient to stage, internal features have to be included. Distinct finger rays present and, in older stage 18 embryos, elbow is present.

Nasal Passages/Respiratory

Nasal tip and frontonasal angle appear. Columella and septum identifiable; the latter has the vomernasal organ. Nasal passages end blindly in early embryos, but break down later to enter the pharynx. Vomeronasal organ present. Choanae result from breakdown of buccopharyngeal membrane.

Carnegie Stage 18

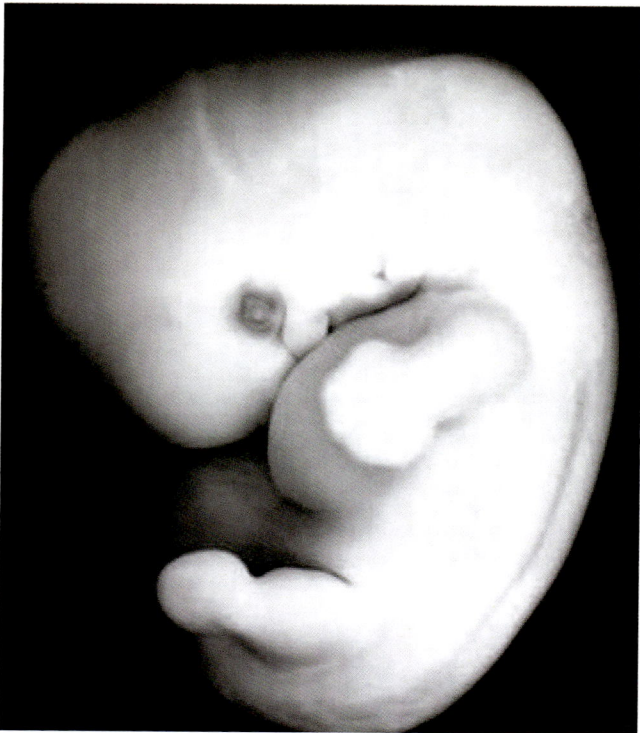

Fig. 3.39 Stage 18: 44–48 days, 13–17 mm. (1) Eyelid. (2) Semicircular canals × 3. (3) Footplate. (Reprinted from The Kyoto Collection, Kyoto University Graduate School of Medicine. Courtesy of Prof. Kohei Shiota and Shigehito Yamada)

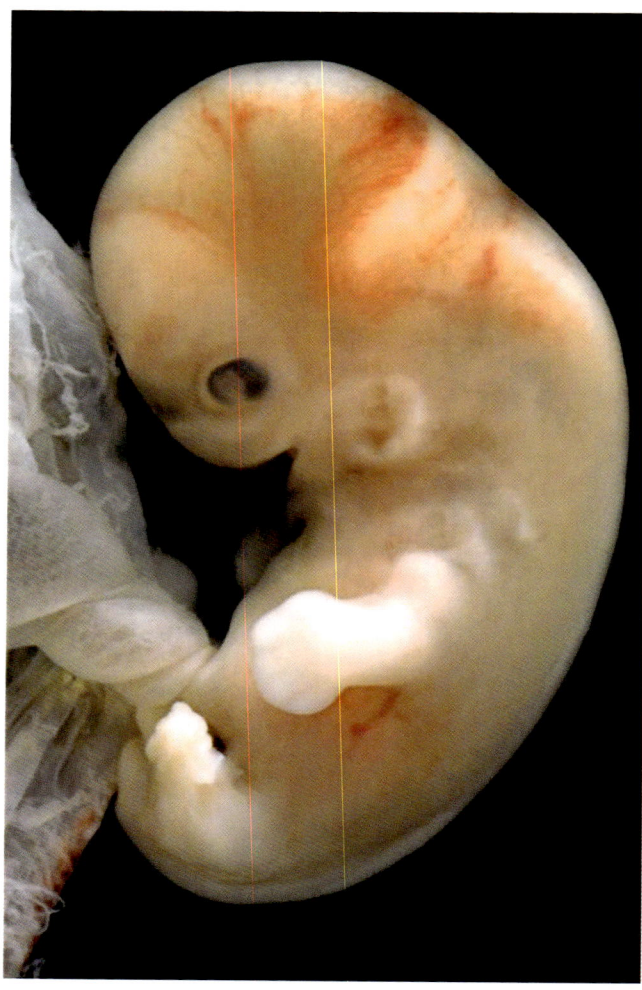

Fig. 3.40 Stage 18 bright field. (Courtesy of Stephen O'Connor, MD)

Trachea structure and epithelium radically differ from the esophagus. Larynx becomes specialized.

Mouth

Upper lip not differentiated. Auricular hillocks are starting to fuse with those rostral to the external auditory canal (from the first arch) and more advanced than those caudal to it (from the second arch). Premaxillary and maxillary fusion for primary hard palate. Secondary hard palate appears as lateral elevations. Submandibular gland first appears thicken epithelium between tongue and jaw.

Nervous System: Future Corpus Striatum, Inferior Cerebellar Peducles, Dentate Nucleus

Hindbrain is most developed with motor nuclei more mature than sensory nuclei. Choroid plexus is present in advanced stage 18. Outer swelling of the cerebellus marks the flocculus. Adenohypophysis now isolated from the mouth. Olfactory bulb developed.

Eyes: Upper eyelid fold present. Grooves mark future conjunctival sacs. Lots of eyelid pigment. The retinas are polygonal. Hyaloid system developed. Scleral thickenings appear with attachment sites for muscles. MNC invades between the lens epithelium and the surface ectoderm. Lens cavity filled in with primary lens fibers.

Ear: Hillocks are fusing. Hillocks rostral to auditory cleft (first arch) are more mature than those caudal to it (second arch). and ventral dorsal. 1–3 semicircular canals emerging from the epithelium of the membranous labyrinth. Cochlear duct is L-shaped. Malleus and incus are chondrifying and stapes now identified with stapedium muscle.

Stage 19 (46–48 Days, 16–18 mm) (Figs. 3.41 and 3.42)

External form.

Limbs stick out straight. Toe rays are prominent but no interdigital notches as yet.

Carnegie Stage 19

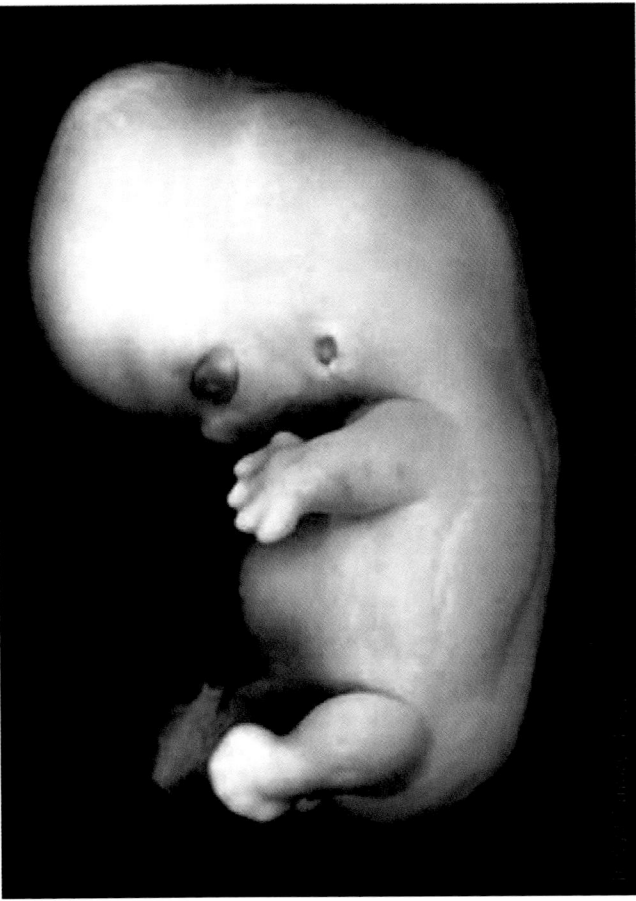

Fig. 3.41 Stage 19: 48–51 days, 16–18 mm. (1) Eye. (2) Auricle. (3) Ossification continues. (4) Foot rays. (5) Straightening of the trunk. (Reprinted from The Kyoto Collection, Kyoto University Graduate School of Medicine. Courtesy of Prof. Kohei Shiota and Shigehito Yamada)

Nervous System: Choroid Plexus of the Fourth Ventricle, Medial Accessory Olivary Nucleus

Eye: Lowe eyelid formed. Eyelid folds meet at lateral canthus.

Optic nerve fibers are passing from retina backward, but have not reached the mid-point of the stalk. Submandibular gland mesenchyme is condensing. Will be invaded by epithelium. Half of the diencephalon is covered by hemispheres.

Carnegie Stage 19

Carnegie Stage 20

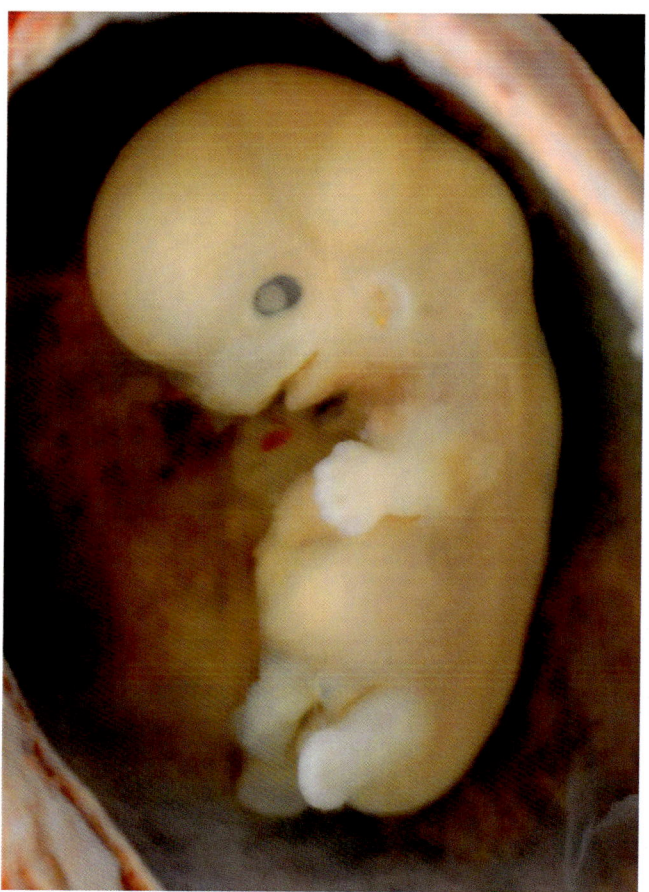

Fig. 3.42 Stage 19 bright field. (Courtesy of Stephen O'Connor, MD)

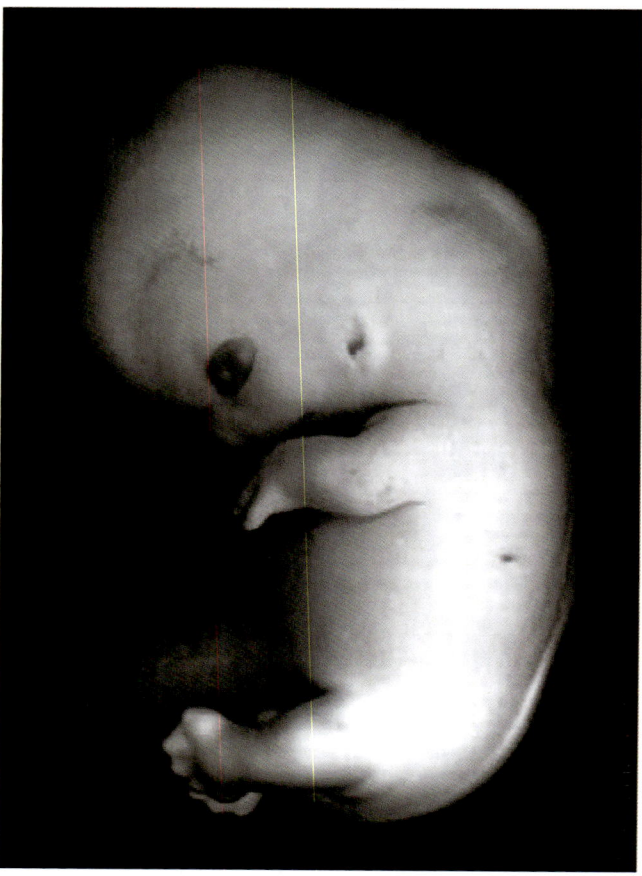

Fig. 3.43 Stage 20: 49–50 days,18–22 mm. (1) Scalp vascular plexus stage 1. (2) Nose. (3) Flexion at elbow and knee. (Reprinted from The Kyoto Collection, Kyoto University Graduate School of Medicine. Courtesy of Prof. Kohei Shiota and Shigehito Yamada)

Ear: Malleus and incus are present. Otic capsule chondrifies but not unified with skull base. Tip of cochlear begins to curl.

Stage 20 (49–50 Days, 22–24 mm) (Figs. 3.43, 3.44, and 3.45)

Scalp vascular plexus, stage 1.
  Elbow flexion present. Fingers distinct but webbed.
  Vomeronasal organ well-defined.
  Submandibular gland had duct.

Nervous System: Choroid Plexus of Lateral Ventricles, Medial Accessory Olivary Nucleus
Hemispheres cover over 2/3 of the diencephalon. Inferior colliculus present.

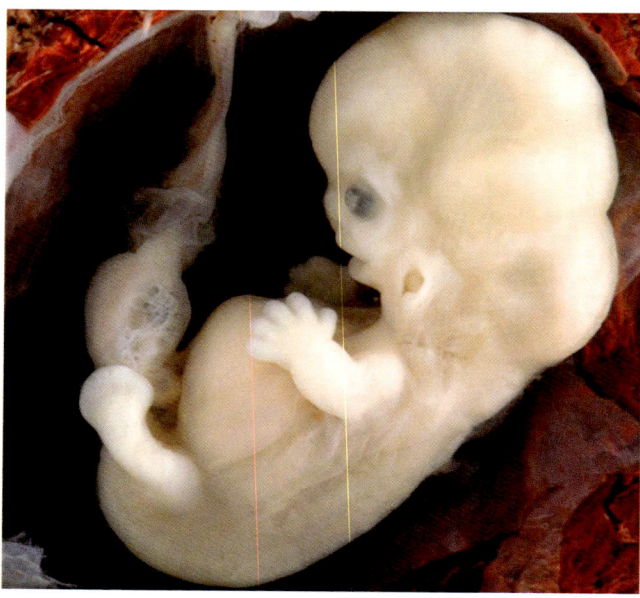

Fig. 3.44 Stage 20 bright field. (Courtesy of Stephen O'Connor, MD)

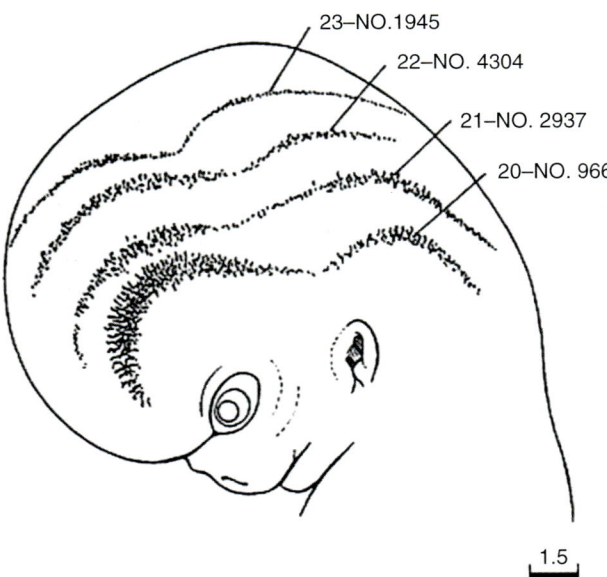

Fig. 3.45 Scalp vascular plexus has four distinct developmental levels that define stages 20–23. Stages determined with respect to the eye-ear line (EEL) and the vertex. Stage 1: less than half the distance between EEL and vertex. Stage 2: half the distance between EEL and vertex. Stage 3: three quarters of the distance between EEL and vertex. Stage 4: plexus greater than three quarters of the distance between EEL and vertex. (Reprinted from Finlay Contributions to Embryology Carnegie Institution 1923; 71: 155–161)

Eye: Upper eyelid formed. Eyelid folds meet at medial canthus. Lens cavity obliterated.

Optic nerve fibers reach the chiasma. Cornea has anterior epithelium, intermediate acellular layer, and posterior epithelium. Trochlea for superior oblique forms stage 20–23.

Ear: Tensor tympani. Otic capsule consolidated with basioccipital and exoccipitals. Cochlea elongates.

Stage 21 (51–52 Days, 2–24 mm) (Figs. 3.46 and 3.47)

Scalp vascular plexus, stage 2.
Fingers are free, toes webbed.
Submandibular gland duct is branching.

Nervous System: Cerebral Hemispheres Have Cortical Plate

Hemispheres cover 3/4 of the diencephalon.
Superior colliculus.

Eye: Levator palpebbrae superioris deliminates from superior rectus. Cells invade intermediate layer of cornea to form substantia propia.

Carnegie Stage 21

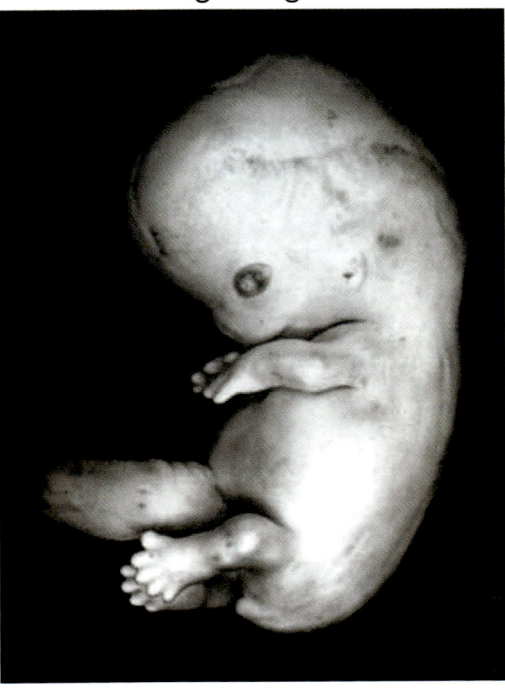

Fig. 3.46 Stage 21: 53–54 days, 22–24 mm. (1) Scalp vascular plexus, stage 2: half way between eye-ear line and vertex. (2) Philtrum. (3) Hand plate with digital webs. (4) Footplate with rays. (Reprinted from The Kyoto Collection, Kyoto University Graduate School of Medicine. Courtesy of Prof. Kohei Shiota and Shigehito Yamada)

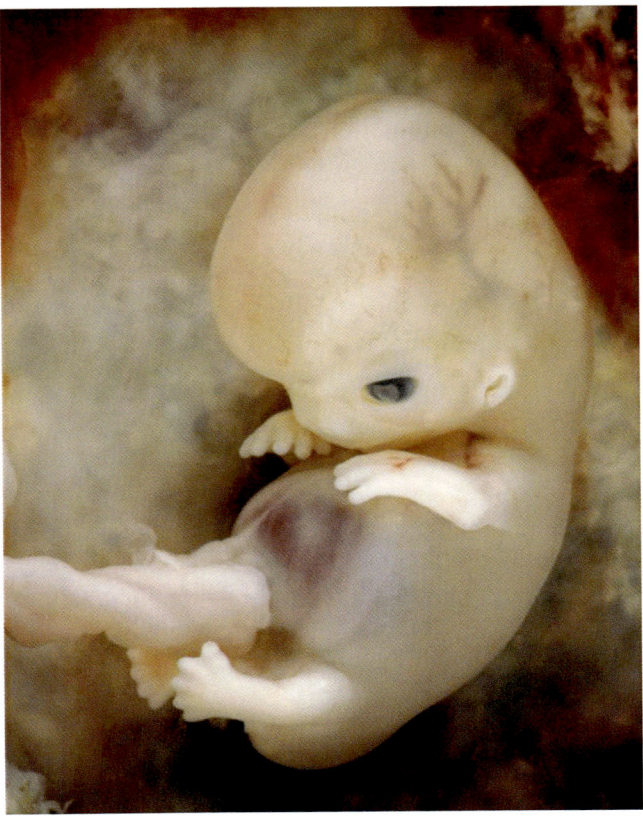

Fig. 3.47 Stage 21 bright field. (Courtesy of Stephen O'Connor, MD)

Stage 22 (53–55 Days, 23–28 mm) (Fig. 3.48)

Scalp vascular plexus, stage 3.

Nervous System: Olfactory Capsule Complete, Internal Capsule Complete

Eye: Eyelids almost closed. Sclera is now definite. Optic nerve now ensheathed.
Ear: Cochlear spiral incomplete.

Stage 23 (56+ Days, 27–31 mm) (Fig. 3.49)

Scalp vascular plexus, stage 4.
    Submandibular gland ensheathed in mesoderm.

Eye: Eyelids closed. Retina complete.
Ear: Cochlea has nearly completed 2.5 turns (tip points down). Labyrinth complete.

Carnegie Stage 22

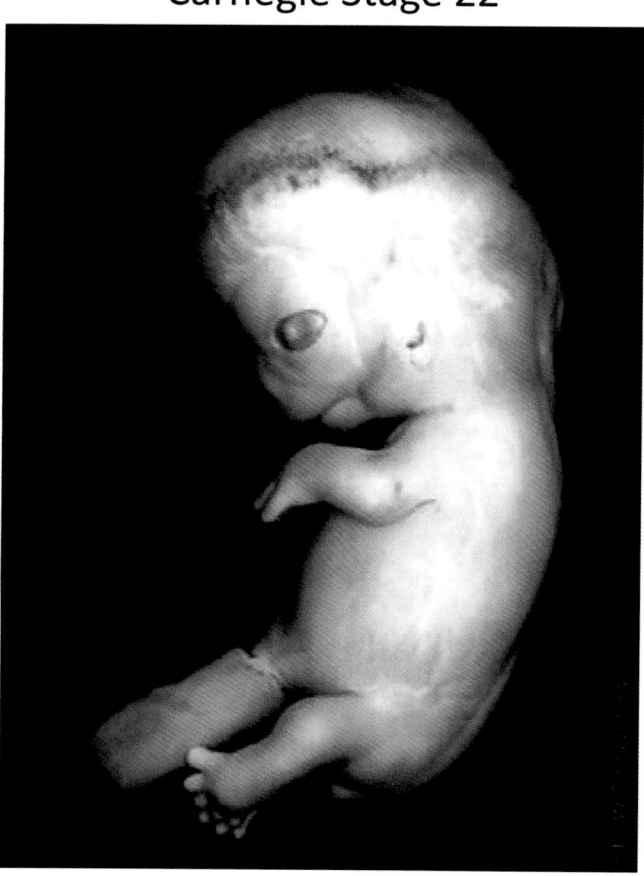

Fig. 3.48 Stage 22: 54–56 days, 23–28 mm. (1) Scalp vascular plexus stage 3: three quarters way between eye-ear line and vertex. (2) Hand plate with separated digits. (3) Foot plate with webbed digits. (Reprinted from The Kyoto Collection, Kyoto University Graduate School of Medicine. Courtesy of Prof. Kohei Shiota and Shigehito Yamada)

Carnegie Stage 23

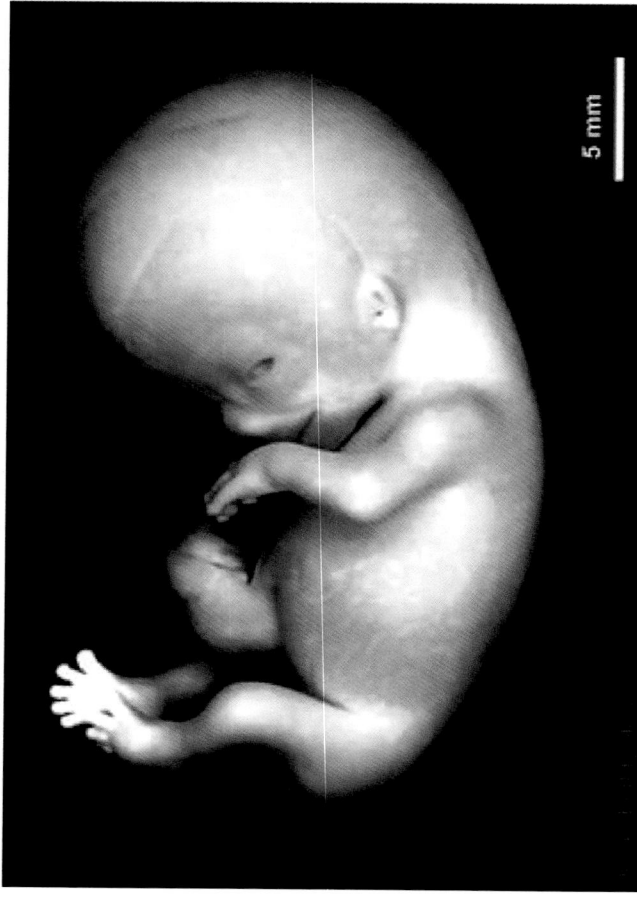

Fig. 3.49 Stage 23: 56–60 days, 27–31 mm. (1) Rounded head. (2) Scalp vascular plexus stage 4: at vertex. (3) Eyelids. (4) Shoulder. (5) Toes separated. (6) Sole of the foot. (Reprinted from The Kyoto Collection, Kyoto University Graduate School of Medicine. Courtesy of Prof. Kohei Shiota and Shigehito Yamada)

Neurovascular Organization and Assembly of the Face

4

Michael H. Carstens

> *Intellectual excellence lies in having faith in the observation of apparently nontranscendental and unimportant facts. To observe an anatomic element calmly, with an open, analytical spirit, and with a spiritual freedom, can lead to an explosive vortex of new knowledge* [1].—Miguel Orticochea, M.D.

Introduction

As outgrowths of the forebrain, the specialized sensory systems of olfaction and vision occupy the center of the face in the form of the letter T, the vertical limb of which is the nose, columella, and philtrum. Late-arriving pharyngeal arch structures, like paired bricks of tissue, clasp from the sides and support the horizontal limbs of the T; however, this spatial arrangement was not always so.

If the face is considered as a composite of neuroangiosomes under tight sequential genetic control, careful analysis of its blood supply, neuroanatomy, and neuroendocrinology can be combined with the neuromeric theory of brain development to construct a useful model of midline facial formation [2, 3].

A most striking aspect of the vascular anatomy of the nose, nasal cavity, and orbit is the existence of sharply defined watershed zones between the internal carotid, internal carotid-derived stapedial, and external carotid systems (ICA, St, and ECA, respectively) exhibited by these structures [4, 5]. The developmental differences between these systems are also striking. The internal carotid per se is primordial, midline, and exclusively dedicated to intracranial structures and the optic apparatus. The stapedial system represents a later embryologic event. It arises as the exclusive extracranial stem of ICA and supplies extradural structures, the orbit, and face. During its development, the stapedial loses its original connection with ICA and forms anastomo-

ses, with the primitive ophthalmic and external carotid. Stapedial induced by V1 connects with primitive ophthalmic to supply the ethmoid complex, internal nose, all extraocular structures, and the entire fronto-naso-orbital soft tissue envelope. Stapedial induced by V2 and V3 connects with ECA maxillary to supply, via the pterygopalatine fossa, the apparatus of zygoma and jaws. The external carotid is also a later event, being formed by a process of rearrangement among the preexisting aortic arch vessels and the primitive dorsal aortae. ECA supplies the entire pharyngeal arch apparatus of the face, *with the exception of the zygoma and jaws* [6–8].

When considering the normal anatomy of the nasal envelope, this author noted the presence of V1 stapedial, the anterior ethmoid arteries (and nerves) hugging the midline above the septum. These vessels continued down through the columella and philtrum about 4–5 mm apart, until dead-ending at the vermilion border where they anastomosed with the ECA labial arcade. Previously reported angiographic studies and surgical dissections [9–12] supported the concept that *the columella and philtrum were constructed from these paired StV1 neuroangiosomes that could be separated from their surroundings.* I found that the nasolabial complex could be conceptualized as a series of independent *developmental fields* (neuroangiosomes) with separate blood supply and embryonic history [13–15].

Let us define three sets of developmental fields. **A fields** are supplied by StV1 and include the frontal bone, the ethmoid complex, nose, and fronto-naso-orbital skin. **B fields** are neural crest bones and supporting structures of the first arch, including the StV2-supplied maxilla-zygomatic complexes, inferomedial and inferolateral walls of the nasal cavity, and the StV3-supplied neural crest mandible. In this chapter, we will be concentrating on the upper face and midface, so the mandible will get short shrift. **C fields** supplied by external carotid are those of the pharyngeal arches. These include all non-jaw-related fields of the first arch. They participate in the soft tissue coverage of the nose and mouth, midface, and lower face. Calvaria bones supplied by external carotid are C fields (Fig. 4.1 A–B–C fields Tessier skull).

M. H. Carstens (✉)
Wake Forest Institute of Regenerative Medicine, Wake Forest University, Winston-Salem, NC, USA
e-mail: mcarsten@wakehealth.edu

Fig. 4.1 Neurovascular fields of the face. V1 stapedial (orange) supplies all MCC derivatives. V2 stapedial (green) and V3 stapedial (white) supply RNC fields of the first arch and second arch, V2 external carotid (not seen), and V3 external carotid (yellow) supply all remaining structures of first arch and second arches. Number in circles refer to Tessier cleft zones (referenced in a separate chapter). (Courtesy of Michael Carstens, MD)

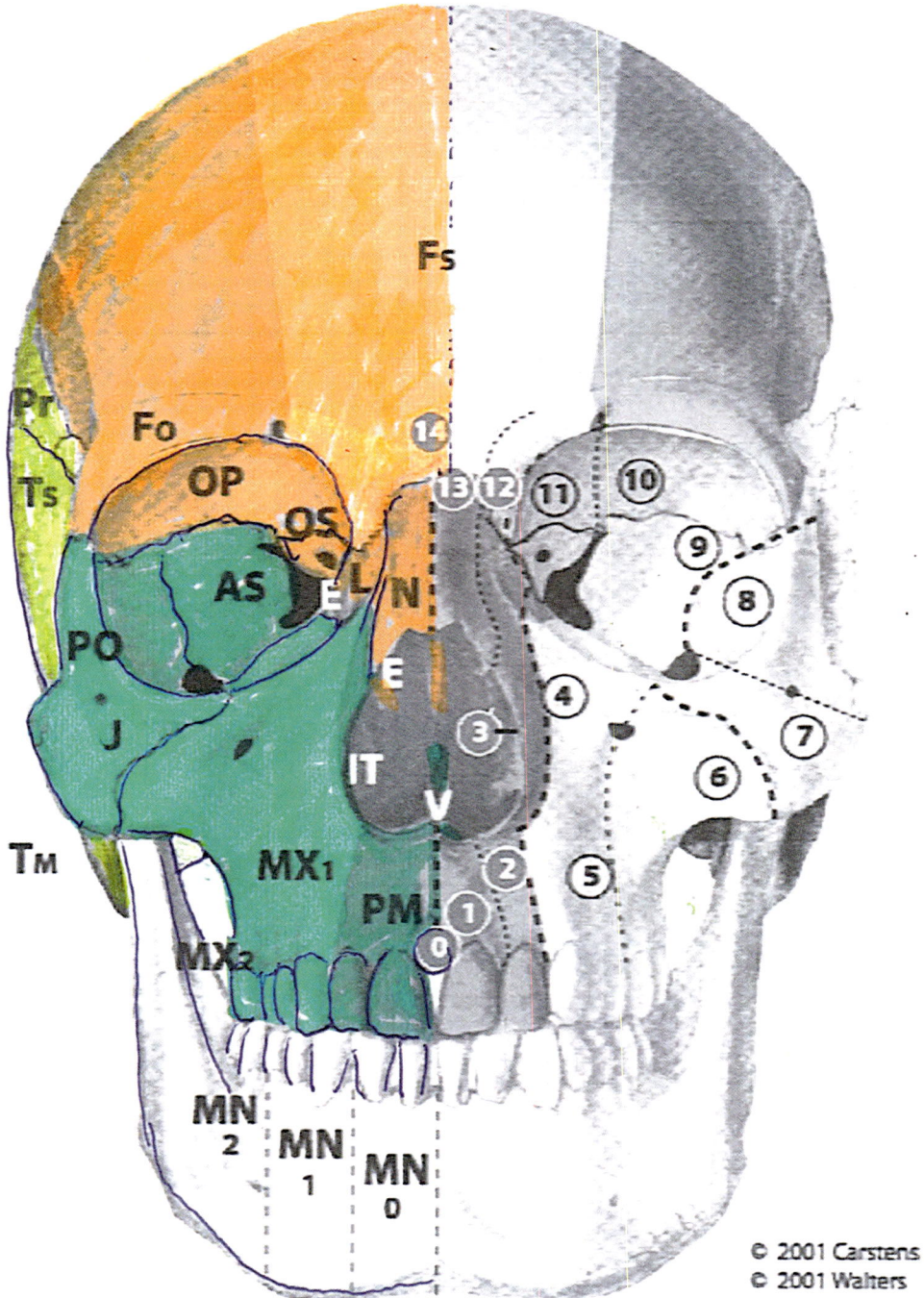

Embryonic developmental fields correspond to blocks of tissue that interact in a programmed fashion to produce the topology of the face. Like tectonic plates, these units possess autonomous innervation and blood supply and move along developmental "slippage planes" according to their mass, velocity, directionality, and timing. Facial development can therefore be considered as the *formation, migration, coalescence,* and *interaction* of separate genetically based fields. Adjacent tissue units required to form a structure (such as the nostril) can be termed *partner fields.* Their interface is a *field boundary*; any process that disrupts the physical integrity of a field or mismatch. This leads to physical distortion of field geometry with-

out any change in the field boundary and will mechanically disrupt normal field interactions leading to *field mismatch.*

The concept that clefting per se leads to subsequent field mismatch has direct clinical relevance for ongoing facial growth and will, over time, further exacerbate the degree of field surface area or volume of the field. An example is the distortion of the lower lateral nasal cartilage in a cleft. It is tethered, pulled down, and dysfunctional but otherwise anatomically intact. The bilateral cleft lip represents a model of midline fusion of paired A fields that includes the ethmoid complex, septum, columella, and philtrum as soft tissue components that interface with B fields vomer, premaxilla, and the hard palate shelves of the maxilla and palatine bones [13]. When bilateral clefting occurs, the prolabium becomes a zone of *cleft-induced field mismatch* between A and B. This model provides a unifying anatomic rationale for the success of surgical strategies such as those of Trott, Mulliken, and Cutting [12, 16, 17]. Cleft lip and palate repairs should be sequentially performed to respect the vascular territories that supply the osteogenic mucoperiosteum of the alveolus [18–20]. **The common denominator of successful cleft repair is the complete restoration of embryonic fields to their proper relationships**.

The interface between the three vascular systems not only defines field boundaries, it also marks the boundary between three general types of gene expression. Contemporary developmental neuroanatomy describes the embryonic first as a tri-partite structure: *prosencephalon* (forebrain), m*esencephalon* (midbrain), and *rhombencephalon* (hindbrain). Subsequently, subdivision into five distinct parts takes place. Forebrain becomes *telencephalon* (cerebral hemispheres) and *diencephalon* (optic apparatus and thalamic structures).

Furthermore, the brain and spinal cord are composed of developmental units known as neuromeres, which are located in cranial-caudal order along the neuraxis. Each neuromere represents a discrete anteroposterior box and includes all parts of that box: floor plate, basal plate, alar plate, and roof plate. The anatomic boundaries of each neuromere are defined by expression patterns of genes of the homeobox family. Each neuromere has its own unique hox "barcode." Neuromeric zones allow the structures of the brain to fit into a genetic "map." Hindbrain consists of 12 *rhombomeres* (r0–r11). A subcomponent of hindbrain, the *isthmus*, separates it from mid brain and is encoded by r0. Midbrain is defined by two *mesomeres* (m1 and m2). Forebrain is defined by six *prosomeres*. Caudal forebrain (diencephalon) consists of four prosomeres (p4–p1). Rostral forebrain (telencephalon) is more complex, but its structures can be lumped into two overall prosomeres (p6 and p5) (Fig. 4.2 The prosomeric map with somitomeres and somites MC).

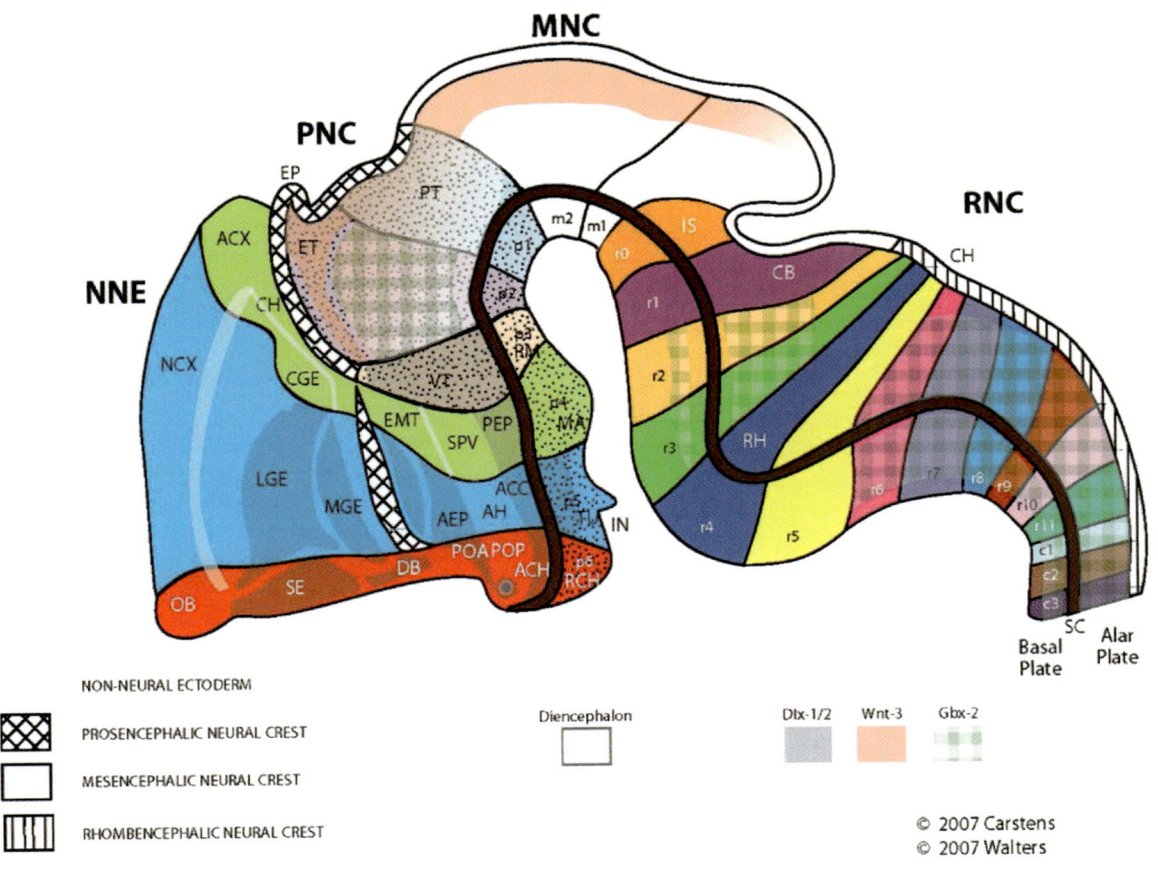

Fig. 4.2 Neuromeric system Puelles version 2001 MHC drawing p1 (aqua), p2 (salmon), p3 (flesh), p4 (light green), p5 (light blue), p6 (red). More recent mapping by Puelles reassigns fields of the secondary prosencephalon—see Chap. 5—so that p6–p4 are no longer considered vertical columns. Nonetheless, mapping of the epithelial derivatives produced by the anterior neural folds over secondary prosencephalon falls into three analogous sectors: nasal epithelium (red), nasal skin (blue), and forehead skin (light green), all of which are supplied by V1 stapedial branches. (Courtesy of Michael Carstens, MD)

Homeotic (hox) genes are among the most primitive to be activated in embryogenesis. Humans have 38 distributed with some redundancy among four chromosomes. Each hox gene has in common a 60 amino acid "loop," or homeobox, that unlocks DNA in a position-specific manner. All organisms use hox genes for axis specification. Humans are 78% equal to *Drosophila* in the same anterior-posterior sequence. Hox genes, as traditionally defined, encode all neuromeres from r3 backwards to specify all segments of the organism down to the tail.

Discovery of additional homeotic genes not in the original hox family has permitted mapping from r2 forward. Genes such as sonic hedgehog (Shh), wingless (Wnt), lim-1, goosecoid, and engrailed allow for the construction of a complex Cartesian map of the forebrain into prosomeric units. These are most discrete in the basal plate of the diencephalon. The same homeotic code for p3–p1 is applied in the model proposed here to describe the populations of neural crest that exist in the neural folds overlying the diencephalon. Less obvious is the coding for the anterior neural folds above the telencephalon. These do not possess neural crest, but are specialized zone of nonneural ectoderm including specialized placodes (adenohypophyseal, nasal, and optic). For our purposes, we shall refer to the territory of the first two as p6 and that of the optic as p5 because inductive genes responsible for the development of the anterior forebrain are also important for the development of the fronto-naso-orbital skin envelope. Finally, the most caudal zone along the anterior neural folds is responsible for the calvarial ectoderm. We shall continue to refer to this a p4.

Although the neuromeric model has undergone several iterations, clinical observations support the existence of genetic factors common to both the frontonasal tissues and underlying brain structures, specifically those anterior to the hypophysis. This is exemplified by holoprosencephaly in which a spectrum of worsening pathologies affects both the CNS and its soft tissue cover. Opposite to the hypothesis of DeMeyer, *the brain predicts the face*.

That vertebrates share a common segmental plan seems intuitively obvious. For more than a century and a half, comparative anatomists agonized over models of segmentation based upon the spinal cord and the gill arches which did not apply themselves well to the mammalian head and neck. The advent of the homeotic gene concept, coupled with new knowledge of the prosomeres and the genes that code for them, constitutes a great "rosetta stone" for developmental biology, a tool for deciphering the hitherto enigmatic bauplan uniting nervous system and craniofacial structure.

The use of neuromeric coding is extremely useful for defining the mesenchymal structures of the face and indirectly its epithelial coverage. With the exception of striated muscles, all craniofacial mesenchyme is derived from neural crest. Neural crest cells developing from the neural folds above a given neuromere bear the same hox "tattoo." Furthermore, the neuroanatomic content of each neuromere is known; and all cranial nerves are ensheathed by neural crest. For example, rhombomere 3 contains the nucleus of trigeminal V3. All neural crest structures innervated by sensory V3 originate from the neural folds above r3 and bear during their embryogenesis the homeotic code of r3. Thus, the fascia of all first arch muscles of mastication and the mandible itself are r3 neural crest derivatives. Using the neuromeric system, it is possible to map out all craniofacial tissues according to their site of origin as indicated by their neuroanatomy.

As has been discussed previously, the paraxial mesoderm resulting from gastrulation begins at level r1—the rostral tip of the notochord. This cranial mesoderm becomes incompletely divided into seven somitomeres—these are responsible for all craniofacial muscles of the first three pharyngeal arches. One can deduce what neural crest is assigned to these somitomeres in terms of fascia and muscle by virtue of their neuroanatomy (Table 4.1).

The facial midline constitutes a unique "window of understanding" through which clinicians and basic scientists can glimpse, analyze, and test the relative contributions of neuromeres to the facial development. Isolated injection of the internal carotid is a useful means to test this interface because it fills branches of ICA and those of StV1 ophthalmic. This sets up a sharp demarcation between the A fields, the B fields, and the C fields. Knowing the origin of these field complexes becomes vitally important. *By putting the "developmental videotape" in reverse, the sequence of migration of the various fields can be tentatively be deduced*. To the extent that this process is better understood, the surgical management of craniofacial deformities can move from geometric maneuvers, such as z-plasties—to biologically based strategies designed to correct defined pathologic mechanisms of development.

Table 4.1 Somitomeres and their muscle derivatives

Somitomere	Innervation	Neuromere	Derivative muscles
Sm1	III	m1	Inferior rectus, inferior oblique, medial rectus
Sm2	III	m2	Super rectus, levator palpebrae superioris
Sm3	IV	r0	Superior oblique
Sm4	V3	r3	Mastication
Sm5	VI, VII	r4	Lat rectus
Sm6	VI, VII	r5	Lat rectus, mastication, facial expression
Sm7	IX/X	r6–r7	Palate, pharynx
Sm8/S1	X, XI, XII	r8	Pharynx, larynx/SCM-trapezius/tongue
Sm9/S2	X, XI, XII	r9	Pharynx, larynx/SCM-trapezius/tongue
Sm10/S3	X, XI, XII	r10	SCM-trapezius/tongue
Sm11/S4	X, XI, XII	r11	SCM-trapezius/tongue

Useful Terminology: Read These First

Developmental field/neuroangiosome: composite block of tissues supplied by a single neurovascular pedicle.

A fields: All structures innervated by V1 and supplied by branches of the stapedial ophthalmic artery (StV1).

B fields: All structures innervated by V2 or V3 and supplied by branches of the stapedial maxillomandibular artery (StV2/StV3) via the pterygopalatine fossa.

C fields: All structures innervated by V1 or V3 and supplied by branches of the external carotid artery.

Stapedial Artery

Bilateral stems representing the dorsal remnants of the second aortic arch *hyoid artery* are translocated backward along the dorsal aortae to junction of third aortic arch artery. These represent the extracranial portion of internal carotid artery. The stapedial is derived from hyoid artery at stage 13. It represents the exclusive extracranial branch of ICA. It promptly enters the skull through the tympanic cavity, passing through the stapes.

Stapedial ophthalmic system (StV1): Following trigeminal V1 to the orbit, stapedial forms multiple branches to all extraocular structures of the orbit. These supply the ethmoid sinuses and upper nasal walls. Terminal branches of StV1 exit to the face to supply all fronto-naso-orbital soft tissues.

Stapedial maxillo-mandibular system (StV2/StV3): Extracranial stapedial follows chorda tympani to the sensory root of V3 where it anastomoses with the maxillary branch of external carotid. This forms a hybrid maxilla-mandibular artery with three segment. Proximal or mandibular segment of MMA has four epaxial stapedial branches directed into the skull, 1 hypaxial stapedial inferior alveolar, and 1 ECA branch, mylohyoid. Middle, or pterygoid segment has no stapedial branches, 2 ECA epaxial branches to masseter and 2 branches to the pterygoid muscles. Distal, or pterygopalatine MMA has exclusively stapedial branches; ECA is not represented.

Stapedial intracranial system (StV1, StV2, StV3): After entering the skull, stapedial branches anteriorly and posteriorly. It forms individual dural arteries in company with sensory branches of trigeminal nerve. It sends out a separate descending branch that exists the skull in company with chorda tympani (Fig. 4.3 Innervation of the dura showing, in particular, the meningeal branches of V1 and V2, V3).

Neuromere

Developmental compartment of the brain, extending from the floor to the roof of the neural tube. Each neuromere is defined anatomically by a unique expression pattern of homeotic genes that specify its anterior-posterior location along the neuraxis.

Embryonic Brain

Begins with three parts and then separates into five parts.

- *Prosencephalon* (forebrain): six prosomeres (p6–p1)
  - Telencephalon (cerebrum): p6 and p5

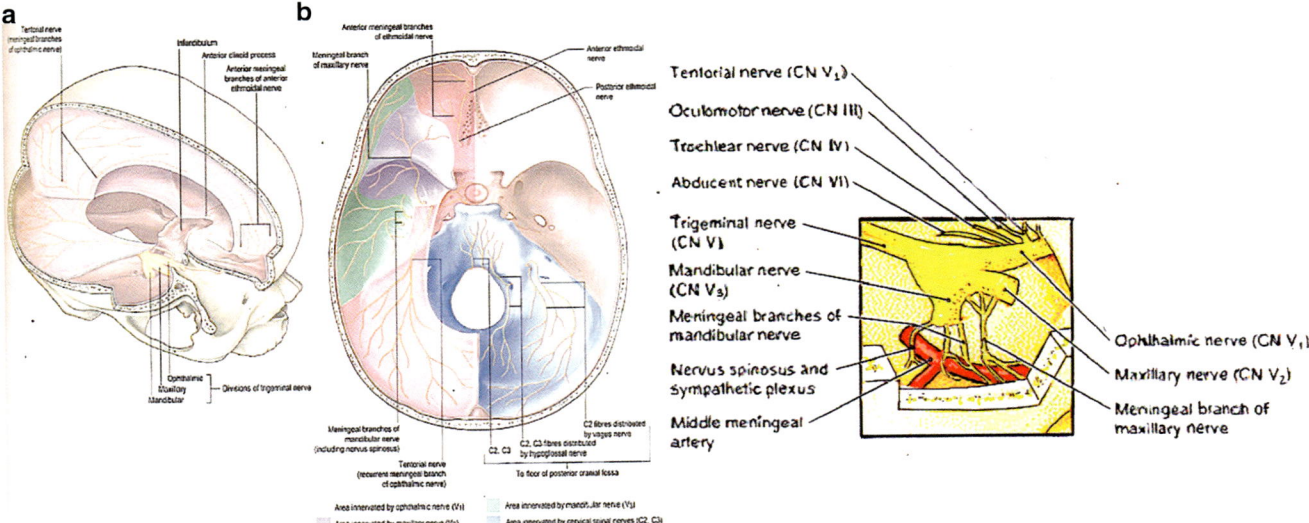

Fig. 4.3 Immediately after emerging from the trigeminal ganglion, the very first branch of each division is dedicated to dura. V1 supplies the anterior cranial fossa and tentorium cerebelli (the latter, because cerebellum arises from r1). After V1 enters orbit (accompanied by branch of stapedial artery), it supplies dura over the ethmoid. V2 and V3 supply the middle fossa. Posterior fossa (below tentorium cerebelli supplied from IX via X) entering at jugular foramen with contributions from C2, C3 entering at foramen magnum. (Reprinted from Standring S (ed). Gray's Anatomy 40th ed. New York, NY: Churchill Livingstone; 2008. With permission from Elsevier)

– Diencephalon (optic apparatus, epithalamus, thalamus, hypothalamus): p4–p1
- *Mesencephalon* (midbrain): two mesomeres (m1 and m2)
- *Rhombencephalon* (hindbrain): 12 rhombomeres (r0–r11)
  – Metaencephalon (isthmus and pons, cerebellum): r0–r7
  – Myencephalon (medulla): r8–r11

Injection Studies

Five fetuses under the age of 22 weeks were presented for necropsy at the Neonatal Pathology Section, Department of Pathology, Kaiser Hospital Oakland, under the direction of Dr. Geoffrey Machin. The common carotid arteries were cannulated with a #25 catheter. Ligation of the external carotid, subclavian, and vertebral arteries was performed. Blue dye was injected. 6–10 cc were sufficient per case. Photographs were taken with a Nikon 105 mm lens at f22 and f32 using Kodak ASA 100 Ektachrome film.

After examination of the face, cerebrectomy was performed and the skull base examined. In selected cases, all midline structures were resected en bloc. Each of these specimens consisted of the following: cribriform plates, crista galli, ethmoid sinuses, perpendicular plate of ethmoid and the nasal column, the lateral wall of the nose, the entire palate, the nose—including nasal bones, columella, philtrum, vomer, and premaxilla.

Case 1: Normal Fetus

Injection of right common carotid resulted in a blush seen first at the right upper eyelid but then spreading to right supratrochlear skin. As more injection was administered, pinpoint stains appeared on the scalp, indicating perfusion of the brain. Similar findings appeared on the left side to a lesser degree, indicating cross-over of the dye. The next finding was a faint blush in the philtrum of the upper lip reaching the vermilion from StV1 anterior ethmoid arteries and then spreading laterally into the ECA labial arteries. No dye reached the surface of the nasal dorsum skin, the remainder of the upper lip or cheek.

The nasal skin was split down the midline, revealing paired anterior ethmoid arteries at the upper border of the upper lateral cartilages. These vessels ran in parallel along the upper border of the septum, then turned caudally to pass through the columella into the philtrum where they formed anastomoses with the ECA labial arcade. Collateral flow from StV1 infratrochlear into ECA lateral nasal and upper labial branches was seen.

Reflection of the anterior scalp flap revealed an expected perfusion via supraorbital and supratrochlear vessels (Tessier cleft zones 10–11 and 12–13, respectively). Although the staining pattern was bilateral, it was more intense on the (right) injected side. Dye was seen in isolated vessels running in parallel with supraorbital on the interior surface of frontal bone. Intense staining of the cribriform plate terminated abruptly at the frontoethmoid suture. At cerebretomy, two normal olfactory nerves were present.

En bloc resection of the midline resulted in a rectangular box of tissue with the cribriform plate and cristae galli constituting its superior margin. The inferior (oral) margin of the specimen disclosed dye running to the anterior nasal spine via the StV2 medial nasopalatine. [Communication was later found in the back of the upper nasopharynx descending from the face of the StV1 sphenoid into the nasopalatine.] The entire maxillary hard palate and alveolus was unstained. When the hard palate was resected away from vomer and premaxilla, no dye was found at the margins, indicating a sharp demarcation of lateral nasal StV2 from medial nasal StV2 (again, to the anastomosis at the sphenoid). Midline section of the lip leaving the nose intact revealed continuity of dye between the lip and columella, between the lip and premaxilla, and between the premaxilla and the base of the nose—all of these findings consistent with anastomoses between StV1 and StV2. Of great importance: (1) directly beneath the philtral skin was a layer of formless mesenchyme of MNC origin; (2) no superficial orbicularis oris was present in the philtrum; (3) deep orbicularis oris was present beneath the philtral mesenchyme; and (4) dye from the philtrum did not perfuse the deep orbicularis oris.

Staining of the lateral nasal wall was limited to its upper half, extending from the orbital floor to the middle turbinate (the distribution of the lateral ethmoid complex). The lower half of the lateral nasal wall consisting of inferior turbinate and maxillary palate was resected; it has no perfusion.

Along the lateral wall of the nose, dye was noted in the anterior and posterior ethmoid arteries (lateral nasal arteries). The former was seen penetrating the anterior ethmoid air cells, subsequently to reemerge beneath the nasal bones, running its course astride the septum to penetrate the columella and philtrum, as previously described. Resection of the lateral wall permitted examination of the perpendicular ethmoid plate and septum. Dye was observed in continuity with the nasal side of the cribriform plate, running down the upper border of the septum. In this specimen, staining over the septum as a whole was not seen.

Case 2: Normal Fetus

Ten cc of dye divided between the two common carotids produced the same sequence of blush production but of equal intensity on both sides. Dye was seen transcutaneously along the nasal dorsum, perhaps due to greater volume of the injectate or the symmetry of its application. Two vessels were observed running vertically through the columella. A blush appeared in the center of the vermilion and spread laterally through the labial arcade to reach the commissures. Although the paired vessels were not seen transcutaneously in the philtrum, removal of skin from the tip of the nose, columella, philtrum, and lip demonstrated the paired vessels running from the columella to the vermilion border. Backflow of dye from the medial canthus into the angular artery and from the nasal dorsum into the alae via anastomoses with the external carotid system was consistent with standard anatomic description. Vertically oriented labial arteries were seen in the late injection phase.

En bloc resection of the A fields was performed. Once again, both the lower lateral nasal walls and palate were not perfused. Upon resection of these structures, a diffuse blush was seen involving the entire septum. Along its caudal border, the posteroanterior course of the medial nasopalatine artery was noted running with the vomer.

Case 3: Unilateral Cleft Lip and Alveolus (Secondary Hard Palate Intact)

Having the two A fields intact, in this otherwise normal specimen, two vessels ran from the columella into the philtrum and thence into the labial arcade on the non-cleft side. Perfusion of the septum was intact. Via the posterior sphenoid anastomosis, dye was seen in the vomer and premaxilla

as well. The brain was normal with a well-defined and perfused cribriform plate. Two olfactory bulbs of equal size were present.

Case 4: Holoprosencephaly with Unilateral Cleft Lip and Midline Cleft Palate; Right Fields Hypoplastic, Left A Fields Aplastic (Fig. 4.4 Holoprosencephaly)

The fetus was hypoteloric with gross lack of nasal projection (best seen on submental-vertex, SMV, view) and an extremely wide cleft of the upper lip. Seen from the SMV, an attempt to form a right naris was present, consisting of a lateral rim and a less well-defined medial rim. The right naris proceeded inward about 4 mm, ending blindly in a small laterally oriented pinhole. The termination of the lateral rim of the naris resembled the nostril sill. The distance from nasal midline to right canthus was twice that from the midline to the left canthus. The left supraorbital ridge was slightly higher than that of the right. A lack of bony definition of the left medial orbit, corresponding in location to the lacrimal bone (which was missing), created asymmetry when compared to the right orbit. The AP profile of the dorsal nasal skin envelope was beak-shaped, with the midline "tip" rolled underneath. This V-shaped "flightless gull" connected the abortive right naris with a blind pitted skin mass where the left naris should have been.

The right lip element had a well-defined orbicularis mass. Its medial border was grossly of normal height; its cephalic

Fig. 4.4 Spectrum of holoprosencephaly. (Reprinted from Raam MS, Solomon BD, Muenke M. Holoprosencephaly: A guide to diagnosis and clinical management. *Indian J Pediatr* 2011; 48:457–466. With permission from Springer Nature)

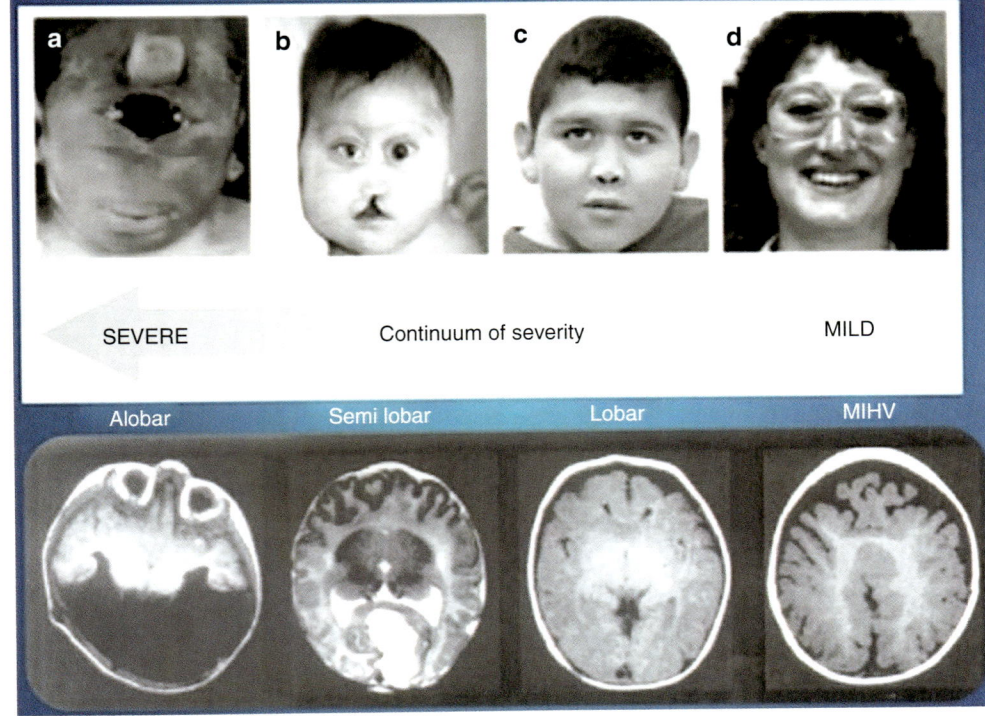

termination was at the medial margin of the lateral nasal process. No muscle fibers invaded the medial rim of the naris. The medial border of the left lip element (defined by the termination of the orbicularis) bore a different relation to the nose. A well-defined groove separated it from the lateral alar "rim."

Intraoral findings were equally striking. A firm midline structure corresponding to the ethmoid complex and septum originated from the most posterior aspect of the roof of the nasopharynx. This ridge descended and increased in vertical height in a posterior to anterior manner, terminating in a vestigial prominence corresponding in location to the premaxilla. The right palatal shelf was cleft from the incisive foramen posteriorly, the cleft beginning precisely at the site where the midline vomer became vertically insufficient to reach the palatal plane.

The left palate shelf was smaller and more poorly defined. It terminated anteriorly into the nasal complex. It had no contact with the midline vomer-septum-ethmoid plate and therefore displayed a complete cleft. The inferior turbinate was small (being built upon a faulty lateral ethmoid complex). The cutaneous mass of what should have been the left naris was displaced posteriorly and cephalically into the cleft site.

The ascending arch of the aorta was then cannulated, permitting use of a #23 gauge catheter and a larger volume of injection. Initially, no cutaneous dye was seen; the amount of the dye was doubled to 20 cc. Eventually, a small blush occurred in the right lip vermilion to the right of the "midline." This reflected an ipsilateral anterior ethmoid vessel. The absence of the left-sided vessel can be attributed to the ipsilateral A field aplasia. The skin of the forehead, nose, lips, and cheeks was removed. A striking asymmetry of the nasal dorsum was observed, with complete absence of the nasal bones and cartilages on the left side. The right side of the nose was bounded by a swatch of yellow-orange fatty tissue. On the left side, this abnormal tissue was more prominent; it occupied the place of the absent nasal capsule.

Although both supraorbital vessels filled equally, no supratrochlear or infratrochlear vessels were present on the left side, indicating localization of vascular pathology to Tessier zones 13 and 12. A single anterior ethmoid appeared from beneath the singular right nasal bone. Upon reaching the upper border of the alar cartilage, this vessel anastomosed with the ECA upper lateral nasal artery which was coursing in a normal manner medially from the angular artery. A caudally directed branch from the midportion of the upper lateral nasal arcade ran along the medial aspect of intermediate crus of the right alar cartilage. Along the caudal border of the severely hypoplastic right alar cartilage, a lower lateral nasal arcade was present.

At cerebrectomy, reflection of the scalp flap showed the left supraorbital vessels to be normal in size and configuration, but closer to the midline than on the right side. This was a reflection of bilateral hypotelorism, left worse than right. Holoprosencephaly was noted with neither cribriform plate, nor cristae galli, nor olfactory bulbs present. A 4 mm recess below the plane of the frontal lobes contained, in its depths, a single vessel passing anteriorly along its right lateral wall.

The intraoral injection pattern was most remarkable. The widely cleft palate was unstained. Its left size was hypoplastic compared with the right. Using 4.5× loupe magnification, the diminutive premaxilla had dye stippling. This would be expected given that sphenoid was intact. This contrasted with the obvious dye blush seen in the normal premaxilla (case 3). It reflects the gross decrease in the StV1 vasculature descending on both sides as well as the nature of the StV2 nasopalatine axis (presumably normal). But vomer was reduced in size as well. Thus, the origin of the palate cleft was appreciated to occur precisely at the take-off of the vomer away from the palatal plane.

Case 5: Cebocephaly (Fig. 4.5 Cebocephaly)

The fetus displayed severe hypotelorism and a complete absence of midline structures including the perpendicular ethmoid plate, septum, and columella, vomer, and premaxilla. Both maxillary shelves were retrodisplaced, causing flattening of the midface. A narrow inverted U-shaped skin envelope provided cover for a vertically deficient single-chamber nasopharynx characterized by total lack of turbinates, both upper and middle from the StV1 ethmoid complex as well at the StV2 inferior turbinate. The entire palate was completely cleft in the midline. A labial orbicularis mass defined a medial border for each lateral lip element. This border did not pass beneath the lateral margin of the nasal envelope. It was demarcated from the lateral nasal prominence by a cutaneous groove.

Injection of 10 cc of dye into the aortic arch perfused both orbits as manifested by upper eyelid staining. No dye entered the dorsal midline. The lateral nasal walls were unstained. This indicates knock-out of the most medial branches of StV1 ophthalmic such as the ethmoids and supra/infra trochlear with an intact supraorbital. The only manifestation of dye in the mouth was in the roof of the nasopharynx, perfusing the midline from posterior-to-anterior fashion. Cerebrectomy disclose alobar holoprosencephaly. No cribriform plate, cristae galli or olfactory bulbs were present. No staining in the midline was noted.

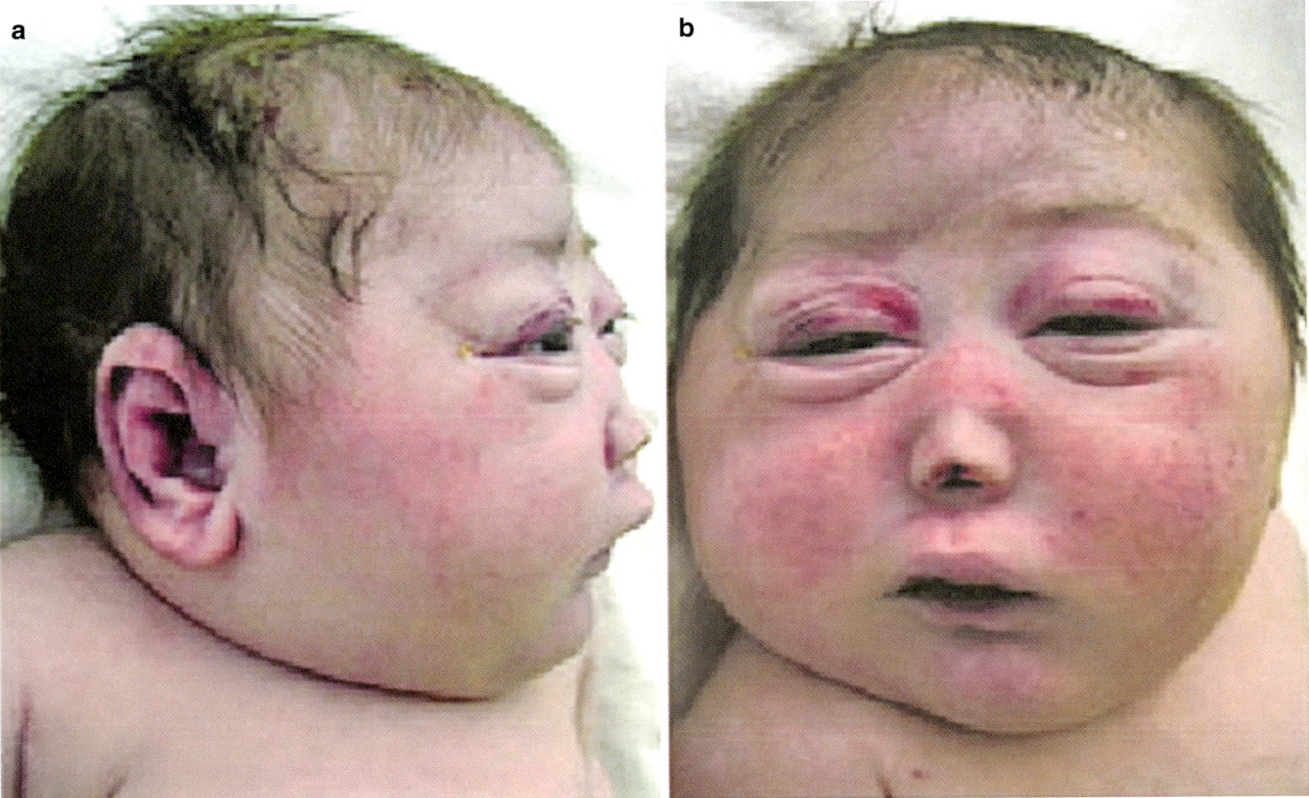

Fig. 4.5 Cebocephaly defined by single chamber nasal passage. (Reprinted from Ritivoiu M, Brezan F, Codreanu I, Stamate M, Anca I. Ultrasound diagnosis of two cases of severe raniofacial anomalies. *Med Ultrason* 2013; 15(4):330–332. With permission from Medical Ultrasonography)

Timetable of Oronasal Field Development

Embryology is the analysis of relationships between temporal order and form. The timing with which anatomic structures make their appearance is all-important to understand the overall process. The use of several different parameters to describe the order of development in the first 8 weeks of life had caused considerable confusion in the literature. These parameters are: (1) age, (2) crown-rump length in millimeters, (3) number of somite pairs, and (4) Carnegie stage. Each system has its own degree of precision and overlap. The Carnegie model for embryonic staging is the most comprehensive. It is discussed in detail in Chap. 3.

Lumping the events of embryogenesis into weeks is imprecise and misleading. Critical events take place within the space of a single day. The fusion of the LNP to the MNP may occur in as little as 6 h [21–23]. Embryo size (C-R length) is a more accurate measurement of maturation, but is not correlated to specific features. The staging system of the Carnegie Institution, described by Streeter [24–26] and refined by O'Rahilly [27–31], categorizes development into a sequence of stages defined by the formation and maturation of key anatomic structures. Each stage is tightly liked to C-R length, intervals of which are as small as 2 mm through stage 15 (33–36 days). The Carnegie staging system is presented in detail in Chap. 3.

Morphologic correlating developmental stage with the formation and number of somites is limited to a certain window of embryogenesis. The first 1–3 pairs appear at Carnegie stage 9 (21–22 days). Recall that the first seven somitomeres remain unsegmented. New somitomeres are produced every 4 h until a total number of about 43. Transformation to somites begins with Sm8. The total number of somites in humans is approximately 36: 4 occipital, 8 cervical, 12 thoracic, 5 lumbar, 5 sacral, and 2 coccygeal. Note that up to 10 coccygeal somites can develop, but these quickly involute back to 2–3; failure of this process results in a persistent "tail." Formation of the 30th pair (observed between days 29–31) defines Carnegie stage 13. Although the number of somite pairs ceases to be of importance for subsequent stages, these entities play a vital role in the segmentation of the spine and the organization of the trunk. This process occurs in register with the neuromeres; these genetic unites determine the cryptic segmentation of the brain and face [32–34].

Concepts of Developmental Fields

The anatomic material in the chapter is a bit dense. To facilitate understanding, I would like to redefine some terms familiar to many readers and introduce some new ones. The goal is to provide a purely descriptive framework that presages the detailed review of experimental work that follows.

At the turn of the nineteenth century, the great German embryologist Wilhelm His defined for the first time the features of the embryonic face using observations under the microscope. He opposed the concept that materials in the nucleus could be of importance in the formation of the organism. Instead, he viewed development (in the face) as the assembly of formed structures he termed *processes*. This idea, repeated in countless drawings based on Wilhelm His' work, continues to adorn the pages of even contemporary plastic surgery textbooks.

Unfortunately, the term "process" has no basis in developmental anatomy. (1) A process is not a singular anatomic unit—it may be made up by several fields. (2) It is not necessarily singular in origin—it may be induced by another field. (3) It does not have autonomous functional significance. Many authors have suggested the substitution of the term prominence for process, but this is also flawed in that it only describes what one sees as the external morphology. Whether seen by light microscopy or SEM, prominences are just the external forms of a much more complex arrangement of fields under central genetic control.

Fields, on the other hand, represent real anatomic structures based on neuroangiosomes. If the face is considered the assembly of such units, its component bones and neurovascular planes take on new significance. For the sake of simplicity, all facial fields can be divided into general categories that reflect the relative origins of coding for the endoderm, mesoderm, and neural crest that make them up. All structures composed of MNC and supplied by the V1 stapedial system can be referred to as A fields. They are in register with prosomeres p6 for the nasal cavity p5 for the nose and orbit, and p4 for the forehead.

The component MNC mesenchyme of the nose and orbit is divided into roughly quadrantic units, stacked up like shoeboxes. Those of the nose are divided into medial and lateral laminae presence of a nasal chamber created by the penetration of the nasal placode from the surface of the embryo straight back into the substance of MNC in communication with Rathke's pouch. The lateral lamina of the nose becomes the medial lamina of the orbit, the lateral wall of which will be provide by B field zygoma and maxilla. Because the bones and cartilages are readily visualized, they constitute convenient "markers" for individual developmental fields. Careful study of the centers of ossification for the membranous bones—as described in older editions of Gray's Anatomy—will demonstrate a one-to-one correspondence between the number of these centers and the number of ancestral bones in the original tetrapod skull! (Figs. 4.6 and 4.7 Tetrapod skull).

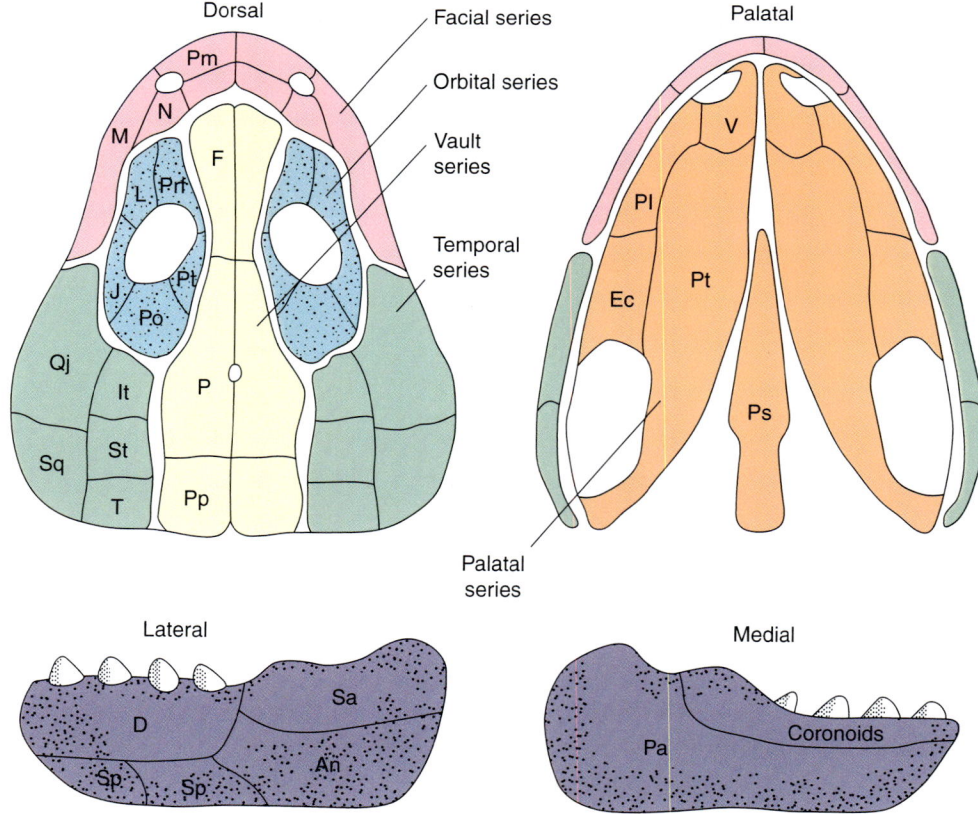

Fig. 4.6 Frontal series (yellow): frontal, parietal, postparietal. Orbital series (blue): prefrontal, postfrontal, postorbital, lacrimal, jugal. Temporal series (green): intertemporal, supratemporal, tabular, quadratojugal, squamosal. Palatal series (orange): vomer, parasphenoid pterygoid, palatine, ectopterygoid. Mandibular, lingual series: coronoids–lingual dentition, postangular. Mandibular buccal series: dentary splenials, angular, surangular. (Reprinted from Kardong K. Vertebrates: Comparative Anatomy, Function, Evolution, 7th ed. New York: McGraw Hill, 2015. With permission from McGraw-Hill Education)

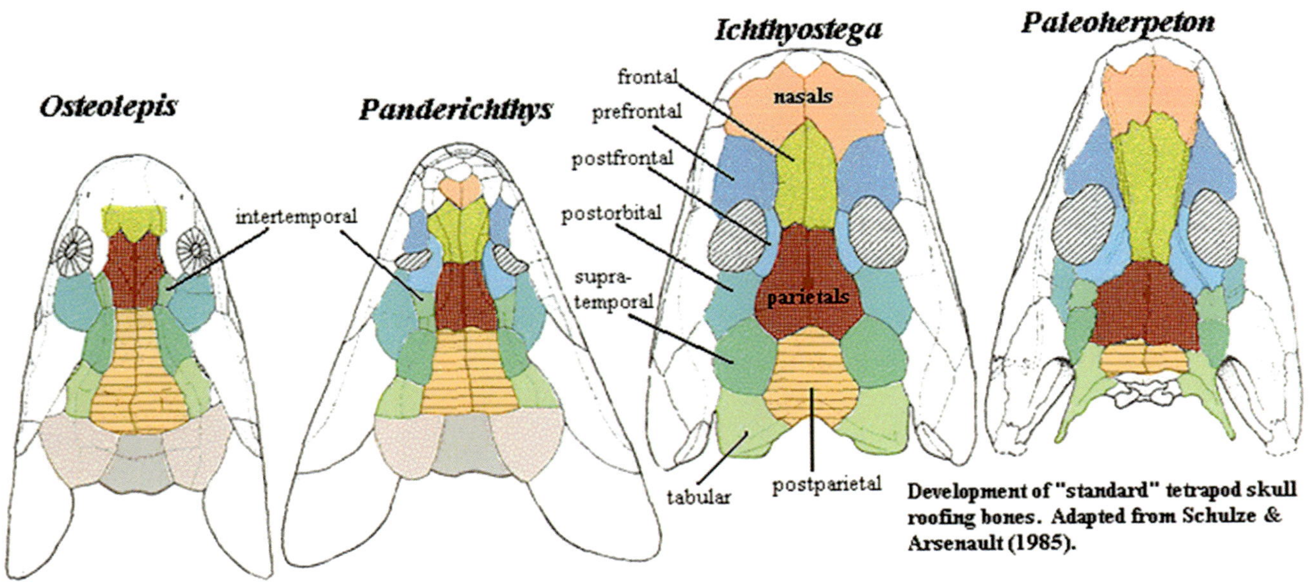

Fig. 4.7 Evolution of tetrapod skull. *Osteolepis* Early aquatic sarcoptyergian. Has extrascapulars and *Panderichthys* Late aquatic sarcoptyergian, adapted for shallow water *Ichthyostega* Mostly aquatic sarcoptyergian, could come onto land ("fish roof"). *Paleoherpeton* Full tetrapod. Note: Only the fully aquatic sarcoptyergians had operculars (cover the gills) and extrascapulars (lock the head to the pectoral gir-dle). Front of the skull gets longer. Parietals shift backward. (Reprinted from Clack JA. A revised reconstruction of the dermal skull roof of *Acanthostega gunnari*, an early tetrapod from the Late Devonian. Earth and Environmental Science Transactions of The Royal Society of Edinburgh 2007;93(2): 163–165. With permission from Cambridge University Press)

The fundamental theme of A field development is the interaction between midbrain-derived neural crest (MNC) that has migrated over the surface of the brain to arrive at the midline as early as stage 10 and a small but critical signaling zone of *prechordal mesoderm* (PCM) located at the base of the forebrain. PCM represents the very first cells to exit at gastrulation. It precedes by a day the exit of paraxial mesoderm during gastrulation—the cephalic terminus of PAM is basisphenoid where the anterior terminus of the notochord resides. All cranial-based structures anterior to basisphenoid are derived from midbrain neural crest.

The fate of MNC is to be penetrated from without by the invagination of the nasal placode. At the same time, MNC is penetrated from within by the outward thrust of the primitive optic apparatus. Later in time the optic placode will likewise invaginate into the optic apparatus. Placodes are specialized epithelial structures often possessing neurosensory capabilities. They are necessary for the development of the final neurosensory organ.

On SEM, it appears that the nasal placodes burrow into the MNC, but in reality, it is the proliferation of MNC around the placodes—which remain genetically connected by fibers in the MNC to the receptor site in the brain. Thus, as the nasal placodes "sink in," they drag p6 epithelium along inside with them to form nasal choanae. The walls of the nasal chamber are formed by the medial ethmoid complex and lateral ethmoid upper and middle turbinates. In so doing, on sagittal view p5, nasal skin lies externally and p6 vestibu-lar lining internally. Sandwiched in between is MNC. The "program" for the nasal cartilages is derived from the intranasal lining. Upper lateral cartilage is programmed by p6 and lower lateral cartilage is programmed by p5.

This nasal chamber is not in communication with the future mouth. In non-primates, the floor of the chamber or *primarer nasenböden* is a continuous flange of cartilage, but in primates the floor is unstable. It opens up by apoptosis along an anteroposterior seam that represents the boundary between the medial and lateral nasoplatine neuroangiosomes. This anatomy will be discussed in detail a bit later. This opening up into the oral cavity constitutes the *primary choana*.

Although neural crest and placode both share a likely neuroepithelial ancestry, *neural crest never enters placodal territory*. MNC constitutes the subepithelial tissues of the nasal chambers. The neural folds of the mammalian forebrain close in a bidirectional manner; from the adenohypophysis/Rathke's pouch cranialward (distal to proximal) and from isthmus caudal (proximal to distal). These two fusion processes terminate at the foramen cecum. The former process represents the future midface from the mouth to the nasofrontal junction, while the latter process represents the closure over the cranium and the upper face. This closure of the *cranial neuropore* takes place at stage 11.

Neural crest under epithelial signals will produce membranous bone directly or sometimes via a cartilage intermediate. Depending upon the signals, it can also produce cartilage as in the septum. In the case of the orbit, MNC receives signals from the p5 neuroepithelium of the globe and from anterior cranial base dura to produce the membranous walls of the orbit. In similar fashion, the p6 vestibular lining stimulates the lateral bony wall of the nose which is medial bony wall of the ethmoid sinuses. From the orbital side, p5 neuroepithelium stimulates the medial orbital wall which is the lateral wall of the ethmoid sinus. Separation of the potential space between these two walls, and its invasion by nasal epithelium, results in the ethmoid sinus. *All sinuses represent the separation of distinct developmental fields and the coverage of the resultant space by epithelium.*

[Note: It has been a dictum that PAM can only produce chondral bone, as in the cranial base. We shall see in the chapter on the neck that PAM can produce membranous bone in the clavicle.]

At this juncture, we must anticipate the section on facial development by Carnegie stage. The first aortic arch arteries appear at stage 9 and are immediately surrounded by neural crest of the corresponding pharyngeal arch. Although they are present one stage earlier, the first and second arches become fully visible at stages 10 and 11. By stage 12, they are confluent and have moved into position adjacent to the A field. This brings in two new sets of tissues and vascular systems. StV2 bone fields of the maxilla are now forming the boundaries of the oronasopharynx, while those of the zygoma will define the lateral orbit. StV3 mandible initiates the lower half of the oral cavity. Surrounding the jaws are skin, mucosa soft tissues, and muscles of first and second arch proper with blood supply from ECA. This configuration of tissues will permit us to describe the nasopharynx in its entirety.

The geometry of the nasoethmoid complex on either side of the midline is like a stack of eight rectangular boxes, the axis of each is anteroposterior. The lateral zone of boxes we can call collectively A2. It consists of an upper deck of midbrain neural crest derivatives from r1: ethmoid sinuses, superior, and middle turbinates. The blood supply is StV1. The lower deck consists of hindbrain neural crest derivatives from r2: palatine bone, inferior turbinate, and maxilla. The blood supply is StV2. The medial zone of boxes we can call A1; it follows the same rules. Upper deck MNC from r1 makes the nasal bone, ethmoid perpendicular plate, and septum, while lower deck RNC from r2 makes vomer and premaxilla.

It should be emphasized that the midline fusion of the two A1 field complexes creates an envelope within which a seemingly singular ethmoid plate and septum develop. There are two vomers, two premaxillae, and two paraseptal cartilages. This normal midline migration depends upon a normal degree of apoptosis that reduces the MNC of the midline ethmoid mass. Failure of this to come to completion can result

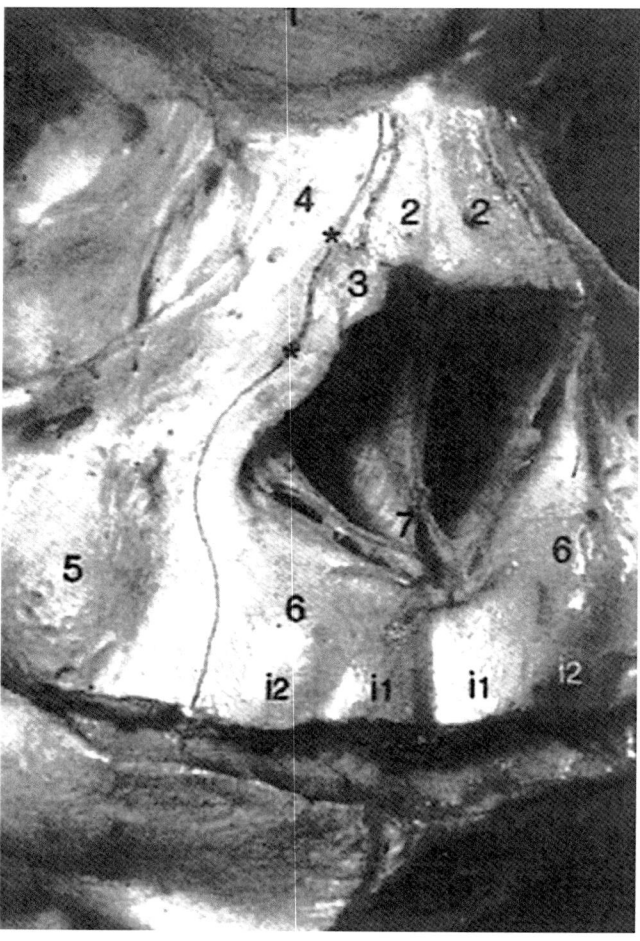

Fig. 4.8 Piriform fossa is bilaminar. Inner wall is frontal process of premaxilla; it arises from lateral incisor field. Outer wall is frontal process of maxilla bearing lateral incisor and canine. (Reprinted from Barteczko K, Jacob M. A reevaluation of the premaxillary bone in humans. *Anat Embryol* 2004; 207:417–437. With permission from Springer Nature)

in enlarged crista galli, bifid septum, or hypertelorism…all bearing witness to the bilateral nature of the midline with unitary MNC separation…the brain is never exposed.

Although most anatomy books consider the piriform rim to be part of maxilla, in truth it is a bilaminar structure. Frontal process of premaxilla (supplied by medial nasopalatine) comes to lie internal to frontal process of maxilla (supplied by medial branch from anterior superior alveolar). Knock-out of the latter results in the Tessier #4 cleft with a fissure running up the lateral wall of the nose but an intact piriform rim (Fig. 4.8 Piriform Barteczko).

What explanation can be given to the coexistence of two histologically distinct structures such as perpendicular ethmoid plate and septum, both from r1 MNC and both developing within the same envelope of p6 vestibular epithelium? The answer lies in the ability of neural crest to form either cartilage or bone depending upon the inductive stimulus. As we shall see the neuroanatomy of the medial and lateral nasal placodes lends different function capabilities to the lining of the nasal chamber. The medial septal wall is right in several distinct neural systems. Thus, genetic differences in the vestibular lining impart different signals to the responding MNC mesenchyme. The posterior nasal chambers are in close contact with prechordal plate mesoderm. It is possible that PCM influences the posterior chamber to become osteogenic. A similar situation could result for nasal bone versus nasal cartilages in terms of the biologic environment deep in the nasal chamber versus out at the surface. This same effect may explain the difference between the membranous presphenoid and chondral basisphenoid, the latter being under the influence of PCM or perhaps even formed by it.

Craniofacial osteology and arterial development are the subjects of their separate chapters. What follows is a brief outline how the developmental fields of the midline are organized. To this point, we have been concentrating on StV1 derivatives. These can be carefully mapped out by following the individual branches of the stapedial ophthalmic to their destinations. Although we have covered the ethmoid complex, it should be noted that mechanism by which the ethmoid sinuses develop is applicable elsewhere to the craniofacial skeleton. The concept of two distinct programming sources inducing opposing osseous laminae is repeated in the orbit and the maxilla.

The posterior orbit consists of the sphenoid complex and the zygoma. When development of the globe takes place, it is in intimate contact with somitomeres 1–3, from which are derived six of the seven extraocular muscles (levator being a common blastemal with superior rectus). PAM from Sm4 is the likely source for chondral basisphenoid (BS). Anterior to BS is membranous presphenoid, an r1 neural crest derivative. BS and PS may represent the ancient trabecular cartilages. They are separated by the sphenoid sinus. This latter has a midline septum which testifies to its bilateral origins. Orbitosphenoid (OS), or lesser sphenoid wing, is an r1 neural crest derivative, but alisphenoid (AS), greater sphenoid wing, is a composite of r1 and r2 neural crest. A foramen for the recurrent branch of meningeal that connects with StV2 middle meningeal penetrates the outer wing of AS. The programming for these bones is the MNC dura of the anterior cranial base.

The contributions of the alisphenoid and pterygoid fields to the sphenoid are as follows. In primitive tetrapods such as *labyrinthodont*, these bones were distinct entities. Their rhombomeric relationships can be deduced from their function. Pterygoid defines the foramen ovale through which passes V3, the master nerve of the caudal first arch. Muscles originating from the lateral lamina control the position of the condyles. From the sphenoid spine, the sphenomandibular ligament descends to the internal aspect of the ramus. Pterygoid is thus an r3 derivative. The juncture of the ali-

sphenoid with the presphenoid defines the foramen rotundum permitting ingress of V2 into the pterygopalatine space and thence forward into the zygomatico-maxillary complex. Thus, AS logically belongs to r2—it is also supplied by StV2 recurrent meningeal that anatomoses with the StV1 lacrimal at the lacrimal gland of the lateral orbit—Tessier cleft zone #9 (Fig. 4.1).

The orbital roof is an extension of the frontal bone complex. It has a sinus cavity which can occupy the entire roof, or just a portion. This results from dual programming. On the orbital side is p5 sclera and on the cranial side is r1 dura. As one leaves the orbit, the bilaminar nature of frontal sinus is continued by induction from the overlying skin envelope and internally from the dura.

Before leaving the orbit, the inductive influence of the optic placode should be emphasized. In the absence of a placode, no cavitation will occur. This can lead to a small or absent orbit. This likely reflects the interactions between the placode and the optic cup to form an intact globe—the prerequisite for an orbit. As seen in cases of ocular phthisis, the dimensions of the orbit are directly molded by the size of the globe. In similar fashion, absence or dysfunction of a nasal placode will lead to absence of a nostril and a complete consolidation of the underlying MNC where choanae should exist.

The maxilla is actually a complex of bones. It arises from primitive palatoquadrate cartilage that is subsequently ensheathed by dermal bones. Like the mandible, it has a tooth-bearing dentary component and a bilaminar support structure anchoring it to the cranial base. Because the medial lamina facing the nasal cavity is incomplete, maxilla is a five-sided box. The remaining medial defect is covered over by inferior turbinate anteriorly and perpendicular plate of palatine posteriorly. Nasal mucosa enters the interspace. As maxilla grows, the sinus expands into the lateral lamina to create its characteristic shape. The program for the parts inferior lateral nasal wall is r2 ectoderm from the first arch. Just as inferior alveolar neuroangiosome supplies the mandibular dental units, superior alveolar neuroangiosome is rationally distributed. Its anterior division (infraorbital) has a medial branch to a maxillary lateral incisor (most often suppressed) and canine, while the lateral branch of ASAA supplies the two premolars. Posterior superior alveolar provides for the three adult molars. In evolution, teeth were common on the ancient bones of the palate and vomer. Thus, it is not surprising that neural crest blastemal for the maxillary incisors travel into position via the medial nasopalatine neuroangiosome.

Mandible begins with a primitive cartilaginous core (Meckel's cartilage) which was quickly ensheathed by tetrapods with neural crest membranous bones. Of the nine bones of primitive mandible, only the dentary remains, the others persisting as genetic memories reflecting binding sites for different muscles of mastication. The primitive mandible had three tooth-bearing coronary bones. These have been amalgamated, but the original neurology persists as three separate sensory nerves of the inferior alveolar nerve complex.

Temporal bone contains middle ear structures derived from primitive forebears. Temporal tympanic bone (Tt) comes from the primitive angular (An). Malleus is the fusion of two primitive bone fields: articular (Ar) give the premanubrium and prearticular PAr) the goniale. Incus is homologous with the reduced quadrate bone. Note that all these bones are derivatives of r3. Interestingly, the malleolo-incudial joint recapitulates the reptilian lower jaw (articular) to upper jaw (quadrate) joint. The stapes represent the old hyomandibula, a second arch derivative that joined the hyoid chain of bones to the piscine skull. This became, in amphibians, a single ossicle of hearing—the columella. The most appropriate derivation of stapes would be from the rhombomere representing the anterior portion of second arch, i.e., the upper deck structures in later development. Thus, r4 neural crest is a good candidate for stapes. In early development, chorda tympani is equal in size to facial nerve. The two likely have nuclei in r4 and r5, respectively. Alternatively, facial nerve could have nuclei in both r4 and r5. This explains which facial nerve has an upper division and a lower division.

The zygoma represents the union of two ancestral bone units: jugal forms the lower body and arch, while postorbital forms the back wall of the orbit. The boundaries of these fields are indicated by the zygomaticofacial and zygomaticotemporal foramina. The zygomatic arch is in continuity with the zygomatic process of the temporal bone (Tz) that leads from the squamosal bone (Ts). Note that Ts and Tz are not developmentally part of temporal bone. Their function is to provide the insertion of muscles of mastication that move the mandibular lever arm. The entrance to the external auditory canal, the tympanic bone, relates to the first two bones of the middle ear. Thus, Ts, Tz, and Tt are all r3 neural crest derivatives "pasted on" to the temporal bone proper. This is only one of many examples of craniofacial bones that have multiple developmental sources that can be traced back to their paleontological parents.

The temporal bone is a complex of eight developmental precursors that follow its centers of ossification. Three of the fields are derived from r3. These are Ts, Tz, and Tt. All of these are neural crest derivatives. Mapping the petrous temporal bone depends upon the location of the two nuclei of cranial nerve VIII in rhombomeres 4 and 5 to give a clue as to neural crest components and a source for paraxial mesoderm. The squamosal may be a combination of r3 NC and possibly Sm4. Three r4–r5 temporal fields are named according to their relative positions to the otic placode (the source of the inner ear). In this model, pro-otic (Tpo) runs from the apex to the arcuate eminence. Opisthotic (Tpo) forms the floor of the tympanic cavity. The roof of the tympanic cavity

is formed by pteriotic (Tpt). Rhombomere r5 contributes to the epiotic (Tep) and the styloid process. Styloid is itself two fields: upper tympanohyal and lower stylohyal. Lesser cornu of hyoid would also be an r5 derivative with posterior belly of digastric slung between.

Mesodermal contributions to petrous temporal bone and mastoid are likely from somitomeres 6 and 7 with the latter providing mesenchyme for the mastoid process. This makes sense from a spatial perspective. Immediately behind Sm7, somitomere 8 (somite 1) is the primary insertion of stenocleidomastoid, the mesenchyme of which arises from the occipital somites.

The second arch, immediately after its inception at stage 11, merges into the substance of the first arch. In so doing, it loses its ectodermal representation, save for the middle lamina of the tempanic membrane. Osseous structures are remnants of the original hyoid series, a series of dermal bones suspended from anterior to the otic capsule down to lesser cornu. If one notes the original piscine model of gills, the hyoid apparatus consists of five cartilages: *epibranchial*, *branchial*, *hypobranchial*, *ceratohyal*, and *basihyal*. The latter may be preserved in the tongues of birds as the entoglossal bone as it runs forward from hyoid to make contact with the foramen cecum of the tongue…precisely the juncture between first arch r3 ectoderm and third arch endoderm from r7. The musculature of second arch arising from somitomere 6 is organized in two distinct functional planes. The *deep investing fascia* (DIF) contains the second arch muscles of mastication; these act in concert with those from the first arch; they include buccinator, posterior belly of digastric, stylohyoid, stapedius, and possibly a component of levator veli palatine. The remainder of second arch muscles are enveloped in the *superficial investing fascia* (SIF). SIF spreads subdermally throughout the head and neck until abutting with cervical somatic dermis of the occiput and posterior neck (Fig. 4.9 Branchial arch cartilages and Fig. 4.10 The hyoid chain).

Second arch fields can be named in conformity with branches of the facial nerve in two divisions: superior/cra-

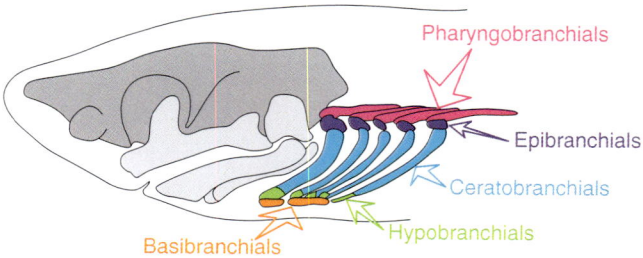

Fig. 4.9 Cartilages of the branchial arches in squalus. The basibranchials unite in the midline. In birds, this forms the *furcula*, or "wishbone." The avian tongue has a midline osseous structure, the entoglossus, formed by fused basibranchials extending forward from fourth arch. (Courtesy of Bret Kuss, PhD)

Fig. 4.10 Cartilages of the branchial arches. The basibranchials unite in the midline. In birds this forms the *furcula*, or "wishbone." The avian tongue has a midline osseous structure, the entoglossus, formed by fused basibranchials extending forward from fourth arch. (Courtesy of Bret Kuss, PhD)

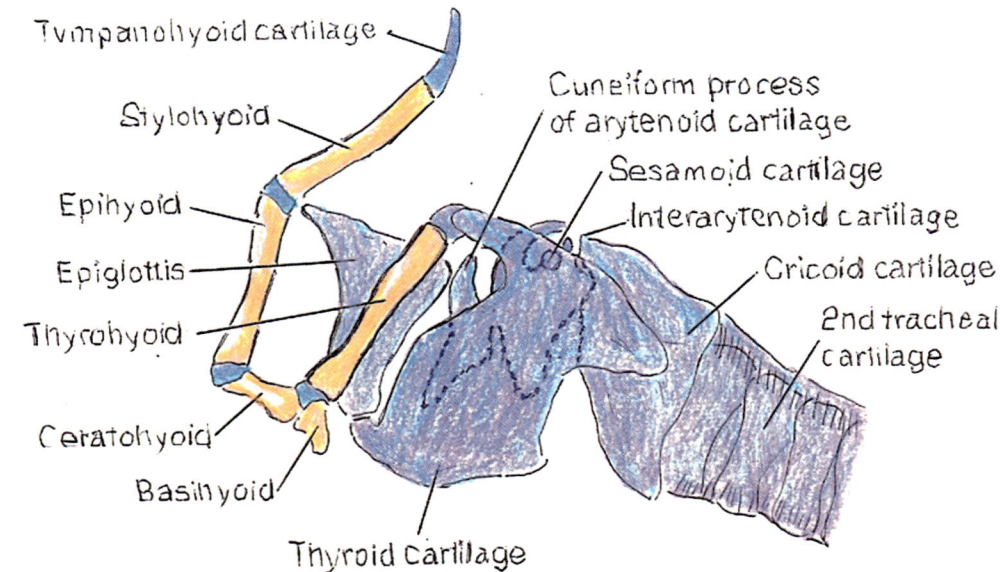

nial (VIIs) and inferior/caudal (VIIi). VII superior covers all muscles in the sensory territories of V1, V2, and epaxial V3. VII inferior covers all muscles in the sensory territories of hypaxial V3 and cervical plexus. For this reason, lateral facial clefts running from oral commissure to the external auditory meatus follow the developmental relationship between epaxial and hypaxial fields with compromising the facial nerve itself. Conditions such as the inanimate facies of Moebius syndrome, when absent represent problems inherent to cranial serves supplying the myoblast populations of somitomeres 5 (lateral rectus) and 6 (muscles of facial expression).

The *basal plan of this neuromeric system* is that the CNS is constructed from a finite number of developmental units arranged along the AP axis and covering all structures within that zone from basal plate to alar plate. Each neuromere has a specific neuroanatomic content. The boundaries of each neuromere are defined by the expression pattern of a combination of homeotic genes. The nomenclature of the neuromeres is as follows: anterior prosencephalon (telencephalon) encoded by prosomeres p6 and p5—distribution is under investigation. Posterior prosencephalon (diencephalon) is encoded by prosomeres p4–p1. Mesencephalon is encoded by mesomeres m1, m2, r0, and r1. Anterior rhombencephalon (pons) is encoded by rhombomeres r2–r7. Posterior rhombencephalon (medulla) is encoded by rhombomeres r8–r11.

The pharyngeal arches are encoded by pairs of rhombomeres "premandibular arch" representing tissues *reassigned from the first and second arches to make the jaws* is encoded by r2 and r3 (Table 4.2)

Table 4.2 Pharyngeal arches: neuroembryologic and functional correlations

Pharyngeal arch	Rhombomere/ nerve	Derivatives
First (mandibular)	r2–r3/V3	Non-jaws, mastication
Second (hyoid)	r4–r5/VII	Mastication, animation, glands
Third (glossopharyngeal)	r6–r7/IX, X	Palate, pharynx, glands
Fourth (pharyngolaryngeal)	r8–r9/X	Constrictors, larynx, glands
Fifth (internal laryngeal):	r10–r11/X	Arytenoids

Anatomy of Facial Fields: A + B + C

Key Concepts

1. Vascular definition: A fields versus B fields versus C fields refers to embryonic source of neurovascular supply. A = V1 stapedial, B = V2 and V3 stapedial, C = V2 and V3 external carotid.
2. B fields are found exclusively in the first arch. All subsequent arches are strictly C fields. Note: second arch has minor representation of StV3.
3. In development, *first and second arches fuse* to form a single functional unit.
4. Therefore, it is understood that unification of face fields will be: A + (B − C).

Note: Vascular anatomy will be covered in the chapters on craniofacial arterial development (Chaps. 6 and 7) (Figs. 4.11 and 4.12 A + B + C).

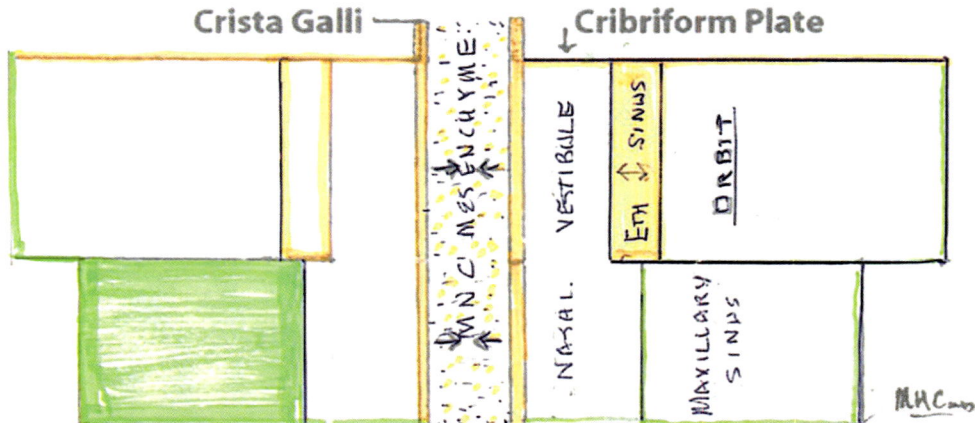

Fig. 4.11 Frontonaso-orbital and maxillary mesenchyme. MNC (orange) is also the source of anterior cranial base frontal bone, ethmoid, and sphenoid. Ethmoid sinuses are separation planes between MNC programmed by r1 sclera (overlying the p5 globe) and p6 vestibu-lar epidermis with r1 submucosa. Naso-orbital-maxillary complex: five-sided box. MNC (orange): ethmoid, sphenoid. RNC (green): maxilla, a five-sided box. (Courtesy of Michael Carstens, MD)

Take-Home Messages

First: Analysis of the neuroanatomic contents of a neuromere allows one to understand the mesenchymal derivatives that it supplies, be they (1) of neural crest that is independent of a pharyngeal arch, (2) of neural crest assigned to a specific pharyngeal arch, or (3) of paraxial mesoderm muscles from a somitomere assigned to a specific pharyngeal arch.

Second: The physical location of a pharyngeal arch-derived neural crest developmental field is a function where its mesenchyme resided within the pharyngeal arch.

Careful study of the trigeminal nerve infers that its three divisions are encoded for by distinct rhombomeres, each with its unique homeotic code. This code is shared with the neural folds corresponding to each rhombomeric zone, and the neural crest cells they give rise to. Thus, all neural crest structures are in register with the neuromere from whence they came. The anatomy of the sensory branches of trigeminal allows us to map out virtually all craniofacial structures.

The trigeminal nerve is a composite structure. Each of its divisions belongs to successive neuromeres. V1 is purely the most extensive. It occupies neuromeres m1, m2, r0, and r1. This is known as the *mesencephalic nucleus of the trigeminal nerve*. V1 is dedicated to the dura and bone of the anterior cranial fossa, the nasal capsule, the orbit, and the fronto-naso-orbital skin envelope. In point of fact, originally V1 was not part of the trigeminal nerve. It represents the ancient *proprius nerve* of the midline snout. A placodal derivative, proprius nerve has a variety of functions in other species, such as thermoreception in snakes. In evolution, with enlargement of the anterior cranial fossa, the midbrain nucleus of proprius extended backward and was captured by the trigeminal nucleus. Thus, V1 is associated with the chondrocranium—the midline support of the brain. Evolutionarily, V1 was never part of the first arch—this fiction is maintained in diagrams depicting pharyngeal arch anatomy.

V2 and V3, on the other hand, have their nuclei in rhombomeres r2 and r3, respectively. They subserve the neural crest bones that add on to the cranial base to support the brain, the so-called viscerocranium, and the surrounding soft tissues. Recall that the original seven gill arches in jawless fishes were all perfused by aortic arches. The invention of jaws by placoderm fishes in the Devonian period involved the sequestration and redefinition of significant chondral elements of the first two gill arches. In the process the first and second branchial arches lost their respiratory function; they became dedicated to food capture. That is why the descendents of the placoderms have five gill arches in chondricthyans (sharks) and four gill arches in teleosts (modern bony fishes).

The invention of jaws required a new means of their metabolic support, the stapedial system. For this reason, both *V2 and V3 serve dual functions*. They induce stapedial arteries supplying the neural crest bone fields of the jaws. All support tissues for the jaws such as skin, fat, fascia, muscle, and glands are supplied by ECA arteries, again accompanied by and induced by trigeminal sensory branches from V2 and V3.

The primary relationship of V1 to the chondrocranium is dramatically revealed by its sensory distribution to the dura of the skull base. The reader will recall that neural crest is the *materia prima* of the meninges. As is well-depicted in standard neuroanatomy texts, V1 not only supplies the dura of the anterior cranial fossa, it also runs backward along the midline of the skull base to the level of the basisphenoid. This marks the boundary of chondrocranium produced by neural crest with that produced from the paraxial mesoderm of the occipital somites. Because cerebellum arises from the isthmus, r0 and from r1, V1 supplies its dura as well.

Lateral to chondrocranium are the supporting bones of the so-called *viscerocranium*. This unfortunate term comes from an association between arches forming the pharynx or digestive system…although the true meaning of viscera refers to the internal organs of the body. Some such as temporal bone

Fig. 4.12 Stapedial V1 (orange) MNC bone/cartilage fields: frontal ethmoid, septum, sphenoid (not seen). V2 stapedial (green): RNC bone fields: maxilla, Vomer (partially obscured by septum), premaxilla. (Courtesy of Michael Carstens, MD)

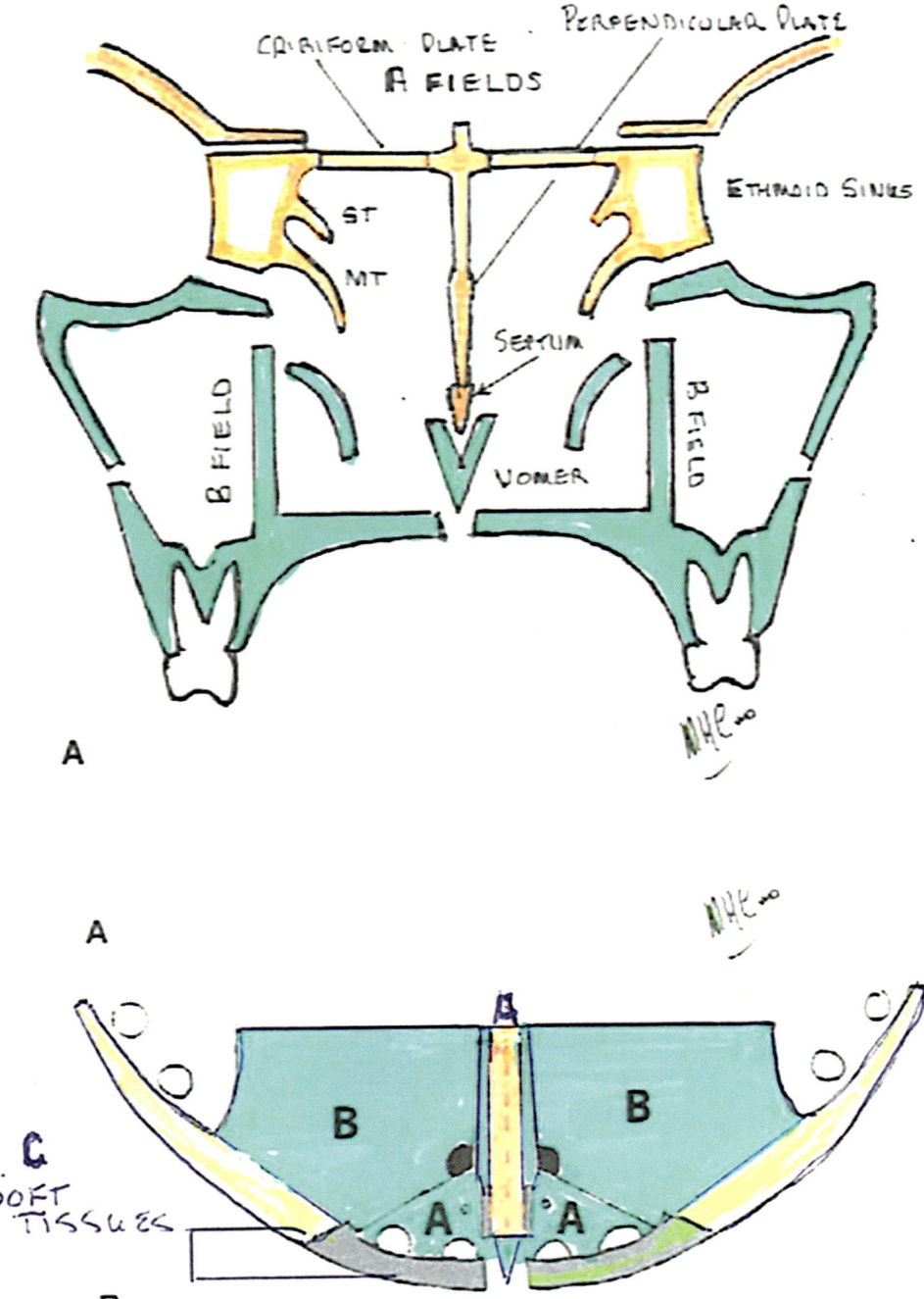

represent a mixture of PAM from somitomeres 6 and 7 plus neural crest from r4 to 5. Others, such as parietal, are formed primarily from r2 to r3 neural crest. The dura underlying the bones of the middle cranial fossa is innervated by V2 and V3; consequently, its neural crest comes from r2 and r3. The neuroanatomic between neural crest derivatives and their neuromeres of origin explains the clinical presentations of referred pain induced by muscle tension or TMJ dysfunction. Knowledge of these embryologic relationships leads to a complete neuromeric mapping of the skull (the subject of a subsequent Chap. 8).

Trigeminal divisions have distinct patterns of egress from the skull. The greater orbital fissure is a boundary between the chondral orbitosphenoid, OS (lesser wing), and the membranous alisphenoid, AS (greater wing). Both are neural crest entities; they simply differ in their means of ossification. But their derivations are different. OS arises from r1, while AS arises from r2, with a contribution from r3 to the pterygoid processes. Thus, V1 enters the orbit through a clear-cut field boundary between r1 and r2. In fact, the fissure is simply a recognition of the preexisting neurovascular structures of V1, III, IV, and VI.

More peripherally, the foramina of the branches of the trigeminal nerve demarcate field boundaries as well. The anterior and posterior ethmoids are distributed to distinct sinus fields. Supraorbital foramen represents the field boundary between the two ancient bone fields of the orbital rim, *pre-frontal medial* and *prefrontal lateral* (PFm and PFl).

V2 and V3 also exit through foramina that represent field boundaries; in which bones form around the neuroangiosomes. The attachment of r2 AS to r1 presphenoid (PS) gives foramen rotundum for V2. The attachment of r3 pterygoid (Pt) to PS defines foramen ovale.

The axis of anterior superior alveolar artery (ASAA)/infraorbital demarcates two zones of dental units each: medially, that of maxillary lateral incisor and canine, and laterally, the premolar zone. Both zones develop as a mesial-distal addition of mesenchyme. Thus, maxillary incisor (when present) always appears before canine. Similarly, first baby molar (and later first premolar) always appears prior to the second dental unit. The zone of the three adult molars develops as a posterior "add-on." Hence, its angiosome, posterior superior alveolar (PSAA), starts out as a single vertical branch down to first adult molar. Later in time two additional branches supply their respective molars. The three branches successively fuse to form a longitudinal arcade such that the stem of the PSAA enters posteriorly.

Zygomatico-facial branch of StV2 constitutes a boundary between the two ancient components of zygoma, a medial jugal field and a lateral postorbital field. Descending palatine separates palatal bone from maxilla. Each of these will have its own dedicated branch.

The neuromeric origins of the sensory components of trigeminal nerve display themselves in *somatotopic* fashion in the neural crest derivatives they supply. Proprius nerve nuclei exist in four locations: m1, m2, r0, and r1. These are distributed among four neurovascular zones of fronto-naso-orbital mesenchyme, clinically seen as Tessier cleft zones #13, #12, #11, and #10. Precise mapping of MNC has not been done to specify exactly what order these populations migrate but, as a hypothesis, it would be logical to consider them following in cranial to caudal order. Evidence for this exists with the somatotopic distribution of anatomic zones to the mesencephalic nucleus with sensation from the most medial and distal tissues in register with m1 and with lateral and proximal tissues "reporting" to r1.

As neural crest migrates into the pharyngeal arches, it is organized into anatomic "compartments" by a position-specifying system of *distal-less* (*Dlx*) genes (DePeuw). The clinical effects of this organization can be seen in the spatio-temporal appearance of second arch facial muscles as detailed by Gasser. Muscle blastema of this system appears in three gradients: deep-to-superficial, lateral to medial, and caudal-to-cranial. This could reflect the position that each blastema has within the substance of the second arch: i.e., "first in, first out."

The same concept of developmental order applies to craniofacial arteries as well. After all, the arteries themselves contain neural crest in their walls. Could it be possible that genetic differences in the NC cells are responsible for specifying the individual branches, for example, the StV1 naso-

ophthalmic? In theory, the vascular tree would develop: (1) in a *spatial order*, supplying the oldest tissues first, from medial to lateral; and (2) as a series of *bifurcations*. Thus, the vascularization of midline structures such as the ethmoid complex would take place before that of subsequent structures, such as the orbital roof. In a bifurcation model, the two zones of the orbital roof, prefrontal and postfrontal, supplied by supraorbital axis would develop simultaneously and independently. Thus, a medial orbital roof defect in Tessier zone #11 would not imply a defect in zone #10: the two zones are biologically autonomous.

Big-picture ideas: (1) The somatotopic representation of tissues in the trigeminal sensory nucleus is a function of their spatial-temporal migration into their final position. Posterior ethmoid sinus is represented in the mesencephalic nucleus cranial to maxilla. Zygoma is posterior to palatine bone. (2) Arterial development is also somatotopic and follows gradients of tissue maturity. (3) A-field development takes place first. B and C fields develop simultaneously. (4) Mesencephalic flexure is required to establish correct spatial relationships among the fields.

A-Fields: StV1 Fronto-Naso-Orbital Arterial System (FNO)

The architecture of the A-fields is defined by the perfusion pattern of the various branches from V1 stapedial after it makes an anastomosis with primitive ophthalmic artery from ICA. Because this developmental anatomy is not widely appreciated, the entire collection of ICA vessels to the globe and non-ICA vessels to extraocular structures has traditionally been lumped together as the ophthalmic arterial system [35, 36]. The anatomy of the stapedial system and the orbital arteries is thoroughly discussed in Chaps. 6 and 7. What follows are the most relevant points.

Development of the StV1 System

1. *There is no such thing as an original ophthalmic artery to the eyeball.* The ophthalmic stem develops in three iterations. The developing forebrain at stage 8 is initially supplied by a *primitive head plexus* which provides nutrition via an ebb and flow mechanism. Primitive internal carotid appears at stage 9. By stage 10 (3 mm), the primitive head plexus is changing. ICA now reaches the site of the optic anlage where it forms two branches. The most distal, *primitive olfactory*, is directed toward the anterior cranial base and has no role in ocular development. The second branch, *primitive maxillary*, supplies the plexus over the optic anlagen. Supply from this artery is temporary until the development of a more formal source of blood supply to the eye.

   At stage 12 (4–5 mm), two critical events take place. First, the ICA bifurcates into a cranial division (the source of both the future anterior and middle cerebral arteries) and a caudal division (the source of posterior cerebral and the posterior communicating branch to the hindbrain system via the circle of Willis). Second, the optic cup forms and folds upon itself to enclose the optic plexus as the *hyaloid artery*. At stage 14, the most proximal branch of the anterior division of ICA gives off primitive dorsal ophthalmic (DOA) which approaches the cup from its caudal and dorsal side. Recall that at this stage the eyeball projects outward from the brain at a 180° angle. DOA connects with hyaloid, making primitive maxillary irrelevant (the latter is subsequently directed medially and backward to supply the hypophysis). At stage 15, a new vessel primitive ventral ophthalmic (VOA) arises from a more distal site along anterior division and supplies the cranial and ventral aspect of the cup. At stage 16, definitive hayloid artery becomes a branch of DOA. At stage 19, the stem supplying DOA and hyaloid shifts backward, becoming immediately proximal to the division point of ICA. This provides more direct blood flow to the globe. DOA and VOA form the temporal and nasal ciliary arteries. They are connected via hyaloid. The adult stem of ophthalmic and its three primary branches are now fully developed.

2. *Nonneural tissue of the orbit requires a separate source of blood supply unrelated to internal carotid.* As we shall see, stapedial stem arises from hyoid artery (a remnant of second aortic arch artery) and, following "programming" by the seventh nerve, proceeds upward through the future tympanic cavity where it bifurcates into intracranial and extracranial divisions. These are programmed, respectively, by superficial petrosal of VII and by chorda tympanic of VII. Upon reaching the trigeminal ganglion, intracranial stapedial follows the branches of V1 and V2 (V3 having exited the skull) to supply dura and the orbit. Inside the orbit, StV2 anastomoses with StV1 as lacrimal artery and subsequently involutes (in most cases). The supraorbital StV1 axis makes and anastomoses with the ophthalmic stem at stage 20. Thus, *StV1 system supplies all nonneural tissues inside the orbit and all frontonaso-ethmoid mesenchyme outside the orbit.*

For the reader, it may seem a trivial point that V1 stapedial be considered a *non-ICA derivative*; after all, it originates from third arch ICA at its juncture with dorsal aortae. But its history reflects that the exclusive source of mesenchyme to the orbit and nasal midline is midbrain neural crest. That of the ocular apparatus is basal forebrain neuroectoderm, but, not being embryologically part of anterior cranial fossa, it is not supplied by anterior cerebral but rather via a stem located proximal to the ICA bifurcation. And the purpose of the StV1 system is to supply non-CNS tissues originating from midbrain neural crest and somitomeric mesoderm.

Organization of the Intracranial StV1 System

The cutaneous manifestations of the A fields are limited to those soft tissues perfused by the terminal external nasal,

upper eyelid, and forehead branches StV1 naso-ophthalmic. The most caudal cutaneous derivatives of anterior ethmoid are columella and philtrum (Cupid's bow). When considered in cross-section, the nasal A fields consist of the septum and ethmoid complex. The latter consists of: cribriform plate, crista galli, ethmoid sinus air cells, and upper and middle turbinates.

The lacrimal bones and lacrimal apparatus constitute the anteromedial boundary of the orbital A fields. The development of the lacrimal duct *precedes* that of the bone. Duct epithelium produces programming signals for surrounding r1 MNC; consequently, the bone ensheaths the duct. Posterior to the lacrimal lies ethmoid, a bilaminar structure that constitutes at once the medial orbital wall and the lateral nasal wall with sinus air cells occupying the space in between. Note that this results from two sources of programming for r1 MNC: nasal epithelium (associated with p6) and sclera/dura (associated with p5).

The various branches of StV1 frontonaso-orbital constitute distinct developmental zones. *Nasociliary branch* runs along the medial boundary of orbital roof and ethmoid complex; it is distributed to three distinct zones. (1) It supplies the ethmoid complex and nasal cavity via the anterior and posterior ethmoids. (2) Its forward continuation as trochlear supplies lacrimal bone and medial forehead. These constitute part of Tessier cleft zones #13 and #12. (3) It also sends a branch to the neural crest of the ciliary body (hence its name). The axis of *supraorbital branch* supplies the bulk of the orbital roof which forms part of Tessier cleft zones #11 and #10. Its forward continuation supplies the lateral forehead. *Lacrimal branch* extends to the lateral recess of the orbit where it anastomoses with StV2 recurrent meningeal branch to supply the orbit and surrounding tissues, including alisphenoid (the greater sphenoid wing).

The structure of the orbital roof is potentially bilaminar. Programming from dura above and sclera below create a potential field separation plane for the *orbital extension of the frontal sinus*. The programming is variable: the sinus may be minimal or may occupy virtually the entire orbital roof. MNC failure in Tessier cleft zones #11 and #10 can result in an orbital roof defect and *encephalocoele*. Encephalocoeles can also occur in Tessier zones #13 and #12; these are directed into the nasopharynx.

StV2 proceeds forward with a dural branch directed to middle cranial fossa and an orbital branch, the *meningo-orbital artery*. At stage 19, it passes through alilsphenoid via cranio-orbital foramen lateral to the superior orbital fissure where it joins StV1 lacrimal branch. It can also enter superior tissue at its lateral corner. Anastomosis with ophthalmic takes place at stage 20 at which time communication with the middle meningeal system becomes irrelevant and, in most instances, meningo-orbital involutes (Fig. 4.13 Intraorbital stapedial system).

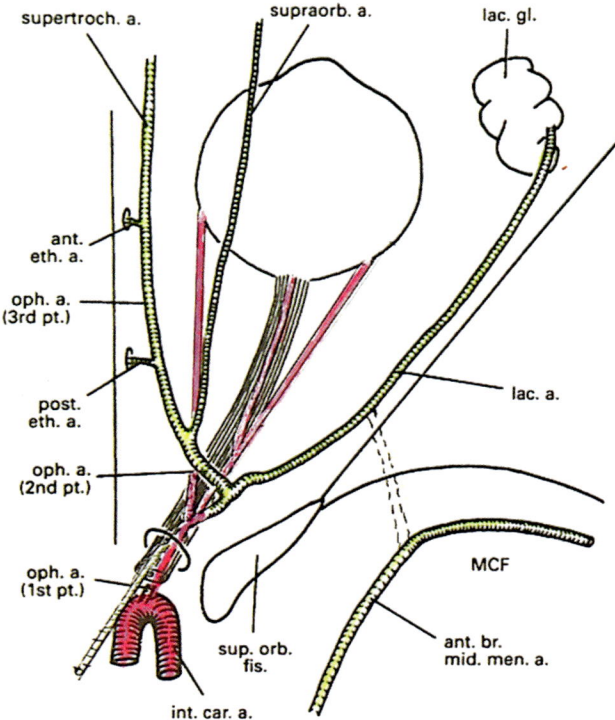

Fig. 4.13 40 mm embryo StV1 (yellow) accompanies optic nerve to the orbit but when anastomosis to ophthalmic takes place, proximal StV1 from orbit to trigeminal ganglio disintegrates, leaving behind StV1 branches to the orbit and anterior meningeal arteries to cranial base above the ethmoid. StV2 (yellow) projects from trigeminal to supply dura of middle fossa and meningo-orbital branch anastomoses with StV1 at what will be lacrimal. In general, the proximal connection, meningo-orbital, disintegrates as well after stapedial and ophthalmic systems have merged. (Reprinted from Diamond MK. Homologies of the stapedial artery in humans, with a reconstruction of the primitive stapedial artery configuration of euprimates. *American Journal of Physical Anthropology* 1991; 84(4): 433–462. With permission from John Wiley & Sons)

B-Fields: StV2 and StV3 Maxillomandibular Arterial System

The architecture of the B-fields is defined by the perfusion pattern of branches from V2 stapedial and V3 stapedial after their common stem (extracranial stapedial) makes an anastomosis with the ECA maxillary artery. Note: this should be called maxilla-mandibular artery because it serves both the lower face and the midface.

Recall that intracranial stapedial gives off stem which proceeds extracranially in company with VII chorda tympani. After joining with ECA at the root of V3, StVII "passes the baton" to branches of the trigeminal nerve, first to V3 and subsequently to V2. Stapedial blood flow proceeds to form the hybrid *maxillo-mandibular artery*. Recall also: (1) all the stapedial derivatives of the proximal sector of MMA are V3-induced, (2) middle sector of MMA has no stapedial derivatives, and (3) distal MMA picks up V2 so all arteries of this sector are StV2.

The anatomy of the pterygopalatine fossa and its StV2 arteries is extensively discussed in the vascular chapter (Chaps. 7, 12 and 13). What follows are the most relevant points.

Evolutionary Considerations

StV2 and StV3 of the MMA supply exclusively neural crest bones involved in forming the jaws and anchoring them to the skull base. The reason for this is evolutionary. Jawless fishes have an external carotid directed toward the "face," but they do now have a stapedial system. The "gnathostome revolution" involved reassigning neural crest tissues dedicated to the first gill arch to another purpose, the creation of jaws and a midface. First aortic arch was destroyed in the process. A new arterial system, the stapedial, came into existence. Those remaining first arch tissues not involved in jaw formation were supplied by the ECA. The ancient trigeminal system with V1 propius nerve to the snout provided programming for orbital stapedial derivatives, while in the remainder of the skull V2 and V3 continued to serve as induction sources for arteries from both stapedial and ECA.

Note that, in evolution, the initial iteration of jaws seen in the early armored fishes (placoderms) required no support structure for the mandible, a condition known as *autosylistic suspension*. Thus, in these agnathans, only first arch was involved. "Soon" thereafter (in paleontologic time), early fishes without body armor connected the jaws to the skull using a second arch derivative, hyomandibula, the amphistylistic suspension. Second arch bones do not depend on stapedial for support.

Organization of the Extracranial StV2/StV3 System

The first (mandibular) sector of MMA gives off four dorsal arteries—all of them StV3—that enter the skull. It gives off two ventral arteries: ECAV3 mylohyoid artery to that muscle and StV3 inferior alveolar artery to the dentary component of mandible. Dentary is the remnant of the original nine-bone structure and contains all the teeth; all the remaining fields of the mandible of C fields. They represent the dermal bones that ensheathed Meckel's cartilage and hence are supplied from external carotid.

The second (temporal) sector of MMA gives off six exclusively ECAV3 arteries, all of which supply muscles of mastication. Three are directed dorsally to the muscles of temporal fossa and three supply the pterygomasseteric sling.

The third (pterygopalatine) sector of MMA gives off exclusively StV2 arteries. Two branches, pharyngeal artery and the artery of the pterygoid canal, are directed backward to sphenoid sinus and auditory tube. They indicate the first arch component of auditory tube. The remaining branches are directed forward into the maxillary-zygomatic complex.

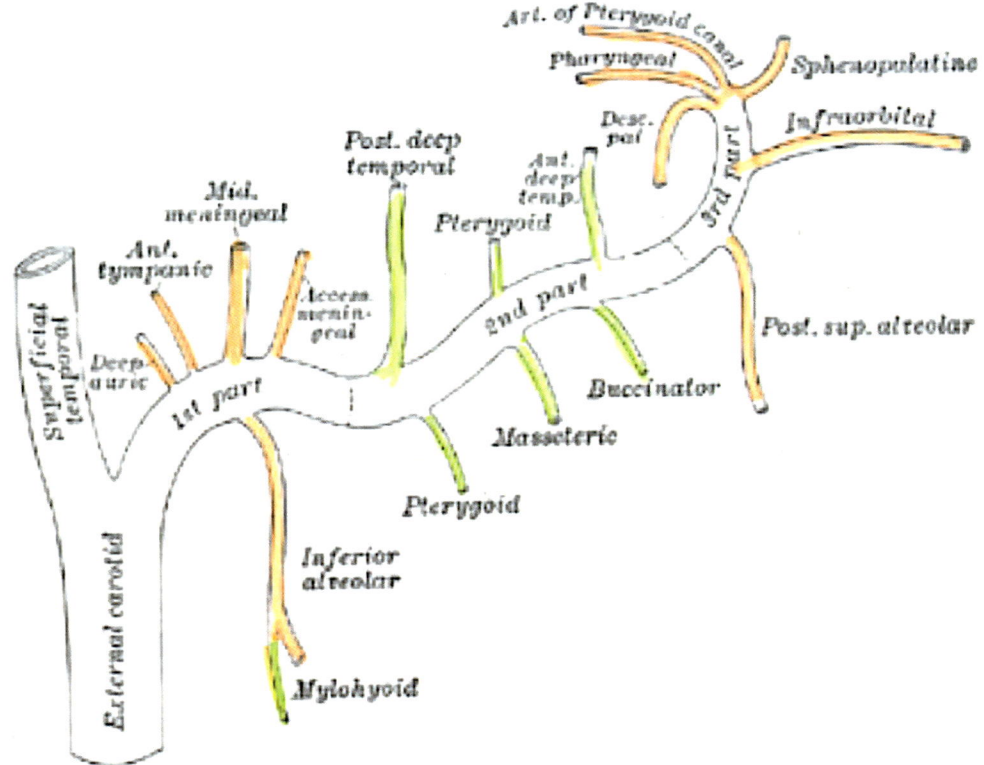

Fig. 4.14 Internal maxillo-mandibular arterial system is a composite of external carotid maxillary branch with extracranial stapedial, conveyed to the root of V3 from tympanic cavity via chorda tympani branch of cranial nerve VII. Stapedial serves neural crest bone reassigned from the first gill arch to form the bones of jaws, the connection of maxilla to the skull base, and dura. These derivatives are in orange. Derivatives directed toward muscles of mastication, also subverted from the first arch, are in yellow. Ever branch is "programmed" by a branch of V2 or V3. Tympanic and deep auricular branches represent communications from stapedial stem outward to form extracranial division. They represent the connection between stapedial and the external carotid system. (Reprinted from Lewis, Warren H (ed). Gray's Anatomy of the Human Body, 20th American Edition. Philadelphia, PA: Lea & Febiger, 1918)

It consists of three distinct systems—all of which subdivide into functional units. (1) *Sphenopalatine artery* is the terminal branch of the StV2 system. Via *nasal branches* supply inferior nasal walls, all r2 mucosa entering the paranasal sinuses, the lower part of septum, and via nasopalatine branches, the medial vomer/premaxilla and the lateral inferior turbinate. (2) *Maxillary artery* also subdivides into two misnamed structures…which I have subsequently renamed. *Posterior maxillary artery* (posterior superior alveolar artery) is proximal and supplies the posterolateral wall of the maxillary sinus as well as molars and the lateral wall of their alveolar housing. *Anterior maxillary artery* (anterior superior alveolar artery) is distal. It supplies the floor of the orbit and inferior oblique muscle. It then bifurcates to supply the anterior wall of maxillary sinus. A medial branch supplies frontal process of maxilla, lateral incisor, and canine, and via a lateral branch, it supplies for the deciduous molars/adult premolars. The dental branches supply lateral wall of the alveolus. (3) *Descending palatine artery* perfuses the palatine bone and subdivides. *Greater palatine* supplies the lingual alveolus and oral lamina of palatine shelf, while *lesser*

palatine supplies the unilaminar horizontal palatine shelf and anterior 1/3 of the soft palate (Fig. 4.14 Maxillomandibular arteries).

C Fields: ECAV2 and ECAV3 Arterial System

The remaining tissues of the face are of exclusively pharyngeal arch origin and are perfused by derivatives of the original aortic arch arteries that provided initial supply of the arches. We shall not consider here the derivatives of the third, fourth, or fifth arches, as they are all posterior to the face. Tissues of interest from the first arch are the skin and mucosa, fat, and muscles of mastication in the deep plane only. Those of relevance from the second arch are glands, a limited number of muscles of mastication in the deep plane, and an extensive number of muscles of animation in the superficial plane.

The behavior of the first and second arches is dramatically different from the remaining three arches. Formed at stages 10 and 11, respectively, by stage 12 they combine into an amalgamated unit, the neural crest for which comes from rhombomeres 2–5 [37, 38]. The consolidation of these two arches reflects the evolutionary innovation of jaws as separate functional units from the original intention of branchial arches as purely respirator/digestive units. It is also seen in the individual fates of the original aortic arches in tetrapods. AA1 and AA2 fall apart and lose their connection with the aortic outflow tract completely. They become dependent on AA3 for survival. AA3 remains intact, becoming common carotid; it gives off the stem of external carotid and continues as the extracranial internal carotid. AA3 and AA4 retain their caudal connection with the aortic outflow tract, but are disconnected dorsally with AA3 directed to the head and AA4 to the remainder of the body. AA5 involves completely, its corresponding pharyngeal arch becoming dependent upon AA4 for survival. AA6 (no corresponding arch) remains attached to outflow tract and is exclusively dedicated to the pulmonary circulation.

The derivatives of the first and second arch complex are supplied by specific branches of the external carotid. First arch is represented by maxillary. Second arch is represented by lingual, facial, and the anterior branches of superficial temporal. Deep plane structures are perfused by the maxillary and lingual arteries, and superficial structures by the remainder. Because the facial muscles of the ear, occiput, and neck (platysma) are not part of the first/second arch complex, they are not included here. In conclusion, we once again return to the concept of somatotopic representation and spatial-temporal development.

The reader is referred to the work of Gasser on stage-specific facial nerve and muscle development. The same rules hold true for deep bone structures as for their superficial soft tissue coverage. Thus, the physical disposition of the maxillary and zygomatic fields from medial to lateral and from deep to superficial is consistent with this type of linear read-out. So too is the timing of their development as reflected in the appearance of calcification with their respective functional matrics. Kjaer makes the interesting observation that all such calcifications are first arrayed around the neurovascular axes of each field [39]. Anatomic work by Ian Taylor defining the territories of *neuro*angiosomes (emphasis mine) [40, 41] is likely to prove homologous!

Conversion of the linear model to three-dimensional reality would be difficult without an appreciation for the role folding in facial development. Participation of the B and C fields in the formation of the oral cavity requires approximation of the first/second arch complex toward the midline. This in turn depends upon the flexure of the embryonic neuraxis. This occurs at the level of the midbrain due to brain expansion. Curving of the embryonic face toward the future chest first occurs at stage 10. The angle between prosencephalon and rhombencephalon decreases from 180° to 150° by

day 22 and to 100° by day 23 [42]. The mesencephalic flexure is responsible for bringing the B and C fields arch into close contact with the A fields. When the tissues are within the critical contact distance, a tightly controlled spatiotemporal sequence of fusions takes place.

Author's note: As will be discussed in detail in the chapter of muscle development, the three-dimensional relationship between the eye and somitomeres bearing extraocular muscles allows for the sequential insertion into the muscles into specific recipient.

The muscle order is as follows: Sm5 lateral rectus[13], Sm3 superior oblique[14], Sm2 superior rectus[14], Sm1 medial rectus[15], Sm1 inferior oblique[16], Sm1 inferior rectus[17], and levator palpebrae superioris[19–21]. How does Sm5 reach the orbit? In this model, development of SM4 preceed that of Sm5. Sm4 "empties out" its contents of myoblasts into first arch, permitting Sm5 direct access to the remaining insertion zone of the globe. From these data, it is evident that these somitomeres are spatially positioned next to the globe which at first projects out 180° from the neuraxis, with Sm1 being closest to its ventromedial aspect. The sclera has a genetic program of insertion sites that "become available" in temporal sequence: dorsal > ventral, lateral > medial. So although myoblasts from the somitomeres are spatially clumped around the globe, the insertion sequence of the muscles reflects the order in which binding sites "turn on their landing lights," permitting the closest muscle available to attach.

Determining Cranial Nerve Geometry in the Face

Important functional consequences arise from the mesencephalic flexure. When the first/second arch complex "spans the gap" to connect with its ipsilateral A-field partners, an internal soft tissue plane results, within which bone and muscle fields develop to complete formation of the orbit, nasal, and mouth. *Recall that there is no endoderm in the mouth until one reaches the posterior aspect of buccopharyngeal membrane.* The BPM forms a ring at the posterior 1/3 of the tongue, the anterior pillar of the tonsillar fossa, and the anterior 1/3 of the soft palate (oral side). All oral mucosa in front of buccopharyngeal membrane is ectoderm, with r2 covering the maxillary and r3 the mandible. The boundary between these neuromeric fields follows a line drawn from the commissure to the external auditory canal.

Big picture idea: Because the biologic "age" of the first arch is greater than that of the second arch, first arch derivatives have spatial priority; second arch derivatives are forced to occupy positions left over by their predecessors.

The melding of first and second arches creates the spatial arrangement of anatomic structures within the complex. Because first arch muscles arise from Sm4 while those of second arch develop within Sm6, the former muscles develop first and come to occupy the deep fascia. Buccinator is opportunistic, being programmed by the oral mucosa while the remaining second arch masticatory muscles attach in series between first arch and third arch derivatives. An example of this is posterior versus anterior digastric and the latter are positioned accordingly. Muscles of facial expression result from reassignment of second arch myoblasts to a new, superficial fascial layer. Beneath the surface of the skin, they fan out over the entire face and head. All sensory afferents from the face are wired to the trigeminal sensory nucleus. Rhombomeres 4 and 5 are not represented. Another example of the melding process is seen in the dual vascular supply of parotid gland, being maxillary from first arch with second arch represented by transverse facial branch of superficial temporal.

The positioning of the facial nerve and the patterning of its division follow a similar design. Because second arch participates in formation of parotid gland, facial nerve must pass through its substance. Note that the duct and the epithelium penetrating into the substance of the gland is of first arch origin. Deep muscles of second arch are laid down first into the deep investing fascia, while the superficial muscles and fascia appear later in time, following the principles of Gasser [43].

The Logic of the Facial Vascular System

Folding has its consequences for the vascular distribution of the external carotid system as well. The ECA arises de novo beginning at stage 12–13 by a mechanism of recombination of the preexistent aortic arch arteries supplying the pharyngeal arches [44]. At the same time, neural crest cells associated with the arches migrate downward into the aortic outflow tract and into the heart. The association of neural crest deficiency states can lead to both cleft palate and cardiac defects.

The developmental fields of the midface and lower face are sandwich of B and C field complexes, respectively. B-field derivatives (the maxillary-zygomatic complex and the mandible) are supplied by StV2 and StV3 branches of the maxillomandibular artery (MMA). C-field derivatives are supplied by arteries of the ECA; these are also induced by sensory branches of V2 and V3. The difference between the two types of tissues is evolutionary. One set of fields has to do with the jaws, the other set of fields are all supporting structures, mucosa, glands, fat, and muscle. Mucoperiosteum over maxilla and mandible is supplied by stapedial. This makes sense because it covers the jaws, whereas the oral mucosa is supplied by lingual and facial.

The fascial layers of the face are arranged in successive layers with distinct sources of blood supply. This is most clearly seen in the temporal fossa. Deepest, of course, is dura. It receives StV3 meningeal vessels from the dorsal aspect of proximal segment MMA. These anastomose in the skull with the intracranial stapedial system programmed by V1 and V2. Structures of the epaxial deep investing fascia

such as temporalis muscle are perfused from the dorsal aspect of MMA. Epaxial structures of the superficial investing fascia such as scalp are supplied by the subcutaneous arteries such as superficial temporal, posterior auricular, and occipital.

Neuraxial Subdivisions of Fields

Developmental fields are the anatomic representation of functional matrices, a concept introduced by Melvin Moss. These are nearly synonymous with Ian Taylor's angiosome concept [40, 41], the difference being the inclusion of an accompanying sensory nerve as an induction agent. Thus, all developmental fields are organized around a unique neuroangiosome. In the face, these are readily appreciated as the neuraxes of the trigeminal nerve. Membranous bones thus form as the "products" of biologic structures supplied by a neurovascular pedicle that preceded the appearance of the bone—therefore the requirement for exit foramina. Histologic study of the infraorbital, incisive, greater palatine, and mental foramina in 26 embryos by Kjaer demonstrated that, in every case, nerve tissue antedated bone [39]. Membranous bone, cartilage, and muscle are interlopers that develop within the confines of a more primitive neutrally organized envelope. There are two important anatomic consequences of this: (1) the boundaries of fields are represented by sutures; and (2) neural foramina are formed by two (or more) contiguous fields…bones form around *preexisting soft tissue structures.*

Facial Clefts as Field Markers: The Neuromeric Basis of the Tessier Cleft Classification

Unusual facial cleft and a variety of craniofacial malformations are manifestations of the neurodevelopmental system gone awry [45]. Facial clefting is, in all probability, a biologically incorrect term. What have been called clefts are varying degrees of mesenchymal (neural crest) states leading to field fusion failure. Field failure is oriented along its neuraxis with involvement beginning distally and, with greater severity, proceeding proximally. When minimal, distal zone deficiency creates a simple boundary zone separation that may exceed the critical contact distance required for union with its partner field. As the severity of deficiency increases, the entire mesenchymal bone field may be wiped out, but the *unfused skin envelope will remain intact.* This is because *the bone defect involved compromise of a specific neurovascular axis from the stapedial system, whereas the skin envelope is supported by external carotid.* Even in grotesque facial clefts, bone is never exposed.

The key to understanding fusion failure resides in interaction between underlying mesenchymal synthesis of bone producing bone morphogenetic protein-4 (*BMP4*) and the expression of Sonic hedgehog (*Shh*) in the epithelium. The principle role of Shh is to maintain the integrity of an epithelial surface. When Shh is fully active, adjacent tissues cannot fuse. When inhibited by BMP4, epithelial surfaces break down and can merge. Normal bone synthesis produces BMP4 which diffuses through the soft tissue to reach the epithelial surface, block Shh, and permit fusion. Thus, any disturbance that reduced the mesenchymal volume of the deep tissue fields will also reduce the amount of BMP4 available to block Shh. In sum, neural crest deficiency states in bone fields negatively affect ability of their overlying epithelial surfaces to achieve physiologic union [46].

Tessier's numeric system is quite rational from this perspective [47]. Each cleft zone involves a deficiency state in one or more bone fields along a specific stapedial neuroangiosome. Orbital cleft zones (10–13) are caused by deficits of the StV1 branches, while maxillary cleft zones (1–8) represent deficits of StV2 branches. The number 9 zone (alisphenoid) is an amalgam among StV1 and StV2 (Figs. 4.1 and 13.21).

The details of the neuromeric model for Tessier clefts have been previously published [48] and will be discussed in this book in a separate chapter. For our purposes, clinical examples of these clefts are presented to give some shape to this concept. Zone 3 cleft involves the loss of both inferior turbinate and maxillary palatal shelf (lateral nasopalatine and greater palatine angiosomes), thus creating wide-open access to the maxillary sinus through a massive defect in the lateral nasal wall. Zone 4 clefts involve the frontal process of the maxilla (medial branch of anterior superior alveolar), but because the frontal process of premaxilla is unaffected, the piriform rim remains intact. Zone 5 involves loss of the medial anterior face of maxilla up to the infraorbital foramen, while zone 6 affects the lateral face of maxilla and buttress. Dental losses in these conditions correspond to the distribution of the neuroangiosome involved.

The zygoma is composed of two zoologic units, jugal (J) and postorbital (PO), defined by the axis of the zygomaticofacial neuroangiosome. Cleft zones 7 and 8 represent deficiency states of these two fields. When hypoplasia strike both J and PO, the malar flattening and downturned canthi seen in Treacher-Collins syndrome result.

In the midline, the Tessier concept of a 0–14 cleft zone is not correct. This zone represents the MNC of the nasoethmoid complex intervening between the orbits and is subject to a normal process of apoptosis that will cause physiologic approximation of the orbits and unification of the nasal passages. Failure of this creates hypertelorism albeit with a normal brain.

Zone 1–13 defects are based on StV1 branches and involve the upper midline. These affect the medial nasal lamina and may be associated with loss of a central incisor. Anterior ethmoid defects affecting the perpendicular ethmoid plate may negatively affect development of an otherwise normal vomer and premaxilla—they cannot descend into the midline. This can lead to a situation midline lip without hypotelorism. Deeper involvement may affect the upper ethmoid complex or cribriform plate. Zone 13 involves defective (or excessive) frontonasal mesenchyme over the midline.

At times, defects in the nasoethmoid arteries are accompanied by defects in the branches of the anterior cerebral artery, causing defective forebrain division (holoprosencephaly) and hypoteloric states, up to and including cyclopia. Cyclopia is often described as a failure of separation of the orbital fields in development. But this condition also involved concomitant failure of MNC structures in the midline with hypotelorism. Here the same type of median cleft lip is now associated with HPE.

Other forms of StV1 knock-outs can have purely frontonaso-orbital effects in zones 13–10, ranging from bone deficit states with consequent encephalocoeles to bone excess states and orbital dystopia.

Zone 2 defects are based on the StV2 medial nasopalatine axis and affect the lower midline. Vomer and premaxilla are hit. If associated with zone 12, a deficit in the p5 nasal lining can cause a "notch" in the alar rim. A normal zone 3 can be affected by a deficiency state in the soft tissues of zone 11 that pulls the alar base upward toward the canthus. We shall discuss these pathologies at length in another chapter.

Assembly of the Face: Developmental Sequence

As the reader will note, understanding the process of midline facial development requires visualizing the process by which fields form, move, and interact. Thorough analysis of a time-dependent series of three-dimensional images will permit their conversion into a form of "embryologic videotape" which can be played forward or backward in the mind as required. Events in the formation of the frontonasal complex are described below at each of the Carnegie stages, described in detail in Chap. 3. To best create such a mental videotape, the reader may find the following readily available resources valuable: (1) a contemporary text of embryology using a molecular approach such as by Carlson or Larson [49], (2) familiarity with Hinrichen's monograph [50], (3) a review of cranial nerve anatomy and functional classification [51, 52], and (4) a skull (preferably one that opens sagittally).

Principles of Placodes

All special somatic afferent systems (the special sense organs of smell, sight, and hearing) employ similar mechanisms in their development. All require the presence of a *placode*, a specialized area of epithelium seen on the surface of the embryonic face as a disc-like structure [53], and they share many common features. (1) All placodes are organized into zones containing tissues of functional importance for the eventual completion of the sensory structure. (2) In all three systems, the placodes remodel and sink into the face, in this way they internalize functional tissue. (3) Once internalized, all placodes undergo structural modifications. (4) Placodes have an inductive influence on the surrounding epithelium and mesenchyme. They are capable of transforming these tissues into separate structures (the nares, the lens, the otocyst). (5) In each case, specialized tissue of the sense organ must make contact with the brain via cranial nerves in order for sensory perception to occur.

Unique to the olfactory placode is the presence of growth hormone-releasing hormone GnRH neurons which originate in the placode, migrate via MNC pathways to the CNS, and terminate in the hypothalamus (a topic that will be discussed in depth in part 2 of the chapter). Okabe showed in human embryos that the GnRH neurons were immunoreactive with antibodies to cytokeratin and cell adhesion molecule 5.2 (CAM5.2), but that after migration, reactivity was extinguished in the early fetal period. Cephalic neural crest cells were not reactive, consistent with the unique field character of the olfactory placode [54].

Before we proceed further, let's remind ourselves of the anatomy of the neural folds of the forebrain. Recall that the posterior prosencephalic folds contain neural crest and that this PNC population is in register with prosomeres p1–p3. Recall as well that the anterior folds overlying the secondary prosencephalon do not contain neural crest, but they do contain placodes of importance, from distal to proximal adenohypophyseal, nasal and optic. These folds can be considered to occupy three specific genetic zones. For the sake of simplicity, we can consider them as p6, p5, and p4. Just outside the neural folds is frontonasal skin in genetic register with the folds. Thus, we can consider the epithelial lining associated with the AP and nasal placodes as p6, that of the nasal skin as p5 and the forehead as p4. Although the precise genetic identity of these fields is still unclear, their embryological and clinical behavior permits us to make use of this simple mapping system.

Early Development of the Forebrain and Stomodeum: Stages 8–11

Stage 8 (days, 18 somitomeres) is defined by the appearance of the nervous system. From its very inception, the future prosencephalon projects itself above the plane of the neural plate. *Stage 9* (20–21 days, 1–3 somites) is marked by the appearance of the first three (occipital) somites. At this time, the earliest segmentation of the brain into its three primary parts is present. Prosencephalon is projected forward by the flexion of the embryo when the primitive heart assumes a ventral position below the pharyngeal arches. *Stage 10* (22–23 days, 4–12 somites) is entered upon the appearance of the fourth somite. All the remainder of the eight cervical somites are produced at this stage. It is here that nasal and optic fields begin to be assembled. Although scanning electron microscopy makes this appear to be a simultaneous event, thin section micrographs at stage 13 demonstrate that the nasal field thickens before the optic field [49]. It may be hypothesized that their genetic coding follows the same sequence (Figs. 4.15, 4.16, and 4.17).

Stage 11 (22–26 days, 13–20 somites) is defined by closure of the cranial aspect of the neural tube—the process

Fig. 4.15 Stage 8 marked by critical events. Germ layers produced during gastrulation (stages 6–7) are organized neuromerically from r1 backward. Nervous system erupts from the neural plate. Growth of the midbrain and forebrain acts as a tissue expander, dragging r1 ectoderm and mesoderm forward like a mantle to provide cover for its side walls. As frontonasal skin expands, it displaces the r1 ectoderm laterally. Residual mesoderm from r1 provides the substance for the primitive head plexus. Although neural crest cells from p1 to p3 provide the dermis for the frontonasal skin, they are biologically incapable of forming angioblasts. Thus, the blood supply over the side walls of prosencephalon is critically dependent on the presence of r1 mesoblast that was "tractioned" into position. Final disposition of r1 tissues is determined by the stage 9. (Reprinted with permission from Johns Hopkins University Department of Medical Art—Gary Lees. Carnegie Institution)

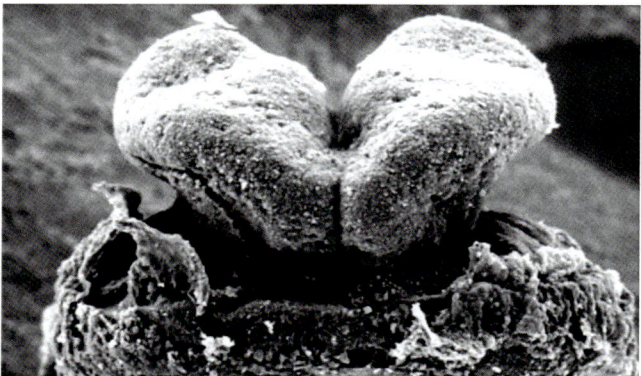

Fig. 4.16 Stage 9 shows prosencephalon rounding into a sausage-like projection. Its base, initially broad at stage 8, pinches inward and the neural folds begin to close, like a zipper, from the most distal zone (adenohypophyseal placode, p6) backward. The undersurface of the forebrain is covered by residual r1 ectoderm and mesoderm. These tissues constitute the temporary coverage for the roof of stomodeum and simultaneously provide the initial blood supply to the developing anterior cranial base. (Courtesy of Prof. Kathleen K. Sulik, University of North Carolina)

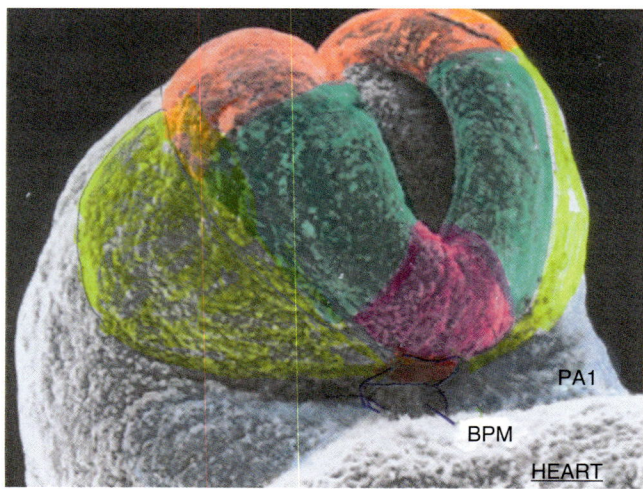

Fig. 4.18 Stage 11 showing the exterior of the anterior neural folds. Adenohypophyseal zone (brown) is pulled inward and abuts buccopharyngeal membrane. Nasal epithelium (red) contains the nasal placodes and will be brought inside the nasal cavities. External nasal epithelium (green) forms frontal nasal skin from the nasofrontal junction forward. Roof of stomodeum is r1 (yellow). Remainder of r1 territory is overtaken by frontonasal skin produced by the neural folds of the forebrain. Maxillary process unfused to forebrain so side walls of prosencephalon are not intact. (Courtesy of Prof. Kathleen K. Sulik, University of North Carolina)

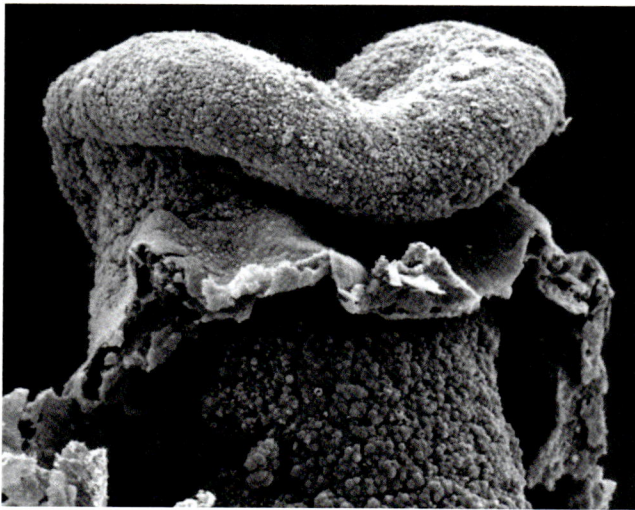

Fig. 4.17 Stage 10 demonstrates the mass of cardiac mesoderm now beneath the future pharynx. Buccopharyngeal membrane being drawn into the stomodeum. Side walls of first arch can be seen at this stage, but unfused to the base of the forebrain. (Courtesy of Prof. Kathleen K. Sulik, University of North Carolina)

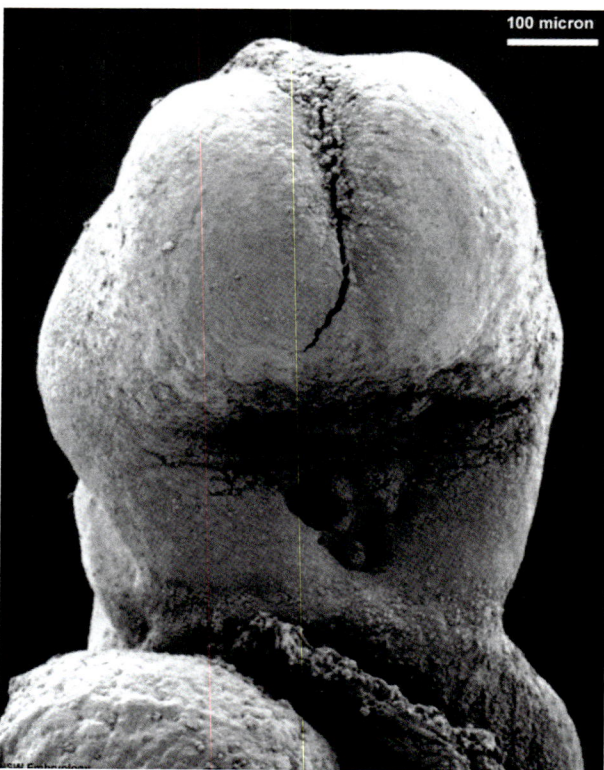

Fig. 4.19 Stage 11 showing the exterior of the anterior neural folds. Adenohypophyseal zone (brown) is pulled inward and abuts buccopharyngeal membrane. Nasal epithelium (red) contains the nasal placodes and will be brought inside the nasal cavities. External nasal epithelium (green) forms frontal nasal skin from the nasofrontal junction forward. Roof of stomodeum is r1 (yellow). Remainder of r1 territory is overtaken by frontonasal skin produced by the neural folds of the forebrain. Maxillary process unfused to forebrain, so side walls of prosencephalon are not intact. (Courtesy of Prof. Kathleen K. Sulik, University of North Carolina)

being more advanced at its caudal margin. At this time, neural crest cells have already migrated: (1) from the midbrain to form frontonasal mesenchyme and (2) from the forebrain neural folds downward to produce frontonasal skin. The forebrain is now protected from the amniotic environment. Epiblastic thickening occurs in the nasal fields [55] (Figs. 4.18, 4.19, 4.20, and 4.21).

Stage 11 is a critical time for mapping of the neuromeric fields of the stomodeum. SEM allow us to look inside the aperture. In its depths, the buccopharyngeal membrane (BPM) can be seen in the process of disintegration. FNP projects forward above the cavity like a large fist. Coming in

Fig. 4.20 Stage 11 is defined by closure of the anterior neuropore. This process is bidirectional. To avoid confusion, it is best to recall the neural fold map. The most rostral folds, from p6 (adenohypophyseal placode), close backwards. Remaining neural folds from r11 close forward. The two processes meet at the *foramen cecum*, marked by the junction of frontal bone with ethmoid bone. Here, emissary veins connecting frontonasal soft tissues and the ethmoid meninges to superior sagittal sinus constitute a danger zone for meningitis and thrombosis of the sinus. The undersurface of prosencephalon is the roof of stomodeum. Its neuromeric coverage is provided by r1. (Courtesy of Prof. Kathleen K. Sulik, University of North Carolina)

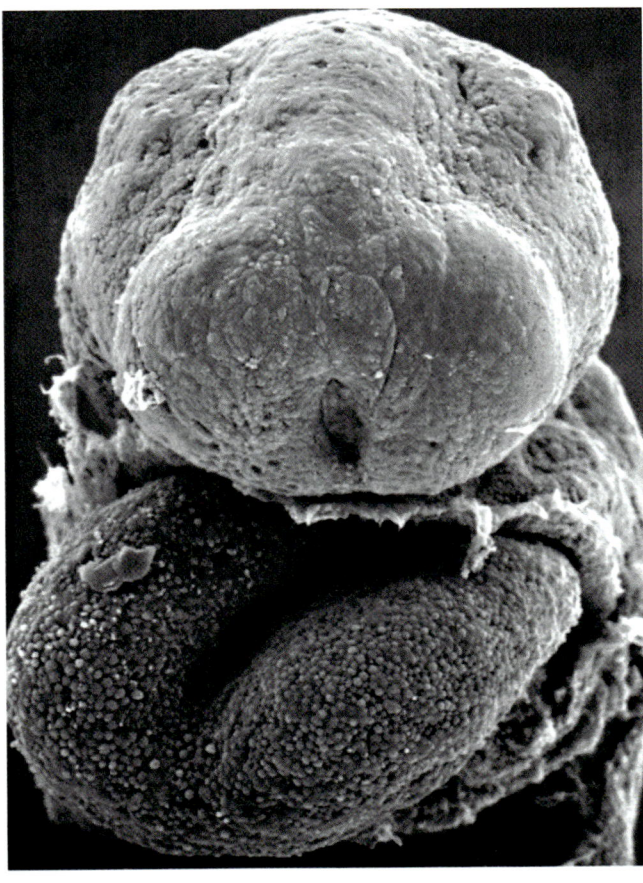

Fig. 4.21 Foramen cecum marks the anterior neuropore. (Reprinted from Wikimedia. Retrieved from: https://commons.wikimedia.org/wiki/File:Gray193.png)

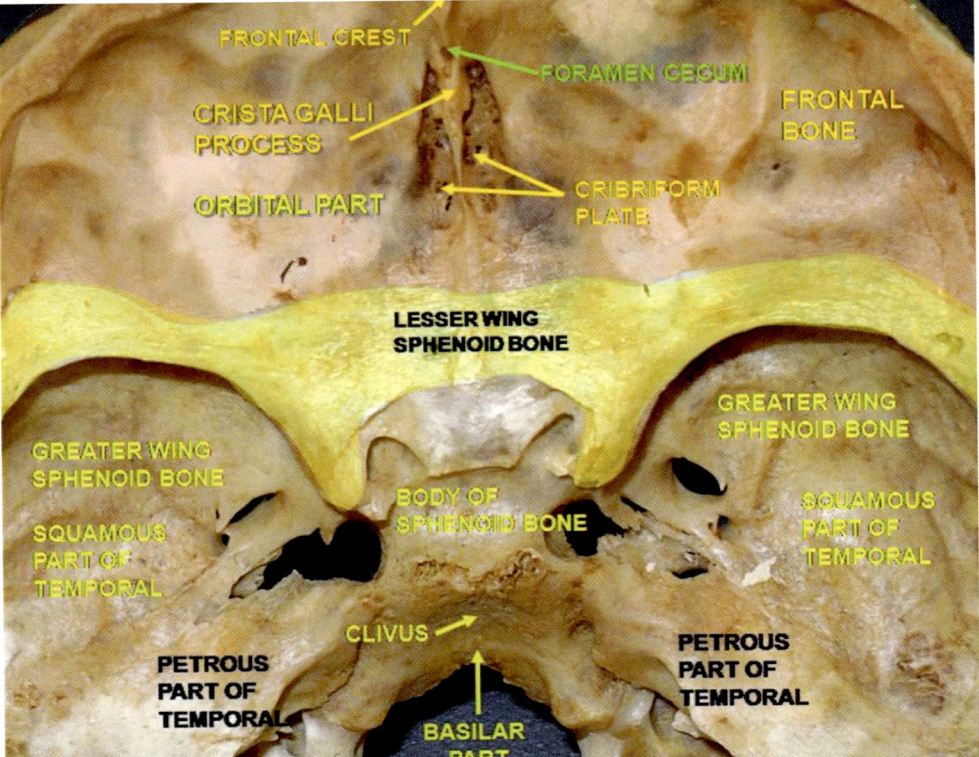

from the sides are the first two pharyngeal arches, hanging downward from the embryo like Michelin tires. They emerge in continuity with the hindbrain and therefore are posterior to the FNP. At this stage, the arches are not in contact with FNP. With growth, they will project forward and fuse with FNP to seal off the lateral boundary of stomodeum.

While the anterior neural tube is still open, we can look into its interior to see how the folds are closing from the adenohypophyseal placode forward to complete the floor of the forebrain. *This is simultaneously the roof of the stomodeum.* What tissue lies on the other side? What provides protection for forebrain from amniotic fluid?

Quail-chick neural crest mapping by Couly and Le Dourain demonstrates that lateral aspects of the primitive forebrain are initially covered by the r1 ectomere. At stage 7, the neural plate from which forebrain develops is surrounded by r1 ectoderm. As forebrain emerges at stage 8, rising out of the planar embryo, its neural folds synthesize additional skin cover.

Let' quickly review the territory of forebrain skin. Recall that its epidermis arises from the nonneural ectoderm of the anterior prosencephalic folds. The dermis is neural crest from the posterior prosencephalic folds. The NNE maps into three zones, nasal epidermis (NE), upper beak ectoderm (UBE), and calvarial ectoderm (CE). In mammals, UBE produces (via the so-called medial and lateral nasal processes): the external nasal skin envelope and the skin of the columella and prolabium. The calvarial field is T-shaped. The transverse limb of the T is frontal bone and the vertical limb is a strip directed straight backward to reach posterior fontanelle. The latter is rather narrow, being encroached upon by the parietal bone fields.

This expansion of frontonasal skin, specifically the CE and UBE fields, forces the r1 ectoderm laterally where it remains as the epithelial coverage over the optic anlage. It will eventually provide the corneal epithelium, the entire conjunctiva, and the skin of the upper eyelid.

The r1 ectomere has yet another role to play. Topologically, it is interposed between the first arch and the forebrain. The tissue expansion effect of the forebrain causes the r1 ectomere to be stretched downward and inward to provide the coverage of the ventral aspect of forebrain, *except in the midline where the folds are in continuity all the way back—past the buccopharyngea membrane—to Rathke's pouch.*

This makes eminent neuro-ophthalogic sense. Recall that at the base of the embryonic forebrain the eye fields appear as a single unit is located in midline of the basal plate corresponding to prosomere p5. Under the influence of sonic hedgehog from the r0, the eye fields split and lateralize to occupy a position just behind the wall of the forebrain neural tube. Midbrain neural crest from r1 is seen beneath the r1 ectomere and surrounding the optic vesicle. This must surely have been present at an earlier stage, before the eyefields

separated. Thus, the r1 ectomere provides the skin cover of the ventral aspect of embryonic forebrain.

What happens to the r1 ectoderm within the stomodeum? Recall that there is no V1 representation within the oral cavity; it is confined to within the nasal cavity. Stage 11 provides a preview of coming attractions. As we shall see, midline apoptosis of r1 mesenchyme drastically reduces the volume of ethmoid mesenchyme interposed between the nasal fields. So too the stomodeum r1 also involutes. At the same time, first arch and second arches fuse to the lateral border of FNP. Tissue from this complex will come to take over the roof of the stomodeum, a process well seen at stage 13.

Early Development of the Nasal Placode and Nasal Fields: Stages 12–15

Stage 12 (26–30 days, 21–29 somites) This stage is marked by the closure of the caudal neuropore. Pharyngeal arch 4 arch forming. The primitive oral opening is bounded above by initial fusion of first arch swelling. Maxillary and mandibular processes are clearly distinct. first arch flanked posteriorly by swellings that represent second arch (Fig. 4.22).

PA1 and PA2 are now in continuity and developing a common mesenchyme. Layers of neural crest within them are laying down the future fascial planes. Within deep investing fascia (DIF), muscles of mastication develop from both SM4 (first arch) and Sm6 (second arch). All of the remaining second arch muscles are "reassigned" to facial animation. They will develop with the superficial investing fascia (SIF). Note that second arch SIF originating from r4 to r5 *spreads over the entire head* and abuts backward against the territory of cervical somite 1–4. This lamination is a defining feature of mammals; the superficial muscles in continuity with the lips are a prerequisite for suckling…something difficult to do with a beak!

With respect to the stomodeum, fusion of the maxillary process brings r2 ectoderm into contact with r1 ectoderm. Why refer only to the first arch? Let's review the following. (1) Quail-chick mapping of ectomeres discloses that the lining of the future oral cavity is strictly ectoderm. (2) First and second arches are distinct but confluent. Within the mouth, second arch ectoderm from r4 dorsal and r5 ventral is obliterated by the expansion of first arch r2 dorsal and r3 ventral. (3) The former site of the buccopharyngeal membrane is located ventrally in the tongue at the junction of its posterior third with the anterior two thirds of the tongue. Dorsally, it is found in the soft palate at boundary between the tensor and levator muscles (again a 1:2 ratio). Here, first arch ectoderm is directly apposed with third arch endoderm: stomodeum meets foregut.

A midline furrow over the developing forebrain appears for the first time at stage 12.

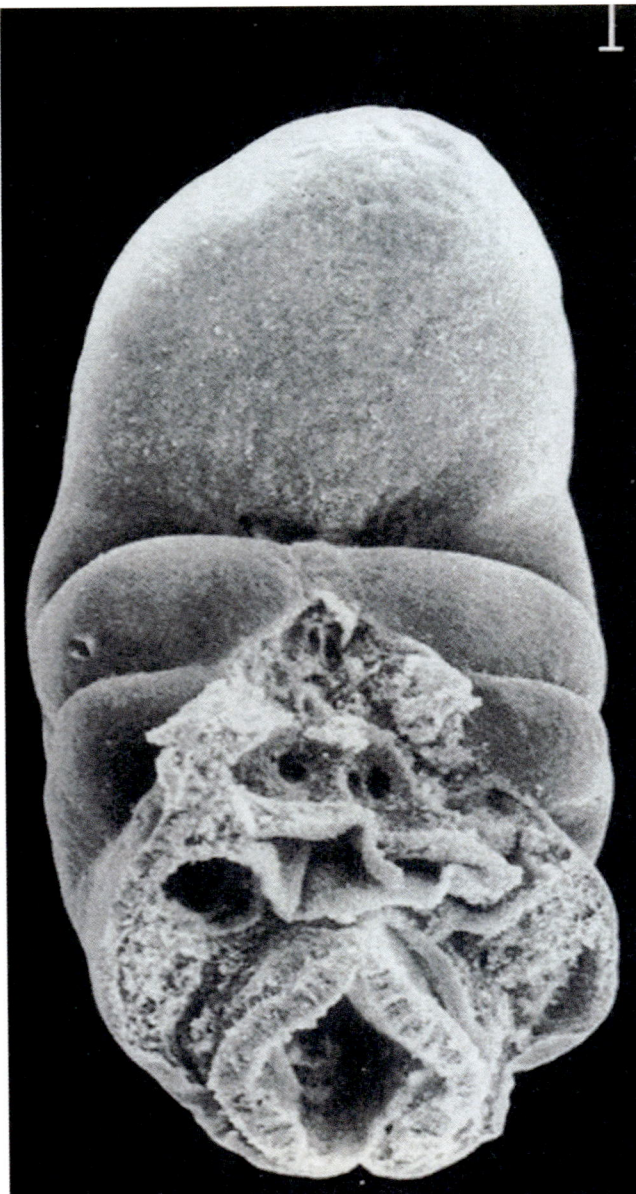

Fig. 4.22 Stage 12 defined by closure of the posterior neuropore. All four arches are present but not seen here due to flexion of the embryo. First and second arches seen; third and fourth arches hidden by embryonic folding. First arch shows signs of developing maxillary process. (Courtesy of Klaus Hinrichsen Collection, Göttingen University; SEM curator Dra Beate Brand-Saberi)

Stage 13 (31–32 days, 30+ somites) is associated with rapid changes in placode morphology. Frontal section micrographs taken at 5 mm show thickening of the nasal placode epithelium (4–5 cell layers), but its surface appearance is flat. The axis of the nasal placodes to the center of the brain is now about 110°, the initial position being 180°. This is due to MNC apoptosis (Figs. 4.23, 4.24, 4.25, and 4.26).

The epithelium overlying the optic vesicle also remains flat. Later in this stage, the vesicle begins to form the optic

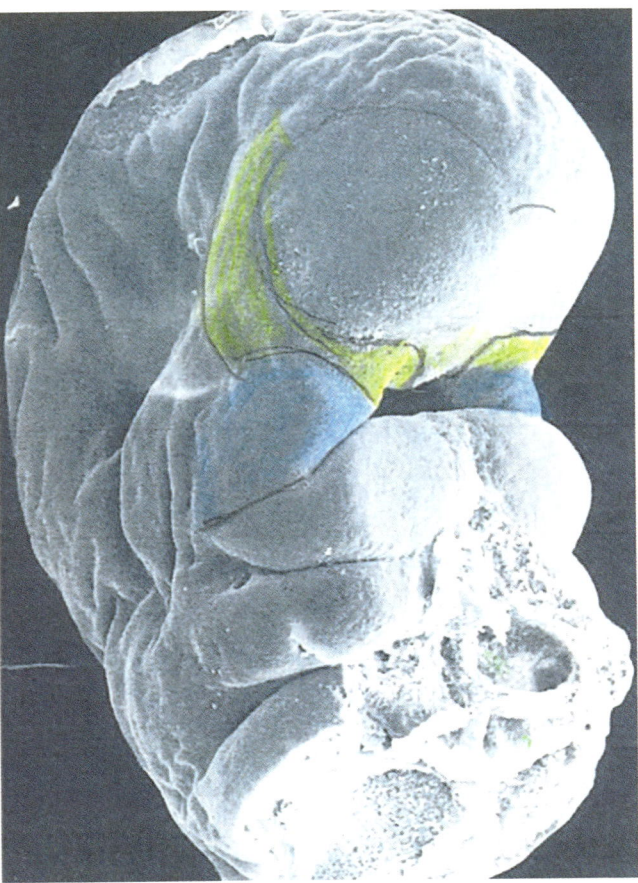

Fig. 4.23 Stage 13 shows nasal placode for the first time. It is flat. Mesenchymal proliferation of MNC around its borders has not yet occurred. Maxillary process developed, but fusion has not yet advanced as far forward as the nasal placode. (Courtesy of Klaus Hinrichsen Collection, Göttingen University; SEM curator Dra Beate Brand-Saberi)

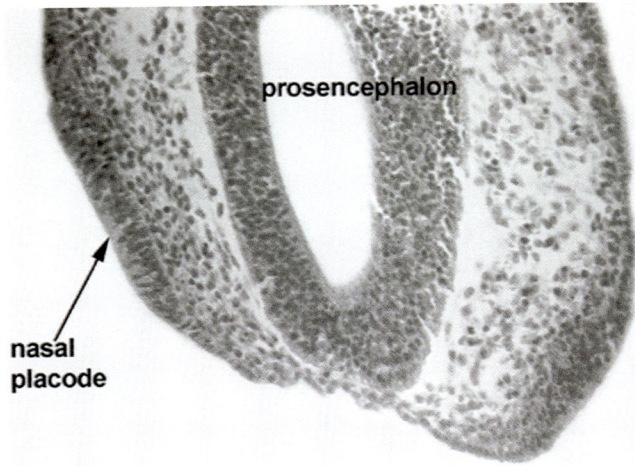

Fig. 4.24 Stage 13 nasal placode is 5–6 cell layers thick. Note MNC intervening between brain and the placode. (Courtesy of Klaus Hinrichsen Collection, Göttingen University; SEM curator Dra Beate Brand-Saberi)

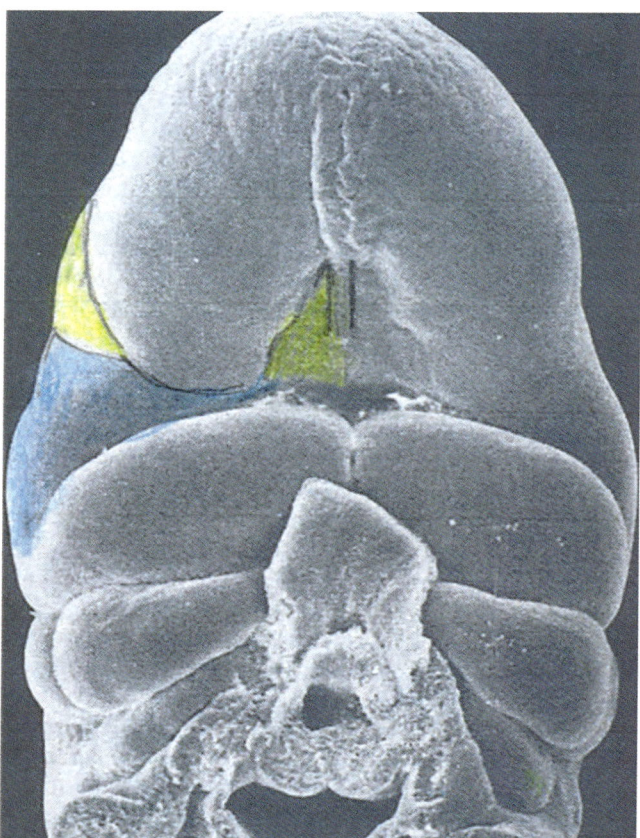

Fig. 4.25 Stage 13 oral view shows, in the roof of stomodeum, the contraction of the MNC A-fields. Epithelium and mesenchyme from r2 are invading forward and medially. (Courtesy of Klaus Hinrichsen Collection, Göttingen University; SEM curator Dra Beate Brand-Saberi)

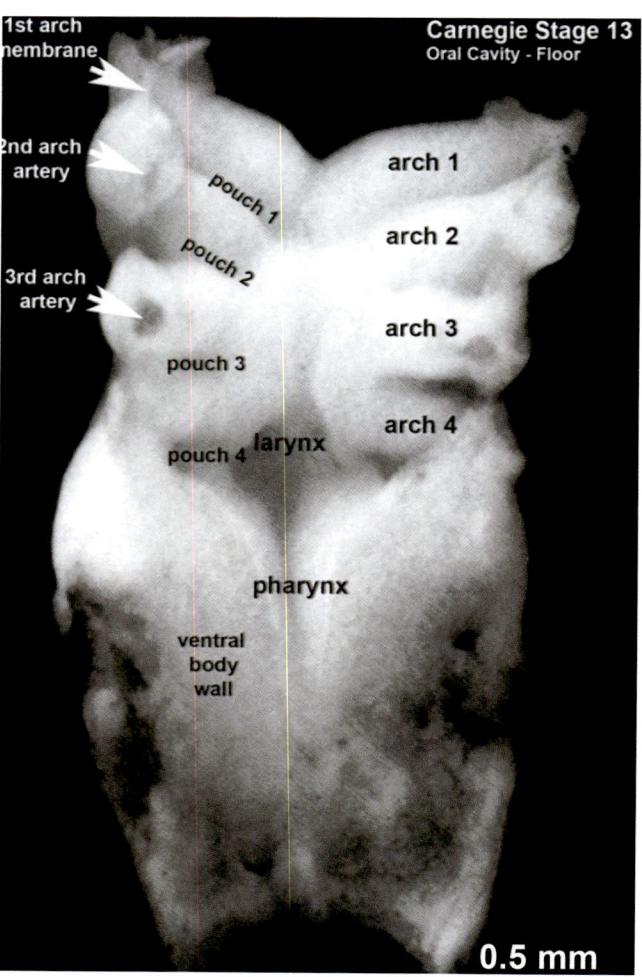

Fig. 4.26 Stage 13 floor of mouth shows clearly the mesenchymal masses of the four pharyngeal arches. Ectoderm from first arch overtakes that of second arch. Buccopharyngeal membrane (disintegrated at stage 11) located at border of second arch with third arch. (Courtesy of Klaus Hinrichsen Collection, Göttingen University; SEM curator Dra Beate Brand-Saberi)

cup, while on the surface a definite rounded optic placode is seen. The first arch is still small. Frontonasal process is almost three times as large and projects forward from it considerably.

Significant changes occur in the rostral corpus as well. The future forehead is initially cone-shaped and planar with the pharyngeal arches. Reflecting the rapidity of brain growth, it is transformed into a bullet-shaped structure projecting far over the level of the arches. A vertical midline furrow appears, making the beginnings of prosencephalic division.

Development of nasal placodes is associated with an expansion of the underlying MNC mesenchyme just beneath them, resulting in forward projection. At the same time, the central mass of the ethmoid complexes causes the nasal fields to diverge. Anterior cerebral axis develops to support the forebrain. Its most distal branch, primitive olfactory artery, is dedicated to the nasal fields, while temporary support of the eye fields is provided by primitive maxillary artery. Later in time the nasoethmoid complex will be supplied by stapedial system along its nasoethomoid axis. These

two systems are closely related, perhaps with common genetic factors. For this reason, holoprosencephaly in its more severe forms will manifest itself externally in the A fields, i.e., in the MNC mesenchyme [56–60]. A-field hypoplasia/aplasia occurs in direct proportion to the degree of cerebral dysgenesis.

The maxillary process is now in direct contact with stomodeum, but sits well behind the nasal placode, leaving intact the continuity of r1 ectoderm between the eye fields and the roof of the stomodeum.

Mapping Out the Mouth

At the time of BPM dissolution, first arch ectoderm comes into contact with third arch endoderm. This is demonstrated by the sensory map of the mouth. In the floor, anterior 2/3 of tongue is supplied by V1 while posterior 1/3 is IX. This transition zone extends laterally to the anterior pillar of tonsillar fossa. In the dorsal oropharynx, the V2 to IX transition is located at the border of the anterior palatine aponeurosis with the posterior 2/3 of the soft palate. Mesenchymal merger in oral pharynx is a slower process. SEM of the floor of mouth at stage 13 shows the four pharyngeal arches as quite distinct, even as the overlying mucosal boundaries have been set up.

Expansion of Midbrain Neural Crest: Mesenchyme of the A Fields

The nasal placode is now disc-like. Intervening between the placodes and the brain is ethmoid MNC. This will subsequently form the nasal chamber. At this time, RNC mesenchyme of the future vomer and premaxilla invades the floor of the nasal fields. RNC comes in from behind, just in front of sphenoid, and tracks forward along the oral margin of the fields from posterior to anterior. This causes them to bulge downward into the stomodeum.

At stage 13, the oral cavity is largely occupied with r2 maxillary tissues. The nasal fields are now separated from the oral cavity by a column of RNC vomer/premaxilla. In the center, mucosa underlying the sphenoid complex can be seen leading back to Rathke's pouch. The ventral oral stomodeum is now the floor of mouth. The developing tongue has well-developed masses representing the contributions of each pharyngeal arch. Note that second arch mesenchyme of the tongue explains the taste receptors in the anterior two thirds, even though the surface ectoderm no longer has second arch representation.

What is the neuromeric composition of the roof of stomodeum? Probably, this lining is a form of nonneural ectoderm with MNC mesenchyme supplied by V1. As we shall see, as the nasal chambers form in subsequent stages, r2 mesenchyme from the medial nasopalatine axis will track forward from just in front of Rathke's pouch, displacing the r1 stomodeal epithelium. When the nasal sacs unite in the midline, and when their temporary floor, the *primärer naseböden*, break down, the walls of the nasopharynx will have two types of epithelium. Dorsally, septum and ethmoid turbinates will be covered by V1 innervated p6 vestibular mucosa. Ventrally, the vomer, premaxilla, inferior turbinate, and the maxilla bear V2 mucosa. The reader is referred to the color-coded figures for this section.

Stage 14 (33–34 days) marks the end of the pharyngeal arch period. The sixth arch dedicated to the pulmonary plexus is in formation. third and fourth arches are still joined together by a segment of the dorsal aortae, but they will separate in the next stage. Amalgamation of the arches is not yet present. But growth of the first arch causes it to project to the level of the frontonasal process (Figs. 4.27 and 4.28).

The stage 14 nasal placodes are large discs. Cross section of the placode shows MNC proliferation resulting thickening at both medial and lateral borders, making the placodes con-

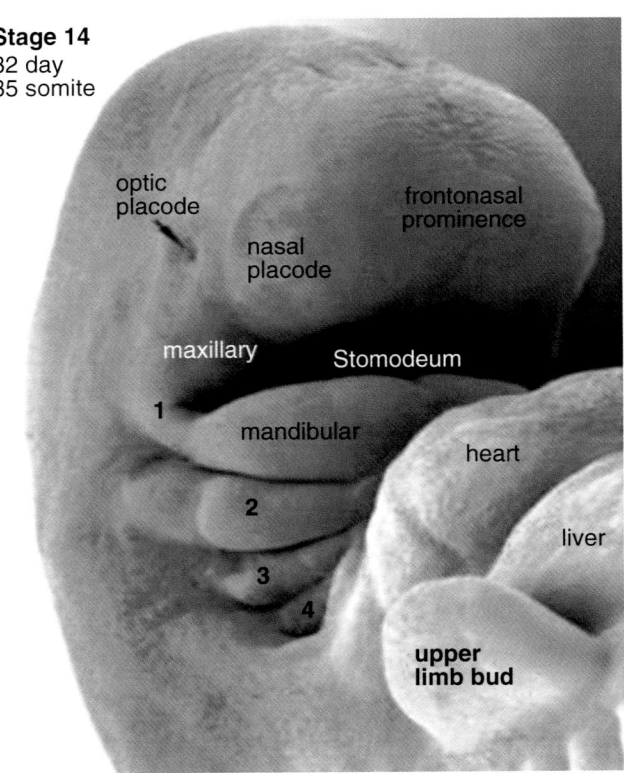

Stage 14
32 day
35 somite

Fig. 4.27 Stage 14 shows nasal placode becoming concave. *Lateral nasal prominence* is present. Some swelling for medial nasal prominence at the next stage. *Maxillary process* now separates from mandibular process, but is positioned behind (posterior to) the nasal placode. (Courtesy of Klaus Hinrichsen Collection, Göttingen University; SEM curator Dra Beate Brand-Saberi)

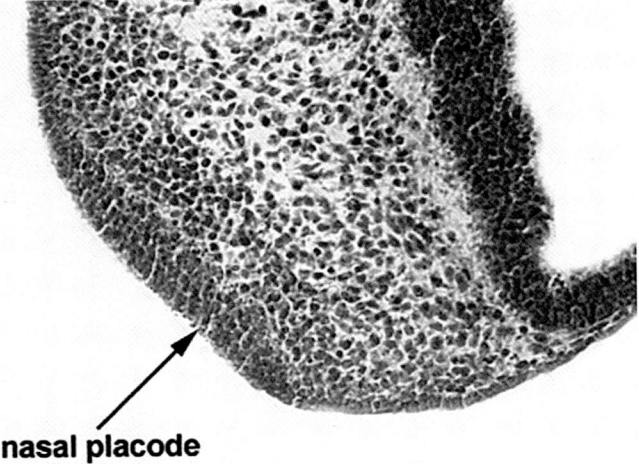

nasal placode

Fig. 4.28 Stage 14 nasal placode demonstrates concavity. Proliferation of MNC at lateral border represents precursor of lateral nasal prominence. (Courtesy of Klaus Hinrichsen Collection, Göttingen University; SEM curator Dra Beate Brand-Saberi)

cave. This will produce subsequently form the *lateral and medial nasal prominences*. These are not yet distinct. The process of midline MNC apoptosis is under way. Therefore, the discs are not as widely separated.

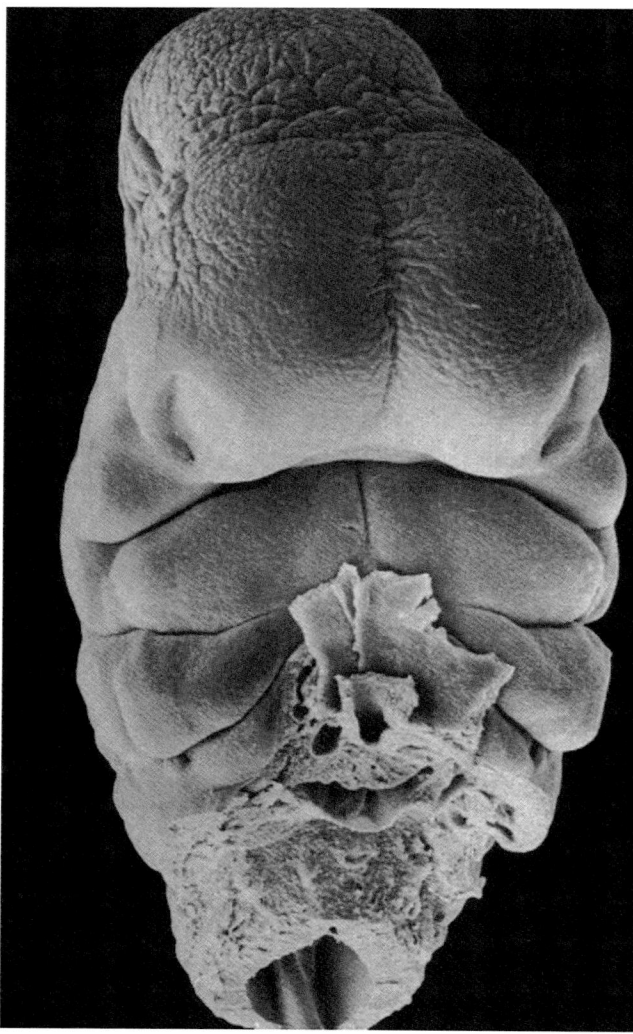

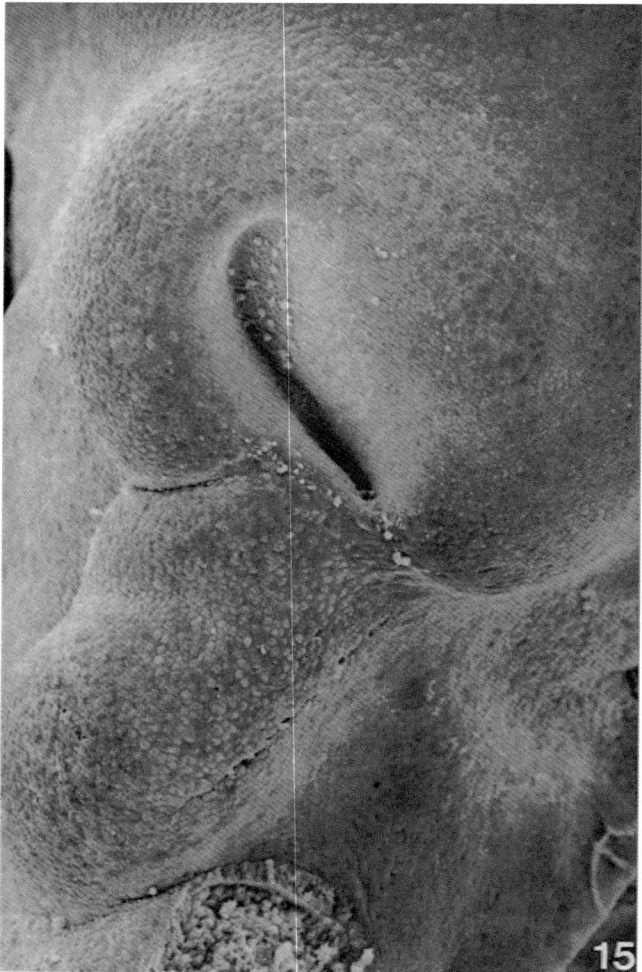

Fig. 4.29 Stage 15 *Medial nasal prominence* now present. *Nasal pit* develops—at this stage it is laterally directed. Fusion has occurred between MxP and LNP. Tissue continuity between LNP and the developing MNP is seen just below the nasal pit. Note lower lip fusing at this stage. (Courtesy of Klaus Hinrichsen Collection, Göttingen University; SEM curator Dra Beate Brand-Saberi)

Fig. 4.30 Stage 15 (late) Medial nasal prominence is well-defined. Note tissue bridge, the *nasal fin* (*band of Simonart*) connects from MNP with LNP. This will permit lateral lip element to advance medially. MxP has joined with LNP. During the course of stage 15, it advances along the nasal fin. Note soft tissue prominence leading from future MNP backward and medially toward Rathke's pouch. This contains ethmoid and vomerine mesenchyme. Neither prolabium nor premaxilla have appeared. Clinical note: If Simonart's band fails to form, nostril floor does not form and a complete soft tissue lip cleft results. Cases exist whereby Simonart's band is formed but MxP fails to "cross the bridge." The result is a unified nostril floor but complete lip cleft. (Courtesy of Klaus Hinrichsen Collection, Göttingen University; SEM curator Dra Beate Brand-Saberi)

Just posterior to the nasal placodes, the optic placode has formed the lens pit.

The frontal midline furrow broadens dramatically.

Maxillary process is now clearly defined. It is now *in contact with the nasal placode*. This closes off the communication between r1 ectoderm of the eye field and stomodeum. From stage 14 onward, the ectodermal cover of the roof of the mouth will change as r2 encroaches from lateral to medial and r1 is reduced in the midline.

Stage 15 (36–37 days) reflects the development of the telencephalon. The nasal placodes have changed from depressed discoid structures. *Lateral nasal processes* (LNPs) are now present. The medial aspect of the placode is thickening (Figs. 4.29, 4.30, and 4.31).

The nasal fields at stage 15 stand on the brink of a structural revolution. Two well-sculpted, laterally directed *nasal pits* appear, tunnelling into underlying MNC. In so doing, they drag their own epithelium inward as lining for the nasal chamber. The two nasal anlagen project away from the midline like spyglasses, giving the facial corpus a definite, paired appearance. Although externally these projections appear to diverge at 120°, on transverse section *the axes of the nasal pits are parallel*. A cellulous strand is seen between the nasal pit and the olfactory field of the brain [55].

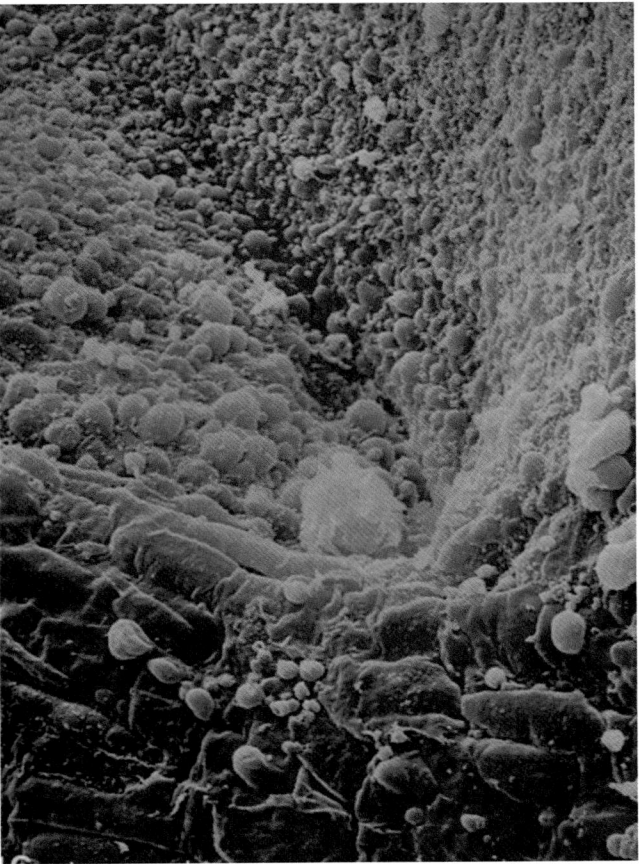

Fig. 4.31 Stage 15 *primare nasenboden* (primary nasal floor) intact but furrow developing at site of future apoptosis. Anterior to primarer nasenboden is the arc of the nasal fin, or Simonart's band. Note that nasal fin is intact and does not show a furrow. Medial indentation to the right just inside the fin marks site of future vomeronasal organ. (Courtesy of Klaus Hinrichsen Collection, Göttingen University; SEM curator Dra Beate Brand-Saberi)

As we shall see, nasal placodes give birth to a variety of neurons. Chemical markers in the MNC mesenchyme provide the "roadmap" by which neurons of the olfactory, accessory olfactory, and GnRH systems find their way from the placodal epithelium to their respective targets in rhinencephalon, the limbic system, and the hypothalamus.

Concomitant changes in surrounding tissues accompany nasal pit formation. On either side of the pit, explosive expansion of extracranial r1 neural crest MNC causes the LNPs and MNPs to assume as semicircular form. They appear on either side of the placodal tunnels. These soft tissues, facing each other in parenthesis-fashion, are termed the *lateral and medial nasal prominences*. LNP appears at stage 15 and MNP at stage 16. Although both are derived from MNC mesenchyme, their structural differences are due to differing signals from the lateral and medial zones of the nasal placode. LNP forms the nasal ala, while MNP contributes to the columella and prolabium. They are the most external manifestation of the lateral A-field lamina [61].

Important: These primary nasal chambers have no communication with the future mouth, either ventrally or posteriorly. They are dead-end against a wall of ethmoid mesenchyme. This will eventually break down to form the *secondary nasal chloanae*. Failure of this to happen causes choanal atresia. With the MNC walls surrounding the nasal chambers, important changes are taking place. *Formation of StV1 and StV2 neuroangiosomes is now complete* [62]. These subdivide into the developmental fields of bone and cartilages that make up the walls of the nasal passages.

At the same time, the maxillary prominence, MxP, is swinging into position. *MxP is now in contact with LNP.* MxP "tracks" forward by a progressive process of epithelial breakdown between it and the frontonasal process. As has been previously discussed, this epithelial instability is due to BMP4 inhibition of SHH. Recall that the BMP4 is released during bone synthesis and that SHH resides within epithelium, where it maintains epithelial stability. The anterior cranial base matures from posterior to anterior; therefore, its development causes instability of its epithelial borders, a process that directs the maxillary process forward.

Despite the fact that LNP and MNP project forward from the primary nasal floor, they are in continuous contact with each other due to the *nasal fin* that unites them. This tissue bridge will conduct MxP from distal to mesial, from LMP to the nasal fin to MNP, where it will finally make contact with prolabium and premaxilla.

Maxillary process is really an amalgam of tissues from the r2 of first arch and r4 of second arch. It contains two types of tissue. B field tissues are supplied by branches of stapedial V2. These are directed to the bone fields of the maxillary-zygomatic complex that were "expropriated" from the first gill arch; the same is true for mandible. Recall that the extracranial stapedial arteries are evolutionarily distinct from the external carotid system. C fields are those fields of the 1st/second arch complex supplied by the ECA. They are derivatives from "original" soft tissues belonging to the first and second arch gill arches.

SEM studies of facial morphology reveal a process of abutment between frontonasal fields and the composite of first and second arch tissues commonly known as the maxillary process. This abutment is marked by deep grooves representing the different developmental origins of the mesenchymal structures (PNC vs. MNC vs. RNC) and of the epidermis (forebrain neural crest versus hindbrain ectoderm). Fusion occurs, followed by a "smoothing out" of the grooves and mesenchymal tissues flow together.

In other situations, external grooves deepen to enclose new structures. At stage 14, MxP bordering the contact zone between MNP and LNP appeared fusiform. By stage 15, a distinct furrow appears between these tissues—the future nasolacrimal duct.

In summarizing intraoral development during stage 15, we can refer to the process of folding as "embryonic origami." The oropharynx is constructed from five arches stacked together like colored Lifesaver® candies. Due to overgrowth of the ectoderm of first and third arches, all ectodermal surfaces of the second arch are obliterated. Thus, there is no endoderm in the mouth…it's all first arch ectoderm back to the posterior 1/3 of the tongue and the tonsillar fossa, the site of the buccopharyngeal membrane (BPM). Here, topologically, the forward-facing layer of BPM is ectoderm and the pharyngeally directed layer of BMP is endoderm. Breakdown of BMP positions first arch ectoderm against third arch endoderm; the Lifesavers® of third arch and fourth arches for the remainder the pharyngeal wall. fourth arch makes the larynx and fifth arch the cricoid. Trachea and esophagus are assembled from lateral plate mesoderm.

Development of the Primary Nasal Chambers: Stages 16–18

Formation of a primary nasal chamber from the A fields occurs in complete isolation from the primitive oral cavity. To understand this process, one must realize that the human nasal chamber does not have the same floor throughout its development. Two different floors, formed from two distinct embryonic fields, are present at two different stages.

The first form of nasal floor comes exclusively from the A fields on either side of the facial midline. Described in 1911 by Peter, the primary nasal chambers, "Primäre Nasenkammern," have primary nasal floors, the "Primäre Nasenböden" [63]. These consist of: vestibular epithelium from NNE, submucosa/dermis from PNC, an intervening layer of r1 MNC mesenchyme, and an oral layer of r2 ectoderm. Recall that r2 overtakes r1 inside the stomodeum as far as the undersurface of the nasal fields. The remaining central area of r1 mucosa involutes and is replaced when the nasal fields fuse in the midline.

Present as well, running in parallel in the lower side walls of each nasal chamber, are two StV1 neuroangiosomes: medial and lateral nasopalatine axes will support the first arch structures that make up the nasal floor. This RNC mesenchyme came from the ptygopalatine fossa forward underneath ethmoid MNC. Its resultant fields develop below those of MNC. Inferior turbinate is a separate bone underneath the ethmoid turbinates, while vomer runs below perpendicular ethmoid plate.

The formation of the primary choanae (*vide infra*) will change this topology completely. Once the placode neuroectodermal tunnels are complete from the exterior to Rathke's pouch, an instability develops in floor of the chamber exactly in the midline between the two nasopalatine neuroangio-somes. This is surely a case of apoptosis (programmed cell death). In any case, a new type of embryonic field relationship is created. For the first time, along the lateral wall of the inverted U-shaped nasal chamber, PNC vestibular lining is now in direct contact with first mucosa. This is seen clinically at cleft palate surgery as a color difference nasal and oral sides of the hard palatal mucosa.

With further growth, MNC apoptosis brings the nasal chambers toward each other in the midline. At the same time, growth of the perpendicular ethmoid plate and septum causes the vault of the "inverted U" cavities to increase in height. The ethmoid plate and septum from r1 MNC will become "sandwiched" between the chambers. Formation of turbinates adds to the vertical dimension of the lateral wall.

This topology is only a passing phase. Once the palatal shelves bud, translate medially, and fuse from incisive foramen backward, a process that is complete by stage 22, the nasal cavities are once again isolated…but with an important difference. The walls of the nasopharynx now include V2-innervated ectoderm as the covering of the vomer and the internal fields of the maxilla. The neuromeric coding of the cranial nasal cavity is StV1 anterior ethmoid and of the caudal nasal cavity StV2 nasopalatine.

Stage 16 (37–40 days) The nasal pits have deepened into *nasal sacs* separated from the oral cavity by the primarer Nasenböden. This floor is also termed the *nasal fin*. The process of nasal pit formation is one of invagination by which the placodal epithelium becomes the internal lining of the future nasal cavity. Those cells occupying the medial and lateral sectors of the nasal placode are localized to the medial and lateral walls of the nasal pit. Accessory olfactory neurons thus become spatially separate from the olfactory neurons. Their "home base," the vomeronasal organ of Jacobson (VNO), lies along the *caudal aspect of the inner wall* of the nasal pit. The VNO will be readily visualized on SEM by Carnegie stage 16 (Figs. 4.32, 4.33, and 4.34).

As the two nasal fields move medially, fusion of these medial walls will create a unified functional matrix "envelope" or sandwich, within which entrapped r1 MNC will form the septum. This is flanked on either side by paired VNOs, paraseptal cartilages, and vomerine ossification centers. Under certain conditions, bifidity of the septum can occur with loose mesenchyme sandwiched in between. This can present a pathway for development of dermoids or even tumors.

The future nostril is defined by a clear-cut inverted U-shaped ring of tissue (made up from MNP) and LNP. Along the caudal border of the nostril, a thin tissue "bridge" between these two structure can be seen distinct from the maxillary prominence. On closer inspection, it can be seen that, by this stage, Mx has moved more medially and is *now in full contact with LMP*. Recall that Mx first makes contact with FNP at stage 14. It tracks medially along the "bridge" during stage 15.

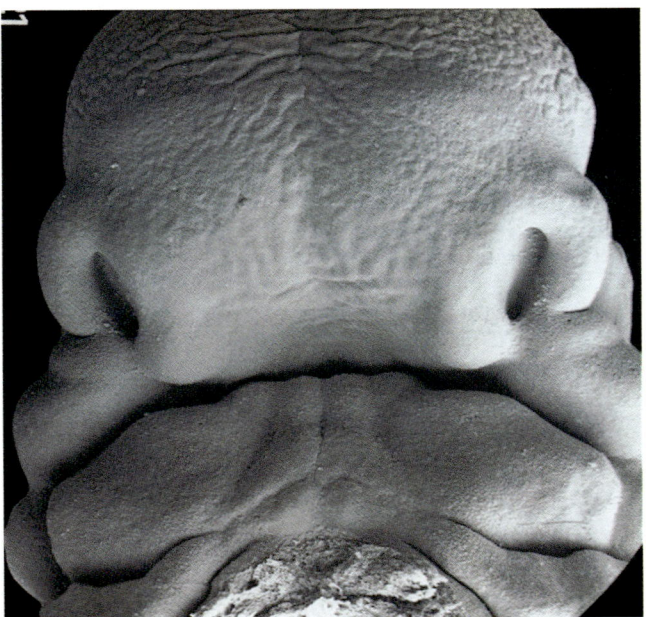

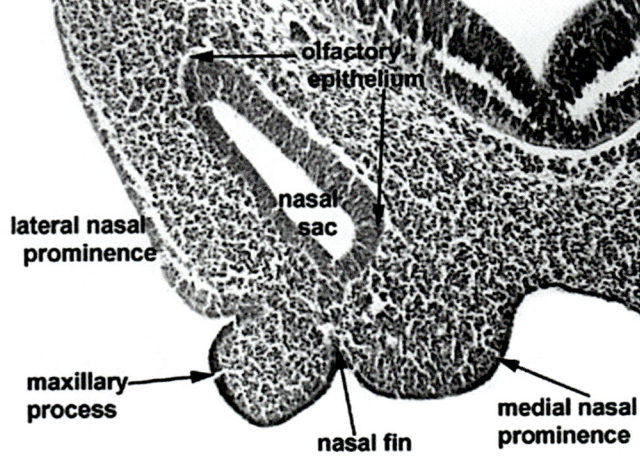

Fig. 4.32 Stage 16 *Nasal pits now opening medially as well as later-ally.* A caudal expansion of MNP, *the globular process of His*, represents prolabium. It can be seen projecting into the stomodeum. Just behind it lies the *vomerine expansion* which will give rise to a premaxilla. The nasal pits deepen to form *nasal sacs* which are separated from the future oral cavity by the Premäre Nasenböden. In front of the nasal sac, the *nasal fin* provides mesenchymal continuity between LNP and MNP. This mesenchymal "bridge," (or ban of Simonart) plays an all-important role in the medial migration of the lateral lip element. MxP now in direct contact with MNP but not yet with prolabium. (Courtesy of Klaus Hinrichsen Collection, Göttingen University; SEM curator Dra Beate Brand-Saberi)

Fig. 4.34 Stage 16 The nasal pit deepens to form a *nasal sac*. In front of the nasal sac, a mesenchymal connection develops between lateral and medial nasal processes, the *nasal fin*. Although floor of the nasal sac disintegrates during stage 17, continuity between the nasal processes is maintained. "Once a nostril, always a nostril." The nasal fin provides a means for MxP to medially and make contact with MNP. In the subsequent stage, this bridge guides the fusion process of the lateral lip element to prolabium. (Courtesy of Klaus Hinrichsen Collection, Göttingen University; SEM curator Dra Beate Brand-Saberi)

Fig. 4.33 Stage 16 intraoral view shows Rathke's pouch and union of LNP and MNP. MxP is making contact with MNP. Frontonasal tissues between the nasal cavities show bulges just medial to MNP. These are the earliest manifestation of the prolabium. Just behind the MNPs, intraoral prominences lead backward toward Rathke's pouch. These represent the future vomeropremaxillary fields. The oronasal floor, primarer nasenboden, remains intact separating nasal and oral cavities. Palatal shelves from the maxillae have not emerged. (Courtesy of Klaus Hinrichsen Collection, Göttingen University; SEM curator Dra Beate Brand-Saberi)

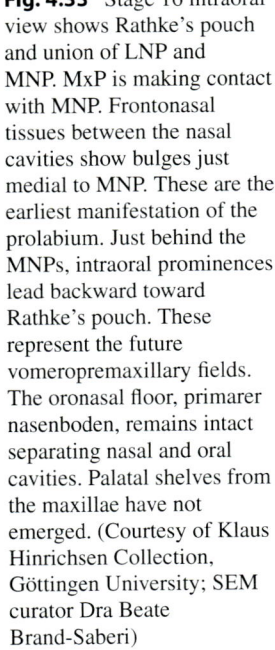

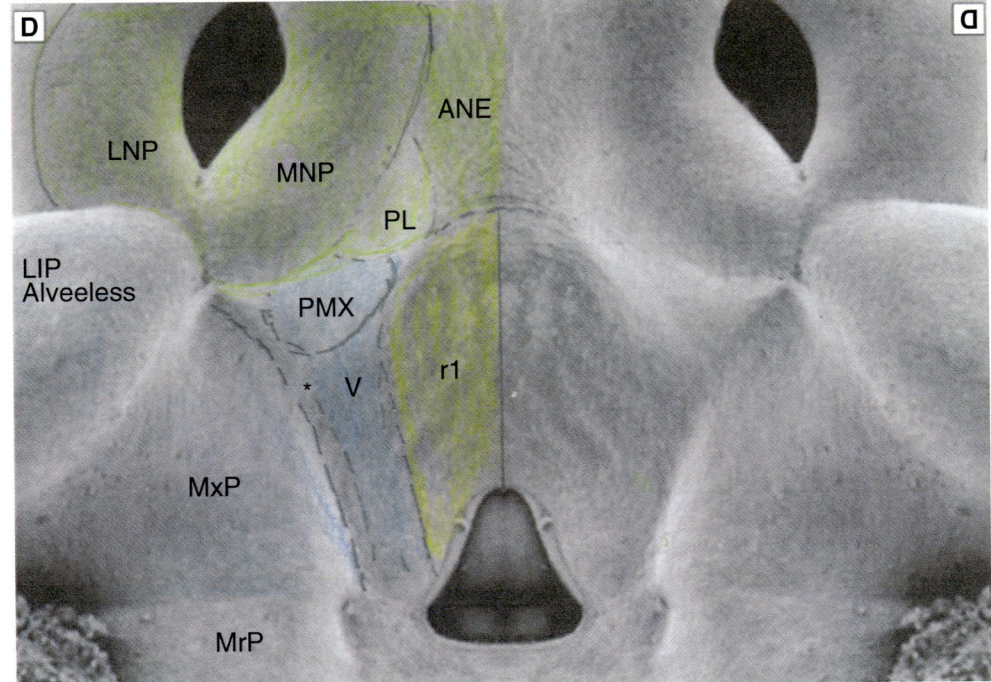

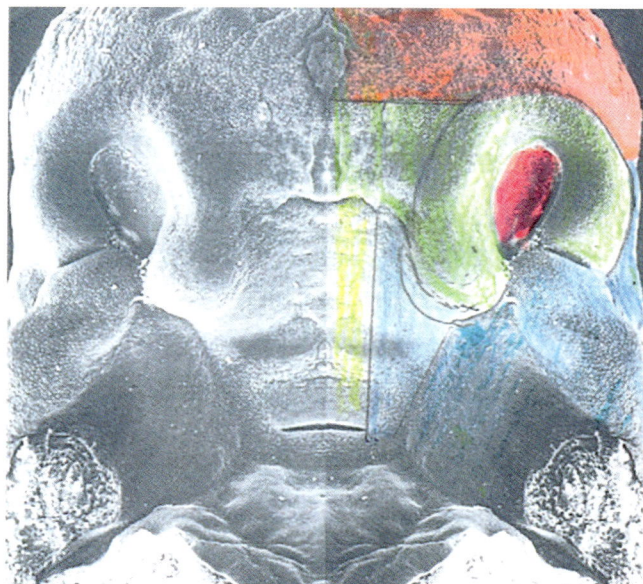

Fig. 4.35 Stage 17 Breakdown of the oronasal membrane has taken place. For the first time, definitive nasal and oral cavities are established. The choanae lead backward toward Rathke's pouch (seen here). Externally, prolabium is expanding downward like a ripe grape from MNP. During this stage, the lip element crosses beneath MNP to achieve contact with prolabium. Mesenchymal flow fills out the upper lip. Breakdown of the primary nasal floor is a required step for the development of the lateral palatine shelves which first appear at this stage. (Courtesy of Klaus Hinrichsen Collection, Göttingen University; SEM curator Dra Beate Brand-Saberi)

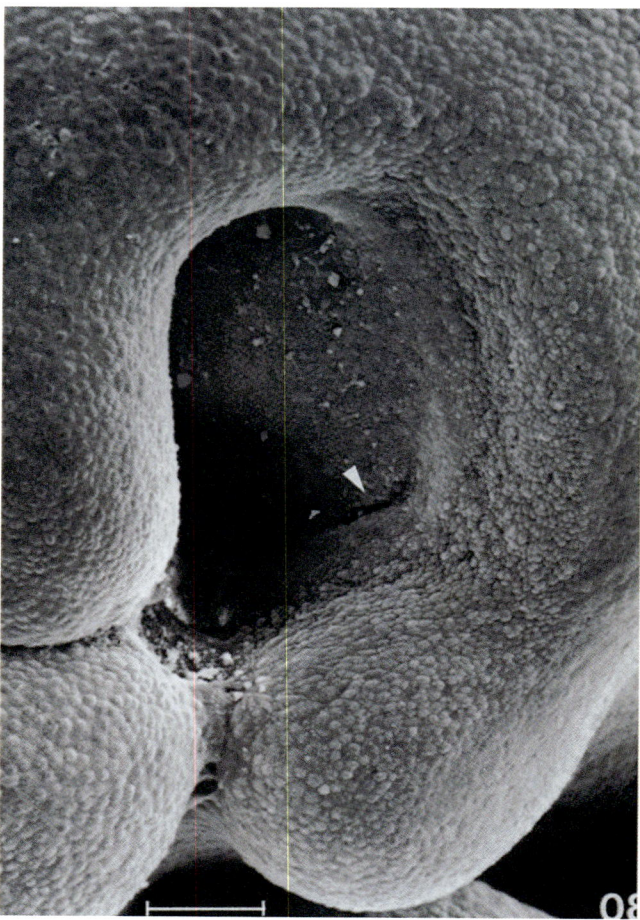

Fig. 4.36 Stage 17 Vomeronasal organ (arrow) contains chemoreceptors. Vomeronasal nerves transmit non-pheromones and terminal nerves convey pheromones. (Courtesy of Klaus Hinrichsen Collection, Göttingen University; SEM curator Dra Beate Brand-Saberi)

The roof of the stomodeum at the beginning of this stage presents as a smooth surface accentuated by two swellings representing the unfused medial A-field walls running posteriorly like the letter V toward Rathke's pouch [64, 65]. The coverage of these fields along their oral aspect is now r2. The center of the stomodeal roof, sandwiched between the A fields, remains r1.

Stage 17 (41–43 days) At this stage, although anteriorly the A-B-C soft tissue complexes are united, immediately posterior to the nostril rim, primary floor of the nose falls apart (Figs. 4.35, 4.36, 4.37 and 4.38).

Distinct furrows representing the apoptotic process in the primary nasal floors are readily seen. The primary nasal choanae are closed. The A fields can be seen on the roof of the primitive mouth as V-shaped longitudinal bulges moving together in the midline from posterior to anterior. These can be envisioned as the *primitive vomers* coming together. The anterior aspects of the A fields are yet unfused, but they now show a vertical projection, the premaxilla. The premaxillary masses of the A fields are in continuity with swellings projecting downward from the medial nasal processes, the so-called *globular process of His.* Each represents the *primitive prolabium.*

Coronal sections show in the roof of the nasal sac thickened *olfactory epithelium*, the receptors for the *primary olfactory system*. The olfactory nerve has two separate plexuses and the terminal nerve is identifiable. The receptors for the *accessory olfactory system* appear in the next stage.

The midline MNC tissues, the so-called "intermaxillary segment," hang down from the cranial base like an upside-down Y. Development of StV2 fields appears to predate that of StV1 fields; the splayed legs of which demonstrate thickened mesenchyme representing the StV2 vomerine-premaxillary fields. A condensation at inferiolateral aspect of the nasal chambers represents the inferior turbinates. No bone or cartilage manifestations of StV1 fields is present at stage 17.

Each bulge contains two types of mesenchyme: dorsal (and unseen) are the StV1 ethmoid fields, ventral (and causing the intraoral projection) are the StV2 premaxilla-vomer fields. No labiobuccal sulcus exists at stage 17. No differentiation toward separate alveolar processes (either from the

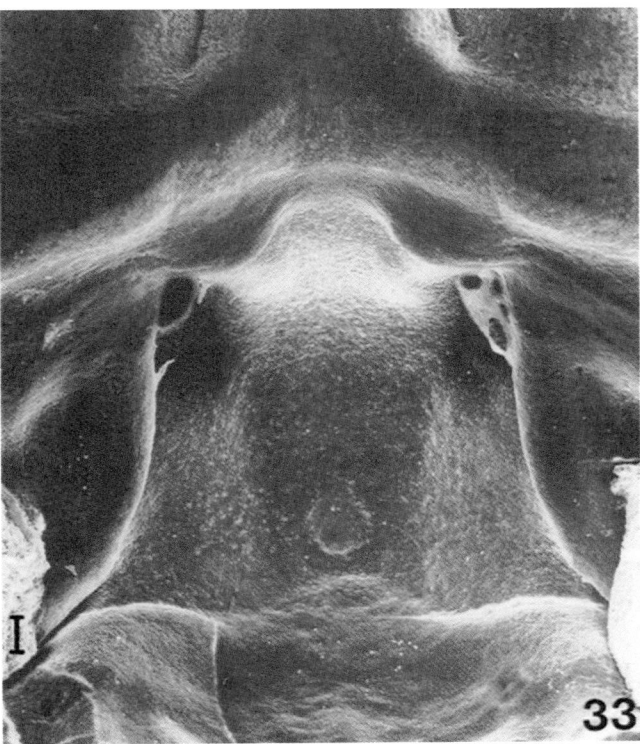

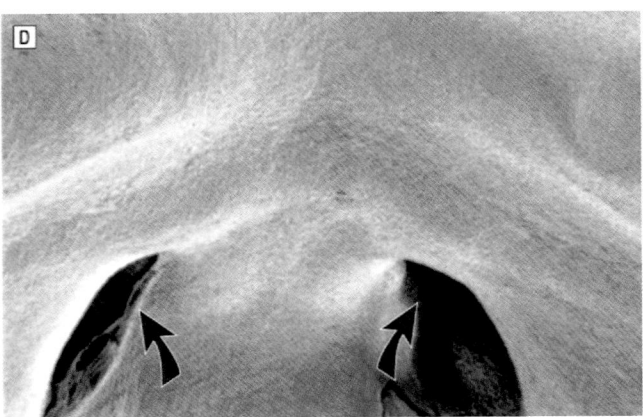

Fig. 4.39 Stage 18 Eyes and nose medializing. Nostrils are vertical. Prolabium and lateral lip elements fusing. Furrow still persists, indicating mesenchymal flow not complete. Paired prolabia remain widely separate. Premaxillary swelling seen. (Courtesy of Klaus Hinrichsen Collection, Göttingen University; SEM curator Dra Beate Brand-Saberi)

Fig. 4.37 Stage 17 Intraoral view shows disintegration of oronasal membrane. This open communication between nasal and oral cavities is known as the *primary choana*. At the same time, lateral palatal shelves have developed from maxilla. They hang vertically downward into the oral cavity along either side of the tongue. (Courtesy of Klaus Hinrichsen Collection, Göttingen University; SEM curator Dra Beate Brand-Saberi)

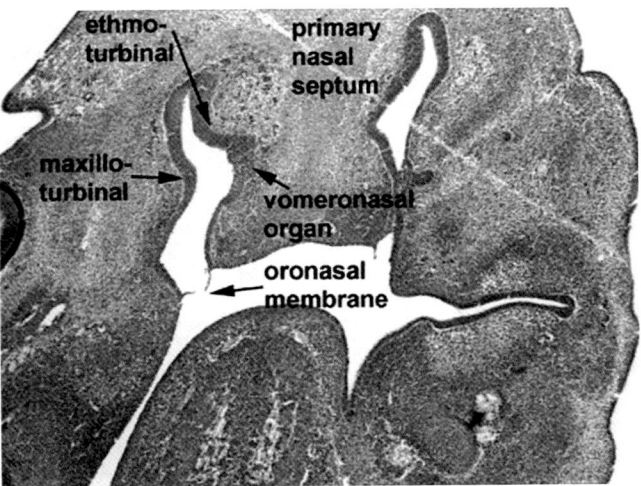

Fig. 4.40 Stage 18 Primary choanae wide open. Primary nasal septum present but not yet projected, so no nasal tip. Ethmoturbinals now seen. (Courtesy of Klaus Hinrichsen Collection, Göttingen University; SEM curator Dra Beate Brand-Saberi)

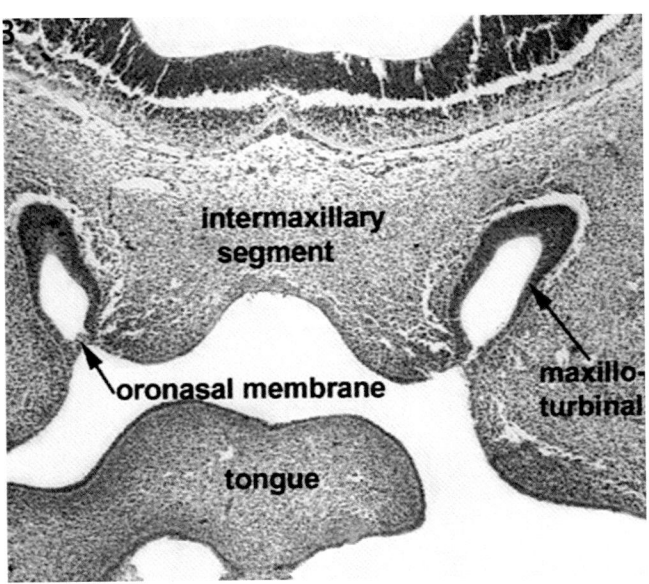

Fig. 4.38 Stage 17 *Oronasal membrane* separates oral and nasal cavities. When it breaks down, primary choana results. Maxillary turbinal tissue represents future inferior turbinate. It develops earlier than the ethmoid turbinal mesenchyme. This is in keeping with developmental primary of extracranial division of stapedial over its intracranial division. Note thickening of the olfactory epithelium which is localizing in the roof of the nose. (Courtesy of Klaus Hinrichsen Collection, Göttingen University; SEM curator Dra Beate Brand-Saberi)

maxilla or the premaxilla) is present. In the brain, this is the last stage at which physical evidence of neuromeres can be seen [66].

Stage 18 (44–47 days) is even more dramatic. Medialization of the nasal chambers and frontalization of the eyes lend for the first time a human semblance to the embryonic face (Figs. 4.39 and 4.40).

The inferior border of the anterior nares is complete. Maxillary process soft tissues have fused with prolabium (globular process of His). The two prolabia and the two premaxillae have not achieved midline fusion. Within the nasal chamber, a medial condensation at the site of the future septo-vomerine suture constitutes the vomeronasal organ (VNO). This represents the receptors for the accessory olfactory system. Olfactory bulbs are now forming to receive the neurons developing in the olfactory epithelia.

At this stage, PNC skin and MNC mesenchyme have formed a definitive midline anlage. As we have previously discussed, lying beneath the forebrain neural crest-derived frontonasal skin, a *solid wall of midbrain neural crest covers the forebrain*, forming dura, anterior cranial base, the frontal bone, ant, and subcutaneous tissue. The surface manifestations of this anlage—termed for convenience A0—lie in *bas relief* between the two medial nasal prominences. A0 represents a paired set of fields supplied by the distal branches of anterior ethmoid. The A0 fields mark the site of the future nasal tip, columella, and philtrum. Septal formation responsible for the projection of these tissues has not yet begun because midline fusion of the A1 fields is not complete. The persistent, albeit-reduced, divergence of these fields at this stage is best seen on intraoral view. The stage 18 prolabium is thus a bifid structure in the form of an upside-down Y, the vertical component of which is columella. Fusion between the two halves of the prolabium with the lateral maxillary complex is demarcated by a faint groove.

Coronal section shows further development of the ethmoid mesenchyme. The *septum makes its first appearance at stage 18.* Due to the large size of the maxillary contribution to the lateral nasal wall, condensations representing the future ethmoid turbinates develop superiorly next to the septum. With growth, these translocate to the superolateral aspect of the nasal chamber. The floor of the nose is gone. The primary choanae represent complete continuity between the nasal chamber and the oral cavity.

Intraorally, at stage 18 clear-cut boundaries exist, separating the StV2 maxillary/premaxillary fields and StV3 mandibular fields (B fields) from the overlying ECA facial soft tissues (C fields). The alveolus (B) is separated by a sulcus from the future soft tissue elements of the lip (C). The two premaxillae B lie deep to the paired A-field prolabial elements. Although the prolabia are fusing the *subjacent premaxillae remain separated* by a prominent midline notch, the lateral margin of each premaxillary alveolar process is coplanar with, but not yet fused with, its neighboring maxillary alveolar process, i.e., the boundaries of primary palate remain unclosed.

The distinct furrows that formerly separated the facial swelling are largely gone, resulting in a "planning" of the facial relief. (This may have to do with development of adipose tissue, but stage-specific documentation is lacking.) This planning is not yet seen at the level of the zygoma or the developing ear. The zygoma lags behind the development of the maxilla. Both are r2 neural crest structures, but the neurovascular pedicles supplying zygoma are more lateral and posterior; thus, they are read out later in time. The ear is in register with more posterior rhombomeres r4–r5 and is therefore also read out later. This explains the persistence of furrows demarcating maxilla from zygoma and the external ear from the rest of the face.

Formation of the External Nasolabial Relief: Stages 19–23

Stage 19 (48–50 days) displays fusion of the prolabia save for a caudal notch.

The growing nasal septum is now forming, causing a deepening of nasal frontal angle, but it has not achieved tip projection; therefore, no nasolabial angle is as yet present. The medial and lateral nasal prominences are not yet sculpted into a nostril rim. Three areas of nasal epithelium (lateral wall, upper nasal septum, and vomeronasal organ) are connected, each with its own neurons (olfactory, terminal, and accessory olfactory).

Coronal section shows further development of the ethmoid mesenchyme. The septum makes its first appearance at stage 18. Due to the large size of the maxillary contribution to the lateral nasal wall, condensations representing the future ethmoid turbinates develop superiorly next to the septum. With growth, these translocate to the superolateral aspect of the nasal chamber (Fig. 4.41).

Intraorally, midline fusion of the ethmo-vomerine complexes is now seen. The premaxillary processes are not yet fused. They nestle together but are still separated by a narrow groove. Definitive maxillary alveolar ridges are present and are in contact with their premaxillary partners. The deep grooves that had so marked the boundary zone between the premaxillary and maxillary alveolar process are more shallow now, as mesenchymal "in-filling" takes place. Palatine processes now project from the aspect of maxilla. They protrude downward on either side of the tongue.

Stage 20 (51 days, webbed fingers) marks the completion of prolabial descent. The premaxilla have fused; all that remains of the original bifid state is an *interincisive suture*. The surface of the premaxillae is flat, the anterior nasal spine having not yet developed. Tip projection secondary to septal growth creates a nasolabial angle, albeit obtuse. Well-defined nostrils are present. Palatal shelves projecting from the border between inferior turbinate and maxillary alveolus are fully developed, but have not ascended into position as they remain blocked by the tongue (Fig. 4.42).

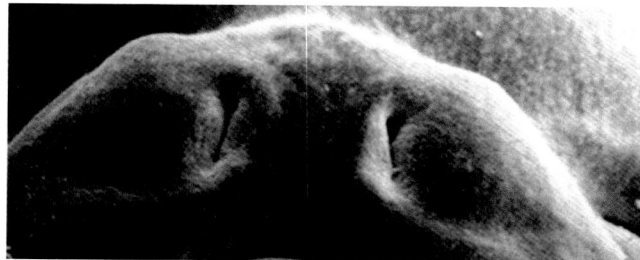

Fig. 4.41 Stage 19 Upper lip continuity established. Prolabial processes and premaxillae descending and medializing. Nostrils are vertical. Primary nasal septal growth creating nasal tip. Columella defined. Palatine processes now cover one half the choanae. (Courtesy of Klaus Hinrichsen Collection, Göttingen University; SEM curator Dra Beate Brand-Saberi)

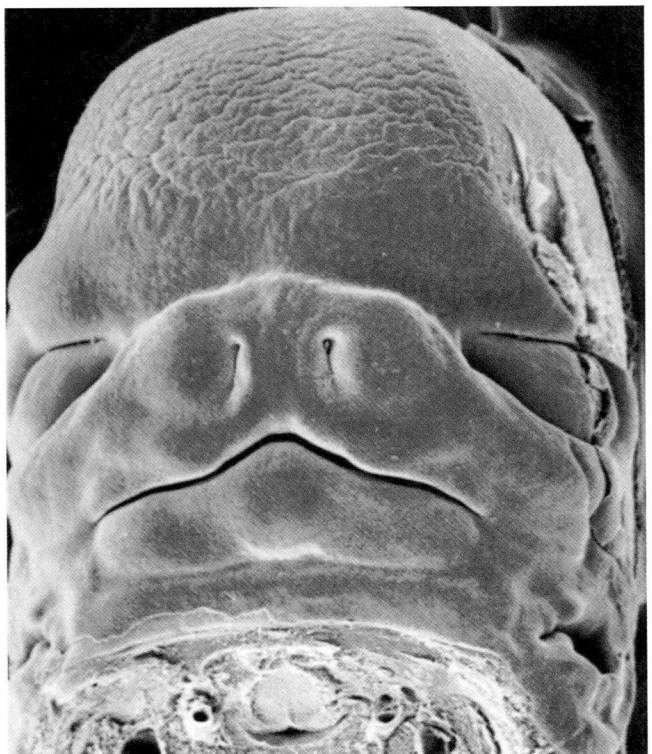

Fig. 4.42 Stage 20 Eyes and nose medialization nearly complete. Nostrils are vertical. Primary nasal septal growth creating nasal tip. Columella defined. The paired prolabia are still separated by a notch. Septum/vomer descending into palatal plane. Palatal processes cover hoanae entirely. Mesenchymal condensation seen in the palatal processes. Palatal shelves achieve upright position at stage 23. (Courtesy of Klaus Hinrichsen Collection, Göttingen University; SEM curator Dra Beate Brand-Saberi)

Coronal section shows epithelial plugs in the anterior nares. Cartilagenous nasal capsules surround the lateral nasal walls, but are not in continuity with the septum. This possibly represents the developmental autonomy of ethmoid zones 13 vs. 12.

Stage 21 (52–53 days) is characterized by an accentuation of the facial profile seen at stage 20. Septal development causes greater nasal tip projection and an accentuated naso-frontal angle. Traction between the caudal septal margin to the premaxillae, mediated by epressor septi nasi, results in *anterior nasal spine*. Despite this, the nasolabial angle remains obtuse. Both maxilla and mandible remain retrusive, as seen in the acuity of both angles SNA and SNB (Fig. 4.43).

Coronal section demonstrates localization of olfactory epithelium to the dorsum. Formerly, it was diffusely distributed at stage 16. Connection now exists between the septum and nasal capsule. Ethmoturbinal fields move into lateral position. All-important in this stage are intraoral changes preparatory to palatal closure. Palatal shelves are now elevating, but not in complete position. They display a prominence just posterior to the future incisive foramen. Fusion will begin here at the next stage. Closure of the alveolar processes of premaxilla and maxilla anterior to the incisive foramen is now complete, creating a united dental arch.

Mesenchymal expansion of the premaxilla follows a dorsal-ventral and lingual-buccal sequence. Defects in this process lead to an isolated alveolar cleft. The worse the cleft, the more incomplete is the closure, and the deeper the notch becomes. Note that the frontal process of premaxilla arises from the lateral incisor field. PMxF ascends as the inner lam-

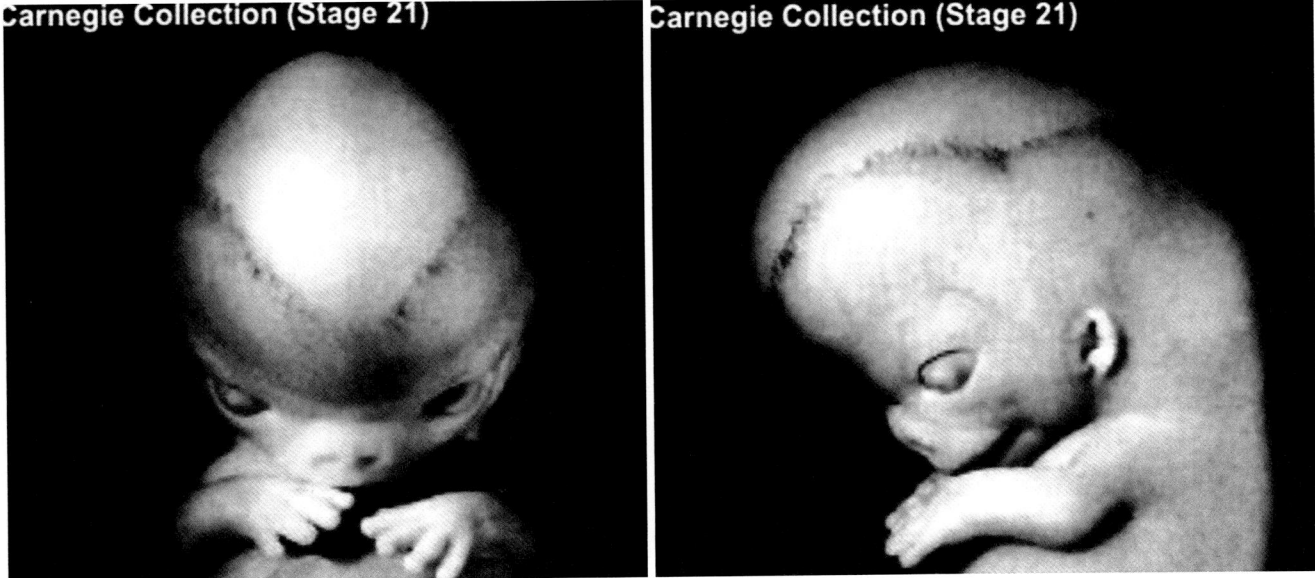

Fig. 4.43 Stage 21 Nose shows tip projection and naso-frontal angle. Nasolabial angle requires further septal growth. (Reprinted with permission from Johns Hopkins University Department of Medical Art—Gary Lees. Carnegie Institution)

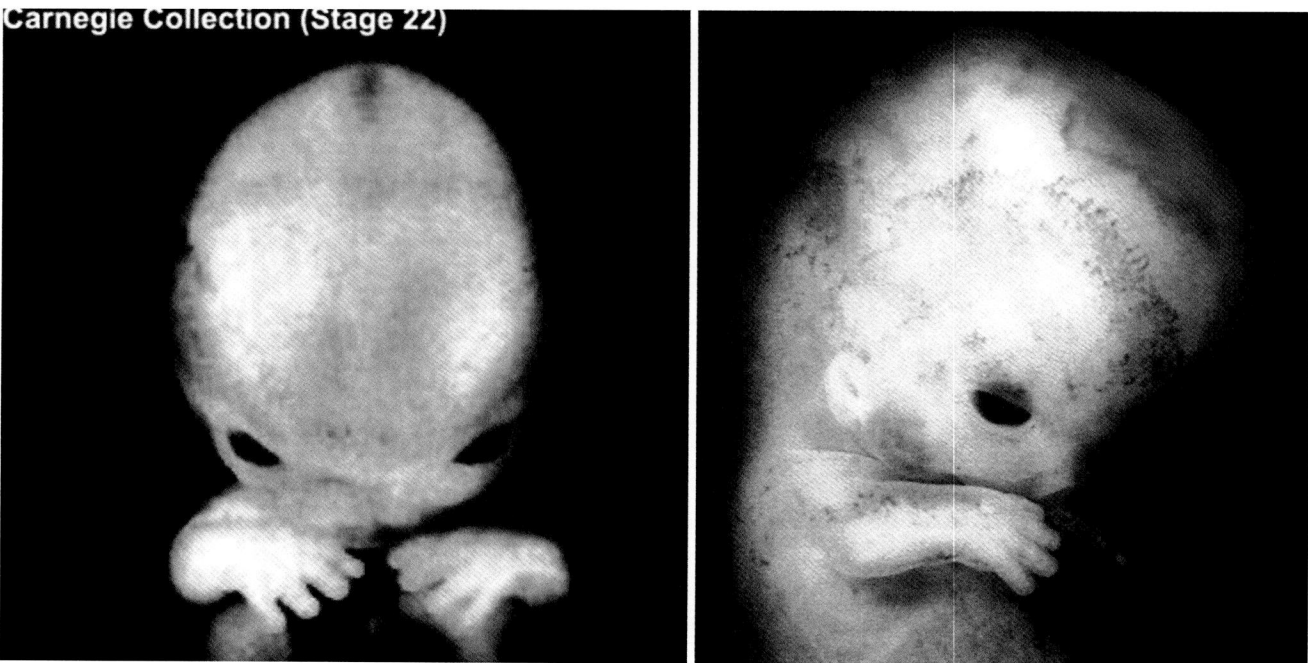

Fig. 4.44 Stage 22 The eyelids are partially closed. Note the zygomatic growth and the pounding out of the skull. (Reprinted with permission from Johns Hopkins University Department of Medical Art—Gary Lees. Carnegie Institution)

ina of the piriform rim. Any reduction in the mesenchymal volumes of PMxB, even in the presence of a lateral incisor, will lead to a deficiency of frontal process and deformity of the piriform fossa. Olfactory bulb formation is now complete.

Stage 22 (54–56 days) marks the closure of the palatal shelves to the vomer. This process is anterior to posterior and represents a unification between the medial and lateral nasopalatine neuroangiosomes (Fig. 4.44).

Middle meatus is now seen between the ethmoturbinates and the maxilloturbinate. This is the first sign of the future maxillary sinus.

Palatal shelf fusion is timed to happen immediately after completion of alveolar fusion. The presence of the nasopalatine bundle at the time of closure generates the incisive foramen. The so-called "incisive suture" has been discovered to be incomplete laterally; although it is directed toward the canine, it never reaches it [67]. Nijo and Kjaer interpret this as a fissure, rather than a suture [68].

Why is fusion of the primary palate an AP process whereas that of the secondary palate take place PA? This phenomenon may depend upon which field is aggregating tissue to itself. (1) In the case of primary palate, the required step involves the development of the lateral incisor field; PMxB involves a progressive filling out of the potential space to accommodate the dental unit. We can imagine mesenchyme flowing out from proximal to distal along the medial nasopalatine axis. It begins from a *fixed point* at the medial incisor field, PMxA, with PMxB being filled out like inflating a tire. Dental development is the final step. Unification of PMx B to maxilla will take place

with or without the lateral incisor. The critical factor is that both blocks of tissue are within the critical contact distance that permits epithelial contact to take place. (2) In the case of the secondary palate, the palatal shelf is a secondary field of the maxilla. The behavior of greater palatine is analogous to medial nasopalatine. Tissue is progressively deposited at its distal tip, from a *fixed point* just behind the lateral incisor or canine fields of the maxilla, the mechanism of inflation being the same. As the palatal shelf expands, it must maintain the proper dimensions in order to maintain contact with vomer. If the tire is underinflated at any point, and contact lost, fusion will fail from that point posterior.

The alveolar ridge is smooth and continuous. Although the blood supply of premaxilla is nasopalatine artery, through the continuity of labial mucosa there is collateral flow from ECA. The face at stage 22 has a less obtuse nasolabial angle due to downward and forward growth of the maxilla and mandible. Both angles SNA and ANB are less acute. A definite distinction can be made between columella and philtrum, although the latter is, at yet, undefined by philtral columns.

In the brain, the lateral olfactory tract is well-defined.

Stage 23 (37+ days) concludes the embryonic period of development.

Breakdown of the ethmoid barrier between the posterior nasal sac and the pharynx establishes the *secondary chonane*. Failure of this process creates *choanal atresia*. Middle meatus has deepened. Perpendicular plate and septum can be seen inserting into vomer. Closure of the hard palate is complete. The mouth and nasal cavity are now separated. Closure of the soft palate extends halfway from the posterior nasal spine to the free margin of uvulus.

Completion of established organ systems occurs during the ensuing fetal period. Full closure of the soft palate is achieved by 9 weeks. The nasal bridge appears at 11 weeks. Development of the second arch facial muscles takes place within the superficial fascial plane and extends throughout the head and face to interface with the cervical somatic neck. Recall that second arch muscles of mastication form earlier within the deep investing fascia plane. Muscles of facial expression surrounding the oral cavity develop in two planes. The deep layer of orbicularis associated with the oral mucosa acts as a sphincter and the superficial layer associated with the skin serves as a dilator. This repeats the overall pattern of branchial arch muscles. At 12 week, superficial orbicularis oris inserts into the MNC mesenchyme underlying the philtrum to form Cupid's bow [69–72]. At 13 week, definitive nares are present [73].

Olfactory and terminal-vomeronasal nerves are clearly seen. A terminal-vomeronasal ganglion is present with autonomic fibers distributed to the terminal component, while chemoreceptor information is presented to the vomeronasal component. Within the brain, connections are completed between olfactory tracts and the cortex. The olfactory area now is definitively related to the claustrum. In keeping with their slightly later development, the optic tract radiations have now reached the lateral geniculate bodies [66, 74].

The Facial Midline: A Vascular Watershed

The perfusion pattern seen in this series reveals the existence of vascular boundary zones involving three embryonic sources of mesenchyme: (1) MNC-derived fronto-naso-ethmoid fields supplied by the V1 stapedial system, (2) neo-pharyngeal arch RNC fields of the jaws and zygoma supplied by V2 and V3 stapedial system, and (3) RNC-derived fields of the first and second pharyngeal arches supplied by the external carotid. In other words, clear-cut neurovascular boundaries exist between the nasoethmoid capsule, the vomer-premaxilla and maxilla, and the overlying soft tissues. Note that the term neo-arch refers to the fact that evolution of jaws sequestered neural crest tissues away from the first two gill arches—a situation that required a new source of blood supply, the maxilla-mandibular stapedial system. All remaining non-jaw components of first and second arches such as muscles of mastication, skin, and mucosa retain their original perfusion source from external carotid.

A midline raphe separating the nasal and oral blood supplies to the maxillary palatal shelves was demonstrated by Maher [18]. Injection of the facial artery and the greater palatine artery perfused neither septum nor vomer [19]. We know now of course that the septum is StV1 and the vomer is StV2 separate from descending palatine. Along the inner aspect of lateral nasal wall, inclusion of the upper two turbinates as StV1 fields (lateral nasal br. of ethmoid) and inferior turbinate as StV2 (lateral nasopalatine) reflects the difference in the neural crest source material: MNC vs. neo-arch RNC [75, 76].

The relative contributions of anterior and posterior ethmoids arteries (AEA and PEA) to the fronto-naso-ethmoid (FNE) envelope are worthy of comment. Multiple perforating vessels through the cribriform plate are seen to create an intense and diffuse staining pattern in the 3 ethmoid sinuses and medial nasal wall [77, 78]. AEA perfuses the anterior and middle ethmoid sinuses, while PEA perfuses the posterior member of the trio.

And because these sinuses drain beneath the upper and middle turbinates, these latter structures are perfused by PEA and AEA as well.

The ethmoid sinuses have medial and lateral walls that represent a separation plane between MNC dedicated to the lateral nasal wall and MNC assigned to the medial orbital wall. These two walls differ in epithelial programming. The former is associated with nasal vestibular lining (p6), while the latter is in contact with sclera. Thus, medial branches from AEA and PEA perfuse the orbital wall, while the nasal wall, constructed from the upper and middle turbinates, is supplied by distinct lateral nasal branches of AEA/PEA that run in the nasal mucosa.

As previously stated, when the paper that forms the original basis of the chapter was written (in 2002), I was unaware of Dorcas Padget's work on the embryogenesis of the cerebral circulation and the existence of the stapedial system. The following passage from the original reveals this misconception which is so readily explained by the vascular embryology as we now understand it: "The perfusion of the septovomerine axis (sic) from the internal carotid raises doubt as to the internal maxillary artery as the primary source of the nasopalatine artery. The existence of anastomoses between NPA and PEA along the septum and lateral nasal wall is well documented" [78].

Now that we know about the stapedial system, the boundaries of these fields all make sense, as do the presence of anastomoses between the systems. A good example of this is *presphenoid bone*, an MNC derivative supplied directly from ICA prior to the take-off of primitive ophthalmic. (Note: The vessel in question, primitive maxillary artery, serves as temporary supply for the optic plexus and thence becomes redirected to the hypophyseal region.) An anastomotic twig exists connecting the bone with the exact junction of nasopalatine into its medial and lateral branches. But this does not mean they are supplied by ICA...one simply has to proceed more posteriorly into the pterygopalatine fossa to realize that the nasopalatine axis is one of the principle branches of StV2.

As will be discussed further, the organization of the A fields is oriented around its neurovascular tissue units, neuroangiosomes. Thus, the nasopalatine arteries, accompanied by nasopalatine nerves, occupy paired grooves on either side of the superior margin of the vomer. Together, these define the vomerine neuroangiosomes.

The embryogenesis of the vomer takes place in a membranous "sling" between paired anterior paraseptal cartilages (APC) and posterior paraseptal cartilages (PPC) [79–82]. [*Author's note*: Non-primate mammals, such as the rabbit and the vole—both possessing snouts—have a single, continuous paraseptal cartilage. The verticality of the human nasal chamber along with the formation of the primary choanae requires a discontinuity between the paraseptal cartilages.] Ossification of the vomer begins at the APCs and moves posterior toward the PPCs [83–86]. This times precisely after the posterior-anterior closure of the A1 medial nasal walls—due to MNC apoptosis—is completed at Carnegie stage 17. The form of the midline at this stage is a scalene triangle with its apex in the roof of the nasopharynx at Rathke's pouch, abutting the sphenoid. But the nasopalatine axes are undoubtedly in place. Vertical and anterior growth of the r1 MNC ethmoid complex is accompanied by an accumulation of r2 RNC mesenchyme along its caudal border...the swellings of the vomers begin to become visible as they expand and fuse. And at the very distal tip of the vomerine axis, the anlage of the premaxilla begins to develop (Fig. 4.45 Ossification pattern of the septum and ethmoid versus the vomer and premaxilla and Fig. 4.46 Vomerine growth with respect to the palatal plane).

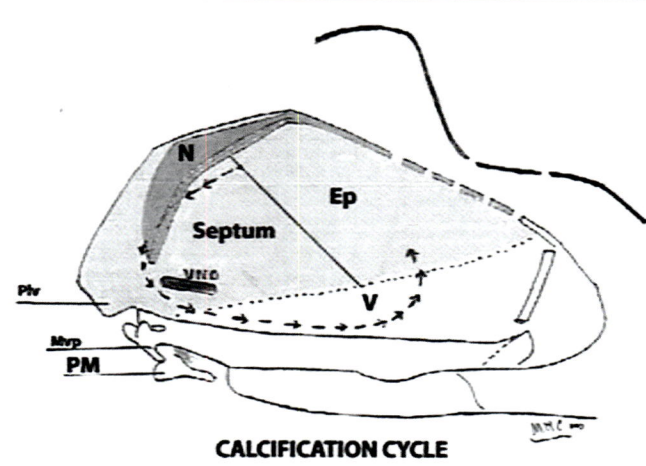

CALCIFICATION CYCLE

Fig. 4.45 Posterior ethmoid artery fields (PEA) develop prior to anterior ethmoid artery fields (AEA). Therefore, the pattern of chondrification of the septum is back-to-front and dorsal-ventral. Distribution of the AEA is more extensive and reaches the vomerine mesenchyme. AEA then becomes *additive* to flow from the medial nasopalatine artery (mNPA). PEA has a greater distance to descend to reach the vomer; therefore, its contribution to mNPA lags behind that of AEA. Therefore, the pattern of ossification of the vomer extends font-to-back from the nasopalatine foramen. These facts are consistent with the development of the vomer as it descends into the palatine plane. (Courtesy of Michael Carstens, MD)

Fig. 4.46 Vascular relationships and growth patterns of the septo-vomerine complex. V1 stapedial from ophthalmic (orange), V2 stapedial from internal maxillary (green), External carotid (pink). Normal vomer descends into midline by stage 18. Palatal shelves develop late stage 17 and are vertically oriented. They achieve horizontal transformation at stage 23. Secondary palate closure 9–12 weeks. Developmental failure of vomer begins posteriorly. If the palatal shelves are of normal volume, the palate cleft will be narrow (right side figures). If palatal shelf is also deficient, the cleft will be wide (left side figures). (Courtesy of Michael Carstens, MD)

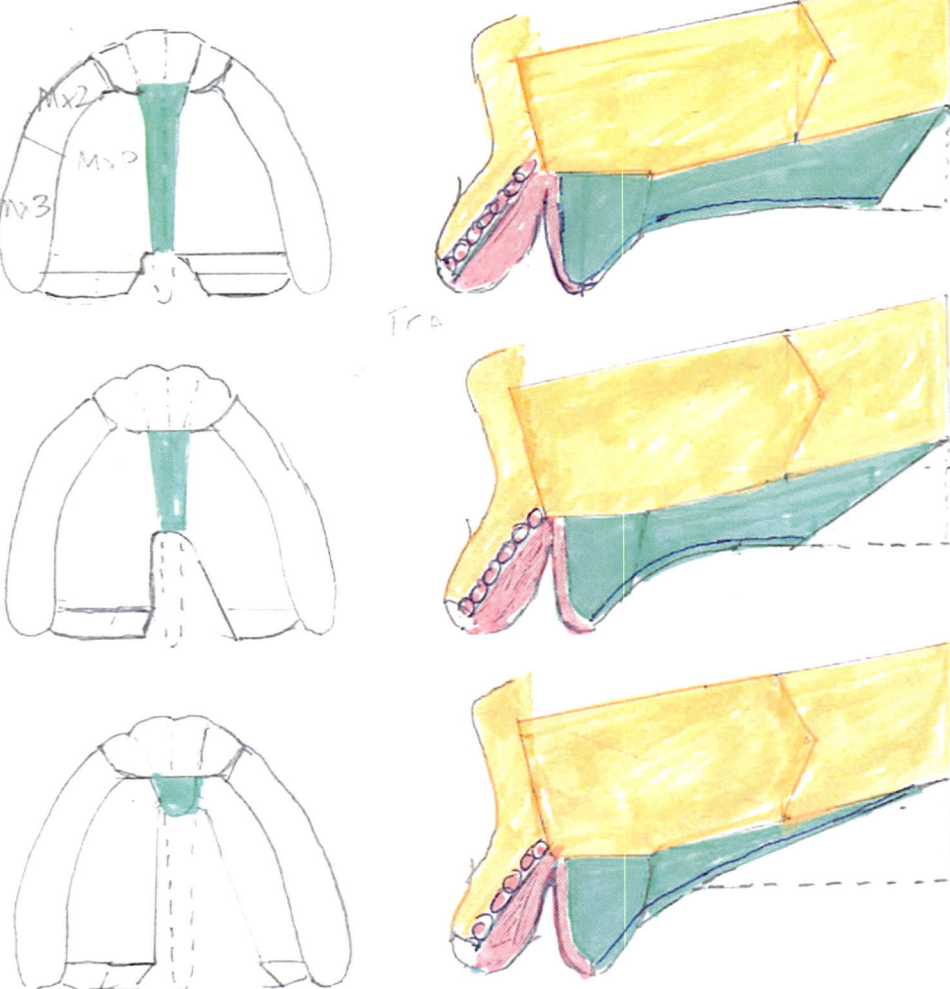

In conclusion, the vascular supply to the facial midline represents the interface with three neurovascular systems. The deep tissues of the medial nose have an upper medial quadrant supplied by V1 stapedial ophthalmic and innervated by V1 medial internal nasal nerve. The lower medial quadrant is supplied by V2 stapedial medial nasopalatine artery and nerve. The upper lateral quadrant is supplied by V1 stapedial ophthalmic and innervated by V1 lateral internal nasal nerve. The lower medial quadrant is supplied by V2 stapedial lateral nasopalatine and descending palatine artery and nerve. Overlying the deep bone-bearing fields medial supply is V1 stapedial ophthalmic and lateral supply is facial from external carotid. The lines of division are straight backward from the piriform fossa—medially along the septovomerine axis and laterally above the inferior turbinate.

Sequence of A Field Development Determines the Dimensions of the Midline (Figs. 4.11 and 4.12)

The three-dimensional pattern of the A fields begins with the proliferation of MNC mesenchyme between the brain and frontonasal skin. Recall that MNC migration over the surface of the forebrain begins as early as stage 9. By stage 13, this tissue expands and begins to project the soft tissues outward. This is seen earliest in the vicinity of the nasal placodes. Because the nasal placodes are anchored, they become internalized. Furthermore, because the MNC cannot penetrate the placodes, peri-placodal MNC proliferation throws up the borders into parenthesis-shaped lateral and medial nasal processes.

The role of prechordal mesoderm in facial development remains poorly defined. It is important in establishing the positioning of the orbits. Evidence exists to suggest that PCM contributes to the blastema of the striated extraocular muscles [87]. This mesoderm is therefore the likely source for somitomeres 1–3 which contain all EOMs except lateral rectus. Note that the mesoderm of somitomere 5, containing lateral rectus, probably originates at level r2–r3, but its abducens nucleus (CN VI) gets displaced further backwards to r4–r5 where it shares a neuromeric origin with facial nerve. Lateral rectus PAM does NOT arise from second arch levels r4–r5. For this reason, Moebius syndrome, characterized by developmental failure of the seventh nerve, is also characterized by failure of the sixth nerve.

Neural plate mapping of the forebrain region in chick embryos shows prior ectoderm to nasal placode [88, 89]. With the formation of the neural folds, the most mesial tissues assume a basal, midline position. Therefore, the olfactory placodes are in position. A unifying statement about prechordal mesoderm is that it originates at gastrulation from levels r0 to r1 and subsequently breaks up into the first three somitomeres, thus providing all but one of the extraocular muscles.

Formation of the septum at stage 18 is a key event for midface projection. It takes place once the process of unification of the medial nasal walls is complete. The septum develops sandwiched between the two A1 fields. Median clefts result from a failure of this apoptotic process. In this condition, *septal widening* and even *bifidity* can be observed anteriorly. The earlier in time that apoptosis is interrupted, the more posterior are the septal defects. Autoradiographic studies by Searles [90–93] using labeled chondrocytes in rats demonstrate that the formation of nasal septal cartilage is an anterior-posterior process, not a vertical downgrowth, as popularly imagined. The posterior branches of the nasoethmoid perhaps develop later than the anterior branches.

The stereotypic sequence in which case ossification centers appear is a recognized means of demonstrating the vascular maturity of a bone field. Work by Kjaer and others [94–96] has elucidated a similar sequence in the developing cranium. On sagittal section, the first evidence of ossification begins in the frontal bone at the nasofrontal junction. Ethmoid complex ossifies later. This sequence may reflect the anatomy of anterior division of internal carotid. ACA has several stems, the fronto-polar stem being the most proximal. Primitive olfactory is the terminal branch of anterior division ICA and is therefore distal to ACA. Primitive olfactory is directed to the olfactory fields and ethmoid. Thus, we observe that ossification of the cranial base posterior to pituitary proceeds from posterior-to-anterior from basioccipital forward to basisphenoid.

These findings are consistent with the development of three vascular systems. (1) Intracranial stapedial develops first with StV1 and StV2. The system of dural arteries is subsequently completed with the ascent of V3 stapedial (middle meningeal) from external carotid into the skull. StV1 bone derivatives may mature prior to those, such as alisphenoid supplied by StV2 or parietal, the dura of which is supplied by StV3. (2) The longitudinal neural arteries fused together beneath pons as the so-called basilar, but then remain separate to supply medulla. Thus, the supply of the hindbrain develops in cranial to caudal sequence. (3) The vertebral contribution plugs into the LNAs just inside foramen magnum. Since the blood supply to basioccipital depends on the extracranial contribution, it makes sense that the ossification should proceed from basioccipital forward. The last membranous bone to ossify is the most medial, the ethmoid complex.

When Granstrom analyzed enzymes active in cartilage and bone formation in rats, a similar pattern was noted [97]. Septal chondrogenesis was followed in sequence by premaxillary ossification and later, that of the vomer. This ossification sequence is carried downward into the alveolus as well [98]. After septal cartilage differentiation is completed, endochondral ossification in the septo-presphenoid zone forms the ethmoid bone. This is yet another example of how cranial neural crest can form a membranous bone via a cartilage intermediate (Fig. 4.45 Septal chondrogenesis and vomer ossification and Fig. 4.46 Descent of the vomer).

Although the closure pattern of the ethmo-vomerine fields due to apoptosis is posterior-anterior, the ossification pattern of the vomer is the reverse. There is no contradiction in this, as these are separate processes. Vomerine mesenchyme is most mature at its distal aspect; as the bone grows forward, new mesenchyme is added *on the back end*. This is exactly how the maxillary palatal shelf grows as well. Premaxilla, on the other hand, has its most mature cells medially and then adds out laterally. It is as if the nasopalatine axis spilled paint at the incisive foramen, creating the housing for the central incisor; it then flows outward to establish the housing for lateral incisor and the frontal process. At any rate, once the vomers are established at stage 18, they are ready to receive and fuse with the palatal shelves as these latter elevate.

Despite a cleft alveolus—due to a deficiency state in the premaxillary field of medial nasopalatine—if the tissue volumes of vomer and palatal shelf are normal, the two fields remain within the critical contact distance, and the anterior-posterior "zipper" mechanism will be activated. This is because events of premaxilla-maxilla contact take place independently anterior to the incisive foramen. Any tissue deficit posterior to incisive foramen may cause a situation in which the bone fields exceed the critical contact distance… wherever this takes place, a palatal cleft will result.

Forebrain Development and Facial Structure: The Brain Predicts the Face

Importance of Timing

The spatial organization of field both those of the CNS and those of the face is a reflection of the timing of their development. Each can be profitably studied to learn about the other. At 15 days' gestation (stage 6), the oropharyngeal membrane and the anteriorly advancing primitive streak define the polarity of the embryo. By 18 days, a prechordal plate is interposed between the oropharyngeal membrane and Hensen's node. The prechordal plate is a small aggregation of mesodermal cells tightly bound to the endoderm (hence the term, mesendoderm). Caudal regression of the Hensen's node and the gastrulation process leave behind prechordal plate and notochord. As this happens, under scanning electron microscopy, tiny paired segmentations of paraxial mesoderm appear. These somitomeres continue to form along Hensen's node as it moves posteriorly [99]. Beginning with the eighth somitomere, the PAM reorganized further into recognizable somites. The first seven somitomeres produce craniofacial muscles for the pharyngeal arches and then disappear.

When the number of somitomeres has reached 19, the first pair of somites replaces the eighth pair of somitomeres. The remaining paraxial mesoderm follows the same sequence of somitomere-to-somite transformation. For a while, an equilibrium exists between somite conversion and the appearance within previously amorphous PAM of a new posterior somitomere. This keeps the number of caudal somitomeres at 11.

But eventually, all are replaced [100]. Humans' embryos produce a total number of 36 somites (4 occipital, 8 cervical, 12 thoracic, 5 lumbar, 5 sacral, 10 coccygeal—only 2 of which survive) at a rate of approximately 4 per day until the process is complete. The process of mesodermal organization is in genetic register with adjacent neuromeres of the midbrain and hindbrain [101]. Organization of the embryonic brain is under control of genes expressed in each neuromere.

Mammalian neurogenesis demonstrates a remarkable degree of order. The segmental expression of genes such as Krox-20 and the Hox family creates within the rhombomeres an architecture in which cranial nerves beginning with III are arranged in dorso-ventral and medial-lateral groups that reflect their function (GSE, SVE, SSE, etc.) [2, 102]. The temporal appearance of the cranial nerves and their pharyngeal arch targets also follows the cranio-caudal pattern of their respective rhombomeres of origin [100–102].

At the same time, signals from the prechordal plate and mesencephalon help to induce the forebrain (prosencephalon). Work by Puelles, Rubenstein, and others [89] has been worked out the developmental units of the forebrain into prosomeres. The system has undergone several iterations. The original prosencephalon of the three-part brain subdivides into a secondary prosencephalon (SP), containing telencephalon (cerebrum), the eye fields and hypothalamus, and a posterior diencephalon (containing the thalamic structures). The anterior forebrain neural folds contain tissue for fronto-naso-orbital skin and the placodes of the nose and eye. The diencephalic neural folds contain neural crest that will migrate forward to complete the skin cover of the upper face. Diencephalon has three prosomeres, p3–p1. Rubenstein's original mouse model postulates the presence of p5 and p6 in the midline of telencephalon as essential for orofacial development [32]. The definition of developmental units for the telencephalon is more complex and is the subject of Chap. 5.

Olfactory and optic nerve development also follows the prosomeric model. Both these structures require in-growth of placodally derived neural tracts, rather than cranial nerves per se. Cranial nerve I, being more rostral and midline, originates from the basal SP. It is therefore associated with the most anterior neuromere, p6 (in O'Rahilly and Müller's terminology) neuromere T [101]. The cranial nerve II originates from basal SP, prosomere p5. (Note that in the O'Rahilly model, it is considered to arise from the diencephalon and is associated with the second neuromere, D1.) As special sense organs, cranial nerves I and II both develop in association with placodes. Shh is associated with epithelial placode formation—the temporal order of its expression should dictate that nasal placode develops before optic.

Hinrichsen's SEM studies [49] demonstrate that such is exactly the case. Prior to their closure, the nasal fields are derived from the anterolateral edge of the neural folds at stage 10 (22 days). This is under control of the same gene (*Pax*-6) as is formation of the optic fields [101–103]. Nasal placodes are seen at stage 13. As these deepen into nasal pits, the MNC surrounding the pits proliferates wildly forming the well-known medial and lateral nasal "processes." But the deep surface of the nasal pit, originally an ectodermal placode, has nothing to do whatsoever with the pharyngeal arch apparatus. It is an invagination of placodal epithelium. The nasal pit appears first at stage 14 and is laterally directed under the influence of lateral nasal process. At stage 15, the medial nasal process appears and the nasal pit ingression extends medially as well. By stage 16, a well-formed nasal sac is present, but is not yet open posteriorly.

The epithelium of the invaginating nasal pit is associated with the developing olfactory tract—ultimately it forms the olfactory epithelium. As the nasal vestibule deepens, it "pushes into" the surrounding MNC…in reality, the process is the reverse: Initially, each nasal vestibule is entirely separate from the oral cavity. When the caudal vestibular floor breaks down, the resultant *primary choanae* constitute a communication with the nasal cavities and the primitive mouth. Although the entire lateral wall of the primitive causes them to swing into position below the orbit to "add on" to the caudal lateral nasal wall, the StV2 neo-arch tissues increase the vertical dimensions of the nasal chambers.

The developing embryonic face demonstrates mesial migration of the nares followed by forward positioning of the eyes at the seventh week. In order for this to happen, the medial walls of the widely separated nasal cavities must fuse

[104]. Internal fusion is a posterior-to-anterior process which begins at Rathke's pouch at stage 14 (33 days). The process is largely complete by 6 weeks by which time a nasal tip is first noted. Fusion of the nasal fields creates a single midline nasal structure into which bone and cartilage will develop. It is the precondition for proper orbital positioning!

Eye fields, on the other hand, originate as a midline fusion of two fields. These must separate and lateralize, a process under tight genetic control. Humans express an equivalent for the Drosophila *eyes absent* (*eya*) gene known as *Eya* [105]. In chickens, Eya2 protein (which is 82% conserved compared with its human form) is expressed in the nasal pit at avian stage 13, but does not appear in the retina until avian stage 24. Fibroblast growth factor 8 (FGF8) exerts differential effects on craniofacial development in Xenopus larvae depending upon the stage at which it is applied [106]. FGF8 beads soaked in heparin at Xenopus stage 13 surpress forebrain, eyes, and midbrain. Application at stage 15 spares the forebrain, but surpresses the eyes. Since development of the nasal placode precedes that of the ophthalmic placode, inhibition of the former will affect the ultimate position and/or form of the latter [107–109]. Cyclopia, with attendant absence of all A fields, is the ultimate manifestation of this principle [110–115]. In summation, A field development, associated with the most anterior of prosomeres, p6, precedes and determines the development of the p5 optic fields (Fig. 4.47 HPE cyclopia).

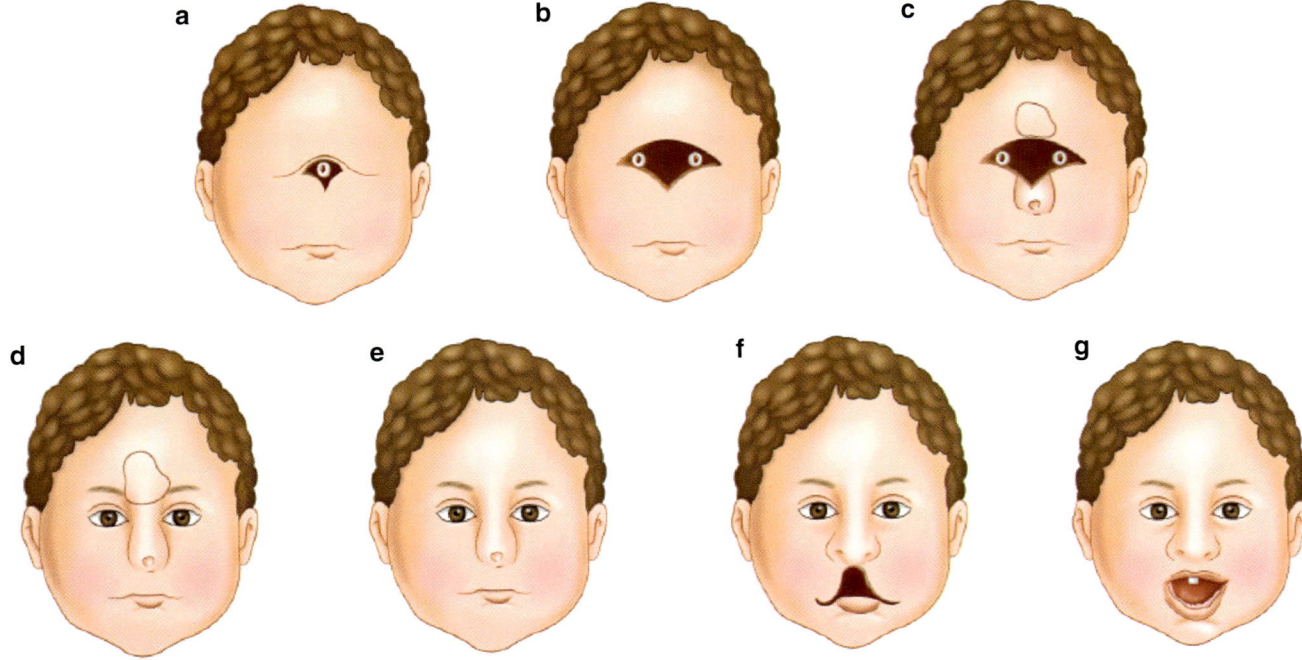

Fig. 4.47 Spectrum of holoprosencephaly. (**g**) Premaxillary insufficiency with single tooth and mild hypotelorism; (**f**) total loss of premaxilla (and vomer); (**e**) cebocephaly with single nostril in correct position; (**d**) cebocephaly/ethmocephaly with double orbits; (**c**) ethmocephaly with single orbit; (**b**) synophthalmia; (**a**) cyclopia. (Reprinted from Bianchi DW, Crombleholme TM, D'Alton ME, Malone FD. Holoprosencephaly. In: Fetology: Diagnosis and Management of the Fetal Patient, 2nd ed. The McGraw-Hill Companies. 2010: 121–128. With permission from McGraw-Hill Education)

How the Prosomeres Relate to Craniofacial Structure

The Forebrain Is Patterned Under New Rules

Central nervous system from the isthmus/midbrain backwards depends upon the induction of a neural plate by non-neural tissues via signaling molecules from the notochord. Studies of gene expression and cell fate maps of the neural plate in vertebrate embryos demonstrate that the spatial origin of these signals—correlated by the homeotic system—determines the 3-D architecture of the CNS.

But what about a forebrain that develops anterior to the notochord? Signals from the r1 and the anterior visceral endoderm are implicated (Fig. 4.48 Carlson—brain induction). Primordial organization can be seen in the layout of the prosencephalic neural plate into domains prior to folding. These genetic domains are described as prosomeres. Mediolateral domains specify longitudinal columns of neural plate, clearly seen in the hindbrain and spinal cord as floor, basal, alar, and roof plates [112]. Anterior-posterior domains specify transverse regionalization in the neural plate. In the midline of the three-somite chick embryo, the adenohypoph-

ysis, lamina terminalis, preoptic area, and hypothalamus are all successive A/P laminations [87, 88]. Along the lateral ridge are located nasal ectoderm, olfactory placode, upper beak ectoderm (nasal bone analog), and calvarial ectoderm. A third layer of complexity is generated by local patterning centers. These include the specialized zones of ectoderm within the olfactory placodes and Rathke's pouch.

Neural tube formation requires that the orderly world of neural plate domains undergoes a critical topologic transformation. In this process, the lateral-most edges roll upward toward each other, much like a handle bar moustache, to form and close up a dorsal midline. The resulting cigarette-like neuraxis now has dorso-ventral domain distribution. Embryonic folding creates flexures in the neuraxis. These, together with differential growth of the neural contents, establish the final three-dimensional configuration of the brain.

Determination of these domains comes from signals produced by adjacent tissues. Nonneural ectoderm specifies the cell fates of the *lateral domains* using members of the transforming growth factor beta (TGF-*B*) superfamily known as bone morphogenetic proteins (BMPs) [116–118]. *Medial*

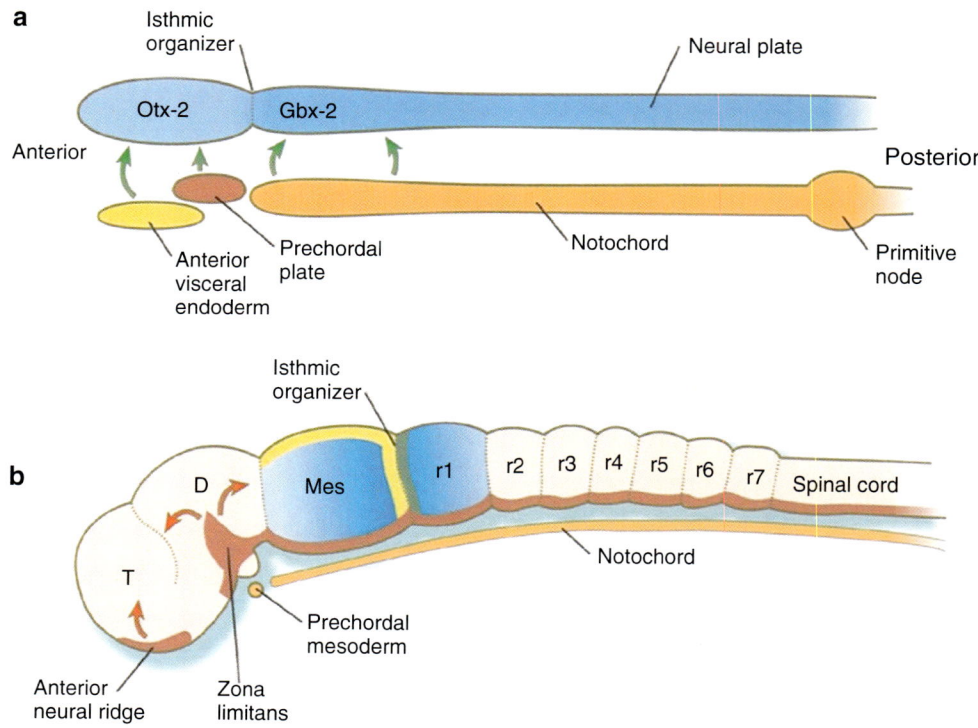

Fig. 4.48 Schematic representation of signaling centers acting on and within the early embryonic brain. (**a**) In response to signals (greens arrows) from the anterior visceral endoderm, the prechordal plate, and the notochord, the neural tube express Oyx-2 in the future forebrain/midbrain and Gbx-2 in the hindbrain/spinal cord. (**b**) Later in development, signals fFGF-8 (green) and Wnt-1 (yellow) from the isthmic organizer induce decreasing gradients of En-1 and En-2 (blue) on either side. Sonic hedgehog (red) is secreted from the other two organizers, the anterior neural ridge and the zona limitans, as well as the ventral part (floor plate) of the neural tube. *D* diencephalon, *Mes* mesencephalon, *r* rhombomere, *T* telencephalon. (Reprinted from Carlson BM. Human Embryology and Developmental Biology, 6th edition. St. Louis, MO: Elsevier; 2019. With permission from Elsevier)

domains are specified by underlying midline tissues, the notochord and prechordal plate. The former determines the midline of the hindbrain and spinal cord [119–121] through its principle specifierm sonic hedgehog (Shh). Although Shh is first seen in the notochord, it is later expressed in the medial neural plate [121, 122]. Shh genetic expression, along with hepatic nuclear factor 3 beta (HNF3b) and Nkx2.1, occurs along the entire midline of the embryonic neural plate [123, 124]. Because the *prechordal plate is the functional anterior equivalent of the notochord*, it is logical that Shh should be produced there.

Shh expression in the dorsal foregut underlying the forebrain takes on great significance. Shh protein is the signal for *Nkx2.1*, the expression of which is required for ventral forebrain development [125]. Shh mutant mice demonstrate failure of midline organization along the axis of the entire CNS, *including cyclopia* [126]. Expression of *Otx2* and *Lim1* genes (also essential for forebrain development) [127–129] occurs in the tissue subjacent to the anterior neural plate as well [130].

Scanning electron microscopy of the notochord and prechordal plate demonstrates both structures to be analogous in structure and origin [131–133]. Both tissues express regulatory genes such as *chordin*, *HNFb*, and *Shh* [134]. Similar gene products may have different roles depending upon their site of origin. The expression of *Lim-1* in prechordal mesoderm suggests that this small structure is important in stimulating the formation of forebrain and cranial structures, whereas its expression in the notochord plays a vital role in stimulating the more posterior parts of the brain and spinal cord [100].

Prechordal plate mesoderm has a specific role in vertebrate forebrain organization. Explant models demonstrate that the prechordal plate and dorsal foregut induce medial properties and repress lateral properties in prosencephalic explants. Furthermore, other tissues associated with the anterior neural plate in 0–1 somite embryos such as lateral foregut or cranial neural crest do not regulate *Nkx2.1* in the explant assays. This result provides evidence that the prechordal plate functions alone in the initial specification of the medial prosencephalon [32]. This patterning model appears to be strongly conserved in vertebrates from the chick [135] to the mouse [136].

Local patterning centers such as the anterior neural ridge (ANR) produce fibroblast growth factor (FGF8). This induces the production of Bf1 (a critical transcription factor for telencephalic regionalization) from the anterior neural plate. Neural plate explants fail to make Bf1 when the ANR is excised. Mouse mutants lacking Bf1 have small forebrains and cannot express *Dlx2*, a basal forebrain marker. The role of the *Dlx* gene family in patterning the pharyngeal arches has been discussed previously.

Local patterning is important in the development of the naso-optic fields in which specialized olfactory and optic placodal epithelia play critical roles. Ophthalmic development involves the following processes: (1) specification of a singular, dedicated eye field in the anterior neural plate; (2) A/P segregation of the optic field from adjacent preoptic and hypothalamic territories; (3) subdivision of the retinal fields via Bf1 and Bf2 into nasal and temporal domains; (4) formation of the optic chiasm; and (5) separation of the eyes by MNC nasoethmoid derivatives [136–138].

Competence of the singular ophthalmic field within the neural plate depends on expression of the *Pax6* gene prior to interaction between the eye field and the lens placode [139]. Subsequent repression of *Pax6* by *Pax2* causes separation at the center of the common eye field and formation of the optic chiasm [140]. Prechordal midline *Shh* induces *Pax2* [141], therefore *Shh* deficiency states are associated with cyclopia in mice [127], zebrafish [142], and humans [143].

Intracranial development of the olfactory bulb depends upon the trophic effect of neurons that develop peripheral to the CNS in the olfactory bulb [62, 144, 145]. Development of neurons within the placode is a precondition for olfactory bulb development [146–148]. *Pax6* is expressed in the olfactory placode. Murine *Pax6* mutants cannot achieve placodal differentiation. Despite nasal pit formation, no olfactory epithelium results; therefore no olfactory bulbs develop [148]. Consequently, these animals are anosmic. Since olfactory neurons come from the lateral aspect of the placode, the Pax6 deficiency state invites comparison with Kallmann's syndrome (anosmia/hypogonadotropic hypogonadism) [149, 150]. A wealth of evidence localizes the origin of GnRH neurons to the medial olfactory placode (*vide infra*). If Pax6 expression involves the entire placode, such mutants should also show endocrine deficiencies and would thus constitute a clinical model for the study of Kallman's syndrome.

A Field Instability: A Clinical Model of Neuromeric Failure

Note: The A fields are defined as those tissues supplied by neuroangiosomes of the StV1 system. Recall that, in development, what were purely MNC fields become invaded from posterior to anterior along the oral border by RNC fields supplied by StV2. Thus, midline fields are properly referred to as the ethmo-vomerine complex.

Stability of the facial midline depends upon fusion of the medial walls of the nasal capsules (A fields). Failure to achieve this process due to failure of physiologic MNC apoptosis is the basis for hypertelorism. Aplasia of one or both nasal capsules leads to inadequate interorbital volume and hypotelorism [111, 127, 151, 152]. Outright failure of A field development results in cyclopia. That interorbital structures (nose, columella, philtrum) contribute a relatively constant proportion to the total width of the face has been known for some time [153, 154]. Burdi [104] divided the embryonic/fetal midface into three cephalometric zones, evaluating their relative contributions of facial width in specimens ranging from 7 to 26 weeks gestational age. Zone 1 was defined as the region between two vertical axes through the septum and the inner-most point of the medial orbital wall, i.e., the nasal capsule/A field.

A field growth occurs earlier in development than in any other zone of the face. In only 2 weeks time, the central zone decreases from 60% of total facial width to a constant 20% by 40 mm CRL. Proper field migration is permissive for the achievement of normal interorbital distance. These findings are also consistent with the stabilization of the optic axes. Eye field migration results in a radical decrease in the interorbital axis from 180° at 5 weeks to 72° at 40 mm CRL. Subsequently, there is little change in the angle. At birth, the angle is 71° with the final adult configuration as 68°.

Clinical Defects

Clinical expression of A field failure involves a diminution of the rhinencephalon, i.e., the p6 zone. The old term of arhinencephaly, introduced by Kundrat in 1882 [155], is not appropriate for this condition, as Yankolev [156] demonstrated that some portions of the limbic lobe and olfactory cortex always persist. The intracranial findings of holoprosencephaly are well known [56, 157–159]. Despite an intact external nasal structure, such cases will demonstrate a diminutive crista galli, an unperforated cribriform plate (no nerves were present at the time of bone formation), and absent olfactory bulbs. These findings result from: (1) diminished volume of the MNC nasal capsule; (2) failure of neuron formation at the placodal level; (3) inadequate trophic influence on the CNS by the peripheral neurons secondary to aplasia of the MNC migratory pathways. Inadequate MNC development is intrinsic to all cases of holoprosencephaly.

MNC produces both intracranial and extracranial structures. The entire frontal bone and anterior cranial base are r1 MNC derivatives as is the dura mater covering the frontal lobes. An intimate vascular relationship exists between the dura and the cerebral cortex. [This is discussed in detail in Chap. 12.] As previously mentioned, the same developmental signals that affect midline brain development and the optic fields also affect the external structures of the nose and orbit. Thus, it is not surprising that a proportionality exists between the facial features of holoprosencephaly and its underlying cerebral anatomy [160–162]. A variety of nasal forms attend this condition ranging from (1) complete arhinia [163–167], (2) arhinia with dual proboscis [168, 169], (3) single nostril with contralateral proboscis [170–173], to complete external nasal structure without midline structures such as the septum [174–176].

Bilateral cleft lip is frequently seen in holoprosencephaly. In the case of holoprosencephalies with "median cleft," the premaxilla can also be absent due to failure of the MNC to permit infraseptal tracking of the StV2 nasopalatine axis. The premaxilla may have no teeth; rarely central teeth are missing but lateral incisors are present [177]. This raises the possibility that each A field has two subunits corresponding to Tessier cleft zones 13 and 12.

In the case of a bilateral cleft, the presence of a lateral incisor is explained by the redundancy of lateral incisor synthesis between medial nasopalatine and the medial branch of anterior superior alveolar artery. Specifically, medial branch of ASAA can potentially produce two dental fields, an "ectopic" incisor as well as the canine. Lateral branch of ASAA supplies the deciduous molars. Thus, *there is no such thing as an "ectopic" lateral incisor*. It is a perfectly normal condition…simply a case of derepression. This explains why, in HPE clefts, nostril formation can take place with a lateral incisor present on the maxillary side of the cleft.

All cases of nasal aplasia demonstrate diminished interorbital distance. Scanning EM of C57B1/6J embryos treated with ethanol during gastrulation demonstrates the effect of midline reduction of MNC within 24 h of injection [178]. Olfactory placodes are more closely spaced and, in some instances, they appear to fuse. Interorbital distance is diminished correspondingly. Mouse embryos exposed to maternally administered ethanol suffer cellular loss from anterior neural ridge. The resulting cell debris is seen on SEM to be concentrated in the mesenchyme of the ventral midline. Furthermore, Sulik's work demonstrates the existence of two distinct populations of prechordal mesoderm, medial and lateral [179]. This suggests that: (1) damage to prechordal mesoderm leads to failure of MNC development, (2) the medial population of prechordal mesoderm is more susceptible.

Experimental Production of Cyclopia and Nasal Capsule Disruption

Along the continuum of A field hypoplasia, cyclopia constitutes its end-stage orbital manifestation [157, 180–183]. During the twentieth century, laboratory production of cyclopic states has been induced by a variety of chemical, surgical, and ultimately, genetic manipulations. In 1910, Stockard used 3% ethanol to create mono or anophthalmic states in 90% of *Fundulus* embryos [184]. Exposure during gastrulation to solutions of magnesium chloride produces cyclopia in fish [185–188]. The same effect is obtained in chicks using lithium chloride [189, 190]. Amphibian cyclopia induced by high concentrations of salt water was shown by Hofstreter to result from incomplete development of prechordal mesoderm [191].

Surgical manipulations in early embryos produce cyclopia when sites associated with A field development are disturbed. Removal of the nasal pits from *Amblystoma* embryos by HS Burr in 1906 had remarkable consequences. The first visible effect of the removal of the nasal sac is a sinking-in of the skin in the nasal region. In the case of bilateral extirpation, this causes a veritable "pug dog" effect which worsens with age. Unilaterally operated larvae show absence of the capsular structures associated with the olfactory sac. The ethmoid column, tectum nasi, and the medial nasal process are all missing [192]. Striking changes in the brain were also noted. The hemisphere of the operated side is markedly smaller than its counterpart. This reduction in size is related largely to the anterior regions of the telencephalon. The olfactory bulb is greatly reduced in size, but in the regions of the hippocampus, the hemispheres are equal [192].

Further experimentation was carried out in more advanced models. In six-somite chicks, Rogers removed the neural folds between the optic vesicles, producing cyclopia [193]. Adelmann targeted prechordal plate mesoderm in amphibians, having previously noticed a surpression of its differentiation in cases of cyclopia [194–197]; Cyclopia was produced.

Genetic surpression and/or deletions experiments which target the MNC A fields (nasal capsule) or their precursors offer definitive proof of the relationships that those fields have with their ophthalmic and StV2 non-pharyngeal and ECA pharyngeal arch partners. Cyclopia in *Shh*-deficient mice has already mentioned [126] the role of *Pax6* gene products [136]. Downregulation of *Shh* by retinoic acid in mouse embryos results in holoprosencephaly [198]. Olfactory capsule development depends upon the ectodermal expression of homeobox gene *Dlx5* [199, 200]. The role of the olfactory placode in the production of the nasal capsule is reviewed by Webb [53]. Mice homozygous for the targeted deletion of *Dlx5* cannot form a nasal capsule [201]. Severe mutants have symmetric aplasia of the capsule, the sidewalls being mere cartilaginous spicules without turbinates. The midline is without septum, paraseptal cartilages, vomeronasal organs, or nasal epithelium.

The findings reported in mice by Depeuw [201] in less than total mutations are of the greatest importance in establishing which structures within the A fields are most sensitive to *Dlx5*; in case of intermediate severity, a pronounced asymmetry, with the right side being more hypoplastic, is seen. The trabecular plate-nasal septum, which is compressed dorsoventrally, deviates to the right to occupy the position of the right cribriform plate and posterior capsule wall. No foramina cribosa are seen on the right and some on the left. The tectum nasi and solum and the nasal prominences are hypoplastic. In these less-than-severe cases, rudimentary branches of nasal epithelium are present, as are hypoplastic rostral turbinate cartilages…indicating deficiency of the lateral eth-

moid capsule. Cartilages in the floor of the nasal capsule, such as lamina transversus anterior, and the paraseptal cartilages, are hypoplastic and sometimes associated with a rudimentary VNO. The dermal bones encasing the nasal capsule (nasal, premaxilla, vomer, lacrimal) are dysmorphic and small.

Neuroendocrine Anatomy of the Forebrain: Cranial Nerves 0 and 1

Introduction

The construction of the olfactory system is ancient in design and highly conserved. Through neurons that arise in specific locations within the nasal placode, chemosensory information is provided to the brain, which is vital for the acquisition of substances, assessment of the environment, and sexual behavior. Unique to the nervous system is the migration of gonadotropin-releasing hormones (GnRH) along cellular pathways from the olfactory placodes to the brain. The wiring of these substances upon completion of development from the lateralized placodes to the midline can only be explained by the existence of paired developmental fields, the medial walls of which fuse in the midline. Externally these consist of PNC skin; internally they are formed by MNC fronto-naso-orbital mesenchyme. Let's refer to these tissues collectively as the A fields. [Tissues provided by V1 stapedial neurovascular axes are knowns as the B fields. Pharyngeal arch tissues are known collectively as C fields.] This chapter presents anatomic evidence to support this model and provides an explanation for Kallman's syndrome.

Functional Anatomy of the A Fields

The A fields contain three populations of neurons, each with a separate location and function. These are: (1) the neurons of the olfactory system; (2) the neurons of the accessory olfactory system associated with complex autonomic responses to chemical stimuli such as pheromones; and (3) gonadotropin-releasing hormone neurons destined to populate the hypothalamus. Each of these populations must make contact with specific target areas of the brain. Each takes its origin from the olfactory placodes (Fig. 4.49 Nasa neuron populations, sagittal; Fig. 4.50 Nasal neuron populations cross-section; and Fig. 4.51 Neuroanatomy of the olfactory and accessory olfactory system, wiring diagram).

The first two populations are chemoreceptors that occupy spatially separate sites within the fetal nasal cavity. On the other hand, GnRH neurons actually migrate from a position near the vomeronasal, a central destination in the hypothala-

Fig. 4.49 Neuron populations of the nasal midline. General sensation: V1 anterior and posterior ethmoid nerves and arteries to septum; (2) V2 nasopalatine nerves and arteries to vomer. Olfactory system: olfactory nerves (circles). Accessory olfactory system: vomeronasal nerves (line). Pheromonal system (GnRH populate the nasal placode): Terminal nerve (dashed line). (Courtesy of Michael Carstens, MD)

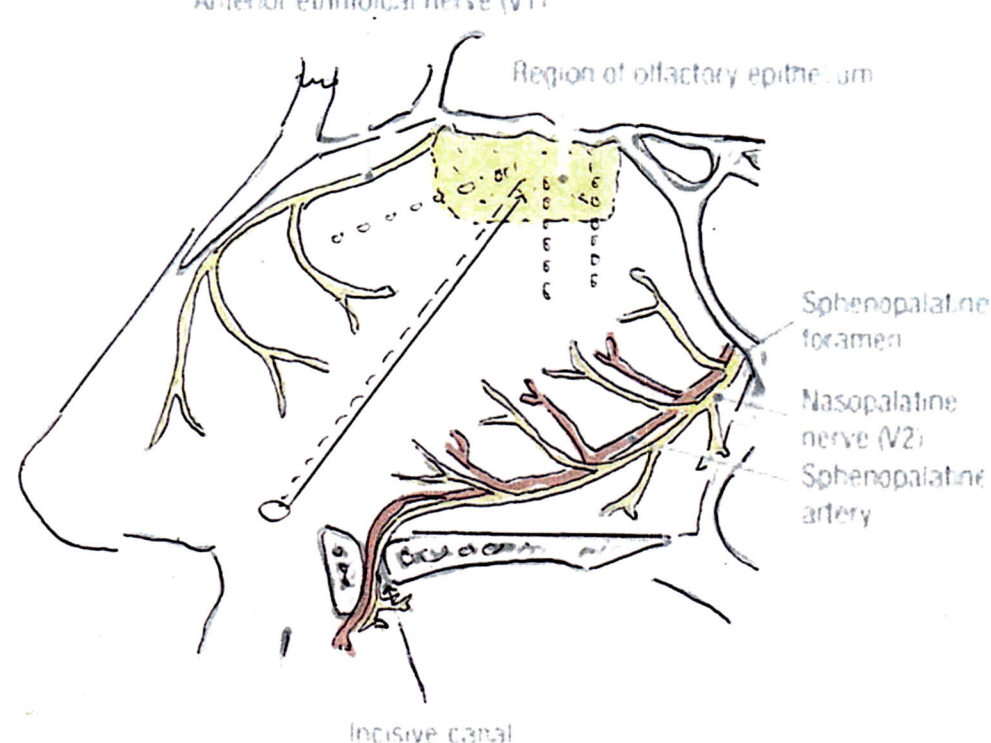

Fig. 4.50 Accessory olfactory system (AOS) arises from vomeronasal organ medial and caudal to olfactory system (OS). VNO is sometimes marked by paraseptal cartilages. Olfactory nerves (gray), vomeronasal nerves (green), terminal nerves (red). PNC p6 (pink), MNC r1 (yellow), RNC r2 (blue). (Courtesy of Michael Carstens, MD)

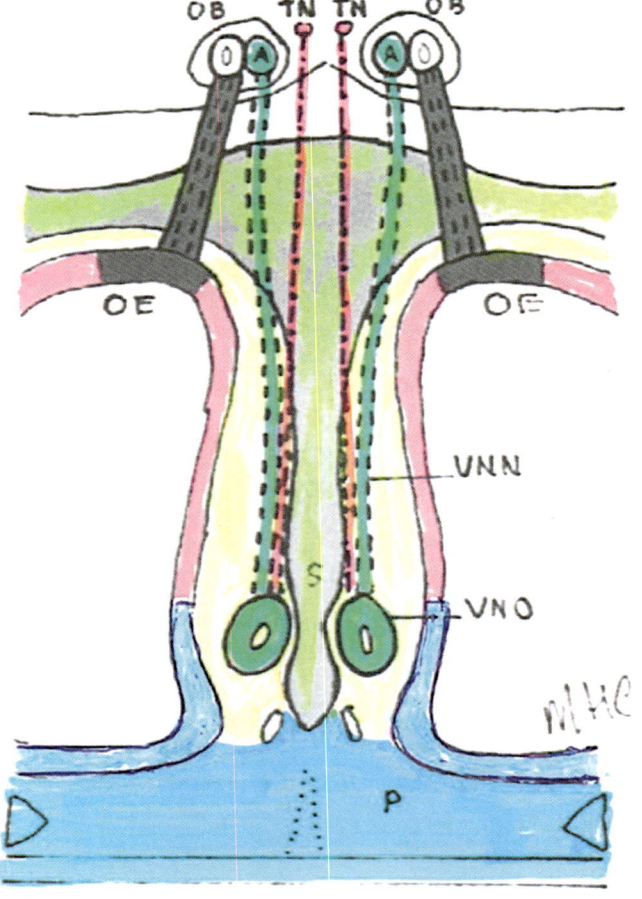

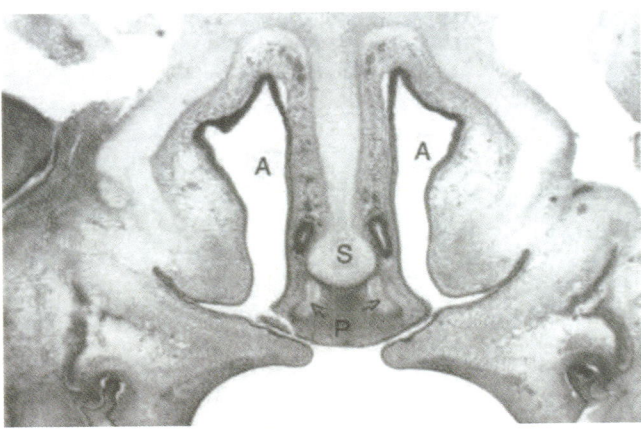

Fig. 4.51 Basic wiring of olfactory system. Neurons arise in olfactory epithelium, travel through cribriform plate, and synaps in main olfactory bulb. Olfactory nerve lies in sulcus at base of forebrain > piriform cortex (PC) > amygdala > entorhinal cortex > hippocampus (memory) > thalamus > insula > orbitofrontal cortex for final cortical integration. Unique characteristics: (1) ipsilateral wiring, (2) profound limbic overlap (emotions) Note this diagram does not show final projections from thalamus to insula and orbitofrontal cortex. (Reprinted from Mooney MP, Siegel MI, Kimes KR, Todhunter JS, Smith TD. Anterior paraseptal cartilage development in normal and cleft lip and palate human specimens. Cleft Pal Craniofac J 1994; 31(4):239–245. With permission from SAGE Publications)

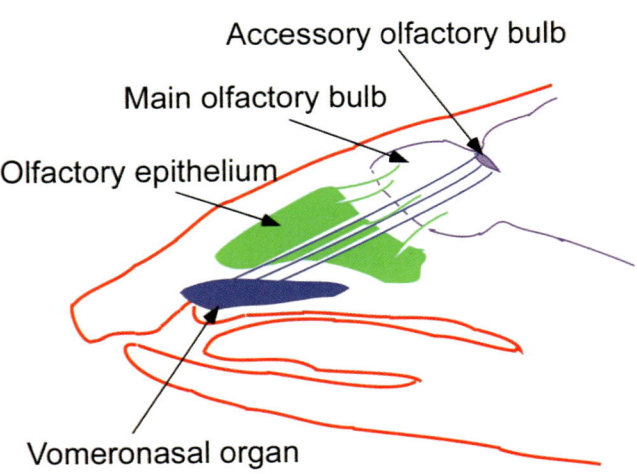

Fig. 4.52 Olfactory sensory neurons (OSNs). OSNs, red dots within the olfactory epithelium (EO), project to the olfactory bulb (OB), forming the olfactory nerve. The projection neurons from the OB send their axons (the lateral olfactory tract (ßLOT)) to the different structures of the olfactory cortex, among them the anterior olfactory nucleus (AON), the taenia tecta, the olfactory tubercule, the piriform cortex (PC), the amygdaline complex (A) and the entorhinal cortex (EC), and the nucleus of the lateral olfactory tract (nLOT; red asterisk). In both cases, the olfactory nerve and the LOT are represented with gross red curved arrows. Neurons from the olfactory cortex and the nLOT project back to the OB (gross orange curved arrows). Main intracortical connections (violet dotted arrows) are toward parts of the neocortex, hippocampus, thalamus, and the hypothalamus, as well as to the neocortex and the contralateral olfactory cortex (forming the anterior commissure, not represented). Vomeronasal chemosensory neurons arise from the vomeronasal organ (VNSO) and project to the accessory olfactory bulb (AOB). The projection neurons from this send their axons (as part of the LOT) to the AON and the nLOT (blue asterisk), and to A (gross blue curved arrows), while axons from neurons from these three cortical structures project back to the AOB (light blue dotted arrows). The projections from the accessory olfactory cortex reach the hypothalamus (turquoise dotted arrow) and the nucleus of the lateral olfactory tract. (Reprinted from Halpern M, Daniels Y, Zuri I. The role of the vomeronasal system in food preferences of the gray short-tailed opossum, Monodelphis domestica. Nutr Metab (Lond) 2005;2:6. With permission from Springer Nature)

mus. When these systems go awry, the clinical manifestations, as in Kallman's syndrome (anosmia with hypogonadic hypogonadism), are produced. This condition affords the opportunity to localize specific neuroanatomic deficits to the nasal placodes.

The developmental field model explains four principle features of this neuroanatomy. (1) Spatial relationships of the neuron populations in the placodes are conserved as the A fields undergo transformation from placodes to pits to chambers to the nasal cavities. (2) Basal midline neuroanatomy is achieved from systems originating in spatially separated embryonic placodes—this requires medial wall fusion of the nasal A fields. (3) Afferent pathways of these systems to the CNS mirror their physical location within the nasal cavity. (4) The nasal systems must grow along an afferent pathway toward the target zones in the brain for which they are wired—the MNC provides the matrix by which this neuronal migration is guided.

Anatomy of the Olfactory System

The common denominator of the special sensory systems of smell, sight, and hearing is that receptor organ neurons must be connected to the brain. Placode development is crucial to the formation of these peripheral sensory receptors. Olfaction is unique in its requirement that two populations of neurons arising from a placode itself must grow backward toward the

brain to make contact with the CNS "mother ship," the olfactory bulb. The latter possesses an intrinsic zonation for each population. As will be seen in the discussion of Kallman's syndrome (vide infra), placode-derived neurons are trophic for the olfactory bulbs [202] (Figs. 4.52 and 4.53).

When midline development is complete, the olfactory epithelium is limited in size to an area 2.5 cm² centered at the apex of the nasal roof [147]. It has some limited spread to the lateral wall above the superior turbinate and along the most cephalic aspect of the midline. Within this specialized epithelium, bipolar olfactory neurons serve two functions: chemoreception and transmission. Once this information is passed to neurons within the olfactory bulbs, it is conducted by olfactory tracts to a staging area, the *olfactory trigone*, out of which emerge medial and lateral *olfactory striae* carrying it to the cortex [203–205].

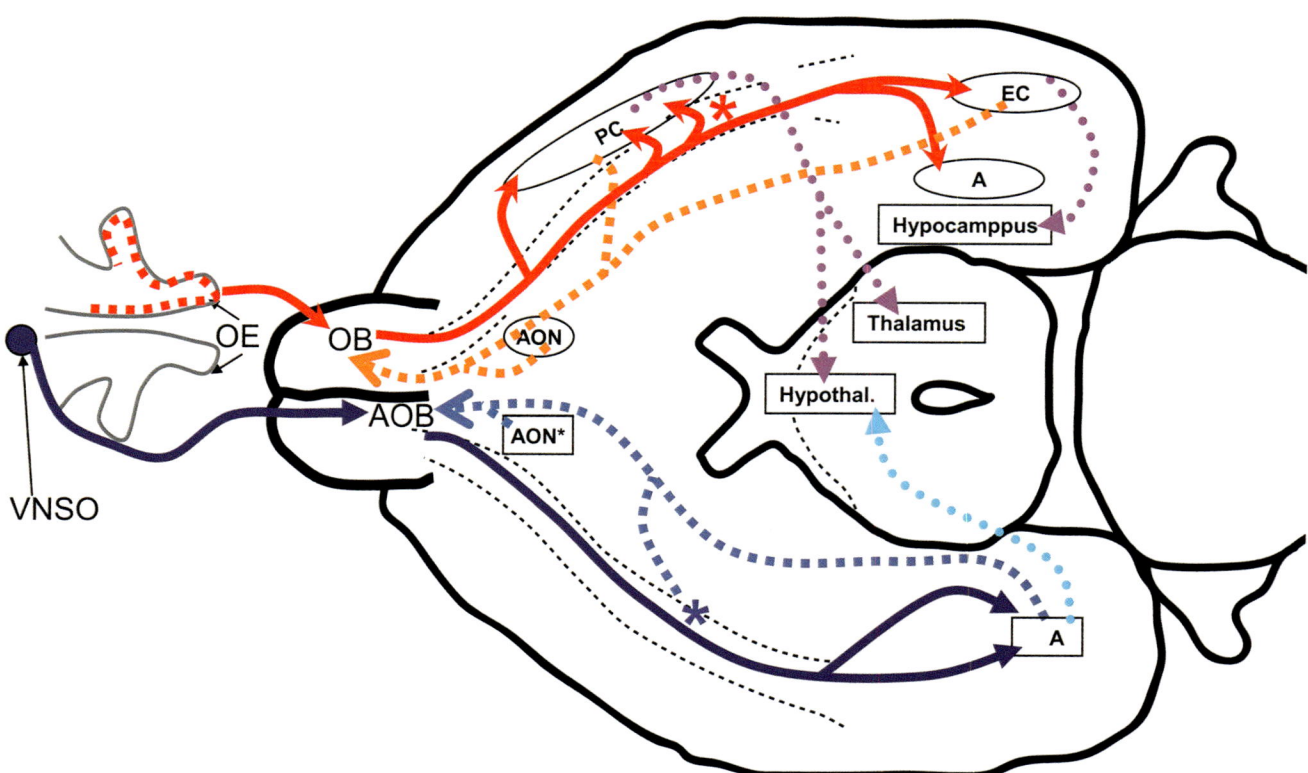

Fig. 4.53 <u>Vomeronasal nerve</u> (VNO) and <u>terminal nerve</u> (TN) travel together. VNO is chemosensory nonsexual and TN detects pheromones. TN also provides pathway for GnRH neurons to reach the hypothalamus. AOM > amygdala > hypothalamus. (Reprinted from DeCastro F. Wiring olfaction: the cellular and molecular mechanisms that guide the development of synaptic connections from the nose to the cortex. Front Neurosci 2009; 3:article 52. With permission from Frontiers Research Foundation)

Appreciation of smell takes place when afferent impulses from the lateral olfactory stria reach the pear-shaped olfactory area. This includes the limen insulae (threshold of the insula), the uncus (hook) of the temporal lobe, and the amygdaloidal (almond-shaped) nucleus. The latter structure is actually a nuclear complex with a dual nature. Its dorsomedial aspect (related to the lateral olfactory area) deals with smell while ventromedial aspect relates to the limbic system. Awareness of odors occurs in the three-layer *uncal lateral olfactory area*. This is passed to the adjacent six-layer parahippocampal gyrus where associations are made. The entorhinal area connects widely to the neocortex, thus integrating smell with other sensory information [206, 207].

Anatomy of the Accessory Olfactory System

A second afferent pathway exists within the olfactory tracts for impulses arising in an anatomically separate region of the septum, the *vomeronasal organ* (VNO). First described in 1777 by Ruysch [208] and later popularized by Jacobson [209], these paired structures exist just posterior and cephalad to the incisive foramen. They are present in 100% of subjects [210, 211]. (Postmortem shrinkage in cadavers and cleft-induced deformations of the septum leading to coronal section artifacts explain variations in VNO incidence previously reported [212].) Each VNO opens into the lumen nasi at its posterior-superior margin. The ultrastructure of the VNO demonstrates a cellular gradient toward the lumen with the most active cells located toward the surface (Figs. 4.54, 4.55, and 4.56 Anatomy of AOS).

Each VNO conveys sensory data derived from pheromones to the CNS via the vomeronasal nerve (VNN) and the posterior branch of the nervus terminalis (NT). Terminal nerve is described in greater detail below. Taking their origin from the septum rather that from the olfactory epithelium, these nerves are spatially medial to the olfactory fibers [144, 213]. When their fibers are traced through the cribriform plate, they are encountered medial to the filia olfactoria. Not surprisingly, accessory olfactory fibers occupy a medial position within the olfactory bulb as well.

At the olfactory trigone, afferent data from the septum are conveyed via the medial olfactory stria to the medial olfactory area. Each area is located along the medial aspect of the frontal lobe just in front of the lamina terminalis and immediately beneath the genu of corpus callosum. Groups of cells here (called septal nuclei due to their proximity to the septum pellucidum) are wired to (1) the main olfactory areas via

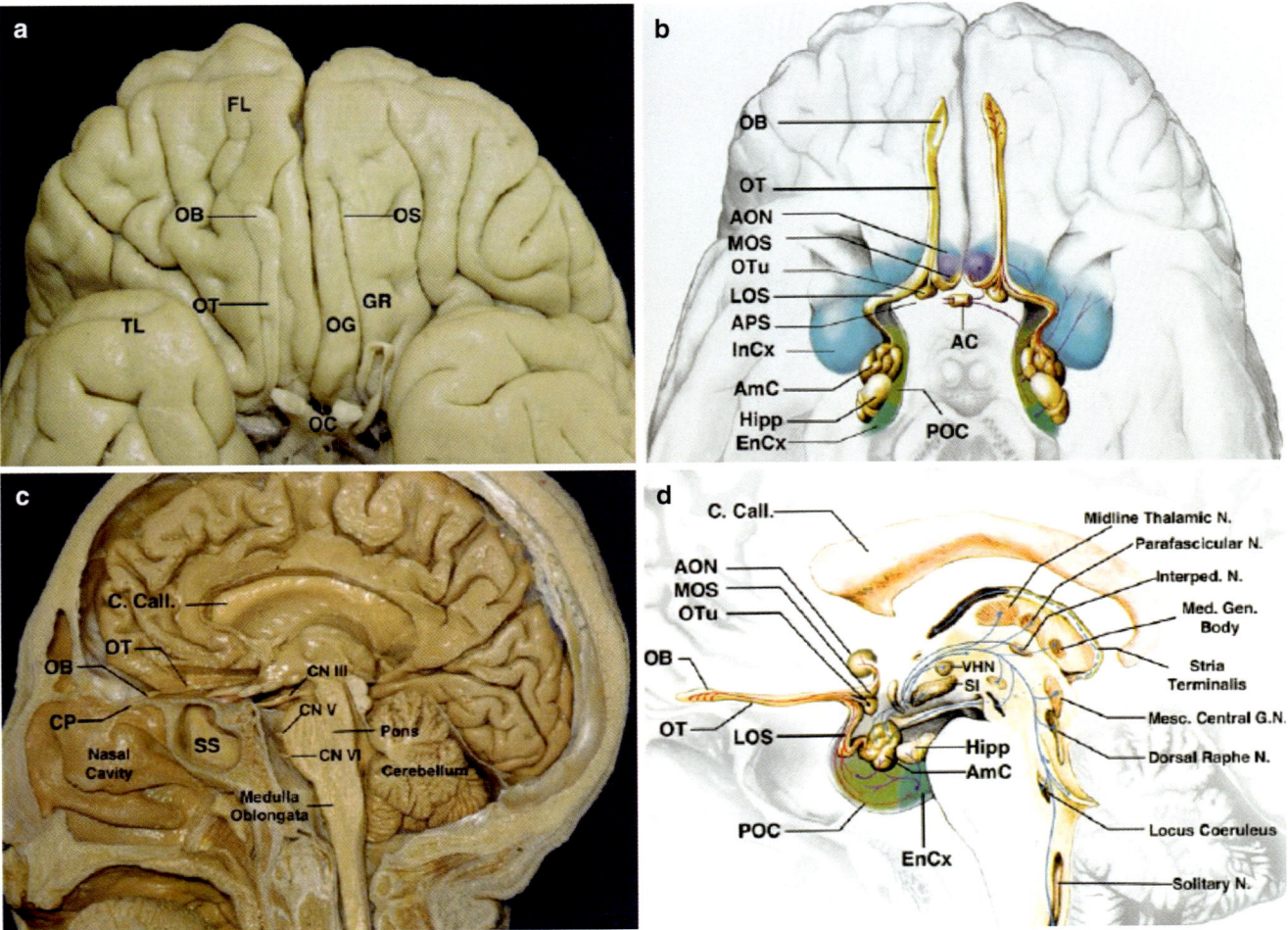

Fig. 4.54 Paraseptal cartilages (P) on either side of the vomer make the site of the vomeronasal organs, seen just lateral to septum (S). (Reprinted from Mooney MP, Siegel MI, Kimes KR, Todhunter JS, Smith TD. Anterior paraseptal cartilage development in normal and cleft lip and palate human specimens. *Cleft Pal Craniofac J* 1994; 31(4):239–245. With permission from SAGE Publications)

the *diagonal band of Broca* and (2) the limbic system. The contribution of the septal nuclei to the sense of smell proper is minor, being likely unconscious. The principle role of AOS input via the septal area is the emotional function of the limbic system.

The medial olfactory area is not primarily concerned with the identification or processing of odors; instead it is wired to the limbic system. The principle areas of this complex, poorly understood system are: (1) the cingulate and parahippocampal gyri collectively known as the limbic lobe; (2) the hippocampus; (3) the amydaloid nucleus;(4) the mammillary bodies of the hypothalamus; and (5) the anterior nucleus of the thalamus. Key functions of the limbic system with reference to olfaction include emotional response to odors, visceral responses, such as nausea, and associations with memory. Seizure activity of the limbic system "the uncinate fit" is characterized by the triad of a panic state, disagreeable odor/taste, and primitive motor acts such as smelling or licking [203].

The relationship of the AOS to sexual function offers insight into the complexities of the limbic system. Direct stimulation of the medial olfactory tract can lead to sperm release [214]. Reproductive behavior triggers by the AOS (such as investigation of urinary or anogenital scent, maternal-infant recognition) have been extensively documented in animal models [214].

In humans, mass receptor potentials from the VNO have been recorded using the electrovomerogram (EVG) [215]. Locally delivered vomeropherins act at the VNO epithelium-nasal mucosa chemoreceptors. Olfactory epithelium and trigeminal nerve are not affected. Topical application of lidocaine or atropine does not block EVG potentials. Responses are determined by the molecular structure of the vomeropherin and the gender of the subject. Males respond more to estratetrenol than females; whereas the female EVG is more active in the presence of androstendione [215]. Behavior changes in randomized, double-blind female subjects when administered topical adrostenedione versus pla-

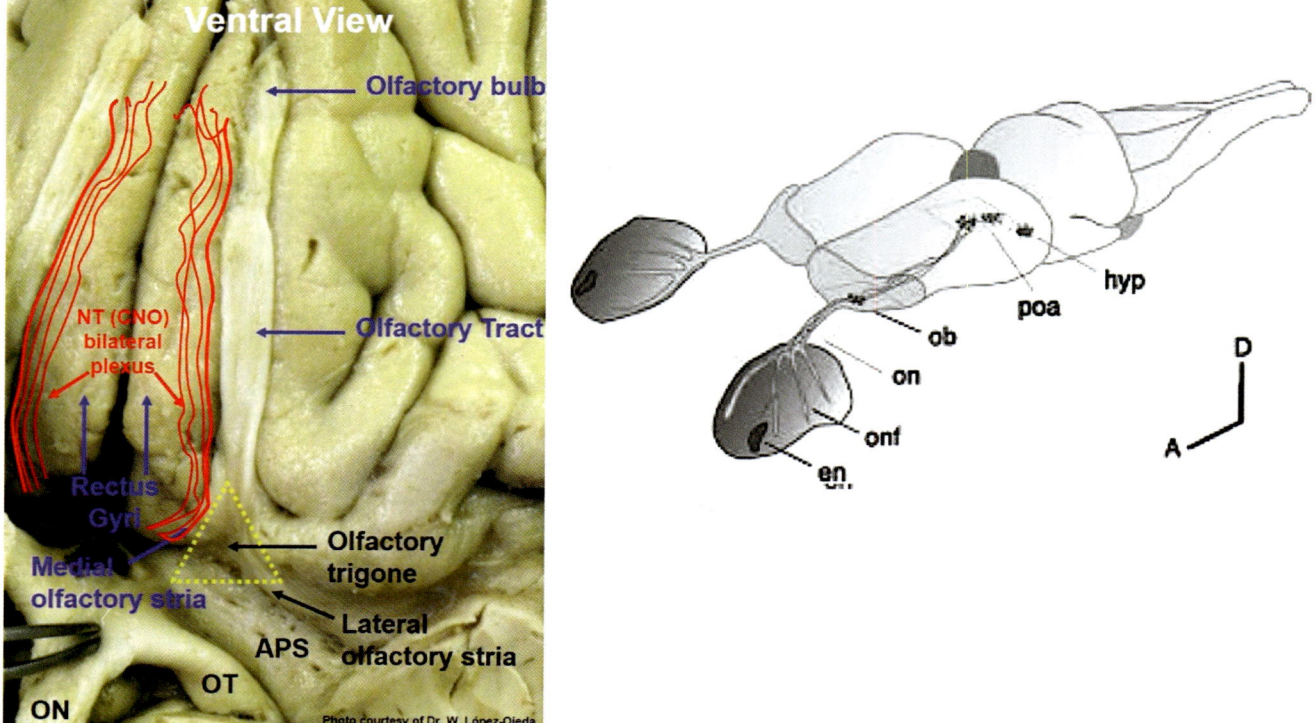

Fig. 4.55 Cranial nerve 0 is also known as nervus terminalis, because of its initial termination in the lamina terminais. The fibers are unmyleinated and intermingle with those of vomeronasal nerve. They pass through cribriform plate media and dorsal l to olfactory nerve fibers. They are a TN that travels in the subpial space over the straight gyrus. Terminate in the septal and preoptic nuclei from there, they have direct access to the limbic system. TN is associated with sexual behavior neurons. TN considered, in contradistinction to accessory olfactory system, as the primary chemosensory system responding to sexual pheromones, with AOS serving a nonsexual chemosensory function. TN serves as a conduit for GnRH cells to migrate to the basal forebrain. This is consistent with LHRH immunoreactivity not seen in either OS or AOS. (Left: Reprinted from Sonne J, Reddy V, López-Ojeda W. Neuroanatomy: Cranial nerve 0 (Terminal Nerve) Stat Pearls (Internet). Treasure Island, FL: STatPearls NCBI Bookshelf, National Library of Medicine, National Institutes of Health; 2021. With permission from Creative Commons License 4.0: https://creativecommons.org/licenses/by/4.0/. Right: Reprinted from Mousley A, Polese G, Marks NJ, Eisthen HL. Terminal Nerve-Derived Neuropeptide Y Modulates Physiological Responses in the Olfactory Epithelium of Hungry Axolotls (*Ambystoma mexicanum*). J Neurosci 2006; 26(29): 7707–7717. https://www.jneurosci.org/content/26/29/7707. Copyright © 2006 Society for Neuroscience)

Fig. 4.56 GnRh neurons are neural crest cells that populate medial nasal placode. They are likely of MNC derivation due to associated secondary findings in Kallman's syndrome such as eye movement disorders. (Reprinted from Sykiotis GP, Pitteloud N, Seminara SB, et al. Deciphering Genetic Disease in the Genomic Era: The Model of GnRH Deficiency. Science Translational Medicine 2010; 2(32): 32rv2. With permission from John Wiley & Sons)

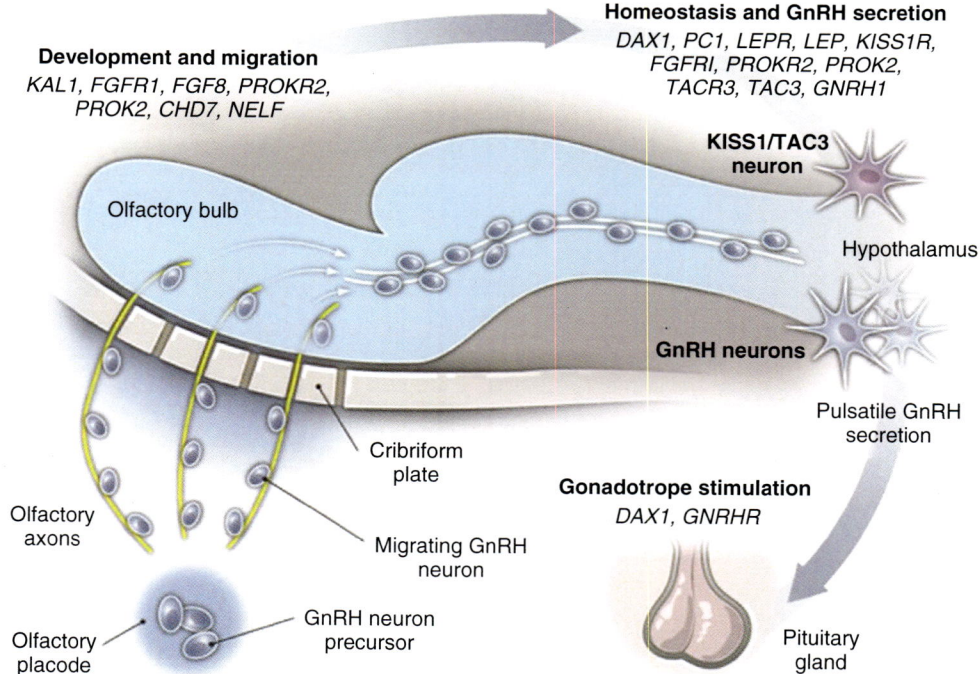

cebo included relaxation and a state of contentment [216]. Picogram quantities of vomeropherins elicit autonomic effects such as bradycardia, parasympathetic-mediated sinus arrhythmia, and increased core temperature [217–219].

Anatomy of Terminal and Vomeronasal Nerves

Each of the midline A field walls contains two neural systems. General sensation from, and autonomic control for, the septal mucosa is supplied by V1 via the anterior and posterior ethmoid nerves. A second system is chemosensory and endocrine in nature and is based upon two distinct nerves: the *terminal nerve* (TN) and the *vomeronasal nerve* (VNN). These define, respectively, internal and external laminae within each A field. The spatial relationships between these two chemosensory systems—preserved all the way from the origins in the septum to their destinations in the brain—are the key to understanding midline pathologies such as Kallman's syndrome and Tessier craniofacial cleft zones #0, #1, and #2 (Fig. 4.55).

The terminal nerve was first described by Fritch in 1878 and by Pinkus in 1895 [220, 221]. The independence of this nerve from the olfactory nerve is well-seen in the porpoise, which lacks olfactory bulbs and pedicle. In 1914, Johnston [222] distinguished NT from VNN and noted that the ganglionated TN has roots that entered telencephalon medial to the olfactory bulb—i.e., it was neuroanatomically distinct. Terminal nerve runs along the cephalic margin of the septum; it has two main branches. The anterior branch (TNa) serves the mucosa forward from the VNO. It anastomoses with the internal nasal branch of V1 *anterior ethmoid nerve*. The posterior branch (TNp) originates at VNO and runs a cephalad course to join with TNa.

Vomeronasal nerve pursues a more oblique course toward the cribriform plate [223]. Its several branches are confluent with the terminal nerve at about the midportion of the septum. Here VN and TN are also confluent with branches of V1 *posterior ethmoid nerve*. The sensory supply of all A fields is V1 and the source of mesenchyme is MNC—with programming provided by p6 vestibular epithelium. Roughly in parallel, below the septum, the vomer, constructed from r2 neural crest, is innervated by StV2 medial nasopalatine.

On cross-section, the sensory and chemoreceptor fibers are layered. Vomeronasal nerve runs an external, superficial (subepithelial) course to terminal nerve. VNN fibers are internal to those from the olfactory epithelium. These relationships are maintained at the cribriform plate. Filaments of the TN from the septum coalesce to enter cribriform plate as a single amalgamated structure anterior to the vomeronasal nerve(s). Once access is gained into the cranium, the fibers of TN and VNN diverge. The separated fibers of TN lie at some distance dorsal to the cribriform plate of the ethmoid bone,

instead of lying directly on its upper surface as the fila olfatoria. The height to which the nerve attains on the lateral surface of the crista galli depends apparently upon the development of the crista galli [224].

Between the olfactory bulb (OB) and the olfactory tract (OT), TN fibers again merge. On the medial surface of olfactory bulb, a close plexus of fibers is intimately associated with the fila olfatoria. In its course over the medial surface of the olfactory tract, the nerve forms a compact bundle of nerve fibers [224]. Within the cranium, layering continues. The olfactory bulbs contain AOS fibers medially and OS fibers laterally, while the terminal nerve runs a midline course outside of and rostral to the olfactory bulb. This runs medially to the olfactory tract along the orbital surface of the frontal lobe (gyrus rectus). The distribution of TN to the brain is likewise independent from the olfactory striae, following the anterior perforated substance in an extremely midline position. TN fibers enter the brain by curving round the convexity of the gyrus rectus just rostral to the optic chiasma.

The nervus terminalis is the most medial neural structure along the base of the anterior forebrain. In monkey and in humans, TN fibers have been demonstrated in a relation medial and inferior to the anterior cerebral artery. In 14 adult brains, McCotter found the nerves were either within or very near the median fissure and embedded in pia [225]. For this reason, the *terminal nerve is considered by some as cranial nerve 0*.

Defining the central connections of the TN proved an elusive task for early anatomists. Greater understanding of its neurophysiology would have to await advances in ultrastructural imaging and immunocytochemistry. Recent work with these modalities indicates that the terminal nerve may have both neurosecretory and hormonal functions. Indeed, the TN may play a significant role in the coordination of gonadal development and reproductive behaviors.

EM studies of TN fibers in both stingrays and mice demonstrate them to be unmyelinated fibers containing particles consistent with the storage of neuropeptides [225, 226]. That TN fibers and ganglia adjacent to blood vessels display dense-cored vesicles (DCVs) and support the idea that TN unloads DCV contents into the circulation [227, 228]. Immunoreactivity to substance P has been in both the TN plexus of the septum and following the TN location along the olfactory nerves of rats [229]. Movement of horseradish peroxidase (HRP tracers from the nasal epithelium to the brain) suggested that TN fibers project to septal and preoptic nuclei [230]. More definitive information would come from the mapping of pathways by which labeled GnRH neurons read the central nervous system.

In summation, terminal nerve has the following functional components: (1) general sensory fibers—in conjunction with the anterior ethmoid nerve—to the anterior septal mucosa; (2) vasomotor fibers providing autonomic control of mucosal blood vessels; (3) secretomotor fibers to Bowman's glands; (4) vascular-associated neurosecretory fibers; (5) neuromodulatory fibers carrying peptides to diffuse sites in the brain—these link gonadal development with reproductive behavior [231].

Anatomy of GnRH Neurons: From Nasal Placode to Hypothalamus

The hypothalamus contains gonadotropin-releasing hormone (GnRH) neurons (also called LHRH—leuteinizing hormone—release hormone); these have a unique peripheral origin. Multiple animal experiments have shown that GnRH neurons originate from the nasal placodes [232–237], specifically, from their medial aspect [238–241]. Migration to the CNS takes place rapidly in the septum via the vomeronasal and terminal nerves. The three-dimensional anatomy of this transformation provides compelling evidence for the existence of paired A fields, the medial walls of which approximate in the midline to form the functional matrix within which the septum lateral develops (Fig. 4.56 Anatomy of GnRH).

Histologic evidence of the effects of olfactory placode ablation was described in rats by Daikoko [242]. Care was taken to operate on day 15.5 gestation so as to predate olfactory nerve ingrowth into the brains. Six days later, the ipsilateral olfactory bulb had failed to develop. The telencephalon was smaller; reductions were noted in the primary olfactory cortex, cortical plate, and hippocampus. Ipsilateral striatum and substantia nigra had diminished substance P and tyrosine hydroxylase fibers. The hypothalamus, in contrast, had a normal complement of TH fibers. Neurons from the olfactory placode are essential for proper forebrain development.

Breakthroughs in tissue preparation and more sensitive antibodies to GnRH have permitted better definition of its location outside and within the central nervous system. GnRH immunoreactivity was first demonstrated in the guinea pig by Schwanzel-Fukuda and Silverman [243]. Other vertebrate studies showed reactivity peripherally in nasal epithelium and olfactory bulbs and centrally in the septal and preoptic nuclei [244]. In mammals, GnRH neurons have been located within the diagonal band of Broca, the stria terminalis bed nucleus, the supraoptic nucleus and the hypothalamic anterior, lateral, and basal nuclei. The consistency of these findings may have evolutionary significance. The data reveal a remarkable similarity across species in the distribution of GnRH neurons, suggesting that, like the peptides themselves, GnRH neuronal structures are phylogenetically ancient, conserved structures [230].

The time sequence of neuronal migration was studied by Kjaer [244] using immunohistochemical labeling of GnRH in 49 fetuses from 15 to 146 mm CRL. Five distinct stages of nasal cavity development were defined. These are consistent in every detail with the A field model.

NAS I: Carnegie stage 18 (44–45 days). Controlled apoptosis leading to midline fusion of the A fields is not yet complete. Nerve fibers are seen running from the mucosa of the medial wall of each nasal chamber to the brain—the chambers having traversed the MNC in the process. No r1 septal development is present—this requires medial wall fusion. Vomeronasal organ and GnRH label are absent.

NAS II: Carnegie stage 20 (51 days) and continuing to 11 weeks. The secondary nasal floor ("zweiterer Nasenboden"), i.e., the r2 palatal shelves, has not elevated from the B fields. The processes are shown hanging vertically along the sides of the tongue. In the midline, fusion of the medial walls is taking place and septal cartilage formation is seen. Vomeronasal organs present bilaterally and connected to the nasal mucosa. GnRH-labeled nerve tissue is seen along the cephalic aspect of the VNO.

NAS III: 10–12 weeks. The palatal shelves have fused. In keeping with the caudal spread of maturation, paraseptal cartilages flank a fully developed septum. No ossification of the palate is seen.

NAS IV: 11–16 weeks. Ossification of the palate is present. Septum is growing. Vomeronasal organs and GvRH label are diminished. GnRH is seen at the level of the cribriform plate. Remnants of the paraseptal cartilages are present and have given way to bilateral vomer precursors.

NAS V: 17–19 weeks. Fusion of vomerine elements beneath the septum gives rise to a single vomer. No GnRH label.

Because 19-week fetak brains display GnRH with the label present as well in terminal nerve ganglia of the nasal cavity, a clear-cut migration has taken place. From its origin in the medial nasal placode, GnRH neurons localize to the vomeronasal organ and then find their way to the hypothalamus via the pathways of the accessory olfactory (vomeronasal) nerve and terminal nerve [234]. This time sequence has been confirmed by Kim [245, 246]. (A cautionary note: Kjaer's conclusion—derived from fixed tissues on coronal section—that the vomeronasal organ is transient at odds with the near 100% incidence in living adult subjects).

How Do Neurons Find Their Way Home? N-CAMs and A-Field Domains

Certain chemicals are important in development based upon their ability to guide the growth of cells. These have been classified into three general categories. (1) Cell adhesion molecules (CAMs) mediate contacts between like and non-like cells. (2) Substrate adhesion molecules SAMs create a tissue environment for cellular movement and attachment.

(3) Junctional adhesion molecules (JAMs) permit mechanical, electrical, or chemical signaling between cells to occur.

Cloning and sequencing of/cams have defined the members of large gene families [247]. N-CAM (neural cell adhesion molecule) is part of an immunoglobulin superfamily including T cell receptors and HLA antigens, growth factor receptors, and neutrophil receptors. Another CAM family is that of membrane glycoproteins known as cadherins [248].

Cells vary in their cadherin specificity—hence within a substrate containing cadherins, the cells can sort themselves out. The SAM family includes familiar substances such as collagen, laminin, and fibronectin belonging to the integrin family. Tenascin (cytotaxin) is a hexameric protein found in the extracellular matrix with domains homologous with fibrinogen, fibronectin II, and epidermal growth factor [249, 250].

It is not surprising that NCAM and tenascin demonstrate diverse effects including: neural aggregation, cell proliferation, cell migration, aggregation, outgrowth of neurites, and regeneration [251]. But NCAM plays a vital role at induction sites outside the nervous system as well. All germ layers are involved, including ectoderm (lens and otic placodes) [251], mesoderm (kidney), and the endodermal buds of the lung and liver.

Abundant evidence suggests the important role of NCAMs. For example, NCAM content in neural crest cells decreases as they move away from their origin and increases as they aggregate to form ganglia. In the cerebellum, glial-cell mediation of granule cell migration requires NCAM for binding of the neuron to the glial cells. Tesascin is involved in the subsequent migration of the neuron along the glial cells to reach its destination. Chick retinotectal projection requires NCAMs [252]. NCAMs occur along the GnRH migration route to the brain and are undoubtedly involved in the process. Extensive work by Schwanzel-Fukuda at the Rockefeller Institute was carried out using a mouse model in which GnRH and NCSM were selectively labeled [62, 253]. At all stages, no "double lable" was detected, meaning that the labeled cells were distinct. In this murine model, antibodies to NCAM block GnRH neuron migration [254]. Given the genetic conservatism displayed by the GnRH system, it is not surprising that similar results have been obtained in rats, primates, and humans [255–257]. The detailed analysis of the mouse data which follows below offers a *specific model for the development and function of A fields.*

At day 9, no immunoreactivity for GnRH is present whatsoever. NCAM is not seen on the placode but Schwanzel-Fukuda made a critical observation: The most rostral localization of NCAM immunoreactivity at this age was found in the anlage of the diencephalon, just caudal to the closure site of the anterior neuropore [253]. The significance of this finding will be discussed later. At day 10, placodal invagination occurs, forming nasal pits. A few hours later, at day 10.5 a small number of NCAM immunoreactive cells are seen between the thickened epithelium of the newly formed nasal pit and the mesenchyme of the developing nasal walls. Traced through serial sections by late day 10, the aggregate of NCAM+ cells and olfactory nerve axons are seen in contact with the rostral tip of the forebrain, forming a thin cap along its ventromedial surface. No GnRH immunoreactivity was seen in the epithelium of the olfactory pit or in the brain at this age [253].

Day 11 in the mouse embryo marks the advent of the vomeronasal organ anlagen. These appear as invaginations in the medial wall of each nasal pit. NCAM is seen at the medial aspect of the VNO. Actual cords of NCAM are seen streaming from the medial pit epithelium into the nasal mesenchyme and thence into the brain. These become NCAM+ vomeronasal and terminal nerves…the migratory pathways for GnRH. The NCAM immunoreactive cells and axons from both the dorsal and medial parts of the olfactory pit appear to form a scaffolding, anchoring the rostral tip of the forebrain to the epithelium of the olfactory pit [253].

GnRH is first seen in the medial olfactory pit at the VNO anlage on day 11, and by day 11.5, GnRH cells have moved into the medial nasal wall along the septum; they then, within hours, follow the course previously laid down by the NCAM+ vomeronasal and terminal nerves. NCAM cells outnumber GnRH 50:1 at this stage. In all sections, GnRH cells are never observed outside the confines of the NCAM pathways.

At days 12–13, label is seen entering the brain. GnRH neurons were observed in thick cores on the nasal septum in association with NCAM-immunoreactive branches of the vomeronasal and terminal nerves, and in the medial part of the NCAM cellular aggregate below the rostral forebrain. At the point where the medial part of the aggregate was in contact with the forebrain, the basement membrane of the ventromedial surface of the brain appeared to have been breached; cords of GnRH neurons coursed through it into brain. Many blood vessels were present in this region [253].

In the 14 days-old mouse embryo, the rostral end of the forebrain is forming an olfactory bulb; the olfactory nerves are organizing as well, medial and caudal to the olfactory bulb. GnRH cells now outnumber NCAM cells by 40:1. The largest proportion of GnRH-labeled cells is now in the area between the nose and forebrain. Antiserum to olfactory marker protein is now reactive for the first time, permitting separation of the olfactory nerves from the vomeronasal and terminal nerves. The entire pathway can be visualized by day 16. At the age, the GnRH neurons form an arch, from their site of entry into the ventromedial forebrain, through the septal and preoptic areas, to the hypothalamus. This pattern of GnRH cell migration is very orderly and consistent from animal to animal [253].

The pathway for olfactory neurons is as follows. Primary olfactory axons are amassed outside the brain in company with heparin sulfate proteoglycan (HSPG) and laminin. They then enter the brain through gaps in the basement membrane at day E14.5 [257]. Immunofluorescence mapping shows that laminin and HSPH both accompany the olfactory axons into the telencephalon. By E16.5, regional basement membrane degradation is complete and direct continuity exists between the CNS and the olfactory pathway [257]. In situ DNS hybridization for neuron-specific markers GAP-43 and class II beta tubulin shows them to appear on immunocytochemistry in the medial forebrain in anticipation of olfactory neuron entry. This suggests that although peripherao olfactory tract input is trophic, specification of intracerebral olfactory tracts occurs as an independent process within the brain [258].

The role of the *Otx-2* gene in mice is intimately associated with the CNS midline. Chimeric mice that are *Otx−/−* fail to initiate regulatory genes such as *Engrailed* in the midline of the neural plate [259]. Failure of forebrain specification ensues. The boundaries for the olfactory tracts are established in the brain by transcription factors such as OTX2. This protein is first expressed in the pre-streak embryo, but by day 8 in the mouse it is confined to the neuroectoderm of the future forebrain. Early dedication of commitment for forebrain tissue has been also shown in the zebrafish model to occur at the shield stage of gastrulation [260]. *Otx2* mutants demonstrate forebrain delation [128]. Immunoblot studies demonstrate that OTX2 products cause upregulation of important cell adhesion molecules such as tenascin-C, NCAM, and DSD-1-PG distributed along the future target zones for the olfactory neurons [261]. Placodal expression of Otx2 may well set up the CAMs along the A fields which will eventually help guide the placodal neurons into the CNS [262].

It will be noted that the molecular environment accompanying the olfactory neurons varies with location. The neurons respond by changes in axon trajectory and the morphology of their growth cones [263]. Before synapsing with mitral cells within the olfactory bulb, axons projecting from the nasal neuroepithelium are ensheathed in cells positive for heparin-binding growth factor FGF-1 [264]. Mitral cells express FGF-1 late in embryogenesis in anticipation of neuroepithelial ingrowth. On laminated portions of the brain (such as olfactory cortex), extracellular proteins such as reelin are expressed strongly in development. But once tract formation is completed, the reelin transcripts disappear [265]. Some systems are expressed in both a spatial and temporal manner. Antibodies to neuronal intermediate filament proteins show that peripherin and alph-internexin are coexpressed in both the olfactory bulbs early in development. Later in time, they separate out, with peripherin being expressed exclusively in the olfactory nerves [266].

On the other hand, signaling molecules such as the semaphorins can alter growth cone behavior. In the presence of soluble collaspin-1, primary olfactory growth cones are halted [267]. Hence, the cell adhesion molecules and the composition of the extracellular matrix determine the assembly and projection of the olfactory neurons, possibly through GABA receptors in the growth cones [268], both outside the cranium and within.

Kallmann's Syndrome (KS): Neuropathology of the A Fields

Although the syndrome of hypogonadism and anosmia that bears his name was described in three families by Kallmann in 1943, credit for its discovery goes to the Spanish pathologist, Maestre de San Juan, who described autopsy findings as "total lack of olfactory nerves with anosmia in an individual in whom existed congenital atrophy of the testicles and the virle member" [269]. The clinical features of Kallman's syndrome are reviewed by Lieblich and others [150, 270, 271]. Because rhino-olfactory and GnRH neurons originate on the respective lateral and medial regions of the olfactory placode, this syndrome permits clinic-pathologic correlation with the embryology [271, 272]. Patients with an intact sense of smell and hypogonadism have only the medial half of the placode involved, whereas those with anosmia have affectation of the lateral placode as well [272] (Figs. 4.57 and 4.58 Kallman).

Kjaer [141] used immunohistochemical labels to separate out the neuronal pathways. Avidin biotin complex/horseradish peroxidase (ABC/HRP) was used to trace GnRH neurons, while olfactory neurons were labeled with protein gene product 9.5 (PGP 9.5). The study confirmed the separateness of the nerve fiber bundles in the young fetus. Since the olfactory nerves are trophic for the olfactory bulb and since MRI scans of patients with Kallmann's demonstrate complete absence of olfactory bulbs and tracts [273, 274], it is reasonable to conclude that such patients have involvement of their lateral nasal placodes.

The limitation of this study to the clinical situation is its failure to separate out the possible roles of terminal nerve versus accessory olfactory pathways. It is interesting to note that certain patients with KS have relatives with anosmia, but not hypothalamic hypogonadism, apparently due to normal development of the medial part of olfactory bulb [275]. Such patients may well have intact terminal nerves capable of linking GnRH to the NCAM scaffold.

The Kallmann defect has been considered by some to be an arrested migration of olfactory axons and GnRH neurons. These are found dead-ended in the meninges, just above the cribriform plate [202]. If NCAM is a product of the A field mesenchyme, i.e., of MNC, then failure of migration could certainly result from a global aplasia of the field or a specific defect in the chemical production of NCAM. The genetic

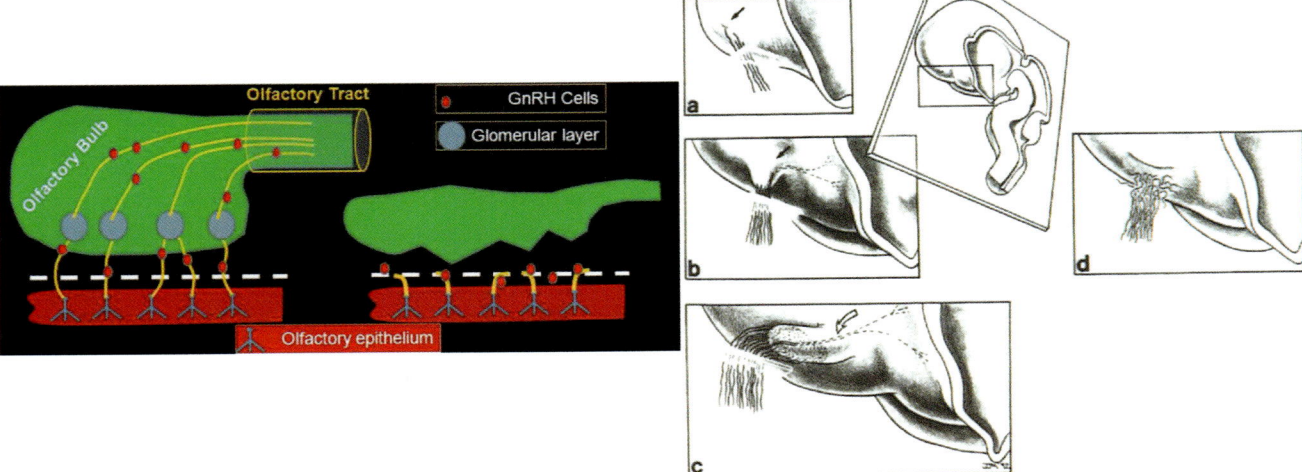

Fig. 4.57 Kallman's syndrome. GnRH cells (neural crest) populate medial olfactory placode. GnRH migrate along vomeronasal axons and gain access to the olfactory bulb. Synapse is in the outermost layer, the glomerular layer, and then travel via olfactory tract to the hypothalamus. Migration block leads to Kallman's syndrome (anosmia and hypogonadism). (Reprinted from Truwit CL, Barkovich AJ, Grumbach MM, Martini JJ. MRI imaging of Kallmann's syndrome: a genetic disorder of neuronal migration affecting the olfactory and genital systems. AJNR 1993; 14:827–838. With permission from American Society of Neuroradiology)

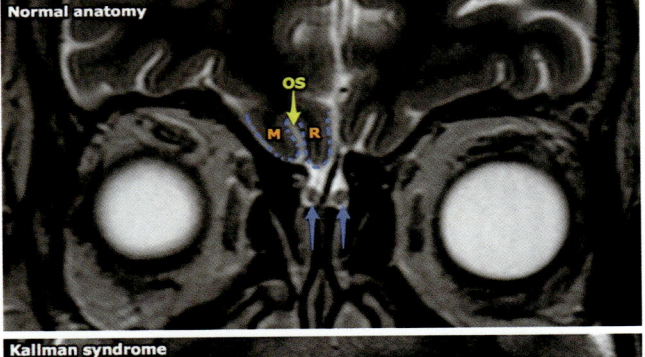

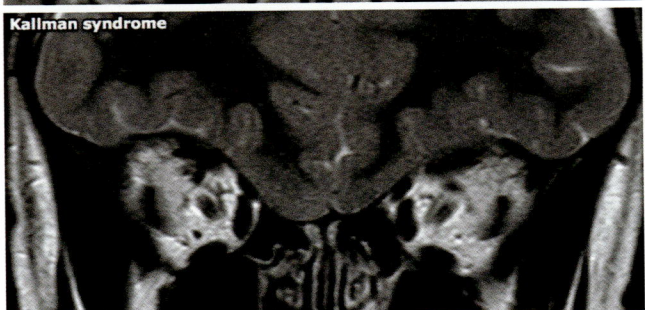

Fig. 4.58 The normal anatomy of the region consists of the olfactory bulbs (blue arrows) located in the olfactory grooves of the anterior cranial fossa. The inferior surface of the frontal lobes usually consists gyrus rectus (aka straight gyrus) (R) separated from the medial orbital gyrus (M) by the olfactory sulcus (yellow arrow). These are absent in Kallman syndrome. Gyrus rectus and medial orbital gyrus are fused. (Case courtesy of Assoc. Prof. Frank Gaillard, Radiopaedia.org, rID: 6083)

basis for the X-linked form of Kallmann syndrome (KAL) has been described and cloned [275]. KAL is a glycosylated membrane protein shown to produce a diffusible component by proteolytic cleavage. Gene expression of KAL has been found in the olfactory bulb, independent of correct innervation by correct innervation by olfactory axons. KAL shares sequence homology with serine proteases of the kind associated with nerve growth cones. Such proteases enable growing nerves to migrate through the extracellular matrix [276]. The KAL molecule may be a potential chemoattractant for olfactory axons and GnRH neurons. This mechanism for olfactory development is highly conserved in vertebrates. The function of GnRH is so important that its absence would not be compatible with species propagation. Along this same line of observation, no mutations of other hypothalamic hormones (CRC, GHRH, TRH) have been described. KAL protein appears to provide a scaffold for GnRH neurons and olfactory nerves to migrate from the olfactory mucosa across the cribriform plate to synapse with mitral cells of the olfactory bulb [277].

Comparison of Kallmann's syndrome with holoprosencephaly is very useful in understanding the function of the A fields. In both conditions, olfactory bulbs and olfactory tracts are missing or hypoplastic. OB and OT development depends upon the trophic influence of fila olfactoria from the nasal cavity. Failure of first cranial nerve development will occur in any condition in which the peripheral tracts fail to penetrate the cribriform plate. KS patients have fully formed noses and no midline deficiencies, such as clefts. In KS failure of peripheral neuron, induction of the OB/OT complex results in either (1) faulty production within the olfactory placode itself or (2) failure of those neurons to reach their destination. In the latter case, cell adhesion molecules along intact A-field "pathways could prove to be the problem."

Outright nasal A-field aplastia (not seen in KS) is intrinsic to the pathology of holoprosencephaly (HPE). Olfactory bulb and tracts are absent in this condition. So too are the crista galli and cribriform plate, both of which are derivatives of r1 MNC. A vestigial hemi-columella and unilateral cleft lip can rarely occur (as in absence of a single A field complex). The more usual state for HPE involves loss of all midline structures, including ethmoid plate, septum, columella, vomer, and premaxilla. In other words, a mixture of midline StV1 and StV2 derivatives from medial nasopalatine axis. Cleft palate results from failure of B field shelves (greater palatine angiosome) to achieve midline contact with vomer and close the nasal floor. In sum, when A field structures are absent, primary neuronal failure precludes CNS olfactory neuronal formation.

Origin and Composition of the A Fields

What emerges from this myriad plethora of evidence is a model of the A fields as a mass of mesenchymal "highways" interacting interposed between brain and frontonasal skin within which are located the nasal placodes. Interposed in the median basal forebrain is a small but important zone of prechordal mesoderm. NCAM reactivity is a marker for this mesenchyme. The observations of Schwanzel-Fukuda in the early mouse embryo demonstrate that NCSM first arises just caudal to the anterior neuropore *prior* to its appearance in the olfactory placode [236]. NCAM reactivity is detected either concomitant with or immediately after the organization of the MNC mesenchyme.

At stage 7, the notochord appears, and by stage 8, it is producing sonic hedgehog (Shh) to induce the formation of the ectodermal neural plate immediately above it [125, 126]. At stage 9, prosomeres are present within forebrain tissue at the time of the first neural folds and the appearance of the first somites. MNC is migrating into position over the forebrain and orbit at stage 9 as well. The migration of MNC from the mesomere, r0 and r1, is truly massive covering the medial midface and anterior cranial fossa. By stage 10, the pharyngeal arch phase begins; B fields supplied by the stapedial system come into position below the orbit and lateral to the nasal passages. More laterally, the C fields from the first arch proper are providing soft tissue coverage over the neural crest bone fields. Because Shh is involved in placode formation, A field development from MNC may probably begin as soon as the neural crest cells are in place, i.e., at Carnegie stages 9–10. As we have seen, by stage 13, MNC in situ around the nasal placodes begins to proliferate.

Prechordal plate mesoderm is the very first product of gastrulation. It is derived from neuromeric levels r0–r1. Since the mesodermal derivatives of Sm4 arise from r2 to r3, PCM is the likely source for the mesoderm of somitomeres 1–3. PCM has a significant role as a signaling center. In the presence of notochord-produced induction molecules such as *noggin* and *chordin*, the prechordal plate is active in forebrain induction through the actions of *Lim-1* and *cerebrus*. Thus, PCM is more reactive with the forebrain, while notochord stimulates more posterior parts of the neuraxis [100]. It is possible that PCM is responsible for the induction of the nasal placodes and the production of NCAMs—or the stimulation of MNC to produce the NCAMs.

Significance of Neuroendocrine Tracts Originating in the Nasal Placodes

Contact between the developing forebrain and the ingrowing, central processes of the olfactory, vomeronasal, and terminal nerves is preceded by a migration of neural crest adhesion molecule (NCAM)—immunoreactive cells from the ventral forebrain out to the epithelium of the olfactory pit and the formation of an NCAM-reactive cellular aggregate in the mesenchyme between the olfactory pit and the forebrain. This NCAM+ cellular aggregate defines the mesenchymal boundaries of the A fields, the blood supply to which will be the V1 stapedial branches with an anastomotic stem at the primitive ophthalmic artery.

Immunoreactive axons of the olfactory, vomeronasal, and terminal nerves grow into the MNC aggregate which is continuous with the rostral tip of the forebrain. The lateral and more rostral part of this cellular aggregate receives the ingrowing olfactory axons that form the olfactory nerve layer of olfactory bulb. This in turn connects via the lateral olfactory stria to multiple areas of the olfactory cortex. The medial, more caudal part of the cellular aggregate receives the central process of vomeronasal and terminal nerves. Vomeronasal nerve terminates in accessory olfactory bulb; this connects via medial olfactory stria to the limbic system. On the other hand, the central processes of terminal nerve end in the medial forebrain.

Gonadotrophin-releasing hormone (GnRH)—immunoreactive neurons also originate from the medial half of the olfactory pit. These GnRH cells migrate into the brain along and within a scaffolding formed by the axons of VNN and TN. They are never seen independent of the NCAM scaffold as they track across the nasal lamina propria.

Vomeronasal and terminal nerves are detectable in the median nasal mucosa of adult humans. They maintain anatomic and functional contact with the vomeronasal organ on each side of the septum. The physical origin of these tissues is superficial and lateral to their final position in the nasal midline. From the bilateral nasal placodes, they become internalized within the substance of MNC in the nasal midline. The following conclusions must be reached: (1) The two laminae of the septum develop from r1 MNC. They do not grow downward from the roof of the nasal cavity. They project forward from the fused r1 perpendicular ethmoid plate(s), are ensheathed in p6 epithelium. (2) Midline fusion of the medial walls of the nasal pits results in a two-layered functional matrix within which paired laminae of perpendicular ethmoid plate fuse in the midline. The nasal midline above the vomer is supplied by V1 stapedial vessels. (3) Paired paraseptal cartilages, vomerine bone, and premaxillae arise as V2 stapedial derivatives moving forward from the caudal ethmoid plate to form the medial structures of the future nasal floor.

The Prosomere Model of Forebrain Development: Its Relevance to the Development of the Midface

The fields of the human face can be conceptualized as belonging to three fundamental areas of genetic expression: (1) those under control of the prosomeres of the forebrain and supplied by V1 stapedial, (2) those under control of the mesomere and rhombomeres r0–r1–r2 and supplied by V2 stapedial, and (3) those under control of rhombomeres r2 and r11 and supplied by external carotid artery. Note the principle contributions to the pharyngeal arches: PA1 (r2–r3), PA2 (r4–r5), PA3 (r6–r7), PA4 (r8–r9), and PA5 (r10–r11). Each pharyngeal arch is supplied by sensory and motor nerves (sensory is the more important for mapping) which reside in specific neuromeres. Neural crest issues innervated by these nerves can be trace to specific sites along the neural tube associated with a specific homeotic code.

Brain structure from Rathke's pouch forward develops as follows. Midbrain develops from mesomere m1–m2, or as a single large mesomere. Diencephalon is coded by prosomeres p1–p3/p4. Telencephalon has two zones with p5 related to the orbit and optic placode and p6 related to the nose and nasal placodes. Although the anatomic relationship of p4 is unclear, it defines structures and tracts in the posterior cerebrum. Cranial nerves follow the same pattern with p6 and p5 setting up, respectively, olfactory and optic nerves.

Author's note: Interesting patterns of cranial nerve nuclei follow their functional classification. This is especially true of the basal plate (home of all the motor nuclei). Here the GSE (general somatic efferent) nuclei are motor to non-pharyngeal arch striated muscles of the eye and the tongue. The midbrain contains CN III. The isthmus (the transition between midbrain and hindbrain) contains IV. In the rhombencephalon, CN VI occupy r4–r5 while r8–r11 contain CN XII nuclei controlling the tongue. As seen in the chart below, SVE (special visceral efferent—a complete misnomer) nuclei innervate the pharyngeal arch muscles away from the midline. Therefore, the SVE column is lateral to GVE. More lateral still are the GVE (general visceral efferent) autonomic nuclei near the sulcus limitans. The patterns demonstrate the spatial relationships between tissues innervated and the neural plate.

The reader should be advised that the functional classification of cranial nerves that has been taught over the last century is inaccurate. First, there are more than 12 cranial nerves. The terminal nerve has already been discussed. Although three great placodal sensory systems involve smell, sight, and hearing, lateral line placodes occur in standardized locations in association with the posterior member of neuromeric dimers (two per arch). In fishes these are of great importance, involving the detection of motion or pressure in the aqueous environment. In humans, these are associated with taste and explain differences in cranial sensory nuclei. Neural crest component of a ganglion always occupies the cephalic position. Thus, SVA nuclei with inferior and superior components will have an inferior nucleus of mixed placodal and neural crest origin while the superior nucleus is neural crest alone.

Second, the distinction between somatic (GSE) and visceral motor (SVE) nuclei is flat-out incorrect. There is nothing "visceral" (relating to smooth muscle) about the pharyngeal arch muscles. All are striated; they all come from paraxial mesoderm. What distinguishes the two columns of motor nuclei is their relative position within the nervous system.

Third, a layered sequence of innervation is also dictated by the physical position of the cranial nerve nuclei. Within a functional classification, expression of the more cranial nucleus occurs first. Tissues that form later are innervated by the more posterior nucleus. Innervation of the salivary glands is an example. Superior salivatory nucleus supplies the sublingual and submandibular glands. The more superficial parotid is supplied by the more caudal inferior salivatory nucleus.

Future work will establish increasing close correlations between individual neuromeres and the developmental fields of the face. A clear-cut pattern of segmentation emerges, albeit not with the same linearity as seen in the hindbrain. The tissues of the fronto-naso-orbital complex are supplied by a series of StV1 angiosomes that play out in a logical order from medial-to-lateral. The walls of craniofacial arteries are constructed from neural crest that originates from specific locations along the neural folds (each of which has its determining hox genetic code). The fundamental topology of the pharyngeal arches is anterior-posterior and external-to-internal, with PA3–PA5 folded up inside on another like a Japanese fan.

Knowledge of the genes that control these spatial relationships is swiftly increasing. Already, anterior and posterior determinations of pharyngeal arch morphology can be characterized by combinations of the *Dlx* gene system [278, 279]. Gene knock-out experiments are helping to provide further details of this intricate system. The ultimate result will be an accurate genetic map that relates neuromere and the neural crest associated with them to peripheral tissue derivatives. Craniofacial dysmorphology can be put on more precise basis, providing clinicians with a more rational approach to diagnosis and therapy.

Conclusion: Toward a New Understanding of the Midline

For over a century, embryologists have conceptualized oronasal development in terms of "processes," yet no clear neuroanatomic basis has been proffered to rationalize their existence. Theories based upon fusion of medial and lateral nasal processes—together with a poorly defined singular globular process—cannot adequately explain the production of the extraoral midline because their assumptions are flawed from the start.

Another intellectual dead-end is the idea of intraoral midline formation as caused by "descent" of a septum. If this were true, the neuron populations and vomeronasal organ which lie on either side of the septum would have to arise from the undersurface of the embryo, not from the nasal placodes. In craniofacial development, bone and cartilage arise within previously formed functional matricies. Why should the septum (which clearly forms at a later stage in midline development) violate this pattern and be presumed to be the architect of its own functional matrix? Human beings are bilateral creatures. How could a singular septum or frontonasal process be topologically possible? And from whence would they originate? How could such conditions as duplicated septum or hemi-nose be explained?

Field theory of midline facial development is predicated upon the following observations:

1. The human face is constructed from neurovascular units called developmental fields.
2. The timing of appearance of these fields is under strict sequential genetic control.
3. Because the development of the internal carotid system precedes that of the stapedial system and the external carotid system, the anatomic distribution (internal to external) of these vessels is a reflection of this timing.
4. Injection of the ICA in isolation defines its territories to include extracranial structures by virtue of the anastomosis of V1 stapedial with primitive ophthalmic.
5. The neuromeric model accurately predicts the organization of the prosencephalon from prosomeres, mesencephalon from mesomeres, and rhombencephalon from rhombomeres.
6. Naso-optic fields develop in intimate synchrony, the former under control of p5 and the latter under control of p6.

7. the internal carotid extends forward from the junction where the dorsal aortae bend downward and backward toward the heart in stage 9. Its first branch is primitive ophthalmic.
8. The ophthalmic system is a hybrid. Primitive ophthalmic from ICA supplies the globe and nothing more. Anatomosis with V1 stapedial is responsible for all extraocular, nasal, and facial derivative.
9. The stapedial system originates from extracranial internal carotid (AA3), migrates intracranially where, following cranial nerves, it gives off the dural arteries, and arrives at the orbit under the inductive aegis of V1. Stapedial loses its connection with the proximal ICA and is subsequently supplied via anastomoses from the external carotid artery. Thus, nonneural orbital, nasal, and facial fields supplied by the hybrid ophthalmic system are actually related to midbrain neural crest.
10. Supporting structures of the caudal nasal cavity and the mouth (B fields) are supplied by branches of the V2 stapedial axis.
11. Although the final configuration of the nose and the eye are very different, both arise from placodes that follow an analogous pattern of development.
12. Fusion of the medial A field nasal walls takes place by a controlled apoptosis of intervening MNC. It results in paired structures (the nasal choanae) and fused structures (the septum).
13. The neuroanatomy of the nasal placodes establishes a polarity within each A field.
14. The MNC of the A fields provides a pathway by which neural tracts reach their destinations within the brain.
15. Correct positioning of the orbits depends upon midline fusion of the A field walls. Failure of this process results in hypertelorism. Hypoplasia/aplasia of MNC mesenchyme interacting with prechordal plate causes hypotelorism or even cyclopia.
16. Functional abnormalities within intact A fields (defective CAMs?) in Kallman's syndrome allow structural and functional differences within the nasal placodes to be observed separately.
17. The origin of the A fields from MNC places them both spatially and temporally ahead of pharyngeal arch derivatives; they appear earlier in time and are in the midline.
18. Co-participation with pharyngeal arch structures is dependent upon extrinsic factors such as the correct development of the various ECA fields.

Development of the facial midline is a bilateral process involving the formation and coalescence of embryonic units called developmental fields. Each field is genetically unique and is supplied by a single neurovascular pedicle. The formation and structure of each field is ultimately in register with one or more neuromeres. A complex pattern of segmentation based on positional genes of the homeobox family is imposed on the neural plate and its cranial nerves, neural crest, paraxial mesoderm, pharyngeal arches, and aortic arch arteries. Domains of the neural plate are set up that, with folding, establish the regionalization of the forebrain.

Two placodal systems lead to development of the olfactory and ophthalmic fields. These have been demonstrated by dye injection to be supplied exclusively by the V1-induced stapedial system which originated as a unique stem from the extracranial internal carotid, spread into the cranium, and (after dissolution of the stem) reattached to the primitive ophthalmic artery. The spatial and temporal control of these two fields is under direction of the anterior basal forebrain from prosomeres p6 and p5 (or their analogs). Development of these fields is independent of that of the StV2 and StV3-supplied jaws and of the external carotid-supplied first and second arch complex. The unique source of mesenchyme for these fields is mesencephalic neural crest from neuromeres m1–m2, and r0–r1.

Genetic signals from prechordal plate are critical for the development of fronto-nasoethmoid-orbital tissues. The muscles of the eye are derived from somitomeres 1–3 and 5. The contribution of prechordal mesoderm to these muscles remains unclear. It may constitute the *materia prima* from which Sm1–Sm3 are formed.

A model of nasal development-based MNC apoptosis causing midline merger of the nasal capsules explains multiple levels of the developmental anatomy of paired midline structures such as the vomernasal organs, paraseptal cartilages, the septum as a bilaminar structure, and the vomerine bones before coalescence. Growth of the septum takes place within the bilaminar envelope of medial nasal fields. Neurovascular failure leading to mesenchymal deficiency and based upon individual axes of the stapedial system (StV1, StV2, and StV3) explains the variations of craniofacial clefting described by Paul Tessier. The neuromeric model is of great value in organizing and understanding a wide variety of congenital craniofacial conditions.

Epilog: A Prediction

As further work in molecular developmental biology and genetics expands our understanding of facial development, the resultant picture will prove to be one of unbearable simplicity, based upon the principles of neuroanatomy, that unifies the painstaking observations across time of all those who have contemplated the internal logic of the human visage.

Acknowledgment The author wishes to acknowledge the outstanding contributions to the work and previous related papers by Alexandra Cabri and Jo Walters. Ms. Cabri contributed clinical illustration depicting surgical procedures. Ms. Walters provided scientific observations and computer graphics. Without their patience, dedication, and insight, these ideas would never have taken form.

References

1. Orticochea M. History of the discovery of the reversal of blood flow through arteries and veins. Letter to Editor. Plast Reconstr Surg. 1997;97:249–52.
2. Rubenstein JLR, Puellas L. Homeobox gene expression during development of the vertebrate brain. Curr Topics Dev Biol. 1994;29:1–63.
3. Puellas L, Rubenstein JLR. Expression patterns of homeobox and other putative regulatory genes suggest a neuromeric organization. Trends Neurosci. 1993;16:472–9.
4. Anderson JE. Grant's atlas of anatomy. 8th ed. Baltimore: Williams & Wilkins; 1983.
5. Gray H, Pick TP, Howden R. Gray's anatomy. 16th ed. Philadelphia: Lea & Febiger; 1985.
6. Congdon ED. Transformation of the aortic arch system during the development of the embryo. Carnegie Contrib Embryol. 1922;14:47–110.
7. Noden DM. Development of craniofacial blood vessels. In: Fernberg RN, Silver GK, Auerbach R, editors. The development of the vascular system. Basel: Karger; 1991. p. 1–24.
8. Sperber GH. Craniofacial embryology, Dental practitioner series. 2nd ed. New York: Year Book Medical Publishers; 1976. London: Oxford; 1983.
9. Rohrich RJ, Gunter JP, Friedman R. Nasal tip blood supply an anatomic study validating the safety of the transcolumellar incision in rhinoplasty. Plast Reconstr Surg. 1995;25:795–7.
10. Ritter FN. Vasculature of the nose. Ann Otol Rhinol Laryngol. 1970;79:468.
11. Song R, Liu C, Zhao Y. A new principle for unilateral complete cleft lip repair: the lateral columellar flap method. Plast Reconstr Surg. 1998;102:1848–52.
12. Trott JA, Mohan JN. A preliminary report on one stage open tip rhinoplasty at the time of lip repair in bilateral cleft lip and palate: the Alor Setar experience. Br J Plast Surg. 1993;46:215–22.
13. Carstens MH. Correction of the bilateral cleft lip using sliding sulcus technique. J Craniofac Surg. 2000;11(2):1–29.
14. Carstens MH. Correction of unilateral cleft lip nasal deformity using the sliding sulcus procedure. J Craniofac Surg. 1999;10:346–64.
15. Carstens MH. The sliding sulcus procedure: simultaneous repair of unilateral clefts of the lip and primary palate–a new technique. J Craniofac Surg. 1999;10:414–29.
16. Mulliken JB. Bilateral complete cleft lip and nasal repair: an anthropomorphic analysis of staged through synchronous repair. Plast Reconstr Surg. 1996;96:9–22.
17. Cutting CB, Grayson B, Brecht L, et al. Presurgical columellar elongation and primary retrograde nasal reconstruction in one-stage bilateral cleft lip and nose repair. Plast Reconstr Surg. 1998;101:630–8.
18. Maher WP. Artery distribution in the prenatal human palate. Cleft Palate J. 1981;18(1):51–8.
19. Maher WP, Swindle PF. Submucosal palate. Dent Progr. 1962;2:167–80.
20. Carstens MH. Sequential cleft management with the sliding sulcus technique and alveolar extension palatoplasty. J Craniofac Surg. 1999;10:503–18.
21. Gui T, Osama-Yamashita N, Eto K. Proliferation of nasal epithelia and mesenchymal cells during primary palate formation. J Craniofac Genet Dev Biol. 1993;13:250–8.
22. Millicovsky G, Johnson MC. Active role of embryonic facial epithelium: new evidence of cellular events in morphogenesis. J Embryol Exp Morphol. 1989;63:53–66.
23. Figueroa AA, Pratt RM. Autoradiographic study of macromolecular synthesis in the fusion epithlium of the developing rat primary palate in vitro. J Embryol Exp Morphol. 1979;50:145–54.

24. Streeter GL. Developmental horizons in human embryos: description of age group XI, 13 to 20 somites and age group XII, 21 to 29 somites. Contrib Embryol Carnegie Inst. 1942;30:211.

25. Streeter GL. Developmental horizons in human embryos: description of age group XIII, embryos of 4 or 5 millimeters long, and age group XIV, period of identification of the lens vesicle. Contrib Embryol Carnegie Inst. 1945;31:27.

26. Streeter GL. Developmental horizons in human embryos: description of age groups XV, XVI, XVII, and XVIII, being the third issue of a survey of the Carnegie Collection. Contrib Embryol Carnegie Inst. 1948;32:13–203.

27. O'Rahilly R, Gardner E. The initial appearance of ossification in staged human embryos. Am J Anat. 1972;134:291–308.

28. O'Rahilly R. The timing and sequence of events in the development of the human eye and ear during the embryonic period proper. Acta Embryol. 1983;168:87–99.

29. O'Rahilly R. Developmental stages in human embryos. Part A: embryos of the three weeks (stages 1–9). Washington, DC: Carnegie Institution; 1973.

30. O'Rahilly R. Early human development and chief sources of information on stages human embryos. Eur J Obstet Gynecol Reprod Biol. 1978;9:273–80.

31. O'Rahilly R, Muller F. Developmental stages in human embryos, Publication 637. Washington, DC: Carnegie Institution of Washington; 1987. p. 175–302.

32. Rubenstein JLR, Martinez S, Shimamura K, Puellas L. The embryonic vertebrate forebrain: the prosomeric model. Science. 1994;266:578–80.

33. Puelles L, Amat JA, Martinez de la Torre M. Segment-related, mosaic neurogenetic pattern in the forebrain and mesencephalon of early chick embryos: I. Topography of AChE-positive neuroblasts up to stage HH18. J Comp Neurol. 1987;266:247–68.

34. Bufone A, Puelles L, Porteus MH, Frohman MA, Martin GR, Rubenstein JLR. Spatially restricted expression of Dlx-1, Dlx-2(Tes-1), Gbx-2, and Wnt-3 in the embryonic day 12.5 mouse forebrain defines potential transverse and longitudinal segment boundaries. J Neurosci. 1993;13:3155–72.

35. Moore KL. Clinical anatomy. Philadelphia: Williams & Wilkins; 1980.

36. Zide BM, Jelks GW. Surgical anatomy of the orbit. New York: Raven Press; 1985.

37. Noden DW. Origins and patterning of craniofacial mesenchymal tissues. J Craniofac Genet Dev Biol Suppl. 1986;2:15–31.

38. Noden DM. Cell movements and control of patterned tissue assembly during craniofacial development. J Craniofac Genet Dev Biol. 1991;11:192–213.

39. Kjaer I. Correlated appearance of ossification and nerve tissue in human fetal jaws. J Craniofac Genet Dev Biol. 1990;10:329–36.

40. Taylor GI, Palmer JH. The vascular territories (angiosomes) of the body: experimental study and clinical applications. Br J Plast Surg. 1987;46:113–41.

41. Taylor GI, Gianoutsos MP, Morris SF. The neurovascular territories of the skin and muscles: anatomic study and clinical implications. Plast Reconstr Surg. 1994;94:1–36.

42. O'Rahilly R, Gardner E. The timing and sequence of events in the development of the human nervous system during the embryonic period proper. Z Anat Entweckl-Gesch. 1994;134:1–12.

43. Gasser RF. The development of the facial muscles in man. Am J Anat. 1967;20:357.

44. Arey LB. Developmental anatomy. 6th ed. Philadelphia: WB Saunders; 1954. p. 364–73.

45. Kawamoto HK Jr. The kaleidoscopic world of rare craniofacial clefts: order out of chaos (Tessier Classification). Clin Plast Surg. 1976;3:529–72.

46. Tessier P, Rougier J, Hervouet F, Woillez M, Lekeffre M, Derome P. Chirugie plastique orbito-palpebrale. Rapport de la Soc Fr Ophthalmol. 1977;15:191–237.

47. Moore KL, Persaud TVN. The developing human: clinically oriented embryology. 6th ed. Philadelphia: WB Saunders; 1998.

48. Larsen WJ. Human embryology. New York: Churchill Livingstone; 1997.

49. Hinrichsen K. The early development of morphology and patterns of development of the face in the human embryo, Advances in anatomy, embryology and cell biology, vol. 98. New York: Springer Verlag; 1985.

50. DeMyer W. Technique of the neurologic examination: a programmed text. 2nd ed. New York: McGraw Hill; 1974.

51. Patten J. Neurologic differential diagnosis. New York: Springer-Verlag; 1977.

52. Langbartel DA. The anatomical primer: an embryological explanation of human gross morphology. Baltimore: University Park Press; 1977.

53. Webb JF, Noden DM. Ectodermal placodes: contributions to the development of the vertebrate head. Am Zool. 1993;33:434–47.

54. Okabe H, Okubo T, Adachi H, Ishikawa T, Ochi Y. Immunohistochemical demonstration of cytokeratin in human embryonic neurons arising from placodes. Brain Dev. 1997;19:347–52.

55. Bossy J. Development of olfactory and related structures in staged human embryos. Anat Embryol (Berl). 1980;161:225–36.

56. Cohen MM, Sulik KK. Perspectives on holoprosencephaly: part II. Central nervous system, craniofacial anatomy, syndrome commentary, diagnostic approach, and experimental studies. J Craniofac Genet Dev Biol. 1992;12:196–244.

57. Gilbert-Barness E, editor. Potter's pathology of the fetus and infant. 3rd ed. St. Louis: Mosby; 1999.

58. Mazzola RF. Congenital malformations in the frontonasal area: their pathogenesis and classification. Clin Plast Surg. 1976;3:573–609.

59. Hengerer AS, Yanofsky SD. Congenital malformations of the nose and paranasal sinuses. In: Stool S, Bluestone E, editors. Pediatric otolaryngology. Philadelphia: Saunders; 1998. p. 831–42.

60. Norman MG, McGillivray BC, Kalousek DK, Hill A, Poskitt. Congenital malformations of the brain: pathologic, embryologic, clinical, radiologic and genetic aspects. New York: Oxford University Press; 1995.

61. Newman MH, Burdi AR. Congenital alar field defects: clinical and embryological observations. Cleft Palate J. 1981;18:188–92.

62. Schwanzel-Fukuda M, Crossin KL, Pfaff DW, Bouloux PMG, Hardelin JP, Petit C. Migration of luteinizing hormone-releasing hormone (LHRH) neurons in early human embryos: association with neural cell adhesion molecules. J Comp Neurol. 1996;366:547–57.

63. Peter K. Atlas der Entwicklung der Nase und des Gaumenb beim Menschen, mit Einschluss der Entwicklungsstorungen. Jena: Gustav Fischer; 1913.

64. Kraus BS. Prenatal growth and morphology of the human bony palate. J Dent Res. 1960;39:1177–99.

65. Kraus BS, Kitamura H, Latham RA. Atlas of developmental anatomy of the face. New York: Harper & Row; 1966.

66. Muller F, O'Rahilly R. The human brain at stages 21–23, with particular reference in the cerebral cortical plate and to the development of the cerebellum. Anat Embryol (Berl). 1990;182:375–400.

67. Sjersen B, Kjaer I, Jacobson J. The human incisal suture and premaxillary area studied on archeologic material. Acta Odontol Scand. 1993;51:143–54.

68. Njio BJ, Kjaer I. The development and morphology of the incisive fissure and the transverse palatine suture in the human fetal palate. J Craniofac Genet Dev Biol. 1993;13:24–34.

69. Park CG, Ha B. The importance of accurate repair of the orbicularis oris muscle in the correction of unilateral cleft lip. Plast Reconstr Surg. 1995;96:780–8.

70. Nicolau PJ. The orbicularis oris muscle: a functional approach to its repair in the cleft lip. Br J Plast Surg. 1983;36:141–53.

71. Monie W, Cacciatore A. Development of the philtrum. Plast Reconstr Surg. 1962;30:313–21.

72. Nammoun JD, Hisley KC, Graepel S, Hutchins GN, VanderKolk CA. Three dimensional reconstruction of the human philtrum. Ann Plast Surg. 1997;38:202–8.

73. England M. Life before birth. Chicago: Year Book Medical Publishers; 1996.

74. Norman MG, McGillivary BC, Kalousek DK, Hill A, Poskin KJ. Ch. 2 Embryology of the central nervous system. In: Congenital malformations of the brain. New York: Oxford; 1995. p. 9–51.

75. Kjaer I. Prenatal skeletal maturation of the human maxilla. J Craniofac Genet Dev Biol. 1989;9:257–64.

76. Kjaer I. Prenatal human cranial development evaluated on coronal plane radiographs. J Craniofac Genet Dev Biol. 1990;10:339–51.

77. Brash JC. Cunningham's textbook of anatomy. 9th ed. London: Oxford University Press; 1951. p. 1264.

78. Anson BJ, McVay CB. Surgical anatomy. 6th ed. Philadelphia: WB Saunders; 1984. p. 90–106.

79. Fawcett E. The development of the human maxilla, vomer and paraseptal cartilage. J Anat Phys. 1911;45:378–405.

80. Wang HG, Kwok P, Hawke M. The embryonic development of the human paraseptal cartilage. J Otol. 1988;17:150–4.

81. Mooney MP, Siegel MI, Kimes KR, Todhunter JS, Smith TD. Anterior paraseptal cartilage development in normal and cleft lip and palate human fetal specimens. Cleft Palate Craniofac J. 1994;31:239–45.

82. Eloff FC. On the relations of the human vomer to the anterior paraseptal cartilages. J Anat. 1952;86:16–9.

83. Kjaer I. Ossification of the human fetal basicranium. J Craniofac Genet Dev Biol. 1990;10:29–38.

84. Sandikcioglu M, Molsted K, Kjaer I. The prenatal development of the human nasal and vomeral bones. J Craniofac Genet Dev Biol. 1994;14:124–34.

85. Couly GF, NM LD. Mapping of the early neural primordium in quail-chick chimeras I. Developmental relationships between placodes, facial ectoderm and prosencephalon. Dev Biol. 1985;110:422–39.

86. Kjaer I. Human prenatal palatal shelf evaluation related to craniofacial skeletal maturation. Eur J Orthod. 1992;14:26–30.

87. Couly GF, LeDouarin NM. Mapping of the early neural primordium in quail-chick chimeras I. Developmental relationships between placodes, facial ectoderm and prosencephalon. Dev Biol. 1985;110:422–39.

88. Couly GF, LeDouarin NM. Mapping of the early neural primordium in quail-chick chimeras II. The prosencephalic neural plate and neural folds: implications for the genesis of cephalic human congenital abnormalities. Dev Biol. 1987;120:198–214.

89. Rubenstein JLR, Shimamura K, Martinez S, Puellas L. Regionalization of the prosencephalic neural plate. Annu Rev Neurosci. 1998;21:445–77. Quotation, pp. 464–465.

90. Searles JC. Radioautographic study of chondrocyte proliferation in the nasal septal cartilage of the 5-day old rat. Cleft Palate J. 1975;12:291–8.

91. Searles JC. Radioautographic study of chondrocyte proliferation in the new born rat. Am J Anat. 1977;150:659–63.

92. Searles JC. A radioautographic study of chondrocyte proliferation in nasal septal cartilage of the prenatal rat. Am J Anat. 1977;150:159–63.

93. Searles JC, Kinser DD. Radioautographic study of chondrocyte proliferation in nasal septal cartilage of the 10-day old rat. J Dent Res. 1972;51:812–28.

94. Shepherd WM, McCarthy MD. Observations on the appearance and ossification of the premaxilla and maxilla in the human embryo. Anat Rec. 1955;121:13–28.

95. Silau M, Njio B, Solow B, Kjaer I. Prenatal sagittal growth of the osseous components of the human palate. J Craniofac Genet Dev Biol. 1992;14:252–6.

96. Kvinnsland S. Observations on the early ossification of the upper jaw. Acta Odont Scand. 1964;27:649–54.

97. Granstrom G, Magnusson BC. Histochemical analysis of enzymes involved in the formation and metabolism of the nasal septal cartilage. Acta Otolaryngol Suppl. 1992;492:15–21.

98. Kjaer I. Prenatal development of the primary maxillary incisors related to maturation of the surrounding bone and to the postnatal eruption. In: Davidovitch Z, editor. Biological mechanisms of tooth eruption and root resorption. Birmingham, AL: EBSCO Media; 1988. p. 233–6.

99. Lumsden A, Krumlauf R. Patterning the vertebrate neuraxis. Science. 1996;274:1109–15.

100. Carlson BM. Human embryology and developmental biology. St. Louis: Mosby; 1999.

101. Muller F, O'Rahilly R. The timing and sequence of appearance of neuromeres and their derivatives in staged human embryos. Acta Anat. 1997;158:83–99.

102. Lumsden A, Keynes R. Segmental patterns of neuronal development in the chick hindbrain. Nature. 1989;337:424–8.

103. Lopez-Mascaraque L, Garcia C, Valverde F, De Carlos JA. Central olfactory structures in Pax-6 mutant mice. Ann N Y Acad Sci. 1998;855:83–94.

104. Burdi AR, Lawton TJ, Grosslight J. Prenatal pattern emergence in early human facial development. Cleft Palate Craniofac J. 1988;25:8–15.

105. Mishima N, Tomarev S. Chicken eyes absent 2 gene: isolation and expression pattern during development. Int J Dev Biol. 1998;42:1109–15.

106. Lombardo A, Slack JM. Postgastrulation effects of fibroblast growth factor on Xenopus development. Dev Dyn. 1998;212:75–85.

107. McDonald R, Barth KA, Xu Q, Holder N, Mikkola I, Wilson S. Midline signalling is required for Pax gene regulation and patterning of the eyes. Development. 1995;121:3267–78.

108. Stoykova A, Walther C, Fritsch R, Gruss P. Forebrain patterning defects in Pax6/Small eye mutant mice. Development. 1996;122:3453–65.

109. Gruss PC, Walther C. Pax in development. Cell. 1992;69:719–22.

110. Mall FP. Cyclopia in the human embryo. Contrib Embryol Carnegie Inst. 1917;6(15):7.

111. Torczynski E, Jacobiec FA. Ch. 6: Cyclopia and synophthalmia: a model of embryologic interactions. In: Duane TD, Jaeger EA. Biomedical foundations of ophthalmology, vol. I. Jakobiec FA, editor. Ocular anatomy, embryology, and teratology. Philadelphia: JB Lippincott; 1982.

112. Lemire RJ, Cohen M, Beckwith JB, Kokich VG, Siebert JR. The facial features of holoprosencephaly in anencephalic human specimens. I. Historical review and associated malformations. Teratology. 1981;23:297–303.

113. Siebert JR, Kokich VG, Beckwith JB, Cochen MM, Lemire RJ. The facial features of holoprosencephaly in anencephalic human fetuses. II. Craniofacial anatomy. Teratology. 1981;23:305–15.

114. Siebert JR, Kokich VG, Warkany J, Lemire RJ. Atelencephalic microcephaly: craniofacial anatomy and morphologic comparisons with holoprosencephaly and anencephaly. Teratology. 1987;36:279–85.

115. Shimamura K, Hartigan DJ, Martinez S, Puellas L, Rubenstein JLR. Longitudinal organization of the anterior neural plate and neuronal tube. Development. 1995;121:3923–33.

116. Liem KF, Tremml G, Roelink H, Jessel TM. Dorsal differentiation of neural plate cells induced by BMP-linked signals from epidermal ectoderm. Cell. 1995;82:969–79.

117. Dickinson ME, Selleck MA, McMahon AP, Bonner-Fraser M. Dorsalization of the neural tube by the non-neural ectoderm. Development. 1995;121:2099–106.

118. Tanabe Y, Roelink H, Jessel TM. Induction of motor neurons by Sonic hedgehog is independent of floor plate differentiation. Curr Biol. 1995;5:651–8.

119. Basler K, Edlund T, Jessel TM, Yamada T. Control of cell pattern in the neural tube: regulation of cell differentiation by dorsalin-1, a novel TGFb family members. Cell. 1993;73:687–702.

120. Tanabe Y, Jessell TM. Diversity and pattern in the developing spinal cord. Science. 1996;274:1115–23.

121. Placzek M. The role of the notochord and floor plate in inductive interactions. Curr Opin Genet Dev. 1995;5:499–506.

122. Placzek M, Jessell TM, Dodd J. Induction of floor plate differentiation by contact-dependent homeogenetic signals. Development. 1993;117:205–18.

123. Marti E, Bumcrot DA, Takada R, McMahon AP. Requirements of 19K form of Sonic hedgehog for induction of distinct ventral cell types in CNS explants. Nature. 1995;375:322–5.

124. Roelink H, Augsberger A, Heemskerk J, Korzh V, Norlins. Floor plate and motor neuron induction by vhh-1, a vertebrate homolog of hedgehog expressed by the notochord. Cell. 1994;776:445–55.

125. Echelard Y, Epstein DJ, St-Jacques B, Shen L, Mohler J. Sonic hedgehog, a member of a family of putative signalling molecules, is implicated in the regulation of CNS polarity. Cell. 1993;75:1417–30.

126. Erickson J. Sonic hedgehog induces the differentiation of ventral forebrain neurons, a common signal for ventral patterning within the neural tube. Cell. 1995;81:747–56.

127. Chiang C, Litingtung Y, Lee E, Young KY, Corden JL. Cyclopia and defective axial patterning in mice lacking Sonic hedgehog gene function. Science. 1996;383:407.

128. Acampora D, Mazan S, Lallemand Y, Avantaggiato V, Maury M. Forebrain and midbrain regions are deleted in Otx-2 mutants due to a defective anterior neuroectoderm specification during gastrulation. Development. 1996;121:3279–90.

129. Matsuo I, Kuratani S, Kimura C, Takada N, Aizawa S. Mouse Otx2 functions in the formation and patterning of rostral head. Genes Dev. 1995;9:1–13.

130. Shawlot W, Behringer R. Requirements for Lim-1 in head organizer function. Nature. 1994;4:425–30.

131. Ang S-L, Jin O, Rhinn N, Daigle N, Stevenson l, Rossant J. A targeted mouse Otx2 mutation leads to severe defects in gastrulation and formation of axial mesoderm and to deletion of rostral brain. Development. 1996;122:243–52.

132. Sulik K, Debhart DB, Inagaki T, Carson JL, Vrablic T. Morphogenesis of the murine node and notochordal plate. Dev Dyn. 1994;201:260–78.

133. Psychoyos D, Stern C. Fates and migratory routes of primitive streak cells in the chick embryo. Development. 1996;122:1523–34.

134. Sasaki H, Hogan LM. Differential expression of multiple fork head related genes during gastrulation and axial pattern formation in the mouse embryo. Development. 1993;188:47–59.

135. Pera EM, Kess M. Patterning of the chick forebrain anlage by the prechordal plate. Development. 1997;124:4153–62.

136. Grindley JC, Davidson DR, Hill RE. The role of Pax-6 in eye and nasal development. Development. 1995;121:1433–42.

137. Saha MS, Servetnick M, Grainger RM. Vertebrate eye development. Curr Opin Genet Dev. 1992;2:582–8.

138. Ozanics V, Jacobiec FA. Ch. 6: Prenatal development of the eye and adnexa. In: Duane TD, Jaeger EA, editors. Biomedical foundations of ophthalmology, vol. I. Jakobiec FA, editor. Ocular anatomy, embryology, and teratology. Philadelphia: JB Lippincott; 1982.

139. Hanson I, Van Heyningen V. *Pax6:* more than meets the eye. Trends Genet. 1995;11:268.

140. Torres M, Gomez-Pardo E, Gross P. *Pax2* contributes to inner ear patterning and optic nerve trajectory. Development. 1996;122:3381–91.

141. Ekker SC, Ungar AK, Greenstein P, von Kessler DP, Porter JA. Patterning activities of vertebrate hedgehog proteins in the developing eye and brain. Curr Biol. 1995;5:944–55.

142. Barth KA, Wilson SW. Expression of zebrafish nk2.2 is influenced by sonic hedgehog/vertebrate hedgehog-1 and demarcates a zone of neuronal differentiation in the embryonic forebrain. Development. 1995;121:1755–68.

143. Roessler E, Belloni E, Gauden K, Jay P, Berta P. Mutations in the human sonic hedgehog gene cause holoprosencephaly. Nat Genet. 1996;14:357–9.

144. Kjaer I, Fischer-Hansen B. Luteinizing hormone releasing hormone and innervation pathways in human prenatal nasal submucosa: factors of importance in evaluating Kallman's syndrome. APMIS. 1996;104:680–8.

145. Graziadei PPC, Monti-Graziadei AG. The development of the olfactory placode on the development of the telencephalon in Xenopus laevis. Neuroscience. 1992;46:617–29.

146. Farbman AI. Developmental neurobiology of the olfactory system. In: Getchell TV, editor. Smell and taste in health and disease. New York: Raven; 1991. p. 19–33.

147. Greer CA. Structural organization of the olfactory system. In: Getchell TV, editor. Smell and taste in health and disease. New York: Raven; 1991. p. 65–79.

148. Schmal W, Knoedelseder M, Favor J, Davidson D. Defects of neuronal migration and pathogenesis of cortical malformations are associated with small eye (Sey) in the mouse, a point mutation at the Pax-6 locus. Acta Neuropathol. 1993;86:126–35.

149. Kallman FJ, Schoenfeld WA, Barrera SE. The genetic aspects of primary eunuchoidism. Am J Ment Defic. 1944;48:203.

150. Lieblich JM, Rogol AD, White BJ, Rosen SW. Syndrome of anaosmia with hypogonadotropic hypogonadism (Kallman Syndrome): clinical and laboratory studies in 23 cases. Am J Med. 1982;73:506–19.

151. Jacobson AG. The determination and positioning of the nose, lens, and ear. I. Interactions within the ectoderm, and between the ectoderm and underlying tissues. J Exp Zool. 1963;154:273–84.

152. Van Eeden FJM, et al. Genes involved in forebrain development in the zebrafish Danio rerio. Development. 1996;123:255–62.

153. Laestadius N, Aase J, Smith E. Normal inner canthal and outer orbital dimensions. J Pediatr. 1969;74:465–8.

154. Mehes K, Kitzveger E. Inner canthal and inter-mammillary indices in the newborn infant. J Pediatr. 1974;85:90–1.

155. Kundrat H. Arhinepncephalie als typishe Art von Missbildung. Wein Med Bild. 1882;5:1395–7.

156. Yankolev PI. Pathoarchitectonic studies of cerebral malformations III Arrhinencephalies (Holoprosencephalies). J Exp Neurol. 1959;18:22–54.

157. Cohen MM Jr, Jirasek JE, Guzman RT, Gorlin RJ, Peterson MQ. Holoprosencephaly and facial dysmorphia: nosology, etiology, and pathogenesis. Birth Defects Orig Artic Ser. 1971;7:125–235.

158. Muenke M. Holoprosencephaly: defects of the mediobasal prosencephalon. In: Norman MG, McGillivary BC, Kalousek DK, et al., editors. Congenital malformations of the brain: pathologic, embryologic, clinical, radiologic, and genetic aspects. London: Oxford University Press; 1995. p. 187–221.

159. Probst FP. The prosencephalies: morphology, neuroradiological appearances, differential diagnosis. New York: Springer-Verlag; 1979.

160. DeMyer WB, Zeman W. Alobar holoprosencephaly (arhinencephaly) with median cleft lip and palate: clinical and electroencephalographic and nosologic considerations. Confin Neurol (Basel). 1963;23:1–36.

161. DeMyer W, Zeman W, Palmer CG. The face predicts the brain: diagnostic significance of median facial anomalies for holoprosencephaly (arhinenecephaly). Pediatrics. 1964;34:256–63.

162. DeMyer WB. Median facial malformations and their implications for brain malformations. Birth Defects Orig Artic Ser. 1975;11:155–81.

163. Lutolf U. Bilateral aplasia of the nose. Maxillofac Surg. 1976;4:245–9.

164. Cohen D, Goitein K. Arhinia. Rhinology. 1986;24:287–92.

165. LaTrenta GS, Choi HW, Ward RF, Hoffman L, Neidich JA. Complete nasal agenesis with bilateral microophthalmia and unilateral duplication of the thumb. Plast Reconstr Surg. 1995;95:1101–4.

166. Meyer R. Total external and internal construction in arhinia. Plast Reconstr Surg. 1997;99:534–42.

167. Wexler A, Neuman A, Benmeir P, Lusthaus S, Wexler MR. A rare case of arhina with severe airway obstruction: case review of the literature. Plast Reconstr Surg. 1993;91:146–9.

168. Gitlin G, Behar AJ. Meningeal angiomatosis, arhinencephaly, agensis of the corpus callosum and large hamartoma of the brain, with neoplasia, in an infant having bilateral nasal proboscis. Acta Anat. 1960;41:56–79.

169. Francesconi G, Fortunato G. Median dysraphia of the face. Plast Reconstr Surg. 1963;43:481–91. See cases 14 and 15, p. 487.

170. McLaren LR. A case of cleft lip and palate with a polypoid nasal tubercle. Br J Plast Surg. 1956;8:57–9.

171. Mahindra S, Daljit R, Jamwal N. Lateral nasal proboscis. J Laryngol Otol. 1973;2:177–81.

172. Boo-Chai K. The proboscis lateralis—a 14-year follow-up. Plast Reconstr Surg. 1985;75:569–77.

173. Poe LB, Hochhauser L, Bryke C, Streeten BS, Sloan J. Proboscis lateralis with associated orbital cyst: detailer MR and CT imaging and correlative embryopathy. Am J Neurorad. 1992;13:1471–6.

174. DeMyer WB. The median cleft face syndrome: differential diagnosis of cranium bifidum occultum, hypertelorism, and median cleft nose, lip, and palate. Neurology. 1967;17:961–71.

175. DeMyer WB. Familial alobar holoprosencephaly arhinencepnaly with median cleft lip and palate. Neurology. 1963;13:913–8.

176. Millard DR, Williams S. Median clefts of the upper lip. Plast Reconstr Surg. 1968;42:4–10.

177. Geoffrey Machin MD. Personal communication, 1999.

178. Sulik KK, Johnston MC. Embryonic origin of holoprosencephaly: interrelationship of the developing brain and face. Scan Electron Microsc. 1982;(Pt 1):309–22.

179. Sulik KK, Johnston MC. Sequence of developmental observations following acute ethanol exposure in mice: craniofacial features of the fetal alcohol syndrome. Am J Anat. 1983;166:257–69.

180. Adelmann HB. The problem of cyclopia. Q Rev Biol. 1936;11:161–82, 284–304.

181. Howard RO, Boue J, Deluchat C, Albert DM, LeHav M. The eyes of embryos with chromosome abnormalities. Am J Ophthalmol. 1974;78:167–88.

182. Torczynski E, Jakobiec FA, Madewell J, Font R, Johnston M. Synophtahlamia: a histological, organogenetic and radiographic analysis. Doc Ophthalmol. 1977;44:311–78.

183. Karseras AG, Laurence KM. Eyes in arhinencephalic syndromes. Br J Ophthalmol. 1973;59:462–73.

184. Stockard CR. The influence of alcohol and other anesthetics on embryonic development. Am J Anat. 1910;10:369–92.

185. Stockard CR. The artificial production of a single, median cyclopic eye in the fish embryo by means of sea water solutions of magnesium chloride. Arch Entmech Bd. 1907;23:S249–58.

186. Stockard CR. The influence of external factors, chemical and physical, on the development of Fundulus heteroclitus. J Exp Zool. 1907;4:165–201.

187. Stockard CR. The artificial production of one-eyed monsters and other defects, which occur in nature, by the use of chemicals. Anat Rec. 1909;3:167–73.

188. Stockard CR. The development of artificially produced cyclopean fish—the magnesium embryo. J Exp Zool. 1909;6:285–337.

189. Rogers KT. Experimental production of perfect cyclopia in the chick by means of LiCl, with a survey of the literature on cyclopia produced experimentally by various means. Dev Biol. 1963;8:129–50.

190. Rogers KT. Radioautographic analysis of the incorporation of protein and nucleic acid precursors into various tissues of early chick embryos cultured in toto on medium containing LiCl. Dev Biol. 1964;9:176–96.

191. Holtfreter J, Hamburger V. In: Willyer DH, Weiss PA, Hamburger V, editors. Embryogenesis: progressive differentiation. Philadelphia: WB Saunders; 1955. p. 230–96.

192. Burr HS. The effects of the removal of the nasal pits in amblystoma embryos. J Exp Zool. 1916;20:27–51.

193. Rogers KT. Experimental production of perfect cyclopia by removal of the telencephalon and reversal of bilateralization in somite-stage chicks. Am J Anat. 1964;115:487–508.

194. Adelmann HB. Experimental studies on the development of the eye. I. The effect of removal of the medial and lateral areas of the anterior end of the urodeliam neural plate on the development of the eyes (Tritan teniatus and Amblyostoma punctatum). J Exp Zool. 1929;45:249–90.

195. Adelmann HB. Experimental studies on the development of the eye. II. The eye-forming potencies of the median portions of urodelian neural plate. J Exp Zool. 1929;54:291–317.

196. Adelmann HB. Experimental studies on the development of the eye. III. The effect of the substrate on the heterotopic development of median and lateral strips of the anterior end of the neural tube of amblyostoma. J Exp Zool. 1930;57:233–81.

197. Adelmann HB. Experimental studies on the development of the eye. IV. The effect of the partial and complete excision of the prechordal substrates of the eyes of Amblyostoma punctatum. J Exp Zool. 1937;75:199–237.

198. Helms JA, Kim CH, Hu D, Minkoff R, Thaller C, Eichele G. Sonic hedgehog participates in craniofacial morphogenesis and is down-regulated by teratogenic doses of retinoic acid. Dev Biol. 1997;187:25–35.

199. Simeone A, Acampora D, Pannese M, D'Esposito M, Stornaiulo A, Gulisano M, Mallamaci A, Kasturi K, Druk T, Huebner K. Cloning and characterization of two members of the vertebrate Dlx gene family. Proc Natl Acad Sci U S A. 1994;91:2250–4.

200. Yang I, Zhang H, Hu G, Wang H, Abate-Shen C, Shen MM. An early phase of embryonic Dlx5 expression defines the rostral boundary of the neural plate. J Neurosci. 1998;18:8322–30.

201. Depeuw MJ, Lui JK, Long JE, Presley R, Menese JJ, Pederson RA, Rubenstein JLR. Dlx5 regulates regional development of the branchial arches and sensory capsules. Development. 1999;128:3831–46.

202. Schwanzel-Fukuda M, Bick D, Pfaff DW. Luteinizing hormone releasing hormone (LHRH)-expressing cells do not migrate normally in an inherited hypogonadal (Kallman) syndrome. Mol Brain Res. 1989;6:311–25.

203. Angevine JB Jr, Cotman CW. Principles of neuroanatomy. New York: Oxford; 1981. p. 150–65, 253–283.

204. Barr ML. The human nervous system: an anatomical viewpoint. New York: Harper; 1974. p. 252–68.

205. Ranson SW, Clark SL. The anatomy of the nervous system: its development and function. 10th ed. Philadelphia: WB Saunders; 1959. p. 327–46.

206. Kimmelman CP. Clinical review of olfaction. Am J Otolaryngol. 1993;14:227.

207. Schiffmann SS. Taste and smell in disease. N Engl J Med. 1983;308(1275):1337.

208. Ruysch F. Thesaurus anatomicus tertius, 70, plate IV, Fig. 48–49. Amsterdam: Wolters; 1703.

209. Jacobson L. Description anatomique d'une organe observe dans les mammiferes. Ann Mus Hist Nam Paris. 1811;18:412–42.

210. Garcia-Velasco J, Mondragon M. The incidence of the vomeronasal organ in 1000 human subjects and its possible clinical significance. J Steroid Biochem Mol Biol. 1991;39:561–3.

211. Johnson A, Josephson R, Hawke M. Clinical and histological evidence for the presence of the vomeronasal (Jacobson's) organ in adult humans. J Otolaryngol. 1985;14:71–9.

212. Smith TD, Siegel MI, Mooney MP, Burdi AR, Todhunter J. Vomeronasal organ growth and development in normal and cleft lip and palate fetuses. Cleft Palate Craniofac. 1996;J33:385–94.

213. Brookover C. The nervus terminalis in adult man. J Comp Neurol. 1914;24:131–5.

214. Demski LS. Terminal nerve complex. Acta Anat. 1993;148:81–95.

215. Halpern M. The organization and function of the vomeronasal system. Ann Rev Neurosci. 1987;10:325–62.

216. Monti-Bloch L, Jennins-White C, Berliner DL. The human vomeronasal system: a review. Ann N Y Acad Sci. 1998;855:373–89.

217. Monti-Bloch L, Jennings-White C, Dolberg DS, Berliner DL. The human vomeronasal system. Psychoneuroendocrinology. 1994;19:673–86.

218. Monti-Bloch L, Grosser BI. The effect of putative pheromones on the electrical activity of the human vomeronasal organ and olfactory epithelium. J Steroid Biochem Mol Biol. 1991;39:573–82.

219. Moran DTL, Monti-Bloch L, Stensaas LJ, Berliner DE. Structure and function of the human vomeronasal organ. In: Doty RL, editor. Handbook of olfaction and gustation, vol. 36. New York: Marcel Dekker; 1995. p. 793–820.

220. Berliner DL, Monti-Bloch L, Jennings-White C, Diaz-Sanchez V. The functionality of the human vomeronasal organ (VNO): evidence for steroid receptors. J Steroid Biochem Mol Biol. 1996;58:259–65.

221. Silver WL, Finger TE. The trigeminal system. In: Getchell TV, editor. Smell and taste in health and disease. New York: Raven; 1991. p. 97–108.

222. Fritsch G. Untersuchungen uber den ferineren Bau des Fischgehirns mit besonderer Berucksichtung der Homologien be anderen Wirbelnecklasses. Berlin: Verlag Guttmansche; 1978.

223. Pinkus VF. Uber einen noch nict beschreiben Hirnnerv des Protopterus annectens. Anat Anz. 1894;9:562–6.

224. Johnston JB. The nervus terminalis in man and mammals. Anat Rec. 1914;8:185–98.

225. McCotter RE. A note on the course and distribution of the nervus terminalis in man. Anat Rec. 1916;9:243–6.

226. Pearson AA. Development of the nervus terminalis in man. J Comp Neurol. 1941;75:39–66. See Fig. 10, p. 50.

227. Demski LS, Fields RD, Bullock TH, Schreibman MP, Margolis-Nunno H. The terminal nerve of sharks and rays. Electron microscopic, immunocytochemical, and electrophysiological studies. Ann N Y Acad Sci. 1987;519:15–32.

228. Jennes L. The nervus terminalis in the mouse: light and electron microscopic immunocytochemical studies. Ann N Y Acad Sci. 1987;519:165–73.

229. Muske LE, Moore FL. Luteinizing hormone-releasing hormone immunoreactive neurons in the amphibian brain are distributed along the course of the nervus terminalis. Ann N Y Acad Sci. 1987;519:433–46.

230. Muske LE. Evolution of gonadotropin-releasing hormone (GnRH) neuronal systems. Brain Behav Evol. 1993;42:215–30.

231. Demski LS, Wright DF. GnRH immunoreactivity in the brain of the brown anole, Anolix sagrei (abstr). Soc Neurosci Abstr. 1992;143:13.

232. Demski LS, Northcutt RG. The terminal nerve: a new chemosensory system in vertebrates? Science. 1983;220:435–7.

233. Demski LS. Terminal nerve complex. Acat Anat. 1993;148:81–95.

234. Schwanzel-Fukuda M, Pfaff DW. Origin of luteinizing hormone-releasing hormone neurons. Nature. 1989;338:161–4.

235. Schwanzel-Fukuda M, Pfaff DW. Migration of LHRH immunoreactive neurons from the olfactory placode rationalizes olfacto-hormonal relationships. J Steroid Biochem Mol Biol. 1991;39:565–72.

236. Schwanzel-Fukuda M, Dellavade TL. Ultrastructure of the nasal mesenchyme during vasculogenesis and formation of the migration route of LHRH neurons originating from the olfactory placode of embryonic mice. Soc Neurosci Abstr. 1998;24:1279.

237. Schwanzel-Fukuda M. Origin and migration of luteinizing hormone-releasing hormone neurons in mammals. Microsc Res Tech. 1999;44:2–10.

238. Wray S, Grant P, Gainer H. Evidence that cells expressing luteinizing hormone-releasing hormone mRNA are derived from progenitor cells in the olfactory placode. Proc Natl Acad Sci USA. 1989;86:8132–6.

239. Wray S, Nieburgs A, Elkabes S. Spatiotemporal cell expression of luteinizing hormone-releasing hormone in the prenatal mouse. Evidence for an embryonic origin in the olfactory placode. Dev Brain Res. 1989;46:309–18.

240. Zheng L-M, Pfaff DW, Schwanzel-Fukuda M. Electron microscopic identification of luteinizing hormone-releasing hormone-immunoreactive neurons in the medial olfactory placode and basal forebrain of embryonic mice. Neuroscience. 1992;46:407–8.

241. Kjaer I, Fischer-Hansen B. Luteinizing hormone releasing hormone and innervation pathways in human prematal nasal submucosa: factors of importance in evaluating Kallman's syndrome. APMIS. 1996;104:680–8.

242. Daikoko S, Koide I. Destruction of olfactory inputs affects the morphogenesis of the telencephalon in rats. Arch Histol Cytol. 1997;60:329–45.

243. Schwanzel-Fukuda M, Silverman NJ. The nervus terminalis in the guinea pig: a new luteinizing hormone-releasing hormone (LHRH) neuronal system. J Comp Neurol. 1980;191:213–25.

244. Demski LS, Schwanzel-Fukuda M. The terminal nerve (nervus terminalis): structure, function, and evolution. Ann N Y Acad Sci. 1987;519:469.

245. Kjaer I, Fischer-Hansen B. The human vomeronasal organ: prenatal developmental stages and distribution of luteinizing hormone-releasing hormone. Eur J Oral Sci. 1996;104:34–40.

246. Kim KH, Patel I, Tobet SA, King JC, Rubin BS, Stopa EG. Gonadotropin-releasing hormone immunoreactivity in the adult and fetal human olfactory system. Brain Res. 1999;826:220–9.

247. Nishizuka M, Arai Y. Glycosoaminoglycans in the olfactory epithelium and nerve of chick embryos: an immunocytological study. Neurosci Res. 1996;24:165–73.

248. Tacheichi M. The cadherins: cell-cell adhesion molecules controlling animal morphogenesis. Development. 1988;102:639–55.

249. Chuong C-M. Adhesion molecules (N-CAM and Tenascin) in embryonic development and tissue regeneration. J Craniofac Genet Dev Biol. 1990;10:147–61.

250. Cowan WM. In: Garrod DR, editor. The development of the vertebrate nervous system. London: Cambridge University Press; 1981. p. 3–33.

251. Richardson C, Crossin KL, Chong CM, Edelman GM. Expression of cell adhesions molecules in embryonic induction. III. Development of the otic placode. Dev Biol. 1987;119:217–30.

252. Silver J, Rutischauer U. Guidance of optic axons in vivo by a preformed adhesive pathway on neuroepithelial endfeet. Dev Biol. 1984;106:486–99.

253. Schwanzel-Fukuda M, Abraham S, Crossin KL, Edelman GM, Pfaff DW. Immunocytochemical demonstration of neural cell adhesion molecule (NCAM) along the migration route of luteinizing hormone-releasing hormone (LHRH) neurons in mice. J Comp Neurol. 1992;32:1–18.

254. Schwanzel-Fukuda M, Reihard GR, Abraham S, Crossin KL, Gm E, Pfaff DW. Antibody to neural cell adhesion molecule can disrupt the migration of luteinizing hormone-releasing hormone-releasing hormone neurons into the mouse brain. J Comp Neurol. 1994;342:174–85.

255. Daikoko-Ishido H, Okamura Y, Yanaihara N, Daikoko S. Development of the hypothalamic luteinizing hormone-releasing hormone-containing neuron system in the rat. In vivo and in transplantation studies. Dev Biol. 1990;140:374–87.

256. Ronnekliev O, Resko JA. Ontogeny of gonadotropin hormone-releasing hormone-containing neurons in early fetal development of rhesus macaques. Endocrinology. 1990;126:498–511.

257. Treloar HB, Nurcomb V, Key B. Expression of extracellular matrix molecules in the embryonic rat olfactory pathway. J Neurobiol. 1996;31:41–55.

258. Hatanaka Y, Jones EG. Early region-specific gene expression during tract formation in the embryonic forebrain. J Comp Neurol. 1998;395:296–309.

259. Rhinn M, Dierich A, Shawlot W, Behringer RR, LeMeur M, Ang SI. Sequential roles for Otx2 in visceral endoderm and neuroectoderm for forebrain and midbrain induction and specification. Development. 1998;125:845–56.

260. Grinblat Y, Gamse J, Patel M, Sive H. Determination of the zebrafish forebrain: induction and patterning. Development. 1998;125:4403–16.

261. Bacharvet KT, von Boxberg Y, Guazzi S, Boncinelli E, Godement P. A potential role for the OTX2 homeoprotein in creating early "superhighways" for axon extension in the rostral forebrain. Development. 1998;125:4273–82.

262. Mallamaci A, DiBlas E, Briat P, Bonicelli E, Corte G. OTX2 homeoprotein in t he developing central nervous system and migratory cells of the olfactory system. Mech Dev. 1996;58:165–78.

263. Whitesides JG, La Mantia AS. Differential adhesion and initial assembly of the mammalian olfactory nerve. J Comp Neurol. 1996;373:240–54.

264. Key B, Treloar HB, Wangerek L, Ford MD, Nurcomb V. Expression and localization of FGF-1 in the developing rat olfactory system. J Comp Neurol. 1996;366:197–206.

265. Alcantara S, Ruiz M, D'Arcangelo G, Ezan F, De Lecca L, Curran T, Sotelo C, Soriano E. Regional and cellular patterns of reelin mRNA expression in the forebrain of the developing an adult mouse. J Neurosci. 1998;19:7779–99.

266. Chien CL, Lee TH, Lu KS. Distribution of neuronal intermediate filament proteins in the developing mouse olfactory system. J Neurosci Res. 1998;54:353–68.

267. Kobayashi M, Toyama R, Takeda H, Dawid IB, Kawakami K. Overexpression of the forebrain-specific homeobox gene six3 induces rostral forebrain enlargement in zebrafish. Development. 1998;125:2973–82.

268. Fukura H, Komiya Y, Igarashi M. J Neurochem. 1996;67:1426–34.

269. Maestre de San Juan A. Teratologia: falta total de los nervios olfatorios con anosmia en un individuo en quien existia una atrofia congenita de los testiculos y el miembro viril. El Siglo Medico. 1856;3:211.

270. Molsted K, Kjaer I, Giwercman A, Vesterhauge S, Skakkbarek NE. Craniofacial morphology in patients with Kallman's syndrome with and without cleft lip. Cleft Palate Craniofac J. 1997;34:417–24.

271. Gauthier G. La dysplasie olfacto-genitale: agenesie des lobes olfactifs avec absence de developpement gonadique a la puberte. Acta Neuroveg. 1960;21(4):345–94.

272. Rugarli EI, Ballabio A. Kallmann syndrome: from genetics to neurobiology. JAMA. 1993;270:2713–6.

273. Truwit CL, Barkovich AJ, Grumbach MM, Martini JJ. MRI imaging of Kallmann's syndrome: a genetic disorder of neuronal migration affecting the olfactory and genital systems. AJNR. 1993;14:827–38.

274. Knorr JR, Ragland RL, Brown RS, Geiber N. AJNR. 1993;14:845–51.

275. Hardelin J-P, Petit C. A molecular approach to the pathophysiology of the x chromosome-linked Kallmann's syndrome. Clin Endocrinol Metab. 1995;9:489–507.

276. Letourneau PC, Condie ML, Snow DM. Extracellular matrix and neurite growth. Curr Opin Genet Dev. 1992;2:625–34.

277. Layman LC. Mutations in human gonadotropin genes and their physiologic significance in puberty and reproduction. Fertil Steril. 1999;71:201–18.

278. Trumpp A, Depew MJ, Rubenstein JLR, Bishop JM, Martin GR. Cre-mediated gene inactivation demonstrates that FGF8 is required for cell survival and patterning of the first branchial arch. Genes Dev. 1999;13:3136–48.

279. Rubenstein JLR. Personal communication, 2000.

The Neuromeric System: Segmentation of the Neural Tube

5

Michael H. Carstens and Harvey B. Sarnat

Introduction

The assembly of the head and neck involves anatomic sub-units, *developmental fields*, that are in genetic register with identifiable zones of the embryonic CNS. These zones, termed neuromeres, have been described using two distinct methodologies, ***descriptive neuroembryology (morphogenesis)*** and ***molecular genetics***. The anatomical approach is represented by the work of Fabiola Müller and Ronan O'Rahilly at the University of Fribourg, Switzerland and, more recently, at the University of California Davis. The genetic approach has many contributors but, because of their unique contributions to the prosomeric system and the elucidation of hindbrain subunits, work by Luis Puelles at the Universidad de Murcia, Spain, in conjunction with John Rubenstein at University of California, San Francisco, deserves particular attention. Although these authorities may differ in their definition of boundaries and neuroanatomical content of individual neuromeres, there is remarkable confluence in the configuration of the neuromeric system as a whole.

This chapter begins with the descriptive neuroembryology of the wife and husband team of Fabiola Müller and Ronan O'Rahilly (FMROR). It then proceeds to consider the prosomeric model of Luis Puelles and John Rubenstein (LPJR). We will look at how the neuromeric system arises, stage-by-stage. In the process, we shall develop a visual picture of the neuromeres and review their key neuroanatomical structures. Neuromeres are morphologically identifiable

transverse subdivisions perpendicular to the long axis of the brain with extensions onto both sides of the body. The Puelles-Rubenstein model has an additional requirement. A neuromere by definition also has a ventral-dorsal dimension such that it forms a box extending all the way from the floor plate to the roof plate. As we shall see, this strict definition proved a stumbling block for understanding the neuromeric map of the telencephalon but, after two decades of research, the issues have been worked out, making the prosomeric model simple, internally consistent, and clinically useful.

In the FMRO model, *primary neuromeres* (6) are present at stage 9; by stage 14, at total of 16 *secondary neuromeres* are described. Using genetic expression patterns, Puelles et al. have expanded the previous large zone of medulla, previously termed rhombomere 8, into 4 *pseudo-rhombomeres*. In this prosomeric model, we arrive at a grand total of 15 distinct neuromeric units (12 rhombomeres, 2 mesomeres, 3 diencephalic prosomeres, and 2 hypothalamic prosomeres). Relationships between the neuromeres, neural crest, paraxial mesoderm, and pharyngeal arches will be discussed.

For the uninitiated, neuroanatomical terminology may seem formidable. The reader is advised that a relatively limited number of structures will be used throughout this text, all of which have been selected from the work of O'Rahilly and Müller [1] and of Puelles [2]. These consistently identifiable structures give us landmarks along the neuraxis, both longitudinal and transverse. Appendices at the back of the chapter define abbreviations and neuroembyological terms. Key resources in neuroscience, neuroanatomy, and clinical neurology by Adams, Blumenfeld, DeMeyer, Patten, Nolte, and Purves are in the reference section.

As a methodology, you may find it helpful to look carefully at the diagrams of the prosomeric system, identifying each structure. It may seem painful at first, requiring some minutes of your time. But, if you study carefully the figures from the initial model of 1993 and familiarize yourself with

M. H. Carstens (✉)
Wake Forest Institute of Regenerative Medicine, Wake Forest University, Winston-Salem, NC, USA
e-mail: mcarsten@wakehealth.edu

H. B. Sarnat
Pediatrics, Pathology and Neurology, Emeritus
University of Calgary, Calgary, AB, Canada

© The Author(s), under exclusive license to Springer Nature Switzerland AG 2023
M. H. Carstens (ed.), *The Embryologic Basis of Craniofacial Structure*, https://doi.org/10.1007/978-3-031-15636-6_5

the terminology, you will appreciate how the reiterations of the prosomeric model in 2003 and 2013 achieve a hands-on mastery of a system that will make sense and become readily accessible. Once you are familiar with the prosomeric model, you will be better equipped to appreciate the work by Múller and O'Rahilly as you visualize how the brain develops between stages 9 and 17.

Historical Background

Historical aspects of brain segmentation are important because, for more than a century, many concepts were reiterated that we now recognize as erroneous based upon modern genetic and neuroembryological data not available to early investigators. Unfortunately, because these old concepts have been taught to generation after generation of neuroscientists and physicians, they are so ingrained that they are difficult to refute or, for some individuals, to reconsider with an open mind. Other old principles of neural tube segmentation remain sound and do not require revision. Thus, selective revision, not total repudiation, is what is required.

As early as 1861, von Höchstetter illustrated the floor of the fourth ventricle of a 5-week human fetus showing 8 transverse ridges alternating with transverse grooves that represent the first depiction of hindbrain segmentation. The term "*neuromere*" was coined by Orr in 1867 to describe

transverse neural units he observed in lizard embryos. He demonstrated these units histologically in hindbrain, midbrain, and forebrain. He also analyzed longitudinal zonation and bending of the brain axis. Orr's study is the historical root of the prosomeric model. Wilhelm His described an alternative neural model [3]. It included the concept of the neural tube as a collection of "plates": floor, basal, alar, and roof. His also described the sulcus limitans as the alar-basal boundary, the concept of the isthmus, and the idea that axial bending could cause morphogenetic deformations of the neural tube. Professor His chaired the committee for the first *Nomina Anatomica* in 1895; his concepts of longitudinal zonation and transverse boundaries (especially those of the isthmus) became widely accepted. These were clearly consistent with neuromeric models. By 1923, it was well-accepted that the hindbrain had subdivisions defined by ridges and sulci, with continuing further documentation [4, 5] (Figs. 5.1, 5.2, and 5.3).

During the early twentieth century, the concept of functional columns (floor plate, basal plate, alar plate, and roof plate) forming a common *bauplan* extending from spinal cord through brainstem was worked out. Alar plate and basal plate were found to have distinct functional zones (GVE, SVE, GVE, GVA, GSA, SSA), now well-known to all readers. At multiple junctures of this text, we endeavored to distinguish the separation between PANS fibers to nonvoluntary smooth muscle and glands coming from *dorsal*

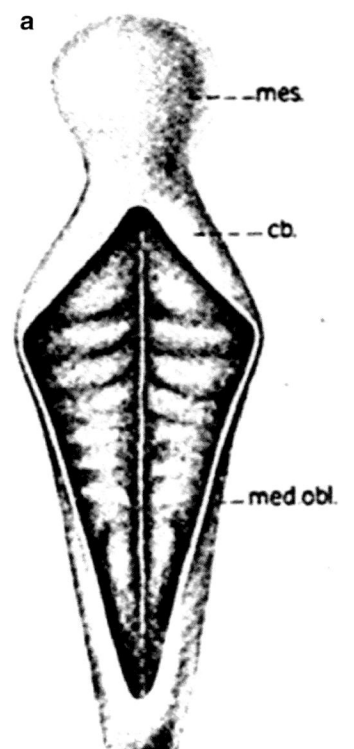
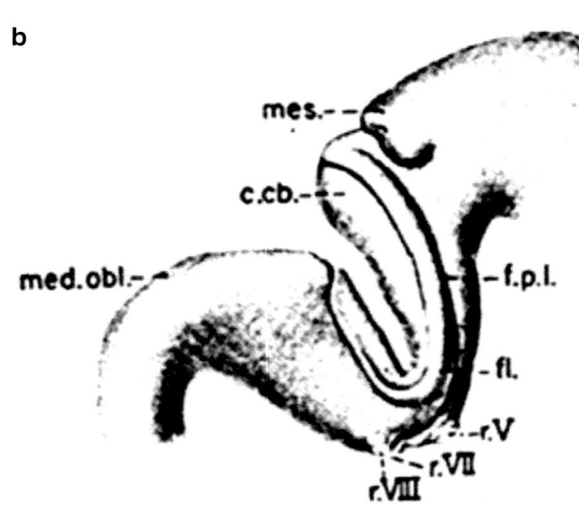

Fig. 5.1 Early discovery of neuromeres. Left: Reproduction of original drawing by von Hochstetter (1861) showing (**a**) alternating transverse ridges and grooves in the floor of the fourth ventricle in dorsal view. Right: The dorsal ridge bordering the rostral and anterolateral margins of the rhombencephalic-shaped fourth ventricle is the rhombic lip or primordial cerebellum elaborately described by His (1867). In (**b**), the cephalic flexure at the mesencephalic level and the slightly more caudal pontine flexure, as well as a cervical flexure at the transition of medulla oblongata (med.obl.) to cervical spinal cord, are clearly illustrated. (Courtesy of Harvey B. Sarnat, MD)

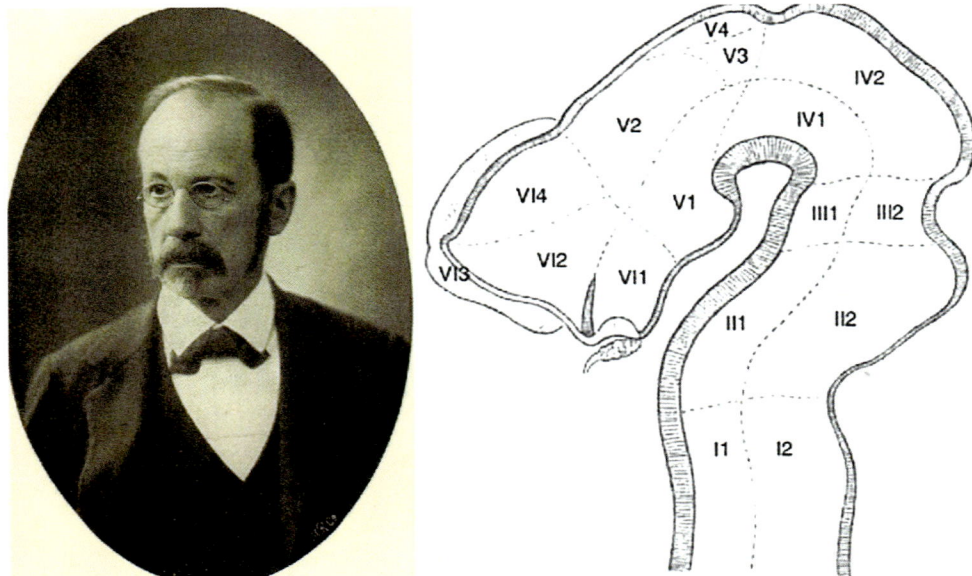

Fig. 5.2 His model 1895 anticipated prosomeric model of the forebrain. His (1895) recognized six transversal units (I–VI) along a bent neural tube. The alar-basal boundary was depicted parallel to the roof and floor plates. Extending to the entire lateral wall. Segmentation was as follows: domain I, myelencephalon; domain II, metaencephalon; domain III, isthmus; domain IV, midbrain; domain V, diencephalon; domain VI, hypothalamus, eye and telencephalon. Basal plates indicated by 1 (I.1–VI.1) and alar plates by 2 (I.2–VI.1). Domain VI.3 is olfactory bulb; V.3, metathalamus; and V.4, habenula. Notes The overall course of sulcus limitans indicates the alar-basal boundary.

Hypothalamus is limited to VI.1 (optic) and V.1 (mamillary) regions. As we now understand it: (1) the optic part is really the tuberal basal part plus a section of alar plate, including suprachiasmatic primordium; (2) optic part is continuous with telencephalic region VI.2 and VI.4. Mamillary pouch (V.1) relates dorsally to thalamus (V.2). V.3 represents pretectum and V.4 the epithalamus. The term neuromere was first coined by Orr in 1887. (Left: Reprinted from Wikimedia. Retrieved from: https://commons.wikimedia.org/wiki/File:Wilhelm_His.jpeg. Right: Reprinted from His W. Die Anatomische Nomenclatur. Nomina Anatomica. Arch. Anat. Entwickelungsges 1895 (Suppl.), 1–180)

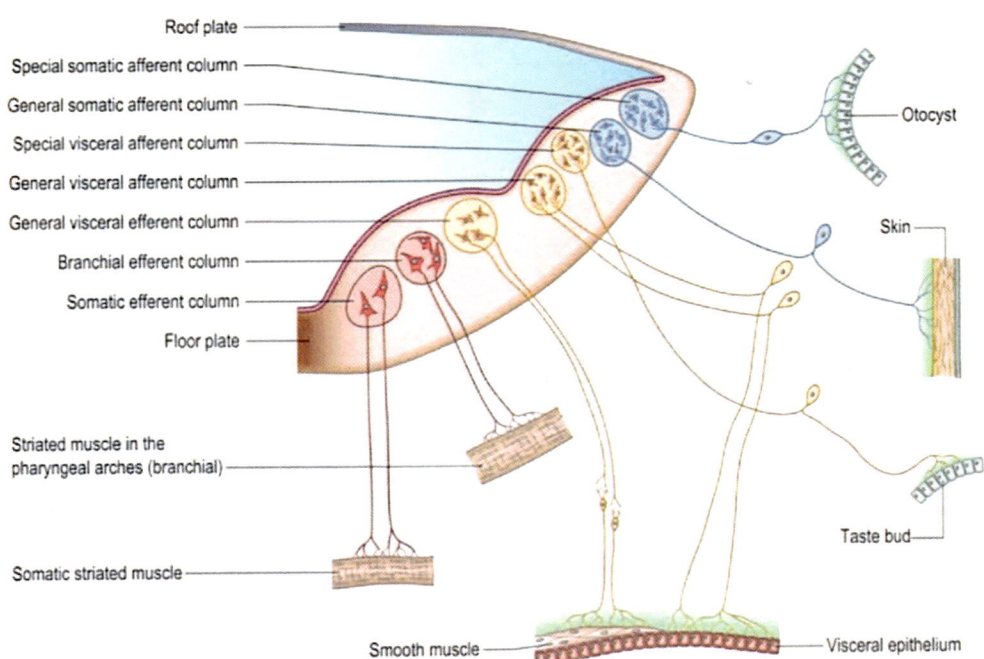

Fig. 5.3 Functional columns of the brainstem. These will be continued into the spinal cord. Note that general visceral efferent, when applied to pharyngeal arch muscles, is a misnomer. Traditional use of the term GVE in the brainstem lumps together branchiomotor and PANS motor (vagus) to the smoot muscle and glands of gut all in one ambiguous term, *nucleus ambiguus*. The striated nature of branchiomeric muscle is

well-established. Here the two functions are clearly separated. In spinal cord neuromeres c1–c4 (and maybe down to c6), GVE is the extension of nucleus ambiguus, r6–r11. Dorsal motor nucleus terminates in the brainstem. (Reprinted from Lewis, Warren H (ed). Gray's Anatomy of the Human Body, 20th American Edition. Philadelphia, PA: Lea & Febiger, 1918) *motor nucleus* r8–r11 and motor fibers for voluntary striated branchiomeric muscle of arches 3–5 coming *from nucleus ambiguus* that continue into *spinal accessory nucleus* c1–c4. What is not widely appreciated are the efforts of neuro-

anatomists to determine if this developmental plan extended further forward into the forebrain, and, if so, in what way (Figs. 5.4 and 5.5).

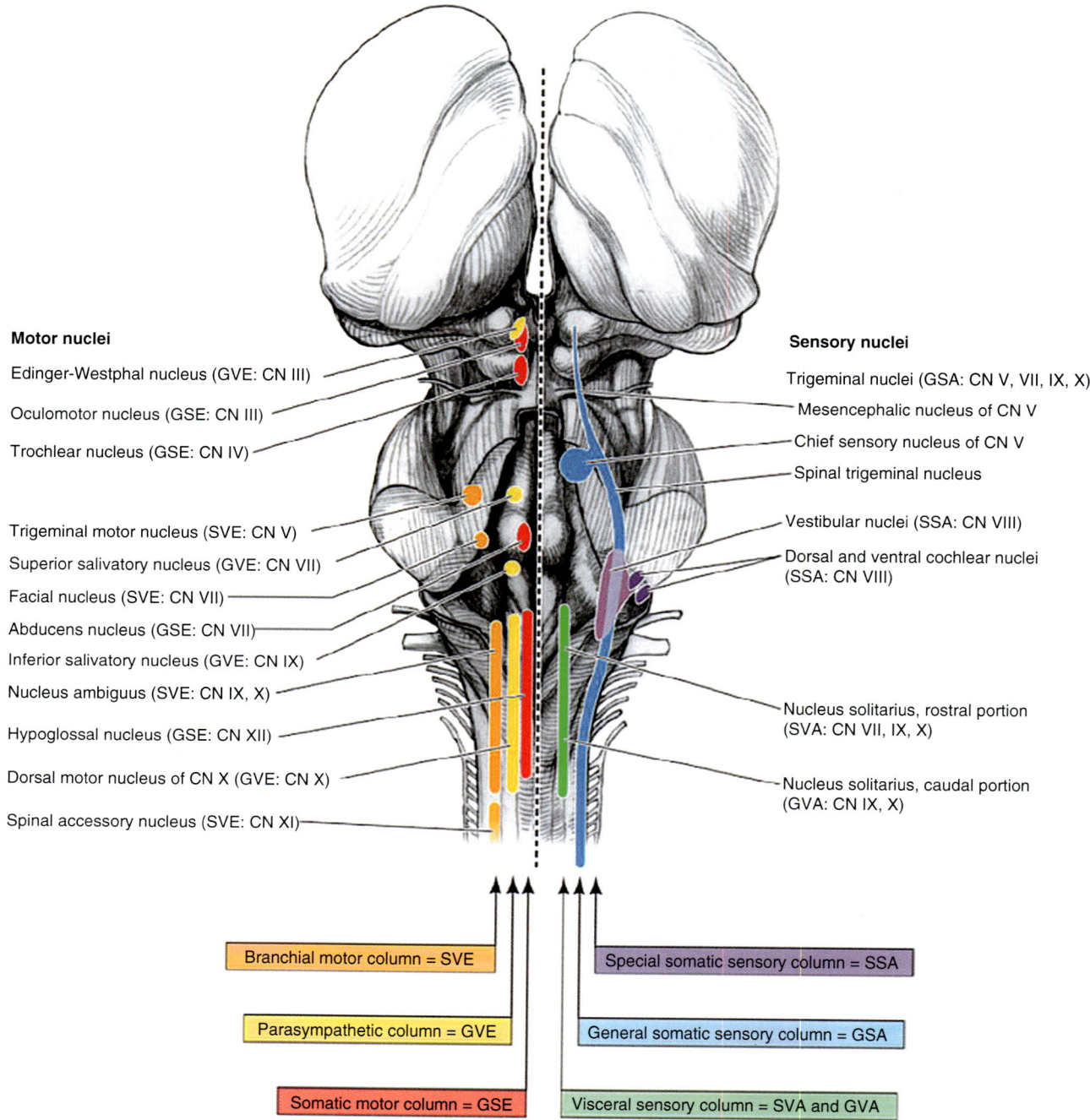

Motor nuclei

Edinger-Westphal nucleus (GVE: CN III)

Oculomotor nucleus (GSE: CN III)

Trochlear nucleus (GSE: CN IV)

Trigeminal motor nucleus (SVE: CN V)

Superior salivatory nucleus (GVE: CN VII)

Facial nucleus (SVE: CN VII)

Abducens nucleus (GSE: CN VII)

Inferior salivatory nucleus (GVE: CN IX)

Nucleus ambiguus (SVE: CN IX, X)

Hypoglossal nucleus (GSE: CN XII)

Dorsal motor nucleus of CN X (GVE: CN X)

Spinal accessory nucleus (SVE: CN XI)

Sensory nuclei

Trigeminal nuclei (GSA: CN V, VII, IX, X):

Mesencephalic nucleus of CN V

Chief sensory nucleus of CN V

Spinal trigeminal nucleus

Vestibular nuclei (SSA: CN VIII)

Dorsal and ventral cochlear nuclei (SSA: CN VIII)

Nucleus solitarius, rostral portion (SVA: CN VII, IX, X)

Nucleus solitarius, caudal portion (GVA: CN IX, X)

Branchial motor column = SVE

Parasympathetic column = GVE

Somatic motor column = GSE

Special somatic sensory column = SSA

General somatic sensory column = GSA

Visceral sensory column = SVA and GVA

Fig. 5.4 Functional columns of the brainstem in situ. In the spinal cord branchiomotor and parasympathetic motor functions are clearly separated. Neuromeres c1–c4 (and maybe down to c6) are GVE; they represent the extension of *nucleus ambiguus* (orange). PANS (vagus) to the viscera comes from *dorsal motor nucleus* (yellow), which is interposed between *hypoglossal nucleus* (red) and nucleus ambiguus. (Reprinted from StackExchange. Biology: What are cranial nerve nuclei? Retrieved from: https://biology.stackexchange.com/questions/44599/what-are-cranial-nerve-nuclei. With permission from Creative Commons License 4.0: https://creativecommons.org/licenses/by-sa/4.0/)

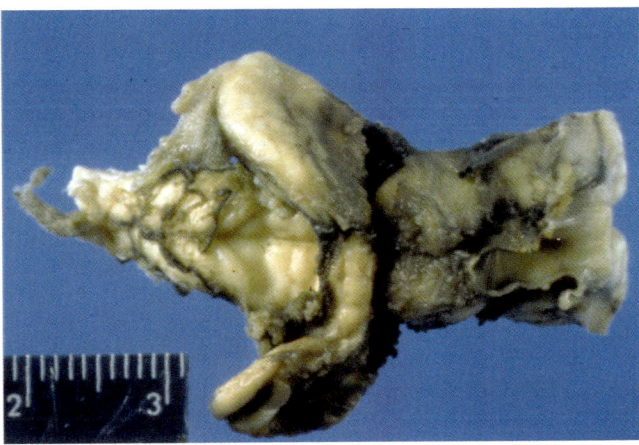

Fig. 5.5 Rhombomeres at autopsy. Dorsal view of the brainstem of a 5-month-old infant born at term exhibits maturational arrest corresponding to about 5–6 weeks gestation. The exposed floor of the fourth ventricle (left) exhibits 8 transverse ridges and grooves denoting embryonic rhombomeres. The rhombic lip (rh.lip) also is evident at the dorsal rim of the rostral fourth ventricle and the anterior medullary velum has failed to form. The collicular plate (col) is forming in the mesencephalic tectum. Swine influenza vaccine administered to the mother in the early first trimester of pregnancy was implicated as the etiology. (Courtesy of Harvey B. Sarnat, MD)

Just as neuromeric brain theory reached its zenith, unrelated findings regarding the functional analysis of motor and sensory nerves cast doubt upon its premises. It was discovered that the components of nerves (either motor or sensory) either originate from, or project on a distinct longitudinal (or columnar) domain of hindbrain or spinal cord. These ascending or descending branches did not appear to respect neuromeric boundaries. The basic model of the hindbrain and spinal cord thus seemed to be columnar, not segmental. After a half century of attention, the transverse model of the brain segmentation fell into disuse. Instead, researchers in the US, such as J.B. Johnston and C.J. Herrick, focused on the functional connotations of "longitudinal columns of the brain." In 1910, Herrick postulated that the diencephalon contained columnar subdivisions [6]. These were considered to be continuous with those of the brainstem and would extend rostrally into the telencephalon, causing confusion by assigning the hypothalamus as a part of the diencephalic basal plate.

The columnar model became widely accepted after World War II. Herrick's work, reiterated in his 1948 book, published almost 40 years after his initial and unchanged 1910 version, remains today as the dogma of diencephalic structural subdivisions. Until the mid-1980s, techniques of experimental neuroscience such as axonal degeneration, axonal transport, and electron microscopy-produced data seemed compatible with the columnar model. All students of neuroanatomy have learned (erroneously) that epithalamus, thalamus, and hypothalamus are longitudinal columns of the forebrain. In the face of this dogma, segmental observations

(consistent with neuromeres) were explained away as transient developmental phenomena, irrelevant for understanding the mature, functioning brain.

The columnar model had remarkable longevity, being reiterated by Swanson as late as 2003, but it proved to be an obstacle to progress because it was too simplistic and did not address causal mechanisms. Columnar theory would not be debunked until the advent of molecular genetics. Due to its entrenched position, we shall contrast it with the neuromeric model further on.

Further interest in neuromere reappears in the premolecular era in studies which defined neuromeres by (1) *neuroanatomical content* (e.g., grey matter nuclei; white matter tracts); (2) patterns of *cell population expansion and differentiation*; (3) reduced *neuroepithelial mitotic activity* at boundaries between segments relative to the mitoses in the middle of segments [7]; and (4) *lineage restriction* that concentrated neuroblasts of a single type and prevented their migration in the longitudinal axis of the neural tube to enable the formation of brainstem nuclei.

The real revolution in our understanding of neuromeres began in the late 1980s and throughout the 1990s as *gene expression domains* began to be mapped out, first in the hindbrain, and later, in the midbrain and forebrain [4, 5, 8–17]. This work has culminated in the prosomeric model of Puelles and Rubenstein, first proposed in 1994, revised in 2003, and further refined in 2013. Two decades of molecular research have established the neuromeric model as a morphological paradigm based on developmental mechanisms which systematize causally the anatomy of all brain structures, based on *genetico-architectural patterns*, throughout all vertebrates.

What criteria can be used to define the physical boundaries of a neuromere? In the 1990s, neuromeric segmentation of the hindbrain was even better defined with the integration of molecular genetics with morphogenesis [11, 18]. The advent of scanning electron microscopy enabled documentation using newer neuroanatomical techniques. With advancing fetal age, ridges and grooves no longer are evident upon gross inspection of the fourth ventricular floor, though they may persist pathologically in cases of maturational arrest (Fig. 5.5). Nevertheless, residua of the initial embryonic neuromeric segmentation persist even in the mature adult brain.

How do neuromeres work? Neuromeres provide compartmentalization that enables neuroblasts of similar lineage to cluster together as neuroanatomical nuclei by providing a *barrier to cellular migration* in the longitudinal axis of the neural tube. The barrier in the grooves between neuromeres is partly physical and partly chemical. A transitory thin transverse sheet of primitive progenitor/glial cells forms from neuroepithelium, its cells similar to radial glial cells that appear at a later stage of development for a different purpose of guiding migratory neuroblasts [19, 20]. The cells

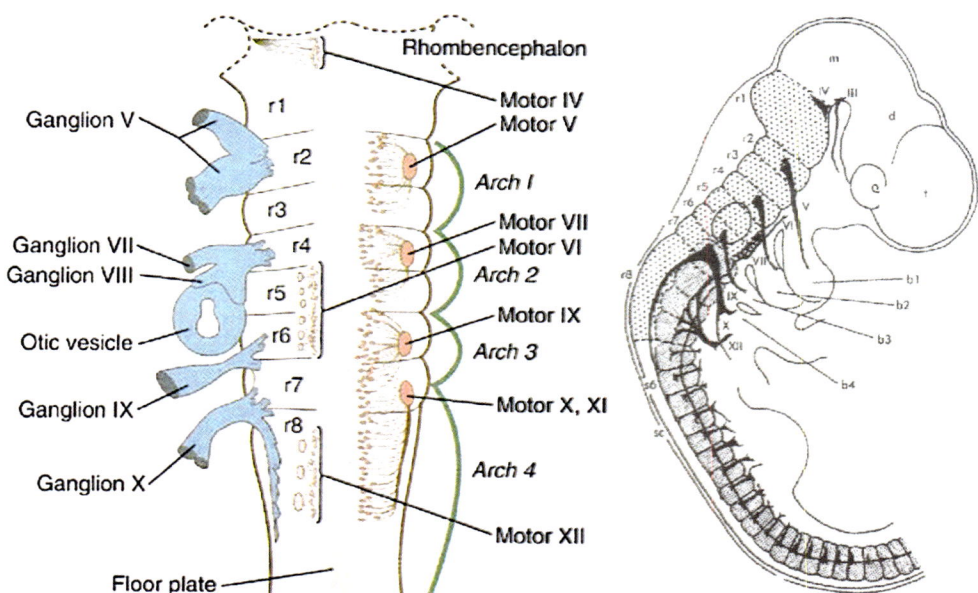

Fig. 5.6 Initial model divided medulla into eight rhombomeres. Drawing of segmentation of the neural tube in the sagittal plane in a stage 18 chick embryo (Hamburger and Hamilton, 1951), with superimposed rhombomeres r1–r8 [11]. The spinal cord (sc) is seen as an extension of r8 rather than as an intrinsically segmented structure. Neuromeric segmentation of the diencephalon (d) and telencephalon (t) is not indicated in this drawing. (Reprinted from Lumsden A, Keynes R. Segmental pattern of neuronal development in the chick hindbrain. Nature 1989;337:424–428. With permission from Springer Nature)

of this thin glial sheet secrete glycoproteins into the extracellular space that serve to repel or at least impede cellular migration through this barrier, but it is not a total physical barrier because it enables growth cones of longitudinal axons to pass through unimpeded.

This same principle of a selective chemical barrier is again used, at a slightly more advanced gestational age, in the dorsal and ventral median septa of the spinal cord and brainstem to prevent ascending and descending axonal growth cones of developing longitudinal tracts from aberrantly decussating, while simultaneously facilitating the crossing of local commissural axons [21, 22]. Mitotic activity in the neuroepithelium is less at the site of the barrier than in the center of the neuromere, hence rhombomeres are centers of cellular proliferation, whereas boundaries contain populations of static cells [7]. The neuromeres also are compartments for the expression of some, but not all, developmental genes that are intrinsically involved with segmentation of the neural tube, particularly rhombomeric homeobox genes of the families of *Hox*, *Wnt*, *En*, *Pax*, and others [23, 24]. Rhythmical electrical activity in cranial nerves V, VII, IX, X, and XII is generated at stages 24–36 in the chick embryo, an electrophysiological correlate of hindbrain segmentation [25] (Figs. 5.6 and 5.7).

The concept of neuromeres thus evolved from a strictly anatomical embryonic segmentation of the neural tube (which remains valid) to the supplementary demonstration that neuromeric morphological units also had highly specific profiles of predictable gene expression. In many instances,

the distinction between one neuromere and its neighbor is not necessarily structural (specific neuroanatomic landmarks), but one of homeotic gene expressions. For example, when we look at cross sections of spinal cord between T4 and T5, how can we differentiate one from the other? The ascending and descending tracts extending through these neuromeres are virtually the same. So what sort of cue determines when a new peripheral nerve unit should emerge? The answer is a switch in gene expression. T5 is distinct from T4 by virtue of a new homeotic pattern. Furthermore, genetic information contained within a neuromere is shared with outlying target structures such as muscle and bone. For this reason, a neuromeric model enables surgeons interested in the structural organization of the head and neck to organize and conceptualize anatomy that otherwise seems confusing at best and at times utterly irrational.

The Number of Neuromeres Is Specific for Each Region of the Neural Tube

The numerical count of neuromeres at various regions of the neural tube also has been an evolving and controversial concept. In the original anatomical model and throughout the twentieth century, it was an accepted concept that there were eight rhombomeres, the eighth incorporating the entire spinal cord. More recent evidence, mainly from molecular genetic studies, enables reassessment of the number of neuromeres in each part of the neural tube. The number of rhom-

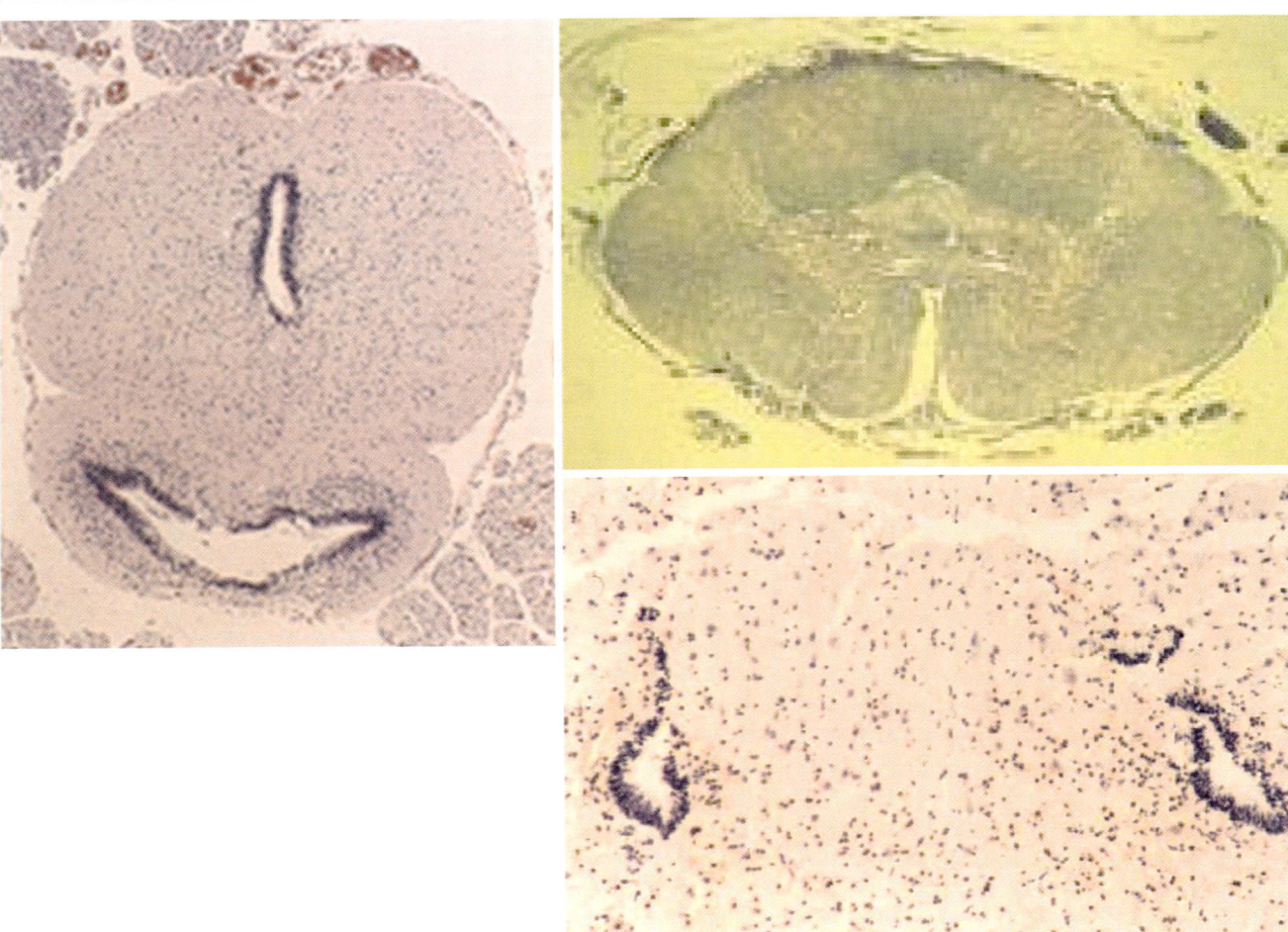

Fig. 5.7 Definition of the midbrain. Left: Transverse section of normal human fetus of 20 week gestation, at the level of the rostral end of the conus medullaris. Two ependymal-lined central canals are seen in the midline. The dorsal is the caudal extension of the canal that extends from the cervical spinal cord; the ventral central canal forms independently within the solid core of secondary neurulation caudal to the closing posterior neuropore. The two canals only occasionally meet for continuity, and both are normal transitory fetal structures. H&E. Right: Transverse sections of cervical spinal cord of a 6-year-old boy with cerebral cortical lissencephaly type II (Walker-Warburg syndrome) and dysplasias of the brainstem and cerebellum as well. There are two cen-tral canals in the horizontal plane, side by side. Both are dysplastic though patent, and both canals lack a roof plate and dorsal median septum, as confirmed by vimentin and nestin immunoreactivities (not shown here). This duplication is not secondary neurulation, which is never seen at the cervical level in any case. In normal secondary neurulation, the two central canals are always one above the other in the vertical axis. Side-by-side canals in the horizontal axis, by contrast, are due to upregulation of a dorsalizing gene (e.g., BMP, PAX) and are pathological; this morphology of duplication is not encountered in fetuses or children with a normal CNS or at any segmental level of the spinal cord. (Courtesy of Harvey B. Sarnat, MD)

bomeres is now 12. The spinal cord consists of 31 myelomeres. The forebrain, however, has been the most problematic. Four to six prosomeres have been described that include segments destined to become diencephalic and telencephalic structures of the mature brain but more important than the exact count, is that the original concept of sequential segments in the longitudinal axis, as with the midbrain, hindbrain, and spinal cord is now incorporated into a new model that encompasses vertical forebrain segments perpendicular to the longitudinal axis.

Despite this perspective, various authors expressed minority views that transverse subdivisions were not only compatible with (and in fact required) for longitudinal segmentation.

Likewise, interpretation of longitudinal columns cannot be done without distinguishing transverse components.

The motor neurons of the brainstem represent an excellent example of the interplay between both systems. All medical students are familiar with the functional classification of the cranial nerves into categories such as general somatic efferent (GSE) and general visceral efferent (GVE). As discussed elsewhere in this text, striated muscles of the pharyngeal arches are longer considered visceral. The older terms have yielded to medial and lateral motor columns. However, we might ask ourselves, why are cranial nerves III and IV discontinuous with VI? What is it about rhombomeres 2 and 3 such that no axial motor column is present? The genetic interplay between target myoblasts in the outlying somitomeres is a factor. The motor nuclei of the second arch reside in r4–r5. Lateral rectus and accessory lateral rectus have absolutely no relationship to the second pharyngeal arch. Thus, it is not surprising that we find motor nuclei for these muscles in r5–r6. But the coincidence of lateral rectus palsy with Möbius syndrome makes perfect sense as r5 is common to both, with the axons of facial motor neurons looping dorsally around the abducens nucleus so that small tegmental watershed infarcts of the fetal brainstem produce typical Möbius syndrome [26] apart from genetic mutations that can cause the same phenotype [27].

Spinal Cord: 31 Myelomeres

The spinal cord no longer is considered a single caudalmost rhombomere, but is demonstrated to be intrinsically segmented despite lack of segmentation of the notochord that induces differentiation of the floor plate ependyma and motor neurons, and despite that in the mature spinal cord parasagittal sections fail to show periodic clustering of motor neurons within their column in the ventral horn and a single continuous unsegmented fetal central canal is seen in most of the spinal cord that is derived from primary neurulation. The periodic clustering of nerve roots and vessels is secondarily imposed by the strongly segmental somites from the paraxial mesoderm. The earliest intrinsic spinal cord motor neurons show no segmental patterns of differentiation, unlike the hindbrain [28].

Nevertheless, anatomical evidence of regional specialization of the spinal cord has long been recognized in the autonomic nervous system, with cervico-sacral regions corresponding to the parasympathetic system and thoraco-lumbar regions subserving the sympathetic system including its preganglionic neurons.

The C4 spinal level has several unique properties that also support intrinsic spinal cord segmentation: (1) It is the site of more than 90% of axons of the phrenic nerves to innervate the diaphragm. (2) Neural crest from C4 produces the spine of the scapula and C4 lateral plate mesoderm contributes to the body of scapula. (3) Level C4 is the breakpoint for the transition between muscles attaching the pectoral girdle to the head (post-hyoid strap muscles) and those connecting it with the upper extremity.

Further evidence that the spinal cord is not derived from a single neuromere is that the conus medullaris that forms caudal to the point of closure of the posterior neuromere is *secondary* neurulation. Rather than folding from a flat plate or placode to form a tubular structure surrounding a cavity that becomes the central canal, in secondary neurulation there is no folding but rather a solid cone of primitive neural tissue and a central cavity canalizes in its center. At the point where primary and secondary neurulation meet, the central canals from each do not always join to form a continuous canal. In

transverse histological sections at this level in the normal fetus in the late first and second trimesters of gestation, two ependymal-lined central canals overlap in the midline vertical axis, the dorsal canal being derived from primary neurulation rostral to that level and the ventral from secondary neurulation caudal to that level. Thus, two central canals in the vertical axis are normal at the lower sacral level, but two canals situated side-by-side in the horizontal axis are always pathological as duplication of structures from upregulation of a dorsalizing gene, nearly always associated with dysgenesis of the CNS at various other levels including the prosencephalon and rhombencephalon; such horizontal plane duplication of the central canal may be found even in the cervical region (Fig. 5.7).

But the strongest evidence of true spinal cord segmentation is the nonuniform differential expression of genes. The homeotic code expressed in each distinct somite level reflects an original pattern affecting all tissues outlying the spinal cord, neural crest, mesoderm, and endoderm. As such, the *HOX* code of any given neuromeric level imposes and modulates the neuroanatomical structure at that level, imposing production of motor and sensory neurons and nuclei that innervate the peripheral tissues of that neuromere (Fig. 5.8).

Hindbrain: 12 Rhombomeres

The neuromeric model by Puelles [29, 30] subdivides the eighth rhombomere of Lumsden and Keynes [4] into four distinct units, r8–r11, one for each nucleus of vagus and hypoglossus. The various vagal nuclei are situated at slightly different levels of the mature medulla oblongata though with considerable overlap. The hypoglossal nucleus is a paired column in the medullary tegmentum, but mainly in the caudal end. Segmentation segregates groups of progenitor cells early in hindbrain development [31]. Hox genes are then involved in regional specification [32]. Motor neurons of the glossopharyngeal and hypoglossal complexes or other more rostral motor nuclei do not project to the periphery before rhombomeres have formed [33].

Boundary zones between rhombomeres are demarcated not only grossly by transverse grooves, but microscopically by the formation of a thin loose transverse membrane of cellular processes, similar to radial glial fibers, that exhibits immunoreactivity to both glial (e.g., glial fibrillary acidic protein; CD-15) and some neuronal (e.g., neurofilament) proteins as well as to primitive cell intermediate filaments such as vimentin, nestin, and actin [7, 11, 20]. Furthermore, mitotic proliferation of the neuroepithelium at these boundary zones is markedly reduced compared with the center of the rhombomeres [7]. The boundary zones limit the longitudinal migration of neuroblasts, to concentrate neurons of the same type for forming cranial nerve nuclei, while not impeding growth cones of long ascending and descending axons. Longitudinal columns of neurons of the same type can form

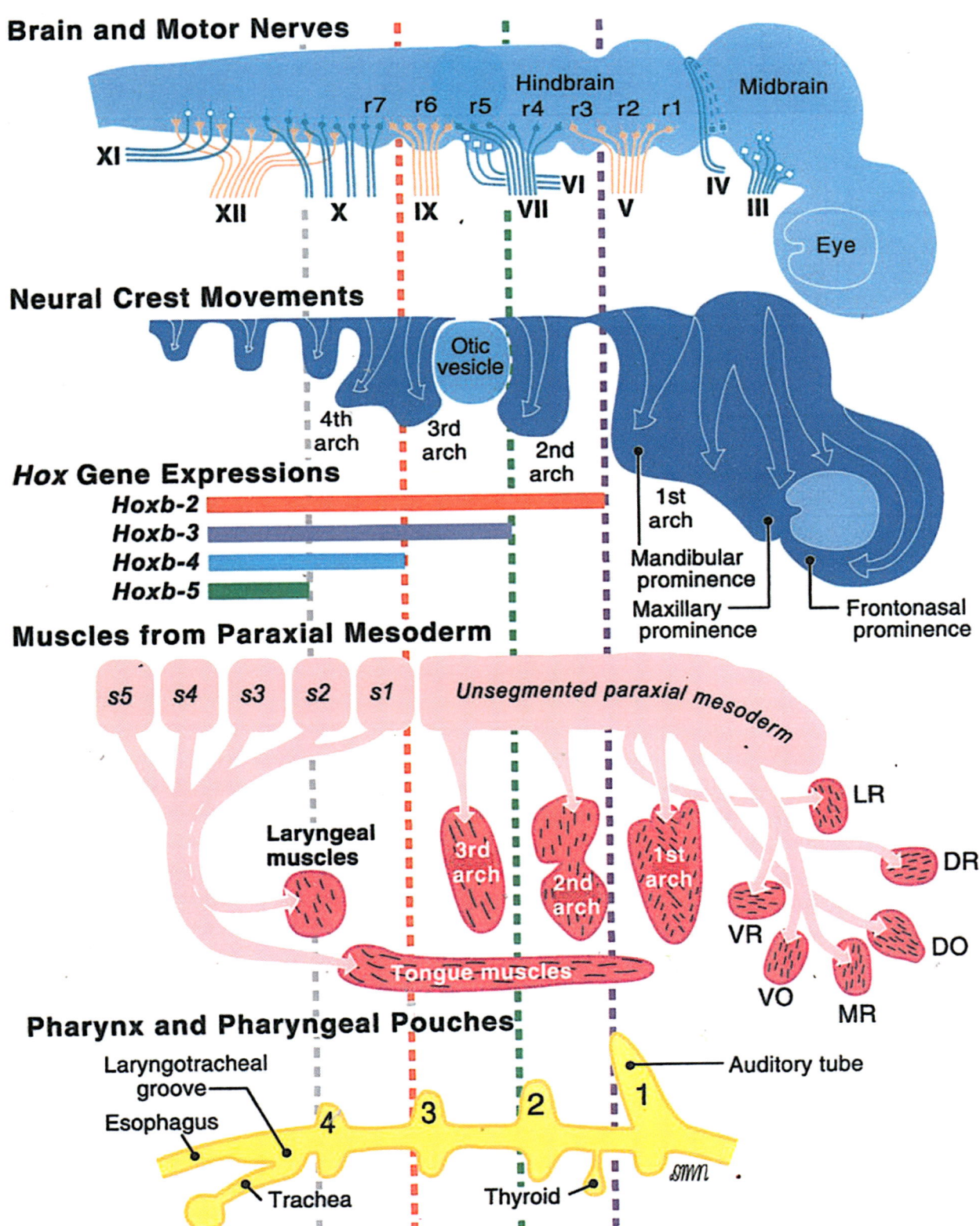

Fig. 5.8 Homeotic mapping of the rhombomeres to outlying tissues. Pharyngeal arches are coded by pairs of neuromeres. Staggered, lateral views of all internal tissue layers in an early avian embryo. These illustrate the changes in locations of each population and the spatial relations among them. Neural crest progenitors, cranial nerves, and myogenic primordia for each branchial arch all arise at the same axial level and maintain this close registration throughout their dorsoventral movements. For example, crest cells that will populate the second branchial arch arise from the same axial location (rhombomere 4) as the seventh cranial nerve and the second arch muscles it will innervate. By contrast, the periocular neural crest, extraocular muscles and the motor nerves that innervate them all arise at separate axial locations and do not establish stable relations until all have reached their sites of terminal differentiation. (Reprinted from Noden DM, Trainor PA. Relations and interactions between craniofacial mesoderm and neural crest populations. J Anat 2005; 207(5): 575–601. With permission from John Wiley & Sons)

in some lower cranial nerve nuclei (e.g., nuclei solitarius and ambiguus of the vagus) because these neurons are generated in two or more adjacent rhombomeres and their nuclei become continuous with maturational regression of the boundary zones between rhombomeres. Such nuclear columns developing in anatomical proximity facilitate reciprocal synaptic connections forming at a later stage of maturation, as exist between the nucleus solitarius and the nucleus ambiguus. Motor centers in the hindbrain develop within pairs of rhombomeres and correspond to branchiomeres [4]. Rostrocaudal migration across rhombomeric boundaries also can contribute to the creation of longitudinal columns of similar motor neurons that comprise mature motor nuclei [34]. The spinal accessory nerve (cranial nerve XI) is likely derived initially from the vagus complex (cranial nerve X) and differs greatly in different classes of vertebrates; the spinal portion may actually have evolved independently from the bulbar portion [35]. In mammals, it is located at the level of neuromeres c1–c4. Immunocytochemical expression of enzymes of neurotransmitter biosynthesis and degradation and of gene expression also define individual rhombomeres [36, 37]. Rhombomeric organization is transitory in the embryo, but does not resolve as early as previously thought and persists in the ventricular zone neuroepithelium, thus restricting longitudinal expansion of the neuroepithelium [38], but rhombomeres do not persist into late fetal, neonatal, or adult life.

Rhombomeres 3 and 5 are the only ones from which motor nerves do not emerge because they are suppressed but not destroyed by the gene *Krox-20* (mouse, =*EGR2*, human), so that the nerve roots from trigeminal sensory neurons in r2 and r3 exit together from r2 and facial motor neurons from r5 exit from r4. This explains why the maxillary (V2) and mandibular (V3) divisions of the trigeminal nerve emerge from the brainstem as a single trunk which then divides peripherally into separate V2 and V3 branches. Host transplantation in chick embryos that reverse the polarity of r3 results in misprojection of a continent of trigeminal axons via the facial nerve exit point; segmental mechanisms intrinsic to the brainstem specify the innervation pattern of motor neurons, not specificity of early aberrant projections in the periphery [39].

An important clarification of the morphological concept of the "isthmus" at the mesencephalic-rhombencephalic boundary was promoted by the leading developmental neuroanatomists of the mid-twentieth century, particularly Elizabeth Crosby [40, 41]. These papers recognized the isthmus not only as a boundary between midbrain (mesencephalic) and pontine (metencephalic) rhombomeres, but as an organizing center of morphogenesis and also important for hypothalamic connections with brainstem centers [41]. This body of work would prove prescient for understanding isthmic signals important for the induction of the midbrain and forebrain. Note that the neural folds above the isthmus do not produce neural crest. For this reason, peripheral derivatives such as zones of V1-innervated dura or bone fields cannot be mapped to the isthmus.

From a phylogenetic perspective, it was noted early that neuromeres are more conspicuous in the embryos of higher (reptiles, birds, mammals) than in lower (lamprey, hagfish) vertebrates in the floor of the fourth ventricle [42, 43]. The protochordate *Amphioxus* does not have evident rhombomeres or may possess only a single primordial rhombomere.

Midbrain: 1 or 2 Mesomeres

It has long been debated whether the midbrain region corresponded to one or to two mesomeres and this issue still is not definitively resolved; within the rostral half the superior (visual system) colliculi and red nuclei form; in the caudal half the inferior (auditory system) colliculi and brachium conjuntivum (decussation of the superior cerebellar peduncles) form. Some structures extend through in both: oculomotor nuclei, substantia nigra, and interpeduncular nucleus. Descending corticospinal, corticobulbar, and corticopontine tracts traverse the entire midbrain in the ventrally located cerebral peduncles. Strictly neuroanatomical criteria alone suggest that there are two mesomeres. By considering genetic expression as well, however, Puelles [29] considered the mesencephalic neuromere (m) to be a single neuromere.

The transition from the caudal mesencephalon (r0) to r1 is sometimes designated the *isthmus*, particularly in older literature before the molecular genetic era [40, 41]. The caudal half of the mesomere (formerly sometimes designated rhombomere 0 or r0) and r1 are the origin of the cerebellum and anterior medullary velum. However, other parts of the cerebellar network are derived from diverse neuromeres: in the *Guillain-Mollaret triangle*, the deep cerebellar nuclei are derived from r2; the inferior olivary nuclei from r6; the red nucleus from the rostral mesencephalon. The pontine tegmental vestibular nuclei are derived from r3 and r4; the ventrolateral thalamic nucleus is of prosomeric origin. The relation of gene expression in the isthmus or mesomere to cerebellar development was shown in the late 1990s [8, 9, 16, 17].

A rare human congenital malformation is selective deletion of the entire mesencephalic neuromere, not just the rostral or caudal half, accompanied by cerebellar hypoplasia and suspected to be secondary to a mutation in the (*Engrailed*) *EN2* gene [44]. This neuromeric deletion is similar to a murine model of *En2* or *Wnt* deletions [10, 45–47]. Thus, alternation of this gene destroys the homeotic coding of m1–m2, essential for production of midbrain and the cerebellus that arises from m1.

Diencephalon/Secondary Prosencephalon: Five Prosomeres (p1–p3, hp1–hp2)

Terminology trap: Traditional neuroanatomy considers the prosencephalon to have two parts: caudal diencephalon and rostral telencephalon. This model was proven to be too simplistic. In the current model, prosencephalon is reiterated as diencephalon with three prosomeres, and in front, a secondary prosencephalonconsisting of hypothalamus divided into two prosomeres: caudal hp1, peducular hypothalamus (PHy), and rostral hp2, terminal hypothalamus (Thy). In this model, the Telencephalon is a gigantic derivative that buds bilaterally from the alar hypothalamus.

The number of diencephalic and prosencephalic neuromeres has been even more controversial than the hindbrain and spinal cord because the anatomical landmarks between putative neuromeres are not as easily demarcated as in the rhombencephalon. The conventional view has been that there were four diencephalic neuromeres that include the epithalamus (pineal), thalamus, hypothalamus, and optic stalk and cup on each side. Secondary prosencephalic neuromeres are the most difficult to distinguish in the embryo. These have been counted between 3 and 5, though the deep telencephalic nuclei (corpus striatum and globus pallidus), hippocampus, amygdala, and cerebral neocortex are easily identified in the mature brain [48]. The traditional view divided the forebrain into three or four caudal (diencephalic) prosomeres and a complex rostral unit containing hypothalamus, eyes, and cerebrum (telencephalon) of variable number according to different authors, ranging from 2 to 4 [49] (Fig. 5.9).

Prosencephalic regions are demarcated by markers at boundary zones, such as CD-15 radial glia, evidence that *neuromeric segmentation exists in the forebrain exactly as it does in the hindbrain* [20].

The problem of lack of consensus among neuroembryologists in specifying the precise number of forebrain neuromeres relates back to the 1910 model proposed by Herrick in which he conceptually simply extended the segmentation of the neural tube associated with underlying notochord to prechordal levels rostral to the mesencephalon where the notochord ends, assuming that some other embryonic structure would substitute for the notochord at the rostral end of the neural tube. The columnar model held that the hypothalamus was (1) the ventralmost longitudinal column of the diencephalon; and (2) that it was intercalated in between midbrain and telencephalon. Herrick continued to assert into the mid-twentieth century that the longitudinal axis of the neural tube was straight, running from the olfactory bulb caudally to the spinal cord [50], and many other twentieth century neuroembryologists and neuroanatomists supported this simplistic concept. This model was attractive because it seems to extend the functional columns of the brainstem to the cerebrum but, with the advent of molecular mapping, it has been definitively debunked [51] (Fig. 5.10).

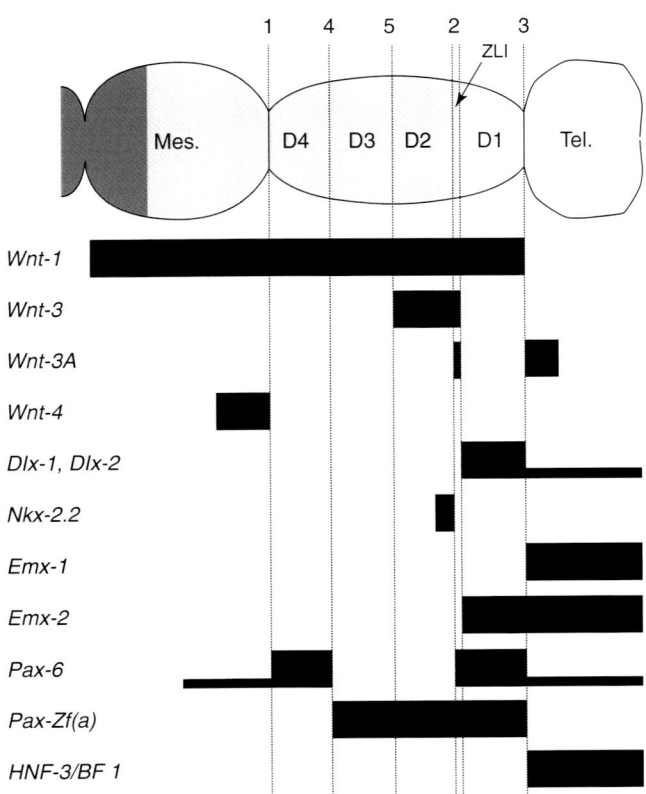

Fig. 5.9 Early neuromeric mapping of diencephalon. Expression domains (black horizontal bars) of 12 putative regulatory genes from mouse frog and zebrafish that demonstrate neuromere boundaries in the diencephalon. Boundaries were defined by morphologic studies and position of axons. In the mouse, neuromeres appear at 9–11 days postcoitum. Light shading shows engrailed homologue, En-2, after grafting of mesencephalic neuroepithelium compared with region of mesencephalon that is normally expressing En-2 (dark shading). Numbers above the neuromere boundaries refer to the order of development: caudal-to-cranial. This compares with hindbrain rhombomere development which is cranial-caudal from r0 to r11. (Reprinted from Figdor MC, SternCD. Segmental organization of embryonic diencephalon. Nature 1993; 363(6430):630–634. With permission from Springer Nature)

The traditional view of classification of mature forebrain structures as either diencephalic or telencephalic also is challenged in reconsideration of the revised concept of prosomeres. The most recent data indicate that the hypothalamus is more closely related to telencephalic structures than to thalamus or epithalamus as diencephalon [53]. Furthermore, the thalamus and olfactory bulb (including its own intrinsic thalamus) are derived from prosomere 2, but the conventional classification assigns the thalamus to the diencephalon and the olfactory bulb to the telencephalon.

The model here proposed does not accept these conventional assumptions. Our concept, initiated by Puelles et al. [54, 2], postulates that in all vertebrates, the neural tube undergoes morphogenetic bending that creates marked flexures of its longitudinal axis. This is most marked at the

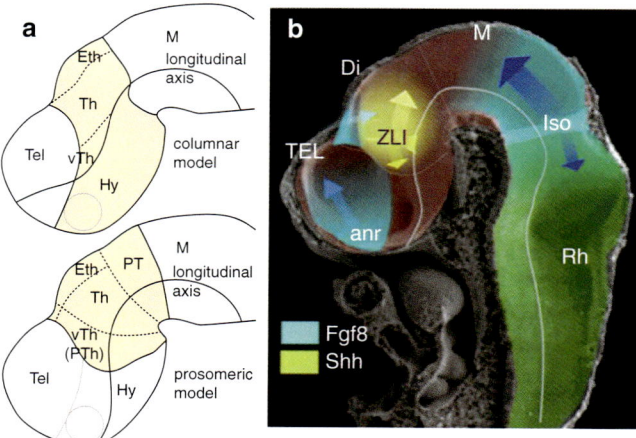

Fig. 5.10 Comparison of columnar vs. prosomeric models of the forebrain. In the columnar model of Johnston, Herrick and Swanson: (i) the hypothalamus is classified together with ventral thalamus as anterior diencephalon and represents the ventral diencephalic column; and (ii) the telencephalon is conceived as the rostral continuation of dorsal diencephalic columns (also classified as caudal diencephalon [52]). In the prosomeric model, the thalami are organized longitudinally, the hypothalamus belongs not to diencephalon but to the secondary prosencephalon,rostral t and completely separate from diencephalon. Schematic representations of a lateral view of E10.5 mouse neural tube where diencephalic territories (yellow) have been represented following columnar (above) and prosomeric (below) models. Scanning electron microscope image showing a lateral view of E10.5 mouse neural tube where the main neural segments and planar secondary organizers are represented. *Abbreviations*: *anr* anterior neural ridge, *Di* diencephalon, *Eth* epithalamus, *Hy* hypothalamus, *IsO* isthmic organizer, *M* mesencephalon, *PT* pretectum, *Rh* rhombencephalon, *Tel* telencephalon, *Th* thalamus, *VTh (PTh)* ventral thalamus (prethalamus), *ZLI* zona limitans intrathalamica. (Reprinted from Martinez S. Molecular regionalization of the diencephalon. Front Neurosci 2012; 6(article 73):1–8. With permission from Creative Commons License 3.0: https://creativecommons.org/licenses/by-nc/3.0/)

cephalic flexure. As a result, the topologically transverse cylindrical sectors of the neural tube are compressed into wedges. From the very beginnings of the neural plate, a common mechanism of dorsoventral (DV) patterning and anteroposterior (AP) patterning takes place. The term *patterning* is defined by gradients of genetic expression along the three axes of the neural tube [55]. All neuromeres share the same DV and AP mechanisms. When the neural plate becomes the neural tube, these result in antagonistic floor plate and neural plate signals that are repeated (under different homeotic environments) throughout the length of the neuraxis. The sequential assembly of the DV/AP "compartments" results in longitudinal zones that are shared along the axia of the brain.

From a genetic standpoint, *the DV pattern is really a medial-lateral sequence* based upon the midline axis of the originally flat neural plate prior to its transformation into a neural tube. The common causal background of the longitudinal zones sets up the property of *serial homology*

(metamery) across all neuromeres. Thus, future tracts will occupy the same spatial position as they descend or ascend along the neuraxis. This occurs irrespective of the fact that individual neuromeres have unique transverse molecular boundaries.

Thus, the prosomeric model visualizes all vertebrate brains as segmented structures constructed along a common *bauplan*, i.e., the same set of DV and AP embryologic units. The neural tube wall thus becomes a checkerboard pattern of domains, each with histogenetic areas that are either unique to that neuromere or shared with others. Thus, the domains of motor nuclei for all pharyngeal arches, trigeminal (r3), facial (r4–r5), nucleus ambiguus (r6–r11), and even the spinal nucleus of XI (c1–c4), occupy a common position in the brainstem and spinal cord.

Segmental identity for each diencephalic unit is specified by a unique set of genes not expressed in the hindbrain and maintained by polyclonal cell lineage restrictions [2, 49, 53, 56–62]. This genetic expression pattern helps define transverse segmental boundaries as well as the differentiation of specific neural structures in each rhombomere.

The prosomeric model has multiple uses. First, it simplifies descriptive neuroanatomy. Nuclei and tracts can be localized into specific sectors. The level at which decussations take place can be specified. Second, it permits comparison of brain structures across taxa (different species). The abducens nucleus in the lamprey is localized to r4–r6, whereas in mice it resides strictly in r5. Third, craniofacial neural crest populations of the forehead orbit and nose can be described in terms of their prosomere of origin. Frontonasal dermis is derived from the neural folds of p3–p1. This creates three potential vertical zones of forehead skin; these conveniently match up with Tessier cleft zones 13–10: zones 13–12 match with the supratrochlear axis, zones 11–10 with the supraorbital axis, and zone 9 with the lacrimal axis.

Unfortunately, traditional neuroanatomy is rife with obsolete paradigms that have become entrenched, camouflaged as "established" nomenclature or as "facts" in textbooks. For example, the use of terms such as epithalamus, dorsal versus ventral thalamus, and hypothalamus is based on the erroneous belief that the forebrain has a straight axis stretching from telencephalon down to the hindbrain. Based on this paradigm, the various parts of the thalamus are parts of a dorsoventral pattern. It is now understood that the vertebrate forebrain axis becomes curved very early in development, at the time of the cephalic flexure. Thus, the thalamic parts arise from AP patterning, although their names are unlikely to be changed. Gene expression patterns, molecular markers, and embryologic observations that were formerly incompatible with the straight-line axis are consistent with the AP model.

The new proposal for prosomeric identity may be applied in all vertebrate classes, but it is noteworthy that there are some major phylogenetic differences in embryonic growth patterns that do not detract from prosomeric units and boundary zones. Primitive or less evolved fishes, such as the jawless hagfish and lamprey, as well as cartilaginous fishes such as sharks and skates, have an inverted forebrain similar to that of sarcopterygians (lungfishes), amphibians, reptiles, birds, and mammals in which the lateral ventricles are enclosed by the forebrain, hence internal. Actinoptyergians, the teleosts or derived bony fishes, by contrast, have an everted forebrain that bends laterally so that the ependymal surface and lateral ventricular homologue are outside at the surface of the brain [63].

Descriptive Neuroembryology (O'Rahilly and Müller)

Stages of Neuroembryonic Development

In 1912, Ernst Keibel and Franklin P. Mall introduced the concept of staging, including a description of early stages 1–9. Mall's successor at the Carnegie Institution, G.L. Streeter, defined later stages 11–23, which he designated "horizons" rather than "stages". In 1987, Fabiola Müller and Ronan O'Rahilly published the first systematic account of the appearance and morphology of human embryos based upon serial sections of 215 embryos of the Carnegie Collection. These authors described the 23 Carnegie stages in 1987, with revisions in 1997, 2006, and 2010. Other neuroembryologists also contributed to the description of the embryonic phenotype at each stage [64]. This system was described in detail in Chap. 3. The Carnegie stages, as defined by FMROR, apply well to the development of the CNS. These are depicted in beautiful detail in: O'Rahilly R, Müller F. The Embryonic Human Brain, 3rd ed. New York: Wiley-Liss, 2006 (Figs. 5.11, 5.12, 5.13, 5.14, 5.15, 5.16, 5.17, 5.18, and 5.19).

In overview, the FMROR classification begins at stage 9 with six primary neuromeres: one each for prosencephalon (P) and mesencephalon (M); rhombencephalon has four (RhA-RhD). In stage 10, P divides into T, D1, and D2. In stage 11, RhA-RhD subdivide into Rh1–Rh8. RhD becomes Rh8. [Note that from our perspective Rh8 contains the four psuedorhombomeres as described by Puelles. Thus the four hypoglossal nuclei are found in r8–r11.] In stage 12, M divides into M1 and M2. In stage 13, D2 subdivides into parencephalon and syncephalon and Rh1 displays a distinct Isthmic rhombomere. In stage 14, parencephalon subdivides into rostral and caudal sectors. Thus, based strictly on *observable neuroanatomic criteria*, the FMROR system describes five forebrain neuromeres, two midbrain neuromeres, and nine hindbrain rhombomeres, for a total of 16.

Obviously, the neuromeric "bodycount" differs between the FMROR model and that of Puelles-Rubenstein (P-R). Furthermore, FMROR, although visually detailed, is a columnar model based on the assumptions of Herrick [6] and radically different in certain neuroanatomic fundamentals. Since our emphasis here is on neuromeres, why should we put effort into studying such a system?

- These two models, morphologic vs. genetic, are complementary.
- FMROR helps us visualize the neuromeres and the timing of their appearance. This makes it accessible for clinicians. But it does not translate readily to understanding relationships between gene expression, the development of the neuraxis, and its relationship to outlying mesenchyme with which it is in register.
- As columnar model, FMROR dating back is based on misconceptions of neuroanatomy which persist to this day. These are worth discussing to clarify these issues for the reader.
- Rubenstein and Puelles provide correlations with neural crest populations arising from the various zones. These are clinically relevant in helping map out relationships between mesenchymal derivatives such as bone and muscle, and the neuromeric origin of their precursor cell populations.

We shall approach the morphologic model in two steps: (1) First we shall present the developmental anatomy stage-by-stage to give you, the reader, a visual impression of neuroembryologic development over time. Look carefully at the figures corresponding to this section. (2) Second, we shall discuss differences between the morphologic and molecular models later on, once we have reviewed the Puelles-Rubenstein model.

The embryonic staging of the human embryonic nervous system is here summarized (see Chap. 3 for more details). Embryo age as per UNSW.

Definitions

Neuromere (FMROR): Morphologically identifiable *transverse subdivisions* in the longitudinal axis of the embryonic brain that extend to either side of the body. *Longitudinal subdivisions* take place at/after stage 15 when all neuromeres are present. Note that the Puelles model is based on a different definition of neuromeres.

Primary neuromeres are six morphologically distinct divisions that appear at stage 9 and can be seen at stage 9 before the embryo closes: prosencephalon (P). mesencephalon (M), and rhombencephalon (A-D).

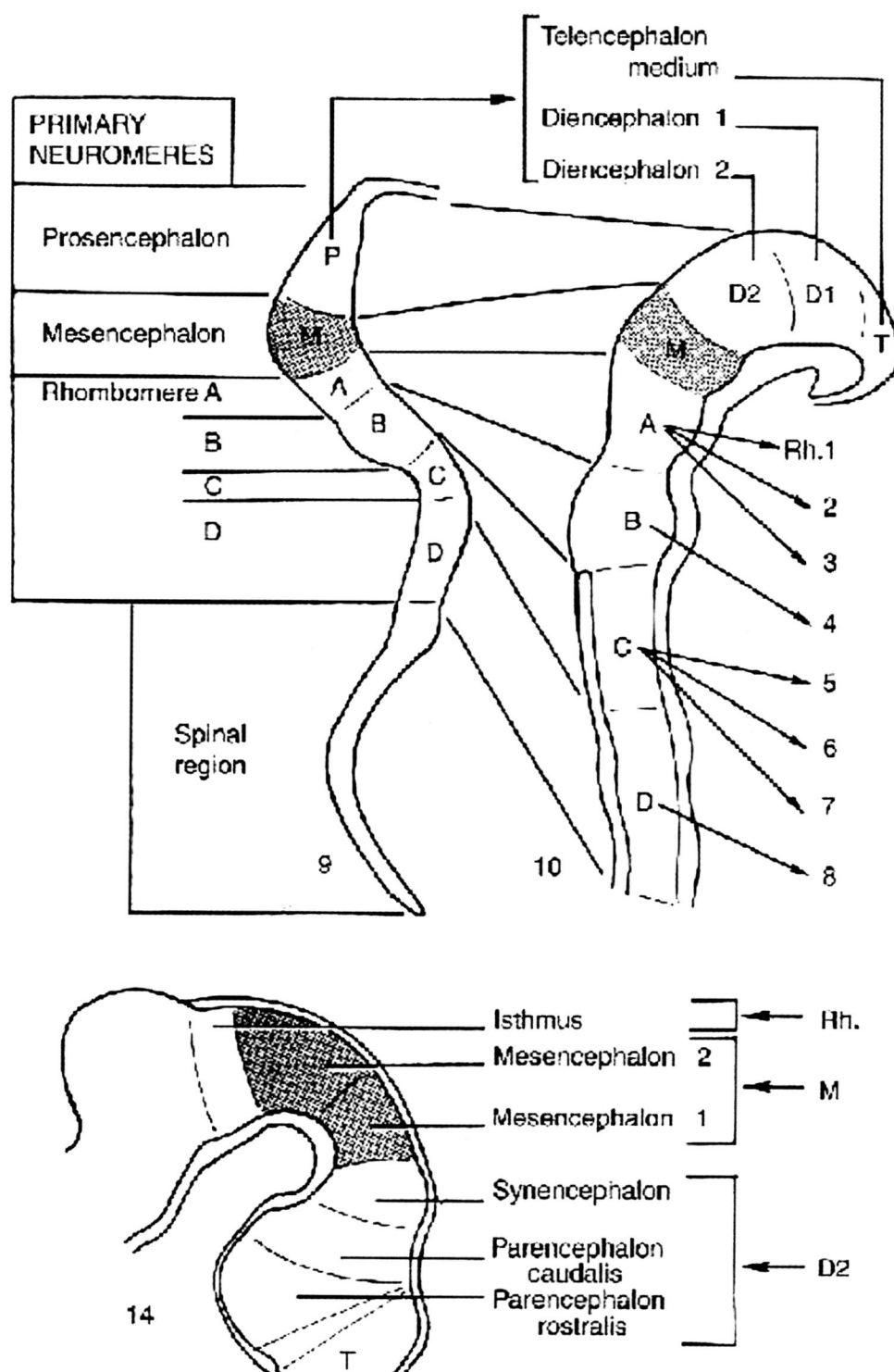

Fig. 5.11 FMROR nomenclature. (Reprinted from O'Rahilly R, Müller F. The Embryonic Human Brain: An Atlas of Developmental Stages, 3rd ed. New York: Wiley-Liss, 2006. With permission from John Wiley & Sons, Inc.)

Fig. 5.12 Primary and secondary neuromeres by stage. (Reprinted from Müller F, O'Rahilly R. Timing and sequence of appearance of neuromeres and their derivatives in stage human embryos. Acta Anat 1997; 158(2):83–90. Copyright © 1997, © 1997 S. Karger AG, Basel)

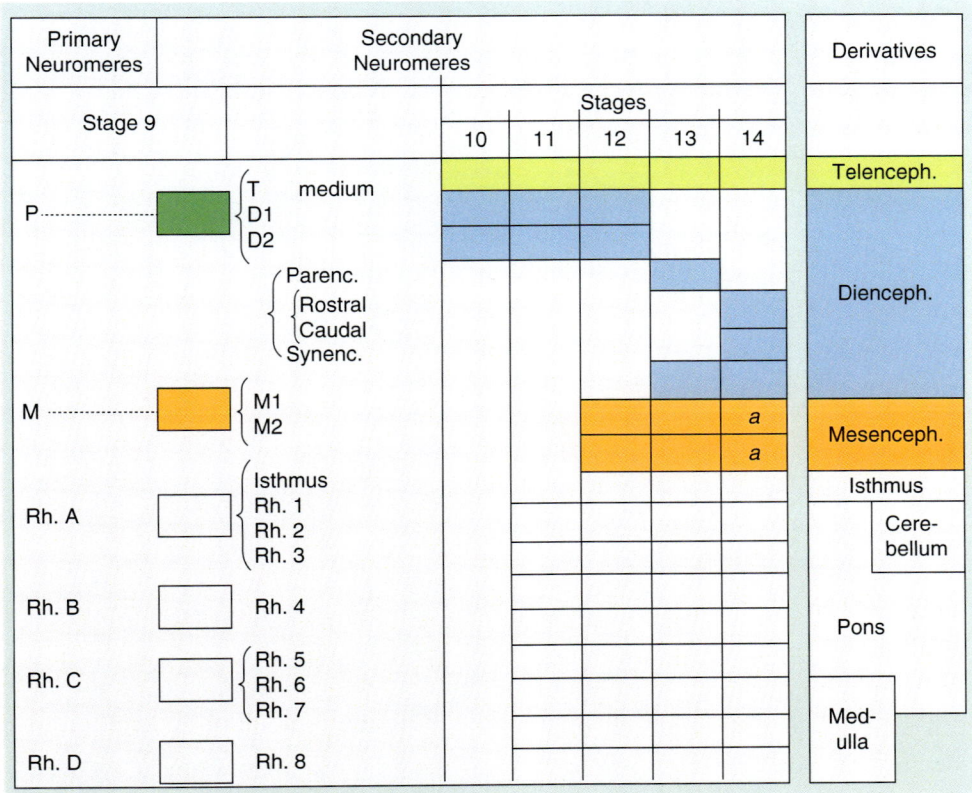

Secondary neuromeres develop during stages 10–14. Note 10–11 prior to neural tube closure and 12–14 after neural tube closure.

Prechordal plate (FMROR): First tissue laid down by primitive streak. Lies ahead of primitive streak ahead of r0 and beneath diencephalon. Diencephalon contains hypothalamus. Telencephalon is considered *prechordal*.

Prechordal plate (LP-JR): Hypothalamus is separate from diencephalon—part of secondary prosencephalon. Prechordal tissue extends all the way forward through hypothalamus.

Anatomical Model for Neuromere Stages 6 Through 17

Stage 6 (13–14 Days)
Primitive streak appears in the two-layer embryonic disc, defining left/right symmetry. Sets the stage for gastrulation.

CNS: no nervous system induction in stage 6

- Stage 6b dorsal primitive node makes the prechordal plate.

Stage 7 (15–17 Days)
Gastrulation takes place. Mesoderm fills the embryo. Segmentation of PAM into somitomeres begins immediately.

CNS: induction taking place but no morphologic sign of the nervous system yet

- From ventral primitive node, a tube is produced, the *notochordal process*. It projects forward beneath the ectoderm until it contacts the prechordal plate. It stops at level of future prosomere **hp2**, the site of the adenophyphysis [65].
- Primitive node retreats backward, laying down a distinct *notochord*.

Stage 8 (17–19 Days, 18 Somitomeres)
The embryonic disc has changed its shape. Originally circular and flat, it is now pear-shaped with a dorsal elevation. The primitive streak and node rapidly achieving their final dimensions. [At stage 7, they occupy 1/2 the length of the disc.] Additional growth during stage 8 results in their rostral-most extension being 1/3 of the disc. Note that in humans the node *included* in the streak; it is not a separate entity. The *neural plate already is present* in stage 7. By the end of stage 8, primitive streak and node are at their maximum length. Subsequently, they will retreat backward, leaving the notochord behind. Gastrulation is complete. By late stage 8, some embryos display a neural groove and neural folds, the first signs of neurulation. The presence of pseudostratified neuroepithelium demarcates the boundary of the neural plate with nonneural surface ectoderm.

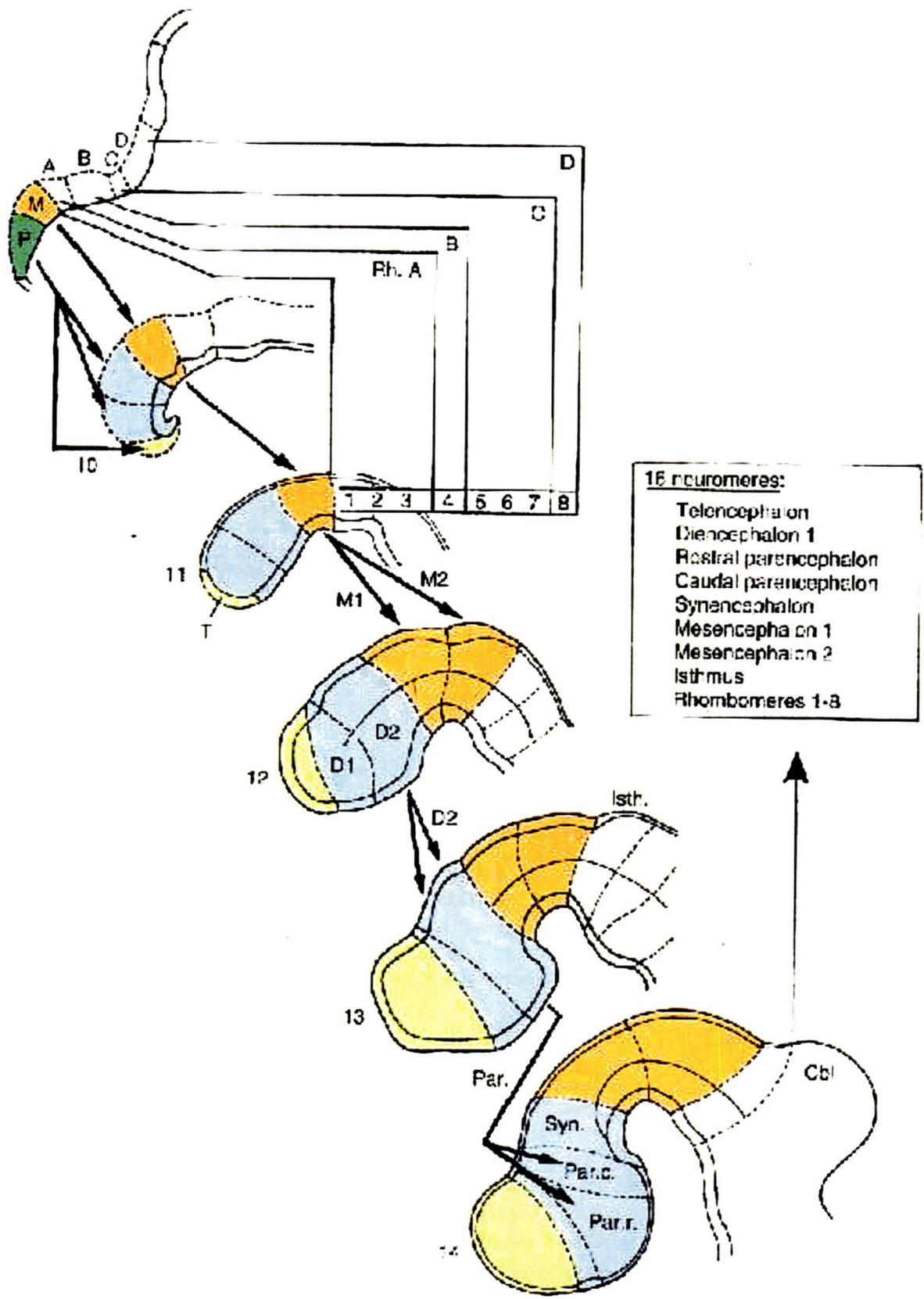

16 neuromeres:

Telencephalon
Diencephalon 1
Rostral parencephalon
Caudal parencephalon
Synencephalon
Mesencephalon 1
Mesencephalon 2
Isthmus
Rhombomeres 1-8

Fig. 5.13 Neuromere development by stages. Compare with chart in Fig. 5.12. Six subdivisions present at stage 9. Secondary neuromeres add to a total of 16. Longitudinal organization in the form of sulci, etc. appears beginning at stage 15. Neuromeres remain visible as late as stage 17. (Reprinted from O'Rahilly R, Müller F. The Embryonic Human Brain: An Atlas of Developmental Stages, 3rd ed. New York, NY: Wiley-Liss, 2006. With permission from John Wiley & Sons)

Fig. 5.14 Stage 9. Neural tube has not formed and vesicles of the brain are not present. Forebrain is nearly exclusive diencephalic. Midbrain marks the cephalic flexure. Hindbrain has four primary rhombomeres. First evidence of neural crest is found over rostral hindbrain and midbrain (r1–m1–m2 complex). Primitive streak is no longer functional; can be seen very posteriorly near cloacal membrane. (Reprinted from O'Rahilly R, Müller F. The Embryonic Human Brain: An Atlas of Developmental Stages, 3rd ed. New York, NY: Wiley-Liss, 2006. With permission from John Wiley & Sons)

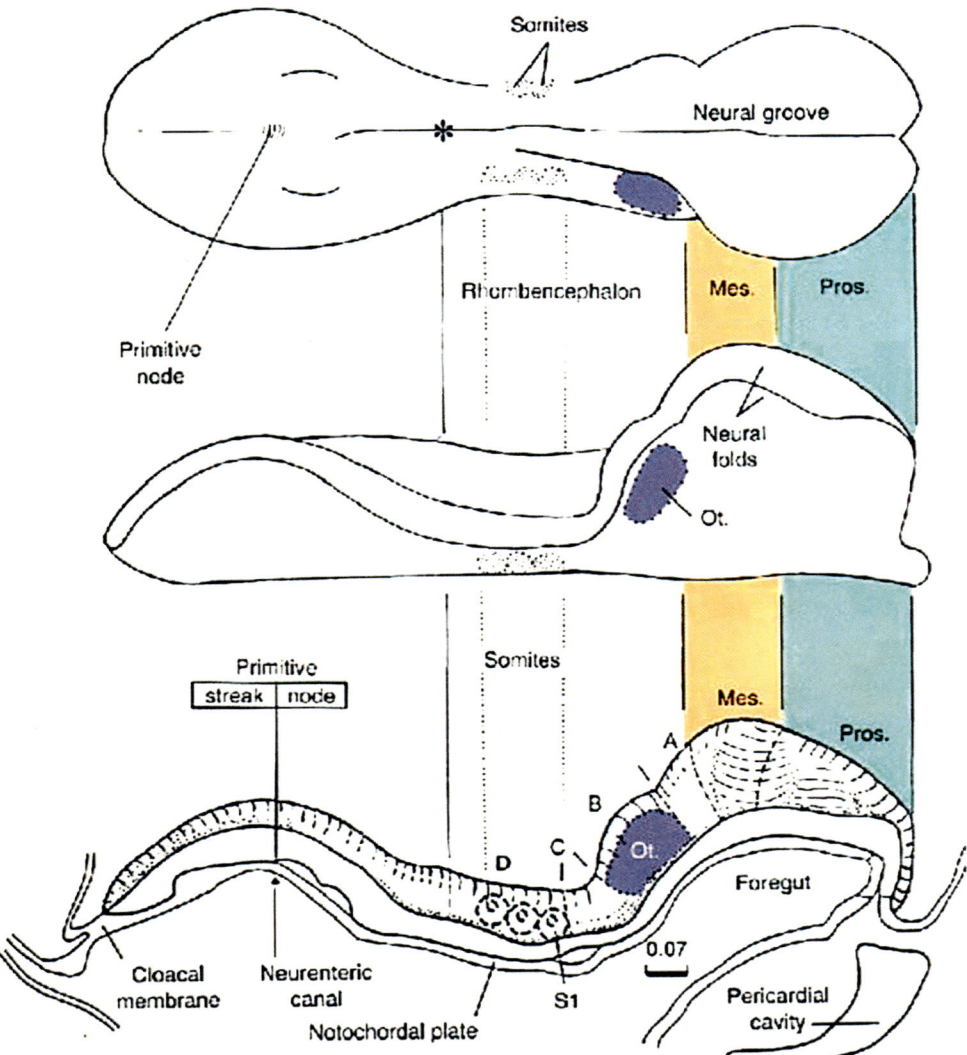

Fig. 5.15 Stage 10.
Hindbrain contains cranial
nerve nuclei. Midbrain flexure
more acute, bring the future
face downward in position.
D2 is the thalamic region and
D1 has the optic anlage,
present here as a sulcus.
Between the two D1 fields is
a tissue bridge between the
optic primordia, the optic
chiasm of torus opticus.
Telencephalon becomes
visible at mid stage 10
between 7 and 12 somites.
Columnar model considers
the notochord to extend as far
as D1, so brain induction is
epinotochordal. D1 and
Telencephalon are considered
prenotochordal. (Reprinted
from O'Rahilly R, Müller
F. The Embryonic Human
Brain: An Atlas of
Developmental Stages, 3rd
ed. New York, NY: Wiley-
Liss, 2006. With permission
from John Wiley & Sons)

Fig. 5.16 Stage 11. At 13
somites, chiasmatic plate
develops. At 14 somites,
lamina terminalis and
commissure develop in the
floor of the telencephalic
vesicle. Optic vesicle is
formed with optic neural
crest, from hp2 neural fold of
telencephalon and from
mesencephalon. Placode for
adenohypophysis is present
and in contact with
neurohypophysis. At 20
somites, anterior neuropore
closes. The process is
unidirectional, from the
anterior commissure
backward, but in the columnar
model it was described as
bidirectional: from D2
forward and Tel. backward.
Note relation of notochord to
the M-D1 border. (Reprinted
from O'Rahilly R, Müller
F. The Embryonic Human
Brain: An Atlas of
Developmental Stages, 3rd
ed. New York, NY: Wiley-
Liss, 2006. With permission
from John Wiley & Sons)

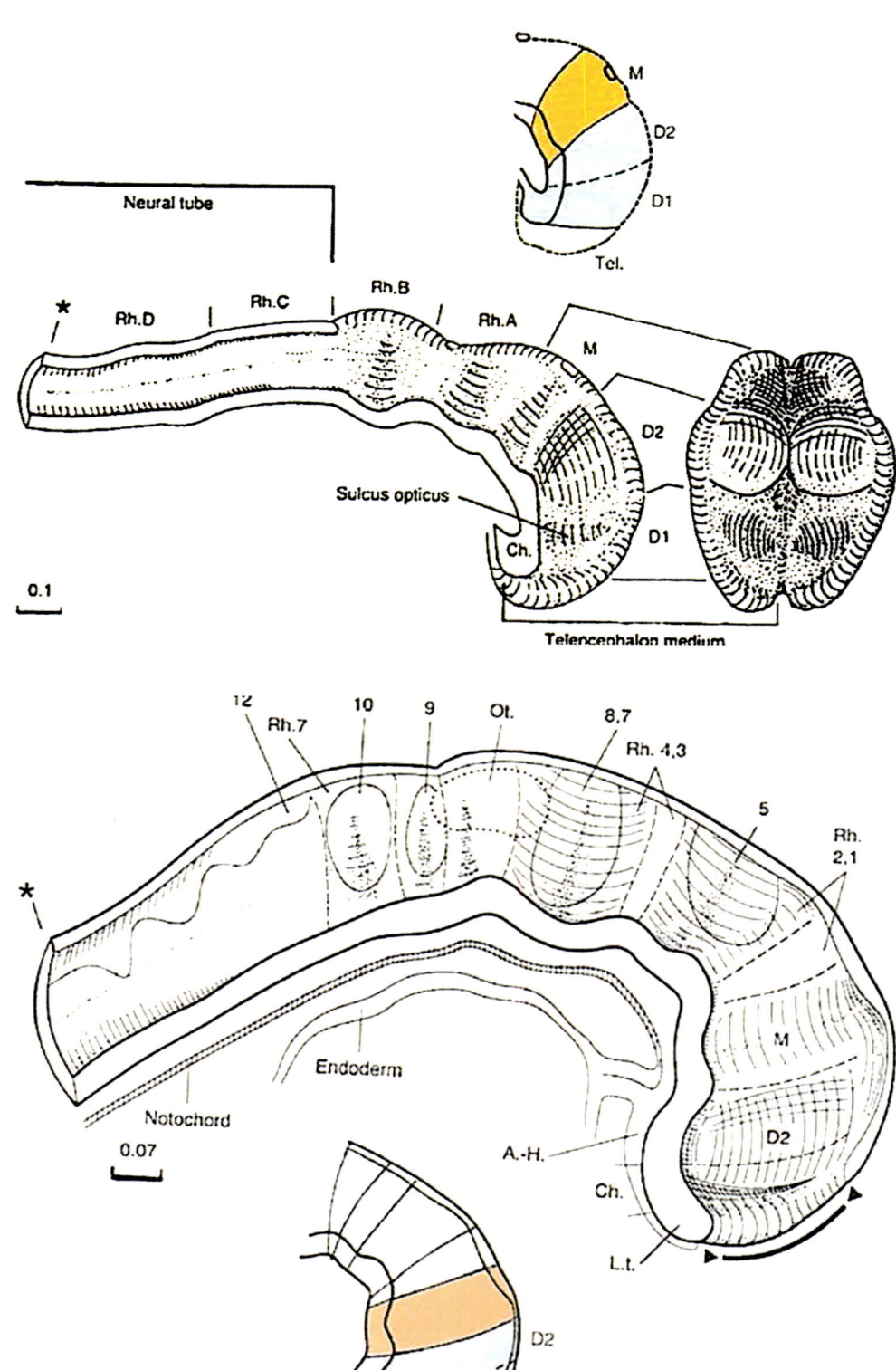

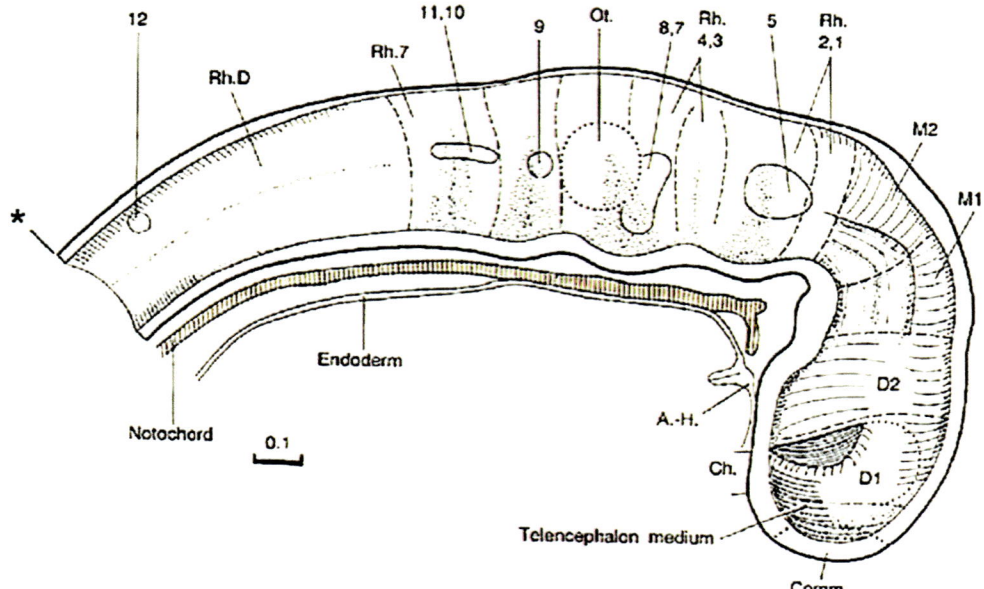

Fig. 5.17 Stage 12. Cranial nerve ganglia develop with neural crest populating epipharyngeal placodes. Hindbrain cranial nerve fibers are developing. Mesencephalon has (caudal-rostral) M2 and M1. D2 is considered related to adenohypophysis and mamillary recess. [In reality, these structures are located further forward in the basal plate of prosomeres hp1 and hp2.] Optic ventricle of D1 opens into the third ventricle. Ependymal fluid is still in contact with amniotic fluid in stage 12. Note that dysgraphia can occur in the place of unilateral terminal lip failure (another manifestation of homeotic bilaterality). (Reprinted from O'Rahilly R, Müller F. The Embryonic Human Brain: An Atlas of Developmental Stages, 3rd ed. New York, NY: Wiley-Liss, 2006. With permission from John Wiley & Sons)

CNS: stage 8a no neural folds, stage 8b neural folds, and neural groove are present

- No neuromeres are present.
- Formation of the definitive notochord
  - 17 days *notochordal process* (*axial notochord*) is a cellular rod in direct contact with neural ectoderm and prechordal plate mesoderm. In mammals, the rod fuses with endoderm to form a *neurenteric canal* that is anteriorly a blind-end, but tracks backward to open into primitive pit (behind the primitive node).
  - 18 days *notochordal plate* results when bottom part of the notochord disintegrates, permitting fluid from yolk sac to pass into amniotic cavity.
  - 19 days solid *definitive notochord* forms when notochordal cells pull away from the endodermal roof of the yolk sac. The result is a solid rod of cells in the midline between endoderm and ectoderm.
- At the end of stage, entire CNS is induced

  - Primitive node, pit, and neurenteric canal are caudal, close to cloacal membrane.

Stage 9 (19–21 Days, 1–3 Somites)

This stage is characterized by sudden transformation from stage 8: (1) the primitive streak is drastically reduced; (2) neuromeres appear for the first time; and (3) neurula-tion—the three major divisions of the embryonic brain can be identified within the folds of the open neural groove.

Mesenchymal organization takes place. The first pharyngeal arch develops. The first three pairs of occipital somites appear. Paraxial mesoderm associated with the head had previously existed as incompletely segmented units, *somitomeres*, lacking the fundamental organization of somites (dermatome, myotome, and sclerotome). The first seven somitomeres maintain mesenchymal continuity, but beginning with caudal aspect of Sm8 somitogenesis begins. Note that in S1 (formerly Sm8), the epithelium is separated from that of S2, but mesenchymal continuity with Sm7 is preserved.

Lateral plate mesoderm vacuolates. Prechordal LPM_V forms the heart.

CNS: major divisions of the brain, mesencephalic flexure, otic disc

- *Prosencephalon* contains prechordal plate that denotes the future neuromere D1. D2 extends backward to the mesencephalic flexure.
- *Mesencephalon* is centered at the apex of the flexure (m1, m2, r0/r1).
- *Rhombencephalon* contains four neuromeres. Rh.A and Rh.B are located on the downslope of the mesencephalic flexure. Rh.A extends to the otic disc (r2–r3), while Rh.B is centered on the otic disc (r4–r5). Rh.C is located in the

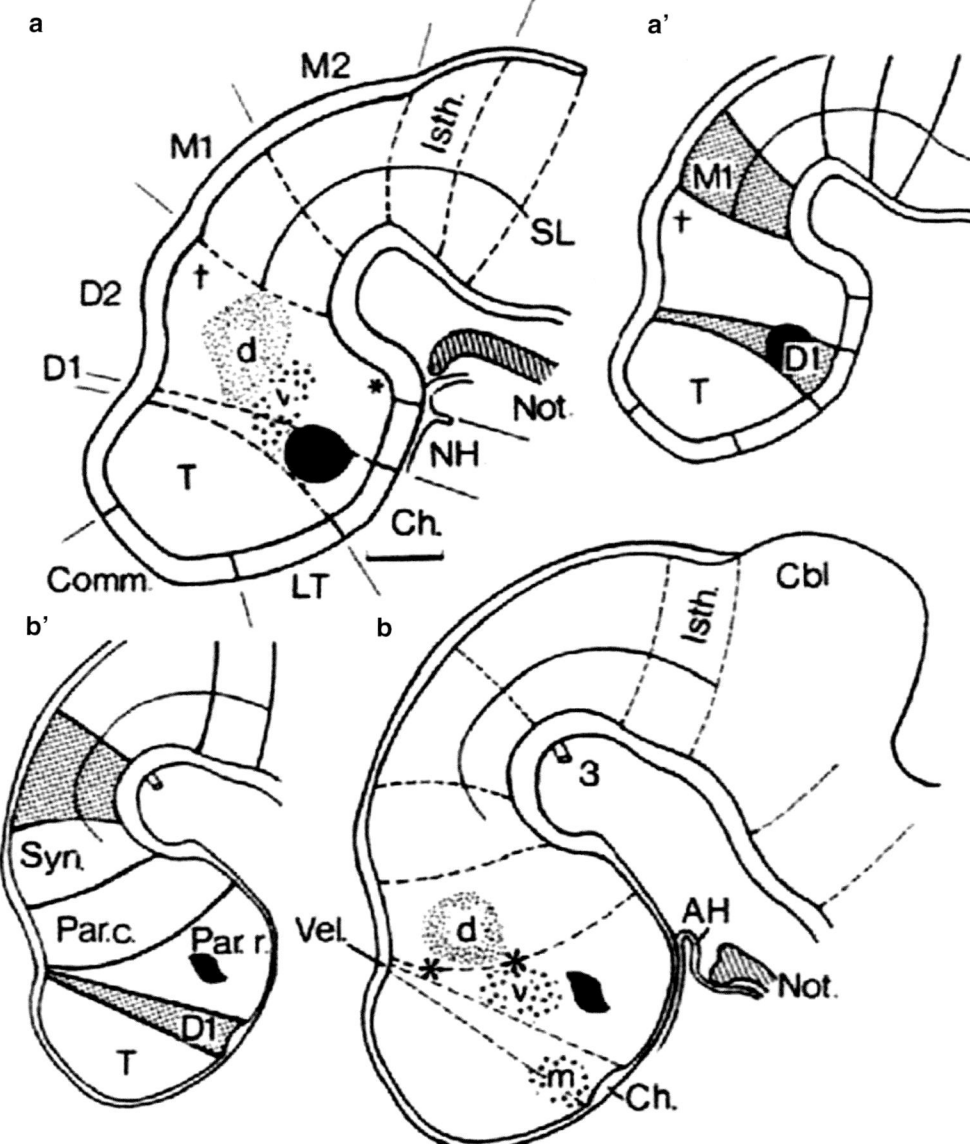

Fig. 5.18 Stages 13–14. With both neuropores closed, future ventricular system not in contact with amniotic fluid. Ependymal fluid produced first by ependymal lining cells and, at a later stage, by the choroid plexuses. <u>Stage 13</u>: Somitogenesis now complete. Cranial nerves motor to pharyngeal arches and tongue are present. Cerebellum arises from r1. Isthmus contains nucleus of trochlear nerve. Oculomotor is not yet formed Neural crest leaves nasal discs to form olfactory, terminal, and vomeronasal nerves. CN0 is pheromonal; it is not connected to the olfactory system, but refers to the medial/lateral septal nuclei and to preoptic areas, all involved in mammalian sexual behavior. Indentation of the lens disc and retina occurs. <u>Stage 14</u>: All neuromeres are complete at this stage. Caudal parencephalon contains dorsal thalamus (future prosomere 2) and rostral parencephalon has ventral thalamus (future prosomere3). Basioccipital and exoccipital bone fields are forming. Cranial nerves II and IV develop. *Torus hemisphericus* divides diencephalon from telencephalon. At its base is found median eminence. Four areas that are prominent in telencephalon during this stage are: (1) hypothalamus, (2) amygdala, (3) hippocampus, and (4) olfactory region. Nasal discs are prominent. The optic cup and lens pit are seen. <u>Note</u>: Blood vessels enter the walls of hindbrain and midbrain. Only place they can access the forebrain is at the sulcus between diencephalon and telencephalon. At this juncture, marked by amygdala, the primary plexiform layer first gets it start. **Key**: Optic sulcus, black; single asterisk, mamillary area; dagger, future syncephalon; double asterisks, sulcus medius. (Reprinted from Müller F, O'Rahilly R. Timing and sequence of appearance of neuromeres and their derivatives in stage human embryos. Acta Anat 1997; 158(2):83–90. Copyright © 1997, © 1997 S. Karger AG, Basel)

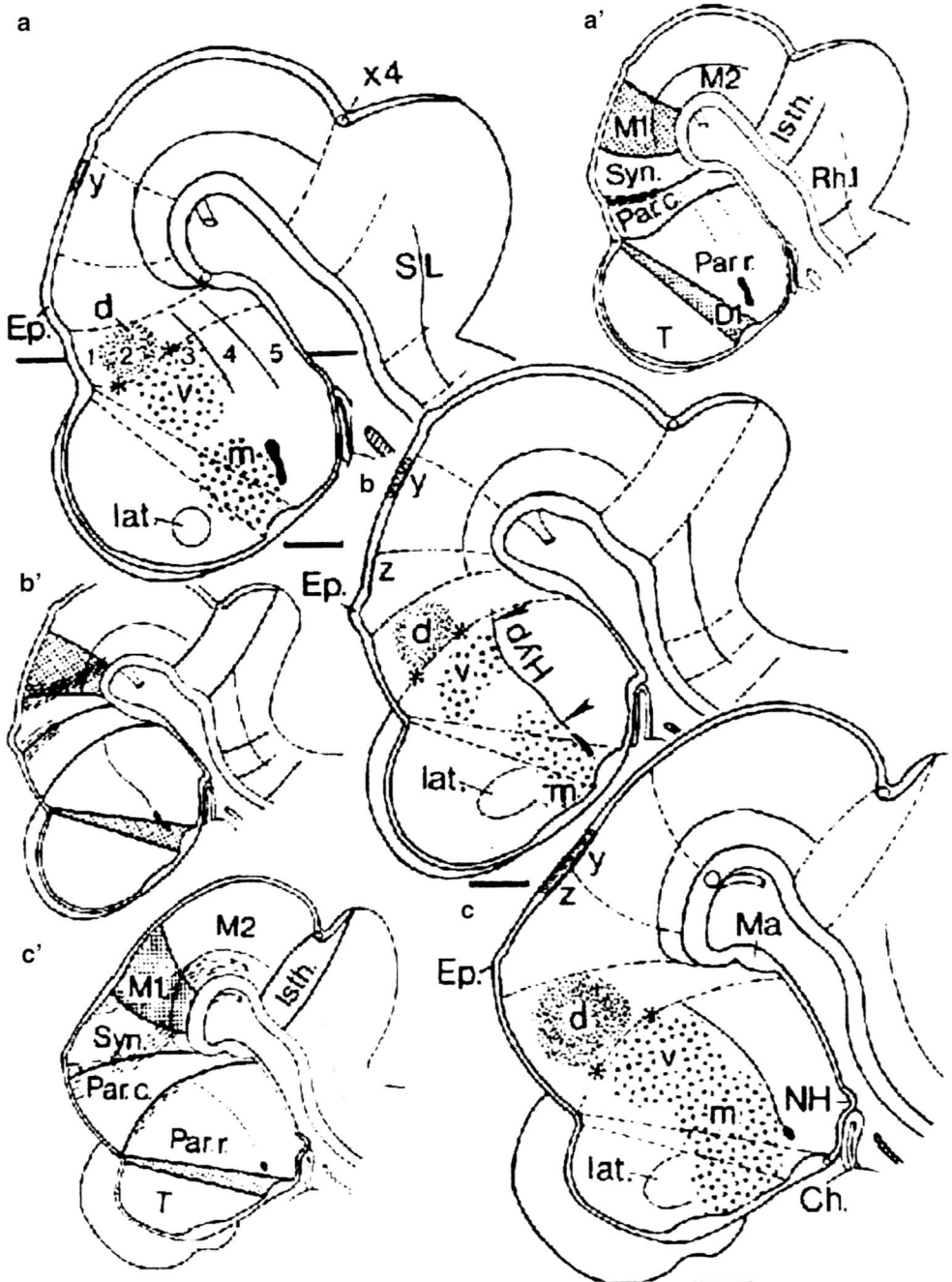

Fig. 5.19 Stages 15–17. Somitogenesis now complete. All cranial nerves motor to pharyngeal arches and tongue are present. Cerebellum arises from r1. Isthmus contains nucleus of trochlear nerve. Oculomotor is not yet formed Neural crest leaves nasal discs to form olfactory, terminal and vomeronasal nerves. CN0 is pheromonal; it is not connected to the olfactory system but refers to the medial/lateral septal nuclei and to preoptic areas, all involved in mammalian sexual behavior. Indentation of the lens disc and retina occurs. A new division arises from D2. Syncephalon contains in its base the interstitial nucleus of Cajal, the forerunner of the medial longitudinal fasciculus. Parencephalon contains the ventral and dorsal thalami. **Key**: Optic vesicle, black spot; double asterisks, sulcus medius; ridge (d) separates syncephalon from cauda parencephalon; Habenulo-interpedicular tract (interrupted lines) separates Par C and Syn in a′. Caudal to that, posterior commissure tract is second set of interrupted lines in b′. Three tracts seen in c′ (caudal-rostral): tract of posterior commissure, habenu-

lopenducular tract, zona limitans interthalamica. Stage 15 (a, a′) Abducens fibers appear. Trochlear decussation appears in the isthmus. Lateral ventricle appears. Hypothalamic cell cord and sulcus limitans of midbrain are signs of formation of five longitudinal zones: epithalamus, dorsal thalamus, ventral thalamus, subthalamus, and hypothalamus [this concept is outdated]. DT and VT are separated by sulcus medius. Stage 16 (b, b′) Neurohypophysis develops. Fiber tracts in diencephalon define the syncephalon which is bracketed by caudally by posterior commissure and rostrally by habenulo-interpedicular tract. Stage 17 (c, c′) Zona limitans interthalamica appears as a ridge between dorsal and ventral thalami. Supramamillary recess receives the termination of sulcus limitans (sulcus medius). Stage 17 is the final stage at which rhombomeres can be seen at transverse ridges in the ventricle. (Reprinted from Müller F, O'Rahilly R. Timing and sequence of appearance of neuromeres and their derivatives in stage human embryos. Acta Anat 1997; 158(2):83–90. Copyright © 1997, © 1997 S. Karger AG, Basel)

flat zone between the posterior margin of the flexure and the occipital somites (r6–r7). Rh.D lies adjacent to the occipital somites (i.e., at r8–r10).

- Neurulation causes physical changes in the embryo. At its midpoint, it has a waist-like constriction. And, just as might happen if one suddenly tightened one's belt, both rostral and caudal ends of the embryo are thrown up into folds. The center point of the rostral prominence will become the mesencephalic flexure point. The "waist" contains the neural plate of the medulla, while the caudal prominence contains the remainder of the primitive streak.

Stage 10 (22–23 Days, 4–12 Somites)

CNS: subdivision of forebrain, optic primordium

- *Prosencephalon* is still wide open. It contains telencephalon and two diencephalic prosomeres: D1, the optic sulcus, and D1, the thalamus. (Note FMROR consider the eye to originate from diencephalon rather than as secondary prosencephalon).
- *Mesencephalon* consists of a single mesomere, M.
- *Rhombencephalon* contains cranial nerve nuclei
  - The otic disc is direct relationship to Rh.B. Neural fold fusion begins at this stage at the cerebrospinal junction.

Stage 11 (24–25 Days, 13–20 Somites)

CNS: rostral neuropore closes, skin coverage now complete over brain

- *Prosencephalon* contains *nasal placode*, the rostral wall of telencephalon. The embryo has between 14 and 20 somite pairs; medial growth and fusion of the neural folds close up the rostral neuropore. The *optic vesicles* are emerging from D1 which is aligned with the *chiasmatic plate*. The future pituitary can be predicted because D2, containing the neurohypophysis, is aligned with adenohypophyseal epithelium which is located just rostral to the buccopharyngeal membrane.
- *Spinal cord*: Sulcus limitans, an important landmark, is present in spinal cord and RhD (caudal medulla, r8–r11), but has not yet extended further forwards.

Cranial nerves: Neural crest populations for IX and X are present.

Calvarial skin: Brain development requires protection from the environment. In its earliest form, this consists of an epithelim derived from the nonneural preplacodal ectoderm (PPE). When diencephalic neural crest from prosomeres p1–p3 arrives, dermis develops. In turn, PPE is converted to epidermis.

Stage 12 (26–27 Days, 21–29 Somites)

CNS: caudal neuropore closes, adenhypophyseal pouch begins, secondary neurulation

- *Prosencephalon*: The optic vesicles remain the major landmark of D1. The floor of D2 begins to display the mamillary recess. D2 is "invaded" by tissues extending forward from the midbrain.
- *Mesencephalon*: Tissues from mesencephalic tegmentum thicken the floor of D2, while those from tectum bulk up the roof. The mesencephalic flexure is approximately 90 degrees. Sulcus limitans is now visible in the midbrain.
- *Rhombencephalon*: Horizontal section shows pharyngeal arch 1 aligned with Rh2 and Rh3, while pharyngeal arch 2 is aligned with Rh4 and Rh5. The otic vesicle is spatially associated with Rh5.
- *Cranial nerves:* Neurons of XII present in floorplate of medulla. Ganglia of V and VII consist of both NC and placodes which have not separated. Ganglia of IX and X are separated (superior = neural crest, inferior = epibranchial placode). Accessory nerve XI is present. Neural crest populations start to line up opposite rhombomeres. Cardiac crest concentrates at r7 and migrates toward heart. Vagal/accessory complex is present. NC for XI is organizing parallel to axis of medulla. Hypoglossal cord contains neural crest from r8 to r110 and PAM from occipital myotomes.

Stage 13 (28–31 Days, 30+ Somites)

CNS: neural tube now closed, isthmus appears, cerebellum develops from r1,

- *Prosencephalon*: D2 subdivides. Its caudal region contains two dorsolateral projections, the syncephalon. The basal part of syncephalon contains the fibers of future *medial longitudinal sulcus*, critical for the coordination of extraocular movements. The rostral region of D2 is parencephalon. It contains the dorsal and ventral thalamic nuclei. In neuromeric terms, these correspond to prosomeres p2 and p3. Paraencephalon grows forward from its basal zone into D1 and makes contact with telencephalon. Median section at stage 13 shows the tip of the notochord abutting up against the adenohypopyseal zone. The AP placodes are located at the extreme anterior limit of the prosencephalic neural folds. This correlates very well with the neural fold map of Couly-LeDouarinmap of ectomeres.
- *Rhombencephalon*: Isthmus appears at this stage at the midbrain-hindbrain junction. It contains the nucleus of cranial nerve IV. Rhombomere 1 becomes metencephalon: its alar lamina gives rise to cerebellum, while its

basal lamina will form pons. The remainder of hindbrain becomes myelencephalon (medulla). It now contains all motor nuclei. Efferent tracts of cranial nerves V, VII, IX–XII can be traced to the external aspect of their respective rhombomeres.

- *Cranial nerves*: Midbrain has III nucleus in m1, isthmus appears containing IV nucleus. Afferent cells appearing in the ganglia. Trigeminal ganglion starts to partition, XI reaches the first cervical spine neural crest. Accessory complex XI reaches first cervical level.

Stage 14 (32 Days)

CNS: cerebral hemispheres, torus hemisphericus, and medial ventricular eminence; optic cup; lens vesicle, pontine flexure

- *Prosencephalon*: Cerebral hemispheres start to invaginate. Medial ventricular eminence expanding. Parencephalon divides into rostral and caudal sectors. All 16 neuromeres are now present.
  - Caudal parencephalon contains dorsal thalamus (p2)
  - Rostral parencephalon contains ventral thalamus (p3)
- *Cranial nerves*: Mesencephalic V present in pontine tectum (r1–r3). Accessory XI connected to remaining cervical ganglia.

Stage 15 (33–36 Days)

CNS: longitudinal zoning of the diencephalon (marginal ridge), di-telencephalic sulcus; lateral ventricular eminence; hippocampal thickening; medial forebrain bundle, nasal pit

- *Prosencephalon*: Lateral ventricular eminence appears in telencephalon. Dorsal and ventral thalami are separated by *sulcus medius* (in reality, this marks the p2/p3 boundary, ZLI).
- *Mesencephalon*: commissure/decussation of superior colliculi. Trochelar decussation appears.
- *Cranial nerves*: Mesencephalic V extends up to isthmus, IV decussates and exits from roof of isthmus, V has three divisions, VI has short axons, VII has branches, VIII fibers are entering vestibular apparatus (cochlear fibers lag behind), XII reaches the tongue.

Stage 16 (37–40 Days)

CNS: evagination of the neurohypophysis, olfactory tubercle, dorsal thalamus, epiphysis cerebrii

- *Rhombencephalon* subdivision of the rhombomeres.
- *Cranial nerves* All CN identifiable (including cochlear), mesencephalic nucleus extends forward into isthmus. VI fibers reach the cavernous sinus. Tympanic br. of IX and superior laryngeal of X present.

Stage 17 (41–43 Days)

Note: This is the last stage with neuromeres that are delineated by ridges on the ventricular surface.

CNS: olfactory bulbs, amygdaloid nuclei, pioneer axons of corpus callosum, forebrain septum; subthalamic nucleus, posterior commissure; red nucleus; internal and external cerebellar swellings, future corpus striatum, inferior cerebellar peduncles, C-shaped hippocampus; inferior and superior cerebellar peduncles, vomeronasal organ

- *Cranial nerves*: External signs of rhombomeres disappear. All ganglia are now present and separated. Oculomotor decussation, V mandibular division joins VII chorda tympani giving the route for the extracranial stapedial artery (motor branches of VII do not appear until after stage 23), VII gives off greater petrosal nerve for the lacrimal gland, glossopharyngeal component of nucleus ambiguus appears, as does ansa cervicalis.

Stage 18 (44–47 Days)

CNS: inferior cerebellar peduncles, corpus striatum, dentate nucleus

- *Cranial nerves*: accessory abducens nucleus develops (VI is now located in both r4 and in r5), nucleus ambiguous shows separation of IX and X (previously comingled).
- Staged development of interneuromeric boundaries.

Stage 19 (17 mm, 46 Days)

CNS: Choroid plexus of fourth ventricle and medial accessory olivary nucleus: dentate nucleus; olfactory bulb; nucleus accumbens; globus pallidus; substantia nigra.

Stage 20 (20 mm, 49 Days)

CNS: Choroid plexus of lateral ventricles: interpeduncular and septal nuclei; optic fibers in chiasmatic plate; habenular commissures; telencephalic flexure initiated.

Stage 21 (23 mm, 51 Days)

CNS: Cortical plate in area of future insula (cerebral hemispheres); optic tract and lateral geniculate body (m1).

Stage 22 (26 mm, 53 Days)

CNS: Internal capsule and olfactory bulbs: olfactory striae; claustrum; internal capsule; adenohypophyseal stalk incomplete.

Stage 23 (29 mm, 56 Days)

CNS: Insula indented, external capsule; caudate nucleus and putamen visible; anterior commissure begins; mesencephalic *Blindsacke*; external germinal layer in cerebellum (Table 5.1, columnar model as seen in Figs. 5.11, 5.12, 5.13, 5.14, 5.15, 5.16, 5.17, 5.18, and 5.19).

Table 5.1 Abbreviations of FMROR columnar model

AH—Adenohypophysial primordium or pouch	m—Medial ventricular eminence	S—Somite
Ca—Cardiac region	Ma—Mamillary region	Scler—Sclerotome
Cbl—Cerebellum	Marg ridge—Marginal ridge	SL—Sulcus limitans
Ch—Chiasmatic plate	Meten—Metencephalon	Sp.cord—Spinal cord
Comm—Commissural plate	Nas—Nasal disc	Sulcus.med—Sulcus medius
d—Dorsal diencephalic elevation or dorsal thalamus	NF—Neural fold	Syn—Syncephalon
Dienceph—Diencephalon	Not—Notochord	Tel—Telencephalon
Ep—Epiphysis cerebri	O Ph—Oropharyngeal membrane	Torus.hem—Torus hemisphericus
Hab-interped—Habenulo-interpeduncula	Ot—Otic disc	V—Ventral thalamuus
Hem—Cerebral hemisphere	P—Prosencephalon	Vel—Velum transversum
Hyp—Hypothalamic sulcus	Par.c—Parencephalon caudalis	X4—Trochlear decussation
Hyp.c—Hypothalamic cord	Par.r—Parencephalong rostralis	Y—Commissure of superior colliculi
Isth—Isthmic neuromere (r0)	PN—Primitive node	Z—Posterior commissure
LT—Lamina terminalis	Post.X—Posterior commissure	
M—Mesencephalon	PP—Prechordal plate	

In summation: Review of the FMROR model gives us a visual picture of how the component parts of the CNS form. Because it is based upon anatomic findings such as fissures and bulges, we cannot assume that cell populations are confined to those locations. After all, the mere process of cellular proliferation in a clone generates topological disturbances. In addition, we note that forebrain neuromere T is added on to the terminus of the neural tube like a large yellow ring without connection as to how it is induced. A more accurate understanding of forebrain development would come with genetic mapping, a model that proves universal across all vertebrates, dating back to the cephalochordate *amphioxus*.

The Columnar Model of the Forebrain: Fundamental Flaws

The original nomenclature of neuroanatomy was based on the morphology of adult brains, often using concepts that are historical and often obsolete. By the end of the nineteenth century, advances in microscopy, evolution, cell theory, embryology, and comparative anatomy led to an initial model of brain development that became widely accepted. The CNS was considered the product of a "segmented neural tube." This theory emerged hand in hand with a segmental model of the vertebrate body. The vertebral axis was considered as being a series of metameric units based upon somites and vertebrae. The cranial base and sacrum represented fusions of these units. Branchial and pharyngeal arches seemed segmented. The brain and spinal cord were postulated to consist of segmental units connected to the periphery with cranial nerves and spinal nerves (Fig. 5.20).

The application of molecular biology to the study of patterning of the nervous system is producing data at an exponential rate. Experimental results revealing how genes function in neural development directly impact our concepts of how brain assembly takes place. In general, these emergent data have contradicted the standard columnar paradigm used to explain forebrain anatomy. Instead, these data are consistent with the older segmental paradigm dating back to His.

Historically, the columnar model was derived from studies of amphibians only. Recall that phylogenetically modern amphibians represent a bifurcation from basal tetrapods that deviates fundamentally from the line leading reptilomorphs, and ultimately to mammals. It should be borne in mind as well that the most primitive amphibians (i.e., those least changed from the ancestral stem amphibian), such as salamanders, have much less neuroblast migration in their brainstem development than do reptiles, birds, and mammals or even more evolved amphibians such as frogs; most neuronal clusters of the adult salamander brainstem remain in the subventricular zone beneath the ependyma of the fourth ventricle, where they were first generated [50, 63].

The fundamental idea of the basis of columnar theory holds that the length axis of the neural tube ends beyond dicencephalon in telencephalon, i.e., that telencephalon is simply a straight line projection of diencephalon. Puelles points out that the principle source of mismatch between the columnar model and molecular genetics does not reside in the concept of longitudinal columns of neurons (e.g., hypoglossal nucleus; nucleus/tractus solitarius) or long tracts (e.g., corticospinal or pyramidal tract), but rather in a mistaken notion of what the longitudinal axis really is. "The fundamental failure of the columnar forebrain model was that it redefined the observable forebrain axis, negating its curvature and substituting (without any specific data) an idea straight line axis." This axis travels forward from the pons into the "caudal" hypothalamus. After traversing the hypothalamus and preoptic area "longitudinally," the axis enters the telencephalon, where it terminates in the olfactory bulb. This concept obviously is at odds with the paired nature of the olfactory nerves and telencephalic hemispheres.

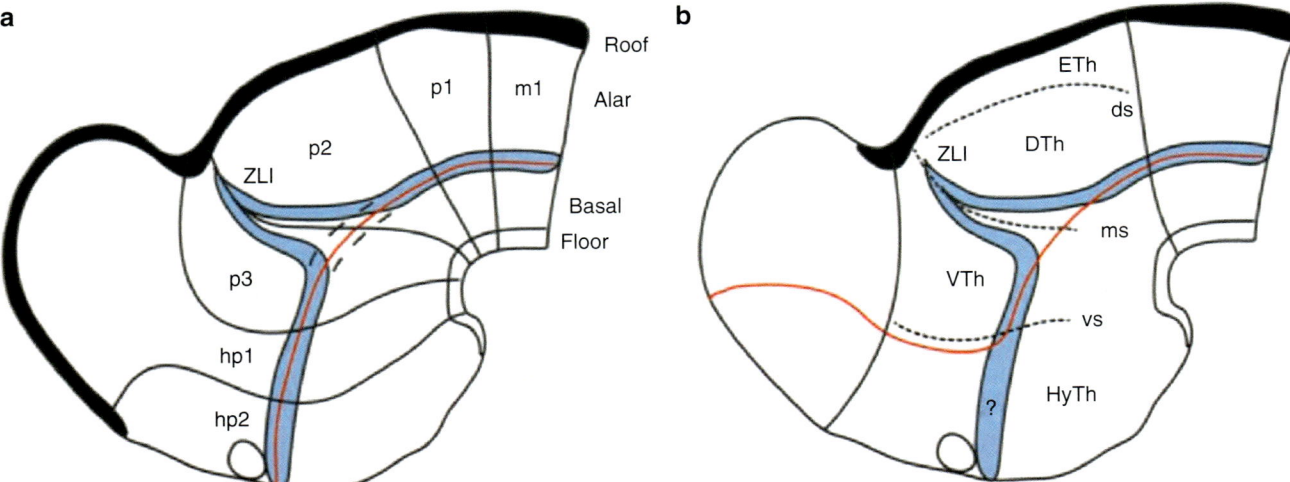

Fig. 5.20 Prosomeric model (**a**) versus columnar model (**b**). Critical differences between the two models are readily seen by domain of expression of Nkx2.2 (in blue) that relates to the alar-basal boundary (red line). The neuromeres are marked for reference in (**a**), as well as the dorsal/middle/ventral diencephalic limiting sulci (ds, ms, vs) in (**b**). Note the transverse spike of the Nkx2.2 domain at the zona limitans interthalamica (ZLI) that separates thalamus and prethalamus is a secondary feature, due to the induction of this gene adjacent to the border of Shh expression, which is ectopically activated at the core of the ZLI. During the early stages of the neural plate and neural tube, the expression band of Nkx2.2 is strictly longitudinal (marked by dashes in **a**). In its development, the boundary is distorted into a spike-like projection by the eruption of SHH from the floorplate. In the columnar model (**b**), the correspondence of the boundary with the gene band is disrupted at the arbitrary deviation of the former into the telencephalon. Moreover, note this model cannot explain why the gene band extends into the hypothalamus, cutting it into two halves, which cannot be understood as alar and basal parts of the hypothalamus, as in (**a**). These zones are indicated in (**b**) by a question mark. The neuromeres are marked for reference in (**a**), as well as the dorsal/middle/ventral diencephalic limiting sulci (ds, ms, vs) in (**b**). Note the transverse ZLI spike of the Nkx2.2 domain that separates thalamus and prethalamus is a secondary feature, due to the induction of this gene adjacent to the border of Shh expression, which is ectopically activated at the core of the ZLI. At neural plate and early neural tube stages, the expression band is strictly longitudinal (marked by dashes in **a**). In the columnar model (**b**), the correspondence of the boundary with the gene band is disrupted at the arbitrary deviation of the former into the telencephalon. Moreover, note this model cannot explain why the gene band extends into the hypothalamus, cutting it into two halves, which cannot be understood as alar and basal parts of the hypothalamus, as in (**a**) (question mark in **b**). (Reprinted from Puelles L. A new scenario of hypothalamic organization: rationale of new hypotheses introduced in the updated prosomeric model. Front Neuroanat 2015; 9:1–23. With permission from Creative Commons License 4.0: http://creativecommons.org/licenses/by/4.0/)

Columnar landmarks are defined by ventricular sulci. *These do not coincide with gene expression patterns.* Furthermore, molecular boundaries are *conserved through evolution*, implying localized control of all histogenesis within that field: proliferation of cell populations, neurogenesis, and the assembly of sequential layer of the cortex. On the other hand, sulci can readily be explained as *tertiary phenomena*: proliferation within one field throws it into prominence with respect to its neighbors. They are not encoded, though the sulci and gyri of the human brain form in a predictable sequence in the second half of gestation and are so constant in the adult brain that each gyrus can be identified and named by macroscopic (gross) examination of the brain at autopsy or in the living patient by neuroimaging or at surgery if craniectomy is performed.

Columns were assumed to be functional and neuromeres not functional. This misconception caused confusion for nearly 100 years. Subdivisions of the forebrain such as epithalamus, dorsal thalamus, ventral thalamus, and hypothalamus are assumed in the columnar model to be associated with the sensori-motor and viscero-somatic organization of the brainstem. They were considered to be vertically stacked, one atop the other. In reality, the thalami, rather than being organized vertically, are distributed longitudinally in distinct neuromeres. Hypothalamus is not the ventral part of diencephalon. It is the ventral part of secondary prosencephalon.

Another example of functional mismatch pertains to the eye. In all vertebrates investigated to date, genetic mapping of *Nkx2.2* in the brain shows that it forms a ribbon-like band extending longitudinally through midbrain, diencephalon, and hypothalamus. This represents the *basal-alar boundary*. The eye develops from hypothalamus above *Nkx2.2*, i.e., from its alar segment. But according to the columnar model, hypothalamus is strictly basal and cannot have an alar plate. Since basal plate is strictly motor, how can hypothalamus give rise to the eye?

A final discrepancy with the columnar model concerns the mapping of the hypothalamus. To date, molecular mapping demonstrates 33 distinct areas of histogenesis. By contrast, columns are supposed to be homogenous.

Molecular Neuroembryology (Puelles and Rubenstein)

Development of the Forebrain

Amniote vertebrates are depicted by development on land; these include reptiles, birds, and mammals. For the uninitiated reader, the ancestral forms for all present amniotes were *reptilomorphs* having a common ancestor from the original tetrapod amphibians. Amniotes then bifurcated from an unknown stem. The *diapsid* line produced dinosaurs (leading to birds) and modern reptiles. The *synapsid* line lead to mammals. Data on brain morphology support the idea that turtles are closest to both stem reptiles and stem mammals. This area of paleobiology is currently in flux, with recent discoveries of feathered dinosaurs adding an exciting twist to our knowledge of the evolutionary record.

The amniote forebrain is a complex structure, including its development. Customarily, discussions of neuroembryology begin with topics such as neural induction, early regional patterning, and the growth of clones immediately prior to neurulation. These fundamental processes are similar across the vertebrate kingdom. They are well-reviewed and will not be further discussed here.

Our discussion of forebrain development and evolution will make use of three approaches: (1) *structural analysis*, (2) *cellular events*, and (3) *molecular mechanisms*. Contemporary developmental biology literature has an abundance of references regarding the latter two. Topologic (structural) analysis has received less attention, but is very important because it gives a positional meaning to cell interactions and to molecular mechanisms. Construction of a three-dimensional brain takes place through modifications of the neural tube. These take place in different degrees and in different directions throughout the ontogeny and phylogeny of an organism.

As the flat neural plate becomes a hollow neural tube, each cell primordium gives rise to larger derivatives (or clones), but the topology does not change. This is referred to as a *one-many transform*. Later in time, other site-specific topological transformations take place, such as the evagination of telencephalic hemispheres. In essence, brain development takes place as a series of three-dimensional expansions that occur at multiple distinct fields of the neural tube wall. Each regionalized field gives rise to a tridimensional radial complex of neural structures extending outward from the ventricular wall toward the pia. Over time cell populations from one radial histogenetic complex migrate tangentially to mingle another such complex. This mixing leads to new neural connections and more complex circuits. In the forebrain, such migrations affect the structure of the brain wall, more layers, and greater sophistica-

tion, but do not change the overall shape. In the hindbrain, some tangential migrations do cause alterations in shape (i.e., the rhombic lip).

Cell groups in the wall of the neural tube share three-dimensional relationships: dorsoventral, DV (i.e., medial to lateral in the neural plate), anteroposterior, AP (cranial to caudal…or vice-versa), and ventriculo-pial, VP (radial). *DV patterning* sets up longitudinal zones along the cylindrical brain wall. *AP patterning* creates the major brain parts (midbrain, metencephalon, etc.) and leads to segmentation (neuromeres) and transverse regionalization. *VP (radial) patterning* permits stratification of different neuronal populations at differing levels within the neural tube wall. In the fourth dimension, *tangential patterning* takes place within a given level connecting adjacent circuits with one another.

Forebrains undergo morphological changes due to a number of well-known processes: cell migration, axonal navigation, formation of neuropil such as neurites and glial processes (especially at synapses), and cell growth versus cell death. To these, we must superimpose regulated proliferation, *clonal patterning*, of specific populations. A simple comparison between the brains of anamniotes (fishes and amphibians) and amniotes demonstrates this principle. Postmitotic periventricular neurons migrate radially outward. In reptilomorphs, these neurons are found entering the outer strata of the mantle zone. Birds and mammals display more extreme degrees of migration. Thus, seemingly strange anatomic relationships between species make total sense when this differential trend is understood. Thus, homologous neural centers can occupy different radial (or tangential) positions depending upon their degree of migration.

Phylogeny of Forebrain Development

In order to appreciate the neuromeric model, we need to get oriented to some principal structures we shall encounter. Let's briefly review this process of forebrain development and see how it changes with evolution. We start seemingly familiar traditional diagram of embryonic development. When the five-part brain develops, it is not obvious that diencephalon gives rise to a secondary prosencephalon consisting of hypothalamus from which project laterally two cerebral hemispheres (telencephalon). Furthermore, hypothalamus, by its very name, was always considered to lie ventral to thalamus…this is simply not the case. As we shall see in the prosomeric model, hypothalamus lies *anterior* to diencephalon. Cerebellum, not present in agnathans, arises from the isthmus (not shown here). Cerebrum, initially small and quite anterior to cerebellum in fishes, enlarges with evolution and expands backwards to cover over the cerebellum (Figs. 5.21 and 5.22).

Fig. 5.21 Embryonic brain development. Note the backward projection of the cortex. Secondary prosencephalon not shown here. Hypothalamus is depicted as ventral to diencephalon. (Reprinted from Puelles l, Harrison M, Paxinos G, Watson C. A developmental ontology for the mammalian brain based on the prosomeric model. Trends Neurosci 2013; 36(10):570–578. With permission from Elsevier)

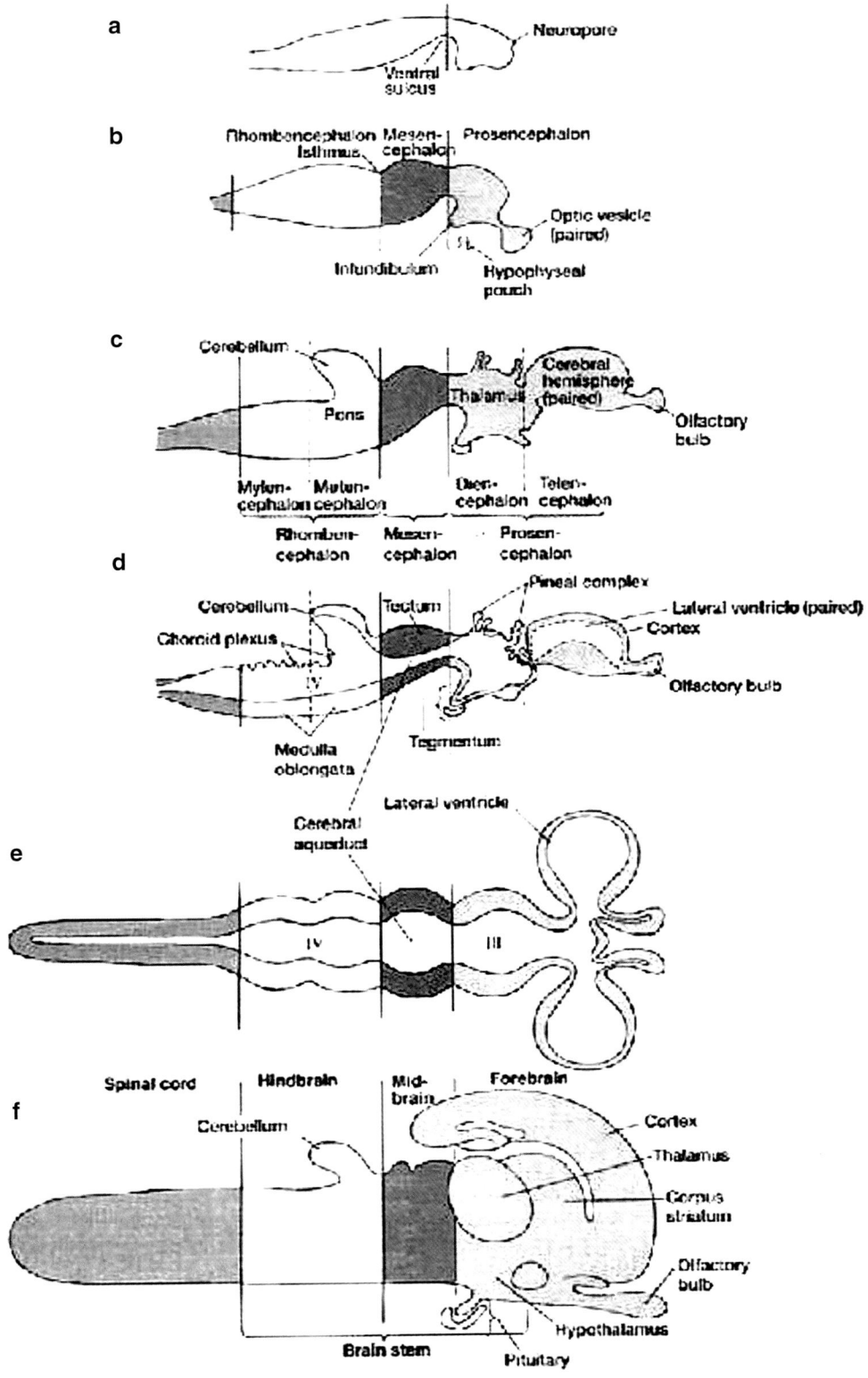

We can better appreciate these relationships in serial sagittal sections of developing mammalian brain with neuromeric labelling. Of note is the boundary between the basal and alar plates (black line) which extends all the way from the spinal cord forward through the hypothalamus. Thus, the two-layer structure of diencephalon is analogous to the ventral *tegmentum* and dorsal *tectum* of midbrain. Note that the preoptic telencephalon (POTel) is the continuation of the alar

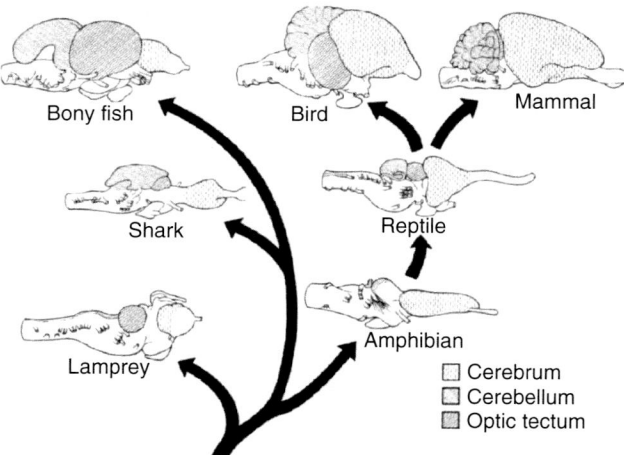

Fig. 5.22 Phylogenetic trends with amniotes. Cerebrum is large (proportionally) in both mammals and birds, but only mammals have gyri, thus increasing complexity while maintaining size. (Reprinted from Puelles l, Harrison M, Paxinos G, Watson C. A developmental ontology for the mammalian brain based on the prosomeric model. Trends Neurosci 2013; 36(10):570–578. With permission from Elsevier)

plate of hp1, the terminal hypothalamus (Thy), formerly known as prepreduncular hypothalamus (PPHy). Telencephalon has additional components, as we shall see later (Fig. 5.23).

The walls of telencephalon have two major zones, pallium and subpallium. The term pallium (pale cloak) refers to the white woolen band with pendants draped around the neck of high ecclesiastics (bishops, pope) in full regalia. From this episcopal ornament hangs a vertical strip, the lappet. In basal vertebrates, pallium contains the olfactory bulb and claustroamygdaloid nuclei. In mammals, pallium forms six-layer isocortex. At its periphery is a more primitive three-layer allocortex shared with hippocampus and the olfactory system. Pallium has three functional parts. Medial and dorsal pallium form hippocampus; lateral pallium forms cerebral cortex.

The subpallium repeats the lappet, hanging beneath pallium. In basal vertebrates is a 3–4 layer structure containing the cortical referral site for olfactory bulb, striate nuclei and preoptic nuclei. In mammals, subpallium has three critical sectors, from dorsal to ventral: striatum (caudate nucleus and putamen), pallidum (globus pallidus), and septum (amygdala). In all vertebrates except derived fishes, the ray-finned actinopterygians, medial pallium rolls inward, as does septum. Changes in the medial pallium in mammals lead to the development of hippocampus, dentate gyrus, and parahippocampus. These structures are not specific to mammals; homologues exist in reptiles and birds (Fig. 5.24).

We can see this mechanism played out over the phylogeny of mammals. In diapsids, striatum remains externalized. Both reptiles and birds demonstrate a midline union between septum and medial pallium. In both the lateral pallium

enlarges, projecting inward in reptiles as the dorsal ventricular ridge and outward in birds as the wulst for processing of stereoscopic vision. In mammals, striatum loses contact with septum and lateral pallium to become internalized. The medial pallium and septum are discontinuous being separate by crossing commissural fibers that span between the hemispheres (Fig. 5.24).

Comments on the Reptilian and Avian Dorsal Ventricular Ridge (DVR)

The historical development of the mammalian brain is prefigured by evolutionary changes seen in reptiles and birds. The telencephalon of these animals contains an important structure called the *dorsal ventricular ridge* (DVR). This refers to a voluminous area of the pallium interposed between the conventional cortex and the striatum. It is so large that it projects into the ventricle. In contrast, the subpallium lying just beneath the DVR does not display ventricular protrusion. The DVR receives afferent neurons from various sensory nuclei of the dorsal thalamus yet its structure is non-cortical. The functional and morphological equivalence of the sauropsidian DVR, as compared to the parts of the mammalian telencephalon, has long been a matter of debate. Karten proposed an evolutionary migratory transformation of sauropsidian DVR centers into site-specific layers of the mammalian isocortex (recall that it has six layers). To understand how this happened, a four-part pallial model works well. Holmgren and later Puelles [29, 66] have proposed that the DVR homolog in the mammalian brain lies within the claustro-amygdaloid complex.

Ancestral tetrapods already possessed a four-part pallium (as evidenced by frogs). Later, reptilomorphs gave rise to diapsid sauropsidians (and later birds). Both developed enormously expanded DVRs, mainly in the lateral and ventral pallial zones. Thus, crocodiles and birds have large DVRs. Synapsids diverged from the reptilomorphs early in evolution. Hence the mammalian DVR did not develop as extensively. Furthermore, the radial migration of neurons (so characteristics of the mammalian brain) tended to transfer DVR derivatives into new locations superficial to the internal capsule. In this way, construction of the claustro-amygdaloidal complex and the olfacto-amygdaloid cortex occurs at the expense of the DVR; thus, mammals have no ventricular bulge. At the same time, expansion of the mammalian striatum causes its own distinctive ventricular bulge.

Expansion of the dorsal pallium in mammals led to the augmentation of the surface layer; it became thrown into folds. A six-layer isocortex now appears. This structure is completely different from submammals. However, throughout the first half of gestation in humans, a radial microcolumnar histological pattern of cortical architecture prevails from radial neuroblastic migration, with the horizontal lamination superimposed from about mid-gestation and eventu-

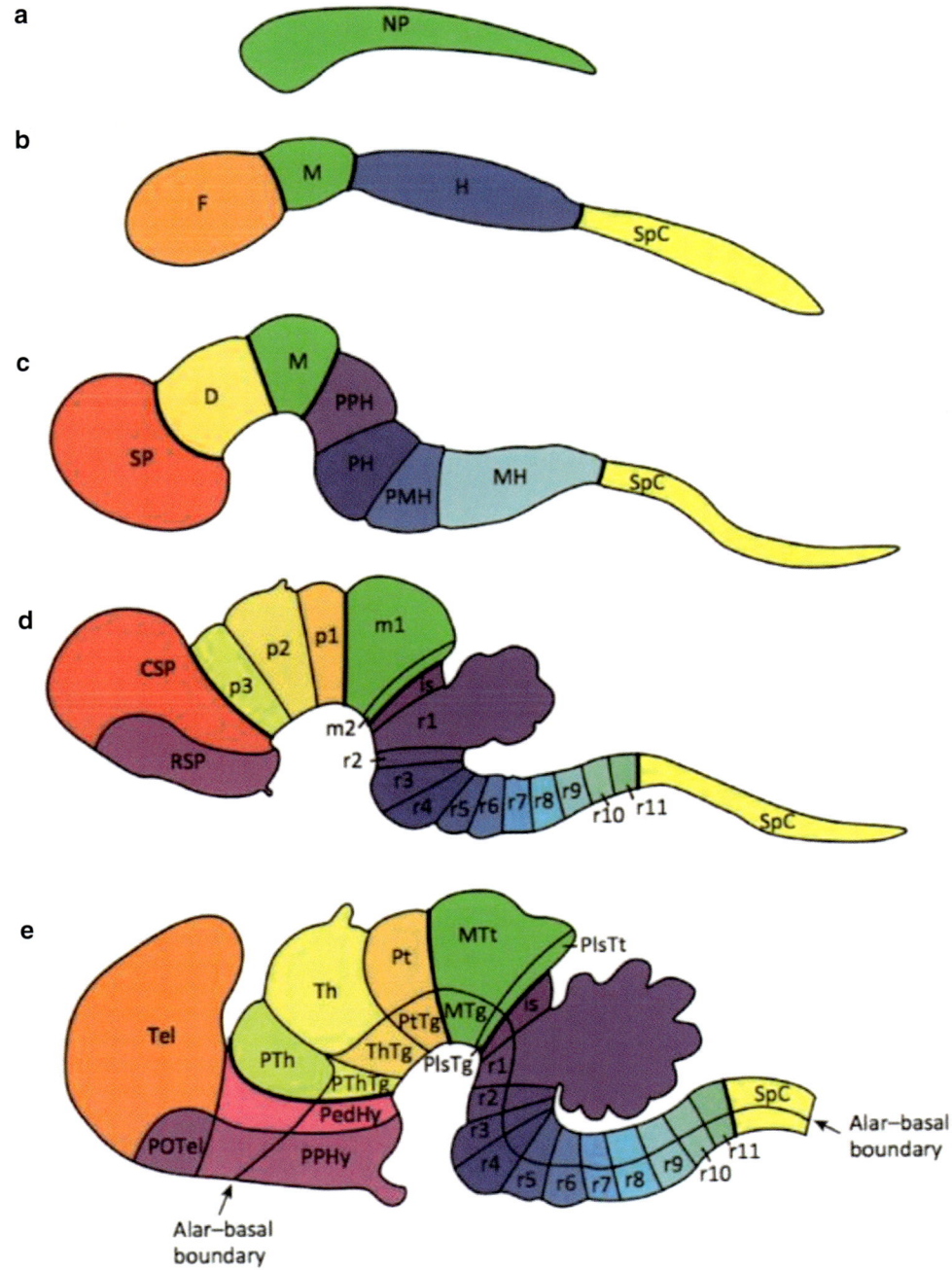

Fig. 5.23 Prosomeric model 2013: developing mouse brain. (**a**) The neural primordium (NP), which is a hollow tube with no subdivisions. (**b**) The rostral (left) part of the neural tube shows the appearance of the forebrain (F), midbrain (M), and hindbrain vesicles (H), with the developing spinal cord (SpC) on the right. (**c**) The forebrain vesicle has two divisions, the secondary prosencephalon (SP) and the diencephalon (D), and the hindbrain is divided into four regions: the prepontine hindbrain (PPH), the pontine hindbrain (PH), the pontomedullary hindbrain (PMH), and the medullary hindbrain (MH). (**d**) From the top, more subdivisions appear in the forebrain [caudal secondary prosencephalon (CSP or hp1); rostral secondary prosencephalon (RSP or hp2); and prosomeres 1–3 of the diencephalon (p1, p2, and p3)], midbrain [mesomere 1 and 2 (m1 and m2)], and hindbrain [isthmus (is) and rhombomeres 1–11 (r1 to r11)]. (**e**) Some parts of the forebrain have become further differentiated: the caudal prosencephalon has formed the main part of the telencephalon; the rostral secondary prosencephalon has formed the preoptic telencephalon (POTel), the terminal hypothalamus (THy), and the peduncular hypothalamus (PedHy); and prosomeres 1–3 have formed the pretectum (Pt), thalamus (Th), and prethalamus (PTh), respectively. Further abbreviations, the diencephalon and midbrain are further subdivided by the alar–basal boundary, which bounds distinct tegmental regions [prethalamic tegmentum (PThTg); thalamic tegmentum (ThTg); pretectal tegmentum (PtTg); midbrain tegmentum (MTg); and preisthmic tegmentum (PIsTg)]. The dorsal part of the midbrain is divided into the main midbrain tectum (MTt) and smaller preisthmic tectum (PIsTt). (Reprinted from Puelles l, Harrison M, Paxinos G, Watson C. A developmental ontology for the mammalian brain based on the prosomeric model. Trends Neurosci 2013; 36(10):570–578. With permission from Elsevier)

Fig. 5.24 Topology of the telencephalon. In all vertebrates except derived bony fishes actinopterygians, pallium inverts and walls of the hemisphere evaginate Septum in prosomeric model is dorsal. Medial pallium, rhippocampus; dorsal pallium and lateral pallium, cerebral cortex; subpallium striatum, caudate nucleus and putamen; subpallium pallidum, globus pallidus; subpallium septum, septal nuclei. (Reprinted from Kardong K. Vertebrates: Comparative Anatomy, Function, Evolution, 7th ed. New York, NY: McGraw-Hill; 2015. With permission from McGraw-Hill Education)

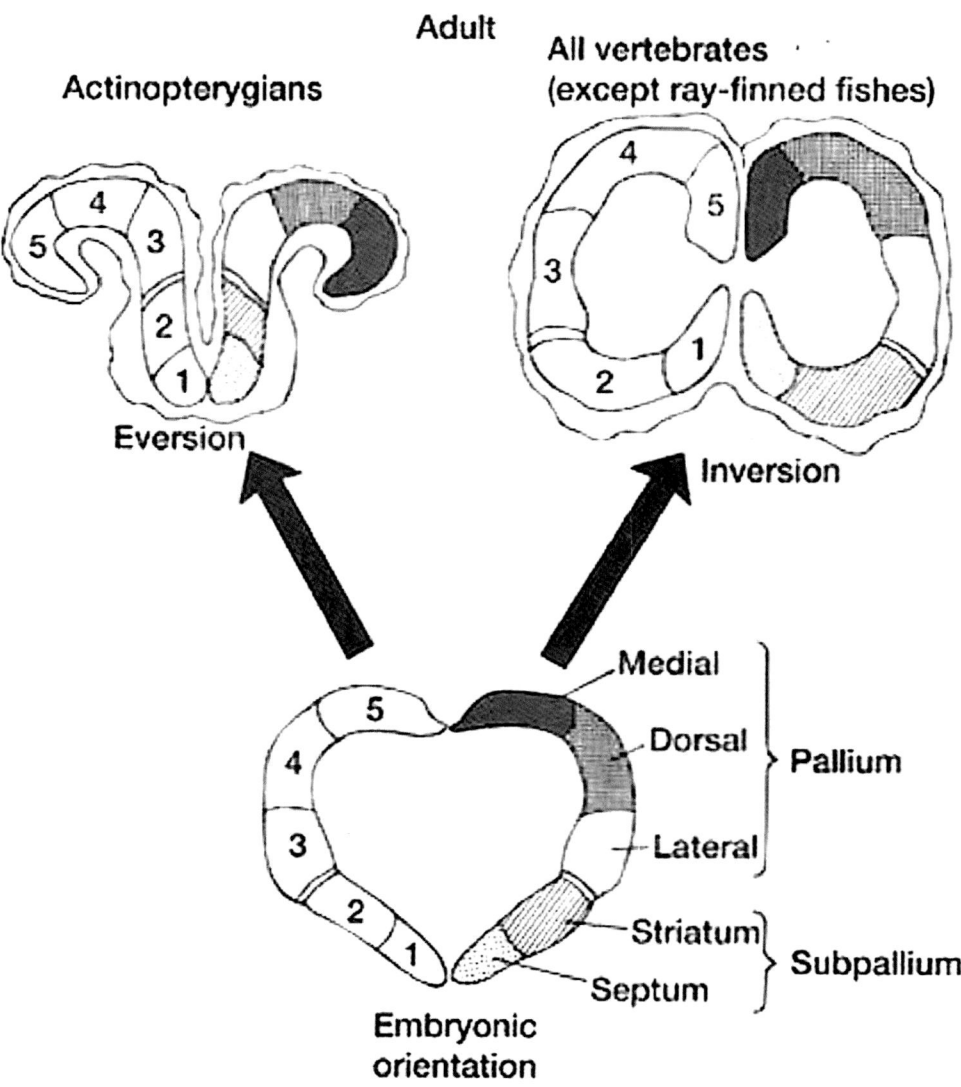

ally predominates; nevertheless, histological evidence of radial lamination is still evident in the adult cerebral neocortex in places where the convolutions bend: the crowns of gyri and depths of sulci [67].

Synaptogenesis in the human cerebral cortex is not initiated until 22 weeks gestation [68]. In normal development, synaptic layers of the cerebral cortex develop in laminae parallel to the pial surface; in cortical dysplasias in which the early radial micro-columnar architecture persists, the layering of synapses is in vertical sheets between the neuronal columns and perpendicular to the pial surface [67], even though electrophysiologically functional radial synaptic sequences are demonstrated even in normal mature human cerebral cortex.

The Prosomeric Model: Iterations

We shall now review in detail the anatomic contents of the individual neuromeres making up the rostral hindbrain, midbrain, and forebrain. Over the past two decades, the prosomeric model has several reiterations. At its inception in 1993, the forebrain was conceived to have six prosomeres: four for diencephalon and two for telencephalon. Isthmus was recognized. The eighth rhombomere was considered a single unit covering the entire medulla (Figs. 5.25, 5.26, and 5.27). The initial topology of the alar plate was (in retrospect) quite simple. By 2001, additional mapping of the hindbrain showed r8 to have four "pseudorhombomeres," r8–r11 (Fig. 5.28). Major changes took place in 2003 (Figs. 5.29 and 5.30). Zona limi-

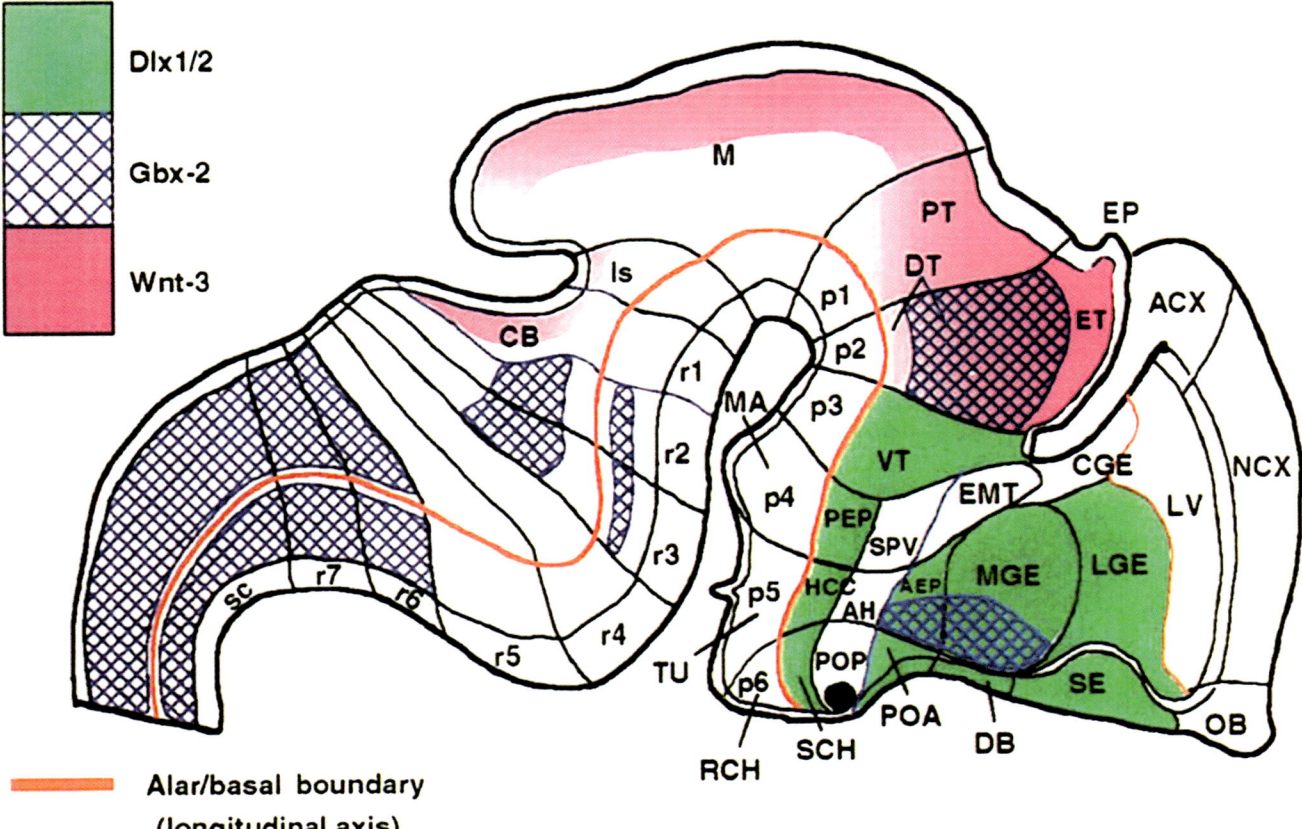

Fig. 5.25 Bulfone Puelles 1993 anatomic map. Realistic map of the expression patterns of the Dlx-1, Dlx-2, Gbx-2, and Writ-3 genes in the brain of an E12.5 mouse. This figure shows a schematic medial view of the brain of an E125 mouse. The medial wall of the telencephalon is opened to show the internal ganglionic eminences. The transverse (neuromeric) subdivisions are delineated by solid lines that are perpendicular to the principal longitudinal subdivision that divides the alar and basal zones and defines the longitudinal axis of the brain (shown by a red line). Other longitudinal zones are delimited by black lines that are parallel to the longitudinal axis. Four longitudinal zones are shown in the spinal cord; from dorsal to ventral they are the roof plate, alar plate, basal plate, and floor plate. These four zones extend rostrally to the absolute forward limit of prosencephalon. The rhombomeres (r1–r7) and theoretical prosomeres (p1–6) are labeled in the floor plate domains of their neuromeres. The expression patterns of the genes studied in this article are shown in the following colors: Dlx-1 and -2, green; Gbx-2, purple; Writ-3, magenta. The Writ-3 expression is shown as a D-V gradient. The site of the optic stalk is indicated by the black circle in the POP domain. Key to abbreviations for Fig. 5.21 included with Fig. 5.22. (Reprinted with permission from Bulfone A, Puelles L, Porteus MH, et al. Spatially restricted expression of Dlx-1, Dlx-2 (Tes-1), Gbx-2, and Wnt-3 in the embryonic day 12.5 mouse forebrain defines potential transverse and longitudinal segmental boundaries. J Neurosci 1993; 13(8):3155–3172. Copyright ©1993 Society for Neuroscience)

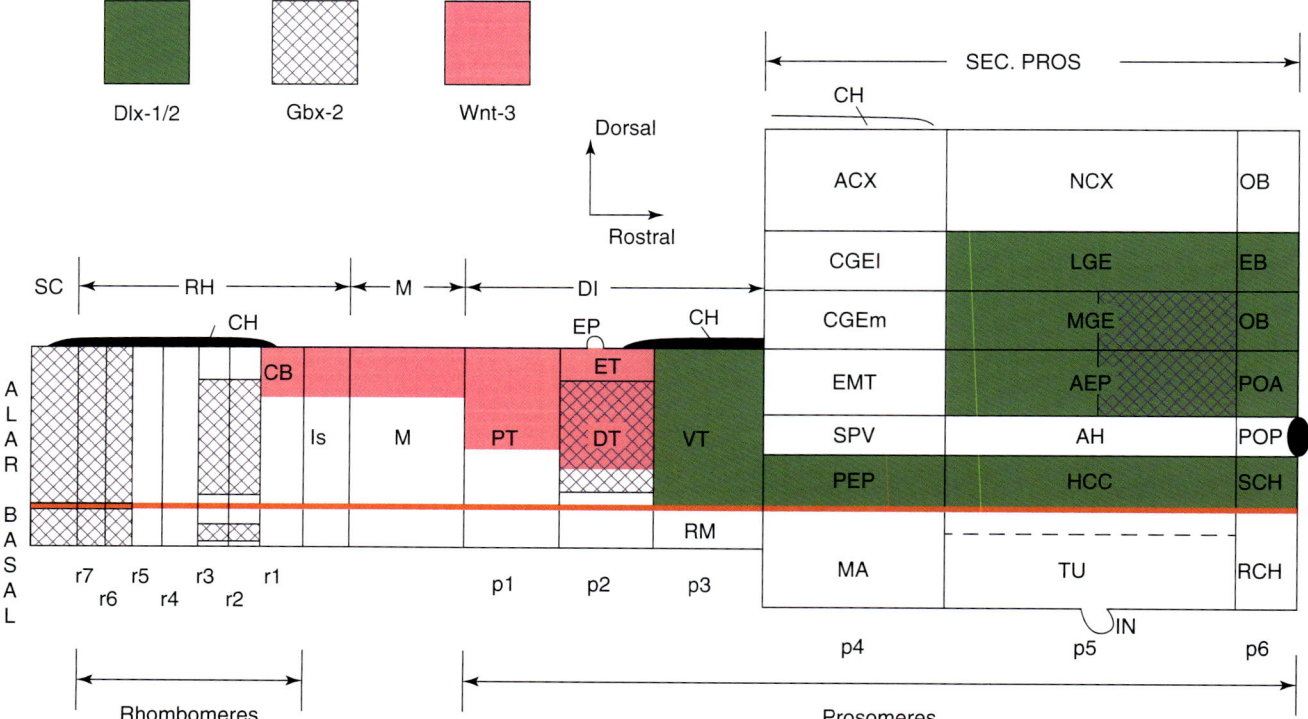

Fig. 5.26 Bulfone Puelles 1993 topologic map. It is a value to compare this rather simplistic and geometric map with the later versions of the prosomeric model, 2003, 2013 iterations. Topologic map of the expression patterns of the Dlx-1, Dlx-2, Gbx-2, and Wnt-3 genes in the brain of an E12.5 mouse. In this figure, the longitudinal axis of the brain is deconvoluted, and the transverse and longitudinal domains are delineated by black lines (the organization of the region rostral to the rhombencephalon is theoretical). The longitudinal boundary that separates the alar and basal regions is shown as a red horizontal line that extends from the spinal cord to the end of the secondary prosencephalon. Additional longitudinal subdivisions within transverse domains are illustrated. For instance, p2 shows four longitudinal domains, based upon the expression patterns of the Wnt-3 and Gbx-2. Midline structures are shown. The expression pattern of the genes studied in this article are shown in the following colors: Dlx-1 and Dlx-2, green; Gbx-2, purple; Wnt-3, magenta. The D-V gradient of Writ-3 expression is approximated. The positions of the rhombencephalon, mesencephalon, diencephalon, and secondary prosencephalon are indicated above the diagram. Telencephalon contains eminentia thalami (EMT), anterior entopeduncular area (AEP), and anterior preoptic area (POA). The locations of the rhombomeres and theoretical prosomeres are shown below the diagram. The optic stalk is represented by a black oval within the POP domain. The position of the CH, EP and IN are shown. The horizontal dotted line in tuberal hypothalamus TU represents the observed Dlx expression in lateral parts of TU; in this region we are not certain whether Dlx expression extends into the basal domain. In later models TU was relocated forward. The caudal boundary of the Gbx-2 expression in p5 was not known with precision. Key (caudal-rostral): *CH* choroid plexus, *CB* cerebellum, *Is* isthmus, *M* mesomere 1 (now m1–m2), *PT* pretectum, *DT* dorsal thalamus, *ET* epithalamus, *EP* epiphysis, *RM* retromamillary area (now located in prosomere hp1), *VT* ventral thalamus (now called prethalamus), *MA* mamillary region, *PEP* posterior entopeduncular area, *SPV* supraoptico-paraventricular region, *EMT* eminentia thalami, *CGEm* caudal ganglionic eminence medial, *CGEl* caudal ganglionic eminence lateral, *ACX* archicortex, *TU* tuberal region, *HCC* hypothalamic cell cord, *AH* anterior hypothalamus, *MGE* medial ganglionic eminence, *LGE* lateral ganglionic eminence, *NCX* neocortex, *RCH* retrochiasmatic area, *SCH* suprachiasmatic area, *POP* preoptic area posterior, *POA* preoptic area anterior, *DB* diagonal band (very important in current model), *SE* septum, *OB* olfactory bulb. (Reprinted with permission from Bulfone A, Puelles L, Porteus MH, et al. Spatially restricted expression of Dlx-1, Dlx-2 (Tes-1), Gbx-2, and Wnt-3 in the embryonic day 12.5 mouse forebrain defines potential transverse and longitudinal segmental boundaries. J Neurosci 1993; 13(8):3155–3172. Copyright ©1993 Society for Neuroscience)

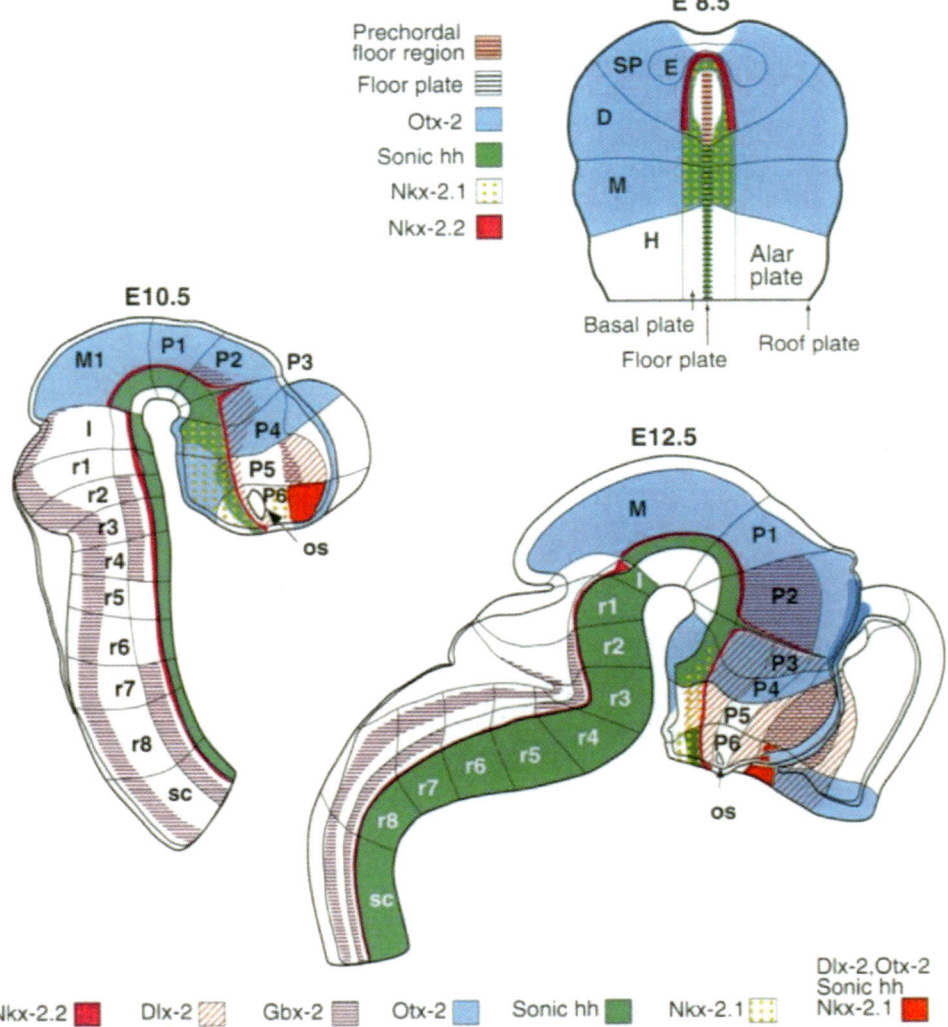

Fig. 5.27 Carnegie stages of prosomere model 1994. Note Shh from notochord to floor plate expressed all the way to anterior p6, the future hypothalamic prosomere hp2. 6 prosomeres, 1 mesomere, r8 not subdivided. E8.5 = Carnegie stage 8.5 neural plate leading to neurulation. E10.5 = Carnegie stage 12 4 pharyngeal arches. The expression of six genes-Dlx-2 14, Gbx-2 14, Nkx-2.1 19, Nkx-2.2 19, Otx-2 18, and sonic hedgehog (sonic hh) (31)-in the neural plate (E8.5) and the neural tube (E10.5 and E12.5) of the embryonic mouse brain. The fate map of the neural plate is based on the studies of other workers and its relation to the expression patterns is hypothetical. The provisional transverse and longitudinal boundaries are indicated as thin black lines. *D* diencephalon, *E* eyes, *H* rhombencephalon-hindbrain, *I* isthmus, *M* mesencephalon-midbrain, *os* optic stalk, *p* prosomere, *r* rhombomere, *sc* spinal cord, *SP* secondary prosencephalon. (Reprinted from Rubenstein JLR, Shimamura K, Martínez S. Regionalization of the prosencephalic neural plate. Ann Rev Neurosci 1998; 21:445–477. With permission from Annual Reviews)

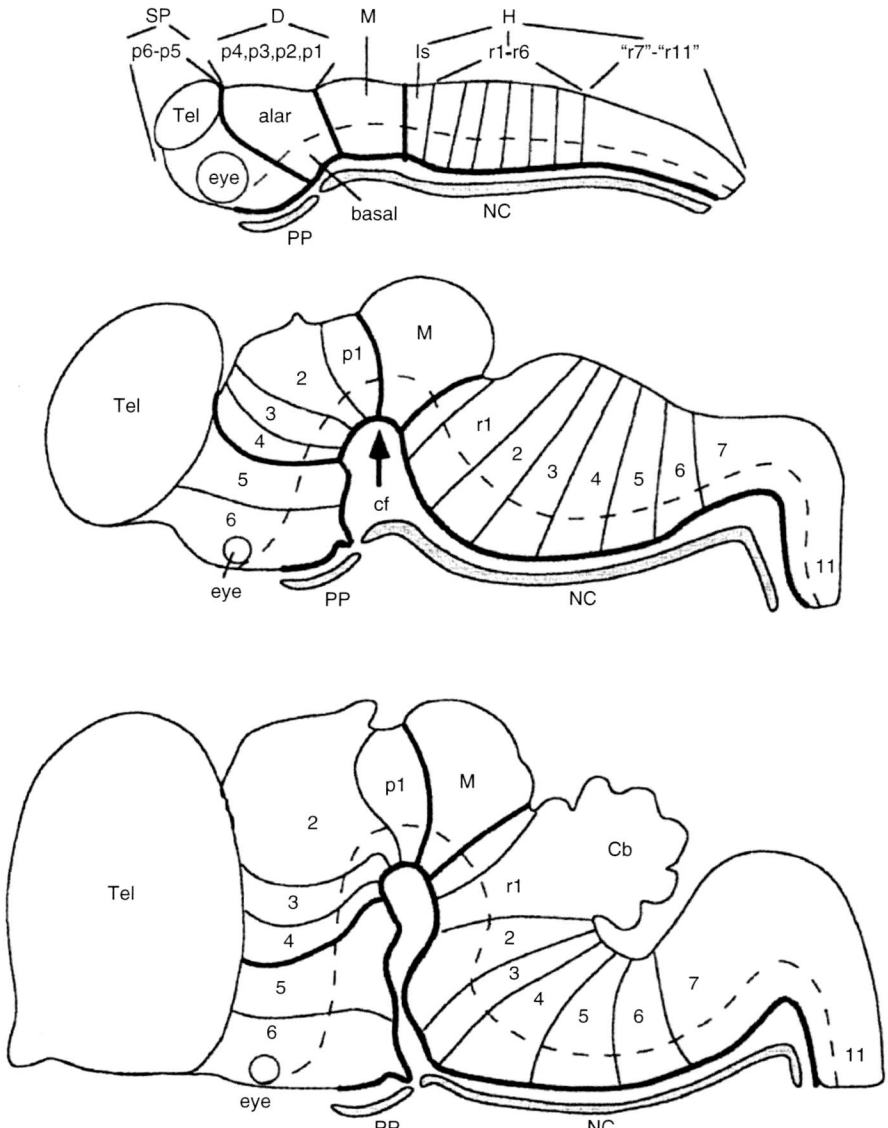

Fig. 5.28 Prosomeric model 2001 at different embryonic stages. New neuromeric segmentation of the neural tube by Puelles [29] at different stages of embryonic maturation. Model maintains six prosomeres but prosomere p4 is now moved back into the diencephalon. Hypothalamus as yet well-mapped. Still only 1 mesomere but medullary r8 has been expanded to rhombomeres r8–r11. Note the cephalic flexion (cf) at the level of the mesomere/p1. In 2018, m2 is recognized as preisthmus and contains inferior colliculi while m1 contains oculuomotor nuclei. Isthmus, r0, contains trochlear nucleus. Schemata illustrating axial incurvation and transversal segmentation of the rostral neural tube in relation to the underlying parts of the axial mesoderm (prechordal plate and notochord). The floor plate induced and maintained by the axial mesoderm is marked as a thicker black line in the floor of the neural tube. The alar-basal boundary appears as a dash line. The thicker transverse lines represent the intertagmatic boundaries (compare Fig. 5.3). Thinner transverse lines mark interneuromeric boundaries in the hindbrain (rhombomeres 1–6 and pseudorhombomeres 7–11) and forebrain (prosomeres 1–6). Note the isthmus (Is) builds the rostral-most distinct part of the hindbrain. The cephalic flexure (arrow marked cf in the middle drawing) progressively separates the midbrain and diencephalon from the rigid axial mesoderm, causing a corresponding deformation of all the longitudinal and transverse boundaries in the area. The relative topology of the eye and that a topologically transverse section through prosomeres 5 or 6 would give the schematic section depicted in

Fig. 5.1d. Note the cephalic flexion (cf) at the level of the mesomere/p1. Prosencephalon is considered prechordal and diencephalon epichordal and caudal to it. Prosomere p4 is reassigned backwards to diencephalon. Hypothalamus is now separated from diencephalon, shifted anterior as the floor of secondary prosencephalon. It is divided into caudal p5 and rostral p6. The eye fields arise from alar p6, but relationships of p5–p6 to telencephalon were not yet mapped out. The eye and telencephalon were recognized as separate bulges. Interthalamic p2/p3, the zona limitans interthalamica, is a well-recognized spike of Shh from the basal plate. *Abbreviations* (proceeding forward from r1 and from ventral to dorsal): r1 cerebellum; r0 isthmus, Is; m esencephalon with tegmentum and tectum; p1 pretectum; p2 (dorsal) thalamus, epithalamus; p3 retromamillary area, (ventral) thalamus, choroid plexus; p4 mamillary area (basal), posterior entopeducular area, supraoptic-paraventricular area, eminentia thalamica, caudal ganglionic eminence (medial), caudal ganglionic eminence (lateral), archicortex, choroid plexus; p5 tuberal hypothalamus (basal), hypothalamic cell cord, anterior hypothalamus, anterior entopeducular area, medial ganglionic eminence, lateral ganglionic eminence, neocortex; p6 retrochiasmatic area (basal), suprachiasmatic area, posterior preoptic area, anterior preoptic area, diagonal band, septum, olfactory bulb. (Reprinted from Puelles L. Brain segmentation and forebrain development in amniotes. Brain Res Bulletin 2001; 55(6):695–710. With permission from Elsevier)

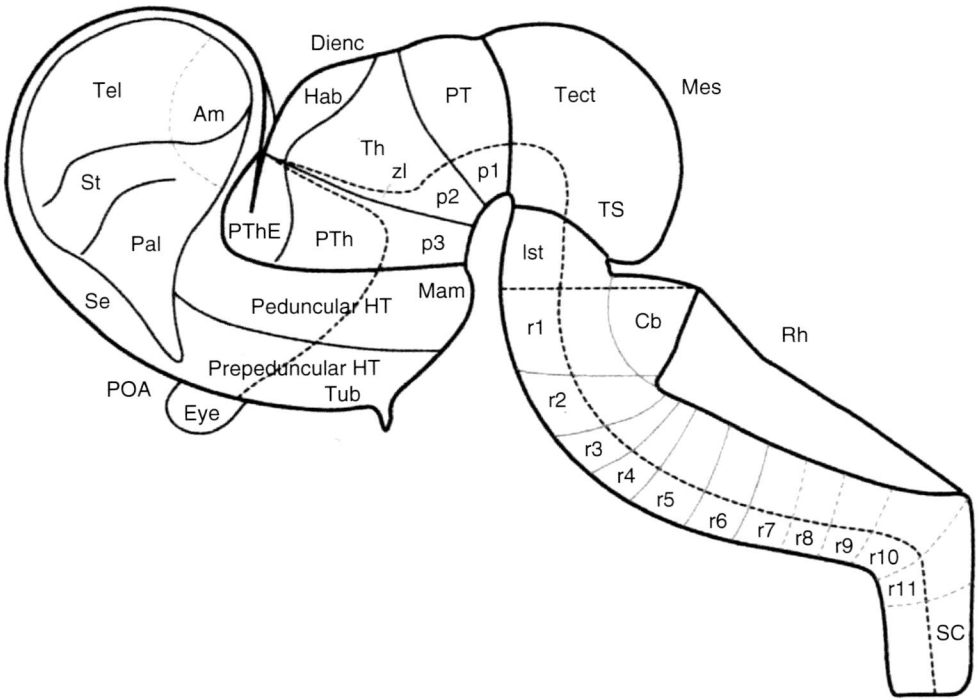

Fig. 5.29 Prosomeric model (2003). In 2003, we will see the following changes to the 2001 model. Isthmus as r0 gives rise along with r1 to cerebellum. Midbrain has two mesomeres with m1 containing oculuomotor nuclei and the superior colliculus, while m2 contains inferior colliculi. Fourth prosomere p4 eliminated. Hypothalamic prosomeres are recognized + telencephalon: p5 becomes peducular hypothalamus (PHy) and p6 becomes prepeducular hypothalamus (PPHy), later termed terminal hypothalamus (THy). Phy in loosely defined continuity with telencephalon and Thy is in continuity with preoptic area (POA). The forebrain lies to the left. Note axial bending at cephalic flexure. The longitudinal alar/basal boundary (dotted line) is present throughout the lateral wall of the neural tube, symbolizing all longitudinal components. [Note: the floor and roof plates not represented.] Zona limitans interthalamica (zl) is a transversal spike-like deviation of the general alar-basal boundary caused by production of Shh in the floor plate. The secondary prosencephalon (Sec Pros) is the rostralmost and most complex unit, consisting of telencephalon (Tel), eye, and hypothalamus (HT: divided in two parts). Septum (Se), striatum (St), pallidum (Pal), preoptic area (POA), and amygdala (Am) regions are identified within the walls of telencephalon; the pallium lies just under the label Tel. Tuberal (Tub) and mamillary (Mam) subregions of the hypothalamus are marked. The caudal forebrain or diencephalon consists of three prosomeres (p1–p3), whose alar regions include the pretectum (PT), the thalamus and habenula (Th–Hab), and the prethalamus (PTh) and prethalamic eminence (PThE); a specific tegmental domain corresponds to each of them (under p1–p3 labels). A simplified view of the large mesencephalic alar plate (Mes) divides it into superior colliculus or tectum (Tect) and inferior colliculus or torus semicircularis (TS); Note 'colliculi' are mammalian terms. The hindbrain or rhombencephalon (Rh) contains 12 neuromeric units, from the isthmus (ist) and rhombomere 1 (rl) down to rhombomere 11 (r11), which limits with the spinal cord (SC). Note the cerebellum (Cb) forms mainly across isthmus and r1. (Reprinted from Puelles L. Forebrain development: Prosomeric model. Puelles. In Lemke G (ed). Developmental Neurobiology. London, UK: Academic Press; 2008: 95–99. With permission from Elsevier)

tans (ZLI) was found to deform the basal-alar junction at the p2–p3 boundary. The fourth prosomere p4 was eliminated. Secondary prosencephalon was recognized. Prepeduncular hypothalamus hp2 (PPHy) and peducular hypothalamus and hp1 (PHy) were mapped out, but their relationships to telencephalon were unclear. To appreciate the differences between the original iteration and the 2003 version, study carefully to comparison figures (Figs. 5.27 and 5.30). The 2013 iteration has additional mapping of midbrain, hypothalamus, and septal roof plate. Prepeduncular hypothalamus was simplified to terminal hypothalamus (Thy) (Figs. 5.31, 5.32, and 5.33). By 2015, the developmental relationships between the hp1 and telencephalon and the individual fields of pallium and subpallium were defined. This resulted in a definitive unified model of secondary prosencephalon with hypothalamic prosomeres hp1 and hp2 extending all the way from the floor plate of hypothalamus to the roof plate of telencephalon, dividing it into evaginated telencephalic vesicle (hp1) and a non-evaginated preoptic area (hp2).

Anatomic Content of Neuromeres, from Caudal-Rostral and Ventral-Dorsal

Contents within each neuromere are listed, by convention, from caudal-rostral and from ventral-dorsal (Fig. 5.30).

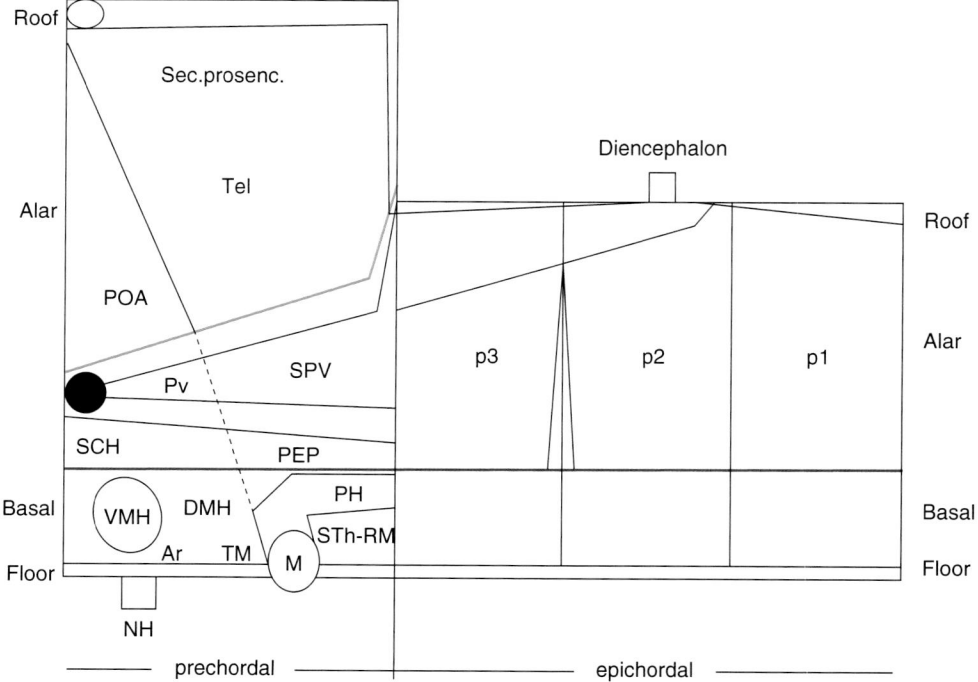

Fig. 5.30 Prosomeric model 2003 topologic map. The primary antero-posterior (AP) divisions, including epichordal diencephalon and pre-chordal secondary prosencephalon, are indicated. Primary dorsoventral (DV) zones are marked at the caudal end (right margin). The alar–basal boundary appears as large dashes superposed on a longitudinal thin line. Note the roof longitudinal zone ends beyond the choroid plexus (ch, gray shading) and commissural septal areas at the anterior commis-sure (ac). Telencephalic subdivisions agree with gene patterns and the avian fate map of Cobos, 2001. The pallial–subpallial boundary is drawn as a red line. The septal (se), amygdaloid (amyg), and stria ter-minalis (ST) regions are histogenetic complexes (small dashes). They get more complex in adults. The optoeminential domain under the tel-encephalic stalk shows tentative internal subdivisions and has at its cau-dal end spike-like dorsal expansions into the amygdala. The hypothalamus appears secondarily divided into caudal (CHy) and ros-tral RHy) marked as medium-large dashes. Note peduncular tract (ped) and fornix tract (fx) are indicated by yellow arrows, Ped connects pos-terior hypothalamus (PH) with ventral pallium (VP). Fornix connects basal stria terminalis (Bst) to mamillary body (M). Retromamillary area is repositioned where the subthalamic nucleus arises and redefined as the posterior hypothalamus. Two AP subdivisions of the prethalamus are indicated (i.e., with higher Dlx-5 signal caudally and higher Pax6 signal rostrally, forming rostral and caudal parts of the zona incerta (RZI, CZI)). Rostral prethalamus has the reticular nucleus (Rt) forming dorsal to the rostral zona incerta, whereas the ventral geniculate nucleus (VG) lies in caudal prethalamus, dorsal to the caudal zona incerta. The p3 tegmentum corresponds to the classic Forel fields (FF). Subdivisions shown in the thalamus are limited to its main three histogenetic tiers (dorsal, intermediate, and ventral) and the lateral geniculate (LG) and medial geniculate MG) primordia. Core and shell domains of the zona limitans intrathalamica are not indicated. The pretectum shows three AP subdivisions, described as precommissural (p), juxtacommissural (j), and commissural (c) domains. Tegmental structures of p1 and p2

include iterated parts of the substantia nigra (SN) and the ventral teg-mental area (VTA), a dopaminergic complex that clearly extends beyond the diencephalon–mesencephalon boundary. *Abbreviations: ac* anterior commissure, *AEP* anterior entopeduncular area, *AHA* anterior hypothalamus, anterior area, *AHC* anterior hypothalamus, central part, *AHP* anterior hypothalamus, posterior area, *Amyg* amygdala, *Ar* arcuate nucleus, *av* anteroventral area of thalamus, *Bst* bed nucleus of stria ter-minalis, *c* commissural pretectum, *ch* choroidal tela, *CHy* caudal hypo-thalamus, *CZI* caudal zona incerta, *D* nucleus of Darkschewitsch, *d* dorsal tier of thalamus, *DMH* dorsomedial hypothalamic nucleus, *DP* dorsal pallium, *Em* eminentia thalami, *FF* Forel fields, *fx* fornix tract, *Hb* habenula (epithalamus), *i* intermediate tier of thalamus, *IC* intersti-tial nucleus of Cajal, *j* juxtacommissural pretectum, *LG* lateral genicu-late nucleus, *LP* lateral pallium, *M* mamillary complex, *MG* medial geniculate nucleus, *MP* medial pallium, *NH* neurohypophysis, *OB* olfactory bulb, *p* precommissural pretectum, *p1–p3* prosomeres 1–3, *Pal* pallidum, *pc* posterior commissure, *ped* telencephalic peduncle, *PEP* posterior entopeduncular area, *PH* posterior hypothalamus, *POA* preoptic area, *PT* pretectum, *PTh* prethalamus (previously known as ventral thalamus), *Pv* anterior periventricular nucleus, *RHy* rostral hypothalamus, *RI* rostral interstitial nucleus, *RM* retromamillary area, *Rt* reticular nucleus, *RZI* rostral zona incerta, *SCH* suprachiasmatic area, *Se* septum, *sm* stria medullaris, *SN* substantia nigra, *SP* secondary prosencephalon, *SPV* supraopto-paraventricular area, *ST* striatum, *STh* subthalamic nucleus, *TH* thalamus (previously known as dorsal thala-mus), *TM* tuberomamillary area, *v* ventral tier of thalamus, *VG* ventral geniculate nucleus, *VMH* ventromedial hypothalamic nucleus, *VP* ven-tral pallium, *VTA* ventral tegmental area, *ZLI* zona limitans intrathal-amica. (Reprinted from Puelles L, Harrison M, Paxinos G, Watson C. A developmental ontology for the mammalian brain based on the proso-meric model. Trends Neurosci 2013;36(10):570–578. With permission from Elsevier)

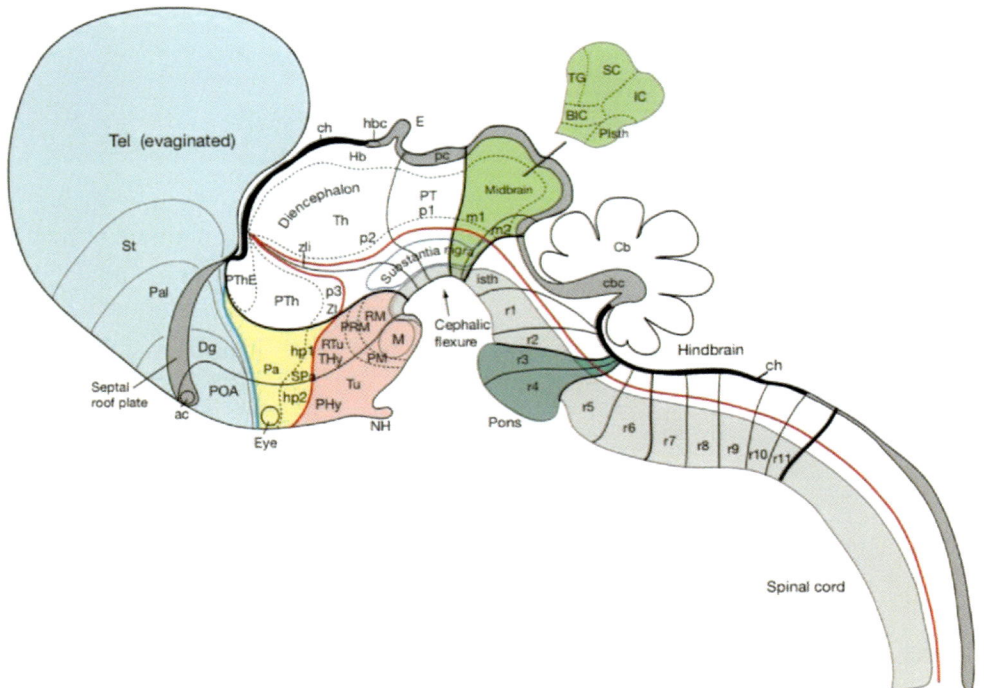

Fig. 5.31 Prosomeric model 2013. The telencephalon is now divided into the pallium and subpallial regions [striatum, pallidum, diagonal domain (Dg), and preoptic area]. The septal roofplate (gray shading) extends from the telencephalic roof to the developing anterior commissure (ac). Within the terminal hypothalamus, the eye vesicle, the neurohypophysis, and the mamillary bodies (M) are differentiating. Within the peduncular hypothalamus, the subthalamic nucleus (STh) is developing. The red line represents the alar–basal boundary, also in the midbrain and diencephalon. In the diencephalon, this molecular boundary is for a short distance pulled to the diencephalic roof as the zona limi- tans (ZLI), which largely separates p2 and p3 at alar plate levels. The gray area above the cephalic flexure represents the most rostral area of Sonic hedgehog (Shh) expression in the floor plate. The diagram shows that the developing substantia nigra extends rostrally from the midbrain into the diencephalon. Other abbreviations are as in Fig. 5.27. (Reprinted from Puelles L, Harrison M, Paxinos G, Watson C. A developmental ontology for the mammalian brain based on the prosomeric model. Trends Neurosci 2013;36(10):570–578. With permission from Elsevier)

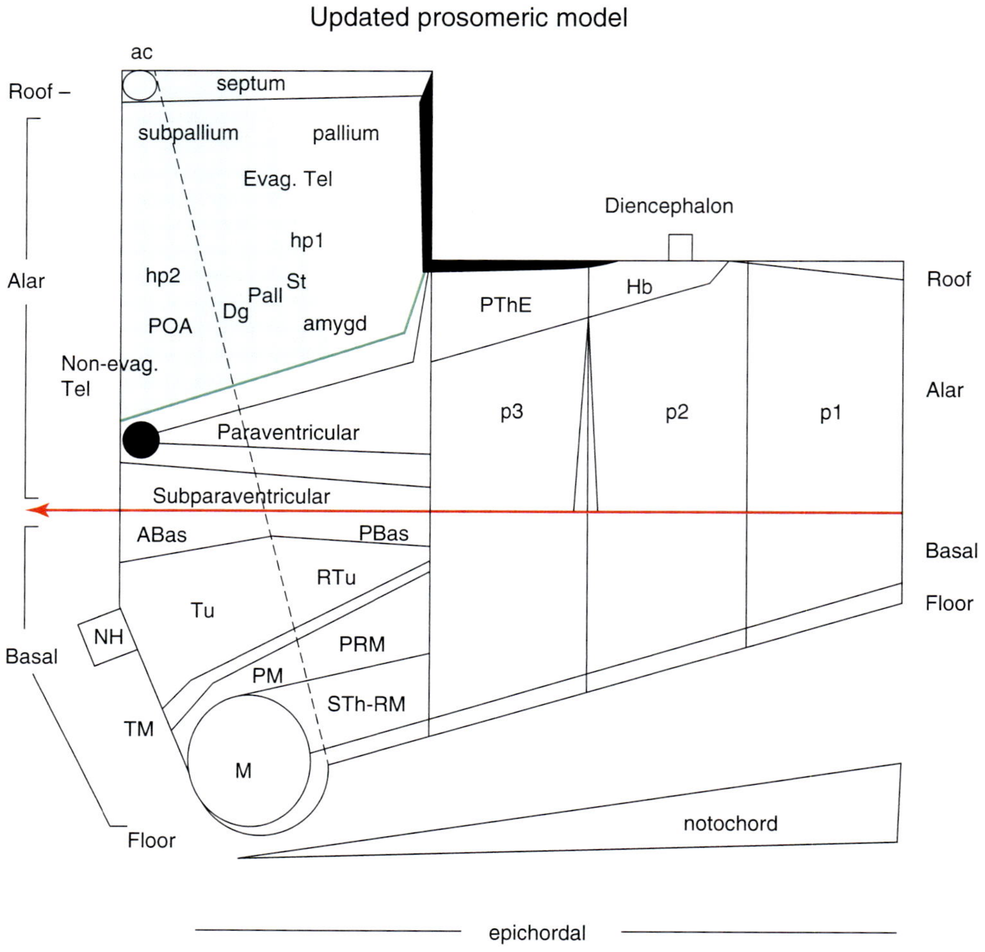

Fig. 5.32 Prosomeric model 2013 topologic map. Intrahypothalamic boundary now extends from the floor plate to the roof plate, distinctly separating the hp1 and hp2 prosomeres and the PHy and THy parts of the hypothalamus. The telencephalic subpallium is identified as a blue field; note its preoptic area (POA), Dg, Pallladium (Pal), and striatus (St) parallel subdivisions. The alar hypothalamus remains essentially unchanged, apart the introduction of the paraventricular and subpara-ventricular area names. The basal hypothalamus is deeply changed, due to our recognizing the mamillary area as occupying an extreme rostral and ventral longitudinal position, consistently with the new floor concept, and the tip of the notochord. This pushes the whole tuberal area, including the median eminence, infundibulum, and neurohypophysis (NH), out of the hypothalamic floor (compare A) and into the rostral end of the basal plate. It represents now a fully longitudinal domain. The novel retrotuberal area (RTu) lies caudally to the tuberal area sensu stricto (Tu) and extends back to the prethalamic (p3) tegmentum, dorsally to the periretromamillary area (PRM). Rostral to PRM lies the perimamillary band (PM). (Reprinted from Puelles L, Harrison M, Paxinos G, Watson C. A developmental ontology for the mammalian brain based on the prosomeric model. Trends Neurosci 2013;36(10):570–578. With permission from Elsevier)

Fig. 5.33 Prosomeric model 2013 neural plate map. (Reprinted from Puelles L, Harrison M, Paxinos G, Watson C. A developmental ontology for the mammalian brain based on the prosomeric model. Trends Neurosci 2013;36(10):570–578. With permission from Elsevier)

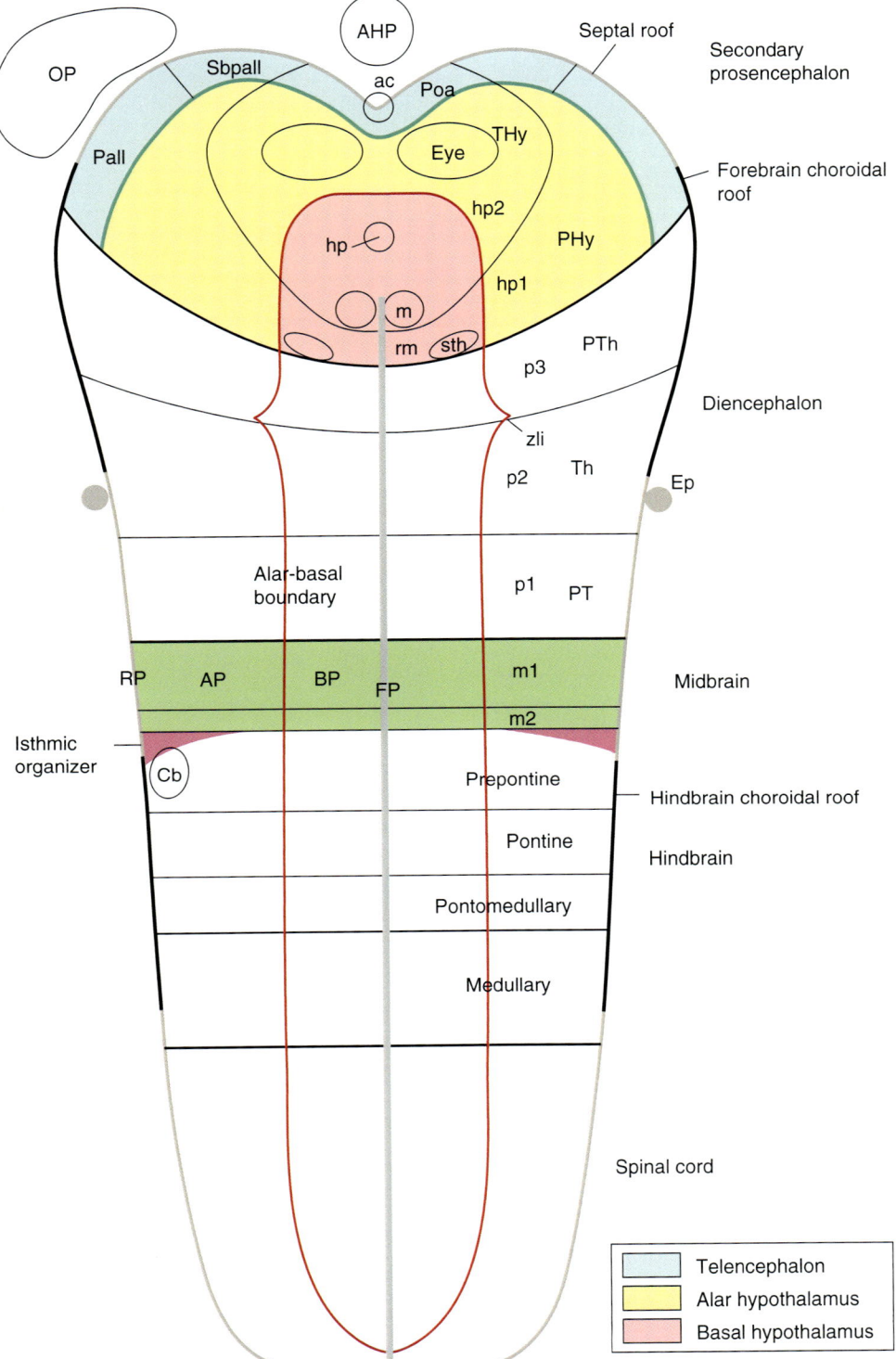

Mesencephalon

- **Mesomere m2 (pre-isthmic mesomere)**: The basal field of m2 has trochlear nucleus. The alar field has the inferior colliculi for auditory system. The ICs receive and process ascending inputs from the auditory system and auditory cortex. They are responsible for the startle response, the vestibulo-ocular reflex and spatial localization from binaural hearing.

- **Mesomere m1**: The basal field of m1 contains oculomotor nuclei (perhaps in rostral and caudal sectors). Alar field has superior colliculi for visual system. In lower vertebrates, the SCs form the optic tectum. In mammals, SCs have a retinotopic maps. They can produce saccadic eye movements and gaze shifts, thus directing movement toward an object.

Diencephalon: The Mysterious "Disappearing" Fourth Prosomere

In the primary iteration of the prosomeric model, diencephalon shared an equal number of prosomeres with secondary prosencephalon. In 2001, p4 was reassigned to the diencephalon. But as the relationships between hypothalamus and telencephalon were clarified, the number of prosomeres in hypothalamus reverted back to p1–p2–p3. Diencephalic prosomeres were found to show differential histogenetic patterns, with signs of mantle layer differentiation and more advanced molecular regionalization. These signals are the prerequisite for the later appearance of specific nuclei with characteristic types of neurons for each prosomeric territory. Prosomeric diversification is more protracted and important in the alar plate than in the basal plate, but the basal plate and the underlying median floor plate also display analogous processes. These are the result of primary gene expressions localized to these particular zones. In this section, the anatomic contents of each prosomere will be described.

The clinical implications of the three-prosomere diencephalon are significant for craniofacial biology. As previously discussed, the dermis of the frontonasal skin zones originates from the caudal neural folds above diencephalon. Recall that epidermis is derived from nonneural crest nonneural ectoderm along the rostral neural folds above telencephalon. Dermal coding determines zones of neurovascular developmental fields. In the forehead and orbit, we observe three primary branches arising the StV1 axis off the ophthalmic: nasociliary and supratrochear, supraorbital, and lacrimal. In this way, we can assign a prosomeric origin to each of these branches.

- **Prosomere 1**
  - Basal p1 contains part of the tegmentum (formerly assigned exclusively to the midbrain). The p1 tegmentum lies behind the mamillary bodies and in front of the red nucleus. The tegmentum caudal to this zone belongs to midbrain proper. Basal p1 shares with basal p2 nucleus of Darschewitsch (D), ventral tegmental area (VTA), and substantia nigra (SN).
  - Alar p1 is the pretectum. Its subfields are: commissural pretectum (c), juxtacommissural pretectum (j), and precommissural pretectum (p).
- **Prosomere 2**
  - Basal p2 shares with basal p1 nucleus of Darschewitsch (D), ventral tegmental area (VTA), and substantia nigra (SN). It also has rostral interstitial nucleus (RI).
  - Alar p2 contains the thalamus (formerly the dorsal thalamus (TH)). TH has ventral, intermediate, and dor-

sal tiers and contains medial geniculate nucleus (MG) and lateral geniculate nucleus (LG). The dorsal zone of alar p2 has posterior half of habenula or epithalamus (Hb). It is thus a summation of epithalamus and dorsal thalamus and is also in the progenitor of the olfactory bulb.
- **Prosomere 3**
  - Basal p3 bears the fields of Forel.
  - Alar p3 contains prethalamus (formerly called ventral thalamus). It has four quadrants. Posterior inferior, caudal zona incerta (czi); posterior superior, ventral geniculate nucleus (VG); anterior inferior rostral zona incerta (RZI); and anterior superior, reticular nucleus (Rt). The dorsal zone of alar p3 has the anterior half of habenula. Together p2–p3 habenular = stria medullaris (sm). Finally, anterior-dorsal eminentia thalmi (Em) and the hemispheric sulcus.

Note: In most vertebrates, the interthalamic transverse boundary between p2/p3 is sharply demarcated by a zone of neurons known as the glial palisade. This boundary, called the *zona limitans interthalamica* (ZLI) contains a "spike" of genetic expression (such as *sonic hedgehog*) that ascends from the basal plate dorsalward into its core. Rostral to this spike, basal genetic markers are more strongly expressed. It is as if the "equilibrium line" separating alar fates from basal fates is strongly displaced dorsalward (toward the roof of the brain).

Secondary Prosencephalon: Hypothalamus

The notochord has an important relationship to the prosencephalon. The entire brain of *Amphioxus* is epichordal. Yet the exact means by which the mammalian forebrain was induced were unclear until recently. In earlier versions of the prosomeric model (2003), Puelles and Rubenstein held that the floorplate of midbrain and diencephalon was induced by signals from the underlying notochord. Thus, mesomeres 1–2 and prosomeres 1–3 were *epichordal*. At the same time, in this construct, secondary prosencephalon, represented ventrally by hypothalamus, was thought to be *precordial*. Its floorplate was not influenced by notochord, but rather prechordal mesoderm. That meant that the floorplate of secondary prosencephalon containing the retromamillary, mamillary, and tuberal fields was entirely prechordal.

In 2012, Puelles demonstrated expression of mouse floorplate genes *Ntn1*, *Shh*, and *Lmx1b* all the way to the rostral tip of hypothalamus, i.e., to mamillary region. Thus, the entire mammalian secondary prosencephalon (and the entire brain) is epichordal. **In sum**: The secondary prosencephalon is that portion of the neural tube that consists of hypothalamus and telencephalon. Hypothalamus continues the structural plan of diencephalon and midbrain: it has basal and alar plates. It gives rise to both the eye and telencephalon. We are indeed the descendants of *Amphioxus*. We shall return to cover anatomic and clinical implications of this point a bit further on.

The hypothalamus has two parts: *caudal hypothalamus* (CHy) and *rostral hypothalamus* (RHy). These have been renamed as hypothalamic neuromeres **hp1** and **hp2**. RHy corresponds to the early expression of *Six3* which extends from septum to neurohypohysis. In the original iteration of the prosomeric model, this was considered the definition of the sixth prosomere. Later experiments with *Six6* knock-out mice demonstrated the loss of the entire secondary prosencephalon acting as a single unit. Basal RHy also expresses *Shh*, critical for forebrain development. Thus, as we shall see, the former prosomere p6, now renamed hp2, is a key element in the pathology of holoprosencephaly.

Hypothalamic neuromeres hp1 and hp2 have basal plates and alar plates exactly analogous to the tegmental and tectal zones of midbrain and diencephalon. Their structures are listed below in ventral-to-dorsal and caudal-to-cranial order.

Basal plates hp1 contains subthalamic nucleus and retro-mamillary area; just dorsal to which lies posterior hypothalamus. The anterior sector of basal hp1 is defined by the mamillary complex which is connected to the limbic system by the fornix tract. Basal plate of hp2 has along it ventral border tuberomamillary area and in front of it neurohypophysis, ringed by arcuate nucleus. Above these structures basal hp2 has dorsomedial hypothalamus and, anteriorly, ventromedial hypothalamus.

Alar plates of hp1 and hp2 have well-defined structures, but each gives rise to an evagination. Alar hp1 contains two structures with long names. Most ventral is posterior-entopeducular area (PEP); above it lies supraopto-paraventricular area. (SPV). The remaining dorsal field is the bed nucleus of stria terminalis (Bst). Alar hp1 gives rise to the telencephalon. Alar hp2 has anterior hypophysis and anterior to it, suprachiasmatic area, positioned mesial to the eye. From the midpoint of anterior alar hp2 is the evagination of the ocular anlage. Directly behind the eye is the complex of anterior hypothalamus posterior and anterior periventricular nucleus (AHP-Pv). Above these we find anterior hypophysis anterior (AHA) and in front of it, the preoptic area (POP).

In sum: hypothalamus, by virtue of being epichordal, and subject to the same types of induction mechanisms, continues the structural *bauplan* of the rest of the brain…with one big difference. Hypothalamus is responsible for the development of the cerebrum. First to emerge are the optic vesicles. These evaginate from the ventral aspect of the alar plate of hp2 (formerly, p6). When this occurs, the surrounding preoptic and postoptic zones of the forebrain wall become topologically distorted. Later in development, the telencephalic vesicles evaginate from the more alar plate of hp1 (formerly

p5). The *telencephalic stalk zone* refers to site where the vesicle crosses the two underlying prosomeres. This is also topologically quite complex.

Note that the alar fields of hp1 (formerly p5) have an extra-telencephalic non-evaginated component. The structures of this region display distinctive patterns of proliferation and histogenesis. It is located strictly dorsal to hypothalamus and contains preoptic, supraoptic-paraventricular, and peducular areas. These domains encroach dorsalward (upward) into the hemispheric stalk, the site of a postulated anterior ectopeduncular area (AEP). The AEP is the site at which oligodendrocyte precursors are manufactured. The AEP is also distinctive because its ventricular zone is the exclusive site of expression for the *Shh* gene.

What we seen then is that anterior growth of telencephalon from hp1 must fold over on top of the optic anlange projecting from hp2. These two zones are entirely separate. *The dural covage is distinct*. The enclosure of the eye by the fronto-orbital "box" distinct from the forebrain results from this development.

- **Hypothalamic prosomere hp1**:
  - Basal hp1 have posterior-inferior subthalamic nucleus and retro-mamillary area (RM). Above it is posterior hypothalamus connected to ventral pallium via the telencephalic pedicle. Anteriorly mamillary area (M) connects to stria terminalis via the fornix (fx).
  - Alar fields have the supraopto-paraventricular area (SPv) in direct contact with posterior basal stria terminalis (Bst).
- **Hypothalamic prosomere hp2**:
  - Basal hp2 has caudally, dorsomedial hypothalamus (DH) and rostrally, ventromedial hypothalamus. Along its inferior border anteriorly is arcuate nucleus (Ar) surrounding the neurohypophysis (NH).
  - Alar hp2 contains anterior hypophysis (AH) and rostral to it the suprachiasmatic area (SCH). Above these structures is anterior hypophysis/central part (AHC) and in front of that, the eye. At the anterior margin of hp2 is the eye and preoptic area.

Telencephalon

Mapping of the telencephalon into distinct molecular and structural zones has been carried out in mammals and birds using genetic markers (transcription factors). These are particularly relevant because these genes are directly involved in the normal development of telencephalon. A good example is the boundary between pallium and subpallium, defined molecularly by *Tbr1, Nkx2.1* and other genes expressed only in the pallium, and the *Dlx* family of genes expressed in the subpallium. The boundary extends all along the telencephalon, from the amygdala caudally to the septum rostromedially. This means that, although septum and amygdala are developmentally more specialized, they share primary genetic markers with the central telencephalon (cortex, claustrum, and basal ganglia) [29].

The pallium has four genetic subdivisions: medial, dorsal, lateral, and ventral. These are based on the expression of *Emx1*, distinct from the traditional view. This four-part pallial model has been observed in mammals, birds, turtles, and frogs.

Subpallium is further divided into three parallel zones: anterior ectopeduncular area (AEP), pallidum, and striatum. Common to all three areas are *Dlx* genes. Pallidum and striatum also express *Nkx2.1,* but only AEP expresses *Shh* in the ventricular zone. The striatum contains diverse specified structures: caudato-putamen and nucleus accumbens, from the distinct central amygdaloidal nucleus. Distinctive markers for these structures are under investigation.

The *lamina terminalis* is an embryonic structure at the traditional junction between diencephalon and telencephalon. Its dorsal part forms the bridge of the corpus callosum and also is the site of origin of prosencephalic neural crest. The anterior commissure forms within a more ventral part of the lamina terminalis. In the mature brain, a remnant of the lamina terminalis persists as a thin transverse membrane extending between the optic chiasm and the anterior commissure.

- **Hypothalamic prosomere hp1**: Above anterior commissure are the following fields in medial-lateral order, prior to invagination. Note the parts of the pallium are named for their final topology after inversion takes place. Note from diagram 10b, rotation hp1 and hp2 90 degrees to the left, positions them as if they were within the skull, over the eye field
  - Anterior entopeduncular area (AEP)
  - Pallidum (Pal)
  - Striatum (ST)
  - Ventral pallium (VP)
  - Lateral pallium (LP)
  - Dorsal pallium (DP)
  - Medial pallium (MP)
- **Hypothalamic prosomere hp2**
  - Preoptic area
  - Anterior commissure

Construction of the Midbrain and Forebrain: Models and Paradigms

Having encountered the previous discussion of neuromeres by O'Rahilly, the reader will note that the prosomeric model provides a new paradigm that allows us to understand neuroembryology in terms of mechanisms. Data from molecular biology, neuroscience, and genetics are assembled into a three-dimensional model that explains how brain development actually might take place. We start a simple set of genes exerting their effects across a flat neural plate. This Cartesian system has genetically defined transverse segments and longitudinal domains. Neurulation rolls the plate into a hollow neural tube. As individual populations expand and interact, topologic distortions of the tubular CNS take place. The end result is a fully functional brain.

Neuromeric theory is *not* in consonance with established views of neuroembryology. The subject is particularly contentious regarding the forebrain. Thus, the reader is entitled to know: Why does this author place such a high value upon an admittedly minority viewpoint?

- First, the neuromeric model, like a Venn diagram, encompasses the existing (longitudinal) model. It emphasizes the importance of transverse segmentation, while not detracting from previous research regarding longitudinal relationships (tracts) within the neuraxis.
- Second, the neuromeric model permits the incorporation of molecular data to explain morphologic events.
- Third, the previous model was largely irrelevant to the larger picture of head and neck development. Knowledge about embryologic processes involved in craniofacial development has expanded but seemingly in isolation to those of brain. By contrast, the neuromeric model applies itself across the board to all aspects of the head and neck, i.e., normal anatomy and pathologies taking place outside the head and neck by virtue of its ability to "map out" mesenchymal populations of neural crest and mesoderm to the anatomic structures they produce.
- Fourth, the prosomeric model is capable of providing clinically relevant explanation for craniofacial anomalies and syndromic relationships.
  - The concept that the homeotic gene expression pattern for a body part peripheral to the head and neck could have a gene "**x**" in common with its local myelomere and also in common with the expression pattern for an intracranial neuromere represents a leap forward for syndromology. The reason for presenting both systems in such great detail is to allow you, the reader, to grasp their similarities and differences.

Structural neurobiology looks at internal brain structures in terms of both form and function. A model is required which permits organization and comparison of structures from the microscopic to the molecular level. Our starting model of the vertebrate brain is one in which differential processes of morphogenesis and histogenesis transform a closed neural tube into an adult structure. What should we expect from such a model?

- It should provide a reasonable explanation about the number of component parts, their anatomic boundaries, and the manner in which they are assembled.
- It should be modular, i.e., capable of being modified and perfected over times.
- It should also be parsimonious, postulating the minimal set of landmarks required to explain the data.
- A segmental model should be able to accommodate unknown or "unfilled" domains, where new data can fit in. The periodic table of the elements is a good example, as it predicts the presence and characteristics of new (undiscovered) elements.
- A good model will predict areas in which meaningful questions can be posed. The existence of distinct genetic zones along the diencephalic neural folds suggests experiments to map out the fate of neural crest derivatives arising from individual zones.
- Finally, a model must be in accordance with clinical data. The occurrence of craniofacial clefts in four vertically oriented zones of forehead dermis and bone skin and frontal bone may correlate with distinct neural crest populations.

Can we understand development without a paradigm? When we look at something as complex as the forebrain, we cannot gather in data without anatomic preconceptions. We need some conceptual structure in order to make sense of things. We normally see what we expect to see. The creative part of the well-prepared mind is to take notice of the unexpected. The segmental paradigm is just set of well thought-out assumptions that can be shared with other colleagues. The tricky part is that the more a paradigm is shared, the less aware we become that we are using it in the first place. In this way, a paradigm can become dogma. As such, it can become a barrier to scientific progress. We get so attached to it that our minds reject thoughts that contradict the dogma. In astronomy, geocentric dogma remained prevalent despite the observations of Gallileo.

In craniofacial surgery, dissection of the prolabium in children with clefts continues to be performed using a design based upon the concept that the prolabium is a single morphologic unit. In reality, the prolabium contains distinct developmental fields that can be separated using incisions along embryologic fusion planes. The outcome of the two surgical approaches is radically different.

Acceptance of a paradigm (morphological or otherwise) depends upon the historical state of scientific knowledge and the reliability of the techniques used to produce that knowledge. Scanning electron microscopy has added greatly to our knowledge of cell structure. When old morphologic concepts can no longer accommodate the accumulation of new data, the paradigm must be modified or abandoned (Figs. 5.34 and 5.35).

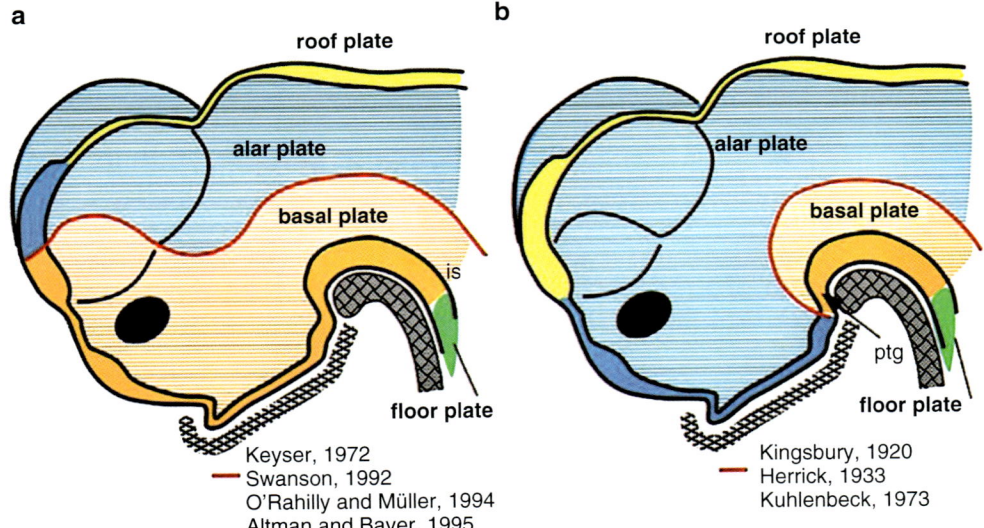

Fig. 5.34 Columnar models: (**a**) Swanson, O'Rahilly and Müller and (**b**) Herrick-Kuhlenbeck. Optic stalk is black oval. Black lines indicate boundary of the telencephalic vesicle and the contour of the ganglionic eminences. Colors: yellow, roof plate; blue, alar plate; red, alar/basal boundary line; orange, basal plate; green, floor plate. Darker shades of same colors indicated where the zones lie across the midline. Floor plate in both (**a**) and ends just rostral to the hindbrain, over the noto- chord. Eye field in (**a**) is in basal plate and in alar plate in (**b**). Alar-basal paradigm is extended into telencephalon in (**a**). (Reprinted from Shimamura K, Hertigan DJ, Martinez S, Puelles L, Rubenstein JLR. Longitudinal organization of the anterior neural plate and neural tube. Development 1995 121:3923–3933. With permission from Company of Biologists)

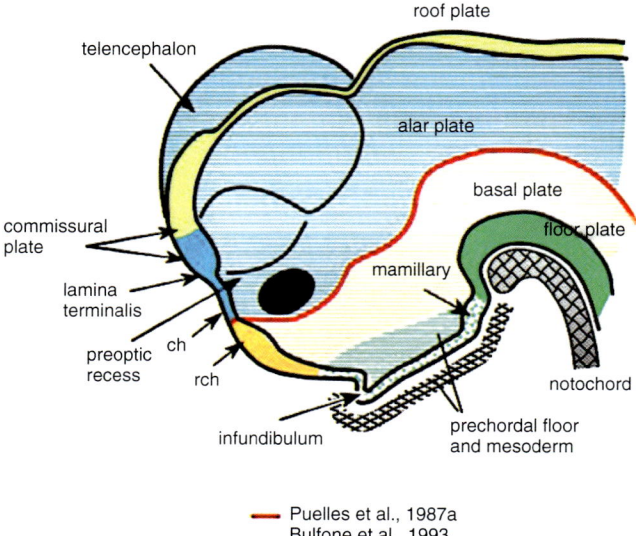

Puelles et al., 1987a
Bulfone et al., 1993

Fig. 5.35 Prosomeric model. Optic tract in alar plate of hp2 in continuity with terminal hypothalamus Thy. (Reprinted from Shimamura K, Hertigan DJ, Martinez S, Puelles L, Rubenstein JLR. Longitudinal organization of the anterior neural plate and neural tube. Development 1995 121:3923–3933. With permission from Company of Biologists)

Evolution of the Prosomeric Model (1993–Present)

Note: To best understand this section, study carefully the original 1993 model, shown in both anatomic and topologic projections. Identify all its component parts. This will enable you to readily appreciate them in subsequent reiterations. Note the changes that take place from 1993, to 2003 and finally in 2013.

The pedigree of the prosomeric model, dating back over 100 years, crediting [69–77] all contributed to the concept of neural tube segmentation.

In 1987, Luis Puelles, using AChE-positive neuroblasts in chick embryos, discovered "mosaic-lie" spatiotemporal patterns of neurogenesis in the midbrain and forebrain that were consistent with the presence of neuromeres in a novel paradigm. Whereas other models considered telencephalon and optic vesicle to represent ring-like first and second neuromeres, stacked one in front of the other like Lifesaver® candies, Puelles described the prosencephalon as dividing into epichordal and prechordal *proneuromeres* called *diencephalon* and *secondary prosencephalon*. The epichordal nature of the diencephalic proneuromeres was previously unreported in the literature. The ACh-label, though suggestive, was too nonspecific to map out neuromeres.

As is so often the case in science, the breakthrough in neuromeric theory came as the result of technology. The biomolecular revolution of the 1970s and 1980s brought about the recognition of homeotic genes and their subsequent application to mapping the organism. This process began with the

hindbrain, but by the early 1990s over 30 *Hox* genes were known to map to the forebrain. Puelles, Rubenstein, and coworkers sought to find expression patterns of regulatory genes in the mammalian forebrain. Their results, reported in 1993, using Dlx-2, Gbx-2, Nkx-2.1, Otx-2, and Sonic hh demonstrated genetically defined boundaries for six forebrain neuromeres. The expression patterns effectively destroyed the previous assumptions of the columnar model. The prosomeric model was born [56] (Figs. 5.25 and 5.26).

The initial model was innovative in several major respects. (1) It recognized the bent longitudinal axis of the forebrain relative to the midbrain and hindbrain. (2) It organized the forebrain into the primary divisions: anteroposterior (AP) transverse neuromeres; and dorsoventral (DV) longitudinal zones. (3) It explored the explicit existence of additional subdivisions of the main AP and DV zones. (4) It recognized the optic and telencephalic vesicles as specialized neural fields with patterning properties that were quite independent from the rest of the neural tube.

In 1994, Rubenstein and Puelles made a clear-cut "declaration of independence" for the neuromeric model *vis a vis* both the revised columnar model [1] and the original columnar model as first proposed by Herrick [50]. The latter conceived of the floor plate as ending at just rostral to hindbrain. Puelles' innovation was to recognize that floor plate continued to extend over the anterior notochordal *all the way to the terminus of the prechordal floor*. Genetic mapping would ultimately confirm this and establish that the entire brain was epichordal (Fig. 5.26 Carnegie stages, Figs. 5.34 and 5.35 Columnar model compared with Fig. 5.25 Prosomeric model).

Puelles' [66] comments on the original model can be summarized as follows. These diagrams of 1993 present an Idealized prosomeric model, by elimination of the gene patterns (for simplicity), thus leaving only longitudinal and transverse subdivisions. Relative sizes of any part are arbitrary. This model only purports to reflect the axial, orthogonal, and neighborhood-invariant relationships obtaining in a mammalian forebrain in a way consistent with available morphological and genetic data, maximizing the capacity to explain regional morphological distinctions at a preliminary, still crude level. Each area has to be imagined in three dimensions, as flower beds growing in a well-organized herbarium and, in so doing, generate the lacking radial dimension. Puelles proposed here that the so-called "paleo-, neo-, and archicortices" might be better represented by successive ventral to dorsal longitudinal domains, possibly ending with a dorsal most septal longitudinal domain below the roof. [The location of the septum in the roof eventually came to pass.] Note that both the 1993 and 1995 iterations maintained the now-debunked idea of an *epichordal floor plate* for diencephalon and a *prechordal floor plate* for the secondary prosencephalon.

The prosomeric model was a paradigm change in neuro-embryology. It recognized, for the first time, the existence of developmental fields within individual neuromeres, making it possible to visualize the topologic changes that are responsible for neural tube shape. Neural tube morphology results from four processes: (1) Patterned cell intercalation, especially in the midline, causes the tube to lengthen and reshape in a medial-to-lateral direction. (2) Neurulation results in neural fold elevation and fusion. (3) Emergence of vesicles for the eyes and telencephalon deforms the forebrain walls. (4) Expansion of individual neurodevelopmental fields (the "flower pot analogy") leads to differential growth of the thickness of the tube and of its surface, as exemplified by the otic placode, causing three-dimensional changes in volume.

2001 brought significant changes to the model while maintaining the prechordal/epichordal construct of forebrain induction. Six prosomeres were preserved, but prosomere p4 as now moved back into the diencephalon. Hypothalamus was not yet well-mapped. Still only one mesomere but medullary r8 has been expanded to rhombomeres r8–r11. Note the cephalic flexion (cf) at the level of the mesomere/p1. In 2018, m2 is recognized as pre-isthmus and contains inferior colliculi, while m1 contains oculuomotor nuclei. Isthmus, r0, contains trochlear nucleus (Fig. 5.28).

In 2003, Puelles and Rubenstein published an updated version of the prosomeric model based on new information regarding gene expression. In a simplification, dorsal thalamus was renamed thalamus and ventral thalamus was renamed prethalamus. They definitively placed hypothalamus as the base of the secondary prosencephalon with two sectors: a caudal *peducular hypothalamus* (PHy) and a rostral *terminal hypothalamus* (Thy). In the process, p4 was eliminated and its contents distributed to PHy, telencephalon, and p3. For example, eminentia thalmi, previously in dorsal p4, were moved dorsal to rethalamus. The previous ventral-dorsal continuity of prosomeres p5 and p6 was interrupted at the boundary between hypothalamus and telencephalon. They redefined telencephalon as an alar plate derivative, probably from PHy. As it evaginates, the telencephalon acquires secondary patterns. The resulting divisions of the pallium and subpallium were *not* found to relate to the underlying patterns of the rostral diencephalon. **In sum**: the secondary prosencephalon was a complex of fields, not neatly subdivided into prosomeres (which should maintain patterning singularities). Prosomeres p1–p3 (pretectum, thalamus, and prethalamus) *are* the caudal forebrain (Figs. 5.29 and 5.30).

By 2013, the prosomeric system underwent its third iteration (Figs. 5.31 and 5.32)

- Anterior commissure, and accordingly, the rostral end of the roof plate, was relocated to dorsomedian preoptic region, which can be conveniently named *septo-commissural preoptic area.*

- Intrahypothalamic boundary was now extended from the floor plate into the roof plate. There are now distinct hypothalamic prosomeres hp1 and hp2 neuromeres extending into telencephalon, creating two parts.
  - *Evaginated telencephalon* (hp1) contains diagonal band, pallidum, striatum, and (laterally) pallium and subpallium.
  - *Non-evaginated telencephalon* (hp2) contains preoptic area and anterior commissure. The PHy (hp1) and Thy (Hp2) parts of the hypothalamus are clearly separated.
- The alar hypothalamus remained essentially *unchanged*, apart from the introduction of the paraventricular and sub-paraventricular area names.
- The basal hypothalamus was *deeply changed*. Mamillary area was shifted from hp1 to hp2, occupying an extreme ventral and rostral position, consistent with the *new concept of the notochord*. New hypothalamic fields were defined: in hp1, the retrotuberal area (RTu) and periretro-mamillary area (PRM), and in hh2, the perimamillary band (PM).

During the first decade of its existence, the prosomeric model laid out a topological organization for gene expression data. Molecularly defined domains of the neural wall could be compared across species. Postulated gene functions were tested serially in terms of regional and cellular fates. The model makes it possible to understand brain embryology more clearly. Each prosomere has its own unique topographic map with its DV and AP divisions being further subdivided into functional compartments, each one capable of producing particular types of neurons and/or glia. Studies of brains from the embryonic to the adult state can be done on a compartment-by-compartment basis, thus documenting how the neuroanatomy of each compartment changes at each developmental stage. This in turn gives a more accurate understanding of the spatiotemporal sequence by which the neural tube is transformed into a functional brain.

The prosomeric model has been duly corroborated in comparative studies of vertebrates as diverse as lampreys and humans. In particular, knowledge of gene expression patterns in nonmammalian vertebrates has expanded greatly in recent years. Genes homologous to those of established molecular boundaries in the mouse forebrain have been documented in the chick, *Xenopus*, and zebrafish. Some of these genes are found in the forebrains of agnathic fishes. Such interspecies comparisons lend themselves to our emerging understanding of the evolutionary origin of the vertebrate neural *bauplan*.

At present, some of the novel premises of prosomeric theory are widely accepted, including: (1) redefinition of the forebrain axis; and (2) the concept of longitudinal zones divided into prechordal and epichordal parts. Previous controversy continuing regarding the number, limits, and neuro-anatomic content of the prosomeres have been resolved.

The Contemporary Neuromeric Model of the Forebrain (2015): Problems and Solutions

Neuromeres are histogenetic segments arranged along the A-P axis. Each neuromere extends the whole vertical length of the neural tube, from floor to roof. Each one has its own floor, basal, alar, and roof longitudinal zones with genetic information specific for that neuromere. All neuromeres share common ventrodorsal patterning mechanisms that originate at the neural plate stage. These represent an antagonism between signals arising downward (medial) from the roof plate and those diffusing upward (lateral) from the floor. Furthermore, the floor plate is induced by the notochord, the source of the homeobox coding specifying each individual neuromere. Regardless of differences in anatomic content, all neuromeres are structurally alike; they are *metameres*.

In the neuromeric model, the midbrain, diencephalon, and secondary prosencephalon are complete rings of the neural tube. Telencephalon is no longer a separate rostral entity. Instead, it is stack above hypothalamus and the eye. Each hemisphere is a gigantic balloon emanating from the dorsal aspect of the alar hypothalamus. The patterning of the telecephalon dates back to the neural plate stage. Subpallium lies rostral to pallium, but with upward and outward growth of the telencephalic vesicle, it comes to occupy a ventral position in the wall. The neural plate map shows clearly the relationship between the two zones of telencephalon: the evaginated cerebral zone arises from hypothalamic prosomere hp1, whereas the non-evaginated preoptic zone arises from hypothalamic prosomere hp2.

The 2015 prosomeric model resolves three critical anatomic issues. We shall describe each of these in detail separately. Taken together, they have implications for a biologic understanding of holoprosencephaly.

- There is no prechordal part of the neural tube. The notochord extends all the way forward below the hypothalamic floorplate. Prechordal plate extends ventro-dosally in front of the terminal (anterior) border of hypothalamus.
- The interprosomeric boundary extends obliquely all the way from basal hypothalamus to telencephalic roof plate. It thus gained a recognition as a complete neuromeric border. Prosomeres hp1 and hp2 no longer stopped arbitrarily at the dorsal margin of alar plate; they were found to continue all the way upward through telencephalon.

- The developmental fields of basal hypothalamus underwent considerable redefinition, making them longitudinal. By organizing the mamillary/retromamillary areas and the tuberal region as longitudinal entities, their various components fit into either the hp1 or hp2 prosomeres.

Problem 1. Determining the Role of the Notochord

In the 2003 prosomeric model, induction of the forebrain was held to take place in two ways. Midbrain and diencephalon were considered *epichordal* part of the neural tube: their floor plate being induced by notochord. Secondary prosencephalon (represented ventrally by hypothalamus) was considered a *prechordal* part of neural tube: its floor plate being induced by prechordal plate mesoderm. By 2013, a series of discoveries led to a unifying concept: *the entire brain* is epichordal. Notochord is responsible for floor plate induction of midbrain, diencephalon and hypothalamus. Prechordal mesoderm cells exert an unexpected vertical influence in forebrain development. We will examine the basis for these findings, but first we must review some definitions.

Developmental Relationships Between Notochord and Prechordal Plate

For many of us, the relationships of primitive node, primitive streak, notochord, and prechordal tissue with one another and their developmental significance may seem lost in the mists of the lecture hall, yet they prove to be critical at this juncture. The best illustration of this at the early stages is by O'Rahilly, although it contains a conceptual flaw. Carlson accurately depicts the morphologic changes in the notochord, but fails to show its full advance to the level of future prosomere hp2.

Staging criteria as per University of New South Wales embryology website, Dr. Mark Hill, director: https://embryology.med.unsw.edu.au/embryology/index.php/Carnegie_stage_"X" (Figs. 5.36, 5.37, 5.38, and 5.39).

Stage 6 (13–14 days)

- The epiblast is a single layer of pseudostratified columnar epithelium. At day 14 and day 17, a distinct population of cells proliferates and moves into the midline. This primitive streak (PS) first becomes visible caudally and then extends forward. This does *not* mean the cells are spreading forward. PS results from an interaction between epiblast and hypoblast from a factor elaborated posteriorly that spreads forward in the midline causing the streak to appear. *Homeotic genes in the midline are turned on in successive potential neuromeric zones.* This is because the process of forming the posterior organizer, *Koller's sickle*, during stage 5c involves cellular movement of epi-

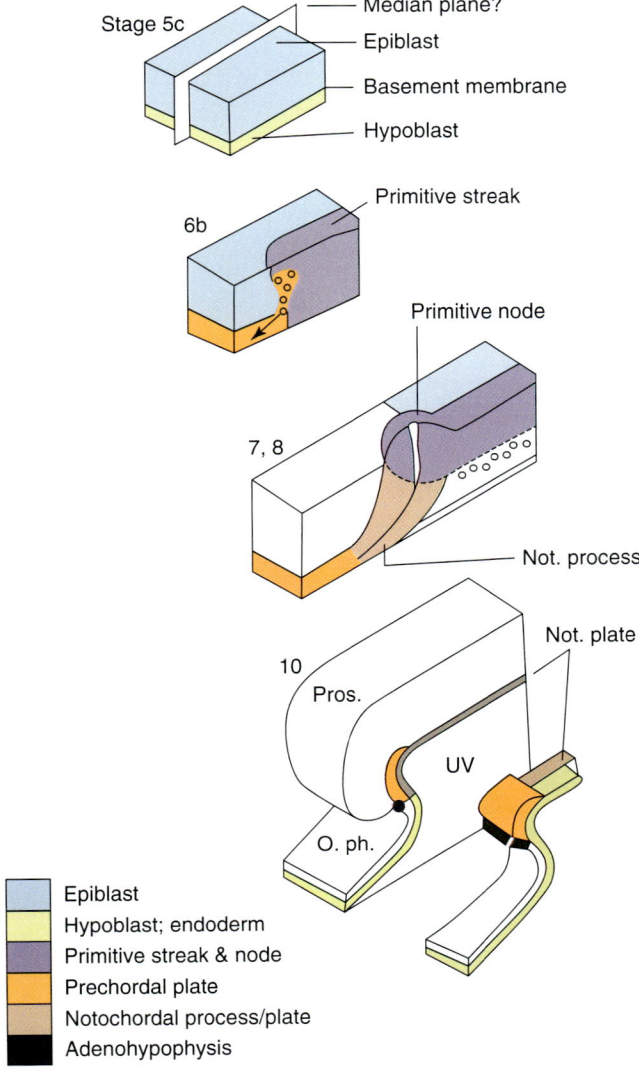

Fig. 5.36 Development of the prechordal plate, stages 5–10 m. Prechordal plate (orange) is the first mesoderm to be produced from the primitive node. It is not born via gastrulation per se, but is synthesized at stage 6b directly from dorsal PN and does not migrate through primitive streak. PCM is extruded forward like toothpaste into the prenodal head region and extends all the way to the heart fields. Although in terms of homeotic coding, PCM would correspond to r0; it is likely influenced by local genes of the endoderm. Notochord is produced as a forward advancing tissue stream from PN at position r0 and time t0. Unlike in the O'Rahilly model, notochord does not stop at the backend of PCM. It penetrates it, arriving at the future prosomeres hp2. (Reprinted from O'Rahilly R, Müller F. The Embryonic Human Brain: An Atlas of Developmental Stages, 3rd ed. New York, NY: Wiley-Liss, 2006. With permission from John Wiley & Sons)

blast such that the embryonic disc is longer random, but instead has electrochemical gradients from A-P and medio-lateral that set up an occult axis in the midline that reveals itself later in development as the sequential expression of homeotic codes unique for each neuromeric level.

- By the end of stage 6, the primitive streak reaches its maximum length at r0. Here it forms the *primitive node* (PN) through which cells egress.
  - The dorsal PN is *proliferative*. It generates the cells of the prechordal plate. Although biologically distinct from epiblast, dorsal PN is indistinguishable from its surroundings.
  - The ventral PN is readily identified. It acts as an *organizer for gastrulation*.

Stage 7 (15–17 days)
- The prechordal plate is visualized. It arises from dorsal PN at level r0, but is *not a product of gastrulation*. Its neuromeric code has been described the mesoderm from levels r0 to r1. The location of PCM is a small zone anterior to r0.
- Primitive node retreats caudally from r0. Just before doing so, it lays down a *notochordal process*. Notochord extends *forward* until it physically encounters prechordal plate. The *anterior notochord* extends from r0 forward to hp2, the future anterior end of terminal hypothalamus.
- As PN retreats, it lays down additional notochord forward of its position.
- Paraxial mesoderm that gastrulated from r1–r2–r3 is dedicated to the vasculature surrounding the brain and to the extraocular muscles. How these populations interact with PCM is uncertain.
- Primitive streak is the site of gastrulation
  - Anterior PS: endoderm, notochord, medial somites.
  - Middle PS: lateral somites, lateral plate (heart, kidney).
  - Posterior PS: Lateral plate, extraembryonic mesoderm.
- As soon as endoderm is produced, PCM incorporates into it, forming a localized thickening.

Stage 8 (17–19 days)
- Prechordal plate in humans is at its maximum in stage 8. It is rectangular and thickened up to eight cells layers. PCM extends forward from r0 to buccopharyngeal membrane and abuts against cardiac mesoderm
- PCM by virtue of binding to endoderm forms *buccopharyngeal membrane* mm.
- PCM becomes the *source of somitomeres Sm1–Sm3*, producing all extraocular muscles except lateral rectus, derived from Sm5.
- *Notochordal canal* develops during 8b. It is continuous with primitive pit. It develops a communication between amniotic cavity and umbilical vesicle, becoming renamed

Fig. 5.37 Prechordal plate, stages 6–10. Notochord advances through the substance of PCM in the midline until reaching the terminus of hp2. Thus, the entire brain is epinotochordal. (Reprinted from O'Rahilly R, Müller F. The Embryonic Human Brain: An Atlas of Developmental Stages, 3rd ed. New York, NY: Wiley-Liss, 2006. With permission from John Wiley & Sons)

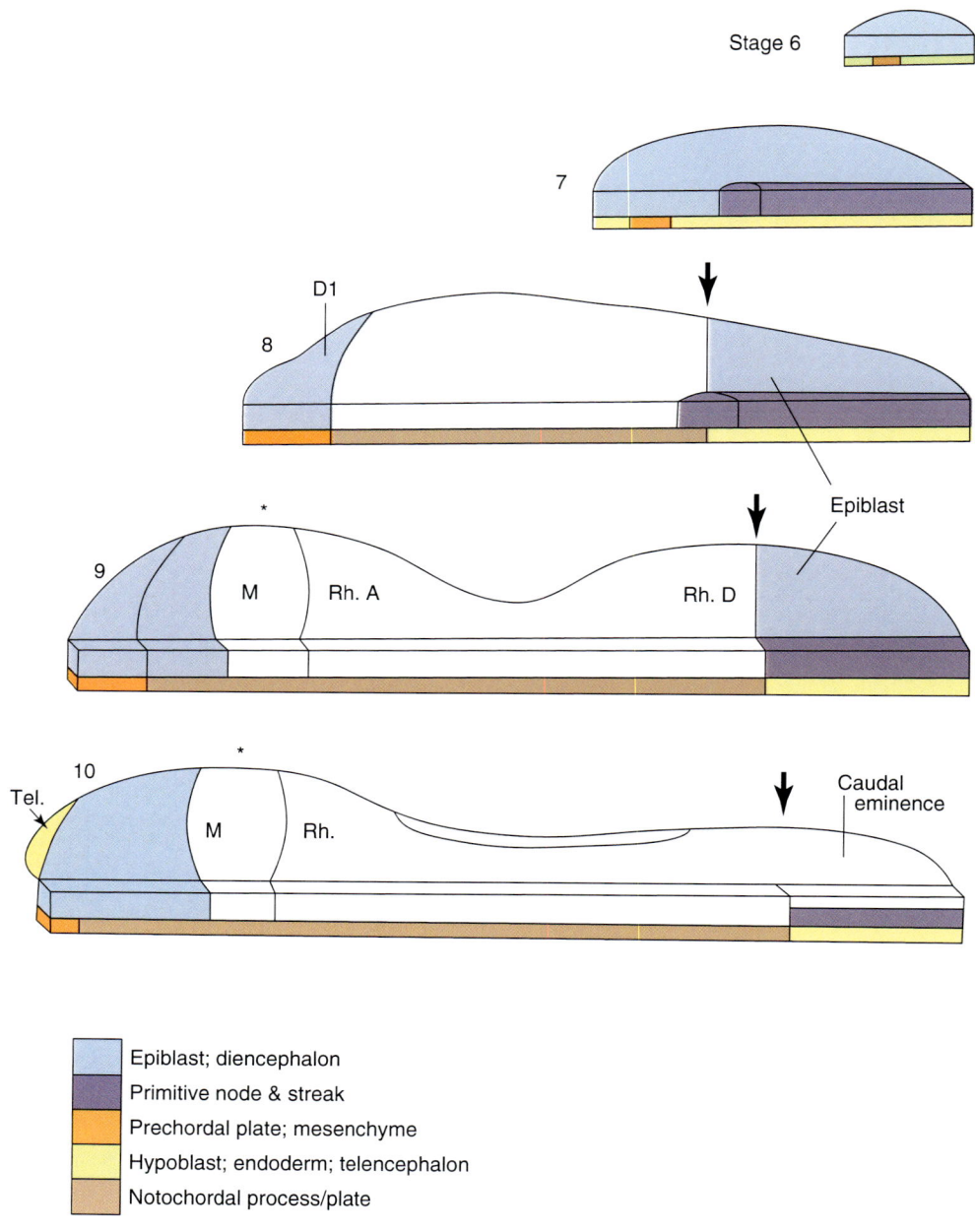

the *neurenteric canal*. The canal moves progressively backward being cervical at stage 9 and at neuromeric level t6 at stage 10.

- *Neurulation* takes place.
- The notochord is laid down like a slime trail behind the retreating primitive node.

Stage 9 (19–21 days, 1–3 somites)

- Prechordal plate is continuous between cardiac mesenchyme and BPM when heart rotates ventrally; so does the PCM. It is thus positioned alongside the first aortic arch.

Stage 10 (22–23 days, 4–12 somites)

- PCM stretched, just like a belt, across the first aortic arches.
- It remains in contact with the notochord at r0.

Stage 11 (23–26 days, 13–20 somites)

- Premandibular condensation—formation of extraocular muscles.

Stage 12 (26–30 days, 21–30 somites)

- Premandibular mesenchyme crosses the midline in front of the first aortic arches. Possibly the source for the r1 sphenoid.

In sum: Prechordal plate mesendoderm, produced at stage 6, is joined at stage 7 by an <u>anterior projection of notochord from primitive node</u>, just before it begins its retreat. Notochord is distributed in the midline forward rom r0 up to hp2. PCM becomes a rectangle beginning at hp2 and proceeding forward as far as the heart field. Notochord will induce the floor plate for the entire midbrain-forebrain complex. Signals from PCM exert a ventral dorsal effect.

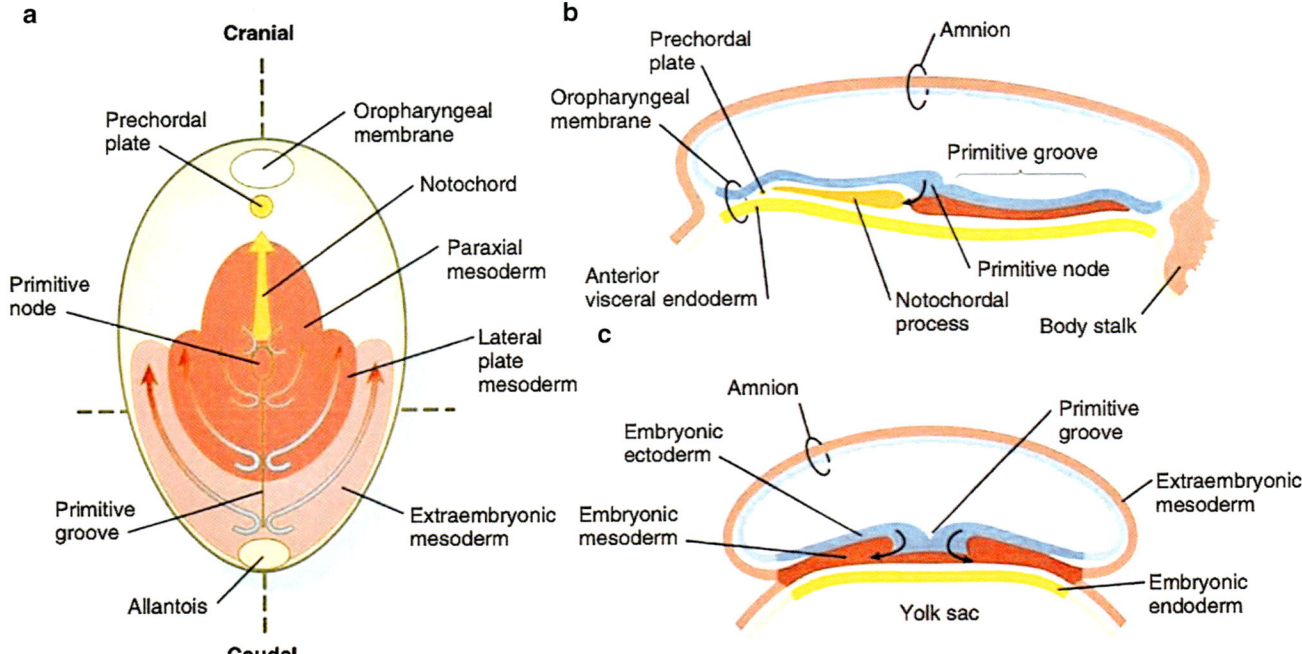

Fig. 5.38 Notochord (orange) migration. The notochord moves first forward from primitive node at time t0 and subsequently, as PN retreats, it is laid down behind much like the slime trail of a banana slug. Note that Carlson as well does not demonstrate the midline penetration of notochord into prechordal plate that is required for brain induction. (Reprinted from Carlson BM. Human Embryology and Developmental Biology, 6th edition. St. Louis, MO: Elsevier; 2019. With permission from Elsevier)

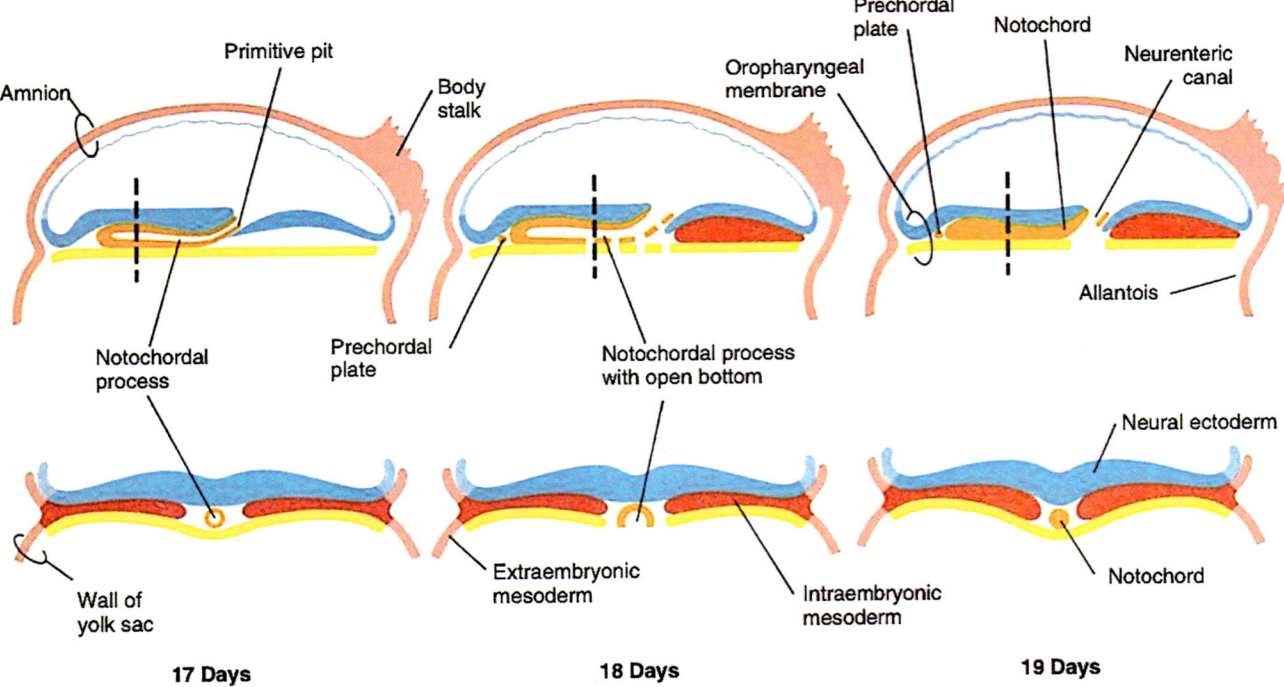

Fig. 5.39 Time course of notochord development during stage 8. By the end of stage 8 primitive pit has arrived at the level of cloacal membrane and induction of the neural system is complete. (Reprinted from Carlson BM. Human Embryology and Developmental Biology, 6th edition. St. Louis, MO: Elsevier; 2019. With permission from Elsevier)

Where Is the Notochord Located with Respect to Hypothalamus?

Notochord is in contact with the neural floor for only a brief period of time through the neural tube stage. The cephalic flexure quickly separates these two structures. No less than [78] described its terminus at or near the mamillary pouch. In 1987, Puelles reported, using ACh-E histochemical labeling, a positive epichordal strip extending from the isthmus all the way to mamillary area. Using differential gene markers, Sánchez-Arrones [79] demonstrated the floor of the forebrain to coincide with the mamillary area in the chick as well. *Shh* is well-known to be directly induced in the floor plate by notochord. Conclusive proof in mammals was demonstrated by when Puelles [2] used *Shh* along with *Foxa1*, *Lmx1b,* and *Ntn* with the label extending to the rostral end of mamillary pouch, exactly as had been predicted by His (Fig. 5.40).

Induction Effects of the Notochord and Prechordal Plate

As reported by García-Calero [80], extirpation of notochordal and prechordal tissues in chick embryos produces a step-wise series of malformations with a startling series of interpretations. Induction sequence: (1) Throughout the CNS, floor plate is induced by notochord (axial homeotic code). Excision of anterior tip of notochord causes deletion of forebrain floor plate and failure of forebrain development. (2) Basal plate results from the combination notochord plus prechordal plate. PCP excision does not affect floorplate. (3) Alar plate develops from basal plate.

- Observation: Prechordal tissue deletion at varying times after its formation causes varying degrees of tissue dedifferentiation, in reverse order of their formation. The later the deletion, the more time for prechordal signals to do

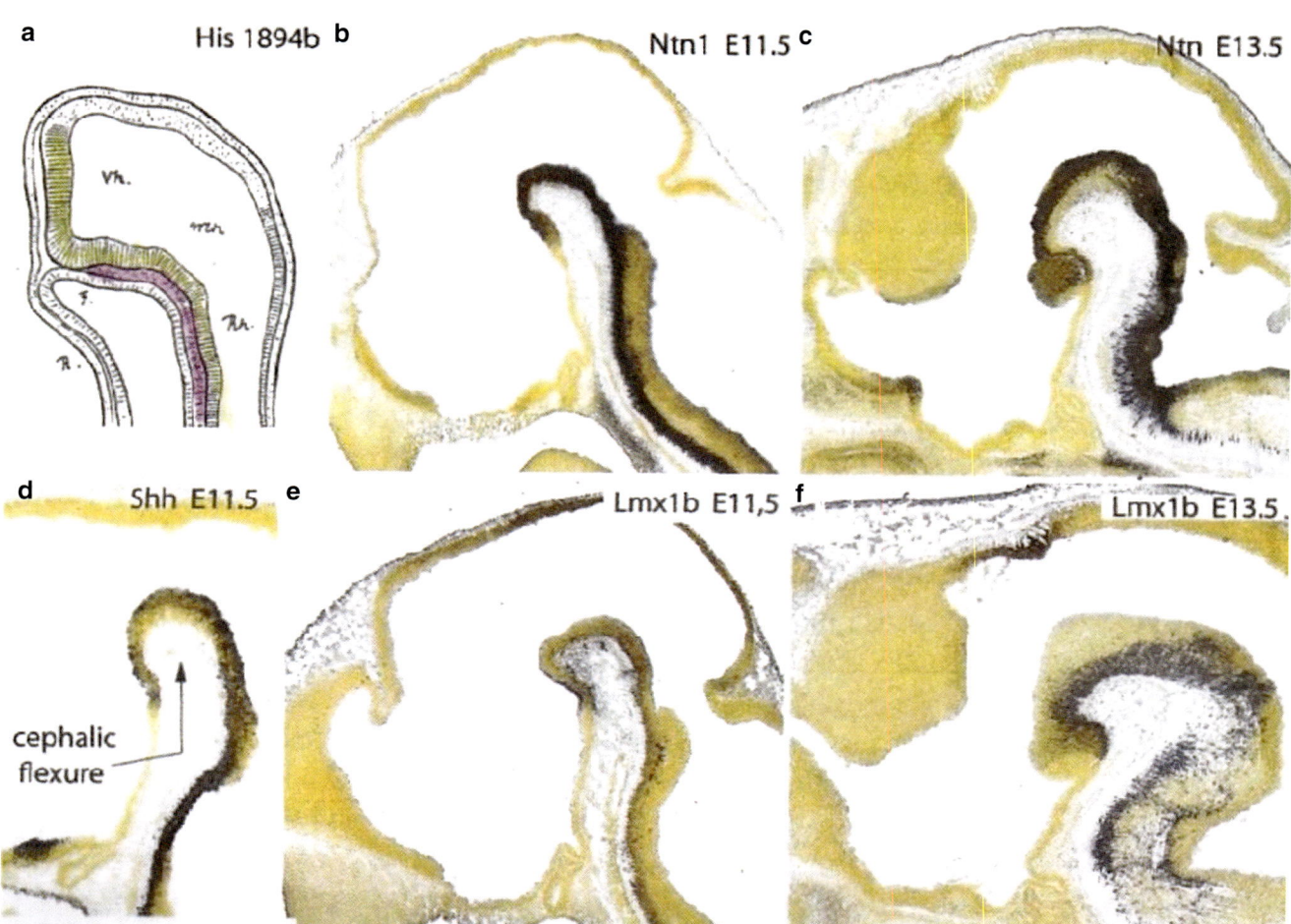

Fig. 5.40 Contact between floor plate and notochord. Gene markers show notochord leading all the way forward to the terminal plate of hypothalamus. Note accuracy of drawing by His [78]. (Reprinted from Puelles L, Martinez-de-la-Torre, Bardet S, Rubenstein JLR. The Hypothalamus. In: Watson C, Paxinos G, Puelles L (eds). The Mouse Nervous System. New York, NY: Academic Press; 2012: 221–303, with permission from Elsevier)

their job; more structures are preserved. Thus, deletion of PCM as the time of its formation causes complete loss of differentiation in (1) basal plate: midbrain, diencephalon, hypothalamus; and subsequently in (2) alar plate with holoprosencephaly and massive dorsalization of remnant tissues. Removal of PCM at progressively later times (1) preserves basal plate in caudal-rostral order first midbrain, then diencephalon, the hypothalamus; and (2) avoids holoprosencephaly and secondarily the expression of *Shh* in subpallium.

- Observation: Prechordal cells migrate upward *along median terminal wall*, progressing from the tip of the floor plate to the tip of the roof plate. They induce hp2 midline structures in the following ventral-dorsal order: (1) hypothalamic basal plate (junction of tuberal and retrochiasmatic areas), (2) rostromedian alar plate (chiasmatic area leading to separation of eyes), (3) lamina terminalis, and (4) preoptic expression of Shh (required for subdivision of the pallium).

In sum: Prechordal cells are produced from alar PN at stage 6, achieve their position, and await the arrival of notochord. Notochord is produced at stage 7. Notochord cells are *nonmotile*. They add on as the primitive node moves backwards. Notochord cells incorporate into the surrounding prechordal plate. Notochord induces the floorplate in a neuromeric order (p1–p2–p3–hp1–hp2) and determines the length of the forebrain. Notochordal signal imposes these boundaries on outlying prechordal plate tissue. PCM cells are *motile*. They migrate dorsally with their prosomeric boundaries.

- Notochord patterning is AP. It is causal for the floor plate and causal (but not sufficient) for basal plate.
- Prechordal plate patterning is DV. The effects are secondary for basal plate, non-axial and ventral-dorsal.

Problem 2. Determining Interprosomeric Boundaries of the Telencephalon

The following discussion explains the prosomeric boundaries as seen on the anterior view of the brain (compare Figs. 5.30 and 5.32).

The prosomeric model requires that each prosomere should extend all the way from floor plate to the roof plate. This was a problem in 2003. The hypothalamus had two domains, but it was unclear where interprosomeric boundary would be located in the roof. Labeling work in 2012 demonstrated that the preoptic area, like a gigantic dagger, implants its tip into the roof plate at precisely that anterior commissure…right between the right and left halves of the brain. The resulting preopto-diagonal border (PDB) formed a separation between the non-evaginated preoptic area (the

old telencephalon *impar*). Furthermore, the preoptic area in hp2 relates directly to the eye and optic chiasma, whereas hp1 is separated from it by the cerebral peduncle. Finally, the PDB explains the curious course of the fornix. **In sum**: the intersection of PDB with anterior commissure defines two parts of the roof plate: an extensive posterior hemispheric zone to telencephalon and a small anterior preoptic zone.

Fate mapping with quail-chick embryos exploded a number of previous dogmas in neuroanatomy. First, although the literature continues to propose a double-closure process for the anterior neuropore, in reality, the anterior commissures as the rostral limit of the roof plate. In all vertebrates, closure takes place at the *lamina terminalis*. Second, labeling demonstrated the idea of the septum as the ventral-most zone of pallium to be false: instead *the septal midline belongs to the roof plate*. Finally, preoptic lamina terminalis belongs neither floor plate nor roof plate. It occupies a position along with optic chiasm to the area where the telencephalic hemispheres emerge. Thus, roof plate of telencephalon is just an enormous expansion corresponding to the hypothalamus from when the vesicle emerged.

Problem 3. Reorganizing the Basal Hypothalamus

Mapping of the hypothalamus floor plate changed a great deal between 2003 and 2013. In the previous interations of the prosomeric model, the floor plate of hypothalamus was assumed to end at the mamillary area (M). This was problematic because M was located in the caudal half of hypothalamus, in prosomere hp1. So how can the programming of hp2 take place? The solution came from mapping showing the mamillary area to be placed anteriorly, thus displacing the tuberal area, including the neurohypophysis, right out of the floor plate and into the rostral basal plate (Fig. 5.41).

Problem 4. Mapping the Acroterminal Hypothalamic Domain: What Does the Front of the Brain Look Like?

The anterior base of the brain is made up of anterior hypothalamus. It extends from the mamillary region of the floor plate to the anterior commissure of the roof plate. Contained in the basal and alar plates of prosomere hp2 are unique structures assembled in a specific ventral-to-dorsal sequence. [The columnar model considers the patterning to be posterior-to-anterior.]

Organization of the Hypothalamus

Before we start, let's review the contemporary mapping of the hypothalamus because the midline structures related to its fields. This review is boring, but neatly packaged and therefore relatively painless. Hypothalamus consists of four

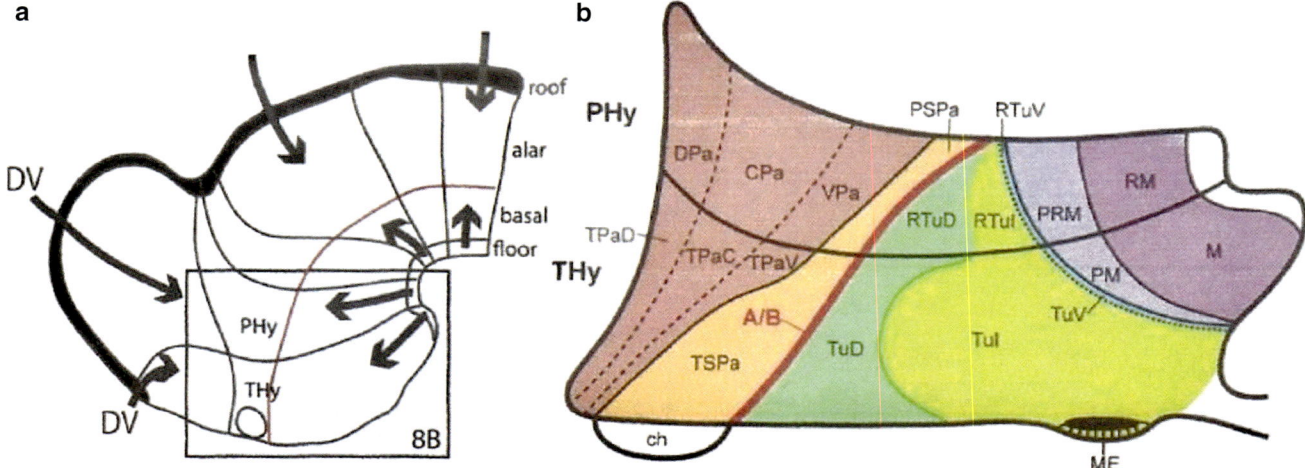

Fig. 5.41 Mapping the hypothalamus. (**a**) Antagonistic dorsoventral patterning signal spread from both the roof plate, including its rostral-most portion at the anterior commissure, and the floor plate, including its rostral hypothalamic sector. These effects presumably establish the alar-basal boundary (red line), as well as the telencephalo-hypothalamic boundary. The blue boxed area is examined in detail in (**b**). (**b**) Map of the known dorsoventral molecular regionalization of the alar and basal hypothalamus, held to result from graded finer interactive effects within the primary dorsoventral pattern. The alar-basal boundary is marked by the thick red line. The alar longitudinal domains are represented by the paraventricular area (subdivided into dorsal, central, and ventral micro-zones) and the subparaventricular area (this relates to the optic chiasm and the initial course of the optic tract). The basal hypothalamus con-sists of similarly dorsoventrally related tuberal and mamillary regions (sensu lato). The updated terminology proposes distinguishing tuberal (Tu) from retrotuberal (RTu) areas, as well as perimamillary and mamillary sensu stricto (PM, M) from periretromamillary and retromamillary sensu stricto areas (PRM, RM), respectively belonging to THy and PHy. Note the Tu/RTu complex can also be subdivided dorsoventrally into dorsal, intermediate, and ventral microzones (TuD, TuI, TuV; RTuD, RTuI, RTuV). (Reprinted from Puelles L, Rubenstein JLR. A new scenario of hypothalamic organization: rationale of hypothesis introduced in the updated prosomere model. Front Neuroanat 2015; 9: article 27. With permission from Creative Commons License 4.0: https://creativecommons.org/licenses/by/4.0/)

quadrants. Basal and alar plates are separated by a heavy red line, whereas hp1, *peduncular hypothalamus* (PHy), and hp2, *terminal hypothalamus* (Thy), are separated by a black line (Fig. 5.41).

- Alar plate is quite orderly. It has four longitudinal fields aligned in parallel:
  - The upper deck <u>relates to the ventricle</u> and consists of three *paraventricular fields* lined up longitudinally (dorsal, central, and ventral).
    Prosomere hp1 (Phy); these are DorsalPa, CentralPa, and Ventral Pa.
    Prosomere hp2; they are TPaDorsal, TPaCentral, and TPaVentral.
  - The lower deck <u>relates to the optic tract</u> and consists of a single *subparaventricular field*. The caudal proso-mere hp1 part is called (predictably) PSPa, while the rostral prosomere hp2 part is TSPa.
- Basal plate is four longitudinal fields, arranged irregularly
  - Tuberal area, dorsal
    Prosomere hp1: retrotuberal dorsal, RTuD
    Prosomere hp2: tuberal, TuD
  - Tuberal area, intermediate
    Prosomere hp1: retrotuberal, RTuI
    hp2: tuberal, TuI
  - Tuberal area, ventral

    Posomere hp1: retrotuberal ventral, RTuV
    Prosomere hp2: tuberal ventral, TuV
  - Paramamillary area
    Prosomere hp1: retroparamamillary, PRM
    Prosomere hp2: paramamillary, PM
  - Mamillary area
    Prosomere hp1: retromamillary, RM
    Prosomere hp2: mamillary, M

Midline Structures of the Anterior Hypothalamus

To understand the developmental fields of the rostromedian domain, we must compare the sagittal anatomy of hypo-thalamus with the coronal anatomy (Figs. 5.32, 5.41, and 5.42). We begin with basal plate of hp2, *Median mamillary region* (MnM), which lies directly dorsal to mamillary floor plate and coextensive with it. Just above it is *median tuber-omamillary recess* area (TM) which sits in front of TuV, the ventral tuberal area. *Median eminence* (ME) contains the critical neurohypophysis surrounded by the arcuate nucleus (Arc). Just behind it is Intermediate tuberal area, TuI. Above median eminence is the horseshoe-shaped *median antero-basal area* (ABasM), the most rostral part of the hypotha-lamic cell cord, formerly known as retrochiasmatic area. ABasM sends out "wings" into dorsal tuberal area (TuD). Nestled dorsally between ABasM and the optic stalks on either side of the midline are the *suprachiasmatic nuclei* (SCH).

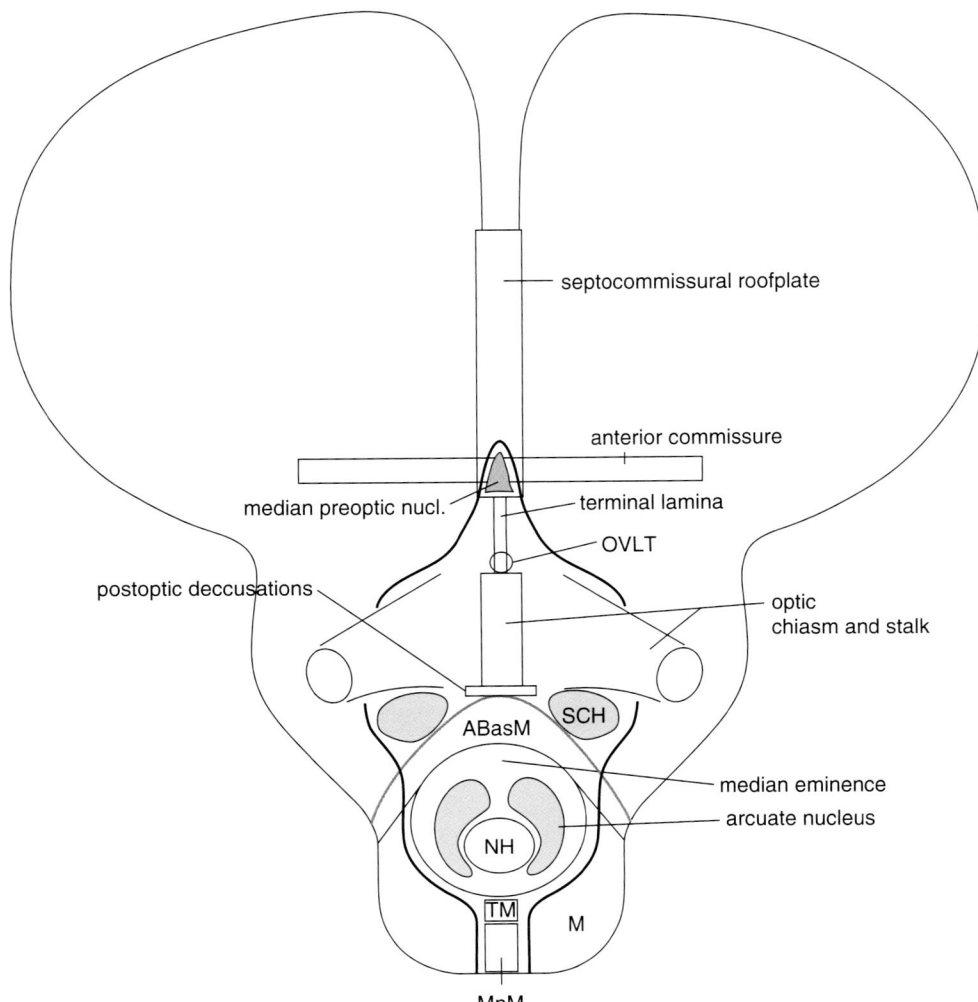

Fig. 5.42 Developmental fields of the acroterminal domain (ATD), i.e. the anterobasal forebrain. The alar-basal boundary is marked in red. The ATD starts at the preoptic roof, encompassing the anterior commissure bed and the median preoptic nucleus (MnPO); further down there is the terminal lamina, and probably also some other neighboring preoptic derivatives, ending with the organum vasculosum laminae terminalis (OVLT), a circumventricular specialization. The alar hypothalamic ATD also includes the optic elements (eyes, stalks, and chiasm) plus the postoptic decussations, and the suprachiasmatic nuclei (SCH) bilater-ally. The basal hypothalamic ATD region includes the precociously differentiating median anterobasal area (ABasM), the median eminence, infundibulum, neurohypophysis (NH), and arcuate nuclei, plus the median tuberomamillary area (TM), finishing with the median mamillary area (MnM). (Reprinted from Puelles L, Rubenstein JLR. A new scenario of hypothalamic organization: rationale of hypothesis introduced in the updated prosomere model. Front Neuroanat 2015; 9: article 27. With permission from Creative Commons License 4.0: https://creativecommons.org/licenses/by/4.0/)

We now enter into the optic zone, the alar plate of hp2. In between ABasM and the optic chiasm is a narrow transverse *postoptic decussation* which makes with *optic chiasm* (OCh) an inverted letter "T". This represents the surface marking of the alar-basal boundary. On the other part of OCh, like a vertical midline extension, is the important landmark, *lamina terminalis* (TL). At the footplate of TL on the OCh is a small circular intensely vascular *median circumventricular organ*, otherwise known by its impossibly long Latin name, *organum vasculosum laminae terminalis* (OVLT). Anteriorly, median preoptic tract is perched like an arrowhead atop terminal lamina. It is crossed by *anterior commissure* (AC).

In sum: acroterminal hypothalamic domain (ATD) lies in prosomeres hp2 and represents the very tip of the terminal wall. It is shared between hypothalamus and telencephalon. It likely develops from ventral dorsal diffusion of morphogens from prechordal plate, but also produces signaling factors of the fibroblast growth factor family that diffuse into hypothalamus to set up the segmentation into hp2 and hp1 zones.

Mechanisms of Patterning

DV Patterning (Dorsoventral Gradient, Vertical Axis)

As discussed clearly in Carlson, this process begins early during formation of the neural plate. Morphogens responsi-

Fig. 5.43 Mechanisms of induction. Key centers in development produce morphogens that diffuse in both dorsoventral and anterior-posterior gradients. (Reprinted from Kiecker C, Lumsden A. The role of organizers in patterning the nervous system. Annu Rev Neurosci 2012; 35:347–367. With permission from Annual Reviews, Inc.)

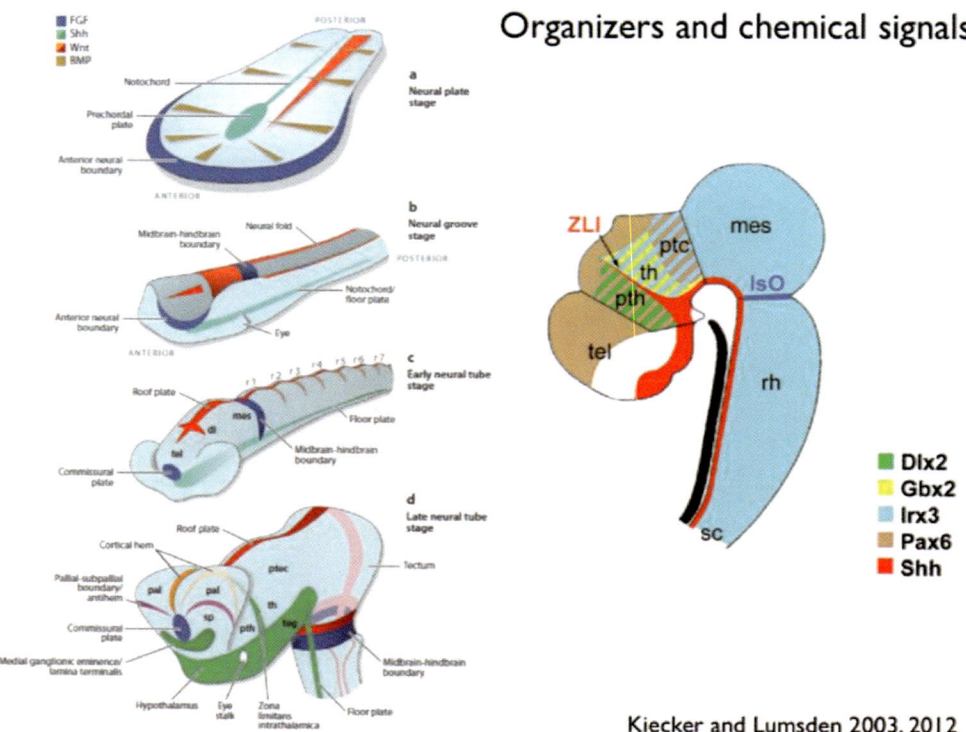

ble for specification of epiblast into neural tissue diffuse outward in the plane of the neural plate from two locations: (1) Hensen's node (the forward terminus of the embryo) and (2) the anterior visceral endoderm (AVE) at the extreme forward margin of the embryo. Recall that once gastrulation has been completed, the original layer responsible for the migrating cells, the epiblast, acquires a new name: ectoderm. Recall further that *the baseline fate of all ectoderm is neural.* The reason peripheral ectoderm becomes epidermis is that these zones are under the repressive influence of *BMP-4.* Otx-2 from the node and *Pit-1* from the AVE repress *BMP-4.* Thus, induction of the epiblast lying between the primary organizer (Hensen's node) and the peripheral boundary with nonneural epiblast creates the neurectoderm of the midbrain and forebrain [81] (Fig. 5.43).

During gastrulation, the node produces prechordal plate and the notochord. The PCP/notochord and the nonneural ectoderm act as mutually opposed sources of signals, medial and lateral, respectively. When the plate rounds up into the neural tube, the *PCP/notochord and NNE signals take on different topologic roles*: ventralizing and dorsalizing, respectively. In addition, paraxial mesoderm lying beneath the floorplate (i.e., beneath the prechordal plate/notochord) provides an ongoing source of *SHH* protein, which produces vertically directed ventralizing signal for the neural tube floor. As neurulation is completed, the neural tube becomes buried beneath nonneural ectoderm. From the NNE and neural crest, dorsalizing proteins such as *WNT* arise.

The prosomeric model postulates that the essential longitudinal organization of the neural tube (floor plate, basal plate, alar plate, and roof plate) results from (1) primary planar and vertical induction processes; followed by (2) secondary ventralizing and dorsalizing effects. Accordingly, the forebrain can be thought of as containing a median floor plate restricted to median eminence, infundibulum, mamillary, retromamillary, and prerubral areas. These structures represent the site of the maximum ventralizing signal concentration. As one moves rostrally, more peripheral (topologically more dorsal) rings of tissue occupy the foremost space (Fig. 5.43 left plate a–d). They "crowd out" the floor plate.

Next comes the basal plate, still ventralized (but less markedly so than the floor plate). It crosses the midline of the rostral forebrain immediately in front of floor plate limit. This takes place at the retrochiasmatic (postoptic) level. It encompasses the eminentia media and infundibular hypothalamus. Crossing the midline more peripherally still are the alar plate and the roof plate. The median alar plate includes the following areas: suprachiasmatic, chiasmatic, and median preoptic (lamina terminalis). Median roof plate contains the anterior commissure. *Thus, the median plane of the rostral forebrain contains a sequence of structures that is topologically and causally comparable to the same DV structure found along the lateral walls of the remaining neural tube.* Clinically, development of the median forebrain depends on PCM and expression of the gene *Cyclops*. Embryologic failure results in the holoprosencephaly sequence…up to, and including, cyclopia.

AP Patterning (Rostrocaudal Gradient, Longitudinal Axis)

As previously discussed, the anatomy of the neural plate consists of longitudinal and transverse genetic zones; these are laid out in a sequential manner. The primary pattern is medial-lateral (i.e., DV); it is laid down along the entire length of the neuraxis. What then follows is the establishment of *metamery*. This term refers to a special case of homology in which separate units exist serially along the axis of the organism (e.g., vertebrae and ribs). These units contain shared morphological features. In the brain, these transverse units (6 prosomeres, the mesomere, and 11 rhombomeres) are structurally compatible or homologous with each other in two fundamental ways: (1) in terms of common longitudinal zonation; and (2) in terms of the mechanisms by which they are laid down. Thus, neuromeres share common DV zones and they develop in cranial-caudal sequence as the result of axis-specifying homeotic genes [81].

In comparative anatomy, homology is a fundamental concept that allows deductions to be drawn between different organisms. *Structural homology* emphasizes site occupied by an anatomic structure within the Cartesian map of the body (Bauplan) and its relationship to neighboring anatomic structures. *Biological homology* (a more modern concept) emphasizes common mechanisms of development (morphogenesis) by which embryonic anatomy is converted to an adult form. Organisms also demonstrate conservation of form over time. This is explained by common mechanisms of constraint (*morphostasis*). Thus, two nasal placodes are required for the formation of the nose, although conditions such as hemi-nose or nasal duplication are occasionally observed. *Historical homology* makes use of taxonomic analysis to formulate an evolutionary account of how the common body plan develops and to explain subsequent variations in this plan. Using such comparisons, disciplines such as paleontology and evolutionary biology, now powerfully reinforced by molecular embryology, seek to construct a "tree of life" that incorporates both vertebrate and invertebrate forms.

Anatomic structures derived from the same metamere (or combination of metameres) may vary greatly in form and function, yet remain homologous. Your nasal bones are (hopefully) shorter than those of a Labrador, although in both instances a finite number of neural crest cells come in contact with p6 vestibular epithelium. The *final size of the nasal bones* is dependent upon the surface area of the epithelial "program" and the size (and growth rate) of the original mesenchymal population. Purposeful use of nasal muscles (such as wrinkling your dorsum in disgust) takes on a whole new dimension when a pachyderm positions itself for a peanut. Thus, a common DV structure in the neural tube establishes from the very beginning a fundamental structural pattern along the entire length of the neuraxis. The only additional structure modification required for a Cartesian system is the serial imposition of transverse limits. Thus, each neuromere shares a common DV layout, while retaining the ability to develop its own unique anatomic identity.

During forebrain development, these anatomic limits appear at the stage when the telencephalic vesicles begin to bulge outward. These become visible in sagittal and horizontal section as localized constrictions where the proliferative activity of the neural is reduced. Such boundaries demarcate strongly proliferating zones of the forebrain wall from one another. These are identified as prosomeres. Note that alar and basal plates have differing rates of proliferation, that of the alar plate being greater. Mammals (humans in particular) have very high proliferation rates. Species with low proliferation rates do not demonstrate clear-cut constrictions and outpouching zones, even though the patterns of subjacent genetic specification may be the same.

Various studies have analyzed the neurohistology of different prosomeres. Each prosomeric field contains its own unique cohorts of basal and alar neurons. These populations develop independently and heterochronically (at different times). Such differential behavior reflects the molecular specificity of each prosomere. Data regarding genes encoding for transcription factors have been accumulating since the 1990s. These are expressed in variously shaped spatial domains of the forebrain wall, usually respecting interprosomeric boundaries. Some genes relate preferentially to transverse zones, while others coincide with longitudinal zones. These expression patterns confirm the previous assumption that alar and basal neuromeric fields in the forebrain display different histogenesis because overlapping transcription factor signals combine into constellations specific for each DV or AP field. The variety of gene patterns demonstrates (1) that certain signals exist all along the neuraxis, i.e., the fundamental unifying role of DV longitudinal zones; and (2) other messages can establish a more restricted positional identity in specific longitudinal or segmental (transverse) territories or create subdivisions with primary alar or basal areas (Fig. 5.44).

Special Note: Olfactory Bulb, Tract, and Intrinsic Olfactory Thalamus

The olfactory bulb is a specialized rostral diverticulum of prosomere 2, beginning as a provisional olfactory sensory field of neuroepithelium (olfactory tubercle) that already receives afferent axons from the olfactory nasal epithelium even before the diverticulum of the primordial olfactory bulb

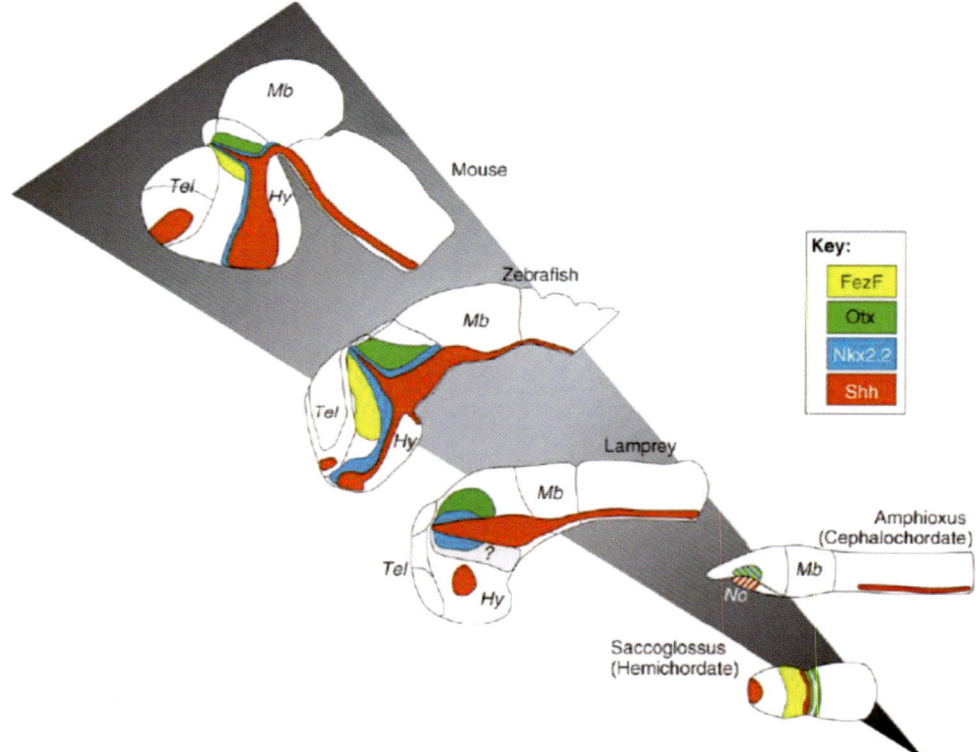

Fig. 5.44 Neuromeres are not added in evolution: evolution of the MDO. Expression of Shh (red) with the flanking expression domain of Nkx2.2 (blue), the prethalamic marker FezF (yellow) and the thalamic marker Otx (green) in the thalamic complex in agnathans (lamprey) and gnathostomes (zebrafish, mouse). Note that Shh is not expressed in the anterior neural tissue in Amphioxus (a cephalochordate), but in the underlying tip of the notochord (No). Data on Fez expression in lamprey are missing (grey). In Saccoglossus (a hemichordate), Shh is expressed in a narrow strip that lies between the Fez and Otx domain, just as in vertebrates, suggesting a basal chordate origin for the MDO. Expression summary in Amphioxus and lamprey is based on different, non-comparative publications. Except for Shh, expression domains are shown only for the thalamic complex. *Hb* hindbrain, *Hy* hypothalamus, *Mb* midbrain, *No* notochord, *Pa* pallium, *SPa* subpallium, *Tel* telencephalon, *MDO* mid-diencephalic organizer. (Reprinted from Scholpp S, Lumsden A. Building a bridal chamber: development of the thalamus. Trends Neurosci 2010: 33(8–2):373–380. with permission from Creative Commons License 4.0: https://creativecommons.org/licenses/by/4.0/)

and tract initiate their growth as anatomically distinctive structures [82]. This axonal connection occurs at Stage 16. As the diverticulum proceeds rostrally from the ventrorostral telencephalon, a thin extension of the lateral ventricle is carried within it to form an olfactory recess which persists until term and then involutes to leave a residue of ependymal clusters and rosettes, much as the spinal central canal involutes in postnatal life [83]. None of the neurons of the olfactory bulb differentiate within the bulb itself, but rather migrate rostrally into the bulb as neuroblasts from the periventricular zone of the main lateral ventricle and then turn radially within the bulb to form the single-cell layer of mitral cells. Mitral dendrites extend into the peripheral synaptic glomeruli on the ventral surface of the olfactory bulb facing the cribriform plate to receive primary axons from the olfactory epithelium, but synaptic contacts are delayed even with the neurites in close proximity. The functional synaptic architecture of the olfactory bulb is essentially radial columnar [84, 85], similar to the functional columns or barrels of neocor-

tex. Primary olfactory axons from the olfactory epithelium (derived from the olfactory placode) are each enclosed by a unique pericyte "ensheathing cell" that is neither a Schwann cell nor oligodendrocyte, but serves a similar function, though olfactory neurons have a continuous turnover and never myelinate even in adult life.

Phylogenetically as well as ontogenetically, olfaction is the earliest special sense to develop. Even protozoans and simple multicellular animals with a nerve net but not a central nervous system, such as medusae (jellyfish) and polyps (hydra; sea anemone), perceive odorous molecules in their ambient water as either attractive (food) or threatening (toxins; predators). In humans, olfactory perception is evident after 28 weeks gestation and the fetus distinguishes strongly aromatic molecules from the maternal diet (e.g., garlic, onion, spices) dissolved in the amniotic fluid that circulates through the nasal passages [86–90] . Reliable olfactory reflexes are demonstrated in preterm and term neonates [89, 91].

The olfactory bulb exhibits essentially the same laminar structure in all vertebrates, except for minor interspecific differences, such as whether the same zone of olfactory epithelium projects to several or only one synaptic glomerulus. The histological structure of the human olfactory bulb is recognized as "mature" as early as 14 weeks gestation [92], but immunocytochemical markers of neuronal maturation indicate that it remains immature at term and into the postnatal period in terms of expression of maturational proteins in all neurons, synaptogenesis, and myelination [83]. The olfactory bulb was the first structure of the human brain to be studied using Golgi impregnations, by Camille Golgi himself [93]. The olfactory tract is not a simple white matter fasciculus because it contains as much grey as white matter, including an extension of the core of granular cells of the olfactory bulb at one end and nodules of the anterior olfactory nucleus at the other [83]. Resident progenitor "stem" cells capable of neuronal differentiation are present in the olfactory bulb and tract, and the bulb remains one of two reservoirs of such cells in the adult brain, the other being the polymorphic zone on the inner surface of the dentate gyrus of the hippocampus. The olfactory epithelium also has many progenitor cells to provide for the rapid turnover of its neurons throughout life. There also is a secondary olfactory system, consisting of the vomeronasal organ and an accessory olfactory bulb that has a similar architecture as the primary olfactory bulb in all vertebrates, as first recognized by Ramón y Cajal [94]. It recedes during the late first trimester and is transitory in humans.

The olfactory is the only primary sensory system of the central nervous system that does not project to the thalamus for relay to the cerebral cortex, because it is the only system that contains its own intrinsic thalamic equivalent [83]. The olfactory bulb is derived from prosomere 2, which mainly gives origin to diencephalic structures including the thalamus which receives all other nonolfactory sensory inputs: somatosensory, auditory, vestibular, and visual. This olfactory thalamus consists of (1) the granule cells forming the core of the olfactory bulb, granule cells being axon-less neurons that form dendrodendritic synapses within the olfactory bulb but not projecting outside; (2) periglomerular inhibitory interneurons that also form dendrodendritic synapses; and (3) the anterior olfactory nucleus which is a group of neuronal aggregates rather than a single nodular structure and a principal terminus of mitral neurons.

Finally, some individuals erroneously deny that the olfactory is a "true" cranial nerve because its peripheral axons do not form a traditional compact nerve bundle, because there is no peripheral ganglion (as with other sensory nerves) and because peripheral olfactory axons do not myelinate throughout life. However, layer 1 of the olfactory bulb is merely a compaction of the separated unmyelinated axons. Layer 2 of the olfactory bulb consists of the synaptic glomeruli on the ventral surface, facing the cribriform plate; this layer w corresponds to a peripheral ganglion that is integrated into the olfactory bulb with primary olfactory nerve fibers forming synapses with dendrites of the mitral neurons in the central olfactory bulb. Finally, most autonomic nerve components of cranial nerves such as the vagal never myelinate. Hence, the olfactory nerve is indeed a true cranial nerve.

Neuromeres and the Phylogeny of Vertebrate Segmentation

Neuromere made their debut at the dawn of vertebrate evolution and have not changed since. The blueprint of the vertebrate brain was present in the last common ancestor of chordates. They have been identified in the hemichordate *Saccoglossus* and in the protochordate *Amphioxus*. The genetic topology of the CNS in *Amphioxus* is shared with higher vertebrate, albeit with further details. The rostral tip of its notochord is homologous to vertebrate prechordal plate. Both the eye vesicles and the telencephalon of *Amphioxus* arise from alar hypothalamus [95]. These relationships are beautifully illustrated in the lamprey. Neuromeres are *not added* in evolution [96], although they may become genetically subdivided and, over time, individuated…as is the case of rhombomere 8 (Fig. 5.44).

The neuromeric system is the final step in a series of evolutionary innovations that gave rise to vertebrate segmentation. Three instances of segmentation are found in extant vertebrates that are conserved with different invertebrate groups. While the ancestor of all Bilateria probably developed with a posterior growth zone expressing *Wnts*, and perhaps oscillating *Notch* signaling along the A/P axis, overt morphological segmentation appears later in the phylogeny.

Pharyngeal segmentation can be dated to the deuterostome ancestor, *Xenambulacracria*, while somitogenesis dates to the chordate ancestor. Rhombomeric organization of the hindbrain appears in the stem vertebrates, seen today in the CNS of the lamprey [97] (Fig. 5.45).

Fundamental to all the above processes in the appearance of homeobox genes in evolution, both *Hox* and *ParaHox* genes have been found in diverse members of *Cnidaria*, including jellyfish, corals, and sea anemones. These creatures along with their fellows, the *Cteriophora*, are virtually basal triploblasts, directly following the invention of three germ layers. Thus, the homeotic system is present in the most basal forms of animal life, from metazoans (Fig. 5.46).

Homeotic genes underwent evolutionary changes in terms of the numbers of chromosomes in which they were located. This involved duplications. From the time of the hemichordates forward through cephalochordates, Hox genes were found in a single cluster. Then with the advent of agnathic fishes, two duplication events took place. At 400 mys (mil-

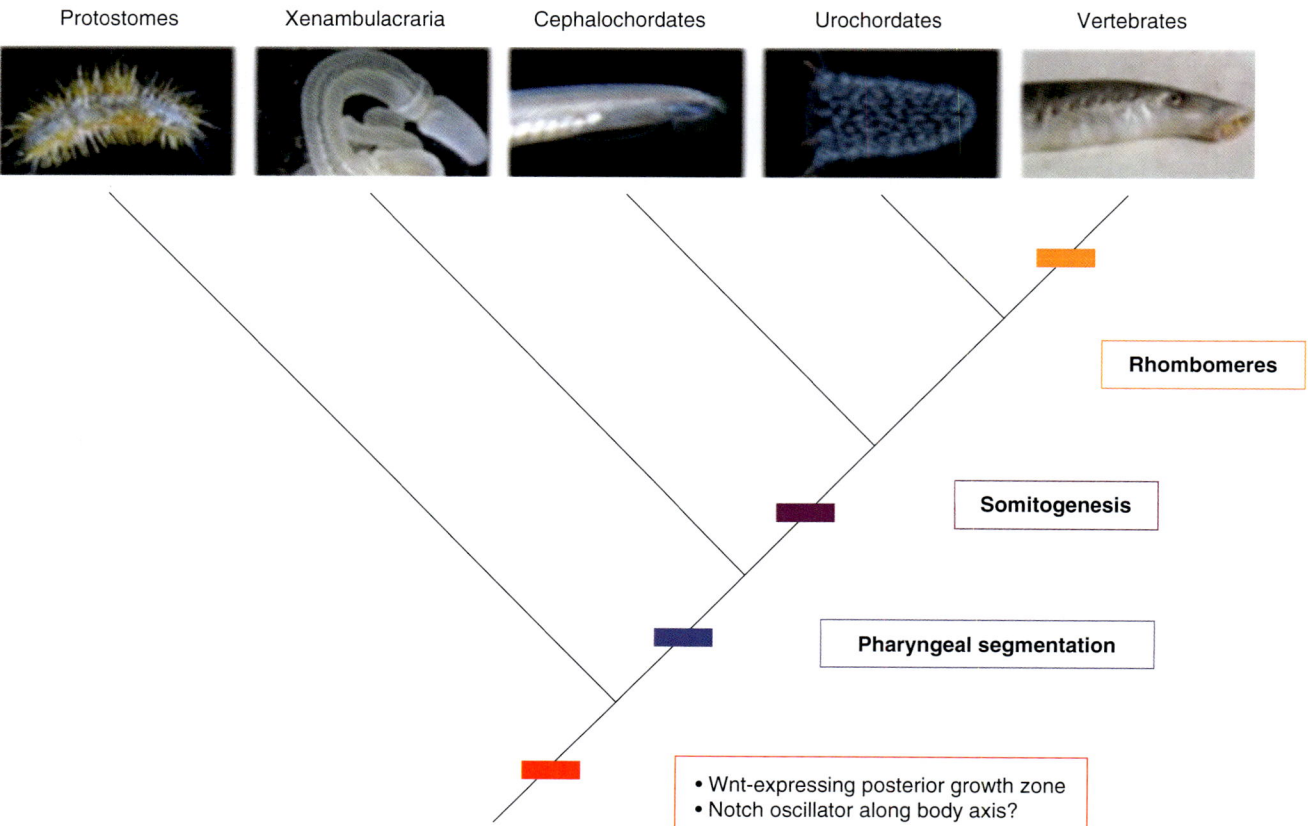

Fig. 5.45 The evolutionary history of segmentation in the vertebrate lineage. Three instances of segmentation are found in extant vertebrates that are conserved with different invertebrate groups. While the ancestor of all Bilateria probably developed with a posterior growth zone expressing Wnts, and perhaps oscillating Notch signalling along the A/P axis, overt morphological segmentation appears later in the phylogeny. Pharyngeal segmentation can be dated to the deuterostome ances-tor, while somitogenesis dates to the chordate ancestor and rhombomeric organization of the hindbrain to the vertebrate stem. (Reprinted from Graham A, Butts T, Lumsden A, Kieker C. What can vertebrates tell us about segmentation? EvoDevo 2014, 5:24–32. with permission from Creative Commons License 4.0: https://creativecommons.org/licenses/by/4.0/)

lion years ago), gnathostomes are divided into the actinopterygian (bony fin) line, leading to derived fishes and the sarcopterygian (fleshy fin) line leading tetrapods. Along the line to fishes, between 400 and 110 mya the genome duplicated, as seen in the zebrafish, leading to 4 hox clusters. Along the other line, the transition between lung fishes and primitive tetrapods involved a similar event leading to 4 hox clusters (Fig. 5.47).

What can we learn about pharyngeal arches from hox genes? Recall that chordates are part of bilateralia that divide into the *protostomes* (worms, mollusks, arthropods) and *deutoerstomes* (echinoderms, hemichordates, and chordates). Chordates have five basic features which constitute the basic body plan: a notochord, a dorsal hollow nerve cord derived from ectoderm, pharyngeal slits, an endostyle or thyroid gland, a postanal tail. *Hox* genes are required for a segmental origin of tissues. Features specific for *Hox* genes are a fixed anteroposterior axis, the dorsalized position of the nerve cord (implying dorsal-ventral patterning), and the notochord, itself a repository for homeotic gene expression.

We have many times remarked on the homeotic pattern of pharyngeal arches in which pairs of neuromeres are responsible for producing a specific arch. Hox genes are embedded in the notochord throughout evolution. We therefore hypothesize that *neuromeres originated about as an expression of segmentation in the nerve cord in response to signals from the underlying notochord.*

Fig. 5.46 Evolutionary history of the Hox gene cluster. Triploblasts are animals with three germ layers. Basal forms of triploblasts are the Cnidarians and the Cteriorphores. Cnidaria have been shown to have both Hox genes and ParaHox genes. Cnidaria are an incredibly rich phylum including coral, jellyfish, and sea anemones that live in both fresh and salt water environments. Characterized by cnidocytes, specialized cells for capturing food. They are basal groups of animal phyla. They have four classes: Anthozoa (anemones, corals), Hydrozoa (hydroids), Scyphozoa (jellyfish), and Cubozoa (the poisonous box jellyfish). Cteriophora (Gk), kteis, 'comb', and pheros 'carry' that use special cilia for swimming. Hox-like genes have been documented in diverse members of Cnidaria. (Reprinted from Ferrier DEK, Holland PWF. Ancient origin of the hox gene cluster. Nat Rev Genet 2001; 2:33–38. With permission from Springer Nature)

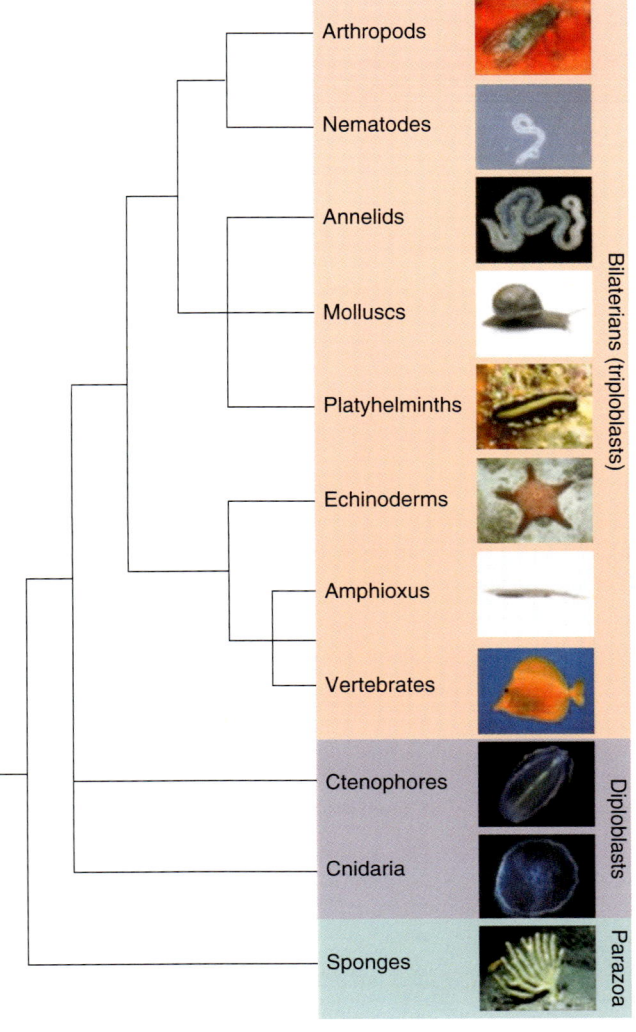

Fig. 5.47 Evolutionary history of hox genes. Duplication events are noted beginning with both derived fishes and tetrapods. (Reprinted from McClintock JM, Robin Carlson, Devon M. Mann, Victoria E. Prince. Consequences of Hox gene duplication in the vertebrates: an investigation of the zebrafish Hox paralogue group 1 genes. Development 2001; 128: 2471–2484. With permission from Company of Biologists)

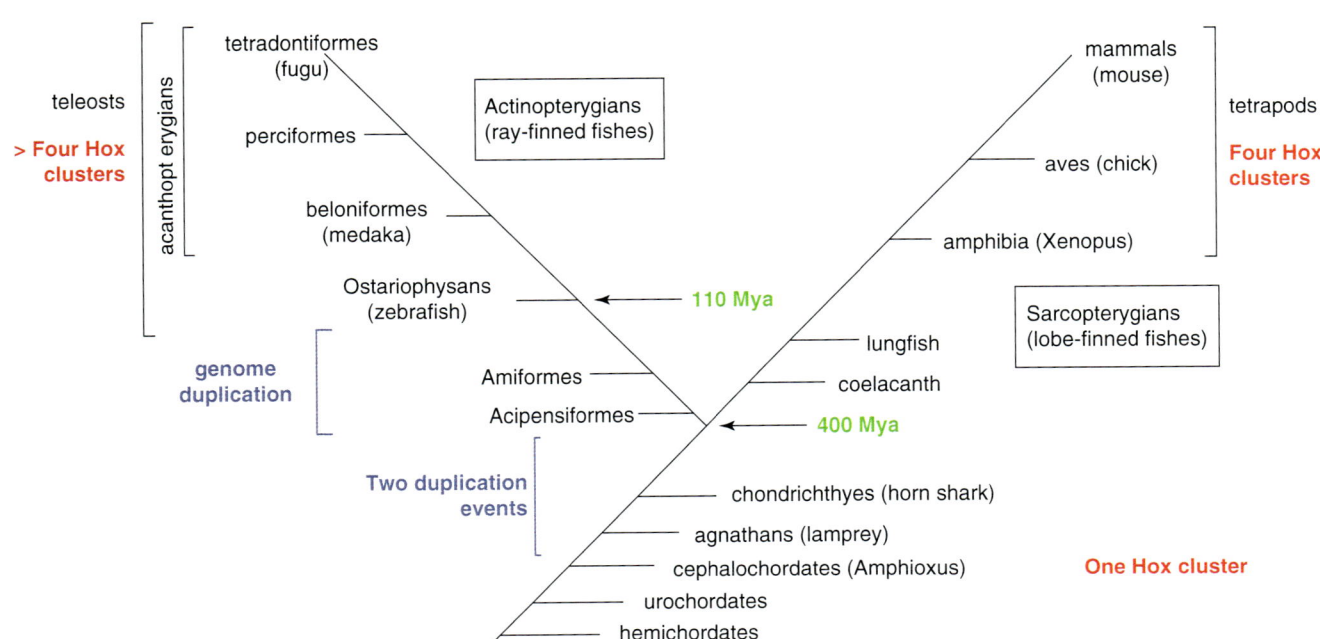

Clinical Significance of the Prosomeric Model

Craniofacial anomalies take many forms. Pathologies can affect mesenchymal structures (bones, muscles, and skin), the CNS, or both. Innovations in surgical treatment depend upon understanding the developmental history of each affected part. Mapping out the origins of non-CNS structures by relating them to a segmental plan of the neural tube makes eminent sense. This approach works equally well for derivatives constructed from neural crest and for those constructed from mesoderm.

Work by O'Rahilly and Müller represents a traditional morphologic description [1]. Their perspective is both challenging and rewarding. We are forced to incorporate a large number of neuroanatomic landmarks, but these can be visualized, owing to the painstaking diagrams. At the same time, we observe how these landmarks relate to the CNS as a whole because they are presented and represented at each successive embryologic stage. The clinical drawback to the morphologic model is its isolation from the rest of craniofacial development. The molecular model of Puelles and Rubenstein builds on the neuranatomical vocabulary. What makes it unique is that it provides us with a genetic reference system connecting each neuromere with the outside world. In the case of the hindbrain, all 12 rhombomeres bear a unique relationship to their respective segment of neural fold. Furthermore, each rhombomere is connected to craniofacial mesoderm as well. Is-r7 supply somitomeres 1–7 while r8–r11 supply somites S1–S4 (formerly somitomeres Sm8–Sm11).

Midbrain neural crest has known sensory innervation and migratory patterns. It is supplied exclusively by V1 and it has been mapped to the periorbital and frontal area. However, the exact derivatives of MNC are unclear. Experimental evidence is not spatially precise. MNC would logically participate, along with RNC from r1, in providing V1-innervated dura mater. Recall that isthmic neural folds do not produce neural crest. Alternatively, the fate of MNC could potentially be osseous; as such, it would respond to dermal signals from PNCc and dural signals from isthmus and r1. Known anatomic information regarding the conjunctiva provides some reference points. The epithelium lining in both the upper and lower eyelids is V1-supplied. This ectoderm is the most rostral product of gastrulation; it represents "r1 skin." Its "dermis" is r1 neural crest. Since MNC is located rostral to r1, it logically migrates earlier in time. It would be more likely to contribute a deeper stratum of mesenchyme, i.e., dura. Maps of the dura based on sensory innervation demonstrate that the territory supplied by V1 is very extensive, including the entire medial wall sweeping dorsally over the corpus callosum all the way to the occiput. When more precise methods of neural crest mapping are applied to the dura, it may be discovered that the V1 territory is comprised of three distinct populations of neural crest: MNC, isthmic RNC, and r1 RNC.

Finally, let us consider the unique anatomy of the forebrain neural folds. It consists of two zones. The rostral folds overlying telencephalon are devoid of neural crest cells per se. These folds are comprised of nonneural ectoderm and placodes. From quail-chick data of Le Douarin et al. [98], the fate of the rostral folds is to produce frontonasal epidermis and sensory apparatus. There are three zones of epidermis: nasal vestibular lining, nasal skin (upper beak epithelium), and forehead (calvarial ectoderm). This epidermis would be useless were it not to have dermal support tissue. The caudal prosencephalic folds overlying diencephalon supply frontonasal dermis by means of neural crest from prosomeres p3–p1. Thus, p3–p1 neural crest cells form a series of dermal "swatches" that provide a biologic identity to successive zones of frontonasal skin. We assume that p1 matures and migrates before p2, such that p1 dermis is biologically "older." Assuming development from the midline laterally, p1 dermis would be supplied by the neuroangiosomes of StV1 nasociliary, p2 dermis would be fed from supraorbital, and p3 dermis would belong to the lacrimal angiosome.

What can be concluded about the prosomeric model? (1) The anatomy of hindbrain neural crest and its migration patterns has been well worked-out. Its migration is segmental, into the pharyngeal arches. Rhombomeric boundaries are based on known expression patterns of homeotic genes. (2) Contributions from midbrain and forebrain neural crest have been previously less accessible in terms how these cells relate to the underlying CNS. The prosomeric model permits us to map out between the brain and the face more precisely. (3) *We postulate that the neural folds above each prosomeric zone share a common genetic "signature" or "tattoo" with underlying neural structures.* Thus, caudal forebrain neural crest consists of a series of genetically distinct populations laid out in linear order over the brain like beads on a string. (4) As these populations migrate forward, they produce frontonasal mesenchymal derivatives of dermis, bone, and dura arranged into precise genetic zones. (5) We postulate that clinical problems can manifest themselves in a specific zone or combination of zones depending upon which neural crest population is affected. Such pathologies fall into reproducible anatomic patterns described by the cleft zones of Dr. Paul Tessier.

Appendix 1: Abbreviations Used in the Text

AH	Adenohypophyseal primordium
Ca	Cardiac region
Cbl	Cerebellum
Ch	Chiasmatic plate
Comm	Commissural plate
d	Dorsal thalamus (dorsal diencephalic elevation)
Dienceph	Diencephalon
Ep	Epiphysis cerebri

Hab-interped	Habenulo-interpedicular
Hem	Hemisphere, cerebral
Hyp	Hypothalamic sulcus
Hyp.c	Hypothalamic cell cord
Isth	Isthmic prosomere
lat	Lateral
LT	Lamina terminalis (embryonic)
M	Mesencephalon
m (VT)	Medial ventricular eminence
Ma	mamillary region
Marg ridge	Marginal ridge
Mes	Mesencephalon (midbrain), superior, inferior colliculi, mesencephalic nucleus V1
Meten	Metencephalon (rostral hindbrain = pons, cerebellar cortex) all structures from r1
My	Myelencephalon (caudal hindbrain) all structures r2–r11
Nas	Nasal disc
NF	Neural folds
NH	Neurohypophyseal primordium
Not	Notochord
O Ph	Oropharyngeal membrane
Ot	Otic disc or vesicle (vestibulocochlear apparatus)
P	Prosencephalon
Par.c.	Parencephalon caudal
Par.r.	Parencephalon rostral
PN	Primitive node
Post.X	Posterior commissure
PP	Prechordal plate
S	Somite (somite series begins at Sm8)
Scler	Sclerotomes
SL	Sulcus limitans
Sm	Somitomere
Sp.cord	Spinal cord
Sulcus.med	Sulcus medius
Syn	Syncephalon
T	Telencephalon
Torus.hem	Torus hemisphericus
V	Ventral thalamus (ventral diencephalic elevation)
Vel	Velum transversum
X4	Trochlear decussation
Y	Commissure of superior colliculi

Appendix 2: Definitions of Neuroembryological Terminology

Compartment. A compartment is a self-contained developmental unit in terms of its cell populations. The primary precursors at the inception of a compartments must generate a spatially restricted polyclone of derivatives. Boundaries are, thus, clonal restriction units. Theoretically, these should be absent inside the compartment; cells should be perfectly free to intermingle. In the 1990s, it was conjectured that brain segments (or neuromeres) might actually be true compartments, at least in terms of neuroepithelial cells. Some authors believe that, if clonal restriction cannot be detected, "other segments" cannot exist. Future research may show that clonal restriction may not prove to be the sine qua non that defines intersegmental boundaries. No one has yet shown that clonal restriction boundaries separating adjacent rhombomeres satisfy the property of completeness. For example, they are absent across the floor plate. Restriction might be absent in some complete transverse limits. It is certainly not uniquely present at all stages of development. What really happens is that a multiplicity of transverse clonal boundaries is set up progressively over time…even inside the segments.

Longitudinal zones. These arise from comparable dorsoventral patterning across the neural primordium.

Metamery. A set of segments of a body form has the topological property of metamery if their fundamental anatomic structure is comparable. They must be serially homologous in terms of basic components. Bilateral symmetry and a length axis are prerequisites. In principle, an animal form can have segments that are nonmetameric. It can also have metameres separated by nonidentical boundaries. Furthermore, continuous metamery throughout the body axis is not obligatory. However, before we can talk about a definition of "fundamental structure," we must first define what common anatomic denominators we will be focusing on.

Neural segment. In this chapter, we apply the general principle of segmentation to the neural tube. Neural segments are distinct and complete parts of the neural tube. The neural tube is bilaterally symmetric and elongated in an A-P manner. Transverse neural subdivisions are already apparent as molecularly distinct part of the neural plate. These *protosegments* already have an anatomic fate. After the neural tube closes up, complete transverse subdivisions known as neuromeres make a transient appearance. They are seen as serial bulges separated by transverse constrictions of the neural wall. The primary cause behind neuromere configuration is differential proliferation. Secondary properties of neuroepithelial cells located at the boundaries also lead to structural separation: (1) clonal restriction, (2) gap junctions cause cell populations to become isolated, (3) differential expression of cell adhesion molecules. Different species bulge in different ways; such bulges are not required for definition of a neural segment. In contrast, differential molecular specification along the AP axis is evolutionarily conserved. Such expression patterning is the primary requirement for segment definition. It gives each segment a unique identity. It also establishes boundaries between segments. Brains change their shape as they develop, but molecular patterns are maintained. Thus, modern molecular techniques allow for a more accurate definition of brain segments (even at early stages) than older, purely morphological methods (see previous chapter). At successive developmental stages, neural segments suffer inevitable morphologic deformation.

Progressive radial expansion of neuronal populations reshapes the neural wall leading to effacement of previously well-defined intersegmental boundaries. This underscores the importance of following cell populations at each stage of embryogenesis. Experimental fate mapping supports the persistence of primitive boundaries, now hidden, which delimit segment-derived domains of the brain wall event in the adult state. These exist, irrespective of "violations" of these limits by tangentially migrating neurons. Thus, the concept that neural segments are "transient" (O'Rahilly) is erroneous. It holds that neuromeres exist only while they can be seen as bulges. From a molecular perspective, neuromeres are described below.

Neuromere. All distinct neural segments postulated by us share a fundamental dorsoventral (DV) structural pattern. The neural tube contains four distinct longitudinal zones: roof, alar plate, basal plate, and floor. These are the consequence of its original Cartesian structure, the neural plate. We therefore see them as *intrinsically metameric*. Thus, the term neuromere refers to the common pattern of fundamental DV zones. The term "neural segment" is less restrictive. It just requires distinct and complete transverse boundaries. Particular regions of the neural tube can be conceived as "tagmata," in which local sets of segments share regional anatomic characteristics. This leads to the classic definition of forebrain, midbrain, hindbrain, and spinal cord neuromeres called prosomeres, mesomeres, rhombomeres, and myelomeres, respectively. Very importantly, please note that, depending on what "fundamental structures" are selected for comparison, prosomeres and rhombomeres might or might not be mutually metameric. Also note that there exist alternative concepts of neuromeres, in which metamery is postulated as the necessary condition for any neural segment. This combines with a definition of segments *according to their boundary properties*, rather than their internal structure. This view often disregards dorsoventral completeness. It accepts only a limited number of "transient brain segments." It leaves other transverse brain parts as morphological nonentities (awaiting some future definition). This point of view (O'Rahilly) is of limited value for causal, comparative, or functional purposes.

Patterning. This term of genetic imprinting is used to describe the programmed morphogenesis of the central nervous system, but it refers more specifically to the gradients of genetic expression in the three axes of the neural tube: dorsoventral or ventrodorsal in the vertical axis; rostrocaudal or caudorostral in the longitudinal axis; mediolateral or lateromedial in the horizontal axis. Most brain and spinal cord malformations can be defined by these gradients (Sarnat 2000).

Segmental or neuromeric subdivisions. These are noncomplete parts distinguished within segments or neuromeres. They can be laid out in various positions relative to the axial dimension of a segment.

Segmentation. For animal forms having bilateral symmetry and a length axis, segmentation involves patterning of the animal into distinct and complete transverse parts (segments) aligned serially along the longitudinal (anteroposterior, or AP) axis. "Distinct" implies that there are detectable intersegmental boundaries, but does not preconceive their nature or how we visualize them. During development, boundaries can change their status from non-overt to overt, and then revert back to a hidden status. Experimental demonstration is thus required at differing stages. "Complete" implies that only a complete cross-sectional (transverse) part of the body form can be considered a segment.

Somitomere. Segmental unit of paraxial mesoderm the first seven of which are incompletely segmented and containing striated muscles of the orbit and first three pharyngeal arches. From the eighth somitomere onward, transformation takes place to somites.

Tagma (*Greek*, meaning a military unit, as in battalion). *Tagma* refers to a specialized grouping of multiple segments, or metameres, into a functional unit. Examples: the thorax of *Drosophila*, the posthyoid hypobranchial muscles with clavicle, the third pharyngeal arch.

Zonal divisions. This term refers to subdivisions that form locally along the DV dimension of a primary longitudinal zone. Advancing DV patterning generates such subdivisions owing to a set-like recruitment of differential gene expression patterns in localized clumps of neural precursors at varying distances from the signal source (e.g., notochord). This leads to the production of specific neuronal or glial cell types along a spatial gradient. The gradient is caused by the differential concentration of molecular signals depending upon the distance from the source. In most cases, these cell populations adopt a stereotypical position in the mantle layer.

Appendix 3: Glossary of Neuroanatomic Terms

Adenohypophysis—The anterior pituitary, derived from the nonneural epithelial part of the Rathke pouch, receives signals from the hypothalamus by blood for secretion of multiple hormones that haematogenously signal endocrine glands in other parts of the body.

Alar plate—The dorsal half of the neural tube above the sulcus limitans.

Amygdala—A mesial basal structure from the subpallium with close relations to the adjacent hippocampus and also the olfactory system; part of the limbic system for mood and emotions; potentially highly epileptogenic non-cortical region of the telencephalon.

Basal plate—The ventral half of the neural tube below the sulcus limitans.

Bauplan—The fundamental architectural plan of the body and of the neural tube as three axes with two opposite gradients of genetic expression and of morphogenesis.

Cephalic flexure—The sagittal bending of the embryonic telencephalic hemispheres and the most rostral area of *Sonic hedgehog* gene expression in the neural tube floor plate.

Cerebral mantle—The prosencephalic neural tissue spanning between the lateral ventricular and pial surfaces of the forebrain.

Claustrum—A thin sheet of grey matter interposed between the external and extreme capsules of the forebrain, lateral to the globus pallidus; it has a structure similar to the thalamus and its reciprocal connections are mainly not with the deep telencephalic nuclei ("basal ganglia"), but rather with the cerebral cortex, principally somatosensory, but also thalamus and hypothalamus. Its function is incompletely understood.

Corpus striatum—The caudate nucleus and putamen, which are really parts of the same nucleus that is divided by the anterior limb of the internal capsule in humans, but is continuous in the rodent brain; the grouping thus is neuroanatomically logical, unlike the "lentiform nucleus" (see below).

Cortex—Any laminated structure of the brain (e.g., cerebral and cerebellar cortices; superior colliculus; olfactory bulb).

Diencephalon—The portion of the prosencephalon from which the thalamus, epithalamus, and hypothalamus form.

Floor plate—The ventral midline of the neural folds and of the early neural tube that is the first ependyma to develop and induces adjacent ependyma, motor neuron differentiation; its ependymal cell basal processes form a barrier to axonal growth cones in the longitudinal axis, but facilitate the midline crossing of commissural axons.

Gastrulation—The formation of a primitive node (of Henson) and primitive streak in the early embryo, which defines three body axes (dorsal/ventral; rostral/caudal; medial/lateral) and gradients along those axes.

Growth cone—The growing tip of an embryonic or fetal axon extending toward its target within the central nervous system, consisting of multiple constantly extending and retracting small processes (i.e., filapodia) seeking signals for the intermediate trajectory of the axon and its end-target neuron with which it will form a synapse.

Hippocampus—A folded structure in the mesial temporal lobe whose grey matter consists of the dentate gyrus (granular neurons) and Ammon's horn (pyramidal neurons); in humans it shifts from a dorsal to a ventral position during fetal life (but remains dorsal in the rodent brain); functions as the principal centre of memory; potentially epileptogenic.

Hodology—The study of connections within the central nervous system (*hodos* = road, Greek).

Hypothalamus—Structure at the base of the prosencephalon traditionally regarded as diencephalic, but more related to telencephalic structures; origin of many molecules that serve to release hormones produced by the anterior pituitary gland (adenohypophysis) or directly via axons to the posterior pituitary (neurohypophysis); also has connections with limbic and olfactory systems.

Induction—The influence of one embryonic tissue upon another so that they differentiate as different mature tissues from each other.

Induseum griseum—A thin strip of neurons on the dorsal aspect of the mature corpus callosum on either side, believed to be remnants of the shift from dorsal to ventral position or "rotation" of the hippocampus during fetal life.

Isthmus—The transitional zone between mesencephalon and rhombencephalon that has a more important role as an organizer of infratentorial neural structures than any neuromere.

Lamina terminalis—A transitory embryonic/fetal structure of the telelencephalic hemispheres through which the major forebrain commissures (corpus callosum and anterior commissure) form; the dorsal part also gives rise to prosencephalic neural crest cells.

Lamination—Horizontal layering of neurons, mostly of the same type, in cortical structures.

Lentiform nucleus—An old artificial grouping of the globus pallidus and putamen because they are close and both are deep telencephalic nuclei, but these two nuclei are quite different and this term is now discouraged because it is neuroanatomically illogical.

Mamillary bodies—Paired ventral bulges at ventral surface of the diencephalic-mesencephalic boundary. They are part of the limbic system, together with the amygdala, hippocampus and hypothalamus, and prefrontal lobes of cortex. (*NOTE:* should be spelled with a single "m" in the middle).

Mesencephalon—The middle of the three dilatations of the early neural tube that will form the midbrain.

Mesomere—Neuromere that forms the mesencephalon (midbrain).

Metamere—(1) embryonic concept: One of a series of homologous segments of the body or of the neural tube that innervate peripheral body structures derived from somites or, in the head region, from pharyngeal (branchial) arches; (2) genetic concept: one of a varying number of common repeating units that make up the repressor segment of a chromosomal locus, the actual number of metameres in a given locus being the degree of repression of the trait in question. (of the neural tube that becomes the pons.

Metencephalon—The rostral part of the embryonic hindbrain (rhombencephalon) that becomes the pons.

Myelencephalon—The caudal part of the embryonic hindbrain (rhombencephalon) that becomes the medulla oblongata.

Myelomere—Neuromere that form the spinal cord.

Neuroepithelium—The portion of the ectodermal germ layer that gives rise to the nervous system; in fetal life the neuroepithelium consists of undifferentiated pre-migratory neural cells that line the ventricular system and have multipotential to form various types of neurons and glial cells, but they are already partially preprogrammed genetically unless an adverse event causes them to change their programme to differentiate as a different type of cell.

Neurohypophysis—The posterior pituitary derived from the neural epithelium of Rathke pouch that receives signals from the hypothalamus through axons within the pituitary stalk (infundibulum) for secretion of vasopressin for renal tubular reuptake of salt excreted in the urine, which also influences systemic blood pressure. It also secretes oxytocin.

Neuromere—Embryonic segments or compartments of the neural tube that differ from adjacent ones in some genetic expressions and also form a barrier to cellular migration in the longitudinal axis to concentrate neurons of the same type to form aggregates for structures.

Neurulation—The bilateral bending dorsally of the primitive neural plate to first form neural folds that come together in the dorsal midline and then fuse to form the neural tube, closing at the rostral neuropore at 24 days and at the caudal neuropore at 28 days gestation in the human embryo.

Nucleus—(1) The membrane-enclosed DNA of the cell; (2) neuroanatomically, an aggregate of neurons of the same type that do not form laminated architecture, though they may be somatotopically arranged (e.g., cranial nerve nuclei of the brainstem; architecture of thalamus and deep telencephalic nuclei).

Ontogeny—Embryonic and fetal development (= *genesis of being*, Greek).

Pallium—The major subdivision of the telencephalon as derivatives of the early prosencephalon at the rostral end of the neural tube, in mammals corresponding to the fetal cerebral neocortex, hippocampus, and several deep structures such as the claustrum and nuclei of the amygdala.

Patterning—The sequence of morphological changes in the development of structures of the central nervous system, generally guided by timing of simultaneous and sequential genetic expressions.

Placode—A plate-like focal thickening of the early embryonic primitive ectoderm from which a sensory organ develops (e.g., olfactory placode; otic placode); the **neural placode** is an earlier region from the primitive streak that forms the neural tube.

Primordium—Undifferentiated cells/tissue programmed to give origin to particular structures with maturation.

Programming—Older synonym for *patterning*.

Prosencephalon—The rostral-most dilatation or vesicle of the early neural tube that will form paired cerebral hemispheres after sagittal cleavage.

Prosomere—Neuromere that forms the forebrain (telencephalic and diencephalic structures).

Rhombencephalon—The caudalmost of the three vesicles at the rostral end of the neural tube that will form the hindbrain (i.e., brainstem and most of cerebellum).

Rhombomere—Neuromere that forms the brainstem and cerebellum.

Roof plate—The ependyma at the dorsal midline of the neural tube; its ependymal basal processes secrete proteoglycans that repel axonal growth cones in the longitudinal axis in the spinal cord and mesencephalon, but permit the midline mesencephalic passage of commissural fibers.

Segmental organization—Architecture derived from repeating units in the longitudinal axis that initially are similar but later differentiate unique structures unlike those of any other unit or segment.

Septal nuclei—Transitory nuclei derived from the ventromedial part of the primitive pallium; the septum atrophies during fetal life to leave only a pair of residual thin glial sheets, the septum pellucidum, with a cavity between them (cavum septi pellucidi) that is not part of the ventricular system.

Somite—One of paired block-like metameric cellular masses of mesoderm arranged segmentally along the embryonic neural tube; somites form in a rostrocaudal gradient for a total of 42.

Somitomere—(1) The paired longitudinal columns of primitive mesoderm on either side of the early neural tube before segmenting into repeating blocks of tissue, the somites; (2) the rostral extension of paraxial mesoderm into the head region of the embryo in relation to pharyngeal arches.

Subpallial region—Telencephalic areas beneath the cerebral cortical plate that include the corpus striatum, globus pallidus, diagonal domain, and preoptic area; may also include origin of amygdalar nuclei.

Substantia nigra—A nucleus in the ventral midbrain and extending into the diencephalon.

Subthalamic nucleus—A small deep telencephalic nucleus ventral to the thalamus.

Sulcus limitans—A symmetrical indentation of the neural tube in the horizontal axis that separates the dorsal alar plate from the ventral basal plate.

Tagma—A high-level unit of biological structure comprising segments that share a general character, for example myelomeres or rhombomeres, by contrast with prosomeres.

Tectum—Dorsalmost part of midbrain, consisting of the superior (visual) and inferior (auditory) colliculi and their commissures, above the cerebral aqueduct.

Tegmentum—Dorsal half of brainstem, above sulcus limitans as the alar plate with grey and white matter.

Telencephalic flexure—The bending of the early telencephalic hemisphere so that the initial posterior pole becomes the temporal, not the occipital pole of the mature cerebral hemisphere; this flexure forms the operculum which, after closure near term, becomes the lateral cerebral (Sylvian) fissure.

Telencephalon—The portion of the prosencephalon that forms the cerebral cortex, deep nuclei (corpus striatum, globus pallidus, subthalamic nucleus, nucleus accumbens), hippocampus and related commissures, and white matter pathways.

Topology—A system for describing relative positions of components within a structure regardless of distortions caused by deformation or malformation.

Zona incerta—A narrow layer of grey matter extending throughout most of the diencephalon, ventral to and separated from the thalamus by the thalamic fasciculus, but laterally continuous with the thalamic reticular nuclei; part of limbic system for mood and emotion.

Zona limitans—A separation between prosomeres 2 and 3 at alar plate levels.

Commentary: Harvey B. Sarnat

I first met Mike Carstens in Los Angeles in 1998 through my uncle, Dr. Bernard G. Sarnat, Professor of Plastic and Craniofacial Surgeon at the UCLA and Cedars-Sinai Hospital. I was impressed not only by Dr. Carstens' fund of knowledge of embryological processes but also by his integration of craniofacial and neural crest development and his ability to conceptualize interactive embryological processes in the manner that Ramón y Cajal had so brilliantly done a century earlier in describing dynamic nervous system ontogenesis. We kept in touch over the years, and I was further inspired by his knowledge of neuroembryology, including modern cutting-edge concepts of segmentation of the neural tube as articulated by Luis Puelles and by J.L.R. Rubenstein.

This was particularly noteworthy since Michael's background is in plastic and craniofacial surgery and not specifically in neuroanatomy and neuropathology. In 2004, as editor of a special issue of the *European Journal of Paediatric Neurology*, I made a point of including his work "Neural tube programming and craniofacial cleft formation" as a chapter-length contribution (*Eur J Paed Neurol* 2004;8(4):181–120). In this article, what I remember most vividly was his ability to reverse, in an ironical way, the order of DeMyers' famous statement in 1964, long considered as dogma in neuroembryology, "The face predicts the brain," into a simple but powerful clinical observation, "The brain predicts the face." This represented for me a fundamental insight into basic mechanisms of ontogeny and the relation of the embryonic nervous system to peripheral structures.

Years later, I was greatly honored and humbled when Michael invited me to work with him on preparing a small, but critical, part to this encyclopedic work. In this chapter, we describe the neuromerical system, not only in its current form, but also historically how this model came into being. Though detailed, I hope that our efforts in this chapter will enable clinicians outside the fields of neuroanatomy, neurology, and neuropathology to conceptualize how this model is structured and how it functions. It is my conviction that neuromerical theory will turn out to be of long-lasting value to explain the origins and development of actual craniofacial in the postnatal state.

Dr. Carstens is a model translational scientist, who has worked for over three decades to achieve a transformation of modern embryological concepts into more rational patient management for improved outcomes. I have borne witness to this process and can appreciate the fruition of these efforts. An example is his complete revision of traditional concepts of congenital cleft lip and palate repair, both primary and secondary. The surgical management of vascular malformations is another practical application of basic embryology. A fundamental contribution is his concept of the neurovascular basis of craniofacial clefts, building upon the pioneering lead of the talented neurosurgical illustrator-turned embryologist at Johns Hopkins and subsequently at the Carnegie Institution, Dorcas Padgett. This work, described in detail in Chaps. 6 and 13, confirms the original classification system of Dr. Paul Tessier, who was a colleague and good friend of my uncle Bernard. It also points toward a specific mechanism to explain these so-called clefts, as failures of formation of vulnerable developmental fields, based on the compromise of specific neurovascular pedicles responsible for their supply.

As a pediatric neurologist, developmental neuroanatomist, and neuropathologist myself, I continued to be awed by Dr. Carstens' grasp in the depth of neuroembryology and the relation of neural tube development to craniofacial structure, as mediated by prosencephalic and mesencephalic neural crest. His neuroanatomical and conceptual contributions to this field represent a new way of mapping out the human embryo based on the relationships of peripheral developmental fields to those parts of the nervous system that supply them. Dr. Michael Carstens' "neuromeric map of the embryo" will be remembered and accredited for generations to come. Dr. Carstens also integrates developmental molecular genetic data but does not fall into the trap of making genetic programming and mutations so exclusive so as to suppress the recognition of mechanisms of neuroembryology and morphogenesis.

Apart from his scientific and anatomical depth of knowledge, Dr. Carstens is a humble, generous, and kind person, with human traits that are equally important and noteworthy.

I also admire his lifetime of service in Latin America and our mutual cultural ties including our spouses. I feel proud and fortunate to know Mike not only as an esteemed and admired colleague, but also as a personal friend.

Harvey B. Sarnat

Calgary, Alberta, Canada

2 August 2022

A Note from Dr. Carstens

Harvey B. Sarnat

In 1998, at Children's Hospital Oakland, Dr. Martin Chin and I were working through the concept of in situ osteogenesis, using the morphogen BMP-2 to convert stem cells in the periosteum into bone-forming cells. In this process, we were in contact with two pioneers in biologic surgery, maxillofacial surgeon Dr. Phil Boyne at Loma Linda and craniofacial surgeon and bone biologist Dr. Bernard Sarnat at the UCLA. I thus met Dr. Harvey Sarnat, a pediatric neurologist and neuropathologist also at the UCLA, and the nephew of the latter. Harvey had written a book about the developmental neuroanatomy, which timed with the appearance of Butler and Hodos' work on comparative vertebrate neuroanatomy. The sources together formed a basis to understand developmental field from the standpoint of Hox genes and segmentation. Later on, during my fellowship year at CHLA with John Reinisch, I got to know Harvey and his wife Laura, also a distinguished pediatric neurologist. Our collaboration grew into two chapters on a new model of craniofacial clefts based on neuroembryology. Over time, Harvey's insights into the system became the structural rib cage for developmental fields, welding together the origins of craniofacial and CNS tissues. Despite my ignorance of the many critical details, Dr. Sarnat gave generously of his time and talent, making this chapter accessible for all readers, regardless of specialty, to have an internal look into how the neuromeric model really works. Without Harvey's contributions, this book would not have been possible.

References

1. O'Rahilly R, Müller F. The embryonic human brain. An atlas of developmental stages. 3rd ed. Hoboken, NJ, Toronto: Wiley-Liss; 2006.
2. Puelles L. Plan of the developing nervous system. In: Rubenstein JLR, et al., editors. Comprehensive developmental neuroscience: patterning and cell type specification in the developing CNS and PNS, vol. 1. Amsterdam: Elsevier; 2013. p. 187–209.
3. His W. Unserer Körperform und das Physiologische Problem ihrer Entstehung. Engelmann: Leipzig; 1874.
4. Lumsden A, Keynes R. Segmental patterns of neuronal development in the chick hindbrain. Nature. 1989;337:424–8.
5. McClure CFW. The segmentation of the primitive vertebrate brain. J Morphol. 1980;4:35–56, 103.
6. Herrick CJ. The morphology of the forebrain in amphibia and reptiles. J Comp Neurol Psychol. 1910;20:413–547.
7. Guthrie S, Butcher M, Lumsden A. Patterns of cell division and interkinetic nuclear migration in the chick embryo hindbrain. J Neurobiol. 1991;22:742–54.
8. Goldowitz D, Hamre K. The cells and molecules that make a cerebellum. Trends Neurosci. 1998;21:375–82.
9. Joyner AL. Engrailed, Wnt and Pax genes regulate midbrain-hindbrain development. Trends Genet. 1996;12:15–20.
10. Joyner AL. Engrailed, Wnt and Pax genes regulate midbrain-hindbrain development. Trends Genet. 1997;124:2923–34.
11. Keynes R, Lumsden A. Segmentation and the origin of regional diversity in the vertebrate central nervous system. Neuron. 1990;2:1–9.
12. McGinnis W, Krumlauf R. Homeobox genes and axial patterning. Cell. 1992;68:283–302.
13. Rowitch DH, Danielian PS, Lee SMK, et al. Cell interactions in patterning the mammalian midbrain. Cold Springs Harbor Symp Quant Biol. 1997;62:535–44.
14. Rowitch DH, McMahon AP. Pax-2 expression in the murine neural plate precedes and encompasses the expression domains of Wnt-1 and En-1. Mech Dev. 1995;52:3–8.
15. Stern CD, Foley AC. Molecular dissection of Hox gene induction and maintenance in the hindbrain. Cell. 1998;94:143–5.
16. Wassef M, Joyner AL. Early mesencephalon/metencephalon patterning and development of the cerebellum. Perspect Dev Neurobiol. 1997;5:3–16.
17. Wassarman KM, Lewandoski M, Campbell J, et al. Specification of the anterior hindbrain organizer is dependent on Gbx2 gene function. Development. 1997;124:2923–34.
18. Rubenstein JLR, Beachy PA. Patterning of the embryonic forebrain. Curr Opin Neurobiol. 1998;8:18–26.
19. Lumsden A. The cellular basis of segmentation in the developing hindbrain. Trends Neurosci. 1990;13:329–35.
20. Mai JK, Andressen C, Ashwell KWS. Demarcation of prosencephalic regions by CD15-positive radial glia. Eur J Neurosci. 1998;10:756–1.
21. Keynes R, Cook GMW. Axonal guidance molecules. Cell. 1995;83:161–9.
22. Snow DM, Steindler DA, Silver J. Molecular and cellular characterization of the glial roof plate of the spinal cord and optic tectum: a possible role for a proteoglycan in the development of an axon barrier. Dev Biol. 1990;138:359–76.
23. Boncinelli E, Somma R, Acampora D, et al. Organization of human homeobox genes. Hum Reprod. 1988;3:880–6.
24. Keynes R, Krumlauf R. Hox genes and regionalization of the nervous system. Annu Rev Neurosci. 1994;17:109–32.
25. Fortin G, Kato F, Lumsden A, Champagnat J. Rhythm generation in the segmented hindbrain of chick embryos. J Physiol. 1995;486(3):735–44.
26. Sarnat HB. Watershed infarcts in the fetal and neonatal brainstem. An aetiology of central hypoventilation, dysphagia, Möbius syndrome and micrognathia. Eur J Paediatr Neurol. 2004;8:71–87.
27. Tomás-Roca L, Tsaabi-Shtvlik A, Jansen JG, Singh MK, Epstein JA, Altunoglu U, et al. De novo mutations in PLXND1 And REV3L cause Möbius syndrome. Nat Commun. 2015;6:7199. https://doi.org/10.1038/ncomm8199.
28. Lim T-M, Jaques KF, Stern CD, Keynes RJ. An evaluation of myomeres and segmentation of the chick embryo spinal cord. Development. 1991;113:227–38.
29. Puelles L. Brain segmentation and forebrain development in amniotes. Brain Res Bull. 2001;55:695–710.

30. Puelles L. Forebrain development: prosomeric model. In: Lemle G, editor. Developmental neurobiology. London: Elsevier/Academic Press; 2009. p. 95–9.
31. Fraser SE, Keynes RJ, Lumsden A. Segmentation in the chick embryo is defined by cell lineage restrictions. Nature. 1990;344:431–5.
32. Krumlauf R. Hox genes in vertebrate development. Cell. 1994;78:191–201.
33. Guthrie S. Rhombomeres and innervation fields. Trends Neurosci. 1995;18:485–488 (response to letter-to-editor).
34. Birgbauer E, Fraser SE. Violation of cell lineage restriction compartments in the chick hindbrain. Development. 1994;120:1347–56.
35. Wake DB. Brainstem organization and branchiomeric nerves. Acta Anat. 1993;148:124–31.
36. Medina L, Reiner A. Distribution of choline acetyltransferase immunoreactivity in the pigeon brain. J Comp Neurol. 1994;342:497–537.
37. Vesque C, Becker N, Seitanidou T, Charnay P. La morphogenèse du cerveau postérieur: vers une analyse moléculaire d'un processus de segmentation chez les vertébrés. Lexique Embryol. 1993;9:975–81.
38. Wingate RJT, Lumsden A. Persistence of rhombomeric organization in the postsegmental hindbrain. Development. 1996;122:2143–52.
39. Warrilow J, Guthrie S. Rhombomere origin plays a role in the specificity of cranial motor axon projections in the chick. Eur J Neurosci. 1999;11:1403–13.
40. Crosby EC, Henderson JW. The mammalian midbrain and isthmus regions. II. Fiber connections of the superior colliculus. B. Pathways concerned in automatic eye movements. J Comp Neurol. 1948;88:53–91.
41. Crosby EC, Woodburne RT. The mammalian midbrain and isthmus regions. Part II. The fiber connections. C. The hypothalamo-tegmental pathways. J Comp Neurol. 1951;94:1–32.
42. Neal HV. Neuromeres and metameres. J Morphol. 1918;31(2):293–315.
43. von Kupffer CR. Heft 4. Kopfentwicklung von Bdellostoma. Lehmann. 1900.
44. Sarnat HB, Rybak G, Kotagal S, Blair JD Jr. Cerebral embryopathy in late first trimester: possible association with swine influenza vaccine. Teratology. 1979;20:93–100.
45. Kuemerle B, Zanjani H, Joyner A, et al. Pattern deformities and cell loss in Engrailed-2 mutant mice suggest two separate patterning events during cerebellar development. J Neurosci. 1997;17:7881–9.
46. McMahon AP, Joyner AL, Bradley A, et al. The midbrain-hindbrain phenotype of Wnt-1−/Wnt-1 mice results from stepwise deletion of engrailed-expressing cells by 9.5 days postcoitum. Cell. 1992;69:581–95.
47. Millen KJ, Hui C-C, Joyner AL. A role for En-2 and other murine homologues of Drosophila segment polarity genes in regulating position information in the developing cerebellum. Development. 1995;121:3935–45.
48. Rubenstein JLR, Shimamura K, Martínez S. Regionalization of the prosencephalic neural plate. Ann Rev Neurosci. 1998;21:445–77.
49. Figdor MC, Stern CD. Segmental organization of embryonic diencephalon. Nature. 1993;363:630–4.
50. Herrick CJ. The brain of the tiger salamander, Ambystoma tigrinum. Chicago: University of Chicago Press; 1948.
51. Puelles L, Rubenstein JLR. A new scenario of hypothalamic organization: rationale of new hypotheses introduced in the updated prosomeric model. Front Neuronal 2015;9(27). https://doi.org/10.3389/fnana.2015.00027.
52. Swanson LW. Brain architecture. New York: Oxford University Press; 2003.
53. Puelles L, Rubenstein JLR. The forebrain. In: Ramachandran VS, editor. Encyclopedia of the human brain. Elsevier; 2002.
54. Ferran JL, Púleles L, Rubenstien JLR. Molecular codes defining rostrocaudal domains in the embryonic mouse hypothalamus. Front Neuronal 2015;8(162): https://doi.org/10.3389/fnana.2014.00162.
55. Sarnat HB, Menkes JH. The new neuroembryology: how to construct a neural tube. J Child Neurol. 2000;21:109–24.
56. Bulfone A, et al. Spatially restricted expression of DLX-1, Dlx-2, (Tes-1), Gbx-2 and Wnt-3 in the embryonic day 12.5 mouse forebrain defines potential transverse and longitudinal segmental boundaries. J Neurosci. 1993;13:3155–72.
57. Krauss S, Johansen T, Korzh V, Fjose A. Expression pattern of zebrafish Pax genes suggest a role in early brain regionalization. Nature. 1991;353:267–70.
58. Porteus MH, Bulfone A, Ciranello RD, Rubenstein JL. Isolation and characterization of a novel cDNA clone encoding a homeodomain that is developmentally regulated in the ventral forebrain. Neuron. 1991;7:221–9.
59. Price M, Lazzaro D, Pohl T, Mattei M-G, Rüther U, Olivo J-C, Duboule D, Di Lauro R. Regional expression of the homeobox gene Nkx-2.2 in the developing mammalian forebrain. Neuron. 1992;8:241–55.
60. Simeone A, Acampora D, Gulisano M, Sotrnaiuolo A, Boncinelli E. Nested expression domains of four homeobox genes in developing rostral brain. Nature. 1992;358:687–90.
61. Tao W, Lai E. Telencephalon-restricted expression of BF-1, a new member of the HNF-3/forkhead gene family, in the developing rat brain. Neuron. 1992;8:957–66.
62. Walther C, Gruss P. Pax-6, a murine paired box gene, is expressed in the developing CNS. Development. 1991;113:1435–49.
63. Sarnat HB, Netsky MG. Evolution of the nervous system. 2nd ed. New York: Oxford University Press; 1981.
64. Heuser CH, Corner GW. Developmental horizons in human embryos. Description of age group 10, 4-12 somites. Contrib Embryol Carnegi Inst. 1957;36:29–39.
65. Puelles L, Harrison M, Paxinos G, Watson C. A developmental ontology for the mammalian brain based on the prosomeric model. Trends Neurosci. 2013;36:570–8.
66. Puelles L. A segmental morphological paradigm for understanding vertebrate forebrains. Brain Behav Evol. 1995;46:319–37.
67. Sarnat HB, Flores-Sarnat L. Radial micro-columnar cortical architecture: maturational arrest or focal cortical dysplasia? Pediatr Neurol. 2013;48:259–70.
68. Sarnat HB, Flores-Sarnat L, Trevenen CL. Synaptophysin immunoreactivity in the human hippocampus and neocortex from 6 to 41 weeks of gestation. J Neuropathol Exp Neurol. 2010;69:234–45.
69. Bergquist H. Formation of the frontal part of the neural tube. Exerpimentia. 1964;20:92–3.
70. Bergquist H, Kallén B. Notes oh the early histogenesis and morphogenesis of the central nervous system in vertebrates. J Comp Neurol. 1954;100:627–60.
71. Gribnau AM, Geijsberts JGM. Morphogenesis of the brain in staged Rhesus monkey embryos. Adv Anat Embryol Cell Biol. 1985;91:1–69.
72. His W. Verschläge sur Eintheilung des Gehirns. Arch Anat Entwickelungsges. 1893;3:172–9.
73. Johnston JB. The morphology of the forebrain vesicle in vertebrates. J Comp Neurol. 1909;19:457–539.
74. Kallén B. Early morphogenesis and pattern formation in the central nervous system. In: RL DH, editor. Organogensis. New York: Holt, Rinehart & Winston; 1965. p. 107–28.
75. Keyser A. Basic aspects of development and maturation of the brain: embryological contributions to neuroendocrinology. Psychoneuroendocrinology. 1983;8:157–81.
76. Tuckett F, Lim L, Morriss-Kay GM. The ontogenesis of cranial neuromeres in the rat embryo. I. A scanning electron microscope study. J Embryol Exp Morphol. 1985;87:215–28.
77. Tuckett F, Morriss-Kay GM. The ontogenesis of cranial neuromeres in the rat embryo. I. A transmission electron microscope study. J Embryol Exp Morphol. 1985;88:231–314.
78. His W, Die Entwickelung des menschlichen Gehirns während der Ersten Monate. Arch Anat Physiol Anat Abt. 1894;1:35–41, 51.

79. Sánchez-Arrones L, Ferrán JL, Rodríguez-Gallardo L, Puelles L. Incipient forebrain boundaries traced by differential gene expression and fate mapping in the chick neural plate. Dev Biol. 2009;335(1):43–65. https://doi.org/10.1016/j.ydbio.2009.08.012. Epub 2009 Aug 19.

80. García-Calero E, Fernández-Garre P, Martínez S, Puelles L. Early mammillary pouch specification in the course of prechordal ventralization of the forebrain tegmentum. Dev Biol. 2008;320(2):366–77. https://doi.org/10.1016/j.ydbio.2008.05.545. Epub 2008 May 29.

81. Kiecker C, Lumsden A. The role of organizers in patterning the nervous system. Annu Rev Neurosci. 2012;35:347–67. https://doi.org/10.1146/annurev-neuro-062111-150543. Epub 2012 Mar 29. Review.

82. Bayer SA, Altman J. The human brain during the early first trimester, Atlas of human central nervous system development, vol. 5. Boca Raton, FL: CRC Press; 2008. p. 450–1.

83. Sarnat HB, Yu W. Maturation and dysgenesis of the olfactory bulb. Brain Pathol. 2016;26:301–18.

84. Kauer JS, Cinelli AR. Are there structural and functional modules in the vertebrate olfactory bulb? Microsc Res Tech. 1993;24:157–67.

85. Willhite DC, Nguyen KT, Masurkar AV, Greer CA, Shepherd GM, Chen WR. Viral tracing identifies distributed columnar organization in the olfactory bulb. Proc Natl Acad Sci U S A. 2006;103:12592–7.

86. Lipsitt LP, Engen T, Kaye H. Developmental changes in the olfactory threshold of the neonate. Child Dev. 1963;34:371–6.

87. Schaal B, Marlier L, Soussignan R. Olfactory function in the human fetus: evidence from selective neonatal responsiveness to the odor of amniotic fluid. Behav Neurosci. 1998;112:1438–49.

88. Marlier L, Gaugler C, Astruc D, Messer J. La sensibilité olfactive du nouveau-né prématuré. Arch Pédiatr (Paris). 2007;14:45–53.

89. Sarnat HB, Flores-Sarnat L. Olfactory development. Part 1. Function from fetal perception to adult wine-tasting. J Child Neurol. 2017;32:566–78.

90. Schall JD. On building a bridge between brain and behavior. Annu Revu Psychol 2004;55:23–50

91. Sarnat HB. Olfactory reflexes in the newborn infant. J Pediatr. 1978;92:624–6.

92. Humphrey T. The development of the olfactory and the accessory olfactory formations in human embryos and fetuses. J Comp Neurol. 1940;73:431–68.

93. Golgi C. Sulli fina struttura dei bulbi olfattorii. Riv Sper Freniat Reggio-Emilia. 1875;1:66–78.

94. de Ramón y Cajal S. La corteza olfativa del cerebro. Trab d Laborat d Investig Biol. 1901:1.

95. Albuixech-Creso B, López-Blanch L, Burguera D, et al. Molecular regionalization of the developing amphioxus neural tube challenges major partitions of the vertebrate. PLoS Biol. 2017;15(4):e2001573. https://doi.org/10.1371/journal.plos2001573.

96. Scholpp S, Lumsden A. Building a bridal chamber: development of the thalamus. Trends Neurosci. 2010;33(8–2):373–80. https://doi.org/10.1016/j.tins.2010.05.003.

97. Graham A, Butts T, Lumsden A, Kieker C. What can vertebrates tell us about segmentation? EvoDevo. 2014;5:24–32. https://doi.org/10.1186/2041-9139-5-2.

98. Le Douarin N, Kalcheim C, Bard J, Barlow PW, Krik DL. The neural crest. 2nd ed. Cambridge: Cambridge University Press; 1999.

Further Reading

Adams RD, Victor M, Ropper AH. Principles of neurology. 9th ed. New York: McGraw Hill; 1997.

Biller J, Greuner G, Brazil PW, editors. DeMyer's technique of the neurologic examination. 6th ed. New York: McGraw Hill; 2011.

Blumenfeld H. Neuroanatomy through clinical cases. 2nd ed. New York: Sinauer/Oxford; 2010.

Bronner-Fraser M. Neural crest formation and migration in the developing embryo. FASEB J. 1994;8:699–706.

Bronner-Fraser M. Origins and developmental potential of the neural crest. Exp Cell Res. 1995;218:405–17.

Butler T. Volume of the human septal forebrain region is a predictor of source memory accuracy. J Int NeuropsychSoc. 2012;18(1):157–61.

Carlson BM. Human embryology and developmental biology. 5th ed. Philadelphia: WB Saunders/Elsevier; 2013.

Cobos L, et al. Fate map of the avian anterior forebrain at the 4-somite stage, based on the analysis of quail-chick chimeras. Dev Biol. 2001;239:46–67.

Eisen JS, Pike SH. The spt-1 mutatin alters segmental arrangement and axonal development of identified neurons in the spinal cord of the embryonic zebrafish. Neuron. 1991;6:767–76.

Ettinger AB, Weisbrot DM, editors. Neurologic differential diagnosis: a case-based approach. New York: Cambridge University Press; 2014.

Ferrier DEK, Holland PWF. Ancient origin of the hox gene cluster. Nat Rev Genet. 2001;2:33–8.

Guthrie S. Patterning the hindbrain. Curr Opin Neurobiol. 1996;6:41–8.

Haines DE. Neuroanatomy in clinical context. 9th ed. Philadelphia, PA: Lippincott, Williams, and Wilkins; 2014.

Herrick CJ. The Brain of the Tiger Salamander, Ambystoma Tigris. Chicago: University of Chicago Press, 1948.

His W. Die Anatomische Nomenclatur. Nomina Anatomica. Arch Anat Entwickelungsges. 1895;(Suppl):1–180.

Karten HJ. Vertebrate brains and evolutionary connectomics: on the origins of the mammalian "neocortex". Philos Trans R Soc Lond B Biol Sci. 2015;370:1684. pii: 20150060. https://doi.org/10.1098/rstb.2015.0060. Review.

Keibel F, Mall F. Manual of human embryology. Philadelphia: Lippincott; 1912.

Kuhlenbeck H. Propaedeutics to comparative neurology. In: The central nervous system of vertebrates, vol. 1. Karger; 1967. p. 159–304.

Louis ED, Stapf C. Unreavelling the neuron jungle: the 1879-1886 publications by Wilhelm His on the embryological development of the human brain. Arch Neurol. 2001;58:1932–5.

Martinez S. Molecular regionalization of the diencephalon. Front Neurosci. 2012;6(article 73):1–8.

Noden DM. Interactions and fates of avian craniofacial mesenchyme. Development. 1988;103:121–40.

Noden DM. Cell movement and control of patterned tissue assembly during craniofacial development. J Craniofac Genet Dev Biol. 1991;11(4):192–213.

Nomes HO, Dressler GR, Knapik EW, et al. Spatially and temporally restricted expression of Pax-2 during neurogenesis. Development. 1990;109:797–809.

O'Rahilly R, Müller F. Developmental stages in human embryos including a revision of streeter's "horizons" and a survey of the carnegie collection. Washington, DC: Carnegie Institution of Washington; 1987. Publication 637.

Orr HA. Contributions to the embryology of the lizard. J Morphol. 1887;2:51–96.

Patten JT. Neurological differential diagnosis. 2nd ed. Berlin: Springer-Verlag; 1998.

Puelles L. A new scenario of hypothalamic organization: rationale of new hypotheses introduced in the updated prosomeric model. Front Neuroanat. 2015;9:1–23.

Puelles L, Rubenstein JLR. Expression patterns of homeobox and other putative regulatory genes in the embryonic mouse forebrain suggest a neuromeric organization. Trends Neurosci. 1993;16:472–9.

Puelles L, Rubenstein JLR. Forebrain gene expression domains and the evolving prosomeric model. Trends Neurosci. 2003;26:469–76.

Purves D, Augustine DJ. Neuroscience. 6th ed. Sinauer; 2017.

Puschel AW, Westerfeld M, Dressler G. Comparative analysis of Pax-2 protein distributions during neurulation in mice and zebrafish. Mech Dev. 1992;38:197–208.

Rhinn M, Dierich A, Shawlot W, et al. Sequential roles for Otx2 in visceral endoderm and neuroectoderm for forebrain and midbrain induction and specification. Development. 1989;125:845–56.

Rubenstein JLR. The prosomeric model: a proposal for the organization of the embryonic forebrain. Science. 1994;266:578–80.

Sarnat HB. Molecular genetic classification of central nervous system malformations. J Child Neurol. 2000;21:675–87.

Sarnat HB, Benjamin DR, Kletter GB, Seibert JR, Cheyette SR. Agenesis of the mesencephalon and metencephalon with cerebellar hypoplasia: putative mutation of the EN2 gene. Report of 2 cases in early infancy. Pediatr Dev Pathol. 2002;5:54–68.

Schaal B, Hummel T, Soussignan R. Olfaction in the fetal and premature infant: functional status and clinical implications. Clin Perinatol. 2004;31:261–85.

Sheehan T, Chambers R, Russell D. Regulation of affect by the lateral septum: implications for neuropsychiatry. Brain Res Rev. 2004;46(1):71–117. https://doi.org/10.1016/j.brainresrev.2004.04.009.

Shimamura K, Hartigan DJ, Martinez S, Puelles L, Rubenstein JLR. Longitudinal organization of the anterior neural plate and neural tube. Development. 1995;123:3923–33.

Swanson LW. Quest for the basic place of nervous system circuitry. Brain Res Rev. 2006;55:356–72. https://doi.org/10.1016/j.brainresrev.2006.12.006.

Tessier P. A new anatomic classification of facial clefts, craniofacial and laterofacial clefts, and their distribution around the orbit. In: Tessier P, Hervouet F, Lekieffre M, Rougier J, Woillez M, Derome P, editors. Plastic surgery of the orbit and eyelids. New York: Masson; 1981.

Vanderah TW, editor. Nolte's the human brain: an introduction to its functional anatomy. 7th ed. Phialdelphia, PA: Elsevier; 2015.

von Baer KE. Über die Entwicklungsgeschichte der Thiere. Konigsberg; 1828.

Wurst W, Auerback AB, Joyner AL. Multiple developmental defects in Engrailed-1 mutant mice: an early mid-hindbrain deletion and patterning defects in forelimbs and sternum. Development. 1994;120:2065–75.

Development of the Craniofacial Blood Supply: Intracranial System

6

Michael H. Carstens

Introduction

Neurovascular anatomy of the head and neck presents a bewildering tangle of structures. In previous chapters, we have focused on the genetically based neuromeric organization of the neural plate, the relationship between the CNS and outlying tissues, and on the neuromeric "coding" of mesenchymal derivatives, be they of mesodermal or neural crest origin. We now turn our attention to the development of arterial supply to the head and neck. Intuitively, we know that this seemingly complex story must possess an underlying simplicity, a logic based upon developmental principles. This chapter will explore the vascular system in a way that makes surgical sense. Craniofacial blood supply in its adult configuration is a rearrangement of intermediate structures such as the aortic arches, the primitive carotids, and the stapedial system. Understanding the sequence in which these vascular "placeholders" produce structures, we are familiar with is an exercise that will provide crucial insights into the fate of the pharyngeal arches and the final anatomy of the face.

All good stories have their beginnings. Any new understanding of vascular development must recognize the contributions of the past and the limitations of present-day discussion. Although all mammals share a common anatomic baseline, humans have a unique sequence of rearrangements from that *bauplan* leading to the final configuration. This is particularly true in the orbit. Hence, we shall concentrate our discussion on human studies, in particular those of Padget (1948) on cerebral circulation and of Condon (1928) on the aortic arches.

Dorcas Padget was a talented art student at Vassar with an interest in medicine. She left college after her junior year to work as a medical illustrator for the department of neurosurgery at Johns Hopkins University. Under the direction of Walter Dandy, she illustrated all the papers and book chapters for that prolific group. In later life she moved to Washington, DC where she was employed at the Carnegie Institution for the Study of Human Development. Taking advantage of the large collection of human embryos at Carnegie, Dorcas Padget carefully studied the vascular supply to the head and neck at various stages of development. Her findings, including painstaking illustrations, were published by the Carnegie Institution and established her as a pioneer in neuroembryology. A short biographical sketch of Dorcas Padget is included at the end of this chapter (Figs. 6.1, 6.2, and 6.3).

Technical aspects of publishing detracted from the effectiveness of Padget's contribution. Her drawings had to conform to the limitation of black and white. To convey three

M. H. Carstens (✉)
Wake Forest Institute of Regenerative Medicine, Wake Forest University, Winston-Salem, NC, USA
e-mail: mcarsten@wakehealth.edu

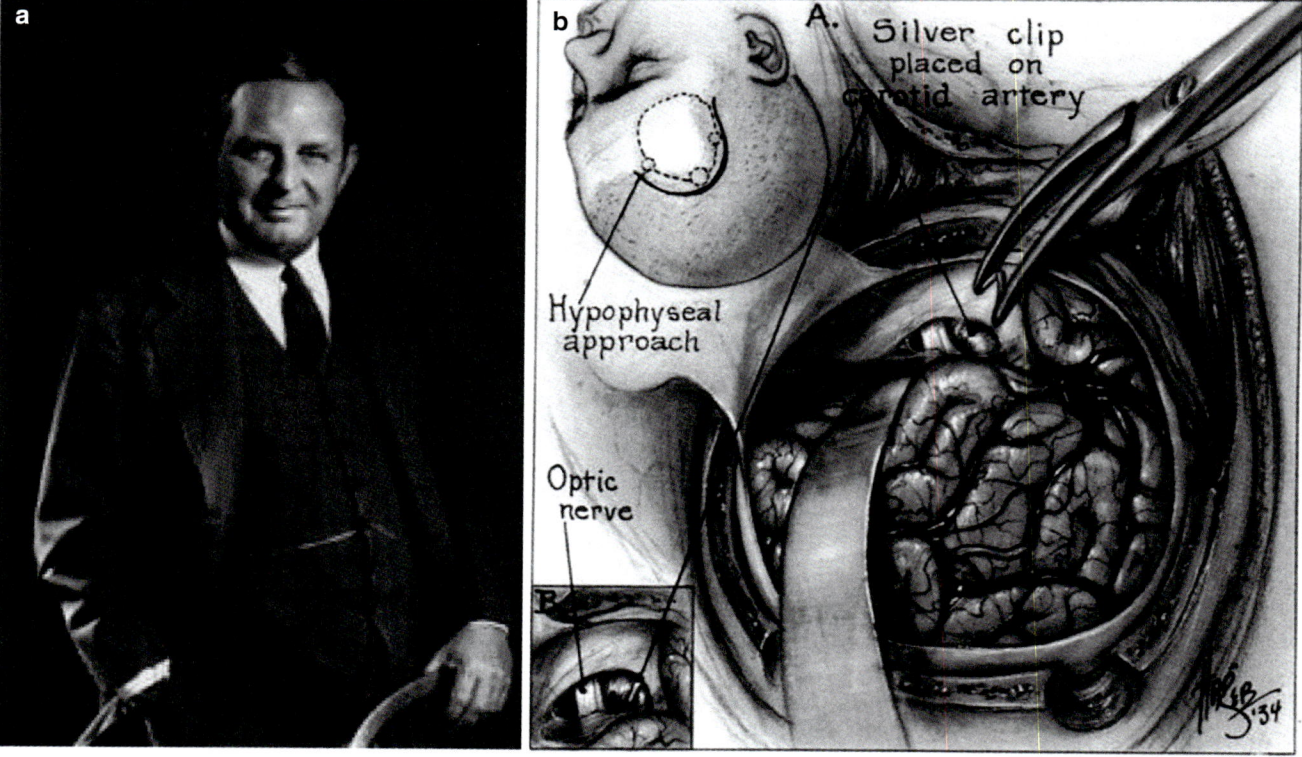

Fig. 6.1 (**a**, **b**) Max Brödel and Dorcas Padget. (Reprinted with permission from Johns Hopkins University Department of Medical Art—Gary Lees. Carnegie Institution—Sonya Bajwa)

Fig. 6.2 (**a**, **b**) Walter Dandy, MD and Padget drawing of a procedure. She did all the artwork for the neurosurgery under Dandy's direction. (Reprinted with permission from Johns Hopkins University Department of Medical Art—Gary Lees. Carnegie Institution—Sonya Bajwa)

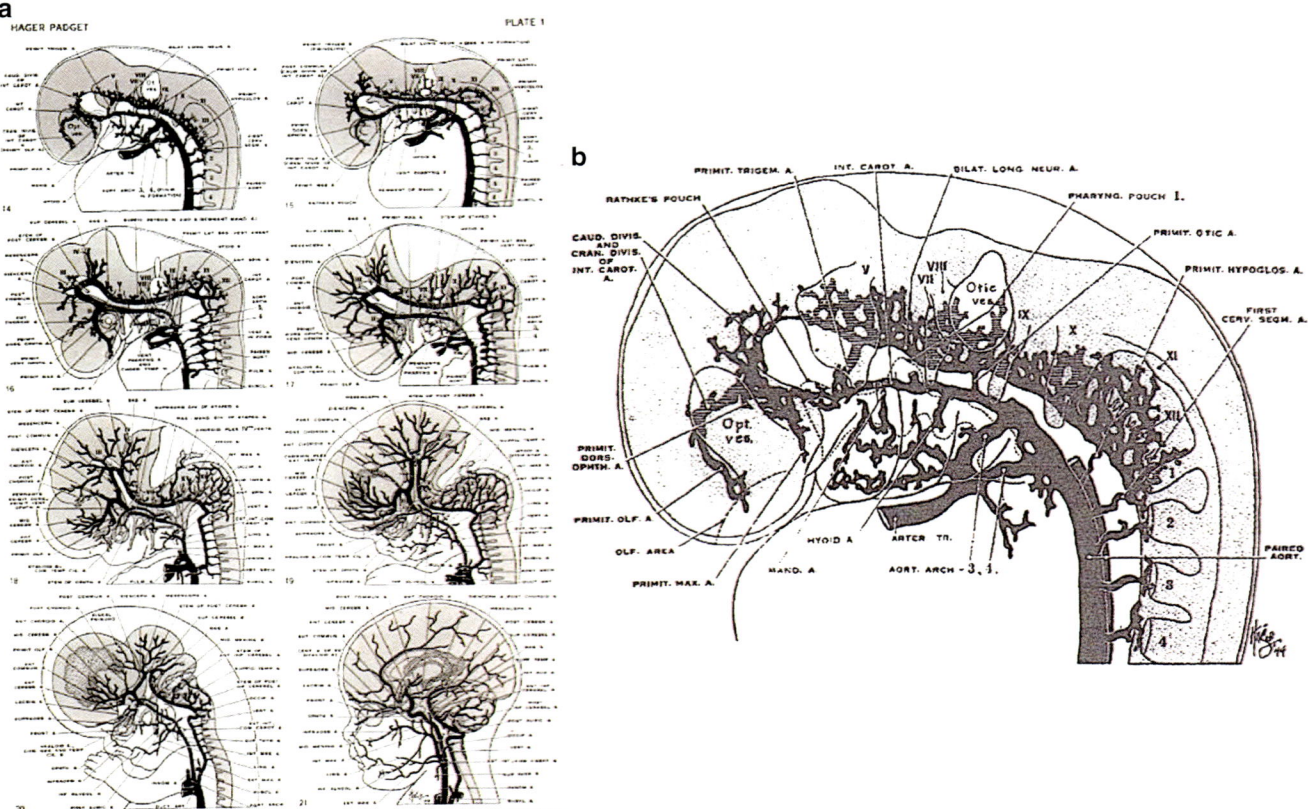

Fig. 6.3 (**a, b**) Dorcas Padget's pioneering work documented craniofacial arteries in embryos from the Carnegie collection. (**a**) Shows serial drawing cerebrovascular development. (**b**) Shows an embryo at Carnegie stage 13, with her dissections shown in exquisite three-dimensional detail. Padget's staging system predated that of Streeter and O'Rahilly. (Reprinted with permission from Johns Hopkins University Department of Medical Art—Gary Lees. Carnegie Institution—Sonya Bajwa)

dimensionalities, she made use of convention such as cross-hatching and dots. Each specimen was depicted in multiple views (lateral, dorsal, ventral) to show spatial relationships between multiple vessels. These constraints make it difficult to readily visualize the developmental process. Nonetheless, her work is of such scientific relevance that readers will be provided direct access to it via the following internet link: www.carstens.embryology.com/vascular/Padget. I have accordingly taken the liberty of reworking her drawing using a color code for the various arterial systems to make the story of arterial development easier to understand. These will be cross-referenced to Padget. Details of this color code are to be found in the illustrations section.

As contemporary readers, we face additional problems in the interpretation of Padget's work. The anatomic system for staging human embryos (elaborated at the Carnegie Institution) was not available to Padget. Consequently, her description makes use empiric stages based on the embryos at hand. Her earliest specimen is 3 mm long and has 20 somites, placing it at stage 11. Thus, we are bequeathed a story that is incomplete. There are times gaps in the sequence that must be filled in. Fortunately, molecular embryology

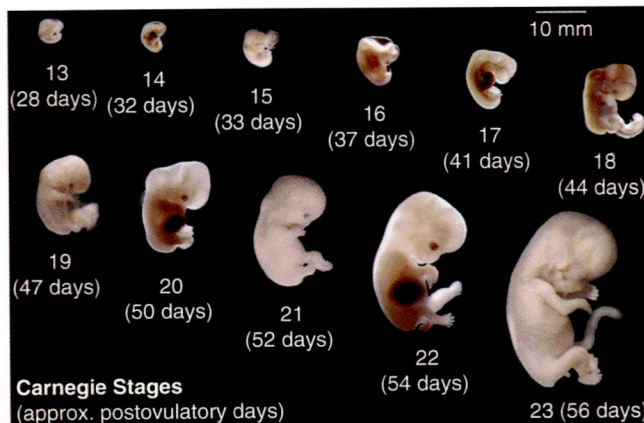

Fig. 6.4 Carnegie stages of embryonic development summarized. (Courtesy of Brad Smith, University of Michigan (brdsmith@umich.edu). NIH award N01-HD-6-3257 P/G F003637)

allows us to work backward to the earliest steps in vascular development (Figs. 6.4 and 6.5).

Recall that at stage 7 the neural plate is induced, with folding beginning at stage 8. The appearance of the first (3)

Fig. 6.5 Staging systems: Carnegie (left) versus Padget stages (right) right. (Courtesy of Michael Carstens, MD)

Relationship between vascular phases of Dorcas Padget and Carnegie stages		
Arterial phase	Venous phase	Carnegie stage
-	1	13
1	-	13-14
2	2	14
-	3	15-16
3	-	16
4	-	17
-	4	17-18
5	-	18-19
-	5	19
6	6	20-21
7	7	40 mm
-	7a	60-80 nmm

somites at stage 9 means that the first 7 somitomeres are *already in place*. Nutrition for the rapidly expanding CNS and mesenchyme requires the presence of vascular channels that predate future aortic arch system at stage 10. ***Nature does not produce a tissue without making provisions for its metabolic support***.

During these early stages, survival of the embryonic CNS depends upon transition through successive vascular designs: (1) ebb-and-flow type of circulation; (2) primordial supply system using separate parallel longitudinal vessels: primitive hindbrain channels and dorsal aortae; (3) functional heart with interconnection of previously independent vessels, separation of supply and drainage, head and neck primordial supplied by aortic arches; and (4) a post-arch remodeling into the adult state. In this chapter we will cover each step of this sequence to understand how blood supply evolves in response to the increasing demands of the embryo.

Traditional anatomy texts make craniofacial circulation difficult to understand. Abbreviated treatments of this subject are unable to show the developmental sequence. Textbooks illustrations can be misleading. Early anatomists did not have appropriate methods for study of embryonic specimens. Dorcas Padget's dissections were not published until 1948. Thus, it is not surprising that earlier editions of Gray's Anatomy depicting a 3-mm embryo with aortic arches and dorsal aorta make no reference to the more dorsal parallel system supplying the CNS. Such misconceptions continue into the latest editions of *Gray's Anatomy*. Some contemporary embryology texts present circuit diagrams of facial circulation that leave the brain out altogether! Small wonder that our understanding of how pharyngeal arches morph into adult structures is vague. And yet…this is exactly what becomes clinically relevant.

As we shall see, neuromeres, somitomeres/somites, aortic arches, pharyngeal arches, and aortic arches are precisely related with each other. Careful study of the stepwise vascular remodeling leading to the final system pays handsome dividends, for it reflects the temporal-spatio reorganization

of craniofacial anatomy. Combining older work with contemporary methods of study (microangiography, transgenic LacZ mice, immunocytochemical labeling, latex casting) proves fruitful because we can "connect the dots" and make sense of the system as a whole. We shall also see that craniofacial vascular development, at every turn, obeys the dictates of the CNS.

Introduction: How to Use This Chapter

Because this material is complex with new concepts, we will cover it from several different steps.

Overview of the System
Our first task will be to evaluate a schematic drawing of the embryonic vascular system. This will provide a quick visual orientation to the principle arterial systems in the head and neck. The following structures will be introduced; each component will be color-coded: deep and superficial head plexus, primitive hindbrain channels, dorsal aortae, segmental arteries, longitudinal hindbrain channels, aortic arches, primitive internal carotid, ventral pharyngeal artery, basilar artery, external carotid artery, stapedial arterial system, and vertebral artery. This color system will be used throughout this chapter.

Mechanisms of Blood Vessel Construction
We begin with the basics of circulation. Where do the progenitor cells of the vascular system arise? How are arteries formed? What is the relationship between nerves and arteries; between arteries and veins? What causes arteries to involute or remodel? How are arteries "assigned" to individual bones, muscles…and to the brain?

Craniofacial Arterial Development by Carnegie Stages
This section provides a narrative description of events in a stage-by-stage manner using the Carnegie system.

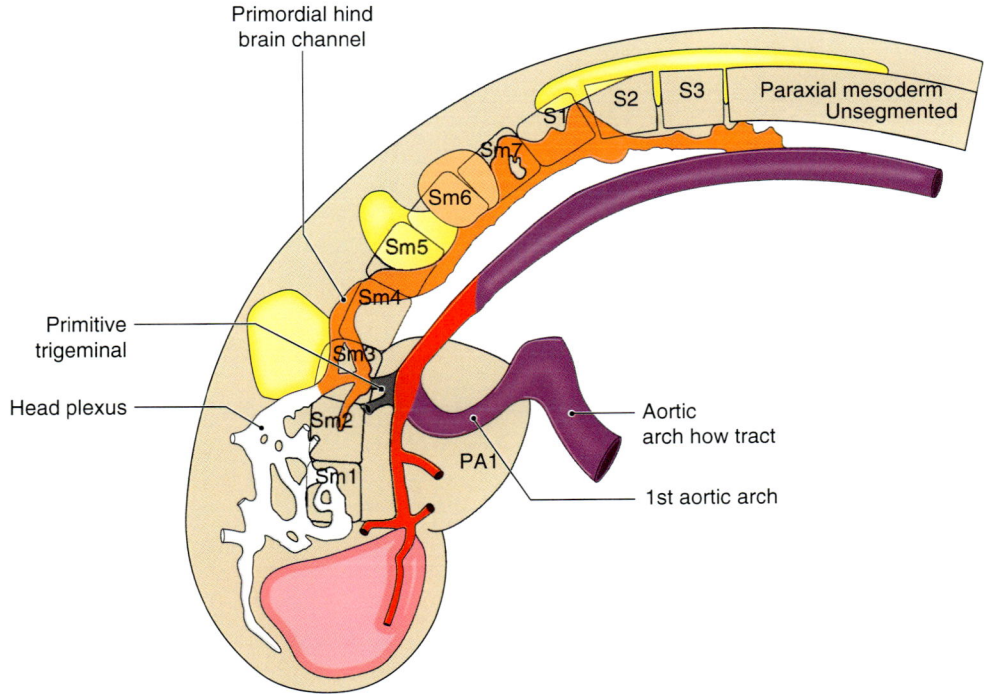

Fig. 6.6 DHP-MC stage 9. First arch: paired dorsal aortae flex at r0/r1, elongate, and ventralize. Primitive trigeminal connects dorsal aortae with primordial hindbrain channels. Primitive internal carotid applied along walls of midbrain and forebrain

Beginning at stage 5, morphogenesis and rearrangement of embryonic tissues "sets the stage" for cranial vessels. Long before the dorsal aortae and heart appear, the embryo has to make yolk sac vessels and blood cells. These important preliminary steps are well illustrated in Carlson; appropriate references are provided. Stage 10 is the earliest embryo studied by Padget. The head plexus and the primitive hindbrain channels are depicted in Fig. 6.6 (corresponding in Padget's monolog to plate 4, Fig. 37). From this point onward, our narrative will follow that of Padget. For the sake of simplification, I have taken license to reproduce Padget's black and white drawings in a color-coded format. This will enable you to follow the development of each arterial systems. Nonetheless, you would be well served to compare my color plates with her originals. The idea of this section is to create a step-by-step visual picture of arterial development.

Blood Supply to the Forebrain and Midbrain
- Timeline
- Components
  - Dorsal aortae
  - Internal
- Descriptive anatomy

- Eye and orbit include derivatives of the stapedial system

Blood Supply to the Hindbrain
- Timeline
- Components
  - The longitudinal neural arteries
  - The vertebral arteries as collateral circulation to the hindbrain
- Descriptive anatomy

Blood Supply to the Pharyngeal Arches
- Timeline
- Components
  - External carotid system
  - Extracranial stapedial
- Descriptive anatomy

Blood Supply to the Orbit and Frontonasal Face
- Timeline of the eye and orbit
- Components
  - Ophthalmic artery
  - Intracranial stapedial
- Descriptive anatomy

Pathologies of the anterior cerebral and intraorbital stapedial systems

Holoprosencephaly vs. orbitofacial clefts (Tessier system).

Do not be daunted as you tackle this material. The goal of this chapter is to help you develop a four-dimensional appreciation of arterial development. As the embryo becomes more complex, its blood supply undergoes predictable changes. Illustrations are key here. Study them carefully. Read the legends and compare them with the narrative as a contextual guide. You will develop a videotape in your mind of when each component of arterial system makes its debut… and of how each contributes to the final anatomy.

Defects in vascular development are vital for understanding the spectrum of craniofacial pathology. The section describing individual arterial components, such as the aortic arches or the many derivatives of the stapedial system, will give you powerful insights into the assembly of the face and how trouble can arise in the process.

I am going to beg your indulgence for a fair amount of repetition in this narrative. You are going to encounter many details at once. Don't worry…presenting the anatomy from multiple perspectives will prevent you from getting lost as you work toward seeing the "big picture."

Craniofacial Arterial Development: An Overview

The development of the arterial supply to the head and neck is a fascinating story involving transformations and interactions among six distinct vascular systems. The best way to tell this story is by describing the arterial anatomy of the embryo at each one of its 23 developmental stages as described by Streeter (and subsequently refined by O'Rahilly and Muller) at the Carnegie Institution in Washington, D.C. The events may seem complex at first but, with a little patience, we shall discover an underlying conceptual simplicity. So let's take a deep breath and get oriented. We shall do this in two ways. First, we are going to *introduce some vocabulary* using three broad phases of arterial development. As we shall see, these closely parallel the six clinical stages described by Padget. Second, let's get a *visual preview* of what these structures look like in a hypothetical embryo.

In the *germ layer phase* (stages 5–8), the embryo is converted from a monolayer to a trilaminar structure. Prior to gastrulation, it survives by simple diffusion. However, as soon as mesoderm appears, specific mesenchymal vessels, hemangioblasts, produce the first primitive vessels via the process of *vasculogenesis*. Hemangioblasts are ubiquitous throughout mesoderm and proliferate wildly. They are not found in prechordal plate mesoderm nor in neural crest. The initial vessels interconnect and expand via angiogenesis. Flow begins from the heart at stage 9 circulation via ebb-

and-flow using three sets of vessels. Nutrients are conveyed to it via the *vitelline veins*. Paired *primitive hindbrain channels* (PHCs) supply the primitive CNS (the hindbrain). In the region of the embryonic "head," the PHCs are known as the *superficial and deep head plexus*. The remainder of the embryo is supported by paired outlying *dorsal aortae* (DAs). These run in parallel to the PHCs; they unite just in front of the future forebrain to form a horseshoe-shaped structure, the anterior "loop" of which will become the future heart tubes.

During the *aortic arch phase* (stages 9–14), the brain becomes more complex with the induction of the midbrain and forebrain; embryonic folding takes place; the primitive heart is formed; and heartbeat is initiated. In stage 9, embryonic folding forces the dorsal aortae directly behind the heart to flex ventrally 90°. This fold is located at the rostral hindbrain, near the trigeminal nucleus. The zone of each newly flexed dorsal aorta is renamed the *first aortic arch* (AA1). Note that the first arches (right and left) do not arise de novo…they are merely the result of stretching and deformation of preexisting structures. Subsequent to this, three now sets of vessels arise. (1) *Segmental arteries* interconnect the dorsal aortae and the primitive hindbrain channels. A common circuit results; the hindbrain is now directly supplied by the heart. Because the segmental vessels are associated with cranial nerve nuclei, they are named accordingly (trigeminal, otic, hypoglossal, etc.). (2) *Primitive internal carotid* arteries now emerge to supply the newly induced midbrain and forebrain. These arise from the "knuckle" of dorsal aortae just below the rostral terminus of the hindbrain, i.e., at Hensen's node (the site of Rathke's pouch). (3) The *"true" aortic arches*, AA2–AA6, now arise sequentially from the outflow tract of the heart. Each stage marks the appearance of a new arch, with the previously formed arches having distinct fates (Fig. 6.7).

Here we must interject some clarifying terminology right up front. Although *six aortic arch arteries* (AA1–AA6) are seen in tetrapod embryos, there are only *five pharyngeal arches* (PA1–PA5), each of which is supplied by an individual aortic arch artery. Furthermore, only four aortic arch arteries supply the developing head and neck. AA5 is a transient structure which involutes, *leaving the fifth pharyngeal arch wholly upon AA4-dependent for its blood supply*. For this reason, as we shall see in our subsequent discussion on the development of the external carotid artery, the first and most caudal branch of the ECA, the superior thyroid artery, represents the intended blood supply to the fourth pharyngeal arch; it also supplies the interior structures of the larynx, i.e., the fifth pharyngeal arch. Paired sixth aortic arch arteries appear at during stage 14; they are exclusively dedicated to the pulmonary circulation.

During the *post arch phase* (stages 15–18), four sets of vessels arise. Stages 15–16 are marked by the appearance of the *vertebral artery* and the reorganization of precursors to

Fig. 6.7 Model embryo at Stage 14 showing different vascular systems. At no time do all the aortic arches appear together. **Primitive ICA** to forebrain and midbrain, **Longitudinal neural** (future **basilar**) to hindbrain, connecting **presegmentals** (trigeminal, hyoid, otic, hypoglossa), **aortic arches** (future **external carotid**), **dorsal aorta** giving off **dorsal intersegmentals** (these will later unite longitudinally as the **vertebral artery**). (Reprinted from O'Rahilly R, Müller F. The Embryonic Human Brain: An Atlas of Developmental Stages. Wiley-Liss, 2005. With permission from John Wiley & Sons, Inc.)

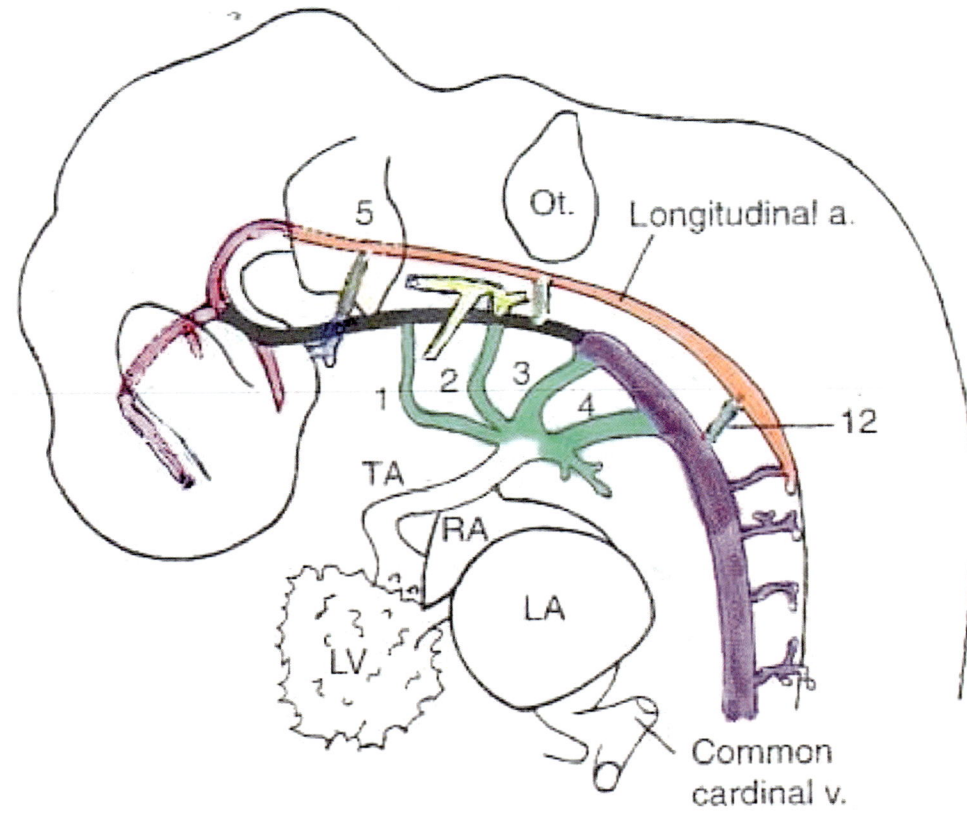

Fig. 6.8 Stage 17 Longitudinal neural arteries fusing into basilar. Vertebral system (blue) develops from vertical anastomoses between intersegmental arteries 1–6. Extracranial stapedial has reached V3 and connects with the ventral pharyngeal artery precursor of the external carotid system, maxillary branch. (Reprinted from O'Rahilly R, Müller F. The Embryonic Human Brain: An Atlas of Developmental Stages. Wiley-Liss, 2005. With permission from John Wiley & Sons, Inc.)

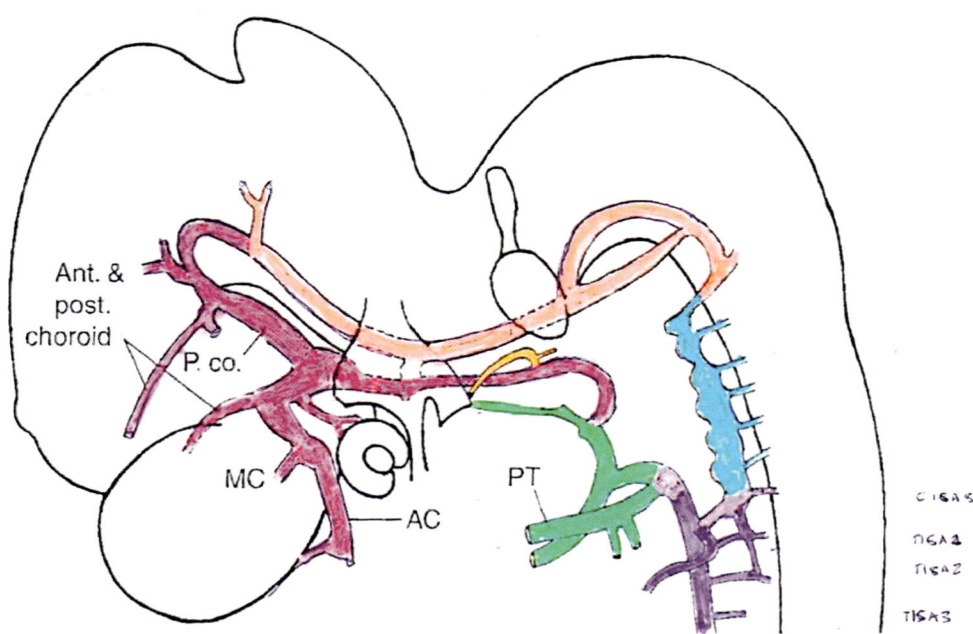

the external carotid system, the *ventral pharyngeal artery*. The *stapedial artery* arises as the continuation of the sole extracranial branch of the internal carotid artery, the *hyoid artery*. It enters the skull via the tympanic cavity where it follows sensory branches of the cranial nerves to form the dural arteries. In particular, following V1, it enters the orbit where it will anastomose with the primitive ophthalmic artery to form the blood supply of the extraocular tissues and frontonasal midline. Stapedial then exits the skull, following chorda tympani and, at stage 17, the *stapedial artery* extends outward along chorda tympani to connect via V3 with the VPA (Fig. 6.8).

Fig. 6.9 Stage 21 External carotid system fully formed. Intracranial stapedial connects with ophthalmic. (Reprinted from O'Rahilly R, Müller F. The Embryonic Human Brain: An Atlas of Developmental Stages. Wiley-Liss, 2005. With permission from John Wiley & Sons, Inc.)

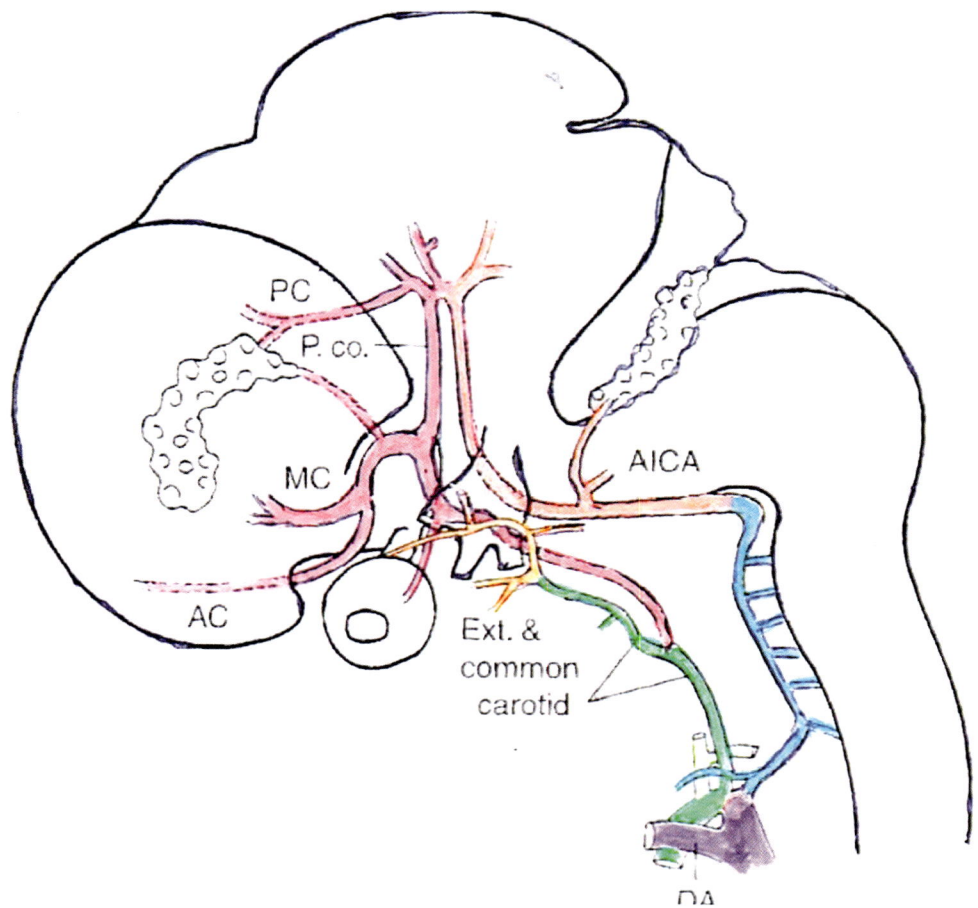

By stage 21, we see composite structures between extra-cranial stapedial and maxillary artery, the first arch remnant of the external carotid. This "hybrid" maxillo-mandibular artery (MMA) will supply two types of derivative structures: (1) muscles of mastication via branches of the maxillary artery and (2) the jaws and zygomatico-orbital-nasal complex via branches of V2 and V3 stapedial. With the formation of MMA, the external carotid system is complete. We also see a similar process with intracranial stapedial extending into the orbit to join with ophthalmic to form a composite system (Fig. 6.9).

The results of the "stapedial revolution" are crucial for understanding the final configuration of vascular anatomy. The stapedial system constitutes an anastomotic system between the internal and external carotid circulations. (1) A forward extension of intracranial stapedial supplies all extra-ocular structures via its anastomosis with primitive ophthalmic. In so doing, stapedial loses its original ICA identity and assumes that of ophthalmic. Thus, all non-ocular and front-nasal-orbital branches of the internal carotid-derived primitive ophthalmic are, in reality, stapedial. As an ICA derivative itself, *stapedial is the "extension cord" of the intracranial internal carotid*, permitting it to supply nonneural tissues. (2) In like manner, a series of anastomoses develop between the intracranial meningeal system and extracranial branches of the external carotid. (3) Finally, anastomosis between the

ECA which is exclusively dedicated to pharyngeal arches and extracranial stapedial supplies those facial tissues which have been diverted away from first arch to form the jaws, mouth, and nose. In so doing, stapedial loses its original ICA identity and assumes that of ECA. *Stapedial is the "extension cord" of the external carotid*, permitting it to supply non-pharyngeal arch tissues.

To get "the big picture," we examine in detail a composite reconstruction of the embryonic vascular system taken at three stages, 14, 17, and 21. Of course, this "master circuit diagram" is artificial because it includes many of the key structures and anastomoses as if they existed all at the same time. However, it does give us a bird's eye view of the critical connections. Here's our color code: (1) dorsal aortae (purple); (2) primitive hindbrain channel, longitudinal neural artery, and basilar artery (magenta); segmental vessels (white); internal carotid artery (red); aortic arches, ventral pharyngeal artery, and external carotid artery (green); and stapedial artery (orange) (Figs. 6.7, 6.8, and 6.9).

All five aortic arches are depicted in place together. In real life, of course, the aortic arches develop (and involute) in a cranio-caudal sequence from stage 9 to stage 14 (roughly one aortic arch per stage), so they are not seen simultaneously. Let's get some terminology straight right off the bat. The arches represent connections between the aortic sac and dorsal aortae. The primitive internal carotid sprouts forward from the "knuckle" of dorsal aorta and the first aortic arch. It then divides. The cranial branch of ICA supplies the developing eye and the telencephalon. The caudal branch supplies the diencephalon and midbrain. Its terminal branch, *posterior communicating artery*, is directed backward and connects to the longitudinal neural artery. Note that the LNA supplies the entire brain from r0 back to r11; *it does not supply the spinal cord*. The dorsal aortae sweep backward, providing segmental arteries to: (1) the occipital somites *but not the hindbrain*; and (2) all post-cranial somites and the spinal cord. Via the first cervical branch of the dorsal aorta, the posterior limit of LNA receives collateral supply.

Each aortic arch also gets into the act. As they connect with the ICA-DA conduit, a sprout from the cranial end of each aortic arch forms a temporary connection between the ICA and the LNA. These pre*segmental arteries* (trigeminal, hyoid, primitive otic, and hypoglossal) will later form important secondary derivatives. The fate of the paired LNAs is, of course, to form a united singular basilar artery. The presegmental vessels have an important, but temporary, role as *carotico-basilar anastomoses* that supply the hindbrain from the heart prior to connection of the establishment of an intact intracranial circuit. Once the caudal division of internal carotid joins with the LNAs at the r0 level, blood flow to the hindbrain comes from the forebrain and from collateral flow via the vertebral system. Consequently, the presegmental arteries involute.

We complete our overview by considering the final disposition of the LNAs and the formation of the vertebral-basilar system. Beginning at stage 13, the LNAs begin to merge in the midline, dragging their carotid anastomoses with them. By stage 16, this process is complete from r1 down to r8: a singular midline *basilar artery* results. Caudal to r8, the LNAs remain separate. They terminate by fusing with the laterally located dorsal aortae via the c1 segmental arteries. Thus, from r8 to r11, the distal LNAs assume the form of an inverted V. They anastomose with the vertebrals at level c1.

In addition, peripheral to the LNAs, an additional system of transverse vessels becomes longitudinally cross-linked to form a "short-circuit" between the level of the otic capsule (r4–r5) and the vertebrals at terminus of the hindbrain, thus forming the *basilar-vertebral anastomoses*.

Note that traditional anatomy texts describe the vertebrals as extending up to the otic capsule, i.e., where LNA fusion in the midline ceases. This is not embryologically correct. In reality, this "bifurcating" arteres are the continuation of the LNAs, all the way down to the vertebrals.

At the same time, the segmental arteries from c1 to c8 undergo two transformations. At stage 15, they become linked together by a "stair-step" longitudinal fusion By stage 16, the axial connections break down, producing the *vertebral artery*. The vertebral-basilar connection is retained via the primordial c1-LNA anastomosis. Not depicted in the diagram are similar intersegmental processes that result in the components of the thyrocervical trunk. The anatomic ratio-

nales of the ascending cervical, inferior thyroid, and transverse cervical arteries will be discussed in detail later.

From a circulatory standpoint, the face is subordinate to the brain. The external carotid system makes its debut only after all four aortic arches are in place. By the time that AAs3 and 4 develop[12–13], involution of the first two arches has occurred, leaving behind them important remnants. (1) Dorsal stumps connected to ICA (the so-called "mandibular" from AA1 and "hyoid" from AA2) will supply future structures. (2) Despite the loss of their aortic arch "cores", nutrition for the first and second pharyngeal arches is maintained. The individual vessels within each arch amalgamate into a single vascular unit (pharyngeal arch plexus). This plexus is, in turn, derived from the common stem of AA3 and AA4 by what is called the *ventral pharyngeal artery* (VPA). When VPA travels outward and plugs into the plexus of the pharyngeal arches, it then morphs into the stem of the external carotid artery system. The ECA proper consists of branches from the original VPA which are arch-specific, and spatially organized: 4th–fifth arch (superior thyroid), third arch (glossopharyngeal), second arch (occipital, posterior auricular, superficial temporal, facial/external maxilla-mandibular), and first arch (lingual). Conspicuously missing from this list is the internal maxillo-mandibular artery (IMMA). As we shall see, IMMA is a hybrid system of arteries between ECA and the stapedial system.

The distal AA2-derived hyoid artery, dangling from the ICA, now gives rise to distally directed branch, the *stapedial artery*. The stapedial develops at the same time as the sprouting of cranial nerves. It is destined to form a system of arteries, each one of which is associated with/induced by a cranial nerve. After entering the middle ear, stapedial splits into a superior ramus that remains intracranial/transorbital and an inferior ramus that exits the skull.

Intracranial stapedial gives rise to three important sets of vessels: *meningeal arteries* (V1, V2 and V3, VII, IX, and X), *orbital arteries* (**V1**), and *fronto-nasal arteries* (V1). The neurovascular axes of StV1 supply **Tessier cleft zones 10–13**. *Extracranial* stapedial gives rise to all branches of the *internal maxilla-mandibular artery*. The neurovascular axes of StV2 supply **Tessier cleft zones 1–8**. The combined axes of StV1 and StV2 supply **Tessier cleft zone 9**. The neurovascular axes of StV3 supply **developmental zones of the mandible**.

For the moment, let's briefly sketch out how stapedial and VPA form the external carotid system. From within the tympanic cavity, common stapedial traverses the ring of the stapes and, remaining intracranial, divides into two branches. The *inferior ramus*, travelling in company (induced by) with chorda tympani, comes to rest at the foot of V3. Chorda tympani and lingual nerve of V3 become chummy (thus taste and somatic sensation arrive together at the anterior 2/3 s of the tongue). VPA, having previously arrived, anastomoses with inferior ramus of stapedial, uniting the external carotid system with the extracranial stapedial axis.

In the trigeminal system, the very first branch of each division is dedicated to supplying dura.

Within the skull, the course of V1 and V2 is as follows. The stem artery follows greater petrosal nerve forward to the trigeminal ganglion. Here, it picks up cranial nerves V1 and V2 as they run forward in the lateral wall of cavernous sinus. Immediately upon leaving the sinus, both V1 and V2 give off *dural branches* prior to exiting the skull. Upper division stapedial artery bifurcates to provide vascular support for all branches of V1 and V2. These arteries supply their respective areas of dura in the anterior and posterior cranial fossae. Orbital branch of V1 passes through superior orbital fissure. Meningeal branch of V1 gains access lateral to the fissure.

V1 stapedial artery follows a similar time course by stage 19 and enters the orbit as *two distinct vessels*. StV1 meningeal branch follows the nerve into lateral orbit via a separate *meningo-orbital foramen* in the greater wing of sphenoid. Sometimes this foramen is merely a lateral extension of the superior orbital fissure. StV1 orbital branch goes through the fissure heading directly for the stem of ophthalmic artery. When lacrimal artery forms, both arteries connect to it, but the flow changes that occur when orbital branch connects with ophthalmic make it dominant. The lateral STV1 usually involutes but sometimes it persists as the recurrent meningeal artery (RMA). The dual anastomosis to lacrimal explains when, in some cases, no connection with ophthalmic transpires. In this case, the entire system of non-ocular orbital arteries is supplied by the StV1, connecting backward to the middle meningeal system. In any case, two dual branches of StV1 explain why the clefts can occur in the lateral-most zone of the orbit (Tessier zone 9).

StV2 follows its nerve as well to supply the dura of anterolateral temporal fosa, including the calvarial lamina of alisphenoid. Once its dural branch is given off, StV2 follows along with the nerve to supply it as its traverses foramen rotumdum. At no time does either V2 or StV2 enter the orbit.

V3, in contradistinction, does not give off a dural branch right away. It exits the skull immediately through foramen ovale, but once in the infratemporal fossa, its first branch is dural. V3 to the dura follows a recurrent trajectory to enter the skull in company with StV3, thus forming the *middle meningeal nerve and artery*. At the trigeminal ganglion, vascular anastomosis takes place between the "newcomer" StV3 and common origin of StV1 and StV2.

This anastomosis causes a sudden change in blood flow. Altered hemodynamic leads to the involution and ultimate disintegration of proximal stapedial. This occurs from tympanic ring (1) forward to the junction of VPA with inferior ramus and (2) forward to the trigeminal ganglion. The union of VPA and stapedial constitutes the "take-off point" of the hybrid internal maxilla-mandibular system. As V2 and V3 grow into their target territories, they are likely to induce from surrounding mesenchyme (by means of a VEGF-type mechanism) the vessels that eventually constitute the branches of IMMA. The same mechanism takes place in the orbit to produce the non-ocular V1 stapedial system.

Disintegration of the proximal embryonic stapedial artery causes shifts in blood flow in its distal derivatives. New anastomoses take place, resulting in the familiar textbook configuration. These will be later detailed, but for now, here is a quick summary to stay oriented. The meningeal system, originally a product of the anterior and divisions of superior ramus, is now supplied by the initial segment of internal maxilla-mandibular artery. The supraorbital branch of anterior division supplies all the adnexal structures of the orbit. All extraocular vessels are programmed by V1. The major named branches follow sensory branches leading out of the orbit. The fasciae of the extraocular muscles arise from midbrain neural crest and is supplied by V1; the arterial supply to all muscles follows the fasciae. Thus, extraocular fasciae are supplied by the StV1 system, while the muscle themselves are supplied from the ophthalmic axis (but the arteries are guided into position by the V1 fascia). Note that the motor nerves to the extraocular muscles certainly induce their own vascular supply (again from surrounding mesenchyme), but this does *not* correspond to muscular arteries, all of which arise from the stapedial ophthalmic.

The proximal stem of the supraorbital stapedial reanastomoses with vessels to the globe derived from the primitive ophthalmic. The final version of ophthalmic artery is thus an amalgam of vessels from primitive internal carotid with those from a "new-comer," the V1 stapedial. We shall see that the clinical spectrum of holoprosencephaly involves the entire axis of PrICA, that of micro-ophthalmia involves the primitive arteries of the optic anlage, while excesses and deficiencies of Tessier zones 10–13 involve V1-induced stapedial arteries.

The reader is advised at this point to study with care the summary diagram of cerebral circulation (Fig. 6.7). We have not made any preliminary comments regarding the developmental basis for the blood supply to the muscles of facial expression. Nor have we discussed the circulation involving the extradural branches of internal carotid, specifically, the petrous, sphenoid (cavernous), and orbitofrontal (ophthalmic). Details of these systems are best left to our final narrative.

Mechanisms of Blood Vessel Construction

Embryonic blood vessels develop from precursor *angioblasts* by three mechanisms.

1. *Coalescence of angioblasts.* This process occurs in situ. It explains the formation of the primordial large vessels: the omphalomesenteric (vitelline) vessels, the dorsal aortae, and the primordial hindbrain channels. In the process of forming tubles, angioblasts recruit local mesenchymal cells to form pericytes which ensheath the primary structures to form a stable tubular vessel. The pericytes function to control porosity and diameter of the vessels.

2. *Angioblast migration.* Other large blood channels, such as the endocardium, arise when angioblasts from a distal site migrate toward the vessel under construction.
3. *Vascular sprouting.* Segmental arteries such as those of the central nervous system and the intersegmental vessels of the body (vertebral, thyrocervical trunk, costocervical trunk, internal thoracic, and epigastric) arise as buds given off periodically by a parent artery.

We will first examine the cell sources of angioblasts in greater detail. Since these arise from mesoderm, the various compartments of mesoderm within the embryo will be discussed. This will include three important patterns of reorganization: (1) segmentation of paraxial mesoderm into somitomeres and somites; (2) informational flow from the CNS into mesoderm via the neural crest migrations; and (3) parasegmentation of the somites and its relationship to intersegmental arteries. Because development of cardiac tissues is concomitant with those dedicated to the vascular system, some remarks regarding the origins of the heart are required. We will finish by discussing the actual mechanisms of vessel formation.

Tissue Origins of Embryonic Blood Vessels

Blood vessels are mesenchymal structures. They cannot be made from ectoderm. The vertebrate cardiovascular system arises from a mesodermal cell source interacting in conjunction with endoderm. Embryonic vessels develop from four sources of mesoderm (Fig. 6.10).

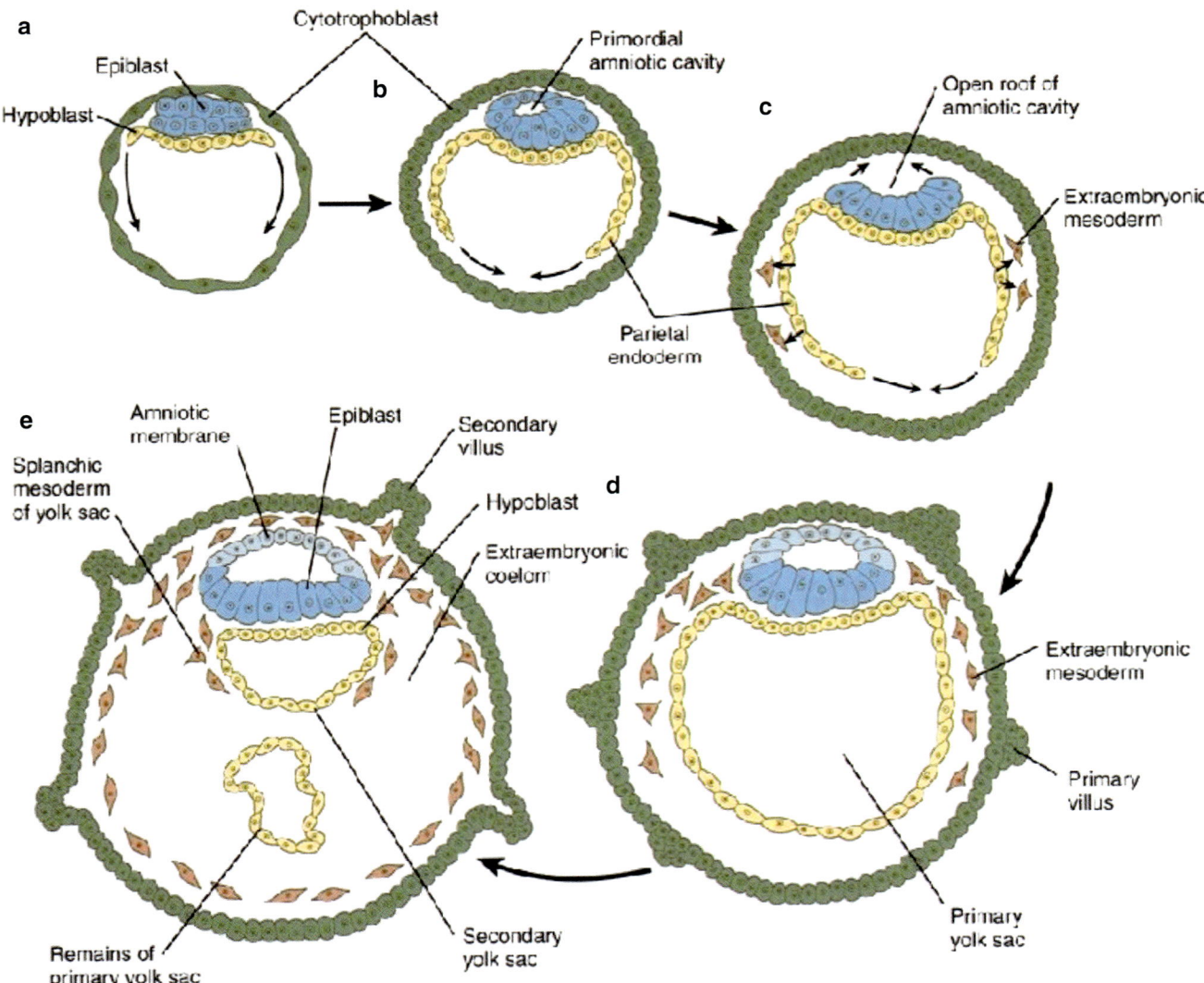

Fig. 6.10 Extraembryonic mesoderm spreads out to form hemangioblasts. (Syncytiotrophoblast not shown here). **A**, 6d = implantation starts; **B**, 7.5d = implantation complete, **C**, 8d = extraembryonic mesoderm; **D**, 9d = extrembryonic ectoderm = amnion; **E**, 2 weeks = EEM invades villi; surrounds primitive endoderm remnant (secondary yolk sac) > vitelline vessels. (Reprinted from Carlson BM. Human Embryology and Developmental Biology, 6th edition. St. Louis, MO: Elsevier; 2019. With permission from Elsevier)

1. *Primitive extraembryonic vessels* arise in situ at stage 5 from extraembryonic mesoderm that surrounds the entire conceptus (amniotic sac, the embryo proper, and the yolk sac). These vessels connect the developing embryo with the yolk sac and future placenta. They serve as a temporary source of nutrition until the heartbeat is initiated.

2. *Primitive intraembryonic vessels* arise with gastrulation at stage 7. The head plexus and primitive hindbrain channels supply the incipient central nervous system. The dorsal aortae supply the remainder of the embryonic body. These vessels combine endothelium from visceral lateral plate mesoderm with an external coat of paraxial mesoderm.

3. *Definitive extraembryonic vessels* result when gastrulation produces additional mesoderm that is extruded from the embryo. Within the body stalk, paired umbilical veins and arteries develop (the right umbilical vein disintegrates). This definitive extraembryonic mesoderm makes contact with preexisting primitive extraembryonic mesoderm. Furthermore, at the level of the body stalk (umbilical cord), *intraembryonic mesoderm becomes continuous with extraembryonic mesoderm*. The original body stalk is broadly based, but it narrows down as pregnancy proceeds. Maternal-fetal circulation (the communication between intra- and extraembryonic mesoderm) becomes confined to the umbilical vessels, with a single umbilical vein carrying oxygenated blood to the sinus venosus.

4. *Definitive intraembryonic vessels* (the cardiovascular system as we know it) arise diffusely within mesoderm. In particular, those of the visceral lamina of lateral plate mesoderm carry blood supply out to the viscera while those of the somatic laminae supply the extremities.

Anatomy of Mesoderm

In previous chapters, we have encountered many times the process of gastrulation: separation of the germ layers. Recall that after gastrulation mesoderm exists in three parallel swatches flanking the neural tube. Closest to the neuraxis is *paraxial mesoderm*, the source of the axial skeleton, dermis of the back, and all striated muscles. *Intermediate mesoderm* is the source of the genitourinary system. *Lateral plate mesoderm* splits horizontally into two layers, each of which relates to an adjacent epithelium. The dorsal layer, *somatic (parietal) lateral plate mesoderm* (LPMs), lies beneath the ectoderm. It generates the appendicular skeleton, participates with neural crest to form the somatic connective tissue system, and forms all remaining dermis of the body. The ventral layer, *splanchnic (visceral) lateral plate mesoderm* (LPMv), lies above the endoderm. From it arise the cardiovascular system, the smooth muscle of the gut and (again with neural crest) the visceral connective tissue system. *Extraembryonic mesoderm* is a technicality. It represents "extra" mesoderm that is pushed out from the embryo into the periphery and is the source for umbilical vessels (Figs. 6.11, 6.12, and 6.13).

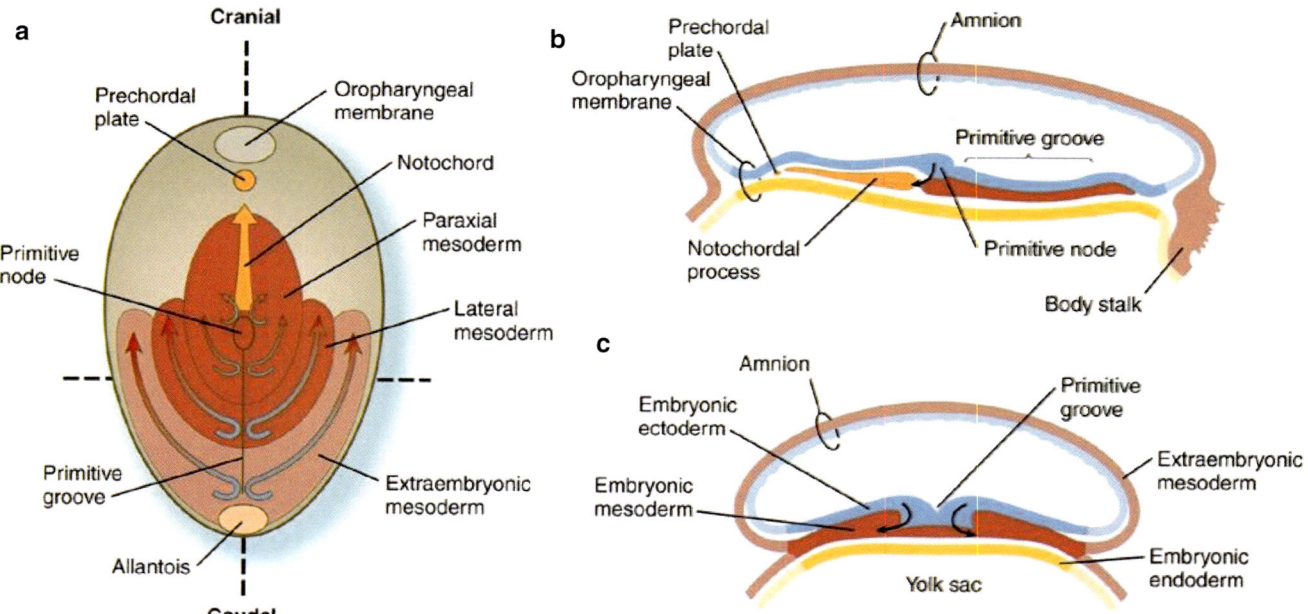

Fig. 6.11 Mesodermal fate determined by its migration pattern through the primitive streak. (**a**) Order of cell entry: (1) endoderm displaces hypoblast; (2) mesoderm cell fates: mesendoderm (notochord), PAM, Intermediate, LPM, Extraembryonic mesoderm. (**b**) Cells passing through primitive node into notochord. (**c**) XS through primitive streak: note displacement of **hypoblast** (light yellow) by **endoderm** (dark yellow). (Courtesy of Bruce M. Carlson, MD, PhD)

Fig. 6.12 Developmental fate of mesoderm. Blood vessels form in each of the mesodermal compartments. Ectoderm and endoderm are incapable of forming blood vessels. (Reprinted from Gilbert SF, Barressi MJF. Developmental Biology, 11th ed. Sunderland, MA: Sinauer; 2016. Copyright © 2016. Oxford Publishing Limited. Reproduced with permission of the Licensor through PLSclear)

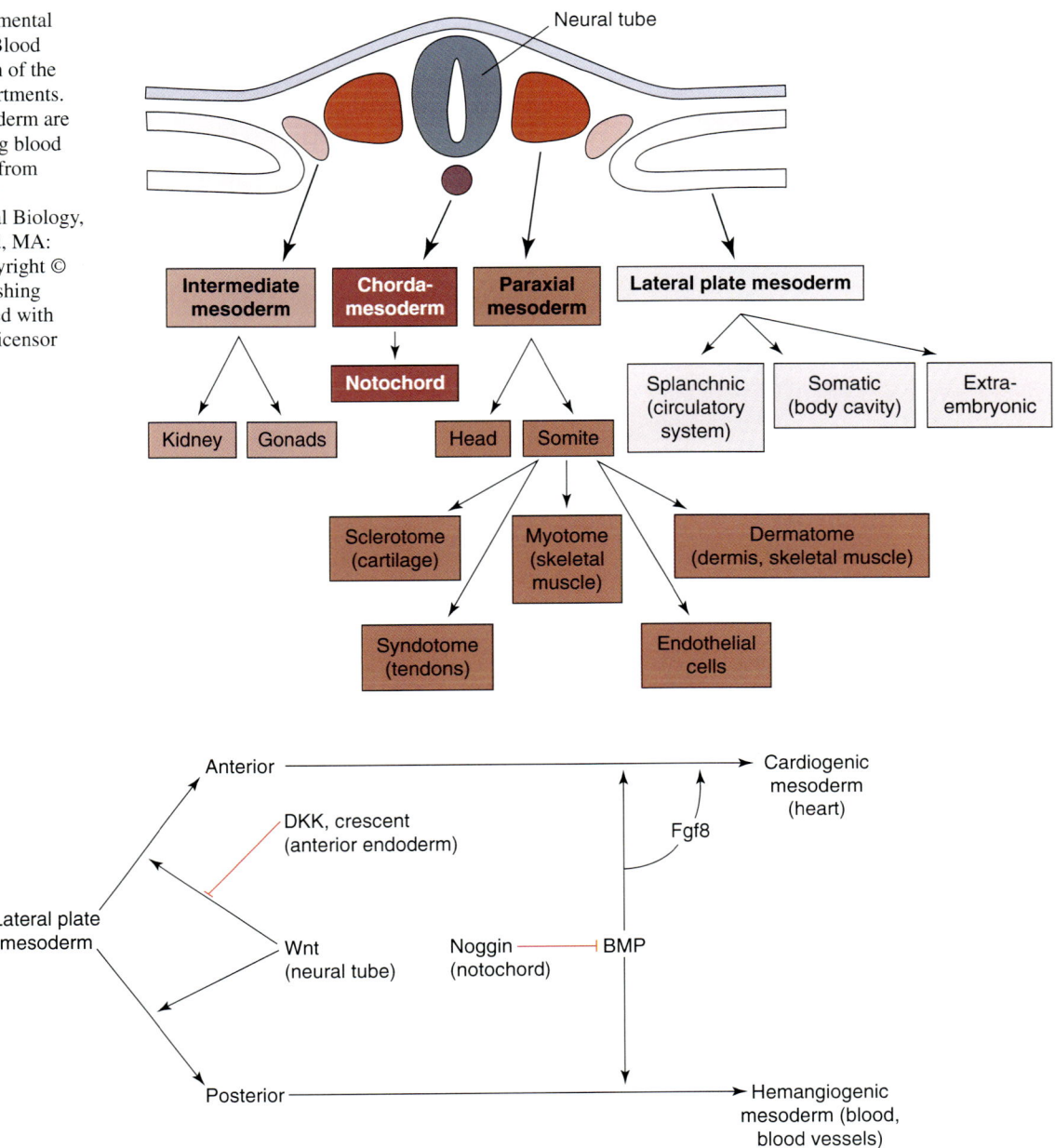

Fig. 6.13 Lateral plate mesoderm: cardiac vs. hemangiogenic. Pharyngeal endoderm induces the heart. Neural tube induces the vascular system. Model of inductive interactions involving the BMP and Wnt pathways that form the boundaries of the cardiogenic mesoderm. Wnt signals from the neural tube instruct LPM to become precursors of the blood and the blood vessels. In the anterior portion of the body, however, Wnt inhibitors (Dickkopf, Crescent) arising from the pharyngeal endoderm prevent Wnt from functioning. This allows later signals (BMP, Fg8) to convert LPM into cardiogenic mesoderm. BMP signals will also be important for the differentiation of hemangiogenic (blood, blood vessel) mesoderm. In the center of the embryo, Noggin signals from the notochord block BMPs. Thus, the cardiac and blood-forming fields do *not* form in the center of the embryo. (Reprinted from Gilbert SF, Barressi MJF. Developmental Biology, 11th ed. Sunderland, MA: Sinauer; 2016. Copyright © 2016. Oxford Publishing Limited. Reproduced with permission of the Licensor through PLSclear)

Let's review the physical extent of these mesodermal "swatches." At gastrulation[6–7], all arise from Hensen's node, starting at level r0 and moving backward. PAM exists in two forms: somitomeric and somitic. The first seven *somitomeres* are partially segmented hollow balls of mesoderm that contain muscles of the face and head. Once produced, they flank the brain from the optic placode to the otic placode (from the eye to the ear). They are innervated by cranial nerves II through IX. The eighth somitomere and all subsequent somitomeres are transformed into fully segmented *somites* having differentiated subunits (dermatome, myotome, and sclerotome). The first four *occipital somites* (formed from somitomeres 8 to 11) are innervated by cranial nerves X through XII. All remaining somites innervated by the peripheral nervous system.

Intermediate mesoderm produces the kidney in three successive iterations (pronephros, mesonephros, and metanephros), the details of which are beyond the scope of this chapter. What is relevant for us is the following. In humans at day 22, beginning with the second cervical somite, signals from paraxial mesoderm induce intermediate mesoderm to form segmental *pronephric tubules*. These consolidate longitudinally into a *pronephric duct*. Thus, the most primitive kidney in the human embryo has a neuromeric relationship to the neck.

Lateral plate mesoderm in the unfolded embryo is shaped like a horseshoe. Although its most anterior cells arise from Hensen's node at r0, LPM expands outward to limits of the embryo. Thus, it sweeps forward around the forebrain, avoiding the "no-man's land" of the buccal-pharyngeal membrane. Once the splitting of lateral plate mesoderm is complete and has split, two layers are formed. *Somatopleure* refers to the combination of ectoderm + LPMs. *Spanchnopleure* (viscero-pleure) refers to the combination of endoderm + LPMv. Between these two layers, a space exists, the *coelom*, which extends from the future neck down through the body cavity. In the head region, an unnamed potential space also exists between the layers. With embryonic folding, the right and left coeloms fuse into a common *intraembryonic coelom* within which the viscera develop. Subsequently, somatic LPM sends tissue extensions inward that subdivide the intraembryonic coelom into pericardial, pleura, and peritoneal cavities. This mechanism has been conserved throughout vertebrate evolution.

Fig. 6.14 Nicole LeDouarin. (Reprinted from Wikimedia. Retrieved from: https://commons.wikimedia.org/wiki/File:Nicole_Marthe_Le_Douarin,_2013_(cropped).jpg. With permission from Creative Commons License 2.0: https://creativecommons.org/licenses/by-sa/2.0/deed.en)

Anatomy of Cranial Neural Crest

Three decades of work by Le Douarin using the quail-chick chimera model have provided a wealth of information about the behavior of neural crest from various locations along the neuraxis. By exploiting interspecies differences in chromatin, sophisticated fate maps have been constructed demonstrating what derivatives are formed from what neural crest population. Contributions of craniofacial mesoderm to vessels have also been assessed. Neural crest migration in mammals takes place very early, well before closure of the neural tube. At stage 8, presomitic mesoderm (the somitomeres) is populated by NCCs. By stage 9, when the first somites are visible, NCCs have moved into the substance of the first pharyngeal arch (Figs. 6.14 and 6.15).

Mesoderm and endothelial cells become separate lineages very early in time, by stage 8–9. Within cephalic mesoderm (somitomeres), a subset of cells express VEGFR2 (a tyrosine kinase receptor for vascular endothelial growth factor). These are fated to become endothelial cells—the building blocks of blood vessel walls. The derivation of pericytes from neural crest has been conclusively demonstrated. Thus, the mesenchyme that will make up the face and pericranial structures contains an admixture of cell populations. Clonal culture studies suggest that, in vitro, migrating cranial NCCs represent, at first, a heterogeneous population of precursor cells that will, under the influence of local cues, become progressively restricted to form distinct lineages. Although it is commonly held that NC caudal to the CNS is more tightly restricted than cranial NC, more recent data show that it can become unrestricted, producing expected bone and cartilage. This may explain pathologic conditions in which these tissues appear ectopically. Such "gain of function" or derepression is also the mechanism for neurofibromatosis.

Craniofacial anatomy is remarkable for the paucity of mesodermal derivatives. Apart from the bones of the cranial base and the striated muscles, the bulk craniofacial mesenchyme is neural crest. Thus, NCCs make up all the epidermis innervated by V1, all dermis supplied by V1–V3, all connective tissues including membranous bone, cartilage, fascia, fat, and the meninges.

Fate mapping experiments in avian embryos demonstrate that the smooth muscle walls of cephalic blood vessels, the aortic arch arteries, and the great vessels are made from neural crest. These findings are equally applicable in mammals. More recent work by Etchevers using the chimera system

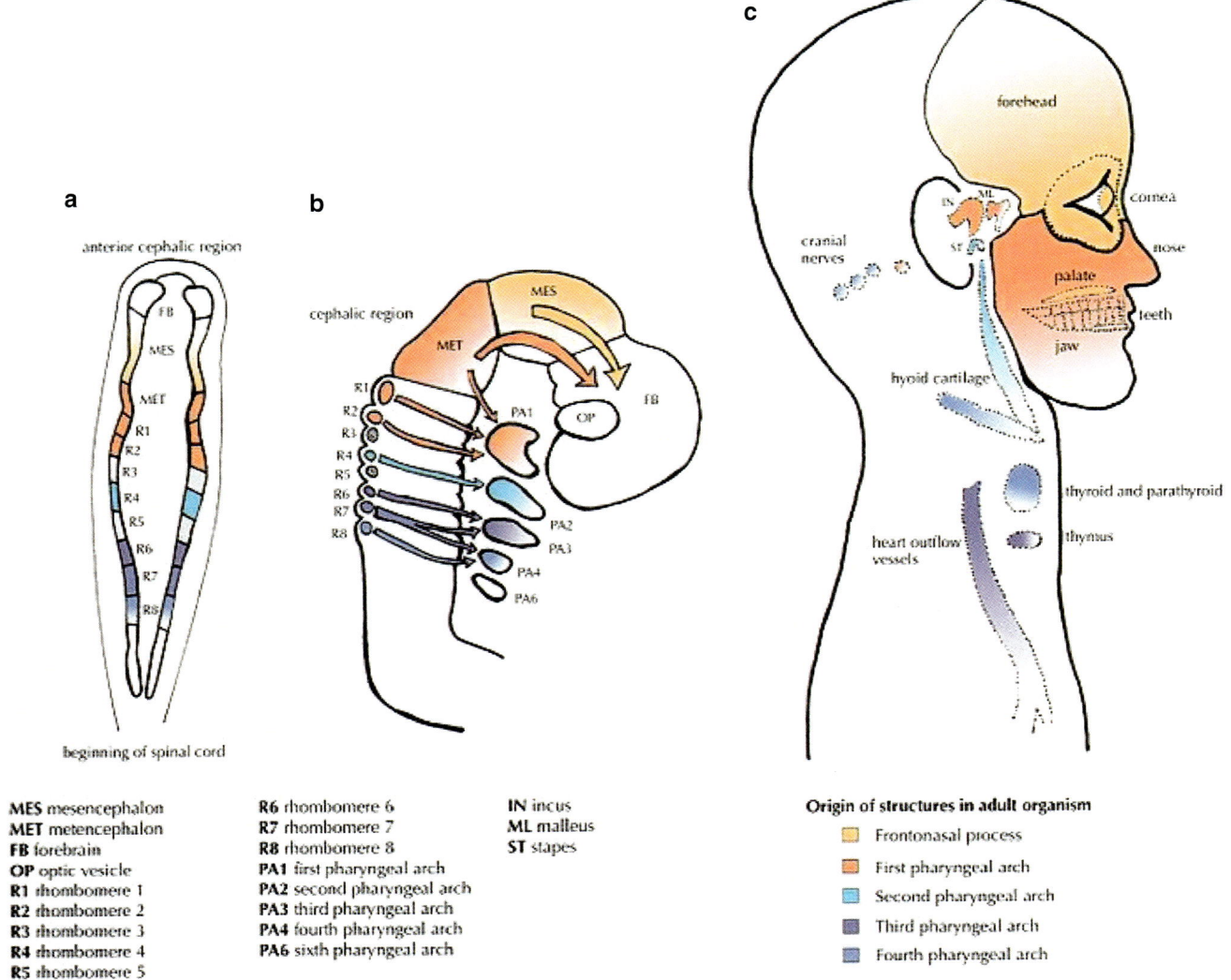

MES mesencephalon	**R6** rhombomere 6	**IN** incus
MET metencephalon	**R7** rhombomere 7	**ML** malleus
FB forebrain	**R8** rhombomere 8	**ST** stapes
OP optic vesicle	**PA1** first pharyngeal arch	
R1 rhombomere 1	**PA2** second pharyngeal arch	
R2 rhombomere 2	**PA3** third pharyngeal arch	
R3 rhombomere 3	**PA4** fourth pharyngeal arch	
R4 rhombomere 4	**PA6** sixth pharyngeal arch	
R5 rhombomere 5		

Origin of structures in adult organism

- Frontonasal process
- First pharyngeal arch
- Second pharyngeal arch
- Third pharyngeal arch
- Fourth pharyngeal arch

Fig. 6.15 Etchevers Neural crest contributions to the craniofacial skeleton. NC migration into craniofacial tissues. (Reprinted from Etchevers HC, Couly GF, Le Dourain NM. Morphogenesis of the branchial vascular sector. Trends Cardiovasc Med 2002; 12(7): 299–304.with permission from Elsevier)

and markers for alpha smooth muscle actin showed parenchymal cells of forebrain meninges to contain pericytes. Interaction between endothelial cells and pericytes is critical for the formation of an intact blood vessel lamina. Outer layers such as tunica media and adventitia contain additional cell types. Graft analysis has shown that (in addition to pericytes) smooth muscle per se and connective tissue cells of the major cephalic arteries are derivatives of cephalic NCCs. Mapping of craniofacial vasculature discloses a striking pattern: the original anteroposterior origin of a NCC from the neural folds corresponds to its final distal-proximal distribution in a defined subset of cephalic blood vessels. ... The "pharyngeal arch vascular sector" is a distinct circuit of blood vessels connecting the ventral aortic outflow tract and pharyngeal arches. The vessels correspond to the forebrain, face, and jaws.

Careful study of Fig. 6.14 demonstrates the above observation. The vessel walls of all arteries (and veins) corresponding to (1) the aortic arch system or (2) the stapedial artery system contain cephalic neural crest. Those vessels not belonging to those systems are entirely constructed from mesoderm. This sets up a stark contrast: forebrain, midbrain, and craniofacial structures are supplied by neural crest arteries, while cerebellum, hindbrain, and occipital structures are supplied by arteries made from mesoderm.

Mapping by Etchevers et al. shows the relationship of NCCs to their corresponding neuromeres. Let's review the data. Diencephalic (posterior forebrain) arteries arise from p4/p3 to p1. Midbrain arteries arise from m1 to m2. Arteries of rostral hindbrain arise from r2 to r7, while those supplying caudal hindbrain originate from r8 to r11 (Fig. 6.16).

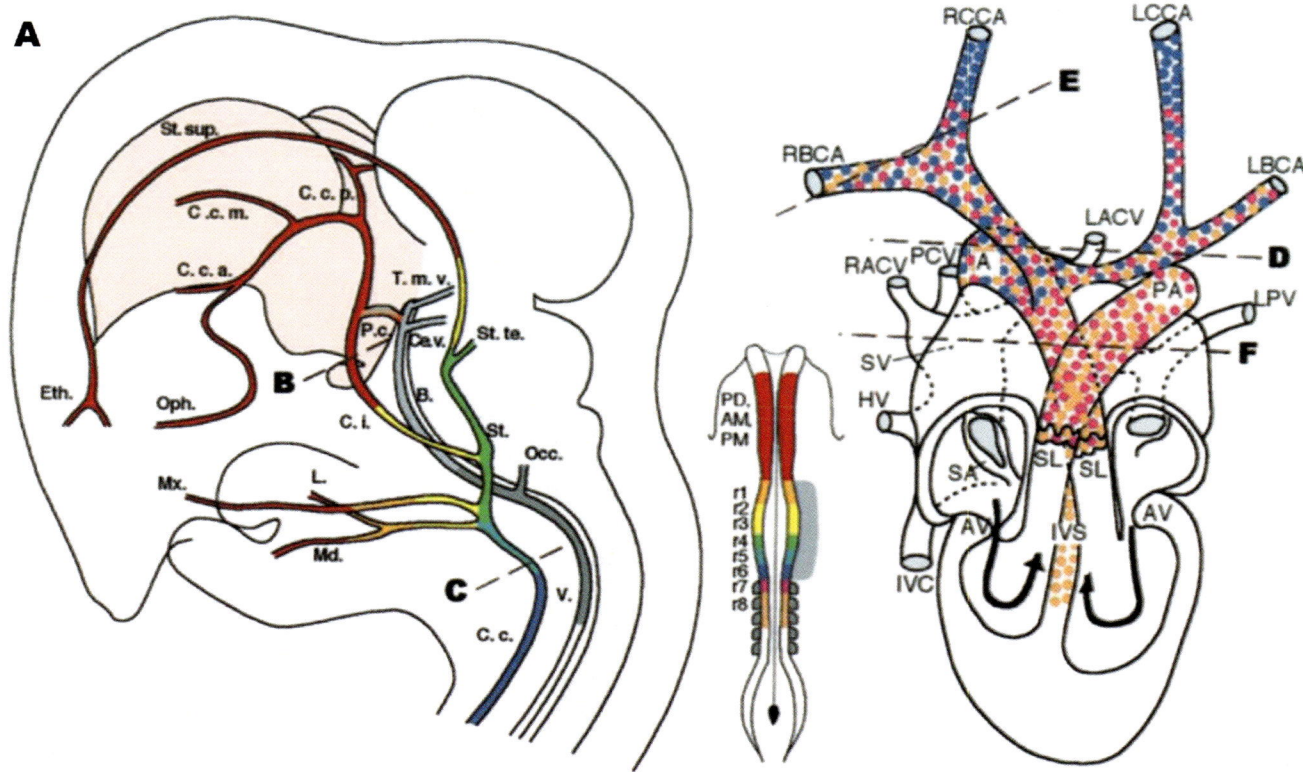

Fig. 6.16 Etchevers' work showing incorporation of NC into the arterial walks. (Reprinted from Etchevers HC, Couly GF, LeDoarin NM. Morphogenesis of the branchial vascular sector. Trends Cardiovasc Med 2002; 12(7): 299–304. With permission from Elsevier)

Although neural crest cells in register with these neuromeres synthesize many structures, the actual material of cephalic arteries follows an entirely different pattern. Ablation experiments by the same group demonstrate that removal of neural crest from p4 back to r3 destroys all frontonasal structures, including the forebrain, medial oculomotor muscles, and anterior pituitary gland (r0). Dermal structures of this entire zone arise from this neural crest. Such dermis constitutes the termination of all individual branches of the extracrania stapedial system: StV1, StV2, and StV3. Dura mater originating from r0 to r3 is supplied by meningeal vessels arising from the same zones.

Segmentation: The Mathematical Basis of Intersegmental Arteries

We now turn our attention to the process of *resegmentation* because this explains the relationship between somites, vertebrae, nerves, arteries, and muscles. We shall refer to these concepts again and again, so now is the best time to get them clarified.

Anterior sclerotome permits the passage of neurons, whereas posterior sclerotome blocks them. The sclerotomes then split apart and recombine to produce vertebrae using the following formula: Vertebra N is derived from posterior sclerotome of N-1 plus the anterior sclerotome of N. Thus, somite 5 combines with somite 6 to produce the body of atlas (Fig. 6.17).

The nerve that runs under atlas in C2 is *segmental*. It supplies muscles having a primary insertion at C1. The artery associated first with C1 is *intersegmental*. It connects spinal neuromeres c1 and c2 with the aorta. Consequently, first cervical intersegmental passes through the interspace cervical somites 1 and 2 (S5–S6). Each intersegmental artery has dorsal and ventral branches.

Recall that mammals have 8 cervical somites from which are produced 8 vertebrae, *not* 7. Cervical somite C1 recombines with occipital somite O4 to produce the **proatlas**, the original articulating element between the skull and the vertebral column. Proatlas contributes the condyles to the exoccipital bone, a ligament uniting basioccipital with the tip of the dens and the upper 1/3 of the dens. Thus, the artery running above the atlas is the true first intersegmental artery. Because its morphology is different, its origins have been lost (Figs. 6.18 and 6.19).

One could legitimately ask: "Wait a minute…I thought the intersegmentals came from the dorsal aortae…and later the aorta. But when I look in the neck, all the intersegmentals are coming off the vertebra. Obviously, there is no aorta in

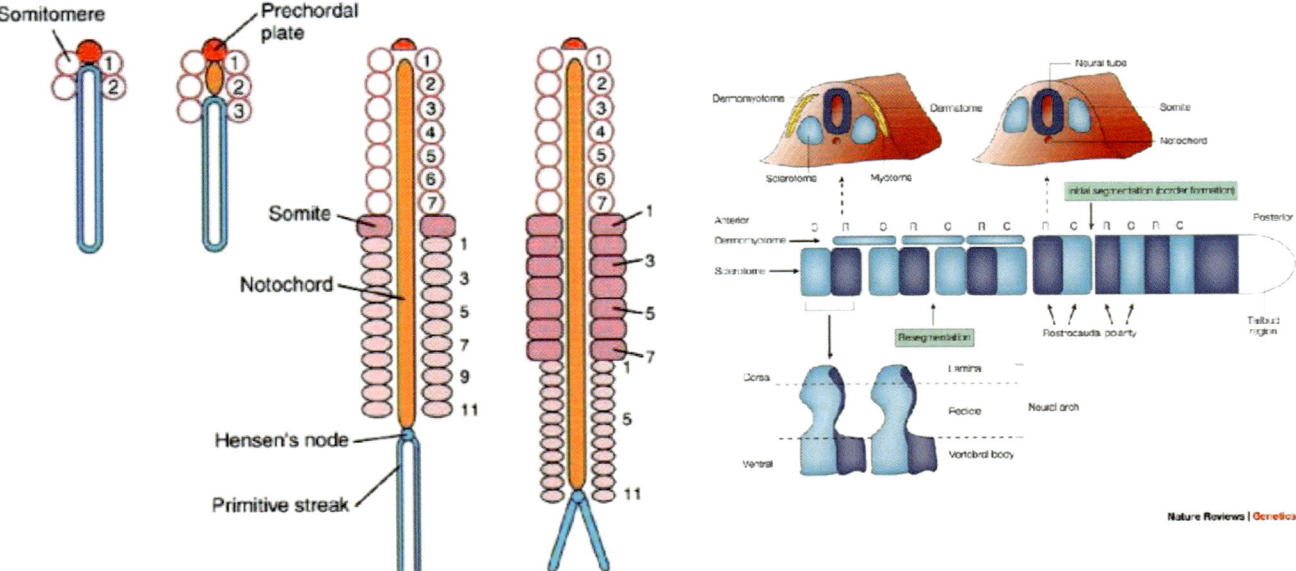

Fig. 6.17 Transformation from somitomere and somites takes place at Sm8 > S1. (1) Sm1–Sm7: muscles to eye and pharyngeal arches 1–3 (including superior constrictor). (2) Sm8–Sm9: pharynx and larynx. (3) S1–S4: posterior fossa and tongue muscles. Note process of PARASEGMENTATION. (Reprinted from Jacobson AG. Somitomeres. Principles of Medical Biology. JAI Press Greenwich, CT; 1998: 209–228. With permission from Elsevier)

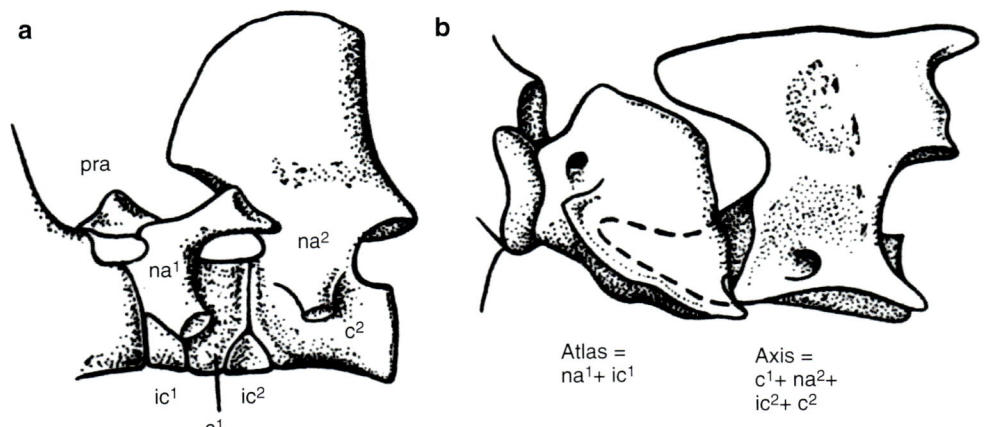

Fig. 6.18 Consolidation of the cranio-vertebral joint. Primitive reptile *Ophiocodon* compared with mammal: c^1, center of atlas; c^2, centra of axis; ic^1 and ic^2, intercentra of the same; na^1 and na^2, neural arches of the same; *pra*, proatlas. In the higher reptilian line, and all mammals, pro- atlas becomes incorporated into the skull as foramen magnum, the occipital condyles, and the cranial 1/3 of the dens. (Reprinted from Romer AS. Vertebrate Paleontology, 3rd ed. Chicago, IL: Univ of Chicago; 1966. With permission from University of Chicago Press)

the neck. What's going on here?" To answer this, we must anticipate our discussion of how the vertebrals develop in the first place. Suffice it to say that, for a time prior to descent of the heart, paired dorsal aortae are sent off to each neuromere. The vertebrals were constructed as *intersegmental branches* intermediate longitudinal anatomoses among the cross-links. When the dorsal aortae are stripped away from the neck to follow the heart down into the chest, the ventral links disappear, leaving the vertebrals and their proximal links untouched.

Some of the intersegmental arteries assume very significant roles. Seventh cervical intersegmental artery runs between the seventh cervical and the first thoracic vertebrae. It then runs undercover of the clavicle becoming the subclavian artery. Curiously enough, the clavicle is the first bone of the body to undergo ossification. Fifth lumbar intersegmental artery runs between the fifth lumbar vertebra and the sacrum. It will give rise (in combination with the umbilical artery) to the common iliac arteries.

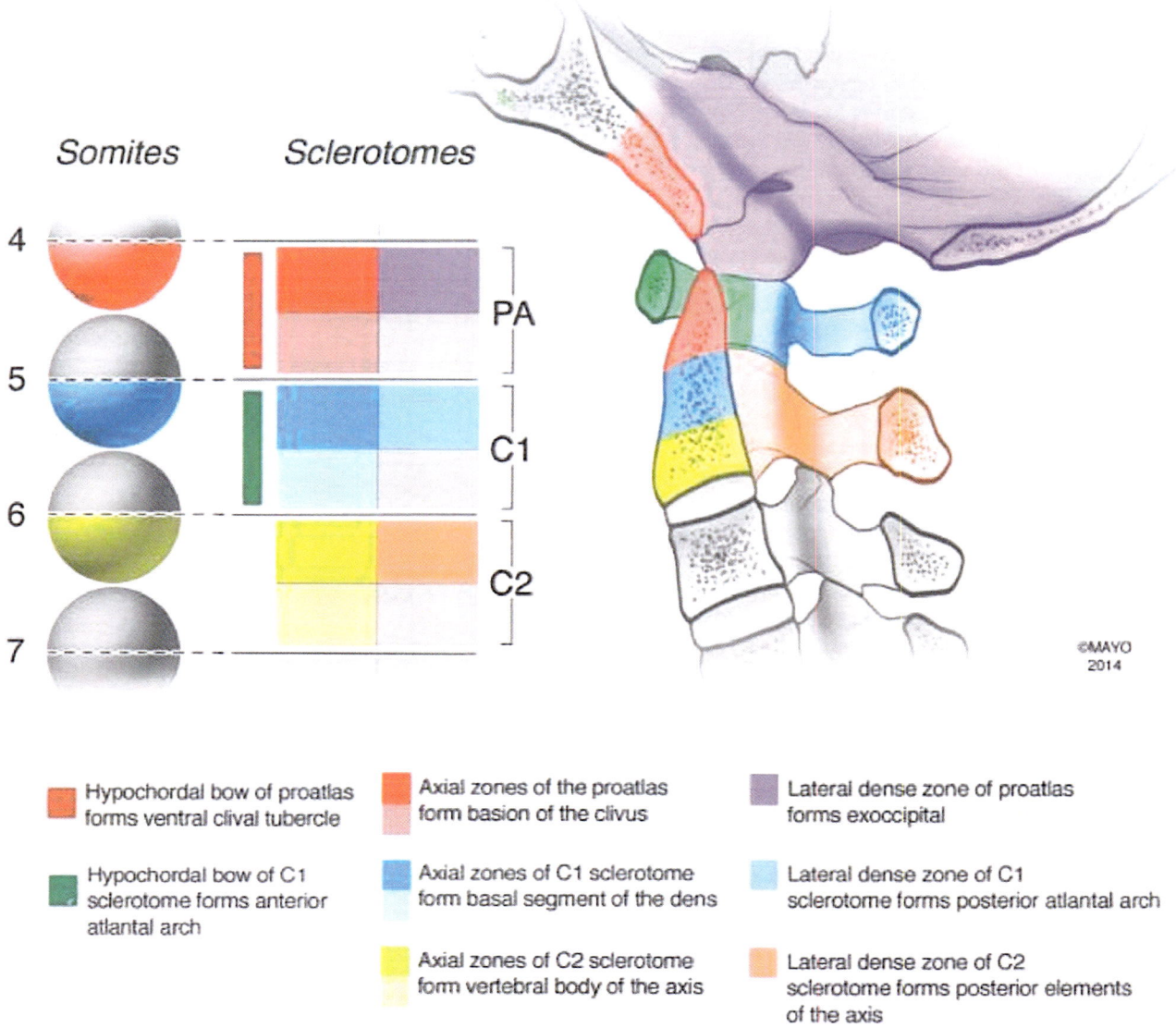

Hypochordal bow of proatlas forms ventral clival tubercle

Axial zones of the proatlas form basion of the clivus

Lateral dense zone of proatlas forms exoccipital

Hypochordal bow of C1 sclerotome forms anterior atlantal arch

Axial zones of C1 sclerotome form basal segment of the dens

Lateral dense zone of C1 sclerotome forms posterior atlantal arch

Axial zones of C2 sclerotome form vertebral body of the axis

Lateral dense zone of C2 sclerotome forms posterior elements of the axis

Fig. 6.19 Parasegmentation creates vertebra body (but not neural arch) from the cranial half of a somite and the caudal half of somite preceeding it. First three occipital somites are fused with the cranial fourth somite. First true cervical vertebra, the *proatlas* is caudal S4 and cranial S5. (Reprinted from Pang D, Thompson DNP. Embryology and bony malformations of the craniovertebral junction. Childs Nerv Syst 2011; 27(4): 523–564. With permission from Springer Nature)

What happens to the stems of arteries retained in the hind-brain? Fusion of the four occipital vertebrae (discussed in the chapter on bone) makes this difficult to ascertain. Furthermore, the neurovascular anatomy of the early embryo, i.e., between the longitudinal neural arteries and the rhombo-encelphalon, is undocumented. Vessels to individual neuromeres and cranial nerve nuclei exist without a doubt. Are these segmental or intersegmental? Does it matter? These questions await further research.

Summary of the Circulations: Two Types of Circulation

Extraembryonic Circulation

The development of the cardiovascular system in mammals cannot be understood without taking into account the formation of the extraembryonic circulation…those blood vessels connecting the embryo with the mother. How this takes place will be discussed in detail in the next section of this chapter, but a few details are relevant at this juncture. First, development of extraembryonic mesoderm takes place one or two stages prior to that of the intraembryonic mesoderm. This makes total sense as a mechanism of gas exchange with the mother and cells to transport oxygen must precede heartbeat.

At stage 5, prior to gastrulation, the two-layer embryo has epiblast and hypoblast, otherwise known as the primitive endoderm. Hypoblast quickly expands to completely line the bastocoele cavity. It subsequently produces an additional layer, interposed between it and the external cytotrophoblast wall: *extraembryonic mesoderm*. This spreads in all directions, eventually intervening between the roof of the amniotic sac and the cytotrophoblast. At stage 6, extraembryonic mesoderm eventually invades the previously solid core of primary cytotrophoblastic villi, converting them into *secondary villi*. Blood vessels form within the *tertiary villi*. At stage 7, the cytotrophoblastic columns make contact with the uterine decidua basalis; these are known as *anchoring villi*. Two important conditions are now met: (1) contact is established with maternal circulation; (2) yolk sac vessels are now formed (the vitelline aa. and vv. in birds and omphlomesenteric aa. and vv. in mammals) (Figs. 6.20 and 6.21).

An *additional source of extraembryonic mesoderm* arises from the embryo itself. During gastrulation at stage 7, epiblast cells ingressing the posterior-most region of the primitive streat produce *intrinsic* extraembryonic mesoderm. This is pushed outward into the body stalk by the lateral plate mesoderm where it makes contact with the extrinsic extraembryonic mesoderm.

Intraembryonic Circulation
Paraxial versus lateral plate derivatives.

Prior to the development of a functional cardiovascular system, the conceptus survives on an ebb-and-flow exchange of nutrients and gases through two primitive sets of vessels. The central nervous system is supplied via a *head plexus* to the prosencephalon and mesencephalon and *primordial hindbrain channels* to the rhomboencephalon. The remainder of the embryonic body is supplied by paired *dorsal aortae*. Although both sets of vessels will be replaced by definitive arterial structures, their importance cannot be overemphasized.

As a general rule, ectoderm cannot form vessels. CNS circulation depends upon an intimate contact between mesenchyme and neural tissue. The meninges consist of a deep layer (pia mater and arachnoid) derived from neural crest and a superficial layer (dura mater) derived from paraxial mesoderm admixed with neural crest. Head plexus and the primordial hindbrain channels lie beneath the pia matere and arachnoid (vide infra). Homeotic signals shared between each neuromere and its corresponding mesenchyme ensure that each zone of neural tissue receives blood supply. Expansion of cerebrum and cerebellum is accompanied by complex in-foldings that carry the vasculature deep into each

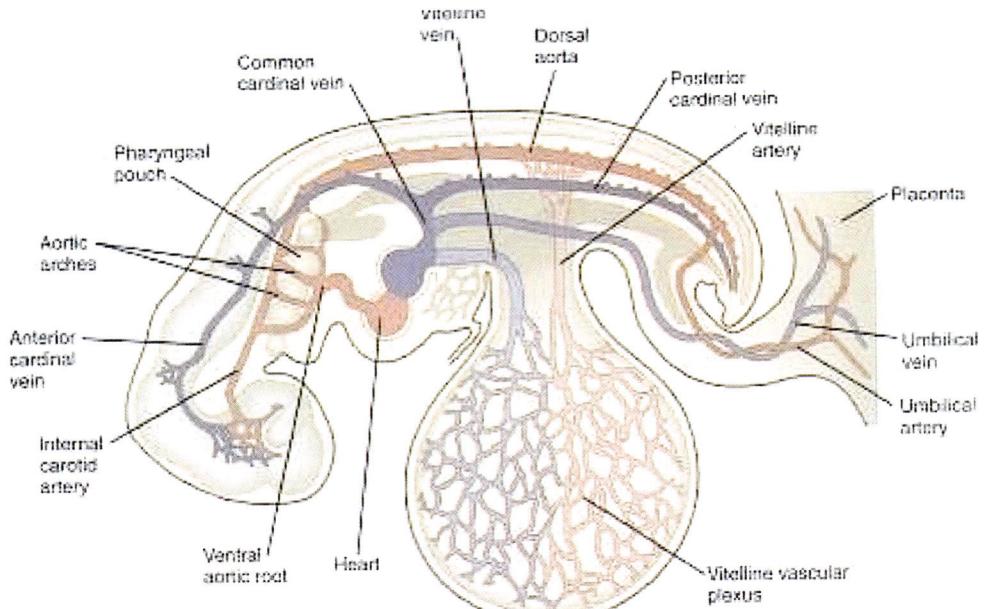

Fig. 6.20 Composite diagram of embryonic circulation at 4 weeks showing: three sources of inflow into the primitive atria: from the embryo (anterior and posterior cardinal veins), from the yolk sac (vitelline veins), and from the placenta (umbilical; veins). Note: the vitelline veins are quickly superceded by the umbilical veins. Schematic drawing of embryonic circulations: BMC 6.26. Dual supply from kolk sac (vitelline vessels) and placenta (umbilical vessels). Vitelline system quickly superceded by umbilical system veins. (Reprinted from Carlson BM. Human Embryology and Developmental Biology, 6th edition. St. Louis, MO: Elsevier; 2019. With permission from Elsevier)

Fig. 6.21 Extraembryonic mesoderm surrounds embryo and is incorporated into the placenta. Between weeks 2 and 3, extraembryonic mesoderm invades into the trophoblast to create villi, the presence of which at 3.5 weeks defines the *chorion*. (Reprinted from Carlson BM. Human Embryology and Developmental Biology, 6th edition. St. Louis, MO: Elsevier; 2019. With permission from Elsevier)

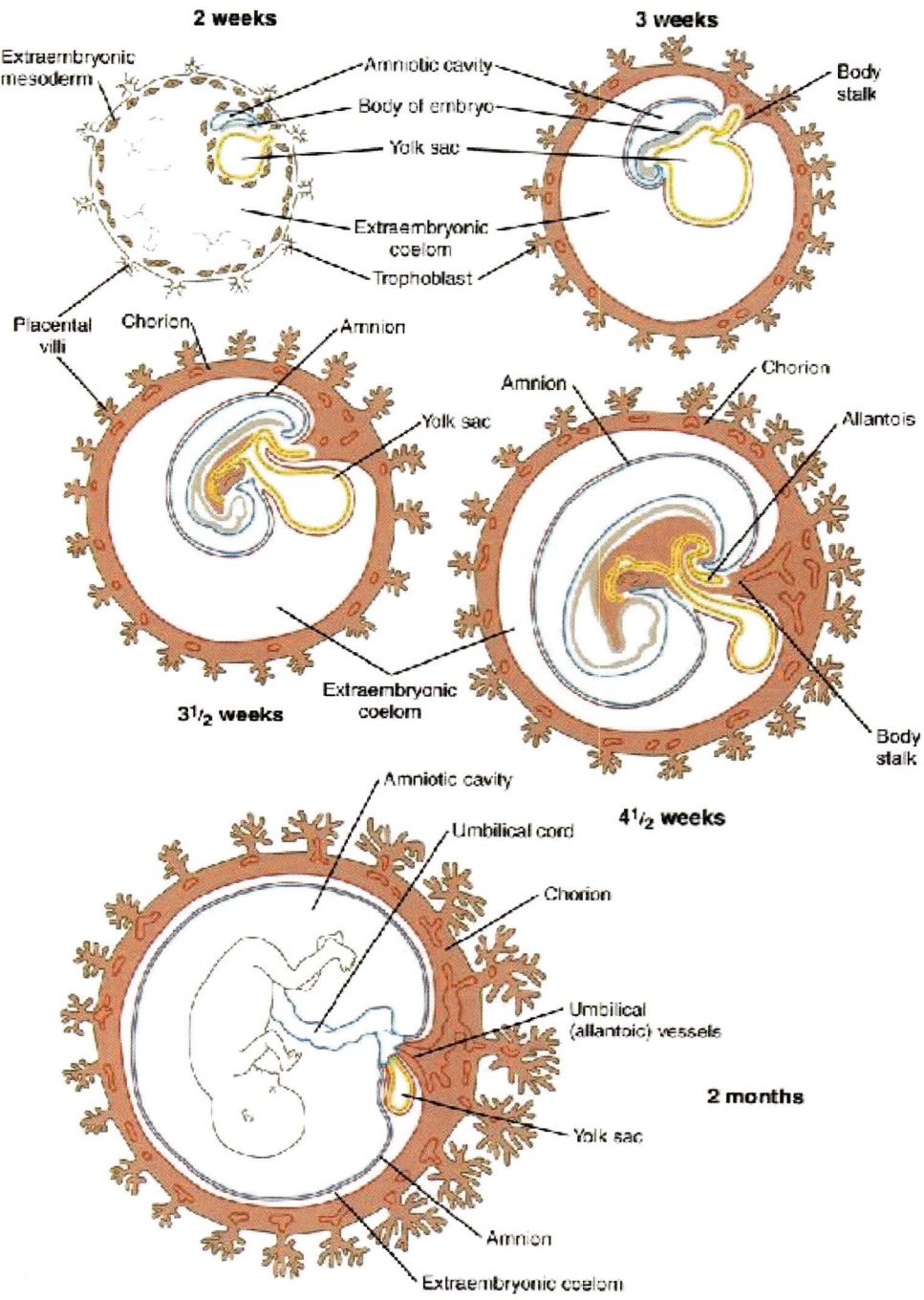

crevice of the brain. Ultimately, these "primitive" arterial systems connect with the "modern" circulatory structures derived purely from lateral plate.

Primitive vessels (head plexus, primordial hindbrain channels, and dorsal aortae) are unique in that their endothelial lining comes from visceral lateral plate mesoderm, but their walls arise paraxial mesoderm. As we shall see, dorsal aorta forward from the third aortic arch become the common carotid, while those segments backward from the fourth aortic arch will fuse in the midline. Two cell sources make up the definitive aorta: paraxial mesoderm is dorsal and lateral plate mesoderm is ventral.

Derivatives of visceral lateral plate mesoderm also arise from distinct regions. In early amniote gastrula, the heart arises from about 50 precursor cells located in the epiblast next to the rostral primitive streak. These cells will migrate early in gastrulation to take up a position in the anterior LPMv. The remaining vessels arise from posterior LPMv (Fig. 6.22).

Vitelline vs. Umbilical Vessels

Review of Carlson's circuit diagram (Fig. 6.20) shows vitelline vein (light blue) connecting over the liver to supply the atria (dark blue). The vitelline artery descends from the

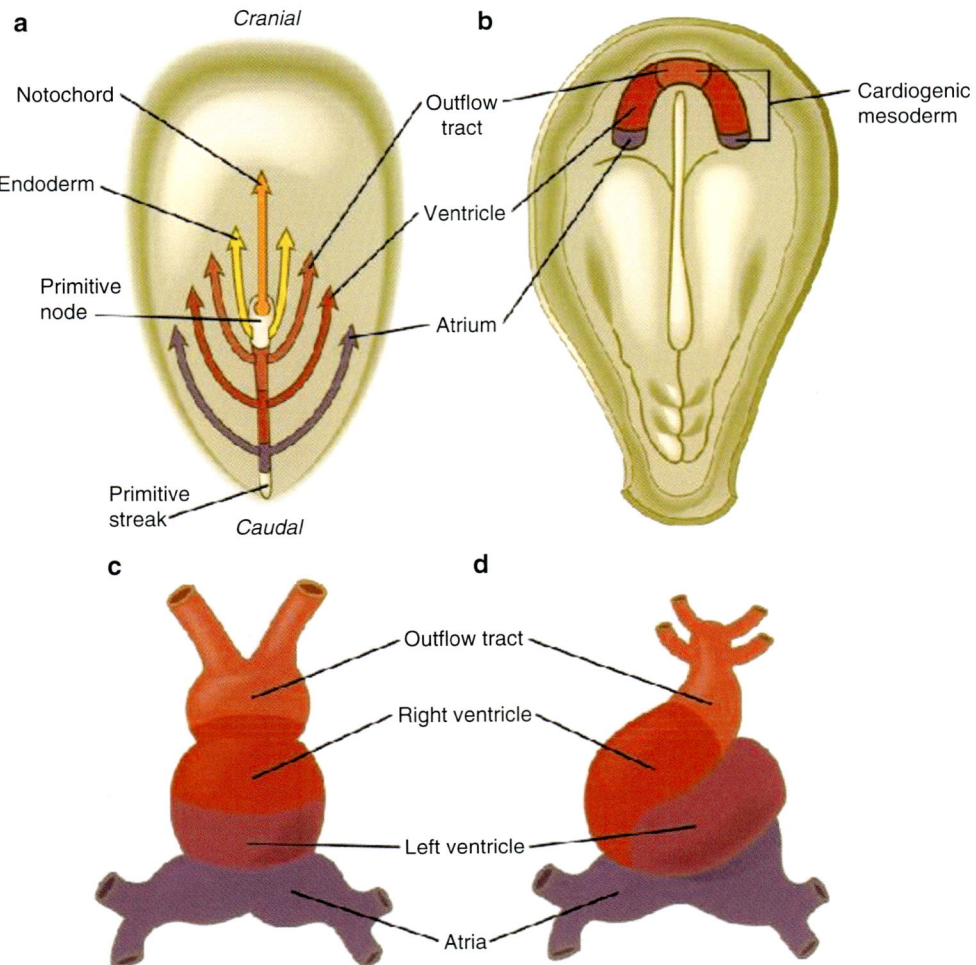

Fig. 6.22 Cardiovascular mesoderm forms in visceral lateral plate mesoderm. Cardiogenic cells of the *primary heart fields* flow forward from a specific location to form the cardiac crescent. Those from the primitive node go the most anterior zone and will form the outflow tract (orange); those of the midsreak will become the ventricles; posterior cells form the atria. *Secondary heart field develops* within the crescent. The anterior portion (green) completes the outflow tract. It causes primary outflow tract to be modified as the right ventricle. The posterior part of the secondary heart field contributes to the atria. Although car-diogenic mesoderm is considered lateral plate in front of the brain, the atria are in continuity with the dorsal aortae from head mesoderm (PAM and LPM). Note: Dorsal aortae running up embryonic axis and either side of the midline plug into the outflow tracts (orange). Atria (purple) receive venous circulation. A note about lateral plate mesoderm: posterior to Hensen's node (r0) LPM is *not split*. Separation of LPM begins again at level c1. (Reprinted from Carlson BM. Human Embryology and Developmental Biology, 6th edition. St. Louis, MO: Elsevier; 2019. With permission from Elsevier.)

future abdominal section of dorsal aorta to send venous drainage from the embryo to the yolk sac. At the same time, a second set of vessels connects the mammalian embryo to the placenta via the body stalk. Umbilical vein (light blue) brings in oxygenated blood to the common cardinal vein (dark blue). Once again, this takes place just above the liver. Umbilical artery connects dorsal aorta caudal to the allantois, i.e., in the tail fold.

Note that the allantois (light green) is diverticulum of the hindgut. It projects into the umbilical cord per se. Note also the physical presence of the tail fold, the tip of which pushes down on the cord. The "notch" between the cord and the tail fold contains the termination of the hindgut. It represents the future cloacal membrane.

So what happens to the structures in the fetus? We conclude our section by consideration of embryonic folding. At stage 7 (gastrulation), folding has not occurred because the germ layers are coming into being. At stage 8, tissue proliferation within the germ layers sets the process of folding in motion. The yolk sac becomes constricted. Its "neck" tightens in purse-string like fashion to form the midgut. The 18-day embryo demonstrates anterior and posterior intestinal portals. At 30 days, the tail fold has pushed the umbilical cord into intimate contact with the yolk sac. Note that these structures remain physically separate. The amniontic membrane is shown here cut…but this structure will turn out to be very important (Fig. 6.23).

Figure 6.21 shows the fetal period. Expansion of fluid within the amniotic cavity will eliminate the extraembryonic coelom. The amniotic membrane fuses to the call of the chorion and envelops the combined structures of yolk sac and umbilical cord. Contact between the yolk sac and the gut is

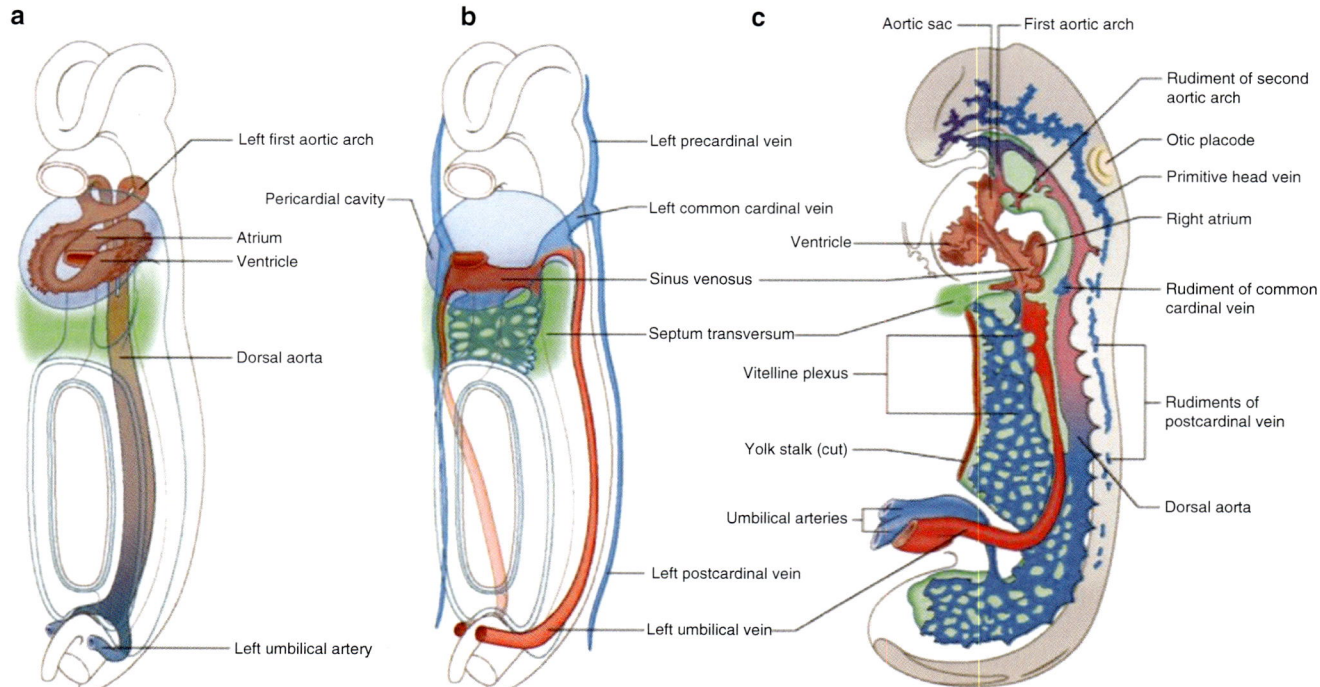

Fig. 6.23 Primitive embryonic circulation at stage 9. (**a**) Continuity of first aortic arches with truncus arteriosus. (**b**) Atria receive three sources of blood (1) vitelline circulation that break up in the liver and create the portal system; (2) common cardinal veins from the body, and (3) oxy- genated blood from umbilical veins—right one involutes. (Reprinted from Standring S (ed). Gray's Anatomy 40th ed. New York, NY: Churchill Livingstone; 2008. With permission from Elsevier)

Fig. 6.24 Three sets of embryonic veins empty into the sinus venosus. (Reprinted from Carlson BM. Human Embryology and Developmental Biology, 6th edition. St. Louis, MO: Elsevier; 2019. With permission from Elsevier)

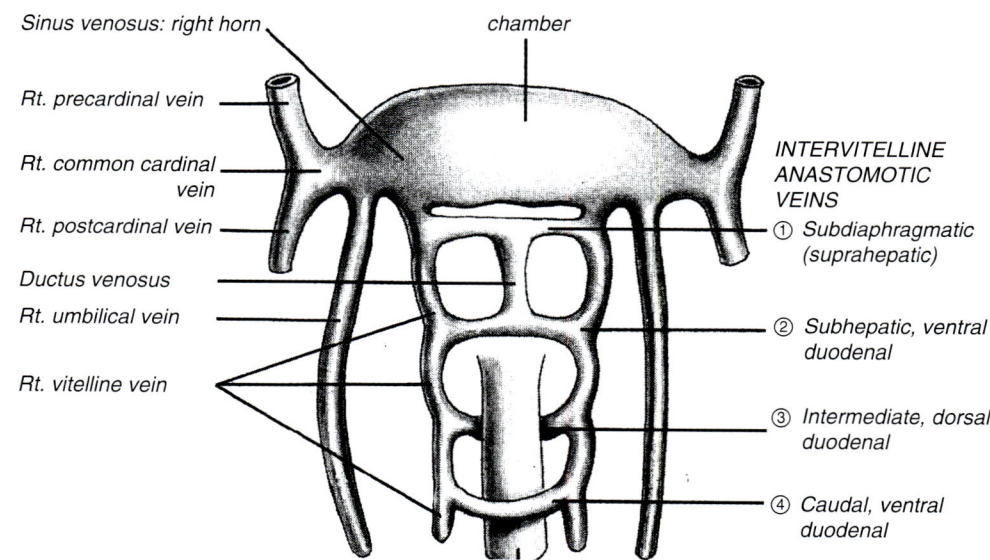

eventually lost. In some cases, it persists as a fibrous cord or outpouching from the small intestine known as *Meckel's diverticulum*. The proximal portion of vitelline vessels yolk sac persists as vessels of the midgut.

The final disposition of the vitelline veins is as follows. They all drain into the sinus venosus of the heart, just above the rapidly growing liver come together into the sinus veno- sus just above the liver. Vitelline veins form anastomosing networks within the substance of the liver. Outside the liver they are associated with the duodenum. The hepatic vitel- line plexus becomes a capillary network that becomes right and left hepatic veins. These drain first into the sinus veno- sus and, ultimately, in association with the portal vein, the inferior vena cava. Proximally, the vitelline veins form an anastomotic network around the duodenum and remodel. In the final configuration, the splenic vein (left) and superior mesenteric vein (right) form the portal vein (Figs. 6.24, 6.25, and 6.26).

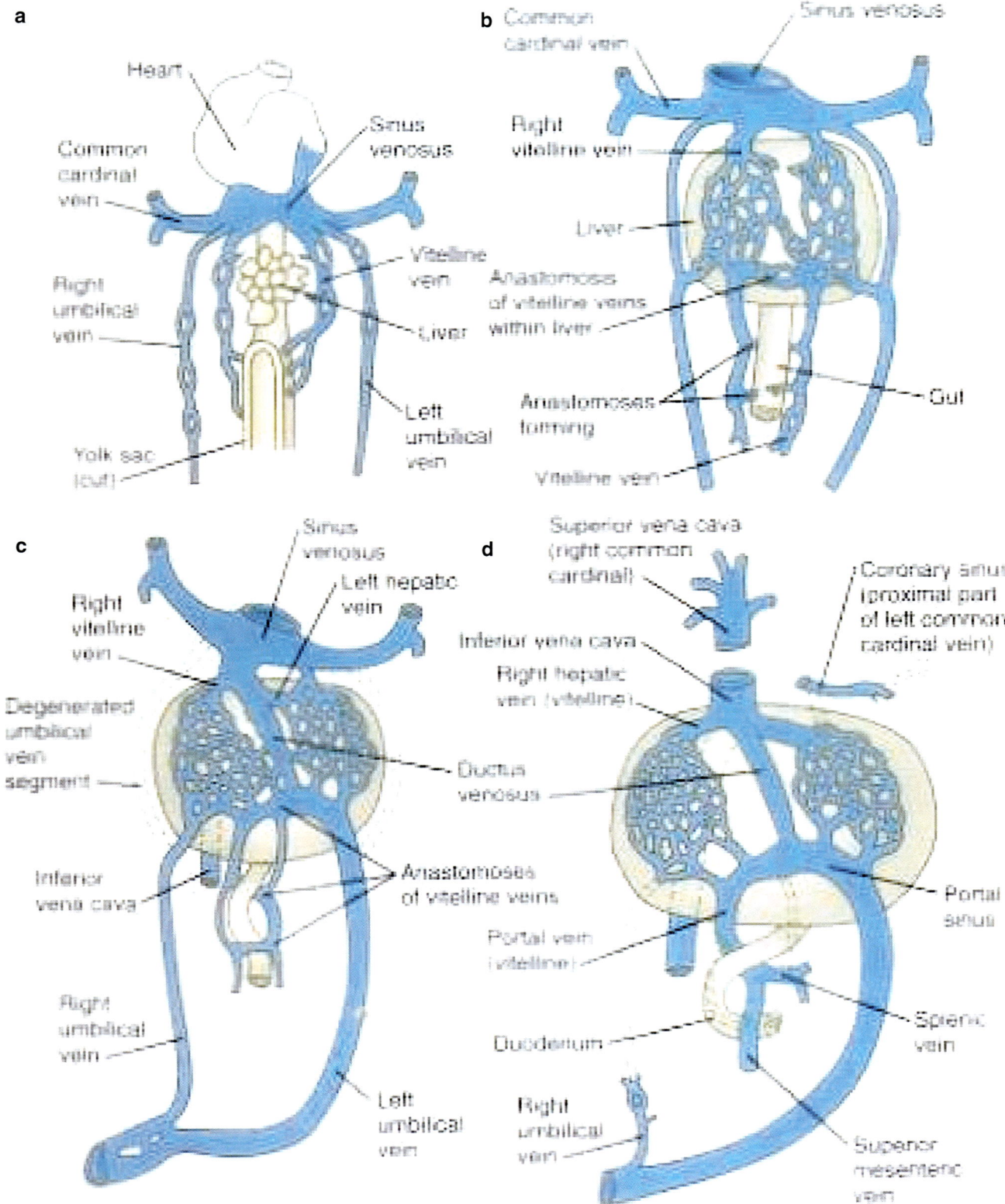

Fig. 6.25 Fate of the embryonic veins. Just as with arteries, longitudinal arteries with transverse anastomoses, embryonic break up creating new longitudinal structures. (Reprinted from Carlson BM. Human Embryology and Developmental Biology, 6th edition. St. Louis, MO: Elsevier; 2019. With permission from Elsevier)

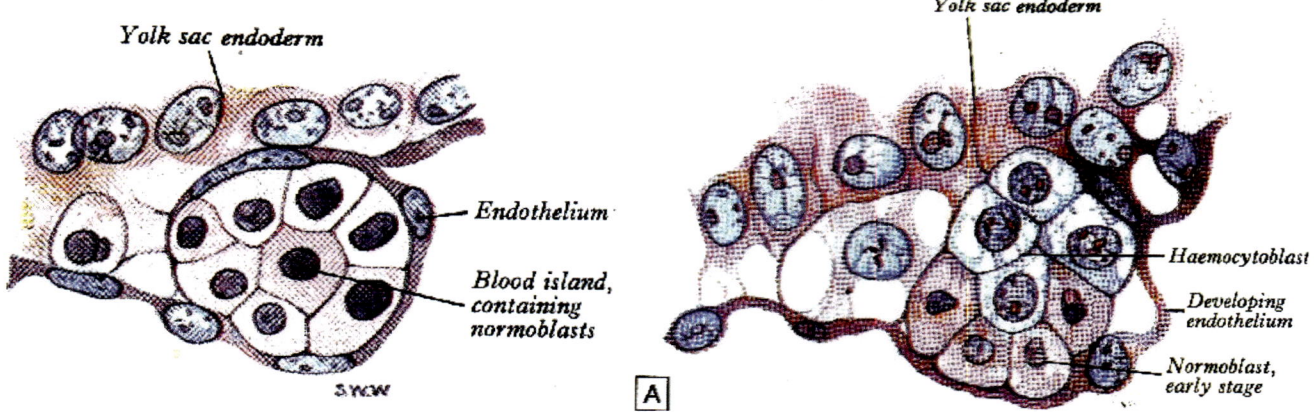

Fig. 6.26 Vasculogenesis: blood islands. (**a**) Blood vessel formation takes place initially in the wall of the yolk sac. Here, undifferentiated mesenchymal cells cluster together to form **blood islands**. These have two functional zones: the inner cells of these clusters (hematocyto-blasts) form blood cells, while outer cells (endothelial cells) become blood vessel endothelial cells. (**b**) Human blood island seen within the mesoderm surrounding the yolk sac. (Reprinted from Gilbert SF, Barressi MJF. Developmental Biology, 11th ed. Sunderland, MA: Sinauer; 2016. Copyright © 2016. Oxford Publishing Limited. Reproduced with permission of the Licensor through PLSclear)

The vitelline arteries retain their developmental association with the primordial dorsal aortae, now converted into a single, fused structure in the abdomen. In recognition of the intimate relationship between the gut and the yolk sac, these arteries supply the entire gut. Thus, the aorta gives off ventral, unpaired branches: the celiac, superior mesenteric, and inferior mesenteric arteries.

At first, the umbilical veins bypass the liver, but their proximal segments soon form connections with the vitelline networks, while the distal segments to the sinus venosus are obliterated. A major channel, the ductus venosus, arises from the left umbilical vein. It shunts oxygenated placental blood through the liver (bypassing the hepatic capillary beds) and delivers into the inferior vena cava. The right umbilical vein, being now useless, involutes.

The umbilical arteries represent the direct continuation of the paired dorsal aortae. These structures exist in the body stalk prior to the appearance of the vitelline vessels. This is proof of the dominance of the allantoic circulation over that from the yolk sac. After the dorsal aortae fuse, the paired umbilical arteries have a ventrolateral take-off. They descend medial to the primary excretory duct. Later on, two new vessels, the fifth lumbar intersegmental arteries, arise from the aorta, pass lateral to the primary excretory duct, and join up with the proximal umbilical arteries. Thus, the umbilical artery has a do*rsal root* (from the fifth intersegmental) and a *ventral root* (the original stem vessel). The dorsal root gives off the common iliac with branches to the pelvic viscera, the external iliac, and finally, the internal iliac artery. The ventral root disappears. What remains of the umbilical artery now arises from the dorsal root distal to the external iliac artery… the internal iliac artery.

Heart Development: Its Relevance to Craniofacial Circulation

Survival and growth of vertebrate embryos depends on a circulatory system consisting of a heart, blood cells, and blood vessels. It is the first functional unit to develop and the heart is the first organ. Nonetheless, the heart does not begin to pump until the embryo has already created its initial circulatory loops. Blood vessels form independently; later they connect up with the heart. Although our focus here is on how craniofacial vessels develop, some remarks regarding the heart will be helpful to understand the overall process (Fig. 6.22).

Cardiovascular fields within the visceral lateral plate mesoderm bear responsibility for the entire circulatory system. Under one set of conditions (in the posterior LPMv), they become *hemangioblasts*. In another set of conditions (in the anterior LPMv), they produce multipotent cardiac precursors. Thus, all different populations of the heart—myocytes of the ventricles and of the atria, smooth muscles of the vasculature, the endothelium of the heart and valves, and the epicardium all are derived from a common precursor population. We shall see later that additional structures of the heart—valve leaflets, the conduction system, and the aortico-pulmonary septum—come from cardiac neural crest. Also known as the *circumpharyngeal neural crest* [1, 2], these cells are associated with pharyngeal arches 3–5, they arise from rhombomeres r6 to r9, and they gain access to the heart via the third and fourth aortic arch arteries.

Four processes are required to make a heart: specification, determination, patterning, and differentiation. We shall examine specification because it references noncardiac vascular tissues as well. The 50 cardiac precursor cells of the epiblast migrate through the rostral primitive streak in a spatiotemporal order. The resultant cardiogenic mesoderm has a medial-lateral organization that reflects the eventual rostral-caudal axis of the heart tube. The earliest cells to leave the epiblast via the streak occupy the most distal positions in the cardiogenic mesoderm. They form the *primary heart field*; it consists of the in*flow tract precursors* (atria and left ventricle). Cells egressing later remain in cardiogenic mesoderm more proximal to the streak. These form the *secondary heart field*. It consists of the *outflow tract precursors* (right ventricle, conus arteriosus, and truncus arteriosus).

To quote Napolean, "Geography is destiny." Specification heart versus non-heart is a question of where the cells reside within the visceral lateral plate mesoderm. First off, let's understand that the "default state" of the vascular system is to form blood vessels and blood cells. This is because all cells egressing is exposed to Wnt, a signal produced by the neural tube. What happens next depends on geography.

Anterior LPMv lies directly above the endoderm of the future pharynx. Wnt inhibitors (Dickkopf, Crescent) from the pharyngeal endoderm prevent the cells from assuming a noncardiac fate. Later in time, BMP and Fgf8 signals convert the anterior LPMv into heart derivatives. Note that this effect is specific to pharyngeal endoderm. Substitution of posterior endoderm will not work.

Posterior LPMv lies directly below the notochord. Although initially these cells are" under the influence" of Wnt—and now on their way to making blood cells and vessels—they are also exposed to cardiogenic signals, such as BMP. Notochord produces Noggin, a BMP inhibitor. It therefore "rescues" noncardiac vascular LPMv from being suborned into heart. As a consequence: (1) Blood cell and blood vessel formation take place within the embryo posterior earlier in time and in a location that is spatially distinct (not connected) from the heart. (2) Because Noggin is produced by a midline structure, cardiovascular fields are not found in the center of the embryo.

Formation of Blood Vessels

Vertebrate Cardiovascular Systems: Common Factors

1. *Physiologic*: All vertebrate embryos must continue to function while, at the same time, developing. Prior to the existence of intestines, lungs, and kidneys, the organism must obtain nutrition, oxygenate its tissues, and excrete. Thus, yolk or placenta supplies nutrition, the extraembryonic or chorionic membranes provide oxygenation, and the allantois receives wastes.
2. *Evolutionary*: Vertebrate circulation is conservative. Strategies previously successful for one group are often reworked, rather than discarded. Although the mammalian yolk sac no longer contains yolk, it receives blood vessels. The avian vitelline system persists in mammals as the omphalomesenteric veins and arteries. Pharyngeal arches in mammals have nothing to do with respiration, yet they follow the same pattern as the branchial arches in fish gills. The transition to lungs will be discussed in a later chapter on the evolution.
3. *Physical*: The design of vertebrate circulation must resolve contradictory demands in order for oxygen to be delivered to the tissue efficiently. Flow through the vessels must be rapid and not consume much energy. At the periphery, oxygen must be easily transported across the vessels into the tissues. By ***Poiseuille's law,*** resistance to flow is proportional to the inverse of the radius to the fourth power, r-4. The best transport is through large diameter vessels. At the same time, gas diffusion is best via small diameter vessels. The problem is resolved using a hierarchical system in which large diameter vessels are connected with a network of small diameter vessels constituting a large cross-sectional area. Fluid dynamics must also be taken into account. The rational behind between the cross-sectional area is to reduce the velocity of flow in the tissues to permit optimal gas exchange. If a large diameter hose is connected to a small diameter hose, flow will increase in the latter. ***Murray's law*** states that, to maintain constant flow, the cube of the radius of the parent vessel must approximate the sum of the cubes of the radii of the recipient vessels. This enables the flow to slow down. In sum, fluid dynamics play an important role in understanding the mechanisms by which vessels branch, form anastomoses, and involute.

Vasculogenesis and Angiogenesis

The process of *vasculogenesis* involves the de novo creation of blood vessels from paraxial mesoderm and lateral plate mesoderm. Later on, this network will be remodeled via a separate process, angiogenesis, into a definitive system of arteries, veins, and intervening capillaries. Recall that when vascular precursor cells leave the posterior primitive streak at stage 7, they become specified as *hemangioblasts*. These cells are wildly invasive; they rapidly populate all sectors of mesoderm except the notochord and prechordal mesoderm. In visceral lateral plate, they form *blood islands*; these exhibit two populations. The inner cells of the blood islands become blood cell precursors, while the outer cells become *angioblasts*, the precursors of blood vessels. Angioblasts then differentiate into *endothelial cells*, the lining of the blood vessels.

The endothelial cells form the *primary capillary plexus*. These delicate, thin-walled structures are ideal for the initial conditions of ebb-and-flow circulation. Subsequently, the angioblasts recruit surrounding mesenchymal cells to

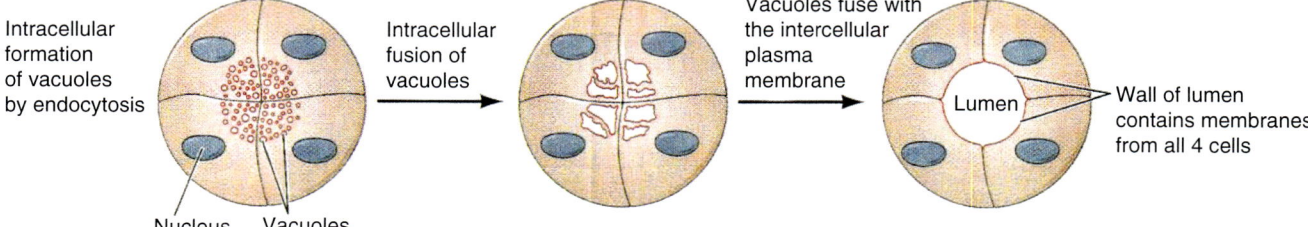

Fig. 6.27 Vasculogenesis: vacuolization. Intracellular vacuolation produces a lumen. The vacuoles form by endocytosis. These create a lumen within a cord, or within an aggregate of cells. Vessel walls thus formed must be stabilized with pericytes. (Reprinted from Gilbert SF, Barressi MJF. Developmental Biology, 11th ed. Sunderland, MA: Sinauer; 2016. Copyright © 2016. Oxford Publishing Limited. Reproduced with permission of the Licensor through PLSclear)

become *pericytes*. These latter cells surround the endothelial cells and create tubular structures. Pericytes are putative precursors of mesenchymal stem cells.

Amniote vasculogenesis takes place in two separate locations and two different points in time. *Extraembryonic vasculogenesis* takes place first in the mesodermal wall of the yolk sac and in the wall of the chorion. Here the hemangioblasts respond to *Indian hedgehog* produced by the yolk sac endoderm. Blood islands form the early vasculature and red cells. *Intraembryonic vasculogenesis* takes place in mesoderm shortly after gastrulation. In paraxial mesoderm, larger vessels such as the head plexus, primordial hindbrain channels, and dorsal aortae are quickly linked with the capillary plexus in all organs.

How does mesoderm appear in the yolk sac prior to the advent of gastrulation? Recall that, at stage 5, hypoblast spreads outward from the bilaminar embryo, flowing along the internal surface of the cytotrophoblast. Hypoblast now lines the entire blastocyst cavity and acquires a new name: primitive endoderm. A new layer promptly appears; it is interposed between primitive endoderm (from which it is derived) and cytotrophoblast. Blood vessels form in the yolk sac lining provide nutrition and gas transport at the early stages prior to initiation of heartbeat at stage 10. As the blood islands grow, they coalesce into a capillary network draining into paired omphalomesenteric veins (vitelline veins in chicks).

Intraembryonic vessels arise in and around organs. They are not initially extensions of the larger vessels. Organs secrete paracrine factors that determine the characteristics of their individual circulations. The capillary network surrounding the brain responds to CNS-derived Wnt to produce the blood brain barrier and proteins for glucose transport.

Several growth factors are involved in vasculogenesis. *Basic fibroblast growth factor*, Fgf2, helps generate hemangioblasts from visceral lateral plate mesoderm. In the placenta, *vascular endothelial growth factor* (VEGF) and *placental growth factor* (PlGF) control vessel development. In particular, VEGF allows endothelial cells to form tubes. *Angiopoietins* control interaction between endothelial cells and pericytes. *Pericytes* are derived from neural crest. They are recruited by the endothelial cells to provide smooth muscle coverage pericytes also enable the vessels to "find their way" through the dura.

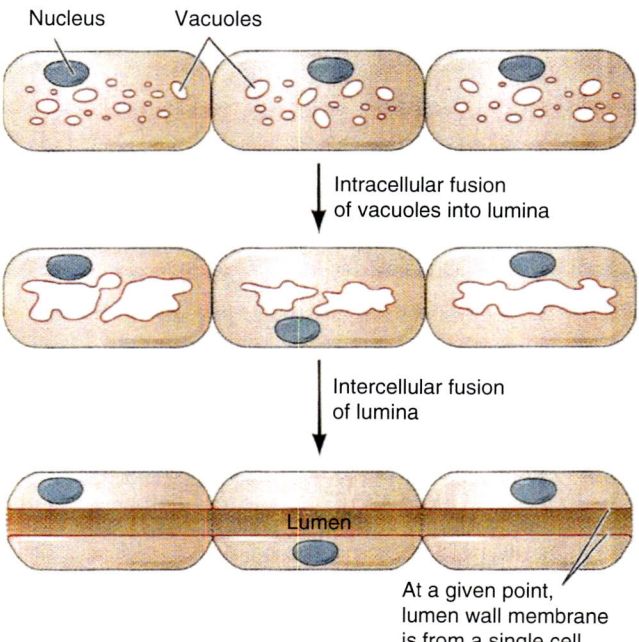

Fig. 6.28 Vasculogenesis: lumen formation. Alternatively, vacuoles in each cell link together with the lining of the lumen arising. (Reprinted from Gilbert SF, Barressi MJF. Developmental Biology, 11th ed. Sunderland, MA: Sinauer; 2016. Copyright © 2016. Oxford Publishing Limited. Reproduced with permission of the Licensor through PLSclear)

The process of angiogensis refers to the remodeling of primary capillary plexus into arteries and veins. In many instances, VEGF-A secreted by an organ causes endothelial cells to migrate from an external position to form capillary networks within the organ. VEGF-A induces a unique group of *tip cells* to produce filopodia that contain densely packed VEGF receptors causing the vessel to extend toward the source of VEGF. Tip cells also have a suppressor mechanism to prevent other neighboring cells from competing. They express a ligand for Notch called Delta-like-4 (Dll4) which causes Notch to be activated in adjacent cells, thus preventing them from responding to VEGF cues. Thus, the location of buds along an axial artery has to do with the distribution of tip cells (Figs. 6.26, 6.27, and 6.28).

Arteries Versus Veins

Arteries and veins arise from the same endothelial precursor cells. They differ in terms of the thickness of their muscular coats, the presence of valves (in veins), and in chemical nature of their endothelial cells. The cell membranes of arteries contain ephrin B2.

Veins have a receptor for Ephrin B4 kinase. Vascular beds develop following a simple model. (1) New arteries form from VEGF produced by the Schwann cells. (2) Using ephrin-Eph interactions, the arteries form the veins that will complete the circulation. This mechanism ensures end-to-end connections that become capillaries. This hypothesis fits with long standing observations in the chick in which arteries appear first within the capillary network followed by vitelline veins.

Although VEGF is the common denominator, the opposite can hold true as well. In vascular beds that have formed, de novo VEGF secreted by vascular smoot muscle cells also be secreted from vascular smooth muscles attracts nerve ingrowth. This explains how an arterial network in an organ can attract the ingrowth of sensory nerves.

Arterial Supply to Tissues

Nerves and Fascia

It has long been known that blood vessels accompany peripheral nerves. The benefits are mutual. Nerves require oxygen; vessels receive autonomic control for vasodilation and vasoconstriction. The mechanisms of induction are also mutual. Nerves produce an angiogenesis factor such as VEGF. Vessels produce nerve growth factor. Peripheral nerves induce arteries near them but not veins [3]. Failure of skin neurons to form leads failure of dermal blood vessels. This may explain the localized pathology of *cutis aplasia* (Figs. 6.29 and 6.30).

Fascia, tendons, periosteum, joint capsules, perichondrum, and fat are all part of a single connective tissue framework dispersed throughout the body. As previously described, the connective tissue system (CTS) develops from different sources depending upon location. Craniofacial connective tissues are exclusively neural crest. Outside the head, connective tissues result from the interaction of neural crest with mesoderm. In the case of the axial skeleton and skeletal muscles, paraxial mesoderm is the target. The appendicular skeleton is constructed from somatic lateral plate mesoderm. Within the body cavity, fascia surrounds the smooth muscles of visceral lateral plate mesoderm. Sympathetic autonomic nerves accompany the arterial system, most likely as an induction from nerve growth factor produced by the arterial walls. What is relevant for our discussion is the concept that connective tissues serve as an alternative sensory peripheral nervous system conveying information back to the brain regarding spatial position, stress/strain status of the musculoskeleton system, and muscle tension. It is exquisitely sensitive to stretch. The CTS is laid out in exact somatotopic relationship with the posterior columns. Coordination of

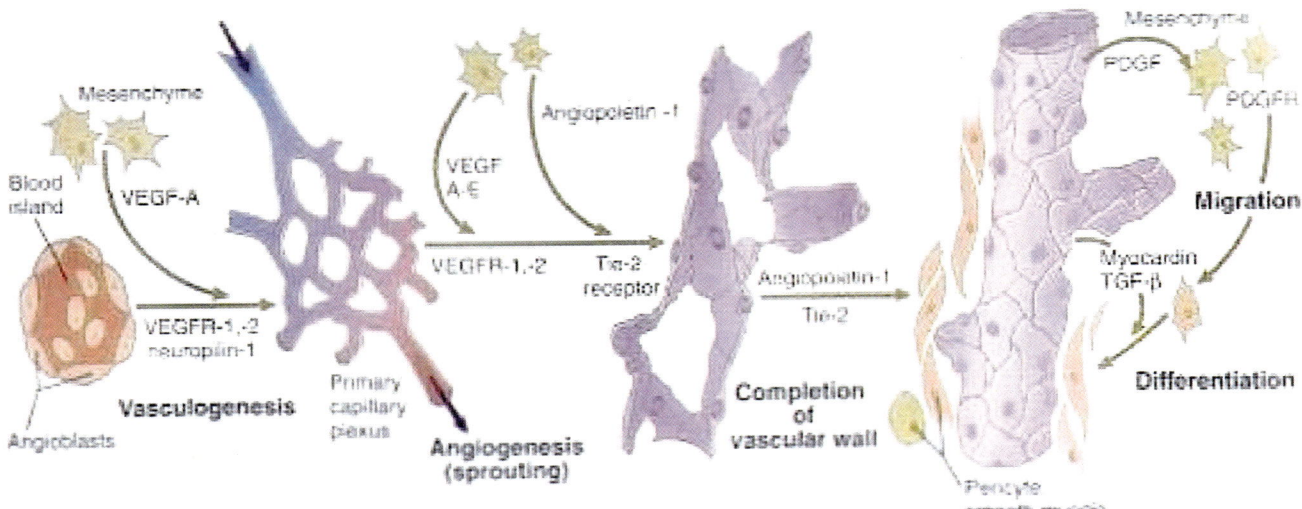

Fig. 6.29 Vasculogenesis and angiogensis. <u>Vasculogenesis</u>: Angioblasts with receptors for vascular endothelial growth factor (VGEFR-2) as stimulated by local VEGF to make primary capillary plexus. <u>Angiogenesis</u>: Endothelial cells form vascular sprouts. Platelet-derived growth factor (PDGF) recruits mesenchymal celles (pericytes). Angioblasts, initially expressing vascular endothelia growth factor receptor (VEGFR-2), are stimulated by vascular endothelial growth factor (VEGF-A) secreted by the surrounding mesenchyme, to form the primary capillary plexus by the process of vasculogenesis. Under additional stimulation by growth factors, competent endothelial cells of the primary capillary plexus form vascular sprouts in the earliest stages of angiogenesis. This is followed by the recruitment of surrounding mesenchymal cells to form the cellular elements of the vascular wall. PDGFm plate-derived growth factor; PDGFR platelet-derived growth factor receptor; TGF-B, transforming growth factor beta. (Courtesy of Bruce M. Carlson, MD, PhD)

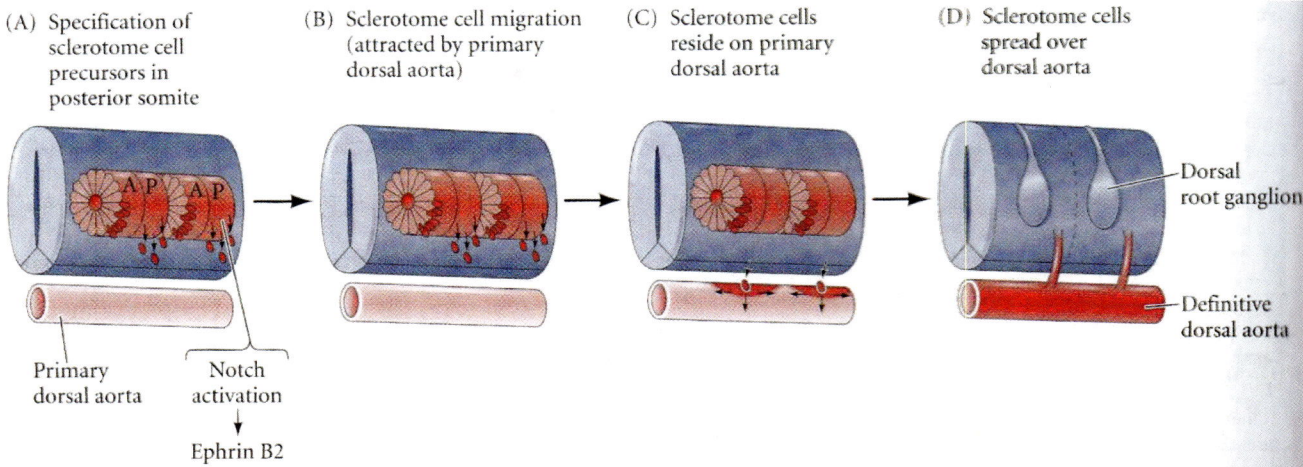

(A) Specification of sclerotome cell precursors in posterior somite

(B) Sclerotome cell migration (attracted by primary dorsal aorta)

(C) Sclerotome cells reside on primary dorsal aorta

(D) Sclerotome cells spread over dorsal aorta

Dorsal root ganglion

Definitive dorsal aorta

Primary dorsal aorta

Notch activation
↓
Ephrin B2

Fig. 6.30 Embryology of the dorsal aorta. The dorsal aortae develop as a composite of paraxial mesoderm (dorsal) and lateral plate mesoderm (ventral). The primary dorsal aorta is lateral plate mesoderm (pink). Sclerotome cells in the posterior somite are activated and migrate to populate the dorsal aorta. The endothelial cells remaining from the primary dorsal aorta become blood cell precursors. Recall that the embryo has folded so what was lateral is now ventral. (Reprinted from Gilbert SF, Barressi MJF. Developmental Biology, 11th ed. Sunderland, MA: Sinauer; 2016. Copyright © 2016. Oxford Publishing Limited. Reproduced with permission of the Licensor through PLSclear)

information takes place in the cerebellum. For this reason, *the CTS is responsible for inducing the arterial supply to target tissues.* Once the stem artery is in place, signal from the tissue determines where the artery enters and how it proliferates within the tissue. Let's now see how this concept applies to various tissues.

Membranous Bone

Membranous ossification produces flat bones directly within a connective tissue envelope; no cartilaginous intermediate is required. The main types of membranous bone are, *periosteal bone* (which envelops and adds thickness to the shafts of long bones like and sleeve and adds thickness to the their shaftes), *patella*, and *dermal bones* formed from dermis and dura. The mesenchyme of the first two bones is somatic lateral plate mesoderm. Dermal bones are found in the skull and face; they develop from neural crest. What are the common denominators among these bones and what sensory nerves are involved?

Neural crest cells respond to high concentrations of BMP produced by dermis or dura makes induction sensory nerve to form compact nodules. The chemistry of BMP is dose-dependent. High concentrations cause the nodules to become cartilage, whereas lower levels induce pre-osteoblast progenitor cells with mRNAs for collagen II and IX. Later on, they express *osteopontin* similar to developing chondrocytes. At this stage, the bone cells are called *chondrocyte-like*

osteoblasts. They subsequently produce Indian hedgehog which commits them as *mature osteoblasts.* These secrete a surrounding collagen proteoglycan matrix called *osteoid.* They are now bone fide bone cells, termed *osteocytes.* Calcification proceeds within the matrix, and at its periphery, calcified spicules arise. Mesenchymal cells surround the spicules and become periosteum. The inner layers of periosteum contain bone precursor cells and these lay down bone in parallel. Bone formed in this way is flat and multilayered bone (Fig. 6.31).

The key question here is the nature of the *pre-periosteal mesenchyme.* In the head and neck, this is without question, neural crest. The sensory innervation of these bones comes predominantly from the trigeminal nerve. Subperiosteal forehead lift can be performed under local anesthesia alone using regional blocks. The periosteum is vascularized by a vessel network originating from overlying local arteries. In the case of the frontal bone, these are the supraorbital axis and the frontonasal axis (supratrochlear). This, the connective tissue-fascia system, is neural in nature and provides coverage from the membranous bones of the skull and face.

It is critical to reiterate that the same neural crest regions responsible for synthesizing the membranous frontonasal bones are also involved with production of the anterior brain (telencephalon). Where the zone of neural crest above p6 and p5 is removed, facial skeleton fails to form, as does the cerebrum. Two possible mechanisms are involved (and may be

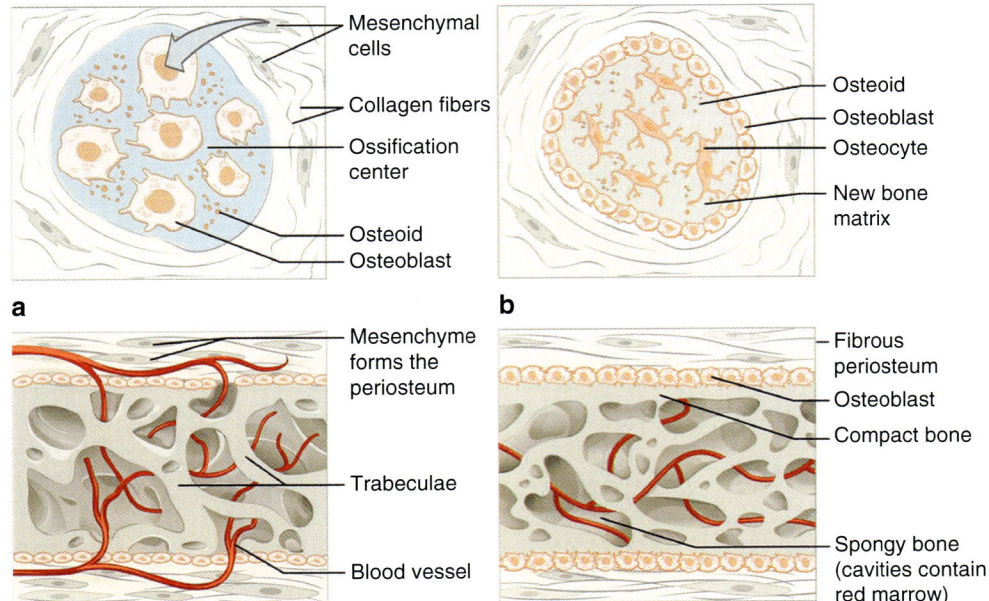

Fig. 6.31 Vascularization of membranous Bone. Intramembranous ossification follows four steps. (**a**) Mesenchymal cells group into clusters, differentiate into osteoblasts, and ossification centers form. (**b**) Secreted osteoid traps osteoblasts, which then become osteocytes. (**c**) Trabecular matrix and periosteum form. (**d**) Compact bone develops superficial to the trabecular bone, and crowded blood vessels condense into red bone marrow. (Reprinted from Biga LM, Dawson S, Harwell A, Hopkins R, Kaufmann J, LeMaster M, Matern P, Morrison-Graham K, Quick D, Runyeon J. Anatomy and Physiology. OpenStax/Oregon State University. Retrieved from: https://open.oregonstate.education/aandp/. With permission from Creative Commons License 4.0: https://creativecommons.org/licenses/by-sa/4.0/)

related). (1) Forebrain meninges arise from this zone. These are critical for setting up the blood supply to the cerebral cortex. (2) Pericytes are also neural crest derivatives. These cells "guide" blood vessels through the meninges and into intimate contact with the brain.

For these reasons, we conclude that the neural crest connective tissue system in head and neck provides all structures related to support and interconnection of bone, cartilage, muscle, dermis, and brain: fascia, tendons, ligaments, periosteum, perichondrium, connective tissue septae, and fat, and meninges.

Chondral Bone

Chondral ossification is responsible for synthesis of all remaining bones of the axial skeleton and the entire appendicular skeleton. Cranial base bones formed this way are the basisphenoid, basioccipital, exoccipital, and supraoccipital. Recall that the infamous invisible proatlas is absorbed into components of the exoccipital, connecting ligament from dens to occipital, and the upper 1/3 of the dens. Each chondral bone is ensheathed in periosteum…this layer not only adds additional bone, but its existence explains the origin of all primary and secondary nutrient arteries. The periosteum

is just an extension of the connective tissue and fascia. As such, it is sensate…part of great "alternative nervous system." Chondral bones form in five steps (Figs. 6.32, 6.33, and 6.34).

1. *Commitment.* A specific population of mesenchymal cells receives signals such as Sonic hedgehog to become cartilage. At this moment, they still appear mesenchymal, but their fate is sealed. The signals likely arise from a surrounding connective tissue "envelope" that defines the location and shape of the bone. Although not depicted in textbook illustrations…the envelope is present nonetheless. This future periosteum is the likely source for the signals that determine commitment…and ultimately the size and shape of the bone.

2. *Compaction.* Mesenchymal cells condense into compact nodules under the influence of BMPs that induce adhesion factors. N-cadherin initiates the compaction and NCAM maintains it. Sox9 produces transcription factors for collage 2 that is vital for cartilage development. Defects in Sox-2 result in camptomelic dysplasia with underdeveloped ribs and tracheal cartilages leading to respiratory failure.

Fig. 6.32 Vascularization of chondral bone: Endochondral ossification follows five steps. (**a**) Mesenchymal cells differentiate into chondrocytes that produce a cartilage model of the future bony skeleton. (**b**) Blood vessels on the edge of the cartilage model bring osteoblasts that deposit a bony collar. (**c**) Capillaries penetrate cartilage and deposit bone inside cartilage model, forming primary ossification center. (**d**) Cartilage and chondrocytes continue to grow at ends of the bone, while medullary cavity expands and remodels. (**e**) Secondary ossification centers develop after birth. (**f**) Hyaline cartilage remains at epiphyseal (growth) plate and at joint surface as articular cartilage. (Reprinted from Biga LM, Dawson S, Harwell A, Hopkins R, Kaufmann J, LeMaster M, Matern P, Morrison-Graham K, Quick D, Runyeon J. Anatomy and Physiology. OpenStax/Oregon State University. Retrieved from: https://open.oregonstate.education/aandp/. With permission from Creative Commons License 4.0: https://creativecommons.org/licenses/by-sa/4.0/)

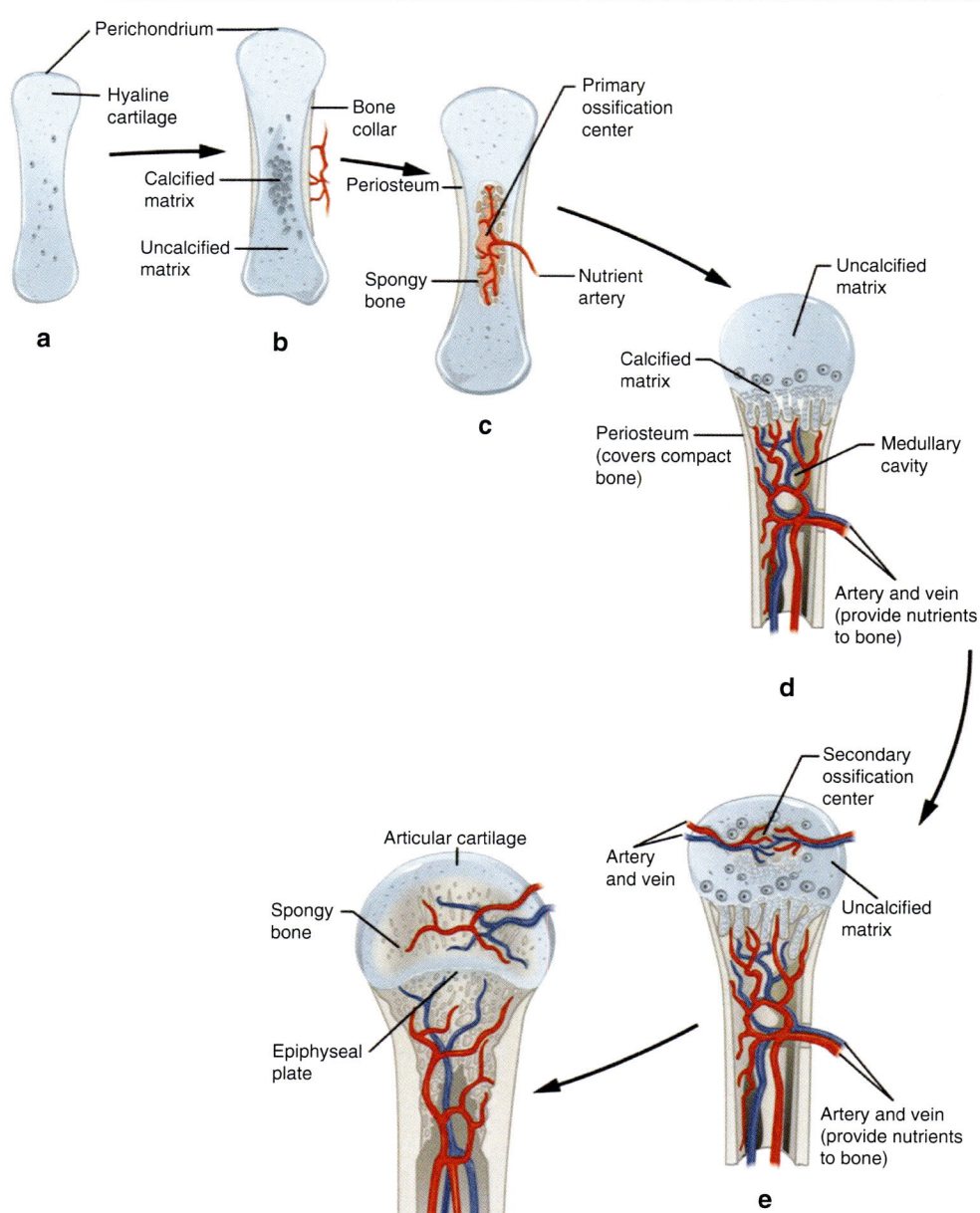

3. *Proliferation*. Nodules differentiate into chondrocytes; these secrete extracellular matrix. The form of the bone appears. Surrounding mesenchyme becomes organized into a definite perimeter…the future periosteum.

4. *Hypertrophic chondrocytes*. Under the influence of Runx2, the chondrocytes stop multiplying and increase in volume. This process is under the control of histone deacetylase 4, an enzyme expressed solely in prehypertrophic cartilage. Mice with HDAC4 knock-outs show premature ossification of limbs and ribs. The hypertrophic chondrocytes become mineralized by calcium phosphate. They also secrete VEGF. This is the critical point. The location of the hypertrophic chondrocytes (in the center of the bone shaft) defines the *primary ossification*

center. VEGF from this site attracts the entry of the primary nutrient artery from the surrounding CTS. The hypertrophic chondrocytes now secrete vesicles containing mineralizing enzymes; then they die. Hydroxyapatite forms in the extracellular matrix. Remodelling by bone cells is dependent upon this mineralization.

5. *Vascularization*. As the hypertrophic chondrocytes die off, blood vessels emanating from the nutrient artery are allowed to enter. They proliferate within the spaces left behind. Blood-bourn chondroclasts and osteoclasts (derived from macrophages) now gain access to the developing bone. A transformation of mesenchymal precursor cells to a bone lineage now takes place. Recall that osteocytes and chondrocytes arise from a common mes-

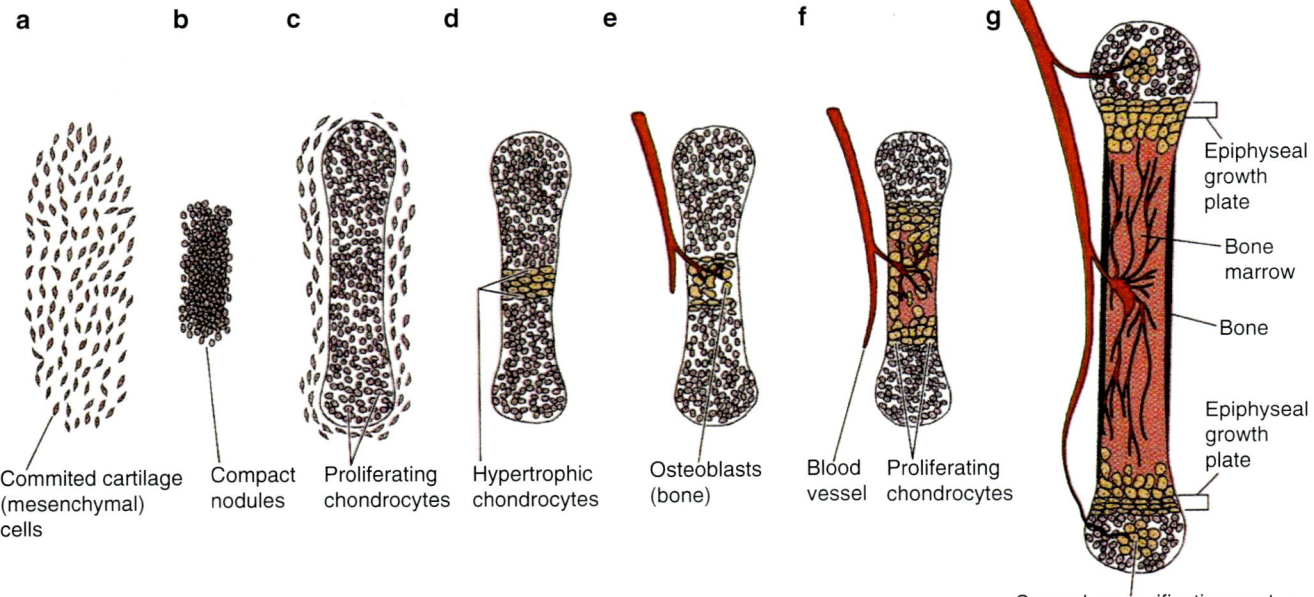

Fig. 6.33 Osteochondral ossification 1. (**a–c**) In chondral ossification, mesenchyme commits to cartilage, and then to compaction, note unseen sheath of fascia surrounding the developing bonel. (**d**) Chondrocyte atrophy leads to apoptosis followed by mineralization. (**e**) Nerves/vessels enter zone of apoptosis. (**f**) Blood vessels bring in osteoblasts, cen- trifugal development. (**g**) Epiphyseal apoptosis > neurovasc >secondary ossification. (Reprinted from Gilbert SF, Barressi MJF. Developmental Biology, 11th ed. Sunderland, MA: Sinauer; 2016. Copyright © 2016. Oxford Publishing Limited. Reproduced with permission of the Licensor through PLSclear)

enchymal precursor, the pre-chondrocyte. In developing bone, a pre-chondrocytes population persists. These cells are acted upon by Indian hedgehog secreted by prehyper-trophic chondrocytes; they now become pre-osteoblasts. In two additional steps, Wnt signals create osteoblasts and Osterix instructs the osteoblasts to form bone. As bone cells replace the chondrocytes, more VEGF is released and more blood vessels are brought in.

6. *Centrifugal ossification.* The central zone of ossification is now surrounded by two flanking zones of hypertrophic chondrocytes. The process of apoptosis, vascularization, osteoblast transformation, and mineralization spreads both proximally and distally away from the central area of ossification. Osteoclasts cause a hollowing out of the internal bone to form a bone marrow cavity. This process will set the stage for paired *secondary ossification centers* at either end of the bone. The same mechanisms take place. Hypertrophic chondrocytes produce VEGF and attract the entry of secondary nutrient arteries from the peripheral CTS to supply the epiphysis. At the same time, new bone forms at the periphery. The CTS surrounding the bone becomes *periosteum.* It contains blood vessels, sensory nerves, and bone precursor cells. Bone forms in this layer by membranous ossification.

Striated Muscle
Previous chapters have presented the concept of the connective tissue system as gigantic alternative nervous system. The CTS encompasses all mesenchymal structures, both somatic and visceral, and extends to the periphery of the body, both external (dermis) and internal (gut). Somatic (striated) muscles occupy prefabricated potential spaces within the fascial network. In a distorted and less obvious way, the same rules apply to intestinal smooth muscle. How does this come about and what relevance does it have for blood supply?

After gastrulation, neural crest cells penetrate and interact with all three columns of mesoderm (paraxial, intermediate, and lateral plate). Within each somite, in response to genetic signals, neural crest subdivides and organizes the myotome into fascial compartments. Each striated muscle then exists within a sensate envelope. Blood vessels induced by the sensory nerves of the fascia are capable of amalgamating into one or more pedicles.

Mathes and Nahai classified muscles into five types, depending upon the anatomy of the vascular pedicle. Type 1 muscles have a single dominant artery. Type 2 muscles have a dominant artery and a subordinate artery. Type 3 muscles have two codominant arteries. Type 4 muscles have multiple arteries, none of which is dominant. Type 5 muscles have a dominant artery and multiple subordinate arteries. A key point is: *the location of the vascular pedicle to a muscle exists independent of its motor nerve.* Motor nerves enter muscle proximally, whereas pedicles may enter at the mid point, or at multiple points. Motor nerves certainly induce their own accompanying vessels, but these are not necessarily the dominant source of blood supply.

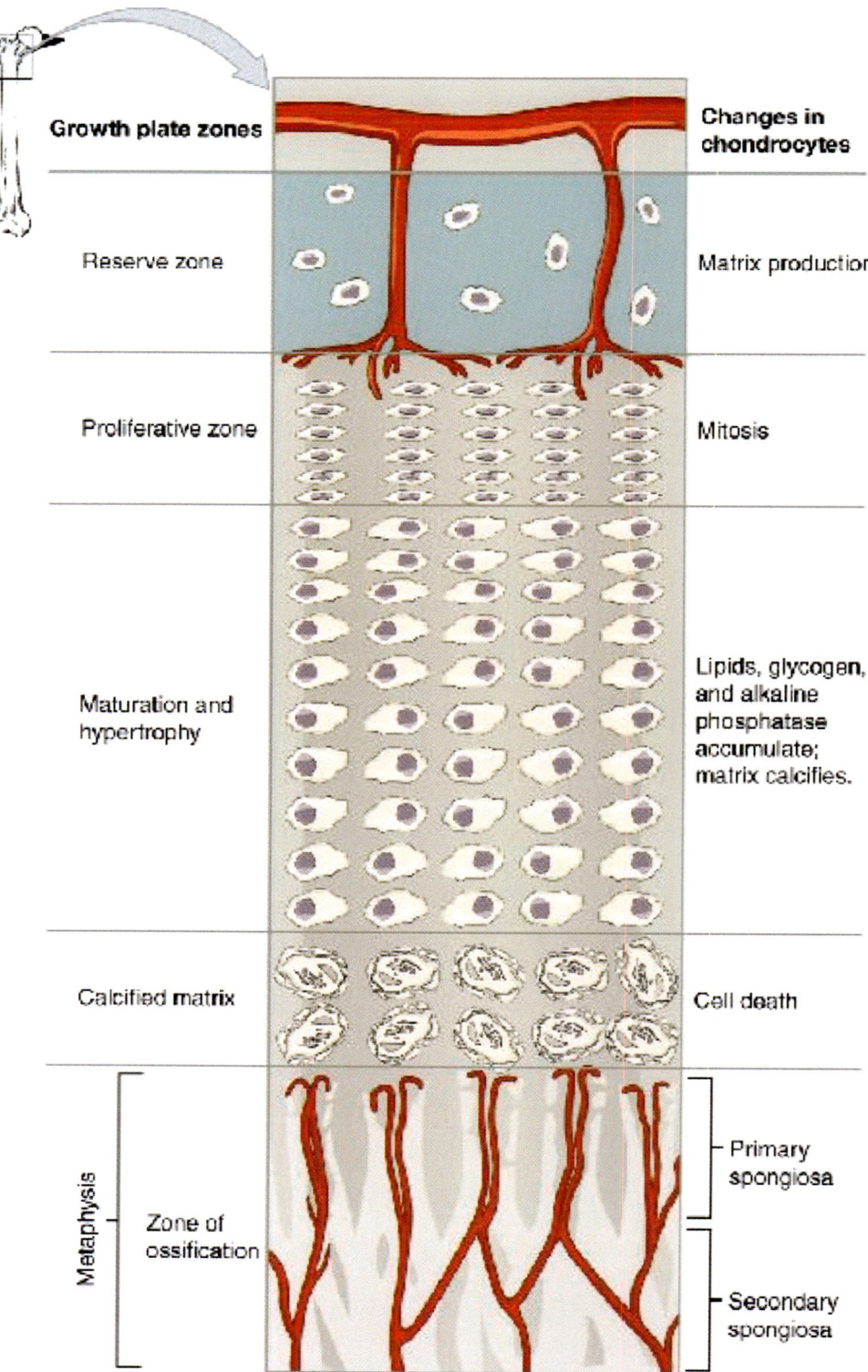

Fig. 6.34 Osteochondral ossification 2. This diagram demonstrates the watershed zone between the blood supply to the epiphysis and that of the medullary cavity. (Reprinted from Betts JG, Young KA, Wise JA, Johnson E, Poe B, Kruse DH, et al. Anatomy and Physiology. OpenStax; March 6, 2013. Retrieved from: https://opentextbc.ca/anatomyand-physiologyopenstax/. With permission from Creative Commons License 4.0: https://creativecommons.org/licenses/by/4.0/)

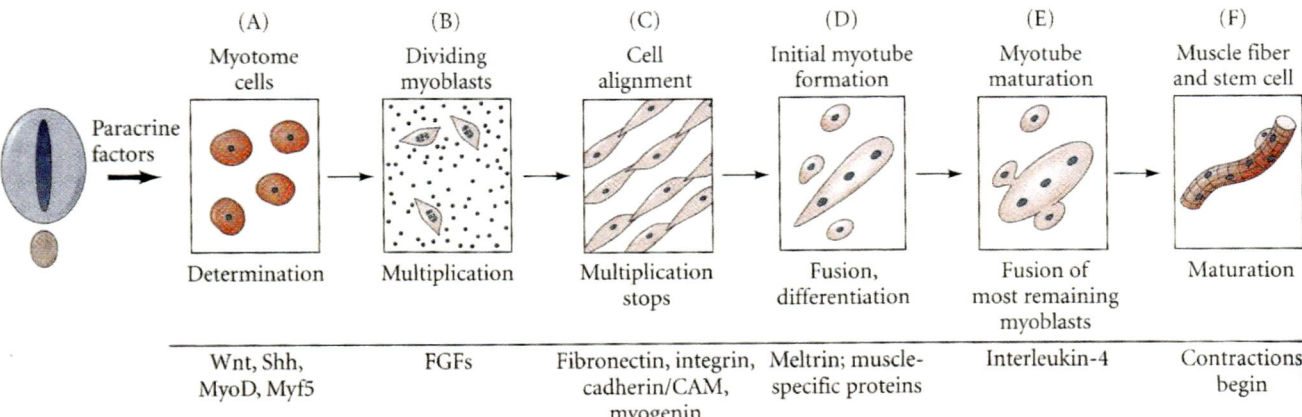

Fig. 6.35 Muscle development. Blood vessels and muscle: entry takes place via the fascia. (a) Paracrine factors commit myotome cells to muscle *and* CTS. (b) FGF growth factors but no muscle proteins. (c–e) Alignment into myotubules (surrounded by CTS). (f) Contractile muscle fibers. Step 1 myoblasts leave cell cycle. Step 2 alignment into chains. Step 3 cell fusion. (Reprinted from Gilbert SF, Barressi MJF. Developmental Biology, 11th ed. Sunderland, MA: Sinauer; 2016. Copyright © 2016. Oxford Publishing Limited. Reproduced with permission of the Licensor through PLSclear)

The concept of perforator flaps has direct relevance to our discussion. Taylor's group [4] built upon the pioneering work of Manchot (1889, translated 1983 [5]) and Salmon [6, 7] to define territories of blood supply. Using injection studies, they discovered that single arteries could supply multiple tissues from bone up to skin. Furthermore, each angiosome was connected by anastomotic arteries to adjacent angiosomes. In 1989, Nakajima showed that a single artery from the deep inferior epigastric artery was transmitted via the deep fascia, so supply an isolated skin flap. Three hundred and seventy four cutaneous perforating vessels have been identified. Some pass through muscle to the skin, while other procede directly to skin. In 2001, the Gent conference distinguished between musculocutaneous and septocutaneous flaps. This has relevance because the surgical dissection required to isolate the flap is quite different, depending upon whether one must violate the muscle.

Perforating vessels extend up through fascial septae and that the pedicles are encased in fat. Perforators are always accompanied by sensory nerves. We hypothesize that the perforator pathways represent blocks of tissue predetermined at the time of embryonic assembly. Thus, *the vascular pedicle for any muscle is conveyed to it via its fascia*, not necessarily by its motor nerve (Fig. 6.35).

Smooth Muscle

Blood supply to the visceral muscle follows an anatomic pattern different from that of somatic muscles. Body and limb muscles are segmentally organized. The origin of a striated muscle is exactly equal to the somitomeres or somites from which its myoblasts arise. This, in turn, can be determined by the roots of its motor nerve. Single somite muscles tend to be small and remain close to the axis. The more complex the motor root, the more extensive is the territory of muscle insertion. In contrast, the neurovascular supply to the gut is midline, not paired. It comes from three stems: celiac axis,

superior mesenteric axis, and inferior mesenteric axis that correspond to the three great plexuses of the sympathetic autonomic nervous system. Vessels flow outward to the segments of muscle in sheets of fatty connective tissue called mesenteries. Gut organization is segmental; it depends upon a homeotic code. (See Carlson [8], chapter on gastrointestinal development). Gut muscle arises from visceral lateral plate mesoderm. There are no small muscles. Layers cover large swatches of territory consistent with peristaltic action.

Gut vessels are programmed by the SANS. This makes eminent sense. Neural crest is the cell source for the autonomic neurons and the myenteric plexus. Thus, although the mesenteric "fascia" may not have sensory nerves with Schwann cells, one nevertheless finds neural tissue capable of inducing vessels. And, just as for striated muscles, the vascular anatomy of the gut does not follow anatomic pattern of its motor nerve (the vagus).

Brain: How Do Arteries Get into the CNS?

(Figs. 6.36, 6.37, and 6.38)

Blood vessels are made from mesenchyme, not epithelium. Neural tissue per se cannot manufacture its own blood supply. During the process of neurulation, the neural plate rounds up in U-shaped fashion to form the neural tube. In the process, paraxial mesoderm surrounds the neural tube wrapping around it like a cloak. PAM contains angioplasts, thus building blocks for vessels are intimately applied to the walls of the neural tube from the very beginning. But the CNS places unique constraints upon it blood supply: (1) vessels not intrinsic to neural tissue must achieve intimate contact with neurons; and (2) a barrier must be constructed between the circulation at large and cerebrospinal fluid.

Both of these challenges are solved by the meninges. These structural elements of these tissues represent yet another iteration of the connective tissue system. Pia mater and arachnoid are constructed at all levels of the CNS from

Fig. 6.36 Vascularization of the brain—meninges elevated and subdural vessels seen. Primitive head plexus on the surface, fed from primitive trigeminal, temporarily supports the brain until internal circulation develops from internal carotid and sends centripetal branches to the surface. (Reprinted from Williams PL (ed). Gray's Anatomy, 38th ed. New York, NY: Churchill Livingstone; 1995. With permission from Elsevier)

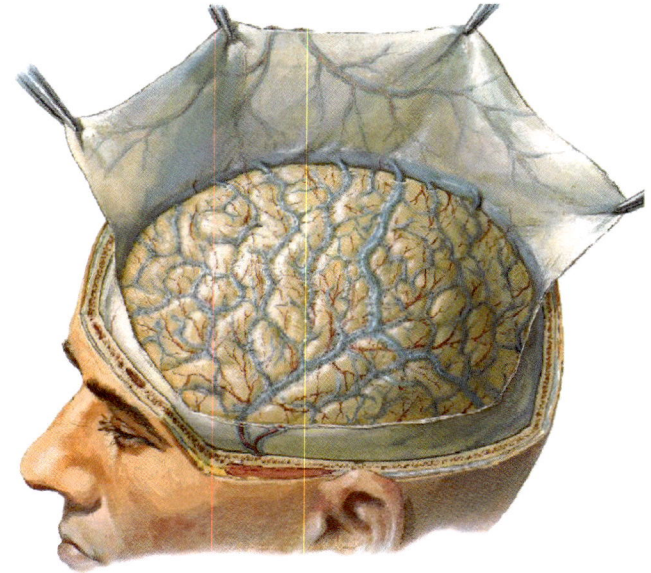

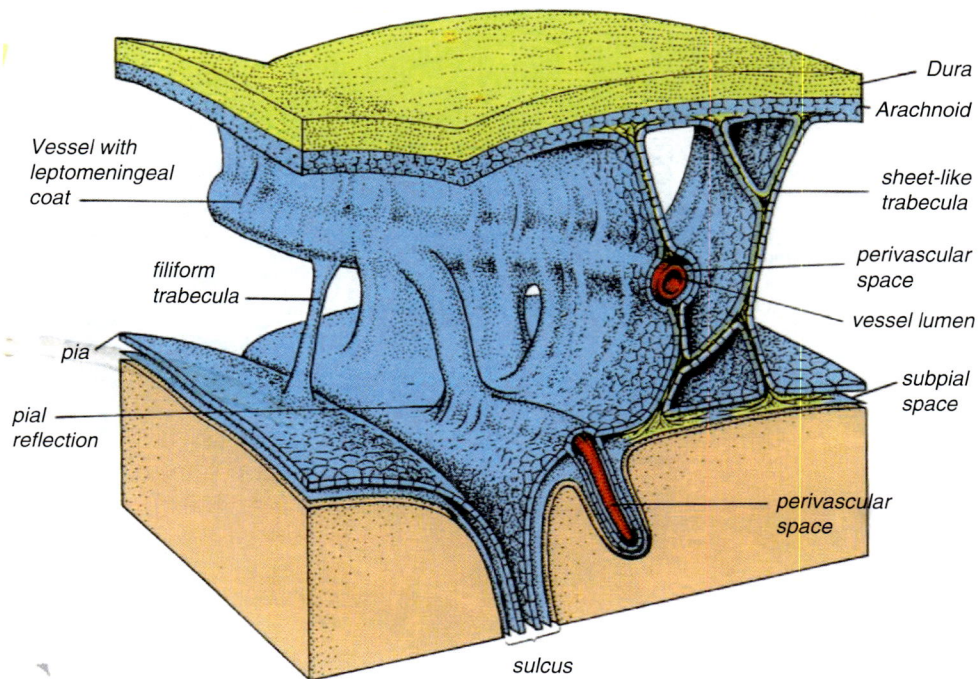

Fig. 6.37 Steady State Meningeal Anatomy and Immune Composition. The meninges covering the central nervous system (CNS) parenchyma consist of the dura mater (adjacent to the skull which is not shown), arachnoid mater, and pia mater. The dura mater is similar to a peripheral tissue with its lymphatic vessels and fenestrated vasculature that allow entry of large molecules (e.g., 45 kDa) from the blood. The dura mater has a large repertoire of immune sentinels, such as dendritic cells (DC), mast cells (MC), innate lymphoid cells (ILCs), meningeal macrophages, T cells, and B cells. The arachnoid matter represents the first impermeable barrier to the CNS parenchyma, as cells within this layer are joined by tight junctions. Therefore, molecules cannot diffuse freely from the dura mater into the subarachnoid space. The subarachnoid space is filled with cerebrospinal fluid, which removes cells and compounds unless they are attached to the stromal matrix (purple). The brain surface is traversed by nonfenestrated veins and arteries encased in the pial sheaths (not shown). Although not surrounded by astrocytic endfeet, vascular endothelial cells in the pia mater meninges are impermeable and joined by tight junctions. The pia matter also contains immune sentinels, although less numerous than in the dura, such as mast cells (not depicted), macrophages, and dendritic cells. The pia mater and glia limitans (formed by surface-associated astrocytes and astrocytic endfeet) represent the final barrier before the parenchyma. Diffusion of small molecules (3 kDa) is restricted to the superficial layers of the parenchyma, and larger molecules cannot cross this barrier. (Reprinted from Rua R, McGAvern DB. Advances in meningeal immunity. Trends Molecular Med 2018; 24(6): 542–559. With permission from Elsevier)

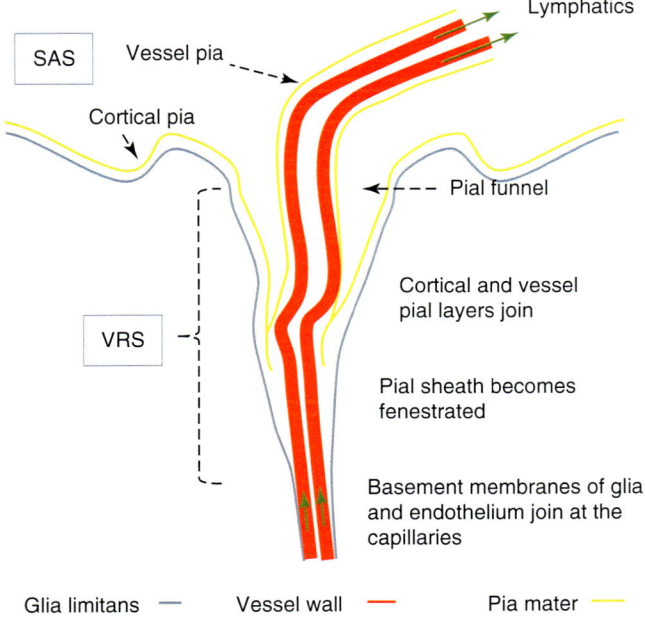

Fig. 6.38 Morphology of Virchow Robin and perivascular spaces. The pial artery ensheathed in pia mater penetrates the brain. Delineated by basal membranes of glia, pia, and endothelium, the Virchow Robin space (VRS) depicts the space surrounding vessels penetrating into the parenchyma. The VRS is obliterated at the capillaries where the basement membranes of glia and endothelium join. The complex pial architecture may be understood as an invagination of both cortical and vessel pia into the VRS. The pial funnel is not a regular finding. The pial sheath around arteries extends into the VRS, but becomes more fenestrated and eventually disappears at the precapillary section of the vessel. Unlike arteries (as shown in this figure), veins do not possess a pial sheath inside the VRS. ISF may drain by way of an intramural pathway along the basement membranes of capillaries and arterioles into the lymphatics at the base of the skull (green arrows). It should be noted that the figure does not depict the recently suggested periarterial flow from the SAS into the parenchyma and an outward flow into the cervical lymphatics along the veins (for discussion see text "Current research"). Also, it is still a matter of debate whether the Virchow Robin space, extending between the outer basement membrane of the vessel and the glia, represents a fluid-filled open space (see text). *VRS* Virchow Robin space, *SAS* subarachnoid space. (Reprinted from Brinker T, Stopa E, Morrison J, KlingeP. A new look at the cerebrospinal fluid circulation. Fluids Barriers CNS. 2014 May 1;11:10. doi: https://doi.org/10.1186/2045-8118-11-10. With permission from Creative Commons License 2.0: https://creativecommons.org/licenses/by/2.0/)

neural crest. When admixed with mesoderm, the neural crest provides structural elements and the mesenchymal cells create the blood vessels. The mixture that will give rise to pia and ararchnoid is known as *meninx primitiva.*

Dura mater covering the cerebrum arises from neural crest admixed with paraxial mesoderm from somitomeres 4 to 7, and possibly somites 1–4. Dura of the midbrain, hindbrain, and spinal cord arises from paraxial mesoderm starting with somite 5. Dura is exquisitely sensate, being supplied by cranial nerves V, VII, IX, X and by peripheral nerves. Since all peripheral nerves contain neural crest components, these contribute to dura mater as well. This explains why dural pathologies such as meningitis can cause symptoms ranging from somatic pain, to vertigo, nausea, and vomiting.

The format of the meninges is simple, based on migration patterns of neural crest and proximity to paraxial mesoderm:

Perineural migration refers to flow of cells downward along the sidewalls of the neural tube, providing a circumferential cover. These cells form the *pia mater* and the *arachnoid*. They engulf the primitive vessels supplying to the brain and spinal cord. For this reason, the internal carotid and longitudinal neural (basilar) arteries are located deep to the dura mater. Vessels associated with the pia mater penetrate into the brain…they bring a coat of pia mater along with them.

Peripheral migration follows two pathways. (1) Ventral migration takes place between the somites and the neural tube. These cells collect in the midline ventral to the notochord. They will form the ganglia of the sympathetic autonomic nervous system. (2) Lateral migration flows outward into all populations of mesoderm: paraxial, intermediate, and lateral plate. These cells form (in conjunction with paraxial mesoderm) the dura mater. Thus, with later developing vascular systems such as aortic arches, the external carotid and stapedial are either extradural or supply the dura proper.

Neural tissue requires 20% of total blood flow. Blood vessels must remain in intimate contact with the brain. How does this occur? The answer lies in the topology of the CNS. A series of subdivisions, flexures, and the growth pattern of cerebrum and cerebellum ensure that the vasculature is carried into every recess of the brain (Fig. 6.39).

Let's look first at the forebrain and midbrain. These structures develop from prosomeres p6 to p1 and mesomeres m1 to m2. In the transition from a three-part brain to a five-part brain, prosencephalon subdivides into the telecephalon (the cerebrum) and the diencephalon (epithalamus, thalamus, and hypothalamus). The cerebrum is constructed from the upper (alar) tier of prosomeres p6 and p5. It undergoes massive growth, spreading out like a set of dumbbells to enclose all the remaining prosomeres. As a consequence, diencephalon is trapped within the cerebral hemispheres. This creates the physical conditions whereby continuous tracts can be established between the cerebral cortex and the underlying thalamic nuclei. Cerebrum also comes to sit atop the midbrain. Infolding of cerebrum into gyri not only adds surface area… it also brings pia and arachnoid into the depths of the brain. These mechanisms bring the internal carotid system to all parts of the forebrain and midbrain.

The hindbrain also undergoes topological changes. It develops from rhombomeres r0 to r11. Rhombencephalon subdivides into metaencephalon (cerebellum and pons) and mylencephalon (medulla). Cerebellum and pons are constructed from r0 and r1. The pons is small and smooth, but the cerebellum has extensive infoldings (folia). Note that these structures are not covered by dura. Like the rest of the brain, their blood supply lies beneath the arachnoid and extends into all potential spaces.

Details of the blood supply to the brain are beyond the scope of this chapter. The key structure to bear in mind is pia

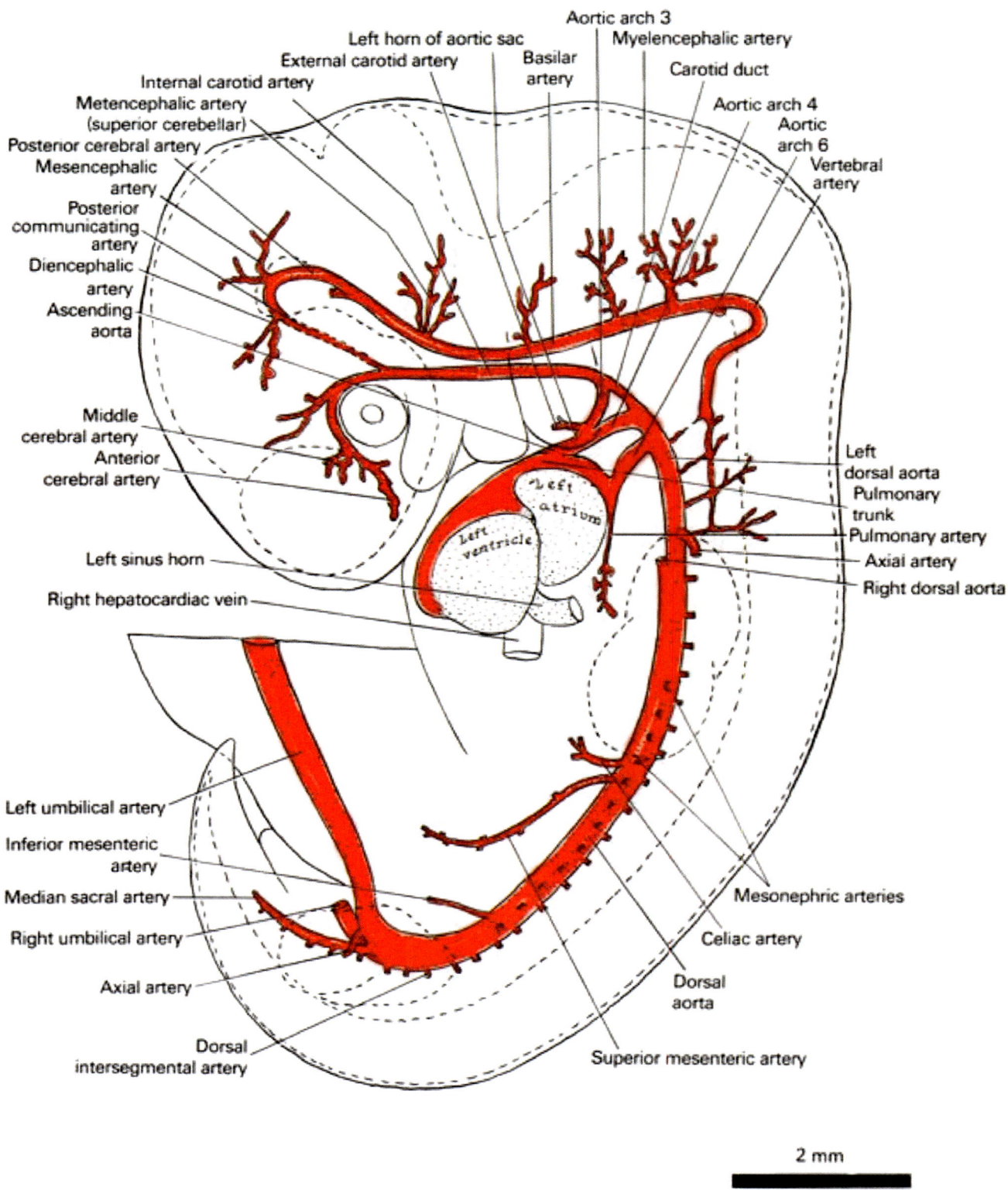

Fig. 6.39 Orientation to cerebral circulation, circa stage 14. Note regular development of vascular axes along the neural tube. (Reprinted from O'Rahilly R, Müller F. Human Embryology and Teratology, 3d. ed. Wiley Liss, 2001. With permission from John Wiley & Sons, Inc.)

mater. All penetrating vessels into the brain are encased by pia mater. Arteries of the carotid and basilar systems run in the subarachnoid space. The arachnoid itself is impermeable, but penetrating vessels from the dural arteries penetrate into the space. The spatial position of the dural vessels reflects the timing of the stapedial system. These arteries arise well after internal carotid supply is established. The dural vessels insinuate themselves between the two layers of the dura mater.

Blood vessel behavior. How do individual vessels "know" what to do?

Assignment. As previously discussed, neural tissue is certainly capable of producing vessel-inducing signals, but repertoire of such signals is likely to be quite limited. It does not make sense to have a different signal for each neuromere, let alone for substructures. Neuromeric identity through a homeotic "tattoo" is a more parsimonious concept. Neural crest from, say, the fourth prosomere interacts with paraxial mesoderm to "label" a certain segment of the primitive internal carotid artery. Branch formation is thus neuromeric. Subsequent budding could be as simple as a binary process of successive divisions, each taking the vessel deeper into the substance of the neuromere.

Connection

A related issue is the relationship between preexisting arteries those developing later in time. How can we explain connections between seemingly unrelated systems? Primitive hindbrain channels develop in paraxial mesoderm immediately after gastrulation (stages 7–8). During the aortic arch phase (stages 9–14), a series of presegmental arteries connect the dorsal aortae with the PHCs, converting them into latter longitudinal neural arteries. Although the absolute number is uncertain, named branches include the primitive trigeminal, primitive hyoid, primitive otic, and primitive hypoglossal. These vessels arise from sites along the dorsal aortae that likely contain neural crest cells having the same homeotic code as their corresponding "target sites" along the LNAs.

How do aortic arches connect the outflow tract of the heart with the dorsal aortae? Let's look at the second aortic arch. Once again, the answer is neuromeric. During stages 7 and 8, neural crest cells from rhombomeres r4 and r5 flow outward, engulfing (and defining) the paraxial mesoderm that makes up their respective segment of dorsal aorta. They subsequently move out into lateral plate mesoderm. This segment of LPMv will become the actual second arch vessel. At stage 9, folding of the embryo brings the heart below the pharyngeal arches. The dorsal aortae at the level of the trigeminal ganglion remain connected to the outflow tract—they become stretched into a downward arc as the first aortic arch arteries. These are immediately surrounded by the neural crest mesenchyme as the first pharyngeal arch aortic A dorsal bud appears from r4 to r5 dorsal aorta simultaneous with a ventral bud from the outflow tract. Intervening r4–r5 mesenchyme becomes the substance of second pharyngeal arch, at the core of which lies the aortic arch artery. Thus, an aortic arch is a "chain link fence" connecting neuroepithelium, paraxial mesoderm, pharyngeal arch, and lateral plate mesoderm. Neural crest migration patterns are faithfully reproduced.

A final example is the distribution pattern of the superficial temporal artery. How does it "know" where it should go? STA is one of two terminal branches of the external carotid system. As we shall see later, all branches of ECA distal to STA and facial artery are derived from the stapedial system.

ECA develops as a remodeling of fragments left over from the aortic arch arteries. We know that first arch and second arch become quickly amalgamated. Vessels encoded by r2–r5 make up arterial network that will become the superficial maxilla-mandibular (facial) artery and the STA. Superficial temporal is programmed in association with dermal sensory nerves from V2 and V3. Recall that dermis does not come from the second arch. STA therefore "seeks out" a region of dura that contains homeotic signals from r2 to r3.

In conclusion, the three examples given demonstrate that vessels develop in consonance with the neuromeric code. Old ones and new ones from a common source connect with one another. Vessels from different sources (PAM vs. LPM) connect. Vessels associated with one tissue (dermis) will seek out tissues from another source (dura).

Remodeling

In the narrative that follows, you will find numerous examples of precursor arteries that disappear or become subsumed by more recent iterations. In other situations, arteries will make an acute bend in their course. How do these changes come about?

Fluid dynamics are an important conceptual tool. Prior to the initiation of heartbeat at stage 10, a connection persists between two primitive systems, the head plexus and the primordial hindbrain channels. Both of these systems provide nutrition to the embryo function by ebb-and-flow diffusion. Stage 9 marks the physical connection between the heart, the dorsal aortae, and the PHCs. New vessels sprout forward to serve the midbrain and forebrain, the primitive internal carotid. The embryo demands better perfusion. Once heartbeat commences, the head plexus becomes disconnected…it will later serve as a template for venous drainage. In analogous fashion, once the internal carotid system connects up with the hindbrain circulation, the need for presegmental arteries between the dorsal aortae and longitudinal hindbrain channels is eliminated. These vessels involute…with exception of the trigeminal artery.

Craniofacial vascular anatomy at term makes no sense without considering the role of anastomoses. As we shall see, the anatomy of the stapedial system is very extensive. Although it arises from a single precursor vessel, it ramifies inside the skull to produce the entire meningeal system. It makes multiple connections with an extracranial system that develops concomitantly, the external carotid. Following induction cues from the intracranial sensory nerves, stapedial reaches the orbit as a supraorbital branch associated with V1. Then, perhaps for reasons involving a change in flow, the stapedial stem involutes. Anastomoses such as the occipital and the internal maxilla-mandibular (AKA internal maxillary) are now considered the source vessels for the meningeal system. The stapedial vessels supplying orbital structures now depend upon a new anastomosis with the primitive ophthalmic artery. With these concepts in mind, the anatomy of craniofacial arteries in the adult becomes more rational and mechanistic.

Part 2. Craniofacial Arterial Development by Stage (Figs. 6.39, 6.40, 6.41, 6.42, 6.43, 6.44, 6.45, and 6.46)

We shall now take up the subject at hand. Since most readers will find this material unfamiliar, let's arm ourselves with two references to stay afloat. As most readers now know, the embryonic period (the first 8 weeks of development) is divided into 23 Carnegie stages, as described by Streeter and later revised by O'Rahilly at the Carnegie Institution in Washington, DC. In 1948, D.H. Padget (also at Carnegie) performed dissections and drawings of human embryos to categorize craniofacial arterial development into six stages (not correlated with the Carnegie system). Her monumental

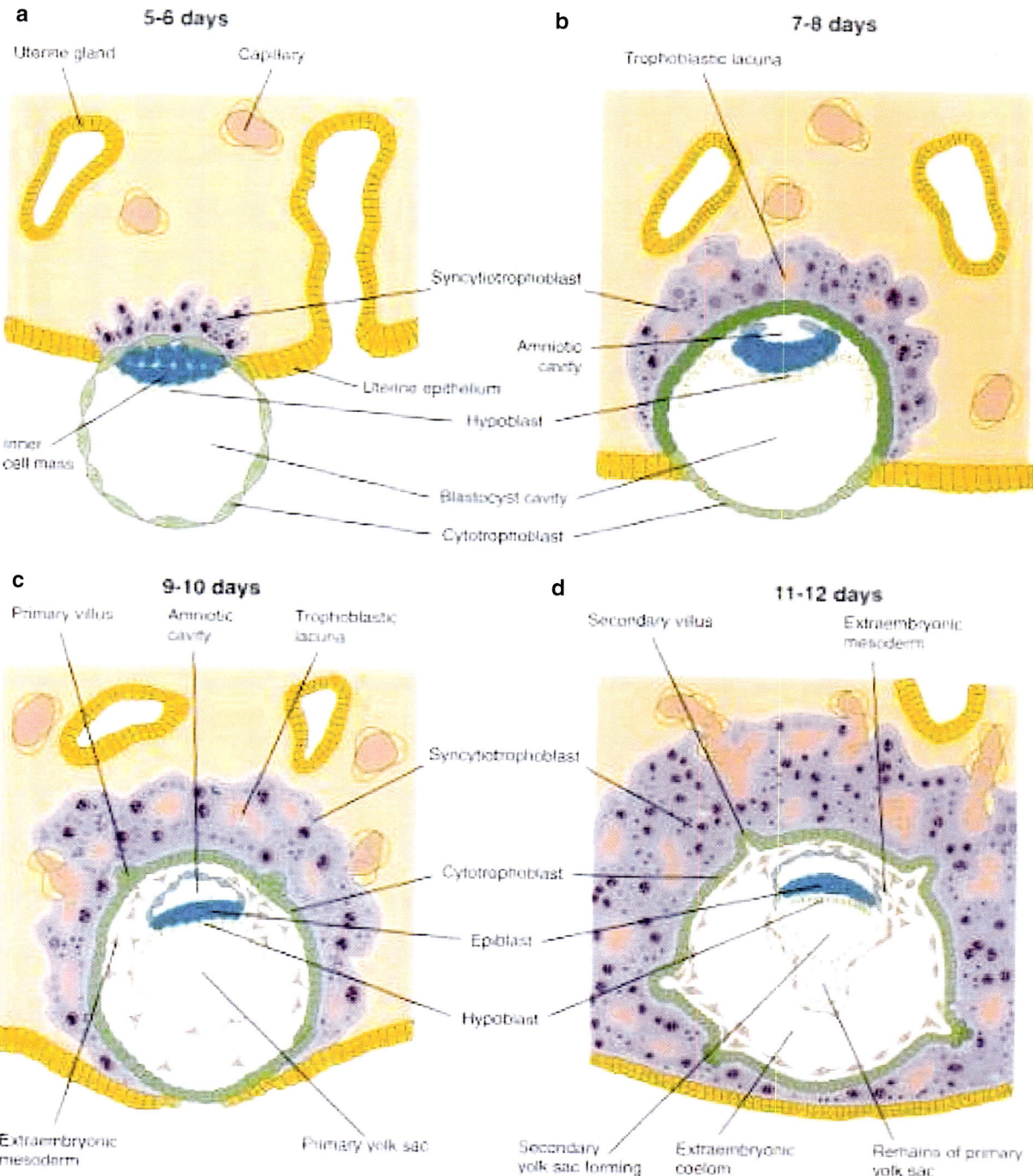

Fig. 6.40 Stage 5. BMC 5–1: Tissue derivatives (color schemes maintained in all illustrations. BMC 5–2: Origins of extraembryonic tissues. (**a**) Implantation, day 6; (**b**) blastocyst at 7.5 days; (**c**) fully implanted blastocyst at 8 days; (**d**) embryo at 9 days; (**e**) late second week. BMC 3–18: Interactions between embryo and uterine lining. Syncytiotrophoblast invades endometrial stroma whereas cytotrophoblast makes the villi. Extraembryonic blood vessels synthesized from extraembryonic mesoderm, itself a product of extraembryonic hypoblast. (Courtesy of Bruce M. Carlson, MD, PhD)

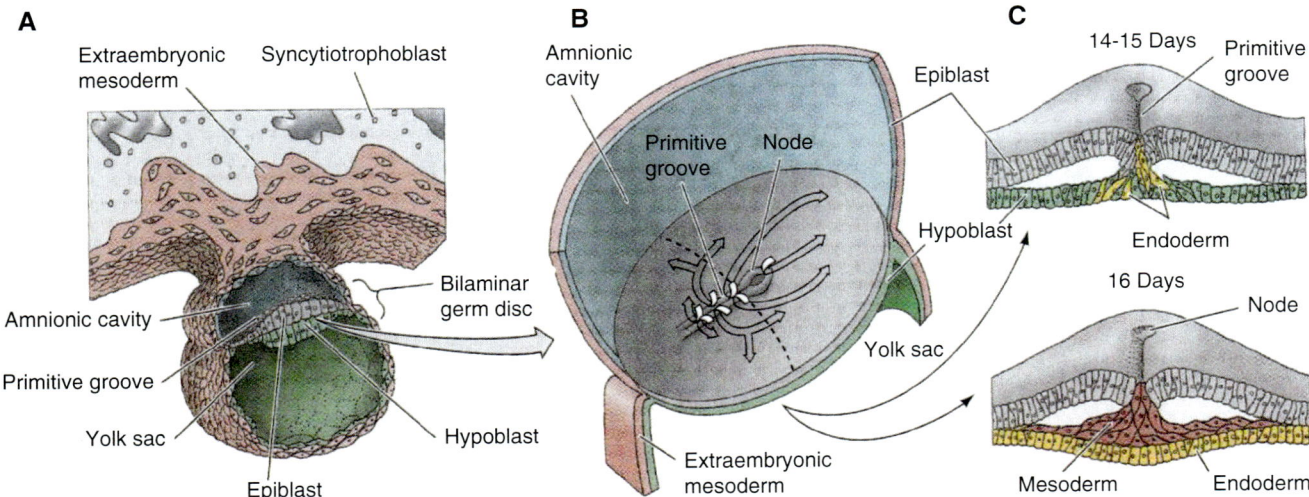

Fig. 6.41 Stages 6–7. Gastrulation is a *bidirectional* process: (1) Prechordal mesoderm produced from the anterior lip of primitive node flows forward from r0 caudal-to-cranial. (2) Just behind that zone, conventional gastrulation is cranial-to-caudal process. This mesoderm surrounds PCM like pincers. At each level, by day 14, true intraembryonic endoderm replaces hypoblast. By days 15–16, mesoderm develops. (Reprinted from Schoenwolf G, Bleyl S, Bauer P, Phillipa-West F. Third Week: Becoming Trilaminar and Establishing Body Axes. In: Larsen's Human Embryology, 5th ed. Philadelphia, PA: Churchill Livingstone; 2014: 57–80. With permission from Elsevier)

Fig. 6.42 Stages 7–8. (Courtesy of Prof. Kathleen K. Sulik, University of North Carolina)

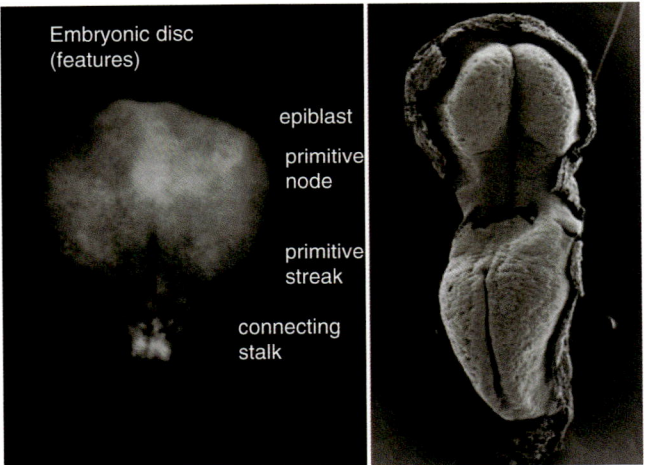

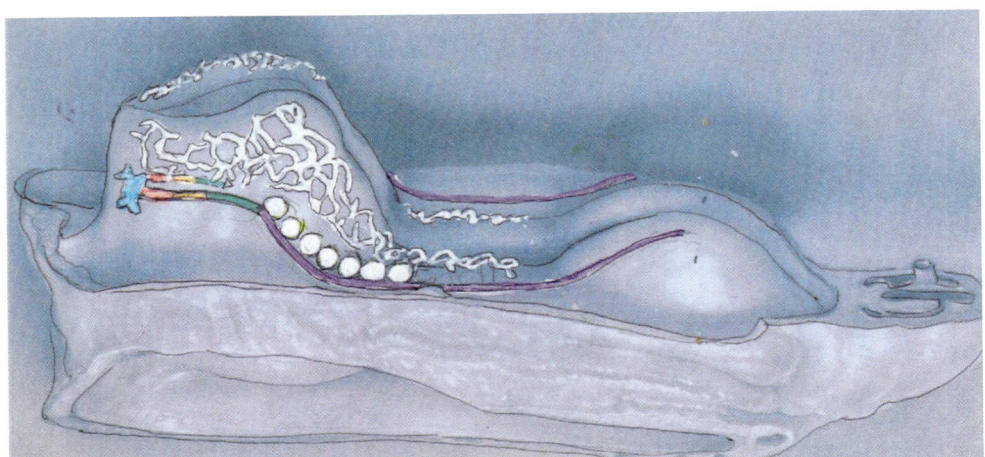

Fig. 6.43 Stage 8. Neural folds are elevating. The sides are covered with primitive vascular plexus (white). Seven somitomeres (circles) with craniofacial mesoderm are shown. At the end of stage 8, a total of 18 have been organized from the paraxial mesoderm. Dorsal aortae (purple) flank the embryo all the way from the outflow tracts (orange) to the first aortic arches (green) to its posterior terminus. These connect with tubular heart fields: outflow tracts (orange), ventricles (pink), and atria (blue). Note posterior extension of the hindbrain plexus over the as-yet undifferentiated cranial mesoderm. (Reprinted with permission from Johns Hopkins University Department of Medical Art—Gary Lees. Carnegie Institution—Sonya Bajwa)

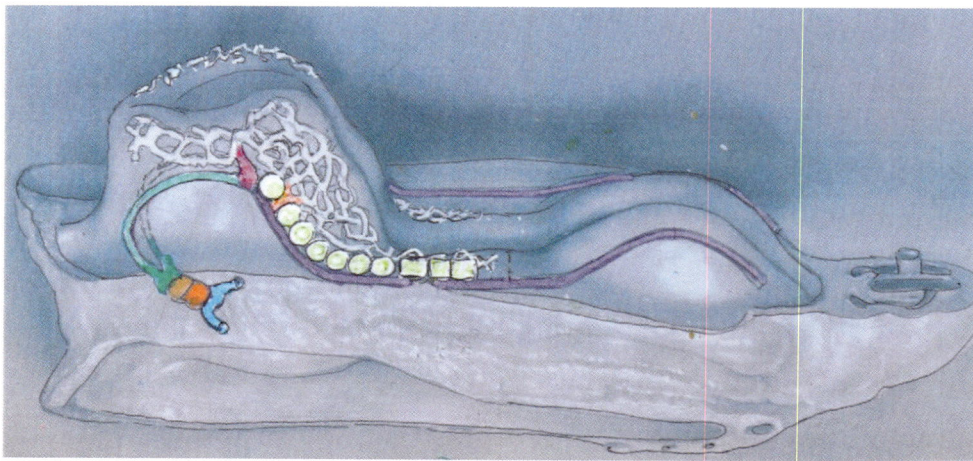

Fig. 6.44 Stage 9. At the time of the 19th somitomere appears, somitomere 8 will undergo transition to first occipital somite. Three somites (squares) are produced during stage 9. Heart has folded beneath the head retreating to a position underneath the future pharyngeal arches. Primitive maxillary artery connects to the head plexus, while primitive trigeminal supplies the primitive hindbrain plexus. (Reprinted with permission from Johns Hopkins University Department of Medical Art—Gary Lees. Carnegie Institution—Sonya Bajwa)

work is dense but well written. I have endeavored to simplify Padget's drawing using a color code, but the reader is referred herein to consult her illustrations directly. To this end, I have included references to Padget's figures. Her work, along with that of ED Conden on the aortic arches, was published by the Carnegie Institution and now available to all reader on internet. The reader is also referred to Carlson's text at hand for background information on various stages. To make things a bit easier, I have used bold suprascript numbers[123456789] to indicate the Carnegie stage[N] where indicated.

Carnegie Stage 5 (0.1–0.2 mm, 7–12 Days)
(Fig. 6.40)

General: Bilaminar embryo (epiblast + hypoblast), primary yolk sac.

CNS: Implantation, and the "invention" of extraembryonic circulation.

CF circulation: Formation of extraembryonic mesoderm as source for placenta, umbilical cord, and primordial hematopoiesis (in the yolk sac).

We begin with some preliminary remarks. Simple diffusion cannot supply the metabolic needs of the rapidly growing mammalian conceptus. A circulatory system consisting of a pump, vessels and blood cells, and a connection with an extraembryonic source of nutrition and oxygen must quickly be assembled. By the completion of stage 7, the embryo has primordial vessels supporting unsegmented mesoderm and an incipient nervous system. At stage 8, connections with the heart are established with an ebb-and-flow circulation. This makes sense because the CNS also appears at this time and

its tissues have high metabolic requirements. The embryo is now sufficiently complex that its survival demands a maternal life support system. This is a case of *supply or die*. This requires simultaneous elaboration of vessels by both the embryo and by extraembryonic tissues. The heart is *plexiform* at stage 9, becomes *straight tubular* at stage 10, and assumes its *S-shaped loop* by stage 11. At stage 11, circulation into the sinus venosus is established from three sources. Cardinal veins from the body and vitelline veins from the yolk sac carry desaturated blood, while the left umbilical vein brings oxygenated blood from the placenta.

Blood vessels and blood cells are constructed from mesoderm. The process of gastrulation[6–7] produces *intraembryonic mesoderm* (IEM). Thus, embryos prior to gastrulation don't have vessels. During this same time period, vascular events are taking place outside the embryo. Hematatopoiesis and a vascular connection with the mother (placentation) must be initiated. These processes make use of *extraembryonic mesoderm* (EEM). EEM is *not* produced by gastrulation. Stem cells destined for this task must be *physically relocated from the embryo to the yolk sac prior to the onset of gastrulation*.

Stage 5 is concerned with the precursor steps required to position stem cells where they will be needed to make this system work. Attachment of the bilaminar blastocyst to the uterine epithelium[4] takes place on day 5–6. (Recall that the periphery of the blastocyst is defined by single-cell-layer cytotrophoblast). Implantation during stage 5 has three 48-h events leading to creation of extraembryonic vascular tissues. (1) The hypoblast spreads outward to line the inner wall of the blastocyst cavity[5a]. (2) The potential space between the primary yolk sac and the cytotrophoblast becomes populated by a new layer, the extraembryonic mesoderm (EEM)[5b]. (3)

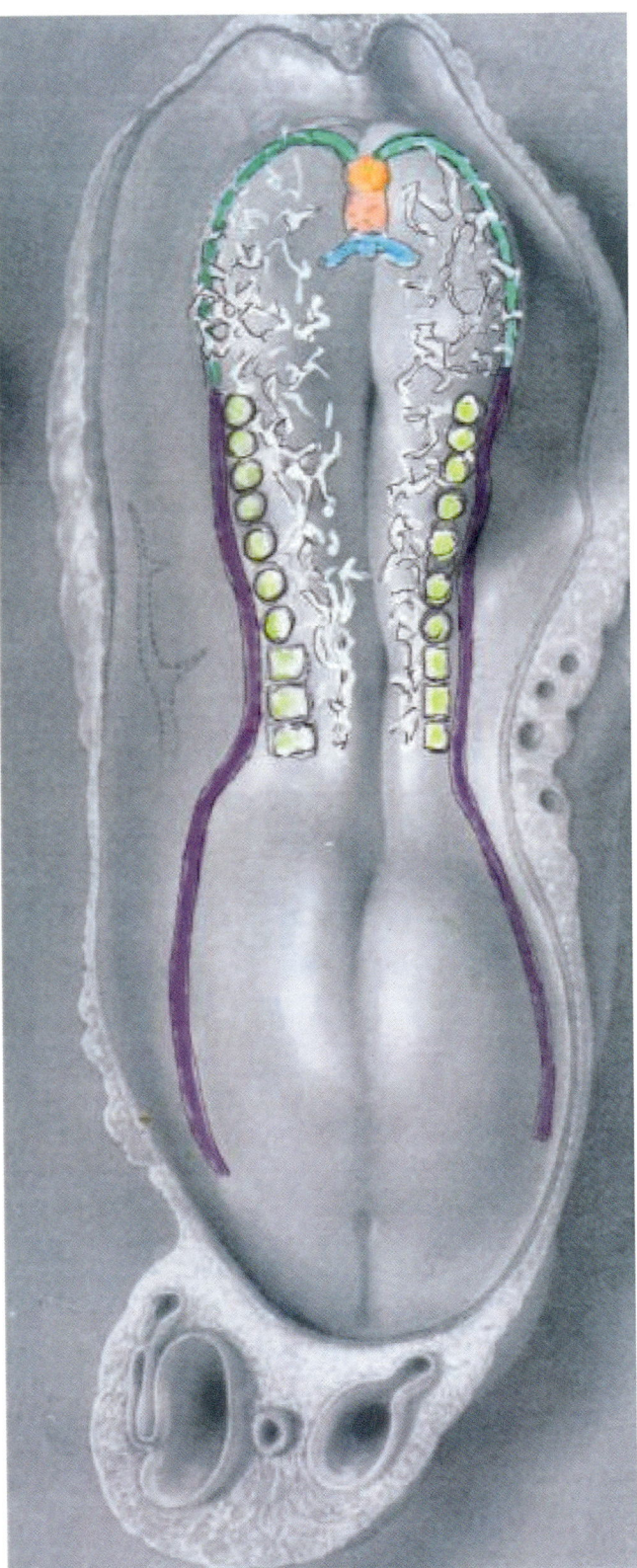

Fig. 6.45 Stage 9. Axial view shows CNS plexus (white) medial to mesoderm and dorsal aortae (purple) lateral to it; they are in continuity with the outflow tracts (orange). (Reprinted with permission from Johns Hopkins University Department of Medical Art—Gary Lees. Carnegie Institution—Sonya Bajwa)

At the completion of implantation, the EEM forms two linings: *extraembryonic somatic endoderm* (EEM$_S$) lies *internal* to cytotrophoblast, while *extraembryonic visceral mesoderm* (EEM$_V$) lies *external* to the hypoblast primary yolk sac. The space between EEM$_S$ and EEM$_V$ is the *extraembryonic coelom*[5c].

These events will prove significant for the vascular system. EEM$_V$ will ultimately become organized into *blood islands*[8]. These fabricate blood-forming cells and endothelial cells; these contain vascular endothelial growth factor receptor (VEGFR-2). Synthesis by surrounding mesenchyme of vascular endothelial growth factor (VEGF-A) causes blood vessel formation in the yolk sac. The hematopoietic duties of the yolk sac are turned over to the liver at 6 weeks.

Due to the "invention" of the placenta, the yolk sac in mammals is small and has little yolk. As the embryo grows and folds, the ventrally located yolk sac becomes associated with the developing gut in a progressively attenuated umbilical cord. It ultimately loses contact with the gut (a remnant can cause a Meckel's diverticulum). Blood vessels that developed to supply the yolk sac persist and remain in contact with the gut. The paired vitelline veins enter the substance of the liver where they become plexiform and ultimately form the hepatic portal vein that drains the intestine and distributes nutrients from the gut throughout the liver.

We turn our attention to the maternal interface. When the blastocyst starts its penetration into the uterine epithelium, the endometrium, "under the influence" of progesterone, has been preparing itself with increase in biosynthetic capacity. The endometrial cells appear epithelioid and full of glycogen[5a]. Implantation of the blastocyst sets off further changes, known as the *decidual reaction*. The cells now stuff themselves with lipid and begin to produce prolactin and insulin-like growth factor. They also acquire a new title: decidual cells[5b]. But decidualization, like beauty, is a superficial, affecting only the stratus compactum. The deeper region, the stratus spongiosum has little stroma (but a whole lot of dilated glands).

"It takes two to tango." Like two dance partners, the uterus and embryo interact in stage 5 to set up vascular connections leading to formation of the placenta[5b–7]. Upon contact with the uterine lining, the single-layer trophoblast divides into an inner *cytotrophoblast* (CTB) and outer *syncytiotrophoblast* (STB)[5a]. The latter is a fusion of mononucleated cytotrophoblastic cells into a specialized mass of multinucleated cytoplasm that is highly invasive of maternal tissue. The burrowing action of STB literally "pulls" the embryo into the uterine wall. STB is initially not continuous around the embryo; it develops at the contact site with the uterus. Once implantation is completed, STB flows amoeba-like along the outer surface of CTB, surrounding it completely. STB produces a number of important hormones, including steroids, human chorionic gonadotropin and

Fig. 6.46 Stage 9. The CNS at this stage occupies about half the total length of the embryo. Note the disparately large size of midbrain. Forebrain is almost exclusively diencephalon. Neural crest migration begins from midbrain and rostral hindbrain. (Reprinted from O'Rahilly R, Müller F. The Embryonic Human Brain: An Atlas of Developmental Stages. Wiley-Liss, 2005. With permission from John Wiley & Sons, Inc.)

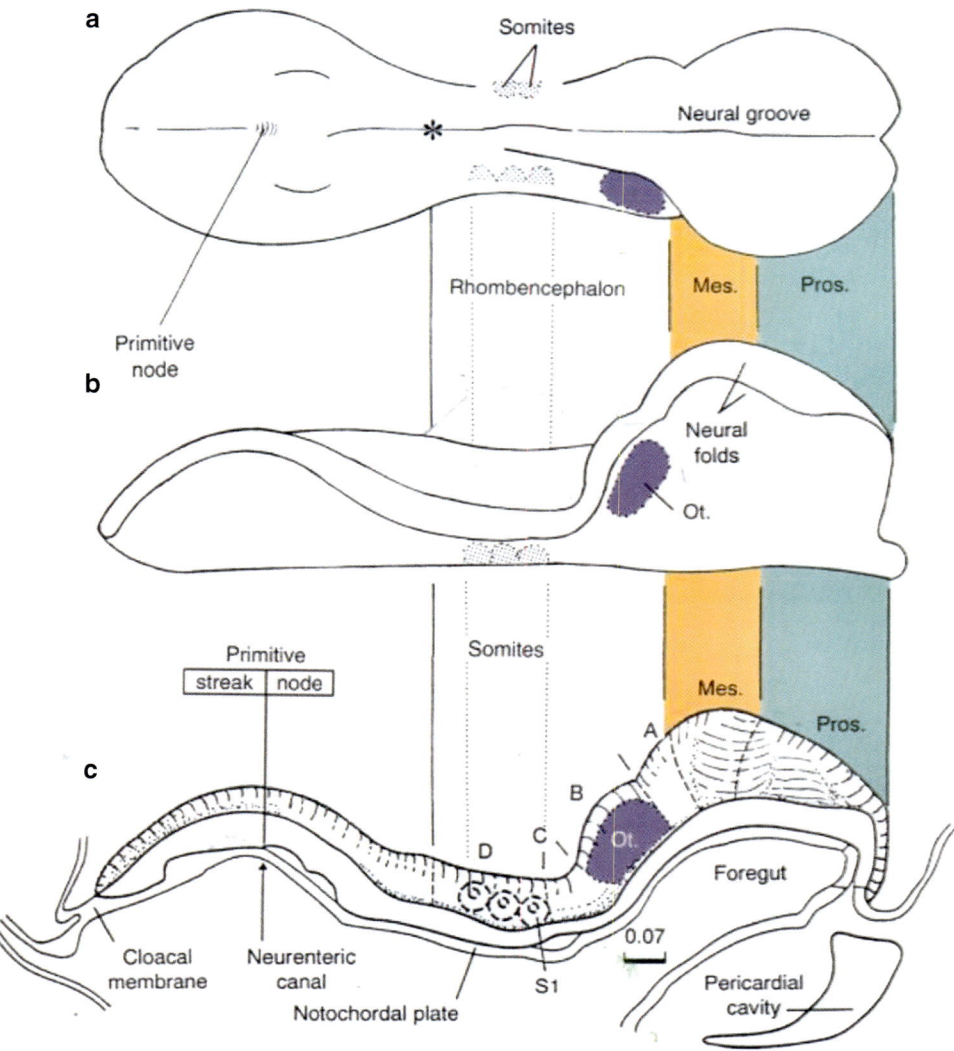

somatotropin (hCH, hCS), and human placental lactogen (hPL). At about 1 week, STB-produced hCG is detectable in maternal urine and plasma. Fluid-filled lacunae form within STB. These come into play when STB erodes uterine blood vessels. The endometrial microvasculature becomes replaced by trophoblastic lacunae.

While this takes place, CTB is not an idle observer. It forms finger-like projections, *stem villi*, into STB[5b]. Each villus has a central core filled with stem cells[5c]. In the subsequent stage, distinct fates will await the CTB villi. Villous CTB also produces somatostatin, gonadotropin-releasing hormone (GnRH), hCG, and histocompatibility antigens designed to shield the embryo from rejection.

Carnegie Stage 6 (0.2–0.3 mm, 13–15 Days)
(Fig. 6.41)

General features: Primitive streak, chorionic villi.
CNS: Axis determination, neuromeric patterning, and placentation.

CF circulation: Several theaters of operation are involved: (1) extraembryonic mesoderm and trophoblast layers interact to form the placenta; (2) extraembryonic mesoderm produces vessels in the yolk sac and connecting stalk; and (3) the embryonic disc undergoes cellular shifts in preparation for gastrulation (Fig. 6.20).

Note: Extraembryonic mesoderm (EEM) originates from two different sources at two different stages of development. Stage 5 EEM differentiates from the hypoblast that has spread out to line the internal surface of the trophoblast. Stage 6 EEM arises from the embryo itself during gastrulation as mesoderm is squeezed out of the embryo like toothpaste to form the body stalk. In both cases, angioplasts within the EEM quickly form blood vessels. Only the yolk sac produces blood cells.

The basic feature of the placenta is the presence of *chorionic villi*. They originate from cytotrophoblast clumps or *primary villi*[5b]. *Secondary villi* arise when extraembryonic mesoderm invades the cores[5c] and, when vascularization occurs[6] they are termed *tertiary villi*. Initially, their tips are not free. Most invaginate directly into preexisting lacunae within which they can wag about like fronds; these are called *floating villi*. Other villi develop an invasive "finger tip," or *cell column* with which they bore all the way through STB, making direct contact with the endometrium. Interaction between underlying EEM and CTB involves homeobox genes *Msx-2* and *Dlx-4*. These cell columns then spread out laterally and link up with each other, forming a *cytotrophoblastic shell* or *basal plate*. Finally, stem cell-containing EEM within the core of the villus can work its way upward, breaking through the cell column to make direct contact with the decidua. *Anchoring villi* represent the mesenchymal fusion between embryo and uterus.

During this stage, the conceptus (amnion, embryo, and yolk sac) is attached to the chorion via the connecting (body) stalk. Right at the time amnion appears, the yolk sac (YS) makes its debut[5]. Initially, it is a single-layer structure of extraembryonic hypoblast (primitive endoderm). But YS stem cells quickly produce extraembryonic mesoderm between it and CTB. The final configuration of the *primary yolk sac* is bilaminar. The "neck" of the sac is quite large, occupying nearly the entire ventral aspect of the bilaminar embryo. As such the "roof" of the YS is bounded by the intraembryonic hypoblast. But that's not the end of the story. Primitive intraembryonic endoderm gets restless again. In a second wave of migration, "hypoblast 2.0" spreads outward away from the bilaminar disc, shoving hypoblast 1.0 out of the way. The original yolk sac is unceremoniously ejected into the extraembryonic coelom where it withers away, never to be heard from again. Left behind is a brand-new *secondary yolk sac*.

Genes determining formation of primitive blood vessels by EEM stem cells are switched on at the same time in both YS 2.0 and the placenta. A spatial dilemma results. The secondary yolk sac dangles off the ventral surface of the embryo into the coelom like the pendulum of a clock. Although YS 2.0 and the placenta share a common layer of EEM, they are spatially quite removed from one another. How do yolk sac vessels connect up with their placental counterparts?

The embryo resolves this topologic problem in a very clever way. Growth is an asymmetrical process characterized by regional differences in cell volume and points of fixation to extraembryonic structures. These physical factors cause folding along the periphery of the embryo. This causes the connection between the developing gut and the yolk sac to narrow down to a stalk. YS 2.0 gets smaller. The spherical remnant of the secondary yolk sac is literally "milked" down the stalk, occupying a distal position. Its attenuated stalk is incorporated into the steadily lengthening umbilical cord. The secondary yolk sac is now adjacent to the chorionic plate. The two EEMs now fuse. The multiplicity of yolk sac blood vessels is now connected to the placenta. As one moves back toward the embryo, efferent flow from-and-to the placenta must be "funneled down" to fit inside the umbilical cord. Two umbilical veins and two arteries (one of which is usually resorbed) do the job.

Yolk sac extraembryonic mesoderm thus functions as a "vascular conduit" between the maternal life support apparatus and the embryo. But that's not all. It is also responsible for hematopoiesis via blood islands. Furthermore, primordial germ cells, formed here, eventually migrate into the gut and, via the dorsal mesentery, populate the gonads. But all three functions (nutrition, blood cells, and germs cells) require a physical connection with the embryo itself. How does this occur?

Blood vessels in the embryo per se appear one stage later than those of the yolk sac and placenta. The mechanisms are the same. The paired dorsal aortae and primordial hindbrain channels develop within intraembryonic mesoderm. All three regions of IEM (paraxial, intermediate, and lateral plate) are capable of forming vessels. Prechordal plate mesoderm and neural crest do not have the capacity to form blood vessels. Obviously, this process cannot occur until gastrulation[7] is initiated. Intraembryonic events during stage 6 prepare the bilaminar embryo for its Houdini-like transformation into a 3-g layer sandwich. The end point of stage 6 is the elaboration of a primitive streak, the mechanical means by which birds and mammals gastrulate. But the streak has an intrinsic significance as well: it represents the confirmation of axis formation. Not only does the embryo have dorsal-ventral layers (epiblast and hypoblast), but the streak, arising from the posterior pole, sets up A-P and left-right axes as well. Finally, because the streak extends forward to a genetically defined point about 3/4 of the way along the epiblast, it defines the future dimensions of the embryonic CNS. Let's take a look at these steps in greater detail.

At first glance, the epiblast appears to be a monotonous layer of pseudostratified epithelium. In theory, all epiblast cells are totipotent; they can produce a full complement of homeotic genes. Looks can be deceiving. By the end of implantation, slight changes in shape[5e] have taken place, making the bilaminar embryo ovoid with the sperm entry point posterior. With our previous model of HNK-1 positive and negative cells in mind, HNK-1 accumulation at the posterior pole creates a *primitive gastrula organizer* (PGO). The consequent mathematical cellular redistribution within the remaining epiblast sets up an anteroposterior and medial-lateral electrochemical gradients that influence via epigenetic means the expression of homeotic genes by the cells of the epiblast.

The physical chemistry of epiblast rearrangement thus creates a *homeotic gradient* whereby individual zones of the epiblast contain unique combinations of homeotic genes that will be expressed further on in development. This, in turn, is responsible for the anteroposterior and medio-lateral axis specification.

Four molecules elaborated by PGO (chordal, noggin, crypto, and Vg1) induce the primitive streak. The streak is characterized by interaction of epiblast cells with hypoblast (primitive endoderm) in the midline; it proceeds forward from the posterior "pole" to a predetermined terminus. At the rostral end of PS, a proliferative population of epiblast piles up to produce the *primitive node*. The location of the node could be determined by (1) the distal extent of diffusion of PGO signals or by (2) the interaction of a signal with a particular homeotic zone. The dorsal part of the node is indistinguishable from epiblast, but the ventral part is an *organizer* from which projects a *notochordal process*. During stage 7 cells contributed at this point will form: (1) foregut endoderm, (2) neural tube floorplate, and (3) paraxial mesoderm.

Although this stage predates gastrulation, gene expression by a zone of the most-anterior midline hypoblast represents the *first step toward induction of the forebrain*. This zone bears the illogical name *anterior visceral endoderm* (AVE). Being a single layer, how can the epiblast be visceral (versus somatic)? ***The native fate of ectoderm is neural unless repressed***. Gastrulation results in the interposition of definitive endoderm and mesoderm rostral to the primitive node. BMPs and Wnt produced by mesoderm block transformation of ectoderm to neural plate. Signals diffusing outward and backward from the AVE antagonize BMP and Wnt. Thus, forebrain induction occurs, thanks to the anti-repressor function of AVE.

In conclusion, two critical events in the stage 6 embryo proper set the stage for its further development. (1) Elaboration of a primitive streak provides the physical mechanism of epiblast migration. The extent of the streak defines the entire body and hindbrain. (2) Specialization of the primitive endoderm/hypoblast creates a forebrain induction center.

Carnegie Stage 7 (0.4–1.0 mm, 16–17 Days)
(Figs. 6.41 and 6.42)

General features: Gastrulation: trilaminar embryo completes formation of germ layers. Notochordal process laid down which determines neuromeric induction of the CNS.

CNS: Germ layers and the notochord preinduction of CNS.

CF circulation: (1) Primordial vessels (primordial head plexus, primordial hindbrain channels, and future dorsal aortae) develop parallel to the neural plate. (2) Vitelline veins develop. (3) Dorsal aortae grow forward and unite as horseshoe-shaped primary heart field. (4) BMP-2 induces secondary heart field. No heartbeat, no blood cells, no folding, and no aortic arches.

At stage 7 *chorionic villi* containing blood vessels within their mesodermal cores are in contact with uterine circulation. 100 spiral arteries surround the embryo. When specialized *invasive CTB cells* migrate out from anchoring villi, the spiral arteries are eroded. Their walls are also modified and widened by CTB, creating increased flow at lower pressures.

Maternal blood that enters the intervillous spaces and bathes the villi has the following characteristics: it is relatively acellular, it is under low pressure, and has a low pO_2. Initial embryonic erythroblasts contain hemoglobin adapted for these conditions. At 12 weeks, when maternal blood has a higher pO_2, fetal hemoglobin undergoes an isoform switch with a more efficient binding curve.

Normal umbilical cords contain two arteries and a single vein (the right umbilical vein degenerates). Venous drainage from the placenta provides oxygenated blood to the embryo via umbilical veins originally paired but the right one degenerates. Two umbilical arteries persist.

Although gastrulation has been discussed in a previous chapter, key aspects need to be repeated. Once an epiblast cell enters the primitive streak, its ultimate fate is determined by two sets of factors: (1) the time of entry into the streak and (2) the spatial position of the cell within the epiblast with respect to the primitive streak. [The positional effect has to do with the expression of homeotic proteins by the notochord, discussed below.]

Gastrulation involves medial displacement of successive groups of epiblast cells into the primitive streak. *The physical position of a cell within the epiblast determines its fate.* The first group of cells to enter the primitive streak are those cells positioned closest to it. After they enter the streak, they displace the primitive endoderm cells laterally. The hypoblast is literally pushed out until it becomes completely displaced into the yolk sac as *extraembryonic endoderm*. The newcomer cells beneath the hypoblast layer become the definitive *intraembryonic endoderm*. Note that a specific population of IEE in the midline axis of the embryo becomes the notochord.

The next group of cells to enter the primitive streak are those originally positioned more laterally within the epiblast. These cells produce mucopolysaccharides, causing an accumulation of fluid in the space between the intraembryonic endoderm and the epiblast. This process causes the two layers to be physically "pried apart", permitting the formation of an intermediate layer. As the cells continue to stream into the space, the "pioneers" are displaced outside the embryo to form the *body stalk* (the future umbilical cord). "Newcomer" cells now fill the space between endoderm and epiblast. This will become the definitive *intraembryonic mesoderm*.

Note: ***Cells entering the streak do so at a specific homeotic level. At gastrulation, they are exposed to the local electrochemical environment and thus acquire the hox code of their entry site.***

The following point is so important that it merits a paragraph all its own. Mesoderm produced by gastrulation at stage 7 becomes continuous from mesoderm produced by the extraembryonic endoderm at stage 5. ***The intraembryonic mesoderm is continuous with the extraembryonic mesoderm.*** Thus, blood vessels developing outside the embryo can now make contact with blood vessels originating within the embryo. This is the basis of materno-fetal circulation.

Recall that mesoderm comes in three "flavors" (paraxial, intermediate, and lateral plate) depending upon physical location of the cells with respect to the neural tube. The "coordinates" of a future mesodermal cell within the epiblast determine when and where will it undergo mesenchymal transformation and make its escape. The site of entry at gastrulation (primitive node, vs. primitive streak) depends upon the craniocaudal level of the migrating cell. Entry into the primitive node produces first, notochord and second, paraxial mesoderm. Entry into the streak layer beneath the epiblast produces from more cranial cells, PAM, from intermediate cells, LPM, and from caudal cells, extraembryonic mesoderm.

When the disc achieves a length of 0.4 mm, an axial structure known as the *notochordal process* (NP) appears. No trace of the CNS is as yet present, but the neural plate of stage 8 will develop immediately dorsal and lateral to the neural plate. Note that immediately rostral to NP is the *prechordal plate*. Understanding the anatomy and functions of the notochord is absolutely key for anyone wishing to make sense of craniofacial anatomy. It is the "master inducer" of vertebrate segmentation; *it sets up the neuromeric axis of the embryo*. The notochord is the product of the very first wave of epiblast cells to enter the primitive node. It produces chemical signals (noggin and chordin) that: (1) convert overlying ectoderm into neural tissue, (2) specify the identity of regions of the neural floorplate; and (3) using BMPs it converts certain regions of somites into vertebrae. The notochord is "the backbone of the backbone."

Just in front of the notochord, the prechordal plate produces signals that induce the forebrain. Its cells are structurally and functionally integrated with the anterior endoderm. Prior to gastrulation, hypoblast anterior to the primitive node begins induction of the head. Later on, induction from prechordal plate mesoderm occurs via a homeobox transcription factor, Lim-1 which blocks signals from BMP-4 and Wnt, thus permitting head formation to proceed.

As gastrulation proceeds, the primitive streak regresses, leaving behind a hollow rod-like *notochordal process* consisting of the cellular precursors of prechordal plate and notochord. Thus, the notochordal process is left behind by the retreating streak, like the slime trail of a banana slug. In mammals, a transitory fusion involving the ventral midline of the notochordal process and the endoderm creates communication between the amniotic cavity above the embryo with the yolk sac below it. The function of this *neurenteric canal* is obscure but may permit passage of chemical signals. In any case, a short time later the endoderm seals up once again leaving behind the final version of the notochord, this time a solid rod running the length of the embryonic axis

from the basisphenoid through the centra of all successive vertebra. Notochordal dysfunction is disastrous: "spare the rod and spoil the child."

The notochord is relevant for craniofacial blood supply because it specifies neuromeric levels of the CNS and the subsequent gene patterns they will express. Recall our axioms: (1) *No tissue is formed without ensuring its own survival, i.e., blood supply*; (2) *When it comes to blood supply or bone, support for the brain comes first*. The developing brain is surrounded by paraxial mesoderm. Angioblasts exist within paraxial and lateral plate mesoderm, but not in notochord or prechordal plate. Thus, at the same time as blood islands are forming in the extraembryonic mesoderm of the yolk sac[8], networks of small vessels probably are already in place during somitomere formation[8] and certainly by the appearance of the first somites[9]. These primitive vessels develop from distinct sources. Alongside the somitomeres, paraxial mesoderm gives rise to a *primordial head plexus* continuous with *primordial hindbrain channels*. Further away from the neuraxis and more ventral, lateral plate mesoderm gives rise to the primitive dorsal aortae (DAs).

Cardiac embryogenesis is a chapter unto itself but a number of features provide correlations with the craniofacial development. *Homeotic definition of distinct heart zones is intrinsic to the gastrulation process*. Migration of heart-forming epiblast cells through the primitive streak occurs in a tightly controlled sequence. Primitive node forms the outflow tract; the mid-streak level forms the ventricles; while the posterior streak forms the atria. Dorsal aortae and the primary heart fields are composed of lateral plate mesoderm and neural crest. The neuromeric coding of primary heart fields extends from r0 back to r11. For conceptual purposes, the atria and ventricles receive neural crest in cranio-caudal order from r0 to r7. The outflow tracts are supplied by neural crest from r8 to r11. Finally, specific regions of the heart are anatomically defined by unique homeotic gene expression patterns.

Blood vessels supplying the CNS and the intersegmental vessels of the trunk arise as sprouts from a main trunk. The basis for sprouting is a factor, angiopoietin-1, that binds to Tie-2 receptors on endothelial cell at sites where a sprout is required. Formation of vessels walls takes place in two steps. (1) "Activated" endothelial cells recruit their own ensheathment by producing PDGF that causes mesenchymal cells to migrate toward the sprout. (2) Subsequent synthesis of TGF-B converts the mesenchyme to vascular smooth muscle and adventitia. In the head and neck, neural crest is a major source of vessel walls, but the endothelial cells are always of mesodermal origin. Not all endothelial cells possess a Tie-2 receptor. Spatial distribution of sprouts serving individual neuromeric levels depends on distribution of Tie-2 + endothelial cells in molecular register with each neuromere. Thus, *the physical act of gastrulation, where the key endothelial cells arise, may determine the anatomic basis for its subsequent vascular supply.*

Visualizing the human embryonic vascular system during stage 7 requires a bit of detective work. Padget's specimen #508 at stage 11 demonstrates the conversion of the primordial hindbrain channels to paired longitudinal neural arteries supplied cranially by presegmental branches from the dorsal aortae and first cervical segmental artery caudally. This configuration results from a series of staged transformations. Working backwards from known steps, the hypothetical configuration stage 7 blood supply follows below.

At the end of stage 7, the embryo is a three-layer planar disc. Mesoderm produced at gastrulation has fanned out in concentric arcs from the primitive node and streak. The most cranial neuromeric level of mesoderm is defined by the homeotic code of r1. Mesoderm fills the entire middle layer of the disc with the exception of two zones. Just rostral to the prechordal mesoderm, adherence of ectoderm and enDoderm to each other creates a "protected zone" into which mesoderm cannot insinuate itself. This zone, known as the *buccal-pharyngeal membrane*, has a caudal counterpart, the *cloacal membrane.*

Mesoderm acquires identities (paraxial, intermediate, and lateral plate) depending upon its distance from the midline. PAM is as yet unsegmented; the craniocaudal separation process that creates the first seven somitomeres is not initiated until stage 8. Cardiogenic mesoderm (originating from r1 to r11) starts out as two blocks of lateral plate mesoderm dorsolateral to the primitive node. These develop forward and unify in the midline, where, under the influence of local induction factors, they form the horseshoe-shaped *primary heart fields*. The posterior ends of the heart fields are *continuous with the primitive dorsal aortae.*

In ensuing stages, the cardiac anlage will undergo a series of reconfigurations. The heart fields differentiate into components (endocardium, myocardium, and epicardium), the expanding forebrain causes the heart to fold under face, and *heartbeat begins*[8]. The paired heart tubes then pass through plexiform[9] straight tubular[10] and looped phases[11]. The distal (sino-atrial) ends of endocardial tubes remain unfused; each side receives three veins[12]. From lateral to medial, these are: (1) common cardinal from the embryo, (2) umbilical veins from the chorionic villi, and (3) vitelline veins from the wall of the yolk sac (umbilical vesicle).

With this overview of the heart out of the way, we can now concentrate on earliest forms of vasculogenesis within the stage 7 embryo. Synthesis of the notochord (the future inducer of the CNS) involves the setting up of cranio-caudal genetic axis. This lays the groundwork for neural plate induction[8] with future segments (neuromeres) organized by homeotic genes. It makes sense that future arterial supply of mesoderm and the neural plate should also be in register with the CNS. Precursor endothelial cells of two parallel longitudinal systems, the primitive dorsal aortae (PrDA) and primordial hindbrain channels, arise (PrHC) within PAM and

LPM. These probably develop as *segmental endothelial units*[7] that subsequently *link up longitudinally*[8]. Later in embryogenesis, this mechanism is used repeatedly to create vessels such as the vertebral artery, thyrocervical trunk, and internal mammary supply to the sternum. Thus, at stage 7, cellular elements in homeotic register with the notochord constitute the precursors for simultaneous development of a segmented neural plate, segmented outlying mesoderm, and a vascular system to supply them both.

The definition of neural crest cells at this stage is critical to understanding cerebral vasculogenesis. As a general assumption, ectoderm is incapable of forming blood vessels. But neural crest is a form of ectomesenchme. It bears a genetic relationship with all levels of the neural tube. Cells of NC derivation form pia and arachnoid that envelope and extend into all sectors of the prosencephalon and mesencephalon. They also produce pericytes, "guide cells" that accompany, direct, and may induce blood vessels in spaces within the brain. As we shall see, the migration of neural crest cells throughout the body is intimately related to a development of fascia and forms the Schwann cells of all peripheral nerves. The concept that the body fasciae, tendons, joint capsules, and periosteum is all part of a common system raises the question of an "alternative nervous system."

Carnegie Stage 8 (1–1.5 mm, 18–19 Days) (Figs. 6.42 and 6.43)

General features: (Hensen's node and primitive pit, notochord and neurenteric canal, appearance of neural plate and neural folds, formation of somitomeres (no somites yet present), blood islands, definition of future neural crest, and migration into the body axis).

CNS: Neuromeric segmentation becomes established, primitive circulation to the brain.

CF circulation: (1) Heart fields fuse as heart tubes. (2) Heart beat initiated. (3) Erythroblasts produced in blood islands available for circulation. (4) primitive head plexus (PHP) supplies the midbrain and forebrain. (5) Primitive hindbrain channel (PHC) serves as both artery and temporary vein for hindbrain. (5) Primitive hindbrain channel receives communicating presegmental branches from dorsal aorta (DA) (Fig. 6.43).

Critical transitions take place in the stage 8 embryo. (1) Having acquired three germ layers, it is preoccupied with organizing its neural plate and mesoderm. (2) Formerly dependent upon a passive ebb-and-flow circulation, the embryo takes control of its own metabolism by means of a functional pump, the availability of O_2-carrying red cells for gas exchange, and a definitive vascular system to transport them.

The neural plate now makes its debut. Neuromeres are not recognizable anatomic entities at stage 7. We postulate that segmentation of the neural plate probably takes place concomitant with segmentation of PAM into somitomeres[8]. O'Rahilly documents their overt physical appearance at stage 9.

Metabolic support for CNS development is so crucial that, from the very moment of induction, its future blood supply is being organized along segmental, homeotic lines. Each individual neuromere maintains a previously established genetic connection[7] with its corresponding segment of PAM within which develops the primitive hindbrain channels. The system is like a ladder; PHC represents the vertical "legs," while the transverse connections to the rhombomeres are the "rungs."

The alert reader will now ask: At stage 8, what is the blood supply to the rest of the brain? Forward from r0, the primitive hindbrain channels are connected to a network of vessels that flank the developing midbrain and hindbrain known as the *primitive head plexus*. Just like the PHCs, the head plexus provides nutrition via ebb-and-flow. It has both arterial and venous functions. Beginning at stage 9, the arterial function of the head plexus is replaced by the invading primitive internal carotid. The vessels remodel into the future venous drainage of this region: anterior, middle and posterior dural plexuses will eventually morph into the dural sinuses.

Stages 7 and 8 last approximately 5 days. During this time, paraxial mesoderm laid down at the previously undergoes cranio-caudal segmentation, producing 18 somitomeres (a rate of approximately 4 somitomeres/day). As discussed previously, the segmentation process is related to homeotic signals from both the nochord and from laterally flanking mesoderm. At the stage 9, as soon as the 19th somitomere is defined, true somites make their first appearance at the level of the eighth somitomere. Recall again that somites are separated from each other by epithelium. They possess distinct zones for future bone, muscle, and dermis. The first seven somitomeres do not undergo this transition. They remain interconnected spheres of PAM admixed with neural crest.

Bone and muscle derivatives of the 7 somitomeres have been mapped out in the avian model with very little variation in mammalian embryos. *Bone derivatives are all epaxial.* Basisphenoid (BS) receives PAM most likely from Sm3 because it is adjacent to the rostral hindbrain. Parietal and squamous temporal bones are covered with temporalis muscle innervated by V3 and therefore have Sm3 mesoderm. The otic capsule has two major components. Prootic develops from Sm5 to Sm6, while opisthotic comes from Sm6 and Sm7. Somitomeric muscles are both epaxial (extraocular series) and hypaxial (muscles of the pharyngeal arches). Each pharyngeal arch receives muscles from a single somitomere. These relationships are summarized below (See Table 6.1).

Primitive blood supply to the brain comes from the head plexus and primitive hindbrain channels (Figs. 6.6, 6.44, and 6.45) All of these structures are synthesized from paraxial mesoderm. Scanning EM studies show a zone of mesoderm

Table 6.1 Muscle derivatives of the somitomeres

Somitomere/somite	Derivatives
Sm1	Superior rectus, imedial rectus, inferior rectus, levator palpebrae superioris
Sm2	Inferior oblique (possibly medial rectus)
Sm3	Superior oblique/Basisphenoid
Sm4	First pharyngeal arch: muscles of mastication
Sm5	Lateral rectus
Sm6	Second pharyngeal arch: muscles of facial expression
Sm7	Third arch: soft palate, superior/middle constrictor
Sm8/Somite 1	Fourth arch: external larynx, middle constrictor
Sm9/Somite 2	Fifth arch: internal larynx, inferior constrictor
Sm10/Somite 3	Tongue, SCM/trapezius
Sm11/Somite 4	Tongue, SCM/trapezius
Sm12/Somite 5	Tongue, SCM/trapezius, C1 muscles
Sm13/Somite 6	Tongue, SCM/trapezius, C2 muscles
Sm14/Somite 7	SCM/trapezius, C3 muscles

flanking the midbrain and forebrain: Sm1–Sm3 provide the mesenchyme of the head plexus. The hindbrain from r0 to r11 is flanked by Sm4 to Sm11; these give rise to the primitive hindbrain channels. As we shall see, these primitive structures become interconnected to the dorsal aortae which provide their blood supply. The PHCs do not continue further downward to supply the spinal cord. Instead, the dorsal aortae provide segmental branches to each neuromeric level.

We now turn our attention to the development of the dorsal aortae. At the same time as blood islands become functional[8], the *cardiogenic plate* forms in visceral lateral plate mesoderm rostral to oropharyngeal membrane. Note: Traditional embryology texts depict a split in the lateral plate mesoderm between an external somatic layer (LPMs) and an internal visceral layer (LPMv). The space between them is called the *intraembryonic coelom*. This split *does not exist* in the lateal plate mesoderm associated with the brain. Anterior to the brain, the split occurs once again; the cardiac system arises from the visceral (or splanchnic) layer of LPM.

Lateral to the brain and spinal cord, endothelial cells within the LPM aggregate with segmental neural crest cell to produce a horseshoe-shaped vascular structure. This structure has three functional zones: (1) A posterior section from r0 backwards will become the paired *dorsal aortae proper.* (2) An intermediate section from r0 forward to p6 (flanking the developing midbrain and forebrain) is the *primitive aortic arch* (the future AA1). (3) An anterior-most section forward from the brain is inverted U-shaped loop called the *primary heart field* or cardiac crescent. Anterior cells of the crescent do not express retinoic acid and assume a ventricular identity. Retinoic acid expressed by the cells of the posterior crescent creates the atria. To complete our story, a *secondary heart field* is formed by forward migration of pharyngeal arch mesoderm. first arch creates the right ventricle[9] and second arch mesoderm forms the outflow tract.

It is assembled into paired *heart tubes*. These have an anterior-posterior axis and consist of developmental zones that are neuromerically defined.

The system at this time resides within the embryonic disc, folding has not yet occurred. A side-to-side fusion between the tubes occurs anteriorly; the future cardiovascular system is H-shaped. The anterior-directed "legs" of the H will receive future venous inflow. The "crossbar" of the H is the future *sino-atrial zone*. The posteriorly directed "legs" have a *ventricular zone* followed by the *outflow tract*. The distal (most anterior) part of the outflow tract, closest to the heart, is known as the *aortic sac*; the aortic arches develop from here. The outflow tracts (OT) are in turn connected with the primitive dorsal aortae.

When head folds arise and when heart flexion creates the first aortic arch[9], a large dorsally directed *primitive trigeminal artery* arises from the flexion site and connects DA with PHC. Additional intersegmental vessels are clearly identifiable at 3 mm[10]. In the following stage, named primitive transverse connections (*trigeminal, otic,* and *hypoglossal*) can be identified[11]. These correspond neuromerically to r0–r3, r4–r7, and r8–r11, respectively. It is not known exactly when these communications are established, but they likely develop in a cranio-caudal sequence. In any case, at stage 8, the PHCs and DAs are *not as yet plugged into the cardiac system*. They simply function as conduits for ebb-and-flow. Because the PHP-PHCs are the exclusive supply of the primitive CNS, their physical extent begins at future p6 and terminates at r11. The DAs will run the full course of embryo, supplying all 36 somites (43 somitomeres).

Stage 8 vascular anatomy plays an important mechanical role in the subsequent events, leading to formation of the first aortic arch. Forward bending of the embryo induced by head growth causes the cardiac loop to be carried ventral to the foregut at stage 9. Scanning electron micrographs of the embryonic face reveal a prosencephalic bulge flanked on either side by the mesenchyme of somitomeric mesoderm. Immediately caudal to forebrain, a 90° bend occurs at the junction of the dorsal aortae and the outflow tracts. Two important arteries project from DA at the take-off of AA1: (1) a medially directed *primitive trigeminal artery* (PrTA) interconnects DA with LNA and supplies the trigeminal ganglion; (2) an anteriorly directed *primitive internal carotid artery* (PrICA) supplies the forebrain and midbrain. This will include fronto-naso-orbital mesenchyme.

Why should a soft tissue system such as this permit such an acute angle bend without itself becoming distorted? When the head flexes, what holds the rest of the embryonic axis in place? What explains the location of the bend at r0/r1? A theoretical answer lies in the series of vascular connections initiated during stage 8. First, the PHP-PHC system is firmly attached to the embryonic brain via a series of vessels (in theory, one for each neuromere). Second, PHP-

PHC are likewise segmentally attached to the seven somitomeres. Third, PHP-PHC are become physically connected to dorsal aortae by three or four *presegmental arteries* which provide blood supply to the neural plate as follows: *primitive trigeminal* serves r0–r3, a putative *primitive hyoid* serves r4–r5, *primitive otic* serves r6–r7, and *primitive hypoglossal* r8–r11. Note that the primitive trigeminal was likely "hardwired" at stage 7. Forward from r0, the dorsal aortae are disconnected from somitomeres. These physical connections ensure that, when mesencephalic flexion creates a right-angle bend at the OT-PDA junction, the *original r0–r1 PDA site stays put*.

We come now to an all-important point: **neural crest migration into adjacent structures**. This is a very early event. [For a good overall description of the process, see the relevant chapter in Carlson.] Neural crest invasion assumes three forms depending on the location. Hindbrain neural crest from r2 backwards is *segmental* into somitomeres and pharyngeal arches. Midbrain neural crest from m1 to m2 and r0 to r1 migrates as *streams* into the anterior cranial base and midface. Forebrain neural crest travels as a *glacier-like sheet* from the posterior neural folds of the diencephalon forward to populate the anterior neural folds. Our emphasis here is the hindbrain. Crest cells surround somitomeres. As the aortic arches develop from LPMv starting in stage 9, the vessels are surrounded by neural crest mesenchyme to create the physical structures known as pharyngeal arches.

Imagine the pharyngeal arches as expansile sacs filled with neural crest with an aortic arch artery running up the center of the sac. Crest cells create the future connective tissue system of the head and neck: fascia, tendons, and periosteum. Responding to the "genetic map" of the epithelial cover of each arch, neural crest creates subdivisions, "pockets" with which future myoblasts will eventually fill. Rapid multiplication of craniofacial neural crest cells drives physical expansion of mesenchymal structures that project outward ("grow") from the neural axis. Five pharyngeal arches result.

As we shall see, the subsequent 6 stages are characterized by the sequential appearance of six aortic arches. AA1 results from the passive bending of the vascular circuit. AA2–AA5 arise as stems from the aortic sac. The first five aortic arches are associated with pharyngeal arches, while the AA6 contributes to the pulmonary circulation. The first four aortic arches share common anatomic features. Each artery becomes engulfed by (mostly) neural crest and (some) paraxial mesoderm to create the pharyngeal arch system. Aortic arches fulfill three functions: (1) they conduct blood from the cardiac pump into the embryonic body via the DA-LNA system; (2) they constitute a vascular "core" around which mesenchyme organizes itself into a pharyngeal arch; and (3) after a remodeling process, remnants of the arches produce derivative structures: common carotid, external carotid, the outflow tract, and great vessels.

The aortic arch system is more than a vascular conduit away from the heart. It also represents a physical means by which stem cells critical for cardiac development can enter the interior of the heart, creating structures such as the semilunar valves. Well-known associations exist between congenital heart anomalies and pharyngeal arch dysmorphology. These can result from defects in *secondary proliferation/migration* of neural crest. They can also result from defects in target tissue. Molecular studies of gene expression within the developing heart and great vessels demonstrate homeotic-based segmentation. Problems in a common homeotic zone may have craniofacial, as well as cardiac, manifestations, e.g., velo-cardio-facial syndrome. We shall now turn our attention to the relationships that develop at stage 9 among the dorsal aortae, the aortic arches, and the somitomere-somite system.

Carnegie Stage 9 (1.5–2.5 mm, 20–21 Days, 1–3 Somites) (Figs. 6.44, 6.45, and 6.46)

General: Appearance of first three somites, deep neural groove, cranial neural folds elevate. Early heart tubes. first aortic arch artery, first pharyngeal arch. Splitting of the eye field.

CNS: Major divisions of the brain.

CF circulation: Folding brings heart ventral to developing forebrain. First aortic arch and pharyngeal arch develop. Primitive internal carotid and primitive trigeminal arteries arise. PrTA supplies anterior end of PHC. Connection between AA1 and the primitive hindbrain channels via trigeminal artery supplies the developing hindbrain (MC Fig. 6.47).

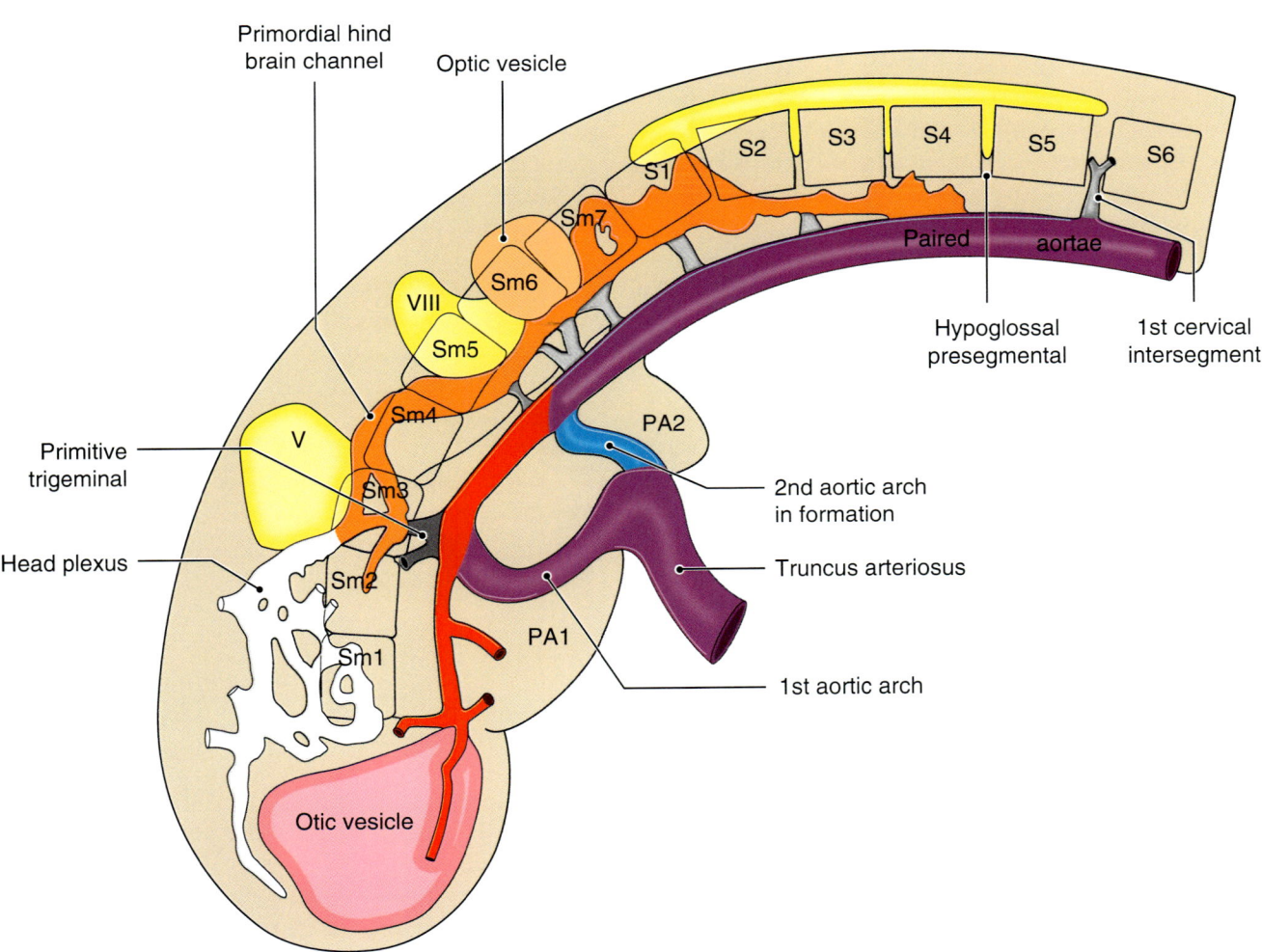

Fig. 6.47 Stage 10. Youngest embryo in the DHP series…may be as late Primitive hindbrain channel is in continuity with the superficial and deep head plexus. Together these act like a vein, until multiple primitive presegmental plug in at various cranial nerve levels (trigeminal, otic, parotic, and hypoglossal) causing arterialization. The longitudinal neural arteries form as a low-resistance "superhighway" to accommodate the flow. Preferential growth of the LNAs causes the plexus to involute. Note PrTrigeminal (PrTG) is the largest presegmental artery. Primitive internal carotid (PrICA) directed toward the optic vesicle which appears at stage 10

Pharyngeal arch formation is initiated during this stage. Scanning electron micrographs of the embryonic face shows the mushrooming prosencephalon flanked on either side by the mesenchymal mass of neural crest mesenchyme. Immediately caudal, the bulges of the first pharyngeal arches are visible. At their core is a fusion of LPM of the dorsal aortas originating at levels r2–r3. AA1 is surrounded by r2–r3 neural crest. Filling out the sac of the first pharyngeal arch are myoblasts originating from Sm4. Thus, although the components of the first pharyngeal arch are present at stage 8, the core artery is identifiably visible at stage 9. By stage 10, PA1 is readily appreciated. In fact, the presence of PA1 is a defining characteristic of stage 9.

The common conception of aortic arches is that they develop upwardly directed (dorsal) from the aortic sac until they come into contact with five ventrally directed mesenchymal "bags," the pharyngeal arches. The reality is quite different. Aortic arches 2–5 result from the union of two distinct sources that meet one another within the substance of the pharyngeal arch: the dorsal aortae give off ventrally directed stems, while the aortic sac produces dorsally directed stems. In the case of the fifth arch, the union fails, although occasional remnants can cause cardiac anomalies.

Development of the first aortic arch is quite different. It is located just beneath the trigeminal ganglion, i.e., below r1–r3. Let's paint this ring-like zone of dorsal aorta fluorescent yellow. Forebrain growth and cardiac displacement induce a 90° kink at this site. With the heart tucked under the face and the DAs strapped onto the neuraxis, *something* has to give…. the yellow zone, is now "stretched" downward like an elastic sock, becoming an elongated yellow tube with a new name: AA1.

Why should the take-off of AA1 be located where it is? Two events occur *prior to head folding*…and *prior to AA1 formation*. Labelling studies demonstrate that paraxial mesoderm from Sm1 to Sm3 remains distinct from the first arch. They are assigned to the orbit and are positioned forward from r0. Embryos always protect newly formed tissues by ensuring adequate blood supply. From the r1 junction of DA and with first aortic arch, two critical vessels develop. (1) *Primitive internal carotid artery* (PrICA) is directed forward. It supplies the forebrain and future orbito-nasal mesenchyme. It probably does so by following anatomic "clues" from the receding primitive head plexus. (2) *Primitive trigeminal artery* (PrTA) interconnects dorsal aorta with longitudinal neural artery; it supplies the trigeminal ganglion. Of the several *presegmental branches* linking dorsal aorta with longitudinal artery, PrTA is the first, largest, and most constant. It is the main conduit from the heart to the hindbrain. All other presegmental branches are unstable, PrTA persists as a remnant.

For mechanical reasons, *fixation of dorsal aorta to the brain* by PrICA and PrTA causes the "hinge point" of the first aortic arch to take level of r0 and r1…just opposite the cranial end of the notochord…and just opposite the future

pituitary. Downward descent of the heart stretches out the intermediate segment of dorsal aorta into the first aortic arch.

Just cranial to its origin, primitive internal carotid produces branches destined to supply the cranial base. The first is *primitive maxillary artery (*PrMxA), which is originally assigned to the anterior forebrain and optic anlage. These duties are temporary. As soon as primitive ICA reaches the optic vesicle, maxillary is diverted to supply the posterior hypophysis. The anterior hypophysis will be supplied by pial vessels. In conclusion, *blood supply is a response to the needs of mesenchyme*. The presence of these two vessels, sprouting at the level of r0–r1, ensures that downward bending of dorsal aortae giving rise to the AA1 will occur adjacent to trigeminal ganglion/somitomere 4 and nowhere else.

In actuality, primitive maxillary artery is a misnomer, since it has nothing to do with the maxilla. However, the zone in which maxillary develops is quickly populated by neural crest from r0 to r2 which will ultimately be supplied by the various stapedial arteries associated with V1 and V2 sensory branches. The back wall of the orbit, alisphenoid (AS), demonstrates this mesenchymal mix. AS is synthesized from two sources: r1 (from midbrain) medially and the r2 lateral wall of temporal fossa. Abutting against it is hindbrain neural crest. For this reason, Tessier cleft zone 9 (the alisphenoid field) is notoriously resistant to clefting.

For the irrepressibly curious alisphenoid, two branches of StV1 pass through alisphenoid. Lateral meningeal branch of StV1 traverses what is cranio-orbital foramen lateral to superior orbital tissue, while orbital branch passes through the tissue proper. Lacrimal artery is induced by orbital branch of StV1, but meningeal branch joins it as well. Prior to anastomosis with ophthalmic, both the StV1 meningeal and StV1 orbital are supplied from behind by supraorbital stapedial. More on all this will be found in the section on the stapedial system.

Carnegie Stage 10 (2.0–3.5 mm, 22–23 Days, 4–12 Somites) (Fig. 6.47)

General: Beginning fusion of neural folds, curving of the embryo, beginning heartbeat. second aortic arch artery, second pharyngeal arch.
CNS: Neural tube and optic primordium.
CF circulation: Second aortic arch artery arises de novo from truncus arteriosus. (1) First stage in differentiation of supply and drainage vessels. (2) Primordial hindbrain channels persist; these are precursors of the longitudinal neural arteries (LNAs) (Fig. 6.6). (3) Primitive internal carotid arrives at the basal forebrain.

The next event in the "circulatory cascade" is the *formation of the second aortic arch*. The scenario is different from its predecessor, because AA2 arises *de novo* as a dorsally directed "sprout" from the aortic sac. Its take-off is just caudal to AA1. The stimulus is unclear, but would reasonably involve signaling between the sac and the r4–r5 segment of dorsal aorta. At the dorsal aspect of the arch, AA2 anastomoses with the dorsal aorta. Nearby, on the opposite side of dorsal aorta, an unnamed presegmental stem interconnects DA with primitive hindbrain channels and the future facial-acoustic nucleus.

From the lateral walls of diencephalon, optic primordia appear. This calls for immediate vascular support which is dutifully supplied by primitive maxillary artery, the first branch of the incipient primitive internal carotid. The sequence of events preceeding this is as follows. At late gastrulation[7–8], under the control of a homeotic gene RAX (retina and anterior homeobox), a single eye field develops just ahead of prechordal plate. The eye field (and the entire forebrain) is split under the influence of sonic hedgehog (shh) secreted by prechordal plate and ventral diencephalic floorplate (basal plate of neuromere p5)[8–9]. Rax and Pax-6 induce the optic primordium[10].

This process of anterior forebrain development does not always go to completion. It can be potentially inhibited by the posteriorizing action of locally produced Wnt. The expression of transcription factor Six-3 blocks Wnt and permits formation of separate eye fields. Six-3 failure leads to loss of shh and failure of the eye fields to split. This causes the clinical spectrum of cyclopia.

Anterior cranial base and the base of the optic primordium are supplied by primitive maxillary. Primitive internal carotid has arrived at the caudal margin of the optic primor-

dium, but has not passed forward dorsally. Recall that the optic anlage is originally at a 180° angle from the brain, so its arriving vessels and neural crest encounter its "backwall." Furthermore, of the four midbrain neural crest populations, r0 and r1 migrate prior to m2 and m1. Thus, r0–r1 cover the anterior cranial base and forebrain, while m1–m2 coat the eyeball, with m1 theoretically assigned to the dorsolateral sector and m2 to the ventromedial sector.

Segmental vessels to the hindbrain and cervical somites are present but not yet prominent.

Carnegie Stage 11 (2.5–4.5 mm, 24–25 Days, 13–20 Somites) (Fig. 6.48)

General: Optic vesicles, rupture of buccopharyngeal membrane.
CNS: Closure of the rostral neuropore.
CF circulation: Third aortic arch artery, third pharyngeal arch, primitive longitudinal neural arteries are plexiform (Fig. 6.48).

Formation of the third aortic arch takes place much the same as that of its predecessor. Its "insertion site" is the r6–r7 segment of dorsal aorta.

At this juncture, a critical event happens. Aortic arch dwindles and eventually falls apart, leaving behind a ventral stump dangling forlornly from the DA, the *mandibular artery*. The reasons behind the involution of AA1 are likely

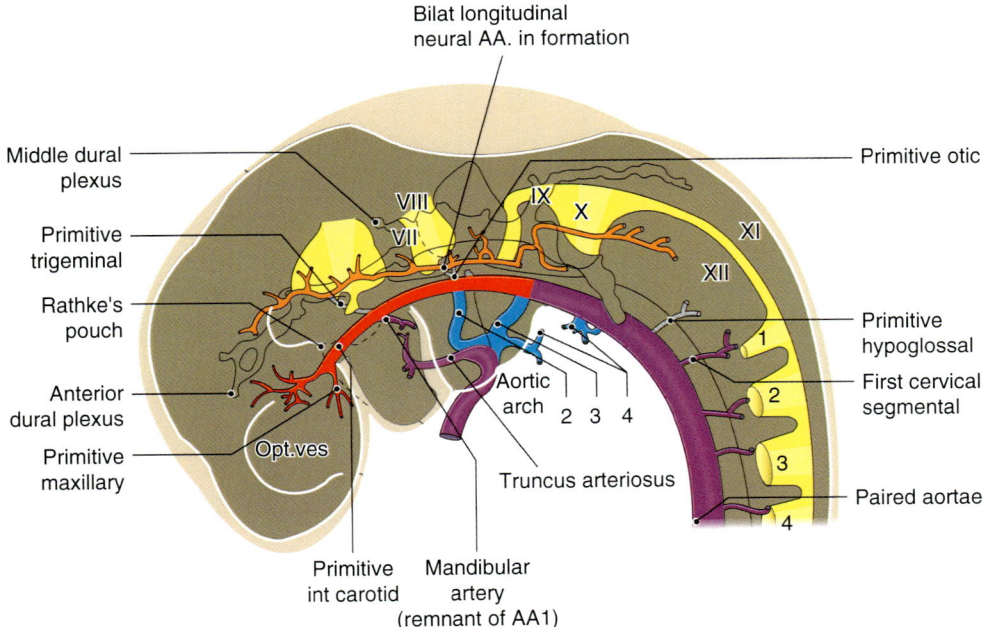

Fig. 6.48 Stage 11. AA3 has formed. Fourth arch is in process formation. Increased flow has converted primitive hindbrain plexus to longitudinal neural arteries. These lie against the wall of the hindbrain. Note primary head vein is fully formed. First aortic arch is disintegrating leaving remnant mandibular artery (future vidian artery). As LNAs move to midline to form the basilar, presegmental otic artery involutes.

PrICA now supplies forebrain and midbrain. Has not yet bifurcated. Now that AA1 is falling apart, the segment of the dorsal aorta associated with it, that gave off PrTG, is now formally considered internal carotid. Primitive internal carotid is defined as that segment between the first aortic arch and the bifurcation

related to fluid hydraulics. Blood flow from the heart is preferential for AA2...it is shunted away from AA1.

What prevents the death of the first arch? Due to free communication of mesenchyme among the pharyngeal arches, *PA2 and PA1 share a common vasculature*. With AA1 now defunct, the combined arches are perfused from AA2. The vessels of each pharyngeal arch retain their original homeotic coding. As, we shall see, construction of the external carotid will assign a branch or branches to each pharyngeal arch.

Melding of the pharyngeal arches is taking place. second arch is confluent with deep and superficial planes of the first arch and with the superficial plane of the third arch. Intraoral expansion of the first arch reduces representation of second arch r4–r5 mucosa and blood supply to the tonsillar fossa.

At stage 11, the optic vesicle is specified. In its distal part, Vsx-2 induces the neural retina[11], while in its proximal part the transcription factor Miif induces retinal pigment epithelium. The optic anlage is supported by maxillary artery, while primitive internal carotid is now advancing over the caudal and dorsal aspect of the vesicle. Primitive internal carotid has not yet produced the caudal division.

The primordial hindbrain channel is now involuting. Plexiform fragments of a vascular tract (the future LNA) lie on the wall of the hindbrain, fed by regional presegmental branches (trigeminal, otic, hypoglossal). Segmental vessels from the paired aortae to the cervical somites are present. No anastomosis between these segments and the LNA is present. The most caudal vessel, the seventh segmental artery, will eventually form the subclavian.

Carnegie Stage 12 (3–5 mm, 26–27 Days, 21–29 Somites) (Fig. 6.49)

General: Fourth aortic arch artery, fourth pharyngeal arch, optic neural crest, otic vesicle, upper limb buds, ICA divides.
CNS: Closure of the caudal neuropore.
CF circulation: Establishment of future supply to carotids and descending aortae, establishment of cranial and caudal divisions of internal carotid, definitive longitudinal neural arteries.

Note: PADGET STAGE 1 was based on embryos 4–5 mm: (please see Padget, Figs. 6.49, 6.50, and 6.51).

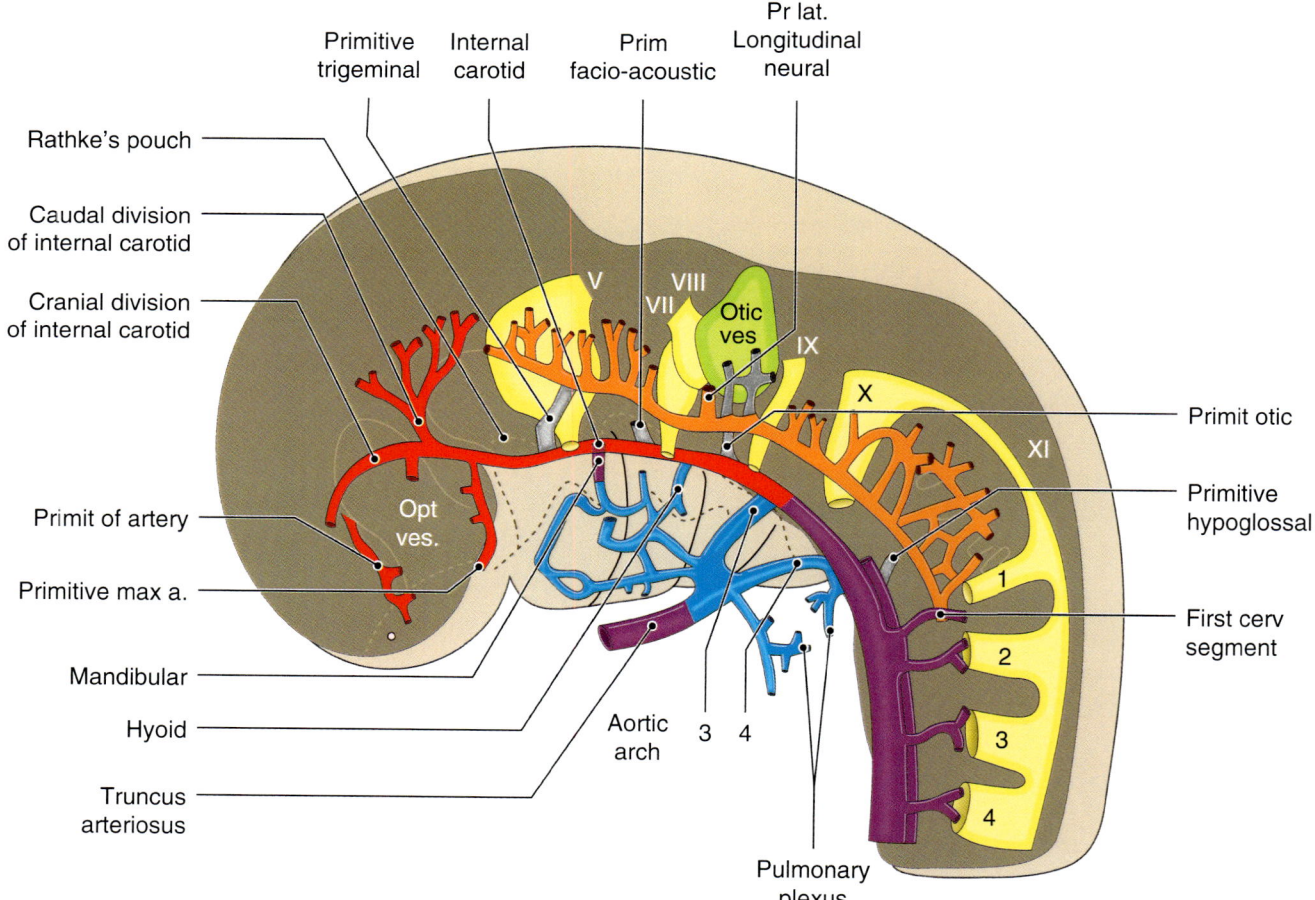

Fig. 6.49 Stage 12. third aortic arch present. Primitive internal carotid has now bifurcated. Its caudal division has not yet connected with the hindbrain circulation. Cranial division has advanced over the optic vesicle to the future nasal fields. Primitive maxillary is beginning to redirect toward the floor of the forebrain. The longitudinal neural artery of the hindbrain is supplied at its cranial aspect by the primitive trigeminal artery and at its caudal aspect by the primitive hypoglossal artery and the first cervical segmental

Fig. 6.50 Stages 11–12. Primitive internal carotid supplies the forebrain via prosencephalic stem (with cranial and caudal divisions) and via the primitive maxillary directed toward the floor of the forebrain. It now has cranial and caudal divisions. The longitudinal neural artery of the hindbrain is supplied at its cranial aspect by the primitive trigeminal artery and at its caudal aspect by the primitive hypoglossal artery and the first cervical segmental

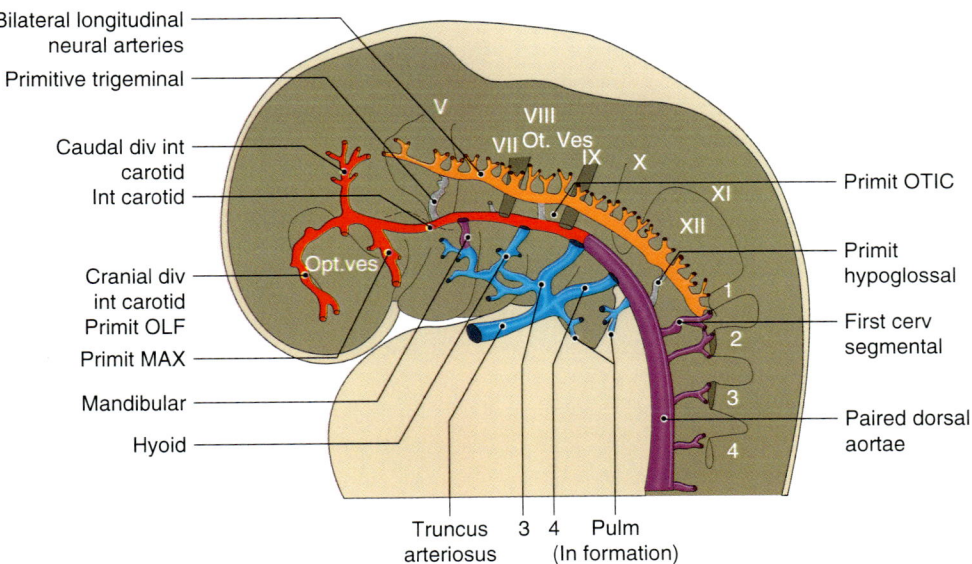

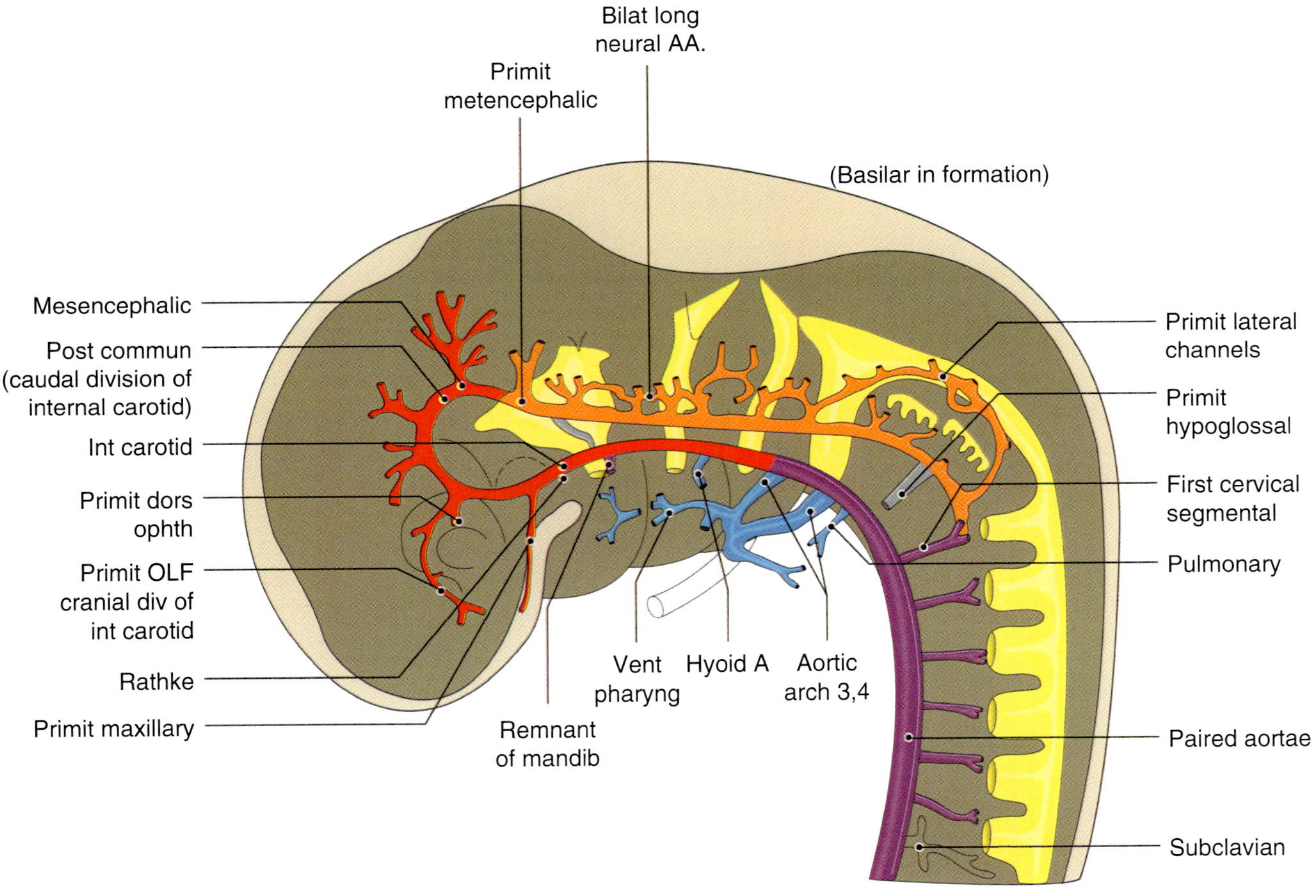

Fig. 6.51 Stages 13–14. Caudal division of PrICA now joins the neural artery. This becomes the posterior communicating artery. LNAs are drifting together to form what will be the basilar. Hindbrain segmental arteries (primitive trigeminal and primitive ophthalmic) are dwindling, reflecting changes in blood flow. Aortic arteries to the first two pharyngeal arches have fallen apart. Vessels within the first and second arches are reorganizing as ventral pharyngeal artery (the precursor to the external carotid system)

The fourth aortic arch artery connects the truncus arteriosus with the r8–r9 segment of dorsal aorta. Development of AA4 is not accompanied by changes in AA3. Fourth pharyngeal arch constructs the hypopharynx and larynx. Neural crest from r8 to r9 contributes to fascia, while PAM from somitomere 8/somite produces the thyroid cartilage, cricothyroid muscle, and middle constrictor. Motor control of these muscles takes place via the superior laryngeal branch of the vagus nerve, the nucleus of which resides in r8. Future development of AA4 is asymmetric with the left side constituting the arch of the aorta (the dorsal aortae running distally into the body having fused). The right side become incorporated into the subclavian artery.

The terminal end of the dorsal aorta, extending forward from the third arch, represents the *primitive internal carotid* and is now, by default, longer. Its "origin" extends backward to the third arch. An *intermediate segment of dorsa aorta* between AA3 and AA4 persists until Carnegie stage 17 (considered the transition between the "pharyngeal arch period" and the "post pharyngeal arch period").

Embryos like to keep their blood supply tidy. The advent of PA4 is a good excuse for the vascular consolidation of the first three arches. As soon as AA4 is established, AA2 obligingly begins to disintegrate apart (right on time). Left behind is an equally forlorn ventral remnant of DA, the *hyoid artery*. This small artery will have a big future…it will eventually produce the stapedial system.

Blood supply to the first two arches is unaffected. Vascular continuity between PA3 and the PA1- and PA2 complex involves the usual tangle of vessels. However, in stage 12, this network becomes simplified. A rostrally directed sprout from the ventral base of AA3, the *ventral pharyngeal artery* (VPA), extends forward like a taproot, seeking out the PA1–PA2 plexus and supplying it. These vascular linkages, formed previously in arches 1 and 2, will later connect up with the vascular plexus of arches 3 and 4. Thus, VPA serves as the "template" for the external carotid system.

The trigeminal ganglion is like Grand Central Station; it is the site of many important departures. First off, r0/r1 marks the right-angle downturn of the outflow tract away from the dorsal aortae. It is neurologically related to the Hensen's node (the rostral terminus of the notochord) and to the adenohypophysis of the pituitary. All vessels forward from this point are derivatives of the primitive internal carotid artery. Medial from ro/r1, the primitive trigeminal artery supplies the ganglion.

Stage 12 marks the highpoint of primitive maxillary supply to the future eye and ventral forebrain. As internal carotid develops, maxillary will become reassigned to the hypophysis. Internal carotid has now bifurcated. The forebrain and the optic vesicle in the 4 mm embryo are supplied by the cranial division of the primitive internal carotid artery. The cranial division prICA curves around the optic vesicle and then terminates in the olfactory zone. The caudal division of PrICA loops backward over the midbrain.

The optic vesicle is surrounded by a vascular plexus supplied up to this point by maxillary artery. At stage 12 the situation changes. Just at the "split point" between its cranial and caudal divisions, PrICA gives off a small branch that travels along the dorsocaudal aspect of the optic vesicle. This marks the *primitive dorsal ophthalmic artery* (PrDOA). Its ventral counterpart will develop in the ensuing stage.

Note carefully that, at this stage, *no circulatory connection exists between the forebrain and hindbrain*. At this stage, bilateral longitudinal neural arteries on the hindbrain wall are fully developed. They are supplied by the carotid via the primitive trigeminal artery and by additional presegmental branches. Note that the caudal end of LNA receives blood from two sources: the primitive hypoglossal artery and the first cervical segmental artery.

Note: PADGET STAGE 2 was based on embryos 5–6 mm: (please see Padget, Figs. 6.50, 6.52, 6.53, 6.54, and 6.55).

Carnegie Stage 13 (4–6 mm, 28–31 Days, 30+ Somites) (Figs. 6.52 and 6.53)

General: Fifth aortic arch artery (fails), fifth pharyngeal arch. Lower limb buds, lens, retina and lens disc, separation of otic vesicle.
CNS: Closed neural tube, cerebellum appears.
CF circulation: Stage 13 is characterized by the appearance of brief, abortive appearance of AA5 and the development of the fifth pharyngeal arch. The fifth aortic arch artery attempts to form but involutes, leaving the diminutive fifth pharyngeal arch dependent for its survival on the blood supply of the fourth arch. This explains why superior thyroid is distributed throughout the larynx.

The fifth arch is composed of mesoderm from somitomere 9 and neural crest from rhombomeres 10–11. From it develop the inferior constrictor, the intrinsic muscles of the larynx, the arytenoid cartilages, and the cricoid cartilage. Motor control of these muscles takes place via the inferior laryngeal branch of the vagus nerve, the nucleus of which resides in r9. The vascular axis starts out as the fifth aortic arch with a dorsal bud from the r10 to r11 dorsal aorta and a ventral bud from the aortic outflow tract. In fishes, AA5 is a normal component of the fifth branchial (gill) arch. Tetrapods modify this system to make lungs so that AA5 degenerates. Remnants of the fifth aortic arch at times persist in certain anomalies of the heart; when present these represent the ventral moiety.

Confusion is no stranger for students of embryology. Does the fifth pharyngeal arch exist or does it not? Many textbook descriptions describe the "disappearance" of a fifth

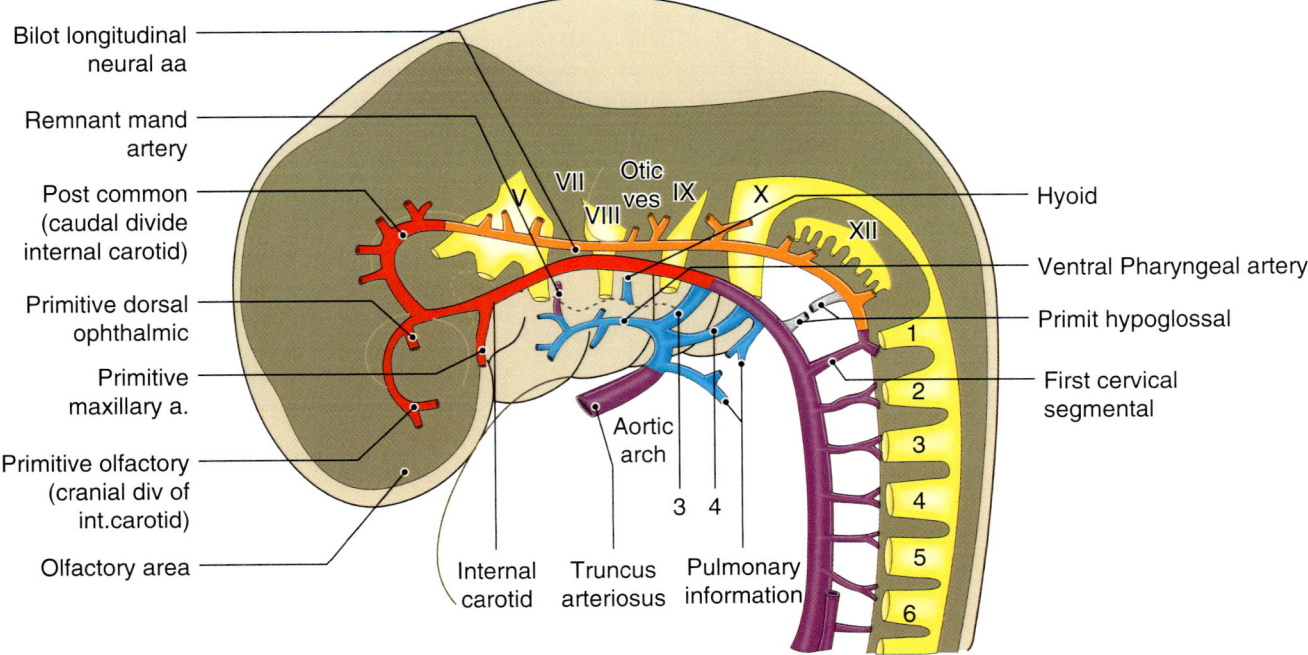

Fig. 6.52 Stage 13a. Fourth aortic arch present. Aortic arches 1 and 2 have involuted. Left behind are two dorsal remnants, respectively, mandibular artery and hyoid artery. The paired dorsal aortae have retreated to the level of aortic arch 3. Forward from AA3 is the internal carotid. Forebrain and optic vesicle are supplied by two arteries: primitive maxillary and the primary cranial division of internal carotid. Primary caudal division of internal carotid runs posteriorly over the midbrain, but is *not* connected to the hindbrain. Longitudinal neural arteries run alongside the wall of the hindbrain. The LNAs are connected to the ICA via presegmental arteries: primitive trigeminal, primitive otic, primitive hypoglossal. Caudal LNAs fed dorsal aortae via first cervical segmental arteries. Note optic vesicle receives the primitive dorsal ophthalmic artery at this time

Fig. 6.53 Stage 13b. Axial view. Neural arteries appearing on either side of the hindbrain wall supplemented by primitive trigeminal and other presegmental arteries

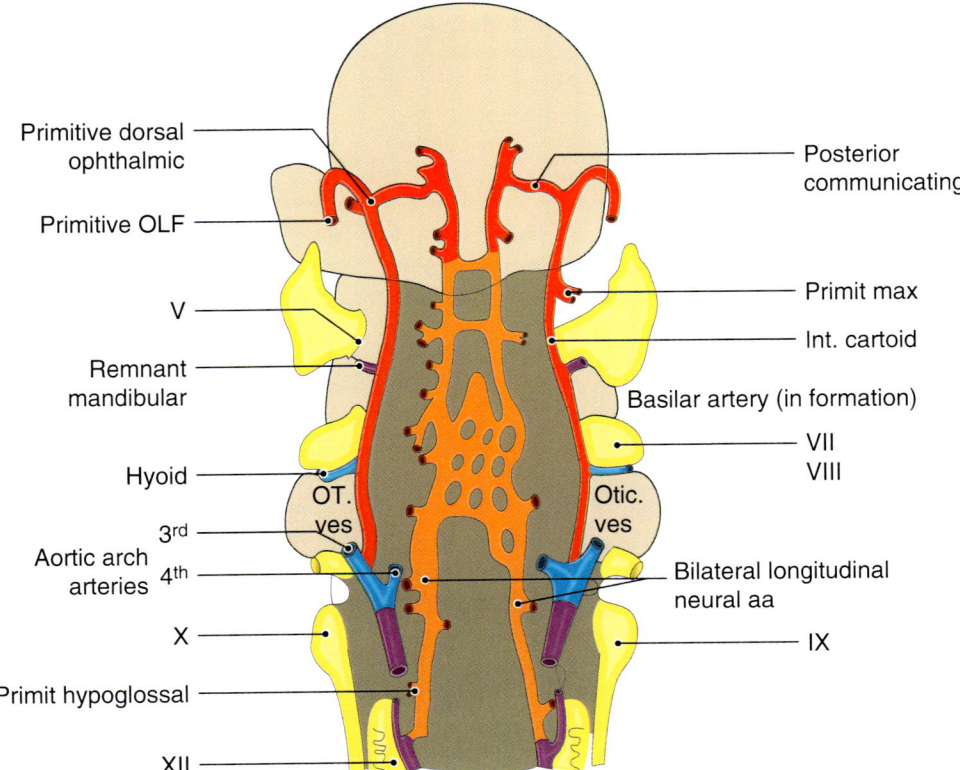

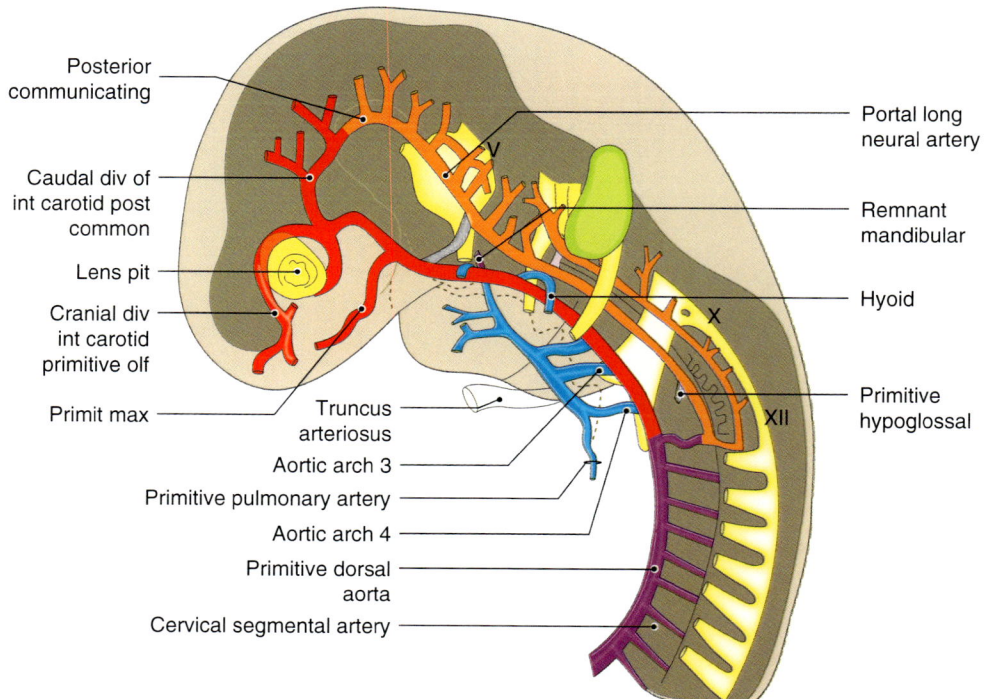

Fig. 6.54 Stage 14. Caudal division of PrICA now joins the neural artery. This communicating segment becomes the posterior communicating artery. Primitive metencephalic (future superior cerebellar-SCbA) is first major branch of the hindbrain. Stem of posterior cerebral not yet seen (it will be given off just before SCb). LNAs are drifting together beneath metencephalon to form the basilar artery. Hindbrain segmental arteries (primitive trigeminal and primitive ophthalmic) are dwindling, reflecting changes in blood flow. Aortic arteries to the first two pharyngeal arches have fallen apart. Vessels within the first and second aortic arches completely disintegrated. Pharyngeal arch plexus supplied by ventral pharyngeal artery (the precursor to the external carotid system)

Fig. 6.55 Stage 14. Primitive maxillary is at the height of its development. With the connection of PrDOA to the optic plexus, it quickly recedes and devotes itself to the hypothalamus and cranial base. Lateral view shows upper cervical segmentals connected to the vertebral and disconnected from dorsal aortae. Lower cervical segmentals intact to the DAs with vertical vertebral interposed. seventh nerve has swung caudally with the hyoid following. Tympanic nerve seen projecting forward from ninth nerve. VPA now in contact with chorda tympani. Stapedial not yet formed. Its extracranial course follows chorda tympani. In this way, it contacts VPA. Eye now supplied by both primitive dorsal and ventral ophthalmics. Nasal fossa now present and supplied by primitive olfactory. Anterior and posterior choroidals present. Basilar artery complete

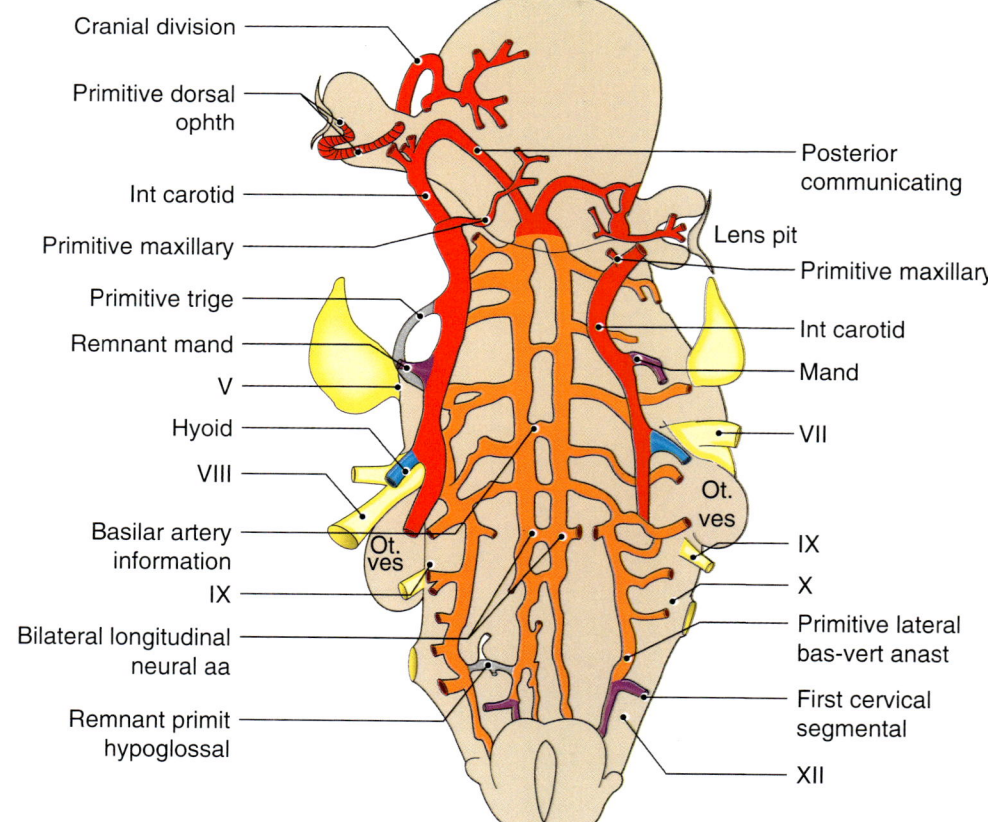

arch, followed by development of a definitive sixth pharyngeal arch. Some contemporary authors do the math differently, describing in mammals, *five and only five*, pharyngeal arches. This discrepancy is due to misconceptions about aortic arch development. The issue is indeed perplexing. Why should a perfectly regular, genetically programmed system of somites and arches become disrupted? Vertebrate neuroembryology is very conservative in its evolution. What would be the consequences for cranial nerves if an entire arch could be summarily wiped out? In 35 years of study, I have never found a satisfactory explanation for this dogma.

Let's cut right to the chase with our conclusions; then we will backtrack to explain how this viewpoint was arrived at. All pharyngeal arches have an aortic arch at their core. *Mammals have five (count 'em) pharyngeal arches*. These are supplied by *four aortic arches*. Mammals do indeed have six aortic arches.

This makes eminent neuroembryologic sense. Crosstalk between the neural plate and outlying paraxial mesoderm determines the anatomic content of the CNS. When a somitomere or somite lacks appropriate muscle targets, no motor neurons will be "assigned" to that site. Loss of an entire pharyngeal arch during evolution would *ipso facto* result in disruption or dislocation of hindbrain nuclei and tracts. Evolution does not support this. Comparative neuroanatomy of the vertebrate medulla demonstrates great conservatism between fishes and tetrapods. In mammals, the same pattern of connections between vagus and spinal accessory nerve and with their respective targets is maintained.

Somitomeres contain myoblasts of epaxial muscles and/or hypaxial muscles. Pharyngeal arches are ventral structures. *Each pharyngeal arch is constructed using paraxial mesoderm from single somitomere*. Thus: PA1 = Sm4, PA2 = Sm6, PA3 = Sm7, PA4 = Sm8, and PA5 = Sm9.

Termination of the pharyngeal arch system and its derivatives takes place at the thyroid and cricoid. Paraxial muscles supplying the voicebox arise from the first and second occipital somites (Sm8–Sm9). The nucleus of superior laryngeal resides in r8, while that of the recurrent laryngeal dwells in r9. Thus, by level r9, all cranial nerves supplying the pharyngeal arch derivatives are accounted for. There is no need to invoke for evidence structural shifts that would result from deletion of a pharyngeal arch.

The pharyngeal arches begin to develop a common vascular system. The mandibular and hyoid branches are about the same size. Hyoid later becomes the ICA source for an all-important intermediate structure, the stapedial artery. At the base of the fourth arch (the *common segment* between third arch and fourth arch), a rostrally direct sprout, *the ventral pharyngeal artery,* connects with the previously established PA1–PA2 network. The trajectory of ventral pharyngeal artery is directed toward the tip of trigeminal nerve V3 which is growing into the future face. Along its course, VPA will give off individual branches to each of the five pharyngeal

arches (please see Padget, Figs. 6.52 and 6.53). These depict VPA *in intimate association with primitive trigeminal artery*. Later in our story, we shall see how this relationship helps to explain why anterior tympanic and middle meningeal possess distinct stems along the proximal third of maxillomandibular artery.

Previously, the relationship of the primitive maxillary artery to r0/r1-derived, V1-innervated orbital derivatives was noted. At stage 13, the vessels of the future eye are highly plexiform, receiving contributions from both PrDOA and PrMxA. With the advent of PrDOA, primitive maxillary begins to redirect itself ventromesially along the margin of Rathke's pouch until it terminates as a terminal plexus at the mesial tip of the prosencephalon. It will supply the posterior pituitary (the anterior pituitary is supplied by intracranial pial branches).

Ocular development at this stage demonstrates a long lateral branch passing over the dorsal and temporal aspect of the optic cup to reach the primordium of the lens. This is the *primitive dorsal (temporal) ophthalmic artery*. The bones of the orbit will develop much later. We shall see that all adnexal structures surrounding the globe will be supplied by a completely distinct vascular system emanating from intracranial stapedial, anterior division, and supraorbital branch. More about that subject anon.

The absolute terminus of PrICA remains *primitive olfactory artery*, the absolute terminus of PrICA. Anterior cerebral develops proximal to PrOlfA; it subsequently gives off lateral olfactory nerve artery (LONA). The spatial order of vessels along the arterial axis of PrICA is clinically relevant. *Holoprosencephaly involves the PrICa axis*. HPE follows a medial-lateral gradient of severity: PrOlfA, LONA, PrVOA, and PrDOA.

The caudal division of PrICA serves diencephalon and posterior cerebrum. At this time, it forms a secondary anastomosis with cranial terminus of LNA; it therefore bears the name *posterior communicating artery* (PCmA). Although at this time the main branches of PCmA are indistinct, let's name them for future reference. In cranial-caudal order, the branches of PCmA are: posterior choroidal, diencephalic, mesencephalic, and posterior cerebral. *Note that PCA belongs to original ICA system*; it does *not* originate from basilar.

"Seduced and abandoned," the primitive trigeminal is now eclipsed by posterior communicating artery as the blood supply to the LNA-basilar artery complex. Sites of fusion between the LNAs, particularly in the otic region, represent early formation of the basilar artery. In this process, the presegmental arteries between DA and LNA are "pulled" medially and become "stretched out." Attenuation of the primitive otic artery is observed. The primitive hypoglossal is replaced by the first cervical segmental as the primary source of blood for the caudal LNA. At about 5 mm, characteristic changes occur in the LNAs that presage the basilar artery. The arteries develop multiple areas of coalescence along the ventral surface of rostral hindbrain.

In the hindbrain, a parallel channel to the caudal LNA develops. Although temporary, it later serves as an accessory anastomosis between the basilar and vertebral arteries. Let's call it VBA. It runs from r4 to r11 between the roots of the lateral motor column (VII, IX, X, XI) and the medial motor column (VI, XIII). Cranially, VBA is fed by lateral branches from the basilar artery. These are located at the trigeminal ganglion; they take the form of an inverted V. Caudally, VBA is supplied by the first cervical vessels.

Carnegie Stage 14 (Figs. 6.54 and 6.55)

General: Sixth aortic arch (pulmonary) plexus, fusion of the conotrucus into a single spiral, optic cup and retinal fissure, nasal pits,
CNS: Closed neural tube, cerebellum appears.
CF circulation: Pulmonary arch. Formation of a complete cerebral circuit between internal carotids and longitudinal neural arteries.

Although the ventral stem of the sixth aortic arch artery is depicted as in discrete vessel arising from the conotruncus, in reality it is a plexus associated with laryngeal and pulmonary buds arising from the base of the fourth aortic arch and connected with ventrally directed stems from the c1 to c2 segment of dorsal aortae. As the respiratory diverticulum and pulmonary apparatus extend, this plexus consolidates into a pair of pulmonary arteries. These structures are in no way analogous to the previous five aortic arches; hence, the term sixth aortic arch should be scrapped in favor of pulmonary arch. Pulmonary arch development, like AA4, is also asymmetric. On the left side, its original connection to c1–c2 dorsal aortae, it persists as ductus arteriosus, a shunt between left pulmonary artery and descending aorta to protect lung vasculature from excessive pressure during intrauterine development.

The caudal division of ICA now connects with LNA via the posterior communicating artery. Presegmentals consequently involute. First branch of the cranial ICA, the primitive DOA, is now in contact with the dorsotemporal eye and is positioned to fall into the retinal fissure.

PADGET STAGE 3 is based on embryos 7–12 mm:

Carnegie Stage 15 (Fig. 6.56)

General: Cerebral hemispheres, nasal pits, hand plates, urogenital sinus.
CNS: Longitudinal zoning of the diencephalon.

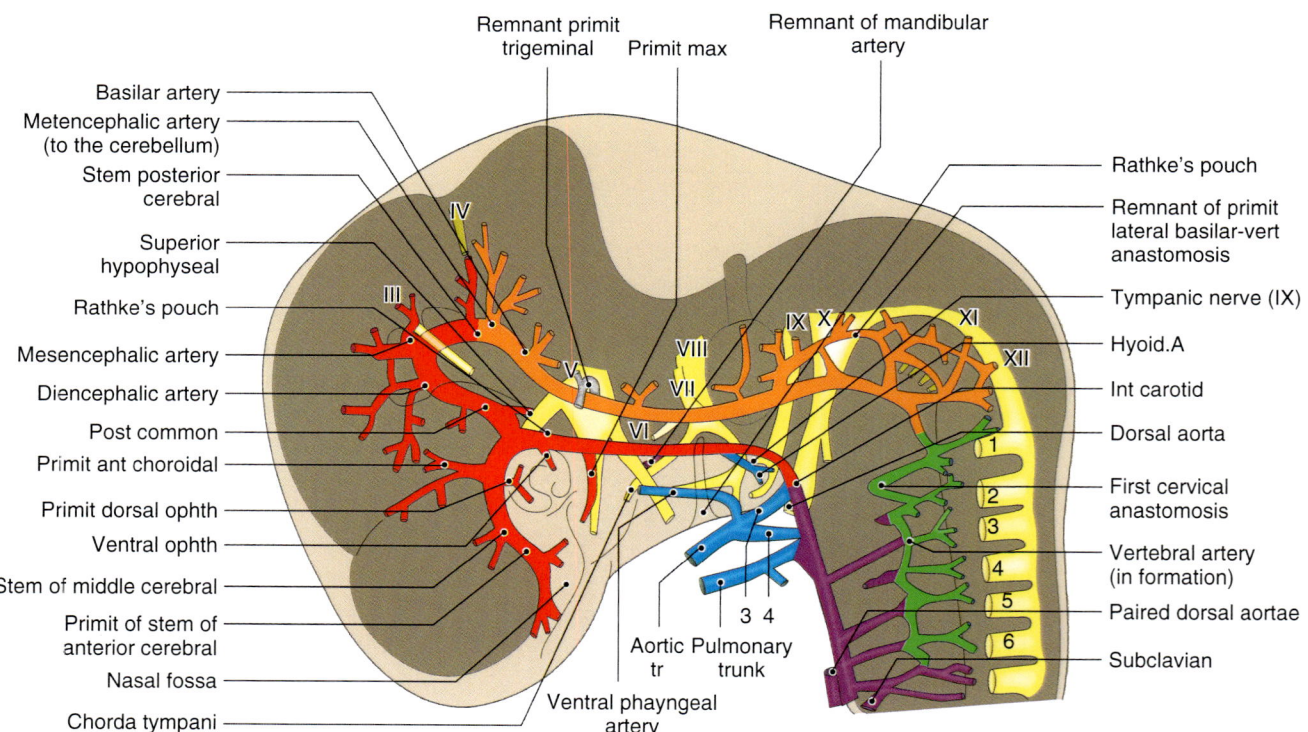

Fig. 6.56 Stage 15. Lateral view shows upper cervical segmentals connected to the vertebral and disconnected from dorsal aortae. Connections of lower cervical segmentals to the dorsal aortae remain intact, but with newly formed vertical vertebral interposed.. seventh nerve has swung caudally with the hyoid following. Tympanic nerve seen projecting forward from ninth nerve. VPA now in contact with chorda tympani. Stapedial not yet formed. Its extracranial course follows that of chorda tympani to make contact with VPA. The eye is now supplied by both primitive dorsal and ventral opthamics. Nasal fossa now present and supplied by primitive olfactory. Anterior and posterior choroidals present. Basilar artery complete. Primitive ventral ophthalmic is seen

Carnegie Stage 16 (Fig. 6.57)

General: Retinal pigment, foot plates, auricular hillocks, upper lip.

CNS: Evagination of the neurohypophysis.

CF circulation: (1) Vertebral artery begins. (2) Cerebral arteries differentiate. (3) Cranial div. of ICA is primitive = primitive olfactory. (4) Optic supply is still plexiform, but has preexistent common temporal ciliary and newly formed common nasal ciliary arteries. (5) Hyoid stem now reaches cranial end of second arch, i.e., PA1–PA2 "overgrowth." (6) Ventral pharyngeal artery follows chorda tympani and arrives at V3.

The internal carotid extends forward from AA3. The remnant of mandibular artery is small (it has not yet been put to use as the vidian artery). The hyoid artery remnant is conspicuous, but has not yet given off the stapedial stem. VPA now supplies the tip of V3. Primitive maxillary leaves the PrICA just lateral to Rathke's pouch and courses ventrally on its way to Rathke's pouch.

Cerebral development is characterized by changes in the cranial division of the ICA. Just distal is the *primitive dorsal (temporal) ophthalmic artery* supplying an extensive network of vessels over the caudal-dorsal optic cup. At stage 15, just distal along the cranial division of ICA, its first cranial branch, *primitive anterior choroidal artery,* supplies diencephalon as it courses toward the choroid fissure. From the opposite side at the same spot, ICA gives off the *primitive ventral (nasal) ophthalmic artery*. It supplies the more cranial and ventral optic plexus.

Note that, when the eye moves away from the brain at a later stage, the globe will have two developmental hemispheres: the nasal globe, being supplied by a more distal VOA, is developmentally distinct than the temporal globe. Nasal sclera receives attachments of Sm1 muscles. Temporal sclera is controlled by muscles from Sm2, Sm3, and Sm5. Note that the blood supply to the eye is largely plexiform at this stage. *No single branch of ICA enters the ocular cleft.* Distal to anterior choroid, middle cerebral emerges. The cranial div. of ICA (*primitive olfactory artery*) terminates at the nasal pit; just beforehand, it gives off the future *anterior cerebral artery*.

Important changes also take place in the caudal division of the ICA. The mesencephalon contains the two nuclei of the oculomotor nerve. IIIa resides in m1 and is responsible for innervating the muscles originating from Sm: inferior rectus and medial rectus. IIIb resides in m2 and innervates Sm2 muscles: inferior oblique, superior rectus, levator palpebrae superioris, and Müller's. Sm3 and Sm5 provide, respectively: superior oblique and lateral rectus. The muscles attach to neural sclera binding sites in two main zones, dorsotemporal and ventral nasal. These correspond to two distinct populations of midbrain neural crest which can be

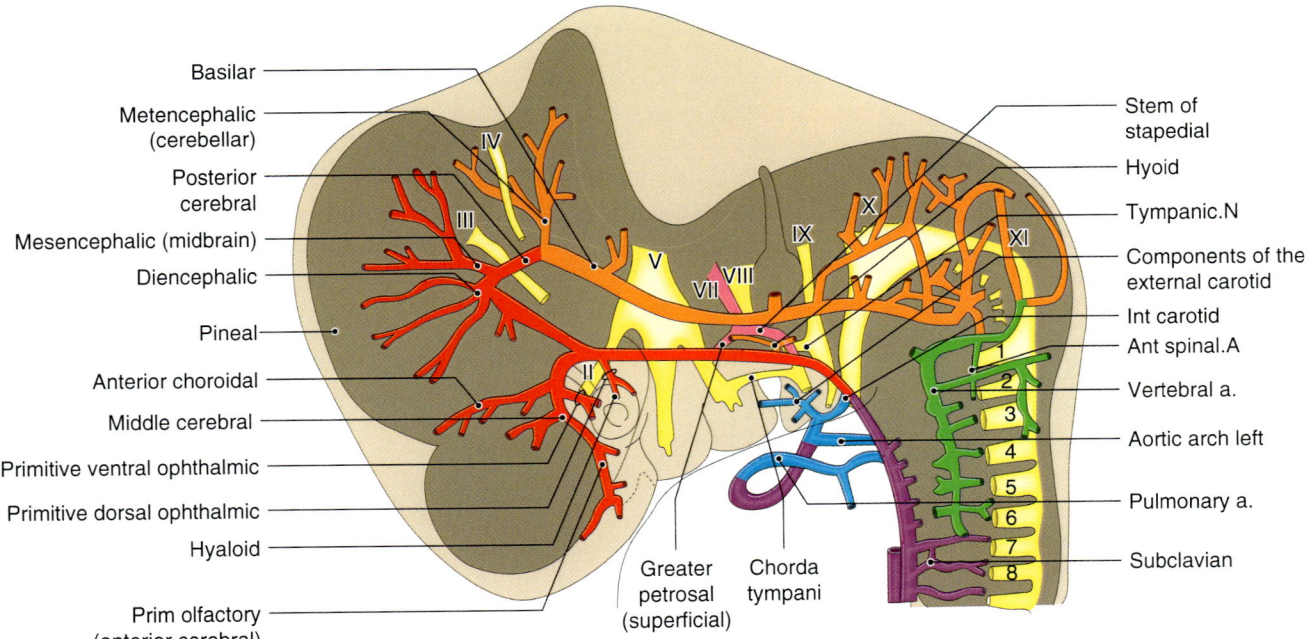

Fig. 6.57 Stage 16–17. Upper seven cervical segmental arteries connecting dorsal aortae with spinal cord and somites have now formed a series of vertical anastomoses that create a distinct vessel linking them all together—the vertebral artery. At this stage, first cervical segmental to the posterior LNA is small and hence all cerebral arteries still depend on internal carotid. PrDOA now has two distinct branches: hyaloid and common temporal ciliary. Hyoid artery (dorsal remnant of AA2) lies at the cranial corner of the second arch. The third arch melding with the first two arches. Vessels within the third arch are included with the ventral pharyngeal artery. VPA now arrives at the root of V3

determined by the relative embryonic age of their vascular supply. The temporal-dorsal zone is the more primitive, being supplied by DOA and is likely in register with the posterior mesomere, m2. The naso-ventral zone is more recent, is supplied by VOA, and receives the muscles of Sm1. Existence of additional binding sites in the DOA zone permits eventual insertion of muscles from Sm3 and Sm5.

Just anterior to the oculomotor nuclei, the caudal end of the posterior communicating artery produces two branches. The first branch supplies diencephalon. It gives rise to a *posterior choroidal artery* projecting upward to join forces with the anterior choroidal artery. The second branch supplies mesencephalon. At the anastomosis of posterior communicating with basilar, the *metencephalic artery* (future superior cerebellar artery) supplies the nucleus of IV and the emerging metaencephalon (cerebellum and pons). Note that before posterior-basilar union takes place, posterior communicating will give rise to posterior cerebral artery. Thus, unlike descriptions in traditional anatomy texts, *PCA is not a derivative of basilar.* All arteries supplying prosencephalon (including the entire cerebrum) and mesencephalon originate from the internal carotid system. All arteries to rhombencephalon trace their lineage back to the primitive hindbrain channel-longitudinal neural artery ancestors.

A stark distinction is seen between the branches destined for midbrain and hindbrain. Undoubtedly, this related to their ontological relationship. Recall that the genes such as *Otx* from rhombencephalon are required for induction of midbrain. These signals work in concert with genes expressed by the primitive anterior visceral endoderm in stages 6 and 7. Subsequently, hindbrain differentiates into the r0–r1 metaencephalon (cerebellum and pons) and myelencephalon (medulla) containing r2–r11. Thus, *hindbrain is the primary developmental unit of the CNS.* Forebrain and midbrain are secondary induction products. The initial disconnection between their respective blood supplies bears testimony to these primitive genetic relationships.

It must be emphasized that, although r0–r1 is structurally part of hindbrain, neural crest from r0 to r1 cooperates extensively with that from midbrain and caudal forebrain. *All non-neural structures (including dura) are supplied by StV1 (intracranial supraorbital stapedial). All neural and cranial base structures are supplied by PrICA.*

At this stage, union of the LNA into basilar is almost complete. However, caudal to otic capsule, they remain apart; thus, basilar union causes the r8–r11 segments to remain splayed out into an inverted-Y. At the same time, collateral channels parallel to LNA described previously become an accessory *basilar-vertebral anastomosis* running from otocyst to first cervical segmental artery. Its embryologic derivation is from the LNAs, not the vertebrals.

Vertebral artery development is initiated at this stage. Previously we have observed transverse segmental branches connecting the dorsal aortae with each level of the spinal cord. At this juncture, the segmental linkages become themselves interconnected via new longitudinal anastomoses. A new vessel, the primitive vertebral artery (PrVA), is now interposed in parallel between the dorsal aortae and the spinal cord. Growth of the neck displaces the dorsal aortae and creates additional "freeway" space beneath the spinal cord. The segmental vessels become "stretched out." Remember that the DAs are fated to move into the midline, fuse, and descend into the chest as the definitive aorta. Disruption of the transverse segmental vessels connecting DA and spinal cord is inevitable. The "weak links" are the ventral transverse vessels spanning between DA and the PrVA. These attenuate and disappear. The dorsal transverse vessels between PrVA and spinal cord remain intact. Completion of this process results in the definitive vertebral arteries.

VPA is now united with plexiform vessels of all four pharyngeal arches and is ready for transition into the definitive external carotid system.

Note: PADGET STAGE 4 was based on embryos 12–14 mm.

Carnegie Stage 17 (Fig. 6.58)

General: Rapid head enlargement, six auricular hillocks, nasolacrimal groove, finger rays.
CNS: Olfactory bulbs, amygdaloid nuclei.
CF circulation: (1) Vertebral artery is irregular but complete; it still originates from the dorsal aorta. (2) Stapedial artery makes anastomosis with VPA at the root of V3. (3) Anterior cerebral emerges from primitive olfactory. (4) Ophthalmic artery (5) Definitive hyaloid artery enters fetal ocular cleft. (6) Components of external carotid now develop.

Rotation of the heart causes a 180° clockwise transposition of AA6 (the future pulmonary arch) with AA4 (the aortic arch). An attenuated segment of dorsal aorta still connects AA3–AA4, but this will be eliminated by the following stage.

Hyoid artery enters the tympanic cavity giving off *stapedial artery*, which courses toward tympanic membrane. After traversing stapedial ring, *common stapedial* runs forward beneath tympanic roof. It then bifurcates. A posterior division exits tympanic cavity and will supply dura vessels of the posterior fossa. The anterior division subdivides into a superior ramus and an inferior ramus (Fig. 6.6).

Inferior ramus develops <u>before</u> superior ramus. It escapes tympanic cavity via a small *canal of Huguier* and emerges medial to the spine of sphenoid…immediately next to foramen spinosum. It follows the pathway of chorda tympani. This brings inferior ramus into direct contact with the medial

Fig. 6.58 Stage 17 (12.5 mm). The "post-branchial phase" is marked by the rapid development of stapedial stem. By the end of stage 17, it has advanced to V3 and annexed the distal end of VPA. The system can now be referred to as external carotid. Cerebral arteries are well-defined. The vertebrals show now established origin from the subclavians. Dorsal aorta segment between AA3 and AA4 is involuting. The ventral connection between AA3 and AA4 along the outflow tract is incorporated into the base of AA3 as the common carotid stem

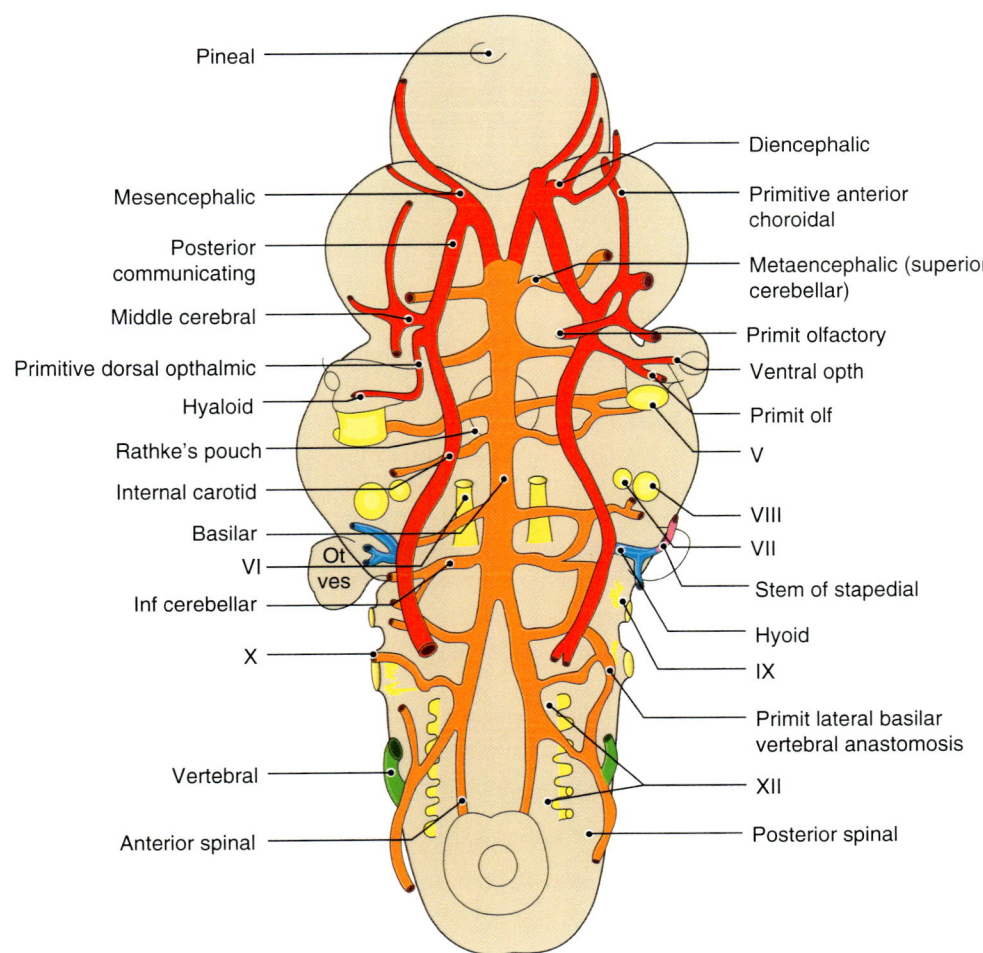

side of V3. This makes sense because the destination of CT is the lingual branch of V3. It is at this point the unification of inferior ramus with VPA will take place.

Superior ramus remains intracranial. It produces dural branches associated with cranial nerves VII, IX, and X. It then runs forward following greater petrosal and comes into association with trigeminal ganglion. There it forms *anterior and posterior divisions* that supply the membranous bones and dura of the middle and posterior fossae. The arterial pathways of the meningeal system reflect dural sensory branches from V2 and V3. Note that stapedial arteries to the membranous supraoccipital bone connect with posterior auricular and occipital branches from VPA when the main stem involutes.

VPA constitutes the root of the external carotid system up to, but not including, internal maxillo-mandibular. The organization of its branches follows the physical arrangement of the pharyngeal arches. Proximal branches supply the "internalized" third, fourth, and fifth arches. Thus, the following pattern is seen: fourth and fifth arch (superior thyroid), third arch (ascending pharyngeal), second arch, postotic (posterior auricular, occipital), second arch preotic (superficial tempo-

ral, external maxillo-mandibular/facial, and lingual), and first arch (lingual and internal maxillo-mandibular). Note the IMMA is joined by stapedial.

Why is the distribution of arteries to the second arch so extensive? The reason has to do with the ability of PA2 muscle and fascia to access the subcutaneous plane throughout the head and anterior neck. The existence of a superficial investing fascia for these muscles of facial expression is unique to mammals.

Union between the VPA and the stapedial artery takes place at V3… a "new addition" to the external carotid stem results. Under guidance of V2 and V3, distal outgrowth of this stem bears the unwieldy title: *maxillo-mandibular division* of the stapedial artery. Anatomic nomenclature can both clarify and obfuscate. This is a case in point. It is said that the stapedial artery grows out to supply its target structures. Later in development, stapedial proximal to VPA disintegrates, leaving behind a small artery internal to stapes. If stapedial is a true anatomic entity, how does it "know" where to go? And why should it later fall apart? Neural induction provides a common unifying hypothesis. Inferior ramus of stapedial develops initially under influence of chorda tympani.

Once anastomosis with the proximal external carotid (VPA) is achieved, V2 and V3 take over the role as "architects" of the internal maxillary system.

Obliteration of the aortic segment between AA3 and AA4 takes place during stage 17. As this process comes to completion, the common carotid artery, exclusively derived from AA3, is now distinct. The definitive aortic arch develops at stage 18. We shall now detail the components of the internal carotid system.

A definitive description of the primary cranial division of the internal carotid is now possible. The first, most proximal, branch is the *primitive anterior choroidal.* The second branch is the *middle cerebral.* The distal internal carotid, previously called the *primitive olfactory artery,* now has two branches. The original one goes to the nasal fossa. A secondary, more mesial branch, now dives inward toward the developing olfactory nerve root (VEGF again?). This will represent the future continuation of *stem of the anterior cerebral.* This seems paradoxical because, in adults, the MCA seems the direct continuation of its ICA "parent" with ACA seemingly a collateral "foundling." Continued mesial growth of the two ACAs results in a series of anastomoses in the midline, the future *anterior communicating artery.*

Similar developments take place in primary caudal division of internal carotid. The posterior communicating artery is much more robust. The *diencephalic artery* (Figs. 6.57 and 6.58) demonstrates four distinct branches. These could represent blood supply to genetic subdivisions of prosomeres 1–3 (p3 being most rostral). The posterior choroidal, emerging from the diencephalic courses dorsally and rostrally to "seek out" the preexisting anterior choroidal, itself a caudal branch of anterior division ICA. These two vessels could logically anastomose at the neuromeric interface between p5 and p4.

Ocular blood supply also advances in this period. The final position of the adult ophthalmic stem is not yet present. The *primitive dorsal ophthalmic artery* (DOA) now divides into two optic branches: the previous common temporal ciliary supplying the caudal-dorsal (temporal) optic cup, and a hyaloid branch entering the ocular cleft. The *primitive ventral ophthalmic artery* (PrVOA) is stretched out in length due to (1) growth of the stalk and (2) ventrolateral movement of the eye away from the brain wall. PrVOA will become the common nasal ciliary artery.

DOA and VOA are now annexed into an ophthalmic that is located more distal along the internal carotid than will be its final position.

The trigeminal may (or may not) persist as a large carotico-basilar anastomosis. If so, the junction between the posterior communicating and the basilar is small. Attenuation of the trigeminal leads to full development between the forebrain/midbrain and hindbrain systems. The theme at this stage is the transition of hindbrain circulation from a supply exclusively by the carotids to a situation in which the hindbrain is supplied by the vertebrals.

Vertebral artery formation, initiated at stage 16, continues to develop. Ventral displacement of dorsal aortae stretches out the segmental vessels and creates additional "freeway" space beneath the spinal cord. Intersegmental anastomoses develop longitudinally, creating an intermediate vessel, the future vertebral artery, in parallel with the dorsal aortae. In a cranio-caudal process, the ventral connections between the DA and VA fall apart. The cranial part of the vertebral is clearly more advanced in DHP 5 (Figs. 6.57 and 6.58). In its final configuration, VAs (via the first cervical segmental) connect the basilar artery with the subclavian.

Note: PADGET STAGE 5 was based on embryos 16–18 mm:

Carnegie Stage 18 (13–17 mm, 44–47 Days) (Fig. 6.59)

General 18: Eyelids, nasal tip, nipples, elbow, toe rays, aortic arch forms.
CNS: Olfactory bulbs, amygdaloid nuclei.
CF circulation: ophthalmic stem reaches final position.

Carnegie Stage 19 (16–18 mm, 48–50 Days) (Figs. 6.60 and 6.61)

General: Trunk lengthens and straightens out, midgut herniation, StV1 reaches the orbit.
CNS
CF circulation, stages 18–19: (1) Vertebral origin moves to level of ductus arteriosus. (2) Stapedial stem beyond stapes trifurcates: (a) superior ramus (anterior division) gives meningeal branches to middle fossa + supraorbital branches to orbit and anterior cranial fossa; (b) superior ramus (posterior division) gives meningeal branches to membranous posterior fossa; and (c) inferior ramus gives internal maxillo-mandibular. (3) Ophthalmic permanent stem + primary optic branches (V1 stapedial remains separate). (4) Most cerebral arteries are defined, especially middle and anterior cerebrals. (5) Olfactory attenuating. (6) Anterior communicating is a plexus. (7) External carotid has 4–5 branches corresponding to development of V2–V3/"stapedial" component.

Descent of the heart into the thorax causes elongation of the common carotid. The vertebral is also straightened, being stretched out by the final positioning of the subclavian.

In Padget stage 4, the ventral tip of V3 marked the site of an intimate rendezvous between inferior ramus of stapedial and VPA. Recall that VPA is not a de novo "real artery," but

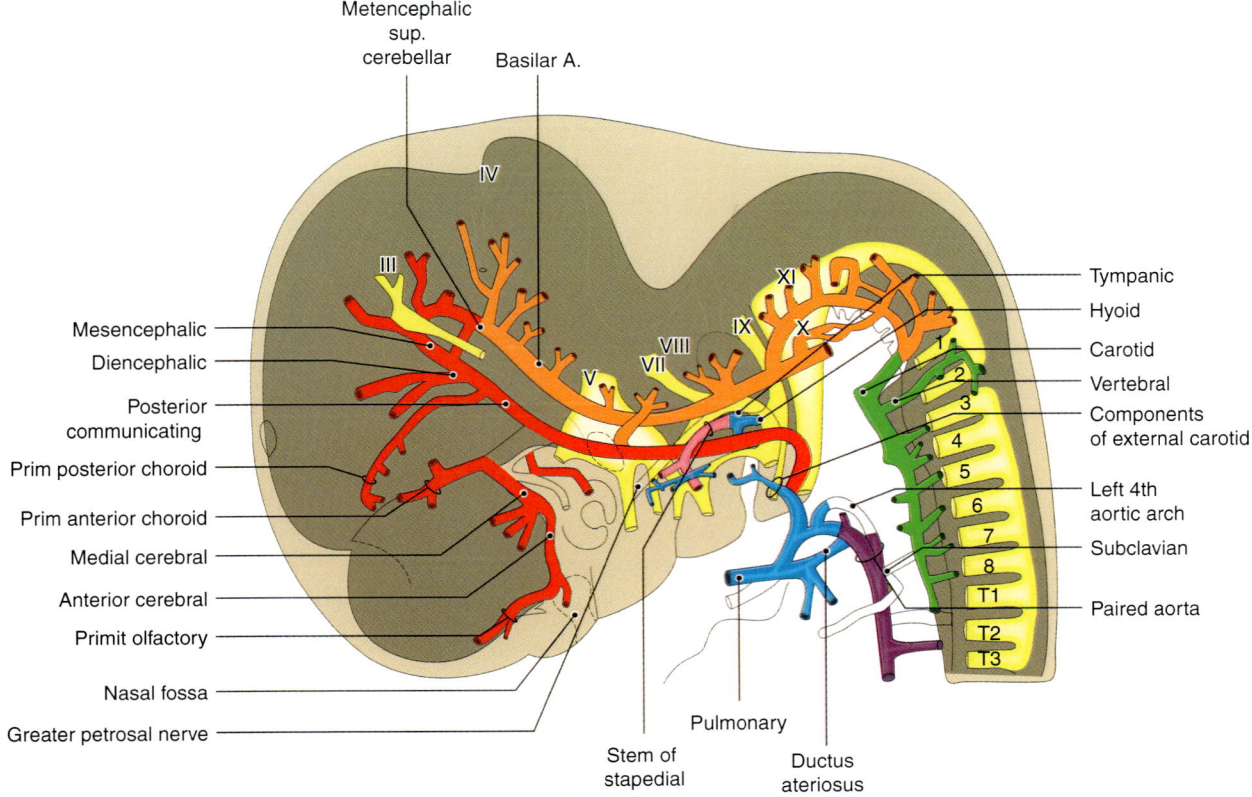

Fig. 6.59 Stage 18 (14 mm). In this embryo, the cerebral circulation appears less refined. Forward extension of the external carotid in as the maxillo-mandibular stem is seen. Intracranial stapedial stem is advancing to the orbit. Ophthalmic stem is now present in its adult location

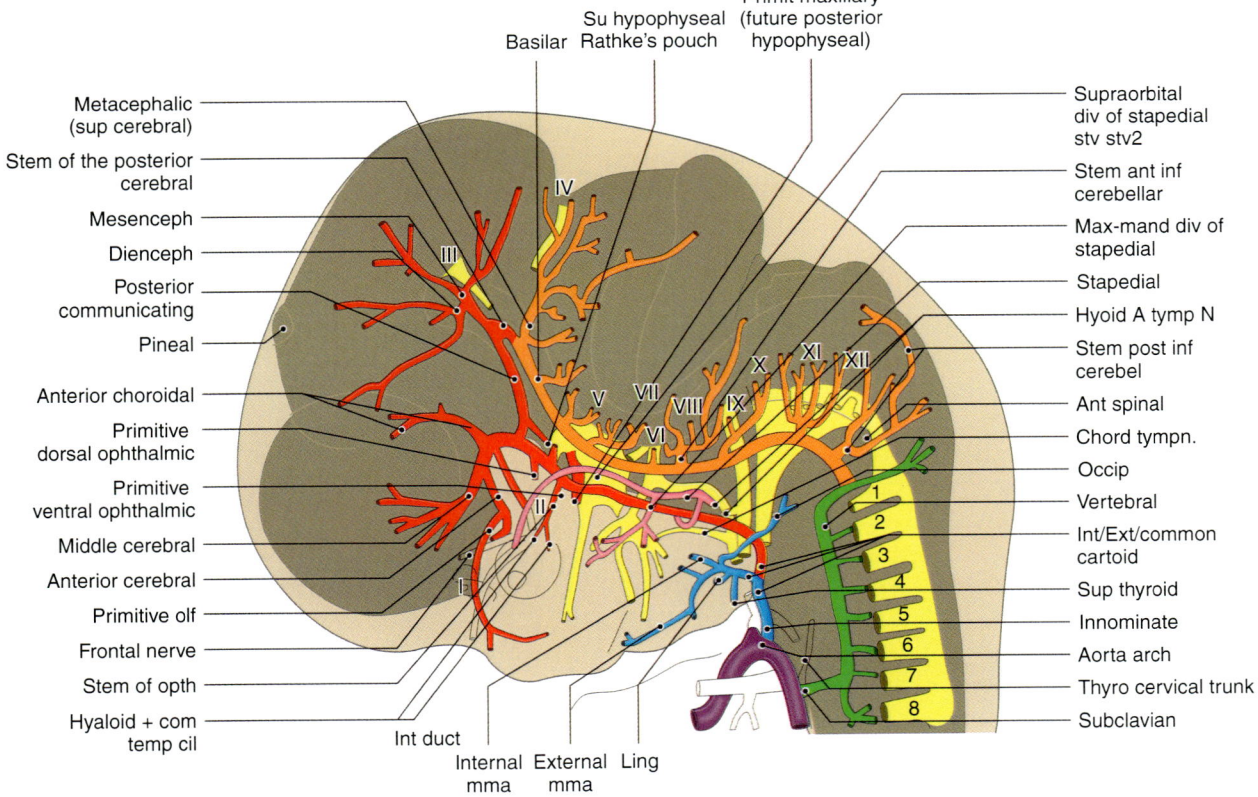

Fig. 6.60 Stages 19–20 (18 mm). Cerebral arteries achieve adult configuration. Intracranial stapedial enters the orbit and reaches the ophthalmic stem. In the face, extracranial produces all branches for the jaws, zygoma, and oronasal cavity

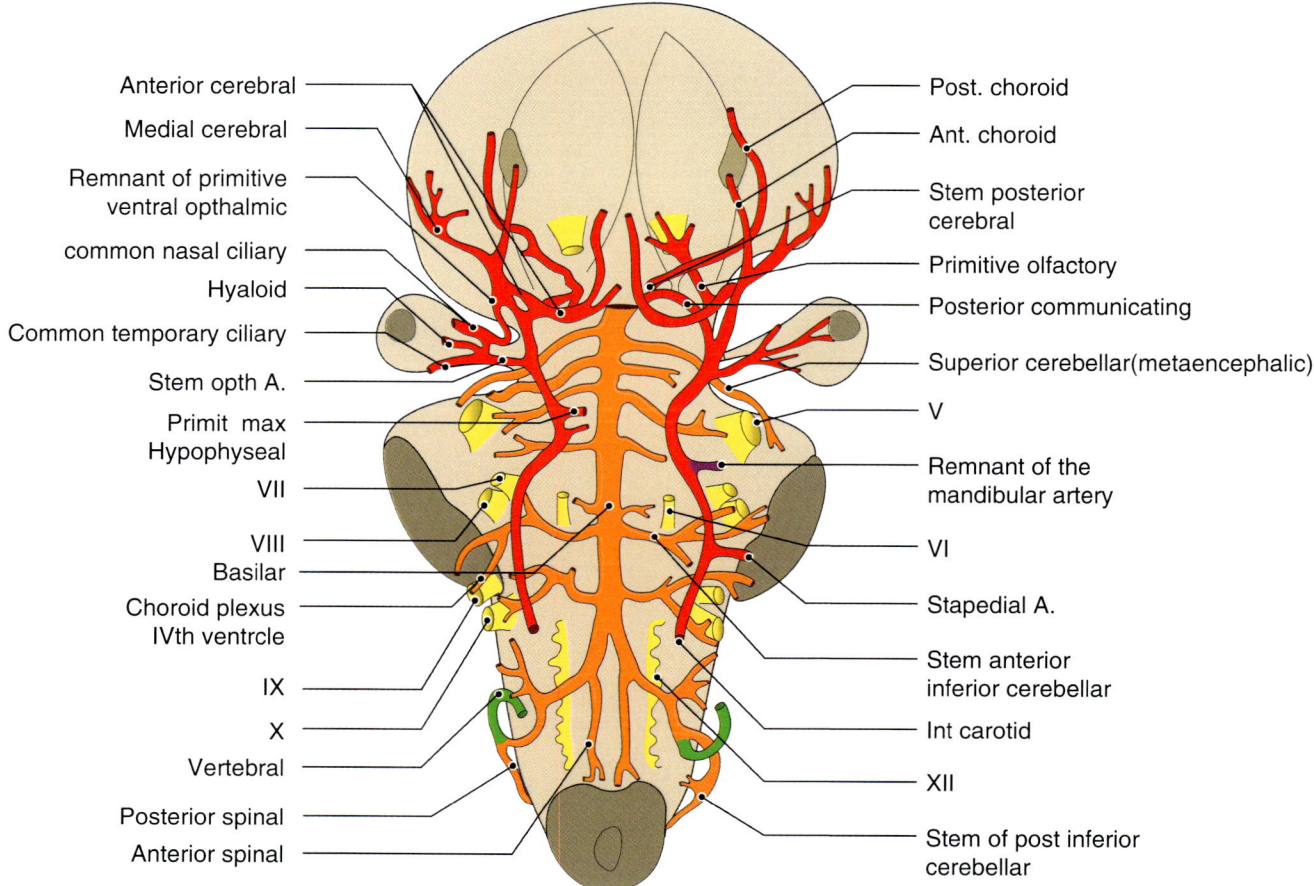

Fig. 6.61 Stages 19–20 (18 mm)

rather the amalgamation of a preexisting network of vessels interconnecting pharyngeal arches 1–3. VPA, originating at AA3, simply reorganizes this network. VPA forms the entire external carotid *not associated with trigeminal neural induction*. The various branches of the non-stapedial ECA may arise by neural induction: superior laryngeal from X., ascending pharyngeal from IX, occipital, posterior auricular, superficial temporal, and facial from VII, lingual from motor V3.

In sum at stage 19, external carotid branches are now visualized for the first time, coincident with the appearance of the common carotid. These include superior thyroid, ascending pharyngeal, occipital, posterior auricular, superior auricular, lingual, and facial (external maxillary). Internal maxillary branches will not appear until a later stage.

Padget's stage 5 marks the final configuration of external carotid. ECA has two embryologic components. (1) <u>VPA-ECA</u> forms all arteries up to the anastomosis with extracranial division of stapedial. (2) The <u>internal maxillo-mandibular system</u> results from the union of VPA-ECA with extracranial stapedial. It has two functional kinds of arteries. Those supplying nonneural crest derivatives, such as muscles of mastication are programmed by the motor branches of V3 and (in the case of buccinator) VII. Those arteries supplying the

neural crest derivatives of the jaws, structural fields connecting the jaws with the skull, and dura are programmed by sensory branches of V2 and V3.

What about the remaining components of superior division? From trigeminal ganglion, an orbital ramus tracks along V1 to supply the orbit and the frontonasal mesenchyme. All dermis, submucosa, bones, sinuses, and dura innervated by V1 will be supported by the supraorbital stapedial system. At the same time, a meningeal ramus tracks along V2 to supply the dura of the anterior temporal fossa.

The alert reader will have deduced that the anatomy described here appears at variance with traditional descriptions and illustrations. The "trifurcation" of stapedial stem, originally intracranial in the euprimate condition, is altered in humans, being dragged forward and downward to a position just behind foramen ovale. The embryonic situation looks like the letter Y with inferior ramus constituting the vertical limb. Inferior ramus develops one stage prior to superior rams. The superior ramus constitutes the angular limbs. In the next stage, changes in blood flow within this system occasion breakdown of the proximal stapedial stem which leads annexation events with ophthalmic and with internal maxillary.

Primitive branches of the internal carotid meet their fate as the ICA begins to achieve a permanent configuration. Two widely separated components of ICA combine to create the adult ophthalmic artery. (1) From the proximal (r4–r5) level of ICA, the orbital division of stapedial arises. *All orbital arteries* originate from this vessel. Note that the nuclei of lateral rectus and accessory lateral rectus reside in r4–r5 as well (we'll discuss the significance of this fact a bit later). From the distal level (p5) of ICA, the primitive *dorsal ophthalmic* and *ventral ophthalmic* arteries now become branches off the ophthalmic stem. They supply the choroid and sclera via *common temporal ciliary* and *common nasal ciliary* arteries. A series of anastomotic "loops" along the carotid appear to permit the primitive PrDOA to migrate *backward* from p5 to r1. It is tempting to speculate that extraocular muscle migrating *forward* into the orbit could pick up blood supply from the developing ophthalmic system using the very same mechanism.

The *primitive maxillary artery* (sic) is directed mesially toward the craniopharyngeal canal. It likely participates in blood supply to the sphenoid; ultimately it contributes to the *inferior hypophyseal artery*. It was likely the source for Sm1 PAM (a mesenchymal source of basisphenoid). The dorsal remnant of the *mandibular artery* follows the vidian nerve (carrying the combined superficial and inferior petrosal branches of VII) as it creeps its way along the lateral wall of the carotid artery.

Forebrain arteries become more defined. Cranial division of internal carotid has the following branches from proximal-to-distal: anterior choroidal, middle cerebral, and anterior cerebral. Its terminus, primitive olfactory artery, now manifests two branches. The original one is directed to nasal fossa. It now gives off a more proximal *olfactory nerve artery* that likely serves the trabecular cartilages and their midline derivative cribriform plates. This artery is described as a continuation of anterior cerebral. In the adult state, this anatomy is deceptive. From a purely topographic standpoint, the anterior cerebral appears to be the direct continuation of the parent ICA, while the more lateral olfactory artery appears to be a branch. This is yet another example in which the relationship between primary and secondary arteries undergoes a reversal.

Middle cerebral artery becomes very prominent. In the roof of the diencephalon, choridal tissue is supplied by the anterior choroidal and posterior choroidal arteries from the respective cranial and caudal divisions of the ICA. Note that the anterior choroidal is more dorsal, supplying the pineal gland, while the posterior choroidal stays more ventral. Mesencephalon is supplied by two prominent branches, one for each mesomere (m1 and m2). It is important to note that diencephalic and mesencephalic supply can at times originate from a common stem of the posterior communicating artery. This anatomy reflects (1) the sharp genetic division between hindbrain and midbrain, (2) the commonality of genetic induction from r1 forward affecting all the way forward to p4, and (3) the separate nature of p5 and p6 which, in the primitive state, received additional induction from signals produced by the AVE (anterior visceral endoderm).

Posterior cerebral is not prominent at stage five. The proximal segment of PCA is seen only at the most caudal segment of the posterior communicating artery. This sets up a paradox. *Posterior cerebrum is developmentally cranial to midbrain.* How can its major artery develop from a segment of posterior communicating that is *caudal* to the mesencephalic stem? Late in the fetal stage, the remainder of the PCA develops as a *collateral* of this stump. To picture this, recall that the midbrain and diencephalon are destined to become encircled by the posterior poles of the ever-enlarging cerebral hemispheres. In the adult state, PCA is connected with the posterior choroidal via multiple small diencephalic and mesencephalic branches. PCA results from a proliferation of these branches. *PCA is ultimately a derivative of the posterior communicating, via the posterior choridal.* As such, it develops prior to, i.e., cranial to, the emergence of the midbrain vascular system from posterior communicating. Thus, the "layout" of the posterior division of ICA has an unappreciated bifurcation of posterior communicating into (1) posterior choridal-diencephalic that gives rise to PCA; and (2) mesencephalic.

Hindbrain differentiates into two subdivisions: metaencephalon (cerebellum and pons) arises from r0/r1; myelencephalon (medulla) arises from r2 to r11. Branches of basilar artery develop in parallel with events in r0–r1 and the fourth ventricle. *Superior cerebellar artery*, established in stage 4, now has a mesial division just caudal to the trochlear nerve (r0–r1) and a caudal division coursing over the lip of the cerebellum. *Anterior inferior cerebellar artery* arises from level of cranial nerve VIII (r4–r5) and terminates in the fourth ventricle choroid plexus. Caudal to the otic placode, bifurcation of the basilar into the vertebrals takes place. Here, branches of vertebral arteries run between the multiple roots of cranial nerves X and XI. From this melange, *posterior inferior cerebellar artery* arises as a single vertebral branch. It runs along the medulla in a cranial direction to seek out the choroid plexus.

Note: PADGET STAGE 6 was based on embryos 20–24 mm long:

Carnegie Stage 20 (18–22 mm, 51 Days)
(Figs. 6.60 and 6.61)

General: Scalp vascular plexus, stage 1. Bending of arms at elbows, fingers distinct but webbed. Anal and urogenital membranes disintegrate.

CNS: Choroid plexus of the lateral ventricles, optic and habenular commissures, interpeduncular and septal nuclei.

Carnegie Stage 21 (22–24 mm, 52–53 Days)
(Figs. 6.62 and 6.63)

General: Fingers free, toes distinct but webbed, external genitalia indifferent.

CNS: First appearance of the cortical plate.

CF circulation, stages 20–21: (1) Circle of Willis completed by anterior communicating (2) Anterior cerebrals extend between hemispheres. (3) Primitive olfactory remnants create medial striate arteries. (4) Ophthalmic stem unites with supraorbital StpV1. (5) Middle meningeal stem. (6) External carotid branches.

At this juncture, the face becomes recognizably human. The head unflexes from the chest and expansion of the cerebral hemispheres creates the normal curvatures of the brain. Appearance of the Circle of Willis marks the maturation of cerebral circulation. We shall see the reconfiguration of the temporary stapedial artery as it undergoes annexation by the ophthalmic and internal maxillary arteries.

The anatomy of the ventral pharyngeal artery and its eventual transformation into the external carotid system follows directly from the above. We have seen that formation of the first four aortic arches creates a series of vascular axes between aortic sac and dorsal aorta, around each of which genetically similar mesenchyme organizes itself into a pharyngeal arch. As each aortic arch is produced in succession, transformations occur in its predecessor arches. Development of AA3 is accompanied by breakdown of AA1–2, leaving behind a *common plexus of vessels* that unites the mesenchyme of PA1–3 into a single unit. AA3 and AA4 do not disappear, but the original relationship they once had as core vessels within pharyngeal arches 3 and 4 undergoes a spatial transformation. AA5 involutes and the mesenchyme it was intended to supply is taken over by circulation from AA4.

As the pharyngeal arches 1–5 amalgamate, overgrowth of the PA1–PA2 complex forces the mesenchymal mass of PA3, PA 4, and PA5 into a progressively more internalized position. The arches fold up upon each other like a Japanese fan. Third arch derivatives of Sm7 such as superior constrictor lie internal to the carotid system. Fourth arch laryngeal muscles derived from Sm8 are deep to the superior thyroid artery, the most proximal branch of the external carotid system. Fifth arch muscles lie tucked within the larynx.

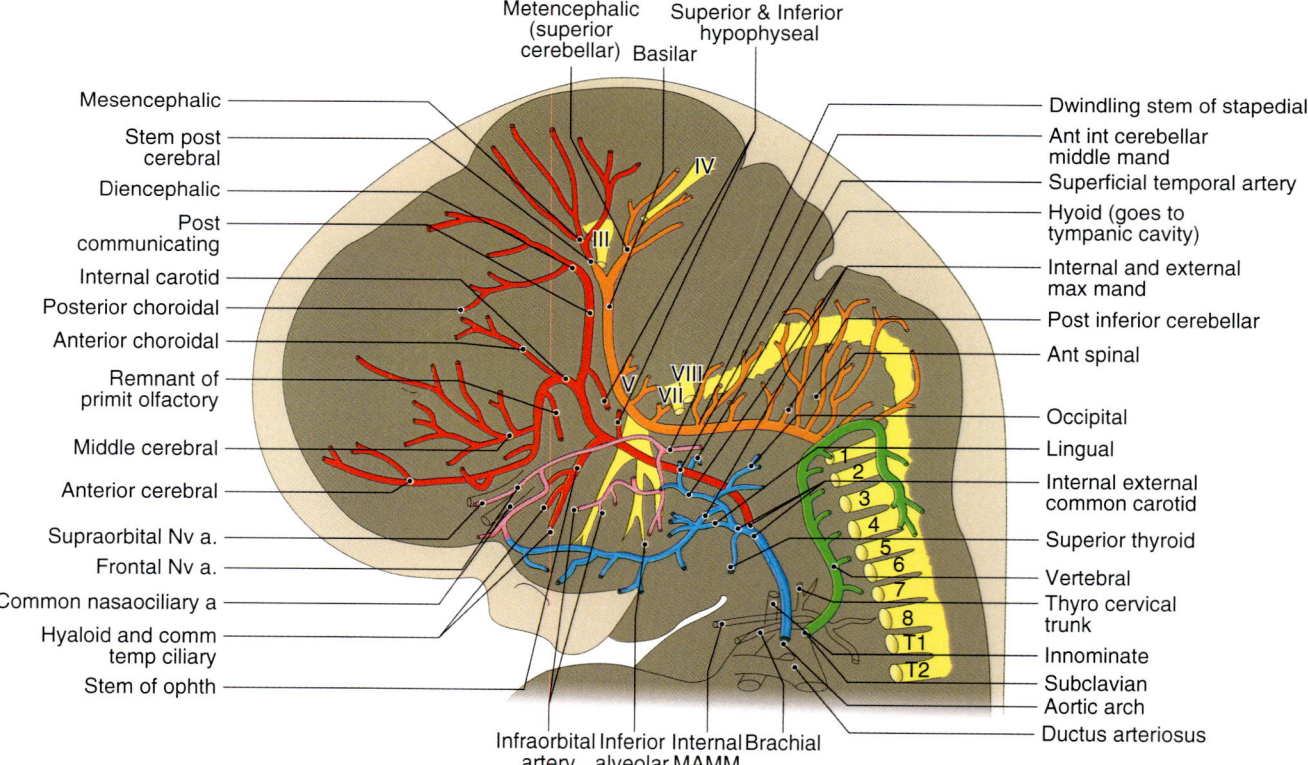

Fig. 6.62 Stages 21–22 (24 mm). Stapedial stem involuting. Anastomosis with ophthalmic stem and formation of orbital vessels. Union anterior communicating arteries complete the circle of Willis. External carotid system fully developed

Fig. 6.63 Stages 21–22
(24 mm)

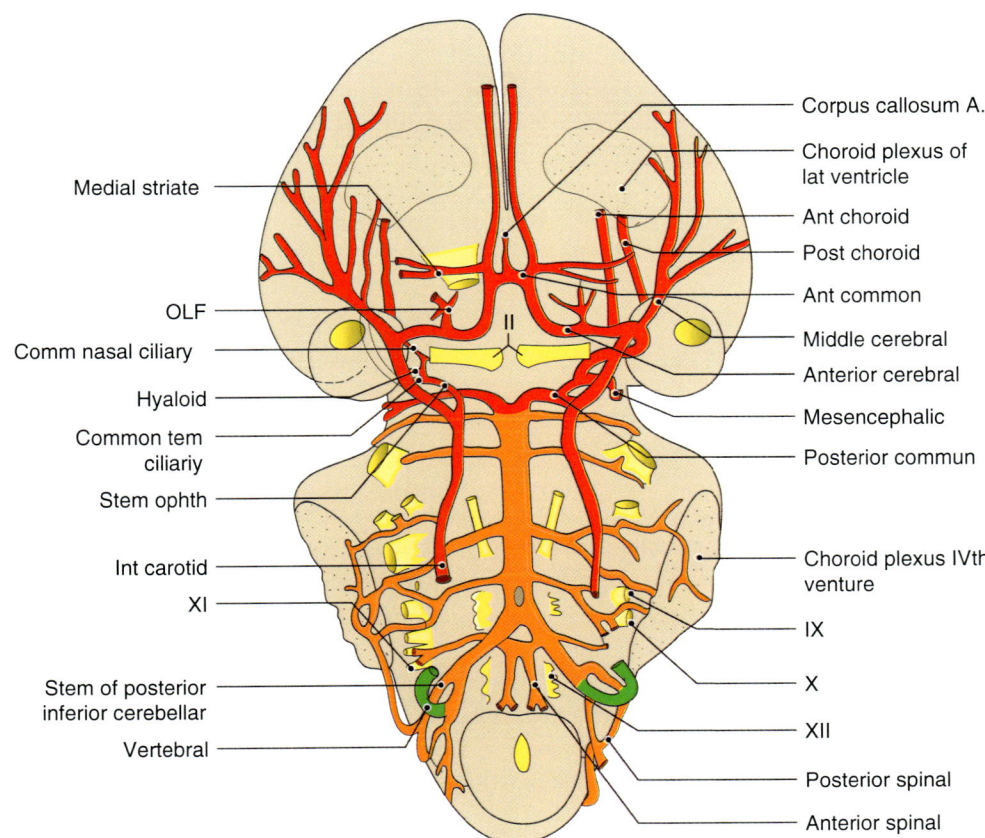

The story of the external carotid system is one of reorganization and simplification. The disparate tangle of vessels shared by pharyngeal arches 1–4 becomes the VPA stretching obliquely across the ventral zones of the arches to reach invading trigeminal system. Transformation of VPA into ECA is characterized by a series of branches arranged in a meaningful proximal-distal order. Each branch is assigned to its corresponding pharyngeal arch. PA4 and PA5 = superior thyroid; PA3 = ascending pharyngeal; PA2 = occipital, posterior auricular, superficial temporal, and facial; and PA1 = internal maxillo-mandibular system.

The rearrangement of the stapedial system at stages 20–21 results from four factors: (1) the anatomic arrangement of the trigeminal system, (2) the requirement that stapedial arteries follow the sensory branches of the trigeminal system, (3) critical anastomoses, and (4) consequent alterations of blood flow. Let's follow the sequence. When the three divisions of trigeminal are given off, each one sends its first branch to supply the dura. In the case of V1 and V2, dural branches are immediately given off prior to the nerves exit via superior orbital fissure and foramen rotundum. V3 makes a very "fast get-away" from the skull via foramen ovale before a dural branch can be produced. V3 is patient however. No sooner does it enter the infratemporal fossa than it produces a recurrent branch, the *middle meningeal nerve*, that tracks upward to reenter the skull to supply the dura of the middle and posterior cranial fossae.

Extracranial stapedial is now unified with external carotid as the external maxilla-mandibular system at the tip of V3. All arteries of IMMA are programmed by trigeminal nerves. Thus, the very first branch of IMMA follows the recurrent V3 nerve to the dura as is the StV3 *middle meningeal artery*. When this artery reaches the trigeminal ganglion, it anastomoses with intracranial STV1 and StV2 to the temporal fossa. A drastic alteration of hemodynamics ensues causing the original anatomy of the intracranial stapedial circuit to fall apart.

By stage 19, StV1 and StV2 had established contact with their respective zones of dura. In addition, both had gained access to the developing orbit at two different sites, STV1 enters through the medial aspect of the superior orbital fissure. StV2 enters as the so-called recurrent meningeal artery through the meningo-orbital foramen in the lateral wing of alisphenoid. StV3 anastomosis between intracranial and extracranial stapedial systems causes the immediate involution of the proximal stem leading from tympanic cavity to the trigeminal ganglion. At the same time, the proximal stem from orbital StV1 back to the trigeminal ganglion disappears.

As a consequence, during stage 20, StV1 to the orbit is annexed by ophthalmic forming a hybrid system just as occurred with IMMA at stage 19. All dural branches of StV1, StV2, and StV3 now depend on the external carotid. In intracranial locations, distal branches of the meningeal arteries, deprived of their original supply, now connect to the branches of the external carotid system. Thus, the various "origins" of dural arteries cited by Gray's Anatomy and other sources are, in reality, anastomotic compensations for the dissolution of the original stapedial system.

With the establishment of blood flow through ECA, changes in blood flow affect the entire system. Proximal stapedial now involutes from just distal to stapes forward to the trifurcation, including a segment proximal posterior division. Hyoid stem remains as the *caroticotympanic artery*. Three critical anastomoses take place, involving connections of distal branches of the stapedial system to each other or connections between the stapedial system and arteries external to the system. (1) A "shunt" connects distal external carotid artery (VPA) to ramus inferioris. Anterior tympanic artery and deep auricular artery are remnants connecting the shunt with the middle ear. (2) Ophthalmic artery annexes the supraorbital division of superior ramus. (3) A "ramus anastomoticus" arising from the shunt enters the cranial cavity via foramen ovale and annexes the meningeal branches of superior ramus (anterior branch). This probably involves a prior connection between primitive trigeminal artery and the meningeal system. The middle meningeal now "originates" from the maxillary division of IMMA. (4) The neomorphic axis of middle meningeal is disconnected from posterior branch of superior ramus. Thus, meningeal branches to posterior fossa arising from posterior branch now appear to "originate" from posterior auricular and occipital.

From the lateral aspect of Gasserian ganglion, two branches of supraorbital take origin. Medially, StV1 follows V1 into the orbit. Laterally, from StV2s meningeal, an *orbital branch* extends forward passing through alisphenoid via a separate *orbito-sphenoid foramen* in the alisphenoid. It supplies the #9 zone of lateral orbit and there anastomoses with StV1 lacrimal artery. When this communication persists, it is known (incorrectly) as is *recurrent meningeal branch* of lacrimal artery.

Lateral orbit is quite distinct from the globe and adnexa. Lacrimal gland likely represents an amalgam of r1 and r2 mesenchyme. This could perhaps account for its division into superior and inferior lobes. Its predominant supply is StV1, supported from below by StV2 infraorbital. Alisphenoid and zygoma are constructed from "add-on" hindbrain mesenchyme (Sm3 PAM and r2 neural crest). AS is thus penetrated by StV2 orbital branch, but its predominant supply is StV2 deep temporal. Zygoma is purely r2 in derivation. Zygomatic branch from infraorbital supplies jugal and postorbital via zygomatiofacial and zygomaticotemporal branches. These in turn anastomose with transverse facial and frontal branches from superficial temporal artery.

Definition of the anterior communicating and anterior cerebral arteries takes place at this stage. The mesial surface of the cerebral hemispheres is now supplied by paired ACAs. Just like the other cerebral arteries, ACA makes its own contribution to the choroid plexus. This branch, noted at the foramen of Monro, precedes the development of corpus callosum. The *anterior communicating artery* at first is plexiform, but later becomes a single channel. It provides a significant branch to the commisural plate (from which the corpus callosum is derived).

In Padget stage 5, the adult stem of ophthalmic consisting of hyaloid and common temporal ciliary (formerly PrDOA) annexes PrVOA to form the common nasal ciliary branch. This anastomosis takes place below the plane of optic nerve.

Primitive olfactory artery completes its development stage 6. Original PrOlfA persists, in its original iteration, as a mesial branch accompanying olfactory nerve. As such, its ultimate destination of *medial olfactory nerve artery* (MOA) are those tissue derivatives of olfactory placode. Within the nasal cavity, it hugs the roof along. Laterally, it forms anastomoses with the anterior and posterior ethmoid branches of StV1 which supply ethmoid labyrinth. A larger lateral branch now appears. *Lateral olfactory artery* (LOA) enters the medial perforated substance of the rhinencephalon; from here it goes on to supply the basal ganglia. Described first in 1909 as the *recurrent artery of Heubner* and more recently as the *medial striate artery*, LOA is formed by anastomosis between elements of PrOlfA and anterior cerebral. Indeed, it can be considered the first branch of anterior cerebral, departing immediately next to anterior communicating artery.

We can make sense of these arteries by recalling that the humans have two olfactory systems. *Accessory olfactory system* is chemosensory. Its neurons originate in medial nasal placode and migrate via cranial nerve 0, nervus terminalis, to midline rhinencephalon. This system is pheromonal. Additional neurons involved in the gonadotropin hormone-releasing hormone (GnRH) system are also located in medial nasal placode. *Olfactory system* represents the sense of smell proper. Its neurons originate in lateral nasal placode; their target is anterior perforated substance. Because MOA occupies the terminal position along the axis of primitive internal carotid, it is tempting to consider it the "oldest" artery, serving a chemosensory system more primitive than olfaction. The ability of sharks to detect blood in the water is a good example of chemo-olfaction.

The circle of Willis is completed but difficult to recognize at this stage. The cerebellar lobes have not formed. In their place is a plexiform configuration of vessels from which both anterior and posterior cerebellar arteries are hard to identify from.

Note: PADGET STAGE **7** is based on embryos 40+ mm (Figs. 6.64, 6.65, 6.66, and 6.67).

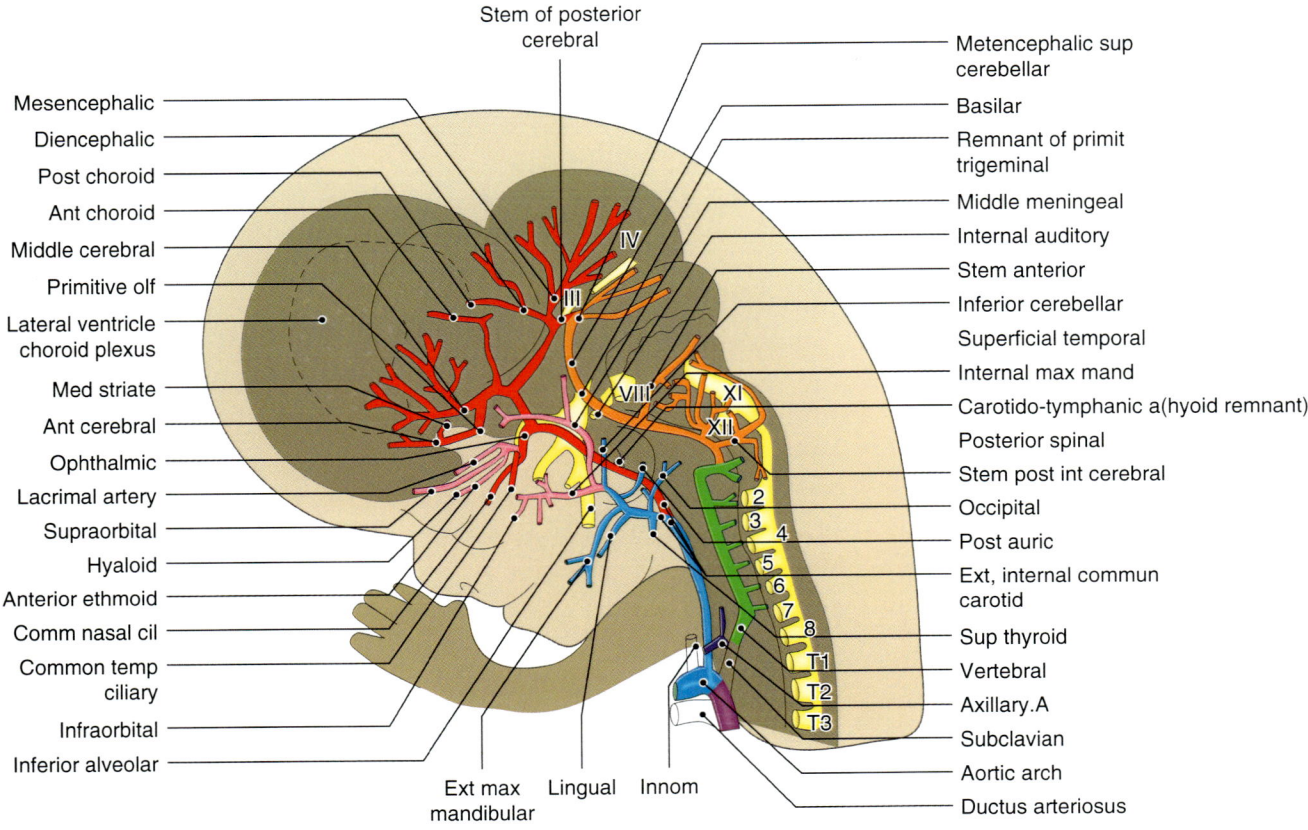

Mesencephalic
Diencephalic
Post choroid
Ant choroid
Middle cerebral
Primitive olf
Lateral ventricle choroid plexus
Med striate
Ant cerebral
Ophthalmic
Lacrimal artery
Supraorbital
Hyaloid
Anterior ethmoid
Comm nasal cil
Common temp ciliary
Infraorbital
Inferior alveolar

Stem of posterior cerebral

Ext max mandibular   Lingual   Innom

Metencephalic sup cerebellar
Basilar
Remnant of primit trigeminal
Middle meningeal
Internal auditory
Stem anterior
Inferior cerebellar
Superficial temporal
Internal max mand
Carotido-tymphanic a(hyoid remnant)
Posterior spinal
Stem post int cerebral
Occipital
Post auric
Ext, internal commun carotid
Sup thyroid
Vertebral
Axillary.A
Subclavian
Aortic arch
Ductus arteriosus

Fig. 6.64 Stage 23+, fetus (43 mm). Cerebellar arteries and posterior cerebral are not yet out to length—awaiting further brain growth. Ocular and orbital system complete

Fig. 6.65 Stage 23+, fetus (43 mm)

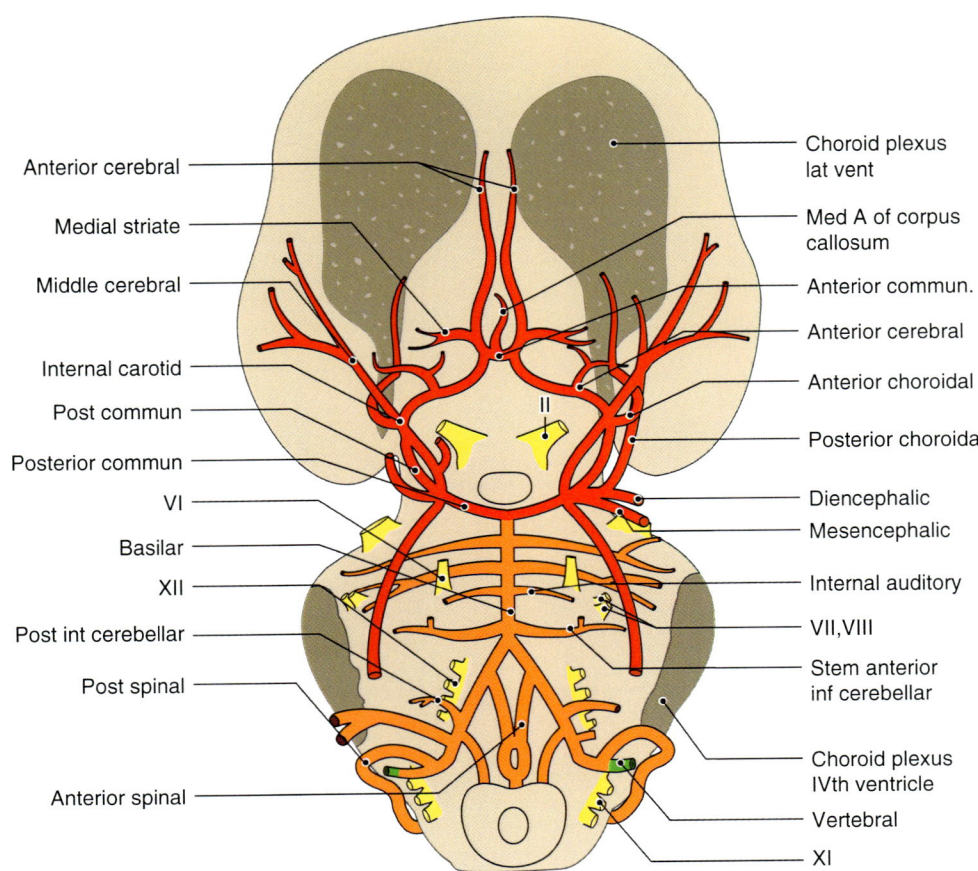

Anterior cerebral
Medial striate
Middle cerebral
Internal carotid
Post commun
Posterior commun
VI
Basilar
XII
Post int cerebellar
Post spinal
Anterior spinal

Choroid plexus lat vent
Med A of corpus callosum
Anterior commun.
Anterior cerebral
Anterior choroidal
Posterior choroidal
Diencephalic
Mesencephalic
Internal auditory
VII,VIII
Stem anterior inf cerebellar
Choroid plexus IVth ventricle
Vertebral
XI

Fig. 6.66 Newborn. Note the sizes of the cerebral arteries and the circle of Willis are still comparable at this stage. The proportions will change with brain growth

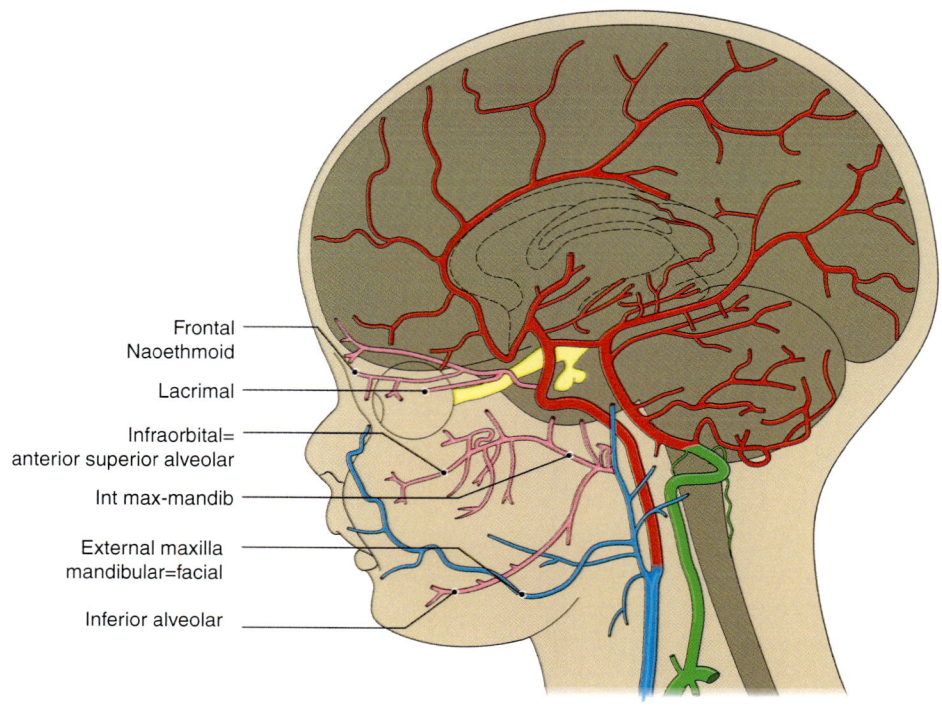

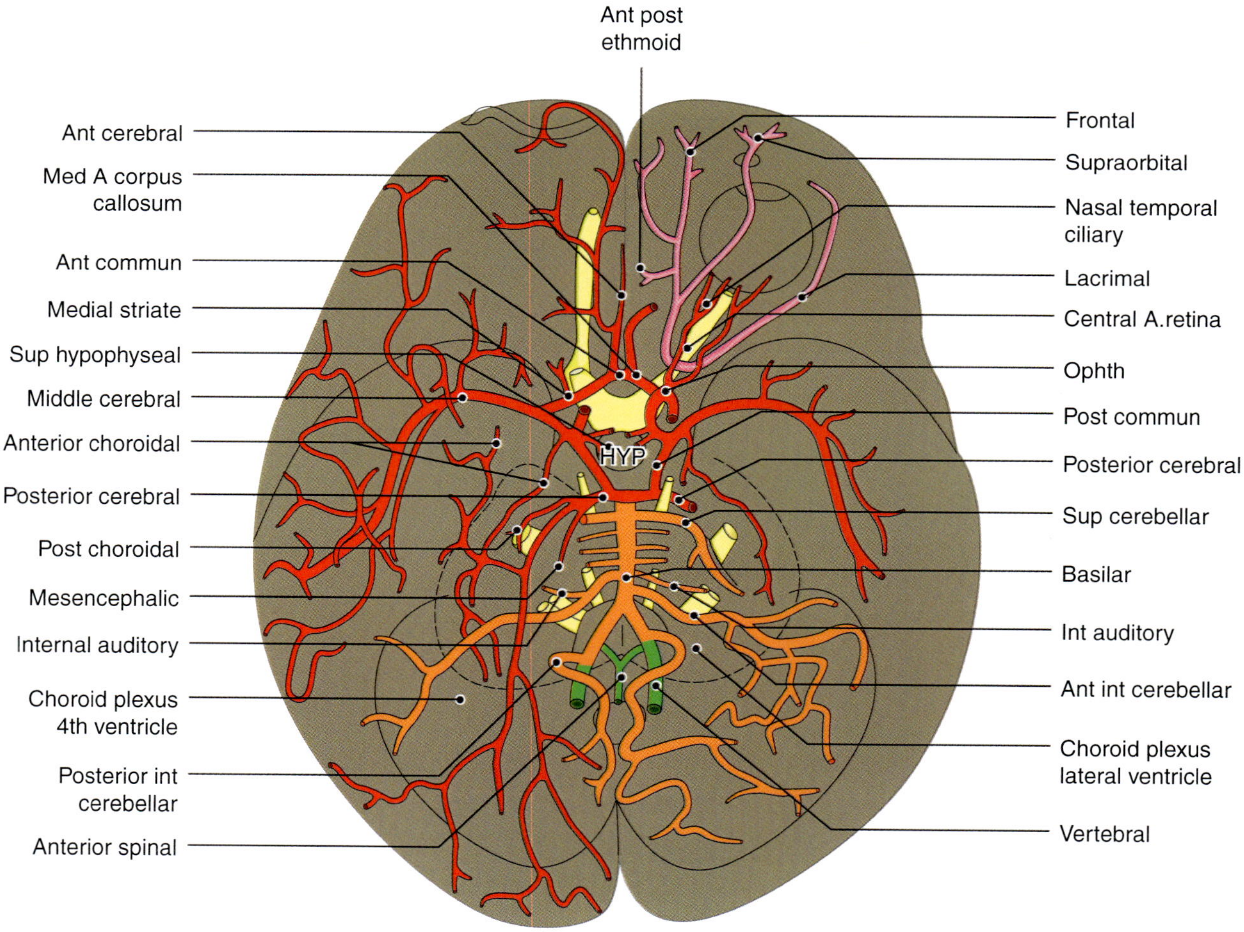

Frontal
Naoethmoid

Lacrimal

Infraorbital=
anterior superior alveolar

Int max-mandib

External maxilla
mandibular=facial

Inferior alveolar

Ant post
ethmoid

Ant cerebral

Med A corpus
callosum

Ant commun

Medial striate

Sup hypophyseal

Middle cerebral

Anterior choroidal

Posterior cerebral

Post choroidal

Mesencephalic

Internal auditory

Choroid plexus
4th ventricle

Posterior int
cerebellar

Anterior spinal

Frontal

Supraorbital

Nasal temporal
ciliary

Lacrimal

Central A.retina

Ophth

Post commun

Posterior cerebral

Sup cerebellar

Basilar

Int auditory

Ant int cerebellar

Choroid plexus
lateral ventricle

Vertebral

HYP

Fig. 6.67 Newborn. Note change in orientation of the posterior communicating artery

This is the adult configuration. The stems of origin of all head arteries are identifiable. Predominance of the cerebral and cerebellar hemispheres over pons and medulla determines the size and direction of the posterior cerebral and inferior cerebellar arteries. A caudal swing of the posterior communicating brings the circle of Willis into view along the ventral brain.

The ophthalmic artery is now complete. The lacrimal branch completes the orbital branches. Its proximal end is the incorporated supraorbital division of stapedial artery. In stage 6, the proximal StV2 intracranial extraorbital stapedial, under influence of V2, becomes the middle meningeal. The distal segment of stapedial becomes the lacrimal.

Part 3. Blood Supply to the Forebrain and Midbrain

Cerebral series, **sagittal**: Figures 6.65, 6.66, 6.67, 6.68, 6.69, 6.70, 6.71, 6.72, and 6.73.
Cerebral series, **axial**: Figures 6.74, 6.75, 6.76, 6.77, 6.78, 6.79, 6.80, 6.81, 6.82 (anterior cerebral), and 6.83.

Dorsal Aortae

For figures, see Chap. 7.

In human embryos up to stage 9, the dorsal aortae are the sole source of blood supply to all nonneural tissues. The organization of these vessels dates back to gastrulation[7] itself, when the mesoderm makes its first appearance. Recall that the mesoderm spreads throughout the embryonic disc with two exceptions. Just anterior to the neuromere r0, i.e., anterior to Hensen's node, a zone exists to which ectoderm and endoderm are adherent, preventing mesodermal penetration. This is the so-called *bucco-pharyngeal membrane*. At the caudal terminus of the embryo, a similar zone of adherence creates the *cloacal membrane*.

The existence of the buccopharyngeal membrane has practical consequence for mesodermal organization. Recall that gastrulation is a cranio-caudal process that begins at Hensen's node. The mesoderm it creates is organized neuromerically. One can visualize it as a series of swatches fanning outward from the embryonic axis, beginning at r0. Anterior to Hensen's node, the swatches from either side become confluent, forming a series of arcs. These skirt around the buccopharyngeal membrane. Dorsal aortae develop within the lateral plate mesoderm, its boundary with paraxial mesoderm.

Note: Intermediate mesoderm destined to form the genito-urinary system lies interposed between lateral plate mesoderm and paraxial mesoderm. But intermediate mesoderm is not present in the head region. It first appears at neuromeric level c1.

The dorsal aortae are quickly invested by an adventitial layer derived from neural crest. Thus, neural crest from each neuromeric level imparts to a particular segment of dorsal aorta an identity in exact register with that particular neuromere. *In sum, we can say that the dorsal aortae, prior to embryonic folding, are derived from neural crest and lateral plate mesoderm in register with rhombomeres r0–r11 and from all 38–40 subsequent myelomeres*. Such segmental coding has multiple consequences.

We cannot overemphasize the importance of neural crest spread into the lateral plate mesoderm. This explains how the autonomic nervous system accompanies the vessels. It is helpful to think of adventitia and fascia as both part of the great connective tissue envelope extending from the CNS to paraxial mesoderm to lateral plate mesoderm. Recall that lateral plate mesoderm splits into a somatic (body) layer and a splanchnic (visceral) layer. Within the somatic lateral plate, neural crest spreads and interacts with mesoderm to produce adventitia, periosteum, ligaments, joint capsules, and ultimately to dermis. Within the visceral lateral plate, neural crest and mesoderm provide connective tissue support for all organs of the thorax and abdomen. The common denominator for all components of the connective tissue system (CTS) is that it provides sensory feedback to the brain. *CTS is an alternative nervous system*.

Very important note: Lateral plate mesoderm *does not split between r0 and r11*. Intraembryonic coelom first appears at level c1. Mesoderm in the region bordering the forebrain and midbrain represents the forward spread of mesoderm from the hindbrain. In the zone anterior to the buccopharyngeal membrane, the LPM once again splits to form the cardiogenic layers.

Recall that in the primordial embryo, the CNS is supplied by paired concomitant vessels, the primitive head plexus and the primitive hindbrain channels. These vessels are well shown in Fig. 6.6. In the earliest stages, no connection exists between the primitive head plexus or primitive hindbrain channels and the dorsal aortae. Circulation is by ebb-and-flow through stage 8.

At stage 9, the two systems become united by a series of anastomotic *segmental arteries* that develop at r0, r4–r5, and r11 (there are perhaps more). The growing brain now has direct access to blood flow from the heart. The PHCs are now renamed: *longitudinal neural arteries*. Segmental arteries also provide blood supply to the somites. The appearance of the first three somites is a defining characteristic of stage 9. Each somite (on either side of the neural tube) receives a nutrient artery from the ipsilateral dorsal aorta. Beginning with the fifth somite (first cervical somite), these arterial stems are in a 1:1 ratio with the somites. The first four (occipital) somites are supplied by the amalgamation of four stems into a single vessel, the *hypoglossal artery*. This vessel also anastomoses with the caudal terminus of the LNA.

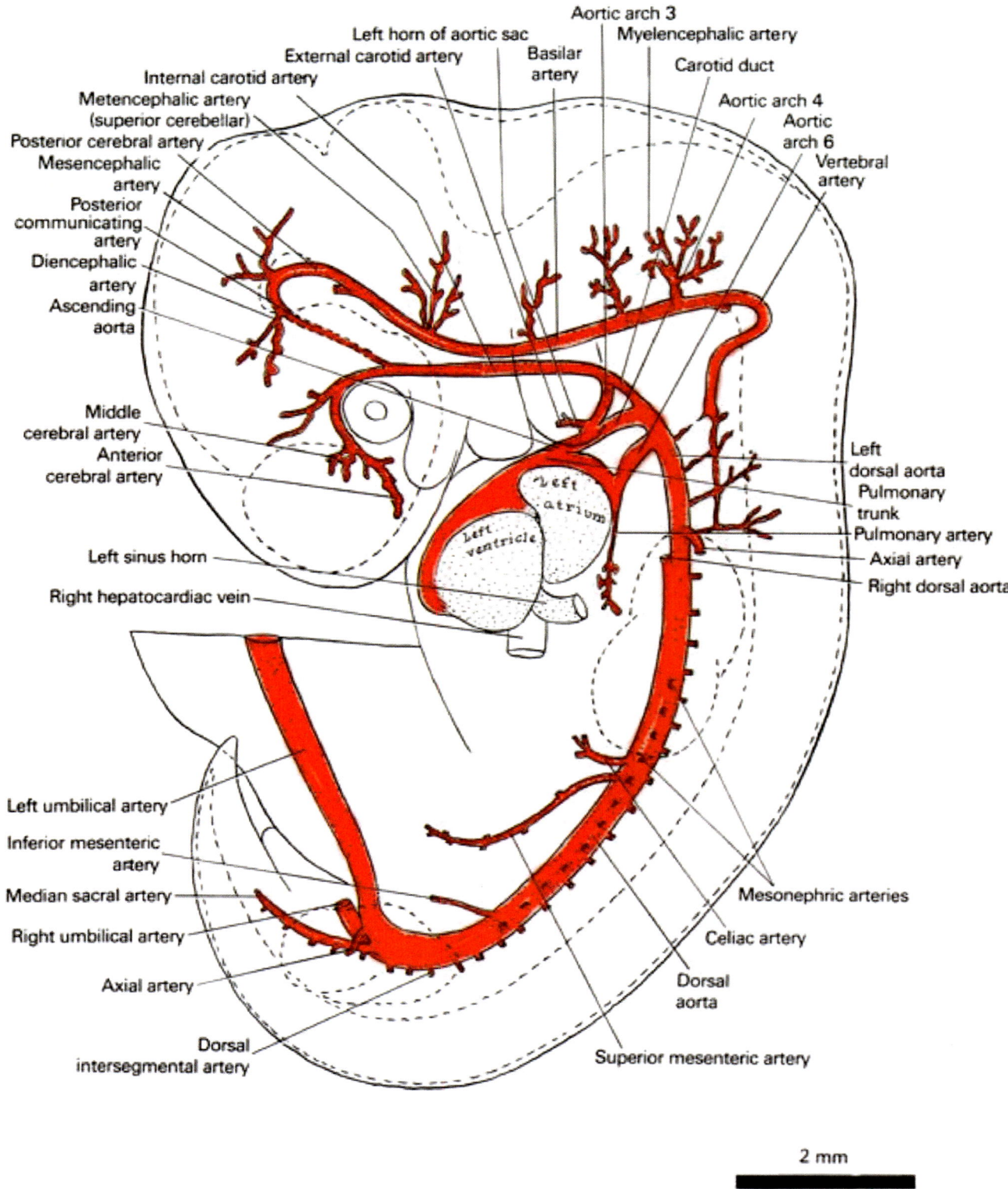

Fig. 6.68 Cerebral circulation at stage 14 emphasizing showing major branches. Posterior cerebral is represented as a small stem, anteriorly directed, form posterior communicating. 25 segmental arteries from the aorta represent 12 + 5 + 5 + 3 somites. 12 mesonephric segmental arter- ies are shown. (Adapted from Gasser RF. Atlas of Human Embryos. Harper & Row; 1975. And used by permission of the Virtual Human Embryo project (https://virtualhumanembryo.lsuhsc.edu))

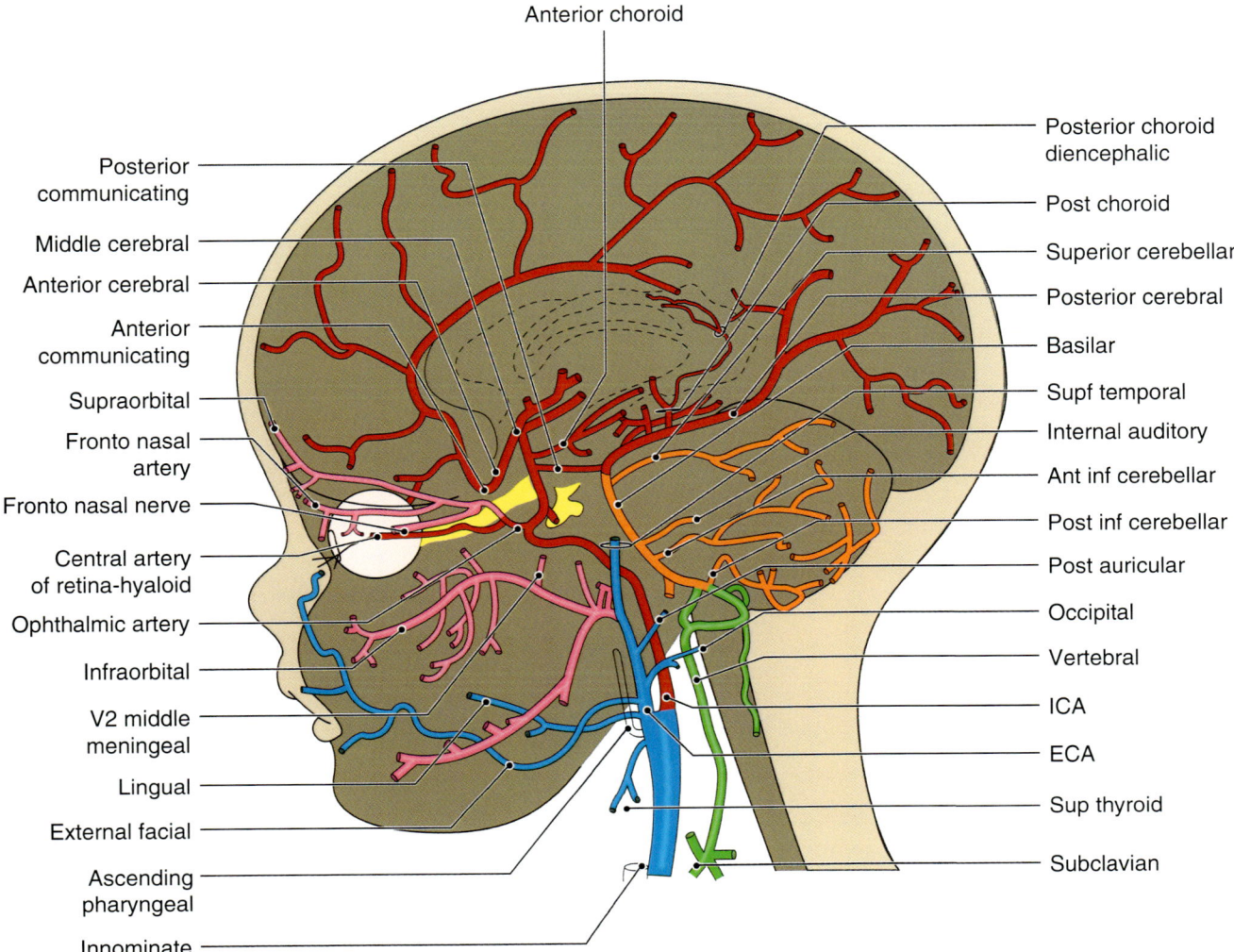

Anterior choroid

Posterior communicating

Middle cerebral

Anterior cerebral

Anterior communicating

Supraorbital

Fronto nasal artery

Fronto nasal nerve

Central artery of retina-hyaloid

Ophthalmic artery

Infraorbital

V2 middle meningeal

Lingual

External facial

Ascending pharyngeal

Innominate

Posterior choroid diencephalic

Post choroid

Superior cerebellar

Posterior cerebral

Basilar

Supf temporal

Internal auditory

Ant inf cerebellar

Post inf cerebellar

Post auricular

Occipital

Vertebral

ICA

ECA

Sup thyroid

Subclavian

Fig. 6.69 Fetal stage. Predominance of cerebral and cerebellar hemispheres stretches the arteries into the adult configuration. Anterior cerebral artery no longer supplies choroid plexus because corpus callosum and fornix separate cerebrum from diencephalon. Posterior cerebral now prominent derivative of mesencephalic branch (which is now reduced). AICA and PICA now very dominant to the cerebellum. They supply the cerebellar choroid plexus as an afterthought

Fig. 6.70 Stages 11–12.
Internal carotid gives off two
branches before reaching the
forebrain. Primitive trigeminal
is the first and largest of the
presegmentals that connect
dorsal aortae with the
longitudinal neural arteries.
Primitive maxillary was
originally directed to the optic
vesicle at stage 9–10, but
becomes redirected away
from it to Rathke's pouch, the
future hypophyseal fossa

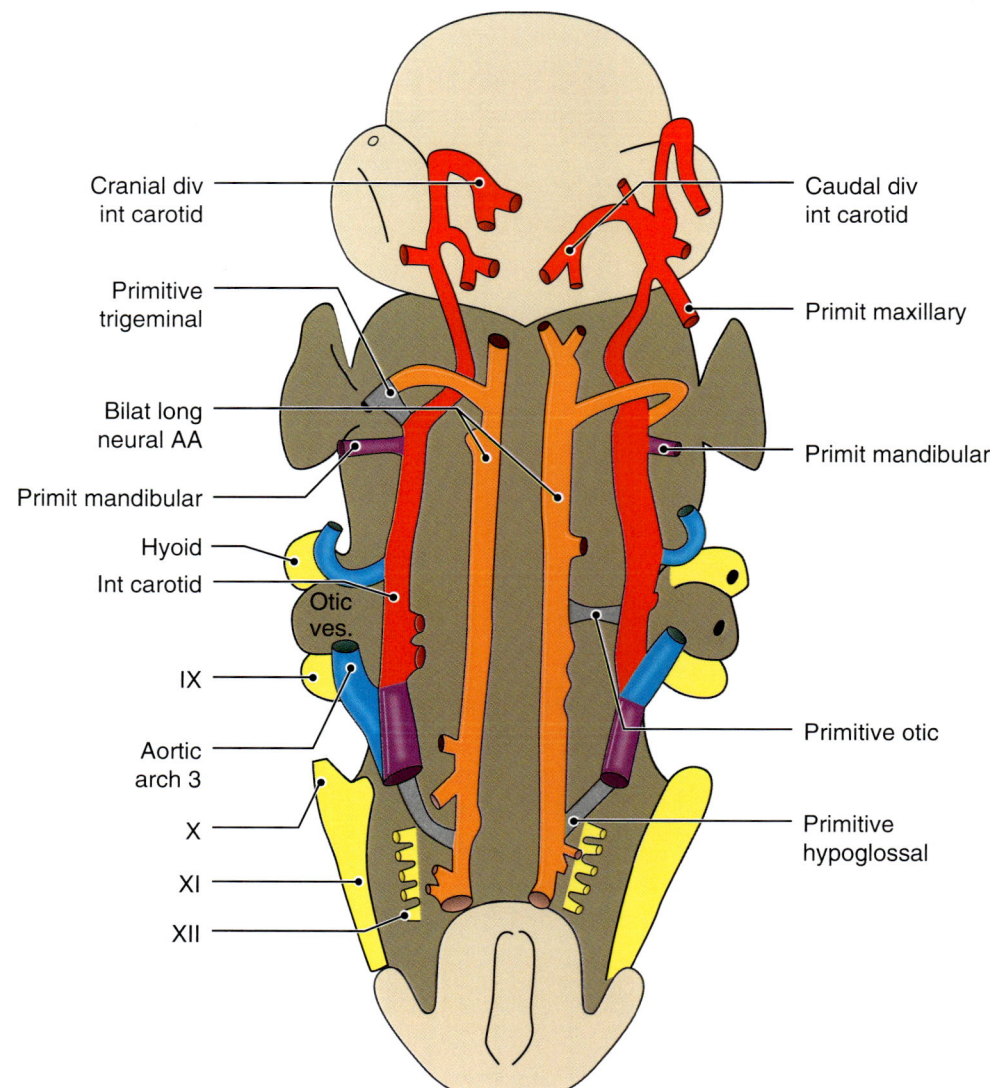

Fig. 6.71 Stages 13–14. The LNAs are fusing in the ventral midline beneath pons to make form basilar. ICA has bifurcated. The terminal segment of the caudal division, just beyond the mesencephalic stem, becomes posterior communicating. The two PComms form a Y-shaped junction and connect with the basilar (here seen in formation). Just before doing so, they give off the posterior cerebrals. The first branches from the basilar are those of metencephalic (superior cerebellar). Direct flow from ICA to LNA makes primitive trigeminal irrelevant; it involutes

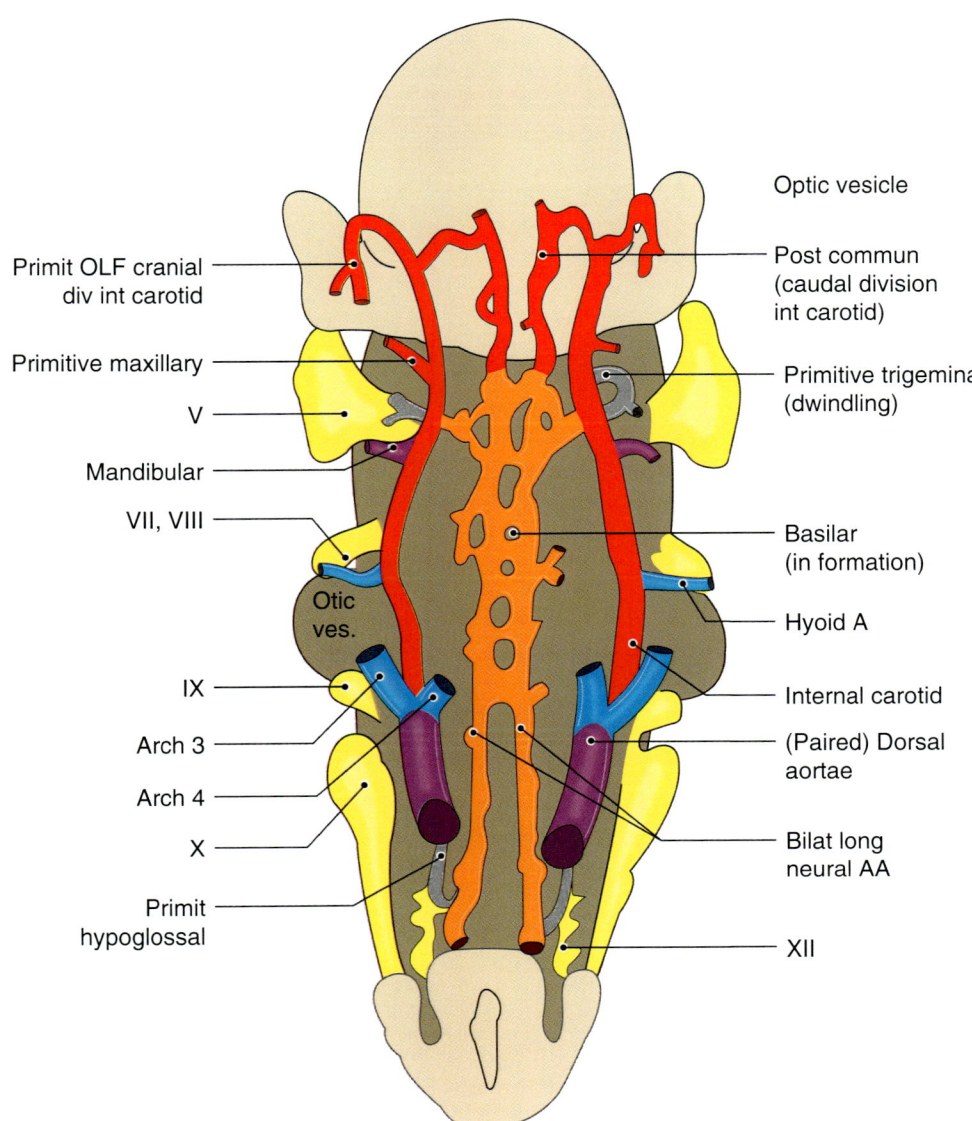

Optic vesicle

Primit OLF cranial div int carotid

Post commun (caudal division int carotid)

Primitive maxillary

Primitive trigeminal (dwindling)

V

Mandibular

Basilar (in formation)

VII, VIII

Hyoid A

Otic ves.

Internal carotid

IX

(Paired) Dorsal aortae

Arch 3

Arch 4

X

Bilat long neural AA

Primit hypoglossal

XII

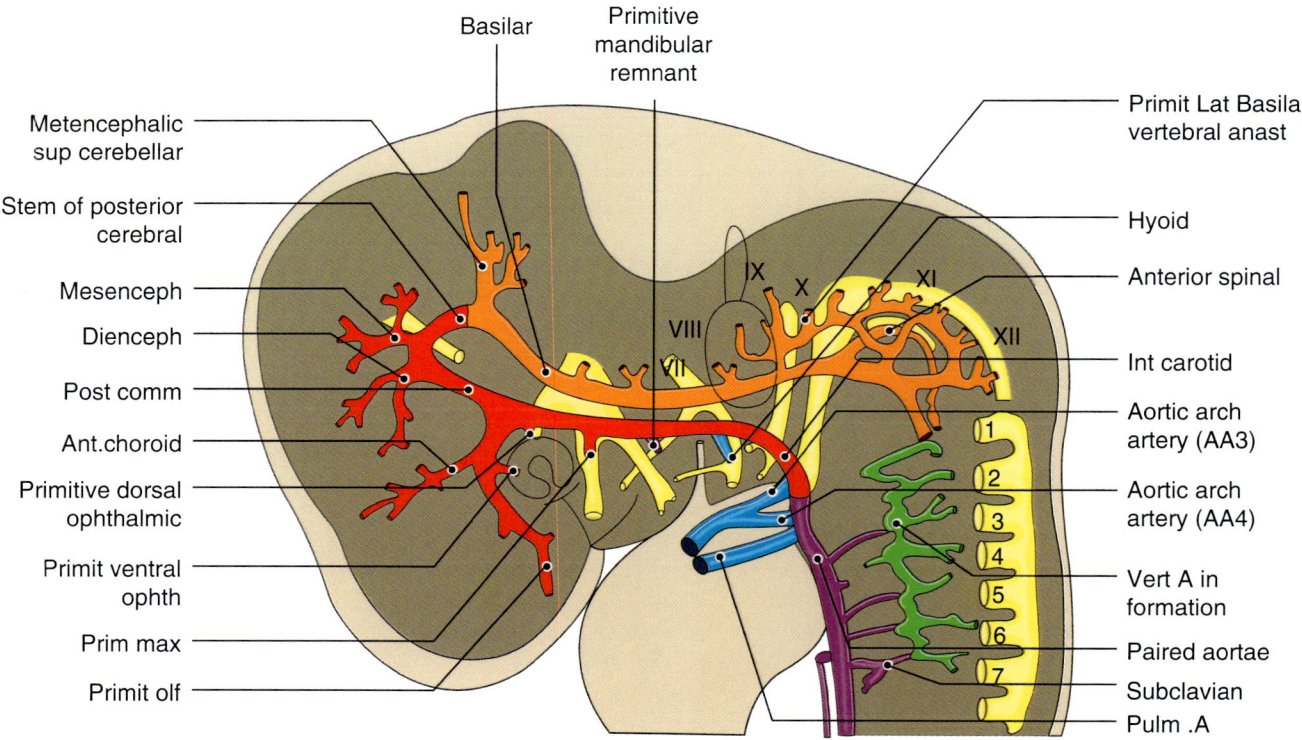

Fig. 6.72 Stages 15–16 (9 mm). Upper seven cervical segmental arteries connecting dorsal aortae with spinal cord and somites now form a vertical chain of anastomoses that link them together. All cerebral arteries still depend on internal carotid. They are not yet connected with the hindbrain circulation. Hyoid artery (dorsal remnant of AA2) lies at the cranial corner of the second arch. The third arch melding with the first two arches. Vessels within the third arch are included with the ventral pharyngeal artery. VPA now arrives at the root of V3. Optic cup now has plexus from the primitive ventral ophthalmic artery

Fig. 6.73 Stages 15–16. Basilar is almost complete. Running on either side of it are parallel vessels that extend from cranial nerve six to the hypoglossal to create the basilo-vertebral anastomosis. Posterior cerebrals can be seen emerging from PCom. Note emergence of diencephalic and mesencephalic arteries from the posterior division

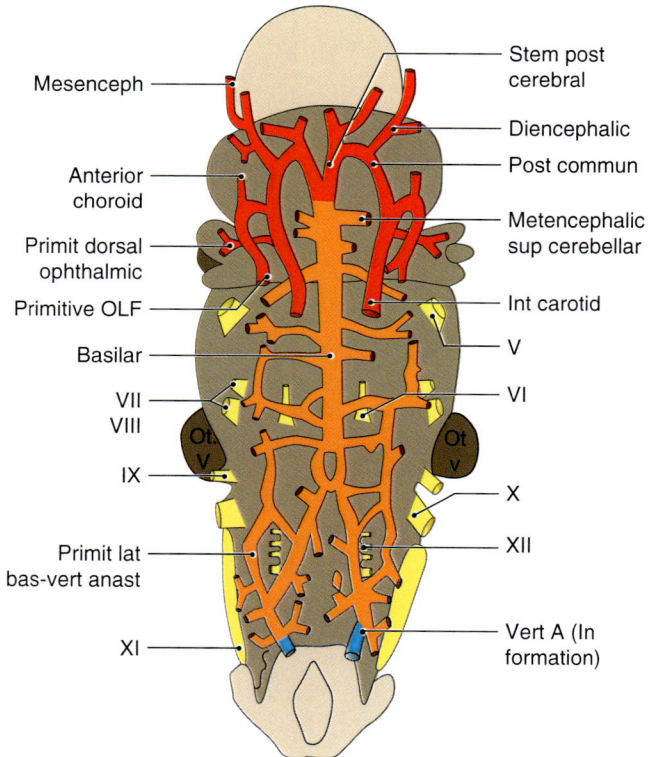

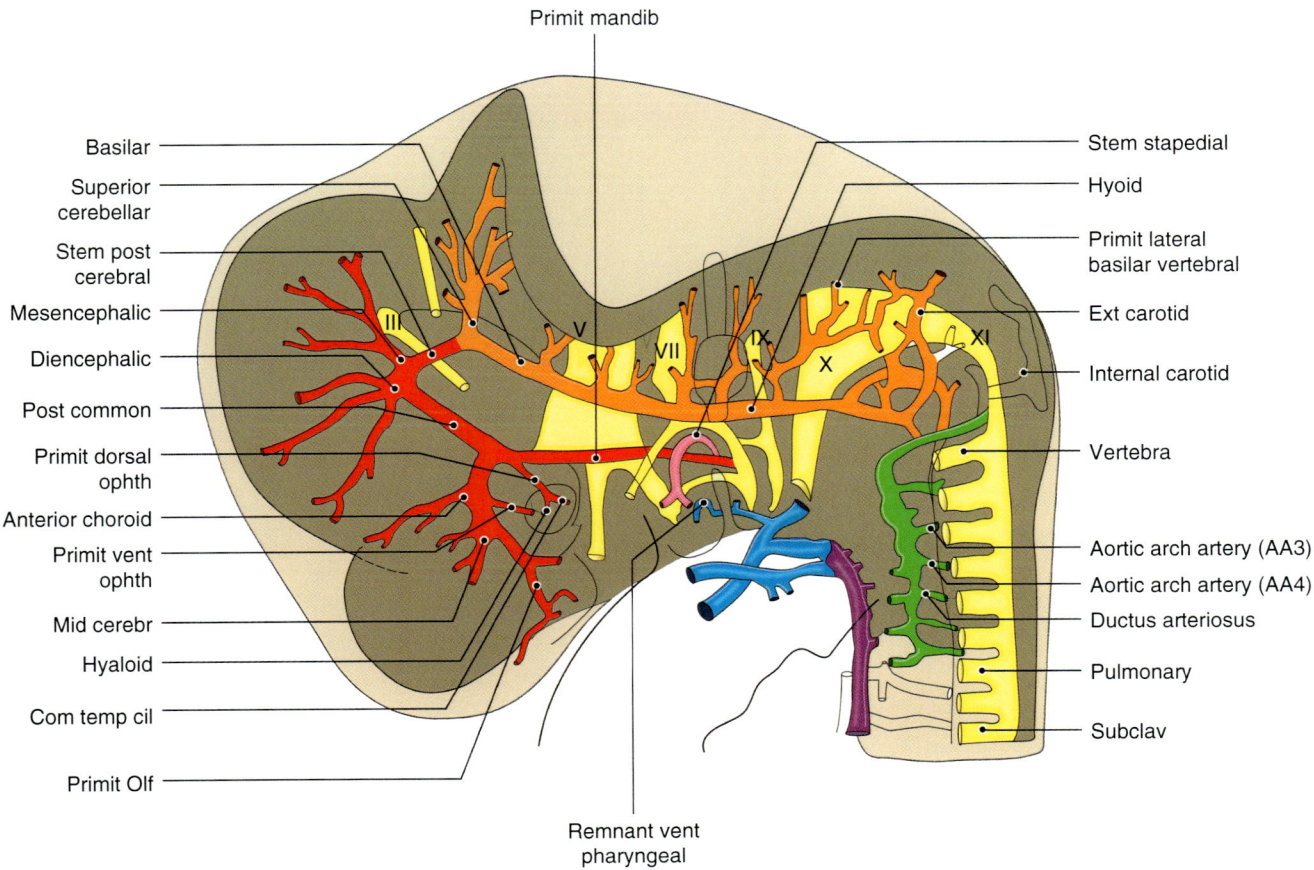

Fig. 6.74 Stages 17–18. Early "postbranchial" phase. The hyoid artery, no longer part of second arch, gives off stapedial artery. Stapedial directed intracranial (not shown) and extracranial, where it joins distal VPA to complete the external carotid artery. Note that stage 17 is characterized by unification of facial processes. It is at this time that craniofacial clefts can become manifest. Dorsal aortae are still paired. The segment of dorsal aorta between AA3 and AA4 begins to dwindle. Common carotid, the segment of outflow tract between AA3 and AA4 becomes the common carotid. Subclavian origin of vertebral artery well defined. Primitive dorsal ophthalmic artery now has two definitive branches: hyaloid and temporal ciliary

Fig. 6.75 Stages 17–18.
Note the two branches of
primitive olfactory artery. The
stem continues to the
*olfactory pit in the nasal
placode*; a medial branch
supplies the *olfactory nerve
root*. From the basilar-
vertebral channel, multiple
branches emerge. These
create a vascular ring around
cranial nerve VI (abducens)

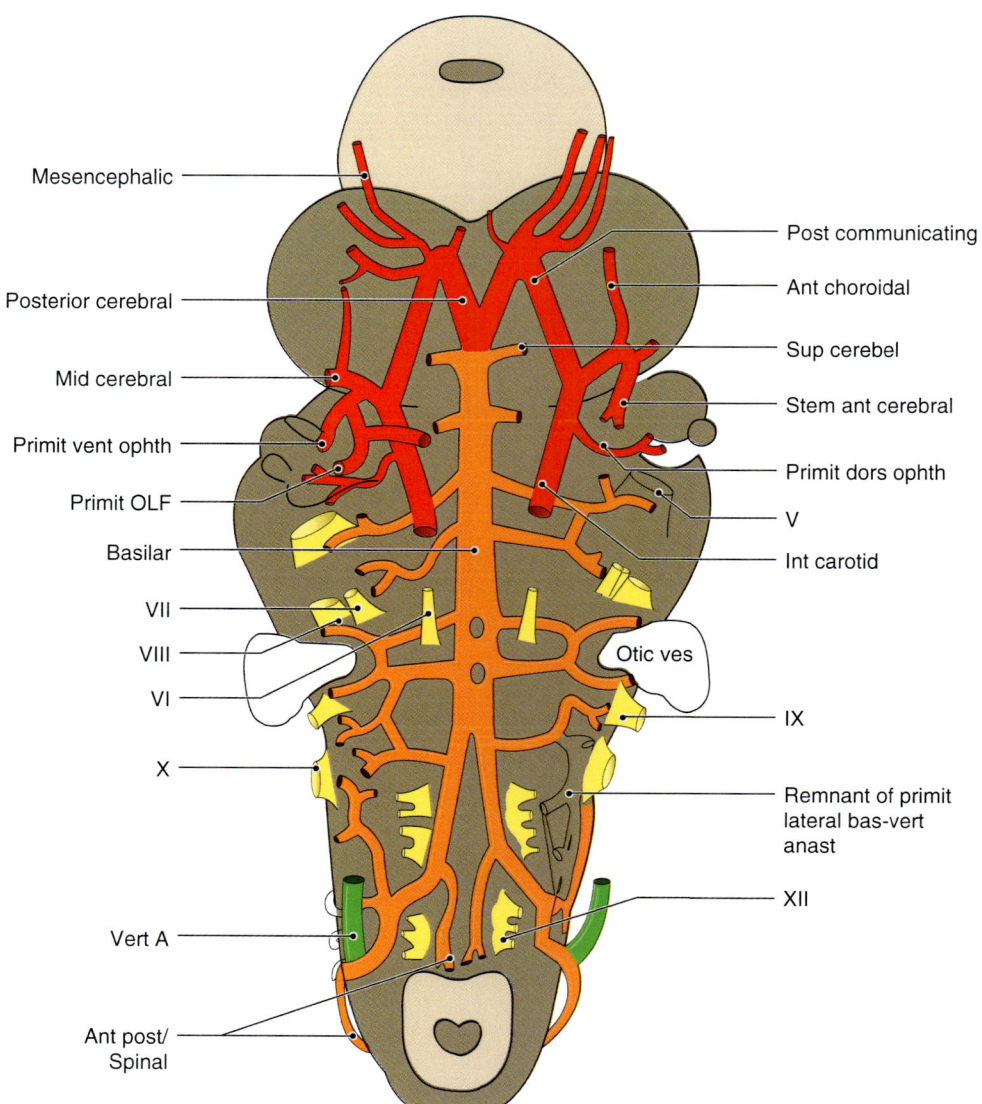

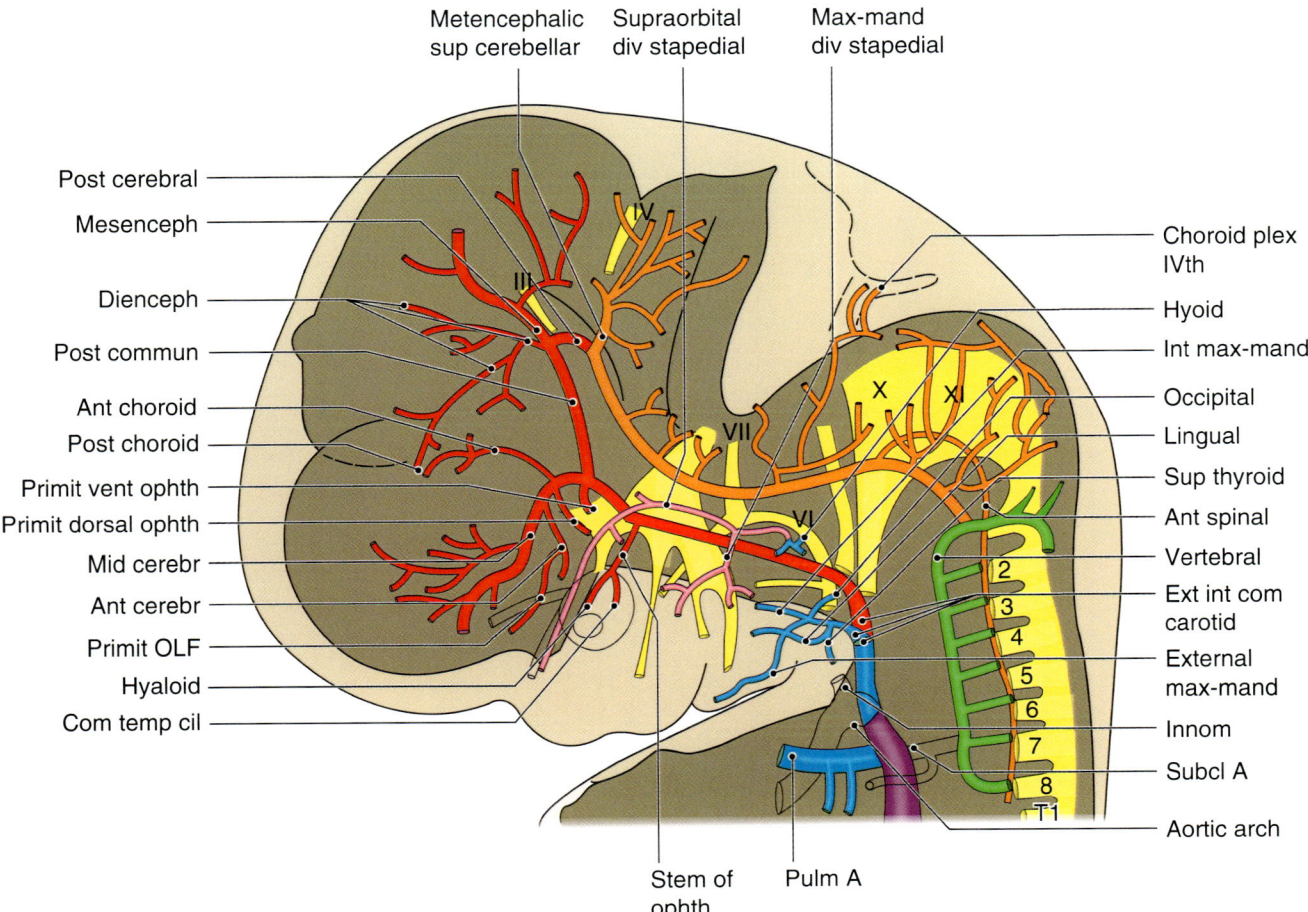

Fig. 6.76 Stages 19–20. Stems are well identified for two systems: cerebral and external carotid. Primitive olfactory artery (formerly the terminal end of the cranial division of ICA) is shriveling in favor of the anterior cerebral. Stapedial artery now has identified branches associated with the trigeminal nerve: orbital, maxillary, and mandibular. Stapedial stem still present; thus, no anastomosis as yet with primitive ophthalmic. Primitive ophthalmic now annexes the primitive dorsal ophthalmic and primitive ventral ophthalamic to create the adult ophthalmic stem. Both PrDOA and PrVOA are V1 inductions from lateral plate mesoderm; they do not involve the stapedial system. Subclavian artery now arises opposite the pulmonary artery. Vertebral artery is now complete

Fig. 6.77 Stages 19–20. Olfactory fossa is supplied by primitive olfactory, but its position as the leading edge of anterior division is becoming obscured in the growth of anterior cerebral. Note variable origin of stems of anterior inferior cerebellar and of internal auditory. Ophthalmic stem has now completed annexation of primitive ventral ophthalmic. The eye has its full complement of vessels. Anastomosis with stapedial comes next

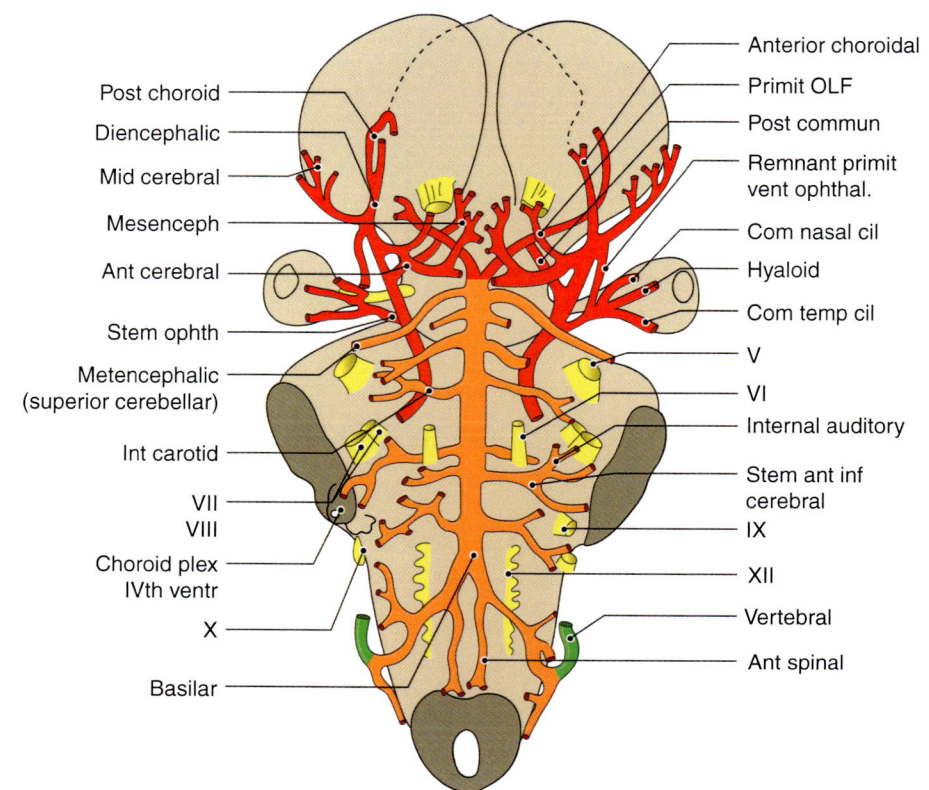

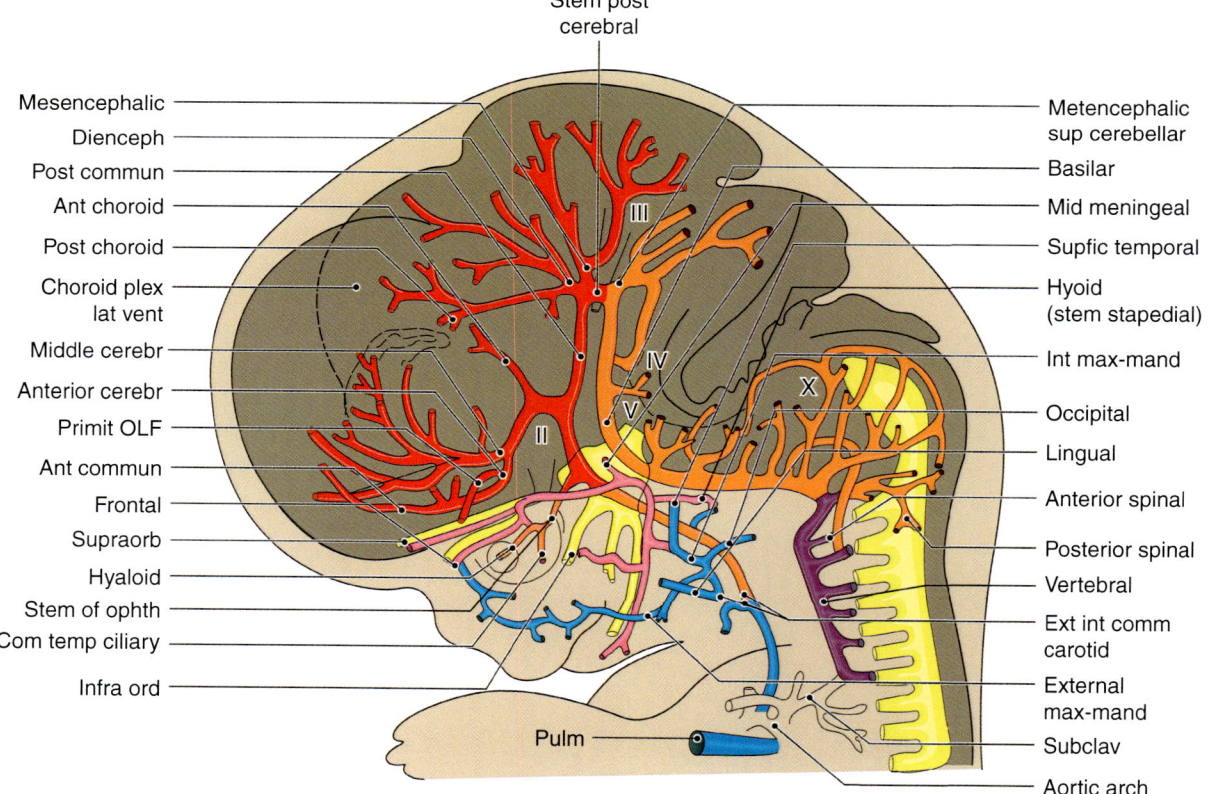

Fig. 6.78 Stages 21–22. CNS tissue cannot form arteries. Thus, all internal carotid and longitudinal neural/basilar vessels lie *external* to the brain. Hindbrain growth is not topologically complex. It is essentially an expansion of the neural tube. Thus, hindbrain vessels remain ventral. Cerebral growth involves expansion and folding which entraps and internalizes internal carotid branches. Anterior cerebral arteries now have a communicating branch. These are stretched medially by the forebrain. Stapedial stem involutes simultaneous with, or as a consequence of, two anastomoses: (1) ophthalmic + supraorbital division and (2) external carotid + maxillomandibular division

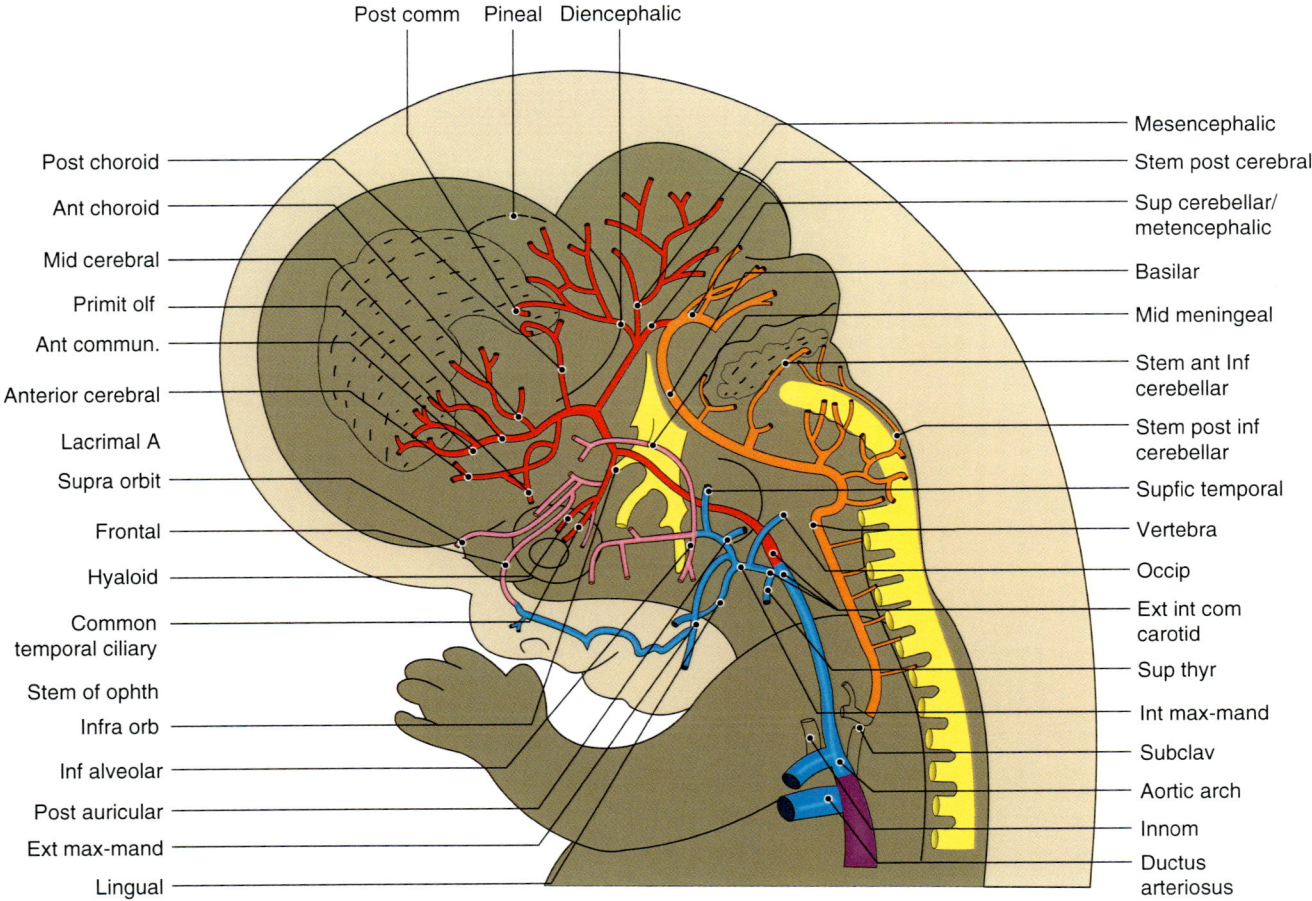

Fig. 6.79 Stage 23+. Posterior communicating originally was directed caudocranially. It now reverses course, becoming craniocaudal. Blood supply to choroid plexus is an important landmark. Two arteries, anterior inferior cerebellar and posterior inferior cerebellar arteries, were previously hidden within the primitive anastomoses between basilar and vertebral arteries. AICA and PICA are now well seen on their way to supply the choroid plexus of the fourth ventricle. Choroid plexus of lateral ventricle receives anterior choroidal from cranial division of ICA and posterior choroidal from caudal division of ICA. At the foramen of Munro, one finds a temporary artery (the primitive branch of anterior cerebral). Ophthalmic artery acquires the STV1 lacrimal artery. The supraorbital (intracranial) division of stapedial remains as the middle meningeal artery

Fig. 6.80 Stages 21–22.
Primitive olfactory is
disintegrating and
incorporated into striate
branches of segment A1 of the
anterior cerebral artery.
Proximal development of
ACA is complete with
emergence of the paired
anterior communicating
branches. They unite in the
midline to complete the circle
of Willis

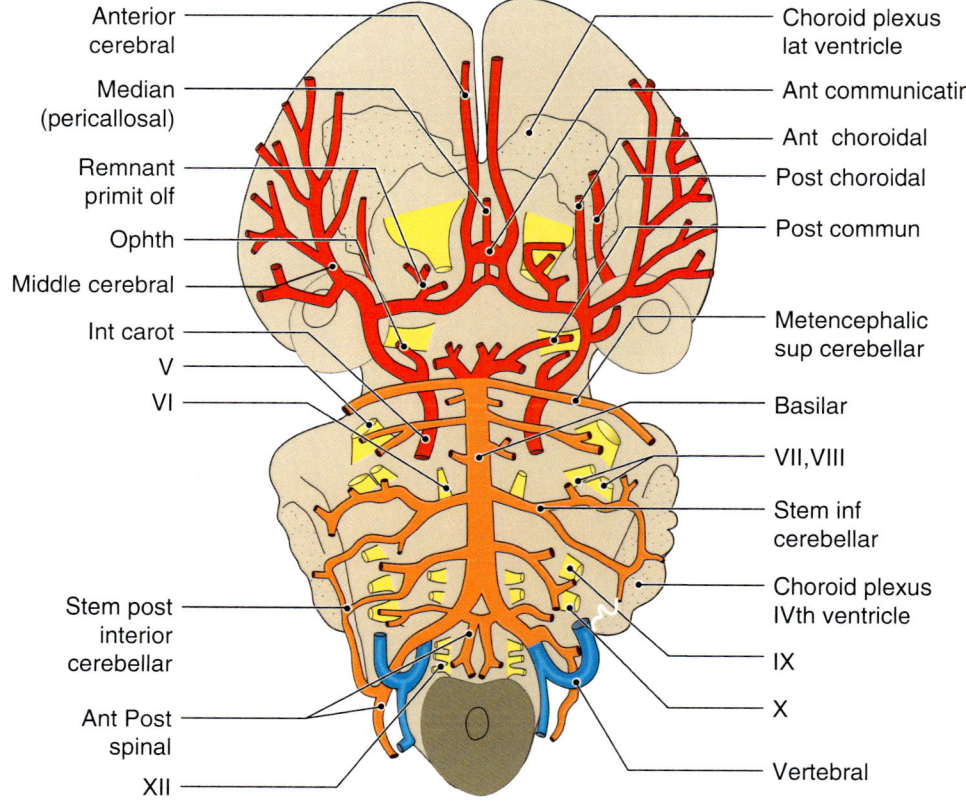

Fig. 6.81 Stage 23+. At
7 weeks, the ventral branches
from the circle of Willis are
well seen. Brain expansion
causes the posterior
communicating artery to
swing backward to supply the
occipital lobe. At this time,
PComm overshadows the
posterior cerebral, but the
latter will expand
dramatically with brain
growth. Note that the blood
flow to PCA, though
embryologically a part of the
internal carotid system,
eventually becomes
dependent more on basilar
than ICA circulation. Despite
the fact that occipital lobe is a
forebrain deliberative, PCA is
confusedly considered by
some to be developmentally
related to the hindbrain
system

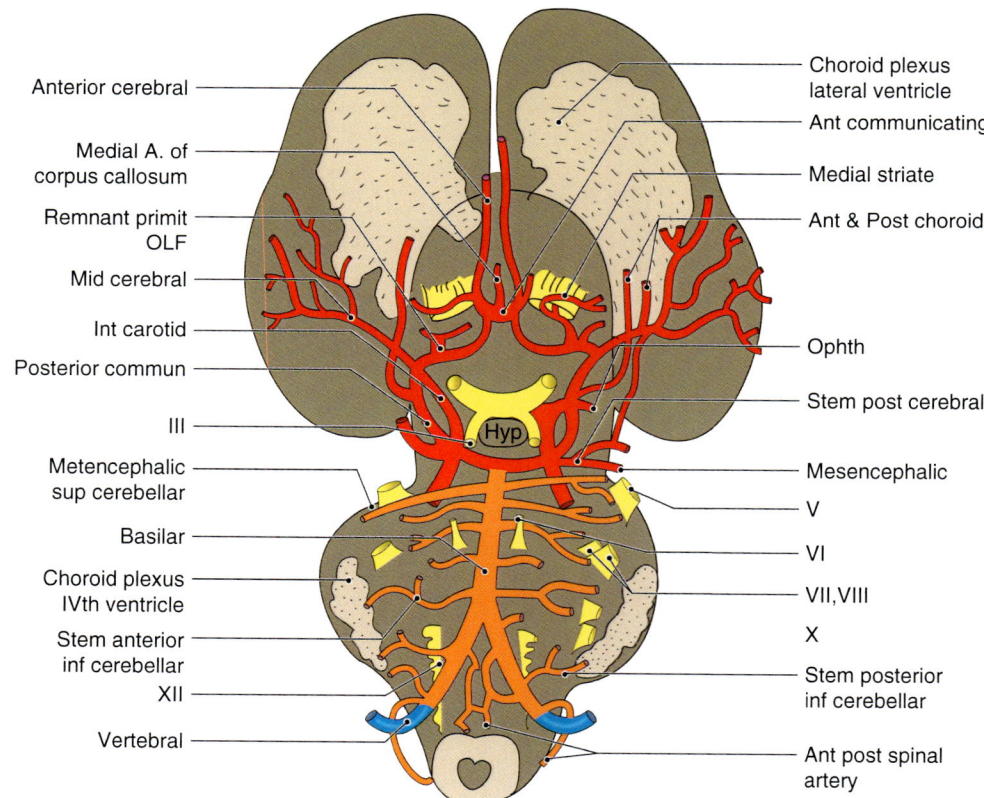

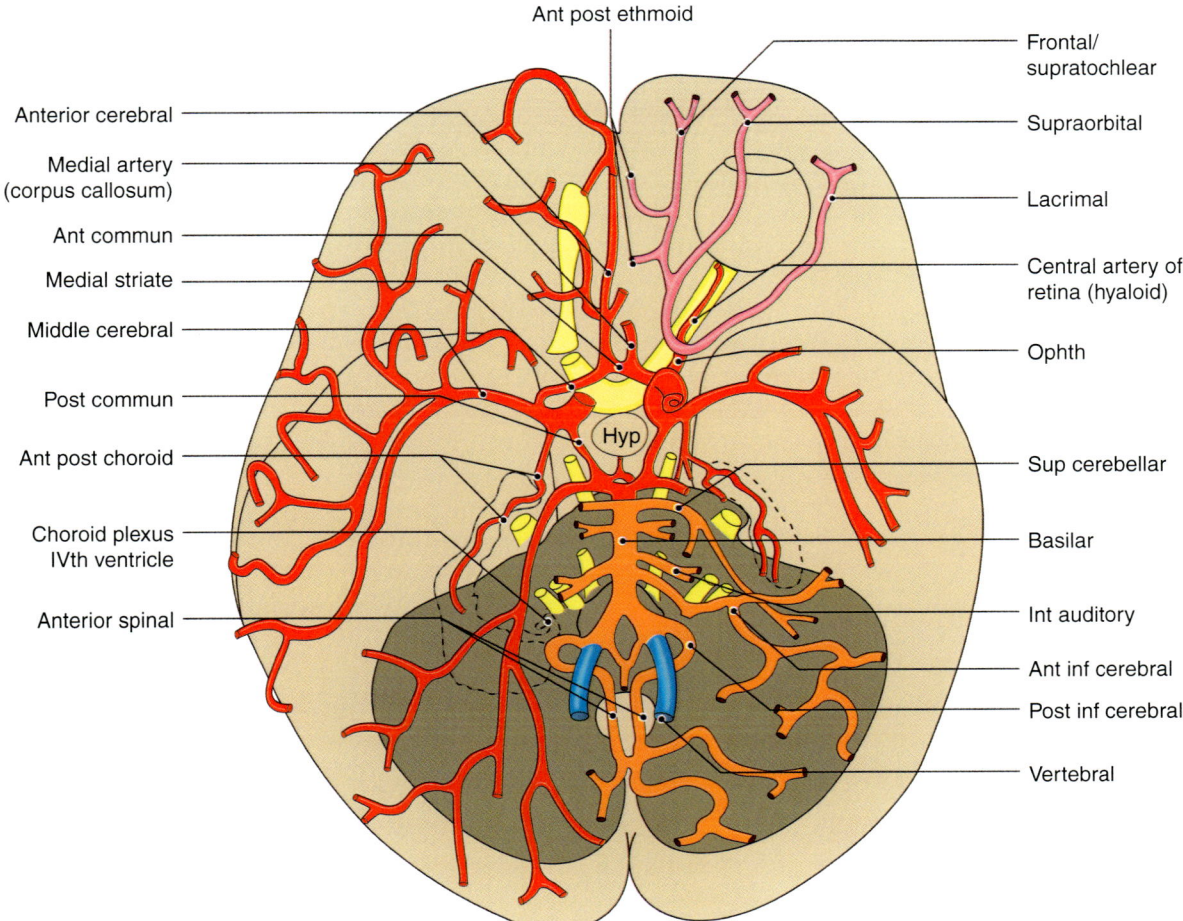

Fig. 6.82 Newborn. Anatomy of ACA is better seen at this time than in the adult state. Note that it is distal to anterior choroid. Variations in the arterial anatomy of the circle of Willis are common in its posterior half. This is due to the transition between dominance of the ICA-PCA relationship in the newborn and that of the basilar-PCA relationship seen in the adult

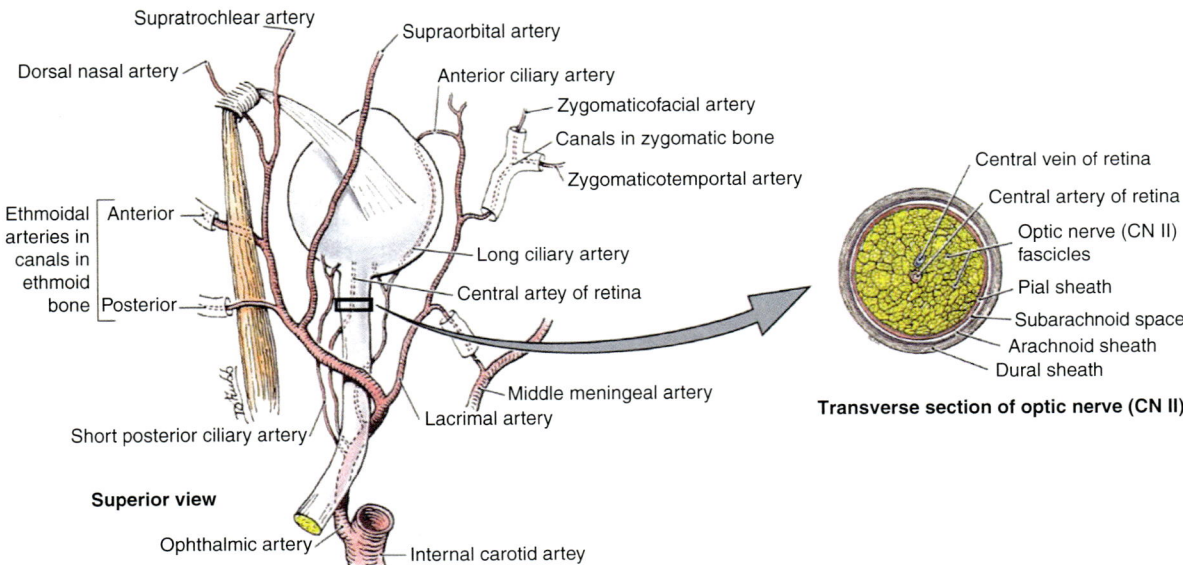

Fig. 6.83 The Ophthalmic System. Supraclinoid internal carotid artery: PrDOA and PrVOA are V1 inductions of PAM. Supraorbital stapedial artery: medial orbital br., lateral meningeal br. Variant of blood supply shows persistence of the lateral meningeal branch of StV1 being fed from middle meningeal. Media orbital branch was lost. The two branches come together just behind the orbit to make a common StV1 stem back to the split with StV2. Portion of middle meningeal proximal the bifurcation of StV1 and StV2 is derived from the ancient supraorbital division of stapedial. See figure on stapedial system for further clarification. (Reprinted from Standring S (ed). Gray's Anatomy 40th ed. New York, NY: Churchill Livingstone; 2008. With permission from Elsevier)

Development of segmental arteries is not random. They appear to "target" major cranial nerve nuclei. How does the system recognize where stems should emerge? If the process is similar to that of aortic arch formation, segmental arteries are the result of fusion of opposing "buds" or stems, one descending from the primitive hindbrain channel and the other ascending from the dorsal aortae. We can postulate the following: *segmental arteries result from fusion of stems having a common neuromeric sequence.* We shall see this mechanism repeated in other scenarios as well.

During the aortic arch phase[9–14], the segments of dorsal aortae corresponding to levels r0 to r11 reorganize. From r0, a forward-directed stem will become the primitive internal carotid, supplying the developing midbrain and forebrain. At each successive stage, as additional aortic arches develop, the definition of the "takeoff" of internal carotid from dorsal aorta moves *backwards*, terminating with the insertion of the third aortic arch. We shall consider the mechanism of aortic arch formation subsequently, in Chap. 7.

The next step in the reorganization of the dorsal aortae involves the involution of the dorsal artery segment connecting the third and fourth aortic arches[15]. When this falls apart, cardiac output is now divided between *flow directed to the head* via third arch and *flow directed to the body* via fourth arch. Subsequent to this, selective involution of the right dorsal aortic root[16] creates the final configuration of the great vessels: the left dorsal aortic root became the arch of the aorta. Caudal to the definitive aortic arch, the dorsal aortae move toward each other and fuse to form the definitive descending aorta. Individual arterial stems supply each level of the vertebral column and spinal cord.

Internal Carotid System

At stage 9, when the ventral folding of the heart drags the dorsal aortae downward, a new arterial stem projects forward from the arterial fold at r0. That is, from the "knuckle" where the dorsal aortae are flexed downward (ventrally) 90 degrees. The primitive internal carotid opposite primitive trigeminal within the lateral plate mesoderm surrounding the "new brain" (mesencephalon and prosencephalon). It thus lies lateral to the primitive head plexus, but connects to it and reorganizes according to developmental cues from the local environment. For this reason, although the early forms of ICA appear plexiform, they quickly become well-defined vessels with locational specificity.

Useful Caveats

Descriptions of the internal carotid artery in traditional anatomic texts make use of *directional terminology* that is misleading and, from a developmental standpoint, backward. For example, internal carotid is described as "entering" the carotid canal and then "ascending." These verbs give an impression of internal carotid artery as something that is "growing." In reality, the development of internal carotid *involves the **reorganization** of specific segments of preexistent dorsal aorta.* Thus, dorsal aorta destined to become internal carotid is composed of mesodermal elements and neural crest corresponding to hindbrain levels r0 through r11. It is critical to understand that the DA is not a distinct object "growing through" the mesenchyme. It is an ancient structure, being established directly after gastrulation. It is also a *genetically segmental structure*, formed by successive contributions from mesenchyme of each neuromeric level. The vessel demonstrates great anatomic constancy. Arterial induction within each somitomere occurs as an interaction between competing signals, an *overlapping of gene at a specific spatial location.*

Recall that the first sign of the primitive internal carotid is a sprout from the dorsal aortae at level r0. Primitive internal carotid grows forward to serve the developing midbrain and forebrain. It is closely applied to the CNS. In contrast, dorsal aorta lies more lateral, being separated from the hindbrain by intervening paraxial mesoderm. As the first two aortic arches develop and disintegrate, *the definition of primitive internal carotid (versus dorsal aorta) keeps moving **backwards**.* So too does the cranial limit of dorsal aorta. Thus, prior to development of third arch, primitive internal carotid interacts with rapidly developing paraxial mesoderm. Recall that basisphenoid bone comes from Sm4, while petrous temporal bone comes from Sm5 to Sm6. Internal carotid becomes *entrapped* within substance of the latter.

The third aortic and fourth aortic arches are constructed from mesoderm and neural crest corresponding to rhombomeric levels r6–r9. The "insertion" of third arch into dorsal aorta marks the caudal limit of the primitive internal carotid. When fourth arch appears, it remains connected to third arch by an intervening segment of dorsal aorta. Eventually, this "bridge" falls apart: third arch is left behind as the common carotid, while fourth arch goes on to form the arch of the aorta. The term *primitive internal carotid* is dropped; the *definitive internal carotid system* is born.

Subsequent events necessary for assembly of the external carotid system [such as the emergence of ventral pharyngeal artery from the third arch to fourth arch junction and of development of the stapedial system from hyoid artery] now proceed in a proximal-to-distal direction. This "maturation sequence" recalls that seen in the development of the vertebrae in which different structures at the same level are assembled following genetic gradients operating in opposite directions.

Developmental Timeline of the Intracranial System

A brief recap is helpful. The reader should review these revised modified drawings based on the staging of Padget and the Carnegie Institution.

Stages 9–10 (1–2 mm)

First and second aortic arch. *Forebrain*: At trigeminal ganglion, internal carotid divides. (1) Dorsolateral primitive trigeminal supplies the primitive hindbrain channel. (2) Medioventral branch makes a plexus at Rathke's pouch. *Hindbrain*: Primitive hindbrain channel connected to each neuromeric level (cf Fig. 6.47).

Stage 11 (Figs. 6.68 and 6.69)

Third aortic arch. *Forebrain*: Primitive internal carotid supplies optic vesicle, forebrain and midbrain; no cranio-caudal division as yet. *Hindbrain*: Disappearance of primitive hindbrain channel, emergence of longitudinal neural artery.

Stage 12/Padget 1 (Figs. 6.68 and 6.69)

Fourth aortic arch. *Forebrain*: Divisions of cerebral internal carotid. Cranial division terminates in olfactory area (newest). Caudal division forms plexus at midbrain. *Hindbrain*: Longitudinal neural arteries fully formed; they are supplied by trigeminal and first cervical.

Stages 13–14/Padget 2 (Figs. 6.50 and 6.70)

Forebrain: Cranial division of internal constitutes primitive olfactory artery terminating at olfactory plate. *Eye*: Primitive dorsal ophthalmic (PrDOA) appears. It supplies temporal globe. Caudal division of internal produces posterior communicating artery that contacts longitudinal neural (thus replacing primitive trigeminal). *Hindbrain*: Longitudinal neural arteries approximate and amalgamate into a midline plexus: the primordial basilar.

Stages 15–16/Padget 3 (Figs. 6.51 and 6.71)

"Post-branchial phase." *Forebrain*: Cranial division forms primitive branches (proximal-distal): (1) prim. anterior choroidal to diencephalon, (2) prim. middle cerebral, (3) prim. olfactory (future stem of anterior cerebral). *Eye*: PrDOA now connects to hyaloid artery. Primitive ventral ophthalmic (PrVOA) appears; it supplies nasal globe. Caudal posterior communication has two large branches at midbrain, just in front of cranial nerve III. (1) Diencephalic artery has four branches (p4–p1). Gives off posterior choroidal. (2) Mesencephalic artery has two branches (m1–m2). *Hindbrain*: Basilar artery gives off superior cerebellar at r1. Nerve roots receive branches of basilar. Vertebral artery being formed by cervical segmentals.

Stages 17–18/Padget 4 (Figs. 6.72 and 6.73)

"Cerebral transition phase." *Forebrain*: (1) Cranial division now has adult parts. The proximal stem gives off anterior choroidal and then middle cerebral. Distal stem has parent primitive olfactory and a secondary mesial branch, the anterior cerebral artery. Caudal division posterior choroidal now grows toward anterior choroidal. (2) Posterior Ophthalmic stem not yet present. Primitive eye arteries become elongated: (1) PrDOA (common temporal ciliary + hyaloid) and (2) PrVOA. *Hindbrain*: Exclusive carotid supply to all cerebral arteries gives way to divided system in which hindbrain is now fed by vertebrals.

Stages 19–20/Padget 5 (Figs. 6.74 and 6.75)

Forebrain: Middle cerebral becomes prominent. Anterior/posterior choroidals anastomose. Posterior cerebral arising from the caudal posterior communicating. *Eye*: Permanent ophthalmic *Hindbrain*: Definitive common carotid artery. Primitive connecting branches (trigeminal, hyoid) now reassigned.

Stages 21–22/Padget 6 (Figs. 6.76 and 6.77)

Forebrain: Circle of Willis completed.

Stage 23+ (Figs. 6.78, 6.79, 6.80, and 6.81)

Embryologic organization of the internal carotid artery is quite different from traditional descriptions. Textbooks rely on nomenclature based on dissections of adult model. Centuries would pass before the embryonic *bauplan* could be properly studied. Technical advances in methods of preservation, vascular injection materials, and magnification have given us a picture that is quite different, but much more conceptual.

Internal Carotid and Its Branches: Neuromeric Model

Anatomic Sectors of the Internal Carotid Artery
(Figs. 6.82, 6.83, 6.84, 6.85, and 6.86)

The ***cervical sector*** of ICA has no branches. At the level of parotid gland, it passes deep to the investing fascia of first and second arch muscles; it runs superficial to the investing fascia of third and fourth arch muscles. From the superior cervical ganglion (neuromeric level r11–c1), SANS fibers climb aboard the ICA and follow its course into the skull.

The ***petrous sector*** of ICA carotid forms within the substance of Sm5–Sm6 petrous temporal bone. After entry into the bone, it makes two successive "L-shaped" turns at 90° to each other. The first is forward and medial; the second is upward and forward. It then enters the skull between two backward-directed projections of the sphenoid: the lingula is lateral and the petrosal process is medial. Internal carotid lies anterior to the cochlea and to the tympanic cavity. It is separated from these structures by a thin sheet of bone. More anteriorly, internal carotid is again separated from trigeminal ganglion by another thin sheet of bone that forms the floor of the ganglion and the roof of the carotid canal. A plexus of veins draining the tympanic cavity accompanies ICA and drains into cavernous sinus.

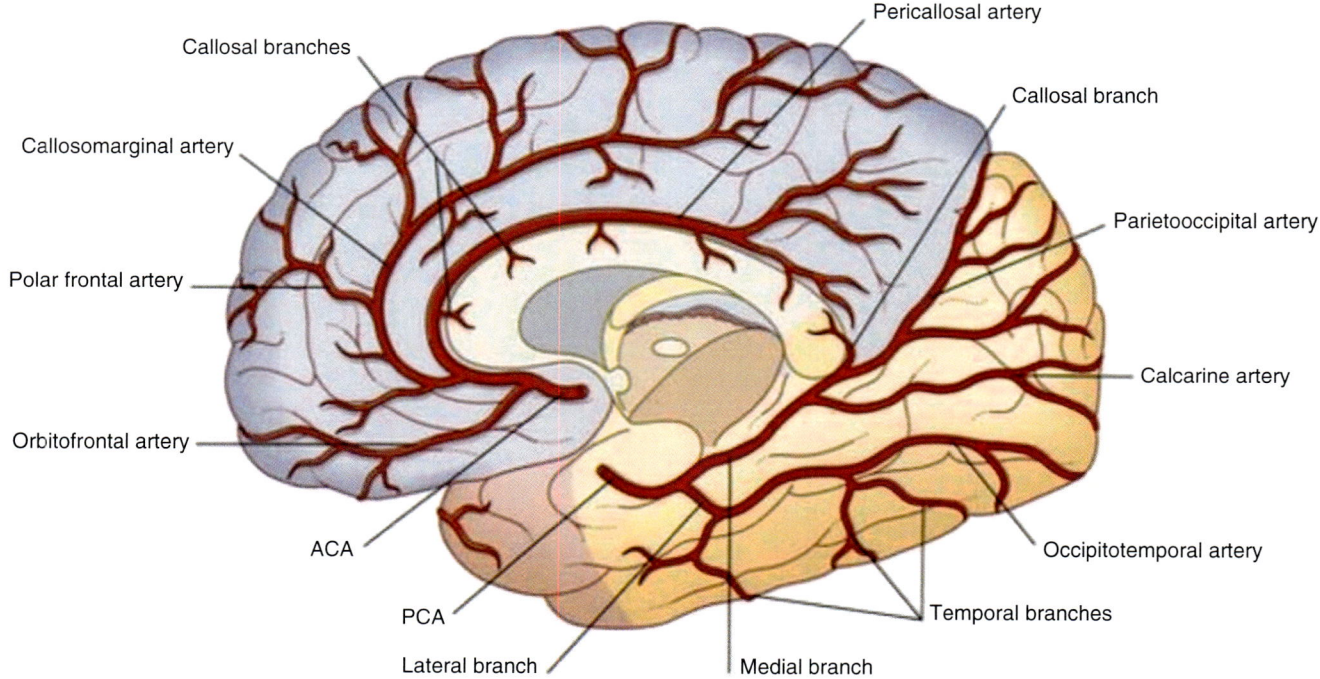

Fig. 6.84 Cerebral ICA territories, sagittal lateral. The three systems are color-coded: ACA (yellow), MCA (red), and PCA (green). ACA is associated at all times with V1; note how far posteriorly it extends. (Reprinted from Clemente C (ed). Gray's Anatomy, 30th American Edition. Philadelphia, PA: Lea & Febiger. With permission from Wolters Kluwer Health)

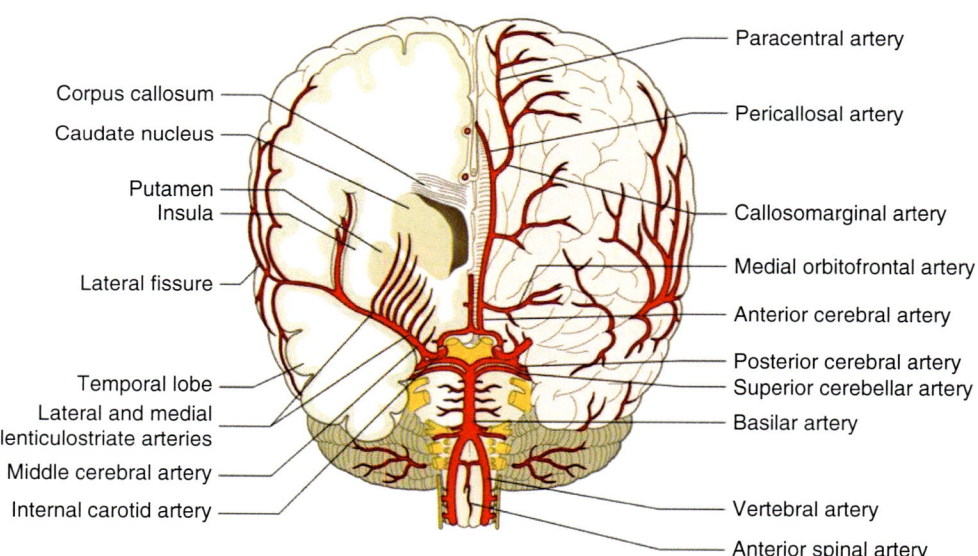

Fig. 6.85 Cerebral ICA territories, sagittal medial. The three systems are color-coded: ACA (yellow), MCA (red), and PCA (green). Midline structures are supplied mainly by branches of ACA and PCA. Note the distinction between marginal callosal artery and pericallosal artery. ACA has three segments. A3 bifurcates into pericallosal and marginal callosal (paracentral) with implications for the spectrum of the minimal form of holoprosencephaly. Note that segments A4 and A5 are described in some classifications but not in others. Note the overlap between MCA from the Sylvian fissure and the ACA coming over the vertex. (Reprinted from Clemente C (ed). Gray's Anatomy, 30th American Edition. Philadelphia, PA: Lea & Febiger. With permission from Wolters Kluwer Health)

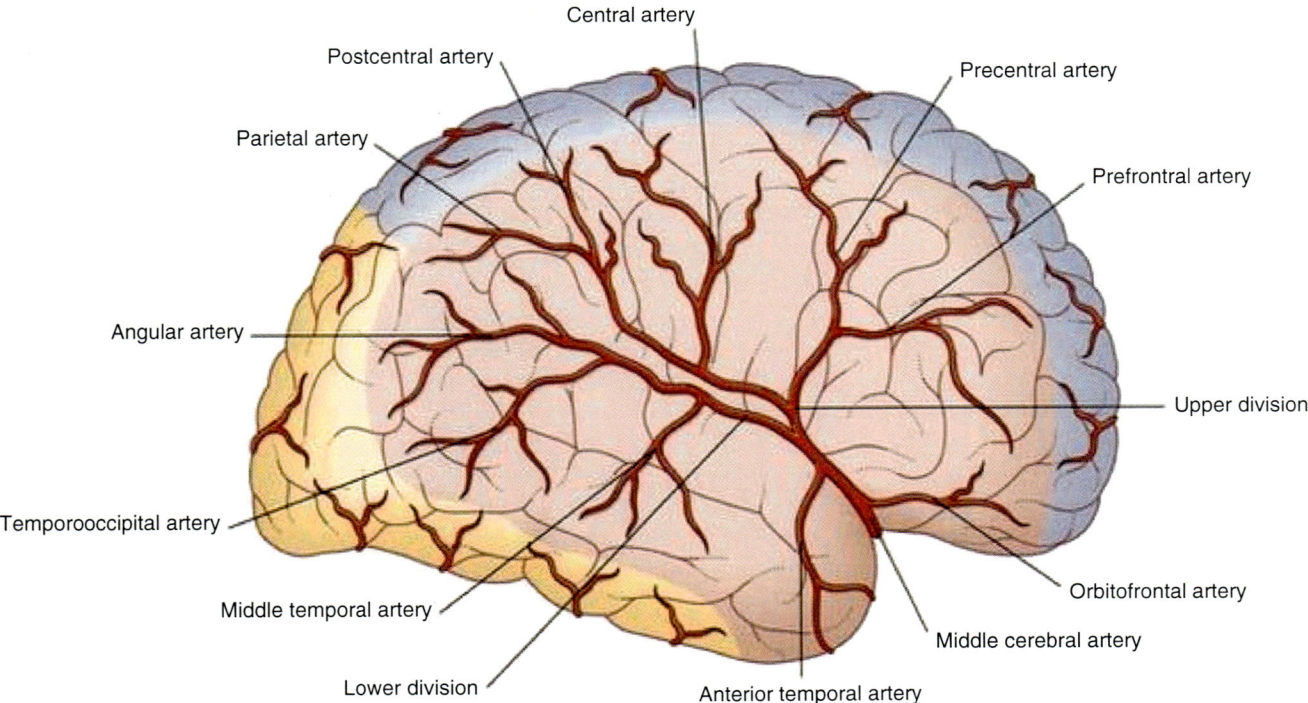

Fig. 6.86 Watershed zones are sites of anatomoses between territories of the internal carotid artery. They represent areas of vulnerability to ischemia. (**a, b**) Shows cortical surface, whereas (**c**) is a coronal section. Frontal cortex supplied by ACA which tracks backward in the midline over corpus callosum all the way to the parieto-occipital sulcus. Base of the brain showing distribution of ACA, MCA, and PCA. Note MCA supplies the lateral basal cortex in anterior cranial fossa. Posterior cere- bral artery (sagittal). Traditional diagram showing the apparent "origin" of PCA from basilar when, in reality, it is the last stem of caudal division before posterior communicating. (Reprinted from Hanes DE. Fundamental Neuroscience for Basic and Clinical Applications, 3rd ed. Philadelphia, PA: Churchill Livingstone; 2006. With permission from Elsevier)

Petrous internal carotid has two branches. *Caroticotympanic branch* is the remnant of the primitive hyoid artery. It passes through the posterior wall of canal and gains access to the promontory where it makes anastomoses with anterior tympanic (from the internal maxillary and with the stylomastoid branch of posterior auricular artery). *Pterygoid branch* passes forward into the pterygoid canal. There it makes an anastomosis with the pterygoid branch of internal maxillary artery that is passing backward from the pterygopalatine fossa.

The *cavernous sector* of the internal carotid is covered by lining membrane of the cavernous sinus. After passing lateral to the basisphenoid (Sm4), the artery turns upward and pierces the dura mater of the sinus roof. In this location, the abducens nerves lie against its lateral aspect—where it can be affected by the development of an aneurysm.

Cavernous internal carotid has four derivatives. The walls of cavernous and inferior petrosal sinuses are supplied by *cavernous branches*. These can form anastomoses with the middle meningeal artery. The primitive maxillary artery (vide infra) gives rise to superior and inferior *hypophyseal branches* to the anterior pituitary. Trigeminal ganglion receives a *ganglionic branch* that represents the primitive tri-geminal artery. Dura of the anterior cranial fossa is inner-vated by V1 (Tessier zones 10–13) and is supplied by the *anterior meningeal branch* that passes above orbitosphenoid (lesser wing). The posterior location of this artery allows it to anastomose with the meningeal branch of the posterior eth-moid. Although posterior ethmoid appears to arise from oph-thalmic, it is really a derivative of the StV1 stapedial system.

The *cerebral sector* of internal carotid perforates dura just medial to the anterior clinoid process. It then passes between cranial nerves III and II and runs forward toward the anterior perforated substance located at the medial extremity of the lateral cerebral sulcus. It has five major branches and a glar-ing addition, the posterior cerebral artery.

Just as internal carotid emerges from cavernous sinus at the medial aspect of anterior clinoid process (i.e., from r1 orbitosphenoid), it enters the optic canal inferior and lateral to optic nerve. In point of fact, these two structures *predate the canal* that forms around them from r1 neural crest bones. Medial presphenoid and lateral orbitosphenoid close down around the neurovascular unit like two jaws of a pair of pli-ers. Internal carotid is going to give off all its branches from medial to lateral, whereas optic nerve has to follow the axis of the globe (directed outward at about 45°). Thus, internal carotid *crosses over* cranial nerve II.

In humans, what are the boundaries of the internal carotid system? Internal carotid bifurcates into a cranial division and a caudal division. The anterior terminus of the cranial division is *olfactory artery* (OlfA). This is deceptive because in the adult state, anterioc cerebral appears to be terminal, but, in reality, it is the third branch. In the embryo, the "leading edge of anterior division of ICA is *olfactory*—followed by ophthalmic and then anterior cerebral.

The posterior terminus of ICA is *posterior communicating artery* (PCmA), not *posterior cerebral artery* (PCA). Again, in the adult state, the anatomy is deceptive because textbooks describe PCA as a branch of the basilar artery…a contention that makes no embryologic sense. We know that the basilar artery represents from a site-specific midline fusion of the longitudinal neural arteries over the metencephalon (r0–5)…and we know that the LNAs provide the unique blood supply to the hindbrain. Recall that in the early stages, hindbrain circulation is completely disconnected from that of the forebrain and midbrain. So it makes no sense for a forebrain structure to be supplied by the hindbrain circulation. Instead, PCA is given off by PCmA immediately before the latter makes anastomosis with the LNA.

Does the Internal Carotid Artery Supply the Face?

The answer to this is both *yes and no*. In the adult state, all the branches to the sinuses, orbital contents upper eyelids, forehead, and nose emanate from the ophthalmic artery… which, as we all know from medical school, are derived from the internal carotid artery. We also "know" that ophthalmic artery supplies the structures of the eye itself. Again, this makes no embryologic sense. Internal carotid artery, in all other instances, maintains a strict alliance with the forebrain and midbrain. Why should it engage in a dalliance with tissues unrelated to the brain or the anterior cranial base?

The resolution to this puzzle lies in the fact that ophthalmic artery in the adult state is a composite of two separate embryonic systems. (1) The eye is an extension of the forebrain. It is supplied by derivatives of an original hyaloid artery plus primitive dorsal and ventral ophthalmic arteries…all of these structures come from the cerebral sector of internal carotid artery. (2) The adnexal structures of the orbit, forehead, and nose are derivatives of r0–r1 midbrain neural crest with extraocular muscles coming from the PAM of somitomeres 1, 2, 3, and 5. These structures are all supplied by V1-induced branches of the supraorbital stapedial artery. We shall tell the story of the stapedial anon…several times over until you are thoroughly sick of hearing it…but suffice it to say that the stem of the entire stapedial system to dura, orbit, and face dies off. By the time "the great stapedial slough" takes place, all stapedial derivatives have already made connections with companion arterial systems. For this

reason, in the adult state, StV1 is joined with the original ocular axis to make up the ophthalmic artery, as we know it.

The same holds true for the rest of the face. Neural crest tissues of the midface and the lower face are supplied by branches of StV2 and StV3 that are connected with the external carotid system. Although intracranial, the dura is not a CNS structure. The stapedial arteries that supply it are ultimately connected to the external carotid system. Internal carotid artery remains faithful to the CNS.

Herein we shall describe the anatomy of the internal carotid system, including that of the eye and orbit, but shall reserve a detailed description of the stapedial system and its contributions to the face and orbit for a subsequent chapter.

The Petrous Sector

Carotico-tympanic artery: This vessel is the remnant of hyoid artery. It is genetically related to r4–r5. It passes backward via a small foramen in the carotid canal and gains access to the promontory (Sm5–Sm6). There it makes connections with the PA1-related *anterior tympanic branch* of IMA and with PA3-related *stylomastoid branch* of ascending pharyngeal.

Pterygoid artery: This branch relates to the tympanic sector of temporal bone. It follows fibers of deep petrosal nerve (VII) into the pterygoid canal, where it runs head-long into the V2 pterygoid branch of internal maxillary artery. Remember: cranial nerves cannot live without a blood supply. This is yet another example of how VII and V interact to induce blood supply and how seemingly irrational neural connections through otic and pterygopalatine ganglia make sense.

The Cavernous Sector

Cavernous: Sinuses do not fly solo. The cavernous arteries provide blood supply to inferior petrosal and cavernous sinus. These branches anastomose with branches of the middle meningeal system. This is because, in its construction, intracranial stapedial follows cranial nerve VII from the tympanic cavity forward via greater petrosal nerve until reaching trigeminal ganglion and the cavernous sinus where it picks up the V1 and V2 branches. Dura forms around these structures to cover the frontal and temporal lobes.

Hypophyseal: Superior and inferior hypophyseal arteries supply the all-important pituitary gland. Pial arteries supply the superior hypophyseal artery, while primitive maxillary artery gives rise to the inferior hypophyseal artery. Recall that somitomeres are produced prior to the existence of the primitive internal carotid artery. They are initially by diffusion from the primitive head plexus and the primitive hindbrain channels. With the advent of primitive internal carotid, this situation changes because the head plexus regresses. Primitive maxillary is the expedient candidate providing temporary segmental supply to the base of the forebrain and future optic anlage.

In her description of "stage 2," Padget observed that primitive maxillary skims past Rathke's pouch and is distributed to ventromedial tip of prosencephalon. This trajectory positions it perfectly to supply frontonasal mesenchyme plus neural crest of ethmoid. An alternative hypothesis is that an unnamed, more anterior companion branch to primitive maxillary is ultimately responsible for superior hypophyseal artery. Evidence for this is suggested by dissections in the stage 13 4 mm embryo. The plexiform nature of ICA around Rathke's pouch would suggest the existence of a companion vessel to PrMxA, but this has not been identified.

As previously stated, the ultimate fate of mesoderm from Sm1 to Sm3 and Sm5 is to form the extraocular muscles. These latter myoblasts will populate fascial "spaces" from r0/r1. Subsequently, the V1 sensory supply to these fasciae will induce individual muscular arteries to develop from the StV1 stapedial system.

Ganglionic: Trigeminal ganglion lies in intimate contact with cavernous sinus. Just as external carotid provides supply to the lateral aspect of ganglion via the primitive trigeminal artery, so too does internal carotid supply the medial aspect of ganglion. Recall as well that stapedial (the "turncoat" artery that switches allegiance from VII to V at the ganglion) follows a course along the medial aspect of V1.

Anterior meningeal: This vessel passes over obitosphenoid (lesser wing to supply dura of the posteromedial anterior cranial fossa). Its distribution can be deduced from its *known anastomosis with StV1 posterior ethmoid* at posterior cribriform plate. It may represent the companion blood supply of the very first branch of V1 which is distributed to the dura of the anterior cranial fossa.

Ophthalmic: The ophthalmic stem arises from the ICA immediately after it extends through the dura of cavernous sinus. The stem is medial to the anterior clinoid process. It runs in company with the optic nerve below and lateral to that structure. Passage of optic nerve and ophthalmic artery into the orbit is demarcated by the optic canal which represents the interface between the orbitosphenoid and alisphenoid bones.

The Eye and Orbit: A Hybrid System (See Figures in Chap. 7)

The adult anatomy of the ophthalmic axis and its various branches to the eye and orbit can only be understood based upon its developmental history because it represents the crossroads of four arterial sources serving distinct functional systems.

- *Internal carotid stem* artery from internal carotid provides support for the visual apparatus. Optic nerves and chiasm are supported from the anterior communicating branches of anterior cerebral.

- *V1 intracranial stapedial* provides vascular support for the orbit by dividing into two branches at the backwall of the alisphenoid: a medial *orbital branch* via superior orbital fissure and a lateral *meningeal branch* via cranio-orbital foramen.

- *V2 intracranial stapedial* (recurrent meningeal artery) supplies the dura of anteriolateral temporal fosss and the alisphenoid. It terminates with a small branch that accompanies V2 out of the foramen rotundum.

- *V2 extracranial stapedial* ascends from the pterygopalatine fossa into the lateral orbit via zygomatic artery to supply the lateral orbital wall via its zygomaticofacial and zygomatico temporal branches.

Because most of the components of the ophthalmic axis are not derived from the internal carotid, we shall reserve discussion about their development for an ensuing section on eye, orbit, and frontonasal face. For our immediate purposes, we shall simply name the component arteries of the combined ophthalmic system and describe their anatomy.

We begin with two statements: (1) *all arteries to the eye and orbit arise as induction products with a corresponding nerve*; (2) the globe itself is a neural unit, being composed of neural ectoderm and covered with neural crest; therefore, *the eye induces its own blood supply.*

Ophthalmic artery is located ventral to the optic nerve. It gives off two main branches: central artery of the retina and paired long posterior ciliary arteries. From these latter arise the short posterior ciliary arteries. In the process, a vascular ring is created that surrounds the optic nerve and is connected with the ophthalmic axis. V1 stapedial and V2 stapedial unite and form a continuous arc which makes anastomoses with ophthalmic via the vascular ring. The stapedial stem winds around the lateral aspect of the optic nerve, becoming dorsal to it. Stapedial then immediately breaks into two major divisions: a laterally directed lacrimal artery and a medially directed frontal artery. Subsequently, the frontal artery subdivides into a lateral supraorbital axis and a nasociliary axis. The latter has two neuromeric zones, proximal and distal.

The **ocular group** of arteries arises from the internal carotid. They supply the globe. The branches are, in proximal-to-distal order, follows:

- Central artery of the retina
- Long posterior ciliary arteries, nasal, and temporal
- Short posterior ciliary arteries 12–15

The **orbital group** of arteries arises in conjunction with the three branches of intraorbital stapedial. These supply the nasoethmoid complex, orbital surround, and extraorbital/

fronto-nasal-orbital soft tissues. **NB**: *Traditional anatomic classification is embryologically incorrect*. Intraorbital stapedial is traditionally described as having three separate branches, but in reality, there are only two, with frontal artery subdividing. In the subsequent section, the embryologic rationale for this classification will be discussed. Our purpose here is to provide the descriptive anatomy. The arteries are discussed in order of their appearance, medial-lateral and proximal-distal.

1. Lacrimal artery (Tessier zone 9)
    (a) Anastomosis of V2 zygomatico-temporal nerve to lacrimal nerve provides programming for "pseudolacrimal" ZT and SF arteries (Tessier zones 8, 7)
2. Frontal artery (Tessier zones 10, 11)
    (a) Lateral division induced by lateral br. of frontal nerve (Tessier zones 10)
       • Supraorbital
    (b) Medial division induced by medial branch of frontal nerve (Tessier zones 11)
       • Proximal sector—induced by nasociliary nerve
    (c) Ethmoid (Tesier zones 12, 13)
       • Intraorbital
          – Anterior ciliary to supply neural crest intraocular muscles
       • Transethmoid/anterior cranial fossa
          – Posterior ethmoid
          – Anterior ethmoid
       • Supratrochlear to glabella of forehead
       • Infratrochlear to lacrimal sac and dorsal nasal envelope

Arteries of the Optic Apparatus—Ophthalmic System

Central artery of the retina: Primitive hyaloid artery is the embryonic precursor of central artery. It is the first branch of ophthalmic artery and is located *below* the optic nerve. About 1 cm behind the globe, it penetrates the dura to run within the sheath of the optic nerve (itself an extension of the dura) and is distributed to the retina. A second branch is more external. We shall see that the conversion of the optic vesicle to the optic cup involves an incomplete folding that results in an inferior invagination, the *optic fissure*. Hyaloid artery becomes entrapped within this fissure.

Long posterior ciliary arteries: These paired vessels are induced by the long ciliary branches of nasociliary nerve. The arteries follow trajectories established by their embryonic precursors; they define the globe into two distinct territories. Temporal LPCA is derived from *primitive dorsal ophthalmic*; it supplies dorsotemporal sclera. Nasal LPCA is derived from *primitive ventral ophthalmic*; it supplies ventronasal sclera. The two arteries penetrate the sclera just beyond the optic nerve and run forward between sclera

and choroid to reach the iris where they join up with the *circulus arteriosus major* around the periphery of iris. Small penetrating branches form a ring around the pupil, the *ciruculus arteriosus minor*.

Short posterior ciliary arteries: Parasympathic autonomic nervous system (PANS) supplies the globe by means of six to ten postganglionic short posterior ciliary nerves. These, in turn, induce 6–10 arteries that develop from the long posterior ciliary arteries. These short arteries penetrate the sclera to supply the choroid and ciliary body. They have posterior branches that anastomose with the central artery of the retina and can play a protective function in cases of central ischemia to the disc.

Arteries of the Orbit: Stapedial System

These arteries supply periocular structures: fat, fascia, muscles, and bones. *They are all derivatives of the stapedial system and arise via neuroinduction.* We will cover them in additional detail in a subsequent section on the eye and orbit.

It is worth recalling that the anastomosis of V1 stapedial with ophthalmic takes place ventral to the optic nerve, but all the subsequent branches will be dorsal. The stem ascends around the lateral aspect of optic nerve to get into position.

The stapedial system is programmed by branches of V1 with some assistance of the cranial nerves III, IV, and VI for the muscular branches. There are three main branches of V1 in the orbit. ***Lacrimal nerve*** travels to the lateral border of the orbit where it transects lacrimal gland and reaches the lateral upper eyelid. ***Frontal nerve*** passes over levator palpebrae superioris and subdivides into *supraorbital* and *supratrochlear* branches, both of which ascend onto the forehead. The supratrochlear nerve is responsible for inducing the extraorbital branches relating to the eyelid and forehead of the nasociliary artery. ***Nasociliary nerve*** runs deep to optic nerve and follows the medial wall of the orbit giving off successively *posterior ethmoid nerve* to the sphenoid and ethmoid sinuses, *anterior ethmoid nerve* to the cribriform plate and nasal cavity, and *infratrochlear nerve* to the lacrimal zone and nasal dorsum. Finally, the ciliary part of nasociliary nerve gives off *long and short ciliary nerves* which program the corresponding branches from the ophthalmic system. Nasociliary nerve is responsible for inducing the intraorbital branches and the extraorbital branches to the lacrimal apparatus and nose.

Lacrimal Artery (Zone 9)

This vessel contributes in induction by the lacrimal nerve. Upon entry into the orbit, lacrimal artery is the first branch off ophthalmic, but it has also been observed to exit posterior to the orbit. It courses laterally *beneath* optic nerve and follows along superior border of lateral rectus to reach lacrimal gland. *Lacrimal artery contributes to lateral rectus.* It traverses lacrimal gland, dividing it into orbital and palpebral lobes; logically supplies the orbital lobe. After exiting gland, it supplies conjunctiva and drops down to connect with the *lateral palpebral arteries* of the upper and lower eyelids.

Lacrimal artery makes a critical anastomosis with the middle meningeal system, one that creates a great deal of confusion regarding pathologies of the posterior and lateral orbital walls. Recall that supraorbital stapedial runs forward with both V1 and V2 until reaching the back wall of the orbit, where it divides. StV2 supply temporal fossa dura and the calvarial lamina of alisphenoid sends a branch to the orbit through a foramen in the alisphenoid. StV1 supplies anterior cranial fossa dura and the orbit with a lateral meningeal branch and a medial orbital branch. These access the orbit through the canio-orbital foramen and superior orbital fissure, respectively.

In stage 19, medial orbital branch programs the orbital vessels, including lacrimal. Lateral meningeal branch attaches to distal lacrimal and then ascends through the orbital plate to supply dura of the anterior cranial fossa. During stage 19, both branches remain in continuity with supraorbital for their blood supply. In stage 20, anatostosis with ophthalmic takes place, as describe below, *vide infra*. Now irrelevant, supraorbital involutes backward to the StV1–StV2 bifurcation.

Sometimes the lateral StV1 branch persists. Because it retains its original connection, and because the former supraorbital is taken over by middle meningeal, it appears as if this StV1 meningeal is part of that system. It is called the *communicating branch*, or *recurrent meningeal artery* (RMA). Sometimes it is quite large and actually replaces the lacrimal within the orbit.

Key fact: sometimes this anastomosis persists, in which case you can trace the RMA forward into lacrimal gland where it connects to rise to *Zygomatico-temporal artery* (ZT) ascending up the inner wall of the lateral orbit. Of note, lacrimal artery is directed to the <u>orbital</u> part of the gland, but ZT supplied the <u>palpebral</u> part. In any case, ZF exits the gland and terminates in the postorbital bone field, Tessier zone 8. 7.

Herein lies a paradox: in many instances the original embryonic continuity between RMA and lacrimal disappears, leaving StV1 lacrimal artery as an apparently the sole source for the ZT and ZF arteries. We know that the zygoma complex is strictly of r2 neural crest derivation. How can the blood supply for these bone fields come from a V1 source?

Neuroanatomy Saves the Day

Zygoma (jugal and postorbital) and maxilla constitute a common set of developmental fields. From pterygopalatine fossa, StV2 sends a sensory to supply all three of them. Posterior superior alveolar is given off just before entering the infraorbital fissure. Once inside, zygomatic nerve is given off and anterior superior alveolar continues forward.

Zygomatic nerve now follows the floor of the orbit laterally and then give offs *zygomaticofacial nerve* (ZF) which

exits the jugal through the ZF foramen. Zygomatic continues upward along lateral orbital wall until it reaches lacrimal gland where it terminates into two branches. Lacrimal branch joins with lacrimal nerve *de* jure to supply the gland with PANS motor fibers. *Zygomaticotemporal nerve* (ZT) zots laterally to exit postorbital bone through the ZT foramen. In the meantime, *in the absence of the RMA anastomosis*, blood supply from ophthalmic flows outward from lacrimal artery, pursuing the two V2 branches: (1) it follows laterally to become *ZT artery* to **zone 8**; (2) it descends, following ZT nerve inside the lateral orbital wall until it encounters ZF nerve, and there, makes its escape through the malar eminence as *ZF artery* to **zone 7**.

In sum: Although blood supply to zones 7 and 8 *appears* to come from lacrimal (implying that they are r1 neural crest derivatives), in reality, these r2 bone fields are supplied as expected. The key is the neuroanatomy. The ZF and ZT nerves do the programming…the artery, departing from its lacrimal blood source, simply plays "follow-the-leader."

Frontal Artery (Zones 10–11)

Supraorbital artery travels with and supplies superior rectus and levator palpebrae superioris. It follows its companion V1 nerve laterally upward and out of the orbit via the supraorbital foramen. Thereupon, it makes a clinically significant division.

A deep *muscular branch* supplies frontalis and subgaleal fascia. Medially, it anastomoses with supratrochlear.

A superficial *cutaneous branch* is distributed to galea and forehead skin corresponding roughly to the dimensions of the eyebrow.

Ethmoid Artery

Transethmoid/intracranial

Meningeal branch is most proximal. It passes backward through superior orbital fissure to reach middle cranial fossa where it anastomoses with accessory meningeal that ascends from internal maxilla-mandibular upward through foramen ovale to reach trigeminal ganglion.

Posterior ethmoid: This artery supplies the *posterior intracranial and intranasal tissues* sector of Tessier zones 12 and 13. It passes through orbital lamina to supply posterior ethmoid sinus. From here, it enters the cranium through the *spheno-ethmoid suture*. Thus, it supplies the ethmoid lateral to cribriform plate, i.e., the roof of the ethmoid sinuses. Small branches descending via cribriform plate anastomose in front of sphenoid with medial nasal branches of medial sphenopalatine. Thus, the neural crest presphenoid bone lies in the posterior nasopharynx. It represents the vascular anastomosis between StV1 and StV2.

Anterior ethmoid: After penetrating the orbital lamina, the artery enters "sinus city." Recall that anterior ethmoid sinus, middle ethmoid sinus, and frontal sinus are all continuous. Via the frontal sinus, the artery pursues a *midline* course to enter cranium, where it divides. *Meningeal branch* supplies dura of cribriform plate. This is intracranial zone 13 of Tessier. Just lateral to crista galli, *nasal branch* dives into the roof of nose where it immediately divides. *Lateral anterior nasal branches* are distributed to superior and middle turbinates, while *medial anterior nasal branches* ramify over septum. From the latter, a branch runs along the undersurface of nasal bone, supplying it. At its inferior border, the artery encounters the upper lateral (triangular) cartilage, the cephalic margin of which is neatly tucked underneath the caudal margin of the nasal bone. Via this anatomic "escape route," the artery achieves a subcutaneous location. Paired *anterior ethmoid arteries* travel together in the midline, following along the upper border of septum. They then descend through columella into the philtrum. The vascular anatomy of the philtral prolabium is the basis of developmental field reassignment cleft surgery.

- Intraorbital
  - *Muscular arteries*: Developmentally, these arteries belong to the stapedial system. Muscular arteries arise from various sites along the frontal artery. They are distributed in two groups; these mirror the geometry of the original embryonic vascular structures. <u>Superior branches</u> supply muscles assigned to the sclera of primitive dorsal ophthalmic artery: superior oblique, superior rectus, and levator palpebrae superioris. <u>Inferior branches</u> supply muscles assigned to the sclera of primitive ventral ophthalmic artery: medial rectus, inferior oblique, and inferior rectus. Lateral rectus is also supplied by lacrimal artery (an important detail to which we shall return later). Muscular arteries accompany the motor nerves. The muscles are enclosed within *V1 supplied fascial envelopes* of the individual extraocular muscles. All sensation from the muscle units required for both nocioception and proprioception is conveyed back to by branches of V1 to the somatotopic map within the mesencephalic nucleus.
  - *Anterior ciliary arteries*: These vessels arise from the inferior group of muscular branches of ophthalmic artery. Recall that these are stapedial branches. Travelling along rectus tendons, the anterior ciliary vessels pierce the sclera just behind the corneo-scleral junction. There they contribute to the *circulus arteriosus* around the attached iris. This is a neural crest zone; it is responsible for supplying the intrinsic muscles of the eye.
- Extracranial
  - *Supratrochlear artery*: This vessel supplies the *medial orbital and frontal tissues* of **Tessier zones 12–13**. A *deep muscular branch* supplies procerus and corrugator. Laterally, it anastomoses with supraorbital. A superficial *cutaneous branch* is distributed to glabella and medial forehead.
  - *Infratrochlear artery (dorsal nasal artery)*: This artery supplies the *nasal sector* of Tessier zones 12–13. It appears above medial canthus, sends a side-branch to superior lacrimal sac, and then proceeds to divide again.
    Dorsal nasal artery
    - The lateral branch supplies the nasal sector of Tessier zone 12, i.e., *lateral nasal wall* as a triangle bounded by nasal dorsum medially, lacrimal duct laterally, and medial canthus superiorly anastomoses with angular artery of facial.
    - The medial branch supplies nasal sector of Tessier zone 13, i.e., nasal dorsum down to tip.

  - *Medial palpebral*: This artery supplies the *canthus* and *lacrimal sac* of Tessier zone 11. This importance of this vessel is out of proportion to its size. Its trajectory is schizophrenic. It arises deep to the pulley of superior oblique, descends, and then splits around the medial canthus. The paired arteries join with their confrères from the lacrimal artery to encircle the eyelids. They form a double arcade corresponding to anatomic components of orbicularis. A *marginal arcade* just above tarsus supplies tarsal orbicularis. A *peripheral arcade* half way between tarsus and orbit supplies palpebral orbicularis. Just medial to the origin of lateral canthus, the medial palpebral arteries make a connecting loop with paired lateral palpebral arteries derived from lacrimal artery. Colobomas of the eyelid are defects in this system.

The Cerebral Sector (Figs. 6.82, 6.83, 6.84, 6.85, 6.86, 6.87, and 6.88)

Textbook descriptions of internal carotid artery are, from an embryologic standpoint, inherently inaccurate because they are based on dissections of adult (or full term) dissections. Centuries would pass before the embryonic *bauplan* could be properly studied. Technical advances in methods of preservation, vascular injection materials, and magnification have given us a picture that is quite different, but much more conceptual. At stage 12, the primitive ICA divides into cranial and caudal divisions; we shall describe the vessels of each division in a proximal-distal sequence.

Cranial Division (From Proximal to Distal)

Anterior choroidal arises just cranial to the division point of ICA. It supplies: (1) optic tract, (2) cerebral peduncle, (3) lateral geniculate body, (4) optic radiation, (5) tail of caudate nucleus, (6) posterior internal capsule, (7) globus pallidus, and (8) choroid plexus.

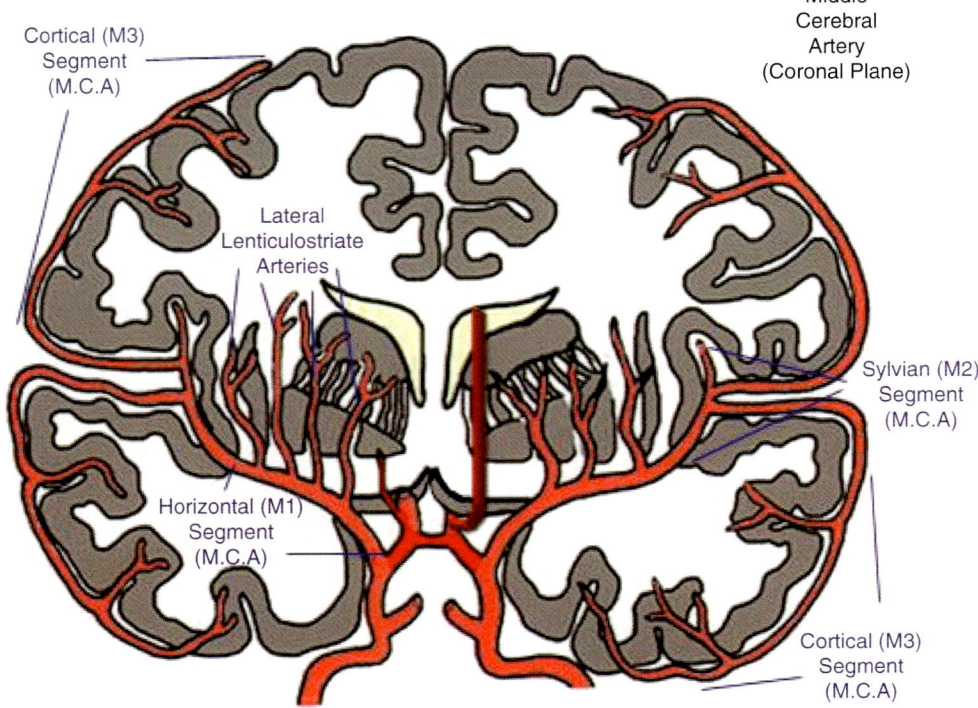

Fig. 6.87 ACA and PCA segments are neuromeric. <u>Anterior cerebral</u> (A1–A5) follows neuromeric coding. Note: anterior communicating segment (A1) over the optic chiasm and supplies anterior hypothalamus. Orbital segment (A2) gives off fronto-polar and orbito-frontal branches. Genu segment (A3) supplies internal frontal br. and the calloso-marginal artery. Marginal segments (A4–A5) continue the callosomargina, the paracentral, and the internal parietal arteries. <u>Posterior</u> <u>cerebral</u> (P1–P4) follows neuromeric coding as well. Note: posterior communicating (P1), thalamic branches (P2), medial and lateral choroidal br. (P3), calcarine and parieto-occipital br (P4). (Reprinted from Hanes DE. Fundamental Neuroscience for Basic and Clinical Applications, 3rd ed. Philadelphia, PA: Churchill Livingstone; 2006. With permission from Elsevier)

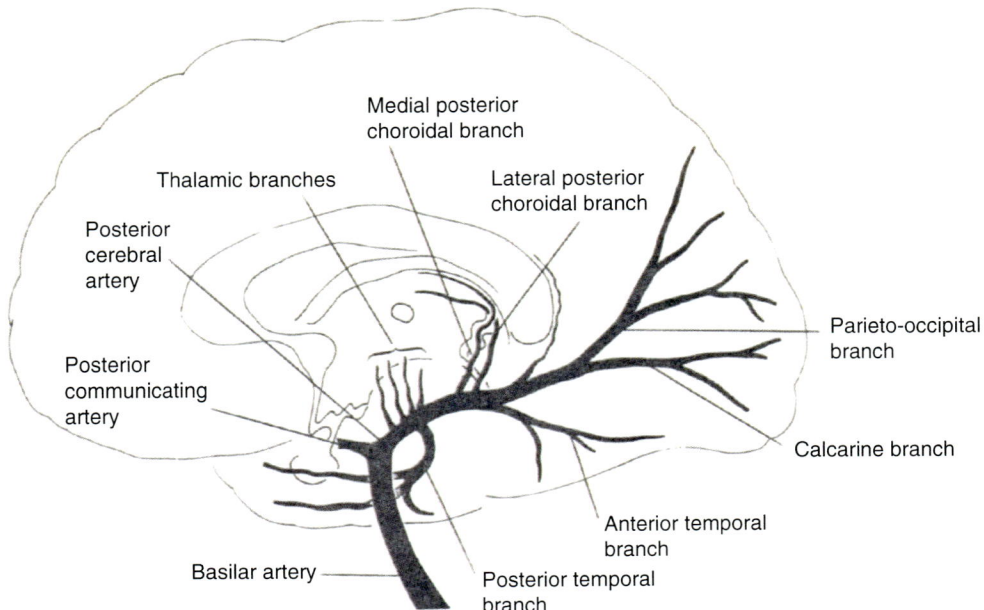

Fig. 6.88 MCA segments are neuromeric. <u>Coronal section</u> showing supply to the deep gray matter from lenticulostriate arteries. Middle cerebral artery distribution to cortex. The lenticulostriate branches supply deep brain segments. Those of segment M1 are important in severe cases of holoprosenecephaly. ACA with three segments. A3 bifurcates into pericallosal and marginal callosal (paracentral) with implications for the spectrum of the minimal form of holoprosencephaly. In HPE pathology works its way inward from the cortex. (Reprinted from Hanes DE. Fundamental Neuroscience for Basic and Clinical Applications, 3rd ed. Philadelphia, PA: Churchill Livingstone; 2006. With permission from Elsevier)

Middle cerebral gives off branches to internal capsule and cortex. It has four surgical segments. **M1,** the sphenoidal segment, runs straight lateral and gives off lenticulostriate arteries. **M2**, the insular segment, runs in the lateral (Sylvian) fissure. **M3**, the opercular segment, emerges from the sylvian fissure. **M4** is the external cortical portion. The branches of MCA are of two types. (1) *Central branches* arise just after MCA enters lateral sulcus. 10–15 lenticulostriate arteries supply head of caudate nucleus and, above globus pallidus, lentiform nucleus. (2) *Cortical branches* of MCA are very cooperative. They share gyri of frontal and parietal lobes with anterior cerebral. As an "equal opportunity neighbor," MCA shares temporal lobe with posterior cerebral. Dura over MCA-supplied cortex distribution is innervated by V2 and V3. An interesting correlation is the difference between the precentral and postcentral gyri. The cotex of the latter territory is V3 and (surprise, surprise) is sensory, while the V2 precentral gyrus is motor.

Anterior Cerebral Artery

Three anatomic segments are described.

Segment **A1** extends from the internal carotid through anterior communicating. Penetrating *anteromedial central branches* from *anterior communicating* supply the optic nerve, caudate nucleus, and anterior portions of internal capsule.

Segment **A2** extends distal from anterior communicating to the stem of callosal arteries. Its first branch, *orbitofrontal*, arises immediately distal to anterior communicating. It supplies the nasal and orbital surfaces of frontal lobe. These may be involved in the clinical spectrum of HPE, particularly when anosmia is present. The second branch, *frontopolar,* supplies the superior medial frontal gyrus, extending to its external surface. Angiosome determines interhemispheric separation of the frontal lobes.

Segment **A3** is the *precallosal artery* and represents the termination of ACA. It begins distal to frontopolar artery and hugs the splenium of corpus callosum before giving off two branches that run in parallel with one another. The proximal branch of A3, *callosomarginal artery*, follows a course that is dorsal and parallel to the corpus callosum, but not in contact with it. It gives off a series of internal frontal, paracentral, and internal parietal branches that are all directed upward along the medial cortex. At the interhemispheric fissure, they access the external surface. The anatomic integrity of these branches is essential for interhemispheric separation. *Internal frontal* branches complete the supply to the posterior frontal lobe and parietal lobe. *Paracentral branches* supply the precentral and postcentral gyri. *Internal parietal branches* reach back to occipital lobe and anastomose with the terminus of pericallosal. The distal branch of A3, *pericallosal artery*, begins just distal to the curvature of splenium and remains closely applied to corpus callosum throughout its length.

At its terminus, it makes an anastomosis with posterior cerebral. This artery is distributed to "midline sharing" structures and represents the exclusive supply to corpus callosum.

Note: Traditional sources consider anterior cerebral to be the terminal branch of anterior division. This is embryologically incorrect. The true terminus of anterior division ICA is the primitive olfactory which is seen as soon as stage 9. PrICA divides at stage12 with subsequent development of anterior cerebral, middle cerebral, and anterior choroidal.

Olfactory artery has two divisions. The lateral branch supplies the nasal field for olfaction; the medial branch supplies the basal ganglia (striatum).

Caudal Division (Figs. 8.37, 8.38, and 11.168)

The caudal division has four important segments. From proximal to distal, these are: (1) diencephalic, (2) mesencephalic, (3) posterior cerebral, and (4) posterior communicating.

Diencephalic

This artery arises just caudal to the division point of ICA, *prior to posterior cerebral*. It follows the course of posterior cerebral around brain stem. It gives off *thalamic perforating* branches to posterior thalamus. *Lateral posterior choroidal* sends one branch laterally to supply parahippocampal gyrus and inferomedial temporal lobe plus (of course) temporal choroid plexus. It sends another branch medially behind pulvinar to lateral ventricle choroid plexus. Anastomoses between lateral and medial posterior choroidal vessels are to be expected. These unite the diecephalic and mesencephalic circulations.

Mesencephalic

This artery is embryologically distinct from diencephalic. Many texts ignore the mesencephalic artery by considering it as a mere branch off posterior cerebral. Obviously, the anatomic zones of midbrain and posterior lobe are developmentally distinct. It has two main branches, anterior and posterior, which supply the quadrigeminal plate of the midbrain tectum. The superior colliculi of the plate integrate eye-head movement; the inferior colliculi integrate auditory reflexes. It gains access to the roof of the third ventricle where it supplies choroid plexus and can be called medial posterior choroidal. As development proceeds towards maturity, the mesencephalic is subsumed into the structure of, can rightfully be called, medial posterior choroidal.

Posterior Cerebral

This misunderstood vessel appears late in development along with the posterior expansion of forebrain. Although the reader is well aware that PCA originates from supratentorial internal carotid artery, *adult anatomy texts depict it as the terminal branch of infratentorial basila artery*. In the fetal period, the stem of PCA shifts caudally; it becomes vertebrobasilar-dependent. The anatomic order of structures appears to be reversed.

In the adult form, the PCA incorporated mesencephalic artery.

The overall *raison d'etre* of PCA is to supply regions dedicated to visual perception and visual motor control. The functions of seeing and looking supplied by PCA are as follows. (1) Midbrain and pontine centers control pupillary reflexes, convergence, positioning of the eyes, and various reflexes, including vestibulo-ocular (balance), corneal (tearing), and palpebral (protective blinking). (2) Lateral geniculate body and pulvinar process signal from retina. (3) Visual information must be transmitted from one side of the brain to another. The bulk of this transfer takes place at splenium of corpus callosum. (4) Memory of visual data is stored in PCA-supplied temporal lobe. (5) Striate cortex—area 17—receives visual input. (6) Peristriate cortex—areas 18–19—comments on the input, transfers it, and coordinates with the oculomotor systems of the midbrain.

Posterior cerebral artery departs from the posterior communicating stem and winds around *cerebral peduncle* to reach the tentorial surface of occipital lobe. It divides into two sets of branches. (1) Central branches go to midbrain and thalamus. *Mesencephalic branches* depart from the posterior aspect of PCA and pass through posterior perforated substance to supply the cerebral pedicles and interpeducular region of midbrain. Critical structures of the visual system are served, including: motor nuclei for III, midbrain reticular formation, substantia nigra, and the midbrain segment of corticospinal tract. *Thalamic branches* hit multiple diencephalic structures. Anterior branches go to mammillary bodies, the optic tract on its way to the orbits, and anterior thalamic nuclei. Some posterior branches penetrate posterior perforated substance to reach thalamic nuclei, while others travel upward to supply geniculate bodies. (2) Cortical branches of posterior cerebral are four; these are perfectly rational in name and function. *Anterior temporal branch* goes to the "memory zone" of uncus and parahippocampal gyrus. *Posterior temporal branch* pursues parahippocampal gyrus to reach temporal-occipital gyri. *Parieto-occipital branch* hugs the medial surface of posterior cerebrum to access the precuneate and cuneate gyri.

Occipital branch supplies the primary visual cortex (lingual and cuneate gyri).

Gray's anatomy divides PCA into three segments *based on nondevelopmental criteria*. The terminology is misleading, so let's clear it up. (1) When traditional texts describe posterior cerebral artery as "originating" from basilar, the terminal segment of posterior communicating a. is termed **precommunical** posterior cerebral. This segment of PCA is considered to "start" at its *false* "point of origin" at the basilar bifurcation. It then is described as proceeding lateral to its junction with posterior communicating. In the process, it gives off *interpeducular branches* which pass through posterior perforated substance to access upper midbrain (cranial

nerves III and IV, red nucleus, and substantia nigra) and hypothalamus. *Quadrigeminal branches* supply superior (visual) colliculus. Medial posterior choroidal artery "originates" here, although it can also arise proximal to take-off of PCA. (2) The **postcommunical** PCA runs between posterior communicating and runs around brain stem to access cortex. The lateral posterior choroidal artery "originates" from this segment. (3) **Terminal PCA** gives off lateral and medial occipital arteries.

Posterior Communicating

The developmental approach to this vessel requires that we give it an importance completely out of proportion with its length. This short vessel is the terminal segment of the caudal division of ICA. After giving off posterior cerebral, it jogs medially and caudally at a 45° angle to anastomose with basilar. These three arteries form a Y-shaped anastomosis, like a referee signaling for a touchdown. Just beneath his "armpits," the oculomotor nerves emerge and run forward. When traditional texts describe posterior cerebral artery as "originating from basilar," the terminal segment of posterior communicating is termed *precommunical* posterior cerebral. In any case, *interpeducular branches* pass through posterior perforated substance to access upper midbrain (cranial nerves III and IV, red nucleus, and substantia nigra) and hypothalamus. *Quadrigeminal branches* supply superior (visual) colliculus. Medial posterior choroidal artery "originates" here, although it can also arise proximal to take-off of PCA.

Part 4. Blood Supply to the Hindbrain
(Figs 6.82, 6.83, 6.84, 6.85, 6.86, 6.87, 6.88, 6.89, 6.90, and 6.91)

We have completed our survey of how the internal carotid, stapedial, and external carotid systems provide the blood supply of the forebrain, midbrain, and face. But that's not the end of our story. A glaring omission has been a detailed discussion of blood supply to the hindbrain. There are several reasons why this topic should be treated as a separate section.

Time and again, we have seen how the hindbrain has demonstrated "developmental primacy." In the first place, it *predates* forebrain. The induction of forebrain itself occurs from genes elaborated at the r0 level (*Otx-2*) in cooperation with those of the AVE. Second, after production of the germ layers, mesodermal organization proceeds in a craniocaudal direction from r0 backward. Vessel formation requires mesoderm. It is not surprising that the primitive dorsal aortae and the primitive hindbrain channels flanking the hindbrain are the first to appear. *The entire internal carotid axis supplying forebrain and midbrain is a forward extension from this hindbrain vascular "platform."*

Development of the heart and great vessels depends on contributions from distinct levels of hindbrain neural crest.

Fig. 6.89 PCA segments are neuromeric: basilar section. Posterior cerebral artery <u>derived from posterior communicating of the ICA,</u> but dominant circulation in the adult is basilar. It anastomoses around posterior genu of corpus callosum with pericallosal artery. In this view, segments P1–P4 previously seen on median sagittal section are readily identified. (Reprinted from Hanes DE. Fundamental Neuroscience for Basic and Clinical Applications, 3rd ed. Philadelphia, PA: Churchill Livingstone; 2006. With permission from Elsevier)

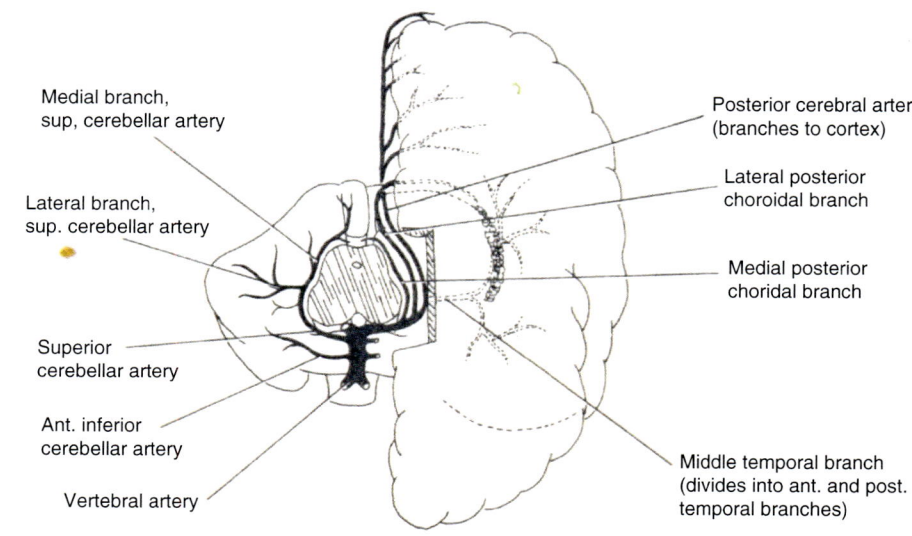

Medial branch, sup, cerebellar artery

Lateral branch, sup. cerebellar artery

Posterior cerebral artery (branches to cortex)

Lateral posterior choroidal branch

Medial posterior choroidal branch

Superior cerebellar artery

Ant. inferior cerebellar artery

Vertebral artery

Middle temporal branch (divides into ant. and post. temporal branches)

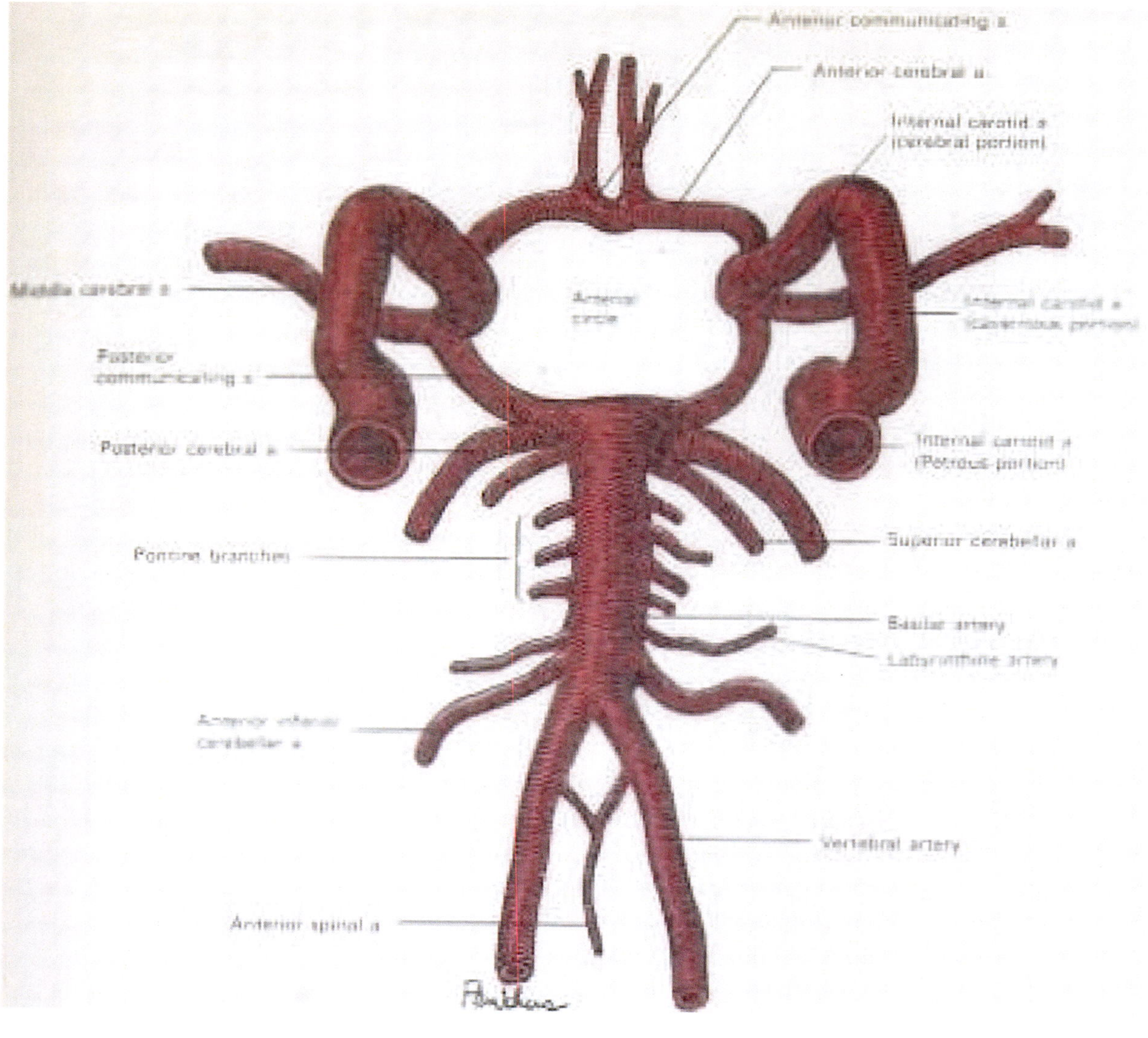

Fig. 6.90 Basilar artery showing extension to neuromeric level r4–r5 and then resumes the divergent pattern of the LNAs. Labyrinthine artery supplies structures innervated by cranial nerves VII and VIII (and thus, from r4 to r5). These are well illustrated in the sagittal cerebral series previously presented. (Reprinted from Hanes DE. Fundamental Neuroscience for Basic and Clinical Applications, 3rd ed. Philadelphia, PA: Churchill Livingstone; 2006. With permission from Elsevier)

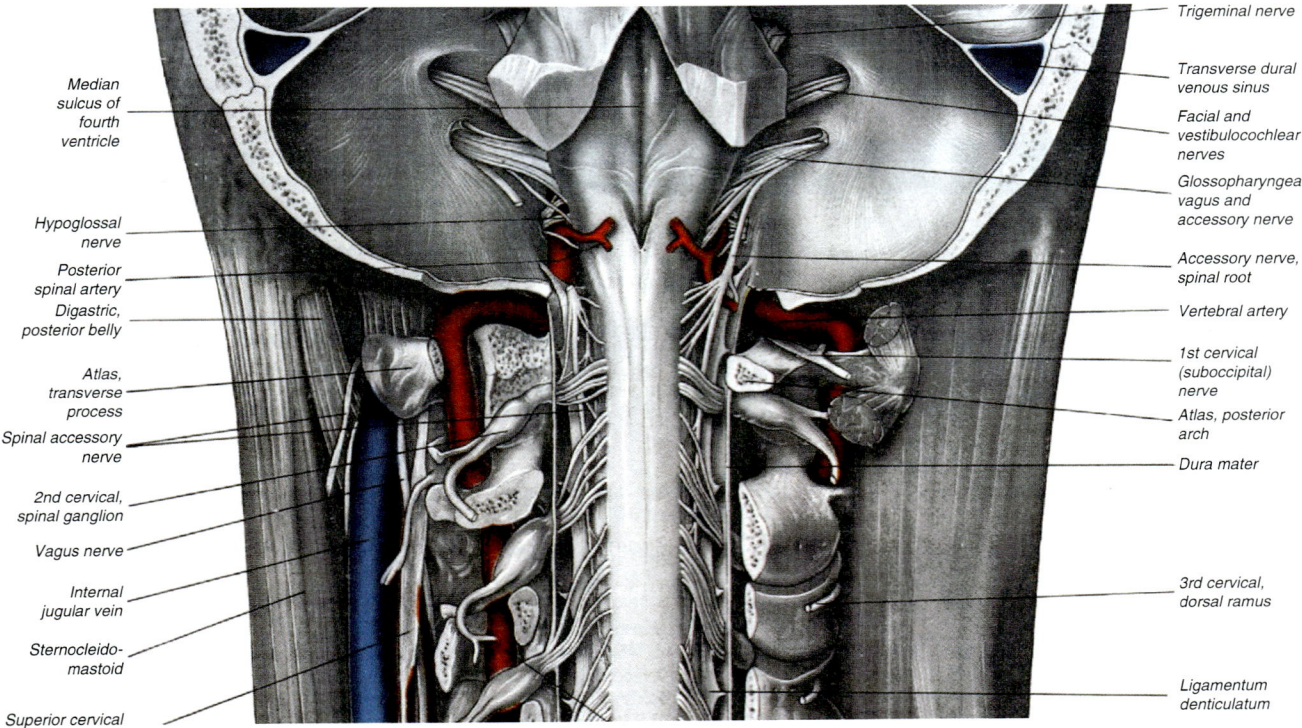

Fig. 6.91 Vertebral artery—note vascularization of the hypoglossal nerve root marks the transition zone to the bilateral longitudinal neural arteries. (Reprinted from Williams PL (ed). Gray's Anatomy, 38th ed. New York, NY: Churchill Livingstone; 1995. With permission from Elsevier)

Rostral hindbrain (r2–r5) mesenchyme is assigned to the heart, but not great vessels. Cardiac neural crest arising from the otic capsule to r11 enters the heart via AA3 and AA4 and the lungs via the sixth arch pulmonary arteries. This neural crest in important for creating structures such as valves.

An interesting speculation: how do stems from the aortic arches "know" where to match the stem from the overlying dorsal aortae? Is it possible that genetic identity (homeotic genes) exists between the neuromeric zones of the dorsal aortae and those of the outflow tract—perhaps in the form of a homeotic *anteroom-posterior palindrome*? In this model, the secondary heart field which gives rise to the outflow tract is constructed from lateral plate mesoderm having a relationship established at gastrulation with the dorsal aortae. The outflow tract could have neuromeric zones proximal to distal r11–r2 versus those of the dorsal aortae r2–r11. Growth and folding of the head folding tuck the heart ventrally. The sequences are now in parallel. They remain connected at level r2 via the first aortic arch.

Mesoderm associated with the caudal hindbrain (r8–r11) has a number of distinct fates. Paraxial mesoderm from somitomeres 8–9 contribute to the fourth and fifth pharyngeal arches. Somites from these same levels (r8–r11) produce tongue muscles and form the posterior cranial base. The origin of sternocleidomastoid and trapezius muscles comes from these somites and possibly from lateral plate mesoderm in register with r8–r11. Thus, blood supply to the hindbrain must interact with distally based blood supply from the external carotids and subclavian to support these structures.

Developmental Timeline of Blood Supply to the Hindbrain

The overall story of the arterial supply to the hindbrain is one of a "peaceful transfer of power" in which the predominance of blood flow shifts from the internal carotid system to the vertebral system. This process goes through several phases. These are summarized below with details to follow in the subsequent narrative

Stages 7–8: Primordial hindbrain channels provide nutrition to the hindbrain via ebb-and-flow, disconnected from the dorsal aortae.

Stage 9: Primitive trigeminal artery provides the first connection between the hindbrain plexus and the heart.

Stages 10–11: Additional presegmental arteries connect with the PHCs at the hyoid, otic, hypoglossal, and first cervical levels. LNAs replace the PHCs and supply the neuromere of the hindbrain with segmental branches. Primitive internal carotid to forebrain and midbrain has not yet developed a caudal division, leaving hindbrain "out of the loop."

Stage 12: Caudal division of ICA develops.

<u>Stage 13</u>: Posterior communicating represents the final branch of caudal ICA.

<u>Stage 14</u>: PComm unites with LNAs. Segmental fusion of the LNAs over the pons, but not the medulla.

<u>Stage 15</u>: Longitudinal union of the transverse cervical segmental arteries creates vertebro-basilar system.

<u>Stage 16</u>: Heart descends into thorax taking origin of vertebral arteries with it.

<u>Stage 17</u>: Segment of dorsal aorta connecting AA3 with AA4 disappears.

Development of the neck and chest takes place concomitantly. The heart descends into the chest. The arch of the aorta gives rise to distinct forms of arterial supply to head and neck, upper extremities, and trunk. The paired dorsal aortae must fuse into a single structure. Establishment of this adult system poses anatomic challenges for the embryo that require radical solutions. These include: (1) separation of the third and fourth aortic arch arteries, (2) elimination of unnecessary vessels to properly direct arterial flow from the heart to the body and avoid shunts, (3) disconnection of dorsal aortae from the hindbrain, cervical spinal cord, (4) maintenance of blood supply to the hindbrain and developing neck even as the heart and aortic arch sink into the chest.

The final iteration of the vascular supply to the hindbrain cannot be understood without an appreciation of the overall architecture of blood supply to the neck. But that subject is inconceivable without understanding the overall vascular bauplan of the somite-based vertebrate trunk. Thus, our discussion of the carotid system closes by considering this system as a highly evolved series of evolutionary innovations designed to support the head as an "add-on" to a more primitive segmental body plan. The "tetrapod revolution" resulted in a separation of head and body by a moveable craniovertebral joint of remarkable evolutionary advantage. The "mammalian masterplan" interposed additional repeating segmental units allowing unique 3-D positioning of the head, while further distancing the brain from its source of metabolic support. As we shall see, the arterial anatomy of the mammalian neck demonstrates the sensitive manner in which variations in neuromeric coding responded to the challenges using variations on preexisting patterns to create a unique solution set.

Longitudinal Neural Arteries

From the moment of its inception, mesoderm begins to produce blood vessels. Gastrulation precedes neurulation. Thus, as the neural plate rounds up into the neural tube, the walls of the embryonic brain are surrounded by paraxial mesoderm containing a primitive vascular network develops. The *primordial hindbrain channels* (PHCs) and the *primitive head*

plexus (PHP) are continuous with each other and supply the brain via an ebb-and-flow mechanism. The organization of the plexus is not segmentally organized. It functions as a low-pressure system.

Beginning at stage 9, the CNS plexus connects with the dorsal aortae via a series of primitive *pre-segmental arteries*, the first and most consequential being the *primitive trigeminal artery*. The stem of TgA arises from the r2 to r3 level of the dorsal aorta and it connects with the plexus at the level of the future trigeminal ganglion. Additional segmental arteries (trigeminal, otic, and hypoglossal—others likely exist) connect the PHCs with the dorsal aortae. These reach the first cervical level by stage 10.

The exposure of the PHCs to increased arterial flow causes a preferential remodeling to take place. Longitudinal neural arteries develop in parallel with the dorsal aorta. These bear a linear array of clearly identified stem vessels distributed to each neuromeric segment. It is likely that these arterial branches sprout along the axis of the LNAs in response to signaling factors elaborated and transmitted from each neuromere. By stage 11, the transformation of the primordial hindbrain channels to LNAs is largely complete.

In the meantime, development of the internal carotid system continues forward from the take-off of the primitive trigeminal arteries to supply the midbrain and forebrain. The early ICA is not initially connected to the hindbrain. Stage 12 and 13 are marked, respectively, by the bifurcation of the ICA producing a caudal division giving rise to a posterior communicating segment. Connection between the ICA and the LNAs takes place at stage 14, providing significantly increased blood supply and taking the stress away from the trigeminal artery.

A concomitant event at stage 14 is the localized fusion of the LNAs at level of the pons to form a singular basilar artery. This fusion does not continue beyond the pons so that, below the ponto-medullary junctions, the LNAs remain divergent. They remain tethered caudally by the first cervical segmental. The resultant configuration takes the form of an upside-down letter "Y." In addition, a parallel circuit develops running peripheral to the LNAs from level of the otic capsule to the junction with the vertebral arteries at the level of neuromere c1.

LNA Branches of the Metencephalon (r0–r5): The Basilar Artery

Superior Cerebellar
The junction between the outstretched "arms" of PCm and LNAs is Y-shaped, with the fused vertical limb representing the fusion of the LNSs into the basilar artery. Just before connecting with the LNAs, posterior communicating arteries give off the posterior cerebral artery. Just caudal to this, cranial nerve III appears beneath each "armpit" of the "Y." The

oculomotor nerves run forward ventral to PCm. Just caudal to III arise the superior cerebellar arteries. These guys cannot give up their orbital fixation. As they wind around the *cerebral peduncle,* they "keep an eye on" trochlear nerve (probably receiving induction signals). Within pia, superior cerebellar (SCbA) ramifies and connects caudally with both anterior inferior cerebellar (AICbA) and posterior inferior cerebellar (PICbA). In addition to its primary target, SCbA is distributed to anterior medullary velum and tela choidea of third ventricle.

In this regard, a few words are in order concerning the significance of the tela choroidea. Recall that, in the broadest terms, telencephalon arises from alar p6–p5, while diencephalon arises from alar p4–p1 plus basal p6–p1. *Neopallium,* the nonolfactory part of cerebral cortex, is the developmental source of most of the cerebral hemispheres. This is most likely associated with alar p5. The primitive cerebral hemispheres project laterally like giant dumbbells, at the center of each is a cavity, the *primitive lateral ventricle.* The thin neural walls of this cavity contain neural stem cells, the proliferation from which forms the cerebral hemispheres. They expand rapidly in a horseshoe fashion. The dorsal and caudal growth of the cerebral hemispheres overtakes and encloses the diencephalon by month 3, the midbrain by month 6, and the hindbrain by month 8.

Not every part of the cerebrum enjoys the benefits of this "inflation." The median strip of tissue uniting the hemispheres does not grow. It stays thin and remains as the *roof of the third ventricle.* The restraining effect of this structure explains the development of a deep groove between the two balloons, the longitudinal cerebral fissure. Recall that mesoderm envelops the brain. Over the midline, this mesoderm is supplemented by r0–r1 midbrain neural crest to form the *falx cerebri.* Further growth of the hemispheres places the ventricles on stretch. Like good tissue expanders, the initially globular ventricles are stretched asymmetrically in response to changes in their surroundings. This explains how they give origin to the *anterior, inferior,* and *posterior horns.*

Just over the primitive interventricular foramen, lateral ventricle along with roof plate of diencephalon maintains a unique zone of epithelial tissue. This ependymal tissue pushes its way into the medial wall of lateral ventricle, forming a groove, the *choroid fissure.* This provides the perfect opportunity for angioblast-containing external mesoderm to insinuate itself between the two layers of ependyma, forming the *tela choroidea.* One can just hear the blood vessels growing! The resulting choroid plexus mushrooms into the third ventricle, during the fourth month almost filling it. Although it eventually abandons its expansionist tendencies, choroid plexus remains as the major site of CSF production.

Blood supply to this important zone, literally at the "heart of the brain," reflects its functional importance. When primitive internal carotid to forebrain transitions into cranial and caudal divisions, the very first branch formed by either one is an artery directed toward future choroid plexus. Thus, cranial ICA gives off anterior choroidal and caudal ICA produces posterior choroidal. But why should ICA do all the work? It would seem fitting that the LNAs should make a contribution as well. This "ambassadorial" function is fulfilled by superior cerebellar artery (1) because it is the most anterior r0–r1 vessel available and (2) it is physically closer to choroid plexus than anterior inferior cerebellar artery.

Pontine

These vessels supply r0–r5 pons and cross over to make anastomoses with vessels supplying midbrain. These latter are derivatives of the mesencephalic branch of caudal division of ICA. *Medial branches* enter pons directly, while *lateral branches* travel a more circuitous route around brain stem penetrating posterolaterally.

Labyrinthine

This important artery is dedicated to the inner ear. It departs from the midpoint of basilar at level r4–r5 and follows VII and VIII into the internal auditory canal. It does not supply middle ear. Because it can also arise from AICA, one can postulate neuroembryonic relationships between the vestibulocochlear unit and anterior cerebellum.

Anterior Inferior Cerebellar

This vessel arises in lower third of basilar. It is always in contact with cranial nerve VI, being dorsal 1/3 and ventral 2/3 of the time. Recall that the nuclei of VI reside in r4–r5. Its target, lateral rectus, develops in Sm5. As it travels downward, it is ventral to VII and VIII but, as these enter temporal bone, it sometimes divides around them. At cerebellopontine angle, AICbA divides to serve medial and lateral aspects of the inferior cerebellum.

The discerning reader might ask why inferior cerebellum merits to two vessels, but superior only one.

LNA Branches of the Mylencephalon (r6–r11)

Posterior Inferior Cerebellar Artery

PICbA is the largest branch of the LNAs and arises 1.5 cm caudal to the ponto-medullary junction just below the rootlets hypoglossal. Its courses posteriorly (dorsally) around the olive. It runs interspersed between the rootlets of IX and X and move on to the surface of inferior cerebellum where it divides. *Medial branch* travels in the groove between superior and inferior hemisphere. *Lateral branch* supplies the "cupped palm" surface of inferior cerebellum.

Anterior Spinal Artery (Fig. 11.33)

Two branches descend along the medulla from the ponto-medullary junction to 2 cm above the foramen magnum where they unify. The singular vessel then runs in the *anterior median fissure* of the spinal cord. Just like PSA, at each neuromeric level, ASA is reinforced by paired segmental arteries that enter the vertebral canal from the sides and travel anteriorly. These are not spinal branches per se. The sources of these branches are regionally distinct: neck (vertebral, ascending cervical branch of inferior thyroid), thorax (intercostal), and posterior abdominal wall/pelvis (lumbar, iliolumbar, and lateral sacral). All these vessels are joined together by vertical anastomoses to form a single *anterior median artery*. It courses beneath pia mater, continuing into filum terminale. The biologic significance of this system (probably the unification of embryonic paired vessels) is that it illustrates the uniform neuromeric basis of blood supply to the CNS.

Posterior Spinal Artery

These vessels appear along medulla (they can also arise from PICA). Near the origin, PSA supplies the lateral wall of fourth ventricle. The vessels travel dorsally and, beyond the medullary-spinal junction, descend in the midline. The PSA runs on the dorsolateral surface of spinal cord, ventral to the dorsal spinal nerves, and supplies dorsal spinal cord. At each neuromeric level, it is reinforced by spinal branches.

Meningeal Branches

These small vessels bide their time and take off from vertebral just as it enters foramen magnum, i.e., at the c1 neuromeric level. They provide critical support *between bone and dura* of the endochondral posterior fossa. They also supply falx cerebelli, but *not* r1 tentorium cerebelli. Please note: ectomeninx (dura mater) is composed of nonneural tissue (extracranial mesoderm plus neural crest that has migrated out into PAM), whereas endomeninx is a neural structure made from the local neural crest. *Ectomeninx must be supplied by extracranial vessels.* The StV2/StV3-based middle meningeal system provides for the membranous posterior fossa. Vascular derivatives from rhombomeres r8–r11 do not contribute to the carotid system. Occipital somites are supplied by LNA-cervical segmental system, fed by the first cervical branch. Thus, dura lining the bone derivatives of these somites in posterior fossa can rightfully expect blood flow from the c1 neuromeric level.

Vertebral Artery

This vessel is the first branch of the subclavian artery. It arises from the mesenchyme of cervical somites 1–8. Recall that vertebrae result from parasegmental contributions of somites (vertebra x represents the caudal half of somite $x - 1$ + the cranial half of somite x). Spinal nerves always exit caudal to their respective vertebral body. The fourth tho-

racic segmental nerves pass deep to vertebra T4. In the neck, integration of proatlas (r11 + c1) into the skull results in only 7 cervical vertebrae. C1 exits caudal to the (unseen) proatlas and cranial to what is called "first cervical vertebra" (Figs. 6.91 and 6.92).

Vertebral artery arises from the posterior aspect of subclavian and runs along the side of the vertebra C7 without any attachment. As it ascends, vertebral artery is encircled by the transverse processes of vertebra C6 to C1. After passing through the atlas, it flips medially over 90° and follows superior articular process to achieve the cranial aspect of posterior arch, where it leaves behind a tell-tale groove (Figs. 5.11 and 5.14) It now must enter the skull. The atlanto-occipital membrane has formed an arch over the preexisting vessel, which passes into the skull through foramen magnum.

The backward jog of vertebral at level C6 is highly functional. C7–T1 is the neck-trunk pivot point. If vertebral were tethered here, it would be constantly subjected to mechanical stress. Fortunately, the insertion of the first rib to C7–T1 modifies the "need" for a transverse process in the first place. Other cervical vertebra can have their puny cervical ribs… C7 gets the real thoracic thing!

Anatomic variation in ribs is a likely result of modification in the Hox gene code. In the case of C7, failure of this modification leads to the presence of an extra rib at C7, above the normal insertion of first rib. This results in a well-known *thoracic outlet syndrome*, the clinical characteristics of which involve compression of the subclavian artery and/or brachial plexus are entrapped between the cervical rib and the scalenus muscle.

In the previous chapter on the osteology of the neck, the reader will recall that discussion regarding why mammals have only seven vertebrae. Regulatory *Hox* genes control the identity of regions along the length of animals, i.e., they specify cell fates. The Galis-Metz hypothesis is that variations in the Hox sequence associated with the neck would lead to increase embryonal cancers [9]. C7 is apparently more susceptible than its cervical colleagues to genetic instability. Cervical ribs occur with an incidence of 1:500, but the frequency of cancer in these patients is 125× that of the general population.

Sectors of the vertebral artery

Let's discuss the parts of vertebral artery before tackling its branches.

Part 1: In its initial course, vertebral lies in the embrace of two vertical muscles, longus colli and scalenus anterior. Inferior thyroid crosses anteriorly on its way to its hypaxial destination. Posterior to vertebral are (logically) transverse process of C7 and the SANS, including stellate ganglion. Like climbing ivy, fibers of SANS take advantage of the vertebral arteries to gain access to the skull.

Fig. 6.92 Vertebral artery—anterior meningeal branch belongs to the LNAs. May represent flow of paraxial mesoderm from cervical somites 1–2 into the posterior fossa where it contributes to periosteum and dura. (Reprinted Cagnie B, Barbaix E, Vinck E. Atherosclerosis in the vertebral artery: an intrinsic risk factor in the use of spinal manipulation Surg Radiol Anat 2006;28(2):129–34. with permission from Springer Nature)

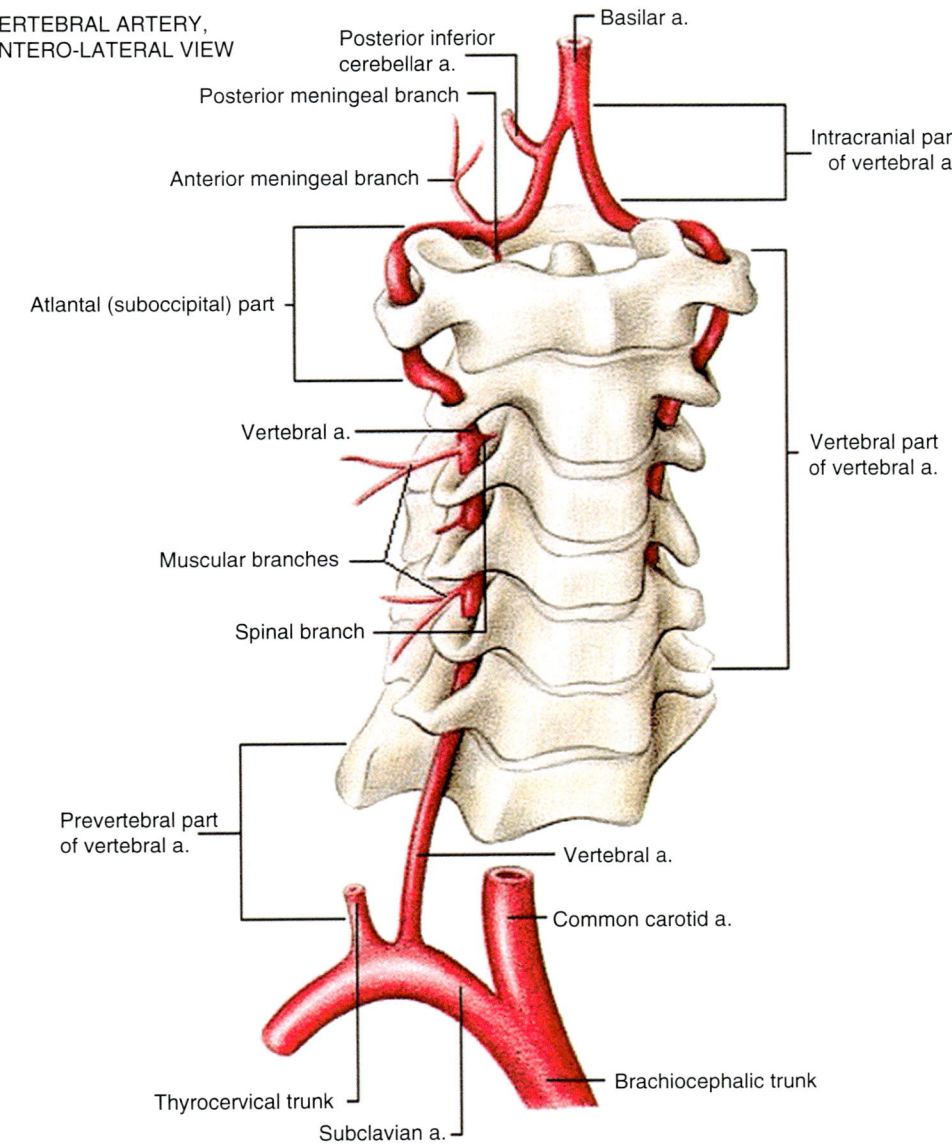

VERTEBRAL ARTERY, ANTERO-LATERAL VIEW

Basilar a.

Posterior inferior cerebellar a.

Posterior meningeal branch

Anterior meningeal branch

Intracranial part of vertebral a.

Atlantal (suboccipital) part

Vertebral a.

Vertebral part of vertebral a.

Muscular branches

Spinal branch

Prevertebral part of vertebral a.

Vertebral a.

Common carotid a.

Brachiocephalic trunk

Thyrocervical trunk

Subclavian a.

Part 2: This zone involves "transforaminal transit" from C6 to C1. Vertebral is now surrounded by cervico-thoracic SANS ganglia. Cervical nerves send out vertical rami along the inferior border of the transverse processes; naturally, these travel behind vertebral artery.

Part 3: This zone marks the exodus of vertebral out from the neck into the head. It exits through a narrow pass medial to rectus capitis lateralis, the muscle that helps to turn the head by steering the transverse process "tiller." Like a caravan, it then embarks on a tortuous trek over the Atlas mountains to reach the medullary Mediterranean. As it enters the vertebral canal, it is contained within the suboccipital triangle. This is bounded by rectus capitis posterior major and two obliques, superior and inferior. Suboccipital nerve (neuromere c1) runs between posterior arch and artery.

Part 4: The dura mater of the foramen magnum constitutes embryologic boundary between the longitudinal neural

arteries of the medulla and the vertebral arteries of the neck. The point of penetration is located between the anterior root of first cervical and hypoglossal nerve and the first cervical nerve. Recall that rhombencephalon subdivides into r0–r5 metaencephalon (cerebellum and pons) and r6–r11 myelencephalon (medulla). Basilar represents the confluence of the longitudinal neural arteries along ventral pons. At the pontomedullary junction, the separation of the LNAs is reinstated. They splay outward and downward, terminating at the first cervical segmentals to become vertebrals.

The LNAs and vertebrals differ in several important ways (1) developmental mechanism, (2) time sequence, (3) mesenchymal composition, and (4) relationship with the brain. **_For the reasons given below, it is anatomically incorrect to refer to the intracranial vessels supplying the medulla as vertebrals_**.

Development of the vertebrals parallels that of other arterial systems, such as the thyrocervical trunk and the internal thoracic artery. They arise from a vertical cross-linking of segmental arteries connecting the dorsal aortae with each level of the neck from neuromeres c1 to c7. Once the cross-links form, the ventral segments between the dorsal aortae and the vertebral chain disintegrate leaving the vertebrals arising from the seventh intersegmental arteries (future subclavians) and connected directly to the neck. In contrast, the LNAs develop as a simple remodeling of a previously existing system, the primordial hindbrain channels.

The LNAs are fully formed by stage 12, whereas the vertebrals do not appear until stage 15.

The mesodermal building blocks of these two system are neuromerically distinct. LNAs are constructed from somitomeres 4–7 and somites 1–4. Vertebrals are constructed from somites 5–10. The time course of development differs by several stages. Vertebrals develop from lneuromeric levels c1 to c7 and have no biologic relationship with the brain.

LNAs are in genetic register with rhombomeres r0–r11, whereas the vertebrals bear components from cervical neuromeres c1–c7. One system serves the brain, the other, the spinal cord.

Branches of the Vertebral Artery

Traditionally (and un-embryologically), the vertebral artery is described as having two groups of branches, cervical and cranial. As we have pointed out, once penetrating the dura, these vessels are rightfully called longitudinal neural arteries.

Muscular Branches

Deep layer neck muscles between atlas and occiput are anatomically distinct from the interconnecting muscles of the cervical vertebrae. These latter have relatively monotonous "targets." The insertion sites for atlanto-occipital muscles are spatially more sophisticated, but they originate from a common embryologic principle, the law of parasegmentation. Thus, those muscles responsible for flexion-extension and rotation of the head at the atlanto-occipital joint originate from the first cervical somite (Sm12 = somite 5). They seek insertion at remnants of the somite 4–5 *proatlas* and with the fused concentric-ring derivatives of 1–4 (supraoccital, exocipital, and basioccipital). These muscles are all supplied by vertebral artery as it snakes its way around the lateral mass of the atlas.

The anastomoses of these muscular branches have embryologic significance. (1) Occipital artery represents external carotid supply to the superficial investing fascia covering posterior skull. This supplies second arch muscles, but (remembering the stapedial) it also receives a trans-calvarial anastomosis from StV3 middle meningeal to posterior fossa. (2, 3) Anterior neck muscles are supplied by the ascending and deep cervical arterial chains (more on these later). These anastomoses represent ancient cross links between the three longitudinal vessel systems: hypaxial, lateral, and epaxial.

Spinal Branches

Each neuromeric level is supplied by a spinal branch which gains access to the intervertebral canal and divides. (1) *CNS branches* pass along nerve roots to supply meninges and cord of each neuromere. (2) *Vertebral branches* form lateral anastomotic chains that run upward and downward on the dorsal surface of the vertebrae, at the junction between body and pedicle.

Meningeal Branches

These small vessels bide their time and take off from vertebral just as it enters foramen magnum, i.e., at the c1 neuromeric level. They provide critical support *between bone and dura* of the endochondral posterior fossa. They also supply falx cerebelli, but *not* r1 tentorium cerebelli. Please note: ectomeninx (dura mater) is composed of nonneural tissue (extracranial mesoderm plus neural crest that has migrated out into PAM), whereas endomeninx is a neural structure made from the local neural crest. *Ectomeninx must be supplied by extracranial vessels.* The StV2/StV3-based middle meningeal system provides for the membranous posterior fossa. Vascular derivatives from rhombomeres r8 to r11 do not contribute to the carotid system. Occipital somites are supplied by LNA-cervical segmental system, fed by the first cervical branch. Thus, dura lining the bone derivatives of these somites in posterior fossa can rightfully expect blood flow from the c1 neuromeric level.

Commentary: Rolf Ewers

Over the years, Dr. Rolf Ewers has been not only a friend and trusted colleague, but also a key contributor to this work, although he himself is unaware of this fact. When I visited his unit in Vienna, I was working out the vascular embryology of the face in an attempt to understand the Tessier classification of rare facial cleft. Rolf sent me over to the Federal Pathological-Anatomical Museum of the University of Vienna. This collection is housed in the so-called Narrenturm, a former hospital for the mentally ill founded in 1784. In addition to many anatomical oddities collected from all over Europe over the centuries are skeletal clefts of all types and configurations. Here, before my eyes, there was proof positive that Tessier's system was based on focal neurovascular abnormalities. I returned from Vienna convinced that Dorcas Padget's observations were correct; this chapter and Chap. 7 describing the vascular embryology of both the intracranial and extracranial arterial systems are the result. I am informing Rolf of his contribution via this means but look forward to thanking him in person.

Michael Carstens

This book from Dr. Michael Carstens is for me, as a cranio-maxillofacial and cleft surgeon, a seminal work. I have known Michael for many years and have had him to our Unit at the University Hospital of Vienna as a visiting professor. He has been researching and operating in this very special field for decades, and his work has been controversial, as it has challenged many previously accepted opinions. His cyanotype, his lodestone, the basic formulation for his theories, has always been the Tessier classification. This model explores in depth for years, seeking to find an explanation for the processes of facial development and the failures of same. The understanding gained in this search guided him to formulate the technical principles of developmental field reassignment surgery.

Michael has always thought about the head and neck in terms of neuromeric terminology. This book presents the relationships between the processes of neurulation and gastrulation to demonstrate how the neuromeric "map" is established in the embryo. In my opinion, this map is the core concept of this book.

Certainly, this work will engender controversial discussions, but the anatomy which he has detailed in this book is the result of a rigorous and continuous line of thought and consequences. If the face is considered as a composite of neuroangiosomes under tight sequential genetic control, careful analysis of its blood supply, neuroanatomy, and neuroendocrinology can be combined with the neuromeric theory of brain development to construct a useful model of midline facial formation.

This idea of common neuromeric definition provides us with a new understanding of the anatomic rationale behind the bones, muscles, and fascia of the craniofacial skeleton. The neuromeric system offers a unique perspective on craniofacial deformities. Pathologic states involving a particular neuromeric level can affect one or all its derivatives. Deformities involving seemingly unrelated bones or muscles can be understood in terms of common neuromeric levels of origin. The neuromeric theory of brain development permits you to construct a useful model of midline facial formation and, when disturbed, deformation. In every instance, neuromeric coding is useful in understanding tissue origins and assembly.

Although this book raises many more questions than answers, the pathway forward to understand development will surely involve the neuromeric mapping of its component parts and the tracing of the final derivatives back to their beginnings.

The concept of fields—each with a specific neurovascular axis, each susceptible to failure of formation vs. disruption based upon a growth cone dysfunction—has been presented in detail in this book. The surgical procedures which follow out of this knowledge are explained and shown.

Michael's final statement is very important: the fundamental goal should be a process-oriented cleft lip and palate repair with the conversion of the dysfunctional bone and soft-tissue matrix into a functional matrix that develops over time in a natural way.

It is my sincere hope that this book should be widely read by students and more importantly by all those training as cranio-maxillofacial cleft surgeons.

Vienna, August 2022

Rolf Ewers

A Note from Dr. Carstens

Rolf Ewers

Over the years, Dr. Rolf Ewers has been not only a friend and trusted colleague, but also a key contributor to this work, although he himself is unaware of this fact. When I visited his unit in Vienna, Rolf was (and is) an outstanding thought leader in cranio-maxillofacial surgery. At the time, I was working out the vascular embryology of the face as a means of understanding the Tessier classification of rare facial clefts. Rolf and Hildegund welcomed me to their home and subsequently directed me over to the Federal Pathological-Anatomical Museum of the University of Vienna. This collection is housed in the so-called Narrenturm, a former hospital for the mentally ill founded in 1784. In addition to many anatomical oddities collected from all over Europe over the centuries are skeletal clefts of all types and configurations. Here, before my eyes, there was proof positive that Tessier's system was based on focal neurovascular abnormalities. I returned from Vienna convinced that Dorcas Padget's observations were correct; this chapter and Chap. 7 describing the vascular embryology of both the intracranial and extracranial arterial systems are the result. I am indebted to Rolf for his seminal contribution; these chapters are but a very inadequate way to express my appreciation and to acknowledge the historical role of Vienna in the development of science and medicine.

References

1. Kuratani S. Is the vertebrate head segmented? Evolutionary and developmental and considerations. Integr Comp Biol. 2008;48(5):647–57. https://doi.org/10.1093/icb/icn015.
2. Kuratani S, Kusakabe R, Hirasawa T. The neural crest and evolution of the head trunk interface in vertebrates. Dev Biol. 2018; https://doi.org/10.1016/j.ydbio.2018.01.017.
3. Mukouyama YS, Shin D, Britsch S, Taniguchi M, Anderson DJ. Sensory nerves determine the pattern of arterial differentiation and blood vessel branching in the skin. Cell. 2002;109(6):693–705.
4. Taylor GI, Pen W-R. The angiosome concept and tissue transfer. St. Louis: Quality Medical Publishers; 2014.
5. Manchot C. The cutaneous arteries of the human body (translation of Hautarterien des menschlichen Körpers by Ristic J, Morain WD). 2nd ed. New York: Springer; 1983.
6. Salmon M, Taylor GI, Tempest MN. Arteries of the skin. New York: Churchill Livingstone; 1988.

7. Salmon M. Arteries of the muscles of the extremities and the trunk: arterial anastomotic pathways of the extremities. CRC Press; 1994.

8. Carlson BM. Human embryology and developmental biology. 6th ed. St. Louis, MO: Elsevier; 2018.

9. Galis F, Kundrát M, Metz JA. Hox genes, digit identitites and the bird/therapod transition. J Exp Zool B Mol Dev Evol. 2005;304(3):198–205. Review

Further Reading

Namba K. Carotid-vertebrobasilar anastomoses with reference to their segmental properties. Neurol Med Chir (Tokyo). 2017;57(6):267–77. https://doi.org/10.2176/nmc.ra.2017-0050.

Padget DH. The development of the cranial arteries in the human embryo. Contrib Embryol Carnegie Inst. 1948;32:205–61.

Michael H. Carstens

Big Picture Idea Branches of the trigeminal and facial nerves accompany arteries of the external carotid system and stapedial system to supply the face, orbit, and dura. The resultant neuroangiosomes delineate the migration pathways of neural crest.

From a developmental standpoint, the structures supplied by the external carotid artery system (ECA) *exclude* the jaws (mandible and the zygomaticomaxillary complex) and the dura. These latter structures are supplied by the extracranial division of the stapedial artery system. The reasons for this date back to the invention of jaws in the chondrichthyan fishes of the Devonian period. Agnathic fishes had seven respiratory gill arches. The "gnathic revolution" involved a reassignment of the cartilage structures in the first two gills into the palatoquadrate, mandibular, and hyoid cartilage precursors of the jaws. This gave them a 2 + 5 configuration. The remaining muscles of the first two arches, formerly involved in respiration, were also reassigned. The external carotid in early gnathic fishes came off the ventral aorta.

In more advanced fishes, a 2 + 4 pattern appears, the seventh arch being assigned to the swim bladder. The external carotid supply to the lower head becomes relocated to the efferent branch of the remnant of the second aortic arch artery, the hyoid artery. This is very important. As tetrapods evolved, the remaining gill arches were replaced with five pharyngeal arches. These became supplied by the external carotid system, now with its stem moved backward to the third aortic arch artery. As we shall see, the hyoid becomes the stem for a new system, the stapedial.

It should be noted that the dura mater represents a third layer of meninges that is not present in non-mammals. This topic is discussed in depth in Chap. 12.

Blood Supply to the Face

(Figures 7.1, 7.2, 7.3, 7.4, 7.5, 7.6, 7.7, 7.8, 7.9, 7.10, 7.11, 7.12, 7.13, 7.14, 7.15, 7.16, 7.17, 7.18, 7.19, 7.20, 7.21, 7.22, and 7.23)

In the section that follows, we shall give a brief description of the events involved in the breakdown of the aortic arches and their reorganization into the external carotid system. This timeline requires that we cover the initial stages of stapedial formation as well, but the details of the stapedial system will be covered in a subsequent section.

Let's conclude with two "sound bites." The extracranial stapedial, born of the ICA, constitutes an *extension cord for the ECA* permitting perfusion of the deep structures of the jaws and midface. Finally, it will be seen that *each of the branches of the ECA is assigned to a specific pharyngeal arch.*

M. H. Carstens (✉)
Wake Forest Institute of Regenerative Medicine, Wake Forest University, Winston-Salem, NC, USA
e-mail: mcarsten@wakehealth.edu

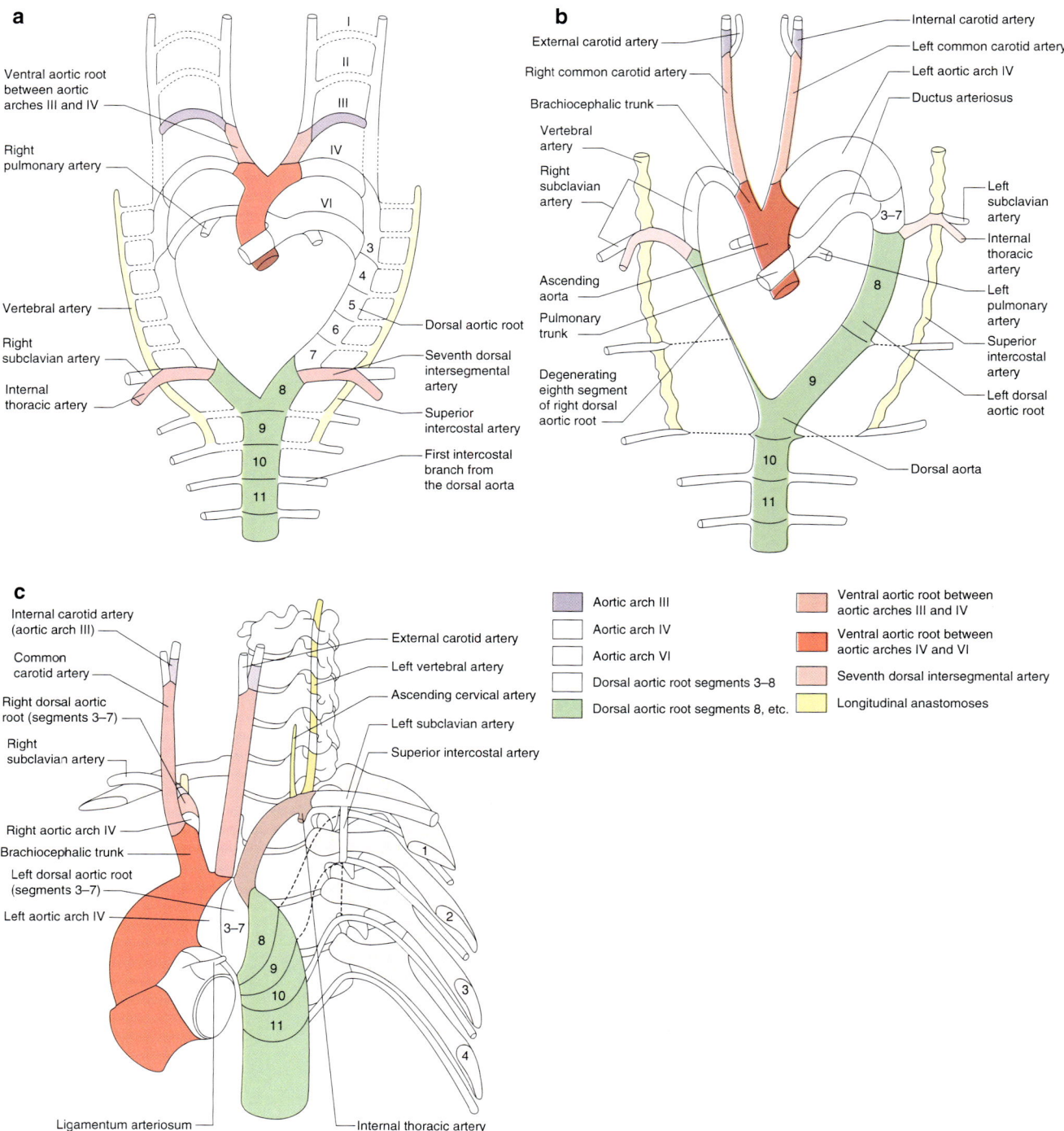

Fig. 7.1 Aortic arch color code. (Reprinted from Lewis, Warren H (ed). Gray's Anatomy of the Human Body, 20th American Edition. Philadelphia, PA: Lea & Febiger, 1918)

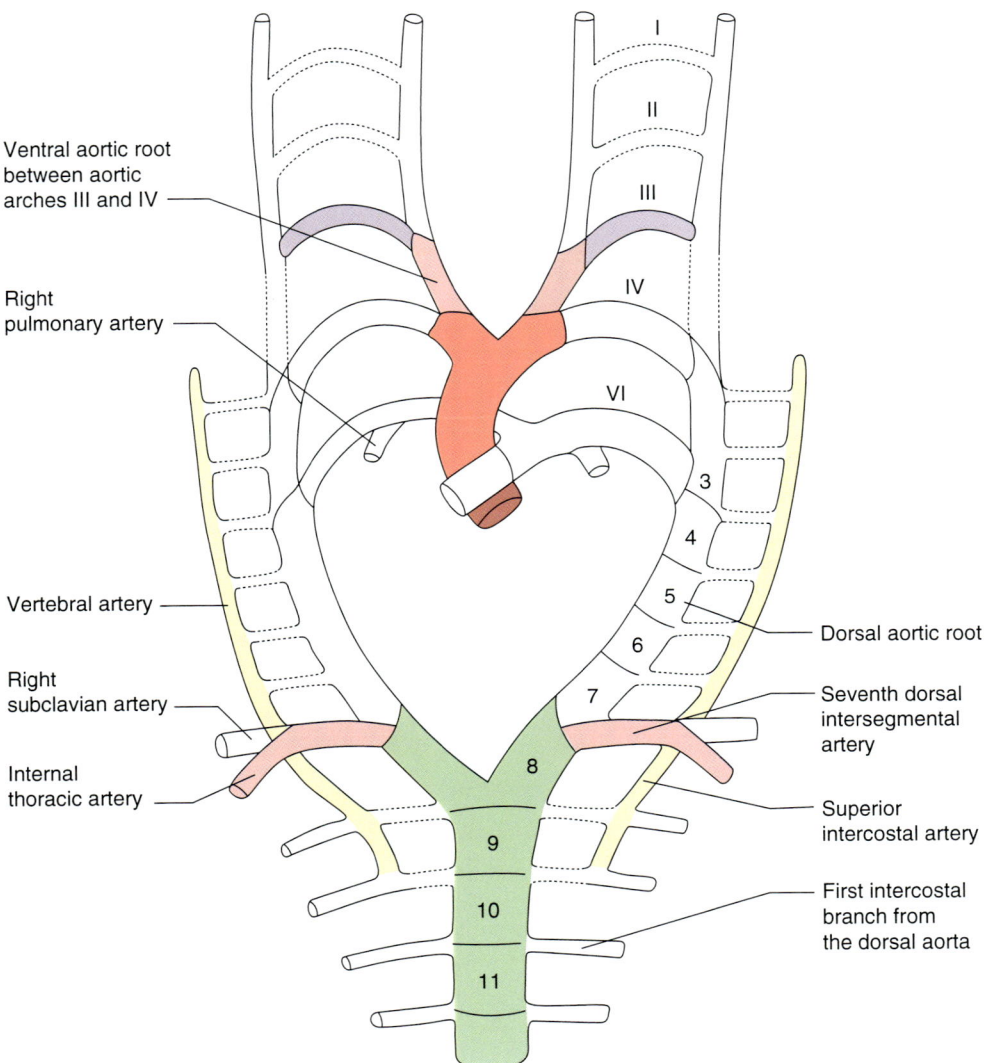

Fig. 7.2 Aortic arches at initial formation. This diagram is very accurate and, given the neuromeric model, has a great deal to teach us Proatlas (S4–S5) is vertebra C0 and the transverse segment of vertebral artery running between it and atlas is the *analog of intersegmental artery*. Let's call it ISA1. ICA0 supplies somites S4 and S5 and therefore gives off meningeal branch into foramen magnum. First ISA beneath atlas C1 supplies S5–S6. Second ISA beneath axis C2 supplies S6–S7. Third ISA beneath C3 supplies S7–S8. Fourth ISA beneath C4 supplies S8–S9. Fifth ISA beneath C5 supplies S9–S10. Sixth ISA beneath C6 supplies S10–S11. Seventh ISA (subclavian) beneath C7 supplies S11–S12. Eighth ISA (via supreme thoracic) in the first intercostal space supplies S12–S13. Ninth ICS (via supreme thoracic) in the second intercostal space supplies S13–S14. Tenth ICA beneath T (third intercostal space) supplies S14–S15 and is called the first intercostal artery. Important: The diagram ignores proatlas therefore segment 1 is really segment 2. Third thoracic segment is really segment 11 (8 + 3). (Reprinted from Lewis, Warren H (ed). Gray's Anatomy of the Human Body, 20th American Edition. Philadelphia, PA: Lea & Febiger, 1918)

Fig. 7.3 Connecting segment between dorsal aorta and left pulmonary plexus is the ductus arteriosus. Left brachiocephalic trunk (brown) is being absorbed into the aortic root. (Reprinted from Lewis, Warren H (ed). Gray's Anatomy of the Human Body, 20th American Edition. Philadelphia, PA: Lea & Febiger, 1918)

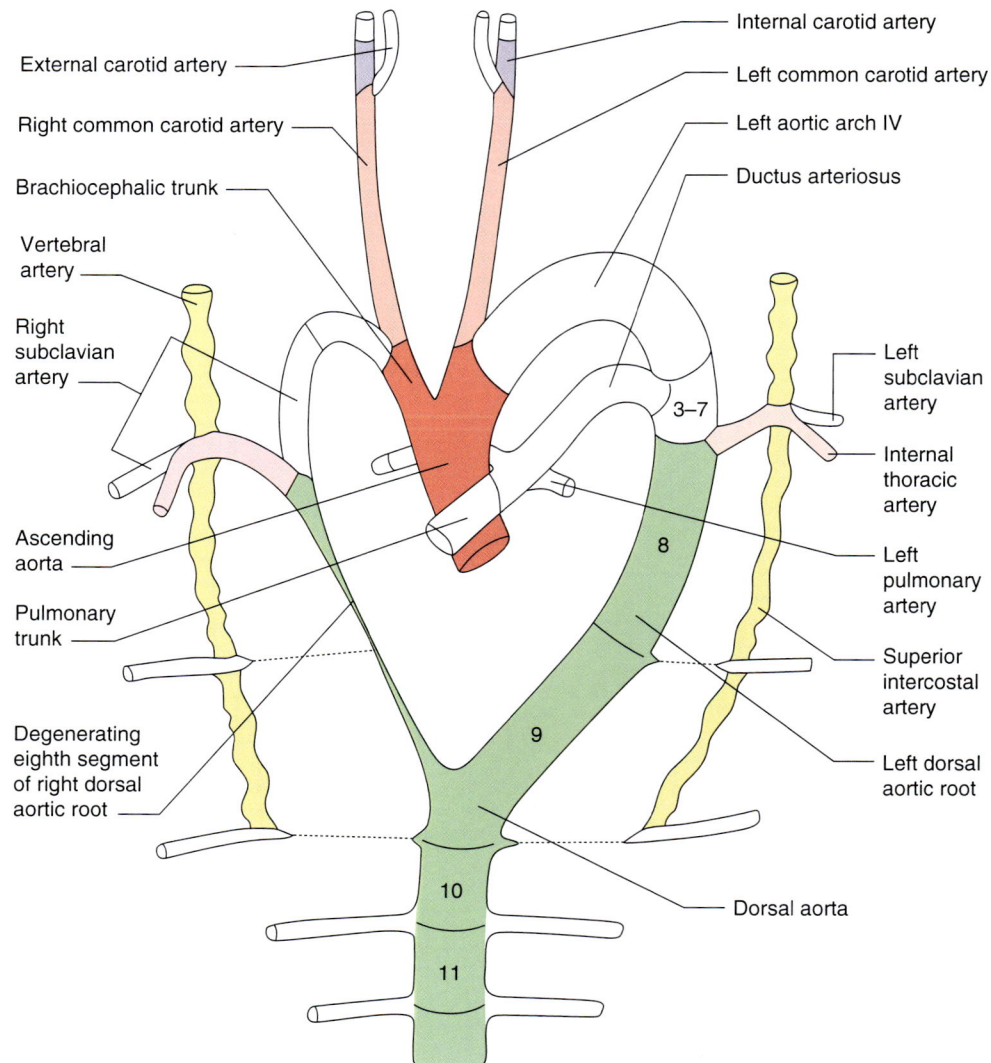

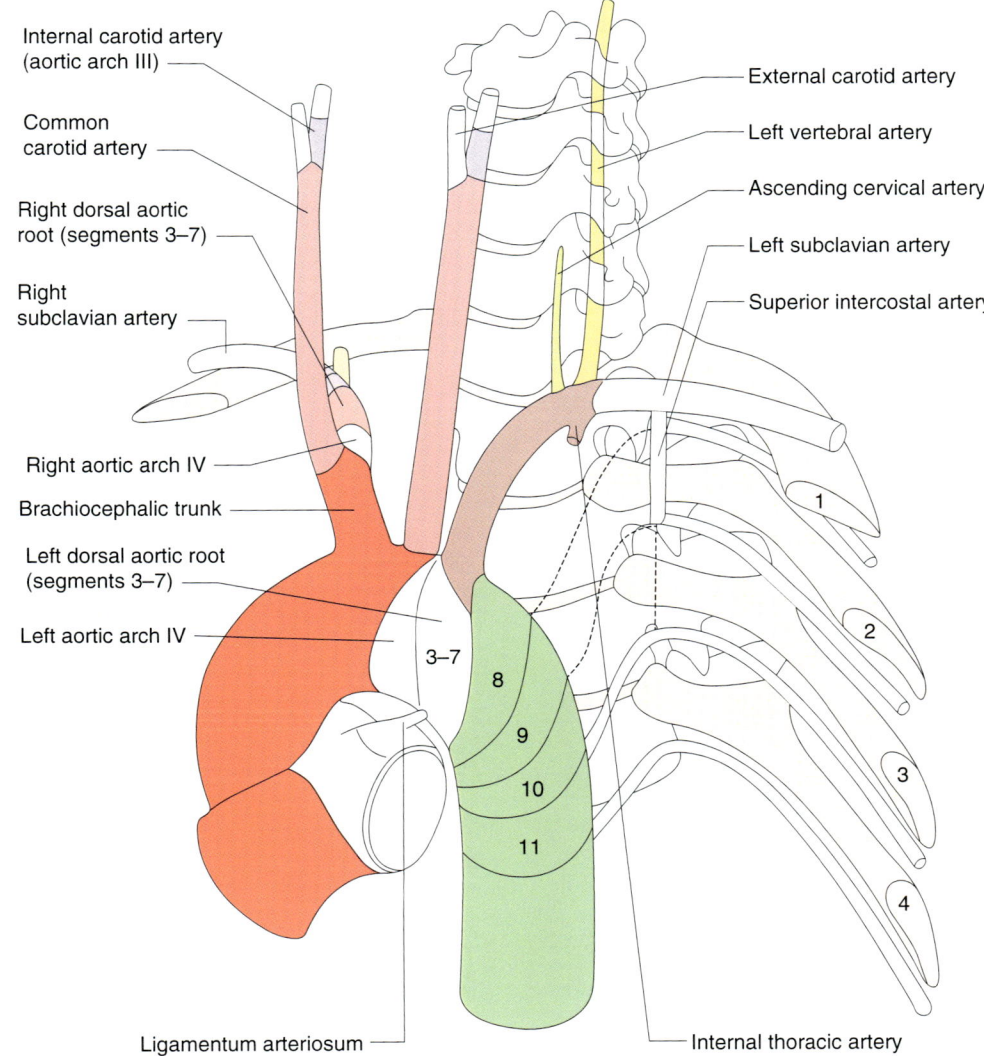

Fig. 7.4 Cervical segments are located in two different sites. On the right side, they are an independent segment located between brachiocephalic and the take-off of the subclavian. On the left side, they are fully incorporated into the wall of the aortic outflow tract just prior to take-off of the subclavian. (Reprinted from Lewis, Warren H (ed). Gray's Anatomy of the Human Body, 20th American Edition. Philadelphia, PA: Lea & Febiger, 1918)

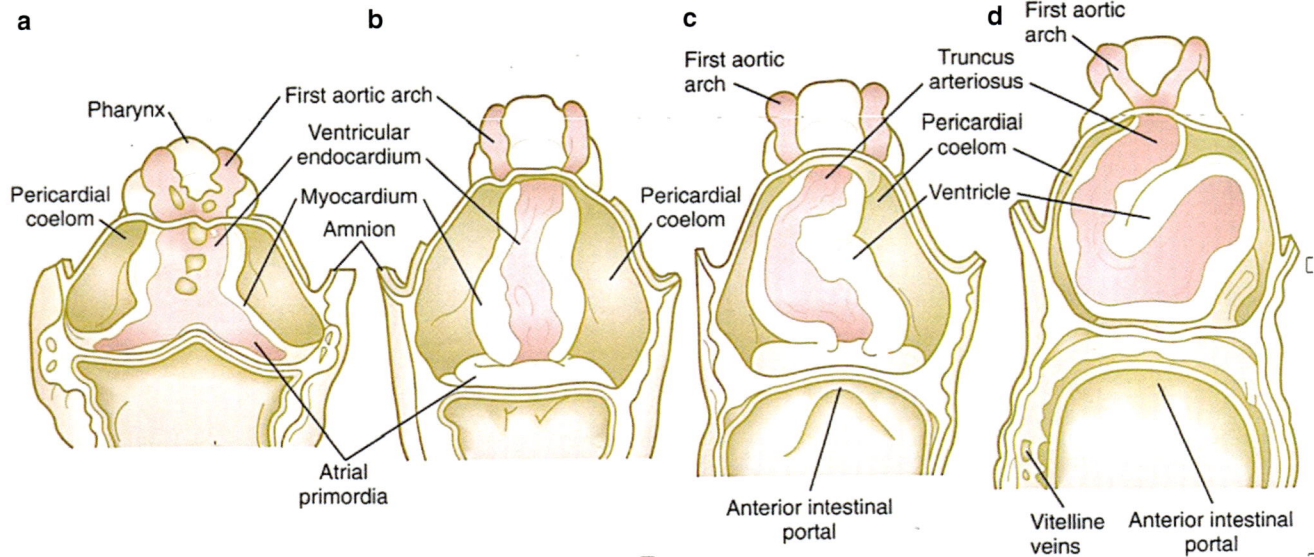

Fig. 7.5 Folding of the heart. Creates first aortic arches as the connecting segment between dorsal aortae and the outflow tract. First aortic arches in stage 9 are in continuity with the paired heart tubes. By the end of stage 9, these have fused, folded into an S-shaped configuration and are tuck directly beneath the pharynx, from which position, second aortic arches will have direct access the dorsal aortae. (Courtesy of Bruce M. Carlson, MD, PhD)

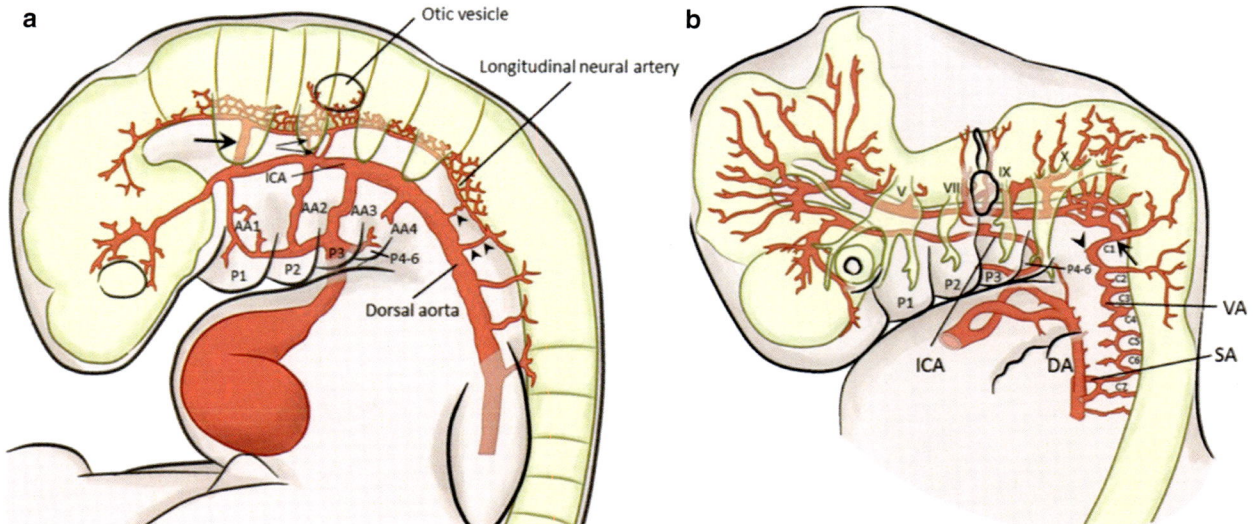

Fig. 7.6 Schematic of aortic arch system and the systemic circulation. Vessels in an embryo of 4 mm. The primitive trigeminal artery (arrow) is arising from the primitive internal carotid artery (ICA). A few primitive otic arteries arise opposite to the second aortic arch (small arrows). The caudal end of the hindbrain is supplied by the primitive hypoglossal (arrowhead) and proatlantal intersegmental (double arrowheads) arteries. The primitive carotid-vertebrobasilar anastomoses communicate the primitive ICA and dorsal aorta with the longitudinal neural artery and accompany their corresponding nerves. *AA* aortic arch, *P* pharyngeal arch. Vessels in an embryo of 7–12 mm. Fusion of the first to sixth arteries (C1–C7) results in the formation of the vertebral artery. Note that the distal portion of the proatlantal intersegmental artery PIA remains as the horizontal portion of the vertebral artery (arrow). Persistence of the proximal portion of the PIA results in the persistent PIA seen in adults (arrowhead). The muscular components of each pharyngeal arc are associated with a single cranial nerve. *DA* dorsal aorta, *ICA* internal carotid artery, *P* pharyngeal arch, *SA* subclavian artery, *VA* vertebral artery. (Reprinted from Namba K. Carotid-vertebro-basilar anastomoses with reference to their segmental properties. Neurol Med Chir (Tokyo) 2017; 57(6): 267–277. With permission from Creative Commons License 4.0: https://creativecommons.org/licenses/by-nc-nd/4.0/)

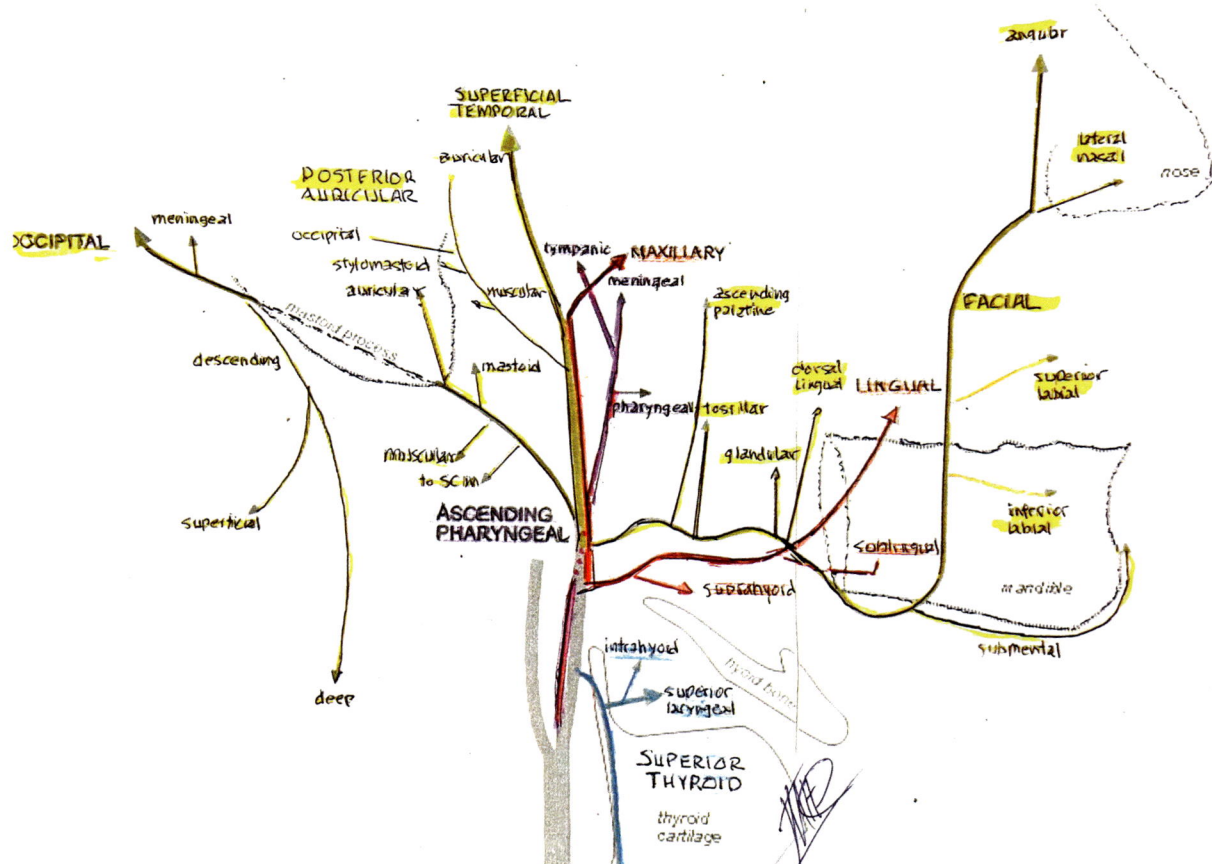

Fig. 7.7 Organization of the external carotid system. (Courtesy of Michael Carstens, MD)

Fig. 7.8 Fourth arch superior thyroid. STA is the first branch from the external carotid artery and represents the fourth and fourth arch components of ECA. Derivatives of the fourth arch include thyroid cartilage and muscles; it extends to the cricothyroid membrane. Derivatives of fifth arch (r10–r11) are, putatively, cricoid cartilage and inferior constrictor so superior thyroid supplies these zones. (Reprinted from Lewis, Warren H (ed). Gray's Anatomy of the Human Body, 20th American Edition. Philadelphia, PA: Lea & Febiger, 1918)

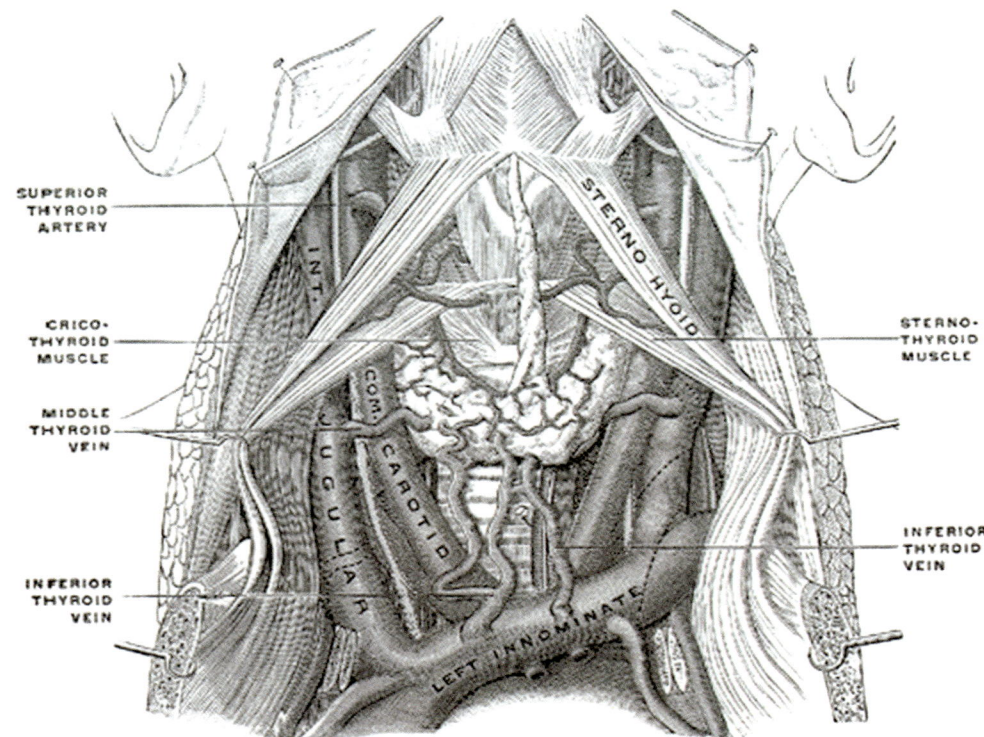

Fig. 7.9 Fourth arch superior thyroid, schematic. It extends to the cricothyroid membrane. Derivatives of fifth arch (r10–r11) are, putatively, cricoid cartilage and inferior constrictor so superior thyroid supplies these zones. Left lobe of thyroid gland (orange) showing two branches of superior thyroid. In 3–10% of the population, the thyroid ima artery ascends to supply the lower gland. It originates in the mediastinum, usually from brachiocephalic, and ascends in front of the trachea, usually accompanied by a rudimentary inferior thyroid artery. If the ima is sectioned at thyroid surgery, bleeding is profuse and the vessel can retract into the chest. STA has the following branches:

- Infrahyoid: running along the lower margin of hyoid at the boundary of rr7–r8
- Superior laryngeal pierces the thyrohyoid membrane to enter the larynx
- Sternomastoid: supplies the upper one-third of SCM (myoblasts in register with r8–r1)
- Cricothyroid: runs transversely across the membrane
- Anterior superior thyroid: supplies trachea (thyroidea ima a.)
- Posterior superior thyroid: terminal axis of STA; contacts inferior thyroid artery
(Courtesy of Michael Carstens, MD)

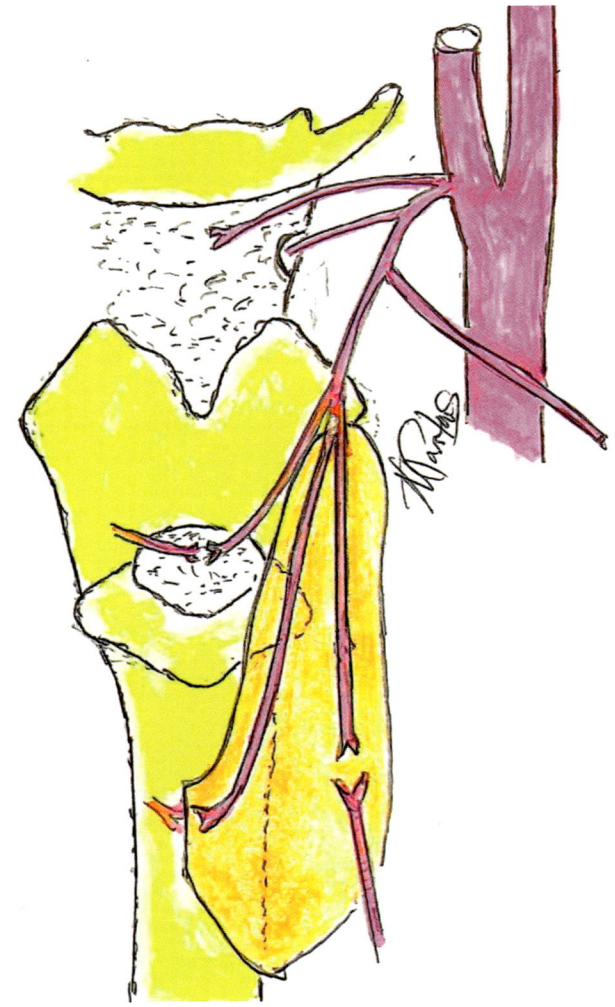

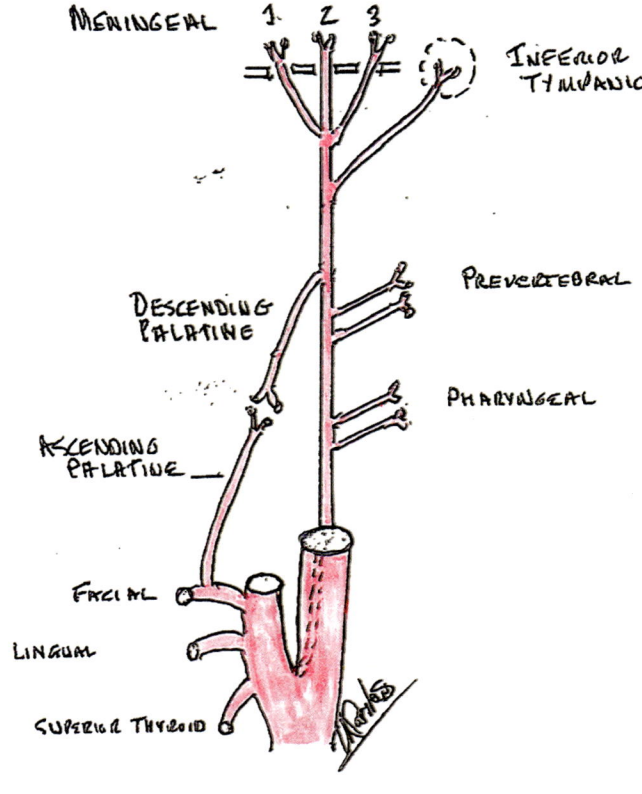

Fig. 7.10 Third arch ascending pharyngeal has a **pharyngeal trunk** and a **neuro-meningeal trunk**. The stem of ascending pharyngeal comes off the internal aspect of external carotid, just above the stem for the inferior thyroid. The pharyngeal trunk supplies two groups of muscles: prevertebral and the pharyngeal constrictors. The neuromeningeal trunk has three branches: (1) clival, (2) jugular, (3) hypoglossal. (Courtesy of Michael Carstens, MD)

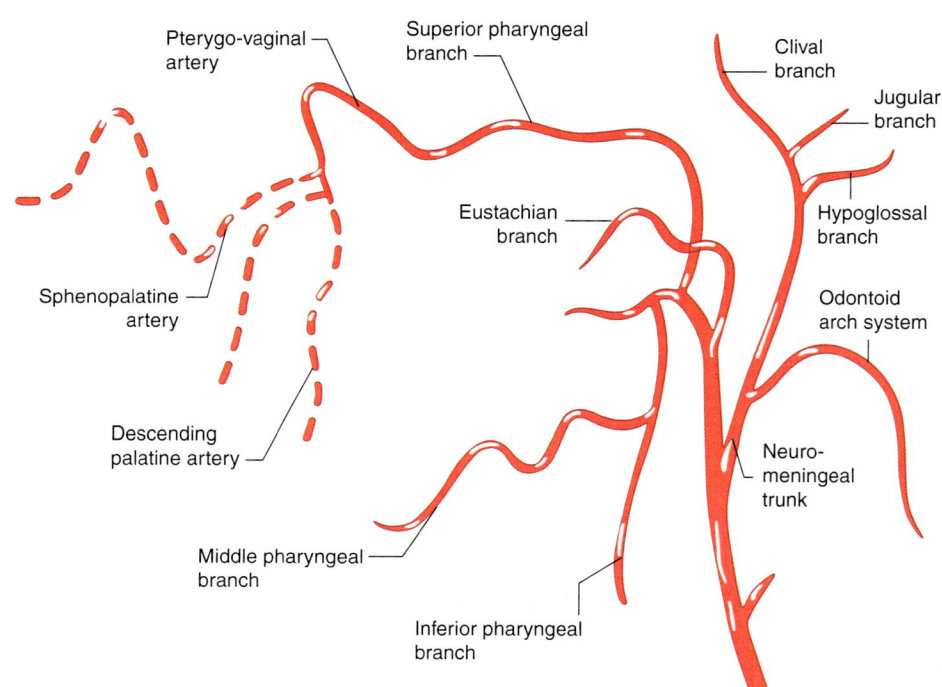

Fig. 7.11 Third arch ascending pharyngeal makes anastomoses with first arch V3 stapedial at the tympanic membrane and with V2 stapedial in the substance of the palate

- Ascending pharyngeal artery
- Posterior pharyngeal branches to the pharyngeal wall
- Posterior muscular branches to prevertebral muscles; anastomose with branches of ascending cervical and vertebral arteries
- Inferior tympanic supplies the middle ear
- Meningeal branches pass through: foramen lacerum, jugular foramen, hypoglossal canal

- Descending pharyngeal branch to the soft palate; anastomoses with asc. pharyngeal of facial
- Ascending pharyngeal branch from the facial artery
- Lingual artery
- Superior thyroid artery

(Reprinted from Spiotta AM, Hughes G, Masaryk T J, Hui FK. Balloon-augmented Onyx embolization of a dural arteriovenous fistula arising from the neuromeningeal trunk of the ascending pharyngeal artery: technical report. J Neurointerven Surgery 2011; 3(3):300–303. With permission from BMJ Publishing, Ltd.)

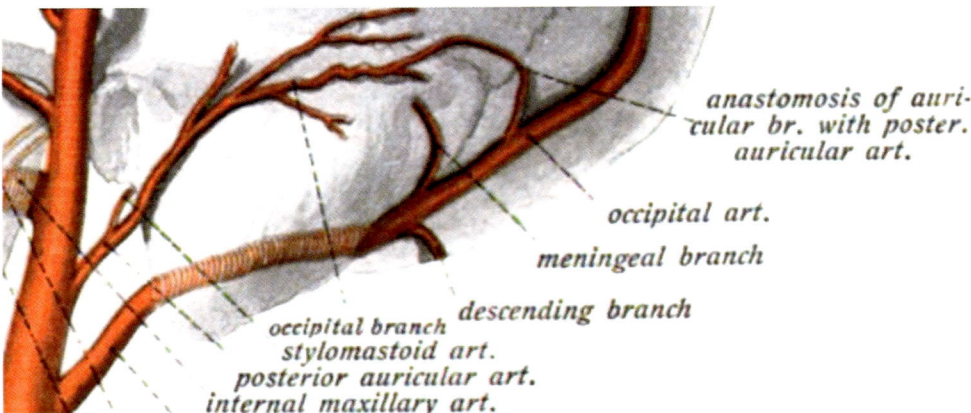

Fig. 7.12 Posterior auricular artery is dedicated to the second arch. Note that it gives off two muscular branches: internally to stylomastoid branch and externally to the second arch muscles on the back of the ear. Note anastomosis with occipital artery which serves the second arch occipitalis muscle. Branches of external carotid artery. Second arch occipital artery supplies the SMAS of the posterior scalp. Provides communicating meningeal branches to underlying V3 stapedial system of the dura. Note connection with stylomastoid artery. (Reprinted from Dr. Johannes Sobotta (1909). Updated nomenclature by Mikael Häggström—Original publication: Atlas and Textbook of Human Anatomy Volume III Vascular System, Lymphatic system, Nervous system and Sense Organs. 1909)

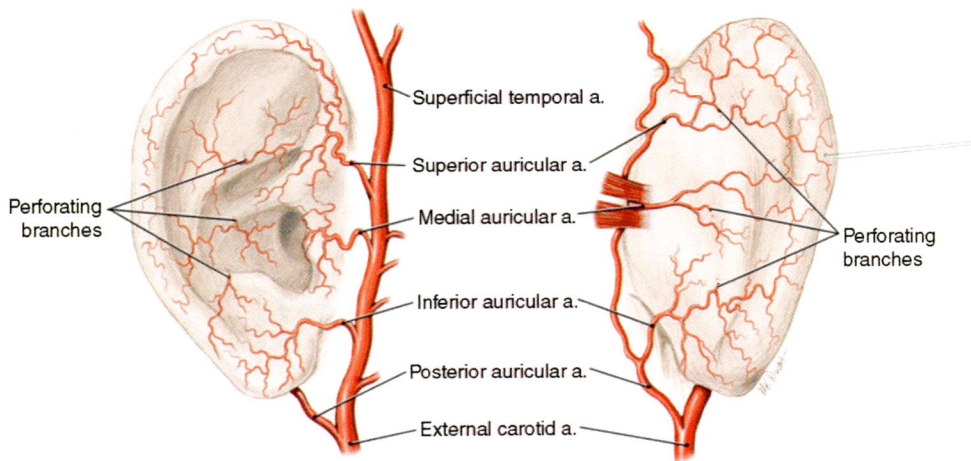

Fig. 7.13 Second arch posterior auricular is in register with second arch and supplies the muscles, both posteriorly and, via penetrating branches, to intrinsic muscles on the anterior surface of the ear. Note that second arch posterior auricular and first arch anterior auricular anastomose over the anterior surface of the pinna. Because the muscles are second arch derivatives anterior auricular supplies skin only. Perforating branches from the posterior auricular are dedicated to intrinsic muscles of SMAS system. (Reprinted from Prendergast PM. Anatomy of the External Ear. In: Shiffman M. (eds) Advanced Cosmetic Otoplasty. Berlin, Heidelberg: Springer; 2013: 5–21. With permission from Springer Nature)

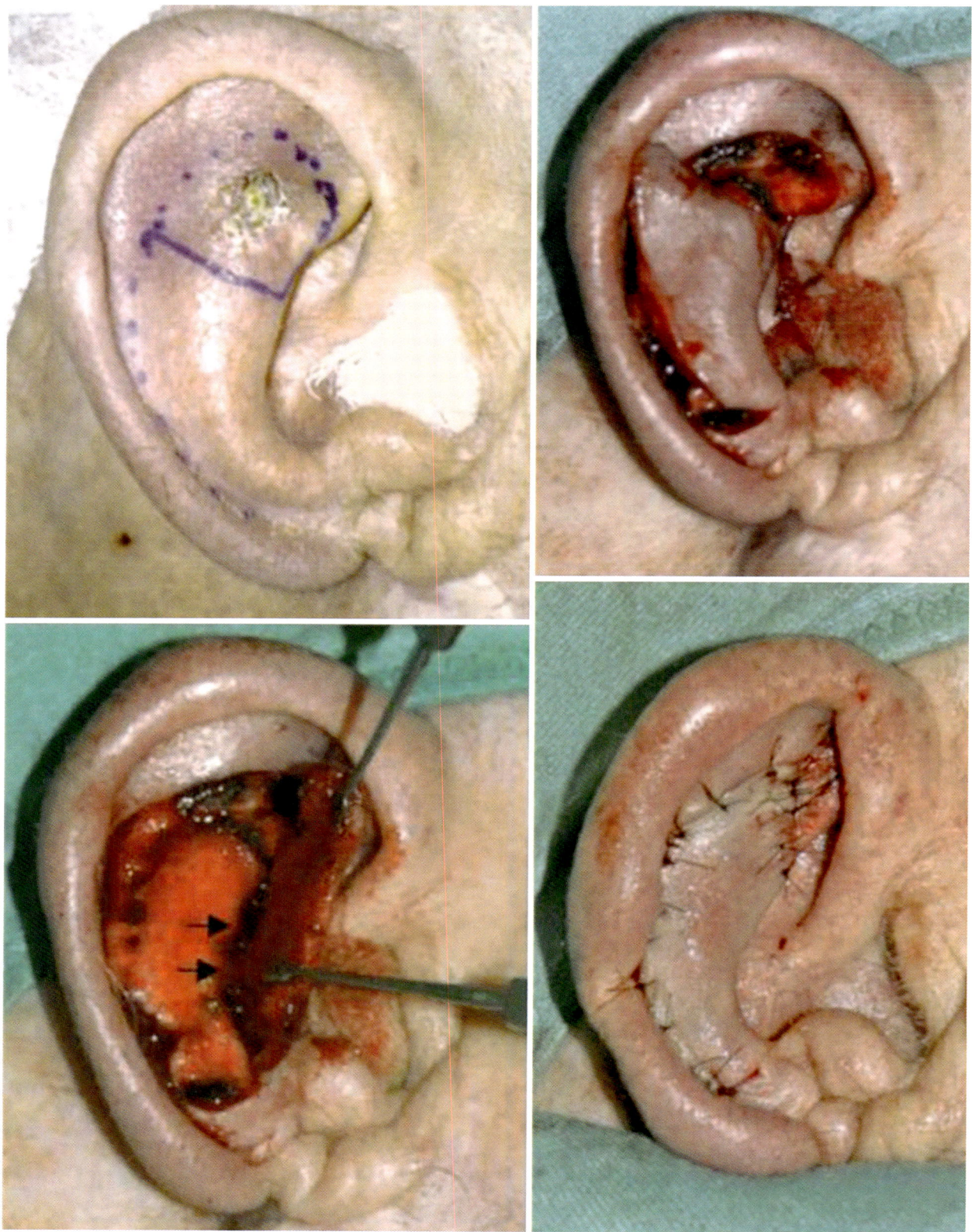

Fig. 7.14 Ear reconstruction using the posterior auricular artery. Perforating branches from the posterior auricular artery power skin of the external (anterior) pinna overlying the anti-helix. (Upper left): Resection defect marked. Reconstructive flap based on remaining skin over lower half of the pinna which is (upper right) elevated away from the anti-helix. (Lower left) The flap is reflected forward to demonstrate two branches from the posterior auricular artery penetrating externally (arrows) to sup- ply the skin flap. (Lower right) The skin flap is now rotated into position, thus completing the reconstruction. (Reprinted from Bacarrani A, Pedone A, Petrella, et al. A new approach in the management of triangular fossa auricular defects: the posterior auricular artery perforator antihelix-conchal flap. Open Reconstructive and Cosmetic Surgery 2010; 3(1): 17–20. With permission from Creative Commons License 3.0: https://creative-commons.org/licenses/by-nc/3.0/)

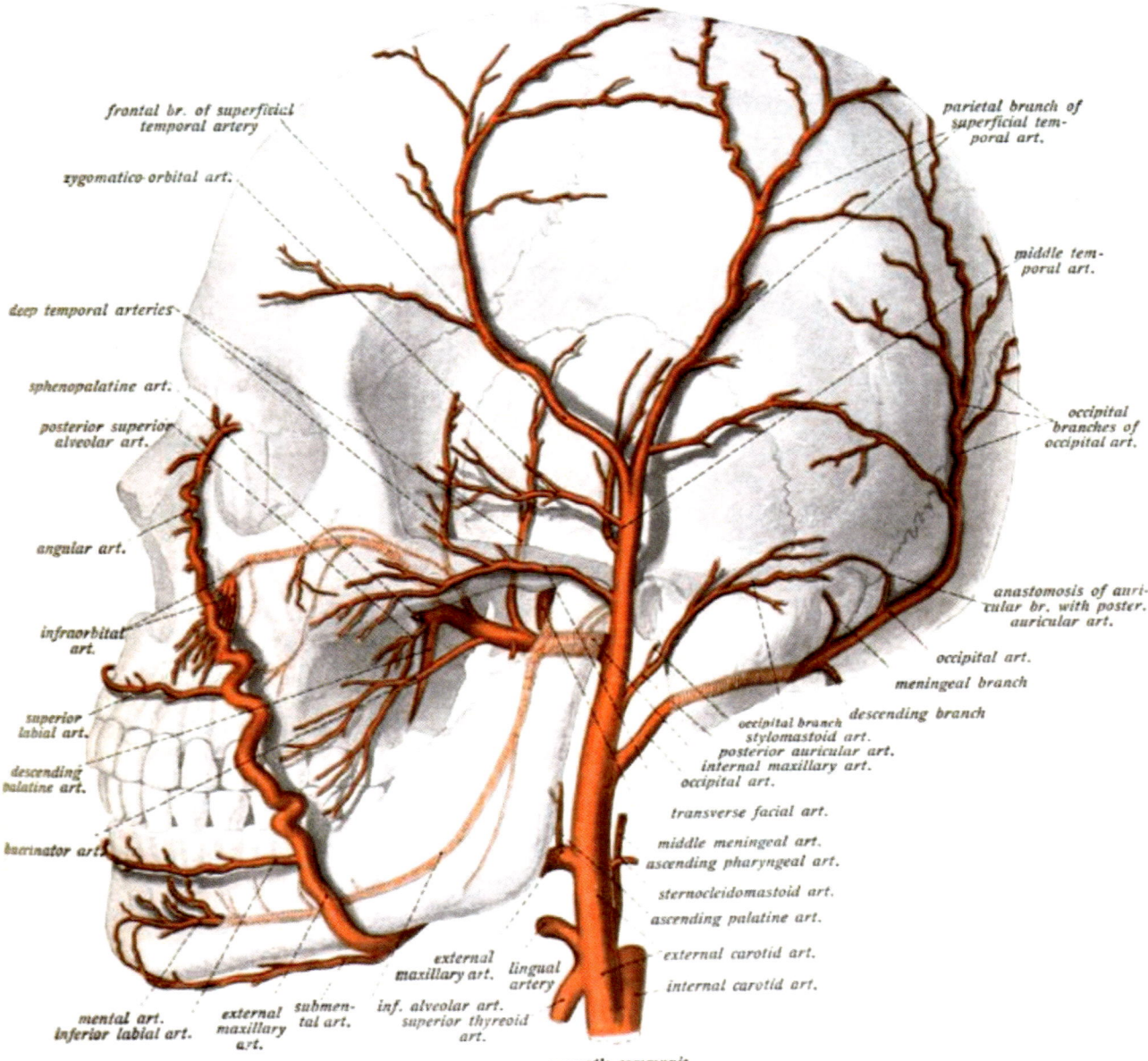

Fig. 7.15 Second arch facial artery supplies the SMAS of the face and fronto-temporo-parietal scalp—It supports the frontalis. Note parietal branch anastomosis with occipital (not shown). Supports scalp over the membranous bones (squamous temporal, frontal, parietal). (Reprinted from Dr. Johannes Sobotta (1909). Updated nomenclature by Mikael Häggström—Original publication: Atlas and Textbook of Human Anatomy Volume III Vascular System, Lymphatic system, Nervous system and Sense Organs. 1909)

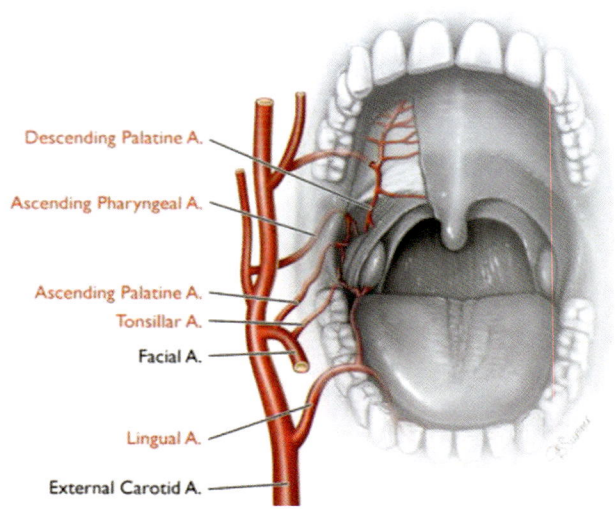

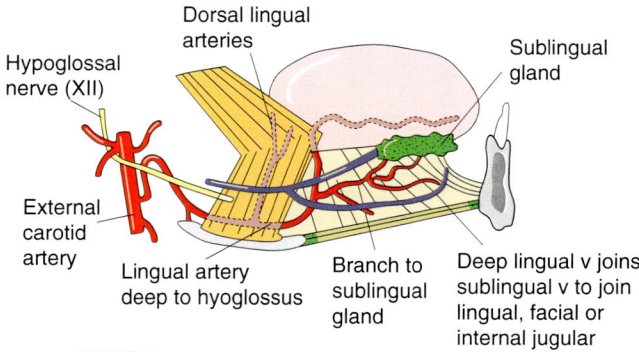

TONGUE - LINGUAL ARTERY

LYMPH
- Tip to submental gland bilaterally
- Dorsum to submandibular mostly unilaterally
- Posterior to jugulo-omohyoid & deep cervical

Fig. 7.16 Second arch tonsillar artery demonstrating the tonsillar fossa as a confluence of first, second, and third arches. (Reprinted from Castagnin LA, Goyal M, Ongkasuwan J. Tonsillitis and Peritonsillar Abscess. In: Valdez T, Vallejo J. (eds). Infectious Diseases in Pediatric Otolaryngology. Cham, Switzerland: Springer International Publishing: 137–150. With permission from Springer Nature)

Fig. 7.17 Second arch lingual supplies muscles of mastication and the deep part of tongue that represents the second arch component that is subsumed by first and third arches. Glands are innervated by PANS from r4 to r5. (Courtesy of Instant Anatomy (www.instantanatomy.net) Retrieved from: http://www.instantanatomy.net/diagrams/HN086b.png)

Fig. 7.18 Second arch lingual. The tongue has three neuromeric zones defined by sensory mucosa. First arch is innervated by V3 and supplied by profundal linguae. Second arch is innervated by VII, contains the circumvallate papillae, and is supplied by anterior branch of dorsal lingual. Third arch base of tongue is innervated by IX and supplied by the posterior branch of dorsal lingual and tonsillar branch of the facial. (Reprinted from Lewis, Warren H (ed). Gray's Anatomy of the Human Body, 20th American Edition. Philadelphia, PA: Lea & Febiger, 1918)

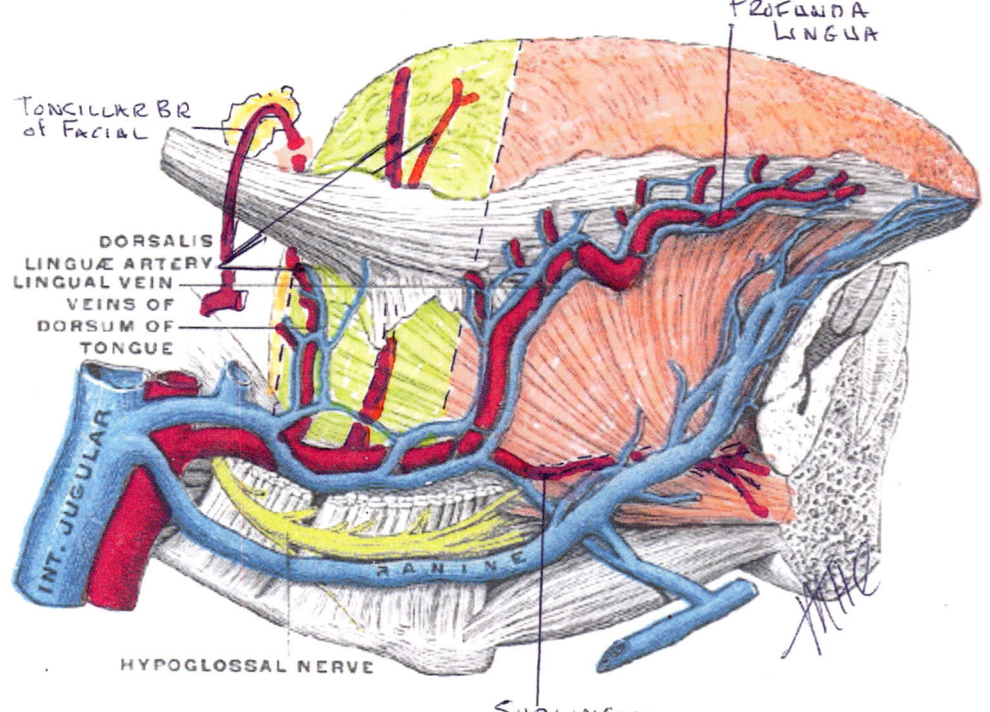

Fig. 7.19 Superficial temporal. Note parietal branch anastomosis with occipital (not shown). Supports scalp over the membranous bones (squamous temporal, frontal, parietal). (Reprinted from Barral JP, Croibier A. Superficial Temporal Artery. In: Visceral Vascular Manipulations. Philadelphia, PA: Churchill-Livingstone; 2011: 1556–1559. With permission from Elsevier)

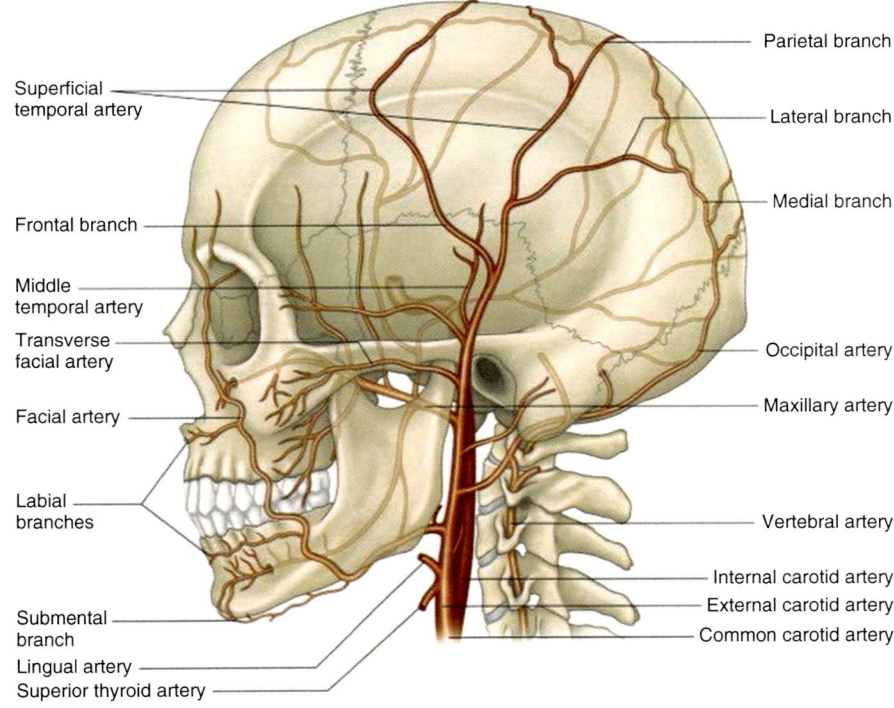

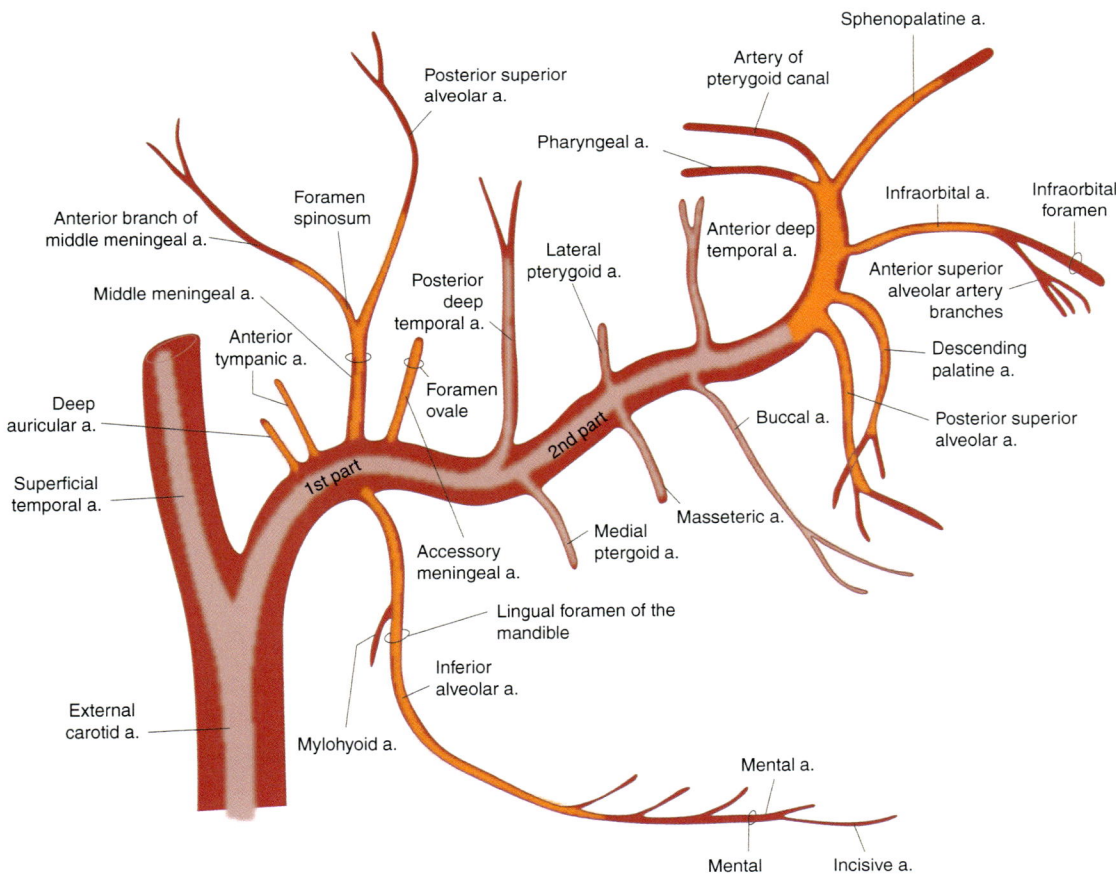

Fig. 7.20 First arch Internal maxillary. Blood supply to the teeth is sharply defined. Mandibular teeth are supplied by posterior-most branch of internal maxillary, the inferior alveolar artery, indicating its biologically "older status" with development of the posterior zone of the first arch preceding that of the anterior zone. Complexities of maxil-lary teeth require major source of the dental arcade supplied by superior alveolar artery with special supply to the premaxillary teeth from the nasopalatine. (Reprinted from Lewis, Warren H (ed). Gray's Anatomy of the Human Body, 20th American Edition. Philadelphia, PA: Lea & Febiger, 1918)

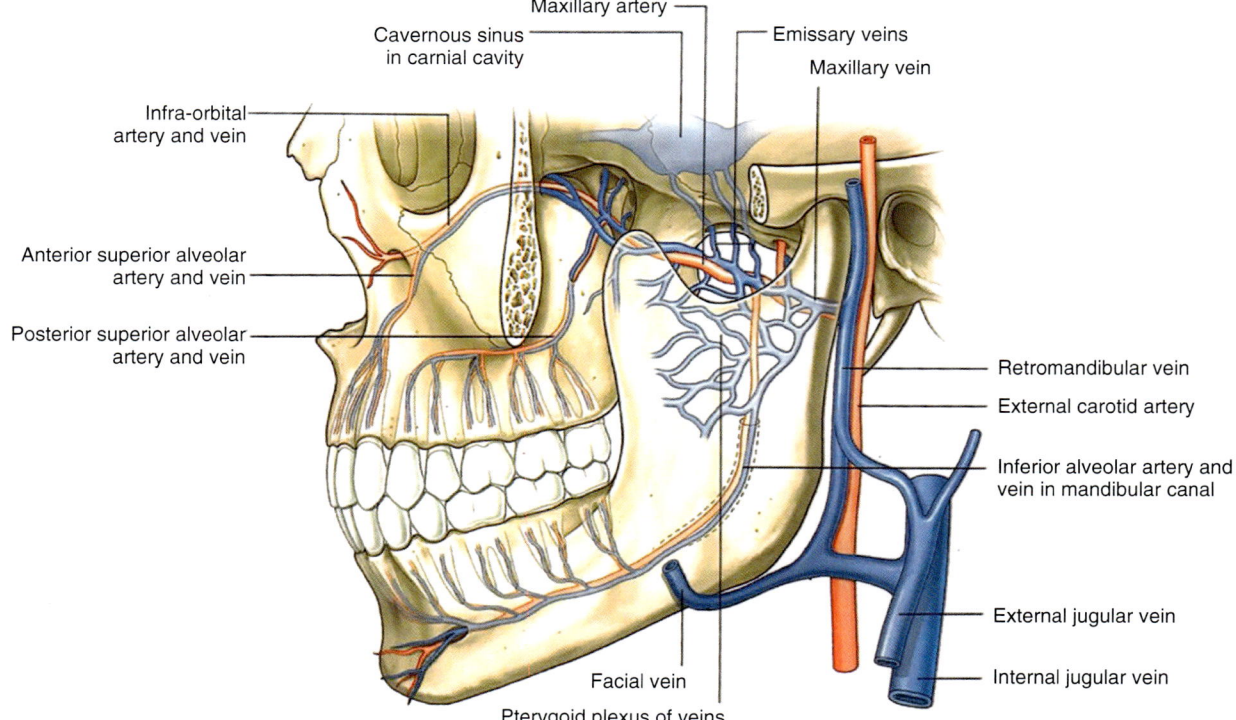

Fig. 7.21 Blood supply to the teeth is sharply defined. Mandibular teeth are supplied by posterior-most branch of internal maxillary, the inferior alveolar artery, indicating its biologically "older status" with development of the posterior zone of the first arch preceding that of the anterior zone. Complexities of maxillary teeth require major source of the dental arcade supplied by superior alveolar artery with special supply to the premaxillary teeth from the nasopalatine. Anterior superior alveolar has two branches: the medial branch supplies the maxillary incisor and canine and the lateral branch of anterior superior alveolar supplies the bicuspids. Note as well that the sensory nerves to the individual teeth and lingual gingiva are given off before the superior alveolar nerve exits the infraorbital foramen. The buccal lamina of gingival tissue is supplied by StV2 after its exit from the foramen. Thus, infraorbital blocks are not sufficient to anesthetize the maxillary teeth. (Reprinted from Drake RL, Vogl AW, Mitchell AWM. Gray's Anatomy for Students, 3rd ed. Philadelphia, PA: Churchill Livingstone, 2015. With permission from Elsevier)

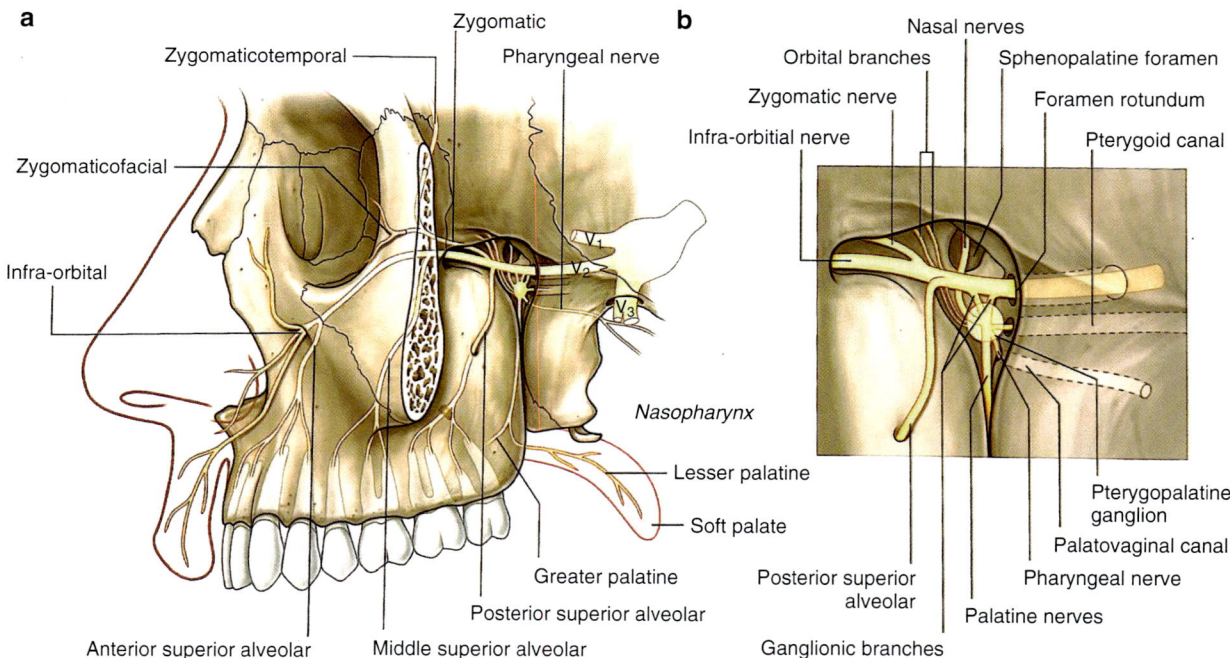

Fig. 7.22 Sphenopalatine fossa—organization. Schematic of arterial branches at the sphenopalatine (pterygopalatine) fossa. Diagram is incomplete as it does not show anterior superior alveolar (infraorbital) artery branches nor the zygomatic artery. (Reprinted from Drake RL, Vogl AW, Mitchell AWM. Gray's Anatomy for Students, 3rd ed. Philadelphia, PA: Churchill Livingstone, 2015. With permission from Elsevier)

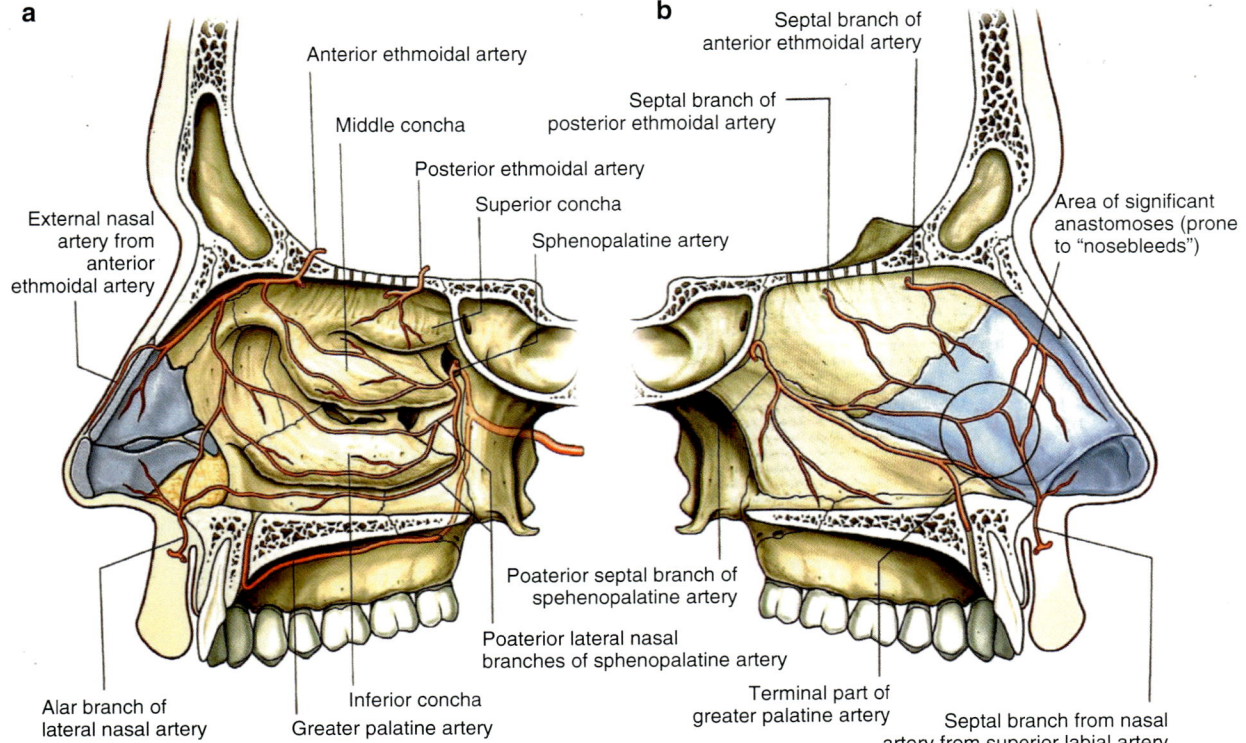

a

External nasal artery from anterior ethmoidal artery

Anterior ethmoidal artery

Middle concha

Posterior ethmoidal artery

Superior concha

Sphenopalatine artery

Alar branch of lateral nasal artery

Inferior concha

Greater palatine artery

Poaterior septal branch of spehenopalatine artery

Poaterior lateral nasal branches of sphenopalatine artery

b

Septal branch of anterior ethmoidal artery

Septal branch of posterior ethmoidal artery

Area of significant anastomoses (prone to "nosebleeds")

Terminal part of greater palatine artery

Septal branch from nasal artery from superior labial artery

Fig. 7.23 Nasopalatine is distributed to extracranial zones 1 and 2. It is analogous to the anterior and posterior ethmoid axis which supply intracranial zones 13 and 12. A common lesion in the stapedial system can affect structures on either side of the skull. (Reprinted from Drake RL, Vogl AW, Mitchell AWM. Gray's Anatomy for Students, 3rd ed. Philadelphia, PA: Churchill Livingstone, 2015. With permission from Elsevier)

Timeline of the External Carotid System

Stage 7: gastrulation produces mesodermal substrate

Stage 8: primitive head plexus, primitive hindbrain channels, dorsal aortae

Stage 9: first aortic arch, embryonic flexion

Stage 10: second aortic arch

Stage 11: third aortic arch

Stage 12: fourth aortic arch

Stage 13: fifth aortic arch—fails, ventral pharyngeal artery stem

Stage 14: pulmonary plexus (sixth aortic arch), definitive common and internal carotid

Stage 15: end of aortic arch period, VPA arrives at V3

Stage 16: branches of ECA forming, anastomosis with extracranial stapedial

Stage 17: external carotid system complete, StV3 middle meningeal intracranial

Aortic Arch Arteries: Precursors of Two Systems

(Figures 7.1, 7.2, 7.3, and 7.4)

Human embryos possess five pharyngeal arches (PAs) with which craniofacial structures are assembled. Running through the core of each arch is a vascular axis, *aortic arch artery* (AA), that spans from the cardiac outflow tract tucked below the pharynx upward to paired dorsal aortae. The first four pharyngeal arches are important. PA5 is diminutive, producing the arytenoids, cricoids, and associated muscles. For this reason, the fifth aortic arch artery involutes almost immediately, making PA5 dependent upon the blood supply of PA4 for its survival. *There is no sixth pharyngeal arch.* Instead, the fate of the sixth and final aortic arch artery is to form pulmonary circulation.

Note: Textbook illustrations of the aortic arches are misleading. We can easily understand how they work by means of a little *embryonic origami*. Fact: the first aortic arch artery does *not* "sprout forth" from the cardiac outflow tract. At stage 8, the embryo is still a flat trilaminar disk. The embryonic heart field is located in front of the future brain and is connected to the two dorsal aortae that run passively backward along the length of the embryo. At stage 9, explosive growth of the forebrain forces the embryonic head to flex downward almost 150°. The hinge point, located at the midbrain, is called the *mesencephalic flexure*. The heart is now positioned under the neuraxis, specifically, directly below posterior pharynx.

Because the dorsal aortae are attached to the ventricles of the heart, in the process of folding, they become stretched downward and backward, forming bilateral arcs connecting the heart below with the embryo above. They are now termed the *first aortic arch arteries*. The remaining five aortic arch arteries develop, one per stage. Each arch represents the union of two distinct stems. From above, the dorsal aortae send out ventrally directed stems while, from below, the outflow tract produces paired dorsally directed stems. The two sets of stems unite to produce the aortic arch artery. Failure of this anastomotic process explains: (1) the failure of AA5 and (2) the breakdown and reorganization of the system.

The six aortic arches appear in craniocaudal sequence in stages 9–14. These mesodermal structures are surrounded by neural crest mesenchyme, running vertically through the core of their corresponding pharyngeal arches. The pharyngeal arch becomes readily visualized about one stage later on scanning electron microscopy. After their formation in stages 10–11, the first and second pharyngeal arches merge together in stages 12–13, thus creating the tissue masses that will complete the construction of the face. In the process, their aortic arch arteries disintegrate, leaving behind arterial remnants dangling from the dorsal aortae. *Mandibular artery*, the dorsal remnant of AA1, will become the (Vidian) artery of the pterygoid canal. *Hyoid artery*, the dorsal remnant of AA2, will become the source of the stapedial system.

At stages 11–12, the plexus of arches 1–2 is clearly connected with a stem connected to the segment of the *ventral aortic root* (VAR) that gives rise to AA3 and AA4. This stem is known as ventral pharyngeal artery (VPA). AA3 and AA4 are both plugged into the dorsal aortae by this interconnecting segment disintegrates at stage 14. By stage 18, AA3 and AA4 are separated, the former supplying the head and the latter the body. The segment of ventral aortic root that was between AA3 and AA4 remains attached to AA3, becoming the common carotid artery. AA3 now becomes the extracranial internal carotid artery. The stem of the external carotid system is located at the transition between ventral aortic root and AA3. AA4 produces paired subclavians and, on the left side, the aortic arch. The two dorsal aortae into which AA4 inserts have merged into a single descending aorta. As a result of embryonic growth and spatial rearrangement of the pharyngeal arches, the hyoid artery becomes the *first branch of the internal carotid artery* just below the otic capsule. Hyoid artery will morph into the stem of the stapedial system.

The reader will note that, at this point, the skull has not yet developed. Osteogenic mesenchyme subsequently condenses around pre-existing soft tissue structures. The various fissures and foramina of the skull are a result of this process.

Development of the Facial Fields

Refer to the cerebral series in Chap. 6, Figs. 6.50, 6.51, 6.67, 6.68, 6.72, 6.74, 6.76, and 6.78.

A Caveat

At stage 9, with the folding of the heart and the formation of the first aortic arch, the distal tip of the primitive internal carotid will become dedicated to the ventral forebrain. The very first branch seen beyond the mandibular is primitive maxillary artery. It should be stressed that this is a misnomer as this vessel never supplies maxilla. However, its area of distribution to the midbrain neural crest of the ventral forebrain is exactly the zone to which the maxillary fields of the first arch will eventually approximate and fuse.

At stage 11, at the same time as third aortic arch artery develops, first and second aortic arch arteries disintegrate and are replaced by a plexus. By at least stage 12, this plexus receives blood flow from AA3 via the *ventral pharyngeal artery* (VPA). By stage 13, this plexus includes tissues of the third and fourth pharyngeal arches. The hyoid stem is directed posteriorly. At stage 14, blood supply to the four arches is no longer plexiform and VPA is readily identifiable, extending outward from the ventral aortic root. Individual pharyngeal arches are seen on scanning EM. Cranial nerves are beginning to invade the future face.

At stage 14, the "post branchial phase," all pharyngeal arches have formed and the pulmonary precursors are present. The pharyngeal arches have undergone a topological shift. The first and second arches are amalgamated, and their backward extension buries arch 3. Arches 4 and 5 are tucked inside the pharynx (Figs. 6.52, 6.53, and 6.56). In the merger of the first three arches, second arch muscles are divided into two functional groups. The deep plane will contain as limited number of muscles of mastication working in synch with those of the first arch. A superficial plane will contain the muscles of animation. This myofascial plane flows like toothpaste into the interstices of the entire face, neck, and scalp. Note that ventral pharyngeal artery is conspicuous at this stage. Hyoid artery gives off the stapedial stem. Cranial nerve nuclei VII–X become more crowded. Chorda tympani reaches the tip of V3. VPA grows outward to meet it.

By at least stages 15–16, individual branches to the pharyngeal arches are identifiable and the modified VPA becomes external carotid. The hyoid artery-derived stapedial stem is directed upward into the mesenchyme of somitomeres 6–7 that will form the tympanic cavity. It splits into an upper cranial division and lower facial division; the latter develops more quickly and follows chorda tympanic to meet and join with the distal branch of external carotid at stage 17. The resulting maxilla-mandibular artery is a hybrid system which supplies all derivatives of the first arch. ECA branches support those structures arising evolutionarily archaic derivatives of the original first arch. Stapedial branches the evolutionarily new derivatives of the mandible and zygomaticomaxillary complex. *We shall give further consideration to the development of the stapedial system in the subsequent section.*

Organization of the External Carotid System

From the stem of external carotid, individual branches are assigned to each of the pharyngeal arches.

We consider Table 7.1 with three comments in mind. First, the second arch has the largest number of branches, befitting the extensive distribution of its mesenchyme. Second, the facial artery supplies two planes. The deep plane involves a limited number of muscles of mastication (buccinator, posterior digastric, stylohyoid, and stapedius). The superficial plane encompasses the mimetic muscle system, which extends over the entire head, face, and anterior neck. Third, the branches of the first arch have a very limited distribution, covering the muscles of mastication (all of which attach to the mandible) and the oral mucosa of the caudal half of the mouth, but this could be possibly assigned to V3 stapedial. Blood supply to the oral mucosa of the cranial half of the mouth, the zygomaticomaxillary complex, and the lower midface is all assigned to a separate system, the V2 stapedial arteries.

Table 7.1 External carotid system

Branch	Arch	Derivatives supplied
Superior thyroid	4th, 5th	Inferior pharynx, larynx
Ascend. pharyngeal	3rd	Superior pharynx, palate
Occipital	2nd	Facial muscles
Posterior auricular	2nd	Facial muscles
Superficial temporal	2nd	Facial muscles
Facial	2nd	Mastication, facial muscles
Lingual	1st	Mucosa floor of mouth and mandible
Maxillomandibular	1st	Mastication

The areas of distribution for each branch of ECA vary widely and are progressively more extensive. Fifth arch is the most diminutive; it is the putative source of the arytenoids and cricoid. It does not merit its own branch but relies on AA4. The territory of fourth arch is larger, as it supports the inferior constrictor portion of the pharynx as well as the laryngeal cartilages. Third arch is still larger, spanning from the upper two constrictors to soft palate and the hyoid apparatus. AA4 and AA3 are each supplied by but a single branch of ECA. The second arch is the most extensive of all, with derivative muscles scattered over the entire cranium supplied by four distinct branches of ECA. The first arch is supplied by only two branches of ECA (lingual and maxillomandibular) but its vascular is the most complex of all the arches. Muscles of mastication are supplied by lingual and the ECA-derived branches of maxillomandibular artery. Bones are supplied by the extracranial stapedial system. The skin cover of the first arch is supplied by second arch branches of the facial and superficial temporal.

Stapedial System: Face

(Figures 7.24, 7.25, 7.26, 7.27, 7.28, 7.29, 7.30, 7.31, 7.32 and 7.33 (orbital), 7.34, 7.35, 7.36, 7.37, 7.38, 7.39, 7.40, 7.41, 7.42, 7.43, 7.44, 7.45, 7.46, 7.47, 7.48, 7.49, and 7.50 (stapedial system))

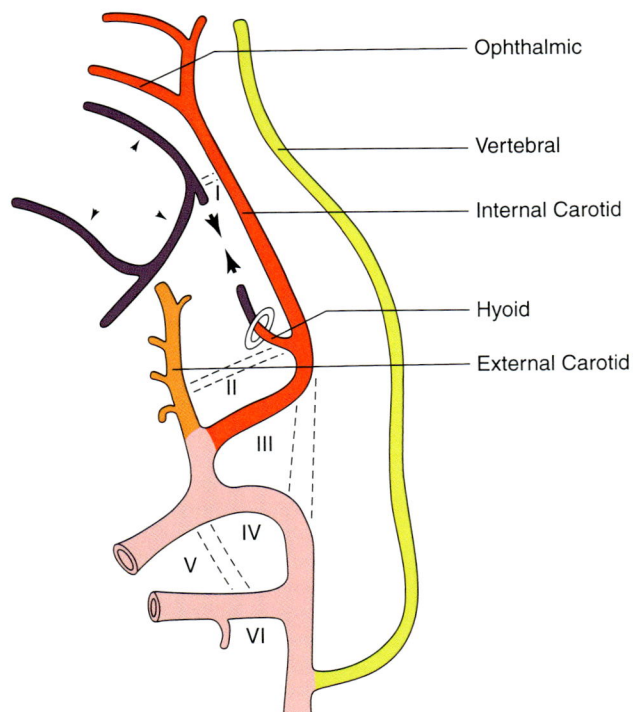

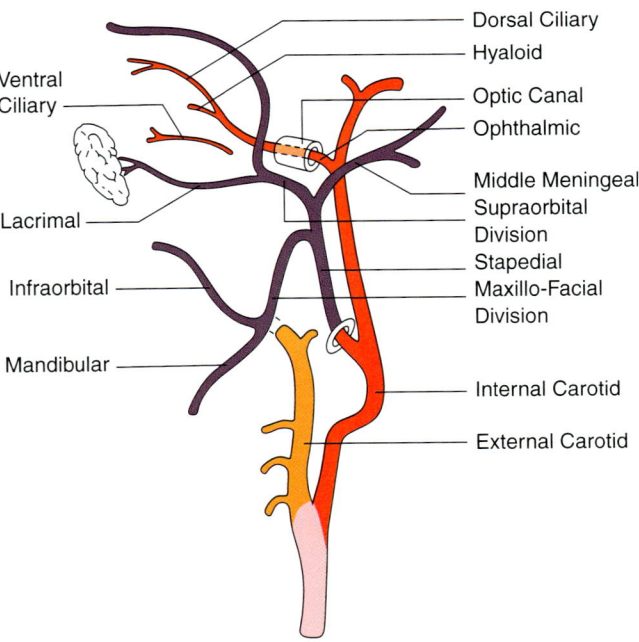

Fig. 7.24 Arteries of the eye and orbit—showing the standard insertion of stapedial into the lateral margin of ophthalmic (via the dorsal arch of the arterial arcade surrounding optic nerve. (Reprinted from O'Rahilly R, Müller F. The Embryonic Human Brain: An Atlas of Developmental Stages, 3rd ed. Wiley-Liss; 2006. With permission from John Wiley & Sons, Inc.)

Fig. 7.25 Organization of arterial supply to the eye and orbit. Vessels of ICS origin (pink) appear in the following order, proximal to distal along the ophthalmic stem: central retinal artery, lateral posterior ciliary, medial posterior ciliary. The order reflects relative order of their appearance during development with CRA as the plexus at stage 12, primitive DOA at stage 14, and primitive VOA at stage 15. Ophthalmic branches from ICA (pink) are jointed branches from stapedial (orange). The remnant of the StV2–StV1 supraorbital arc becomes lacrimal. Stapedial bifurcates into frontal and nasociliary axis. The latter is programmed by two nerves, nasociliary and (at its extreme distal margin) medial branch of frontal. (Reprinted from Oyster C. The Human: Structure and Function. Sinauer Associates 1999. With permission from Oxford University Press)

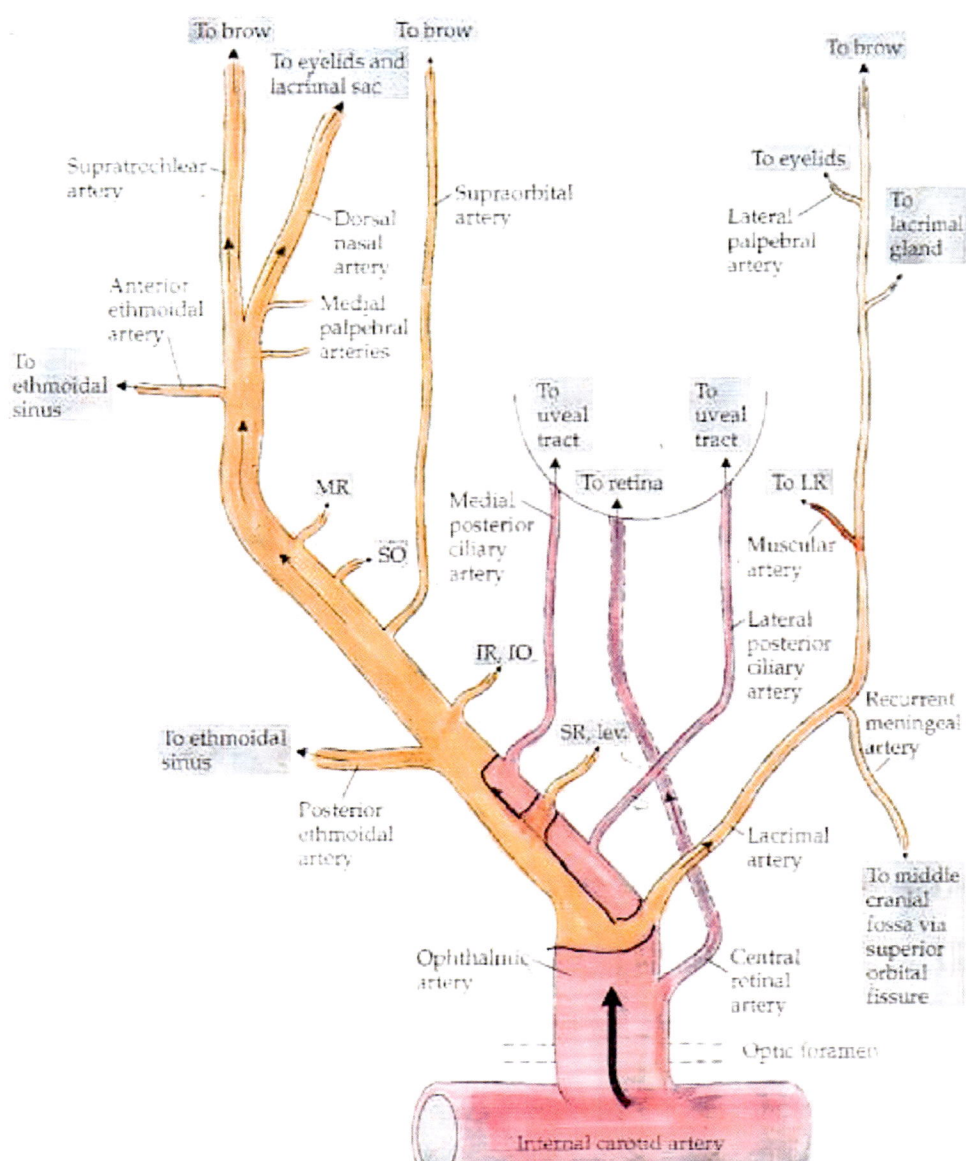

Fig. 7.26 Nerves of the orbit showing the extreme medial trajectory followed by supratrochlear branch of frontal. This determines the final branch of nasociliary as the supraorbital to zones 12 and 13. As it diverges from zones 10–11, it serves to program the nasociliary arterial pedicle. 210214–002288. (Reprinted from Standring S (ed). Gray's Anatomy 40th ed. New York, NY: Churchill Livingstone; 2008. With permission from Elsevier)

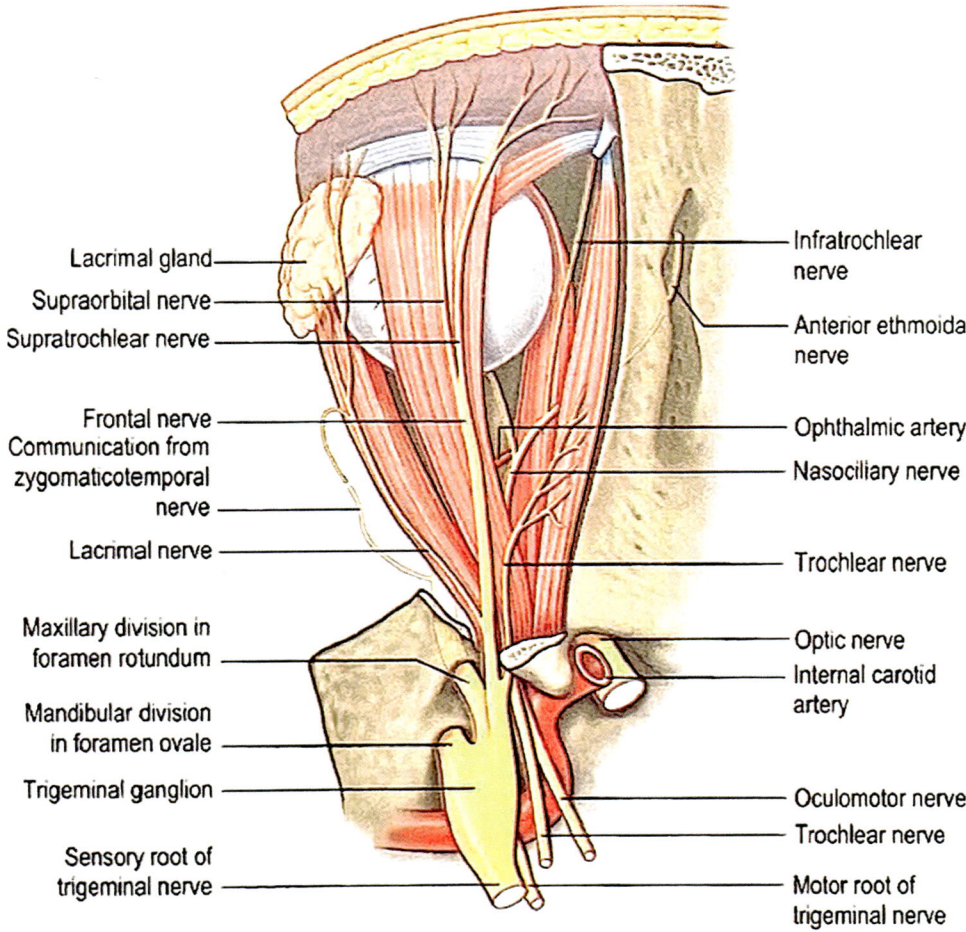

Fig. 7.27 Annulus of Zinn encloses the motor nerves for all muscles except superior oblique beneath the axis of the optic nerve. Note that cranial nerve IV is the only one to cross the midline before leaving the brain. (Case courtesy of Assoc. Prof. Craig Hacking, Radiopaedia. org, rID: 52363)

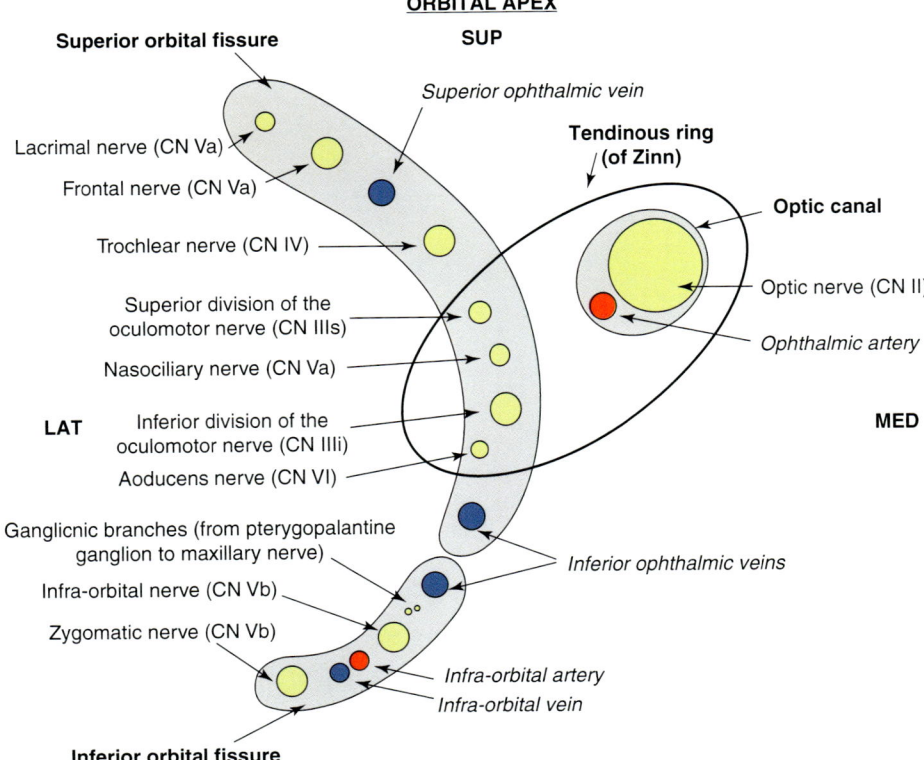

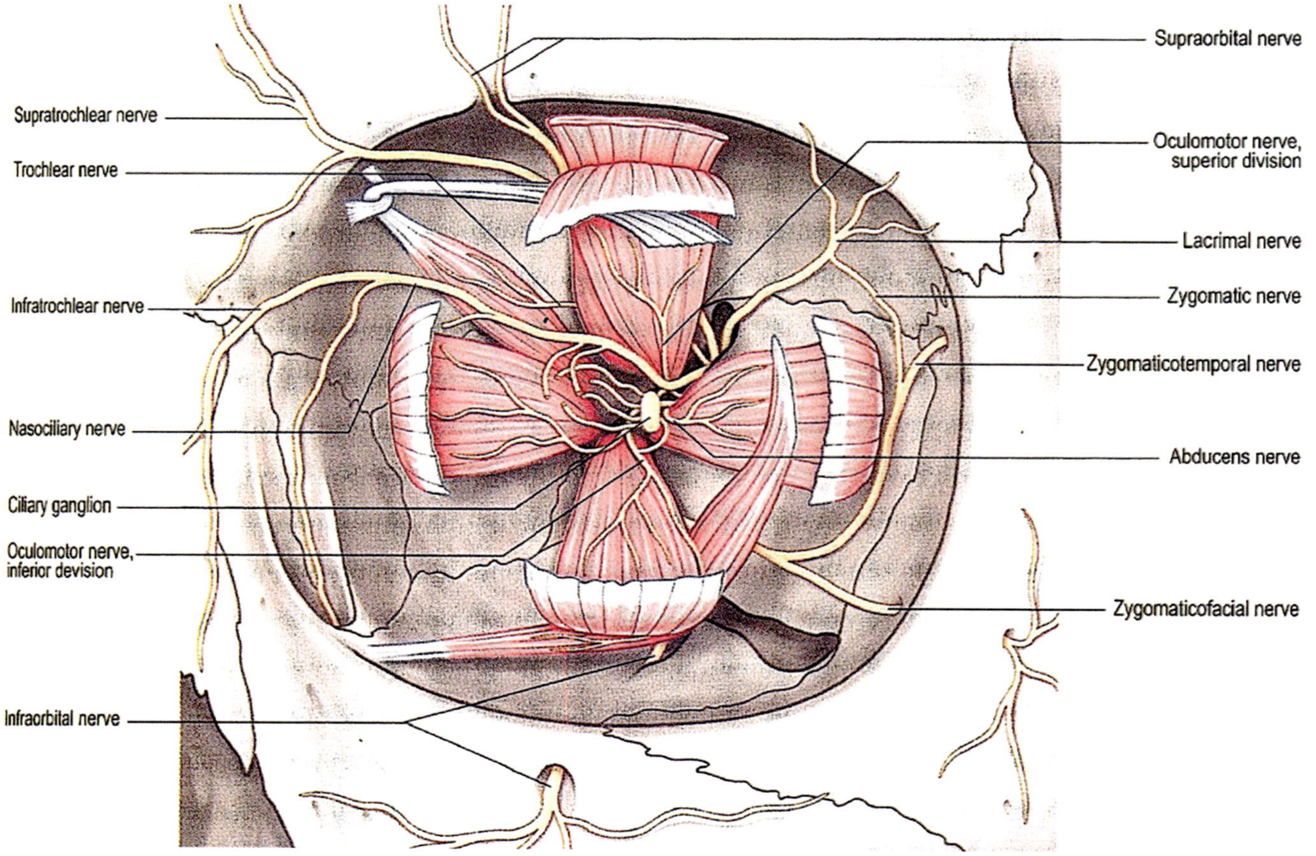

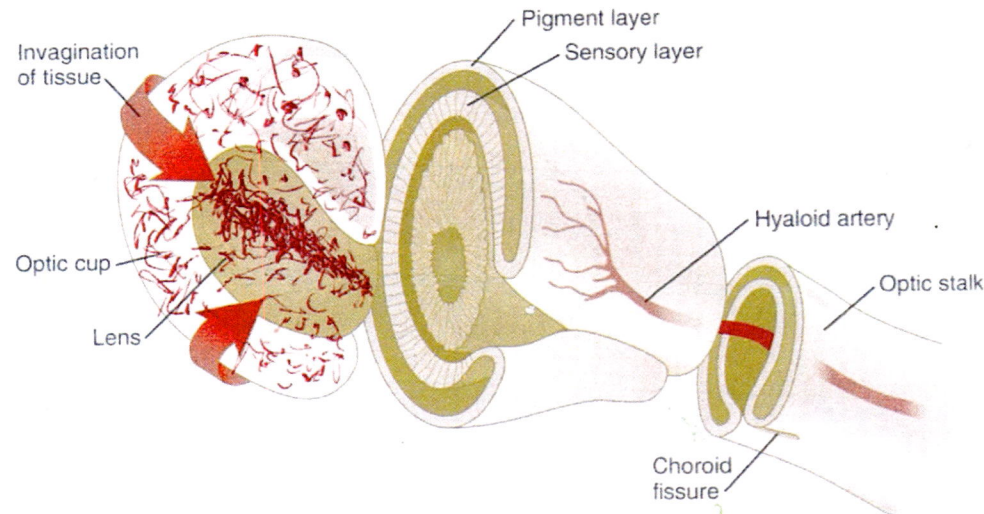

Supratrochlear nerve

Trochlear nerve

Infratrochlear nerve

Nasociliary nerve

Ciliary ganglion

Oculomotor nerve, inferior devision

Infraorbital nerve

Supraorbital nerve

Oculomotor nerve, superior division

Lacrimal nerve

Zygomatic nerve

Zygomaticotemporal nerve

Abducens nerve

Zygomaticofacial nerve

Fig. 7.28 Organization of cranial nerve V1. (Reprinted from Standring S (ed). Gray's Anatomy 40th ed. New York, NY: Churchill Livingstone; 2008. With permission from Elsevier)

Fig. 7.29 Folding of the optic vesicle into the optic cup leaves a incorporates the plexus that overlies it into the hyaloid artery. Plexus is initially supplied by primitive maxillary but is taken over at stage 14 by primitive dorsal ophthalmic. (Courtesy of Bruce M. Carlson, MD, PhD)

Invagination of tissue

Optic cup

Lens

Pigment layer

Sensory layer

Hyaloid artery

Optic stalk

Choroid fissure

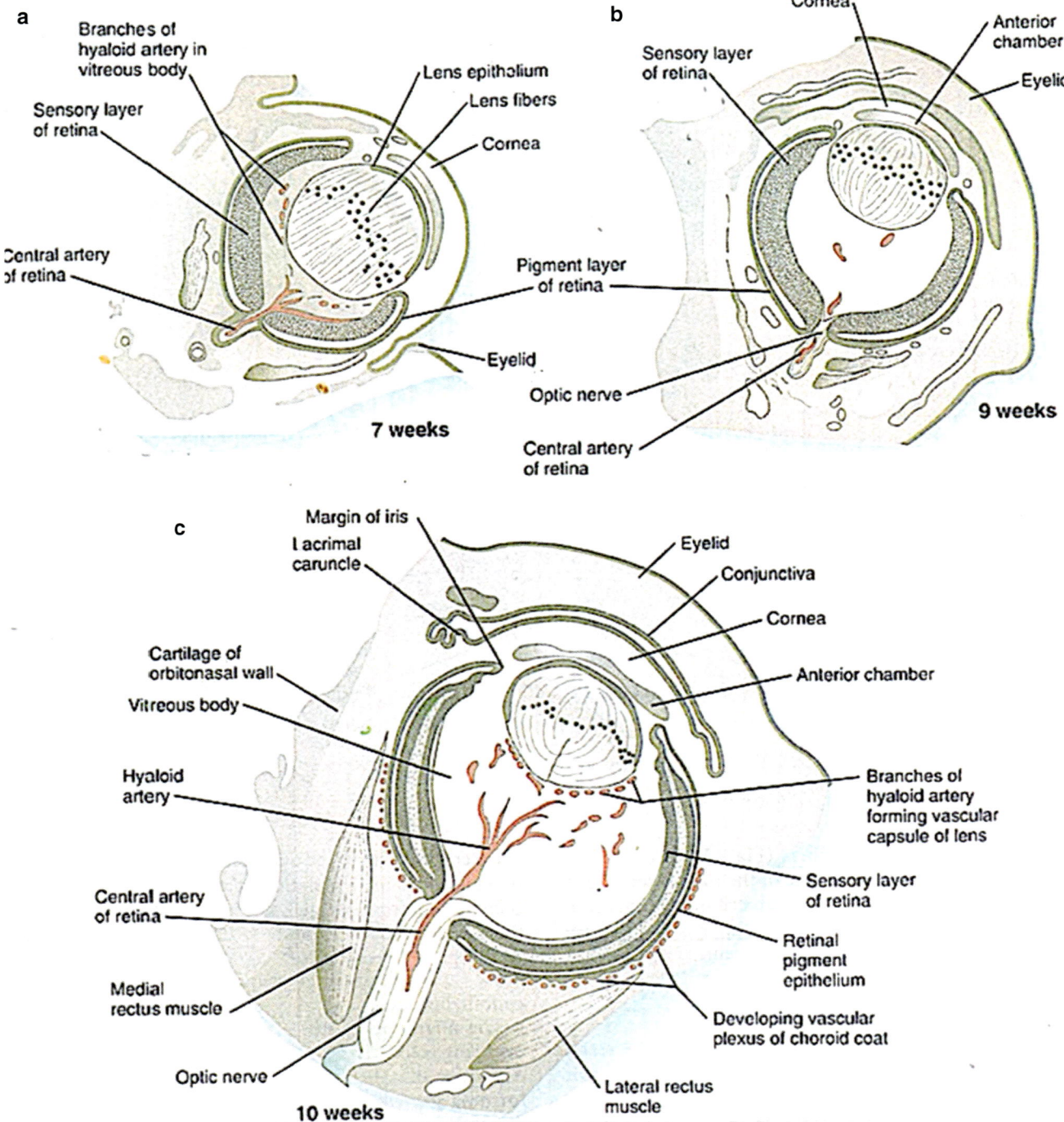

Fig. 7.30 Fate of the hyaloid: distal segment in the vitreous disintegrates. Proximal segment becomes the central artery of the retina. (Courtesy of Bruce M. Carlson, MD, PhD)

Fig. 7.31 Lacrimal gland has two sources of blood supply: lacrimal to the orbital lobe and conjunctival from the anterior ciliaries to the palpebral lobe. (Reprinted from Jones LT, Reech MJ, Wirtshafter JD. Manual of Clinical Ophthalmology. Rochester, American Academy of Ophthalmology and Otolaryngology, 1970. With permission from American Academy of Ophthalmology and Otolaryngology)

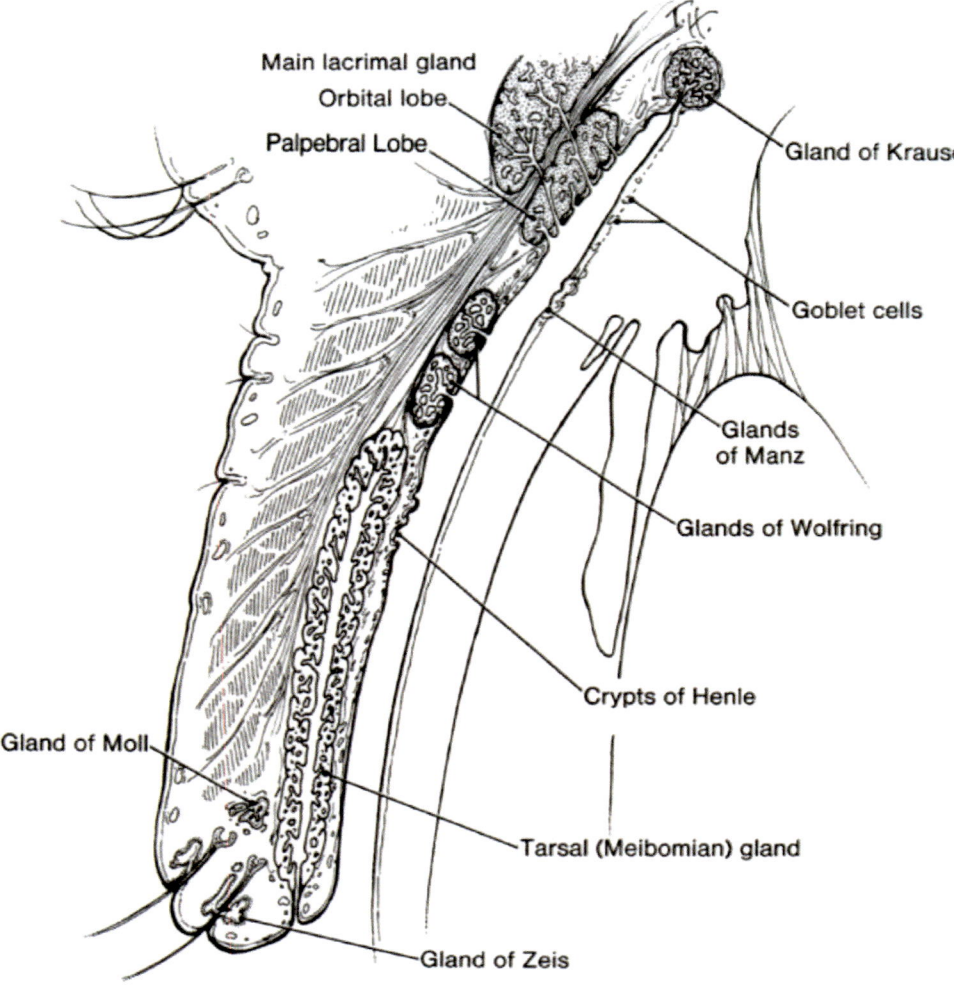

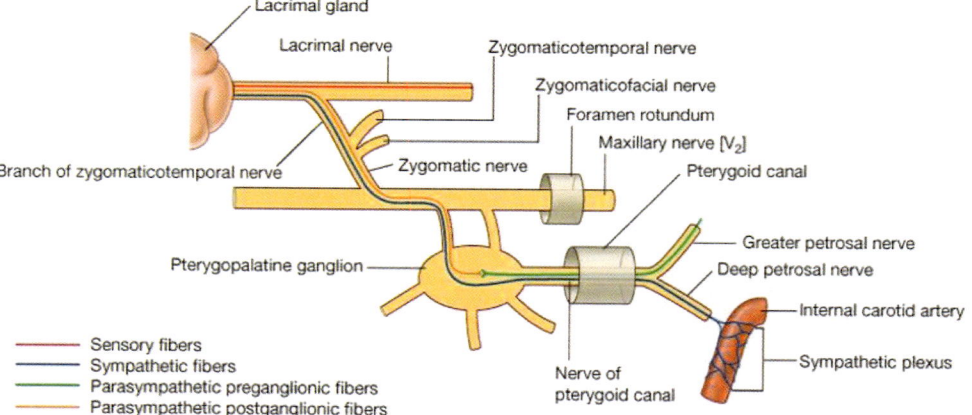

Fig. 7.32 Blood supply and Innervation of the lacrimal gland relate to embryology:
- Capsule and stroma of orbital lobe: r1 neural crest: V1 and lacrimal a
- Capsule and stroma of palpebral lobe: r1 neural crest: V1 and lacrimal a

- Glandular elements r4 neural crest: V2 and zygomaticotemporal a
- Ductal epithelium from r1 conjunctiva: V1 and lacrimal a
  (Reprinted from Drake RL, Vogl AW, Mitchell AWM. Gray's Anatomy for Students, 3rd ed. Philadelphia, PA: Churchill Livingstone, 2015. With permission from Elsevier)

Fig. 7.33 Vidian nerve combines PANS fibers from greater petrosal with SANS fibers from superior cervical ganglion (via carotid). (Reprinted from Martin RG, Grant JL, Peace D, et al. Microsurgical relationships of the anterior inferior cerebellar artery and the facial-vestibulocochlear complex. *Neurosurgery* 1980; 6(5):483–507. With permission from Oxford University Press)

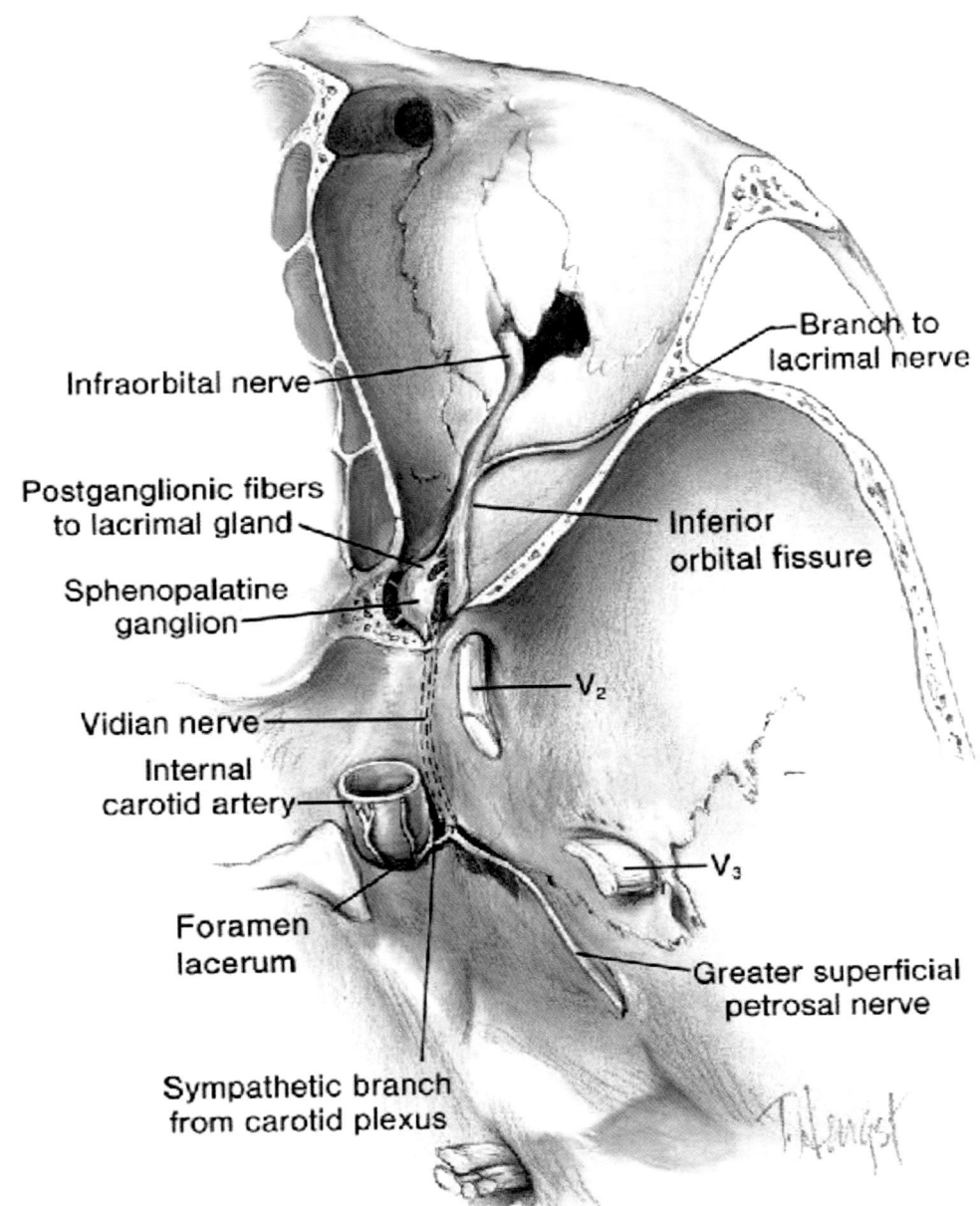

Fig. 7.34 Branches of the Stapedial system. Stem of stapedial
- Ramus superioris:
  - Posterior intracranial to dura
  - Anterior intracranial (supraorbital) to temporal fossa and orbit
- Ramus inferioris
  - Maxillomandibular
(Reprinted from Diamond MK. Homologies of the stapedial artery in humans, with a reconstruction of the primitive stapedial artery configuration in primates. Am J Phys Anthro 1991; 84(4):433–462. With permission from John Wiley & Sons.)

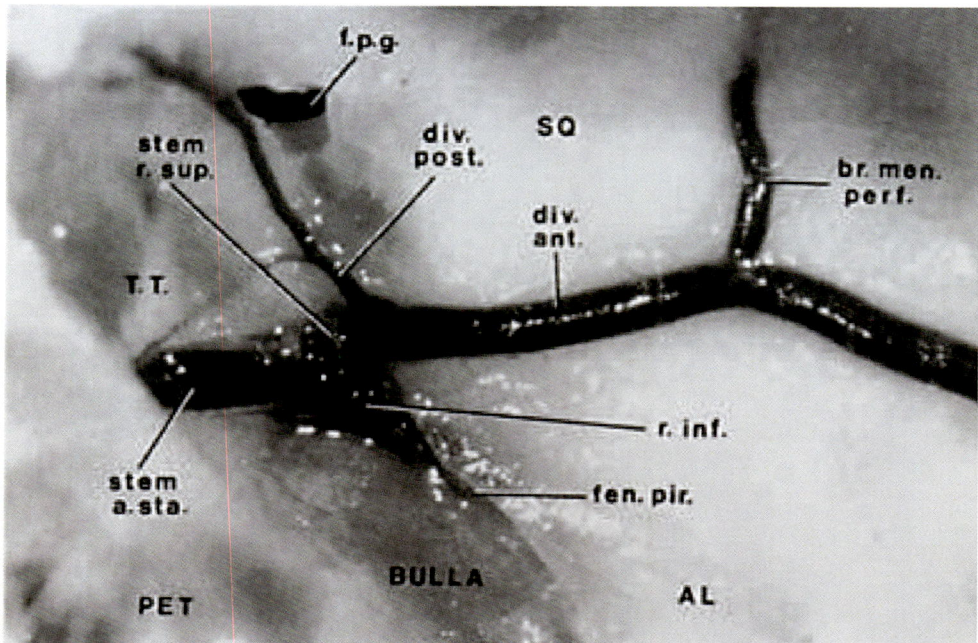

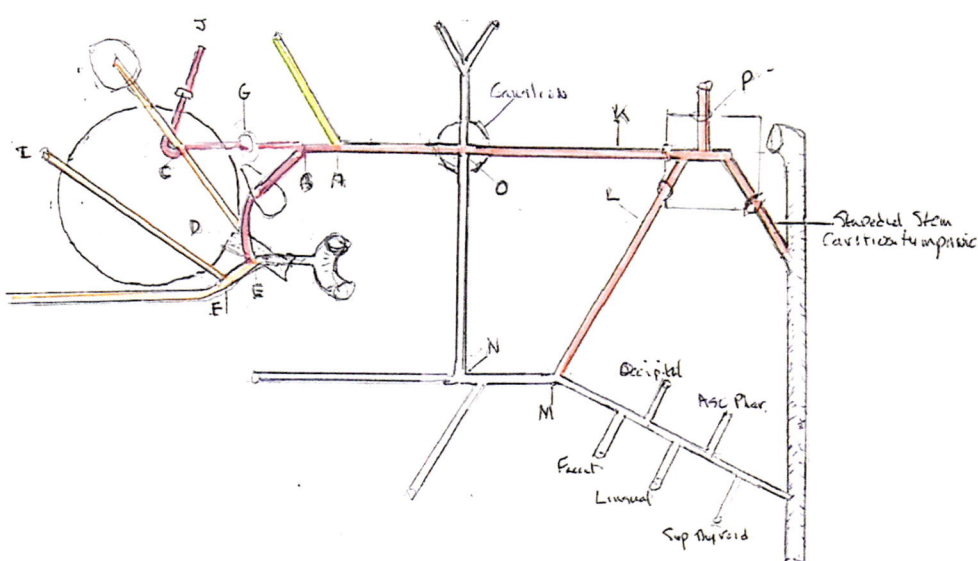

Fig. 7.35 Stapedial artery development 1. Stage 16 Stapedial stem (pink) comes from derivative of AA3 common carotid into tympanic cavity (T)
- TP Upper division, posterior (pink) is programmed by branches of VII supplies posterior dural mesenchyme
- TKA Upper division, anterior (pink) is programmed by VII (greater petrosal) until it reaches V ganglion. It supplies anterior dural mesenchyme
  - Although terminology is outdated we will refer to anterior branch of upper division as the supraorbital artery (pink)
- TL is lower division. It is approached by ventral pharyngeal artery
Stage 17 supraorbital approaches the orbit. Anastomosis of VPA with lower division at M forms internal maxillomandibular MN which gives:
- StV2 to pterygopalatine fossa, StV3 to mandible, and StV3 middle meningeal which re-enters the cranial cavity with recurrent meningeal nerve

Stage 18 Supraorbital splits at point **A** to give off StV2 meningeal branch (yellow) to anterior temporal fossa and common StV1 **AB** (magenta) to orbit. Middle meningeal is united with supraorbital at **O**

Stage 19 StV1 enters the orbit: meningeal branch **BC** and orbital branch **BD**. Three orbital branches (orange) **EF** and **EH** develop; they are still supplied from behind by supraorbital. StV3 developing to the dura (frontal branch and parietal branch) causing proximal involution of supraorbital **OK**. **MN** will persist as anterior tympanic

Stage 20 anastomosis between StV1 system and ophthalmic artery
Stage 21 Supraorbital **AB** involutes

Establishment of anastomosis between the lower division of the stapedial artery (future IMAX) and the ventral pharyngeal artery

(Courtesy of Michael Carstens, MD)

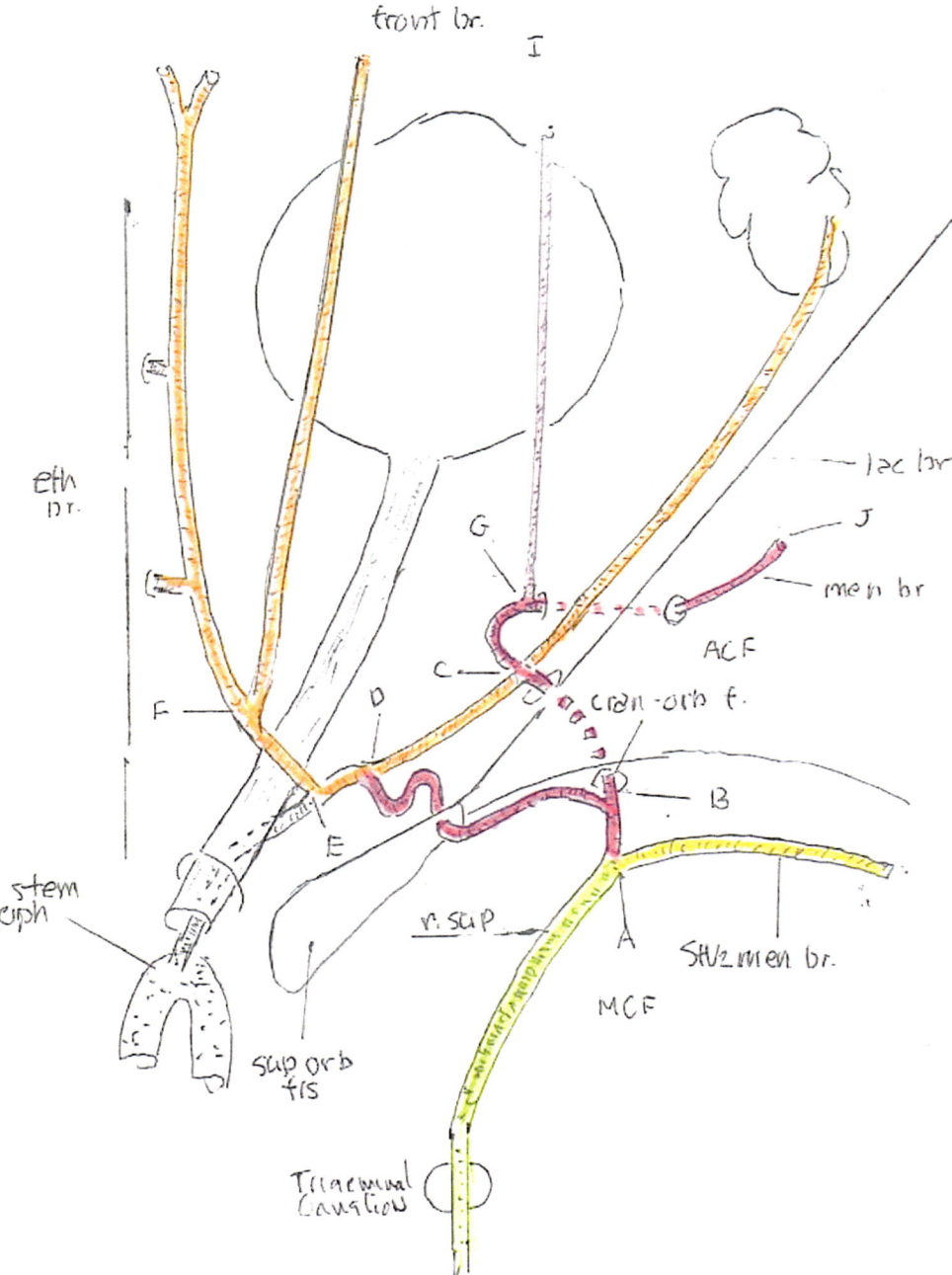

Fig. 7.36 Embryonic stapedial system to the orbit. Stapedial stem given off by internal carotid via the tympanic cavity; it then divides into (1) extracranial stapedial (StE) to join the external carotid system and (2) intracranial stapedial (StI) which supplies the entire dura, the individual meningeal branches being programmed sensory branches of trigeminal nerve. Anterior branch of StI (pink) proceeds from trigeminal ganglion through middle cranial fossa (MCF) toward anterior cranial fossa

- At point A, StI divides: (1) AB follows V1 toward the orbit—this is StV! (magenta); (2) the remainder of StI follows V2 branches to the dura of anterior middle cranial fossa. It also supplies the calvarial lamina of alisphenoid
- At point B, StV1 divides into two V1-programmed branches. (1) Medial orbital branch BDE through great orbital fissure into the orbit. (2) Lateral meningeal branch BCGJ passes through meningo-orbital foramen in alisphenoid to enter the lateral orbit; it then exits via an unnamed foramen to enter the floor of the anterior cranial fossa where it supplies the dura

- At point D, cranial nerve V1 divides into three sensory: the large common branch continues medially, then splits to become ethmoid n. and frontal n.; a small lateral branch proceeds laterally as lacrimal nerve
- At point E, StV1 makes an anastomosis with stem ophthalmic. It then follows V1 sensory nerves. StV1 proceeds medially as EF to supply zones 13–12 via the ethmoid branch (nasociliary) and zones 11–10 via the frontal branch (supraorbital). StV1 proceeds laterally as EDCH to supply zone 9 via
- Involution of embryonic stapedial supply has alternative consequences
  - ABD involutes leaving the intraorbital stapedial system connected to ophthalmic. This is the most common situation
  - BD can involute with ABC intact with recurrent meningeal supplying the intraorbital stapedial system
  - Minor variants have persistent ABD ± BC

(Courtesy of Michael Carstens, MD)

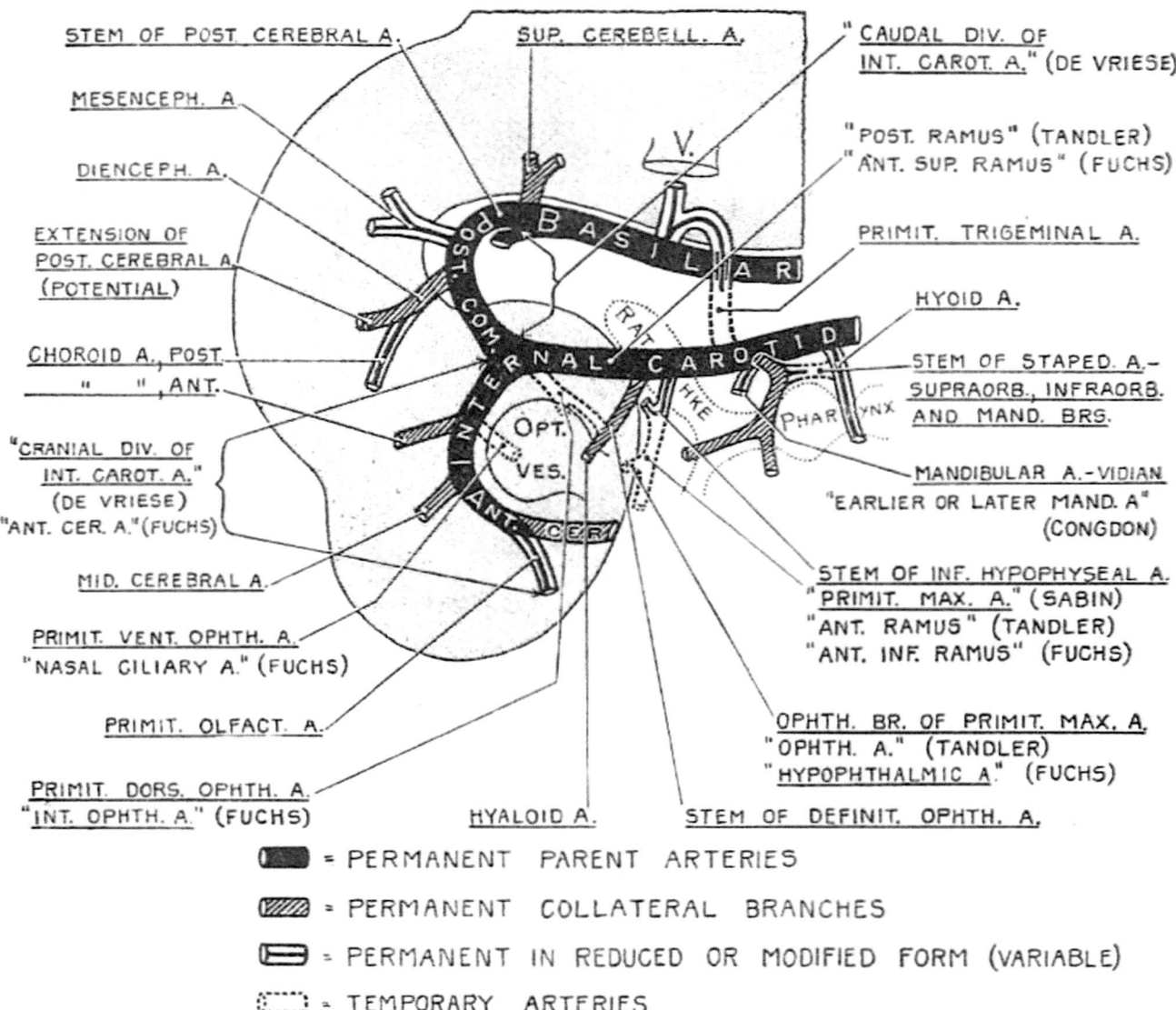

Fig. 7.37 Composite diagram of arteries situated around the embryonic forebrain. Variations in terminology are presented. (Reprinted with permission from Johns Hopkins University Department of Medical Art—Gary Lees. Carnegie Institution—Sonya Bajwa)

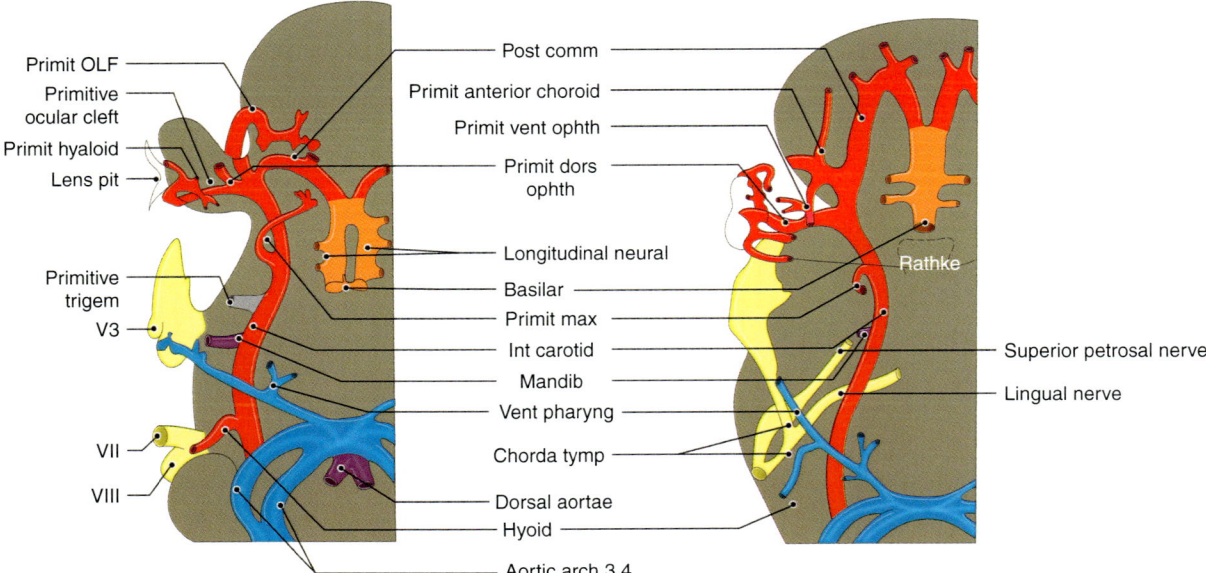

Fig. 7.38 (Left) Stage 14 (4–5 mm). The primordial cup and lens supplied by primitive dorsal ophthalmic (prDOA) have arrived. It supplies: (1) primordial cup and lens; (2) caudal and dorsal zone of the optic vesicle. Temporary support for base of the vesicle at this time comes from a transient branch of primitive maxillary. PrMx is possibly a support for Sm1 and Sm2 as well. (prDOA). At stage 14, hyoid arises as a remnant of AA2 and is associated with VII. VPA arises from AA3 to AA4. It supplies all three arches and arrives at V3. *Multiple sources nourish CN II.* (Right) Stage 15–16 (9 mm). Optic cup has a plexus representing branching from prDOA. Neural crest has invaded; pigment present. PrDOA inserts into annular vessel at lens cup and also into *distal* hyaloid. Primitive ventral ophthalmic artery (prVOA) arises from more proximal PrICA prior to bifurcation into caudal branch. Cranial nerve VII is branching: lingual, superficial petrosal, and chorda tympani (CT now associated with V3). Hyoid artery physically associated w/ proximal chorda tympani. This maxillomandibular division of stapedial out of the tympanic cavity and connect it with V3. VPA lies medial to distal CT. V1 sensory nerve lies *dorsal* to mid-section of eye. The V1 stapedial arteries have not yet arrived. The prDOA supplied plexus ensheaths the optic cup. New on the scene is primitive ventral ophthalmic (prVOA) which contributes as well; it is *not* the primary supply for the plexus. Surrounding the lens at its margin is an *annular artery* which is in continuity with the *primitive hyaloid* which is running in the ocular cleft. PrDOA is inserted into its distal aspect. Note hyoid artery and ventral pharyngeal arteries (green) have shifted

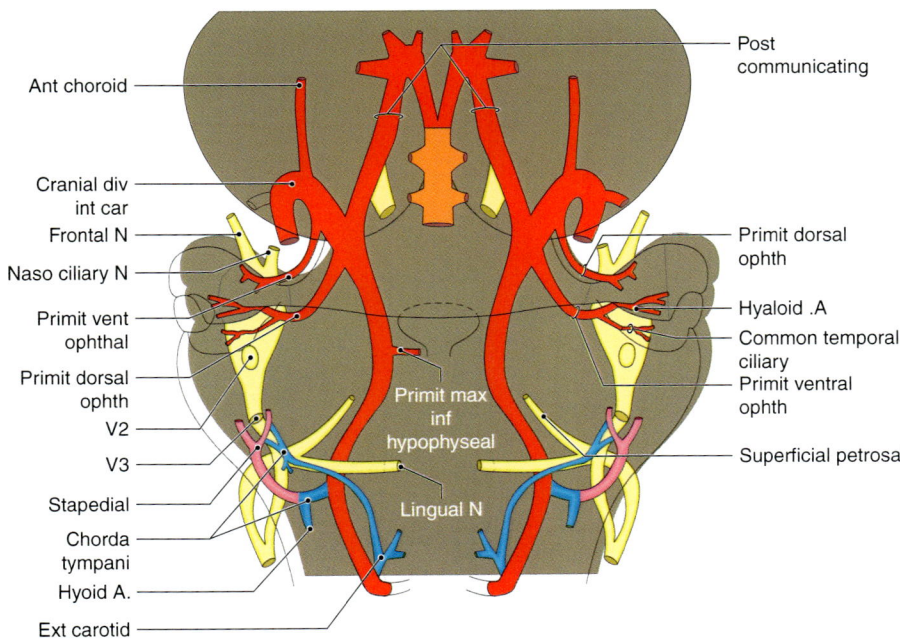

Fig. 7.39 Stage 17–18 (14 mm). PrDOA and prVOA are both much longer. PrDOA sends a direct branch to the lens, this vessel becomes the *definitive hyaloid*. Another branch of prDOA is *temporal long ciliary*, induced by temporal V1. Temporal development precedes nasal. PrVOA much longer Stapedial lies lateral to distal CT and lateral to V3. Proximal VPA = external carotid. V1 moves forward and is now dorsal to junction of globe and optic nerve. At this point, it has two branches, nasociliary and frontal (lacrimal comes later). In the face, the distal end of stapedial max-mand. Branch has anastomosed with ventral pharyngeal artery. It is now supplied by both stapedial stem in tympanic cavity and by external carotid. This persists as anterior tympanic artery

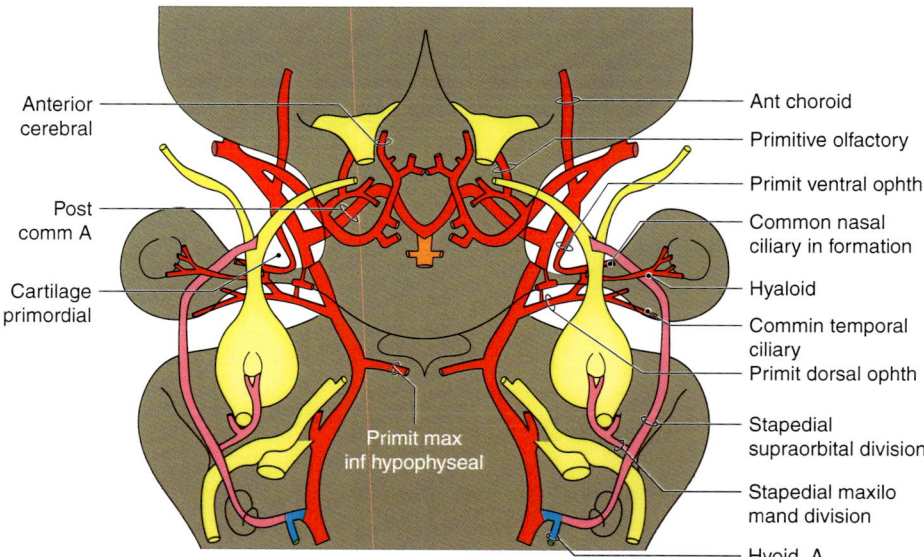

Fig. 7.40 Stages 19–20 (18 mm). <u>All ocular branches from ophthalmic stem are now in place</u>. The long and tortuous stem of PrDOA is replaced by a more direct permanent *dorsal ophthalmic stem*. ANS fibers associated with the primitive ophthalmic induce *short ciliary aa*. V1 located dorsal to optic nerve. Hyaloid branch of prDOA now makes secondary anastomosis with PrVOA to make *nasal long ciliary aa*, induced by nasal V1. (Ethmoid and frontal) branches have now extended outside the orbit. Lacrimal is not mentioned specifically but is depicted as a branch off of frontal. (Lacrimal may be the first orbital branch). Stapedial branches are dorsal to globe, ventral to frontal n. and dorsal to nasociliary n. StV1 meningeal and orbital branches enter the orbit. Distal lacrimal is accessed by StV1 meningeal. Newly formed maxillomandibular system lies medial to V3, and grows forward to make contact with V2

- *V1 induces all orbital branches of stapedial*
- *CN II induces primitive ophthalmic*

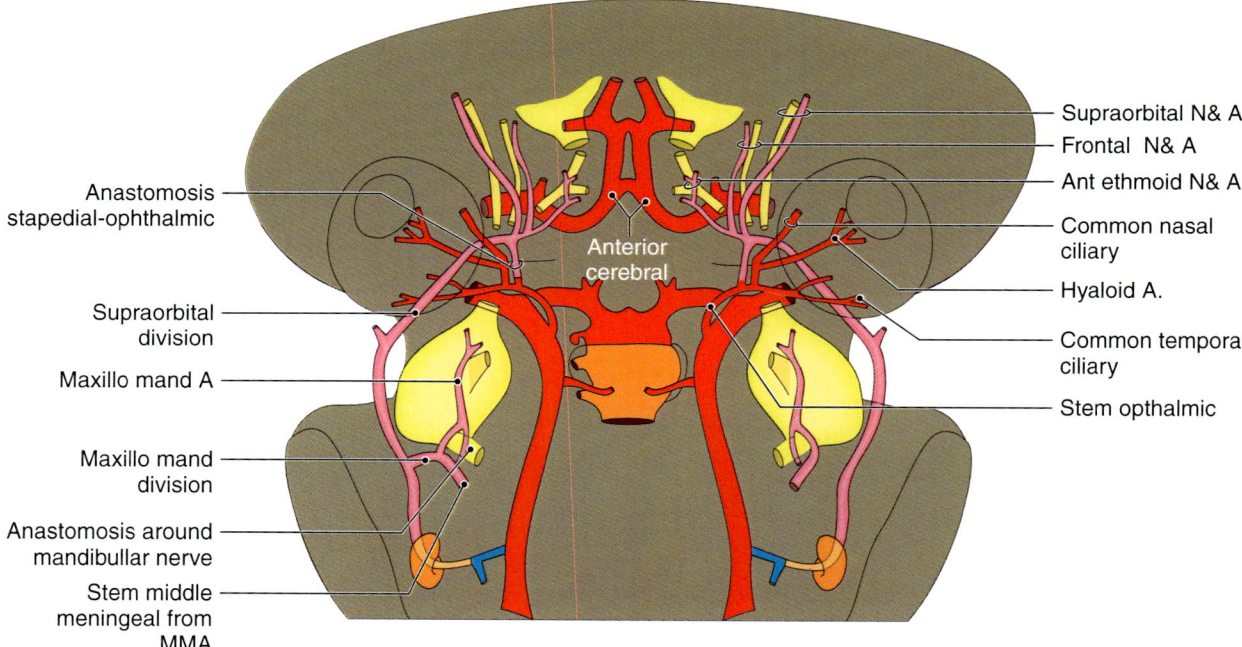

Fig. 7.41 Stage 20 (20 mm). <u>All orbital branches from V1 stapedial are now in place</u>. In stage 20, StV1 orbital system is annexed by ophthalmic The parent stem, supraorbital stapedial, is now dwindling. Note that stapedial makes a secondary anastomosis (lateral) around V3. External carotid branches induced. Branches in individual muscles follow trigeminal sensory supply to their fasciae

- Arterial "ring" is mechanism by which ophthalmic artery passes from the ventral side to the dorsal side of optic nerve. <u>This positions it to annex StV1</u>

- Internal maxillary uses a "ring" mechanism to reach external surface of mandibular nerve
- Lacrimal gland: (1) fascia and stroma, r1 neural crest supplied by lacrimal; (2) glandular elements, r4 neural crest* supplied by zygomaticotemporal; (3) epithelium of ducts as ingrowth of conjunctiva, r1 ectoderm supplied by lacrimal PANS secretomotor fibers from the pterygopalatine ganglion follow the zygomatic nerve up to the lacrimal gland. See Chap. 13 for more details

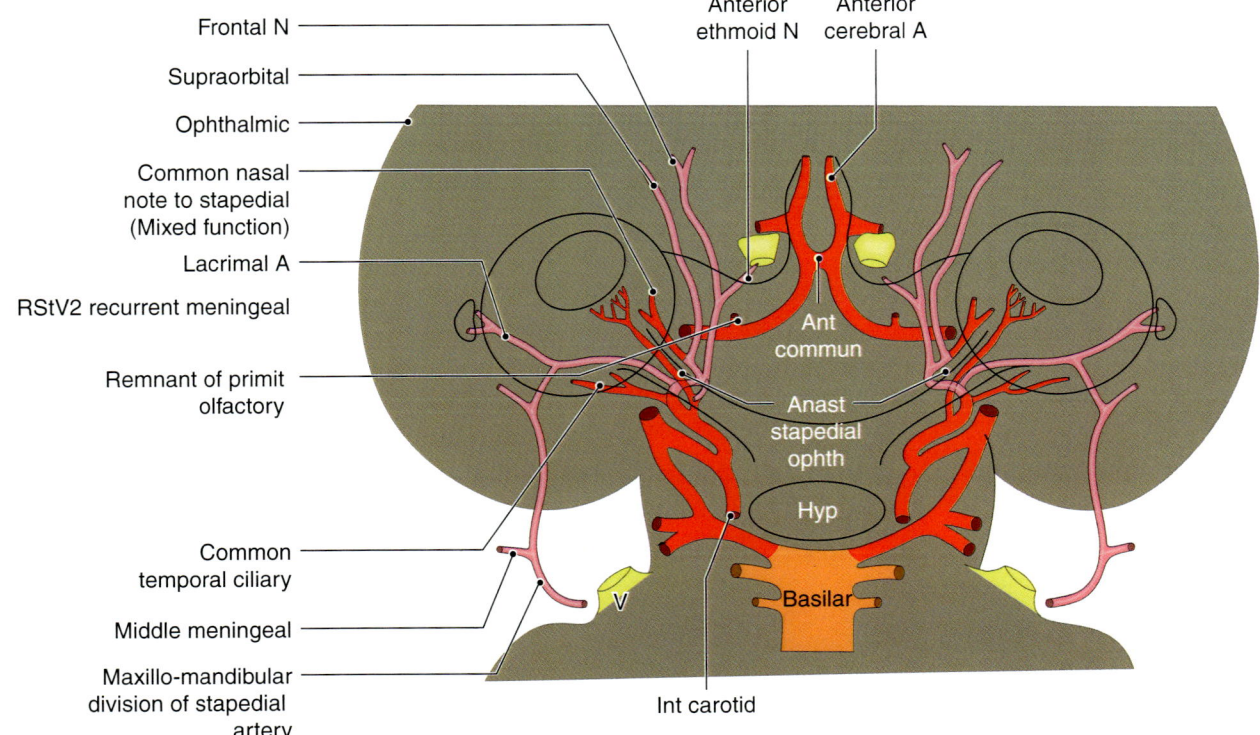

Frontal N
Supraorbital
Ophthalmic
Common nasal note to stapedial (Mixed function)
Lacrimal A
RStV2 recurrent meningeal
Remnant of primit olfactory
Common temporal ciliary
Middle meningeal
Maxillo-mandibular division of stapedial artery

Anterior ethmoid N
Anterior cerebral A
Ant commun
Anast stapedial ophth
Hyp
Basilar
Int carotid
V

Fig. 7.42 Stage 23+ (40 mm). In this stage, two embryonic vessels disintegrate. (1) Supraorbital stapedial stem dies from the orbit backward to the StV2 bifurcation. The remainder persists as *anterior branch of middle meningeal*. (2) Cranioventral half of the optic "ring" dies; the caudal-dorsal half is the anastomosis running from ophthalmic up to stapedial.

Recall that ophthalmic arose from ICA at Rathke's pouch caudoventral to optic nerve. This is very far away from posterior communicating, the origin of PrDOA. This was close to r1 early on but with growth it stretches out. This elicits backward vascular "leapfrog" which terminates at r1. The ocular cleft is thus supplied from the caudo-ventral aspect. See text

Fig. 7.43 Stages 13–14 (5.3 mm). During the "branchial" phase, cranial nerves and aortic arch arteries supply each individual pharyngeal arch and pharyngeal pouch. The system then undergoes revisions. Recall that first and second arch merge into PA1–2 causing a dissolution of the aortic arches; these are replaced by the ventral pharyngeal artery: During stage 13–14, two embryonic vessels cross through the territory of PA1–2: *Ventral pharyngeal artery* is seen heading toward V3. The second arch remnant *hyoid artery* appears in the vicinity of facial ganglion. First arch leaves behind a remnant, mandibular artery. This will become the *artery of the pterygoid canal*. A dense capillary plexus covers the lens vesicle

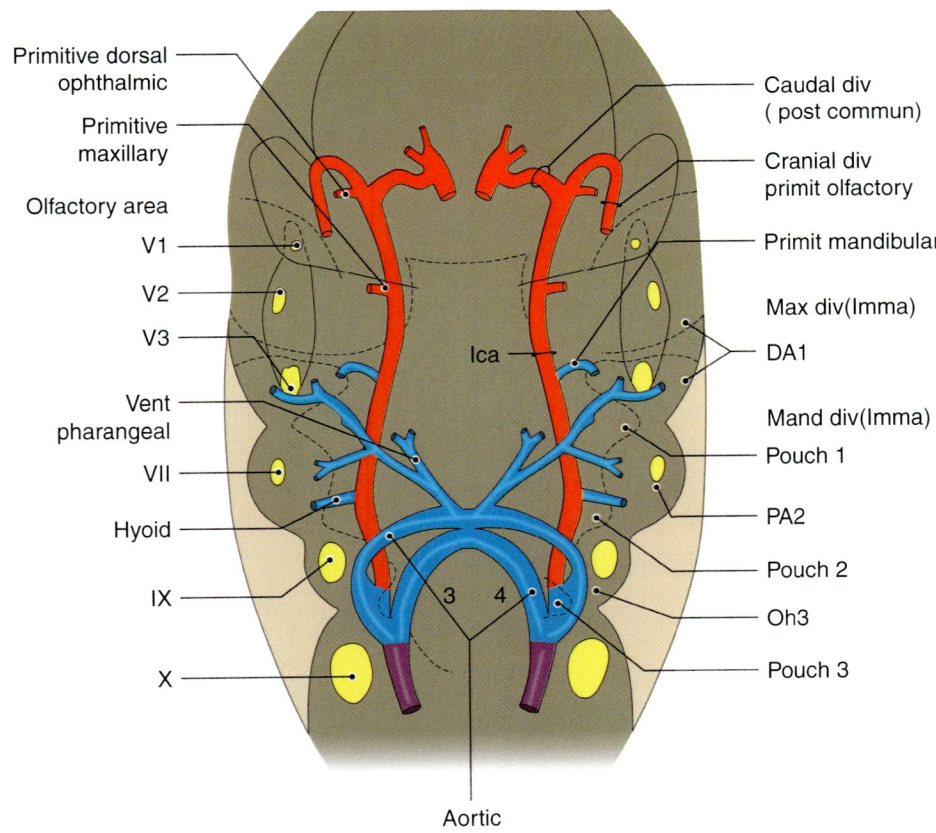

Primitive dorsal ophthalmic
Primitive maxillary
Olfactory area
V1
V2
V3
Vent pharangeal
VII
Hyoid
IX
X

Caudal div (post commun)
Cranial div primit olfactory
Primit mandibular
Max div(Imma)
DA1
Mand div(Imma)
Pouch 1
PA2
Pouch 2
Oh3
Pouch 3

Ica
3   4

Aortic arch

Fig. 7.44 Stages 15–16
(7 mm). Third pharyngeal
arch is no longer
recognizable. It has been
internalized, i.e., the
"Japanese fan" model, as seen
in the "bunching together" of
cranial nerves V–X. Cranial
nerve VII is redirected
posteriorly. The *carotid origin
of the hyoid* shifts as well
from the caudal border of the
second arch backward
following the translocation of
the carotid to just opposite
CN X. Thus hyoid is relocated
to the caudal border of the
third arch / cranial border of
the fourth arch. In the context
of embryonic folding this
places the hyoid artery
directly beneath the future
temporal bone complex

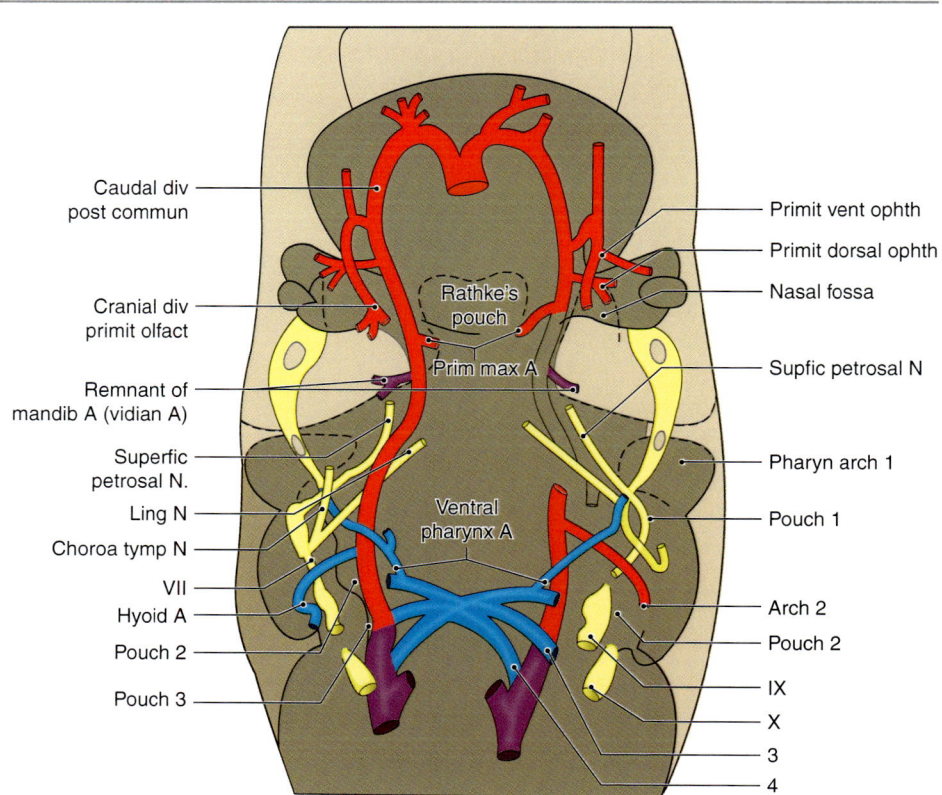

Fig. 7.45 Stages 17–18
(14 mm). Origin of hyoid
stem now further forward, in
front of the first pharyngeal
cleft, i.e., between the PA1–2
sandwich and PA3. This
translocates extracranial
stapedial into proximity with
V3 and with the VPA. The
stapedial along chorda
tympani is accordingly
renamed *maxillomandibular
division*. Note the V1 has
crossed over the ophthalmic
stem and two branches are
seen

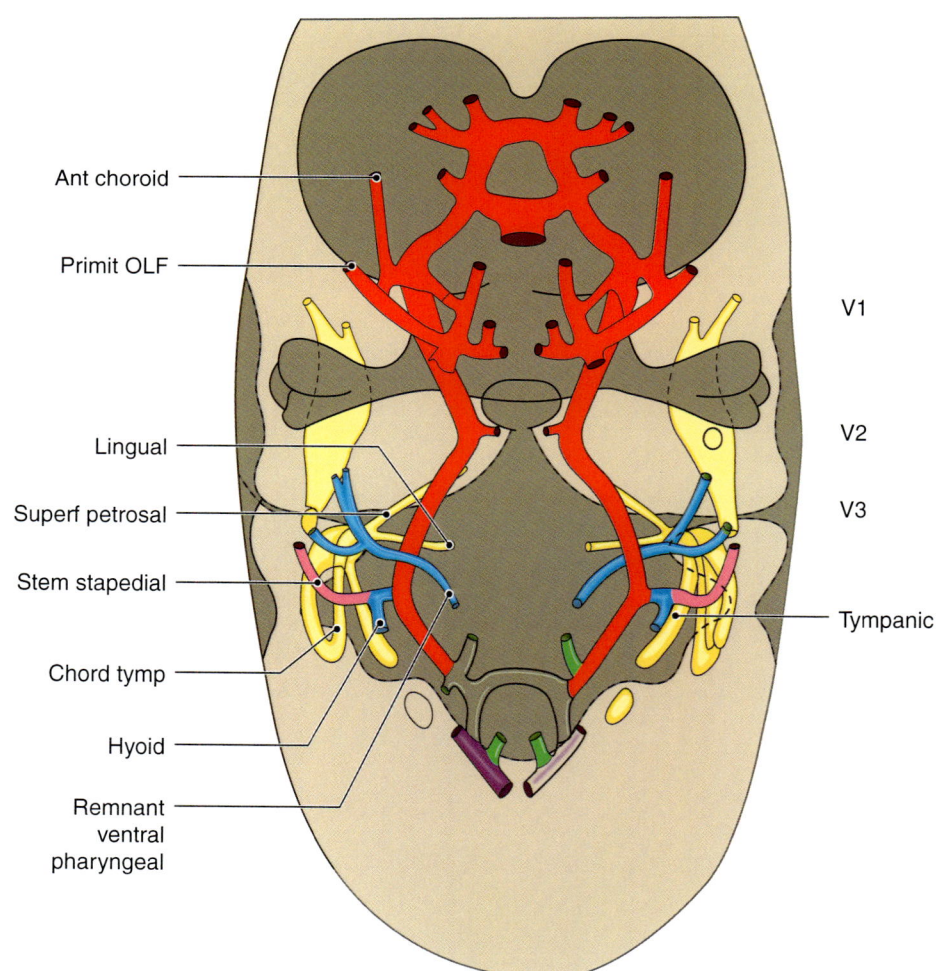

Fig. 7.46 Stages 19–20 (18 mm). StV1 meningeal and orbital branches now seen entering the orbit from lateral to medial. The three orbital artery branches are programmed by medial StV1 just inside the superior orbital fissure. Definitive ophthalmic in place. Gives off 2 of the 3 ocular branches: *hyaloid artery* and *common temporal long ciliary* (previously, the PrDOA). In the process of annexing the third ocular artery, *common nasal long ciliary* from PrVOA, the future, seen here approaching the optic nerve from the medial side

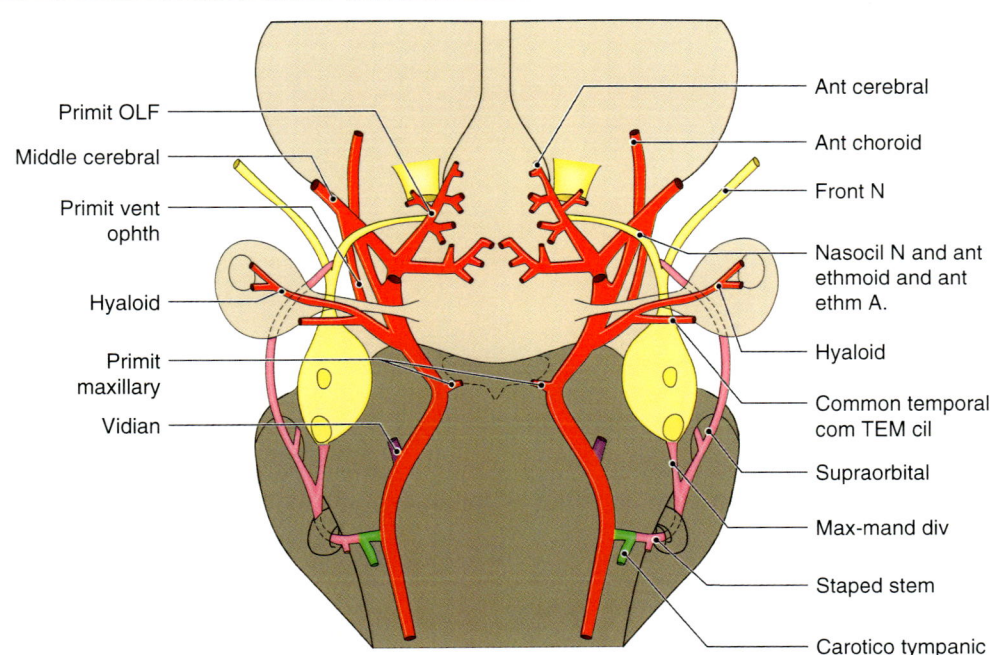

Fig. 7.47 Blow-up of DHC-MC 33

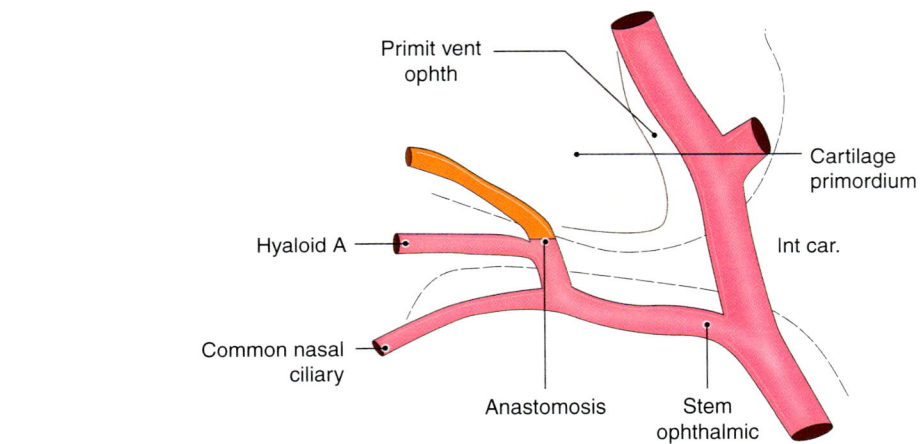

Fig. 7.48 Stage 20 (21 mm). All three ocular arteries are present. Nasociliary and frontal nerves are dorsal to the optic nerve. Right side is more primitive, with persistence connection between StV2 from meningeal into the orbit with STV1. On the left side, StV2 disconnected from orbit, leaving meningo-orbital artery remnant. Note that the ophthalmic stem is now *dorsal* to optic nerve, meaning that the transfer via the vascular "ring" is complete and anastomosis with StV1 has taken place

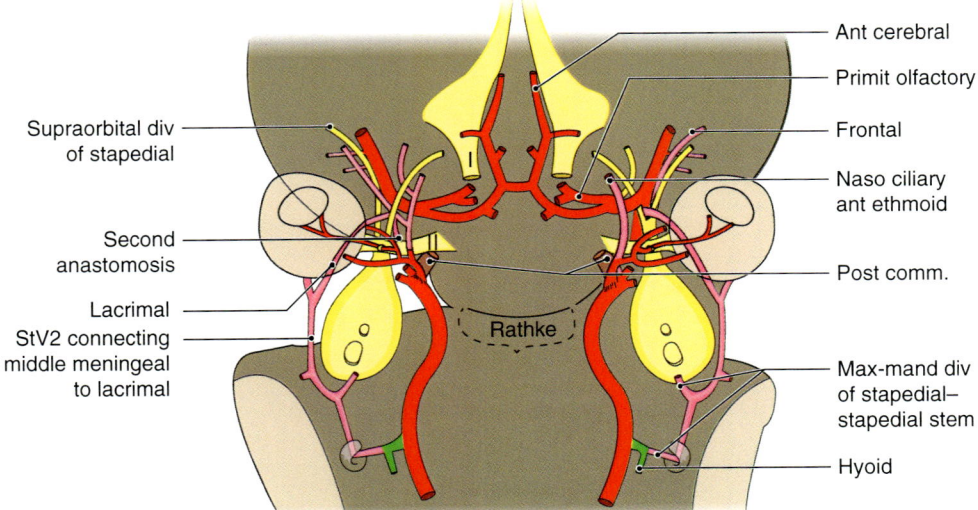

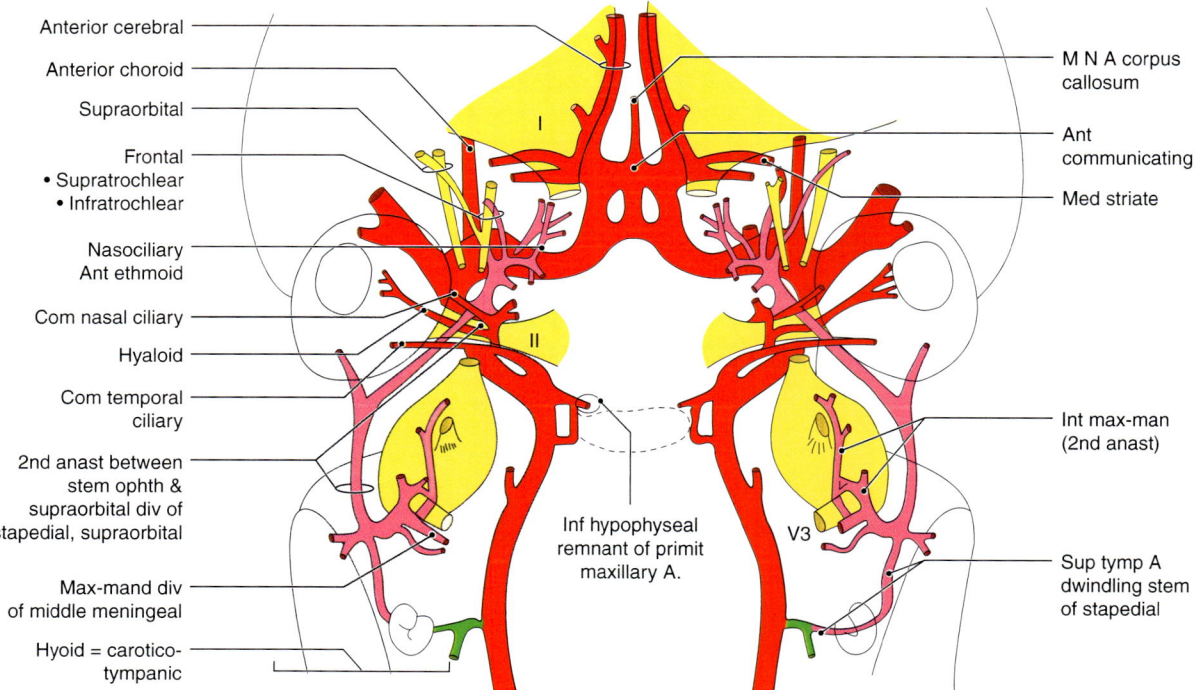

Fig. 7.49 Stages 21–22 (24 mm). For a lateral view, refer to DHP-MC 8a. The purpose here is to show how ophthalmic artery migrates from beneath optic nerve to a dorsal position where it can annex the StV1 orbital system previously synthesized during stage 9. Presence of arterial ring around the optic nerve permits the ophthalmic artery to wrap around the nerve and gain a dorsal position. Just proximal to this stem, stapedial makes it anastomosis with ophthalmic *below* the nerve. On the left side, lacrimal artery (unlabeled) is just beginning to emerge. Note that the stem of stapedial from hyoid artery, i.e., supraorbital stapedial from tympanic cavity forward to middle meningeal, has disintegrated. When this happens, maxillomandibular ECA (internal maxillary branch) becomes the sole source for the middle meningeal system

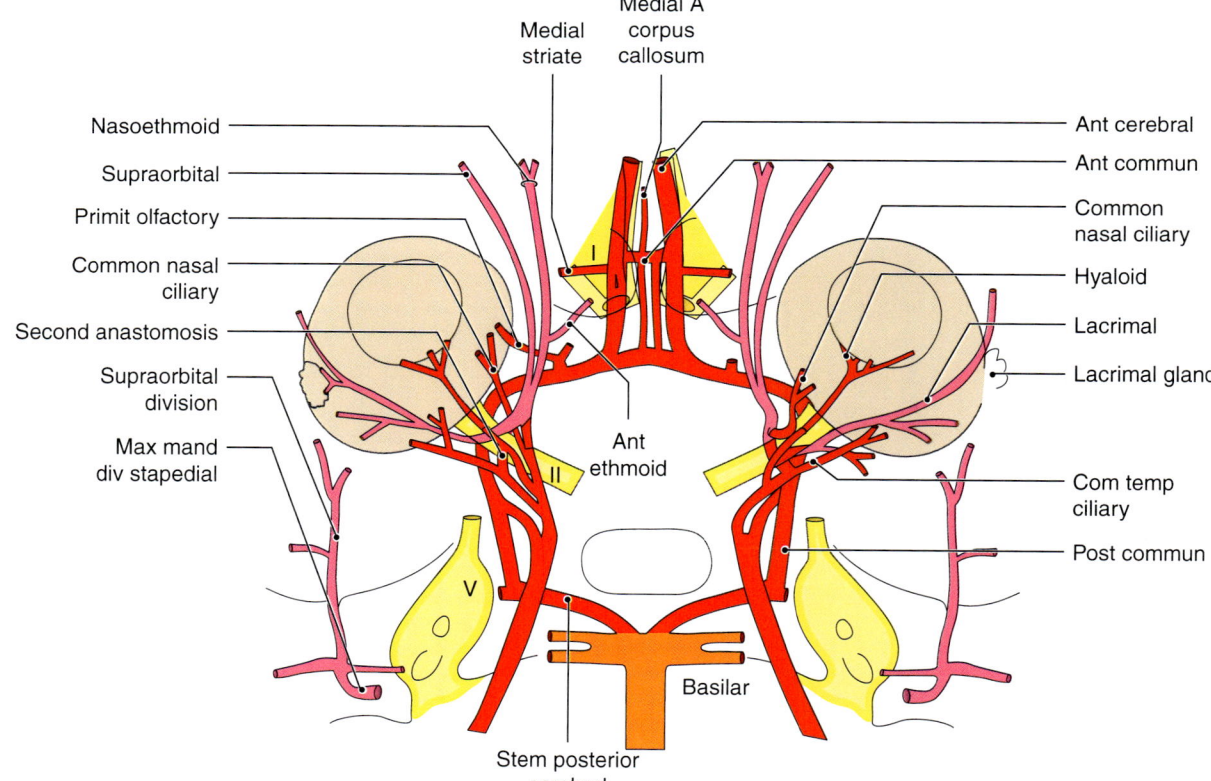

Fig. 7.50 Stage 23+ (39 mm). Complete union of stapedial and ophthalmic systems. Left side shows the adult state. On the right side, persistent arterial ring keeps ophthalmic in ventral position to optic nerve, whereas on the left side it is dorsal. Lacrimal artery present. On the right side, lacrimal-meningeal anastomosis remnant is seen

The stapedial system represents the vascular response to the evolution of jaws. As previously described, the "gnathostome revolution" involved a radical reassignment of neural crest tissue from respiratory derivatives of the first and second gill arches to cartilaginous structures for predation. With the advent of jaws, the first and second arches lost all relationship to respiration; they became a single functional unit. In this process, the first gill slit is eliminated although retained in sharks as the *spiracle*. The representation of the second arch within the pharynx is virtually eliminated. Thus, the tonsil, a second arch derivative, sits within mucosal folds innervated by V3 (first arch) and IX (third arch). In the process of tissue folding and reorganization, *endodermal second arch representation is all but eliminated*. In similar fashion, a cutaneous representation of the second arch does not develop owing to the absence or elimination of dermal elements derived from r4 to r5 neural crest. First arch dermis takes over. To support the lower jaw, second arch neural crest was reassigned to make hyoid bone. All these changes required a new iteration of vascular anatomy.

Let's consider the fate of the primitive precursor arteries that supplied the first and second arches. Each remains attached to the internal carotid. The mandibular artery becomes the *artery of the pterygoid canal*. The remnant of hyoid artery is *caroticotympanic artery*. Its spans through a tiny foramen in the posterior wall of the carotid canal and climbs over the promontory. There it makes an anastomosis with anterior tympanic of the internal facial (internal maxilla-mandibular artery). This represents the remnant of the disintegrated stapedial within the tympanic cavity.

The various components of the stapedial system are programmed by cranial nerves V, VII, and possibly IX. Before going further, we should review the anatomy of sensory branches of the trigeminal nerve. *Of particular importance is the anatomic pathway for trigeminal supply to the dura.* After leaving the trigeminal ganglion, V1 and V2 travel forward in the lateral wall of the cavernous sinus while V3 descends directly out of the skull. Note: ***the most proximal sensory branch of each part of trigeminal is dedicated to the dura.***

V1, upon its exit from the cavernous sinus and immediately prior to its entry into the orbit, gives rise to the lateral ***anterior meningeal nerve*** supplying the dura of the anterior cranial fossa, ***orbital nerve*** then enters the superior orbital fissure. Once in the orbit, V1 again supplies the dura via a separate *dural branch of anterior ethmoid nerve* that re-enters the cranial cavity via the cribriform fossa.

V2, before exiting the skull via foramen rotundum, gives off the ***middle meningeal nerve*** to the postorbital dura of anterolateral frontal lobe behind alisphenoid.

V3, after exiting the skull via foramen ovale, gives off the ***recurrent meningeal nerve*** supplying the temporo-parietal-occipital lobes. This requires that it tracks upward with the middle meningeal artery via the foramen spinosum.

Developmental Anatomy

When the second aortic arch to PA2 disintegrates at stage 12, it leaves behind a dorsal remnant, the hyoid artery, dangling from the dorsal aorta. With the growth of the embryo by stage 13, the hyoid is repositioned backward toward the otic capsule.

At stage 14, the definitive circulation of the internal carotid is established. The hyoid artery enters the middle ear at the postero-medial angle. It travels forward to the promontory where it divides into the *promontory artery* and the proximal *stapedial artery*. It then ascends the promontory and passes anterior to fenestra cochleae. The stapedial system then undergoes two critical divisions within the tympanic cavity. (1) Just prior to accessing the stapes, it divides into anterior and posterior branches, with the anterior branch passing through the stapes while the posterior branch does not. (2) Upon exiting the stapes, the anterior branch subdivides into superior and inferior divisions, both of which subsequently exist the tympanic cavity.

Posterior Branch

Just proximal to fenestra vestibuli, stapedial stem gives off a *posterior branch* that ascends, traveling in proximity (but in the opposite direction) with horizontal and descending facial nerve. Within the tympanic cavity, the artery sends a tiny branch along motor branch of VII to supply the stapedius muscle and then supplies the posterior wall. It subsequently travels backward and upward in the sulcus between the petrous temporal bone and the squamous temporal bone, the *petrosquamous sinus*. Occasionally, it can pass through into the mastoid region. Posterior branch then moves on to the posterior fossa. It gives off *meningeal branches*, *temporal branches* that pierce the bone to enter temporalis, and supplies the occipital area via a large trans-osseous *arteria diploetica magna* temporarily.

The dura of the posterior fossa served by this primitive posterior stapedial system develops from neural crest from three distinct sources. When the original stapedial stem disintegrates, resulting in anastomoses with the external carotid system reveal the underlying embryology. Temporal-parietal dura comes from r3 and is supplied by first arch-related middle meningeal artery. Posterior fossa dura is partially derived from second arch neural crest and is supplied by meningeal branches of the occipital that access the skull base via the condyloid canal and jugular foramen (along with the internal jugular vein). Third arch neural crest also contributes to the dura of posterior fossa. This zone receives posterior meningeal artery of the ascending pharyngeal via jugular foramen, foramen lacerum, and the hypoglossal canal. When proximal stem of stapedial disintegrates, the dural arteries of these zones are described as originating from the external carotid when in fact this is a secondary system.

Recall that the brain is surrounded by mesenchyme, both mesoderm and neural crest, so that the dura and its arterial supply develop synchronously. As discussed in the chapter on the meninges, dura becomes identifiable later in the embryonic period. It follows a developmental sequence from the hindbrain forward and from the cranial base upward toward the vertex of the cerebrum. This is discussed in greater detail in Chap. 12.

Anterior Branch

The *anterior branch* of stapedial stem ascends, passing "through" the intercrural foramen of stapes. (Recall that stapes is a secondary condensation of neural crest around the artery). Stapedial travels antero-laterally underneath the horizontal portion of VII. It then swerves laterally along the undersurface of tegmen tympani. There it splits, with each ramus exiting the tympanic cavity via its anterior wall. (1) The superior ramus exits the roof of the tympanic cavity and remains intracranial and follows the greater pretrosal branch of VII forward to the level of trigeminal ganglion. (2) The inferior ramus exits the anterior wall of tympanic cavity and pursues an extracranial course following the chorda tympani branch of VII outward to make contact with V3 which has descended via foramen ovale into the infratemporal fossa.

Inferior Ramus

Note: The inferior ramus is referred to as the *maxillomandibular stapedial*. Development of inferior ramus anticipates that of superior ramus by one stage. It runs along the inferior surface of tympanic roof, exits the middle ear via the *canal of Huvier*, and joins with the distal external carotid at the tip of V3 to form the composite internal maxilla-mandibular system. In the proximal one-third of the IMMA, it will supply mandible. In the distal one-third of IMMA, it produces all the arterial branches to neural crest tissues of the zygomaticomaxillary complex. The anastomosis between StVII and external carotid takes place around stage 16–17 *well before that for the orbit.*

Superior Ramus

Note: The superior ramus is frequently referred to in the literature as the *supraorbital stapedial*. Unfortunately, this term is also applied to the frontal artery seen at term. Thus, an early embryonic structure is confused with another much later branch of V1 stapedial. Furthermore, in the embryo, supraorbital is programmed by both StV1 and StV2 but only branches from STV1 enter the orbit. The V1 branch is ignored. Here, we shall apply the general term supraorbital stapedial to the combined V1–V2 branches. When the three individual arteries of the orbit are described, the names of branches will be unambiguous: ethmoid, frontal, and lacrimal.

Superior ramus takes leave of its inferior colleague, making a sudden jog straight upward through the tegmen tympani to enter the cranial cavity. The foramen through which this occurs (the "tegmen trapdoor") is, of course, a field boundary during skull development.

Superior ramus travels forward in company with the superior petrosal branch of VII along what will be the petrosal ridge (the bone has not yet formed). Upon reaching the trigeminal ganglion the fickle superior ramus "jumps ship from superior petrosal; shifting its allegiance to follow along the branches of V1 and V2. Both axes immediately give off *meningeal branches* to supply the dura of the middle and anterior cranial fossae. Subsequently, StV1 and StV2 skirt along the anterior half of the squamous temporal as far as the alisphenoid. There, StV1 enters orbit in two locations whereas StV2 terminates in the middle cranial fossa. StV1 orbital gains access via what will become the superior orbital fissure. StV1 meningeal enters via the orbit via the extreme lateral corner of the superior fissure or via a poorly appreciated *cranio-orbital foramen (Hyrtl's canal)* located at the center of the greater wing.

Note: No mention has been made yet of V3. This is because the V3 root immediately leaves the ganglion to descend into the future temporal fossa and is thus not physically available to interact with the superior ramus. Not to worry, it will meet up with stapedial via the inferior ramus at the tip of V3.

StV1: A Straight Shot to the Orbit

In the anatomic literature, StV1 destined for the orbit has been given a confusing name, the *supraorbital branch*. We are going to scrap this terminology completely (1) because it ignores the inductive role of V2, it professes ignorance of the dural branches, and (3) this name is commonly applied elsewhere to one of the major arteries of the orbit.

StV1 becomes identifiable at stage 18; at stage 29, it connects with StV2 just over the bifurcation of nasociliary and frontal nerves. During stage 19 StV1attaches to primitive ophthalmic artery; the anastomosis is complete at stage 20. The "annexation" of stapedial takes place due to the involution of the stapedial stem. Thus, *the final product, "ophthalmic artery," is a hybrid system*. Primitive ophthalmic, being derived from ICA, supplies CNS tissue exclusively, i.e., the optic nerve and globe. All remaining tissues of the orbit innervated by V1 are supplied by branches from the derivatives of the original stapedial system. The StV1 divides into ethmoid, frontal, and lacrimal arteries. All subsequent branches of the system are inductions. The various branches were discussed in Chap. 6.

StV2: A "Two-Timer"

From trigeminal ganglion, StV2 travels forward to two destinations. The *dural branch* supplies the temporal fossa. The *maxillary nerve branch* is clinically insignificant. It accompanies V2 as it exits the skull via the foramen rotundum and ducks into the pterygopalatine fossa. The artery itself is clinically insignificant.

Just behind the bifurcation point of StV1 and StV2, the supraorbital becomes a target for the middle meningeal artery derivative of lower division stapedial arising from outside the skull and programmed by V3. Extracranial stapedial (see below) passes up from below via the foramen spinosum as the middle meningeal nerve and artery. It provides high-pressure flow that supports the forward extension of StV1 and StV2 into the anterior cranial fossa; they arrive at the orbit at stage 18. During stage 19, StV1 enters the lateral orbit via the superior orbital fissure or via a separate meningo-orbital foramen in the lateral wing of the sphenoid. Sometimes this foramen is merely a lateral extension of the superior orbital fissure. At the same time, it connects medially with the developing StV1 lacrimal artery. At stage 20, if StV1 meningeal persists, after the supraorbital segment disappears this artery will remain connected with middle meningeal system, now is converted into an extension of StV2, the *anterior branch of the middle meningeal artery*.

StV3: Exile and Return

As previously stated, V3 exits from trigeminal ganglion directly out of the skull. Its sensory root makes contact with chorda tympani of VII to produce lingual nerve. In so doing, it is paired up with extracranial stapedial which follows chorda tympani. At stage 16, the distal "maxillary" branch of ECA is present at the tip of V3. At stage 17, two arteries merge to form the axis of maxillomandibular artery (IMMA). All subsequent branches of IMMA are induced by V3 proximally and V2 distally. StV3 accordingly gives off a hypaxial branch, inferior alveolar, to mandible; directly opposite it gives off an epaxial branch, middle meningeal, to the dura. The anatomic course of middle meningeal fulfills the duty of V3 to give off its very first branch to the dura—thus its recurrent course upward. At stage 18, an anastomosis takes place at the level of trigeminal ganglion between middle meningeal superior stapedial ramus, one that drastically alters the pattern of blood flow and results in a radical reconfiguration of the intracranial system.

Reunification, Disintegration, and Fate of the Stapedial System (See Diagrams)

Toward the end of the embryonic period, the anatomy of the stapedial system is drastically altered, making it virtually unrecognizable. Three anastomoses are responsible for these changes.

1. At stage 16–17, inferior division stapedial joins external carotid just lateral to the sensory root of V3 immediately below its exit from the skull. This anastomosis is the most distal branch of the external carotid system, the hybrid *maxillomandibular artery* (MMA). Middle meningeal is immediately produced.

2. From MMA, middle meningeal artery follows sensory V3 middle meningeal nerve upward to supply the dura. At stage 18, middle meningeal annexes StV2. It does not annex StV1 proximal to the orbit because this segment has already undergone involution. Thus, intracranial middle meningeal becomes a hybrid system: its anterior branch arises from StV2 and its posterior branch arises from StV3.

3. At stage 20, within the orbit, The StV1–StV2 arc is annexed by ophthalmic. The proximal StV2 segment connecting orbit with the middle meningeal system dies backward (in most cases) to trigeminal ganglion.

4. By stages 20–21, all traces if the proximal intracranial stapedial are gone. The meningeal arterial system is complete albeit the dura mater is still in the process of formation.

The common denominator of these anastomoses is the *exposure of the distal vessel to higher flow*; the proximal segment therefore involutes. Superior division of stapedial is eliminated completely back to stapes. Inferior division of stapedial distal to stapes persists as the *anterior tympanic artery*. Proximal to stapes, the stapedial stem persists as the *caroticotympanic artery*. These changes force StV1 to seek annexation with ophthalmic. StV2 becomes wholly dependent for survival on StV3 middle meningeal.

Arteries of the Pharyngeal Arches and Jaws

Let's start our revisionism by considering the neuromeric design of the external carotid at its terminal branches. Then, we will consider the ways neural induction by the trigeminal system affects the blood supply to the middle fossa. Finally, we will take on the facial nerve and how its various branches explain important vascular relationships.

Note on figures: These structures are readily accessed in Gray's anatomy. Images can be pulled up in seconds on Wikipedia as well.

Fifth Arch and Fourth Arch: Superior Thyroid Artery

(Figures 7.8 and 7.9)

The most caudal branch of ECA is the sole arterial axis for derivatives of the fourth pharyngeal arch (r8–r9) and the diminutive fifth pharyngeal arch (r10–r11). It arises just caudal to greater cornu of hyoid bone (i.e., the border of the third and fourth arches). It proceeds inferiorly under cover of omohyoid, sternohyoid, and sternothyroid muscles. It is deep to the internal branch of r8 superior laryngeal nerve (SLNi) from which it is programmed. Both nerve and artery pass through the space between middle and inferior constrictors, i.e., between Sm8 and Sm9.

Superior thyroid's topology faithfully reenacts the repositioning of arches 4–5 caudal and internal to the third arch. The fascial plane of the larynx descends from the base of the tongue. One muscle, cricothyroid, arises from the first somite (Sm8) and is attached to external surface of the larynx. Another Sm8 muscle, middle constrictor, attaches to the posterior border of the larynx. Muscles of the second somite (Sm9) are also assigned to the fourth arch. They track downward to their attachment sites within the larynx itself and between larynx and the arytenoids. The fifth arch is the putative source for arytenoids and cricoid. The spatial position of arches 4–5 terminates with the trachea and the esophagus. Remaining structures originating from the occipital somites are not associated with the pharyngeal arch system. Therefore, they must occupy a position external to the trachea-esophageal "core." Thus, the strap muscles, originating from somites 6–8, develop later and occupy a position superficial to superior thyroid artery.

Infrahyoid artery follows thyrohyoid between PA3 hyoid and PA4 thyroid cartilage. *Sternocleidomastoid artery* supplies the middle part of sternocleidomastoid, i.e., that originating from Sm8. *Superior laryngeal artery* follows internal laryngeal branch of SLN. It supplies the laryngeal cartilage and intrinsic muscles of the larynx: cricothyroid, posterior cricothyroid, lateral cricothyroid, arytenoid, and thyroarytenoid. Recall that external branch of SLN is "mono-muscle-motor," innervating cricothyroid only. Internal SLN is exclusively sensory. It divides into three branches: epiglottis, aryepiglottic folds, and mucous membrane of posterior larynx. The latter branch is the most inferior; it communicates with ILN.

The muscles of the larynx are supplied by both the superior thyroid artery and the inferior thyroid artery. This is understandable because their territories overlap. However, we can separate these groups out by neurologic criteria. Cricothyroid is supplied by SLN, the nucleus of which resides in r8; it therefore arises from Sm8, i.e., the hypaxial sector of the first occipital somite. ILN maintains base of operation in r9; thus, all remaining laryngeal muscles come from Sm9, i.e., the hypaxial sector of the second occipital somite.

Cricothyroid artery Because the cricothyroid muscle is the sole object of superior laryngeal nerve's desire, it merits it very own muscular branch.

Third Pharyngeal Arch: Ascending Pharyngeal Artery

(Figures 7.10 and 7.11)

This is one of two arterial axes dedicated to derivatives of the third pharyngeal arch (r6–r7). It arises from the posterior aspect of ECA and ascends along the pharyngeal wall in the plane anterior to longus capitis. Recall that longus capitis is *not* a pharyngeal arch derivative. It is segmental muscle originating from cervical somites 2–4. Due to its "border zone" status, its ventral fascia receives neural crest contributions from r6 to r7. These lead to sensory innervation from cranial nerve IX. Thus, ascending pharyngeal provides anterior branches to the prevertebral musculature.

Pharyngeal Arteries The intrinsic muscles of the pharynx (superior and middle constrictors) receive 3–4 branches from the ascending pharyngeal. The extrinsic stylopharyngeus gets a single branch (note that palatopharyngeus is supplied by APA).

Palatine Arteries This vessel is developmentally significant. At times, it replaces a PA2 derivative, the *ascending palatine branch of facial* artery. The interchangeability of these two vessels sheds light on a seemingly logical aspect of pharyngeal anatomy. Tonsil lies between PA1 mucosa (over palatoglossus) and PA3 mucosa (over palatopharyngeus). If tonsil is truly a second pouch derivative, how can it be in direct contact with first arch mucosa? Whatever happened to first pouch? The dilemma is resolved by recalling that amalgamation of the first two pharyngeal arches causes "disappearance" of the ventral sector of first pouch (dorsal sector persists as eustachian tube). Palatine branch arcs over the upper border of superior constrictor such that it supplies oral side of soft palate and auditory tube. Recall that auditory tube is itself an amalgamation of lateral r3 and medial r6 (r7 mastoid lying directly behind).

Prevertebral Arteries These supply posterior pharyngeal wall structures: muscles (longus capitis and longus colli) and nerves (X, XII, and sympathetic chain). The muscles are segmental, arising from the ventral myotomes of the cervical somites. Motor control is provided by the ventral rami of C1–C4, not from the cervical plexus. Hence, their motor innervation enters from the dorsal side. The dorsal fascia arises from segmental neural crest; it is supplied by ventral sensory nerves. The ventral fascia is admixed with third arch neural crest from r6 to r7. Prevertebral branches of ascending pharyngeal penetrate the ventral aspect whereas branches from the ascending cervical branch of inferior thyroid artery enter dorsally. These are the two arterial systems, one dedicated to PA3 and the other, a part of the segmental supply to the neck, anastomose within the muscles.

Inferior Tympanic Artery Cranial nerve IX has a tympanic branch which is sensory to the medial wall of the tympanic cavity. This artery, induced by the nerve, enters via the *tympanic canaliculus* of temporal bone. Note that mucosa lining the promontory, along with that of medial eustachian tube, is in continuity with third arch pharyngeal mucosa.

Meningeal Arteries Three vessels stem from APA supply dura; these enter the skull via three foramina. Jugular foramen, containing IX, X, and XI, and (occasionally) hypoglossal foramen supply posterior fossa. A second branch traverses foramen lacerum to gain access to middle fossa.

Second Pharyngeal Arch: Occipital Artery

(Figure 7.12)

The posterior analog of linguofacial, occipital arises deep to two second arch muscles, posterior belly of digastric and stylohyoid. XII winds around from behind and crosses over the occipital artery which, in turn, crosses internal carotid, X and XI. It ascends in the space between the mastoid process and the transverse process of C1, i.e., between bone formed from Sm7 or the first somite versus bone derived from the seventh and eighth somites. Thereafter, it is directly applied to the occipital groove of the mastoid. It is covered by posterior digastric and sternocleidomastoid in company with the "capitis chums" (splenius and longissimus). The occipital "floor" consists of rectus capitis lateralis, obliquus superior, and finally, semispinalis capitis. After being entrapped in the confines of SCM and trapezius, occipital artery "escapes" upward, slithering its way through the fascia to unite with branches from posterior auricular and superficial temporal. It follows greater occipital nerve (c2–c3).

Muscular Arteries Posterior digastric^{Sm6} and stylohyoid^{Sm6} are supplied by occipital. It also sends collaterals to their nearby "capitus chums," splenius capitis, and longissimus capitis.

Sternocleidomastoid Artery This vessel passes posteriorly and downward *over* hypoglossal nerve to enter the substance of SCM in its upper one-third. Within the substance of the muscle, it joins with sternocleidomastoid branch of superior thyroid to the middle one-third. Sternocleidomastoid, being supplied by spinal accessory nerve, has likely origins from somites 1–4.

Auricular Artery This branch supplies mastoid process and the skin in back of the concha. These zones are composed from a mixture of r7 neural crest and Sm7 PAM. The dermis is supplied by epaxial sensory branches of C2–C3; these provide the induction.

Mastoid Artery This branch forms a transcranial anastomosis with middle meningeal. Recall that MMA to the membranous supraoccipital bone develops from intracranial division of stapedial (posterior branch). Externally, this vessel can arise from either occipital directly or via its auricular branch. This artery supplies the mastoid air cells, diploic space, and dura. Recall that membranous supraoccipital is bilaminar. Its internal lamina is supplied by ICA stapedial (which originates from r4 to r5). The external lamina is supplied by ECA occipital, the neuromeric coding of which is likely r6–r7.

Meningeal Arteries The original supply to the periosteal lining of posterior fossa (membranous bones) was from the intracranial stapedial. Meningeal branches from the occipital represent an anastomosis with that system. They enter the skull via the jugular foramen and condyloid canal. Induction is by cervical sensory branches (C2–C3) to the local fascia and extracranial periosteum.

Descending Branch Perhaps the most intriguing branch of occipital is that which descends away from the skull, giving off two branches to the neck. The *superficial branch* erupts upward through splenius capitis and spills into sternocleidomastoid, within which it anastomoses with superficial br. of transverse cervical. Once again, we see the neuromeric complexity of SCM. Its upper one-third is supplied by the SCM branch of occipital. The middle one-third receives contributions from descending branch and transverse cervical branch. The lower one-third is supplied by inferior thyroid. The significance of this is discussed in Chap. 9 and 10 cf on myology. The deep branch travels between the paired longitudinal "semispinalis sisters" (capitis and cervicis). It makes anastomoses with vertebral artery and with deep cervical artery from the thyrocervical trunk. Recall that vertebral, inferior thyroid, and transverse cervical share a common mechanism of formation (vide infra).

Terminal Occipital Branch The blood supply of occipitofrontalis supplies the overlying scalp. An additional transcranial branch accesses the dura via the *parietal foramen*. Again, the induction if C2–C3.

Second Pharyngeal Arch: Posterior Auricular Artery

(Figures 7.13 and 7.14)

Just opposite the apex of styloid process, this not-sizeable artery makes its appearance above posterior digastric and stylohyoid. If one elevates parotid gland, it can be followed along the styloid process. It then follows a groove between ear cartilage and mastoid bone and then divides.

Stylomastoid Branch Supplying mucous membrane of the tympanic cavity, mastoid antrum, and cells as well as the semicircular canals (all of which are r6–r7 structures), this artery is induced by branches of glossopharyngeal nerve. It forms the *posterior tympanic artery* that constitutes the posterior half of a *tympanic vascular ring* supplying internal surface of tympanic membrane. The other component of the ring, *anterior tympanic artery*, is apparently a branch of IMMA but, in reality, is a derivative of the stapedial within the tympanic cavity. The tympanic ring represents yet another example of pharyngeal arch amalgamation as it applies to the middle ear. Stylomastoid makes an anastomosis with the *petrosal branch of the middle meningeal artery*, itself derived from IMA.

Auricular Branch This branch ascends between auricularis posterior and auricular cartilage. It is induced by cranial nerve IX sensory branches to the dermis and fascia. Some branches extend outward to the rim; others penetrate the posterior cartilages (derived from second arch neural crest) to reach the external surface of the posterior pinna. This identifies the posterior pinna as an epaxial structure associated with, but not derived from, the third arch.

Occipital Branch Unlike the occipital proper, occipital branch of posterior auricular does not travel beneath SCM. Instead, it passes over SCM to supply scalp behind and above the ear. Quite logically, it anastomoses with occipital artery.

Second Pharyngeal Arch: External Facial Artery

(Figure 7.15)

The fascinating, mysterious anatomy of the facial artery begins with its misleading name. All craniofacial surgeons are aware of the important VII-innervated muscles of facial animation and the SMAS fascia from which they originate (we addressed this subject at length previously). These peripatetic wanderers take up residence in diverse sites of the head and neck. Since they all arise from Sm6, it would be anatomically gratifying if we could somehow unify them around a common arterial source. Alas, this simplistic approach is not possible. The facial artery serves a wide variety of structures including the lateral eyebrow, levators of the lip, anterior body of the mandible, posterior digastric, and the mastoid process. These targets arise from the first three pharyngeal arches plus non-arch mesenchyme.

The branches of the facial artery are *not* distributed according to the motor branches of cranial nerve VII. The key to understanding the vascular anatomy of the facial system is the sensory supply to the fasciae of second arch structures. As previously discussed in Chap. 5, sensory nerves to the fascia are responsible for induction of arterial supply. Facial fasciae are derived from neural crest mostly from r1 to r3 but with some contributions from r4 to r5 and from r6 to r7. Sensory nerves to these fasciae come from V1 to V3, with additional supply from VII, and IX. Using this model, the branching pattern of the facial artery makes sense.

Ascending Palatine Artery

This vessel originates just past the take-off of facial artery and ascends, hugging the pharynx between stylopharynge-

us^{Sm6} and the more external styloglossus^{Sm8}. It continues its pharyngeal pathway, being sandwiched between superior constrictor^{Sm7} and PA1 medial pterygoid^{Sm4}. This positions it between PA3 and PA1. Upon reaching the skull base, it divides at levator veli palatini^{Sm7}. (1) A palatal branch follows the course of LVP over the *upper border* of superior constrictor. This brings it to the *nasal aspect* of the soft palate. There it anastomoses with the StV2-induced greater palatine artery. (2) An auditory branch passes through superior constrictor, supplies the palatine tonsil, and terminates on the medial lamina of eustachian tube. It has anastomotic relationships with its neighbor, tonsillar branch of facial, and, significantly, with ascending pharyngeal. This vessel is third arch all the way. It takes care of all Sm7 muscles to palate. Its relationship to LVP tags it to the PA3 side of auditory tube.

Tonsillar Artery

(Figure 7.16)

This branch treks upward between deeply placed medial pterygoid^{Sm4} and, more lateral, styloglossus^{Sm11}. It lies directly along lateral aspect of superior constrictor^{Sm7}, which it perforates. Its distribution is eminently PA3; its branches spread out into the root of the tongue and the tonsillar bed. It enters inferior aspect of pharyngeal tonsil as its main artery. Traditional embryology classifies tonsil as a derivative of the second pharyngeal pouch. Our revisionist view is that pouches 1 and 2 are amalgamated. The tonsillar crypt is an interface zone between three pharyngeal arches, laminated together from medial to lateral. Lateral r3 mucosa is V3-innervated, while medial r7 mucosa is IX-innervated. Neuromeric coding of the underlying muscles surrounding tonsil is *spatial*. Palatopharyngeus runs beneath medial r7 mucosa; its innervation (both sensory and motor) is cranial nerve IX, a perfectly sensible arrangement.

Laterally, palatoglossus is a different story. Like all tongue muscles, palatoglossus is a perfect opportunist. It arises from tongue base and sweeps upward, exploiting a vacant pathway beneath r3 mucosa to access the soft palate from the oral side. The fascia of palatoglossus is likewise derived from r3 neural crest. Sensory input from this muscle is processed by the trigeminal. Thus, *sensory nerves* to the fascia are responsible for induction of the tonsillar branches from the facial artery. Although hypoglossal nerve is motor to palatoglossus (an induces its own accompanying perineural arteries), it is not responsible for the vascular supply to the muscle.

Glandular Artery

This segment of facial artery travels external to submandibular gland. Recall that lingual artery, the previous external carotid derivative, pursues a course internal to hyoglossus (and therefore internal to submandibular gland). Three to four branches supply the gland, with a distinct branch following along Wharton's duct itself. The duct runs between mylohyoid[Sm4] and hyoglossus[Sm9]. So too, does XII. Hypoglossal nerve is a purist. It stays external to hyoglossus. IX runs external to middle constrictor; it stays internal to the plane of hyoglossus and styloglossus. This makes sense because *tongue muscles cannot violate previously established pharyngeal arch neurovascular planes.*

V3-lingual nerve has no such constraints. Its task is to run from just inside mandible all the way forward to supply the mucosa lining floor of mouth and the anterior two-third of tongue. To accomplish this goal, lingual nerve unscrupulously crosses *over* hyoglossus and *under* Wharton's duct (and, perforce, the ductal branch of glandular artery).

What significance can we attach to the glandular branch? Let's focus on its most distal component, the *ductal branch.* Recall that the embryology of the major salivary glands involves invagination of epithelial duct into a mesenchymal mass. Interactions cause formation of glandular structures, all connected back to the ductal apparatus. In essence, the gland develops in an "r3 submandibular sanctuary" with mandible above, anterior belly of digastric[Sm4] in front, stylomandibular ligament (linking r3 mandible to r6 styloid process) and mylohyoid[Sm4] are deep to it. Thus, pathway of ductal branch recapitulates development of the gland itself.

Submental Artery

The last (and largest) of the cervical branches is dedicated to the muscles of two triangles, submental and submandibular. It lies sandwiched between two Sm4 muscles: mylohyoid (deep) and anterior digastric (superficial). After supplying these, it makes two anastomoses: (1) sublingual branch of sublingual and (2) mylohyoid branch of inferior alveolar. *These represent a mediation between a deep system supporting tongue and a superficial system belonging to StV3.* At the symphysis, submental takes an upward course and splits. A *superficial branch* runs external to the second arch levator labii inferioris; it supplies V3 skin of the lower lip. Here, it encounters and anastomoses with the other artery dedicated to this zone: inferior labial. A *deep branch* runs internal to the muscle, supplying it and the underlying bone. Anastomoses with inferior labial and mental arteries complete its course. What submental artery demonstrates is that sensory innervation from V3 induces blood supply for muscles from two distinct arches: PA1 (mastication) and PA2 (facial animation).

Inferior Labial Artery

Near the oral commissure, this artery dives deep to depressor anguli oris. Since the latter is inserted on orbicularis oris, the artery must penetrate the latter. It runs along the margin of muscle beneath the mucosa. Recall that deep orbicularis oris (DOO) is programmed by r3 mucosa while superficial orbicularis oris (SOO) is programmed by r3 dermis. These two planes meet at the white roll (the boundary between mucosa and skin). The sheet-like DOO follows the curl of lower lip vermillion until reaching skin. It thus takes the form of an inverted letter J until it makes contact with SOO. Tucked within the crotch of the J, one finds the inferior labial artery. Mental branch of inferior alveolar artery connects to it.

Superior Labial Artery

Slavishly imitative of its inferior colleague, superior labial pursues a similar anatomic strategy. Diving deep to zygomaticus major, it gains access to the same plane between DOO and SOO of upper lip. Superior and inferior labial, afraid to part company, remain connected at the lateral commissure. Vertical branches (termed septal and alar) ascend in the fat plane between DOO and SOO until reaching their respective nasal targets.

Lateral Nasal Artery

Ascending along the side of the nose terminal, facial artery overlies the lacrimal duct, supplying the skin of Tessier zone 4, ultimately making anastomoses with ECA infraorbital and ICA dorsal nasal branch (infraorbital branch) of ophthalmic. Lateral nasal branches supply intrinsic nasal muscles. Nasal skin medial to the lateral nasal artery is of prosencephalic origin and contains additional blood supply from anterior ethmoid branches of ophthalmic.

Angular artery This represents the termination of the facial artery. Running beneath levator labii et alaeque nasi, it supplies orbicularis oculi[Sm6] and lacrimal sac. SMAS fascia in this region is admixed with r2 neural crest. Sensory fibers from V2 are responsible for induction of the angular artery.

Muscular arteries Numerous branches from facial artery support the muscles of the region. For example, the cervical division supplies medial pterygoid[Sm4] and stylohyoid[Sm6] while the facial division supplies masseter[Sm4], buccinator[Sm6], and the remaining muscles of facial expression[Sm6].

The take-home message of external facial is that it provides blood supply to skin, mucosa, and muscles belonging to the first two pharyngeal arches. Working in tandem with internal maxillary artery, EMA supplies every structure of the PA1–PA2 region. Naturally, a number of third arch structures abut the posterior border of this zone. Like a good soldier, EMA discharges its duty and takes care of them as well.

There is one glaring exception to this rosy picture, the tongue. The presence of a ***common linguofacial artery*** explains the situation neatly because it neatly ties together several intriguing aspects of blood supply to the face. Sm6 produces both muscles of facial expression in the superficial plane and muscles of mastication involving the deep plane. The migration pathways of both groups of muscles follow the motor branches of VII. All sources of blood supply to the superficial plane muscles of facia follow branches genetically related to r4–r5. The blood supply to second arch muscles of mastication has arisen from an anterior, more distal vessel just proximal to the internal facial, the lingual artery.

Second Pharyngeal Arch: Lingual Artery (Linguofacial)

(Figures 7.17 and 7.18)

Frequently, the lingual and facial arteries emerge as a *common vessel* that subsequently divides into its respective branches. The developmental significance is that these two arteries are assigned to first pharyngeal arch structures and save the neural crest bone derivatives associated with the jaws, supporting bones, and the oropharynx.

Examination of the relationship between lingual artery and the muscles it serves adds to our understanding of how occipital somite derivatives contribute to the head and neck.

The tongue is a muco-epidermal sack slung between r3 mandible and r5 hyoid. The neuromeric origins coding of this envelope can be deduced from the sensory supply of its submucosa because, just like dermis, it originates from neural crest. The anterior two-third of tongue receives general sensation from V3 and special sensation (taste) from VII. Thus, the submucosa of this zone arises from r3 neural crest while the population of taste organelles likely originates from r5 neural crest. This pattern is entirely consistent with pharyngeal arch development. After all, PA1 and PA2 are completely integrated; all second arch derivatives are enclosed within the confines of first arch. The posterior one-third of tongue receives both general sensation and taste from IX. PA3 is self-contained unit sandwiched between the PA1–PA2 amalgam anteriorly and the fourth arch epiglottis.

The epithelial envelope of the tongue is a potential space. It is invaded by the hypoglossal fascia, a neural crest structure originating from r8 to r11. This fascia is populated by myoblasts originating from the hypaxial myotomes of occipital somites 1–4. The migratory pathway of occipital fascia is recapitulated by the spatial organization of the hypoglossal nerve. Lingual muscles bear no relationship to the pharyngeal arch system. Nonetheless, they insert themselves into that system by exploiting tissue planes that give them access to the interior of the sack. In so doing, they negotiate their way around pre-existing muscles, in a constant search for unoccupied osseous binding sites. For this reason, the muscles of somites 1–4 are known collectively as the *anterior hypobranchial cord*. Note that the cervical strap muscles, arising from somites 5–7, are derived from the ancient cora-

comandibularis, span from the c1 level of pectoral girdle beneath all the arches to reach the mandible, and are known collectively as the *posterior hypobranchial cord*.

When these myoblasts migrate, they are never far from metabolic support, taking advantage of the unorganized plexus of PA1–PA4. The blood supply of these muscles is a function of the target to which they are assigned. Once in place, they are surrounded by local fasciae, from which they gain their blood supply. As you review the distribution of branches of the lingual, note that they are directed, not toward muscles per se, but to attachment sites, periosteum, and submucosa in other words, to the fascia. Local sensory innervation to their respective fasciae determines the final entry point of the arteries into the muscle bellies.

The migratory pathway of tongue muscles is indicated by hypoglossal nerve. Tongue muscles are also instructive because their blood supply has no anatomic relationship to that of occipital somites. Posterior hindbrain is supplied by derivatives of the longitudinal neural arteries. The LNAs are supplied antegrade from the circle of Willis and retrograde by the vertebrals. By way of contrast, laryngeal muscles arise from Sm8 to Sm9 (somites 1–2) and behave in somitomeric fashion, i.e., they follow a pharyngeal arch pathway. Their arterial supply is that of the combined source for the fourth and fifth arches, i.e., superior thyroid.

Before considering the individual branches of the lingual artery, for the sake of orientation, let's review the muscles of the tongue. Details of these muscles are discussed in the chapter on myology. Four extrinsic and four intrinsic muscles make up the tongue. The myoblasts arrive at the future floor of mouth forming a mound-like structure, the *tuberculum impar*. From this origin they fan out, seeking insertion sites. They are arranged in two groups. *Extrinsic muscles* position the tongue in space. These arise, one per somite, from somites 1–4 and migrate in cranial-caudal sequence. Their primary insertion sites are proximal to distal: lateral to medial from the mucosal surface inward. The muscles migrate and insert in the following order: styloglossus (r5 styloid process), hyoglossus (r5 greater cornu), chondroglossus (r5 medial hyoid body), and genioglossus (r3 mandible). The secondary insertion sites have a lateral to medial pattern: mucosa bears styloglossus, intermediate layer is hyo- and chondroglossus, and finally genioglossus gains the tubercles of the mandibular midline. *Intrinsic muscles* insert themselves into the interstices in a deep-to-superficial, lateral-medial pattern.

The pathology of tongue clefts is instructive because it is always in the midline. Thus, defects arise in the most distal derivative, the one that arrives last.

Suprahyoid Artery This small branch follows superior border of hyoid bone supplying its muscle attachments. Its relative position as the "last" (most proximal) lingual vessel is confirmed by anastomoses with the PA4–PA5 superior thyroid artery.

Dorsal Lingual Artery Arising beneath hyoglossus (tongue inserting into second arch hyoid), these branches are dedicated to posterior dorsum of tongue, i.e., the posterior one-third or r7 zone. They also supply mucosa up to r7 soft palate and r8 epiglottis. Anastomoses with tonsillar arteries of facial have been documented.

Sublingual Artery This vessel exploits the space between genioglossus lateral and mylohyoid (connecting r3 mandible with r4–r5 lesser cornu of hyoid). It supplies mucoperiosteum of mandible. It connects with submental branch of facial.

Deep Lingual Artery The "latest" (most distal derivative of lingual), this vessel lies lateral to genioglossus, the "latest" muscle, which inserts into r3 genial tubercle. Lateral to the artery is r3 mucous membrane (anterior two-third).

The arterial sequence of the tongue follows the same proximal-distal and medial-lateral pattern. Suprahyoid supplies the initial muscles arising from the third arch zone hyoid. Dorsal lingual goes further to the third arch mucosa of the tongue. Sublingual and deep lingual both seek out first arch tissues of the r3 mandible in a medial-lateral order. This sequence demonstrates the principle that the earliest muscle of a group will bind to the first available site. It acts as the "pathfinder" for subsequent migrations. it anastomoses at the tip of the tongue with its colleague from the opposite side. Deep lingual demonstrates, once again, *the migratory "history" of subsequent muscles will be reflected in the spatial positioning of their arterial branch with respect to the pathfinder*. In the case of the tongue, more proximal branches serve muscles that migrated later, were forced to use a more internal route, and inserted more distally.

Second Pharyngeal Arch: Superficial Temporal Artery

(Figure 7.19)

Considered by some as the continuation of the ECA, STA travels under cover of auricularis anterior muscle. Both temporal and zygomatic branches of cranial nerve VII cross over this vessel, while V3 auriculotemporal nerves travel posterior to it. STA crosses the posterior aspect of zygomatic bone; 5 cm later it divides into frontal and parietal branches.

Transverse Facial Branch This is the largest branch of STA. It emerges from parotid gland prior to exit of STA itself. It treks forward obliquely between zygomatic arch and parotid duct. This can be a clinically useful relation in locating the duct deep within the gland. The vessel rests

directly on masseter that benefits from its largesse, along with parotid and overlying skin. Branches of facial nerve are fellow travelers.

Middle Temporal Branch This artery supplies **Tessier zone 9** (transverse facial may also contribute). Appearing immediately above zygomatic arch, it plunges through the fascia and follows the course of squamous temporal bone[r3 and Sm4], supplying both it and temporalis muscle[Sm4]. Not surprisingly, it encounters anastomotic branches of *deep temporal artery* from IMA.

Zygomatico-Orbital Branch This artery supplies **Tessier zones 8 and 7** (postorbital and jugal). This vessel is induced by V2 lies superficial to temporalis muscle fascia and deep to subgaleal fascia. It travels forward *along zygomatic arch* until it reaches the orbit. At this juncture, it supplies orbicularis oculi. It also makes an important anastomosis with two branches of ICA ophthalmic: lacrimal and palpebral.

Anterior Auricular Branch This artery supplies the r3 skin of tragus, anterior external auditory canal, and anterior auricle. Anastomoses with posterior auricular signify the amalgamation of PA1, PA2, and PA3.

Frontal Branch Supplying VII-innervated muscles of the forehead, this branch is induced by V2. It sweeps forward over Tessier zone 9 where it cooperates with both supraorbital and supratrochlear to supply the skin of Tessier zones 10–13.

Parietal Branch As the "big Kahuna" to the temporo-parietal scalp, this artery provides vital supply for the clinically relevant fascia of this region. Induction is by V3.

First Pharyngeal Arch: Internal Facial (Maxillomandibular) Artery

(Figures 7.20, 7.21, 7.22, and 7.23)

Superficial temporal artery and internal maxillary artery are often considered the "terminal branches" of the external carotid system. This is developmentally incorrect. Recall that the stem vessels of ECA represent a rearrangement of the older aortic arch system. They supply all structures of the first and second pharyngeal with the exception of the mandible, zygomaticomaxillary complex, and dura. ECA of the IMMA supplies V2–V3 dermis, fascia, glands, muscles of mastication, and the fascia of anterior preauricular muscles associated with r3 neural crest cartilages. The final stem for the first arch is internal maxillary which immediately fuses with the stapedial to form internal facial or internal maxillomandibular. This hybrid we shall call MMA and it is traditionally divided into three sectors: mandibular, pterygoid, and pterygopalatine. Those branches serving oral mucosa or muscles of mastication are supplied by the ECA. *All remaining arteries are* *derived from the **stapedial system** and induced by V2 and V3*. In our description, all stapedial arteries will be highlighted in red.

Mandibular Sector (Proximal One-Third)

This sector is a hybrid of the stapedial maxilla-mandibular artery and the ECA internal maxillary. It gives off two ventral branches which are fused initially fused, one, mylohyoid, from ECA and the other, inferior alveolar, from StV3. There are four cranial branches of stapedial derivation. They connect ECA maxillary stem vessel with (1) the tympanic cavity and (2) the dura; the StV3 inferior alveolar supplies the neural crest components of the mandible.

Mylohyoid Artery This vessel travels with inferior alveolar as the first ventral branch of MMA. It then diverges away to supply mylohyoid, a first arch muscle of mastication (part of original first arch).

Inferior Alveolar Artery Great clinical significance pertains to this artery as it supplies four out of five mandibular developmental zones. The vessel diverges from IMA as the latter winds around the neck of the mandible and the sphenomandibular ligament. Recall that the SP ligament unites r2 alisphenoid with r3 mandible immediately behind mandibular foramen (the entry point for inferior alveolar nerve and artery). Inferior alveolar nerve and artery descend along medial ramus, supplying it as a separate bone field. Immediately prior to the foramen, the mylohyoid nerve and the lingual nerve are given off. These induce companion arteries to muscle and mucous membrane of the lingual sulcus. More anteriorly, the sulcus is supplied by the sublingual branch of lingual artery proper. Within the mandibular canal, it supplies the individual dental units.

Distal-to-proximal organization of inferior alveolar reflects the maturation sequence of teeth. The distal branches are biologically the "oldest." *Mental branch* emerges from mental foramen. There it encounters two components of the facial system. It makes a proximal anastomosis with *submental* and a distal anastomosis with *inferior labial*. Within the substance of mandible, the incisors are supplied by an *incisor branch*. Dental development distinguishes between the incisor zone, the cuspid or premolar zone (canine and bicuspids), and tricuspid or molar zone. No gross distinction of IAA anatomy pertains to these zones, proximal IAA simply consists of branches distributed one by one to the tooth roots and thence to the dental pulp. It is not surprising that teeth arise from the inner core of PAM enshrouded by a covering mantle of neural crest enamel.

Mandibular evolution, as previously discussed, is a perfect demonstration of Herring's law, in which the number of craniofacial bones for a structure becomes simplified with evolution. The original tetrapod mandible had nine bones, four of which translocated into the skull as components of tympanic bone, malleus, and incus. Mammalian mandibles are made of an apparently single bone, the dentary. However, from a neurovascular standpoint, mandible retains five developmental zones. Condyle is represented by deep auricular. Inferior alveolar supplies the non-tooth-bearing ramus and three dental zones (each with distinct sensory supply).

Deep Auricular Artery One of two derivatives of the stapedial stem within the tympanic cavity, this misunderstood artery is traditionally described as emerging from the parotid behind the TMJ, to supply the anterior wall of the external canal and the external surface of tympanic membrane. Following V3 sensory nerves are distributed between two r3 neural crest derivatives: tympanic temporal bone and TMJ (condylar head). This demonstrates the developmental unity between condyle and posterior joint space.

Mandibular hypoplasia, seen in craniofacial microsomia and otomandibular (Goldenhar) syndrome, demonstrates a clinical spectrum revolving around these fields, presenting as three grades of severity as described by Pruzansky (modified by Kaban et al. [1] and Bergfeld and Heike [2]). In class I, the mandible is small, with normal musculature bone volume in the glenoid fossa. IIa demonstrates a hypoplasia and malformation of glenoid fossa, TMJ, and ramus. The musculature is, like class I, essentially normal. Despite this, the joint is functional. In IIb inferior and medial malposition of the joint prevents symmetric opening of the mandible. The musculature is frankly hypoplastic. Class III presents with agenesis of the ramus. Obviously, glenoid fossa is also absent.

Anterior Tympanic Artery This artery follows V3, inserting itself into the tympanic cavity via the petrotympanic fissure. It promptly forms a vascular ring around tympanic membrane; it represent an anastomosis between V3 anteriorly and VII sensory posteriorly. As such, it represents a derivative of the original stapedial within the tympanic cavity. Anterior tympanic makes two other anastomoses of developmental important within the tympanic cavity. The first is with *caroticotympanic* branch from the ICA (the remnant of the embryonic hyoid artery). The second is with the *artery of the pterygoid canal* (the remnant of embryonic mandibular). The physical association of these two arteries mirrors the genetic relationship between the TMJ field and the middle ear field. Ossicular abnormalities are often seen in conjunction with mandibular hypoplasia. Simultaneous "hits" to deep auricular and anterior tympanic fields would explain this spectrum of pathology.

Middle Meningeal Artery Dural arteries represent not only blood supply but pathways of mesenchymal migration that contribute to the formation of the dural layers themselves. The inner layers of meninges, pia and arachnoid, arise from neural crest cells that flow from the neural folds downward around the neural tube-like honey poured over an apple. The outer layer, dura, originates from neural crest associated with surrounding somitomeres. In the parietal zone, PAM from Sm4 contributes as well. Dural arteries are primarily distributed to bone but also contribute perforating branches into underlying meningeal layers.

The middle meningeal artery (obviously from ECA) is the largest arterial source to the dura. It supplies membranous bone of the entire middle fossa, part of anterior fossa, and part of membranous posterior cranial fossa. These bones are (with exception of parietal) of exclusive neural crest derivation. Difficulty arises when we consider additional arterial sources: (1) *Anterior meningeal arteries* from ICA anterior ethmoid supply the anterior fossa. (2) *Perforating dural arteries* from occipital supply membranous posterior fossa. (3) *Posterior meningeal arteries* from the vertebral supply chondral posterior fossa. What sense can we make of this system? As we shall see, the answer lies in the original wiring diagram of extracranial and intracranial stapedial system.

Let's consider first the gross anatomy of MMA and then concentrate on its individual branches. MMA arises proximal to mandibular in 94% cases. Three percent cases are co-extent and 3% are distal. In general, MMA follows the upward trajectory of V3 to the dura. Why this recurrent pathway? The answer is that from trigeminal ganglion, V3 immediately exits the skull before having a chance to give off a dural sensory branch. Once it arrives at the level of trigeminal, StV3 MMA connects with the superior division of stapedial which had been supplying StV2 to the middle fossa and StV1 to the anterior fossa.

Prior to this anastomosis, superior division was a relatively low-flow system due to a small caliber stem from the tympanic cavity. StV3 MMA brings high flow from external carotid. The altered hemodynamics cause/contribute to a drastic remodeling of the system. From the tympanic cavity forward, proximal stem of StVII involutes making StV2 entirely dependent on StV3 MMA. At the same time, the segment of StV1 between trigeminal ganglion and the orbit also degenerates. This leads to annexation of the STV1 orbital system by ophthalmic. At times, this annexation does not happen. In such cases, the entire orbital system assumes its supply from middle meningeal via the StV2 orbit.

The target tissue of the meningeal system consists of neural crest from r1 to r3 that encompasses the entire brain. Although MMA is itself an StV3 artery, its anastomoses permit it to supply the dura over the entire cerebrum.

The physical trajectory of middle meningeal is as follows. It travels between lateral pterygoid[Sm4] and the neural crest sphenomandibular ligament[r2-r3]. Auriculotemporal V3 splits around MMA; the vessel then enters the skull via *foramen spinosum*. Several important branches are given off just after entry into the skull. We shall describe these later. It then follows a groove within PAM of the derivative alisphenoid[Sm3] which it supplies and then divides.

The *anterior (frontal) branch* passes over greater wing of sphenoid. It then runs in a canal at the *alisphenoid angle* of parietal bone where it spreads out. The posterior (parietal) branch follows squamous temporal bone and reaches parietal bone some distance ahead of the *parieto-mastoid angle*.

Immediately upon entering foramen spinosum, middle meningeal gives off significant *non-dural branches*. (1) *Ganglionic branches* spread over the trigeminal ganglion, the roots of V1–V3, and a bit of nearby dura. (2) *Petrosal branch* enters the hiatus for greater pretrosal nerve and (apparently) runs *backward* to reach facial nerve. It supplies VII and anastomoses with stylomastoid branch of posterior auricular. The alert reader will detect that *stylomastoid supplies VII as it transits from the skull*. Recall that the disintegration of distal aspect of stapedial stem occurs inferolateral to foramen ovale. Foramen spinosum arises at that site. VII and its companion stylomastoid artery are surrounded by alisphenoid mesenchyme. Facial nerve is the likely induction agent for these two arteries. (3) *Superior tympanic branch* follows VII within the canal for tensor tympani to supply the mucosa and muscle. (4) *Temporal branches* represent connections between the intracranial stapedial system and the extracranial stapedial system. These vessels pass downward through the horizontal portion of greater wing. Upon entering the infratemporal fossa, they connect with branches of deep temporal artery. (5) The most distal component of MMA is the *lacrimal anastomotic branch*. This passes into the orbit through either greater orbital fissure or via a separate canal in the greater wing. Here, it encounters recurrent meningeal branch from the lacrimal itself.

Accessory Meningeal Artery A remnant of *primitive trigeminal artery*, this vessel arises from either the maxillary stem or as part of middle meningeal. It follows V3 into the skull via the foramen ovale. It supplies the semilunar ganglion and surrounding dura.

Our study of the mandibular artery teaches us important lessons. (1) Laterofacial microsomias (craniofacial microsomia, Treacher-Collins-Franceschetti, and Goldenhar syndrome) all affect the mandible, TMJ, and middle ear in varying ways. Neuromeric relationship involving fields supplied by deep auricular, anterior tympanic, and inferior alveolar arteries provides a unifying explanation as to the various forms of pathology encountered. The common denominator among the three fields is that they all involve interaction with V3. Disturbances in r3 neural crest and Sm4 PAM at a very early stage of development, up to gastrulation itself, may constitute the fundamental pathology in the syndromic states. (2) Stapedial artery embryogenesis reflects relationships r0–r5, both epaxial as in the vascular supply of the dura mater, and hypaxial as in the amalgamated state of pharyngeal arches 1 and 2. The system is programmed in sequence by VII and V. *All dural arteries (save those from vertebral to the chondral posterior fossa) originate from the stapedial system.*

Temporal Sector (Middle One-Third)

These branches are strictly muscular. All of them represent the maxillary stem of ECA. In these situations, V3 is both somatic sensory and motor; it provides the induction.

Deep Temporal Artery This artery has two branches, both of which run in the plane between temporalis muscle and membranous squamous temporal and parietal bones. Anterior deep temporal sends communicating branches through greater wing of sphenoid and through zygoma to connect with lacrimal artery in the orbit. Because these vessels run low rather than high, they are likely to anastomoses with the remnant of StV2, the original artery to lacrimal gland, rather than with StV1, the lacrimal artery associated with ophthalmic stability.

Pterygoid Arteries These branches supply the pterygoids.

Masseteric Artery Like mandibular, masseteric artery arises between neck of mandible and sphenomandibular ligament. It travels through mandibular notch in company with V3 masseteric nerve. It anastomoses with (1) masseteric branch of facial and (2) transverse facial branch of STA.

Buccal Artery This small vessel follows the course of the buccal nerve. Its supplies posterior buccinator (facial artery supplies anterior buccinator). It also supplies subjacent mucosa.

Sphenopalatine Sector (Distal One-Third)

All components of MMA in the distal sector are strictly V2 stapedial derivatives (vide infra). These arteries follow sensory branches of V2. All of them traverse foramina or canals.

Posterior Superior Alveolar Artery This artery supplies **Tessier zone 6**. Just as maxillary artery enters the pterygopalatine fossa, it gives off posterior superior alveolar, often along with infraorbital as a common stem. It travels down the tuberosity of maxilla, giving branches both external mucoperiosteum and the internal lining of maxillary sinus. It is the predominant supply of sinus mucosa. PSA is distributed over the buccal surface of the gum as far as canine. Recall that bucco-labial gum of premaxilla is supplied by medial sphenopalatine. Thus, PSA is the predominant source of blood flow to mucosa of maxillary sinus and buccal mucoperiosteum of gum.

Anterior Superior Alveolar Artery This vessel, also called infraorbital artery, runs along the infraorbital groove of the orbital floor and then traverses through the infraorbital rim via a short canal to enter the deep tissues of the anterior face, while in the sinus, it gives off arteries to three zones.

Orbital floor branches These arteries supply **Tessier zones 4 and 5**, specifically the inferior orbital rim. They arise while in the infraorbital groove. Medial and lateral twigs interact with their respective superior alveolar branches to supply These provide additional supply to inferior rectus and inferior oblique; a lateral twig goes up to lacrimal. Extracranial StV2 enters lacrimal gland from below, inferior to the plane of the globe. Intracranial StV2 enters lacrimal gland from above, superior to the plane of the globe. Recall that lacrimal gland has two lobes. Infraorbital StV2 supplies the orbital lobe and middle meningeal StV2 supplies the palpebral lobe. As previously described, in formation of the definitive orbital vessels from the ophthalmic stem, StV1 connects with MMA StV2. With the demise of the proximal stapedial, StV1 remains the dominant source of blood flow to lacrimal gland.

Medial anterior superior alveolar artery This vessel supplies **Tessier zone 4**. It targets maxillary wall medial to infraorbital foramen and lateral to nasolacrimal duct, medial one-third of infraorbital rim, and a dental branch to the canine. It also provides collateral support to frontal process of maxilla and therefore the incisors. Traditional texts pay little attention to premaxilla as a separate bone; they ascribe blood supply of incisors solely to ASAA.

Lateral anterior superior alveolar artery This artery supplies **Tessier zone 5**. It targets maxillary wall lateral to infraorbital foramen and middle one-third of orbital rim, and dental branches to the premolars.

Facial branches analogous to ASAA and MSAA supply mucoperiosteum of zones 4 and 5. Ascending branches supply lacrimal sac and meet angular branch of facial. Medial branches anastomose with dorsal nasal branch of ophthalmic Tessier zone 11. Descending branches pass in the space between levator labii superioris and levator labii et alaeque nasi where they encounter buccal artery and the extreme terminus of transverse facial artery.

Descending Palatine Artery *Greater palatine nerve* combines V2 with secretomotor fibers from pterygopalatine ganglion. It induces descending palatine artery. The neurovascular bundle runs in a canal of ascending process of palatine bone. *Greater palatine artery* runs forward to supply the entire lingual surface of the hard palate and alveolus up to the incisive foramen. *Lesser palatine artery* exits via a separate foramen at the junction of maxilla and horizontal palatine shelf prior to reaching greater palatine foramen. It travels posteriorly to supply the predominantly r3 *oral surface of anterior soft palate* and palatine aponeurosis. Recall that *nasal surface of soft palate* is supplied by r7 palatine branch of ascending pharyngeal.

Artery of Pterygoid Canal The pterygoid or Vidian canal extends medially from the roof of the pterygopalatine fossa straight back into middle cranial fossa via the medial pterygoid plate to foramen lacerum. It develops as the inverted-U-shaped intersection of transverse pterygoid body[r1] with medial pterygoid plate[r1] and lateral pterygoid plate[r2].

Artery of pterygoid canal is induced by its companion nerve. It is distributed to mucosa of upper pharynx and lateral auditory tube[r7], where it quite logically anastomoses with the supply of media auditory tube[r7], the ascending pharyngeal[PA3]. Its final destination is lateral sphenoid sinus, where it picks up SANS fibers from the cavernous sinus plexus. There, it also anastomoses with tympanic branch of ascending pharyngeal. The induction relationship between artery and nerve works in reverse, efferent nerve directed forward and efferent blood flow directed backward. This pattern is reminiscent of the relationship between facial nerve and stylomastoid artery.

Foramen lacerum is a cartilaginous plug marking the confluence of basisphenoid, basioccipital, and the apex of petrous temporal bones. It forms around several pre-established structures: lesser petrosal nerve (LPN), greater superficial petrosal nerve (GSPN), nerve and artery of pterygoid canal, and meningeal branch of ascending pharyngeal artery.

Neurology of the Pterygopalatine Fossa

Nerve of Pterygoid Canal

This small, but complex structure is formed by the union of superficial and deep petrosal nerves at the foramen lacerum. As such, nerve of the pterygoid canal (NPC) can be conceptualized as "*the autonomic distributor to the pterygopalatine ganglion.*" Deep petrosal nerve carries postganglionic SANS fibers to PPG. Superficial petrosal nerves carry preganglionic PANS fibers to PPG. Note that pterygopalatine and sphenopalatine are synonymous.

1. *Greater deep petrosal nerve* is "the sympathetic root" of pterygopalatine ganglion. SANS to the head is predominantly supplied when pre-ganglionic fibers from cervical roots C1–C4 synapse at the *superior cervical ganglion.* Postganglionic fibers glom on to the carotid artery as the *carotid plexus.* The para-sphenoid *cavernous sinus plexus* provides the take-off of GDPN. The nerve runs a short distance in foramen lacerum until it joins fibers of SPN. Thus, the nerve of pterygoid canal and its precursors come to be encased in cartilage.

   To avoid confusion, note the existence of a *lesser deep petrosal nerve.* It conveys secretomotor PANS fibers originating from IX to parotid. The *tympanic nerve* (of Jacobson) runs upward from the neural crest *inferior ganglion* of IX (neural crest) to tympanic plexus. Lesser petrosal nerve (the continuation of tympanic) penetrates petrous bone just medial to tensor tympani[Sm4], climbing its way upward to achieve an intracranial position on petrous bone *immediately lateral to the hiatus of the facial canal.* At this point, it picks up fibers of VII. LDPN exits middle cranial fossa through an opening in alisphenoid. It "dead-ends" in the otic ganglion, a 4-mm structure located immediately outside of foramen ovale, just medial to V3. It is medial to V3, lateral to auditory tube, posterior to tensor veli palatini[Sm4], and anterior to middle meningeal artery. A small *sphenoidal branch* connects it to the nerve of pterygoid canal. Otic ganglion, being physically next to V3, connects to auriculotemporal nerve which carries PANS fibers to parotid gland.

2. *Greater superficial petrosal nerve* is the parasympathetic "root" of pterygopalatine ganglion. GPN comes originally from the geniculate ganglion of VII. Although famous as a PANS provider it is largely sensory, GPN distributes sensory fibers via V2 lesser palatine nerve to soft palate and auditory tube. PANS fibers to pterygopalatine ganglion from which they distribute themselves in ingenious ways to salivary glands of palate and the lacrimal gland.

Pharyngeal

Long and narrow, medial pterygoid plate[r1] is more delicate than its lateral partner. Its lateral surface is part of pterygopalatine fossa while its medial surface forms the lateral boundary of the posterior nasal cavity. Its inferior terminus has a posterior extension, the *pterygoid hamulus,* that provides a "pulley" for the tendon of tensor veli palatini. At its superior margin, medial pterygoid plate projects two delicate laminae at a 90° angle beneath sphenoid body, as if to offer it support. These *vaginal processes* have two articulations: anteriorly, with *sphenoid process of palatine bone*[r2] and posteriorly, with the *alae of vomer bones*[r2]. Thus, the base of sphenoid is an interface zone between r1 and r2 fields.

The anterior margin of medial pterygoid plate articulates with perpendicular plate of palatine bone. Its posterior margin provides insertion for pharyngobasilar fascia and, along lower one-third, superior constrictor[Sm7]. Auditory tube is supported at its midpoint by a projection, the *processus tubarius*.

The undersurface of the vaginal process had a groove. Where this intersects sphenoid process of palatine bone, the groove becomes the *palatovaginal canal*. Through this space are transmitted *pharyngeal branch of maxillary artery* and pharyngeal branch of pterygopalatine nerve. The structures provide support for r2 nasopharyngeal mucosa posterior to auditory tube. Recall that r7 mucosa medial to tube is supplied by IX.

Sphenopalatine (Nasopalatine)

Internal maxillary artery terminates with this clinically important artery. It gains access to the superolateral corner nasal cavity via the sphenopalatine foramen just behind the superior turbinate of ethmoid[r1]. It then divides into medial and lateral branches. The interface between the ECA sphenopalatine axis and the ICA ophthalmic (via ethmoid) axis is critical for understanding pathology in Tessier zones 1–3.

Lateral sphenopalatine This artery supplies **Tessier zone 3**. It describes an oblique course downward and forward along the lateral nasal wall. *Lateral posterior nasal* branches provide collateral support for mucosa and underlying bone of sphenoid and perpendicular plate of palatine; they are the unique supplier of inferior turbinate. Sphenoid has multiple sources of support, while descending palatine is "dedicated" to that bone field. In these zones, medial sphenopalatine plays "second fiddle." For this reason, in Tessier 3 clefts, palatine bone remains intact. Isolated defects of palatine bone can occur independent of the inferior turbinate. Superior and middle turbinates are supplied from lateral nasal branches of the ethmoid arteries.

Medial sphenopalatine This artery supplies **Tessier zones 2 and 1**. Its trajectory arcs over the nasal roof, *medial posterior nasal* branches supplying medial sphenoid and perpendicular plate of ethmoid. It descends along the ipsilateral vomer[rr2] and terminates in ipsilateral premaxilla[r2]. Prior to premaxilla, LSPA gives off a nasopalatine branch that passes through incisive foramen to anastomose with ipsilateral greater palatine artery. Premaxilla receives MSPA at the midpoint of its posterior wall about 5–10 mm anterolateral to incisive foramen. Defects isolated vomer cause Tessier 2 isolated cleft palate. Defects involving premaxilla or premaxilla/vomer cause cleft lip either isolated or in combination with cleft palate.

Blood Supply to the Orbit

(Figures 7.24, 7.25, 7.26, 7.27, 7.28, 7.29, 7.30, 7.31, 7.32, 7.33, 7.34, 7.35, 7.36, 7.37, 7.38, 7.39, 7.40, 7.41, 7.42, 7.43, 7.44, 7.45, 7.46, 7.47, 7.48, 7.49, and 7.50 (stapedial system))

We shall now consider in detail the configuration of primitive arteries supplying the developing eye and orbit. By way of background, an overview of eye and orbit developmental stages will be presented. This will be followed by description of how the ophthalmic artery and its branches to the globe are assembled. We next will cover the development of the orbital vessels supplied by the stapedial system. The goal here is to explain wherever possible the developmental rationale of these arteries and the structures they supply. The section will conclude with comments regarding the embryologic rational behind the mesenchymal contents of the orbit: walls, muscles, fasciae, and lacrimal gland. **Note**: suprascripts in bold refer to Carnegie stage.

Blood supply to the orbit can be subdivided between vessels supplying the globe and optic nerve and vessels for everything else. Simply stated, the eye is a gigantic neuron, an extension of the brain itself, suspended in a sea of nonneural mesenchyme. This neural tissue induces the development of *optic vessels* that are exclusively supplied by the cranial division of primitive internal carotid artery. *Orbital vessels* supply the adnexal structures of midbrain neural crest and extraocular muscles from somitomeric paraxial mesoderm. They develop from an intermediate system, the intracranial division of stapedial artery, one branch of which follows V1 from the middle fossa into the orbit (StV1) while a second branch, "under the influence" of V2 (StV2), pursues a lateral course to supply alisphenoid and lacrimal gland.

Timeline of Ophthalmic and Orbital Development

Note that the general steps of ocular development are well described in the respective chapters in Carlson and O'Rahilly and will not be repeated here. Staging is per O'Rahilly, 1987. For those readers consulting the original 1987 paper, embryonic ages were revised between 1987 and 2001. Measurements as reported by Padget were made prior to these classifications; the correlations are not exact.

Stage 10 (2–3.5 mm, 4–12 somites, 28 days)

Appearance of optic primordia and sulcus in neural folds at 8 somites. Primordia meet at chiasmatic ridge (that will become the optic chiasm). MNC invasion.

Stage 11 (2.5–4.5 mm, 13–20 somites, 29 days)

Right and left optic primordia form optic chiasm.

Optic evagination at 14 somites—ventricle continuous with forebrain. Optic evagination contacts surface ectoderm. Neural crest MNC and maybe PNC invades surface mesenchyme at 14–16 somites and interposes itself between surface ectoderm and the evagination. Some NC cells migrate through the underlying basement membrane to form future pigment cells of the uvea.

At 17–19 somites, optic evagination now becomes the *optic vesicle*.

Caudal limiting sulcus between vesicle and forebrain marks future backwall of orbit.

Stage 12 (3–5 mm, 21–29 somites, 30–31 days)

PNC (optic neural crest) invasion. Neural crest completely circumscribes the optic vesicle and covers it—source of future sclera. Vesicle continuous with future third ventricle.

Stage 13 (4–6 mm, 30–31 somites, 28/32 days)

Optic vesicle located just beneath ectodermal surface. Two basement membranes present: one beneath the ectoderm and the other coving the vesicle.

Lens disk becomes thickened and prominent. In contact with retinal disk, the future invaginated layer that will become the neural retina.

Primordia of *lateral rectus*.

Stage 14 (5–7 mm, 32+ somites, 33–35 days)

Uveocapillaris layer supplied by *primitive dorsal ophthalmic artery*.

Retinal disk zone of the vesicle invaginates to create *optic cup* with a double-layer retina. Brings its basement membrane with it. Outer layer in contact with uvea is *pigmented retina*; inner layer is *neural retina* with multiple cell layers. Note basement membrane invaginates as well.

Retinal fissure contains condensation of the plexus = primitive hyaloid artery.

Lens pit indents in placode to make *lens pit*. Optic vesicle becomes *optic cup* that communicates with the surface via optic pore.

Oculomotor nerve appears.

Primordia of *superior oblique, superior rectus*.

Stage 15 (7–9 mm, 36–37 days)

Lens pit disappears to form *lens vesicle*.

Choriocapillaris plexus supplied by *primitive ventral ophthalmic artery* (one stage later).

Retinal pigment appears in outer layer of the cup.

Vitreous forms in the lentiretinal space. Optic cup begins to close and presses against the surface of the ectoderm.

Cornea develops its own basement membrane and its own epithelium *below the surface ectoderm*.

Primary vitreous develops, hyaloid approaches lens.

Three divisions of trigeminal appear.

Stage 16 (8–11 mm, 38–40 days)

Eyelid grooves appear before eyelids.

Lens body grows, creating a D-shaped lens cavity. Blood vessels form two arcades around the lens.

Perilentil blood vessels from hyaloid.

Primordium of *medial rectus*, common primordia of *inferior oblique* and *inferior rectus*.

Stage 17 (11–14 mm, 41–43 days)

Upper eyelid fold.

Optic cup has pentagonal shape.

Retina fissure closes posterior-anterior; notch persists anteriorly.

Proliferation and differentiation of retina.

Annexation of hyaloid by DOA and formation of *common temporal ciliary artery*.

Extracranial stapedial arrives at V3 to complete external carotid, *middle meningeal* forms.

Distinct primordium of *Inferior rectus*.

Stage 18 (13–17 mm, 44–45 days)

Lower eyelid fold, groove of the conjunctival sac.

Neural crest invades space between lens epithelium and surface ectoderm, also invades behind the lens—Forms anterior and posterior epithelium of the lens, intrinsic muscles of the eye.

Stem of PrDOA relocates caudally along internal carotid, forms *stem of ophthalmic artery*.

Stem of PrVOA pulled downward, annexed by ophthalmic, now forms *common nasal ciliary artery*.

Forward growth of supraorbital stapedial toward the orbit, StV2 predominant as V2 is developmentally more mature than V1.

Middle meningeal anastomosis with intracranial stapedial stem.

Stage 19 (16–18 mm, 46–47 days)

Eyelid folds become eyelids, meet at lateral canthus.

Ganglion cells in the neural layer of retina give rise to optic nerve.

Hyaloid within optic nerve becomes *central artery of retina*.

Stapedial artery passes from middle meningeal into future orbit above the globe, from lateral to medial, encounters V1.

Primordium of *levator palpebrae superioris* emerges from common superior rectus.

Stage 20 (18–22 mm, 49–50 days)

Eyelids meet at medial canthus.

Optic nerve fibers reach the brain.

Ophthalmic artery annexes supraorbital stapedial.

Stapedial branches follow V1 nerves and develop medial to lateral: nasoethmoid > frontal > lacrimal.

Trochlea for superior oblique develops stages 20–23.

Stage 21 (22–24 mm, 51–52 days)
Delamination of levator palpebrae superioris (its fascia subdivides lacrimal gland).
Stage 22 (23–28 mm, 53–55 days)
Eyelids cover half the eye.
Scleral condensation.
Stage 23
Eyelid closure complete.
Eight layers of retina.

Eyeball Development: A Global Perspective

Recall that the neural plate[8] induces a zone of ectoderm overlying the future eye to become *lens ectoderm*[9]. In late Carnegie stage 10 (22 days), bulges appear in the lateral walls of basal p5 diencephalon, the *optic grooves*. Within 48 h, these optic vesicles have enlarged outward until they come into near approximation with r1 surface epithelium[11]. Inductive signals from the vesicle transform the epithelium into the *optic (lens) placode*[13]. In the next stage the *lens vesicle*[14]; becomes separated from the overlying epithelium. Simultaneously, the optic vesicle inverts into the *optic cup*[14]; nasal pits are also apparent at this stage. Lens vesicle now assumes an inductive role of its own, causing formation of the *cornea*.

Formation of the optic cup determines the layered blood supply to the globe. First is not symmetrical process. The infolding of the cup results occurs along the ventral midline of the optic vesicle (not the center). The resultant *choroid fissure* is continuous with a ventral groove in the *optic stalk* that runs all the way back to the diencephalon. In the vicinity of the choroid fissure, the network of vessels surrounding the cup becomes entrapped and consolidates, forming the *primitive hyaloid artery* that passes forward through the vitreous and dead-ends at the posterior wall of the lens. The vitreous part subsequently involutes leaving behind the central artery of the retina (please see Gray's Anatomy Fig. 30.13.10).

A second way that formation of the optic cup affects vascular supply is the creation of a double-layer retina. The posterior wall of the optic vesicle becomes a thin outer layer that is populated by neural crest to become the two-layer *pigmented retina*. The anterior wall of the vesicle forms the thick five-layer *neural retina*. Later in development, neuronal processes from the ganglion cells of the neural retina ganglion cells will grow backward toward diencephalon within the tubular space of the external layer (the optic stalk). They will surround the hyaloid artery, pass *through* the optic nerve and chiasm, and enter the contralateral optic tract.

Blood supply to the eye follows these discrete layers. The retina is supplied by perineural arteries running through the optic nerve and the surrounding pia/arachnoid. These arter-

ies are derived from the original plexus of head mesoderm and midbrain neural crest that surrounded the optic vesicle. It is also supplied by extraneural vessels that penetrate sclera and choroid to access the retina. These arise posteriorly from the ophthalmic artery and anteriorly from the V1 stapedial arteries.

Central artery of retina runs in the central canal of the optic nerve. The nerve is ensheathed by pia and arachnoid mater within the latter which *pial arteries* penetrate to supply the outer wall of the nerve. They also supply the surrounding optic nerve head (ONH). *Short posterior ciliary arteries* (SPC) pierce the sclera near the disk and run in the deep layer (*Sattler's layer*) of the choroid, to supply fine vessels of posterior choriocapillaris. *Long posterior ciliary arteries* (LPC) are paired along the dorsolateral and ventromedial axes of the globe. As we shall see, their embryology is significant. They run in the outer layer of choroid (*Haller's layer*) all the way forward to the termination of the choroid at its boundary with the iris at the ora serrata. There, the LPCs give off recurrent branches that dive down into Sattler's layer, supply anterior choriocapillaris, and anastomose with the SPCs about halfway back from the iris to the optic nerve. They also give anterior branches to the *major arterial circle* where they anastomose with *anterior ciliary arteries* that have penetrated sclera via the tendons of the four rectus muscles. Note that these latter are associated with the stapedial system and supply neural crest structures via the major arterial circle with LPCs and exclusively via the *minor arterial circle*.

Finally, let us take a careful look at the adult anatomy of the ophthalmic axis. It exits from internal carotid immediately as the latter emerges from the cavernous sinus, just medial to anterior clinoid process. It passes through optic canal inferior and lateral to optic nerve. Here, it gives off a common stem that supplies *central artery* to the retina and *medial long posterior ciliary artery*. Just beneath the superior rectus muscle, it produces *lateral long posterior ciliary artery*. Fifteen to twenty *short posterior ciliary arteries* arise from both long posterior ciliary counterparts about halfway from the optic canal to the globe.

From this point onward, the ophthalmic system ends and the stapedial system begins. The anastomosis is located directly from ophthalmic as a separate stem located immediately distal to the take-off of lateral LPC. Alternatively ophthalmic can give a common stem to both lateral LPC and stapedial. In any case, stapedial stem vessel loops around optic nerve to achieve a position superior to the nerve where it proceeds to subdivide into its lacrimal branch and frontal branches. In 20% of cases, stapedial gains its medial position by passes inferior to the nerve. This variation is, as we shall see, due to an ancient encirclement of the nerve by a series of anastomoses the result from the unification of long ciliary branches and the stapedial system.

Development of the Ocular Arteries

On the face of it, the above arrangement of ophthalmic and stapedial anatomy is hard to rationalize. It seems intuitive that all branches to the ocular apparatus are given off prior those supplying extraocular structures. But how this comes about can only be appreciated by understanding the intermediate embryonic vessels that support the developing eye and orbit. This involves separate origins of the to the walls of the globe, the complete relocation of the stem of the ophthalmic artery, the creating of an arterial ring around the optic nerve, and the annexation of a newcomer to the orbit, V1 stapedial from the middle cranial fossa, and the unique developmental situation of the lacrimal gland. We shall now proceed to tell story of how these events take place.

The Rise and Fall of Primitive Maxillary Artery

At stage 8, head mesoderm and neural folds are developing. These tissues are supplied by diffusion from dorsal aorta, and the primitive neural plexus, respectively. Initially, these primordial vessels do not communicate with the other but at stage 9, when internal carotid projects forward from level r0, presegmental primitive trigeminal artery connects dorsal aortae with the CNS plexus. At the same time, *primitive maxillary artery* (prMx) arises as the very first branch of internal carotid. It is directed toward the base of the forebrain and optic anlage, as previously described. It supplies the optic plexus until the arrival of primitive dorsal ophthalmic.

Primitive maxillary plays an important, but temporary role in supporting the circulation to the eye. Its ultimate fate is to supply posterior pituitary. A review of the steps of pituitary development is useful here. The gland has two embryonic origins: ectoderm of the adenohypophyseal placodes in the roof of the oral cavity (stomodeum) invaginates via Rathke's pouch to make the adenohypophysis. Infundibulum grows down from the ventral floor of diencephalon to make the neurohypophysis.

Week 4—hypophyseal, Rathke's pouch, diverticulum from the roof

Week 5—elongation, infundibulum, and forebrain diverticulum make contact

Week 6—connecting stalk between oral cavity and hypophyseal pouch degenerates

Week 10—ACTH and growth hormone produced

Week 16—adenohypophysis fully differentiated

Padget's depiction of a 3-mm stage 10 embryo (Fig. 7.37) shows the arrival of primitive ICA at the optic vesicle.[10] A small, unnamed branch is directed toward the base of the vesicle; this vessel is clearly identified one stage later as primitive maxillary. At 4-mm prMx has developed a lateral branch supporting the optic cleft and a medial branch just lateral to the brain at Rathke's pouch, r0, site of the future hypothalamus.[11] At 5–6 mm, the advent of the primitive dorsal ophthalmic artery brings blood flow directly from ICA to the optic vesicle; consequently, the orbital branch of primitive maxillary regresses. The medial branch remains visualized at 14 mm near Rathke's pouch just lateral to trigeminal ganglion.

The 18-mm embryo at stages 17–18 shows narrowing of the neck of Rathke's pouch as the pituitary gland matures.[17–18] The definitive ophthalmic artery, present at this stage, is cranial to prMx and isolates to the intracavernous part of internal carotid. Primitive maxillary becomes the inferior hypophyseal artery with the superior hypophyseal circulation supplied by pia arteries. The significance of this is that prMx represents the most proximal part of the internal carotid system with a non-neural ectoderm involved in producing the hypophyseal placode. This is followed by the optic anlage, again involving a placode for induction. Although both these organs can be considered part of the central nervous system, all subsequent branches of internal carotid are intracranial and dedicated to the brain.

The Iterations of the Ophthalmic System and the Stapedial System

At 3 mm[11] (25 days, 20 somites) primitive carotid dead-ends as a plexus over the emerging optic vesicle (Fig. 6.47). In the next stage, internal carotid is seen more cranially as it climbs over the dorsal aspect of the optic vesicle[12]. Primitive maxillary contributes to the plexus as well (Fig. 6.48). The plexus around the optic vesicle is supplied from these sources. We have previously described how a portion of the plexus becomes entrapped in the optic fissure and subsequently consolidates into the hyaloid artery. The proximal source of hyaloid arises from the plexus but this will change in the two next stages.

At 4–6 mm[13–14], a series of changes occur. ICA splits into a cranial and caudal divisions, the latter being biologically "newer." Infolding of the optic vesicle creates the two-layer retina. A vessel emerges at the exact bifurcation of the internal carotid and takes a lateral course. It encircles the lens pit along its caudal-lateral 180° and drops into the ocular cleft. *Primitive dorsal ophthalmic artery* (PDOA supplying the cup and lens can now be identified[14]. DOA originates from the distal primitive internal carotid immediately before its bifurcation into cranial and caudal divisions. PDOA runs caudally to reach the eye, passing *dorsal to the optic nerve* (this detail becomes important later). There connects with the hyaloid at two locations, anteriorly at the lens and posteriorly at the margin of the plexus. All connection between PrMx and optic vesicle is now lost.

At 9 mm[15], the plexus of the optic vesicle receives a second source of supply. *Primitive ventral ophthalmic* (PVOA) appears exactly opposite of the take-off for the anterior choroidal. In this way, it becomes the most proximal branch of the cranial division. The vessels surrounding the optic cup have become a dense plexus. The lens is encircled by a vessel at the margin of the cup. This is connected posteriorly with the irregular hyaloid vessel. Thus, the lens is supplied anteriorly by a plexus representing the distal ends of PDOA and PVOA, both of which originate directly from internal carotid. It is also supplied from behind by primitive hyaloid, originating from the primitive head plexus.

The spatial relationship of these two arterial axes is very important as it divides the globe into genetically distinct hemispheres. PDOA has a caudo-dorsal relationship to the optic cup. This will correspond the dorsotemporal retina. PVOA has a cranioventral relationship; this will correspond to the ventro-nasal retina. Both these arteries now supply the plexus, which constitutes a physical barrier. Although both arteries have definitive stems, once they reach the plexus their axes are poorly defined. Note that the stem of PDOA along the axis of primitive internal carotid is proximal to its bifurcation, i.e., proximal to the stem of PVOA. Thus, *the temporo superior globe* which it supplies is *more primitive that nasoinferior globe* supplied by PVOA. This has important implications for the pathology of holoprosencephaly.

The dense plexus over the cup now breaks up in favor of two well-defined arterial axes.[16–17] Faced with a dwindling blood supply, primitive hyaloid (which takes its origin from the plexus) needs a more reliable source. Like a fickle lover, hyaloid abandons its allegiance to the plexus and makes an anastomosis with PDOA just before it enters the optic nerve. Thus, PDOA, as it approaches the back of the eye, gives off a medial branch, the *definitive hyaloid artery*, and its lateral branch is now renamed *common temporal ciliary artery*.[17] The term "common" means that this stem eventually gives rise to *both* the long posterior ciliary artery and a number of short posterior ciliary arteries to that half of the globe. The proximal segment of PDOA will undergo a drastic revision.

At 14 mm[17], growth of the eye causes the stems of both PDOA and PVOA to be stretched out. PDOA, under tension, must cross over the dorsal aspect of the optic stalk. This routing is circuitous and clumsy—as a consequence, the stem of the elongated primitive dorsal ophthalmic now undergoes a drastic change.[18] It shifts *caudally and ventrally* along the course of internal carotid through a "leap frog" mechanism we shall discuss later, coming to its final adult position *lateral to and below* the optic nerve as the definitive *ophthalmic artery*. The ophthalmic now bifurcates to supply both hyaloid (the future central artery of the retina) and common temporal ciliary. Note that the original stem of PDOA, positioned dorsal to optic nerve, will eventually supply the optic chiasm.

Advance warning: The situation in which hyaloid artery is associated with common temporal ciliary will undergo a 180° reversal. In the final state, ophthalmic give off medially

directed stem which bifurcates into central artery of the retina (the former hyaloid) and common nasal ciliary. The process by which this transition takes place creates a transient annular ring around the optic nerve and sets the stage for the anastomosis of the ophthalmic and stapedial circulations.

Note that at stage 16 external carotid and extracranial stapedial have united but intracranial (supraorbital) stapedial has not yet extended forward from trigeminal ganglion. Middle meningeal ascension into the cranial cavity is accompanied by rapid forward growth of StV1 and StV2 toward the orbit.[17] Orbital StV1 is smaller in caliber than Orbital StV2. The two intracranial stapedial systems unite in the orbit, where all subsequent branches are induced by V1.[18] Unification of stapedial with ophthalmic takes place at medial aspect of the perineural arterial ring.[19] This process is discussed in further detail in the following section.

During stage 18, a second anastomosis also takes place, one that determines the final anatomy of the ophthalmic circulation. Recall that the origin of primitive VOA is physically located in the anterior cranial fossa, just opposite of anterior choroid artery. In order to access the eye, this VOA must exit the future skull by traversing the developing junction of the orbital roof and ethmoid complex. Since these bones will close, the stem of VOA is preordained for extinction. The initial stem of VOA is quite short and has direct access to the back of the eye.[15–17] But the same process of growth that eventually forces DOA to detach and reconnect with ICA also drags VOA backward. Its course runs parallel along internal carotid from distal-to-proximal, until it encounters the medial border of optic nerve just after the nerve enters the orbit. VOA then follows optic nerve distally to the globe.

From ophthalmic artery, the branch that supplies hyaloid now bifurcates to form an anastomosis with VOA. Primitive VOA distal to the annexation point is now renamed the *common nasal ciliary artery*.[18] For a brief time during stage 18, ophthalmic appears as a simple trifurcation, providing the central artery to the retina and the two common ciliary arteries. Higher flow from internal carotid into nasal ciliary causes the primitive VOA proximal to the annexation point to involute and fall apart.[19] Its stump remains in the anterior cranial fossa to supply optic chiasm.

The neurovascular developmental sequence of the eye and orbit runs from lateral to medial. The final anatomy appears to flout the principle. The ophthalmic axis starts out at stage 18 with temporal ciliary and hyaloid being its primary branches with nasal ciliary as a secondary branch. By stage 19, the situation is reversed, and it appears that hyaloid and nasal ciliary are primarily associated with each other. Here is what happens, and why.

Growth of the eye and the medialization of the orbits cause a physical change in arrangement of the ophthalmic arteries that leads to their final geometry. The stem of common temporal ciliary is stretched laterally to nearly double the length of its nasal counterpart. This pulls it away from the hyaloid so that appears as an independent branch from ophthalmic. Hyaloid and common nasal temporal arise as a separate, stem from ophthalmic. Thus, ophthalmic bifurcates and then the medial stem bifurcates again. We shall see this geometric pattern repeated with the V1 sensory nerves and the stapedial arteries that are programmed by them.

A few comments regarding supply to the retina. It will be noted that the vascular support for hyaloid artery undergoes several iterations. Initially, the future eye consists of a head plexus, fed by primitive maxillary artery and supplemented by MNC invasion.[10] With further PNC invasion, optic evagination is covered by choriocapillaris.[12] With optic cup and retinal fissure, local choriocapillaris is entrapped and organizes into primitive hyaloid.[14] Primitive dorsal ophthalmic arrives at lens and connects primitive hyaloid anteriorly.[14] Dense plexus forms and connects PDOA and PVOA.[15] The retinal fissure closes and PDOA is forced to annex hyaloid proximally, becoming common temporal ciliary.[16] Proximal stem of PDOA is replaced by definitive ophthalmic artery which now supplies hyaloid.[18] The hyaloid is now its maximum development. Backward growth of neurons from retina produces optic nerve which ensheaths hyaloid, now termed central artery of the retina.[19] Hyaloid artery in the vitreous breaks up.[23]

Common temporal ciliary is located posterodorsal to the optic nerve. Primitive ventral ophthalmic is positioned anteroventral to the nerve. When hyaloid is given off by CTA, before it penetrates the substance of optic nerve, it pursues an oblique course that runs distal and ventral. It is therefore ideally situated midway between ophthalmic artery stem and its target, VOA. The subsequent anastomosis either connects VOA to ophthalmic via hyaloid (or directly to ophthalmic). Thus, in the final state hyaloid often appears to originate from nasal ciliary.

Prior to the union of the ophthalmic and stapedial systems, the anastomoses we have described create vascular ring around the optic nerve. The ventral part is more tenuous and consists of the connection that permitted the DOA to join with the newly relocated adult ophthalmic stem. The dorsocaudal half of the ring is larger and more recent. It represents the connection between supraorbital stapedial and ophthalmic. Stapedial approaches optic nerve dorsally, following a lateral to medial course. At the lateral border, it curls downward to make the anastomosis. This occurs in 85% of cases. But in 15% of cases, stapedial tracks completely over the nerve and then winds downward at its medial border to access ophthalmic artery and, in so doing, takes advantage of the inferior ring.

At 20 mm[20], supraorbital develops ethmoid, frontal, and lacrimal branches (in that order). Its stem is involuting.

Stapedial is being annexed by ophthalmic artery (the ocular branches of which were set up previously).

At 40 mm[fetal], ophthalmic artery achieves its adult configuration with the interruption of two vessels: (1) cranioventral segment of the arterial ring around optic nerve; and (2) stapedial. In the anterior cranial fossa, stapedial becomes lacrimal while in the posterior cranial fossa it becomes anterior branch of middle meningeal.

Development of the Optic Arteries: Leapfrog

How can we make sense of these events? First, recall that the ectodermal lens is very far removed from its blood supply. As it sinks beneath the surface, it becomes embraced by the optic cup, the rim of which is vascularized. You can duplicate this by placing your fist (the lens) into the cupped palm of the opposite hand (the optic cup). The cleft between your thenar and hypothenar eminences represents the ventromedially located ocular cleft, which is, (inexplicably?) 180° opposite to its blood supply, PrDOA.

Just as a snake sheds its skin to accommodate growth, PrDOA reinvents itself to fulfill its mission: blood supply to the lens. At 7–8 mm[14], it follows a beeline course from ICA along the posterior (caudal) and dorsal margin of the optic cup. Optic cleft opens caudally. PDOA gains access to it near the lens. There it connects with distal primitive hyaloid. Recall that proximal hyaloid is fed from the optic plexus and ultimately from primitive maxillary artery. At 7–10 mm[16], two events reflect the activity of neural crest within the optic cup: (1) a network of vessels, *primitive choriocapillaris* envelops the cup; and (2) pigment develops in the outer layer of the cup.

Choriocapillaris presents a problem for PDOA. Originally, it travels a straight line over the dorsal eye to enter the outer (lateral) end of the ocular cleft. At 7–12 mm[16], it is forced to negotiate its way through the intermediate plexus of choriocapillaris.

Making use of multiple anastomoses available to it via the plexus, a series of communications between PDOA and hyaloid work their way backward along the cleft margin. This form of "leapfrog" is driven by involution of the plexus and progressively more stable hemodynamics. By 12–14 mm[17], PrDOA is seen entering the inner (mesial) end of the ocular cleft. This marks the *definitive hyaloid artery*.

The companion vessel to PDOA, the naso-ventral PVOA, arises one stage later. At 9 mm[15], it sprouts from the cranial division of ICA just opposite the anterior choroid artery. PVOA is also spatially challenged. Initially, its course is short but in succeeding stages (due to growth of the forebrain and of the eye in opposite directions), the course of PVOA becomes stretched out. At 18 mm[19–20], PrVOA is now directed ventrally and elongated. It runs in parallel with proximal internal carotid.

The common stem of ophthalmic artery is now present. It sends out a clearly defined *common temporal ciliary branch* to the dorsotemporal optic cup. Common nasal ciliary branch does not come about until the next stage. *This sequence of arterial development demonstrates the genetically driven dorsotemporal to ventronasal maturity gradient at work.*

Appearance of cartilage primordium marks the separation of orbit from anterior cranial fossa. PVOA runs between the cartilage and the anterior (future cranial) margin of optic stalk. The proximal stem of PVOA is destined to become obliterated by bone but at this stage is grows downward until it hits the medial edge of optic nerve. Here, it makes a right turn, following the nerve to reach the eye. At 21–24 mm[21], an anastomosis between PVOA and ophthalmic artery takes place *ventral* to the nerve. PVOA distal to the anastomosis is renamed *common nasal ciliary branch*. PVOA proximal to the anastomosis is obliterated.

Recall that the stem of PDOA arises more proximally along ICA, immediately prior to its split into cranial and caudal divisions. The stem of common nasal ciliary arises more distally, moving up into the cranial division just opposite anterior choroidal. In any case, it supplies the developmentally "newer" hemisphere of the eye. The remaining carotid remnants of PVOA and PDOA are retained as minute *branches to the optic chiasm.*

Now it is time to consider actual stem of ophthalmic artery in its adult position. It does not make its appearance until 18 mm[19]. Prior to this time, blood supply to the eye is conveyed exclusively by PDOA which arises (Figs. 6.60 and 6.61) from ICA just lateral to Rathke's pouch, well before the bifurcation. PDOA has two branches: hyaloid and common temporal ciliary artery. At stage 19 the stem, from the common StV1–StV2 branch extending forward from trigeminal ganglion is dwindling. In man, expansion of the cerebral hemispheres causes the arteries projecting outward from the circle of Willis to be drawn out dorsally and caudally. The origin of ophthalmic artery moves caudally along the ICA by anastomotic progression. It comes to its final position caudal and ventral to optic nerve.

Stapedial System: Orbit

The embryonic stapedial system is a critical component of craniofacial vascular development but it defies an easy description. Both intracranial and extracranial divisions become completely incorporated as component part of ophthalmic and external carotid circulation and are not recognized by traditional anatomic texts. This omission, due to the delayed incorporation of modern embryology into gross anatomy, remains a cause for confusion, both in terminology and in concepts. Our purpose here is to deconstruct and simplify the various systems so that we can appreciate the contributions each makes to the final adult state.

Because of the anatomic complexies presented by the stapedial system, we have dealt with it twice before in this text. Refer to the timeline on page 319. First, under the topic of the internal carotid system, in which all its component vessels were described, it was necessary to include the stapedial contributions to the ophthalmic system. This was done without developmental details but it permitted us to flesh out the internal carotid system in its entirety. In the ensuring topic of external carotid system, the overall story of early stapedial development was recounted in detail to create an understanding of how the inferior ramus could be incorporated simultaneously into the maxilla-mandibular arterial axis and, at the same time, return intracranially to create the meningeal system.

For the sake of simplicity and to avoid going back and forth between sections, the timetable for stapedial development is repeated below.

Timeline of Stapedial Development

Stage 12 fourth aortic arch artery.

Stage 13 Hyoid stem.

Stage 14 Hyoid sends out stem of stapedial artery common carotid is assembled from the aortic outlet segment connecting third and fourth aortic arches.

Stage 15 Otic vesicle becomes the otic capsule, stapedial enters future tympanic cavity 2 divisions (1) posterior and anterior, (2) anterior divides into intracranial supraorbital branch and extracranial maxillomandibular branch.

Stage 16 Extracranial stapedial arrives at V3.

Stage 17 Anastomosis between extracranial stapedial and ECA.

Stage 18 Extracranial stapedial follows V3 middle meningeal branch into middle cranial fossa where it unites with intracranial stapedial axis at trigeminal ganglion to form meningeal system. Meningeal StV1 and StV2 branches arrive at medial and lateral corners of superior orbital fissure.

Stage 19 Lacrimal and orbital connect to form *common supraorbital artery* extending medially to the bifurcation of nasociliary and frontal nerves.

Stage 20 Anastomosis of supraorbital with ophthalmic stem.

Stage 21 Lacrimal artery replaces supraorbital as lateral branch of StV1 system, now supplied by ophthalmic StV2 meningo-orbital involutes (sometimes persists).

Stage 22 Lacrimal artery, supplied by ophthalmic, extends to the lacrimal gland.

We are left with several critical questions. Why should the orbit be supplied from two sources, StV1 and StV1? If the perineural arteries supplying the branches of V1 are present earlies and are supplied from V1 meningeal, why it is necessary to make the connection with the StV2 orbital branch? And what might drive the anastomosis between the supraorbital stapedial axis and ophthalmic?

To answer our first question, the orbital walls are made of neural crest. Medially orbitosphenoid is an r1 derivative whereas laterally, alisphenoid comes from r2. These zones are associated with V1 and V2, respectively. Each requires arterial support from its respective neuromere. The fields are innervated by branches of V1 nerve that are given off prior to entry into the orbit. The initial perineural vascular supply of these nerves comes from very small branches of StV1. Since the sensory anatomy of V1 in the orbit is basically a bifurcation between the medial nasociliary and frontal branches and the lateral lacrimal, it is no surprise that lacrimal nerve enters the orbit surrounded by its sheath of dura. The nerves appear in a medial to lateral order, reflecting the arrival times of the neural crest populations they will innervate. From stages 17 to 19, nasoethmoid precedes frontal which precedes lacrimal nerve. The further lacrimal nerve gets from its point of entry into the orbit at lateral superior orbital fissure, the more tenuous its blood supply. Similarly, the distal end of V2 orbital branch is likewise progressively more ischemic. When these vessels encounter one another, they immediately connect.

That blood supply to the lateral orbit should originate from the StV2 orbital branch of meningeal makes perfect sense because it supplies two bone fields constructed from r2 neural crest: alisphenoid and posterior orbital wall of zygoma. (Note that the lateral orbital wall of zygoma is penetrated by branches of zygomatic artery ascending from infraorbital.)

But immediately upon entering through the lateral margin at stage 18 of superior orbital fissure, orbital meningeal *loses its accompanying V2 nerve*. It picks up the local V1 signal from lacrimal nerve and grows medially. In so doing, it crosses over the globe and inserts itself between the frontal and nasociliary branches. It now drops downward to anastomose with ophthalmic artery at stage 20. The anastomosis results from the availability of high flow from the very proximal ophthalmic stem versus the distal flow coming from all the way back at trigeminal ganglion.

Development: A Summation

Arterial development in the orbit demonstrates several important principles. (1) Arteries are induced by signals from neural tissue, specifically sensory nerves. In the case of the globe, varying patterns of gene expression over its surface can stimulate vessel formation at one point in time or lead to vessel breakdown at another. (2) Arteries are fickle. They will switch allegiance from one nerve to another in order to achieve growth. This implies that the signaling mechanism is not nerve-specific. (3) Quantitative production of inducing substance could vary from one nerve to the next. Stapedial "jumps ship" from greater petrosal (facial) to tri-

geminal when it arrives at the distal terminus of the former and encounters a strong signal from the latter. (4) Arteries are sensitive to changes in blood flow. They obey the law of supply and demand. Faced with a decline in supply (but with ongoing obligations to their client tissues), they seek out newer (and more robust) sources of supply. (5) Arterial prefer a simple, direct route to their target tissues to minimize resistance to blood flow. For this reason, when the proximal portion of DOA becomes stretched out, it relocates its stem along ICA to achieve favorable flow. (6) When myoblasts migrate into their destination, they encounter and are surrounded by fascial "compartments." Sensory nerves to these fascial envelopes are responsible for the induction of arteries supplying the muscles residing within those envelopes. (5) Both a muscle and its motor nerve are functional unit; they share a common neuromeric definition. This is the mechanism by the motor nerve "finds" the muscle. Of course, the motor nerve induces its own perineural arterial supply from the surrounding mesenchyme.

Soft Tissues of the Orbit

Notes on the Trigeminal Ganglion

A few words about the trigeminal (semilunar, Gasserian) ganglion are in order. The ganglion of V is a troglodyte, holed up in *Meckel's cave*, a pocket of periosteum at the apex of petrous temporal bone. It is rather sizeable, measuring $2 \text{ cm} \times 1 \text{ cm}$. It is situated lateral to cavernous sinus. Its fibers run beneath the superior petrosal sinus. Ophthalmic division leaves the ganglion and passes directly through superior orbital fissure. But prior to entry into the orbit, it produces lacrimal, frontal, and nasociliary branches. Lacrimal nerve runs within a separate sleeve of dura mater. Within the substance of the gland, it receives secretomotor PANS fibers from zygomatic nerve.

Extraocular Muscles

The origin of craniofacial muscles from somitomeres has been previously discussed in the Introductory chapter of this book. The initial position of the eye fields is 180° apart. The orbits begin the process of medialization at stage 15; although it is largely complete by stage 23, it continues through term. Thus, with embryonic folding, ocular structures are in direct physical proximity to the myoblasts of somitomeres 1–3 and 5. The sclera of the eyeball has two developmental zones, dorsocaudal and cranioventral. These awkward names result from the right-angle location of the eyes.

When development is completed, these two zones are renamed superior (dorso-) temporal and inferior (ventral-) nasal. Genetic markers define these two hemispheres and the retina as well. (Please see Carlson's text, Fig. 13.13) Distinct populations of neural crest likely populate the two hemispheres. Since the extraocular muscles insert into the sclera and since 5 of the 7 arise from somitomeres 1 and 2 (recall that levator palpebrae shares a common blastemal with superior rectus), and since the oculomotor nerve to the recti arises from midbrain, we can assume that: (1) dorsocaudal sclera comes from m1 neural crest; and (2) ventrocranial sclera comes from m2 neural crest.

No specific mapping studies have yet been reported defining the neuromeric zones of the eye and orbit. But based on the above reasoning and the order of appearance of structures in the orbit, we can hypothesize the most of globe as being encoded by m1 and m2. The neural crest origins for the actual structures of the orbit are likely r0 medially, r1 laterally, and in the posterolateral wall, r2. Anterior cranial fossa dura would be encoded from r0 medially (zones 13 and 12) and r1 laterally (zones 11 and 10). Dura of the middle fossa originates from r2 anterolaterally and r3 for everything else.

The order of appearance of extraocular muscles is worth examining and is as follows:

LR[13] (Sm5), SO[14] (Sm3), SR[14] (Sm1), MR[16] (Sm1), IO[16] (Sm2), IR[17] (Sm1), LPS[19–21] (from pre-existent SR). We also note that the oblique muscles insert posteriorly to their rectus counterparts. To do so, they must follow a superficial course. The "read-out" is posterior to anterior, from lateral to medial and from superior to inferior, with the obliques inserted prior to their rectus counterparts. This is because the posterior zone of each hemisphere is *biologically "older"* than anterior zone. And this follows from the *progressive vascularization of the choroid* from posterior to anterior. The DOA hemisphere is vascularized before the VOA hemisphere. **Note**: this involves the first three "early bird" muscles (LR[13], SO[14], and SR[14]). Midbrain development proceeds caudal to cranial so m2 gets matures before m1. LPS is anomalous simply because it "breaks-out" later in development from the common anlage of the superior rectus mesenchyme.

Does this mean that the myoblasts for each muscle migrate into position at different stages? Possibly, but an equally strong argument could be made about the fascia into which they myoblasts find themselves inserted, a fascia that segregates them into individual compartments. And it is this neural crest fascia which does indeed migrate with a spatial temporal order perhaps being attracted by homeotic expression zones of the optic cup. Let's examine the properties of the orbital fasciae in more detail.

Orbital Muscle Fascia Has a Neural Function

All fascia of the orbit arises from MNC neural crest. When myoblast migration takes place, the cells move collectively into a new neighborhood in which they are "assigned" into individual "homes" of orbital fascia in relation to their neural crest insertion site. But the fascia is more than a simple external covering. Craniofacial muscle fascia is intrinsically a sensory neural structure. This point is absolutely critical. Fascia invades the interstices of each muscle and connects

back to centers of command and control based on proprioception and coordination, essential functions which allow extraocular muscles to track objects with pinpoint accuracy.

How does this square with motor supply? Each extraocular muscle has its own innervation but they share a common source of V1 fascia, one which reports back to the coordination centers of the mesencephalic nucleus of V, which is somatotopically organized. Vascular induction of r1 neural crest and prechordal mesenchyme is non-specific for the motor nerve. Thus, the perineural blood supply of motor nerve to superior oblique represents an induction of vessels by cranial nerve IV from the surrounding mesenchyme to create a muscular artery from the stapedial system, *not* directly from the ophthalmic axis.

The same model of packaging exists in the pharyngeal arches.

For example, somitomere 6 contains all precursor myoblasts for the muscles of the second pharyngeal arch. Motor supply is provided by facial nerve, which has two nuclei residing in r4 and r5. The upper division of VII is likely wired to r4 while the lower division of VII resides within r5. Thus, neural crest from r4 flows into the cranial sector of second arch (prior to folding) that will contain upper division muscles, while those of lower division would arise from r5 neural crest in the caudal part of the arch.

Let us repeat: fascia is intrinsically neural. It contains structures that will differentiate into sensory structures of the muscle units and tendons so that accurate proprioceptive information about muscle function can be conveyed to the brain…but how? Recall that VII does not contain a processing center for somatic sensation. How can facial muscle fatigue be conveyed? The answer is the entry into the fascia of trigeminal sensory fibers. These provide the relay for muscles to the brain. Although the motor control of these muscles is from cranial nerve VII, sensory information flows backward via their fascia to the respective center of integration centers in main sensory nucleus of V (nociception) and mesencephalic nucleus (coordination). The fascial envelopes of all craniofacial muscles originating from somitomeres are supplied by somatic sensory fibers from the V1–V3, IX, or X. It is these fibers that induce arterial supply into the muscle bellies. Thus, posterior digastric receives motor supply from VII.

Table ">" refers to induction

Note: the spatio-temporal arrangement of muscles argues for their sequential migration into place, or the sequential maturation of the fascial compartments which they come to occupy. We postulate that the myoblasts move into position and are surrounded by local neural crest that will form their fasciae, all of which is innervated by V1.

Somitomere 1

Muscle: superior rectus[14], medial rectus[16], inferior rectus[16], levator palpebrae superioris[19–]

Motor nerve: rostral oculomotor nucleus in mesomere m1

Somitomere 2

Muscle: inferior oblique[16]

Motor nerve: caudal oculomotor with nucleus in mesomere m2

Somitomere 3

Muscle: superior oblique[14]

Motor nerve: trochlear with nucleus in rhombomere r1

Somitomere 5

Muscle = lateral rectus[13]

Motor nerve: abducens with nuclei in rhombomeres r4 (and r5)

Muscular arteries Developmentally, these muscles belong to the stapedial system. Muscular arteries are distributed in two groups: these mirror the geometry of original embryonic vascular structures. Superior branches supply muscles assigned to the sclera of primitive dorsal ophthalmic artery. Inferior branches supply muscles assigned to the sclera of primitive ventral ophthalmic artery. Muscular arteries are induced by *V1 sensory nerves to the fascial envelopes* of the individual extraocular muscles. These arteries are *not* induced by the motor nerves III, IV, and VI.

Lacrimal Gland

The mesenchymal origins of lacrimal gland can be inferred from the structures to which it relates and to its blood supply and innervation. Lacrimal gland has two lobes, orbital and palpebral. The orbital lobe is an MNC derivative which shares fibers confluent with the periosteum of the orbital roof, likely r1 neural crest. It is separated from the actual globe by the tendons of superior rectus and lateral rectus. Orbital lobe directly receives V1 lacrimal nerve and is supplied by lacrimal artery from the supraorbital stapedial system.

The palpebral lobe lies in direct contact with the posterolateral conjunctiva, an r1 derivative. It is separated from the orbital lobe by a layer of fascia. Six to twelve ducts from the conjunctiva represent invasion of r1 epithelium into the gland. Additional ducts ascend from palpebral gland into the orbital lobe, where tear production takes place. Surgical resection of the palpebral lobe interrupts these communications and can therefore compromise the lacrimation.

Lacrimal gland is approached from below by V2 zygomatic nerve branch derived from infraorbital nerve. This nerve is supplied by perineural vessels from the anterior superior alveolar artery. This raises the question as to a possible alternative source of blood supply to the gland. It also brings the possibility that r2 neural crest may contribute to the mesenchyme of the gland and be involved in the reception of parasympathetic neural input. Would their presence explain why PANS originate at the r4–r5 level of hindbrain

(facial nerve) rather than from ciliary ganglion of midbrain? After all, if lacrimal is derived from midbrain neural crest, would it not be logical to share its PANS fibers with those of the rest of the orbit? Such questions merit further research.

The construction of the orbit from neural crest mesenchyme makes use of two primary pathways. The superior and medial walls are formed by MNC which contributes to the medial posterior wall as well. RNC from r2 forms the inferior and lateral walls plus the lateral zone of the posterior wall (alisphenoid). Lacrimal gland, by its location in the superolateral orbit, occupies a watershed position. It is a safe assumption that r1(?r2) neural crest precursors of lacrimal gland crest arrive at that location and are subsequently enclosed by surrounding structures.

Recall that these signals stimulating tear production (corneal sensation) are relayed from the V1 sector of mesencephalic nucleus to the lacrimal nucleus of facial nerve in the r1 pons. From here, signals pass through facial nerve in r4–r5. Just distal to geniculate ganglion VII gives off the greater pretrosal nerve, the course of which we are now well aware. These pre-ganglionic PANS fibers are joined by SANS fibers internal carried by lesser petrosal nerve. The two nerves unite at the level of foramen lacerum travel via V2 as the *nerve of the pterygoid canal* otherwise known as *vidian nerve*. Recall that this artery is derived from the *primitive mandibular artery*. PANS fibers gain access to the pterygopalatine fossa, which they synapse in its ganglion. Postganglionic PANS fibers follow the *zygomatic nerve* V2, a branch off of infraorbital. From the inferior orbital fissure zygomatic ascends along the lateral orbital wall. Two sensory branches, *zygomaticofacial* and *zygomaticotemporal*, supply malar and postorbital skin. Branches of zygomaticotemporal anastomose with the *lacrimal branch of V1* (the nerve responsible for inducing the lacrimal artery). Note that in some cases, zygomatic simply continues onward after giving off the cutaneous branches and connects with lacrimal. Activation of this pathway produces tears. All land-dwelling vertebrates have a lacrimal system. For obvious reasons, fishes are exempt!

The temporospatial development of lacrimal artery proceeds in parallel with that of the gland (see Table 7.2). Development of the optic cup and its vessels precedes that of the adnexa. PDOA[14] is present before the appearance of permanent ophthalmic stem. Annexation by ophthalmic stem of PVOA[18] precedes that union of StV1 lacrimal and StV2 orbital branches[19]. Annexation by the ophthalmic follows.[20] In the subsequent stage, lacrimal extends lateral beyond the entry point of STV2 and seeks out the lacrimal gland.[22]

Developmental Timetable

Stage 18 Primordial lacrimal tissue—orbital lobe development precedes palpebral

Table 7.2 Timeline of stapedial development

Stage 12	4th aortic arch artery, formation of common carotid
Stage 13	Hyoid remnant
Stage 14	Common carotid, hyoid gives off stapedial stem
Stage 15	Stapedial divides in otic capsule, maxillomandibular division
Stage 16	Supraorbital division, maxillomandibular div. arrives at V3
Stage 17	Anastomosis stapedial with external carotid
Stage 18	StV1 and StV2 bifurcate from supraorbital, MMA stem
Stage 19	StV1 orbital/meningeal brs. enter orbit, MMA joins supraorbital
Stage 20	StV1 orbital anastomosis with ophthalmic
Stage 21	Lacrimal artery Involution of StV1 meningeal
Stage 22	Lacrimal artery reaches gland

Stage 19 Mesenchymal condensation-induced epithelial thickening (palpebral lobe)

Stage 20 Well-defined future "gland area"

Stage 21 Conjunctival epithelial buds condense in the fornix

Stage 22 Gland approached by lacrimal artery and nerve at posterior-medial border

Stage 23 Epithelial buds have lumen, superior rectus inserts [end of the bud phase]

Week 9 Levator palpebrae superioris

Week 10 Levator palpebrae superioris divides the gland—through stage 12

Week 13 Glandular acini appear; zygomatic and lacrimal nerves unite to supply them. Anastomosis takes place within the interior of the gland. Zygomatic is an independent actor—consistent with r2 neural crest precursor mesenchyme? Stroma is not condensed

Week 14 Branching of gland parenchyma

Week 15 Stroma condenses around the acini

Week 16 Union of acini and stroma forming lobes, each lobe gets a vessel

Neurovascular Organization of the Adnexa

We now consider the walls of the orbital box. Each bone is "programmed" from two distinct sides; each has two distinct arterial sources. Induction of superior, posterior, and medial walls is supported on the orbital side by ICA via V1 stapedial and on the cranial/nasal side by ICA proper (non-stapedial).

The floor on the orbital side is supplied by ICA via V1 stapedial and on the maxillary side by ECA from infraorbital. The supply to the lateral wall on the cranial side is ICA via V1 stapedial; on the orbital side, it is ICA/ECA via V2 stapedial/middle meningeal.

Facing the external world, the eyelids have an internal supply to conjunctiva and an external supply to the skin. Furthermore, they are divided into medial and lateral zones.

Skin

Upper eyelid skin The medial third (Tessier zone 12) is supplied by superior palpebral branch of infratrochlear; the middle third (Tessier zones 11–10) is supplied by frontal; and the lateral third is supplied by lacrimal + transverse facial from superficial temporal artery. Thus, medial two-third of upper eyelid dermis comes from r1 neural crest while lateral one-third of upper eyelid dermis comes from r1 and r2 neural crest.

Lower eyelid skin The medial third (Tessier zone 4) is perfused by inferior palpebral branch of infratrochlear; middle third is perfused by infraorbital, and lateral third (Tessier zone 5) is perfused by StV2 zygomatico-orbital of superficial temporal. All dermis originates from r2 neural crest.

Conjunctiva

Develops from neural crest mesenchyme interposed between lens and future eyelid skin. Has three zones: orbital, tarsal, and marginal. Orbital conjunctiva is supplied by the anterior ciliary arteries that enter the limbus via the four rectus muscles. Tarsal and marginal conjunctiva are supplied by penetrating arteries of the tarsal and marginal arcades running in the subcutaneous tissue.

Superior marginal and peripheral arcades between superior medial palpebral br. of infratrochlear and lacrimal.

Inferior marginal and peripheral arcades between inferior medial palpebral br. of infratrochlear and transverse facial of superficial temporal.

Full-thickness congenital defects, colobomas, spare the fornix and orbital conjunctiva. Because the tarsal and marginal conjunctiva is supplied by the external arcades, a vascular defect will be full thickness.

A curious fact, beloved by board examiners in ophthalmology, concerns the blood supply to the lateral rectus. All other extraocular muscles are supplied from V1-named branches of ophthalmic (read stapedial) and from ciliary arteries (primitive dorsal and ventral ophthalmic). Lateral rectus is unique in two respects. (1) It receives no contribution from ciliary. (2) It is supplied by *two sources* stapedial: the lateral muscular branch and the lacrimal branch.

The Eye and Orbit

The adult anatomy of the ophthalmic axis and its various branches to the eye and orbit can only be understood based upon its developmental history because it represents the crossroads of four arterial sources serving distinct functional systems.

- *Internal carotid stem* artery from internal carotid provides support for the visual apparatus. Optic nerves and chiasm are supported from the anterior communicating branches of anterior cerebral.
- *V1 intracranial stapedial* provides vascular support for the orbit by dividing into two branches at the backwall of the alisphenoid: a medial *orbital branch* via superior orbital fissure; and a lateral *meningeal branch* via cranio-orbital foramen.
- *V2 intracranial stapedial* (recurrent meningeal artery) supplies the dura of the anterolateral temporal fossa and the alisphenoid. It terminates with a small branch that accompanies V2 out of the foramen rotundum.
- *V2 extracranial stapedial* ascends from the pterygopalatine fossa into the lateral orbit via zygomatic artery to supply the lateral orbital wall via its zygomaticofacial and zygomaticotemporal branches.

Because most of the components of the ophthalmic axis are not derived from the internal carotid, we shall reserve discussion about their development for an ensuing section on eye, orbit, and frontonasal face. For our immediate purposes, we shall simply name the component arteries of the combined ophthalmic system and describe their anatomy.

We begin with two statements: (1) *all arteries to the eye and orbit arise as induction products with a corresponding nerve*; (2) the globe itself is a neural unit, being composed of neural ectoderm and covered with neural crest; therefore, *the eye induces its own blood supply*.

Ophthalmic artery is located ventral to the optic nerve. It gives off two main branches: central artery of the retina and paired long posterior ciliary arteries. These developed from the primitive dorsal and ventral ophthalmic arteries. From these, latter arise the short posterior ciliary arteries. In the process, a vascular ring is created that surrounds the optic nerve and is connected with the ophthalmic axis. When V1 stapedial enters the orbit, it is annexed by ophthalmic at the vascular ring. The stapedial stem winds around the lateral aspect of the optic nerve, becoming dorsal to it. Stapedial then immediately breaks into three major divisions.

Ocular Arteries

These arteries supply the structures of the globe. They are derivatives of the original primitive internal carotid system and arise via neuroinduction. Recall that the eye is neural tissue. The optic primordium belongs to the diencephalon; it extends outward from the ventral fifth prosomere. Recall that dorsal fifth prosomere is telencephalon. Furthermore, the sheath of the optic nerve is a continuation of the dura. The sensory supply to the globe comes from two branches of V1: dorsolateral and ventromedial. Thus, the eye induces its own blood supply.

Central Artery of the Retina (derived from the primitive hyaloid artery) branches very close to optic foramen. It penetrates dura to run within the sheath of optic nerve (itself an extension of dura). About 1 cm behind the globe, it enters the substance of optic nerve itself. Recall how conversion of optic vesicle to optic cup caused hyaloid artery to became enveloped within optic fissure. Within the retina, it branches into temporal and nasal distributions mirroring those of the long posterior ciliary arteries.

Short Posterior Ciliary Arteries Parasympathetic autonomic nervous system (PANS) supplies the globe by means of six to ten postganglionic short posterior ciliary nerves. These, in turn, induce the arteries. Let's do a quick review of the relevant neuroanatomy. Ciliary ganglion is the relay station for PANS to the eye. It is situated about 1 cm behind the back wall of the orbit. It is lateral to optic nerve and medial to ophthalmic artery and rectus lateralis. Recall that CN III also passes through the lateral recti. Oculomotor nerve has two divisions, the inferior being larger. Pre-ganglionic PANS fibers from Edinger-Westphal nucleus in midbrain travel with inferior division directed toward inferior oblique (*the first muscle to arise from Sm1*). Pre-ganglionic fibers synapse in a specific (posterolateral) sector of ciliary ganglion.

From the anterior face of ciliary ganglion, wraith-like short ciliary nerves run forward in two bundles, above and below optic nerve. They pass anteriorly on the inner surface of sclera. Thus, short ciliary and long ciliary follow parallel courses but travel in distinct planes: the PANS fibers are superficial to the SANS fibers. Short ciliary nerves provide PANS to ciliaris and sphincter pupillae. They also carry SANS to dilator pupillae along with V1 pain fibers from cornea and iris. For a vivid description of their activity, just ask anyone with contact lenses!

Superior division of oculomotor, not to be outdone, picks up postganglionic SANS fibers from cavernous sinus. These innervate the non-striated Muller's muscle attached to levator palpebrae superioris. You can test this out by scaring a friend (the method is left to your discretion) and observing the upper lid. Sudden stimulus of Muller's causes the eyes to open up wide!

Long Posterior Ciliary Arteries These vessels follow trajectories established by their precursors. Temporal LPCA is derived from primitive dorsal ophthalmic; it supplies dorsotemporal sclera. Nasal LPCA is derived from primitive ventral ophthalmic; it supplies ventronasal sclera. These two arterial axes originate from developmentally distinct locations along the cranial division of the primitive internal carotid. PrVOA arises anterior (distal) to PrDOA, i.e., it is closer to the terminus of PrICA. Thus: *PrDOA is more genetically more "primitive" than PrVOA*. Furthermore, both arteries arise proximal to the take-off of anterior cerebral artery. These facts are vital for understanding the pathologic spectrum of holoprosencephaly. As previously discussed, ***HPE worsens in a medial-to-lateral gradient***. Thus, the nasal hemispheres of the globe are damaged before the temporal hemispheres.

The "developmental primacy" of primitive ventral ophthalmic artery might seem, at first glance, in contradiction with the findings of Padget. Her account visualizes PrDOA and hyaloid joining together as the original axis of ophthalmic stem with PrVOA tacked on secondarily. But, long before PrDOA makes its debut, primitive olfactory artery has already arched over the optic disk to reach the ventral midline. This means that the future "sprout points," the future stems of artery to olfactory nerve, anterior cerebral artery, and primitive ventral ophthalmic artery, although not observed, are *genetically present* distal to the stem of primitive dorsal ophthalmic.

Of course, inductive interactions between the optic cup and the arterial plexus play their own role. The topologic changes involved in folding the cup require that the groove for future hyaloid artery be ventral and lateral. (The cup cannot turn in upon itself medially.) In any case, the long posterior ciliary arteries quickly pierce posterior globe and run forward between choroid and sclera. Upon reaching the limbus, each one divides into an upper and lower branch. These unify as the *circulus arteriosus major*. Radial branches in the iris converge until reaching the pupil at which point a second ring is formed, the *circulus arteriosus minor*.

Anterior Ciliary Arteries These vessels arise from the inferior group of muscular branches of ophthalmic artery. Traveling along rectus tendons the anterior ciliary vessels pierce the sclera just behind the corneo-scleral junction. There they contribute to the circulus arteriosus around the attached iris.

Orbital Arteries

These arteries supply periocular structures: fat, fascia, muscles, and bones. *They are all derivatives of the stapedial system and arise via neuroinduction. We will cover them in additional detail in a subsequent section on the eye and orbit.*

It is worth recalling that the anastomosis of V1 stapedial with ophthalmic takes place ventral to the optic nerve but all the subsequent branches will be dorsal. The stem ascends around the lateral aspect of optic nerve to get into position.

The stapedial system is programmed by branches of V1 with some assistance of the cranial nerves III, IV, and VI for the muscular branches. There are three main branches of V1 in the orbit.

- Lacrimal nerve travels to the lateral border of the orbit where it transects lacrimal gland and reaches the lateral upper eyelid.
- Frontal nerve passes over levator palpebrae superioris and subdivides into *supraorbital* and *supratrochlear* branches, both of which ascend onto the forehead. The supratrochlear nerve is responsible for inducing the extraorbital branches relating to the eyelid and forehead of the nasociliary artery.
- Nasociliary nerve runs deep to optic nerve and follows the medial wall of the orbit giving off successively *posterior ethmoid nerve* to the sphenoid and ethmoid sinuses, *anterior ethmoid nerve* to the cribriform plate and nasal cavity and *infratrochlear nerve* to the lacrimal zone and nasal dorsum. Finally, the ciliary part of nasociliary nerve gives off *long and short ciliary nerves* which program the corresponding branches from the ophthalmic system. Nasociliary nerve is responsible for inducing the intraorbital branches and the extraorbital branches to the lacrimal apparatus and nose.

Lacrimal Artery (Zone 9)

This vessel contributes is induced by the lacrimal nerve. Upon entry into the orbit, lacrimal artery is the first branch off ophthalmic but it has also been observed to exit posterior to the orbit. It courses laterally *beneath* optic nerve and follows along superior border of lateral rectus to reach lacrimal gland. *Lacrimal artery contributes to lateral rectus*. It traverses lacrimal gland, dividing it into orbital and palpebral lobes, logically supplies the orbital lobe. After exiting gland, it supplies conjunctiva and drops down to connect with the *lateral palpebral arteries* of the upper and lower eyelids.

Lacrimal artery makes a critical anastomosis with the middle meningeal system, one that creates a great deal of confusion regarding pathologies of the posterior and lateral orbital walls. Recall that supraorbital stapedial runs forward with both V1 and V2 until reaching the back wall of the orbit, where it divides. StV2 supplies temporal fossa dura and the calvarial lamina of alisphenoid sends a branch to the orbit through a foramen in the alisphenoid. StV1 supplies anterior cranial fossa dura and the orbit with a lateral meningeal branch and a medial orbital branch. These access the orbit through the cranio-orbital foramen and superior orbital fissure, respectively.

In stage 19 medial orbital branch programs of the orbital vessels, including lacrimal. Lateral meningeal branch attaches to distal lacrimal and then ascends through the orbital plate to supply dura of the anterior cranial fossa. During stage 19, both branches remain in continuity with supraorbital for their blood supply. In stage 20, anastomosis with the ophthalmic takes place, as described below, *vide infra*. Now irrelevant, supraorbital involutes backward to the StV1–StV2 bifurcation.

Sometimes the lateral StV1 branch persists. Because it retains its original connection, and because the former supraorbital is taken over by middle meningeal, it appears as if this StV1 meningeal is part of that system. It is called the *communicating branch*, or *recurrent meningeal artery* (RMA). Sometimes it is quite large and actually replaces the lacrimal within the orbit.

Key fact: sometimes this anastomosis persists, in which case you can trace the RMA forward into lacrimal gland where it connects to rise to *Zygomaticotemporal artery* (ZT) ascending up the inner wall of the lateral orbit. Of note, lacrimal artery is directed to the <u>orbital</u> part of the gland but ZT supplied the <u>palpebral</u> part. In any case, ZF exits the gland and terminates in the postorbital bone field, Tessier zone 8. 7.

Herein lies a paradox: in many instances, the original embryonic continuity between RMA and lacrimal disappears, leaving StV1 lacrimal artery as an apparently the sole source for the ZT and ZF arteries. We know the zygoma complex is strictly of r2 neural crest derivation. How can the blood supply for these bone fields come from a V1 source?

Neuroanatomy Saves the Day

Zygoma (jugal and postorbital) and maxilla constitute a common set of developmental fields. From pterygopalatine fossa, StV2 sends a sensory to supply all three of them. Posterior superior alveolar is given off just before entering the infraorbital fissure. Once inside zygomatic nerve is given off, anterior superior alveolar continues forward.

Zygomatic nerve now follows the floor of the orbit laterally and then give offs *zygomaticofacial nerve* (ZF) which exits the jugal through the ZF foramen. Zygomatic continues upward along lateral orbital wall until it reaches lacrimal gland where it terminates in two branches. Lacrimal branch joins with lacrimal nerve *de* jure to supply the gland with PANS motor fibers. *Zygomaticotemporal nerve* (ZT) zots laterally to exit postorbital bone through the ZT foramen. In the meantime, *in the absence of the RMA anastomosis*, blood supply from ophthalmic flows outward from lacrimal artery, pursuing the two V2 branches: (1) it follows laterally to become *ZT artery* to **zone 8**; (2) it descends, following ZT nerve inside the lateral orbital wall until it encounters ZF nerve and there, makes its escape through the malar eminence as *ZF artery* to **zone 7**.

- In sum: Although blood supply to zones 7 and 8 *appears* to come from lacrimal (implying that they are r1 neural crest derivatives), in reality, these r2 bone fields are supplied as expected. The key is the neuroanatomy. The ZF

and ZT nerves do the programming, the artery, departing from its lacrimal blood source, simply plays "follow-the-leader."

Frontal Artery (Zones 10–11)

Supraorbital artery travels with and supplies superior rectus and levator palpebrae superioris. It follows its companion V1 nerve laterally upward and out of the orbit via the supraorbital foramen. Thereupon, it makes a clinically significant division.

- A deep *muscular branch* supplies frontalis and subgaleal fascia. Medially it anastomoses with supratrochlear.
- A superficial *cutaneous branch* is distributed to galea and forehead skin corresponding roughly to the dimensions of the eyebrow.

Ethmoid Artery

The ethmoid axis has also referred to nasociliary. Its only intraorbital branch, anterior ciliary, is very distal and supplies the neural crest intraocular muscles of the iris and lens. The remainder of its branches exit the orbit either medially, through the ethmoid complex or forward to supply the tissues of the nasolacrimal system and medial eyelids.

- Transethmoid/intracranial
  - *Meningeal branch* is most proximal. It passes backward through superior orbital fissure to reach middle cranial fossa where it anastomoses with accessory meningeal that ascends from internal maxilla-mandibular upward through foramen ovale to reach trigeminal ganglion.
  - *Posterior ethmoid* This artery supplies the *posterior intracranial and intranasal tissues* sector of Tessier zones 12 and 13. It passes through orbital lamina to supply posterior ethmoid sinus. From here, it enters the cranium through the *spheno-ethmoid suture*. Thus, it supplies the ethmoid lateral to cribriform plate, i.e., the roof of the ethmoid sinuses. Small branches descending via cribriform plate anastomose in front of sphenoid with medial nasal branches of medial sphenopalatine. Thus, the neural crest presphenoid bone lies in the posterior nasopharynx. It represents the vascular anastomosis between StV1 and StV2.
  - *Anterior ethmoid* After penetrating the orbital lamina, the artery enters "sinus city." Recall that anterior ethmoid sinus, middle ethmoid sinus, and frontal sinus are all continuous. Via the frontal sinus, the artery pursues a *midline* course to enter cranium, where it divides. *Meningeal branch* supplies dura of cribriform plate. This is intracranial zone 13 of Tessier. Just lateral to crista galli, *nasal branch* dives into the roof of nose where it immediately divides. *Lateral anterior nasal branches* are distributed to superior and middle

turbinates while *medial anterior nasal branches* ramify over septum. From the latter, a branch runs along the undersurface of nasal bone, supplying it. At its inferior border, the artery encounters the upper lateral (triangular) cartilage, the cephalic margin of which is neatly tucked underneath the caudal margin of the nasal bone. Via this anatomic "escape route," the artery achieves a subcutaneous location. Paired *anterior ethmoid arteries* travel together in the midline and follow along the upper border of septum. They then descend through columella into the philtrum. The vascular anatomy of the philtral prolabium is the basis of developmental field reassignment cleft surgery.

- Intraorbital
  - *Muscular arteries* Developmentally, these arteries belong to the stapedial system. Muscular arteries arise from various sites along the frontal artery. They are distributed in two groups; these mirror the geometry of the original embryonic vascular structures. Superior branches supply muscles assigned to the sclera of primitive dorsal ophthalmic artery: superior oblique, superior rectus, and levator palpebrae superioris. Inferior branches supply muscles assigned to the sclera of primitive ventral ophthalmic artery: medial rectus, inferior oblique, and inferior rectus. Lateral rectus is also supplied by lacrimal artery (an important detail to which we shall return later). Muscular arteries accompany the motor nerves. The muscles are enclosed within *V1 supplied fascial envelopes* of the individual extraocular muscles. All sensation from the muscle units required for both nociception and proprioception is conveyed back to by branches of V1 to the somatotopic map within the mesencephalic nucleus.
  - *Anterior ciliary arteries* These vessels arise from the inferior group of muscular branches of ophthalmic artery. Recall that these are stapedial branches. Traveling along rectus tendons the anterior ciliary vessels pierce the sclera just behind the corneo-scleral junction. There they contribute to the *circulus arteriosus* around the attached iris. This is a neural crest zone; it is responsible for supplying the intrinsic muscles of the eye.
- Extracranial
  - *Supratrochlear artery* This vessel supplies the *medial orbital and frontal tissues* of **Tessier zones 12–13**. A *deep muscular branch* supplies procerus and corrugator. Laterally, it anastomoses with supraorbital. A superficial *cutaneous branch* is distributed to glabella and medial forehead.
  - *Infratrochlear artery (dorsal nasal artery)* This artery supplies the *nasal sector* of Tessier zones 12–13. It appears above medial canthus, sends a side-branch to

superior lacrimal sac, and then proceeds to divide again.

Dorsal nasal artery
- The lateral branch supplies the nasal sector of Tessier zone 12, i.e., *lateral nasal wall* as a triangle bounded by nasal dorsum medially, lacrimal duct laterally, and medial canthus superiorly. Anastomoses with angular artery of facial.
- The medial branch supplies nasal sector of Tessier zone 13, i.e., nasal dorsum down to tip.

Medial palpebral This artery supplies the *canthus* and *lacrimal sac* of Tessier zone 11. This importance of this vessel is out of proportion to its size. Its trajectory is schizophrenic. It arises deep to the pulley of superior oblique, descends, and then splits around the medial canthus. The paired arteries join with their confrères from the lacrimal artery to encircle the eyelids. They form a double arcade corresponding to anatomic components of orbicularis. A *marginal arcade* just above tarsus supplies tarsal orbicularis. A *peripheral arcade* half way between tarsus and orbit supplies palpebral orbicularis. Just medial to the origin of lateral canthus, the medial palpebral arteries make a connecting loop with paired lateral palpebral arteries derived from lacrimal artery. Colobomas of the eyelid are defects in this system.

The Spectrum of Holoprosencephaly

(Figures 7.51, 7.52, 7.53, 7.54, 7.55, 7.56, 7.57, 7.58, 7.59, 7.60, 7.61, 7.62, 7.63, 7.64, and 7.65)

Holoprosencephaly is a complex spectrum of malformations affecting the brain and (variably) the face. It involves incomplete separation of forebrain structures with increasing degrees of disorganization. Facial manifestations are also progressive, ranging from mild hypotelorism to frank cyclopia or anophthalmia. Recent evidence supports the concept that signaling interactions between brain and face are the common denominator of HPE, i.e., "the brain predicts the face."

The general facts of HPE are well known and summarized in reviews (see Further Reading List). The condition is more common than previously thought and occurs in as many as 1/250 conceptuses (see Further Reading List). Newer methods of investigation (MRI) can detect more subtle findings. Our purpose here is to review the anatomic spectrum of HPE and explore its vascular basis to identify a possible sequence of craniofacial field defects found in both the brain and the face.

We shall begin with conventional anatomic classification. Following that, correlations will be made with specific cra-

Fig. 7.51 Branches of the anterior cerebral artery. ACA with three segments. A3 bifurcates into pericallosal and marginal callosal (paracentral) with implications for the spectrum of the minimal form of holoprosencephaly. MCA segment M1 gives the lenticulostriates to deep gray matter structures. (Reprinted from Hendrix P, Griessenauer CJ, Foreman P, et al. Arterial supply of the upper cranial nerves: A comprehensive review. Clinical Anatomy 2014; 2013, 27(1):108–117. With permission from John Wiley & Sons, Inc.)

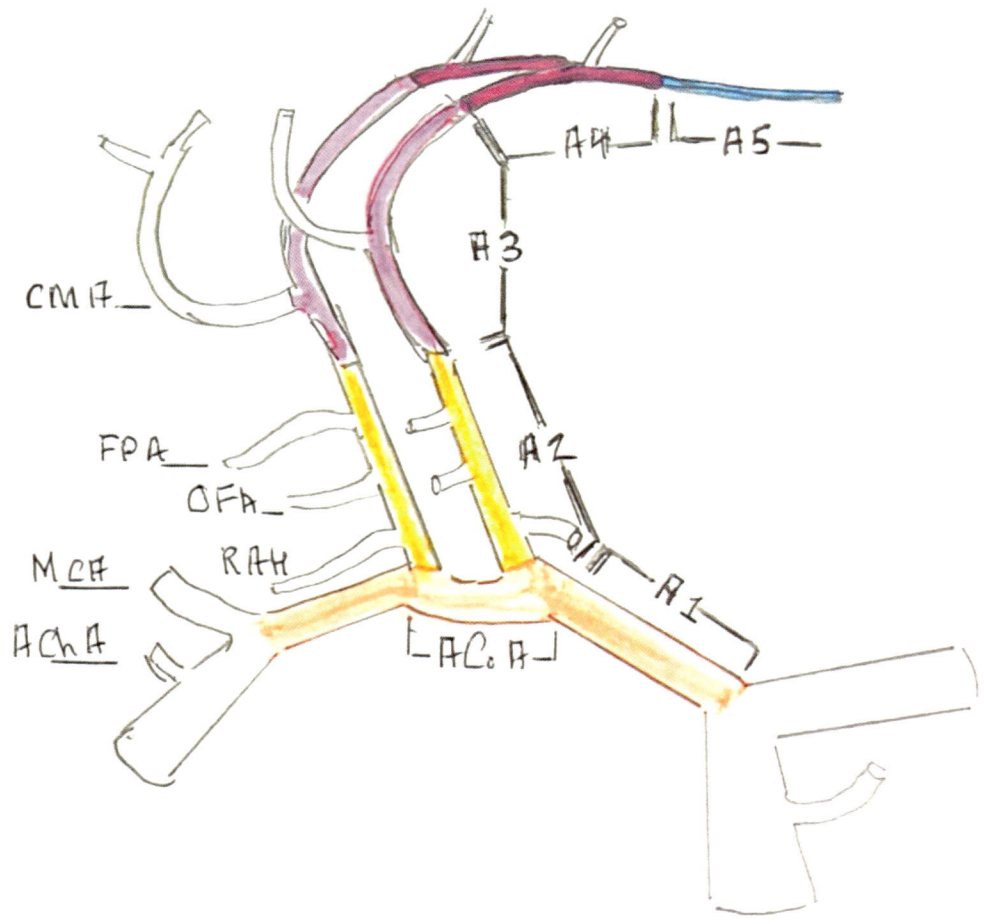

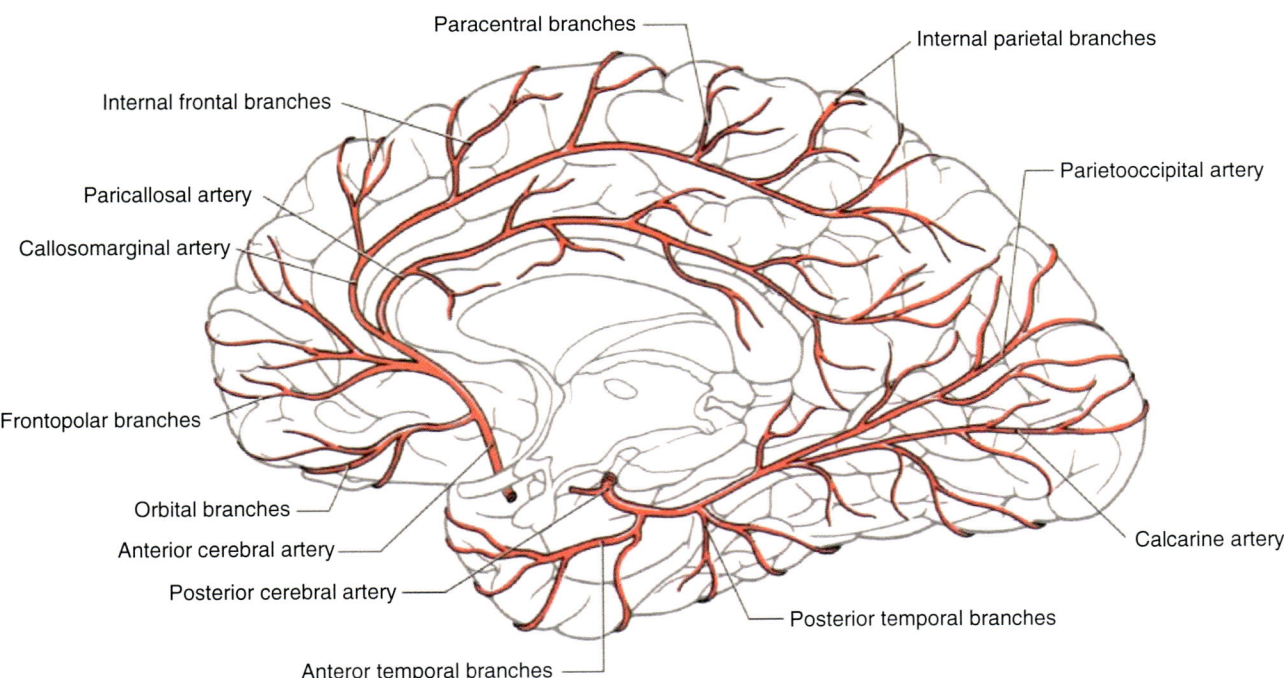

Fig. 7.52 Pericallosal and callosomarginal arteries. Neuromeric segments of cortex are demarcated by branches of the callosomarginal arteries. (Reprinted from Hanes D. Fundamental Neuroscience: Basic and Clinical Applications. Philadelphia, PA: Churchill Livingstone; 2006. With permission from Elsevier)

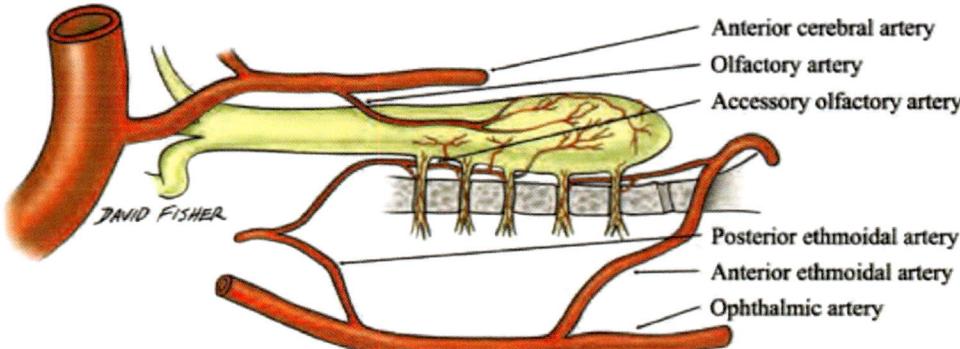

Fig. 7.53 Blood supply of olfactory nerve comes from two sources. (1) ICA-derived primitive olfactory artery is medial and is the main source for the nerve. (2) StV1 anterior ethmoid is lateral; it supplies cribriform plate and provides collaterals to the nerve. These are intracranial manifestations zones 13 and 12. Anosmia without hypotelorism is purely olfactory artery. Any degree of hypotelorism implicates the stapedial system as well. Perhaps the two branches share common genetic controls for their neural crest components. (Reprinted from Hendrix P, Griessenauer CJ, Foreman P, et al. Arterial supply of the upper cranial nerves: a comprehensive review. Clin Anat 2014; 27(8): 1159–1166. With permission from John Wiley & Sons)

Fig. 7.54 Development of the craniofacial primordia. (**a**–**d**) A frontal view of the prominences that give rise to the main structures of the face. The frontonasal (or median nasal) prominence (red) contributes to the forehead (**a**), the middle of the nose (**b**), the philtrum of the upper lip (**c**), and the primary palate (**d**), while the lateral nasal prominence (blue) forms the sides of the nose (**b**, **d**). The maxillomandibular prominences (green) give rise to the lower jaw (specifically from the mandibular prominences), to the sides of the middle and lower face, to the lateral borders of the lips, and to the secondary palate (from the maxillary prominences). Note that: Contemporary depictions of development do not include the frontal zones along with those of the midface. External manifestations of zones 12 and 13 involve the nose and the median forehead. Zone 13 (red) related to medial nasal placode and zone 12 (blue) relates to lateral nasal placode. (Reprinted from Helms JA, Cordero D, Tapadia MD. New insights into craniofacial development. Development 2005; 132: 851–861. With permission from The Company of Biologists Ltd.)

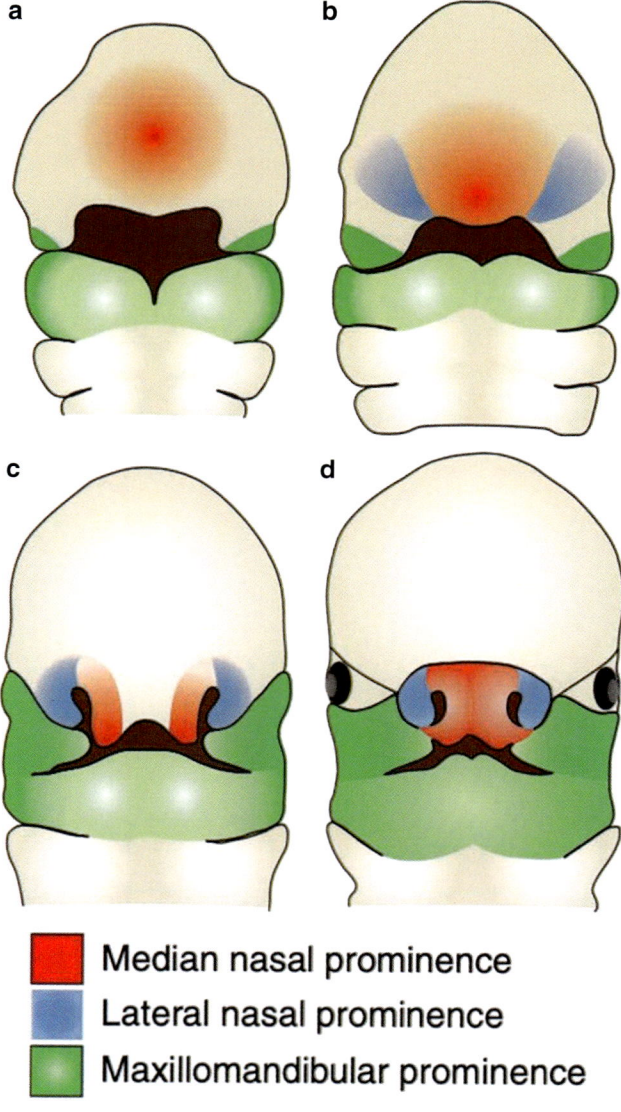

Fig. 7.55 Spectrum of holoprosencephaly. From top to bottom and left to right: (1) rudimentary premaxilla and vomer with bilateral palate cleft and bilateral cleft lip, (2) complete absence of vomerine and premaxillary bones—median cleft, (3) fusion of lip and cebocephaly, (4) ethmocephaly, and (5) cyclopia. (Reprinted from Bianchi D. Fertology: Diagnosis and management of the fetal patient, 2nd ed. Mc-Graw Hill; 2010. With permission from McGraw-Hill Education)

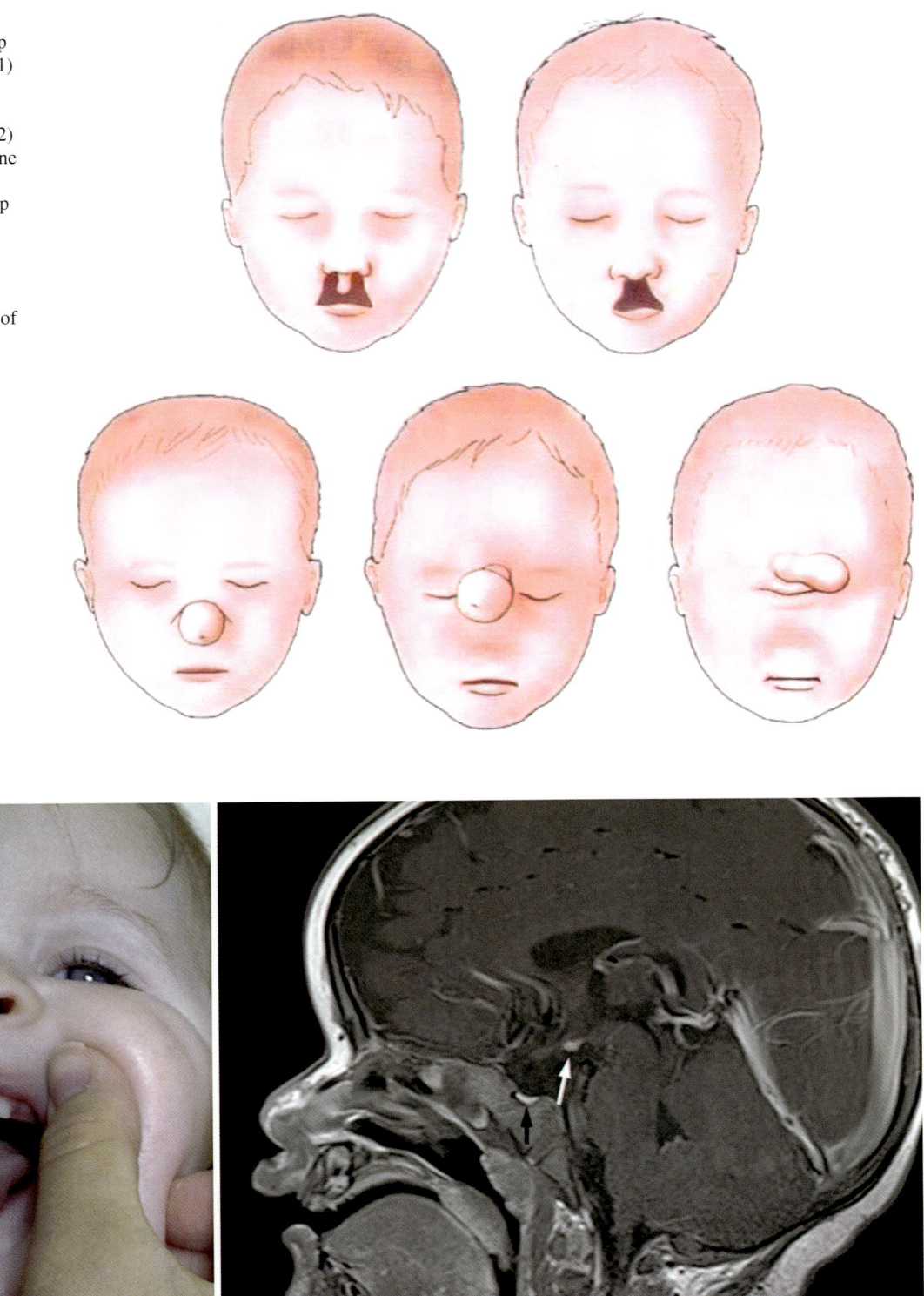

Fig. 7.56 HPE: forme fruste showing fusion of the two central incisors. Total volume of premaxilla is reduced. Lateral incisors are OK. (Courtesy of Michael Carstens, MD)

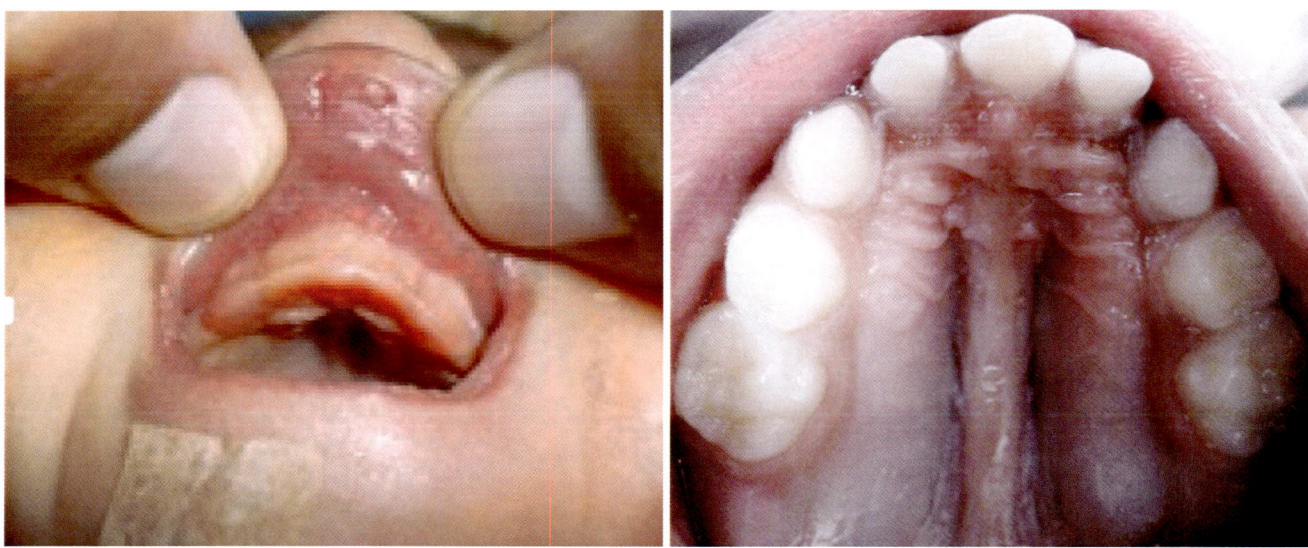

Fig. 7.57 HPE: reduction of vomer—note incomplete bilateral palate clefts on either side of the midline. (Courtesy of Michael Carstens, MD)

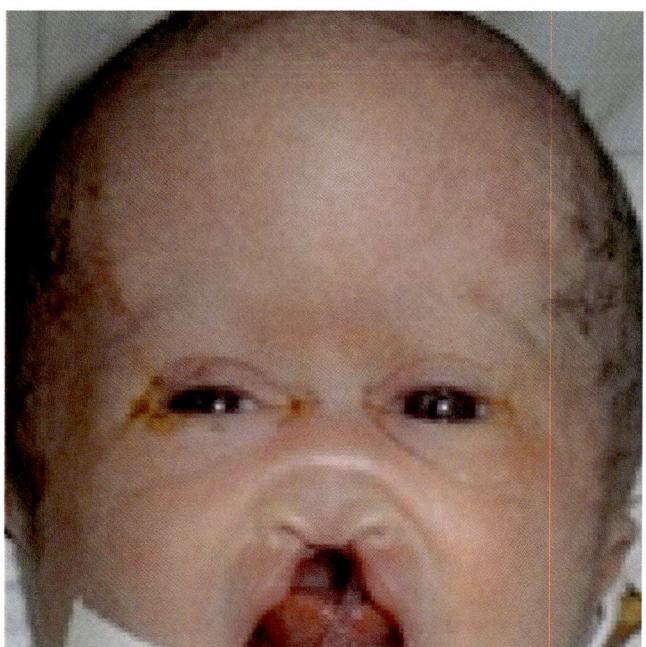

Fig. 7.58 Midline cleft, absent premaxilla and vomer, forme fruste cebocephaly, nasal bones are absent. (Reprinted from ECLAMC Centro Anomalias de Malformaciones Congenital, Professor Ignacio Zarante, Institutode Genetic Humana, Potifica Universidad Javeriana Bogotá, Colombia)

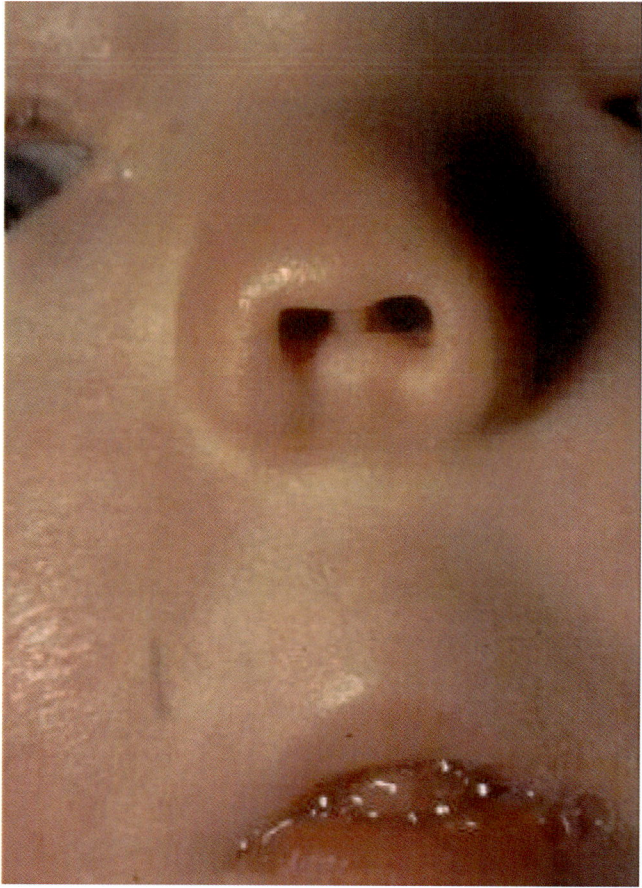

Fig. 7.59 Cebocephaly (low proboscis). (Reprinted with permission from Ritivoiu M, Brezan F, Codreau I. Ultrasound diagnosis in two cases of severe craniofacial anomalies. Medical Ultrasonography 2013;15(4):330–332)

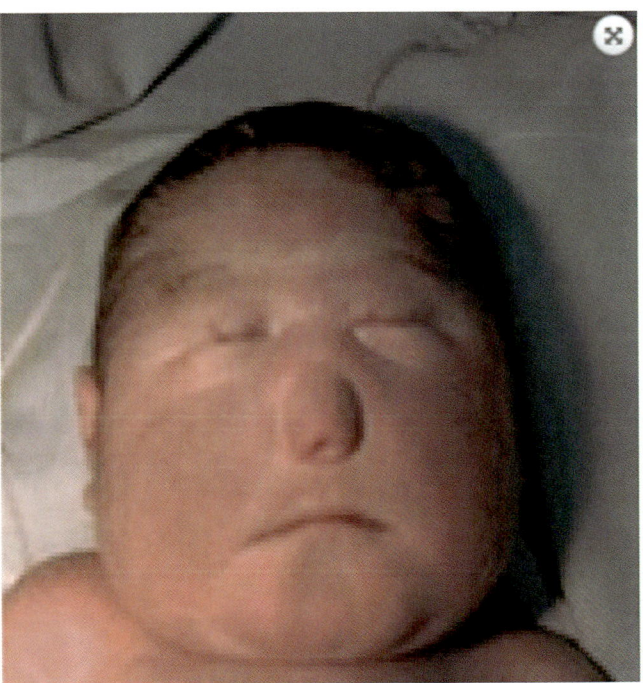

Fig. 7.60 Cebocephaly (high proboscis at nasion). (Reprinted from ECLAMC Centro Anomalias de Malformaciones Congenital, Professor Ignacio Zarante, Institutode Genetic Humana, Potifica Universidad Javeriana Bogotá, Colombia)

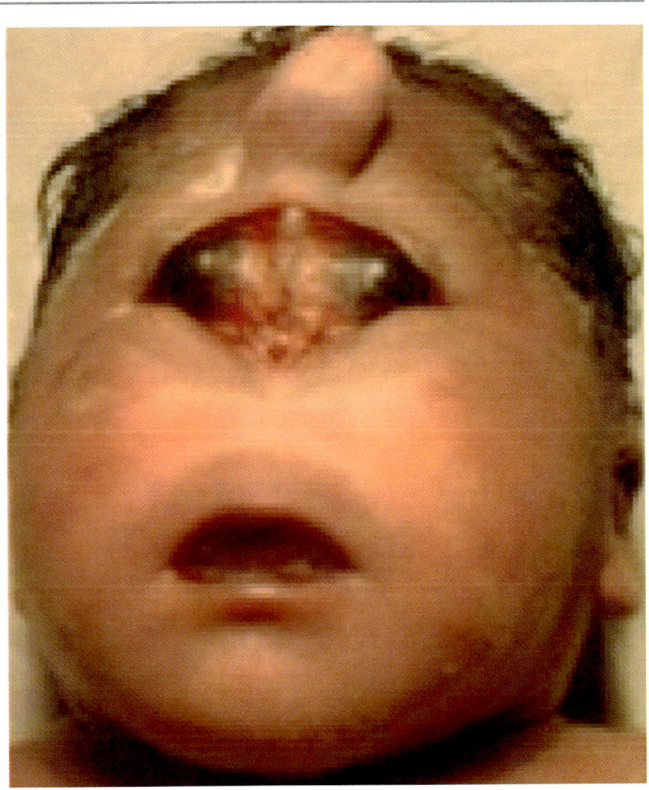

Fig. 7.62 Ethmocephaly/cyclopia—with proboscis. (Reprinted from ECLAMC Centro Anomalias de Malformaciones Congenital, Professor Ignacio Zarante, Institutode Genetic Humana, Potifica Universidad Javeriana Bogotá, Colombia)

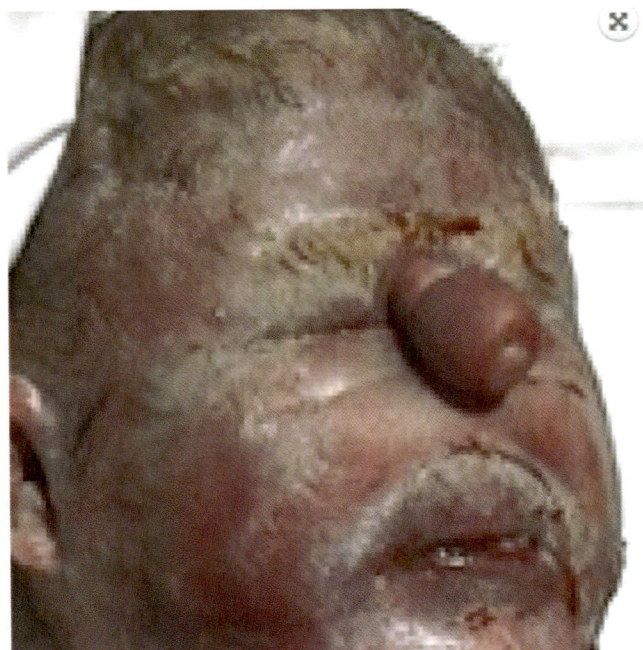

Fig. 7.61 Ethmocephaly (proboscis above the single orbit, eyes separate). (Reprinted from ECLAMC Centro Anomalias de Malformaciones Congenital, Professor Ignacio Zarante, Institutode Genetic Humana, Potifica Universidad Javeriana Bogotá, Colombia)

Fig. 7.63 Pseudo-cyclopia two ocular remnants. (Reprinted from ECLAMC Centro Anomalias de Malformaciones Congenital, Professor Ignacio Zarante, Institutode Genetic Humana, Potifica Universidad Javeriana Bogotá, Colombia)

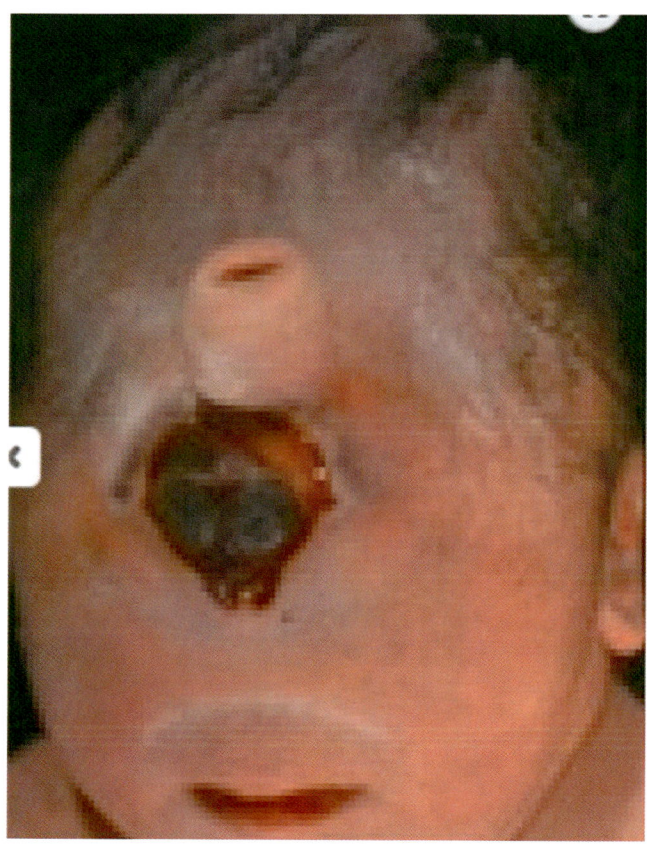

Fig. 7.64 Cyclopia median eye. (Reprinted from ECLAMC Centro Anomalias de Malformaciones Congenital, Professor Ignacio Zarante, Institutode Genetic Humana, Potifica Universidad Javeriana Bogotá, Colombia)

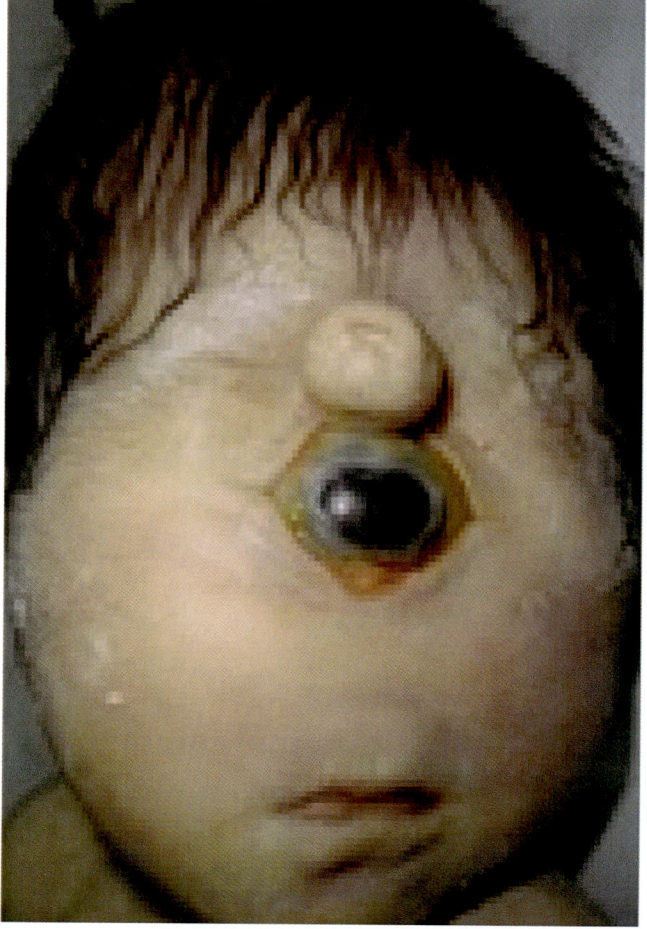

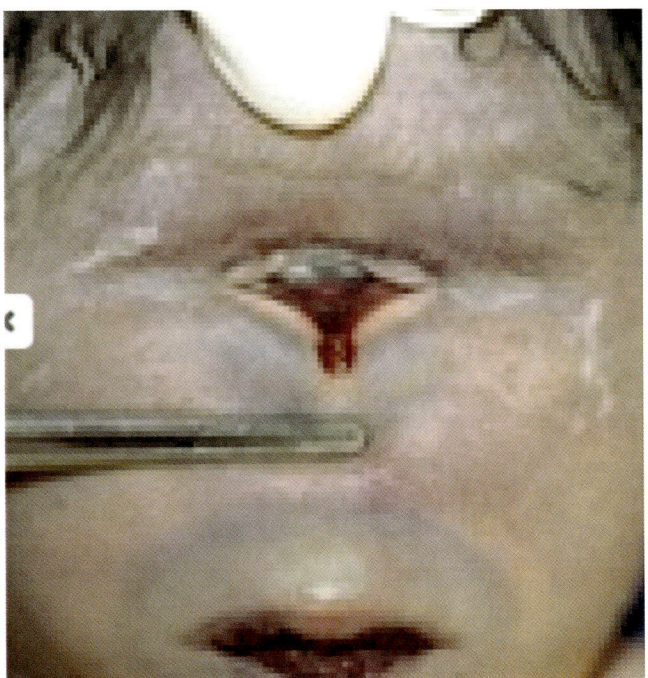

Fig. 7.65 Cyclopia—no eye, no proboscis. (Reprinted from ECLAMC Centro Anomalias de Malformaciones Congenital, Professor Ignacio Zarante, Institutode Genetic Humana, Potifica Universidad Javeriana Bogotá, Colombia)

niofacial angiosomes. We will then consider the sequence of HPE as a progression of vascular deficits arranged along the axis of three systems: for the brain, the anterior division of the internal carotid artery, and for the naso-orbital tissues and eye, the stapedial system and the extracranial internal carotid derived ophthalmic axis. This same type of analysis, cataloging anatomic findings with neurovascular field deficits, will be subsequently applied to other craniofacial anomalies. Note that the degree of deep gray matter separation parallels the degree of hemispheric non-separation. The former indicates lesions along the axis of anterior cerebral artery whereas the latter reflects pathology of the middle cerebral artery axis.

The ocular and orbital manifestations of **holoprosencephaly** involve one or more neurovascular axes from three systems: *primitive internal carotid artery*, *anterior cerebral artery,* and the *medial branch of stapedial*. The first two system affects the eye while the stapedial affects the orbit. If we consider the arterial axes from their spatial position along the course of the internal carotid, from distal-to-proximal these are (1) primitive ventral ophthalmic artery, (2) primitive dorsal ophthalmic artery, and (3) from the new ophthalmic stem, the medial stapedial branch to frontal, and then to nasociliary. The developmental course of these arteries also follows the same time course. As we shall see, HPE follows a progression of increasing vascular compromise.

Anterior cerebral artery is the most distal major branch of the anterior division of ICA. It is responsible for supplying the eye.

Craniofacial Manifestation of HPE

HPE has concomitant craniofacial manifestations. Let's review the scenarios from best-case to worst-case scenario. (1) Mild cases involve a reduction in the r0 neural crest mesenchyme of the ethmoid complex and frontal bone. These zones are supplied by nasociliary. This leads to a minimal hypotelorism and reduced frontal dimensions without affecting the globe. Artery only will spare the artery to olfactory nerve, maintain a relatively normal cribriform plate, and have minimal anomalies such as a deficient corpus callosum. (2) Cases not involving the globe but with involvement of anterior cerebral always have forebrain abnormalities. (3) Cyclopia adds a knock-out of PrVOA. The cyclops presents with a midline fused globe consisting of two nasal retinas and a common "orbit." Hypotelorism is profound. (4) Anophthalmia involves additional knock-out of PrDOA. An orbit, if present, is vestigial. Isolated microphthalmia/anophthalmia can occur without significant craniofacial involvement (other than a small orbit). It likely involves the original hyaloid artery.

CNS Manifestations of HPE

Minimal Interhemispheric Form (MIHF)
- Callosal genu and splenium normal
- Absence/defect of corpus callosum (posterior sector)
- Parietal and posterior frontal neocortex fused
- Interhemispheric fissure division complete
- Hypothalamus and basal ganglia normal separation
- Deep gray matter normal
- Cerebral lobes normal

Lobar Form
- Corpus callosum from normal to absent
- Frontal neocortex fused in midline (entire neocortex is fused, except occipital)
- Interhemispheric fissure division complete
- Olfactory tract hypoplasia
- Deep gray matter separated
- Cerebral lobes normal

Semi-Lobar Form
- Corpus callosum absent
- Complete medial cortical fusion (anterior frontal lobe now involved)
- Interhemispheric division incomplete
- Absence/hypoplasia of olfactory tracts
- Deep gray matter shows varying degrees of non-separation
- Rudimentary cerebral lobes

Alobar Form
- Absent corpus callosum
- Fused forebrain U with single ventricle

- No interhemispheric fissure
- Absent olfactory bulbs and tracts
- Deep gray matter non-separated
- Cerebral lobes

Neuroangiosome Analysis

Minimal Interhemispheric Form (MIHF)

This presentation begins with a lesion in the distal axis of corpus callosum. Here, a watershed zone exists between the terminus of pericallosal artery and posterior cerebral artery. That the genu and splenium are normal indicating that there is defect in the proximal branches from *pericallosal* into the tracts. Thus, corpus callosum can demonstrate distal hypoplasia. Fusion of parietal neocortex indicates a lesion in the distal branches of *marginal callosal* artery. As this defect worsens, the fusion moves progressively more anterior: from medial parietal, to paracentral, and finally to medial frontal. Occipital neocortex remains unaffected as it lies in the territory of posterior cerebral. No fusion of anterior frontal lobe because the frontopolar branch proximal to the callosals remains intact.

Recall that anterior cerebral distribution to the medial cortex and corpus callosum can take two forms. If marginal callosal exists as a separate branch from the A3 stem, then its lesions are independent of pericallosal. Parietal and posterior frontal fusion can occur in the presence of an intact corpus callosum. If marginal callosal is not present, ascending medial cortical branches arise from pericallosal. Common defects will arise in both structures in a posterior-to-anterior progression.

The interhemispheric fissure remains intact because, as these branches ascend to the surface they are in continuity with one another. The blood supply for the developing basal ganglia comes from branches of the anterior communicating artery. These supply the cingulate gyrus, fornix, optic chiasm, parolfactory areas, and hypothalamus. Cerebral lobes are normal due to overall intact nature of anterior and middle cerebral circulation to the cortex.

Lobar Form

Corpus callosum can be normal to absent depending upon the extent of compromise to pericallosal stem. Fusion of the anterior frontal neocortex (in addition to previous posterior fusion) indicates a new defect involving *frontopolar artery*, the stem of which is located in segment A2, proximal to common callosal. Under these circumstances, if corpus callosum is intact, marginal callosal and pericallosal are co-equal and the main axis of A2 and A3 is normal. If A2 itself were compromised, corpus callosum could not exist. Note that neocortical fusion does not yet affect the surface of the interhemispheric fissure. Development of deep brain structures and the cerebral lobes remains unaffected.

Semi-Lobar Form

The axis of pericallosal is completely compromised as indicated by absence of corpus callosum. Fusion of the medial cortex is complete. The process has extended more proximally along segment A2 with compromise of the *naso-orbital branch*, as seen by hypoplasia or absence of the olfactory bulbs and tracts. Non-separation of deep gray matter indicates involvement of segment A1, most likely at the level of central branches of anterior communicating. Partial absence of Interhemispheric division is seen. The cerebral lobes are hypoplastic. The defect may be spreading out to the more proximal stem from anterior division of ICA, with involvement of distal branches of middle cerebral artery in segments M4 (cortical), M3 (opercular), and M2 (insular). Deformed ventricles indicate involvement of anterior choroid, the most proximal branch of anterior division.

Lobar Form

All previous findings indicating diffuse involvement of the entire ACA. Definite compromise of proximal A1 segment in wipe-out of the both nasal and orbital circulation. Recall, however, that the primitive DOA and VOA arise more proximally. The stem of PVOA is located at the most proximal position within the anterior division and while that of PDOA arises from ICA just proximal to the division itself. Complete absence of interhemispheric division is seen with a single ventricle. Deep gray nuclei are fused indicating involvement of lenticulostriate arteries of segment M1. In its final form, the "pancake brain" demonstrates diffuse failure of both the ACA and MCA. Note that posterior cerebral artery (PCA) embryologically is the penultimate branch of occipital development and remains protected due to collateral supply to the PCA from the basilar artery (proximal fusion of the longitudinal neural arteries).

Craniofacial Manifestations of HPE

DeMyer classification, modified by Kawamoto

Type VB: Normal brain
- Hypotelorism (ethmoid fields)
- No cleft lip (normal premaxilla)
- Midface hypoplasia
- Normal nose and piriform fossa

Type VA: Minimal Form (Zone A3)
- Forme fruste brain findings
- Hypotelorism (frontal and ethmoid fields)

- Median cleft lip (rudimentary premaxilla)
- High arched palate without cleft
- Midface hypoplasia
- Flat nose (hypoplastic nasal bones, normal piriform fossa)

Type IVB: Lobar Form (Zone A2 Distal)
- Hypotelorism
- Median cleft lip (no vomer or premaxilla)
- Cleft palate
- Flat nose (hypoplastic nasal bones, piriform fossa narrow)

Type IVA: Semi-Lobar Form (Zone A2 Proximal, Zone A1)
- Hypotelorism
- Median cleft lip (no vomer or premaxilla)
- Flat nose—loss of septum, flat nasal bones
- Single nostril

Type III: Alobar Form—Cebocephaly (Entire ACA, Zone M1)
- Hypotelorism
- Median cleft lip
- Proboscis in situ

Type II: Alobar Form (Ethmocephaly)
- Extreme hypotelorism—separate orbits
- Median cleft lip
- Proboscis located above the orbits

Type I: Alobar Form (Cyclopia)
- Single orbit
- Divided eye vs. single eye
- Proboscis may be absent

Neuroangiosome Analysis

Type VB represents a mild form of ethmoid hypoplasia in the nasociliary axis, with partial compromise of both the posterior and anterior ethmoid branches. This reduces the transverse dimensions of the interorbital MNC nasoethmoid complex. Volume of the ethmoid sinuses should be reduced, causing mild hypotelorism. This is sufficient to retard forward projection of the r2 maxillary-zygomatic complex.

Type VA Midline hemispheric HPE. Supratrochlear branch of nasociliary affects volume of Tessier zones 13 and 12. More extensive involvement of the ethmoid fields making up the lateral nasal wall reduces dimensions of the piriform fossa and further limits maxillary development. Perpendicular plate preserved by septal hypoplasia results from medial nasal of anterior ethmoid. Posterior ethmoid supply support perpendicular plate. Development of vomer and premaxilla is restricted due to poor transmission of the medial nasopalatine axis. Supratrochlear branch involvement from nasociliary axis begins to progressively affect Tessier zones 13 and 12, beginning with zone 13. This can be seen as median approximation of the eyebrows and on CT.

Type IVB Lobar HPE brain findings. Progressive hypotelorism. Involvement of posterior ethmoid to perpendicular ethmoid plate blocks development premaxilla, causing a median cleft. Intact by hypoplastic vomer causes high arched palate but no cleft palate. Midline ethmoid deficiency reduces nasal tip projection. Lateral branches of anterior ethmoid branches to internal and external nasal envelope cause hypoplasia of nasal bones. Medial nasal placode is dysfunctional to hypoplastic causing defects in accessory olfactory system and vomeronasal organ with absence of terminal nerve. This is the earliest manifestation of pathology in the olfactory axis. Medial olfactory placode contains growth hormone-releasing hormone (GnRH) and relates to pituitary. Perpendicular plate and septal reduction cause high arched palate but no cleft.

Type IVA Semi-lobar HPE brain findings manifest. The nasal deficiency is more advanced with absent nasal bones, no septum, and complete loss of projection. Absent vomer causes midline cleft palate with normal maxillary shelves unable to fuse. Medial nasal placode hypoplasia and severe reduction in the nasal aperture. Endocrine disturbances can occur at the level of the hypothalamus. The condition progresses backwards along the axis of the anterior division of ICA. Cerebral lobes are affected indicating MCA involvement. Involvement of orbito-nasal Iris coloboma can reflect defective development of PVOA which is the most proximal stem from the anterior division of ICA.

Type III Alobar (cebocephaly) This condition represents a complete loss of the lateral fields of the nasal placode. Olfactory tracts are compromised meaning generalized involvement of fields by primitive olfactory artery. Severe hypotelorism with downward slanting of the medial canthi toward midline deficiency zone. Frontal deficiency is seen as midline approximation fusion of eyebrows. *Proboscis* results when the two nasal placodes fuse in the midline, a process involving complete aplasia of the medial half of each nasal placode. Since the nasal placodes are supplied by primitive olfactory arteries, it involves wipe-out of the medial branch of PrOlf.

Type II Alobar (ethmocephaly) The orbits are still separated by residual ethmoid mesenchyme. Eyes not normal with findings affecting the nasal retina. PVOA affected. Defects in the sclera of the nasal hemisphere can cause medial rectus malinsertion and strabismus. Blood supply can affect development of the lens. Abnormal lens. Approximation of the orbits forces the proboscis to relocate above the orbits where it is interposed between upward-slanting eyebrows.

Type I Alobar (cyclopia) Medial walls of the orbit are gone. Entire nasociliary axis failure. Lateral retinas are abnormal but PDOA still intact. Proboscis is absent indicating complete failure of primitive olfactory artery. Eyebrows fused in arc above the orbits. Position of the eyebrows determined by the two bone fields of the orbital rim: prefrontal and postfrontal. These appear to be drawn upward toward a deficiency state of frontal in the midline with zone 13 more affected than zone 12.

Developmental Field Sequence of HPE

Embedded in the classification of HPE is a progression of embryologic deficits.

CNS Manifestations of HPE

HPE follows a clear-cut distal-to-proximal progression involving the axis of anterior division of internal carotid artery. Neuropathology along each of the stems of the cranial division of ICA proceeds from minor superficial findings to more central disorganization and aplasia.

Nasal field finding involving the olfactory zone is always present, indicating involvement of primitive olfactory artery. Clinical manifestations of anosmia are noted as the lobar form but have not been studied in the minimal form. These are well described in Kallman's syndrome, a condition involving defects of the nasal placode. As the condition worsens in the alobar form and the nasal placodes come to the midline, primitive olfactory failure causes proboscis. ACA and MCA findings have been cataloged. Ventricular findings in semi-lobar HPE implicate compromise of the most proximal stem, anterior choroid. This localizes the lesion at the level of PrVOA, compromise is accompanied by ocular findings involving the naso-ventral hemisphere of the eye.

Compromise of PrDOA indicates disease proximal to the bifurcation itself. [Recall that primitive internal carotid artery refers to the segment between the first aortic arch and the bifurcation. Once the bifurcation is complete, it is no longer considered "primitive" so PrICA is renamed the ICA.] In failure of PrDOA, the condition falls outside the anatomic zone of HPE and is arguably not part of the syndrome.

Craniofacial Manifestation of HPE

HPE in the orbit and face is caused by lesions along the axis of the nasociliary artery of the orbital stapedial system. Nasociliary has three branches: posterior ethmoid, anterior ethmoid, and common trochlear. The territories of distribution have been described for each. All three stems appear to be affected concomitantly. A distal-proximal progression is logical but is more difficult to piece together.

Both posterior ethmoid and anterior ethmoid supply zones 13 and 12, within the intracranial cavity as well as the nose. They not only support dura but also the olfactory fibers coming through the cribriform plate. Hypoplasia of the olfactory apparatus represents a negative synergy between posterior ethmoid and primitive olfactory from the ACA. Anosmia is present in the lobar but studies have not been done to document olfactory dysfunction in the minimal form of HPE. True placode dysgenesis is characteristic of the alobar form.

Within the nasal cavity, both lateral wall sinus involvement (zone 12) and/or midline ethmoid/septal hypoplastia (zone 13) occur very early in HPE. These are responsible for the hypotelorism and median cleft seen from the very beginning of the sequence.

Note: Pathologies of the nasal placodes and of the eye, although they are manifested externally, are the result of the intracranial HPE sequence. This makes sense because both structures are anatomic extensions of the brain, the eye as an extension of basal diencephalon and the olfactory placode as containing neurons that will migrate into the brain to form the olfactory and accessory olfactory systems.

Relationship of HPE to Tessier Craniofacial Clefts

Craniofacial clefts of the orbitofrontal zone involve disruption of normal development between developmental fields, some of which are bone-forming and others of which produce soft tissue structures only. This spectrum of clefts described by Tessier is discussed in a separate chapter. It is important however, to make some very simple distinctions between these entities and HPE.

Four Mesenchymal Units Are the Anatomic Basis of Tessier Cleft Zones 10–13

All four involve the arteries of the orbital stapedial system. From medial to lateral, the nasociliary axis supplies zones 13 and 12 while supraorbital supplies zones 11 and 10. Defects in this system do not directly affect the CNS but deficits in the orbit or frontal bone can permit formation of encephaloceles.

Each artery of the stapedial system has a unique genetic definition. Thus, the biologic behavior of artery is independent of the others. Orbitofacial clefts do not show a "progression" of severity from one axis to another. Multiple fields can be affected at the same time.

In all established forms of HPE, deficits of neuroangiosomes in the anterior division of ICA are combined with deficits along a single axis of the stapedial system, nasociliary artery.

The common denominator of both Tessier clefts and HPE has to do with (1) the individual genetic identities of distinct branches of anterior division of internal carotid and those of stapedial nasociliary; and (2) the common genetic definition shared by all these branches. The pathology underlying HPE is distinct from that which produces craniofacial clefts of the Tessier variety. In the latter, a specific "hit" affects the mesenchyme within the homeotic distribution of a particular neurovascular axis (e.g., the dorsal nasal artery). The deficits are localized and are not additive.

In HPE, the "hit" must affect a genetic component common to the entire length of anterior division of internal carotid from its distal terminus backward to its take-off. There may be subcategories of genetic definition for each branch of the division. The defect manifests itself at the olfactory branch of ACA. As the deficit worsens, its clinical manifestations are seen more distally, i.e., further away from the midline. Furthermore, the defect has a quantitative "dose-dependent" effect on development of the forebrain supplied by anterior division. The worse the defect, the more profound the cerebral insult.

Finally, it should be noted that in HPE there are no frank bone deficits between the craniofacial fields. The fields are confluent and progressively more hypoplastic. But encephaloceles are not seen.

It should be noted that some syndromic forms of craniofacial clefts have neurologic manifestation. Thus, when the genetic definition of each arterial system is worked out, and when anatomic zones of the brain and face are analyzed, it will be found that they share a common genetic basis linking their vascular supplies together.

Where is the pathology located? Here are some possibilities. (1) *Local environment hypothesis:* Vessel sprouting at specific sites results from interaction between ICA mesenchyme and the local environment. Problems in the induction of a specific branch could occur while zones remain normal. (2) *Genetic gradient hypothesis:* Elaboration of a gene product could be required from proper differentiation of the vascular axis. These could be expressed quantitatively such that a progressive deficit of the gene product would express itself as a proximal-to-distal gradient. (3) *Failure of neural induction hypothesis:* In craniofacial clefts, inadequate development of the sensory or motor nerve responsible for inducing its companion artery would lead to a tissue deficit. (4) *Premigratory mesenchymal hypothesis:* Neural crest arising from specific sectors of neuromeres m1–m2, and r1–r0 could be affected by local pathology prior to migration.

What is important to keep in mind is that common pathologies such as clefts and HPE can be rationally explained by looking at the neurovascular axis as the "end organ" of the pathology. In sum: clefts are independent entities involving multiple independent vascular units. HPE in all but its minimal forms requires the addition of an extracranial vascular axis, the most medial of the stapedial arteries, nasociliary. The two conditions share an overlapping Venn diagram. Research that defines the molecular basis of craniofacial arterial development can provide a common denominator to better understand these clinical spectrum of HPE and orbitofacial clefts.

Concluding Remarks

In this chapter, we have explored in depth the sequential events involved in the development of craniofacial arterial supply. This has been done by examining each of its component parts (external carotid, stapedial system, etc.) and how they relate to one another. Two areas remain for discussion which will be deferred to separate chapters altogether. The development of the meninges and their blood supply involves the concept of concentric layers of integument and fascia. Once again, the stapedial system will be examined in depth to explain the anatomy of the dural arteries. Although we have concluded with the vertebral system and how it contributes to the hindbrain circulation, the overall topic of neck development is also a separate chapter. It is my hope that you can take out if this chapter is a firm sense of the timeline in which the craniofacial vascular systems develop, interact, and reorganize into their final adult form. This information will prove of great value as we move forward to further discussions of craniofacial pathologies in terms of neuromeric failure to produce mesenchymal and neurologic structures.

References

1. Kaban LB, Padwa B, Mulliken JB. Mandibular deformity in hemifacial microsomia: a reassessment of the Pruzansky and Kaban classification. Plast Reconstr Surg. 2014;134(4):657e–8e.
2. Bergfeld CB, Heike C. Craniofacial microsomia. Semin Plast Surg. 2012;26(2):91–104. https://doi.org/10.1055/s-0032-1320067.

Further Reading

General

Carlson BM. Human embryology and developmental biology. 6th ed. Elsevier; 2018.
Liem KF, Bemis WB, Walker W, Grande L. Functional anatomy of the vertebrates: an evolutionary perspective. 3rd ed. Cengage Learning; 2000.
Padget DH. The development of the cranial arteries in the human embryo. Contrib Embryol Carnegie Inst. 1948;32:205–61.
Schoenwolf GC, Bleyi SB, Brauer PR, Francis-West PH. Larsen's human embryology, Chapter 13. 5th ed. Elsevier; 2014.
Taylor GI. The angiosome concept and tissue transfer. St. Louis: Quality Medical Publishers; 2014.

Holoprosencephaly

Artman HG, Boyden E. Microphthalmia with single central incisor and hypopituitarism. J Med Genet. 1990;27:192–3.

Cohen MM. Perspectives on holoprosencephaly L Part III. Spectra, distinctions, continuities, and discontinuities. Am J Med Genet. 1981;34:271–68.

Cohen MM. Malformations of the craniofacial region: evolutionary, embryonic, genetic, and clinical perspectives. Am J Med Genet. 2002;115:245–68.

Cohen MM. Holoprosencephaly: clinical, anatomic and molecular dimensions. Birth Defects Res A. 2006;76:658–73.

Dubourg C, Bendavid C, Pasquier L, Henry H, Odent S, David V. Holoprosencephaly. Orphanet J Rare Dis. 2007;2:8. https://doi.org/10.1186/1750-2-8.

Elias DL, Kawamoto HK, Wilson LF. Holoprosencephaly and midline facial anomalies: redefining classification and management. Plast Reconstr Surg. 1992;90(6):951–8.

Hahn JZS, Barnes PD, Clegg NJ, Stashinko EE. Septopreoptic holoprosencephaly: a mild subtype associate with midline craniofacial anomalies. Am J Neuroradiol. 2010;31:1596–601.

Hall RK. Solitary median maxillary central incisor (SMMCI) syndrome. Orphanet J Rare Dis. 2006;1:12.

Hattori H, Okuno T, Momoi T, et al. Brief clinical report: Single central maxillary incisor and holoprosencephaly. Am J Med Genet. 1987;28:483–7.

Johnson VP. Holoprosencephaly: a developmental field defect. Am J Med Genet. 1981;34:258–64.

Kjaer I, Keeling JW, Graem N. The midline craniofacial skeleton in holoprosencephalic fetuses. J Med Genet. 1991;28:846–55.

Machado E, Machado P, Grehs B, Grehs RA. Síndrome do incisivo central superior solitário: relato do caso. Dental Press J Orthod. 2010;15(4):55–61.

Mallick S, Ray PSS, et al. Semilobar holoprosencephaly with 21q22 deletion: an autopsy report. BMJ Case Rep. 2014; https://doi.org/10.1136/bcr-2014-203597.

Pineda-Álvarez DE, Solomon BD, Muenke M, et al. A broad range of ophthalmological anomalies is part of the holoprosencephaly spectrum. Am J Med Genet. 2011;155(11):2713–20.

Petryk A, Graf D, Marcuclo R. Holoprosencephaly: signaling interactions between the brain and the face, the environment and the genes, and the phenotypic variability in animal models and humans. Wiley Interdiscip Rev Dev Biol. 2015;4(1):17–32.

Raam MS, Solomon BD, Muenke M. Holoprosencephaly: a guide to diagnosis and clinical management. Indian J Pediatr. 2001;48(6):457–66.

Sevastano CP, Bernardi P, Seuánez HN, Martins-Moreira MA, Orioli LM. Rare nasal cleft in a patient with holoprosencephaly due to mutation in the ZIC2 gene. Birth Defects Res A. 2014;100:300–6.

Siebert JR, Kokich VG, Beckwith JB, Cohen MM, Lemire RJ. The facial features of holoprosencephaly in anencephalic human specimens, II. Craniofacial anatomy. Teratology. 1981;23:305–15.

Souza JP, Siebert JR, Beckwith BD. An anatomic comparison of cebocephaly and ethmocephaly. Teratology. 1990;42(40):347–57.

Solomon BD, Muenke M. Holoprosencephaly overview. GeneReviews NCBI Bookshelf. www.ncbi.nlm.nih.gov/books/NBK1530

Tessier P, Ciminello FS, Wolfe SA. The arrhinias. Scand J Plast Reconstr Surg Hand Surg. 2009;43(4):177–96. https://doi.org/10.1080/02844310802517259. PMID:19401938

Developmental Anatomy of the Craniofacial Bones

Michael H. Carstens

Introduction

How to Make This Chapter Work for You

The purpose of this chapter is to discuss in detail the bones of the craniofacial skeleton in such a way as to make each one "come alive." Each bone has a unique developmental history that determines its location within the skull as a whole, its shape, its foramina and grooves, and its function. How can we accomplish this? Some new tools are required. Let us keep the following five caveats in mind.

First, applying our new conceptual model to the calvarium requires us is to make a change in our linguistics. How we go about describing something determines how we shall visualize it. Most of us think of the skull in textbook terms, either as a whole or as individual parts. But these images are all based on the final state. We open a model of the brain and skull; we observe the superior orbital fissure. It contains neurovascular structures (e.g. ophthalmic vein and cranial nerves such as III and V1). These are described as "passing through" superior orbital fissure when, in reality, the sequence is the opposite. Pre-existing neurovascular structures "assigned" to the globe develop lateral to the mesoderm of the future basisphenoid. Later in time, r1 neural crest migrating above them is deposited medially and anteriorly as presphenoid and lesser wing. Finally, r2 neural crest comes in laterally to form greater wing. The result is the superior orbital fissure.

With the exception of the jaws and pharyngeal arch bones, the calvarium is protective of pre-existing soft tissue structures. Every foramen, fissure, and groove has a story to tell… based largely upon those neural and vascular structures that will become surrounded by bone-forming mesenchyme. This is an entirely different mindset than that used by traditional texts and requires that we use our anatomic language in a new way. For example, consider a phrase "the carotid artery passes through the petrous bone on its way to cavernous sinus." Obviously, arteries are not augers. They cannot "drill" their way through bone. Neurovascular development is primal. Nature never produces a structure without first ensuring for its survival. Thus, each bone of the craniofacial skeleton develops in the context of its vascular "surround."

Second, we need to view neuromeric terminology as a means to unify and simplify development. After all, bone formation is a rather simplistic. In order to classify bone development, we need to know two key pieces of information: (1) the identity of its mesenchyme (neural crest vs. mesoderm) and its neuromeric source (where it comes from); and (2) the identity of any participating epithelium (ectoderm vs. endoderm). We must also bear in mind two caveats: (1) some bones are composites consisting of several mesenchymal "units" and these do not necessarily come from the same neuromeric location. (2) Some membranous bones develop from two distinct epithelial programs. Such bones will be bilaminar, the potential space being occupied with marrow or a paranasal sinus.

Third, our model is inherently neuroanatomic. Follow the nerves and you cannot get lost. This explains the apparently illogical coding of epidermis and oral mucosa. Coding mesenchyme is easy. Every neuromeric level is represented. But the neuromeric identity of ectodermal epidermis or endodermal mucosa is not fixed by gastrulation. Instead, it is determined by signals from the subjacent dermis/submucosa. In craniofacial anatomy, these layers are neural crest derivatives while in the rest of the body, they come from mesoderm. This makes the neuromeric "count" of ecto/endoderm *discontinuous* because neural crest mesenchyme for dermis/submucosa comes from limited sources. *Frontonasal dermis* (including the upper half of the nasal chambers) comes from caudal forebrain neural crest (p3–p1) and nasal submucosa comes from midbrain neural crest (r1): both are innervated by V1. *Periorbital dermis* from the upper eyelid to the eyebrow comes from r1 and is innervated by V1. *Facial and scalp dermis* is supplied by hindbrain neural crest (r2–r3), most of which is innervated by V2–V3. No further dermis is

M. H. Carstens (✉)
Wake Forest Institute of Regenerative Medicine, Wake Forest University, Winston-Salem, NC, USA
e-mail: mcarsten@wakehealth.edu

© The Author(s), under exclusive license to Springer Nature Switzerland AG 2023
M. H. Carstens (ed.), *The Embryologic Basis of Craniofacial Structure*, https://doi.org/10.1007/978-3-031-15636-6_8

produced by r4–r11 nor does the first cervical somite have a dermatome. Thus, to all intents and purposes, facial epidermis is dermis defined. The sensory map of the head and neck reveals that the junction between craniofacial neural crest dermis and neck/trunk mesodermal dermis is the boundary between V3 and C2.

Fourth, bone coding is conceptually simple. Forebrain neural crest does not produce bone. Hindbrain neural crest from levels r2 to r11 produces the calvarial bone and the bone/cartilage structures of the first to fifth pharyngeal arches. Midbrain neural crest is responsible for the synthesis of the anterior cranial base and fronto-naso-orbital complex.

Fifth, the developmental sequence of craniofacial bones depends on the *physical location of osteogenic mesenchyme* and its *vascular support*. It cannot reliably be deduced by ossification pattern. Given the right environment, paraxial mesoderm and neural crest can develop via either membranous or chondral mechanisms. This makes eminent biologic sense. The function of the cranium is brain protection. The cranial base forms first; the lateral calvarial walls are peripheral "add-on" bones. Thus, paraxial mesoderm from the fourth somitomere backwards is immediately available to support the neuraxis. Dermal bones of the cranial suprastructure (including the fronto-orbito-nasal complex) arise later from a *direct migration* of neural crest populations from the midbrain and rostral hindbrain (r0–r3). Remaining craniofacial bones develop from an *indirect migration* hindbrain neural crest via intermediary structures, the pharyngeal arches.

Phylogeny of Craniofacial Bones

Each bone field has a story to tell: a story about its evolutionary history. In this chapter, we will see many illustrations of the bone fields as they appear in the line leading to humans: prehistoric fishes, tetrapods, reptilomorphs, and pre-mammals. Note how bone fields change in relationship to one another but maintain their neuromeric identity. Careful study of these images will give you a dynamic picture of how the components of the human skull were derived.

Craniofacial Bone fields

Neurocranium (r0–r3)
- Sphenoid
  - Basisphenoid
  - Orbitosphenoid
  - Alisphenoid (epitpterygoid)
  - Pterygoid process
- Ethmoid
- Nasal

- Lacrimal
- Frontal
- Parietal
- Interparietal
- Squamosal
- Tympanic

Neurocranium (r4–r11)
- Temporal bone complex
- Occipital bone complex

Splanchnocranium (r2–r11)
- Vomer (parasphenoid)
- Inferior turbinate
- Palatine (pterygoid, ectopterygoid)
- Premaxilla
- Maxilla
- Zygoma
- Mandible
- Hyoid

Classification of Craniofacial Bones

Bone development can be classified into three types of categories: (1) mesenchymal source, (2) ossification mechanism, and (3) neuromeric level.

Source of Mesenchyme
This first category recognizes that skull bones arise from neural crest, paraxial mesoderm or a combination of the two. The second category distinguishes bones that form in membrane, those that form in cartilage, and those that start out as cartilage but become membranous. Finally, the neuromeric system permits anatomic localization of each derivative. Neural crest of the maxilla arises from a different location (r2) along the neural folds than that of the hyoid (r4–r5). Paraxial mesoderm of the petrous temporal arises from somitomeres 5–6 (in register with r4–r5) whereas that of the supraoccipital arises from the fused sclerotomes of somites 1–4 (in register with r8–r11) (Fig. 8.1).

Mechanism of Ossification
Membranous bones involve an interaction between mesenchyme and a signaling source, generally an epithelium, that provides spatial cues as to the physical dimension of the bone. For example, pharyngeal endoderm "programs" the hyoid. Experimental excision and 180° rotation of this zone in chick embryos results in a complete inversion of the hyoid bone. Calvarial vault synthesis involves interaction between overlying skin and underlying dura with an intervening mesenchymal population made up of neural crest dermis and, in some locations, rhombencephalic paraxial mesoderm (PAM).

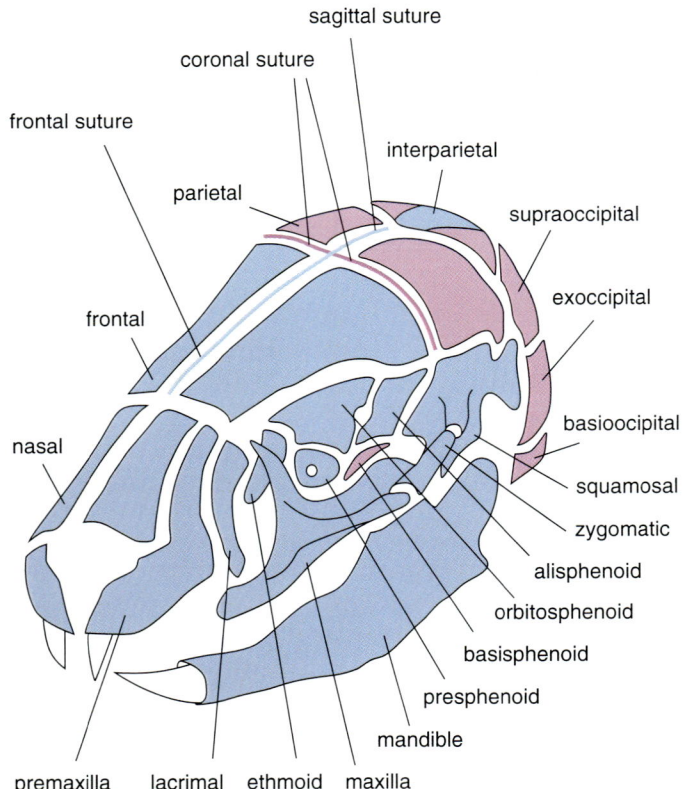
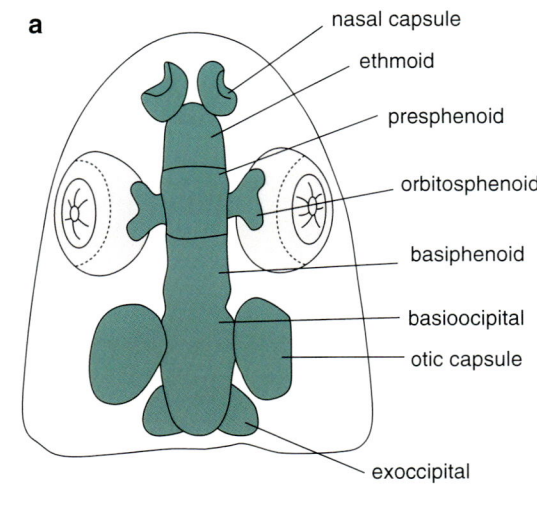

Fig. 8.1 Mesenchymal components of the skull. Neural crest mapping in mammals discloses basisphenoid to be neural crest as well. Small paraxial mesoderm acrochordal cartilages separate it from PAM basioccipital. Notochord is surrounded by PAM. Neural crest bones either ossify directly in membrane or form a cartilaginous intermediate (e.g., the parietal cartilage) which subsequently undergoes membranous ossification. Skull base is neural crest from basisphenoid forward and PAM from basioccipital backward. Otic capsule (future petromastoif and squamosal) is non-somitic PAM. Entire occipital complex is somitc PAM. [Left: (Reprinted from Mishina Y, Snider TN. Neural crest cell signaling pathways critical to cranial bone development and pathology. Exp Cell Res 2014; 325(2):138–147. With permission from Elsevier). Right: (Reprinted from Morriss-Kay GM Derivation of the mammalian skull vault. *J Anat* 2001; 199:143–151. With permission from John Wiley & Sons)]

Forebrain is regionally organized into prosomeres; the subjacent neuroanatomy undoubtedly contributes to regionalization of the dura as well (Fig. 8.2).

Ossification pattern may depend upon (1) mesenchymal cell type and (2) the nature of the biologic "surround" (neighboring tissues of similar or different type). Neural crest bone can be either membranous, chondral, or membranous via a cartilage intermediate. PAM bone from somites is chondral. PAM bone from somitomeres is chondral but can be admixed with neural crest (as in parts of the petrous temporal bone). In all fairness to somatic PAM, these cells are internalized within the organism; they are not exposed to an epithelium. As a first-pass assumption, the decisive effect upon what type of ossification mechanism a mesenchymal responder population might pursue probably has more to do with signals it receives from the biologic surround. A possible mechanism may be the relative amounts of bone morphogenetic protein (BMPs) to which the mesenchyme is exposed. For example, BMP-2 derivatives are dose-dependent: low dose gives fat, medium dose cartilage, and high dose bone.

The shoulder girdle constitutes an exception to the above generalizations. Clavicle and scapula originally spanned the distance between skull and trunk of fishes. The tissue map of these two bones shows that they are composite structures in which postotic neural crest (r8–r11) forming chondral bone and mesoderm from neuromeric levels c2–c4 undergoes membranous ossification. This fascinating subject is covered in detail in our discussion of the neck that follows.

Neuromeric Level

Note: The genetic definition of rhombomeres r4–r11 and all subsequent neuromeres of the spinal cord follow the original Hox series of homeotic genes (Fig. 8.3).

Rhombomeres r8–r11: The Myelencephalon Is a Transition Zone

Skeletal derivatives originating at the level of the medulla come from two sources. Here, four occipital somites in register with r8–r11 form the cranial base of the posterior fossa and the hypobranchial muscles dedicated to the tongue. At

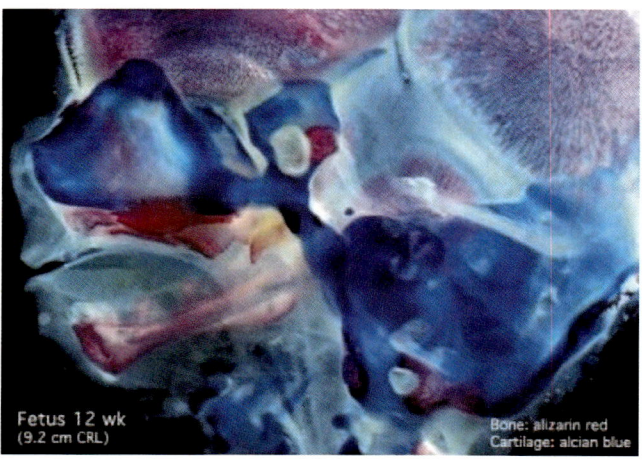

Fig. 8.2 Fetal skull at 12 weeks. Cranial base (dark blue): (1) anterior cranial fossa is anterior to notochord (basisphenoid) and develops from r1 neural crest ossifying as either cartilage intermediate to membrane or as chondral bone; (2) middle cranial fossa is chondral bone from r4 to r7 paraxial mesoderm via somitomeres; posterior cranial fossa is chondral bone from r8 to r11 paraxial mesoderm via occipital somites S1–S4. (Reprinted from UNSW Embryology. Retrieved from: https://embryology.med.unsw.edu.au/embryology/index.php/File:Fetal_head_lateral.jpg. With permission from Prof. Virginia M. Diewert). Neuromeric system. <u>Forebrain, anterior</u> (formerly p6–p4) under review; <u>forebrain, posterior</u> (p3–1); <u>midbrain</u> (m2–m1); hindbrain, non-pharyngeal arch (s forr0–r1); hindbrain, pharyngeal arch (r2–r11)

the same time, these four rhombomeres as the source of neural crest for pharyngeal arches 4 and 5 as a source for the structures of the larynx and hypopharynx.

Occipital somites play an important developmental role which is poorly understood. From the first cervical somite (S5) backward, the formation of the axial skeleton is straightforward, with neuromeres, spinal nerves and somites in perfect register. This 1:1 relationship seems apply as well to the first four occipital somites (S1–S4). Key to this argument is the subdivision of r8 into four distinct genetic units (pseudo-rhombomeres), one for each component of the cranial nerve nuclei that reside within the caudal hindbrain (medulla).

Understanding the rationale behind a 1:1 pairing system of the caudal medulla with the occipital somites is very important. At risk of being repetitive, the reader will recall from our discussion of neuromeres the two contrasting models of the hindbrain neuroanatomy. The *O'Rahilly model* is morphologic, the medulla is conceived as being a single unit. The *Puelles-Rubenstein model* is derived from genetic mapping. Although rhombomeres r8–r11 are not visually distinct, they do pair up individual branches of motor nerves nicely with their targets myoblast populations in the four outlying myotomes of occipital somites 1–4. Furthermore, the existence of a genetic segmentation system aligned with

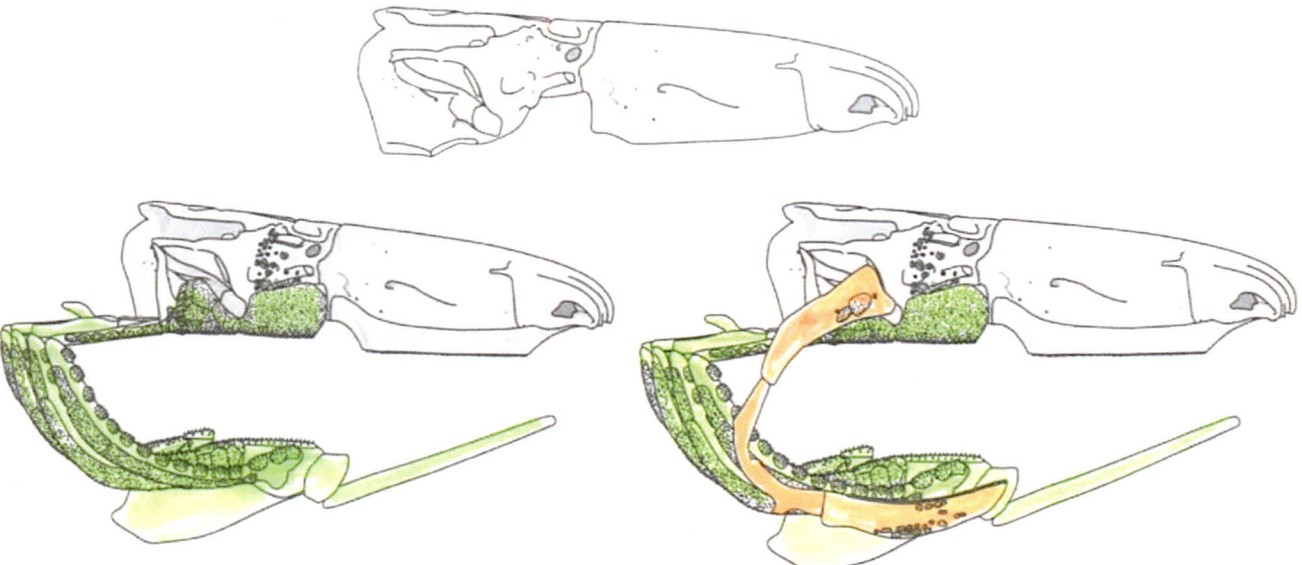

Fig. 8.3 Vertebrate skull construction 1. The basic vertebrate chassis as demonstrated in *Eusthenopteron*, a bony fish of the early Devonian period typifying the structural components of the gnathostome skull. We shall see a great deal of this character in the course of this chapter. Here, the skull base or chondrocranium is seen in isolation. The skull is hinged at the terminus of the notochord, between basioccipital and basisphenoid (neural crest-mesodermal border). We next add branchial arches (yellow). The basic model has a total of seven branchial arches (in some species six). BA3–BA6 shown here are all tooth-bearing. Next, we add the hyoid arch, BA2 (orange). Its dental units are scanty. (Reprinted with permission from: GEOL431—Vertebrate Paleobiology course web resources. Thomas Holtz and John Merck, Department of Geology, University of Maryland, College Park. https://www.geol.umd.edu/~jmerck/geol431/)

somites is consistent with that observed along the remainder of the neuraxis. For this reason, the original designation of r8–r11 as "psuedorhombomeres" has given way to formal recognition of their existence.

Let us test this out by looking more closely at neuromuscular anatomy of the occipital somites. The target muscles are those of the larynx (X), the sternocleidomastoid-trapezius derivatives of the ancient branchial arch dilator muscle complex, the cucullaris (XI), and the tongue muscles (XII). Tetrapods, unlike sharks, have undergone complete loss of the sixth and seventh arches but retain the original neuroanatomy. Thus, vagus and spinal accessory occupy the *nucleus ambiguus* that begins with r6–r7 and runs downward through caudal medulla and into spinal cord. Vagus territory terminates at r11 but accessory runs all the way down to neuromeres c3–c4. Furthermore, within the same zone, beginning with r9, three nuclei of the mammalian hypoglossal innervate tongue muscles originating from S2 to S4. Some models postulate the participation of somite 1 as well. One again, our conclusion is that the region of r8 is really a set of four genetically distinct pseudorhombomeres, each of which is in register with a specific occipital somite.

How well does this model apply itself to bone mapping? Huang et al. demonstrated in the avian model that the basioccipital, exoccipital, and chondral supraoccipital are assembled as concentric laminations from the first five occipital sclerotomes [1]. Although mammalian systems have four occipital somites, the rules should be the same. Furthermore, when one compares the spatial positions of each of these bones, a homology can be readily observed with the ancient fish vertebrae (minus the hemal arches, lost in evolution). The concentric lamination of the centra produces the basioccipital, that of the lateral masses gives the exoccipital/condyle, and the neural arches are assembled into the chondral exoccipital bone. We will cover the specifics of the posterior fossa later on.

Pharyngeal arches 4 and 5 are in register, respectively, with neural crest from r8 to r9 and with r10 to r11. Recall that these two structures develop in concentric fashion, like Russian dolls. Fourth arch gets a lop-sided proportion of neural crest as it forms the entire larynx. Fifth arch is relegated to the production of the cricoid and possibly the arytenoids. Motor control of speech arises from two levels and two somitomeres. Cricothyroid is supplied by external laryngeal of vagus (r8); its myoblasts arise from Sm8. All remaining muscles are supplied by internal laryngeal nerve of vagus (r9); these form from somitomere 9 and are the exclusive domain of paraxial mesoderm. Middle and inferior constrictors are supplied via vagus, most likely from levels r8 to r11. Although conclusive mapping has not been done, these likely arise from Sm8 to Sm9 and from Sm10 to Sm11, respectively.

You might ask, why all the niceties about somitomeres versus somites? The reason is that not solid evidence maps these muscles to occipital somites. The neuroanatomy of the larynx is well known; for this reason, Noden has mapped these muscles to levels r8 and r9. Esophagus and trachea are derivatives of the lateral plate (not PAM) so they cannot be somite derivatives. Furthermore, LPM splits in the neck at level c1. The caudal extent of inferior constrictor therefore defines the boundary between the neuromeric territories of r11 and c1.

Rhombomeres r4–r7: A Very Orderly Place

Mapping becomes relatively straightforward as one proceeds forward into the caudal metencephalon, i.e., the lower pons. The second pharyngeal arch is supplied by r4–5. The embryonic chorda tympani nerve is surprisingly large, nearly equal in size to the facial motor nerve! The nuclei of these two structures reside within r4 and r5, respectively, along with the nuclei of the vestibulo-cochlear nerve.

Neural crest products of the second arch are the inheritors of the original cartilage chain of hyoid arch. In the original chondrichthyan fishes, these were part of a branchial arc. With the evolution of the bony fishes, the second arch was subverted into a suspension system anchoring the mandible to the cranium. With the advent of the temporomandibular joint, mandible achieved an autonomous articulation with the skull. The hyoid series remained attached to skull via stylohyoid ligament and descended down to hyoid bone in the neck, attached to mandible and mouth via muscles.

Somitomeres 5–6–7 are the source for paraxial mesoderm that forms the parachordal cartilages of the cranial base of middle cranial fossa and the optic capsules.

Somitomere 5 contains the myoblasts of lateral rectus and palpebral depressor while somitomere 6 gives rise to the muscles of facial animation. Sm6 shares with Sm7 muscles originally intended for jaw opening (posterior digastric, stylohyoid, and possibly a component of levator veli palatini).

A note for the idle minds: The nucleus of lateral rectus was originally designed to fit into position r2 but it travels back to a final position at r4. It therefore commandeers a somitomere at a more posterior level. The original lay-out of oculomotor nuclei was m1, m2, r1, and r2.

The third pharyngeal arch is supplied by r6–r7. Its neural crest is dedicated almost exclusively to the fascial suspension of the upper and midpharynx. The lower half of hyoid bone, including greater cornu, is an afterthought.

Motor neurons from r6 to r7 travel via both glossopharyngeal and vagus to supply all muscles of Sm7: soft palate (except tensor tympani) and both superior and middle constrictors are supplied via IX or X. Recall that IX and X share the same position in nucleus ambiguus. *It is not important if the nerve supply to the palate travels physically in one nerve or the other.* What is important is the origin of the neurons from r6 to r7. Sm7 has a remaining hypobranchial muscle, stylopharyngeus, which spans between the second arch above and the middle constrictor below.

Given the above, it makes perfect sense that the pathways of VII into both the skull and the face should involve a *field separation planes*. Petrous temporal bone is essentially bilaminar box, with Sm5–Sm6 providing the antero-lateral laminae and Sm7 the posterior medial laminae. The mastoid temporal bone would arise from Sm7, possibly admixed with PAM from Sm8 (the first occipital somite).

Note: All neuromeres anterior to r4 are defined by non-Hox homeotic genes. This results in three functional populations in terms of bone formation. (1) Hindbrain neural crest (r2–r3) produces first arch bones and parieto-occipital membranous bones. (2) Midbrain neural crest (m1–m2 and r0–r1) behaves as single unit functional unit with r0–r1 producing frontonasal bones. (3) Forebrain neural crest does not produce bone, only frontonasal skin.

Recall that *first arch is evolutionarily unique*. Its original configuration in jawless fishes was that of a humdrum branchial arch; the skeleton support of which of neural crest cartilage. The agnathic first arch shared with its fellows a common genetic definition, with its AP positioning determined by hox genes and its distal-proximal axis via Dlx genes. As stated before, the "gnathostome revolution" was characterized by a reassignment of neural crest from gill structures to jaws. In so doing, the original Hox code specifying first arch and held in common with the remaining arches was replaced by a new non-Hox homeotic system. Thus, neural crest cells arising from diencephalon posterior to r3 express non-Hox homeotic genes, with the tradition Hox system beginning at r4 and continuing caudally throughout the length of the organism.

Rhombomeres r2–r3: A Schizophrenic Situation
Neural Crest Derivatives
The r2–r3 neural folds of the rostral metencephalon (upper pons) produce neural crest with two distinct destinies: (1) mesenchymal supply for the first pharyngeal arch; (2) coverage for the forebrain. This produces a schizophrenic crisis. Neural crest from r0 to r1 is not neurotic. These levels are not in register with the first arch. They produce neural crest that migrates immediately forward over the surface of prosencephalon. There it produces the dura and skeletal structures of the anterior cranial base from presphenoid forward.

Neural crest from r2 to r3 behaves like Hamlet. One population remains outside the first arch. It follows the migration pathway of r1 and pursues an epaxial course over the caudal prosencephalon. It produces all dura and dermis innervated by V2/V3 and contributes to the calvarial sidewalls. Another population assumes a hypaxial fate. It dives downward to fill up the first arch with the mesenchyme.

Inside the arch, r2–r3 NC has two alternative fates. (1) It can form traditional structures such as dermis and fascia for the muscles of mastication. (2) It can also opt to serve as the *materia prima* for synthesis of the mandible and zygomatico-maxillary complex. In the first instance, r2–r3 neural crest derivatives will be supplied by the traditional external carotid system that arose with the jawless fishes. But in the new iteration, r2–r3 membranous bone, be it incarnated in the calvarium, the jaws, or the dura, has a new source of blood supply: the stapedial system.

Paraxial Mesoderm Derivatives
Epaxial populations of paraxial mesoderm from levels r2 to r3 have a very limited role. These cells, in cooperation with PAM from level r1, provide immediate coverage of the forebrain with angioblasts that make the *primitive head plexus*. This plexus keeps the forebrain alive until the internal carotid system has the time to develop its requisite branches.

Hypaxial populations of PAM from r2 to r3 have several functions. This mesenchyme is concentrated in the fourth somitomere. Sm4 populates the parietal bone, where it is admixed with r2–r3 neural crest. It ossifies in two distinct mechanisms. The lower zone of *parietal cartilage* is chondral while the upper zone is membranous. As motor nucleus of V3 resides in r3, Sm4 provides myoblasts for muscles of mastication. The source of the acrochordal cartilage that forms the dorsum sella is uncertain. This PAM is admixed with the neural crest of posterior trabeculae. It lies anterior to the tip of notochord in basiocciptal and therefore cannot arise from occipital somites. Because Sm4 lies immediately lateral to the tip of notochord, it is a possible source of this mesoderm.

Mesomeres m1–m2 and Rhombomeres r0–r1: Married to the Midline
The neural folds of m1, m2, r0, and r1 migrate forward to provide the structures of the anterior cranial fossa, nasal complex and orbit. These four neuromeres are referred to collectively as *midbrain neural crest* (MNC). An appreciation for the commonalities and differences between their derivatives is critical to understanding the skull.

In the early brain development, mesencephalon is a huge structure, nearly equal in size to the forebrain. It has a single neuromere r1 midbrain which abuts against r1 of the hindbrain. Recall that r1 differentiates into an anterior isthmus, r0, which gives rise to cerebellum. At the same time, the midbrain tectum (roof plate) also subdivides into a posterior *pre-isthmic mesomere*, m2. Although remains to be defined about the mesomeres, m1 and m2 are contain anatomically distinct structures: two oculomotor nuclei and the superior/inferior colliculi. At the midbrain-hindbrain junction, rm2 and r0 abut one another.

Important Points of Simplification Regarding Neural Crest from the "Midbrain"

1. As we proceed in mapping out the derivatives from these two zones, for the sake of simplicity, let us refer to those in register with the junctional (anterior-most) hindbrain as r1 and to those in register with the midbrain as m1.
2. Gastrulation starts at levels r0–r1.
   - This process is initiated at stage 6b when primitive node produces prechordal mesoderm which is extruded forward from primitive node at r0.
   - From stage 7, gastrulation proceeds backwards beginning with a broad swatch of r1 ectoderm, beneath which is r1 paraxial mesoderm.
   - The neural folds produce r1 neural crest.
3. Levels m1–m2 lie anterior to the gastrulation process; their sole product is m1 neural crest.
4. Prechordal mesoderm from r1 induces midbrain. Neural fold closure at level r1 proceeds that of m1. Thus, the migration of r1 neural crest arrives at the cranial base first and relegates Johnny-come-lately m1 neural crest to the orbit.
5. MNC migration to the face and orbit takes place during stages 9–10. The neural crest cells follow the pathway of PAM from level r1 previously deposited over surface of the forebrain as it arose in stage 8.

The Triple Roles of r1 Paraxial Mesoderm

At all levels of the vertebrate embryo, CNS development takes place such that it ensures its own survival. From the hindbrain backward, the process of neural tube formation pulls in adjacent paraxial mesoderm, thus providing: (1) a nearly instantaneous form of primitive vasculature; and (2) a support system for the notochord/body axis. The overlying non-neural ectoderm forms an epithelial to prevent the exposure to amniotic fluid.

Development of the forebrain/midbrain accomplishes the same goals using a slightly different form of topology. Recall that these tissues are induced anterior to the notochord. They arise much as a balloon is inflated…and consequently must push through and incorporate the pre-existent tissues from r1 to r3 that lie in front of the notochord.

Gastrulation from levels r1 to r3 produces three germ layers arranged in three concentric arcs which flow outward and forward to fill in the space between notochord and buccopharyngeal membrane. Outside of them, and further forward, is a broad swatch of lateral plate mesoderm shaped like a horseshoe and directed backward. Close forward like pincers to fill in the remainder [2, 3].

This flat Cartesian sea is distorted by the development of forebrain arising like a primeval volcano from the surface of the ocean. In so doing, it enshrouds itself with the surrounding tissues. Thus, its anterior most zone is protected by r1 ectoderm. With the subsequent expansion of frontonasal skin, this temporary r1 "skin" over the brain will be replaced. It persists as the epidermis of the upper eyelid. As telencephalon expands, it appropriates surrounding territories of epaxial r2 and r3 ectoderm which will form the epidermis of the remainder of the scalp. And of course, beneath the ectoderm lies paraxial mesoderm.

What are the roles of subjacent paraxial mesoderm for r1–r3? Recall that craniofacial PAM is *biologically incapable of making dermis*. This property first appears at level c2. Instead, it provides angioblasts to rapidly construct the primitive head plexus (ectodermal tissue cannot form blood vessels on its own). The pre-existence of PAM over the walls of the forebrain solves an additional problem: how to guide neural crest cells to their appropriate targets. PAM provides a neuromere-specific lattice for orderly migration: r1 NC follows r1 PAM, etc. Subsequent neural crest migration takes place in the plane between ectoderm and the primitive vasculature. Thus, the craniofacial dermis is supported by an underlying subdermal plexus.

PAM from Levels r1 to r3 Plays No Role in Cranial Base Protection

This is accomplished using neural crest. A singular exception is mesoderm of the acrochordal cartilage that is in association with posterior basisphenoid. This could originate PAM at level Sm4.

The primordial duty of PAM is to provide striated muscles, first to the axial skeleton and second to the appendicular skeleton. Furthermore, *every segmental unit of paraxial mesoderm, be it one of the seven cranial somitomeres or any subsequent somite is defined by a motor nerve supplying striated muscle.*

Pharyngeal arch (r2–r3) contains myoblasts from Sm4. So why is it that three somitomeres lie anterior to PA1? The math is simple to understand. Motor nuclei for oculomotor muscles exist in m2, m1 and in r1. Each nucleus demands its own somitomere. Level r1 produces a large amount of PAM. This mesenchyme is subsequently divided to produce somitomeres 1–3.

Upper oculomotor (m1), lower oculomotor (m2), and trochlear (r1) nerves are constructed from the local neural crest. These supply, respectively: Sm1 (medial rectus, inferior oblique, and inferior rectus), Sm2 (superior rectus, and levator palpebrae superioris), and Sm3 (superior oblique).

For these reasons, we map midbrain neural crest into the orbit as a source for extraocular muscle fascia and ligaments. Genetic mapping of the sclera follows from that of the globe proper and demonstrates two distinct zones, inferonasal and superotemporal. This explains the insertion sites and sequence of extraocular muscles—see Chap. 9—and may involve distinct populations of m1 and m2 but this remains a future research question.

Epithelial Substrates: Neural Ectoderm and Dura Mater

In our study of gastrulation, we noted that transformation of single-layer epiblast into a trilaminar system does not take place all at once, throughout the embryo. Rather, each new layer appears in a deliberate spatio–temporal sequence, one neuromere at a time. It proceeds via a "cellular clock" beginning at the cranial end of the primitive streak and continuing caudally backward. The sequence is as follows: (1) epiblast, (2) epiblast + hypoblast (primitive endoderm) = bilaminar embryo, (3) primary wave of migration from epiblast through primitive streak forms true endoderm (with hypoblast pushed out laterally), and (4) secondary wave of migration from remaining epiblast through primitive streak creates first an axial zone of mesendoderm (the future notochord) and second, the definitive mesoderm. At this juncture, the epiblast cells that did not participate in gastrulation now constitute the ectoderm. The trilaminar embryo is established.

Hindbrain mucosa and skin Let us now examine how the spatial layout of embryo is programmed. Recall that axis specification begins at the posterior pole at the sperm entry point. The primitive epiblast thus possesses invisible antero-posterior and medial-lateral homeotic gradients (even though the primitive streak has not yet appeared). The advent of the hypoblast defines the dorsal–ventral axis. Reorganization of the hypoblast is non-random. It takes its cues from the epiblast. It subsequently interacts with the overlying epiblast which acquires a *mirror image homeotic program*. Let cells next to the streak have program hox_M (medial) and cells at the periphery have program hox_L (lateral). When hox_M enters the streak it immediately binds to the first available site. Subsequently hox_L enters and moves past hox_M to assume a more peripheral position. Thus, *the hox pattern of the epiblast is duplicated within the endoderm*. Recall that the formation of the definitive endoderm rudely shoves the hypoblast out of the way—no longer part of the embryo, it becomes extra embryonic endoderm. Mesoderm that enters subsequently the intervening space between epiblast and endoderm encounters the same homeotic program from both sides...but with different induction effects.

Definitive endoderm provides one set of signals for the mesenchyme to obey; the other set of signals originates from definitive ectoderm (the remaining epiblast). In this context, endoderm has the likely priority. Neural crest population now rapidly invade this space. In the head and face, NC mesenchyme responds to the dual-signal system to produce submucosa and dermis. The next set of derivatives will be the membranous bones and connective tissues. Finally, invading myoblasts seek out binding sites along the newly created bones.

Recall that neural crest populations migrate in spatially distinct patterns. *Rostral hindbrain* goes first, followed by *midbrain and caudal hindbrain*, with *forebrain* neural crest being the last to complete the process.

Why is there no epithelial representation of cranial nerve VII? First, the trigeminal system is designed to report craniofacial sensation. Second, first and second arches function together as a composite. Mesenchyme from the second arch exploits the subdermal plane of all skin supplied by trigeminal. It also exploits the subdermal plane of hypaxial c2–c4 skin reaching the clavicle. In this course, its overlying ectoderm becomes irrelevant. The same process takes place in the oral cavity where the first arch mucosa and the third arch mucosa abut one another.

In sum, oral mucosa is ectodermal until the buccopharyngeal membrane. In the pharynx, hindbrain endoderm and neural crest submucosa in register with r6–r7 patterns hyoid and that from r8 to r11 patterns the fourth arch laryngeal cartilages.

Midbrain Orbital mapping of neural crest is still uncharted territory. Neural crest from m1 and m2 forms the oculomotor nerves of the 7; it is reasonable to assume that m1–m2 forms fascia of the orbit as well and the V1 innervated conjunctiva (both upper and lower). Since the optic anlage forms as an extension of the brain we will refer to sclera as an r0–r1 structure. The skin and subcutaneous tissues of the upper eyelid definitely come from r1. For the sake of simplicity, we shall refer to intraorbital and extraorbital neural crest innervated by V1 as MNC.

Forebrain nasal lining and skin Gastrulation does not take place anywhere anterior to r1 and therefore cannot provide for skin cover over the frontonasal mesenchyme. As discussed in Chap. 2, the developmental basis of frontonasal skin is unique. Its epidermis originates, not from non-neural epiblast but from non-neural ectoderm of the rostral forebrain neural folds. We shall discuss the development of craniofacial skin in a later chapter.

The developmental fields of frontonasal skin are laid out in a Cartesian grid. Its epidermis arises from the rostral prosencephalic neural folds, p6, p5, and p4. The zone from the midline to the nasal placodes belongs to p6. Upper beak ectoderm comes from p5. Calvarial ectoderm from glabella and eyebrows backwards comes from p4. Migrating forward beneath this layer, caudal prosencephalic neural crest from p3 to p1 slides forward between neural crest epidermis and brain like a hand thrust into a glove. This secondary layer will form the frontonasal dermis. Successive lamination from p3 to p1 could theoretically set up the potential for a genetic gradient of frontonasal dermis in vertical "stripes" from medial (p43) to lateral (p1) but this remains to be investigated.

A clinical note: Tessier's description of supraorbital facial clefts recognizes four vertical relatively vertical zones within the forehead skin and frontal bone. Deficiency or absence in these zones produces a characteristic cleft. The presence of such genetic zones could explain the vascular anatomy of the forehead skin. Supratrochlear would be "assigned" to p3. Supraorbital would be assigned to p2. Lacrimal would belong to p1. A lateral–medial gradient would also explain the successive pattern of forehead muscles: procerus, corrugator, and frontalis.

Frontonasal mesenchyme from the forebrain has no blood supply of its own. Rather is fed by a series of arteries derived from the stapedial system and programmed by the individual branches of V1. These arteries all belong the naso-orbital-ethmoid neural crest mesenchyme that comes from r0 to r1. Extracranial StV1 supplies the outer table of the frontal bone, plus the nasal and lacrimal bones and the frontonasal skin. We shall see that inner table of the frontal bone is supplied by dural vessels of the intracranial stapedial system.

The non-neural ectoderm (NNE) of the anterior forebrain contains specialized placodes that retain a relationship to the CNS. The adenohypophyseal and nasal placodes reside in the p6 zone while the optic placodes are located posteriorly in the p5 zone. These are anchored firmly to the brain or the developing eye. As we have seen in Chap. 4, p6 nasal placode is surrounding by proliferating p5 neural crest. This causes an apparent invagination of the nasal placodes backward into the underlying MNC mesenchyme, thus creating a chamber, the walls of which have vestibular lining of p6 derivation.

Dura and brain Development of the dura and its intimate relationship to the stapedial artery system was discussed in Chap. 8. Dura mater together with overlying dermis provides stem cells for the membranous calvarium. Neural crest from r1 to r3 is the predominant component of dura mater although there are small contributions corresponding the cranial nerves IX via X through the jugular foramen. None of these are significant for membranous ossification. V3 extends itself backward to provide stem cells for the membranous supraoccipital.

The important concept is that epithelial programming of the skull involves brain. Dura mater is just a mesenchymal interloper. A myriad of vessels interconnect dura with cerebral cortex. Although the vessel walls are mesodermal, penetration into the brain requires neural crest pericytes. Thus, signals between brain and dura shape signal arising from dura to bone.

Final Comments and Summary

Thus far, we have made a clear-cut distinction between the neuromeric source of mesenchyme from which a bone forms and the source or sources of epithelia that provide instructions to the mesenchyme. We have done so for three important reasons. (1) *The terminology used to describe a bone will be based on the mesenchyme, not the epithelium.* (2) *Bones that develop between two epithelial surfaces will be*

bilaminar, the potential space becomes filled with *marrow* or, when in continuity with the oronasal cavity, enlarges to form a *sinus*. A good example of these principles is the ethmoid complex. It develops from the r1 *trabecular cartilages*. All components of the ethmoid are r1 bones. This mesenchyme that forms the labyrinth intervenes between a medial program from p6 nasal vestibular lining and the lateral program from r1 MNC sclera. Under these conditions, a field separation results in the anterior and posterior ethmoid sinuses. (3) *Genes located at the rostral end (r1) of the most primitive part of the brain, rhombencephalon, are responsible of the induction of the midbrain and forebrain. OTX-2,* expressed at the anterior part of the primitive streak, interacts with genes located in the extreme anterior pole of the pregastrulation embryonic disc, a zone of hypoblast known as the anterior visceral endoderm (AVE).

The foremost priority of skull bone is brain protection. Forebrain/midbrain, as induction products of hindbrain. Neural crest arising from forebrain and midbrain is incapable of producing bone. *All bones of the craniofacial skeleton, be they of neural crest of PAM origin, arise from mesenchyme in register with hindbrain (r0–r11).*

Mesenchymal Sources of Craniofacial Bones

Neural Crest Bones

r1: basisphenoid and presphenoid, ethmoid, frontal, nasal, lacrimal

r2: premaxilla, maxilla (epipterygoid, quadrate), alisphenoid, jugal/postorbital, inferior turbinate, palatine, ectopterygoid

r3: squamosal/quadratojugal, parietal, mandible, tympanic, malleus, incus

r4: petrous temporal (with Sm5) prootic

r5: petrous temporal (with Sm6) opisthotic, hyoid lesser cornu

r6: mastoid (with Sm7)

r7: mastoid (with Sm7), hyoid greater cornu

r8–r11: core of the clavicle, inferior border clavicle, designated zones of scapula

Paraxial Mesoderm Bones

Undefined paraxial mesoderm (Sm1–Sm3): orbital cartilage (orbitosphenoid, lesser wing) and acrochordal cartilage (posterior basisphenoid).

Sm4: parietal

Sm5: prootic temporal

Sm6: epiotic temporal

Sm7: opisthotic temporal

Sm8–Sm11 (occipital somites 1–4): basioccipital, exoccipital, chondral supraoccipital

Organization of Craniofacial Bones Based on Blood Supply

V1 stapedial: anterior cranial fossa, frontoethmoid

V2 stapedial: the maxillary–zygomatic complex

V3 stapedial and external carotid: middle cranial fossa

Internal carotid: petrous temporal complex

Vertebrobasilar: posterior cranial fossa

Embryology of the Cranial Base: A Quick Review

This section is designed to be short and sweet. We have previously discussed these concepts along with general introduction to the skull and its history so what follows is by way of review. We shall first approach this subject by organizing and clarifying terminology regarding models of the skull that are often confusing. We will then consider development in terms of four general processes, followed by a stage-specific narration of events involving chondrocranium.

Models of the Skull

The craniofacial skeleton can be classified in three ways: (1) by function, (2) by type of ossification, (3) by type of tissues, and (4) by neuromeric origin of tissue. In terms of function, *neurocranium* protects the brain while the *splanchnocranium* protects everything else. The skull ossifies in two ways: bones that form in cartilage, *chondrocranium*, and bones that form in membrane. *Dermatocranium. All craniofacial cartilages, except those of the oto-occipital region (otic capsule and occipital somites) are preformed in neural crest.* All dermal bones are formed directly from neural crest. With the exception of parts of the clavicle and scapula, all remaining bones of the body are mesodermal. Classification by tissue type (somitomeres vs. somites) is very useful for discussing craniofacial muscles but it is not as relevant for categorizing bones. Although the origins of cranial bones from somitomeres has been worked out in the avian model, reliable information for mammals is not yet available for mammals. Furthermore, paraxial mesoderm is physically adjacent to craniofacial bone targets, postulating an intermediate step of organization into somitomeres may be an unnecessary step (which nature, in her wisdom, will eschew). For our purposes in this book, we shall refer to bone structures their neuromeric source, using sensory innervation as our reference point.

Processes of Cranial Base Development

From a common assembly of mesenchyme cranial base develops in successive iterations, each with functional sig-nificance. The order in which cartilages appear has nothing to do with when their mesenchyme is produced. All the relevant actors are present on stage at the start of the play. Gastrulation happens so quickly that all the mesoderm and neural crest is in place by stage 8. Its rather a matter of when the *blood supply* is sufficiently mature to support the organization of mesenchyme into cartilages and subsequently into bone. Development of the chondrocranium accomplishes three goals: (1) establishment of the protective chondrocranium, (2) assembly of sensory capsules for the organs of hearing and balance, and (3) construction of protective sidewalls.

Making a Secure Platform

The first iteration is dedicated to making a secure platform and consists of four stages. The first parachordal stage sees the appearance of *parachordal cartilages*. These extend from the tip of the notochord at dorsum sella all the way back to the primordia of the occipital somites. They are produced by PAM from r1 to r7. The parachordals appear first because the longitudinal neural arteries serving the hindbrain mature before the circulation to the midbrain and forebrain. Furthermore, the LNA mature backwards. As a consequence, in the basal plate stage, otic capsule and four occipital somites appear. These represent physical transformations of the 5th through 11th somitomeres and are produced by PAM from r8 to r11. During the prechordal stage, r1 neural crest forms pairs *trabeculae crani* extending from the tip of the notochord, the source of basisphenoid and presphenoid. Finally, in the ethmoid stage, the trabeculae extend further forward to produce the ethmoid complex. Note that the third and fourth stages take place when the stapedial system has matured sufficiently to support them.

Sensory Capsules

The second iteration involves cartilage capsules surrounding the organs of sensory perception. They appear in the following order. Neural (r1) crest forms the *nasal capsule* and is appended onto the ethmoid capsule. It manifests multiple foramina representing pre-existing vessels and nerves for olfaction. *Otic capsule* is made from PAM plus some neural crest (r4–r7) which engulfs the otocyst and surrounds multiple cranial nerves. The *optic capsule* forms last and is more complex. Midbrain neural crest (m1–m2) forms the sclera while r1 neural crest produces the frontal bone, including the orbital root.

Sidewalls and Picayune Details

In the third iteration, the cranial base becomes more complex, and more recognizable. Distinct cartilages arise from the original substrates, fusion of components takes place, and additional bone fields are added on. This is especially true for the sphenoid complex. *Ala orbitalis cartilage* (r1) produces orbitosphenoid, i.e., the lesser wing. *Ala tempora-*

lis cartilage (r2) produces the medial part of alisphenoid while r2 and r3 neural crest cells adds on to the lateral orbit and lateral pterygoid plate. *Acrochordal cartilage* (r1) is interposed between basisphenoid and basioccipital. It constitutes the posterior part of sella turcica.

Fusions take place between previously cartilages, often around neurovascular structures. Union of the olfactory capsule with the central trabeculae produces the lateral zones of the ethmoid, the midline cristae galli and perpendicular plates respectively. The basal plates extend laterally to join with otic capsule. Jugular foramen is a good example of the intersection of r7 with r8 fields respecting three cranial nerves and the jugular vein. The occipital complex fuses with the parachordals. Finally, support for the internal carotid and hypophysis is established.

New components appear. The ancient pterygoid finds its retirement home as the medial pterygoid plate. It is joined posteriorly by ectopterygoid directed downward from basisphenoid to a rather ignominious fate as the Hamulus. Alisphenoid sends an r3 envoy downward as lateral pterygoid plate and alisphenoid. Occipital somites fuse to form the basioccipital bone while their lateral and dorsal fields produce exoccipital and supraoccipital bones.

P.S. picayune (Fr. *picaillon* = a small, worthless copper coin)

Figures 8.3 and 8.4 demonstrate the assembly of the prototypical vertebrate skull.

Development of the Chondrocranium by Carnegie Stages

Recall that posterior chondrocranium is paraxial mesoderm, anterior chondrocranium is neural crest (with exception of the hypochiasmatic cartilage). The cochlear portion of the auditory capsule is neural crest and the canalicular part is PAM. Finally, basisphenoid is initially neural crest, but is subsequently populated by paraxial mesoderm (Figs. 8.5, 8.6, 8.7, 8.8, and 8.9).

The mammalian chondrocranium consists of 14 pairs of cartilages. The sequence of mammalian chondrocranial development has been documented in the mouse by McBratney-Owen and the murine stages (ES) can be extrapolated to the Carnegie system.

- **Stage 8** (E11): development begins in paraxial mesoderm when parachordal cartilages are fused in the midline. The notochord is positioned atop the parachordal cartilages; it has reached its final forward projection to level r1 as a precondition for gastrula.
- **Stage 9** (E12) is marked early on by the initial organization of mesoderm from levels r5 to r11 into recognizable cartilage structures: the occipital arches and the canalicular part of the auditory capsule. Prechordal neural crest mesenchyme is still an anlage.
- **Stage 10** (early E13): the rostral trabecular cartilages organize as nasal capsule. Later in the same stage, hypophyseal cartilages appear in the dead-center of the cranial base, just anterior to notochord. Paranasal cartilages develop in the lateral walls of the nasal capsule.
- **Stage 11** (E14): Protection of the eye and brain are the themes here. Hypochiasmatic (orbitosphenoid) cartilage appears medial to the optic nerve, orbital cartilage develops lateral to it, while frontal cartilage forms the roof. Trabecular cartilages are now chondrifying caudally.
- **Stage 12** (E15): Parietal cartilages form laterally and fuse with the cranial base. In the remaining, cartilages of the

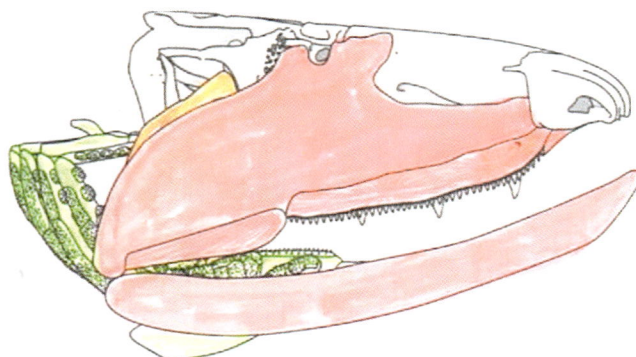
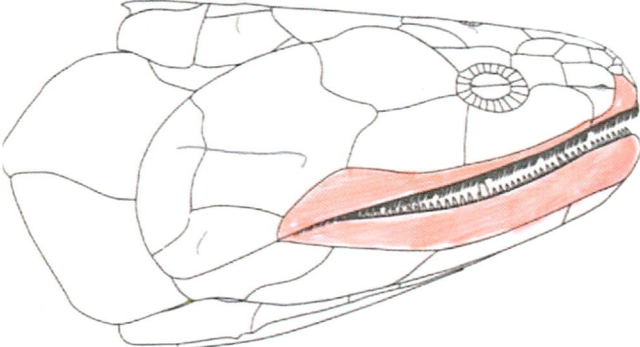

Fig. 8.4 Vertebrae skull construction 2. The penultimate step is the addition of the maxillo-mandibular arch, BA1. It consists of a palato-quadrate cartilage as the future maxilla and a dentary cartilage as the future mandible. In the final iteration, dermal bones are added to the skull. They serve several functions: (1) ensheathment, both intraoral and extraoral for the jaws, (2) protection for the brain sensory organs,

(3) coverage of the gills—the opercular series, and (4) connection between the skull and the pectoral girdle—the extrascapular series. We shall cover all these elements in due time. (Reprinted with permission from: GEOL431—Vertebrate Paleobiology course web resources. Thomas Holtz and John Merck, Department of Geology, University of Maryland, College Park, https://www.geol.umd.edu/~jmerck/geol431/)

Fig. 8.5 Cartilages of the cranial base. Mature chondrocranium, color-coded. Extensive contribution of the frontal cartilage to the orbit. Pterygoid cartilages on ventral surface are not shown. Newborn cranial base with bones in red and cartilages in blue. Newborn cranial base, color-coded. Bones shown in dark gray ossify in membrane. (Reprinted from McBratney-Owen B, Iseki S, Bamforth SD, Olsen BR, Morriss-Kay G. Development and tissue origins of the mammalian cranial base. 2008; *Developmental Biology* 322:121–132. With permission from Elsevier)

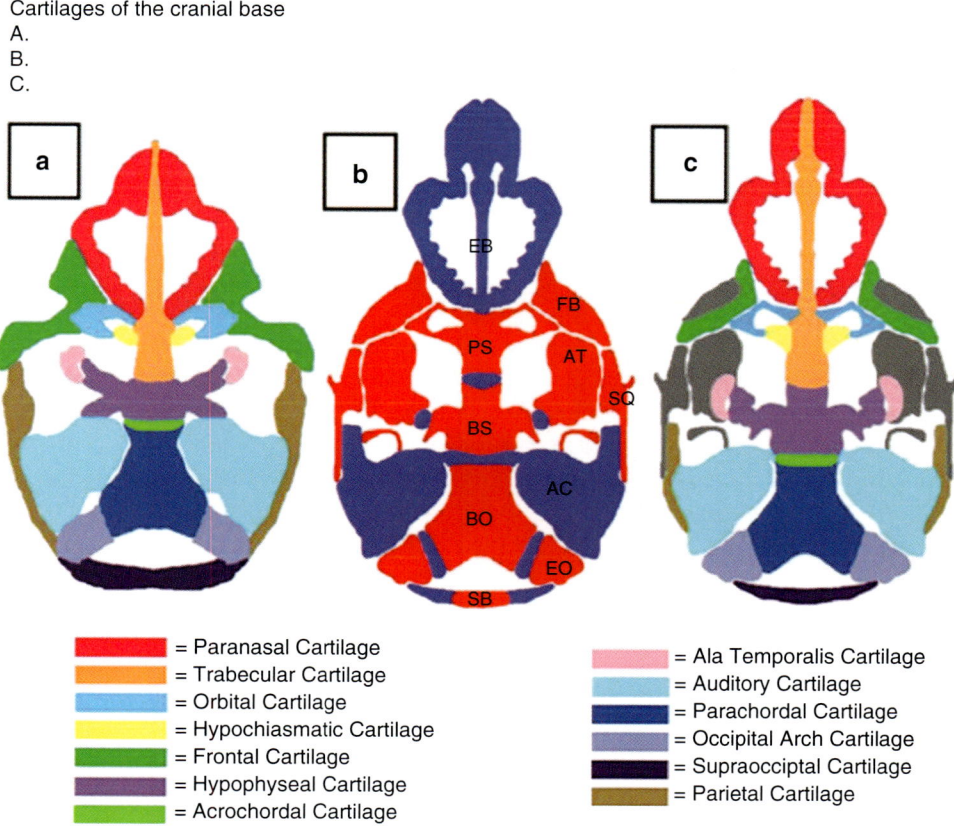

Cartilages of the cranial base
A.
B.
C.

= Paranasal Cartilage
= Trabecular Cartilage
= Orbital Cartilage
= Hypochiasmatic Cartilage
= Frontal Cartilage
= Hypophyseal Cartilage
= Acrochordal Cartilage

= Ala Temporalis Cartilage
= Auditory Cartilage
= Parachordal Cartilage
= Occipital Arch Cartilage
= Supraocciptal Cartilage
= Parietal Cartilage

sphenoid complex develop. Basitrabecular cartilage (lesser wing) ala temporalis (greater wing) appear lateral to the now-fused hypophyseal cartilages. Acrochordal cartilages (PAM) straddle the border between basisphenoid and basioccipital; they will constitute dorsum sellae. They are located just over basicranial fenestra, site of the ancient intracranial joint. Meanwhile, in the back room, occipital somites form supraorbital cartilages to roof over foramen magnum and the cochlear part of auditory capsule chondrifies.

- **Stage 13**: The chondrocranium is considered complete by and has the following characteristics. Midline fusion is complete in the anterior cranial fossa. Fusion between frontal and parietal stabilizes the nasal structures to the cranial base. Finally, at an unspecified later stage, the trabecular cartilages split into a rostral nasal septum and caudal presphenoid. Paired pterygoid cartilages appear on the ventral surface of hypophyseal cartilages. The hypochiasmatic cartilages unite with presphenoid as the lesser wings while orbital cartilages make a lateral parenthesis that encloses the optic canal.

Chondrification in the Basal Tetrapod

Critical sectors of the craniofacial skeleton develop in cartilage. With the exception of the oto-occipital region, all cartilage precursors arise from neural crest (Figs. 8.10, 8.11, 8.12, 8.13, and 8.14).

Neurocranium: cranial support and sensory capsules
First arch: jaw elements

- Palatoquadrate cartilage:
  - Epipterygoid (**epi**): dorsal margin of pq, medial alisphenoid (**as**)
  - Quadrate (**q**): posterior margin of pq, articulates with mandible
- Meckel's cartilage—programs dermal units of lower jaw

Second arch: bones supporting the jaw, in tetrapods, incorporated in occiput

- Pharyngeal (hyomandibula) cartilage suspends jaw joint
- Epi-, cerato-, and hypohyal cartilages support mandible
- Basihyals unify as hyoid body

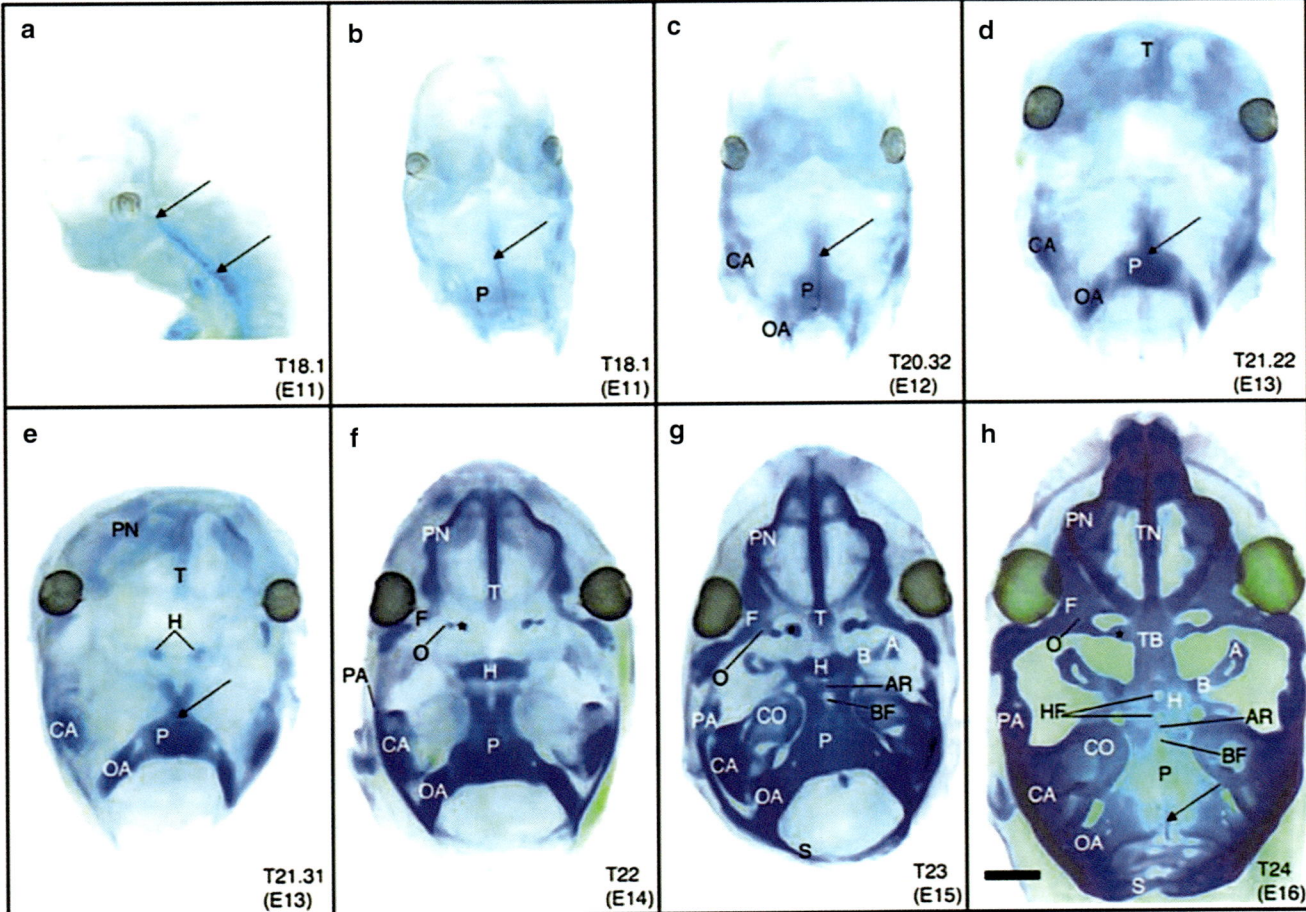

Fig. 8.6 Development of mouse chondrocranium. As shown by Alcian blue stain from E11 to E16, equivalent to Carnegie stages 8–13, all images are dorsal, except **a** (lateral). Parachordal cartilage (P) appears in association with the notochord (arrows). At E16, all cartilages have fused and ossification is beginning to take in posterior cranial fossa. **Key**: *A*, ala temporalis cartilage; *ACO*, alicocochlear commissure; *AR*, acrochordal cartilage; *AT*, ala temporalis (greater wing) of basisphenoid bone; *A*, *AC* auditory capsule; *ACO*, alicochlear commissure; *BS*, basitrabecular process, basisphenoid; *CA*, canicular part of auditory capsule; *CM*, cranial mesenchyme; *CO*, cochlear part of auditory capsule; *DI*, diencephalon; *EB* ethmoid bone; *EO*, exoccipital bone (will eventually fuse to basioccipital bone); *F*, frontal cartilage; *FB*, frontal bone; *FV*, fourth ventricle of hindbrain; *GW*, greater wing (ala temporalis) of the basisphenoid bone; *H*, hypophyseal cartilage; *HF*, hypophyseal fenestra; *LW* lesser wing of the presphenoid bone; *MNP* medial nasal process; *MV*, mesencephalic vesicle of midbrain; *MX*, maxillary process; *NL*, orbitonasal lamina of the paranasal cartilage; *O*, orbital cartilage; *OA*, occipital arch cartilage; *OF*, optic foramen; *ON*, optic nerve; *OR* optic recess; *P*, parachordal cartilage; *PA*, parietal cartilage; *PI*, developing pituitary gland; *SO*, paranasal cartilage presphenoid bone presphenoidal synchondrosis supraoccipital cartilage supraoccipital bone (will eventually fuse to basioccipital bone); *SOS*, spheno-occipital synchondrosis; *SQ*, squamosal bone; *T*, trabecular cartilage (trabecular plate); *TB*, basal portion of trabecular plate; *TG*, trigeminal nerve; *TN*, nasal portion of trabecular plate; *TO*, tongue; *TV*, third ventricle of forebrain; *Y*, hypochiasmatic cartilage. (Reprinted from McBratney-Owen B, Iseki S, Bamforth SD, Olsen BR, Morriss-Kay G. Development and tissue origins of the mammalian cranial base. 2008; *Developmental Biology* 322:121–132. With permission from Elsevier)

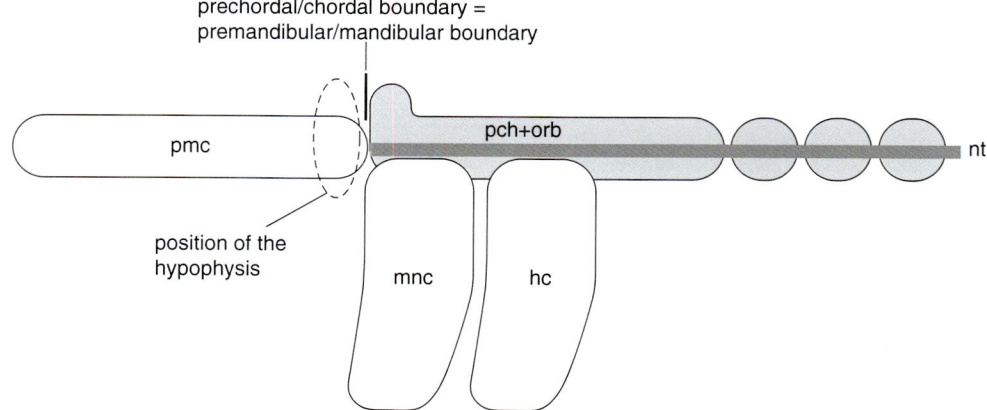

Fig. 8.7 Prechordal neurocranium. It is generally accepted that the rostral, prechordal part of the neurocranium is formed of premandibular ectomesenchyme. Thus, the boundary between the premandibular and the mandibular ectomesenchyme is expected to be found at the same level as the boundary between the prechordal and the chordal cranium, and the polar cartilage is simply understood as the posterior part of the trabecula. In this scheme, the hypophysis is found between the caudal portions of trabeculae (the prechordal cranial element) and the parachordal cartilages—boundary indicated by acrochordal cartilage. *hc* hyoid arch crest cells, *mnc* mandibular arch crest cells, *nt* notochord, *pch+orb* parachordal and orbital cartilages, *pmc* premandibular crest. (Reprinted from Kuratani S, Adachi N, Wada N, Osi Y, Sugahara F. Developmental and evolutionary significance of the mandibular arch and prechordal/premandibular cranium in vertebrates: revising the heterotopy scenario of gnathostome jaw evolution. *J Anat* 2013; 222:41–55. With permission from John Wiley & Sons)

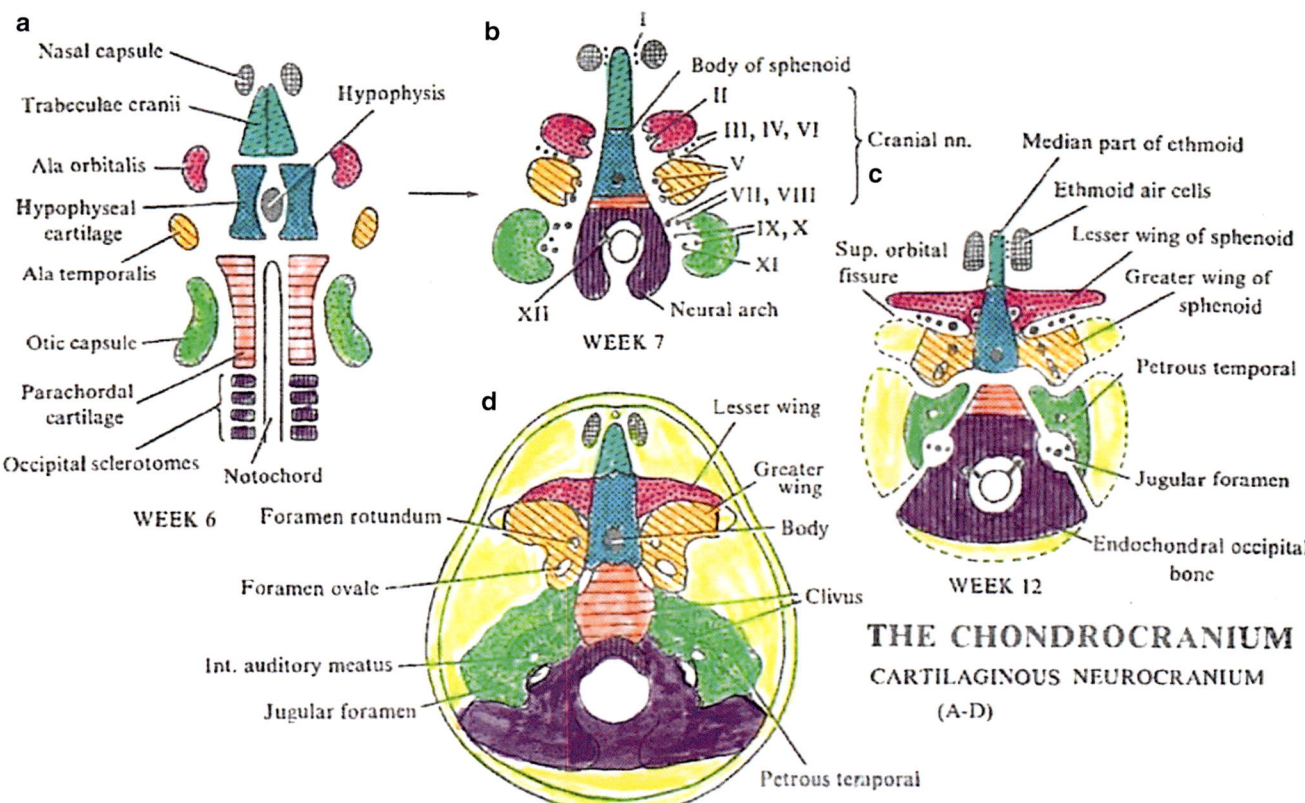

Fig. 8.8 Cranial base development: 6 weeks to newborn. Cartilage bones in colors, neural crest (yellow). Steps in the formation: (1) parachordal, (2) basal plate/occipital, (3) prechordal, (4) ethmoid, (5) sensory capsules, (6) integration. (Reprinted with permission from Pansky B. Review of Medical Embryology. New York: Macmillan, 1982. https://discovery.lifemapsc.com/library/review-of-medical-embryology/chapter-67-appendicular-skeleton-and-skull-development. © 1982)

a
skull base : vertebrate

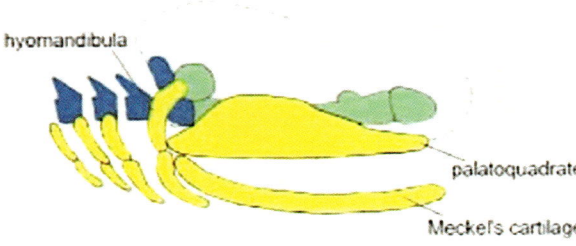

c
tetrapod viscerocranium and occiput

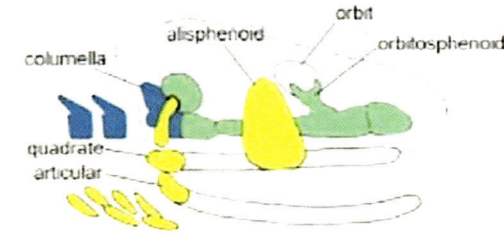

b
skull base and cartilaginous viscerocranium

d
skull components: mammals

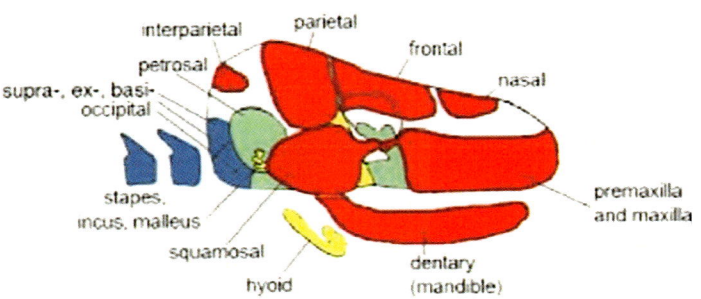

skull base / sensory capsules (green)
pharyngeal arches or viscercranium (yellow)
dermatocranium (red)
occipital somites (ble)

Fig. 8.9 Skull development by embryologic components. **Note errors** Components of the skull change with evolution. Old terminology at odds with mesenchymal mapping. (1) Anterior cranial fossa from basisphenoid forward, is neural crest; otic capsule (petrosal squamosal) is non-somitic PAM, occipital complex is somitic PAM. (2) Pharyngeal arches are neural crest; (3) dermatocranium (red) ensheaths the brain and jaws and is neural crest. bIn A–C dermal bones not shown. Basioccipital in **a** should be colored blue. (Reprinted from Morriss-Kay GM Derivation of the mammalian skull vault. *J Anat* 2001; 199:143–151. With permission from John Wiley & Sons)

Third arch: lower half of hyoid bone

Fourth and fifth arches: thyroid and cricoid cartilages do not ossify (normally)

Dermatocranium

Order Out of Chaos

What can explain the seemingly random pattern of facial bones? In this chapter, we shall see that they are derived from precursors, may represent mergers, or demonstrate changes in position. Nevertheless, a basal tetrapod configuration exists. If we turn the pages backward to the time of the prehistoric placoderms, we find fishes with large skull bones (in fact with armor all over the body) that have no relationship to the basal bony fish condition. The exception occurs with a Chinese placoderm Qilinyu dating back to the Silurian. This fish has three dermal bones around the mouth (premaxilla, maxilla, and dentary) but nowhere else. This marked the start of the dermal demographic expansion. These bones apparently replaced pre-extent oral plates, the infragnathic (dentary), posterior supragnathic (maxilla) and anterior supragnathic (premaxilla). But what rationale can we find for other facial bones without an obvious precursor?

Craniofacial bones clearly articulate with one another and this must be coordinated based on a developmental rationale. But the multiple bones seen in early fishes does not appear to have any sort of control. How was order established out of chaos?

A Mechanism for Genetic Selection

Let us consider the facial series. These bones provide the definition of the anterior skull, define the nares, and included anterior teeth. When we look at early bony fishes before *Cheirolepis* we see a welter of small bones, varying from species to species. These different dermal ossifications grew and impacted on each other. They would simply grow until the hit another bone field and then quit. During the late Silurian period, various fishes hit upon a combination that maximized the mobility and control for a premaxilla studded

Fig. 8.10 Median series: nasal (**n**), frontal (**f**), parietal (**p**), postparietal (**pp**). Circumorbital series: lacrimal (**l**), prefrontal (**prf**), postfrontal (**pf**), postorbital (**po**). (Reprinted with permission from: GEOL431—Vertebrate Paleobiology course web resources. Thomas Holtz and John Merck, Department of Geology, University of Maryland, College Park, https://www.geol.umd.edu/~jmerck/geol431/)

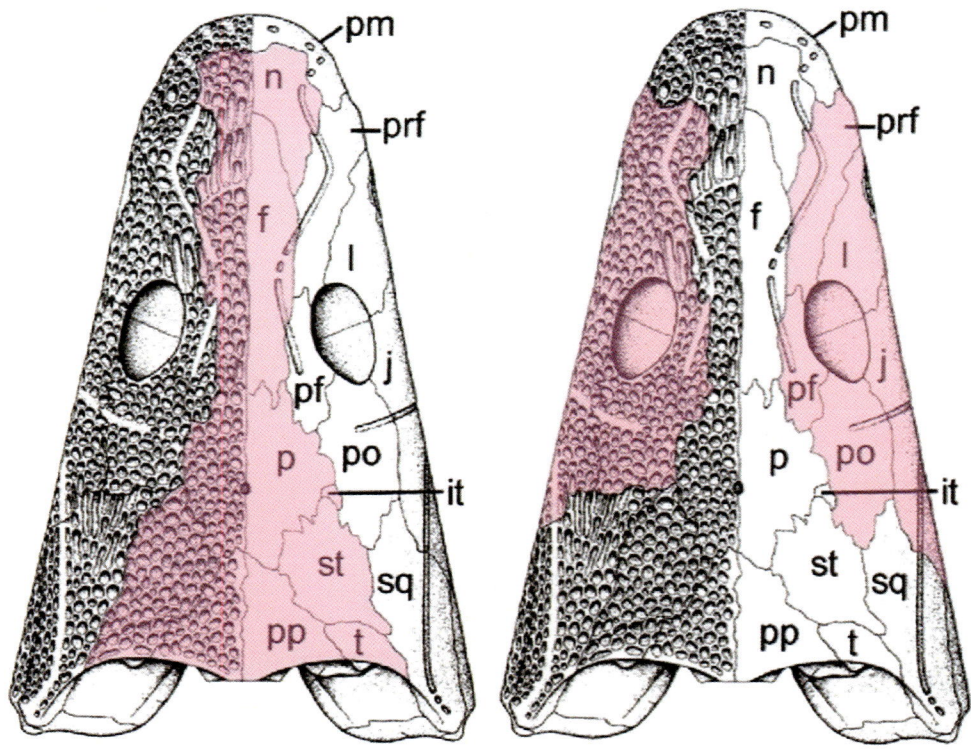

Fig. 8.11 Marginal series premaxilla (**prm**), maxilla (**m**) is a barely seen; it covers ("papers over") the palatoquadrate cartilage. Cheek series squamosal (**sq**), quadratojugal (**qj**). (Reprinted with permission from: GEOL431—Vertebrate Paleobiology course web resources. Thomas Holtz and John Merck, Department of Geology, University of Maryland, College Park, https://www.geol.umd.edu/~jmerck/geol431/)

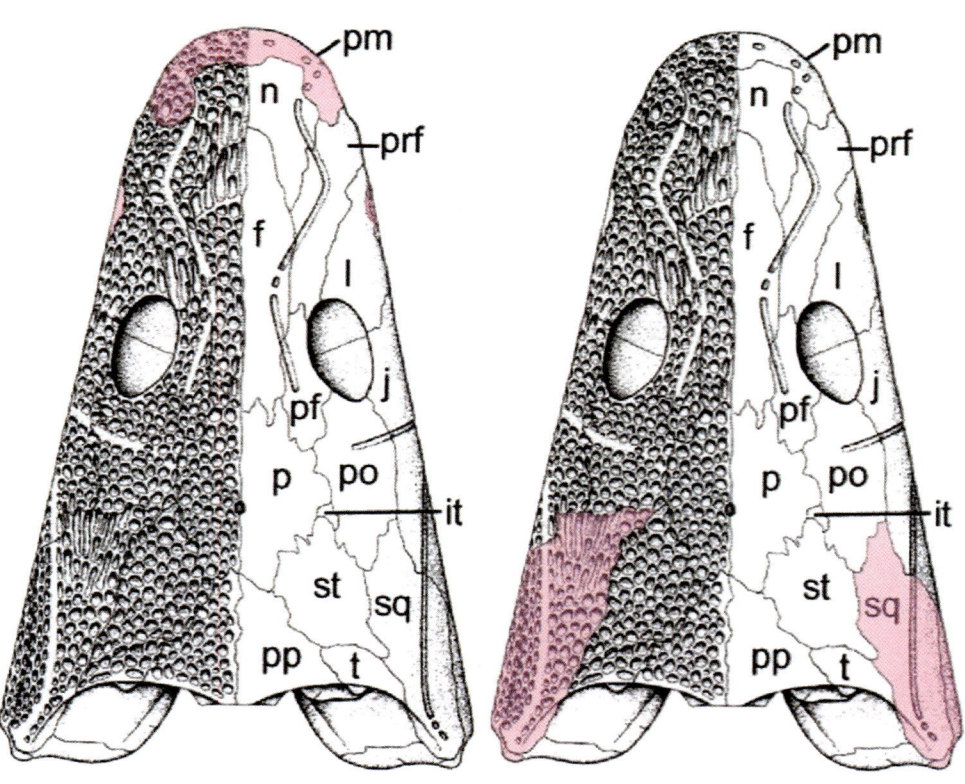

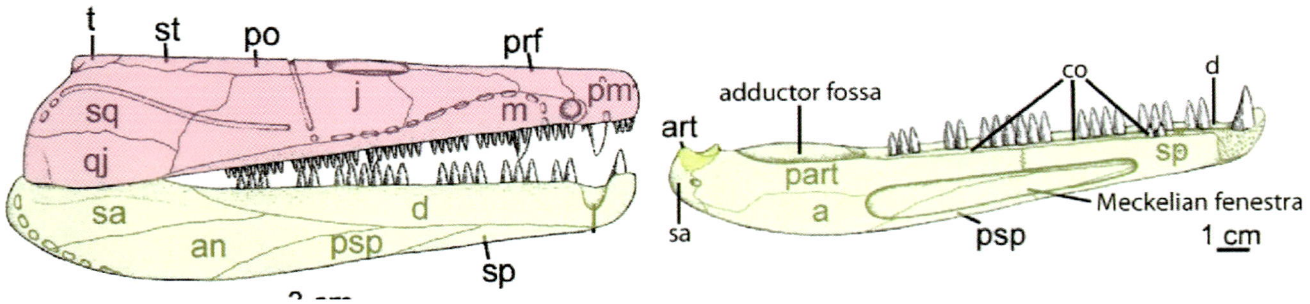

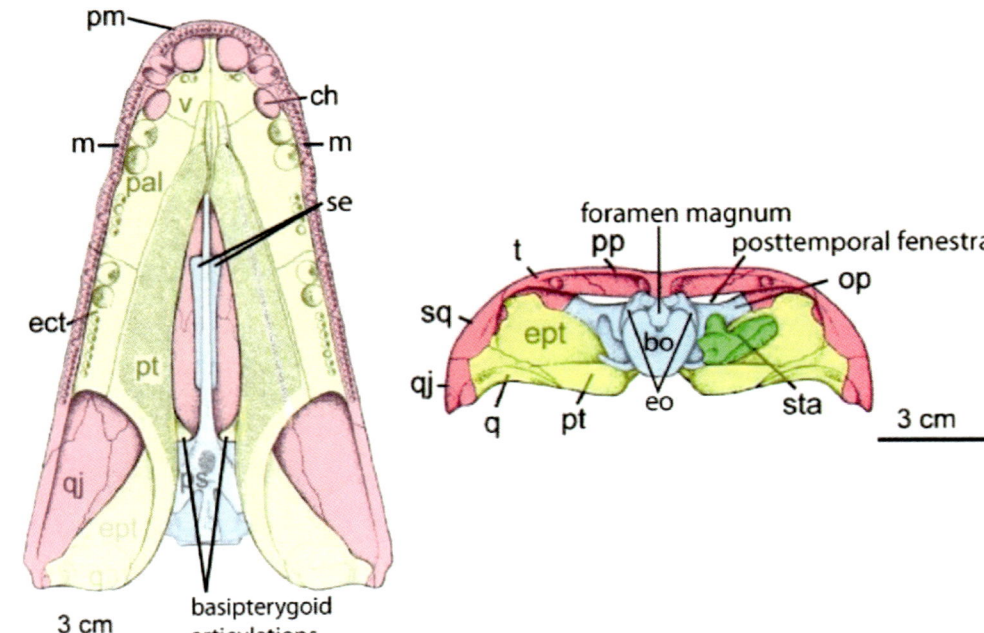

Fig. 8.12 Meckel's series, lateral angular (**a**), dentary (**d**), post splenial (**psp**), splenial (**sp**), surangular (**sur**). Meckel's series, medial articular (**art**), coronoids (**co**), prearticular (**part**). (Reprinted with permission from: GEOL431—Vertebrate Paleobiology course web resources.

Thomas Holtz and John Merck, Department of Geology, University of Maryland, College Park, https://www.geol.umd.edu/~jmerck/geol431/)

Fig. 8.13 Palatoquadrate series vomer (**v**), palatine (**pal**), ectopterygoid (**ect**), pterygoid (**pt**), sphenethmoid (**se**), choana (**ch**), maxilla (**m**). Posterior neurocranium series stapes (**st**), opisthotic (**op**), exoccipital (**eo**), note "encirclement" by pterygoid (**pt**), quadrate (**q**), quadratojugal (**qj**), squamosal (**sq**), tabular (**t**), and postparietal (**pp**). (Reprinted with permission from: GEOL431—Vertebrate Paleobiology course web resources. Thomas Holtz and John Merck, Department of Geology, University of Maryland, College Park, https://www.geol.umd.edu/~jmerck/geol431/)

with teeth. This necessitated a higher degree of genetic control. Dermal bones could not just grow willy-nilly in response to an epidermal signal. This favored the selection of a single line of stem cells to arrive in position, execute their program, and therefore become "dependable." This forced surrounding bones to respect the position of these "pioneers" and therefore assume their own position. In so doing, the secondary cells would acquire a genetic code to specify in what position they should remain.

Thus, functional changes in a single bone field, could via a "domino effect," create order in neighboring bone fields.

The Premaxilla as "the Enforcer"

As a thought experiment, let us consider how the premaxilla could have become an instrument of evolutionary change.

This primordial tooth-bearing bone, in the dead center of the snout, was designed to bite. For this to be effective, the stability was needed. By the late Silurian, a number of different fish species happened on a consolidation pattern centered on the premaxilla that was useful for predation. This became genetically dominant. The bones surrounding premaxilla, instead of forming from a random dermal program, came under stringent genetic control as well. A specified population of neural crest cells now arrives in situ and executes a program. In sum, *once one dermal bone becomes "reliable," surrounding bones become "reliable" as well, since they must stop growing at the boundaries of the fixed bone.* Thus, by virtue of food capture, premaxilla became a Mafiosi "enforcer" exerting order in the snout…a process that spread to the rest of the face.

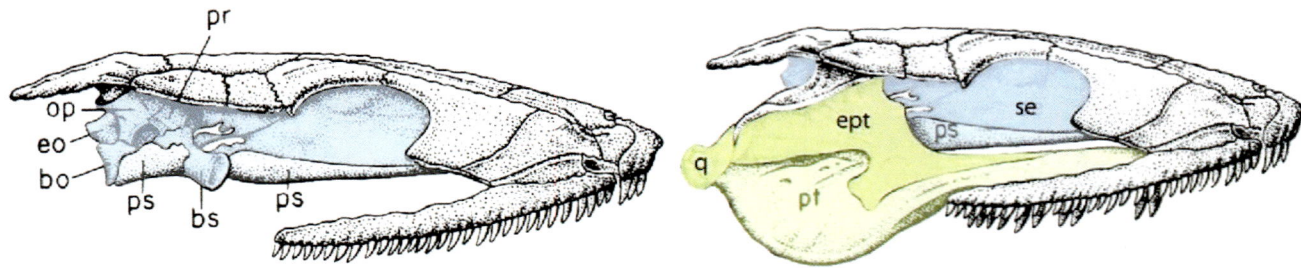

Fig. 8.14 Sagittal neurocranium parasphenoid (**psp**), basisphenoid (**bs**), basioccipital (**bo**), exoccipital (**eo**), opisthotic (**op**), prootic (**pr**). Bone fields added to neurocranium from palatoquadrate (yellow) epipterygoid (**ept**), pterygoid (**pt**), quadrate (**q**), note cranial base (blue) with sphenethmoid (**sp**) and parasphenoid (**ps**). (Reprinted with permission from: GEOL431—Vertebrate Paleobiology course web resources. Thomas Holtz and John Merck, Department of Geology, University of Maryland, College Park, https://www.geol.umd.edu/~jmerck/geol431/)

Dermal Bones: Basal Tetrapods

Median series
- Midline
- Nasal (**n**)
- Frontal (**f**)
- Parietal (**p**)
- Postparietal (**pp**)

Temporal
- Intertemporal (**it**)
- Supratemporal (**st**)
- Tabular (**t**)

Circumorbital series
- lacrimal (**l**)
- Prefrontal (**prf**)
- Postfrontal (**pf**)
- Preorbital (**prf**)

Marginal series
- Premaxilla (**pm**)
- Maxilla (**m**); lateral bone covering palatoquadrate

Cheek series
- Squamosal (**sq**)
- Quadratojugal (**qj**)

Palatoquadrate series
- Ectopterygoid (**ect**)
- Palatine (**pal**)
- Pterygoid (**pt**)
- Vomer (**v**)

Meckel's series, lateral
- Angular (**an**)
- Dentary (**d**): the alveolus, tooth-bearing bone
- Postsplenial (**psp**) continues splenial series posteriorly
- Splenial (**sp**): beneath dentary, ? program for dental zones
- Surangular (**sa**): above and in back of dentary, source of coronoid

Meckel's series, medial
- Articular (**art**)
- Coronoids (**co**): tooth-bearing bone fields medial to dentary
- Prearticular (**part**)

Neurocranium: Summary of the Braincase

Sagittal neurocranium, midline—these bone fields are bilateral but fused
- Basisphenoid (**bo**)
- Basioccipital (**bs**)
- Parasphenoid (**ps**) dermal, covers floor of neurocranium, not in mammals

Sagittal neurocranium, paramedian
- Exoccipital (**eo**)
- Opisthotic (**op**)
- Prootic (**pr**)

Posterior neurocranium
- Stapes (**st**)
- Landmarks

Neurocranium (r0–r3): Neural Crest, Epaxial Stapedial System

Sphenoid Complex

Descriptive Anatomy

The sphenoid deserves the to be called a complex for three reasons. (1) It is assembled from distinct cartilages and has discrete components: basisphenoid, presphenoid, orbitosphenoid, alisphenoid, and pterygoid processes. (2) It develops from distinct sources: paraxial mesoderm and neural crest. (3) Its various components ossify via three distinct mechanisms: chondral, membranous via a chondral intermediate, and membranous (Figs. 8.15, 8.16, 8.17, 8.18, 8.19, 8.20, 8.21, 8.22, 8.23, 8.24, 8.25, 8.26, 8.27, 8.28, and 8.29).

Sphenoid has two forms of ossification and two sources of mesenchyme. Basi*pre*sphenoid (known, more simply, as presphenoid, PS) and basi*post*sphenoid (more commonly termed basisphenoid, BS) are both r1 neural crest derivatives that form chondral bone. The former articulates with ethmoid and contains the sphenoid sinus and carotid sinus. The latter contains the pituitary which constitutes an ectodermal connection between the buccal cavity and the basal forebrain. This craniopharyngeal canal is an embryologic boundary zone representing two sources of mesenchyme. Posterior to pituitary, acrochodral cartilage is a focal contribution of paraxial mesoderm to basisphenoid. The craniopharyngeal canal can give rise to *cysts* with epithelial lining or to a *craniopharyngioma*. Note that these tumors do not erupt via the sphenoid sinus. On the other hand, both *encephaloceles* can explain the sutural boundaries of the sphenoid complex to access the oronasal cavity.

A vivid way of visualizing sphenoid anatomy is to compare it to a silent sphinx, standing guard in the center of the skull. The rostral terminus of the notochord is within basisphenoid. Enclosed within its BS/PS center, in the *hypophyseal festestra*, is the pituitary gland. Its shape is described by most textbooks as being bat-like, but this appearance is deceiving. In reality, the lateral "wings" of the sphenoid belong to a different bone, the *alisphenoid*, an r2 neural crest derivative. When these alisphenoid wings are clipped off, the "stripped down" version of sphenoid indeed resembles a sphinx. Its leonine "body" is made up of a solid posterior *basisphenoid* (BS), a chondral bone derived from r1 neural crest. Its human "face" is fashioned from a hollow anterior *presphenoid* (PS) derived from r1 neural crest. Two paws dangle down from basisphenoid, the *medial pterygoid laminae*. These are termed medial because the "alisphenoid add-on" contributes its own separate set of *lateral pterygoid laminae*. The pterygoid laminae arise from r3 neural crest. Our sphinx needs a magnificent crown. This is represented by flanking r1 neural crest *orbitosphenoid cartilages*. These

r1 neural crest derivatives are transformed into the membranous *lesser wings*. Attachment of the orbitosphenoids to presphenoid respects pre-existing cranial nerves and arteries; the optic foramen and the pterygoid canal result.

Recall that the pituitary occupies the interface between presphenoid and basisphenoid. Mapping studies of skull mesenchyme demonstrate conclusively that all craniofacial bones anterior to pituitary are derived from neural crest. Some of these may begin with a cartilage intermediate (lesser wing), while others form directly in the membrane (greater wing). When these bone fields share a potential space with the oronasopharynx, mucosa will form in that interface. This results in cavitation; air sinuses result. Thus, presphenoid contains paired air cells separated by a midline lamina. Surgical access to the pituitary via transsphenoidal hypophysectomy depends upon this embryology.

The intracranial surface of presphenoid (PS) contains several landmarks. Directly anterior to PS lies yet another r1 neural crest bone, the ethmoid. A stubby rectangular projection of the presphenoid, the e*thmoid process*, articulates in a tongue-in-groove manner with the cribriform plate, enhanced mechanical stability of the anterior cranial fossa results. In rare instances, field defect at this site permits egress of *sphenoethmoidal encephalocele*. Posteriorly, the smooth surface of *jugum sphenoidale* bears longitudinal grooves created by the olfactory nerves as they project forward from rhinencephalon over sphenoid to terminate as the olfactory bulbs resting atop the cribriform plate. Further posterior, the crossing of the optic nerves is marked by a transverse *chiasmatic groove* leading to the optic canals. Proceeding backward is the sella turcica, covered with dura, in which resides the pituitary. This marks the boundary between presphenoid and basisphenoid.

Three clinoid processes serve as important landmarks and provide attachment for the dura. The *anterior clinoid process* is formed by a scroll of orbitosphenoid as it makes the lateral wall of the optic canal. The *middle clinoid processes* define the anterior limits of sella turcia. The *posterior clinoid processes* fulfill a similar function but also anchor the dura of tentorium cerebelli. They arise as outgrowths of the achrochordal cartilages. Cranial nerve VI carves a longitudinal groove on either side of the dorsum sellae, while just below it, another "stabilizer," the *petrosal process*, articulates with the apex of the petrous temporal bone. We shall see that petrous bone arises from Sm4 to Sm5 paraxial mesoderm and r4–r5 neural crest. Like the prow of a ship, the r4 petrous apex comes nosing in medially as it attempts to "dock" with r3 component of the sphenoid. This maritime marriage creates the *foramen lacerum*. The terminal portion of dorsal basisphenoid is the *clivus*. During development, a transient cartilaginous *sphenoccipital synchondrosis* defines the articulation between Sm4 basisphenoid and Sm5–Sm11 basioccipital.

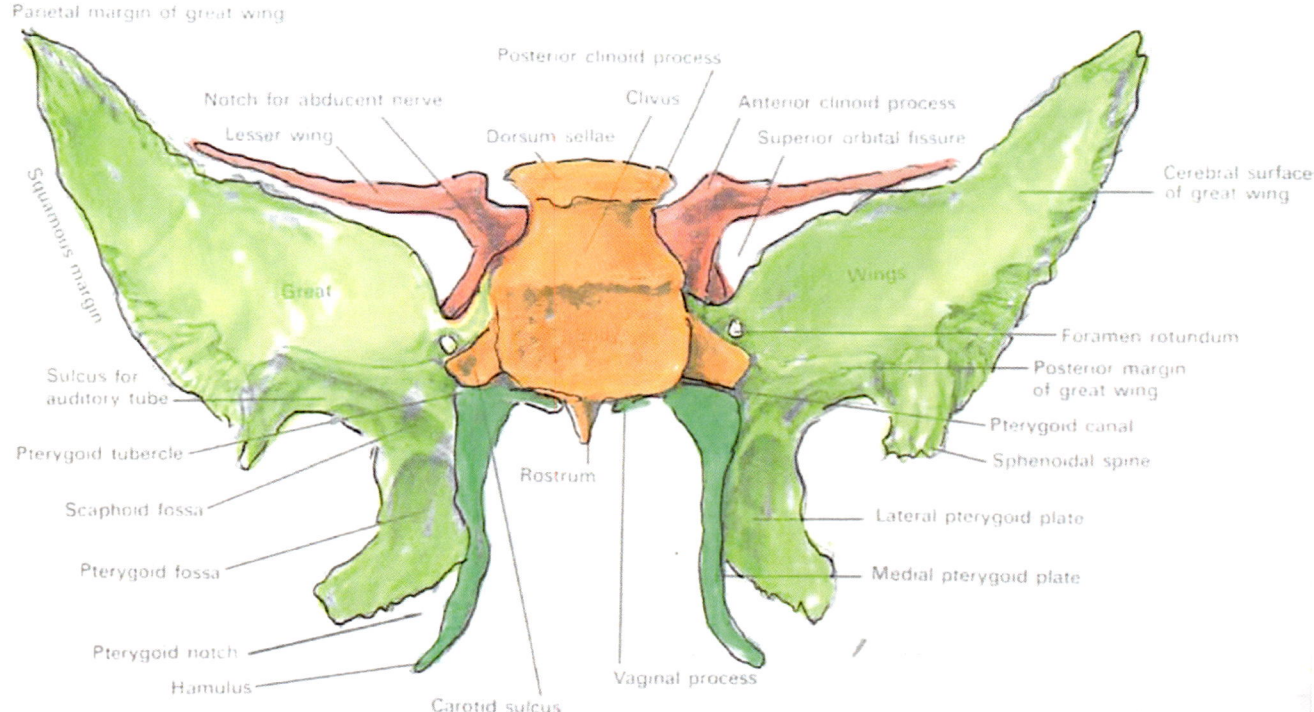

Fig. 8.15 Sphenoid, posterior aspect. Medial pterygoid process = pterygoid bone; lateral pterygoid process = epipterygoid (alisphenoid). Lesser wing originates from r1 orbitosphenoid cartilage. Foramen rotundum is breakpoint between orbitosphenoid and alisphenoid, i.e., between r1 and r2. Alisphenoid has two parts: greater wing (r2) and lateral pterygoid plate (r3). (Reprinted from Lewis, Warren H (ed). Gray's Anatomy of the Human Body, 20th American Edition. Philadelphia, PA: Lea & Febiger, 1918)

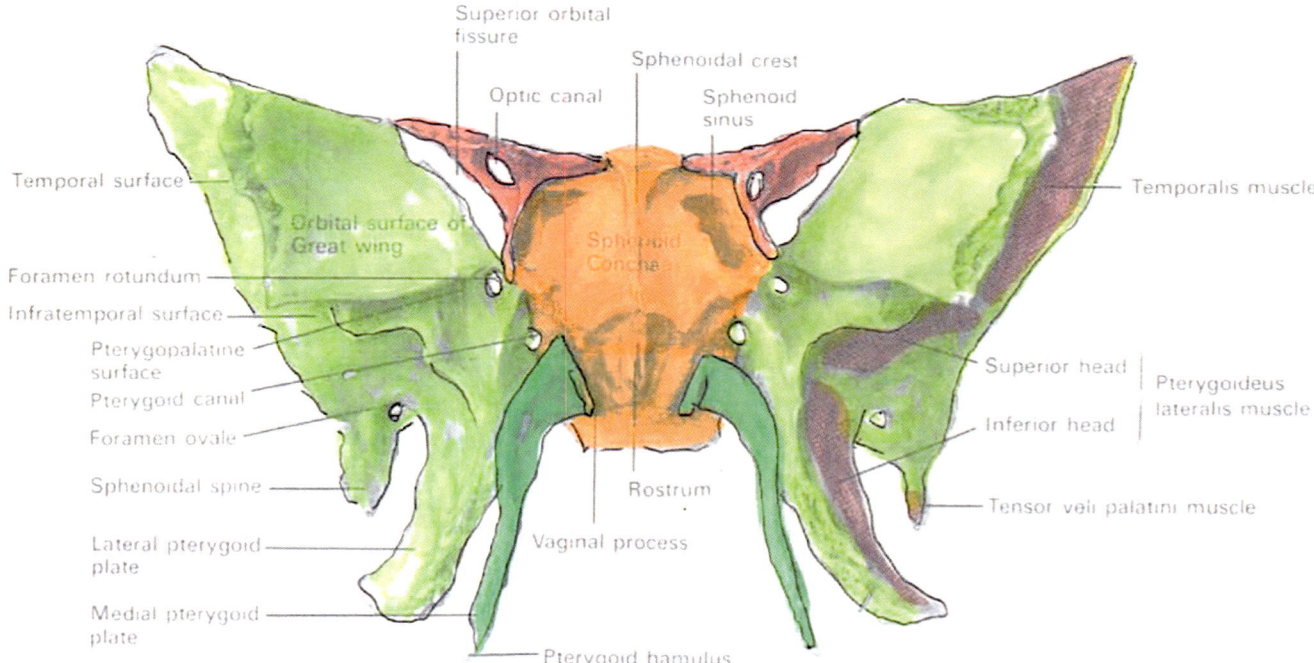

Fig. 8.16 Sphenoid, anterior. Note insertions of Sm4 muscles of mastication of epaxial (r2) and hypaxial (r3) alisphenoid. Medial pterygoid plate derived as remnant of pterygoid bone. (Reprinted from Lewis, Warren H (ed). Gray's Anatomy of the Human Body, 20th American Edition. Philadelphia, PA: Lea & Febiger, 1918)

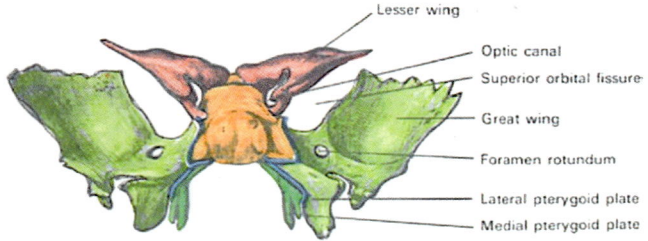

Fig. 8.17 Sphenoid at 4 months. Alisphenoids extend to the entire combined body of PS and BS. Greater wing is r2 membranous lateral and chondral (lateral orbital cartilage) medially. The r2 zone surrounds foramen rotundum. The r3 zone surrounds foramen ovale and gives off lateral pterygoid plate. Note supraorbital cartilage of r1 frontal bone not seen. Note separate components for medial pterygoid plate (pterygoid bone) and hamulus (ectopterygoid bone). (Reprinted from Lewis, Warren H (ed). Gray's Anatomy of the Human Body, 20th American Edition. Philadelphia, PA: Lea & Febiger, 1918)

The anterior surface of the presphenoid participates in forming the posterior nasal chambers. A midline *sphenoidal crest* articulates with the paired laminae of the perpendicular ethmoid plate. On either side of the crest, paired concave *sphenoid conchae* make up the anterior face of the presphenoid and the back wall of the nasopharynx.

Paired openings into the presphenoid sinus are located on either side of the sphenoidal conchae. Thus, the sinuses drain into the *sphenoethmoidal recesses*. It is of interest that occasional penetration of the sphenoid air sinuses has projected backward through the substance of basioccipital bone to the foramen magnum. This must represent a *midline basioccipital cleft*, better considered a failure of midline fusion, which allowed posterior extension of the sphenoid sinus to occur. Inferolaterally are articulation sites for the sphenoid processes of the r2 palatine bones.

Fig. 8.18 Cranial base cartilages contributing to sphenoid complex. Not shown here: acochordal (postsphenoid polar), Medial pterygoid plated (pterygoid) and lateral pterygoid plate (alisphenoid) are below the plane of the brain. (Reprinted from Nemzek WR, Brodie HA, Hecht ST, et al. MR, CT, and Plain Film Imaging of the Developing Skull Base in Fetal Specimens. American Journal of Neuroradiology 2000; 21(9): 1699–1706. with permission from American Society of Neuroradiology)

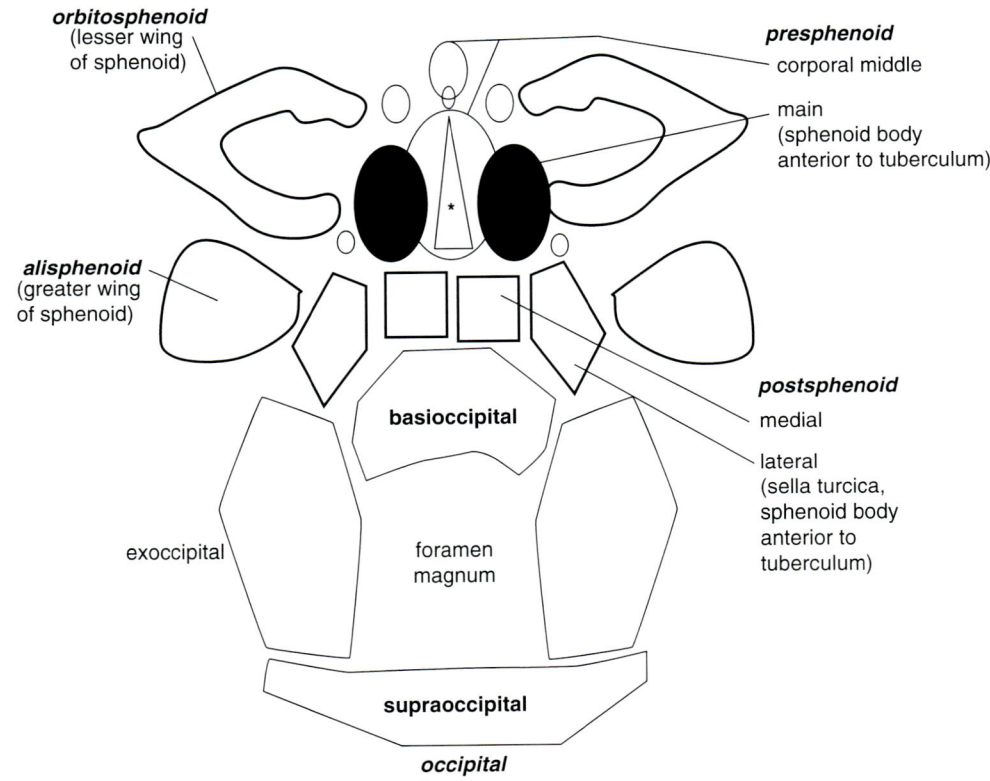

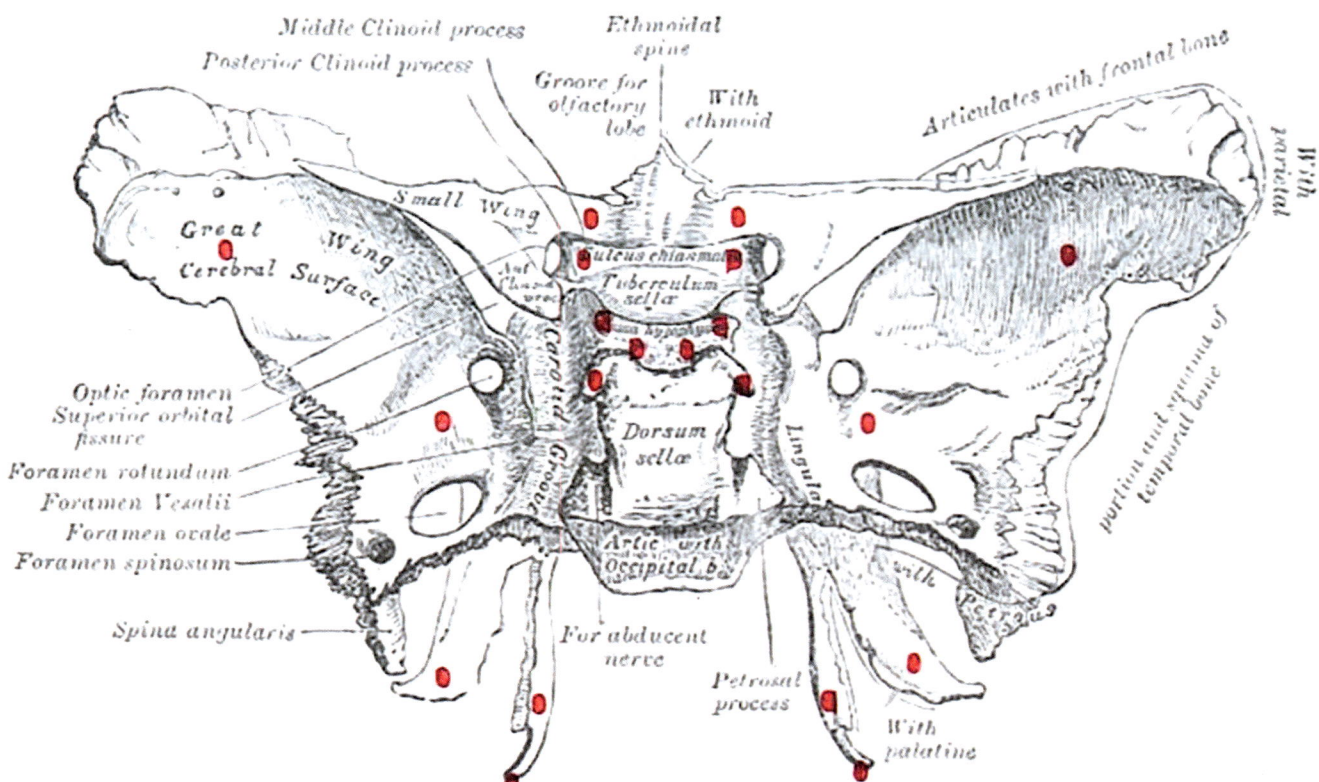

Fig. 8.19 Sphenoid ossification centers both chondral and membranous number 20. Basisphenoid cartilage (4), acrochordal/dorsum sella (2), presphenoid (2), orbitosphenoid (2), r2–r3 alisphenoid, medial chondral and lateral membranous (4), r3 alisphenoid lateral pterygoid plate (2), r2 medial pterygoid plate, and (2) pterygoid hamulus. (Reprinted from Lewis, Warren H (ed). Gray's Anatomy of the Human Body, 20th American Edition. Philadelphia, PA: Lea & Febiger, 1918)

The inferior surface of the presphenoid has a midline *sphenoid rostrum* that is in continuity with the sphenoid crest. This central projection receives proximal articulations with the paired r2 vomer bones. On either side of the rostrum, *vaginal processes* represent the superomedial extension of the r3 medial pterygoid plates (the presphenoid "add-ons").

Blood Supply and Mesenchyme

Arterial supply to the sphenoid complex arises from five distinct sources: (1) internal carotid artery, cavernous portion, pterygopalatine sector, (2) intracranial stapedial system (meningeal arteries), and (3) extracranial stapedial system (maxillo-mandibular artery), (4) external carotid to mucoperiosteum, and (5) external carotid via muscle attachments.

Internal Carotid

Recall that the internal carotid artery extends forward from the third aortic arch at levels r6–r7 as far as the terminus of the notochord at levels r0–r1. It arises from the dorsal aortae and its wall is made from the same paraxial mesoderm that also forms the surrounding petrous temporal bone and basi-sphenoid, i.e., Sm7, Sm6, Sm5, and Sm4. The petrous portion of ICA begins with carotid canal that takes ICA forward anterior to cochlea and tympanic cavity. [Since we postulate the mastoid process to be a derivative of Sm7 this PAM may possibly not be involved with ICA.] At the level of trigeminal ganglion (r1–r3) petrous ICA is separated from the ganglion by only a membrane.

The cavernous portion of ICA passes forward along the body of the basisphenoid. The blood supply through this dura supplies the lateral walls of basisphenoid. At the level of the anterior clinoid process, ICA perforates the dura of the roof of sphenoid sinus, i.e., at the boundary between basisphenoid and pre-sphenoid; this defines the beginning of the *intracerebral portion of ICA*. Its very first branch, *primitive maxillary artery*, arises before the ICA. Of transient importance as a supply of the eye, the primitive maxillary is reassigned to supply the posterior hypophysis. After the ICA bifurcation, arteries from the A1 sector of the anterior cerebral artery supply anterior hypophysis and neighboring bone.

Cavernous ICA has four main branches. (1) *Cavernous branch* supplies two sinuses: inferior petrosal and cavernous. The former is likely induced by a branch of VII. (2) Superior and inferior *hypophyseal arteries* supply the pituitary. The

a

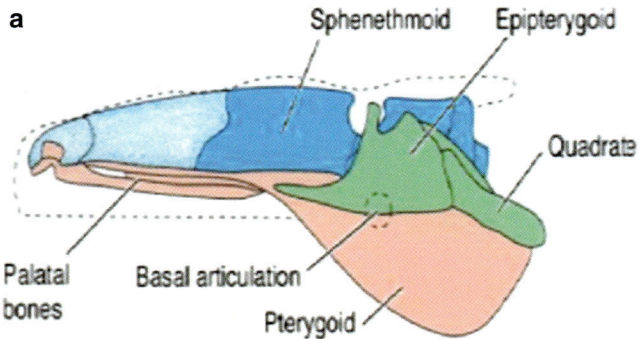

Lateral view with dermatocranium removed

b

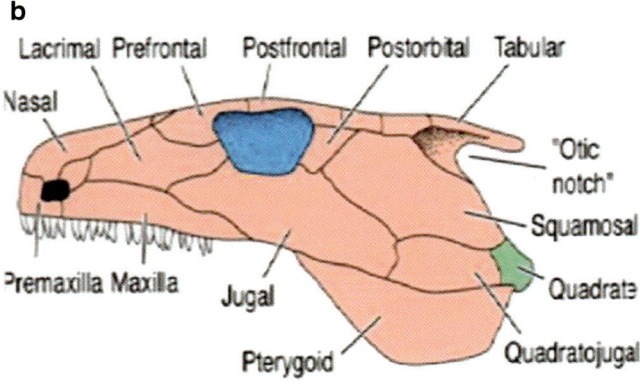

Lateral view of dermatocranium

Fig. 8.20 Chondrified derivatives of the palatoquadrate cartilage. Epipterygoid (Ept) is part of palatoquadrate cartilage; it has basal articulation with pterygoid. It articulates with the skull roof. Ept becomes alisphenoid. It maintains the connection with the orbit. Even though pterygoid gets absorbed into basisphenoid alisphenoid will produce the lateral pterygoid process. (Courtesy of William E. Bemis)

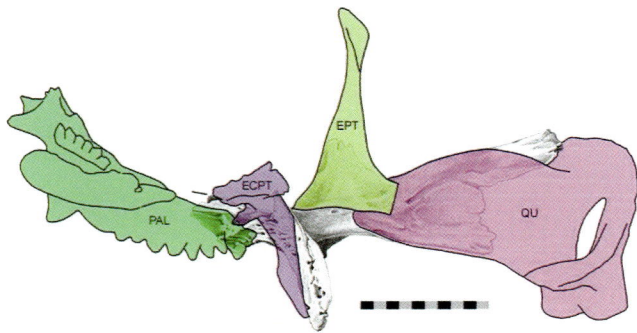

Fig. 8.21 Basic osteology: palatoquadrate cartilage bone complex. Seen from left side Pal = palatine bone series *Ectopterygoid*, Ecpt (green), is really just the most posterior of the palatine bones. It laminates on to the medial pterygoid process as the hamulus. *Epipterygoid*, Ept (yellow), morphs into alisphenoid and is hypaxial extension becomes the lateral pterygoid plate. Note epipterygoid is like royalty. It is an authentic extension of posterior PQ cartilage and represents the original articulation between braincase and palatoquadrate. Pterygoid is just a routine dermal bone that is produced from r2 in response to PQ. BS is covered ventrally by the dermal parasphenoids. Basipterygoid process connects the cranial base to the palate. Up to basal tetrapods it was mobile but becomes solidified. (Reprinted from: www.palaeo-electronica.org. Retrieved from: http://palaeo-electronica.org/2011_2/251/images/figure_71.jpg)

latter is a remnant of the embryonic maxillary artery. (3) *Ganglionic* branches supply the trigeminal ganglion. These represent the *primitive trigeminal artery* connecting the dorsal aorta with the longitudinal neural artery. (4) *Anterior meningeal* is of interest. It arises via induction from extraorbital StV1 as the latter ascends into the orbit. The course of anterior meningeal thus traverses over the lesser wing to supply dura at the posterior-medial aspect of the anterior cranial fossa. It is at this juncture that it anastomoses with posterior ethmoidal from intraorbital StV1. Laterally, over the greater wing, dural anastomoses exist with StV2 meningo-orbital branches.

Intracranial Stapedial System: Meningeal Arteries

Dura covering the anterior cranial base arises from the r1–r3 neural crest and is supplied by meningeal arteries associated with trigeminal sensory nerves. As previously discussed, these are all derivatives of the stapedial system. StV1 supplies both presphenoid and orbitosphenoid.

The vascular supply for r2 alisphenoid arises from two distinct sources.

- The *basal segment* receives branches from the anterior deep temporal, a derivative of the first arch external carotid component of the second portion of the maxillo-mandibular artery.
- The *alar segment* receives two different sources of support.
  - Dura lining its temporal side is supplied by a forward-directed branch of from the intracranial StV2 middle meningeal artery.
  - The periosteum lining the orbital side of the alisphenoid is supplied from StV1 lacrimal.
  - A third artery, the *meningeal branch of StV1*, penetrates AS via a separate cranio-orbital foramen lateral to superior orbital fissure. This vessel, also known as the *recurrent meningeal artery*, connects the common stem of StV2 and StV1 running forward from the middle meningeal artery with intraorbital StV1 lacrimal artery. It later involutes (see Chap. 8) and lacrimal proceeds forward on its own.
- This seemingly irrational picture is a product of the phylogeny of the alisphenoid bone. The upshot is that the dual source of blood supply explains why Tessier zone #9 affecting alisphenoid are so very rare. The presence of these two distinct vascular zones explains why alisphenoid is considered to have two distinct centers of ossification.

Fig. 8.22 Amniote epipterygoid (medium green) connects pterygoid (brown) with the backside of orbital cartilage, pila antotica (magenta). Note that epipterygoid articulates with inferior border of parietal. It is also in contact with periotic neurocranium (light blue). (Courtesy of Lawrence M. Witmer, PhD)

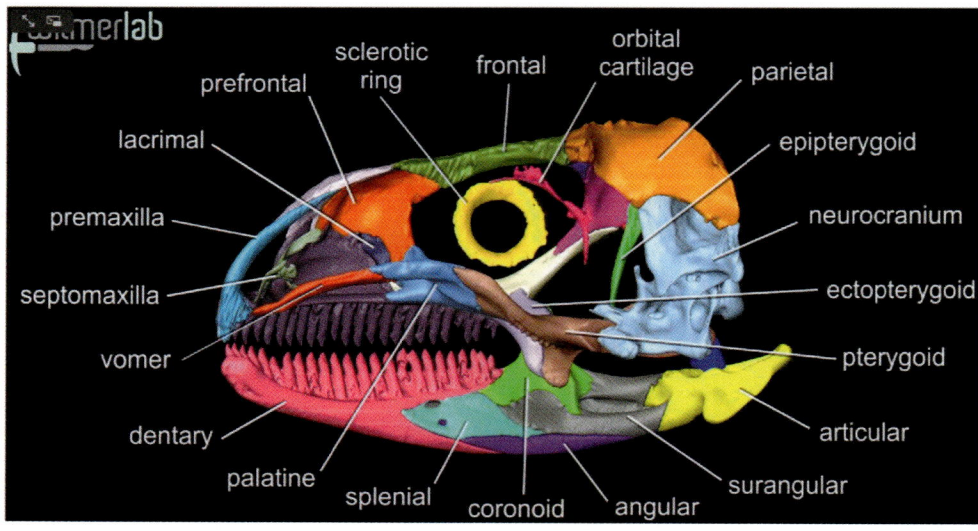

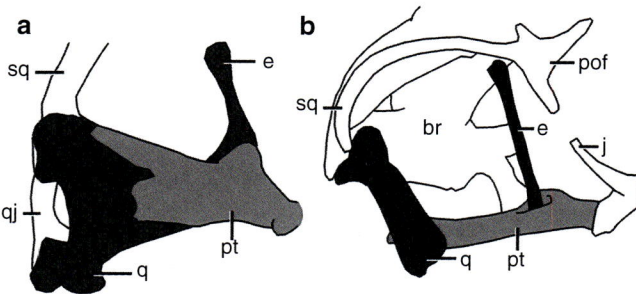

Fig. 8.23 Radical changes in quadrate (future stapes) from basic tetrapod to pre-reptile. Relationship between the quadrate (q), pterygoid (pt), and epipterygoid (e) in *Sphenodon*, medial view (8) (**a**), and *Varanus*, lateral view (original) (**b**). Not to scale. *br* braincase, *j* jugal, *pof* postorbitofrontal, *qj* quadratojugal, *sq* squamosal. In *Sphenodon* and basal taxa, the epipterygoid is welded onto pterygoid. In the lizard *Varanus*, a kinetic joint develops (the *streptostylistic* condition). Quadrate became stressed with biting and development of a lower temporal bar which quadrate attaches to the jugal (the future zygomatic arch) provided additional support. What's more, the advent of the joint between epi and pt means that pterygoid began to move forward to sphenoid with brain growth the attachment of epi could be disarticulated and transferred. (Reprinted from Moazen M, O'Higgins PO, Evans SE, Fagan MJ. Biomechanical assessment of evolutionary changes in the lepidosaurian skull. PNAS May 19, 2009;106(20): 8273–8277. With permission from PNAS)

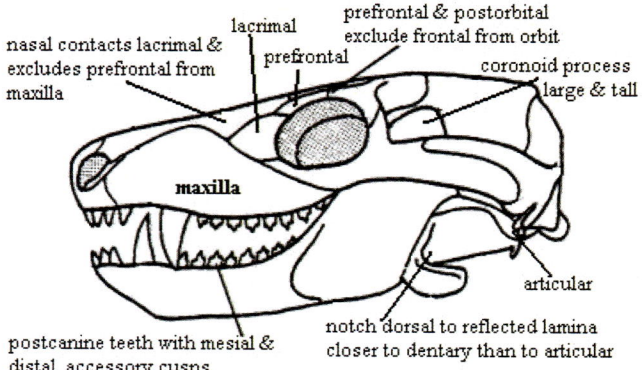

Fig. 8.24 Cynodont apomorphias 1. Note postcanine teeth beginning to specialize. Premolars and molars are present with *Morganucodon*. (Reprinted from Carroll RL. Vertebrate Palaeontology. New York, NY: WH Freeman and Company; 1988)

Extracranial Stapedial System: Third Part of Internal Maxillo-Mandibular Axis

IMMA in its distal third gives off two arteries departing backward from the pterygopalatine fossa. These are programmed from sensory branches of V3 that supply the cranial base. *Artery of the pterygoid canal* passes backward to the upper pharynx where it anastomoses with StV2 nasopalatine branches and ascending pharyngeal arteries representing the third arch component of external carotid. It supplies sphenoid sinus and via ascending pharyngeal, the auditory tube, and

tympanic cavity. Thus, the artery of pterygoid canal is an anastomosis between the first arch and third arch derivatives. The smallest branch of IMMA is the *pharyngeal artery*. It runs backward in company with the pharyngeal branch of V3 pterygopalatine nerve via the palatovaginal canal to supply sphenoid sinus, upper pharynx, and auditory tube.

External Carotid System: Second Part of Internal Maxillo-Mandibular Axis

IMMA in its middle third gives off muscular branches supplying muscles of mastication which attach to various zones of the sphenoid complex. Anterior temporalis covers the posterior aspect of the greater sphenoid wing and its undersurface, the anterior roof of infratemporal fossa. The medial and lateral pterygoids plates provide attachment sites for their respective muscles, supplied by a named branch descending from the second part of IMMA.

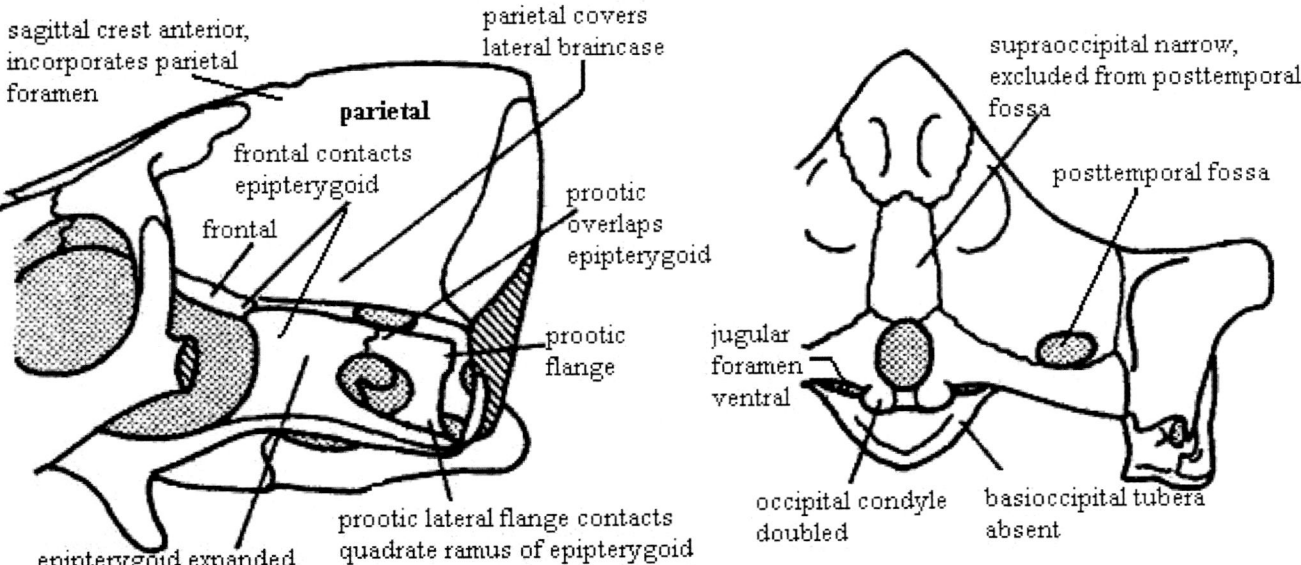

Fig. 8.25 Cynodont apomorphies 2. Note contact between alisphenoid and frontal bone. Not quadrate ramus of AS overlapping pro-otic externally. At the same time, pro-otic overlaps alisphenoid with *anterior* *lamina*, the future squamous temporal bone. (Reprinted from Carroll RL. Vertebrate Palaeontology. New York, NY: WH Freeman and Company; 1988)

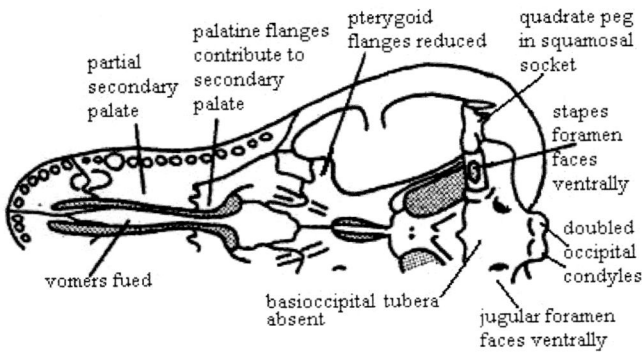

Fig. 8.26 Cynodont apomorphies 3. Note Pl and Pl are coming in as the "maxillary shelf" to fuse with vomer. P3 is forming the palatine horizontal shelf. Pterygoid flanges reduces as pterygoid prepares to disconnect in meld with basisphenoid. (Reprinted from Carroll RL. Vertebrate Palaeontology. New York, NY: WH Freeman and Company; 1988)

External Carotid System: Pharyngeal Branches of Ascending Pharyngeal

These arteries fan out over the posterior pharyngeal wall as high as the skull base. As such, they provide cover over the nasopharynx behind the palate and, perforce, up to the level of Rathke's pouch and the ventral aspect of presphenoid.

Significance of the blood supply. One notes that four of the five sources supply neural crest-derived bone sectors of the sphenoid complex.

Development

During the mid-stage 16 (early E13), trabecular cartilages appear in the nasal capsule and extend backward. By stage 17 (late E13), hypophyseal cartilages appear around the developing neurohypophysis. During stages 18–19, three structures appear as the optic capsule: ala orbitalis, frontal, and hypochiasmatic cartilages. Ala orbitalis and hypochiasmatic will fuse to form orbitosphenoid which subsequently fuses with presphenoid. During stages 20–21(E15), basisphenoid (now fused) adds on new lateral components, the *basitrabecular* cartilage. Alisphenoid now appears. *Acrochordal* cartilages fuse in the midline to the rostral terminus of the parachordal cartilages. They become the posterior "backing" of the dorsum sellae and constitute the boundary between the sphenoid and the basioccipital.

Ossification

The raw materials for the entire chondral skull base are in place with gastrulation by stage 8. Neural crest migration into the area is accomplished between stages 9 and 10. As we have stated before, the ossification sequence of a bone is does not document bone "age" but provides a *relative guide to the development of vasculature sufficient to support bone development.*

Ossification of the sphenoid complex takes place first in the postsphenoid zone. It is seen in the basal alisphenoid between foramen ovale and foramen rotundum at 8 weeks. Body of basisphenoid appears shortly thereafter. Medial pterygoid plate descends from the sphenoid body and ossifies in the membrane at 9–10 weeks. At 12 weeks, hamulus appears from cartilage. At 6 months, fusion between the medial and lateral pterygoid plates takes place.

The presphenoid zone develops a week later, with ossification in membrane at 9 weeks from orbitosphenoid carti-

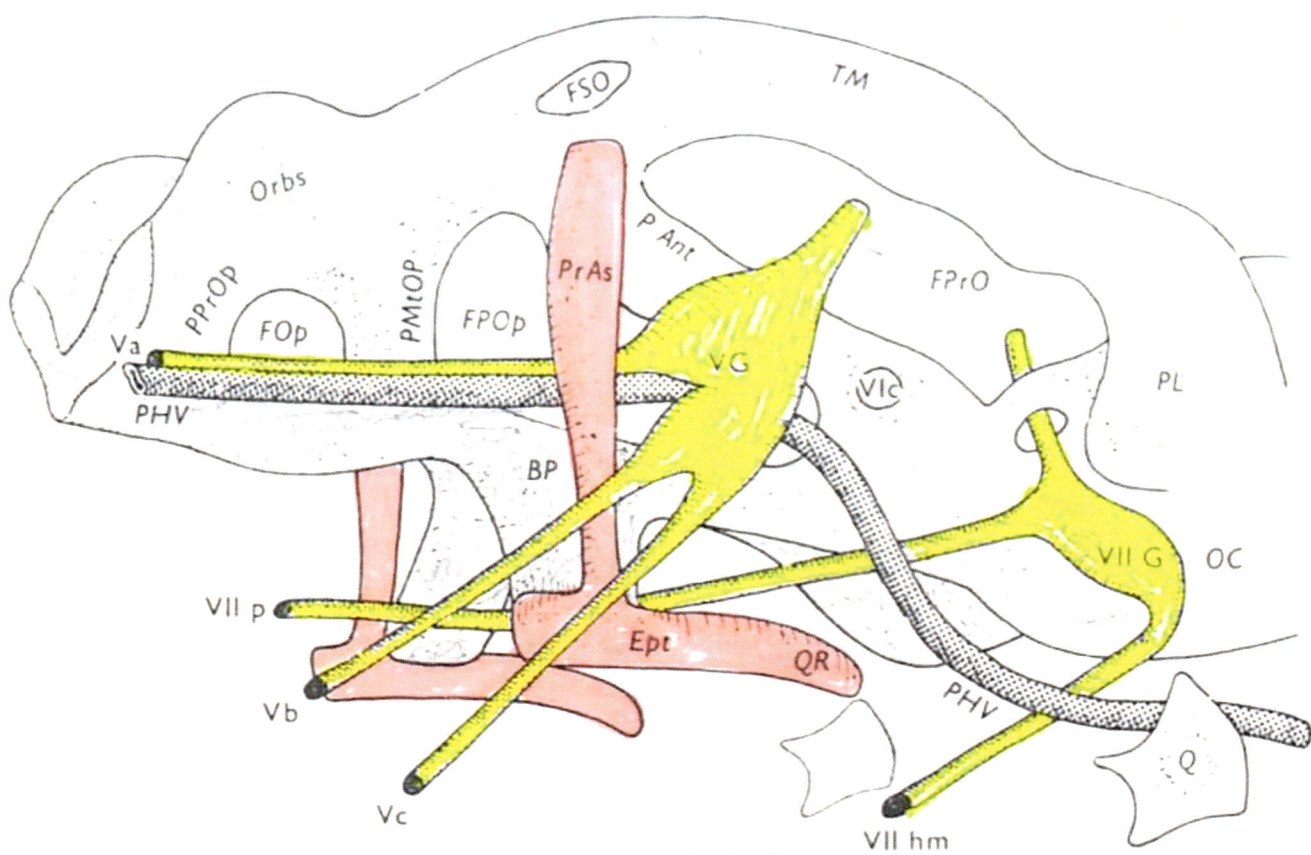

Fig. 8.27 Alisphenoid: ectopterygoid–orbital relationships. Note how the processus ascendens passes between V1 and V1. *BP* basitrabecular process, *Ept* epipterygoid, *FOp* foramen opticus, *FPOp* fenestra postoptic, *FPrO* fenestra pro-otica, *FSO* fenestra supraorbitale, *OC* optic capsule, *Orbs* orbitosphenoid, *Pant.* Pila antotica, *PL* parietal lamina, *PHV* primary head vein, *PMtOp* pila metoptica, *PPrOp* pila preoptica,

PrAs procesus ascendens of epipterygoid, *Q* quadrate, *QR* quadrate ramus of epipterygoid, *TM* taenia marginalis, *VG* trigeminal ganglion, *Vic* abducens canal, *VIIG* facial ganglion, *VII hm* facial nerve, *VIIp* palatine (vidian) nerve. (Reprinted from Presley R, Steel FLD. On the homology of the alisphenoid. *J Anat* 1976; 121(3):441–459. With permission from John Wiley & Sons)

lage appearing just lateral to the optic canal. Presphenoid body appears at 11 weeks. Finally, the sphenoidal conchae develop at prenatal month 5.

At birth, the sphenoid complex is tripartite. Pre- and post-sphenoid fuse to form a common body at 8 months. At 1 year, the alisphenoids fuse to the body. The lesser wings approximate one another in the midline to form a smooth, yoke-shaped structure, jugum sphenoidale. Fusion of the sphenoid conchae with the ethmoid labyrinth occurs at 4 years and ossifies by age 12.

Phylogeny

The core components of sphenoid, or better stated, the sphenethmoid are primordial dating back to the agnathic fishes. Alisphenoid and the pterygoid plates are newer. Alisphenoid is derived from the original palatal series connecting the oral cavity with the skull. The biggest changes in this complex occurred with the transition to later therapsids, the immediate precursors to mammals. Changes in feeding mechanisms permitting complex chewing made these ani-

mals more energy-efficient. These included the invention of a temporomandibular joint and increasing specialization of teeth. The palate, formerly kinetic, became welded to the braincase. Dermal *pterygoid bones*, previously very large structures, became relocated to the ventral sphenoid as medial pterygoid laminae while the adjacent tooth-bearing *ectopterygoids* as the likely source of the lateral pterygoid laminae. These structures provided for new muscle attachments to the mandible permitting complex motion. *Epipterygoid*, in the pre-mammalian synapsid condition, was attached to the root of the pterygoid process below and suspended by soft tissue from the parietal bone. In mammals, it became anchored to orbit and the fronto-parietal junction, pterion, as the alisphenoid. *Orbitosphenoid* appears as a distinct bone in mammals.

The sphenoid complex *sensu strictu* arises from 16 distinct ossification centers. If one includes the alisphenoid bones and the pterygoid plates, each with two ossification centers, the number of ossification centers in the complex rises to 24. Sphenoid development should be considered in

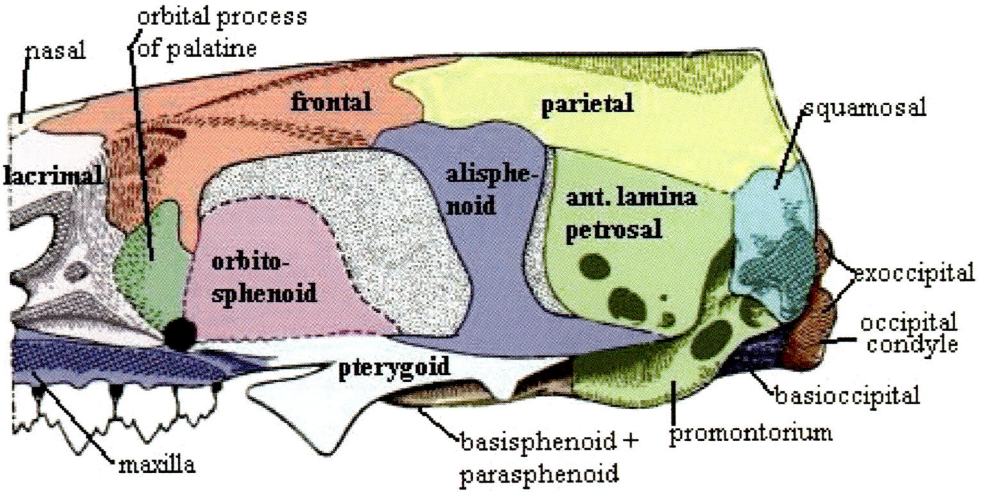

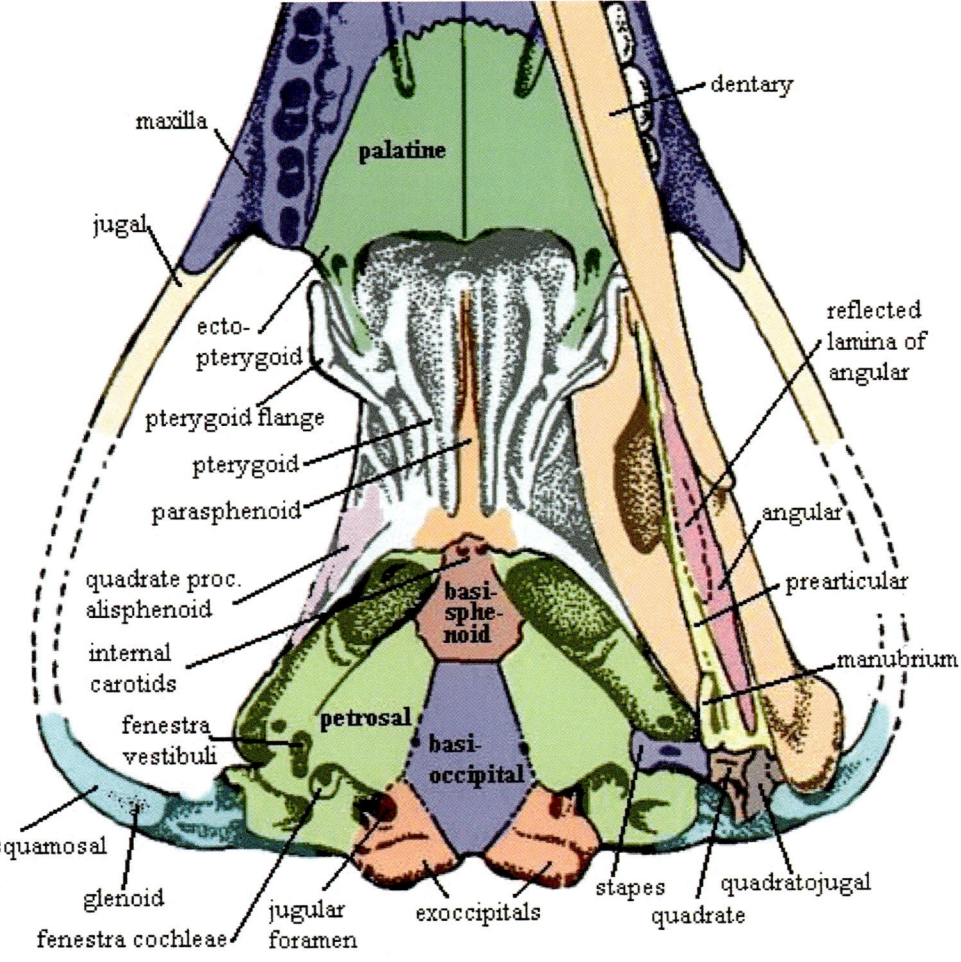

Fig. 8.28 Alisphenoid in mammals. Alisphenoid and orbitosphenoid are mammalian inventions. In *Morganucodon*, lateral wing of alisphenoid is connected at the corner of the frontal and parietal. It retains this position as pterion. Orbitosphenoid relates to orbital process of palatine bone. As the entire orbits rotates forward, OS and AS bone fields will rotate outward and forward 90°, while palatine retains its sagittal posi- tion. Pterygoid (pale blue) will be absorbed, except basipterygoid pro- cess (seen posteriorly), as the medial plate. Alisphenoid will give off a descending process as lateral plate. (Reprinted from Kermack KA, Mussett F, Rigney HW. The skull of Morganucodon. Zool. J. Linn. Soc. 1981;71:1–158. With permission from Oxford University Press)

Fig. 8.29 Ectopterygoid and the medial pterygoid plate. Fate of pterygoid (pale blue) and ectopterygoid (green) in mammals, as seen in *Morgaunucodon*. Pterygoid is incorporated into basisphenoid, leaving behind the pterygoid flange as medial pterygoid plate. Ectopterygoid laminates the posterior aspect of medial pterygoid plate. It remains in humans as the humulus. These changes are occasioned by brain growth, shortening of the snout, and changes in the orbits. Basisphenoid, here very posterior, will move to the center of the skull, directly behind the orbits. (Reprinted from Kermack KA, Mussett F, Rigney HW. The skull of Morganucodon. Zool. J. Linn. Soc. 1981;71:1–158. With permission from Oxford University Press)

four ways: anatomic components, the initial appearance of its individual components, the ossification sequence, and the final fusion pattern.

Components of the Sphenoid Complex

Basipostsphenoid (BS) consists of four cartilages that arise from r1 neural crest. Two medial *hypophyseal* (or polar) cartilages flank the tip of the notochord and enclose the pituitary. They abut against the basioccipital bone. A pair of lateral *basitrabecular* cartilages form the sella turcica and sphenoid body posterior to tuberculum. Paired *posterior clinoid processes* arise from separate acrochodral cartilages and constitute two additional ossification centers. They provide attachment for tentorium cerebelli; abducens nerve runs below and lateral to them. Thus, basisphenoid has six ossification centers.

Basipresphenoid (PS) develops from six centers of r1 neural crest. Paired *main presphenoid cartilages* form the body of sphenoid anterior to tuberculum sellae. They will hollow out to form the sphenoid sinus. Anterior to sphenoid body, a pair of *corporal middle cartilages* fuse in the midline. They fill in the space between the lesser wings, giving rise to midline structures such as sphenoidal crest. The anterior wall of presphenoid that abuts against the ethmoid cartilage consists of paired concave discs, the *sphenoidal conchae*. Laterally, between the conchae and the lesser wings, the membranous bone is incomplete, giving rise to open communication between the sphenoid sinus and the upper nasopharynx.

Orbitosphenoid (OS) appears to arise from the presphenoid but represents separate *ala orbitalis* cartilages. This r1 bone develops from four ossification centers. An initial pair appear just lateral to the optic foremen. Along its posterior and medial surface, two separate *hypochiasmatic* cartilages form the *anterior clinoid processes*. These provide insertion for the r1 dura of tentorium cerebelli and constitute additional ossification centers.

Alisphenoid (AS) forms the greater wing. Although it appears to arise from, basisphenoid, Alisphenoid is a distinct bone field. Its precursor is the *ala temporalis* cartilage. The true source of AS is r2 *epipterygoid bone*, a dermal bone representing an outgrowth from the palatoquadrate cartilage. Each AS has two distinct zones. Each alisphenoid can thus be considered to have two distinct ossification centers. This critical bone deserves a section unto itself, *vide infra*.

- The <u>alar segment</u> (r2) ossifies in the membrane and forms the back wall of the orbit membrane, *ala orbitalis* and the side wall of the braincase, *ala temporalis*. Its arterial supply is StV2.
- The <u>basal segment</u> (r3) ossifies in cartilage. It contains foramen ovale for transmission of V3 out of the skull

and forms the roof of the anterior infratemporal fossa. It is supplied by StV3.

Medial pterygoid laminae (MPt) appears to project downward from basisphenoid but it is really an "add-on." The true source of MPt is r2 *pterygoid bone* that was evolutionarily absorbed into BS. Intervening between MPt and presphenoid is the *pterygoid canal* to transmit the nerve and artery of the same name. Each MPt has two ossification centers corresponding to the ancient *pharyngobranchial* and *epibranchial* cartilages of the first arch. The lamina per se is membranous while the hamulus is chondral.

Lateral pterygoid lamina (LPT) arises from the undersurface of the alisphenoid. LPT arises from the P4 bone field of the palatine series, the *ectopterygoid*. LPt forms in the membrane and has two ossification centers.

Alisphenoid the Transformation of Epipterygoid

When we look at alisphenoid in humans, it is depicted as a monolithic structure. But we already know that this is not so. The lateral pterygoid is a separate bone field, ectopterygoid. Medial pterygoid (and its famous hamulus) is the remnant of the dermal roofing bone of the palate, pterygoid. The transition of the epipterygoid into the mammalian alisphenoid and the associated changes in surrounding bone fields highlight two major trends in the mammalian skull evolution: (1) the need for increasing skull strength; and (2) expansion of the braincase.

Increasing Skull Strength

Expansion of the brain and skull permitted the development of larger muscles with greater biting power. Strength and stability were needed.

- Multiple bone fields united. The supraorbital bones were annexed into frontal. The temporal series gave supratemporal and intertemporal to parietal, converting it into a four-quadrant complex. The most posterior of the temporal series, tabular, became incorporated into post(inter) parietal. Bones fields surrounding mandible were incorporated into the dentary. Jugal and postorbital unite.
- Kinesis between chondrocranium and the palate was lost. In this process, the pterygoid–epipterygoid joint is eliminated. As shown in the figures, this was convenient, because it allowed alisphenoid.

Expansion of the Braincase

In ancestral amniotes, the lateral surface of the brain was protected by chondrocranium but it proved incapable of adapting to brain growth. For this reason, the mammalian skull has chondrocranium in a limited zone, the cranial base, from orbit to floor and the back end of the braincase. In all other locations, dermal bone fields developed to cover the sides and roof of the brain. Of interest for us is the development of lateral coverage with reference to parietal and epipterygoid, seen best on coronal section (Fig. 8.30).

- The anapsid condition shows epipterygoids resting on, but not united with pterygoid. Neither are they united with basisphenoid or parasphenoid. [Note: BS is chondral while presphenoid is membranous. We will discuss it with the vomer.] An intervening fascia connects the Epi with the overlying parietal. Note that the chewing muscle, adductor mandibulae is not superficial; it lies deep to an external lamina of parietal, postorbital, and zygoma.
- The synapsid condition (pre-mammal) shows a freeing up of parietals. They are disconnected from postorbitals and have moved forward into the orbit. Zygomatic arch is bowed outward. This creates the *temporal fenestra*. Epipterygoids are now fused medially to basisphenoid

and pterygoid is lost, leaving behind pterygoid processes, the medial pterygoid plates. Parietal is expanding downward o almost, but not quite, contact Epi. Note a small portion of postorbital keeps them separate.
- The mammalian condition shows partials and alisphenoids sutured together. Adductor mandibulae has subdivided into temporalis and masseter. Internal masseter will give of the pterygoid muscles.

In sagittal view, we appreciate in ancestral amniotes that epipterygoid is embedded in adductor fascia, lies lateral to basisphenoid, and rests on pterygoid. From its inferior border, Epi sends out a posteriorly directed *periotic process* which grasps the lateral law of the otic capsule. A membranous anterior lamina buds forth from otic capsule, the future squamous temporal bone. The same section of the basal mammalian skull shows alisphenoid in contact with the orbit. It is incorporated in the side wall. Squamous temporal bone is well shown in Fig. 8.31.

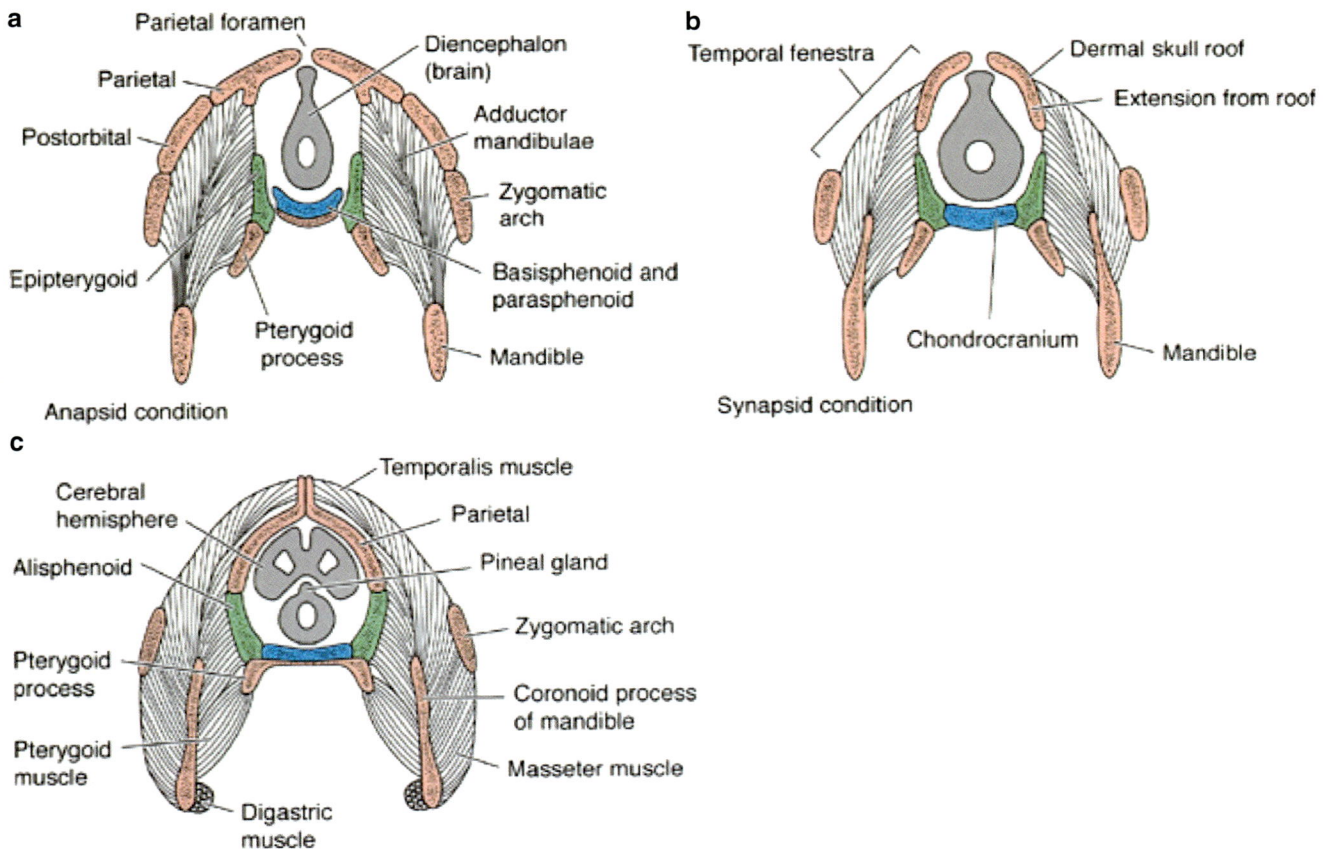

Fig. 8.30 Coronal sections of the temporal region showing the anapsid (**a**), synapsid (**b**), and mammalian (**c**) configurations. Development of temporal fenestra involved a separation of the postorbital from the parietal and the reduction of the postorbital to the orbital moiety of the zygomatic arch. This permitted a differentiation of the adductor mandibulae into distinct functional components for greater mandibular control. With the evolution of pterygoid muscles, mammals had independent control of mandible and TMJ permitting an enhanced form of mastication suitable for a wider variety of food sources. Note that the epitpterygoid (green) and the pterygoid process remain in contact. The latter always related to the inner mass of the adductor mandibulae. With differentiation, this muscle becomes the separate pterygoideus. (Courtesy of William E. Bemis)

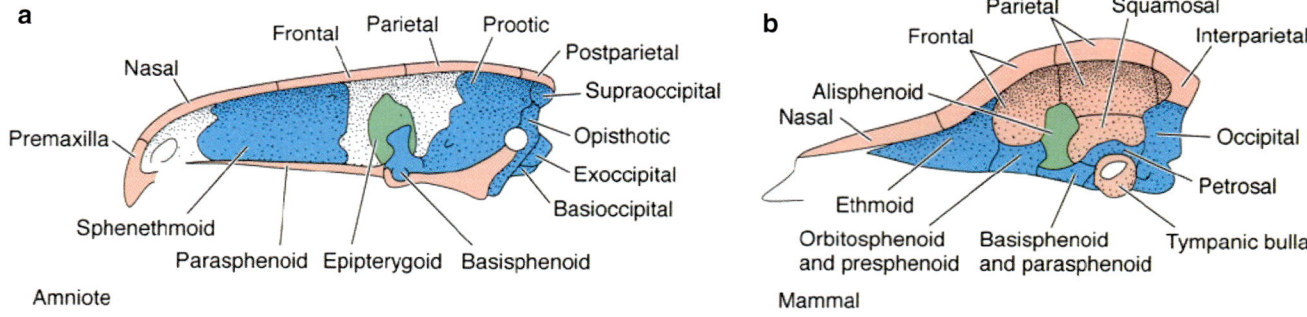

Fig. 8.31 Evolution of the epipterygoid (green) in basal amniote into the mammalian alisphenoid. Whereas in the amniote state, it was surrounded by membrane, it comes to be flanked by dermal bones, the frontal and the parietal. The origin of squamous temporal bone is well defined. It represents the forward projection of the squamosal, coming to articulate with the frontal bone. (Courtesy of William E. Bemis)

In the amniote *Lacertus*, the original function of epipterygoid is to serve as a strut between the mobile palate and the cranium. We see in the lizard how it appears as simple strut.

Critical changes in epipterygoid take place in cynodont-mammal transition. The early Triassic cynodont, *Thrinaxodon*, shows in the sagittal section that epipterygoid has gotten much large, becoming a lamina in front of periotic. Directly above it is parietal. It has foramina for V2 and V3. From periotic, a membranous extension, anterior lamina, projects forward. This will be part of the future anterior squamous temporal. It is a forward extension of the original squamosal which remains in continuity with the jaw joint.

As one progresses from morganucodontids to triconodontids (closer to real mammals) the epipterygoid separates from periotic completely and anterior lamina fills in the space. In marsupials, one sees foramina in alisphenoid for a space between it and orbitosphenoid for V1 (future orbital fissure), and foramina for V2 and V3. In true mammals, it becomes the anterior inferior wall of the lateral skull but is not yet in the orbit. It articulates extensively with frontal bone.

The final stage of alisphenoid is driven by forebrain growth and frontalization of the orbits. As frontal bone expands laterally over the eyes, it drags alisphenoid with it. Contact is made with the zygomatic complex.

Ethmoid Complex

Descriptive Anatomy

The ethmoid bone spans the midline of the anterior cranial fossa between the presphenoid bone and the frontal bone. As such, it is part of an unbroken chain of r1 neural crest bone fields that provide support for the frontal lobe. Its initial derivative, cribriform plate, develops around *pre-existing neural fibers connecting the p6 nasal placode with receptors for smell in the brain*, hence the story of its development is inseparable from that of the olfactory and accessory olfactory system (Figs. 8.32, 8.33, 8.34, 8.35, 8.36, 8.37, 8.38, 8.39, 8.40, 8.41, 8.42, 8.43, 8.44, 8.45, 8.46, 8.47, and 8.48).

The fronto-ethmoid complex is tightly interlocked. Cribriform plate is inserted into *ethmoid notch* of the frontal bone. In its center, paired cristae galli serve as attachments for falx cerebri. In hyperplasias of zone 13, widening, and even bifidity, of the cristae have been documented. Dura mater of prosencephalon develops from r1 to r3 neural crest; that of anterior cranial fossa is r1. Falx cerebri, being innervated by V1, is logically attached to the cristae and the interior vertical midline of frontal bone. Dural insertions have an internal logic based upon the mesenchymal composition of the target bones. From the anterior edge of cristae, galli paired triangular *alae* fit into the frontal bone, thus creating the *foramen cecum*. On either side of the anterior cristae is the *foramen of nasociliary nerve*.

Quadrilateral perpendicular plate articulates with sphenoidal crest, vomer, septum, and, anteriorly, with both nasal bone and (just above it) frontal bone. Along the surface of superior perpendicular plate are numerous canals bearing neural filaments of the accessory olfactory system.

The labyrinth, or lateral mass contains three groups of air cells. Let us think of it as a box with four sides.

1. The lateral surface of the labyrinth consists of a smooth structure, orbital plate of ethmoid (OPE). Orbital plate of ethmoid has four important articulations. Above it is *orbital plate of frontal bone*. The anterior and posterior ethmoid foramina are located along this intersection. Below it are the *orbital process of palatine bone* and maxilla. Posteriorly it abuts against sphenoid while anteriorly it articulates with lacrimal. OPE constitutes the medial orbital wall and covers over the middle and posterior ethmoid sinuses. The anterior ethmoid sinus group is located anterior to the orbital plate but does not go naked. These anterior air cells are overlapped by frontal process of maxilla and lacrimal bone.

Fig. 8.32 Alisphenoid in man. AS is divided into epaxial and hypaxial zones by infratemporal crest. In the temporal fossa *m. temporalis* inserts broadly up to, but not including postorbital. Infratemporal fossa bears *m. pterygoideus lateralis* which inserts into the TMJ capsule. It has pterygoid hamulus, which bears tendon of TVP, and sphenoid spine, which carries chorda tympani along its medial surface and connects to mandible via sphenomandibular ligament (SPL), itself a derivative of Meckel's cartilage. This establishes hypaxial alisphenoid as r3. Meckel's derivatives: anterior, dentary; middle, SPL; posterior, tympanic, malleus, incus. (Reprinted from Boileau JC. Grant Atlas of Human Anatomy Philadelphia: Williams & Wilkins 1943)

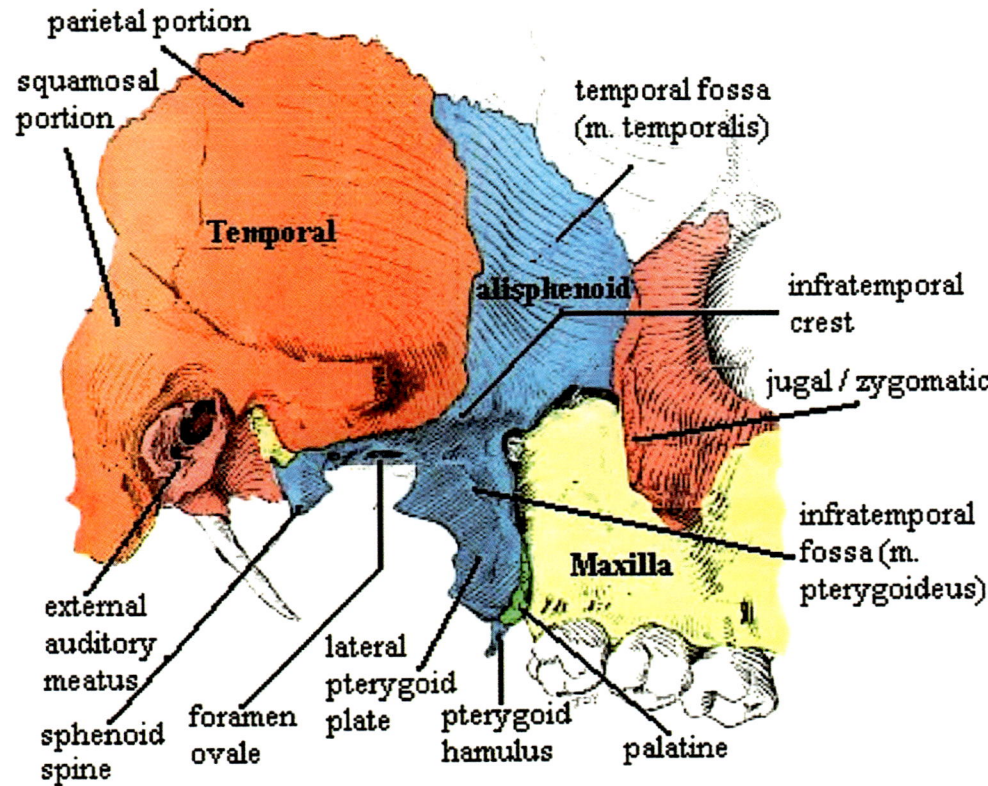

2. The medial surface of the ethmoid labyrinth projects two conchae (turbinates) into the nasal chamber. A downward projection of laminar bone from cribriform plate creates middle turbinate. Its mucous membrane contains many olfactory nerves. Within the middle conchae, air cells create a swelling, bulla ethmoidalis. Underneath the scroll-like middle conchae the *middle meatus* leads anteriorly to a drainage site, the infundibulum. The back part drains the maxillary sinus; the middle part drains the anterior ethmoid air cells and the frontonasal duct while the back part drains the middle ethmoid air cells. Sometimes the meatus posterior to middle conchae remains unclosed, providing wide-open access to the maxillary sinus. Finally, superior meatus separates the middle concha from the superior concha. The latter surely must have a "superiority complex" because it is about half the size of middle neighbor. Under cover of superior conchae, *superior meatus* extends backwards and provides draining for posterior ethmoid air cells.

The ethmoid complex resembles a man standing with his arms outstretched, each hand bearing an empty rectangular sinus box. So as to not tire out, the bottom of each box rests upon a larger maxillary box extending upward from the floor. Just lateral, an orbital box provides support; it shares a common wall with the ethmoid sinus box.

The best way to understand the functional anatomy of ethmoid complex is to break it down into four component fields. The medial fields consist of paired *perpendicular plates* projecting downward (caudally) and paired *cristae galli* projecting upward (cranially). The lateral fields develop into vertically oriented *labyrinths*. These two vertical blocks of bone are connected by thin transverse laminae, the *cribriform plates*. Initially the fields appear as a single anlage but, in reality, they are genetically distinct blocs. These are subdivided by neural crest tissue connecting the nasal placodes with the rhinencephalon. With growth, apoptosis of the fields along the mid-placodal plane results in development of potential spaces between the midline perpendicular plates and the lateral labyrinths, the *nasal choanae*. As this process occurs, a horizontal plate of bone develops around the olfactory nerve fibers. This is known as the cribriform plate. Failure of canalization results in *choanal atresia*.

Nasal epithelium contains neurons that can detect pain, temperature, and pressure and communicate with V1 to the trigeminal sensory nucleus. It also contains olfactory receptor neurons which need to gain access to the brain. Cribriform plate represents an adaptation by which these connect with mitral neurons of the olfactory bulb. Here, some 2000 glomeruli are specialized around specific smells. They may also have a binary code capable of detecting 16×10^6 smells. As such, the glomeruli are the primary level of olfactory processing.

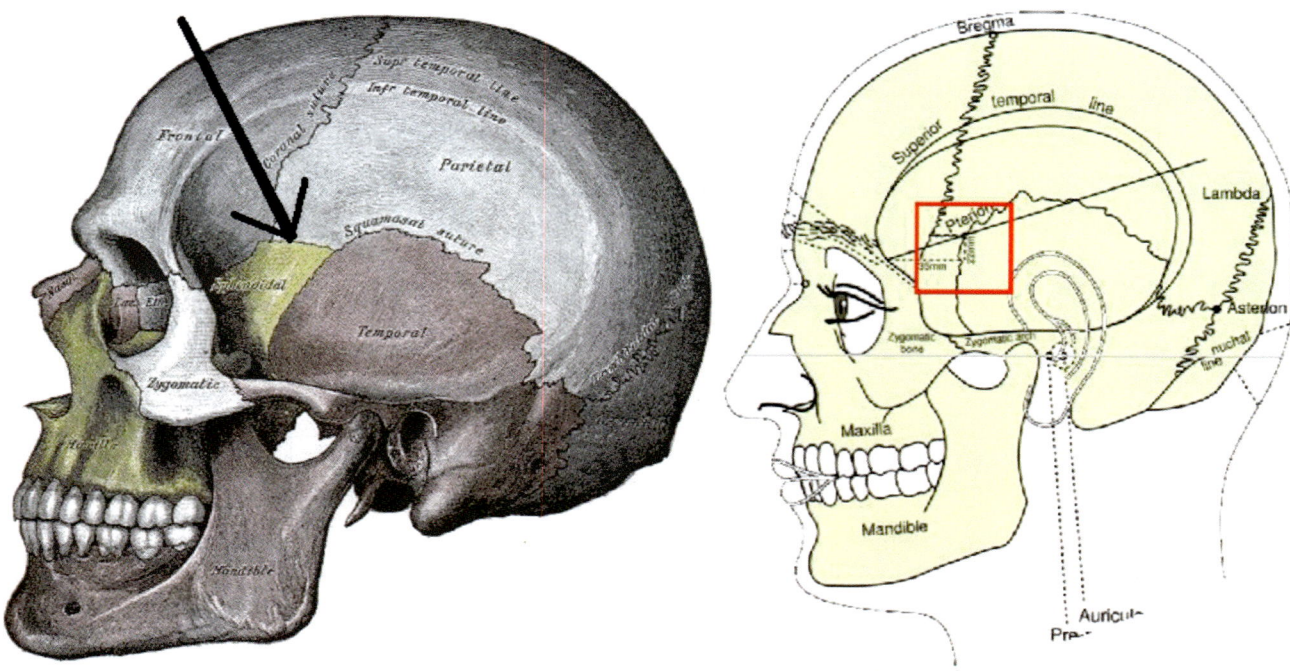

Fig. 8.33 Pterion represents the final position of alisphenoid behind the primate orbit. [Left: (Reprinted from Lewis, Warren H (ed). Gray's Anatomy of the Human Body, 20th American Edition. Philadelphia, PA: Lea & Febiger, 1918). Right: (Reprinted with permission from Wikimedia. Retrieved from: https://commons.wikimedia.org/wiki/File:Pterion.PNG)]

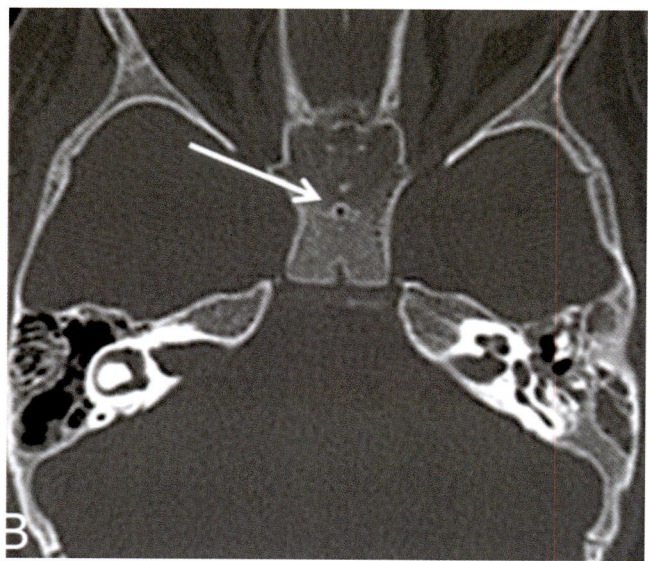

Fig. 8.34 Craniopharyngeal canal. (Reprinted from Abele TA, Salzman KL, Harnsberger HR, Glastonbury CM. Craniopharyngeal canal and its spectrum of pathology. *Am J Neuroradiol* April 2014, 35(4):772–777. With permission from American Society of Neuroradiology)

It should be noted that with development, the plane of the cribriform plate can be substantially lower than that of the orbits. This is marked within the orbit by the frontoethmoid suture. Upon entry through the suture for dissection of the superior portion of the ethmoid sinuses, one can inadvertently penetrate the dura.

Mammals have two types of turbinals, scroll-like projections from the lateral nasal wall. *Ethmoturbinals* (the superior and middle conchae) are associated with the cribriform plate and are covered with olfactory epithelium. They augment the surface area for contact between these neurons and respiratory air. *Maxilloturbinals* are separate bones from the ethmoid complex. They do not play a role in smell. Their function is to maintain the environment of the respiratory passages and will be discussed under inferior turbinate.

Blood Supply and Mesenchyme

The neurovascular anatomy of the ethmoid complex involves distinct systems supplying the nasal chamber and the intracranial cristae galli/cribriform plate. Recall that the p6 nasal placode related functionally to the rhinencephalon. Neurons that originate in the placode will penetrate the brain: those of the medial half connect to the *accessory olfactory cortex* while those of the lateral half are directed to the *primary olfactory cortex*. Meanwhile, all around the borders of the placode MNC mesenchyme is proliferating, causing the skin surrounding the placode to project forward. This causes the nasal placode apparently "sink in" to the ethmoid mesenchyme. It forms a backward-directed tunnel which terminates at the adenohypophysis, the eventual target of it gnRH neurons.

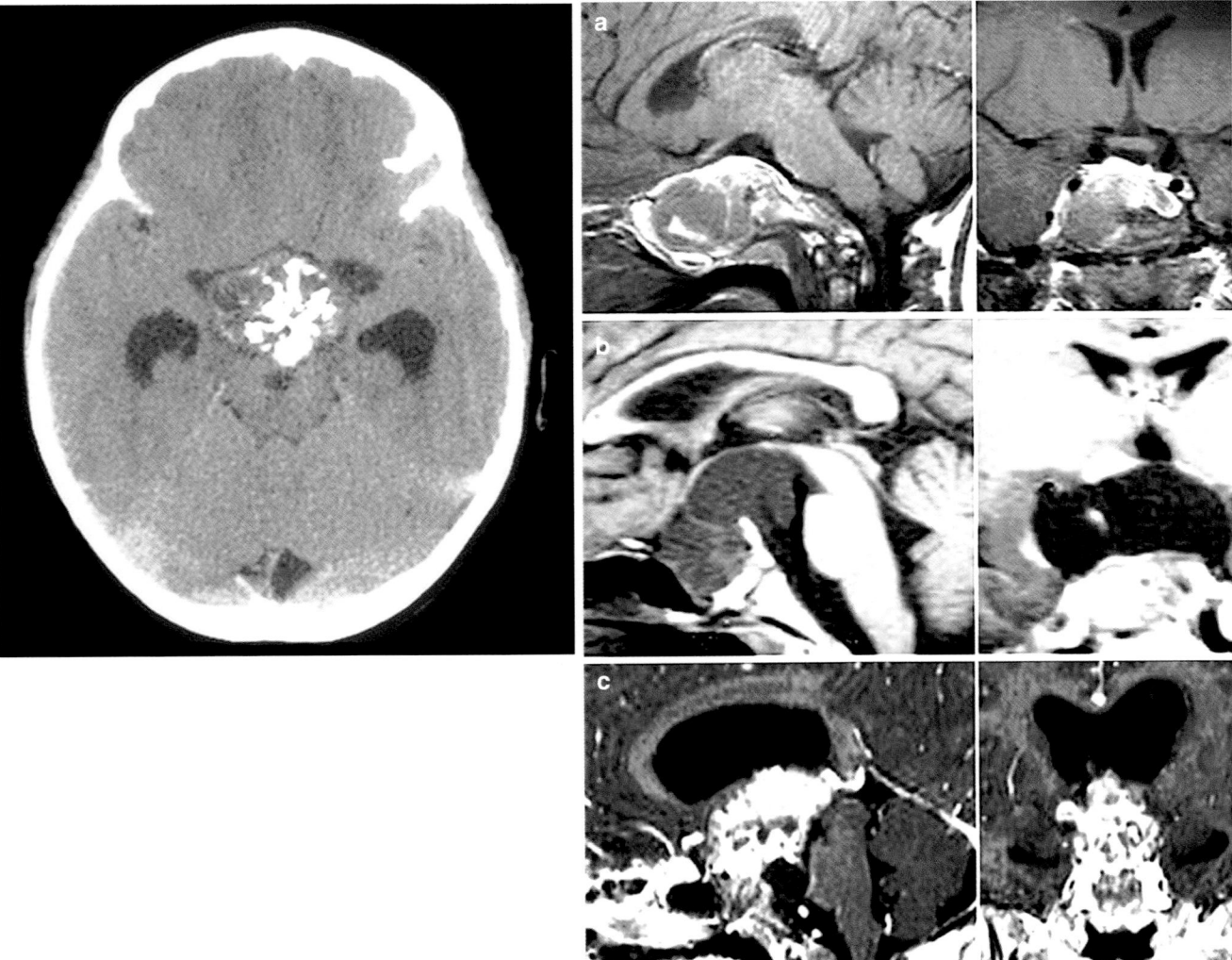

Fig. 8.35 Craniopharyngioma extending into the pharynx. Left: Unenhanced CT shows a calcified cystic structure in the supra sellar region; hydrocephalus is present. Right: Enhanced T1 weighted MRI scans. (**a**) Grade 1 olid sellar/supra sellar tumor—discrete from the hypothalamus (grade 0). (**b**) Grade 1 cystic sellar/suprasellar tumor that distorts the hypothalamus (grade 1). (**c**) Grade 2 tumor invasive of the hypothalamus (not seen). (Reprinted from Garnett MR, Puget S, Grill J, Sainte-Rose G. Craniopharyngioma. *Orphanet J Rare Dis* 2007; 2,18: https://doi.org/10.1186/1750-1172-2-18. With permission from Springer Nature)

Within the primitive nasal cavity, the upper epithelium is p6 PNC, the lower epithelium is r2, and the subepithelial tissue is MNC. Blood supply for the nasal cavity comes from two sources. The roof and sides are supplied by medial and lateral nasal branches from the StV1 anterior ethmoid artery. After giving off an important *meningeal branch*, the artery sends *descending branches* into the nose via a slit just lateral to crista galli. *Medial descending branches* extend downward from the roof to the vomerine border while *lateral descending branches* course over superior and middle turbinates. Paired distal branches emerge from under the nasal bone at its junction with the triangular (upper lateral) cartilage; these run forward to supply the columella and philtrum.

Recall from Chap. 4 that the separation between the primitive nasal cavity and the mouth, *primärer nasenböden*, is destined to disintegrate. At stage 17, the bottom falls out like a trap door from beneath each nasal choana. Beginning at stage 14, first pharyngeal arch has already invaded from lateral-to-medial. The ectoderm on both sides of the trapdoor is r2. When one inspects the nasal cavity, transition between neural crest epithelium and oral ectoderm, derm is marked by a color change. This readily appreciated at cleft palate surgery.

V2-innervated vestibular lining derives its epithelium from r2 endoderm and its dermis from r2 neural crest. It is supplied by the StV2 sphenopalatine axis. A *medial naso-palatine branch* supplies vomer and premaxilla. On the lat-

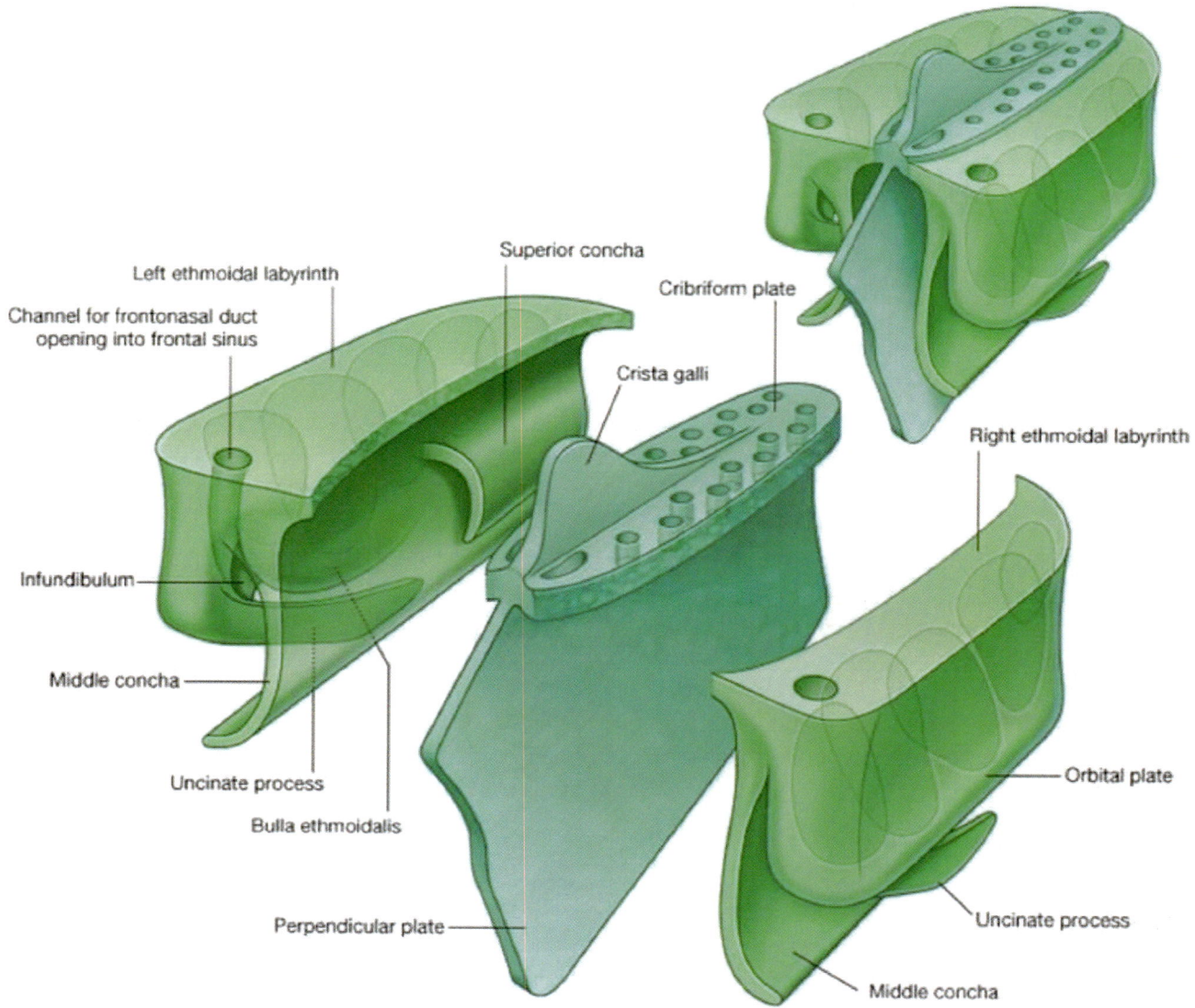

Fig. 8.36 Ethmoid bone complex. The ethmoid consists of paired sets of fields of r1 neural crest: (1) zone 13 medial fields consist of cristae galli and cribriform plate; (2) zone 12 lateral fields have labyrinth with nasal lamina and orbital lamina programmed separately. (Reprinted from Drake RL, Vogl A, Mitchell A. Head and Neck. In: Gray's Anatomy for Medical Students, 3rd edition. Philadelphia, PA: Churchill Livingstone; 2014: 875–959. With permission from Elsevier)

eral side, inferior turbinate and the nasal mucosa of the palatal shelf are supplied by the *lateral branch of nasopalatine*. Frontal process/canine zone of maxilla flanks the lateral entrance to piriform fossa; it is supplied by StV2 anterior superior alveolar artery—also known as infraorbital artery.

Within the nasal cavity, the lateral walls bear the scroll-like projections of the upper and middle turbinates. Genetic programming intrinsic to the epithelium is responsible for the boundaries of these fields. Indeed, throughout the pharyngeal arch system, oropharyngeal epithelium fulfills this function. The neuromeric identity of nasal and nasopharyngeal lining represents the interface between two sources of

neural crest: forebrain NC yields inferiorly to r2 at the level of inferior turbinate.

Sinuses represent potential separation planes between adjacent field. At times, the fields arise from distinct neuromeric sources. The petrous temporal bone, synthesized from Sm5 (prootic) and Sm6 (opisthotic), forms a sinus with the Sm7 mastoid. In the case of the ethmoid sinuses, the mesenchyme arises from a single source MNC but receives instructions from distinct epithelial sources. Upper nasal vestibular lining induces the (medial) nasal wall while the sclera (an extension of dura) induces the (lateral) orbital wall. The potential space between the two laminae is exploited by mucosa in continuity with the oral cavity.

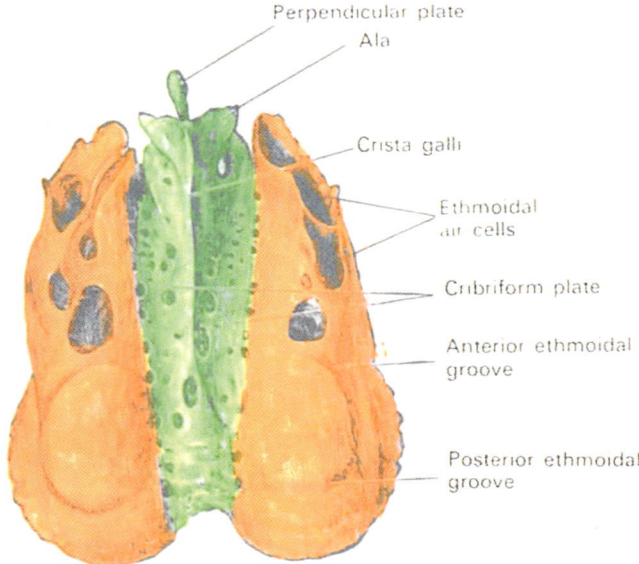

Fig. 8.37 Ethmoid dorsal view. Posterior and anterior ethmoid arteries send branches through the cribriform plate to supply r1 dura of anterior cranial base. Zone 13 is distinguished from zone 12 by parallel rows of perforations. (Reprinted from Lewis, Warren H (ed). Gray's Anatomy of the Human Body, 20th American Edition. Philadelphia, PA: Lea & Febiger, 1918)

Finally, it should be noted that the intracranial component of ethmoid has two sources of supply. Anterior and posterior ethmoids send up meningeal branches that supply the lateral zones. The primitive olfactory arteries constitute the midline supply to crista galli; they provide dedicated arteries to the olfactory nerve.

Development

In mammals, each ethmoid complex consists of two distinct zones. The <u>central zone</u> contains a perpendicular plate and horizontal tissue with multiple perforations for nerves originating from the olfactory placode. It has a single source of programming for p6 nasal epithelium. The <u>lateral zone</u> has two laminae. The nasal side lamina contains the superior and middle conchae and is programmed by p6 nasal epithelium. The orbital side lamina is smooth and is programmed by neuroepithelium from r1/p5.

The entire ethmoid complex originates from paired trabecular cartilages derived from r1 neural crest. Recall that mapping of these populations has not been done. It has four components: cribriform plate, paired perpendicular plates, and paired labyrinths. Several sources of epithelium provide programming for the ethmoid mesenchyme. These are r1/p5

Fig. 8.38 Perpendicular plate, right nasal side (zones 13). Perpendicular plate receives the ethmoid spine of sphenoid. It also articulates with the midline sphenoidal crest. The vaginal processes of sphenoid connect with vomers. (Reprinted from Lewis, Warren H (ed). Gray's Anatomy of the Human Body, 20th American Edition. Philadelphia, PA: Lea & Febiger, 1918)

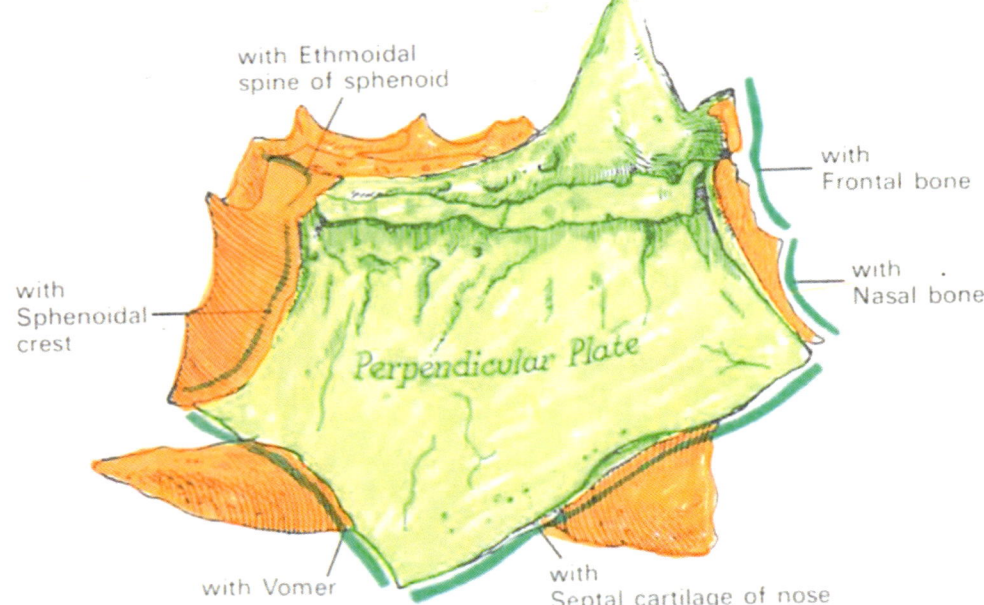

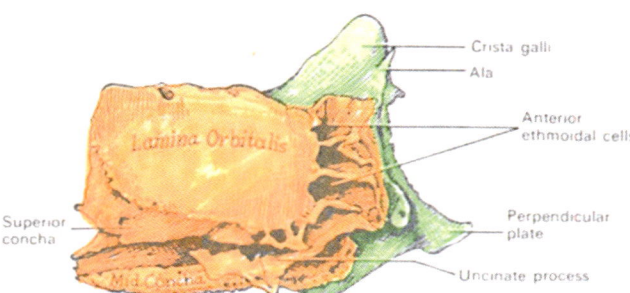

Fig. 8.39 Ethmoid labyrinth, right orbital side (zone). Middle and posterior ethmoidal cells are hidden by the lamina orbitalis. Removal of the lachrymal bone exposes the anterior cells. (Reprinted from Lewis, Warren H (ed). Gray's Anatomy of the Human Body, 20th American Edition. Philadelphia, PA: Lea & Febiger, 1918)

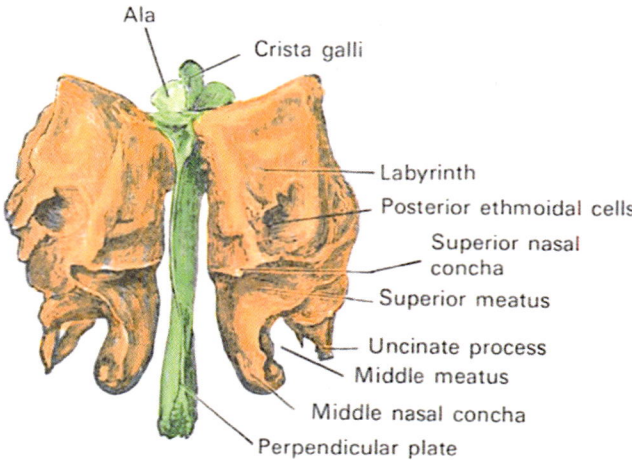

Fig. 8.40 Anterior view of the ethmoid complex. Note how uncinate process is suspended into space from the medial concha. (Reprinted from Lewis, Warren H (ed). Gray's Anatomy of the Human Body, 20th American Edition. Philadelphia, PA: Lea & Febiger, 1918)

sclera, p6 nasal vestibular lining, and r1 dura of the anterior cranial fossa. The lateral and medial laminae of the r1 labyrinth are in contact with r1 and p6 epithelia, respectively. This creates a *field separation plane* leading to sinus formation. The medial nasal walls contain the perpendicular plates and, later on, the septum sandwiched between paired p6 lining. Programming of the cribriform takes place from the p6 nasopharynx below and r1 dura above. When intervening MNC undergoes involution, fusion of the medial fields forms the four-part ethmoid complex.

Ossification of the future labyrinthine fields appears at 4–5 months. It follows a cranio-caudal and lateral-medial gradient from the lamina orbitalis downward into the conchae of the ethmoid sinuses. At about a year, the perpendicular plates and cristae galli originate from paired ossification centers. Because these are contiguous in the midline they

have been described as a single ossification center. Ossification of the perpendicular plate and cristae galli takes place within the first year; fusion with the labyrinth occurs during the second year. The ethmoid air cells appear during the first year of life; their development is complete by the third year.

Developmental analysis of the nasal capsule demonstrates that the r2 components (maxillary turbinates) mature one stage prior to the ethmoturbinates. This reflects, once again, that the StV2 arteries from pterygopalatine fossa develop prior to those descending from orbit.

Phylogeny

This bone is the anterior part of the sphenethoid complex which is laid down in cartilage in all vertebrate skulls. There is considerable evolutionary variation. In non-mammals, the ethmoid region may not ossify. Some osteoglossomorph teleost fishes (having bone-bearing tongues which they bite against) do not have an ethmoid cartilage at all. Eels present a single cartilaginous fusion of ethmoids to the vomers.

It is difficult to pin down the phylogeny of the turbinals because they are only ossified in birds (diapsids) and mammals (synapsids). Extinct species that might give us a clue had turbinals made from cartilage or very thin laminae and hence are not preserved. Turbinals appear to arise with amniotes. A lateral nasal capsule with two conchae is found in the New Zealand tuatara lizard *Sphenodon* is known as a living fossil as it constitutes the split of the amniotes into the diapsid line leading to dinosaurs, crocodiles and birds. These have three conchae and some mammals can have up to six. This variation in number likely reflects the reiteration of agenetic program rather than representing individual structures with separate evolutionary origins.

The *maxilloturbinal* (inferior turbinate) bone appears in the lateral nasal wall at stage 17. It is distinct from the ethmoturbinals and will be discussed separately.

The *ethmo turbinals* appear at stage 18 in the medial and superior aspect of the nasal capsule. They are covered with olfactory epithelium and likely function to expand the surface area available for olfaction. Their dorsal position ensured a neurologic relationship with V1 and with the arterial supply from StV1 ophthalmic.

During next three stages, growth of the nasal capsule surpasses that of inferior turbinate. Hence, IT tends to take up less space in the lateral nasal capsule. This provides an opportunity for the ethmoturbinals to expand. At stage 21, the ETs shift their position laterally, separating themselves from the septum. By stage 22, they assume their final position in the superolateral nasal wall. At stage 23, middle meatus appear, providing a well-defined separation between the two groups of turbinals.

Fig. 8.41 Ethmoid sinus air cells. The ethmoidal air cells are medial to the removed lacrimal bone, and may extend anteriorly to pneumatize the maxillary bone of the lacrimal sac fossa...*because it is bilaminar*, consisting of frontal process of maxilla and frontal process of premaxilla. (Reprinted from Burkat CN, Lucarelli MJ. Anatomy of the lacrimal system. In: Cohen AJ, Mercadetti M, Brazzo BG (eds). The Lacrimal System. New York, NY: Springer; 2006: 3–19. With permission from Springer Nature) .

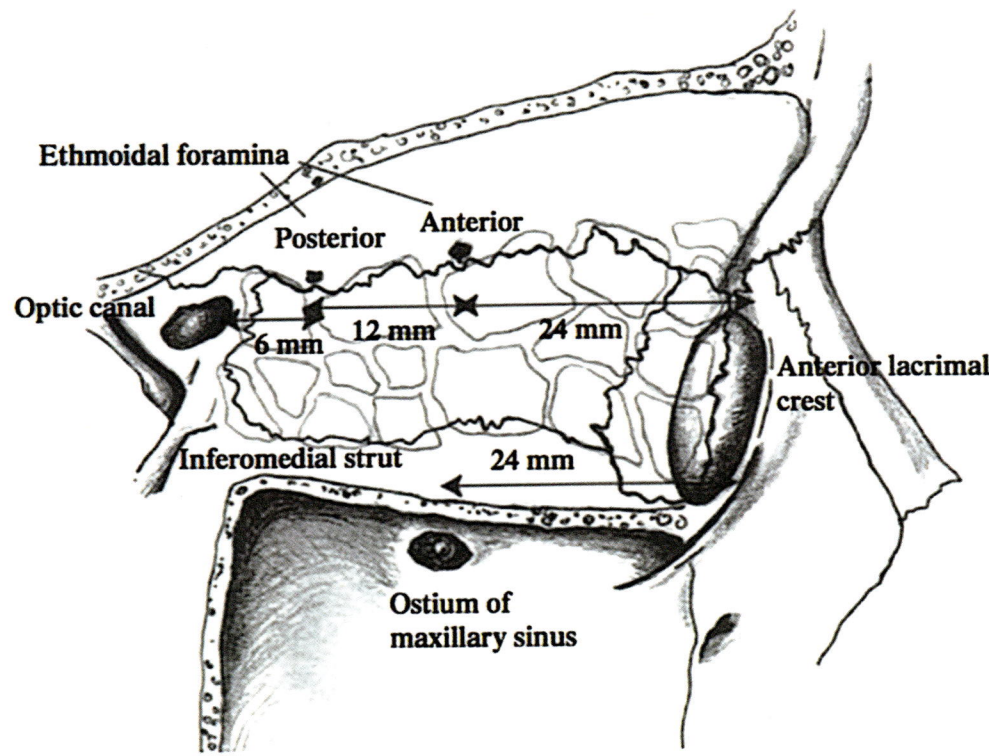

NOSE, NASAL CAVITY AND PARANASAL SINUSES

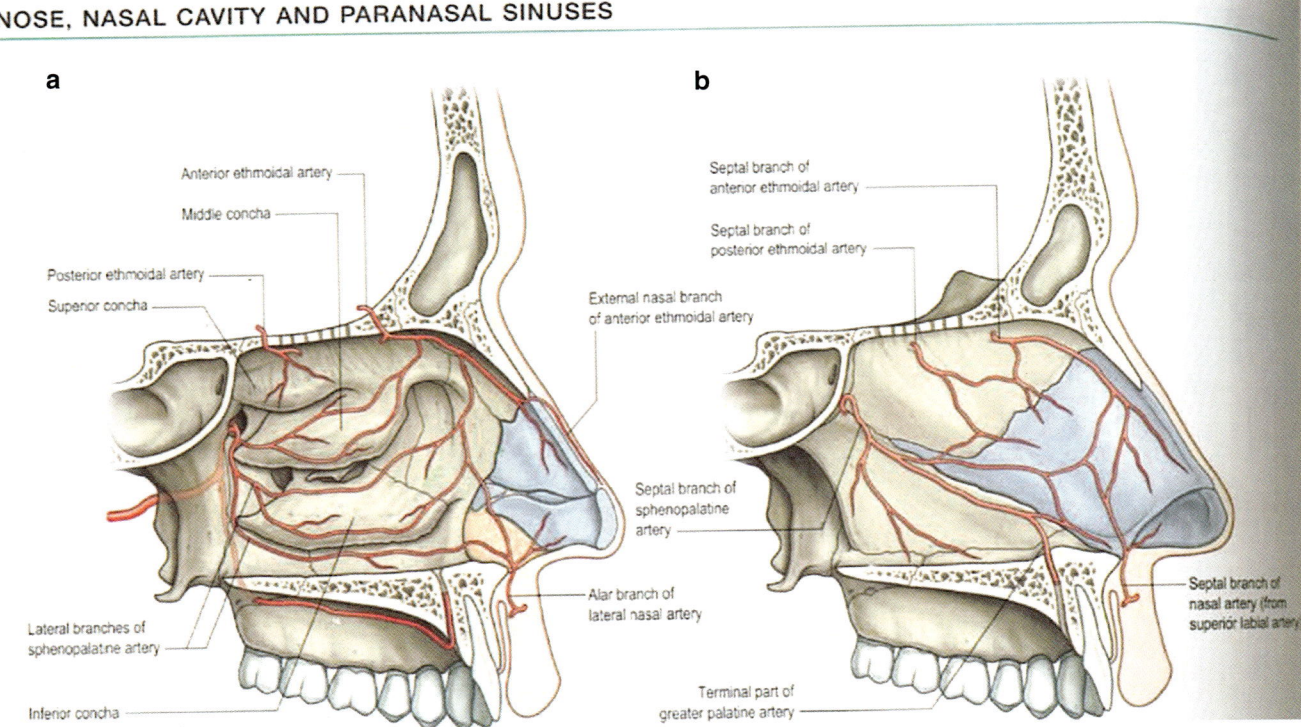

Fig. 8.42 Blood supply to turbinates. Medial and lateral nasal walls are supplied by angiosomes from StV1 and StV2. Blood supply over middle turbinate is redundant from both anterior and posterior ethmoid (StV2). Medial nasopalatine (StV1) is exclusive supply to inferior turbinate. (Reprinted from Drake RL, Vogl A, Mitchell A. Head and Neck. In: Gray's Anatomy for Medical Students, 3rd edition. Philadelphia, PA: Churchill Livingstone; 2014: 875–959. With permission from Elsevier)

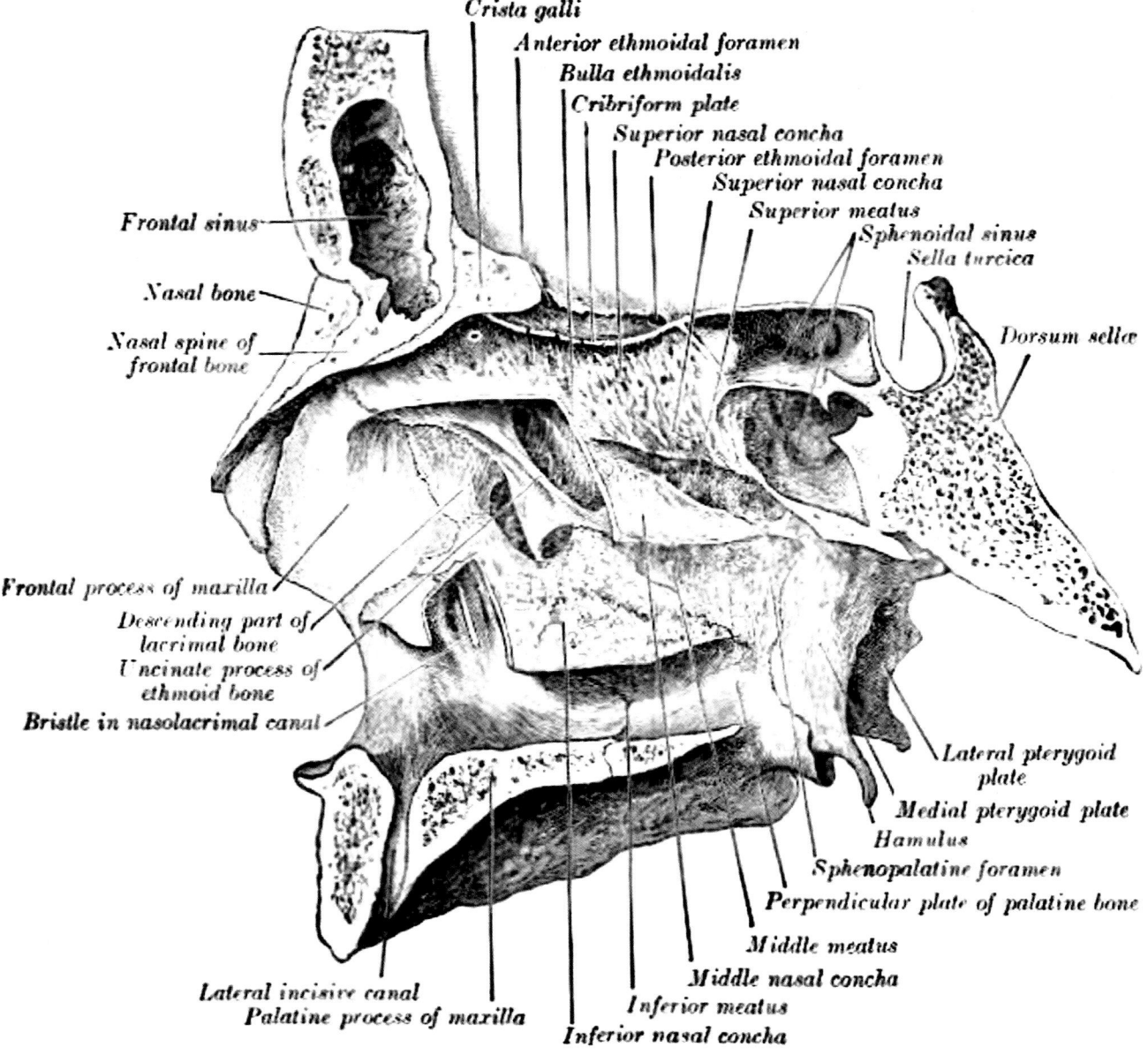

Crista galli
Anterior ethmoidal foramen
Bulla ethmoidalis
Cribriform plate
Superior nasal concha
Posterior ethmoidal foramen
Superior nasal concha
Superior meatus
Sphenoidal sinus
Sella turcica

Frontal sinus

Dorsum sellæ

Nasal bone

Nasal spine of frontal bone

Frontal process of maxilla

Descending part of lacrimal bone

Uncinate process of ethmoid bone

Bristle in nasolacrimal canal

Lateral pterygoid plate

Medial pterygoid plate

Hamulus

Sphenopalatine foramen

Perpendicular plate of palatine bone

Middle meatus

Middle nasal concha

Lateral incisive canal
Palatine process of maxilla

Inferior meatus

Inferior nasal concha

Fig. 8.43 Uncinate process. The mucosa has been stripped off from the conchae. Articulations of uncinate and lacrimal with inferior turbinate seen. (Reprinted from Lewis, Warren H (ed). Gray's Anatomy of the Human Body, 20th American Edition. Philadelphia, PA: Lea & Febiger, 1918)

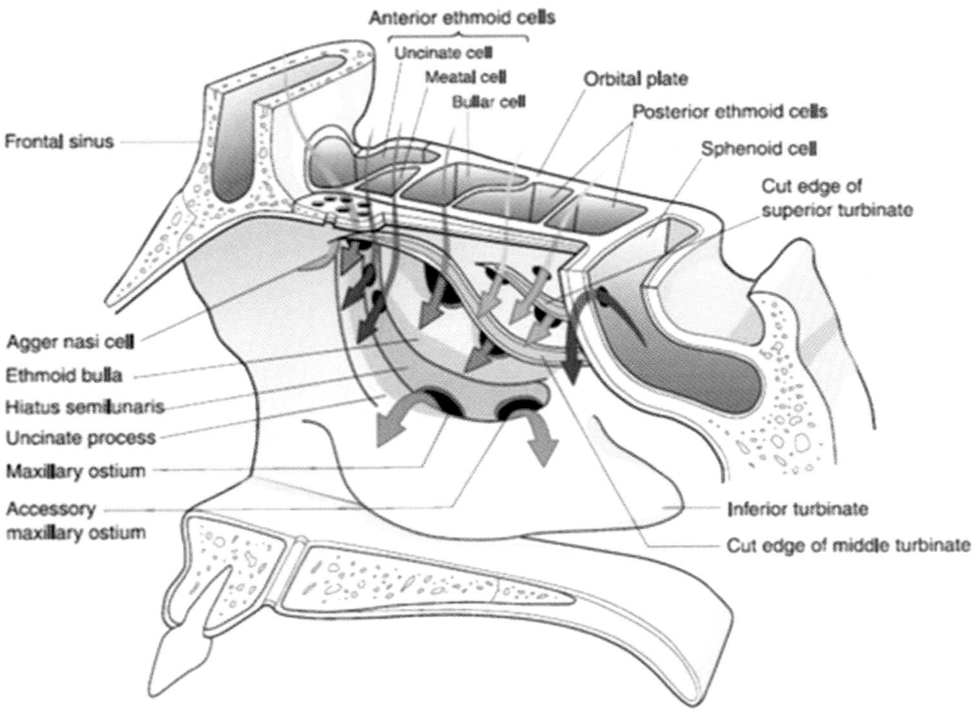

Fig. 8.44 Upper turbinal ventilation system. *Uncinate process* (Hook like bony extension of medial wall). *Hiatus semilunaris* (crescent passage between uncinate process and ethmoid bulla through which middle meatus communicates with ethmoidal infundibulum). Frontal recess (drainage channel of frontal sinus). *Bulla ethmoidalis* (most constant and largest anterior ethmoid air cell that projects inferomedially over hiatus semilunaris). *Ethmoidal infundibulum* (funnel-shaped passage through which anterior ethmoid cells and maxillary sinus drains into middle meatus). *Maxillary sinus ostium* (drainage channel of maxillary sinus). (Reprinted from Hou D, Fang L, Zhao Z, Zhou C, Yang M. Angular vessels as a new vascular pedicle of an island nasal chondromucosal flap: Anatomical study and clinical application. Experimental and Therapeutic Medicine 2013; 5:751–756. With permission from Spandidos Publications)

Clinical Correlations: Craniofacial Clefts

The ethmoid is the site of several craniofacial pathologies. Its fields comprise zones 13 and 12 in the cleft classification system of Tessier.

Normal development involves a centralization of the orbits. This requires two possible mechanisms in zone 13. *Apoptosis* refers to programmed cell death within the zone. *Differential growth* favors lateral fields at the expanse of #13. The former seems more logical. When these mechanisms fail, hypertrophy of zone 12 and/or 13 takes place leading to hypertelorism and frontonasal dysplasia, and outright midline facial clefts can result.

Canalization of the potential space between the central fields and the lateral fields takes place, respecting the olfactory systems. Nasal placode dysfunction can occur with normal choanal development. Medial placode failure leads to *anosomia*, whereas lateral placode failure produces *hypogonadic hypogonadism*. These features are known as *Kallman's syndrome*. In the worst-case scenario, failure of normal canalization to occur leads to choanal atresia in which the airway is totally blocked.

On the other hand, selective pathologic deficits seen in holoprosencephaly are characterized by a reduction or outright absence of the ethmoid complex. The vascular basis of this is explored in Chap. 7. Such deficiency can be unilateral or bilateral. The degree of severity follows a *medial-to-lateral gradient*. Crista galli can be narrowed or absent followed by narrowing of the cribriform plate. In more advanced states, the ethmoid sinus is compromised. In such cases, hypotelorism ipsilateral to the labyrinthine deficit is always present.

In addition, selective deficits of Tessier zones 10–13 can also exist. In such cases, normal surrounding fields will be affected in one of two ways: (1) if the cleft zone is deficiency but intact, surrounding normal fields will collapse inward toward the defect; (2) if the defect involves full thickness communication with the cranium, the resulting encephalocele will enter the defect, causing otherwise normal sur-

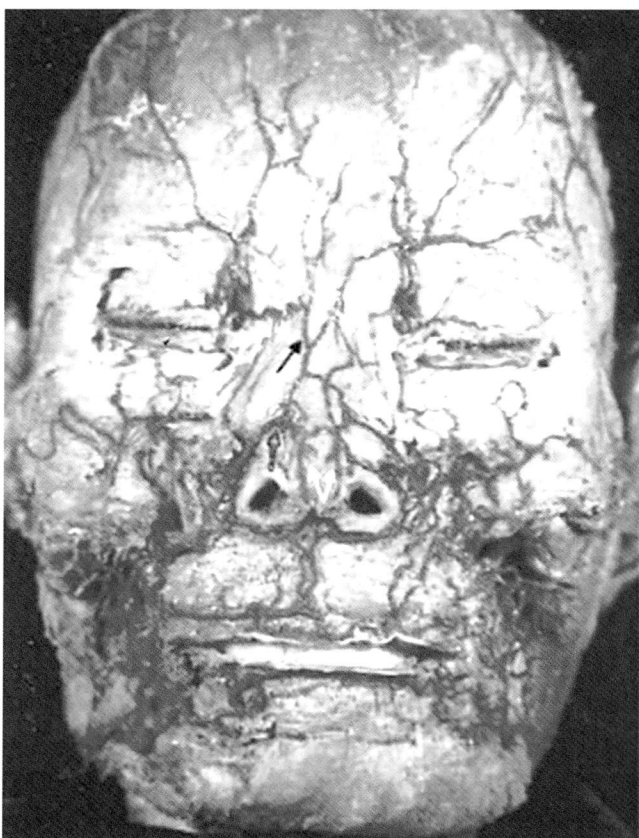

Fig. 8.45 External versus internal nasal blood supply. Blood supply to philtrum uses penetrating arteries that course downward inside columella from deep dorsal nasals not appreciated on this injection which shows the external blood supply with anastomosis between StV1 and StV2 over the alae. (Reprinted from Hou D, Fang L, Zhao Z, Zhou C, Yang M. Angular vessels as a new vascular pedicle of an island nasal chondromucosal flap: Anatomical study and clinical application. Experimental and Therapeutic Medicine 2013; 5:751–756. With permission from Spandidos Publications)

rounding fields to be pushed apart. In such cases, dystopia is often present.

Nasal Bone

Although nasal bone is not truly a part of the neurocranium, due to the commonalities of its blood supply and innervation with the ethmoid complex, we shall discuss it here (Figs. 8.49, 8.50, 8.51, and 8.52).

Descriptive Anatomy

The nasal bone develops in the membrane from r1 neural crest. It is programmed by the lining of the nasal vestibule, the mucosa of which is p6 and the submucosa is MNC. The bone is supplied by the medial and lateral branches of nasal artery from the anterior ethmoid stem.

Paired nasal bones articulate with paired frontal bones. They are sandwiched between a pair of bilaminar pillars. On either side, the external lamina is formed by the frontal process of the maxilla while internal lamina is formed by the frontal process of the premaxilla. The internal nasal surface is grooved by *external nasal branch of anterior ethmoid nerve* (V1). The external surface is convex. Immediately above it, second arch superficial musculoaponeurotic system (SMAS) gives rise to procerus and nasalis. Note that SMAS is *not* attached to either nasal bone or overlying skin. For example, procerus ascends to insert into a zone of dermis just cranial to the nasofrontal junction, while nasalis inserts above and lateral to the incisive and canine fossae of the maxilla.

The superior border of the nasal bone inserts into a notch in frontal bone. The inferior border gives an important attachment to upper lateral cartilage. The saw-tooth lateral border fits into the frontal processes (vide supra). The medial border articulates with nasal septum, perpendicular plate, and the beak-like nasal processes of the frontal bone.

The cutaneous envelope of the nose consists of two distinct zones. (1) The internal vestibular lining is made up of p6 epidermis with medial dermis p4 dermis and lateral dermis p3. (2) The external nasal skin is made up of p5 epidermis and p4 and p3 dermis. Blood supply to the nose (and forehead) begins with the terminal branches of StV1 nasociliary (the most medial stapedial branch of ophthalmic). This arterial axis is induced by nasociliary nerve. As we have seen with ethmoid, the nasociliary axis is distributed to zones 13 medial and 12 lateral. Anterior ethmoid supplies intracranial and intranasal structures (via anterior ethmoid), while supratrochlear supplies extranasal structures.

After giving off posterior ethmoidal, StV1 divides into nasociliary (zones 12–13) and supraorbital (zones 10–11). Nasociliary has two terminal branches.: (1) anterior ethmoid enters the ethmoid labyrinth where it sends a recurrent branch to anastomose with posterior ethmoidal. (2) Posterior ethmoid artery also enters the posterior ethmoid labyrinth and analogous branches. AEA supplies the ethmoid air cells and ethmoid cribriform plate/crista galli. Within the cranium anterior, AEA gives off a *meningeal branch*. This makes an

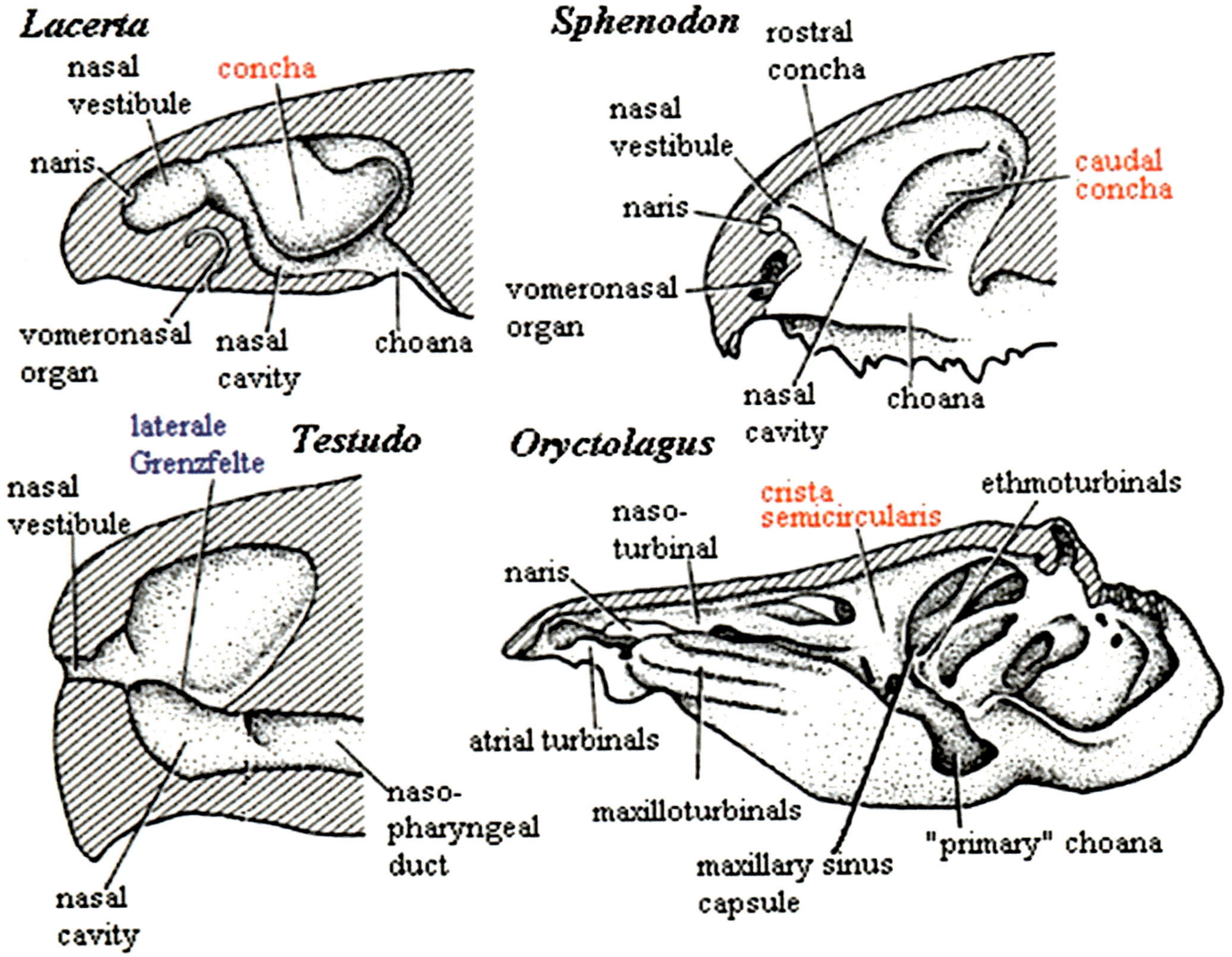

Lacerta

nasal vestibule

concha

naris

vomeronasal organ

nasal cavity

choana

Sphenodon rostral concha

nasal vestibule

naris

vomeronasal organ

caudal concha

nasal cavity

choana

Testudo

laterale Grenzfelte

nasal vestibule

naso-pharyngeal duct

nasal cavity

Oryctolagus

naso-turbinal

crista semicircularis

ethmoturbinals

naris

atrial turbinals

maxilloturbinals

maxillary sinus capsule

"primary" choana

Fig. 8.46 Phylogeny of turbinals. Note relationship between ethmo-turbinals and maxilloturbinal well defined in mammals. (Reprinted from Wittmer LM. Homology of facial structures in extant archosaurs (birds and crocodilians), with special reference to paranasal pneumaticity and nasal conchae. J. Morphol. 1995;225:269–327. With permission from John Wiley & Sons)

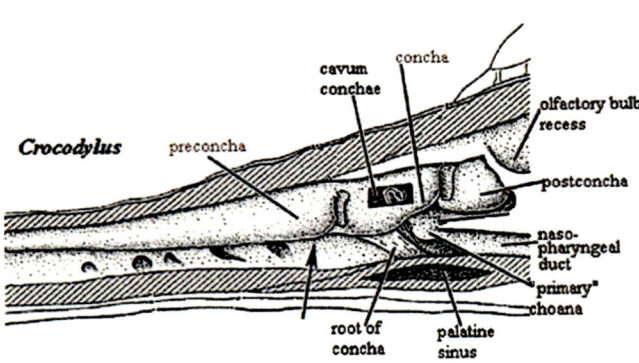

Crocodylus preconcha

cavum conchae

concha

olfactory bulb recess

postconcha

naso-pharyngeal duct

"primary" choana

root of concha

palatine sinus

Fig. 8.47 Redundancy of turbinals. Note: position of olfactory bulb with respect to ethmoturbinals, presence of a sinus in the hard palate, nasopharyngeal duct in the parasphenoid. These can be redundant. (Reprinted from Wittmer LM. Homology of facial structures in extant archosaurs (birds and crocodilians), with special reference to paranasal pneumaticity and nasal conchae. J. Morphol. 1995;225:269–327. With permission from John Wiley & Sons)

anastomosis with the *meningeal branch of cavernous portion of internal carotid*. Ascending along the inner surface of frontal bone, it gives blood supply to the frontal sinus. Just lateral to crista galli, the nasal branch of AEA passes through a slit and descends along the inner surface of nasal bone. It then divides into medial and lateral nasal branches. *Median nasal branch* supplies Tessier zone 13. It courses over the septum and ultimately produces *terminal branch* of AEA to the columella and philtrum. *Lateral nasal branch* supplies Tessier zone 12. It irrigates the upper and middle turbinates, inferior turbinate being supplied by lateral sphenopalatine of the StV2 system.

In sum, the developmental components of the nose are as follows:

1. Vestibular lining: zones 13 and 12 have p6 epithelium and MNC submucosa.

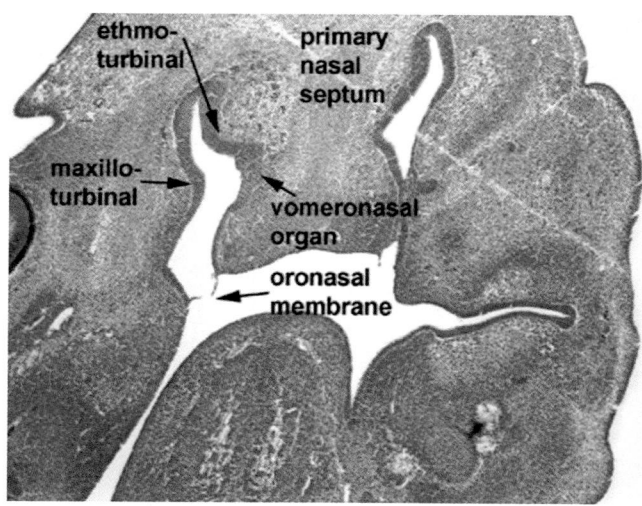

Fig. 8.48 Stage 18 Ethmoturbinal development. Ethmoturbinals appear for the first time in the vicinity of the septum. Large maxilloturbinals (which appear at stage 17) do not leave room for ethmoturbinals at first. (Reprinted from Kim C-H, Park HW, Kim K, Yoon J-H. Early development of the nose in human embryos: a stereoscopic and histologic analysis. *Laryngoscope* 2004; 114:1791–1800. With permission from John Wiley & Sons)

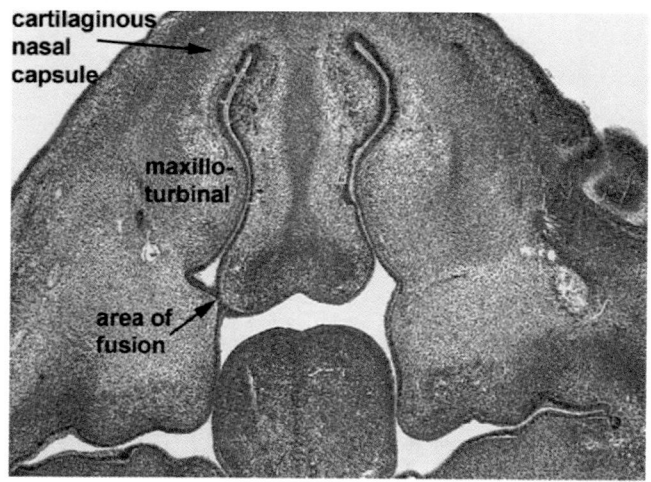

Fig. 8.49 Stage 20 ethmoturbinal development. Chondrification of nasal capsule. Ethmoid turbinals still medial. (Reprinted from Kim C-H, Park HW, Kim K, Yoon J-H. Early development of the nose in human embryos: a stereoscopic and histologic analysis. *Laryngoscope* 2004; 114:1791–1800. With permission from John Wiley & Sons)

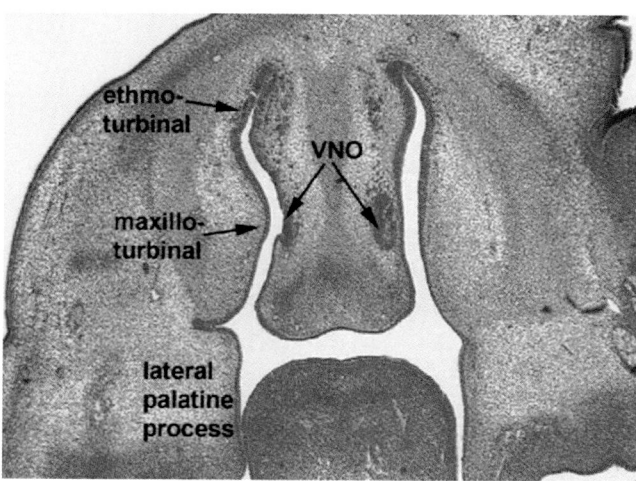

Fig. 8.50 Stage 21 ethmoturbinal development. Note lateral palatine processes on side of tongue, not yet elevated. (Reprinted from Kim C-H, Park HW, Kim K, Yoon J-H. Early development of the nose in human embryos: a stereoscopic and histologic analysis. *Laryngoscope* 2004; 114:1791–1800. With permission from John Wiley & Sons)

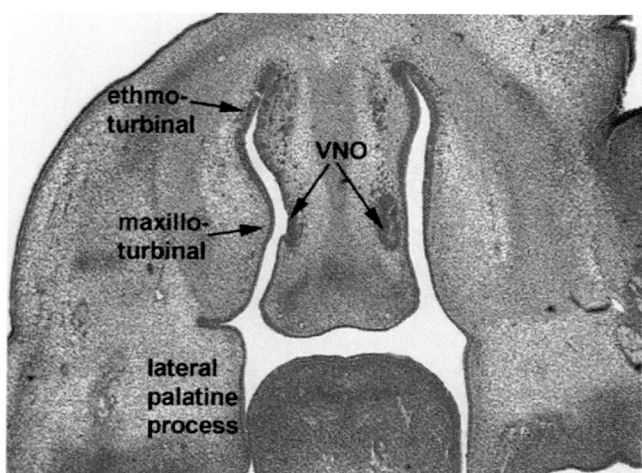

Fig. 8.51 Stage 22 ethmoturbinal development. Growth of the nasal capsule is dorsal, and forward, giving space for the ethmoturbinals. They shift away from septum into the lateral capsule. (Reprinted from Kim C-H, Park HW, Kim K, Yoon J-H. Early development of the nose in human embryos: a stereoscopic and histologic analysis. *Laryngoscope* 2004; 114:1791–1800. With permission from John Wiley & Sons)

2. Nasal skin (extends from the nasal tip into the nostril up to the external nasal valve): both zones 13 and 12 have p5 epidermis. Dermis could theoretically be p3. Note nasal skin does *not* program nasal bone; it *does* program lower lateral cartilage.
3. Blood supply to nasal bone and vestibular lining: naso-ciliary (StV1) > *anterior ethmoid* = medial nasal branch (zone 13) + lateral nasal branch (zone 12).

4. Blood supply to nasal skin: *supratrochlear* = (frontal branch) to zones 13 and 12 of forehead skin + *infratrochlear* (dorsal nasal branch) to nasal dorsum.

This r1 neural crest bone forms in direct contact with p6 vestibular lining. It is not programmed by p5 nasal skin. *Were this the case, the nasal bone would have two sources of programming, would be bilaminar, and no potential space would exist for SMAS.* Nasal bone ossifies from a single center at the eighth to ninth week. Although its officiation is directly membranous, nasal bone is in continuity with eth-

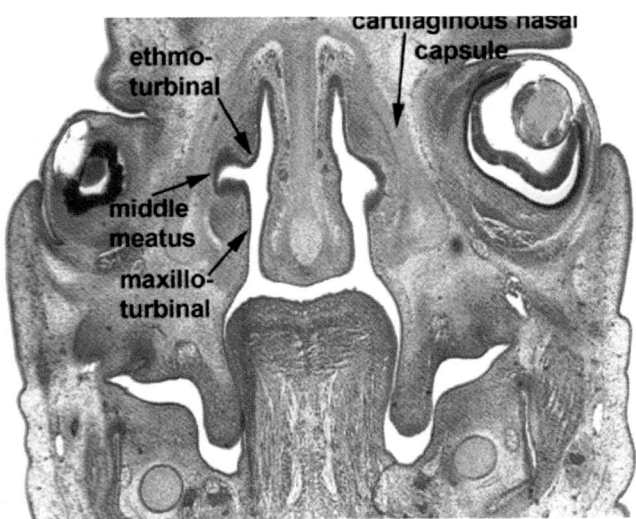

Fig. 8.52 Stage 23 ethmoturbinal development shows development of middle meatus. (Reprinted from Kim C-H, Park HW, Kim K, Yoon J-H. Early development of the nose in human embryos: a stereoscopic and histologic analysis. *Laryngoscope* 2004; 114:1791–1800. With permission from John Wiley & Sons)

moid capsule, a neural crest derivative formed via a cartilaginous intermediate.

Development

The mesenchymal source of the nasal bones is MNC and they are programmed by the p6 nasal vestibular lining. Ossification takes place in membrane at the eighth week. The bones lengthen over time but their width remains essentially the same. This correlates with the fixed dimensions established between the orbits at birth.

Phylogeny

Nasal bones evolution parallels that of the frontal bone. In primitive sarcopterygians of the early Devonian, such as the osteolepiforms (having thick and enlarged scales at the base of their fins), the snout was a haphazard series of multiple small bones. In the late Devonian, advanced rhipidistians such as tristichopterid, *Eusthenopteron*, have a linear array of two nasal bones (anterior and posterior) and a frontal bone on either side of large *median rostral bone* that occupies the midline in front of parietal. Lacrimal is separated from the nasals by prefrontal. In the terminal clade of elpistostegalians just before the tetrapod, median rostral is lost and centralization occurs. *Pandericthys* shows a clear-cut series of midline bones extending back to the occiput. Intervening bones persist between the premaxillae and the nasals. The original pairing of nasal bones has shifted from an anterior–posterior relationship to medial lateral. The two nasal bones are paired side-to-side. This may be reflected in the medial and lateral cleft zones, 13 and 12. Thus, although each nasal bone appears as a unitary structure, it may have two distinct fields.

A Digression on the Phylogeny of the Choanae

Early bony fishes (~420 Mya) had two pair of nostrils, one pair for incoming water (known as the anterior or incurrent nostrils) and a second pair for outgoing water (the posterior or excurrent nostrils), with the olfactory apparatus (for sense of smell) in between. In the first tetrapodomorphs, porolepiforms (~415 Mya), the excurrent nostrils migrated to the edge of the mouth, occupying a position between the maxillary and premaxillary bones, directly below the lateral rostral (a bone that vanished in early tetrapods) ([4], p. 74).

In all but the most basal (primitive) tetrapodomorphs, the excurrent nostrils have migrated from the edge of the mouth to the interior of the mouth. In tetrapods that lack a secondary palate (basal tetrapods and amphibians), the choanae are located forward in the roof of the mouth, just inside the upper jaw. These internal nasal passages evolved while the vertebrates still lived in water [5]. In animals with complete secondary palates (mammals, crocodilians, most skinks), the space between the primary and secondary palates contains the nasal passages, with the choanae located above the posterior end of the secondary palate.

Clinical Correlations

Craniofacial Cleft Zones

Zone 13 = perpendicular plate and medial ½ of nasal roof (corresponding to crista galli). Zone 12 = lateral nasal wall/medial wall of labyrinth + frontal process of premaxilla. Zone 11 = medial wall of orbit/lateral wall of labyrinth + lacrimal + frontal process of maxilla

Tessier zone 13 pathologies can have dual manifestation: extracranial (intranasal, extranasal) and intracranial (ethmoid and frontal). The medial zone of nasal bone and/or septum can be affected. Recall that upper lateral cartilage is programmed by p6 vestibular epithelium, whereas lower lateral cartilage is programmed by p5 nasal epithelium. A defect involving the medial zone can therefore create a notch in the domes. Isolated absence of columella has been described. In the latter case, prolabium can remain unaffected because it receives collateral blood supply through premaxilla from medial sphenopalatine, through lingual mucoperiosteum from greater palatine and through buccolabial mucoperiosteum from posterior superior alveolar. Intracranial defects can involve deficiency states permitting escape of encephaloceles. Field excess of the crista galli and/or frontonasal skin in the supratrochlear distribution can be seen.

Tessier zone 12 pathologies have a similar scope of distribution. Deficiency of the lateral zone of nasal bone can create a notch "cleft" to appear where the nasal bone normally articulates with the piriform rim (PMxF and MxF) and involves the posterolateral nasal wall. Deficient ethmoid mesenchyme with foreshortening of the upper and middle turbinates will distort the position of nasal bone—itself an r1 structure.

The association between zones 13-1 and 12-2 involves spatial similarities between StV1 anterior ethmoid and StV2 medial sphenopalatine. The central axis of AEA supplies proximal structures and then, after giving off lateral nasal, continues in the midline down septum. These branches anastomose at the septo-premaxillary junction with medial naso-palatine (AKA nasopalatine).

Septum, Columella, and Prolabium

Although the skin of the columella–philtrum is in continuity with dorsal nasal skin, its blood supply is embryologically distinct. The columella and the philtrum are perfused by the *terminal branches* of the *medial nasal arteries*. The anterior ethmoidal stem produces both meningeal and nasal branches. These latter penetrate the cribriform place at a performation next to crista galli. The nasal artery stem descends along the inner surface of nasal bone, leaving a groove (a paleontologic "footprint" of the original neurovascular unit). It then produces medial and lateral nasal branches. It then forced to emerge between nasal bone and upper lateral cartilage. This is because the cartilage is attached to its epithelial "program," the underlying p6 vestibular lining. There is no place for the artery; it must be displaced above and medial. It continues downward beneath p6 mucosa at the upper border of septum, supplying vertically directed branches to septal mucosa. The arteries emerge in the columella and travel downward in front of medial crura. In the philtrum, they are located about 1–2 mm medial to the philtral column, i.e., *at a distance equal to the width of the columella.* Upon reaching the white roll, the philtral branches of medial nasal anastomose with the labial branches of the facial. This takes place between the intersection between the caudal border of superficial orbicularis (SOO) and the terminus of the vertical portion of deep orbicularis oris (DOO). The program for DOO is oral mucosa. It therefore descends to the vermillion border and then curls upward, much like the letter "J." *The white roll is produced exactly at the interface of DOO with SOO.*

The r1 septal cartilage anlage descends and grows forward in the nasal midline sandwiched between the two zones of p6 vestibular lining in zone 13. It receives programming instructions from both sides. In situations where r1 mesenchymal excess is present, the septum will be wider, even taking on a dorsal groove giving a semblance of bifidity. Under normal circumstances, septal development accompanies that of premaxilla and vomer. Recall that the pathway by which r2 mesenchyme of PMx and V descend into the midline is entirely dependent upon proper development of the r1 perpendicular ethmoid plates.

The developmental sequence of the columella-philtrum is as follows. (1) Anterior to sphenoid, the trabecular cartilages produce paired, vertically directed laminae. These descend along the anterior surface of sphenoid, articulating with sphenoidal crest. They subsequently fuse as the perpendicular plate. (2) From the inferior surface of sphenoid, paired rostral processes of the sphenoid provide articulation for the paired streams of the medial sphenopalatine mesenchyme, the contents of which are premaxilla and vomer. PMx is the more primitive. Its descent precedes that of vomer in space and time. The forward position of PMx is determined by the growth of septum and vomer. *A fully-formed premaxilla can exist in the presence of a vertically reduced vomer.* (3) As PMx-V develops, it carries forward with it an island of dedicated skin. This r2 skin is in absolute continuity with that of p5 nasal skin. It marks the cutaneous junction of V1 and V2. Thus, in the normal prolabium with an intact lip, the "Cupid's Bow" is the p5 skin with r2 mesenchyme beneath it. The floor of the nose, as its introitus, has an r2 "shoulder" of the columella articulating just internal to the nostril sill.

Bear in mind that the p6 skin is actually broken up into two vertical zones by the genetic coding of the dermis. Zone 13 = p6 epithelium + p4 dermis. Zone 12 = p6 epidermis + p3 dermis. Under influence of the nasal placode, the epithelium of these zones in the vestibule is mucosa, while outside the vestibule, it remains skin.

The topology of the columella–philtrum is quite straightforward. The entire complex originates at the sphenopalatine (pterygopalatine) fossa. At this point, r2 mesenchyme sweeps in posteriorly from the hindbrain, bearing with it separate complexes of developmental fields, each of which is organized around its own StV2 neurovascular axis. First to arrive is medial sphenopalatine mesenchyme, i.e., premaxilla-vomer. It bears its own V2-innervated skin. It will come into direct contact with prosencephalic skin immediately lateral to the adenohypophyseal placodes (Rathke's pouch).

SEMs of the embryonic face demonstrate this vividly (Hinrichsen). Looking from below, the roof of the mouth displays two columns of mesenchyme running from Rathke's pouch forward to the nasal placodes; these diverge almost 90°. Differential growth of zone 12 is much great than that of zone 13; the latter shrinks by comparison. In reality, the mechanism is physiologic apoptosis. If we look at the embryonic face straight on, the nasal placodes approximate each other, the skin between the placode disappears, and the medial walls of each nasal chamber fuse. This is a process of programmed ethmoid apoptosis.

Again, from below, the primitive anlage of the trabecular complex, vomer and premaxilla appear as distinct ridges leading backward to Rathke's pouch. Initially they appear as V-shaped but soon become parallel ridges. With growth of the septovomerine complex, the premaxilla and the skin covering it are positioned downward and forward. They drag their neurovascular supply along with them. Thus, medial nasal arteries from the anterior ethmoid stem find themselves stretched along the upper border of septum and downward into premaxilla.

One can imitate this process by making a fist with the arm held horizontally. Observe the radial aspect of the index finger: metacarpal = frontal skin, proximal phalanx = nasal dorsum, middle phalanx = columella, and distal phalanx = prolabium. The thenar eminence = vestibular lining leading back toward the sphenoid. The base of the thumb metacarpal = Rathke's pouch. Straightening of the finger = growth of the septo-vomerine-premaxillary complex. In cases of severe holoprosencephaly, there is absence or a rudiment of septum. A nasal cavity is a single chamber. Tuck up inside its midline is a primitive anlage of skin…in continuity with that of the nasal dorsum. This represents columella–philtrum. The anlagen may contain bone; a tooth may be present.

Other pathologies can be explained by this model. Recall that prolabium receives collateral blood supply via premaxilla. Columella contains the terminal medial nasal arteries (plus collaterals from more posterior branches of medial nasal). Thus, columella is vulnerable to congenital absence or infarction, leaving behind an apparently normal prolabium and premaxilla. Septum remains intact as well due to its more proximal supply. However, congenital absence or loss of the common medial nasal artery can affect septum alone or both structures (depending upon the extent to which columella is supported by collateral circulation).

Lacrimal Bone: Eyelids

Descriptive Anatomy

Osteology

The smallest of the facial bones, lacrimal bone, has an academic importance disproportionate to its size because its articulations, and the manner in which these are conditioned by the pre-existence of the lacrimal system, tell us a great deal about how developmental fields interact. It is unique in that it houses a soft tissue conduit for tears to reach the nasal cavity (Figs. 8.53, 8.54, 8.55, 8.56, 8.57, 8.58, 8.59, 8.60, and 8.61).

The lacrimal bone develops in the membrane from r1 mesencephalic neural crest. It is unilaminar, and therefore has a single source of epithelial programming, the p6 nasal lining. On the orbital side, the soft tissue coverage is r1 MNC. Overlying the lacrimal bone is the medial canthus, the junction of r1 upper lid skin with r2 lower lid skin. The bone is supplied by the lacrimal branch of infratrochlear artery (otherwise known as medial palpebral branch).

The medial orbital rim is formed by locking together of the r2 *maxillary process of the frontal bone* and the r1 *frontal process of the maxilla*. Just inside the orbital rim is a depres-

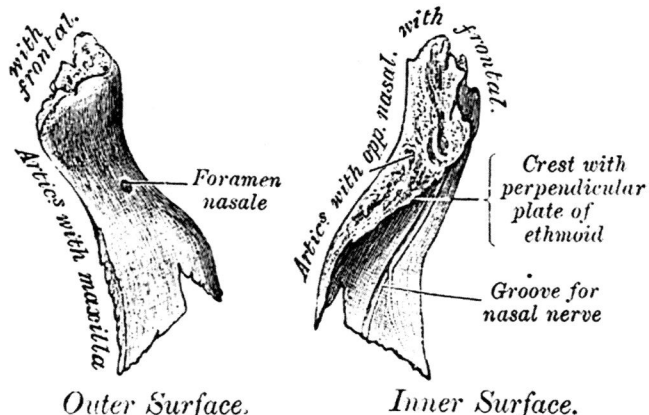

Fig. 8.53 Gray 30.4.100–111 nasal bones. The nasal bones demonstrate a groove for V1 and the common lateral nasal branch of StV1 anterior ethmoid. At the inferior border of nasal bone this splits into a submucosal branch and a deep subcutaneous branch. The latter follows along the dorsal septum and descends into the columella and philtrum. (Reprinted from Lewis, Warren H (ed). Gray's Anatomy of the Human Body, 20th American Edition. Philadelphia, PA: Lea & Febiger, 1918)

sion, the *fossa of the lacrimal sac*. The fossa represents the junction between the maxillary and frontal bones and is defined by two sharp bony ridges: the anterior lacrimal crest of the maxillary bone (ALC) and the posterior lacrimal crest of the lacrimal bone. The fossa is wider at its base; it constitutes a passage into the nose for the nasolacrimal duct. In its depths runs the maxillary-lacrimal suture.

The orbital aspect of lacrimal bone has two distinct zones defined by the *posterior lacrimal crest* (PCL). Posterior to the PCL, its smooth surface joins that the orbital lamina of ethmoid to make up the medial orbital wall. Anterior to the PCL, a vertical depression exists between lacrimal and frontal process of maxilla, the *lacrimal sulcus*, marked by the *maxillary-lacrimal suture*. The upper portion of the lacrimal sulcus house the lacrimal sac; the lacrimal duct resides in the lower portion. At the lower terminus of the lacrimal sulcus, the two ridges of PLC and ALC encircle the lacrimal duct along its orbital aspect. They do so by means of a small, but important projection from the inferior terminus of PLC, the *lacrimal hamulus*. This articulates with *lacrimal tubercle* of r2 maxilla. Thus, as lacrimal and maxilla develop, a bony canal surrounds the nasolacrimal duct. This explains the upper part of the nasolacrimal duct.

Attached to the posterior crest is the *lacrimal part of orbicularis*. Just below it is the posterior limb of the medial canthal ligament. Just opposite this, on the anterior crest, anterior limb of the medial canthal ligament gains attachment. As these two structures straddle the lacrimal sac, contraction of the lids and "milk" the lacrimal sac. A final structure of importance is a medially-directed branch from infraorbital artery. It runs along the ALC and penetrates the bone to supply the frontal process of maxilla. Compromise of the blood

Dermal Head Skeleton

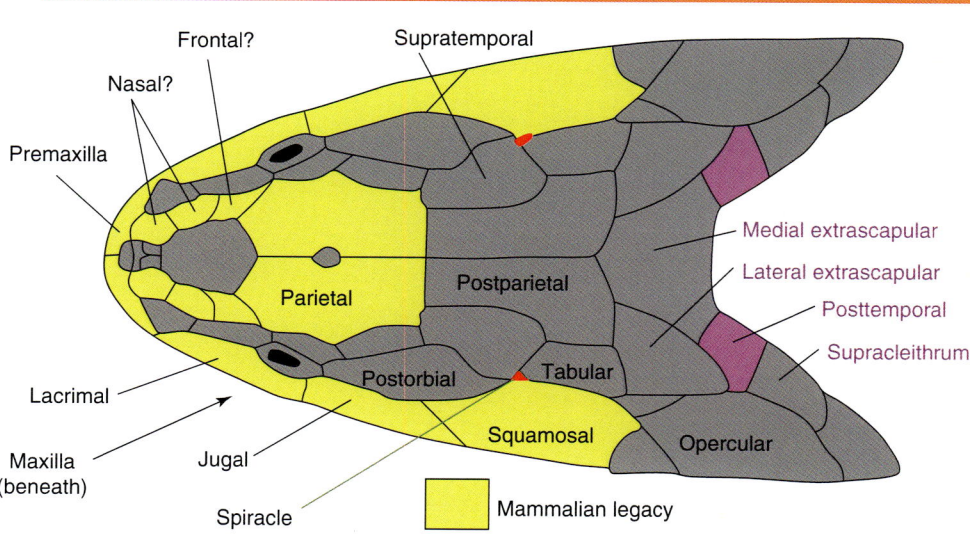

Eusthenopteron

Fig. 8.54 Nasal bones *Eusthenopteron*. Primitive *Eusthenopteron* has paired nasal bones aligned front to back. Nasolacrimal duct is present. Note how the nasals marginalize the frontals. Recall that sarcopterygians divide into coelacanths, lungfishes, and our ancestors, the tetrapodomorphs. These latter are defined by the presence of an internal nostril (choana) and the presence of a humerus. *Eusthenopteron* is an

early tetrapodomorph belonging to the Osteolepdiform order. (Reprinted from Andrews SM, Westoll TS. The Postcranial Skeleton of Ensthenopteron foordi Whiteaves. Earth and Environmental Science Transactions of The Royal Society of Edinburgh 2012; 68(9):207–329. With permission from Cambridge University Press)

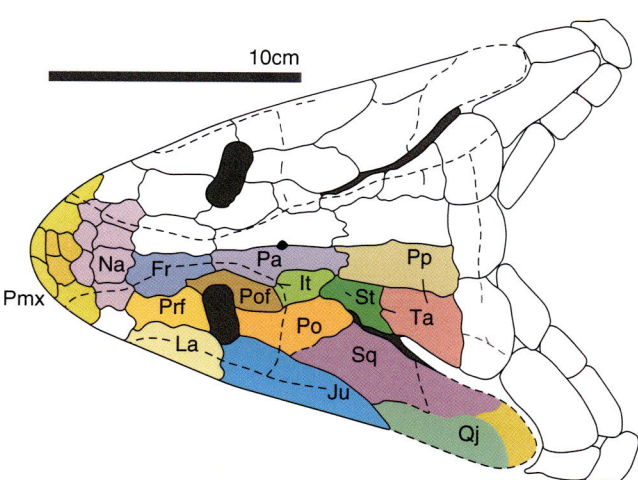

Fig. 8.55 Nasal bones *Panderichthys*. *Panderichthys* is a late tetrapodomorph belonging to the precursor to the Elpistostegalian order, immediately proximal to tetrapods. Here, the nasals are paired side by side. In contrast to *Eusthenopteron*, frontals in *Panderichthys* have moved into the midline but are excluded from the orbit by prefrontal and postfrontal. (Reprinted with permission from: GEOL431—Vertebrate Paleobiology course web resources. Thomas Holtz and John Merck, Department of Geology, University of Maryland, College Park, https://www.geol.umd.edu/~jmerck/geol431/)

supply is part of the Tessier #4 cleft in which the lacrimal system, located in cleft zone #3 remains intact.

The nasal aspect of lacrimal bone is smooth and concave. It has two zones defined by a vertical furrow. Anteriorly, it forms part of middle meatus. Posteriorly, it articulates with ethmoid. The zone anterior and inferior to the furrow is the *middle meatus* while the zone posterior and superior to the furrow blends in with the *anterior ethmoid air cells*. The *uncinate process* of middle turbinate descends from this common juncture. Thus, upper lacrimal and uncinate process form a cul-de-sac with the nasal cavity. Note that anterior ethmoid air cells can expand forward deep to lacrimal and invade the frontal process of maxilla.

The articulations of lacrimal bone are as follows: anterior border with frontal process of maxilla, posterior border with ethmoid, superior border with frontal bone, and (importantly) inferior border, with two key articulations (once again defined by posterior lacrimal crest). Posterior to the PLC, lacrimal attaches to *orbital plate of maxilla*. Anterior, the *descending process of lacrimal* joins with the *lacrimal process of inferior turbinate*. This defines the lowest part of the lacrimal canal.

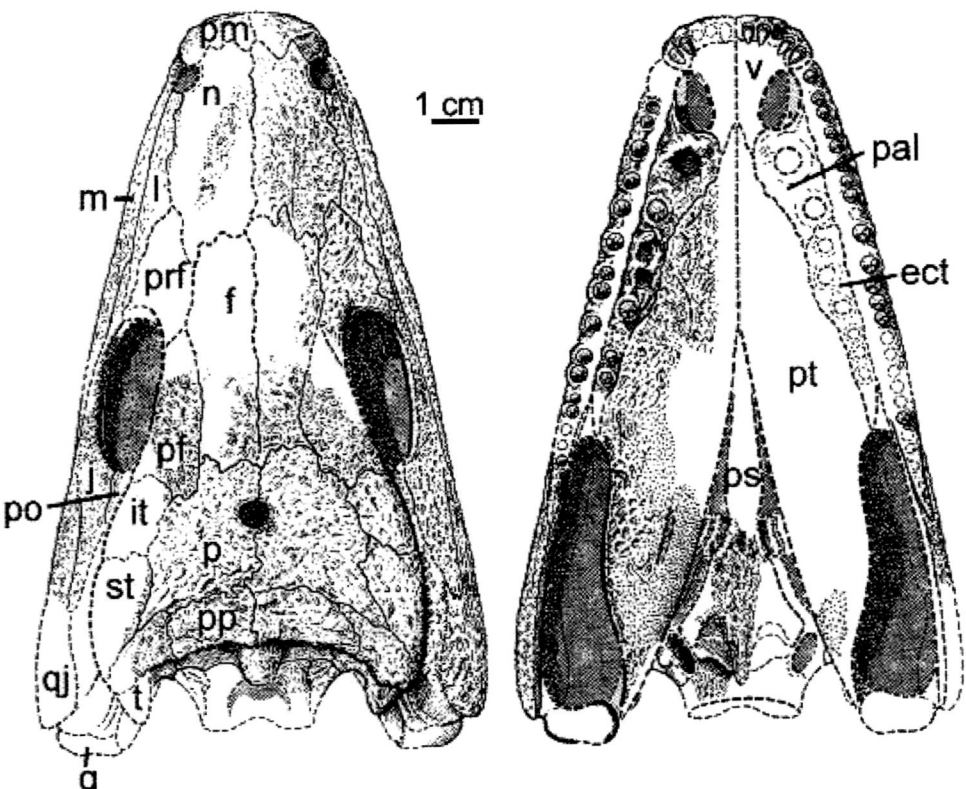

Fig. 8.56 Nasal bones, *Eoherpeton*. Stem tetrapods are problematic, as they combined an aquatic existence with air-breathing, the capacity to hear airborne sound and partial ability to walk. Acanthstega, Greerepeton, Watcheeria, and Megalophalus all have bone remnants of gills. *Eoherpeton*, a crown tetrapod of the Carboniferous, shows a single pair of nasal bones. These remain stable throughout vertebrate evolution. The tetrapods subsequently split into the amphibians and the reptilomorphs. *Eoherpeton* is a stem <u>anthracosaur</u>. Anthracosaurs, so named for their fossils being discovered in coal deposits, are the clade of tetrapods leading to mammals and is defined as amniotes plus all other tetrapods more closely related to amniotes than they are to amphibians. (Reprinted with permission from: GEOL431—Vertebrate Paleobiology course web resources. Thomas Holtz and John Merck, Department of Geology, University of Maryland, College Park, https://www.geol.umd.edu/~jmerck/geol431/)

The Lacrimal Duct and Sac

The derivation of the soft tissue contents of lacrimal fossa is readily understood by its formation. At stage 14 (7 mm), the frontonasal process and the upper maxillary process begin to fuse in a craniocaudal direction. The sulcus represents the boundary between the midbrain neural crest associated with (p5) frontonasal skin and the hindbrain neural crest with r2 skin coverage. The two epithelia flow together in the depths of the sulcus running from the medial boundary of the orbit to the interior of the nose beneath inferior turbinate. As additional mesenchymal tissue flows inward, the sulcus becomes progressively more shallow. At between stage 17 and 18 (43 days), the skin surfaces fuse over the sulcus. This entraps in its depths a rod-like column of p5-r2 epithelium and surrounding with r1 MNC and r2 RNC. The rod remains connected only with the surface at its orbital and nasal ends. It extends two finger-like columns of cells upward, one for each lid, as the canaliculi. This makes use of the embryological "fault line" between skin and conjunctiva, the so-called "gray-line."

The ducts terminate prior to the medial margin of the tarsal plate.

Canalization of the rod takes place during the early fetal period at 4 months via a process of apoptosis. It begins in the upper system (canaliculi and sac) and then proceed caudally into the nose. In so doing, it reiterates the original closure gradient between the maxillary and frontonasal processes. For this reason, the superior membrane at the puncta becomes patent at 7 months gestation whereas the inferior membrane in the nasal fossa can remain closed resulting in congenital obstruction.

Blood Supply and Mesenchyme

Distal to the stem of anterior ethmoidal artery, the nasociliary vascular axis gives rise to a *trochlear stem*; it bifurcates around the pulley of superior oblique. (1) *Infratrochlear artery* arises just below the pulley and runs forward to emerge at the medial canthus, where it gives rise to paired *palpebral arteries*. (2) *Supratrochlear artery* passes above the pulley and subsequently splits. Dorsally, its named

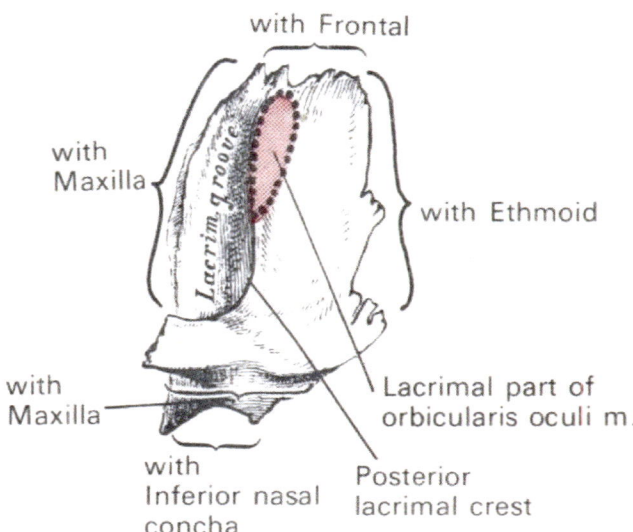

Fig. 8.57 Gray 30.4.127 left lacrimal, orbital (lateral) side. Note articulations with four bones. Orbicularis oculi has a secondary insertion (red). (Reprinted from Lewis, Warren H (ed). Gray's Anatomy of the Human Body, 20th American Edition. Philadelphia, PA: Lea & Febiger, 1918)

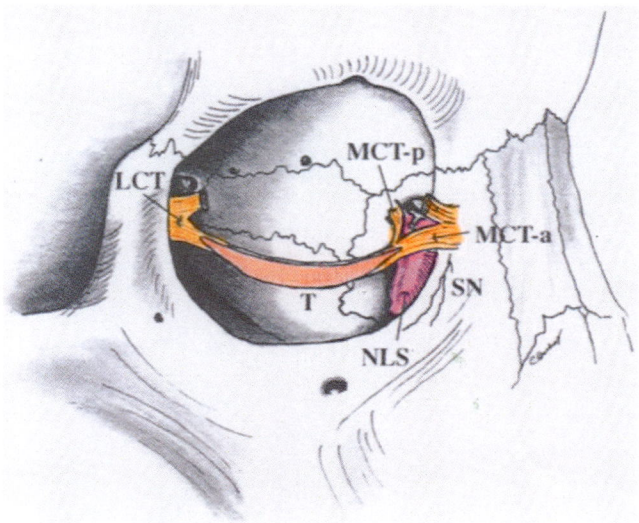

Fig. 8.59 Lacrimal pump. Contraction of the tarsal component of the three-part orbicularis tenses two pronged medial canthal tendon and squeezes down on the sac. *LCT* lateral canthal tendon, *MCT-a* medial canthal tendon-anterior limb, *MCT-p* medial canthal tendon-posterior limb, *NLS* nasolacrimal sac, *T* tarsus. (Reprinted from Burkat CN, Lucarelli MJ. Anatomy of the lacrimal system. In: Cohen AJ, Mercadetti M, Brazzo BG (eds). The Lacrimal System. New York, NY: Springer Verlag; 2006: 3–19. With permission from Springer Nature)

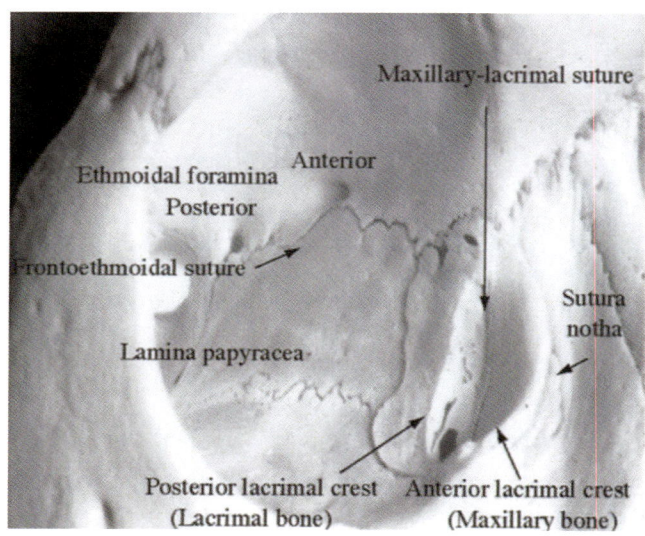

Fig. 8.58 Lacrimal fossa. Maxillary lacrimal fossa represents boundary between r1 lacrimal and r2 frontal process of maxilla. (Reprinted from Burkat CN, Lucarelli MJ. Anatomy of the lacrimal system. In: Cohen AJ, Mercadetti M, Brazzo BG (eds). The Lacrimal System. New York, NY: Springer Verlag; 2006: 3–19. With permission from Springer Nature)

continuation, or *frontal branch*, supplies zones 13 and 12 of the forehead. Ventrally, the *dorsal nasal branch* supplies zones 13 and 12 of the external nasal envelope.

Although the skin of the columella-philtrum is in continuity with dorsal nasal skin, its blood supply is embryologically distinct. The columella and the philtrum are perfused by the terminal branches of the anterior ethmoid axis, the *medial nasal arteries*. These run beneath the nasal bones,

leaving a groove on their undersurface, and emerge from beneath the inferior margin to run down the midline, just hugging the septum until they descend downward through the columella and philtrum. The developmental origins of this skin can be traced back to the original neuromeric anatomy of the postgastrulation embryo at stage 8, immediately before folding begins. This fascinating subject is discussed with the vomer bone.

The *superior palpebral artery* courses laterally along the upper eyelid in two parallel arcs: the *inferior arch* runs along just above the tarsus while the *superior arch* is halfway between the margins of the lid and the orbit. The arches are interconnected by vertical branches to form an arcade. The upper lid is divided into three zones. *Zone 11* (medial third) is supplied by the stem of superior palpebral. *Zone 10* (middle third) contains anastomoses between the upper arch and supraorbital. In *zone 9* (lateral third), the two arches coalesce; they subsequently anastomose with the StV1 lacrimal. *Zone 8* (postorbital zone) makes a contribution from anterior branch of superficial temporal. In summation, the skin of the eyelids is embryologically distinct from frontonasal skin. In the upper eyelid, epidermis is r1 ectoderm and the dermis is r1 MNC. It is supplied by V1 stapedial. In the lower eyelid, the epidermis is r2 ectoderm and the dermis is r2 RNC.

The *inferior palpebral artery* runs laterally over the tarsus. It does not have a parallel arch. The lower lid is also divided into three zones. Zone 11 is supplied by the stem of the inferior palpebral. Zone 10 contains anastomoses

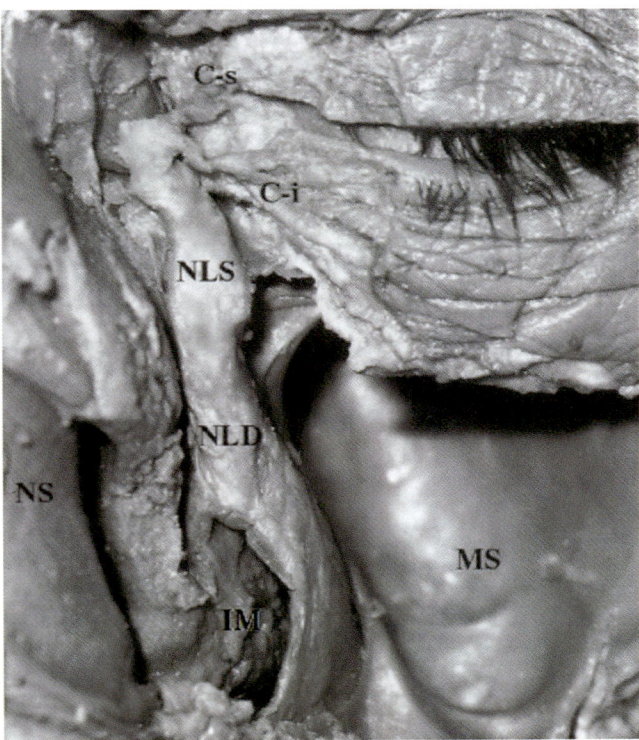

Fig. 8.60 Nasolacrimal sac is intraorbital while the duct is covered medially by the inferior shelf of lacrimal bone. *NS* nasal septum, *NLD* nasolacrimal duct, *NLS* nasolacrimal sac, *C-I* inferior canaliculus, *C-s* superior canaliculus. (Reprinted from Burkat CN, Lucarelli MJ. Anatomy of the lacrimal system. In: Cohen AJ, Mercadetti M, Brazzo BG (eds). The Lacrimal System. New York, NY: Springer Verlag; 2006: 3–19. With permission from Springer Nature)

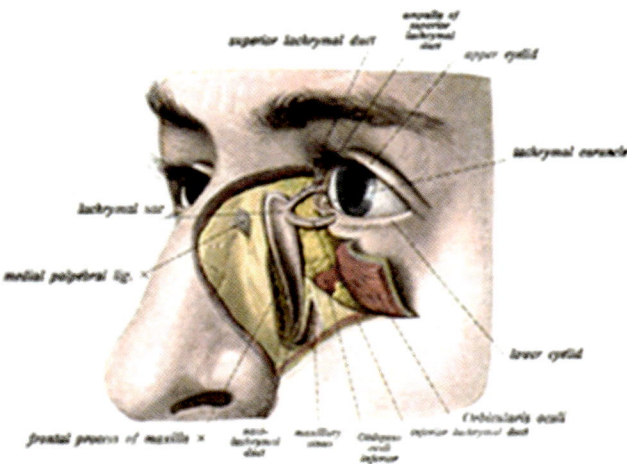

Fig. 8.61 Nasolacrimal duct. NLD follows neuromeric pathway between r2 maxilla and r1 ethmoid medial concha. Valve of Hausner located at the r2–r2 boundary of maxilla with inferior turbinate. This marks the embryonic pathway of solid rod of mesenchyme which underwent cavitation via apoptosis and was subsequently populated by r1 nasal epithelium. (Reprinted from Burkat CN, Lucarelli MJ. Anatomy of the lacrimal system. In: Cohen AJ, Mercadetti M, Brazzo BG (eds). The Lacrimal System. New York, NY: Springer Verlag; 2006: 3–19. With permission from Springer Nature)

between the arch and infraorbital. In the lateral third, the arch makes two connections. It continues laterally beneath the lateral canthal tendon to anastomose with StV2 zygomatico-orbital. It also turns upward to join with the inferior arch of the upper lid. Inferior eyelid, like its partner, has three segments but *the transition between StV1 and STV2 takes place in the middle, not lateral, sector*. Thus, the medial zone of lower eyelid is StV1, while the lateral two zones are StV2.

Both eyelids are supported by collateral circulation from the external carotid system: angular artery of the facial is medial while transverse facial is lateral.

In sum, lacrimal bone *per se* is supplied by both medial superior and medial inferior palpebral arteries. Specific distributions include medial canthus (superior) and lacrimal sac (inferior).

Development

Ossification of the lacrimal bone takes place around the 12th week. Two sites are involved: the centrum and the hamulus. The latter could be expected to develop later in time.

Phylogeny

The lacrimal duct and the lacrimal bone first make their appearance in the sarcopterygian line from the Osteichthyes. It has been documented in *Eusthenopteron*, a primitive tetrapodomorph. This represented a useful adaptation maintaining eye moisture when emerging from water. It was located beneath the eye between the jugal and the nares. It was bounded by prefrontal medially and maxilla laterally. With the reptilomorphs, the lacrimal bone was relocated backwards with nasal bone filling in the space. It remained outside the orbital rim, being flanked by prefrontal above and maxilla below. In mammals, it become internalized in the orbit.

All tetrapods have evolved eyelids as a means of coping with a dry environment. The advent of a moveable eyelid occurs with reptiles, although the upper lid remains static. Many tetrapods have possess a third eyelid, the *nictating membranes*. These structures fulfill a lubricating function because they possess *Harderian glands* which produce an oily secretion. In many tetrapods, the basicranial muscle has been retained as a retractor of the globe. When the eyeball is brought into the socket, it allows the lids to close and lubrication to ensue.

Clinical Correlations: Lacrimal System and Eyelids

Lacrimal duct stenosis is a common congenital condition. Its can occur in three forms: (1) canalicular; (2) lacrimal sac, and (3) at the outflow tract. The latter condition most likely reflects a malformation (possible an excess) of the hamulus.

The Frontal Bone Complex

Descriptive Anatomy

The frontal bone is a misnomer. It is actually a complex of six r1 neural crest bones. The <u>forehead</u> is composed of two separate squamous bones separated by the metopic suture. The inferior terminus of the squamae is above the orbital rim, just above the eyebrow. The vertical squamae develop in membrane. The <u>roof of the orbit</u> is made from paired *prefrontal* (PrF) and *postfrontal* (PF) bones, separated by the supraorbital neurovascular axis. They project forward and make up the orbital rim. Laterally, postfrontal abuts against the postorbital bone field, i.e., the frontal process of zygoma, the latter being made of two separate fields, jugal and postorbital. Medially, the two prefrontals are separated by paired vertical islands of squama that proceed downward to terminate with the nasal bones. This central zone is known as the *glabella*. The horizontal squama develop from the *frontal cartilages* (Figs. 8.61, 8.62, 8.63, 8.64, 8.65, and 8.66).

The *pars frontalis* present a smooth external surface punctuated by paired *tuber frontale* (frontal eminences) located 3 cm lateral to the metopic suture. Below each eminence, a trough-like depression accentuates the supraciliary

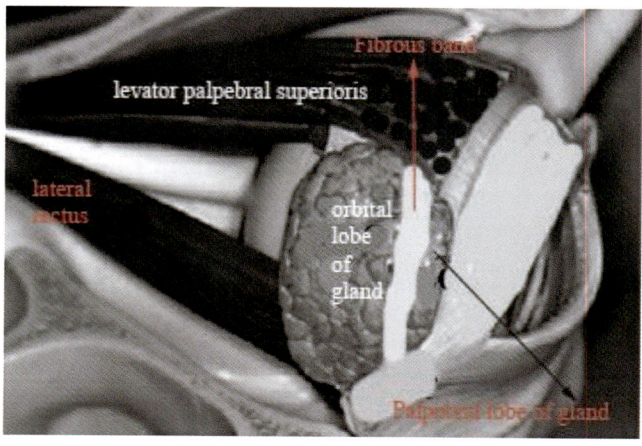

Fig. 8.62 Lacrimal gland. The gland has two lobes divided by r1 fascia from levator palpebrae superioris. LPS mesenchyme originate from Sm6 at the level of r4–r5. This could also convey r4–r5 neural crest mesenchyme into the gland. Lacrimal gland can be thought of a r1 stroma with r4–r5 glandular elements. This could be the basis for its dual innervation with V1 providing sensation and VII PANS motor control (recall that the entire ANS system is of neural crest derivation). Thus, the presence of r4–r5 tissue calls forth PANS fibers present in pterygopalatine ganglion to take advantage of V2 zygomatic nerve destine to penetrate the r2 bone through the zygomatico-facial and zygomatico-temporal foramina to reach the surface of the skin. V1 lacrimal branch lies nearby thus an anastomosis takes place that carries PANS fibers from V2 to V1 and its target mesenchyme. For this reason, pain from the V1 cornea elicits a reflex stimulation of tear production. (Reprinted from Burkat CN, Lucarelli MJ. Anatomy of the lacrimal system. In: Cohen AJ, Mercadetti M, Brazzo BG (eds). The Lacrimal System. New York, NY: Springer Verlag; 2006: 3–19. With permission from Springer Nature)

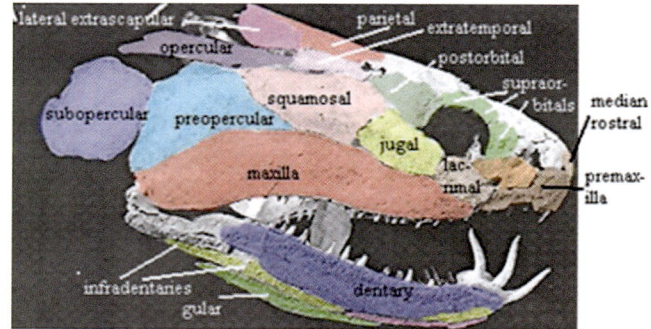

Fig. 8.63 Lacrimal bone in *Onychodus*. Sarcoptyerygians produced two lines: first were the Onychodontiforms ("claw-teeth"), *Onychodus*, followed by the Rhipidistians that were (in order): dipnoans, rhizodontiforms, osteolepiforms, elpistostegalians, and TETRAPODS. *Onychodus* demonstrates **lacrimal bone and duct** indicative of exposure of the eye to air. Note the strange premaxillary teeth, designed to skewer prey. These actually were received in separate "pocket" of the palate. (Reprinted from Long JA. *On the relationships of Psarolepis and the onychodontiform fishes*. J. Vert. Paleontol. 21(4):815–820. With permission from Taylor & Francis)

arches, the size of which is proportional to the frontal sinuses. The arches terminate medially as the glabella and laterally as the zygomatic process. From these latter, a *temporal line* sweeps upward and backward. It quickly subdivides into a *superior temporal line* and an *inferior temporal line*; these are continuous with their named counterparts over the parietal bone. The developmental significance of the STL is that is marks the cephalic border of the deep investing fascia covering the temporalis muscle. The anterior aspect of temporalis DIF is r2 neural crest and so too is the overlying dermis. The posterior aspects of these structures are r3 neural crest. STL is thus an important site of fascial fusion between r1 frontonasal skin and epaxial facial skin from r2 and r3. Anteriorly this is marked by a prominent vein. Takedown of this fascial insertion is an important component in certain aesthetic procedures to optimize browlift.

The internal surface of the squamae bears a trough-like *sulcus of superior sagittal sinus*, the edges of which provide attachment for the paired *falx cerebri*. Continuing caudally the two ridges fuse as the *frontal crest*; it articulates with the ethmoid bone. The frontoethmoid suture contains a potential opening, the *foramen cecum*, site of the embryonic *anterior neuropore*. This aperture is usually closed but occasionally permits a vein from sagittal sinus to the nasal roof. It also can be a portal of exit for encephaloceles.

The *pars orbitalis* consists of paired bilaminar orbital plates. On the orbital side, one notes two depressions, laterally for the lacrimal gland and medially, a fovea trochlearis into which fits the cartilage pulley of superior oblique. This structure can be located by following the medial sweep of the orbital rim half-way between supraorbital foramen and the frontal process of maxilla. The fovea is located about a cm inside. Continuing along the medial wall of the orbit, the

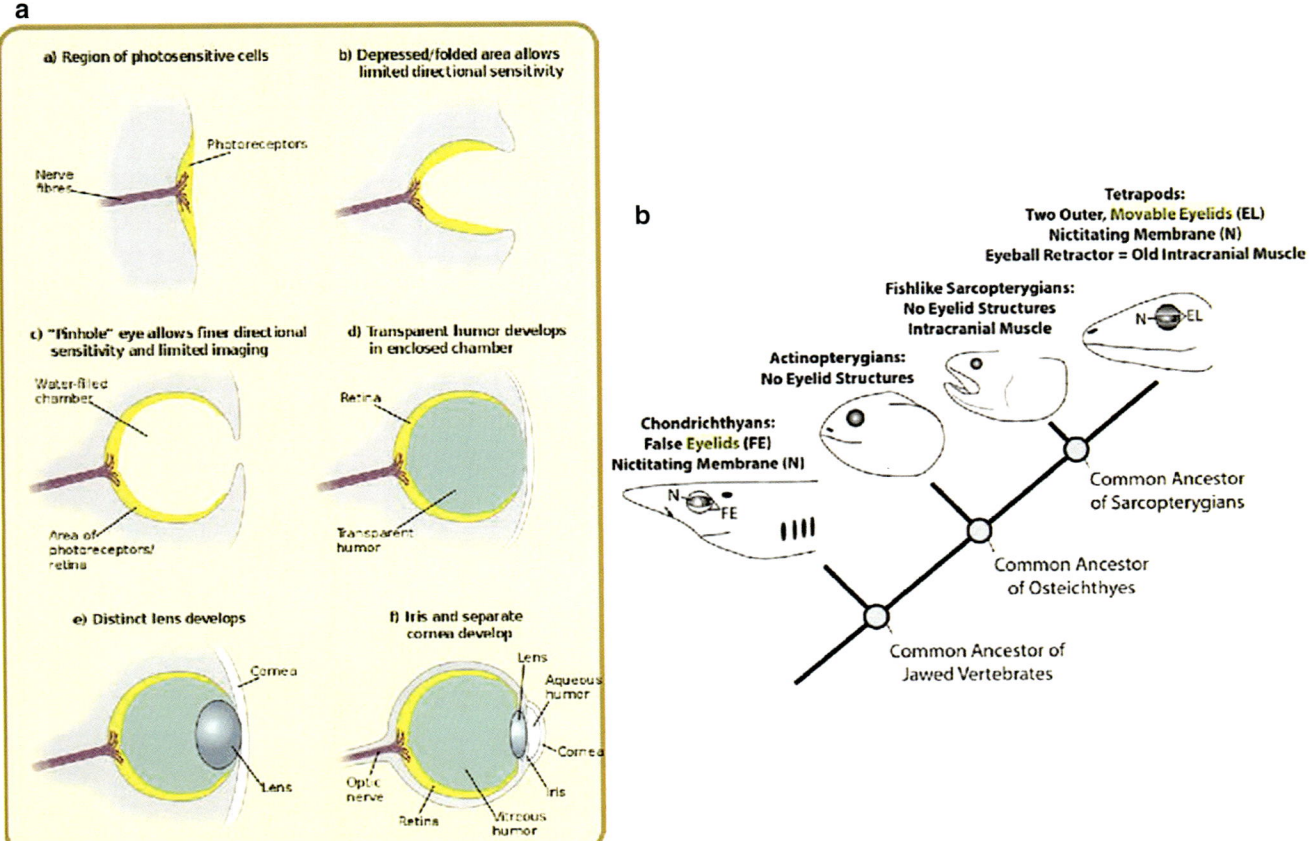

Fig. 8.64 Eyelid phylogeny. Eyelids appear in advanced sarcopterygians. Nictating membranes disappear in haplorrhine primates. Note that the haplorrhine upper lip is disconnected from both the nose and the gum. (Reprinted from Nilsson D-E. Eye evolution and its functional basis. Vis Neurosci. 2013 Mar; 30(1–2):5–20. With permission from Cambridge University Press)

lateral ethmoid plate has two perforations permitting the anterior and posterior ethmoid arteries to enter the ethmoid labyrinth.

Frontal sinuses occupy the interlaminar space in both the squama and the pars orbitalis. These drain into the nose via the middle meatus located beneath the middle turbinate. The course of the frontonasal duct curves inferior and posterior. In 50% of cases, it is confluent with *infundibulum*, the drainage point of the anterior ethmoid air cells. Recall that the posterior ethmoid air cells drain via the superior meatus beneath the superior turbinate.

The boundaries of frontal bone are represented by the following sutures. Recall that the space between the bone fields at each suture is composed of neural crest cells from underlying dura that are insinuated to reach the external layer of periosteum. The *fronto-parietal* (coronal) suture extends down to pterion. Below this point it continues as the alispheno-temporal suture. (2) *Fronto-alisphenoid* suture runs horizontally forward from the pterion and terminates against the backwall of the orbit. (3) *Fronto-ethmoid* suture separates the orbital roof plate of prefrontal bone from lateral wall of the ethmoid sinuses. It connects the anterior and pos-

terior ethmoid foramina, extending from the lesser wing forward as the *fronto-lacrimal* suture, the *fronto-maxillary* suture demarcating the upper limit of the frontal process of maxilla, and terminates with the *fronto-nasal* suture.

Blood Supply and Mesenchyme

The frontal lamina of each squama is supplied by two arterial axes of StV1. Nasociliary gives rise to the *frontal branch of supratrochlear*; its territory is defined by p3 and encompasses Tessier cleft zones 13 and 12. The *dorsal nasal branch of supratrochlear* is assigned to p3 of glabella and nasal process. Supraorbital supplies p2, i.e., Tessier zones 11 and 10. In like fashion it contributes to the upper eyelid. The dura covering of the cranial lamina of squama and the cranial lamina of orbital plate receive *meningeal branches of anterior ethmoid* (a branch of nasociliary *proximal* to the common infraorbital axis). The orbital lamina of orbital plate is supplied by supraorbital, i.e., zones 11 and 10. Thus, prefrontal constitutes zone 12 and post frontal zone 1211. Finally, the most lateral zone of the orbital roof is supplied by *lacrimal artery*; it corresponds to Tessier cleft zone 9 and includes soft tissue of the lateral eyebrow and upper eyelid.

Fig. 8.65 Gray 30.4.90 frontal bone fields exterior. Prefrontal and postfrontal bone fields (yellow) make up the orbital roof. Corrugator and orbicularis oculi have secondary insertions (red) specifically into the profrontal bone field. (Reprinted from Lewis, Warren H (ed). Gray's Anatomy of the Human Body, 20th American Edition. Philadelphia, PA: Lea & Febiger, 1918)

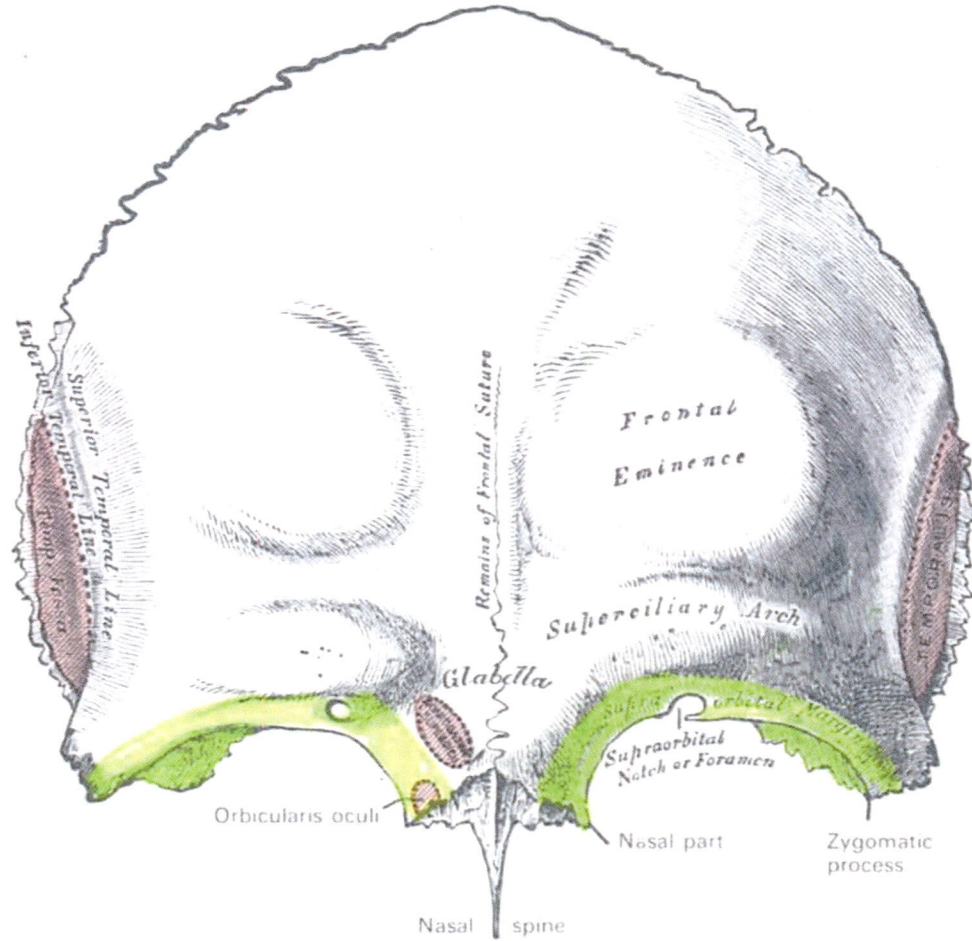

Development

The bones of the frontal complex develop by via membranous ossification in response to a unique *dual programming system*. The deep component is the *neuroectoderm* of the developing secondary prosencephalon. The forebrain is covered with a mixture of *blood vessels* from r1 paraxial mesoderm and the *primordial meninx* from r1 MNC. The superficial component is *frontonasal skin* composed of nonneural ectoderm epidermis of the anterior prosencephalic neural folds and neural crest dermis from the posterior prosencephalic neural folds, p3–p1. This is theoretically arranged as three vertical populations, from medial to lateral, p3 to p1. These would correspond to the territories of nasociliary (or supratrochlear), supraorbital, and lacrimal. Tracking into this system are the above-mentioned vascular axes.

Note: Vertical zonation of the forehead is not just an academic concern. The only factor that could explain the necessity of three separate neurovascular axes is the presence of genetically different environments associated with the type of underlying neural crest. PNC comes from prosomeres p1–p3, each of which is genetically distinct; therefore, it is the

logical candidate. Maturation of the posterior neural folds is from back-to-front. Hence it would appear that p1 neural crest would migrate first to the midline creating a p1–p3 gradient from medial to lateral but the order could also be reverse. The important point is that this region has defined zones of cleft formation that correspond to the neurovascular axis. We are thus obligated to explain why this is so. Furthermore, if there is a commonality between the homeotic code for a prosomere, and underlying neurologic fields, CNS deficits seen with craniofacial clefts become logical. Indeed, the genetic mapping of craniofacial defects becomes a way of understanding the relationships between deep and superficial structures.

Later in development, an additional layer from second arch tracks beneath the skin from lateral to medial. This new layer is composed of *superficial investing fascia* (SIF) containing *facial muscles* from somitomere 6. SIF is constructed with neural crest from r4 to r5 assigned to upper and lower division facial muscles. Therefore, frontonasal SIF is an r4 structure. This combined system is also known as SMAS. Although in the rest of the face, the SMAS is supplied by external carotid, in this region it territories overlap

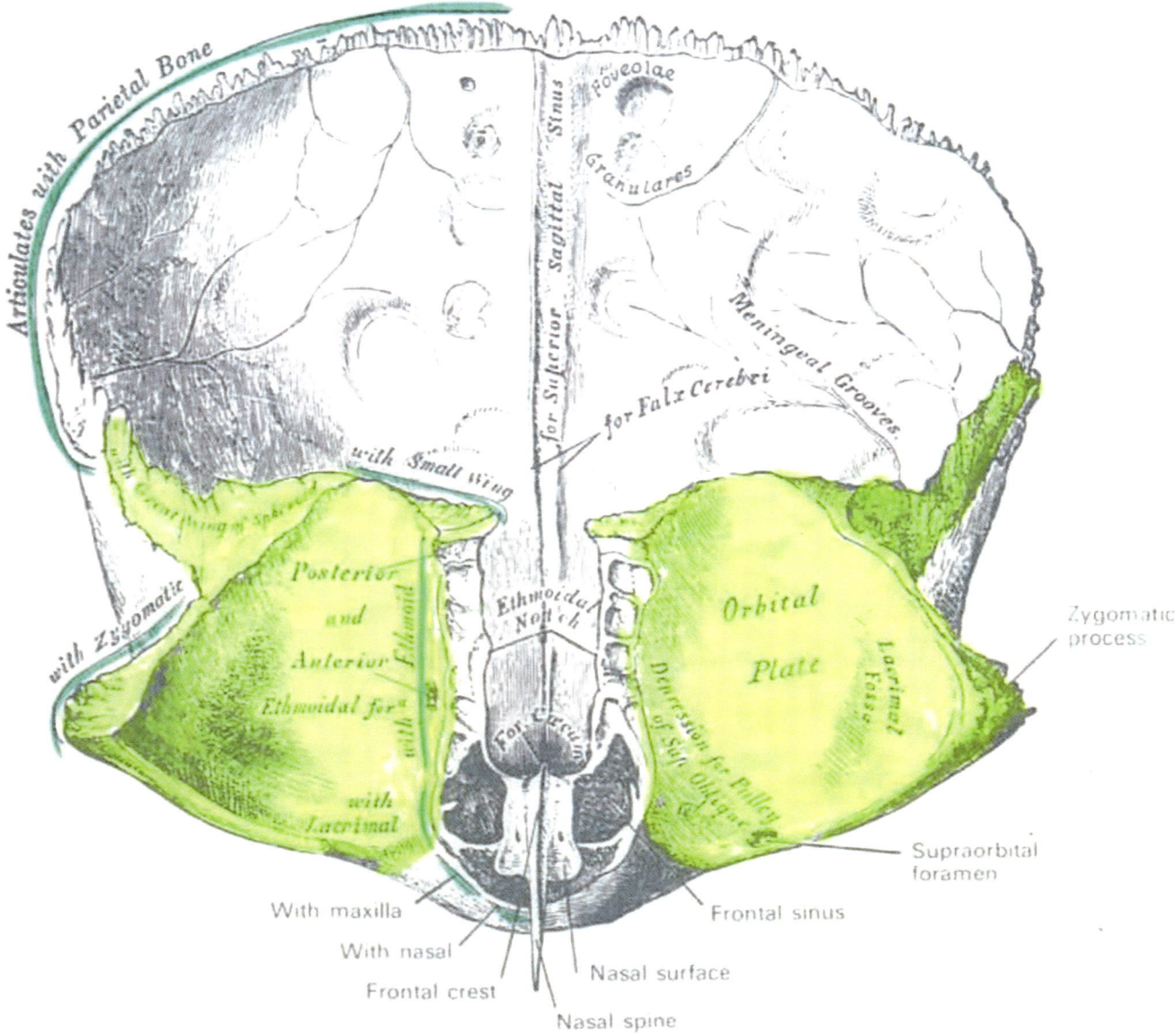

Fig. 8.66 Gray 30.4.91 frontal bone fields internal. Orbital roof initiated with r1 frontal cartilage intermediate(s) that undergoes membranous ossification. These are the likely sources of the prefrontal and postfrontal bone fields. (Reprinted from Lewis, Warren H (ed). Gray's Anatomy of the Human Body, 20th American Edition. Philadelphia, PA: Lea & Febiger, 1918)

with those of StV1. Thus, frontalis has a dual blood supply from ECA frontal branch of superficial temporal and STV1 supraorbital, as does corrugator. Procerus shares a blood supply between StV1 infratrochlear and ECA facial artery.

For this reason, the vascular axes, as they exit from the orbit, demonstrate deep and superficial branches over and under the SMAS. These StV1 stapedial vessels will actually penetrate the SMAS to reach the periosteum of the outer table.

Recall that dura is not an epithelium. Nonetheless, it has a vascular relationship with the underlying cerebral cortex. *The genetic composition and activity of dura are directly and indirectly related to the neuromeric zones of the brain it covers and supplies.* Thus, r1 dura over p6 olfactory cortex is distinct from r1 dura covering the p5 frontal lobe.

Phylogeny

Agnathic fishes had skull with an open roof and an unprotected brain. Evolution quickly corrected this situation. Cartilaginous fishes achieved brain protection with a fused cartilaginous cap. Bony fishes (osteoichthyces) accomplished the same goal using dermal bones. In early choanate fishes, such as *Eusthenopteron*, the frontals were one of many small bones surrounding the nares. They were separated from the midline by an intervening postrostral bone. These fishes faced a difficult problem. The anterior skull was short, so the size of the jaws was limited. For better predation, a longer jaw was needed but it required mechanical bracing against the braincase.

Recall that our lineage can be traced back sarcopterygian (fleshy-fin) fishes in the mid-Devonian period with fins joined to the body via a single bone for greater mobility. From these, a clade (subgroup) Rhipidistia were directly ancestral to lung fishes and tetrapodomorphs. In this transition, the bones of the anterior snout begin to organize into a "normal" pattern. A major theme was the development of a central column of bones from snout to occiput. In the basal osteolepidid, *Osteolepis*, the parietals are located between the orbits; the anterior bones are small and jumbled. In *Panderichthys*, the frontals move posteriorly to orbits but the nasal bones remain multiple.

All this changed with the tetrapod line. *Ichthystega* had the first feet. With the loss of the extrascapular series, the forelimb girdle was disconnected from the skull and became mobile. With tetrapods, the frontals became centralized as part of the nasal series (premaxilla, nasal, frontal, parietal). In Paleoherpeton, we can see a reduction in the snout and the explosive growth of the forebrain brought the frontals over the mouth and nose. The orbital series demonstrates from the very beginning the primal relationship prefrontal and post frontal to the orbit as frontal bone moves backward into its final position.

Our paleohistory demonstrates that frontal bone is indeed a complex and that the supraorbital bones fields predate the presence of the squamous forehead. Although they are neural crest, they form via a cartilage intermediate, succeeded by membranous ossification.

Two primary centers of ossification are noted at the end of the second month of life. These are located at the supraciliary arch. They probably represent bilateral condensation of r1 neural crest mesenchyme as the frontal cartilages. These divided into the prefrontal and postfrontal bone fields of the frontal. From this center, bone spreads in three directions: upward as aquama, downward toward the nose and horizontally as the orbital plate. Two secondary centers of ossification appear at the inferior margin of the metopic suture. Studies have not been done regarding the intracranial ossification of the orbital plate.

Clinical Correlations: Anencephaly, Tessier Cleft Zones 13, 12, 11, and 10

It should be noted that the orbital cartilages that form the sphenoid complex fulfill the primitive necessity for coverage over the eyes (i.e., the brain). The so-called "flattening" of the frontal bone seen in anencephaly likely results from the preservation of prefrontal and postfrontal, in the face of the total loss of membranous frontal bone.

Parietal Bone

The parietal bone develops from paraxial mesoderm arising from r2 and r3 probably via the fourth somitomere [6]. It is bilaminar structure; the bone can be split though the intervening marrow. Parietal bone is programmed from below by the dura and from above by the dermis. SMAS becomes interposed between dermis and bone—but not until the fundamental programming has taken place. The parietal dura interacts with precentral gyrus of frontal lobe and postcentral gyrus of parietal lobe. These may be involved in programming the dural arteries that supply their external layer. Parietal bone is surrounded by four sutures, three of which are potential sites of premature closure, craniosynostosis.

Although parietal bone has a single source of mesenchyme, each of its two laminae can be divided into four biologically distinct quadrants, based upon the neurovascular anatomy of adjacent programming tissues, either dura or dermis. Common forms of craniosynostosis can be understood as faulty interactions between neighboring zones (Figs. 8.67, 8.68, 8.69, 8.70, 8.71, and 8.72).

Descriptive Anatomy

The convex external lamina of parietal presents a rounded *tuber parietale* (parietal eminence). This marks the epicenter at which ossifications are first noted. The eminence is covered by a three-layer mesenchymal "sandwich." *Dermis* and *subgaleal fascia-periosteum* constitute the "bread." These layers originate from r3 neural crest. A second intervening layer, the *galea aponeurotica*, is the "cheese." It arises from the second arch neural crest, likely in register with r5. Galea aponeurotica over the parietal bone does not itself contain muscle. Instead, it spans between muscle-bearing fasciae of the forehead and posterior skull, *galea frontalis* and *galea*

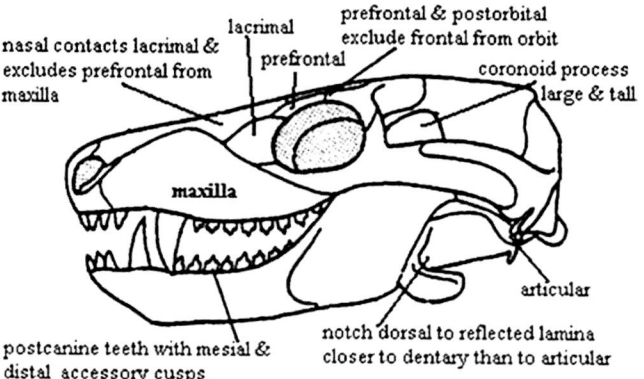

Fig. 8.67 Cynodont skull synapomorphies: frontal continues to be excluded from the orbit. Prefrontal and postfrontal fields are incorporated by the primate line. Primates achieve postorbital enclosure. The primate lacrimal bone is fully incorporated into the orbit. (Reprinted from Carroll RL. Vertebrate Palaeontology. New York, NY: WH Freeman and Company; 1988)

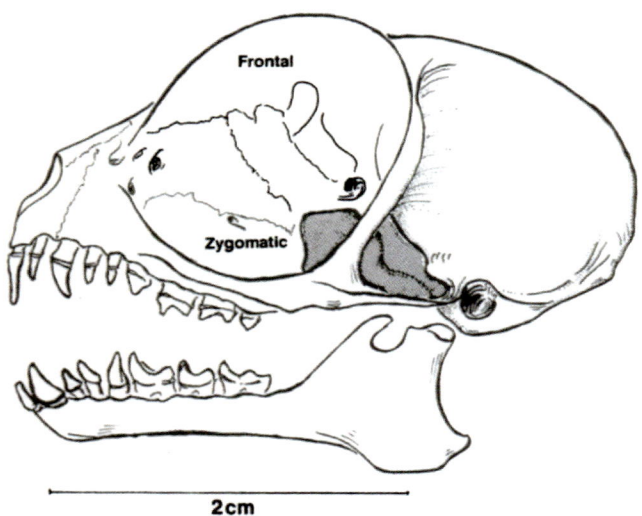

Fig. 8.68 Tarsier skull. Floor of the orbit is entirely zygomatic. Maxilla remains excluded. Primate evolution is characterized by binocular vision and increased visual acuity, adaptations for arboreal existence. This involves (1) medial positioning of the orbits, (2) increased axial length of the eye, (3) addition of a fovea to the retina for improved night vision, and (4) increased color perception. and splits into two lines. The addition of the fovea meant Primates split into two groups; *Strepsirrhini* (Gr. strepsi "curved nose") (lemurs and lorises) have crescent-shaped (wet) noses and two types of cones in the retina. They have no posterior orbital wall. *Haplorhini* (gr. Haplo, "simple nose") have flattened (dry) noses and have three types of cones for full color vision. About 55 million years ago haplorrhines divided into the tarsiidae (represented today only by tarsiers) and the anthropoidae. Tarsiers have enlarged orbits, a posterior orbital wall and integration of the frontal, prefrontal and postfrontal bone fields. (Reprinted from Schütz P, Ibrahim HHH, Rajab B. (April 22nd 2015). Contemporary Management of Frontal Sinus Injuries and Frontal Bone Fractures. In: Mohammad Hosein Kalantar Motamedi (ed). A Textbook of Advanced Oral and Maxillofacial Surgery Volume 2, IntechOpen; 2015. https://doi.org/10.5772/59096. With permission from Creative Commons License 3.0: https://creativecommons.org/licenses/by/3.0/)

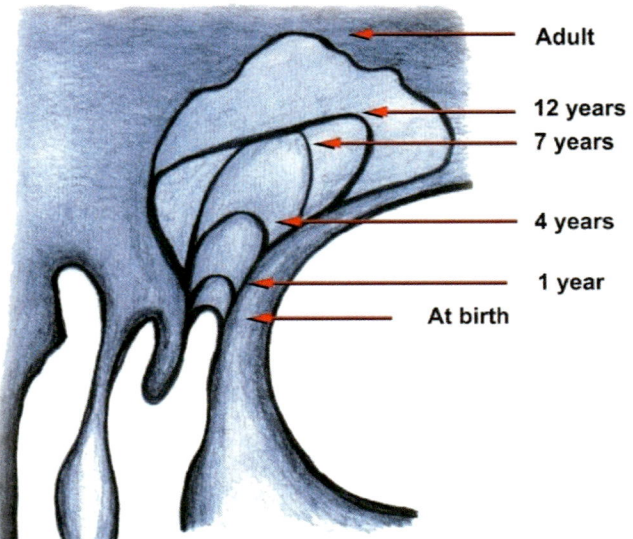

Fig. 8.69 Frontal sinus development. As bilaminar bone fields grow, the mucosal dead space increases with expansion into both frontal bone fields and the orbital roof. (Reprinted from Schütz P, Ibrahim HHH, Rajab B. (April 22nd 2015). Contemporary Management of Frontal Sinus Injuries and Frontal Bone Fractures. In: Mohammad Hosein Kalantar Motamedi (ed). A Textbook of Advanced Oral and Maxillofacial Surgery Volume 2, IntechOpen; 2015. https://doi.org/10.5772/59096. With permission from Creative Commons License 3.0: https://creativecommons.org/licenses/by/3.0/)

occipitalis. It is also continuous with the SMAS of the face. Because this layer envelopes the entire head, it is properly called the *epicranial fascia*. A third layer, temporalis muscle, constitutes the third layer of the sandwich. It covers the lower half of parietal bone. It arises from somitomere 4.

Characteristics of the epicranius are: (1) it contains muscles of facial animation arising from Sm6; (2) it insinuates itself between precursor layers; (3) it does not provide sensory nerves to the overlying skin; and (4) it does not participate in the programming of bone fields. Four branches of external carotid artery (facial, superficial temporal, posterior auricular, and occipital) trace out the migration patterns of the facial muscles. All branches are derived from the segment of external carotid originating from the second aortic arch artery. These arteries are located in a distinct plane from the deep investing fascia. They relate to the *superficial investing fascia* or SMAS. Sensory branches of V3 induce these arteries, either alone, or in company with VII. Blood supply to the epicranius enters its various muscles from below.

Below the parietal eminence, new layers displace the epicranius to a more superficial position. At *superior temporal line*, temporalis muscle fascia is attached. Just below that, a parallel *inferior temporal line* marks the uppermost boundary of temporalis muscle. Like all pharyngeal arch muscles, temporalis is enclosed by *deep investing fascia* (DIF). The upper layer of DIF is straightforward fascia. The lower layer *doubles as periosteum*. Superficial to the temporalis muscle fascia lies the spongy, but vascular *subgaleal fascia* (SGF). Superficial to the SGF the second arch mimetic muscles are enclosed by *superficial investing fascia* (SIF). Cephalic to the superior temporal line, subgaleal fascia continues, uniting with periosteum and providing it with blood supply. This structure, common parlance termed *pericranium*, extends over the entire calvarium.

The concave internal lamina of parital bone presents deep grooves represent the pathways of the middle meningeal vessels. As one ascends toward the midline a groove in noted. This provides attachment for the falx cerebri. Running between the two attachments in the midline is superior sagittal sinus.

The parietal articulates with six bones: interparietal, occipital, mastoid temporal, squamous temporal, alisphenoid, frontal, and with the opposite parietal. It bears two

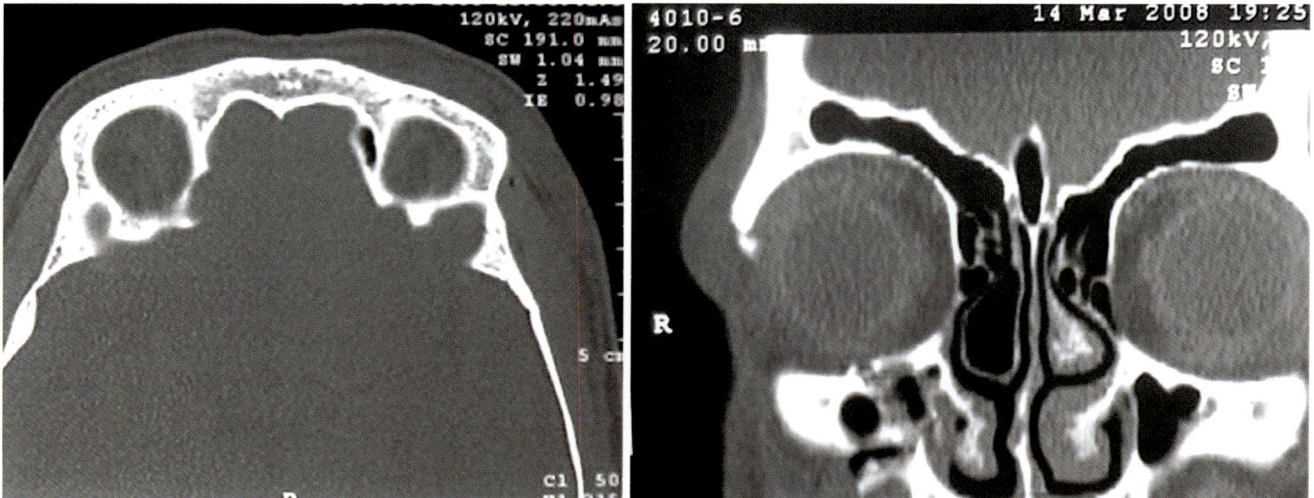

Fig. 8.70 Frontal sinus aplasia versus hyperplasia. (Reprinted from Schütz P, Ibrahim HHH, Rajab B. (April 22nd 2015). Contemporary Management of Frontal Sinus Injuries and Frontal Bone Fractures. In: Mohammad Hosein Kalantar Motamedi (ed). A Textbook of Advanced Oral and Maxillofacial Surgery Volume 2, IntechOpen; 2015. https://doi.org/10.5772/59096. With permission from Creative Commons License 3.0: https://creativecommons.org/licenses/by/3.0/)

Fig. 8.71 Parietal bone, external surface showing lower two r2–r3 quadrants (yellow) into which Sm4 temporalis has its primary insertion. Recall that in pre-mammals temporalis and masseter were once a single unit (no pterygoids existed). (Reprinted from Lewis, Warren H (ed). Gray's Anatomy of the Human Body, 20th American Edition. Philadelphia, PA: Lea & Febiger, 1918)

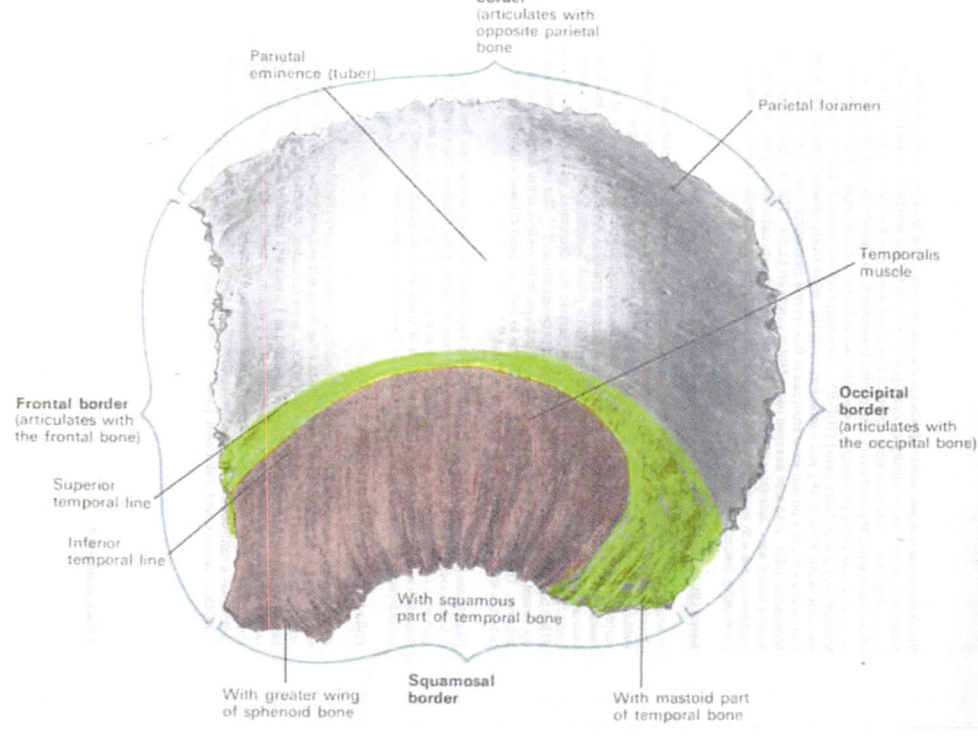

Fig. 8.72 Parietal, internal surface. Frontal and parietal branches of middle meningeal vessels indicate border between r2 and r3 neural crest. In most specimens, frontal is less widely represented. Parietal bone forms in cartilage and is predominantly paraxial mesoderm. (Reprinted from Lewis, Warren H (ed). Gray's Anatomy of the Human Body, 20th American Edition. Philadelphia, PA: Lea & Febiger, 1918)

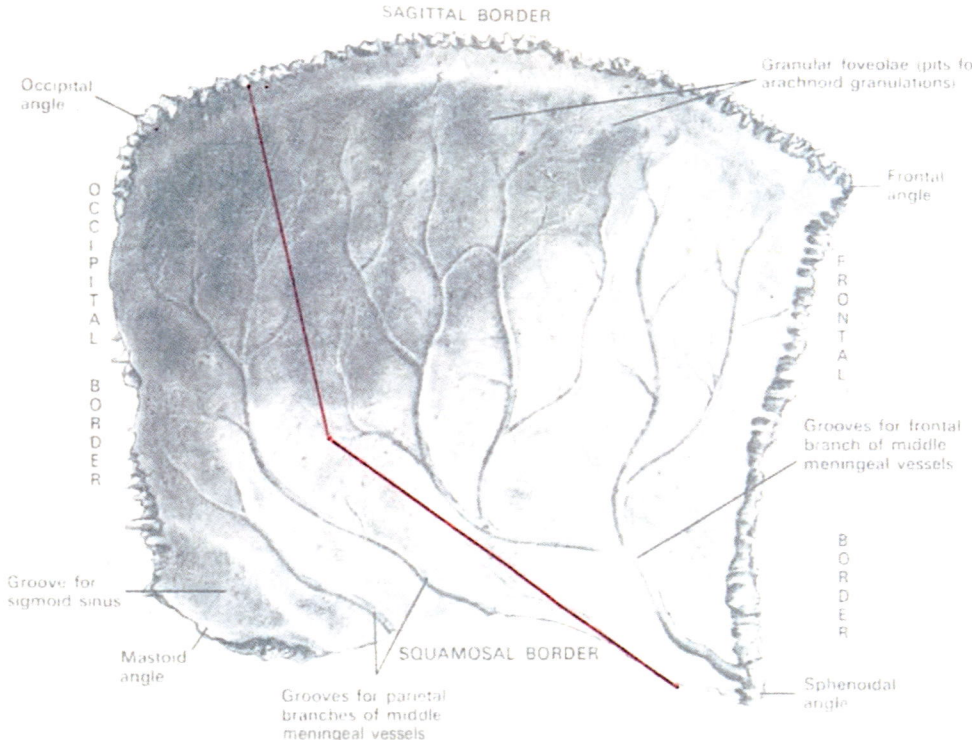

points of neurosurgical importance. *Pterion* represents the intersection of frontal, alisphenoid, squamous temporal and parietal. *Asterion* represents the transition point between of the transverse sinus to the sigmoid sinus. This takes place at the *mastoid angle*, a shelf along the inner surface of the parietal where it meets the mastoid temporal bone.

Blood Supply and Mesenchyme

Inner Table

The internal lamina of parietal bone is divided into four quadrants based on neurovascular supply of the underlying dura. The lower 80% of parietal dura is perfused by the arteries of the middle meningeal system, themselves derived from the stapedial axis. Recall that the stem of stapedial arises from the dorsal remnant of second aortic arch artery. Recall further that middle meningeal artery splits into two roughly equal branches, frontal and parietal. Although these appear to bifurcate from a common source, they represent an anastomosis between the pre-existent StV2 from the intracranial division of stapedial and the vertical ascending branch of StV3 arising from the extracranial axis of internal maxilla-mandibular.

StV2 and StV3 divide the parietal dura into anterior and posterior zones. The mesenchyme supplied by the MMA is neural crest, r2 (anterior) and r3 (posterior), admixed with paraxial mesoderm from Sm4. An additional zone of parietal dura runs longitudinally along the upper 1/5 of parietal bone, the midline border containing sagittal sinus. It consists of r1 neural crest.

Nota bene As discussed in the vascular section, when posterior forebrain erupts upward it requires immediately vascular support. This is provided by PAM from levels r2 and r3. These cells are likely to pass directly over the brain but they could also proceed through the somitomere stage...in which case they can be referred to coming from as Sm4 (Table 8.1).

Outer Table

The external lamina of parietal bone is divided into four quadrants based on neurovascular supply of the overlying dermis, fascia, and muscle. The entire skin/fascia cover is supported by the superficial temporal artery, itself a product of the external carotid system. Recall that STA is one of four branches (facial, STA, posterior auricular, occipital) that are derived from the stem of ECA assigned to supply second arch structures. The STA divides into frontal branch and

Table 8.1 Developmental fields of parietal bone, inner table

	Anterior sector	Posterior sector
Upper 1/5		
Neurovascular	StV2 MMA, frontal br.	StV3 MMA, parietal br.
Mesenchyme, dura	Neural crest r1	Neural crest r1
Lower 4/5		
Neurovascular	StV2 MMA, frontal br.	StV3 MMA, parietal br.
Mesenchyme, dura	Neural crest r2, PAM r2	Neural crest r3 + PAM r3

parietal branch; these are programmed by V2 and V3 respectively.

The major axis of STA runs in the galeal (SIF superficial investing fascia). Recall that SIF is a neural crest structure arising from levels (r4–r5). It gives off branches to overlying neural crest dermis: r2 (anteriorly) and r3 posteriorly. STA also produces penetrating branches that penetrate the outer table of parietal bone…but only in the upper 50% of the bone.

STA cannot supply the lower 50% of parietal bone because of an intervening muscle layer. Outer table differs from inner table in that temporalis muscle, a derivative of Sm4, is inserted into its lower 50%. This muscle receives the anterior and posterior deep temporal arteries, both of which are ECA derivatives from the second part of maxillo-mandibular axis (Table 8.2).

Development

Development of mesenchyme over the zone of future parietal takes place in three successive planes defined by three different vascular networks. Gastrulation, completed stage 7–8 pre-dates the development of the forebrain in stage 9 at which time mesoderm from levels r2 to r3 is immediately available to provide the blood vessels for the lateral cerebrum. Neural crest moves in from the neural folds, travelling over the surface of the brain. The deep layer of PAM and NC makes the *pial plexus*. An intermediate layer forms the parietal dura from which arises the parietal bone arises At stage 16–17, the *stapedial system of meningeal arteries* develops to support the dura and bone. The superficial layer consists of scalp with its own layer of neural crest dermis. At stages 20–23, *superficial temporal system* develops, from below upward.

Given the maturation of the STA vessels, it is not surprising that ossification of the parietal bone is observed at the eighth week of life, i.e., at or after stage 23. It takes place takes from two centers positioned one above the other. These

Table 8.2 Developmental fields of parietal bone, outer table

	Anterior sector	Posterior sector
Upper 1/2		
Neurovascular	V2 STA, frontal br.	V3 STA, parietal br.
Mesenchyme, dermis	Neural crest r2	Neural crest r3
Mesenchyme, SIF fascia	Neural crest (r4)	Neural crest (r4)
Lower 1/2		
Neurovascular, superficial	V2 STA, frontal branch	V3 STA, parietal branch
Mesenchyme, dermis	Neural crest r2	Neural crest r3
Mesenchyme, SIF fascia	Neural crest r4–r5	Neural crest r4–r5
Neurovascular, deep	Ant deep temporal	Post deep temporal
Mesenchyme, deep	Temporalis, ant, Sm4	Temporalis, post, Sm4

likely represent the *temporal zone* and the *supratemporal zone*. Parietal mesenchyme demonstrates chondral ossification in the lower zone and membranous ossification in the upper zone.

The developmental sequence of parietal bone likely results from temporal differences in its tissue coverage. Temporalis muscle with its rich blood supply has greater flow to the bone than do the small vessels penetrating the bone from the scalp. At this time, the STA vessels are advancing steadily upward. Thus, parietal ossifies first over its temporal zone and subsequently over the calvarium. All surgeons dissecting over the upper lateral cranium note the paucity of sub-scalp tissue overlying the vault.

Phylogeny

Parietal bone belongs to the midline series of tetrapods: nasal, frontal, parietal, and interparietal (squamous occipital). Its initial position in *osteolepis*, is quite anterior, directly between the orbits. Growth of the jaws (and consequently the face) provided enhanced biting surface for predation. This combined with brain caused expansion and posterior repositioning of the frontals while the orbits remained stationary. The parietals shifted backwards. Throughout evolution the parietals remained bounded posteriorly by the postparietals. As well shall see subsequently, these combined with other bone fields to produce the membranous supraoccipital bone complex. This situation is seen in the labyrinthodont *Paleoherpeton*. Here, parietals are bounded, from front to back, by frontal, postfrontal, intertemporal, supratemporal, temporal and postparietal by squamosals laterally and postparietals posteriorly.

Clinical Correlation

No other calvarial bone has the mesenchymal complexity of parietal. The most common forms of craniosynostosis occur along three of the four boundary zones of parietal bone: coronal, sagittal, and lambdoid. The pathology underlying the synostosis, be it partial or complete, reflects and a deficiency state or dysfunction in one or more quadrants.

Furthermore, the syndromic variations that accompany some forms of synostosis can be understood on the basis of simultaneous involvement of structures sharing a common neuromeric relationship. Homeotic analysis of tissues on either side of a suture, both in the normal and pathologic state can be a great value in factors maintain suture patency or premature shutdown.

The source material for the parietal bone, like that of frontal and interparietal, arises from paraxial mesoderm at the level of the dura. No dura is purely a neural crest structure. It contains PAM to make its blood vessels. Parietal PAM arises at gastrulation from levels r2 and r3. These are translated into the overlying bone. The biologic map of parietal dura is neuromeric and affects neural crest populations as well. Thus, parietal is a rectangle of mesodermal bone with a genetic signature different at each of its boundaries and different from those of its neighbors.

The sutural boundaries of parietal bone are characterized by the interposition of dura arising from below and becoming continuous with the outer periosteum. In the *coronal suture* contains r1 from the frontal dura and r2 from the anterior parietal dura. The *sagittal suture* combines r1 with r2 anteriorly ad r3 posteriorly. The *lambdoid suture* mixes r3 with C2 inferiorly and C3 medially. This has implications for this development of synostosis. *Parietal-squamosal* is simple: on either side the neural crest populations are the same r2 versus r2 and r3 versus r3.

Interparietal Bone Complex

Interparietal bone complex forms in membrane under the influence of dura. It has multiple fields and arises from both mesoderm and neural crest: it is non-somitic. Its structural design is almost a carbon copy of parietal. It is biplanar. Its external lamina arises from two sources, with medial and lateral fields composed of C3 and C2 mesoderm, respectively. The mesenchyme of the internal lamina is much simpler as its arises from r3 neural crest dura. Interparietal complex in

mammals is associated with muscles of fascial expression, the occipitofrontalis (Figs. 8.73, 8.74, 8.75, 8.76, 8.77, 8.78, 8.79, 8.80, 8.81, 8.82, and 8.83).

Before we begin, a caveat

Traditional anatomic terminology regarding the posterior skull is confusing and wrong from both a developmental and a paleontological perspective. The occipital bone portrayed in all texts considers to be a single occipital bone consists of two distinct sets of bone fields. The occipital bone complex—described (appropriately) at the terminus of this chapter—belongs to the cranial base. It is composed of paired basioccipital, exoccipital, and supraoccipital bones, all of which form in cartilage derived from the occipital somites. Furthermore, we shall see that supraoccipital itself has eight distinct centers of ossification.

Confusion arises for two reasons. (1) The individual components of the chondral occipital complex are lumped together with no explanation as to how they form. (2) The supraoccipital bone is described as having two distinct zones. *Planum occipitale* (the true interparietal) is membranous and gives insertion for a single second arch muscle, occipitalis. *Planum nuchalae* (supraoccipital bone proper) is chondral and gives insertion for head extensors.

It is impossible that planum occipitale belongs to supraoccipital. It contains non-neural crest mesenchyme,

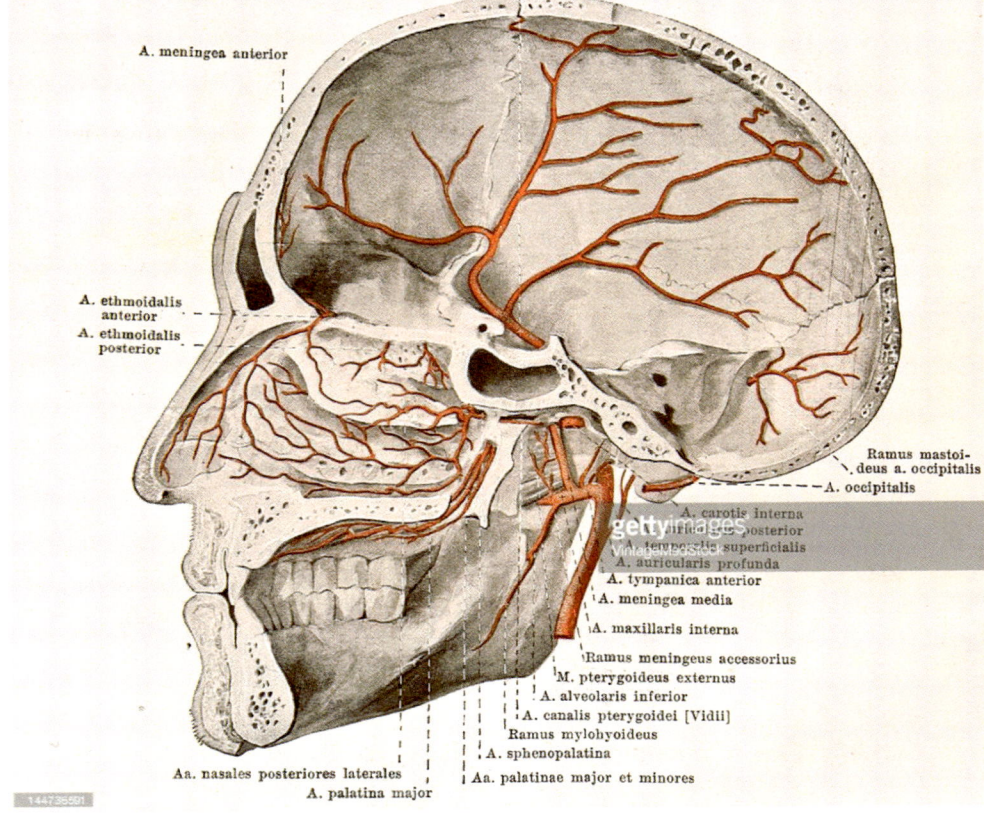

Fig. 8.73 Middle meningeal artery by Werner Spalteholtz. Note dissection of StV2 meningo-orbital artery (recurrent meningeal). Note division of parietal into two distinct zones, StV2 and StV3. (Reprinted from Spalteholtz W. Hand atlas of Human Anatomy. Philadelphia, PA:, Lippincott; 1903. Available from: https://babel.hathitrust.org/cgi/pt?id=uc1.31822000936377)

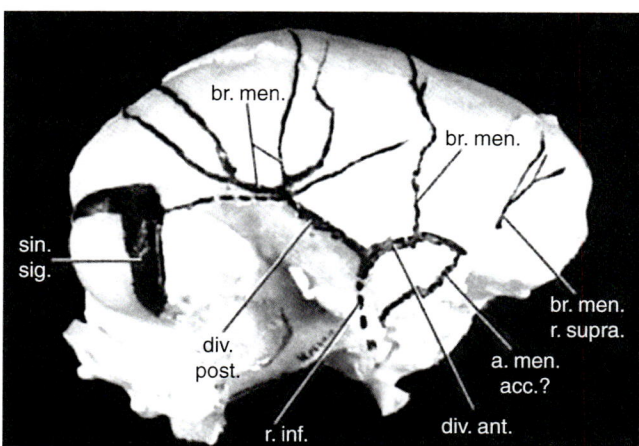

Fig. 8.74 Stapedial system endocast. Endocast of primate skull showing trifurcation of the stapedial system. The stapedial stem has involuted making the entire system dependent on the external carotid via StV3 middle meningeal. StV2 meningo-orbital (recurrent middle meningeal) branch seen. Orbit removed, therefore, StV1 orbital branches not present. In the orbit, StV2 connects to StV1 lacrimal branch. StV1 anastomosis to ophthalmic changes flow so recurrent meningeal involutes. (Reprinted from Diamond MK. Homologies of the Stapedial Artery in Humans, With a Reconstruction of the Primitive Stapedial Artery Configuration of Euprimates. *Am J Phys Anthro* 1991; 84:433–462. With permission from John Wiley & Sons)

incompatible with the somite sclerotomes. Its associated muscle, occipitalis, is somitomeric remains calvarial, whereas all muscles inserted into supraoccipital arise from somite dermatomes and are connected to the postcranial skeleton.

Descriptive Anatomy

The convex external lamina of interparietal extends from the *external occipital protuberance* upward to the lambdoid suture separating it from parietal bone. Two lines of importance arch away from the protuberance. These are not widely separated but each has its own anatomical significance. *Superior nuchal line* (SNL) is the more caudal. It marks the upper boundary of supraoccipital bone and provides insertion for the extensor muscles connecting the posterior skull to the body. Running just below this line, from medial to lateral, are trapezius and sternocleidomastoid, two unique muscles originating (in part) from occipital somites. These are discussed in greater detail in Chap. 10. All muscles below originate from cervical somites. *Highest nuchal line* (HNL) is cranial to SNL. The membranous bone between HNL and SNL is a developmentally obscure but important structure,

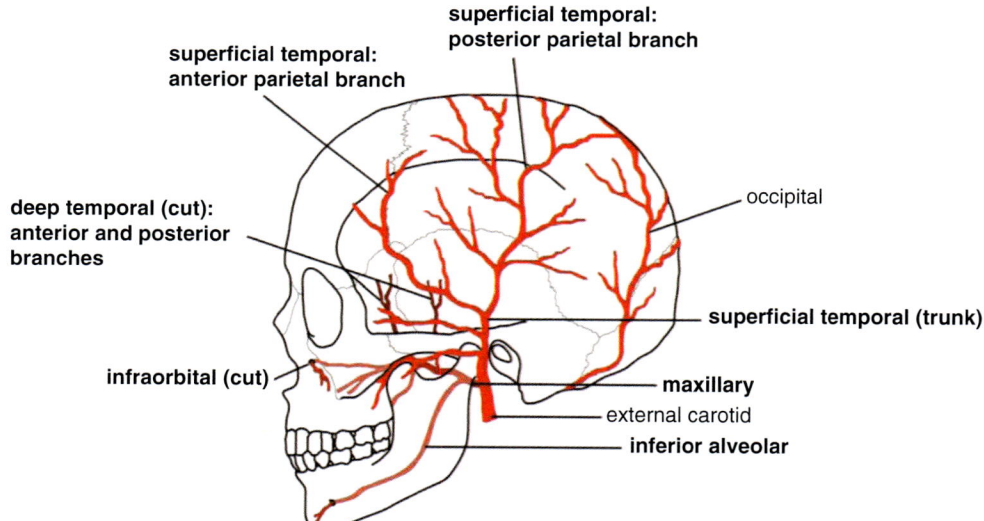

Fig. 8.75 Superficial temporal artery and parietal bone. External vascular fields of the temporo-parietal region. Squamosal bone is vascularized by temporal vessels to temporalis muscle. V2-innervated territory does not extend beyond superior temporal line. Parietal bone is supplied internally by StV2 frontal and StV3 parietal branches of middle meningeal and externally by anterior and posterior branches of superficial temporal artery, both of which are programmed by V3. (Reprinted from Snoddy, AME, Buckley HR, Elliott GE, Standen VG, Arriaza BT, Halcrow SE. Macroscopic features of scurvy in human skeletal remains: A literature synthesis and diagnostic guide. Am J Phys Anth 2018; 167(4):876–895. With permission from Creative Commons License 4.0: http://creativecommons.org/licenses/by/4.0/)

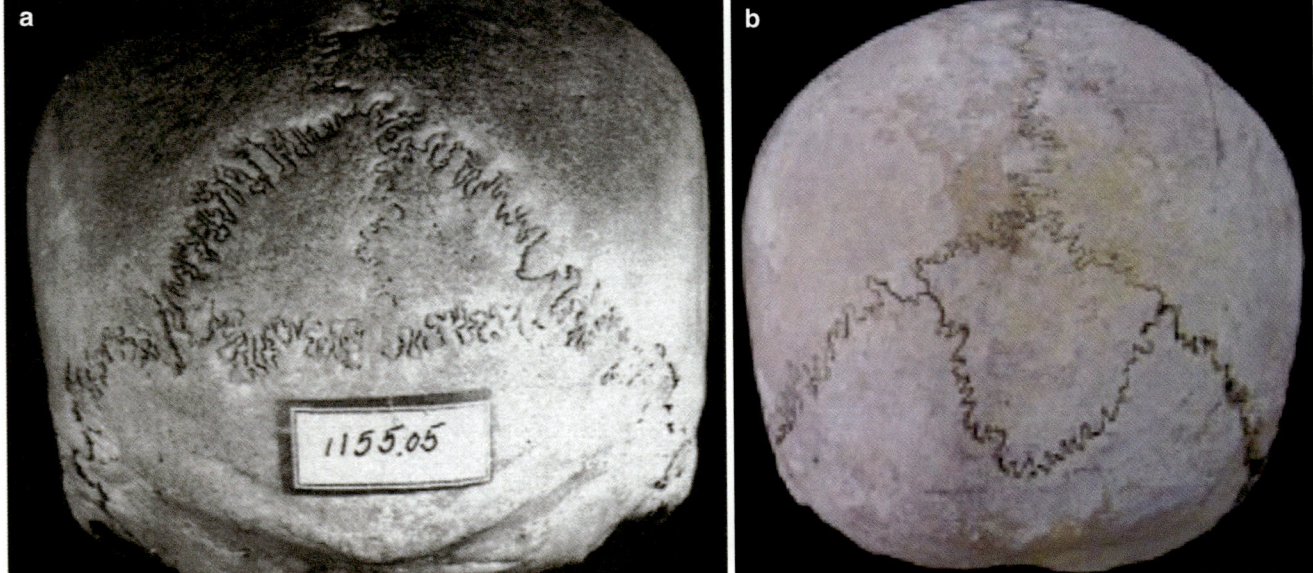

Fig. 8.76 Parietal bone evolution. Backward repositioning of the parietal and interparietal bone fields. *Osteolepis*: small nasal bones but midline frontals. *Panderichthys*: an advanced Elpigostegalian fish. Fin bone fit pattern: humerus, radius, ulna Braincase is tetrapod-like *Ichthyostega*. First tetrapod with feet, with loss of extrascapulars, postparietals arrive at occiput. *Paleoherpeton*: early tetrapods so not have a supraoccipital bone. (Reprinted from Schulze, H-P & M Arsenault. The panderichthyid fish Elpistostege: a close relative of tetrapods? Palaeontology 1985;28:293–309. With permission from The Paleontological Association)

Fig. 8.77 Interparietal bone ossifications can have many variations (**a**): Pal (India); (**b**): Nikolova (Bulgaria). (**a**) The central interparietal bone is divided into two equal fields. (**b**) Variations of the interparietal contribute to a complex map of the occipital fields—see Fig. 8.76. (**a**: [Reprinted from Pal GP, Tamankar BP, Routal RV, Bhagwat SS (1984) The ossification of the membranous part of the squamous occipital bone in man. J Anat 1984;138:259–266. See also: Pal GP. Variations of the interparietal bone in man. J Anat 1987;152:205–208.with permission from John Wiley & Sons). **b**: (Reprinted from Nikolova S, Toneva D, Yordanov Y, Lazarov N. Variations in the squamous part of the occipital bone in medieval and contemporary cranial series from Bulgaria. Folia Morphol (Warsz). 2014 Nov;73(4):429–38. With permission from Creative Commons License 4.0: https://creativecommons.org/licenses/by-nc/4.0/)

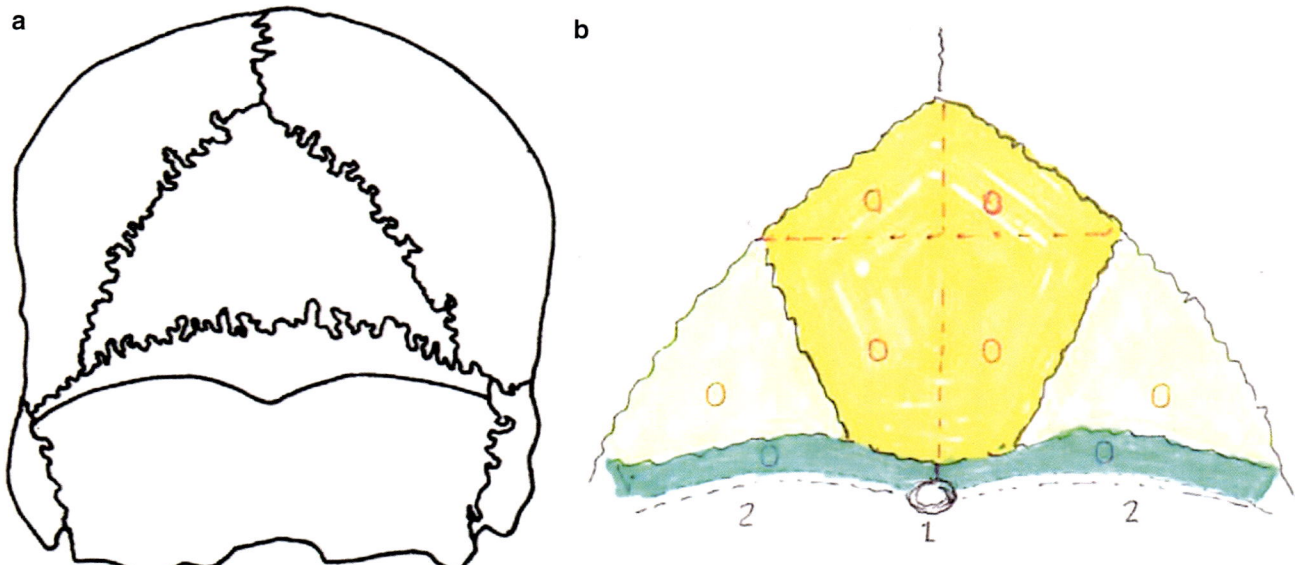

Fig. 8.78 Interparietal bone ossifications can have many variations. Interparietal bone seen (left) intact with no subdivisions and (right) with eight potential ossification centers. Dark yellow = interparietal, Light yellow = tabular, Green = torus occipitalis lateralis or lamella triangularis, 1 = superior occipital protuberance, 2 = superior nuchal line. Torus occipitalis lateralis has two ossification centers. Note: Srivastava considers possibility of additional lateral centers in each tabular field.

[Left: (Reprinted from Srivastava HC. Development of the ossification centers in the squamous portion of the occipital bone in man. *J. Anat.*, 1977;124:643–9. With permission from Blackwell Publishing Ltd). Right: (Reprinted from Srivastava, H C. Ossification of the membranous portion of the squamous part of the occipital bone in man. *J. Anat.* 1992; 180:219–24. With permission from John Wiley & Sons)]

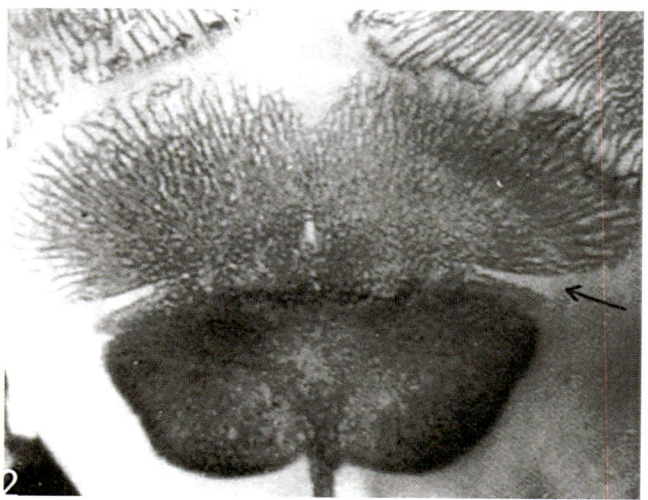

Fig. 8.79 Torus occipitalis lateralis. Photograph of 16 week fetus showing the *lamina triangularis* (intermediate segment) fused with the supraoccipital bone below and medial portion of the lateral plate above. Its lateral portion is separated from the lateral plate by the lateral fissure (arrow). The lateral plate is fused with the medial plate. The median fissure separating the two medial plates is obliterated except in its small lower portion. (Reprinted from Srivastava, H C. Ossification of the membranous portion of the squamous part of the occipital bone in man. *J. Anat.*, 1992;180:219–24. With permission from John Wiley & Sons)

torus occipitalis lateralis. It represents the point of attachment for occipitalis. This muscle is the most posterior component of the SMAS, i.e., the superficial investing fascia originating from neural crest associated with the second arch. Thus, r4–r5 SIF envelops the entire head as the epicranius.

The internal surface of interparietal is smooth, being subdivided by the posterior terminus of superior sagittal sinus. Its inferior border with the chondral basioccipital bone is marked by the *transverse sinus*. Attached to these structures are the falx cerebri and the tentorium cerebelli. Nestled into the interparietal fossae are the posterior cerebral lobes while the hemispheres of the cerebellum occupy the two supraoccipital fossae.

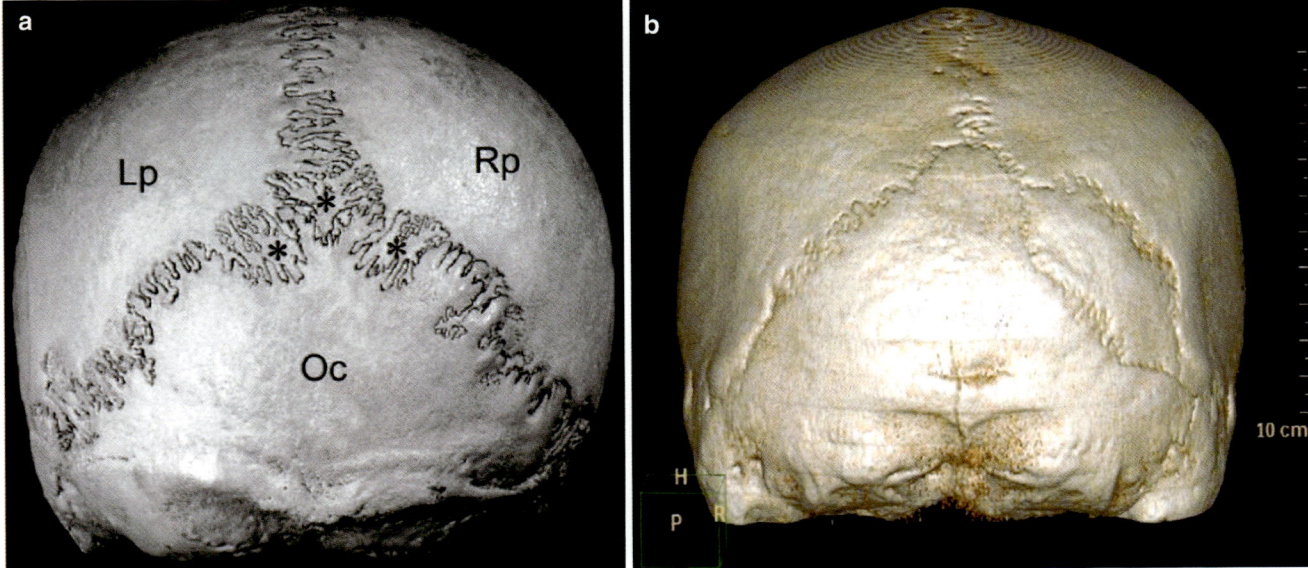

Fig. 8.80 Wormian bones marked by asterisks. #-D CT demonstrates microstructure of incomplete trabeculae. (**a**) Small wormian bones seen within the interstices of the lambdoid sutures. (**b**) Isolated right tabular bone. Interparietal fields can be misinterpreted as sutural bone. [**a**: (Reprinted from Magdalena Kozerska, Janusz Skrat, Jerry Walocha, Andrzej Wrobel, Krzystof Piech. Imaging of the Wormian bones using microcomputer tomography. Folia Med Cracov 2013 53(4):21–28. With permission from Folia Medica Cracoviensia). **b**: (Reprinted from Wikimedia. Retrieved from: https://commons.wikimedia.org/wiki/File:Wormian_bone_lambda_VRT.jpg. With permission from Creative Commons License 3.0: https://creativecommons.org/licenses/by-sa/3.0/deed.en)]

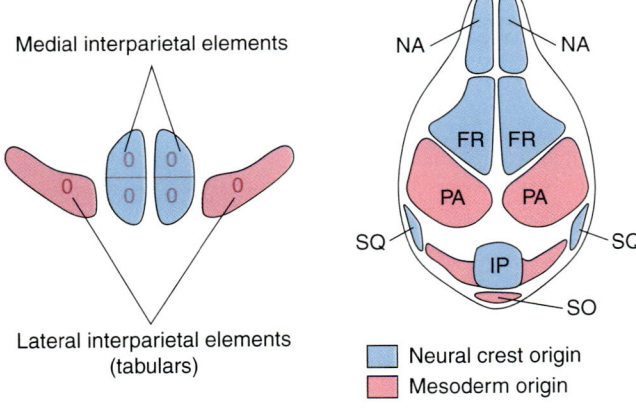

Fig. 8.81 Mesenchymal mapping: interparietal and tabular. Ossification centers of interparietal represent combination of interparietal bones (possibly with retained extrascapulars with inclusion of tabular bone fields. (Reprinted from Koyabu D, Maier W, Sánchez-Villagra MR. Paleontological and developmental evidence resolve the homology and dual embryonic origin of a mammalian skull bone, the interparietal. *PNAS* 2012; 109(35):14075–14080. With permission from PNAS)

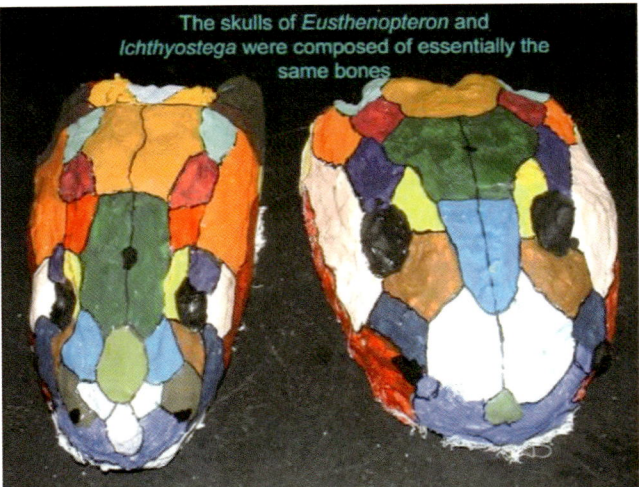

Fig. 8.82 Disconnection of pectoral girdle from skull. Sarcopterygiann *Esthenopteron* versus basal tetrapod *Ichthyostega*. Color code: white (nasal), blue (frontal), dark green (parietal), brown (postparietal), red (posttemporal), aqua (tabular), pale blue (medial extrascapular), tan (lateral extrascapular). Note loss of extrascapulars, posterior displacement of midline skull series with postparietal now at the occipital border, and posteromedial repositioning of the tabulars. Supraoccipital does not appear in evolution until the amniotes. (Courtesy of Dr. Walter Jahn, Dept. of Biology, SUNY Orange Campus)

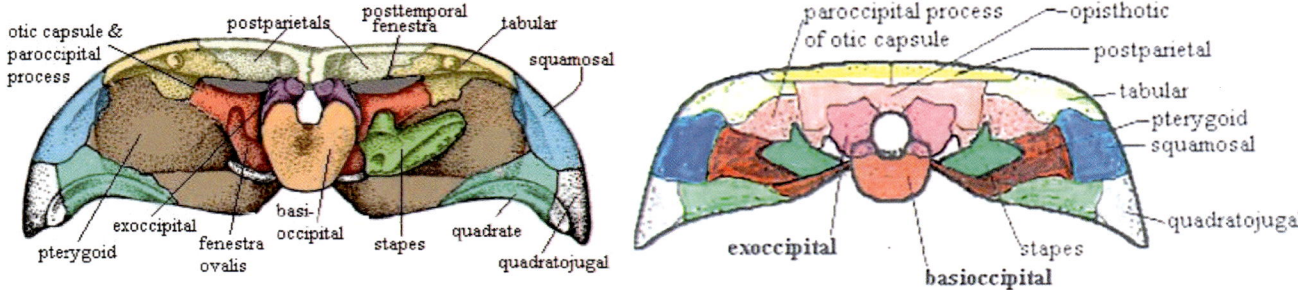

Fig. 8.83 Tetrapod *Greererpeton* shows postparietals flanked by tabulars. The space between the otic capsule and the skull roof, *posttemporal fenestra*, is filled in with unspecified mesenchyme. This will be the future lateral supraoccipital zone. Skull roof provides the cover for foramen magnum. Basal amniote *Paleothyris* had an anapsid skull. Otic capsules (opisthotics) spanned between exoccipital to cover foramen magnum. This will mark the beginnin of supraoccipital. [Left: (Reprinted from Carroll RL. Vertebrate Palaeontology. New York, NY: WH Freeman and Company; 1988). Right: (Reprinted from Berman DS. Origin and evolution of the amniote occiput. J Paleontol 2016; 74(5):938–956. With permission from Cambirdge University Press)]

Blood Supply and Mesenchyme

The Internal Lamina

The internal lamina of interparietal bone is in contact with the dura of posterior fossa above transverse sinus. Neural crest from r3 is admixed with mesoderm from cervical somites S7–S7 (C2–C3) (Table 8.3).

The external lamina of interparietal bone is divided into four quadrants based on neurovascular supply of the overlying dermis, fascia, and muscle. The entire skin/fascia cover is supported by the occipital artery medially and the posterior auricular artery laterally. These arteries run with greater occipital nerve C3 medially and lesser occipital nerve C2 laterally. Penetrating branches supply the external table of interparietal directly in its upper 50%. Occipitalis muscle, also supplied by the same two arteries, provides supply directly in the lower 50% (Table 8.4).

Occipital Artery

The occipital artery is one of the four branches supplying SIF fascia and the tissues of the second arch. It originates from posterior aspect of external carotid and passes posteriorly, pursuing a course beneath and parallel to the posterior digastric muscle and stylohyoid. Hypoglossal nerve skirts around it on its way to the tongue. Occipital crosses both the ICA, internal jugular and cranial nerves X and XI. It then passes upward in a furrow medial to mastoid process. At external occipital prominence, it ascends into the scalp. To do so, it must bore a hole in the dense fascia of sternocleidomastoid and trapezius. In the scalp, occipital artery *pursues a medial course upward*, making an anastomosis with posterior auricular laterally and, at the vertex, with parietal branch of superficial temporal.

Branches of occipital artery are as follows:

1. *Muscular branches*: supply medial aspect of occipitalis muscle and posterior digastric (both second arch muscles

Table 8.3 Developmental fields of interparietal bone, inner table

	Medial sector	Lateral sector
Neurovascular	V3 MMA	V3 MMA
Mesenchyme, dura	Neural crest r3 + PAM C3	Neural crest r3 + PAM C2

Table 8.4 Developmental fields of interparietal bone, outer table

	Medial sector	Lateral sector
Upper 1/2		
Neurovascular	Greater occipital C3 / Occipital artery	Lesser occipital C2 / Posterior auricular
Mesenchyme, dermis	PAM C3	PAM C2
Mesenchyme, SIF fascia	Neural crest r4–r5	Neural crest r4–r5
Mesenchyme, bone	Neural crest r3	Neural crest r3
Lower 1/2		
Neurovascular	Greater occipital C3 / Occipital artery	Lesser occipital C2 / Posterior auricular
Mesenchyme, dermis	PAM C3/no SIF	PAM C2/no SIF
Mesenchyme, SIF fascia	Neural crest r4–5	Neural crest r4–5
Mesenchyme, deep	Occipitalis (Sm6)	Occipitalis (Sm6)
Mesenchyme, bone	Neural crest r3	PAM Sm6–7

from Sm6). It also provides collaterals to three of the four muscles below superior nuchal line: sternocleidomastoid, splenius capitis, and semispinalis capitis of the head

2. *Stylomastoid artery*: follows facial nerve to supply tympanic membrane, second arch components and mastoid cavities, as well as the r4–r5 semicircular canals

3. *Auricular branch*: back of the ear and mastoid

4. *Meningeal artery*: penetrates the mastoid foramen to supply dura mater of the mastoid and the diploe

5. *Descending branch* in the neck (a) stays superficial to supply trapezius and anastomose with ascending branch

transverse cervical; and (b) goes deep to anastomose with vertebral

Posterior auricular artery arises just above the take-off of occipital and directly opposite styloid process. It ascends more lateral than occipital, pursuing a groove between mastoid process and the external ear.

It has three main branches:

1. *Stylomastoid branch* supplied tympanic cavity and the tympanic membrane, semicircular canals and mastoid process, all structures related to r4–r5.
2. *Auricular branch* supplies ear muscles from the back and perforates to anterior surface.
3. *Occipital branch* pursues a lateral course upward; it supplies occipitalis and lateral aspect of the supraoccipital scalp.

In summation, the external table of postparietal bone has two zones of perfusion corresponding medially with C3 and occipital artery and laterally with C2 and posterior auricular artery. Both arteries represent branches of the external carotid system in register with the second aortic arch, both supply muscles of facial expression originating from Sm6. They supply not only the superficial investing fascia that encloses the second arch muscles also the overlying skin of the scalp.

This raises an interesting and unanswered question. Occipital scalp skin, being innervated by C2–3, arises from somites 6 and 7. These levels obviously has epaxial branches supplying skin over the nape of the neck. Why should the blood supply for this skin be dependent upon another source, i.e., the external carotid? Recall that neuromeres r4–r11 and c1 are not capable of producing skin. (with the exception of a small amount of the auricle). No somatic sensory nerves exist at these levels. Their mesenchyme is diverted away from dermis. Any ectoderm produced at these levels during gastrulation seems to have been overtaken, perhaps through a process of apoptosis, and replaced by dermis and epidermis beginning with the second cervical neuromere.

We can speculate that this interface of skin, between the back of the skull and the neck, represents a transition zone in which, with the expansion of the brain and the occiput, skin from neuromeres c2–c3 that is in contact pulled forward through a process of tissue expansion, much as the cowl pulled forward over the head of a monk.

At any rate, two sets of vascular territories result and this has developmental implications, as we shall see, in accounting for the ossification centers of what was originally a complex of bones, now fused into an apparently single unit.

Development

Development of the planum occipital has traditionally been considered to proceeds from paired ossification centers near the midline at 8 weeks. Several months later, about the fifth month, an additional pair of centers appears *inferolaterally*.

The anatomic literature documents the existence of multiple ossification centers. Srivastava considers the membranous bone to have 8–10 such centers distributed in three zones. The most caudal zone (confusingly termed the *intermediate segment*) lies between superior nuchal line and highest nuchal line. This zone was previously identified by Ranke as the *lamina terminalis* or *torus occipitalis traversus*. It contains two centers. Lying above lamina triangularis/torus is the true interparietal bone complex. It is triangular in shape and has four bone fields. Paired central interparietal contain four ossification centers while two lateral interparietal fields have one to two center(s) each.

Fusion of various centers creates variations in the shape of the bone complex. In some models [7, 8], the bases of the triangular lateral fields meet in the midline, pushing the central fields upward. What is important here is not the exact number of ossification centers but the precise number of constituent bone fields.

Koyabu et al. demonstrate that the interparietal ossification centers are homologous to the postparietals while the lateral interparietal ossification centers are homologous to the tabulars [9]. The mesenchymal origins of these bone fields are different. The postparietals arise in register with dura innervated by V3 and therefore are of r3 neural crest derivation. The torus and tabulars originate from non-somitic paraxial mesoderm, most probably from the sixth somitomere. The reason for this assertion is that torus bears the insertion of occipitalis, a Sm6 muscle which it shares with mastoid, also non-somitic.

Phylogeny

How can we make sense of this scenario? A brief survey of the early braincase gives us some clues. Recall that bony fishes are divided into Actinopterygii (spiny fins) and Sarcopterygii (fleshy fins), the latter being the evolutionary branch leading to tetrapods. Basal tetrapodomorph fishes emerged in the early Devonian. The best-known specimens are from China and demonstrate changes of the brain and lower jaw that are unique to tetrapods. Primitive clades of sarcopterygians (porolepiforms, onychodontids, rhizodontids, osteolepiforms) had disorganized snouts covered with variable numbers of bones. In the basal form, *Osteolepis*, anterior to the parietals, the dermal bones are highly variable. But beginning with the parietals, order was established. Just behind the parietals are found the postparietals.

In subsequent tristchopterids such as the *Eusthenopteron*, the head, gills, and shoulder girdle attached to the skull all became covered with dermal bones. *The back of the skull consisted of medial and lateral extrascapulars.* The postero-lateral corners of the postparietals was in contact with the *tabular bones.*

Next on the stem lineage to tetrapods were the panderich-thyids (also known as the elpistosegids) in which the snout begins to change. In *Panderichthys*, previously lateralized nasal bones consolidate and move into the midline, as do the frontal bones. Parietal is pushed backwards but the back of the skull remains the same. Parietal and postparietals are *unsutured*, permitting flexion of the intracranial joint. What we now have, however, is an unbroken chain of bones from snout to occiput.

In true tetrapods, dramatic changes occur synchronously. In *Ichthyostega*, (Gr. "fish-limb"), a primitive neck appears, with a true occiput and a new relationship at the cranial ver-tebral junction. The forelimb girdle, previously welded to the skull becomes disconnected, permitting mobility. This requires elimination of the extrascapular series.

Loss of the extrascapular bones in *ichthyostega* causes the back of the skull to look shorter, but the biting apparatus has advanced in front of the orbits. The absent medial extrascap-ulars are replaced by the post parietals which moved back-ward in the midline to arrive at the posterior border of the skull. The loss of the lateral extrascapulars is compensated for translation of the tabulars which have swung around to crowd in on the posterior lateral margins of the postparietals. The supratemporals have shifted back to replace the tabulars. But as we shall see, perhaps the genetic "idea" of the medial and lateral extrascapulars was not lost after all.

Torus Occipitalis: The Mysterious Membranous Bone

Thus far we have been able to account for the history of the interparietal bone complex as six ossification centers. The two lateral fields are remnants of the tabular bones. They are in register with C2 and are supplied by posterior auricular. The central four fields represent the original paired postpari-etal bones. These are in register with C3 and are supplied by occipital artery. Alternatively, the two upper central zones could also represent separate, *pre-interparietal* fields, a con-troversial term. But can we make of the lowest zone, lamina triangularis? This obscure anatomic structure turns out to play an important role in understanding how the occiput is constructed and how its component parts evolved over time.

This topic leads us inevitably into a consideration of the occipital bone complex. Supraoccipital is somatic and chon-dral. There exists above it a transverse zone of membranous non-somitic bone unrelated to either the postparietals or the tabulars. Where does it come from?

Fishes have no need for a craniovertebral joint. They function quite nicely in their watery environment with an unrestricted notochord which is inserted directly into the otic capsule, the otoccipital skull. The occiput is an invention of

tetrapods. In *Acanthostega*, it is a very simple affair. Foramen magnum is bounded by basioccipital below, exoccipitals at the sides, the otic capsules making up the roofline. There is no supraoccipital bone. (Note: the piscine "supraoccipital bone" is a completely unrelated affair designed to confuse those of us that have to deal with tetrapods anatomy.). *Hidden in the connective tissue spanning between the two otic cap-sules is the source material for the membranous bone, the future torus occipitalis.* Perhaps this bone fields in this zone represent genetic holdovers of the medial and lateral extrascapulars. Gene mapping will potentially clarify the situation.

During the Carboniferous period, the next iteration of tet-rapods evolved and eventually split into the batrachomorphs (frog forms) and the reptilomorphs (precursors of amniotes). In this transition, reptilomorph *Greererpeton* demonstrates the backward translocation of the postparietals into the "occipital slot." We can see them resting almost directly on the top of the otic capsule. Almost, but not quite… The space in between, the *post-temporal fenestra*, will be one that is occupied by the torus.

The environmental catastrophes of the late Carboniferous marked the end of the basal tetrapods. In the line toward amniota, the temnospondyls diverged first, leading to the Lissamphibia (frogs, salamanders and the like). The reptilo-morph line produced the lepospondyls (microsaurs), an evo-lutionary dead-end. By the early Permian period, stem amniotes, embolomeres (often termed anthracosaurs) emerged.

Changes in the amniote occiput took place above fora-men. The otic capsules become internalized and are no lon-ger seen the by the time of the temnospondyls such as *Lapillospsis* (Fig. 8.80). The skull shows optisthotics extend-ing two lateral processes to connect with the tabulars. At the same time, two medial processes extend straight upward linking the exoccipitals with the postparietals. This sets up four potential spaces for bone formation. These are candi-dates for new supraoccipital bone fields. Do these spaces represent genetic "footprints" of the ancient medial and lat-eral extrascapulars?

Moving into the synapsids a gross expansion of supraoc-cipital has taken place. The theriodont Gorgonopsids demon-strate all four components of the interparietal bone complex. Characteristic of mammals they had incipient ear bones but they not fully pelagic (hair-bearing) and so probably did not have the mammalian fronto-occipitalis. Therefore, the exis-tence of a lamina triangularis is doubtful at this evolutionary level.

The synapsid line continued on to the cynodonts. The cynodonts had greatly expanded brains with lateral expan-sion of the jaw joint, a zygomatic arch with enhanced mus-cles of mastication, and the occiput was affected as well. *Galeasaurus planiciceps* was a basal pre-mammilloform cynodont in South Africa (Fig. 8.81). A thick rim of bone is seen interposed between foramen magnum and supraoccipital. As we shall see, supraoccipital bone also has multiple ossification centers (8). This zone is the likely source of the lower tier of occipital bone fields.

In summation, dorsal braincase covering the occiput between the lambdoid suture and the superior nuchal line consists of four bones fields of differing mesenchymal composition. The lateral fields are composed of Sm6 mesoderm and are homologous with the tabular bones. Paired medial fields are composed of r3 neural crest and are homologous with the postparietal bones. Interposed between these four fields and the chondral supraoccipital bone complex is a zone of membranous bone of somitomeric mesoderm that is unrelated to the occipital somites. It likely arises de novo as a condensation of mesenchyme connecting the two opisthotic bones prior to their incorporation into the posterior cranial fossa. Although the source of this bone field has not been mapped out, it is likely to arise from Sm7.

Squamosal/Quadratojugal Complex (sq-qj)

Descriptive Anatomy

The proper zoologic name for the squamous temporal bone is the *squamosal quadratojugal bone complex*. This has nothing to do with the temporal bone. In evolution, these bones originally appeared as a dermal bones of the skull with no relationship to the otic capsule. Squamosal develops in membrane from r3 neural crest. The degree to which Sm4 PAM contributes to its mesenchyme is unknown. The bone is

thin and unilaminar. The internal surface of the squama bears the trench of the posterior branch of *middle meningeal artery*. At the inferio-medial border of Sq is the petrosquamous suture. It borders anteriorly with alisphenoid greater wing, and postero-superiorly with parietal bone. The external surface of Sq provides for the insertion of the temporalis muscle which passed downward to insert into the coronoid process and anterior ramus. The tendinous portion temporalis corresponds to the surface area of Sq whereas its muscular part extends upward over the lower half of parietal bone. Just above the external auditory canal the bone is grooved by the *middle temporal artery* (Figs. 8.84, 8.85, 8.86, 8.87, 8.88, 8.89, 8.90 and 8.91).

The upper border of squama bears an attachment of *temporalis muscle fascia*. Caudal to this line the muscle has two fasciae. The deep lamina of temporalis muscle fascia is directly associated with the muscle itself and attaches to the deep margin of zygomatic arch. The superficial lamina attaches to the external margin of zygomatic arch. Between the two laminae is a fat pad. This is an important surgical landmark, the details of which will be discussed in the following chapter on craniofacial muscles and fasciae. Temporalis muscle fascia attaches above it. Its lower and lateral borders bear the secondary attachment of the masseter muscle.

Projecting forward from the squama is the *zygomatic process*. Zygomatic process originates from two separate zones. Posterior root lies just above external auditory meatus; it is a continuation of the upper border of quadratojugal bone and connects with the temporal line. Anterior root is a continuation of the lower border of jugal bone and terminates with the

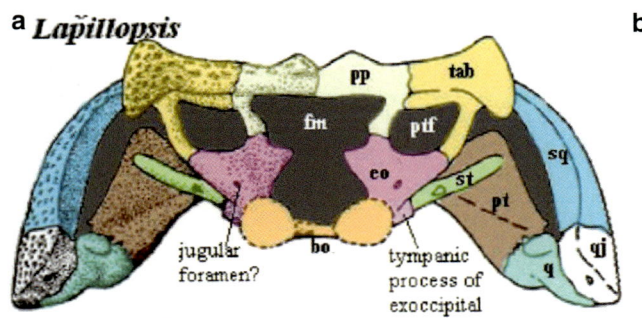
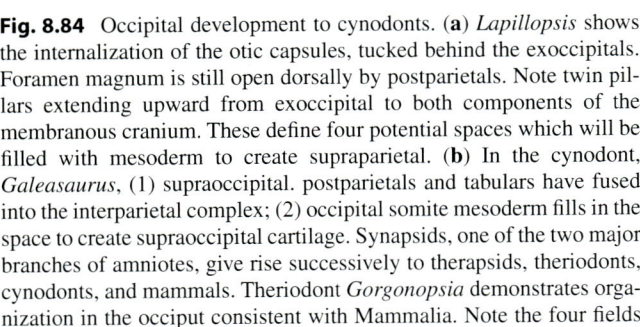
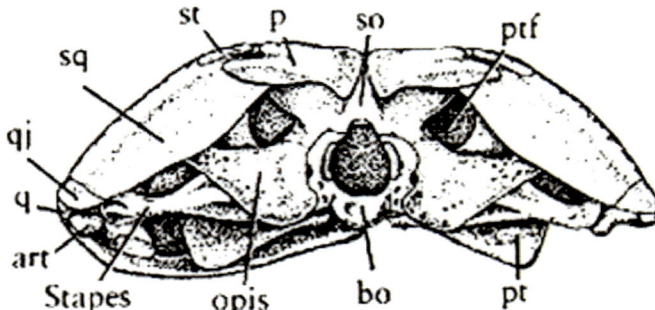

Fig. 8.84 Occipital development to cynodonts. (**a**) *Lapillopsis* shows the internalization of the otic capsules, tucked behind the exoccipitals. Foramen magnum is still open dorsally by postparietals. Note twin pillars extending upward from exoccipital to both components of the membranous cranium. These define four potential spaces which will be filled with mesoderm to create supraparietal. (**b**) In the cynodont, *Galeasaurus*, (1) supraoccipital. postparietals and tabulars have fused into the interparietal complex; (2) occipital somite mesoderm fills in the space to create supraoccipital cartilage. Synapsids, one of the two major branches of amniotes, give rise successively to therapsids, theriodonts, cynodonts, and mammals. Theriodont *Gorgonopsia* demonstrates organization in the occiput consistent with Mammalia. Note the four fields

of the interparietals (formerly postparietals) flanked by the tabulars. Supraoccipital is massive. *bocc* basioccipital, *exocc* exoccipital, *ip* interparietal, *op* opisthotic, *ptf* posttemporal fenestra, *socc* supraoccipital, *sq* squamosal, *t* tabular. [**a**: (Reprinted from Yates AM. The Lapillopsidae: a new family of small temnospondyls from the Early Triassic of Australia. J. Vert. Paleontol. 1999;19:302–320. With permission from Taylor & Francis). **b**: (Reprinted from Araujo R, Fernandez V, Polcyn MJ, Fröbisch J, Martins RMS. Aspects of gorgonopsian paleobiology and evolution: insights from the basicranium, occiput, osseous labyrinth, vasculature, and neuroanatomy. *Peer J* 5:e3119. With permission from Creative Commons License 4.0: https://creativecommons. org/licenses/by/4.0/)]

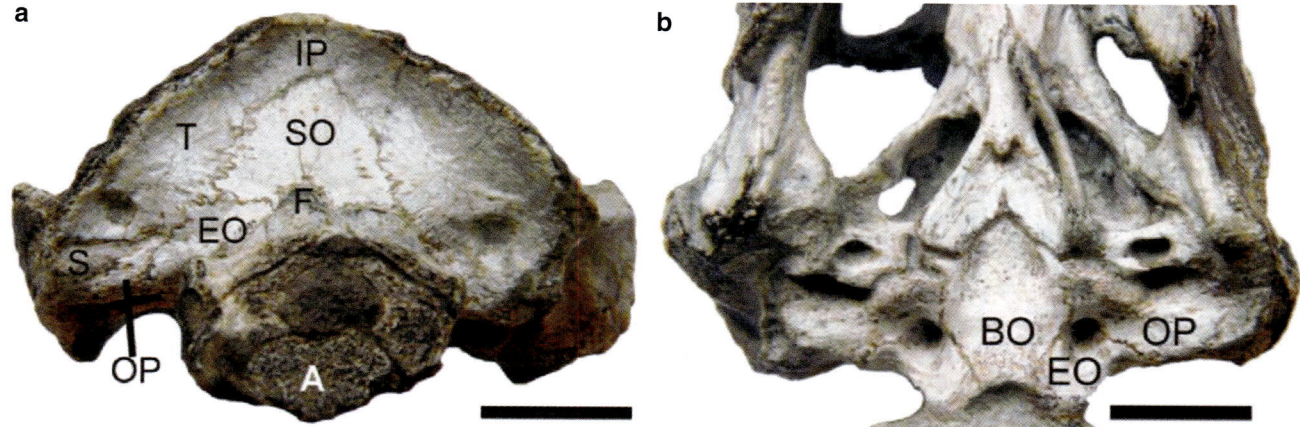

Fig. 8.85 *Galeasaurus planiciceps* was a basal pre-mammilloform cynodont in South Africa. *A* atlas centrum, *BO* basioccipital, *EO* exoccipital, *F* foramen magnum, *IP* interparietal, *OP* opisthotic, *S* squamosal, *SO* supraoccipital, *T* tabular. Note for cross-reference in our discussion of the temporal opisthotic bone and the occipital bone that opisthotic is posterior and exposed. It has not been internalized. It bridges between exoccipital and squamosal. Note on left the zone of

bone interposed between foramen magnum and SO. (Reprinted from Araujo R, Fernandez V, Polcyn MJ, Fröbisch J, Martins RMS. Aspects of gorgonopsian paleobiology and evolution: insights from the basicranium, occiput, osseous labyrinth, vasculature, and neuroanatomy. *Peer J* 5:e3119. With permission from Creative Commons License 4.0: https://creativecommons.org/licenses/by/4.0/)

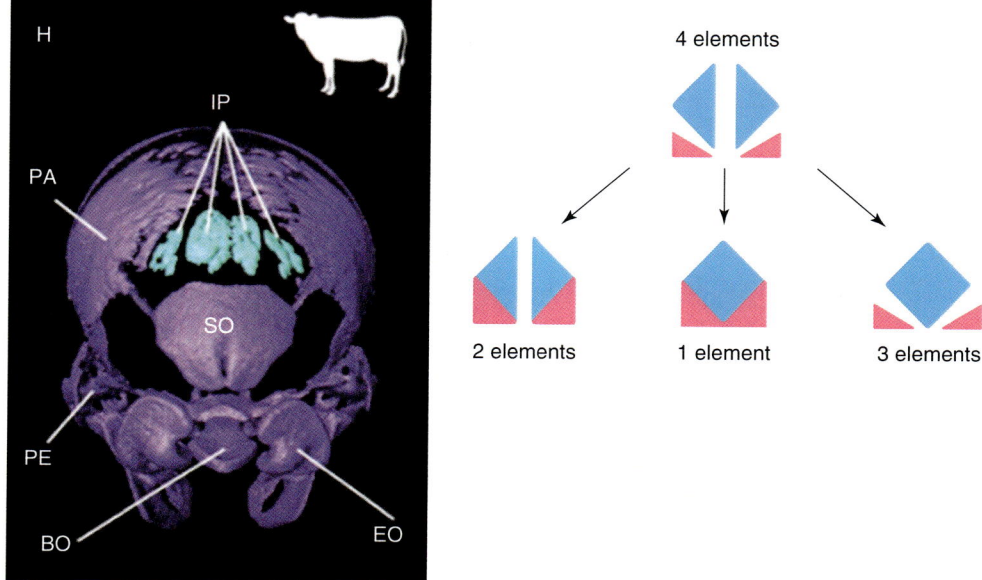

Fig. 8.86 Developmental fields of interparietal. Interparietal is universal in mammals but appears in a variety of forms. The four unfused elements of interparietal (postparietals and tabulars), seen here in the ungulate (cow), can combine in three different ways. Added to this is the contribution of lamina triangularis which may or may not be fused with upper fields (blue, postparietal; red, tabular). The supraoccipital

cartilage is seen completely distinct from exoccipitals. (Reprinted from Koyabu D, Maier W, Sánchez-Villagra MR. Paleontological and developmental evidence resolve the homology and dual embryonic origin of a mammalian skull bone, the interparietal. *PNAS* 2012; 109(35):14075–14080. With permission from PNAS)

Fig. 8.87 Blood supply to interparietal fields. Lower muscular zone has two angiosomes: posterior auricular (lateral) and ocipital (medial). Upper fascial zone supplied by occipital but in two divisions, lateral and medial, corresponding to the tabular and postparietal bone fields. (Reprinted from Vanelderen P, Lataster A, Levy R, et al. Evidence-based medicine: occipital neuralgia. Pain Practice 2010; 10(2):137–144. With permission from John Wiley and Sons, Inc.)

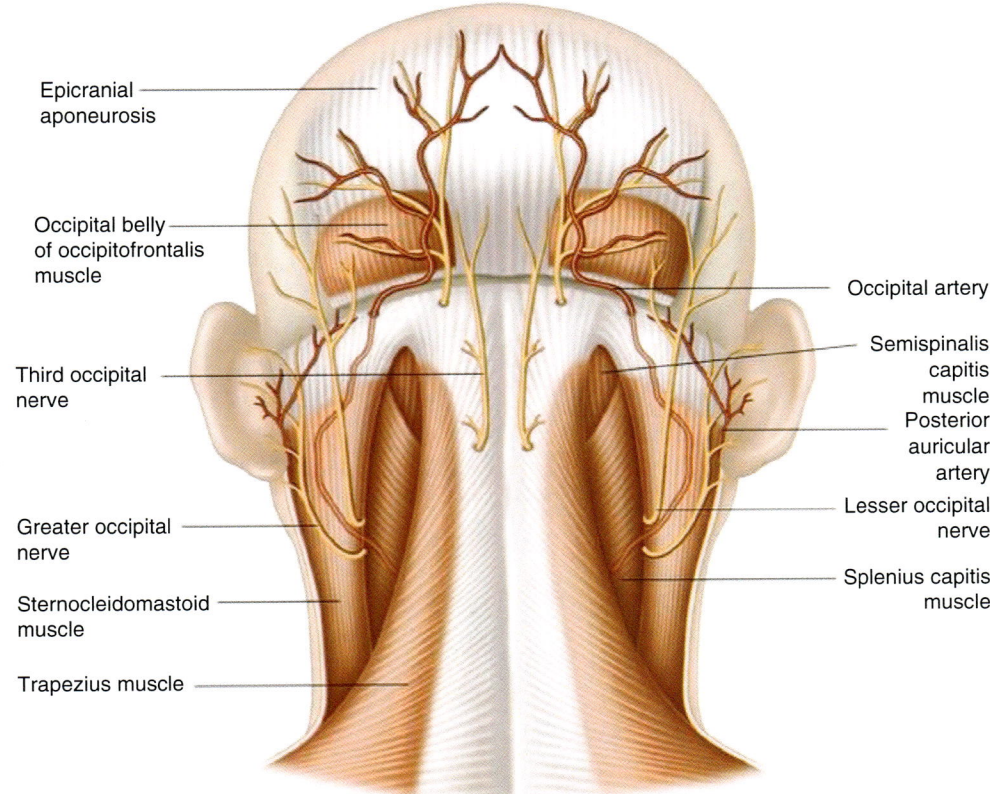

Epicranial aponeurosis

Occipital belly of occipitofrontalis muscle

Third occipital nerve

Greater occipital nerve

Sternocleidomastoid muscle

Trapezius muscle

Occipital artery

Semispinalis capitis muscle

Posterior auricular artery

Lesser occipital nerve

Splenius capitis muscle

articular tubercle which is the anterior boundary of mandibular fossa.

The glenoid fossa is defined anteriorly by articular tubercle and posteriorly by external acoustic meatus. It is lined with cartilage. The anterior fossa belongs to the squama; it receives the condyle. Behind, a *post-glenoid process* prevents retrodisplacement. The posterior fossa is non-articular. It is part of the tympanic temporal bone. These two halves are separated by the *squamosal-tympanic suture* that separates the two bones. Proceeding inward is another field separation boundary, the *petrotympanic (Glaserian) suture*. In the middle ear, this suture transmits *anterior tympanic* branch of IMMA.

The *anterior canaliculus of chorda tympani* is located along the outside of the Eustachian tube in a small triangle between the squama and the petrous temporal bone. Otherwise termed the *canal of Huguier*, this represents the "escape route" of the primitive stapedial, which, as it exits, induces the anterior tympanic artery.

In coronal section, one can appreciate that Sq flows medially over the tympanic cavity. It provides support for the lateral occipito-temporal gyrus of temporal lobe. As such this lamina may constitute the primitive *epiotic bone field*.

Craniofacial microsomia presents a picture of progressive osseous deficiency centered around the TMJ. The spectrum of CFM is represented by the classification system of Kaban et al. devised a classification system 1, the fossa is reduced but the joint remains intact. In Posnick 2 the joint is absent but the condyle is present. In Posnick 3, the condyle itself is absent. These finding represent a progressive defect of the superficial temporal axis beginning with the middle temporal artery.

Blood Supply and Mesenchyme

The mesenchyme of sq-qj is neural crest. At the level of zygomatic arch, it thickens. The internal lamina is derived from r3 dura, which is supplied by posterior division of *middle meningeal artery*, running along its longitudinal axis. This dura interacts with the three lateral gyri of temporal lobe (all are longitudinally-oriented with the arterial axis). The mesenchyme of the external lamina is also r3 neural crest. Blood supply to Sm4 temporalis muscle is also dual, thus creating two distinct vertically-oriented sectors. The anterior sector is supplied by the *posterior branch of deep temporal artery* which originates from the middle sector of IMMA. The posterior sector of temporalis is *middle temporal artery*, a branch of superficial temporal. This vessel arises just above zygomatic arch. The posterior zygomatic arch that belongs to qj is supplied by superficial temporal artery, in particular, its *transverse facial branch*.

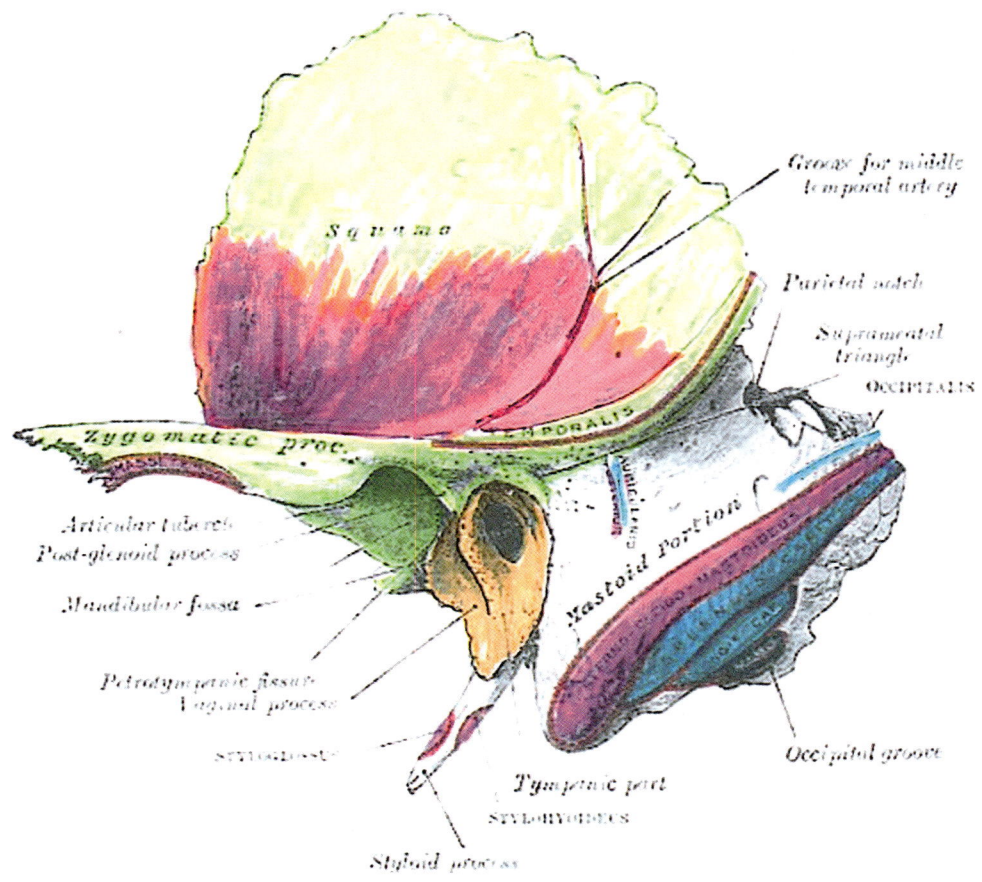

Fig. 8.88 Squamosal-quadratojugal complex (yellow) give primary insertion for muscles of mastication. Note that masseter does not insert into the angerior (jugal) half of zygomatic arch. Tympanic (orange) has vaginal process that splits around styloid process. Tympanosquamous fissure traces up to stylomasoid foramen and facial nerve. Second arch styloid gives rise to Sm6 stylohyoid and styloglossus (lower half) inserting into third arch thyoid and posterolateral tongue and Sm7 sty-lopharyngeus inserting into third arch pharynx. Mastoid from Sm7 opisthotic has two zones. Cranial zone (infrasquamous) gives secondary insertion to Sm6 facial muscles. Caudal zone (mastoid process) gives primary insertion to sternocleidomastoid from occipital somites (Sm8–Sm11). (Reprinted from Lewis, Warren H (ed). Gray's Anatomy of the Human Body, 20th American Edition. Philadelphia, PA: Lea & Febiger, 1918)

Development

Squamosal-quadratojugal develops from r3 neural crest partially in membrane. The sequence for the temporal (qj) part of the arch is undocumented but likely membranous. Ossification begins at the root of the arch and spreads in both directions.

Phylogeny

The bones of the squamosal-palatoquadrate complex are derivatives of r3 neural crest. They ossify in membrane beginning at the posterior root of zygomatic arch. This marks the epicenter between the two bone fields. Ossification takes place at the second month of post-natal life.

Neither squamosal nor quadratojugal bone fields exist in osteichthyes. In *Amia*, a faithful replica of the basal condition, the immediate zone behind hyomandibula is occupied by the four bones of the *opercular series* (recall our discussion of this basic anatomy in the previous chapter). The origin of squamosal is unclear but metapterygoid, a derivative of palatoquadrate cartilage in *Amia* is a possible candidate. In primitive sarcopterygians, the dinomorph lungfish *Strunius*, sq and qj are paired antero-posteriorly and appear concomitantly with subopercular and interopercular. In advanced tetrapodomorphs, the operculars are "lost" in the transition from gills and the development of upper limbs. In the latter process, the extrascapular series must detach from the back of the skull, and from the operculars as well. Interopercular is considered to be the source of quadratojugal. The prototetrapod, *Panderichthyes*, shows loss of opercular and replacement by squamosal.

The situation stabilizes with tetrapods. *Eoherpeton* demonstrates a 90° rotation, with squamosal and quadratojugal

Fig. 8.89 Internal view of squamosal (yellow) is deceptive. The outer lamination that leads to mandibular fossa cannot be seen, lying just outside the subarcuate fossa. This is a weak zone where the head of the condyle can be driven upward into the middle cranial fossa. Anterior border articulates with alisphenoid. (Reprinted from Lewis, Warren H (ed). Gray's Anatomy of the Human Body, 20th American Edition. Philadelphia, PA: Lea & Febiger, 1918)

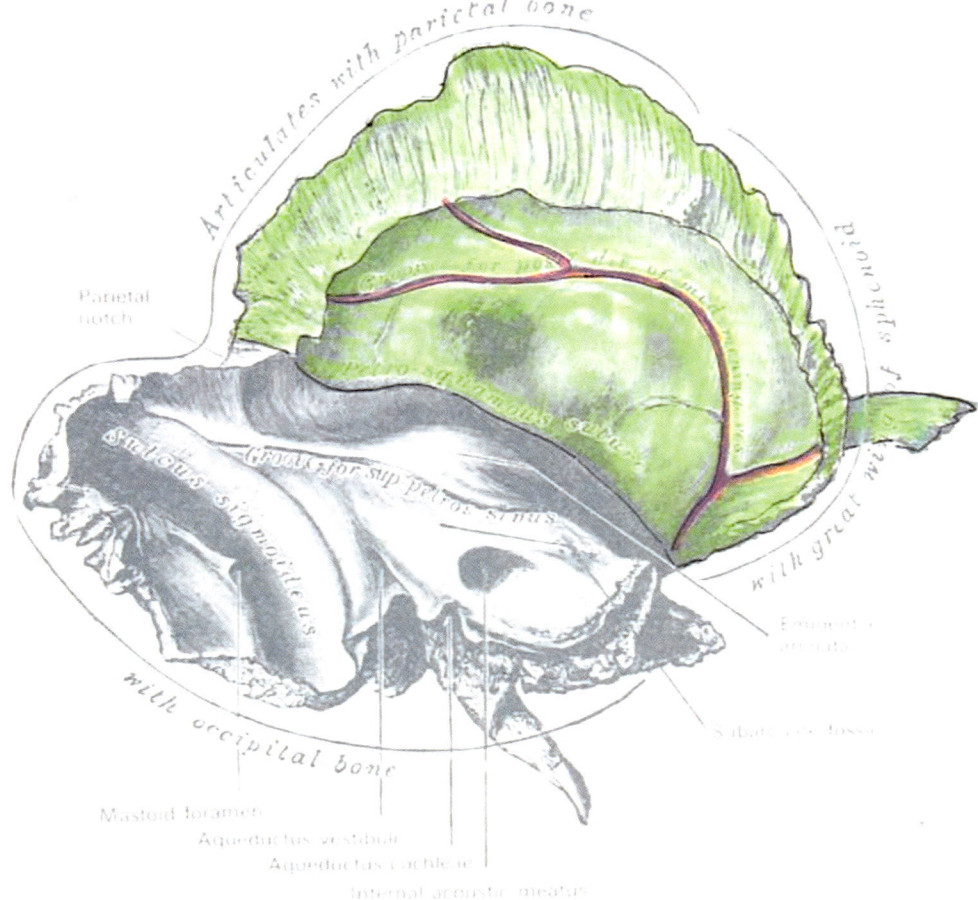

Fig. 8.90 *Strunius.* Squamosal makes its debut with the lungfish, as in *Strunius.* Here squamosal (gray) occupies a position anterior to an associated quadratojugal field (light green). (Reprinted from Long JA. On the relationship of psarolepsis and onychodontiform fishes. J Vert Paleontol 2001; 21(4):815–820. With permission from Taylor & Francis)

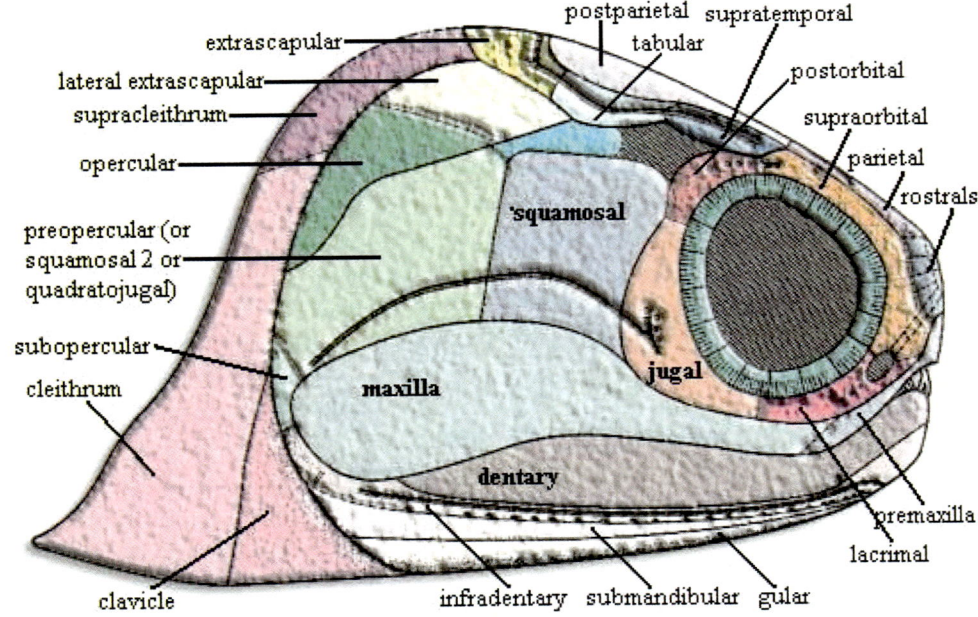

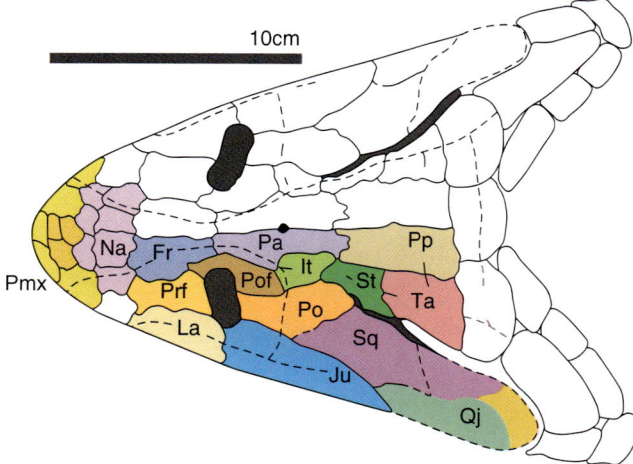

Fig. 8.91 Squamosal *Panderichthyes* shows interconnectedness between squamosal (violet), quadratojugal (green) and tiny (but important) quadrate (tan). All three bones are r3 neural crest. Occipital margin covered with medial and lateral extrascapulars. Outrigger bone series connects the skull with the anterior fin. (Courtesy of Mr. David Peters)

now aligned one above the other, a condition which remains through evolution. In therapsids, squamosal is in contact with the tympanic temporal and prootic capsule such that an external auditory canal develops. Cynodonts continue to modify the temporal fossa. Basal cynodont *Procynosuchus* has a two-component zygomatic arch formed between jugal and quadratojugal for the attachment of powerful muscles. *Probainognathus* develops a unique double joint for the mandible, which proved to be the penultimate step before the invention of the mammalian temporomandibular joint. The posterior margin of surangular develops an articular process, the future condyle. At the same time, a fossa develops in squamosal to receive the condyle, the future TMJ. Probainognathus leaves behind unfinished business. The squamosal still lies posterior to parietal and has not swung beneath it. At the apex of the cynodonts is *Morganucodon*, arguably the first true mammals because it fully incorporated incus and malleus into the middle ear. In Mammalia, ptergyoids develop with new insertions compatible with grinding up foodstuffs.

The story of squamosal terminates with the evolution of euprimates and the growth of the brain case. Frontoparietal expansion is accompanied by flattening of the face, reduction in jaw size (especially the mandible). Squamosal expands as well, coming forward beneath parietal and into contact with alisphenoid. The median sagittal crest into which masseter was inserted recedes and the muscle retreats to a lower, more modest, position in keeping with change in diet. Formerly devoid of muscle, squamosal accepts the duties of insertion. Zygomatic unchanged in design since the cynodonts now bears the masseter muscle, displaced from the skull.

Tympanic Bone (tym)

Descriptive Anatomy

This small bone, found only in mammals, is intimately associated with the squamosal bone. Its mesenchyme is likely derived from the first arch lining (r3) of the anterior external auditory meatus. The most likely mechanism of ossification is membranous via a chondral intermediate (Figs. 8.92, 8.93, 8.94, and 8.95).

The tympanic bone cannot be appreciated without a consideration of the bony anatomy of the ear in general. The external auditory canal is a five-sided box 24 mm long. The sixth side, the lateral wall, is empty, being open to the air; the medial wall is the tympanic membrane. The remaining four walls are made up of two opposing L-shaped bones. Mastoid is posterior and superior. Tympanic is anterior and inferior. The external auditory meatus consists of V3-innervated (r3) skin anteriorly and IX-innervated (r7) skin posteriorly. The cartilaginous EAC is the first 8 mm. Although it starts out as a complete ring attached to the ear, once inside, it becomes discontinuous. It is firmly attached to the *auditory process* of tympanic bone only. Thus, the cartilage is deficient cranially and posteriorly. The osseous EAC is 16 mm long. Its anterior and inferior parts belong to tympanic temporal. In the fetus, this is seen as a completely separate *annulus tympanicus*.

Middle ear has six walls: tegmental (roof), jugular (floor), membranous (lateral), labyrinthic (medial), mastoid (posterior), and carotid (anterior). We shall consider three of these here and the remainder in our discussion of the petromastoid temporal bone. *Tegman tympani* is a thin lamina located at the junction of anterior petrous with squamous. It provides a semicircular trough for *tensor tympani*. Its lateral margin abuts petrosquamous suture. Embryologically, this could represent a membranous lamina from squamous temporal. The *jugular wall* is thin, separating middle ear from jugular fossa. Derived from tympanic, it abuts the labyrinthic wall (likely r7) where an interface suture permits entry of tympanic branch of glossopharyngeal nerve. The *membranous wall* is formed by the tympanic membrane and by the ring-shaped tympanic bone into which it is inserted. A hiatus between tympanic and mastoid leaves a gap in the upper ring, the *notch of Rinvinus*. This forms a landmark for three small, but very important openings (see below).

- *Posterior canal of chorda tympani* (*iter chordae posterius*) opens into tympanic cavity at the fissure between mastoid and membranous wall. This is just behind tympanic membrane, dead level with manubrium of malleus. Recall that chorda tympani arises from r4 while motor VII has its nucleus in r5. It is not surprising that chorda tympani is contained in a separate canal running parallel to facial canal. The two nerves finally unite near (but not at) stylomastoid foramen.

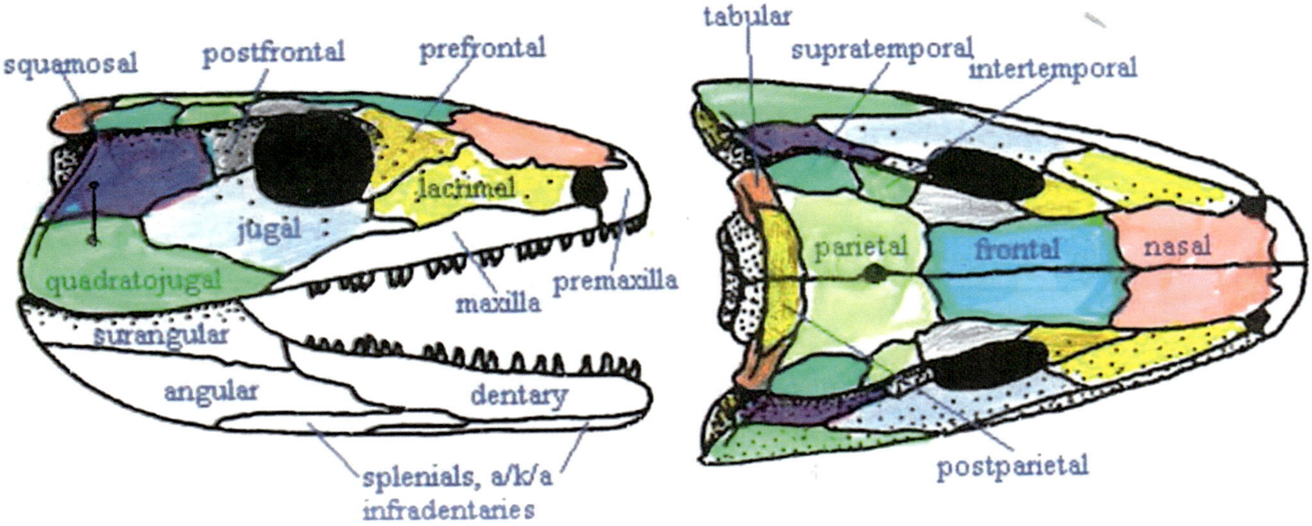

Fig. 8.92 Basal tetrapod *Eoherpeton* shows loss of extrascapulars, disconnection of the pectoral girdle from the head, and verticalization of the squamosal-quadratojugal fields. Anapsid skull has no temporal fossa and no zygomatic arch. Quadratojugal relates to small quadrate (tan) posteriorly as seen from above. (Reprinted from Smithson TR. *The morphology and relationships of the Carboniferous amphibian Eoherpeton watsoni Panchen*. Zool. J. Linn. Soc. 1985;85(4):317–410. With permission from Oxford University Press)

- The *petrotympanic (Glaserian) fissure* is a narrow 2 mm slit, located just anterior and superior to the bony ring housing the tympanic membrane, it contains the anterior ligament of the malleus and the anterior tympanic artery.
- *Anterior canal of chorda tympani* (*iter chordae anterius*) is located at the medial end of the Glaserian fissure. Note that the nerve runs lateral to tympanic membrane, neck of the malleus, but medial to tensor tympani muscle.

Here's how tympanic bone fits into the equation. Posterior surface forms three walls of the external auditory canal: anterior, floor, and posterior (partly). Anterior surface is quadrilateral and forms the back wall of the mandibular fossa below (ventral) to the temporomandibular joint. Superior border abuts medially with squamosal, the tympanosquamous fissure, and laterally with postglenoid fossa of the temporomandibular joint. Inferior border forms a sharp *sheath of the styloid process.*

Styloid process is a remnant of the second arch. It therefore has two ossification centers, tympanohyal is proximal and stylohyal is distal. Attached to it are two ligaments (stylomastoid and stylohyoid) and three muscles (stylohyoid, stylopharyngeus, and styloglossus). Stylohyoid represents a connection between second arch and second arch; stylopharyngeus connects second arch with third arch, and styloglossus with tongue that is at the neuromeric level of fourth arch.

Blood Supply and Mesenchyme

The best way to understand tympanic bone is to consider the anatomy of the tympanic membrane ™ that is attached to it. TM has three layers. The outer cutaneous stratum has a superficial layer of squamous cells beneath which is an epidermal prickle cell layer. No dermal papillae are present. An intermediate stratus consists of neural crest fibers projecting from the handle of malleus. We can map these layers to r3. The inner stratum is a single layer of non-ciliated mucosa in continuity with the lining of tympanic cavity. Its origin is from the first and second arch mucosa of nasopharynx.

Innervation of TM is key to understanding blood supply. V3 provides pain fibers. Fibers from inferior ganglion of IX and superior ganglion of X combine as the *auricular branch of vagus* (Arnold's nerve). It travels posterior to internal jugular until it reaches the mastoid canaliculus. Just in front of stylomastoid foramen, it joins the *auricular branch of VII* which ascends gains entry through tympanomastoid fissure and gives off two branches. One branch connects to posterior auricular branch of V3 (somatic sensory). The other branch supplies a small area of auricular skin (cranial aspect), the posterior and inferior walls of external canal, and the outer tympanic membrane. These connections explain the reactions of heartrate and pain to ear irrigation. In sum, the majority of TM has V3 innervation and first arch mesenchyme but a small, posterior component has a source from second arch.

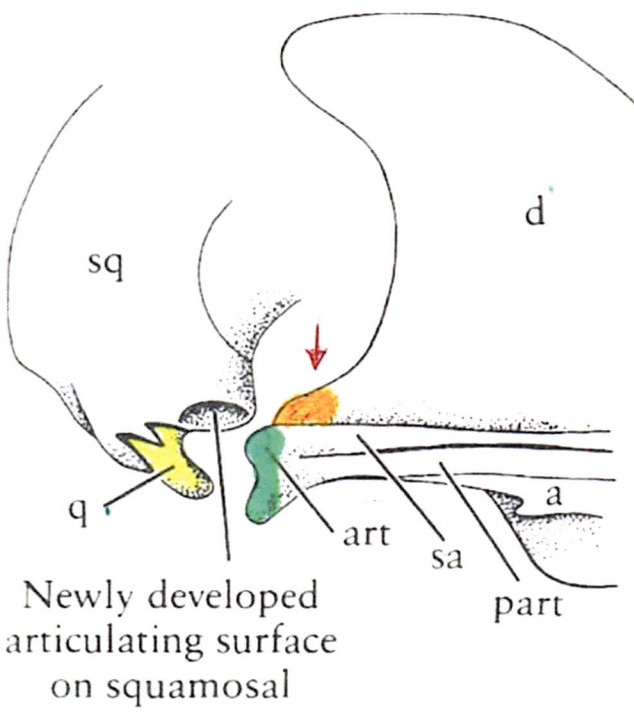

Newly developed
articulating surface
on squamosal

Fig. 8.93 Epicynodont transition to the TMJ. Transition to the TMJ takes place with an articulatory fossa develops in the squamosal-quadratojugal in epicynodont *Thinaxodon*. The r2 quadrate moves into direct contact with stapes. A large coronoid process projects upward from the dentary, perhaps representing a posteriorization of the third coronoid. It does *not* form an articulation with squamosal. Later on, in *Probaingnathus* the postdentary rod develops. The PDR is still in contact with surangular. Note the orange zone of the posterior surangular/dentary. This is the *first sign of a condylar process*; it relates directly with the new glenoid fossa. A transient second jaw joint between dentary and squamosal appears in Probainognathus and Morganucodon. In mammals (see *Sinocodon*), the PDR breaks off, allowing surangular to fill in the posterior mandible. Brain growth (1) dislocates the coronoid from squamosal and (2) acts as a tissue expander to bring the glenoid-condyle articulation upward and forward. (Reprinted from Benton MJ. Vertebrate Paleontology, 4th ed. Blackwell; 2015. With permission from John Wiley & Sons)

It is supplied by the two most posterior branches of mandibular sector of IMMA: *anterior tympanic artery* and *deep auricular artery*. These vessels often arise from a common stem. The represent forward projections of the primitive stapedial stem, known (in the adult state) as the *caroticotympanic branch of internal carotid*. Passing forward, ATA ramifies along the internal lamina of tympanic membrane. Here, it shares a connection with the external carotid artery, the *stylomastoid branch of posterior auricular artery*. This is yet another example of the connections between ECA and stapedial systems. DEA passes behind temporomandibular joint which it supplies. It supplies the cartilaginous part of EAC and the external lamina of tympanic membrane. Thus, we observe yet another example of outside–inside lamination. The most proximal artery serves the outer r3 TM; the more distal artery is distributed to the inner r3–r5 TM.

Development

Tympanic bone unites with squamosal just before birth. Tympanic bone is an incomplete right with the aperture directed superiorly. Ossification takes place at the week 9–10 of fetal life.

Phylogeny

Tympanic bone is a transformation of initial bone field of caudal Meckel's cartilage. Recall that Meckel's has two sectors, mandibular and tympanic. The latter has four fields: angular, prearticular, articular and quadrate. Angular is responsible for producing tympanic; prearticular and articular form the reflected lamina and body of malleus; and quadrate becomes the incus. Tympanic bone is an r3 neural crest derivative that forms in cartilage. The phylogeny of tympanic bone is discussed in the section of the bones of the middle ear.

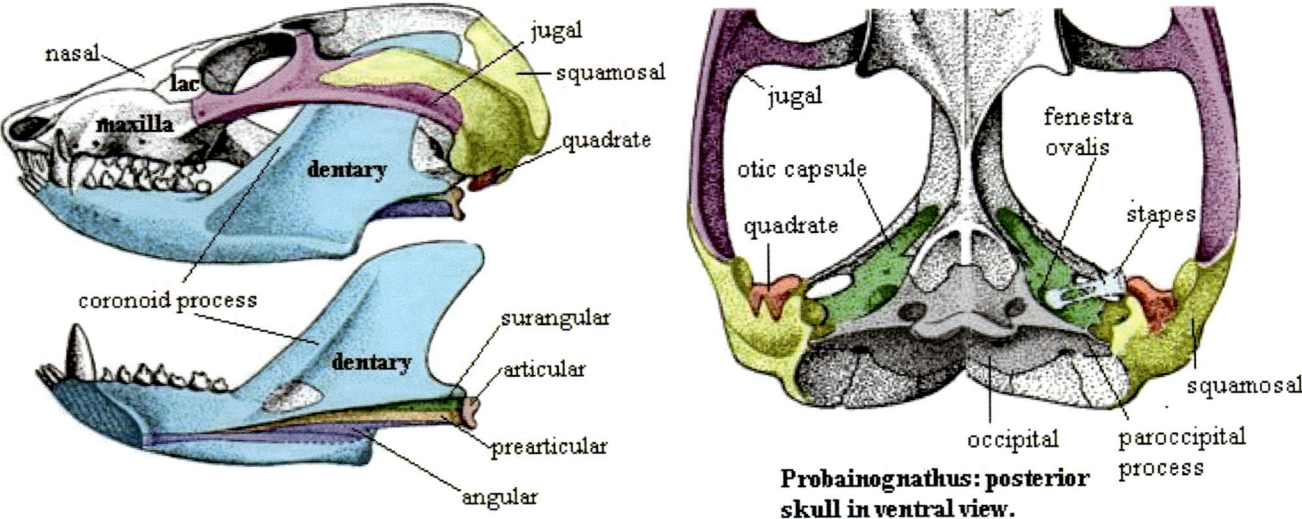

Fig. 8.94 *Probainognathus*. Left shows two-component mandibular joint and arch connecting zygomatic arch connecting jugal (pink) with quadratojugal (yellow). Prearticular (tympanic), articular (malleus), and quadrate (incus) remain in situ, having not transitioned into the ear. Amalgamation of articular and prearticular, followed by angular forms the postdentary rod (PDR). Right: Note joint between quadrate-squamosal and its articulation with stapes. (Reprinted from Carroll RL. Vertebrate Palaeontology. New York, NY: WH Freeman and Company; 1988)

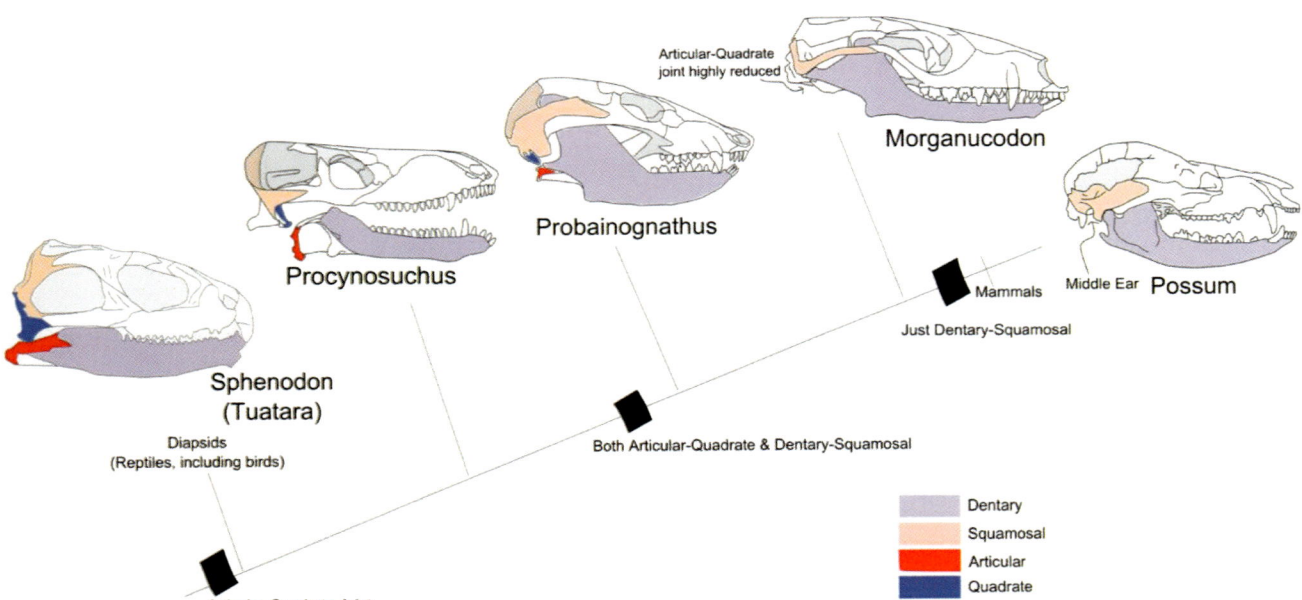

Fig. 8.95 Reassignment of ear bones: evolution of the TMJ. The double joint exists in *Probainognathus* and *Morganucodon* but in the latter the PDR is dragged backwards by quadrate and the coronoid process is both reduced in size and positioned forward. True mammals have a 3-ossicle ear and a TMJ. The position of basal mammal *Sinocodon* is marked. Specialized pterygoid muscles split off from the temporalis blastema in mammals. Basal cynodonts *Dvinia* (not shown) and *Procynosuchus* develop reflected lamina or articular (the future ecto-tympanic). In epicynodonts, *Galeosaurus* (not shown) and *Thrinaxodon* (not shown). Here, the articular is up-righted to form the postdentary rod (PDR). Eucynodont *Probainognathus* shows quadratojugal component now transferred forward to incorporate into zygomatic arch. A double joint exists between posterior surangular and quadrato-jugal as well as between coronoid and parietal. Although mammiliform *Morganucodon* continues the double joint the posterior (articular-quadrate) joint is greatly reduced. Proto-mammal *Sinocodon* has the first true TMJ. Mammalia has final position of squamosal-qj and a verticalized ramus. (Used with permission from: Understanding Evolution. 2020. University of California Museum of Paleontology. 22 August 2008 http://evolution.berkeley.edu/)

Neurocranium (r4–r7)

Introductory Remarks

This section and the one that follows are quite complex as they present two mesodermal bone complexes which are distinct in mesenchymal source, development and function yet structurally related. The <u>petrous temporal complex</u> forms from somitomeric mesoderm produced at rhombomeric levels r4–r7. Its three bone fields (prootic, opisthotic, and the diminutive epiotic) are chondral and fit together in a complex way to (1) complete the middle cranial fossa and (2) encase the apparatus for the systems of hearing and balance. The <u>occipital complex</u> (basioccipital, exoccipital, and supraoccipital) forms from somitic mesoderm produced at rhombomeric levels r8–r11. Its bone fields represent a lamination from the four occipital somites which can only be discerned by neuromeric mapping. These function to (1) complete the posterior cranial fossa and (2) articulate with the trunk. Finally, these two bone complexes articulate with one another to complete the skull base. Although the petrous temporal complex is hidden from view in mammals, it was (along with stapes) an integral part of the occiput in pre-mammals. Thus, the contemporary anatomic relationships among these bone fields can best be understood on the basis of their paleontology.

Our subject at hand seems messy, at first, from a neuromeric point of view because the petrous temporal complex houses neural crest bones from the first and second arches. Furthermore, the boundaries between somitomeres 4 5 and 6 physically surround the first and second pharyngeal clefts. These relationships are critical for understanding the anatomy of the external auditory canal, the tympanic cavity, and the Eustachian tube. Finally, the evolution of the mammalian invention of the three-ossicle ear is critically dependent upon the formation of a new and functionally sophisticated temporomandibular joint. For all these reasons, our discussion will admix anatomic concepts seen previously with the squamosal and tympanic bones; and it will continue forward when we discuss the mandible.

Regarding content and illustrations, our intention is not to repeat the excellent work of contemporary embryology texts regarding the microanatomy of the ear. The goal is to provide concepts and visual insights into critical steps in this process.

Finally, a recommendation for the reader. The temporal complex has many details. The "big picture" resides in the description of its component parts and its evolution, with emphasis on the prootic and opisthotic bones. Regarding development of the internal, middle, and external ear, I do recommend the section dealing with the assembly of the middle ear because it has to do with temporomandibular joint evolution. The reader can decide *peri passu* what is of interest.

Petrous Temporal Bone Complex

The temporal bone complex is an assembly of bones that represent the transition between the membranous neurocranium and the chondral neurocranium. In continuity with the bones of the first, second, and third arches, the temporal provides functional attachments for the mandible and hyoid apparatus; it houses the organs of hearing and balance (Figs. 8.96, 8.97, 8.98, 8.99, 8.100, 8.101, 8.102, 8.103, 8.104, 8.105, 8.106, 8.107, 8.108, 8.109, 8.110, 8.111, 8.112, 8.113, 8.114, 8.115, 8.116, 8.117, 8.118, 8.119, 8.120, and 8.121).

The bones of the complex can be conceptualized in terms of their relationships to the otic placode and its derivatives. An external lamina, the *squamosal* bone, provides protection for the lateral skull and connects with the mandible via condylar fossa. The ring-like *tympanic bone* provides a framework for the externa auditory canal, and houses the tympanic membrane. Deep to these structures, the *petro-mastoid bone* consists of three parts: an anterior *pro-otic bone* extending forward into the middle cranial fossa, a posterior *opisthotic* bone extending backward toward the occipital complex, and (in mammals) an external *epiotic* bone field that forms the superior and lateral walls of the middle ear and contains the *mastoid sinus*. Housed within the confines of the tympanic cavity are r2 malleus, r3 incus, and r4 stapes. Because these component parts are fused it is difficult to see them as separate units; for this reason, the embryology can be best understood on the basis of the phylogenetic relationships established in evolution.

We start with two disclaimers. Two of these components have already been discussed. Why? Both squamosal and tympanic arise from different mesenchymal sources and have distinct histories from the petromastoid complex. The external lamina, the *squama*, properly known as *squamosal–quadratojugal* arises elsewhere and is translocated into position in the therapsid line. It is an r3 neural crest bone. Thus, sq-qj is a "new-comer" to the temporal complex. It provides external protection, forms the posterior half of zygomatic arch, and connects with the mandible via condylar fossa. Tympanic bone, also an r3 neural crest derivative, arises from the posterior sector of Meckel's cartilage. Through complex process it detaches from mandible and becomes incorporated into temporal territory. For these reasons, our discussions of the squamosal–quadratojugal and of the tympanic precede those dealing with the temporal bone complex proper.

The blood supply to this complex is an amalgam as well. Temporal vessels external carotid maxillary supply first arch mastication apparatus. Stapedial derivatives perfuse the

Fig. 8.96 Tympanic bone (orange) in situ. Tympanic bone is derived from r3 angular bone. It forms the posterior wall of the TMJ (mandibular fossa). Within the ear canal it is U-shaped. Lower border splits to include the styloid process, a second arch derivative, which inserts into petromastoid, indicating its derivation from Sm4 to Sm5. Note: primary insertion of temporalis (red) not colored. (Reprinted from Lewis, Warren H (ed). Gray's Anatomy of the Human Body, 20th American Edition. Philadelphia, PA: Lea & Febiger, 1918)

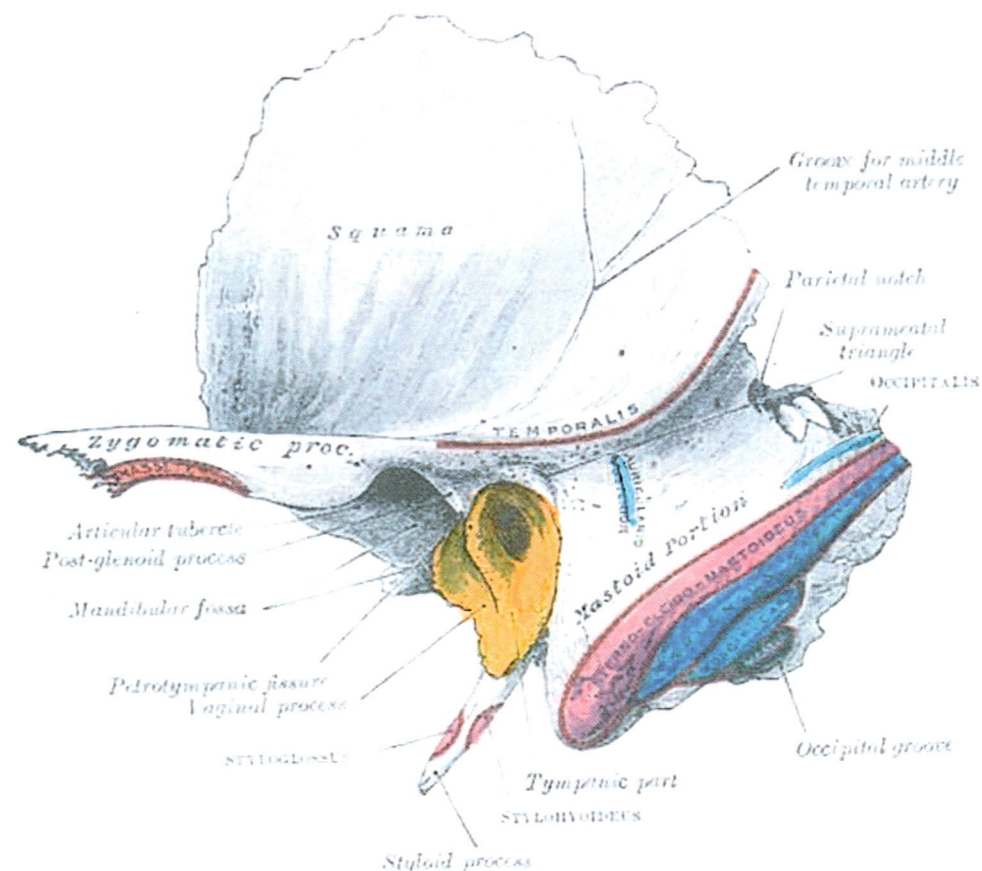

squamous temporal, tympanic, and middle ear bones and represent contributions associated with second aortic arch contributions to the PA1/PA2 complex. Petrous carotid supplies the periotic/opisthotic zone. Branches from the longitudinal neural system such as internal auditory supply the neural structures of the inner ear. Epiotic (mastoic) receives supply from the external carotid and (via muscle attachments) the vertebral system.

Finally, this zone represents a watershed among two forms of mesenchyme. Neural crest from first and second arch forms the ear bones and the squamous–tympanic complex. The otic capsule develops from somitomeric paraxial mesoderm. Mastoid bone contains bone marrow cells.

We shall discuss the temporal bone complex properties in the following terms. First, its component parts will be described with specific attention to blood supply and mesenchyme. Next, the evolution and development of the bone complex will be detailed. Finally, we shall look at the specific developmental processes involving the inner ear, middle ear, and external ear.

Descriptive Anatomy: Petrosal Bone

The petromastoid bone is frequently referred to simply as the periotic bone, indicating its relationship to the structures derived from the otocyst. The mesenchymal sources for the petromastoid complex are PAM from Sm5, Sm6, and Sm7 plus neural crest from r4 to r7. This bone is unique because, encased within it, are soft tissue structures of vital importance: the internal carotid artery, the facial nerve, and the vestibulo-cochlear system.

Four centers of ossification are traditionally described for the embryonic petromastoid. These centers (*pro-otic*, *opisthotic*, *epiotic*, and *pteriotic*) are named in accordance with their relation to the otic capsule. Bast describes *14 centers of ossification*, all of which are related to nerve terminations, the internal auditory meatus, and the semicircular canals. Only when an individual part of the otic capsule has reached, its final size will bone formation take place.

The resulting structure is spatially complex. Its anatomic details are described in several monographs. For a quick and visually appealing overview of the anatomy and embryology, the reader is referred to Moore and Carlson, respectively. Our goal is to understand the origin of these components and how they are assembled.

Note As we shall see, paleontological evidence proves very helpful in understanding the derivation of structures related to the otic capsule. In point of fact, the mastoid, found only in mammals, turns out to be just a consequence of the opisthotic. But traditional descriptions of the temporal complex constantly refer to petrous and mastoid as if they were

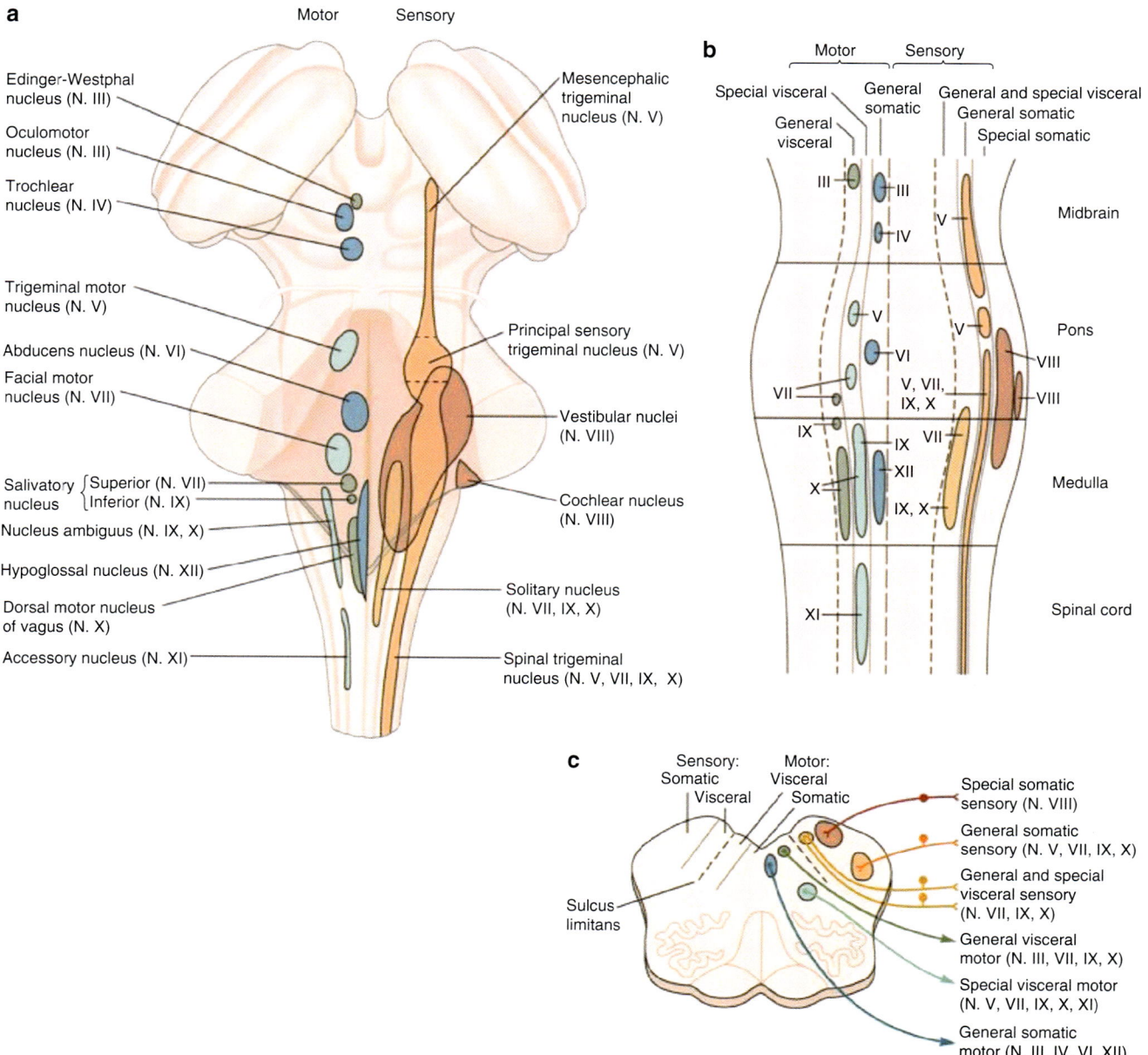

Fig. 8.97 Grand generalization (in three parts). (**a**) All pharyngeal arches are in register with pairs of rhombomeres. (**b**) Motor control and general sensory perception for pharyngeal arch structures are in strict neuromeric register with the brainstem via the lateral motor column and the lateral sensory column. Neurons dedicated to a specific arch may arrive at their desination either directly or indirectly, using a different nerve to gain access. (**c**) *Pharyngeal arch structures should be classified, not by the individual nerves that supply them, but by the neuromeres which supply the neurons in the first place.* (Reprinted from Kandel ER, Schwartz JH, Jessell TM, Siegelbaum SA, Hudspeth AJ, Mack S (eds). Principles of Neural Science, 5th ed. McGraw Hill; 2013. With permission McGraw-Hill Education)

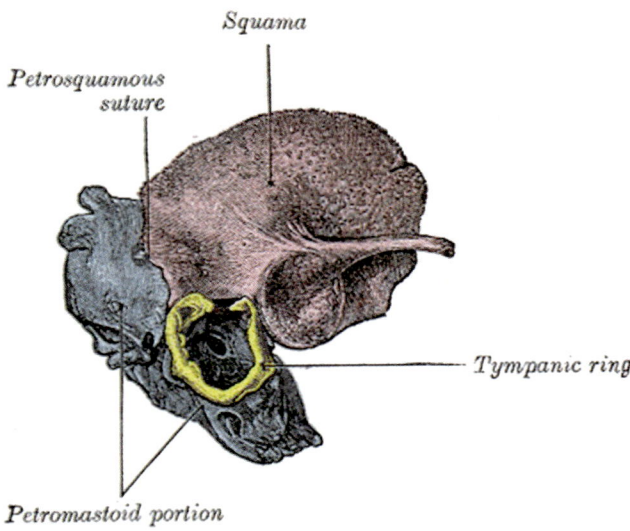

Fig. 8.98 Tympanic bone at birth in situ. The tympanic bone comprises the entrance to the tympanic cavity save for a superior portion. (Reprinted from Lewis, Warren H (ed). Gray's Anatomy of the Human Body, 20th American Edition. Philadelphia, PA: Lea & Febiger, 1918)

discrete entities, so we shall stick to that paradigm for the moment.

The geometry of petrous temporal is that of a pyramid sandwiched between the occipital and sphenoid bones. It thus has a base, an apex, and three walls. Its flat <u>base</u> is plastered against the mastoid and squamous temporal bones while its <u>apex</u> projects forward into the angle between basi-occipital bone and greater wing of sphenoid, i.e., into the interface between r0–r1 and r2. The <u>anterior wall</u> borders with the inner table (cerebral surface) of squamosal at the *petrosquamous suture*. The <u>posterior wall</u> constitutes the anterior wall of the posterior cranial fossa. Being a true pyramid, the anterior and posterior walls come together as a knife-like *petrosal ridge* that projects upward against the undersurface of temporal lobe. The <u>inferior surface</u> of petrous is faces outside the skull and constitutes the posterior roof of infratemporal fossa.

The surfaces of petrosal bone bear anatomic features worth reviewing so we can stay oriented to this complex structure.

Fig. 8.99 Tympanic cavity with bones 3-D. (Reprinted from Buytaert JAN, Aerts J, Salih WHM, et al. Visualizing middle ear structures with CT—Studying morphology (bone and soft tissue) & dynamics. J Morphol 2015; 276(9):1025–46. With permission from John Wiley & Sons)

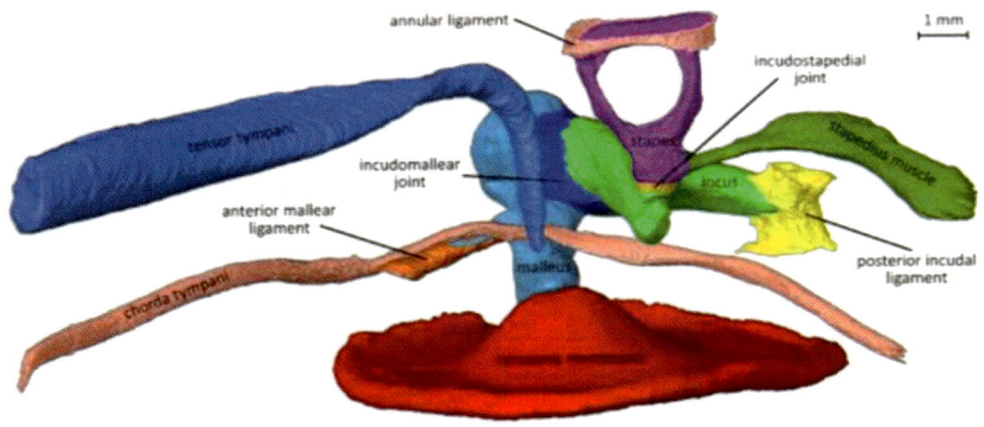

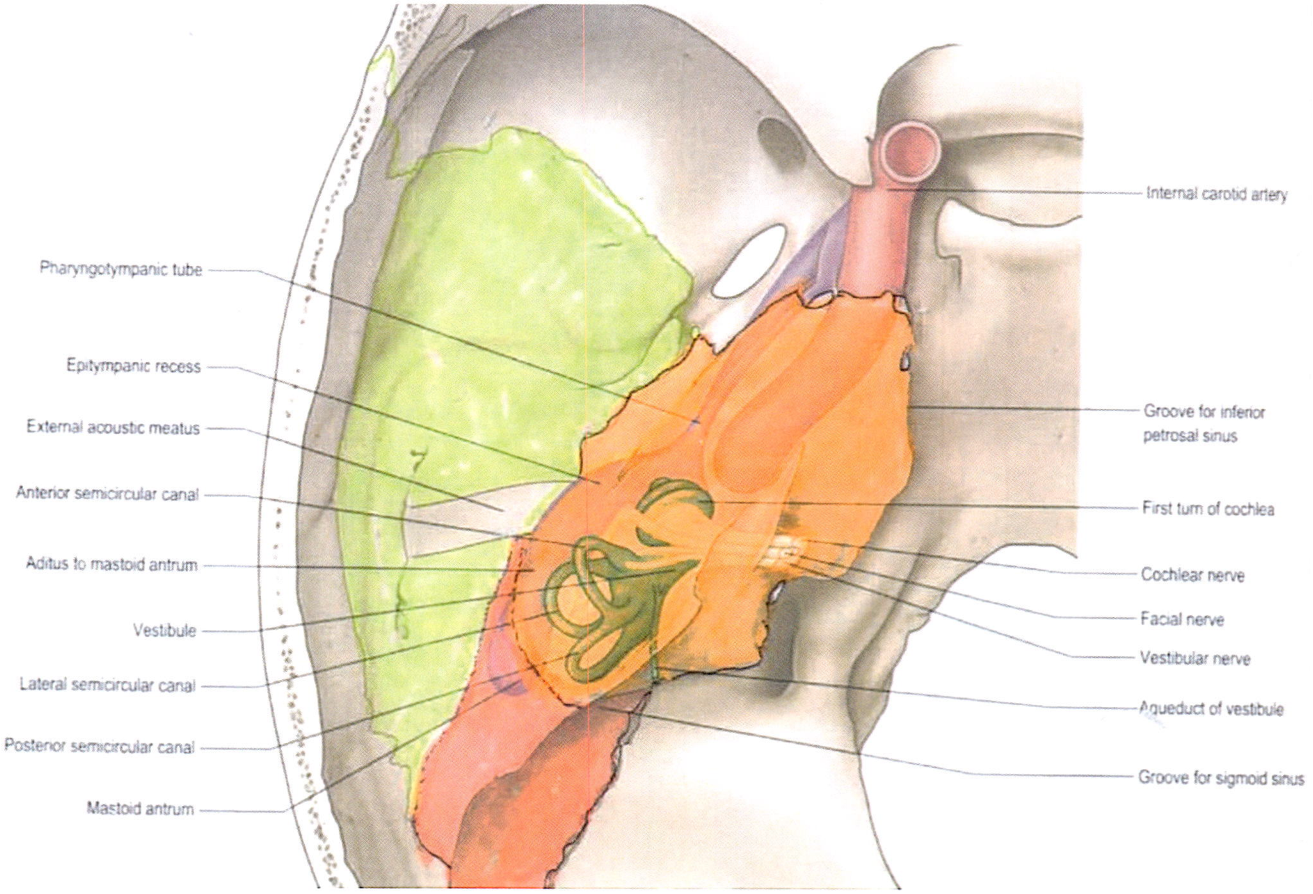
Internal carotid artery
Pharyngotympanic tube
Epitympanic recess
External acoustic meatus
Anterior semicircular canal
Aditus to mastoid antrum
Vestibule
Lateral semicircular canal
Posterior semicircular canal
Mastoid antrum
Groove for inferior petrosal sinus
First turn of cochlea
Cochlear nerve
Facial nerve
Vestibular nerve
Aqueduct of vestibule
Groove for sigmoid sinus

Fig. 8.100 Temporal bone fields, superior view. The external acoustic meatus belongs to tympanic bone. Opisthotic is poorly represented in this view, being positioned inferiorly and posteriorly. As such, it makes up the floor of the tympanic cavity but it contains mastoid antrum. Opisthoic and prootic flank the external auditory, with opistothic forming the backwall of tympanic cavity. (Reprinted from Lewis, Warren H (ed). Gray's Anatomy of the Human Body, 20th American Edition. Philadelphia, PA: Lea & Febiger, 1918)

Anterior wall: At its apex lies, the exit site for the *carotid canal*, just above which is a depression marking the fossa for trigeminal ganglion. *Arcuate eminence* lies in the epicenter of anterior wall and marks *anterior semicircular canal. Tympanic cavity* lies just anterior and lateral to it and is covered over by a paper-thin lamina, the *tegmen tympani*. The canal for PANS-bearing *greater petrosal nerve* is conveniently marked by a groove caused by the overlying petrosal branch of meningeal artery. Just lateral to the hiatus for the GPN is the passageway for the SANS-bearing *lesser petrosal nerve*.

Posterior wall: A large hole in its center marks the *internal auditory meatus* which transmits cranial nerves VII and VIII, nerves intermedius, and internal auditory artery into the inner ear. Just behind it is a tiny opening for endolymphatic duct; it is known the *aqueduct of the vestibule*. The passageway for the facial nerve starts at the internal acoustic meatus and terminates at stylomastoid foramen. It heads straight out laterally, until reaching the medial wall of tympanic cavity. In so doing, it threads a needle between cochlea in front and the semicircular canals to the back. Just before the middle ear, it takes two right-angle turn, first going posteriorly over the fenestra vestibule and then, with the semicircular canal still located posteriorly, plunging downward to exit the skull.

Inferior wall: The surface is very busy. Let us proceed from anterior to posterior. The roughened surface of the apex is a rectangle to which is *attached levator veli palatini*, just lateral to which is the cartilaginous auditory tube with attached fibers of LVP that serve to open it. Just behind the tube is the entry point for the *carotid canal*. Just behind and medial to the canal the *cochlear canaliculus* transmits (1) a tube of dura mater that connects the subarachnoid space with the perilymphatic space, and (2) a venous connection between internal jugular and the cochlea. Behind the canaliculus is located the *jugular fossa* in which is seated the bulb of jugular vein. Lateral to that is the *canaliculus for the tympanic branch of IX* into the middle ear cavity. This, the *nerve of Jacobson*, provides (1) sensation for the tympanic membrane, (2) PANS fibers from the inferior salivatory nucleus via lesser petrosal nerve and thence to parotid gland, and (3)

Fig. 8.101 Temporal bone fields, medial view. per MC: prootic petrous (orange), Opisthotic petrous (pink), squamosal (yellow). Opisthotic is cup-like, being directed posterior, inferior, and medial to prootic. It includes the epiotic field. Mastoid process borders with posterior (retroauricular) squamosal. Prootic is a pyramid directed anteriorly and superiorly. It borders with squamosal, except the retroauricular part. (Reprinted from Lewis, Warren H (ed). Gray's Anatomy of the Human Body, 20th American Edition. Philadelphia, PA: Lea & Febiger, 1918)

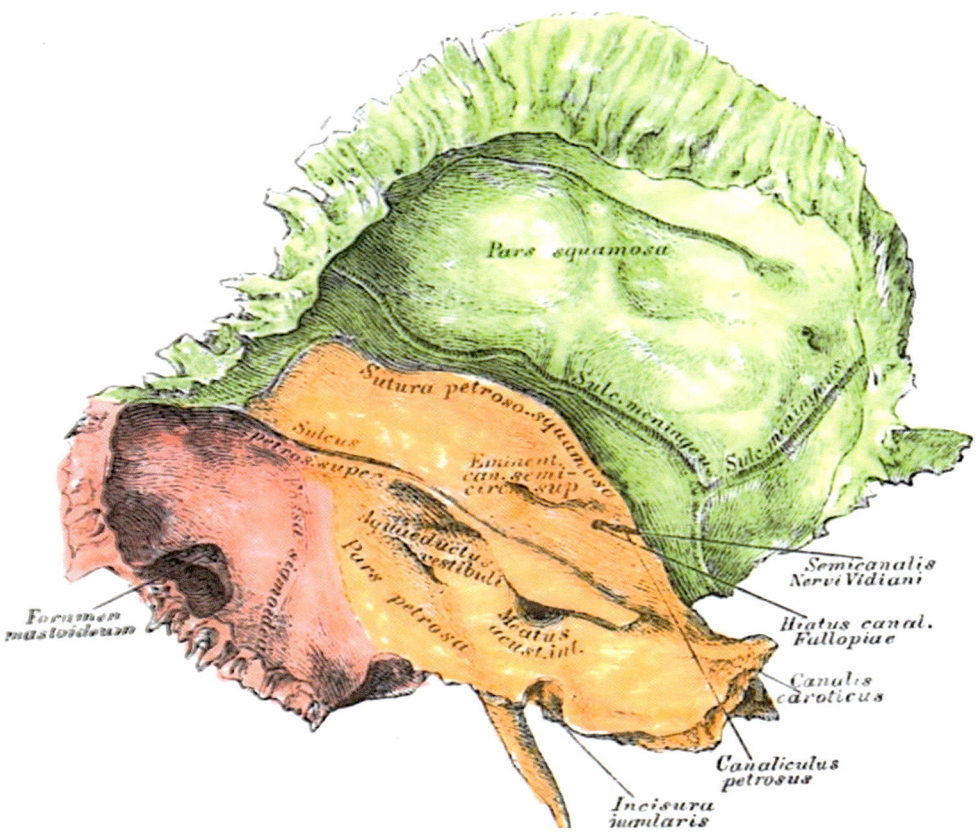

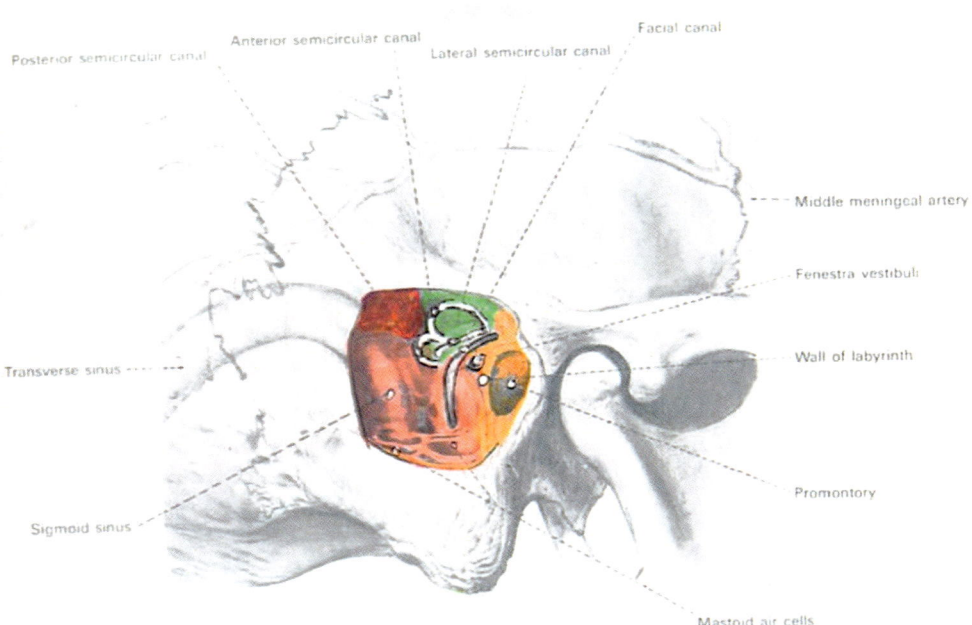

Fig. 8.102 Medial tympanic cavity (my best-guess mapping). Tympanic cavity with tympanic bone removed. Posterior and medial walls cover the semicircular canals. Note the opisthotic and prootic meet at the oval window (fenestra vestibuli). Note: epiotic is incorporated into opisthotic and pteriotic is part of prootic. Prootic (yellow): superior semicircular canal and forward to the apex, anterior roof, anterior medial wall, and anterior wall. Opisthotic (orange): promontory, cochlear window, floor of tympanic cavity, goes medially below the internal auditory meatus, posterior medial wall, back wall and floor. Pteriotic (green): lateral semicircular canal, roof of tympanic cavity and up into the antrum. Epiotic (pink): posterior semicircular canal, posterior roof and into the mastoid process. (Reprinted from Lewis, Warren H (ed). Gray's Anatomy of the Human Body, 20th American Edition. Philadelphia, PA: Lea & Febiger, 1918)

Fig. 8.103 Gray 30.4.95 tympanic bone fields, saggital showing opisthotic acting like a posterior cap behind, below prootic. (Reprinted from Lewis, Warren H (ed). Gray's Anatomy of the Human Body, 20th American Edition. Philadelphia, PA: Lea & Febiger, 1918)

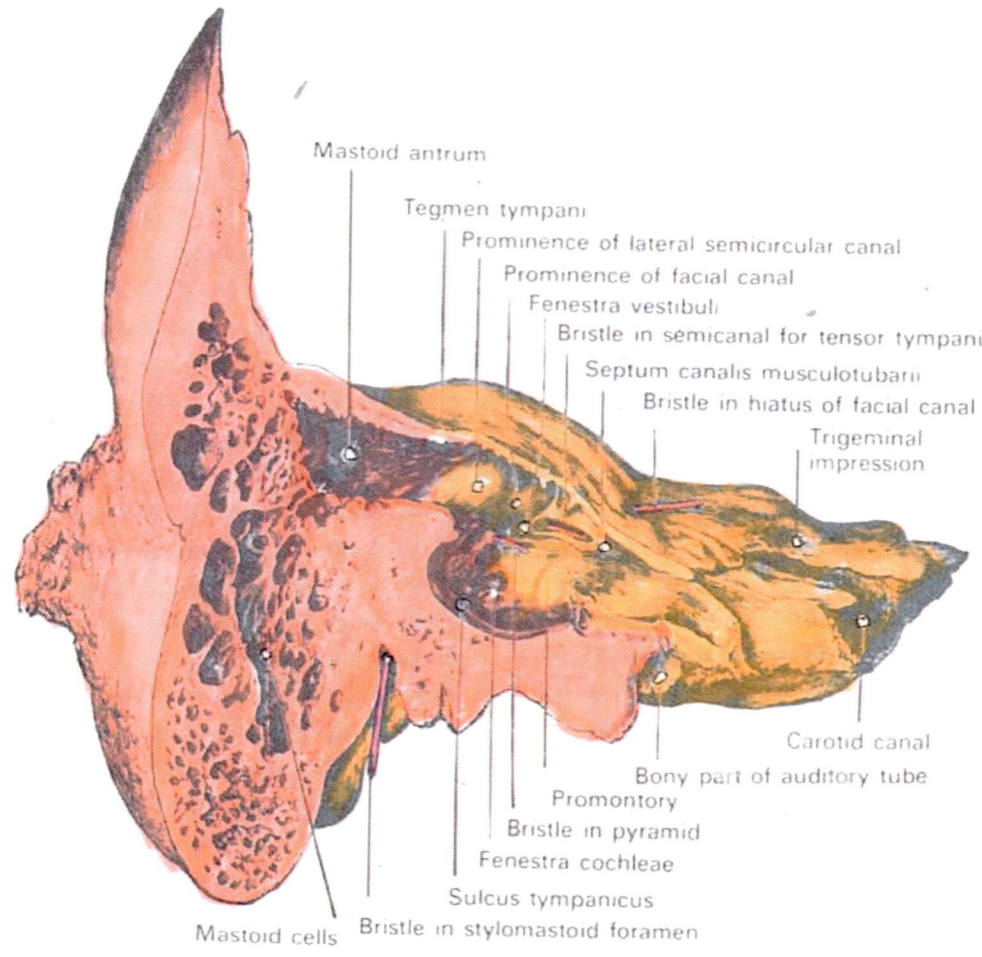

Mastoid antrum
Tegmen tympani
Prominence of lateral semicircular canal
Prominence of facial canal
Fenestra vestibuli
Bristle in semicanal for tensor tympani
Septum canalis musculotubarii
Bristle in hiatus of facial canal
Trigeminal impression
Carotid canal
Bony part of auditory tube
Promontory
Bristle in pyramid
Fenestra cochleae
Sulcus tympanicus
Bristle in stylomastoid foramen
Mastoid cells

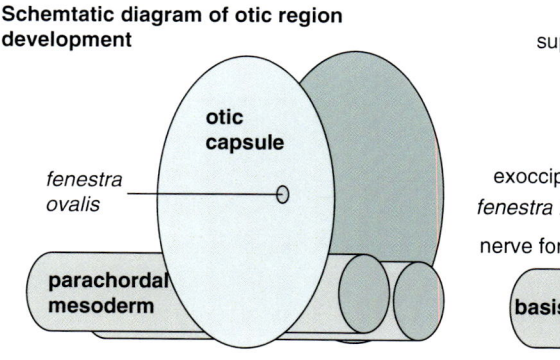

Schemtatic diagram of otic region development

otic capsule

fenestra ovalis

parachordal mesoderm

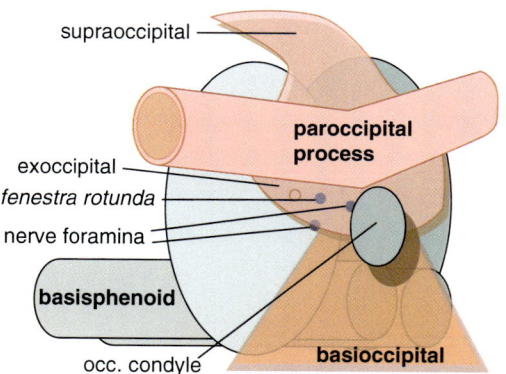

supraoccipital

paroccipital process

exoccipital
fenestra rotunda
nerve foramina

basisphenoid

occ. condyle

basioccipital

Fig. 8.104 Parachordal mesoderm is the source for both prootic and opisthotic bone fields. Together they form the otic capsule. The otocyst is housed in the anterior prootic. Otic capsule is flanked posteriorly by exoccipitals. Supraoccipitals develop from cartilages produced by occipital somites. Posterior expansion of the brain pushes the entire bone complex downward 90° such that occipital condyles (paired in mammals) face downward. (Courtesy of Augustus T. White, Palaeos. com. Retrieved from http://palaeos.com/vertebrates/bones/ear/overview02.html)

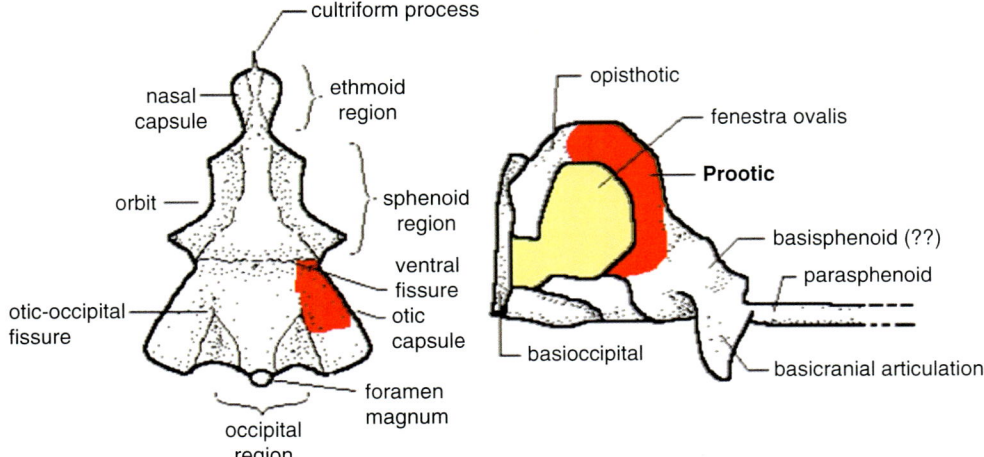

Fig. 8.105 Prootic bone. Prootic contains the cochlea and semicircular canals in the anterior part of the petrous complex. Note its position relative to opisthotic. Otic capsule was originally designed to protect the vestibular apparatus which was housed in prootic but as tetrapods advanced, the capsule became dedicated to hearing as well. Hyomandibula becomes stapes which by virtue of its attachment to otic capsule served a bracing function for prootic. As stapes were reduced to become a sound transducer, the roles of these two bones were reversed, with prootic now serving as the support for stapes. Note that pootic and opisthotic come together to share the oval window. This is seen in the field diagrams of the tympanic cavity. (Courtesy of Augustus T. White, Palaeos.com. Retrieved from http://palaeos.com/vertebrates/bones/braincase/prootic.html)

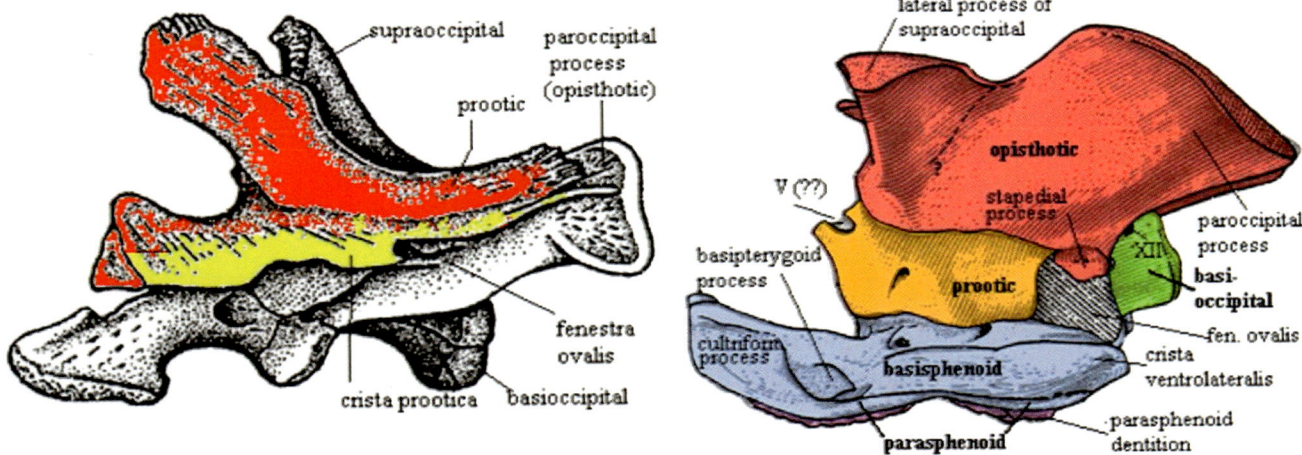

Fig. 8.106 New Zealand monitor lizard *Varanus*, an evolutionary throw-back. Prootic (red) and opisthotic (yellow) are shown; they are sheltered by *crista prootica*. Crista one of the more reliable landmarks on the prootic. It is a ridge or ledge of the prootic which runs anteroposteriorly along the side of the braincase above the otic capsule. In reptiles, crista prootica can be elaborated into a major component of the braincase as shown in Varanus where it represents and merges with opisthotic or exoccipital to form the paroccipital process. This process reinforces the back of the skull like a horizontal cross-bar. Anteriorly, the crista forms a scroll that protects into a kind of cranial nerve foramina. Crista prootic it shelters fenestra ovalis and serves as a reinforcing protection for the nerves. *Aerosaurus* ("copper lizard") shows position of fenestra ovalis and articulation for stapes. Note how posterior the otic capsule is located. Paroccipital process supports the basioccipital. [Left: (Reprinted from Rieppel O, Zaher H. *The braincases of mosasaurs and Varanus, and the relationships of snakes.* Zool. J. Linn. Soc. 2000;129:489–514. With permission from Oxford University Press). Right: (Reprinted with permission from Texas Memorial Museum. The University of Texas at Austin)]

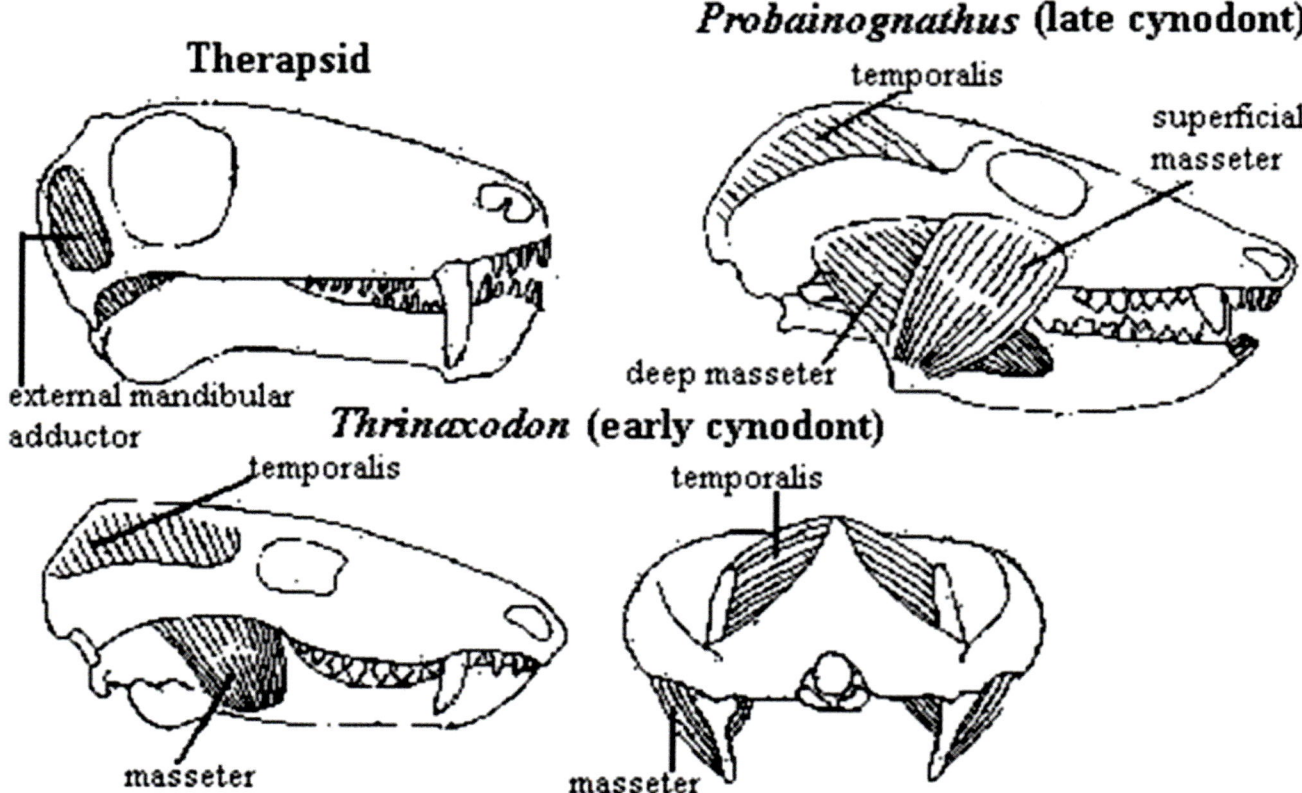

Fig. 8.107 Muscle evolution causes changes in the posterior skull. Therapsid has synapsid skull with a small temporal fossa and a single mandibular adductor blastema. Early cynodont *Thrinaxodon* has differentiated into temporalis and a masseter attached to the zygomatic arch. *Probainognathus* has bi-planar masseter with deep layer that will become pterygoids. It also has dentary squamosal joint. With the forward positioning of the joint, brain growth was directed posteriorly with consequent changes in the position of prootic and opisthotic. (Reprinted from Carroll RL. Vertebrate Palaeontology. New York, NY: WH Freeman and Company; 1988)

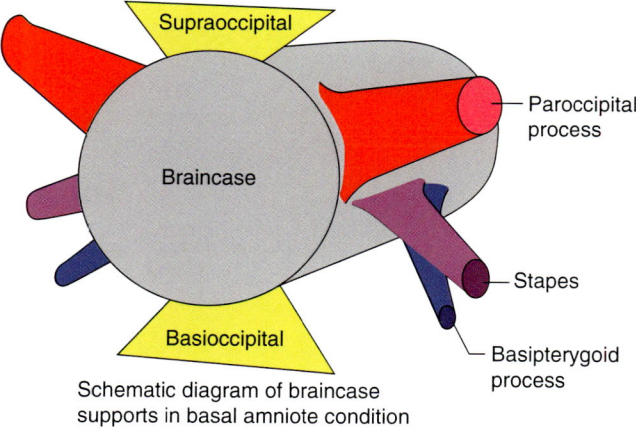

Fig. 8.108 Braincase in amniotes runs from the occiput forward to basisphenoid. It is supported ventro-dorsally by basioccipital and (in advanced amniotes) supraoccipital. It has three side struts (basipterygoid, stapes, and paraoccipital process) that brace it to the rest of the cranium. Paraoccipital process is an extension of opisthotic. (Courtesy of Augustus T. White, Palaeos.com. Retrieved from: http://palaeos.com/vertebrates/bones/braincase/opisthotic.html)

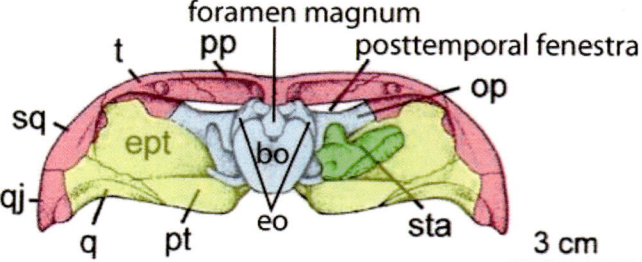

Fig. 8.109 Post temporal fenestra in *Greererpeton*. This represents the occiput of a generalized primitive tetrapod. Mesoderm is present between the otic capsules and the postparietal. *Non-somitic PAM from levels r8 to r9 flows into this space to produce the supraoccipital cartilage.* Stapes is attached to opisthotic but has not yet reached laterally to squamosal. Posttemporal fenestra remains unfilled laterally. Exoccipitals sending braces in two directions: outward to tabulars and upward to post parietals. The space between postparietals and exoccipitals has not yet been filled in by supraoccipital. (Reprinted from Godfrey SJ. The postcranial skeletal anatomy of the Carboniferous tetrapod Greererpeton burkemorani Romer, 1969. Phil. Trans. R. Soc (Lond.), 1989;B323:75–133. With permission from The Royal Society (U.K.))

Fig. 8.110 Amniote occiput. Basal amniote Paleothyris. (Reprinted from Laurin M, Reisz RR. A reevaluation of early amniote phylogeny. Zoological Journal of the Linnean Society 1995;113:165–223. With permission from Oxford University Press)

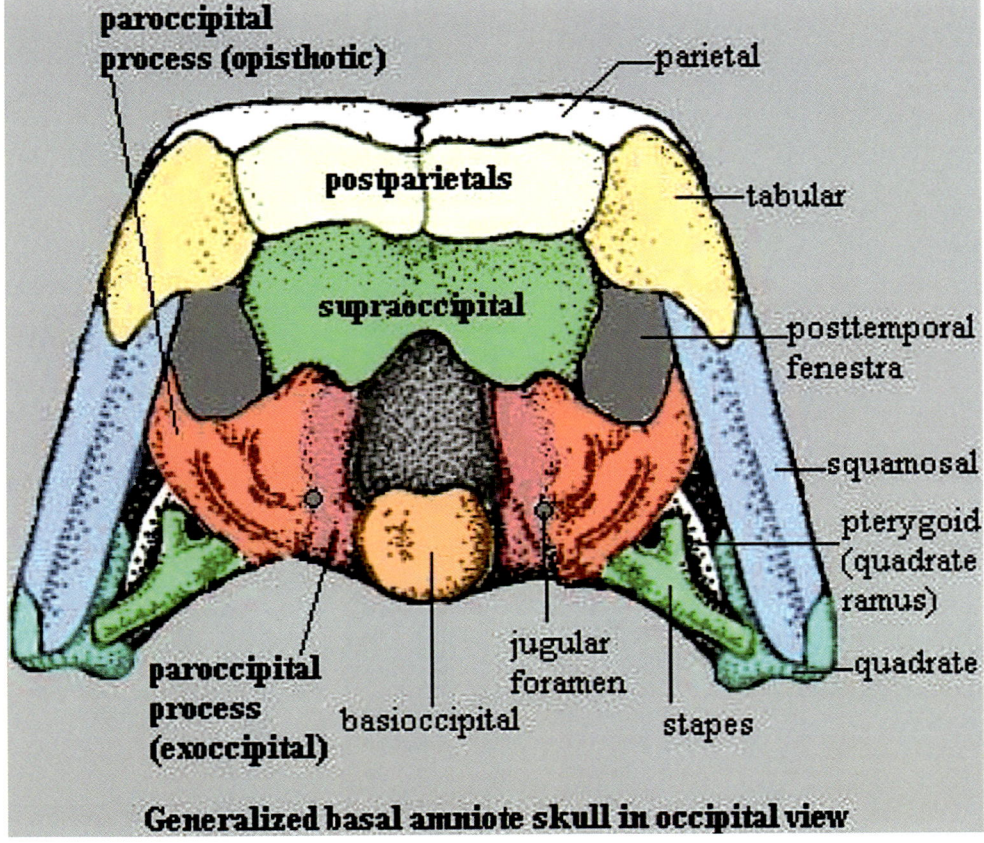

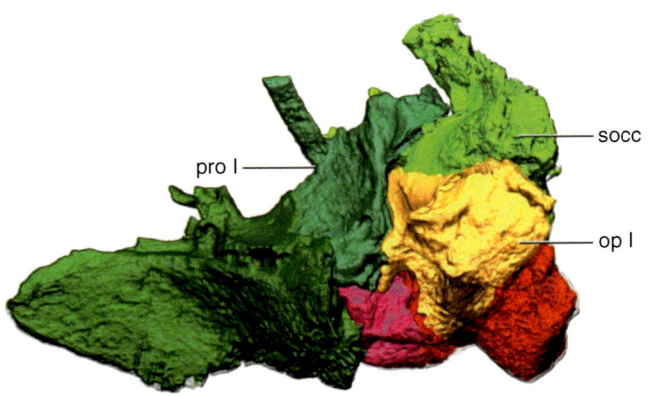

Fig. 8.111 *Captorhinus* sagittal cut of braincase. Premammalian relationship of prootic and opisthotic. Brain expansion will push the opisthotics remain fixed so they are drawn into the skull. At the same time, exoccipitals will expand to complete coverage of the posterior skull. *bocc* basioccipital, *bspost* basipostsphenoid, *op l* left opisthotic, *pro l* left prootic, *pro mp* prootic medial process, *ps+bspre* parasphenoid + basipresphenoid, *socc* supraoccipital. (Reprinted from Araujo R, Fernandez V, Polcyn MJ, Fröbisch J, Martins RMS. Aspects of gorgonopsian paleobiology and evolution: insights from the basicranium, occiput, osseous labyrinth, vasculature, and neuroanatomy. *Peer J* 5:e3119. With permission from Creative Commons License 4.0: https://creativecommons.org/licenses/by/4.0/)

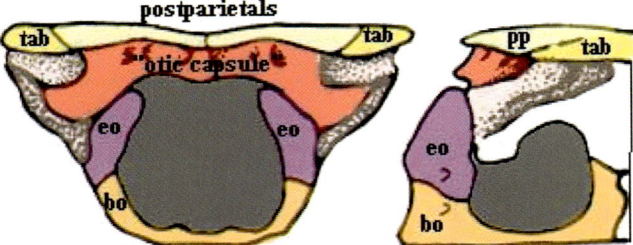

Fig. 8.112 *Acanthostega* shows the position of opisthotics in basal anamniote tetrapod. The capsules are joined together by a common mesoderm but it has yet been converted into distinct bone fields. (Reprinted from Berman DS. Origin and evolution of the amniote occiput. J Paleontol 2016; 74(5):938–956. With permission from Cambirdge University Press)

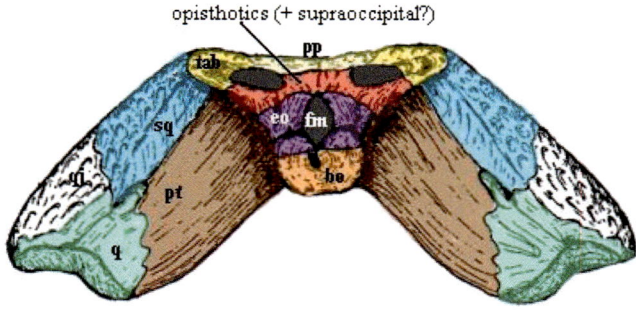

opisthotics (+ supraoccipital?)

Fig. 8.113 *Megalocephalus* occiput. Here the inter-opisthotic mesoderm is organizing itself as the future supraoccipital. (Reprinted from Carroll RL. Vertebrate Palaeontology. New York, NY: WH Freeman and Company; 1988)

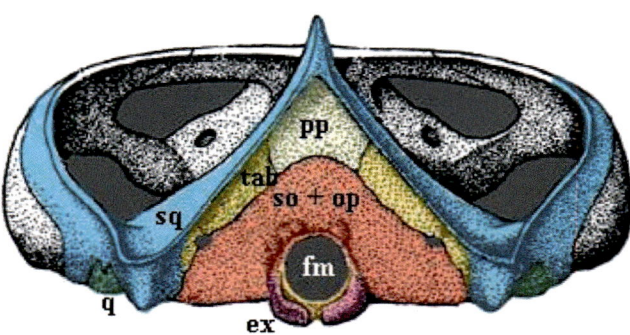

Fig. 8.114 *Probainognathus* occiput shows clear-cut relationships between interparietal complex and the occipital complex. Supraoccipital (so) and opisthotic (op) are draped widely around the exoccipitals. (Reprinted from Kermack, KA, Mussett F, Rigney HW. The skull of Morganucodon. Zool. J. Linn. Soc. 1981; 71:1–158. With permission from Oxford University Press)

SANS fibers via deep petrosal nerve to carotid plexus. These latter gain access to pterygopalatine fossa via the vidian nerve. From the back surface of carotid canal, a bony extends downward as two laminae; the lateral lamina being continuous with squamosal. Medial to it lies the *styloid process* and posterior to that, the mastoid. In the center of these two structures is *stylomastoid foramen*. Thus, dissection which follows structures inserting into styloid process will automatically lead to the main trunk of facial nerve. Our final structure is the *tympanomastoid fissure* which transmits the *auricular branch of X*.

Descriptive Anatomy: Mastoid Temporal Bone

The mastoid forms the posterior margin of the temporal bone complex. Its <u>external surface</u> provides the secondary insertion for occipitalis. *Mastoid foramen* constitutes a communication between the occipital artery and the meningeal system of the posterior fossa mastoid foramen. Inserted into the <u>inferior surface</u> has a prominent projection, the *mastoid process*, into which are inserted sternocleidomastoid and splenius capitus (which are shared with the superior nuchal line of occipital bone) and splenius capitus. Medial to the mastoid process is a fossa which bears the insertion of posterior belly of digastric. The <u>internal surface</u> of mastoid has groove for the sigmoid sinus which is separated from the mastoid air cells by a tiny bony lamina. For this reason, mastoid infections can cause meningitis and/or thrombosis of the sinus.

The mastoid process is in continuity with the mucous membrane of the tympanic cavity. Therefore, it contains the air cells of the *mastoid sinus*. Bone marrow can be found in the mastoid process. The *mastoid antrum* is located in the

Fig. 8.115 Mammiliform *Morganucodon* showing **fusion of the prootic and opisthotic.** (Reprinted from Kermack KA, Mussett F, Rigney HW. The skull of Morganucodon. Zool. J. Linn. Soc. 1981; 71:1–158. With permission from Oxford University Press)

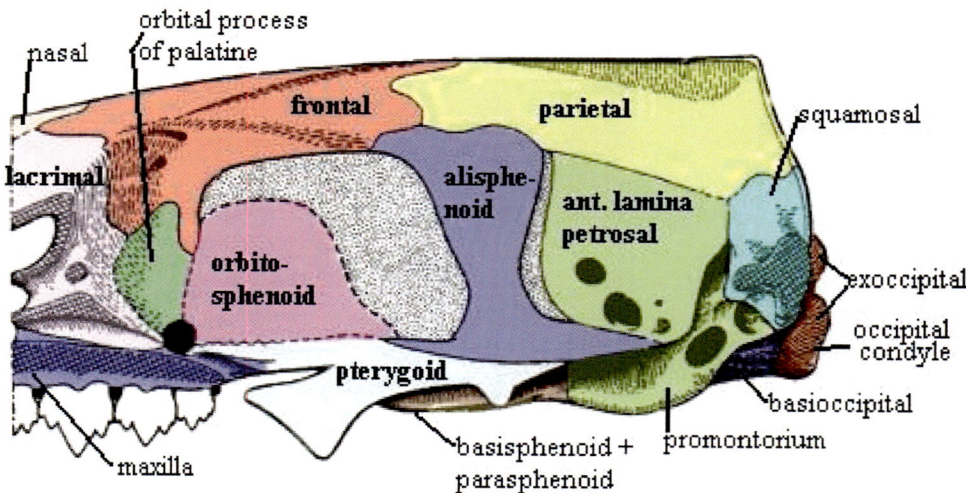

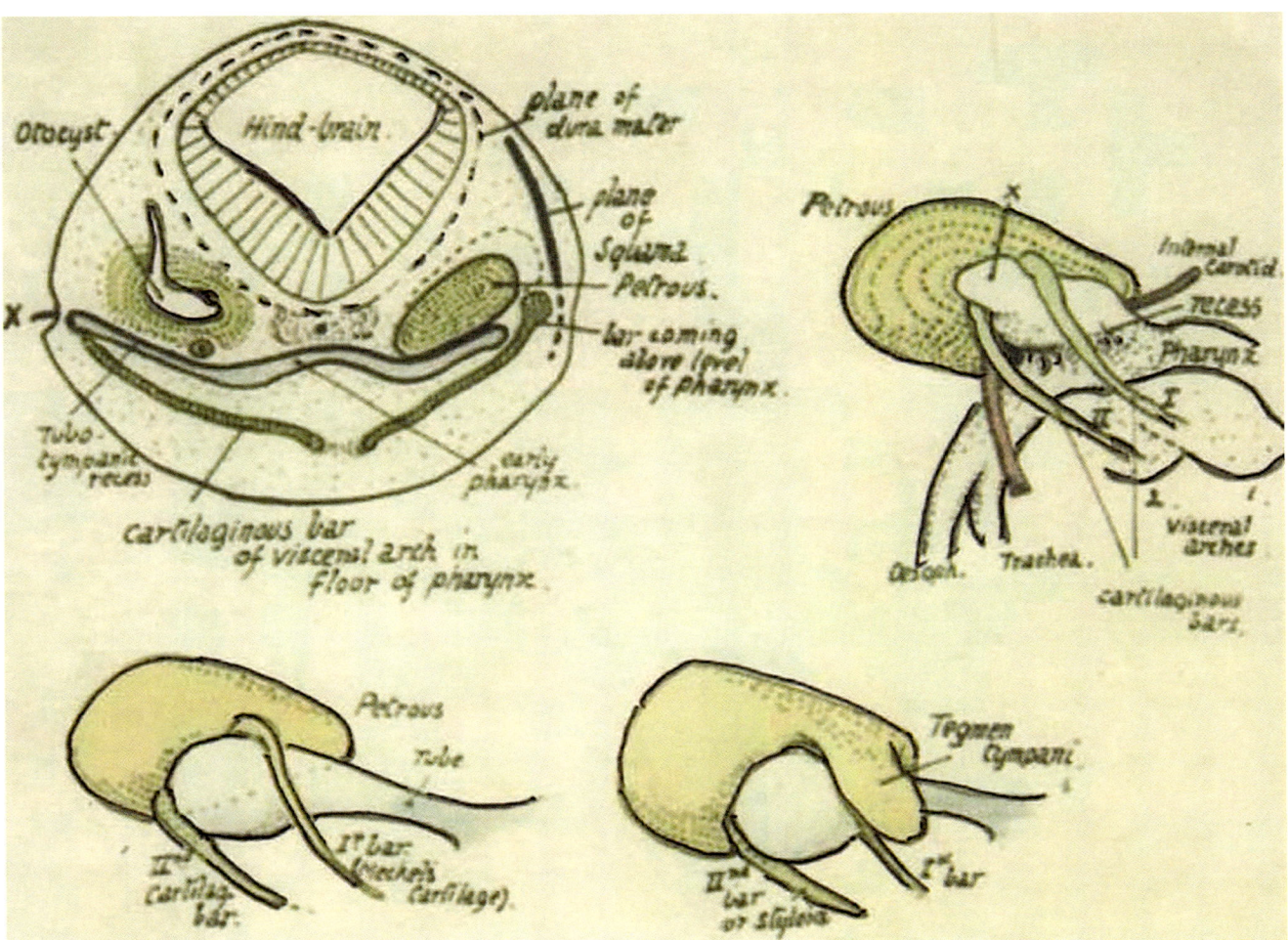

Fig. 8.116 Petromastoid development 1. <u>Upper left</u>: Cross section shows otocyst in the petrous with outlying squamous. Quadrate (stapes) is braced to opisthotic which consequently covers *fenestra ovalis*. <u>Upper right</u>: Pharyngeal diverticulum inserted between cartilagenous bars of the first and second arches is the Eustachian tube. <u>Lower left and</u> <u>right</u>: Petrous growth, indented by the tube, subsequently sends out a diverticum over the tube, *tegment tympani*. (Reprinted from Breathnach AS (ed). Frazer's Anatomy of the Human Skeleton, 6th ed. London: J&A Churchill; 1907)

anterior and the superior part of mastoid process. It communicates with the air cells but contains free air and is in continuity with the middle ear. Its roof is a thin wafer of bone, the *tegmen tympani* which separates the sinus from middle cranial fossa and constitutes another weak spot for perforating infection causing meningitis. Medial to the antrum is a bulge caused by the lateral semicircular canal.

Blood Supply and Mesenchyme

The anatomy of the internal carotid follows the classification system of Bouthillier as having seven segments: cervical (C1), petrous (C2), lacerum (C3), cavernous (C4), clinoid (5), ophthalmic (C6), and terminal (C7) [10]. The petrous segment is housed within a canal which results when the mesenchyme of the bone-forming fields surrounds the artery

and then proceeds to ossify. The *ascending segment* passes in front of cochlea and tympanic cavity, separated from it by only a thin bony wall. It then bends forward and passes underneath the trigeminal ganglion from which it is also separated by a delicate lamina of bone. The *lacerum segment* begins above foramen lacerum (through which ICA does not pass) and terminates at the petrous apex. It then travels along the upper lateral shelf of the sphenoid sinus of presphenoid bone. The *meningo (inferior) hypophyseal branch* localizes this *petrosal segment* to the level in front of neuromeric level r0. It exits the sinus at proximal dural ring as the clinoid segment until reaching distal dural ring which is part of the falx where it runs briefly with cranial nerve II as the *ophthalmic segment*. When it passed between cranial nerves II and III, the ICA forms its anterior and posterior divisions. The petro-

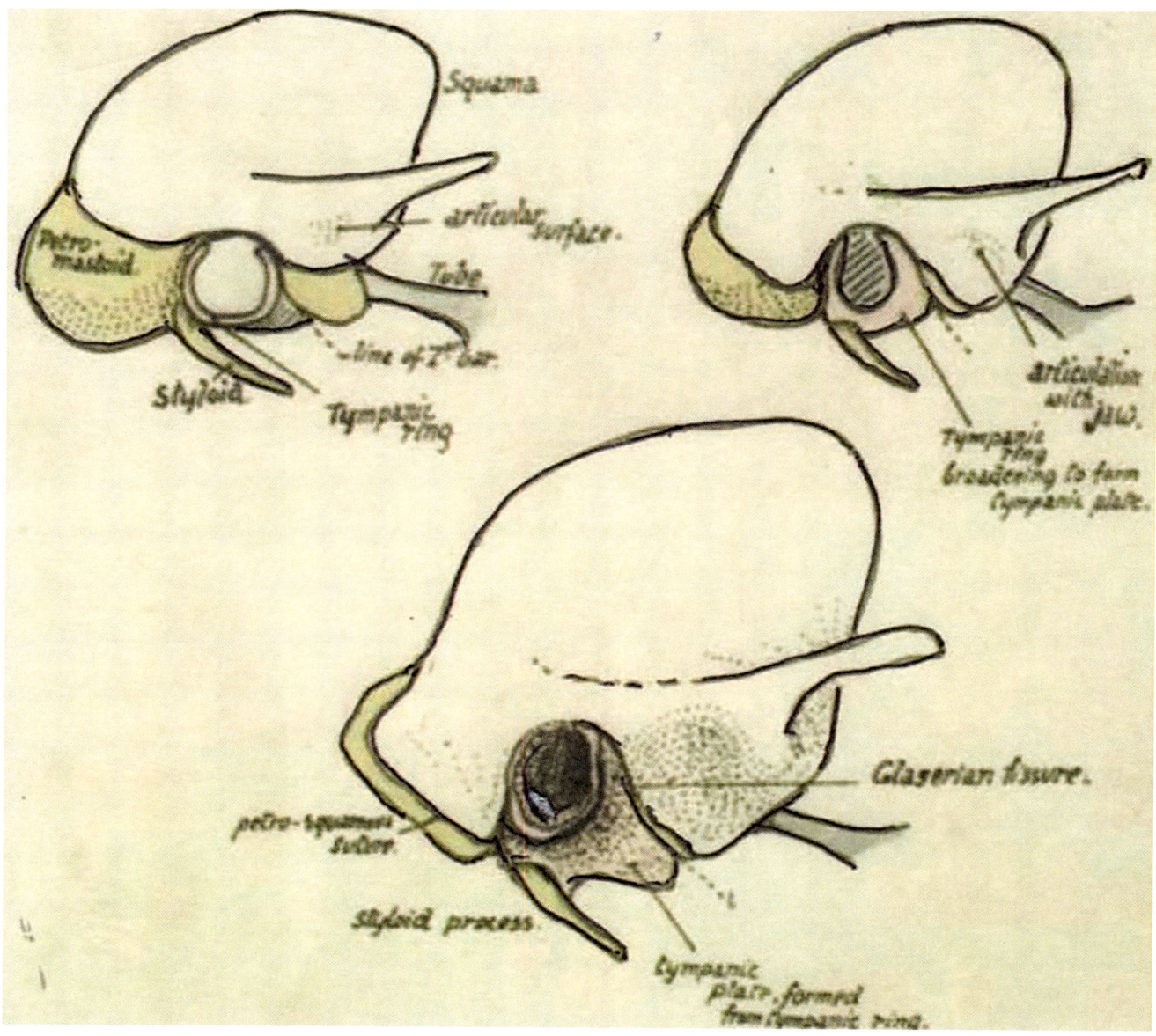

Fig. 8.117 Petromastoid development 2 (Frazer). Upper left: tympanic ring now incorporated from the tympanic portion of Meckel's into petromastoid. Styloid process projects from Sm4 to Sm5 boundary. Sm4 and Sm5 are putative precursors of prootic and opisthotic, respectively. Upper right: squamosal forming indentation to accommodate the mandibular condyle. Petrous prootic lies deep to the joint. Tympanic ring expanding, still medial to the styloid. Lower center: petrosquamous fissure marks posterior and inferior expansion into the mastoid process. Tympanic plates extend both external and internal to styloid, embracing it as the vaginal process. (Reprinted from Breathnach AS (ed). Frazer's Anatomy of the Human Skeleton, 6th ed. London: J&A Churchill; 1907)

sal and lacerum segments have a few named branches. Petrous bone is supplied by multiple un-named small vessels.

Longitudinal neural artery, on the other hands, gives off multiple branches directed laterally. These may well be in a 1:1 relationship with neuromeres r2–r7.

Both the ICA and the paraxial mesoderm that constructs the walls of both vessels arise from levels r2 to r7. Because traditional hox genes come into play at level r3, the role of r2 may be unclear. Certainly somitomeres 1–3, the PAM of

which arises from level r0 to r1, are not available for ICA construction. Instead they are directed toward supplying the mesenchymal coverage of the forebrain and orbit.

The petrous temporal bone contains within it the internal carotid derived from the embryonic dorsal aorta. Housed within petrous temporal is the inner ear; this is constructed from three components.

(1) Somatic ectoderm from the otic disc forms the *membranous labyrinth*, consisting of three semicircular ducts, an utricle and saccule, and the cochlear duct. (2) Neurosensory

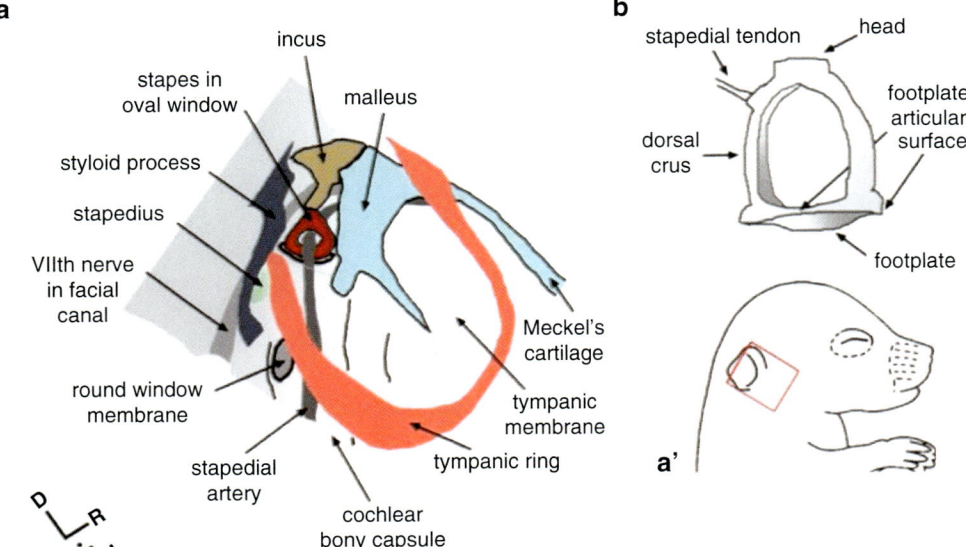

Fig. 8.118 Tympanic ring. The body of tympanic ring from reflected lamina of angular is first to ossify. Nine weeks later, prearticular adds on the had of the tympanic ring. Articular (malleus) and quadrate (incus) are beginning to ossify. Ossification of all three bones is complete at birth. Prootic and opisthotic are divided by the facial nerve. Note Meckel's cartilage extending backward into the posterior dentary extension backward. (Reprinted from Raft R, Coate TM, Kelly M, et al. Pou3f4-Mediated Regulation of Ephrin-B2 Controls Temporal Bone Development in the Mouse. October 2014 PLoS ONE 9(10):e109043. With Creative Commons License 1.0: https://creativecommons.org/publicdomain/zero/1.0/)

areas derived from the otocyst form three *ampullary crests*, two *maculae*, and the *spiral organ*. These structures are directly related to the brain in both innervation (the eighth cranial nerve) and blood supply (the basilar artery being derived from the longitudinal neural artery). (3) Paraxial mesoderm belonging to the epaxial half of somitomeres 5–7 is induced by the otocyst to form a surrounding *osseous labyrinth* consisting of three semicircular canals, a vestibule, and a cochlea.

The middle ear is formed by neural crest associated with the first, second, and third pharyngeal arches. These tissues about the petrous bone to form the *tympanic cavity and the ossicular chain*. Innervation is provided by cranial nerves V, VII, and IX (via X). Blood supply is primarily from the stapedial system and the external carotid. A small, internal component comes from the hindbrain basilar system via *subarcuate artery*, a derivative of the internal auditory artery.

Neural crest from r3 produces *tympanic temporal bone*. It also combines with Sm4 PAM to form *squamous temporal bone*. The *mastoid process* is an "add-on" structure; it does not appear until the 29th week. It develops from epaxial Sm7; it directly abuts Sm8 somitic PAM (i.e., the first occipital somite). Its blood supply (the mastoid branch of occipital artery) relates with the second arch occipitalis muscle. This explains its relation to the external carotid system and its likely somitomeric source of mesenchyme.

Phylogeny and Development of the Temporal Bone

Because temporal bone embryology involves so many actors, we should now set out a developmental timeline so we can keep oriented. This is a three-step process.

1. We must get acquainted with the embryonic components of the cranial base. To visualize them, the reader is advised to review the description and diagrams at the beginning of this chapter and in Carlson.
2. The next step is to review these events the Carnegie-O'Rahilly embryologic staging system reviewed in Chap. 3.
3. The development of each anatomic unit is described separately.

The bone fields dating back to the dawn of the skull base, the *pro-otic bone* and the *opisthotic bone* fit together as like two distorted bookends to form the otic capsule. A separate *epioccipital bone* of the fish model which "disappears" with tetrapods may prove to be a genetic idea that is resuscitated in mammals as the mastoid process. These bones all arise from non-somitic paraxial mesoderm in register with rhombomeres r4–r7. The posterior margin of mastoid interfaces with chondral supraoccipital is a product of the occipital somites r8–r11. As many features of the occiput can only be

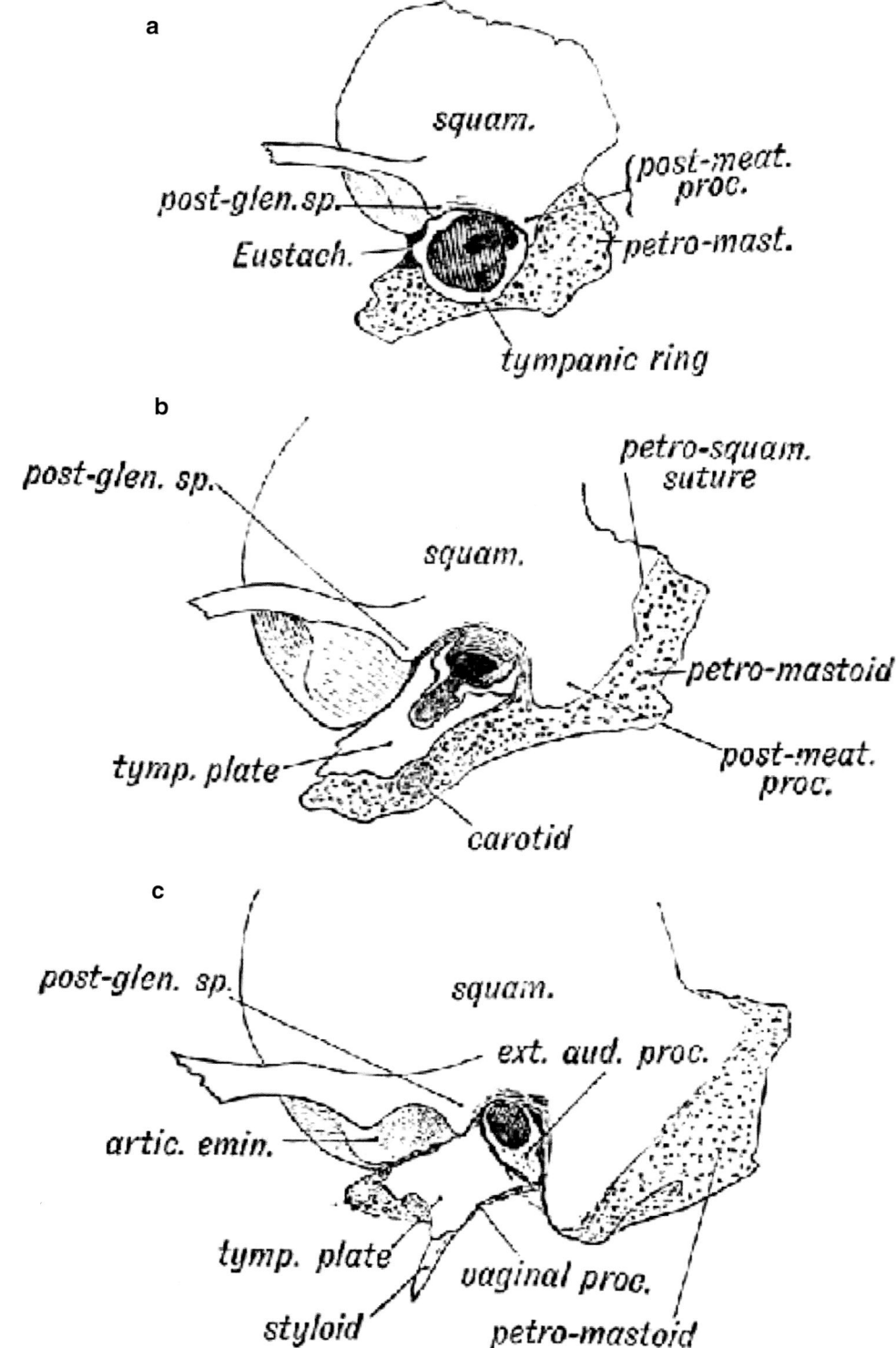

Fig. 8.119 Development of tympanic bone. (**a**) Tympanic bone in situ. (**b**) Mastoid growing downward and tympanic plate appears. (**c**) Styloid process appears. (Reprinted from Keith A. Human Embryology and Morphology. London: Edward Arnold; 1902)

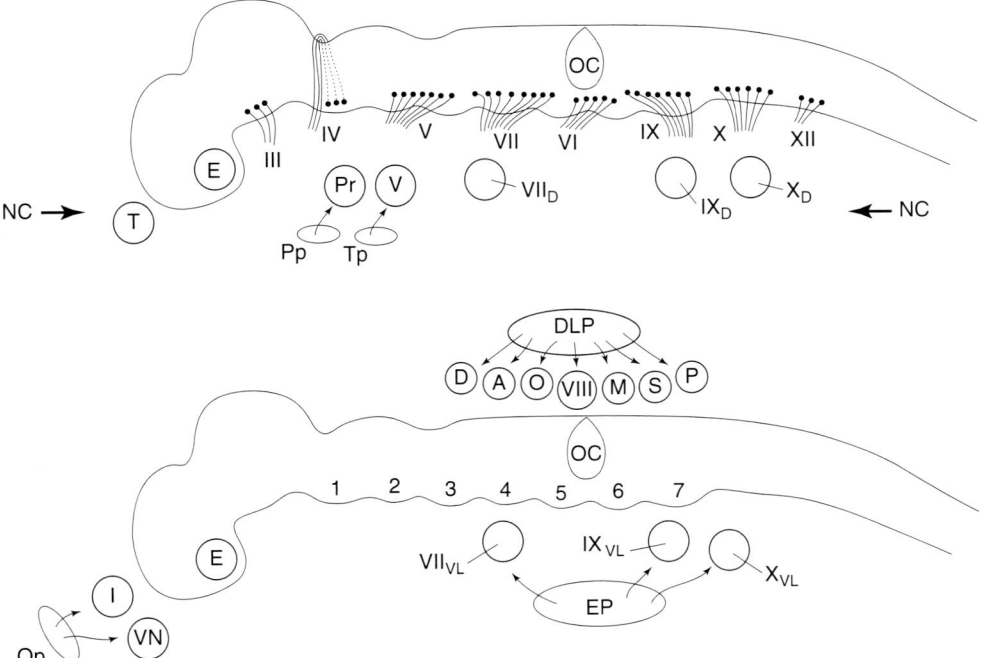

Fig. 8.120 Placodes, neural crest and cranial nerves. Parasaggital section of neural tube and somitomeric mesoderm. Above: Derivation of cranial nerves and ganglia from neural crest and placode contributions from profundus placode (P$_P$) profundus nerve P and trigeminal placode (T$_P$) to trigeminal nerve V. Below: Derivation of cranial nerves and ganglia from dorsolateral series, ventrolateral series, epibranchial series and rostral olfactory placodes. *A* anteroventral lateral line, *D* anterodorsal lateral line, *M* middle lateral line, *O* otic lateral line, *OC* otic capsule, *P* posterior lateral line, *Pr* profundus, *S* supratemporal lateral line, *T* terminal, *VN* vomeronasal. (Reprinted from Butler AB, Hodos, W. Comparative Vertebrate Neuroanatomy, 2nd ed. John Wiley, 2005:157–182, fig. 9.8. With permission from John Wiley & Sons)

understood in terms of these precursor bones, this discussion will seguy into the next part of this chapter into the occipital bone complex.

Prootic Bone

Descriptive Anatomy

The prootic bone forms the anterior and the dorsal part of the otic capsule. It merges without sutures with opisthoic to house the cochlea and labyrinth. Prootic articulates with dermal skull bones (squamosal and tympanic) whereas opisthotic does not. Specifically, prootic articulates with quadrate, so that, in pre-tetrapod basal vertebrates, it serves as a lateral strut or brace for the skull. It can also receive processes that connect it with the sphenoid complex.

In tetrapods, the physical relationship of prootic to opisthotic is like that of two bookends surrounding the otocyst. Prooic forms anterior wall, parts of the roof and floor, and the lateral wall. Opisthotic is the mirror image and completes the box. Along their medial aspects, the two bone fields surround the fenestra ovalis (Fig. 8.101).

Prootic is constructed from paraxial mesoderm in register with rhombomeres 4–5. It is likely to pass through an intermediate somitomeric form because non-somitomeric PAM is already plastered up against the wall of the neural tube.

1. *Protection for the labyrinth and inner ear.* Semicircular canals (SCC) are an essential invention for mobility and predation permitting an orientation to one's environment. They first appear in the primitive agnathic fishes. A single posterior semicircular canal is found in cyclostomes of order Myxinoidea ("slime producer") and seen in modern-day hagfishes. Going one step up to petromyzontiformes ("stone lickers"), lampreys and later, in the conodonts, a second, vertical semicircular appeared. Gnathostome fishes added the third horizontal canal, making the more effective predators.

2. *Braincase support.* Prootic is a fundamental component of the cranial base. A prominent ridge, *crista prootica* runs anteroposteriorly above the otic capsule. In reptiles, it joins either with opisthotic bone or with the exoccipital bone to form a *paraoccipital process*, a reinforcing crossbar spanning across the back of the skull (more on this later). In mammals, the crista may be represented by the *ridge of the petrous apex*. Cranial nerves such as VII and VIII enter the skull via foramina under cover of crista prootica.

Fig. 8.121 Inner ear neurosensory structures. Prootic: <u>anterior semicircular canal</u>, anterior root of tympanic cavity, and forward to the apex. Opisthotic: promontory, cochlear window, floor of tympanic cavity, goes posterommedially below the internal auditory meatus. Pteriotic: <u>lateral semicircular canal</u>, posterior roof of tympanic cavity and up into the antrum—part of prootic. Epiotic: <u>posterior semicircular canal</u> and into the mastoid process—part of opisthotic. (Reprinted from Blausen.com staff (2014). "Medical gallery of Blausen Medical 2014". WikiJournal of Medicine 1 (2). https://doi.org/10.15347/wjm/2014.010. ISSN 2002-4436. With permission from Creative Commons License 3.0: https://creativecommons.org/licenses/by/3.0/deed.en)

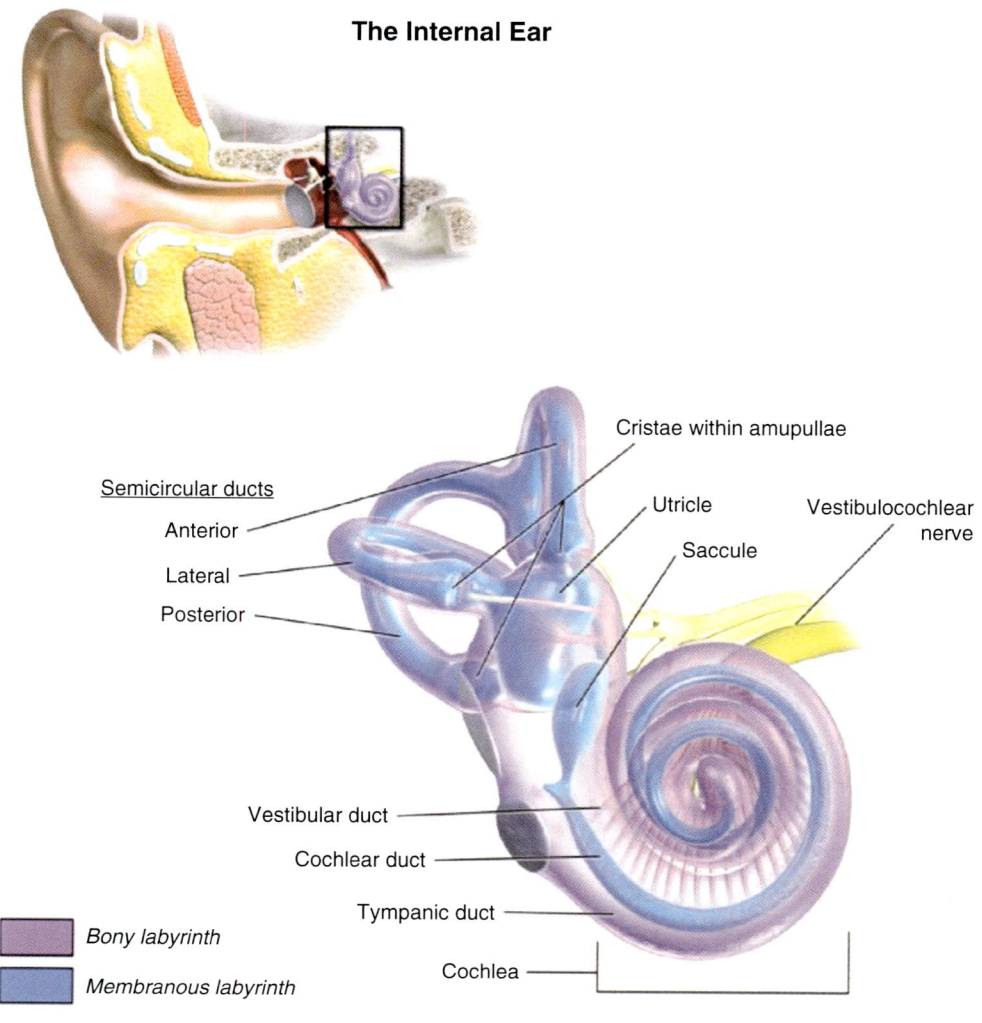

The Internal Ear

Semicircular ducts
Anterior
Lateral
Posterior

Cristae within amupullae
Utricle
Saccule
Vestibulocochlear nerve

Vestibular duct
Cochlear duct
Tympanic duct
Cochlea

Bony labyrinth
Membranous labyrinth

3. *Articulation with stapes.* With the appearance of tetrapods, mandibular suspension from the skull involving the hyomandibula of the second arch underwent a change becoming autostylistic. In this condition, maxilla is fused to the skull and mandible articulates between two first arch bones, articular from the mandible and quadrate, formerly a sheathing bone of palatoquadrate cartilage (the template for upper jaw). Primitively, stapes was a large heavy second arch bone connecting the otic capsule with surrounding dermal bones such as squamosal and tabular. As such, it carried vibrations to the otic apparatus. With evolution, the support function became unnecessary and stapes evolved into a delicate bone for sound transmission. It is now supported by prootic rather than the reverse.

Phylogeny

The evolution of a vestibular system in jawless fishes up to and including the osteostracans took place without an otic capsule and hence, without prootic. In the placoderms, the braincase presents with a single mass bearing a ridge, the crista prootica (a very ancient structure, indeed). Early bony fishes and the sarcopterygians had a well-defined but undivided otic mass. A prootic exists in modern teleosts and tetrapods but it is not clear if this these structures are homologous. The fundamental relationship between prootic and opisthotic becomes evident in basal tetrapods (Fig. 8.102).

Amniotes link the occiput to the otic capsule via a supraoccipital of the braincase. This frees stapes of additional stress, permitting its evolution as a transducer. In the cynodont synapsids, the immediate precursors to mammals, anterior positioning of the jaw joint was accompanied by a *reassignment of chewing muscles* along the external surface of dermal bones. The original masseteric mass split, resulting a "new" temporalis, now inside zygomatic arch. This *removed constraints on brain growth* and the position of the otic capsule. Prootic was now free to associate more with squamosal. The growth of the middle ear apparatus favored a lighter bone with more space. In mammals, prootic merges with opisthotic (Fig. 8.103).

Opisthotic Bone

Descriptive Anatomy

This bone is literally "the other half" of the petrous complex. It sits behind prootic. On the one hand, it completes the coverage of the vestibulocochlear apparatus. On the other hand, it forms multiple articulations. Because the otic capsule is internalized in mammals, it is not visible externally but in basal amniotes it plays an essential role in bracing the occipital skull.

Opisthotic is constructed from paraxial mesoderm in register with rhombomeres 6–7. In the transition from mammaliaforms to mammals, a fusion takes place between prootic and opisthotic. However, by looking at anatomic landmarks such as internal auditory meatus and jugular canal, by review of the neural structures encased by these bones, and by a consideration of ossification centers, it is possible to reconstruct how opisthotic and prootic fit together.

Let us look at the standard condition of a basal amniote (Figs. 8.104 and 8.105). We see the braincase from behind as a cylinder directed forward toward the basisphenoid. It is supported from below by basisphenoid which articulates with the pterygoids via the *basipterygoid process*. Recall that the pterygoids are very large at this stage of evolution and extend all the way to the back of the skull. The dorsal braincase terminates with the supraoccipital bone, which, as discussed previously, arises from somitomeric mesoderm and has no relationship with the underlying occipital somites.

The braincase is connected with the rest of the skull by reinforcing struts. From its position at the rear of the otic capsule, opisthotic projects outward a lateral *paraoccipital process* that articulates with squamosal bone. Stapes (recently modified from myomandibula) is a truly massive bone located at the junction between prootic and opisthotic. It runs laterally and ventrally to connect the otic capsule with the quadrate bone, located in the extreme ventrolateral corner of the occiput.

Let us examine opisthoic in situ as part of the tetrapod occiput. It flanks exoccipital. Jugular foramen marks the interface of these two fields. Each opisthotic sends out a *paraoccipital processes* which connects laterally to squamosal. Together these act as a sling suspending from the center of the cranial base from the side wall. Superomedially, opisthotic joins with supraoccipital. The tabulars frequently produce an *opisthotic process* that drops down like a window shade to contact opisthotic and thereby close off the *posttemporal fenestra*.

The intercalation of opisthotic is well seen in the skull of *Eocaptorhinus*, a therapsid in the line leading to mammals (Fig. 8.106). Of note is the lateral view demonstrating the central position of stapes between prootic and opisthotic. Also the supraoccipital reaches down from the dorsal cal-varium to grip the skull from above. Definitely this bone is not a derivative of the occipital somites.

More efficient food-gathering is the name of the game in evolution. For carnivores, this meant developing stronger jaw muscles. The temporal fenestra provided a greater attachment surface. In mammals, this becomes externalized to house a new temporalis muscle. The strong stapes acted as a reinforcing strut at the back of the skull permitting the muscles to shift forward. Note evolution of the stapes will be discussed under the section entitled "Development of the Middle Ear."

Mesenchymal source material for opisthotic is PAM from rhomberic levels r6–r7. It probably arises posterior to that for the r4–r5 prootic. Because opisthotic forms jugular foramen with exoccipital, and because the foramen in constructed around IX and X, it is logical to consider its mesoderm to arise from a zone more posterior to that of the otic capsule.

Phylogeny

The panderichthyids (AKA elpistostegalids) were the most evolved tetrapod-like fishes before limbs appeared with the first tetraopods, *Acanthostega*. Externally, its skull had many similarities with tetrapods but internally it retained the piscine platform. The braincase of *Panderichthys* maintained the intracranial joint running from between basisphenoid and basioccipital to otic capsule. Hyomandibula retained its original function. Recall that with the advent of jaws, second arch gill structures disappeared to produce a chain of bones, hyomandibula was recruited to suspend the jaws from the skull base at the otic capsule. Recall further that the epibranchial cartilage of hyomandibula is located just behind the *spiracle*, the former first gill pouch between the first and second branchial arches of jawless fishes. With the transition to land, hyomandibula becomes the stapes. The remaining epibranchial, ceratobranchial, hypobranchial, and basibranchial cartilages form a chain from styloid process to lesser cornu and body of the hyoid bone. Spiracle morphs into Eustachian tube and lies accordingly anterior to oval window where stapes ultimately inserts.

Transformation of the tetrapod braincase took place rapidly with many innovations, a new occiput being one of them braincase was rapid. That of *Acanthostega* is rather rudimentary with basioccipitals and exoccipitals forming a U-shaped sling around foramen magnum (Fig. 8.107). A mesodermal mass connects the otic capsules, being slung from one side to the other like pizza dough. The opisthotic sling roofs over foramen magnum. Sitting atop the otic capsule like flat mosaic tiles are the postparietals and tabulars. On the lateral view, a space is present between exoccipitals and the otic capsule. This is filled in with connective tissue within which will develop the dorsal supraoccipital bone.

In *Greererpeton*, paraoccipital process connects with tabular and extends a toe towards squamosal (Fig. 8.108). Stapes lies posterior to it. At this level of evolution, the division between preootic and opisthotic is not well defined so it looks as though paraoccipital process could be arising from either the former or the latter. Further up the line in *Lapillopsis* the opisthotic on the move. It is no longer seen next to exoccipitals. Instead, in the baphetid *Megalocephalus*, we see it translocated above foramen magnum (Fig. 8.109) (viz interparietal section figure). This anticipates the condition that will be obtained in the later amniotes in which this region produces a true dorsal supraoccipital bone.

In the evolution of the diapsid amniotes (reptilomorphs), the tabular, supratemporal, and postparietal bones diminish and are officially declared dead. But in synapsids, amniotes such as *Paleothyris*, the tabulars and postparietals are retained (Fig. 8.107). The opisthotic and supraoccipital complex is a solid mass. The subsequent therapsids are ancestral of mammals. They have the newly developed musculature occupying the expanded temporal fenestra (temporal fossa). In the therapsid cynodont clade (dog teeth), powerful muscles inserted into a sagittal crest compress the membranous occiput into a triangle. *Probainognthus* shows opisthotics flanking the exoccipitals, as expected, but above foramen magnum, the paraoccipital plate, i.e., membranous supraoccipital bone is seen (Fig. 8.110). It contains all the bone fields previously mentioned. Double condyles are present but they remain ventral. They will migrate to the lateral position in true mammals.

In mammaliaforms, such a *Morganucodon*, prootic and opisthotic undergo fusion into an (apparently) singular petrosal. Persistent paraoccipital processes between exoccipital and squamosal are the only remaining evidence of opisthotic. A posterolateral bulge marks the promontorium of the petrosal. It contains cochlea. In early mammals, the cochlear was not coiled up, as it is today. Expansion of opisthotic with *mechanical assistance from sternocleidomastoid* explains the mastoid prominence seen in humans.

Development (Figs. 8.112, 8.113, 8.114, and 8.115)

The mesenchyme for the bones of temporal complex is laid down very early in development. Organization of the otic capsule follows immediately after gastrulation at stages 8–9. Neural crest contributions follow soon thereafter with first arch components in place by stages 9–10 and second arch components in place by stages 10–11.

Ossification centers for the temporal bone complex are as follows: (1) middle ear bones, (2) membranous labyrinth, (3) squama, (4) zygomatic process, (5) tympanic bone, (6) prootic, (7) opisthotic, and (9) styloid process. Since the component parts are assembled essentially at birth, we will deal with them from simple to complex, i.e., from outside to inside. The middle ear bones will be described later in that section. Note that within petrous, the

neurosensory organs have separate ossification centers for each component.

Styloid process has two ossification centers. *Tympanohyal* is proximal and ossifies just before birth. *Stylohyal* is distal and ossified later, at around 2 years of life. The component cartilages in mammals have not been well-traced. The current accepted sequence of branchial arch cartilages recognizes 5 units from proximal to distal: pharyngobranchial, epibranchial, ceratobranchial, hypobranchial, and basibranchial. Older literature follows the same sequence but uses confusing terminology. Here is a reasonable approximation: pharyngobranchial cartilage (tympanohyal bone), epibranchial cartilage (stylohyal bone), ceratobranchial cartilage (stylohyoid ligament), hypobranchial cartilage (lesser cornu of hyoid), and basibranchial cartilage (upper border of hyoid bone).

Tympanic ring arises from the angular bone, the first one of the post dentary sequence. It has a single ossification center. It unites with squamosal after birth.

Squamosal and zygomatic process. These bone fields have one ossification center each. Squamosal starts ossifying at the base of the future zygomatic arch. Development spreads upwards in sequence with the blood supply. Later, a second and more distal ossification center marks the development of zygomatic arch proper. Congenital defects can affect the development of the arch causing hypoplasia or absence. Zygomatic development is contingent upon that of squamosal. Asymmetry of squamosal size can occur.

Otic capsule develops from four centers at 4–5 months gestation. (1) *prootic* ossification is located at the arcuate eminence. The spread of bone is superior over the auditory canal and lows forward to the apex of the bone. It forms the medial wall, and front wall of tympanic cavity, covers part of the cochlea and vestibule, particularly *superior semicircular canal*. (2) The *opisthotic* ossification is located over the promontory surrounding the cochlear window. It forms the back wall and floor of tympanic cavity. It completes coverage of the cochlea. (3) The *pteriotic* ossification center likely belongs to opisthotic bone. It appears over the *lateral semicircular canal* and fills in the roof of the tympanic cavity, and spreads backwards to the antrum. (4) The *epiotic* ossification center makes its debut over *posterior semicircular canal*. It flows backwards to complete the mastoid. This bone field may represent the *epioccipital bone field* present in teleost fishes because of its association with the posterior semicircular canal.

In the immediate pre- and postnatal period, the temporal bone complex pursues the following directions of growth. Before birth, the tympanic ring and squama unite, making access to the braincase water tight. During year one, petromastoid and squamosal fuse. Also, the proximal bone field of styloid process, tympanohyal, fuses to the skull base but distal bone field, stylohyal, lingers until puberty or even later.

The <u>tympanic bone</u> cannot advance against the petrosal. Hence its only options for growth are lateral and posterior. The <u>zygomatic arch</u> starts out with a very shallow fossa for the TMJ which actually faces outward. This is because squama lies inferior to the arch but with changes in the skull base the lower squama extends inwardly. At the same time, the root of the arch grows outward, thus the fossa deepens and comes to faces inferiorly. <u>Mastoid process</u> responds to growth and traction by the sternocleidomastoid by projecting downward and inferiorly. The foramen for facial nerve, initially rather shallow, becomes tucked under the skull base. It also affects tympanic bone, pushing its floor anteriorly until the floor of the meatus becomes relocated to the anterior wall of the canal.

Developmental Timeline of the Ear: Carnegie/O'Rahilly Stages

Stage 9 (20 days, 1–3 somites)
> CNS: Major divisions of the brain.
> General: folding of the head causes passive formation of the first aortic arch.
> Ear: otic placode.

Stage 10 (22 days, 4–12 somites)
> CNS: Neural tube and optic primordium.
> General: pharyngeal arch 1.
> Ear: Otic disc at S10 stage; cells migrate from otic disc at S12. Acousticofacial crest.

Stage 11 (24 days, 13–20 somites)
> CNS: Rostral neuropore.
> General: pharyngeal arch 2.
> Ear: Otic pit. Caudal migration to final position dorsal to second pharyngeal cleft at S16.

Stage 12 (21–29 somites; 26 days)
> CNS: Caudal neuropore, secondary neurulation.
> General: pharyngeal arch 3.
> Ear: Otic vesicle connected to surface via a pore. Ventral wall of otic vesicle > vestibulocochlear neural crest.

Stage 13 (28 days, 30+ somites)
> CNS: Closed neural tube, cerebellum.
> General: pharyngeal arch 4.
> Ear: Basement membrane of otic disc now surrounds the otic vesicle. Capillary network. Beginnings of mesodermal condensation into otic capsule. Vesicle now closed off from surface. Dorsomedial otic vesicle > endolymphatic appendage. Vestibular part of the vestibulocochlear ganglion and vestibular nerve fibers are now formed.

Stage 14 (32 days)
> CNS: Cerebral hemispheres.

> Ear: Cochlear duct elongates from ventral otic vesicle elongates into cochlear duct.

Stage 15 (33 days)
> CNS: Longitudinal diencephalon zones.
> Ear: Otic capsule condensed. Utriculo-endolymphatic fold. Vestibular fibers reach otocyst epithelium. Auricular hillocks visible. Most ventral second arch (the hyoid bar) = antitragus.

Stage 16 (37–39 days)
> CNS: Evagination of neurohypophysis.
> Ear: Semi-circular canals presaged by thickened walls of otic vesicle. Utriculosaccular diverticulum. Spiral ganglion. Stapedial artery and stapes.

Stage 17 (41 days)
> CNS: Olfactory bulbs and amygdaloid nuclei.
> Ear: Otic capsule near chondrification. No semicircular canals as yet. Geniculate ganglion of VII. Tubotympanic recess and chorda tympani. Ossicles. Six hillocks: (1) tragus, (2) and (3) crus helicis, (4) and (5) helix, and (6) antitragus.

Stage 18 (44 days)
> CNS: Corpus striatum, inferior cerebellar peduncles, dentate nucleus.
> Precartilagenous otic capsule in contact with epithelial labyrinth (patterning). Semicircular ducts appear: anterior > posterior > lateral. Cochlea L-shaped. Stapes identified. Auricular hillocks merge.

Stage 19 (44 days)
> CNS: Choroid plexus of fourth ventricle.
> Ear: Otic capsule now cartilaginous (unconnected to basal plate). Cochlear tip is curled. Malleus and incus identified.

Stage 20 (47 days)
> CNS: Choroid plexus of lateral ventricles, optic, and habenular commissures.
> Ear: Parietal lamina. Otic capsule connects to basal plate and to exoccipitals. Tip of cochlea is curled. Tensor tympani and stapedius visible.

Stage 21 (50 days)
> CNS: Cortical plate.
> Ear: Cochlea first spiral.

Stage 22 (52 days)
> CNS: Internal capsule olfactory bulbs.
> Ear: Cochlea second spiral.

Stage 23 (56 days)
> CNS: End of embryonic period.
> Ear: Cochlea final ½ spiral. Labyrinth now complete. Ductus reuniens well seen.

<u>Nota Bene</u>: The anatomy of the temporal bone is particularly challenging. Readers are referred to the following references for definitive three-dimensional drawings: Anson, Gulya, Nadol, and Wolff.

The Inner Ear

The inner ear is constructed from the *otocyst*, an epithelial placode containing the neurons of hearing and balance. Recall that placodes are specialized zones of ectoderm that are functionally tied to the brain. Let us review this concept briefly. The otocyst is a *dorsolateral placode*, as such it is related to the primitive *lateral line system* in fishes (Fig. 8.116).

Dorsolateral "epibranchial" placodes have four sensory categories: (1) the motion detection, (2) electric currents, (3) hearing, and (4) balance. The *mechanosensory lateral line* detects motion in the water at a distance, suppressing currents generated by the animal's own movements. It functions in a wide range of environments, from clear surface water to utter darkness. These placodes have a long evolutionary history and are widely distributed, being identified in agnathic fishes, cartilaginous fishes, bony fishes, lungfishes, and water-dwelling amphibians. The *electrosensory lateral line* can identify prey, even when buried under the sand. It also plays a role in social interactions. It was present in ancestral vertebrates (at least the lamprey), was lost at the teleost level, and has re-evolved independently several times since. The *octaval system* consists of the auditory and vestibular rami of cranial nerve VII and their associated central structures. These rami are present in all vertebrates but some parts of the peripheral auditory apparatus have evolved independently in fishes and tetrapods. Dorsolateral placodes make use of common sensory detection structures such as hair cells and neuromasts.

To complete the picture, *ventrolateral placodes* deserve a brief word. The terminal nerve, *nervus terminalis*, is the most rostral cranial nerve; it conveys GnRH neurons and relates to the medial half of the olfactory placode. Neurons from the same location form the chemosensory *vomeronasal nerve*. Fishes have a distinct profundus nerve in the same distribution as V1 in mammals.

Comparative neuroanatomy suggests that trigeminal ganglion is a composite structure, *V1 being placodal whereas the remainder of the ganglion arises from neural crest*. Ventrolateral "epibranchial" placodes carry taste information to the gustatory nucleus from VII_{VL}, IX_{VL}, and X_{VL}.

The blood supply to the membranous labyrinth, the *internal auditory artery*, arises from the LNA either directly from basilar segment or indirectly via the anterior inferior cerebellar artery. Despite the intimate relationship between cranial nerves VII and VIII within the internal auditory canal, the labyrinth is exclusively supplied by the vestibulocochlear nerve and is strictly in register with r4–r5. Facial nerve is merely in transit. Neural crest from r4 has been traced to the vestibular part of the vestibulo-cochlear ganglion. It is unclear if the cochlear part is supplied by r4 or r5. In general, the developmental sequence favors vestibular over cochlear

thus these two organs are organized in an anterior–posterior manner. In any case, the otocyst can be considered to be in genetic register with both rhombomeres 4 and 5.

What pathway does the otocyst follow as it "migrates" from the surface ectoderm into PAM? Recall that the physical location of the otocyst is posterior to that of the future Eustachian tube and auditory canal. These structures are located at the level of the *spiracle*, a passageway (seen in jawed fishes) from the dorsal aspect of the first branchial pouch to the dorsal first branchial cleft. In the transition to jaws, the ventral aspects of the pouch/cleft are lost; only the spiracle remains.

The otic placode, on the other hand, is hard-wired to r4 and r5, i.e., to the second arch. It is physically located directly above the second pharyngeal pouch, the boundary between second arch and third arch. Recall that these two arches receive paraxial mesoderm from rhombomeres r4–r7. Inside the braincase the internal acoustic canal lies between r4 and r5. Indeed, the fibers of VII and VIII travel together into the internal auditory canal. They are not split into rostral and caudal segments. They can likely be mapped to a boundary between prootic and opisthotic bone fields to r4–r5 versus r6–r7.

The petrous complex as part of the skull base is an epaxial structure. Although the pharyngeal arches lying beneath it are hypaxial, these two sets of tissues bear have common genetic and neuromeric relationships. Mapping of the mammalian petrous complex has not been done as yet but we can draw inferences from the anatomy. We do know that jugular foramen, permitting exit of IX, X, and XI out of the posterior fossa, is located at the interface between opisthotic and exoccipital. This places somitomeric mesoderm from levels r7 (i.e., the third arch) against somitic mesoderm from r8.

The semicircular ducts develop from three flat pouches extending outward in three right angle planes from the otocyst. The center of each pouch collapses into itself and involutes, leaving behind an arciform tube. The part of the otocyst where these three ducts are confluent is known as the *utriculus*. Located at one end of each semicircular duct is a special ampulla patches of specialized hair cells (ampullary crests) are housed. Each ampulla is connected to the utriculus. Hanging off from each it is the bag-like *sacculus*. Both utriculus and sacculus have a hair cell-bearing macula as well, bringing the total of dedicated neuroepithelial areas in the vestibular system to five. Recall the developmental relationship between motion-sensing lateral line placodes and hair cells.

The cochlear duct is physically distinct from the utriculus. Its only point of anatomic continuity with the vestibular system is via a small *ductus reuniens*; it connects the cochlea to the sacculus. The cochlea has its own dedicated neuroepithelial area, the *spiral organ of Corti*, which runs the entire length of the duct.

In sum, the membranous labyrinth is (topographically speaking) "A Tale of Two Cities." It starts out as a simple cystic structure, only to morph into three arches and a spiral. If the vestibular system were elastic, one could envision placing it on stretch, applying a few snips, and a large circular tube would emerge. Similarly, if one were to unite the proximal and distal ends of the cochlear duct and stretch it out, a similar circular tube would result. Thus, *the final shapes assumed by the components of otocyst may very well depend upon the mesenchymal microenvironment into which they are expanding.* The genetic make-up of the mesodermal fields of r4–r7 creates the conditions necessary to accommodate two such disparate, yet similar systems.

We shall leave this subject on a provocative note. *Sm6 contains the cochlear space into which the cochlear duct forms. The Sm6–7 interspace contains the vestibule into which the utriculus and sacculus develop. Sm7 contains the semicircular canals that come to occupy the semicircular ducts.*

There are two interpretations of this scenario. First, the otocyst expands into the mesenchyme via genetically determined embryonic fusion planes (programmed cell death), passively accommodates, and then stimulates its own bony encasement. Second, the otocyst induces the periotic mesenchyme using factors that break down or remodel the mesenchyme to assume its final configurations. The bottom line: the three-dimensional anatomy of the membranous labyrinth is not a complex program. It consists of a sequence of signals emanating from neuroepithelial sites on the otocyst (up to 14) that interacts with the mesenchymal environment to construct three different configurations, one for each somitomere.

As we move forward with this discussion, we shall examine the neuromeric coding implicit in the developmental anatomy, neurology, and blood supply of the membranous labyrinth and its housing, the petrous temporal bone. This leads us to the following hypothesis. (1) The spatio-temporal development of the otocyst parallels that of cranial nerve VIII. (2) The anatomic distribution of vestibular and cochlear fibers in the medulla is a somatotopic representation of the otocyst. Thus, *the development of the petrous bone reflects a program inherent to the neural plate.*

Inner Ear: Functional Components (Figs. 8.117 and 8.118)

Ears are designed to carry out four functions: (1) angular acceleration, (2) linear acceleration–gravity, (3) hearing, and (4) low-frequency vibration separate from hearing. The ear is an evolutionary "add-on" to the old lateral line system of fishes. All its functions involve *mechanosensation*: the detection of mechanical displacements caused by sound, pressure, or changes in orientation using hair cells. Let us look at the two devices used by the ear for transduction and then at the

four anatomic structures in which this takes place. In so doing, I am shamelessly adapting the comments in Palaos for simplicity and elegance.

Specialized *hair cells* have microvilli oriented in a fixed orientation in space. This makes hair cells directional. When deformed, they fire using an "on-off" mechanism. These microvilli are reinforced by actin fibers which make them more or less "stiff." This modulates the strength of their response to the deforming source.

Hair cells exist in an enclosed environment, a *neuromast organ*, in which they are supported by a gel-like cupula and can be deformed by fluid waves. In the semicircular canals and cochlear, they respond to movements of fluid. The cupula in acceleration-detecting chambers called *ampullae* contain tiny mineralized particles of calcium carbonate, *otoliths*, which reside in situ until perturbed by a force, in which cased they are dispersed through the liquid like as sandstorm. The otoliths have weight sufficient to deform the cupula and thus detect acceleration (an airplane taking off) and a change in position (essential for Olympic gymnasts).

Endolymphatic duct is a pressure equalizer.

The labyrinth has three fluid-filled tubes that function as slosh receptors. Depending on one's orientation in space, one of the three tubes will respond. At the bottom of each tube is an *ampulla* containing a neuromast organ that responds to shear on it cupula but firing signals to the brain. Since these signals have gradations of intensity, change in space and its rate are calculated and communicated.

The maculae are sensitive both linear acceleration (gravity) and low-frequency vibrations (distant thunder, an earthquake). For some reasons, only sharks have three of them, the rest of us have to function with two, although a vestigial diverticulum exists in some species, the *macula neglecta*, to remind us of what we are missing. They are found in both the utricle and saccule but we do not have any functional correlation.

The lagena or cochlea is coiled in mammals, whereas in the rest of nature it is gently curved. It provides a more structured environment for the hair cells.

Inner Ear: Development

At this point, we need to take a deep breath and not faint. Yes, the inner ear is spatially complex. Most accounts of its development are confusing, to say the least. The pitfall lies when the embryonic events are presented in a linear fashion with all structures described simultaneously. Gulya's sequence is much easier to follow because it looks at each component in four time periods: 0–4 weeks, 4–8 weeks, 8–16 weeks, and 16+ weeks. What we shall discover is that spatio-temporal order of development of the membranous labyrinth and its subsequent ossification pattern very much follows that of its neurology [11, 12].

Membranous Labyrinth

The membranous labyrinth is an interconnected system of ducts and chambers that is lined by epithelium and filled with endolymph. Its components are the cochlear duct containing the organ of Corti, the utricle and saccule with their respective maculae, the semicircular ducts with their attendant cistae ampullares, and the endolymphatic duct/sac. Signals from these structures interact with the surrounding periotic mesenchyme to create the otic capsule (which eventually is converted into bone).

The labyrinth originates from a localized zone of surface ectoderm (*not neural crest*) at the end of week 3. This plaque-like structure is located just lateral to the neural fold dorsal to the first pharyngeal groove, i.e., between r3 and r4. In point of fact, its lining basement membrane is continuous with the r3–r4 rhombencephalic groove. Within a matter of days, these placode cells demonstrate a ciliary brush border. This presages the morphogenesis of movement and sound detection apparatus.

The otic placode is in intimate contact with a collection of neural crest cells, incorrectly termed the "acousticofacial ganglion." (Recent work has shown that derivatives of this population relate to facial nerve only.) From the wall of the neural groove, ganglion cells delaminate from the neural groove and pursue a ventrolateral migration. Immediately prior to the fourth week, the placode sinks into the mesenchyme, creating the *otic pit*. The fourth week per se consists of four main events. The pit dilates into the *otic sac*. When the mouth of the sac fuses, the *otic vesicle* or otocyst is born. Soon thereafter the dorsomedial otocyst gives rise to a diverticulum, the *endolymphatic appendage*. Finally, the induction of the surrounding mesenchyme begins create the cartilaginous *otic capsule*.

During weeks 4–5, dorsoventral elongation of the otocyst takes place. Three folds create three subdivisions of the vesicle: (1) the endolymphatic duct and sac, (2) the saccule with its projecting cochlear duct, and (3) the utricle with its semicircular ducts. The first "partitioning fold" is vertically directed into the dorsal utricular part of the vesicle. It "pinches off" a dorso-medial projection, the endolymphatic sac. Next, the cochlear duct projects ventromedially. Nerve fibers are beginning to grow in from the statoacoustic ganglion. In response, the medial wall of the vesicle thickens into the primary *macula comunis*. This subsequently subdivides. The superior segment of the common macula becomes the macula of the utricle and the ampullary of the superior and lateral semicircular ducts. The inferior segment becomes the macula of the saccule and the ampullary crest of the posterior semicircular canal.

The above sequence is not random. It is based on that of the vestibular division of VIII. Recall that this nerve has five epithelial targets: three ampullary crests (one for each semicircular canal) and two maculae (utricle and saccule). The spatiotemporal growth of vestibular VIII depends upon the spatial localization of nuclear cells for each target in the brainstem. (Recall that this is medial to that of cochlear VIII.) *The vestibular apparatus has a representation in r4–r5 brainstem that is somatotopic.*

The second partitioning fold is horizontal; it is located along the medial wall of the otocyst. It separates endolymphatic duct from saccule. The third partitioning fold is also horizontal but is located along the lateral wall of the otocyst. It separates the utricle above from the saccule below.

Between weeks 8 and 9, the three partitioning folds approximate one another. As a consequence, the utricle and saccule are nearly separate. Their only means of communication is via the utriculo-saccular duct. The free edge of the vertical fold becomes known as the *utriculo-endolymphatic valve of Bast*. At weeks 10–12, the growth of the membranous labyrinth is complete.

Cochlear Duct

The cochlear duct first appears as a ventral projection from the saccule at 6 weeks. It coils medially, with the first turn being complete by the end of the sixth week. The remaining 2½ turns are complete by the end of the eighth week. The wide-mouth communication with saccule shrinks down to a delicate *ductus reuniens*. Within the cochlear ducts, two epithelial ridges differentiate. The larger lateral ridge becomes the *spiral limbus*. The medial ridge becomes the *organ of Corti*. All aspects of development proceed from proximal to distal, i.e., from the basal turn to the apex. During the eighth week modiolus, tympanic scala and vestibular scala, and the surrounding otic capsule have all begun to differentiate.

During weeks 8–16, further growth of the cochlea is all in diameter and shape. At week 11, its shape changes from round to oval. At week, 16 it is triangular. The epithelium of the posterior wall (basilar membrane) becomes the organ of Corti and tectorial membrane. Differentiation is always proximal–distal (basal–apical) and parallels that of the cochlear nerve in the spiral lamina.

From 16 to 20 weeks, the various cellular components of the organ of Corti make their debut. The otic capsule is now a bony shell. At the 19th week, definitive supply by the labyrinthine artery to the cochlear is present. The vessels traverse the cochlear modiolus, spiral laminae, and the walls of the scalae. This is a true end-vessel system without anastomoses. Note that the blood supply to the otic capsule is distinct from that of the membrane, it does not reach the inner periosteal layer. The stria vascularis is fully developed at the 20th week. By the 22nd week, all histologic elements of the organ of Corti are present from base to apex. SEM studies of innervation to the organ of Corti at 18–20 weeks demonstrate afferent fibers develop earlier than efferent fibers. This suggests that the developmental gradient of sensory cells depends upon afferent innervation and blood supply.

Utricle and Saccule

The pars superior of the membranous labyrinth is phylogenetically older. It consists of the utricle and semicircular ducts. The pars inferior is younger. It consists of the saccule plus the cochlear duct. Note that a constriction during the seventh week separates the utricle from the saccule, creating the delicate ductus reuniens. During the eighth week, the three partitioning folds create the definitive utricle, saccule, and endolymphatic duct. The macula of the utricle is derived from the labyrinth at the entry point of VIII. By the end of the eighth week, this epithelium has changed from a simple to a pseudostratified histology.

During weeks 8–16, the vertically-oriented fold deepens creating a utriculo-endolymphatic valve. Between the 10th and 12th week, the macular sensory cells have tufts free borders. A gel-like cushion, the otolithic membrane, contains crystals of $CaCO_3$, the *otoconia*. By the 14th to 16th weeks, the maculae are fully differentiated.

Semicircular Ducts

By week 4, two paddle-like projections from the utricle represent the canalicular divisions of the otic capsule. During the sixth week, the two paddles have partial resorption of their cavities. The epithelial walls in the center collapse together and fuse, leaving the peripheral aspect of the paddles as a U-shaped tube. The sequence is superior, posterior, and lateral. At one end of each of the three ducts, a dilated ampulla exists where the duct enters the utricle. The superior and lateral canals have ampullae located at their anterior crura. The posterior canal ampulla is at the inferior crus. The non-ampullated crura also communicate with the utricle. During the seventh week, a ridge of neurosensory epithelial cells, the cristae ampullares appears precisely where the vestibular nerve fibers enter. The cristae have a perpendicular orientation to the direction of endolymph. Note that fusion of the posterior and superior ducts fuse (crus commune) means that the utricle receives 5 crura not 6.

Between 8 and 16 weeks, the ducts grow. Between the 10th and 12th weeks, the epithelium of the cristae starts as simple squamous and then differentiates into sensory hair cells and their supporting cell populations. The cristae are fully developed by the 15th week. The superior semicircular duct achieves adult size at the 20th week. This is followed by the posterior and the lateral ducts (according to their phylogenetic order).

Endolymphatic Duct

This structure is the first diverticulum to appear from the otic capsule. It appears at the sixth week. Near the vestibular terminus it is narrow, but broadens out distally into a spade-like sac with an epithelial lining.

Otic Capsule

The otic capsule is a cartilaginous mass containing the otic capsule. By the end of week 4, this is discernable as an increased density in the mesenchyme surrounding the otic vesicle. It is located above the lateral aspect of the tubotympanic recess. It will be fated to form the petrous temporal bone. At the fifth week, mesenchyme is condensing everywhere except in the vicinity of the endolymphatic duct. At the sixth week, it has surrounding the membranous labyrinth and is becoming cartilaginous. At the site of the future internal acoustic meatus, an absence of condensation is noted. This is due to the presence of the neural structures. At the close of the sixth week, precartilage differentiates into true cartilage. This process requires an additional 2 weeks to complete. The membranous labyrinth at 8 weeks has the full adult configuration but has not completed growth. Although the outer precartilage to cartilage transition is largely complete at 8 weeks, *events in the mesenchyme in immediate contact with the labyrinthine epithelium are very different.* Here, a retrogressive process of de-differentiation of the cartilage into a loose vascular reticulum will favor the expansion and remodeling of the labyrinth.

At 9 weeks, de-differentiation of precartilage surrounding the membranous labyrinth permits expansion of the labyrinth into loose mesenchyme. This pattern of dedifferentiation is seen along the outer (leading) edges of the semicircular canals, probably reflecting epithelial-mesenchymal interaction. The mesenchyme facing the inner (trailing) edges of the canals re-differentiates into cartilage. The otic labyrinth now undergoes three phases of growth. In the first phase, the inner precartilage is transformed into three zones. The epithelium of the labyrinth is enveloped by dense areolar tissue, the *membrana propia*. Just outside this is an arachnoid-like tissue containing fluid. Finally, one encounters a dense outer zone, the *perichondrium of the otic capsule*. In the second phase, from 9 to 10 weeks, the center zone coalesces to create the true *periotic space*. In the third phase (12–16 weeks) is characterized by the complete conversion to cartilage and the beginnings of ossification. The formation of petrous temporal bone thus starts at the basal cochlea during the 16th week and continues up shortly before birth.

At 16+ weeks, the ossification process of the otic capsule in both canalicular and cochlear regions accelerates. This has several unique features. Growth is rapid, especially between 15 and 21 weeks. Despite the small space, a large number of ossification centers are involved (14 in all). Fusion of ossification centers takes place without intervening epiphyseal bone. The otic capsule has a trilaminar histology. Fetal architecture persists in both periosteal and endosteal areas without remodeling. Each ossification center and each layer appear and ossify independently.

Perilymphatic Space

A strange tissue exists between the membranous labyrinth and the inner periosteum of the otic capsule. This issue is mesodermal, from Sm6 and Sm7. The de-differentiation process takes place between weeks 8 and 24. Just beneath round window, a site of rarefaction indicates the future scala tympani.

The periotic reticulum undergoes a process of vacuolization in which it "hollows itself out" to accommodate the perilymphatic system. Even at the tenth week, the process is far advanced, especially over the lateral aspect of the utricle and saccule. (Note: the more medial canalicular region develops more slowly.) This creates a mesothelial lining that separates the saccule from the footplate of stapes. Support fibers and neurovascular structures are immediately laid down. By the 12th week, the reticulum surrounding the cochlear duct and spiral laminae has broken down extensively. *Interfibrillar spaces coalesce together to form the scalae.* Definitive periotic spaces appear late in the 12th week. The first site is the perilymphatic cistern just adjacent to the round window. This is followed by scala tympani near round window. Scala vestibuli presents as the perilymphatic cistern next to the oval window. Within the connective tissue between the periotic scalae, two membranous bones develop, the *spiral laminae* and the *modiolus*. Both are likely of neural crest origin. Note that spiral lamina extends itself distally to connect with modiolus. By the 16th week, the canalicular periotic labyrinth is vacuolated.

At 16+ weeks, the development of the perilymphatic space lags behind that of the scalae. Full development of the space between the semicircular canals and the cistern of the vestibule is complete by 20 weeks. In its final iteration, the reticulum thins out to create a single uninterrupted space between the membranous labyrinth and the periosteum.

Capsular Channels

At 7 weeks, the precartilage is observed to break down over the medial wall of the basal turn of cochlea. This will form the primordial cochlear aqueduct. It contains the periotic duct connecting scala tympani with subarachnoid space. Anatomic continuity exists between the reticular tissue of the periotic duct and cranial nerve IX, dura and inferior petrosal sinus. Thus, round window is continuous with posterior fossa via cochlear aqueduct.

At 16+ weeks, two definitions need to be kept in mind. *Cochlear aqueduct* is the channel in the otic capsule. The *periotic duct* refers to an enclosed membranous duct that communicates between the perilymphatic space and the subarachnoid space. At 16–18 weeks, the cochlear aqueduct consists of three structures: the tympanomeningeal fissure, the periotic duct, and the inferior cochlear duct. At 20 weeks, the growth of the petrous apex places the inferior cochlear vein into a separate compartment. At 24 weeks, a fusion between the promontory and the round window eliminates the tympanomeningeal fissure. At 32 weeks, the cochlear aqueduct (and its contained periotic duct) elongates along the medial aspect of the otic capsule. At 40 weeks, arachnoid tissue grows into the cochlear aqueduct.

The Arteries

In humans, five aortic arch arteries appear from stages 9 to 13. These vessels connect the aortic sac with the dorsal aortae; in so doing, they traverse the mesenchyme of five pharyngeal arches. Recall that, with the dissolution of AA5, blood supply to the fifth pharyngeal arch is provided by AA4. Recall as well the sixth aortic arch artery is non-pharyngeal; it forms the pulmonary circulation. Prior to embryonic flexion, the hindbrain is already being supplied by the longitudinal neural arteries. The dorsal aortae support all non-neural tissues; thus, the mesodermal fields (r4–r7) that will receive the auditory vesicle are segmentally supplied from the DAs. At the time of embryonic flexion[9], the dorsal aortae and LNAs communicate via the primitive trigeminal arteries, sprouts from the DAs grow forward from level r3 as primitive internal carotid arteries to supply the midbrain and forebrain. The labyrinth is now supplied from the LNAs via primitive otic arteries. As the LNAs fuse to in the metencephalic zone to form the basilar system, the transient primitive otic arteries involute.

At 4 weeks, the internal carotid artery has three developmental segments. (1) *Common carotid* artery is the segment of aortic sac connecting AA4 and AA3. External carotid artery arises just distal to common carotid. It sends out branches dedicated to each pharyngeal arch. (2) *Extracranial internal carotid* is exclusively derived from AA3 prior to its insertion into the dorsal aortae. (3) *Petrous internal carotid* is the span of the dorsal aorta from the entry of AA3 into temporal bone at level r7–r8 its exit point just beyond insertion site of AA1 at level r3. This is represented by a dorsal remnant vessel, the *primitive mandibular artery*. (4) *Cavernous internal carotid* is the span of DA from levels r3 to r0–r1.

At 4–5 weeks, each pharyngeal arch and pouch receives an aortic arch artery and a cranial nerve, the mesenchyme of which is in register with the neuromeric level of the arch. An exception to this is the fifth pharyngeal arch. It receives anastomotic supply from AA4, which takes over when AA5 disintegrates. By the time PA3 develops, AA1 and AA2 have both involuted, leaving behind two dorsal remnants: primitive mandibular and hyoid, respectively. Embryonic survival depends upon a new alternative circulation to the first and second pharyngeal arches, the well-discussed stapedial system.

As previously discussed, the hyoid gives rise to a new branch, the stapedial artery. Stapedial passes from the posterior wall into the middle ear where it bifurcates. Intracranial stapedial gives rise to the intracranial meningeal system. It also supplies the orbit via the StV2 recurrent meningeal artery and the StV1 supraorbital artery. Extracranial stapedial joins forces with the primitive external carotid axis at V3, just distal to the stem of maxillo-mandibular artery. Branches from the 1/3 of MMA refer back to the tympanic cavity and represent the original pathways of stapedial as it exited tympanic cavity.

The longitudinal neural system is simple and segmental. The term *labyrinthine artery* is preferable to the more common internal auditory artery because it supplies both vestibular *and* cochlear structures. By the sixth week, the LNAs have fused into a common basilar artery from which multiple branches arise, most likely under induction from cranial nerves. Recall that during the fifth week, the ganglion of VIII is already subdividing. Labyrinthine and anterior inferior cerebellar both arise from basilar. They run outward in parallel, with cranial nerve VI being interposed between. Depending upon involution pattern, labyrinthine thus can arise from either basilar or AICA. At the end of the eighth week, subarcuate artery arises from either labyrinthine artery or AICA. It supplies part of the labyrinthine capsule and part of mastoid.

At the beginning of the tenth week, all branches to the ear are identifiable. We can see that initial contributions from the dorsal aortae important for survival of the somitomeric mesoderm are no longer relevant. The dependence of the labyrinth upon *labyrinthine artery*, a branch of LNA, relates the otocyst (and its petrous housing) to rhombencephalon. *Subarcuate artery* represents r4–r7 bone. It supplies part of the labyrinth and part of mastoid. Intracranial branches of stapedial (e.g. greater petrosal) contribute to petrosal as well (most likely to parts derived from neural crest). Stapedial *meningeal branches* from r8 to r11 are induced by cranial nerves IX, X, XI, and XII. They communicate via jugular foramen and condylar canal with non-stapedial ECA occipital artery.

Non-stapedial external carotid also contributes. (1) *Ascending pharyngeal* supplies PA3. It gives off *inferior tympanic* branch (induced by glossopharyngeal nerve) that supplies the medial wall of the tympanic cavity. *Meningeal branches* communicate with intracranial stapedial via foramen lacerum, jugular foramen and hypoglossal foramen. At foramen lacerum, the deep petrosal nerve meets superficial petrosal nerve. Thus, ascending pharyngeal may contribute to petrous apex. (2) *Occipital artery* arises from VPA. Its mesenchyme is in register with from r4 to r7. It supplies the proximal portion of sternocleidomastoid (derived from somitomeres 8 and 9). Anastomoses with intracranial stapedial have been mentioned. (3) *Posterior auricular artery* has two branches of relevance to the ear. The *stylomastoid artery* supplies facial nerve, tympanic cavity antrum, mastoid air cells, and semicircular canals and endolymphatic duct. In this regard, it represents an ECA anastomosis with the LNA

labyrinthine system. The *auricular branch* supplies hypaxial structures of the external ear (the deep/cranial portion of auricle).

Focus in-Depth: Development of the Petrous Internal Carotid Artery

One of the most striking features of petrous temporal bone is the manner in which it ensheaths the internal carotid artery. This will be the story of how its *materia prima*, the dorsal aorta, with a few additions and subtractions, achieves its final anatomic state. Next, we will examine the "organizer" of the inner ear, the otocyst. The step-wise development of the vestibulocochlear apparatus will deform the previously linear otic mesenchyme, forcing the ICA to assume a tortuous course. Next, we will consider the development of the facial nerve and vestibulocochlear nerves. Its final configuration will result from its embryologic origins in the brain stem and how it must accommodate itself to the otocyst.

The petrous portion of internal carotid has a long embryonic history. This segment develops from the *dorsal aortae extending caudal to primitive trigeminal artery backward to, and including, the insertion of the third aortic arch*. Recall that at the time of embryonic folding,[8–9] each dorsal aorta rostral to primitive trigeminal doubles back 180° upon itself and fuses, producing the *primitive pharyngeal arch artery (PrPA)*. This artery supplies the mesenchyme cranial to the first arch (specifically: somitomeres 1–3 and neural crest from m1, m2, and r1). It eventually becomes the stem of the primitive ICA. Thus, the embryonic definition of PrPA is that it refers to that segment of DA which lies between PrICA and primitive trigeminal.

Directly opposite primitive trigeminal is the ventrally directed AA1. Over the next several stages, as the aortic arches form and disintegrate, the territory of dorsal aorta that will be progressively "re-defined" as carotid artery keeps extending backwards until it reaches AA3. Recall that the segment of dorsal aorta between AA3 and AA4 eventually disintegrates after formation of AA6. In its final form, common carotid is likely constructed from mesenchyme in genetic register with r6–r7. The segment of DA that becomes petrous ICA is made up of r4–r7 mesenchyme.

Mesoderm used to construct the petrous ICA comes from paraxial mesoderm of rhombomeres r4–r7. Somitomeres 5–6 are in approximate genetic register with rhombomeres 4–5; Sm7 is in register with rhombomeres 6–7. Recall that, by definition, *somitomeres are not discontinuous*. Physical separation by a distinct epithelium that is characteristic of somites does not take place until the caudal border of Sm8, i.e., the Sm9–Sm9 junction. Whether or not somitomeres represent discrete intermediate entities in the assembly of structures such as the ICA or the petromastoic complex is unknown and likely irrelevant. What matters is that clumps of r4–r7 mesoderm, assembled in linear fashion, form a continuous tube.

What about the construction of ICA forward from the petrous apex? Where does its paraxial mesoderm come from? And how does this related to the synthesis of the lNA system or other arteries supplying the CNS? The best answer to this is to realize that the forebrain/midbrain "eruption" drags with its PAM from levels r0 to r3. As we have seen these cells make the *primitive plexuses* that are later remodeled into mature arterial systems. Hindbrain neurulation accomplishes the same thing. It is not clear whether the neural tube is cloaked with PAM immediately, as gastrulation has just been completed, or whether the PAM organizes into somitomeres first and then disperses. In any case, the source material for the entire ICA system supplying the forebrain and midbrain comes from this early mesodermal source.

Recall that traditional descriptions of ICA use terminology that implies direction. For example, the ICA is said to "enter" the skull base via the *carotid canal* just medial to stylomastoid foramen. It is positioned lateral to cranial nerves IX, X, XI, and XII and lateral to internal jugular. ICA then "ascends," pursuing a tortuous course through the PTMC until "entering" the skull between the petrous process of the sphenoid and the lingual. Just anterior to middle ear ICA takes a sharp bend and runs forward and medially beneath eustachian tube to the petrous apex. The walls of this canal are extremely thin (0.5 mm); at times they are but a membrane. There are two main "weak spots." The first is just anterior to tympanic cavity and the cochlea. The second is next to trigeminal ganglion. *These sites may represent neural crest contributions formed by induction from the neighboring neural structure.* Recall that investing fibers of SANS neurons surround the petrous ICA.

This model is misleading; the ICA develops in exactly the *opposite* direction. In point of fact, the dorsal aortae appear during somitogenesis (just after gastrulation). The arterial walls of the dorsal aortae are constructed in cranial–caudal sequence as a series of ring-like donations of PAM. At the physical location r6–r7 and r8–r9, the third and fourth aortic arches plug into the DAs. Next, the connecting segment of DA between AA3 and AA4 is obliterated, and the common carotid artery is formed. The remaining caudal portions of the dorsal aortae participate with the eventual development of the heart and great vessels. They eventually fuse as the descending aorta.

The primitive anlage comprising periotic (petromastoid) temporal bone can be conceived as *four rhombomeric mesodermal "beads" on an internal carotid "string."* Contained within this linear array are cranial nerves VII and VIII. As these develop, particularly, the vestibulocochlear apparatus, the temporal bone becomes physically perturbed, differential growth takes place, and the somitomeric string morphs into its final shape.

The Inner Ear: Molecular Mechanisms

Let us now revisit these events using a flow chart of inductive events and tissue transformations (Figs. 8.119 and 8.120). The primitive ectoderm overlying the PAM of the future ear undergoes five successive inductions. Signals from chorda-mesoderm and later from the PAM itself prepare the periotic ectoderm to receive a third signal, *FGF-3* from the hindbrain.[9] This induces the *otic field*; it expresses *Pax-2*. Note that the medial aspect of otic disc adheres directly to neural plate. Hindbrain *Wnt* signals now induces the Pax-2-positive cells directly overhead to become the *otic placode*.[10–11] A fifth signal, additional hindbrain *FGF-3*, converts this into the *otic vesicle* or *otocyst*, which sinks into the periotic mesenchyme.[12–13] The location of the otocyst is dorsal to the second pharyngeal pouch, that is How does this help us to understand composition of petromastoid? When we look at the components of the middle ear, we see somitomere 4 represented anteriorly as tensor tympani and somitomere 6 as stapedius. Tensor has a primary insertion into the *bony roof of the pharyngotympanic tube* and greater wing of sphenoid. It has a secondary insertion into the *manubrium of the malleus*. Stapedius has a primary insertion into the *pyramidal process of the descending facial canal* and a secondary insertion into the *posterior neck of the stapes*. Thus, the tympanic cavity stands as the dorsal pouch between the first and second pharyngeal arches, that is to say, between rhombomeres 5 and 6.

The next step in the formation of the inner ear is the molecular patterning of the otocyst. The dorsal neural tube, via *Wnts*, specifies the vestibular system. It is further developed by *Dlx-5* and *Gbx-2*. The ventral neural tube via *Shh* specifies the cochlea. *Pax-2* causes development of the endolymphatic duct and cochlea. Note that *FGF-3* produced by r5 and r6 is required for the duct. Semicircular canals development depends upon a series of homeobox genes *Nkx-1* in general, *Otx-1* for the lateral semicircular canal, and *Dlx-5* for the anterior and posterior canals.

Development of the vestibular part of the vestibulocochlear ganglion[13] predates the spiral ganglion to the cochlea[16]. In a similar fashion, the semicircular ducts are fully formed[18] well prior to the final revolution of the cochlea[23]. As we shall see, this pattern is a result of a neuroembryologic "maturity gradient" determined by the physical location of the cell columns within the neural plate. Because the acoustic tract is more lateral than the vestibular tract, it is subject to different induction signal strengths depending upon whether the gene product is elaborated at the neural midline or in peripheral mesenchyme.

Encasement of the membranous labyrinth in bone involves the induction of the surrounding mesenchyme by BMP-4 produced by the otocyst. Condensation starts at stage 15, is complete at stage 19, and the petrous bone becomes anchored to exoccipital at stage 20 weeks.

Neurology of the Inner Ear

We next direct our attention to the anatomy of cranial nerves VII, VIII and IX. How does each behave and what does this have to do with petromastoid assembly? Let us first review where their nuclei reside. Both VII and VIII occupy r4–r5 while IX occupies r6–r7 PANS neurons controlling the salivary glands of the mouth (general visceral efferent) is anatomically split between those directed to the palate (r4) and those directed floor of the mouth (r5). VII taste fibers (special visceral efferent) spatially mimic the above, with palate reporting to level r4 and tongue to level r5. It is unclear if VII motor control of striated muscles (special visceral efferent) is split in the same way (with upper facial division in r4 and the lower facial division in r5) or if all motor neurons arise in r5 **Statoacoustic nerve** VIII occupies r4–r5 as well. Its vestibular fibers are located *medial* to the acoustic fibers.

The brainstem nuclei of both VII and VIII reside in rhombomeres 4 and 5 yet these two nerves exit at distinct points. Why? Recall that brainstem nuclei exist in four functional columns, arranged from medial to lateral. (1) *Somatic motor* neurons are divided into a medial column for non-pharyngeal arch muscles (extraocular, epibranchial, and hypobranchial) and a lateral column for pharyngeal arch muscles. (2) *Visceral motor* neurons are autonomic. (3) *Visceral sensory* neurons are divided into medial *general* afferents from the gut and lateral *special* afferents from taste buds. (4) *Somatic sensory* neurons are divided into medial *general* afferents from skin, muscle, etc. and lateral *special* afferents dedicated to the vestibulocochlear system. The cochlear neurons are the most lateral of all.

The "floorplan" of the neural plate has important implications. Development of VII–VIII neurons follows a medial–lateral pattern that mimics the functional gradient: motor nerve to facial muscles (r4–r5), nervus intermedius-taste (r4), vestibular (r4–r5), and cochlear (r5). Because neurons originate from the neural tube and migrate outward, the medial population develops first. Thus, VII exits *anterior* to VIII. As the otocyst develops, VII drapes *over* it. Furthermore, as we study the otocyst, the vestibular development will *precede* that of the cochlea.

Cranial Nerve VII
Development

At the third week of gestation, a neural crest mass, the *facial primordium* arises in close proximity to the otic placode. At 4 weeks, this population, containing general somatic sensory fibers and somatic motor fibers, is seen attached to the rhombencephalon *rostral to the placode*. This makes sense because the placode is located between r5 and r6 whereas the facial nerve nuclei reside in r4–r5. It contains general somatic sensory fibers and somatic motor fibers. At 4–5 weeks, neural crest at the level of the epibranchial placode is transformed into the *geniculate* ganglion (this is independent of motor fibers).

The facial primordium now splits. The caudal division gives rise to the main *motor trunk* of VII. The rostral division, *chorda tympani*, is equal in size. It travels ventral to first pharyngeal pouch to access the first pharyngeal arch. At 6.5 weeks, chorda tympani reaches submandibular ganglion. At 7 weeks chorda tympani and lingual nerve (V3) have united immediately proximal to the ganglion. Two additional branches develop from geniculate ganglion. *Greater superficial petrosal nerve* arises at 5 weeks from the ventral ganglion. It contains PANS fibers to the palate. *Nervus intermedius* arises from geniculate ganglion at 7 weeks and travels forward between the VII and VIII.

At 8 weeks, the still-cartilaginous posterior otic capsule demonstrates a groove on its tympanic wall, the future *facial canal*. A distinct motor branch to stapedius muscle is present. At 12 weeks, nervus intermedius communicates with both motor VII and the cochlear nerve. The dorsomedial aspect of facial nerve gives off two branches that fuse and subsequently hook up with both the superior ganglion of IX and the superior ganglion of X. The resulting nerve supplies the dermis and subcutaneous tissue of the external auditory canal. By week 17, all neural connections of VII are established.

At 15 weeks, the geniculate ganglion is full-size. It lies sandwiched in membranous bone between the squamous temporal bone and the middle fossa plate. By week 26, with the progressive ossification of the otic capsule, the facial sulcus becomes the true facial canal. The bone of the deep portion is periosteal, likely representing osteoinduction by the nerve. At 35 weeks the geniculate ganglion rests upon a thin bony lamina that separates it from the epitympanum. Note that the intracranial surface of facial canal remains open. Thus dura is in direct contact with the perineural tissue of geniculate ganglion.

Anatomic Considerations of Facial Nerve for the Temporal Bone

Note that facial nerve has two separate components. Nervus intermedius resides in r4. It contains visceral sensory (taste) fibers returning from the mouth and tongue. It also conveys PANS motor fibers from *superior salivatory nucleus* to be distributed via the pterygopalatine ganglion to target glands. Somatic motor VII resides in r5. Both join together at the internal auditory meatus. Just before reaching the tympanic cavity, they "shake hands and part company." This bulge is called the genu. At this point, r5 VII dives backward and downward through the facial canal, *taking along with it r4 fibers to the tongue*. These shall eventually pursue a separate pathway, the chorda tympani. Nervus intermedius continues forward into the petrosal bone *taking along with it r4 fibers to the V1–V3 distribution*.

Motor VII gives off nerve to stapedius muscle and then chorda tympani just prior to exiting stylomastoid. CT enters the tympanic cavity via the *posterior canaliculus*. It runs along the internal fibrous surface of tympanic membrane and is then draped over by mucosa. It exits tympanic cavity via the *anterior canaliculus*. It conveys PANS fibers to submandibular and sublingual glands (via the submandibular ganglion), and of course, supplies taste to the anterior 2/3 of tongue.

We are now going to talk about three distinct petrosal nerves, the first two of which relate to VII. These are *greater petrosal nerve* (GPT), *deep petrosal nerve* (DPT), and *lesser petrosal nerve* (LPT). The major branch of nervus intermedius runs forward from the genu as the greater petrosal nerve. It runs along the knife-edge of the petrous temporal bone, probably synthesizing a small wafer of protective bone via VEGF from the dura lying above it. Thus, what appears to be a tunnel through the bone is really just a canal. Just below greater petrosal lies the throbbing ICA, covered with a "fish-net stocking" of autonomic fibers. These coalesce as *deep petrosal nerve*, running parallel to greater petrosal. As GPT exits the anterior petrous bone, it drops down beneath trigeminal ganglion. At this point, it joins up with DPT to create the *nerve of the pterygoid canal (Vidian nerve)*. This nerve dead-ends at the pterygopalatine ganglion to which it brings three kinds of fibers: (1) taste afferents, (2) PANS motor to palatine salivary glands (GPT); and (3) autonomic motor SANS from the carotid (DPT).

What about lesser petrosal? PANS fibers to the parotid originate in the *inferior salivatory nucleus* connected functionally to glossopharyngeal nerve. The lesser petrosal nerve conveys these fibers over the substance of petrous bone until reaching foramen ovale. There, LPT exits and connects to the otic ganglion lying medial to V3. After synapse, postganglionic fibers travel via auriculotemporal nerve to parotid gland. In so doing, it surrounds the motor branch of V3 to medial pterygoid. So what we have here is a split of VII around the substance of tympanic cavity between two branches, an anterior r4 sensory branch to mouth and a posterior r4 sensory branch to the tongue + r5 motor.

This anatomy makes evolutionary sense. In the primitive gnathostome condition, all cranial nerves associated with branchial pouches have three branches/pouch. The mucosa of the pouch itself was supplied by a *pharyngeal ramus*. Anterior to the slit was a *pretrematic ramus*. Both of these rami contained only visceral sensory fibers. The *posttrematic ramus* runs posterior to the slit and contains both motor and sensory fibers (both visceral and somatic). Cranial nerve VII follows the same rules: its pretrematic ramus is nervus intermedius and its posttrematic ramus is chorda tympani + the motor nerve to the muscles of the sixth somitomere (Fig. 8.121).

We next direct our attention to the anatomy of cranial nerves VII, VIII, and IX. How does each behave and what does this have to do with petromastoid assembly? Let us first review where their nuclei reside. Both VII and VIII occupy r4–r5 while IX occupies r6–r7 PANS neurons controlling the salivary glands of the mouth (general visceral efferent) and are anatomically split between those directed to the palate (r4) and those directed floor of the mouth (r5). VII motor control of striated muscles (special visceral efferent) is split in the same way, with upper facial division in r4 and the lower facial division in r5. VII taste fibers (special visceral efferent) spatially mimic the above, with palate reporting to level r4 and tongue to level r5. **Statoacoustic nerve** VIII occupies r4–r5 as well. Its vestibular fibers are located medial to the acoustic fibers.

The brainstem nuclei of both VII and VIII reside in rhombomeres 4 and 5 yet the two nerves exit at distinct points. Why? Recall that brainstem nuclei exist in four functional columns, arranged from medial to lateral. (1) *Somatic motor* neurons are divided into a medial column for non-pharyngeal arch muscles (extraocular, epibranchial and hypobranchial) and a lateral column for pharyngeal arch muscles. (2) *Visceral motor* neurons are autonomic. (3) *Visceral sensory* neurons are divided into medial *general* afferents from the gut and lateral *special* afferents from taste buds. (4) *Somatic sensory* neurons are divided into medial *general* afferents from skin, muscle, etc. and lateral *special* afferents dedicated to the vestibulocochlear system. The cochlear neurons are most lateral of all.

The "floorplan" of the neural plate has important implications. Development of VII–VIII neurons follows a medial–lateral pattern that mimics the functional gradient: motor nerve to facial muscles (r4–r5), nervus intermedius-taste (r4), vestibular (r4–r5), and cochlear (r5). Because neurons originate from the neural tube and migrate outward, the medial population develops first. Thus, VII exits *anterior* to VIII. As the otocyst develops, VII drapes *over* it. Furthermore, as we study the otocyst, the vestibular development will precede that of the cochlea.

How does this help us to understand composition of petromastoid? When we look at the components of the middle ear, we see somitomere 4 anteriorly as tensor tympani and somitomere 6 as stapedius. Tensor has a primary insertion into the *bony roof of the pharyngotympanic tube* and greater wing of sphenoid. It has a secondary insertion into the *manubrium of the malleus*. Stapedius has a primary insertion into the *pyramidal process of the descending facial canal* and a secondary insertion into the *posterior neck of the stapes*. Thus, the tympanic cavity stands as the dorsal pouch between the first and second pharyngeal arches.

Cranial Nerve VIII

Development

The primordium of statoacoustic nerve arises from cells of the anteromedial otic placode during the third week. In the fourth week, these cells migrate through the epithelium and the basement membrane to reach the region where the VIIIth nerve ganglia will develop. Between weeks 4 and 5, the ganglion divides into superior and inferior segments directed to the otic vesicle. The *pars superior* supplies the macula of the utricle and the cristae corresponding to the superior and lateral semicircular ducts. At 5–6 weeks, the *pars inferior* subdivides. The upper segment supplies the macula of the saccule and the third crista corresponding to the posterior semicircular duct. The lower segment supplies the organ of Corti. At week 6, the multiple fibers to the posterior ampulla condense into an apparently single nerve. After 7 weeks, the pars superior enlarges; by 8 weeks it has reached adult size, with the upper and lower divisions extending out of the ganglio. The cochlear nerve is also adult size at 8 weeks. Differentiation of sensory epithelium and support cells requires interactions between the nerve and the otic capsule.

Anatomic Considerations of Statoacoustic Nerve for the Temporal Bone

Vestibulocochlear nerve has two distinct sets of fibers that arise as bipolar ganglion cells and connect disparate parts of the inner ear with distinct central neuroanatomic connections. The neuroanatomy of these two systems is complex and well described in Grays's 40th. We include a few details herein.

Cochlear nerve arises from bipolar cells in the *spiral ganglion of the cochlea* which is located just at the edge of the bony spiral lamina. It connects the hair cells of the organ of Corti to the cochlear nucleus in r5–r6. Cochlear nerve contains both afferent and efferent fibers. *Afferent* fibers pass from the hair cells via organ of Corti to the anterior zone of the vestibulocochlear nucleus which is tonotopic. *Efferent fibers* arise from the olivo-cochlear system in the brainstem. Within the cochlea they modulate the response of the hair cells through frequency selection and control of sensitivity. This is how we "listen in" to certain melodies in within a musical score or "zero in" on a particular conversation at a noisy party. Autonomic fibers are strictly sympathetic and, by influencing blood flow, may control the metabolism of various cell types—interestingly avoiding the organ of Corti. Within the internal auditory canal, cochlear nerve passes beneath facial nerve, both being anterior to vestibular nerve. The auditory nerve supplies the cochlear nucleus, which is tonotopic, spanning the pontomedullary junction. Thus, auditory nerve and its central connections distinguish between r5 and r6.

Vestibular nerve also arises from bipolar cells in the *ganglion of Scarpa* which is located in the superolateral corner of the internal auditory meatus. Three sets of fibers innervate the maculae of utricle and saccule and the ampullae of the semicircular canals. These exit via tiny foramina, are assembled, and pass posterior to cochlear nerve en route to the medulla. Vestibular nerve enters at the cerebellopontine junction and the splits into (1) ascending branches connected to the vestibular nuclei and cerebellum; and (2) descending branches forming the spinal root of vestibular nerve.

The take-home point of cranial nerve VIII is that along with VII it established the anterior territory of temporal bone as mesenchyme in register both anatomically and functionally with r4–r5. This is likely the neuromeric definition of the prootic field.

Cranial Nerve IX

Anatomic Considerations: First, a Grand Generalization

Third pharyngeal arch and its main cranial nerve, glossopharyngeal, are directly relevant for understanding the embryologic construction of the pharynx and indirectly relevant for mapping the temporal bone complex. Although we have visited this topic before (and will do so again) specific points need to be reiterated here. First, the functional anatomy of a pharyngeal arch is determined, not by which particular cranial nerve innervates it, but from which neuromeric level(s) and functional column(s) are responsible for the innervation. Second, the same neuromeric level and column may choose to send out more than one motor or sensory nerve to a given zone. Third, nerves always seek out tissues from the same neuromeric register.

Third arch is in register with rhombomeres 6 and 7. These contain lateral motor column neurons supplying the branchiomeric muscles of the palate and pharynx via both IX and X. Recall, under the old terminology, that these are considered to the *general visceral efferent*. Recall further that it now known that pharyngeal arch muscles (so-called branchiomeric) are *not* visceral. The only reason for maintaining this out-of-date terminology is that GSE and GVE apply to two distinct functional motor columns: medial motor (GSE) referring to muscles of the eye and tongue, and lateral motor (GVE) referring to all the rest.

Thus, although all third arch muscles originate in Sm7, only one, stylopharyngeus, receives r6–r7 motor fibers via the specific physical entity of glossopharyngeal nerve. Palatal and pharyngeal muscles are all supplied by r6–r7 via the pharyngeal plexus of vagus. For this reason, the literature, by clinging to a mistaken "one arch-one cranial nerve" model, muddles up the identities of the embryologic components of the third and fourth arches.

What comes through here is a single unifying plan for the voluntary control of pharyngeal arch muscles. What lateral motor column means is that *every level is represented* from r3 down to c3, no matter what specific nerve bears the efferent fibers. Trigeminal and facial have their own motor nuclei.

Nucleus ambiguus extends from r6 to c3 and is somatotopic. It contains glossopharyngeal (r6–r7), vagus (r7–r11), and spinal accessory (r10–c3). Why the latter? Did not we consider the pharyngeal arches to end at r11? In the chapter on the neck, we will discuss the derivation of the muscles of XI, sternocleidomastoid, and trapezius. We shall see that their organization may hearken back to the times of sharks, when there were seven arches, extending backward four additional neuromeric units.

The same confusion applies when we consider pain sensation from the third arch. General visceral afferent fibers from r6 to r11 all refer to yet another mis-named structure, the *trigeminal sensory nucleus* (TSN). This structure extends all the way from r0 to the cervical spinal cord. It spatially distinct from the medial sensory column *nucleus solitarius* for taste. We can therefore refer to TSN as the **lateral sensory column**.

It is described as having three regions: mesencephalic, principle sensory, and spinal. Fibers from V1 to V3 enter the ganglion and split, some ascending into the brainstem and some making local synapse. Mesencephalic nucleus is *proprioceptive* for oculogyrics muscles, the muscles of mastication, the teeth, and the muscles of facial expression. Principle sensory nucleus is somatotopic and handles general sensation from all territories supplied by V1–V3, i.e., rhombomeres r0–r5. Spinal nucleus is somatotopic as well. It receives all inputs from IX, X, and XI, i.e., rhombomeres r6–r11. Thus, the sensory territory of glossopharyngeal (upper pharynx and palate) connects at levels r6–r7. Vagus (mid and lower pharynx, larynx, and upper esophagus) connects levels r8–r11. Spinal accessory (sternocleidomastoid and trapezius) connects to levels r10–r11 and c1–c3, although these muscles may originate as far forward as r8.

Here, as promised, is the **grand generalization** (in two parts). (1) *Motor control and general sensory perception for pharyngeal arch structures are in strict neuromeric register with the brainstem via the lateral motor column and the lateral sensory column.* (2) *Pharyngeal arch structures should be classified, not by the individual nerves that supply them, but by the neuromeres which supply the neurons in the first place.*

Anatomic Considerations of Glossopharyngeal Nerve for the Temporal Bone

Glossopharyngeal nerve exits the brainstem via four rootlets, in conjunction with those of vagus, spanning from the first four rhombomeres of the medulla. Its origins are as follows. (1) Motor fibers arise from r6 to r7 of *nucleus ambiguus* (lateral motor column) versus r7 to r11 for vagus. (3) General sensory fibers arise from r6 to r9 of *spinal nucleus* (lateral sensory column) versus r8–r11 for vagus. Thus, some degree motor and sensory overlap affects soft palate and the upper two constrictor. (3) Special visceral motor fibers for parotid

gland arise from r6 to r7 inferior salivatory nucleus. (4) Special sensory fibers for taste refer to levels r6–r7 of nucleus solitarius.

Glossopharyngeal passes through jugular foramen anterolateral to X and XI. It is closest to the opisthotic, being separated from the other nerves by a sleeve of fascia. This indicates a *separation plane between opisthotic and exoccipital*, i.e., between bone derivatives of r7 versus r8. Lodged within jugular foramen is petrous ganglion which houses cell bodies of sensory nerves.

Just as it exits the skull glossopharyngeal receives an important anastomosis from facial nerve carrying PANS fibers from the r4 to r5 superior salivatory nucleus. IX then runs anteriorly just superficial to internal carotid until it reaches the r5 styloid process. Here it tracks inward, giving off a single motor nerve to stylopharyngeus. Recall that stylohyoideus, supplied by VII, is inserted more lateral on the styloid. Thus, third arch is folded inside of second arch.

Glossopharyngeal nerve has multiple functions. We shall list them and then concentrate on the recurrent relationships that exists between IX and the tympanic cavity. Note incorrect use of "visceral" below:

- Somatic afferents supply the upper pharynx, soft palate and posterior 1/3 of tongue.
- Visceral afferents (special) supply taste to posterior 1/3 of tongue.
- Visceral afferents (general) secretomotor for parotid, innervate the carotid sinus.
- Visceral efferents (general) r6 direct to stylopharyngeus, r6–r7 indirect via vagus to palate and pharynx.
- Visceral efferents (special) carotid sinus.

Tympanic nerve carries PANS fibers from the r6 to r7 *inferior salivatory nucleus*. It arises from the ganglion in the jugular fossa and gains access to tympanic cavity from below penetrating its floor just next to the medial wall. Both floor and medial wall arise from the r6 to r7 opisthotic. Once inside, tympanic nerve climbs over the promontory of the medial wall and contributes to the posterior aspect of *tympanic plexus*. Thus, IX provides sensory supply for the three opisthotic walls of the cavity and the posterior tympanic membrane.

Meanwhile SANS fibers from carotid plexus perforate the wall of carotid canal, enter tympanic cavity, and join with tympanic plexus as well. The tympanic nerve, now with both kinds of autonomic fibers, exits the cavity and terminates in the otic ganglion.

Otic ganglion is located just medial to the root of V3. It receives fibers from two cranial nerves IX and (to a lesser extent) VII because facial nerve sends an anastomotic branch to glossopharyngeal as soon as it exits the skull. Thus, otic ganglion gets PANS fibers from two sources: *superior sali-*

vatory nucleus (r4–r5) and *inferior salivatory nucleus* (r6–r7). Although IX is more important than VII, *the embryologic source of parotid gland itself must be from r4 to r5* because there is no known instance in which third arch tissues are not located deep to the plane of the second arch.

Final Points of Clarification

The petrous temporal bone complex carries three distinct nerves, each with autonomic fibers, and with confusingly similar names.

Greater petrosal nerve (GPN) carries PANS fibers from geniculate ganglion and conveys them to pterygopalatine ganglion.

Deep petrosal nerve (DPN) carries SANS fibers from superior cervical ganglion via carotid plexus and conveys them to greater petrosal nerve.

Vidian nerve (VN) originates at foramen lacerum as GPN + DPN.

Lesser petrosal nerve (LPN), the continuation of tympanic nerve, carries PANS and SANS fibers from tympanic plexus and conveys to otic ganglion.

The Middle Ear

The middle ear of modern amniotes (reptiles, birds, and mammals) is a mechanical sound transduction system interposed between the outside world and neurosensory apparatus buried deep within somitomeric mesenchyme. It is an assembly neural crest derivative laminated together from PA1 to PA3 to form a box and an ossicular chain. The tympanic cavity communicates forward to pharynx via the eustachian tube and backward to the mastoid air cells via the aditus ad antrum (Figs. 8.119, 8.120, and 8.121).

In diapsids (reptiles and birds), the tympanic cavity contains a single bone known as the stapes in reptiles and the columella in birds. It spans the gap between the tympanic membrane and the inner ear. In contrast, mammals have three ossicles, their tympanic membranes are different in location and more complex is derivation, their external auditory canals form via a different mechanism, and their Eustachian tubes are anatomically and physiologically different.

What amniote middle ears share in common is the mesenchymal composition of the stapes which comes from two sources. The footplate of stapes that sits in the oval window is derived from r4 to r5 mesoderm of the otic capsule while the remaining components come from r3 neural crest.

Stem amniotes did not have a tympanic middle ear. Their fossils show a large stapes, derived from the second arch hyomandibula. Its role was to stabilize the skull during mastication because, as we shall see, the palate was mobile. New connections freed the stapes from this role and it became more delicate, capable of transmitting vibrations. During the early Triassic, the stapes was transformed in all amniotes, with exception of mammals, into a single-ossicle ear. Mammals pursued an independent course to form a three-ossicle ear, due to the evolution of the temporomandibular joint. This enabled better transmission of high frequencies, and undoubted evolutionary advantage…just ask any bat.

We shall now discuss the structural components of the middle ear, including their evolutionary history. This section concludes with a synopsis of how they are constructed together as a system.

Middle Ear: Anatomic Components and Development
The Ossicular Chain

The middle ear contains three bones, each of which comes from two different sources. Arches. Sound waves cause positional changes in the tympanic membrane which are amplified by this ossicular chain. The membrane itself is trilaminar with an external layer of first arch ectoderm, a middle layer of r3 neural crest, and a posterior epithelium derived from a combination of first and third arch neural crest, *not* endoderm. The mucosal lining of the Eustachian tube is derived from the first arch and third arch.

- Malleus comes from two sources: the body is derived from *articular* and the manubrium from *retroarticular process of articular*. The head of malleus bears has a fact to articulate with incus; it then narrows to a small neck, gives off its two processes, and sends manubrium out to engage with the eardrum. The anterior process is connected embryonically to Meckel's cartilage; this is retained as a vestigial ligament to the petrotympanic fissure. The lateral process inserts into upper tympanic membrane (pars flaccida) and sends out two *malleolar folds* to the *notch of Rivinius*. This refers to a defect in the tympanic ring just above the petrotympanic fissure in the back wall.

- Incus also comes from two sources. The body comes from first arch *quadrate* while the lenticular process comes from the second arch, likely the *symplectic*. The body has a facet for malleus and gives off two crua. The short crus connects via a ligament to the posterior epitympanic recess in a site called *fossa incudis*. The long crus terminates in a small laterally directed lenticular process connected to stapes.

- Stapes comes from two sources as well. Almost the entire bone is a second arch derivative of the old hyomandibular but its footplate is mesoderm from the otic capsule. The head of stapes articulates with lenticular process of incus, then its neck splits into two crura which connect to an oval plate articulating with oval window. If we hearken back to our paleontology, we recall that hyomandibula gives off and anteriorly directed *symplectic* which is connected to quadrate throughout evolution and posteriorly directed interhyal which connects downward through the cartilage chain to styloid process and hyoid bone.

Blood Supply and Innervation

Tympanic membrane is supplied by two main arteries. *Anterior tympanic artery* (tympanic branch of maxillary) arises from the first segment of IMMA, just next to deep auricular. It passes behind TMJ (but does not supply it) and enters the cavity via petrotympanic fissure. It supplies the inner aspect of TM and anastomoses with stylomastoid branch of posterior auricular. *Deep auricular artery* arises just proximal to ATA. It also passes behind TMJ, to which is sends a branch, and then perforates the cartilage of the EAC to supply its anterior wall. It makes anastomosis with ATA.

The tympanum is innervated by V3, auriculotemporal branch. It also receives tympanic branches of IX and the auricular branch of vagus. Recall that r6–r7 mucosa enters the tympanic cavity via the posterior wall of the Eustachian tube.

Mucosa of tympanic cavity of the tympanic cavity receives six arteries. Anterior tympanic and deep auricular, as listed above. Posterior auricular from second arch supplies the posterior tympanic cavity and also perfuses the mastoid. Stapedial V3 had a *petrosal branch* associated with middle meningeal artery and anastomoses with *tympanic branch of ascending pharyngeal*. Finally, following the course of the tympanic tube is *artery of pterygoid canal*. Innervation of the tympanic cavity is mixed. Many components of the tympanic plexus have been described elsewhere. Seen as a whole it comprises: (1) tympanic branch of IX, (2) caroticotympanic plexus bearing SANS fibers through perforations in the wall of carotid canal, and (3) lesser petrosal nerve carries preganglionic PANS fibers to otic ganglion. *Chorda tympani* exits facial canal must 6 mm before the main nerve reaches stylomastoid foramen. It follows its own separate canal. Recall that this is the means by which extracranial stapedial artery exits the skull (Fig. 8.122).

Eustachian Tube and Lining

The eustachian tube arises as a *fusion of the first and second pharyngeal pouches*. Because of its laminar nature, it presents embryonic planes in continuity with the mucosa of the oronasal cavity. When the primitive mesenchyme cavitates, these potential spaces are exploited by the mucosa to create pneumatization. The mucosa lining the eustachian tube and the middle ear cavity is endoderm. Recall that with the elimination of second arch mucosa, IAC comes to be an admixture of first second, and third arch mesenchyme. It is not surprising that the blood supply and innervation of the posterior wall come from ascending pharyngeal (AA3), and tonsillar/pharyngeal branches of facial artery (AA2).

The fusion of pharyngeal pouches 1 and 2 draws together muscles from the first, second, and third arches. Muscles

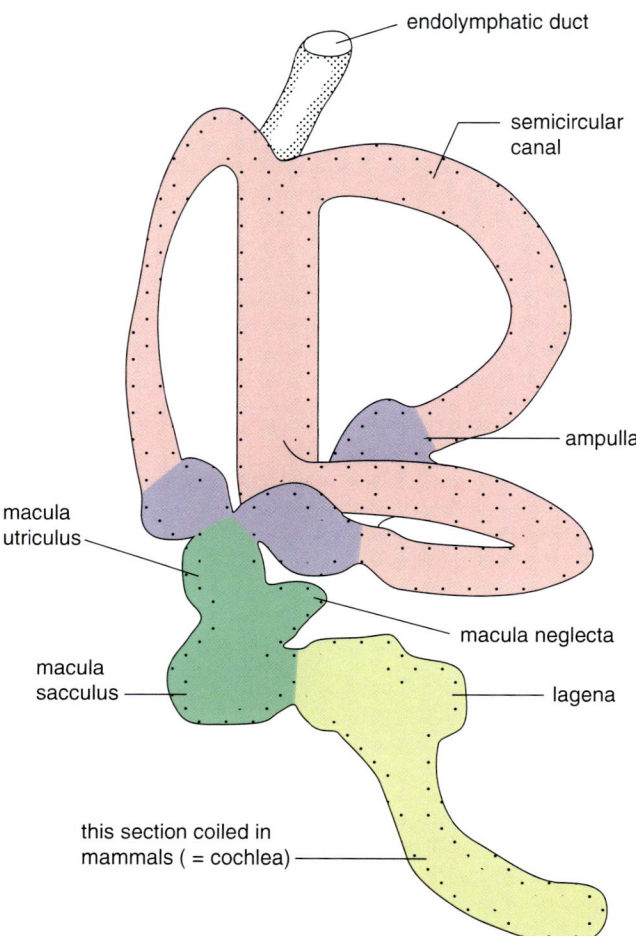

Fig. 8.122 Mechanosensory transduction takes place in three sites. Neuromast organs are found in the ampulla, macula, and lagena. The endolymphatic duct is just a pressure equalizer. (Courtesy of Augustus T. White, Palaeos.com. Retrieved from: http://palaeos.com/vertebrates/bones/ear/overview.html)

attached to the tube are (laterally) Sm4 tensor veli palatini and (medially) Sm7 levator veli palatini. This same geometric arrangement is maintained within the middle ear: tensor tympani (Sm4) is situated external to stapedius (Sm6). The bony walls of the middle ear are derivatives of neural crest from the first and second arches. Geometrically they form five sides of a box. The sixth side, i.e., the medial wall, is the petrous promontory. As we shall see, neurovascular "mapping" of the medial and posterior walls (opisthotic bone) indicates a contribution from r6 to r7 mesenchyme.

The lining of the middle ear "box" comes from two different sources. Endoderm from the Eustachian tube covers the inner half of the box, whereas the outer half, including the inner lining of the tympanic membrane neural crest, results from a mesenchymal to epithelial transformation. This process, contrary to traditional descriptions, is detailed at the end of this section.

Tympanomastoid

During the third week, the pharyngeal arch system is laid down in cranio-caudal fashion. At appearance of the fourth pharyngeal arch, second arch migration has taken place. The first and second pharyngeal pouches have coalesced in a ventral-to-dorsal sequence, creating a dorsally directed endodermal tube. The lateral wall is constructed with PA1 neural crest and Sm4 mesoderm (tensor veli palatini). The medial wall is constructed from PA3 neural crest with Sm7 mesoderm (levator). The dorsal extension of the tube will become the future tympanic cavity. *In the dorsal tube, pharyngeal pouch coalescence is partial.* PA2 neural crest is present (the future stapes) as are myoblasts from Sm6 (stapedius). Some PA2 endoderm is theoretically present as well, providing coverage over the stapes (Figs. 8.123 and 8.124).

In the fourth week, the terminal end of dorsal tube expands and flattens dorso-ventrally, becoming a slit-like tunnel shaped like a spade. It comes into direct contact with the infolded first pharyngeal groove. This situation is temporary. PA1 mesenchyme inserts itself between the endodermal (pouch) layer and the ectodermal (groove) layer. This will become a portion of tympanic membrane, the *tunica propia*, and the manubrium of the malleus.

The fifth and sixth weeks are marked by expansion of the tympanic cavity. This is in synch with growth of the otic capsule. In the seventh week, remaining mesenchyme of the second arch proliferates, causing a ring-like constriction. Lateral to this site is the *tubotympanic recess*. Medial to the site is the *primordial eustachian tube*. The definitive pars propia of tympanic membrane and manubrium of malleus develop at the end of the eighth week. A process of cavitation begins to take place in the inferior (ventral) middle ear. It spreads slowly upward. From the otic capsule, a projection of mesoderm, the *superior otic process*, extends outward over the ossicles. It becomes *lateral tegmen tympani*. Medial tegmen tympani is neural crest mesenchyme.

By the 12th week, tympanic cavity now extends over the medial surface of the inferior tympanic membrane. Further cavitation has now created the epitympanic recess. Mesenchymal dissolution accompanies development of the ossicles and muscles. Endodermal mucosa within the cavity is quick to exploit the situation; it ensheaths the bones. At the 16th week, the boundaries of the middle ear are established. An outgrowth of the squamous temporal bone, the *tympanic process*, makes two contributions to the middle ear: (1) the anterior wall of the epitympanic cavity; and (2) the lateral wall of tympanic cavity proper. A separate lamella from petrous pyramid creates the floor. The lateral boundary is now tympanic membrane and the tympanic ring. In sum: anterior wall = tympanic and prootic; medial wall = prootic, posterior wall = opisthotic; floor = opisthotic; roof = epiotic/opisthotic.

From the 16th to the 20th weeks, Eustachian tube elongates and fibrocartilage appears. At the 21st week, connective tissue within epitympanic recess proliferates and expands laterally; this will become antrum. Ossification of the tegmen tympani starts at 23 weeks and continues onward until birth. At the 24th week, just after ossification of the otic capsule is complete, air cells appear within the petrous pyramid. Air cells also develop in the wall of tympanic cavity, near the genicular ganglion (supracochlear cells) and in pericarotid bone.

The mastoid process now makes its debut. Periosteum from the otic capsule spreads over antral connective tissue. It

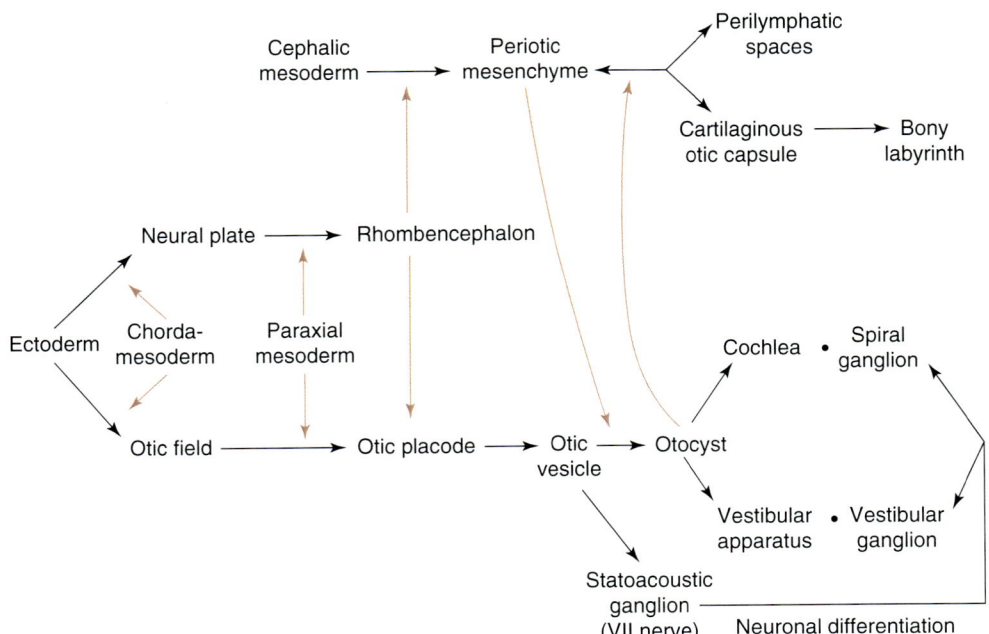

Fig. 8.123 Inductive events and tissue transformations in the developing ear. Colored arrows refer to inductive events. (Reprinted from Carlson BM. Human Embryology and Developmental Biology, 5th ed. Elsevier; 2014. With permission from Elsevier)

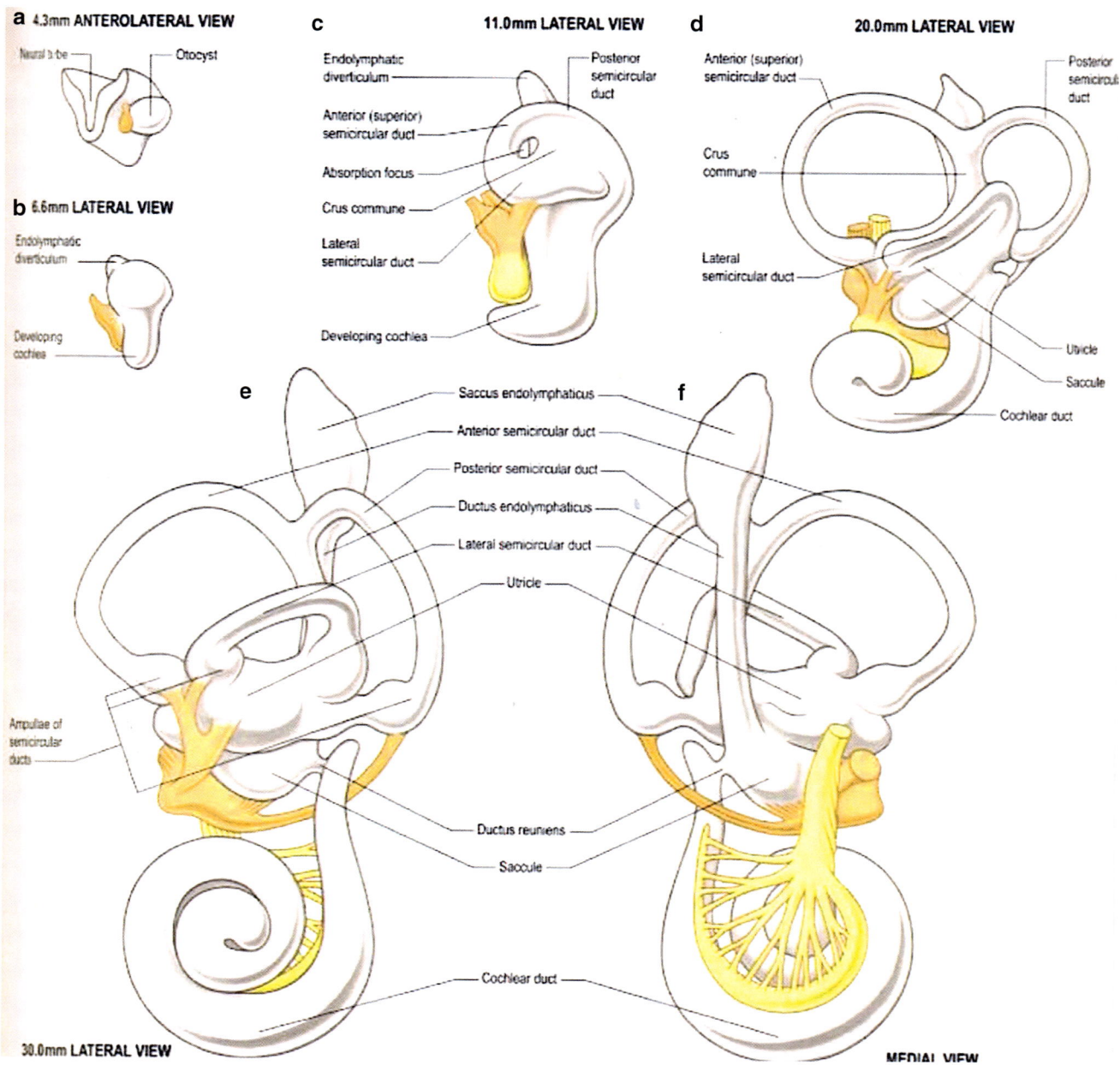

Fig. 8.124 Inner ear development. Cranial nerve VIII has two nuclei: vestibular nerve in r4 and auditory nerve in r5. Neural crest from r4 to r5 makes up the soft tissue structures and is enclosed by mesoderm of the same derivation. (Reprinted from Lewis, Warren H (ed). Gray's Anatomy of the Human Body, 20th American Edition. Philadelphia, PA: Lea & Febiger, 1918)

subsequently fuses with tympanic process of squamous temporal bone. Into this tissue mass, the epitympanic recess extends posteriorly at 30th week to form the antrum. By the 35th week, antral development is complete. Mastoid growth continues after birth until 5–10 years of age.

Malleus and Incus
Neural crest blastema corresponding to the future malleus and incus appear between 4.5 and 6 weeks as extensions of Meckel's cartilage. They in close association with the chorda

tympanic nerve. CT sweeps around the ventral aspect of the malleal and incudal primordia; it constitutes a point of physically anchorage for these tiny bones. It also acts to induce blood supply. In the sixth week, CT separates the malleal–incudal blastemae from the hyoid (second arch) bar. Cartilage now is the ossicles as well as in Meckel's and Reichert's. In the seventh week, conversion to bone appears. Short process of incus comes into contact with otic capsule. By the end of the eighth week, malleus has its adult form (but it is still connected to Meckel's. This membranous connection becomes the anterior process of malleus, *processus Folianus*. Incus is different. It is separate from Reichert's because CT severed the connection. In compensation, the long process of incus extends toward the head of stapes. It maintains a persistent connection with hyoid arch. In contrast, incus remains separated from malleus by mesenchyme.

Between 8th and 10th week, the ossicles grow rapidly, whereas anterior malleal process grows more slowly. By the 15th week, adult size is achieved. Bone formation takes place at 16th week. Meckel's cartilage begins to degenerate, leaving behind the anterior ligament of malleus.

At the 17th week, perichondrial bone extends over the neck of malleus so that it becomes encircled. Vascular buds invade and marrow form. The ossicles are fully mature at the 20th week. By the 27th week, the ossicles are enveloped in mucous membrane and are fully vascularized.

Stapes

At 0–4 weeks, stapes has a dual origin. The ring of stapes arises from second arch neural crest in the of the *symplectic bone*. It has the following derivatives: a *capitulum* that articulates with incus, two *crura*, and the lateral (tympanic) surface of *footplate*. The inner footplate, *lamina stapedialis*, arises from otic capsule mesoderm. Its derivatives are the annular ligament and the medial (labyrinthine) surface of footplate. At the fourth week, these structures are as yet unrecognizable. What is observed is a blastema just dorsolateral to the hyoid bar immediately next to VII and its induction product, the stapedial artery.

At the fifth week, interaction between the growing blastema and pre-existing neurovascular structures results in three physical changes in the stapedial mesenchyme. (1) The solid blastema is now "grooved" by the stapedial artery, i.e., the blastema is growing around the arterial structure. As the mass encircles the artery, the stapedial ring forms. The central defect created by the stapedial artery is the obturator foramen. (2) Blastemal growth around the facial nerve separates stapedial primordium into three parts: *stapes proper*, the *laterolyale*, and an intermediate "bridge," the *interhyale*. (3) The two poles of stapes grow away from each other. Stapes proper expands upward and forward while laterohyale pursues a downward and backward course. As this happens, the blastema rotates around the fixed axis of facial nerve so that laterohyale winds up posterior to stapes.

The seventh week marks the "docking" of stapes with the otic capsule. Laterohyale meets the otic capsule. Interhyale condenses. Stapedial ring enlarges to accommodate to the oval window. Long process of incus makes the contact with head of stapes. Differentiation into cartilage now begins, concomitant with chondrification of the otic capsule. It would seem that all these neural crest derivatives are responding to common genetic signals.

In the eighth week, the otic capsule continues to chondrify save for one exception: the bases of future crura and footplate become fibrous, forming the *lamina stapedialis of the otic capsule*. From this arises the medial (vestibular) surface of stapedial footplate. When the footplate has grown to its final size, the tissue of its rim will condense into the *annular ligament*.

The ninth week marks the appearance of stapedius muscle the tendon of which arises from interhyale. Laterohyale and otic capsule now fuse. In so doing, two bony structures are formed: the anterior wall of facial nerve and the pyramidal eminence supporting stapedius muscle. Because differentiation of laterohyale predates that of the stapedial pyramid it can be argued that stapedial tendon is the primary insertion while stapedial pyramid represents the secondary insertion. Note that anterior wall of facial canal distal to laterohyale develops as a membranous ossification associated with Reichert's cartilage.

The cartilaginous phase of stapedial development lasts from the 9th to the 15th week. In the 16th week, blood vessels are observed running from the primordial facial nerve canal to footplate and head of stapes. These arise from the stylomastoid branch of posterior auricular artery, exclusively supplying derivatives of the second pharyngeal arch. Also supported by this circulation is mastoid antrum and tympanic cavity bone overlying the semicircular canals.

At the 18th week, ossification of the stapes takes place. It starts at the footplate, spreads over the crura and eventually reaches the head. In the 20th week, the footplate widens, causing a dedifferentiation of the mesenchyme from cartilage to mesenchyme. AT maturity the fenestral rim redifferentiates into the fibrous annular ligament that encircles the footplate, providing anchorage to the oval window. Stapes achieves its final form at the 32nd week. Unlike the long bones, it does not thicken; the fetal bone is retained.

Middle Ear: Phylogeny

Among the defining characteristic of mammals, is a middle ear with three ossicles. The system increases amplification of sound through air. Two muscles, tensor and stapedius, modulate the ear drum. The evolutionary history of these bones has fascinated paleontologists and comparative anatomists for 200 years. Good accounts are readily available [13, 14]. Our purpose here is to streamline important highlights from various sources to provide an integrated picture of how these bones can to be, taking the following topics one-by-one: (1) review of the cartilaginous components of the pre-tetrapod branchial arch; (2) the history of the second arch bone, stapes; (3) the history of the first arch bones, tympanic, malleus, and incus; (4) how middle ear evolution related to the development of the TMJ (Figs. 8.125, 8.126, 8.127, 8.128, 8.129, and 8.130).

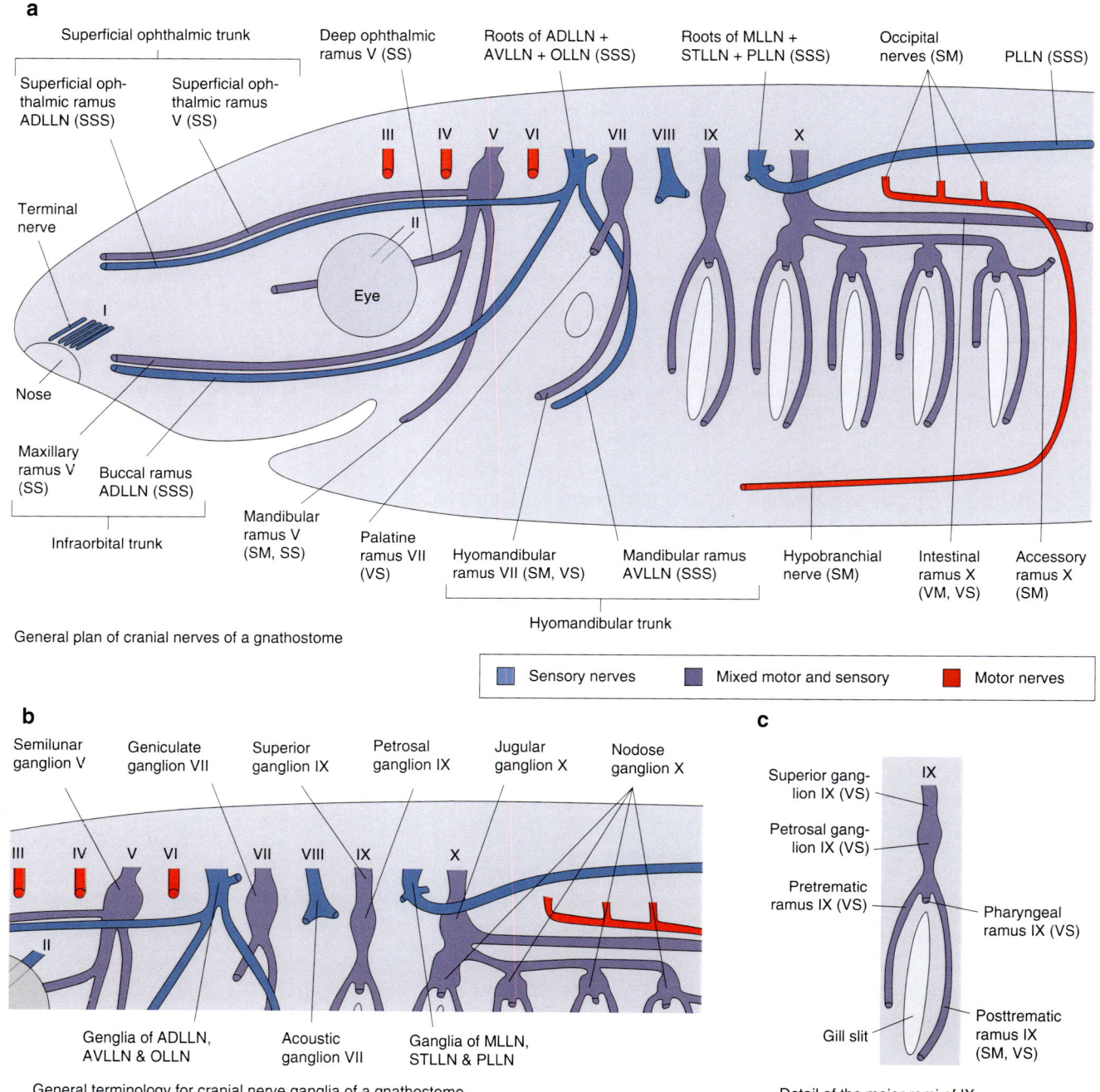

Fig. 8.125 Cranial nerves *Latimeria*. Cranial nerves based on the sarcoptyergian coelacanth *Latimeria*. Note that the six lateral line nerves (ADLLN, AVLLN, OLLN, MLLN, STLLN, and PLNN) are considered separate cranial nerves. Older texts lump them together as parts of other cranial nerves. Note that pretrematic branches are sensory whereas posttrematic branches are both sensory and motor. Note the pharyngeal ramus is visceral sensory to the entire arch. *SM* somatic motor, *SS* somatic sensory, *SSS* special somatic sensory (lateral line nerves and statoacoustic nerve), *VM* visceral motor (parasympathetic), *VS* visceral sensory. (Courtesy of William E. Bemis)

Fig. 8.126 Right tympanic membrane, viewed from the inside. It passes into Eustachian tube, a sandwich of first arch and third arch mesenchyme. Pharyngeal endoderm begins with third arch (r6–r7) and extends into the middle ear providing the lining of the medial, internal, and posterior walls with sensory innervation of IX/X. This explains the wide distribution of referred pain from middle ear otalgia. (Reprinted from Lewis, Warren H (ed). Gray's Anatomy of the Human Body, 20th American Edition. Philadelphia, PA: Lea & Febiger, 1918)

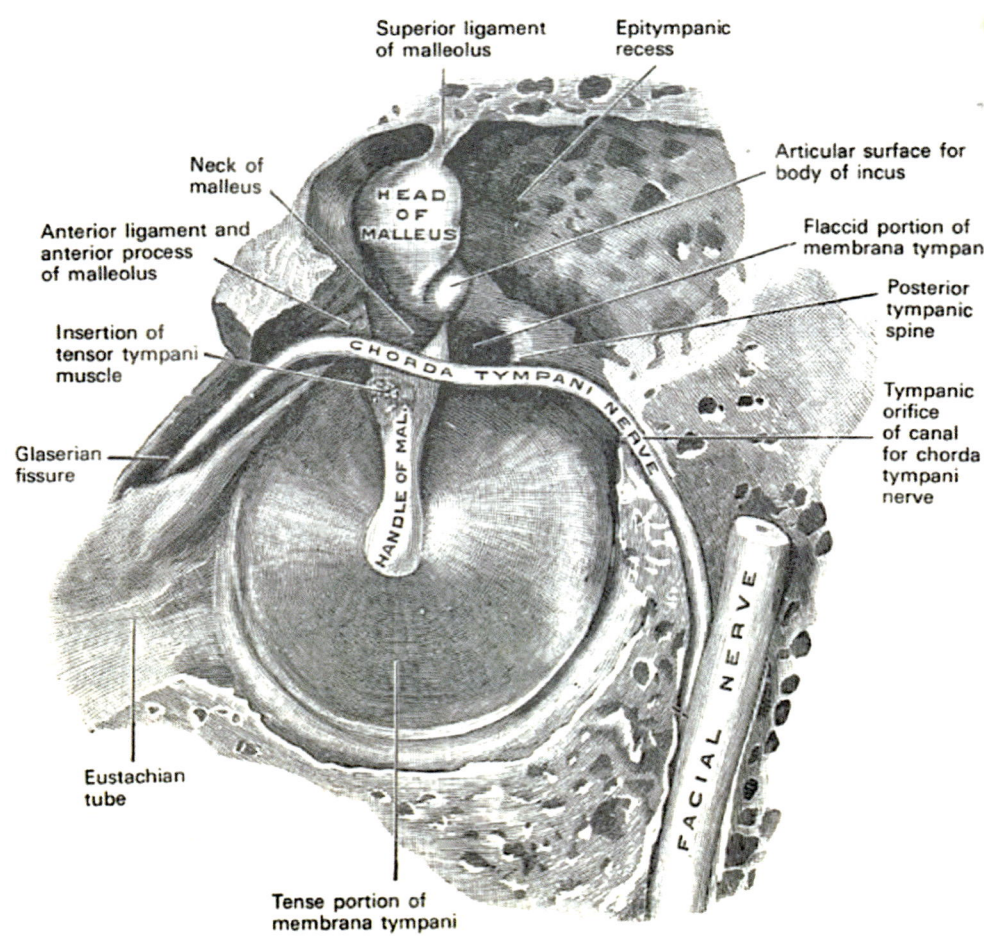

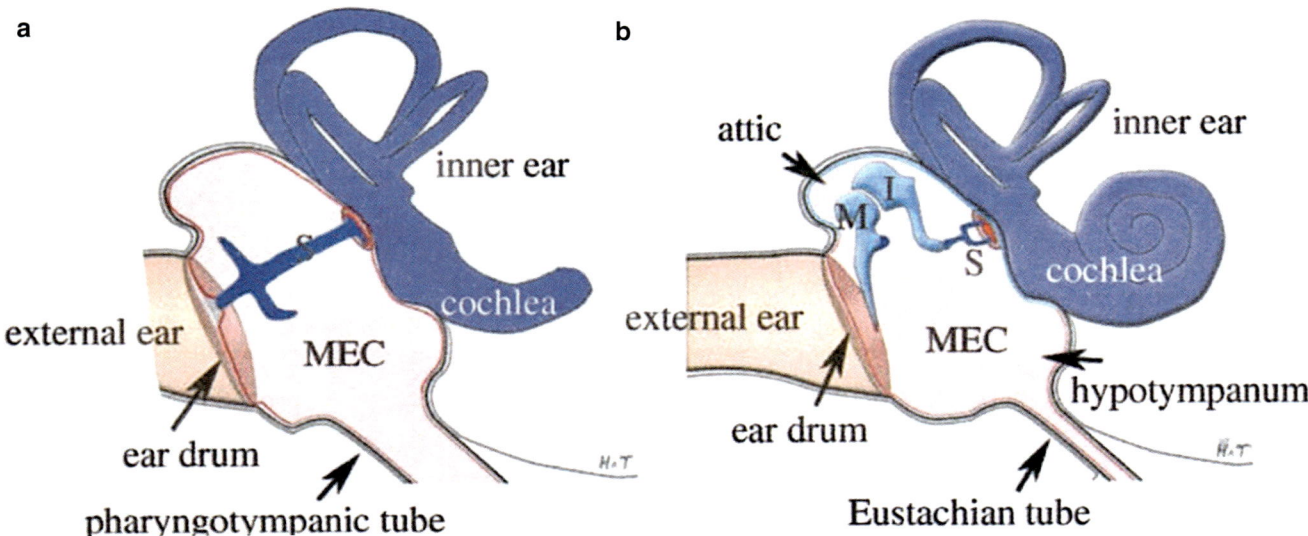

Fig. 8.127 Single versus triple ossicle ear. Sauropsid middle ear showing single ossicle (stapes) spanning from tympanic membrane to oval window compared to mammalian middle ear with three ossicles within the cavity. Origin of ossicles: Light blue denotes first arch neural crest derived tissue. Dark blue denotes second arch neural crest derived tissue. Red denotes mesoderm-derived tissue (stapes footplate). *S* stapes, *M* malleus, *I* incus, *MEC* middle ear cavity. (Reprinted from Tucker AS. Major evolutionary transitions and innovations: the tympanic middle ear Phil. Trans. R. Soc. B 2017;372(1713): https://doi.org/10.1098/rstb.2015.0483. With permission from Creative Commons License 4.0: https://creativecommons.org/licenses/by/4.0/)

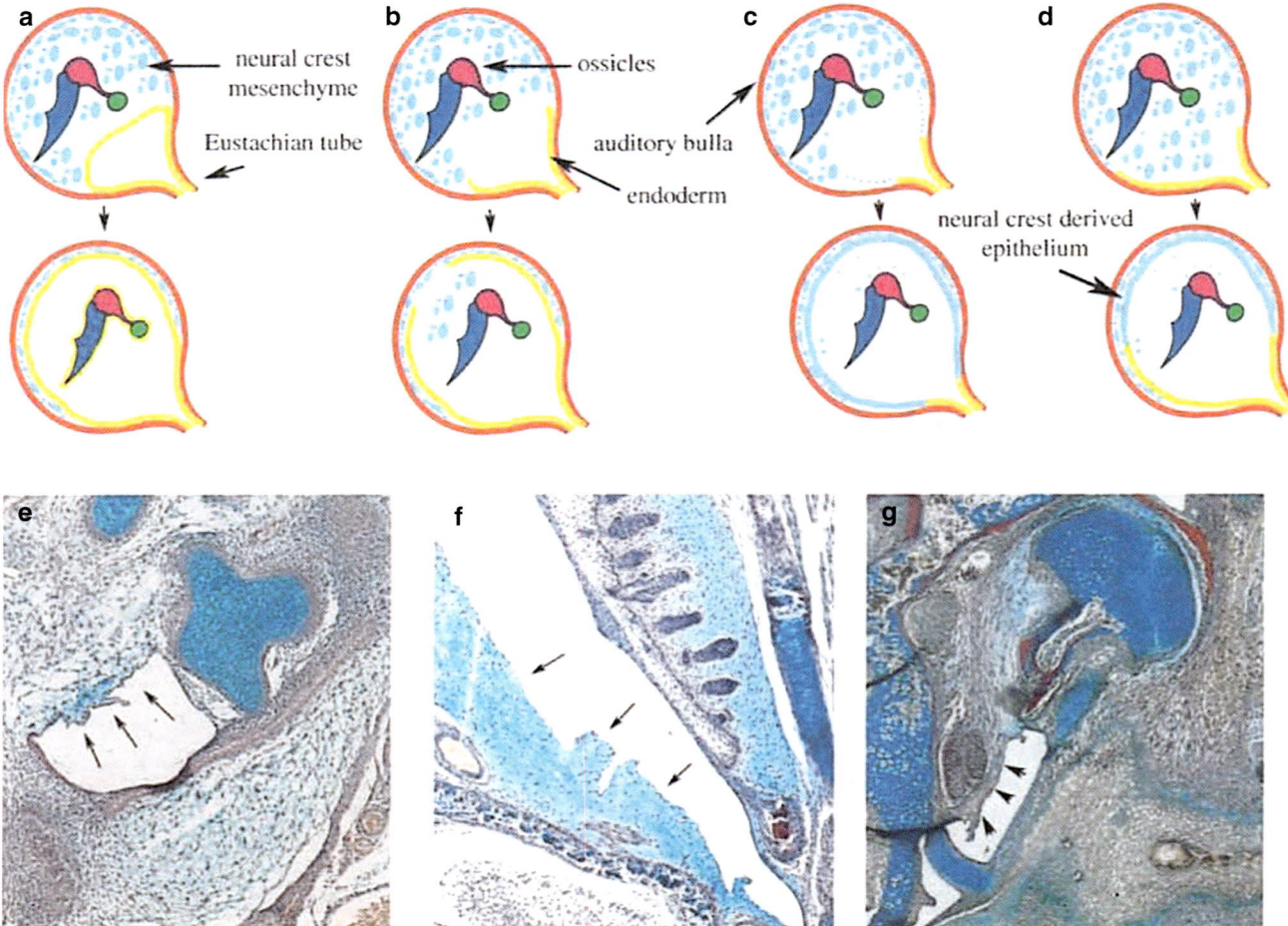

Fig. 8.128 Cavitation of the ear. (**a–c**) Proposed processes of creating an air-filled space. (**a**) Invasion of the endoderm as a sheet of tissue wrapping around the ossicles. (**b**) Break of the endoderm to allow the tissue to move around the ossicles. (**c**) No invasion of the endoderm but creation of a cavity by retraction and transformation of the mesenchyme. (**d**) Process based on lineage tracing showing a dual origin of the middle ear lining incorporating some of the ideas from the previous three models. (**e–g**) Histology sections through the middle ear during retraction of the mesenchyme: (**e**) mouse E18.5; (**f**) shrew P5; (**g**) postnatal opossum. Arrows represent the mesenchyme retracting back from the forming tympanic membrane creating a cavity. Endonderm, yellow; neural crest, blue; mesoderm, red. (Reprinted from Tucker AS. Major evolutionary transitions and innovations: the tympanic middle ear Phil. Trans. R. Soc. B 2017;372(1713): https://doi.org/10.1098/rstb.2015.0483. With permission from Creative Commons License 4.0: https://creativecommons.org/licenses/by/4.0/)

Branchial Arch Cartilages: Building Blocks of Jaws

In bony fishes and in the tetrapods, jaws develop as ensheathments of first arch cartilages by dermal bones. Pterygopalatoquadrate cartilage (PQ) is covered (partially) by premaxilla and maxilla. The quadrate protruded out the back end (we shall hear more about it later). Meckel's cartilage is laminated on the outside by dentary, 2–3 splenials, angular and surangular; and on the inside by coronoids preangular, and angular (Fig. 8.125).

The primitive cartilages themselves represent a *chain of bone fields* that reproduce the original vertically-oriented *branchial arch series*: pharyngo-, epi-, cerato-, hypo-, and basibranchial. The upper two are dedicated to the upper jaw while the remainder are assigned to the lower jaw. Palatoquate is a modification of the first arch pharyngo- and epibranchial cartilages. Meckel's is derived from the first arch ceratobranchial cartilage. The second arch is subverted in a similar way. The upper cartilages become *hyomandibula* which is attached to the otic capsule. The lower cartilages form the hyoid. The remaining five arches form the gills (Figs. 8.126).

The jaws of the earliest fishes were simple conversions of the first gill arch, *leaving the second gill arch intact for respiration*. The basal jaw joint involved an articulation between the *quadrate* bone of the upper jaw and the *articular* bone of the lower jaw. The relationship between these two bones has remained unchanged throughout evolution. Jaw suspension in placoderms was *autostylic* in which the upper jaw itself was jointed anteriorly with the cranium but the lower jaw was unsupported. The hyomandibula was not involved.

This situation proved unstable, so when the evolutionary line split into cartilaginous fishes and bony fishes, a more functional hinge evolved, making use of the second gill arch.

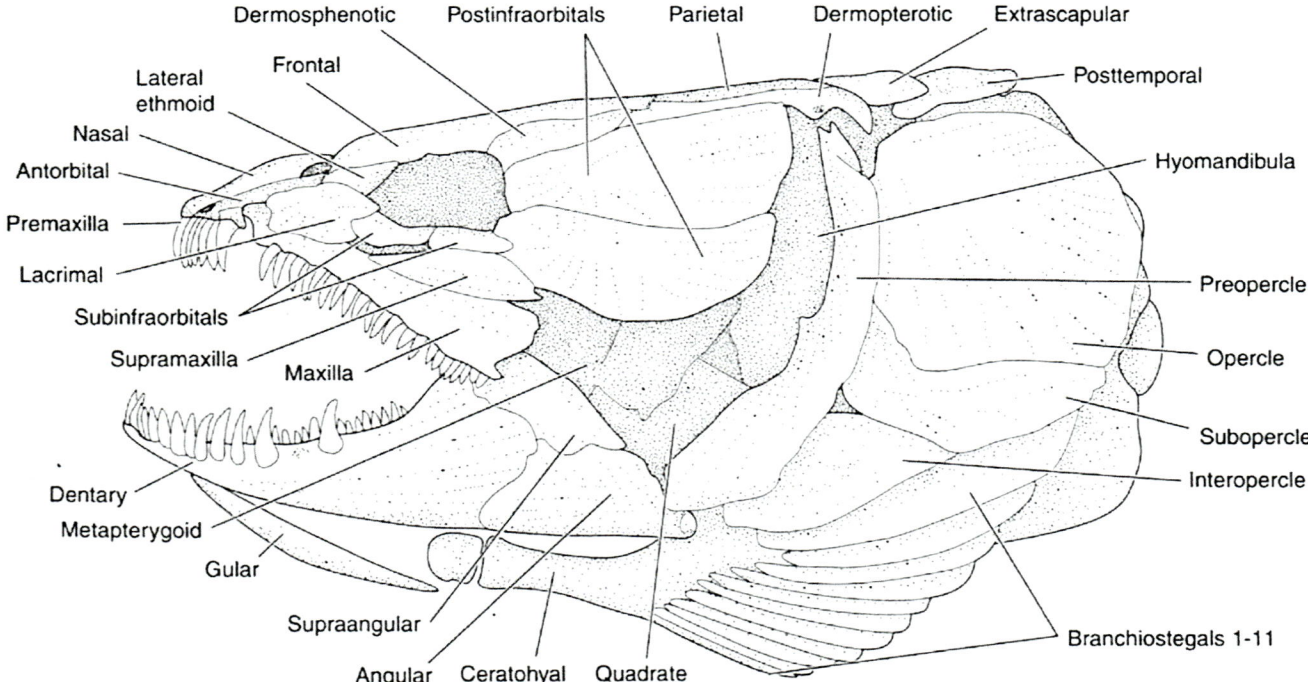

Fig. 8.129 Cranial bones of the basal (primitive) bony fish *Amia calva*. From dorsal to ventral note the following. Extrascapulars border the pectoral girdle but do not attached directly to it. Posttemporal connects the skull to the pectoral girdle. The tetrapod transition in charac-terized by loss of the extrascapular and posttemporal. Opercular bones cover the gills. Branchiostegals (usually 10) connect the skull to the second arch in the hypopharynx. (Courtesy of William E. Bemis)

Fig. 8.130 Banchial arch cartilages. The first two arches are no longer respiratory. Palatoquadrate = pharyngo- and epibranchial cartilages. Meckle's = ceratobranchial cartilage. (Reprinted from Kardong K. Vertebrates: Comparative Anatomy, Function, Evolution, 7th ed. New York, NY: McGraw-Hill; 2015. With permission from McGraw-Hill Education)

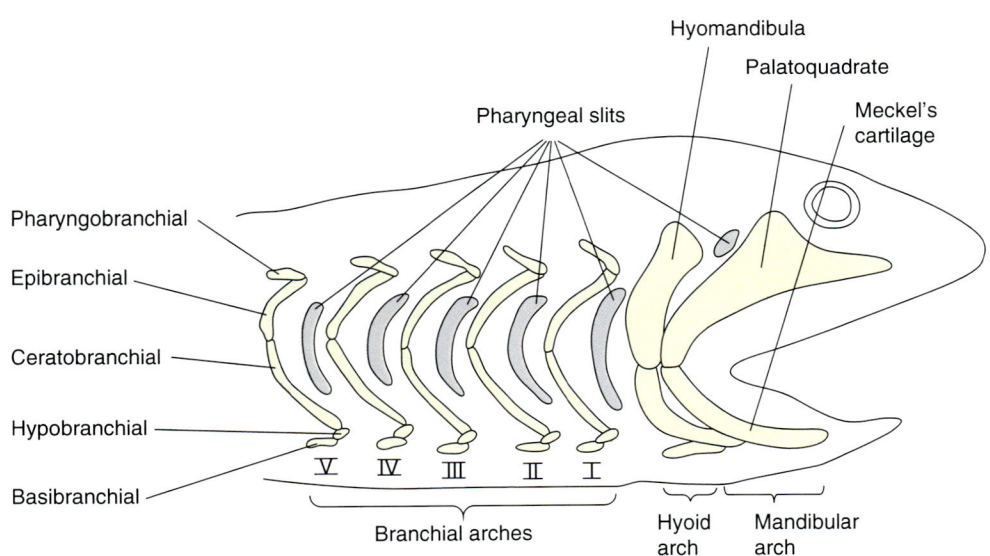

In these derived fishes, the upper cartilages of the branchial chain were converted into a single bone, the *hyomandibula*, which attached to the otic capsule rotates downward to contact quadrate. Projecting forward from its caudal margin was a new bone element, the *symplectic*, that connects with the first arch quadrate. This structure will turn out to be of importance in the later assembly of the mammalian middle ear. Dangling down from hyomandibular, are the three remaining cartilages: ceratohyal, hypohyal, and basihyal. These create the fish hyoid bone. These relationships can be readily seen in *Amia* (Figs. 8.127 and 8.128).

History of Stapes: Present at the Beginning

Second arch stapes evolves long before its companions. The transition to therapods severed the link between the head and forefin. A number of bones are lost that change the anatomic environment of hyomandibula. Four bones covering the gills and providing support for the lateral skull (preopercular, interopercular, subopercular, and opercular) lost. With the loss of the *opercular series*, hyomandibula takes up the slack by becoming a laterally directed brace for the occiput. It also gains a new name, stapes. These changes are seen well in *Greererpeton* (Fig. 8.105).

As we move up the scale, the basal amniote, such as *Paleothryis*, shows a truly massive stapes spanning between optic capsule and quadrate (Fig. 8.106). At the diapsid/synapsis split, the reptile *Claudiosaurus* shows two bones come to the rescue of stapes: (1) paraoccipital process from opsthotic joins occiputs to lateral skull squamosal; (2) pterygoid lends support to quadrate as well. No longer needed for skull stabilization stapes becomes more delicate. At this time, a *tympanic membrane is developing* behind the quadrate. Stapes transmits sound from the TM to oval window providing hearing for reptiles and birds. Note that the relationship between quadrate bone and stapes is never lost. It is this connection, the symplectic, that will guide quadrate and its partner articular into an expanded tympanic cavity as the incus and malleus. These ossicles will insert themselves between tympanum and stapes, thereby dramatically expanded the range of hearing.

Incus, Malleolus, and Tympanic: New Kids in Town

The addition of new elements to the middle ear cavity came about from several concomitant processes. The embryonic anatomy of the Meckel's cartilage was substantially altered as small bone fields moved into a new position. Quadrate bone remained relatively stable in the process. The evolution of the mammalian ear took place in synapsid amniotes from therapsida to theriodontia to cynondontia to probainognathia to Mammalia (Figs. 8.129 and 8.130).

Meckel's cartilage occupies a position within the mandible that is ventral and lingual. It has two zones, mandibular and tympanic. The latter has from proximal-to-distal, four bone fields with their subsequent derivatives. *Angular* and *reflected lamina of angular* form the tympanic ring; *prearticular* and *articular* form malleus. The *quadrate* belongs to the palatoquadrate cartilage of upper jaw and forms incus. Reflected lamina of angular is an outgrowth of that bone the posteroventral limb of tympanic membrane while body of the angular forms the anterodorsal limb. Slung between them is pars tensa. *Retro- (pre) articular process* of articular forms neck of malleus and the base of manubrium, whereas articular forms the manubrium proper.

The therapsids developed an external auditory canal which is seen in squamosal and the jaw joint sits very posteriorly at the back of the skull. The tympanic process of Meckel's shows ear bone fields held in a canal and connected with mandible. In the therian clade of cynodonts, just proximal to mammals, three things happened. (1) The bone bridge broke away completely from the mandible, the bones being held together with cartilage and fibrous tissue. (2) The surangular bone expanded to form a *coronoid process*. (3) Posterior dentary evolved a new vertically directed field, the *ramus*, which made the contact with squamosal.

The transformation begins with the cynodonts with *Sinoconodon* being a basal form and Morganucodon considered the first true mammal. *Sinoconodon* is a truly transitional species. It was polyphydont, meaning its teeth were constantly being replaced. Specialized premolars did not exist. It had a deep trough from which the postdentary rod resided.

Thrixanodon is an example of an early cynodont with nasal and lacrimal finally in contact, and a double occipital condyle. A reflected lamina is seen on the surface of angular. The dentary shows the first signs of invasion of muscles of mastication. Coronoid process is well inside the zygomatic arch. Although articular-quadrate joint persists, the quadrate is much reduced in size. Stapes now contacts quadrate. The surangular bone field projects upward, just behind the coronoid process of dentary; it sits forward of squamosal.

Probainognathus is where the switch takes place. Surangular is now in contact with squamosal in a hollowed-out glenoid fossa. Articular becomes rod-like and is held by fibrous tissue to a groove along the medial side of mandible. We see a brief experiment with a double jaw joint.

Morganucodon is considered the first true mammal. It was a diphyodont, having but two sets of teeth. Premolars now exist. The double jaw joint is further refined. Quadrate-articular joint persists. Surangular now loses contact with squamosal. From posterior dentary, ramus/condyle develop and articulate in the glenoid fossa. As a consequence, the bones of the old reptilian jaw joint, articular and quadrate, are much reduced. Ventral view of the skull shows the dentary condyle in direct articulation with squamosal. Stapes and quadrate are seen in contact with petroa.

Mammaliaforms split into the monotremes and eutherians; in each line, full separation of the middle ear from the jaw takes place. This involves (1) an apoptosis of Meckel's cartilage, and (2) differential growth of the mandible forward, (3) expansion of the petrous complex to include the ear bones.

Tympanic Membrane and Tympanic Cavity

Although small in size, the tympanic membrane is of critical importance to under stance the evolution of the ear. It had been thought by many that since reptiles, birds, and mammals evolved from a common sauropsid ancestor that somehow that mammals somehow added two additional bones to the sauropsid single-ossicle ear. But there are multiple lines of evidence against this including the anatomic course of chorda tympani and the physical disposition of the tympanum with respect to Meckel's cartilage (again review Frazer). Presumably the external ear canal and the tympanic membrane would form in the same way. But not only EAC is located in a different site in mammals, but the suspension of the TM is completely different. Mammals use the reflected lamina of angular from the lower jaw to create an encircling ectotympanic bone while reptiles and birds suspend their TMs from the quadrate bone of the upper jaw. This could represent a simple signal switch between r2 and r3 fields (Fig. 8.114).

The mammalian tympanic membrane is unique in that it forms from two parts. The upper part corresponding to the sauropsid TM become the *pars tensa*. The lower part, a mammalian neomorph, becomes *pars flaccida*. In sauropsids, the angular and prearticular continue their roles within the lower jaw whereas in mammals all bone fields are subsumed by the dentary. In mammals, the additional mobility of the TMJ allowed these bone fields to detach from the mandible and assume a new role for tympanic membrane support.

The mammalian external auditory canal forms when a local accumulation of mesenchyme superficial to tympanic membrane disintegrates via apoptosis, permitting surface ectoderm to invade. In birds, this process involves an invagination of surface ectoderm. The location of TM in mammals is deeper than in sauropsids as well. The mammalian and sauropsid ear canals form on opposite sides of the jaw joint.

Tympanic membrane is traditionally, but incorrectly, described as having three layers: and outer ectoderm, a middle mesenchyme (mesoderm), and a lining endoderm, the source of which comes as an invasion from the eustachian tube. In this endodermal model, the pharyngeal pouch invades the middle ear, creating a cavity lined by endoderm. This is topologically possible in non-mammals with only a single ossicle. But how can this occur, if the tympanic cavity has stapedial artery branches, chorda tympani, and sensory nerves?

In 1959, Schwartzbart showed in human cadavers that the endoderm stops short. The endodermal diverticulum runs between what we consider somitomeres 5 and 6 and then ruptures. Why? Recall that the somitomeres, unlike somites, are not separated by epithelium, they are confluent. Eustachian tube runs into a solid wall of mesoderm which it cannot penetrate. The future tympanic cavity is an admixture of neural crest mesoderm which is internally mapped out. Mesodermal apoptosis takes place leaving the neural crest ossicles behind. Via a mesenchymal-to-epithelial transformation (MET) neural crest forms the lining of the cavity all the way to the "doorway," outside of which Eustachian tube was loitering. Endoderm now proceeds to line the medial wall and medial aspects of the other four walls. This dual origin of middle ear epithelium in mammals was proven by Thompson and Tucker using labels of *Sox17-icre* for endoderm and *Wnt1cre* for neural crest. Neural crest also provides epithelial cover around the ossicles and the cochlea (Figs. 8.123 and 8.124).

The Eustachian Tube: Passage to the Pharynx

The mammalian pharyngotympanic tube differs in anatomy and function, being more sophisticated than that of sauropsids. In reptiles, it is bony and wide open to the pharynx whereas in mammals it is cartilaginous, small and collapsible. We resolve this problem with two muscles located on either side of the tube: Sm4 tensor veli palatine on the outer aspect and Sm6 levator veli palatine on the inner aspect. This is anatomically consistent with the spatial "internalization" of the third arch inside the second and first arches. Note that TVP has been identified as part of the lateral pterygoid muscle blastema and is innervated by V3, there identifying it as a muscle of mastication. LVP is part of the superior constrictor blastemal and is supplied by neuromere r6–r7 conveyed by vagus.

Nota bene: in the neuromeric model, the mapping of a structure to an individual arch depends on the neuromeric source of the neurons, not the identity the nerve that conveys them. Thus, although vagus is identified with the fourth arch, LVP (being supplied from r6 to r7) remains a third arch muscle.

It is thought that the more narrow and flexible Eustachian tube in mammals is a response to brain growth. The presence of cilia in mammals and a muscular "valve" system are compensatory innovations. But Eustachian anatomy poses clinical challenges as well. Serous otitis media can be occasioned by high altitude travel. Infectious otitis can result from dysfunctional motility or dysplastic muscle, as in patients with cleft palate.

How Do the Middle Ear Bones Get into the Ear?

Development of more complex dentition and a new form of jaw joint were innovations permitting mammals complex chewing and the ability to handle new foodstuff as well as to maximize nutrition for available sources. The basal mammaliaform *Morganucodon* continued to have the old reptilian joint between articular and quadrate as well as a new, improved joint between the squamosal and a posterior expansion of dentary. This "proto-TMJ" permitted shearing and gridding motions. It allows for greater movement between the teeth while blocking dislocation. These changes were seen in an improved occlusion pattern. As this TMJ took over, the bones of the existing joint lost their functional role and underwent a process of involution, but in the process, they did *not* lose their anatomic connection. Articular (malleus) remained attached to quadrate (incus) and the latter remained attached via *symplectic* to stapes. And herein lies an important developmental relationship. Stapes via *interhyale* retained contact with ceratohyal, the source for the styloid process.

Styloid process gives rise to two ligaments of interest, stylohyoid and stylomandibular. The latter ligament inserts exactly into the former site of angular bone. The ligament is straddled laterally by StV3 inferior alveolar nerve/artery and medially by chorda tympani of VII. In evolution, a breakdown occurs between the dentary and tympanic parts of Meckel's cartilage, *precisely at the insertion point of stylomandibular ligament*. The bone fields, being disconnected from the mandible, are repositioned upward by differential growth, their course being guided by stylomandibular ligament. They are adsorbed or engulfed by the growth of r4–r5 mesenchyme and brought into contact with the otic capsule, within which are developing concomitantly, the vestibular and cochlear neurosensory organs. Stapes have never relinquished its relationship to the otic capsule, remaining attached to the oval window. Thus, the incus and malleus are dragged into and enveloped by second arch neural crest perioticmesenchyme which forms a chamber about them, the potential tympanic cavity. Tympanic ring, being angular origin, is the last link in this chain. It remains on the outside guarding the entrance to the tympanum. Just superficial to it a mesodermal mass, the auditory meatal plug, is programmed in precisely the same dimensions as the ectotympanic ring. Subsequent breakdown of the cord, followed by epithelial ingrowth results in the external auditory canal.

Between cynodont-type double jaw joint in which the standard reptilian joining between articular and palatoquadrate was combined with a demonstrated improved occlusion (more on this later with the mandible) and the subject is best relegated to our discussion of the mandible but some comments relating to the middle ear are important.

We should not here the relationship between the stapedial artery and these ear bones. Recall that the initial two branches of the proximal IMMA both connect with tympanic cavity, deep auricular artery and anterior tympanic artery. They likely represent the pathway by which the extracranial divi-sion of stapedial departed from tympanic cavity to access the "maxillary" branch of external carotid in *the* infratemporal fossa. It is not surprising what these two arteries supply. *Deep auricular artery* is directed to TMJ, the lining of external auditory meatus and the outer lining of tympanic membrane, all representing first arch neural crest structures. *Anterior tympanic artery* supplies the inner surface of tympanic membrane, once again, r3 neural crest. It forms and anastomotic arcade around TM, connecting with stylomastoid branch of posterior auricular artery representing the second arch. Note that anterior tympanic also anastomoses with caroticotympanic branch of internal carotid; in so doing, *it recapitulates the anatomic course of the embryonic stapedial stem*.

The Middle Ear and the TMJ: Functional Significance

After a number of synapsid experiments lead to evolutionary dead ends the Therapsid line emerged. These had an external acoustic meatus and a jaw joint in line with the back of the skull, exactly the location of stapes. The tympanic bone fields began to separate away from the mandible as a distinct *postdentary rod* (PDR). *Symplectic remained connected with quadrate.* Just at the split between monotremes (platypus) and the therians leading to marsupials and mammals, the bones broke into units held together with ligaments. In therians, the ear bone complex became a bridge of ligaments and cartilage between the dentary and the future ear cavity.

This process involved experimentation with different forms of skulls. In the earliest amniotes, such as *Paleothryis*, the anapsid skull was smooth with no new insertion sites. The line quickly demonstrated experimentation with two new forms, both of which placed muscles *outside* the skull, permitting subsequent brain expansion and more efficient degrees of efficiency for foot processing. The *diapsid* skull, seen in reptiles and bird, has two fenestrae. The synapsid condition, unique to mammals, has a single lower temporal fenestra surrounded by postorbital, squamosal and jugal. Recall that, in humans, postorbital is the upper process of zygoma and jugal is the anterior zygomatic arch. Recall further that posterior zygomatic arch is a forward extension of squamosal. A lower one is the same as in mammals whereas an upper fenestra is bounded by postorbital, squamosal, and parietal (Fig. 8.131). As we shall see, the *synapsid condition is a prerequisite for the development of the TMJ*.

It should be obvious by now that jaw and ear development are integrated processes. Much of this story has to do with development of more sophisticated means of food gathering and processing. **In evolution, it's not that you become *what* you eat…you become *how* you eat**. So, in the ensuing section on the mandible, we will discuss the developments in the musculoskeletal system created new and more efficient means of food acquisition and processing…and that these innovations had selective advantage. In concluding this sec-

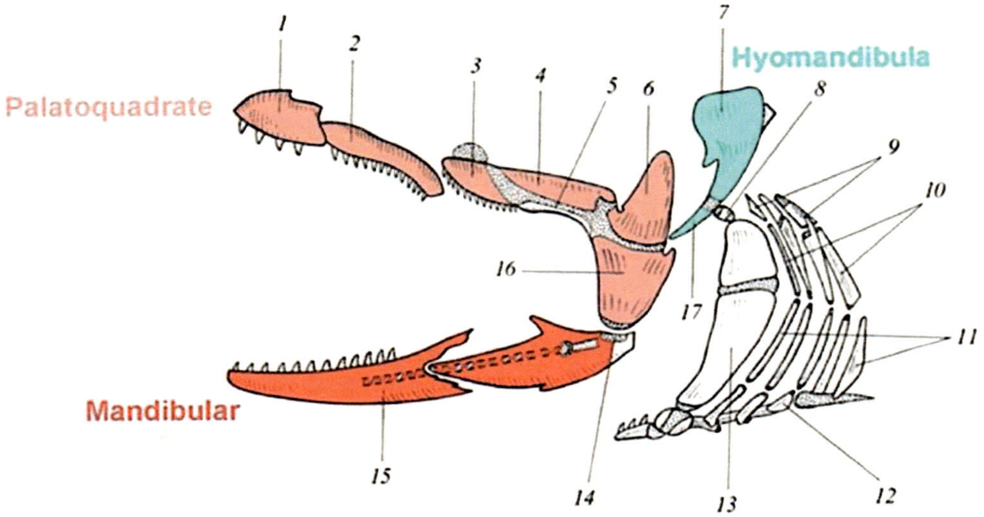

Fig. 8.131 Branchial arch cartilages in basal gnathostome fish. The original agnathic fish model has seven branchial arches. This became modified twice: initially, in placoderms, bon conversion of the gill structures of first branchial arch into jaws; and later, in derived fishes, by the "invention" of the hyomandibula from second branchial arch to support the jaws via direct contact with quadrate. Note symplecticum (blue) directly opposite interhyal and connected to quadrate (incus). This represents the future neural crest component of stapes. 1, premax- illa; 2, maxilla; 3, palatine; 4, entopterygoid (palatine series); 5, ectopterygoid (last palatine); 6, Metapterygoid; 7, Hyomandibula; 8, interhyal; 9, pharyngobranchials; 10, epibranchials; 11, ceratobran- chial; 12, hypobranchials/basibranchials; 13, ceratohyal; 14, tympanic part of Meckel's cartilage (articular = malleus); 15, dentary part of Meckel's; 16, quadrate (incus); 17, symplecticum (stapes). (Reprinted from Goodrich ES. Studies on the Structure and Development of Vertebrates. Macmillan; 1930)

tion on ear development, we shall see that forward position- ing of the masticatory complex created spatial conditions for ear bone migration. And, in return, increased auditory acuity proved to be a significant advance for mammals.

The External Ear (Figs. 8.132 and 8.133)

Mesenchyme and Blood Supply

Pharyngeal arch lamination is an extremely important con- cept. Recall that the first and second pharyngeal arches are anatomically and functionally fused. Let us explore this from a vascular, a neurologic, and a fascial standpoint. The evolu- tion of jaws made first and second pharyngeal arches co- dependent. Their respective aortic arches are quickly lost; they survive by means of a vascular network that is eventu- ally shared with the third and fourth arches: the *ventral pha- ryngeal artery.* VPA becomes the proximal (non-stapedial) external carotid artery. Extracranial stapedial links up with VPA at V3. Thus, ECA distal to the external maxillo- mandibular (facial) artery is the StV2–StV3 internal maxillo- mandibular artery. In sum, the bone and soft tissues of the first and second arches share a common blood supply.

Facial dermis comes almost exclusively from r1 to r3 neu- ral crest. Rhombomeres r4 and r5 have very little cutaneous representation, save some innervation to the posterior exter- nal auditory canal. Muscles of facial animation migrate widely, finding their secondary insertion sites subjacent to dermis supplied by V1–V3 and C2–C4. Because the devel-

opment of prosencephalic skin and PA1 skin *predates* inva- sion by PA2 myoblasts, the muscles of facial expression always respect the neurovascular bundles from V2 to V3. These muscles are interlopers. They exploit a potential space between the *superficial* (subcutaneous) *fascia* (SF) and either the *deep investing fascia* (DIF) containing either muscles of mastication or facial bones programmed by r1–r3. Sm6 facial muscles have a common primary insertion in a distinct structure, the *superficial investing fascia* (SIF). Thus, SIF is sandwiched between SF and DIF. From the SIF, facial muscles seek out secondary insertions either superficially into the overlying dermis or deep into the periosteum of maxilla and mandible.

When we study embryos during in the so-called branchial phase dramatic differences are seen in the physical appear- ance of mesenchyme, depending upon whether it is epaxial or hypaxial. Epaxial PAM and neural crest are confluent (even though genetically based neuromeric zones are pres- ent). Hypaxial extensions of neural crest create five succes- sive pharyngeal arches. These are partially separated externally by ectodermal *pharyngeal grooves* and internally by *pharyngeal pouches.* Thus, the vertical first pharyngeal groove and pouch lie between PA1 and PA2. Note that dor- sally, these separations terminate at the level of the neuraxis, i.e., pharyngeal "clefts" are strictly hypaxial creatures. *As the second pharyngeal arch undergoes its diaspora forward into first arch and backward into third arch backward, PA2 loses virtually all its ectodermal and endodermal representation.*

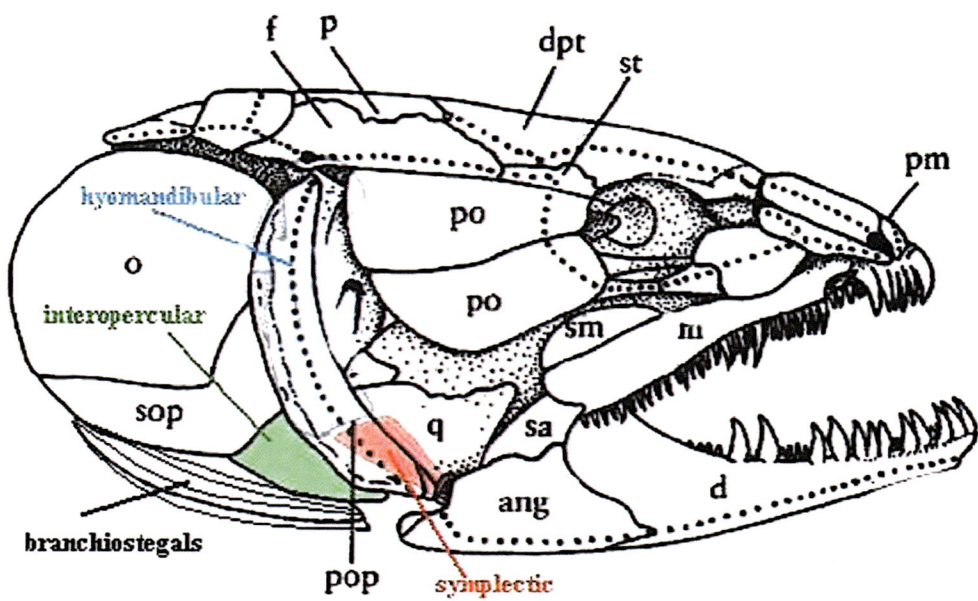

Fig. 8.132 Dermal bones of *Amia*. *Amia* showing **symplectic** (red) connecting hyomandibula with the first arch quadrate. The articulation of the jaw is quadrate-articular, which is partially hidden by angular. Preopercular (removed) covers over most of hyomandibula and symplectic. Note: *POP* preopercular, *IO* interopercular, *SOP* subopercular, *O* opercular. (Reprinted from Grande L, Bemis W. A comprehensive phylogenetic study of Amiid fishes based on comparative skeletal anatomy: an empirical search for interconnected patterns of natural history. J Vert Paleontology 1998; 18(suppl):1–681. With permission from Taylor & Francis)

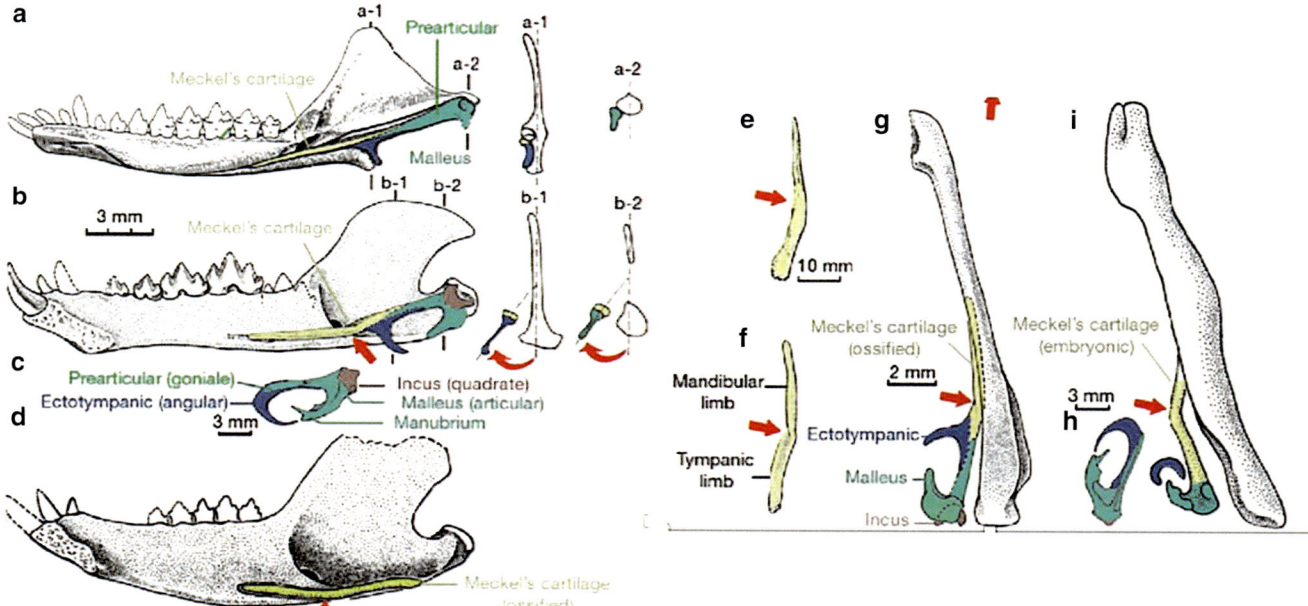

Fig. 8.133 Tympanic part of Meckel's cartilage (yellow) has distinct bone fields. Angular bone (blue) give off a reflected lamina which forms, the body of tympanic ring, the *ectotympanic*. Prearticular bone (green) forms anterior process of malleus. Articular bone (green) forms body of malleus. The retroarticular process of articular (green) forms the manubrium of mallius. Quadrate bone (tan) arises from posterior palatoquadrate cartilage and forms the incus. (Reprinted from Luo Z-X, Li G, Chen P, Chen M. A new tricodont mammal and evolutionary development in early mammals. Nature 2007; 446(7133):288–293. With permission from Springer Nature)

This "sandwich phenomenon" is seen in the lamination structures between Eustachian tube, tympanic cavity, and external auditory canal. Intraoral mucosa from r4 to r5 is absent in both the tube and the bony cavity. The petrosal bone complex represents r4–r5 prootic docking around the otocyst with final coverage by r6–r7 opisthotic. Note that the real "business end" of the inner ear is all in prootic territory. Cranial nerve IX has no derivatives there. Finally, external auditory canal is a straight-forward interface between first and second arch derivatives. But because epithelial derivative of r4–r5 have no sensory representation, pain in the ear canal is V3 auriculotemporal with involvement of IX internally.

The mesenchyme of the external ear is complex: skin cover and muscles come from three completely different sources with a blood supply specific to each: first arch skin and cartilage, second arch cartilage and muscles, and skin from cervical somites two and three.

First arch skin covers the anterior three cartilage hillocks, the anterior external auditory canal and the outer surface of tympanic membrane. It supplies the posterior *Anterior auricular artery* arises four branches from superficial temporal supplies the skin. It supplies anterior auricular skin, the tragus, and anterior external acoustic meatus. It also supplies anterior auricularis.

Second arch makes *almost no contribution to ear skin*, saving posterior external auditory canal. Sm6 does supply the three extrinsic auricular muscles and six intrinsic muscles. Of the three extrinsic muscles (anterior auricular, posterior auricular, and superior auricular), the latter two are supplied by *posterior auricular artery* which winds around the posterior ear deep to auricularis posterior. It ramifies sending perforating branches through the cartilage to supply the anterior intrinsic muscles. Motor supply of the second arch pinna is bilaminar. Two extrinsics, auricularis anterior, and auricularis superior plus all four of the anterior intrinsics are supplied by temporal branch of VII. Auricularis posterior and all the posterior intrinsics are innervated by posterior auricular branch of VII.

Auriculotemporal of V3 supplies the anterior skin including tragus and a small anterior zone external skin. The skin of the remainder of the auricle is supplied as follows: *greater auricular* from C2 to C3 supplies the lower internal (cranial) surface and the back half of external surface. *Lesser occipital* C2 has a more lateral distribution. It covers the upper cranial surface. *Auricular vagus* innervates concha.

The innervation tells the embryologic story. When second arch hillocks are laid down, they are deposited at the boundary of cervical somite skin. The hillocks expand posteriorly underneath this skin cover. This explains the sensory innervation from C2 and C3. However, the rapid invasion of second arch SMAS muscles beneath the surface quickly brings in blood supply from the external carotid system which takes over the area.

Developmental Anatomy

The Osseous Platform (a Quick Review)

Traditional descriptions of the adult temporal list five components: squama, petrous, tympanic, mastoid, and styloid process. The latter two develop postnatally. Squama and tympanic are membranous (neural crest) while the others are chondral. Petrous remains cartilaginous until the 20th week when its 14 ossification centers begin a sequential process of bone formation. The eighth week is marks the bidirectional development of squama forward into zygoma and backward over the skull. The ninth week demonstrates the zygomatic process. The middle ear appears at this time as a projection of otic capsule, the *periotic process*. This shelf extends itself over the ossicles, becoming *lateral tegmen tympani*. Medial segment tympani remains fibrous. Tympanic bone appears between the ninth and tenth week.

Sixteenth week the *postauditory process of squama* extends backwards, posterior to tympanic ring. It flows superficial to underlying bone to produce the smooth *anterosuperior mastoid* bone. During the 20th–24th weeks, the cartilaginous otic capsule begins to calcify from multiple centers. Inner ear (the labyrinth) and all parts of middle ear (tympanic cavity) except the floor are now adult size. Because mastoid is pneumatized it continues to grow. The 25th week marks the completion of middle ear; the floor is either an independent bone or a bony projection of the petrous pyramid. In the 29th week, tympanic process of squama joins up with the antral part of periosteal otic capsule. The result is lateral wall of antrum. At birth, an ossification center located proximal (dorsal) in Reichert's cartilage fuses with otic capsule. The resulting bony lamina provides anterior coverage for the distal facial nerve canal. It produces a visible swelling within the floor of the tympanic cavity, the *styloid eminence*.

The mastoid antrum at birth is large and thin-walled. Recall that it has received external coverage from the postauditory squama. Further growth medializes both the antrum as well as the facial nerve. Reichert's cartilage also completes growth in the postnatal period. Its proximal segment becomes styloid process. Its distal segment produces lesser cornu and upper body of the hyoid bone.

The Pinnae

At fifth and sixth weeks, the periotic mesenchyme demonstrates condensations into cartilaginous mounds, the *hillocks of His*. Pre-otic hillocks 1–3 arise from first arch r3 neural crest; they are supplied by V3 auriculotemporal nerve. Post-otic cartilages 4–6 are more controversial. The classical view postulates a second arch origin; they arise from r7 neural crest. They receive from the rostral side a small cutaneous branch of VII. From the caudal side, they are supplied by cervical nerves. Greater auricular (ventral C2–C3) supplies the more ventral posterior pinna while lesser occipital (C2) is directed dorsally. This view fails to account for the innervation of posterior EAC skin by IX via X (nerve of Arnold). Levine (332) depicts a 6-week embryo with a second pharyngeal cleft bearing two hillocks as well. The latter structure directly abuts cervical territory posteriorly. Recall that arches 4–5 lie internal to PA3 (the "Japanese fan" model). Fusion of the second cleft will literally "drag" cervical innervation forward onto the posterior aspect of the ear. It would also account for the sensory presence of glossopharyngeal nerve to posterior EAC dermis.

Accessory Structures: Auditory Canal, Tympanic Membrane and Tympanic Ring

As soon as first and second pharyngeal arches have formed,[10–11] the future EAC appears as a funnel-like invagination in the dorsal aspect of first pharyngeal cleft. Between the fourth and fifth weeks, the ectoderm of the PA1–PA2 cleft approximates the endoderm of the PA1–PA2 pouch. Recall that, just ventral to this site, melding of the second arch creates a fusion between the first and second pouches. This will produce the eustachian tube. The sixth week marks the invasion of mesoderm between the epithelial laminae. At the eighth week, mesenchymal growth on either side deepens the tunnel creating the fibrocartilaginous EAC.

In the ninth week, a mass of ectodermal cells, the meatal plug, extends into the lower wall of the middle ear cavity. The epithelial cells (endoderm) of the tubotympanic recess cavity will become the internal lamina of TM. Between these two epithelial layers, interposed mesoderm becomes the *fibrous layer of lamina propia*. Within this plane, blood vessel form to supply the TM. Four ossification centers in the tympanic membrane coalesce at the 12th week; the bony tympanic ring is complete at the 16th week. An open area at the superior rim, the notch of Rivinus, permits direct attachment of TM to petrous bone. At all other levels, TM is inserted into a sulcus in the tympanic ring.

At the 21st week, the previously solid meatal plate begins to cavitate. This process starts deep and works its way externally. Disintegration of the ectodermal plate at the 28th week results in the formation of a canal. Remnants of the ectoderm plate form the inner bony segment of IAC and the superficial layer of tympanic membrane. Fixation between the tympanic ring and the otic capsule takes place at the 34th week. This starts posteriorly and works its way anteriorly. It is not complete until birth. The genetic "drive" for this may reside more in PA2.

The walls of the outer EAC are cartilaginous. They are not derived from internal sources; rather, they represent an inward extension of the auricular cartilage. Communications between EAC and parotid gland via the cartilage take place

as fissures of Santorini. Horizontal plate of temporal squama contributes the superior wall. Tympanic ring gives rise to the floor, the mandibular fossa, and the styloid sheath.

Coda

We have come to the end of a long and complex section. Let us recap the main points. The assembly of the temporal bone complex demonstrates how individual mesenchymal fields are programmed to receive and shelter invaluable neurosensory structures critical for survival. We have endeavored to reconstruct how these units are put together; each foramen, every space providing clues about their geometry. The neuroanatomy shows us how complex controls arising from rhombomeres 4 to 7 are relayed throughout the head and neck. Nerves are dedicated to tissue units based on neuromeric relationships. Nerves and arteries co-exist as mutually programming units. Bone fields form around them; hence, neurovascular axes are a way of them mapping out. Temporal complex is primeval, as old as mobile vertebrate life itself, but the anatomy we see today make sense in the context of its evolutionary history.

Part 5: Neurocranium (r8–r11)

The Occipital Complex

Introduction: A Thrice-Told Tale

This occipital bone complex of the posterior cranial base is anything but boring. It differs in mesenchymal composition and mechanism of development from the rest of the skull and demonstrates dramatic changes with evolution. In fishes, the occipital results from the lamination of three occipital somites that produces two pairs of chondral bone fields: basicoccipital and exoccipital. Non-amniote tetrapods add a fourth occipital somite. S4 breaks apart and its rostral fields join the cranial base via a process of parasegmentation. The loss of extrascapulars in early tetrapods amniotes brings the postparietal bone fields backward to form the dermal roof of the occiput. Amniotes present an entirely new pair of supraoccipital bone fields arising from non-somitic mesoderm. These are interposed between the postparietals and the exocipitals. Supraoccipital mesenchyme ossifies as chondral bone. In advanced synapsids the sharing of bone fields, coupled with the break-up of the occipital condyle result in a more mobile neck.

Together the bone fields of the occipital complex perform four basic functions.

- Posterior cranial fossa protects the hindbrain and cerebellum.

- Exit points for spinal cord (foramen magnum) and cranial nerves (foramina).
- Craniovertebral articulation between an occipital condyle (synapsids have paired condyles) and atlas.
- Structural support for the posterior skull, in which they interact with surrounding bone fields.

Achtung! The occiput is so important that it is discussed three times. We first encountered it in relation to the somitomeric prootic and opsithotic bone fields of the temporal complex. We now consider the braincase as a product of the occipital somites. We shall meet the occiput again in our discussion of the neck.

Anatomic Components

The basic structure of the occiput in anamniotes (which we briefly review under phylogeny) consists of dermal skull bones that roof over a foramen magnum this surrounded by a U-shaped complex of exoccipitals on the sides and basioccipitals below. Amniotes add additional mesenchyme to the situation and create an intervening zone of paraxial mesoderm, which will become the supraoccipital cartilage.

Since our primary interest (as hominid narcissists) is in our future ancestors we begin our discussion of occipital anatomy *in media res*, with Bob the Basic Amniote (Figs. 8.134, 8.135, and 8.136). The dermal skull roof consisting of supratemporals, parietals, and interparietals (not shown) sits like the crossbar over paired supraoccipitals. These are supported from below by exoccipitals which are braced laterally by opisthotics which in turn articulate with

dermal bones such as quadrate, squamosal, or pterygoid. The supraoccipitals and exoccipitals form the roof and sides of foramen magnum. Its floor comes from paired basioccipitals. In amniotes, these form the occipital condyle with contributions from the exoccipitals. On either side of the primitive single condyle are depressions to receive the articulation with the first cervical vertebrae. As we shall see, this initial arrangement inhibits motion. Synapsids resolve this problem by splitting the condyle into two parts, each with additional contributions from proatlas. Basioccipital is structurally critical because it shares weight-bearing with basisphenoid in front for the skull and with pterygoid below to receive stresses from the palate.

Let us consider Bob's braincase a bit further, from two perspectives: behind and lateral. Note that in basal amniotes, the anterior braincase remains cartilaginous.

- The supraoccipitals are chondral bone fields found only in tetrapods. *They do not develop from occipital somites.* Instead, they result when r8–r9 paraxial mesoderm is stretched across the back of the skull, like a suspension bridge between the opisthotic bones. They form *palanum nuchale*. Recall that *planum occipitale* arises from r6 to r7 neural crest. These concepts clarify what is otherwise a confusing situation. The supraoccipital complex is really two distinct embryologic parts: four dorsal dermal fields (interparietals and tabulars) and four ventral chondral fields, the "true" supraoccipital. Supraoccipital is supported from below by the exoccipitals.

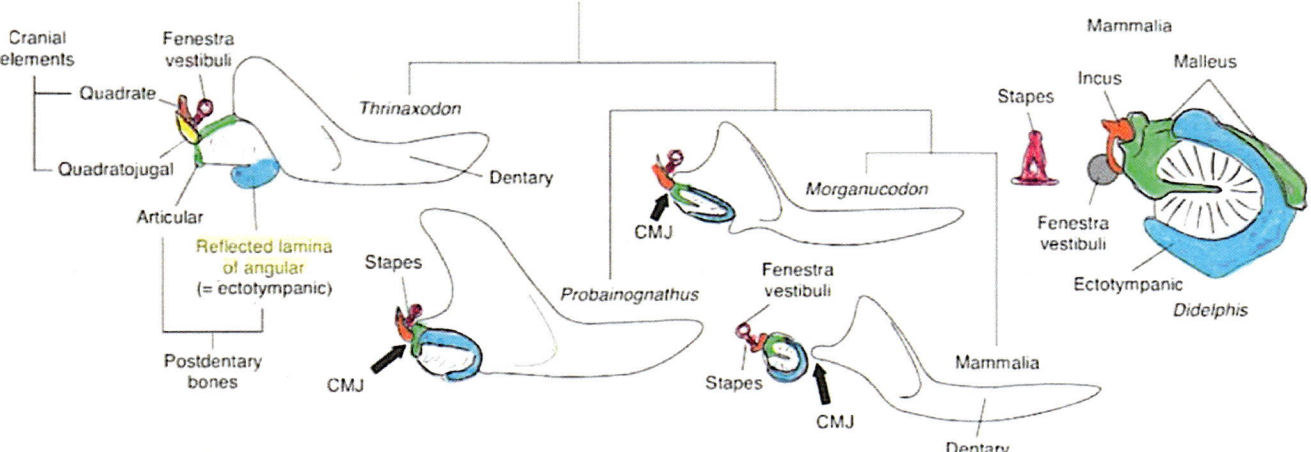

Fig. 8.134 Detachment of ear bones from Meckel's cartilage. *Procynosuchus* (not shown), the prototype cynodont, had a reflected lamina of the angular but maintained the reptilian jaw joint (articular-quadrate). *Thrinaxodon* experimented with first postdentary bones (but not true stapes): tympanic ring enclosing the tympanic membrane, quadrate and quadratojugal remained connected. *Probainognathus* and *Morganucodon* both had two jaw joints: the old reptilian and a new

dentary-squamosal. But appearance of a defined stapes. *Probaingnathus* lacks a manubrium (green) of malleus; this appears in *Morganucodon* and is transmitted to mammals. In the mammalian condition, the bones are freed from the back of mandible. (Reprinted from Benton MJ. Vertebrate Paleontology, 4th ed. Blackwell; 2015. With permission from John Wiley & Sons)

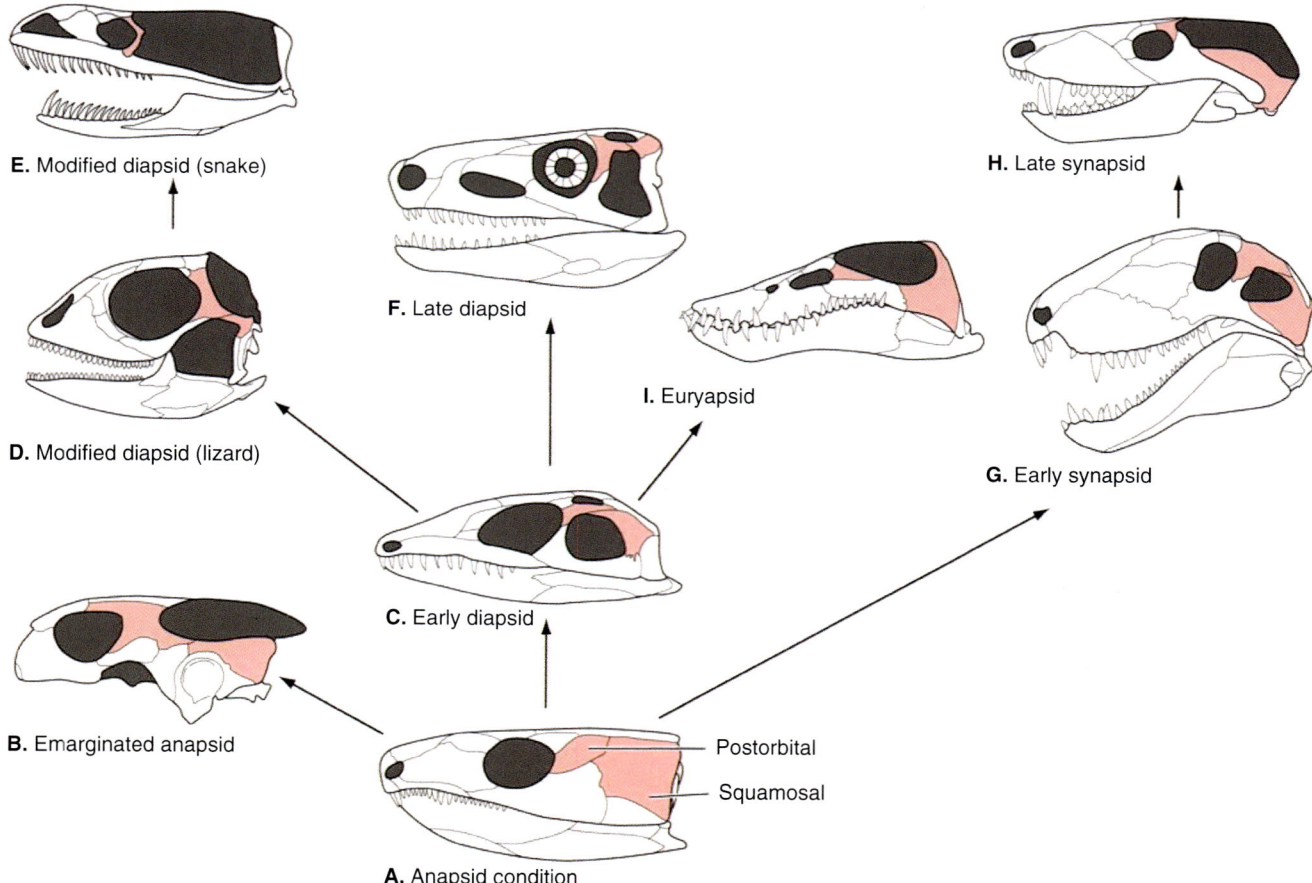

Fig. 8.135 Phylogeny of amniote skulls beginning with reptilomorphs. <u>Anapsid condition</u>: the skull of basal amniote *Paleothyris* was similar to the ancestral anamniote tetrapod *Paleoherpeton*, retaining the parietal foramen for median eye, but it lacked an otic notch. The occipital condyle is single. From fenestra ovalis, a large columella buttresses against quadrate. No tympanic membrane. Bones shift around depending upon forces exerted on them. <u>Diapsid condition</u>: basal condition in *Sphenodon* with subsequent loss of dermal bones in the temporal region. Tympanic

ear develops with a slender columella but disappears in snakes. Birds are also diapsids. They arise from the therapod group of dinosaurs. <u>Synapsid condition</u>: single temporal fenestra initially defined by postorbital and squamosal bones. In late synapsids postorbital incorporates into jugal to form zygoma leaving postfrontal and parietal as boundaries of the fenestra. In mammals, postfrontal is incorporated into orbital rim. Fenestra communication with orbit blocked by alisphenoid. (Courtesy of William E. Bemis)

- The <u>exoccipitals</u> develop from the neural arches of the occipital somites. These are present in all vertebrates from primitive fishes on. They contribute in basal amniotes a single occipital condyle.
- <u>Opisthotic bones</u> are the posterior part of the otic capsule and are fused anteriorly to the prootic bones. These endochondral bone fields were discussed in the temporal bone section. Jutting laterally outward from the opisthotics are <u>paraoccipital processes</u>. They attach to the r2 dermal quadrate bones, the cranial half of the jaw joint and connected to squamosal on the side of the skull. Thus, the jaw joint in all amniotes except later synapsids relates directly to the otic capsule. In cynodont synapsids, it moves forward to become the TMJ.
- <u>Fenestra ovalis</u> is located anterior the paraoccipital process. It is a membrane between middle ear and inner ear to which is attached the stapes.

- <u>Stapes</u> in its original iteration was a supporting cartilage of the second arch in agnathans. In derived fishes, it became the hyomandibula. The basal tetrapod iteration of stapes was a strong support element between quadrate and the braincase. As paraoccipital process became more robust, stapes was reduced to a slender rod, the <u>columella</u>, the single ossicle of reptilian hearing. Stapes achieves its final form in mammals as part of the three-bone ossicular ear.
- <u>Occipital condyle</u> in amniotes is a ball-and-socket design formed from basioccipital, with help from exoccipital. It becomes double in cynodonts, acquiring the last vestigages of proatlas.
- <u>Basioccipital</u> is derived from paired central of all four occipital somites. It unites on either side with posterior otic capsule to make the floor of the posterior fossa. Projecting from below in basal amniotes but not mam-

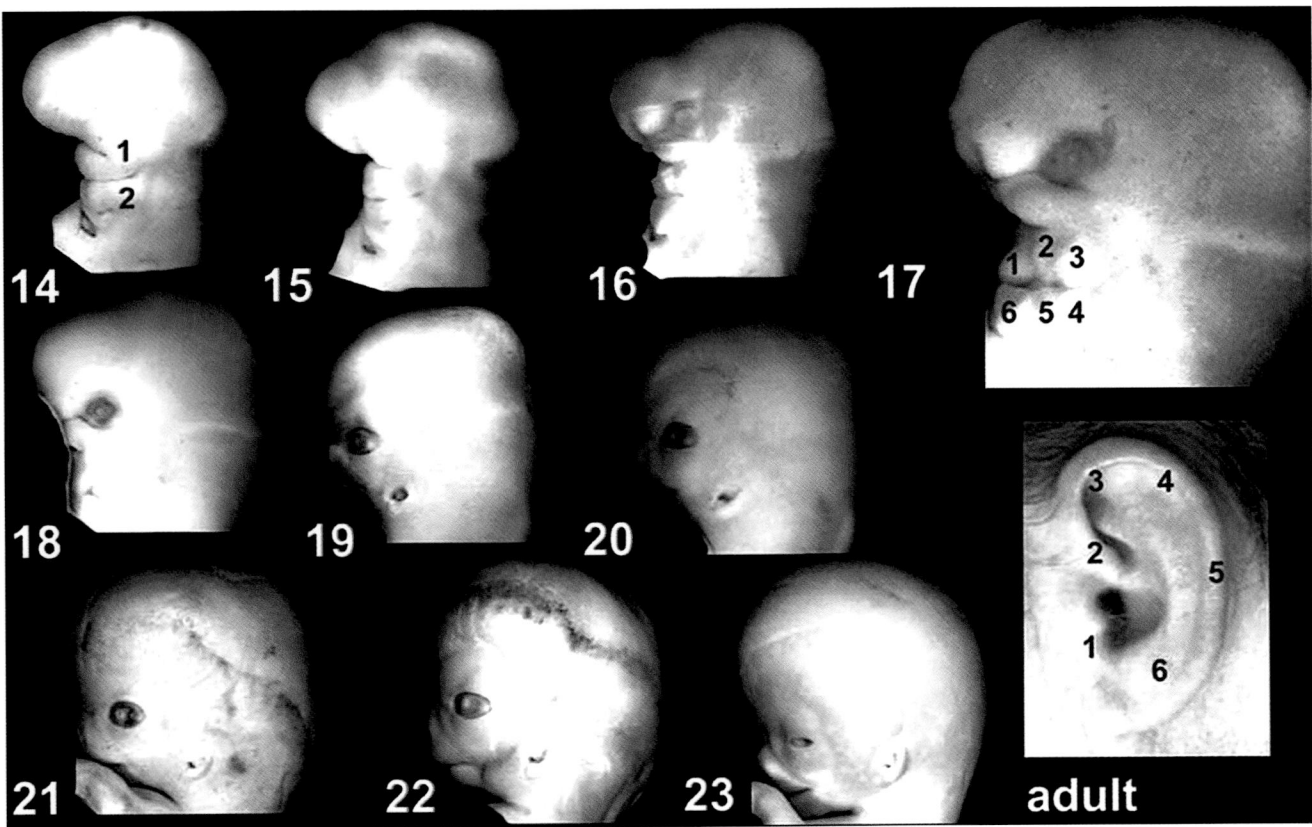

Fig. 8.136 External ear development, Carnegie stages. Hillocks 1 (tragus), 2 (helix), 3 (cymba concha), 4 (cavum concha), 5 (antihelix), 6 (antitragus). (Reprinted from Streeter GL. Development of the auricle in the human embryo. Contrib Embryol Carnegie Inst. 1922;14:111–138)

mals is <u>basioccipital tubers</u> that bear attachments for the ligaments between the skull and neck. Watch out! <u>Basipterygoid processes</u> (sic) are a misnomer. They belong to basisphenoid and attach it to the palate.

- <u>Pterygoid</u> is one of the dermal bones that line the inner surface of palatoquadrate.
- <u>Basisphenoid</u> is covered on its palatal surface by parasphenoid. It gives rise to the poorly named "basipterygoid" processes bind to the palate.
- <u>Parasphenoid</u> is welded to the bottom of basisphenoid. It extends forward in the midline of the palate. Olfactory tracts run along its upper (cranial) surface.
- <u>Epitpterygoid</u> (sic) is another mis-named bone field. Although it appears to arise from pterygoid, it is really an upward projection from the original palatoquadrate (pterygoid is just a dermal new-comer). Epitpterygoid will articulate with orbit.
- *Foramina* Every tetrapod has (1) *basioccipital foramina* for the hypoglossal nerve, (2) a *jugular foramen* (or analogous structure) between opisthotic (r6–r7) and the exoccipital (r8–r11), *fenestra ovalis* between oposthotic and prootic admits cranial nerve VII.

Descriptive Anatomy, Ossification Centers, and Bone Fields (Figs. 8.137, 8.138, 8.139, 8.140, 8.141, 8.142, 8.143, 8.144, 8.145, 8.146, 8.147, 8.148, and 8.149)

The ***basioccipital*** is a rectangle extending from the foramen magnum to the sphenoid. A cartilage plate, the *spheno-occipital synchondrosis*, separates the two bones at birth but becomes ossified by age 25. Along its external (inferior) surface about 1 cm in front of the foramen magnum is the *pharyngeal tubercle*. Attached to it, anterior to foramen magnum, is the anterior atlanto-occipital membrane. Its lateral aspects bear the secondary insertions of rectus capitus anterior and longus capitus.

The ***exoccipital*** sits on either side of foramen magnum. From its external surface, paired occipital condyles project downward. These will articulate with the superior facets of the atlas. Along the medial side of each condyle is a tubercle that bears the alar ligament. The posterior margins lead to the *condyloid canal*; the anterior margin bears the hypoglossal canal through which a communication between the ascending pharyngeal artery (the axial vessel of the third arch) and the intracranial stapedial system. Projecting laterally from

Fig. 8.137 External ear development, hillock formation. (Reprinted from Wood-Jones F, I-Chen W. Development of the external ear. J Anat 1934; 68(4):525–533)

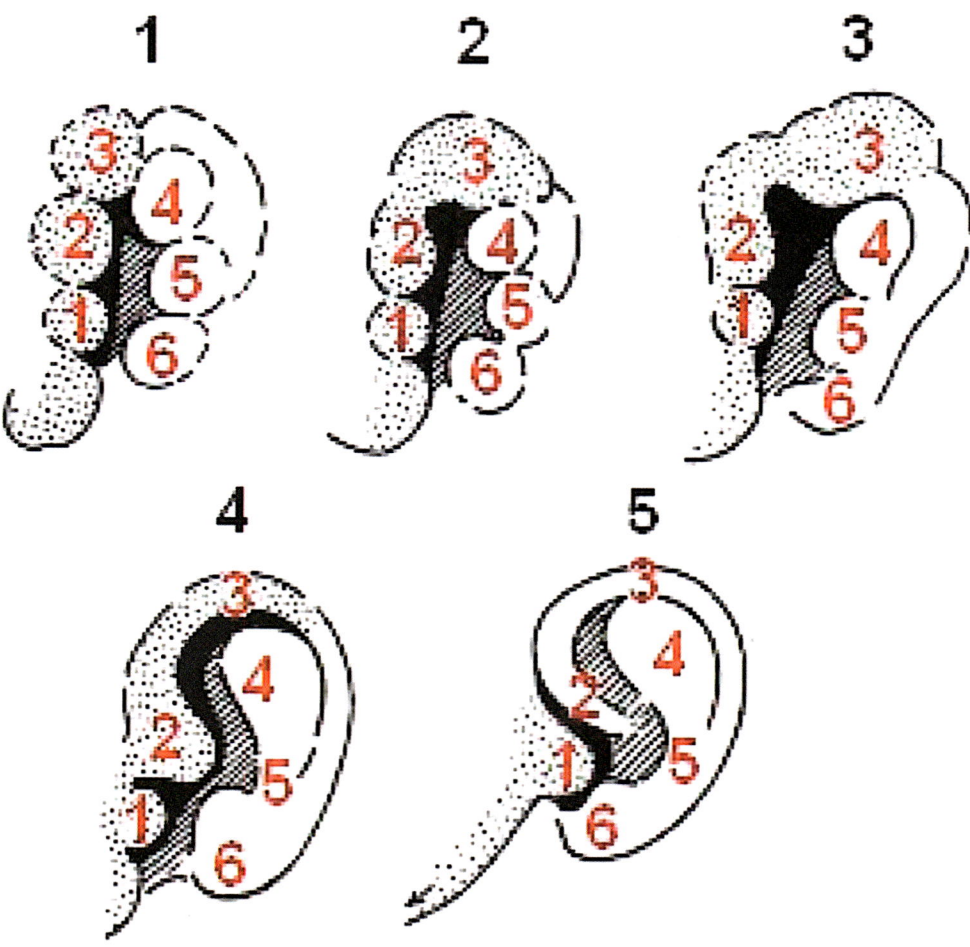

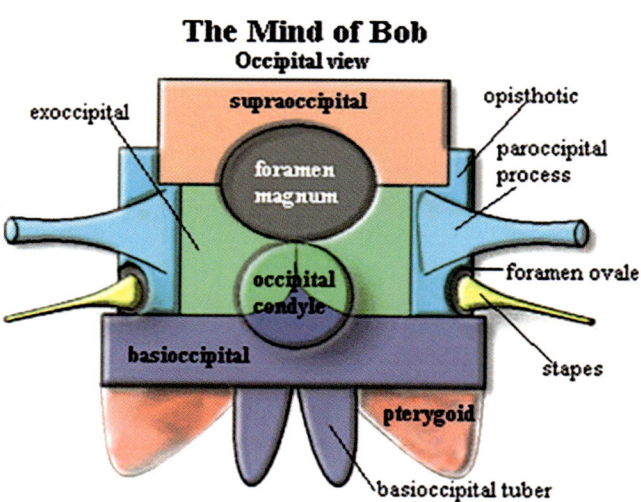

Fig. 8.138 Cranial base osteology, occipital, in Bob the Basic Amniote. The dermal skull roof of anamniotes is now physically separated from foramen magnum by the interposition of non-somitic PAM supraoccipital fields straddling between the two opisthotic bones. (Courtesy of Augustus T. White, Palaeos.com. Retrieved from: http:// palaeos.com/vertebrates/bones/braincase/overview. html#Dermosphenotic)

each condyle is a quadrangular bony shelf, the *jugular process*. This bears the secondary attachment of rectus capitus lateralis and lateral atlanto-occipital ligament. The anterior surface of the jugular process bears an indentation, the *jugular notch*. The internal surface of pars lateralis is distinguished by the *jugular tubercle*, a marker for the hypoglossal canal. It is grooved by cranial nerves IX, X, and XI. Sigmoid sinus leaves an impression as well.

The ***supraoccipital*** is a misnomer. It is really a bone complex consisting of two mesenchymal populations. As we consider Gray's anatomy, we immediately run into the problems with the neural crest interparietal bones fields previously mentioned. We shall refer to these using the term *planum occipitale*. The lower zone is derived from somitomeric mesoderm; it forms via chondral ossification; it is referred to as *planum nuchale*.

Note: Fifth occipital somite has an unstable dermatome: although it forms, it quickly involutes. Therefore, we find no C1 nerve fibers in occipital scalp. Fifth somite produces a myotome but it is strictly hypaxial, supplying muscles anterior to the spine as flexors for the skull base.

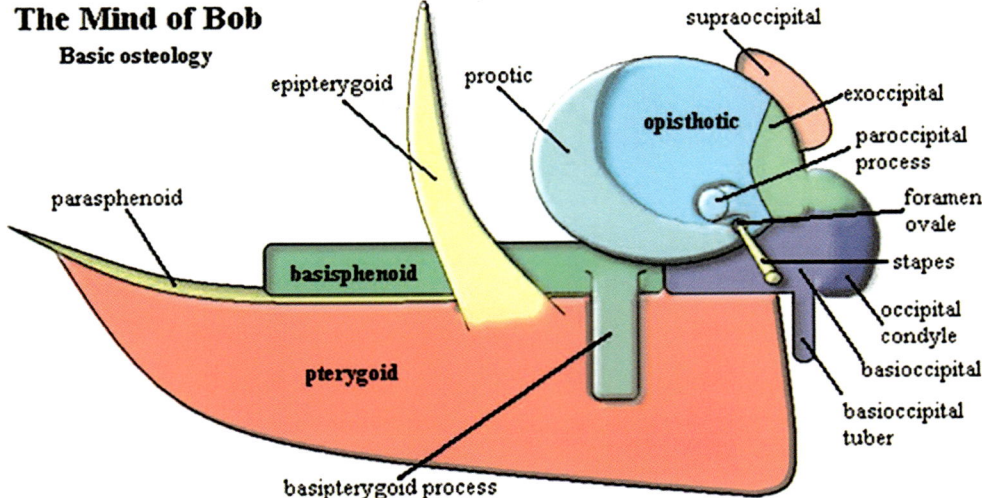

Fig. 8.139 Cranial base osteology, lateral. Suproccipital (light pink) is non-somitic PAM from levels r8 to r9. Exoccipital (light green) sits atop basioccipital (purple). Otic capsule consists of posterior opisthotic (blue) and anterior prootic (blue-green). Opisthotic projects paraoccipital process (blue) and the foramen ovale which bears stapes (yellow). Basisphenoid (green) is covered on its ventral surface by parasphenoid (yellow-green). Pterygoid (red) flares off to the side but does not arise with cranial base. In reality, it is a dermal bone lining the inner surface of palatoquadrate. Basipterygoid process (sic) has nothing to do with pterygoid. It originates from basisphenoid (therefore it is colored green) and projects downward to secure the palate to the braincase. Eppterygoid (lemon) arises from pterygoid and extends upward to the orbit. It will become in mammals alisphenoid. (Courtesy of Augustus T. White, Palaeos.com. Retrieved from: http://palaeos.com/vertebrates/bones/braincase/overview.html#Dermosphenotic)

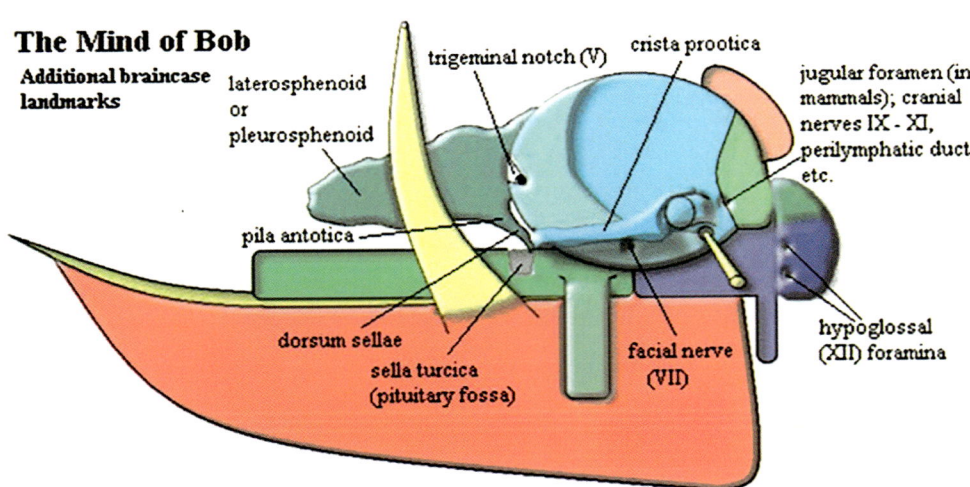

Fig. 8.140 Cranial base osteology, lateral, with additional landmarks. *Jugular foramen* admits IX and X. Remember that XI is really the r10–r11 roots of X, the *palato-pharyngo-laryngeal nerve*. Hypoglossal foramina are two in fishes and three in tetrapods (three occipital somites vs. four). Facial nerve VII excapes at the interface between prootic and opisthotic. Trigeminal V passes through primitive trigeminal notch between prootic and pleurosphenoid (the future lesser wing of sphenoid). Pila antoica connects orbitosphenoid with basisphenoid just behind the sella turcica. (Courtesy of Augustus T. White, Palaeos.com. Retrieved from: http://palaeos.com/vertebrates/bones/braincase/overview.html#Dermosphenotic)

Fig. 8.141 The occipital bone complex in situ. Complex is shown in toto. Basioccipital and exoccipital separated by synchondrosis. Exoccipital and supraoccipital separated by synchondrosis, the inferior horizontal suture or line (I), seen here as a thick ridge separating cerebellar fossa from condylar zone. Exoccipital contains jugular foramen. (Reprinted from O'Rahilly R, Müller F. Human Embryology and Teratology, 3rd ed. Wiley-Liss, 1981. With permission from John Wiley & Sons)

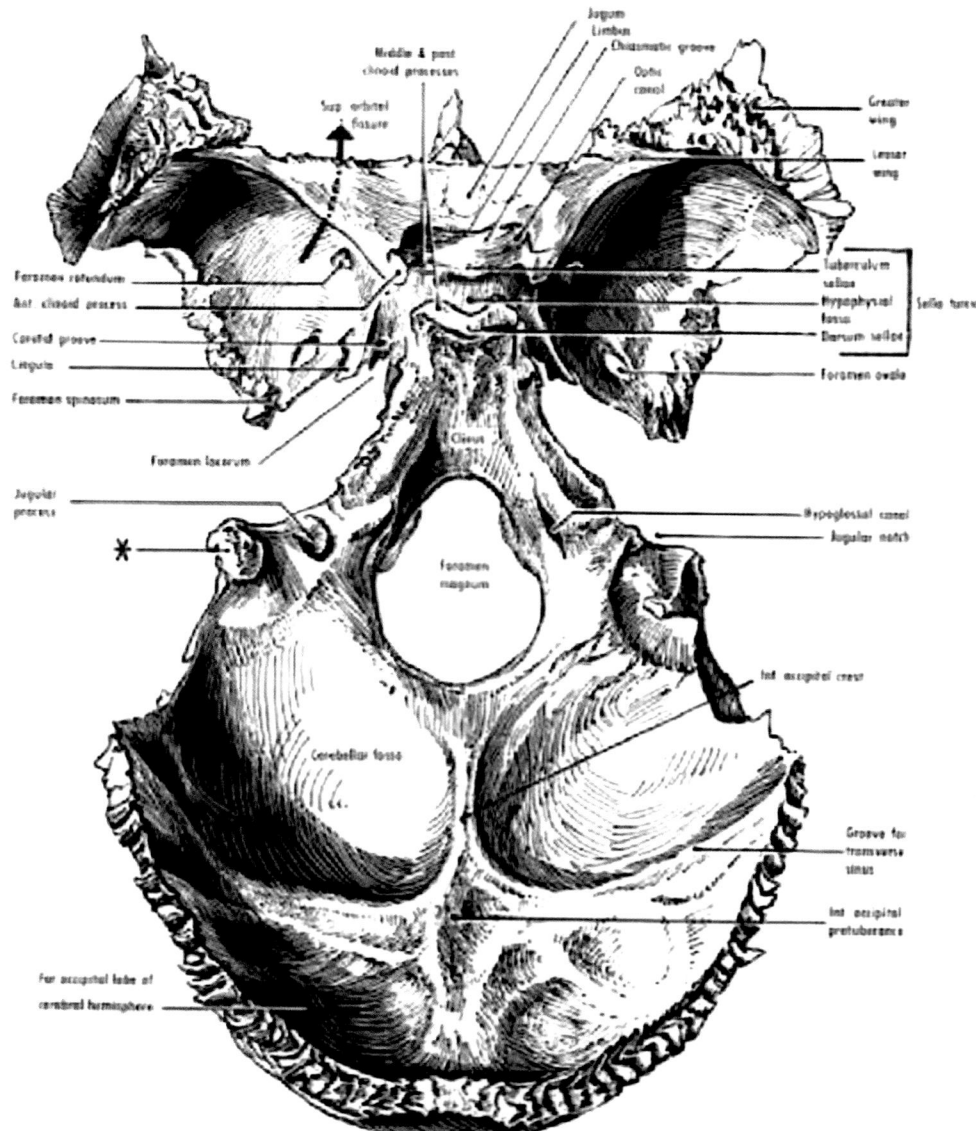

Fig. 8.142 Exoccipital and basioccipital bone fields CT. Proximal boundaries of exoccipital are readily seen internally and these correspond externally to the inferior nuchal line. Left: Basioccipital 3D. Right: Exoccipital 3D. [Left: (Reprinted from Wikimedia. Retrieved from: https://commons.wikimedia.org/wiki/File:Basilar_part_of_occipital_bone14.png. With permission from Creative Commons License 2.1: https://creativecommons.org/licenses/by-sa/2.1/jp/deed.en). Right: (Reprinted from Wikimedia. Retrieved from: https://commons.wikimedia.org/wiki/File:Lateral_parts_of_occipital_bone08.png. With permission from Creative Commons License 2.1: https://creativecommons.org/licenses/by-sa/2.1/jp/deed.en)]

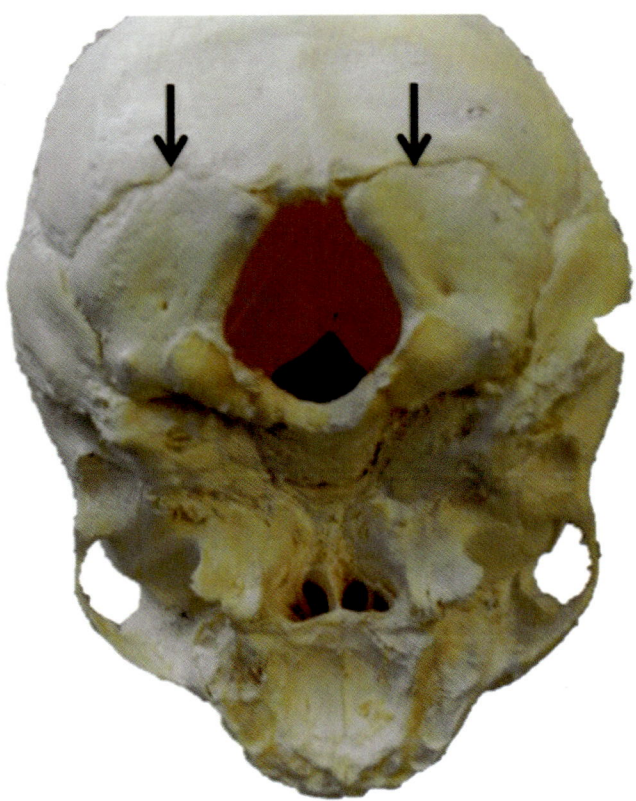

Fig. 8.143 Synchondroses of exterior occiput: (1) basisphenoid and basioccipital, (2) basioccipital and exoccipital, (3) exoccipital and supraoccipital, i.e., *planum nuchale*. Inferior nuchal lines are the boundary between exoccipital and supraoccipital. (Reprinted from Tubbs RS, Bosmia AN, Cohen-Gadol AA: The human calvarium: a review of embryology, anatomy, pathology, and molecular development. *Childs Nerv Syst* 2012; 28:23–31. With permission from Springer Nature)

The chondral *planum nuchale* is monolamellar. In some zones, such as the cerebellar-supporting inferior fossae, the bone is quite thin, nearly translucent. It lacks a diploic space. At other sites, such as the occipital condyles and the pars basilaris, the bone is thick. As we shall see, four occipital somites laminate together to produce all parts of planum nuchale. As the axial skeleton transitions from cranium to cervical the manner in which individual somites merge together to produce the bony elements changes from a segmental to a parasegmental pattern. Evolutionary changes involving somites 4–7 help explain the configuration of paired occipital condyles and the atlas–axis complex.

Planum Occipitale

This zone is smooth. It lies between the lambdoid suture above and the *superior nuchal line*. The membranous planum occipitale is r3 neural crest derived from the dura covering the posterior cerebral hemisphere. It interacts two additional sources of mesenchyme, an overlying dermis from the second and third cervical somites (S6–S7) and intervening second arch superficial investing fascia. Under the dual influences planum occipital, behaving similar to other calvarial bones has two layers. Enclosed within the compact inner and outer tables is the diploic space filled with cancellous bone.

The muscle attached to planum occipitale is of somitomeric origin (Sm6), has its primary insertion in the SMAS, and its secondary insertion into the cranial margin of SNL. Running in parallel to SNL (and slightly higher) is the *highest nuchal line* to which galeal aponeurotic is attached. Note both HNL and SNL terminate in a midline knob-like

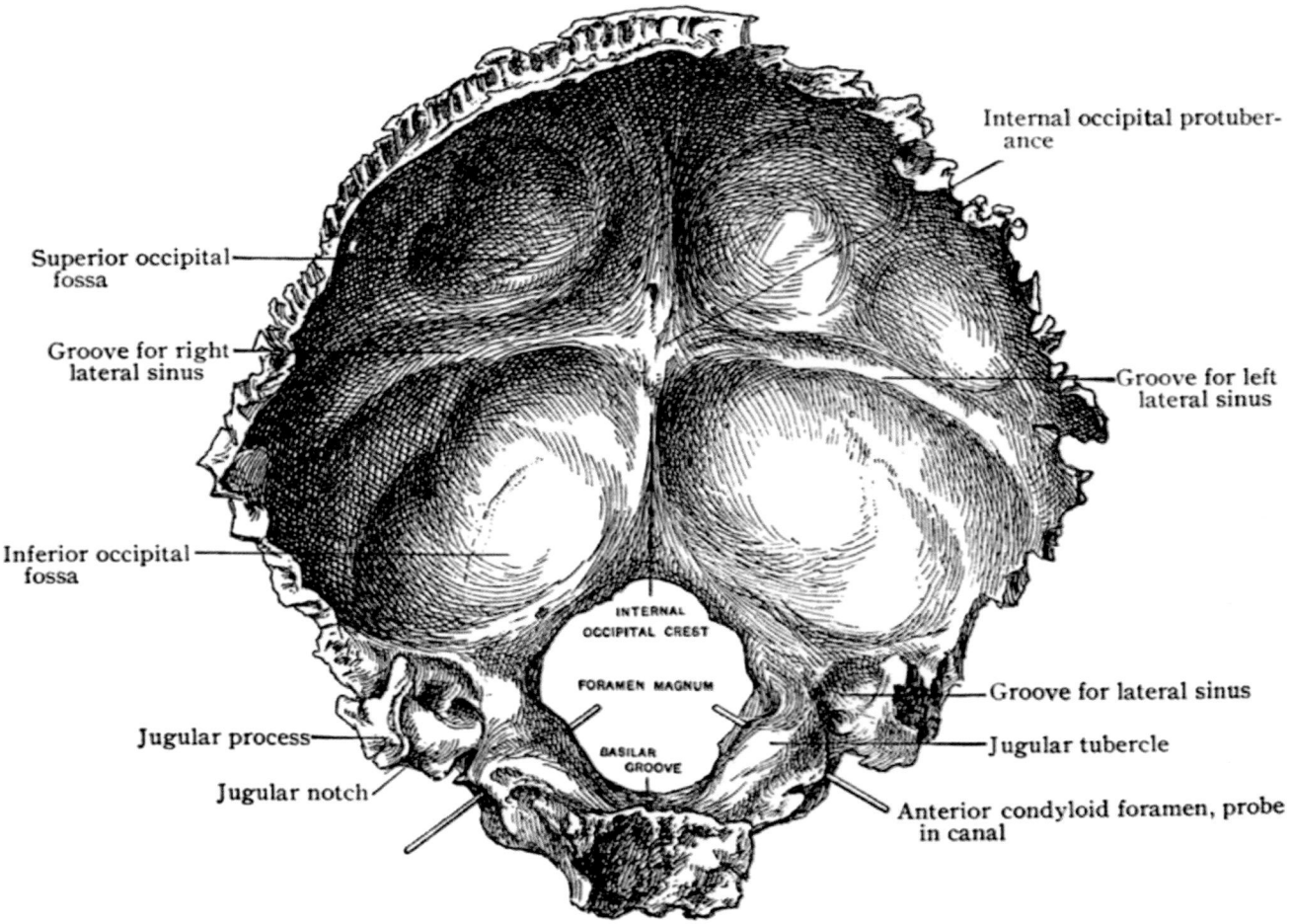

Fig. 8.144 Synchondroses of interior occiput demonstrates thickened ridge of inferior horizontal suture defining border of exoccipital with supra-occipital. Inferior nuchal line. (Reprinted from Sobotta J. Atlas and Textbook of Human Anatomy Philadelphia: WB Saunders; 1909)

Fig. 8.145 Occipital ossifications (redrawn). <u>Mendosal suture</u> = *superior nuchal line*: separates interparietal complex from supraorbital complex. <u>Superior horizontal suture</u> = *inferior nuchal line*: defines four centers including Kerckring's centers. <u>Inferior horizontal suture</u> defines exoccipitals—here seen as synchondrosis. <u>Vertical suture</u>: not shown. (Reprinted from Lewis, Warren H (ed). Gray's Anatomy of the Human Body, 20th American Edition. Philadelphia, PA: Lea & Febiger, 1918)

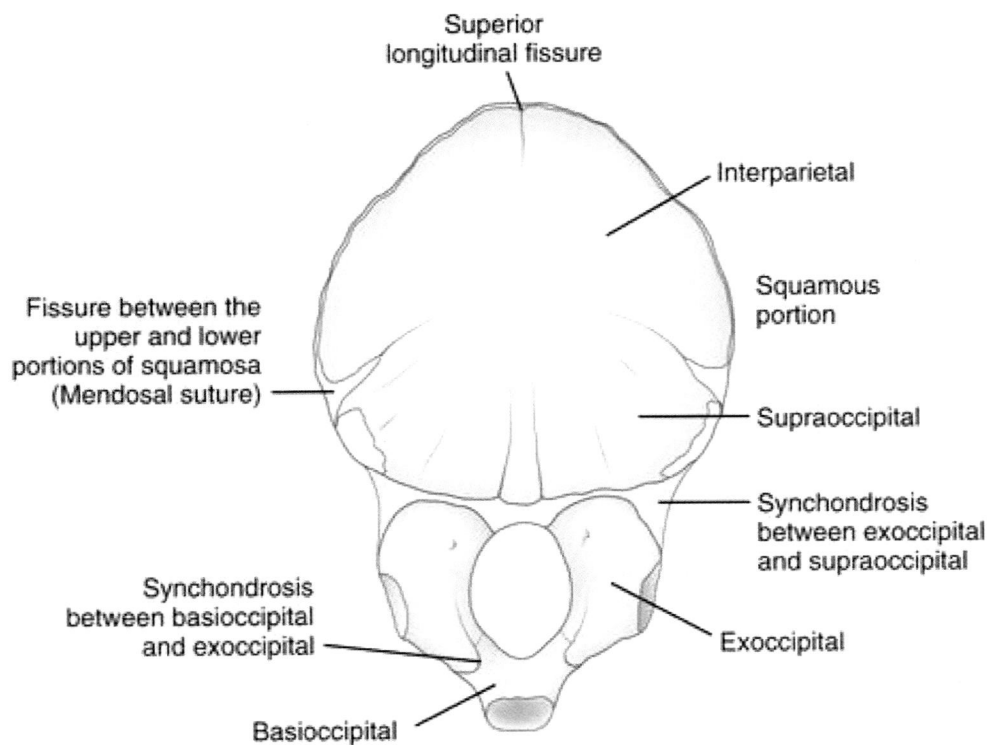

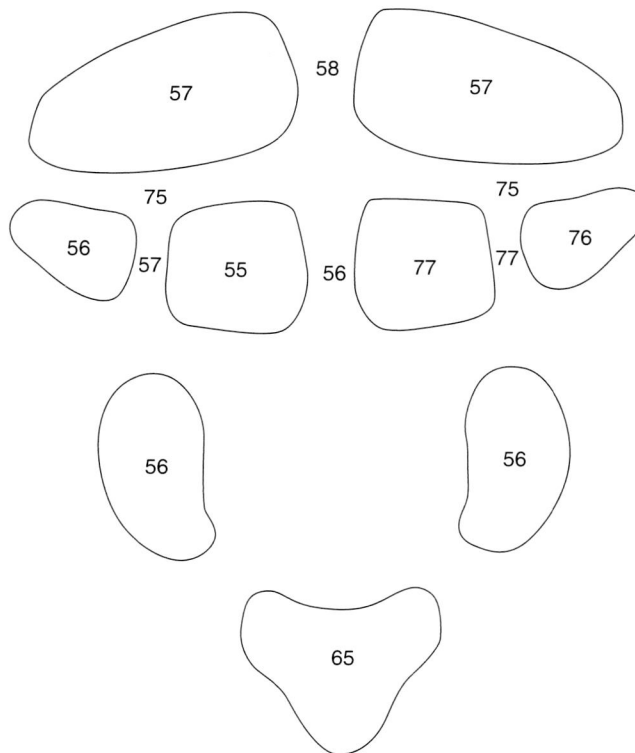

Fig. 8.146 Ossification of the supraoccipital bone complex. Planum nuchale has eight quadrants, each with specific muscle insertions. Exoccipital is more extensive than shown. Medial zone is the condyle. The lateral zone is jugal process bears *m. rectus capitis lateralis*. Each one has separate ossification center. Day 55, two cartilages above FM; day 56, lateral centers, exoccipitals, and inferomedial merger; day 57, unification of inferior zone SO, appearance of upper zone fields; day 58, unification of upper zone; day 65, basioccipital centers. (Reprinted from Mall FP. On ossification centers in human embryos less than one hundred days old. Amer J Anat 1906; 5:433–458)

projection, the *external occipital protuberance*. It is of interest that no somitic muscle attaches to neural crest planum occipitale, only the second arch SMAS system, the Sm6 *occipital belly of galeofrontalis*, the superficial investing fascia of which is neural crest.

Having previously discussed these interparietal bone fields, we shall turn our entire attention towards the neuraxis of the skull, foramen magnum and the supraoccipital bone that provides its roof.

Planum Nuchale

This zone refers to the chondral (mesodermal) occipital bone. It lies below superior nuchal line. Its roughened surface offers attachment to muscles that flex and extend the head. The caudal margin of SNL bears the primary insertion sites for two muscles derived from somite 1–4: trapezius (medially) and sternocleidomastoid (laterally). Just below and nearly in parallel is *inferior nuchal line*. In between SNL and INL are located semispinalis and obliquus. Further down, below INL and posterior to the occipital condyle are the deep extensors of the head: rectus capitus posterior minor and rectus capitus superioris. From the posterolateral border of foramen magnum, the posterior atlanto-occipital membrane descends, anchoring the skull base to the posterior ring of atlas. All of these muscles arise from somites 1 to 3.

Muscles attached to planum nuchale are of somitic origin, with primary insertions into either axis or atlas and secondary insertions into the skull. They are all supplied by branches of *suboccipital nerve*, the dorsal primary ramus of the first cervical nerve. The dorsal primary ramus of second cervical nerve gives off a medial branch, the *greater occipital nerve*.

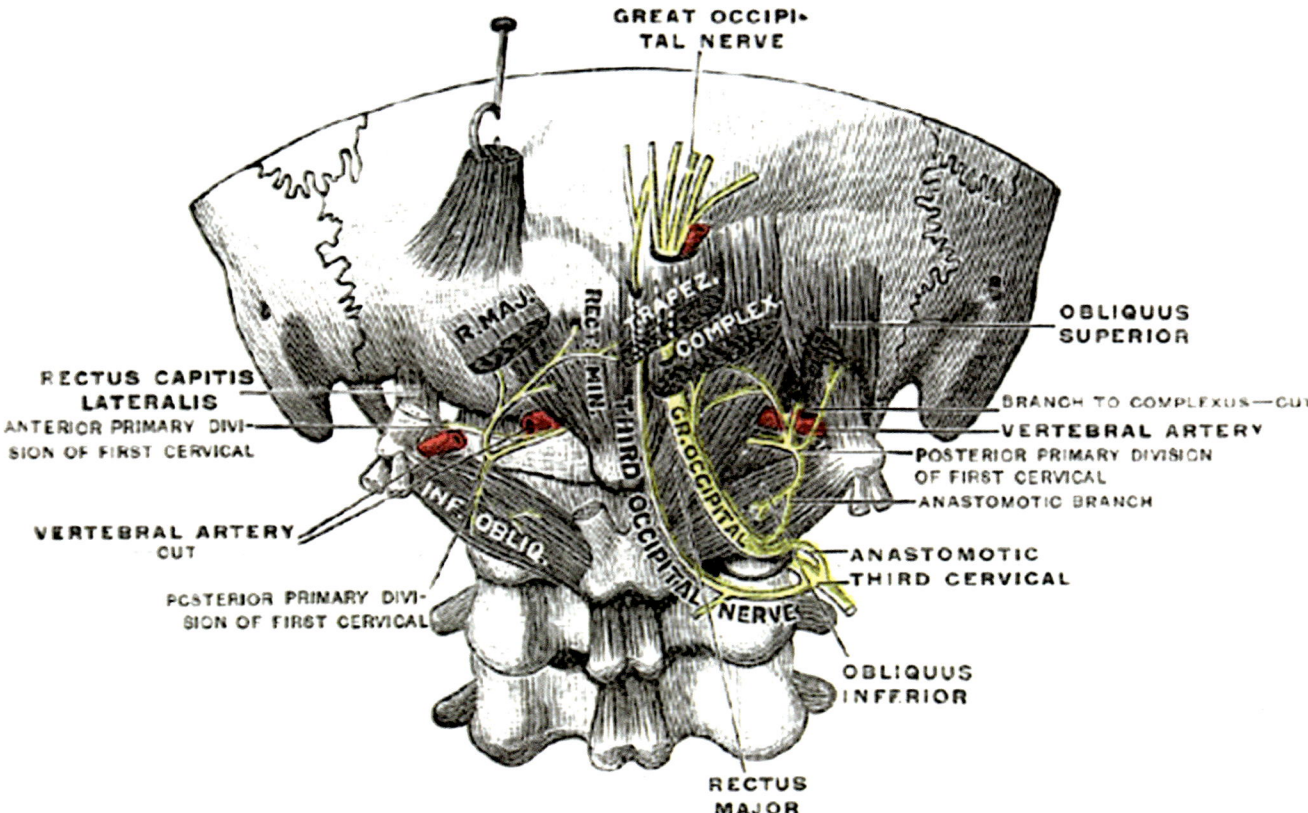

Fig. 8.147 Supraoccipital bone fields (after Srivastava). Exoccipital (dense stipple) has two centers: occipital condyle and juglar process. Supraoccipital (light stipple) has four centers below inferior nuchal line, four centers above inferior nuchal line. Interparietal (gray) has two centers between SNL and HNL which bears occipitalis belly of *m. occipitofrontalis*. Note: suture between exoccipital and supraoccipital is not indicated here (but it exists). (Reprinted from Lewis, Warren H (ed). Gray's Anatomy of the Human Body, 20th American Edition. Philadelphia, PA: Lea & Febiger, 1918)

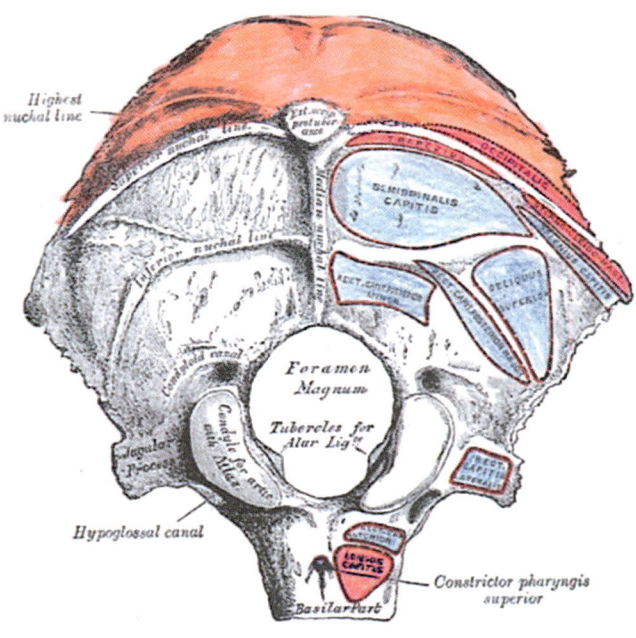

Fig. 8.148 Ossification centers of the occipital bone. These are 14 in number (BO, 2; XO, 4; SO, 8)

Basioccipital has two centers

Exoccipital has two centers: condylar and jugular. The latters which bears *m. rectus capitus lateralis*

Supraoccipital *planum nuchale* is divided into two zones by *Inferior tranverse line* ITL, *inferior nuchal line* (INL) and *superior nuchal line* (SNL). Inferior transverse line (ITL) separates exoccipital from planum nuchale. Lower zone has four ossification centers

Upper zone

First occipital somite

Lower zone (between ITL and INL), four centers: insertions of c1 muscles. Medial: *m. rectus capitus posterior minor* (c1), lateral: *m. rectus capitus posterior major* (c1) and *m. obliquus superior* (c1).

Second and third occipital somites

Upper zone (between INL and SNL), four centers: insertions of c2–c3 muscles. Medial center bears *m. semispinalis capitis* (greater occipital n. c2–3); lateral center bears *m. splenius capitis* (c3). This zone develops from the second and third occipital somites

Interparietal *planum occipital* is divided into two zones by superior nuchal line (SNL) and highest nuchal line (HNL). Between SNL and HIL is a narrow zone with a one or two ossification centers bearing the insertions of trapezius and sternocleidomastoid

(Reprinted from Lewis, Warren H (ed). Gray's Anatomy of the Human Body, 20th American Edition. Philadelphia, PA: Lea & Febiger, 1918)

The dorsal primary ramus of third cervical nerve also contributes to greater occipital nerve.

The internal surface of the squama bears cruciate bony ridges that mark the venous sinuses. These divide the squama into four distinct fossae. The upper two provide shelter for the posterior cerebral hemispheres while the lower two bear the cerebellar hemispheres. The overall structure of the squama is bilaminar with an intervening marrow space, save in the inferior fossae. Here the thinned-out bone is nearly translucent. This is consistent with a single programming source.

Development of the Occipital Complex

The assembly of the occipital complex is radically different from all the other bones of the skull. It is the product of serial contributions from the four occipital somites, laminated one upon the other. This information, derived from molecular tracing experiments, has not been incorporated into standard texts. Each somite makes specific contributions producing three successive pairs of bone fields: basioccipital, exoccipital, and supraoccipital. In the final anatomy, the boundaries between exoccipital and supraoccipital are particularly vague. We will consider occipital development from two points of view (Figs. 8.150, 8.151, 8.152, 8.153, 8.154, and 8.155).

Traditional Model

This model is based upon ossification centers. It does not consider where the mesenchyme comes from. There is no consideration of occipital somites.

Occipital bone in the traditional sense is a misnomer. It is not a single bone. It has multiple components from four occipital somites; all of them ossifying in cartilage.

- Basioccipital or *pars basilaris* is described as having two midline ossification centers. These appear at the sixth week. Most likely these represent a fusion of four paramedian centers.

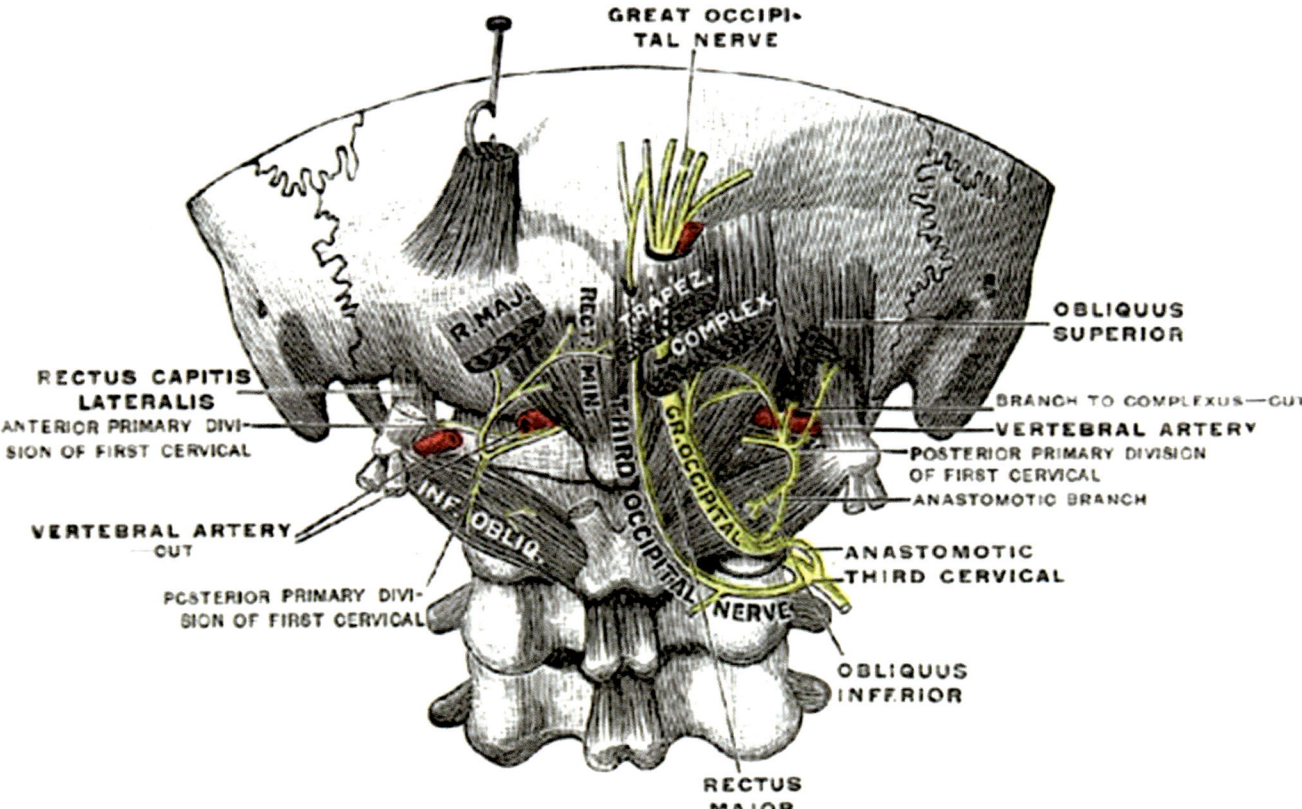

Fig. 8.149 Occipital muscles. Note that supraoccipital receives secondary insertions exclusively from axis or atlas. Recall that these structures develop an amalgam of cervical somites 1–3 (proatlas have been incorporated into the skull based) thus all are supplied by dorsal primary rami of cervical nerves 1–3. Ventral rami c1–c3 are directed to the cervical plexus. Recall as well that, due to the incorporation of proatlas into the skull base, the first cervical nerve emerges *above* the atlas. The first four somites of the neck constitute the original tetrapod neck, to which mammals added an additional four neuromeres and somites giving a total of seven vertebra but region of the neckmuscles from axis upward. (Reprinted from Lewis, Warren H (ed). Gray's Anatomy of the Human Body, 20th American Edition. Philadelphia, PA: Lea & Febiger, 1918)

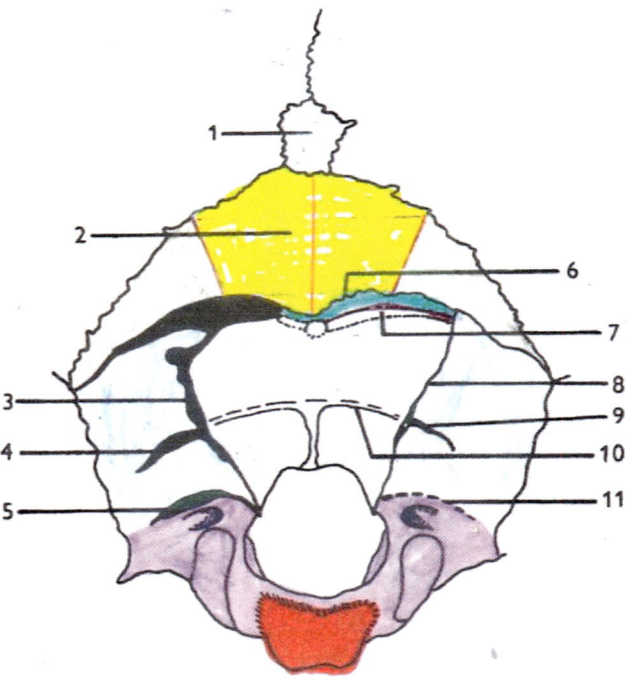

Fig. 8.150 Occipital bone fields: sutures of the supraoccipital complex as per Prof. HC Srivastava. Diagram showing the occipital bone sutures. 1, Pre-interparietal bone; 2, interparietal; 3, left vertical suture; 4, left superior horizontal suture corresponds to inferior nuchal line; 5, left inferior horizontal suture; 6, interparietal-supraoccipital suture; 7, superior nuchal line = Mendosal suture; 8, right vertical suture; 9, right superior horizontal suture; 10, inferiornuchal line; 11, right inferior horizontal suture. *Torus occipitalis transversus* lies between 6 and 7. Internal occipital crest is vertical and midline. Not labeled but seen as a vertical separation in specimen by GP Pal. *Inferior horizontal suture* (HIS) defines exoccipitals. *Vertical suture* (VS) is poorly appreciated/unrecognized. It separates medial and lateral fields of the supraoccipital. *Superior horizontal suture* (SHS) is commonly termed superior nuchal line. Supraoccipital is bounded by SHS and VHS; laterally, it abuts with mastoid. (Reprinted from Srivastava HC. Development of the ossification centres in the squamous portion of the occipital bone in man. *J. Anat.*, 1977;124:643–9. With permission from Blackwell Publishing Ltd.)

- Exoccipital *or pars lateralis* or appears on either side of foramen magnum from single ossification centers at the eighth week. These fields interface with opisthotic and form the jugular foramen. In mammals, the occipital condyles are formed exclusively from the exoccipitals.
- Supraoccipital or *planum nuchale* forms in cartilage from somitomeric somitic mesoderm probably in register with Sm8 and Sm9 (neuromeric levels r8–r9). During the seventh week, eight ossification centers appear. At the third month, planum occipitale and planum nuchale start to fuse.

Nota bene Interparietal and tabular bone fields make up amniote *planum occipitale*. This is not part of the supraoccipital bone. These fields develop in the membrane from non-somitic mesenchyme, likely from levels r6 to r7. Planum occipital is triangular and consists of from six to eight fields.

Four bone fields border the lambdoid suture. In the center, paired *interparietal* bones with 2 or 4 ossification centers that appear near the midline at the eighth week. On either side of the central fields, lateral fields representing the *tabular* bones have one center each. Beneath this tetrarchy is a narrow strip of membranous bone sitting atop the superior nuchal line somatic supraoccipal bone, the *torus occipitalis transversus*, with two centers on either side of the midline. As previously described, TOT arises as an ossification of transverse mesenchyme connecting the opisthotic bone fields of otic capsule.

Order of ossification of the occipital complex is basioccipital (6 weeks) > supraoccipital (7 weeks) > exoccipital (and interparietal complex) (8 weeks) [15, 16].

At birth, the occipital bone consists of four pieces (squama, two lateral parts, and the basilar segment) connected together by cartilaginous zones (Fig. 8.147). This represents a process of chondrification that is spreading outward asymmetrically from central nucleation zones. The unossified cartilage strips are simply awaiting their turn. Unification of the squama and the exoccipitals takes place at 4 years. Pars basilaris joins at 6 years. The cartilage zone between basioccipital and basisphenoid, the *spheno-occipital synchondrosis*, ossified at age 18–25 years.

Frustrating as it may be, the development of the occipital complex remains poorly described in the medical literature. Contemporary resources do not give us a clear picture of how the posterior fossa relates to surrounding mesenchyme. The anatomy and functional significance of the occipital somites has remained the province of developmental biology and comparative anatomy. We shall now turn our attention to this subject.

Neuromeric Model

Let us start out with a few simple premises. Craniofacial bones resulting from membranous ossification are *segmental*; they are discrete entities, separated by sutures. Bones of the anterior cranial fossa sitting ahead of the notochord are synthesized from neural crest with a small prechordal component to posterior basisphenoid to form the sella turcica (Figs. 8.152, 8.153, 8.154, and 8.155).

Transformation of somitomeres into somites begins at the level of Sm8. Thus, occipital somites S1–S4 are derived from Sm8 to Sm11; cervical somites S5–S12 are derived from Sm12 to Sm19. Vertebra and ribs develop via *parasegmentation*: the caudal ½ of somite *n-1* combines with the cranial ½ of somite *n* to produce vertebral number *N*. This process is fully operative at the level of the third cervical vertebra. Thus, C3 = caudal 7th somite + cranial 8th somite. C7 = caudal 11th somite + cranial 12th somite. T1 = caudal 12th somite + cranial 13th somite.

Evolutionary "reshuffling" of the occipital–cervical junction in tetrapods having an atlas–axis complex means that *parasegmentation between caudal S4 and cranial S7 is irreg-*

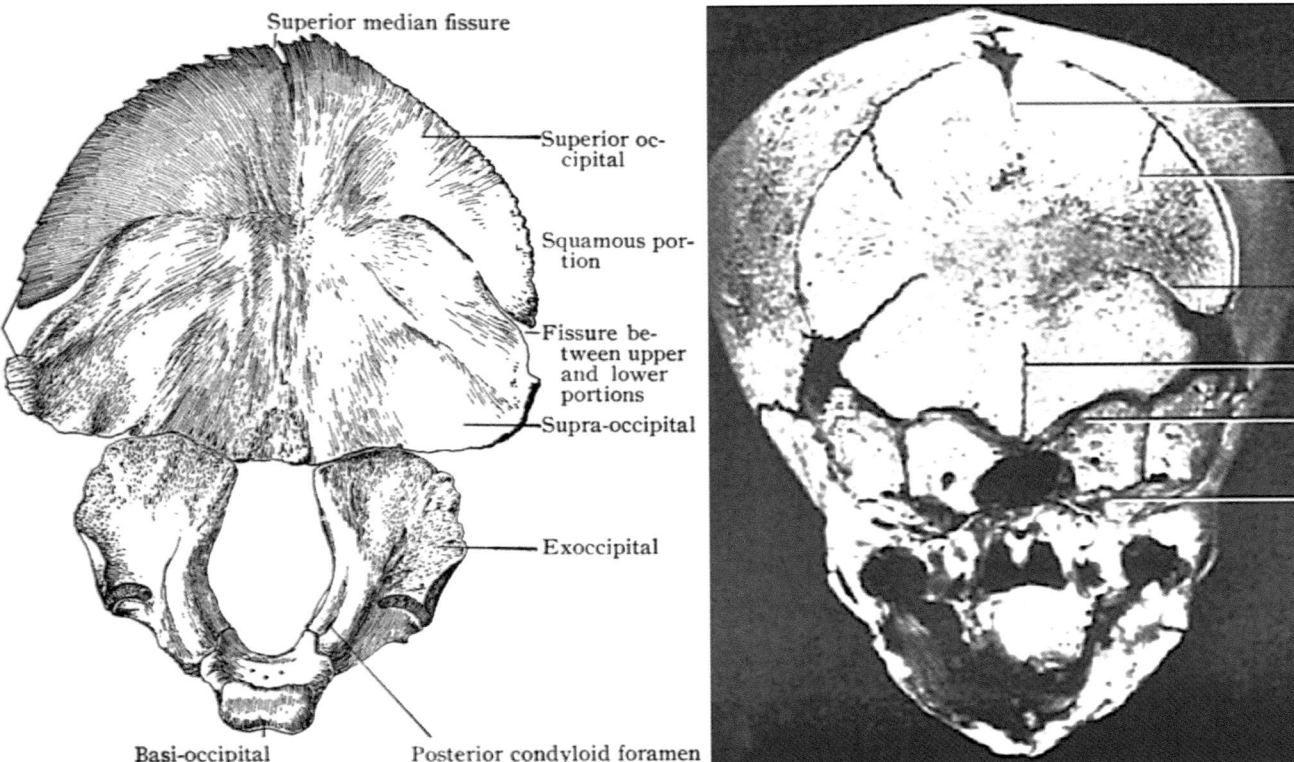

Superior median fissure

Superior oc-
cipital

Squamous por-
tion

Fissure be-
tween upper
and lower
portions

Supra-occipital

Exoccipital

Basi-occipital   Posterior condyloid foramen

Fig. 8.151 Occipital complex at birth. Clearly shown is the suture between exoccipital and supraoccipital. Note that the anatomist recognized the interparietal complex as the "superior occipital portion," separated from the supraoccipital complex by superior nuchal line (SNL). Supraoccipital lies between SNL and the inferior horizontal suture. Lower tier of supraoccipital is defined by superior and inferior horizontal sutures; it has four distinct fields. [Left: (Reprinted from Pierson GA, Dwight T. Piersol's Human Anatomy, including structure and development and Practical Considerations. Philadelphia, PA: Lippincott; 1919). Right: (Reprinted from Pal GP, Tamankar BP, RoutalL RV, Bhagwat SS. The ossification of the membranous part of the squamous occipital bone in man. *Journal of* Anatomy 1984;138:259–266. With permission from John Wiley & Sons)]

ular. We shall discuss this in greater detail under the heading of occipital somites and segmentation.

What's important to keep in mind is that the development of the posterior fossa is altogether different. Four sclerotomes (S1–S4) fuse together to produce basioccipital and exoccipital *by a mechanism that is neither segmental nor parasegmental.* To see how this works, let us make a simple model of how the occipital sclerotomes fit together based on (1) the anatomy fish vertebra and (2) mapping studies of avian occipital somites.

Vertebrae and ribs are not present in agnathic fishes (hagfish and lamprey). The sole source of skeletal support for the hagfish is the notochord. Once at the level of chondrichthyan fishes some elements of vertebrae and ribs appear.

Actinopterygians have ribs in seemingly odd places. Such intermuscular and subperitoneal ribs are part of the locomotor apparatus. They play no role in respiration in fishes because gas exchange takes place in the gills. A baseline vertebra will have a *neural arch* to protect the spinal cord. Between the bases of successive neural arches spinal nerves pass outward through *intervertebral foramina.* Each neural arch is attached to a *centrum* from which projects downward a *hemal arch.*

The function of the hemal arch in the tail segment of fishes is to protect the caudal vascular bundle from compression. Beginning with choanate fishes ancestral to tetrapods such as the osteoleptiform[1] and *Eusthenopteron*[2], lateral projections from the neural arch-centra, *parapophyses*, appear to support the ribs. Such projections presage the future *pedicle.*

Having no tails to swim with, the hemal arches were lost in tetrapods. Lumbar ribs were lost, gaining greater mobility. The basic vertebra contained three paired components, each with an ossification center: centra, two lateral masses, and two neural arches.

Here's how the occipital sclerotomes are assembled to make an occipital complex. The sclerotomes of each somite migrate medially to click together around the neuraxis like a pair of double-opposing U-shaped magnets. The **centra** form the basioccipitals; the lateral masses and **neural arches** form the exoccipitals. It is not clear if occipital somites have **lateral masses** or, if so, what their contribution might be. Note that the occipital somites decrease in size. Let us now

[1] extinct species.

[2] extinct species.

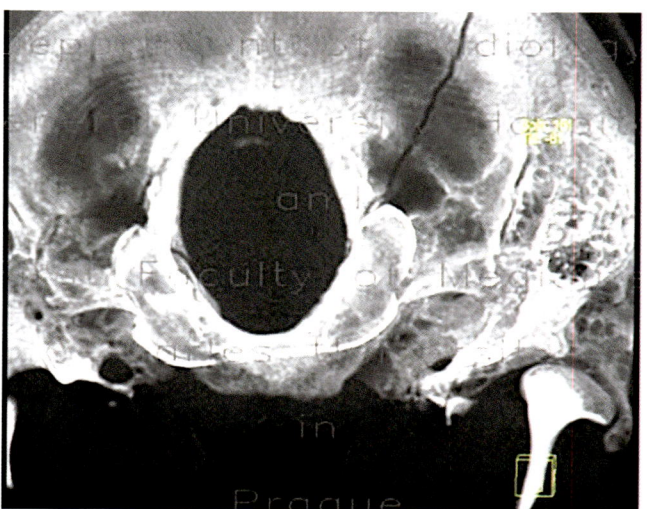

Fig. 8.152 Vertical suture of supraoccipital. Right side shows clearly the vertical suture. Note the distinction of the inferior nuchal line dividing right supraoccipital into upper and lower zones. Left side show the inferior horizontal suture following the posterior base of the condyle. (Source: Fissures of occipital bone, terminating in foramen magnum. *Atlas of radiological images* 2017; v.1. Charles University of Prague. http://atlas.mudr.org/Case-images-Fissure-of-occipital-bone-terminating-in-foramen-magnum-524; http://atlas.mudr.org/img-Fissure-of-occipital-bone-terminating-in-foramen-magnum-1752)

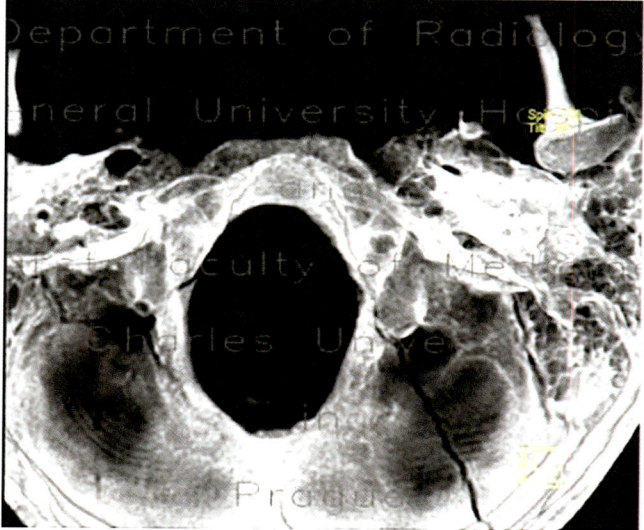

Fig. 8.153 Inferior horizontal suture of the supraoccipital. From the interior of the skull, left inferior horizontal suture is seen as well as the right vertical suture. (Source: http://atlas.mudr.org/Case-images-Fissure-of-occipital-bone-terminating-in-foramen-magnum-524; http://atlas.mudr.org/img-Fissure-of-occipital-bone-terminating-in-foramen-magnum-1752)

fuse these sclerotomes together such that their various zones merge together concentrically, like Russian dolls. This sequence is played out in real time…ossification proceeds from posterior to anterior so the somites can laminate one on the other (Fig. 8.152).

Now take the neuraxis and bend it downward 90°. The spinal cord will now be draped on top of the basioccipital, the exoccipitals will lie lateral to foramen magnum and the supraoccipital will extend upward from the posterior margin of foramen magnum (Fig. 8.153).

Labeling studies of avian occipital somites by Huang et al. demonstrate that the posterior fossa is indeed assembled in concentric rings [1]. Birds are considered to have five occipital somites; mammals have four. This view is contested by O'Rahilly who considers birds to have only four. In the latter model, resegmentation is considered to take place between somites 4 and 7.

If we consolidate the Huang and O'Rahilly models, somites 1–4 each contribute to the posterior fossa as concentric strips of mesoderm. The hypoglossal nerve is an important landmark because it demarcates the third and fourth somites. This is manifested by the hypoglossal foramen in the exoccipital bone. Embryonic staging by O'Rhailly–Carnegie system distinguishes stages 9 and 10 by the appearance of the fourth occipital somite (Figs. 8.154 and 8.155).

Somite four becomes highly specialized because *it is the first somite forming a motion segment*. It spans from the lip of foramen magnum to the upper ½ of dens, a product of S5. Recall that dens projects upward from the body of axis. S4 also produces most of caudal exoccipital and the occipital condyle. As such, it articulates with Atlantic ring of which it forms the cranial ½ of Atlantic ring (as well as the cranial ½ of Atlantic spinous process). Caudal ½ of dens and cranial ½ of axis body are derived from S5 while the caudal ½ comes from cranial S6. Neural arch of axis is likewise S6–S7. Third cervical vertebra is S7–S8.

Note The topic of the craniovertebral junction is very important from an evolutionary standpoint because of its functional implications. We shall discuss this in further detail in Chap. 10.

Phylogeny of the Occipital Complex

The story of the occipital complex is its transition from fishes to mammals to primates is one of brain expansion, of filling a hole, and of adjustments to the craniovertebral junction accommodate life with a neck and eventual arboreal/bipedal posture. Here is a bird's-eye view. The cranial base in the midline of the posterior fossa can be considered like a sliding drawer which was pulled out backward as it incorporated a fourth somite and responded to the expansion of the membranous calvarium. At the same time, the otic capsules remained fixed in position to the basisphenoid at center of the skull. Brain growth forced the post-temporal dermal bones to enlarge and shift in position. This, plus the appearance of a neck between the skull and the trunk created post-temporal fenestrae behind the otic capsules. In amniotes, these spaces were filled in by somitomeric mesenchyme to create new supraoccipital bone fields. Finally, the anatomic

Fig. 8.154 Development of supraoccipital bone: the supraoccipital cartilage. Occipital cartilage (dark gray) forms from somitomeric PAM (r6–r7) early at E12/Carnegie stage 9–10. Note frontal cartilage (green) forms from neural crest at same time. (Reprinted from McBratney-Owen B, Iseki S, Bamforth SD, Olsen BR, Morriss-Kay G. Development and tissue origins of the mammalian cranial base. 2008; *Developmental Biology* 322:121–132. With permission from Elsevier)

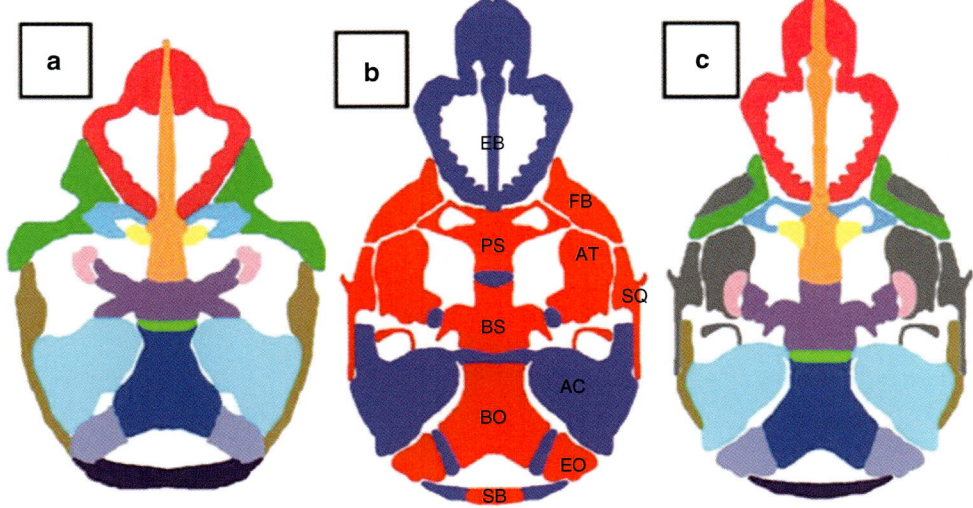

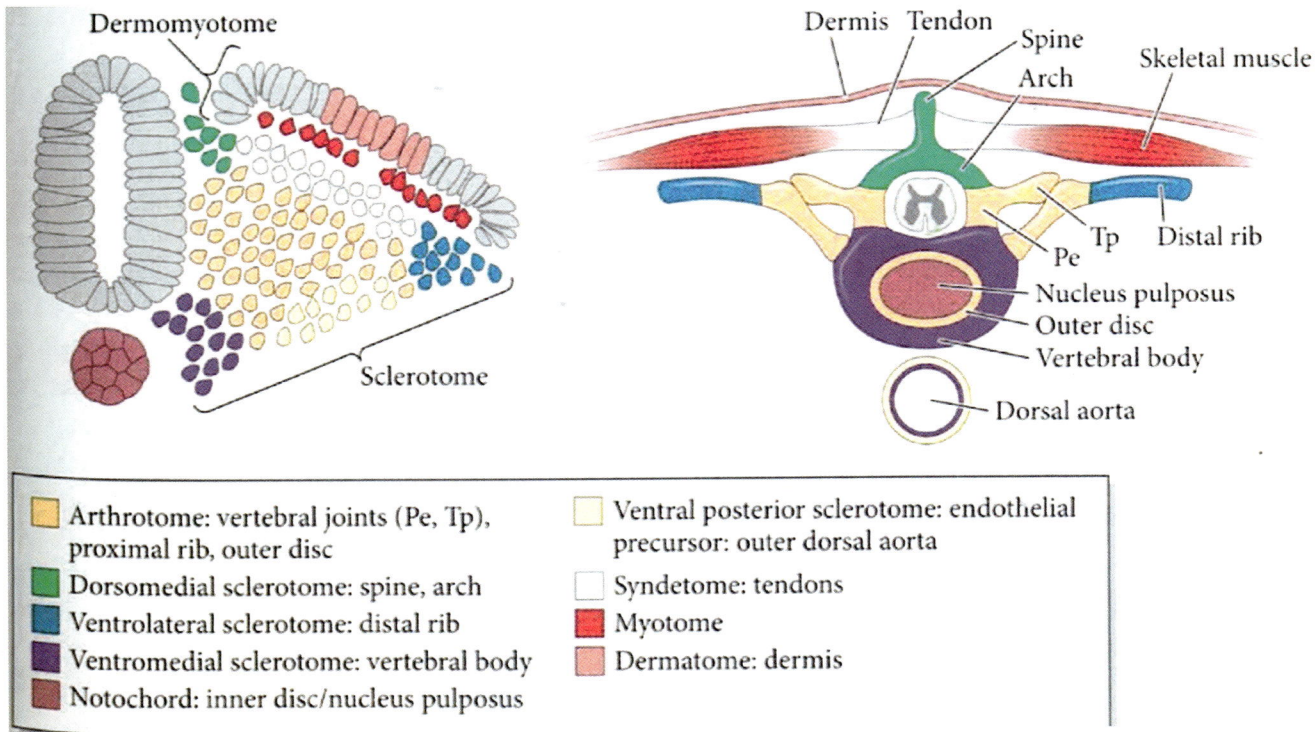

Fig. 8.155 Somite structures. The four occipital somites are in register with r8–r11. S1–S3 appear at stage 9. The "somitomere-somite transformation" takes place at Sm8 as soon as Sm19 is synthesized. The appearance of S4 defines stage 10. The developmental repertoire of occipital somites is reduced to bone and hypaxial muscle. They cannot make dermis. S1–S4 are fused together in concentric rings. The ventro-medial sclerotome forms basioccipital; the lateral sclerotome, exoccipital; and the dorsomedial sclerotome forms exoccipital and supraoccipital. (Reprinted from Gilbert SF, Barresi M. Developmental Biology, 11th ed. Sinauer: Sunderland, MA, 2016. Reproduced with permission of the Licensor through PLSclear)

composition around foramen magnum was dynamic, with new contributions made from the atlas to the cranial base, the appearance of an atlas–axis complex, and a more sophisticated craniocervical joint. So, let us begin.

We start off by considering the changes in the mesenchymal map that takes place in the transition from the sarcopterygian fish *Eusthenopteron*, a transitional tetrapod, and a full tetrapod, our focus being the foramen magnum (Fig. 8.156). In our fish ancestors, foramen magnum was ringed by somitic field (basioccipital and exoccipital). Otic capsule made of cephalic (somitomeric) mesoderm was excluded. In *Acanthostega*, foramen magnum gets much larger in comparison to the rest of the occiput. The somitomeric "ring" opens up dorsally; an expanded otic capsule now forms the roof over the spinal cord. Hyomandibular has been converted into stapes the footplate of which is firmly planted into the fenestra vestibule of the otic capsule. In the crown state prior to amniotes, exemplified by Greerepeton, expanded exoccipitals reclaim territory around the dorsolateral foramen magnum and trade otic capsules are pushed out of the way to the sides.

Let us turn back the pages to our basal actinopterygian, the amiable *Amia* (Fig. 8.157). Here exoccipitals form a triangular A-frame house to shelter the spinal cord. Lateral to the exoccipitals, intercalars are intercalated with the posttemporals and the braincase. Between the roof of the skull and the occiput is a large rectangular space, the *posttemporal foramen*. Spanning it are paired *epioccipitals*. These are notable because, in future iterations, these twin pillars will be replaced by identical arms extending upward from the exoccipitals, this time from exoccipitals. It is this mesodermal subdivision of a space into four zones that is intriguing as a source of mesenchyme for the supraoccipital that will eventually cover over this space.

Not much happens in fish evolution so let us turn to two early tetrapods at the very shift of locomotory dominance, in which the operculogular bone series are lost and the pectoral girdle becomes disconnected from the skull. As we have seen previously, the 8-digit *Acanthostega* was primarily an aquatic animal. Lacking wrists its forelimbs were probably not useful. The 7-digit *Ichthyostega* (Gr. "fish-roof") had a stronger pectoral girdle and probably came out of the water. It had a long, narrow braincase flanked by opisthotics. These reached the back of the skull with flange-like processes that provided reinforcement for the exoccipitals. In both cases, a true neck-like occipital condyle had not evolved. The notochord was flexible but springy so these animals probably could not move their heads much. Strong muscle connections to support the back of the skull out of water were lacking. What is significant in *Acanthostega* is the somitomeric mesodermal "hammock" slung across the back of the skull form the opisthotics. It provides the roof for foramen magnum (Fig. 8.108).

The neuromeric origin of this mesenchyme is likely as follows. Prootic and opisthtoic are respectively r4–r5 and r6–r7. The mesenchyme that connects them is likely to arise from r8 to r9 as it will form the future membranous interparietal bone complex.

Greererpeton is the next stem up from *Acanthostega* in the line toward amniotes (Fig. 8.79). This is important because *Greererpeton* is probably the most primitive well-known occiput without fish features. *Greererpeton* was aquatic organism and may have had gills, but no opercular bones; and its pectoral girdle was largely free of the skull. The posttemporal fenestra is narrow such that exoccipitals are almost in contact with the postparietals. Opisthotics are prominent and send paraoccipital processes out to contact the tabular. But at this point, osseous connections have yet been established between the posterior dermal roof and the cranial base.

Greererpeton has a well-developed arch of the axis and is adapted to occipital muscles. The centra of axis and atlas in are similar in size. In the line leading to amniotes, the following changes take place: (1) atlas loses the neural spine—an impediment to motion; (2) pleurocentrum of atlas fuses with intercentrum of axis as the beginnings of the dens; (3) neural arch of axis fuses with the pleurocentra lying just beneath it.

The next step up from *Greererpeton* toward the amniotes in *Megalocephalus* (Gk. "big head"). These large alligator-like tetrapods (called baphetids) had enormous keyhole orbits caused by reduction in the lacrimals and replacement with a pterygoideus muscle (an early experiment for better chewing). A bony structure which is clearly related to the amniote opisthotic appears on the surface. It is W-shaped, with two lateral processes that it sends outward to join with the tabulars; it also sends two fused central processes straight up to postparietals. Thus, the opisthotics cover the top of the foramen magnum. In amniotes, this central region is filled in with loose mesenchyme. But this will form cartilaginous bands (Fig. 8.109).

We now arrive to the reptilomorph *Paleothyris*, our basal amniote, with an anapsid skull (Fig. 8.106). Here, the posttemporal fenestrae remained open but they would be filled in subsequently. Two landmarks appear with exoccipital, both of which it shares with opisthotic, the *jugular foramen* and the *fenestra ovalis*, or fenestra vestibula. These are shared between the otic capsule and exoccipital for the rest of vertebrate evolution. Note that paraoccipital processes extend from opisthotic out to squamosal; these two bone fields will also remain connected as the petrosquamosal suture.

Note: postparietal bone fields posterior to the parietals and in contact with the exoccipitals. These dermal bones (plus the tabulars) form the <u>dermal interparietal bone complex</u> *planum occipital* that sits atop the true <u>chondral supraoccipital bone</u>.

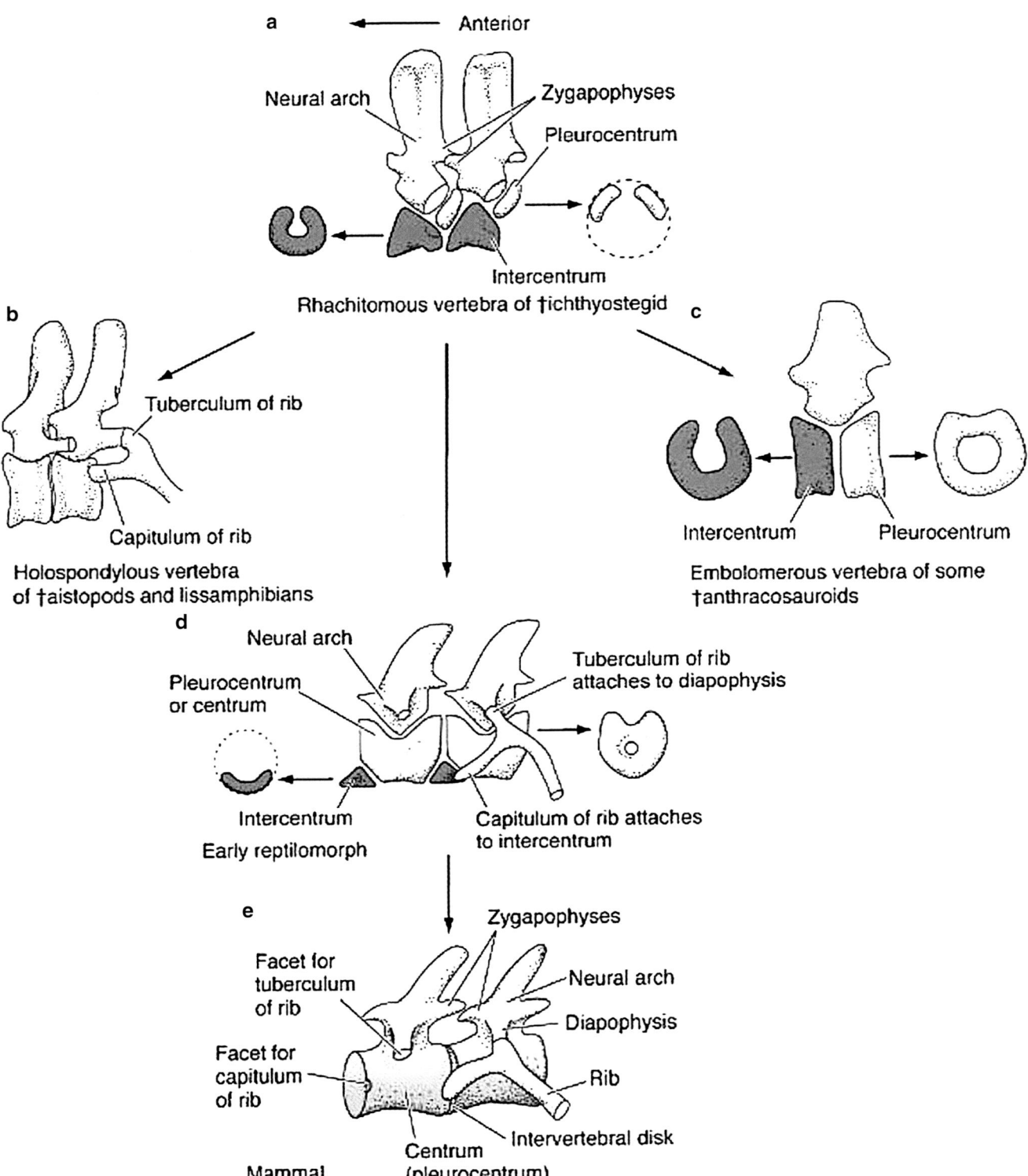

Fig. 8.156 Evolution of vertebra. Basioccipital cartilage comes from the centra. Exoccipital cartilage is derived from the neural processes. Supraorbital cartilage is not a derivative of occipital somite neural process. It develops from somitomeric mesenchyme in continuity with the oposthotic bones of otic capsule. As such, supraoccipital PAM it is probably coded r6–r7. *Rachitomous* vertebra (Gr. *Rhaschis* = spine, *tomos* = cut) seen basal tetrapod, *Icthyostega*, in which centrum is divided. Intercentrum (dark gray) dominant and two pleurocentra light (gray). *Holospondylous* vertebra (amphibians) has complete fusion of both units. In extinct marine reptiles, the two units are equally fused. *Embolomerous* vertebra (extinct aquatic anthracosaur) as intercentrum and pleurocentrum equally represented. Early reptilomophs continue the rachitomous condition but with pleurocentra are dominant and fused together. Mammals lose the intercentrum. Capitulum inserts onto the previous centrum. Zygapophyses add stability. (Reprinted from Kardong K. Vertebrates: Comparative Anatomy, Function, Evolution, 7th ed. New York, NY: McGraw-Hill; 2015. With permission from McGraw-Hill Education)

Fig. 8.157 Neuromeric development of occipital bone 1. Russian doll model of occipital somite development to form posterior cranial fossa. At stage somites are in place and organizing into occipital complex. Final form is achieved with occipitocervical flexion. (Adapted from Puelles L, Harrison M, Paxinos G, Watson C. A developmental ontology for the mammalian brain based on the prosomeric model. Trends in Neurosciences 2013; 36(10):570–578. With permission from Elsevier)

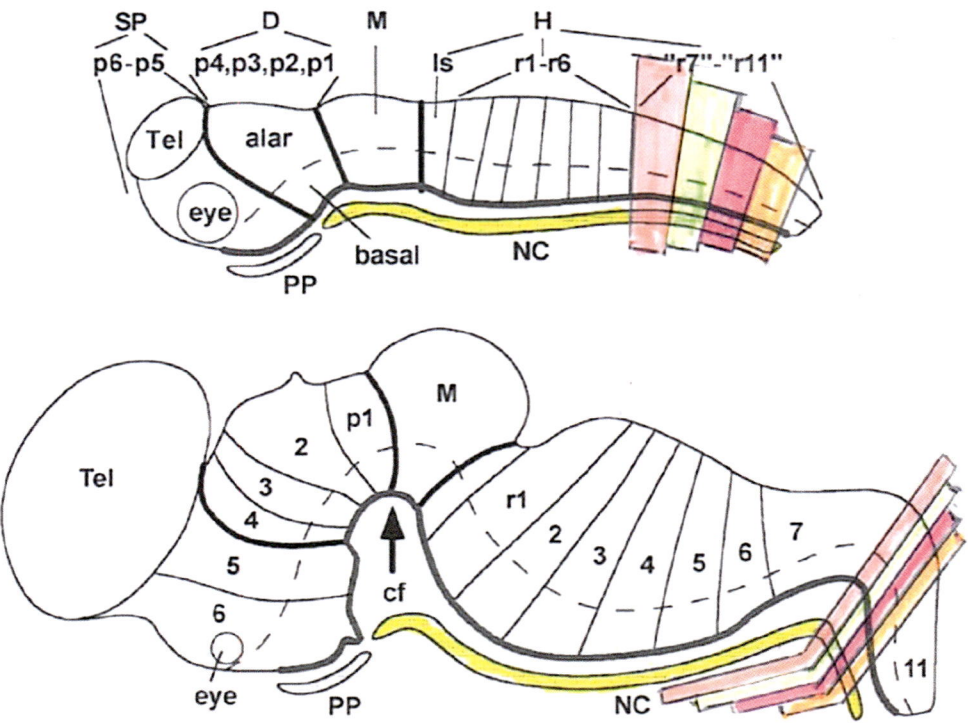

The triangular configuration of the three dorsal elements is easy to understand. The supraoccipital ties the occiput to the dermal skull roof, while the exoccipitals are braced, directly or through the opisthotics, to the dermal elements that make up the posterolateral walls of the skull. Together, the triangle of supraoccipitals and exoccipitals make up the top and sides of the foramen magnum.

The basic amniote craniovertebral joint is a ball-in-socket articulation. The two moieties of basioccipital form a midline occipital condyle. Flanking the condyle are two "pockets" excavated from basioccipital. These receive elements of the first cervical vertebra. This arrangement limits lateral and/or dorsoventral rotation of the skull with respect to the spinal column and resist hyperextension of the neck. These brakes on rotation protect the spinal cord from bending stresses.

The amniote basioccipital is braced strongly to resist compression by articulating with cranial base elements such

as the basisphenoid and pterygoid. The design problem is acute, because the basioccipital is performing its structural duties dorsally, and there is no direct ventral support. Recall that the basal amniote spine extends straight backwards.

The reptilomorphs quickly gave rise to the synapsid line leading to mammals and lateral to the diapsid line of reptiles and birds. In the cynodont skull, the posttemporal fossae are replaced with bone and the supraoccipital is established. As we have seen, *Probainognathus* (Gk. "progressive jaw") was quite an innovator, the author of the "double jaw joint" experiment, and proud possessor of a secondary palate (Fig. 8.110). It developed a double condyle, characteristic of mammals. The supraoccipitals wrapped all the way around the exoccipitals. The relationship with more cephalic interparietal bone fields was very clear. By the time we reach the therapsids, such as *Galeosaurus*, the supraoccipital, as we know it today was fully constructed (Fig. 8.158).

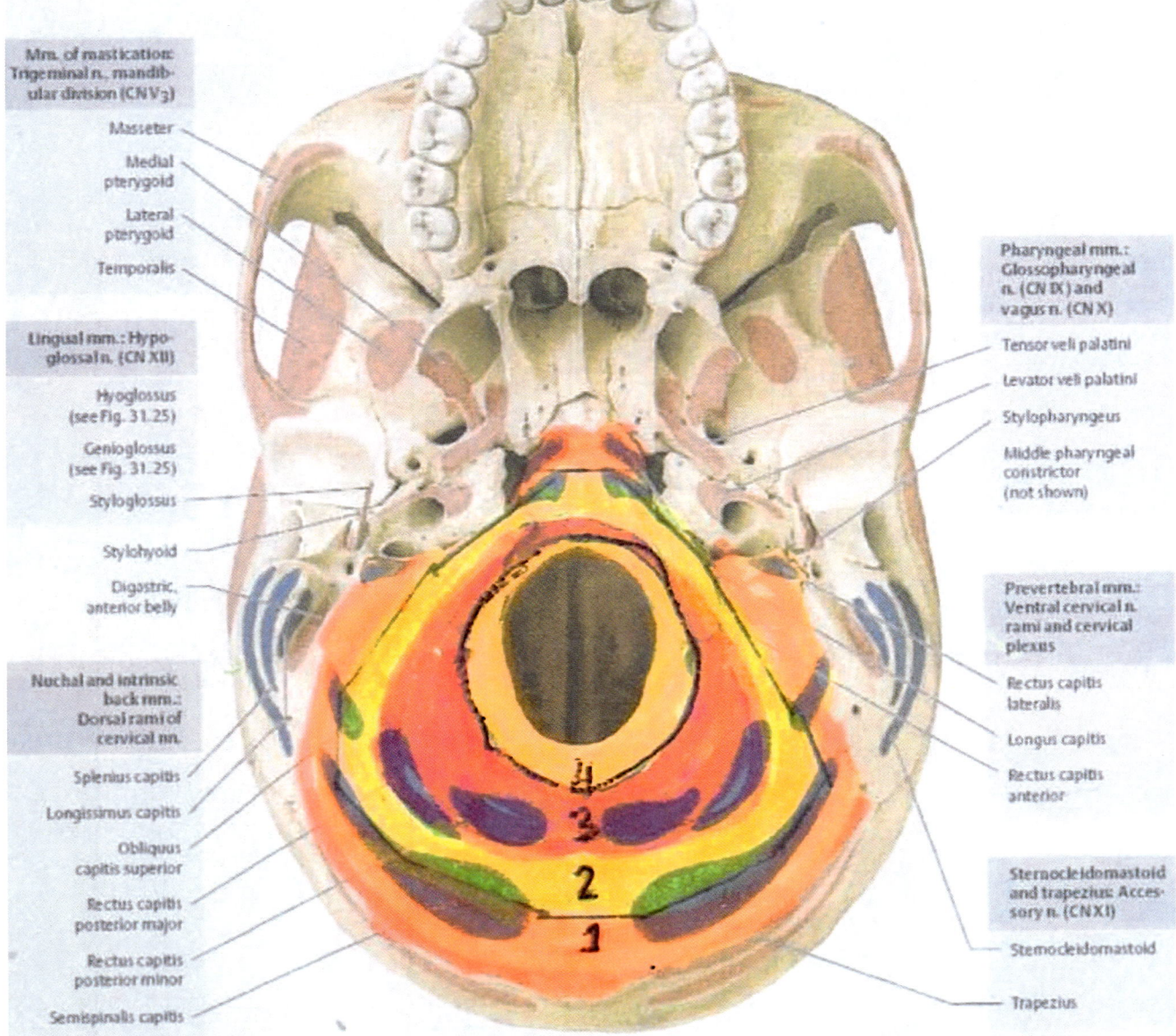

Fig. 8.158 Neuromeric development of occipital bone 2. Lamination of the occipital vertebrae forms concentric rings of developmental fields, shown here projected on the skull base. S1 (r8) is outermost and S4 (r11) rings the foramen magnum. S2 (r9) and S3 (r10) share the occipital condyle. (Reprinted from Lewis, Warren H (ed). Gray's Anatomy of the Human Body, 20th American Edition. Philadelphia, PA: Lea & Febiger, 1918)

Splanchnocranium (r2–r11)

The Vomer

Descriptive Anatomy

The thin bony vomer is really an amalgam of two separate embryonic laminae. Each lateral surface side bears a groove for the medial sphenopalatine (nasopalatine) vessels and nerve. The vomer has four borders, three of which bear articulations. (1) Along the superior border is a groove that receives an articulation with *sphenoid rostrum*. Projecting laterally from the superior border two horizontal *alae* provide two sets of articulations: posteriorly with the medial pterygoid plate *vaginal processes* and anteriorly with the palatine bone *sphenoidal processes*. (2) The inferior border articulates with a both the maxillae and the palatine bones via the *nasal crest*. (3) The anterior bears two articulations, posteriorly with the perpendicular ethmoid plate, and anteriorly with septum. (4) The posterior border of vomer is free (Figs. 8.159, 8.160, 8.161, 8.162, 8.163, and 8.164).

Fig. 8.159 Neuromeric development of occipital bone 3. Mammals have four occipital somites, but principle is the same. (Adapted from Huang R, Zhi Q, Patel K, Wilting J, Christ B. Contribution of single somites to the skeleton and muscles of the occipital and cervical regions in avian embryos. *Anat Embryol* (2000) 202:375–383. With permission from Springer Nature)

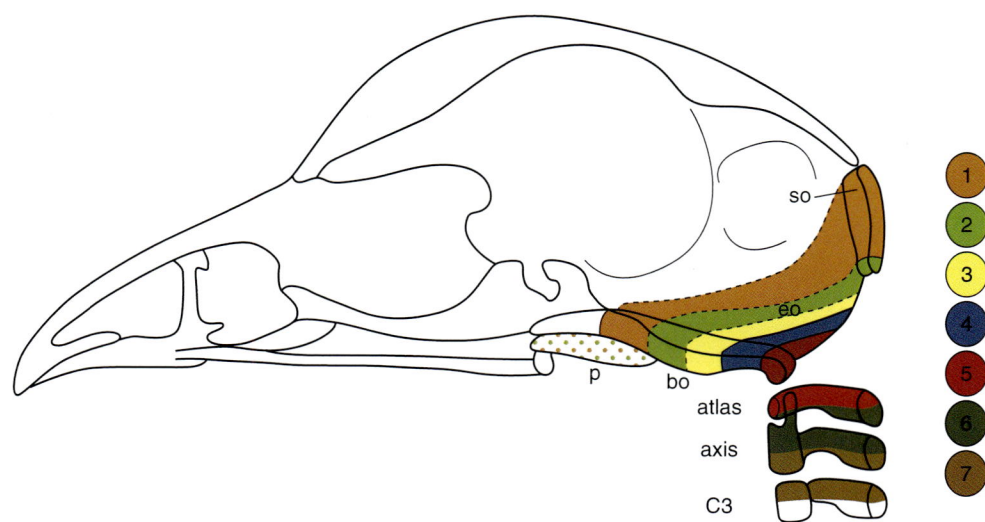

Blood Supply and Mesenchyme

The vomerine complex develops in the membrane from r2 neural crest. It is programmed by r2 oral ectoderm. Each vomerine lamina is supplied by the proximal segment of medial nasopalatine artery, a StV2 induction product.

Nasopalatine artery proceeds through the sphenopalatine notch to enter the nasal cavity whereupon it divides into medial and lateral axes. The former descends to supply the vomer and, upon reaching the incisive foramen, makes an anastomosis with of the incisive foramen with greater palatine. Lateral NPA courses over and below the surface of inferior turbinate and spreads out to supply the mucoperiosteum covering the nasal aspect of the maxillary palatal shelf. It also supplies a limited zone of posterior.

Development of the vomer is inseparable from that of nasal septum and perpendicular plate of ethmoid. Nasal septum develops from r1 mesenchyme in a biphasic manner. It starts out as a cartilaginous plate. Posteriorly this will become osseous perpendicular plate of ethmoid. *Further anterior growth is cartilaginous*. This is likely to results from a dose-dependent relationship of a growth factor with the epicenter located posteriorly. A likely mechanism is one in which the concentration of growth factor decreases anteriorly…cartilage develops as a result.

Vomerine mesenchyme tracks down into position by following the ethmoid template. Conditions where the ethmoid template is reduced vertically thus draw the vomer upward toward the roof of the mouth. The results is the high arched palate. Another variation involves a deficiency in the vomerine field itself.

Development

Formation of the vomerine bones develop takes place in membrane. A pre-existing neural crest field forms a chondral intermediate in which each bone has a single ossification center; the paired laminae appear at the eighth week of fetal life (week 20). During the third month of fetal life (24 weeks), these unite beneath the septum. Note that growth of the vomer takes place backwards and downwards.

Phylogeny

In osteichthyans dating back to the Devonian period, each vomer develops beneath the primary palatoquadrate commissure. It is therefore in sequence with the dermopalatine series. The vomers of *Cheirolepis* are separated by the midline parasphenoids (plural because they are fused, of course). By the time of the sarcopterygian Eustheopteron, they have fused anteriorly in the midline with *parasphenoid* continuing forward up to the future incisive foramen. Note that such primitive species, parasphenoid, contain within it a *buccohypophyseal canal* containing the stalk of Rathke's pouch. In tetrapods, the cranial base sets up a barrier between the oral cavity and the pituitary.

The histories of the vomer and the parasphenoid are intertwined. They are physically co-extent, and homologies have been proposed. Parasphenoid is present in all vertebra taxa, except placental mammals. It is described as a singular bone but most certainly represent the fusion of two fields just off the midline. Parasphenoid extends from below the ventral ethmoid (at the region of the future incisive foramen) backwards to the temporal capsule, i.e., from r0 to r4–r5. Because it reaches the level of the ethmoid, the posterior portion of parasphenoid provides a coverage to the brain from the oral cavity. Sitting in front of it are the vomer bone fields, and further forward, paired prevomers. For this reason, homologies in this bone series have been proposed. We shall consider parasphenoid first, with respect to its structure and evolution and second, with respect to its homology (or lack thereof) to the vomer.

Parasphenoid is initially complex but becomes simplified into a simple midline structure. It is first seen in the Silurian agnathan placoderms about 445 Mya. The ancestors of agna-

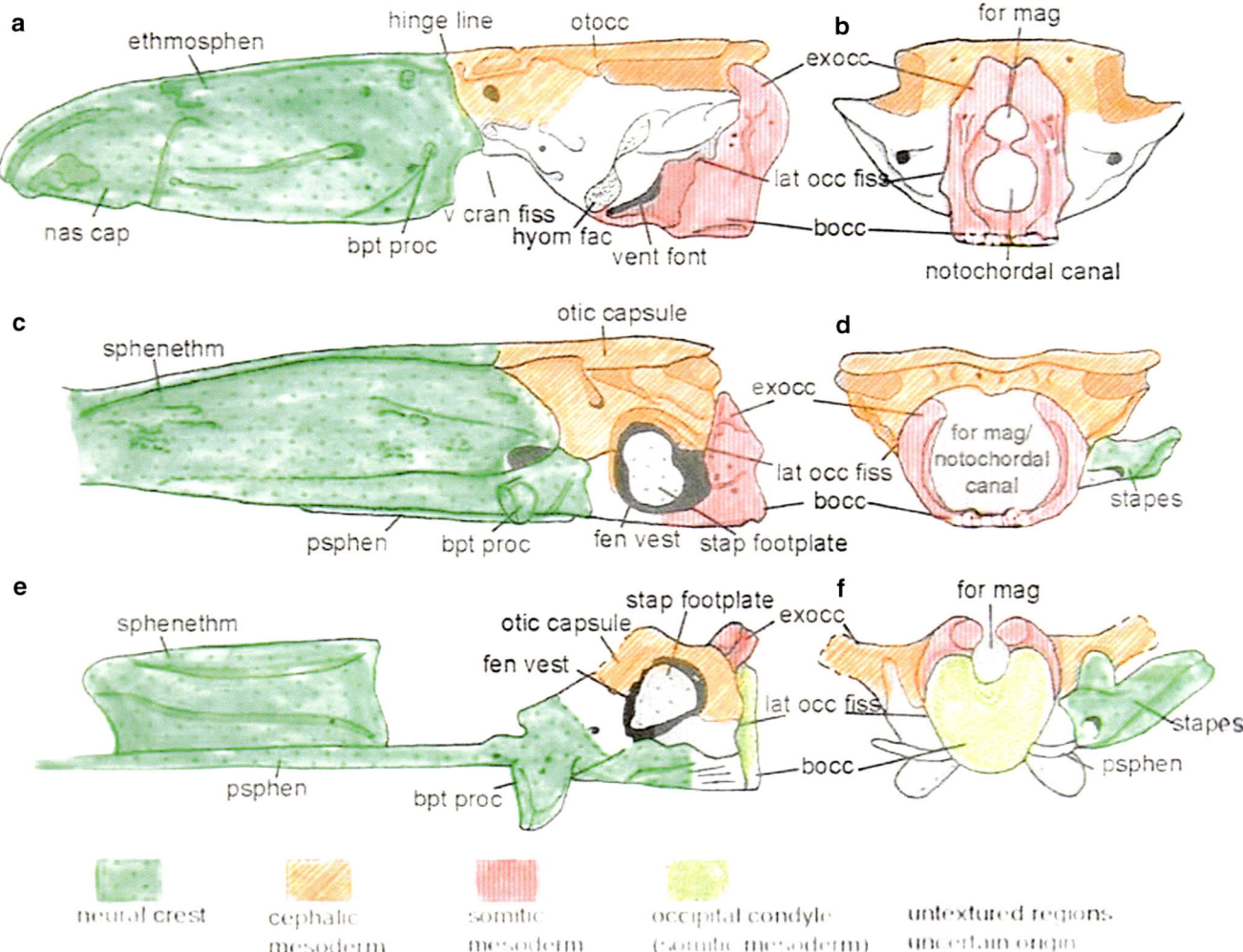

Fig. 8.160 Braincase mesenchyme. Braincase of sarcopterygian, transitional tetrapod, full tetrapod. *Eusthenopteron* shows notochordal canal and foramen magnum enclosed by basioccipitals. Otic capsule does not have separable prootic and opisthotic fields. Note facet in otic capsule to receive hyomandibula. Otic capsule is somitomeric mesoderm. Initial relationship of basioccipital and exoccipital is **vertical**. This could reflect the primordial shape of the occipital somites. *Acanthostega* has a combined canal for notochord and foramen magnum. Hyomandibula has now morphed into stapes which articulates with fenestra vestibuli. Basioccipial has folded 90° with spinal cord running on top of it. No neck articulation, limited head mobility due to flexible notochord. In *Greererpeton*, basioccipital makes a single the condyle. (Reprinted from Clack JA. Gaining Ground: the Origin and Evolution of Tetrapods, 2nd ed. Indianapolis, IN, University of Indiana Press, 2012. With permission from Indiana University Press)

thans had multiple dermal plates covering the cranial floor and basal visceral arches. The fusion of the plates is considered the origin of the parasphenoid (and potentially other oral bone fields). The anterior 1/3 of parasphenoid is denticle-bearing. Details about parasphenoid in its initial iteration such as ossification and potential muscle attachments are unknown, sitting interposed and behind the vomers.

Between the placoderms and the teleost fishes, parasphenoid elaborated two lateral wings. These move posteriorly in teleosts. Prior to amniotes, muscles that were attached to parasphenoid shift to the pterygoid plates, thereby creating the conditions for the eventual revolution in mastication that occurs with the synapsid skull. The lateral wings enlarged to form articulations with basioccipital. In the extinct cynodonts that are ancestral to mammals the parasphenoids actually fused with basioccipital to form part of the basisphenoid itself. In the mammalian line, the parasphenoids continued with the marsupials and monotreme but are considered to disappear in the placentals. In point of fact, Kardong considers them to persist in human embryos.

Since the nineteenth century, the reptilian parasphenoid has been considered homologous with the mammalian vomer. But a problem exists in non-mammals due to the co-existence of "vomers" and parasphenoids. If parasphenoid becomes the true vomer, what is the meaning of these original vomers? De Beer tried to resolve this by renaming the

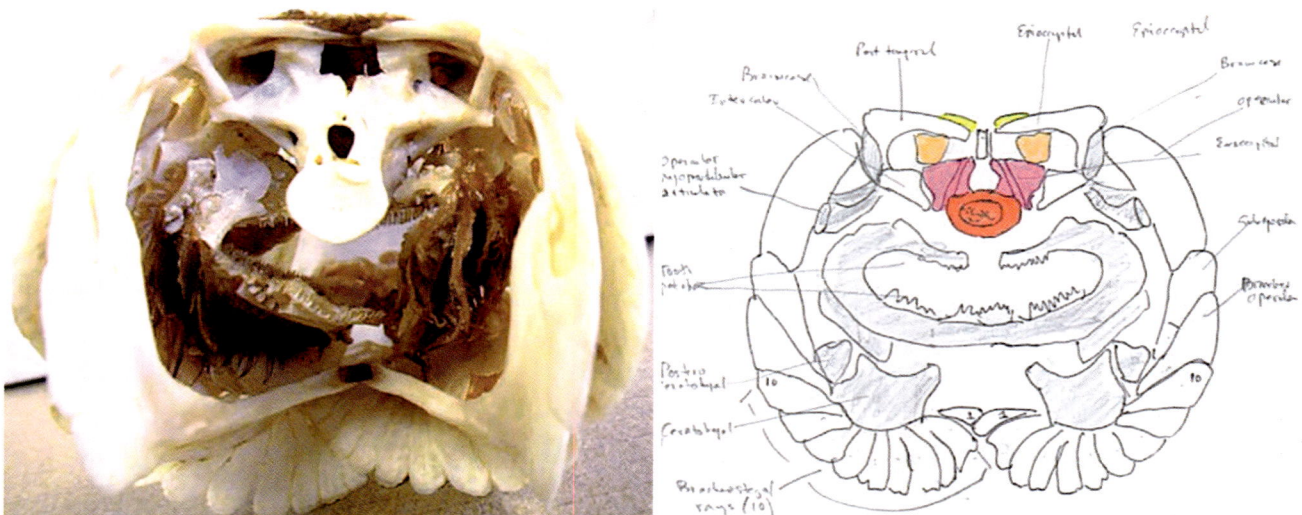

Fig. 8.161 Amia occiput: note epioccipital centers breaking up the posttemporal fenestar. Note also the absence of participation of the opisthotics in support of the basioccipital. BO is really quite unsupported. (Courtesy of Michael Carstens, MD)

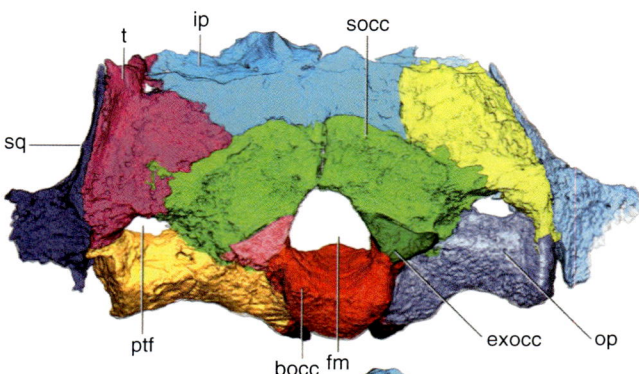

Fig. 8.162 Therapsid Captorhinus occiput demonstrating supraoccipital fields. Note that the posttemporal fenestra (ptf) are still open. When these fill in with bone, supraoccipital (socc) will have medial and lateral fields. *bocc* basioccipital, *exocc* ex-occipital, *fm* foramen magnum, *ip* interparietal, *op* opisthotic, *ptf* posttemporal fenestra, *socc* supraoccipital, *sq* squamosal, *t* tabular. (Reprinted from Araujo R, Fernandez V, Polcyn MJ, Fröbisch J, Martins RMS. Aspects of gorgonopsian paleobiology and evolution: insights from the basicranium, occiput, osseous labyrinth, vasculature, and neuroanatomy. *Peer J* 5:e3119. With permission from Creative Commons License 4.0: https://creativecommons.org/licenses/by/4.0/)

reptilian vomers as *prevomers* which he considered ancestral for the *palatine process of premaxillae* in mammals, making presphenoid ancestral for the true vomer.

There are problems with this model. From tetrapods onward, the amniote line via reptiles leads to Mammalia. In the split between marsupials and placentals, parasphenoid persists in the former group in which it articulates with vomer and is fused with basisphenoid. Furthermore, the mammalian vomer lacks many features of parasphenoid. It has no lateral wings. It articulates with many more bone fields (sphenoid, ethmoid, palatine) than does parasphenoid.

The case for or against homology between parasphenoid and vomer can be tested using Patterson's criteria. *Similarity*: homologous structures must have the same 1:1 topographical relationship to other structures, i.e., the pterygoid. *Congruence*: the homologous character must fit in with the other characteristics of its clade (a group of organisms sharing a common ancestor). *Conjunction*: If two structures are homologous, they cannot co-exist in the same organism. Vomer fails to meet the same relationship with pterygoids does parasphenoid. Unlike parasphenoid, it has no history of muscle attachment. The two bone fields co-exist. In sum, by all three criteria, vomer fails to be the homolog of parasphenoid.

An alternative is presented by Atkins in which vomer is not homologous with parasphenoid. Instead, parasphenoid is lost or absorbed and vomer takes over its role to support the nasal septum.

Clinical Correlations

Deficiency of the vomer is noted at the inferior and posterior border. As this worsens, it recapitulates in reverse, the developmental sequence of the bone. The defect moves superiorly and anteriorly. The effect is to lift the vomer out of its proper horizontal fusion plane with the palatal shelves. Isolated deficiency of the vomer leads to a cleft of the secondary hard palate. As the vomer is lifted out of the plane of the hard palate, closure cannot be achieved. Diagnosis of an isolated case occurs when the maxillary palatal shelfs are close apposed, but unfused; and the vomer is recessed. Combined cases of vomerine deficiency and horizontal maxillary shelf deficiency are common.

Fig. 8.163 Vomer in situ. Paired vomer bones fuse in the midline and are cradled in bilateral crests running the length of the hard palate from incisive foramen to just posterior border of palatine bone. They articulate with the presphenoid, delimineated by sphenoid sinus. (Reprinted from Lewis, Warren H (ed). Gray's Anatomy of the Human Body, 20th American Edition. Philadelphia, PA: Lea & Febiger, 1918)

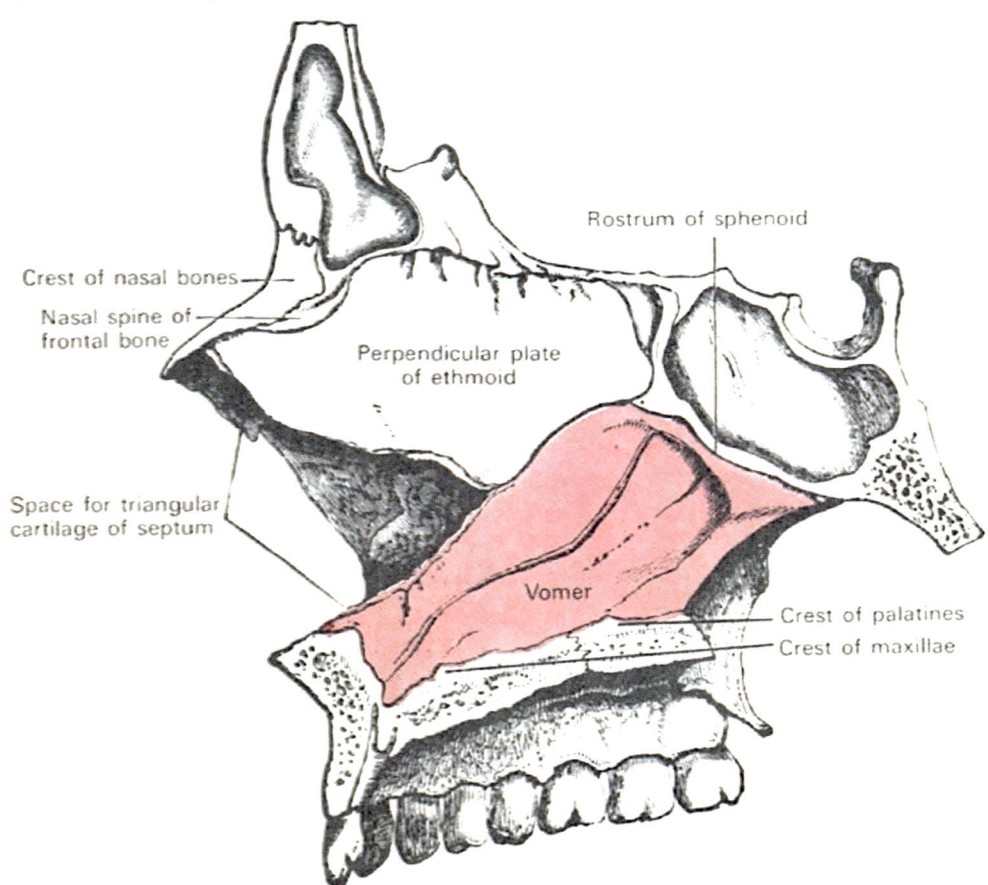

The Premaxillary Complex

Premaxillae are paired dermal bones in series with the nasals and the maxillae. They define part of the nares. In paleontology when a structure has multiple functions, it tends to be stable over time. This is true for premaxilla. (1) It bears teeth essential for biting and holding prey. (2) It supports the internal nares choanae. (3) It defines the chemosensory accessory olfactory system. (4) Premaxilla can form a keratinous beak. (5) As the most anterior bone, it is small and robust, capable of absorbing trauma while protecting the rest of the rostrum. (6) Premaxilla in fishes is mobile and jointed to gape open to receive take in large morsels (Figs. 8.165, 8.166, 8.167, 8.168, 8.169, 8.170, 8.171, 8.172, 8.173, 8.174, 8.175, 8.176, 8.177, 8.178, 8.179, and 8.180).

Both premaxilla and maxilla were tooth-bearing from the start, with multiple units, Dinosaurs, such as *Tyrannosaurus*, had four sharp teeth per hemipremaxilla! Despite its presence in all non-homid tetrapod skulls some authorities, such as Ashley-Montague, have denied its existence in man. This is simply not the case. The developmental history of the premaxilla is definitively reviewed by Barteczko. Furthermore, premaxillae are among the dermal jaw bones documented in the first stem gnathostomes 419 million years ago.

Descriptive Anatomy

Premaxilla articulates between vomer posteriorly and maxilla laterally. It has three distinct subfields, defined on the basis of the deciduous teeth. These are from medial to lateral: central incisor (PMxA), lateral incisor (PMxB), and frontal process (PMxF), an upward extension from the lateral incisor zone. Frontal process articulates laterally with the frontal process of the maxilla (MxF) and superiorly with nasal bone. In fetal life, these two units are separated by a visible suture. With time MxF growth exceeds that of PMxF, the latter becomes internalized within the piriform fossa, and the suture between them disappears before birth. As a result of this lamination, the piriform rim is bilaminar. The extra-rigid piriform is considered a "buttress" of the face (ideal for placement of surgical reconstruction plates).

Blood Supply and Mesenchyme

The premaxillary complex develops in the membrane from r2 neural crest. It is programmed by r2 oral ectoderm. Each premaxillary lamina is supplied by the distal segment of medial nasopalatine artery, a stapedial V2 induction product. Because arterial blood flow is proximal to distal, the biologic "age gradient" is: PMxA > PmxB > PMxF. For this reason, central incisor erupts prior to lateral incisor. Dental develop-

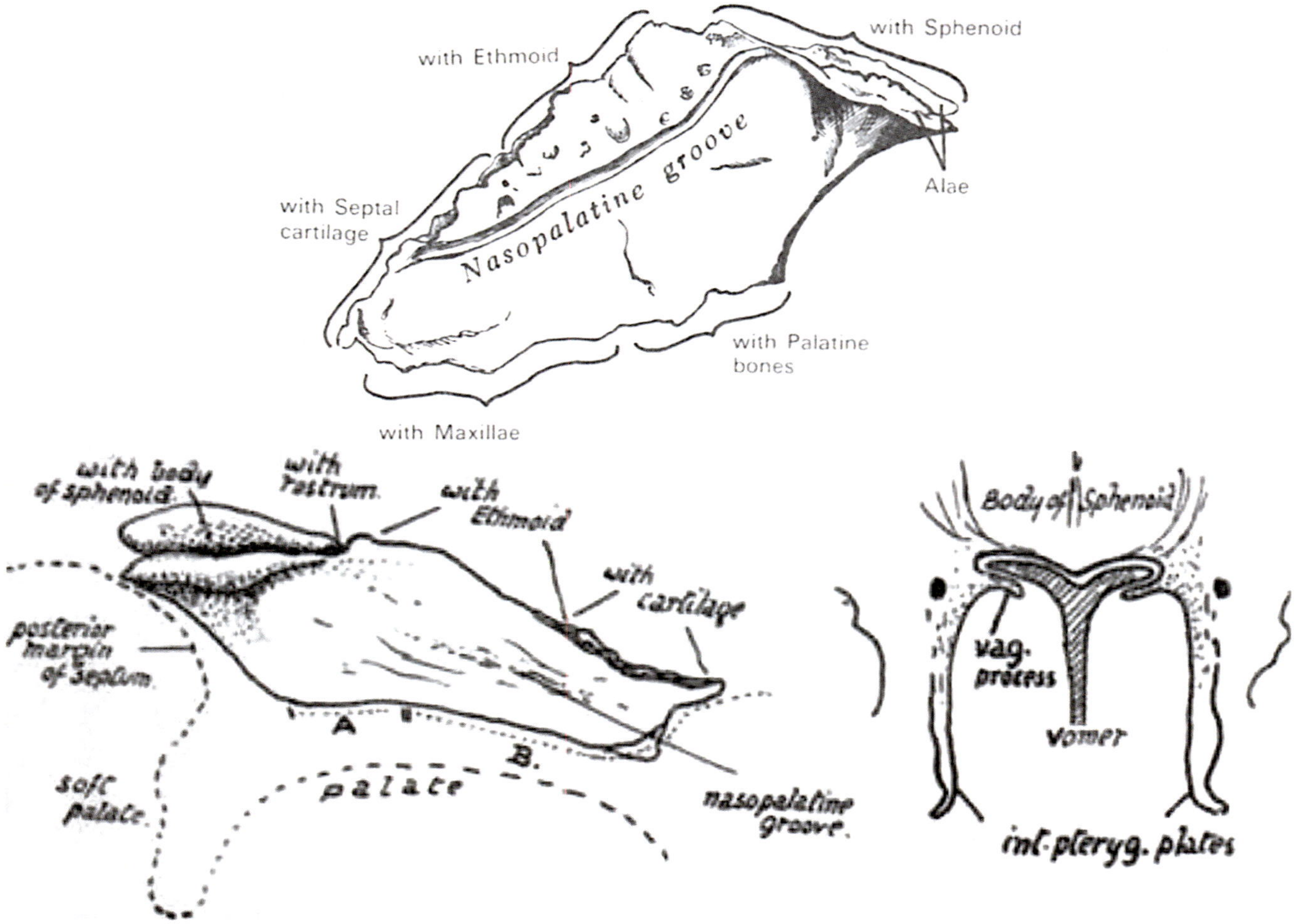

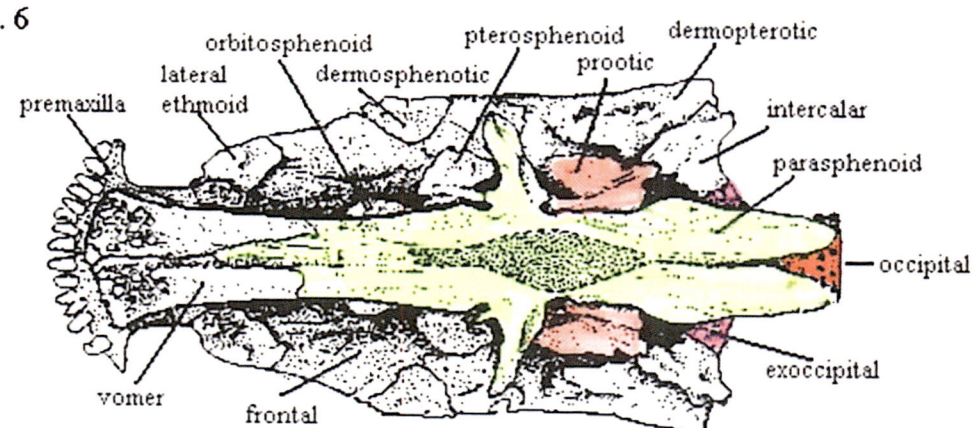

Fig. 8.164 Vomer (Frazer). Vomer is the fusion of paired fields. Nasopalatine neurovascular pedicle on either side. The cephalic half above the groove thickens and flares out posteriorly as the alae. These actually interdigitate with presphenoid. (Reprinted from Breathnach AS (ed). Frazer's Anatomy of the Human Skeleton, 6th ed. London: J&A Churchill; 1907)

Fig. 8.165 Parasphenoid covers the cranial base from the vomers to the occiput. Its coexistence with vomers rules it out as the homolog for the vomer. (Reprinted from Grande L, Bemis W. A comprehensive phylogenetic study of Amiid fishes based on comparative skeletal anatomy: an empirical search for interconnected patterns of natural history. J Vert Paleontology 1998; 18(suppl):1–681. With permission from Taylor & Francis)

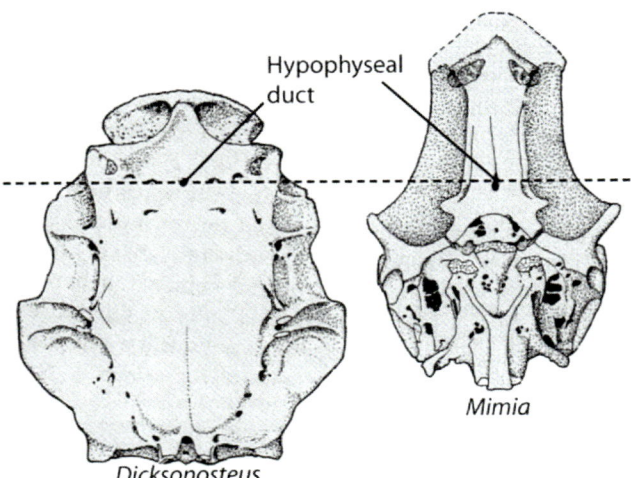

Fig. 8.166 Prehistory of parasphenoid. Palate of placoderm versus fish showing duct leading to pituitary. *Dicksonosteus* is a primitives arhrodire placoderm of the early Devonian. As such, it has primitive scissor-like jaws and no palatine bones so the duct is a direct passage from adenohyphyseal placodes upward through the cranial base to the brain. Parasphenoid is seen in the prehistoric actinopterygian fish *Mimi*. Here hypophyseal duct runs through the parasphenoid which has emerged as a dermal ensheathment covering the forebrain all the way back to basisphenoid. Note the lateral flanges in contact with the otic capsule. In mammals, the parasphenoid is absorbed and the duct is known as Rathke's pouch. (Reprinted from Grande L, Bemis W. A comprehensive phylogenetic study of Amiid fishes based on comparative skeletal anatomy: an empirical search for interconnected patterns of natural history. J Vert Paleontology 1998; 18(suppl):1–681. With permission from Taylor & Francis)

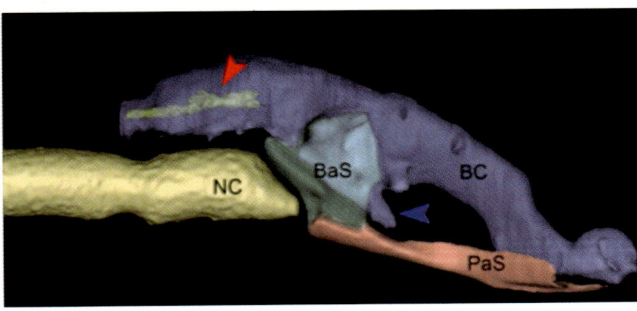

Fig. 8.167 Buccophyophyseal canal. Lateral sagittal view showing the notochord (yellow), the cranial cavity (purple) containing a very small and posteriorly placed brain (red arrowhead), the basisphenoid (green) and the parasphenoid (red). Same structures in closer view; the red arrowhead indicates the brain; the much elongated pituitary gland extends from the brain towards the skull base in a groove formed by the basisphenoid (blue arrow). *BaS* basisphenoid, *BC* cranial cavity, *NC* notochord, *PaS* parasphenoid. (Reprinted from Khonsari RH, et al. The buccophyphyseal canao is an ancestral vertebrate trait maintained by modulation in sonic hedgehog signaling. *BMC Biology* 2013; 11:27. With permission from Creative Comons License 2.0: http://creative-commons.org/licenses/by/2.0)

ment follows the *law of vascular priority*. The eruption sequence is determined by the vascularization sequence.

As we shall see, premaxilla and maxilla share with mandible a common dental developmental design; otherwise, we should have no occlusion. This is seen in their vascular design. Mandible is supplied by inferior alveolar artery accompanied by three sensory branches of V3, one for each dental group (incisors/canine, bicuspids, tricuspids). In the upper jaw, the necessity of an airway causes a 3-way distribution of blood supply from the sphenopalatine axis with nasopalatine to premaxilla, anterior superior alveolar to ectopic incisor/canine/bicuspid and posterior superior alveolar to the tricuspids. From sphenopalatine fossa, a common stem passes over the palatine notch and divides with lateral nasopalatine serving inferior turbinate and the inferolateral nasal wall above the palatal shelf. The medial nasopalatine swerve over behind the vaginal process of sphenoid where it encounters the vomerine ala and descends, supplying it with collaterals to lower septum. Medial nasal process (MNP) enters the premaxilla supplying central incisor field followed by those of lateral incisor and frontal process.

A final note is regarding the lability of the lateral incisor. The so-called "ectopic" incisor is a common occurrence but, in reality, it merely reflects the inherent ability of maxilla to produce such a unit. For this reason, the medial division of anterior superior alveolar artery sends a separate medial branch to the frontal process of maxilla, the potential housing of the ectopic incisor, and a lateral branch to canine. *When present, ectopic maxillary incisor always erupts before canine.* There are clinical circumstances in which the maxillary lateral incisor erupts but the premaxillary lateral incisor fails to develop as part of a cleft sequence.

Development

The arrival of distal neurovascular pedicle brings neural crest mesenchyme into the zone. Because this is the most distal of all the StV2 arteries, it makes sense that it would follow that of the maxilla, i.e., the lateral sphenopalatine axis. For that reason, ossification of the maxilla is noted first at stages 16–17. Premaxillary ossification, as described by Mall, first appears at the incisor region at stages 17–18 (week 7), *one full stage after the appearance of ossification over the canine.* Barteczko notes three ossification centers, one over each incisor and one at the lateral piriform process, i.e., PMxF. Recall that when this takes place, the *primarer nasenboden* is still intact, that is, there is absolute mesenchymal continuity between both sets of fields. In point of fact, the "fault line" that develops in the primary nasal floor is repre-

Fig. 8.168 Parasphenoid phylogeny. The parasphenoid bone forms dermal lining beneath the entire length of the cranial base. In mammals, it fuses with basisphenoid, remaining as an articulation. It is not homologous to vomer. (Reprinted from Franz-Odendaal TA, Atkins JB. The evolutionary and morphological history of the parasphenoid bone in vertebrates. Acta Zoologica 2016; 97(20):255–263. With permission from John Wiley & Sons)

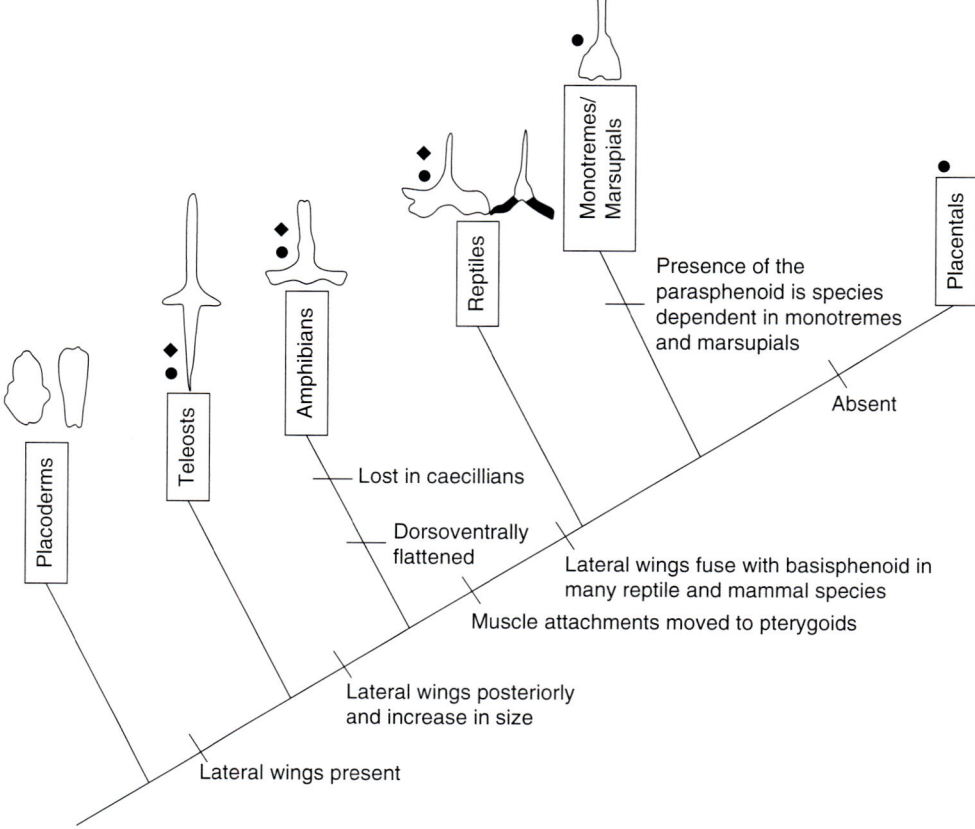

Fig. 8.169 Dinosaur premaxilla. (Reprinted from Keith A. Human Embryology and Morphology. London: Edward Arnold; 1902)

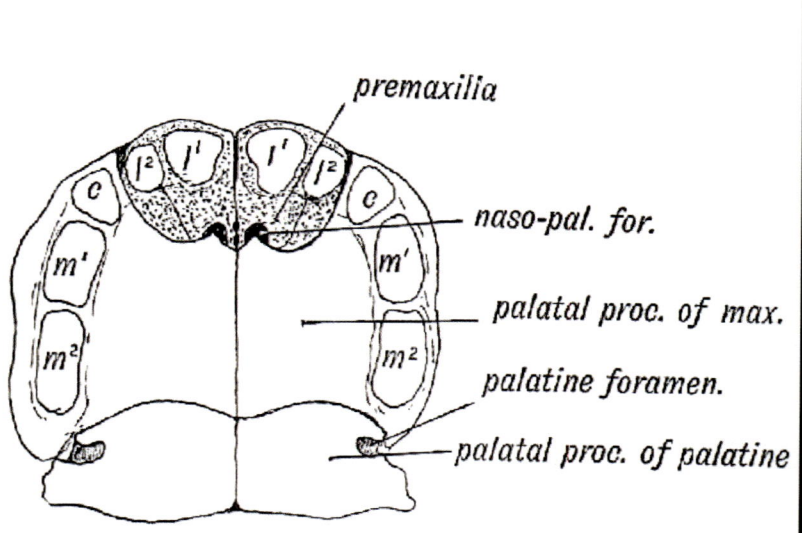

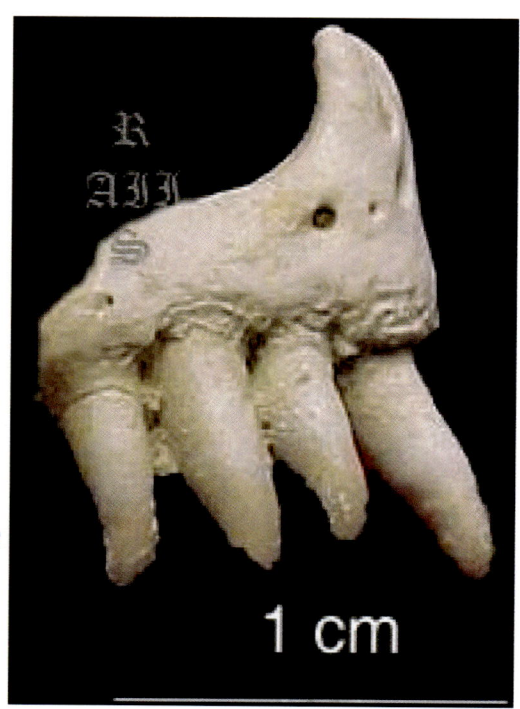

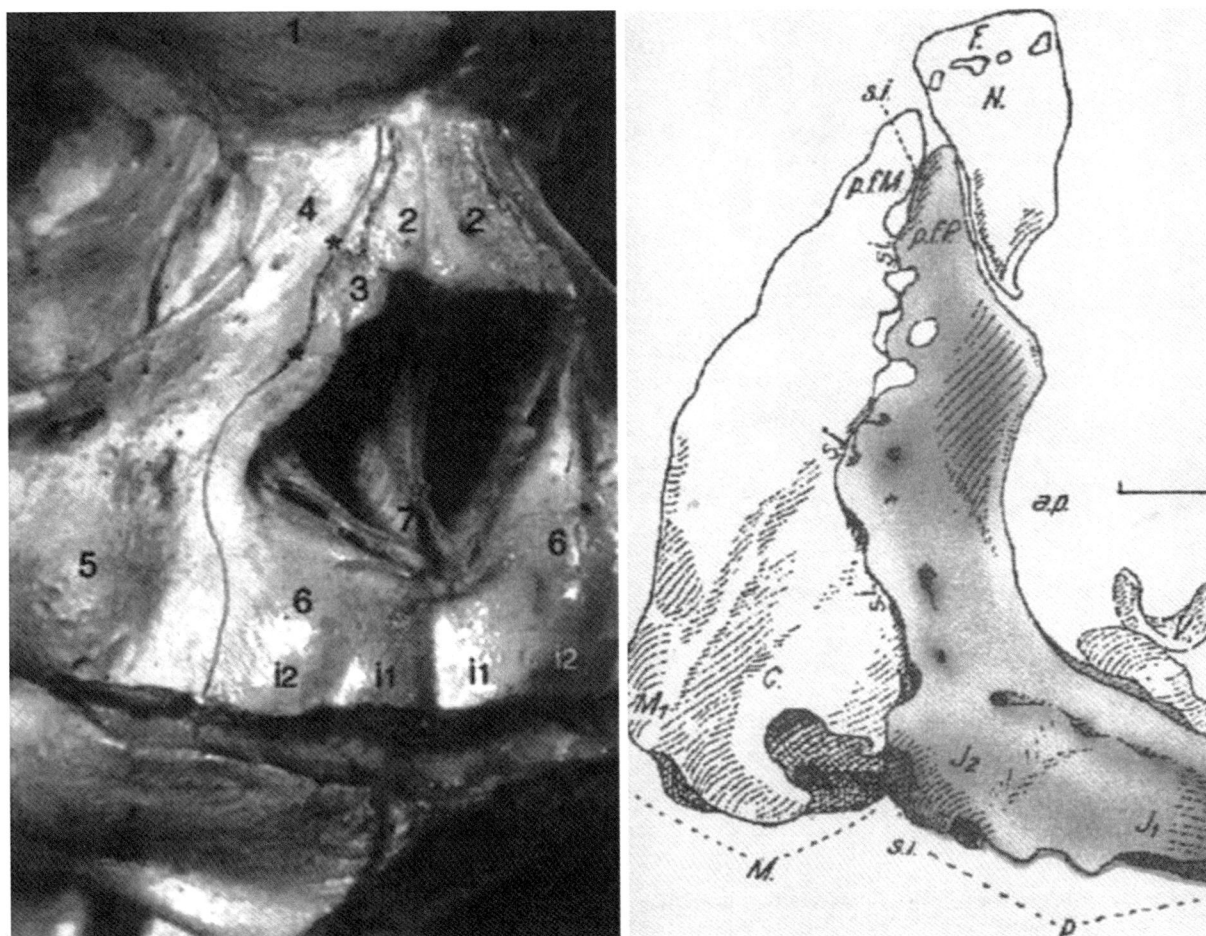

Fig. 8.170 Premaxilla in fetus at 7 months. Sutura incisiva is well seen. 1, frontal bone; 2, nasal bone; 3, frontal process of maxilla; 4, frontal process of premaxilla; 5, maxilla; 6, premaxilla; 7, nasal spine (processus Stenonianus nasalis). (**b**) Reconstruction by Felber at 5 months showing side-by-side position of the frontal processes. Superimposition of MxF over PMxF creates a bilaminar piriform rim.

[Left: (Reprinted from Barteczko K, Jacob M. A re-evaluation of the premaxillary bone in humans. *Anat Embrol* 2004; 207:417–437. With permission from Springer Nature). Right: (Reprinted from Felber P. Anlage und Entwecklung des Maxilllare und praemaxillare beim Menschen. *Gegenbaurs Mophol Jb* 1919; 50:451–499)]

sents a programmed cell death along the interface between the medial and lateral nasopalatine angiosomes. Ossification appears much later, after the primary nasal floor has split open. Development of the premaxilla from the septo-premaxillary junction is downward and forward. From PMxB, a vertical process develops which ascends in tandem with MxF until the frontal bone is reached. From the posterior aspect of premaxilla, a palatal shelf (PMxP) develops in medio-lateral fashion. This remains temporarily separated from the anterior margin of maxillary palatal shelf (MxP) by the *sutura incisive lateralis*.

The developmental gradient for the premaxillary fields follows the arrival of their mesenchymal population, thus lateral incisor (PMxA) > media incisor (PMxB) > frontal process PMxF. Vascular insufficiency hits these populations in reverse order. Thus, all cleft lip situations have a deficient or absent premaxillary frontal process defect

causing a "scooping out" of the bone stock of the piriform fossa. As we shall see, the reduction in bone stock has quantitative consequences for the production of local biochemical signals essential for normal soft and hard tissue closure.

Stage 16: 5.5 weeks
- Prolabium begins to descend from medial nasal process.

Stage 17: 6 weeks

- Prolabium contacts first the medial nasal process, then the lateral lip element: lip fusion begins.
- Breakdown of primarere nasenboden.
- PMx swellings appears.
- Maxillary palatal processes (MxP) appear, hanging vertically.

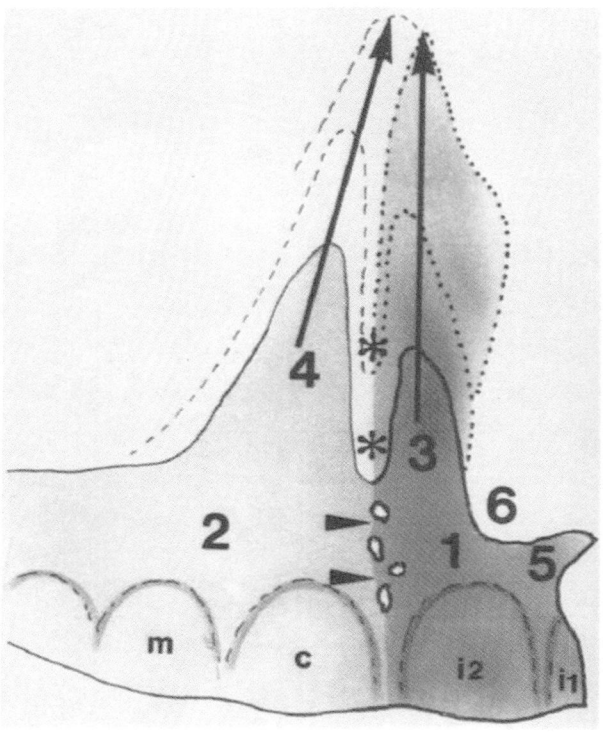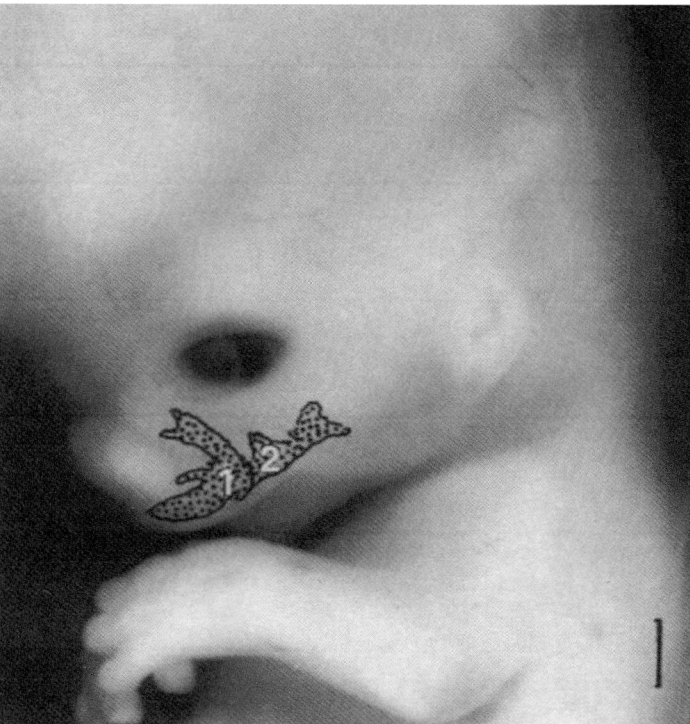

Fig. 8.171 Growth pattern of frontal processes of premaxilla and maxilla, line of fusion marked by asterisks. Embryo, stage 23, showing premaxilla (1) and maxilla (2) in situ. Note the size of the premaxillary frontal process, PMxF. (Reprinted from Barteczko K, Jacob M. A re-evaluation of the premaxillary bone in humans. *Anat Embrol* 2004; 207:417–437. With permission from Springer Nature)

Stage 18: 6+ weeks

- Lip closure complete.
- Prominent premaxillary swelling seen on SEM.
- Maxillary ossification center separates from dentary (16 mm).

Stage 19: 7 weeks
- Premaxillary ossification now present (18 mm).
Stage 20: 7+ weeks
- Premaxilla and maxilla have extended to length over the dentary and PMxF is developing (20 mm).
Stage 21: 7++ weeks
- 3-D expansion (23 mm).
Stage 22: 8 weeks

- Premaxillary alveolar process fuses to maxillary alveolar process, primary palate closes proximal–distal, dorsal–ventral, and medial–lateral (25 mm).
- MxPs elevating.

Stage 23

- PMx alveolae are still separate, palatine process of PMx appears, nasal septum fused to PMx (29 mm).

- Prenoverine process present with nasopalatine artery in incisive canal (32 mm).
- MxPs now level.

9 weeks and beyond

- Palatal shelves of PMx bone develop medial-to-lateral.
- MxPs moving forward to close the *sutura incisive lateralis* (41 mm).
- Palatal part of premaxillary bone more prominent (45 mm).
- Incisors surrounded by bone (55 mm).
- Sutura incisiva still separates PMxP from maxillary palatal process MxP.
- Central incisor appears, followed by lateral incisor (68 mm).

Phylogeny

Ancient History: In Media Res
The premaxilla is a very ancient bone. It made its debut with the first appearance of dermal bones in prehistoric armored fishes, the extinct Placodermi. As a matter of fact, the presence of dermal bones around the mouth is diagnostic of ver-

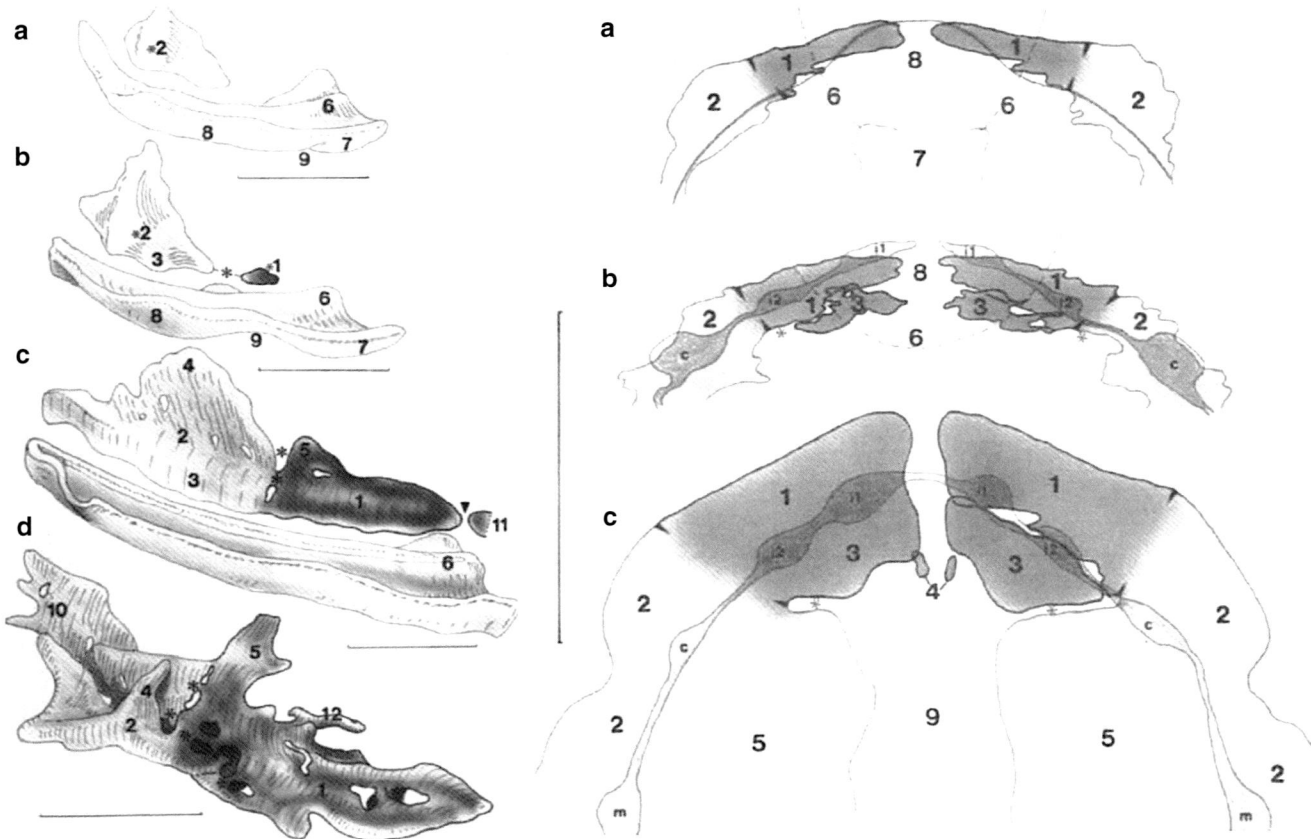

Fig. 8.172 Development of right upper jaw: (**a**) 6 weeks (stage 18); (**b**) 7 weeks-early (stage 19); (**c**) 7 weeks-middle (stage 20); (**d**) 9 weeks. 1, Premaxilla; 1, premaxillary ossification center; 2, maxilla; 2, maxillary ossification center; 3, facial alveolar wall; 4, frontal process of maxilla; 5, frontal process of premaxilla; 6, dental ridge; 7, outer contour of medial nasal process; 8, contour of maxilla; 9, nasolacrimal sulcus (sic); 10, palatine bone; 11, part of left premaxilla; 12, early formation of piriform aperture. (Reprinted from Barteczko K, Jacob M. A re-evaluation of the premaxillary bone in humans. *Anat Embrol* 2004; 207:417–437. With permission from Springer Nature)

tebrate line leading ultimately to humans. Let us take a closer look at the history of the premaxilla (Figs. 8.170 and 8.171).

Jaw were invented in early in the Paleozoic. The armored fishes, the extinct Placodermi, had a macromeric head skeleton made up of large plate-like bones. These gave rise to (1) chondrichthyans with purely cartilaginous skeletons (and so not of interest) and to (2) osteichthyans (bony fishes) which divided into two clades, according to the type of skeletal support for their fins: actinopterygians (ray-finned fishes) and our ancestors, the sarcopterygians (fleshy-finned fishes).

Although placoderms and osteichthyans both have macromeric dermal skeletons, it was thought that that these bones had no homologies and two completely different nomenclatures arose. Furthermore, two lines produced by the placoderms produced two lines with a micromeric dermal skeleton: sharks with complete loss of bone and the extinct acanthodians, a sub-group of ray-finned fishes. How did one condition lead to the other? The prevailing micromeric crown model required that macromeric placoderm skull was replaced by a micromeric condition in an unknown

ancestor. Subsequently the osteichthyans would re-acquire the macromeric condition de nova, with consequent development of new and non-homologous bone. Yet even in the earliest osteichthyans, certain placoderm features persisted, such as a multi-component shoulder girdle with dermal elements.

In sum, the beak-like skull of placoderms seemed to bear no relationship to ours. Until 2013, paleontologists thought that the common ancestor of living jawed vertebrates had no recognizable facial bones. The placoderm condition degenerated into the shark with a cartilaginous skeleton and a covering of small bony plates. Somehow bony fishes evolved an independent of large facial bones and modern jaws were re-invented.

All this changed in 2013, with the discovery of *Entelognathus primordialis*, a 419 million years old jawed fish of the Silurian was discovered in China. This fossil combines osteichthyan-like dermal bone of the first and second arches, specifically identifiable premaxilla, maxilla, and dentary along the jaw margins, with a placoderm-like braincase, skull roof, and pectoral girdle. As this

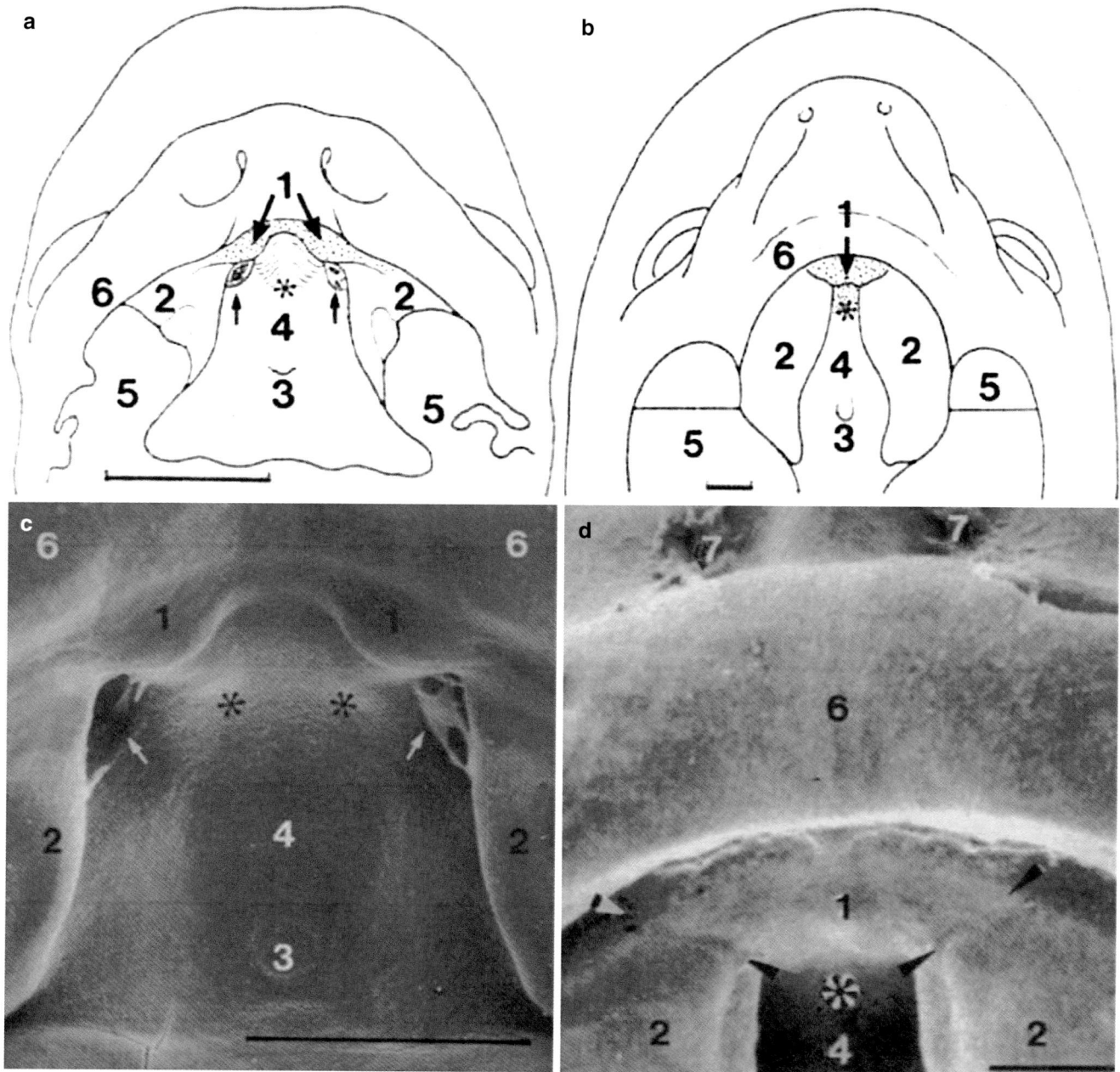

Fig. 8.173 SEMs of oral cavity at stages 19 and 22 (**c** and **d**) with accompanying drawings (**a** and **b**). Note mandible removed. 1, Premaxilla; intermaxillary bulge; behind premaxilla 2, secondary palate; 3, Rathke's pouch; 4, roof of mouth; 5, removed mandible; 6, upper lip; 7, nasal plug. Stage 19 shows the opening of the choanae. Early alveolar ridge can be appreciated. Pemaxillae still unfused. Note at stage 22 sutura incisiva transversa as a depression between palatal process of premaxilla (1) and palatal process of maxilla (2). Alveolar ridge will be defined. (Reprinted with permission from Hinrichsen K. Early development of morphology and patterns of the face in the human embryo. *Adv Anat Embryol Cel Biol* 1985;98:1–79)

Fig. 8.174 Placoderm skull 1—before the appearance of premaxilla. Jaw bones were unitary constructs. They had several oral plates: <u>supragnathic anterior</u>, <u>supragnathic posterior</u>, and <u>infragnathic</u> These became in the maxillate pacoderms (*Entelognathus* and *Qilinyu*) the premaxilla, maxilla, and dentary bones. (Reprinted from Wikimedia. Retrieved from: https://commons.wikimedia.org/wiki/File:Dunkleosteus_terelli_-_placoderm_fish_skull_-_Smithsonian_Museum_of_Natural_History_-_2012-05-17.jpg. With permission from Creative Commons License 2.0: https://creativecommons.org/licenses/by-sa/2.0/deed.en)

combination had never been seen before in stem gnathostomes so the description uses osteichthyan terms for the bones of the mandibular and hyoid arches and placoderm terms for the armored skull and trunk (Figs. 8.172, 8.173, 8.174, and 8.175).

The snout of *Entelognathus* is short and pointed. It is made up of premaxillae covered by premedian and then rostrals, all fused in the midline. The nares are triangular, sandwiched between the rostrals and the sclerotic ring around the eye. (The eyes of *Entelognathus* are immobile). The premaxilla has a palatal lamina directed posteriorly to contacts a palatal lamina from the maxilla. Premaxilla also has a facial lamina that extends outward to touch the lacrimal. Within the mouth, the supragnathal bone characteristic of placoderms is not seen. No vomer is present. *Thus, the mesenchyme of premaxilla evolved separately from that of vomer.* Recall that the stapedial system that supplies both bone fields is an invention concomitant with the incorporation of the second arch into the jaws as seen in true osteichthyans. It is possible that the stapedial system did not exist at the time of Entelognathus. The latter is a derived bone field taking the advantage of a common blood supply.

Cheek bones are present in *Entelognathus* as well. These include maxilla, lacrimal, jugal, and possibly quadratojugal.

Fig. 8.175 Placoderm skull 2. Note the three primitive bones of the jaws: anterior superognathal, premaxilla; posterior superognathal, maxilla; and inferognathal, mandible. Note that cucullaris spans from anterior dorsolateral to paranuchal. It acts as an extensor of the head. (Reprinted with permission from: GEOL431—Vertebrate Paleobiology course web resources. Thomas Holtz and John Merck, Department of Geology, University of Maryland, College Park, https://www.geol.umd.edu/~jmerck/geol431/)

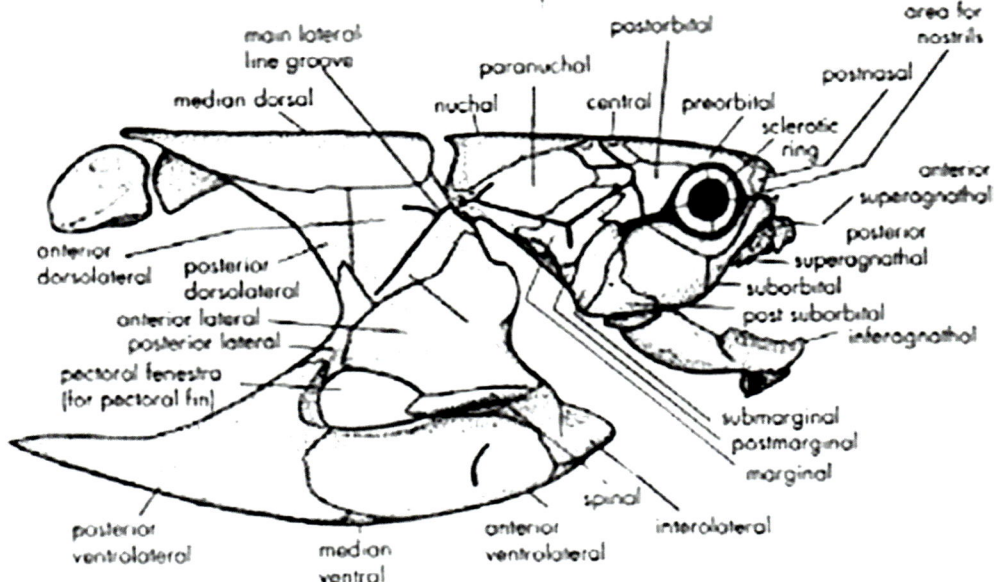

Entelognathus   Qilinyu

Fig. 8.176 *Entelognathus* and *Qilinyu*. Maxillate placoderms of the Silurian, 419 million years ago. [Left: (Reprinted from Zhu M, Yu X, Ahlberg PE, et al. A Silurian placoderm with osteichthyan-like marginal jaw bones. Nature. 2013 Oct 10;502(7470):188–93. With permission from Springer Nature). Right: (Reprinted from Zhu M, Ahlberg PE, Pan Z, et al. A Silurian maxillate placoderm illuminates jaw evolution. Science 2016; 354:6310. With permission from The American Association for the Advancement of Science)]

 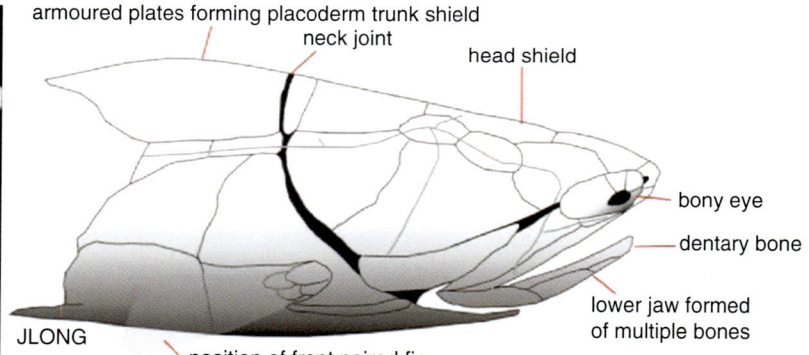

Fig. 8.177 *Entelognathus principalis* showing paired premaxillae—compare with diagram. Premaxillae (yellow) are roofed over by rostrals (pink) and flanked by maxillae (green). The skull is jointed. The eye is immobile. Mandible had dentary and three infradentary bones. Lacrimal, jugal, and qudratojugal also appeared. [Top: (Reprinted from Zhu M, Ahlberg PE, Pan Z, et al. A Silurian maxillate placoderm illuminates jaw evolution. Science 2016; 354:6310. With permission from The American Association for the Advancement of Science). Bottom: (Reprinted from The Conversation. Retrieved from: https://theconversation.com/extraordinary-missing-link-fossil-fish-found-in-china-18461. Courtesy of Prof. John A. Long)]

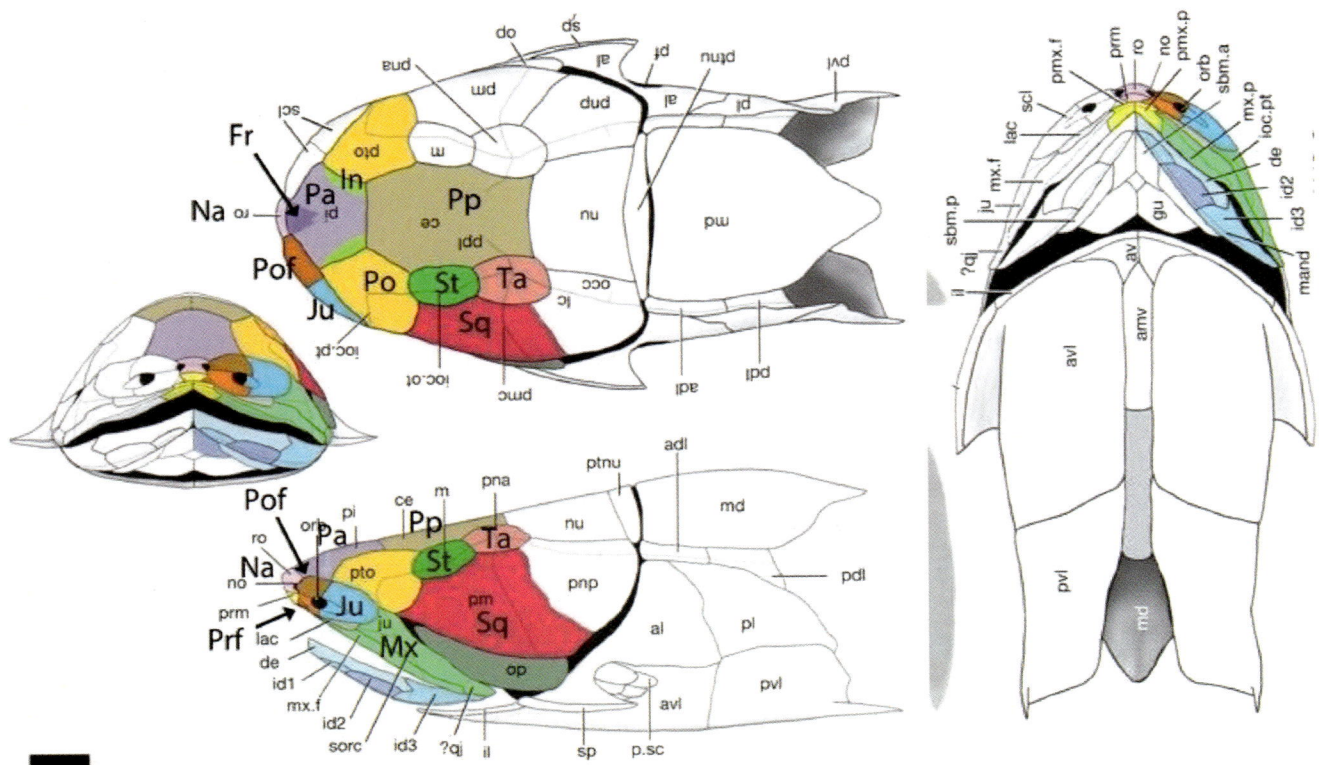

Fig. 8.178 *Entelognathus principalis.* Bone homologies. Premaxillae have three fields: **pmx.f** facial lamina of premaxilla, and **pmx.p**, palatal lamina of premaxilla. Premaxillae are roofed over by **ro** rostral bone (pink). Existence of pmx.f explains the source of a second ossification center and represents the future frontal process of premaxilla, referred to in comparative anatomy as the septomaxilla. Note premaxillae are not physically related to nostrils. Note three infradentary bones, id1–3. **Bone chart** shows color coding for bone fields homologous with derived fishes and tetrapods. Facial bones are in color and skull/trunk bones in white. Bone abbreviations: *adl* anterior dorsolateral plate, *al* anterior lateral plate, *avl* anterior ventrolateral plate, *amv* anterior medioventral plate, *av* anteroventral plate, *avl* anterior ventrolateral plate, *ce* central plate, *de* dentary, *gu* principal gular, *id1–3* infradentary first to third, *il* interolateral plate, *ioc.ot* otic branch infraorbital line groove, *ioc.pt* postorbital branch of infraorbital line groove, *ju* jugal, *lac* lacrimal, *lc* main line lateral groove, *m* marginal plate, *mand* mandibular line groove, *md* median dorsal plate, *mx.f* facial lamina of maxilla, *mx.p* palatal lamina, *no* nostril, *nu* nuchal, *occ* occipital cross commissure, *op* opercular, *orb* orbital fenestra, *pdl* posterior dorsolateral plate, *pf* pectoral fenestra, *pi* pineal plate, *pl* posterior lateral plate, *pm* postmarginal plate, *pmc* postmarginal line groove, *pmx.f* facial lamina of premaxilla, *pmx.p* palatal lamina of premaxilla, *pna* anterior paranaauchal plate, *pnp* posterior paranuchal plate, *ppl* posterior pitline, *prm* premedian, *ptnu* postnuchal plate, *pto* postorbital plate, *pvl* posterior ventrolateral plate, *p.sc* pectoral fin scales, *qj* quadratojugal, *ro* rostral plate, *sbm.a* anterior submandibular, *sbm* posterior submandibular, *scl* sclerotic plate, *sorc* supraoral line groove, *sp* spinal plate. (Courtesy of Mr. David Peters)

Maxilla presents an insertion for adductor mandibulae. The horizontal shelf of maxilla contacts that of premaxilla inside the margin of the upper jaw, constituting a biting surface for mandible. There are not teeth along the oral margin. Dermal bones medial to the maxilla (palatine, ectopterygoid, or the placodermal supragnathal) are not found. Quadratojugal is welded to maxilla and jugal; it is similar to postsuborbital in placoderms.

The mandible of Entelognathus also represents a step forward from placoderms. It has a dentary and three infradentaries but no teeth. Dermal ensheathing bones (coronoids, prearticular, or the placodermal infragnathal) are not present.

Entelognathus represents a combination of new osteichthyan dermal bones along the jaw margins with the conservation of the placoderm braincase and body armor. It shares

characteristics of its palatoquadrate with crown gnathostomes. In these latter, the metapterygoid portion is tall making the palatoquadrate in a "cleaver shape" that accommodates the adductor muscle medially. However, in *Entelognathus*, metapterygoid remains low, as in placoderms. This suggests that changes in the muscle anatomy and/or biting surfaced preceded the development in higher gnathostomes of a more advantageous palatoquadrate.

Let us summarize the importance of *Entelognathus*, and the premaxilla. First, until 2013, dermal marginal bones (premaxilla, maxilla, mandible) were unknown in placoderms; they were considered key synapomorphies of Osteichthyes. This discovery established the macromeric crown model of evolution. In this, homologies established with Osteichthyes remain, and further dermal bones continue to evolve. The micromeric condition seen in living chon-

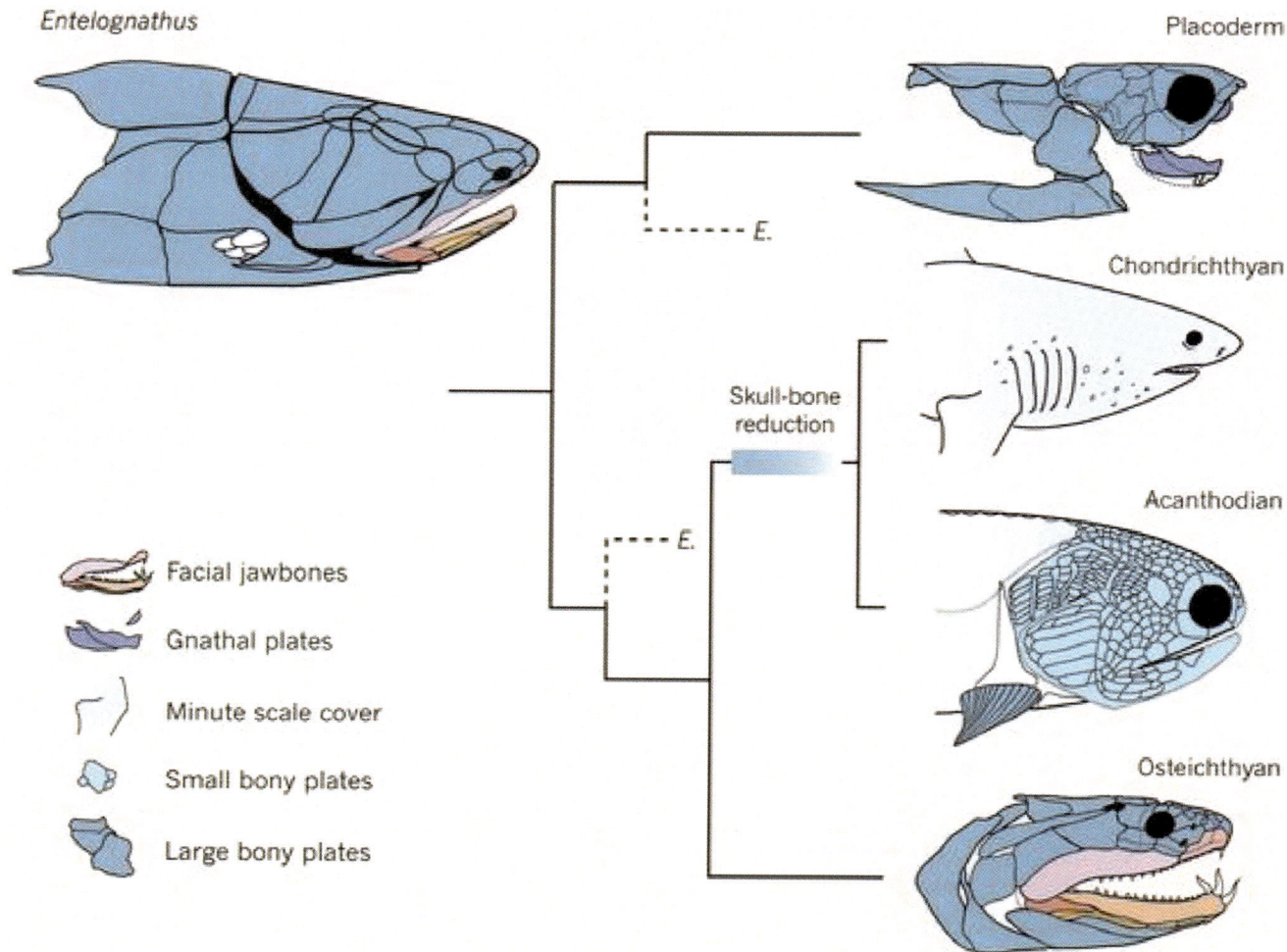

Fig. 8.179 Skeletal evolution. Chondrichthyans and acanthodian bony fishes have micromeric dermal skeletons made of scales. Osteichthyans and subsequent vertebrates have macromeric dermal bones. **Old hypothesis**: Placoderms were thought to have an exclusively macromeric dermal skeleton, characterized by plates. In crown gnathostomes, this was replaced by a micromeric condition which was perpetuated by the sharks and acanthodians. Derived bony fishes were thought to acquire a macromeric skeleton de novo. **New hypothesis**: *Entelognathus principalis* ("true jaw") combines a placoderm-like dermal skull roof, braincase, and shoulder girdle with a palatoquadrate (characteristic of gnathostomes) covered with dermal bones of the first and second arches. For this reason, nomenclature for *Entelognathus* combines osteichthyan terms for the branchial arch bones and placoderm terms for the rest of the skull roof and trunk. At critical stem point: (1) non-osteichthyans undergo skull bone reduction to a 100% micromere pattern; and (2) osteichthyans under skull expands to a 100% macromere pattern. (Reprinted from Friedman M, Brazeau MD. A jaw-dropping fish. Nature 2013; 502:175–177. With permission from Springer Nature)

drichthyes and the extinct acanthodians represents an acquired condition. Second, premaxilla is associated with an external contact with lacrimal which presages the frontal process. Third, premaxilla has an intraoral contact with maxilla, presaging the primary palate. Fourth, vomer appears later in evolution, certainly in the sarcopterygian line.

Ancient History: A La Recherce du Temps Perdu

Until the discover of Entelognathus, it had been assumed that the evolution of the tri-partite jaw (premaxilla, maxilla, and mandible) in modern bony fishes was an event independent from the bone structure of placoderms. But this posed a question for evolutionary biology. Were the simple blade-like jaws of the placoders the most primitive condition of the "first jaws" or were they more complex structures with multiple parts?

A second Silurian maxillate placoderm again from China reported in 2016 provides clarification. *Qilinyu rostrata* is considered <u>one step earlier</u> than *Entelognathus* and leads to the proposition that: (1) premaxilla, maxilla, and mandible are all homologous to the gnathal plates seen in placoderms, (2) all these bones represent a common dental arcade, and (3) the gnathal-maxillate transformation happened at the same time in both jaws (Figs. 8.172 and 8.176).

The body habitus of *Qilinyu* is quite different from *Entelognathus*. It had a face like a dolphin. It was a bottom-

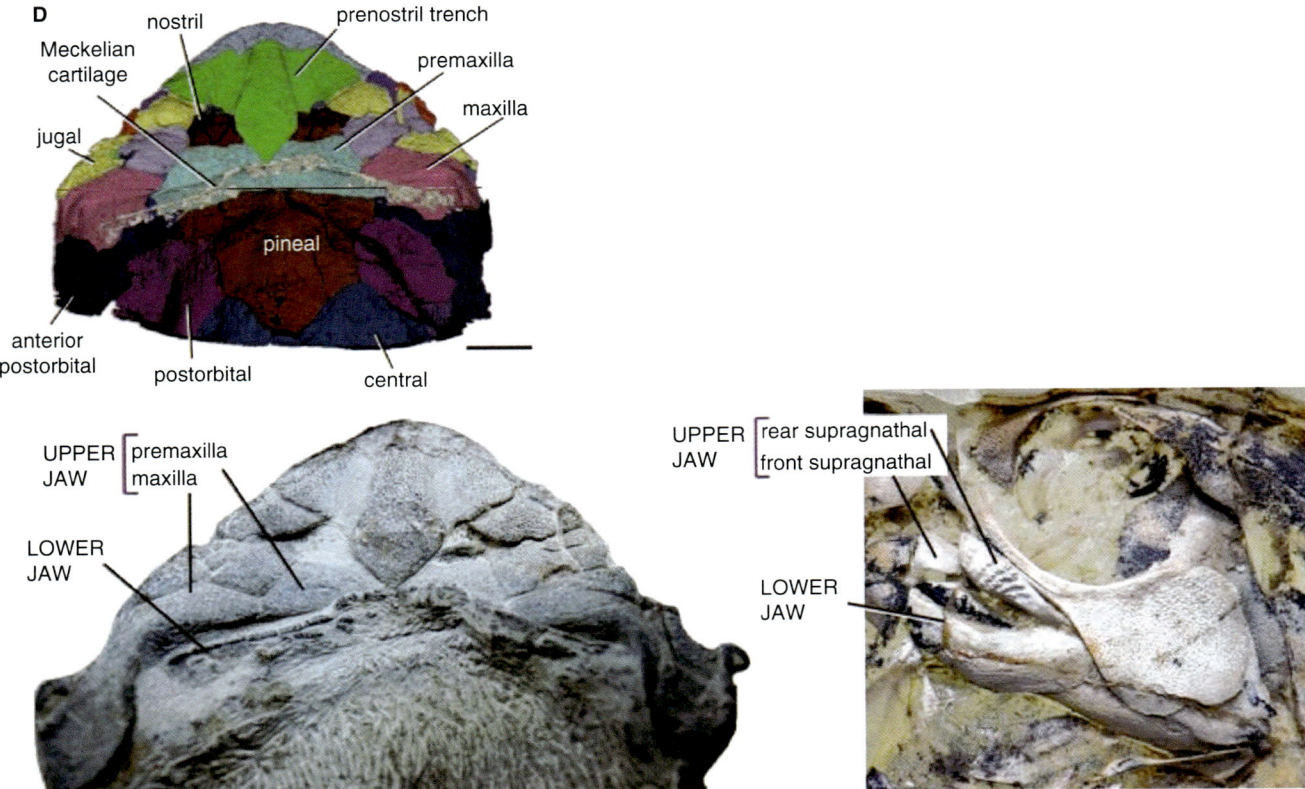

Fig. 8.180 *Qilinyu*. Premaxilla is located immediately below nostrils in Qilinyu. Its jaws are reinforced by dermal bones, a finding seen more extensively in Entelognthus. *Qilinyu* (left) has upper and lower jaws outside the palate, combined with many teeth inside the mouth as well. The upper jaws of *Bruntonichthys*, a conventional placoderm, have supragnathals and wholly within the palate. It has no external dermal bones. Unlike *Entelognathus*, the lower jaw of *Qilinyu* does *not* have external dermal bones. The jugal bone appears here as well and incorporated the upper jaw into the cheek. Extensive dermal bones connecting the palatoquadrate cartilage to the skull base develop later in the osteichthyans. (Reprinted from Zhu M, Yu X, Ahlberg PE, et al. A Silurian placoderm with osteichthyan-like marginal jaw bones. Nature. 2013 Oct 10;502(7470):188–93. With permission from Springer Nature)

feeder, as its mouth is ventrally positioned. Its eyes were mobile. The premaxillae and maxillae of both are edentulous, with broad palatal laminae. The dentary biting surface of *Qilinyu* is broad and flat whereas that of *Entelognathus* is narrow and sharp. Neither fish has dermal bones internal to the jaws.

Placoderms have three types of intraoral bones: the *infragnathal* on Meckel's cartilage, the *posterior supragnathal* arising from palatoquadrate, and the *anterior supragnathal* on the ethmoid (Fig. 8.171). In the past, these bones have been considered homologous with the inner dental arcades of maxilla (vomers, dermopalatines, ectopterygoids) and of the mandible (coronoids) as seen in ostheichthyans. In both number and position, the new model considers them as precursors to premaxilla, maxilla, and mandible. The major difference of the marginal jaw bones in these two species is that they have external facial laminae (they are visible externally), whereas in placoderms, they are not.

The implication of *Qilinyu* is that its dental arcade conserves the gnathals of the placoderms but that they acquire facial laminae. Somewhere between Entelognathus and crown ostetichthyan, the palatal laminae of premaxilla and maxilla were "lost" and replaced by a new series of palatal bone, such as vomers, dermopalatines ectopterygoids, pterygoids, coronoids, and the like. For a review of the lineage of the premaxilla, see Fig. 8.177.

The Septomaxillary Bone: Precursor of Frontal Process?

The original location of the nares was lateral between palatine, maxilla, and vomer but in sarcopterygian fishes, the brain expanded forward and forced the nares toward the midline where they became bounded by the premaxillae. In this regard, the origin of the frontal process of premaxilla is of interest. What relationship does this bone field have to lateral incisor? The answer lies with the *septomaxillary* a, small dermal bone just inside the naries. It has three possible configurations. Type A septomaxilla can be found either attached to, or as extension of lacrimal bone. It is involved in maintaining lachrymal duct. Type B septomaxilla is tucked inside of the naris in contact with vomer to form the floor of the nose. It extends over to the vomeronasal organ. Type C septomaxilla is a garden variety dermal bone that roofs over the naris, acting as a *ventral "add-on" to nasal bone*. It is almost <u>outside</u> naris and partially covers it. As such, type C is consistent with frontal process of premaxilla in humans. Muscles can be added on to all these variants as either constrictors or dilators of the nostril (Figs. 8.178, 8.179, and 8.180).

Septomaxilla appears in the Devonian period as the result of a reorganization of previously chaotic rostral bones that takes place in our ancestors, the tetrapodomorph fishes, resulting in sets of bones that regularized over time. Of interest are the *tectal series*, the forerunners of nasal and lacrimal, and the *rostral series* that around the edge of the nostril. Recall that in *Panderichthys*, the nasal bones are lined up medial–lateral. In the latter species, the nares are ringed by four bones: maxilla, premaxilla, lateral rostral, and anterior tectal. The anterior tectal bone lies dorsal to lateral rostral bone. The latter two bones are considered potential ancestors of septomaxilla. Note that the lateral rostral bone is interposed between premaxilla and maxilla (Fig. 8.178). During the course of our evolutionary line, *premaxilla slips underneath anterior tectal and lateral rostral to make the contact with maxilla*. That puts anterior tectal as the candidate for type C septomaxilla—and for the frontal process of maxilla.

Type C septomaxillae are found in basal tetrapods *Acanthostega* and *Ichthyostega*. After some experimentation, we find in early (anapsid) amniotes that type A becomes plesiomorphic, i.e., the ancestral trait. This makes sense because terrestrial animals need a lacrimal system. In the early anapsid, *Procolophon* septomaxilla is found sitting along the lateral rim of the piriform fossa (Fig. 8.179). The type A condition is continued into the synapsids, bridging between premaxilla and nasal bone in the basal therapsid *Haptodus* (Fig. 8.180). By the cynodonts, septomaxilla seems to have been incorporated into premaxilla as seen in *Morganucodon*, although it persists as a separate bone in monotremes, as in your favorite, the platypus.

Clinical Correlations

Tessier described a cleft between the medial and lateral incisors, terming it cleft zone 1. The tooth germs for lateral incisor are highly variable. Often, a supranumerary tooth is present on the maxillary side. Absence of the lateral premaxillary dental unit can occur in the absence or in the presence of a maxillary supranumerary. This produces distinct clinical pictures. (1) When the mesenchymal housing of B and the frontal process are intact, and no supranumerary is present, an intact arch remains with a space. (2) When the housing of B and PMxF are intact and a supranumerary is present when an intact arch remains with a small or malpositioned false "lateral incisor." (3) When PMxF is deficient or absent and B is present, lateral piriform volumes is reduced, leading to a clinical gradient varying between an incomplete cleft of the lip, primary palate or both. (4) When PMxF is absent, PMxB is grossly reduced, a complete cleft of the primary palate, accompanied by a cleft of the lip results. NB: This biology is covered extensively in Chaps. 15–17.

Inferior Turbinate (Maxillary Turbinal)

Descriptive Anatomy

The inferior turbinate (inferior choncha) develops in the membrane from r2 neural crest. It is positioned at the junction of two sectors of the nasal capsule: parietotectal cartilage for the roof and upper sidewall and paranasal cartilage for the lower sidewall. IT is programmed by r2 vestibular lining. It is supplied by the lateral nasopalatine artery and nerve (LNPA). A terminal branch of MNPA represents the medial division of the pterygopalatine segment of the maxillary artery (Figs. 8.181, 8.182, and 8.183).

The bone is spongy and is rolled up upon itself. If one flexes the wrist downward and curls the fingers into the palm, the dorsum of hand will represent the nasal surface of inferior turbinate, and the fingers scroll. It should be emphasized that inferior turbinate is not part of the r1 ethmoid complex. Slung between ethmoid above and palatine below, inferior turbinate is the center-piece of a three-bone complex making up the lateral nasal wall. A unique feature is that its base is unsupported. Like a Post-It® attached to a bulletin board with pins, inferior turbinate is suspended in mid-air from its neighbors via four processes.

Let us consider how IT achieves this in greater detail. Middle meatus, an aperture sheltered by the middle turbinate of the ethmoid, lies above inferior turbinate. The convex *medial surface* of IT faces the nasal cavity, as the "main humidifier of the nose," the spongy bone contains longitudinal grooves for blood vessels. The *lateral surface* of IT (i.e., the inner surface of the scroll) faces the lateral nasal wall. Anteriorly, the lateral surface is in contact with the lacrimal footplate, this aperture being the inferior meatus. Thus, *superior border* contains four articulatory processes, all of them facing laterally. The *inferior border* (the bottom of the scroll) hangs free.

The attachment sites of IT can be visualized by dividing the superior border into thirds. The anterior 1/3 abuts the *conchal crest of the maxilla*. The "copy cat" posterior 1/3 abuts the *conchal crest of the palatine bone*. The middle 1/3 bears three distinct processes. The lacrimal process is the most anterior. It articulates with the *descending process of the lacrimal bone*. It also articulates with the *nasolacrimal groove of maxilla* located just behind MxF (but lateral to PMxF). This interface represents a field boundary between the r1 lacrimal bone and the r2 inferior turbinate. As such, it creates the canal for the nasolacrimal duct. *The lacrimal sac is just like a sinus*, it represents the communication of oronasal mucosa into a potential space created between the r1

Dermal jaw bones from fish to human

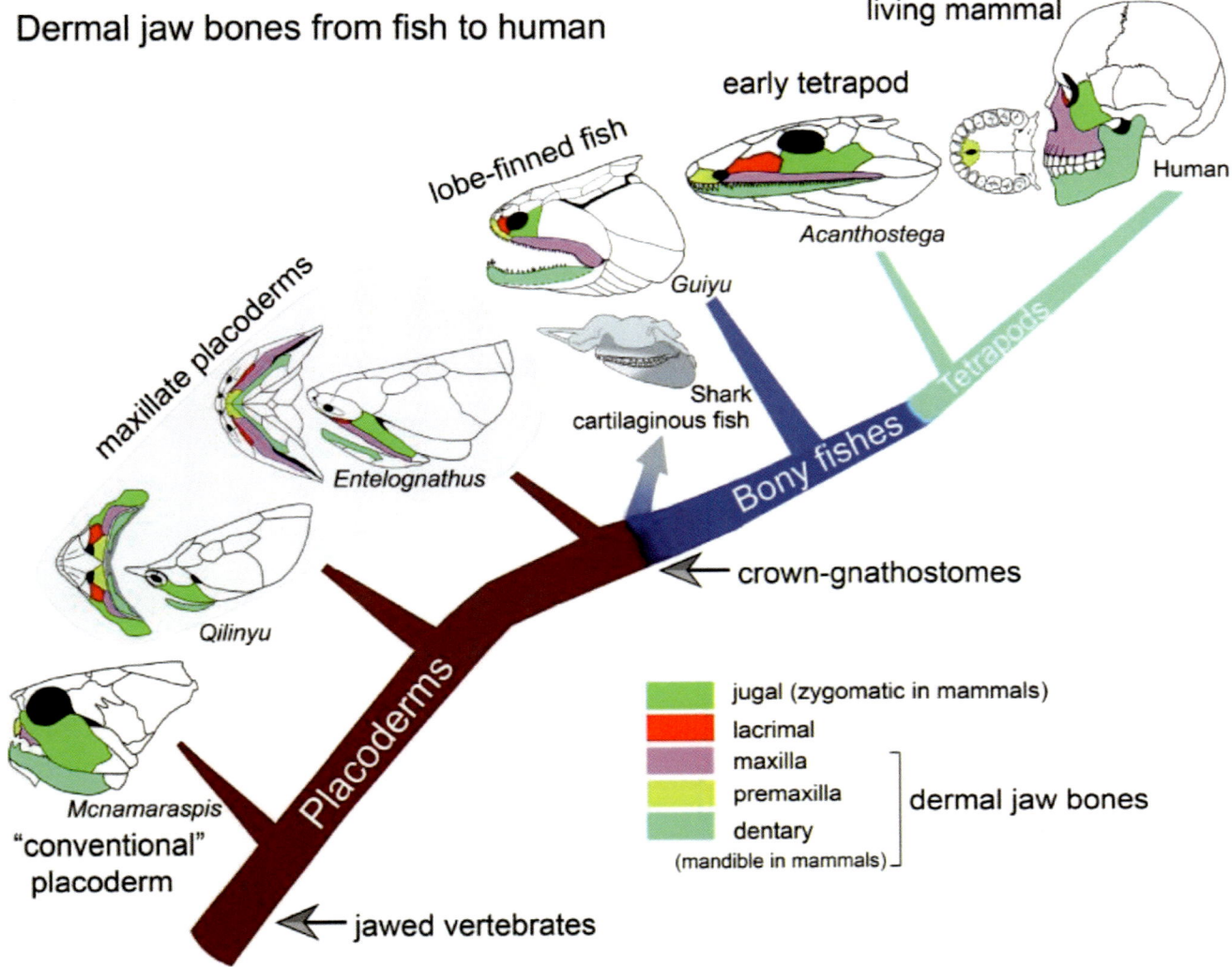

Fig. 8.181 Premaxillary phylogeny. Dermal jaw bones (premaxilla, maxilla, dentary) can be traced back prior to the advent of true osteichthyans (derived bony fishes). Ancient placoderms such as *Mcnamaraspis* and *Dunkleosteus* had no teeth, using gnathal bony plate for biting. These became the homologs of marginal jaw bones (premaxilla, maxilla, and mandible). (Courtesy of Science X)

lacrimal and the r2 frontal process of maxilla. Just behind the lacrimal process, IT sprouts an *ethmoid process*. This unites with the most dependent portion of ethmoid, the *uncinate process*. Extending from the ethmoid process like an afterthought, the *maxillary process* forms part of the medial wall of the maxillary sinus.

Blood Supply and Mesenchyme
Inferior turbinate is assembled from r2 neural crest. It is supplied by the lateral branch of nasopalatine artery, itself the medial branch of the StV2 arteries departing from IMM in the pterypalatine fossa. The artery snake over the top of the sphenopalatine notch where it divides. Lateral nasopalatine artery (LNPA) hugs the wall, as the exclusive supply of inferior turbinate and the mucoperiosteum covering the nasal surface of palatal maxillary palatal shelf. Distal branches of

LNPA reach the soft tissue of the nasal introitus where they anastose with ascending branches from superior labial.

The mesenchyme of inferior turbinate is r2 neural crest. It probably corresponds to a precursor structure. The homolog of the inferior turbinate was described by Wittmer.

Development
Inferior turbinate first appears in the lateral nasal capsule at stage 17. Its mesenchyme occupies most of the lateral wall, forcing the ethmoturbinals to develop above nasal septum. These subsequently move into a superior position in the lateral nasal capsule at stage 21. By stage 23, the middle meatus has formed creating a definitive separation between the turbinals. This is all consistent with the two-part phylogeny of the nasal capsule, consisting parietotectal cartilage (roof and upper sidewall) and paranasal cartilages (lower sidewall).

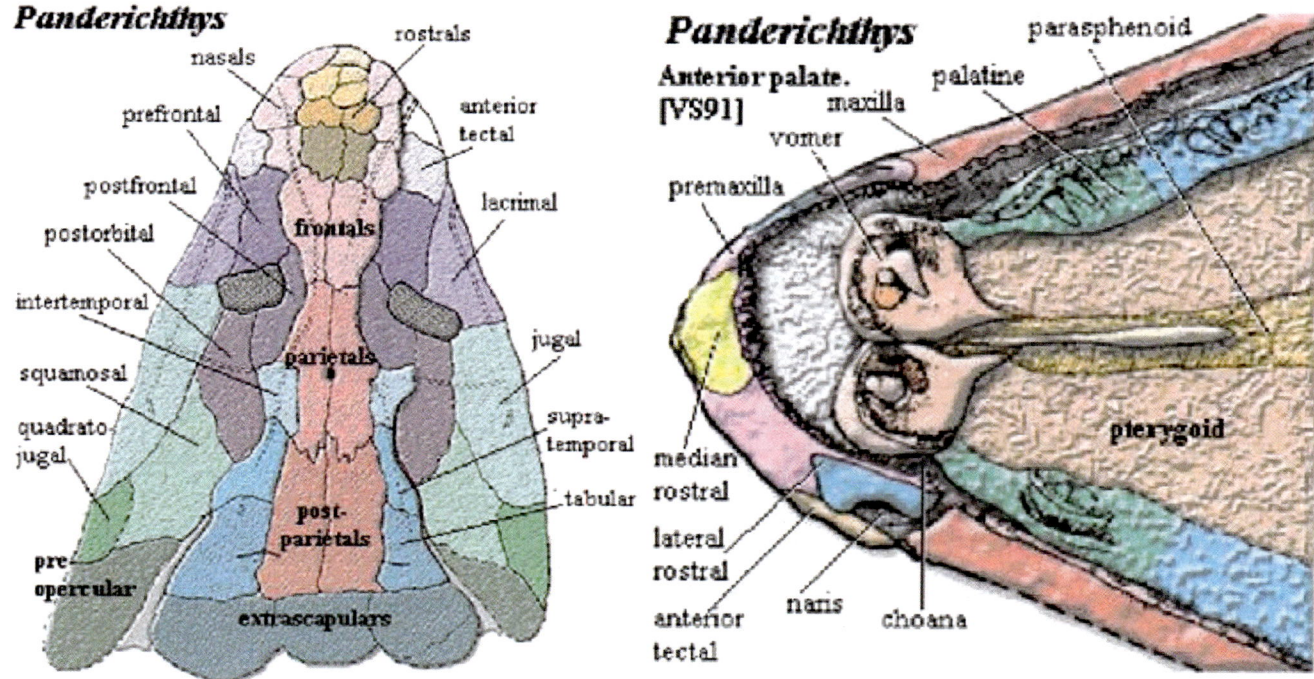

Fig. 8.182 Septomaxilla 1: Nares are ringed by maxilla, premaxilla (pink), anterior tectal (flesh), lateral rostral (blue). Note nasal bones paired anterior and posterior. Maxilla is separated from premaxilla by lateral rostral. Premaxilla *Panderichthys*. (Reprinted from Vorobyeva EI. Observations on two Rhipidistian fishes from the Upper Devonion of Lode, Lativa. Zool J Linnean Soc 1980; 70(2):191–201. with permission from Oxford University Press)

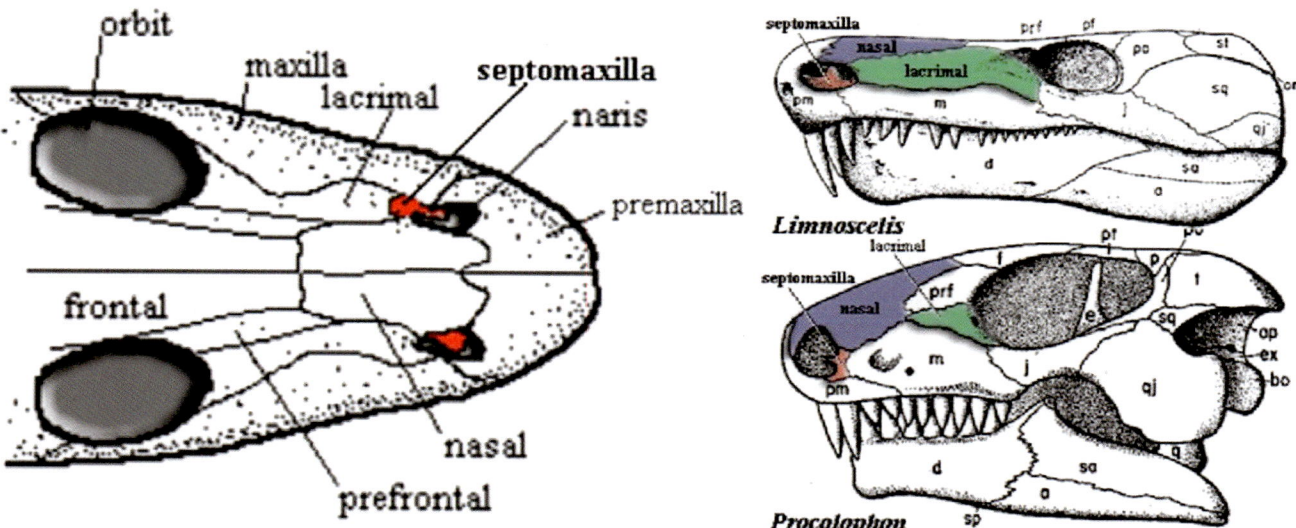

Fig. 8.183 Septomaxilla 2: In amniotes, type A septomaxilla is plesiomorphic (ancestral) and relates to the lacrimal system, essential for terrestrial life. (Reprinted from Gregory WK (ed). SW Williston's Osteology of the Reptiles. Harvard University Press; 1925)

At the fifth month of fetal life (32 weeks), ossification at the caudal margin of the cartilaginous lateral nasal capsule marks the first appearance of IT. The mesenchyme spreads anteroposteriorly to complete its articulations; then it projects into the nasal cavity to form the definitive scroll.

Phylogeny

The phylogeny of the inferior turbinate is difficult to ascertain in paleohistory. IT, if it existed before amniotes, was cartilaginous and delicate, impossible to fossilize. It is likely an apomorphy of amniotes. Wittmer describes IT as the homolog of the *primary concha*. As seen in lizards, reptiles, turtles, and mammals, all have the same structure. Primary concha appears in the nasal capsule at the junction of parietotectal cartilage and paranasal cartilage. It appears in development prior to the ethmoturbinals, it is near to (but *not* innervated by) V1, and it articulates with the lacrimal bone.

Primary turbinate exhibits an unusual behavior it that it can be multiple. Whereas in mammals, it is always singular, in the tuatara reptile *Sphenodon* it is doubled, and crocodiles it appears in triplicate. This is not a trivial issue and it teaches us a lesson that applies to other craniofacial bones, such as palatine. IT apparently violates the criterion of *conjunction* which holds that if two structures are homologous, they cannot appear in the same organism. We previously saw how Patterson's criteria applied to issue of the vomer versus parasphenoid. But in the case of the multiple turbinals, it is not that they are different structures, but rather a reiteration of a genetic program within the various points in the ethmoid and nasal capsule.

The Palatine Bone Complex

The palatine bone plays a key role in connecting the nasal wall, maxilla, and palate with the orbit. Its history is intimately connected with the evolution of the dermal bones of the palate. Over time the ancient joint between the skull base and the palate becomes consolidated. In this process, the history of the palatine bone is inseparable from that of the *pterygoid bone*, the "components" of which fragment undergo a diaspora (Figs. 8.184, 8.185, 8.186, 8.187, 8.188, 8.189, 8.190, and 8.191).

Note up front confusing terminology regarding two bones that have no biologic relationship whatsoever to pterygoid. *Epipterygoid* originates from palatoquadrate and remains connected to the cranium. *Ectopterygoid* originates from the

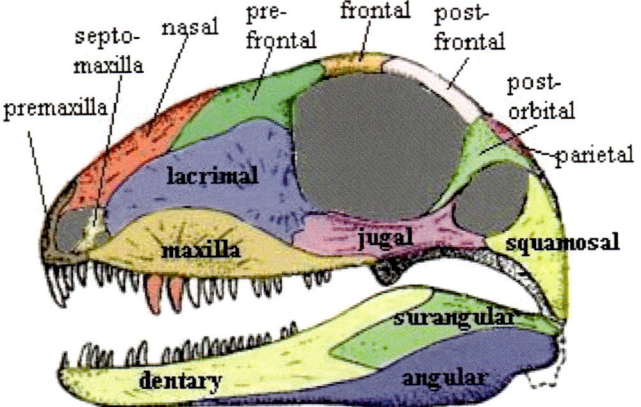

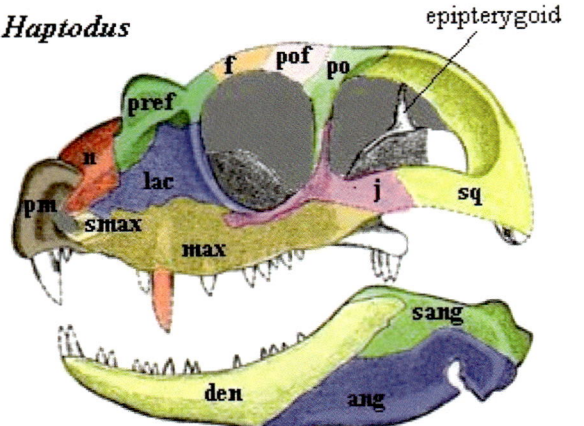

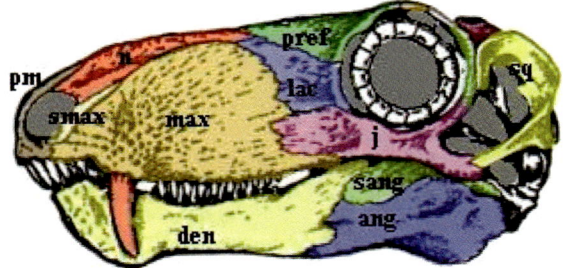

Fig. 8.184 Septomaxilla 3: Septomaxilla in Therapsida. Basal therapsid *Haptodus* show septomaxilla slung between premaxilla and nasal bone. [Top: (Modified from Lewis, Warren H (ed). Gray's Anatomy of the Human Body, 20th American Edition. Philadelphia, PA: Lea & Febiger, 1918). Middle: (Reprinted from Ivakhnenko MF. Biarmosuches from the Ocher Faunal Assemblage of Eastern Europe. Paleontological Journal 1999; 33(3): 289–296. With permission from Pleiades Publishing, Ltd.). Bottom: (Reprinted from Laurin M, Reisz RR. A reevaluation of early amniote phylogeny. Zool J Linnean Soc 1995; 113(2):165–223. with permission from Oxford University Press)]

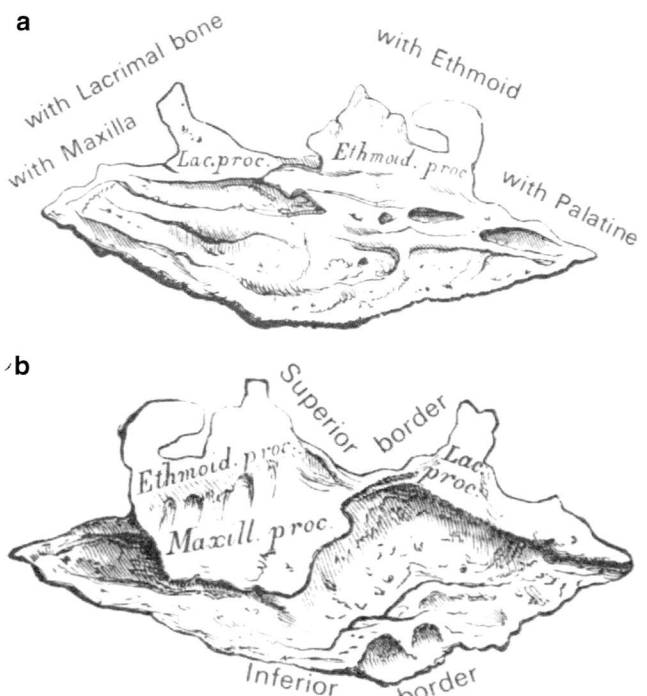

Fig. 8.185 Inferior turbinate landmarks. Medial (nasal) aspect shows scroll-like bone projecting into the nasal cavity. Behind that is located nasolacrimal duct. Articulatory processes all project from its lateral aspect. Lateral (maxillary) aspect shows free inferior border and two articulatory processes: anteriorly to lacrimal bone, and posteriorly to uncinate process. Vomer is tucked into a U-shaped cradle formed by maxilla anteriorly and palatine posteriorly. (Reprinted from Lewis, Warren H (ed). Gray's Anatomy of the Human Body, 20th American Edition. Philadelphia, PA: Lea & Febiger, 1918)

palatine bone complex and retains a connection with ptery-goid. We shall go into these details anon, but we get ahead of ourselves…

Descriptive Anatomy

The palatine bone joins the orbital floor, and perforce the skull base, with the upper jaw. It has a characteristic L-shape which results from the intersection of two plates, horizontal and perpendicular, at a slightly acute 80°–85° angle. As we shall see, this geometry will have surgical implications for cleft palate repair. It also bears three significant processes.

Horizontal plate: The dimensions of the horizontal plate are determined by those of the horizontal plate of the pre-palatine. If PPl is reduced in width, the horizontal plate of Pl will likewise be foreshortened. The nasal surface is concave. The oral surface bears near the midline a transverse eleva-tion, the *palatine crest*, which serves as an attachment for the aponeurosis of tensor veli palatini. The anterior border is irregular, forming a rough articulation with the accessory palatine bone. The posterior border provides attachment for the palatine aponeurosis. The lateral border forms an acute

angle (80°–85°) with the perpendicular plate. Two bony prominences project from the fused medial borders. Musculus uvulae inserts into *posterior nasal spine*. The *nasal crest* articulates with vomer.

Perpendicular plate is a thin lamina with medial and lat-eral sides plus four borders. The medial side is scooped-out, forming the posterior part of *inferior meatus*. Just above that, a transverse ridge, the *conchal crest*, articulated with the palatine process of inferior turbinate. Higher up, a second transverse ridge, the *ethmoid crest*, provides and articulation for middle turbinate. Tucked between the two crests is the *middle meatus*. Still higher up, above ethmoid crest, is a third depression, for *superior meatus*. The lateral side is rough and irregular; it articulates with maxilla. Along its posterior aspect, a well-chiseled vertical greater palatine groove marks where the maxilla and perpendicular plate enclose the neuro-vascular bundle.

The anterior border is punctuated by a forward projection, the *maxillary process*, at the same level as the conchal crest. The rough posterior border articulates with the r1 medial pterygoid plate. This expands into the pyramidal process into which medial pterygoid is inserted. Superior border contains a forward-facing *orbital process* and a backward-directed *sphenoidal process*.

Pyramidal process projects backward from the intersec-tion between horizontal and perpendicular plates. Its poste-rior surface insinuates itself into a crevice between the medial and lateral pterygoid plates. An important landmark is the lateral surface and is a roughened zone along its ante-rior aspect. This will abut against the tuberosity of the alveo-lus. The *lesser palatine foramen* is located at the intersection between pyramidal process and horizontal plate.

Orbital process contains an air cell. It has two non-articulatory and three articulatory surfaces. Superior surface forms the posterior orbital floor. Lateral surface contributes to inferior orbital fissure. Anterior surface articulates with maxilla. Posterior surface contains the opening of the air sinus, which can communicate with the sphenoid sinus. Medial surface articulates with ethmoid labyrinth. Sometimes, this can serves as an alternative escape route for the palatine air sinus, which communicates with the poste-rior ethmoid cells.

Sphenoidal process has three surfaces and three borders. Superior surface connects with medial pterygoid plate and abuts against the sphenoid concha. Medial surface pokes its tiny nose into the nasal cavity. Lateral surface connects with lateral pterygoid plate. Anterior border forms the posterior margin of the sphenopalatine foramen. The medial border of sphenoid process, by virtue of its proximity to the sphenoid bulla, can reach as far as the ipsilateral ala of the vomer where it receives the sphenoid rostrum. Not to be left out, medial pterygoid plate has a vaginal process that articulates with posterior border.

Fig. 8.186 Inferior turbinate (maxilloturbinal) in situ. Maxilloturbinal/inferior turbinate articulates front-to-back with maxilla, lacrimal (duct passes between), ethmoid uncinate process, and palatine. Entry into maxillary sinus can be either in front of or behind the uncinate process. Note the vascular distribution of StV2 lateral sphenopalatine artery (blue) includes inferior turbinate, the lateral nasal wall between IT and the hard palate, and forward up the internal aspect of piriform fossa, i.e., posterior to the frontal process of premaxilla. StV1 anterior and posterior ethmoids (red) supply the upper lateral nasal vault. (Reprinted from Lewis, Warren H (ed). Gray's Anatomy of the Human Body, 20th American Edition. Philadelphia, PA: Lea & Febiger, 1918)

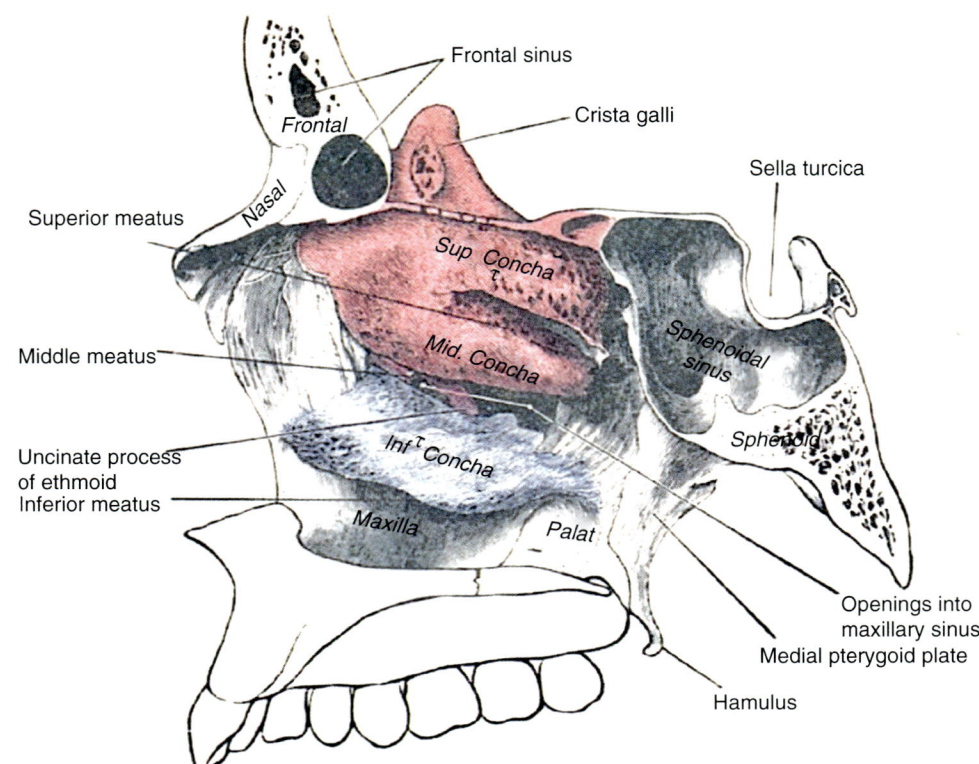

Blood Supply and Mesenchyme

The palatine bone develops in the membrane from r2 neural crest. It is programmed by the r2 nasal and oral mucosa. Descending palatine artery (DPA) provides its blood supply.

Descending palatine artery enters the palatine bone at the notch and plunges downward through the perpendicular plate, giving off multiple small branches to the bone. Just above the junction of the perpendicular and horizontal plates, DPA axis gives of a lesser palatine branch to the soft palate. LPA supplies the entire horizontal palatine shelf and the anterior 1/3 of the palatine aponeurosis. The axis of DPA then continues, emerging onto the oral surface of the maxillary shelf via a foramen at the boundary between horizontal palatine lamina and the maxillary lamina. Its terminus is at either the canine or the lateral incisor dental fields. GPA supplies the entire oral mucoperiosteal envelope of the maxillary palatine shelf. On the oral side of the shelf, GPA anastomoses with the lateral nasopalatine axis.

Development

The palatine bone field shows ossification at the eighth week of fetal life (20 weeks). The epicenter is between the horizontal plate and the perpendicular plate. It spreads out in three directions. The medially directed horizontal plate and the posteriorly directed pyramidal process develop first, followed 2 weeks later by the superiorly directed perpendicular plate. At birth, the horizontal and perpendicular plates are

equal in size. Over time, increasing depth of the nasal chamber increases the relative size of perpendicular plate. As the perpendicular plate extends upward, it encounters the pre-existing sphenopalatine neurovascular axis. The osseous mesenchyme splits around the pedicle. The anterior orbital and posterior sphenoid processes results. In point of fact, the complexity of the various processes of palatine bone stems from its relatively insertion into previously established bone fields such as sphenoid, ethmoid, and maxilla. The palatine mesenchyme just has to fit in where it can.

Phylogeny

In mammals, the hard palate is firmly welded to the skull base. In the evolutionary record, this was not always the case.

All vertebrates need some means to capture and process food. Agnathans are represented today by two survivors, the cyclostomes hagfish (slime eel) and lamprey. The former species has horny teeth on its tongue to rasp away tissue into the mouth. The latter arranges teeth around a sectorial mouth to open blood vessels and lap up its meal with its tongue. Extinct lines such as the ostracoderms began to experiment with dentin and fragments of bone.

The placoderms, armored fishes of the early Devonian, cruised the bottom of the sea like giant vacuum cleaners. These earliest gnathostomes had body armor using plates of dermal bone including a cluster of small bones around the

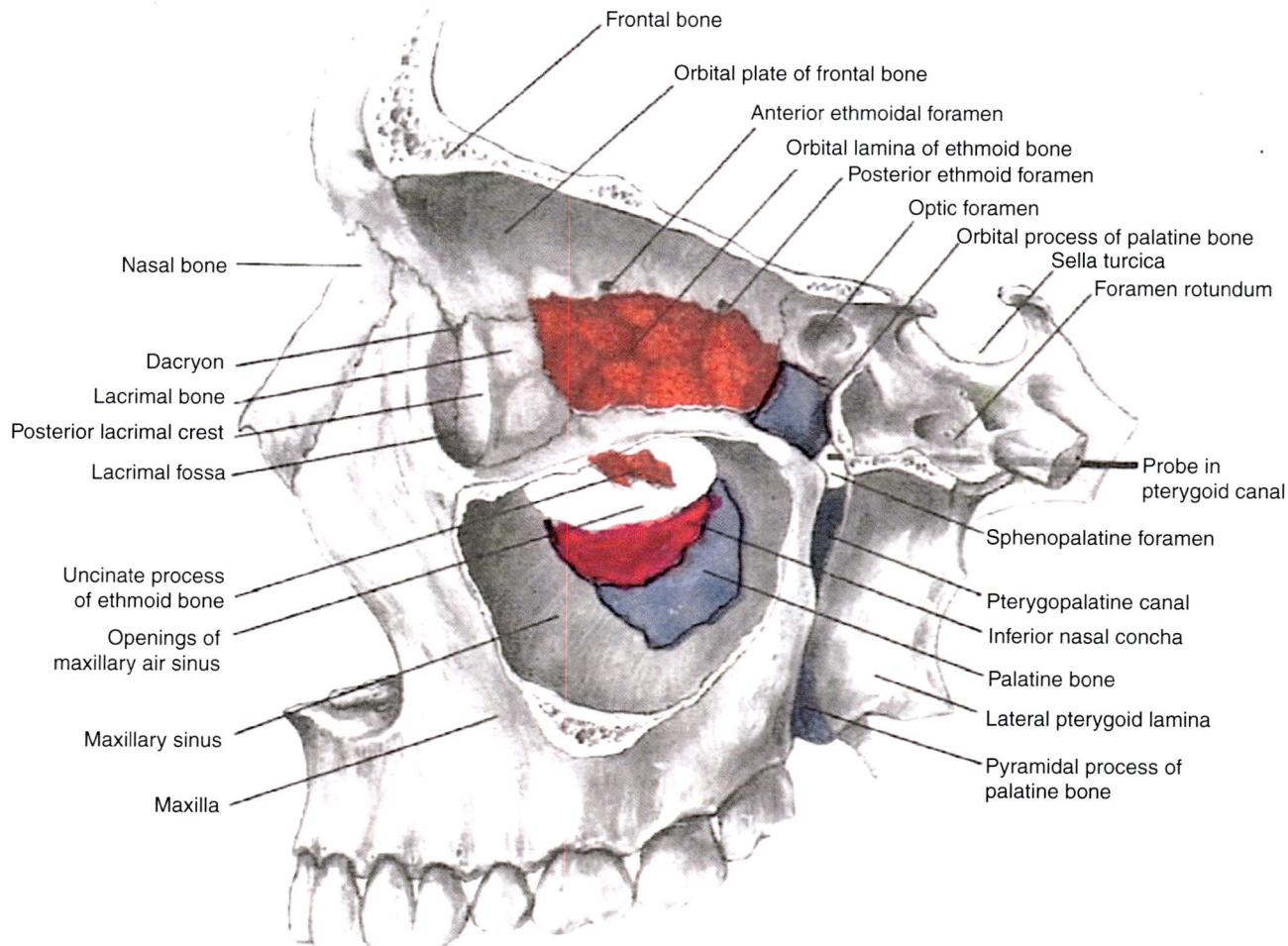

Fig. 8.187 Inferior turbinate is attached to lateral nasal wall supported along its inferior border by maxilla (gray), uncinate process (red) and palatine bone (blue). Note two distinct introituses to maxillary sinus.

(Reprinted from Lewis, Warren H (ed). Gray's Anatomy of the Human Body, 20th American Edition. Philadelphia, PA: Lea & Febiger, 1918)

mouth. Formerly considered edentulous, the placoderms had shearing bone plates arranged on the outside of the jaw; these functioned as teeth. Recent CT evidence of the arthrodire *Campagopiscis* showed intraoral teeth with bone, dentin, and pulp cavities. The placoderm, *Entelognathus*, considered a bridge to the basal gnathostomes, has well-defined premaxilla, maxilla, and dentary.

In the subsequent split into chondrichthyes and osteoichthyes, a definitive palate appears. The earliest well-documented palate appears in actinopterygians *Mimia* and *Cheirolepsis*. The entire palate is studded with teeth. An external row of bones (rostrals, premaxillae, maxillae) is lined internally by four palatines on either side. These are not genetically distinct; they are serial homologues. In the midline behind the premaxillae are paired vomers, followed by fused parasphenoids articulating with basioccipital. Pterygoids are positioned between the midline and the palatines. Ectopterygoids are posterolateral. These dermal bones are all tooth-bearing.

The anatomy becomes simplified in sarcopterygian line. The tetrapodomorph *Eusthenopteron* consolidates the anterior palatines into a single bone in the same position. The ectopterygoid is likely derived from the posterior serial palatine homologue(s). Basal tetrapods maintain the marginal dentition in the palatine. But, with retreat of the parasphenoid, the pterygoids become the dominant structures of the palate. Amniotes show in the reptilomorphs a reduction and fusion of the ectopterygoids with palatine. The ectopterygoids make the contact with transverse processes of the pterygoid, a connection that will remain intact throughout ontogeny. The vomers have now moved into the midline to complete coverage of the anterior palate.

In the evolution of the amniotes, a secondary hard palate occurs twice. In the diapsid line, it is seen in crocodiles. In the synapsid line, it becomes ubiquitous. Let us look at this process in greater detail. Each step involves modifications in the palate. The synapsids arise in the mid Carboniferous. They have loss of teeth from the pterygoids as a synapomor-

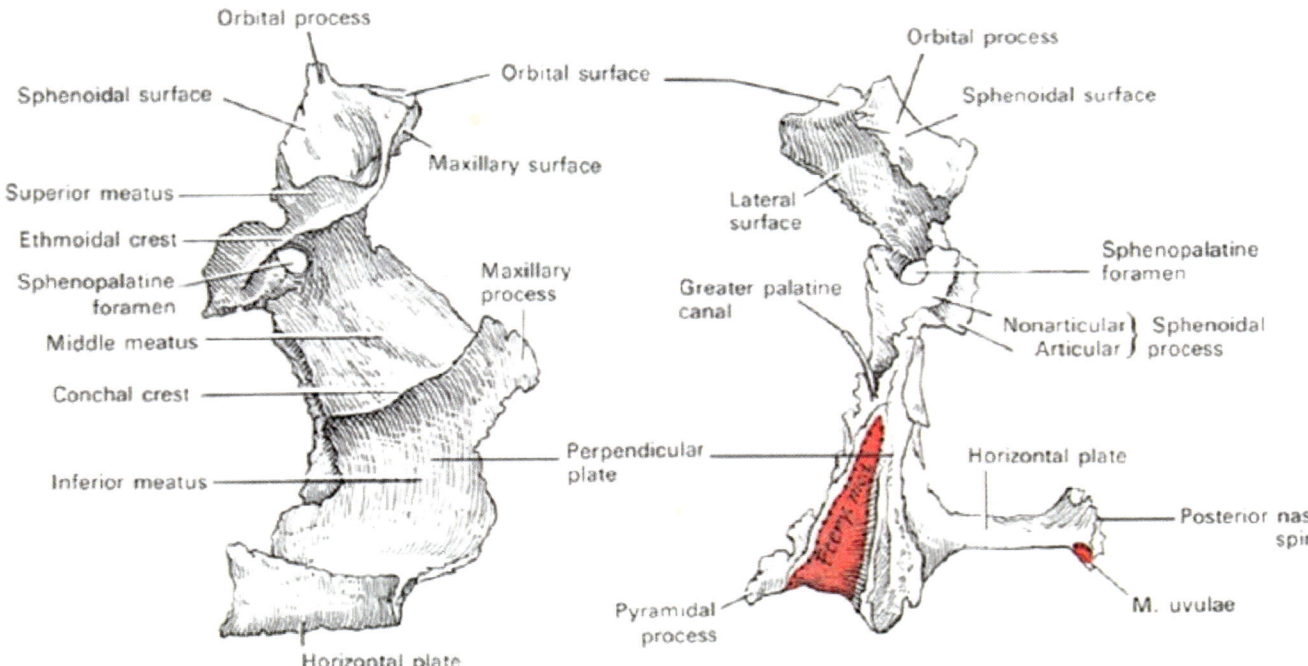

Fig. 8.188 Palatine bones. Note primary insertion of medial pterygoid muscle attaching to the r3 lateral pterygoid plate derived from hypaxial alisphenoid. The orbital process extends considerably above (cranial to) the sphenopalatine foramen. (Reprinted from Lewis, Warren H (ed). Gray's Anatomy of the Human Body, 20th American Edition. Philadelphia, PA: Lea & Febiger, 1918)

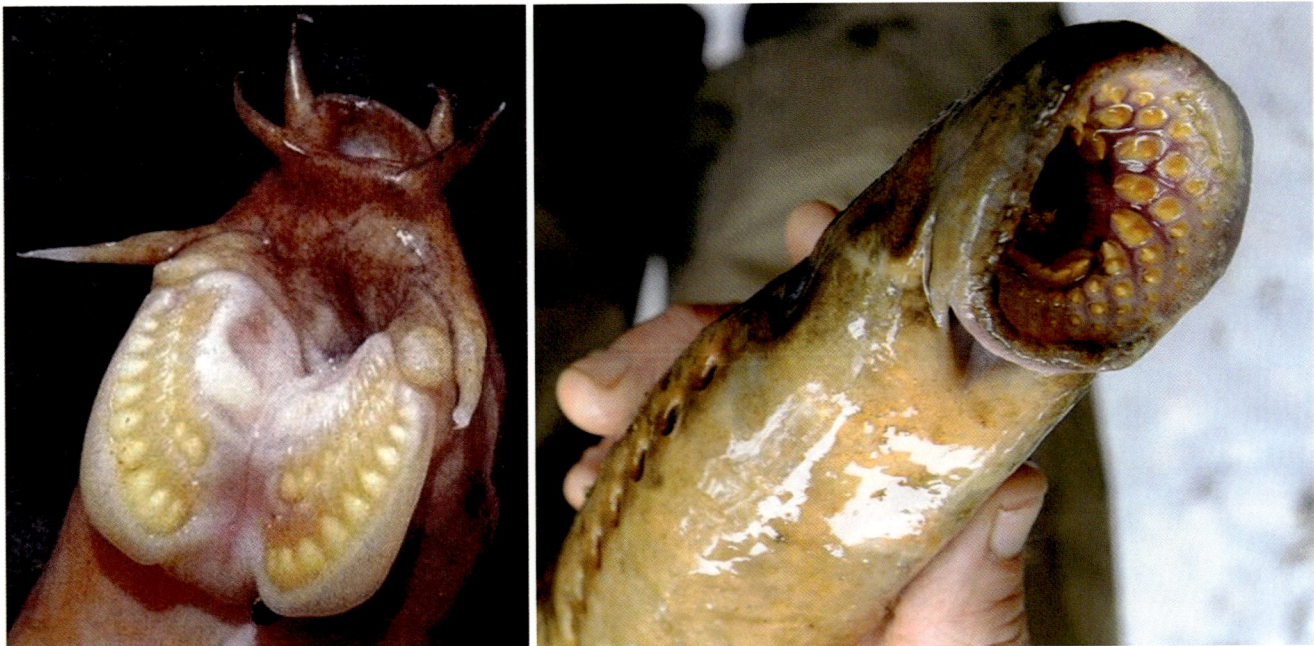

Fig. 8.189 Oral teeth without jaws: hagfish (left) and lamprey (right). The latter has the eight branchial baskets. (Courtesy of Critterscience.com)

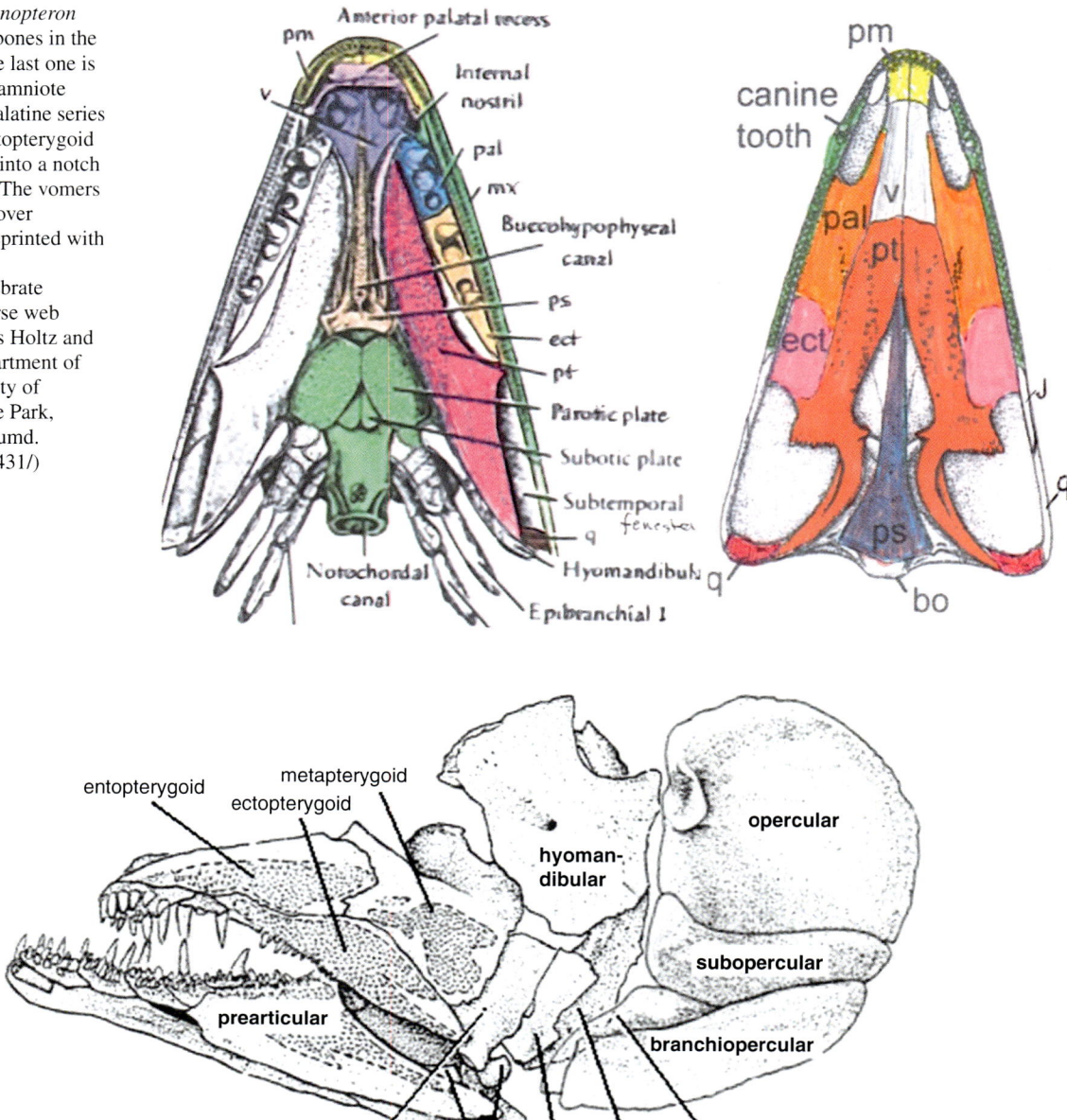

Fig. 8.190 *Eustenopteron* palatine has four bones in the palatine series; the last one is ectopterygoid. In amniote *Paleothyris*, the palatine series has coalesced. Ectopterygoid (last palatine) fits into a notch in the pterygoids. The vomers extend backward over parasphenoid. (Reprinted with permission from: GEOL431—Vertebrate Paleobiology course web resources. Thomas Holtz and John Merck, Department of Geology, University of Maryland, College Park, https://www.geol.umd. edu/~jmerck/geol431/)

Fig. 8.191 Palatal series: entopterygoid, true pterygoid bone; ectopterygoid, last member of palatine bone series; metapterygoid, "potential candidate for squamosal." *Epipterygoid* is a chondral extension of the palatoquadrate. It is external and unrelated to the pterygoid. It runs just inside the lateral dermal wall, maintaining an articulation with the cranium. In mammals, it becomes alisphenoid and the articulation persists to the frontal bone at the level of the orbit. (Reprinted from Grande L, Bemis W. A comprehensive phylogenetic study of Amiid fishes based on comparative skeletal anatomy: an empirical search for interconnected patterns of natural history. J Vert Paleontology 1998; 18(suppl):1–681. With permission from Taylor & Francis)

phy. Evolution then took several false starts. The next iteration, therapsida (mammaliaform reptiles), appears in the early Permian. They show loss of teeth from ectopterygoid. In the penultimate clade before mammals, theriodontia (wild animal teeth) teeth are lost from the palatine bones. Thus, just before final pre-mammal clade of cynodontian (dog teeth), the bones of the palate are endentulous.

This is clinically relevant because aberrant dental development can occur in a variety of sites within the oral cavity. This is simply an "unmasking" of the ancient program.

In the final "push" from cynodonts to mammals, three transformations take place which center around the palatine bone, are associated with mastication, and result in the definitive hard palate. (1) Stabilization of the palate involves the welding of the palatine bone to the skull base. (2) Musculature formerly attached to the pterygoid bone shifts forward to the pterygoid processes of the sphenoid, enabling complex movements at the temporomandibular joint. (3) A secondary hard palate separates the airway from the pharynx for more efficient coordination of mastication and respiration (Fig. 8.131).

Epipterygoid is a dermal bone derived from the r2 palato-quadrate and is readily seen in the lateral skull of derived fishes such as *Amia* in which it articulates with the cranium, the so-called *kinetic skull*. It relates internally to presphenoid. Epipterygoid is the homolog of alisphenoid and its epaxial component is the likely source of the *perpendicular plate* because it maintains the relationship to both the orbit and the sphenoid bone.

Alisphenoid is unique to mammals. In cynodonts epipterygoid morphs from being a slender columellar supporting strut to a flat quadrangular plate which articulates with the lower border of parietal and laminates inside the periotic bone. Anteriorly it articulates with frontal. With brain growth, parietal expands outward and alisphenoid is pushed forward away from periotic, like a sliding shower door. It thus becomes the inferolateral wall of the temporal fossa. We still see this again in Chap. 13 when we explore the orbit.

Ectopterygoid is the fourth field of the palatine bone series. It relates to the posterolateral corner of pterygoid; when the latter bone moves forward so does ectopterygoid.

In the anapsid skull of the early amniotes, pterygoid bone becomes reduced and moves up into the plane of the sphenoid, with which it will eventually fuse. In so doing, pterygoid leaves behind a souvenir of its time in the oral cavity, ventral extension which will be *medial pterygoid process*. Ectopterygoid fuses to it and is the source of the *lateral pterygoid process*. Anapsids have chewing muscles that span from mandible to the *inside* of the postorbital and parietal bones and to the outside of the pterygoid process. These

muscles are not very efficient. The anapsid design also limits brain expansion.

In the synapsid skull of non-mammals, a temporal fenestra opens up, offering the masseteric complex new, and more mechanically advantageous, attachments to the outside of the postorbital and parietal bones. On the mandibular side, the development of a coronoid process gives more extensive muscle insertions for generation of greater force. Epipterygoid fuses with basisphenoid thus uniting the palate, the cranial base and the cranial fossa as the *akinetic skull*.

The synapsid skull of mammals has a well-developed fusion of pterygoid and ectopterygoid elements as the medial and lateral pterygoid processes. These provide insertion of neomorphic muscles of mastication, the medial and lateral pterygoids. These muscles have new insertion sites into the mandibular angle and the TMJ. They constitute a crucial innovation for grinding up food stuffs in a more energy-efficient manner. The epaxial part of ectopterygoid, as the alisphenoid, is fully integrated into the anterior cranial fossa at both the level of the orbit and of basisphenoid.

Development of the secondary hard palate can be readily traced from the theriodonts to mammals. *Eutheriodont* shows a wide-open communication between the oral cavity and the primary palate, i.e., the skull base, made from sphenoid and vomer. The ectopterygoids attach posteriorly to the transverse processes of pterygoids. *Procynosuchus*, an entry-level cynodont, shows shelf-like extensions of premaxilla, maxilla, and palate. The maxillary shelves are expanding in size and moving toward the midline, in the process displacing the palatal shelves of the palatine bones backward.

Thrinaxodon represents a quantum leap. This so-called secondary palate lies ventral to the original roof of the mouth. Simultaneously, a regression of primary palate opens up the nasal cavity. The maxillary and palatine shelves meet in the midline. In *Probainognathus*, the posterior palatal margin continues backwards, a process completed by *Morganucodon*, considered as the first "true" mammal. Ectopterygoid remains as a small afterthought, part of the hamular process.

Clinical Correlations

Recall that vomerine development proceeds posteriorly and downward. The very last site of articulation is into the palatine horizontal plate. A posterior notch could reflect absence of the vomerine terminus, defective medial extension of the horizontal shelf, of both. The common denominator is absence of posterior nasal spine with concomitant disorganization of the soft palate midline.

Multiple aspects of palatine bone biology are clinically relevant and these are discussed in the respective Chaps. 14 and 16 on cleft palate biology.

The Maxillary Complex

The maxillary complex consists of the maxilla proper, a five-sided box with an absent wall facing the nasal cavity. This space is covered over by ethmoid middle turbinate, a separate inferior turbinate and the palatine bone. The embryonic anlage of the upper jaw, once a single structure, produces two independent bone fields, ectopterygoid and quadrate, which become alisphenoid and incus, respectively. It also serves as the template for dermal bone fields, assembled on either side: a tooth-bearing dentary which become maxilla proper, and bones of the hard palate. Although our focus here is just on the maxilla, the other bone fields are described separately (Figs. 8.192, 8.193, 8.194, 8.195, 8.196, 8.197, and 8.198).

The best way to understand the structure of the upper jaw is to compare it with that of the lower jaw. Although they appear very different, their functions are identical, as is their basic design. Both are organized around an embryonic cartilage core: *palatopterygoquadrate* (PQ) made from maxilla and *Meckel's cartilage* (MC) for the mandible. Both are *bilaminar* and consist of *multiple dermal bone fields* assembled into three functional components: a *single housing field for dental units* and *supporting fields bearing muscle insertions* and *articulatory fields* connecting the jaws with each other and with the skull.

The mandible of choanate fishes and early tetrapods consists of a tooth-bearing dentary bone supported from below by two splenial bones and a more posterior articular bone that constituted the primitive connection to the skull. These structures have all fused (Herring's law) to make a unitary bone but both sides of the body and mandible are perfused by distinct arteries supplying the attached muscles which bear witness to its more complex past. The potential space between the laminae of the dentary bone contains the dental units while that of the supportive lever arm is filled with marrow.

On the opposite side of the bite plane, the original maxillary bone has a tooth-bearing component analogous to the dentary of mandible. The upper jaw was not always fixed to the cranial base. In early bony fishes, the tooth-bearing palatoquadrate cartilage (the precursor of the maxilla) had a moveable articulation with the chondrocranium. In its current tetrapod iteration, the support structure of the mammalian maxilla is fixed to the skull, but the overall design remains (just like mandible) a six-sided box, programmed on the intraoral side by endoderm and externally by ectoderm. But as this box grows, its intrinsic structure will lead to a deformation with a sinus cavity in the center.

Descriptive Anatomy: How the Maxilla Gets Its Shape

The medial wall of the box (the lateral wall of the nose) is discontinuous. When all the non-maxillary bones are disarticulated from the base, the sinus presents a wide-open aperture into the nasal cavity, the middle meatus (Fig. 8.196). The lateral nasal wall in the articulated skull is made up of five non-maxillary bones: frontal process of premaxilla and palatine bone articulate with each other like bookends. They "hold hands" with each other beneath the inferior turbinate. Coming down from above is uncinate process of ethmoid. It extends a timid toe downward to touch inferior turbinate. Finally, inferior turbinate is the "Great Communicator," it connects with all its neighbors. Just as the components of a camera diaphragm close down its aperture, so the non-maxillary bones largely "cover up" the middle meatus, leaving behind two bony discontinuities, one on either side of inferior turbinate. It is precisely these openings, permitting oral mucous membranous membrane to enter the potential space between the maxillary laminae that give rise to the maxillary sinus.

To see how this happens, let us take a quick look at the topology of our five-sided box, the entire blood supply of which comes from the lateral division of the sphenopalatine axis. Posterior wall and lateral wall are supplied by posterior superior alveolar artery and nerve. Anterior wall and roof are supplied by anterior superior alveolar artery and nerve. The floor narrows down to a bilaminar tooth-bearing alveolus the dental units of which are supplied by the superior alveolar axis...in exactly the same way as mandible is supplied by an inferior alveolar axis.

The missing sixth wall of the box corresponds to the pentagonal bony "diaphragm" of the lateral nasal wall—a large hole in the center that is partially covered over by the uncinate process of the ethmoid and the inferior turbinate. This is a critical point because the *nasal epithelium gains access to the interface between the maxillary laminae* and *prevents them from fusing*. Thus, as growth of the maxillary complex takes place, this potential space expands to form the maxillary sinus. Because the medial lamina of the maxilla is fixed to the cranial base by the ethmoid and palatine bones, expansion of the maxillary sinus can only occur laterally and forward projecting outward beneath the orbit. This explains the unique shape of the maxillary sinus.

Blood Supply and Mesenchyme

The bones of the maxillary complex all develop from r2 neural crest distributed through the superior alveolar axis. It has an anterior component supplying the alveolar and dental

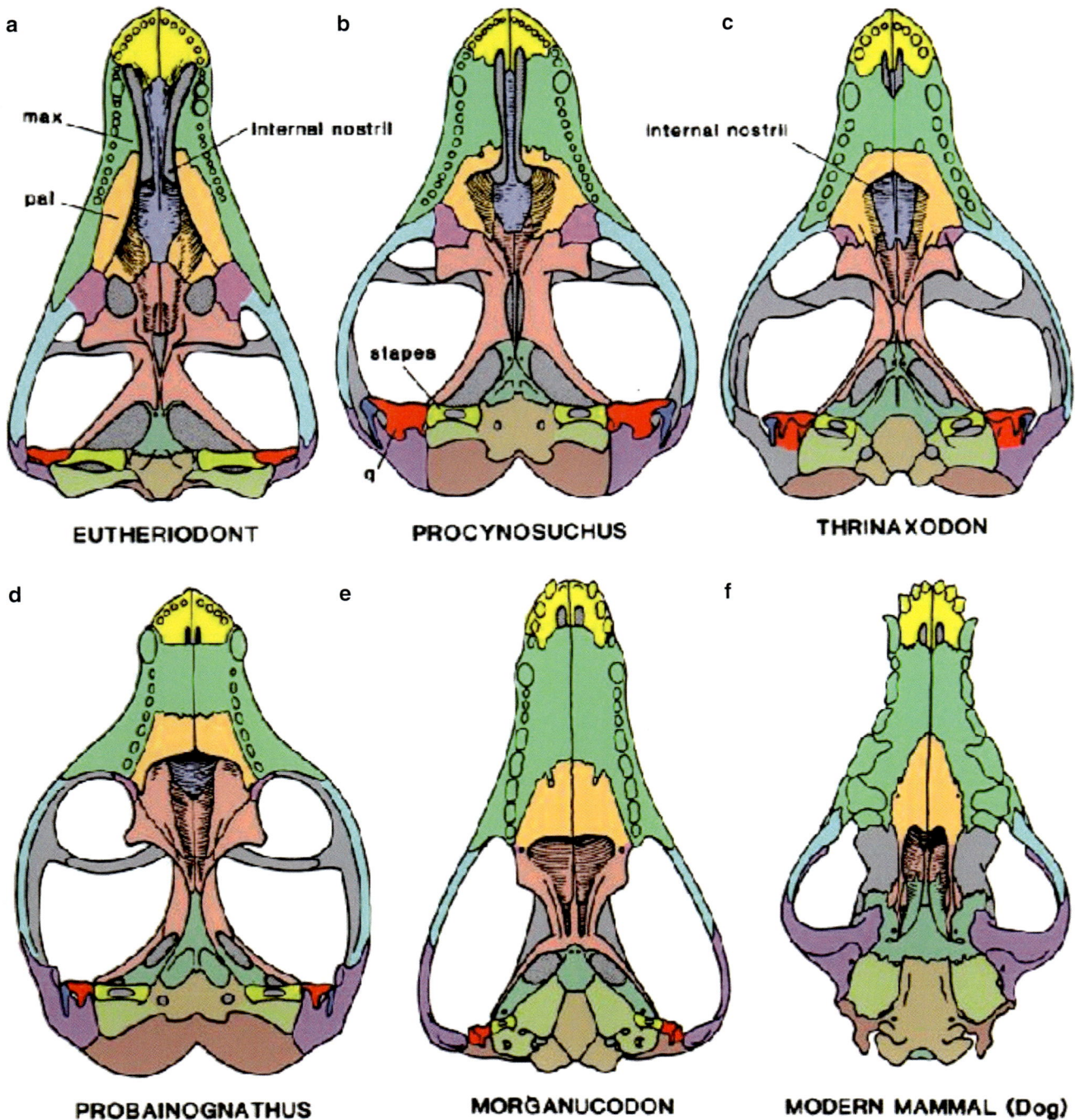

Fig. 8.192 Palatine phylogeny. Color code: premaxilla, yellow; vomer, gray; parasphenoid, lilac; maxilla, green; palatine, tan; pterygoid (medial plate), flesh; ectopterygoid (hamulus), magenta; articular (malleus), blue; quadrate (incus), red; columella (stapes), lemon; prootic/opisthotic (petrosal), purple. Note the extension of pterygoid (brown) in *Eutheriodont* all the way to the occiput. Ectopterygoid (lilac) is just the most posterior of the four palatines. In *Procynosuchus*, the pterygoid moves out of the occiput and insinuates a flange behind ectopterygoid. In *Probainognathus*, pterygoid is incorporating into basisphenoid and ectopterygoid becomes part of future medial pterygoid plate hanging down from sphenoid. Alisphenoid (the future lateral pterygoid plate) does not appear until Mammalia. It is accompanied by another new orbital element, orbitosphenoid. (Courtesy of Mr. David Peters)

Fig. 8.193 Ectopterygoid is P4, the last bone of the palatatine complex. (Reprinted from: www.paleo-electronica.org. Retrieved from: http://palaeo-electronica.org/2011_2/251/images/figure_71.jpg)

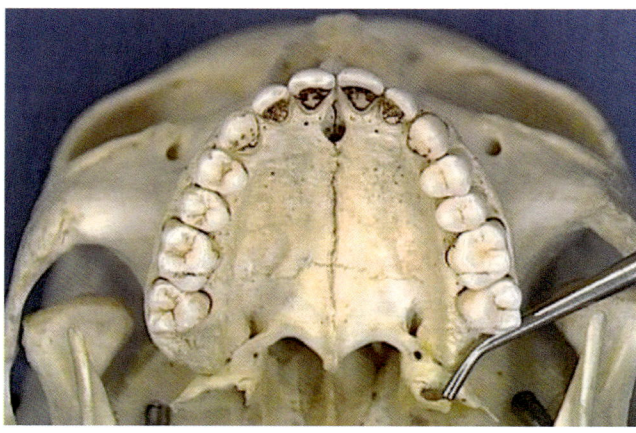

Fig. 8.195 Hamulus: fusion line with the pterygoid plate can be appreciated in this specimen. (Reprinted from Hamulus Barchetta NF, de Oliveira RLB, Silveira AVS, Faig-Leite H. Clinical and morphofunctional aspects of pterygoid hamulus: literature review. Braz Dental Sci 2015; 18(4):5–11. https://doi.org/10.14295/bds.2015.v18i4.1078, with permission from Creative Commons License 4.0: https://creativecommons.org/licenses/by/4.0/deed.en)

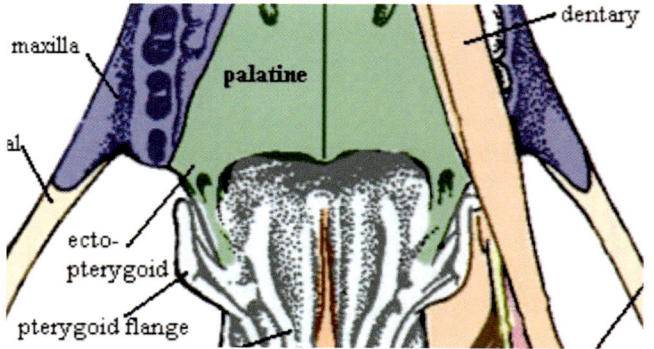

Fig. 8.194 Origin of the hamulus in mammals, as seen in Morgaunucodon. Pterygoid (pale blue) is incorporated into basisphenoid, leaving behind the pterygoid flange as medial pterygoid plate. Ectopterygoid (green) laminates the posterior aspect of medial pterygoid plate. It remains in humans as the hamulus. (Reprinted from Kermack KA, Mussett F, Rigney HW. The skull of Morganucodon. Zool. J. Linn. Soc. 1981;71:1–158. With permission from Oxford University Press)

fields of the lateral incisor, canine and premolars via medial and lateral branches. It has a posterior component supplying the alveolar and dental fields of the molars. The anterior and posterior superior alveolar axes perfuse the membranous bone connecting the alveolar arch with the rest of the skull. As such, the neurovascular plan for the maxilla is exactly the same as that for the mandible. In the lower jaw, a single neurovascular pedicle (consisting of three subunits) enters the mandibular alveolus posteriorly and runs along its entire length supplying all three functional groups of teeth: (uni) cuspid, bicuspids, and tricuspids. In the case of the maxilla, the dental units are the same but two pedicles are required; a third one is designated for the premaxilla.

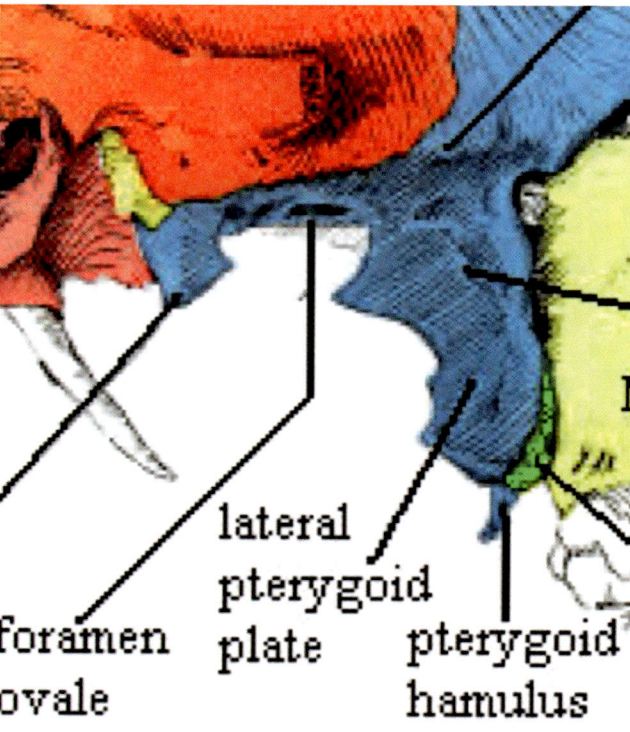

Fig. 8.196 Alisphenoid arises from epipterygoid, a dermal bone lining the interior surface of palatoquadrate cartilage. It is divided into epaxial and hypaxial zones by infratemporal crest. In the temporal fossa, *m. temporalis* inserts broadly up to, but not including postorbital. Infratemporal fossa bears *m. pterygoideus lateralis* which inserts into the TMJ capsule. It has pterygoid hamulus, which bears tendon of TVP, and shpenoid spine, which carries chorda tympani along its medial surface and connects to mandible via sphenomandibular ligament (SPL), itself a derivative of Meckel's cartilage. This establishes hypaxial alisphenoid as r3. Meckel's derivatives: anterior, dentary; middle, SPL; posterior, tympanic, malleus, incus. (Reprinted from Boileau JC. Grant Atlas of Human Anatomy Philadelphia: Williams & Wilkins 1943)

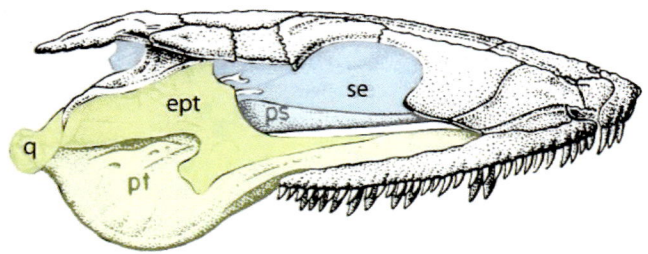

Fig. 8.197 Palatoquadrate (pq) in Eusthenopteron: derivatives of lateral maxilla (white) ensheaths the outer aspect of PQ. It is dentary bone, as in the mandible. PQ produces two chondral extensions that beome epipterygoid (ept) and quadrate (q) bone. Medial to pq is pterygoid bone (pt) and the palatine series (not shown). (Reprinted with permission from: GEOL431—Vertebrate Paleobiology course web resources. Thomas Holtz and John Merck, Department of Geology, University of Maryland, College Park, https://www.geol.umd.edu/~jmerck/geol431/)

The maxilla proper consists of a four-walled body (containing the sinus) and four bony processes: zygomatic, alveolar, frontal, and palatine. The maxillary walls provide the labio-buccal lamina of the alveolus. They contribute to the infratemporal fossa and the pterygopalatine fosse.

The *posterior* or *infratemporal* wall bulges outward, forming the anterior wall of infratemporal fossa. At the tuberosity (for the wisdom tooth), it wraps around medially to articulate with the pyramidal process of the palatine bone. The interface between this nasal extension of posterior wall and palatine bone houses the *canal of the descending palatine neurovascular pedicle.*

The *lateral wall* provides the lateral lamina of the alveolus housing the molars. At this juncture, three foramina provide alveolar canals for the posterior superior alveolar nerves and arteries to the molars. It is separated from posterior wall by the tuberosity. It is also from the anterior wall by a ridge running upward from the first molar, the *maxillary buttress.*

The *superior* or *orbital* wall is a triangle, the medial border of which articulates with lacrimal bone, orbital lamina of ethmoid, and orbital process of palatine bone. Note that the lacrimal sac predates bone formation; it is nested in the embrace of lacrimal bone and maxilla, creating the *lacrimal notch.* At its posterolateral margin, orbital plate articulates with the *orbital surface of alisphenoid.* Anteromedial margin of orbital plate bears the primary insertion of *inferior oblique.*

Down the center of the orbital plate runs the inferior orbital fissure containing the pedicle of infraorbital nerve and artery (IONA), more accurately named the anterior superior alveolar axis (ASA). *Thus, medial orbital plate is Tessier zone 4 while lateral orbital plate is Tessier zone 5.* Note the existence of two canals at the terminus of the fissure: the infraorbital canal proper and a canal for the medial branch of ASA, directed to the canine and incisors. On occasion, a more laterally situated canal conveys the lateral branch of ASA to the premolars. Superior alveolar nerve and artery is

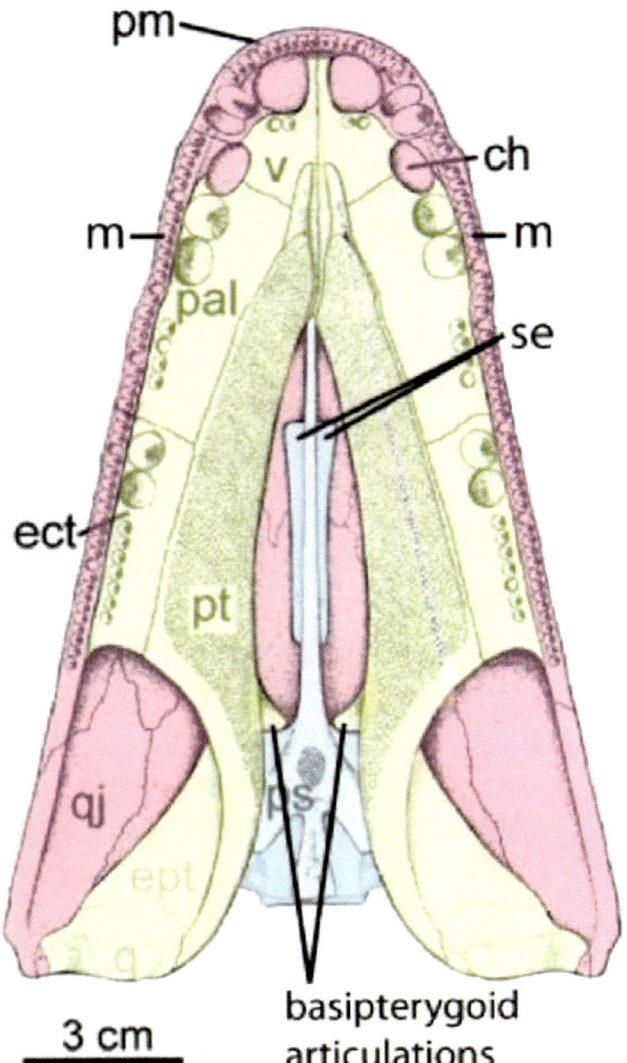

3 cm

basipterygoid articulations

Fig. 8.198 Palatoquadrate (pq) in Eusthenopteron: oral derivatives of palatoquadrate, external series (pink) and internal series (yellow). (Reprinted with permission from: GEOL431—Vertebrate Paleobiology course web resources. Thomas Holtz and John Merck, Department of Geology, University of Maryland, College Park, https://www.geol.umd. edu/~jmerck/geol431/)

described as being distributed to all teeth but is certainly reserved for the molars.

In fetal life, the *anterior* or *facial* wall of maxilla presents an articulation with that of the premaxilla. Along the alveolus, dental eminences correspond to tooth roots, the most prominent being that of the cuspid. Just lateral to the *canine eminence* is the *canine fossa,* the periosteum above which bears the insertion of *levator anguli oris.* Directly above the canine fossa is the infraorbital foramen and above that, the inferior orbital rim, the periosteum of which bears the secondary insertion of *levator labii superioris.*

Immediately upon exiting the infraorbital neurovascular pedicle divides into descending medial and lateral branches.

The former (also termed anterior superior alveolar) supply the *external alveolar wall of the canine and incisors*. This is *zone 4 of Tessier*. Note that development of the external alveolar lamina is an overgrowth of the premaxillary internal alveolar lamina. Recall that, under normal circumstance maxilla donates mesenchyme to the premaxilla. This includes the above, plus lateral incisor, either as a primarily or as a supranumerary unit. The lateral branch of ASA supplies the external alveolar wall of the premolars. This is *zone 5 of Tessier*.

The Maxillary Hard Palate

The maxillary hard palate develops in the membrane from r2 neural crest. The palatal shelf of the maxilla represents a replacement for the pterygoid bone which was derived from the ancient palatoquadrate. It arises from the junction between the internal maxillary lamina and the maxillary alveolus (derivative of the ancient palatoquadrate). Its blood supply reflects these two distinct sources. Lateral nasopalatine artery from the internal lamina supplies the nasal side. Greater palatine artery from the alveolus provides its blood supply on the oral side. The interface between these two neurovascular zones, between the lingual alveolar wall below and the inferior turbinate above, is a biologic "signal" for the palatal shelf to develop. As we shall see, a similar interface zone exists posteriorly that signals the emergence of the soft palate envelope.

The hard palate makes its first appearance in crocodiles. It is unclear whether it represents an evolutionary innovation that buds off from the medial wall of alveolus or whether it is the manifestation of the genetic reappearance of a previous ancient palatal bone field.

The Palatal Process of the Maxilla Is Bilaminar

Alveolar bone in both maxilla and mandible is a *bilaminar* structure with dental units housed between distinct buccolabial and lingual bones. The *buccal wall of the alveolus* belongs to the maxilla per se and is supplied by the medial and lateral branches of the infraorbital artery, better named the anterior superior alveolar artery, and by the posterior superior alveolar artery. The *lingual wall of the alveolus and the palatal shelf* is a single functional unit supplied by greater palatine branch of the descending palatine artery. These structures appear to be a single unit although, based on comparative anatomy, the shelf of the hard palate may be the evolutionary descendent of a pre-existent bone field.

The concept of the hard palate shelf as a bilaminar structure may come as a shock to many readers but comparison of the maxillary alveolus with the mandibular alveolus is most reassuring. Both structures develop within the first pharyngeal arch in direct opposition to each other (thereby explaining occlusion). Both structures are tooth-bearing—recall that teeth are always housed in bilaminar bones. Both structures, in the evolutionary record, are composed of distinct bone fields. Both structures have distinct arterial supplies to their labio-buccal versus lingual walls, and both have distinct arterial supply to the dental units. The posterior superior alveolar artery is the primary supply to the teeth. It is the analog of the inferior alveolar artery of the mandible. Thus, the functional design of the maxilla is virtually the same as that of the mandible...the structures have a different shape and different names but the functional significance remains the same.

There are four lines of evidence for the bilaminar nature of the palatal shelf. First, like other membranous bones such as the frontal bone, it has a dual blood supply (and a dual source of programming). Second, the palatal shelf contains marrow (characteristic of flat bones of the skull). Third, formation of a sinus within the palatal shelf has been documented in primates. Fourth, this bone can be tooth bearing. This is not surprising. Primitive vertebrates had multiple rows of teeth and multiple tooth-bearing bones in the oropharynx. Over the course of time the number of bones is reduced and simplified, a phenomenon known as *Herring's law*. Like many calvarial bones, the maxillary shelf (prepalatine) is bilaminar because it has two sources of programming. The distinct color difference between the oral mucoperiosteum and the nasal periosteum show that its two neurovascular axes are biologically different.

Development

Ossification of the maxillary hard palate starts at the incisive foramen and progressively sweeps backwards as more mesenchyme to added on from proximal-to-distal (anterior-to-posterior) and from medial-to-lateral. The additional mesenchyme demands that new arterioles sprout off from the main axis. For this reason, *clefts of the secondary hard palate begin posteriorly and medially...in the most recent and most vulnerable zone of bone formation*. As the mesenchymal deficiency worsens, as the deficit hits progressively "older" (previously formed) zones of mesenchyme, the cleft extends forward until it terminates at the incisive foramen. In exactly the same way, the neurovascular axis to the lingual lamina of the premaxilla adds mesenchyme from mesial-to-distal, that is, from proximal-to-distal. On the buccolabial side, the same pattern is seen. The medial branch of anterior superior alveolar ossifies and sustains eruption before the lateral branch of anterior superior alveolar artery (ASAA).

Ossification of the maxilla occurs in the same progression with all neuroangiosomes. Documentation exists for anterior superior alveolar and greater palatine. Gradient of bone formation is the same as for the palatine bone: anterior-to-posterior and medial-to-lateral.

Phylogeny

Understanding the evolutionary biology of the upper jaw and palate begins with the *palatopterygoquadrate cartilage*. This represents a conversion of the upper, epibranchial cartilage of the first gill arch in agnathans into a deceptively simple unitary structure, albeit one with built-in genetic subunits. The functional structure of PQ is exactly the same as the of Meckel's cartilage. Both have an anterior tooth-bearing *dentary zone* and a posterior *articulation zone* dedicated to the jaw joint. Thus, we shall refer to anterior PQ as the *maxillary dentary*.

The various component bones of the maxillary complex arise in response to genetic cues from developmental zones of PQ by ossifying in one of two ways. We shall describe the basal situation in tetrapods and modify it for mammals.

Endochondral ossification takes place *directly on the surface* the PQ. It "recognizes" two intrinsic bonefields: (1) *Epipterygoid* (**ept**) develops along the upper (dorsal) margin of PQ and connects with the roof of the skull. It will eventually form alisphenoid. (2) *Quadrate* (**q**) forms a joint with articular bone of Meckel's cartilage.

Membranous ossification involves the formation of distinct dermal bones along the internal surface of PQ in response to cues from PQ. These bones all project medially. Two bones form a long narrow strip just inside PQ. (1) *Palatine bone* (**pal, pl**) has both a vertical and a horizontal component and projects all the way forward to touch the backside of the palatal extension of premaxilla. (2) *Ectopterygoid* (**ect**) sits behind palatine (falsely) and projects all the way backward to the terminus of the dentition. (3) Medial to, and attached to both palatine and ectopterygoid lies the very substantial *pterygoid* (**pt**). The midline palate in the basal tetrapod state has a parasphenoid (**ps**) which extends from behind premaxillae and between the vomers all the way to basisphenoid.

As we have seen previously, the original dermal pterygoid formed a ventral lining "strut" beneath the sphenopalatine cartilage until dead-ending against the lateral "wings" of basisphenoid, the basipterygoid articulation. It between PS and the pterygoid shelves, the palate was wide open up to the cranial base, the so-called *interpterygoid vacuity*. In evolution of the mammalian hard palate, the pterygoid is absorbed into basisphenoid, its only trace being two ventrally directed pterygoid processes. Meanwhile, a maxillary shelf (MxP) encroaches into the oral cavity between the premaxillary shelf and the palatine bone. This maxillary self along with the displace palatine shelf will form the secondary hard palate. Since the original configuration of the vertebrate palate has multiple palatine bones, and since the ectopterygoid is considered just a modification of the posterior palatine, so mxp could well arise from the same mechanism or, alternatively, as *replacement for* the anterior zone of the mysterious involuting pterygoid.

Pterygoid is identified in the palate of sarcopterygian fishes as early as *Cheirolepsis*. As the amniotes split into the synapsids and diapsids, pterygoid bone becomes toothless. Palatine teeth is lost in therapsids and in the later theriodonts, pterygoid is lost (being absorbed into the sphenoid as the medial pterygoid plate). A secondary palate emerges at this time. The maxillary hard palate occurs at exactly the same location as its predecessor. Moreover, its shape is very similar to that of pterygoid. Thus, although it does not represent a transformation of the pterygoid bone per se, it likely develops according to the same positional cues.

Vomer is considered by some a derivative of PQ. This flies in the face of its blood supply and mesenchyme. Vomer and premaxilla are both r2 bones to be sure but they share a common vascular axis distinct from the vessels supplying maxilla. Premaxilla and maxilla are on the periphery of the palate. In point of face, up to the time of sarcopterygians, maxilla lies posterior to the naris and in not even in physical contact with premaxilla, let alone vomer. True physical contact between maxilla and vomer do not occur until the pre-mammalian cynodonts.

By now, having discussed the history of the palatine bone, and, perforce, the evolution of the hard palate as well as the contributions making up the zygomatic complex, a great deal about the phylogeny of maxilla has been covered. We should mention the late expansion of the maxillary sinus that takes place in anthropoidea. It should also be emphasized that the unusual shape of mammalian maxilla is due in part to the multiple bone fields that bind to the calvarium, to the palate and to the skull base. Over time, these fixed points constrain the directions in which maxillary growth can take place.

A second misconception concerns the relationship of maxilla to the orbit. For a considerable period of evolution, jugal excludes maxilla. The orbital floor is jugal with the infraorbital fissure drawn between jugal and ethmoid. Not until jugal moves posteriorly can maxilla participate in the orbit.

What is worthy of our attention is the question of dental specialization. Are there developmental zones in maxillary dentary that might explain this phenomenon? Our old friend *Amia* demonstrated that bony fishes can have two distinct *supramaxillary* bones located along the upper border of the maxilla. The existence of the two units suggest that genetic boundaries exist within maxilla that determine dental identity. These would logically consist of a medial sector for maxillary lateral incisor and canine, as well as a lateral sector for the premolars.

Evolutionary Highlights of Maxillary Phylogeny

Neural crest appears in the first vertebrate of the Epoch 1 of Cambrian and by Epoch 4, end of the Cambrian, a calcified dermal skeleton appears. The armored awless fishes of the early Paleozoic, collectively known as ostracoderms, had a jumble of small bones of great variation. The advent of jaws caused this situation to normalize. Palatoquadrate cartilage and Meckel's cartilage can be identified in the earliest placoderms at the beginning of the Silurian but there was no hyoid support. Subsequent evolution of Osteichthyes saw a dermal skull roof. These dermal bones plated PQ creating maxilla and mandible with teeth ankylosed in place. Enamel shows up in the subsequent Sarcopterygians. They developed sensory lines linking up maxilla forward with nasals and backward to jugal, presumably a valuable innovation to detect motion and potential predators. Just before the Devonian, in Eotetrapodiforms, the vomer expands posteriorly and begins to overlap the parasphenoid. As this process continues, PS will retreat backward and be absorbed with no known derivative.

In tetrapods, the lateral line system disappears as they spent their time out of the water. The palate retains teeth but vomer and palatine bones lose their fangs. The vomers become long and narrow. Amniote do not show much change until one reaches Synapsida in the mid Carboniferous. A maxillary buttress appears in *Dimetrodon* and progressive loss of teeth takes place in the bones of the palate beginning with pterygoid and proceeding with ectopterygoid and palatine. Development of the edentulous palatine bone times with the appearance of a palatal shelf from maxilla in secondary hard palate begin to form in the pre-cynodont *Eurtheriodont*.

Rapid changes in the hard palate take place in Cynodontia, setting the stage for the final mammalian condition. Maxillary shelves appear in Procynosuchus and achieve midline fusion in *Thrixandon*, although they remain short. At this time, premaxilla has four incisors. Arguably the first mammal, Morganucodon, achieves full palate length. True mammals have three incisors per premaxilla. As described for zygoma, the flaring of the malar prominence occurs with the expansion of the mammalian maxillary sinus.

Clinical Correlations

Our model of the maxilla differs substantially from that presented by traditional texts for several reasons. (1) The blood supply to the maxillary walls is from the primary (lateral) division of sphenopalatine stem. That of the inferior wall and medial wall is from the primary (medial) division. These two arterial axes are spatially and genetically distinct. (2) Premaxilla has a frontal process that laminates internally with that of the maxilla. Thus, PMxF is the visible anterior surface of medial nasal wall. The developmental relationships between maxilla and premaxilla are of the utmost importance for the understanding of cleft formation. Hence, our discussion must cross reference that of premaxilla (*vide supra*).

Dental anesthesia provides a good example of neurovascular zones of the alveolus. Blocks of the infraorbital nerve are highly effective for the soft tissues of the lip, cheek, and labio-buccal alveolar mucosa but may not be helpful as stand-alone anesthesia for extraction. This is because branches of IONA external to the fissure are readily blocked whereas nerve branches internal to the maxilla can be reached only if the drug diffuses back along the nerve and/or into the bone. Two reliable solutions are available: (1) central V2 block at the pterygopalatine fossa and (2) gingival infiltration. Because the internal (lingual) alveolar lamina is supplied by the greater palatine nerve both sides of the gums must be injected.

The Zygomatic Complex (Jugal-Postorbital)

Descriptive Anatomy

The zygoma complex consists of two zoologic bone fields: *jugal* (inferior) and *postorbital* (superior). These bones are separated by a horizontal plane connecting the zygomatico-orbital foramina. The two components of the jugal–postorbital complex (JPO) fuse together at the time of the basal synapsids. Their position and structural roles are conservative and inseparable from the lateral orbit. Only very late in evolution, under duress for binocular vision in the anthropoids, does JPO grudgingly admits two new members to the "inner circle" of the orbit: (1) maxilla participates in the orbital floor and (2) alisphenoid joins the back wall. The malar process provides insertion for two second arch muscles of facial expression while the zygomatic arch supports first arch masseter for heavier duties. Postorbital (the orbital process of zygoma) receives an insertion from temporalis (Figs. 8.199, 8.200, 8.201, 8.202, 8.203, 8.204, 8.205, 8.206, 8.207, and 8.208).

The lateral surface of zygoma is punctuated by *zygomatico-facial foramen*. This is the vascular axis for craniofacial *Tessier cleft zone 7*. Just below the foramen is the periosteum into which the zygomaticus minor and levator labii superior muscles are inserted. The lower postero-medial surface bears the attachment with the maxilla while the upper postero-medial surface is attached to greater wing of sphenoid. The superior surface forms the orbital floor along with that of maxilla. The *zygomatico-orbital foramina* are located here. These bear V2 nerves outward, zygomatico-facial and zygomatico-temporal supply zones 7 and 8 of the face. Frontal process has two serrated articulations: superiorly, for frontal bone and posteriorly, for alisphenoid. 1 cm below the Z-F suture and just within the orbital margin, the tuberculum marginal provides attachments for four structures. Most internal is a lateral extension from the levator palpebrae superioris aponeurosis. There follow three ligaments in succession: ocular suspensory ligament, lateral palpebral ligament (lateral canthus), and lateral cheek ligament.

Temporal process of zygoma articulates obliquely with zygomatic process of quadratojugal. Note that from tempo-

Fig. 8.199 Maxilla without zygoma. The zygoma has been removed, demonstrating that the inferior orbital fissure is a field separation point with maxilla making the medial part. Thus, anterior superior alveolar does not split the maxilla, it merely follows the zygomatico-maxillary boundary. Once it reaches the external surface, it fans out both medially and laterally to supply zones 4 and 5. Thus maxilla, when considered from above, is L-shaped. (Reprinted from Lewis, Warren H (ed). Gray's Anatomy of the Human Body, 20th American Edition. Philadelphia, PA: Lea & Febiger, 1918)

ralis fascia attaches firmly from Z-T suture back to T-Z. Masseter gains an insertion along the posteroinferior border.

Ossification of zygoma occurs at the midline along the plane dividing postorbital from jugal. This division can at times be observed in the newborn. It is readily appreciated in texts of comparative anatomy and paleontology.

Innervation to the Zygomatic Complex

Recall that in mammals, the maxilla and jugal form a developmental complex consisting of three bones: maxilla proper, jugal, and postorbital. Therefore, from pterygopalatine fossa, a common StV2 zygomatico-maxillary branch (ZM) ascends to supply them all. Just before entering infraorbital canal, ZM gives off posterior superior alveolar nerve. Once it enters the canal it has two more branches: anterior superior alveolar nerve and zygomatic nerve.

The zygomaticofacial nerve runs along the lower outer aspect of the orbit and arrives onto the surface of the face through a foramen in the zygomatic bone. It then passes through the orbicularis oculi and innervates the skin over the prominence of the cheek. It ramifies with the zygomatic branch of the facial nerve and also the palpebral branches from the maxillary nerve of the trigeminal nerve.

Fig. 8.200 Maxillary sinus entrance. Three bone fields at together like the shutter of a camera to partially close off the entrance into the maxillary sinus. (Reprinted from Lewis, Warren H (ed). Gray's Anatomy of the Human Body, 20th American Edition. Philadelphia, PA: Lea & Febiger, 1918)

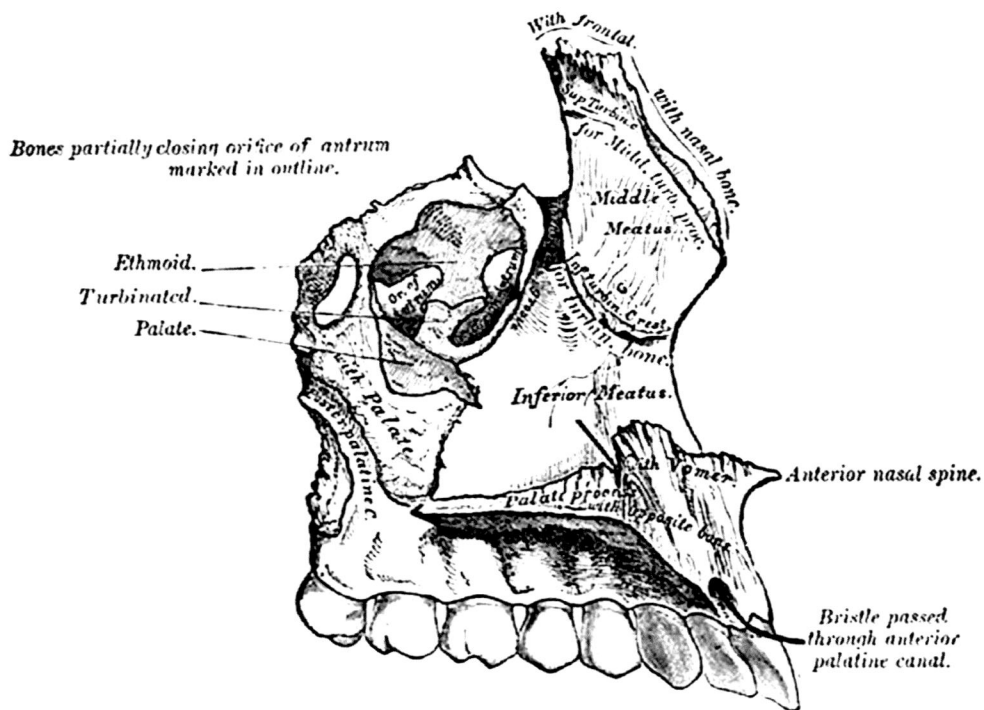

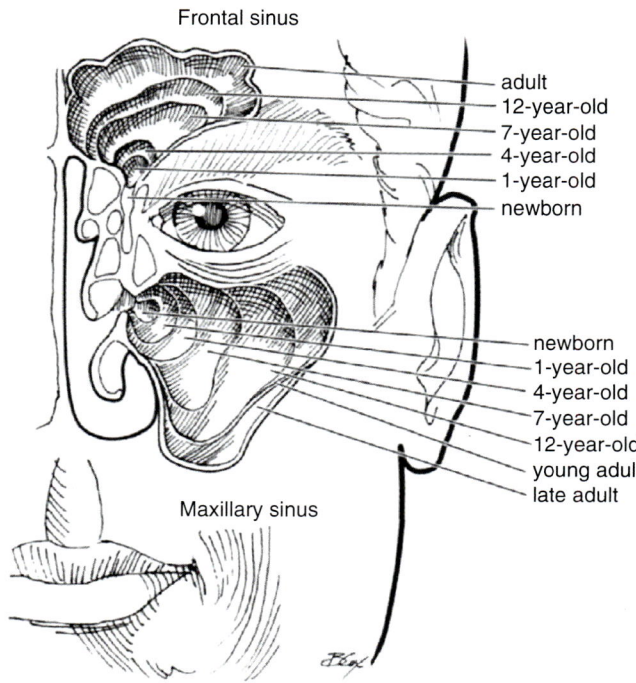

Frontal sinus

adult
12-year-old
7-year-old
4-year-old
1-year-old
newborn

newborn
1-year-old
4-year-old
7-year-old
12-year-old
young adult
late adult

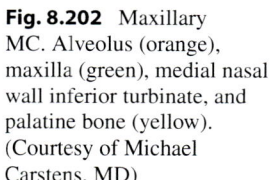

Maxillary sinus

Fig. 8.201 Maxillary sinus growth. Maxillary growth represents field expansion in permissible vectors: lateral and anterior. (Reprinted from Zalzal HG, O'Brien DC, Zalzal GH. Pediatric Anatomy: Nose and Sinus. Operative Techniques in Otolaryngology-Head and Neck Surgery 2018;29(2):44–50. With permission from Elsevier)

Fig. 8.202 Maxillary MC. Alveolus (orange), maxilla (green), medial nasal wall inferior turbinate, and palatine bone (yellow). (Courtesy of Michael Carstens, MD)

The zygomaticotemporal nerve runs along the lower outer aspect of the orbit and provides a branch to the lacrimal nerve. It then passes through a small canal in the zygomatic bone and then arrives into the temporal fossa. It passes superiorly between the bone and the temporalis muscle. It will then go through the temporal fascia just a couple of centimeters superior to the zygomatic arch and innervates the skin of the temple region. It anastomoses with the facial nerve and also with the auriculotemporal nerve forms the mandibular division of the trigeminal nerve. When it pierces the temporal fascia, it also sends a small branch between the two layers of temporal fascia to reach the outer aspect of the eye.

Blood Supply and Mesenchyme

Both jugal bone and postorbital bone develop in membrane from r2 neural crest. The program resides in V2-innervated skin. The blood supply for these bones comes from three different sources. Within the orbit, StV2 gives rise to the zygomatico-facial axis and to the zygomatico-temporal axis. The two pedicles are transmitted by the bone through foramina. They are dedicated to the jugal and postorbital bone fields, respectively. *Anterior deep temporal artery* also contributes. It is an ECA derivative of the second part of MMA (all three branches of which supply muscles from Sm4 and Sm6 (buccinators).

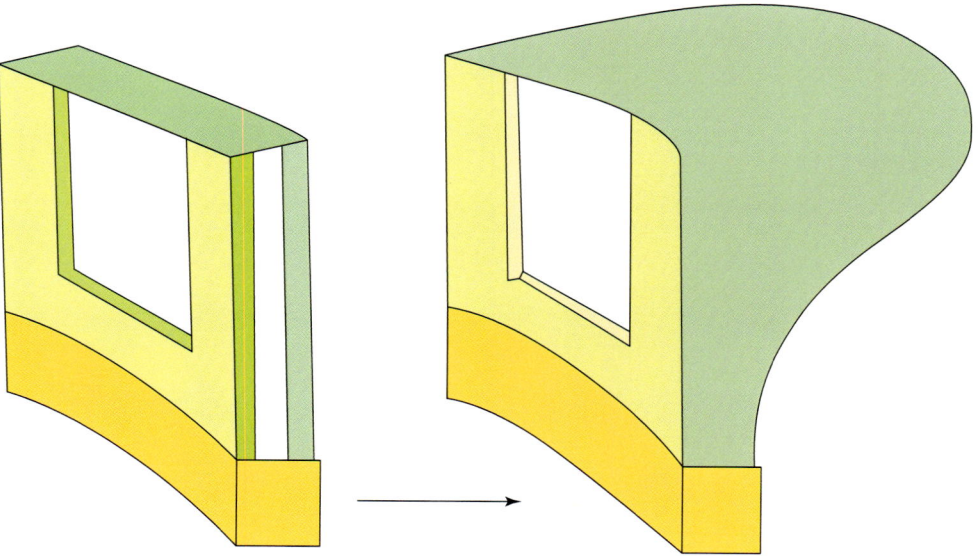

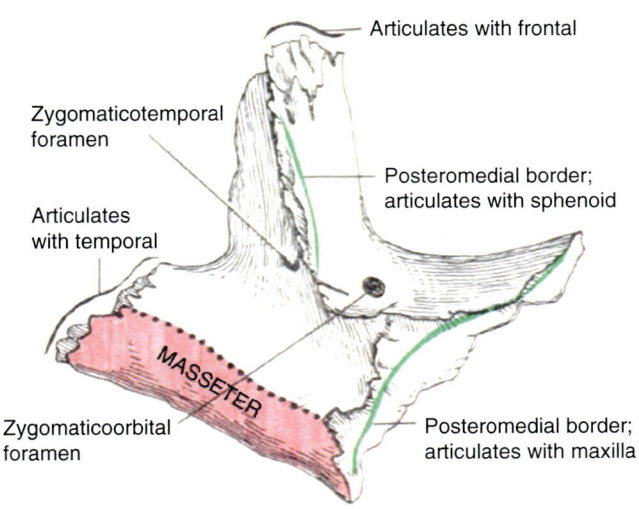

Fig. 8.203 Gray zygoma, internal 30.4.129. Masseter has a secondary insertion. Zygomatico-temporal is the axis of postorbital bone field. (Reprinted from Lewis, Warren H (ed). Gray's Anatomy of the Human Body, 20th American Edition. Philadelphia, PA: Lea & Febiger, 1918)

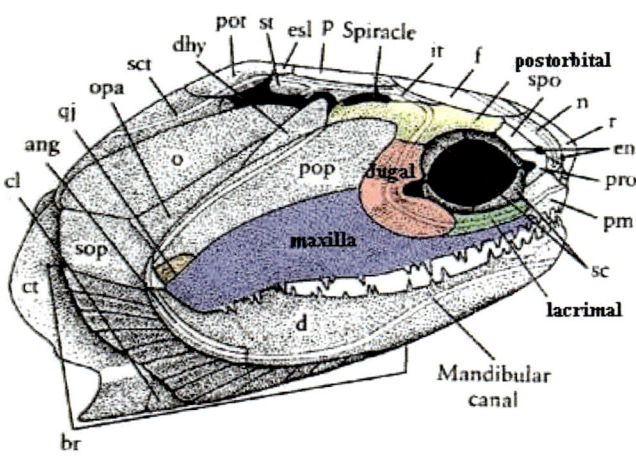

Fig. 8.205 Jugal, *Cheirolepsis*. In the early sarcopterygian, *Cheirolepsis*, jugal is associated with postorbital. Physically separated from quadratojugal, with which it will eventually connect. Jugal and lacrimal exclude maxilla from the orbit. (Reprinted with permission from: GEOL431—Vertebrate Paleobiology course web resources. Thomas Holtz and John Merck, Department of Geology, University of Maryland, College Park, https://www.geol.umd.edu/~jmerck/geol431/)

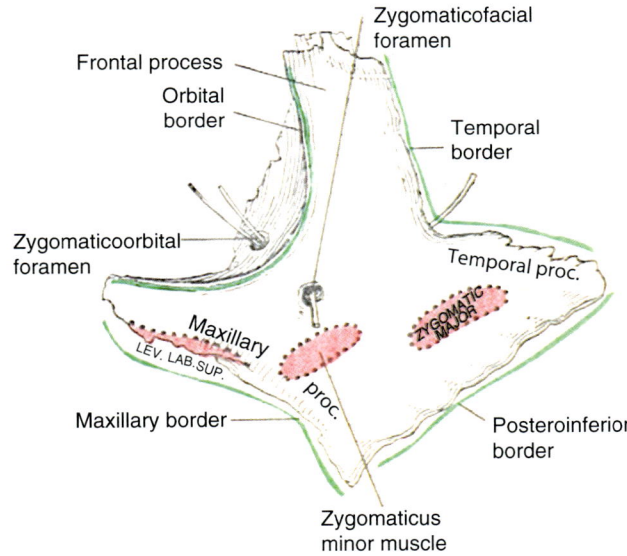

Fig. 8.204 Zygoma, external facial muscles all have secondary insertions (red). Zygomatico-orbital is the neurovascular axis of jugal bone field. (Reprinted from Lewis, Warren H (ed). Gray's Anatomy of the Human Body, 20th American Edition. Philadelphia, PA: Lea & Febiger, 1918)

Development

Ossification of zygoma is seen at 8 weeks, on or just after stage 23. It occurs at the midline along the plane dividing postorbital from jugal. This division can at times be observed in the newborn. It is readily appreciated in texts of comparative anatomy and paleontology.

Clinical correlations: craniofacial cleft zones 7 and 8.

Phylogeny

Jugal and postorbital begin their lives as members of the circumorbital series, a ring of bones that guard the eye. These bones begin in primitive fishes as a series of sclerotic ossicles but become dermal bones in osteoichthyes. *Amia* reflects the primitive condition as follows (left orbit, counterclockwise): lacrimal (SO$_1$), jugal (SO$_2$), suborbitals (SO$_3$, SO$_4$, and SO$_5$), dermosphenotic, postfrontal, supraorbital, and prefrontal. Jugal plays a very conservative role over time. Examination of its relationship to maxilla proves helpful in understanding the anatomic design of the human orbit.

In the sarcopterygian fishes, jugal undergoes a transition from an ossicle to a dermal bone and shifts from directly beneath the orbit to the posterolateral position. In the clade

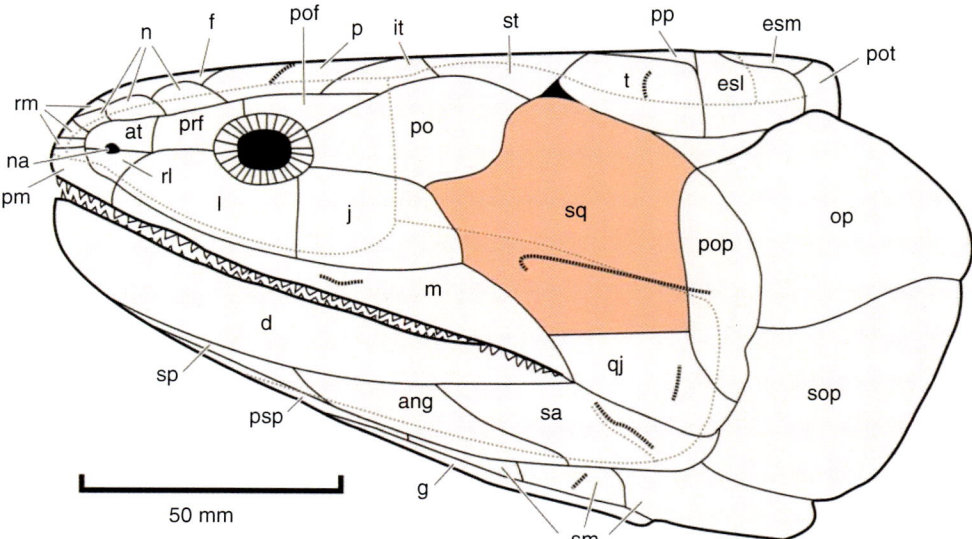

Fig. 8.206 Jugal-squamosal, *Eusthenopteron*. Eusthenopteron has jugal in contact with squamosal, a possible conversion of metapterygoid. As the snout shortens, maxilla is translocated forward but the orbit remains the same so jugal maintains its position. At the maxilla slides, forward jugal will come into contact with quadratojugal. (Reprinted

from Wikimedia. Retrieved from: https://commons.wikimedia.org/wiki/File:Eusthenopteron_postorbital.png. With permission from Creative Commons License 4.0: https://creativecommons.org/licenses/by-sa/4.0/deed.en)

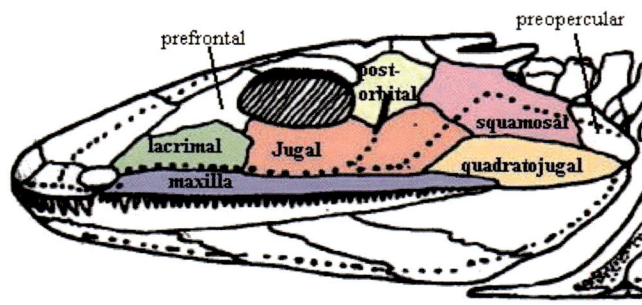

Fig. 8.207 Jugal, *Acanthostega*. Tetrapods show final contact established between jugal and quadratod jugal. No zygomatic arch is yet present. With brain expansion into cynodonts, squamosal will stretch downward and insinuate itself pushing j-qj outward into an arch form. (Reprinted with permission from Clack J Gaining Ground: The Origin and Early Evolution of Tetrapods, 2nd ed. Bloomington, IN: University of Indiana Press; 2012)

immediately prior to the tetrapodomorphs, the osteolepid *Cheirolepsis* shows jugal *below the sclerotic ring* (an important detail to which we will return later). It is in contact with lacrimal anteriorly, thereby *excluding maxilla from the orbit*. Jugal is in continuity posterosuperiorly with postorbital. This constant relationship is preserved in the human zygomatic complex (Fig. 8.201). Note in *Cheirolepsis* that premaxilla, lacrimal, jugal, and postorbital constitute a common neurologic sector, threaded together by the *lateral line system*. A neurosensory structure unique to fishes (but of relevance for us), the lateral line, consists of open pores arranged in regular lines over the head and body of the fish. These detect changes in pressure or waves of motion in the environment.

Fishes can triangulate this information to detect the source (and hence, a predator). The lateral line system functions as hair cells and is homologous with our vestibular system (Fig. 8.116).

Jugal assumes its standard condition in the crown sarcopterygians and early tetrapods. The operculars have been "eliminated" with appearance of the squamosal as a replacement bone. Note the persistence of the latera line system in both tetrapodomorph *Eusthenopteron* and in tetrapod *Acanthostega*. In the latter, zygoma has expanded posteriorly and passes beneath squamosal (in the same plane) to make and oblique contact with quadratojugal. This contact between jugal and qj persists through evolution. This sets the stage for the temporal fossa and zygomatic arch (Figs. 8.202 and 8.203).

We now trace the relationship between jugal, maxilla, and the orbit. Synapomorphies of the amniote skull include direct contact between the frontal bone and the orbit. This implies that the pre-existing prefrontal and postfrontal bones 201 are incorporated with frontal squama to form the frontal bone complex. The initial anapsid skull split between the synapsids leading to mammals and reptilia (they can be either diapsids or eurapsids). In all cases, maxilla remains excluded. Another key finding is the skull of *Dimetrodon*, a transitional reptilomorph that stands just before the true synapsids, both maxilla and jugal are massive (Fig. 8.204). The latter extends a *postorbital process upward*. This evidences the eventual fusion between the jugal and postorbital fields. Note that *the entire orbital floor is an extension of the jugal bone*. Maxilla

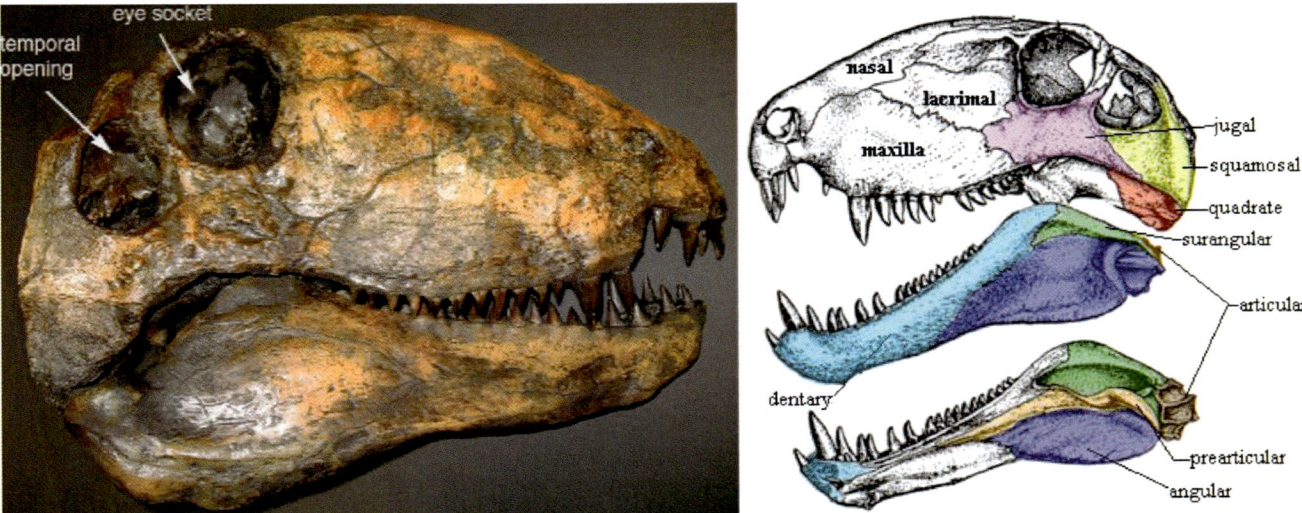

Fig. 8.208 Jugal in *Dimetrodon*: The pelycosaur ("mammal-like reptile") *Dimetrodon* ("two shapes of teeth") had massive lacrimal excluding maxilla from orbit. The orbital floor is made up by the medial extension of the jugal bone. [Left: (Courtesy of Prof. Richard L. Squires). Right: (Reprinted from Carroll RL. Vertebrate Palaeontology. New York, NY: WH Freeman and Company; 1988)]

has no contact, being excluded from the orbit by lacrimal bone. Thus, in humans, the orbital floor lateral to the axis of infraorbital (anterior superior alveolar) retains this ancient basal condition.

Changes in jaw muscles, development of the temporal fossa, and expansion of the brain were major themes in cynodont evolution. The massive adductors are reduced and spatially specialized muscles appear. Neomorph pterygoids insert into the pterygoid wings and controlled the future TMJ. The common adductor breaks into a deep temporalis with a much-reduced insertion and a superficial masseter, capable of lateralizing motion. These changes produced a much more sophisticated mastication and ability to handle previously hard to digest substances such as nuts and grasses (Fig. 8.205).

The mammalian braincase expanded. Instead of letting itself be confined to the basisphenoid–basipterygoid–occipital skull base, expansion of membranous parietal, occipital, squamosal, and frontal bones permitted the cerebrum to grow in all directions and particularly behind the eyes. A consequence of this was the lateral positioning of jaw muscles, formerly inserted quite posteriorly. The epipterygoid evolved into *alisphenoiid, greater wing of sphenoid*. This was always connected to frontal but it now occupies space between postorbital and parietal. The downward extension of alisphenoid united with pterygoid process, forming lateral pterygoid plate, and thus anchors the palate to the braincase. Alisphenoid uses temporal process of zygoma outward. Parietal expands downward as does the squamosal–quadratojugal complex. The jugal process of qj is also pushed outward. A definitive zygomatic arch emerges with the cynodonts. Let us look at these trends in sequence (Fig. 8.205).

The first mammals appeared in the late Triassic. *Adelobasileus* is the oldest, most basal mammal. Its orbit bears two diagnostic mammalian features. An entirely new bone field appears for the first time: *obitosphenoid* connects with frontal bone with ascending/orbital process of palatine bone to create a solid medial wall of the orbit. In so doing, orbitosphenoid becomes the *lesser wing of sphenoid*. At the same time, the amniote epipterygoid, a slender column of bone, becomes a broad sheet, *alisphenoid*, which sits behind orbitosphenoid. The brain expands posteriorly and in encased in bone. The lateral skull changes as well. The brain in basal amniotes is only partially covered by epipterygoid, prootic and opisthotic. Adelobasileus achieves lateral coverage by sending from prootic a forward-projecting *anterior lamina of periotic* which marries up in front with alisphenoid and above with parietal. This is the source of the squamous temporal bone. Cheek teeth developed two roots.

Adelobasieleus is followed in the Jurassic by *Sinocodon* (from China) which elaborated a well-developed dentary/surangular-squamosal joint (the future TMJ). Occlusion was not complete. Its petrosal promontorium provided housing for cochlear which would permit a full spiral. Expansion of the squamosal and parietal bones was accompanied by absorption of the tabular bones into the interparietal bone complex. Bone field of the skull in *Morganucodon* is the same as the previous taxa but it presents refinements elsewhere. The squamosal-dentary joint is now the main hinge for the jaw. The reptiloform quadrate-articular joint persists but its elements are now incorporated into the middle ear. Dentary is now dominant in the mandible and coronoid is present. Posterior tympanic bone fields of Meckel's cartilage remain as a consolidated rod composed of surangular, prearticular, angular, and articular. Cheek teeth are now in distinguished between premolars and molars. A precise occlusion is achieved.

Definitive arrangement of the human zygoma is achieved, perhaps unsurprisingly, with the development of binocular vision in primates. As discussed elsewhere, our immediate ancestory begins with the "almost but not quite" pleasiadapiforms in the Paleocene. *Plesiadapis* was a tree-climber with large eyes that still faced sideways. To accomplish this, it developed an opposable thumb with a fingernail, but all remaining digits had claws. It did not have binocular vision. Euprimates (the real deal) radiated in the early Eocene, had binocular vision (among other innovations), and are divided into two groups. The *Strepsirhini* (Gr. Strepsi, "turning inward") are the lemurs and lorises, characterized by inward-slanting nostrils and wet noses (like your dog). Haplorhini (tarsiers, monkeys, apes, and us) have rounded nostrils and dry noses. Living tarsiers are island-dwellers in Southeast Asia with huge eyes for a nocturnal existence and an orbital floor shared with minimally maxilla (Fig. 8.206).

To complete our story, Anthropoidea (human-like) is higher on the evolutionary scale than tarsiers and has many innovations. It is divided into the *platyrrhines* (flat, broad nose) of the New World and the *catarrhines* ("hook nose") of the Old World. Anthropoid orbits are regular in size, indicating a diurnal habitus. They have a smaller cornea and longer focal length for distance vision, and color perception. Color vision differs by geography. The catarrhines have only two types of cones whereas the catarrhines have three…so we are fortunate to come from Africa. On the other hand, catarrhines lose significant volume of the olfactory cortex, so our sense of smell is not as refined. The first catarrhine anthropoid clade, *Catopithecus*, comes from Egypt in the early Oligocene. Maxilla has entered the orbit (Fig. 8.207). The later Egyptian hominid *Parapithecus* shows a flaring of the malar eminence with a visible zygomatico-maxillary suture (Fig. 8.208). (1) This change is likely driven by normalization of the ratio between the maxilla and orbit and the development in anthropoids of maxillary and sphenoid sinus cavities.

The Mandibular Bone Complex

Understanding the evolutionary biology of the mandibular complex begins with *Meckel's cartilage* (MC) (Fig. 8.209). The mesenchyme of this important bone is exclusively r3 neural crest. It is a very ancient structure, dating back to the Precambrian. Once in position as a lower jaw, MC, surrounded by supporting bone fields became a bone complex, the history of which demonstrates anatomic changes that constituted advantages for food acquisition and processing. You are not just *what* you eat but *how* you eat. Changes in

length, muscle insertions, mechanical leverage, and innovations in dentition all played a role in the survival of species.

Traditional descriptions consider each hemi-mandible to be a single structure consisting of a tooth-bearing alveolus supported by a body in continuity with a vertically oriented ramus from which project a coronoid process and a condylar process. But looks can be deceiving. Like the maxilla, the mandible is more complex than it would appear. It is *bilaminar*; the external and internal laminae of the mandible can be surgically split. *Bone marrow* fills the potential space between the laminae. Multiple arterial sources supply its cortex. This implies that is it comprise of *multiple developmental fields*. Indeed, the original bauplan of the tetrapod mandible (as seen in *Labyrinthodont*) consisted of ten bones surrounding Meckel's cartilage. We shall explore the evolutionary fate of these bones later. Finally, mandible *articulates with the cranial base*. In mammals, this relationship is defined by a temporo-mandibular joint but various other configurations have been explored by evolution over time.

Descriptive Anatomy (Figs. 8.210, 8.211, and 8.212)

Projecting from the ramus are two processes of developmental significance. The *coronoid process* is triangular. Its anterior surface is continuous with that of the ramus. The secondary insertion of temporalis muscle occupies the entire coronoid process and spills down over the internal lamina of anterior ramus all the way down to the third molar. The *condylar process* articulates with the tympanic temporal bone. The head is covered with fibrocartilage. Interposed between is articular cartilage of the TMJ. The head bears a small lateral tubercle to which is attached the *temporomandibular ligament*. The neck bears a medial depression, the *pterygoid fovea*, which bears the primary insertion of lateral pterygoid muscle.

Muscles of mastication (all of one of them being Sm4 derivatives) have primary insertions to the mandible. The external ramus harbors masseter, while the coronoid process contains temporalis. Note that *no Sm4 muscle is associated with the external body*. The internal surface of the body has Sm4 muscles anterior digastric and mylohyoid; that of the ramus has medial pterygoid, the coronoid bears temporalis, and the condyle has lateral pterygoid. Each muscle attachment brings blood supply to the periosteum. Sm6 contributes a single muscle to the masticatory apparatus, posterior digastric, which attaches at the internal aspect of the mandibular angle.

Muscles of facial expression (Sm6 derivatives) have secondary attachments to the lateral mandibular body: mentalis, depressor labii inferioris, depressor anguli oris, platysma, and buccinator. Note that *no Sm6 muscle is associated with the internal body*. The medial mandibular body supports genioglossus (from the first occipital somite) and attaches

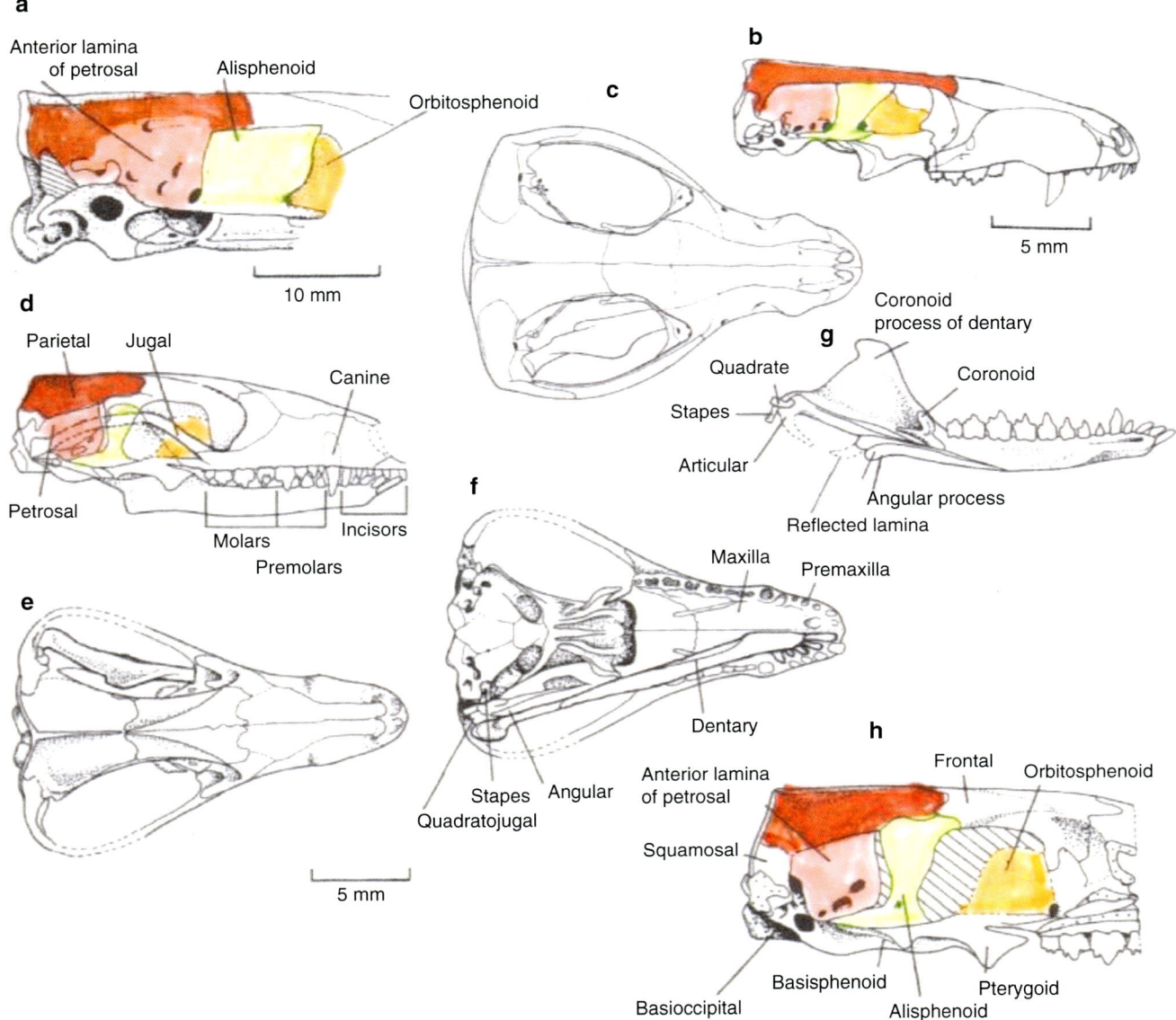

Fig. 8.209 Zygoma/Orbit in basal mammals. Adelobasileus (**a**) has innovations of new *orbitosphenoid* (orange) and a modified epipterygoid, the *alisphenoid* (yellow). Note enhanced lateral skull coverage with the anterior lamina of prootic/petrosal (pink) extending forward to contact alisphenoid and upward to contact frontal (brown). Note in *Morganucodon* the persistent tympanic rod makes up of four bone fields (surangular, prearticular, angular, and articular). (Reprinted from Luo Z-X, Ji Q, Yuan C-X. Convergent dental adaptations in pseudotribosphenic and tribosphenic mammals. Nature 2007;450:93–97. With permission from Springer Nature)

distally while a slip from Sm7 superior constrictor attaches behind the tuberosity.

It can be hypothesized that Sm4 muscles in register with r3 migrate prior to those supplied by more posterior neuromeres. These finding suggest that *the bones fields of the internal lamina may be synthesized before the external lamina.*

Facial artery makes a groove along the lateral body just distal to the ramus. Proximal to this landmark is Sm4 territory; distal to it is Sm6 territory. We postulate this boundary to be a remnant of the precursor bones: body and ramus are

genetically distinct We further postulate that the developmental "maturity gradient" of both sectors is distal-to-mesial. For this reason, pathology of the ramus worsens as one proceeds from condyle to gonion. If body is also involved, it spreads medially from the molars forward.

The inferior alveolar nerve provides sensation for the mandible and teeth. This nerve is actually a collection of three distinct sensory nerves bundled together and housed within a U-shaped trough along the medial mandible. Later in development, this will seal up, enclosing the nerves in the alveolar canal [17].

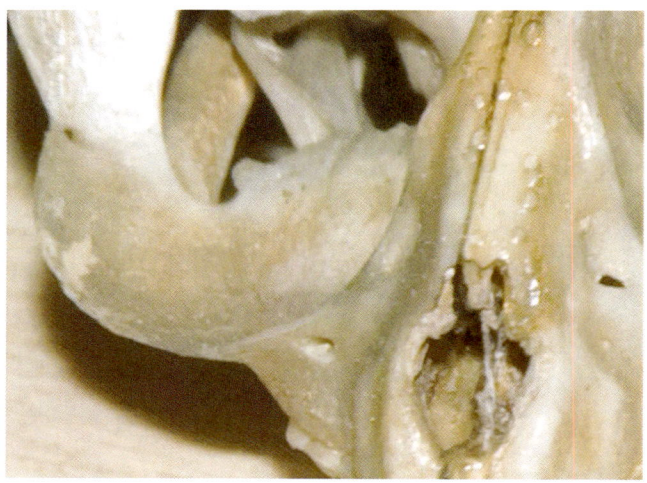

Fig. 8.210 Jugal evolution in primates. In tarsiers, jugal continues to form almost the entirety of the orbital floor. (Reprinted from Wikimedia. Retrieved from: https://commons.wikimedia.org/wiki/File:Tarsier_skull.jpg. with permission from Creative Commons License 2.0: https://creativecommons.org/licenses/by-sa/2.0/deed.en)

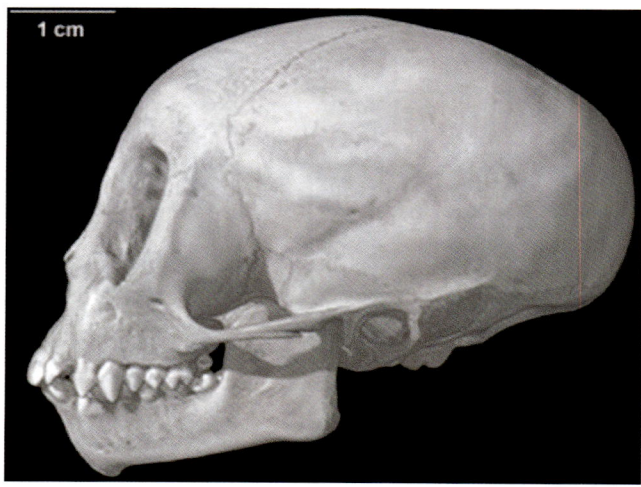

Fig. 8.211 *Catopithecus* Gr. "*cato*" = "below" and "*pithekos*" "one who plays tricks" (an ape plays trick); thus, "one below ((prior to) the ape"). Early haplorhini found in Egypt. (Reprinted with permission from Dr. James Rossie, 2002, "Saguinus oedipus" (On-line), Digital Morphology. Accessed January 8, 2019 at http://digimorph.org/specimens/Saguinus_oedipus/306845/. Dr. James Rossie. Stony Brook University, NSF digital library, University of Texas Austin)

Multiple sources supply the mandible. We will start with the internal lamina and marrow space. Inferior alveolar artery originates from IMMA just as that vessel runs between the mandibular neck and the sphenomandibular ligament. Almost immediately it gives off *lingual artery* that is programmed by it accompanying StV3 lingual nerve. The lingual pedicle supplies the lingual sulcus and lateral floor of mouth. Inferior alveolar artery descends with its companion StV3 nerve [17].

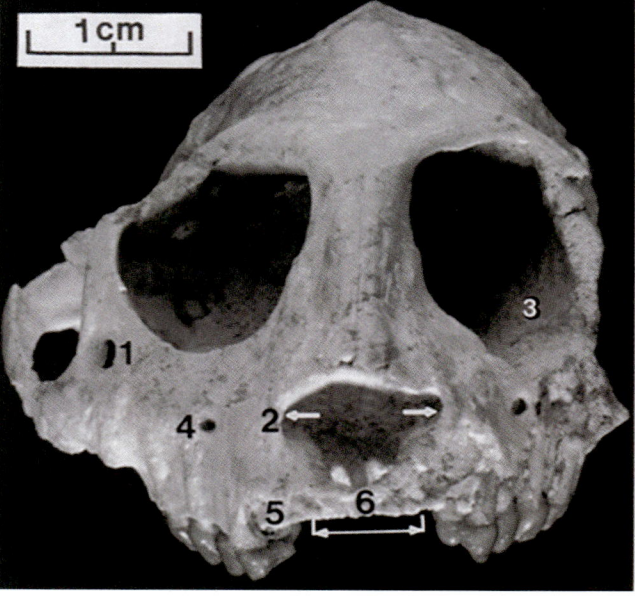

Fig. 8.212 *Parapithecus* was an early Egyptian hominid showing the flaring of the jugal process. (Reprinted from Simons El. The cranium of *Parapithecus grangeri*, an Egyptian Oligocene anthropoidean primate. Proc Nat Acad Sci U.S.A. 2001; Jul 3; 98(14):7892–7897, with permission from PNAS)

As StV3 approximates the mandibular canal, it gives off yet another pedicle *mylohyoid artery* and nerve. The vessel runs along the medial lamina and supplies the muscle. In both form and function, mylohyoid artery is analogous to greater palatine artery. Having entered the ramus, inferior alveolar passes into the body where it provides dental branches to all the teeth. At the adult first premolar tooth (the deciduous first molar), it divides. The *mental artery* exits the bone while the inaccurately named *incisor artery* stays in the bony canal to supply canine, lateral incisor, and central incisor. In sum, inferior alveolar artery is analogous to superior alveolar artery. Both vessels penetrate the bone, both pursue a course within the interlaminar space and supply all the dental units. Note that in the incisor region of premaxilla, SAA makes an anastomosis with medial sphenopalatine artery whereas IAA simply anastomoses with its contralateral vessel.

Blood Supply and Mesenchyme
Conclusions from Tables 8.5 and 8.6.

1. Dentary bone fields are defined by types of dentition. They are supplied by three branches of a single arterial axis, the inferior alveolar artery which is StV3 from the proximal segment of MMA.
2. Non-dentary bone fields are defined by muscle insertions. These are supplied by multiple branches from the ECA.
3. The various arterial fields can be understood on the basis of paleontology.

Table 8.5 Lingual laminae

Bone field	Muscle insertion	Arterial supply
Condyle	TMJ	Superficial temporal
	Lateral pterygoid	MMA/pterygoid, lat. br.
Coronoid	Temporalis	MMA/posterior deep temporal
Ramus	Medial pterygoid	MMA/pterygoid, medial br.
Body, proximal	Mylohyoid	MMA/mylohyoid
	Platysma	Facial/cervical br.
Body, distal	Anterior digastric	Facial/cervical/submental
Symphysis	Genioglossus	Lingual/deep lingual
	Geniohyoid	Lingual/deep lingual

Table 8.6 Buccolabial laminae

Bone field	Muscle insertion	Arterial supply
Condyle	TMJ	Superficial temporal
Coronoid	Temporalis	MMA/posterior deep temporal
Ramus	Masseter	MMA/masseteric
Body, proximal	Buccinator	MMA/buccal
	Platysma	Facial/cervical/submental
Body, distal	Depressor anguli oris	Facial/cervical/submental
	Depressor labii inferior	Facial/cervical/submental
Symphysis	Mentalis	Facial/cervical/submental

4. Recall that the primitive *Labyrinthodont* mandible had multiple bone fields, of which only the *dentary* bone persists in mammals. Their fates are as follows:

- **Coronoids**: There were three splenials along the cephalic border. Do these now reveals programming of three dental fields embedded within the epithelium?
- **Dentary**: Origin of the symphysis.
- **Splenials**: Two splenials on either side of caudal border of mandible. Do these presages the mesial and distal bone fields of the mandibular body?
- **Surangular**: The likely precursor of ramus.
- **Angular**: Precursor of tympanic temporal.
- **Prearticular**: Precursor of anterior process of malleus.
- **Articular**: Precursor of malleus.

Arterial Axes of the Mandible

Two distinct arterial axes supply the body of the mandible: facial artery and lingual artery. Facial artery is composed of four proximal *cervical branches* and four distal *facial branches*. Let us proceed from distal-to-proximal. First off, all along the course of its course unnamed *muscular branches* are assigned to individual muscles. The cervical segment provides branches to medial pterygoid and stylohyoid. The facial segment gives rise to masseter and the buccinator/mimetic muscle group. Astute readers will note that facial

artery is an "equal opportunity lender." It provides for muscle derivatives of both the first and second pharyngeal arches (Sm4/V3 and Sm6/VIII respectively). Via the muscles of mastication blood supply is conveyed via the periosteum to their respective regions of the mandible.

The branches of facial artery associated with the soft tissues of the nose and midface (*angular, lateral nasal,* and *superior labial*) are irrelevant for the mandible. The *inferior labial* artery helps to support the external alveolar cortex from canine to midline. It makes an anastomosis with mental branch of inferior alveolar.

The cervical branches are so-named because they arise in the neck and travel forward. Contrary to their name, they do not supply the neck. Most distal is *submental artery*. It runs along the inferior border of mandible superficial to mylohyoid and deep to anterior belly of digastric. It supplies these muscles and their respective zones of external mandibular cortex. Quite logically, it is connected with the other two extramedullary vascular axes of mandible: mylohyoid branch of inferior alveolar and sublingual branch of lingual artery. The final destination of submental is the chin, where it divides. The deep branch supplies levator labii inferioris and bone. Again, it makes a logical anastomosis with mental branch of inferior alveolar artery. The superficial branch ascends into the lower lip where it connects with the inferior labial arcade. The role of submental artery is to supply the inferior border of non-alveolar body.

More proximal on the cervical segment are *glandular branches* distributed to the submandibular gland, lymph nodes, and masseter. The facial notch of the mandible reflects this supply to caudal border of the body-ramus junction, i.e., the gonial angle. The *tonsillar branch* passes between medial pterygoid and styloglossus. It supplies the palatine tonsil and the root of the tongue. As such it supplies the cranial border of the body–ramus junction, i.e., it is the mirror-image of gonion. *Ascending palatine artery* is a bit-player regarding the mandible. One branch perforates the Sm7 superior constrictor en route to palatine tonsil and Eustachian tube. As a consequence, it connects with ascending pharyngeal, the arterial axis of the third arch (to which superior constrictor belongs). The other branch tracks along the course of another Sm7 muscle, levator veli palatini, ascends over the upper border of superior constrictor, and supplies soft palate, where it connects with greater palatine.

We come now to the other main source of blood supply to the mandibular body, the lingual artery. This axis also has four named branches. (1) *Deep lingual artery* is the most distal. It runs between the mucosa and the most peripheral tongue muscle (inferior longitudinal). It is also situated on the lateral aspect of genioglossus which it comes to supply at the genial tubercle of the symphysis. (2) *Sublingual artery* has its stem at the anterior border of hyoglossus. It then runs forward between genioglossus and mylohyoid, which it supplies. This explains the vasculature to the bulk of internal lamina of mandibular body. It also supplies the mucoperiosteum of the lingual alveolus. In this respect, *sublingual artery fulfills the same function as the greater palatine artery*. A final anastomosis is made with submental artery. (3) *Dorsal lingual branches* are irrelevant to mandible. (4) *Suprahyoid artery* supplies the r3 moiety of the hyoid bone and the mylohyoid muscle.

The principle sources of blood supply to the coronoid process and condyle are *inferior alveolar artery* and *superficial temporal artery*. This vessel begins in the deep portion of parotid posterior to the neck of the condyle. It gives unnamed branches to the posterior sector of masseter muscle, the medial pterygoid, and temporomandibular joint. It subsequently has six named branches. *Transverse facial artery* arises within the substance of parotid gland prior to emergence of the STA from the parotid. Transverse facial supplies both gland and duct. It follows the course of masseter to which it contributes. Adjacent to it are one or two branches of facial nerve accompany. This makes biologic sense. Because VII is not a somatic sensory nerve, it must be accompanied by sensory fibers of V3. This explains the mechanism whereby parotid malignancy can cause facial pain. *The induction of transverse facial may depend on these V3 fibers*. Just above zygomatic arch is *middle temporal artery*. This vessel supplies temporalis muscle and the external lamina of squamous temporal bone. It forms an anastomosis with posterior deep temporal branch of IMMA.

Development

Development of the mandible is analogous to that of the maxilla. In a nutshell, a central cartilaginous core, Meckel's cartilage, becomes ensheathed with membranous bones. It consists of an anterior tooth-bearing *dentary zone* and a posterior *tympanic zone* which contains the bone fields of the future middle ear (Figs. 8.213, 8.214, 8.215, 8.216, 8.217, 8.218, 8.219, 8.220, 8.221, 8.222, and 8.223).

Meckel's cartilage extends downward from the otic capsule to the midline below the mouth. The largest segment of Meckel's represents the ceratobranchial cartilage of BA1. It forms all mandibular fields distal (posterior) to the exit point of V3 mental nerve and has accessory cartilages. Meckel's also has accessory cartilages which represents the hypobranchial and basibranchial cartilages of BA1. These form the symphysis fields mesial (anterior) to the V3 mental nerve.

The fate of Meckel's cartilage is as follows. The extreme posterior aspect retains a connection with the skull base as the *sphenomandibular ligament*. Subsequently, *accessory cartilages* appear: (1) the condylar nucleus extends downward along the postero-lateral ramus; (2) the coronoid nucleus along anterior ramus; (3) discrete rami along both sides of the distal alveolar walls; and (4) along the anterior–inferior border. Separate ossification centers are not seen in

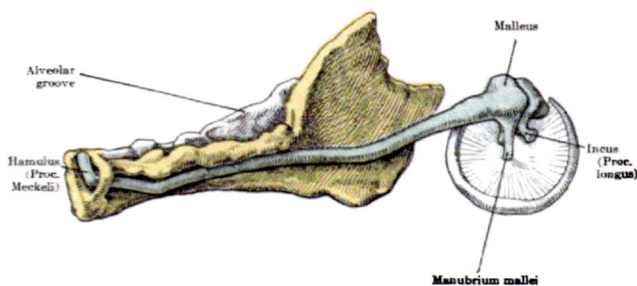

Fig. 8.213 Meckel's cartilage (after Kollman) approximately 95 mm. Meckel's is a neural crest structure which has two sectors defined by the lingula. The dentary sector (anterior) provides the template for flanking bone fields that develop in membrane. Note that the anterior zone, the hamulus or *processus Meckeli*, gives rise to the chondral symphysis. The tympanic sector (posterior) contains the bone fields of the ear. (Reprinted from Kollman JKE. Handatlas der entwicklungsgeschichte des menschen (Atlas of the Development of Man), Jena, Gustav Fischer; 1907)

these nuclei. They are subsequently invaded (and absorbed) by surrounding membranous bone. The inner alveolar border arises from a *splenial center* that arises from the bony mass of the body [18–20].

Radlanski et al. report the existence of chondral ossification centers next to the dental primordia in humans beginning as early as 95 mm (12 weeks) [21]. These provide structural evidence of discrete developmental units within the bone. Such sites could indicate discrete developmental cues for the organization of neural crest to form individual dental units.

These observations will become very important when we discuss the evolutionary history of the mandible because *the number of membranous bones in the basal tetrapod state is nine*. Only one of these, the dentary, is thought to persist, an assumption we shall challenge. Although preexisting bone fields have been incorporated into the dentary, their genetic markers persist, providing the definition for the individual developmental fields of the mandible.

Embryonic Period

- Stage 9 (day 20–21, 1.5–2.5 mm). The mesenchyme of the first pharyngeal arch is first assembled.
- Stage 13 (28–32 days, 4–6 mm). Mesenchymal condensation of Meckel's cartilage is seen.
- Stage 17 (41–43 days, 11–14 mm). Extracranial stapedial has united with maxillary branch of ECA to produce the internal maxillomandibular arterial axis, from the first sector of which inferior alveolar artery forms in conjunction with V3. Thus, at 41 days, primary ossification appears along the lower border of Meckel's cartilage. The ossification center appears precisely at the bifurcation of V3 inferior alveolar into the *incisor branch* to the dental units of the future symphysis and the *mental branch* to

Fig. 8.214 Mandible, lateral. (Reprinted from Lewis, Warren H (ed). Gray's Anatomy of the Human Body, 20th American Edition. Philadelphia, PA: Lea & Febiger, 1918)

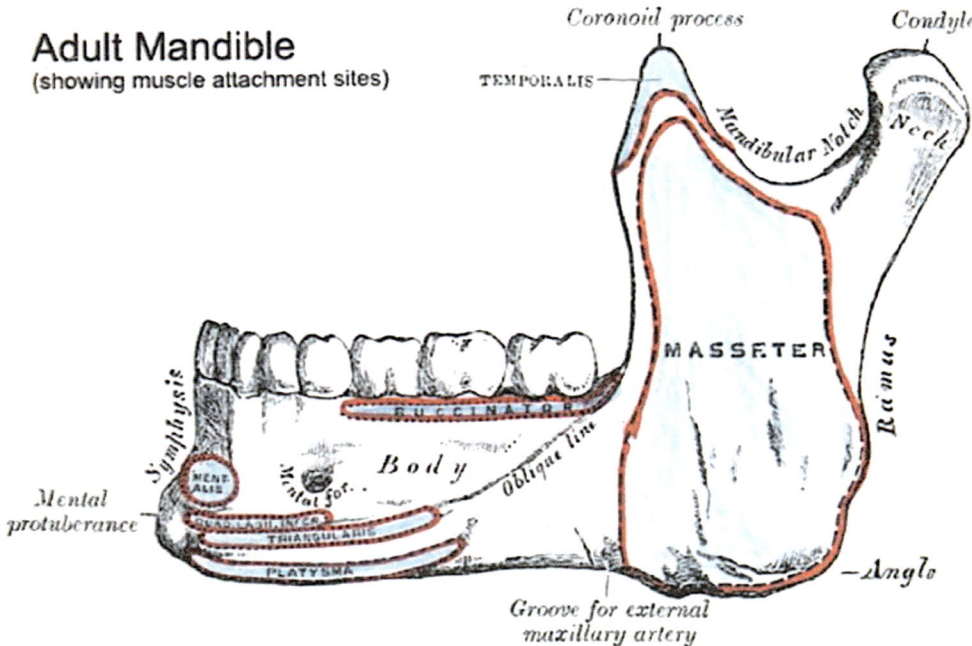

Fig. 8.215 Mandible, medial. (Reprinted from Lewis, Warren H (ed). Gray's Anatomy of the Human Body, 20th American Edition. Philadelphia, PA: Lea & Febiger, 1918)

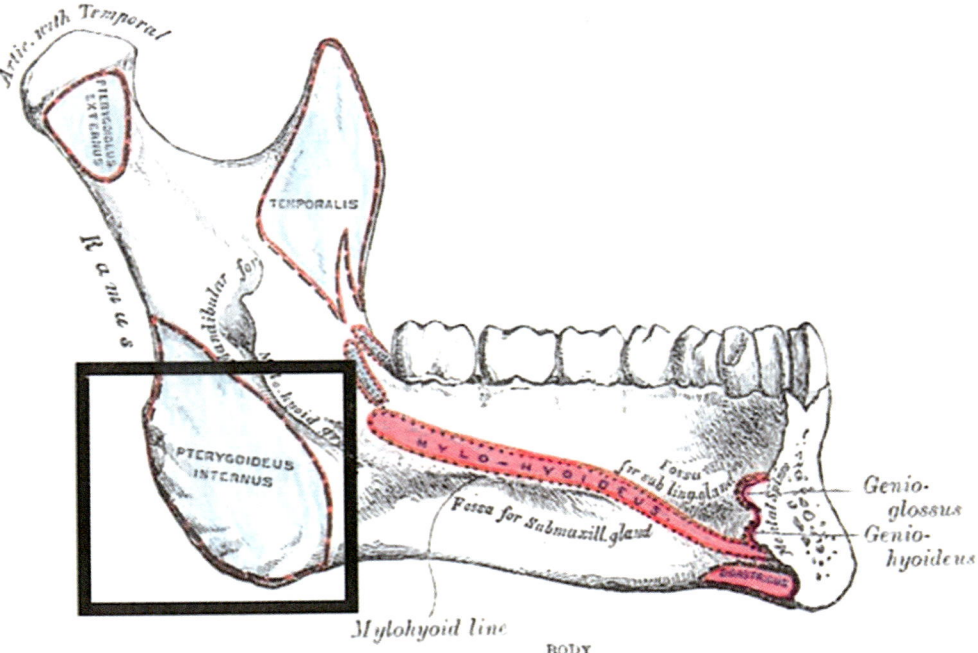

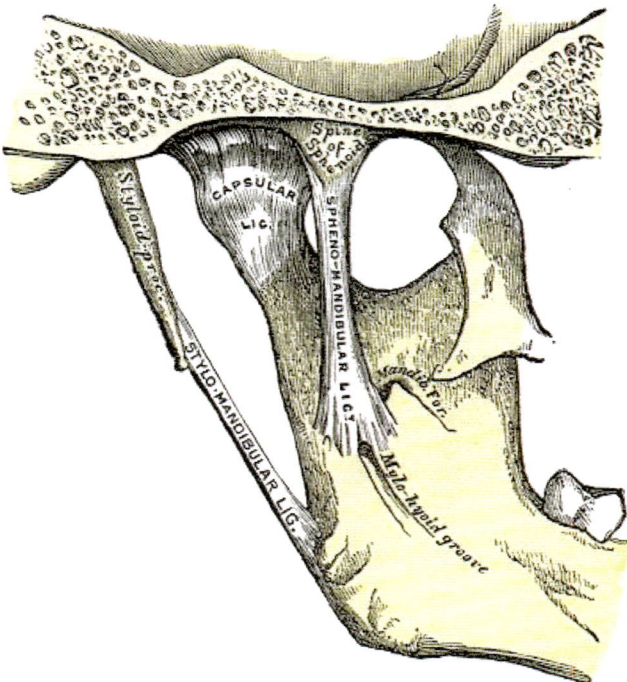

Fig. 8.216 Ligaments. Sphenomandibular ligament (SpM) runs from spina angularis of r3 alisphenoid to just below the lingula. Medial to it runs chorda tympani br of VII. Lateral to it runs inferior alveolar branch of V3 Stylomandibular ligament (StM) from r5 styloid process of petrous to the angle. Note that condyle arises from between SpM and StM. (Reprinted from Lewis, Warren H (ed). Gray's Anatomy of the Human Body, 20th American Edition. Philadelphia, PA: Lea & Febiger, 1918)

overlying soft tissues. Muscle insertions are observed at stage 18 (44 days, 13–17 mm).

- Stage 19 (48–50 days, 16–18 mm). The seventh week shows ossification at the location of the future canine (i.e., the mental foramen). From this initial site, bone spreads out over the external aspect of Meckel's cartilage in radial fashion. *This implies that the "maturity gradient" of the mandibular fields will be lateral-to-medial.* Both nerve to mylohyoid and inferior alveolar nerve are internal to the bone but eventually the latter emerges laterally as mental nerve. Condyle begins to form as a secondary cartilage. It is distinct from primary cartilages, such as cranial base and septum. It predates coronoid [19, 22] (Figs. 8.216 and 8.217).

- Stage 21 (51–52 days, 24 mm). Unification of the symphysis with lateral body creates mental foramen. V3 straddles Meckel's, with mylohyoid and lingual nerves running along its medial surface while inferior alveolar nerve remains applied laterally in association with the membranous bone. The course of lingual nerve defines the tympanic sector of Meckel's cartilage. Chorda tympani enters V3 at this point. This is also the entry point of St3 inferior alveolar artery. Malleus and incus develop from the tympanic sector of Meckel's. Proximally, at the level of first and second arch interbranchial segments, Meckel's cartilage and Reichert's cartilage come together. Reichert's provides stapes. Common facial nerve descends along the medial aspect of malleus and incus. Internal to

Fig. 8.217 Ossification centers of mandible: outer lamina (dentary), inner lamina (splenials), coronoid (surangular), ramus, condyle. (Reprinted from Keith A. Human Embryology and Morphology. London: Edward Arnold; 1902)

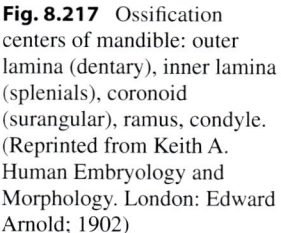

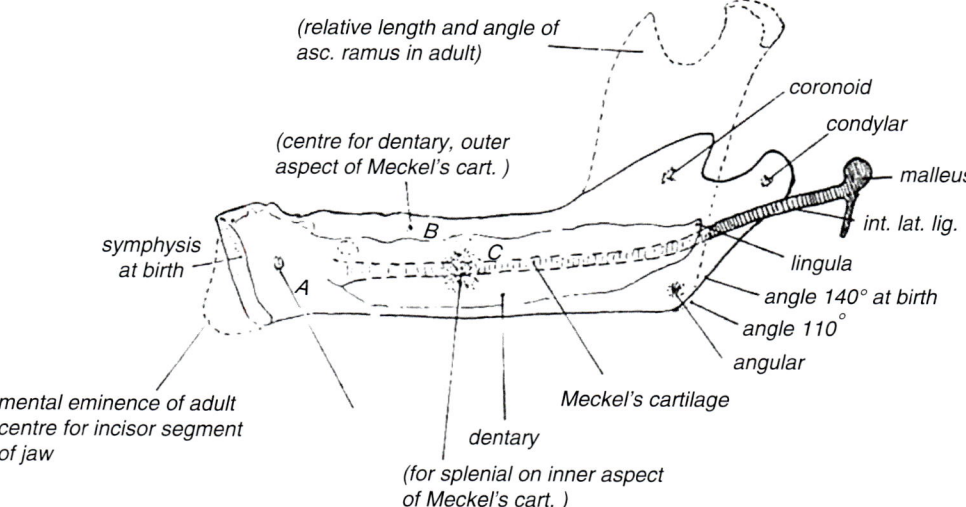

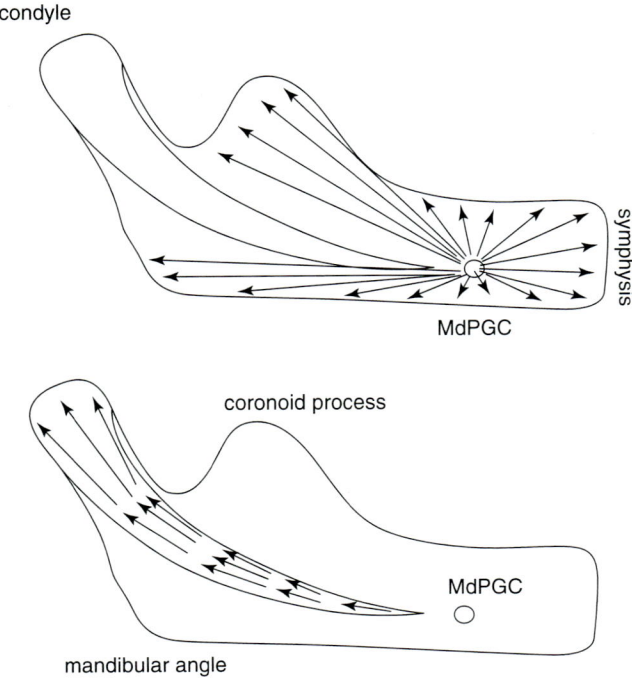

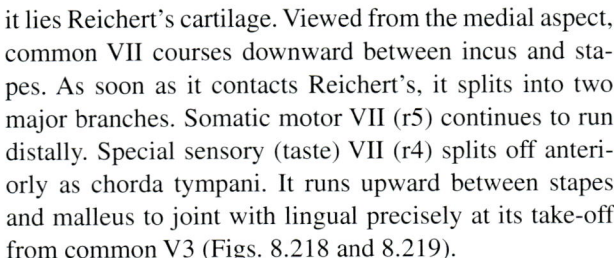

Fig. 8.218 Mandibular growth vectors. <u>Intramembranous</u>: (1) body, except menton, (2) ramus up to mandibular foramen. <u>Endochondral</u>: (1) symphysis, (2) ramus above mandibular foramen, (3) coronoid process, (4) condyle. (Reprinted from Lee, SK, Kim YS, OH HS, Yang KH, Kim EC, Chi JG. Prenatal development of the human mandible. *Anat Rec* 2001; 263:314–32. With permission from John Wiley & Sons)

it lies Reichert's cartilage. Viewed from the medial aspect, common VII courses downward between incus and stapes. As soon as it contacts Reichert's, it splits into two major branches. Somatic motor VII (r5) continues to run distally. Special sensory (taste) VII (r4) splits off anteriorly as chorda tympani. It runs upward between stapes and malleus to joint with lingual precisely at its take-off from common V3 (Figs. 8.218 and 8.219).

- <u>Stage 23</u> (57 days+, 27–31 mm). During the eighth week, this process has proceeded backward (cranially) to cover the entire external surface of Meckel's cartilage. But the external mandibular lamina ossifies in two ways. *Mesial to mental foramen it is chondral; distal to the foramen it is membranous.* Muscle attachments for pterygoids, masseter, and temporalis are present. Condyle blastema are attached to pterygoid muscle. Ramus expands and endochondral ossification occurs. Coronoid process develops.

Fetal Period

- <u>8.5 weeks</u>: Chondral ossification spreads from the mental foramen medially beneath the inferior margin of Meckel's to reach its internal (lingual) surface. From here bone, formation proceeds forward towards the midline (but not backward). This forms the internal (lingual) wall of the symphyseal fields. These fields are organized from the hyobranchial and basibranchial cartilages of BA1.

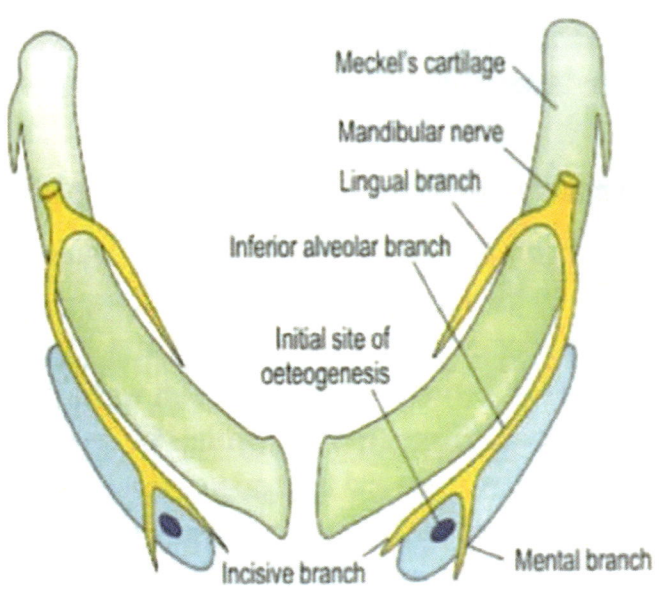

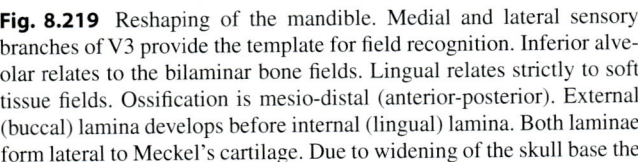

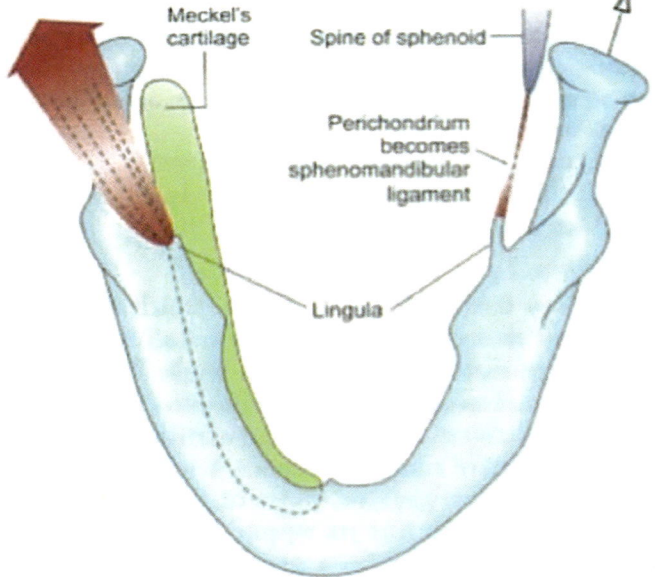

Fig. 8.219 Reshaping of the mandible. Medial and lateral sensory branches of V3 provide the template for field recognition. Inferior alveolar relates to the bilaminar bone fields. Lingual relates strictly to soft tissue fields. Ossification is mesio-distal (anterior-posterior). External (buccal) lamina develops before internal (lingual) lamina. Both laminae form <u>lateral</u> to Meckel's cartilage. Due to widening of the skull base the sphenomandibular ligament and the TMJ distract the ramus laterally. This causes bone development to deviate away from Meckel's cartilage at lingual: the boundary of mandibular versus tympanic sectors of Meckel's. (Reprinted from Nanci A. Embryology of the Head, Face, and Oral Cavity. In: Ten Cate's Oral Histology, 9th edition. St. Louis, MO: Elsevier, Inc.; 2007: 23–40. With permission from Elsevier)

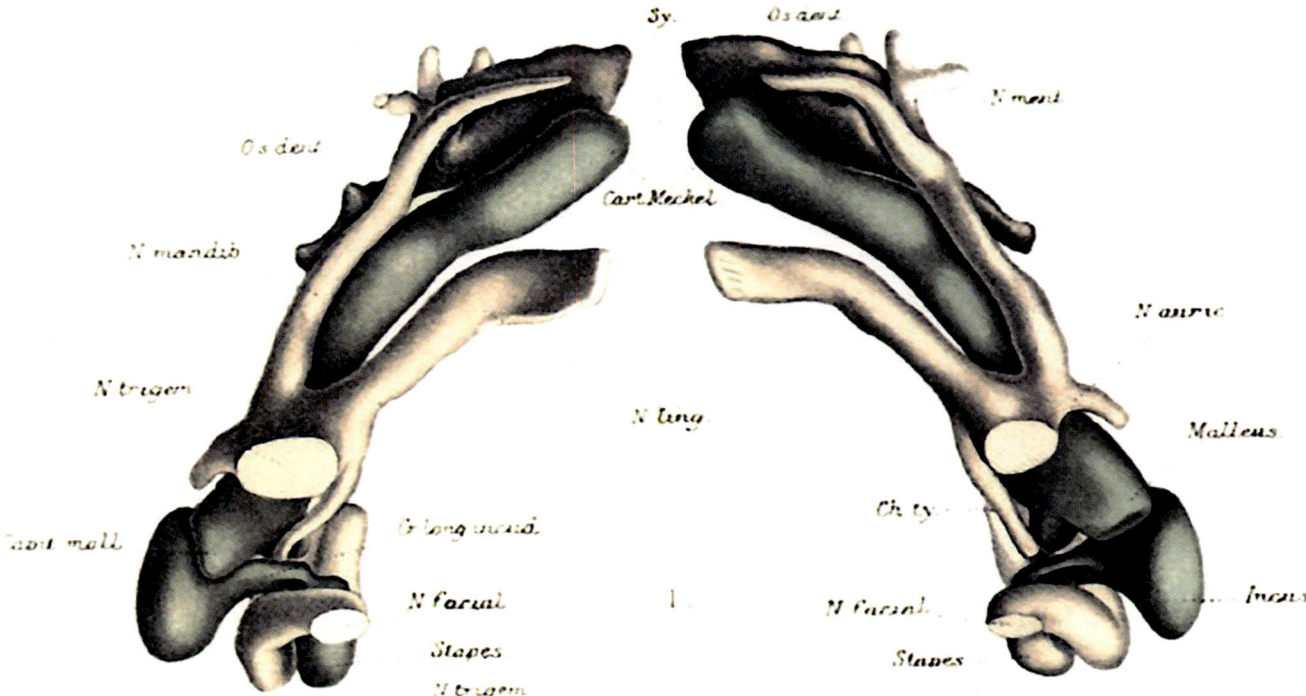

Fig. 8.220 Mandible development: 18 mm from above. Mesenchymal pathways on either side of Meckel's cartilage are defined by sensory branches of V3. Bone deposition begins distally and laterally; it then proceeds proximally. (Reprinted from Fanz Keibel, Kurt Elze (Freiborg). Nomentafel zur Entwicklungsgeschichte des Menschen. Jena: Verlag von Gustaf Fischer, 1908)

Fig. 8.221 Mandible development: 18 mm, right side, lateral. Body ossified in membrane lateral to mental nerve. Symphysis ossifies in cartilage medial to mental nerve. It is initially separate from the body. (Reprinted from Fanz Keibel, Kurt Elze (Freiborg). Nomentafel zur Entwicklungsgeschichte des Menschen. Jena: Verlag von Gustaf Fischer, 1908)

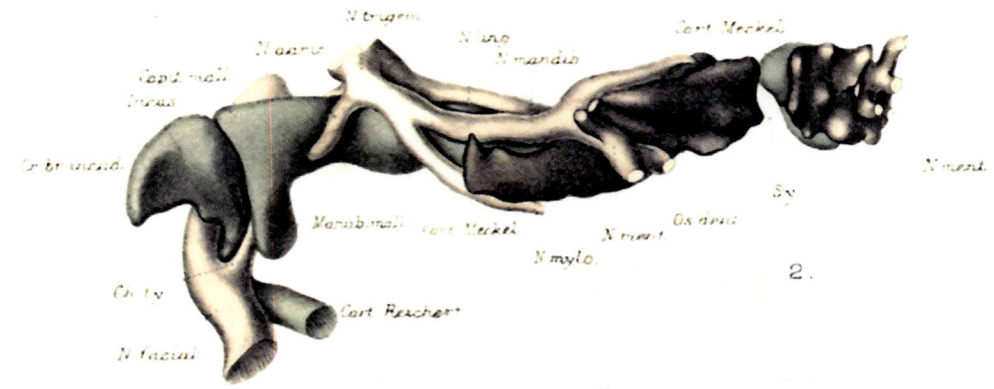

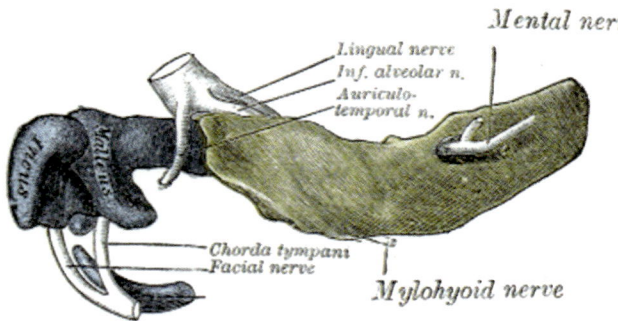

Fig. 8.222 Mandible development: 24 mm, (56 days) right lateral. Unification of the symphysis of with lateral body has taken place, surrounding the mental nerve. (Reprinted from Fanz Keibel, Kurt Elze (Freiborg). Nomentafel zur Entwicklungsgeschichte des Menschen. Jena: Verlag von Gustaf Fischer, 1908)

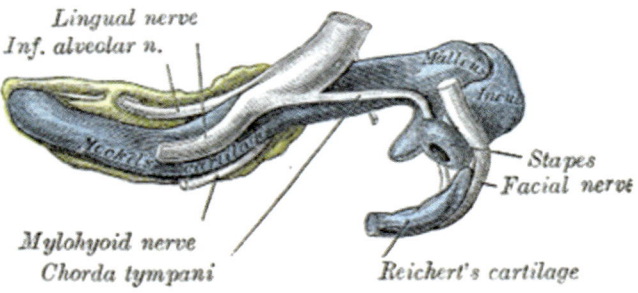

Fig. 8.223 Mandible, 24 mm (56 days), right medial. Lingula defines the tympanic sector of Meckel's cartilage. Chorda tympani enters V3 at this point. This is the entry point of St3 inferior alveolar artery. Ear bones develop in the tympanic sector. Proximally, at the level of first and second arch interbranchial segments, Meckel's cartilage and Reichert's cartilage come together. Reichert's provides stapes. The anlagen of these cartilages and the manner is which they drape around the pharyngotympanic tube is illustrated by Frazer. (Reprinted from Fanz Keibel, Kurt Elze (Freiborg). Nomentafel zur Entwicklungsgeschichte des Menschen. Jena: Verlag von Gustaf Fischer, 1908)

- 10 weeks: The entire incisor region of Meckel's cartilage is encased in chondral bone. This will become the symphysis. Condylar cartilage (CC) presents as an inverted cone within the ramus. The apex of the cone points toward the mandibular foramen. Vascular canals (VC) develop at junction between CC and Meckel's cartilage. This site is marked by the auriculotemporal branch of V3. Posteromedial VSx are 1–3 in number. A single posterolateral VC exists at upper outer pole of the condyle.

- 11 weeks: We turn our attention next to the ossification of the internal mandibular lamina. With symphyseal ossification complete, the mandible begins to change its shape. Transverse growth of the skull base pushes the middle ear (to which Meckel's cartilage is fixed) laterally. At the same time, ligaments and muscle attachments at the posterior mandibular restrain it: the formerly linear mandible

becomes angulated. Meckel's cartilage becomes *S-shaped*. A consequence of this movement is that Meckel's is now pulled away (internally) from the outer membranous lamina. Into this gap, a new layer of membranous bone forms, cranial to mental foramen and, once again, lateral to Meckel's cartilage. The mandible is now bilaminar. The internal (lingual) lamina arises in distinct two phases. (1) Mesial to canine, and prior to mandibular displacement, the lingual lamina develops as chondral bone along the internal (lingual) perichondrium of Meckel's. (2) Distal to canine, and subsequent to mandibular displacement, the lingual lamina develops as membranous bone along the external (buccal) lamina of Meckel's. Local signals determine the ossification pattern of the neural crest anlage. Intramembranous ossification of the condyle extends upward and medial from the insertion of lateral pterygoid.

- 12 weeks: In the early fetal period (90 days, 95 mm) proximal growth of lateral membranous bone produces the *ramus*. It splits along the plane of V3 to produce coronoid notch. *Accessory cartilages* are shown in stippled blue. *Coronoid process* arises from dentary in apparent spatial isolation from Meckel's. In contrast, the *condylar process* ascends as a separate entity from the lateral aspect of Meckel's cartilage *per se*. For a time, it can be seen along the lateral aspect of the ramus. This implies that as the membranous ramus forms, it first ascends on either side of the condylar process. Secondarily, the condylar process is converted into membranous bone. This is traditionally described as an "invasion" but it is more likely just a transformation of the neural crest cartilage intermediate (similar to lesser wing of sphenoid). A medial view of the mandible at this stage shows that membranous bone has crept inward and upward below the inferior alveolar nerve to encase it. This creates a zone of medial bone above (anterior to) Meckel's. Virtually the entire medial ramus arises from this anlage. However, a separate U-shaped sector of membranous bone wraps around the condylar accessory cartilage from lateral to medial to form the gonial angle. The developmental implication here is that zone of membranous bone in association with inferior alveolar nerve and above Meckel's will house the dentition. Below Meckel's a zone of dense bone will support the alveolar process. Vascular invasion of the condylar cartilage begins. Endochondral ossification of the condyle proper takes places between weeks 12 and 15 (Figs. 8.220 and 8.221).

- 14 weeks: Ossification of the posterior half of ramus takes place from condylar process.

- 16 weeks: Ossification of the anterior half of ramus takes place from the coronoid process. Alveolar process develops.

Post-natal Growth and Development (Figs. 8.222 and 8.223)

- At birth, the body of the mandible is a shallow trough containing the sockets for two incisors, the canine, and two deciduous molars. The mandibular canal is sizeable and runs close to the lower border of the bone. The mental foramen in located beneath the first molar. The angle is obtuse (175°), and the condyloid portion is nearly in line with the body. The coronoid process is of comparatively large size, and projects *above* the level of the condyle.

- Early childhood events are as follows. The two segments of mandible unite at the symphysis. The process takes place from below upward, mimicking the initial spread of mesenchyme with respect to Meckel's cartilage. The alveolar margin may maintain a separation into the second year of life. The body becomes elongates in its whole length, but more especially behind the mental foramen, to provide space for the three additional teeth developed in this part. The depth of the body increases owing to increased growth of the alveolar part, to afford room for the roots of the teeth, and by thickening of the subdental portion which enables the jaw to withstand the powerful action of the masticatory muscles; but the alveolar portion is the deeper of the two, and, consequently, the chief part of the body lies above the oblique line. The mandibular canal, after the second dentition, is situated just above the level of the mylohyoid line; and the mental foramen occupies its final adult position. The angle becomes less obtuse, owing to the separation of the jaws by the teeth; about the fourth year it is 140°.

- During adulthood, the alveolar and subdental portions of the body are usually of equal depth. The volume of the alveolar portion is maintained by the stresses on the bone due to mastication. The mental foramen opens midway between the upper and lower borders of the bone, and the mandibular canal runs nearly parallel with the mylohyoid line. The ramus is almost vertical in direction, the angle measuring from 110° to 120°.

- The process of aging affects the mandible as follows. If dental loss occurs the alveolar process is absorbed and the bone becomes greatly reduced in volume. Consequently, the main part of the bone is below the oblique line. The mandibular canal, with the mental foramen opening from it, now runs close to the alveolar border. The ramus is oblique in direction, the angle measures about 140°, and the neck of the condyle is more or less bent backward.

Phylogeny (Figs. 8.224, 8.225, 8.226, 8.227, 8.228, 8.229, 8.230, 8.231, 8.232, 8.233, 8.234, 8.235, 8.236, 8.237, 8.238, 8.239, 8.240, 8.241, 8.242, 8.243, 8.244, 8.245, 8.246, 8.247, 8.248, and 8.249)

History of Meckel's Cartilage (MC)

The origins of Meckel's cartilage date back to early chordates and related to feeding, not respiration. The earliest creatures were small enough to survive on of trans-epidermal absorption of O_2. Protochordates has dorsal pharyngeal slits which with chordates flipped into a ventral position. These slits were lined with mucous for the entrapment of foot particles. Cephalochordates introduced supportive cartilage rods into the pharyngeal bars forming a *branchial basket*.

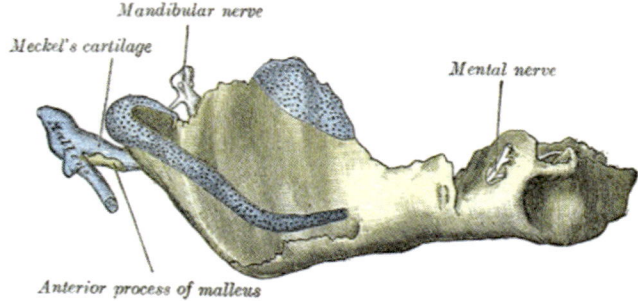

Fig. 8.224 Mandible 95 m, (90 days) right lateral. Dentary has taken over the entire surface. Coronoid appears. Reflected lamina is seen prior to forming ectotympanic. (Reprinted from Lewis, Warren H (ed). Gray's Anatomy of the Human Body, 20th American Edition. Philadelphia, PA: Lea & Febiger, 1918)

Fig. 8.225 Mandible 95 mm (90 days) right medial. Note relationship of Meckel's cartilage to the symphysis. (Reprinted from Lewis, Warren H (ed). Gray's Anatomy of the Human Body, 20th American Edition. Philadelphia, PA: Lea & Febiger, 1918)

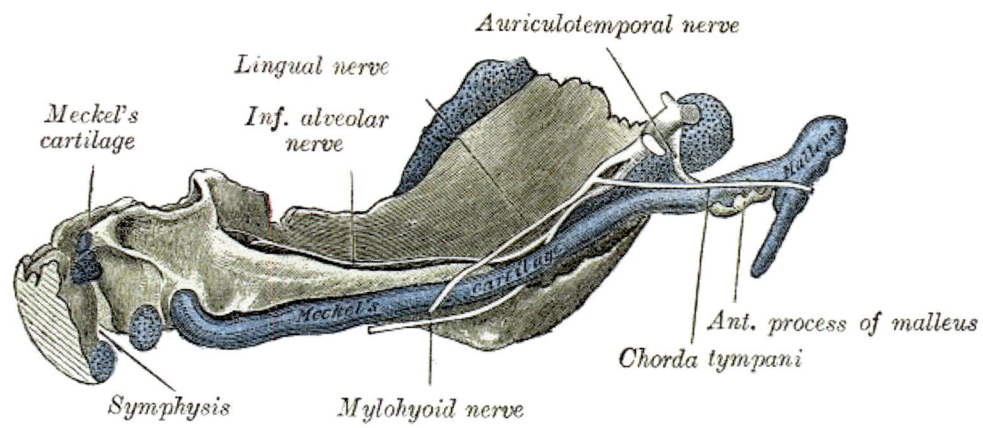

Fig. 8.226 Mandible at birth and childhood. Note the change of the mandibular angle. (Courtesy of Michael Carstens, MD)

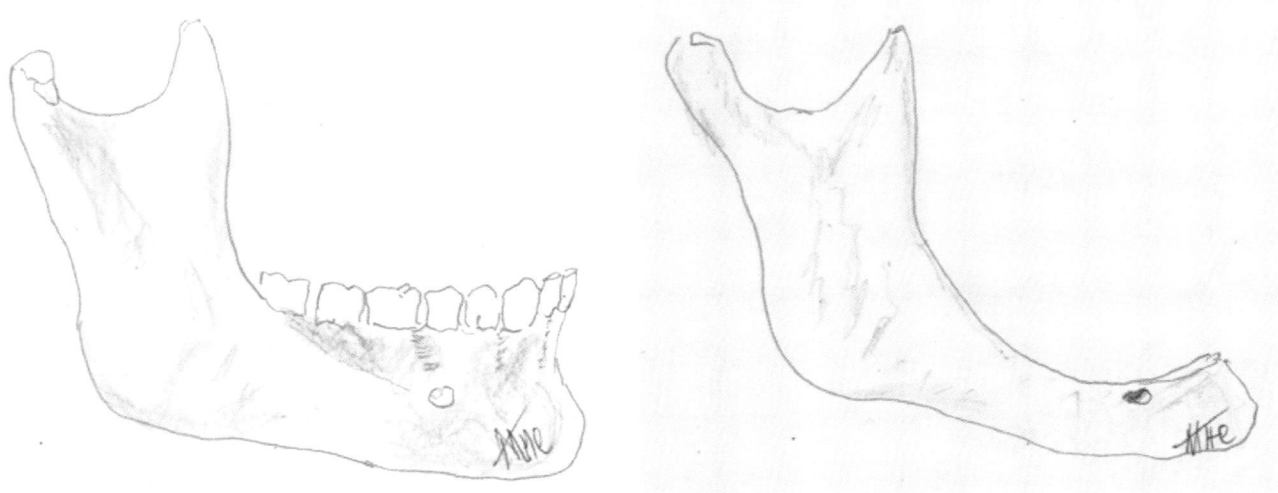

Fig. 8.227 Mandible of early and late adulthood. Figures assume edentulous condition in old age. Posterior resorption of the ramus places the condyle in an apparently retroclined position. (Courtesy of Michael Carstens, MD)

Branchiostoma (Amphioxus) attached muscles to branchial basket to create one-way valves for water intake to augment nutrient capture. Basal jawless craniates such as *Petromyzon* (lamprey) have eight visceral arches, located superficial to gill pouches. Experimentally they can the latent capacity to produce a mineralized skeleton within the dermis. Thus, <u>an externalized branchial skeleton is primitive for chordates</u>. Some jawless taxa invented a *velum*, a sail-like flap of tissue controlled by the branchial basket for bringing water into the pharynx. In osteostracans, the velum/first gill/first arch was innervated by trigeminal nerve. With increased size and mobility, the branchial basket assumed a respiratory role 222–247 and the seven-arch system with five component cartilages emerged (Figs. 8.224 and 8.225).

Meckel's cartilage (MC) arises from r3 neural crest that forms the ceratobranchial, hypobranchial, and basibranchial cartilages of the first branchial arch. It makes its debut as a jaw with teeth with chondrichthyans. In the osteichthyans, it is rapidly ensheathed with dermal bones: dentary, surangular, the splenials, and the coronoids (Figs. 8.3 and 8.4). Adductors establish control. Although MC remains cartilaginous, its proximal "exposed" end ossifies as quadrate bone, the leading element in a series of bone fields (angular, goniale, prearticular articular and quadrate) that will form middle ear. Despite many changes in dentition, covering bones, and reorientation to the skull, MC persists as an embryologic element in the jaw.

Milestones in Mandibular Design

Jaw development proceeded in two iterations. The earliest placoderms re-engineered the first arch only. Palatoquadrate was firmly attached to chondrocranium but the actual joint between the upper and lower jaws was unsuspended, the so-called primitive *autostylistic* condition. Bony teeth erupts in

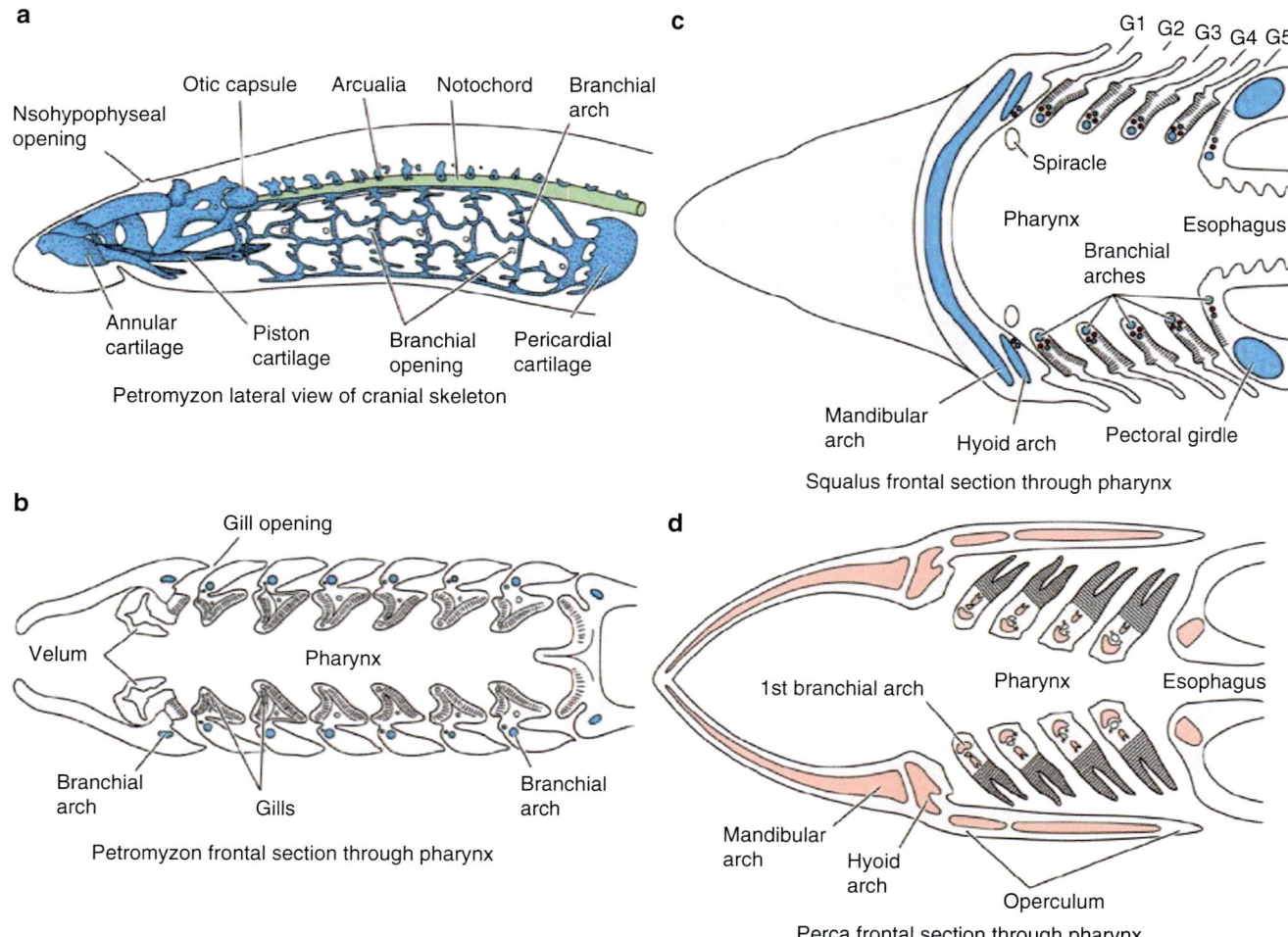

a

Nsohypophyseal opening

Otic capsule   Arcualia   Notochord   Branchial arch

Annular cartilage   Piston cartilage   Branchial opening   Pericardial cartilage

Petromyzon lateral view of cranial skeleton

b

Gill opening

Velum   Pharynx

Branchial arch   Gills   Branchial arch

Petromyzon frontal section through pharynx

c

G1 G2 G3 G4 G5

Spiracle

Pharynx   Esophagus

Branchial arches

Mandibular arch   Hyoid arch   Pectoral girdle

Squalus frontal section through pharynx

d

1st branchial arch   Pharynx   Esophagus

Mandibular arch   Hyoid arch   Operculum

Perca frontal section through pharynx

Fig. 8.228 Origins of Meckel's cartilage. Hagfishes and lampreys are often lumped together as surviving jawless craniates. As hagfishes are not true vertebrates, although they have a cartilaginous skull and cartilaginous visceral arches. These were positioned superficial to the gill pouches. Jawless vertebrates had dermal bone in the cranium but in evolution was lost in evolution. Hence the cranium of sea lamprey *Petromyzon marinus*, the sole survivor is entirely cartilaginous. Lamprey has eight visceral arches that are arranged as baskets and lying deep to the gills in apposition to the walls of the pharynx. Primitive fishes such as sharks evolved a 2 + 5 system of jaws plus five visceral arches. In derived fishes, this may be reduced to a 2 + 4 system. The neural crest origin of the annular and piston cartilages gave rise to the concept of 1–2 *premandibular arches*, as espoused by Kuratani. Whether these entities existed or not, the neuromeric identity of facial mesenchyme can be assigned to rhombomeres r0 and r1. (Courtesy of William E. Bemis)

the early Silurian period marking the immediate transition from placoderms to the bifurcation point of chondrichthyans and osteichthyans. Greater stability was needed for more effective predation. Early cartilaginous and bony fishes quickly improved the situation by modifying second arch into a suspensory bone, the hyomandibula, spanning from the otic capsule to the caudal tip of palatoquadrate cartilage (soon known as the quadrate bone). This is known as *amphistylistic suspension*. Five respiratory arches remained behind (Fig. 8.226, types of jaws suspension).

With evolution, jaw support took two final directions. Derived chondrichthyes and all bony fishes disconnect the maxilla from the cranium as a means of gaping open the jaws (remember the movie). The resulting *hyostylistic suspension* depends strictly upon the second arch hyomandibular bone.

All tetrapods and all remaining fishes have the maxilla fused to the cranium. The mandible is hinged from the cranium. In this secondary autostylistic suspension, hyomandibula is "relieved of its duties" and takes on a new role (as stapes) for hearing.

How did this revolution take place? Several explanations exist. (1) *Serial theory* or the classic theory of Alfred Romer (1983) considered an ancestral fish to have nine gills and eight gill slits. The three anterior gill arches were modified with a premandibular arch becoming, the second mandibular arch the jaws, and the third arch the hyomandibula. Mapping of non-pharyngeal arch midbrain neural crest was unknown at the time. (2) *Composite theory* was put forth by Swedish Paleontologist Erik Jarvik (1978) denied the concept of "one arch, one mandible." Mandible was thought to result from a

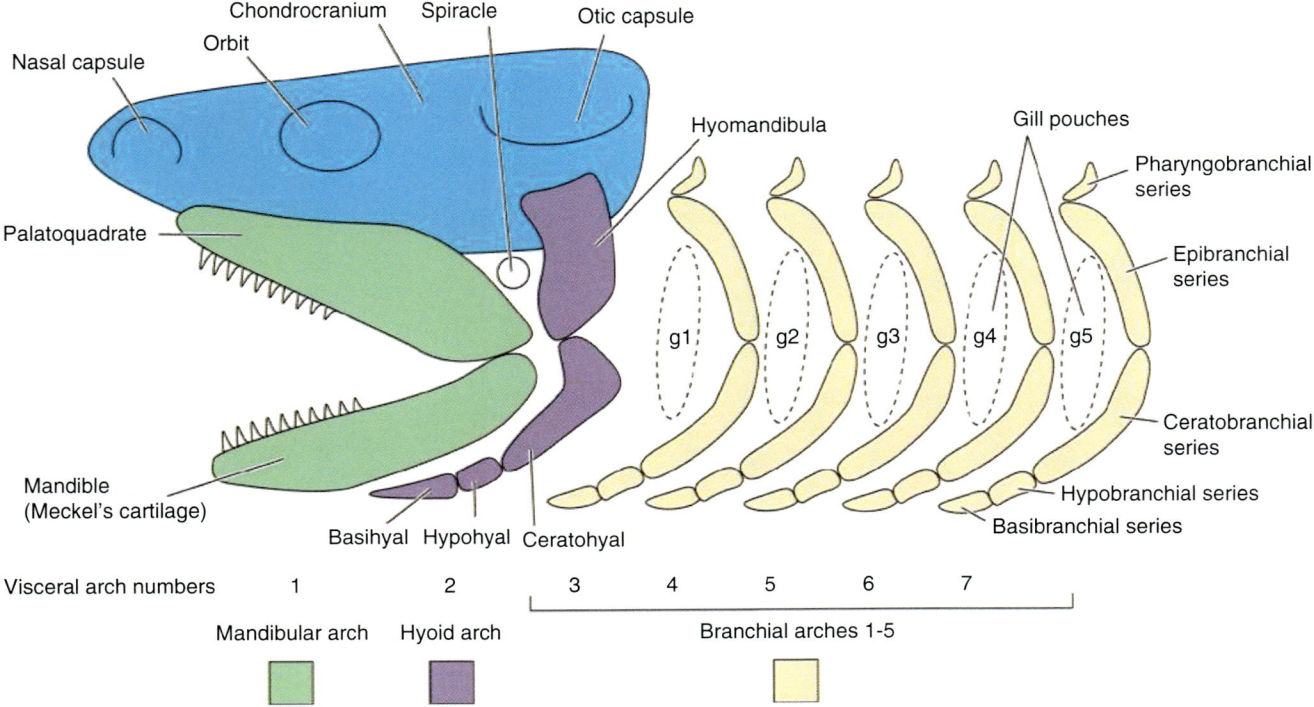

Fig. 8.229 Spanchnocranium in basal gnathostome. The 7-arch system in primitive gnathostomes. The original gnathostome condition, as seen in placoderms, did not include the hyoid arch. Derived fishes, both chondrichthyans and osteichthyans, incorporated BA2 into the jaws, creating the 2 + 5 system, seen here. Palatoquadratea cartilage = pha-ryngobranchial + epibranchail cartilages of BA1. Meckel's cartilage complex = ceratobranchail + hypobranchial + basibranchial cartilages of BA1. More advanced bony fishes (teleosts) have a 2 + 4 system. (Courtesy of William E. Bemis)

complex fusion from at least three different arches. Again, this flies in the face of homeotic coding. (3) Ventilation theory of Mallett (2008) considers the jaws to be primarily modifications for respiration. Whereas lampreys have their gills internal to the skeleton and gnathostomes place them externally. (4) *Heterotopic theory* of Shigatani (2002) is more genetic and takes into consideration that the same neural crest cells that make up the lips and velum in the lamprey become the jaws. This concept recognizes that one cannot find predecessors of jaws in jawless vertebrates (Fig. 8.227 and 8.228 theories).

This text has made use of *neuromeric theory*, a modernized version of the serial model backed up with neuromeric coding. It is based on a seven-arch model for agnathans in which the first arch r2 epibranchial cartilage produced PQ and r3 ceratobranchial cartilage (r3) produced Meckel's. Second arch makes r4 hyomandibula and r5 hyoid apparatus. The remaining five arches remained respiratory; all derived fishes, be they cartilaginous or bony, have five gill arches (or sometimes four) fishes. In tetrapods, the third to the fifth arche lost their respiratory function but retain their role as pharyngeal arches. The sixth and seventh arches disappeared as physical entities but were neuromerically preserved mesenchymal levels c1–c4 dedicated to the neck.

Neuromeric theory is integrative. It can be experimentally tested using homeotic mapping. Given that r0 and r1 exist in agnathans, these neural crest populations localize directly to the anterior cranial base. It is not necessary to postulate that they are "sidetracked" into producing a premandibular arch. Genetic mapping shows that r2 and r3 neural crest in perioral soft tissues made from r2 to r3 neural crest become "reassigned" to make bone.

The Standard Condition

The Standard Condition of the mandible appears very early in evolution, before tetrapods in the choanate sarcopterygian fishes. From onward, it varies little. *Dentary* covers the front half of mandible. It contains the outer row of teeth. Just below dentary two *splenial* bones make up the lower border. Medial to dentary are two to three *coronoid* bones contain the inner row of teeth. Behind dentary lies the *surangular* bone. Drapes over both sides and extends two fingers forward on either side of the posterior dentary. *Angular* covers the lower border and lateral side of the posterior half of mandible, just behind the splenials. The posterior and superior border contains a small bone, articular, which is destined to

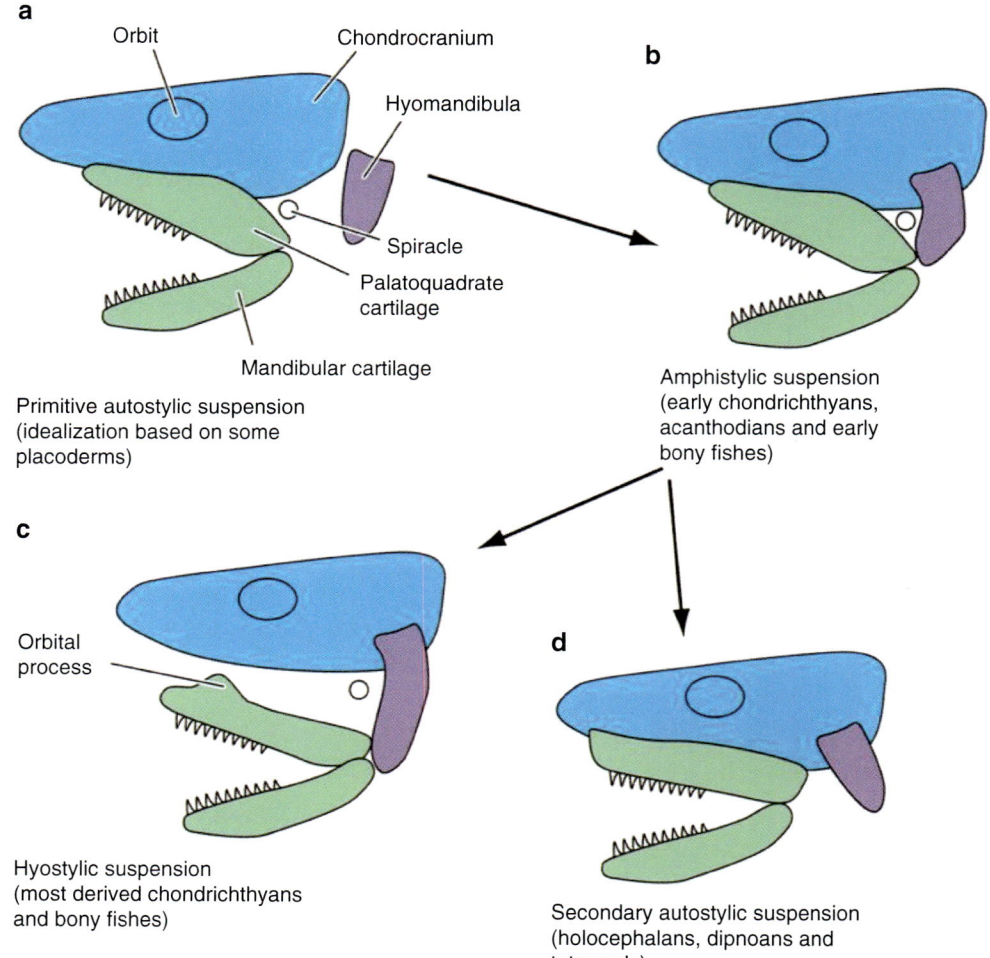

Fig. 8.230 Phylogeny of mandibular suspension. The articulation between the skull and the mandible changes with evolution. *Paloestyly* appears in early agnathans. Here no connection exists between the arches and the skull. The earlies gnathic fishes (placoderms) had *euautostyly*, a simple connection between the mandible and the skull without assistance from the second arch. Internediate fishes (both chondrichthyans and osteichthyans) presented with an additional element, the second arch hyomandibula (yellow). This dual *amphistylistic* condition used a ligament from palatoquadrate to the skull and a separate connection with hyomandibular. It is continued in modern sharks. More advanced bony fishes developed first a *hyostylistic* suspension with mandible attached to the skull via the intermediate hyomandibular. Subdivision of the hyoid arch created an intermediate *symplectic bone* between the hyomandiular fixed to the skull and the quadrate. Tetrapods (except mammals) use meta *autostylistic* suspension with quadrate bone attached to the skull. Hyomandibula is no longer necessary and becomes stapes. Mammals evolved the TMJ, the *craniostylstic* condition. (Courtesy of William E. Bemis)

form malleus. Finally, *prearticular* is slung across postero-medial mandible like a slack rope in a gentle curve connecting the last coronoid with articular; it is sandwiched between which is clasped between dentary and angular (Fig. 8.229).

Dentary and Surangular Bones

Over the course of time, dentary bone has been involved in a low-grade guerilla war with its neighbors, eventually winning out by "replacing" them. However, jas we shall it, in this process, dentary does not eliminate the genetic program of its opponents. Rather, it incorporates these fields, such that they continue to exert their influence internally, creating signals that determine the insertion of muscle groups. By the time of tetrapods, it has taken over nearly the entire lateral half of jaw and is slowly encroaching on the coronoids' turf. Territory formerly occupied by splenial and surangular has been overrun by the dentarian legion.

Mandible is unlike maxilla in that its dental zones are restricted to the dentary and the coronoids. All remaining dermal bones are endentulous. In the upper jaw, bones of arising long the medial zone of PQ are all tooth-bearing. Over time, of course, premaxilla and maxilla both win out. The lateral-most row of teeth provides the best mechanical advantage and the greatest degree of precision for occlusion. Warm blooded animals require calories for maintenance of body temperature. Food stuffs need to be processed in the mouth not by means of a gizzard. The complex motions of the molars available to mammals

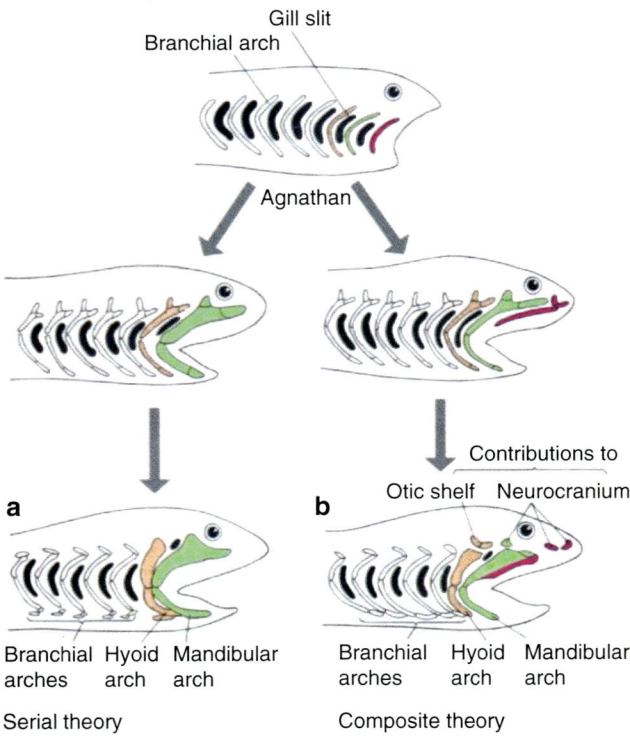

Fig. 8.231 Theories of mandible development: serial and composite. (Reprinted from Kardong K. Vertebrates: Comparative Anatomy, Function, Evolution, 7th ed. New York, NY: McGraw-Hill; 2015. With permission from McGraw-Hill Education)

through the TMJ, pterygoids, and masseter-enabled maximal nutrition.

Surangular drapes over dorsolateral surface of Meckel's cartilage like a towel draped sideways behind the dentary. The former is tooth-bearing, surangular is muscle-bearing. It proffers a posteriorly directed helping hand to support the articular bone of the joint. In turn, it receives back-up from the rather substantial angular bone extending upward from the lower border of the jaw. It is aligned on the other side of the mandibular demilitarized zone with quadrato-jugal. The surangular bone field plays a critical, yet unappreciated role in the evolution of the mammalian mandible. During its existence as an identifiable bone in evolution surangular serves to: (1) support the jaw joint by reinforcing articular, (2) strengthen the jaw, permitting to lengthen, and therefore become more effective at snapping up prey, and (3) create a secondary jaw joint which plays a temporary, but important, role in the translocation of jaw muscles. As we shall see, in mammals, the surangular bone field is apparently "lost", being covered over by dentary. Nonetheless, its persistence explains the development of both the coronoid process and condyle.

Surangular first appears at the nodal point of fish evolution in the early Silurian period when osteichthyans developed a craniofacial skeleton, gaining a roof over the brain and reinforcing the facial cartilages. Shortly thereafter, at the

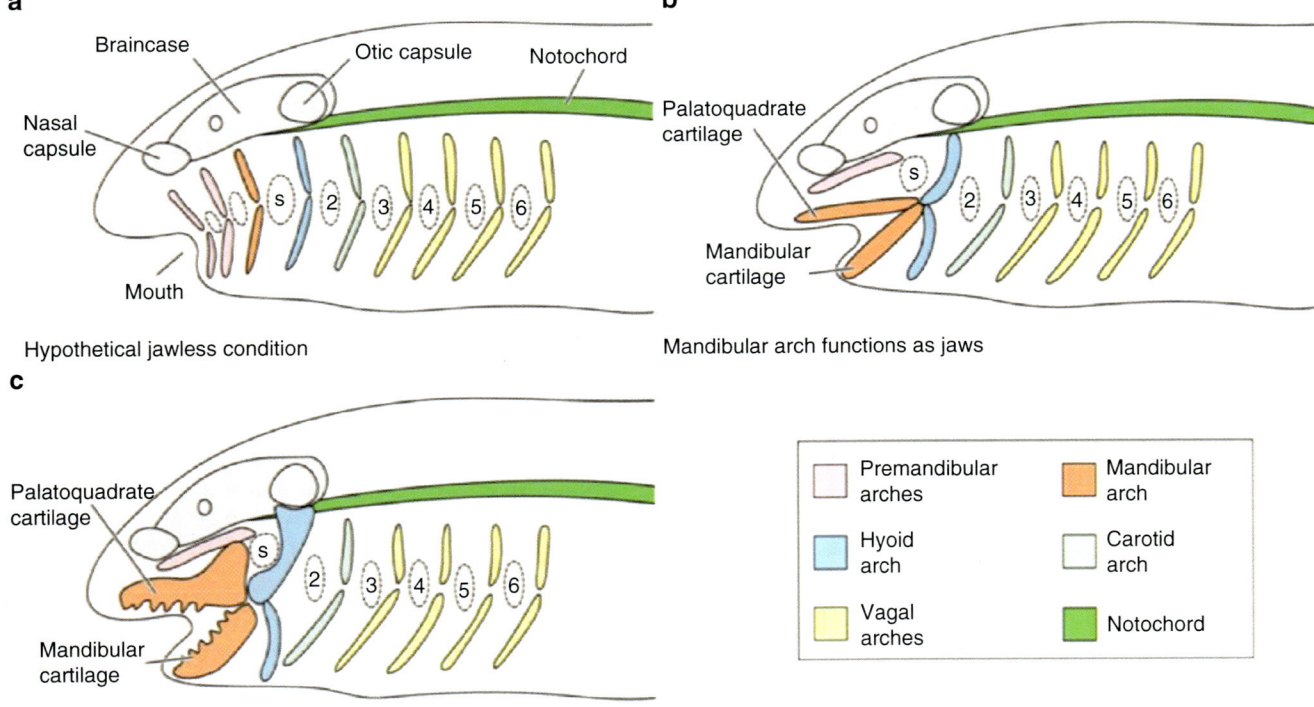

Fig. 8.232 Kuratani and neuromeric models differ in source material of neural crest from r0 to r1 anterior to the first arch. (Courtesy of William E. Bemis)

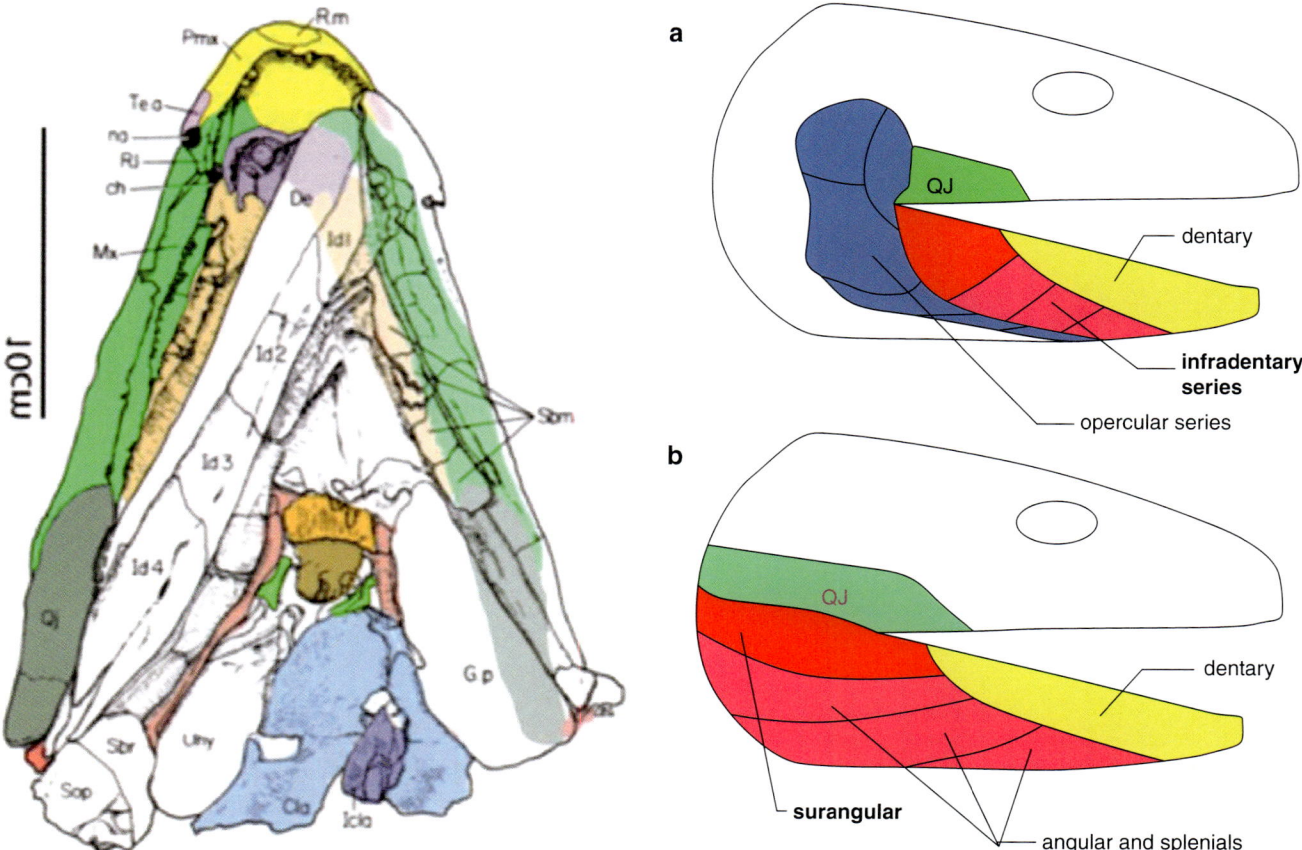

Fig. 8.233 Mandible in *Panderichthys*. Four internal dental segments, id1–id4, identified below the dentary. Internal to these are the bones of the palate. *Rm* median rostral, *Pma* premaxilla, *Tea* anterior tectal, *na* nasal, *Mx* maxillary, *De* dentary. (Reprinted from Carroll RL. Vertebrate Palaeontology. New York, NY: WH Freeman and Company; 1988)

Fig. 8.234 Derivatives of internal dental series: hypothetical (a) sarcopterygian and (b) early tetrapod skulls demonstrating homologies between the bone fields of the infradentary series and those of the angular series. (Courtesy of Augustus T. White, Palaeos.com. Retrieved from: http://palaeos.com/vertebrates/bones/dermal/mandibular-surangular.html)

branch point between actinopterygians and sarcopterygians, these latter developed enamel on their teeth, and *splenial bones*.

Our ancestor, the sarcopterygian *Panderichthys* demonstrates a series of four *infradentary* bones that are diagonally oriented, like ribs (Fig. 8.229). These infradentaries are a transitional form of the gill-covering opercular apparatus. When the operculars are lost in tetrapods, the infradentaries change their orientation, becoming horizontal; they form the remaining dermal bones: *The lowest pair, id1–id2 become the splenials, id3 forms the angular, and the uppermost, id4, becomes the surangular*. Note how surangular now comes into extensive contact with quadratojugal (Fig. 8.230).

The fundamental mission of surangular is to serve as an insertion site for muscle, whereas that of mission of dentary is to provide housing for teeth. As we move into the early Devonian period, tetrapodomorph fishes developed a *precoronoid fossa* (for jaw adductors). At the same time, an organized army of teeth marches backwards along the jaw in successive coronoid bones. This reaches its posterior extent

in the elpistostegalians, as in *Panderichthys*. A boundary is thus created between the tooth-bearing sector of mandible and the muscle-bearing sector.

In early tetrapods, surangular occupies both the buccal and lingual sides of the posterior mandible. The dentition is not regionally specialized. It is arranged in two rows. Lingual dental zones are organized by three coronoid bones whereas the buccal teeth lie within a singular dentary bone. The basal bone of the mandibular body has two splenials corresponding to id1 and id2, followed by angular, di3 and surangular id4. Later in evolution, with the departure of the posterior bone fields into the ear, surangular will constitute the sole bone field of the posterior mandible.

Surangular assumes new importance in the synapsid line of amniotes. Cynodonts develop a unique but temporary secondary jaw joint between a secondary cartilage, the coronoid process, arising from surangular and squamosal. With mammalian brain growth, deepening of the temporal fossa, diversification of muscles, and the dominance of the TMJ eliminates the need for this joint. The more sophisticated

Fig. 8.235 Mandibular bone fields in derived choanate fish and basal tetrapod. (Courtesy of William E. Bemis)

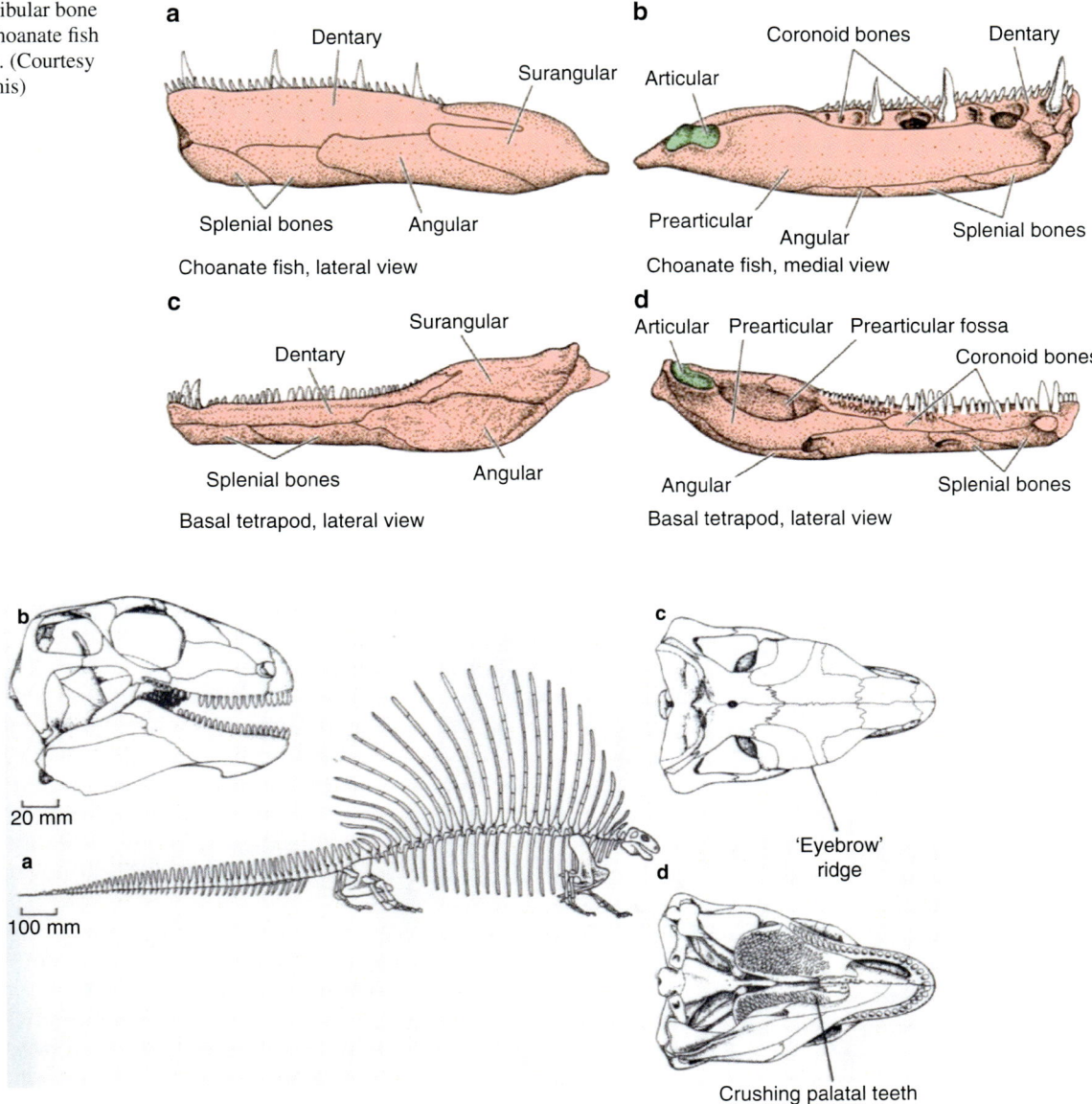

a

Dentary

Surangular

Splenial bones

Angular

Choanate fish, lateral view

b

Coronoid bones

Dentary

Articular

Prearticular

Angular

Splenial bones

Choanate fish, medial view

c

Surangular

Dentary

Splenial bones

Angular

Basal tetrapod, lateral view

d

Articular   Prearticular   Prearticular fossa

Coronoid bones

Angular

Splenial bones

Basal tetrapod, lateral view

b

20 mm

a

100 mm

c

'Eyebrow' ridge

d

Crushing palatal teeth

Fig. 8.236 Mandible *Edaphosaurus*. Pelycosaurs were herbivores, adapted for grinding food using "pavement-like" teeth and a sliding jaw joint. The coronoid process shifts anteriorly and the lateral temporal fenestra was enlarged for larger musculature. Note the persistence of crushing palatal teeth. These are not lost until the cynodonts. (Reprinted from Modesto SP. The Lower Permian Synapsid Glaucosaurus from Texas. *Palaeontology* 37:51–60 http://www.reptileevolution.com/edaphosaurus.htm. With permission from the Palaeontological Association)

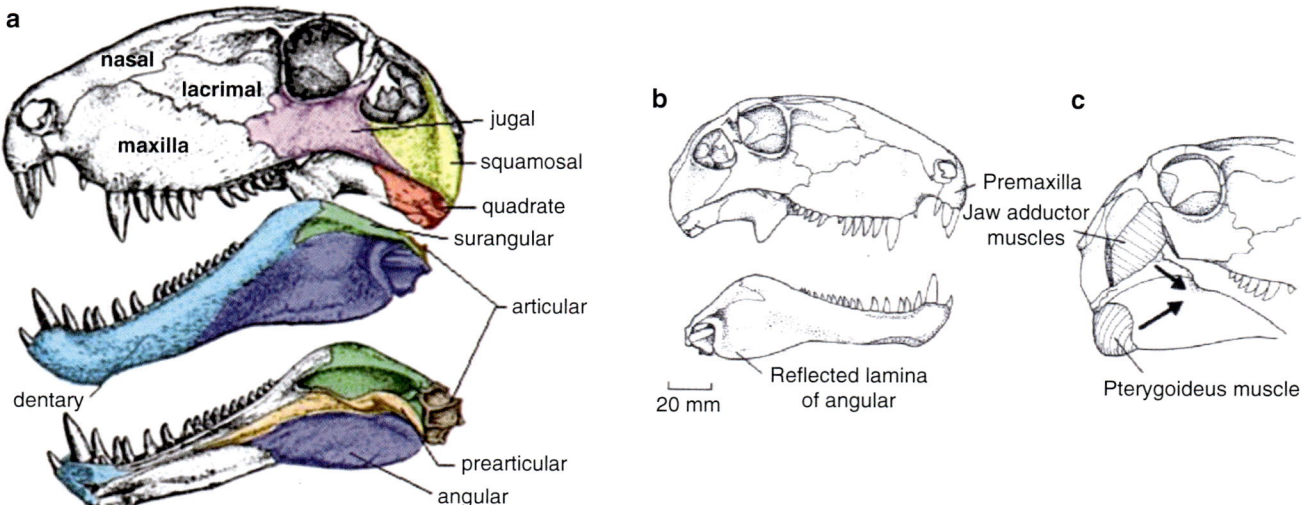

Fig. 8.237 Mandible *Dimetrodon*. Dentary is "climbing over" surangular produces a coronoid to which will be attached temporalis. [Left: (Reprinted from Carroll RL. Vertebrate Palaeontology. New York, NY: WH Freeman and Company; 1988). Right: (Reprinted from Benton MJ. Vertebrate Paleontology, 4th ed. Blackwell; 2015. With permission from John Wiley & Sons)

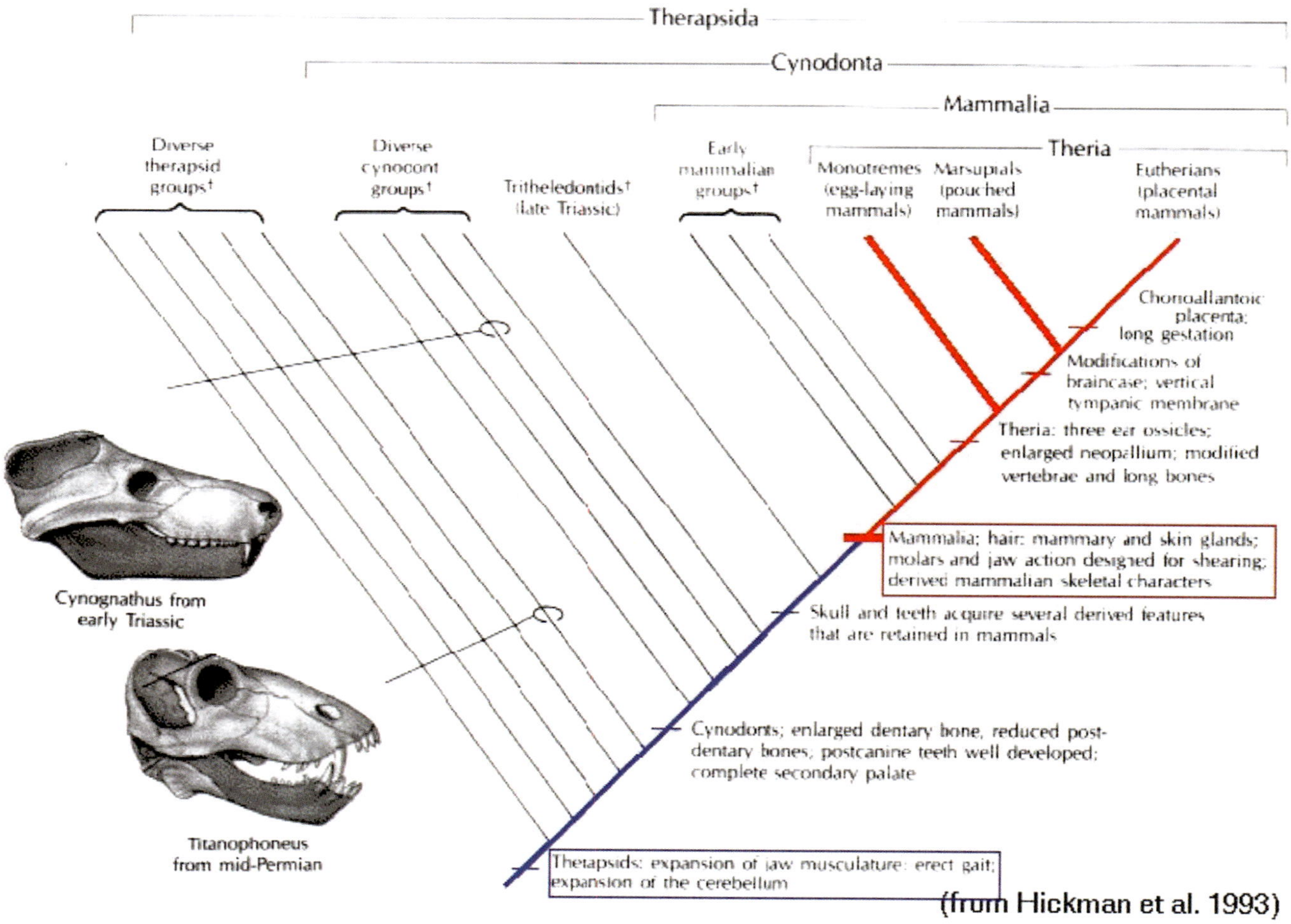

Fig. 8.238 Phylogeny of Therapsida. (Courtesy of Prof. Stephen M. Carr, Department of Biology Memorial University of Newfoundland, St. John's NF, Canada)

Fig. 8.239 Tetraceratops. Fangs project from the primordial placoderm bone fields. (Courtesy of Dimitry Bogdanov)

masticatory system wins out. <u>Surangular fuses with dentary</u> and the <u>remnant of the cynodont secondary joint</u> becomes the temporalis-bearing <u>coronoid process</u>. At the same time, with the flight of bone fields from the tympanic sector of Meckel's into the middle ear, the surangular bone field is the only player left standing in posterior mandible. Although it is covered over by dentary, the surangular field produces a secondary cartilage which expands posteriorly into the zone vacated by the angular: the <u>condylar process</u>.

These processes present during the seventh and eighth weeks as small eminences along the surface of mandible. At week 9, chondrogenesis takes place. Both cartilages have a cone-like "root" in the surangular and develop outward. The condylar process chondrifies distinctly lateral to Meckel's cartilage—condyle does *not* arise from Meckel's. As the pro-

Fig. 8.240 Mandible *Tetraceratops*. Note improvements for chewing: a larger temporal fossa and an anteriorized jaw joint. Palatal teeth become less important. (Reprinted from Benton MJ. Vertebrate Paleontology, 4th ed. Blackwell; 2015. With permission from John Wiley & Sons)

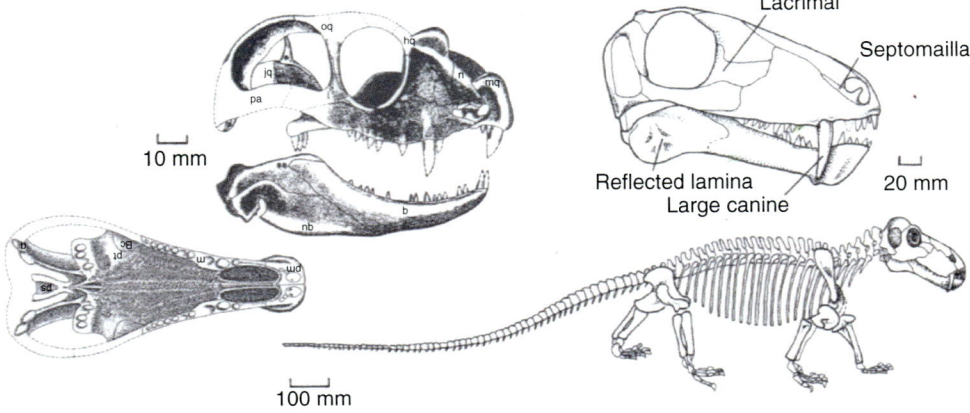

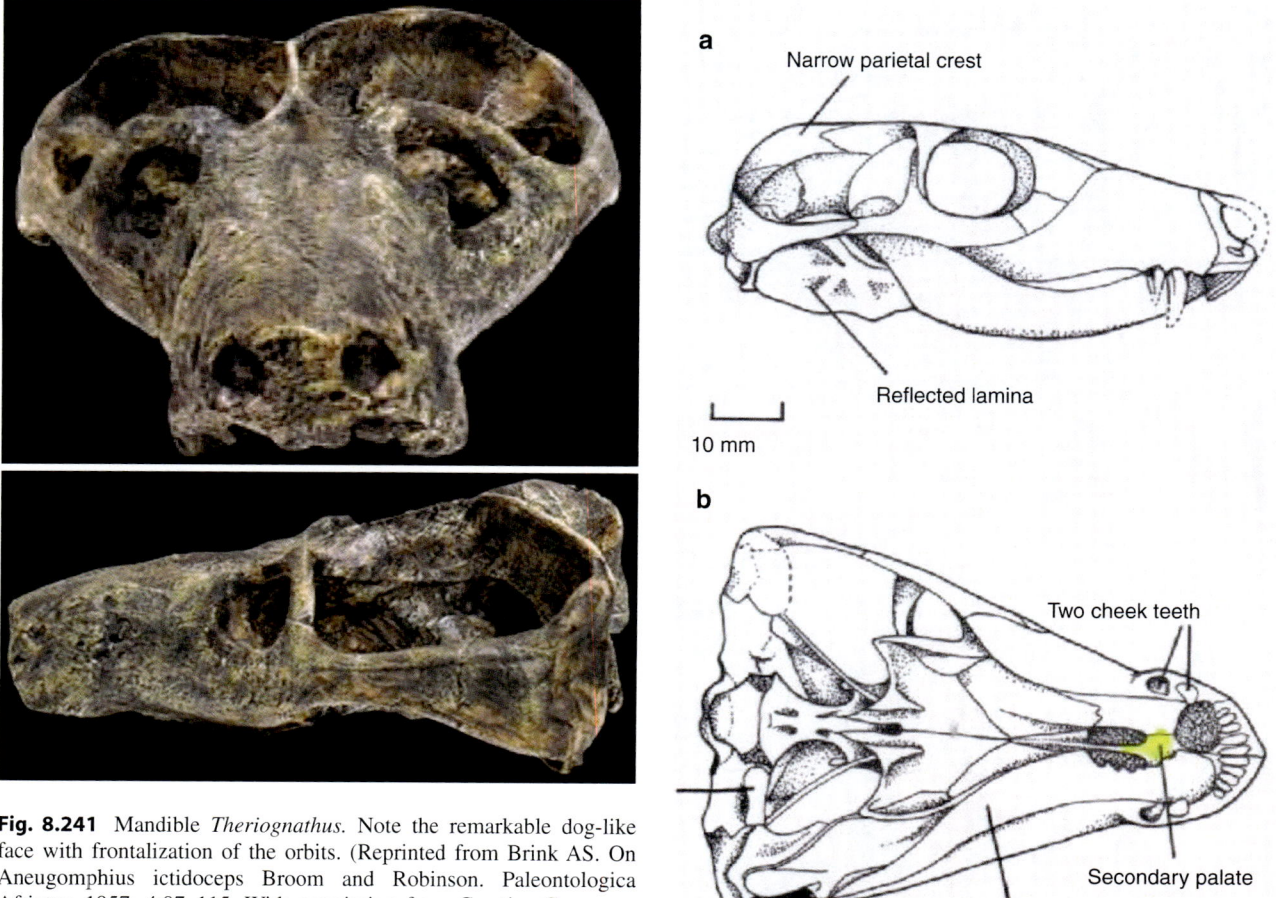

Fig. 8.241 Mandible *Theriognathus*. Note the remarkable dog-like face with frontalization of the orbits. (Reprinted from Brink AS. On Aneugomphius ictidoceps Broom and Robinson. Paleontologica Africana 1957; 4:97–115. With permission from Creative Commons License 4.0: https://creativecommons.org/licenses/by/4.0/)

Fig. 8.242 Feeding mechanism of theriodont, *Theriognathus*. *Theriognathus* had an enlarged coronoid process and initial appearance of secondary hard palate. Line in (**b**) indicates stapes. (Reprinted from Benton MJ. Vertebrate Paleontology, 4th ed. Blackwell; 2015. With permission from John Wiley & Sons)

cesses grow outward, they are surrounded at their base by membranous bone. With further mandibular growth, the ramus uprights itself with its anterior zone biologically determined by coronoid process and the posterior zone by condylar process. Blood supply to the condyle is from the medullary branches of inferior alveolar artery.

Why such detail about the origins of the mandibular bone fields? The significance has to do with the individual genetic signatures of each. Although they become "absorbed" by dentary into a seemingly unitary structure, their influence remains because they determine individual binding sites for the many muscles that inset into the mammalian mandible.

Synapsid Innovations

Let us finish out our tour of mandible by considering the sequence of steps leading from basal synapsids to mammals. Recall that back in the Carboniferous, the amniotes split into Synapsida (our line) and the sister group of Diapsida (Parareptilia and Eureptilia). The basal synapsids were called pretherapsids or pelycosaurs ("wooden bowl lizards"), so-named named for the elongated neural spines projecting upward from their back. The initial group, exemplified by

Edaphosaurus, was herbivorous, combining peg-like teeth with a propalinial (sliding) jaw joint that acted in a saw-like motion. These synapsids continued to have palatal teeth (Fig. 8.232). A later sister iteration, *Dimetrodon*, was carnivorous. It had substantial adductors attached to the inside of the mandible to pull it shut. Its pterygoid bone was edentulous and sent pterygoideus muscle forward to attach outside the angular to pull the jaw backward. The relocation of pterygoideus would prove invaluable when the TMJ was invented because it permitted more complex chewing movements (Fig. 8.233).

In the middle Permian, the pelycosaurs were succeeded by therapsida, the lineage leading to mammals (see schematic Fig. 8.234). Therapsida are characterized by having 12 or fewer teeth behind the canines. These post-canine teeth were more sophisticated. They developed serrations for better cutting up of food. The "true" coronoid process appears. Ectopterygoid teeth are lost. Changes in the relationship of

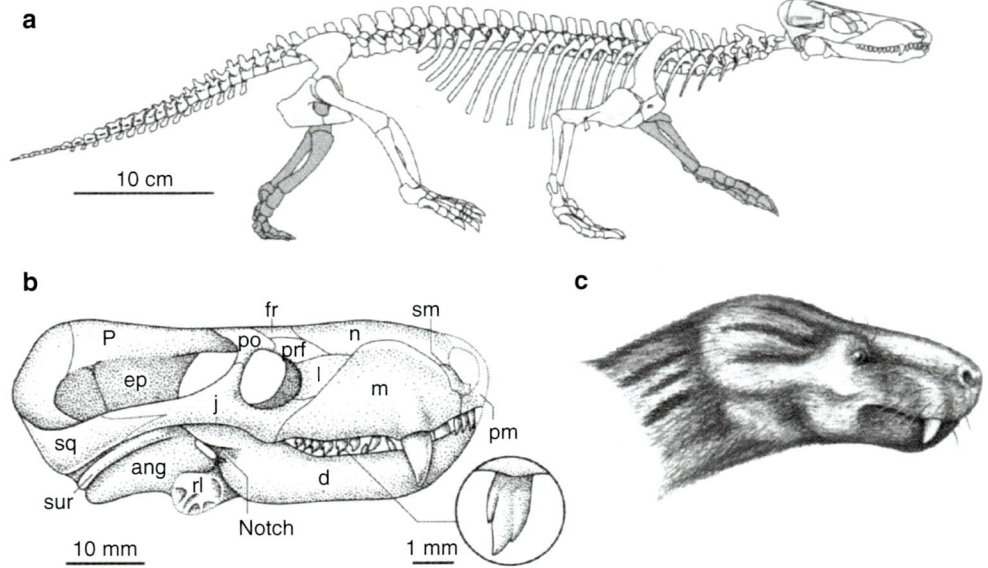

Fig. 8.243 Inventing the TMJ 1: new bone fields. Basal synapsed (left) versus basal cynodont *Dvinia* (right). Dvinia was basal to *Procynosuchus.*. Note "take-over" by dentary of lateral bone fields followed by medial bone fields. Dentary "climbs over" suragular and incorporates it to produce coronoid process, essential for insertion of temporals. Articular begins to point backwards. Articular forms the body of the malleus. Reflected lamina of the articular will become the manubrium of malleus. The reflected lamina of angular will form U-shaped bone directly posteriorly that becomes tympanic. In cynodonts before mammals, the tympanic does not separate from the posterior mandible. (Reprinted from Tartarinov LP. Morphology and systematics of the Northern Dvina cynodonts (Reptilia, Therapsida; Upper Permian). *Postilla* 126, 1968. With permission from The Peabody Museum of Natural History at Yale University)

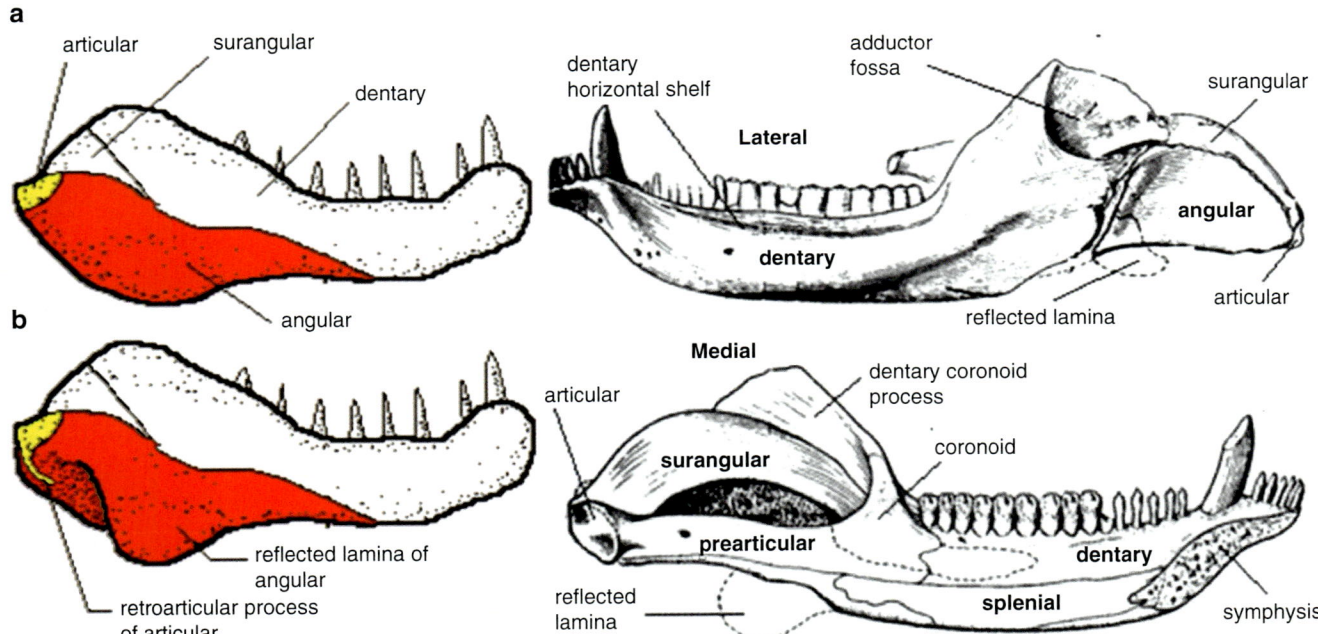

Fig. 8.244 *Procynosuchus.* (Reprinted from Botha J, Abdala F, Smith RMH. The oldest cynodont: new clues on the origin and early diversification of the Cynodontia. *Zoological Journal of the Linnean Society*. 2007; 149:477–492. With permission from Oxford University Press)

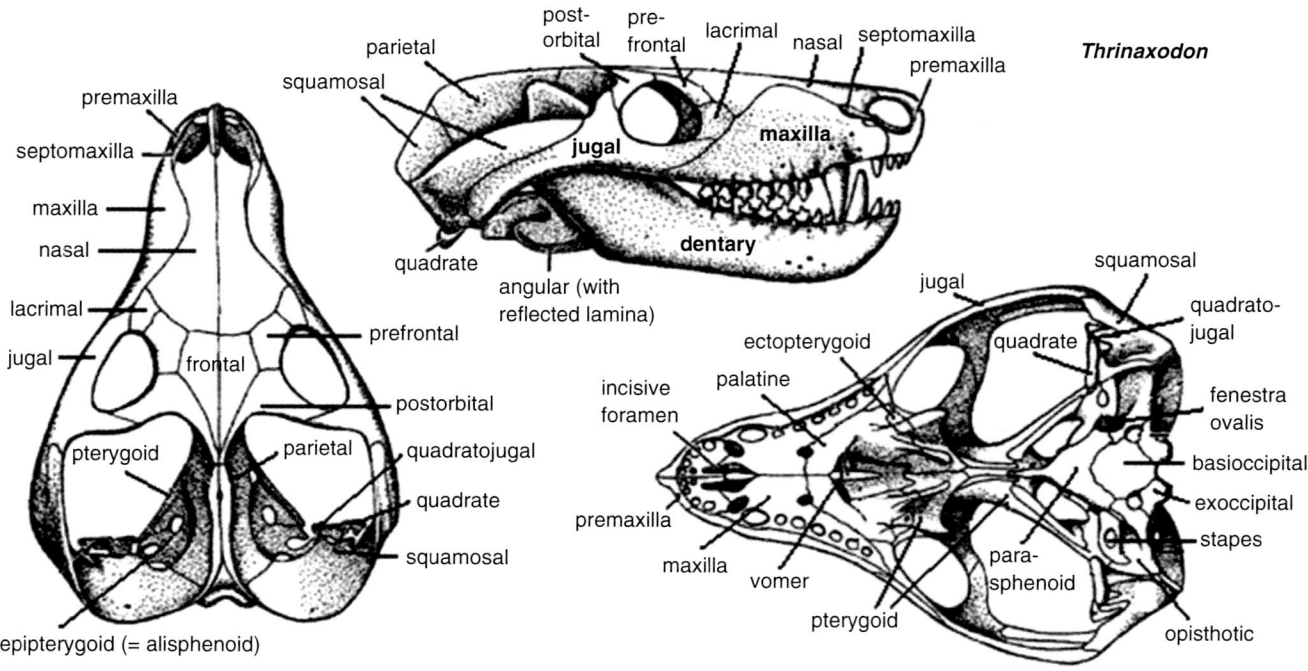

Fig. 8.245 *Thrinaxodon.* Dentary is now completely dominant. Proatlas fragment still present but head rotation was possible. (Reprinted from Carroll RL. Vertebrate Palaeontology. New York, NY: WH Freeman and Company; 1988)

Fig. 8.246 *Thrinaxodon.* Many innovations are shown including the double condyle, interlocking teeth slide together (occlusion was not yet established). (Reprinted from Benton MJ. Vertebrate Paleontology, 4th ed. Blackwell; 2015. With permission from John Wiley & Sons)

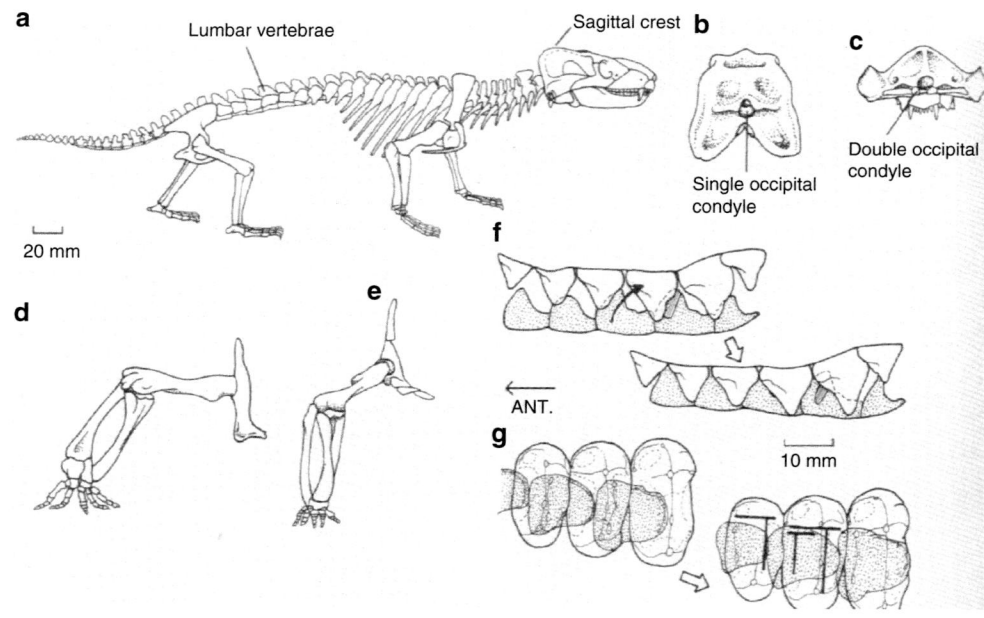

Fig. 8.247 Ectopic third molar in the ramus indicates potential source of the surangular from third coronoid. Upper panorex shows cyst in the center of the right ramus; in the lower panorex, tooth is located high and posterior, at the base of the right condyle. [Left: (Reprinted from Findik Y, Baykul T. Ectopic third molar in the mandibular sigmoid notch: report of a case and literature review. *J Clin Exp Dent* 2015; 7(1):e133–137. With permission from Creative Commons Attribution License). Right: (Reprinted from Wang C-C, Kok S-H, Hou L-T, et al. Ectopic third molar in the ramus region: a report of a case and literature review. *Oral Surg Oral Med Oral Path Oral Radiol* 2008; 105:155–161. With permission from Elsevier)]

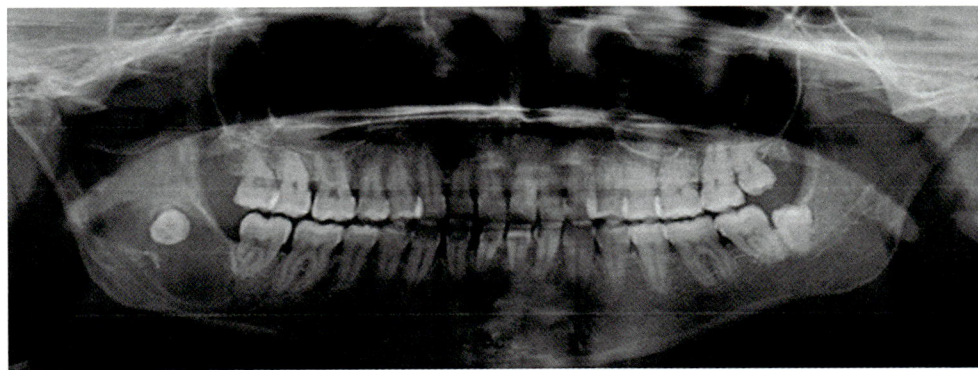

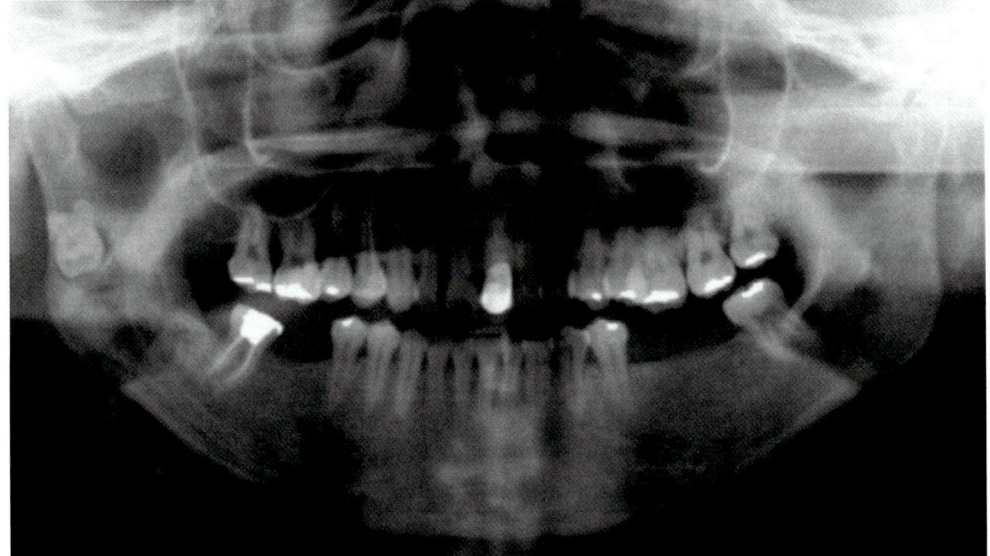

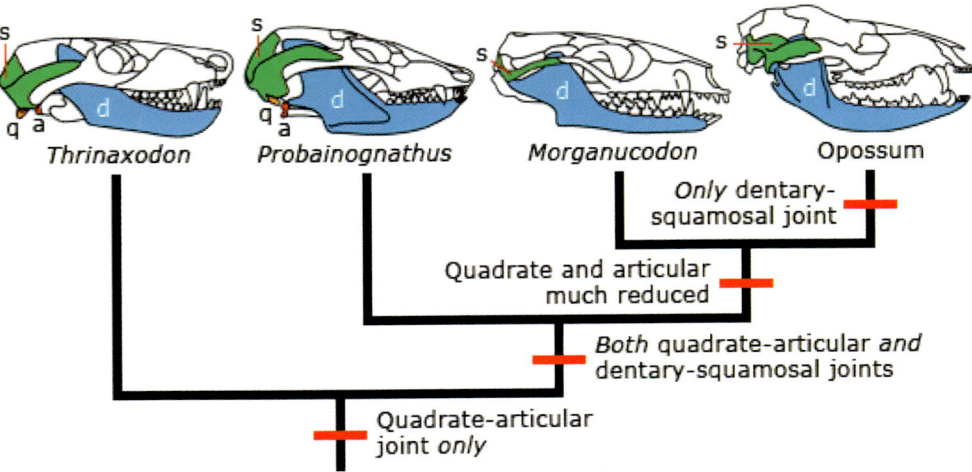

Fig. 8.248 Evolution of the TMJ takes place during the cynodont-mammal transition. Basal cynodont *Procynosuchus* not shown. Epicynodont *Thrinaxodon* has single modified quadrate-articular joint. Eucynodont *Probainognathus* and mammiloform *Morganucodon* have two separate joints, quadrate-articular and dentary-squamosal; and create single dentary squamosal TMJ. Basal mammals after *Morganucodon* have a single TMJ. This is caused by transverse brain expansion which dislocates the dentary-squamosal articulation. *Sinocodon* (not shown) has prototype TMJ in which the condylar process from surangular falls into the glenoid fossa. With subsequent growth of squamosal, the glenoid fossa moves forward and the ramus becomes upright. Mammals reduce the size of the coronoid process. (Copyright 2020 by The University of California Museum of Paleontology, Berkeley, and the Regents of the University of California)

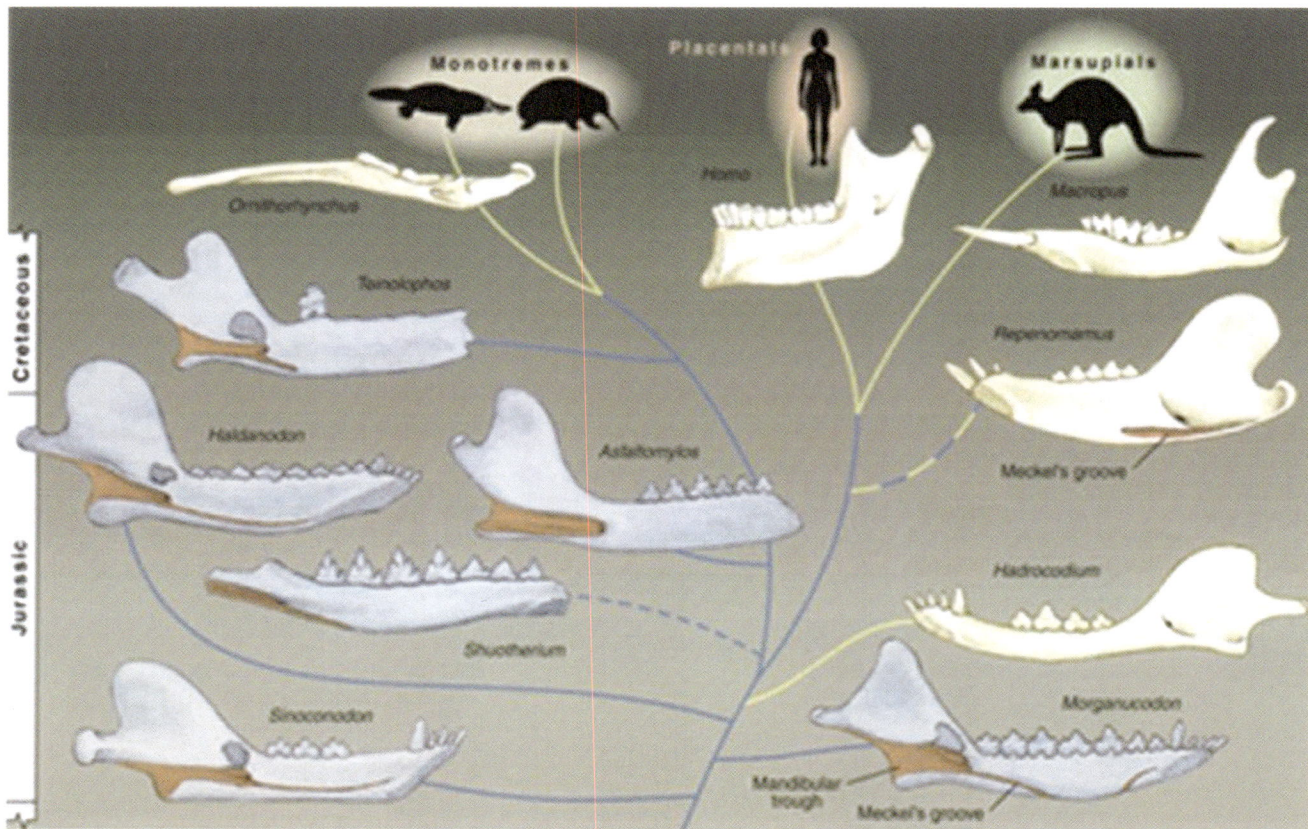

Fig. 8.249 Rise of the mammalian mandible. Extent lines = yellow; extinct lines = blue. *Therapsid* dentary bone: angular (tympanic), articular (malleus), quadrate (incus) directly connected to mandible. Bone separate from dentary as the **postdentary rod** (PDR). *Monotremes*: mandibular trough accomodates PDR by means of ligaments. Rod used for hearing. *Therians*: mandibular trough not present; bones held to jaw by ossified Meckel's cartilage. Bone bridge breaks away from jaw to form ear cavity. Note how posterior process of dentary grows outward as ramus and develops a condyle with cartilage cap (separate ossification center). (Copyright 2020 by The University of California Museum of Paleontology, Berkeley, and the Regents of the University of California)

the jaw to the skull appeared. In pre-therapsids, the angle of the dentary was in line with the occiput. The basal therapsid *Tetraceratops* ("four-horned face emblem") demonstrated a shift in the dentary angle forward to the level of the postorbital bar (Figs. 8.235 and 8.236). In crown therapsids, the Therocephalian, intraoral changes take place. *Theriognathus* (Gr. *therios*, wild beast/mammal "wild animal jaw") had a dog-like face with frontalization of the orbits. It developed mammal-like canines and a vaulted hard palate constructed from premaxilla vomer, maxilla, and palatine bones (Figs. 8.237 and 8.238).

Note: Synapsid evolution is marked by the search for occlusion, a highly desirable train permitting maximum processing of food. This required the loss of that palatal teeth. Basal synapsids lose them from the pterygoid bone. Loss from ectopterygoid takes place in therapsids and then from palatine bone. In theriodonts ("mammals-like teeth"), the palatine bone becomes edentulous. At this point, immediately prior to cynodonts, mandibular dentition can now articulate strictly with maxilla.

Toward the end of the Permian, the Cynodonts arose. Their evolution is characterized by radical experimentation in mastication: specialized teeth for grinding, a reduction in the number of cycles of tooth replacement, a unique temporomandibular joint permitting three-dimensional motion, and new muscles capable of executing complex movements. We shall look at the following examples: basal cynodont *Procynosuchus*, epicynodont *Thrinaxodon*, eucynodont *Probainognathus*, mammiform *Morangucodon*, and protomammal *Sinocodon*.

Basal cynodonts *Procynosuchus* ("before dog crocodile") and *Dvinia* demonstrated changes in bone fields that promoted a dramatic increase in adductor mass: (1) the zygomatic arch flared outward and (2) an adductor fossa appears behind dentary to accommodate insertion. The beginnings of the secondary hard palate occur at this level (see sections of Maxilla and palatine bone). The coronoid process shows evidence of invasion of occlusal muscle (Fig. 8.103).

Important changes in the masticatory apparatus make their appearance. Angle of the dentary, which was in line with the occiput in therapsids, now moves further forward to the level of the postorbital bar. This gave greater mechanical advantages with less muscle mass. Primary insertion of muscle is displaced forward as well. Primitive adductor mandibulae subdivides into masseter and temporalis, as indicated by a notch at the base of the coronoid process. All palatal teeth are lost: mandibular dentition must now articulate strictly with maxilla (Fig. 8.103). *Procynosuchus* has multiple-cusp teeth (precursors of molars) for grinding. These specialized teeth, combined with the need for absolute occlusion, would drive the development of a temporomandibular joint permitting complex movements. Nota bene: Cynodonts have stiffened thoracic rib articulations and reduction in lumbar ribs, raising speculation that a diaphragm might have appeared for the first time in *Procynosuchus*. Certainly, it is present by the time of *Probainognathus* (Figs. 8.239 and 8.240).

The Permian-Triassic extinction wiped out 90% of creatures living in the Permian. Epicynodont **Thrinaxodon** ("three-pointed tooth") was a survivor and presents a truly transitional synapsid with more mammal-like features (Figs. 8.241 and 8.242). The occipital condyle is now double and the atlas–axis joint is complete. The zygomatic arch shows characteristic S-shape curvature. The lower jaw is almost entirely dentary. It projects the coronoid process upward inside true S-shaped zygomatic arch, a fateful innovation as we shall see. The reason for this may be an increase in the mass of the surangular field buried within the dentary. But for the moment, there is no temporal joint. *Thrinaxodon* is the last gasp of the original gnathostome jaw. A single articular-quadrate joint remains, but quadrate is much reduced. It moves into contact with stapes in the middle ear.

In cynodonts before mammals, the tympanic bone fields have not separated from the posterior mandible. But we can observe preparations for this event taking place at the time of *Thrinaxodon* when quadrate shift backward and a new articulating surface developed on squamosal, the future *glenoid fossa*. In response to the backward retreat of quadrate, the tympanic bone fields of Meckel's cartilage consolidate to form the *post dentary rod* (PDR). It sits in horizontal *mandibular trough* and is held in place by ligaments (Fig. 8.89).

Inside the mouth, incisors are reduced in number to four and the secondary hard palate achieves midline closure, but still remains short. Cheek teeth (premolars) with two roots are present. These changes all point to an ability to process greater volumes of food and thus support a greater activity. Although initial occlusion is inaccurate at final jaw closure the teeth are interlocked. Maintenance of this occlusion will drive TMJ evolution. Finally, turbinals appear for the first time, indicating signs of mammalian endothermy.

Eucynodonts arose in the Middle Triassic and divided into two clades: cynognathiae and probainognathiae. The cynognathiae were herbivores adapted for grinding food. They demonstrate a well-developed lateral crest of the dentary, the probable forerunner of the buccal wall of the alveolar process. Their wear facets demonstrate the first evidence of precise occlusion. The probainognathiae, the immediate ancestors of mammals, were carnivores. *Chiniquodon* had a longer secondary palate. They demonstrate absence of lumbar ribs indicative of a diaphragm, an essential element for a higher metabolic rate. Changes in the acetabular fossa allowed for greater motion of the femur and an advance in posture away from the sprawling gait.

Probainognanthus has more derived dentition with additional cusps and an incredible innovation: a double jaw joint! In addition to the original joint, surangular becomes adventurous. It reaches up and articulates with squamosal enabling a rocking movement, but this innovation marks the beginning of the switch to the TMJ. The ear bones are in contact (Fig. 8.243).

Note the following in Probainognathus. Recall that articular, prearticular, and angular amalgamate to form the postdentary rod which is in continuity with the quadrate. Quadrate has relocated further backward. Just in front of it the squamosal forms a *glenoid fossa*. Just above the PDR, surangular produces a projecting coronoid process which lies in immediate proximity to the glenoid (Fig. 8.89).

Along with its dual-joint system, *Probainognthus* developed a more sophisticated masticatory apparatus. Note the further subdivision of masseter into a deep head anchored on the posterior zygomatic arch for retrusion and a superficial head connected anteriorly for protrusion. The neural control for these muscles would prove useful later in evolution for innervating the pterygoids (Fig. 8.103).

Where is dentary bone in all this? In basal cynodonts *Dvinia* and *Procynosuchus* (Fig. 8.240), we have noted dentary beginning to expand posteriorly, arching over the back of surangular. In so doing, it creates an *adductor fossa*, a home base for masseter. As dentary expands astride surangular, it projects a distinct *coronoid process field*. This is seen in mammals as a secondary cartilage. It could well be the remnant of the most posterior of the coronoid fields, driven backwards. This can explain the unusual finding of an *ectopic molar* buried in the walls of the normally edentulous ramus [23] (Fig. 8.244).

The mammilloform *Morganucodon* is a transitional form (Fig. 8.243). It still maintains the double joint but changes are underway. Dentary is in the process of surmounting surangular and produces a secondary cartilage, *coronoid process* which, with brain growth, is displaced out of the squamosal glenoid. A new condylar process now projects backward from the squamosal (covered over by dentary). Quadrate moves into the middle ear. By default, the condylar process of posterior dentary (the future ramus) slips into glenoid fossa, replacing the previous surangular. This forms the definitive mammalian jaw joint. *Morganucodon* shows many changes in its the feeding apparatus as well. The hard palate extends backward to its full extent permitting efficient breathing during mastication. Differentiation of generalized amniote teeth takes place producing mammalian incisors, canines, and multicuspid teeth. Precise occlusion was achieved. The dentition is *diphyodont* meaning it has only milk teeth and adult teeth. It thus is possible that *Morganucodon* nursed its young. For this reason, in some classifications, it is considered as a true mammal (Figs. 8.190 and 8.234).

Temporomandibular Joint: The Ear Bone Switch

The multi-directional TMJ is one of the crown jewels of mammalian evolution. It was preceded in cynodonts by innovations in jaw muscles. A reduction in the adductors is accompanied by three specific muscles, temporalis, masseter, and the pterygoids. These are manifested by the outward projection of the zygomatic arch. The lower jaw could be moved from side to side rather than simply up and down. The transverse distance between the lower jaws became narrower than that up the upper jaws, the *anisognathus condition*. This permitted chewing on one side of the jaw and then the other. At the same time, complex cusp configurations appeared on the molars. In sum, with the ability to move the jaws through circular patterns, grinding up of complex foodstuffs (grasses, shrubs, or tough meat) expanded the nutritional landscape for mammals, and as a consequence. an expansion of the habitat.

Mammaliaforms divides the Jurassic period. One line leads to the monotremes living in Australia. Monotremes maintain the PDR and use it as a hearing device. They have no TMJ. The other line the *Theria* leads to true mammals. The evolution of therians to mammals is described in terms of innovations in dentition, indicating progressive sophistication of the jaw joint. The fundamental change took place between Morganucodon and mammals as the Meckelian tympanic bone fields broke free from mandible and migrated into the middle ear. With the PDR out of the way, the posterior margin of dentary became free to issue a new structure, the condyle (Figs. 8.245 and 8.246).

We can see this in most basal taxa, the *triconoconts*, so-named for having three molar cusps. *Hadroconium* and *Sinocodon* were small creatures (the latter was found in China) that demonstrate a true condyle projecting backward into the glenoid fossa of squamosal. The later taxa *Priocodon* shows the change in the rod-like primitive condye into a structure with a well-defined neck and head. Study Figs. 8.246 and 8.247 to appreciate these changes (Figs. 8.247 and 8.248).

In humans, TMJ development undergoes morphologic changes from a primitive stage not dissimilar from *Sinocodon*. Primordial stage (6 weeks), Carnegie stage 16, shows development of membranous bone fields in register with Meckel's cartilage. A primary joint between r2 quadrate (incus) and r3 articular (malleus) is present but coronoid and condylar fields not ossified. The blastemic stage (7–8 weeks) shows the development of intraosseous growth centers within anterior surangular (coronoid process) and posterior surangular (condylar process) concomitant with temporal bone synthesis. These processes start out at 7 weeks as conical buds with their apices directed inward toward the center of the surangular fields (anterior and posterior). They then erupt from the surface of the bone and project outward. Temporal bone is also growing. The two sets of blastemata approximate one another. In the cavitation stage (9–11 weeks), the articular fossa, discs, and capsule appear and lower joint space forms. This is followed by a maturation stage (12 weeks+) in which the upper joint space develops. As the ramus assumes the upright position, it will have two biologic zones of growth, an anterior zone in association with the coronoid field and a posterior zone in association with the condylar field. These zones can be safely separated by a vertical osteotomy in which the coronoid field remains in continuity with the buccal cortex and the condylar field is kept in continuity with the lingual cortex (vide infra) (Figs. 8.249 and 8.250) [24].

Clinical Correlation

The surgical splitting of the mandible follows a sequence determined by the development. From above downward to the alveolar canal, the process is simple. Fields on either side are distinct. However, below the canal the bone is not bilaminar. It is exclusively of lateral origin as it wraps around inferior alveolar canal. Hence, no natural separation plan exists, it must be cut from above-downward with an osteotome or saw. For mandibular advancement or recession, a sagittal split ins used, either in the manner of Obwegeser or as the Dal modification. To accomplish a simple rotation of the mandible as in the abnormal downward vector of craniofacial macrosomia, a full thickness osteotomy can be safely performed by staying posterior to the nerve (Fig. 8.251).

Final Thoughts

Maxilla and maxilla share common developmental themes. Both are organized around embryonic cartilages that program bilaminar membranous bone fields and teeth.

As development proceeds both palatoquadrate and Meckel's cartilages disintegrate.

Phylogenies are similar: both originate from placoderm bone plates

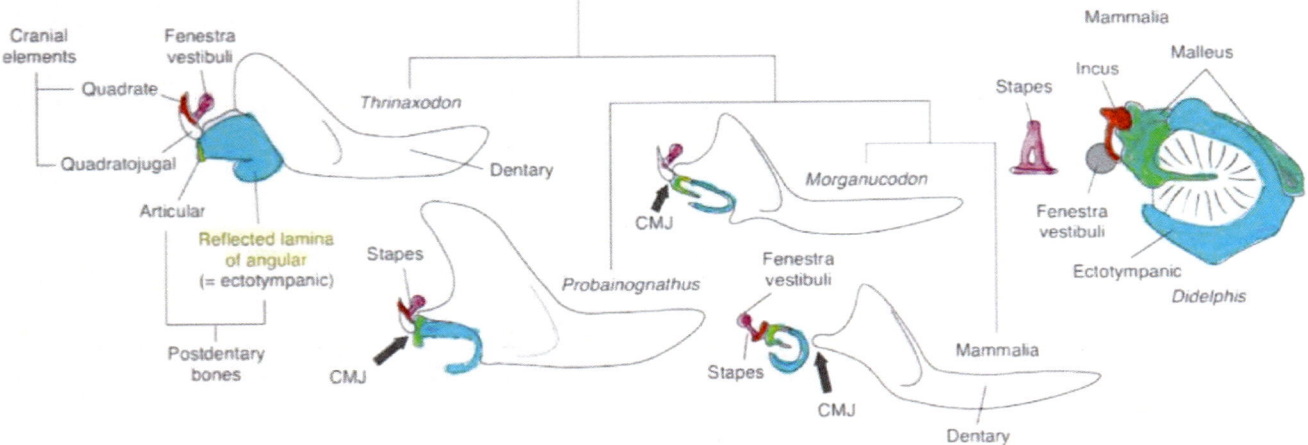

Fig. 8.250 Evolution of the jaw joint and middle ear. Prior to the cynodonts, the Theriodonts had a reptilian jaw joint but developed two innovations: (1) a coronoid process arose that nested in the masseteric fossa. (2) The dentary extended backwards beneath the angular. This would eventually lead to a protruberance that became in cynodonts the condyle. The saga of the TMJ begins with the cynodonts. Epicynodont *Thixanodon* shows postdentary bones appearing from Meckel's cartilage. Experiments in two dsimultanous joints in *Probainognathus* and *Morganucodon*. In the latter, manubrium of malleus appears. But in both cases, the tympanic remains attached to the posterior mandible. In true mammals, tympanic is completely free and is incorporated into temporal bone. (Copyright 2020 by The University of California Museum of Paleontology, Berkeley, and the Regents of the University of California)

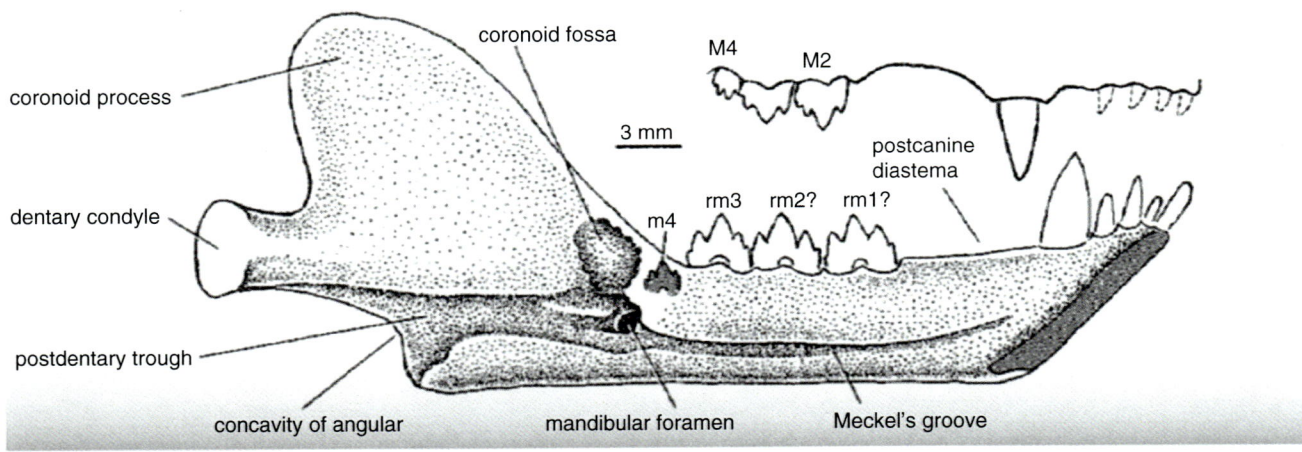

Fig. 8.251 Inventing the TMJ, mammalian iteration 1. Proto mammal *Sinococodon* shows postcanine diastema (no premolars), a coronoid process, a condyle and a postdentary trough. There is a single jaw joint. (Reprinted from Benton MJ. Vertebrate Paleontology, 4th ed. Blackwell; 2015. With permission from John Wiley & Sons)

- Palatoquadrate gives rise to anterior superognathal (premaxilla) and posterior superognathal (maxilla)
- Meckel's gives rise to inferognathal with anterior cartilages (symphysis) and posterior cartilages (body)

Meckel's has two sectors with distinct fates

- Mandibular sector remains within walls of mandibular symphysis and body and eventually involutes
- Tympanic sector is exteriorized lingual to the dermal bones; it eventually breaks off and joins middle ear

The bilaminar nature of the dental housing has distinct structure and function

- Maxillary walls form a five-sided box containing maxillary sinus
- Mandibular walls enclose a marrow space propitious for surgical splitting

All dental units are supplied by a common neurovascular system based on the trigeminal nerve and stapedial arteries

- Identical homeotic neural subdomains
- Palatoquadrate and Meckel's have identical programs to specify dentition occlusion

Dermal bone fields of mandible contain two distinct functional programs

- Dental fields: dentary and coronoids form lateral and medial alveolar laminae that house the teeth
- Muscle-insertion fields: infradentary series consists of four bones, each with specific muscles

  - id1–id2 become splenials (body) = depressors
  - id3 becomes angular (malleus) = tensor tympani
  - id4 becomes surangular (coronoid, condyle) = elevators

Dermal bone fields are dynamic

- Number of incisors in the symphysis and number of post-canine teeth in the body diminish
- Dentary bone subsumes neighboring fields but their homeotic programs remain
- Surangular becomes the source of additional bone fields, despite being covered over by dentary

Developmental fields of the mandible have consequences for pathology

- Symphysis
  - Vertical or horizontal excess/deficiency = malposition
  - Transverse excess/deficiency = diastemata/crowding
- Coronoids
  - Excess/deficiency = hyperdontia/hypodontia, ectopic molar
- Splenials (id1–id2):
  - Excess/deficiency = diastemata/crowding
- Angular (id3)
  - Anomalies of malleus
- Surangular (id4)

  - Hyperplasia/hypoplasia = prognathism/retrognathism
  - Coronoid = excess or deficiency
  - Condyle = ankylosis, Treacher-Collins, loss of ramus

In sum: The textbook description of the human mandible is as a derivative of the dentary bone. This model in incorrect for many reasons. The various developmental field have just been reiterated. The arterial supply to mandible comes from a variety of sources. In addition to the inferior alveolar artery, attached muscles perfuse local areas of cortex. Multiple distinct developmental fields can be identified: the medial and lateral laminae of the alveolus organized around three dental zones, the medial and lateral laminae of the symphysis and body the medial and lateral laminae of the ramus and, within the latter, the coronoid and condylar processes, and finally the tympanic bone fields. Mammalian dentition is characterized by three different morphologies (incisors, cuspids, and molars) all programmed by neural crest in register with V3. These genetically distinct zones also correspond to the primitive coronoids. In sum, the mandibular complex is just that, a complex assortment of individual developmental fields, each of which continues to play a role in development. Embryologic analysis offers an innovative way to understand pathologic states of the mandibular complex.

Hyoid Bone Complex

The hyoid bone develops at the interface of the second and third pharyngeal arches. As such, it serves as a unique reminder of our evolutionary past. In its original design of the piscine, branchial arch has five cartilages arranged either above the axis of the gill or below it. The reader is referred to the phylology of the gill arches. In the transition to jaws, these cartilages are modified: (1) pharyngobranchials are lost; (2) epibranchials becomes palatoquadrate in the first arch, and hyomandibula in the second arch; (3) ceratobranchials becomes mandible in the first arch, ceratohyal in the second arch, the third arch component of hyoid bone, and potential sources of neural crest for structures of the fourth and fifth arches; (4) hypobranchial and basibranchial become hypohyal and basihyal in the second arch and are obliterated in the first arch.

As we shall see, the various components of the hyoid complex fulfill several functions: a cross-link support system running from the sternum to the mandible, depression of the mandible, a midline organizing structure for the tongue, and an embryonic "guidewire" for migration of the thyroid (Fig. 8.252, 8.253, 8.254, 8.255, 8.256, 8.257, 8.258, 8.259, 8.260, 8.261, 8.262, 8.263, 8.264, and 8.265).

Descriptive Anatomy

The hyoid bone complex arises from neural crest contributions from second arch (r5) and third arch (r7). It has six fields, each of which has its own ossification center: a *corpus ossis hyoidei* (fused body), upper *cornua majora* (greater cornua), and a *cornua minora* (lesser cornua) (Fig. 8.252).

Each half has three ossification centers representing a body and two posteriorly directed projections, an upper *lesser cornu* and a lower *greater cornu*. The ossification of the body represents the hypohyal field. The second arch contribution comes from r5 neural crest. It is relatively small, consisting of the upper (cranial) aspect of the hyoid body and the lesser cornu. The third arch contribution comes from r7 neural crest. It forms the bulk of the body and the greater

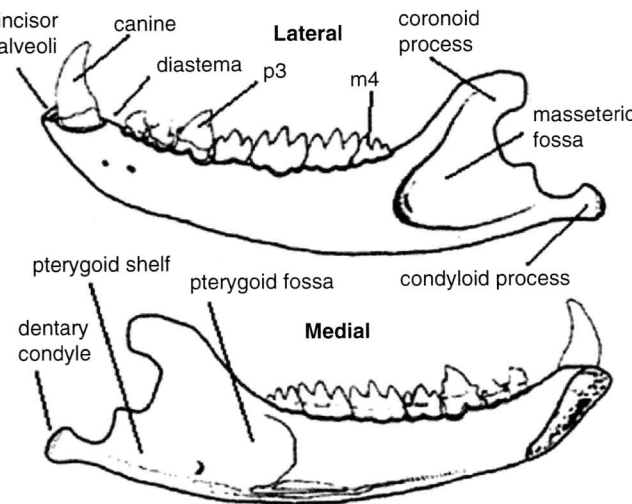

Fig. 8.252 Inventing the TMJ, mammalian iteration 2. *Priocodon* was a Mesozoic mammal of the late Jurassic to early Cretaceous. It appears just beyond the monotremes, and therefore represents the first true mammal. It belongs to the tricondonta, so named for three equal cusps on the molars. The dentary of *Priocodon* shows absorption of the coronoids and splenials. The coronoid process is inclined 135°. Condylar process projects backward from dentary. (Reprinted from Quiang J, Luo Z-X, Yuan C-X, Wible JR, Zhang J-P, Georgi JA. The earliest known eutherian mammal. Nature 2002; 416(6883): 816. With permission from Springer Nature)

cornu. The upper and lower sectors of the hyoid are defined by a well-developed *transverse ridge*. Muscles and ligaments are attached to the hyoid in accordance with its individual developmental fields.

Hyoid is U-shaped and is described as having a compact central body from which is continues straight backward by two great cornua. Lesser cornua are smaller in size and project from the upper border, demarcating the body from the great cornua. Along the medial border of the greater cornua is inserted thyrohyoid membrane, indicating that hyoid projects anteriorly to the plane of the thyroid cartilage. The lesser cornua provide the insertion site for *stylohyoid ligament*, a point of suspension from the skull.

The dorsal surface is confluent with the fourth arch epiglottic cartilage but separated from it by a *bursa*, permitting independent movement. Superior border is connected by *thyrohyoid membrane* to the fourth arch thyroid cartilage. The hyoid body bears the insertions of three groups of muscles (Figs. 8.253 and 8.254).

1. *Pharyngeal arch* muscles function in mastication and the coordination of swallowing.
2. *Suprahyoid* muscles (intrinsic tongue muscles) arise from the occipital somites, S1–S4. They are supplied by hypoglossal nerve (XII). They extend from second arch structures forward into the tongue. The first three of the four

extrinsics (S1–S3) have primary insertions along the upper border of hyoid.

3. *Infrahyoid* (strap) muscles such as sternohyoid, sternothyroid, and thyrohyoid represent muscles connecting pectoral girdle with the pharyngeal floor and ultimately with mandible, making a "pitstop" along the way at the hyoid bone. They represent the ancient *coracomandibularis* muscle (seen today in sharks) that unites sternum with mandible. In so doing, coracomandibularis "strings together" the ventral aspects of all intervening branchial arches. With exception of geniohyoid (see below), the strap muscles all have a secondary insertion into the lower border of the hyoid. Strap muscles are all (again with exception of geniohyoid) polyneuromeric, arising from the first three cervical somites, S5–S7 (C1–C3). They are innervated by the *ansa cervicalis* (neuromeres c1–c3). Their evolutionary significance is discussed separately in Chap. 9 on craniofacial muscles.

Muscle insertions into second arch hyoid (upper border) have two functional laminae: an internal trio of pharyngeal arch muscles of mastication (Sm4, Sm6, Sm7) is flanked by an external trio of extrinsic tongue muscles (S1–S4) supplied by hypoglossal nerve (r8–r11). Of the hypobranchial cord muscles (r8–c3), these arise from the occipital component (r8–r11). The order of attachment is neuromeric and proximal-distal.

Mylohyoid is a derivative of Sm4 and is innervated by V3. It has a very extensive primary insertion into mandible, running along the mylohyoid line all the way from symphysis to the third molar. It inserts again into the anterior zone of lower hyoid border.

Digastric, posterior belly is also a derivative of Sm6 and is innervated by VII. It spans from the *mastoid notch of temporal bone* (r4–r5 derivative) located immediately behind styloid process. It then inserts as an *intermediate tendon* into the posterior zone of lower hyoid border and posterior greater cornu. Note that the r4–r5 mastoid notch is posterior in neuromeric sequence to r4–r5 styloid process. For this reason, posterior digastric inserts behind stylohyoid.

Digastric, anterior belly is a derivative of Sm4 and is supplied by inferior alveolar branch of V3. Digastric is unique among muscles of mastication in that it is polyneuromeric with motor neurons arising from adjacent rhombomeres r3–r4. Its secondary attachment is the last available along the mandible.

Middle constrictor occupies the most medial and posterior insertion site along the greater cornu. It also inserts extensively along the posterior lamina of thyoid. Medial constrictor arises from somitomere 7 and demarcates the boundary between third arch (the hyoid) and fourth arch (thyroid cartilage). It lies internal to the tongue muscles as expected.

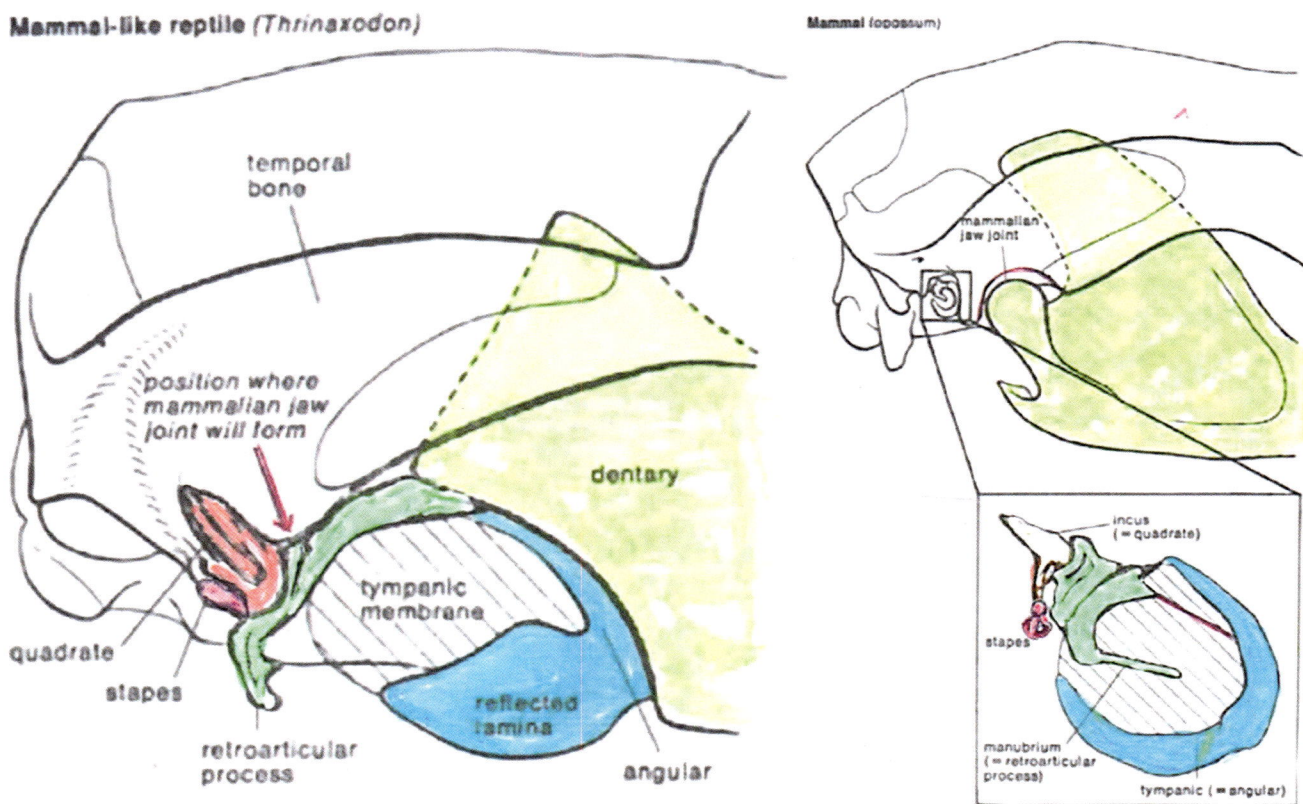

Fig. 8.253 Summary: TMJ evolution between *Thrinaxodon* and Opossum. The basal state of the ear at the node between diapsids (reptiles and birds) and synapsids (mammals) was that of a single bone (stapes) transmitting from a tympanum to the inner ear. The jaw joint continues the articular-quadrate mechanism. The genetic potential of Meckel's cartilage remained unexploited in the diapsid line. Differentiation into three distinct bones is diagnostic for the synapsid line leading to mammals but only in mammals are the three bones trans-ferred to the temporal bone. Evolutionary changes in hearing mechanism begin with *Thrixanodon*. Here the jaw joint bone performs two functions: (1) create a hinge and (2) conduct vibrations from the tympanic membrane which is sheltered in the reflected lamina of angular bone of the stapes. Mammals separate out these two functions creating a much more sensitive mechanism for sound detection. (Reprinted from Benton MJ. Vertebrate Paleontology, 4th ed. Blackwell; 2015. With permission from John Wiley & Sons)

Pharyngeal constrictors are established *prior to* tongue muscles; these latter must navigate external to the pharynx in order to reach the tongue.

Three of the four extrinsic muscles of the tongue have primary insertions into second arch moiety of the hyoid bone. All are located along the superior border. The fourth muscle, styloglossus arises from occipital somite 4 and is attaches to second arch styloid process. The fifth muscle, palatoglossus, is not a true extrinsic tongue muscle. It connects third arch soft palate with the posterolateral base of the tongue, also of third arch derivation. As expected palatoglossus is innervated by r6–r7 pharyngeal plexus (IX via X).

The *secondary insertions* of the extrinsics into the tongue follow a medial-to-lateral sequence: genioglossus, chondroglossus, hyoglossus, and styloglossus. Since all are innervated by four successive roots of hypoglossal nerve (r8–r11), it can be surmised that they originate and migrate from separate somites.

Genioglossus develops from S1. It is a large muscle that migrates forward and downward from the occipital somites, attaching to the anterior–superior aspect of hyoid. Genioglossus is the last muscle to insert into the tongue and occupies the midline, attaching to the *genial process* of mandible.

Chondroglossus likely arises from S2. It is considered the part of hyoglossus. Its primary insertion is a little higher up, from lesser cornu.

Hyoglossus likely arises from S3. It has developmentally priority over chondro- and genioglossus and therefore occupies a more lateral position within the tongue. It has two "heads," internal and external, separated by fibers of genioglossus. Both heads are positioned on the hyoid posterior to genioglossus. The *internal head* bears the confusing name of chondroglossus. It attaches along the upper border of hyoid, just anterior to lesser cornu. These attachments indicate that the "maturity gradient" of r5 body is posterior–anterior. The *external head* of hyoglossus is external to the previous two muscles. Its insertion into hyoid is thus a more superficial, yet extensive. It runs along the transverse ridge itself all the way back to the greater cornu.

a

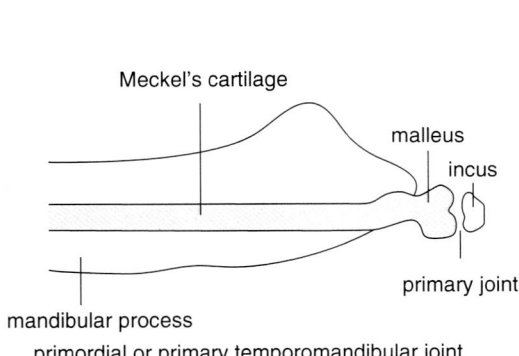

primordial or primary temporomandibular joint

b

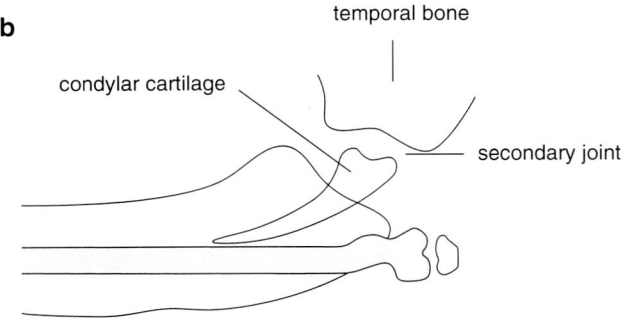

blastemic stage of development of the temporomandibular joint

c

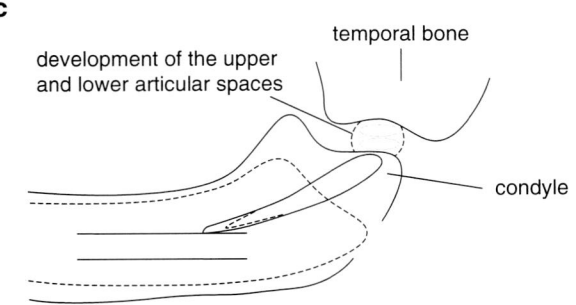

cavitation stage of development of the temporomandibular joint

d

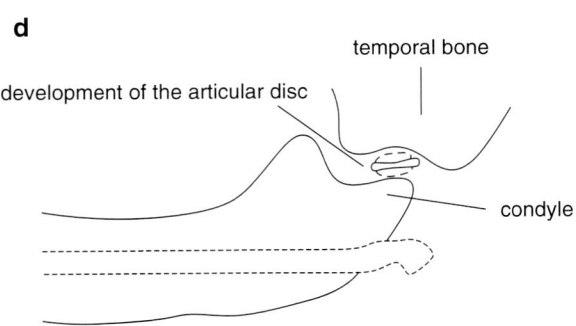

maturation stage of development of the temporomandibular joint

Fig. 8.254 TMJ development. Primordial stage 16, 6 weeks with the development of membranous bone fields along Meckel's and the primary joint between r2 quadrate (incus) and r3 articular (malleus). Coronoid and condylar fields not ossified. Blastemic stage (7–8 weeks): development of intraosseous growth centers within anterior surangular (coronoid process) and posterior surangular (condylar processs) concomitant with temporal bone synthesis. The two sets of blastemata approximate one another. Cavitation stage (9–11 weeks): articular fossa, discs and capsule, and lower joint space. Maturation stage (12 weeks+): development of upper joint space. (Reprinted from Badel T, Savić-Pavačin I, Zadravec D, Marotti M, Krolo I, Grbeša D. Temporomandibular joint development and functional disorders related to clinical otologic symptomatology. Acta Clin Croat 2011; 50:51–60. With permission from Acta Clinica Croatica)

Geniohyoid is the final muscle to gain insertion to dorsal hyoid. It follows the rules as well. Its myoblasts originate from the fifth somite (since somites 1–4 are supplied by hypoglossal). This muscle originates from S5 and is innervated by the first cervical nerve via XII Geniohyoid is a mandibular depressor; as such, it is a continuation of the strap muscles, not an intrinsic tongue muscle. It is thus forced to migrate inferior (beneath) genioglossus. The primary insertion is at the anterior border of the hyoid body, both above and below the transverse ridge, i.e., to anterior border of hyoid body at the r5/r7 junction. It secondarily inserts into the inferior mental spine of r3 mandible.

Muscle insertions into the third arch hyoid (lower border) are a single functional lamina consisting of a quartet of strap muscles (S5–S7, i.e., C1–C3). Of the hypobranchial cord muscles (r8–c3) these arise from the cervical component (c1–c3). Just as in the second arch upper hyoid, strap muscles insert into lower hyoid in a neuromeric anterior–posterior sequence.

Stylohyoid (StH) is a derivative of Sm6 and is innervated by VII. It first attaches to styloid process and thence to mid zone of lower hyoid border. StH has a shorter migration distance to hyoid than does posterior belly of digastric, hence it "wins out" to gain the more anterior insertion site.

Digastric, posterior belly is also a derivative of Sm6 and is innervated by VII. It spans from the mastoid notch of temporal bone (r4–r5 derivative) and then inserts as an *intermediate tendon* into the posterior zone of lower hyoid border and posterior greater cornu. Note that the insertion of digastric into hyoid is posterior to that of StH because its primary insertion at mastoid notch is posterior in neuromeric sequence and physical location to the r4–r5 styloid process.

Digastric, anterior belly is a derivative of Sm4 and is supplied by inferior alveolar branch of V3. Digastric is unique among muscles of mastication in that it is polyneuromeric with motor neurons arising from adjacent rhombomeres r3–r4. Its secondary attachment is the last available at.

Geniohyoid (GH) is the terminal muscle of the strap sequence. It completes the work of coracomandibularis to

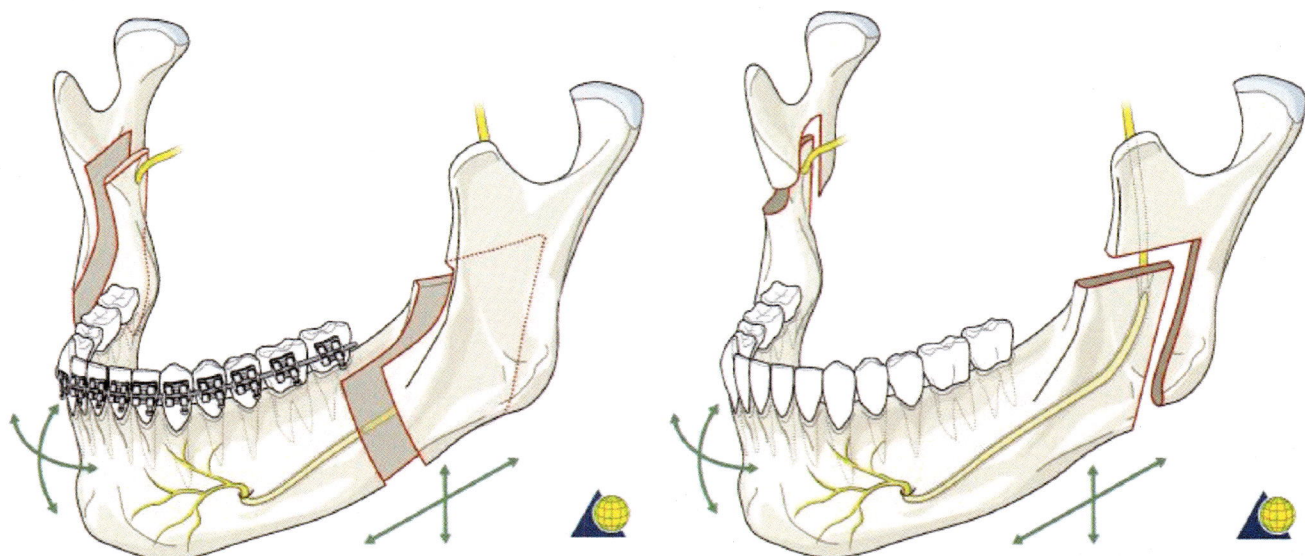

Fig. 8.255 Osteotomies of the mandible. Left: Saggital split osteotomies separate an external ramus and condylar fragment from an internal body fragment bearing the inferior alveolar nerve. These osteotomies permit either forward movement of the mandible in cases of inadequate projection or recession of the mandible in cases of prognathism. The original design by Obwegesser extended the internal osteotomy in a horizontal fashion above the nerve and extended completely backward to the posterior border. The Dal Pont modification shown here descends vertically, preserving the posterior order. It theoretically creates a larger surface area for contact between the bone segments but with a higher risk of iatrogenic nerve injury during the split. Right: a simple full thickness osteotomy through the ramus field allows for rotational movements of the mandible in cases of abnormal growth vectors (usually downward) in cases of craniofacial macrosomia. (Copyright by AO Foundation, Switzerland)

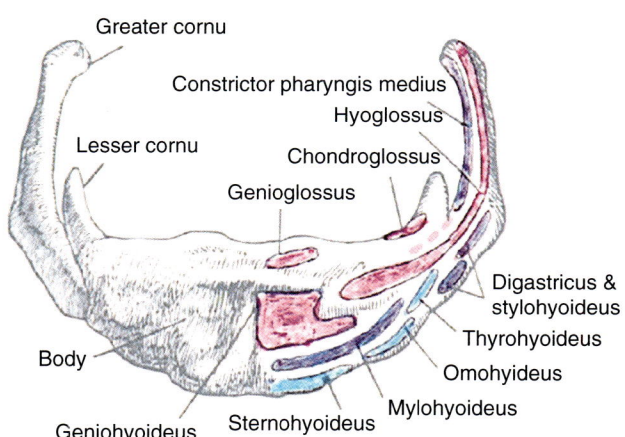

Fig. 8.256 Hyoid. Hyoid bone is a combination of second and third arch neural crest. Primary insertion, suprahyoid (pink): extrinsic tongue muscles, and suprahyoid strap muscle geniohyoid. Secondary insertion, suprahyoid (purple): pharyngeal arch muscles. Secondary insertion, infrahyoid (blue): strap muscles. (Reprinted from Lewis, Warren H (ed). Gray's Anatomy of the Human Body, 20th American Edition. Philadelphia, PA: Lea & Febiger, 1918)

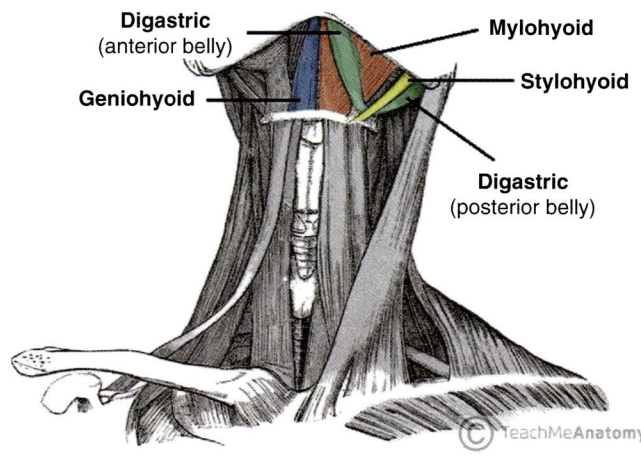

Fig. 8.257 Suprahyoid muscles. (Reprinted from Lewis, Warren H (ed). Gray's Anatomy of the Human Body, 20th American Edition. Philadelphia, PA: Lea & Febiger, 1918)

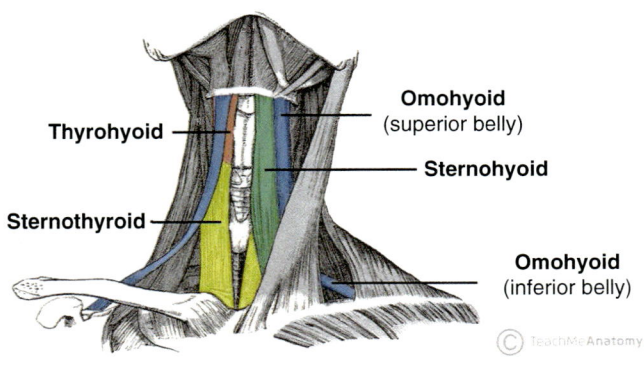

Fig. 8.258 Infrahyoid muscles. (Reprinted from Lewis, Warren H (ed). Gray's Anatomy of the Human Body, 20th American Edition. Philadelphia, PA: Lea & Febiger, 1918)

unite the sternum to the mandible. It is mononeuromeric, arising from the fifth somite and innervated by C1. It is the only strap muscle to connect second arch hyoid to first arch mandible. Its primary attachment at the hyoid body is midline and along its anterior surface. It secondarily attaches to the inferior mental spine inside the symphysis.

Sternohyoid (SH): Like all the strap muscles, the origin of sternohyoid is polynueromeric (c1–c3); therefore, the muscle originates from the first three cervical somites. It subsequently attached to the back of manubrium sterni and medial margin of clavicle. The secondary insertion to the lower margin of hyoid is near the midline.

Sternothyroid (ST) and *thyrohyoid* (TH) form a chain connecting sternum to a more lateral site on hyoid. SH seeks a

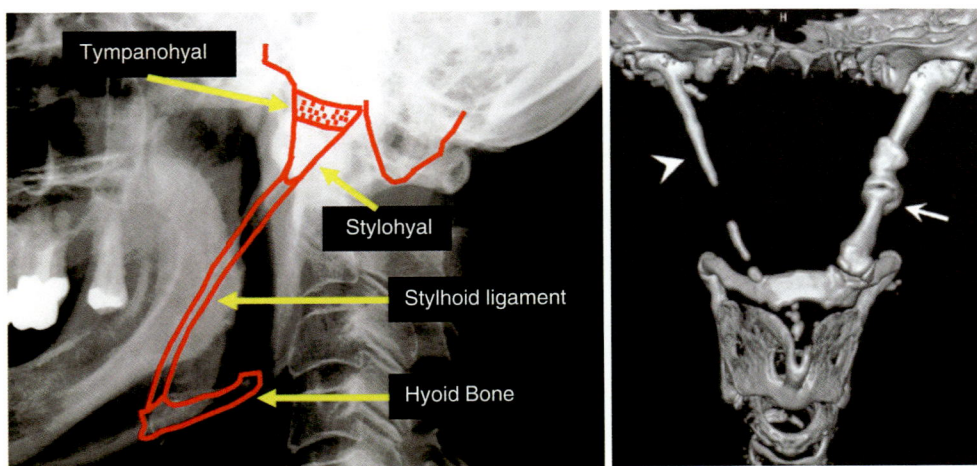

Fig. 8.259 Stylohoid ligament. Hyoid is an embryologically composite bone. Second arch forms lesser cornu and superior part of body; third arch forms greater cornu and inferior half of body. Note that a continuous embryonic signalling pathway exists along the ventral aspects of all five pharyngeal arches and thence to c1–c4 levels of clavicle (see text). This represents the ancient coracomandibularis muscle which breaks into two groups. Hyobranchial infrahyoid (strap) muscles are derived from c1 to c4 and run from sternum/clavicle to lower hyoid. Hypobranchial suprahyoid (extrinsic tongue) muscles are derived from r8 to r11 and run from upper hyoid to tongue. Eagle syndrome (ES) is an aberrant ossification of the stylohyoid apparatus, consisting of the styloid process, the attached stylohyoid ligament, and the lesser cornu of the hyoid bone. Anatomically, the styloid process arises from the temporal bone and passes downwards, forwards, and medially. Embryologically, it represents the ceratohyal cartilage of second pharyngeal arch. Derived from the Reichert's. There are two types of ES as described originally by Eagle: first is the classic styloid process syndrome due to fibrous tissue causing distortion of the cranial nerve endings in the tonsillar bed after tonsil removal; and a second type due to

compression of the sympathetic chain in the carotid sheath Ossification of the styloid process and the stylohyoid ligament leads to an increase in the thickness and length of the styloid process, which then presses on the adjacent structures like the internal jugular vein, carotid artery, facial nerve, vagal nerve, glossopharyngeal nerve, and hypoglossal nerve, resulting in various pressure symptoms. The styloid process normally measures 2.5–3 cm in length; when the length exceeds 3 cm, it is said to be elongated. This syndrome represents a retention by ceratohyal of its embryonic state such that it fails to regress to a fibrotic state. [Left: (Reprinted from Kirchhoff G, Kirchhoff C, Buhmann S, Kanz K-G, Lenz M, Vogel T and Kichhoff RM. A rare differential diagnosis to occupational neck pain: Bilateral stylohyoid syndrome. Journal of Occupational Medicine and Toxicology 2006, 1:14. https://doi.org/10.1186/1745-6673-1-14. With permission from Creative Commons License 2.0: https://creativecommons.org/licenses/by/2.0/). Right: (Reprinted from Raina D, Gothi R, Rajan S. Eagle Syndrome Indian J Radiol Imaging. 2009; 19(2): 107–108. With permission from Wolters Kluwer Health)]

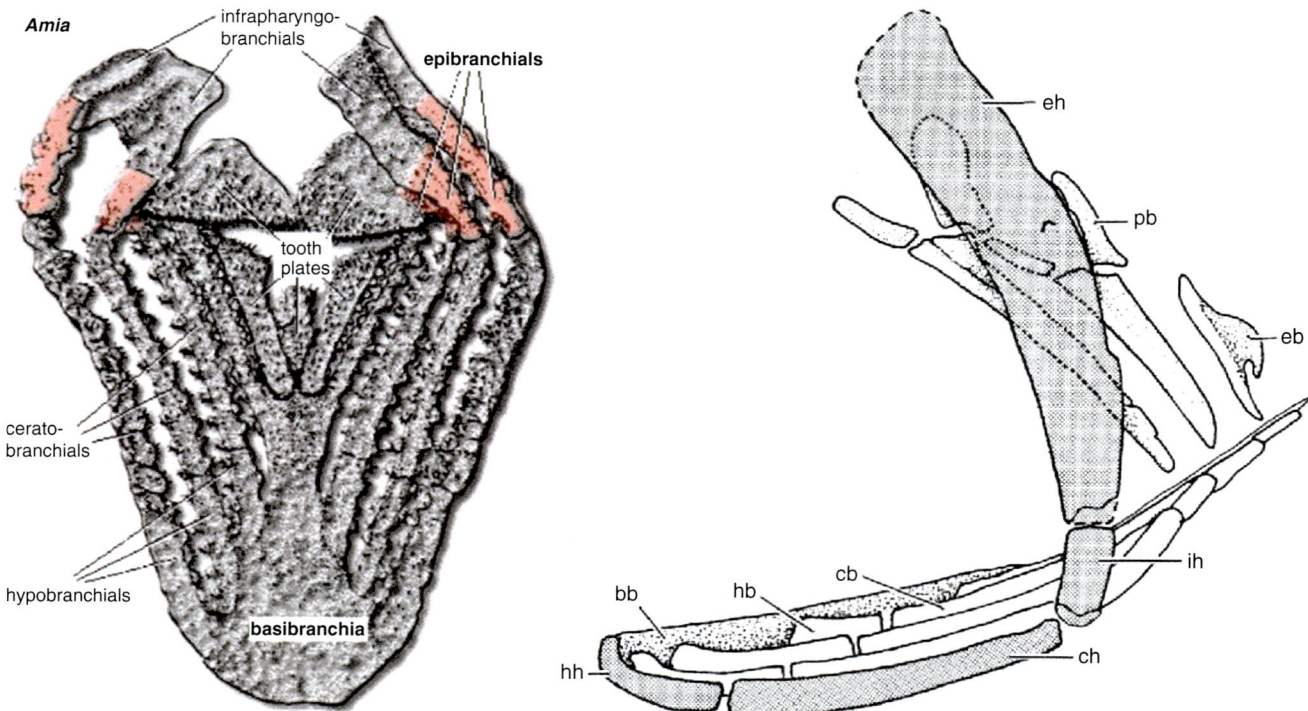

Fig. 8.260 The basal fish *Amia* floor of mouth. Left: Oral cavity bears multiple teeth, including in the pharynx. Note fusion of first two basibranchials. Recall that fishes have *two rows of teeth* in the mandible, the dentary line and the coronoid line. It is possible that these two laminae represent a fusion of the ceratomandibular and ceratohyoid? Right: Synapomorphies of the hyoid arch. Dermal ossifications: epihyal (hyomandibula, unusually big), interhyal, ceratohyal, hypohyal. Of the remaining five branchial arches: pharyngo branchials 1–4 fuse (fifth one is lost). Branchiostegal rays become operculars. Basibanchials from mandibular arch and hyoid arch fuse into a single unit, projected forward, with implications for the tongue. **Key**: *pb* pharyngobranchia, *eb* epibranchial, *cb* ceratobranchial, *hb* hypobranchial, *bb* basibranchial, *eh* epihyal, *ih* interhyal, *ch* ceratohyal, *hh* hypohyal. (Reprinted with permission from: GEOL431—Vertebrate Paleobiology course web resources. Thomas Holtz and John Merck, Department of Geology, University of Maryland, College Park, https://www.geol.umd.edu/~jmerck/geol431/)

primary insertion inferior and lateral to its predecessor, from the back of sternum below SH and also from the cartilage of the first and second ribs. The secondary insertion is into the fourth arch thyroid along the oblique line. From there, sternohyoid ascends to attach to inferior border of cartilage, along the lateral line. TH is the continuation of ST. It begins at the oblique line and gains secondary insertion *directly behind SH* into the lower border of body and greater cornu.

Omohyoid (OH): This is the most lateral of the strap muscles. It has two bellies that form a chain. The *lower belly* begins at upper border of scapula. As it travels medially and upward, it is bound down to clavicle by a fascial loop. It passes behind the sternocleidomastoid (a more derivative of the occipital somites) where an *intermediate tendon* forms. A superior belly passes straight upward, hugging the lateral border of SH and inserts in lower border of hyoid just lateral to the insertion of SH.

Blood Supply and Mesenchyme

Hyoid bone is an interface between second and third pharyngeal arches. Neural crest from r4 to r5 provides the mesenchyme for the upper and lower borders of hyoid bone respectively. Lesser cornu is strictly r4 and receives the second arch stylohyoid ligament.

Lingual artery (first and second arch) supplies hyoid bone via suprahyoid branch and mylohyoid via the sublingual branch.

Facial artery (second arch) supplies mylohyoid and surrounding muscles via submental branch.

Occipital artery (second arch) supplies digastric and stylohyoid muscles.

Ascending pharyngeal artery (third arch) supplies middle constrictor via its pharyngeal branch.

Superior thyroid artery (fourth arch) supplies the strap muscles (along with lingual).

Development

Hyoid bone ossifies from three pairs of centers: lesser cornu represents second arch, greater cornu represents third arch, and body is a combination of both arches.

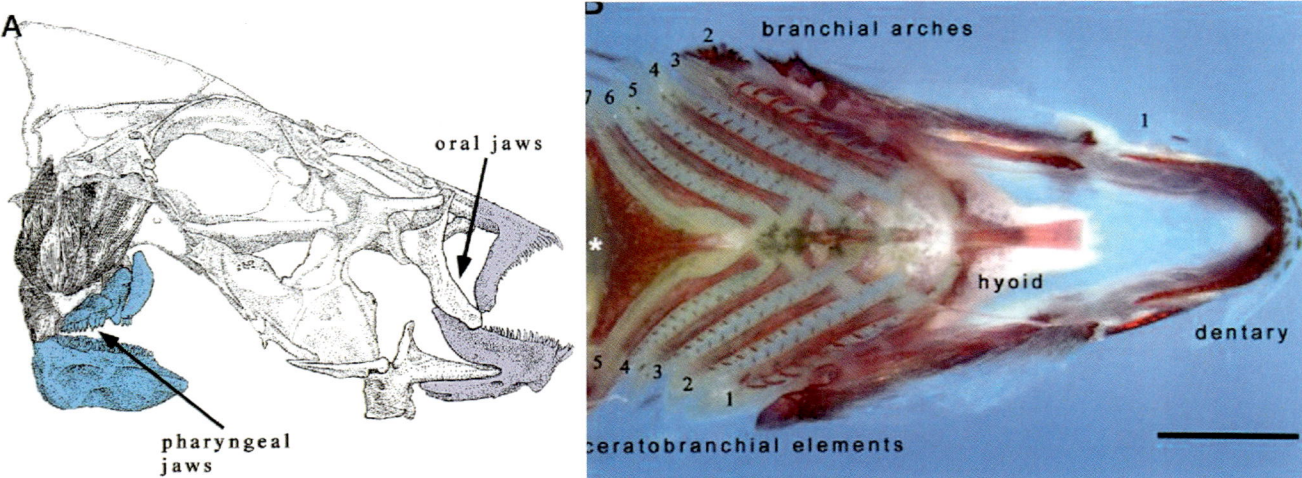

Fig. 8.261 Pharyngeal jaws. During evolution, teeth originated deep in the pharynx of ancient and extinct jawless fishes. Later, with the evolution of bony fish, teeth appeared in the mouth, as in most current vertebrates, although some living fishes retain teeth in the posterior pharynx. Pharyngeal jaws occupy *hox-positive, endodermal* sites, and oral jaws develop in *hox-negative* regions with *ectodermal* cell contributions. Pharyngeal teeth of jawless vertebrates utilized an ancient gene network before the origin of oral jaws, oral teeth, and ectodermal appendages. The first vertebrate dentition likely appeared in a hox-positive, endodermal environment, and expressed a genetic program including **ectodysplasin** pathway genes. This ancient regulatory circuit was co-opted and modified for teeth in oral jaws of the first jawed vertebrate, and subsequently deployed as jaws enveloped teeth on novel pharyngeal jaws. Thus, a core dental gene network is used in the construction of all teeth, regardless of location and lineage. The network is evolutionarily essential: nature appears never to have made a dentition without it. (Reprinted from Fraser GJ, Hulsey CD, Bloomquist RF, Uyesugi K, Manley NR, Streelman JT. An ancient gene network is co-opted for teeth on old and new jaws. *PLoS biology*, 2009; **7**(2):e1000031. With permission from Creative Commons Attribution License)

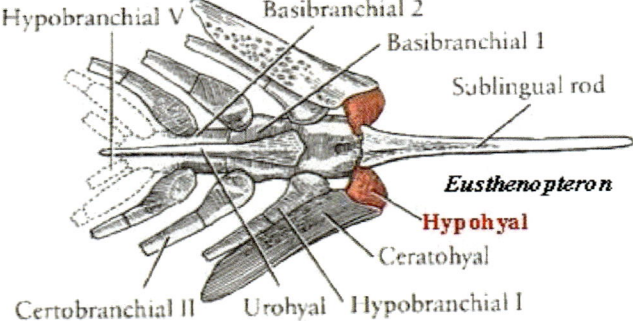

Fig. 8.262 Lingual rod. *Eusthenopternon* showing fusion of basimandibular and basihyal as the sublingual rod. Remaining basibranchials fuse posterior to hyoid. The same genetic signals that program guides the sublingual to connect mandible to hyoid and backward to the pectoral girdle. (Reprinted from Fraser GJ, Hulsey CD, Bloomquist RF, Uyesugi K, Manley NR, Streelman JT. An ancient gene network is co-opted for teeth on old and new jaws. *PLoS biology*, 2009; **7**(2):e1000031. With permission from Creative Commons Attribution License)

Phylogeny

The hyoid bone is a very ancient structure with manifestations in the gill systems of the earliest bony fishes. The details of the gill arch system and the phylogeny of pharyngeal arches has been previously discussed so we will just hit some highlights to show how the derivation of the human hyoid bone.

In the original agnathic fish model, at least, seven gill arches (or more) were present (Fig. 8.127). This ancient system serves as the template for all vertebrates today. These have five components: pharyngobranchial, epibranchial, ceratobranchial, hypobranchial, and basibranchial. Recall that jaw evolution was a two-stage process in which a subversion of the first gill arch gave armored fishes (placoderms) a set of jaws, albeit with loose attachment to the skull. Upper jaw palatoquadrate cartilage came from epibranchial and lower jaw Meckel's cartilage came from cerato branchial.

In the next piscine iteration, both the chondrichthyans and the osteichthyans came up with a new suspension system which transformed epihyoid of the second gill arch into a hyomandibula bone attached to the otic capsule. Those second gill arch cartilages left over (hypohyal and basihyal) created a fish hyoid. Derived fishes such as *Amia* have five branchial arches remain. Fusion of pharyngobranchials 1–5 takes place. Epibranchials 1–2 are also fused dorsally. Epibranchial 3–4 support the tooth plates of the fish jaw. Epibranchial 5 disappears (Fig. 8.128). Ceratobranchials are conserved in form and function. All five remain and they support teeth. Hypobranchials are attached proximally and form a joint distally with basibranchials. These latter tend to fuse in the midline and project (Figs. 8.256 and 8.257).

The pre-tetrapod sarcopterygian *Eusthenopteron* shows basibranchials 1–5 fused in the ventral midline. The basihyals project forward as a *sublingual rod*. In birds and lizards, this structure becomes entoglossal bone. It can be used as a drill for getting morsels out of tree trunks. It can also support a projectile tongue for prey capture (Fig. 8.258).

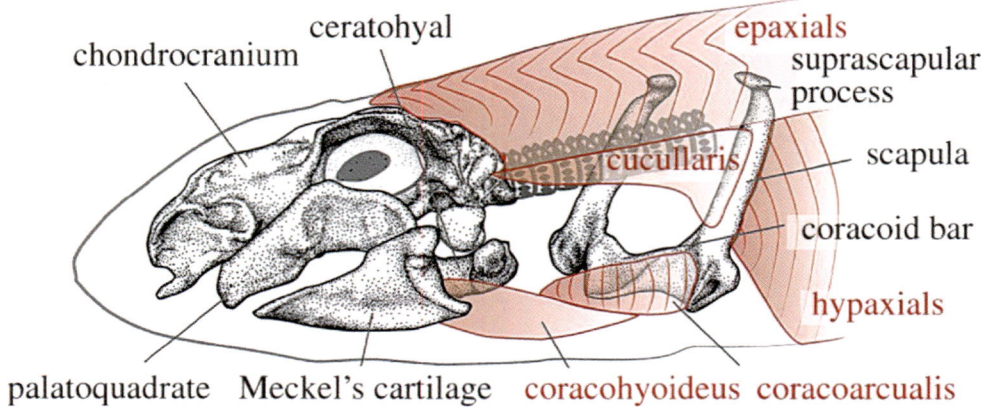

Fig. 8.263 Coracomandibularis. Note the chain of muscles in modern chondrichthyan, *Squalus* (shark) that fixes mandible to the fins. This is reminiscent of the way osteoichthyan fishes use the opercular bones to ratchet open the mouth. Hypobranchial muscles connect to hyoid and thence run beneath adductor mandibulae to the lower jaw. (Reprinted from Camp AL, Scott B, Brainerd EL, Wilga CD. Dual function of the pectoral girdle for feeding and locomotion in white-spotted bamboo sharks. Proc Royal Soc B 2017; 284(1859). https://doi.org/10.1098/rspb.2017.0847. With permission from Royal Society Publishing)

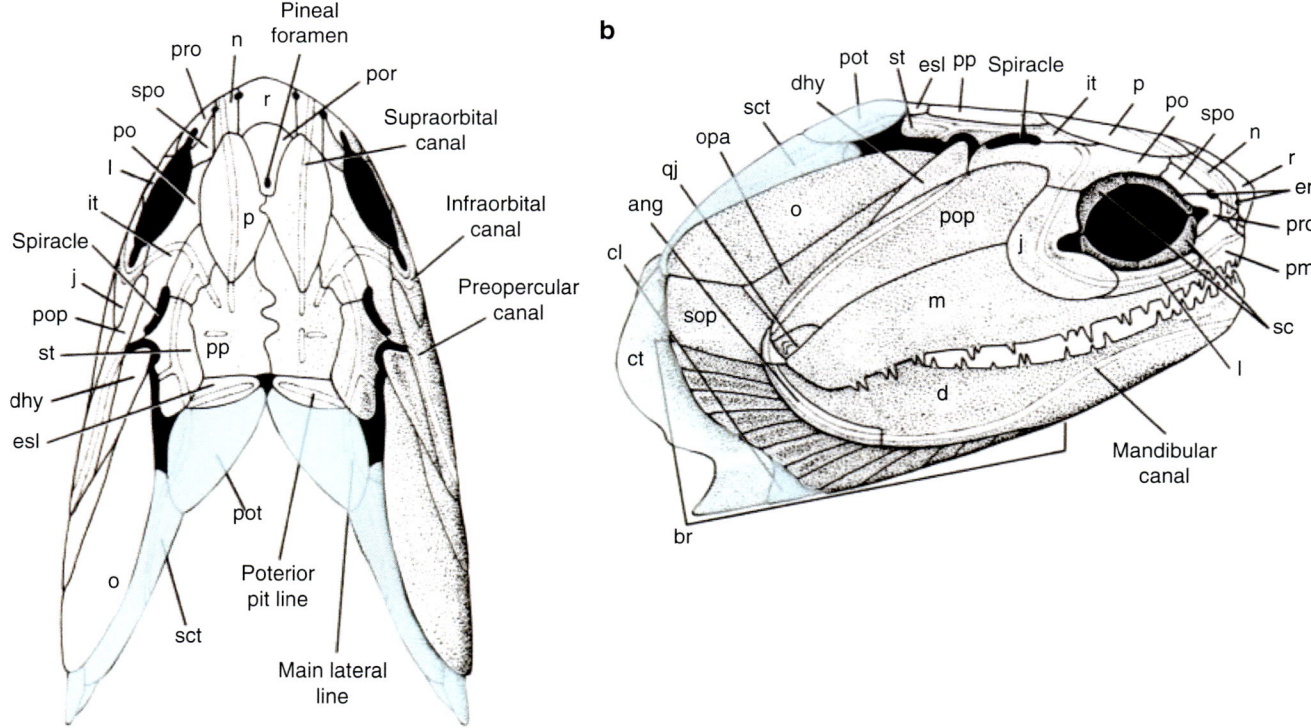

Fig. 8.264 *Cheirolepsis* and the suspension of the pectoral girdle. Behind the skull are two groups of bones (aqua). Extrascapular, medial, and lateral are directly attached to postparietals (**pp**) and supratemporals (**st**). The pectoral line is suspended from extrascapulars as follows: posttemporal (**pt**), supracleithrum (**sct**), postcleithrum (**pct**), cleithrum (**ct**), and finally clavicle. Note ten branchiostegal rays. Above them (the breakpoint being the interbranchial cartilage), the operculars represent modified branchiostegal rays. Thus, with the terminus of the fish skull is r11, this bone series represents the next 4 neuromeres. Pectoral girdle of Polypterus versus human. (Reprinted from Carroll RL. Vertebrate Palaeontology. New York, NY: WH Freeman and Company; 1988)

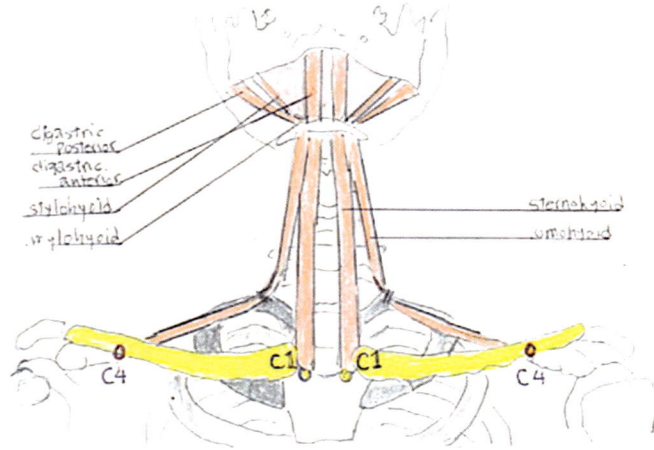

Fig. 8.265 Pectoral girdle of *Polypterus*. Dermal bones are red; replacement bones are black. Posttemporal is actually a piscine skull bone at the terminus of the occiput. This could be the r11–t1 position, referring to the fish trunk and homologous with the c1 position in mam- mals. The pectoral girdle bones are encoded by c1–c4 and provide cor- responding primary insertions form muscles arising from those levels. (Courtesy of Michael Carstens, MD)

The exact derivation of hyomandibula bone is not clear— it could be a combination of pharyngobranchial cartilages from both BA1 and BA2. In any case, its caudal margin of hyomandibular presents as an upside down "Y" with two projecting bones. Anteriorly, *symplectic* connects with first arch quadrate, while posteriorly, *interhyal* connects with epi- hyal. The subsequent derivatives of the second arch are as follows: *epihyal* = styloid process, *ceratohyal* = styloid liga- ment, *hypohyal* = lesser cornu, and *basihyal* = upper half of body. Third arch derivatives are greater cornu (*hypobran- chial* = greater cornu and *basibranchial* = lower half of body (Fig. 8.157).

Relationships with Thyroid Cartilage

Although undocumented, the shape of the upper thyroid car- tilage is homologous with that of the hyoid. It has a superior cornu as well which runs as far forward as the position of lesser hyoid cornu and then drops downward. The contours of the upper border could represent genetic contributions corresponding to the ceratobranchial and hypobranchial car- tilage of the fourth arch.

Correlations: Why Does Thyroid Migrate Through the Hyoid?

In the human tongue, "biologic idea" of the basibranchial projects forward to the boundary between first and second arch mesenchyme, exactly the location of the thyroid gland. The *thyroid anlage* arises as a small population specified from the ventral foregut endoderm—as few as 60 cells in the mouse. These elite commandos proliferate into the surround- ing mesoderm forming a thyroid bud characterized by a unique quartet of transcription factors (e.g. Pax-8) which are required for the development to proceed. The mesodermal bud erupts into the ventral midline of the pharyngeal floor

just at the boundary between first arch and second arch (Figs. 8.259, 8.260, 8.261, and 8.262).

At 4 weeks, thyroid descends in a tract, *represented by pretracheal fascia*, that leads it through the midline of the hyoid body, arriving at the below cricoid cartilages by 7 weeks. This the thyroglossal duct runs from level r3 to r5 to just beyond level r11. The pathway follows above the geniohyoids back to the hyoid body, i.e. to level r6–r7. From there, signals from the hypopharyngeal cord guide further midline migration. This implies that the potential migration route for thyroid may correlate with the midline neuromeric markers as far as the terminus of the strap mus- cles which can relate as far caudal as c8–t2, i.e. to a subster- nal position.

Consider this sequence of events:

Stage 10: thyroid gland begins as an endodermal thickening in the floor of the pharynx

Stage 11: thyroid now visible as a midline pouch between first arch and second arch

Stage 12: third pharyngeal arch, oropharyngeal membrane breaks down

Stage 13: fourth pharyngeal arch, thyroid bilobed, anchored to pharynx by a stalk

Stage 14: swelling appear on the floor of mouth—future tongue

Stage 15: occipital myoblasts migrate, following hypoglos- sal chord into tongue

Stage 16: thyroid detaches from pharynx and migrates downward

Stage 16: third pharyngeal pouch (dorsal) = parathyroid III

Stage 16:third pharyngeal pouch (ventral) = thymus

Stage 17: fourth pharyngeal pouch (dorsal) = parathyroid IV (7 weeks)

Stage 18: Parathyroid in place: IV is cranial and medial to III

At stage 17 (7 weeks), parathyroid III and thymus *maintain a connective tissue bridge*. A thymus plunges downward past thyroid on its way to mediastinum, dragging parathyroid III along with it. The two lobes of thymus are positioned *lateral* and *caudal* to thyroid. Thus, parathyroid III will be positioned in thyroid tissue caudal and lateral to parathyroid IV.

Correlation: Insertion of hypobranchiall muscles into the clavicle

The curious insertion of the strap muscles to sternum and clavicle has to do with the original evolutionary position of the pectoral girdle, a topic discussed further in Chap. 10. For our purposes, recall that that the pectoral girdle consists of a series of dermal bone attached to the skull directly behind the last gill arch: posttemporal, supracleithrum, postcleithrum, cleithrum, and clavicle. This is well seen in the osteolepiform fish *Cheirolepsis*. Assuming two neuromeres per arch, this puts the boundary of the upper fin/limb just at the fourth neuromere of the fish trunk. In humans, this corresponds to level C4. Since clavicle is proximal to c4 scapula, the neuromeric coding of clavicle is c1–c3. The original position of clavicle placed in medial margins together in the ventral midline, precisely where manubrium is eventually located. This was exactly the position of coracomandibularis in ancient chondrichthyans and modern sharks. For these reasons, the muscles of the infrahyoid sector of hypoglossal cord migrate downward in the midline from level c1 and then laterally as far as c3. They then proceed upward to insert on inferior border of hyoid (Figs. 8.263, 8.264, and 8.265).

Commentary: William Bemis

When Michael asked me if I would like to look over the manuscript for this book, he described it to me as a commentary on Gray's Anatomy. I said yes, I would like to see it, and he sent it along to me. Perhaps he said "commentary" in reference to a wonderful sentence in the opening of Chap. 1: "to present the rationale for a new gross anatomy of the embryo." After spending a couple of weeks reviewing his text and figures, I reached the conclusion that Michael had greatly understated the case.

The work goes far beyond the rationale for a new gross anatomy of the embryo and is much more than a commentary. The manuscript is a synthesis that could only have been written by a person deeply curious about developmental and evolutionary morphology and how these fields illuminate human craniofacial anatomy and restorative surgeries. The book treats all cranial organs, their nerve supplies, vasculature, and embryology in due course, just as we did when I taught human anatomy long ago following the approach of many prior generations. But Michael's treatment does not stop there, for he goes on to explore the progress in developmental genetics that helps us understand the underlying patterns and the organization of human cranial components.

Perhaps most surprising in a book ostensibly providing a new gross anatomy of the embryo is the incorporation of new information about the evolutionary history and morphology of vertebrates including sharks, bony fishes, Devonian tetrapods, early amniotes, and mammalian fossils.

I have been privileged to know many excellent vertebrate anatomists, developmental biologists, phylogeneticists, and paleontologists. But none of them could bring all four of those areas into a synthetic treatment of this scope, let alone relate them to the real-world problem of surgically correcting craniofacial anomalies. This manuscript represents thinking so far outside of the box that the book will deserve a place on the shelf of vertebrate biologists everywhere.

A Note from Dr. Carstens

William Bemis

Professor Bemis at Cornell likes to go by "Willy" and that is totally appropriate. Although I have never had the pleasure of walking into his office, the exuberance of his personality is matched only by the length of his beard. He is the quintessential college professor you never forget and with whom you know that the office door will always be open for you. But he is also the author of the single book that changed my professional life and made the one in your hands become a reality. "Comparative Anatomy of the Vertebrates: An Evolutionary Perspective" drove home an important lesson: one cannot understand human adult anatomy without knowing both its development and its prehistory. The final form of our structures and systems must be understood in the context of other vertebrate species extent and of their historical precursors through the passage of time. Consider the humble ossification centers of any bone (the occipital bone, for example), described in small print in Gray's Anatomy. Do these not represent separate, previously autonomous bone fields which have been melded together in the distant past to create the structure before our eyes today? Bemis text was riveting; each chapter brought new insights into structures that I thought I knew … only to find something different more relevant. Although he wears multiple hats, Willy Bemis lives and breathes his subject, be it paleontology, evolutionary biology, genetics, or comparative anatomy. When I discussed this project with him, very much aware of my own ignorance, he straightaway sent me all the slides from his book and threw open his mind (and his time) to help me out. As a scientist, he would be better seen in khakis and a butterfly net than with a lab coat. But I am con-

vinced that any medical curriculum with a serious interest in the developmental biology of the twenty-first century and beyond would do well to make Willy Bemis' text required reading for all premedical curricula. For it is this approach that will make the final interface with genetics and molecular biology. Without Willy Bemis, Chapter 8 (indeed all of them) could never have been imagined. My only regret is not to have had Dr. Bemis as my professor or, nowadays, as that eccentric but delightful neighbor just around the corner.

21. Radlanski RJ, Renz H, Zimmermann C, Schuster FP, Voigt A, Heikinheimo K. Chondral ossification centers next to dental primordia in the human mandible: a study of the prenatal development ranging between 68 to 270 mm CRL. Ann Anat. 2016;208:49–57.
22. Mitzoguchi I, Toriya N, Nakao Y. Growth of the mandible and biological characteristics of the mandibular condylar cartilage. Jpn Dent Sci Rev. 2013;49:139–50.
23. Findik Y, Baykul T. Ectopic 3rd molar in the mandibular sigmoid notch: a case report and literature review. J Clin Exp Dent. 2015;7(1):e133–7. https://doi.org/10.4317/jced5187.
24. Badel T, Savić-Pavačin I, Zadravec D, Marotti M, Krolo I, Grbeša D. Temporomandibular joint development and functional disorders related to clinical otologic symptomatology. Acta Clin Croat. 2011;50:51–60. https://www.researchgate.net/publication/5154239

References

1. Huang R, Zhi Q, Izpisua-Belmonte J-C, Christ B, Patel K. Origin and development of the avian tongue. Anat Embryol. 1999;200:137–52.
2. Carlson BM. Human embryology and developmental biology. 5th ed. Elsevier; 2014.
3. Gilbert SF, Barresi MJF. Developmental biology. 11th ed. New York: Sinauer Associates/Oxford; 2020.
4. Clack JA. Gaining ground: the origin and evolution of tetrapods. 2nd ed. Indiana University Press; 2012. ISBN 0-253-35675-X
5. Linzey DW. Vertebrate biology. JHU Press; 2012. p. 150–1.
6. Morris-Kay GM. Derivation of the mammalian skull vault. J Anat. 2001;199:143–51.
7. Pal GP. Anatomical note: variation of the interparietal bone in man. J Anat. 1987;152:205–8.
8. Srivastava HC. Development of ossification centers in the squamous portion of the occipital bone in man. J Anat. 1977;124(3):643–9.
9. Koyabu D, Maier W, Sánchez-Villagra MR. Paletontological and developmental evidence resolve the homology and dual embryonic origin of a mammalian skull bone, the interparietal. Proc Natl Acad Sci. 2012;109(32):14075–80.
10. Bouthillier A, van Lovern H, Keller J. Segments of the internal carotid artery: a new classification. Neurosurgery. 1996;38(3):425–38.
11. Fritzch B, Baral KF, Lomax MI. Early embryology of the vertebrate ear. In: Rubel WE, Poppoer AN, Fay RR, editors. Development of the auditory system (Springer handbook on auditory research), vol. 9. New York: Springer; 1998. p. 80–145.
12. Gulya AJ. Anatomy of the temporal bone with surgical implications. 3rd ed. New York: CRC Press; 2007.
13. Benton MJ. Vertebrate paleontology. 4th ed. Blackwell; 2015.
14. Kardong K. Vertebrates: comparative anatomy, function, evolution. 7th ed. New York: McGraw-Hill; 2015.
15. Mall FP, Kiebel F. Manual of human embryology, vol. 1. Philadelphia: JB Lippincott; 1910.
16. Mall FP, Kiebel F. Manual of human embryology, vol. 2. Philadelphia: JB Lippincott; 1912.
17. Kjaer I, Keling JW, Fischer-Hansen B. The prenatal human cranium: normal and pathologic development. Copenhagen: Munksgaard; 1999. p. 22, fig 2-3.
18. Lee SK, Kim YS, Oh HS, et al. Penatal development of the human mandible. Anat Rec. 2001;263:314–25.
19. Mérida-Velasco JR, Rodríguez-Vásquez JF, De la Cuadra-Blanco C, Campos-López R, Sánchez M, Mérida-Velasco JA. Development of the mandibular condylar cartilage in human specimens of 10–15 weeks gestation. J Anat. 2009;214:56–64.
20. Spyropoulos MN. The morphogenetic relationship of the temporal muscle to the coronoid process in human embryos and fetuses. Am J Anat. 1972;150:295–410.

Further Reading

Adkins JB, Franz-Ordendaal TA. The evolutionary and morphological history of the parasphenoid bone in vertebrates. Acta Zool (Stockholm). 2016;97:255–63.
Anson BJ, Donaldson JA. The surgical anatomy of the temporal bone and ear. Philadelphia: Saunders; 1976.
Ashley-Montague MF. The premaxilla in the primates. Q Rev Biol. 1935;10:32–59.
Battail B, Surkov MV. Mammals-like reptiles from Russia. In: Benton MJ, Shishkin MA, Unwin DM, Kurochkin EN, editors. The age of dinosaurs in Russia and Mongolia. Cambridge: Cambridge University Press; 2000. p. 86–119.
Baumel J, King AS, Breazile JE, et al. Handbook of avian anatomy. 2nd ed. Cambridge, MA: Nuttall Ornithological Club; 1993.
Bernard S, Loukas M, Rizk E, Oskouian RJ, Delashaw J, Tubbs RS. The human occipital bone: review and update on its embryology and molecular development. Childs Nerv Syst. 2015;31:2217–23.
Brink AS. On Aneugomphius ictidoceps Broom and Robinson. Paleontol Africana. 1956;4:97–115.
Butler AB, Hodos W. Comparative vertebrate neuroanatomy. 2nd ed. Wiley-Liss; 2005.
Chase SW. The early development of the human premaxilla. J Am Dent Assoc. 1942;29:1991–2001.
Clemente C. Gray's anatomy, 30th American edition. Philadelphia: Lea & FEbiger; 1985. p. 182–183, figs. 4-79 to 4-82.
Couly GF, Coltrey PM, Le Douarin NM. The triple origin of skull in higher vertebrates: a study in quail-chick chimeras. Development. 1993;117:409–29.
De Beer G. Development of the vertebrate skull. Oxford: Clarendon Press; 1938.
Fawcett E. The development of the human maxilla, vomer, and paraseptal cartilages. J Anat Physiol. 1911;45:378–405.
Gill FB. Ornithologym. 2nd ed. New York: WH Freeman; 1994.
Jarvik E. Basic structure and function of vertebrates. London: Academic; 1980.
Kaban LB, Mullikan JB, Murray JE. Three-dimensional approach to analysis and treatment of hemifacial microsomia. Cleft Palate J. 1981;18(2):90–9.
Kaban LB, Moses MH, Mullikan JB. Correction of hemifacial microsomia in the growing child: a follow-up study. Cleft Palate J. 1986;23(suppl 1):50–2.
Kaban LB, Moses MH, Mullikan JB. Surgical correction of hemifacial microsomia in the growing child. Plast Reconstr Surg. 1988;82:9–19.
Kardong KV. Vertebrate: comparative anatomy, function, and evolution. 8th ed. New York: McGraw Hill; 2018.

Latham RA. Maxillary development and growth of the septo-premaxillary ligament. J Anat (London). 1970;107:471–8.

Latham RA. Development, structure, and growth pattern of the human mid-palatal suture. J Anat (London). 1971;108:31–41.

Laurin M, Reisz RR. The osteology and relationships of *Tetracdratops insignis*, the oldest known therapsid. J Vertebr Paleontol. 1996;16:95–102.

Liem KF, Bemis WE, Walker WF Jr, Grande L. Functional anatomy of the vertebrates. 3rd ed. Belmont, CA: Brooks Cole; 2001 lateral line system. p. 406–11.

Low A. Further observations on the ossification of the human lower jaw. J Anat Physiol. 1910;44:83–95.

McBratney-Owen B, Iseki S, Bamforth SD, Olsen BR, Morriss-Kay GM. Development and tissue origins of the mammalian cranial base. Dev Biol. 2008;322:121–32.

Merchant SN, Nadol AB. Schuknecht's pathology of the ear. 3rd ed. Shelton, CT: PMPH USA; 2010.

Modesto SP. The lower Permian Synapsid Glaucosaurus from Texas. Palaeontology. 37:51–60.

Northcutt RG. The New Head Hypothesis revisited. J Exp Zool (Mol Dev Evol). 2005;304B:274–97.

O'Rahilly R, Müller F. Human embryology and teratology. 3rd ed. Wiley-Liss; 1981.

Parada C, Han D, Chai Y. Molecular and cellular regulatory mechanisms of tongue myosynthesis. J Dent Res. 2012;91(6):528–35.

Patterson C. Morphological characters and homology. In: Joysey KA, Friday AE, editors. Problems in phylogenetic reconstruction. London: Academic; 1982. p. 21–74.

Presley R, Steel FL. The pterygoid and ectopterygoid in mammals. Acta Embryol (Berlin). 1978;154(1):95–110.

Rodriguez-Vásquez JF, Mérida-Velasco JR, Verdugo-López S, et al. Morphogenesis of the second pharyngeal arch cartilage (Reichert's cartilage) in human embryos. J Anat. 2006;208:179–89.

Rodriguez-Vásquez JF, Kim JH, Verdugo-López S, et al. Human fetal hyoid body origin revisited. J Anat. 2011;219:143–9.

Romer AS, Parsons TS. The vertebrate body. Philadelphia: Saunders; 1977.

Rücklin M, et al. Development of teeth and jaws in the earliest jawed vertebrates. Nature. 2012;491:748–51.

Woo JK. Ossification and growth of the human maxilla, premaxilla, and palatine bone. Anat Rec. 1949;105:737–61.

Michael H. Carstens

Introduction

How to Use This Chapter

We are now going to embark on an exploration of how the soft tissues of the face develop and are organized. Our centerpiece will be the neuromuscular apparatus but we cannot leave out the arterial supply. This is obviously an ambitious project, and we will need to give it some structure. Cranial nerves and the pharyngeal arches are the most obvious places to start but this approach has inherent problems.

First, the face contains three systems of special sensory integration: the nose, eye, and ear. Their development, and its surgical implications, is the subject of separate chapters. Nonetheless, if we are to appreciate the neuromuscular system of the face as a whole, we cannot arbitrarily delete these organs from our discussion. So we shall deal with extraocular and auricular muscles right up front. And this will necessitate that some material will need to be repeated in Chap. 13, and for this, author requests the reader's forbearance in advance.

Second, although pharyngeal arches have long been used as a schema to understand head and neck development, the model has its limitations. Pre-gnathostome vertebrates (the agnathic fishes) had a full complement of seven branchial arches, every one having gills dedicated exclusively to respiration. The invention of jaws was a revolution because it changed the manner in which vertebrates could feed. This necessitated complex rearrangements of the first and second arches. The first two arches in gnathostomes are therefore properly termed *visceral arches*, and the remaining four or five remain *true branchial arches*.

Third, another problem with the traditional pharyngeal arch model is that it does not fully incorporate all craniofacial mesoderm, nor does it integrate neatly with the neuromeric system. As far as the first arch is concerned, its only source of muscles comes from the paraxial mesoderm (PAM) of somitomere 4. How does first arch relate (or not) to somitomeres 1, 2, and 3? We know that the bulk of pharyngeal arch mesenchyme comes from neural crest and that these populations are in genetic register with their source neuromeres. Neural crest experts [1, 2] describe arch one as being populated from three sources, r1, r2, and r3, without specifying what the neurofacial derivatives really are. Furthermore, the original Hox code, applied to the hindbrain and beyond, begins at r3. Neuromeres rostral to r3 are encoded with neo-Hox genes. Bones originating from r2 neural crest, such as the premaxilla and vomer, are not traditionally considered part of the maxillary complex, despite the fact that they share common sources of innervation and blood supply with the maxillary complex. How do they arise? Furthermore, how do we classify bones innervated by V1, such as the ethmoid complex? These questions have led some investigators [3] to postulate a *premandibular arch* (perhaps encoded by r0 and r1) as a means to integrate these mesenchymal tissues and place them in an evolutionary context. We can consider the neural crest structures of the jaws as being quite distinct from the remaining tissues of first arch.

Fourth, the manner in which presomitic craniofacial mesoderm is organized is complex. Our approach has been to make use of the somitomere model of Meier because of its conceptual power. Physical evidence of mesodermal segmentation is unclear as it is the relationship among specific blocks of PAM and the pharyngeal arches. Overt subdivision into identifiable somitomeres may not exist (see review by Noden). However, an unquestionable, *homeotic gene-controlled segregation of mesodermal populations* exists; it is in register with neuromeres of the CNS. Motor and sensory neurons assigned to specific mesodermal population will originate from a neuromere with the same genetic definition. Although the physical model may be too simplistic, neuroembryology provides clinically useful way to understand craniofacial development.

M. H. Carstens (✉)
Wake Forest Institute of Regenerative Medicine, Wake Forest
University, Winston-Salem, NC, USA
e-mail: mcarsten@wakehealth.edu

Fifth, developmental events at the *transition zone between the pharyngeal arches and occipital somites* receive little attention in standard texts. The intention of this book is to describe this region in a consistent, biologically reasonable, and conceptually understandable basis. Here are a few key soundbites. Mammals produce six aortic arch arteries but five pharyngeal arches. Padget's studies of human embryos clearly demonstrate that AA6 is *not* a pharyngeal arch structure. AA5 disintegrates leaving the "orphaned" fifth pharyngeal arch dependent for blood supply upon that of the fourth arch. Now the muscle supply for each arch comes from only one (and only one) somitomere. This 1:1 relationship continues backward along the entire somite system. PA4 and PA5 are thus in genetic register with somitomeres 8–9. These two arches are involved in the formation of the larynx, clearly not a somitic structure. At the same time, Sm8–Sm9 are transformed into the first and second occipital somites. These in turn contribute to the posterior cranial fossa and muscles of the hypoglossal cord, i.e., the tongue.

Finally, some comments are on illustrations and references. This chapter is notable for a relative dearth of images. This is to avoid redundancy with the exquisite (spelling correction) figures depicted in Gray's *Anatomy*. Developmental considerations are covered in depth in Gilbert's Developmental Biology, 11th ed (Chaps. 17 and 19), Kardong, Liem, McLoon, Michailovici, and Noden (see Bibliography).

A Working Agenda

Craniofacial muscles can be functionally categorized based on the spatial organization of their motor nuclei within the brainstem. (1) *Axial muscles* are supplied by the *medial motor column* (MMC). These muscles control the eye and the tongue, structures unrelated to the pharyngeal arch system. Thus, axial muscles are *non-branchiomeric*. (2) *Paraxial muscles* are supplied by the *lateral motor column* (LMC). These muscles exist in two functional classes: (a) *branchiomeric muscles* develop in register with an arch and (b) *presomitic muscles* of soft palate, pharynx, larynx, and upper esophagus are hybrids. Although their origins are branchiomeric, their behavior is somitic. More on this later.

The medial motor column is discontinuous in midbrain and hindbrain. It has two zones. *Rostral MMC* supplies the extraocular muscles. These are controlled by discrete motor nuclei located in m1, m2, r1, and r4–r5. To understand these muscles, we must learn about the developmental "map" of the sclera. Muscle-binding sites for the extraocular muscles result from the sequence by which the sclera is vascularized. This material will cross-reference Chaps. 6 and 7 dealing with the vascular supply to the forebrain. *Caudal MMC* supplies the muscles of the tongue. These are controlled from a single motor nucleus, the hypoglossal nucleus, which spans the entire caudal medulla from r8 to r11.

The lateral motor column is organized in a similar manner. *Rostral LMC* supplies the branchiomeric muscles of the first and second arches. These are controlled by discrete motor nuclei located in r3 and r4–r5. In contrast, *caudal LLC* supplies both branchiomeric and presomitic muscles. These are associated with third, fourth, and fifth arches. Their muscles are controlled by a single motor nucleus, *nucleus ambiguus*. It too spans the caudal medulla from r6 to r11. We shall see that ambiguus sends somatic motor fibers into glossopharyngeal nerve (r6–r7), to vagus nerve (r8–r9), where it travels to pharynx and larynx via X. Ambiguus continues into spinal cord as the spinal accessory nerve. We shall see that the same concepts of motor columns and functional muscle classification are also powerful way to understand the anatomy of the neck.

How to Survive This Chapter

Neurology Not everyone's favorite subject, cranial nerve anatomy is nonetheless critical because it represents the "wiring diagram" for each of the pharyngeal arches. It is key to understand what functions each nerve plays. Pay attention to the layout of the nuclei in the brainstem. Sensory branches define bone fields and skin fields. They also explain arterial anatomy, in particular the stapedial system. Motor branches reflect the "flow pattern" of myoblasts as they stream into position. This information is what determines proximal versus distal insertion.

Myology OK, we *do* have to talk about individual muscles. Although reading one-by-one descriptions of muscles might seem like a cross between cold oatmeal and burnt toast, in reality what we are trying to establish is the temporal–spatial sequence of muscle development. The key to this is (of course) the layout of the motor nerves. As you have been forewarned, our neuromeric model contradicts traditional definitions of anatomy. Origins are strictly defined by the neuraxis. We will try to make sense out of insertions as well: Where are the proximal and distal limits of the muscle? Which part of the muscle inserts first? **And how does it know where to insert in the first place?**

Angiology Pharyngeal arch derivatives have two unifying sources of vascular supply: the external carotid system and the stapedial system. The arteries of the ECA are "holdovers" from the original aortic arches. Pharyngeal arch derivatives "stolen" from the first two arches to make jaws require a new means of perfusion: the stapedial artery system programmed by the sensory branches of V2 and V3. Frontoethmoid mesenchyme from mesencephalic neural crest, not traditionally considered to come from first arch—or perhaps the remnant of the Kuratani's putative premandibular arch—is also supplied by the stapedial system, with branches associated with V1. the development of which is timed with the outgrowth of cranial nerves. For each arch, we shall see how these two types of arteries play "mix and match" to create a final vascular anatomy.

Development In this section, the goal is to create a dynamic picture in your mind of how the muscles of each arch are assembled and how they relate to surrounding bone and soft tissue structures.

So as you can see, you can use each section in two ways. You can start with the static description of how pharyngeal arches are organized emphasizing individual components. You can then follow up this with the dynamic story of how development takes place. Reversing the order is also OK. I have used similar subheadings throughout the chapter so that you can pick and choose or (if you have the patience) read straight on through.

At the end of the day, what we should hope to accomplish is a perspective on craniofacial development in four dimensions. This is not a simple task but it is very much worth the effort. Once we can visualize how structures form in space and time, we can more readily grasp what can go wrong with the various mechanisms, what the anatomic consequences of a breakdown in the system will be, and how these might be surgically addressed.

A Note on Illustrations This chapter works best if you have your trusted copy of Gray's anatomy on hand. We will just concentrate here on giving you an interpretation of how the system works.

Before We Begin: Six Big Picture Ideas for Review

Big Picture Idea #1
Neural crest cells entering the face come from three different sites with three different migration patterns.

Forebrain: *Prosencephalic Neural Crest* (PNC) > Fronto-Orbito-Nasal Skin
The neural folds above the forebrain have two different zones, anterior and posterior. Tissues from these sites create the skin of the forehead, nose, and upper eyelid. Epidermis arises from *non-neural ectoderm* (NNE) of the anterior zone. Dermis arises *neural crest* of the posterior zone (PNC). PNC develops from p1 to p3. It migrates forward underneath NNE as a *single sheet* with distinct genetic zones. These correspond to Tessier cleft zones 10–13. FNO skin is unique; *frontonasal dysplasias* seen in the upper face are a consequence of its development. The explanation for the absence of neural crest in the anterior neural folds has to do with signals elaborated from the underlying prechordal mesoderm.

Midbrain: *Mesencephalic Neural Crest* (MNC) > Upper Face
MNC develops from m1 to m2 from midbrain and r0–r1 from isthmus. We will refer to these neuromeres collectively as MNC or r1. These four populations travel forward in *streams* over the lateral aspect of the forebrain to reach the midline. They produce the entire V1-innervated dura. MNC bones are as follows: the sphenoid complex (except alisphenoid), the ethmoid complex, the lacrimal bone, the frontal bone, and the membranous bones of the orbital series. When the optic cup evaginates from diencephalon, it becomes coated with MNC. The fields of MNC are supplied by V1-induced branches of stapedial ophthalmic axis. These neuroangiosomes are referred to as StV1.

Hindbrain: *Rhombencephalic Neural Crest* (RNC) > Midface
RNC develops from r2 to r11 in the hindbrain and migrates primarily into the five pharyngeal arches (some populations also serve meninges and the otic capsule). Each arch is supplied by a pair of rhombomeres. Blood supply for all these derivatives is external carotid. In the first arch, a separate population of RNC from r2 and r3 is responsible for synthesizing the mandible, zygomaticomaxillary complex, and alisphenoid. These *non-pharyngeal arch jaw fields* are supplied by V2 and V3-induced branches of the stapedial maxillomandibular axis. These neuroangiosomes are referred to as StV2 and StV3.

Big Picture Idea #2
Developmental fields can be lumped into three groups: A + B + C

The face is constructed from three blocks of fields on either side of the midline. We can classify the by their blood supply and innervation. This requires understanding of the stapedial arterial system.

A field neuroangiosomes consist of an *FNO skin coverage* and *bone fields of the anterior cranial fossa and medial-superior orbital walls*. The epidermis contains *nasal and optic placodes*, islands of specialized non-neural ectoderm that form the nasal cavity and complete development of the globe. The MNC bone fields are as follows: sphenoid (except alisphenoid), the ethmoid complex, the frontal bone, and the orbital series (lacrimal, prefrontal and postfrontal). These are innervated by V1. The blood supply is V1-induced branches of the stapedial ophthalmic axis.

B field neuroangiosomes are **non**-*pharyngeal arch bone fields* of the jaws and supporting bones: mandible, vomer, premaxilla, maxilla, palatine, zygoma, and alisphenoid. These are innervated by V2. The blood supply is V2-induced branches of the stapedial maxillomandibular axis. Mandible, supplied by StV3, is not included in our discussion.

<u>C field neuroangiosomes</u>: *pharyngeal arch tissues* (except the jaws). These are innervated by V2. The blood supply is from the branches of external carotid. Muscles of mastication are innervated by V3 and VII. Facial muscles are innervated by VII.

Big Picture Idea #3

The stapedial artery system is the "Rosetta Stone" to understand the classification of craniofacial clefts

The stapedial artery system is one of the great innovations of evolution, dating back to the invention of jaws. [This fascinating history is detailed as an addendum.] Its various branches are programmed by the trigeminal nerve complex. The distribution of stapedial is vast, but very specific; it encompasses all fronto-nasal-orbital structures, the jaws and the dura (see Table 9.1). Deficiencies in this system are the basis for craniofacial cleft states.

Big Picture Idea #4

Subsequent development of the stapedial follows cranial nerve VII

Aortic Arch Arteries: Precursors of the Stapedial System

Human embryos possess five pharyngeal arches (PAs) with which craniofacial structures are assembled. Running through the core of each arch is a vascular axis, *aortic arch artery* (AA), that spans from the cardiac outflow tract tucked below the pharynx upward to paired dorsal aortae. The first four pharyngeal arches are important. PA5 is diminutive, producing the arytenoids, cricoid, and associated muscles. For this reason, the fifth aortic arch artery involutes almost immediately, making PA5 dependent upon the blood supply of PA4 for its survival. *There is no sixth pharyngeal arch.* Instead, the fate of the sixth and final aortic arch artery is to form the pulmonary circulation.

Note: Textbook illustrations of the aortic arches are misleading. We can easily understand how they work by means of a little *embryonic origami*. Fact: the first aortic arch artery does *not* "sprout forth" from the cardiac outflow tract. At stage 8, the embryo is still a flat trilaminar disk. The embry-

onic heart field is located in front of the future brain and is connected to the two dorsal aortae that run passively backward along the length of the embryo. At stage 9, explosive growth of the forebrain forces the embryonic head to flex downward almost 150°. The hinge point, located at the midbrain, is called the *mesencephalic flexure*. The heart is now positioned under the neuraxis, specifically directly below posterior pharynx. Because the dorsal aortae are attached to the ventricles of the heart, in the process of folding they become stretched downward and backward, forming bilateral arcs connecting the heart below with the embryo above. They are now termed the *first aortic arch arteries*. The remaining five aortic arch arteries develop one per stage. Each arch represents the union of two distinct stems. From above, the dorsal aortae send out ventrally directed stems while, from below, the outflow tract produces paired dorsally directed stems. The two sets of stems unite to produce the aortic arch artery. Failure of this anastomotic process explains the following: (1) the failure of AA5 and (2) the breakdown and reorganization of the system.

The six aortic arches appear in craniocaudal sequence in stages 9–14. The pharyngeal arches follow the same sequence but become identifiable one stage later. After their formation in stages 10–11, the first and second pharyngeal arches merge together in stages 12–13, thus creating the tissue masses that will complete the construction of the face. In the process, their aortic arch arteries disintegrate, leaving behind arterial remnants dangling from the dorsal aortae. *Hyoid artery*, the dorsal remnant of AA2, will become the source of the stapedial system.

By the end of the pharyngeal arch period (stage 14), AA3 is transformed into the common carotid; it gives off the external carotid artery and then continues onward as the extracranial internal carotid. AA4 produces paired subclavians and, on the left side, the aortic arch. The two dorsal aortae into which AA4 inserts have merged into a single descending aorta. As a result of embryonic growth and spatial rearrangement of the pharyngeal arches, the hyoid artery becomes the *first branch of the extracranial internal carotid artery* just below the otic capsule. It enters the r4–r5 mesenchyme of what will be tympanic cavity. Hyoid artery morphs into the stem of stapedial system.

The reader will note that, at this point, the skull has not yet developed. Osteogenic mesenchyme subsequently condenses around pre-existing soft tissue structures. The various fissures and foramina of the skull are a result of this process.

Big Picture Idea #4

Sensory nerves of the trigeminal system accompany and program the individual arteries of the stapedial. The neuroangiosomes that result delineate the migratory pathways of neural crest to the face and dura.

Table 9.1 Vascular supply to the CNS by embryonic regions

Field	NCrest	Neuromeres	Blood supply	Derivatives
A	PNC	p3–p1	StV1	Fronto-naso-orbital dermis
A	MNC	m1–m2, r0–r1	StV1	FNO dura, mesenchyme, and bone
B	RNC	r2–r3	StV2/ StV3	Jaws and supporting bone fields
C	RNC	r2–r11	Ext carotid	All other pharyngeal arch structures

How the Trigeminal Nerve Innervates the Dura and Programs the Stapedial System

During stage 14, the cranial nerves develop rapidly. They (especially trigeminal and facial) determine the trajectories of the stapedial system. The innervation pattern is all important for programming the branches of the stapedial system. *Of particular importance is the anatomic pathway for trigeminal supply to the dura.* After leaving the trigeminal ganglion, V1 and V2 travel forward in the lateral wall of the cavernous sinus while V3 descends directly out of the skull. Note: *the most proximal sensory branch of each part of trigeminal is dedicated to the dura.*

The following facts about trigeminal are relevant to understand how these nerves conduct the blood supply to the orbit:

- V1, upon exiting the cavernous sinus, splits into two branches. The anterior meningeal branch enters the orbit lateral to superior orbital fissure. It then passes through the orbital plate via an un-named foramen to supply the dura of the anterior cranial fossa. The orbital branch runs medially to enter the superior orbital fissure; it then proceeds to do two things, in the orbit, V1. It forms the three primary sensory branches which will program the stapedial arteries of the orbit. It supplies the anterior cranial via a separate *dural branch of anterior ethmoid nerve* that re-enters the cranial cavity via the cribriform plate.
- V2, before it exits the skull via foramen rotundum, gives off the *middle meningeal nerve* to the postorbital dura of anterolateral frontal lobe behind alisphenoid.
- V3, after exiting the skull via foramen ovale, gives off the *recurrent meningeal nerve* supplying the temporo-parietal-occipital lobes. This requires that the nerve backtrack upwards in company with the middle meningeal artery, with both structures entering the skull via foramen spinosum.

The Stapedial Artery Stem Divides Inside Tympanic Cavity

The stem of stapedial is the remnant of the primitive *hyoid artery* to the second pharyngeal arch. When the second aortic arch to PA2 disintegrates at stage 12, it leaves behind a dorsal remnant, the hyoid artery, dangling from the dorsal aorta. With growth of the embryo, the hyoid is repositioned backward toward the otic capsule where it eventually is identified as the sole extracranial branch of ICA.

At stage 14, the definitive circulation of the internal carotid is established. The hyoid artery remains extracranial but it gives off the *stapedial stem* which immediately tracks upward into the tympanic cavity where stapes form around it (henceforth the name). Directly beyond stapes, the artery bifurcates.

- The upper division divides again, with one branch directed backward and the other branch directed forward. These branches supply the developing meningeal mesenchyme. Of note, the anterior branch is commonly (and confusingly) called *supraorbital artery*, which is not to be confused with the frontal artery inside the orbit. Supraorbital runs forward to the trigeminal ganglion where it picks up V1 and V2, just as these nerves exit the cavernous sinus. It then heads straight for the orbit.
- The lower division follows chorda tympani out from the tympanic cavity into the face via the petrotympanic fissure. It then picks up V3 sensory nerve just below foramen ovale. All subsequent branches of the stapedial system follow the trajectories of V1, V2, or V3.

Upper Division Stapedial projects forward to the orbit and upward to V supply 1 dura and V2 dura.

The stem artery follows greater petrosal nerve forward to the trigeminal ganglion. Here, it picks up cranial nerves V1 and V2 as they run forward in the lateral wall of cavernous sinus. Immediately upon leaving the sinus, both V1 and V2 give off *dural branches* prior to exiting the skull. Upper division stapedial artery bifurcates to vascular support for all branches of V1 and V2. These arteries supply (1) their respective areas of dura and (2) the orbit.

As mentioned, StV1 enters the orbit in two places. The orbital branch is responsible for programming the ethmoid, frontal, and lacrimal arteries but lacrimal is also joined by meningeal branch in the lateral orbit just next to cranio-orbital foramen. The medial branch pursues V1 nerve into superior orbital fissure. StV1 becomes identifiable at stage 18, enters the orbit during stage 17, and forms the orbital vessels. At stage 20, "annexation" to primitive ophthalmic occurs. Flow is much greater from the ophthalmic stem than either supraorbital stem. Both StV1 branches orbital and meningeal involute backward. Situations exist, however, where the anastomosis to ophthalmic fails. In this case, the entire orbital system is supplied via the lateral branch which retains its connection to middle meningeal as the so-called *recurrent meningeal artery.*

The final product, "ophthalmic artery," *is a hybrid system.* Primitive ophthalmic, being derived from ICA, supplies CNS tissue exclusively, i.e., the optic nerve and globe. All remaining tissues of the orbit innervated by V1 are supplied by branches from the derivatives of the original stapedial system. In the meantime, StV2 supplies the dura to the infero-lateral middle cranial fossa. *At no time does V2 or StV2 enter the orbit.*

- Meanwhile, Back at the Ranch At stage 18, ascending branch from maxillary artery, the middle meningeal artery, connects external carotid to the supra-orbital at the level of the trigeminal ganglion (*vide infra*). By stage 19, the change in flow causes supraorbital to "die backward" from the ganglion to tympanic cavity. StV2 is now converted into the *anterior branch of the middle meningeal artery*. StV3 subsequently defines the *posterior branch of middle meningeal artery*.

Lower Division Stapedial forward to the jaws and upward to the V3 dura

As soon as chorda tympani departs from the tympanic cavity, it makes a beeline for the sensory root of V3, where it seeks out lingual nerve by which to convey itself to the tongue. Lower division stapedial tracks along with the nerve. At the same time, the poorly named maxillary branch from external carotid tracks forward toward V3. At stages 18–19, an anastomosis between lower division stapedial and ECA creates the hybrid *maxillomandibular artery* (MMA).

MMA has two functions and three distinct zones. Branches associated with the external carotid supply all original structures of the first arch, such as the muscles of mastication, fat, and glands. Branches associated with the stapedial system supply those structures *reassigned* from the first arch: jaws, the suspensory bones of the maxilla, and the dura. The zones of MMA are as follows:

- Proximal (*mandibular*) zone gives off stapedial derivatives that re-enter the skull to supply the tympanic cavity and the dura of the middle and posterior cranial fossa. It also sends StV3 inferior alveolar downward to supply the mandible. The sole ECA branch of the proximal zone supplies mylohyoid.
- Middle (*infratemporal*) zone branches are distributed exclusively to muscles of mastication, i.e., to non-neural crest structures.
- Distal (*pterygopalatine*) zone gives off StV2 branches in the pterygopalatine fossa that are subsequently distributed to the maxilla and its suspensory bones.

Big Picture Idea #5

Anastomoses to the stapedial system are responsible for its dissolution

Reunification and Disappearance of the Stapedial System (See Figs. 7.34, 7.35, 7.36, and 7.37)

Toward the end of the embryonic period, the anatomy of the stapedial system is drastically altered, making it virtually unrecognizable. Three anastomoses are responsible for these changes.

- Stage 17 inferior division stapedial joins ECA at V3 ganglion to form hybrid *maxillomandibular artery* (MMA).
- Stage 18–19 middle meningeal connects supraorbital.
- Stage 20 StV1 orbital branch is annexed by ophthalmic.
- By stage 21, supraorbital is completely eliminated.

- Inferior division of stapedial distal to stapes persists as the *anterior tympanic artery*.
- Proximal to stapes, the stapedial stem persists as the *caroticotympanic artery*.

Fate of the Intracranial Stapedial Derivatives

Supraorbital supplies StV1 and StV2.

V1 stapedial first gives off a branch to the dura of the anterior cranial fossa. Once in the orbit, it gives off three branches, each of which supplies structures both within the orbit and outside of the orbit.

- Ethmoid (nasociliary) supplies zones 13–12 composed of the ethmoid complex, the nasal envelope, and the medial forehead between the eyebrows and medial upper eyelid.
- Frontal (supraorbital) supplies zones 11–10 consisting of the orbital roof, the central upper eyelid, and lateral forehead defined by the eyebrows.
- Lacrimal supplies zone 9 which includes lateral orbital plate, the lacrimal gland, and then lateral corner of the upper eyelid.

V2 stapedial supplies trigeminal V2 nerve as it traverses the pterygoid canal en route for pterygopalatine fossa.

- Dural branch supplies zone 9 sphenoid wing originally arose as the *epipterygoid bone*, part of the ancient *palato-quadrate cartilage* from which maxilla evolved. Additional support for alisphenoid comes from temporalis muscle via *anterior deep temporal branch* of the external carotid component of maxillomandibular artery. In sum, Tessier zone 9 is complex, supplied by the confluence of three distinct systems.

Fate of Extracranial Stapedial Derivatives

V2 maxillomandibular artery gives off branches to the maxillary complex and palate all of which are V2 induced. Thus, the arteries supplying the two fields of zygoma, jugal and postorbital (see below), arise in the pterygopalatine fossa.

- Zygomaticofacial and zygomaticotemporal supply Tessier zones 7 and 8.
  - Lacrimal gland receives blood supply from StV2 zygomatic (which brings PANS control as well as from StV1 lacrimal to the capsule and to the ductal epithelium penetrating the gland from the conjunctival recess).
  - All oral salivary glands arise from either r2 or r3 neural crest.
- Superior alveolar supplies Tessier zones 6, 5, and 4.
- Lateral nasopalatine and descending supply anterior and posterior sectors of Tessier zone 3.

- Medial nasopalatine supplies Tessier zones 2 and 1.

  V3 stapedial supplies two functionally related zones.

- Inferior alveolar supplies the dental units and the ramus whereas
- Anterior tympanic refers backward to supply the eardrum, malleus, and incus. In so doing, these reflect the paleohistory of the mandibular bone complex in which its proximal components such as articular were brought backward into the skull as structures of hearing.

Big Picture Idea #6

Two embryonic arterial systems, stapedial and internal carotid, recombine to supply the extra-ocular structures of the orbit. Two arterial systems, stapedial and external carotid, recombine to supply the first pharyngeal arch.

- In the orbit, ICA supplies the globe; stapedial system supplies everything else.
- In PA1, stapedial supplies the jaws and bones connecting the jaws to the skull. ECA supplies everything else (Table 9.2).

Useful Terminology: Read These First

Developmental field/neuroangiosome: composite block of tissues of supplied by a single neurovascular pedicle.
A Fields: all structures innervated by V1 and supplied by branches of the stapedial ophthalmic artery (StV1).
B Fields: all structures innervated by V2 or V3 and supplied by branches of the stapedial maxillomandibular artery (StV2/StV3) via the pterygopalatine fossa.
C Fields: all structures innervated by V1 or V3 and supplied by branches of the external carotid artery.

Table 9.2 Stapedial development by embryonic stages

Stage	Size (mm)	Vascular features
13	4–6	Hyoid stem, primitive maxillary supplies eye
14	5–7	Primitive dorsal (temporal) ophthalmic from distal ICA, stapedial is born
15	7–9	Primitive ventral (nasal) ophthalmic from proximal ICA
16	8–11	Ventral pharyngeal artery (precursor of ECA) arrives at V3
17	11–14	prDOA and prVOA unite to make ophthalmic, stapedial enters tympanic cavity
18	13–17	Tympanic vessels visualized
19	16–18	Supraorbital enters the orbit, maxillomandibular joins ECA
20	18–22	Stapedial stem involution: anastomoses

Stapedial Artery: Bilateral stems representing the dorsal remnants of the second aortic arch *hyoid artery* are translocated backward along the dorsal aortae to junction of third aortic arch artery. These represents the extracranial portion of internal carotid artery. The stapedial is derived from hyoid artery at stage 13. It represents the exclusive extracranial branch of ICA. It promptly enters the skull through the tympanic cavity, passing through the stapes.

Stapedial Ophthalmic System (STV1): Following trigeminal V1 to the orbit, stapedial forms multiple branches to all extraocular structures of the orbit. These supply the ethmoid sinuses and upper nasal walls. Terminal branches of StV1 exit to the face to supply all fronto-naso-orbital soft tissues.

Stapedial Maxillo-Mandibular System (StV2/StV3): Extracranial stapedial follows chorda tympani to the sensory root of V3 where it anastomoses with the maxillary branch of external carotid. This forms a hybrid maxilla-mandibular artery with three segments. Proximal or mandibular segment of MMA has four epaxial stapedial branches directed into the skull, one hypaxial stapedial inferior alveolar, and one ECA branch, mylohyoid. Middle or pterygoid segment has no stapedial branches, two ECA epaxial branches to masseter, and two branches to the pterygoid muscles. Distal or pterygopalatine MMA has exclusively stapedial branches; ECA is not represented.

Stapedial intracranial system (StV1, StV2, StV3): After entering the skull, stapedial branches anteriorly and posteriorly. It forms individual dural arteries in company with sensory branches of trigeminal nerve. It sends out a separate descending branch that exists the skull in company with chorda tympani.

Neuromere: Developmental compartment of the brain, extending from the floor to the roof of the neural tube. Each neuromere is defined anatomically by a unique expression pattern of homeotic genes that specify its anterior–posterior location along the neuraxis.

Embryonic Brain

Begins with three parts and then separates into five parts.
Prosencephalon (forebrain): six prosomeres (p6–p1)
   Secondary prosencephalon (hypothalamic fields + telencephalon): hp1 and hp2
   Diencephalon (optic apparatus, epithalamus, thalamus, hypothalamus): p3–p1
Mesencephalon (midbrain): two mesomeres (m1 and m2)
Rhombencephalon (hindbrain): 12 rhombomeres (r0–r11)
   Metencephalon (isthmus and pos, cerebellum): r0–r7
   Myelencephalon (medulla): r8–r11

A Fields: StV1 Naso-Orbital Arterial System, AKA (sic), StV1 Ophthalmic

The architecture of the A fields is defined by the perfusion pattern of the various branches from V1 stapedial after its anastomosis with primitive ophthalmic from ICA. Because this developmental anatomy is not widely appreciated, the entire collection of ICA vessels to the globe and non-ICA vessels to extraocular structures has traditionally been lumped together as the ophthalmic arterial system. Primitive ophthalmic artery has a misleading name. This nomenclature dates back to Padget's original Figs. 4–6. Coming from the Circle of Willis forward, this branch should probably be labeled naso-ophthalmic artery. It is actually the *terminal branch of intracranial ICA* because all the arteries to the cerebrum have their take-off from a more proximal branch. Furthermore, naso-ophthalmic, after supplying the eye, terminates in the intracranial midline to supply key soft tissue structures such as the olfactory bulbs.

The anatomy of the orbital arteries is extensively discussed in the vascular chapters (Chaps. 6 and 7). What follows are the most relevant points.

For the reader, it may seem a fine point that V1 stapedial be considered a *non-ICA derivative*; after all, it originated from third arch ICA at its juncture with dorsal aortae. But its history reflects the migration of MNC into the orbit and nasal midline which is separate and distinct from the original blood supply to the eye directly off of the most anterior ICA. And the purpose of the StV1 system is to supply non-CNS tissue of MNC and somitomeric origin.

The cutaneous manifestations of the A fields are limited to those soft tissues perfused by the terminal external nasal, upper eyelid, and forehead branches StV1 naso-ophthalmic. The most caudal cutaneous derivatives of anterior ethmoid are columella and philtrum (Cupid's bow). When considered in cross-section, the nasal A fields consist of the septum and ethmoid complex. The latter consists of cribriform plate, crista galli, ethmoid sinus air cells, and upper and middle turbinates.

The lacrimal bones and lacrimal apparatus constitute the anteromedial boundary of the orbital A fields. The development of the lacrimal duct *precedes* that of the bone. Duct epithelium produces programming signals for surrounding r1 MNC; consequently, the bone ensheaths the duct. Posterior to the lacrimal lies ethmoid, a bilaminar structure that constitutes at once the medial orbital wall and the lateral nasal wall with sinus air cells occupying the space in between. Note that this results from two sources of programming for r1 MNC: nasal epithelium (associated with p6) and sclera/dura (associated with p5).

The various branches of StV1 naso-ophthalmic constitute distinct developmental zones. *Ethmoid (nasociliary) branch* runs along the medial boundary of orbital roof and ethmoid complex; it is distributed to three distinct zones. (1) It supplies the ethmoid complex and nasal cavity via the anterior and posterior ethmoids. (2) Its forward continuation as trochlear supplies lacrimal bone and medial forehead. These constitute part of Tessier cleft zones #13 and #12. (3) It also sends a branch to the neural crest of the ciliary body (hence its name). The axis of *frontal branch* supplies the bulk of the orbital roof which forms part of Tessier cleft zones #11 and #10. Its forward continuation supplies the lateral forehead. *Lacrimal branch* extends to the lateral recess of the orbit where it anastomoses with StV1 meningeal branch to supply the orbit and surrounding tissues, including alisphenoid (the greater sphenoid wing).

The structure of the orbital roof is potentially bilaminar. Programming from dura above and sclera below creates a potential field separation plane for the *orbital extension of the frontal sinus*. The programming is variable: the sinus may be minimal or may occupy virtually the entire orbital roof. MNC failure in Tessier cleft zones #11 and #10 can result in an orbital roof defect and *encephalocoele*. Encephaloceles can also occur in Tessier zones #13 and #12; these are directed into the nasopharynx.

B Fields: StV2 and StV3 Maxillomandibular Arterial System

The architecture of the B fields is defined by the perfusion pattern of branches from V2 stapedial and V3 stapedial after their common stem (extracranial stapedial) makes an anastomosis with the ECA maxillary artery. Recall that intracranial stapedial gives off stem which proceeds extracranially in company with VII chorda tympani. After joining ECA, maxillary artery "passes the baton" first to V3 and subsequently to V2 to form the hybrid "*maxillomandibular artery*". Recall also (1) all the stapedial derivatives of the proximal one-third of MMA are V3 induced, (2) middle one-third of MMA has no stapedial derivatives, and (3) distal MMA picks up V2 so all arteries of this sector are StV2.

What follows are the most relevant points regarding the anatomy of the pterygopalatine fossa and its StV2 arteries.

The fields served by StV2 and StV3 are exclusively neural crest bones reassigned from the first and second arches to form jaws. This allows for V2 and V3 to serve a programming function for ECA as well. The first (mandibular) sector of MMA gives off four epaxial arteries—all of them StV3—that enter the skull. It gives off two hypaxial arteries: ECAV3 mylohyoid artery to that muscle and StV3 inferior alveolar artery to the dentary component of mandible. Dentary is the remnant of the original 9-bone structure and contains all the teeth. All the remaining fields of the mandible are C fields. They represent the dermal bones that ensheathed Meckel's cartilage and hence are supplied from external carotid.

The third (pterygopalatine) sector of MMA gives off exclusively StV2 arteries. Its first two branches, pharyngeal artery and the artery of the pterygoid canal and directed backwards to sphenoid sinus and auditory tube. They indicate the first arch component of auditory tube. The remaining branches are directed forward into the maxillary–zygomatic complex. It consists of three distinct systems—all of which subdivide into functional units. (1) *Sphenopalatine artery* is the terminal branch of the StV2 system. Via *nasal branches* it supplies inferior nasal walls, all r2 mucosa entering the paranasal sinuses, the lower part of septum, and via naso-palatine branches the medial vomer/premaxilla and the lateral inferior turbinate. (2) *Maxillary artery* also subdivides into two mis-named structures which I have subsequently re-named. *Posterior maxillary artery* (posterior superior alveolar artery) is proximal and supplies the posterolateral wall of the maxillary sinus as well as molars and the lateral wall of their alveolar housing. *Anterior maxillary artery* (anterior superior alveolar artery) is distal. It supplies the floor of the orbit and inferior oblique muscle. It then bifurcates to supply the anterior wall of maxillary sinus. A medial branch it supplies frontal process of maxilla, lateral incisor, and canine, and via a lateral branch, it supplies for the deciduous molars/adult premolars. The dental branches supply lateral wall of the alveolus. (3) *Descending palatine artery* perfuses the palatine bone and subdivides. *Greater palatine* supplies the lingual alveolus and oral lamina of palatine shelf while *lesser palatine* supplies the unilaminar horizontal palatine shelf and anterior one-third of the soft palate.

C Fields: ECAV2 and ECAV3 Arterial System

The remaining tissues of the face are of exclusively pharyngeal arch origin and are perfused by derivatives of the original aortic arch arteries that provided initial supply of the arches. We shall not consider here derivatives of the third, fourth, or fifth arches as they are all internal to the face. Tissues of interest from the first arch are the skin and mucosa, fat, and muscles of mastication in the deep plane only. Those of relevance from the second arch are glands, a limited number of muscles of mastication in the deep plane and extensive number of muscles of animation in the superficial plane.

The behavior of the first and second arches is dramatically different from the remainder. Formed at stages 10 and 11 respectively, by stage 12 they combine into an amalgamated unit, the neural crest for which comes from rhombomeres 2–5.[37,38] The consolidation of these two arches reflects the evolutionary innovation of jaws as separate functional units from the original intention of branchial arches as purely respirator/digestive units. It is also seen in the individual fates of the original aortic arches in tetrapods. AA1 and AA2 fall apart, lose their connection with the aortic outflow tract completely, and remodel depending on AA3 for survival. AA3 remains intact, becoming common carotid; it gives off the stem of external carotid and continues as the extracranial internal carotid. AA3 and AA4 retain their caudal connection with the aortic outflow tract but are disconnected dorsally with AA3 directed to the head and AA4 to the remainder of the body. AA5 involutes completely, its corresponding pharyngeal arch becoming dependent upon AA4 for survival. AA6 (no corresponding arch) remains attached to outflow tract and is exclusively dedicated to the pulmonary circulation.

The derivatives of the first and second arch complex are supplied by specific branches of the external carotid. First arch is represented by maxillary. Second arch is represented by lingual, facial, and the anterior branches of superficial temporal. Deep plane structures are perfused by the maxillary and lingual arteries, and superficial structures by the remainder. Because the facial muscles of the ear, occiput, and neck (platysma) are not part of the first/second arch complex, they are not included here. In conclusion, we once again return to the concept of somatotopic representation and spatial–temporal development. The reader is referred to the work of Gasser on stage-specific facial nerve and muscle development. The same rules hold true for deep bone structures as for their superficial soft tissue coverage. Thus, the physical disposition of the maxillary and zygomatic fields from medial to lateral and from deep to superficial is consistent with this type of linear read-out. So too is the timing of their development as reflected in the appearance of calcification with their respective functional matrices. Kjaer makes the interesting observation that all such calcifications are first arrayed around the neurovascular axes of each field.[39] Anatomic work by Ian Taylor defining the territories of *neuro*angiosomes (emphasis mine)[40,41] is likely to prove homologous!

Conversion of the linear model to three-dimensional reality would be difficult without an appreciation for the role folding in facial development. Participation of the B and C fields in the formation of the oral cavity requires approximation of the first/second arch complex toward the midline. This in turn depends upon the flexure of the embryonic neuraxis. This occurs at the level of the midbrain due to brain expansion. Curving of the embryonic face toward the future chest first occurs at stage 10. The angle between prosencephalon and rhombencephalon decreases from 180° to 150° by day 22 and to 100° day 23.[42] The mesencephalic flexure is responsible for bringing the B and C fields arch into close contact with the A fields. When the tissues are within the critical contact distance, a tightly controlled spatiotemporal sequence of fusions takes place.

Author's note: As will be discussed in detail in the chapter of muscle development (Chap. 13), the three-dimensional relationship between the eye and somitomeres bearing extraocular muscles allows for the sequential insertion into the muscles into specific recipient zones. The muscle order is as follows: Sm1 inferior rectus, inferior oblique, medial rectus, Sm2 superior rectus, levator palpebrae superioris, Sm3 superior oblique, and Sm5 lateral rectus. Sm4 "empties out" its contents in first arch, permitting Sm5 direct access to the remaining insertion zone of the globe. From these data, it is evident that these somitomeres are spatially positioned next to the globe which at first projects out 180° from the neuraxis. The sclera has a genetic program of insertion sites that "become available" in temporal sequence: inferior > medial > superior > lateral. Because the somitomeres are read out in linear order, their extraocular muscle blastemal comes to occupy the binding sites of the sclera in same sequence.

Development of Craniofacial Muscles

Until the advent of molecular embryology, striated craniofacial muscles were considered to exist in two categories: somatic and visceral. Because the latter are associated with the pharyngeal arches, and because the pharynx is in continuity with the gut, the "visceral" muscles were thought to have a phylogenetic relationship to non-striated gut muscle. Recent work has demonstrated that *all craniofacial striated muscles are alike*; they arise from paraxial mesoderm. Smooth muscle (and cardiac muscle) arises from lateral plate mesoderm. In sum, the concept of "visceral striated muscle" is false.

Traditional neuroanatomy courses describe three distinct motor columns in the brainstem. These were named using vexing acronyms of GSE, VSE, and SVE designed to torture first-year students. Now that visceral striated is passé, what shall we do with these terms? The answer is simple and intellectually satisfying. The existence of brainstem motor columns comes about from the spatiotemporal sequence in which neurons develop within the neural plate. Thus, GSE is closest to the midline and SVE furthest away. *This pattern reflects the spatiotemporal sequence by which the muscle targets develop*. The *medial motor column* (general somatic efferent) supplies all the extraocular muscles and the muscles of the tongue. The *lateral motor column* (general visceral efferent) supplies muscles assigned to the pharyngeal arches. Both muscle groups are PAM-derived; they arise close to the neuraxis. The *sympathetic motor column* (special visceral efferent) is the most lateral because it supplies structures arising from neural crest. [The visceral part is a misnomer sympathetic autonomic nervous system (SANS) may supply structures such as blood vessels and smooth muscle that arise from lateral plate mesoderm, or glands that are endodermal, but the end result of SANS activity affects somatic tissues as well as internal organs.]

Where and how do craniofacial muscles develop? Recall that paraxial mesoderm is segmented into somitomeres and somites; these have quite different structures. The former are hollow balls organized around a hollow chamber, the somitocoele. The latter have three distinct developmental zones (the well-known sclerotome, myotome, and dermatome). Bear in mind that all somites were initially somitomeres. Somitogenesis, the process by which a somitomere is converted into a somite, starts with Sm8. Although each somitomere appears to be a homogenous sphere, within its walls there exists a spatial hierarchy. Cells in one location are destined to become muscles, whereas those in another location will become bone. *The fate of each population is its relative distance (i.e., exposure) to gene products arising from the midline versus those produced within the periphery.*

Let's take this idea a bit further. Somitomeres Sm4–Sm7 are quite simplistic: midline myoblasts become midline muscles whereas lateral myoblasts become associated with pharyngeal arches. Once one comes to Sm8–Sm9 (occipital somites 1–2), a third option appears. A small amount of PAM from these two segments is associated with pharyngeal arches 4 and 5: the external and internal laryngeal muscles. These muscles are supplied by the superior and inferior laryngeal nerves, i.e., by the tenth cranial nerve. They are likely to develop and migrate into position during the somitomere stage, i.e., prior to the reorganization of Sm8 and Sm9 into occipital somites 1 and 2. All remaining myoblasts develop and migrate slightly later, within a myotome proper. Those occupying a medial position will form the pharyngeal walls; those from an intermediate position become tongue muscles; and those in the lateral myotome will contribute to the sternocleidomastoid and trapezius.

Lateral plate mesoderm flanking occipital somites S1–S4 and cervical somites S5–S7 can be considered a backward extension of cephalic mesoderm. It is supplied by neural crest from r8 to r11 and c1 to c3. These muscles, such as trapezius, represent derivatives of the ancient arches 5–7. Thus, although the muscles do not map to a somite myotome per se, their biologic behavior places them in the nucleus ambiguus and its extension into the spinal cord.

We can round out our discussion of craniofacial myoblast groups by re-visiting an old friend gastrulation. Recall that the segmentation of PAM into somitomeres results from repetitive cycles of a "cellular clock" which functions in a rostral–caudal sequence. This mechanism permits ingression of epiblast cells into the primitive streak and "assigns" them to a future neuromeric level. Each somitomere represents a cellular "pocket" associated with a specific level of the nervous system. Epiblast cells entering a given pocket are like "marbles." When the pocket is full, the cellular clock is reset, and a new cycle of gastrulation commences. All subsequent cells will be innervated by the next caudal neuromere.

Another key idea: those cells of the epiblast that enter the primitive streak at a given level do not have an intrinsic specification. Recall that, at each level, gastrulation takes place in "waves." The first cohort of epiblast cells to enter the streak are those located closest to the midline. These will become the notochord and the endoderm. The next cohort is positioned further out; these cells will become the paraxial mesoderm. Finally, the most peripheral cohort will become lateral plate mesoderm and cardiac mesoderm. This simple process goes on from head to tail, wave after wave. Those epiblast cells remaining behind have now earned the right to be called ectoderm. Once the germ layers are established, two additional processes reconfigure the anatomy of PAM. (1) Homeotic gene products establish segmentation into somitomeres and somites. (2) Interaction of gene products from the neural tube and the periphery creates spatial subdivisions into myoblast zones. In sum, spatial position within the single-layer epiblast determines the first level of fate: allocation to germ layers.

Craniofacial muscles arise from four zones of mesenchyme. The classification is based on motor innervation. The **rostral medial motor column** is found is located in m1–m2, r0, and r4–r5. It innervates the *extraocular muscles*. These arise from somitomeres 1, 2, 3, and 5. The **caudal medial motor column** is located in r8–r11. It innervates the *tongue muscles*. These arise from the four occipital somites, S1–S4. The **rostral lateral motor column** is located in r3–r5. It consists of the trigeminal motor nucleus and the facial motor nucleus. These innervate the *muscles of pharyngeal arches 1–2*; these arise from somitomeres 4 and 6. The **caudal lateral motor column** is located in r6–r11. It consists of a single motor nucleus, nucleus ambiguus that supplies two distinct groups of muscles. These are organized based on the segmental boundaries of the original branchial arch system. *Branchiomeric muscles of pharyngeal arches 3–5* retain a structural relationship to their associated arch. These arise from somitomeres 7–9. *Presomitic muscles* are muscles that have undergone extensive modifications. Some, like the muscles of the soft palate, larynx, and pharynx, maintain a physical relationship with the pharynx. They represent gill arch constrictors. *Branchiomeric muscles of pharyngeal arches 6–7* so as the laryngeal muscles, sternocleidomastoid, and trapezius are descendents of external, gill arch dilators but have moved into new anatomic territories as *cranio-appendicular muscles*.

Cervical and upper extremity muscles have a zonal innervation pattern as well. The **rostral medial motor column** is located in neuromeres c1–c4. It innervates the *occipital and upper paraspinous muscles*. These arise from somites 5–8. The **caudal medial motor column** is located in neuromeres c5–c8. It innervates the *lower paraspinous* muscles. These arise from somites 9–12. The **rostral lateral motor column** is located in neuromeres c1–c4. It innervates the *muscles of the cervical plexus*. These arise from somites 5–12. The **caudal lateral motor column** is located in neuromeres c5–c8. It innervates the *muscles of the brachial plexus*.

This chapter will examine each group of craniofacial muscles in turn. The anatomy of the cervical/upper extremity girdle muscles will be discussed in the following chapter. We will begin our discussion with a quick review of neuromeric anatomy of the brain. The four functional groups of muscles (based on motor column) will be covered in turn. Particular attention will be given to the relations each muscle has with surrounding bones. Attention will also be given to the three-dimensional anatomic course of individual cranial nerves. **Note** Muscles of the neck are discussed in Chap. 10.

Neuroembryologic Organization of the Brain: A Review

You are well aware by now that the three-part embryonic brain develops sequentially. Gastrulation itself is initiated at the r0/r1 level of rostral hindbrain and proceeds caudally. Thus, first part of the brain to develop is rostral hindbrain. Next, at the same time as the caudal hindbrain develops, genes from r0 and the prechordal plate induce the midbrain. Finally, a combination of gene products from the prechordal mesenchyme and the anterior visceral endoderm induce the forebrain. Brain–muscle relationships also follow a similar spatial pattern.

Forebrain

Mapping out the developmental units of the prosencephalon has been a major objective of researchers over the past two decades. In Chap. 5, we discussed similarities and differences between the two principal model extent: descriptive neuroanatomy (O'Rahilly and Muller) and molecular genetics (Puelles and Rubenstein). In broad brushstrokes, we conceptualize the 2-part forebrain. The caudal forebrain, or diencephalon, has three *prosomeres* (p1–p3) the major structures of which (thalamus and limbic system) are distributed into distinct basal and alar "decks" which are a continuation of the midbrain. The rostral forebrain, or secondary prosencephalon (SP), has two *hypothalamic prosomeres* (hp1 and hp2). The ventral sector of SP, the hypothalamus, directly continues the basal/alar structure of midbrain and diencephalon. It has two parts: a caudal peduncular hypothalamus (PHy) and rostral terminal hypothalamus (Thy). PHy gives rise to the two cerebral vesicles of telencephalon (Tel). Thy gives rise to the eye and postoptic areas (POA). The two hypothalamic prosomeres of SP run vertically upward from floor plate to base plate. Prosomere hp1 contains PHy and Tel while prosomere hp2 contains Thy and POA. [This system clarifies archaic, confusing terminology regarding the *hypo*thalamus and *epi*thalamus which are neither "hypo" nor "epi."]

Just as a reminder, in the original 1993 iteration of the prosomeric system, the telencephalon was divided into two prosomeres with a rhinencephalon (p6) and a non-olfactory cerebrum (p5). From these zones, two specialized sensory nerves project ventrally. The olfactory nerve (hp1) and the optic nerve (hp2) are not nerves at all; they are actually direct outgrowths of the brain. Cranial nerve I is unique because it accesses the ipsilateral cerebral cortex directly *without a relay station in the thalamus*. Cranial nerve II is likewise notable for its extensive distribution. Visual tracts from each retina project both ipsi- and contralaterally throughout the hp2 cerebrum, beginning ventrally with the eye and terminating dorsally with what will become the visual cortex of posterior cerebral lobe.

No motor nerves or somatic sensory nerved originate from the forebrain. Its structures are involved in the reception, integration, and interpretation of data from the body. Nonetheless, input from the olfactory and optic systems leads to a wide variety of motor responses, both unconscious and volitional. Survival demands complex, fast-acting responses to threats (an approaching fist) and changes in body position (performance on parallel bars). For this reason, eye motions must be tightly synchronized. For this reason, it is not surprising that the hp2 eye becomes an object of intense interest for extraocular muscles controlled by nerves from m1, m2, r0/r1, and r4–r5. We shall see how the neuroanatomic mapping of the globe provides six insertion sites for the EOMs. We shall also see how the very act of embryonic folding "presents" the globe into intimate contact with the somitomeres that contain the EOMs. Like hungry fleas, the EOM myoblasts like hop down upon the globe, each one biting into the most topologically accessible insertion site, dragging their motor nerves along with them.

Midbrain

Although the mesencephalon does not display anatomic subdivisions, O'Rahilly considers it to be composed of two mesomeres, m1 and m2. This is quite reasonable from a functional standpoint. The oculomotor nerve contained within it has two distinct nuclei: rostral and caudal, both of which reside in m1. The rostral nucleus of III supplies the muscles of Sm1: superior rectus, medial rectus, levator palpebrae superioris, and Muller's muscle. The caudal nucleus of III supplies the muscles of Sm2: inferior oblique and inferior rectus. Furthermore, the superior (optic) colliculi are found in m1 whereas the inferior (auditory) colliculi are found in m2.

Intricate relays exist between two sets of primary integration centers in the midbrain. Crosstalk between the superior

(optic) colliculi in m1 and the inferior (vestibule-cochlear) colliculi in m2 drive responses required for coordination of eye movements and the kinesthetic control of the body. In addition, a small bundle of fibers, the *medial longitudinal fasciculus*, runs in the dorsal midline from the colliculi all the way down the floor of the fourth ventricle, i.e., to the termination of the rostral hindbrain. The MLF coordinates among all cranial nerve nuclei connected to the extraocular muscles from m1 to r5.

Rostral Hindbrain

The rostral hindbrain is composed of rhombomeres r0–r5. It contains the nuclei of cranial nerves IV–VII. These nuclei are *discrete*. Note that r0 is otherwise known as the *isthmus*; it contains the *decussation of trochlear nerve*; the nucleus of IV itself resides in r1. Note that the neuroanatomy of isthmus is unclear, but it is considered part of hindbrain. In this text, we shall consider coding of r0 and r1 as co-equal. Rhombomeres r1–r3 contain the nucleus of trochlear nerve and trigeminal nuclei, with r1–r2 being sensory *only*, whereas r3 is both motor and sensory. Within r4–r5 are the exclusively motor nuclei of abducens nerve as well the motor and sensory components of the facial nerve.

The motor nuclei of rostral hindbrain come in two varieties. (1) Nuclei within the *medial motor column* are assigned to axial muscles: the eight extraocular muscles and (in some mammals) an accessory lateral rectus. **NOTE** Within the medulla, cranial nerve VI follows a complex pathway, looping around the tract of the facial nerve. (2) Those residing within the *lateral motor column* are in register with pharyngeal arches 1–3 and with somitomeres 3–7. These motor nuclei are assigned to muscles associated with the pharyngeal arches or (in the case of VII) the mimetic muscles of the face and scalp. Note that the functional pattern shared among neuromeres, arches, and paraxial mesoderm is *irregular*.

Although PA1, PA2, and PA3 all receive innervation from paired neuromeres, each arch is supplied by myoblasts from one (and only one) somitomere. Recall that the first three arches are responsible for the entire oropharyngeal cavity down to the esophagus. Perhaps this morphologic complexity demands extra neurologic control, i.e., two neuromeres instead of one. What is certain from the get-go is that a 1:1 relationship exists between all the pharyngeal arches and the mesoderm assigned to them.

Caudal Hindbrain

The caudal hindbrain is composed of rhombomeres r6–r11. It contains the nuclei of cranial nerves IX–XII. The nuclei are organized into long, multi-level tracts. From this point onward, the registration pattern becomes *highly regular.* Every neuromere is associated with one somitomere. Furthermore, beginning at r8, a new pattern of morphologic complexity is seen as the somitomeres undergo conversion into somites. Neural crest from levels r8 and r9 forms the final two pharyngeal arches PA4 and PA5.

The caudal hindbrain has two nuclei of great importance: hypoglossal nucleus and nucleus ambiguus. Hypoglossal nucleus supplies the axial muscles of the tongue. Because these arise from somites 1–4, hypoglossal is polysomitomeric. Nucleus ambiguous supplies all branchiomeric muscles of the palate, pharynx, and larynx (with the exception of styloglossus). It does so by making connections with the sensory cranial nerves to pharyngeal arches 3–5. These arise from somitomeres 7–11; thus, ambiguous is also polysomitomeric.

Glossopharyngeal nerve deserves a final comment because it has created such confusion. It has its nuclei in r6–r7, is both motor and sensory, and is considered to be the cranial nerve of the third arch. Its sensory distribution is extensive but is credited with supplying only one muscle, stylopharyngeus. All the rest of the third arch muscles are innervated by vagus (pharyngeal plexus). But how can this be you can reasonably ask? I thought X was dedicated the the fourth and fifth arches!

The answer lies in our confusing a neural structure like the vagus nerve with its target. Cranial nerve X co-habits with IX in r6–7. Neurons dedicated to the soft palate, for example, have their motor nuclei in r6-r7; these are the *exclusive neuromeres of the third arch.* They simply choose to travel with X rather than with IX. Thus, the whole concept of a 1:1 relationship between a cranial nerve and a pharyngeal arch is misleading. The true relationship is with the neuromeres, regardless of what physical means the neurons choose to reach their target. Whew, that was a mouthful but keep this principle in mind and you will avoid confusion.

Medial Motor Column (m1–m2, r0–r1, r4–r5) Somitomeres Sm1, Sm2, Sm3, Sm5

Neuromyology of the Eye and Orbit

All vertebrates make use of six extraocular muscles to position the visual axis. Also included in our discussion are two muscles controlling the upper lid, levator palpebrae superioris, and Muller's muscle.

As we shall see, these muscles have many unique features. Our ultimate goal is to better understand the developmental anatomy of these muscles in the context of the overall development of the eye and orbit. This requires a brief sojourn through basic science material germane to non-ophthalmologists.

Next, we will explore in depth the developmental rationale underlying the origins and insertions of the extraocular muscles. EOMs arise without variation from the paraxial mesoderm of four distinct somitomeres: 1, 2, 3, and 5 (please note that some authors attribute the mesenchymal source for these muscles to prechordal mesoderm). Their contents are as follows: Sm1 = SR, MR, and IR. It is the source for LPS. Muller's muscle is non-striated and comes from a transformation of neural crest. Sm2 = IO and possibly MR (they arise from a common blastema). Sm3 = SO, and SM5 = LR. At first glance, this arrangement seems odd. Why are they so distributed? Why should lateral rectus originate so far away from the orbit? Is there any sense to the pattern of primary and secondary insertions observed? By what mechanism and sequence do these myoblasts "know" how to find out their way to the globe in the first place?

Any discussion about EOMs must take into account their blood supply and innervation. We shall see that the orbit and its contents follow a developmental sequence that is recapitulated by its various sources of blood as these are sequentially induced by the optic nerve and supply. This necessitates a review of the dorsal aortae and the sequential development (beginning at stage 9) of three derivative systems: the primitive internal carotid and the aortic arches, and the stapedial system.

Before we begin, let us clarify the terminology of origin and insertion as it applies to the extraocular muscles. All EOMs originate as myoblast populations with the paraxial mesoderm of somitomeres. These are identified by the motor supply. Each population assumes a spatial position with respect to the developing optic vesicle. "Assignment" of which muscle associates with what part of the eyeball is determined by two factors: (1) genetic coding of the scleral insertion sites and (2) physical proximity of between the globe and the somitomeres. Note that the *extraocular myoblasts associate themselves with the globe long before the orbital bones have formed.* As the myoblasts proliferate, the individual muscles extend bi-directionally. Fusion with a scleral target is likely the primary event. The muscles then grow outward from the globe. In the case of the four recti, this will be *backward along the optic nerve axis* to what becomes the annulus of Zinn. Inferior oblique seeks out a secondary insertion site by running *forward to the maxilla.* Superior oblique extends *backward to the orbitosphenoid* (lesser wing).

In summation, extraocular muscle growth is rapid; distal and proximal extensions likely take place simultaneously. But the timing of insertions into the sclera likely precedes those into the bones.

Patterns of Eye Movement

Eyes move constantly but some motion is required for vision. Large, rapid eye movements are used to survey the environment and to place retina images of on the fovea. Slow eye movements trace movements and compensate for position changes of the head/body. Eye movement velocities vary enormously, by 10^5. Overlapping eye fields require coordination of movement. Slow eye movements in opposite directions keep corresponding images on the fovea of both retinas simultaneously. Strabismus is a misalignment of two visual axes under binocular viewing conditions.

Control of Eye Position

The six extraocular muscles are arranged as three reciprocally innervated agonist–antagonist pairs. These have distally inserted into the sclera in two distinct zones. The *anterior zone* near the limbus contains the four recti. The *posterior zone* lies behind the coronal midline and lateral to the sagittal plane. Here are inserted to obliques, with SO above the axial plane and IO below the axial plane. Eyes are stationary when the opposing forces exerted by the extraocular muscles are in balance. Imbalanced forces produce rotations of the eye. Muscle force is related to muscle length. Most of the force required to maintain eye position is passive. Equilibrium muscle length and fore in different gaze positions is a function of innervational command. Different patterns of innervation are required for fast vs. slow eye movements. Extraocular motor neurons are located in three interconnected nuclei. Motor commands result from the interaction between visual and nonvisual inputs to the motor control centers. A copy of the innervational command is used to verify the system's operation. Extraocular motor neurons receive inputs from premotor areas of the brainstem to generate appropriate signals for saccadic eye movements. The pathways for smooth pursuit movements and for vergence go through the cerebellum, but vergences have a separate control center near the oculomotor nucleus.

Extraocular Muscle Structure and Function

Muscle fibers are the units from which muscle are constructed. Striated muscle fibers have a parallel arrangement of contractile proteins that interweave to cause contraction. Striated muscle fibers have different structural, biochemical, and contractile properties. Extraocular muscles contain muscle fiber types not found in skeletal muscle. Thick and thin extraocular muscle fibers have different contractile properties. Different muscle fiber types are not randomly distributed within the muscles. Different muscle fiber types may receive different innervational commands. Extraocular muscles have very small motor areas. Acetylcholine at the neuromuscular junction depolarizes the cell by opening Na^+ channels. The spread of depolarization along the sarcolemma may differ among muscle fiber types, producing different contractile properties. Extraocular muscles are highly sensitive to agents the mimic or block acetylcholine. Extraocular muscles often exhibit early symptoms of myasthenia.

Neurotoxins that interfere with acetylcholine action can be used to alleviated strabismus and blepharospasm.

Sensory Endings in Extraocular Muscles and Tendons

Skeletal muscles have two major types of sensory organs. Human extraocular muscles have anatomically degenerate sensory organs and *exhibit no stretch reflexes*. Passive extraocular muscle stretch is intensely stimulating and may produce *bradycardia*. Sensory endings in extraocular muscles probably *do not* convey information about eye position. Sensory signal from the extraocular muscles may be involved in motor learning, motor plasticity, and development.

Actions of Extraocular Muscles

All the extraocular muscles except the inferior oblique have their primary insertions at the apex of the orbit. The anatomical primary insertion of inferior oblique and the functional primary insertion of the superior oblique are anterior and medial in the orbit. The four rectus muscles are arranged as horizontal and vertical pairs with secondary insertions into the anterior zone of the globe. The horizontal recti rotate the eye around a (sagittal) axis. The vertical recti rotate the eye around a horizontal (axial) axis. The four recti define a muscle cone within the orbital cavity that contains most of the ocular blood vessels and nerves. The oblique muscles constitute a third functional pair; they are distally inserted into the posterior zone of the globe. Extraocular muscle actions cannot be measured directly. The classic description of extraocular muscles is based upon the geometry of their primary and secondary insertions. Boeder diagrams attempt to describe the actions of extraocular muscles completely. The presence of Tenon's capsule and muscle pulleys invalidates the geometric model of extraocular muscle function.

Eye Development by Stages

Eye muscles arise from four somitomeres surrounding the optic cup. Their development involves the following steps: (1) gastrulation, (2) segmentation of PAM into somitomeres, (3) neural crest invasion of somitomeres, creates subcompartments and potential myoblast targets, (4) concomitant development of medial motor column neurons in m1, m1, r0 (isthmus), r1, and r4–r5, (5) homeotic "assignment" of medial motor columns to target somitomeres 1–3 and 5, (6) physical contact between groups of myoblasts and the optic cup, and (7) differentiation of primary insertions into the sclera and secondary insertion into the annulus of Zinn or the orbital bone anlage.

The **somitomeres of origin** for the EOMs are as follows. The superscript written above each muscle refers to the Carnegie stage at which the muscle is first identified: Sm1 (medial rectus[16], inferior rectus[17]), Sm2 (superior rectus[14], inferior oblique[16], levator palpebrae superioris[19-221]), Sm3 (superior oblique[14]), and Sm5 (lateral rectus[13]).

The **order** **of** **insertion** is as follows: $LR^{13} > SO^{14} > SR^{14} > MR^{16} > IO^{16} > IR17 > LPS19\text{--}21$

So that we can keep track of these processes let's summarize of key events in eye development (including blood supply) organized by Carnegie stages. This represents highlights from Chap. 10 with emphasis on the eye and the arterial system of the orbit. NOTA BENE: *Take your time and review this carefully*. We will then make a few generalizations.

Stage 8 (18 days) *nervous system, segmentation, and supply*
> General: Hensen's node, primitive pit, notochord and neurenteric canal, neural plate, neural folds, blood islands. Note that PAM segmentation proceeds caudally, producing somitomeres.

Stage 9 (20 days, 1–3 somites) *major subdivisions of the brain*
> General: three somites visible, representing transformations of Sm8–Sm10. first aortic arch, primitive internal carotid, and primitive trigeminal arteries. Elevation of cranial neural folds. At formation of the 18th somitomere, Sm8 is transformed into the first somite, S1. Embryonic folding positions the heart ventral to face the pharynx. Primitive internal carotid artery (PrICA) dedicated to midbrain/forebrain. Primitive trigeminal artery (PrTgA) connects dorsal aorta to primitive hindbrain channel. Heart outflow now pumps via first aortic arch to dorsal aortae and then split in three directions: (1) via dorsal aorta backward to rest of the body; (2) via PrTgA to hindbrain; and (3) via PrICA to midbrain and forebrain. Note that two brain circulations are *entirely separate*. Also note that first pharyngeal arch is organized around AA1.

Stage 10 (22 days, 4–12 somites) *major divisions of the brain*
> General: 4–12 somites. First pharyngeal arch, neural folds fuse, second aortic arch appears, beginning heart beat.
>
> Eye: optic primordia (optic grooves) appear at eight somites.

Stage 11 (24 days, 13–20 somites) *forebrain subdivides, optic primordium subdivides*
> General: 13–20 somites. Second pharyngeal arch, third aortic arch appears, optic vesicles, buccopharyngeal membrane ruptures.
> Disintegration of AA1 and AA2
> Eye: optic evagination appears at 14 somites and optic ventricle is continuous with the forebrain ventricle. At 16 somites, the wall of the optic evagination provides neural crest to the sheath. This separates the optic epithelium proper from the overlying tissues (PNC+

MNC). At 17–19 somites, the optic vesicle is present. Surrounded by vascular plexus is the terminus of primitive internal carotid.

Stage 12 (26 days, 21–29 somites) *caudal neuropore closes*
> General: 21–29 somites. Third pharyngeal arch, fourth aortic arch appears, assembly of future ventral pharyngeal artery. Cranial and caudal divisions of primitive internal carotid. Primitive hindbrain channels convert to longitudinal neural arteries.
> Eye: Optic vesicle is completely covered by optic neural crest from m1, m2, r0, and r1. Primitive internal carotid "climbs over" the dorsal aspect of vesicle.

Stage 13 (28 days, 30+ somites) *neural tube now closed, cerebellum develops from r1*
> General: 30–31 somites. End of pharyngeal arch period. fourth pharyngeal arch, fifth aortic arch appears and disintegrates, leaving blood supply to fifth pharyngeal arch dependent upon AA4. Primitive internal bifurcates into "older" cranial division and newer caudal division.
> Eye: Optic vesicle is covered by basement membrane. Surface PNC "ectoderm" *also* has a basement membrane. Lateral hemivesicle becomes retinal disk (the future inverted layer of the optic cup) in continuity with a lens disk. Primitive dorsal ophthalmic artery (PrDOA) is now seen emerging at ICA bifurcation. Primordia of **lateral rectus** now appear.

Stage 14 (32 days) *cerebral hemispheres*
> General: fifth pharyngeal arch (blood supply dependent on PA4), lens vesicle, optic cup, nasal pits, sixth aortic arch for pulmonary circuit (*no sixth pharyngeal arch*).
> Eye: Retinal disk invaginates, forming optic cup. Retina ("choroid") fissure *appears*. Lens disk indents to become lens pit. Primordia of **superior oblique** and **superior rectus**. Optic nerve bud appears. Stapedial V1 *primitive dorsal ophthalmic artery* (PrDOA) pursues a lateral course, encircles the lens pit from its caudal–dorsal 180°, and enters the ocular cleft.

Stage 15 (33–36 days) *longitudinal zoning of the diencephalon*
> Eye: Retinal pigment appears. Lentiretinal space contains primary vitreous body and hyaloid artery. Lens surrounded by lens capsule. Restored surface r1 ectoderm is now the anterior epithelium of future cornea. Neural crest is very active: (1) retinal pigment; (2) primitive choriocapillaris network envelopes the optic cup. Stapedial V1 *primitive ventral ophthalmic artery* (PrVOA) appears just opposite anterior choroid artery. Abducens (VI) and trochlear (IV) nerves appear.

Stage 16 (37–40 days) *evagination of the neurohypophysis*
> General: primitive olfactory a. (PrOfA) terminates at the nasal pit having first given off anterior cerebral.

Eye: PrDOA "negotiates" its way through the intermediate choriocapillaris plexus. Eyelid grooves appear (eyelid folds develop during stages 17–19). Ciliary ganglion is present. It provides PANS fibers to the orbit via III. Primordia of *medial rectus* and common primordium of *inferior oblique* and *inferior rectus* are present. IR subsequently delaminates from IO to insert farther forward.[17]

Stage 17 (41–43 days) *olfactory bulbs, amygdaloid nuclei*
General: dorsal aorta segment between AA3 and AA4 disintegrates. Stapedial anastomosis with VPA at V3. All components of external carotid present. Craniocaudal disintegration of the ventral connections between dorsal aortae and vertebrals.

Eye: Definitive hyaloid artery enters ocular cleft. PrDOA and PrVOA elongate. PrDOA reaches mesial end of ocular cleft. There it makes an anastomosis with hyaloid that proves to be clumsy, circuitous, inefficient, and therefore temporary.[17–18] Nasolacrimal groove appears.

Stage 18 (44 days) *inferior cerebellar peduncles, corpus striatum, dentate nucleus*
Eye: V1 stapedial arrives at orbit. Ocular axis is still at 90° angle to midline.

Stage 19 (44–48 days) *choroid plexus fourth ventricle, medial accessory olive*
Eye: Traitorous hyaloid abandons PrDOA and anastomoses with PrVOA. Orbital branches arise as V1 sensory nerves grow into muscles. Recall StV1 to orbit-induced PrDOA at stage 13. These will secondarily join with ophthalmic. Adult stem of ophthalmic arises from ICA lateral to Rathke's ventral and caudal to the optic nerve. PrDOA converts to common temporal ciliary artery.

Stage 20 (47–51 days) *choroid plexus lateral ventricles, optic and habenular commissures, interpeduncular and septal nuclei*
Eye: Three main branches of StV1: ethmoid, frontal, and lacrimal. *Anastomosis* between orbital stapedial and primitive ophthalmic. Stem of stapedial involutes. PrVOA converts to common nasociliary artery supplied from common ophthalmic stem. The trochlear pulley that ensnares SO develops at stages 20–23.

Stage 21 (50–52 days) *cortical plate appears in the cerebral hemispheres*
Eye: Ventral–nasal ciliary artery arises from ophthalmic and annexes PrVOA now shifts *ventral* to optic nerve, thus creating adult common nasociliary artery. **Levator palpebrae superioris** delaminates from superior rectus at stages 21–23.

Key Take-Home Message from the Timeline

(1) Gene products from r0 to r1 diffuse forward to induce midbrain and forebrain in a caudal–rostral. (2) From the moment of its emergence from forebrain, the epithelium of optic vesicle is likewise polarized with the caudal (temporal) aspect more primitive and the rostral (nasal) aspect. (3) The sclera covering the caudal vesicle comes from hindbrain r0–r1 neural crest. (4) The sclera covering the rostral vesicle comes from midbrain m1–m2 neural crest. (5) V1 provides the exclusive sensory supply to sclera; it induces PrDOA and PrVOA, precursors of the long ciliary arteries. (6) The sclera is mapped into genetic zones defined by neurovascular structures. (7) Extraocular myoblast populations cluster around the sclera, occupying distinct genetic zones in a precise spatiotemporal order. (8) The "availability" of scleral insertion sites determines the sequence of EOM development.

Where Do Extraocular Muscles Arise and How Are They Innervated?

Let us begin with some simple observations, based on the "biologic maturity" of the various muscles. Note that suprascripts indicate Carnegie stage. LR is observed at day 27–28[13]. SO and SR both appear at day 32[14]. MR, IO, and IR are present at day 37[16]. Levator palpebrae superior is seen at day 43[19]; this delay is not surprising since the first manifestation of eyelid creases is at stage 16. Turning our attention to the motor nerves, we note the following sequence. Oculomotor nerve has rostral and caudal branches. III_R develops first appears at stage 14; it is assigned to superior rectus and, much later at stage 19 (day 46), to levator palpebrae superioris. Abducens and trochlear nerves both appear at stage 15 but the LR is more mature than SO. III_C does not have available targets until stage 16.

From the above, we can make six deductions of the great anatomic importance.

- Extraocular muscle development has been documented based upon when each set of myoblasts appears in association with the optic vesicle. It follows a strict spatio-temporal sequence: $LR^{13} > SO^{14} > SR^{14} > MR^{16} > IO^{16} > IR^{17} > LPS^{19-21}$.
- Extraocular myoblasts arise from somitomeres 1, 2, 3, and 5 as follows: Sm1 = SR, MR, IR, and LPS; Sm2 = IO (?MR); Sm3 = SO; and Sm5 = LR.
- Somitomeres develop according to the mechanism of formation of their mesoderm.
- All extraocular myoblast populations are biologically "available" for migration to the eye by stage 12, i.e., when the optic vesicle is fully coated with neural crest.

- In theory, if the contents of Sm1 predate those of Sm5, LR should insert last. This prediction is not in accordance with the observed pattern. Sclera has four zones and is completely ready for insertion. Insertion depends up positioning of the globe with respect to somitomeres.
- The neural crest sclera surrounding the optic vesicle possess six binding sites, each of which becomes "available" and will attract myoblasts in a fixed order. Anteriorly, these are as follows: LR, SR, MR, and IR. Posterior globe is developed before anterior globe. Beginning with lateral rectus, an imaginary line joining these insertions spirals *inward* exactly opposite to the previously described *insertion spiral of Tillaux* (White et al. 1989).
- As the binding sites become available, the "choice" as to which somitomeres will provide the available myoblast group to the eye is determined by solely by the physical proximity between that sector of the globe to that particular somitomere.

Why Is the Lateral Rectus Displaced Back to the Fifth Somitomere? (Fig. 9.1)

We come now to a neuroanatomic fact of the greatest importance. *The medial motor column is **continuous** through neuromeres m1, m2, r0, r1 but it is **discontinuous** through the remainder of the hindbrain.* Medial motor column is not present in levels r2–r3 because of the nuclei of V1 and V2, either due to their physical size or due to lack of expression of Hox code for medial motor column. Level r2 is purely sensory. Level r3 contains the nuclei of V3; these are both sensory and motor. V3 is assigned to the muscles of Sm4, i.e., to the first pharyngeal arch.

The medial motor column reappears in levels r4–r5. Many vertebrates possess two muscles here: lateral rectus and accessory lateral rectus. In humans, it is not clear if the motor neurons of VI exist in r4 alone or in r4–r5. In any case, lateral rectus myoblasts reside within Sm5. Levels r4–r5 are also in neurologic register, just next door in the adjacent lateral motor column, with the two nuclei of facial nerve. VII is assigned to the muscles of Sm6, i.e., to the second pharyngeal arch. The juxtaposition of two adjacent motor columns within r4–r5 explains the neuroanatomic pathway pursued by abducens neurons en route to the orbit (VI will make a detour around VII).

No medial motor column is represented at levels r6–r7. MMC makes its appearance again as the hypoglossal nucleus within the caudal hindbrain, r8–r11.

If EOM myoblasts arise from Sm1 to Sm3, why do not we find them in the fourth somitomere as well? Why are they displaced backward? A possible explanation involves *retractor bulbi* muscles documented in the very first jawed fishes,

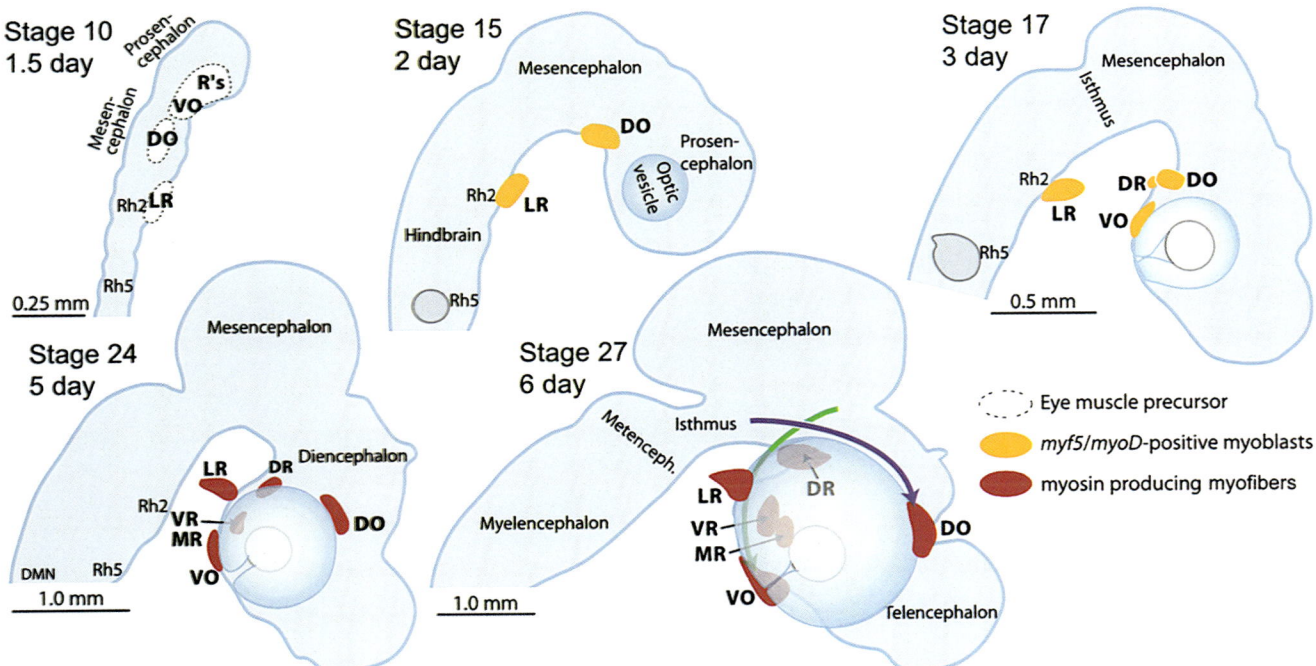

Fig. 9.1 Extraocular muscle migration: chick stage vs. Carnegie stage. All EOMs except LR arise from prechordal mesoderm corresponding to Sm1–Sm3 which lie adjacent to midbrain. SM5 is located in the rostral hindbrain. Muscle origins are as follows: Sm1 = MR, IR; Sm2 = SR, IO. LPS and Müller's; Sm3 = SO; and Sm5 = LR. The order of insertion is as follows: LR[13] > SO[14] > SR[14] > MR[16] > IO[16] > IR[17] > LPS[19–21].

Developmental sequence: Stage 10/1.5 days = stage 10; stage 15/2 days = stage 11; stage 17/3 days = stage 14; stage 23/5 days = stage 17.5; stage 27/6 days = stage 18.5. (Reprinted from Noden DM. The differentiation and morphogenesis of craniofacial muscles. *Dev Dyn* 2006; 235(5):1194–2128. With permission from John Wiley & Sons)

placoderms. These gave rise to two muscles responsible for controlling the nictating membrane of birds, a structure used for corneal protection. Quadratus and pyramidalis are innervated by V3 and therefore develop in somitomere 4.

The mammalian line is divergent from the reptilian-avian line. As we have discussed, many innovations involved chewing, such as specialized teeth, the temporomandibular joint, and a drastic expansion of jaw control muscles permitting chewing of a wide variety of foodstuffs. The neurologic control is far more complex than that required from non-mammals. It is possible that a transfer of mesenchyme to the temporal region required for these innovations reassigned motor nuclei from the medial column to the lateral column. In this scenario, nuclei for the lateral recti were forced backward from r3 into rhombomeres r4–r5.

Two opposing theories exist. Let us take a look at each one and then seek a synthesis.

Theory #1: no motor neuron, no muscle. The segment of rostral hindbrain in genetic register with Sm4 (rhombomere 3) does not contain a medial motor column because there is no room for it. Thus, no corresponding myoblasts exist in Sm4; it is completely dedicated to first arch, i.e., lateral motor column. Development of jaws in vertebrate evolution transformed the functional status of musculature in Sm4. In agnathic fishes, represented today by the lamprey and the hagfish, all visceral arches contain respiratory structures called branches (thus the term branchial arch). In gnathic fishes (and all subsequent vertebrates), the mesenchyme of the first two visceral was reassigned to mastication, not respiration. As a consequence, extensive control of jaw function required an expansion of V3 trigeminal control, leaving no room for a LR motor nucleus. The abducens nuclei were unceremoniously "shoved backward" into the fourth and fifth rhombomeres. In this scenario, in the absence of an appropriate nerve, no myoblasts dedicated to the eye can arise within Sm4.

Theory #2: no muscle, no motor neuron. This model postulates that neurologic content within the CNS is responsive to myoblast groups within somitomeres and somites. Recall that the pharyngeal arches and their respective muscles are hypaxial structures whereas the extraocular muscles are epaxial. LR myoblasts cannot exist within Sm4 without a genetically specified epaxial sector. If r3 and Sm4 are in register, and no epaxial myoblasts exist in Sm4, r3 cannot contain a medial motor column.

We can unite these disparate models with a simple hypothesis. *A somatotopic relationship exists between the motor neurons in the brainstem and their muscle targets.* Muscles develop in somitomeres and somites in spatially distinct zones defined by distinct gene patterns. The genetic "coordinates" for the lateral rectus "compartment" in Sm5 are shared with those of the medial motor column in r4–r5.

Development of the Sclera: Neurovascular Basis of Muscle Insertion

The optic primordium is a purely p5 neuroepithelial structure that appears in the prosencephalic fold[10]. As it projects laterally[11], the wall of the optic evagination picks up neural crest mesenchyme from its surroundings. One stage later, the entire vesicle is covered by *optic neural crest*[12]. Where does this layer come from and what are its functions?

Prosencephalic neural crest is not available to cover the vesicle. It is preoccupied with two tasks: (1) development of frontonasal epidermis and dermis and (2) interaction with prosencephalic placodes. Recall that forebrain neural folds have two zones. The rostral folds contain non-neural epithelium (NNE), the source of FNO epidermis. The caudal folds (p3–p1) provide FNO dermis. Outlying the rostral folds, a specialized zone of *non-neural ectoderm* contains three cranial placodes. The first two are associated with p6. The *hypophyseal placodes* become the primordium of the adenohypophysis. The *olfactory placodes* are precursors of olfactory epithelium. From the lateral nasal placodes arise olfactory neurons; from the medial nasal placodes arise accessory olfactory neurons and neuroendocrine cells (GnRH) that migrate into the brain. Associated with p5 are the *optic placodes*, the precursors of the lens.

PNC flows outward to surround and interact with the placodes but it does not penetrate them. Thus, in the primitive state, frontonasal skin circumscribes the nasal and optic placodes. The ectoderm around the nose is p6 while that of the future periocular skin is eye is p5. Frontonasal dermis is synthesized from p4 to p1 but how the contributions of individual prosomeres are apportioned remains unclear. Suffice it to say that the distribution of frontonasal skin matches exactly that of V1 with one glaring exception: *frontonasal mesenchyme remains extraorbital.*

The distribution of frontonasal skin within and without the future nasal cavities is as follows. The p6 skin forms the vestibular lining inward from the internal valve to Rathke's pouch. This "programs" the septum and upper lateral cartilages. The p5 skin forms the nasal lining outward from the internal valve. It includes the external valve and "programs" the lower lateral cartilages. Proceeding outward from the nasal chamber, p5 skin includes columella, the nasal envelope (except the alae), the upper eyelids, and the entire forehead.

You might wonder why we have strayed seemingly so far from the sclera. The answer lies with the potential conflict that could arise as the p5 optic vesicle grows outward to approaches the overlying p5 optic placode surrounded by PNC dermis. Fusion of these structures would be a developmental disaster. What can be done to prevent this?

In her infinite wisdom, nature ensures that optic vesicle is enveloped by neural crest from an entirely different, non-prosencephalic source. The optic vesicle, as it exits from ventral p5, projects outward into surrounding neural crest mesenchyme like a fist thrust into a sock. In the most likely arrangement, the neural crest comes from the midbrain, MNC: the caudal (temporal) half of the vesicle is coated with m1 while the rostral (nasal) half of the vesicle is coated with m2.

How can we make this approximation? Why MNC and not RNC? Hindbrain neural crest from r0 to r3 migrates before midbrain. It develops in craniocaudal order, so r0/r1 goes first. The immediate goal of this neural crest is to supply the forebrain with a primordial head plexus. Midbrain neural crest migrates later in caudal–cranial order. First to leave is m2; it travels to the most mesial zone of the future orbit. In sequence, m1 follows but does not have to travel so far and remains more proximal. This explains which Sm1 muscles are directed to the dorsal and proximal sector of the globe with Sm2 inserting into the opposite sector.

Making use of neural crest mesenchyme from a variety of sources is a clever arrangement[11–12] that pays handsome dividends. During stage, the p5 epidermis acquires a basement membrane; this binds it to the underlying caudal PNC dermis. Voila! True prosencephalic skin is now formed. At the same time, optic vesicle neuroepithelium acquires its own basement membrane. Like Velcro®, this facilitates contact between it and the overlying neural crest (the future sclera and choroid). We can conceive of the neural crest coating the vesicle as consisting of two hemispheres. The proximal one will become sclera; the distal one will become retinal disk (choroid). Contact now occurs between the retinal disk (the future inverted layer of the optic cup) and the overlying optic disk. These structures *interact, but cannot fuse*. A marginal zone is now detectable around the retinal disk. Into it will insert the rectus muscles in a specific temporal–spatial order.

At stage 14, the retinal disk invaginates to form the optic cup. The lens disk indents and forms a lens pit into which cell remnants of the disk are extruded. During stage 15, the outermost layer of the optic cup acquires pigment and vitreous begins to form between the lens and retina. The hyaloid artery now enters the lenti-retinal space. A lens capsule encloses the lens. Because the lens has lost its cells, the lens pit now closes up and we are left with an acellular lens vesicle. It pinches off and detaches itself from the PNC surface epithelium.

Recall that the p5 PNC skin and the p5 optic placode are biologically distinct; the skin can neither penetrate nor cover the optic placode. That the placode has "sunk beneath the waves" to become the lens, the PNC skin closes up on itself, like the shutter on a camera. Its boundaries will be the set by r1 conjunctiva. The eyelids have not yet developed.

Moreover, the lens now becomes bioactive: it elaborates factors that induce, from surrounding RNC/MNC, and entirely new epithelium with its own basement membrane the cornea. Thus, by the end of stage 15, we have two epithelial structures stacked one above the other: p5 skin overlies

superior half of the cornea and r2 skin covers the inferior half. A potential plane of separation now exists that will prepare the way for the conjunctival sulci from r1 (upper) and r2 (lower). Eye lids develop. This will be exploited in stages 16–19 as eyelid grooves appear[16], deepen into folds[17], the conjunctival sac forms[18], eyelids appear and meet at the lateral canthus[19], the medial canthus forms[20], the eyelids begin to approach the globe[21], but remain open[22].

What Is the Anatomic Rationale for the Insertions of the Extraocular Muscles?

Let us start with a few facts. Extraocular muscles are distributed among four somitomeres. Because paraxial mesoderm segmentation is a cranial-caudal process, the EOMs theoretically originate and mature in a *linear order*: Sm1 > Sm2 > Sm3 > Sm5. Warning: the insertion pattern of the six EOMs does not follow the same sequence. The globe contains six potential insertion sites. These become "activated" in a *sequence determined by the order in which the scleral arteries develop*. From a neuroanatomic standpoint, the optic vesicle is a p5 structure; it lies far forward from the somitomeres. Myoblasts are spatially associated with midbrain and rostral hindbrain. How can they get access to the globe? Head folding forces the optic cup backward. It "offers itself" to the myoblasts, all six populations of which are available and ready to migrate. As previously discussed, EOMs are observed to appear in relation to the globe in the following actual sequence running from day 27 to day 33: **LR[13] > SO[14] > SR[14] > MR[16] > IO[16] > IR[17] > LPS[19–21]**. This involves Sm5 > Sm3 > Sm2 > Sm1. Let us now explore *why* this biology is so very sensible.

Recall that early blood supply to the eye involves two branches from the primitive internal carotid. Primitive dorsal ophthalmic (PrDOA) develops first as a direct branch from the plexus of primitive internal carotid located behind the eye. One stage later, PrVOA buds off the primitive olfactory. Now the optic vesicle per se is a purely neuroepithelial structure per se. *Prosencephalic neural crest is not available* to cover the optic vesicle. PNC is instead preoccupied with the development of frontonasal placode structures and the production of frontonasal skin. *Coverage for the optic vesicle is provided by mesencephalic and rhombencephalic neural crest.* As it exits from p5, the developing eye and optic nerve project outward into surrounding mesenchyme like a fist thrust into a sock. The caudal (temporal) half is coated with RNC from r0 and r1; the rostral (nasal) half of the vesicle is coated with MNC from m1 and m2.

Recall that midbrain arises as a caudal–cranial induction from hindbrain: *MNC is inherently less "mature" than RNC.* If we transpose this to the sclera of the globe, we can see that temporal half is more mature than the nasal half and that the MNC between globes is the least mature. Thus, it is reasonable that PrDOA should develop *prior* to PrVOA. Two developmental hemispheres are thus created: caudal–dorsal and rostral–ventral. Thus, along the limbus of the globe, rectus muscles will develop in a fixed order: lateral > superior > medial > inferior. This is exactly what is observed. LR appears at day 27 (stage 12) while IO and IR are observed at day 33 (stage 15).

To fully appreciate the above sequence, we must also rationalize why the oblique muscles should insert *prior* to their rectus counterparts. The anatomic "progress" of both PrDOA and PrVOA is from proximal to distal. This creates two additional developmental hemispheres: posterior sclera is more mature than its anterior counterpart. Thus, SO develops prior to SR and IO precedes IR.

Extraocular Muscles: Origins, Primary, and Secondary Insertions

All seven EOMs are assembled after the optic cup but long before the orbital bones. Each EOM will have a *primary insertion* into the sclera and a *secondary insertion* to bone/fascia. The maturity sequence of scleral binding sites explains the order of primary insertion. All four recti are inserted near the limbus anterior to the x-axis of the globe. Both oblique muscles are inserted posterior to the x-axis. It is as if the sclera were divided into two sets of genetically and functionally distinct hemispheres. Distinct patterns of blood supply define posterior and anterior hemispheres based on the coronal plane. Distinct patterns of sensory nerve supply define dorsolateral and ventromedial hemispheres based on the axial plane.

Temporary physical relationships between the globe and the somitomeres determine the access routes by which myoblasts migrate onto the surface of the eye.

The cephalic flexure rotates forebrain downward. The caudal wall of optic vesicle is now rotated cephalically as well. By stage 12, the future globe is "presented" to the ventral surface of the rostral hindbrain and midbrain. To better visualize this topology, try this experiment. With your arm forearm and hand in the prone position and parallel to the floor, make a fist and extend your thumb fully. Your forearm represents the neuraxis. The volar wrist crease is r0. Your thumb is the optic vesicle; the thumbnail represents the caudal margin of the optic vesicle. It points backward. Now fully flex your wrist; the thumbnail has now rotated cephalically almost 90°. Note how your thumbnail is "presented" toward the volar surface of your wrist.

Let us also visualize how the somitomeres relate to the optic nerve axis. Sm1 produces the basisphenoid cartilage, Sm2 produces the orbitosphenoid cartilage (lesser wing), and Sm3 contributes to the alisphenoid (greater wing). Recall that the foramen for optic nerve occupies the intersection between basisphenoid and orbitosphenoid. Thus, Sm1 lies

rostal (nasal) to the optic axis while Sm2 is positioned caudal (temporal).

$$LR^{13} > SO^{14} > SR^{14} > MR^{16} > IO^{16} > IR^{17} > LPS^{19-21}$$ thus CN VI > CN IV > CN III

<u>Note</u>: by convention will shall use the symbol ® to indicate a primary insertion, where a muscle is most likely to insert first. Because ossification process of the sphenoid begins take place at 8 weeks, we hypothesize the muscles to insert first into the sclera and secondarily into bone. In some cases of congenital orbital bone defects, the proximal muscle insertions compete with one another but remain in their original configuration along the globe.

From a spatial standpoint, because the LR® is closest to Sm5, it receives these myoblasts. Next up to bat, and more proximal to Sm3, lies SO®. Just distal to that is SO®. Sm2 provides the myoblasts with the rostral division of oculomotor III_1 at the ready. Note that superior rectus must bridge over superior oblique to reach the limbus. Because the upper eyelid is not yet formed, levator palpebrae superioris (supplied by III_R) must wait its turn. Three binding sites remain, all on the rostral (nasal) aspect of the vesicle, all of which will be supplied by Sm1 and innervated by successive branches of III_2. MR® goes first, receiving myoblasts from Sm1. It lies closest to the optic axis. IO® is in the posterior zone so it matures before IR®. Because it has "priority," the next group of Sm1 myoblasts to migrate will seek it out and attach. Perhaps the priority of IO® is also spatial; it is also physically closer to Sm2. IR® lies distal; thus, the myoblasts must pass beneath IO to reach their target.

You, the reader, are right to question why we should bother contradicting terminology dating back (at least) to Henry Gray. What's wrong with "origins" at the back of the orbit and "insertions" into the globe? Unfortunately, these terms are unscientific. In Gray's words, "The origin is *usually* the more fixed and proximal end, while the insertion is the more movable and distal end. For example, the pectoralis arises or has its origin from the sternum, ribs, and clavicle, and its insertion is into the humerus. If the individual were climbing a tree, however, the origin and insertion would be reversed, since the upper extremity would be more fixed and the body more movable. The designations of origin and insertion…are, therefore somewhat *arbitrary* and a *matter of convention among anatomists.*" How can we square these concepts with molecular embryology?

The origin of a muscle is best understood as the somitomere or somite from which the myoblasts emerge. Furthermore, the muscle will be in genetic register with the neuromere or neuromere that provides its motor supply. This definition is simple, based on embryology, and easy to test out. Second, we have previously stated that the primary insertion for a muscle will include a bone, fascia, ligament (or sclera) that originates from same neuromeric level as the motor nerve. The bones of the orbit arise from neural crest r0–r2 and somitomeres 1–3. These have no correlation with lateral rectus. But the sclera contains elements that are held in common with all six muscle groups. Third, one the primary insertions are set, the secondary insertions take place according the sequence of biologic maturation of surrounding bone and fascial fields.

Myology of Somitomere 1

Superior Rectus SR is the first muscle to access the globe from this somitomere, inserting at stage 14. It secondarily inserts into the annulus of Zinn spanning from the superior margin of the optic nerve sheath to the lateral margin of orbitosphenoid.

Medial Rectus MR is the second muscle to access the globe from this somitomere, inserting at stage 16. MR® lies directly medial to the limbus 5.5 mm from the cornea. As the third of the recti, it secondarily inserts into the annulus and is directly medial to optic nerve.

Inferior Rectus IR is the second muscle to access the globe from this somitomere, inserting at stage 17. IR must pass external to its predecessor IO. IR® lies directly below the limbus 6.7 mm from the cornea. As the fourth of the recti, it secondarily inserts in the annulus of Zinn, directly below the optic nerve.

Levator Palpebrae Superioris LPS is the third muscle to access the globe from Sm2, inserting at stages 19–21. LPS® is the tarsal plate, an r1 structure programmed by the conjunctiva. It secondarily inserts into the superior crus of orbitosphenoid, just lateral to superior oblique.

Muller's Muscle Functionally distinct within the orbit, Muller's muscle represents a transformation of local neural crest into smooth muscle innervated strictly by the SANS.

Myology of Somitomere 2

Inferior Oblique IO is the second muscle to access the globe from somitomere 2, inserting at stage 16. It precedes IR because it the posterior sclera matures prior to the anterior limbus and because the distance is shorter. IO® is more lateral that SO®. IO seeks out its secondary insertion on the orbital plate of the maxilla, immediately below naso-lacrimal fossa. This insertion site is directly below the trochlea of the frontal bone. IO is the antagonist of SO; its course runs in parallel to that of SO from the globe to the pulley. The actions of IO are as follows: x-axis = elevation; y-axis = lateral; z-axis = external rotation. It pulls the eye up and out, with ex-torsion.

Oculomotor Nerve

Cranial nerve III arises from two oculomotor nuclei located in both the m1 and m2 zone of the midbrain below the rostral terminus of the aqueduct of Sylvius. Its general somatic efferent fibers supply levator palpebrae and four extraocular muscles except superior oblique and lateral rectus. *It also supplies the intrinsic muscles and saves dilator pupillae* (these muscles are under autonomic control). It conveys PANS fibers to ciliary ganglion.

III emerges from the midbrain into the posterior fossa just medial to the cerebral peduncle. First, it is covered in pia, passing between superior cerebellar artery and posterior cerebral, i.e., precisely at the junction between internal carotid and longitudinal neural systems. Running forward, just lateral to the posterior clinoid, it acquires arachnoid, and then, between posterior and anterior clinoid processes, it pierces the dura. The nerve occupies the most dorsal position within the cavernous sinus.

Just prior to reaching the superior orbital fissure, it divides, entering the apex of the muscle cone as distinct branches. *Superior branch* innervates SR and LPS. It runs forward along the underbelly of SR for its proximal one-third at which point it gives off penetrating fibers that access overlying LPS. Less commonly, a separate branch winds around SR en route to LPS. The *inferior branch* runs below and lateral to optic nerve. It sends motor fibers to MR by passing under optic nerve. Note that this pattern replicates the developmental sequence by which a temporary vascular ring is seen encircling the optic nerve during the transition of ophthalmic artery stem and the V1 stapedial into their final configuration.

Extensive spatial mapping of the nucleus in macaques demonstrates that cells for different muscles are represented in a rostral to caudal sequence. In general, superior rectus neurons are not present in the rostral pars, inferior rectus neurons are not found in the caudal part, inferior oblique neurons extend throughout, and levator neurons are exclusively in the most caudal region. Overall, the sequence for the five muscles is as follows: MR, IO (throughout), IR, SR, and LPS, exactly as predicted from the insertion site sequence. Thus, Sm1 would contribute MR, IO, and IR to the globe while Sm2 would contribute SR and LPS.

One significant finding with oculomotor muscles is concerned with ipsi- versus contralateral. With one exception, the cells of the oculomotor nucleus direct their axons to the *ipsilateral* eye. Neurons supplying superior rectus cross the midline between the oculomotor nuclei and pass into the contralateral nucleus to enter the contralateral oculomotor nerve. Thus, the right superior rectus is innervated by axons originating in the left oculomotor nucleus and vice versa. Fortunately for neuroanatomically beleaguered clinicians, the system is redundant: cutting the oculomotor nerve only affects the ipsilateral eye. But do these findings suggest anything of value? There may indeed be something genetically distinct about the superior zone of the sclera such that SO and SR possess something in common: neurologic control of both muscles requires input from the contralateral midbrain.

Myology of Somitomere 3

Superior Oblique SO arises from Sm3. It inserts at stage 14. SO® is located in the posterior superior sclera just lateral to the vertical (sagittal) midline. It proceeds forward internal to superior rectus and makes a beeline for the *trochlear fovea of the frontal bone*. Here finds itself entrapped by a fibrocartilaginous pulley. From this position, it is forced to pass external to superior rectus, outside the annulus of Zinn. It attaches to the orbitosphenoid at its most superior crus, well medial to the eventual attachment of levator palpebrae superioris. Recall that optic foramen results from when the crescent-shaped OS joins the basisphenoid via two crura, superior and inferior. These grow medially, spanning above and below the optic nerve, and click onto the basisphenoid.

How can we make sense of this muscle? As we shall see, the two obliques follow exactly the same plan. From a functional standpoint, SO and IO are an antagonistic pair. Both begin in the posterior aspect of the eye and run forward to secondary insertion site. IO stops dead in its tracts on the maxilla, whereas SO continues onward. Is there an evolutionary significance to the trochlea? Certainly, if no trochlear fovea were present, SO would wind awkwardly around SR in point of fact SR would wind up serving a pulley function. Not a very good plan. Alternatively, suppose the original designs were dead-end, the SO on the frontal bone with an additional head directly backward by default?

The actions of SO are as follows: x-axis = depression; y-axis = abduction; z-axis = internal rotation. It pulls the eye down and out, with in-torsion.

Trochlear Nerve

Cranial nerve IV has its nucleus in r0/r1, just below that of III_2 in m2. The trochlear nerves cannot exit from the base of the brainstem. They are blocked by the pons. Instead, they pursue a dorsal course on both sides of the aqueduct of Sylvius and proceed headlong toward the inferior colliculi. Here, the axon bundles decussate just behind the inferior colliculi. The trochlear nerves now must hug the sides of the brainstem, pursuing a forward course *in parallel with and below the oculomotor nerves*. Within the cavernous sinus, the trochlear nerves run inside a common bilaminar structure of dura/arachnoid that contains II, IV, V1, and V2. Trochlear nerve is always ventral to III and dorsal to V1.

The trochlear nerves pass into the orbital cone outside the muscle cone of the recti. This is because III and VI are dedicated to the recti and *inferior oblique arises with inferior rectus as a common blastema*. Thus, trochlear nerve serves a muscle outside the cone. Furthermore, from a developmental standpoint, LR is established *one full stage prior* to the advent of SO. Trochlear nerve is simply blocked out from the cone. In any case, it pursues a leisurely course medially, passing *external to* levator palpebrae superioris. It finds its target in the proximal muscle belly. It enters the muscle from

its external aspect; SO is the only extraocular muscle to be so innervated.

Why does trochlear nerve enter its target 180° opposite its colleagues? Perhaps the answer lies in a *somatotopic relationship between the muscle and the motor nucleus* with which it is in register. The physical course of the trochlear nerve upward to the colliculi, around the brainstem and forward, *requires a 180° rotation* of the axon bundles.

Myology of Somitomere 5

Lateral Rectus LR arises from Sm5 (and possibly from Sm6). LR® is directly lateral to the limbus 6.9 mm from the corneal margin. It secondarily inserts into two distinct sites. The larger insertion is into the annulus (common tendon) of Zinn along its most lateral arc. The smaller head attaches to alisphenoid along the lower margin of superior orbital fissure.

Abducens Nerve

Cranial nerve VI originates from one, and possibly two, nuclei located in r4 and r5. These are medial to and slightly rostral to the motor nucleus of VII. The abducens nerve exits just below the caudal margin of cerebellum. Because it lies caudal to oculomotor and trochlear nerves, abducens must run below their trajectories to enter cavernous sinus. Within the sinus, the abducens nerves are *internal to all the other nerves en route to the orbit and immediately external to internal carotid artery*. Here, VI hugs the wall of internal carotid and then enters the superior orbital fissure. Just beyond the annulus of Zinn, it divides to innervate both heads from their *undersurface*.

Angiology of the Eye and Orbit

Central retinal artery This is the first branch of primitive ophthalmic. It begins below the optic nerve but is enveloped within its dural sheath. It enters the actual substance of the nerve about 1 cm behind the eye and thence runs forward into the eye in company with the retinal vein.

Long posterior ciliary arteries are two in number. They are the remnants of internal carotid. PrDOA and PrVOA eventually join up with StV1 derivatives as the ICA part of common temporal ciliary and common nasal ciliary arteries. These posterior arteries run forward to the ora serrata where they break up to supply the arterial axis of the iris.

Short ciliary arteries 15–20 of these small vessels are direct branches from primitive ophthalmic. They pierce the back wall of the eyeball and supply the choroid.

Anterior ciliary arteries These are branches of the muscular arteries; hence, they come from the StV1 system. These follow the muscle insertions of the four recti to form a circular network at the sclerocorneal junction. They bring neural crest cells into the eye that will form the intrinsic muscles of the iris.

Developmental Considerations of the Orbit

Time and again, we have observed the intimate developmental relationship between neurons and arteries. Nerves are metabolically active; they are quite proactive about providing blood supply for themselves. Their neural crest ensheathing cells produce angiogenic signals (such as VEGF) to synthesize companion arteries. In the orbit, V1 is no exception. From the get-go, it supplies the very neural crest tissues that will envelop the optic vesicle. A good example is the choriocapillaris network at stage.

The migration pathways of all EOM myoblasts follow trajectories that cluster around the axis of the optic nerve. They are also in close proximity V1. Each muscle takes up its respective positions on the sclera, *bringing along its own V1 sensory branch*. Recall that at an earlier stage, supraorbital division of intracranial stapedial artery entered the orbit guided by V1. Hence, *each branch of V1 to an EOM induces an accompanying StV1 muscular artery*. Note that the motor branches to the EOMs (III, IV, and VI) do *not* induce the muscular arteries.

Vascular development of the embryonic arteries to the sclera and to the muscles follows a temporal–spatial gradient that explains the anatomy in its final state. As we shall see, the four recti convey blood supply via the tendons into the globe as *anterior ciliary arteries*. These supply conjunctiva, cornea, limbus, ciliary muscle, and iris.

V1 sensory nerves supplying the globe also have a vasoinductive role. Recall that *the initial orientation of the mammalian eye is lateral; it subsequently undergoes a 45° rotation*. The future nasal hemisphere is positioned forward while the future temporal hemisphere faces backward. The *primitive ventral (nasal) ophthalmic artery*, PrVOA, is induced by the anterior (rostral) ophthalmic branch of V1. This branch is likely associated with Sm1. The *primitive dorsal (temporal) ophthalmic artery*, PrDOA, is induced by the posterior (caudal) ophthalmic branch of V1. This branch is likely associated with Sm2.

We next consider the physical relationships that exist between the developing globe and the extraocular muscles. How can myoblasts from somitomeres gain access to the sclera? Let us review some key milestones. Gastrulation begins at stage 6b with the extrusion of prechordal mesoderm from the anterior dorsal lip of the primitive node. During stages 7–8, the process of gastrulation creates in three-layer body embryo. Mesoderm thus established becomes compartmentalized (in human embryos) into seven somitomeres Sm1–Sm2 in prechordal mesoderm and Sm4–Sm7 in paraxial mesoderm. Somitogenesis proceeds in cranial-caudal fashion, starting at the eighth somitomere. The process is controlled by a molecular clock that produces (in chicks) one somite every 90 min. Each cycle involves the expression of *c-hairy* mRNA. This gene is a homolog of the pair-rule segmentation gene, *hairy*, responsible for the segmentation of *Drosophila*. The human molecular "clock" is slower, producing one somite every 6 h. By day 30, the process is complete, with 36–38 somites produced.

Note that in humans, neural crest migration ensues at stage 9 and quickly covers over and penetrates the somitomeres. As each somite is produced, it receives neural crest as this constitutes the predominant source of mesenchyme for all the pharyngeal arches is neural crest. Within the myotome, neural crest cells serve to organize connective tissues and the future fasciae.

During stage 9, several critical events take place. First, explosive growth of the CNS forces a 120° bend in the neuraxis, the *cephalic flexure*. This has two consequences: (1) the paired heart tubes (now fused into a primitive heart) are now repositioned ventral to the future pharynx; and (2) the embryonic face now "faces" upward in direct contact with the mesoderm of the somitomeres.

Second, a *fundamental shift in circulation* from ebb-and-flow (stages 4–8) to a cardiac pump takes place. The heart, connected with the dorsal aortae, begins to beat. The primitive cardiovascular system must now develop new connections so that it can supply the entire embryo. Primary elements of the embryonic arterial system give way to important secondary (definitive) arteries. From the dorsal aortae arises the forward-directed *primitive internal carotid* arteries. These will supply the embryonic midbrain and forebrain. Paired *primitive trigeminal arteries* bud medially to provide a cross-link connection between the ipsilateral dorsal aortae (serving the entire embryonic body) and longitudinal neural arteries (supplying the embryonic hindbrain).

We cannot stress enough the importance of the physical relationships created at stage 9. The embryonic eye projection directly outward from the brain is now folded backward. It lies in nearly direct contact with the overlying somitomeres. As we shall see below, the future globe contains six binding sites for extraocular muscles. These sites are arranged in "gradient of biologic maturity" established previously by neurovascular development of globe. Myoblasts of the future extraocular muscles now have direct access to the eye. They will migrate outward from their respective somitomeres, seeking attachments to the sclera in a strict spatiotemporal order described below.

The physical layout of the circulation at stage 9 explains the development of the pharyngeal arches. Paired dorsal aortae lie in the axis of the body. Directly ventral to the DAs is another parallel system: the aortic outflow tract and the heart. The DAs above, and aortic sac below, are connected by two curved tubes; these are remnants of most proximal heart tubes. They do not fuse. They become surrounded by mesenchyme of the first pharyngeal arch. The initial "connecting tubes" acquire a new name: the *first aortic arches*. Subsequently (in mammals), five more pairs of arteries develop in cranial–caudal sequence between the aortic sac below and the dorsal aortae above. Clumped around each one is the mesenchyme of pharyngeal arches 2–5. Because the fifth pharyngeal arch involutes pharyngeal arches, 4–5 share a common blood supply from the fourth aortic arch artery. Note that the sixth aortic arch bears no relationship whatsoever to the face: it is associated with the lungs.

Traditional embryology texts depict the aortic arches are arising from the aortic sac. 3-D scanning electron micrographs indicate a different mechanism. Arterial budding takes place simultaneously from the dorsal arteries downward and from the aortic sac upwards. The buds connect, running a course directly through the pharyngeal arches. This makes eminent sense. Recall that all arteries require mesoderm for their walls. The pharyngeal arches represent a potential "mesenchymal conduit" through which dorsal and ventral buds connect and acquire a mesodermal sheath.

The following paragraph is critical to understand the mechanism by which the internal carotid develops.

An old adage states: a chain is only as strong as its weakest link. This is also true for the six aortic arches. Because each is constructed from two buds, dorsal and ventral, a *weak spot exists at the midpoint*. AA1 and AA2 fall apart while AA3 and AA4 stay intact. When this happens, *the definition of what is considered internal carotid now extends backward to AA3*. These two arteries share a common segment of aortic outflow tract and a common segment of dorsal aortae, thus forming a circle. What happens next is that the segment of dorsal aorta disintegrates while the ventral connection persists. AA3 will now forward feed the internal carotid while AA4 turns backward as the aortic arch to supply the body. The ventral segment between AA4 and AA3 becomes common carotid artery. The take-off of AA3 is connected to ventral pharyngeal artery; VPA subsequently morphs into the *external carotid system*. The portion of AA3 cranial to ECA is the vertical *extracranial internal carotid*. The first segment of horizontal internal carotid where AA3 enters deep to the otic placode becomes enclosed by paraxial mesoderm from Sm5, Sm6, and Sm7. This becomes *petrous internal carotid*.

Note that AA5 develops one stage after AA4 but it disintegrates *not so the fifth pharyngeal arch*. Blood supply to PA5 thus comes from AA4. Finally, AA6 is not associated with a pharyngeal arch. When its dorsal remnants disappear, its ventral remnants become the pulmonary arteries.

Recall from Chap. 4 that mesencephalon and prosencephalon share a common arterial supply, as a forward extension of the primitive internal carotid. The arterial axis to the midbrain arises from a posteriorly directed stem, the posterior communicating artery. PCom also gives rise to the axis of the

posterior diencephalon and the axis of the posterior cerebrum prior to joining the Circle of Willis. *It is a common neuroanatomic misconception to consider posterior cerebral artery (PCA) as originating from the vertebral-basilar system.* When one looks at the Circle of Willis, this may appear to be the case but, in point of fact, PCA is the final endpoint of the telencephalic arterial circuit.

The eye, arising as it does from the ventral aspect of prosomere 5, is exclusively supplied by branches arising from the primitive internal carotid. Recall that the pharyngeal arches are ventral structures; they are "newer" than the brain. Consequently, the arches (and the muscles assigned to them) are supplied by a newer vascular axis, the external carotid artery. Extraocular muscles are *not* affiliated with the arches. They are assigned to a neural structure, the eye. But the EOMs are "add on" structures to the eye. Not all vertebrates have EOMs; they are not present in agnathic fishes such as the lamprey and hagfish. Thus, the "invention" of EOMs during evolution took place during the development of jaws. For this reason, extraocular muscles require a "newer" blood supply, the stapedial system.

The primary insertions of the extraocular muscles make sense in relation to the developmental zones of the sclera. These are neurovascular. Recall that the blood supply to the anterior hemisphere of the globe is distinct from that of its posterior counterpoint. Branches from the primitive dorsal and ventral arteries pursue a posterior-to-anterior course following medial and lateral sensory branches of V1. Anteriorly, near the limbus, PrDOA and PrVOA give off branches that course backward within the choroid. This makes sense when one considers the tri-phasic topology of the globe. It begins as a vesicle, inverts upon itself as a cup, and finally the "rim" of the cup closes in upon itself. Thus, vessels that initially terminated at the anterior "apex" of the globe become drawn inward, making a U-shaped turn at the limbus with respect to their external axes.

We arrive at a critical point of integration. From a neurovascular standpoint, the globe can be divided into the *x*, *y*, and *z* axes into three sets of hemispheres, each of which has a more "primitive" half. In the coronal plane, posterior > anterior; in the sagittal plane, lateral > medial; and in the axial plane, dorsal > ventral hemisphere of the sclera can be considered to be the more "primitive." NOTA BENE: *the 3-axis model of the eye is constructed the same as that of the semicircular canals.* A nearly exact correspondence exists between the somatotopic organization of the vestibulocochlear system and that of the eye. The semicircular canals lie approximately 45° off horizontal. Thus, axes of the globe are arranged with a similar deviation. We can thus derive a "map" of the sclera with insertion sites arranged as follows: (1) LR, (2) SO, (3) SR, (4) MR, (5) IO, and (6) IR.

During development, the axes of PrDOA and PrVOA will follow a course that is posterior to anterior. Given a choice, an extraocular muscle will preferentially insert posteriorly. Migration will attain primary insertion anterior to the coronal boundary. Muscles with a later onset of migration are forced to accept a more anterior primary insertion site.

Medial Motor Column (r8–r11): Somites S1–S4

Two classes of hindbrain muscles: do not confuse the medial versus lateral motor columns.

Talk about schizophrenic! The developmental anatomy of the caudal medulla and the mesenchymal derivatives it supplies has been driving everyone crazy since time immemorial. Two functional groups of muscles have their neurologic home base in the same zone of caudal brainstem—although both are listed below, in this section we shall only deal with those of the MMC.

- Medial motor column muscles arise from somites Somites S1–S4 or S5 and are supplied by distinct roots of the r8–r11 hypoglossal nucleus via cranial nerve XII plus c1 via XII. These GSE muscles control the tongue. They are all hypaxial.
  - MMC continues downward through the entire length of spinal cord to supply both epaxial and hypaxial muscles controlling the craniovertebral joint and vertebral column.
- Lateral motor column muscles arise from somitomeres Sm7–Sm11 and are supplied by *nucleus ambiguus* via glossopharyngeal (r6–r7) and vagus (r8–r11) nerves. These SVE muscles control the palate, pharynx, and larynx. Thus, motor vagus can be divided into an SVE *palato-pharyngo-laryngeal nerve* and a GVE *viscero-intestinal nerve*. Differences among these groups are deeply rooted in the evolutionary transition to land. In this section, we shall concentrate on the myology of the medial motor column only.
  - LMC also continues downward into cervical spinal cord, becoming the central motor column. Despite the change in name, CMC still represents nucleus ambiguus which originally supplied branchiomeric muscles corresponding to the ancient branchial arches 6–8. Thus, muscles arising from c1 to c6 (sternocleidomastoid and trapezius) are intrinsically branchiomeric. These muscles are supplied by the *transitional nerve*, formerly known as *spinal accessory nerve* (SAN).

As you know, somitomeres 8–11 are transformed into occipital somites, S1–S4. These myoblasts constitute the *hypobranchial muscles* that migrate to the floor of the mouth to produce the segmented muscles of the tongue. They are all supplied by XII. Because the root of the tongue resides in the level of the third arch, the lingual muscles are also known collectively as the *prehyoid hypobranchial* column.

Not bene Medial motor column also extends posteriorly into the cervical spinal cord at neuromeric levels c1–c4. Neuromeres c1–c4 produce the *posthyoid hypobranchial muscles* which span from clavicle/manubrium to the hyoid apparatus.

Hypoglossal Nerve (Fig. 9.2 CN XII)

The 12th nerve emerges from the medulla via a series of rootlets tucked into the ventrolateral sulcus that spans from pyramid to olive. Hypoglossal makes the following connections:

- Vagus—the nerves run tightly together at the skull base
- Pharyngeal plexus—where hypoglossal crosses occipital artery
- SANS—superior cervical ganglion at C1
- Lingual nerve at anterior hyoglossus
- First and second cervical nerves share a common loop

Hypoglossal Nerve Has the Following Branches:

- Meningeal nerves. convey sensory fibers from connecting loop between C1 and C2 via hypoglossal nerve to posterior cranial fossa.
- Geniohyoid/thyrohyoid (conveys fibers from C1 of cervical plexus).
- Tongue.

Myology of Somitomeres 8–11/Somites 1–4

Evolution of the Tongue

All vertebrates must accomplish food transport from the mouth to the gut. In an aquatic environment, this is a straightforward affair: water currents generated by skull bones transport prey to the pharynx. Precursor structures for a tongue in fishes consist of a median ventral structure in the floor of the mouth, the *primary tongue*, that results from fusion of the *basihyal cartilages*, the most distal of the second arch. It has no role in food transport.

A mobile tongue appears for the first-time tetrapods as myoblasts from the occipital somites descend to populate around the *basihyal rod*, localized at the base of the tongue, where it serves as an anchorage point between the tongue mass and the hyoid arch skeleton. This structure dates back to *Eusthenopteron* (see Fig. 8.261 in Chap. 8). Note: The hyoid apparatus consists of a chain of small bones or cartilages suspended from the skull: tympanohyal, stylohyal, epihyal, thyrohyal, ceratohyal, and basihyal. Thus, it is not

Fig. 9.2 Cranial nerve XII, hypoglossal. Four neuromeres are represented (r8, r9, r10, r11), each with a specific muscle from somites S1–S4. Because these are read out in cranial to caudal order, we can appreciate how the tongue muscles are assembled from the midline laterally with insertion sites into r3 mandible (genioglossus), r5 hyoid (hyoglossus), and r7 styloid (styloglossus). (Reprinted from DE Haines. Fundamental Neuroscience for Basic and Clinical Applications, 3rd ed. New York: Elsevier; 2006. With permission from Elsevier)

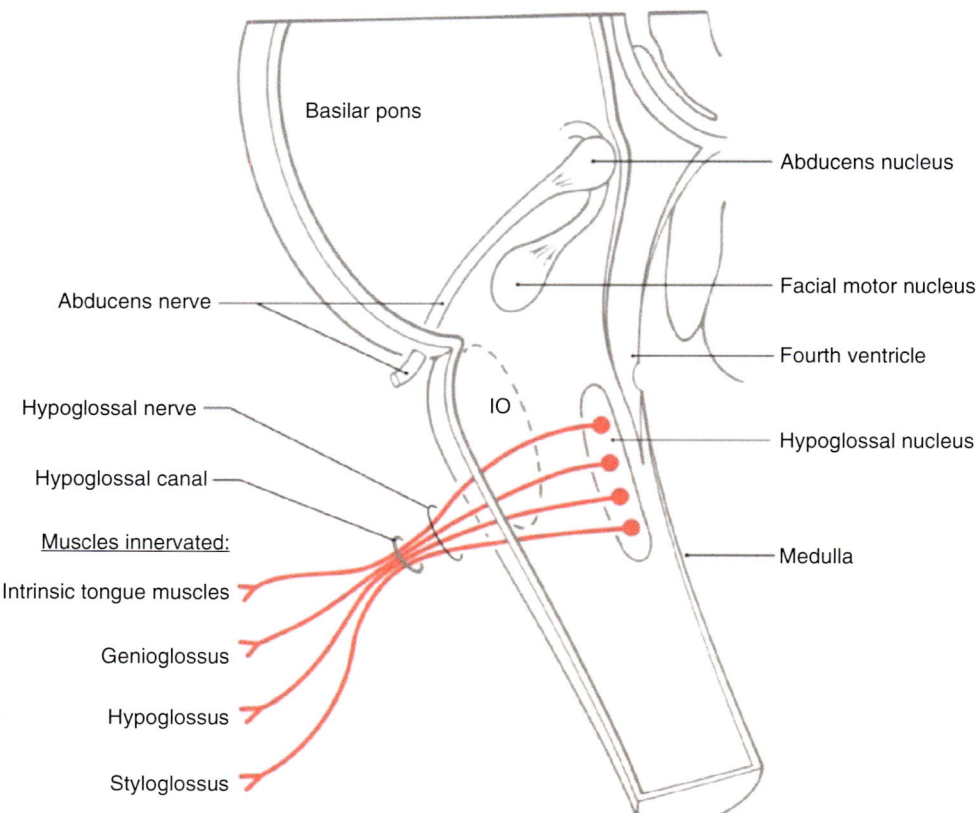

surprising that three of the four extrinsic tongue muscles are inserted into successive bony elements of this series: stylohyoid process, medial hyoid body, and lateral hyoid body.

(Figures 9.3, 9.4, and 9.5, hypobranchial muscles, muscles of the tongue).

The tongue is controlled by *internal series of hypobranchial* muscles, migrating forward in the midline; these are descendents from the chondrichthyan *coracomandibularis*. In tetrapods, this divided into a more superficial *geniohyoideus* running from the mandible back to the second arch and an internally placed *genioglossus* running from the tongue forward to the first arch (genial tubercle). Mammals demonstrate further specialization with styloglossus, chondroglossus, and hyoglossus belonging to the former (geniohyoid) while genioglossus and the intrinsic muscles belong to the latter.

Note: the *external hypobranchial* muscles appeared in chondrichthyans as the rectus cervicis series. These became the strap muscles, spanning between third arch hyoid and the manubrium/clavicle. These muscles originate from somites Sp1–Sp3. Geniohyoid represents the sole forward continuation of this series. For this reason, although it is adjacent to other muscles originating from the hyoid, serving a mastica-

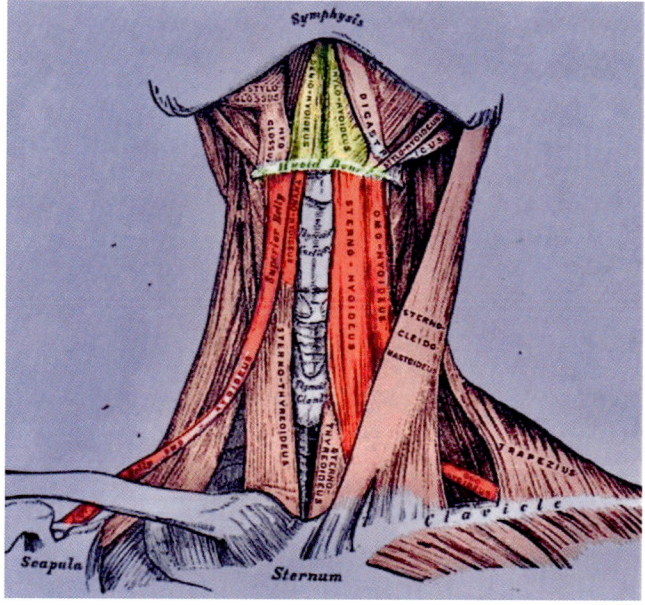

Fig. 9.3 Straps originate from S5 to S8 (C1–C4). S5 produces geniohyoid whereas the rest are polyneuromeric and development from S6 to S8. (Modified from Lewis, Warren H (ed). Gray's Anatomy of the Human Body, 20th American Edition. Philadelphia, PA: Lea & Febiger, 1918)

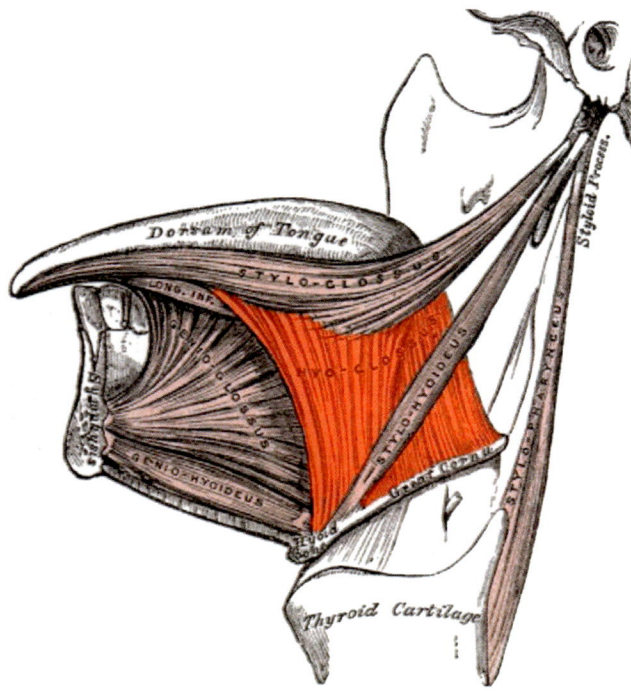

Fig. 9.4 Extrinsic tongue muscles migrate from somite S1 to S4 in medial–lateral order, attaching to progressively more posterior arch structures. This migration takes place internal to the migration for coracomandibularis, i.e., the strap muscles and geniohyoideus. (Reprinted from Lewis, Warren H (ed). Gray's Anatomy of the Human Body, 20th American Edition. Philadelphia, PA: Lea & Febiger, 1918)

tory function, developing from Sm4 and supplied by V3, genioglossus retains its relationship with the first cervical somite and is innervated by a C1 branch coming off of hypoglossal (Fig. 9.3).

Mobility and fine motor control are hallmarks of the tongue. To this end, it has two sets of muscles, four extrinsic and four intrinsic. These differ in their patterns of anatomic distribution and blood supply.

Extrinsic Muscles of the Tongue (Fig. 9.4)

These paired muscles are characterized by *well-defined boundaries*, a *layered arrangement*, and distinct *patterns of blood supply*. They reside within the parenchyma of the tongue proper, arranged in three longitudinal bundles separated by a *median fibrous septum*. From lateral to medial, these are as follows: styloglossus, chondroglossus, hyoglossus, and genioglossus. All four muscles share a common sensory supply (lingual branch of V3) and are innervated by hypoglossal nerve, XII.

Differences in arterial blood supply to the extrinsic muscles provide clues as to their developmental sequence. Genioglossus, chondroglossus, and hyoglossus all have dual arterial sources from two different sectors of the external carotid artery. *Submental branch of lingual* is a derivative of the AA1 sector of ECA. It exits from the first segment of the MMA. *Sublingual branch of facial* is a derivative of the AA2 sector of ECA. These share a common territory and supply

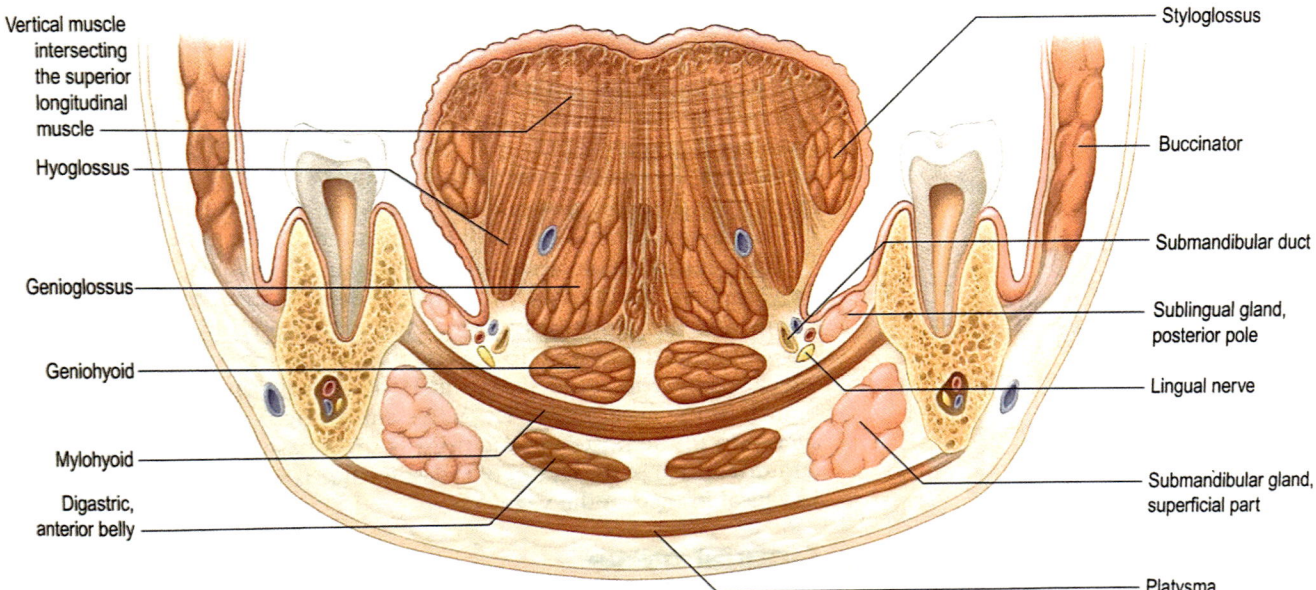

Fig. 9.5 Tongue muscles, intrinsic. (Reprinted from Lewis, Warren H (ed). Gray's Anatomy of the Human Body, 20th American Edition. Philadelphia, PA: Lea & Febiger, 1918)

mucosa of the anterior two-third of tongue. Styloglossus occupies the most lateral position on the second arch styloid process. It is supplied by the *ascending palatine br. of facial artery* (quite naturally) AA2 sector of ECA. As we shall see, this vascular arrangement is an important clue as to the developmental sequence of these muscles.

Styloglossus This muscle migrates downward and forward to insert into two proximal second arch fields: stylohyoid bone and upper stylohyoid ligament. It runs distally into the tongue, where it occupies the most lateral position adjacent to the lateral mucosal border. Note that styloglossus is supplied by ascending palatine branch of facial from the AA2 sector of ECA and also by ascending pharyngeal artery from the AA3 sector of ECA. Thus, it may share a common blastema with palatoglossus (*vide infra*).

Chondroglossus This 2 cm muscle is inserted into the medial aspect of the second arch (r5) upper body and lesser cornu of the hyoid bone. It becomes admixed with hyoglossus.

Hyoglossus Quadrangular in form, hyoglossus also has a broad proximal insertion along the r7 zones of the hyoid, i.e., the third arch lower body and greater cornu. It enters the side of the tongue sandwiched between genioglossus and styloglossus. This corresponds to the proximolateral field of the primitive synapsid mandible, the *angular bone*.

Genioglossus This muscle, triangular in shape, lies in the midline. It has an extensive proximal insertion beginning at anterior upper hyoid (r3). This position is due to the physical presence of hyoglossus, which occupies most available territory of the hyoid body. Fibers of this muscle extend backward to middle constrictor (Sm8). Its distal insertion is into the *superior genial tubercle* behind the symphysis, giving it the midline position. This corresponds to the distolateral field of the primitive synapsid, mandible, the *splenial* bone.

Palatoglossus Positioned like an outrigger, palatoglossus is spatially and developmentally *not a true tongue muscle*. Although this statement flatly contradicts traditional anatomy textbooks, it is supported by multiple embryologic factors. It is proximally inserted into the palatine aponeurosis on the *oral side* of soft palate. Recall that all muscles within this zone arise from Sm7 and consequently all receive sensory supply from cranial nerve IX and motor supply from r6 to r7 vis X pharyngeal plexus. [The nasal zone has one muscle from Sm4, tensor veli palatini, and one from Sm7, levator veli palatini.] Palatoglossus is distally inserted into the *dorsal mucosa* of posterior tongue (again innervated by IX) and into its intrinsic muscles. Unlike true tongue muscles, motor innervation for palatoglossus is not from hypoglossal, but from glossopharyngeal (via the pharyngeal plexus). It receives two sources of blood supply, both of which are hold-overs from the embryonic pharyngeal arch system. *Ascending branch of facial artery* is derived from the second arch; *ascending pharyngeal artery* is derived from the third arch.

Intrinsic Muscles of the Tongue (Fig. 9.5)

These muscles are characterized by *poorly defined boundaries*, a *diffuse arrangement of fibers* interwoven with those of the extrinsic muscles, and a *monotonous neurovascular supply*: hypoglossal nerve and lingual artery. Note that the primary insertions are peripheral to the axial extrinsic group.

Superior longitudinal These fibers lie just under the mucosa of the dorsum. They span forward from deep base of tongue in front of epiglottis and also from the midline septum, fanning outward to lingual margins.

Inferior longitudinal A mirror image to its confrere, these muscle fibers run forward from root of tongue to apex. Laterally, they blend with styloglossus.

Vertical Spanning from ventral to dorsal, these fibers are localized to the tongue apex.

Transverse These fibers pass from median septum outward to the submucosa of the lingual margin. There they intermingle with palatopharyngeus.

Angiology of the Tongue

Recall that the external carotid system develops as the sequential assembly of contributions from pharyngeal arch arteries 1–4. Its most proximal branch, superior thyroid, supplies fourth and fifth arch derivatives. Ascending pharyngeal belongs to the third arch. *Lingual artery* is the principal supply for the tongue. It arises from external carotid between ascending pharyngeal artery and facial artery just opposite the greater cornu of the hyoid bone. Lingual and facial often arises together as *linguofacial artery*; they subsequently separate. The intermediate position between third arch and the amalgamated first/second arch complex means that lingual artery supplies two distinct boundary zones. It extends backward to mucosa innervated by IX such as dorsum of tongue and tonsil. It also supplies mucosa innervated by V and VII. It does not supply the first arch masticatory apparatus nor the facial muscles. Thus, lingual artery is a unique structure spanning adjacent embryologic territories.

Lingual artery arises just opposite the greater cornu of hyoid. It runs along the surface of medial constrictor and then passes deep to hyoid at its posterior border. It then makes a downward loop and is crossed by hypoglossal nerve XII. It dives internally at *posterior border* of hyoglossus and enters the tongue deep to that muscle hyoglossus.

The paramedian position of lingual artery in the tongue has embryologic consequences. After giving off its branches, it runs forward sandwiched between the genioglossus medially and its companion, inferior longitudinal muscle. Defects of this neuroangiosome can affect development of midline and result in a cleft. These could also arise from deficiency of the genioglossus muscle.

Lingual artery has four branches. *Suprahyoid artery* runs along the upper r3 border of hyoid bond. In addition to adjacent structures, it supplies the mucous membranes and gingiva of the proximal internal mandible, formerly the *proximal splenial bone field*. Two to three *dorsal lingual arteries* course (predictably) medial to hyoglossus. They supply mucous membranes: posterior tongue, around to tonsillar pillars (and tonsil), downward to epiglottis, and upward to soft palate. *Sublingual branch* takes off at *anterior border* of hyoglossus. It runs sandwiched between genioglossus medially and mylohyoid. It supplies sublingual gland superficial and proceeds deep to supply the mucous membranes and gingiva of the distal upper internal mandible, *distal splenial bone field*. *Deep lingual artery* is the terminal branch. It can be located at the frenulum. Note that the symphyseal bone fields are chondral derivatives from neural crest, developmentally distinct from the rest of the membranous mandibular bone fields. Therefore, it is not surprising that the blood supply to the chin is also via a separate, submental branch of the cervical division of facial artery.

Phylogeny and Development of the Tongue

As we know, the muscles of the tongue originate in the occipital somites a site very far removed from the floor of the mouth. How do they get there, and in what order? Once in place, what happens? Let us start with some evolutionary history. We will then review the relevant neuroembryology and the role of the hypoglossal cord. This leads to a consideration of mechanism and order of muscle migration and vascular supply. Finally, we shall look at the relationship between the tongue, surrounding pharyngeal arches, and neuromeric coding of the lingual mucosa.

Aquatic and terrestrial vertebrates differ in the mechanism they use for disposition of food within the oral cavity. Once you have that tasty morsel, what to do with it? In aquatic feeding, movements between the skull bones and hyoid apparatus create water currents that food back into the pharynx. Precursor structures for a tongue in fishes consist of a median ventral structure in the floor of the mouth, the *primary tongue*, that results from fusion of the *basihyal cartilages*, the most distal of the second arch. The piscine *primary tongue* is a specialized structure of the oral floor that can have teeth. Although it does not participate in transport, it may serve to hold onto ingested food.

Tetrapods invented the mobile, muscular tongue by combining muscle elements from their occipital myotomes with osseous structures of the first, second, and third pharyngeal arches. Myoblasts from the S1–S4 (some authors add the fifth somite as well) descend to populate zone around the *basihyal rod*, localized at the base of the tongue. The basihyal is a remnant of the fish model and constitutes an anchor-

age point between the tongue mass and the hyoid arch skeleton. Note: The hyoid apparatus consists of a chain of small bones or cartilages suspended from the skull: tympanohyal, stylohyal, epihyal, thyrohyal, ceratohyal, and basihyal. Thus, it is not surprising that three of the four extrinsic tongue muscles are inserted into successive bony elements of this series: stylohyoid process, medial hyoid body, and lateral hyoid body.

Control of a mobile tongue requires two separate muscle systems. (1) Muscles projecting from a bony platform (the hyoid) into the substance of the tongue originate from somites S1 to S4. These are the extrinsic and intrinsic muscles of the tongue. Because these are anchored to the third arch hyoid bone, they are termed *prehyoid hypobranchial* muscles. The tongue muscles migrate first and so occupy an internal position within the floor of mouth. (2) The hyoid is a mobile structure. Muscles that stabilize the hyoid with respect to the pectoral girdle originate from somites S5–S8. These are so-called strap muscles. Because they extend from the original coracoid process forward to hyoid, they are termed *posthyoid hypobranchial* muscles. The strap muscles migrate external to the tongue and span from the pectoral girdle to hyoid and thence forward via geniohyoid to mandible. How does this system arise?

The most primitive version of this system is seen in the Dipnoi, i.e., the lungfishes. Here, four occipital and four spinal somites give rise to a single column of muscle, the *coracomandibularis* that descends beneath the branchial arches to attach into the lower jaw. At the hyoid, it breaks into two components: an anterior *geniohyoideus* and a posterior *rectus cervicis*. In elasmobranchs (sharks), geniohyoideus grows backward internal to rectus cervicis to create single-column genio-coracoideus.

Tetrapods undergo modifications of their hypobranchial muscles in the process of creating a tongue and straps muscles. The prehyoid segment of geniohyoideus produces an additional internal derivative, genioglossus. In mammals, both components produce the definitive elements of the tongue. Genioglossus gives rise to an additional *lingualis*, the source of the intrinsic tongue muscles. Anterior geniohyoideus produces in turn three muscles of the tongue: *styloglossus*, *chondroglossus*, and *hyoglossus*. Posterior geniohyoideus becomes two internal straps, *sternothyroid* and *thyrohyoid* that function in series to unite inner manubrium with fourth arch with third arch. External to them, posthyoid rectus cervicis gives sternohyoid and omohyoid. This explains why the straps exist in two planes.

The tongue has many uses. The distal part of the amphibian tongue, the *gland field*, arises as two ventromedial swellings interposed between the secnd (hyoid) arch and the 1st (mandibular) arch. Amniotes add on to the gland field by means of rostrally placed *lateral lingual swellings*; the combination is known as the *tuberculum impar*. Tetrapods now can position food with respect to the teeth and convey it backward for swallowing. In mammals, the oral cycle is biphasic: (1) protrusion of the tongue and upward extension against the palatal rugae, followed by (2) tongue retraction and shortening away from the palate.

Interesting specializations of the tongue are seen in nature. Frogs, salamanders, and chameleons have a sticky substance on the tip of their tongues with which they can snatch up prey with amazing speed. Woodpeckers store their spiny tongues in a hypobranchial "shed" extending back to the skull base. After using their beaks to drill a hole in a tree, the tongue is inserted to seek out insects. Tongue muscles of anteaters extend caudally to the sternum! Many carnivores have raspy tongues, as cat lovers can attest.

Tongues can bear other sense organs besides taste. Snakes pick up odors with their forked tongues, which they subsequently insert into their vomeronasal organs for analysis. Dogs cool themselves off in the summertime by panting. Mammals make use of the tongue to suckle. Monotremes, like the duckbill platypus, cannot nurse (that horny beak gets in the way) so they lap up milk that is secreted onto hairs.

Tongue muscles originate in the four occipital somites (S1–S4). Geniohyoid arises from S5 and its C1 nerve travels with hypoglossal. Note that although geniohyoid gives rise to three tongue muscles, it remains outside the tongue and functions as a strap muscle. It merely completes the muscular chain leading from pectoral girdle to mandible coracomandibularis.

What mesenchymal products to the occipital somites produce? As we already know, the occipital somites and the first cervical somite (S5) are "incomplete" in their composition as compared with "fully developed" somites. Although initially all somites possess dermatomes, those of the occipital somites disappear. The myotomes of S1–S4 are also anomalous. Their epaxial moieties involute while hypaxial zones remain. Thus, all occipital somite muscles are exclusively ventral. [This will have great significance later on when we examine the peculiar behavior of trapezius.] Beginning with the fifth somite, epaxial muscles make their debut. C1 supplies epaxial rectus capitis and hypaxial longus capitis. With the sixth somite, stable dermatomes appear. C2 innervates epaxial occipital scalp and hypaxial cervical plexus.

Occipital myotomes mature in craniocaudal order. Their growth is directed ventrally. Due to the pre-existence of the posterior arches (in mammals, 3, 4, and 5), the myoblasts must follow a posterolateral course until reaching the floor of the pharynx. At the ventral margin of the third pharyngeal arch, they encounter the developing hyoid bone at which point they will seek out available binding sites. Recall that the base of the tongue is anchored from.

As we recall from Chap. 10, the hyoid bone has multiple binding sites, each of which will become the attachment site of a muscle depending upon the spatiotemporal order of their arrival. By the time the tongue muscles reach the hyoid apparatus, one of these has already been pre-empted and is consequently not available. Medial constrictor is a branchiomeric muscle from r7 to r8. It migrates from Sm8 prior to the conversion of the latter into the first somite. Hence, its development precedes that of the somitic myotome. Because it makes up the pharynx, its course is internal. Its distal attachments are first to hyoid at the superior border of greater cornu (and more distally to posterior thyroid cartilage). The hyoid binding site for middle constrictor is totally logical. This zone is the most posterior and cranial part of hyoid, i.e., the most spatially accessible.

Extrinsic lingual muscles follow the rules. Respecting the position of middle constrictor, they pass externally and distally to find new binding sites. Styloglossus is the lead muscle. Its primary insertion is from styloid process and stylomandibular ligament. Its secondary insertion occupies the most lateral zone of the tongue. Chondroglossus (putatively part of hyoglossus) come next, binding at the posterior superior border of r3 hyoid. Its secondary insertion positions it medial to styloglossus. Hyoglossus is now forced outside. It attaches lateral to its predecessor, at the later-maturing r5 lower body of hyoid. It is admixed with chondroglossus and shares the same intermediate position. Genioglossus is last extrinsic muscle to develop. It is crowded out by hyoglossus so its primary insertion is to the upper border of hyoid body. It attaches to superior mental spine directly above the insertion of geniohyoid. As a consequence, it occupies the midline of the tongue. Bifid tongue abnormalities reflect developmental failure of this muscle.

Blood supply to the tongue is also logical. Recall that chondroglossus, hyoglossus, and genioglossus all have a dual arterial supply. Submental lingual artery represents the actual neurovascular axis of these muscles. Sublingual facial artery is a regional pharyngeal vessel. When the tongue myoblasts achieve their final position, they acquire this additional blood supply "by default." Styloglossus is a straggler, it lives by grace of the lingual artery alone. Note that genioglossus in the midline is supplied by the furthest away from lateral blood supply. The axis of the angiosome runs from proximal to distal.

How do tongue muscles "know" to enter the oral cavity? Once again, the likely mechanism has to do with pre-existing structures. Prior to the arrival of tongue myoblasts, the oral floor consists of bony walls (the mandible) strapped together by a muscle sling, the Sm4 mylohyoid and geniohyoid. Just below lies Sm6 platysma. Just above lies the oral mucosa of the primitive floor of mouth. So given the above, tongue myoblasts simply have no alternative, they flow into the potential space between oral mucosa and mylohyoid.

The submucosa of the oral floor is constructed from three parallel "swatches" of neural crest as indicated by their sensory innervation. V3-innervated neural crest from r3 makes up the anterior three-fourth of the floor. Into the back half of V3 flows taste-bearing mesenchyme innervated by VII, the source of which is likely r5 neural crest. IX-innervated neural crest from r7 makes up the posterior one-fourth of floor and brings in its own complement of taste receptors. Thus, the neuromeric diagram of floor of mouth would like a series of colored stripes, like Lifesaver® candies. As the mesenchymal components of the tongue proliferate, mucosal representation of second arch obliterated. But the contribution of taste glands from second arch is preserved. Thus, we can visualize two different maps of the tongue: one for mucosa and one for submucosa. Given blue for V3, red for VII, and yellow for IX, the mucosa is as follows sublingual mucosa and anterior two-third of tongue in blue with posterior one-third of tongue in yellow. However, if we consider the subcutaneous histologic components, the anterior two-third becomes purple (blue + red), reflecting the second arch elements and innervation for taste by chorda tympani.

Buccopharyngeal membrane is an embryologic barrier at which a change in oral lining takes place. The epithelium of the mouth and upper pharynx is ectoderm while that of the lower pharynx and larynx (and the remainder of the gut) is endoderm. Where does this transition take place? *Waldeyer's ring* is the landmark of the buccopharyngeal membrane. It contains lymphoid tissues which are "assigned" along a biologic line extending from posterior to anterior and from cranial to caudal as follows: adenoid tonsils > peritubal tonsil > soft palate tonsil > lingual tonsil. Embryonic tonsillar tissues are tucked unto recesses formed by the confluent posterior margins of arches 1–3. These sites are all areas of redundant blood supply. Descending palatine (StV2) from internal maxillary relates to first arch, ascending palatine from facial artery relates to second arch, and tonsillar branches from ascending pharyngeal relate to third arch. If tonsils result from lymphoid precursors deposited hematogenously, the blood supply to Waldeyer's ring is propitious indeed!

BPM lies behind Waldeyer's ring. It demarcates ectoderm from endoderm. It also marks a boundary between third arch and fourth arch and between superior constrictor and middle constrictor. Sensory supply to the pharyngeal mucosa also changes. Superior constrictor is purely IX. Fine touch is sharply localized here. A small instrument (such as a dental probe) placed on this mucosa can be detected with pinpoint accuracy a retched (sic) sensation! Pharyngeal plexus is an admixture of glossopharyngeal and vagus. It begins in earnest at middle constrictor and proceeds downward toward the esophageal junction. Vagus sensation is less well localized, although it does report to descending nucleus of V. As we shall see in Chap. 12, sensory supply to the esophagus makes an interesting transition between neuromeric levels c4 and c5.

All this fuss about Waldeyer's ring comes into play when we consider the point at which the lingual muscles swing internally toward hyoid bone and their floor of mouth destination. Let us look for a moment at the boundary between superior constrictor and middle constrictor, i.e., between Sm7 and Sm8. At this location, two third arch structures swing inward to penetrate the pharynx: cranial nerve IX and its muscle, stylopharyngeus. In addition, the constrictor muscles are all laminated. Middle constrictor is *external to* and *overlaps* inferior constrictor. Recall that the distal insertion of middle constrictor spans from r5 stylohyoid forward to the entire r5 greater cornu.

Let us see how these facts fit together to explain the biology of the extrinsic tongue muscles. We know that these muscles arise from the occipital somites but their exact origins are unclear. Let us postulate (1) that the tongue muscles develop in a linear order; (2) that they come from individual somites; (3) that they track inside one another; and (4) that they insert into progressively more anterior arch structures. That gives us the following model: styloglossus (S1) > chondroglossus (S2) > hyoglossus (S3) > genioglossus (S4). Let us further postulate that middle constrictor arises from Sm8 prior to its conversion into S1 and finally bear in mind that Sm7 *palatoglossus is already in place.*

The tongue muscles now find their way into mouth with military precision. The charge of the "lingual light brigade" sweeps past middle constrictor and then "turns its flank." Sensing a weak spot in the defensive line, the tongue muscles breakthrough between Sm7 and Sm8, grab onto the hyoid bone, and thrust themselves forward into the mouth. The space between palatoglossus and middle constrictor is that the "*open door to the oral floor.*"

Finally, a word about geniohyoid, a muscle we shall consider further in Chap. 10. Geniohyoid is the terminal muscle of the "strap series" that spans between the anterior lip of the second arch hyoid body and the first arch genial tubercle of the mandible. Because it is synthesized after the tongue muscles, its trajectory passes beneath them so it forms part of the floor of the mouth that supports the tongue. It shares pretracheal fascia with ventral parts of all five pharyngeal arches (r3, r5, r7, r9, r11).

Lateral Motor Column, Rostral (r3, r4–r5): Somitomeres Sm4 and Sm6

The First Pharyngeal Arch

All muscles of first pharyngeal arch develop from the fourth somitomere. They consist of two distinct functional groups: control of mastication and communication from the auditory tube to soft palate and to middle ear. Recall that the anterior border of tympanic membrane is first arch and innervated by V3. Motor control for these muscles is provided by cranial nerve V3, the nucleus of which resides in r3. Sensory supply to muscles from Sm4 comes from V3 as well. It also innervates all r3 dermis. Our plan is as follows. (1) We consider the innervation design of the first arch. (2) Anatomy of individual muscles will be discussed. (3) We will next examine the blood supply to the first arch in general, and to its muscles in particular. (4) Finally, we will look at the temporal–spatial sequence of the first arch muscles. Evolutionary considerations of the craniofacial muscles will be discussed.

Trigeminal Nerve

The trigeminal nerve is in register with three rhombomeres, r1–r3. It is physically the largest cranial nerve. V has enormous sensory representation: frontonasal skin, the majority of the scalp, facial skin, oral and nasal mucosa, teeth, dura mater, and cerebral blood vessels. It is motor to the muscles of mastication and suprahyoid muscles, anterior digastric and mylohyoid. It is proprioceptive for the above muscles plus the extraocular muscles, and the muscles of facial expression. Note that proprioception from the latter muscles is not as important as in non-craniofacial skeletal muscle. The more important sources of craniofacial proprioception come from the mechanoreceptors of the skin, mucosa, and periodontal membranes. Such information is vital for chewing (biting an apple versus chewing a steak), language production (tongue placement), and the subtleties of facial expression (the broad grin, the arched eyebrow). The smaller motor root lies medial and anterior while the much larger sensory lies posterior and lateral. Because of its more exposed position, sensory V is in contact with three different arteries: superior cerebellar, anterior inferior cerebellar, or pontine branches from the basilar artery. These can all cause trigeminal neuralgia. [Note that basilar results from the fusion of two longitudinal neural arteries, each of which supplies the hindbrain by multiple penetrating vessels.]

The *sensory root* enters the ventrolateral pons. Its axons have the cell bodies in the semilunar (Gasserian) ganglion. *Motor root* exits medial and rostral to sensory root. It also includes proprioceptive fibers from the muscles of mastication; these are connected with the mesencephalic nucleus. A poorly appreciated *intermediate root* is tucked in between the other two roots.

The three roots pursue an intracranial course from the posterior fossa forward to form the semilunar (Gasserian) ganglion. This structure is large by neurologic standards, 2 cm × 1 cm. This lies within a shallow depression (Meckel's cave) located at the ape of the petrous apex. This is located approximately 5 cm deep to the root of zygomatic process of temporal bone. Medial to the ganglion lies the foramen lacerum containing auditory tube cartilage and the internal carotid artery within foramen lacerum. Running beneath the ganglion (and above the petrous bone) are two nerves of

importance: motor root and greater petrosal nerve. While passing under the ganglion, motor root acquires sensory fibers, becoming a mixed nerve, V3.

Ganglion sends branches to tentorium cerebelli and the dura of middle cranial fossa. The floor of the fossa thus has dual programming with r2 neural crest medial and r3 neural crest lateral. A line traced forward from foramen ovale to foramen rotundum demarcates the field boundaries of alisphenoid in middle fossa. From ganglion are formed three divisions with which are associated four important PANS ganglia: *ciliary ganglion* from oculomotor (V1), *pterygopalatine ganglion* from facial (V2), *submandibular ganglion* from facial (V3), and *otic ganglion* from glossopharyngeal (V3).

Sensory Nuclei

Trigeminal sensory fibers, upon entering pons (but prior to reaching the nucleus) split 50:50 into rostral and caudal extensions. Ascending fibers connect with the ***pontine trigeminal nucleus***. Otherwise known as the ***principle sensory nucleus***, this important small but important structure is located in the dorsomedial pons.

Other ascending fibers connect with the ***mesencephalic nucleus*** of V. This nucleus resides in m1, m2, and r0. It contains unipolar cells the peripheral axons of which bring in proprioception. Structures innervated are extraocular muscles, periodontal ligaments and teeth, muscles of mastication, and facial animation muscles. This nucleus is the relay for the only non-spinal monosynaptic reflex: the jaw-jerk. Why should such an important relay center exist in the midbrain? The most obvious reasons are functional; they are related to the visual apparatus and the head. Vital reflexes such as avoidance of a flying object depend on "rapid response system." Visual cues are essential for smiling and biting. Just ask a vampire.

Descending fibers make up the *spinal tract of the trigeminal nucleus*. This extends down to upper cervical spinal cord, neuromeres c1–c4. All along the length of the spinal tract, at multiple levels, fibers plug into the ***spinal trigeminal nucleus***. It, in turn, has three components: *nucleus oralis*, *nucleus interpolaris*, and *nucleus caudalis*. We shall hear more about these later. Note the importance of neuromeric level c4, a key developmental landmark. Spinal trigeminal nucleus and spinal accessory nucleus both terminate here. It also marks the transition between cervical plexus and brachial plexus.

Trigeminospinal tract (and the nucleus it reports to) is the master sensory receptor of the face. It registers somatic sensation (pain, temperature) from all five pharyngeal arches. Embryology is parsimonious. Separate sensory nuclei for each arch would be illogical; information from structures innervated by VII, IX, and X is channeled to a single reference center. Thus, trigeminospinal tract receives *general somatic* fibers (including proprioception) from skin and muscle: extraocular (V1), mastication (V3), facial animation (VII), palate and pharynx (IX), pharynx and larynx (X). It also receives *general visceral* fibers from mucosa: naso-oral (V1–V3), palate-upper pharynx (IX), and lower pharynx-larynx (X). Note that *special visceral* fibers (taste) refer to a separate center, nucleus solitarius.

Not surprisingly, the trigeminospinal tract has a very precise somatotopic organization. Ophthalmic fibers (V1) are always ventrolateral; they extend down to C1. Maxillary fibers (V2) are intermediate; they extend to caudal medulla. Mandibular fibers (V3) are always dorsomedial; they extend past mid-medulla. The dorsal margin of trigeminospinal tract contains general visceral afferent sensory fibers from facial, glossopharyngeal, and vagus nerves. These fibers synapse in the nucleus caudalis.

Neurologists and neurosurgeons treating patients with facial pain have long recognized the clinical importance of the trigeminospinal tract because lesions here can cause dissociated sensory losses in the trigeminal distribution. *Wallenberg's syndrome* involves an occlusion of the posterior inferior cerebellar branch of vertebral artery. This results in loss of pain and temperature in the ipsilateral face but with retention of touch. In the 1890s, medullary tractotomy for paroxysmal trigeminal neuralgia was noted to cause dissociative thermanalgesia of the face. Pain was relieved but touch was preserved.

The organization of fibers in the spinal nucleus is also somatotopic although there are differing views as to its nature. The *rostro-caudal model* is as follows. V1 fibers are ventral but do not descend past c1. Maxillary fibers are intermediate and do not extend beyond r11. Mandibular fibers are dorsal; they terminate at mid-medulla. Neurosurgical ablations of the trigeminospinal tract have been used to map out the nucleus. Operative finding shows the *obex* to be a critical landmark. Just 4 mm below the obex, V1 and V2 areas are analgesic, but V3 is spared. Touch is also preserved, but not "tickle." When section takes place at the obex, V3 is knocked out as well.

Clinical findings also support a *dorsoventral model* of the tract. In this model, divisions V1–V3 all terminate throughout the nucleus. Ophthalmic (V1) does not go as far caudal. The posterior face (V3 next to C2) ends in the most caudal part of the nucleus while upper lip and nose are more rostral. Years ago, when tertiary syphilis was rampant, cases of *syringobulbia* were observed in which segmental parts of the tract were affected, causing sensory loss across all three divisions.

Very important for us to keep in mind is that the spinal tract of trigeminal contains common sensation (general visceral afferent) from territories supplied by facial, glossopharyngeal, and vagus nerves. Fibers from these nerves form a *dorsal column within the tract* that proceeds downward to the most caudal part of the spinal trigeminal, the subnucleus

caudalis. When the dorsal part of the tract is sectioned, the following non-trigeminal areas become anesthetic: tonsillar sinus (VII and IX), posterior tongue and pharyngeal wall (IX), and cutaneous ear (IX via X).

The lower spinal nucleus is accessed by two additional sources. The dorsal roots of upper cervical nerve c1–c3 are significant. Recall that spinal cord representation of non-craniofacial neuromeres begins with c4. The cervical plexus is constructed from the ventral roots of c1–c4. Sensorimotor cortex also provides feedback.

We should be in mind that the spinal nucleus of trigeminal has three subnuclei. *Subnucleus oralis* is the most rostral; it is plastered up against the pontine nucleus (nucleus principalis). This is followed by *subnucleus intermedius* and *subnucleus caudalis*. Trigeminal pain processed by these centers is often vague and hard to localize. This is because they receive a confluence of inputs, both superficial and deep. Thus, during dental procedures, mandibular fibers synapsing in caudalis bring in not only pain from tooth roots (remember how that drill feels) but also stimuli from the tongue and jaw.

Subnucleus caudalis has been extensively studied. It has a different structure, similar to the posterior horn of spinal cord. It is an important relay site for nociception. All rostral trigeminal nuclei project here.

You will note right away that the organization of craniofacial sensation differs from that of the spinal cord. From the neck on down, we see the pathways separate out. Somatic information is conveyed in the *spinothalamic tracts*. These are spatially organized by function. We will use the neural plate model, so from medial to lateral the sequence is: pressure > touch > pain > temperature. The spinothalamic tracts are also laid out by neuromeric segments in a dorsal-to-ventral sequence: c4 > s5. The most dorsal fibers begin with c4. [Once again, we observe the developmental importance of the c3–c4 junction.] The most ventral fibers are sacral. Thus, as one proceeds downward the cord, the spinothalamic tracts get thicker.

Proprioception, exteroception, and vibration are conveyed in the *posterior columns* (dorsal funiculi). The spatial layout of these tracts is diametrically opposite that of their spinothalamic siblings. Functional modalities have a dorsal-to-ventral sequence: touch > pressure > movement > vibration > pressure. The segmental sequence is medial to lateral: c4 > s5. Morphologically, the columns are described as fasculi. The more medial *fasciculus cuneatus* contains mid-thoracic to sacral fibers. The more lateral fasciculus gracilis contains mid-thoracic to sacral fibers. Thus, as one proceeds upward along the cord, the posterior columns get wider.

Motor Nucleus

This nucleus lies in the upper pons. It is just medial to the principle sensory nucleus. It lies within the lateral motor column and supplies branchiomeric striated muscles. It received afferent fibers from both the ipsilateral and contralateral cortex. Mesencephalic nucleus contributes fibers need for proprioceptive control of chewing (not to mention that favorite reflex of neurologists, the jaw-jerk). Medial longitudinal fasciculus conveys information from the visual system. Coordination between mastication and salivation arises from inputs by reticular formation, red nucleus, and nucleus coeruleus.

Trigeminal Nerve, First Division, v1

The nerve exits from the anterosuperior corner of the ganglion. Being directly adjacent to internal carotid, V1 naturally becomes incorporated into the wall of the carotid sinus. Cranial nerves III and IV are given off rostral to (and prior to) V1. Accordingly, they run closer to the undersurface of the brain. V1 has priority over V2. Within the lateral wall of the cavernous sinus, the same dorsa-ventral spatial order is preserved: III > IV > V1 > V2. As passes into the orbit, it is flanked medially by r1 orbitosphenoid (lesser wing) and laterally by r2 alisphenoid (greater wing) while in contact with internal carotid, V1 receives sympathetic fibers.

Tentorial Branch

This nerve comes directly off the ganglion. It crosses over lateral aspect trochlear nerve and accesses tentorium cerebelli between its two laminae. Recall that cerebellum arises from r0 and r1.

Lacrimal Nerve

The smallest of the three branches of V1, this nerve does not deserve its name. Lacrimal nerve's function is purely somatic sensory. It *cannot* per se supply lacrimal gland! So why the misleading name? Yes, it does *passively* convey secretomotor fibers from VII via V2 (see below) but its primary function is to supply dura of lateral orbit, the skin of lateral upper eyelid, and lateral brow. It has its own dural sheath. Entrance into the orbit is via either the extreme corner of superior orbital fissure or via a separate foramen. It runs high up along the roof, just outside of periorbita. V1 precedes lateral rectus so therefore LR lies internal to it. Lacrimal nerve enters lacrimal gland and accompanies lacrimal artery. It then passes outward to supply the skin of the extreme lateral upper eyelid (Tessier zone 9).

Zone 9 is a neurovascular watershed zone. Here, sensory supply from V1 lacrimal is confluent with that of V2. Lacrimal gland shows this embryologic mixture as well. Lacrimal mesenchyme comes from the hindbrain. This takes place when the head is flexed and the eye is "presented" to the ventral surface of the embryo. The blood supply to the gland comes from two vessels. Lacrimal artery is the first branch of StV1. It can be given off prior to entry into the orbit. This spot marks the fusion of the StV1 arterial system with primitive ophthalmic artery. An arterial communication between StV2 meningeal and StV1 lacrimal bears the misleading name of *recurrent meningeal branch of lacrimal artery*.

Innervation of the lacrimal gland takes place via hitchhiking. The original source of preganglionic PANS fibers is facial nerve. These arrive via greater petrosal nerve to the pterygopalatine ganglion. Why don't they simply jump onto V1 right away? Recall that because greater petrosal runs *under* trigeminal ganglion, it is 180° away from V1. Greater petrosal runs in close proximity to internal carotid artery. Here, it acquires additional SANS fibers from deep petrosal nerve. The combined GPN-DPN has the awkward name *nerve of the pterygoid canal*, or vidian nerve.

As we shall see, a lot happens at the pterygopalatine ganglion. Postganglionic PANS fibers jump onto V2; they are carried into the orbit via the zygomatic nerve. This gives off an ascending branch, zygomaticotemporal nerve, which subsequently connects with lacrimal nerve. *Congenital absence of lacrimal nerve* has been reported; in such cases it is replaced by zygomaticotemporal. Thus, zone 9 can be innervated by StV1, STV2, or both.

Dual neurovascular supply may be protective. Craniofacial clefts in zone 9 is extremely rare. Nonetheless, we can focus in on lateral eyebrow pathologies. Dropout of the brows in hypothyroid states is well documented. Eyebrows represent specialized interface zone. They can, at times, extend all the way across the midline, i.e., from zone 13 to zone 9. This would suggest that programming occurs at the junction where the frontal (supratrochlear and supraorbital) and lacrimal neurovascular axes undergo a dorsal-ventral "split." In the majority of individuals, zones 13 and 12 (supratrochlear) are not involved. In other words, eyebrows develop in skin supplied by the supraorbital and lacrimal nerves. In cases of lateral eyebrow "dropout" are we seeing an issue of inadequate programming in zone 9? And why should this zone be susceptible?

Hirsutism (or lack thereof) is closely tied to the endocrine system. Congenital hypothyroidism results in a fine coat of villous hair over the backs of affected newborns. Congenital adrenal hyperplasia (especially the 25-hydroxylase variant) causes hirsutism and, in girls, labial fusion and clitoral enlargement. Virilizing tumors of the adrenal gland (17-oxysteroid) and ovary (testosterone) cause hirsutism. Acromegalic facies have selective prominence of the medial two-third of the brow. Cornelia de Lange patients have confluent eyebrows; entire involvement of the forehead has been recorded. On the other hand, in hypothyroid states, alopecia of the later one-third of the brow is well documented. In this condition, the secretory cells of eccrine sweat glands are noted to PAS-positive granules.

This is the largest branch of V1. It supplies Tessier zones 10–13 of orbital roof, upper eyelid, and forehead. Frontal nerve enters the orbit outside the annulus of Zinn. It runs very high, just below periosteum and above levator palpebrae superioris. This indicates the primary of the nerve, as the muscle develops below it. At the midpoint of the orbit, near LPS, frontal nerve divides.

At this juncture, it is vital that we recall the relationship between STV1 and the extraocular arterial system. As previously discussed, the two axes of supraorbital division of stapedial artery (StV1) and primitive ophthalmic artery are embryologically distinct. With the involution of the stapedial stem, StV1 and prOA become fused into a single structure. Recall that StV1 and all its branches induce companion arteries. Although these have an apparent origin from the ophthalmic branch of internal carotid, they are embryologically distinct. Primitive ophthalmic and the derivatives it supplies are strictly ocular and have nothing whatsoever to do with Tessier cleft zones. True indeed, the eye has its own cohort of developmental pathologies (e.g., microphthalmia). These can occur with a normal orbital surround or with concomitant pathologies involving supraorbital stapedial neurovascular axes.

Supraorbital Nerve

Supraorbital nerve supplies Tessier zones 10 and 11 of the upper eyelid and forehead. It exits the supraorbital foramen (where it is sometimes seen to be duplicated) and divides. The inferior division gives medial and lateral branches directly into upper eyelid skin. Zones 11 and 10 account for virtually the entire upper eyelid, zones 10–11. [A small part of lateral upper eyelid, corresponding to lacrimal gland, is supplied by both StV1 lacrimal nerve and StV2 zygomaticofrontal nerve.] The superior division splits into similar zones but follows a more tortuous course. Its branches run at first deep to frontalis. Because the relationship between V1 and frontal skin predates the arrival of frontalis, the muscle flows around the medial branch. Medial branch innervates scalp up to the parietal bone (i.e., up to coronal suture). The lateral branch strays outward to the extreme margin of frontalis where it penetrates the epicranial fascia (SIF). Its trajectory is very extensive; it innervates scalp all the way back to the lambdoid suture.

Supratrochlear Nerve

Supratrochlear nerve supplies Tessier zones 12 and 13 of the nose and forehead. It pursues a similar course. Because the trochlea is embryologically "newer" than V1, the nerve becomes straddled between periosteum and trochlea. It first sends a small *branch to infratrochlear branch of nasociliary nerve*. It then supplies a small zone of upper eyelid skin and conjunctiva (zone 12). Its final trajectory is similar to supraorbital. First, it goes below corrugator and the medial fibers of frontalis. It then ascends to supply the skin of the midline forehead (zones 12 and 13). This artery provides the axis for

a very useful flap. The forehead flap has excellent color match for the nose and is a workhorse method for the reconstruction for nasal defects from trauma or tumor resection.

Nasociliary Nerve

This nerve is intimately associated with the globe. It thus enters the orbit within the confines of the annulus of Zinn. It is interposed between the inferior oculomotor ramus (to somitomere 1) and the superior oculomotor ramus (to somitomeres 2). It crosses above optic nerve along with ophthalmic artery. It is, of course, accompanied by its induction product, nasociliary artery. It runs between below both superior rectus and superior oblique. We can think of the nerve as having two structural components. Its first four branches are strictly ocular. They have nothing whatsoever to do with the bones or muscles of the orbit. Once medial long ciliary is given off, the nerve is strictly extraocular. It is developmentally associated with the nasal and ethmoid components of Tessier zones 12 and 13. Nasociliary gives off the following nerves *in sequence*.

Lateral long ciliary nerve pierces the posterior sclera near optic nerve. It runs forward external to choroid and deep to sclera until it reaches the limbus. The nerve is sensory to the entire lateral globe and adjacent conjunctiva, iris, and cornea. It carries with its SANS motor neurons from the superior cervical ganglion to dilator pupillae.

Ramus communicans enters ciliary ganglion, which lies lateral to optic nerve 1 cm posterior to the globe. It carries Pans fibers.

Short ciliary nerves 10–15 short ciliary nerves exit the sclera surrounding optic nerve. They connect to the ganglion and consequently to nasociliary nerve.

Medial long ciliary nerve continues in company with the remainder of nasociliary nerve. It is given off *prior* to the ethmoidals. Its sensory and motor functions are the same as its named companion.

Posterior ethmoidal nerve exits the orbit via its named foramen. This is a crucial surgical landmark. Using superior and medial subperiosteal dissection, it is easily identified, lying 1.5 cm anterior to the optic nerve. It supplies the sphenoid sinus and posterior ethmoid air cells.

Infratrochlear nerve is the *penultimate* nerve of V1. It runs above medial rectus but below trochlea. It supplies the skin of the medial canthus and descends down the lateral nasal wall medial to the nasolacrimal duct. Just lateral to the ducts, the skin is supplied by V2. The medial border of this nerve is lateral crus of lower lateral cartilage. The external nasal skin supplied by infratrochlear nerve is Tessier zone 12.

Anterior ethmoidal nerve This is the *terminal* nerve of V1. In human and high primates, foreshortening of the nose changes the topology of the orbit such that the anterior ethmoid appears to be the penultimate nasociliary derivative. The reverse is true. It exits the orbit with its own foramen and enters the anterior cranial fossa where it runs forward beneath the dura in a longitudinal groove on dura and ethmoid bone. This divides cribriform plate into Tessier zones 12 and 13. Just lateral to the crista galli, it encounters a slit and passes out of the cranium and into the nose. It now runs along a groove in the undersurface of the nasal bone (once again splitting the bone into two zones). It now gives off two *internal nasal branches. Medial internal nasal branch* descends along the septum toward incisive foramen where it will meet with V2 medial sphenopalatine nerve. *Lateral internal nasal branch* supplies the ethmoid bones of the lateral nasal wall, i.e., the superior and middle turbinates. Recall that inferior turbinate and the nasal lamina of prepalatine bone are supplied by V2 lateral sphenopalatine. As its final act, anterior ethmoid gives off the *external nasal nerve*. This exits from beneath the lower border of nasal bone. It runs forward to supply p5 vestibule, nasal lobule, columella, and philtrum. Note that paired anterior ethmoid nerves (and the arteries they induce) run about 2–4 mm apart to supply the columella and philtrum. Nasal skin supplied by these nerves is Tessier zone 13.

Trigeminal Nerve, Second Division, v2

The layout of V2 represents the "flow pattern" of neural crest mesenchyme from its origin at the second rhombomere into the face. It will be apparent that the V2 "family" of bone fields consists of several subgroups: zygoma, maxilla, lateral nasal wall (excluding the ethmoid turbinates, middle and upper) and hard palate, and medial nasal wall. Within the subgroups, individual bone fields are biologically independent. The loss of one field will not prevent formation of neighboring fields but it can lead to them to become distorted. From a clinical standpoint, we know that the "epicenter" of facial clefts lies between zones 2 and 3. As one proceeds outward in both directions, the incidence of facial clefts diminishes. Does this imply the existence of a "susceptibility gradient" among the various neurovascular axes? Research in this area may well lead to a common mechanism for all the Tessier cleft zones.

- Meningeal nerve
- Zygomatic nerve
  - Zygomaticotemporal (zone 8)
  - Zygomaticofacial (zone 7)
- Maxillary nerve, proximal (orbital and dentoalveolar)
  - Superior alveolar nerve, molars (zone 6)
  - Middle superior alveolar nerve, cuspids (zone 5)
  - Anterior superior alveolar nerve, incisors (zone 4)
- Maxillary nerve, distal (facial soft tissues)
  - Nasal 9 (alar) branches (ala) (zone 3)

- Labial branches (zones 4–5)
- Palpebral branches (zones 4–5)
- Pterygopalatine ganglion
- Descending palatine nerve
  - Greater palatine nerve: oral lamina of secondary hard palate (zone 3)
  - Lesser palatine nerve: soft palate
- Nasopalatine nerve
  - Lateral nasopalatine: inferior turbinate, nasal lamina secondary hard palate (zone 3)
  - Medial nasopalatine: vomer, premaxilla (zones 2, 1)

Meningeal Nerve

The anatomic function of this nerve follows the ground rules set up by the meningeal branch of V1. *It is the first branch given off by V1*. This follows the theme set up by V1: sensory supply to the dura, induction of accompanying arteries from stapedial stem, and autonomic control of those dural arteries (headache). V2 meningeal nerve is the first order of business of maxillary nerve. It is an intracranial nerve, being given off immediately before exiting foramen rotundum. As such, it supplies the dura mater of the anterior floor of middle cranial fossa and the greater wing of sphenoid. This tells us that these neural crest bone fields are synthesized from mesenchyme from r2.

V2 meningeal picks up SANS fibers from the internal carotid plexus. Vasodilation of StV2 arteries is a mechanism for headache in the temporal–parietal distribution. V2 meningeal induces the frontal branch of middle meningeal artery. Together they provide neurovascular supply to the anterior temporal and anterior parietal dura. V2 meningeal does not reach dura at the vertex of the skull. Recall that a swatch V1 meningeal nerve sweeps backward on either side of the sagittal sinus.

Ganglionic Branches

Having given off the meningeal nerve, V2 now makes two important connections with pterygopalatine ganglion. Understanding this ganglion is crucial. Due to its complexity, we will first describe the branches of V2 and then return to consider the ganglion in greater detail. For the moment, here are the highlights. (1) Certain derivatives of V1 have no direct anatomic pathway to convey sensory data back to the brain. These include extraocular V1 periosteum and V1 mucosa lining the nasopharynx. (2) Lacrimal gland, a hindbrain structure, receives parasympathetic motor fibers from facial nerve via greater petrosal nerve. At the ganglion, these fibers are conveyed to V2 and subsequently its very first branch, zygomatic nerve.

Zygomatic Nerve

Just beyond pterygopalatine ganglion, V2 enters inferior orbital and divides into two branches: zygomatic and infraorbital. Zygomatic runs along the lateral wall of the orbit and gives off two branches. Recall that the zygoma is comprised of two bone fields. The malar eminence represents the jugal bone and the lateral orbital rim is derived from the postorbital bone. Zygomaticofacial supplies Tessier zone 7 (the jugal field) and zygomaticotemporal supplies Tessier zone 8 (the postorbital field).

Zygomaticofacial nerve supplies Tessier zone 7. The nerve exits the orbit via a canal at its inferolateral angle of the orbit and emerges from a foramen in epicenter of the jugal field. It supplies the overlying skin, where it is joined by the palpebral branch from lateral infraorbital. Malar bone develops from neural crest deposited circumferentially around this axis. From it is also derived the malar fat pad. Because this structure tends to sag with age, aesthetic rejuvenation procedure seeks to reposition and anchor it. Later-arriving orbicularis flows around the nerve. Textbooks describe the nerve as "penetrating" the muscle.

Malar clefts can occur in this zone. The zone 6 cleft centers the suture between the maxilla and the jugal field. It is caused by a dropout of mesenchyme from the maxillary buttress. The zone 7 cleft is an outright deficiency of the malar field. In this situation, the maxillary buttress is intact by the jugal field is reduced. Because the jugal field extends backward and the *zygomatic process of the temporal bone*, a zone 7 deficiency results in absence of the anterior arch.

Zygomaticotemporal nerve supplies Tessier zone 8. The nerve climbs up the lateral margin of the orbit and sends a branch to lacrimal nerve. It then turns backward through a separate foramen in the epicenter of the postorbital bone field. It then ascends between bone and temporalis muscle. 2 cm above zygomatic arch, ZT "penetrates" the temporalis muscle and fascia to emerge to supply the skin of the temple. It is joined there by the frontal branch of V3 auriculotemporal nerve.

A number of interesting pathologies take place in this zone. Postorbital defects can range from defective formation of the lateral canthus all the way to outright loss of the lateral orbital. When mesenchymal deficiency involves the skin, the sideburn is dragged forward. This is noted in Treacher–Collins–Franceschetti syndrome.

Lacrimal gland does not develop from mesenchyme in neuromeric register with the midbrain. Its neural crest is likely derived from r1 and possibly r2 neural crest. Epithelial ducts pass into its substance from the r1 conjunctiva of the upper lid. Its primary blood supply comes from two stapedial derivatives: StV1 lacrimal artery and StV2 recurrent meningeal artery. For this reason, lacrimal gland is described as having a palpebral zone (r1) and an orbital zone (r2).

Lacrimal gland has a dual phylogenetic origin from oil glands and serous glands. In non-primates, the serous lacrimal gland is found in the lower eyelid; therefore, it develops strictly from r2 mesenchyme. In primates, the serous gland "migrates" to the upper lid. This represents additional mesenchyme from r1 neural crest. Consequently, the gland is supplied by V1 lacrimal nerve and the StV1 lacrimal artery.

Lacrimal gland, despite its location in the orbit, does *not* get its PANS motor supply from oculomotor nerve (via ciliary ganglion). It is, in point of fact, a hindbrain structure. Its preganglionic PANS fibers arise from a special *lacrimal nucleus in lower pons*. This nucleus is co-linear with, but separate from, salivatory nucleus. These secretomotor fibers have their cell bodies in the geniculate ganglion of facial nerve. Fibers now travel outward as greater petrosal nerve to pterygopalatine ganglion. From here, the pathway is V2 > zygomatic nerve > zygomaticotemporal nerve > lacrimal branch of ZT to V1 lacrimal nerve.

Maxillary Nerve, Proximal (Dentoalveolar)

Posterior Superior Alveolar Nerve

Terminology has made this nerve confusing; it isn't. It supplies Tessier cleft zone 6. Functionally, it innervates the permanent molars, the posterior and lateral walls of the maxilla (including the buttress), and the buccal lamina of the alveolus from the third molar all the way forward to include the first premolar. The nerve arises from the trunk of V2 *prior to* the infraorbital groove. It travels around the maxillary tuberosity, sending branches to the mucoperiosteum of the lateral gum and to adjacent mucous membranes of the cheek. It enters the alveolar bone through multiple canals. From here, it sends branches inward to the mucous membrane of the lateral maxillary sinus. It proceeds forward to supply the apices of the molar teeth. Neural crest mesenchyme flowing along this nerve is responsible for the synthesis of the buccal alveolar lamina, molar region.

Here, the terminology gets confusing. In the mandible, we have a very simple situation. The inferior alveolar artery is the neurovascular supply to the dentary bone. It supplies all the teeth, period. Traditional anatomic texts divide superior nerve into three branches: posterior middle and anterior. This creates confusion. Why is there not a single dental neurovascular pedicle?

From the alveolar bone point of view, PSAN supplies the buccal alveolus all the way to canine. Why is it considered to spare the premolars? After all, the deciduous teeth in this zone were molars, not premolars. We also know that middle superior alveolar nerve is quite variable. Sometimes it is duplicated; sometimes it is absent. So why split hairs?

Probably, the best perspective on this problem is clinical. Craniofacial clefts occur in distinct patterns in zone 4 (medial to infraorbital nerve) and in zone 5 (lateral to infraorbital nerve). These are separate and distinct from the clinical findings in zone 6. Retention of traditional nomenclature continues to distinguish between the three classes of teeth as well.

Middle Superior Alveolar Nerve

This is the nerve of Tessier cleft zone 5. As stated, this branch is variable. It departs quite proximally from the infraorbital groove and runs down the inner aspect of the lateral anterior maxillary wall. On its way, it supplies sinus mucosa and thence innervates the two premolar teeth. Neural crest mesenchyme flowing along this nerve is responsible for the synthesis of the alveolar lamina, cuspid region.

Anterior Superior Alveolar Nerve

This nerve supplies Tessier zone 4. It departs distal to MSAN at the midpoint of the infraorbital groove. Inexplicably, its take-off from the parent nerve occurs from *lateral aspect*, at the midpoint of the infraorbital groove. Of great importance is the *nasal branch of ASAN*. It passes through a hiatus into the nose where it supplies the anterior lateral floor of piriform fossa. It ascends up the lateral wall all the way up to the level of the maxillary sinus meatus (middle turbinate). *Nasal branch of ASAN supplies the frontal process of the maxilla.* It proceeds medially to the anterior nasal spine. This neurovascular field supplies the *external (labial) lamina of the incisor-bearing alveolus.* Deep to this lies the *internal (lingual) lamina of the incisor-bearing alveolus.*

The implications of this anatomy for cleft surgery are simply stunning. Teeth develop within a bilaminar structure, each wall of which has a separate neurovascular supply. Lateral to the piriform fossa, the alveolus consists of two maxillary bone fields. The external lamina is supplied by posterior superior alveolar, while the internal lamina is supplied by greater palatine. The origins of the piriform rim and the alveolar housing of the incisors are as follows. *The maxilla provides the external frontal process and the external alveolar lamina. These are supplied by medial branch of infraorbital and by the nasal branch of ASAN. The premaxilla provides the internal frontal process and the internal alveolar lamina. These are supplied by the medial nasopalatine artery.* Thus, development of both mandible and maxilla follows exactly the same embryologic principles.

Maxillary Nerve, Distal (Facial Soft Tissues)

Infraorbital Upon exit from the foramen of the same name, infraorbital divides into two branches supplying Tessier zones 3, 4, and 5. It innervates the mucoperiosteum of the external (labial) alveolar lamina containing incisors and premolars (zones 4–5); the skin of the entire lower eyelid (zones 4–5); the upper lip, nasal skin of the ala, and anterior (non-vestibular) nasal floor (zone 3). An important medial branch supplies the frontal process of maxilla at the lateral canthus; it joins zygomaticofacial nerve while at the membranous septum it joins with external nasal branch of anterior ethmoidal nerve that supply the nasal wall behind.

Pterygopalatine Ganglion / Sphenopalatine Ganglion

This is the largest PANS ganglion. It lies deep in the pterygopalatine fossa, anterior to foramen rotundum and to the pterygoid canal. From a functional standpoint, this ganglion is an extension of facial nerve. It is the vehicle by which PANS fibers from VII obtain access to the orbit and naso-oropharynx. Because it is so intimately associated with V2, its fibers are all distributed along V2 and, indirectly, to V1.

Nerve of the pterygoid canal is the *parasympathetic motor root* of the ganglion. It contains preganglionic PANS fibers from lacrimal nucleus in the lower pons. These proceed to lacrimal gland as follows: VII greater petrosal > VII n. of pterygoid canal > V2 zygomatic > V2 zygomaticotemporal > V1 lacrimal. Secretomotor PANS fibers from the VII-linked superior salivatory nucleus are distributed via V2 to glands in the nose (via V1 collaterals), pharynx, and palate.

The *sympathetic motor root of the ganglion* is also found in the pterygoid canal. Pre-ganglionic fibers from the carotid plexus synapse in the superior cervical ganglion. From there, they travel via deep petrosal nerve until reaching pterygoid canal.

Orbital branches travel through the posterior ethmoid foramen to supply the sphenoid and ethmoid sinuses.

Descending Palatine Nerve

Posterior nasal nerves are given off by DPN in the canal. They penetrate anteriorly to supply the nasal wall behind the middle and superior turbinates.

Greater palatine nerve runs forward in a groove to the canine tooth where it communicates with medial nasopalatine nerve. It supplies the oral mucoperiosteum of the secondary hard palate. Neural crest mesenchyme flowing along this nerve is responsible for the synthesis of the lingual alveolar lamina, from tuberosity to the canine.

Lesser palatine nerve runs backward to supply horizontal plate of palatine bone, soft palate, uvula, and tonsil (from above). This is also the supply to the anterior palatine aponeurosis.

Nasopalatine Nerve Enters via sphenopalatine foramen formed by the notch between the posterior (sphenoid) process of palatine bone and the anterior (orbital) process of palatine bone.

Lateral nasopalatine nerve sweeps down diagonally to cover the non-ethmoid lateral wall, including inferior turbinate and the nasal mucoperiosteum of hard palate.

Medial nasopalatine nerve passes downward in a groove along the ipsilateral vomer, supplying it, and, eventually, the premaxilla.

Innervation of the Middle Ear

The ear is an embryological *menage a quatre*. (For an inventive turn of phrase, nothing beats the French language). Because it is constructed from multiple sources of neural crest and mesoderm, it is not surprising that its innervation is similarly complex. Although the eye and ear deserve a designated chapter, it makes sense to sketch out some details here. We will continue to reinforce and amplify these ideas elsewhere in this text. Please note that all somatic sensory afferent from the ear, regardless of the cranial nerve involved, all converge to a single processing center in the brainstem to the descending trigeminal nucleus and to nucleus solitarius.

Eustachian tube Lateral wall cartilage and mucosa are first arch structures. *V2 pharyngeal nerve* arises from the ganglion. It induces a companion *pharyngeal branch of maxillary artery*; together, they supply the mucosa behind auditory tube. This explains that the lateral wall also bears tensor veli palatini, the motor supply of which is V3. Medial wall cartilage and mucosa are third arch structures. Jacobson's nerve (the tympanic branch of IX) is sensory. Medial wall bears levator veli palatini, an Sm7 muscle with motor control from nucleus ambiguous via IX.

Tympanic membrane Within the middle ear, tympanic membrane is trilaminar. It has a cutaneous *external stratum* in continuity with the skin of the external auditory canal. It is innervated as follows: anterior 2/3 = auriculotemporal nerve (V3); posterior 1/3 = Arnold's nerve (IX via X). A fibrous neural crest intermediate stratum has radial fibers. The mucosal *internal stratum* is in continuity with the lining of the tympanic cavity—both of which are neural crest derived, *not* endoderm. It is supplied anteriorly by V3 and posteriorly by VII.

Tympanic cavity Medially (the wall of the promontory) and posteriorly (the mastoid sinus), the mucosa is supplied by IX (Jacobson's nerve). The anteromedial walls are supplied by V3. Endoderm from first arch and third arch enters the medial wall of the cavity. But all the rest of walls are covered by neural crest that underwent a mesenchymal-to-epithelial transition.

Trigeminal Nerve, Third Division, v3

Conceptual Plan

Of its three divisions, mandibular nerve is the largest. V3 is a mixed nerve. Its sensory root lies at the inferior corner of the ganglion; the ganglion itself supplies V1–V3. The motor root is smaller. Its only targets are the muscles of Sm4. These two roots travel exit via foramen ovale (motor root lying medial to sensory) and quickly unite to form a 2–3 cm *main trunk of V3*.

Just medial to the point of fusion lies the *otic ganglion*. It surrounds the medial pterygoid nerve. Main trunk then subdivides into anterior and posterior branches. Pterygospinous ligament separates the branches; it is occasionally ossified. If so, the anterior branch will pass through the *foramen of Civinini*.

Recall that alisphenoid is composed of neural crest from two sources. Anterior wall of middle cranial fossa is centered around V2 at foramen rotundum comes from r2. The floor of middle cranial fossa is centered around V3 at foramen ovale. PAM from somitomeres 3 contributes to greater wing. Thus, lateral and medial pterygoid plates are r3 bone.

Please note how V3 is spatially distributed within a muscle "pocket." Its branches run in the space bordered laterally by masseter and lateral pterygoid and medially between medial pterygoid and temporalis.

Prior to Division, Main Trunk of V3 Gives Off Two Named Nerves

Meningeal branch of V3 (nervus spinosus) is responsible for inducing middle meningeal artery. Together, they enter the skull at foramen spinosum. Anterior division connects up with the meningeal branch of V2. Posterior division travels all the way back to posterior fossa where it supplies the air cells of the mastoid sinus. Based on our previous discussion of the stapedial system, the reader will recognize that extracranial stapedial and intracranial stapedial communicate with one another at the trigeminal ganglion.

Medial pterygoid nerve has more functions than one would suspect from its name. It supplies three muscles. *Nerve to medial pterygoid* quickly enters the muscle along its deep surface. *Nerve to tensor veli palatini* penetrates the muscle at its primary insertion at lateral pterygoid plate. Indeed, tensor and medial pterygoid develop from a common blastema. Nerve to tensor tympani enters the cartilage of the auditory tube and therein supplies the muscle. Its trajectory carries it in parallel with lesser superficial petrosal nerve. One can appreciate, in the clinical implications of isolated defects involving the common blastemata of this neuroangiosome. Patients having the submucous form of cleft palate would have deficits in middle ear ventilation as well as soft palate function.

Anterior Division of Mandibular Nerve Is Mixed, but Primarily Motor

It receives all motor fibers *except* medial pterygoid nerve and mylohyoid nerve. Because these muscles will provide sensory feedback, a modest sensory component is also present.

Masseteric nerve follows lateral pterygoid but does not enter it. At the coronoid notch, it enters masseter along the undersurface of zygomatic arch. It is in a good position to innervate TMJ.

Deep temporal nerves run forward in the infratemporal fossa. Anterior deep temporal runs fused with buccal nerve and then branches off between the heads of lateral pterygoid to enter anterior temporalis. Posterior deep temporal may arise with masseteric nerve. It jumps over the upper margin of later pterygoid and skims along temporal bone to enter posterior temporalis.

Lateral pterygoid nerve may also follow buccal nerve on its way to the deep surface of the muscle.

Buccal nerve passes outward through the lateral pterygoid heads to run downward along its external surface. Ramifying along the external surface of buccinators, it joins with motor fibers from VII. It supplies skin overlying buccinators and mucous membranes internal to buccinators.

Posterior Division of Mandibular Nerve Is Mixed, but Primarily Sensory

Auriculotemporal Nerve has two rootlets. These pass backward around the middle meningeal artery. This encirclement is due to the biologic priority of the pre-existent anterior branch. Running along the internal aspect of the mandible at the neck, it turns upward with superficial temporal artery (which it helps to induce). Cloaked by a temporary ensheathment from parotid, it sneaks its way between the condyle and ear. Once over the zygomatic arch, it breaks free and forms superficial temporal branches.

Auriculotemporal nerve makes several *communications and branches* of importance.

1. *Facial nerve* These nerves have a rendezvous in parotid at the posterior border of masseter. Thus, zygomatic, buccal, and mandibular branches of facial nerve carry V3 sensory fibers via target mimetic muscles into overlying skin.

2. *Otic ganglion* Auriculotemporal and ganglion communicate right away. Otic ganglion receives preganglionic PANS from glossopharyngeal nerve. Auriculotemporal faithfully conveys these secretomotor fibers into the substance of parotid. One can conceive of parotid as containing two embryonic cell types. First arch mesenchyme provides the parenchyma while third arch supplies the glandular tissue. Epithelial ducts from r2 and r3 oral mucosa hook up with the glandular tissue. Voila! Saliva is produced and secreted.

3. *Anterior auricular* These branches supply anterior half of external pinna, i.e., tragus and helix. Recall that posterior external pinna is supplied by cranial nerve IX (r7).

4. *External acoustic meatus* Supplies anterior skin and anterior tympanic membrane.

5. *Articular branches* Innervates joint from behind; masseteric nerve supplies front side.

6. *Parotid branches* Carry PANS postganglionic fibers from otic ganglion.

7. *Superficial temporal branches* These nerves accompany (read induce) superficial temporal artery; temporal skin; zygomaticofacial and zygomaticotemporal connections.

Lingual Nerve Starts out on deep surface of lateral pterygoid, running in parallel with inferior alveolar nerve. It lies medial to the latter because it supplies midline structures. Lingua nerve is joined by (what else?) chorda tympani. Skims along surface of mandible underneath medial pterygoid. Crosses over superior constrictor and styloglossus (where it can be identified) and then encounters the tongue. Interposed between hyoglossus and deep submandibular gland surface then swings outward to run along lateral surface of the submandibular gland.

Lingual nerve has an assortment of connections and branches.

1. *Chorda tympani* CT joins lingual nerve 2 cm from foramen ovale. It carries preganglionic PANS motor fibers from VII to the submandibular ganglion. Returning in CT are special sensory fibers for taste. These belong to the ventrolateral facial placode (VII$_{VL}$). They convey taste from the anterior two-third of the tongue to the gustatory nucleus located in the rostral division of nucleus solitarius. Note that rostral division is strictly for taste; it receives VII$_{VL}$, IX$_{VL}$, and X$_{VL}$. The caudal division receives visceral afferent information from the gut, respiratory, and cardiovascular system via dorsal glossopharyngeal IX$_D$ and dorsal vagus X$_D$.

2. *Submandibular ganglion communications* The ganglion appears as if it were "suspended" from the lingual nerve by two branches, each with a distinct function. Proximal branch conveys preganglionic PANS from VII$_{CT}$. Distal branch takes postganglionic fibers to the submandibular gland.

3. *Hypoglossal communications* Tongue muscles perform very complex movements requiring precise proprioceptive feedback (pain as well). A plexus between lingual fibers and hypoglossal nerve exists at the anterior border of hyoglossus. This brings fibers from V3 out to the tongue muscles along the branches of XII. Via these fibers, induction of arterial branches from lingual artery to the muscles and mucosa takes place.

4. *Distribution fibers* These fibers of lingual nerve convey afferent sensation from the mucosa of the anterior two-third of the tongue, adjacent gums, and the sublingual gland. They also connect to the taste buds of the same region.

Inferior Alveolar Nerve

Running with the artery it induced, this nerve hugs the internal surface of lateral pterygoid. It then skims along the medial border of the ramus against which it finds itself entrapped by the later-developing sphenomandibular ligament. Via the mandibular foramen, it enters the mandibular canal passing between the lingual and buccal laminae of the mandible until exiting at the mental foramen. The raison d'etre of the foramen is that it defines the embryologic border between the membranous neural crest zones of the mandible and the chondral neural crest symphyseal zones paired across the midline.

The individual branches of inferior alveolar nerve define various zones of the mandible.

1. *Mylohyoid nerve* Departs from inferior alveolar prior to entry into the lower jaw. It runs along deep surface of ramus to supply first, mylohyoid and second, anterior belly of digastric. These may represent genetic "footprint" of the lingual bone fields seen in the primitive synapsid jaw: angular (proximal) and splenial (distal).
2. *Dental branches* Proximal branches supply premolars and molars. This explains the developmental relationship between deciduous molars and permanent premolars. Distal (incisive) branches supply the canine and incisors. Note that mental nerve exits from inferior alveolar at the boundary between these two nerves.
3. *Mental nerve* Exits beneath depressor anguli oris, where it trifurcates to supply labial gum, sulcus, and lower lip.

Myology of Somitomere 4

Masseter This muscle is primarily inserted into the r2–r3 zygomatic arch. It is secondarily inserted into r3 mandibular ramus via three laminae. (1) *Deep anterior lamina* begins at the ramus. It is secondarily inserted into the inner surface of the anterior two-third of zygomatic arch. (2) *Superficial anterior lamina* begins at the angle and lower half of lateral ramus. It is secondarily inserted into the inferior surface of the anterior two-third of zygomatic arch. (3) *Posterior lamina* begins at superior ramus and lateral coronoid process. Its fibers pass posteriorly and superiorly to gain a secondary insertion along the inner surface of the posterior zygomatic arch, i.e., the r3 zygomatic process of squamous temporal bone.

From our previous discussion of craniofacial bones, mandible has a dentary field supported by four body fields (two on each side). Ramus-condyle is a separate entity. As pathology worsens, one sees loss of the condyle and, eventually, loss of the ramus. Thus, posterior lamina of masseter is the biologically "newest" component.

Masseter is supplied by the first arch masseteric branch of IMMA. It receives collaterals from second arch facial artery proper and second arch transverse facial br. of superficial temporal.

Medial pterygoid This quadrilateral muscle has two primary insertions. (1) The medial head inserts into the internal aspect of the r3 lateral pterygoid plate and the groove of the palatine bone. (2) The lateral head inserts into the r2 pyramidal process of palatine bone and into the r2 maxillary tuberosity. It secondarily inserts into the r3 mandible along the internal surface of the angle and lower half of ramus, extending as high up as the inferior alveolar foramen. Five structures separate upper border of medial pterygoid from the mandible: sphenomandibular ligament, internal maxillary artery, inferior alveolar artery and nerve, and lingual nerve.

Note that medial pterygoid and masseter share common functional characteristics. They form together an anatomic "sling" suspending the mandible from the maxilla and zygoma. As the lower jaw moves, it follows a center of rotation established by this sling and by a ligament running from the spine of the r3 sphenoid of the infratemporal medial cranial fossa to medial ramus: the *sphenomandibular ligament*.

Recall that alisphenoid is composed of neural crest from two sources. Anterior wall of middle cranial fossa is centered around V2 at foramen rotundum and comes from r2. The developmental center of middle cranial fossa floor is centered around V3 at foramen ovale. PAM from somitomere 3 contributes to greater wing. Thus, lateral and medial pterygoid plates are r3 bone.

Blood supply to the medial pterygoid is first arch pterygoid branches of IMMA.

Lateral pterygoid This muscle shares a number of characteristics with its *confrere*. It also has two primary insertions. (1) The inferior head is developmentally "older." It follows a deeper, more internal course from the lateral surface of the r3 lateral pterygoid plate. (2) The superior head, being "newer," projects medially and inferiorly, from the r3 infratemporal crest and also from the inferolateral r3 surface of alisphenoid. The two heads converge into two secondary insertions at developmentally "newer" zones of r3 mandible: the anterior margin of condylar neck and the anterior margin of the TMJ disk.

Blood supply to this muscle is pterygoid branches of IMMA with collaterals from second arch ascending palatine br. of facial artery.

Temporalis Primary insertion of temporalis occupies developmental zones of r3 mandible distinct from those of other muscles of mastication. It is attached to (1) internal surface, apex, and anterior border of coronoid ramus; and (2) anterior border of ramus, extending as far forward as the third molar.

Its insertion is quite broad, involving the deep surface of temporalis fascia and the temporal fossa, spanning from r3 squamous temporal bone forward to r2 alisphenoid and the inferolateral corner of p5 at the frontozygomatic suture. Blood supply to temporalis is provided by first arch deep temporal br. of IMMA.

Tensors

Tensor tympani This muscle connects the pharynx with middle ear, recapitulating the evolutionary relationship between these sites. It has two points of primary insertion: scaphoid fossa is into the r3 lateral wall of the Eustachian tube (recall that the medial wall of the tube is r6–r7). It then passes up through a tunnel formed between r3 auditory tube and r2 greater wing of sphenoid, ending as a tendon that travels backward internal to tympanic membrane. The tensor tendon inserts secondarily into the manubrium directly into the root of its manubrium. Note that chorda tympani courses forward directly above the insertion. The nerve serves a convenient landmark separating the downward-directed manubrium from the neck and head of the malleus. Blood supply to tensor is StV3 superior tympanic branch of middle meningeal artery.

Tensor veli palatini This muscle is also served as a "bridge" between the pharynx and the soft palate. It is triangular with three points of primary insertion: anteriorly from the r3 pterygoid fossa of the sphenoid bone, posteriorly from the r3 medial spine of the sphenoid, and medially into the lateral wall of r3 auditory tube right at the isthmus (where the bony later one-third joins the cartilaginous medial two-third). Indeed, it shares some fibers with tensor tympani but pursues an opposite course. The muscle sends a delicate tendon around the *medial pterygoid hamulus*, passing through buccinators to the palatine process of the r2 *horizontal plate of the palatine bone*. It passes forward and downward, attaching to medial spine of sphenoid.

Let's pay a little attention to functional considerations of these insertion sites. Looking at the skull from below, we note the petrous processes of the temporal bone forming the anterior border of middle cranial fossa. These pass from the lateral skull base forward to the axis of the skull base: the spheno-occipital suture. The abutment between medial petrous apex and lateral basilar occipital bone, the *petro-occipital suture*, is incomplete posteriorly, the space being the jugular foramen. On the lateral side of the petrous apex lies the infratemporal wing of alisphenoid. These two bones are separated by the petrosphenoid suture nestled within the auditory tube. Infratemporal wing has two distinct developmental components. It contains foramen ovale for V3 and (posteriorly) foramen spinosum for the middle meningeal artery. These neurovascular spaces suggest that both lamina

projecting downward from infratemporal alisphenoid are derivatives of r3.

TVP begins its trajectory from the pterygoid fossa, the undersurface of alisphenoid. Alisphenoid is a membranous bone derived from two sources of neural crest: r2 and r3. The former is responsible for synthesis of the greater wing and the anterior wall of middle cranial fossa. Its mesenchyme ossifies outward from foramen rotundum. The floor of middle cranial fossa develops from r3; its ossification spreads outward from foramen ovale. Two pterygoid plates project downward from alisphenoid. They are most likely r3 derivatives. Medial pterygoid plate lies just posterior to r2 palatine bone.

Tensor veli palatini has two sources of blood supply. Primary is StV2 lesser palatine branch of descending palatine to surrounding mucosa. Secondary is second arch ascending palatine br. of facial. This makes developmental sense, because the internal auditory tube, with which TVP is related, constitutes a boundary zone between PA1, PA2, and PA3.

In soft palate development, TVP is the pioneer, followed next by levator veli palatini and thence the remainder of the Sm7 muscles (palatopharyngeus, and palatoglossus and the midline uvulus). Given this insertion mechanism, embyologic defects in the bony targets these muscles are seeking, will throw them off course. Isolated defects of pterygoid hamulus have been described. Quite commonly, when the dimensions of the prepalatine bone are deficient (horizontal, anteroposterior, or both), the horizontal palatine shelf will be foreshortened. This reflects a deficiency state in the lesser palatine neuro-angiosome. As a consequence, the insertion of TVP will be pulled forward. Alternatively, horizontal shelf can itself be reduced. This creates a "notch" in the posterior bony palate. Such r2 defects can occasionally be associated with hypoplasia of TVP or the muscle can be abnormal in the presence of normal bony anatomy. In any case, affectation of TVP will affect middle ear function.

Relations of this thin muscle lie lateral to medial pterygoid plate, the auditory tube and (of course) to levator veli palatini (the latter arising as it does from medial auditory tube). Also medial to TVP are V3, auriculotemporal nerve, chorda tympani, and (last but not least) middle meningeal artery. MMA is also not surprising because it represents the developmental relationship between intracranial and extracranial stapedial, i.e., between the primitive trigeminal artery running along V3 and StVII following chorda tympani downward from middle ear to V3.

Suprahyoid Muscles

Mylohyoid These paired muscles are flat; they make up the muscular floor of the mouth. As such, they lie above anterior border of digastric. The primary insertion occupies the entire length of the lingual mandible along the mylohyoid line from symphysis to the last molar. This is the original territory of both angular and splenial bones. Two sets of fibers seek out two separate secondary insertions. Posterior fibers attach to the r3 lesser cornu of the hyoid (the old ceratohyal) while anterior fibers insert from the symphysis into the r3 body of the hyoid (old basihyal). Note: the pterygoid sling posteriorly and the mylohyoid anteriorly are separated by a "bare area." The developmental sequence of mandible is theoretically: body > ramus > condyle.

In rare congenital conditions, the posterior bone elements are missing, along with their associated muscles. In such cases, floor of mouth remains intact. Absence of the tongue can occur, again with intact floor of mouth. The blood supply of mylohyoid relates to that of tongue. Recall that the programming of the oral epithelium is all-important is transmitting signals that instruct the tongue muscles to find their correct place. Behaves like tongue likely predates these of the mastication muscles. Anterior two-third of the tongue has second arch sensory elements (taste). Thus, it is not surprising that second arch arteries are actively involved in supplying mylohyoid. The main vessel, sublingual branch, first arch lingual artery, pierces the muscle; StV3 inferior alveolar artery, sublingual branch, ramifies along the inferior surface of the muscle. It connects to submental branch of second arch, facial artery.

Anterior digastric This muscle follows mylohyoid in developmental sequence. Its primary insertion is along inferior lingual border of mandible: the original territory of splenial. The secondary insertion is into a tendon that it shares with second arch posterior digastric. The tendon spans from the lateral r2 body of hyoid posteriorly to the r7 greater cornu. It is said that the tendon "perforates" stylohyoid but developmentally this situation is the reverse. Digastric pre-exists when the stylohyoid blastema arrives. At any given location, Sm4 muscles always pre-empt Sm6 muscles. Thus, the second arch stylohyoid must divide and reunite as it flows around anterior *digastric*. Blood supply comes from sublingual branch, second arch facial artery.

Angiology of Sm4 Muscles

Arterial branches to these muscles originate from the middle one-third of the internal maxillomandibular artery (IMMA), the *infratemporal fossa segment*. The proximal one-third of IMMA is called the *mandibular segment* (Fig. 7.20). IMMA is a composite structure. It represents the union of extracranial stapedial descending from tympanic cavity to the sensory root of V3. From the axis of IMMA, it supplies all structures of the jaws and bones connecting the jaws to the skull.

This branch of external carotid represents the contribution from remnants of AA1 that supply the first arch floor of the mouth and muscles of mastication. Recall that *the ECA develops as a rearrangement of the ventral remnants of pha-*

ryngeal arch arteries 1–4. Ventral remnants from the first pharyngeal arch artery are admixed with contributions from the extracranial stapedial system. Deep auricular artery and anterior tympanic artery represent induction products from chorda tympani. Middle meningeal and accessory meningeal arteries are induced by V3. Thus, *IMMA is really two-arteries-in-one*.

We can make the following observations. *Muscles of mastication* are supplied by the more "primitive" first and second pharyngeal arch arteries. The *tensor muscles* are supply structures that are evolutionarily "newer." They thus make use of a "newer" stapedial artery system. Hence, stapedial induction products supply tensor veli (StV2) and tensor tympani (StV3). Mylohyoid is a *hypobranchial muscle* analogous to the tongue. Its mucosal cover is r3 ectoderm. Its principle vessel is lingual artery. It is interesting that branches dedicated to the muscles of mastication are *not* major suppliers of facial skin. This role is taken up by the SMAS, the external carotid branches of which perfuse the skin. Cutaneous vascular anatomy does however contain multiple anastomoses between IMMA and EMMA. These represent the *admixture* of pharyngeal arches 1–2.

Developmental Sequence of First Pharyngeal Arch Muscles

Sm4 muscles fall into three temporo-spatial groups. Mastication: temporalis > masseter > lateral pterygoid > medial pterygoid. Floor of mouth: mylohyoid > anterior digastric. Palato-tympanic: tensor veli palatini > tensor tympani.

The Second Pharyngeal Arch

The fifth and sixth rhombomeres (r5–r6) contain the nuclei of two cranial nerves, abducens and facial. Cranial nerve VI supplies the fifth somitomere while cranial nerve VII innervates the muscles of the sixth somitomere. The sensory supply (nociception and proprioception) for lateral rectus is V1 while V2 and V3 innervate second arch muscles. Neurons for proprioception synapse in the mesencephalic nucleus of V, which extends from r3 forward into the midbrain. Those for nociception proceed to the trigeminal sensory nucleus. The second arch branchiomeric muscles are divided into two functional categories (mastication and facial expression), each with its distinct fascia and blood supply. The blood supply of these muscles originates from the four distal major branches of the ECA proper: occipital, lingual, facial, and superficial temporal.

Another appropriate term for the facial artery is the external maxillomandibular artery (EMMA). This system is the exclusive supply of the superficial investing fascia (SIF), better known to surgeons as the superficial musculoaponeurotic system (SMAS). The SIF is a neural crest derivative that encloses all second arch *muscles of facial expression*. Deep investing fascia (DIF) invests the second arch *muscles of mastication*. EMMA most likely represents a rearrangement of remnants from the second pharyngeal arch artery.

The predominant form of mesenchyme in the pharyngeal arches is neural crest, not mesoderm. Paraxial mesoderm from the somitomeres migrates and is distributed expaxially to the meninges and hypaxially into the arches where it forms blood vessels and muscles. At times, this process fails, as seen in craniofacial microsomia and Moebius syndrome. Recall that, very early in development, pharyngeal arches 1–2 become melded together. Unlike arches 3–5, arches 1 and 2 are uniquely confluent. This process is possible because *the first 7 somitomeres are not separated from one another by epithelium*. In the stage following its inception, second arch melds into the first arch. By stage 12 the two arches are in continuity. First and second arch muscles of mastication form the deep plane of this complex. Muscles from the second arch form a superficial plane which spreads subcutaneously beyond the confines of first arch to enclose the entire face, head, and neck.

Why is the behavior of the first and second pharyngeal arches so distinctive? The answers are inextricably linked to the transition between agnathic (jawless) fishes to gnathostomes. This topic merits a chapter of its own, evolution of jaws. For our purposes, modern-day agnathans such as the lamprey and hagfish demonstrate that the first two arches were originally branchial, i.e., they have respiratory function. These creatures feed via a non-mobile stoma that scoops up food as they swim. The fossil record demonstrates that primitive agnathans had up to 15 branchial arches! The radical anatomic rearrangements required for mobile jaws convert the branchial arches 1–2 into a new feeding apparatus. Modern-day chondrichthyans (sharks) help us get the terminology straight. They have seven visceral arches arranged as a 2 + 5 system: two arches dedicated to jaws + five branchial arches dedicated to respiration. With the transition to land, the respiratory function transitions into a food reception (the pharynx) and respiration. The mammalian system is 2 + 3: two arches for jaws and three arches dedicated to the pharynx.

Mammalian Sm6 muscles are divided into two classes: (1) an original group functionally associated with the jaws; and (2) an evolutionarily newer group concerned with facial animation. The fascial anatomy of these two groups is distinct. Posterior digastric, stylohyoid, and buccinator are synchronized with the muscles of mastication. These are all co-planar and make use of a common *deep investing fascia* (DIF). Mammals "invented" an entirely new form of jaw suspension, the TMJ. No longer required for mastication, a majority of second arch muscles now assumed an entirely new role: facial animation including suckling. As we know, the facial muscles reside within a separate layer, the *superficial investing fascia* (SIF). In common surgical parlance, this fascia is known as the *superficial musculoaponeurotic system* or SMAS. How did a completely new layer of fascia come into existence?

In primitive agnathan fishes, the second arch was truly branchial: it was a functioning gill arch. Although the standard number of gills in the lamprey is 7, some primitive agnathans had up to 12 gills! The original function of the hyoid muscles was to constrict and expand the gills: i.e., pump water through the gills. With the invention of jaws, neural crest cartilage of the hyoid arch gill was reassigned to create supporting bones for the jaws. Hyomandibula is attached to the chondrocranium. It articulates with a ventral ceratohyal. These two bones are "assigned" (respectively) to palatoquadrate cartilage (maxilla) and Meckel's cartilage (mandible). The former constrictors of the second gill arch are now associated with these bones: levator hyomandibula and interhyoideus. The entire system of gill arch muscles and jaw muscles in fishes is co-planar.

In mammals, the muscles of Sm6 develop and migrate at different times. The "primitive" second arch muscles dedicated to mastication (posterior digastric, stylohyoid, and buccinator) develop earlier—concomitant with its first arch "functional fellows." One stage later, they are joined by their third arch "pharyngeal *confrères*." Together, these muscles all share a common *deep investing fascia*, the DIF. The "modern" second arch muscles dedicated to facial expression develop and migrate later in time. As they depart from their somitomeres, they acquire a separate "new" layer of neural crest. This becomes the *superficial investing fascia*, also known by clinicians as the "superficial musculoaponeurotic system," or SMAS.

The second wave of second arch muscles is very opportunistic. They spread far and wide to envelop the entire head as the epicranius. The myoblasts of the muscles of facial expression are lodged within the SIF at predetermined sites. This structure constitutes the *primary insertion site* for all muscles of facial expression. Subsequently, these muscles seek out *secondary insertion sites* in dermis or periosteum.

SIF migrates subcutaneously throughout the head and neck. It crosses the midline of the forehead and the vertex of the calvarial vault. Its boundaries are those between r3 neural crest dermis and c2 paraxial mesoderm dermis over the back of the skull and neck. The caudal extension of SIF with platysma muscle reaches the C2–C5 clavicle. In the pharynx, SIF acts as a sort of "tissue expander" that quickly covers over the cutaneous manifestations of the third, fourth, and fifth arches to form the *Sinus of His*. The ectodermal components of these arches are obliterated and they fold up on each other like a Japanese fan to create the pharynx. At the same time, intraoral expansion of third arch abuts up against first arch, obliterating representation of second arch save for tonsil and the lining of the Eustachian tube (see Chap. 10).

We shall approach the muscles of Sm6 in layers. This is best organized around the functional neuroanatomy of cranial nerve VII. The anatomic distribution of r4–r5 is so widespread that the best way to get a handle on the subject is via its master integrative structure, the facial nerve. This will yield much valuable insights, not only into the organization of its larger motor root, but also the smaller *nervus intermedius* containing PANS visceromotor fibers and special sensory fibers for taste. We next consider the individual muscles supplied by facial nerve. Our discussion will follow the same spatial sequence as that of the facial nerve. (Ear muscles will require a digression.) The vascular supply to the second arch will emphasize the relationship between branches to Sm6 muscles that come from the original external carotid systems. The "new-comer" stapedial system. Finally, we will present a dynamic picture of how muscles of Sm6 spread far and wide to cover the face and anterior neck.

Facial Nerve

Conceptual Plan of the Seventh Cranial Nerve

The dual roots of the facial nerve emerge at the inferior border of pons. They are accompanied by vestibulocochlear nerve. From medial to lateral, these are motor VII, sensory VII, and VIII. By virtue of its interposition, the sensory root is called *nervus intermedius* (nerve of Wrisberg). In general, VII supplies mimetic muscles, buccinator, posterior digastric, the accessory muscles of mastication (buccinators, posterior digastric, and stylohyoid), and platysma. In so doing, it makes use of two divisions. We shall, by convention, consider these two reflect original dorsal and ventral muscles in fishes. We shall also explore a hypothetical (as yet unproven) relationship between rhombomeres r4 and r5 with the upper and lower divisions of motor VII.

The three above-named nerves "enter" the internal acoustic canal. Almost immediately, within the fundus of the canal, VII and VIII part company. Facial nerve now "enters" petrous temporal bone. We are using quotation marks because the very language used to describe this anatomy is deceptive. In all instances, bone forms secondarily around the neural structures. Neural crest swarming down from r4 and r5 along the nerves is quickly deposited around them as mesenchyme. We know for example that somitomeres 5 and 6 contribute PAM to petrous temporal bone and that somitomere 7 is likely to synthesize, in part, the mastoid cavity. Similarly, neural crest cells from r4 to r7 contribute to petrous and mastoid as well.

Facial canal (*aqueductus Fallopii*) now follows a circuitous course. It proceeds laterally between cochlear and semicircular canals. Just near middle ear cavity, it makes a sudden turn backward, running within the medial wall just underneath thin bone directly over oval window. It can be visualized there as a prominence. It then makes a U-shaped bend and the two roots fuse. This marks the site of the geniculate ganglion. From the ganglion, it drops precipitously downward along the mastoid air cells, proceeding headlong toward stylomastoid foramen. After exiting, facial nerve travels within parotid gland. It crosses over external carotid and, at the posterior border of mandible splits into: (1) superior division (temporal, zygomatic, infraorbital) and (2) inferior division (buccal, marginal mandibular, cervical). We shall consider these various components in greater detail.

Geniculate Ganglion

This is the sensory ganglion housing unipolar ganglion cells destined to access the gustatory nucleus. VII_{VL} is a ventrolateral placodal derivative carrying fibers originating via lingual nerve to chorda tympani. Other sources of taste come from soft palate supplied by V2 lesser palatine via greater superficial petrosal. From the posterior wall of external auditory canal and from the mastoid process, general somatic afferent fibers of VII refer pain and temperature back to the auricular branch of vagus. For this reason, ear surgery can produce unexpected nausea.

Contributions of Facial Nerve to Nervus Intermedius

This is sometimes referred to as *glossopalatine nerve*. It has two parts, sensory and PANS. Sensory includes the geniculate ganglion, chorda tympani, and greater petrosal nerve. Its afferent division ends in the nucleus solitarius (gustatory nucleus). PANS includes the pterygopalatine ganglion and submandibular ganglion.

Communications of the Facial Nerve

1. *In the internal auditory canal,* VII contributes fibers to VIII.
2. *Within the geniculate ganglion,* VII talks to otic ganglion via lesser petrosal nerve. It also receives SANS fibers ascending along middle meningeal. These travel via the diminutive external superficial petrosal nerve.
3. *In the facial canal,* VII connects with auricular vagus just above stylomastoid foramen.
4. *After exit from stylomastoid foramen,* VII receives two branches of auriculotemporal V3. These pass behind neck of condyle. Cutaneous sensation is thus conveyed to the skin overlying terminal branches of the cheek.

Branches of VII from Geniculate Ganglion

Greater Superficial Petrosal Nerve GPN is a mixed nerve conveying sensory fibers back to the brain and secretomotor PANS fibers from geniculate ganglion forward. It pursues a short passage in the petrous temporal bone. It then enters the middle cranial fossa via the *hiatus of the facial canal*. It travels in a sulcus along the petrous ridge under cover of the dura. It passes beneath trigeminal ganglion, also subdural within Meckel's cave. Upon reaching foramen lacerum, it encounters the cartilage of auditory tube. GPN crawls over the tube, crosses over the lateral aspect of internal carotid, and unites with the SANS-bearing *deep petrosal nerve*. Together these two nerves form the *Vidian nerve* or the *nerve of the pterygoid canal* which is subsequently distributed to the nose, salivary glands of the palate, and the lacrimal gland.

Nerve of the Pterygoid Canal This nerve first appears at foramen lacerum. It represents the union of the postganglionic PANS *greater petrosal* and the pre-ganglionic SANS *deep petrosal*. The bulk of its fibers are *sensory from the soft palate* via lesser palatine nerves. Vidian nerve is *parasympathetic* root of pterygopalatine ganglion. It conveys PANS secretomotor fibers from geniculate ganglion that will find their targets in the distribution of V1 and V2, i.e., in the nose salivary glands of the palate, and the lacrimal gland. The trajectory of Vidian nerve is short. It enters a bony tunnel (oddly named the pterygoid canal). Not wanting to travel alone (and desirous of blood supply), it induces an artery to accompany it in the canal. Also, within the canal, it receives an additional *ascending sphenoidal branch* from otic ganglion. If one observes the floor of the sphenoid sinus (as with an endoscope), one observes a prominent ridge caused by this canal. It sends off a few tiny branches to the sphenoid sinus mucosa.

Pterygopalatine Ganglion (PPG, *Meckel's ganglion, sphenopalatine ganglion*) PPG is very important despite its puny size (about 5 mm in diameter). It sits deep in the fossa just inferior to the maxillary division of trigeminal nerve. Remember that V3 has already been given off. PPG receives two autonomic inputs. The *parasympathetic root* comes via greater superficial petrosal nerve; the *sympathetic root* comes via deep petrosal nerve.

Here are some ***important connections*** of the pterygopalatine ganglion.

1. *Maxillary nerve* sends two short *pterygopalatine nerves* through PPG without making a synapse. This is the mechanism by which V2 picks up postganglionic fibers from PPG.
2. *Superior cervical ganglion* sends postganglionic SANS fibers along the carotid plexus. These are conveyed to PPG via *deep petrosal nerve*; it is the *sympathetic root* of PPG.
3. Branches from PPG supply *PANS motor fibers to the eye, nose, and mouth*. These are not independent nerves. They "hitchhike" with branches of V2 to reach the glands of the nasal cavity, pharynx, and palate. Via a connection between V2 zygomatic branch with V1 lacrimal branch, PANS fibers reach the lacrimal gland as well. NOTE: this is another example of anastomosis between the V1 and

V2 components of stapedial artery Lacrimal gland, as we have seen, gets a dual blood supply. StV1 lacrimal artery connects with StV2 zygomatic branches in the lateral orbit, i.e., in "Tessier zone 9".

Branches of VII Within Facial Canal, Prior to Stylomastoid Foramen

Nerve to Stapedius Muscle This nerve passes downward and penetrates (is surrounded by) the posterior bony wall. The muscle is accessed via a minute fissure at the base of the tympanum.

Chorda Tympani This runs downward parallel to facial nerve. It enters a foramen in posterior tympanic wall 6 mm proximal to stylomastoid foramen. It travels cranial and forward, parallel to facial nerve but diverges, heading toward lateral wall of tympanum. Chorda tympani now enters the tympanic cavity a foramen, *iter chordae posterius*, located in the posterolateral corner, between the base of the pyramid and the tympanic membrane. It runs horizontally across the tympanic membrane, crossing the manubrium. It exits the tympanic cavity via the *iter chordae anterius* and plunges into a canal located in the fissure between petrous temporal bone and tympanic temporal bone. Via this fissure (*canal of Hugier*), chorda tympani makes its escape from the skull just medial to the sphenoid spine. Here, it receives a branch ganglion. It then unites with lingual nerve, sandwiched between lateral and medial pterygoid muscle.

Most fibers of chorda tympani are special visceral afferents conveying taste from the anterior two-third of tongue. However, it also conveys preganglionic PANS fibers to submandibular ganglion.

Note that chorda tympani is accompanied by anterior tympanic artery from the first segment of maxilla-mandibular artery. This is the remnant of the original lower division of stapedial that exited the tympanic cavity.

Submandibular Ganglion This structure is 5 mm, equivalent in size to pterygopalatine ganglion. It is suspended from lingual nerve by two 5 mm branches. It is located internal to and above submandibular gland on hyoglossus, just behind mylohyoid. The proximal root is parasympathetic conveying PANS fibers from geniculate ganglion via chorda tympani. These fibers are secretomotor to *submandibular* gland and *sublingual* gland.

Branches of VII After Exit from Stylomastoid Foramen

Digastric nerve and *stylohyoid nerve* These nerves are purely motor. They subserve "primitive" second arch muscles associated with mastication. At times, these nerves arise together. Digastric muscle is innervated very close to the stylomastoid foramen whereas stylohyoid receives its motor nerve at its midpoint.

Posterior Auricular Nerve This nerve is motor to the muscles of the posterior ear and to occipitalis. It has clinically important somatic sensory functions that explain patterns of referred pain. Arising just distal to stylomastoid foramen, the nerve runs anterior to mastoid process. From the ear, a branch of auricular vagus wanders out to join it. It has two important developmental connections with the cervical plexus. Recall that all components of the cervical plexus are ventral. (For more details regarding the cervical plexus, see Chap. 12). The reader should be undaunted by the following detail because hidden within it are important clues as to the embryology of the external ear.

The first connection of PAN is to *posterior branch of greater auricular nerve*, the superficial branch of ventral C2–C3. After emerging from posterior border of sternocleidomastoid, it penetrates the deep investing fascia surrounding that muscle to run along the external surface of SCM. Overlying platysma is contained by superficial investing fascia. GAN ascends, sandwiched between these structures, and divides.

1. *Anterior (facial) branch of greater auricular* supplies skin over posterior parotid along its posterior and inferior border. Diving branches connect to facial nerve in the substance of parotid. When facelift interrupts the connections, at times re-anastomoses between GAN and VII take place, *gustatory sweating* can occur.
2. *Posterior (mastoid) branch of greater occipital* supplies the skin of the mastoid. NOTA BENE: it also innervates the ear skin in two locations. (a) lower two-third of posterior auricular skin. This corresponds to hillocks 5 and 6; it also supplies posterior earlobe (b) a penetrating branch tracks outward through the substance of the ear (possibly below hillock 6) and achieve supplies lower concha and the anterior earlobe.

The second connection of PAN is to *lesser occipital nerve*, the superficial branch of ventral C2. Like its co-conspirator, greater auricular, lesser occipital winds around the posterior border of sternocleidomastoid. It emerges higher up. This is because the position of C2 in the plexus is logically more cephalad than that of C2–C3. Right at the insertion of SCM into the mastoid, lesser occipital penetrates through its fascia and ascends behind the ear. It supplies the upper one-third of posterior auricular skin and the rim of the helix. This corresponds to hillock 4.

Clinically, the connections between PAN and the epaxial sensory nerves to mastoid skin, posterior auricular skin, and occipital scalp are of great importance for understanding the diverse patterns of facial pain from herpetic neuralgia. From geniculate ganglion, viral involvement spreads along the individual branches of greater auricular and lesser occipital nerve.

Posterior auricular has two branches supplies two different types of muscles. Its *auricular branch* is smaller. It runs exactly along with the mastoid branch of greater occipital. It therefore is assigned to the zone of cartilage derivatives from hillocks 5 and 6. It is motor to *auricularis posterior*, one of the three extrinsic ear muscles. Recall that the mastoid branch of GAN penetrates the substance of the auricle from its internal (mastoid) surface to its external (facial) surface. The PAN has direct access to all the intrinsic muscles of the ear more about these muscles later. The occipital branch of posterior auricular is larger. It follows the superior nuchal line, i.e., the lower border of occipitalis, which it innervates.

Parotid Plexus

The terminal portion of facial nerve is strictly motor. Within the substance of parotid gland, it splits into a temporofacial division (temporal zygomatic, and buccal) and a cervicofacial division (mandibular and cervical). Without exception, the branches of facial nerve must traverse the parotid fascia en route to their respective targets. Most of these ascend to a position superficial to parotid but deep to the muscle. Buccal nerve sends deep branches below parotid; its target muscles are innervated from above.

Temporal Branches Zygomatic arch is crossed. First encountered are anterior auricular; it corresponds to hillocks 1 and 2. Next is superior auricular, corresponding to hillock 3. Temporal VII communicates with V3 auriculotemporal and V2 zygomaticotemporal. The nerves run forward to supply frontalis, upper orbicularis, and corrugator. Connections are made with V1 supraorbital and lacrimal branches.

Zygomatic Branches These nerves run roughly in parallel with zygomatic arch. They supply lower orbicularis from below. Branches communicate with V2 zygomaticofacial and V1 lacrimal.

Buccal Branches These branches are the largest of the five groups. They run both superficial and deep. Superficial branches run just under the skin; they innervate their targets muscles, including procerus, from above. Deep branches supply the elevators of the upper lip. They pass deep to zygomatic major, zygomaticus minor, and levator labii superior. Continuing forward, they supply the small muscles of the nose from above. Branches of facial artery to the external nose are induced by these nerves. Consequently, the vessel arteries are superficial to the muscles.

Mandibular Branch This is a singular nerve that pursues a trajectory deep to platysma (which it does not supply). It emerges at the depressor anguli oris and supplies branches to lower lip and chin muscles.

Cervical Branch Also singular, this nerve runs beneath platysma, which it *does* supply.

Myology of Somitomere 6

Stapedius Historically, this was a chewing muscle. In fishes, the second arch bone, *hyomandibula*, served to suspend the jaws from the skull. In tetrapods, it became *columella*, the primary bone of the tetrapod middle ear. By connecting tympanum to cochlea, columella became a sound conductor. In birds and mammals, two bones from the first arch lower jaw are added to middle ear. Articular becomes malleus and quadrate becomes incus. Columella becomes bifid and thins out as stapes. Muscle evolution followed the same patterns. *Levator hyomandibula* in sharks became *depressor mandibulae* in reptiles. This muscle is located in close proximity to columella. As columella becomes internalized as stapes to the other two bones, it draws a slip of depressor mandibulae with it. This becomes stapedius.
Stapedius occupies a cavity in the pyramidal eminence along the posterior wall of the tympanic cavity providing the muscle with its proximal attachment. Just behind, it lies the descending segment of facial nerve. Stapedial sends out a delicate tendon that attaches to the backside of the neck of stapes. It is supplied by a designated motor branch that arises from within the facial canal. Blood supply comes primarily from second arch ECA posterior auricular artery with collaterals from StVII anterior tympanic and StV3 middle meningeal. Note that both of these represent "escape routes" for the original stapedial artery as it made its exit forward from the tympanic cavity. Stapedial is the antagonist of tensor tympani (tensor dampens out chewing). It is protective against loud noises; it acts to damp down vibrations. Paralysis produces hyperacusis. Note that vagal fibers conveyed by IX into tympanic cavity travel through the facial sheath of stapedius muscle almost to geniculate ganglion. At that point, they penetrate into lesser petrosal nerve.

Posterior Digastric Digastric implies "two bellies." It shares a common evolutionary history with stylohyoid. Both originated from the *interhyoideus* of sharks. This muscle persists in both amphibian and reptiles. In mammals, it breaks up into digastric and stylohyoid.

The posterior belly is the larger one. Its primary insertion is the *mastoid notch* of temporal bone. It inserts into the same tendon as described for the anterior digastric. Functionally, this muscle belongs to the original hyoid series. It is co-planar with the first arch muscles, with which it shares deep investing fascia. Its motor nerve enters very proximally, close to stylomastoid foramen. It is supplied by the posterior two branches of second arch ECA: posterior auricular artery and occipital artery.

Stylohyoid This muscle develops from interhyoideus as well. It is positioned anterior and superior to posterior digastric. Its primary insertion is r4–r5 styloid process of temporal bone. Developing downward, it seeks out its secondary insertion at the r5 posterior body of the hyoid just above and behind omohyoid. Stylohyoid receives three of the four arteries of second arch ECA: posterior auricular, occipital, and facial.

Taken together, these two muscles constitute a useful *approach to the facial nerve* from a Risdon-type incision. Both are enclosed by DIF. They lie deep to, *and separate from*, the platysma (the latter being enclosed in SIF). Fortunately for surgeons, stylohyoid is immediately seen beneath platysma. Moreover, the fascias of posterior digastric and stylohyoid are separate but fused at their common border. If one follows the posterior margin of stylohyoid backward, one will arrive directly at stylohyoid, the styloid process, and facial nerve!

Buccinator This muscle extends like a quadrilateral sheet just deep to the r2–r3 buccal mucosa between maxilla and mandible. It is attached to the pterygomandibular raphe, which puts it in continuity with superior constrictor. A ligament slung between maxillary tuberosity and pterygoid hamulus admits the tendon of tensor veli palatine, indicating the functional status of TVP as a masticatory muscle. Recall the TVP and medial pterygoid arise from a common blastema. Since TVP is a Sm4 muscle, it is in place *prior* to migration of buccinators from Sm6; hence, the ligament must skirt around the tendon. A plane of separation exists between buccinator attached to r2 buccal mucosa and that attached to r3 buccal mucosa. For this reason, buccinators sends fibers forward into deep orbicularis of the lips without decussation. Lateral facial clefts between the commissure and the ear divide the upper and lower halves of buccinators.

Second Arch Muscles of Facial Expression (by Motor Nerve)

Posterior Auricular Branch: Posterior Ear and Posterior Scalp

As far as second arch myoblasts are concerned, the ear is the center of the universe. As we have already seen, neural crest SIF flows upward around it in the form of a Y. Before listing the individual muscles, let's review the gross anatomy of the auricle.

The rim of the auricle is the *helix*. Where this makes a turn, its apogee, is *Darwin's tubercle*. Running parallel is *antihelix*. These two curves are separated by a "boat-shaped" depression, the *scapha*. Cephalically, antihelix divides into a Y, forming superior and inferior *crura*. The space between the crura is the triangular fossa. The inferior-directed *crus helicis* spills out onto deep cavity, the concha. In so doing, it divided concha into an upper *cymba conchae* and a lower *cavum conchae*. Entrance to the meatus is defined by two cartilaginous prominences. Anteriorly, *tragus* is so names because it bears hairs like a goat's beard! Its opposite fellow is antitragus.

Examination of the cartilage framework from behind is valuable. Auricular cartilage is absent in the earlobe and in the space between tragus and helix. In front of auricle, just where helix begins to curve upwards is spina helicis. Along its posterior aspect, the terminal helix has a separate tail, *cauda helix*, that is not fused with the body of the cartilage. The gap between these structures is *fissura antitragohelicina*.

Extrinsic Auricular Muscles

These muscles all attach to the *medial* (cranial) surface of the ear cartilages. Their action is to move the ear as a whole. Can *you* wiggle your ear? Note that the epicranius fascia is continuous with the SMAS in the face; both are part of the superficial investing fascia (SIF) within which are found muscles derived from Sm6. Anterior nerve supply auricularis anterior and auricularis superior are supplied by temporal branch. Posterior auricularis is supplied by posterior auricular nerve. The extrinsic muscles are supplied exclusively by the second arch ECA posterior auricular artery. They are *all* supplied by posterior auricular nerve.

Auricularis posterior has two heads; these are difficult to identify because they are pale and wispy. They run from hillocks 5 and 6 (posterior concha) back to r7 mastoid. This muscle forms before the other auriculares.

Auricularis superior This is the largest muscle. It spans from hillocks 2 and 3, the upper surface of concha to temporal epicranius.

Auricularis anterior is the smallest of the three muscles. Its fasciculi span from the second hillock at the conchal bowl to preauricular epicranius fascia.

Intrinsic Auricular Muscles

These muscles reside on the external (facial) surface. These link various components of the cartilage framework together. They are *all* supplied by posterior auricular nerve. This penetrates the substance of the ear by following the penetrating sensory branch of greater auricular nerve. Because four of these muscles reside on the external surface, they are supplied by superficial temporal nerve. Posterior auricular nerve supplies transversus and obliquus. Arterial supply is second arch ECA anterior = superficial temporal; posterior = posterior auricular.

Helicis major sits on anterior margin of helix. It spans upward from spina helicis (hillock 1) to just before helix begins its backward flexure (hillock 2).

Helicis minor spans across from posterior spina helicis to concha.

Tragicus spans the lateral aspect of tragus.

Antitragicus runs from antitragus (hillock 1) to antihelix (hillock 6).

Transversus auriculae is located on the *posterior* pinna. It extends from eminentia conchae to posterior prominence of the scapha.

Obliquus auriculae is on *posterior* pinna.

Observations/Hypotheses

1. All intrinsic muscles come from a single source, Sm6, and are innervated by a single motor nerve, posterior auricular. It makes sense that *all auricular cartilages are likewise derived from second arch neural crest.*
2. Clefts of the auricle are observed to extend on an oblique line from between 1 and 2 backward to between 4 and 5. It is hypothesized that represents a *genetic division between cranial hillocks from r4 neural crest and caudal hillocks from r5 neural crest.*
3. The six hillocks observed are the result of programming of this second arch crest by overlying anterior ear skin. Sensory neuroanatomy reveals the neuromeric map of the external ear skin. Anterior to EAC: first arch, r3 dermis. Posterior to EAC: third arch, r7 dermis. Within posterior wall, EAC: second arch r5 dermis.

Occipitalis

Recall that the supraoccipital bone has both chondral and neural crest components; these are demarcated by superior nuchal line. Below SNL, chondral supraoccipital develops from the sclerotomes of the four occipital somites. It provides secondary insertion to four sets of muscles the mesenchyme of which originates from cervical somites. Above SNL, membranous supraoccipital develops as a bilaminar induction of neural crest of uncertain derivation (possibly r3). mesenchyme. Internal induction is provided by r3 dura. Externa induction comes from c2 to c3 dermis. Second arch neural crest SIF inserts along the entirety of the superior nuchal line. SIF and the epicranius are synonymous. Occipitalis has its primary insertion along SNL. It extends forward (upward) where it finds its secondary insertion into a fascial confluence of SIF and periosteum, termed the *highest nuchal line*. The blood supply to occipitalis is from the two posterior branches from the second arch ECA: posterior auricular artery and occipital artery.

Temporofrontal Nerve: Anterior Ear, Anterior Scalp, Circumorbital/Palpebral Group

Auricularis superior This is the largest muscle. It spans from hillocks 2 and 3, the upper surface of concha to temporal epicranius.

Auricularis anterior is the smallest of the three muscles. Its fasciculi span from the second hillock at the conchal bowl to preauricular epicranius fascia.

Frontalis Flanking each other across the midline of the forehead, the two bellies of frontalis have no "traditional" insertion; they are packaged within epicranius fascia. Along its medial inferior margin, frontalis blends with three muscles: medially with procerus, at its midpoint with corrugator supercilii, and laterally with orbicularis oculi. Frontalis develops later than occipitalis. The arterial supply is StV1 supratrochlear and StV1 supraorbital to zones 13–10. Lateral support to zone 9 outside the muscle in zone 9 is provided by the superior termination of second arch ECA, frontal branch of superficial temporal.

Orbicularis oculi The anatomy of this muscle is a series of concentric ellipses that acted in concert as a sphincter. The take-home message is that neural crest approaches the orbit in a pincer-like movement orbit from lateral to medial. The innervation to this muscle is frontal branch to the upper lid and zygomatic branch to the lower lid.

> *Palpebral lamina* This lamina is thin and pale. Its fibers span over the orbital septum from the r1–r2 medial canthus all the way to the r1–r2 lateral palpebral raphe. It gives rise to some delicate fibers behind the eyelashes, the *ciliary bundle of Riolan.*
>
> *Orbital lamina* This lamina is thick and reddish. It arises from the r1–r2 bifurcation of medial palpebral ligament and swoops out peripherally to blend with corrugator supercilii and frontalis. Some fibers insert into eyebrow skin; these are the *depressor supercilii.*
>
> *Lacrimal lamina* This component spans from r1 lacrimal crest behind the sac, using two slips to insert first, into the tarsi just lateral to the canalicular apertures, and second, along the outer aspect of the tarsal plate all the way over to the r2 palpebral raphe.
>
> Arterial supply to orbicularis comes from second arch ECA superficial temporal to lateral canthus, second arch ECA terminal facial to medial canthus, StV1 ophthalmic to upper lid and StV2 infraorbital to lower lid. We shall spend more time with the embryology of the eyelids in Chap. 12.

Corrugator supercilii Pyramidal in shape, this muscle lies the most medial frontalis and orbicularis; hence, it precedes them in development. Its primary insertion is to dermis at the midpoint of the supraorbital margin. It secondarily flows downward to attach to the most medial aspect of supraorbital margin. Because the muscle spans all the way from zone 13 to zone 9, it has a dual arterial supply: StV1 ophthalmic to zones 13–10 and second arch ECA superficial temporal to zone 9.

Zygomatic and Buccal Branches: Nasal Group and Buccolabial Group

With the exception of procerus, the mesenchyme of the nasal muscles closely follows the surface of the maxilla. It comes in low and sweeps upward. The primary points of attachment start above canine and proceed medially. The canine-lateral incisor muscles sweep upward along the lateral aspect of the nose, while the central incisor depressor septi spans straight upwards into septum. The likely order of formation is as follows: procerus, levator labii superioris et alaeque nasi (LLSAN), nasalis, dilator naris anterior, and depressor septi. Again, the exception of procerus, all these muscles receive their primary blood supply from below via terminal branches of the facial artery.

Procerus This muscle has its primary insertion into the p5 fascia overlying inferior nasal bone and upper lateral cartilage. It runs upward to its secondary insertion into p5 skin of the glabella, where it is admixed with the most medial and inferior fibers of frontalis. Innervation of these upper fibers is from superficial temporal; lower fibers are supplied by the *deep branch* of the zygomatic branch of VII. Arterial supply is from second arch terminal facial artery. These indicate that the pathway of the blastema follows two lateral-to-medial pathways: infraorbital and supraorbital. From lateral to medial is infraorbital; it then ascends up the side of the nose.

Levator Labii Superioris Et Alaeque Nasi This muscle has its feet firmly planted on two nasolabial sites: (1) medial slip into alar groove and thence to alar cartilage lateral crus perichondrium; (2) lateral slip confluent with orbicularis and levator labii superioris. It spans upward to a secondary insertion into the most distal (upper) margin of frontal process of maxilla. Motor supply is zygomatic to (1) and superior buccal to (2). The latter is important because it runs in a plane deep to the buccolabial group. Formation of this muscle antedates the orbicularis. Arterial supply comes in from below via terminal angular from facial and laterally, from StV2 infraorbital.

Nasalis This muscle has two parts with different names: transverse and alar. (1) Transverse nasalis, or *compressor naris*, attaches to maxilla superolateral to incisive fossa. Its fibers sweep upward, lateral to the alar part, to the lateral wall of the nose. There they ascend until reaching the dorsum where they form an aponeurosis. This fibrous aponeurosis blends with that of procerus and with LLSAQ. (2) Alar nasalis, or *dilator naris posterior*, attached to maxilla above lateral incisor and canine. This is medial to transverse nasalis but lateral to depressor septi. Its fibers are inserted into the *skin of the ala at the lateral crus* and into the *posterior mobile septum.* Talmant's concept of cleft repair seeks to reestablish aberrant nasal insertion into the maxilla and thereby help open and maintain the airway. Note that dilator is medial to compressor. The muscle is caudal to procerus. Buccal branch is dominant. Thus, arterial supply is terminal facial with collateral supply from StV2 medial infraorbital.

Dilator Naris Anterior Known otherwise *apicis nasi or dilator naris minor*, this muscle works in synch with the alar nasalis (to which it is attached) to prevent collapse of the nostril during inspiration. It spans from upper lateral cartilage and dilator nasalis major down to the caudal edge of lateral crus. In so doing, it encircles the nostril; it functions as the primary dilator of the nostril.

Depressor Septi This small, but aesthetically important muscle lies just deep to the mucosa of the upper lip. It spans from central incisor fossa between the mucosa of the sulcus (where it is admixed with overlying deep orbicularis). It sends fibers upward into the membranous "mobile" septum and the footplate of inferior crus. It is innervated solely by buccal branch. Depressor septi has been confused as part of the nasalis complex but its arterial supply is quite separate, being (logically) perfused by second arch superior labial branch of facial artery.

Buccolabial Muscles

During development, myoblasts for nine separate muscles converge to a point just lateral to the commissure. There at the modiolus, they are arranged in pinwheel fashion, radiating outward. The central body has a cleft through which facial artery is transmitted from deep to superficial. Modiolar muscles exist in two functional groups. *Cruciate modiolar muscles* form an X: zygomaticus major, levator anguli oris, depressor anguli oris, and the pars moldiolaris of platysma. *Transverse modiolar muscles*: buccinators, risorius, incisivus superioris, incisivus inferioris.

Zmajor, Zminor, LAO, and LLS all span vertically outward from the lip into from r2 bones into the lip. Because of this geometry, they have a motor supply from zygomatic or buccal or perhaps both. Zmaj and Zmin below to Tessier zone 5; LAO and LLS belong to Tessier zone 4. In a similar manner, mentalis, DII, and DAO span outward from lower lip to r3 mandible.

The sensory supply of these muscles (proprioception) is V2 infraorbital.

Zygomaticus Major The blastema of this muscle is associated with that of orbicularis complex. Its primary insertion is into the corner of the mouth where it blends with levator anguli oris and depressor anguli oris. The insertion is into malar periosteum just anterior to r2–r3 zygomaticotemporal suture. Because of the muscle's relationship to the lip, it receives blood supply from second arch superior labial branch of facial.

Zygomaticus Minor Like Zmaj, this muscle forms part of orbicularis complex. Its primary insertion is into upper lip medial to Zmaj and lateral to LLS. The secondary insertion is into lateral zygoma just posterior to zygomaticomaxillary suture. It is supplied by superior branch of labial. Because of the muscle's relationship to the lip, its blood supply is second arch ECA, superior labial branch of facial.

Levator Anguli Oris The blastema of this muscle follows close to bone and then extends downward into the lip. Thus, its primary insertion is canine fossa of maxilla below infraorbital foramen. It precedes levator labii superioris in development because it lies deep to it. It inserts into angle of mouth. Some fibers insert into the inferior nasolabial fold. The blood supply of LAO is bidirectional. Primary (from below) is second arch, superior labial br. of facial. Secondary (from above) is StV2 infraorbital. Its sensory supply is V2 (infraorbital medial).

Levator Labii Superioris As another part of modiolus complex, LLS has its primary insertion is into orbicularis of upper lip between LLSAQ and levator anguli oris. The secondary insertion is into periosteum corresponding to maxilla and zygoma above the level of infraorbital foramen. Like LLO, the blood supply is also bidirectional: LLS supplied by second arch terminal facial artery (from below) is primary. StV2 infraorbital (from above) is secondary. Surgical approaches to the region using a subperiosteal approach can distort these relationships if care is not taken to reattach the periosteum to points of fixation such as the inferior orbital rim or temporal fascia.

Buccinator This thin, rectangular muscle is programmed by the buccal mucosa between maxilla and mandible. It is stretched out like a sheet between these two bones. The proximal attachment is at the pterygomandibular raphe. The two proximal "corners" of the sheet are located opposite the molars. A small ligament is noteworthy because it bears proximal buccinator fibers, thus indicating the embryonic pathway of the muscle. It extends from maxillary tuberosity to pterygoid hamulus. A small space in the pharyngeal wall exists here. The tendon of tensor veli palatini makes use of this space to pierce the pharyngeal wall and gain access to pterygoid hamulus. Note that buccinators lie in the same plane as medial pterygoid plate. Anteriorly, buccinators fibers from maxilla and proceed forward to enter the respective lip. But a central group of fibers, direct descendents of the raphe, does something very different. At the modiolus, they decussate.

Sm6 buccinator lies in direct contact with Sm7 superior constrictor. Posteriorly, a fat pad separates the muscle from mandible; this permits passage of the buccal artery. The superficial surface of buccinators is crossed by the expected entourage of facial muscles: zygomaticus, risorius (when present), the anguli oris levator, and depressor. Parotid duct starts externally at the level of third molar and penetrates at second molar.

Buccinator spans between r2 maxilla and r3 mandible. It has a dual blood supply: anteriorly, second arch ECA facial artery; and posteriorly, from StV3 buccal branch of inferior alveolar artery.

Orbicularis Oris. Anatomic revisionism has refashioned our knowledge about this muscle. It is now recognized to consist of two laminae divided into four quadrants, each of which has a larger external lamina, *pars peripheralis*, and smaller internal lamina, *pars marginalis*. Pars peripheralis fibers originate from the modiolus and enter the lips. These fibers are thickest at the white roll. Pars marginalis occupies the core of the vermillion. It is uniquely developed for speech. Non-primate mammals do not have a marginalis; it is rudimentary in non-human primates. *Incisivus labii superioris* starts just above the lateral incisor. It runs deep to pars peripheralis; at modulus it segregates into superficial and deep fibers. *Incisivus labii inferioris* begins lateral below the eminence of the lateral incisor. Blood supply to orbicularis follows similar rules. Primary vessels are all second arch ECA: second arch ECA superior and inferior labial arteries from facial, and the second arch transverse facial artery from superficial temporal. Collateral flow comes from StV2 infraorbital artery and StV3 mental artery.

Mandibular Branch: Muscles of the Lower Lip

Mentalis Arising from a primary SMAS attachment into lower lip skin, this muscle attaches to periosteum of the incisive fossa. It is protrusion and eversion of the lower lip (helpful in drinking). Blood supply is second arch inferior labial branch of facial artery with collateral support of StV3 mental branch of inferior alveolar.

Depressor Labii Oris This muscle has it primary insertion in the skin and mucosa of lower lip. As such, it is fused with lower lip orbicularis. It extends downward to secondarily insert on the oblique line of the mandible. Primary blood supply is second arch inferior labial branch of facial artery with a secondary contribution from StV3 mental branch of inferior alveolar artery. DLO develops prior to DAO.

Depressor Anguli Oris As a member of the modiolus complex, DAO fans outward to find a secondary insertion from the oblique line of the mandible and the mental tubercle. Primary blood supply is second arch inferior labial branch of facial artery with a secondary contribution from StV3 mental branch of inferior alveolar artery. It develops lateral to (and later than) levator anguli oris.

Cervical Branch: An Unrestricted Constrictor

Platysma In the shark, the precursor muscle of platysma (and the facial muscles in general) is ventral constrictor. In tetrapods such as amphibians and reptiles, this becomes sphincter collis. Platysma lives up to its name. This broad, flat muscle has an extremely important embryologic function. Traditional texts place this muscle in company with sternocleidomastoid and trapezius but this is incorrect. Platysma

has its own fascia and its own derivation. It defines the permissible limits for caudal expansion of the SIF down to C3 on the chest (it covers over clavicle). Indirectly, it gives us evidence of the transition between two kinds of necks: the head-related neck (C1–C4) and the trunk-related neck (C5–C8).

The primary insertions of platysma fibers reproduce the forward and downward sweep of its mesenchyme. Posteriorly, *pars facialis* blends with the subcutaneous tissue of the superficial to the posterior jaw. Further forward, *pars modiolaris* blends with modiolar muscles. *Pars mandibularis* is attached to the lower anterior mandibular body. A *pars labialis*, attached to the lateral half of the lower lip, lies deep to depressor anguli oris. Pars anterioris blends with its opposite in the nuchal midline. Distally, platysma terminates where SIF blends with the fascia covering the upper pectoralis and deltoid.

Platysma provides blood supply to overlying skin of anterior dermatomes C2–C4 (the dermatome of the first cervical somite involutes). Thus, it is not surprising that the muscle is supplied from multiple sources. second arch ECA submental branch of facial represents the original pharyngeal arch contribution. Thyrocervical trunk is the second branch of subclavian, after vertebral. Both these vessels develop from the vertical cross-linking of multiple horizontal segmental arteries serving levels c1–c8. Thyrocervical trunk gives off three branches. *Inferior thyroid artery* runs all the way up to the inferior constrictor. *Suprascapular artery* supplies the skin of hypaxial dermatomes (C2–C4) of the anterior triangle. *Superficial cervical artery* is distributed to the posterior triangle. Thus, because platysma interposes itself subcutaneously in anterior triangle, it picks up blood supply intended for those dermatomes.

Angiology of Second Pharyngeal Arch Muscles

Arterial branches to these muscles originate from the middle one-third of the internal maxillomandibular artery (IMMA), the *infratemporal fossa segment*. The proximal one-third of IMMA is called the *mandibular segment* (Fig. 7.20). IMMA is a composite structure. It is the terminal branch of the external carotid system; i.e., its take-off is distal to superficial temporal. Recall that *the ECA develops as a rearrangement of the ventral remnants of pharyngeal arch arteries 1–4*. Ventral remnants from the first pharyngeal arch artery are admixed with contributions from the extracranial stapedial system. Deep auricular and anterior tympanic represent induction products from chorda tympani. Middle meningeal and accessory meningeal arteries are induced by V3. Thus, *IMMA is really two-arteries-in-one*. Note that muscle branches most likely arise from the more "primitive" first pharyngeal arch artery. Branches supplying neural crest bones are stapedial products (StV3 to the mandible and StV2 to the maxillary complex). It is interesting that branches dedicated to the muscles of mastication are *not* major suppliers of facial skin. Cutaneous vascular anatomy does, however, contain multiple anastomoses between IMMA and EMMA. These represent the *admixture* of pharyngeal arches 1–2.

Sm6 muscles come in various categories. The suprahyoid muscles, posterior digastric and stylohyoid, represent the "original" hyoid arch; they are dedicated to the masticatory apparatus. These muscles are supplied exclusively by second arch ECA arteries.

Stapedius is an innovation not seen in fishes. Its blood supply is therefore more complex blood reflecting the evolution of the stapes itself. (1) ECA is represented by posterior auricular artery. (2) Stapedial derivatives include anterior tympanic artery and a posterior branch from middle meningeal. Recall that chorda tympani induce stapedial artery to cross over the neck of stapes. It subsequently sends out two branches that escape the tympanic cavity to connect with proximal one-third of IMMA, plugging into it just proximal to middle meningeal. At times, the two vessels are confluent. After involution of the stapedial stem, the residual anatomy of the branches is as follows. *Anterior tympanic artery* accompanies chorda tympani and supplies mucosal lining of the anterior tympanic cavity. It forms a vascular ring around the tympanic membrane, joining up with posterior tympanic artery (from stylomastoid artery). It then goes on to provide a collateral supply to stapedius.

Evolutionarily "new" muscles from Sm6 are divided into three groups. Posterior scalp and ear muscles are supplied exclusively by second arch ECA. Anterior scalp and orbits are supplied by derivative arteries of StV1. Muscles of facial expression in the nasal and circumorbital groups are always supplied by second arch ECA but, in most cases, they receive collaterals from StV2 or StV3 derivative arteries.

Development of Second Pharyngeal Arch Muscles

Having just spent a great of time with the facial nerve, I had hoped to describe the facial muscles it supplies in a rigorous order following each nerve in sequence. However, this proves unworkable because it puts the cart before the horse. The roadmap of the facial nerve faithful recapitulates the migration patterns of second arch neural crest and the muscles it contains. Gasser's work demonstrates that developmental of facial muscles and the branches of the facial nerve takes place according to three spatiotemporal gradients. These are as follows: *proximal to distal and deep to superficial*. Recall that epaxial is neuroanatomically closer to the midline.

As we have stated before, these neuromuscular gradients appear to be universal for all pharyngeal arches. The biology of these gradients is likely based upon the sequential vascular development of the arches. The vascular axis of each arch is a combination of a ventral stem from the aortic sac and a dorsal stem from the dorsal aortae. The ventral stem devel-

ops first. Moreover, the axis of the pharyngeal (aortic) arch artery is eccentric within each arch. The arterial conduit is not dead center but posterior and medial. Thus, the pharyngeal arches are arranged into metabolic maturity gradients. Mesenchyme from each sector within an arch develops and migrates according to its "status" on the "vascular totem pole." Let's now follow this process.

The second pharyngeal arch is fully formed at stage 11. The bulk of its mesenchyme comes from the neural crest of rhombomeres 4 and 5. As neural crest migrates into the arch, it "spills over" somitomeres 5 and 6, much like syrup poured over scoops of ice cream. The somitomeres are both coated and penetrated by neural crest. Recall that the somitomeres already possess intrinsic spatial organization. Gene signal from both the periphery and the neural tube has created gradients within the paraxial mesoderm. Neural crest invasion further compartmentalizes the somitomeres. In so doing, the contents of each somitomeres obtain further information between the neurons of r4 and r5. Lines of communication are established between the CNS and its prospective myoblast targets.

Somitomere 5 does not participate in the formation of the second arch. It provides mesenchyme for rostral petrous temporal bone, likely housing the cochlear apparatus. In addition, its lone muscle, lateral rectus, proceeds to migrate outward toward the eye on instructions from the medial motor column. Recall that at this stage, head folding has conveniently "presented" the optic apparatus to the overlying somitomeres. Lateral rectus is the earliest muscle to insert, being seen at stage 13.

Somitomere 6 does two things. Its dorsal contents contribute mesenchyme to caudal petrous bone, likely centering on the semicircular canals. At the same time, its ventral contents have merged with the expansion of r4–r5 neural crest to form the ventrally directed second pharyngeal arch. Thus, PA2 hangs down from the embryonic axis like a saddle bag.

We come to some critical facts regarding the cranial "displacement" of the otic placode. We know, at first glance, that the otic placode will give rise to the otocyst, which will, in turn, produce the vestibulo-cochlear apparatus. And we also know that facial nerve runs with VIII into the midline of these structures. What is the relationship between r4–r5 and the placode? How does the position of the placode change over time as the temporal bone and middle cranial fossa are assembled? And most importantly how does this affect the migration routes of r4–r5 neural crest?

The otic placode appears very early in development at stage 9. Its medial is adherent to neural tube and contains neural crest. The lateral half is placodal; it is composed of somatic (non-neural) ectoderm. Neural crest cells from r4 to r5 populate the placode. This correlates with the eventual location of both the facial and vestibulocochlear nuclei. By stage 12, the otocyst forms. Its ventral wall contains r4–r5 neural crest precursors of the vestibulocochlear ganglion. By stage 13, the vestibular component of the ganglion is present. The spiral ganglion of the cochlea is visible at stage 16. As the neurons of cranial nerve VIII grow medially toward the hindbrain, they "return back home" to r4–r5.

Events are now rapidly in motion that will determine the location of the external auditory canal, the external ear, and the permissible migration routes for neural crest. In order to understand the final neuromuscular anatomy, we must consider what happens from the two perspectives. Development of the ear is *epaxial*. It involves a series of relationships running from middle cranial fossa (temporal bone), to inner ear, to the tympanic cavity, and outward to the auricle. Development of the pharyngeal arches is *hypaxial*. It is also characterized by changing relationships between structures.

The physical location of the otic placode is important to appreciate. At stage 9, it lies *directly above the first two epibranchial (sic) placodes, these being facial and glossopharyngeal.* In other words, it would appear that the ear canal should be located between the second and third pharyngeal arches. Such is (initially) not the case. At stage 12, the placode-to-otocyst transformation is accompanied by rapidly multiplying neural crest. This heaped-up mesenchyme makes the otocyst appear prominent. It is now located *directly above the cleft* between pharyngeal arches one and two. It would thus appear that the ear canal should be located between the first and second pharyngeal arches.

Pharyngeal arch development is an exercise in origami. It is characterized by three processes. (1) The first and second arches become confluent, although their neuromuscular structures remain distinct, spatially they form a single functional unit. (2) The third, fourth, and fifth arches fold up inside one another, like a Japanese fan. (3) Neural crest arising from r4 to r5 exploits a newly available plane and flows widely, enveloping both the entire head and the ventral neck down to the clavicles: this creates the superficial investing fascia for the SMAS muscles.

At stage 12, the third pharyngeal arch is present. At stage 13, the otocyst separates from the surface and sinks into the paraxial mesoderm of somitomeres 5–6. The fourth pharyngeal arch is now present. At stage 15, the fifth pharyngeal arch is complete but recall that it is supplied by the fourth pharyngeal (aortic) arch artery. The fifth aortic arch is unstable but, by virtue of anastomosis between its dorsal remnant and the fourth aortic arch, blood flow is shared between the fourth and fifth arches. At stage 16, the pharyngeal arch period is complete. Arches 1 and 2 are now confluent; they lie superficial to the remaining three arches.

The remainder of our story has to do with the prodigious peregrinations of r4–r5 neural crest, from the vertex of the skull to the level c4 on the chest. Note carefully the wording. I did *not* say second arch neural crest *exclusively*. The reason for this is that pharyngeal arches, by their very nature, are hypaxial structures. But recall that neural crest migrates *downward* from the most dorsal site of the embryo, the neural folds. Although some of it passes into the pharyngeal arches, another pathway is to hug the wall of the neural tube. This neural crest remains epaxial, forming the primitive meninx, or pia-arachnoid. Thus, the neural tube of rhombomeres r4–r5 is covered by its corresponding neural crest.

We now arrive at a dramatic moment: the explosive migration of second arch neural crest. It should be obvious that this expansion postdates the formation of the primitive meninx; it postdates that of first arch crest. Of course, some second arch crest will remain in-line and co-planer with its predecessor. Thus, early-migrating DIF fascia provides an avenue for "traditional" hyoid muscles (posterior digastric and stylohyoid) to participate with first arch muscles of mastication in jaw control. But the remainder of second arch crest migration, involving the vast majority of its mesenchyme, takes place later. A new fascial layer, the SIF, spreads outward in all directions. These new migration pathways are defined by pre-existing tissues. Second arch crest needs to be clever; it must exploit potential planes. It will therefore flow wherever permitted.

We have previously discussed how the skin of the face is formed. Recall that functional dermatomes do not appear until somite 6 (level c2). Thus, facial skin has to develop using alternative sources. Frontonasal skin (innervated by V1) gets both its epidermis and dermis from of forebrain neural crest. V1 ectoderm is reserved for the conjunctiva. Facial skin (innervated by V2 and V3) derives its epidermis from r2 to r3 ectoderm and its dermis from r2 to r3 neural crest. *A plane of separation exists between craniofacial skin and underlying structures: epaxial brain and hypaxial first arch.*

So we come now to some important statements. Second arch crest produces two fasciae, DIF and SIF, both of which contain myoblasts in distinct spatial compartments. It is not as if the fascial layers are laid down and the myoblasts find their way into position later. SIF contains discrete populations of second arch myoblasts; it constitutes their *primary insertion*. The SIF is like toothpaste containing colored beads (facial muscles). As the toothpaste spreads in the subcutaneous plane, it carries the beads along with it. Thus, the beads (muscles) maintain their position in the toothpaste (SIF) until such time as they locate an object (bone, muscle, or fascia) to which they are attracted and can make their *secondary insertion*.

The order and location in which facial muscles develop are determined by the migration patterns of neural crest. And so "the great toothpaste squeeze" is on, with SIF taking its myoblasts over the head and down the neck. But, as neural crest mesenchyme departs for better climes, there is a price to be paid. Vitiated by this volume loss, a number of second arch structures involute or fuse with surrounding structures.

For this reason, there are very few second arch bones in the face. First and third arch structures become juxtaposed by default. Waldeyer's ring has mucosa supplied by V2/V3 in apposition with IX. The auditory tube as a lateral first arch lamina and a medial third arch lamina with chorda tympani sandwiched in between.

The same phenomenon affects the ear. The confluence of the three arches is seen in the neuroanatomy. Completely confluent this affects the neuroanatomy of the external pinna. The upper zone is supplied by first arch, auriculotemporal nerve V3. The skin of external auditory canal is supplied by V3 anteriorly and VII posteriorly (as expected) but the concha is supplied by auricular branch of vagus. Recall that *vagus is represented as far forward as r7*. Therefore, it falls in the territory of third arch. We know that second arch (r4–r5) does not produce dermis. Third arch (r6–r7) neural crest produces submucosa in the pharynx. Thus, neural crest origin of auricular skin likely comes from two primary sources: first arch (r3) and third arch (r7). These skin zones cover over second arch external auricular muscles. Note that the lobule and lower helix are products of cervical somites C2 and C3. Second arch myoblasts do not penetrate into these areas.

And so, we come to the migration routes of second arch crest. What explains the distribution of facial muscles and the timing of their development? As we proceed downward along the axis of the facial nerve, we recall Gasser's gradients: *proximal to distal, deep to superficial, and ventral to dorsal*. The epicenter of second arch neural crest fascia is stylomastoid foramen. All muscles have SIF unless otherwise specified. All SIF muscles are *hypaxial* except those supplied by temporal branch.

Let's follow Gasser's observations. By 20 mm deep muscle, condensations have formed epaxial stapedius and hypaxial posterior digastric and stylohyoid. Deep muscles differentiate before superficial muscles. Cervico-mandibular and occipital muscle development precedes that of frontal and midfacial muscles. There is a 1:1 correspondence between the proximal–distal order of the motor branches and the spatial positioning of the target muscles.

Summary: Does the SIF Work?

Frontalis nicely illustrates the biology of the SIF. As we shall see, it is contained within the fascia and it has no traditional "insertions." Superficial investing fascia spreads into all available zones of the head and neck. When this process is complete, myoblast populations within genetically defined boundaries are transported passively into their final position. Each mature muscle has three options. (1) It can stay put within the fascial envelope (2) It can extend out from the fascia to the nearest available binding site: dermis, bone, or cartilage. (3) It can choose two binding sites, taking the SIF with it.

Lateral Motor Column, Caudal (r6–r11): Somitomeres Sm7–Sm11

Terra incognita is a very good way to describe the neuromotor apparatus supplied by the rhombomeres r6–r11. Why is this place so confusing? First, *unclear relationships exist between the neural tube and mesenchyme*. In the rostral medulla, we saw rhombomeres, cranial nerves, and their targets all in 1:1 registration. In caudal medulla, we find six rhombomeres in genetic register with one somitomere (Sm7) and four occipital somites 1–4 (formerly Sm8–Sm11). Pharyngeal arches from paired rhombomeres. Thus, PA3 comes from r6 to r7, PA4 from r8 to r9, and PA5 from r10 to r11. Second, *two major cranial nerves live here and one does not*: Glossopharyngeal nerve IX is in register with r6–r7, vagus nerve X is in register with r8–r11, and spinal accessory nerve XI **does not exist**. Third, the neuromeric locations of hypoglossal nerve and spinal accessory nerve change positions in evolution.

Watch out for an amazing coincidence! Here are four important facts. (1) The motor columns of midbrain and rostral medulla share common characteristics. (2) The same pattern is also seen in the motor columns of the caudal medulla. (3) MMC control of axial muscles has been conserved through evolution all the way down through the spinal cord. (4) LMC relationships get rearranged in the course of mammalian evolution. We shall summarize the highlights below:

Medial motor column muscles of the eye and tongue both exhibit sequential patterns of development and insertion as they attach their target organ directly or indirectly to skeletal structures. Eye muscles arise from Sm1, Sm2, Sm3, and Sm5, each of which is supplied by a single motor nucleus. Oculomotor and trochlear, and abducens motor nuclei are *monosomitomeric*. These relationships change as one proceeds down the neuraxis. Tongue muscles arise as a group from occipital somites 1–4 all of which are supplied by a **common motor nucleus**. The hypoglossal motor nucleus of caudal medulla is *polysomitomeric*. The organization is neat and quite ordinary.

In a very similar manner, muscles innervated by lateral motor column r3–r5 follow a well-defined developmental sequence, arise from distinct somitomeres, and are supplied by two distinct nuclei in the brainstem. There is no crossover. Trigeminal and facial motor nuclei are *monosomitomeric*. Here, we come to a real breakthrough. Muscles innervated by lateral motor column r6–r11 also follow a well-defined developmental sequence, arise from somitomeres 7–11, and are supplied by a **common motor nucleus.** Nonetheless, the muscles supplied by nucleus ambiguus are *monosomitomeric*.

Knowing the above, we can re-interpret the neuromuscular anatomy of r6–r11 in a radical new way. Nucleus ambiguus is exclusively motor and supplies branchiomeric muscles.

<u>Levels r6–r7</u> contain motor nuclei for IX and X, both of which innervate the third arch. Thus, palate and upper pharyngeal musculature is supplied from a common neuromeric source via two physically distinct nerves, IX and the pharyngeal plexus of X. <u>Levels r8–r9</u> contain motor nuclei for X and innervate the fourth arch. They supply the middle pharynx via pharyngeal plexus and larynx as superior laryngeal nerve. <u>Levels r10–r11</u> contain motor nuclei for X and innervate the fifth arch. They supply the lower pharynx via the pharyngeal plexus and the muscles of the internal larynx via recurrently laryngeal nerve. The caudal two roots of nucleus ambiguus, formerly considered as the cranial root of XI, are now known to connect exclusively with vagus.

The origins of the lateral motor column in the brain begin with a series of muscles that served as constrictors and dilators of the branchial arches. The term **"presomitic"** refers to the evolutionary history of these muscles as constrictors and dilators of the branchial apparatus in fishes. At the tetrapod transition, these assume new roles. The constrictors remain internal contributing to palate, pharynx, and larynx. The dilators move backward into the spinal cord to become sternocleidomastoid and trapezius. This evolutionary history is discussed along with cranial nerve XI and in Chap. 10 on the neck.

LMC in the medulla is known as ***nucleus ambiguus***. It is strictly branchiomotor (old terminology special visceral efferent, SVE). Nucleus ambiguus innervates Sm7 to Sm11 which provide muscles for the soft palate, pharynx, and larynx. These are supplied neuromerically. <u>Third arch muscles</u> (r6–r7) include stylopharyngeus; soft palate and superior constrictor are supplied by either IX alone (stylopharyngeus) or via X (palate and superior constrictor). <u>Fourth arch muscles</u> (r8–r9) include external thyroid and middle constrictor and are supplied by vagus. <u>Fifth arch muscles</u> (r10–r11) are those of internal larynx and inferior constrictor; these are innervated by recurrent laryngeal of vagus. Note: it is a mistake to think that third arch muscles are supplied by a motor nerve (vagus) of the fourth arch. It is just that vagus has a motor nucleus in r7 and therefore can supply the palate.

Nucleus ambiguus continues into cervical spinal cord as the <u>central motor nucleus</u> (c1–c6). Its lateral component (c1–c4) supplies ***branchiomeric dilator muscles***.

Nucleus Ambiguus: Master Motor Control for All Branchiomeric Muscles Posterior to the Jaws (Fig. 9.6 Nucleus Ambiguous)

Primitive agnathic fishes had up to 15 branchial arches involving a total of 30 neuromeres. Ten are cranial (r2–r11) and 20 are postcranial. The first and second arches of these creatures were gill-bearing. Each gill arch had the same arcade of cartilages. In crown agnathans, such as the placoderms, the number of gill arches dropped to 8. This raises an interesting conjecture. Let us hypothesize: (1) beginning with r2 these 30 neuromeres possess some intrinsic capability to form branchial arch (body wall) muscles and osseous or bone structures; (2) although the number of neuromeres involved in arch formation decreases with evolution, elements of the primitive homeotic "signature" are retained. This leads to two observations.

- In the primitive state, postcranial segments 1–20 are potentially branchial. This could explain location of the thoraco-abdominal boundary. and the location of the mammalian diaphragm between the 24th and the 25th somites (T12–L21).
- Further along the evolutionary ladder, these same homeotic elements, applied to tetrapods, localize the cervico-thoracic boundary at the terminus of the arch system. Based upon the shark model of seven arches, the initial neck-thorax junction stabilizes between the eighth and ninth somites, C4–C5. This is also, conveniently, the origin of the diaphragm.

In the course of evolution, the number of postjaw branchial arches is steadily reduced. When jaws were "invented," the structures of arches 1–2 ceased being branchial. Gnathostome mesenchyme associated with the first and second arches (r3–r5) was reassigned to the jaws; the separate mesenchymes from r2 to r5 became confluent. In the first jawed fishes, the total number of branchial arches dropped to eight with 6-gill arches. This was further reduced in chondrichthyans (sharks) to 5-gill arches. Modern osteichthyans (teleosts) have 4 gill arches.

Sharks cannot actively control their branchial arches. In order to achieve water flow past their gills, they must swim perpetually or die. The acquisition of improved branchial arch muscles enabled osteichthyes to actively pump water through their gills and oxygenate more efficiently. Branchial arches are essentially biplanar: deep muscles constrict the gills and superficial muscles are dilators. Recall that all branchial and pharyngeal arches are hypaxial structures, as are the muscles assigned to them.

In the transition to land, the gill arches became pharyngeal arches; these were reduced in number the 3. As a consequence, muscles associated with branchial arches 3–5 (r6–r11) were reassigned to new roles. Deep layer muscles, all constrictors, underwent extensive transitions. Some became associated with the *new pharyngeal wall*, performing a constrictor function. *Palatal muscles* set up communication between the tetrapod ear and pharynx; the *velopharyngeal sphincter* is also a constrictor mechanism. Some migrated into new reconfigurations of branchial arch cartilages: the laryngeal and arytenoid cartilages received a series of small *muscles of phonation*.

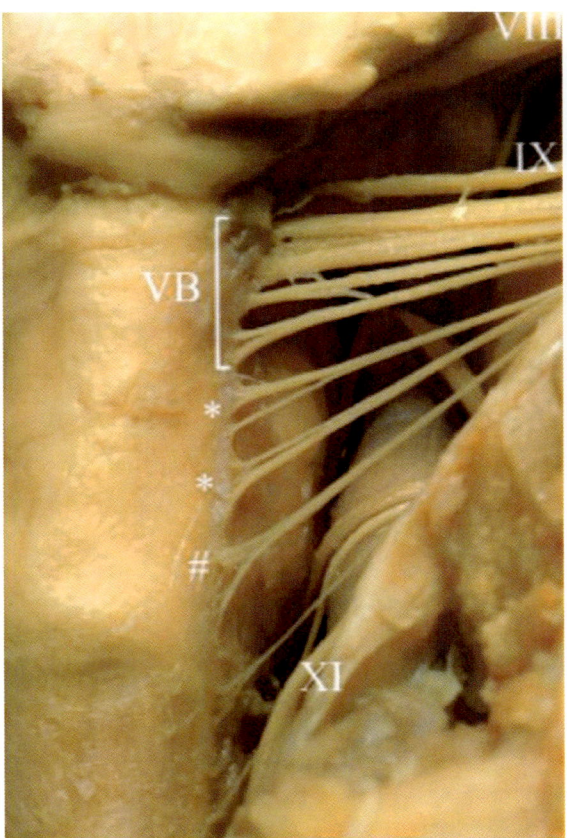
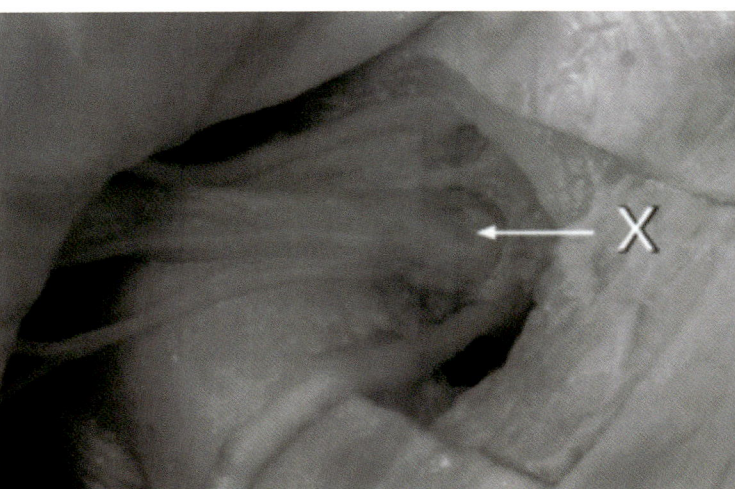

Fig. 9.6 Cranial nerve XI, accessory. Left: Right posterior brainstem. It has been rotated slightly to the left to demonstrate the olivary eminence, the bulbous protrusion seen deep to the rootlets. IX and XI are clearly distinct. Rootlets from multiple levels converge at the jugular foramen. Those from r10 to r11 are spaced out more than those of vagal bundle VB. Right: Dissection of the jugular foramen shows all rootlets between IX and XI remain separate from each other but converge to form the superior ganglion of the vagus, X. [Left: (Reprinted from Wiles CCR, Wrigley B, Greene JRT. Re-examination of the Medullary Rootlets of the Accessory and Vagus Nerves. Clin Anat 2007; 20:19–22. With permission from John Wiley & Sons). Right: (Reprinted from Lachman N, Acland RD, Rosse C. Anatomical evidence for the absence of a morphologically distinct cranial root of the accessory nerve in man. Clin Anat 2002; 15:4–10. With permission from John Wiley & Sons)]

Superficial layer muscles also underwent modifications. The best example of this is the cucullaris muscle seen in shark. This muscle is a series of gill dilators extending along the dorsal aspect of the gills from r6 caudally to the attachment of the pectoral girdle, i.e., from r6 to first spinal nerve (sp1). This tidbit of paleontological trivia becomes relevant to understand the anatomy of trapezius because in mammals, the innervation of trapezius, derived from cucullaris, is a continuation of nucleus ambiguus down to level C4.

What happened in evolution is that, in tetrapods, with the loss of gills, cucullaris control of the branchial arches was no longer needed. Furthermore, with the separation of the pectoral girdle from the skull, cucullaris made a backward "frameshift" of six neuromeres. When before it went from r6 to the back of the skull, in mammals it spans from the back of the skull r11 down to c4–c6 on the pectoral girdle.

This process was proposed by Dutch anatomist Ariens Kappers as Special Visceral Efferent theory. In this model, nucleus ambiguus in medulla r6–r11, being strictly branchiomeric, contains four neuromeres of vagus as r8–r11. In early

vertebrates, what we now call transitional nerve, or accessory nerves were part of vagus but as phylogeny progressed with creation of the neck it migrated caudally to occupy positions c1–c4 in the cervical spinal cord. Thus, the derivative muscles of cucullaris, sternocleidomastoid, and trapezius are branchiomeric and strictly GVE. They are not supplied by a cranial nerve at all, spinal accessory nerve being strictly peripheral from neuromeres c1–c4. We shall consider the history of cranial nerve XI, *including its debunking*, a bit later on.

The Third Pharyngeal Arch

Glossopharyngeal Nerve IX

Conceptual Plan of the Ninth Cranial Nerve
(Fig. 9.7 CN IX)

The glossopharyngeal nerve originates from the sixth and seventh rhombomeres. This nerve just does not get much respect. It is said to supply but a single muscle, stylopharyngeus, but (thanks to connections with the vagus) its distribution is much greater. IX has four types of nerve fibers: (1) Somatic efferents supply striated muscles: stylopharyngeus, all muscles of the soft palate (except tensor), and superior constrictor. (2) Special visceral efferents provide secretomotor fibers to the small mucous glands of pharynx, but *not* soft palate. (3) General visceral efferents connect with the baroreceptors of the carotid sinus. (4) Special visceral efferents bring taste information from the posterior tongue.

Brainstem Nuclei of IX

IX emerges from the medulla in the sulcus between olive and inferior peduncle. It runs along the flocculus of the cerebellum until it reaches jugular foramen. In the foramen, its position is lateral and anterior to vagus and spinal accessory. X is assigned to pharyngeal arches 4–5 which lie caudal to third arch. Recall that arches three- to fivefold into one another like a Japanese fan (or like Russian dolls). Thus, vagus must be positioned internal to glossopharyngeal. Vagus is anterior and lateral to spinal accessory nerve (SAN). As we shall see, the roots from r10 to r1 do not constitute a separate cranial root of XI. Instead, they simply contribute to vagus. So SAN merely represents a peripheral transitional nerve with roots from c1 to c6 which is entrapped by the incorporation of the fourth occipital into the head. This process runs concomitant with the forward repositioning of hypoglossal nucleus in mammals.

IX runs in a groove on the inferior surface of petrous temporal bone. It is kept separate from the bone by its dural sheath (all cranial nerves have them). It passes anterior to, and between, jugular vein and internal carotid. It hugs the lateral surface of ICA. This makes sense because extracranial ICA is the direct descendent of the embryonic third pharyngeal (aortic) arch artery. IX is positioned posterior to styloid process because the latter is a second arch derivative (r5 neural crest). It passes posterior to stylopharyngeus for 2–3 cm and then, after innervating the muscle, passes around its lateral surface to the posterior border of hyoglossus. Note that this exit point takes place at the *hiatus between superior constrictor and inferior constrictor.* These relationships permit identification at surgery. At this point, it disperses into palatine tonsil, pharynx, and posterior tongue.

IX has two ganglia. Within jugular foramen lies *superior (jugular) ganglion.* It is thought to be a detached "rest" from inferior ganglion. For this reason, it is small, or absent. IX runs along inferior petrous bone. A depression in the latter contains *inferior (petrous) ganglion.* It contains cell bodies for sensory fibers. Jugular ganglion, when present, communicates with its petrous patron.

The glossopharyngeal nerve originates from the sixth and seventh rhombomeres within nucleus ambiguus. This nerve just does not get much respect. It is said to supply but a single muscle, stylopharyngeus, but *thanks to connections via nucleus ambiguus* to vagus r6–r7 supply soft palate and upper constrictor as well. The mucosal distribution of IX is very extensive. Cranial nerve IX carries four types of nerve fibers.

- Somatic efferents supply striated muscles: stylopharyngeus. Remaining r6–r7 muscles of the soft palate (except

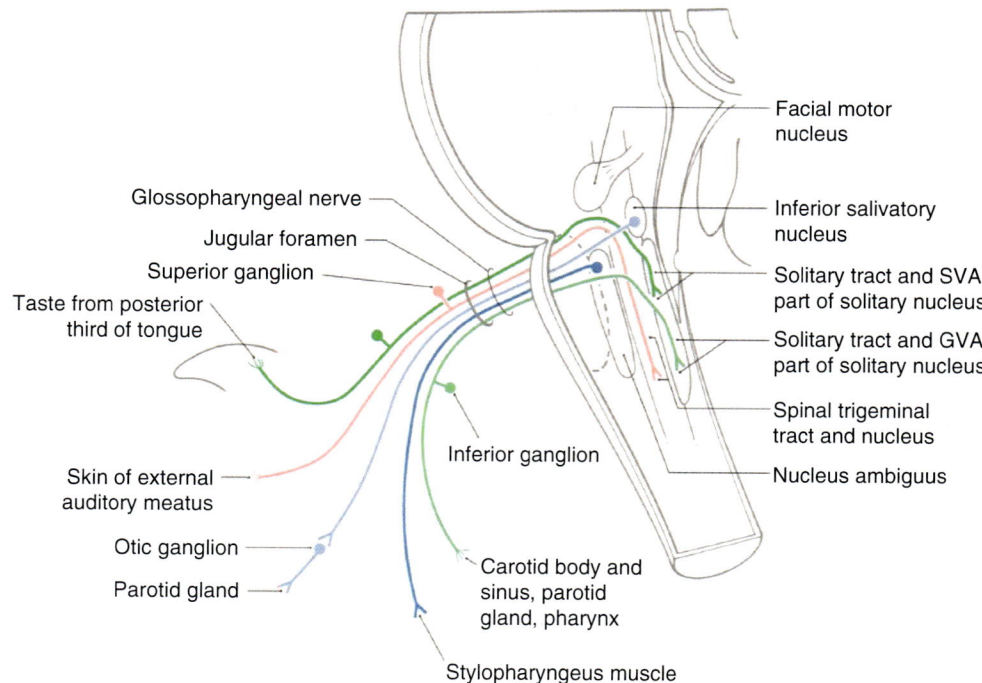

Fig. 9.7 Cranial nerve IX, glossopharyngeal. CN IX is primarily visceral sensory with small motor components. Spinal trigeminal tracts receive general somatic sensory (red) from ear canal and general visceral efferents (light blue) from pharynx. Rostral nucleus ambiguus sends branchiomotor efferent (dark blue) to stylopharyngeus. Rostral nucleus solitarius receives SVA (taste) while caudal nucleus solitarius receives GVA from pharynx and carotid. (Reprinted from DE Haines. Fundamental Neuroscience for Basic and Clinical Applications, 3rd ed. New York: Elsevier; 2006. With permission from Elsevier)

Glossopharyngeal nerve
Jugular foramen
Superior ganglion
Taste from posterior third of tongue
Skin of external auditory meatus
Otic ganglion
Parotid gland
Inferior ganglion
Carotid body and sinus, parotid gland, pharynx
Stylopharyngeus muscle
Facial motor nucleus
Inferior salivatory nucleus
Solitary tract and SVA part of solitary nucleus
Solitary tract and GVA part of solitary nucleus
Spinal trigeminal tract and nucleus
Nucleus ambiguus

tensor), and superior constrictor are supplied via vagus pharyngeal plexus.

- Visceral efferents provide autonomic secretomotor fibers to the small mucous glands of pharynx, but *not* soft palate.
- General visceral afferents connect with the baroreceptors of the carotid sinus. Nociceptive information from the third arch is provided by V3.
- Special visceral afferents bring taste information from the posterior tongue.

Nuclei of Glossopharyngeal Nerve

Nucleus of tractus solitarius is somatotopic. Rostral tract = taste. Middle tract = chemoreceptor data (pO_2) from carotid body and baroreceptor data from carotid sinus. Caudal tract = general visceral afferent (pain, temperature) from palate and pharynx mucosa.

Motor Nucleus Glossopharyngeal motor nucleus represents *rostral nucleus ambiguous*. It is located in the reticular formation just medial to the spinal tract of V and to the nucleus of trigeminal nerve. Recall that spinal tract descends in parallel with nucleus. Both are somatotopic. Thus, the tract sends fibers into nucleus in a craniocaudal order. Recall as well that nucleus has an extensive caudal projection. Textbooks describe glossopharyngeal as supplying stylopharyngeus only. Recent anatomic work demonstrates this zone to innervate all soft palate muscles (save tensor) and superior constrictor (via IX and X). This is accomplished because nucleus ambiguous is the lateral motor column supplying all branchiomeric/presomitic muscles of caudal hindbrain. This encompasses rhombomeres r6–r11, i.e., somitomere 7 and somites 1–4.

Both *superior salivatory nucleus* and *inferior salivatory nucleus* are part of the visceral efferent column in reticular formation. PANS fibers from SSN travel via VII to supply submandibular and sublingual glands while those from ISN supply parotid and lacrimal.

Extramedullary Anatomy

IX emerges from the medulla in the sulcus between olive and inferior peduncle. It runs along the flocculus of the cerebellum until it reaches jugular foramen. In the foramen, its position is lateral and anterior to vagus and spinal accessory. X is assigned to pharyngeal arches 4–5 which lie caudal to third arch. Recall that arches three- to fivefold into one another like a Japanese fan (or like Russian dolls). Thus, vagus must be positioned internal to glossopharyngeal. In turn, vagus is anterior and lateral to spinal accessory. The motor targets of XI (sternocleidomastoid and trapezius) are outside the skull itself. Furthermore, these muscles were originally part of the epaxial cucullaris muscle in shark. All pharyngeal arches are hypaxial. Thus, XI is positioned dorsal and internal to its companions.

IX runs in a groove on the inferior surface of petrous temporal bone. It is kept separate from the bone by its dural sheath (all cranial nerves have them). It passes anterior to, and between, jugular vein and internal carotid. It hugs the lateral surface of ICA. This makes sense because extracranial ICA is the direct descendent of the embryonic third pharyngeal (aortic) arch artery. IX is positioned posterior to styloid process because the latter is a second arch derivative (r5 neural crest). It passes posterior to stylopharyngeus for 2–3 cm and then, after innervating the muscle, passes around its lateral surface to the posterior border of hyoglossus. Note that this exit point takes place at the *hiatus between superior constrictor and inferior constrictor*. These relationships permit identification at surgery. At this point, it disperses into palatine tonsil, pharynx, and posterior tongue.

IX has two ganglia. Within jugular foramen lies *superior (jugular) ganglion*. It is thought to be a detached "rest" from inferior ganglion. For this reason, it is small, or absent. IX runs along inferior petrous bone. A depression in the latter contains *inferior (petrous) ganglion*. It contains cell bodies for sensory fibers. Jugular ganglion, when present, communicates with its petrous patron.

Communications of Glossopharyngeal Nerve

IX joins with *auricular branch of vagus*. This is critical because it represents the embryologic pathway of r6–r7 neural crest to the pinna. This forms the dermis of external pinna, posterior to the external auditory canal. Second arch neural crest provides the mesenchyme for cartilage hillocks 4–6 that make the auricle being remodeled into posterior helix, antihelix, and antitragus. Because first arch skin remains anterior to the meatus, posterior skin and sensory innervation for this zone is provided by cervical plexus. External auditory canal is the exception. Here, third arch lines its posterior wall and the posteromedial tympanic cavity. Sensory innervation for this r6–r7 skin and mucosa is carried back to trigeminal nucleus via X as Arnold's nerve. Note that a communication occurs between the trunk of IX just below petrous ganglion and facial nerve just after the latter exits from stylomastoid foramen.

Branches of the Glossopharyngeal Nerve

Tympanic Nerve of Jacobson

The cell bodies for this nerve lie within petrous ganglion. It is distributed to the tympanic cavity and to the parotid gland via the otic ganglion. The nerve exits petrous ganglion and then pursues a course with a tiny bony canal inside a process that separates the jugular fossa from the carotid canal. It enters into tympanic cavity through the medial floor. It then travels along the promontory of the medial wall. At this point, it participates in *tympanic plexus*. At the processus cochleariformis, it seeks out the canal of tensor tympani and the dives internal to it. At this point, it becomes the *lesser petrosal nerve*.

Tympanic Plexus Two *caroticotympanic nerves* (superior and inferior) penetrate the wall of the carotid canal; these convey SANS fibers into tympanic cavity. *Greater petrosal nerve* contributes PANS fibers via a tiny foramen in the wall of the bony labyrinth just of the vestibular window.

- *Sensory fibers* of this plexus carry pain and temperature from oval window, round window, inner tympanic membrane, Eustachian tube, and the air cells of mastoid sinus.
- *Lesser petrosal nerve* is the termination of tympanic nerve. After encountering tensor tympanic, it burrows into petrous bone medial to tensor, emerging into medial cranial fossa just lateral to facial canal hiatus. Following a space between petrous bone and greater wing of sphenoid, it encounters the otic ganglion.

Jacobson's Nerve (Tympanic Nerve of IX) This nerve is designed to carry PANS salivatory fibers from the inferior salivatory nucleus to parotid gland via the facial nerve. It takes off from the inferior ganglion, enters tympanic cavity, and forms a plexus over the promontory. This plexus connects: (1) to greater petrosal nerve of VII, (2) mucosa of tympanic cavity, (3) medial auditory tube, (4) mastoid air cells, and (5) lesser petrosal nerve. Thus, IX conveys back somatic sensation from the posterior and medial walls of the tympanic cavity and from the mastoid sinus.

Arnold's Nerve (Auricular Branch of X) Arising from superior ganglion of X and is *joined by a branch from inferior ganglion of IX*. The nerve enters the mastoid canaliculus that takes it into the jugular fossa. It then travels over temporal bone and, 4 mm above the stylomastoid foramen, crosses the facial canal. Here, it sends an ascending branch into facial nerve. This small detail is important for understanding herpetic neuralgia because this connection allows herpes virus in the geniculate ganglion to creep out and invade the distribution of Arnold's nerve. See the section below on "Glossopharyngeal Nerve".

Subsequent to its dalliance with facial nerve, Arnold's nerve passes through the tympanomastoid fissure and divides. The *external branch* joins with posterior auricular branch of the facial nerve, which is motor to the auricularis posterior. The *internal branch* of Arnold's nerve supplies: (1) external skin of the posterior pinna; (2) posterior wall and floor of the external auditory meatus; (3) posterior external surface (more like a curvilinear rim) of tympanic membrane.

This is an *extremely important* detail for understanding ear development: the IX-bearing auricular branch of vagus can also be considered the X-bearing auricular branch of glossopharyngeal.

Otic Ganglion This small 4 mm structure likes must medial to V3 just outside foramen ovale. It lies on the lateral aspect of auditory tube just behind the proximal insertion of V3-innervated tensor (recall that TVP is attached to the lateral wall). *LPN is the PANS root of otic ganglion.* It contains preganglionic secretomotor fibers from *inferior salivatory nucleus.* V3 fibers to TVP pass through the ganglion, consistent with the development of trigeminal prior to that of glossopharyngeal. Postganglionic PANS fibers from ganglion communicate with *auriculotemporal nerve.* This branch of V3 subsequently sends branches to the common trunk of facial nerve just behind the neck of the mandible. Thus, PANS fibers from r6 to r7 follow r4–r5 facial nerve fibers throughout the substance of r2–r3 parotid gland. This neurology deftly demonstrates how derivatives of arches 1–3 are "layered."

Carotid Sinus Nerve Also known as the *nerve of Hering*, this nerve is given off just beyond jugular foramen. It immediately shares a communication with pharyngeal branch of vagus. It then runs straight down anterior surface of internal carotid until reaching the bifurcation. At this juncture, we have the beginning of common carotid, formed by the ventral connecting segment between fourth and third aortic arches. This segment represents the embryonic outflow tract. External carotid artery thus contains elements of AA4.

Branches to ***soft palate and pharynx*** These join with the pharyngeal branches of vagus opposite middle constrictor. The three pharyngeal constrictors have not been precisely mapped out; nonetheless, they likely arise originate from Sm6 to Sm7 (superior), Sm8 to Sm9 (middle), and Sm10 to Sm11 (inferior). Palate and superior constrictors are innervated by fibers from r6 to r7 in upper nucleus ambiguus. These eschew the motor branch of IX *sensu stricto* and join X to access the muscles via pharyngeal plexus. Although both nerves cooperate to provide somatic motor control, it is convenient to consider r6–r7 vagus IX as assigned to superior constrictor, leaving r8–r11 vagus to deal with the rest. A difference in the perception of pain takes place as one proceeds from pharynx to esophagus. Upper pharynx and cervical somatic sensation permit one to localize a foreign body whereas within thoracic esophageal pain is diffuse. These findings are explained by neuroanatomic differences between brainstem sensory nuclei to which IX and X report.

Branch to Stylopharyngeus Traditionally described as the sole motor branch of IX, this antiquated idea misses the mark significant contribution of third arch to the pharynx.

Tonsillar Branches These provide sensory supply to palatine tonsil and thence to soft palate where they connect with V2 *lesser palatine nerve.*

Lingual Branches These contain general visceral efferent fibers carrying from the mucosa of the posterior tongue. They also have special visceral efferent fibers for taste from the vallate papillae.

Myology of Somitomere 7

Superior Constrictor This muscle likely arises from a combination of Sm6 and Sm7. It is readily distinguished from the other two constrictors due to its pale color. Its primary insertion is located along the posterior midline from an aponeurosis attached to the pharyngeal spine of the basiooccipital bone. Superior fibers run forward below the Sm6 levator veli palatini and the Eustachian tube, seeking out a secondary insertion along the inferior one-third of the r2 medial pterygoid plate and the r2 hamulus. The upper border of superior constrictor is separated from the skull base by an aponeurosis known as the *sinus of Morgagni.* Middle fibers insert into the r2–r3 connective fascia, the *pterygomandibular raphe.* Inferior fibers insert into the mandibular alveolus just above the posterior terminus of the mylohyoid line. Blood supply is dual. The primary supply is third arch pharyngeal branch of ascending pharyngeal artery. The alternative vessel is second arch tonsillar branch of facial artery. Superior constrictor and levator veli palatini arise from a common blastemal and share the same blood supply. It is possible that superior constrictor may have motor supply from VII thereby coordinating speech with facial nerve function. Innervation of superior constrictor originates from r6 to r7 with fibers transmitted to the muscle via pharyngeal plexus of vagus.

Middle Constrictor This muscle fans out from its primary insertion along the posterior median fibrous raphe. It seeks out sequential secondary insertion sites. (1) It drapes along the inferior border of stylohyoid ligament, running from the r4 to r5 styloid process to the r4–r5 lesser cornu of hyoid. (2) It extends further along the lower border of the lesser cornu. (3) Finally, it follows along the superior border of r6–r7 greater cornu. Note that the rostral fibers of middle constrictor lie external to superior constrictor but the caudal fibers are positioned internal to inferior constrictor. Motor fibers from r6 to r7 originate in the cranial part of accessory nerve, i.e., in nucleus ambiguus, and access the muscle via the pharyngeal plexus of vagus Blood supply is third arch ascending pharyngeal artery but it also receives second arch tonsillar branch of facial artery.

Levator Veli Palatini this muscle body has its proximal attachment to petrous ape and the medial auditory tube. It passes above the concave border of superior constrictor. In cases of unilateral craniofacial microsomia ipsilateral palatal elevation is weak or absent. Blood supply relates to mucosal territories. Third arch pharyngeal branch of ascending pharyngeal artery runs along its medial aspect while second arch ascending palatine branch of facial is lateral.

Uvulus This muscle is attached to posterior nasal spine and the palatal aponeurosis running in parallel with posterior margin of palatine bone. Its blood supply comes via mucosa: the nasal side anteriorly being StV2 lesser palatine and the oral side being third arch pharyngeal branch of ascending pharyngeal.

Palatopharyngeus Proximal insertion in soft palate, oral side. Levator and uvulus divide it into two fasciculi. The two posterior fasciculi are continuous. The anterior fasciculi are more bulky. The paired muscles are also confluent across the midline. Passing down medial to palatine tonsil, the muscle becomes *confluent with stylopharyngeus.* The two muscles are distally inserted into posterior thyroid cartilage. Blood supply demonstrates the physical connection with pharynx. Third arch pharyngeal branch of ascending pharyngeal artery is primary. Second arch ascending palatine branch of facial artery is secondary.

Palatoglossus Proximal insertion in soft palate, oral side. The paired muscles are continuous across the midline. Distal insertion into the tongue, sending some fibers deep into the tongue parenchyma and some to the dorsum. Blood supply is simpler than that of palatopharyngeus, being: third arch ascending pharyngeal artery, second arch ascending palatine branch of facial artery.

Stylopharyngeus This muscle has its primary insertion from the *medial* aspect of styloid process. Recall that stylohyoid is located on the *lateral* and aspect of styloid process. Thus, the third arch derivative is internal to that of the second arch. Styloglossus sweeps downward later and occupies the remaining binding site on the anterior styloid process. Stylopharyngeus travels in company with its motor nerve IX, passing in between superior constrictor and middle constrictor. It seeks out its secondary insertion in the mucosa adjacent to posterior border of thyroid cartilage. Blood supply is third arch pharyngeal branch of ascending pharyngeal.

Angiology of the Third Pharyngeal Arch

The seventh somitomere is innervated by motor and branches of cranial nerve IX. The nuclei of IX reside in rhombomeres r6 and r7. Sensory supply to these muscles is provided by IX. The blood supply to these muscles demonstrates the following patterns.

Third arch ECA: stylopharyngeus.

Third arch ECA + second arch ECA tonsillar branch of facial: superior constrictor.

Third arch ECA + second arch ECA ascending palatine branch of facial: palatoglossus.
Third arch ECA + second arch ECA ascending palatine branch of facial + StV2 greater palatine: palatopharyngeus, salpingopharyngeus.
Second arch ECA ascending palatine br. of facial + StV2 descending palatine: uvulus.

Thus, third arch ECA is involved in all muscles with a physical attachment to the pharynx. All palate muscles are supplied by cranial part of accessory nerve via pharyngeal plexus. Uvulus is located in such a midline position that it cannot be supplied directly from the pharynx. Hence, it must live off the mucosa alone.

At this juncture, some clarification is required regarding the individual development of the ascending pharyngeal vis-à-vis that of the superior thyroid artery. Recall that both the third and fourth pharyngeal arch arteries do not involute. Both arteries plug into the dorsal aorta at separate sites; this intervening segment of DA persists for a short while. During stage 14 (formation of the sixth aortic arch), the *common dorsal connection* between AA3 and AA4 involutes, with the distal fourth arch remaining connected to the systemic aortic circulation. A *common ventral connection* also exists between AA3 and AA4. It connects them to the aortic sac. The combined AA3–AA4 segment becomes the common carotid artery. Thus, the tissue source for ECA includes AA4.

Development of the Third Pharyngeal Arch

The overall theme of the third arch is sphincter formation. Soft palate, in conjunction with superior constrictor, exercises a valve-like action to direct airflow. Working synchronously, the constrictors propel food toward the esophagus. Superior and middle constrictors are rather boring but the functions of inferior constrictor are more complex (see below). PAM mesenchyme for these muscles sweeps downward and forward from somitomere 7. Superior constrictor may be the first, as it must attach to the pterygomandibular raphe (and the backside of buccinators). Thus, nasal surface receives tensor first, and then, levator. The myoblasts next populate the oral side: palatoglossus > palatopharyngeus > uvulus. Superior constrictor comes in a bit later. Superior and middle constrictors are rather boring but the functions of inferior constrictor are more complex. Mesenchymal programming for the constrictors depends on signals from the ectodermal lining of the palate and the endodermal.

The Fourth and Fifth Pharyngeal Arches (r8–r11): Somitomeres Sm8–Sm11

Vagus Nerve

The vagus nerve carries a "full complement" of neurons (Fig. 9.8).

- <u>Somatic afferent</u> fibers carry sensation from *glossopharyngeal-innervated skin* of the posterior external pinna and the external aspect of posterior acoustic meatus.
- <u>Visceral afferent</u> fibers supply mucous membranes of the respiratory system (trachea, bronchi, lungs), the heart, the gastrointestinal system from esophagus to the anal verge, and the kidneys.
- <u>General visceral efferent</u> (PANS) fibers supply the heart as well as the smooth muscle and glands of the above systems.
- <u>Special visceral efferent</u> (lateral motor column) supplies voluntary striated muscle of the pharynx and larynx. Vagus conveys motor fibers from IX to the muscles of soft palate (except tensor).

Nuclei of Vagus Nerve

Dosal nucleus is general visceral efferent. Eighty percent of its neurons are preganglionic PANS to smooth muscle. It sits in dorsomedial medulla next to the floor of fourth ventricle. Its caudal limit is c1. An interposed structure nucleus interculatus separates dorsal nucleus from hypoglossal. The nucleus has topographical representation with the gut rostral, stomach and pancreas in the middle, and heart and lungs caudal.

Nucleus ambiguus is special visceral efferent. It lies directly caudal to a corresponding nucleus for IX. It supplies the striated muscle of the fourth and fifth pharyngeal arches with r8–r9 dedicated to fourth arch and r10–r11 supplying fifth arch. Once again, there is topographic representation, with soft palate located rostral (probably with crossover to IX), pharyngeal constrictors in the middle, and larynx caudal.

Nucleus tractus solitarius receives special visceral afferent (taste) fibers from VII, IX, and X in a topographic pattern. Oral cavity and anterior two-third tongue via chorda tympani and greater superficial petrosal are rostral. Circumvallate and foliate papillae from posterior one-third tongue, palate, and pharynx via lingual branch of IX are distributed rostromedial while those from epiglottis, larynx via superior laryngeal nerve are distributed caudolaterally. Carotid sinus and aortic body fibers terminate here as well.

Fig. 9.8 Cranial nerve X, vagus. Vagus has three sensory tracts and two motor tracts. <u>Spinal trigeminal tract</u> receives GSA from the mucosa of external auditory canal. <u>Rostral solitary nucleus</u> (gustatory) receives SVA fibers (dark green) from the root of tongue and epiglottis. <u>Caudal solitary nucleus</u> (cardiorespiratory) gets GVA input from lungs, gut and viscera. <u>Nucleus ambiguus</u> (r6–r11) supplies "GSE" neurons (dark blue) to branchial muscles of palate (r6–r7), pharynx (r6–r11), and larynx (r8–r11). <u>Dorsal motor nucleus</u> supplies SVE (light blue) PANS neurons to smooth muscle and glands of the viscera. (Reprinted from DE Haines. Fundamental Neuroscience for Basic and Clinical Applications, 3rd ed. New York: Elsevier; 2006. With permission from Elsevier)

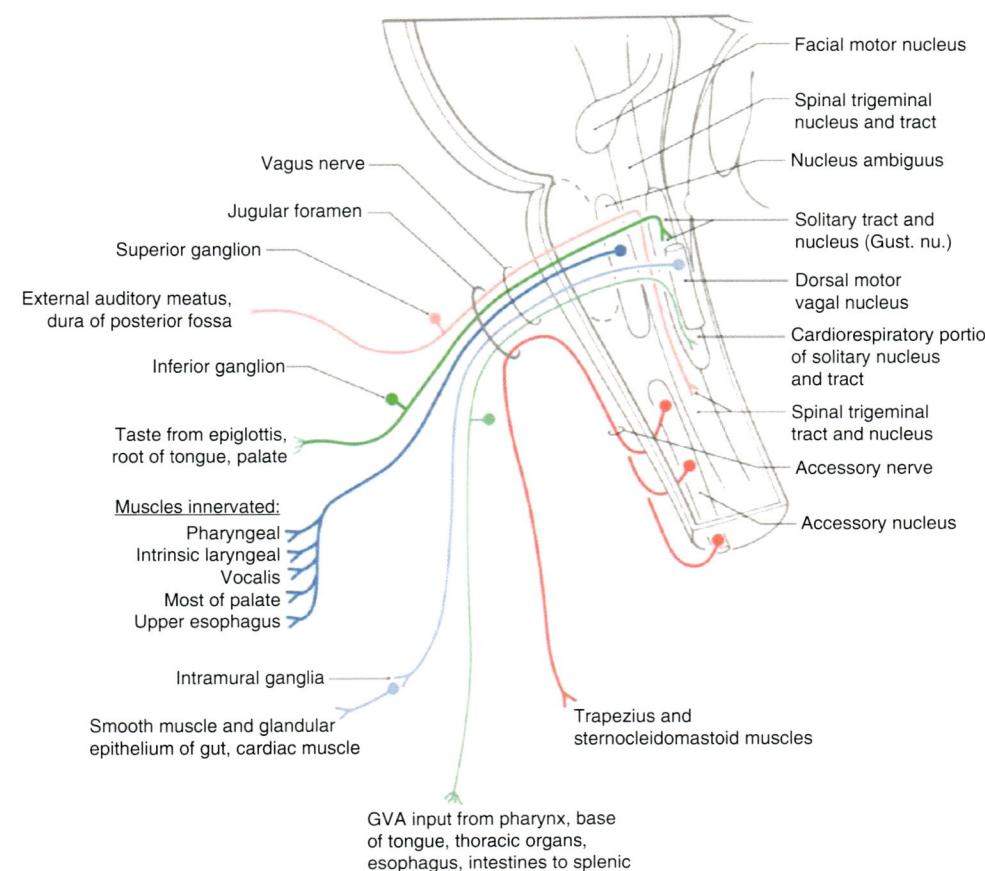

Spinal trigeminal nucleus receives general somatic afferent (pain) fibers conveyed to vagus from skin of concha, posterior external pinna, periosteum of posterior fossa, and foramen magnum.

Extramedullary Course

The nerve makes its exit from medulla via 8–10 small roots in a sulcus spanning from olive to inferior peduncle. It is inferior (medial) to glossopharyngeal nerve and superior to C1–C4 spinal accessory nerve The most caudal rootlets, traditionally attributed to the erroneous cranial root of spinal accessory nerve, are probably dedicated to the fifth arch internal muscles of the larynx. Why? The rootlets combine into a single nerve that passes backward beneath cerebellar flocculus to reach jugular foramen. Here it exits, sharing a *common dural sheath* with accessory nerve. Vagus remains physically separate from IX which exits anterolaterally in its own sheath. At this point, the vagus enlarges into two sensory ganglia.

Superior ganglion (jugular ganglion) lies within the foramen. It is small (4 mm) and contains unipolar ganglion cells, most of which enter the auricular branch. For this reason, surgical procedures of the ear, by virtue of IX-innervated skin, can refer on to vagus, causing intense nausea.

Inferior ganglion (nodose ganglion) is located just outside the jugular canal. It is about 2.5 cm and fusiform. Its unipolar cells have two fates: some comprise the internal root of superior laryngeal near; the rest go on to supply the viscera.

Communications of Vagus

Ganglionic Connections with Craniofacial Structures

1. At *superior ganglion,* vagus nerve communicates with (1) cranial part of spinal accessory nerve; inferior ganglion of IX; via the auricular branch to facial nerve; and to the SANS superior cervical ganglion. Note that the cranial part of XI resides in r8–r11 as well. It provides motor fibers to branches of vagus distributed to pharynx and larynx. This is important for understanding *voluntary control of striated muscle in middle constrictor and inferior constrictor.*

2. At *inferior ganglion*, vagus nerve sends fibers to hypoglossal, to superior cervical ganglion (once again), and to the connecting loop between C1 and C2.

Vagus nerve trunk refers to the union of vagus with cranial part of XI. This travels downward within carotid sheath behind first, internal carotid and later, common carotid. It lies in front of jugular vein. It is deep to both vessels. Recall that common carotid represents a melding of AA4 and AA3 and AA4. By tracking along carotid, vagus is ensured to access the embryonic aortic root. In the neck, vagus diverges.

Right vagus crosses first part of subclavian artery. It follows right side of trachea until reaching lung where it makes the pulmonary plexus. Below this plexus, it continues to dorsal esophagus where it communicates with left vagus. The result is a single trunk, the *posterior vagus nerve*. This disappears down the esophageal hiatus of diaphragm and then proceeds along the lesser curvature of stomach.

Left vagus accesses the thorax between subclavian and left carotid, scoots around the left side of aortic arch, contributes to posterior pulmonary plexus, and reaches esophagus roughly at the level of the tracheal bifurcation. After receiving communications from right vagus, it creates *anterior vagus nerve* which flows over anterior stomach.

Branches of the Vagus in the Jugular Fossa

Meningeal branch arises at superior cervical ganglion. It accesses the skull at jugular foramen and spreads out over the dura of posterior fossa. Kimmel demonstrated that the nerve actually conveys sensory fibers from C2 to C3. Recall that the occipital somites and the first cervical somite lack dermatomes (these were present but involuted); hence, there are no somatic sensory fibers available for posterior cranial fossa bone derivatives of S1–S4.

Auricular branch (*nerve of Arnold*) takes off from superior ganglion and communicates with *inferior ganglion of IX*. It tracks along the internal aspect of jugular vein and enters the skull at mastoid canaliculus. Here are *crosses and communicates with facial nerve* 4 mm above stylomastoid foramen. It passes external at the tympanomastoid fissure. One branch goes to the skin of *posterior external auricle*. The other branch goes to *posterior external auditory canal*. Somatic afferent fibers refer pain to the spinal nucleus of trigeminal and cause nausea to be virtue of referral to.

Branches of Vagus Nerve in the Neck

Pharyngeal Branches Two nerves arise at inferior ganglion. (1) An all-important *motor branch* supplies *branchiomeric muscle* from rhombomere r6–r11 of *nucleus ambiguus*. As such, it is properly called *palato-pharygo-laryngeal nerve*. It also contains sensory fibers from target

mucosa. These branches follow internal carotid to *superior border of middle pharyngeal constrictor*. At the hiatus between superior and middle constrictors, these nerves gain access to the pharyngeal wall where they form the *pharyngeal plexus*. (2) *Nerve to carotid body* supplies chemoreceptors that sense changes in pO_2.

Superior Laryngeal Nerve This nerve has its nuclear source in r8–r9. SLN is the motor and sensory nerve of the fourth pharyngeal arch. As such, its sensory distribution defines those structures that are derivative from the arch.

It takes off from the lower inferior ganglion, heading inward deep to internal carotid, seeking out superior cornu of thyroid cartilage. It receives SANS from superior cervical ganglion.

- *External branch* is purely motor. It runs deep to sternothyroid. It supplies cricothyroid and inferior constrictor. The nerve is closely associated with superior thyroid artery. It is thus at risk during thyroidectomy.
- *Internal branch* is purely sensory. It pierces thyrohyoid membrane along with its artery. Once inside, it divides into three branches; these are organized somatotopically to supply mucosa and glands all the way down to vocal folds.
  - *Superior branch* goes to piriform fossa.
  - *Middle branch* supplies the ventricle (quadrangular membrane). It is the afferent source of the *cough reflex*.
  - *Inferior branch* is distributed to the ventricle and subglottic cavity.

Inferior Laryngeal Nerve The nucleus is located in r10–r11. ILN represents the motor and sensory nerve of the fifth pharyngeal arch. Its sensory distribution defines the structures that are derived from the fifth arch. It is also known as the ramus internus of cranial nerve XI. ILN is well known for its different anatomy.

- *Right ILN* passes in front of first part of subclavian, loops under it, and travels dorsal along trachea and esophagus.
- *Left ILN* arises from vagus as it skirts along the lateral aspect of aortic arch. Here, distal to ligamentum arteriosum, it curves back behind aortic arch to access trach and esophagus. The ILNs lie medial to the hanging saddlebags of the thyroid lobes. At this juncture, they are very close to inferior thyroid artery. Note that ITA has great embryologic significance more on this in Chap. 12. The ILNs run under the inferior border of inferior constrictor. They penetrate cricothyroid membrane and supply *all thyroid muscles except cricothyroid*.

The upper (ascending) part of the nerve is close to the inferior thyroid artery. It passes either deep to (two-third of cases) or superficial to (one-third of cases) cricopharyngeus, which it innervates. At the lower border of inferior constrictor, it divides. Anterior branch is motor. Posterior branch is sensory. Anterior branch ascends posterior to cricothyroid joint and then terminates in thyroarytenoid muscle. Thus, the linear order of muscles supplied by ILN is as follows: cricopharyngeus > posterior cricoarytenoid > interarytenoids > lateral cricoarytenoid > thyroarytenoid. This reproduces the spatiotemporal sequence by which the fifth arch muscles populate the internal larynx.

Superior Cardiac Branches These provide PANS motor control to the heart.

Myology of Somitomeres 8 and 9

Precise mapping of muscles from this zone has not been done but some accurate guesswork is possible. The thyroid, arytenoid, and cricoid cartilages are neural crest derivatives. Thyroid and arytenoid arise from r8 to r9 while cricoid arises from r10 to r11. These correspond to pharyngeal arches 4 and 5, respectively. Muscles assigned to these cartilages arise and migrate prior to tongue development. They could be construed to form during the more primitive somitomere stage.

Inferior Constrictor Inferior constrictor is the largest and thickest of the three constrictors. It spans between fourth arch and fifth arch and extends down to the margin of trachea and esophagus. It arises from somitomeres Sm8–Sm9 because it relates exclusively to the external surface of thyroid cartilage but, due to a component dedicated to cricoid, it is possible that it originates from somitomere 9 as well. Its primary insertion is posterior into the median pharyngeal raphe. Recall that this is ultimately suspended (via superior constrictor) into the pharyngeal tubercle of basioccipital. The muscle has two parts.

Thyropharyngeus has secondary insertions into oblique line of thyroid lamina; a small slip goes into the lesser cornu.

Cricopharyngeus inserts into lateral cricoid cartilage. This takes place between the caudal attachment of thyrocricoid muscle and the articular facet for inferior thyroid cornu. Two sectors of this muscle component are described. An upper *pars oblique* sweeps back to the posterior midline raphe. A lower (and deeper) transverse *pars funiformis* forms a circular band (therefore the term sac) and does *not* attach to the raphe.

The anatomic minutia of cricopharyngeus portion of inferior constrictor create two spaces of clinical importance. They represent *anatomic "weak spots" for diverticula.*

Killian's triangle refers to a small triangular zone <u>located posteriorly</u> between the pars oblique and pars funiformis.

Larimer's triangle lies posteriorly as well, just beneath cricopharyngeus.

Blood supply of inferior constrictor differs from that of middle constrictor. By virtue of being third arch ECA ascending pharyngeal artery and subclavian muscular branches of inferior thyroid artery. As we shall see in the ensuing chapter, inferior thyroid artery develops as a vertical reconstitution of multiple segmental arteries serving each level of pharynx and esophagus from c4 to t2.

The Larynx: A Quick Orientation

Always getting short shrift in anatomy courses, the larynx is a complex structure with multiple functions: it directs airflow, has a mechanism to prevent aspiration, and serves as an organ of phonation. As such, it is a defining characteristic of humanity. For detailed information and excellent illustrations, the reader is referred to the 40th edition of Gray's anatomy. Our purpose here is to explore the formation of this organ and, perforce, the muscles that make it work. This is not an easy task, for little experimental work exists regarding its development. Nonetheless, in order to carry out our tasks we need to summarize the landmark structures of the larynx.

The larynx consists of paired cartilaginous structures, three large and three diminutives. Anatomy texts sometimes consider thyroid, cricoid, and epiglottis to be singular but, in point of fact, they are assembled from bilateral sources of mesoderm.

Thyroid cartilage is the largest structure. It is a derivative of r8–r9, or the fourth pharyngeal arch. At birth, the larynx develops high in the neck, directly opposite C1–C4 vertebrae. By age 6, it has moved to level C4–C7. Thus, the upper border of thyroid cartilage at C4 sits is a landmark for the carotid bifurcation. Thyroid cartilage is suspended from hyoid bone by thyrohyoid membrane. This marks the boundary between third arch and fourth arch. An aperture in the TH membrane designates the pathway of internal branch of superior laryngeal nerve and its companion, the fourth arch ECA superior thyroid artery. A similar membrane connects it with cricoid cartilage below. Four cornua project posteriorly, like stalactites and stalagmites. The lower cornua straddle the posterior cricoid like ice tongs, creating thyrocricoid joints. The upper cornua ascend in the posterior margin of thyrohyoid membrane. Although they do not reach hyoid, they certainly stiffen up the membrane.

Epiglottis is shaped like a tennis racket. Its handle or stalk, the *petiolus*, is attached via a thyroepiglottic ligament to the backside of the thyroid notch. This allows it to bend backward passively under pressure from the tongue and the contraction of aryepiglottic muscles. It is not critical for swallowing. When absent, aspiration is rare. Fourth arch taste buds are located here with innervation from vagus. The base of epiglottis is confluent with that of the arytenoid cartilages. These are connected by the *aryepiglottic folds*.

The upper lingual surface of epiglottis is covered with mucosa (non-keratinized stratified squamous epithelium) and it flows outward, forming epiglottic folds. Those that connect with the arytenoid cartilages contain aryepiglottic muscles. Its anterior surface lies behind thyrohyoid membrane. The *pre-epiglottic space* in between contains fat and is clinically important. The posterior surface of epiglottis is covered with respiratory ciliated mucosa. Multiple perforations in the epiglottis permit passage of internal branch of superior laryngeal nerve and constitute a communication with the pre-epiglottic space.

Arytenoid cartilages are also derivatives of the fifth arch. They form a synovial joint with the posterior superior cricoid laminae. Two movements take place here, both of which cause alterations in the vocal important for speech. Rotation causes the cords, which are attached to the bases of the arytenoids, to swing outward. Gliding allows the apices of the arytenoids to approximate each other or recede.

Cricoid cartilage is positioned level with C6. It can be considered a derivative of r10–r11 or the fifth pharyngeal arch. Its internal surface is lined with mucosa. Congenital malformations can cause subglottic stenosis. The airway here is also vulnerable to acquired stenosis from prolonged intubation.

Corniculate cartilages hide in the posterior aryepiglottic folds. They sit like tiny curved fingers facing one another atop the arytenoids. Sometimes they form a joint with the arytenoids; at other times, the cartilages are fused.

Cuneiform cartilages are teardrop-shaped. They also reside in the aryepiglottic folds anterior and superior to the corniculates.

Several clinically important details should be noted regarding the histology of the larynx. The internal aspect of the larynx is covered by a ciliated pseudostratified respiratory epithelium. For this reason, it shares with the rest of the respiratory tract a mucociliary clearance function. The vocal folds are a different story. Because they are subjected to constant stress, they are covered with non-keratinized stratified squamous epithelium. This same type of epithelium covers the external aspects of the larynx because it is constantly exposed to food, fluids, and (in some cases) marijuana smoke.

Mucous glands provide lubrication for the system but they do not exist over the cords. For this reason, prolonged speaking can lead to hoarseness. Vocal folds consist of five layers, the most superficial of which contains collagen fibers and is poorly attached to the underlying vocal ligament. Edema can accumulate here, Reinke's space, which extends the full length of the cords. Vocal folds meet at the *anterior commissure*. Here, the fibers blend into the perichondrium of the thyroid cartilage. Known as *Broyle's ligament*, the site contains blood vessels and lymphatics. It is an excellent escape route for cancer cells. Differences in the attachments of mucosal folds within the larynx predispose to accumulations of fluid and swelling of the vocal cords, *Reinke's edema*.

Embryology of the Larynx

Discussing this topic is both appropriate and awkward. The development of a larynx is inseparable from the terrestrial transition and the evolution of the neck. For our purposes here, it suffices to make a few broad points. Neural crest provides the mesenchymal substrate for laryngeal cartilages. These represent an evolutionary "rehash" of the branchial cartilage of arches 4 and 5. As an oversimplification, fourth arch is the source for thyroid, epiglottis, cuneiform, and tritiate cartilages, while fifth arch gives rise to cricoid, arytenoid, and corniculate cartilages. It is possible that both arches contribute. On the other hand, fifth arch may be limited to ultimobranchial body. Paraxial mesoderm produces the striated muscles of the larynx; these arise from two sources. Labeling experiments of the laryngeal muscles in birds by Noden localize them to the first two occipital somites but intraoperative stimulation shows laryngeal activation to arise from the posterior zone of nucleus ambiguus. Cricothyroid muscle, the target of superior laryngeal nerve, originates from r8 to r9 and is innervated by external branch of laryngeal nerve of vagus. The internal laryngeal muscles originate from levels r0–r11. They are innervated by internal branch of inferior (recurrent) laryngeal nerve of vagus.

Development of trachea and esophagus begins at neuromeric level c1, i.e., from somite 5. At birth, larynx is located directly opposite C1–C4. It begins to descend at age 2; at age 6, it reaches its final position opposite vertebrae C4–C7. At any rate, from somite 5 downward, paraxial mesoderm assumes new, and alternative, roles. Although it continues to produce the axial skeleton, its striated muscles are exclusively assigned to the trunk and appendicular skeleton. Thus, all cartilage and muscle of the viscera in the neck, i.e., the trachea and esophagus, are synthesized from lateral plate mesoderm.

Muscles of the Larynx

Laryngeal muscles first got their start in sharks as the superficial constrictors and interbranchials. For greater detail on the fish anatomy, see the chapters on muscle evolution in publications by Carlson and Kardong. The tetrapod transition in amphibians shows further specialization into dilator larynges, subarcuals, transversi ventrales, and depressors arcuum. Evolution is largely completed by reptiles with the advent of specific intrinsic laryngeal muscles.

Extrinsic muscles affect the larynx during respiration and phonation. *Infrahyoid strap muscles* (thyrohyoid, sternothyroid, and sternohyoid) move the larynx directly and the hyoid indirectly. We shall spend more time with these muscles in the next chapter. *Inferior constrictor* has an extensive secondary insertion along the oblique line of thyroid cartilage. Two *pharyngeal elevator muscles* (stylopharyngeus and palatopharyngeus) attach to the posterior thyroid cartilage and superior cornua.

Intrinsic muscles do not have any attachments outside the larynx. With the exception of cricothyroid, innervation is exclusively supplied by the inferior (recurrent) laryngeal nerve. The blood supply for these muscles is provided by anastomoses between superior thyroid from fourth arch ECA and inferior thyroid from thyrocervical trunk. To understand the rationale of these muscles is helpful to think about them in two ways.

- Mesenchymal pathway:
- The linear order of muscles supplied by ILN is: cricopharyngeus > posterior cricoarytenoid > interarytenoids > lateral cricoarytenoid > thyroarytenoid. This reproduces the spatiotemporal sequence by which the fifth arch muscles populate the internal larynx.
- Functional groups:
  - **Control of rima glottis**: *posterior crico-arytenoid, lateral cricoarytenoid,* and
  - *Transverse arytenoid*
  - **Vocal ligament tension**: *cricothyroid* and *thyro-arytenoid*
  - **Modify laryngeal inlet**: *oblique arytenoid, aryepiglottic,* and *thyro-epiglottic*

Cricothyroid spans from fourth arch to fifth arch. It can be considered a fourth pharyngeal arch derivative. It is the only intrinsic muscle innervated by r8–r9 superior laryngeal nerve, external branch. It has its proximal (primary) insertion along the undersurface of thyroid cartilage. Two parts are described. *Pars recta* is anterior. It causes rotation at the cricothyroid joint. In turn, this changes the distance between the vocal process of the arytenoid (where the cord attaches) and the posterior lamina of thyroid cartilage. *Pars obliqua* is posterior. It pulls the entire thyroid cartilage forward, thus stretching the vocal folds. Blood supply for cricothyroid is exclusively via superior thyroid artery. It "enters" larynx via the thyrohyoid membrane. The developmental pathway of this muscle is external to thyroid cartilage. Inferior constrictor occupies the entire oblique line. Just below this, the blastema encounters a newly available binding site; it inserts itself into the space.

All the remaining muscle blastemata originate from somitomeres 10–11. Their pathway of descent is posterior and internal to that of cricothyroid and inferior constrictor also originates from Sm9. *The only way for the intrinsic muscles to are forced to access the interior of the larynx due to the pre-existence of inferior constrictor in development.* Inferior laryngeal nerve enters from below upward,

external to esophagus and internal to the thyropharyngeus part of inferior constrictor.

Posterior cricoarytenoid is the only muscle to open the rima glottis. The muscles act by rotating the arytenoids laterally around their vertical axis. This separates the vocal processes and, consequently, the vocal folds. These muscles are active in the production of *unvoiced sounds.*

Lateral cricoarytenoid extends from the upper border of cricoid upward to the lateral footplate of arytenoid, the *muscular process.* It causes the vocal processes to approximate. As such, it constitutes the *antagonist* of posterior cricoarytenoid.

Transverse arytenoid lies deep to oblique interarytenoid, i.e., it develops first. It appears to be solitary (can't be), filling in the space in posterior larynx between the two arytenoids. It attaches to the backside of the muscular processes. This is important for speech because it closes up the posterior intercartilaginous space of rima glottis. This action is required for *whispering.*

Thyroarytenoid runs lateral and parallel to the vocal fold. Its fibers run between the lower internal aspect of thyroid cartilage and the anterolateral surface of arytenoid. Contraction of this muscle tilts the arytenoids forward toward thyroid cartilage. A relative tension is placed upon the posterior aspect of the vocal cords. This *raises the pitch* of the voice. These muscles also affect the bulk of the cords, thus affecting the *timbre* of the voice.

Oblique arytenoid and *aryepiglottic* These muscles are posterior; they lie superficial to the transverse arytenoid. They describe a X-shaped, criss-cross pattern that adducts the aryepiglottic folds, giving them a sphincteric function.

Thryoepiglotticus This muscle is probably an extension of thyroarytenoid that goes all the way forward to epiglottis. These muscles cause a widening of the laryngeal inlet.

Angiology of the Fourth and Fifth Arches

Superior Thyroid Artery This vessel is the most caudal branch of external carotid. It represents the supply to the fourth pharyngeal arch. It arises at the level of hyoid greater cornu and then runs down into the neck under cover of the strap muscles and superficial to inferior constrictor but runs directly to thyroid gland where it divides into an anterior branch to the ventral thyroid gland and a posterior branch deep to the gland.

Infrahyoid branch runs deep to thyrohyoid muscle, a continuation of sternothyroid muscle. It anastomoses with suprahyoid branch of lingual (second arch) because not third arch structures are in the vicinity.

Sternocleidomastoid branch supplies the superior part of the muscle. Recall that SCM has three level-specific sources of blood flow.

Superior laryngeal branch supplies almost the entire contents of the larynx.

Cricothyroid branch represents a separate branch to the membrane. If there is truly a fourth/fifth arch boundary, it would be represented by this territory.

Recall that fifth aortic arch artery involutes, making the mesenchyme which was assigned to it dependent on superior thyroid artery. Recall also that sixth aortic arch artery is dedicated exclusively to the pulmonic circulation. The ultimate fate of left aortic arch artery is to become the aortic outflow tract. For this reason, rare congenital cases of persistent fifth aortic arch artery can result in the so-called "double-barrel" anomaly, consisting of a systemic shunt from the ascending aorta transversely to descending aorta.

Inferior Thyroid Artery This artery is one of three branches of thyrocervical trunk, the others being suprascapular and transverse cervical. As previously described, it develops from the vertical cross-linking of multiple transverse branches arising from the first branch of subclavian artery and therefore runs upward along the border of medial border of scalenus anterior. Inferior thyroid gives of branches that act in a neuromeric manner.

- Muscular branches supply the infrahyoid, scalenus anterior, and, most importantly, inferior constrictor.
- Ascending cervical parallels phrenic nerve to supply deep muscles of the neck and the spinal canal as well.
- Esophageal branches connect with segmental counterparts from aorta.
- Tracheal branches connect with bronchial arteries and anastomoses with superior thyroid.
- Inferior laryngeal branch travels deep to inferior constrictor along with inferior (recurrent) laryngeal nerve it also anastomoses with superior thyroid.

The Spinal Accessory Nerve: Is It a Separate Cranial Nerve? (Figs. 9.2, 9.6, and 9.9)

History of a Controversy

The existence of separate cranial nerves with distinct physiologic roles was first defined in 1664 by Sir Thomas Willis (of the Circle of Willis), whose remarkable work has been translated from Latin to English. Willis described a spinal accessory nerve (SAN) with six to seven cervical roots that exited the jugular foramen in company with vagus and then proceeded to innervate sternocleidomastoid and trapezius. Although he considered it an atypical spinal nerve, since it did not proceed directly to its targets, he did not classify SAN as a cranial nerve. Interestingly, Willis grouped together VII with VIII and IX with X.

Misconceptions about the nature of XI have persisted for more than two centuries. All contemporary anatomy texts from Gray's first edition in 1858 have described SAN as having two separate components: a cranial root consisting of four or more branches admixed in jugular foramen with IX and X and a peripheral root originating from C1 to C6. The bulbar root was described for the first time in 1838 by Friedrich Arnold at the University of Heidelburg. [Arnold also described the auricular branch of vagus supplying ear canal, tragus and auricle, a nerve that today bears his name: Arnold's nerve.] Henry Gray referenced Arnold's work and used it to depict the two roots of cranial nerve XI. The illustrations of Arnold and Gray have been recopied ever since.

The actual work required for full exposition of this region, either at surgery or in the dissection room, is complex and laborious, requiring craniotomy and magnification. In cadavers, the process of embalming causes adhesions of the nerves, making use of fresh tissue mandatory. Even so, filmy adhesions of arachnoid can dry out in the wet lab dragging the nerves together. Putting the nerves on stretch proves helpful.

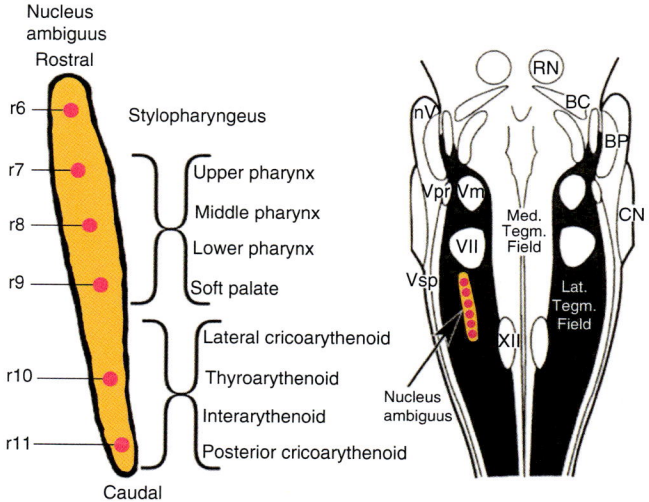

Fig. 9.9 Cranial nerve XI, accessory and somatotopic map of nucleus ambiguus. The anatomy of the nucleus ambiguus is neuromeric and somatotopic is technically part of the glossopharyngeal nucleus. **Neuromeres r6–r7** are organized as follows: r6 cooperates with r7 to give sensory supply to the upper pharynx and posterior two-third of the soft palate. Motor control is dedicated to stylopharyngeus muscle (r6) to the soft palate (except tensor) and superior constrictor (r6–r7). **Neuromeres r8–r9** have sensory supply to the middle pharynx; while it has motor control of middle constrictor and lateral cricothyroid. **Neuromeres r10–r11** have sensory supply the lower pharynx and motor control to the muscles of the larynx (except lateral cricothyroid). Thus, the branchiomotor contributions from nucleus ambiguus to the vagus give it a new name based on its function: the ***palate-pharyngo-laryngeal nerve***. (Reprinted from Holstege G, Subramanian HH. Two different motor systems are needed to generate human speech. J Comp Neurol. 2016 Jun 1;524(8):1558–77. With permission from John Wiley & Sons)

Given these constraints is not surprising that this anatomy has remained refractory to investigation.

The two-root model and the existence of a separate cranial nerve XI were challenged by Black in 1912.

Dutch neuroanatomist Ariens Kappers considered the spinal component of accessory as the true accessory proper and that the "cranial" component of accessory was just a continuation of vagus. Furthermore, Kappers proposed that in phylogeny, vagus and accessory were co-mingled in medulla. In this case, r6–r11 accessory would exit at the cervical junction to supply the cucullaris. With evolution, the nucleus of accessory migrated four to six units backward into the spinal cord and produced an independent nerve. Nevertheless, the exit point of the nerve remained at the r11/c1 junction because, with the acquisition of the fourth occipital somite by tetrapods, nerves IX, X, and accessory were entrapped within jugular. In mammals, accessory is a continuation of the SVE branchiomeric column of nucleus ambiguus.

Lachman and Acland exposed the posterior fossa in 15 fresh cadavers and identified, using a stereomicroscope, identified rootlets emerging from caudal medulla. Instead of merely using foramen magnum, they defined boundary of medulla as the caudal margin of the olive. Arising from medulla, they found a row of densely packed posterior medullary rootlets (PMRs). Just below the PMRs, the rootlets thinned out to form distinct row of four caudal posterior medullary rootlets (CPMRs). As the PRMs and CPMRs approached the jugular foramen, they came together as an aggregate and then abruptly separated into a discrete rostral trunk, glossopharyngeal nerve IX, and a similar caudal trunk, vagus nerve X. Upon entry into jugular foramen, the two nerves shared a common sheath of dura.

About 5 mm below the medulla, a new series of distinct and well-formed cervical roots emerged directly outward from the spinal cord. These joined together as an unambiguous SAN, directed toward jugular foramen but without any connection with the CPMRs whatsoever. In sum, all rootlets from the retro-olivary junction joined with the vagus and none communicated with SAN.

Intradural exposure of nerve roots at the craniocervical junction showed that stimulation of the so-called cranial roots of XI (i.e., the CPMRs) failed to stimulate SCM or Tpz but activated the vocal cords. These authors concluded that the cranial "root" of SAN is merely an extension of vagus into neuromeres r10–r1 [4].

An additional piece of evidence regarding the nature of accessory nerve comes from homeobox knockout experiments. SAN, being a continuation of nucleus ambiguus, supports branchiomeric muscles. In mammalian embryogenesis, homeotic transcription factor of the Nkx family determines regional patterning of the ventral CNS in response to graded Shh signaling. In knockout experiments, Pabst et al. showed in mice that *Nks2.9* mutants had normal hypoglossal and spinal cord neurons whereas spinal accessory nerve was markedly reduced [5]. Similar changes were observed furthermore in glossopharyngeal and vagus nerves.

In mammals, sternocleidomastoid and trapezius are not derived from occipital somites; these are strictly cervical muscles. SCM and Tpz originate from somites S5–S6/C1–C2 and S7–S8/C3–C4, respectively. Trapezius may map further backward to S10/C6. This is at odds with mapping data in birds by Couly in which cucullaris was reported to originate from somites S1 to S6 whereas Huang found it restricted to S1–S2. In both cases, the avian cucullaris is innervated by accessory nerve.

The phylogeny of the SAN is characterized by changing relationships with regard to the nucleus ambiguus and to the hypoglossal nucleus. In particular, XII demonstrates in the mammalian lineage a rostral migration from the spinal cord into medulla. This exemplifies the process of *neurobiotaxis*, described by Kappers as the migration of cell bodies in evolution toward the source of the stimuli that they most frequently receive. This progression is well demonstrated by comparison of axial and sagittal views of organisms representing distinct evolutionary levels. Let us compare the neuromeric positions of vagus, SAN, and hypoglossal nuclei in selected vertebrates: basal gnathostome fish, basal tetrapod, and mammal (human).

The classical elasmobranch, the skate *Raja eglanteria*, considered ancestral to the shark, is considered a living model for the ancestral gnathostome condition. In skates, the SAN nucleus lies ventral in the spinal cord and overlaps other ventral spinal neurons, but rostral to hypoglossal. SAN is also positioned posterior to the branchiomeric vagus, i.e., nucleus ambiguus. In the basal tetrapod, the salamander, hypoglossal nucleus moves forward into medulla to approximately r10 while SAN remains at c1. In mammals, the hypoglossal has relocated fully forward into the r8–r11 position. SAN remains in the c1–c6 position. Repositioning of the mammalian hypoglossal nucleus makes evolutionary sense because mammals, having developed complex a masticatory system with specialized teeth and a temporomandibular joint adapted for grinding, benefit by a highly mobile tongue capable of managing intraoral foodstuffs in good example of neurobiotaxis.

Mapping of the spinal accessory in rats indicates distinct functional columns within the ventral horn. A medial column is located in dorsomedial ventral horn at C1–C3 supplying sternomastoid and cleidomastoid. A lateral column found in dorsolateral ventral horn at C2–C6 supplies trapezius.

In summation, the spinal accessory nerve, cranial nerve XI, may not exist as a separate entity.

- The rootlets of XI emerging from caudal-most medulla emerge from r10 to r11 and are responsible for supplying the laryngeal muscles of the fifth pharyngeal arch.

- Although these roots emerge separately from those of r6–r9, they ultimately come together to form the common vagus, the *palate-pharyngo-laryngeal nerve*.
- The cranial root of XI, whether considered as a separate structure or as a continuation of vagus, represents the caudal end of nucleus ambiguus.
- Nucleus ambiguus is a strictly branchiomotor (SVE) nerve.
- With the exception of glossopharyngeal nerve, which carries selected motor fibers from r6 to r7 to stylopharyngeus, all remaining motor roots from r6 to r11 are SVE and serve a common purpose.
- Since vagus from r6 to r9 supplies branchiomeric muscles of <u>both</u> the third and fourth arches, why should a completely separate cranial nerve exist to supply the fifth arch? And, if so, why does it not pursue an independent course from vagus?
- The continuation of nucleus ambiguus into levels c1–c6 of spinal cord, designated as the central motor column supplies muscles derived from cervical somites S5–S10. Unlike all other somite-derived muscles (the connective tissues of which arise from paraxial mesoderm), the connective tissue of SCM and Tpz are of neural crest derivation. Therefore, these muscles may represent modified branchiomeric muscles related to branchial arches BA6–BA8.

Angiology

At the time of their inception, the fourth and fifth arches are organized around an aortic arch axis. These arteries develop from bidirectional buds originating from the aortic outflow tract below and the dorsal aortae above. In the case of AA4, at stage 13, the dorsal and ventral buds approximate and fuse, creating paired arteries to the fourth arches. At stage 14, the buds of AA 5 attempt to do the same thing but they either <u>fail to connect</u> or connect, but then <u>disintegrate</u>. In either case, the result is the same: the fifth pharyngeal arch becomes dependent on the fourth for its blood supply.

Beginning at stage 15, two events take place that complete the blood supply to arches 4 and 5. First is the *separation of arterial supply into the carotid circulation to the head and neck and the systemic circulation to the body*. This is initiated when the segment of dorsal aorta between AA3 and AA4 falls apart. The outflow of AA4 is now separated from the head; it becomes completely dedicated to the non-head body. Obviously, how the paired fourth aortic arches and the dorsal aorta supply the entire rest of the embryo. And equally obviously, they must remodel into a single left-sided aortic arch and descending aorta. Note, the split of abdominal aorta into paired common iliac arteries reflects the distal persistence of the paired dorsal aortae. The second event affecting blood supply to the fifth arch is the *development of the inferior thyroid artery* as a branch of the thyrocervical trunk, the events of which are recounted below.

Superior thyroid artery AA4 is represented in the external carotid system by the superior thyroid artery, the first branch of ECA. It arises at the level of greater cornu of hyoid bone. It tracks along lateral border of thyrohyoid muscle until reaching the apical lobe of the thyroid gland which it supplies via *glandular branches*. *Infrathyroid artery* runs deep to thyrohyoid and supplies infrahyoid strap muscles. *Superior laryngeal artery* pierces the thyrohyoid membrane along with internal branch of superior laryngeal nerve. It supplies the mucosa of *upper* larynx, *not* the intrinsic muscles. *Cricothyroid artery* supplies the named muscle. *Sternocleidomastoid artery* supplies the middle region of the muscle.

AA5 quickly disintegrates. Pharyngeal arches 4 and 5 are physically located deep in the future throat behind the tongue. The mesenchyme of the fifth arch is now supplied by collateral flow from the fourth arch. Recall that the first branch of the external carotid is superior thyroid artery. In the fetal state, the fifth arch derivatives of larynx will be supplied *from above* by superior thyroid artery.

In the meantime, as AA4 morphs into the aortic arch and subclavian arteries, the neck will receive blood supply from below by three successive arterial systems: vertebral artery, thyrocervical trunk, and costocervical trunk. Originally, the paired dorsal aortae gave off horizontal segmental arteries to each somite, from the first cervical somite (S5) all the way down into the chest. Next, the horizontal vessels formed vertical cross-link; these were plugged into the newly created subclavian arteries. Finally, the vertical systems separated from the dorsal aortae altogether (and from each other as well). Because of this mechanism, these arteries are not accompanied by veins.

Inferior thyroid artery rises all the way from its stem on the first part of subclavian to the inferior constrictor muscle. It has four types of branches. *Muscular branches* march all the way along longus colli to the inferior constrictor. *Ascending cervical artery* supplies the scalene muscles. *Pharyngeal branches* supply trachea, esophagus, and thyroid. *Inferior laryngeal artery* follows the inferior laryngeal (recurrent) branch of vagus and is distributed to the internal laryngeal muscles and mucosa.

Fascial Layers of the Pharyngeal Arches

The craniofacial/cervical region in mammals differs from the rest of the body in having three layers of fasciae: (1) subcutaneous- or superficial-fascia (2) superficial investing fascia, and deep investing fascia.

The *subcutaneous fascia* (SF) runs beneath the skin of the entire body. It has two layers. The external layer, *panniculus adiposus*, contains a variable amount of fat. It ranges in thickness from several centimeters in the overly nourished to virtually nothing in skinny individuals. This fat is absent in certain zones such as nose, eyelids, pinna of the ear, labia minora, and penis/scrotum. The inner layer is very thin and elastic. Running in the interfascial space are the superficial neurovascular supply and lymphatic drainage.

The *superficial investing fascia* (SIF) refers to a fascial zone unique to mammals. It is exclusively associated with the muscles of facial expression, all of which originate from the sixth somitomere. In the head, the territory of SIF envelops the entire skull. Its posterior margin is the superior nuchal line. From the occiput, one can follow it forward over the mastoid process. It thence proceeds downward (with platysma), flowing over the sternocleidomastoid and thence drapes over c1–c4 clavicle until reaching the anterior chest at the intersection of the fourth cervical dermatome and the second thoracic dermatome.

The *deep investing fascia* (DIF) is associated with all remaining muscles of the head and neck. This layer subdivides to include the following muscle groups: Sm4 muscles of mastication, Sm6 infrahyoid muscles, Sm7 muscles of the pharynx, Sm8 external laryngeal muscles, Sm9 internal laryngeal muscles Sm8–Sm9 sternocleidomastoid, Sm8–Sm11 muscles of the tongue, Sm12–Sm19 deep muscles of the neck, and the Sm13–15 (C2–C4) strap muscles.

Conclusion

This chapter, despite its telegraphic nature, represents a framework for understanding the pharyngeal arch system.

- Improved terminology specifies muscle origins and insertions in terms of developmental mechanism. We should note that primary and secondary insertions are not always explained by a simple proximal-distal model. A muscle of the extremity that attaches to a distal primary insertion can "retrace its steps" to a more proximal site where it makes secondary insertion.
- Neuromeric relationships are faithfully reflected in the patterns of innervation. In particular, the evolutionary history of the brachial plexus demonstrates how neuromeric boundaries shift caudally in the construction of the mammalian neuraxis.
- Knowledge of the branchial arch system in agnathans and its morphologic changes in jawed fishes, tetrapods, and mammals gives greater understanding of how muscles evolve from simple units to more complex subunits capable of exerting more precise motor control.

References

1. Hall BK. The neural crest and neural crest cells in vertebrate development and evolution. 2nd ed. New York: Springer; 2009.
2. Le Douarin NM, Kalcheim C. The neural crest. 2nd ed. Cambridge: Cambridge University Press; 1999.
3. Kuratani S, Adachi N, Wada N, Oisi Y, Sugahara F. Development and evolutionary significance of the mandibular arch and prechordal/premandibular cranium in vertebrates: revising the heterotopy scenario of gnathostome jaw evolution. J Anat. 2013;222:41–55.
4. Brinzeu A, Sindou M. Functional anatomy of the accessory nerve studied through intraoperative electrophysiological mapping. J Neurosurg. 2017;126:913–21.
5. Pabst O, Rummelles J, Winter B, Hans-Hemming A. Targeted disruption of the homeobox gene *Nkx2.9* reveals a role in development of the spinal accessory nerve. Development. 2003;139:1193–202.

Further Reading

General Developmental Biology and Extraocular Muscles

Baker CVH, Bronner-Frser M. Vertebrate cranial placodes, I. Embryonic induction. Dev Biol. 2001;232:1–61.
Barishak YR. Embryology of the eye and its adnexa. Basel, Switzerland: Karger; 1992.
Bertelmez GW, Blount MP. The formation of the neural crest from the primary optic vesicle in man. Contrib Embryol Carnegie Inst. 1954;35:55–71.
Bravo H, Insunza O. The oculomotor nucleus, not abducent innervates the muscles of the nictating membrane in birds. Acta Anat (Basel). 1985;122(2):99–104.
Cveckl A, Piatigorski J. Lens development and ocular mesenchyme: new insights from mouse models and human diseases. BioEssays. 2004;26:374–86.
Gilbert PW. The origin and development of the human extrinsic ocular muscles. Contrib Embryol Carneg Inst. 1957;36:59–78.
Gilbert S, Barressi M. Developmental biology. 11th ed. Sunderland, MA: Sinauer; 2016.
Graw J. The genetic and molecular basis of congenital eye defects. Nat Rev Genet. 2003;4:876–88.
Johnston MC, Noden DM, Hazelton RD, Coulombre JL, Coulombre AF. Origins of avian ocular and periocular tissue. Exp Eye Res. 1979;29:27–43.
Kardong K. Vertebrates: comparative anatomy, function, evolution. 7th ed. New York: McGraw Hill; 2015.
Kondoh H. Development of the eye. In: Rossant J, Tam PL, editors. Mouse development: patterning, morphogenesis and organogenesis. San Diego: Academic; 2002. p. 519–39.
Levin AV. Congenital eye anomalies. Pediatr Clin North Am. 2003;50:55–76.
Liem K, Bemis WE, Walker WF, Grande L. Functional anatomy of the vertebrates. 3rd ed. Belmont, CA: Brooks/Cole; 2001.
Mann I. The development of the human eye. 3rd ed. Br Med Assoc; 1964.
McLoon LK, Andrade F. Craniofacial muscles: a new framework for understanding the effector side of craniofacial muscle control. New York: Springer; 2012.
Michailovici I, Eigler T, Tzahor E. Craniofacial muscle development. Curr Top Dev Biol. 2015;115:3–30. https://doi.org/10.1016/bs.ctdb.2015.07.022. Epub 2015 Oct 1. Review. PMID: 26589919
Mui S, et al. The homeodomain protein Vax2 patterns the dorsoventral and nasotemporal axes of the eye. Development. 2002;129:797–804.

Noden DM. Differentiation and morphogenesis of craniofacial muscles. Dev Dyn. 2006;235(5):1194–218.

O'Rahilly R. Early development of the eye in staged human embryos. Contrib Embryol Carneg Inst. 1966;38:1–42.

O'Rahilly R. The prenatal development of the human eye. Exp Eye Res. 1975;21:93–112.

O'Rahilly R. Timing and sequence of events in the development of the human eye and ear during the embryonic period proper. Anat Embryol. 1983;168:87–99.

Padget DH. The development of cranial arteries in the human embryo. Contrib Embryol Carnegie Inst. 1948;32:205–61.

Pearson RH. The development of the eyelids. Part I. External features. J Anat. 1980;130:33–42.

Porter JD, Baker RS. Anatomy and embryology of the ocular motor system. Chapter 25. In: Miller NR, Newman NJ, editors. Walsh & Hoyt's clinical neuro-ophthalmology, vol. I. 5th ed. Baltimore: Williams & Wilkins. p. 1043–99.

Schubert FR, Singh AJ, Afoyalan O, Kioussi C, Dietrich S. To roll the eyes and snap a bite – function, development and evolution of craniofacial muscles. Semin Cell Dev Biol. 2018; https://doi.org/10.1016/j.semcdb.2017.12.013. Epub ahead of print. Review. PMID: 29331210

Sevel D. A reappraisal of the origin of human extraocular muscles. Ophthalmology. 1981;88:1330–8.

Sevel D. Origins and insertions of the extraocular muscles: development, histologic features, and clinical significance. Trans Am Ophthal Soc. 1986;24:488–526.

Standring S, editor. Gray's anatomy. 40th ed. New York: Elsevier.

White MH, Lambert HM, Kincaid MC, et al. Ora serrata and the spiral of Tillaux. Ophthalmology. 1989;96(4):508–11.

Wozniak W, O'Rahilly R. The times of appearance and the developmental sequence of the cranial parasympathetic ganglia in staged human embryos. Anat Rec. 1980;196:255A–6A.

Spinal Accessory Nerve

Benninger B, McNeil J. Transitional nerve: a new and original classification of a peripheral nerve supported by the nature of the accessory nerve (CN XI). Neurol Res Int. 2010; Article ID 476018; https://doi.org/10.1155/2010/476018.

Huang R, Zhi Q, Patel K, Wilting J, Christ B. Contributions of single somites to the skeleton and muscles of the occipital and cervical regions in avian embryos. Anat Embryol. 2000;202:375–83.

Kappers A, Huber C, Crosby E. The comparative anatomy of the nervous system of vertebrates, including man, vol. 2. New York: Hafner Publishers; 1967.

Krammer EB, Lischka MF, Egger TP, Riedel M, Gruber H. The motoneuronal organization of the spinal accessory nuclear complex. Adv Anat Embryol Cell Biol. 1987;103:1–62.

Lachman N, Acland RD, Rosse C. Anatomic evidence for the absence of a morphologically distinct cranial root of the accessory nerve in man. Clin Anat. 2002;205:193–201.

Pearson A. The spinal accessory nerve in human embryos. J Comp Neurol. 1937;68:243–66.

Tada MN, Kuratani S. Evolutionary and developmental understanding of the spinal accessory nerve. Zool Lett. 2015;1:4. https://doi.org/10.1186/s40851-014-0006-8.

Tubbs RS, Benninger BI, Loukas M, Cohen-Gadol AA. Cranial roots of the accessory nerve exist in the majority of adult humans. Clin Anat. 2014;27:102–7.

Wiles CCR, Wrigley B, Greene JRT. Re-examination of the medullary rootlets of the accessory and vagus nerves. Clin Anat. 2007;20:19–22.

The Neck: Development and Evolution

10

Michael H. Carstens

Introduction

A structural analysis of the muscles of "the neck" is somewhat arbitrary; it depends upon one's concept of the region and encompasses a variety of structures, many with evolutionary changes, incurred with the transition to life on land. Cervical vertebrae were initially three which we mammals have expanded to seven. The larynx and trachea were invented while the primitive gut became the esophagus. Glands such as the thyroid were relocated. Cranial nerve XI resulted from the forward incorporation of a multi-segment peripheral nerve. In discussing the muscles of the neck, we are making a neuromeric transition between muscles arising from somitomeres and those arising from somites. With the important exceptions of sternocleidomastoid and trapezius, all muscles in this chapter are innervated by peripheral nerves. For purposes of this discussion, we will confine ourselves to those muscles supplied by XI and cervical nerves 1–8. We therefore bid adieu to the floor of mouth, the suprahyoid series, and to the muscles of the pharynx and larynx. But there will be obligatory cross-talk regarding SCM and trapezius and their chaperone, the platysma.

This chapter, like its predecessor, makes use of the motor column model to organize muscles into functional groups. But, if we hope to understand the neck in a deeper, more conceptual way, we shall have to give up the segmental comforts of cranial nerves and pharyngeal arches. Neck anatomy is the *sine qua non* of tetrapod existence. It is all about change and innovation. New systems of respiration and locomotion get their start here. Dermatomes are stretched out into seemingly irrational patterns. Muscles have new roles: some are organized in a garden-variety fashion around the head and spinal column for axial control. Others act as "bridges" to the outboard pectoral girdle. Still another mus-cle, subcoracoideus, an elevator of coracoid and antagonist of pectoralis, is reincarnated in mammals as the diaphragm.

The evolutionary pressures and mechanisms responsible for this mammalian plan are important to emphasize. Therein, we shall plunge back into the comfortable world of segmentation, based upon the basic vertebrate *bauplan*: the fish body. We will examine in greater detail how the anatomy of the forefin explains the development of the shoulder girdle. We can gain some insight into the important new role played by lateral plate mesoderm in constructing the trachea and esophagus. Interesting correlations about the twin trajectories of thyroid and thymus will be sought. And, of course, all that is terribly relevant for our study of neck muscles. So … the reader is advised to simply get through this chapter … Subsequently we shall discuss neuromuscular evolution. Subsequently we shall discuss neuromuscular evolution. Feel free to bounce back and forth among the sections of this chapter. In this way, the correlations will leap out at you.

The first section of this chapter will examine the osteology of the region. We will discuss the bones of the occipito¬cervical junction (occipital, proatlas, atlas, and axis), as well as typical cervical vertebrae (C7), paying particular attention to their mechanisms of development. Next, we shall look at the development of the clavicle and scapula. When one combines information regarding the origins of individual muscles from their respective somite (s) with the ossification sequence of the individual bones, the spatial arrangement of muscle binding sites follows a quite predictable neuromeric pattern.

The second section is myology. *Five distinct neuromuscular systems* are involved: neck to head, neck to neck, neck to trunk, neck to shoulder girdle, and that most unique mammalian invention, the diaphragm. We will organize our discussion about the muscles on the basis of functional groups and motor columns (medial central and lateral). The vital role of the fascia will be emphasized. The origin of the mammalian.

The third section, neurology, reviews the general layout of the cervical spinal cord (emphasizing the motor column

M. H. Carstens (✉)
Wake Forest Institute of Regenerative Medicine, Wake Forest University, Winston-Salem, NC, USA
e-mail: mcarsten@wakehealth.edu

model) and the spinal nerves. The cervical plexus will be discussed in detail. Owing to our craniofacial focus (and for the sake of brevity), *the brachial plexus will be declared "out of bounds."*

The fourth section, angiology, will explore the rationale behind the arterial supply to the neck. Far from being non-sensical, the vascular system is a logical outcome of segmental developmental processes.

The fifth section, systems analysis, explores important structures that pass through the neck into the chest: trachea, esophagus, endocrine glands, thyroid, and thymus.

As a final note, the overarching theme of the neck is that it represents evolution at work. A craniovertebral joint is created. Vertebral specialization allows for better positioning of the head. Even the length of the neck is dynamic. Central to these changes are shifts in homeotic coding along the cranio-caudal axis. These redefine the anatomic contents at various neuromeric levels. Formerly cranial muscles subdivide, with some remaining in place while others are displaced backward into the "new neck." The brachial plexus relocates and gets larger, that most curious of muscles, the cucullaris, assumes a radical reiteration in mammals, allowing us to oxygenate more efficiently. So keep the hox code in the back of your mind as you enter into the brave new world of the neck.

Drawing the Line: How Do We Define the Neck?

The neck can be conceptualized in various ways. As an expropriation of the rostral trunk, it serves as a simple *passageway* for pre-existing organ systems between head and trunk. The neck is also a functional entity unto itself. It ensures the *positioning of the head and its organs of perception* for survival functions such as finding and grabbing a meal…or avoiding becoming someone else's lunch. Finally, cervical muscles are important for control of the upper extremity: pectoral girdle, arm, forearm, and hand. The list of muscles and bones associated with cervical neuromeres becomes very lengthy. Since our emphasis is craniofacial, where shall we draw the line?

First off, we'll cover all muscles acting on the neck itself. That means that muscles spanning from the neck up to the head and down to the trunk are fair game. The pectoral girdle is also a target in our crosshairs. It provides support for pharyngeal arch structures (the strap muscles) and connects to the skull. Finally, cervical muscles attach the pectoral girdle to the neck and trunk All remaining structures of the upper extremity are out-of-bounds…with one important exception, the diaphragm, which, as we shall see, represents an evolutionary subversion of a pectoral girdle muscle, subcoracoideus.

Rest assured, all muscles of the upper limb follow the same neuromeric rules: they simply are beyond our scope.

One can only hope some reader with a keen interest in the extremity will apply these concepts to better understand the appendicular skeleton and teach the rest of us.

- In sum: the scope of our inquiry includes all structures associated with cervical neuromeres c1 through c8 with the exception of the upper extremity distal to the clavicle and the scapula.

Organization of Cervical Mesenchyme

At multiple occasions, we have discussed how epiblast cells acquire an identity during gastrulation. Those that ingress through either the *lateral aspect of primitive node* or the *rostral zone of primitive streak* will become committed to the somitic cell line. After traversing the streak these cells maintain contact with the basal laminae of both the overlying epiblast and underlying hypoblast. Recall our previous discussion in which cell movements in gastrulation are liken to the flow of toothpaste. The cells nearest the streak dedifferentiate, detach, and dive inward. They attach immediately, pushing the primitive endoderm (hypoblast) laterally. Subsequent cells play "leapfrog," over their predecessors, migrate laterally and attach at the first available position. Recall the first wave of cells to ingress represents the chorda-mesoderm. These cells kick the hypoblast to either side and settle dead center, where they will ultimately form the notochord. The next wave of cells represents true endoderm. It pushes the hypoblast completely out from beneath epibast. The third wave of cells is the embryonic mesoderm.

Once the cells are in place, and the hypoblast has been pushed out of the embryo, the mesoderm closest to the midline, referred to as paraxial mesoderm (PAM), flanking the notochord. These cells acquire the homeotic code of the neural plate with which they are in register. This creates their anterior-to-posterior identity. They also acquire signals that will confer additional transverse spatial identities that become important during somitogenesis. Those that entered through *lateral Hensen's node* become the *medial halves* of the somites, whereas those that entered 200 μm posteriorly through *rostral streak* become the *lateral halves* of the somites. This implies that the first wave of cells occupies the medial position; those that enter later are forced outside. Like students filing into an auditorium, latecomers must file past those already seated. When we later examine the development of the occipitocervical junction, we shall see that medial and lateral zones of the somites have different behaviors.

Intermediate mesoderm exists as a column of extending from neuromere c1 down through the lumbar-sacral neuromeres. IM is the source of the genitourinary system. It bears homeotic segmentation in register with paraxial mesoderm.

This is manifested as a series of nephrotomes, one for each neuromeric level, and each with its segmental artery. At a metaneuromeric level, the embryonic kidney takes three iterations as it "migrates" caudally from the neck: pronephros, mesonephros, and metanephros. Aberrancies in the cervical spine have a correspondingly high incidence of genitourinary anomalies.

Lateral plate mesoderm does not display overt segmentation but its organization is homeotic. It has two layers: the somatic lamina (LPM_S) is associated with ectoderm; the visceral lamina (LPM_V) is associated with endoderm. Lateral plate mesoderm is organized neuromerically into two broad zones.

- *Cardiac LPM* is a single mesenchymal mass in register with r0–r11. Its role is the formation of the heart, great vessels, and the blood supply to the head. Both LPM_S and LPM_V are involved in cardiac formation. LPM_V forms aortic arches organized according to pairs of rhombomeres.

Posterior LPM begins at neuromere c1. Its role is the formation of the appendicular skeleton, the post-cranial vascular system, smooth muscle, and viscera. PLMP splits into two laminae: somatic lateral plate mesoderm LPM_S and visceral lateral plate mesoderm. LPMS and LPMV cannot fuse together. The space between the layers is the *intraembryonic coelom* (IEC).

Lateral plate LPM_V from c1 downward wraps around the GI tract to form its musculature and surrounds the larynx to envelop the respiratory diverticulum as the substance of the developing lungs. This mesenchyme drops down into the chest on either side of the heart, and descends from craniofacial LPM mesenchyme. For this reason, congenital malformations or aplasia of the lungs within the chest can co-exist in the presence of an intact heart.

Lateral plate LPM_S from c1 to t1 relates to the upper extremity. LPM_S from c1 to c6 forms pectoral girdle as the core of clavicle and 100% of scapula. LPM_S from c5 to t1 forms the bones of the upper extremities. Its vascular core is surrounded by LPM_V fascia.

From the above, the intraembryonic coelom begins at level c1 but manifests itself at level c4. How? The upper extremity develops as a diverticulum from levels c4 to t1. The skin envelope has PAM dermis. Just inside the envelope are PAM myoblasts. LPM_S bones are at the center, accompanied by LPM_V blood vessels. PAM muscles and their fascia attach to LPM_S bones but they never fuse with LPM_V fascia. For this reason, vascular pedicles always remain distinct from the muscle units. Furthermore, LPN_V vascular pedicles never fuse with the LPM_S periosteum. Based on these facts we conclude the following:

- The intraembryonic coelom begins at c4 and extends downward into the chest.
- The IEC is not seen in the neck because is *expropriated by the upper limb.*
- The upper boundary of IEC is c4 and is subcutaneous in the neck behind the clavicle as the cupula which receives the apex of the lung.

The innervation of LPM gives us a clue as to its neuromeric anatomy. All lateral plate derivatives are hypaxial and are innervated by ventral roots. Furthermore, these are always organized as plexuses: cervical, brachial, celiac, superior mesenteric, inferior mesenteric, and lumbo-sacral. What that means is that LPM populations from individual neuromeres migrate in complex 3-D ways. They are not organized in a simple linear order, as are those from the somites. The clinical implication of this is that an event at a single neuromeric level might manifest itself within lateral plate mesoderm at several different levels.

A partial exception to this pattern is the innervation to LPM in the neck. Here, dorsal motor nucleus of vagus supplies esophagus but, as yet, individual segments have not been traced back to a specific rhombomere. Sympathetics from c1 to c3 through the superior cervical ganglion, from c4 to c5 through the middle cervical ganglion, and from c6 to c8 through the inferior cervical ganglion are distributed to the myenteric plexuses in their respective regions of the esophagus.

Neural Crest in the Neck   The role of neural crest cells in the neck is much more limited. In the upper neck they contribute to the cartilage structures of the fourth and fifth arches. The fasciae and connective tissues of branchiomeric muscles, including sternocleidomastoid and trapezius, are all neural crest whereas in all remaining cervical muscles these tissues are PAM. Neural crest is fundamental for the sympathetic nervous system and peripheral nerves throughout the body. Cervical dermis is PAM but contains neural crest melanocytes. The axial skeleton is exclusively PAM but the pectoral girdle contains neural crest elements: the cortex of clavicle and the spine of the scapula.

Osteology: Axial Bones

Why begin with the bones? Let us take a deep breath and a step backward. The spinal column represents the very core of the vertebrate body. It is also the world's best place to observe the segmentation process at work. As we begin our story of vertebral assembly, we are going to start at the beginning, at its very first step: the segmentation of paraxial mesoderm. The next step is somitogenesis. Here, we shall see how vertebral bodies are formed from opposing halves of adjacent somite, a process termed *parasegmentation*. We shall then consider how each somite subdivides into functionally distinct units (sclerotome, myotome, and dermatome). Once these concepts are reviewed, we'll look at the actual assembly process of "prototypical" vertebrae. This will be extended to the anomalous atlas and axis. Finally, we will delve into how the mammalian craniofacial junction evolved.

Certain evolutionary questions loom large here. Mammals have eight cervical somites but only seven vertebrae? Why do cervical nerves emerge above their respective vertebral arch, whereas spinal nerves at all other levels must pass underneath? Why on earth should strap muscles supplied by the cervical plexus insert on the sternum and clavicle? Why are the motor targets of spinal accessory nerve so far removed from the head and neck? I will beg the reader's indulgence for any redundancy. The twists and turns of evolution are quite fascinating and terribly relevant. You won't be bored (Figs. 10.1 and 10.2).

The fundamental story of the mammalian cervical spine is the production of 8 vertebrae, Proatlas, the first cervical vertebra, has a complex evolutionary history in which it fragments into component parts to create the craniovertebral joint, that is, the atlas–axis complex. The fourth through the eighth cervical vertebrae repeat a basic developmental plan which is continued all the way to the sacrum. We shall discuss the osteology of the axial spine in terms of its building blocks (paraxial mesoderm and the somites), how these are assembled to form vertebrae, ossification centers of the vertebra as clues to developmental components, the developmental anatomy of the craniovertebral junction, and finally the individual bones beginning with the cranial base.

Fig. 10.1 All mammals, regardless of size, have seven vertebrae. Compare the okapi versus giraffe. (**a**) brontosaurus, (**b**) velociraptor, (**c**) giraffe, (**d**) okapi, (**e**) human. Left: [Reprinted from Vidal D, Mocho P, Páramo A, Sanz JL, Ortega F (2020) Ontogenetic similarities between giraffe and sauropod neck osteological mobility. *PLoS ONE* 15(1): e0227537. https://doi.org/10.1371/journal.pone.0227537. With permission from Creative Commons License 4.0: https://creativecommons.org/licenses/by-sa/4.0/deed.en.] Right: [Reprinted from Wikimedia. Retrieved from https://commons.wikimedia.org/wiki/File:Okapi_Giraffe_Neck.png. With permission from Creative Commons License 4.0: https://creativecommons.org/licenses/by-sa/4.0/deed.en]

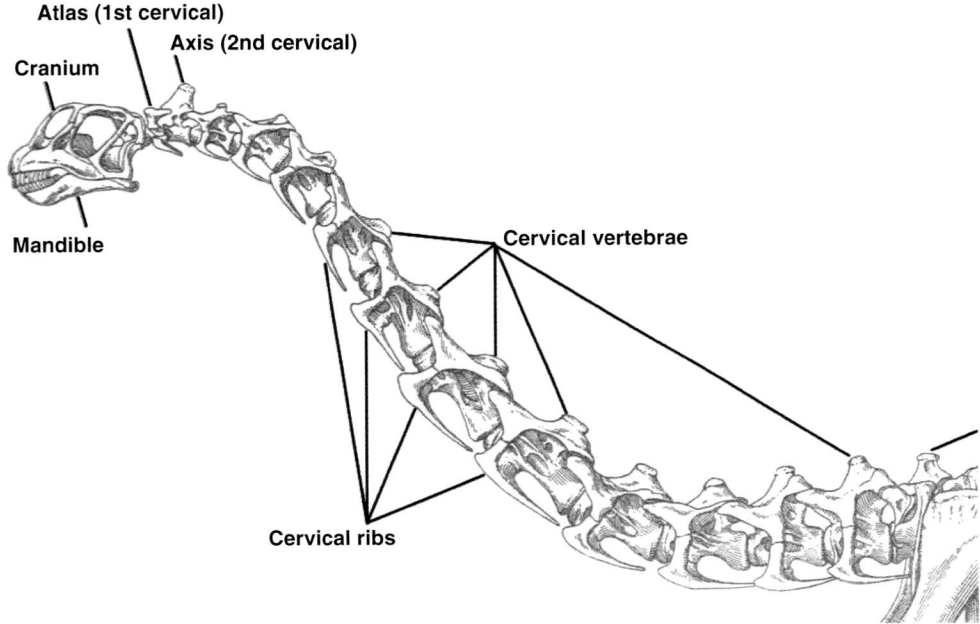

Fig. 10.2 Camarasaurus. Sauropods had 11–19 cervical vertebrae. Backward-pointing cervical ribs are seen here. In mammals, these are bicipetal, fused to the vertebrae as transverse processes through which pass the vertebral artery, a longitudinal fusion of previous segmental cervical arteries. Note: amniotes divide into diapsids (reptiles and birds) and synapsids (mammals). In synapsids the number of cervical vertebrae stabilizes at the level of cynodonts to 7 whereas in diapsids, along the avian line, 8 or more are possible. [Reprinted from M Taylor MP, Wedel MJ, and Cifell RL. A new sauropod dinosaur from the Lower Cretaceous Cedar Mountain Formation, Utah, USA. Acta Palaeontologica Polonica 2011;56(1): 75–98. doi: https://doi.org/10.4202/app.2010.007. With permission from Creative Commons License 4.0: https://creativecommons.org/licenses/by/4.0/]

Somitogenesis

Immediately after gastrulation, paraxial mesoderm is a loose mesenchyme without cellular polarities. This *presomitic mesoderm* (PSM) quickly organizes itself due to the elaboration of cell adhesion molecules, such as fibronective and N-cadherin, a process known as *compaction*. The cells condense into somitomeres consisting of a basal lamina surrounding a single layer of epithelial cells with polarized organelles oriented radially around a central cavity (somitocoele) that contains mesenchymal cells. Craniofacial PSM has two zones with two distinct fates. The anterior zone extends from midbrain back to the otic vesicle and is registered with mesomeres m1–m2 and rhombomeres r0–r7 and is sometimes referred to as *head mesoderm*. It remains in the somitomeric state. The first seven somitomeres produce craniofacial muscles of the first three arches, the posthypophyseal cranial base, and the primary blood vessel systems of the head and face. The posterior zone extends from the otic vesicle to the craniocervical junction, is in register with r8–r11, and undergoes transformation to somites (Figs. 10.3, 10.4, 10.5, 10.6).

Mammalian somitogenesis begins at Sm8 but, in terms of timing, this transformation occurs upon the development of the 19th somitomere. This is interesting because, in humans, these twelve levels correspond to the first 12 somite levels (4 occipital and 8 cervical). The first somite forms directly caudal to the otic vesicle. As the subsequent somitomere develops caudally (at position Sm20) Sm9 becomes the second somite. Thus, new pairs of somites are added on in rostrocaudal order until the number fixed for the species is reached. In humans, segmentation takes place from day 16 to day 28. It produces a total of 7 somitomeres and 42–44 somites at the rate of 4 per day. Recall that of the 8–10 coccygeal somites only 2–3 persist.

In sum, depending upon what stage the gastrulation process is observed, the expanding column of body mesoderm consists of (1) a rostral zone of mature somites that are differentiating into their subcomponents (e.g., sclerotome and dermomyotome); (2) an intermediate zone of pre-differentiated somites, and (3) a caudal zone of somitomeres just rostral to the retreating primitive streak (Fig. 10.7).

The "Clock and Wavefront" Model

The process of metameric transformation is regulated by the "clock and wavefront" model proposed first by Cooke and Zeeman in 1976 and subsequently elaborated by Pourquié in 2003. Presomitic mesoderm cells oscillate between a permissive and a non-permissive state for somite transformation. These changes of state are controlled by an autonomous *seg-*

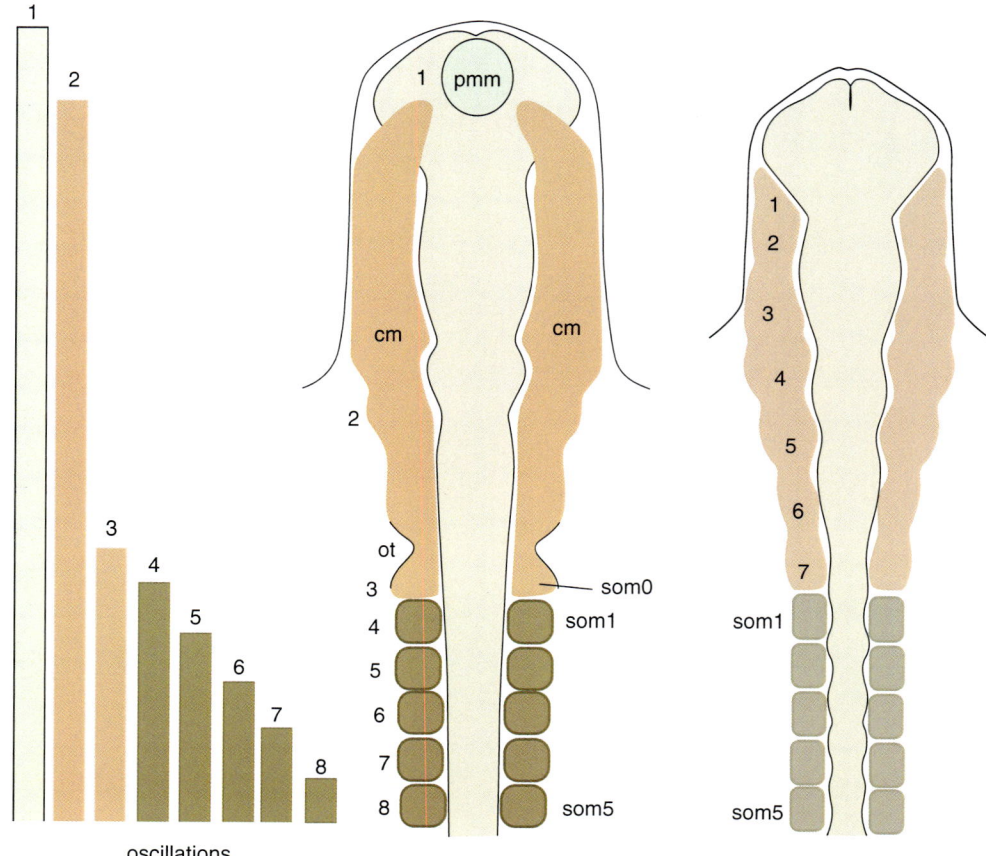

Fig. 10.3 Somitomeres and somites. Somitomeres and somitogenetic genes. The morphological pattern of somitomeres in the chick embryo is shown on the right, as a simplified illustration. Hypothetical somitomeres are numbered. On the left is shown the oscillating expression of a somitogenetic gene, chairy, in the early chick embryo.. Each oscillation is numbered together with the mesodermal part generated after that oscillation. Note that there are only two oscillations in the head mesoderm, one for the premandibular mesoderm, and the other for the rest of the cephalic mesoderm. Birds have 5 occipital somites and mammals have 4. Both birds and mammals have 7 somitomeres. These model, are physically confluent, have no internal compartments, and produce striated muscles only Somitomeres 1–3 and 5 are exclusive to the orbit. Somitomeres 4, 6, and 7 provide pharyngeal arches 1, 2, and 3 with mesoderm. is The fourth and fifth pharyngeal arches, not shown here receive banchiomeric mesoderm from somitomeres 8–11 prior to their reformatting into occipital somites with defined myotomes. Sm1–Sm4 produce the hypobranchial muscles of the tongue. [Reprinted from Kuratani S. Craniofacial development and the evolution of the vertebrates: the old problems on a new background. Zoological Sciences 2005; 22: 1–19. With permission from The Zoological Society of Japan]

Fig. 10.4 Somite distribution is controlled by Hox genes. Chick versus mouse pattern. Like all mammals, mice have 7 cervical vertebrae. Birds have longer necks. The ostrich Cervcal-thoracic boundary is Hox5 versus Hox6. [Reprinted from Gilbert SF, Barressi MJF. Developmental Biology, 11th ed. Sunderland, MA: Sinauer; 2016. Copyright © 2016. Oxford Publishing Limited. Reproduced with permission of the Licensor through PLSclear]

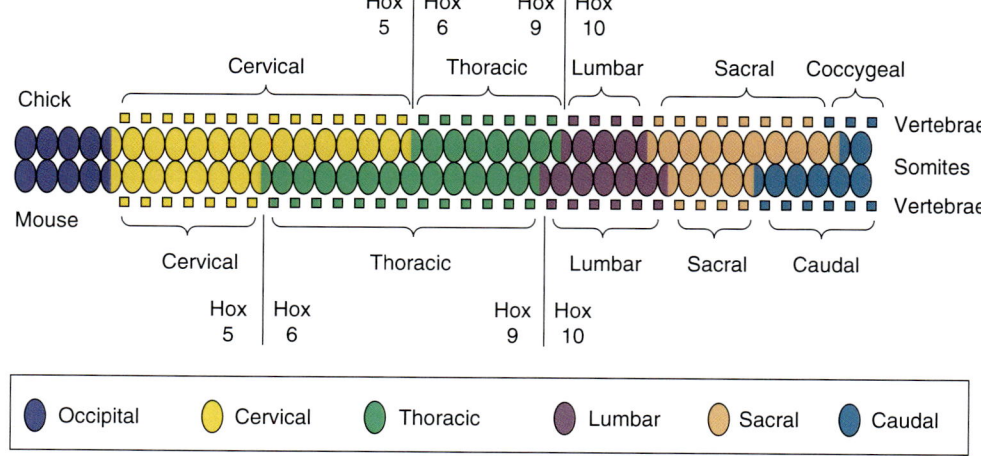

a

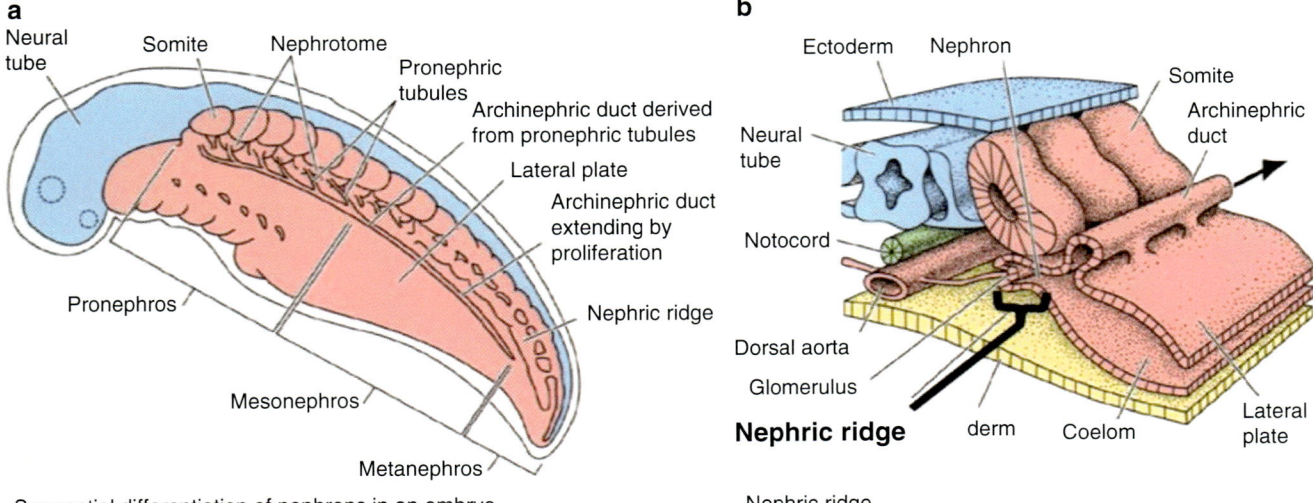

Sequential differentiation of nephrons in an embryo

b

Nephric ridge

Fig. 10.5 Organization of cervical mesoderm. Paraxial mesoderm is organized as somites. Intermediate mesoderm begins at level c1. It rounds up as the *archicnephric duct*. Individual nephrotomes and ductules are neuromeric. Lateral plate mesoderm posterior to cardiac LPM also begins at level c1. Its somatic and visceral layers are separate, creating the coelom. Pleuroperitoneal folds arise from the nephric ridge. Therefore conduction from the neck into the thorax is spatially organized [Courtesy of William E. Bemis]

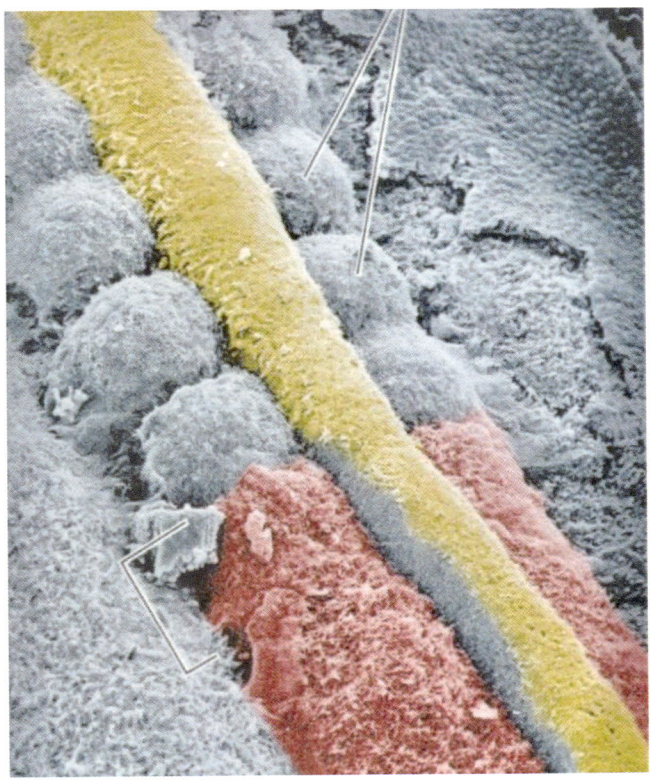

Fig. 10.6 Conversion of presomite mesoderm. Somitomeres after Sm8 are transient structure and immediately transition to somites. [Reprinted from Gilbert SF, Barressi MJF. Developmental Biology, 11th ed. Sunderland, MA: Sinauer; 2016. Copyright © 2016. Oxford Publishing Limited. Reproduced with permission of the Licensor through PLSclear]

mentation clock, present in all vertebrates. It consists of a wavefront of maturation that sweeps caudally along the axis of the embryo, causing physiologic changes wherever it is located. The clock has an intrinsic periodicity (4 hours in humans) and is responsible for generating the segmentation of the somites [1] (Fig. 10.8).

The nuts and bolts are as follows. *Notch* is the putative stimulus for somitomere-somite transformation. It is a transmembrane receptor gene that gets periodically turned on by a number of time-based cycling genes known as the *c-hairy1* family. These cycling genes cause coordinated pulses of *notch* mRNA to appear when a somite is being formed. In somitomeres, factor FGF8 keeps the cells in the mesenchymal state…and therefore prevents transformation into a somite. In any somitomere, the FGF8 gradient decreases toward the rostral axis. Thus, the rostral zone, where somitogenesis is taking place, has very low FGF8. This permits the cells to respond to *Notch*, to differentiate and acquire an epithelial boundary. In the caudal zone, high FGF8 prevents epithelialization of PSM cells. Thus, the wavefront of somitogenesis corresponds to the boundary between two domains of fgf8 expression: high/non-permissive (in the caudal PSM) and low/permissive (in the rostral PSM).

The body axis is growing and elongating caudally with constant addition of new somitomeres. Thus, as new PSM cells with strong fgf8 expression are added to the rear, the overall FGF8 gradient is displaced caudally as well. This guarantees that the metameric boundaries on both sides of each somite are separated by a cellular distance correspond-

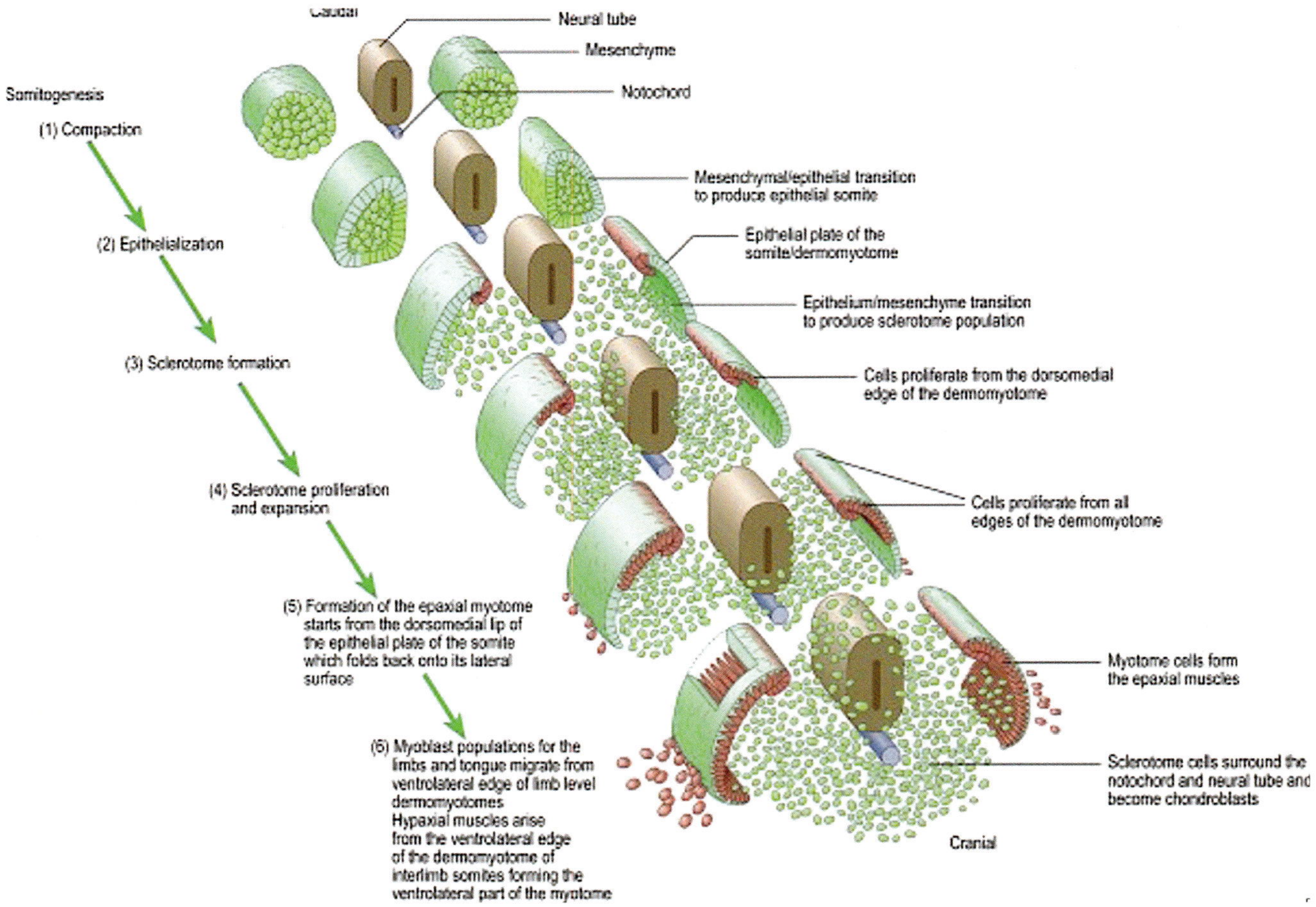

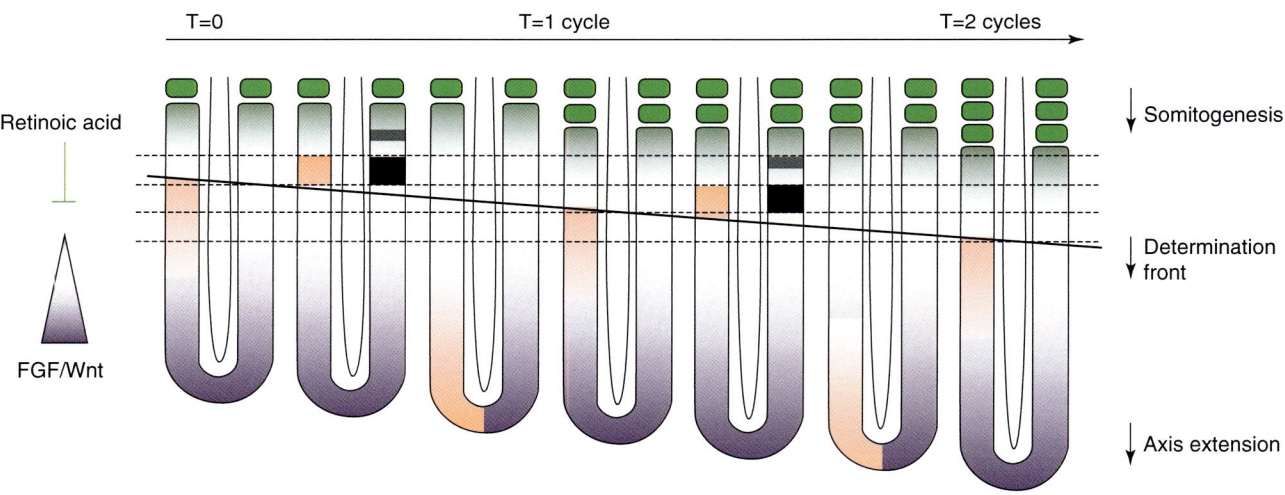

Fig. 10.7 Somitogenesis. Cranial to caudal progression consists of 6 steps. Somite lower right is more mature and therefore are further along in the process. [Reprinted from Lewis, Warren H (ed). Gray's Anatomy of the Human Body, 20th American Edition. Philadelphia, PA: Lea & Febiger, 1918]

In the figure (Fig. 10.7), the following labels appear:

Somitogenesis
(1) Compaction
(2) Epithelialization
(3) Sclerotome formation
(4) Sclerotome proliferation and expansion
(5) Formation of the epaxial myotome starts from the dorsomedial lip of the epithelial plate of the somite which folds back onto its lateral surface
(6) Myoblast populations for the limbs and tongue migrate from ventrolateral edge of limb level dermomyotomes. Hypaxial muscles arise from the ventrolateral edge of the dermomyotome of interlimb somites forming the ventrolateral part of the myotome

Caudal — Neural tube — Mesenchyme — Notochord
Mesenchymal/epithelial transition to produce epithelial somite
Epithelial plate of the somite/dermomyotome
Epithelium/mesenchyme transition to produce sclerotome population
Cells proliferate from the dorsomedial edge of the dermomyotome
Cells proliferate from all edges of the dermomyotome
Myotome cells form the epaxial muscles
Sclerotome cells surround the notochord and neural tube and become chondroblasts
Cranial

Fig. 10.8 Clock and wavefront. The body axis of vertebrates is composed of a serial repetition of similar anatomical modules that are called segments or metameres. This particular mode of organization is especially conspicuous at the level of the periodic arrangement of vertebrae in the spine. The segmental pattern is established during embryogenesis when the somites—the embryonic segments of vertebrates—are rhythmically produced from the paraxial mesoderm. This process involves the segmentation clock, which is a travelling oscillator that interacts with a maturation wave called the wavefront to produce the periodic series of somites. *Segmental patterning of the vertebrate embryonic axis (PDF Download Available).* Available from: https://www.research-gate.net/publication/5439063_Segmental_patterning_of_the_vertebrate_embryonic_axis [accessed Dec 15 2017].

In the figure (Fig. 10.8): T=0, T=1 cycle, T=2 cycles; Retinoic acid; FGF/Wnt; Somitogenesis; Determination front; Axis extension.

Nature Reviews | Genetics

ing to the caudal displacement of the determination wavefront during a single oscillation cycle of the segmentation clock. The velocity of somite production is thus linked to the periodicity of the gene clock which is species-specific: 90 minutes for the chick, 2 hours for the mouse, and 4 hours in humans.

Differentiation of Somites

Somite boundaries are defined by *epithelialization*. This occurs when a mesenchymal > epithelial transformation takes place along the periphery of the somite. The somite is now physically separated from its more cranial neighbor by an epithelial segmental barrier. Within hours, further differentiation takes place along the dorsal–ventral axis. Ventromedial cells lose their epithelial characteristics. They are no longer bound to one another. They migrate medially to surround the notochord to form *sclerotomes*. In the meantime, dorsolateral cells retain some of their epithelial behaviors to form *dermomyotomes*. These immediately break apart. The superficial layer of mesenchyme interacts directly with the epithelium to form the *dermatome* giving rise to dermis and subdermal smooth muscle. The deep layer remains densely packed to form the *myotome*, the future source of axial striated muscle. Note that these changes take place due to external signals from the notochord; they are not intrinsic to the cells of the somite.

Sclerotomal cells differentiate in medial-lateral fashion according to their distance from the signaling centers of the notochord. Axial sclerotome compacts around the notochord to form the future *vertebral body*. Lateral sclerotome appears triangular and will form the remaining vertebral elements,

and also displays a densely packed caudal zone and a loosely packed cranial zone; the zones are separated by the *fissure of von Ebner*. The cranial loose zone never chondrifies. It produces signals that stimulate the neuraxis to send out nerve tissue and neural crest into the space. In contrast, the caudal dense zone of the lateral sclerotome takes the form of a triangle. The base abuts up with the perichordal sclerotome (future vertebral body) to form the *pedicle*. The dorsolateral side produces the *neural arch* while the ventrolateral gives rise to the *costal process* (Fig. 10.9).

The final result is a somite divided into distinct mesenchymal zones. The dermomyotome, responsible for dermis and muscle, requires some clarification. Until very recently, it was believed that somites were responsible for *all* populations of dermis. In reality, their role is limited to the skin overlying the epaxial muscles. In these locations, dermis is formed by a de-epithelialization of the underlying dermomyotome. Blood vessels to this dermis are also derived from dorsomedial dermomyotome. Hypaxial dermis is arranged in dermatomes in a common neuromeric register with its epaxial counterpart but the source material is *somatopleuric lateral plate mesoderm* in register with the overlying ectomere (Fig. 10.10).

The function of the myotome needs clarification. Myotomal cells subjacent to the dermatome behave exactly as expected to produce the epaxial muscles that correspond to their level.

The v*entrolateral* edge of somites produces hypaxial cells that behave in an altogether different manner, depending upon the neuromeric location of the somite. These populations do not aggregate around the spine. Instead, they migrate outward to cover the entire ventrolateral body wall and the extremities. In the case of somites located at the neuromeric levels of the limbs (c5–t1 and l3–s2) the migration deposits

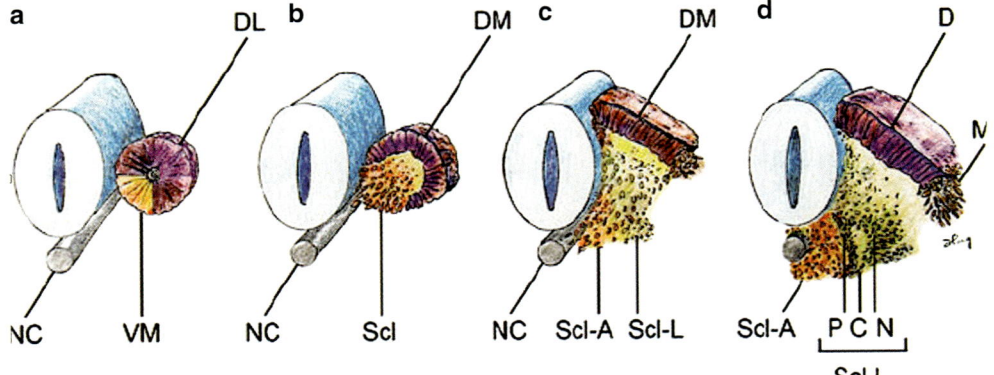

Fig. 10.9 Differentiation of somite. (**a**) Ventromedial cells (VM) form the sclerotome. (**b**) Ventromedial sclerotome cells (Scl) de-epithelize from the somite and migrate towards the ventral notochord (NC). (**c**) Sclerotomal cells further subdivide into an *axial cluster* (Scl-A) surrounding the notochord, and paired *lateral clusters* (Scl-L) flanking the perichordal axial sclerotome. Dorsolateral somite retains its epithelial pattern to become the dermomyotome (DM). (**d**) The lateral sclerotome

(Scl-L) forms a triangle next to the axial sclerotome. The three sides of the triangle become Anlagen for the pedicle (P), neural arch (N), and the costal process (C), respectively. The dermomyotome also subdivides into the dermatome (D) and the lateral migrating myotome (M). [Reprinted from Pang D, Thompson DNP. Embryology and bony malformations of the craniovertebral junction. *Childs Nerv System* 2011; 27:523–564. With permission from Springer Nature]

Fig. 10.10 Somite: component parts. Note internal cells of sclerotome become arthrotome to form joints. Dorsomedial sclerotome also forms dura. [Reprinted from Gilbert SF, Barressi MJF. *Developmental Biology*, 11th ed. Sunderland, MA: Sinauer; 2016. Copyright © 2016. Oxford Publishing Limited. Reproduced with permission of the Licensor through PLSclear]

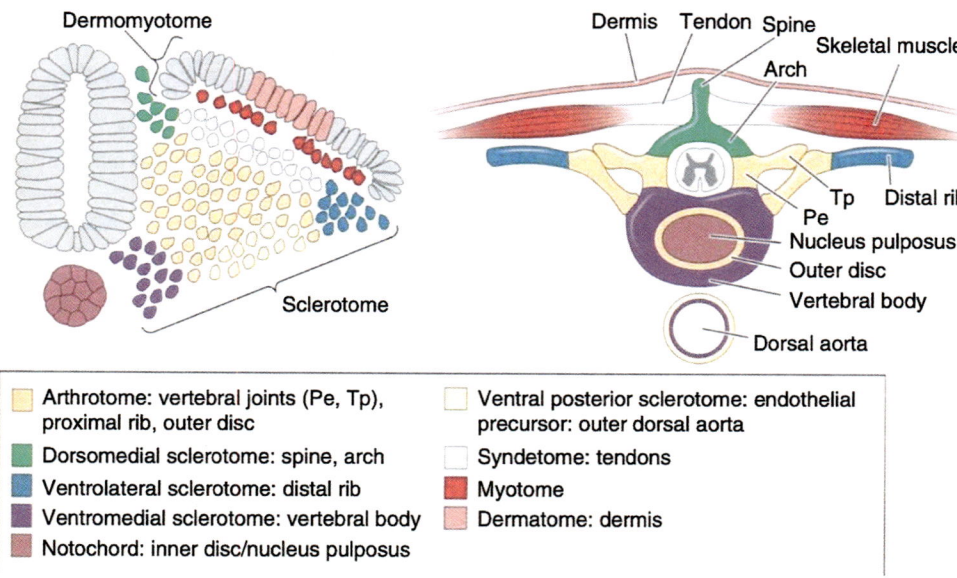

☐ Arthrotome: vertebral joints (Pe, Tp), proximal rib, outer disc
☐ Dorsomedial sclerotome: spine, arch
☐ Ventrolateral sclerotome: distal rib
☐ Ventromedial sclerotome: vertebral body
☐ Notochord: inner disc/nucleus pulposus
☐ Ventral posterior sclerotome: endothelial precursor: outer dorsal aorta
☐ Syndetome: tendons
☐ Myotome
☐ Dermatome: dermis

them into the limb bud. Similarly, somites located along the interlimb neuromeres (t2–l2) produce muscles that migrate outward along the body wall as intercostals and abdominal wall muscles.

Resegmentation of Somites

Remak in 1855 is credited with recognizing a shift in the metameric boundaries of somites at the level of the sclerotome such that the vertebral body develops from two adjacent somites whereas the neural arch arises from a single somite. This he termed *neugleiderung*, or resegmentation. Let's first outline the process and then detail the mechanism [2] (Fig. 10.11).

After the lateral sclerotome develops its *pedicular, arcual,* and *costal* components, the mesenchyme of the axial sclerotome begins to show signs of compartmentalization as well. Its caudal zone receives mesenchyme from the dense zone of more distal somite. In the process, the caudal axial sclerotome *becomes dense.* The result is the unification of the vertebral body with the outlying pedicle. Note that the most medial cells of the axial sclerotome remain loose throughout development as the *nucleus pulposus.* The most cranial part of the axial dense zone is aligned with *von Ebner's fissure* in the lateral sclerotome. It condenses dramatically to produce the *intervertebral boundary zone* (IBZ). The IBZ encircles the central core of nucleus pulposus and goes on to make the

annulus fibrosus of the intervertebral disc. When fusion of the two pedicles to the body and the neural arches to one another is completed the resultant *vertebral ring* provides circumferential protection for the spinal cord. Note that the costal component gives rise to the *transverse process* and, potentially, to rib. Costal formation was secondarily eliminated in vertebrate evolution in the cervical and lumber regions as it gave greater mobility for the neck and trunk, although under certain congenital conditions *cervical ribs* can occur.

Since the two dense zones of axial and lateral sclerotome are lined up with the IBZ it makes sense that the mature pedicles will be attached to the cranial half of the vertebral body, and not to its caudal half. The position of the spinal nerve and intersegmental artery also makes sense because they pass through the loose cranial zone of lateral sclerotome and therefore must cross *above* the neural arch.

Examine Fig. 10.10 with care. Note that the sclerotome has two zones with two types of behavior. Both the axial sclerotome (Scl-A) and lateral sclerotome (Scl-L) develop dense and loose zones but *the former re-segments, whereas the latter does not.* Please bookmark this important point; we shall return to it later when discussing the O'Rahilly Müller model of occipitocervical development. These zones are likely genetic remnants of the original arculalia (Fig. 10.12).

During re-segmentation, the sclerotome is formed from the caudal and rostral halves of two adjacent somites, such

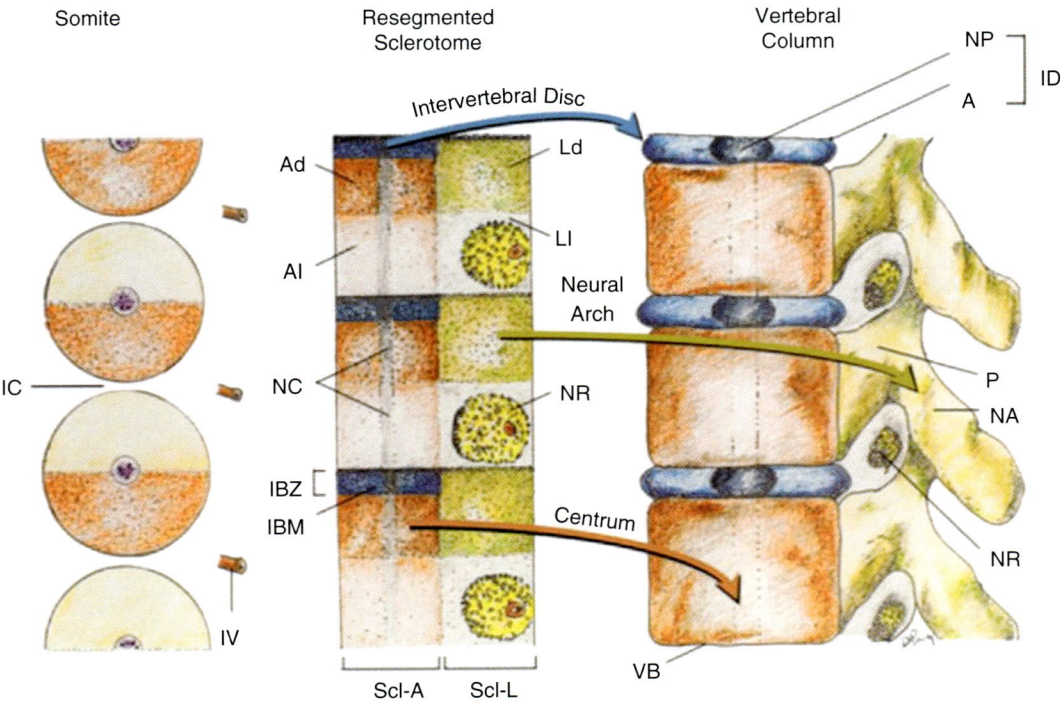

Fig. 10.11 Resegmentation of somites to form sclerotomes and changes of sclerotomal primordia to mature vertebral parts. The somitic and primordial origins of vertebral parts and phenotypic parts are colour-matched, and the locations of the somites, resegmented sclerotomes, and vertebrae along the embryonic axis are approximately counter-registered. During resegmentation, the sclerotome is formed from the caudal and rostral halves of two adjacent somites, such that the middle of the resegmented sclerotome lines up with the intersomitic cleft (IC). Both the axial sclerotome (Scl-A) and lateral sclerotome (Scl-L) develop dense and loose zones. The dense zone of the lateral sclerotome (Ld) becomes the neural arch (NA). Thus the neural arch and is *not* resegmented. Pedicle (P), which is attached to the rostral part of the vertebral body (VB) chondrifies from the recombination of the loose zone of axial sclerotome (Al) and part of the dense zone of axial sclerotome (Ad). The *rostral layer of the dense zone of the axial sclerotome* soon forms the intervertebral boundary zone (IBZ) containing intervertebral boundary mesenchyme (IBM), which ultimately forms the annulus (A) and, together with notochord remnants (NC), the nucleus pulposus (NP) of the intervertebral disc (ID). The loose zone of the lateral sclerotome (Ll) does not form bone but promotes emergence of the nerve roots (NR). Thus, the neural arch is derived from a single somite but the vertebral body receives contributions from two adjacent somites. IV intersomitic vessel. Arrows indicate developmental fates of the sclerotomes. [Reprinted from Pang D, Thompson DNP. Embryology and bony malformations of the craniovertebral junction. *Childs Nerv System* 2011; 27:523–564. With permission from Springer Nature]

that the middle of the resegmented sclerotome lines up with the intersomitic cleft (IC). The dense zone of the lateral sclerotome (Ld) becomes the neural arch (NA). Thus, the neural arch is *not* resegmented. Pedicle (P), which is attached to the rostral part of the vertebral body (VB) chondrifies from the recombination of the loose zone of axial sclerotome (Al) and part of the dense zone of axial sclerotome (Ad). The *rostral layer of the dense zone of the axial sclerotome* soon forms the intervertebral boundary zone (IBZ) containing intervertebral boundary mesenchyme (IBM), which ultimately forms the annulus (A) and, together with notochord remnants (NC), the nucleus pulposus (NP) of the interverte-

bral disc (ID). The loose zone of the lateral sclerotome (Ll) does not form bone but promotes emergence of the nerve roots (NR). Thus, the neural arch is derived from a single somite but the vertebral body receives contributions from two adjacent somites. IV intersomitic vessels travel through the loose zone.

Arrows indicate the developmental fates of the sclerotomes. Intervertebral disc arises from rostral-most layer of the axial sclerotome dense zone. Neural arch develops from a single-level dense zone of the lateral sclerotome; it is not resegmented. Centrum results from resegmentation of two adjacent axial sclerotomes.

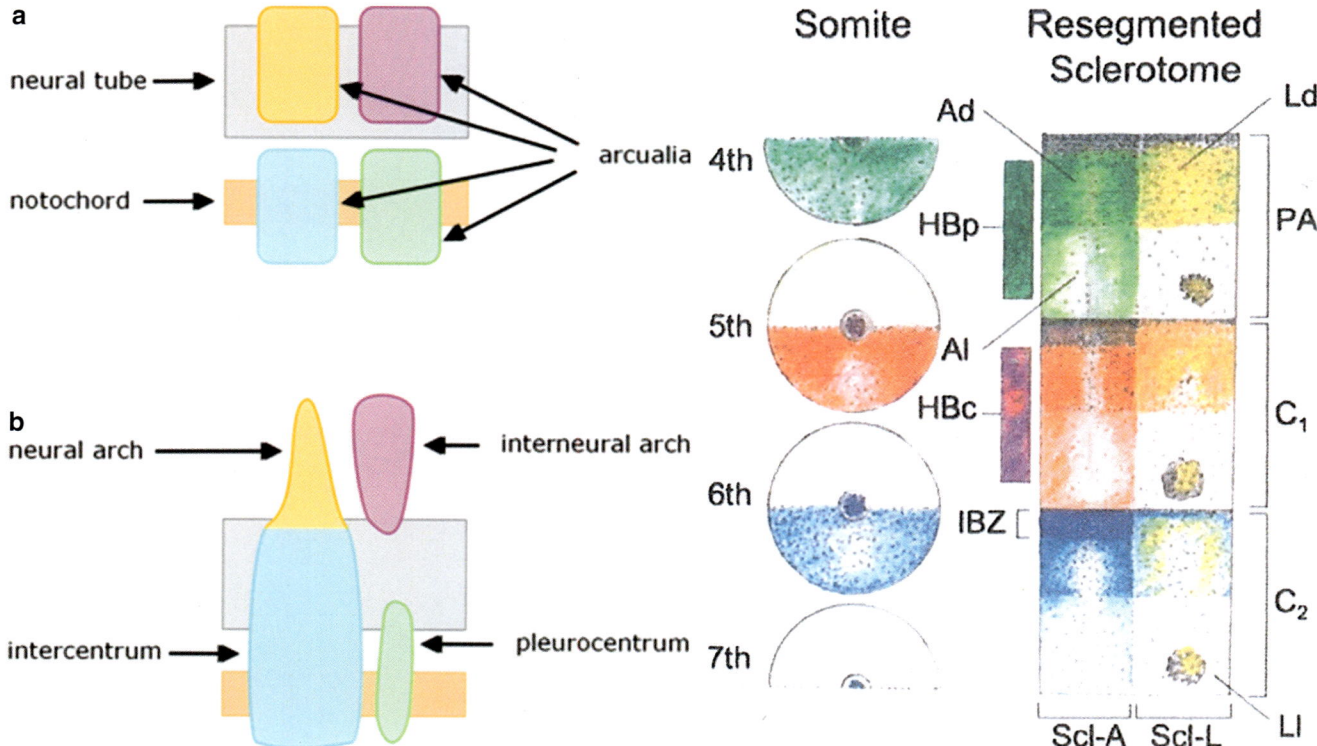

Fig. 10.12 Left Arculalia, embryonic (**a**) versus adult (**b**). Arculalia are seen in fishes as 4 segmented blocks of cartilage per segment, the result of an obvious built-in genetic program. Right These are not present in tetrapods because tetrapod centra result from perichordal mesenchyme. Nonethless, the spatial programming that determines the piscine system may be operative to determine zones of the sclerotome. Basodorsal (tan) = neural arch, Basoventral (blue) = intercentrum,

Interdorsal (pink) = interneural arch, Intervertebral (green) = pleurocentrum. Source: Kardong 8.10 Scl-A—axial sclerotome, Ad = axial dense, Al = axial loose, Scl-L = lateral sclerotome, Ld = lateral dense zone, Ll = loose zone, IBZ = intervertebral boundary zone. [Reprinted from Pang D, Thompson DNP. Embryology and bony malformations of the craniovertebral junction. *Childs Nerv System* 2011; 27:523–564. With permission from Springer Nature]

Sneak Preview: Resegmentation in the Ocicipito-cervical Junction

Prior to resegmentation, the eight cervical somites produce eight cervical vertebrae but the first one, proatlas, also called C_0, is split up between the basioccipital bone and the atlas. In the process, its lateral derivatives (pedicle, neural arch) are lost. Thus, the C_1 root that should have passed below proatlas now passes above the neural arch of atlas. The C_8 nerve root passes below C_7 neural arch but above the T_1 neural arch because the T_1 neural arch develops from the caudal half of the C_8 somite. Note that resegmentation persists in the proximal ribs but the sclerotomes become mixed distally (Fig. 10.13).

Assembly of Vertebra

As we have seen, sclerotomes arise in ventral zone of the somites. Each one quickly divides into a rostral half and a caudal half. This transition is marked by a boundary zone of the extracellular matrix called *Von Ebner's fissure*. The cell

population of the rostral half is loosely packed, while that of the caudal zone is densely packed. We need to make use of the following notation: *rostral loose zone* of somite N = SN_R and the *caudal dense zone* of the same somite = SN_C. Thus, for the first somite, $S1 = S1_R + S1_C$. When a somite subdivides, its outer layer, the *epithelial plate* remains unaffected. *It spans both hemi-somite*s, as does its derivative, the dermomyotome. The *perinotochordal sheath* imparts a chondrogenic fate to those mesenchymal cells with which it comes into contact. Control of segmentation is dual. The *notochord* also controls ventral segmentation. When excised, ventral plate spinal ganglia control dorsal segmentation (Fig. 10.14).

Sclerotomes recombine to form vertebrae; this is termed *resegmentation*. Each vertebra develops as the combination of the caudal (dense) sclerotome combined with the rostral sclerotome of the next-most-caudal somite. For example, the eighth thoracic vertebra develops from $T7_C + T8_R$. Using our numbering system (starting from the first occipital somite, S1), vertebra $T8 = S19_C + S20_R$. The fusion of sclerotomal tissue around the notochord creates the *centrum* of the vertebra. By Carnegie

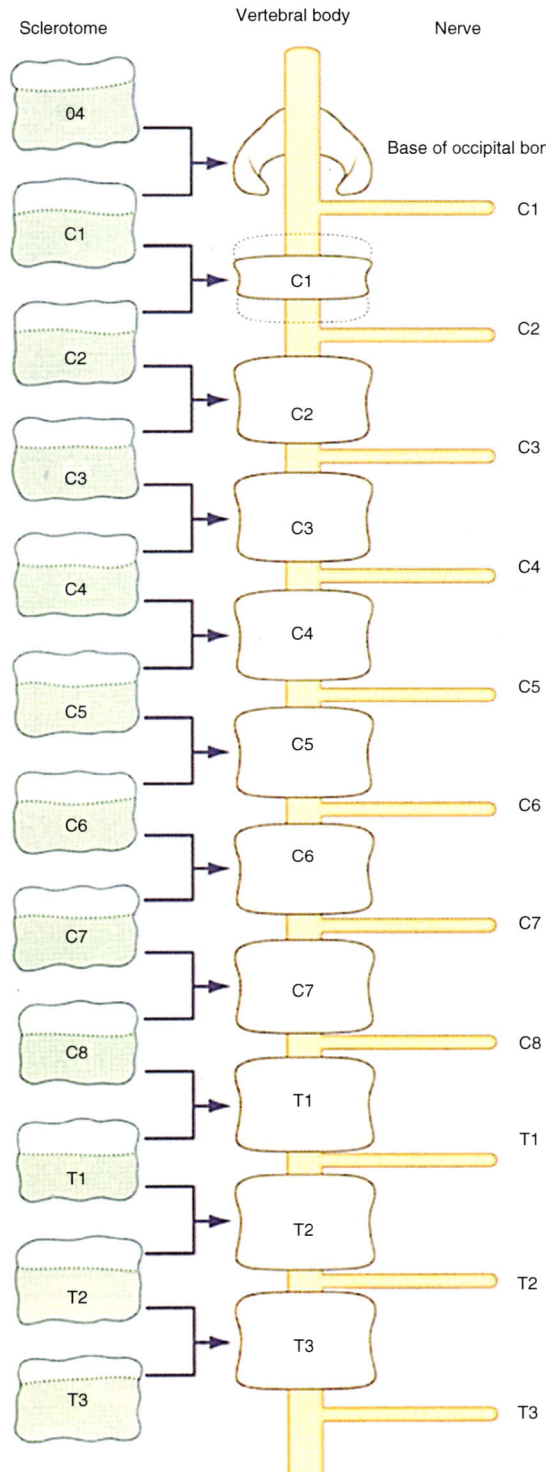

Sclerotome   Vertebral body   Nerve

Base of occipital bone

Fig. 10.13 Results of resegmentation. Proatlas the original first cervical vertebra (S4 + S5) loses the neural arch, donates its ventral body to foramen magnum, and its pedicles to the occipital condyles. The first cervical segmental nerve runs correctly beneath proatlas but upon the incorporation of proatlas into the cranial base, nerve C1 is now positioned atop atlas. All cervical nerve follow this pattern which "normalizes" numerically at neuromeric level t1, the first thoracic vertebra. [Reprinted from Lewis, Warren H (ed). Gray's Anatomy of the Human Body, 20th American Edition. Philadelphia, PA: Lea & Febiger, 1918]

stage 15 human embryos demonstrate condensation around the notochord; *neural arches, or processes,* now enclose the spinal cord. They extend downward and make contact with the dorsolateral centrum. The centrum, located ventral to the spinal cord, encases the notochord. Neural arches have two sets of projections: dorsolateral laminae and ventrolateral pedicles. Note that fishes have entirely separate ventrally-located *hemal arches,* the function of which is to protect the aorta from compression during swimming. The hemal processes project upward to join the pedicles (Fig. 10.15).

To this point, two critical joint structures are missing: vertebra-to-vertebra and vertebra-to-rib. A solution to this problem occurs at the junction between the laminae and pedicles. *Superior and inferior articular processes* (zygapophyses) will create facet joints. The *transverse process* will form the costovertebral joint. What about the ribs? These grow out from the ventral junction of pedicle with centrum. They have their own ossification center. The coastal processes have a primary articulation here but they develop a secondary projection that expands backward and upward to make contact with the transverse process. Remember that the *proximal ribs have a dual somitic origin* (the same as that of its vertebra) whereas the *distal rib comes from a single somite.* Thus proximal fifth rib is constructed from caudal t4 and cranial t5. The distal fifth rib structure is unitary and is exclusively t5 (Figs. 10.16 and 10.17).

Like a suspension bridge, the vertebrate bodies are slung between two sets of vertical pylons (the extremitites). Ribs are struts essential for support. They relate to the vertebrae either by articulation or by fusion. Embryologically, ribs develop from cartilage within *myoseptae,* sheets of connective tissue in the coronal plane that partition off body segments. In fishes, there are two sets of ribs associated with each vertebral segment, a dorsal set, and a ventral set. *Dorsal ribs* form at the intersection of each myoseptum with a horizontal sheet of connective tissue, the *horizontal septum,* that runs longitudinally down the entire length of the fish body. It intersects with the ventral aspect of the vertebral column. *Ventral ribs* form at the intersection between the myoseptae and the coelom (visceral cavity). Ventral ribs of the fish body are serially homologous with the hemal arches of the fish tail. Ventral ribs are lost in tetrapods, but will occasionally make a guest appearance as a vertebral anomaly.

Ribs are named based on their type of distal articulation. *True ribs* articulate with the sternum, *false ribs* articulate with each other, and *floating ribs* have no articulation. True ribs have two proximal articulations: they are *bicipital.* This is best understood from the primitive condition. The *tuberculum* articulates with neural arch via the *diapophysis.* The *capitulum* articulates with the intercentrum via the *paraphysis* (Figs. 10.18 and 10.19).

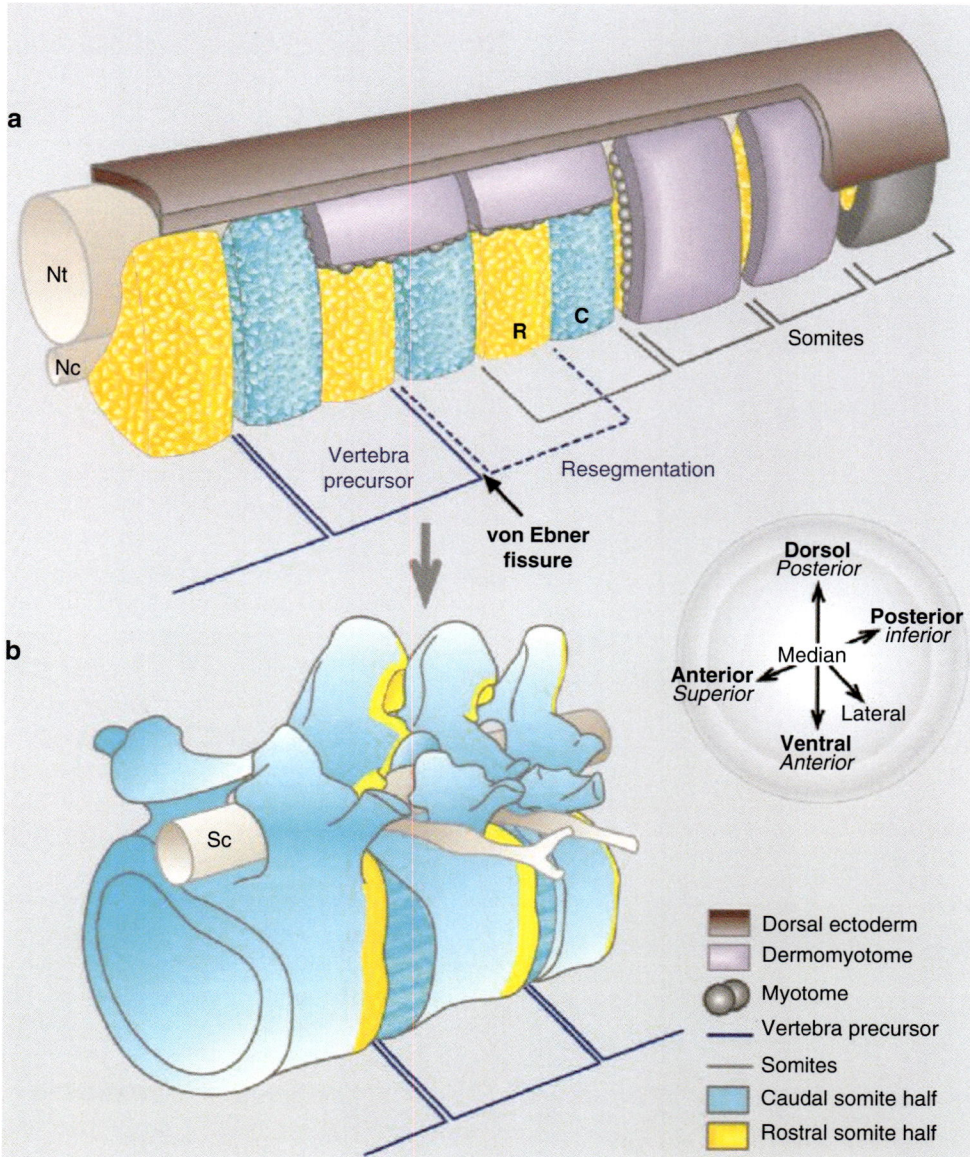

Fig. 10.14 Von Ebner fissure. Resegmentation of the sclerotome and its contribution to vertebrae. Schematic temporal sequence of sclerotome resegmentation (side view). Sclerotome rostral and caudal compartments are separated by the von Ebner fissure. The rostral compartment of one somite/sclerotome (yellow) fuses to the caudal compartment of the consecutive somite/sclerotome (blue) to form one vertebra. Thus, the somites and the vertebrae are out of register by one-half of a segment. The dorsal ectoderm (brown) and dermomyotome that do not resegment (purple) have been removed to visualize the underlying sclerotome. Fate of the rostral and caudal sclerotome compartments projected onto adult human vertebrae. Respective contribution of the somite caudal and rostral compartments is shown. The orientation of the embryonic axes is indicated in black bold in the circle and the corresponding medical terminology is shown in gray italics. (R) Rostral somite sclerotome compartments. (**c**) Caudal somite sclerotome compartments, (Nt) neural tube, (Nc) notochord, (Sc) spinal cord. [Reprinted from Chalupa's J, Pourquié O. Patterning and Differentiation of the Vertebrate Spine. In: Pourquié O (ed) The Skeletal System. Cold Spring Harbor Laboratory Press, 2009, pp. 41–116. Cold Spring Harbor Laboratory Press]

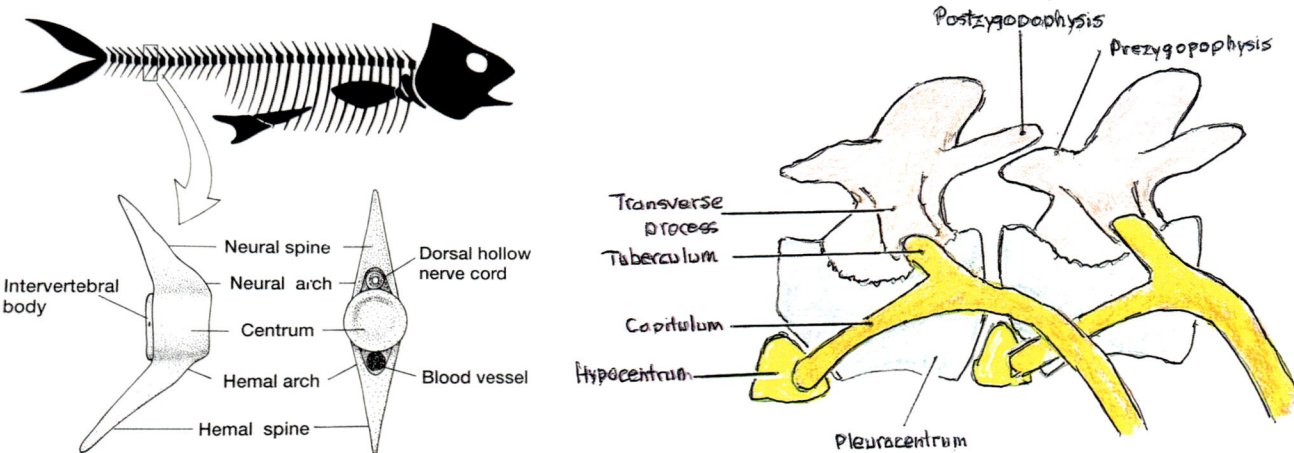

Fig. 10.15 Component parts of vertebra. Hemal arch found in fish for protection of aorta in mobile tail, in most reptiles, some birds and in mammals with long tails. Hypocentrum (intercentrum) is absorbed into the more proximal vertebra body N-1 where it receive the capitulum of rib from vertebra N. Left: [Reprinted from Kardong KV (ed). Vertebrates: comparative anatomy, function, evolution. McGraw-Hill Education; 2015. With permission from McGraw-Hill.] Right: [Courtesy of Michael Carstens, MD]

Intervertebral Discs: The Legacy of the Notochord

The notochord is a very ancient structure. Derived from the very first cell population to undergo gastrulation, it spans the length of the neuraxis, from Hensen's node (r0) to the tail. As such, the notochord is a structural "fossil" that defines the anatomic site where gastrulation was initiated. In mammals, this is conveniently seen as the pituitary fossa. During early embryonic assembly, genes expressed by the notochord play a vital role in organizing the CNS into neuromeric compartments. The notochord runs all the way down the axial cranial base from the pituitary fossa, through the basisphenoid, and down the length of the basioccipital to emerge as the occipito-dental ligament (apical ligament of the dens). It subsequently travels through the centra along the entire length of the vertebral column.

The fate of the notochord depends upon whether it is encased *within* a vertebral body or whether it spans the interspace *between* the vertebrae. Recall that when the sclerotome has four subcomponents. *Ventral sclerotome* surrounds the notochord as the *perinotochordal sheath*. At 6 months of age, notochord cells within vertebral bodies begin to degenerate; by the second decade, they are all gone. Sometimes persistent notochord remnant can degenerate into a tumor known as a *chordoma*. Between the vertebrae, we have a very different story. Here the notochord and its sheath expand to form the *nucleus pulposus*, better known in its adult form as the intervertebral disc.

The original metameres of the body are co-extensive with the sclerotomes. Thus, every fissure of Von Ebner and every disc lie opposite the mid-point of a body segment. Recall that every somite has at its core, a transient hollow center, the somitocoele, a remnant of its original somitomeric form. Von Ebner's fissure transects the somitocoele. When two somites recombine, the more cranial somite donates its caudal dense zone to the newly forming vertebra. *Somitocoele cells that are dragged backward with this caudal half remain at the interface between the vertebra and its cranial predecessor.* The intervertebral disc develops as an interaction between these somitocoele cells and the perinotochordal sheath.

Running through the center of the disc is notochord. Intervertebral discs first appear between the axis and the third cervical vertebra. Recall that the third cervical vertebra, C3 is the product of the caudal dense zone of S7 and the cranial loose zone of S8. Thus, "spacer" genes responsible for discs are not expressed between somites S1 and S7. What we have is a solid basioccipital bone, an occipito-atlantal joint, an atlantoaxial joint, and (finally) a run-of-the-mill intervertebral joint. Obviously, gene expression necessary for interspace anatomy does not take place cranial to axis-C3. But although there are no discs cranial to C3, don't think for a moment that the notochord is not represented. After running downward through the center of the basioccipital bone notochord is continued by the occipitodental, or apical, ligament (a remnant of the loose zone of ancient proatlas), thence onward into the dens; and then finally into the body of axis.

Disc formation starts at stage 23 and thereafter. Centra of the vertebrae expand and compress down the notochord until it is eliminated. Abnormal persistence can lead to a chordoma. But between adjacent vertebrae the notochord persists

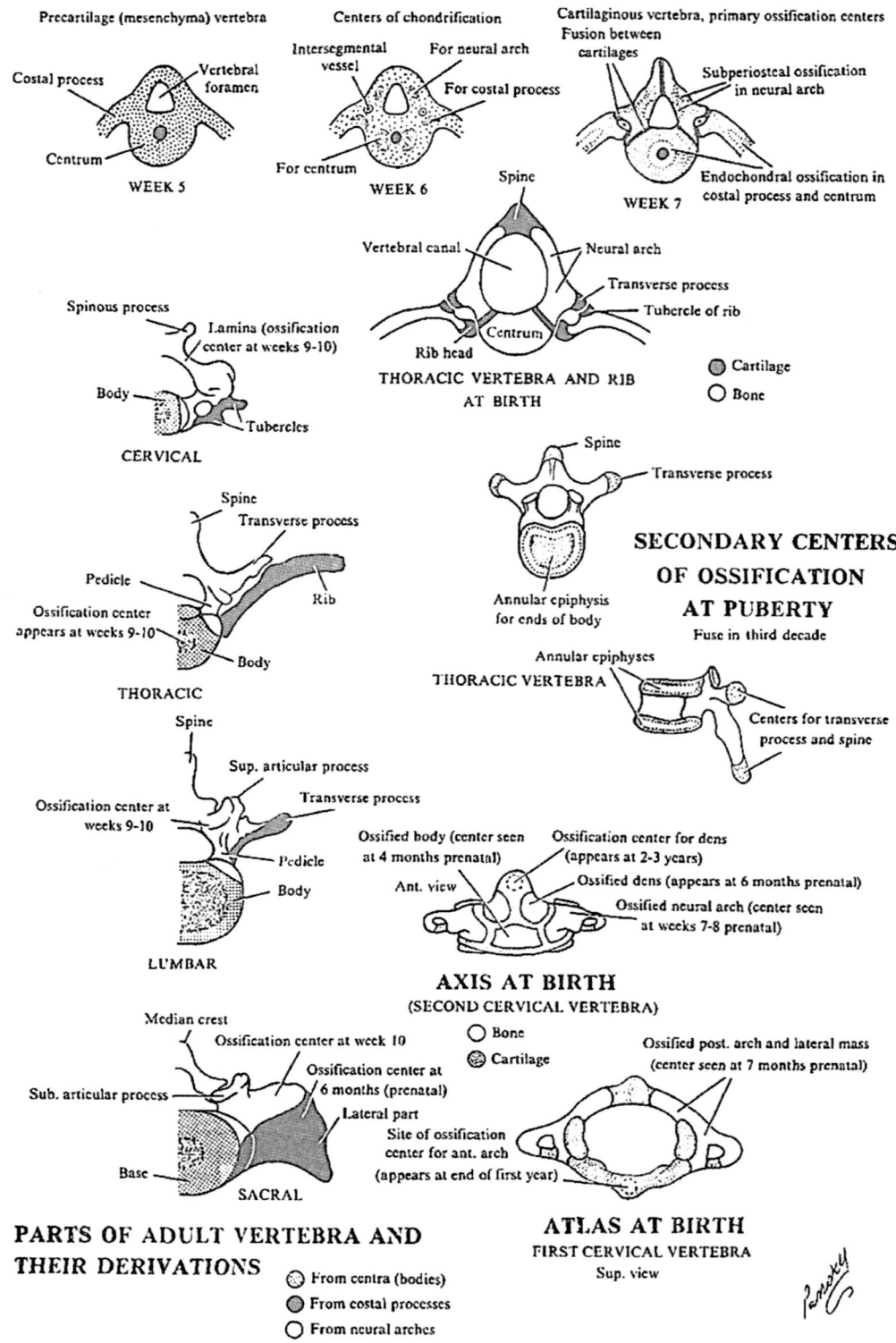

Fig. 10.16 Development of vertebra Note chondrification centers are always paired but later coalesce; they become ossification centers. In early childhood, the centers of each vertebral arch fuse and are joined to the vertebral body by a cartilaginous *neurocentral junction*. This junction allows growth to accommodate the enlarging spinal cord. Fusion of the neurocentral junction usually occurs between the third and sixth years. Anterior notching of the vertebrae is sometimes seen in the infant's or child's vertebrae and shows the site of somite fusion. Secondary ossification centers develop at the ends of the transverse and spinous processes and around the vertebral end plates at puberty. These fuse by age 25 years. Congenital defects are common in the axial sys-

tem. Variations in the lumbar spine occur in about one-third of individuals. Spina bifida occulta is common. Hemivertebrae result from a failure of formation or segmentation. Such lesions are frequently associated with genitourinary abnormalities and less frequently with cardiac, anal, and limb defects; tracheoesophageal fistula; and conductive hearing defects if the cervical spine is involved. [Reprinted with permission from From Pansky B. Review of Medical Embryology. New York: Macmillan, 1982. https://discovery.lifemapsc.com/library/review-of-medical-embryology/chapter-66-development-of-the-axial-skeleton. © 1982]

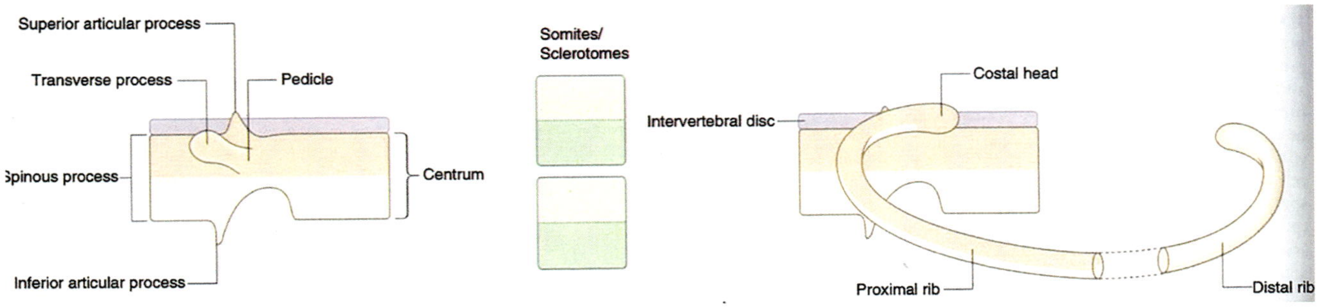

Fig. 10.17 Resegmentation of the rib includes the pedicle and the proximal rib, in which the contributions from cranial and caudal somites halves do not mix. The distal rib demonstrates more mixing. For this reason, sensory blocks for any give rib n must be also be placed into intercostal nerves $n-1$ and $n + 1$. [Reprinted from Lewis, Warren H (ed). Gray's Anatomy of the Human Body, 20th American Edition. Philadelphia, PA: Lea & Febiger, 1918]

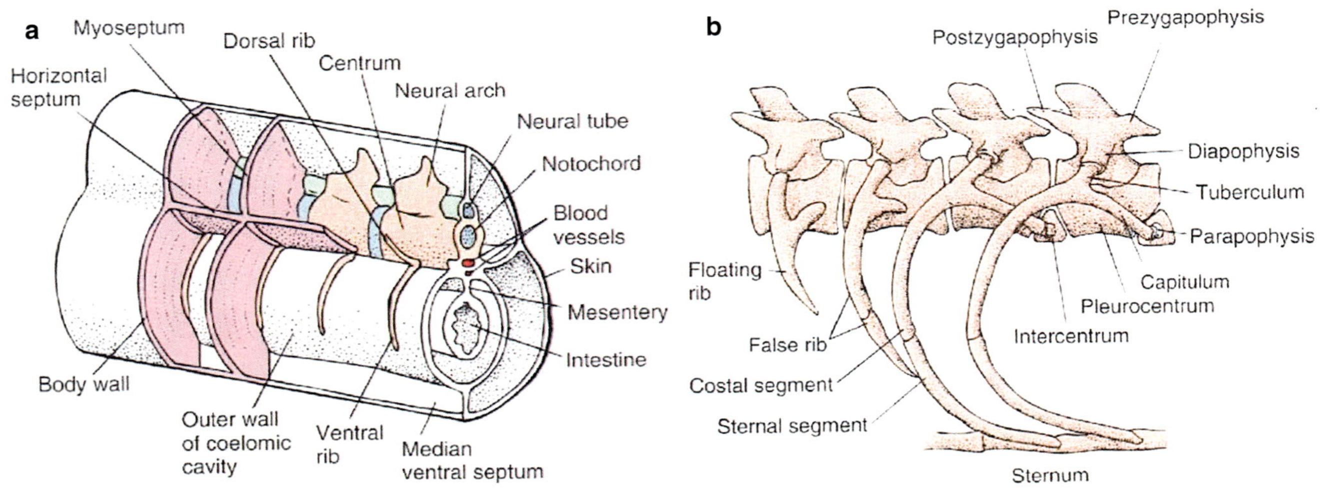

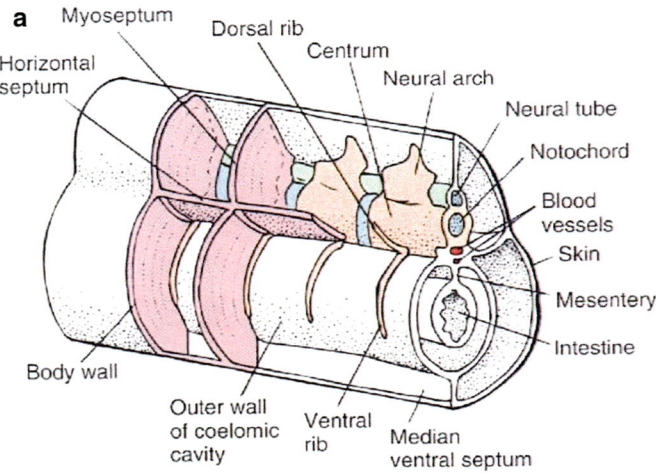

Fig. 10.18 Evolution of ribs Fishes have dorsal and ventral ribs with myoseptum dividing epaxial versus hypaxial muscles. In caudal vertebrae, the ventral ribs close together like pincers to form a hemal arch to protect the aorta from torsion while swimming. The hemal arch is analogous with the dorsal arch which protects the spinal cord. Amniote ribs gain stability by the addition of interlocking zygoapophyses in continuity with the pleurocentrum. Note insertions of each rib with tuberculum articulating with pleurocentrum of the same level and with capitulum articulating with intercentrum (which belongs to the preceeding vertebra). Note that false ribs have both insertions into the same level. [Reprinted from Kardong KV (ed). Vertebrates: comparative anatomy, function, evolution. McGraw-Hill Education; 2015. With permission from McGraw-Hill.]

as the central zone of the developing disc. Surrounding chondroblasts create a fibrous shell around the nucleus pulposus (Fig. 10.20).

Ossification of Vertebrae

Ossification centers might seem, at first glance, yet another example of anatomic minutia…fit only for small print. In reality, they represent centers of genetic activity from which bone formation spreads outwards. As such, we are really looking at the "maturity sequence" of a given bone. As such, we can expect that muscle attachment centers will develop in exactly the same sequence.

Once the sclerotomes surround the neural tube, they begin to express collagen II: conversion into bone follows. Stage 17 marks the onset of chondrification. It proceeds in a fixed sequence. Vertebrae typically ossify from four primary centers: one in each centrum and one in each neural arch. The centra have ossification centers located dorsal to the notochord. These are incorrectly described as being singular. Clear-cut bilateral ossifications can occur. Surpression of one centrum can produce a hemivertebra. During the initial postnatal years each centrum shares with the ipsilateral neural arch a synchondrosis, the *neurocentral joint* (Fig. 10.21).

Ossification of the neural arches starts at the base (the root of the transverse process) and spreads in two directions: backward into the laminae and forward into the pedicles >

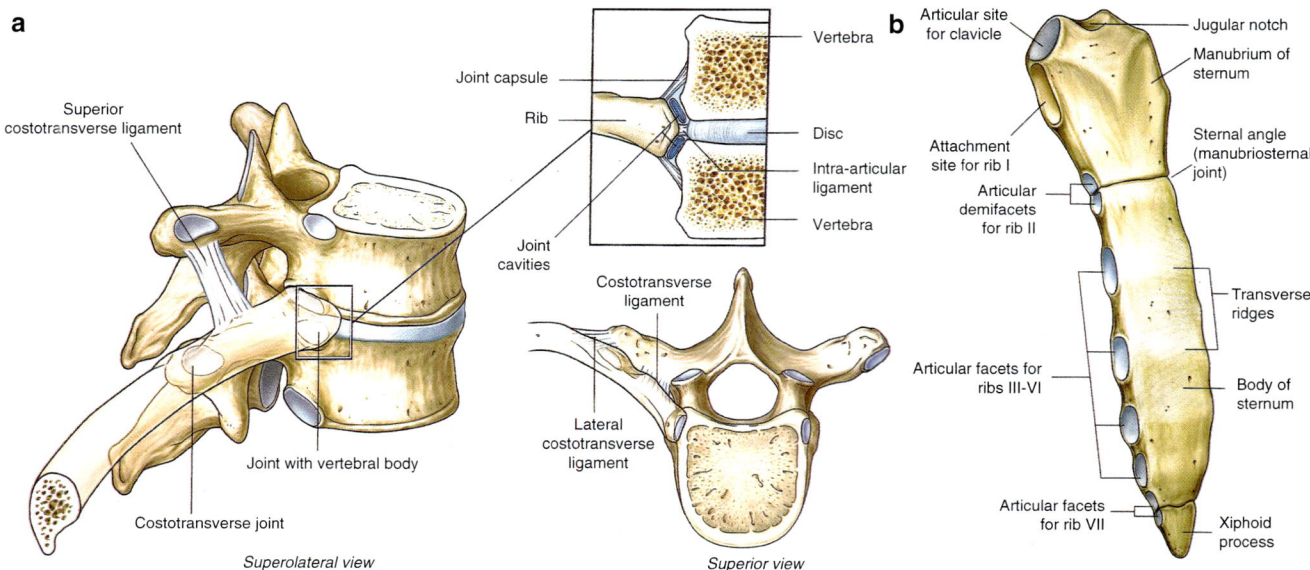

Fig. 10.19 Articulations of the rib. The seventh rib is shown. Head of rib articulates with the tubercle of the zygaphophysis of vertebra of its same number (pleurocentrum) and with the vertebral body preceeding it (a genetic remnant of the intercentrum). Re-segmentation process affects ribs. Upper part of fifth rib comes from T4$_C$ and lower half comes from T5$_R$. This parasegmentation exists in the proximal have of the rib but the mesenchyme of the distal half is more uniformly from the same level. [Reprinted from Drake R, Vogel AW, Mitchell AWM. Gray's Anatomy for Students, third edition. Philadelphia, PA: Churchill-Livingstone. 2015. With permission from Elsevier]

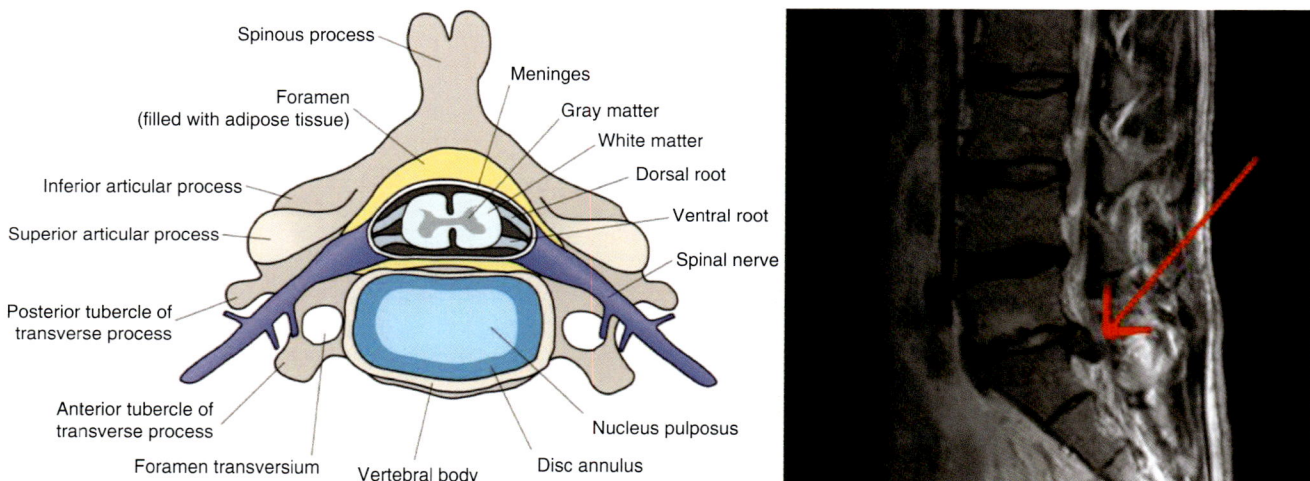

Fig. 10.20 Cross-section of cervical vertebra with disc. Intervertebral discs develop as a fibrous ring around nucleus pulposus, remnant of the notochord. Cervical disc herniation occurs most commonly in the fifth, sixth, and seventh vertebrae. These represent the "newer" vertebrae acquired in amniote evolution. Degeneration leads to Stress over time concentrates at the border between the embryonic pedicle and body, an anatomic weak spot, leading to degeneration and compression. This leads to a defect in the annulus allowing gelatinous nucleus pulposus to protrude outward. Left: [Reprinted from Wikimedia. Retrieved from: https://commons.wikimedia.org/wiki/File:Cervical_vertebra_english. png. With permission from Creative Commons License 3.0: https://creativecommons.org/licenses/by-sa/3.0/deed.en.] Right: [Reprinted from Wikimedia. Retrieved from: https://commons.wikimedia.org/wiki/File:Lagehernia.png]

further forward into the body > lateral into transverse processes > downward into the articular processes. At stage 23, the vertebral column looks like spina bifida: almost all vertebrae are present (32–33) but there are no spinous processes, the "tops" are open. Fusion of the spinous processes takes place much later, during the fourth postnatal month. Vertebral arch centers appear in craniocaudal order (some studies dispute this). Unification of the arches follows a distal–proximal sequence: lumbar > thoracic > cervical. Fusion of the centra with the arches follows a proximal–distal sequence: cervical > thoracic > lumbar.

Certain zones remain cartilaginous for many years. During puberty, *six secondary ossification centers* make their debut. The upper/lower borders of vertebral bodies

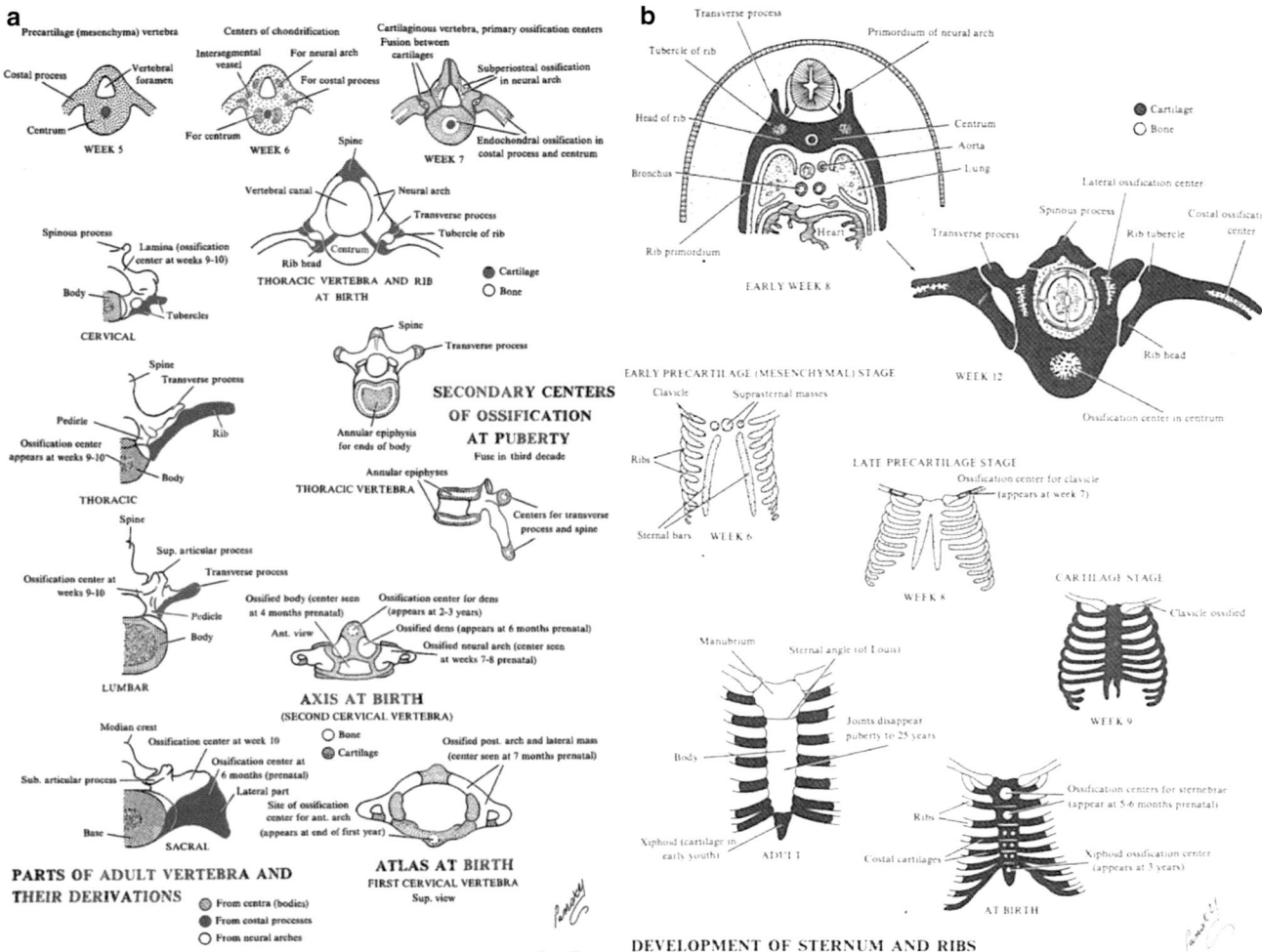

Fig. 10.21 Ossification of vertebrae. Left: Vertebra C1 (atlas) lacks a centrum. Vertebra C2 (axis) has two centra, one of which forms most of the dens. The development of "typical" cervical vertebrae is shown, including the primary ossification centers within the hyaline cartilage, The development of a thoracic vertebrae is shown, including the three primary ossification centers (observe the joints present at this stage), followed by secondary ossification centers at puberty. The development of the lumbar vertebrae is shown, including the primary and secondary ossification centers. Note the anular epiphyses separated from the body, and the anular epiphyses in place. The development of the sacrum is not shown. Note that the ossification and fusion of sacral vertebrae may not be complete until age 35. Right the primary and secondary ossification centers of the sternum (with ribs developed from costal elements. Left: [Reprinted from Pansky B. Review of Medical Embryology Macmillan, 1982 Chapter 66 Axial Skeleton, fig. 31 ISBN: 978-0023906206 https://discovery.lifemapsc.com/library/review-of-medical-embryology/chapter-66-development-of-the-axial-skeleton. With permission from LifeMap Sciences, Inc.] [Reprinted from Pansky B. Review of Medical Embryology Macmillan, 1982 Chapter 67 Appendicular Skeleton fig. 32 ISBN: 978-0023906206 https://discovery.lifemapsc.com/library/review-of-medical-embryology/chapter-67-appendicular-skeleton-and-skull-development. With permission from LifeMap Sciences, Inc.]

form ring-like annular apophyses. From them will project *costal articular facets*. The apices of processes (spinous and transverse) each have a center.

Five Reasons Why Should We Care About Ossification Centers

Reason 1. Somites have a spatial "fate map" which is determined by the site at which epiblast cells ingress during gastrulation. Cells passing through Hensen's node will be assigned to the medial somite. Those passing through the rostral streak will be positioned laterally.

Reason 2. Sclerotomes come from medial somites and these subdivide into four functional parts: *central sclerotome* = ventral neural arch, pedicles, and proximal ribs; *ventral sclerotome* = perinotochordal sheath; *dorsal sclerotome* = dorsal neural arch; and *lateral sclerotome* = distal ribs.

Reason 3. Ossification centers represent genetically distinct developmental fields. Thus, the pattern of vertebral development follows the order in which ossification centers appear. The temporal–spatial sequence is ventral > dorsal and medial > lateral.

Reason 4. Collagen II produced by ossification centers attracts myoblasts. Muscle insertions into the vertebra occur in the same temporal–spatial sequence as the ossification centers. This is particularly well demonstrated by the back muscles.

Pathologies or evolutionary rearrangements can selectively involve one vertebral field versus the rest. Inappropriate fusion of the lower cervical vertebrae (C3–C7) is seen in Klippel-Feil syndrome. A *cuneiform vertebra* caused by a missing centrum can lead to scoliosis. *Differences in segmentation between the centra and the neural processes of somites 4–6 are responsible for the occipitocervical junction*, a revolutionary development in tetrapods (see below).

Phylogeny of the Centrum

The four zones of the sclerotome are genetically distinct developmental fields. Each ossification center represents a single developmental field. An apparently singular ossification center may represent paired centers at the boundary zone of neighboring fields. Under certain conditions *component parts may fail to fuse or display abnormal fusion patterns*. These pathologies reflect the expression of otherwise suppressed developmental fields (Fig. 10.22).

The number of centra per vertebral body is variable, ranging from none (*aspondyly*) to six (*polyspondyly*), seen in Holocephali (octopuses) and Dipnoi (lungfishes). In its primitive form, the tetrapod vertebra is *diplospondylous*; it consists of two paired parts, cranial pleurocentra, and caudal intercentra plus a neural arch. Such vertebrae, in which all three arch elements are discrete, are called *aspidospondylous*. Rhipidistians (lungfishes) had a particular type of *rachitomous* (Gr. *rhachis* = spine + *tomos* = cut) vertebra consisting of a large intercentrum, a small pleurocentrum, and a neural arch. The intercentrum was an incomplete U-shaped ring with its open segment directed dorsally whereas the pleurocentrum was U-shaped, but with its open segment facing ventrally. In the development of the craniovertebral joint, *the intercentrum fragments along these lines*.

The earliest tetrapods, such as *Ichthyostega*, had rachitomous vertebrae but invented a new lateral stabilizing element, the zygapophysis. An immediate split then occurred between the anamniote line leading to amphibians and the amniote line leading to anapsid reptilomorphs. In the former, the intercentrum became the main element of the vertebra body and is associated with the neural arch whereas, in our ancestral line, the pleurocentrum became dominant. The early reptilomorph condition explains the anatomy of modern-day rib articulations. With the appearance of definitive amniotes, the vertebrae become *holospondylous*, that is,

characterized by the fusion of all component parts. The intercentrum, being absorbed into the posterior aspect of the preceding vertebral body, bears the insertion of the capitulum [3].

Vertebrae of the mammalian line are holospondylous, they form the fused centra with the pleurocentra dominant and the intercentra absorbed posteriorly. When intercentra are present, they remain as unossified intervertebral cartilages.

What is the significance of the above? (1) Developmental fields in the lateral zones of vertebrae can behave differently from those in the center. (2) Fields do not necessarily have to fuse with one another. Thus, ventral sclerotome, associated with the notochord, may remain distinct from the other parts of the sclerotome. (3) The diplospondylous condition is consistent with two genetic parts of the sclerotome: a *cranial loose zone* and a *caudal dense zone*. (4) The existence of rachitomous vertebrae proves that neural arches are assembled in a similar manner to the centra. (5) These facts put together will explain the various vertebral forms during evolution. We will sketch these out briefly in this chapter but will deal with them more completely in our discussion of neck evolution.

Rearrangement of Component Parts of the Centrum = A New Joint

Two fundamental events in the adaptation to land were the invention of the neck which granted mobility for predation and the liberation of the pectoral girdle from the skull. As we shall see later, in our discussion of the cucullaris muscle, the development of a moveable joint between the head and trunk was not a new idea. Placoderms had a pivot joint with a flexor/extensor system, the cucullaris, and levator capitis connecting the skull with rigid body armor. In subsequent fish evolution, these ideas were forgotten. When head trunk separation next appeared, a new innovation took place, a joint between the skull base and internal axial vertebral column. The cucullaris remained attached to the pectoral girdle, now displaced to the level of the fifth and sixth cervical vertebrae. The creation of this new joint required the recombination of somites at the cranial–vertebral junction, we shall now examine this process in detail (Figs. 10.23, 10.24, 10.25, 10.26, 10.27).

Frameshift of the Skull Base and First Three Spinal Vertebrae = A New Joint

Note on nomenclature: $SomiteNumber_{CAUDAL/ROSTRAL}$.

Example: fourth somite, caudal zone = $S4_C$; seventh somite, rostral zone = $S7_R$.

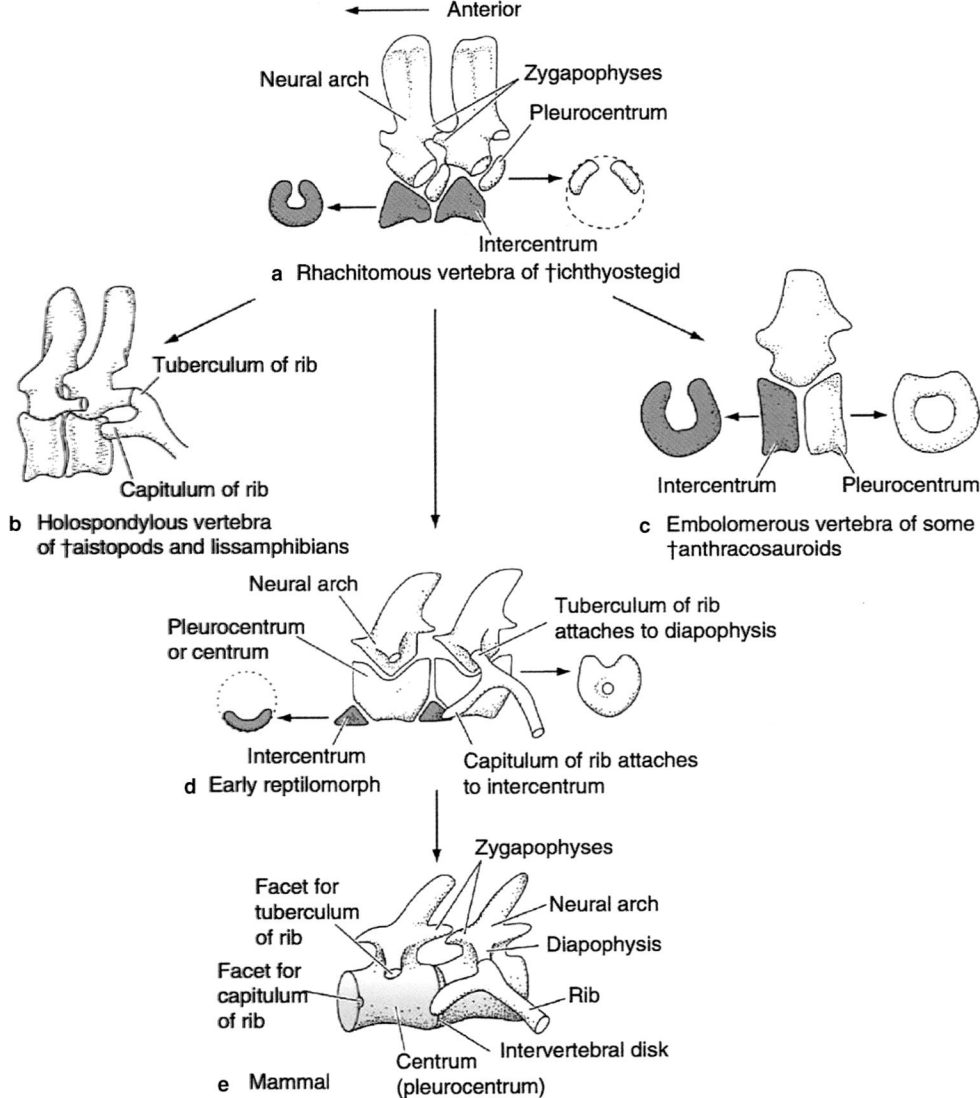

Fig. 10.22 Evolution of vertebrate centrae. (**a**) Basal tetrapod *Ichtyostega* had rhachitomatous vertebrae with U-shaped Intercentrum having a small aperture for a persistent but reduced notochord. NOTE Ancestrally, the neural arches simply protected the spinal cord. Once the the vertebral column began to support the animal's weight, special articulations between the neural arches—the zygapophyses evolved. The first interation was between neural arch and centrum. This was transformed in amniotes to an articulation between the neural arches themselves. Most vertebrae had ribs. Sternum has not been identified at this level. The fusion of holospondylous vertebrae prevents identifying either intercentra or pleurocentra. Neck vertebra can barely be distinguished. (**b**) Amphibian neck consists of a single atlas with very limited mobility. (**c**) †Anthrocosaursn experimented with emolomerous centrae in which intercentrum and pleurocentrum were co-dominant. (**d**) In Reptilomorphs have schizomerous condition: the pleurocentrum is exerting dominance. The reptilomorph OCJ consists of a single occipital condyle that articulates with centrum of the atlas. The ancient condition is seen in rib insertion, in which tuberculum of rib N articulates with the pleurocentrum of vertebra N, while capitulum of rib N articulates with the intercentrum of vertebra N−1. E In synapsids, beginning with therapsids, the condyle divides and shifts dorsally. Mammals lose (1) proatlas entirely, and (2) neural spine of the atlas. Atlas and axis lose their zygapophyses. [Courtesy of William E. Bemis]

Axial Sclerotomes

The fusion planes of S1–S3 have grooves denoting the previous existence of embryonic segmental arteries. The fourth somite is a normal, healthy somite with both loose and dense zones; hence the fusion between the $S4_R$ and its predecessors is marked by *hypoglossal canal* for the hypoglossal artery, the last remaining intersegmental vessel from the ancient longitudinal neural arteries. $S4_C$ recombines with $S5_R$ to form the sclerotome of the original primary cervical vertebra, the proatlas. The cranial region of proatlas ($S4_C$) becomes incorporated as a *ligament into the basion* at the foramen magnum while $S5_R$ becomes the *apical segment of the dens*. The notochord runs downward through basioccipital, through

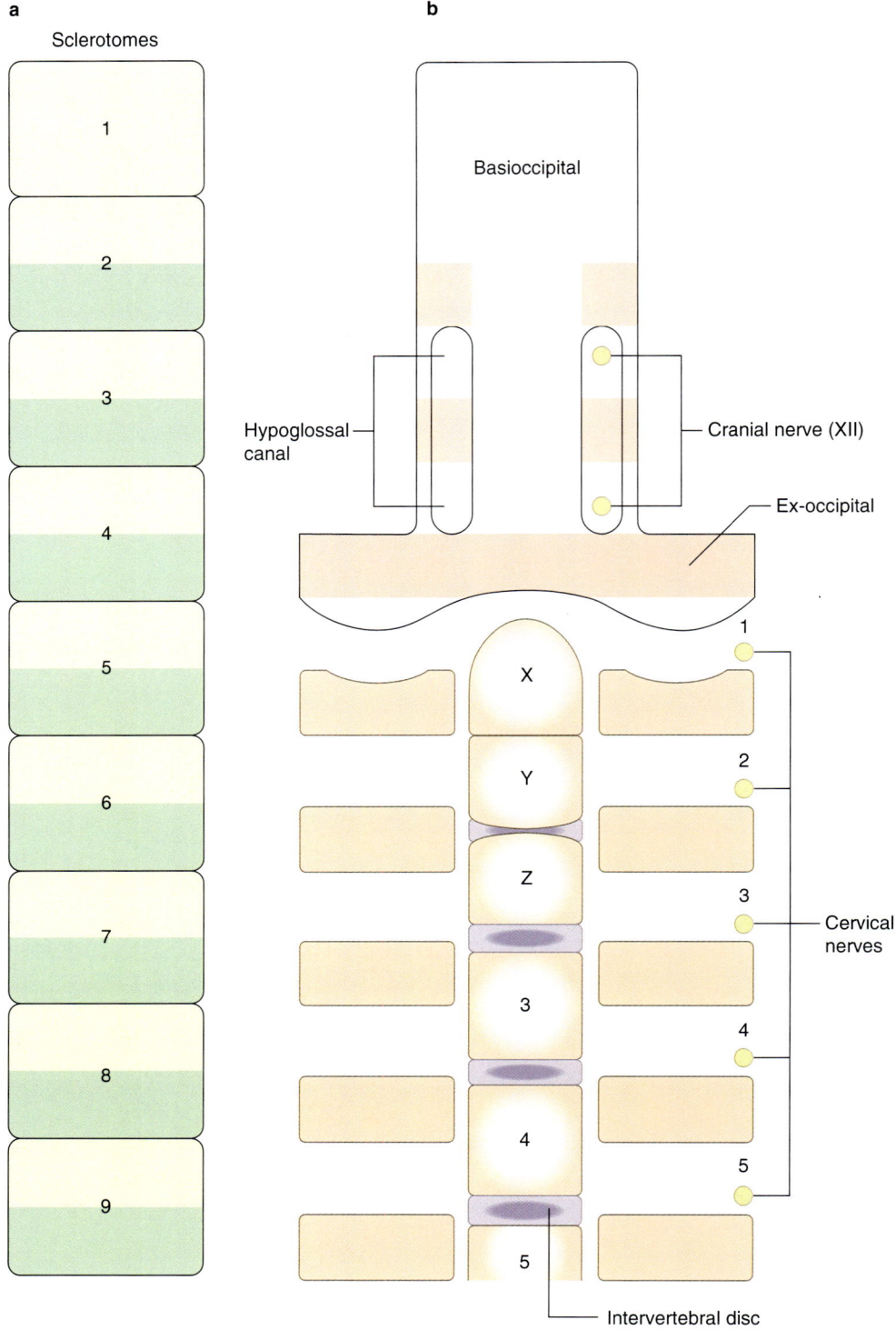

Fig. 10.23 Formation of the human craniovertebral junction. Sclerotomal primordia and their vertebral phenotypes are colour-matched. Proatlas The proatlas resegmented sclerotome (PA) comes from the fourth and fifth somites. Derived from the proatlas are: the axial zones (Ad and Al) which become the basion (B) of the basioccipital or clivus (CL) and the apical segment of the dens (AD); the lateral dense zone (Ld) becomes the exoccipital comprising the occipital condyle (OC), and lateral rim and opisthion (OT) of the foramen magnum; the hypochordal bow of proatlas (HBp) forms the ventral clival tubercle (CT). C1 The C1 resegmented sclerotome (C1) comes from the fifth and sixth somites. Derived from the C1 sclerotome: the axial zones form the basal segment of the dens (BD); the lateral zone forms the posterior atlantal arch (C1P); the hypochordal bow (HBp) forms the anterior atlantal arch (C1A). C2 The C2 resegmented sclerotome (C2) comes from the sixth and seventh somites. Derived from the C2 sclerotome: the axial zone forms the C2 vertebral body (AB); the lateral zone forms the neural arch of C2 vertebra. IBZ (internvertebral boundary zone) between the proatlas and C1 sclerotome forms the upper dental synchondrosis (US) and the IBZ between the C1 and C2 sclerotomes forms the lower dental synchondrosis (LS). [Reprinted from Pang D, Thompson DNP. Embryology and bony malformations of the craniovertebral junction. *Childs Nerv System* 2011; 27:523–564. With permission from Springer Nature]

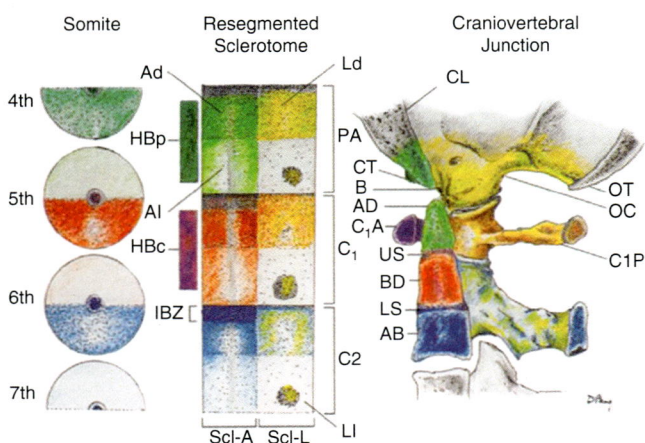

Fig. 10.24 Severance line traces the final separation of the skull from the cervical spine. It runs through the original resegmentation fronts of S4$_C$ and S5$_R$. These are: (1) the junction between the basion and apical segment of the dens in the axial proatlas, and (2) the junction between the exoccipital, or future occipital condyle, and the lateral mass of C1 (derived from the lateral portion of the C1 resegmented sclerotome). [Reprinted from Pang D, Thompson DNP. Embryology and bony malformations of the craniovertebral junction. *Childs Nerv System* 2011; 27:523–564. With permission from Springer Nature]

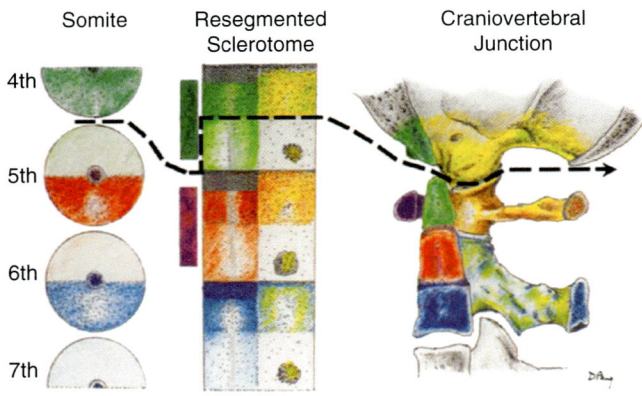

Fig. 10.25 O'Rahilly/Müller model Centra, neural arches, spinal nerves and ganglia, comparative relationships. On left are the somites, prior to resegmentation, beginning with S1. Note that S1 has no dense zone, it is strictly a loose zone. S2$_R$ is loose so the combination of S1$_C$ and S2$_R$ does not convert caudal S1 into a dense zone. S4$_C$ + S5$_R$ becomes the proatlas. On the right shows the lateral masses all as dense zones from a single somite level (no resegmentation). The centra are resegmented. X, Y, and Z from proatlas, atlas, and axis form the tip of dens, body of dens and centrum of the axis. Tip of dens is S5$_C$ + S6$_R$ and the latter forms the upper synchondrosis of the dens. Body of dens is S6$_C$ + S7$_R$ with the latter loose zone persisting as the lower synchondrosis of the dens. Body of the axis is S7$_C$ + S8$_R$. Note that the lateral masses are not resegmented. Body of the atlas comes strictly from S2. [Reprinted from Standring S. The Back. In: Gray's Anatomy, 40th edition. Philadelphia, PA: Churchill-Livingstone; 2008: 763–774. With permission from Elsevier]

this zone, and down into body of axis proper. The articulation of apical denta centrum with the basion creates the pivot point of the new craniovertebral junction.

The *intervertebral boundary zone* of the CVJ between somites 4 and 5 has unique properties. Recall that the IBZ is a tight compaction of cells located in the dense zone of the axial sclerotome in line with von Ebner's fissure of the lateral sclerotome. Under normal conditions, IBZ forms the intervertebral discs. In proatlas IBZ, a physical disruption of cells, mediated by cleavage genes, separates the bony union between the dens and the basiocciput. The dentooccipital ligament and notochord remain behind in the midline. In this manner, the skull gains its independence from the vertebral column.

Dens is strapped into place by ligaments derived from dense proatlas cells ventral to the notochord, the *hypochordal bow*. These give rise to the anterior clival tubercle located on the ventral basioccipital. The lateral sclerotome of proatlas has important derivatives. Its dense regions form paired exoccipital bones. These fuse to the anterolateral border of the foramen magnum to form the two occipital condyles. The additional flexion of this system was an instant evolutionary advantage. The loose zones of the lateral sclerotome stimulate the development of the C$_1$ nerve root.

Proatlas sclerotome, as the product of S4$_C$ and cranial S5$_R$, is quite properly the first cervical vertebra, but it breaks up. Unfortunately, current nomenclature does not recognize the proatlas. Therefore, the product of S5$_C$ and S6$_R$ is falsely termed "the first cervical vertebra." Rather than quibble with anatomic dogma, let's just refer to proatlas as cervical vertebra 0.

Unlike all remaining sclerotomes caudal to S7, in which IBZ morphs into the disc, *the IBZs of somites 6 and 7 disappear*. Instead of having discs, the dens now has two synchondroses which ultimately ossify. In this way, the proatlas apex from S5$_R$ is cemented to the basal dens from S5$_C$/S6$_R$ atlas, and the basal dens is united with the S6$_C$/S7$_R$ body of the axis.

The upshot of resegmentation is to produce three midline components: the *apical dental segment* from caudal proatlas (cervical sclerotome 0), the *basal dental segment* from the first cervical sclerotome, and the *body of axis* from the second cervical sclerotome. Ossification of the midline is a lengthy process that takes place caudal to rostra. The lower synchondrosis between the axis body and the basal segment occurs at birth but the apical and basal dental segments are not united until adolescence. The apical ligament is derived from proatlas dens. The transverse atlantal and alar ligaments securing the CVJ arise from the S6 first cervical sclerotome and the basal dental segments—recall that the axial structures originally came from paired sclerotomes.

Fig. 10.26 Staged
development of occipital
somites. (**a**) Stage 12 four
somites (S1 = D) with
hypoglosal nerves (yellow
circles) and potential
myoblasts arising from S2 to
S4. Note that the occipital
somites increase in size. Only
S4 has a true neural arch,
which is uses to make
posterior foramen magnum
(anterior formen magnum is
basioccipital). Neural crest
gangia (green) indicated. (**b**)
Stage 14 S1–S4 now have
sclerotomes and are
positioned ventrally along the
notochord. Loose zones (light
green) and dense zones (dark
green) are indicated. Nerves
travel through loose zones.
Occipital neural crest
diminished and cervical
neural crest has formed
ganglia. (**c**) Perinotochordal
sheath has reached tip of
notochord (r1). [Reprinted
from Standring S. The Back.
In: Gray's Anatomy, 40th
edition. Philadelphia, PA:
Churchill-Livingstone; 2008:
763–774. With permission
from Elsevier]

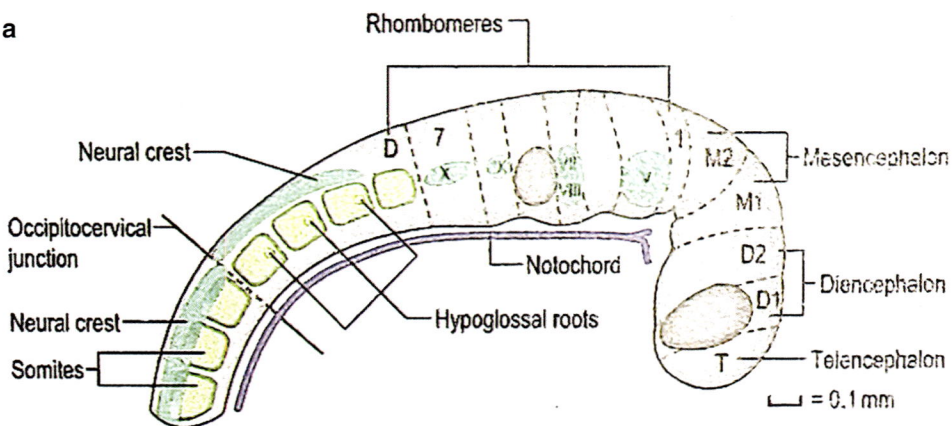

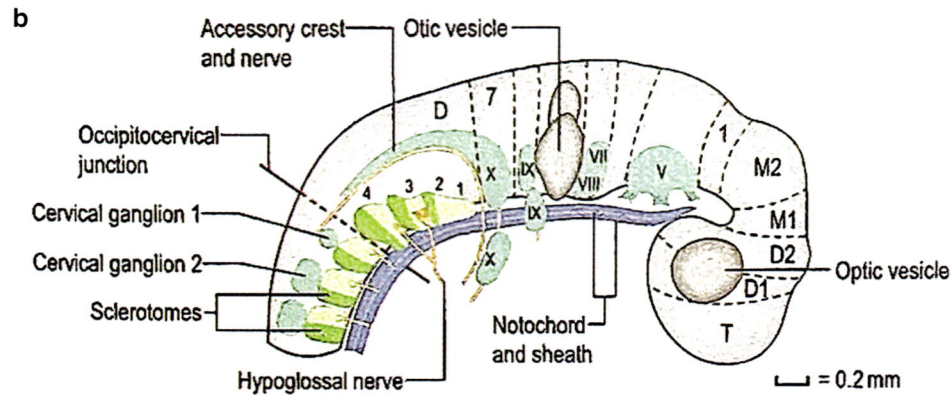

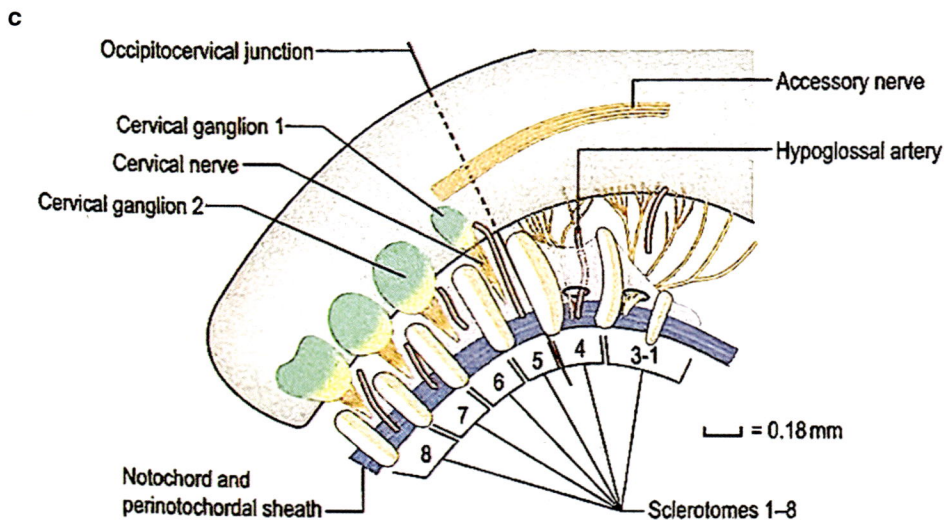

Lateral Sclerotomes

Lateral dense zone of first cervical sclerotome makes the
posterior arch of atlas. Lateral dense zone of the second cervical sclerotome makes the posterior arch of axis. Loose
zones promote second and third cervical nerves. Hypochordal
bow of first cervical sclerotome forms anterior S4 and the
anterior arch of atlas. No further equivalent hypochondral
bows were produced.

To visualize this process, review Figs. 10.23, 10.24,
10.25. The first model is that of O'Rahilly and Müller
(Fig. 10.23) and is based on direct observations of human
embryos. It is schematic but tells the story. The second
model, from Pang and Thompson (Figs. 10.24, 10.25) is
more three-dimensional and presents the final anatomy in a
color-coded lateral view of the occiput.

The osteology of the posterior cranial base is an important
topic with little clarity in the literature (Figs. 10.26, 10.27, 10.28).

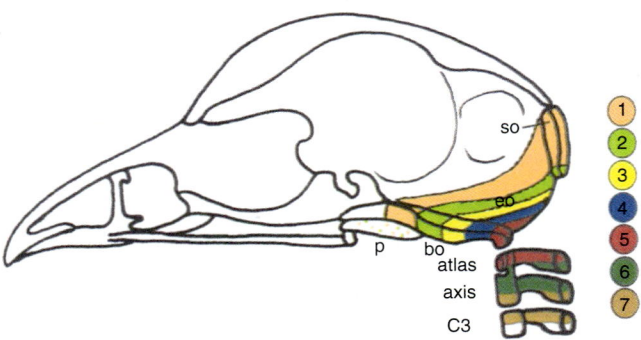

Fig. 10.27 Construction of the posterior fossa from occipital somites in the avian model. Contributions of individual somites to the occipital bone complex have been mapped. Paraxial mesoderm from each somite flows backward in concentric "swatches" which laminate together like Russian dolls. Recall that protoatlas was composed of $S4_C + S5_R$. The fifth somite is split between the foramen magnum and atlas. [Reprinted from Huang R, Quia Z, Patel K, Wilting J, Christ B. Contributions of single somites to the skeleton and muscles of the occipital and cervical regions in avian embryos. *Anat Embryol* 2000; 202(5):375–383. With permission from Springer Nature]

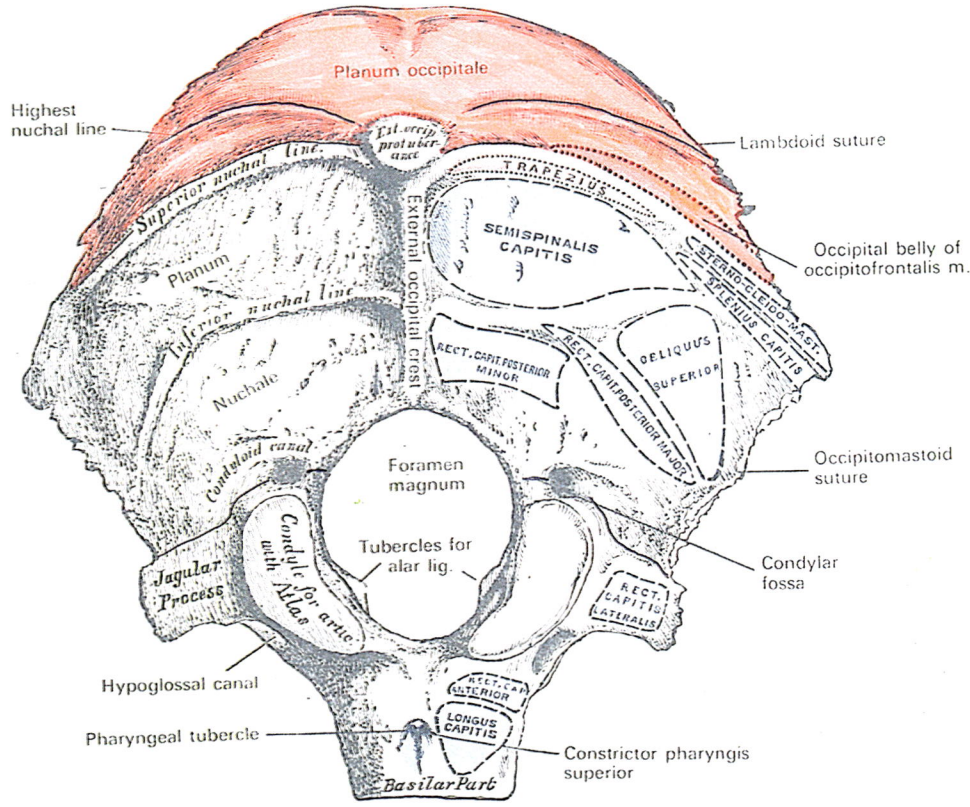

Fig. 10.28 Occipital somites S1–S4 produce only muscles dedicated to the head, either as branchiomeric muscles related to arches 3–5 or as the somatic hypobranchial muscles of the tongue. All muscles attached to the occipital bone are by secondary insertions (blue). Superior nuchal line bears secondary insertions of sternocleidomastod and trapezius, the latter is incorrectly labeled in red. Its fibers descend from C2 to C6 (somites S6–S10) to the scapula and then have return to the axial skeleton. Membranous occipital bone has four interparietal fields (the post-parietals and tabulars) indicated in pink. Superior nuchal line bears the secondary insertion of occipitalis, incorrectly labeled in red. Note the inferior rim of foramen magnum and medial part of occipital condyles in mammals represent contributions from protoatlas. [Reprinted from Lewis, Warren H (ed). Gray's Anatomy of the Human Body, 20th American Edition. Philadelphia, PA: Lea & Febiger, 1918]

We shall discuss it anon. Suffice it to say that the mammalian posterior fossa is made up of four occipital somites. The original three (alpha) fuse together and the fourth one (beta) is added on. Hypoglossal artery and nerve communicate through the interface between alpha and beta. This corresponds to the rostral (loose) zone of S4. During its early development S4 (unlike S1–S3) has proto-neural arches. Its dense zone produces an axial contribution to the terminal part of the basioccipital. The lateral elements of S4, acting as neural arches form the dorsal rim of the foramen magnum and part of the occipital condyles. Rostral S5 (proatlas) donates the rest of the condyles.

The cranial vertebral boundary in fishes lies between S3 and S4 (they have no neck). In diapsid amniotes (dinosaurs, birds, and true reptiles) it is between S4 and S5. Mammals incorporate the proatlas and have a complete reorganization of the atlas/axis complex. The mammalian cranial-vertebral boundary is between somites S5 and S6. Hypoglossal nerve and first cervical nerve in mammals are separated by the incorporated proatlas. The original first cervical vertebrae (proatlas) develop from the caudal half of S4 and the cranial half of S5. The joint that we use to flex our neck lies between proatlas (the occipital condyles) and atlas.

Development of the Occipital Somites

Nota bene Work of O'Rahilly and Müller constitutes the best resource for understanding the occipital region. The drawings are superb (Figs. 10.23, 10.26).

Stage 9 (1.5–2.5 mm, 26 days, 1–3 somite pairs)

First appearance of somites. The first somite relates to the posterior otic capsule and corresponds to rhombomere 8.

Stage 10 (2–3.5 mm, 29 days, 4–12 somites) pharyngeal arch 1

First somite is now the same size as somites 2–4. Somitocoeles still present. They disappear as differentiation takes place. Sclerotomes appear at ventromedial aspect of each somite.

Stage 11 (2.5 to 4.5 mm 30 days, 13–30 somites) pharyngeal arch 2

Rostral neuropore closed. Somitocoeles gone. Somite 1 is now smaller relative to the rest. It has *no* contact with surface ectoderm. Its sclerotomic material is confined to the caudal half. This will ultimately be carried forward as the hypoglossal cord It lies caudal to vagal and accessory nerve neural crest. No evidence of a hypoglossal nerve. Somite 2 in contact with surface ectoderm. Its sclerotome is interposed between the compact zone and the neural tube. Somite 3 has a more "regular" shape resembling somite 4. Its sclerotome closely follows that of the neural tube. *Intersegmental arteries* from the LNA (future basilar-vertebral system) are seen between S2–S3 and S3–S4. This is because intersegment S1–S2 is supplied by the fourth aortic arch.

Stage 12 (3–5 mm 31 days, 21–30 somites) pharyngeal arch 3

Dermatomyotomes 1–4 are still visible. For the first time, occipital somites are anatomically distinguished from cervical somites by two criteria: (1) primordial of hypoglossal nerve; and (2) the appearance of the cervical neural crest. The hypoglossal nerve presents as cellular strands that emerge from basal lamina of hindbrain and terminate in the myotomes of S2–S4. Cervical neural crest flows into somite 5 and further below. This will produce the posthyoid hypopharyngeal cord (strap muscles) c1–c3. These bind to the c1–c3 zones of clavicle and sweep forward under arches 5, 4, and 3.

None of the occipital somites has a somitocoele any longer. Somite 1 continues to have a dermatome but is being transformed into a loose mesenchyme. Although S2–S4 contain dermatome, myotome, and sclerotome, the ventral 2/3 of dermatome is becoming disorganized. The second intersegmental artery is being transformed into a hypoglossal artery between S3 and S4.

Pharyngeal arches 1–3 present with neural crest cells present. Epipharyngeal disc activity is seen (particularly above PA1). Disc cells have joined up with ganglia of V, VII, and X. PA3 contains neural crest from IX and X. Arch 4 contains neural crest destined for the larynx and tongue.

Stage 13 (4–6 mm 30+ somite pairs) pharyngeal arch 4

Dermomyotomes thinned out but are recognizable. Myotomes present in S1–S4 are separated by neural crest septae; they descend as the *hypoglossal cord* followed by XII nerve roots. This marks the early formation of the tongue. The first appearance of dense and loose zones is seen in S4 and in S5–S12.

Stage 14 (5–7 mm) pharyngeal arch 5, end of pharyngeal arch period

Rostral sclerotomes of S1–S3 are fused with perinotochord. Hypoglossal nerve root and hypoglossal artery separate them from the S4 sclerotome.

Stage 15–22 incorporation of S4 into the occiput

At stage 15 perinotochord of S4 has a dense zone that extends to the midline by stage 17. During stages 16–17, the occipital somites form a single unit extending forward toward the pituitary. This marks the consolidation of the basioccipital. In stage 18, cartilaginous exoccipital appears. Central segments X, Y, and Z represent S5–S7. In stage 19 the dense tissue below (ventral) to notochord condenses to form the *hypochordal bow* at the level of S5. This is the precursor of *anterior arch of atlas*. Basioccipital is mainly S3–S4.

At stage 20 the regular pattern of dense and loose zones seen in the central cervical region changes. The dense zones have two derivatives: (1) the central part surrounds the notochord, and; (2) the peripheral part becomes *annulus fibrosus*. Note that between X and Y there is no annulus fibrosus while between Y and Z it is present, but reduced in size. Well into the fetal period central column of axis continues to display these component parts. Thus, body of axis = S5 + S6 + S7. Neural arch of axis = S6 + S7 while anterior arch of atlas originates from S5.

In stage 21, occipital condyles come from two sources: central segment 4 and exoccipital. The central column of the axis originates from central segments 5–7. Neural arch of the axis comes from S6 to S7; accordingly, it is extra-large. Note that the position of the most superior ganglion is *above* (cranial) to neural arch. This is further evidence of the *pre-existence of an original first cervical segment, the proatlas*, constructed from S4 and S5.

In sum: Segmentation in human embryos is a sudden event at stage 9 marked by the appearance of primary rhombomeres [4], the first aortic arch, the otic disc, and three pairs of somites. These latter developed out of the more primitive somitomeres Sm8–Sm10. Neural crest is already present at this stage. Stage 10 is characterized by the appearance of the first pair of pharyngeal arches and the appearance of occipital sclerotomes. Stages 11–12 present with hypoglossal nerve, and spinal myotomes + neural crest. At stage 14 hypoglossal neural crest and the hypoglossal cord populate PA3 and PA4.

The Supraoccipital Bone = S1 to S4$_R$ + S4$_C$

Humans have 37 pairs of somites (up to 44 during development). These are distributed as follows: 4 occipital, 8 cervical, 12 thoracic, 5 lumbar, 5 sacral, and 8–10 coccygeal. In most of us, the coccygeal somites involute, leaving us with 1–3 remnants, but on occasion, persistence of coccygeal somites presents as a stubby tail. Contributions of occipital somites to the occiput have been mapped out by Huang (Fig. 10.27).

Descriptive Anatomy

Bookmark these fundamental evolutionary facts:

- All somites from S4 backward express Hox3B; S1–S3 do not.
- Fishes have three occipital somites; tetrapods have four.
- The original first spinal vertebra in basal tetrapods is proatlas, S4$_C$ + S5$_R$.
- Fragmentation of proatlas contributes to cranial base and is responsible for the formation of the atlas–axis complex.

As previously stated, the composition of occipital bone has changed through the course of evolution. In its original piscine version, it was constructed from three occipital somites. In its current mammalian version, four occipital somites posterior fossa contain important contributions from proatlas, a bone not present in modern tetrapods. The anat-omy of the proatlas is absolutely key to the understanding of the modern occipital bone and OCJ. Proatlas is synthesized from caudal S4 and rostral S5. We will use the following type of shorthand: S4$_C$ and S5$_R$.

We will start with some generalities about the chondral occipital bone, emphasizing its three components: basioccipital, exoccipital, and supraoccipital. For a more in-depth description, including the multiple components of the membranous occipital bone, the reader is referred to Chaps. 8 (bones) and 12 (dura) (Fig. 10.28).

Sclerotomal fusion around the notochord is the earliest manifestation of the vertebral body. The standard pattern involves the unification of the upper half of one sclerotome (caudal) with the lower half of its neighbor (cranial). This pattern begins with S4–S5. We have previously emphasized that the four sclerotomes behave very differently: their *central elements* fuse completely, while their *lateral elements* maintain a separation.

Conceptually, it makes sense that the first priorities of the developing brain are to protect itself. It does so by means of blood supply and an underlying cartilaginous cranial base. At stage 9 (day 20) the first three somites are visible; the appearance of S4 is the definition of stage 10 (day 22). Unification of S1–S4 results in a single midline basioccipital bone, two lateral exoccipital bones, and the posterior rim of the foramen magnum. The anterior rim originates from elsewhere (the proatlas). Vascularization of the occipital bone from the basilar-vertebral system occurs at the same time. The caudal occipital segment is demarcated by the hypoglossal artery and first cervical artery.

Changes in the nervous system are taking place rapidly at this time. Stage 10 is also characterized by the formation of the neural tube, a process involving the fusion of the neural folds. *Neurulation is initiated at the fourth somite;* it proceeds *forward* from the occipitocervical junction. During weeks 5 and 6, the CNS differentiates into its component parts. The fourth ventricle roof thins out, giving rise in the midline to the foramen of Magendie and, laterally, to the foramina of Luschka. The fourth ventricle is now in communication with the subarachnoid space and CSF circulation begins.

Since the primordial function of the skull is the protection of the brain, chondrocranium is constructed first, during stages 8–13. Recall that all cranial base cartilages are neural crests, with exception of the otic capsule and the occipital bone complex. This process is described in Chap. 8.

After creating a supporting floor of cartilage, the brain seeks to protect its sidewalls with membranous bones. The process of membranous ossification occurs quickly because it does not require a cartilaginous intermediate. Membranous bone formation is completed by stages 16–17 (38–40 days). Occipital sclerotomes are in register with the segmental components of hypoglossal nerve. The nerves pass through the lateral loose zone of the sclerotomes as follows: the first root emerges between S1 and S2, the second root emerges between S2 and S3, the third root emerges between S3 and S4, and the last root defines emerges below the occiput.

The membranous and chondral components of the occiput have been *erroneously referred to as a single bone*, supraoccipital. As discussed previously, the upper membranous zone represents the four interparietal bone fields while the lower chondral zone of true supraoccipital also has four zones. IP and SO are divided by several landmarks. *External occipital protuberance* is readily palpable; it lies midway between the foramen magnum and the apex of the lambdoidal sutures. From this central point, two curvilinear lines arch outward (they are only visible on the dry skull). Uppermost is the highest nuchal line and, below it, the superior nuchal line. The importance of these lines has to do with the muscle attachments associated with them. *Highest nuchal line* bears along its lateral margin, the caudal attachment of the epicranial aponeurosis. This marks the most posterior insertion of the second arch superficial investing fascia and the caudal attachment of occipitalis. *Superior nuchal line* bears the proximal attachment of two muscles connecting the skull with the shoulder girdle. Along its medial half is the trapezius while the lateral half bears sternocleidomastoid. These muscles belong to the lateral motor column and develop later in time in comparison with muscles of the suboccipital triangle. They therefore must assume attachments at the distal margin of the chondral supraoccipital bone. Just below the sternocleidomastoid lies the distal attachment of splenius capitis. Running vertically downward from external occipital protuberance is the median external occipital crest. At the midpoint of this vertical line one encounters another transverse landmark, the *inferior nuchal line*. Two shallow fossae lie between superior and inferior nuchal lines. These are filled up with the broad distal attachments of semispinalis capitis and obliquus superior. Between inferior nuchal line and foramen magnum lies another set of fossae. These contain medially, rectus capitus posterior minor and more laterally, rectus capitus posterior major.

What we can glean from this anatomy? Three functional groups of muscles, arising from deep to superficial, achieve distal insertion into the supraoccipital/interparietal bone fields at progressively higher (and more superficial) levels. The four *suboccipital muscles* all have proximal attachments from the atlas or axis. All are innervated by dorsal ramus of C1. Three of the four fan upward and outward, seeking fame and fortune along the "occipital Riviera" *below inferior nuchal line* in the following sequence: rectus capitus posterior minor > rectus capitus posterior major > obliquus capitis superior. The Lone Ranger of this group is obliquus capitis inferior; its spans from the spinous process of the axis to the transverse process of the atlas. (2) Semispinalis capitis is supplied by descending branch of greater occipital nerve,

dorsal C2, and dorsal C3. It is forced to attach *above inferior nuchal line*. A large muscle, it monopolizes the entire fossa all the way up to superior nuchal line. (3) Splenius capitus belongs to the most superficial quartile of the erector spinae muscle and is innervated by dorsal C2–C3. As such, it is forced to attach to the rough surface of occipital bone just below the sternocleidomastoid and, more laterally, to the mastoid process. The attachment of epicranius lies outside the chondral supraoccipital altogether. As a purely neural crest fascia, the SIF naturally relates to the purely neural crest membranous supraoccipital bone.

Exoccipital bones flank foramen magnum. They likely develop from serial contributions from S2 to S3. These quadrilateral flat bones are known as the jugular tubercles because they make up the posterior margin of the jugular foramen. From the ventral neural arches ($S4_C$) of proatlas, the exoccipital bones acquire occipital condyles. These articulate with the superior articular facets of the transverse processes of the atlas. Anatomic confirmation of this is seen in the alar ligaments that unite the medial tubercle of each occipital condyle to the tip of the dens. The roughened area of the jugular process gives a distal insertion to the rectus capitis lateralis (supplied by ventral C1–C2) that spans upward from transverse process of atlas.

The basilar part of occipital bone shares a growth center with basisphenoid, which ossifies at age 25. The inferior margin of the foramen magnum comes from proatlas; it bears a tubercle that provides the suspension for the fibrous pharyngeal raphe. Just anterior to the occipital condyle is rectus capitis anterior (ventral C1–C2). Further forward and lateral to the tubercle is the proximal attachment of longus capitis (ventral C1–C3).

Mechanism of Development

Posterior fossa is constructed in mammals from four occipital somites plus the contribution from the rostral half of the first cervical somite, $S5_R$. Occipital somites have the same component parts as their truncal counterparts but differ in behavior in several ways. Their topology arrangement is different. Instead of being separated by intervertebral discs, they are stacked together in succession, like Russian dolls to surround the spinal cord, as demonstrated by mapping work of Patel and Huang. Their dermatomes, initially present, involute. Their myotomes are strictly hypaxial, producing hypobranchial muscles dedicated to the tongue. Although they have rostral loose zones and caudal dense zones the axial components fuse into a solid block of bone whereas the lateral zones contain foramina that are limited to loose zones, through which cranial nerves pass to the periphery. Jugular foramen contains two cranial nerves, IX and X, and a spinal nerve, representing roots c1–c6, that is mislabeled as cranial nerve XI.

For these reasons, a standard model of parasegmentation does not apply well to the occipitovertebral region. Gray's Anatomy states the following:

- Occipital sclerotomes 3 and 4 are most distinct by stage 14, by which time the first three sclerotomes have fused. Vertebrae are formed from the fifth somite caudally: the first cervical vertebra is formed by the caudal half of occipital somite 4 and the cranial half of cervical somite 1. This shift in somite number accounts for the production of seven cervical vertebrae from eight cervical somites.

Seems plausible, doesn't it? Certainly, it resolves the age-old student's dilemma, *why are eight cervical nerves but only seven cervical vertebrae?* Unfortunately, this neat picture is flawed.

O'Rahilly and Müller to the Rescue

Occipitocervical development is best explained by the embryologic model of O'Rahilly and Müller, which takes into account differences in behavior between the centra and vertebral arches. Their conceptual breakthrough resulted from an analysis of the early development of the occipitocervical region not in terms of sclerotomes alone, but by taking into account differences in genetic expression between the centra and the vertebral arches. These translate into differences in behavior. All somites from *S4 caudal express Hox3B; S1–S3 do not.*

Models have to be predictive. The parasegmentation model fails to account for several anatomic facts. First, the basioccipital bone has *two un-named foramina* through which two of the three neurovascular pedicles of hypoglossal nerve make their exit. The third pedicle emerges through a separate *hypoglossal canal* between basioccipital and exoccipital bones. Second, atlas and axis are obviously different from all other vertebrae. Why is the atlas only a ring? Are the dens and the odontoid process one and the same? (3) What is the significance of the *occipito-dental ligament* (apical ligament of the dens)? Third, the explanation seems to fit mammals fairly well but how can we use it to explain the anatomy of the CVJ in other life forms, that is, fishes and early tetrapods?

O'Rahilly and Muller separated out occipital and cervical somites into their medial and lateral components. From a dorsal perspective, they created a map of the CVJ in which the centra are placed in the middle and the neural processes are placed laterally (cf Fig. 10.13).

Recall that:

- *ventral sclerotome* = perinotochordal sheath, that is, the heart of the centrum;
- *central sclerotome* = ventral neural arch, pedicle, and proximal rib;

- *dorsal sclerotome* (late developing) = dorsal neural arch;
- *lateral sclerotome* = distal rib.

When seen from this new perspective, lateral elements display a normal process of parasegmentation. Central elements, being genetically distinct, are free to behave differently (and do so). For this reason, the axial elements of S1–S3 fuse together. Parasegmentation at the axial zone first begins with S4. When S4 is incorporated into the tetrapod skull, its anterior loose zone combines with S3 which is strictly a loose zone. As a result, a mesenchymal gap demarcates the S3–S4 boundary.

The centra of the rostral 3 cervical somites (S5–S7) were designated by O'Rahilly and Muller as X, Y, and Z. Although equal in size to those of somites 8–12, X–Z display unusual behaviors, most (but not all) are explained by this model. At stage 17, X and Y are fused while an intervertebral disc exists (transiently) between Y and Z. By stage 21, Y and Z are fused. Thus, the dens consist of three centra, all of which belong to the axis. Using our terminology, the vertebral body of axis = c1 + c2 + c3.

In standard parasegmentation theory, the neural arch is composed in just the same way as the body. The cranial part is dense; it comes from the caudal zone of the previous somite. The caudal part is loose; it arises from the cranial zone of the same sclerotome.

Atlas is explained as follows. It has "given up" its body (the pleurocentrum) to the axis. The anterior arch remains unexplained. The posterior arch of atlas is quite narrow. It develops from the dense area of sclerotome 5; there is no contribution from the loose area of sclerotome 6.

Axis develops as follows. Its body consists of cranially projecting dens (X + Y) plus Z, X = pleurocentrum of atlas, Y = intercentrum of axis, and Z, the true centrum of axis = the dense zone of sclerotome 6 + the loose zone of sclerotome 7.

Nota bene these caveats to the FMROR model: (1) Terminology referring to the centra as X, Y, and Z is paleontologically fuzzy. Recall that the "centrum" previously consisted of a rostral intercentrum and a caudal pleurocentrum. It is easy to see how the "body" of the ancient proatlas could be split apart. (2) Soft tissue structures such as ligaments are *not* a part of the FMROR analysis. Paired *alar ligaments* connect dens to the medial aspect of occipital condyles. These are 11 mm long, thick, and stout. They serve a "check rein" function to prevent excessive rotation. The *apical ligament of the dens* starts from the anterior margin of the foramen magnum, its fibers coalesce to attach at the tip of the dens. It is flanked on either side by the alar ligaments. *This ligament represents the continuation of the notochord from the skull base into the vertebral bodies.* It is a direct product of the ventral sclerotome of the first cervical somite.

Now, for a moment, consider two fundamental differences between fishes and tetrapods. The fish skull is connected to the pectoral girdle…the head and trunk are welded together. Axial stability is provided by the notochord. It runs continuously from the skull all the way down the body. We'll explore these ideas later on; suffice it to say that the earliest tetrapods, such as *Acanthostega* and *Ichthyostega*, demonstrate two revolutionary ideas, both of which are crucial for the creation of a neck: (1) disconnection of the head from the body, leading to greater mobility; and (2) a redesign of the first vertebra into a cervical vertebra permitting, for the first time, head rotation (Figs. 10.29, 10.30).

Recent labeling experiments tracking neural crest behavior indicate that the head–trunk interface in mammals does *not* lie at the occipitocervical boundary, but rather lies at the somite 3/4 level. This reflects the more ancient posterior occipital boundary. *In fishes, only the first three somites contribute to the occipital bone.* Furthermore, somites 1–3 develop normally. They all have cranial loose zones and caudal dense zones. They all have the expected proportions of sclerotome and dermomyotome compartments. Neural crest cells arising from rhombomere opposite somites 1–3 contribute to the posterior pharyngeal arches. Despite the fact that S1–S3 all possess posterior somitic "barrier" zones, crest cells seem to ignore them and proceed blithely forward. Neural crest cells arising opposite somites 4–5 do not contribute to the head. Instead, they behave in a "truncal" fashion, migrating ventrally, in a selective fashion, only through the "permissible" anterior loose zones.

Motor nerves, when they encounter occipital somites, behave differently. They proceed through loose zones only which correspond to developmental field boundary zones. Hypoglossal behaves in a "truncal" fashion, just as if it were a spinal nerve. Thus, while hypoglossal neurons distinguish between the anterior and posterior zones of S1–S3, neural crest cells do not.

Why should neural crest behavior change so radically at the somite S2–S3 boundary? All somites from the fourth somite onward, express Hox3B. Thus, differences in migratory behavior of anterior occipital and posterior occipital neural crest are not intrinsic to the crest cells but are dictated by the genetic environment of the individual somite.

In sum: the head-trunk interface lies at the somites 3–4 border. This is aligned with the head-trunk boundary of lower vertebrates. In both teleost fishes and amphibians, only the first three occipital somites make up the occipital bone. Evolution of amniotes involves the recruitment of an additional somite to make up the occipital bone but this incorporation was *not* accompanied by an expansion of the occipital NC domain.

Evidence of Intervertebral Discs in the Cranial Base

Recall those occipital somites have cranial loose zones and caudal dense zones; the boundary between them is Von Ebner's fissure (AKA…you guessed it…the somitocoele). Thus, when somites recombine to make vertebrae, they "split

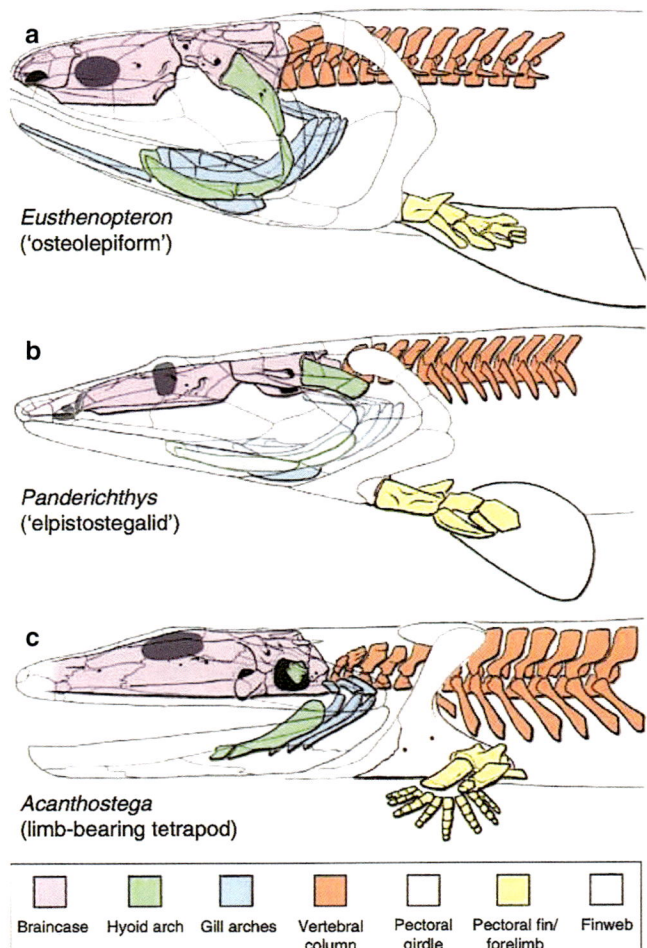

Fig. 10.29 Mid-level sarcopteryigian became tetrapodomorphs. Tristichopterid fish, the oteolepiform *Eustenopteron* has dermal bones of pectoral girdle (white) attached to the skull at post-temporal bone. These arch backward from spinal 1 to spinal 6. Mesodermal pectoral girdle. The primitive chondral scapulocoracoid (pink) connects to the pectoral fine (yellow). The crown level tetrapodomorph, the elpistostegalid, *Panderichthys*, is very close to the earlies tetrapod, Tiktaalik. It has a humerus and the earliest evidence of bone marrow. In Acanthostega the pectoral girdle is separated from the skull, its dermal bones are simplified, and the chondral bones are enlarged. [Reprinted from Benton MJ (ed). Vertebrate Palaeontology, fourth Edition. Oxford, UK: Wiley Blackwell; 2014. With permission from John Wiley & Sons]

apart" along this fissure. The intervening space becomes the disc.

At first glance, occipital somites have all the equipment to make discs but fail to do so. Basioccipital bone appears as a monotonous solid block, with tell-tale lateral hypoglossal foramina. The first attempt to make a disc is between Y and Z of the axis. This disc is unstable; its brief lifetime lasts from stage 17 to stage 21. The first permanent intervertebral disc appears between the axis and the third cervical vertebrae. What's going on in the skull base?

Let's take the Muller–O'Rahilly model and apply it, first to fishes and later to mammals. FMOR established differences in biological behavior between the central elements and lateral elements of sclerotomes. This idea is crucial: it

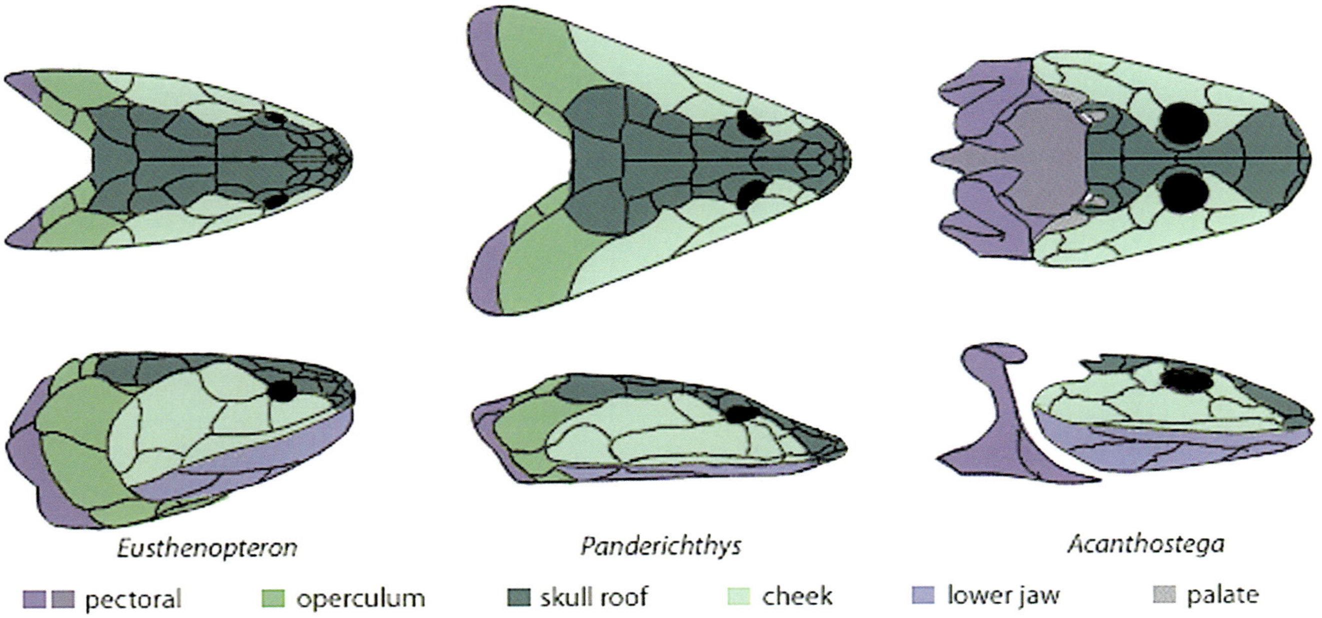

Fig. 10.30 Separation of the pectoral girdle *Eusthenopteron* has opercular series (medium green) attached to cheek series (light green) and to skull roof (dark green). Dorsal view shows, just behind the parietals (in dark green) the post-temporals (medial) and tabulars (lateral). Posterior and internal to the operculars are medial and lateral extrascapulars (medium green). *Panderichthys*: Pectoral series (purple) remains attached to operular series. *Acanthostega*: disconnection of pectoral series from posttemporal. Opercular series is gone. [Courtesy of Dennis C. Murphy]

allows us to reconstruct the anatomy of the basioccipital bone and occipitocervical joint in a way that accounts for all these elements. Recall that the cranial loose zone of the first occipital somite is fused with somitomere 7.

The *central elements of S1–S3 do not have genes permissive for segmentation*. Thus, the bodies and neural arches of $S1 + S2 + S3_R$ are fused. The *lateral elements of S1–S3 do possess genes permissive for segmentation*. Although the occipital sclerotomes do not have facet joints, they interact laterally with spaces that permit passage of hypoglossal neurovascular pedicles. Thus, *genes permissive for segmentation are fully operative in the dorsal and lateral sclerotomes from S1 onward*. Consequently, the pedicles of S1, S2, and cranial S3 are separate. They act like facet joints. The resulting spaces provide exit for the first two branches of the hypoglossal…the remaining two branches are truncal.

Phylogeny of the Occipital Bone

Fishes use the first three somites to construct the braincase. *Basioccipital* = $S1 + S2 + S3_R$. *First truncal vertebra* = dense $S3_C$ + loose $S4_R$ Note that the S1–S3 are Hox3B negative, do not have parasegmentation, and are fused. Escape routes for hypoglossal nerve are occult boundary markers between these somites.

Basal tetrapods incorporate the rostral loose zone of the fourth somite into the braincase. This occurs probably at the level of *Tiktaalik* and is associated with the separation of the pectoral girdle (Fig. 10.31). Note that pectoral separation does not take place *Basioccipital* = $S1 + S2 + S3 + S4_R$. *First truncal vertebra (proatlas)* = dense $S4_C$ + $S5_R$. No axis is present. Transposition of the fourth somite from a trunk to the braincase resulted in (1) a more capacious posterior fossa; and (2) the incorporation of two spinal nerves, accessory and hypoglossal, into the braincase.

Advanced tetrapods (both lines, the temnospondyls and the anthracosaurs) incorporate proatlas *partially*, probably as ventral border of foramen magnum. Atlas does not have a ring. Axis is created but has no dens. The condyle is single and ventral.

Amniote synapsids (therapids) have more extensive incorporation of proatlas elements developing a double condyle from loose S5 (proatlas) lateral to the foramen magnum.

Mammals have fragmentation and remodeling of the primitive axis. Final incorporation of proatlas is occiput is the ligament of dens. The central elements of S1–6 are thus characterized by direct bone-to-bone continuity. Fusions. Dense S1–loose S5 = basioccipital bone. Dense S5 = rim of foramen magnum, medial condyles, and apical ligament of the dens. Loose S6–loose S7 = dens/body of axis. Dense S7–loose S8 = third cervical vertebra:

$$\text{Proatlas} : \text{C0} = S4_C + S5_R$$

Proatlas is first vertebral body in early tetrapods during the Devonian period. No axis exists; *all subsequent cervical vertebrae are alike*. Proatlas was incorporated into the skull in subsequent anthracosaurs, such as Gephyrostegus, by which time the second cervical vertebra is transformed into an identifiable axis (although very different in form from its mam-

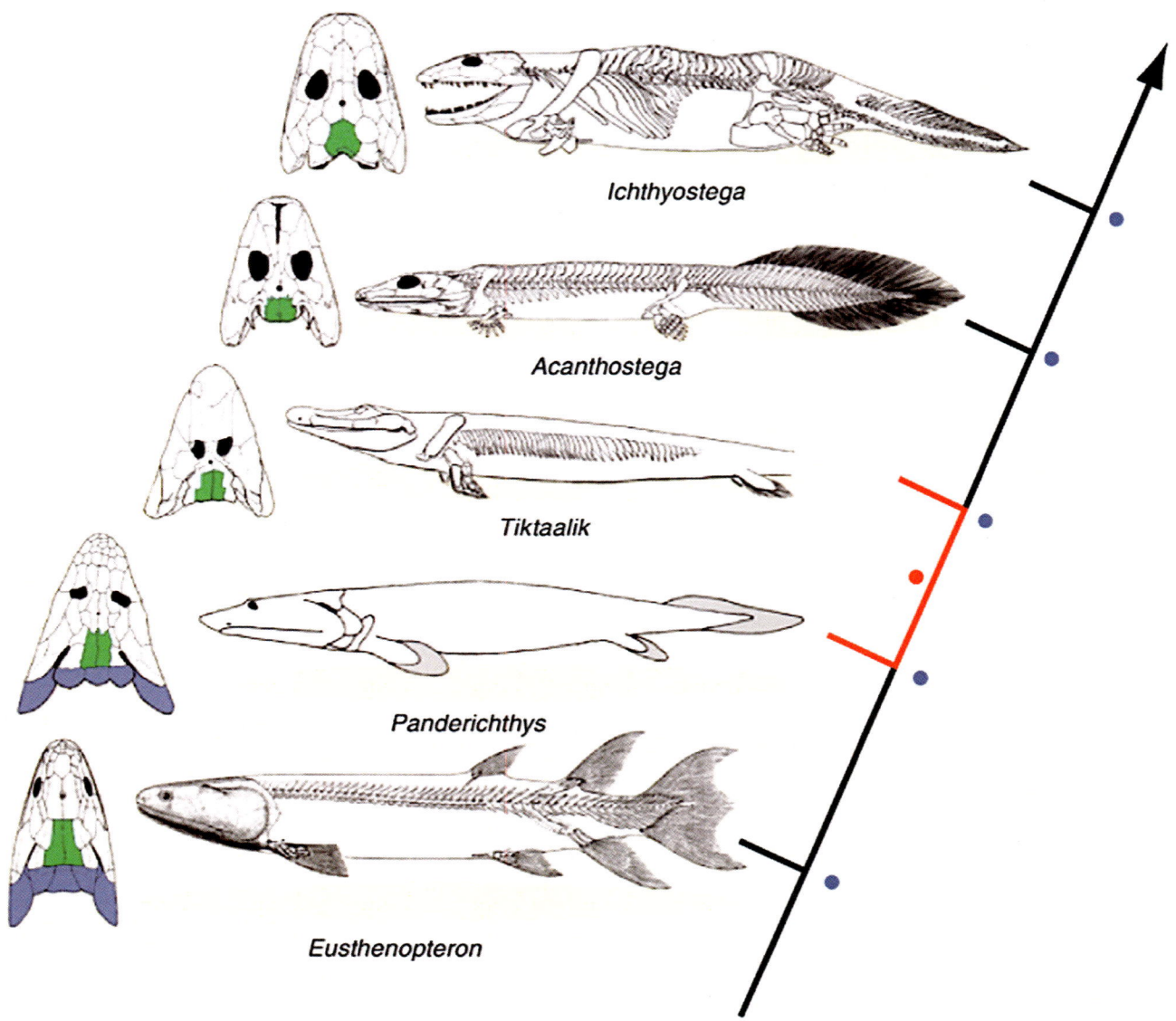

Fig. 10.31 Phylogeny of the pectoral girdle: relation to vision? Note between *Panderichthys* and *Tiktaalik* loss of extrascapular series (blue) on the skull side and on the pectoral girdle, loss of post-temporal, supra-cleithrum and postcleithrum (anoclaiethrum). [Reprinted from Ahlberg PE, Clack J Paleontology: A firm stem from water to land. Nature 2006; 440(7085): 747–749. With permission from Springer Nature]

malian descendent). The ancient proatlas has been seen in mammals in two forms: (1) Hox gene manipulation in mice causing a "frameshift" mutation; and (2) congenital abnormalities. Proatlas has a "body" consisting of 6 elements: two anterior intercentra, two posterior pleurocentra, and paired dorsal neural arches. The intercentra develop from caudal S4 while the pleurocentra come from rostral S5. The dorsal neural arches of proatlas developed in the same manner. Projecting from the base of the neural arch were lateral masses containing posteriorly-directed articulatory surfaces; these formed a "standard" joint with the subsequent vertebra.

Proatlast remnants exist in a variety of forms [5]. The original form of proatlas without axis is reproduced in the Truang–Goehmann malformation [6] (Fig. 10.32).

In the anapsid reptilomorphs. a fragment of proatlas neural arch is left behind, positioned dorsal to the atlas. This situation restricted movement. This remnant of proatlas persisted all the way into synapsids and disappear with mammals. The remaining fragment of proatlas was lost in mammals with a resultant gain in mobility for the neck.

Here is the final disposition of proatlas: it breaks up in three ways. (1) The centra *split off from the body*, leaving behind a ring-like neural arch. (2) Intercentra and pleurocentra separate along the coronal plane. The intercentra become the anterior tubercle of the clivus. The pleurocentra enclose the notochord as the apical ligament of the dens (occipitodental ligament); they also form the tip of dens. The surviving centrum persists as the hypochordal bow. (3) The ventral and dorsal halves of the neural arch and "pedicles" also separate along the coronal plane. The rostral components (S4) are reassigned *forward and ventral* while the caudal components (S5) go *backward and dorsal*. Rostral neural arch forms the

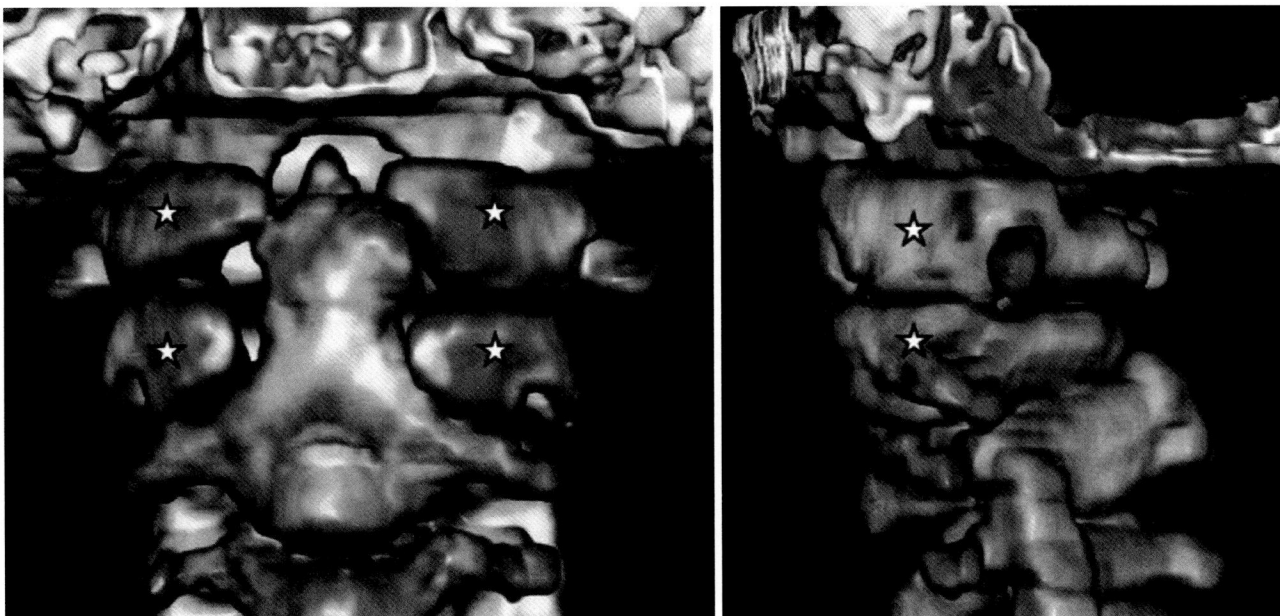

Fig. 10.32 Persistent proatlas. Dens is unfused with the vertebral rings of proatlas (*) and atlas (*). Os dontoideum is normal, presented as triangula nubbin. Lateral lateral view shows two separate bodies (*) [Reprinted with permission from Spittank H, Goehmann U, Hage H, Sacher R. Persistent proatlas with additional segmentation of the cranio-vertebral junction: the Tsuang-Goehmann malformation. Radiol Case 2016; 10(10):15–23. https://www.radiologycases.com/index.php/radiologycases/article/view/2890]

U-shaped anterior rim of foramen magnum. The rostral lateral masses fuse with the exoccipital bones to form the true occipital condyles. Alar ligaments connect these structures with the dens. Caudal neural arch forms the posterior arch of atlas (C1) and the alar and crucial ligaments. The dorsal lateral masses become the lateral masses of the atlas.

The secession of the proatlas centra from its body produces the S5 tip of dens. This will fuse with central elements of atlas and with body of axis to produce the odontoid process. The new "slimmed down" version of the proatlas body is known as the *hypochordal bow*. In general, it disappears along with the rest of proatlas. But at times the hypochordal bow of proatlas will persist. In such cases, it fuses with the anterior arch of the atlas. This produces an abnormal articulation between clivus, anterior arch of atlas, and the odontoid apex (Figs. 10.33, 10.34).

$$\text{Atlas} : \text{C1} = \text{S5}_\text{C} + \text{S6}_\text{R} \text{ a modified ring}$$

Atlas is, in reality, the second vertebral bone of the neck. Designed to permit flexion/extension of the head, atlas is a very "odd duck." Embryologically complex, atlas is synthesized in a regular manner from caudal S5 and rostral S6 but also with contributions from proatlas. The atlas has the forma of an ovoid ring. It consists of two lateral masses connected by anterior and posterior arches. The lateral masses are bulkier anteriorly; they bulge into what would otherwise be a smoothly oval vertebral canal. This makes the anterior arch appear smaller. The inner "shoulders" of the lateral masses are connected by a transverse ligament. This ligament divides the vertebral canal into two compartments. The anterior third is occupied by the dens while the posterior two-thirds contain the spinal cord and its coverings. Both dens and cord are of similar diameter (Fig. 10.35).

Atlas lacks a true body. This odd situation seems more rational if we consider that its genetic components behave similarly to those of somites 1–5. That is, *central elements maintain allegiance to one another and remain fused, while the lateral elements remain separate*. The centra of the atlas behave in exactly the same manner as those of the proatlas: they also secede from the body of C1. Thus S5_C–S6_R body of the atlas fuses above, with S5_R body of proatlas, and below, with the S6_C–S7_R body of the axis. The lateral masses of the atlas are proatlas remnants; these come from S5_R neural arch. They bear two sets of articulations directed above, toward the occipital condyles, and below, toward the axis. The transverse processes of the atlas are long, almost equal to those of C7. They serve as leverage points for muscles controlling head position. The neural arch of the atlas is derived in a very standard way from S5_C and S6_R. In conclusion, we can see that both *proatlas and atlas share very similar anatomy*.

Ossification centers of the atlas are four. Those of the lateral masses appear at the seventh week. It spreads outward and backward along the neural arches; these unite at 3–4 years. Ossification of the anterior arch follows a slower course with paired midline centers (mistaken for a single center) appearing at about a year. At times ossification spreads forward from the lateral masses, another manifestation of the bilateral nature of the anterior arch.

Muscle attachments follow from key structural landmarks of the atlas. As expected, they follow the ossification sequence.

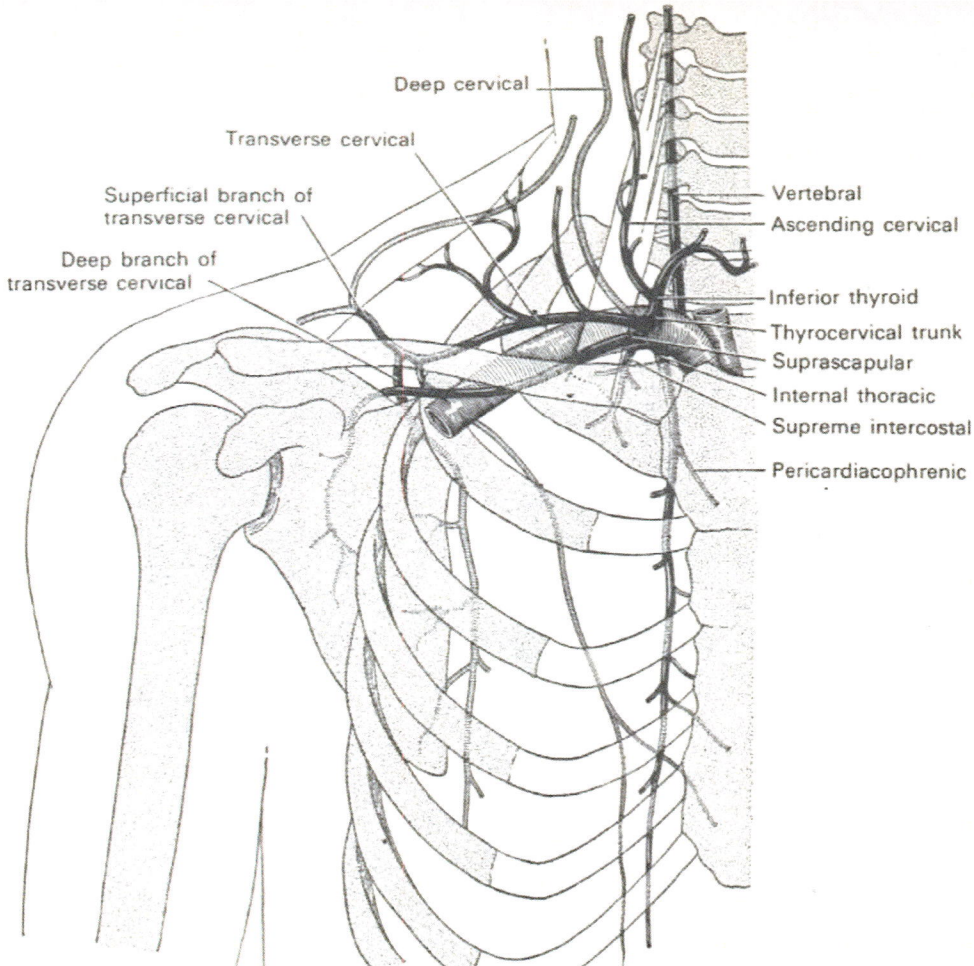

Fig. 10.33 Apical and alar ligaments represent loose zone of proatlas (S5$_R$). Exposure of the cruciform ligament after removal of posterior arches of atlas and axia removed andupward reflection of the tentorial membrane These span from dense zone of proatlas (S5$_C$) to the anterior rim of foramen magnum which may represent S5$_R$ contribution or dense zone of fourth occipital somite (S4$_R$). The centra of proatlas split up in the coronal plane. Intercentrum becomes anterior tubercle of clivus. Pleurocentra become the apical ligment of the dens and the dens apex. The neural arch of proatlas separates from the centra and breaks into to U-shaped segments. Rostral neural arch goes ventral and contributes to anterior rim of foramen magnum. Caudal neural arch goes dorsal to form posterior arch of the atlas. [Reprinted from Lewis, Warren H (ed). Gray's Anatomy of the Human Body, 20th American Edition. Philadelphia, PA: Lea & Febiger, 1918]

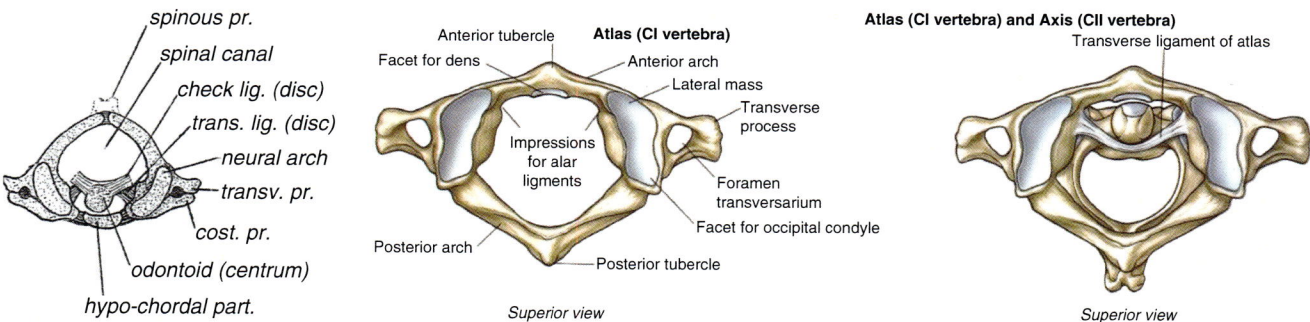

Fig. 10.34 S1–S3 Mammalian atlas contains element of proatlas. The neural arch of proatlas splits in the coronal plane. Centrum breaks into three segments. Rostral intercentrum becomes anterior tubercle of clivus. Caudal pleurocentrum forms dens and the apical liament. The remaining fragment of centrum is the hypochordal bow. Neural arch of proatlas splits as well. Rostral neural arch shifts ventral to form anterior rim of foramen magnum (not seen here). Caudal neural arch shifts dorsal to form posterior arch of atlas. Left: [Reprinted from Keith A. Human Anatomy and Morphology, fourth ed. New York: Longmans, Green & Co; London: Edward Arnold, 1921.] Right: [Reprinted from Drake R, Vogel AW, Mitchell AWM. Gray's Anatomy for Students, third edition. Philadelphia, PA: Churchill-Livingstone. 2015. With permission from Elsevier]

The *anterior tubercle* on the anterior arch gives proximal attachment for longus colli. *Anterior surface of lateral mass* is the primary attachment for the upward-directed rectus capitis anterior. Flanking *posterior tubercle* are rectus capitis posterior minor. *Transverse process* bears rectus capitis lateralis superiorly. Behind that is obliquus capitis superior. The very *apex of tranverse process* bears a muscle going up, obliquus caitis inferior, and three muscles heading downward: levator scapulae, splenius cervicis, and scalenus medius.

Caveat   Recall that in mammalian evolution, the pleurocentra become the dominant component of the vertebral body.

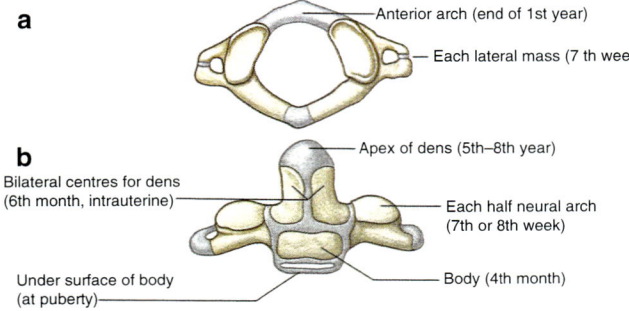

Fig. 10.35 Ossification of atlas and axis. By the same logic Axis has 8 centers with tips of dens and body coming from two sources each. The presence of these centers in the axis [Reprinted from Standring S. In: Gray's Anatomy, 40th edition. Philadelphia, PA: Churchill-Livingstone; 2008. With permission from Elsevier]

Intercentra are relegated to a low-status knob (parapophysis) wcapitulum) of the rib. We have to be careful about our terminology here. Previously, we used the anatomy of primitive tetrapod vertebrae to make a point. These creatures had bodies in which intercentra and pleurocentra were co-equal in size. We took advantage of this to point out the relationship to parasegmentation, in which pleurocentra were derived from the loose zone of the somite in question whereas intercentra originated from the dense zone of the next more cranial somite. In modern mammals, although the body is just pleurocentra that do not mean it arises from a single somite… the separate genetic centers behind parasegmentation are alive and well in twenty-first-century mammals:

$$\text{Axis}: C2 = S6_{C+}\, S7_R \text{ a pivot post}$$

Axis is designed to permit rotation of the head. It develops exclusively from sclerotomal elements of caudal S6 and cranial S7 but includes additions from proatlas and atlas. The body of the axis is the sum of the central sclerotomes (from which project the ventral neural arches) and the ventral sclerotomes (which enclose the notochord). Sitting atop the body is the odontoid process (dens), the formula for which is $S4_C + S5_R + S5_C$. These are completely fused. At birth, a cartilaginous process representing an abortive attempt to create an intervertebral disc is seen between the odontoid process and axis body, that is, between $S5_C$ and $S6_R$. This

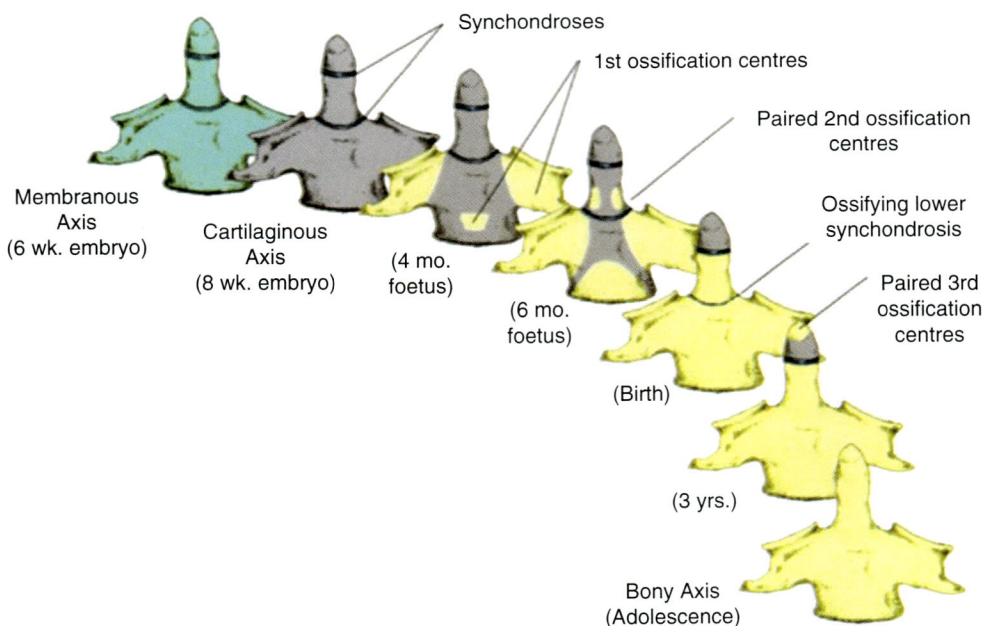

Fig. 10.36 Development of the axis. The axis is a composite structure: $S5_C$ (odontoid process of dens) + $S6_R$ (dens) + $S6_C$ (synchondrosis) + $S7_R$. The three developmental phases of the axis (C2) and the three waves of ossification. The primordia for the dens components are assembled during the membranous phase. Upper and lower dental synchondroses are shown as dense lines. First wave of ossification at fourth foetal month consists of bilateral centres for the neural arches and a single centre for the centrum. Second wave at sixth foetal month consists of bilateral ossi-fication centres for the basal dental segment. At birth, the basal dental centres should have integrated in the midline and begun to be fused to the centrum. Third wave of C2 ossification occurs from 3 to 5 years post-natal life at the apical dental segment, which does not become fused to the basal dens till the 6–ninth year, and fully formed during adolescence. [Reprinted from Pang D, Thompson DNP. Embryology and bony malformations of the craniovertebral junction. *Childs Nerv System* 2011; 27:523–564. With permission from Springer Nature]

synchondrosis is present in most children up to age for but disappears by age 8 (Fig. 10.36).

The axis has primary and secondary ossification centers. Like any cervical vertebra, each vertebra arch has two centers; these appear in the seventh week. In the fourth–fifth month, ossification of the central is observed. This starts at the midline from paired centers (erroneously described as being single). The odontoid components from the axis show up at six months as paired centers. These generally fuse before birth. The odontoid components from proatlas demonstrate ossification between 5 and 8 years with fusion to the remainder of the dens by 12 years. Abnormalities of ossification can involve the apex with dens. If the odontoid process fails to fuse with the body the result is *os odontoideum*.

Anterior body of axis provides attachment for the longus colli. The *tips of the transverse process* are quite busy, containing proximal attachments for ventral scalenus medius and dorsal splenius cervicis and levator scapulae. Spanning between the tips are intertransversus muscles. The projection of the spinous process gives attachment first, to obliquus capitis inferior and further posteriorly, to rectus posterior major.

CVJ Derivatives: A Summary (Fig. 10.37)

Somites 1–3 (alpha unit)

- basioccipital bone,

Somites 2–3

- exoccipital, including jugular tubercles.

Somite 4 (beta unit)

- Centrum ($S4_C$) is *incomplete* = posterior basioccipital, anterior tubercle of clivus.

Somite 5

- Intercentrum ($S5_R$) = apical, transverse ligaments of dens;
- Pleurocentrum ($S5_C$) = dens (apex), apical ligament.
- Hypocentrum (what remains) = hypochordal bow.

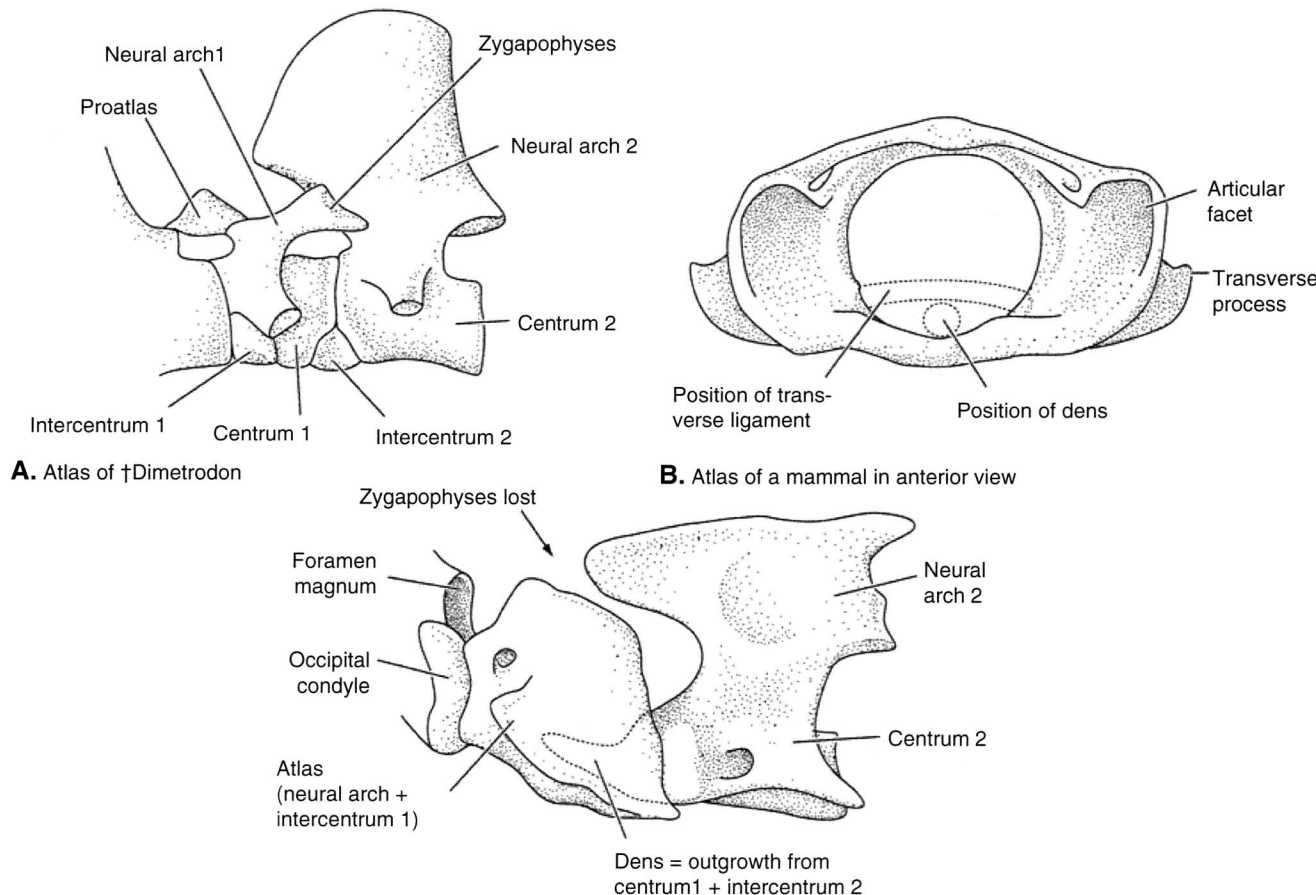

A. Atlas of †Dimetrodon

B. Atlas of a mammal in anterior view

C. Atlas and axis complex of a mammal in lateral view

Fig. 10.37 Assembly of the atlas and axis Anapsid reptilomorph *Dimetrodon* (just before split into diapsid and synapsid lines) had two cervical vertebrae but no atlas and no axis. Centrum 1, intercentrum 1, and neural arch are separate and proatlas persists. Mammals shows fusion between intercentrum 1 and neural arch 1. Centrum 1 intercentrum 1 separate from atlas and fuse. They join with centrum 2 the body of the axis to become the tripartite dens. [Courtesy of William E. Bemis]

- Neural arch rostral (S5$_R$) shifts ventral = anterior U-shape to foramen magnum, occipital condyles, alar, and cruciate ligaments.
- Neural arch caudal (S5$_C$) shifts dorsal = posterior arch of atlas (rostral), lateral atlantal masses.

Somite 6

- Hypocentrum = anterior arch of the axis.
- Centrum = dens (body).
- Neural arch = posterior atlas arch (caudal).

Somite 7

- Hypocentrum = disappears.
- Centrum = (S7$_R$) = body of axis.
- Neural arch = posterior arch of the axis (rostral).

Cervical Vertebrae C3–C7: garden variety

We don't have much to report here. It is as if, from c4 downward, nature simply repeats herself. The benefit for mammals is an increase in mobility. It's not fair to compare mammals with birds because avian cervical vertebrae are *amphicoelous*, that is, they are double, opposing saddle joints… not unlike the human thumb. In humans, the transverse process is well out of the way of the *foramen intertransversarium*…let's give preference to the vertebral arteries. These vertebrae do possess a coastal process but, under normal circumstances, it aborts. The distal parts of the cervical processes do not develop. But we are all aware of the so-called "cervical rib." This is especially common at the level of c8 (the seventh cervical rib). Anomalies of lower cervical ribs occur via inappropriate cervical fusion: *Klippel-Feil syndrome*. The phenotype includes a low hairline, foreshortening of the neck, and reduced movement (Fig. 10.38).

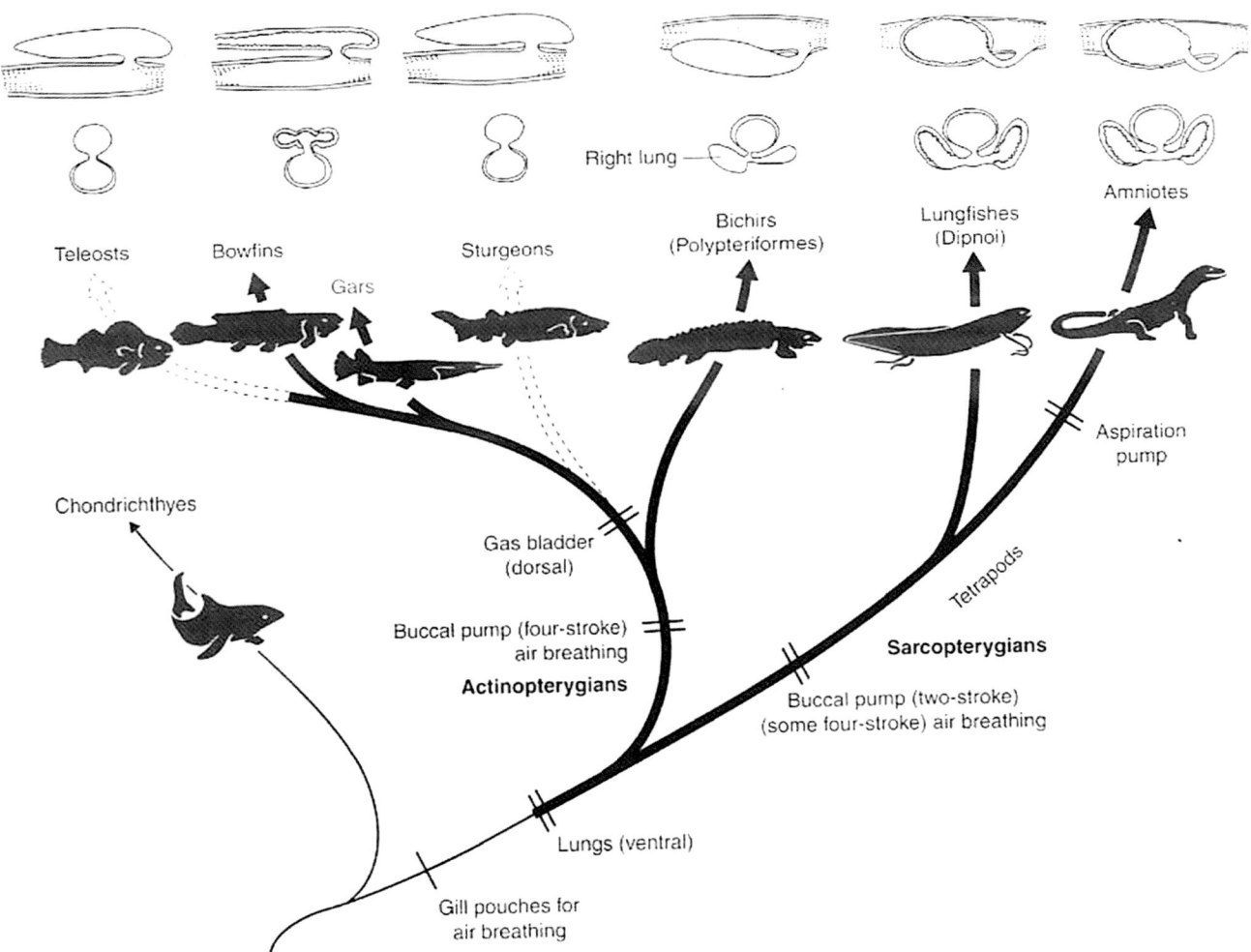

Fig. 10.38 Klippel-Feil Syndrome is defined by congenital fusion of two cervical vertebrae (at any level) with consequent limitation of motion. A short neck and low hairline are associated finidngs along with associated scoliosis and spna bifida. Left: [Reprinted from Ulmer J. Klippel-Feil Syndrome: Ct and MR of acquired and congenital abnormalities of the cervical spine and cord. J Comp Assist Tomo 1993; 17(2): 215–24. With permission from Wolters Kluwer Health, Inc.] Right: [Reprinted from McGaughran JM, Kuna P, Das V. Audiological abnormalties in Klippel-Feil Syndrome. Arch Dis Childhood 1998; 79(4): 352–355. With permission from BMJ Publishing Ltd.]

Cervical vertebra C3–C7 possess proximal muscle attachment sites that are analogous to those found in thoracic and lumbar vertebrae. The reader will note that these sites, whether dorsal or ventral, follow a similar pattern: midline-to-lateral, and from deep-to-superficial. The *dorsal pattern exactly mimics the spread of ossification from the base of the neural arch* (1) outward laterally along the lamina to the transverse process; and (2) outward dorsally along the spinous process.

Distal muscle insertions follow the same spatial pattern. Dorsal musculature is arranged in three layers. These mostly-forgotten muscles are important not for their specifics but for the biological pattern of their attachments. They are described in detail in the myology section.

Intrinsic layer muscles are monosomitic (with exception of; from deep to superficial and from medial to lateral: *multifidus, semispinalis, interspinale,* and *intertransversari.*

Erector spinae are polysomitic muscles arranged in three columns, from medial to lateral: *spinalis cervicis, longissimus cervicis,* and *iliocostalis cervicis.*

The splenii connect the occiput and the OCJ with the trunk: *splenius capitus* and *splenius cervicis.*

C7 does present a few peculiarities. Because of its long spinous process, this bone is often termed the *vertebra prominens. Foramina transversaria* transmits vertebral veins, but *not* the vertebral arteries. It has two very active insertion points. *Anterior tubercle of the transverse process* receives scalenus medius and levator costarum. Posteriorly, the *tubercle of the spinous process* receives the insertions of interspinales, multifidus, semispinalis thoracis, spinalis capitus, and, finally, trapezius.

Phylogeny of the Cervical Vertebrae

We are now going to use what we know from the fossil record to reconstruct the evolution of the cervical spine with emphasis on three specific issues.

- What adaptations took place in tetrapod vertebrae for life on land?
- How did the cervical spine expand in length?
- What specific innovations took place in mammals to enhance control of the upper extremity?

Once upon a time, there was no such thing as a neck…just ask any fish! The terrestrial triumph of tetrapods involved multiple evolutionary advances including gills into lungs, fins into limbs, a novel joint interposed between the head and body, and a flexible neck designed for positioning of the head. As we shall see, the cutaneous envelope of neck and its musculoskeletal infrastructure developed as modifications of the piscine trunk, specifically involving somites 5–12. How the neck came into being will be discussed at the end of this chapter. For our purposes here, we are going to focus on the

following topics: (1) the number of cervical vertebrae, (2) the parasegmentation of the occipitocervical junction, a structure that literally revolutionized life on land, and (3) the broader topic of vertebral evolution to the mammalian line (Fig. 10.39).

The Vertebral Axis in Fishes: Adaption for Swimming

Life in water in the water has its advantages. The axial skeleton of fishes faces few demands. Due to buoyancy, gravity has little influence. Body design must resist "telescoping," contraction while swimming due to muscle contractions. The bauplan of the fish body is seen in the cephalochordate *Branchiostoma,* better known as amphioxus (Gr. *amphi* = both + *oxy* = sharp), so named for its two pointed ends. Amphioxus is supported by a notochord that runs the length of its body and can be stiffened by contraction. Its body wall is constructed from somites and it uses a series of V-shaped myomeres for swimming (Fig. 10.40).

Although axial systems of extinct jawless fishes are as yet an evolutionary black box. Perhaps their body armor precluded the development of a spine. A few examples, such as osteostracans, have random skeletal elements surrounding the notochord and spinal cords. In living jawless descendents, hagfishes have no vertebral elements but lamprey has cartilaginous neural arches, *arculalia,* which partially protect the spinal cord. Jawless fishes do not have ribs.

A true axial skeleton evolved in gnathostomes. Chondrichthyans protect the spinal cord with a continuous jacket of segmented cartilage plates and poorly developed ribs. The vertebral column of bony fishes is divided into a trunk region and a tail region with modifications in the latter for swimming. Although the notochord continues straight into the skull the first trunk vertebra has articulation but essentially no mobility, due to the stiff notochord. The vertebrae in basal choanate fishes, such as the osteolepiform *Eusthenopteron* were rachitomatous with *paired* elements fused in the midline (referred to hereafter in the singular): small dorsal pleurocentra, dominant U-shaped intercentra, and neural arches. In the tail, mirror-image hemal arches remained to protect the aorta during swimming. Due to the low-stress conditions of water, fish vertebrae are connected with ligaments; there is no need for interlocking bony joints or zygapophyes (Figs. 10.41, 10.42).

The Vertebral Axis in Tetrapods: Adaption for Weight-Bearing

Aquatic life placed little demand on the axial skeleton. Not unlike a weightless astronaut, you spend your time swimming around with a buoyance system. But the transition to land meant suspension of the body between limbs. Axial

Fig. 10.39 Phylogeny of vertebrae Neural and hemal arches do not become important until placoderms. Vertebral centra appear late in evolution of both chondrichthyan and osteichthyan fishes. In tetrapods, the centra replace the notochord except for nucleus pulposus. [Reprinted from Kardong KV (ed). Vertebrates: comparative anatomy, function, evolution. McGraw-Hill Education; 2015. With permission from McGraw-Hill.]

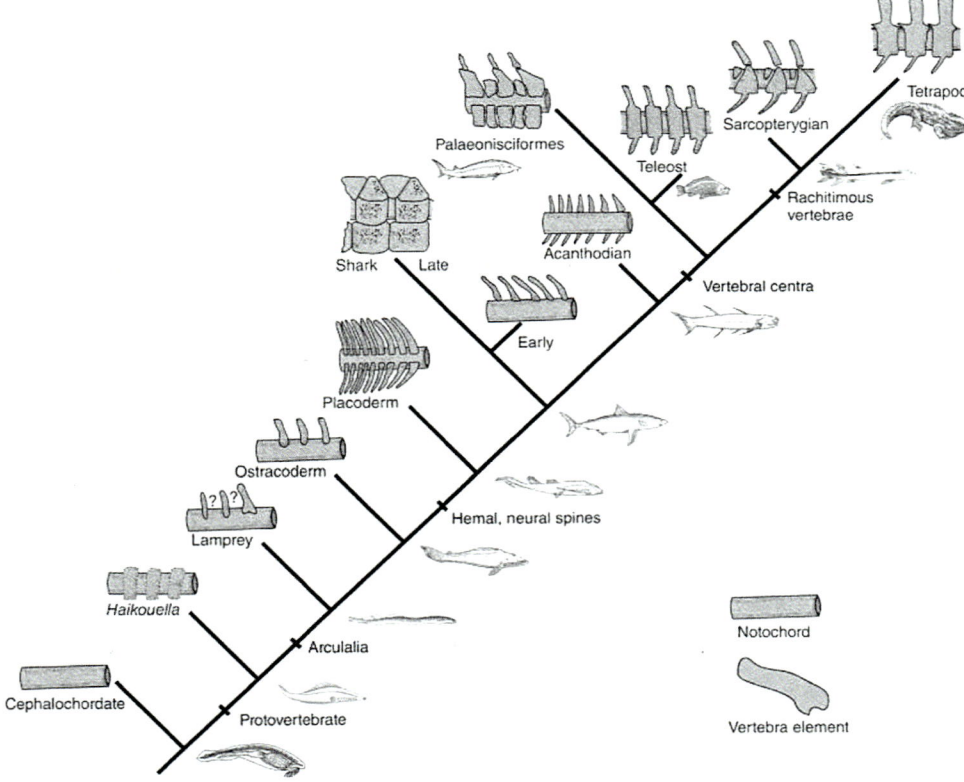

Fig. 10.40 Amphioxus Note the eight cartilaginous branchial baskets for filter feeding. The nerve cord has partial covering with a neural arch or spine. The perioral head has a neural crest annular cartilage. [Reprinted from Kardong KV (ed). Vertebrates: comparative anatomy, function, evolution. McGraw-Hill Education; 2015. With permission from McGraw-Hill.]

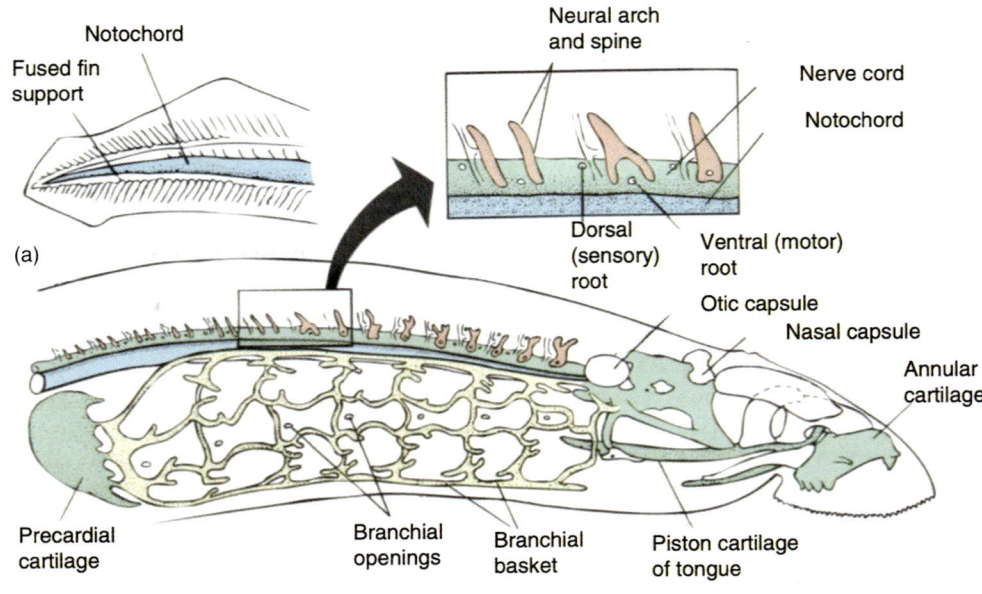

skeletal adaptions to the new mechanical demands resulted in vertebra with fused elements.

Basal tetrapods, the labyrinthodonts evolved directly from rhipidistians. They were ponderous creatures. We do not know if they had a sternum but most of their vertebrae were rib-bearing. Abdominal ribs were useful to present the collapse of their lungs and viscera as their massive bodies lay on the ground. It appears that they had primitive cervical vertebrae, the first of which was the atlas but not

axis. The number of cervical vertebrae is uncertain (Fig. 10.43).

The vertebrae of *Acanthostega* and *Ichthyostega*, were apsidospondylous and intercentrum dominant. This loose conglomeration of bony elements was ill-suited for life on land. They remained notochord dependent. Its elastic sheath provided form mobility which allowed them to incorporate the primitive neural connections used in side-to-side swimming motion of fishes into undulations used

Fig. 10.41 Shark spine [Reprinted from Kardong KV (ed). Vertebrates: comparative anatomy, function, evolution. McGraw-Hill Education; 2015. With permission from McGraw-Hill.]

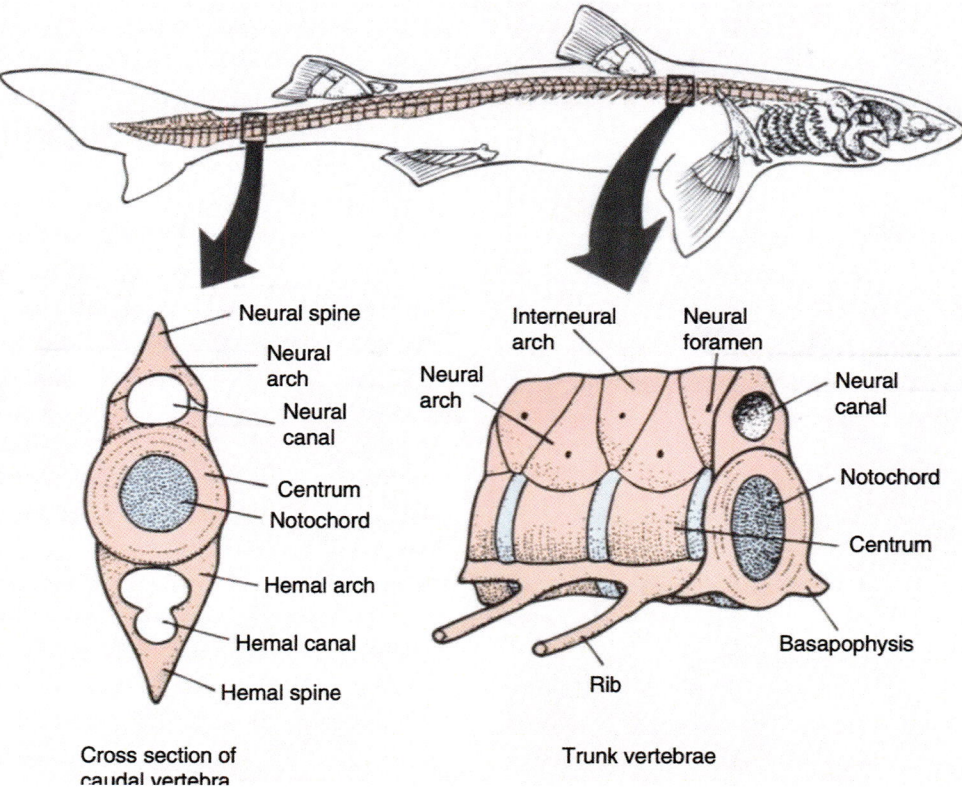

Fig. 10.42 Osteolepiform *Eusthenopteron* vertebrae. The basal condition is intercentrum dominance. [Reprinted from Kardong KV (ed). Vertebrates: comparative anatomy, function, evolution. McGraw-Hill Education; 2015. With permission from McGraw-Hill.]

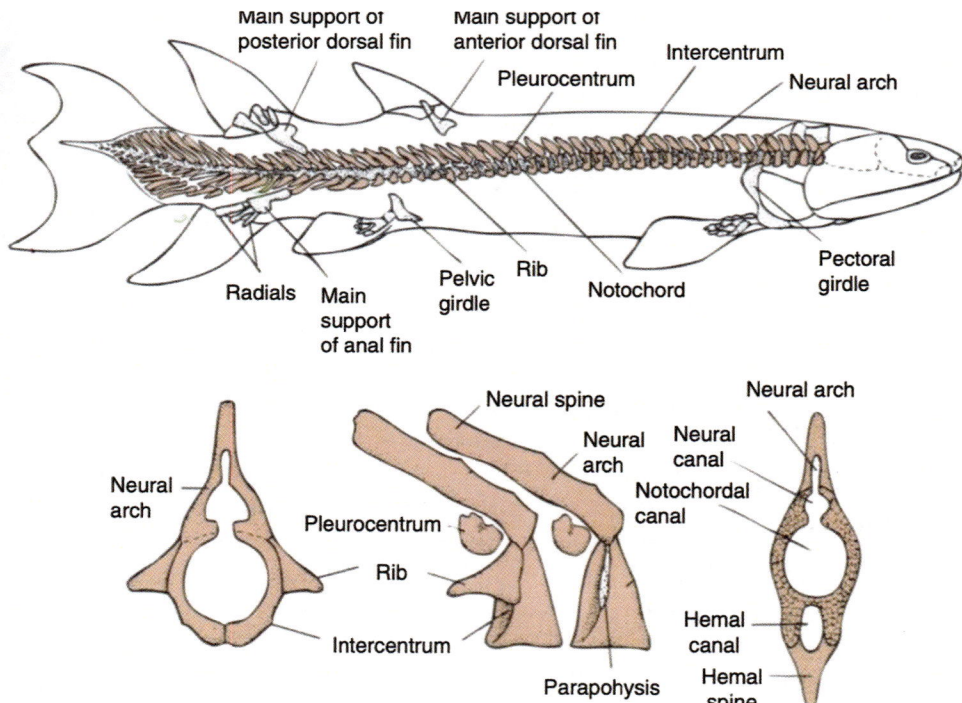

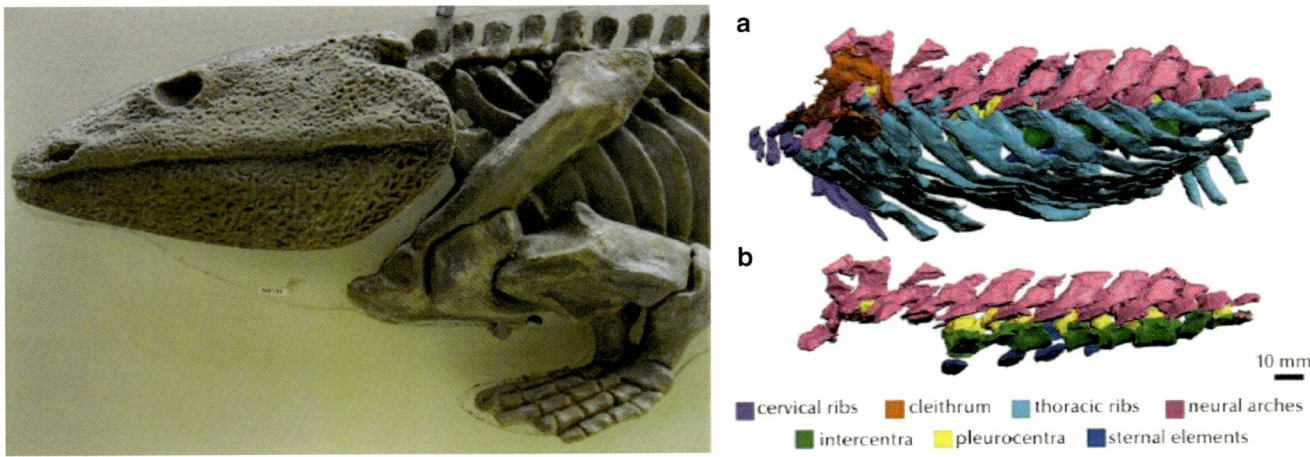

Fig. 10.43 Cervical vertebrae in Ichthyostega. Ribs present at all levels. Number of cervical vertebrae is uncertain…based on cervical ribs the original tetrapod neck had 3 cervical vertebrae. Frogs and salamanders have a single cervical vertebra, the atlas. The frog brachial plexus consists of sp2 and sp3 with sp3 belonging to the first truncal vertebra. Recall that sp1 belongs to the proatlas. [Reprinted from Ahlberg PE, Clack JA, Blom H. The axial skeleton of the Devonian tetrapod †*Ichthyostega*. *Nature; 2005;* 437:137–141. With permission from Springer Nature]

to propel one foot in front of another. Such lateral body swings are preserved today in salamanders. But their ponderous side-to-side pattern of locomotion placed undue torque on the spine; it åthis with two innovations. First, their cervical ribs were drastically reduced and eventually incorporated into the vertebra as *zygapophyses*. The processes reach across intervertebral joints as gliding articulations. They permit bending of the spine but *resist twisting*. The second innovation was the *sacral region*. By providing an attachment of the pelvic girdle to the vertebral column hindlimb, propulsive forces could be directly transferred to the spine.

Labyrinthodonts gave rise to two different lineages characterized by differences in the relative prominence of the centrum elements. Both directions reflect a reduced role for the notochord, with bony elements strengthened for load-bearing. *Temnospondyls* retained dominant intercentra. They are most closely associated with ribs and axial muscles used in swimming. Pleurocentra were not eliminated; they were eventually incorporated into the intercentrum as a single mass, the holospondylous condition, seen today in modern amphibians. The evolutionary consequence of holospondyly is possibly reflected in the foreshortened amphibian neck. Only a single cervical vertebra, the atlas, is present. Elsewhere in evolution, reshuffling the developmental subunits of the vertebra is responsible for the creation of a proatlas–atlas–axis complex. This is possible when these elements are fused (Figs. 10.44, 10.45, 10.46).

The other lineage of Labyrinthodonts was the *Anthracosaurs* (so-called because their skeletal remains were found in coal beds). The vertebrae of these creatures were characterized by expansion of the pleurocentrum at the

expense of the intercentrum. In one direction, embolomeres, the two elements remained unfused but equal in size. In the other direction, intercentrum was reduced; this gave rise to amniotes.

Crown anamniotes are represented by the reptilomorph *anthracosaur* lineage. These creatures were embolomeres, meaning that their intercentra and pleurocentra were equally represented. The anthracosaurs ultimately gave rise to amniotes. Loss of cervical and abdominal ribs led to greater mobility.

In amniotes pleurocentra enlarged at the expense of the intercentra, ultimately eliminating them altogether. Pleurocentra function to support the neural arches. They undergo successive interlocking by virtue of their zygapophyses. This reduces axial flexibility but strengthens the vertebral column for load-bearing. The limbs became less ponderous leading to more speed and agility.

Tetrapod Vertebral Column: Regional Variation and Size

The tetrapod axial skeleton has more regional variation than in fishes due to multiple functions. For example, the stresses and mobility required for the upper extremities are different from the lower extremities. Living anamniotes went in a different direction. As amphibians reinvaded a watery habitat, swimming became important for the pre-adult phase. Subsequently, specializations for leaping were required. These include reduced numbers of vertebrae (they have but 1 cervical vertebra), enlarged centra, and extensive ossification of the vertebrae with the fusion of components: the holo-

Fig. 10.44 Vertebral body evolution from Rhipidistians forward can be classified by composition of vertebral centrum. Temnospondyls are intercentrum dominant and lead to all living land-dwelling anamniotes. Anthracosaurs are pleurocentrum dominante and lead to amniotes. [Reprinted from Kardong KV (ed). Vertebrates: comparative anatomy, function, evolution. McGraw-Hill Education; 2015. With permission from McGraw-Hill.]

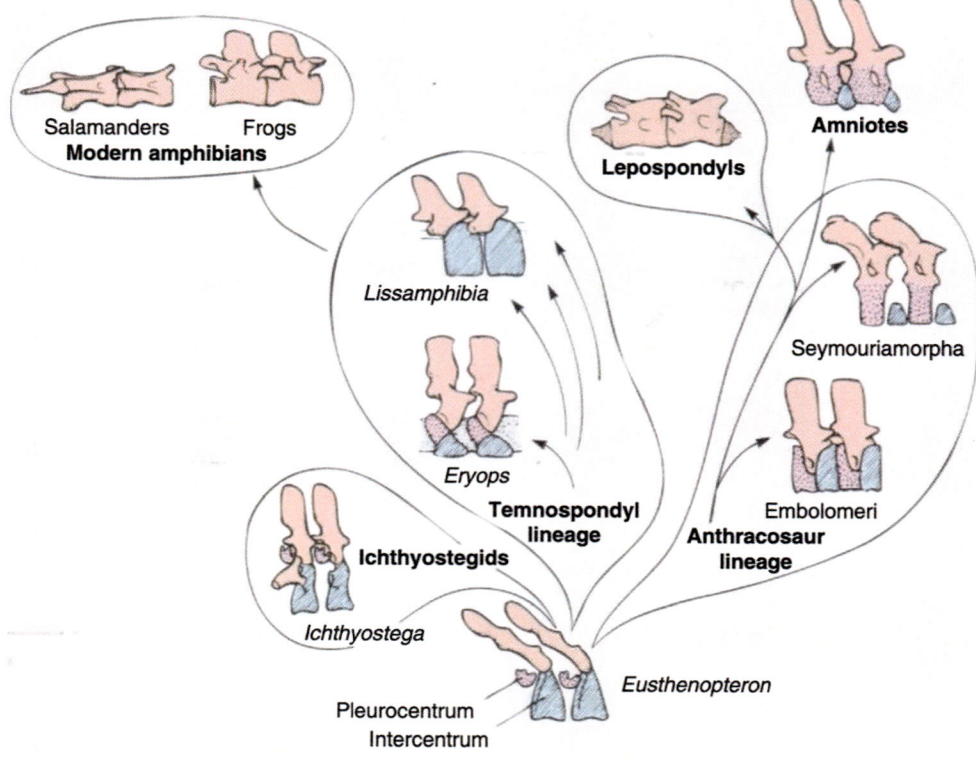

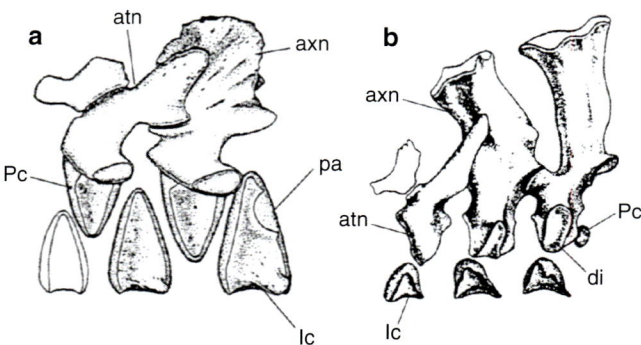

Fig. 10.45 Proatlas partial incorporation Atlas–axis complex appears in the bifurcation of basal tetrapods into temnosondyls and anthraco-saura. Note proatlas fragment dorsal to neural arch of atlas in two temnospondyls *Greererpeton burkemorani* (left) and *Eryops*. Note the reduction of pleurocentra in the latter. Proatlas (C0) incorporation must be a concomitant event with the differentiation of C1 atlas and C2 axis. [Reprinted from Shishkin MA. Evolution of the cervical vertebrae in temnospondyl amphibians and differentiation of the early Tetrapods. *Paleontological Record* 2000; 34(5):534–546. With permission from Springer Nature]

spondylous condition. These small creatures have a sternum but ribs for body support are not important. The sum of these changes strengthens the central core in preparation for propulsion by leaping.

Basal amniotes, the reptilomorphs, developed many innovations. They had anapsid skulls, without a temporal fenes-

tration for specialized chewing muscles. The number of cervical vertebrae in anaspids is 5, including the invention of an axis, thereby allowing for enhanced neck mobility. Neck length of reptilomorphs immediately expands in subsequent lines. Diapsids increase their cervical vertebrae by expropriation from the trunk. Modern reptiles have 6–10 while birds have 11–25. This shifts the cervical–thoracic junction backward while maintaining a brachial plexus of 4 roots (Fig. 10.47).

Reptilomorph needed a stronger skeleton to assist their invasion of the terrestrial environment so they reduced the intercentrum to a small piece and pushed in ventrally between the definitive centra (recall that all these pieces are fusions of bilateral embryonic components). Ribs were embedded in the body wall as important sources of lightweight support, extending all the way from atlas to the caudal (post-pelvic) vertebrae. In the anterior trunk, they developed a sternum, possibly as a fusion of anterior ventral cartilaginous rib elements, *paraphernalia*. Interclavicle appears for the first time and unites, together with coracoid, to the sternum, thereby locking pectoral girdle into a ventral position. Intercentrum, used by reptilomorphs as the attachment for capitulum of the rib, is incorporated into the pleurocentrum of the previous vertebral body so rib attachments now span two vertebrae. Accordingly, the rib from first vertebra of the trunk attaches to the upper sternum.

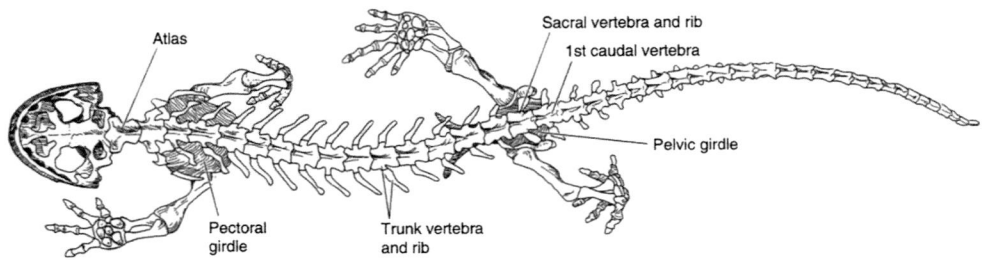

Fig. 10.46 Spine salamander. Note single cervical vertebra, pectoral girdle associated with second spinal vertebra. Note in amphibians, nerve sp1 is autonomous. The salamader brachial plexus is sp2–sp4. The frog brachial plexus is reduced to sp2 (flexors) and sp3 (extensors). This reflects the drastic reduction in its upper extremity. [Courtesy of William E. Bemis]

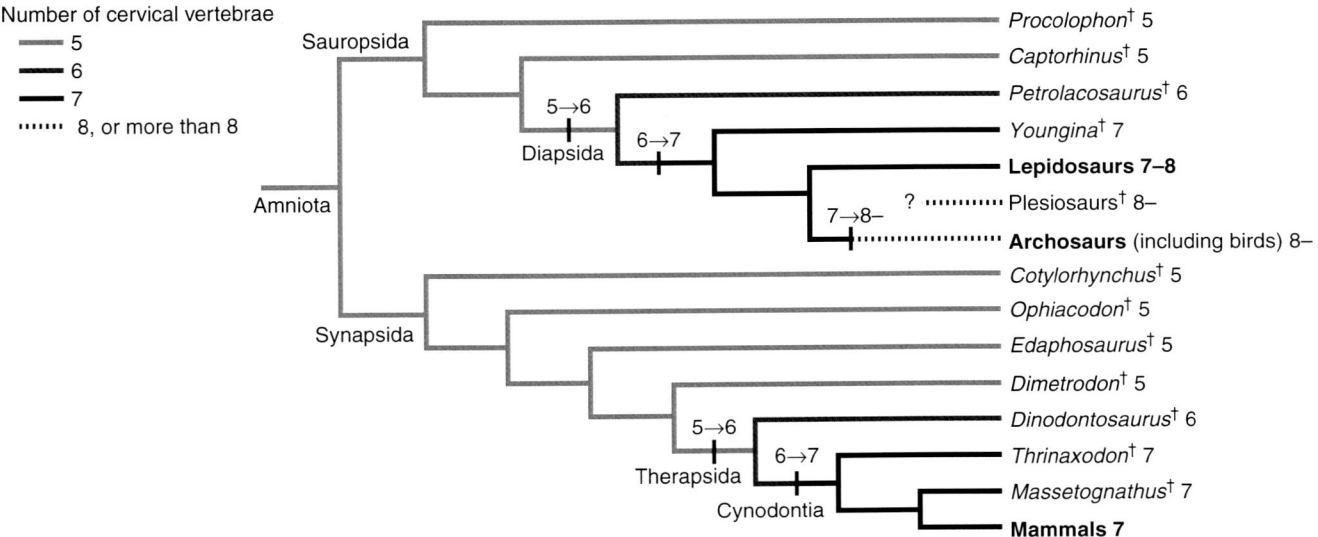

Fig. 10.47 Expansion of cervical vertebrae during amniote evolution through two crown amniote clades. Sauropsida are almost all diapsids. The extant (living) reptiles are Lepidosaurs and Archosaurs—the line leading to birds. Synapsids started out at the level of pelycosaurs with 5 vertebrae which increased to 6 with the therapids *Dimetrodon*† and sta- bilized at 7 with the cynodonts (*Thrinaxodon*† and mammals). [Reprinted from Hirasawa T, Kuratani S. A new scenario of the evolutionary derivation of the diaphragm from shoulder muscle. *J Anat* 2013; 222:504–517. With permission from John Wiley & Sons]

Table 10.1 Variations in the vertetebral column by region

Fishes	None	Trunk	Caudal	
Amphibians	Cervical 1	Trunk 8–22	Sacral 1	
Reptiles	Cervical 6–10	Trunk 10–22	Sacral 2–3	
Birds	Cervical 11–25	Thoracic + Lumbar 14–30	Sacral 10–23	Pygostyle
Mammals	Cervical 7	Thoracic + Lumbar 14–30	Sacral 3–5	

Throughout our discussion of diapsids, we have ignored the avian line because it does not lead to mammals. Some comments regarding avian vertebrae are appropriate for understanding a different form of neck. Despite the usefulness of wings, birds cannot use them for food acquisition. They compensate for this using an elaborate system for head positioning and control. Birds are sauropods, direct descendants of dinosaurs; accordingly, they have both an atlas–axis complex and cervical ribs. But the mobility they require for head positioning is achieved by increased neck length.

In synapsids, cervical vertebrae expand, becoming six in therapsids and seven in cynodonts. They accomplish this by duplicating levels C4–C5 using the same hox code.

and pushing additional codes backward. The resulting 6-root mammalian brachial plexus supplies greater complexity of upper extremity muscles (Table 10.1).

Phylogeny of the Occipital-cervical/ Craniovertebral Articulation

We now turn our attention to the evolution of the vertebral column in general to that of the craniovertebral junction. First, we review the changing definition of the head-trunk interface over time. We then discuss how the anatomy of the joint evolved.

Basioccipital bone in primitive fishes develops from the fusion of three occipital vertebrae; its termination is the loose zone of S3. The first truncal vertebra is dense S3 + loose S4. *The original head–trunk interface for all vertebrates is located at S3–S4.* Fishes have direct contact between the head and pectoral girdle using a series of *opercular bones*. As they are swimming in a watery, low-impact environment, weight bearing is not an issue. "Shock absorber" discs are not needed.

The fish-tetrapod transition is characterized by three well-known organisms: *Tiktaalik, Acanthostega,* and *Ichthyostega. Tiktaalik* is the oldest and most primitive tetrapod. It demonstrates the loss of the opercular bone, the subopercular bone, and the extrascapular series (the two remaining opercular bones are lost later in evolution). The pectoral girdle is liberated from the skull and a neck joint exists. It had synovial joints in the forelimb. The hindlimb was incomplete. Although it had a pubis and ilium with an acetabular joint, ilium was not attached to the spine; no ischium was present. *Acanthostega* demonstrates the five basic elements of the pectoral girdle: anocleithrum, cleithrum, scapulocoracoid, clavicle, and interclavicle. Ischium and the sacrum appear for the first time.

When the upper limb was freed from the skull, the basic cervical plan was one of five vertebrae but fossils do not show us intact original proatlas. Fragments of it do persist up to and including the pre-amniote pelycosaurs. But today we do know what it looks like because of unusual congenital conditions in which this bone reappears (see Fig. 10.32). Genetic manipulation of the homeotic code can also cause it to be produced. The proatlas seen in fossils today is probably the neural arch of the original first cervical vertebra (call it C0) which was produced by neuromeres r11 and c1 in the first tetrapods (without an axis) and subsequently incorporated into the skull. Fossil evidence for an atlas–axis complex dates back to the anthracosaur line but where the incorporation of proatlas occurred is unclear.

At the dawn of our story, we shall start with basal anapsid amniotes. The reptilomorph neck seen in *Paleothyris* consists of five vertebrae. The craniovertebral joint consists of a single occipital condyle lying ventral to foramen magnum which articulates with a depression in the atlantal centrum. Just dorsal to this lies a fragment of proatlas that limits motion. This is followed by axis and by three standard cervical vertebrae. This allows for the existence of the primitive brachial plexus consisting of four roots, C3–C5 + T1. This plexus, exemplified in chameleons, has just two terminal branches: radial nerve and brachial nerve (Figs. 10.48, 10.49).

The craniovertebral joint became more sophisticated in the synapsid transition. Our therapsid ancestor, *Dimetrodon,* showed a split of the occipital condyle, probably back into its two original embryonic parts. These are now relocated dorsally and are situated on either side of the foramen magnum. Concomitant with this was the formation of bilateral facets on the neural arch of atlas to receive the double condyles. Therapsids added a sixth cervical vertebra. All three parts of atlas (neural arch, intercentrum, and pleurocentrum) remained separate. The atlas–axis complex of cynodonts, such as *Thrinaxodon* still retained proatlas. The axis remains a full complement of seven cervical vertebrae that were now present. The penultimate iteration of the cervical spine.

In mammals, the proatlas is lost. The atlas undergoes a partitioning in which intercentrum joins with neural arch to make a ring while pleurocentrum is displaced backward to fuse with intercentrum of axis to form the dens which in turn fuses with pleurocentrum of axis, thus joining dens with the body. To facilitate rotation, the zygapophyses between the neural arches of atlas and axis are eliminated.

Fig. 10.48 Evolution of proximal cervical vertebrae. Cynodonts are the last step before mammals. The transofrmation to mammalian atlas–axis complex, is shown in fig. 36 [Reprinted from Kardong K. Vertebrates: Comparative Anatomy, Function, Evolution, seventh ed. New York, NY: McGraw-Hill; 2015. With permission from McGraw-Hill Education]

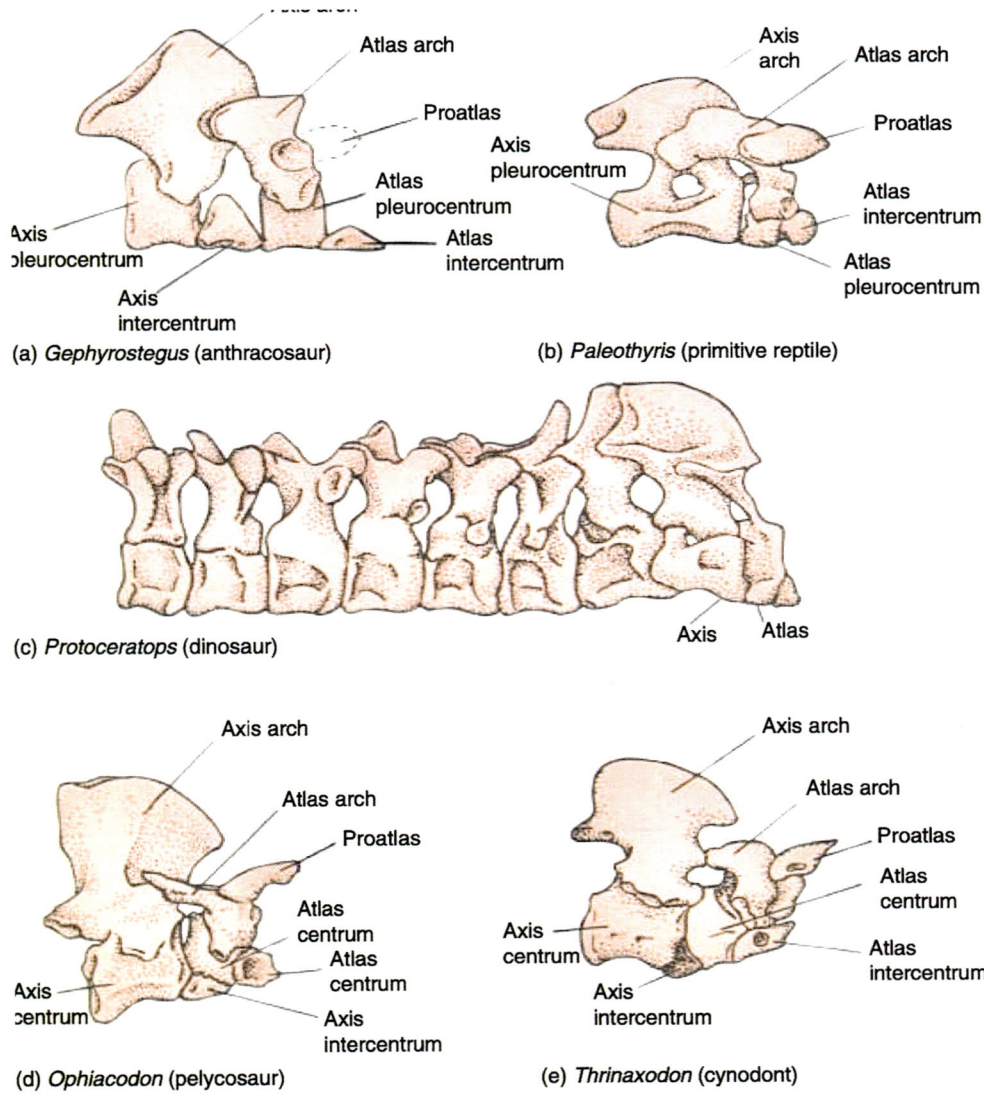

(a) *Gephyrostegus* (anthracosaur)

(b) *Paleothyris* (primitive reptile)

(c) *Protoceratops* (dinosaur)

(d) *Ophiacodon* (pelycosaur)

(e) *Thrinaxodon* (cynodont)

Fig. 10.49 Basic amniote pattern of the brachial plexus. br1–br4 (first through fourth spinal nerves) constitute the basic formula for the brachial plexus; brl, *N. brachialis longus*; rad, *N. radialis*; spc, *N supracorcoideus*. Note the amniote has the basic number of 5 cervical vertebrae. Radial and branchial nerves have two roots each and are connected. [Reprinted from Hirasawa T, Kuratani S. A new scenario of the evolutionary derivation of the diaphragm from shoulder muscle. *J Anat* 2013; 222:504–517. With permission from John Wiley & Sons]

Taxon	Spinal nerve number			
Most reptiles	C6	C7	C8	T1
Chameleons	C3	C4	C5	T1
Varanid lizards	C7	C8	C9	T1
Pigeon	C12	C13	C14	T1
Chicken	C13	C14	C15	T1
Ostrich	C18	C19	C20	T1

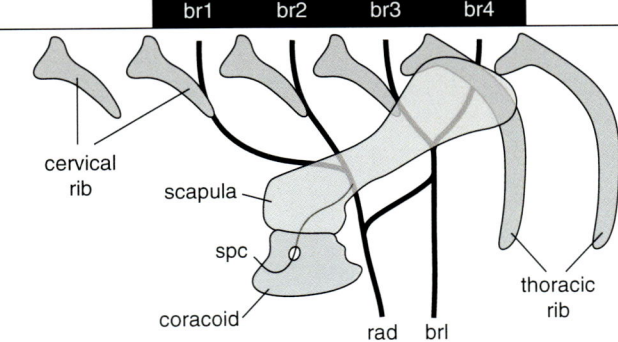

Osteology: Appendicular Bones

Organization of Lateral Plate Mesoderm (LPM)

Without exception, all bones of the head and neck are formed from either neural crest or paraxial mesoderm. Furthermore, all remaining bones of the axial skeleton, the vertebral column, and ribs, are PAM as well. Lateral plate mesoderm is the source of bone for the appendicular skeleton. The pectoral girdle, for evolutionary reasons we shall later explore, is an exception to this rule. The clavicle is a neural crest bone with a core of LPM marrow. Scapula is largely LPM, but specific sites along its periphery and the scapular spine bear neural crest cells.

Lateral plate mesoderm is produced by the same mechanism of gastrulation as its paraxial partner but, unlike its paraxial confrèr, it does not have an intermediate form of organization analogous to somites. First off, recall that the primary role of LPM is the production of the cardiovascular system, smooth muscle (as in the gut) the viscera, and the appendicular skeleton. Forward from neuromere c1 the role of LPM is strictly cardiovascular. Posterior to c1 the "portfolio" of LPM expands to form the bones of the appendicular skeleton and the smooth muscles of the gut and genitourinary system.

LPM has two regions: cardiac mesoderm (CM) and posterior lateral plate mesoderm (PLMP). PLMP is collinear directly behind CM and begins at neuromeric level c1. It is organized along the anterior–posterior axis of the embryo in the same way as somites, using Hox genes to establish positional specification. The separation of PLPM into somatic (LPM_S) and splanchnic (LPM_V) layers does not take place in the agnathic state. It is present in all gnathostomes [7] (Fig. 10.50).

In any case, LPM cell populations remain in register with the neuromeric level of the primitive streak from which they were produced and from which they are innervated. *LPM of the head and neck receives its coding from the neural crest to penetrates it and surrounds it to make the fascia.* Thus, scapula arises from LPM at levels c2–c6. The middle third esophagus has a mixture of PAM striated muscle and LPM smooth muscle, thus retaining somatotopic nocioception, whereas the muscle of distal third is purely LPM.

Phylogeny of Lateral Plate Mesoderm

Paraxial mesoderm with its subdivisions into somites has been present since the dawn of vertebrate evolution. Lateral plate mesoderm, in contrast, has undergone several iterations. The primitive state of LPM is seen in the cephalochordate amphioxus (the lancet), *Branchiostoma*. While its PAM is organized into myotomes. Its lateral plate mesoderm is not regionalized. The *Forkhead* gene *FoxF* is expressed throughout. Kuratani, working with lamprey, a living agnathan, established that its LPM is specified by Hox genes. Furthermore, it is regionalized into an anterior cardiac zone (CLMP) and a posterior truncal zone (PLMP). Lamprey CLMP has separate somatic and visceral layers but PLMP remains unseparated. No genes associate with limb production in gnathostomes are present in lamprey CLMP.

At the stem of gnathostomes, two changes take place in PLMP that are key to the production of fins/limbs. Its gene profile changes, such that it has a somatic layer, a requirement for the appendicular skeleton. Furthermore, it now expresses novel hox genes *Tbx4* and *Tbx5*. TBX4/5 induces *Fgf10* in mesoderm and *Fgf8* in overlying ectoderm, both steps being requisite for limb formation. Nested expression of these genes specifies the "fin field." As we shall see in our discussion of cervical neurology, the expression of Hox genes is involved in a shift from the ancestral pattern of neurons controlling the muscles of the pectoral fin to a new position coinciding with disconnection of the pectoral girdle from the skull.

Fig. 10.50 Lateral plate mesoderm evolution, plate 3. [Reprinted from Onimaru K, Shoguchi E, Kratani S, Tanaka M. Development and evolution of the lateral plate mesoderm: Compariative analysis of amphiosus, and lamprey with implircations for the acquisitions of acquired fins. *Dev Biol* 2011; 359(1):124–136. With permission from Elsevier]

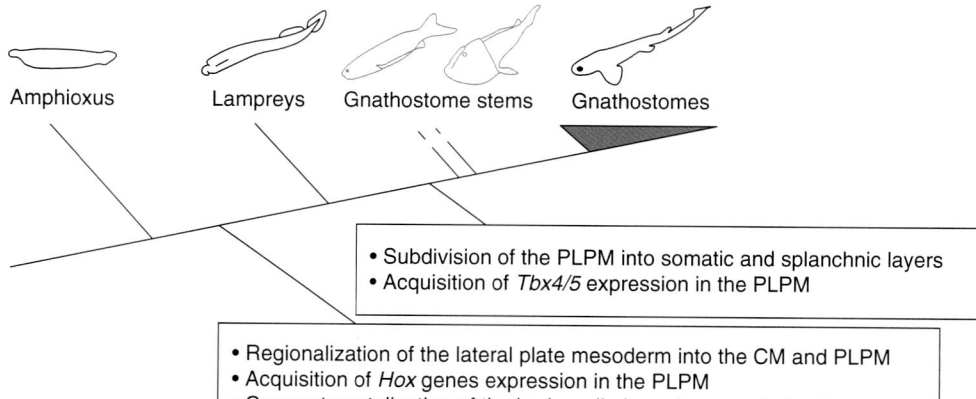

Amphioxus   Lampreys   Gnathostome stems   Gnathostomes

• Subdivision of the PLPM into somatic and splanchnic layers
• Acquisition of *Tbx4/5* expression in the PLPM

• Regionalization of the lateral plate mesoderm into the CM and PLPM
• Acquisition of *Hox* genes expression in the PLPM
• Compartmentalization of the body wall along dorso-ventral axis

Neuromeric Analysis: How to "Code" Lateral Plate Bones

Where does mesenchyme of clavicle and scapula originate? Neuromeric analysis of muscles having insertions/attachments to these bones allows us to "map" the bones into developmental zones. We can also reconstruct the *order* in which the bone fields are assembled. Here's an example. Let's recall (once again) that the origin of a muscle is given by the neuromeric location of its motor nerve. Levator scapulae is supplied by the dorsal scapular nerve (C3–C5), thus its myoblasts arise from the seventh, eighth, and ninth somites (S7–S9). The insertion/attachment of a muscle will be to a bone field that is registered with C3–C5. Coding is *not* determined by the secondary (distal) insertion) of the muscle. Levator scapulae develop from paraxial mesoderm. Levator scapulae make its primary insertion into the medial border of the scapula; its attachment is a triangular zone just medial to the spine. Its secondary insertions are tendinous slips to the transverse processes of atlas, axis, and the third and fourth cervical vertebrae. Thus, the mesenchyme making up this region of scapula originally came from lateral plate mesoderm in register with neuromeric levels C3–C5.

Warning Traditional anatomic texts as misleading. Gray's Anatomy uses the term "origin" in a purely descriptive way that has nothing to do with where the myoblasts of a particular muscle arise. In many instance (levator scapulae is a good example), the attachments of a muscle are exactly the reverse of its development. This is not to bash Gray. He simply described muscles arbitrarily based upon what he considered to be their fixed point (origin) and movement point (insertion). Moore's text uses the terms "proximal attachment" and "distal attachment," but again, uses them in a non-developmental way, leading often to a reversal (levator scapulae again). Such archaic terminology prevents us from seeing the unitary relationships that exist between bones and muscles in their earliest state.

Muscles attach to bone fields in a specified order. This can be determined by the spatial position of their motor nerves. At first glance, the process required to map out the bone fields of the appendicular skeleton (pectoral girdle and upper extremity) appears a bit more complex when compared to the mapping of the axial skeleton (skull and cervical vertebrae). Here's why. Cervical muscles supplied by the medial motor column (spinal nerve) are midline or paramedian. Those specifically from C1 to C3 form the post-hyoid hypobranchial column which extends forward from the pectoral girdle, runs beneath the pharyngeal arches, and attaches via hyoid to the mandible. These muscles act as depressors of the lower jaw. Cervical muscles supplied by the lateral motor column relate to outboard structures. These are all hypaxial/ventral, even though, for evolutionary reasons, muscles associated with the scapula appear to occupy a "dorsal" position.

Embryonic muscles seeking attachment to the shoulder girdle and limb have complicated migration patterns. That is because their bony targets develop in an inboard–outboard, proximal–distal direction. Recall the evolutionary sequence: *stylopodium*, *zeugopodium*, and *autopodium*. Furthermore, the postaxial bone units subdivide, whereas the axial units simply add on linearly. For this reason, the segmental innervation of upper limb joints follows a strictly linear sequence with shoulder abductors and lateral rotators supplied by C5 while the intrinsic muscles of the hand are innervated by C8, T1. Finally, as the upper limb develops, it undergoes a series of movements: internal rotation of the humerus > elbow flexion (stage 20, 51 days) > pronation. These movements cause different zones of the developing bones to be "presented" to the developing muscle masses for attachment. This is particularly true for the humerus.

Because upper extremity myoblasts have complex migration patterns, their motor nerves often coalesce and then separate. This results in a plexus. The game plan of the cervical plexus is the most simplistic. Its superficial branches innervate the skin while its deep branches are motor. Muscles supplied are: (1) the anterior and lateral capitus group; (2) the superficial hypobranchial muscles connecting the ventral midline with hyoid and mandible; (3) the diaphragm; (4) the most medial muscles connecting head and cervical spine to should girdle, and; (5) a muscle (scalenus medius) connecting cervical spine to the first rib (c8–t1).

Brachial plexus is a far more complicated switchyard. Its organization is proximal–distal and also postaxial–preaxial: roots > trunks > divisions > cords > branches. Muscles are spatially organized in the same order. As one goes farther out the chain, more muscles are supplied.

Roots: *dorsal scapular* = levator scapulae + rhomboid; *long thoracic* = serratus.

Trunks: *nerve to subclavius*; *suprascapular* = supraspinatus + infraspinatus.

Divisions: no branches.

Cords: *pectoral nerves*; *subscapular* = teres major; *thoracodorsal* = teres minor + latissumus; and *axillary* = detoid. Note that the axillary nerve (C5–C6) is the last motor nerve of the posterior cord. It marks the endpoint for all muscles with primary (proximal) insertions to the shoulder girdle.

In our next section, we shall describe the embryology of the clavicle and scapula. Detailed information regarding the neuromeric coding of each muscle is presented. Illustrations depicting the developmental sequence are labeled using neuromeric terminology.

Hox Gene Mapping of the Neck Bones: New Rules of Engagement

The shoulder girdle represents an interface zone between the head, neck, and trunk. Our assumptions up until this point regarding what bone comes from neural crest and what bone comes from mesoderm have been based upon a classical *ossification model* in which bones formed by membranous ossification are presumed to arise from neural crest whereas those formed by chondral ossification are assumed to be of mesodermal origin. In the neck, these rules become flexible and give rise to the *muscle scaffold model* in which the mesenchymal origin of a region of bone is determined by the composition of the fascia accompanying its various muscle attachments. All muscle fibers are composed of mesoderm but their connective tissue elements and their fascial envelope can be neural crest or mesoderm. Muscles will attach.

As we shall see, clavicle and scapula are composite structures with certain zones developing from neural crest and others from mesoderm depending upon the muscle inserted into that zone. Furthermore, we find that post-otic neural crest (PONC) from r8 to r11 has multiple fates. (1) It synthesizes the *cartilaginous structures* of the fourth and fifth arches, that is the thyroid, cricoid, and arytenoids cartilage. (2) It contributes to the *walls of the superior thyroid artery and its branches* supplying the fourth and fifth arches. (3) It provides *connective tissue and fascia* for the muscles of the tongue as well as for the sternocleidomastoid and trapezius muscles.

Experimental work fundamental for our understanding of how the clavicle and scapula develop was reported by [8] using Wnt-1 and Sox-10-Cre recombinase-mediated fat mapping to answer three fundamental questions. First, can PONC form the chondral bone? The answer is yes. (2) Are the dermal bones behind the otic capsule strictly neural crest? The answer is no; in this zone, mesoderm can form membranous bone as well. (3) Does the distribution of neural crest and mesoderm into the shoulder girdle correlate with ossification type or muscle insertion sites? The answer is that insertion sites determine bone mesenchymal identity.

Note that, for the purposes of neuromeric coding, *only the primary insertion of a muscle is important* (Figs. 10.51, 10.52).

The sternocleidomastoid is innervated by the spinal accessory nerve from neuromeric levels r8–r11. It has an additional sensory supply from c1 to c4. Thus, although the muscle arises from four somites (occipital 1–4), it receives neural crest from neuromeric levels r8–r11 and c1–c4 which forms the connective tissue and fascia of SCM. These relationships play a role in the primary insertion of SCM to the crest of the occipital bone. Trapezius has a similar formulation.

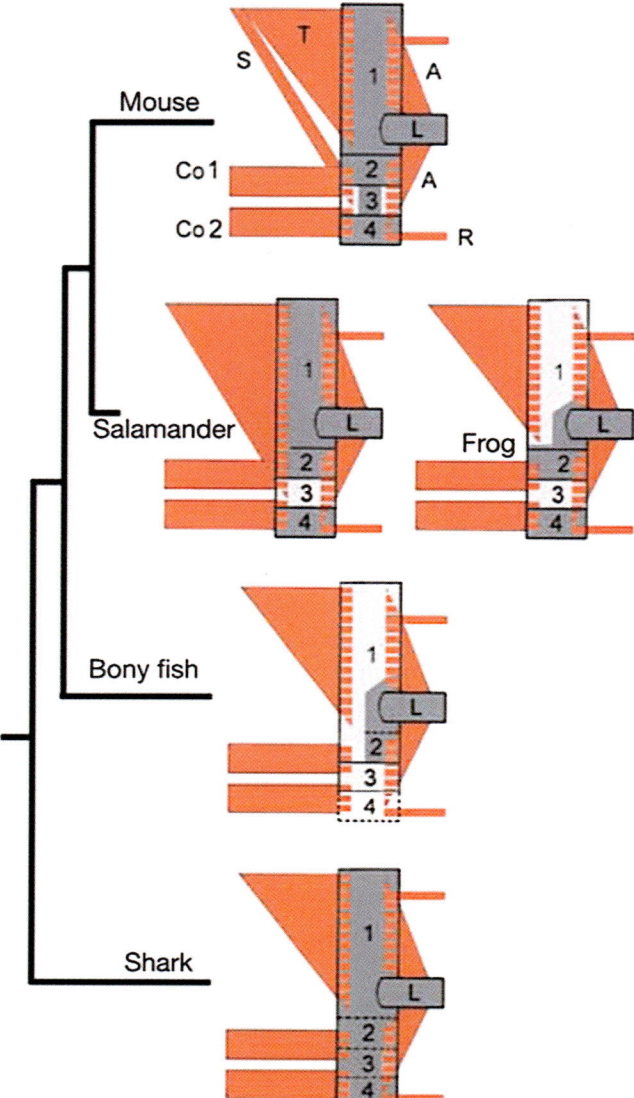

Fig. 10.51 Mesenchymal analysis of plectoral girdle show diversity of ossification. Highly conserved neck muscle scaffolds (red) attach (hatched areas) on a shoulder skeleton (boxes 1–4) that displays variable dermal (light grey) and endochondral (dark grey) ossification type. Attachment regions (hatching) of the gnathostome trapezius muscle (T) are endochondral in sharks, salamanders5and all amniotes but are dermal in fish and frogs. A = limb muscles, Co1 and Co2 (coraco-branchialis and coraco-hyoideus); L = limb skeleton; R = trunk muscles; S = sterno-cleido-mastoid. In the shoulder skeleton, box 1 is the dorsal cleithrum (dermal) in bony fish (Polyodon or Amia) and frog (Rana) and the scapular region (endochondral) in salamander, mouse and living amniotes; box 2 is the acromio-coracoid (endochondral); box 3 is the clavicular region (dermal/dermal+endochondral), although in sharks, bone is absent and its space is taken by part of the scapulo-coracoid (stippled); and box 4 is the sternal region, comprising the sternum (endochondral) or connective tissue (bony fish). [Reprinted from Matsuoka T, Ahlberg P, Kesaris N, Iannerelli P, Dennehy U, Richardson WD, McMahon AP, Koentges G. Neural crest origins of the neck and shoulder. *Nature* 2005; 46:437–355. With permission from Springer Nature]

Numerous pathologies obtain for defects in PONC. Pathological flexibility of PONC differentiation. Connective tissues can be over-expressed entirely leading to

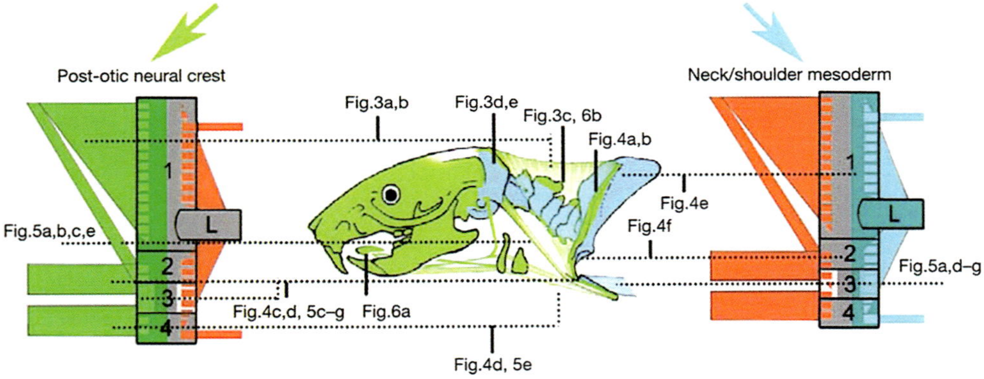

Fig. 10.52 Genetic lineage labelling of PONC and somitic mesoderm. Otic PONC (green) and mesodermal (blue) populations are indicated. Note that the spine of scapula, the former cleithrum, is PONC. [Reprinted from Matsuoka T, Ahlberg P, Kesaris N, Iannerelli P, Dennehy U, Richardson WD, McMahon AP, Koentges G. Neural crest origins of the neck and shoulder. *Nature* 2005; 46:437–355. With permission from Springer Nature]

"pseudo" macroglossia (in patients with trisomies). PONC fate to become bone may change to soft tissue mesenchyme, as in Arnold-Chiari syndrome. PONC fated to become connective tissue may be for ectopic bone, as in Klippel-Feil syndrome (Fig. 10.53).

The Clavicle

Clavicle is a curvilinear bone derived and synthesized from postotic neural crest and lateral plate mesoderm from neuromeric levels c1–c4. It articulates between the manubrium sterni and the acromion of the scapula. Functionally it serves as a fulcrum enabling muscles to extend the upper extremity laterally. Animals using anterior prehension make active use of the clavicle but, in many carnivores, it is simply suspended in the muscle sling. For many reasons, the clavicle is unusual. Despite its position, draped over across the root of the neck, it belongs to the piscine head-trunk interface and is associated with the cucullaris muscle. Recall that cucullaris spans between the primitive pectoral girdle to the dorsal tips of the respiratory branchial arches, as far forward as r6–r7 (Figs. 10.54, 10.55, 10.56).

But the limb migrates in mammals down to the level of c5 with muscles attaching to the scapula from this position. Clavicle follows along with the chondral component, scapulocoracoid, of the pectoral girdle. Thus, the phylogeny of the human clavicle reflects the caudal migration of the limb. Clavicle develops in two ways. Its external surfaces, both anterior and posterior, form in membrane. Clavicle is the *first bone of the body to ossify*, reflecting its relationship to the first post-cranial neuromeres. Recall that the original insertion of the pectoral chain was to the post-temporal bone of the occiput, likely at position r10–r11. Therefore, the most proximal zone of clavicle should code from c1. Recall that ossification of the cranium starts at the occipital somites and sweeps forward.

Clavicle ossifies from three constant centers, two primary and one secondary. At stage 17 (14 mm, 41–42 days), the clavicle is a mesenchymal condensation of mixed neural crest derivation slung between two mesodermal bones: the first rib and the acromion of the scapula. These latter two bones develop from paraxial mesoderm and lateral plate mesoderm, respectively. Within the band, medial and lateral zones of cartilaginous transformation (precartilage) appear. This "core of the clavicle" represents the ability of PONC to form endochondral bone. Next, two *primary dermal ossification centers* appear, one ventral and one dorsal. The ventral ossification center is of PONC derivation while the dorsal ossification center is from the lateral plate mesoderm. *Mesoderm thus forms membranous bone on the back side of the clavicle.* The two membranous centers quickly fuse around the chondral core by day 45. Next, the sternal and acromial zones become true cartilage from which endochondral ossification extends outward from the shaft. Expansion of diameter takes place via subperiosteal deposition [9, 10].

The medial center is responsible for 2/3 of the total length. Two secondary centers appear around age 20. The sternal center is constant and represents the remnant of the procoracoid while the acromial center is variable. The nutrient artery of the clavicle is derived from the suprascapular artery. It enters via a foramen at the lateral end of the subclavius groove.

As discussed previously, the early tetrapod pectoral girdle had a separate bone, the *interclavicle*, positioned medial to the clavicle. In the synapsid line, the interclavicle persists in therapsids and monotremes but in mammals, it is described as absent. We can postulate that the interclavicle may continue to exist as the secondary ossification site at the sternal border. The manubrium is physically different; it is a derivative of the ancient anterior coracoid process. The manubrium is a mammalian invention. We shall see that its original form fuses with interclavicle to create a larger struc-

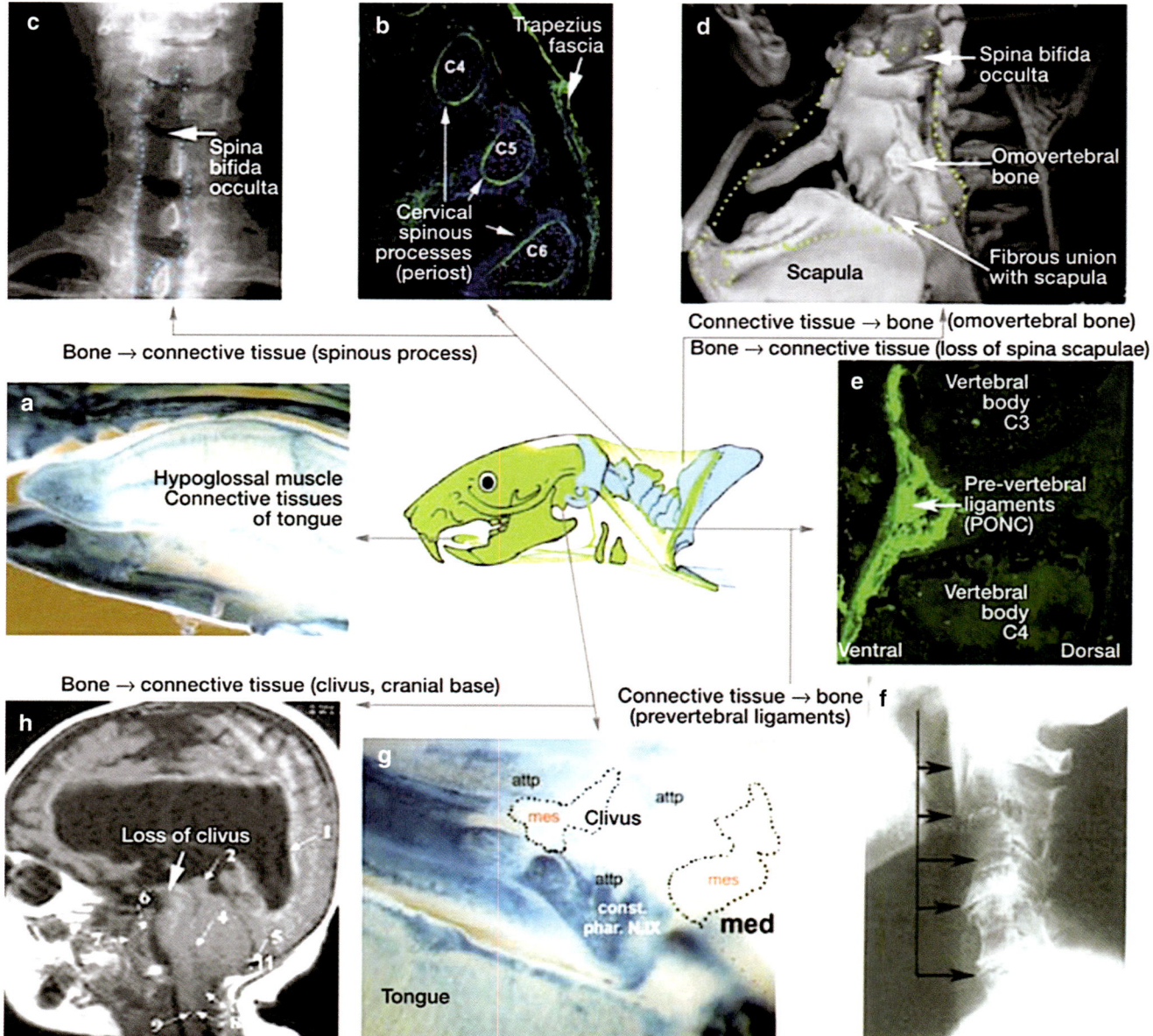

Fig. 10.53 Pathologies of PONC differentiation. (1) Changes in PONC connective tissue *per se. Trisomy pseudo-macroglossia* (**a**). All tongue-muscle connective tissues are entirely derived from neural crest (blue areas) explaining enlarged tongues as neurocristopathic. (2) Changes in PONC fate from bone/periosteum (**b**) into connective tissue can explain localized cervical defects in *cleidocranial dysplasia* (**c**) where neural spines fail to close and *Arnold-Chiari syndrome* I + II (**h**). In the latter, the PONC-derived clivus (blue in **g**) of the otherwise mesodermal (mes) cranial base, which serves as the attachment point (attp) for pharynx constrictor muscles (constr. Phar. N.IX/X.) in front of the medulla (med), fails to form and is replaced by fragile connective tissue. (3) Changes in PONC fate from connective tissues into bone. CT of pharynx constrictor muscles (const.phar. N.IX/X in g that are connected to cervical vertebrae (**e**) can ectopically become a vertical bridge of bone (**f**), leading to neck immobility either in *Klippel-Feil syndrome* (**f**) of as ectopic, 'omovertebral' bones inside trapezius territory (stippled line in **d**) inpatients with *Sprengel's deformity*, a frequent finding in Klippel-Feil syndrome 19, 20. Note also the concomitant loss of the PONC-derived (but not mesodermal) scapular spine in patients with Sprengel's deformity (**d**). [Reprinted from Matsuoka T, Ahlberg P, Kesaris N, Iannerelli P, Dennehy U, Richardson WD, McMahon AP, Koentges G. Neural crest origins of the neck and shoulder. *Nature* 2005; 46:437–355. With permission from Springer Nature]

ture with articulations for both clavicle and firest rib (see the development of manubrium).

Three hypaxial muscles having insertion on the clavicle define the neuromeric origins of the bone and divide the bone into two developmental fields. The connective tissue and insertions of all three muscles are derivatives of paraxial mesoderm The medial field is defined *by sternohyoid (C2–C4), which attaches along the upper and anterior surfaces.* The lateral field is defined inferiorly by *subclavius (C5–C6)* and by *deltoid (C5–C6)* superiorly. Note that the subclavius is innervated from the lateral trunk of the brachial plexus.

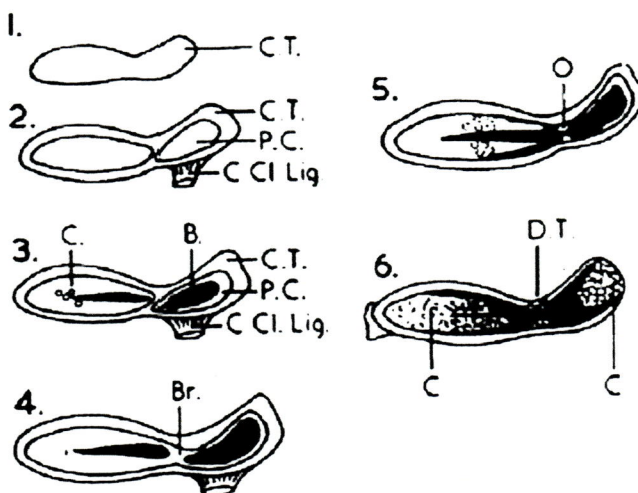

Fig. 10.54 Development of the clavicle. Clavicle consists of a neural crest outer coating and a mesodermal core. Coding is c1– to c4–c5 from medial to lateral, Lateral articulaltion with acromion which is c4–c5. [Reprinted from Fawcett J. The Development and Ossification of the Human Clavicle. Am J Physiol 1913; 47(pt2):225–234]

Deltoid being situated more laterally has a most distal innervation from the posterior cord of the brachial plexus. This indicates that the subclavius gains access to the clavicle first, while the deltoid arrives later and thus is positioned more laterally. All remaining muscles attached to clavicle are hypaxial, with primary insertions from either head or trunk. The upper extremity has no secondary insertions into clavicle.

Thus, clavicle arises from postotic neural crest r8–r11 and lateral plate mesoderm in register with neuromeric levels c2–c6. Its two developmental fields each have a primary ossification center. Clavicular LPM from neuromeric levels c2–c6 associates itself medially with PAM from levels c8–t1 that forms the sternum and first rib and, peripherally, with lateral plate mesoderm assigned to the upper extremity.

Clinical Correlation: Cleidocranial Dysostosis

This neurocristopathy affects membranous bones, in particular causing hypoplasia or absence of the clavicle and an enlarged patent anterior fontanelle. Frontal and parietal bossing are prominent. Cranial suture may display small Wormian bones. Nasal bones may be missing. Hypoplasia of the zygoma is accompanied by faulty maxillary sinus development with chronic sinusitis. The sygomatic arch is thin or even discontinuous at the zygomaticotemporal suture. Mandible has a slime ramus and pencil-like coronoid. The mandibular symphysis is patent. Enamel issues and dental maldevelopment, including delays in eruption, are present. Excessive development of the dental lamina leads to hyperdontia and crowding (Figs. 10.57, 10.58).

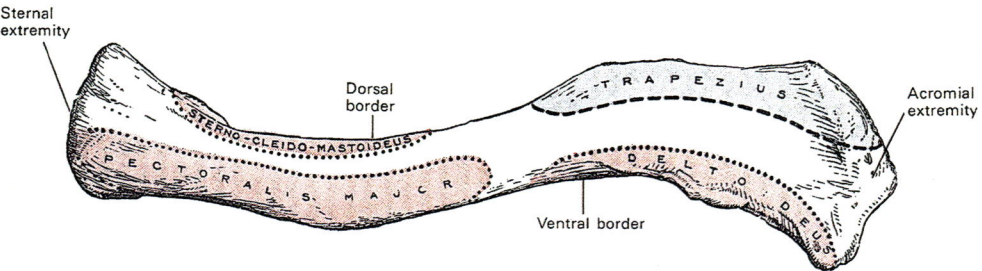

Fig. 10.55 Clavicle, superior. Each muscle has a primary insertion. Trapezius should be colored red. As part of the pectoral girdle, all muscles connecting to head follow an antegrade–retrograde have a primary insertion pattern. All muscles connecting clavicle to humerus are antegrade–antegrade. Trapezius is unusual: upper fibers are antegrade–ret- rograde and lower fibers are antegrade–antegrade. All insertions on clavicle are primary and are colored red. Trapezius has a primary insertion into clavicle and should be colored red. [Reprinted from Lewis, Warren H (ed). Gray's Anatomy of the Human Body, 20th American Edition. Philadelphia, PA: Lea & Febiger, 1918]

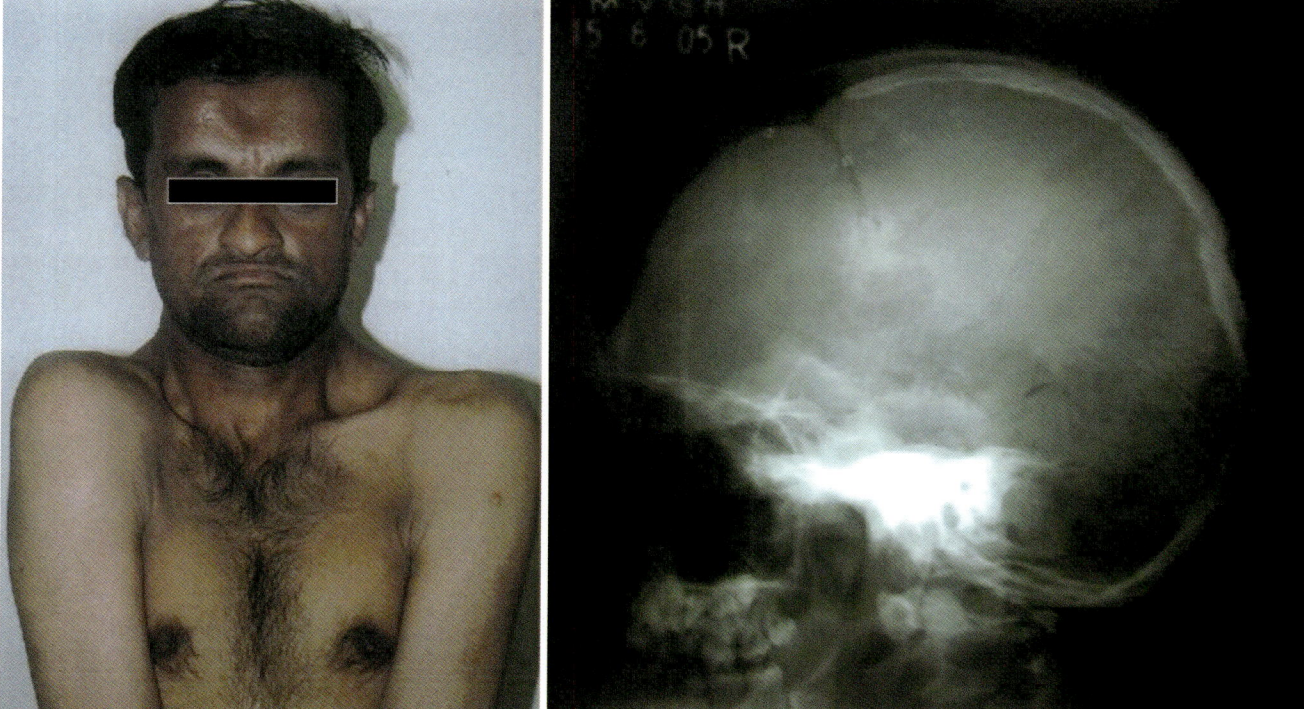

Fig. 10.56 Clavicle, inferior. As part of the pectoral girdle, all muscles connecting to head follow an antegrade–retrograde have a primary insertion pattern. All muscles connecting clavicle to humerus are antegrade–antegrade. Trapezius is unusual: upper fibers are antegrade–retrograde and lower fibers are antegrade-antegrade. All insertions on clavicle are primary and are colored red. [Reprinted from Lewis, Warren H (ed). Gray's Anatomy of the Human Body, 20th American Edition. Philadelphia, PA: Lea & Febiger, 1918]

Fig. 10.57 Cleidocranial dysostosis. Absent collar bones, fforehead is prominent. Maxillary hypoplasia, both vetical and horizontal with mandibular prominence. X-ray shows open fontanelles, underdeveloped sinuses. The atlantoaxial complex is impacted up against the skull base. [Reprinted from Garg RK, Agrawal P. Clinical spectrum of cleidocranial dysplasia: a case report. Cases Journal BMC 2008; 1: 377–381. With permission from Creative Commons License 2.0: http://creativecommons.org/licenses/by/2.0]

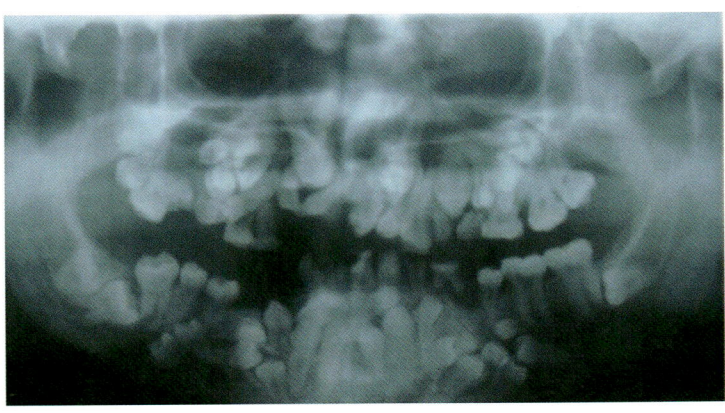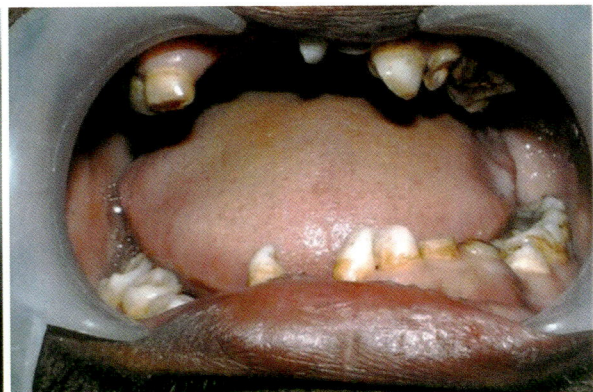

Fig. 10.58 Cleidocranial dysostosis. Multiple ectopic teeth, flattened nasal bridge, hyperteolorism is seen. [Reprinted from Garg RK, Agrawal P. Clinical spectrum of cleidocranial dysplasia: a case report. Cases Journal BMC 2008; 1: 377–381. With permission from Creative Commons License 2.0: http://creativecommons.org/licenses/by/2.0]

Mapping the Clavicle

Manubrium sterni is derived from anterior coracoid process. It codes to C1–C2.

It gives primary attachment to sternothyroid and sternohyoid.

Neural crest contributions to clavicle

Cartilagenous core is postotic neural crest r8–r11.

Stenocleidomastoid brings neural crest from its distal fibers to the medial 1/3 of anterior.

Medial 2/3 of Clavicle: C2–C4

Sternocleidomastoid arises from S5 to S6. Its motor supply is from the C1 to C2 roots of spinal accessory nerve. Its primary attachment to clavicle is at the anterior (neural crest) upper border. This muscle is branchiomeric with neural crest connective tissue likely originating from r8 to r11 leading it to connect it secondarily to the skull.

Pectoralis major arises from S9 to S13. It is supplied by C5, C6, C7, C8, and T1. It has two primary insertions: a clavicular head along the anterior lower border and a sternocostal head. Its secondary attachment is to the humerus.

Sternohyoid (lateral fibers) from C2 to C4 has a primary attachment situated at posterior (mesodermal) medial border, adjacent to manubrium. It ascends as part of the straps to the third arch border of the hyoid bone.

Lateral 1/3 of Clavicle: C5–C6

Subclavius This obscure muscle arises from S9 to S10. It is supplied by the subclavian nerve; that is, the trunks of C5–C6. It has its primary insertion within a groove on the inferior border of clavicle. Its secondary insertion is to first rib (T1 = S12–S13). Subclavius precedes deltoid.

Deltoid This large muscle arises from S9 to S10 as well. It is innervated by the axillary nerve, that is, by C5–C6 from the posterior cord. It has three primary insertion sites. The anterior fibers of this muscle have their primary insertion from anterior and superior lateral border of clavicle. Middle fibers are proximally attached at the acromion of the scapula whereas posterior fibers are located along the posterior border of the scapular spine.

Trapezius (posterior) has a primary insertion into posterior border of clavicle.

The Scapula

The scapula is a flat bone of triangular shape that it lies like a pancake over the dorsal surface of ribs 2–7. It is personified by 3's. Three borders face superiorly toward the neck, medially toward the spinal column and laterally toward the armpit. Three angles make the triangle, superior sharp surface for the scapular elevator muscles, an inferior dagger for teres to pull the scapula toward the humerus and for serratus to retract it toward the chest wall, and a lateral labium to receive with humerus. Three processes projecting outward from the home plate in three different directions anchor the scapula and determine its movements. The *spine* projects dorsally and create two fossae for muscle attachments. Projecting laterally from the spine is an outward-looking peninsula, the *acromion*, designed for articulation with clavicle. The *coracoid process* projects ventrally and gives attachment to the three muscles, pectoralis minor is medial; lateral is biceps brachii and coracobrachialis (Figs. 10.59, 10.60, 10.61).

In conceptual terms, the scapula is a floating platform for the upper extremity that, via multiple muscles attached to the head, spine, and thoracic cage and via its articulation with the clavicle, permits placement of the hand in space. The mesenchymal origins of scapule demonstrate it to be mostly LPM.

Scapula appears shortly after stage 23, that is, during fetal life. It is synthesized from lateral plate mesoderm (LPM) in register with neuromeres c2 through t1. It also bears contribu-

tions from neural crest [11–13]. The overall flow of mesenchyme is from dorsal-to-ventral. This is evidenced by the identical innervation of the supra/infraspinatus twins (C5–C6) from the upper trunk which lie dorsal to the bone plate versus subscapularis (C5–C6) from the more distal posterior cord which occupies the ventral position. Scapular spine represents the ancient coracoid and bears the attachment of the trapezius from neuromeres c2 to c6. Bone formation has an epicenter at the suprascapular notch (zone c1–c2) from which it radiates both medially and laterally, and then downward. As the coracoid process represents an add-on, it is not surprising to find that this zone represented LPM from c5 down to t1.

Scapula ossifies from eight distinct centers: body (1), coracoid process (2), acromion (2), medial border (1), inferior angle (1), and lower glenoid rim (1). Ossification of the body of scapula takes after stage 23. It is initiated at the scapular neck, a collar-like ring of bone holding outward the glenoid fossa like a chalice. From there a flat quadrilateral plate

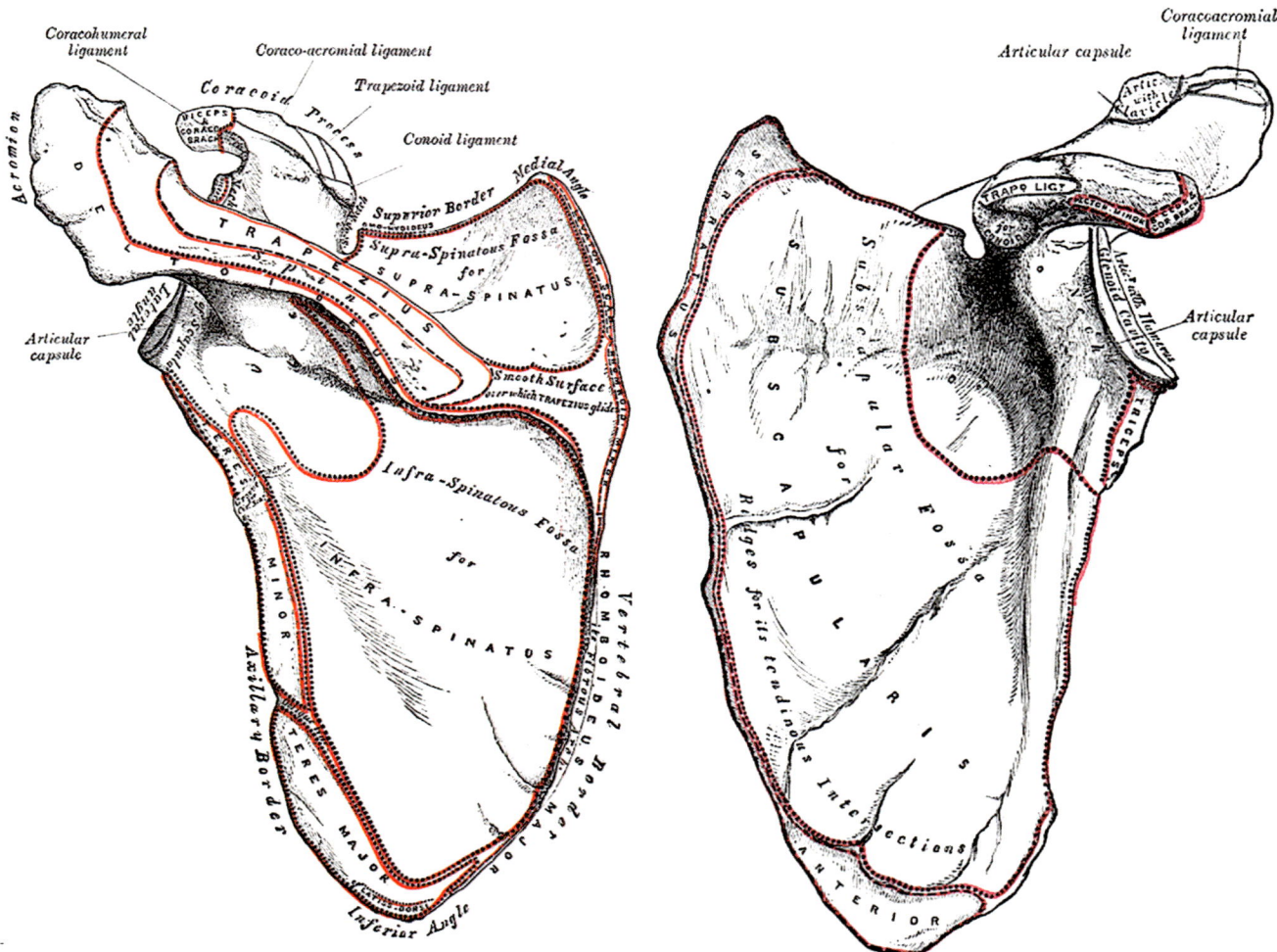

Fig. 10.59 Scapula: neuromeric coding. The scapular plate is created from lateral plate mesoderm via chondral osteogenesis; scapular spine is LPM dermal bone. The bones are in register with neuromeres c3–t1. Connective tissues of pre-scapular muscles down to c3 have neural crest connective tissue. Post scapular muscles have mesodermal connective tissue. Muscles having primary insertion into scapula are in three classes. (1) Intrinsic muscles insert into the surface of the scapula and control internal/external rotation of shoulder joint and humeral abduc-

tion (subscapularis, teres minor and the supra- and infra-spinatus muscles). (2) Extrinspc muscles insert into projections of the scapula, that is, coracoid process, supra- and infra-glenoid tubercle of scapula, and scapular spine (biceps, triceps, and deltoid). (3) Stabilizers and rotators of the scapula muscles (trapezius, serratus, levator scapulae, rhomboids). [Reprinted from Lewis, Warren H (ed). Gray's Anatomy of the Human Body, 20th American Edition. Philadelphia, PA: Lea & Febiger, 1918]

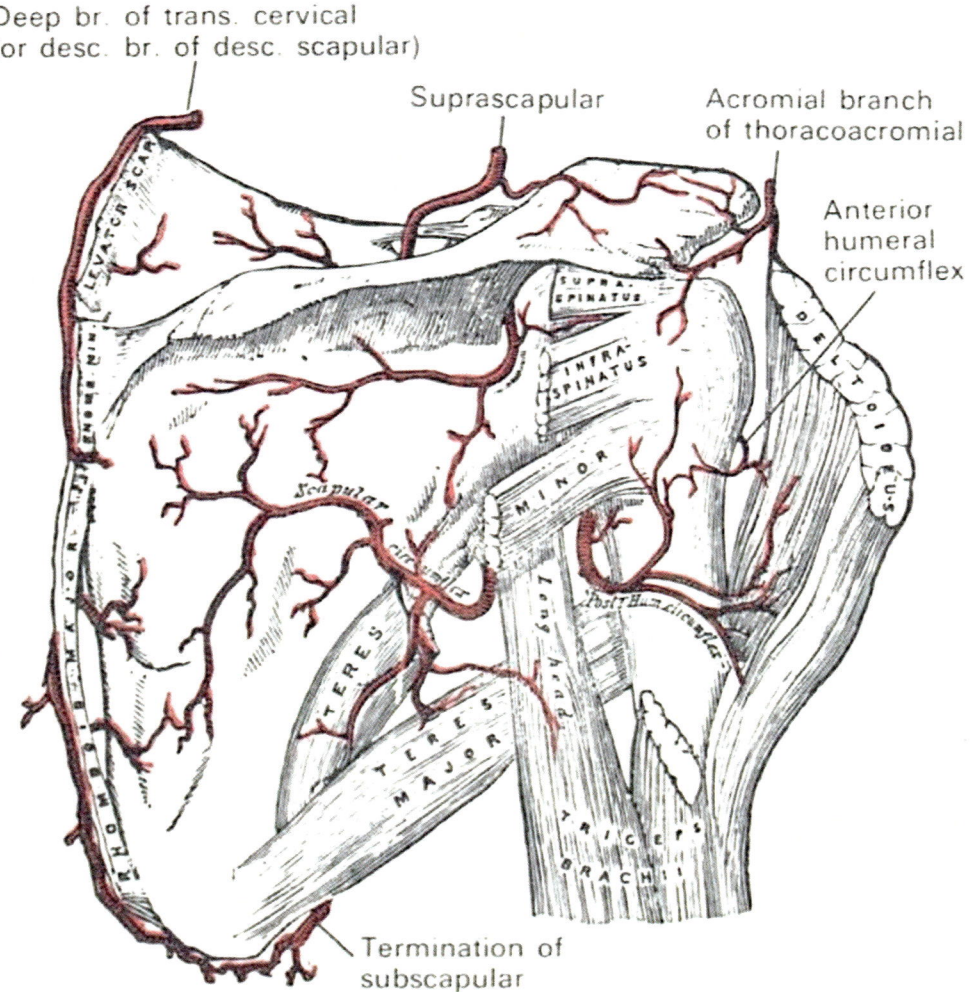

Deep br. of trans. cervical
(or desc. br. of desc. scapular)

Suprascapular

Acromial branch
of thoracoacromial

Anterior
humeral
circumflex

Termination of
subscapular

Fig. 10.60 Vascular supply of the scapula. *Upper (rostral) half of scapula is supplied from the subclavian artery.* Although subclavian has three parts, only thyrocervical trunk is important. First part of subclavian gives rise to thyrocervical trunk with three branches. After *inferior thyroid* goes into the neck *suprascapular artery* supplies structures along the clavicle as far as acromion; it then passes over to supply dorsal scapular muscles (supra- and infra-spinatus). *Transverse cervical artery* supplies the elevators (levator scapulae, rhomboids) and deep plane muscles of scapula to the second part of subclavian produces costocervical trunk. Second part of subclavian lies beneath scalenus anterior and gives rise to costocervical trunk with two branches, neither of which supplies scapula. *Supreme intercostal* goes down to posterior ribs 1–2 and pleura. *Deep cervical* goes up the neck in parallel to inferior thryoid. Third part of subclavian in 50% of cadavers gives rise to *descending scapular artery* which substitutes for deep branch of transverse cervical. *The lower (caudal) half of scapular is supplied by axillary artery.* Axillary artery also has three parts, scapula being supplied from the second and third. First part of axillary artery produces one artery, supreme thoracic (supremely unimportant) to pectorals and anterior chest wall. second part of axillary artery gives rise to two arteries: thoracoacromial trunk and lateral thoracic both perfuse the scapula. (1) Thoracoacromial has 4 branches. *Pectoral branches* go to the muscles attached to clavicle and coracoid process. *Clavicular branch* supplies subclavius. *Acromial branch* goes all the way to distal acromion and the humeral head. *Deltoid branch* supplies the muscle as its insertion into acromion. (2) Lateral thoracic artery descends along the chest wall to supply serratus. Third part of axillary artery has three arteries. (1) Subscapular artery supplies scapula via two branches. *Circumflex scapular* is larger and goes through triangular space to infraspinous fossa. It supplies the teres twins and subscapular. Thoracodorsal is smaller and is the continuation of subscapular to latissimus. (2 and 3) *Anterior and posterior humoral circumflex* are exclusive to the neck of humerus. [Reprinted from Lewis, Warren H (ed). Gray's Anatomy of the Human Body, 20th American Edition. Philadelphia, PA: Lea & Febiger, 1918]

of bone appears. In the third month, the blade of the scapula extends downward and the scapular spine, a derivative of the original anterior coracoid process, projects dorsally. This reflects the rich blood supply around the humerus and thence surrounding the borders of scapula. Failure of ossification at the acromion can create a painful non-union, *os acromiale* (Figs. 10.61, 10.62).

At birth, the scapula is largely osseous but all parts having to do with articulation with the clavicle and humerus remain cartilaginous: the glenoid, coracoid process, and acromion. Those parts having articulation with the thoracic cage, vertebral border, and inferior angle, also remain cartilaginous. The remaining flat surfaces develop via membranous ossification. Final ossification takes place during adolescence in this order: coracoid process, proximal acromion (base), inferior angle and lower vertebral border, distal acromion (apex), and upper vertebral border. These sites reflect vascular growth patterns. Superolateral border represents the axillary

Fig. 10.61 Ossification centers of the scapula. [Reprinted from Anwar I, Amiras D, Khanna M, Walker M. Physes around the shoulder girdle: normaldevelopment and injury pattern. *Clin Radiol 2016; 71(7):702–209.* With permission from Elsevier]

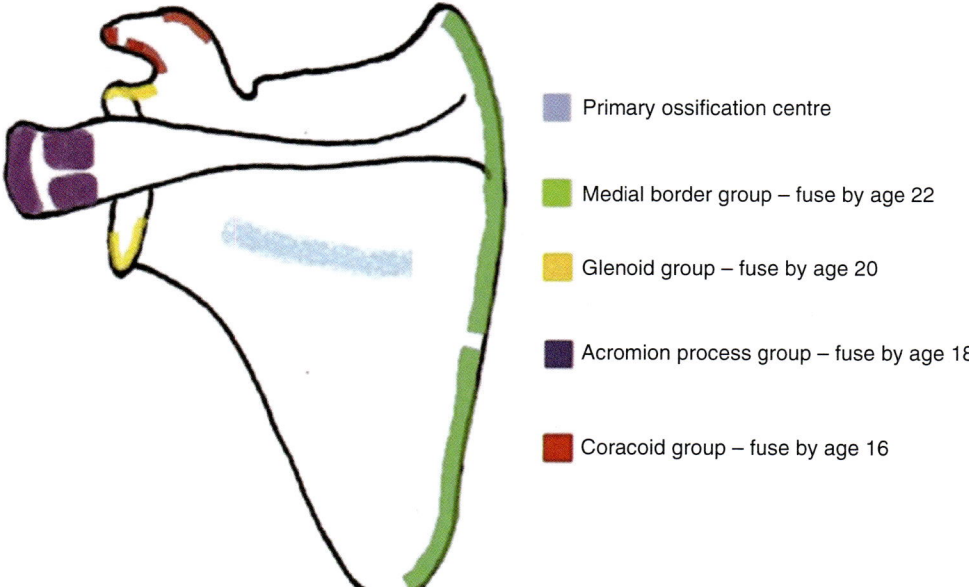

- Primary ossification centre
- Medial border group – fuse by age 22
- Glenoid group – fuse by age 20
- Acromion process group – fuse by age 18
- Coracoid group – fuse by age 16

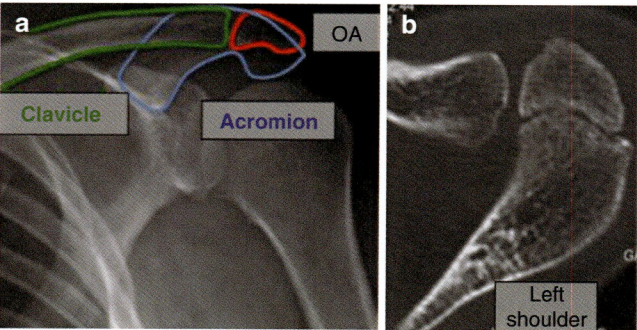

Fig. 10.62 Os acromiale. [Reprinted from Barbier O, Block D, Dezaly C, Sirveaux F, Mole D. Os acromiale, a cause of shoulder pain, not to be overlooked. Orthopaedics & Traumatology: Surgery & Research 2013; 99(4): 465–472. With permission from Elsevier]

artery spreading outward, first ventrally toward coracoid and then dorsally to acromion. Inferomedial border is the scapular arcade: inferior half is subscapular and is followed by superior half, the descending branch of the transverse cervical.

Mapping the scapula: the insertion sequence of appendicular muscles.

Terminology note: At the primary insertion site a muscle can be said to *arise* without implying anything about its origin. Muscles always originate from the somite(s) of their motor nerve(s).

In all instances, appendicular muscles associated with the pectoral girdle have their primary insertions in the LPM bones *with which they are in register*. Bones mature in a fixed biological sequence: their binding sites become progressively available in a fixed spatiotemporal order: deep-to-superficial, distal-to-proximal, and ventral–dorsal. We can make the following generalization: for any given neuromeric level, muscles compete for binding sites according to the "landing site sequence." As myoblasts enter an extremity

they migrate first to the distal limit and then proceed to add up, like water filling up a glass.

Once a primary attachment is accomplished, muscles seek out their secondary insertion site(s). Those supplied by the cervical plexus swerve back to secondary targets at more proximal neuromeres along the midline. Strap muscles (C1–C3) from somites S5–S7, for example, seek out primary insertions into the manubrium and clavicle and thence rebound backward toward the body following the ventral margins of the branchial arches and terminating in the midline geniohyoids. Pectoral girdle muscles supplied by brachial plexus proceed forward to secondary targets at equal or more distal neuromeres along the body wall or upper extremity. Pectoralis minor (C8–T1) attaches first to coracoid process and inserts into third–fifth ribs (T3–T5).

What can insertion patterns reveal about the biological zones of the scapula? One boundary zone seems to occur along the scapular blade (both dorsal and ventral aspects), and along its margins. Levator scapulae (C3–C4) are inserted into the upper zone whereas the rhomboids (C5–C6) from brachial plexus insert below. The order of muscle attachments into specific layers is also useful. As scapula matures, it attracts insertions in the same sequence. Supraspinatus (C5) inserts above the spine; the larger infraspinatus (C5–C6) have a more distal root and inserts caudally. Thus scapular development proceeds, as predicted, from cranial to caudal. This gives us a picture of "mesenchymal flow" over scapula. Subscapularis and supra/infraspinatus attach first. The former grabs onto the chest wall and the latter two claim the highest position along humerus. Lateral border muscles such as teres minor and major come in next. They must attach to humerus more distally. Coracoid process muscles would be laid down in a late sequence. A great "cover-up" then takes place with the antegrade retrograde muscles attaching. On the ventral side and lateral sides, pectoralis

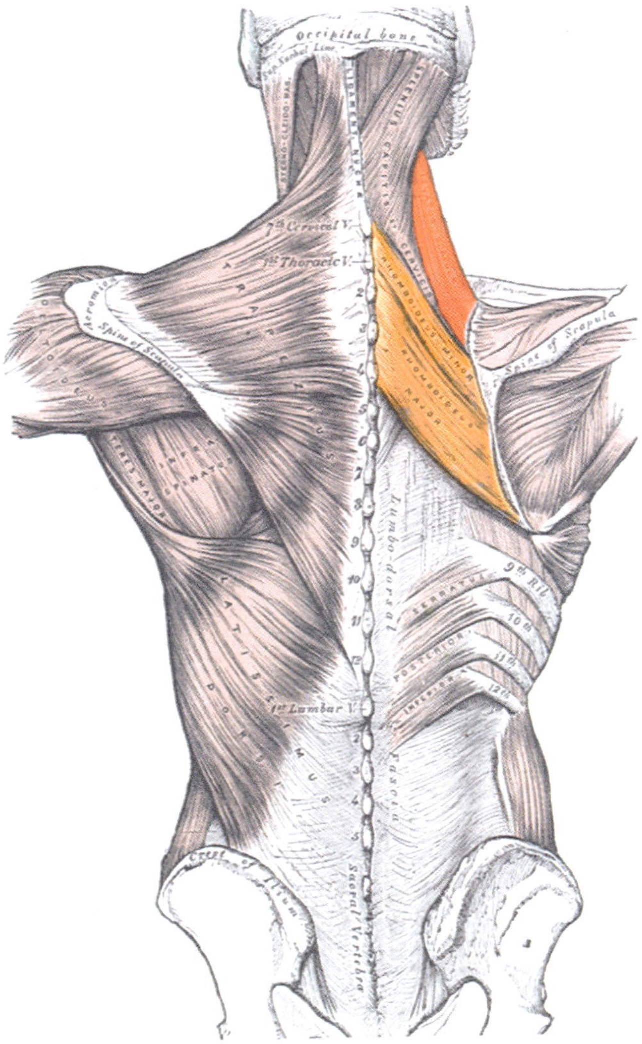

Fig. 10.63 Muscle layering over the scapula. Muscles connecting pectoral girdle to the axial skeleton at or above neuromere t1 represent reassignment of previous muscles and continue to innervated by medial motor column, either directly or as cervical plexus. [Reprinted from Lewis, Warren H (ed). Gray's Anatomy of the Human Body, 20th American Edition. Philadelphia, PA: Lea & Febiger, 1918]

major and deltoid are subcutaneous. Trapezius fulfills the same function over dorsal scapula. It covers over its competitors, supra- and infraspinatus (Fig. 10.63).

Superior Border: LPM from c1 to c4

Omohyoid (inferior belly) arises from S6 (and sometimes S7). It is supplied from cervical plexus C2 (C3). The primary insertion is immediately medial to the suprascapular notch. It joins with an anterior belly that develops from S5 (S6). Together, the muscle sling makes a secondary insertion into hyoid just lateral to sternohyoid. Recall that we are coding manubrium as a c1 PAM bone. Spine of scapula—a c4 derivative—divides muscles connecting with the head from those extending medially toward the body. Blood supply to the upper omohyoid is lingual (second and third arches) and to

lower muscle from superior thyroid (fourth and fifth arches). Thus, the vascular supply betrays the function of this muscle as it connects the pectoral girdle to the pharyngeal arches.

Acromion: LPM from c5 to c6

Deltoid (middle fibers) arises from S9 to S10 and is supplied by axillary nerve (C5–C6, posterior cord). Because it spans from pectoral girdle to humerus its blood supply reflects distinct zones of axillary artery. The proximal muscle receives the thoracoacromial from second part of axillary artery while the distal muscle is supplied by the anterior and posterior humeral circumflex branches from third part of axillary. At the tip os of.

Coracoid Process: LPM from c5 to t1

This process projects forward from the upper border at almost a right angle. It lies below the lateral one-fourth of the clavicle. It has three muscles associated with it. Recall that the coracoid process represents the incorporation of the ancient *posterior coracoid* into scapula. It develops from medial to lateral, that is, short head of biceps > coracobrachialis > pectoralis minor. Because the latter two muscles have a common motor nerve, coracoid has two ossification centers: an inferior center represents the musculocutaneous components while the superior center represents the medial pectoral nerve-supplied pectoralis minor.

Biceps (short head) arises from S9 to S10 and is supplied by musculocutaneous nerve (C5–C6). The primary insertion is at the lateral apex of the coracoid process (in common with coracobrachialis). It terminates at the tuberosity of the radius (posterior side).

Coracobrachialis arises from S10 to S11 and is supplied by the musculocutanous nerve (C6–C7), the terminal branch of lateral cord. The primary insertion is at the medial apex of the coracoid process. It has its secondary insertion on the humerus between the primary insertions of triceps, proximally, and brachialis. Blood supply comes from the third part of axillary as anterior humeral circumflex.

Pectoralis minor arises from S12 to S13 and is supplied by the medial pectoral nerve (C8, T1) from brachial plexus and medial cord. The primary insertion is the superior aspect of the process. It fans out to insert into the chest wall at ribs 3–5. Pectoralis minor defines the three parts of axillary and is supplied by thoracoacromial axis from its second part. Its medial border also receives the superior thoracic artery, the sole branch from first part of axillary artery.

Medial Border, Posterior: LPM from c3 to c5

Levator scapulae arises from S7 to S9 and is supplied by spinal nerves of C3 and C4 plus C5 via dorsal scapular nerve. Its primary attachment is to superior medial border of scapula above the spine. It ascends to make secondary attachments to the transverse processes of atlas and axis, C3, and C4. It is supplied by two vessels. Ascending cervical is from inferior thyroid branch of thyrocervical trunk whereas transverse cervical arises directly from thyrocervical trunk, from first part of subclavian immediately lateral to the vertebral axis.

Rhomboid minor and major both originate from S8 to S9 and are supplied by a branch of dorsal scapular nerve, C4, C5. Its primary insertion is to inferior medial border of scapula below the spine. Blood supply is from transverse cervical artery from thyrocervical trunk. This latter, being proximal to suprascapular, descends under the rhomboids to reach the tip of scapula.

Medial Border, Anterior: LPM from c5 to c7

Serratus anterior arises from S9 to S11 and is supplied by long thoracic nerves C5, C6, and C7. It has its primary insertion to medial border of scapula. Secondary insertions are sought out to ribs 1–8 (and sometimes to ribs 9 and 10). It is supplied by the lateral thoracic artery from second part of the axillary artery.

Lateral Border, Posterior: LPM from c5 to c7

Triceps brachii (long head) arises from S9 and is likely supplied by the upper components (C5–C5) of radial nerve (C5–T1) from the posterior cord. The secondary insertion is at the infraglenoid tuberosity: exactly the location of an ossification center! The primary supply of triceps is distal, away from the muscles associated with scapula, specifically from profunda brachii from superior ulnar collateral. However, the proximal zone of the long head is perfused from posterior humeral circumflex.

Teres minor arises from S9 to S10. It is supplied by axillary nerve, C5, and C6. Its insertion is upper 2/3 of posterior lateral border and the lateral 1/3 of the fossa. It has a dual blood supply, the proximal scapular zone perfused by circumflex scapular, from subscapular artery, the largest branch of the third part of axillary artery. The distal zone is supplied by posterior humeral circumflex.

Teres major arises from S9 to S11. It is supplied by lower subscapular nerves C5, C6, and C7. Its primary insertion is from the lower 1/3 of the posterior lateral border and the lateral 1/3 of the fossa.

Pectoralis minor arises from S12–S13 and is supplied by medial pectoral nerve (C8 and T1) from brachial plexus, medial cord. The primary insertion is superior aspect of the process. It fans out to insert into the chest wall at ribs 3–5. Like teres minor, it also has a dual blood supply with proximal zone from thoracodorsal closer to the chest wall leading to posterior humeral circumflex at the arm.

Dorsal Scapula: LPM from c5 to c6

Supraspinatus arises from S9 to S10; it is supplied by suprascapular nerve (C5 and C6) from the upper trunk. It has an insertion from the medial 2/3 of the fossa. Two arteries perfuse it. Suprascapular artery is a branch of thryocervical trunk, dorsal scapular artery is the deep branch from transverse cervical, and also a branch of thyrocervical trunk.

Infraspinatus has the same characteristics as supraspinatus. Its blood supply comes from suprascapular and circumflex scapular, a descending branch of subscapular.

Ventral Scapula: LPM from c5 to c6

Subscapularis arises from S9 to S10; it is innervated by subscapular nerve (C5, C6), the first branch from the posterior cord. The branch sequence is subscapular > axillary > radial. This large muscle is perfused by branches from three sources: axillary to the upper lateral half, and subscapular to the lower lateral half, with a contribution of suprscapular.

Scapular Spine: LPM from c3 to c6

Trapezius has primary insertion along the entire spine from lateral to medial and is encoded probably c3–c4 to c6. It has three distinct vascular zones. The upper zone is supplied by the transverse muscular branch of occipital artery, with neural crest related to the second arch. This may relate to the embryologic part of the muscle that originally was associated with the hindbrain. The middle zone is supplied by the transverse cervical artery from thyrocervical trunk from the first part of subclavian. The lower zone is supplied by dorsal (descending) scapular artery which arises from the second or third part of subclavian.

Clinical correlations of pectoral girdle coding: the strap muscles demystified.

From neuromeric levels, r8–r11 and c1–c3 somites produce two groups of hypobranchial muscles that follow a curvilinear pathway downward and forward to their targets. They bear the sobriquet hypobranchial because they must migrate around the boundaries of the arches. The first group from S1 to S4 provides the extrinsic muscles of the tongue and has primary attachments from the second arch (styloglossus and hyoglossus) and first arch (genioglossus) into the tongue matrix. S5 produces genioglossus which is mistakenly included with the tongue muscles…really is the anterior extension of the strap muscles.

The second group of hypobranchial muscles comes from S5 to S7 and is known as the strap muscles. These descend downward first to the sternum and clavicle and thence upward beneath the fourth and third arches to attach to the hyoid bone. Their primary insertions are into the sternum, manubrium, and clavicle in medial-to-lateral sequence. Because none of the strap muscles are innervated by c4, the insertion of the omohyoid stops short of distal clavicle.

Note that second arch platysma muscle fibers drape over the clavicle but are separated from it by neural crest SIF fascia whereas the fasciae enclosing muscles with true attachments to the clavicle is DIF and is mesodermal.

Phylogeny of the Pectoral Girdle

Life on land would be unthinkable without locomotion. The transition from fins into limb is a fascinating story, well worth an entire chapter (or even a book). Because the appendicular skeleton is well represented in the fossil record, we can directly observe the structural details of animals long extinct; we can also follow the phylogenetic changes leading to the modern human limb. For our purposes, we shall focus on the relationship of the pectoral girdle (and perforce the upper extremity) to the head and neck. In our previous discussion of the neuromuscular organization of the neck, we noted that the spinal accessory nerve and the eight cervical nerves innervate

several functional classes of muscles: (1) cranium to pectoral girdle, (2) cervical vertebrae to cranium, (3) cervical vertebrae to each other, (4) cervical vertebrae to pectoral girdle, and (5) cervical vertebrae to trunk. Obviously, the development of the clavicle and scapula is inseparable from that of occipital bone, mastoid, and the cervical column…but how? [14] (Figs. 10.64, 10.65, 10.66, 10.67, 10.68, 10.69).

Let's get a quick bird's-eye view of the subject. The appendicular system consists of pectoral and pelvic girdles that support the limbs. The girdles differ radically in composition. The pectoral girdle appears first in evolution with Tiktaalik; it is composed of both dermal and endochondral elements. The pelvic girdle occurs later in time with tetrapods Acanthostega/Ichthyostega; it is strictly endochondral. *The existence of a dual design suggests differing evolutionary mechanisms.*

Endochondral bones arose from the basal supporting structures of the forefin. During evolution, these lateral plate structures expanded outward into the surrounding skin envelope, where dermal bones derived from the original encasing bones were encountered. Thus, coracoid articulates posterior to the neural crest bone series that is suspended between the

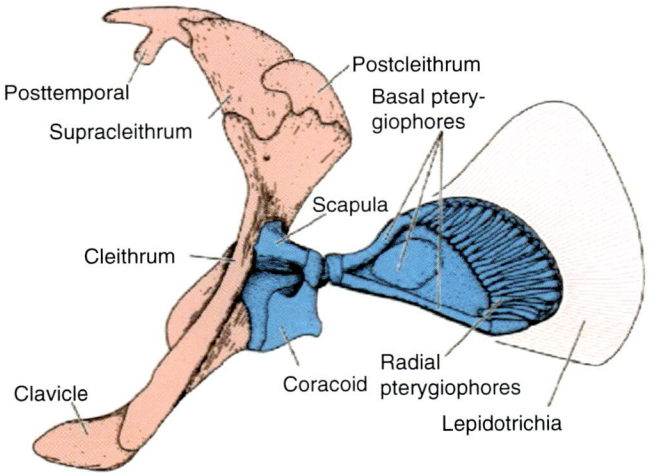

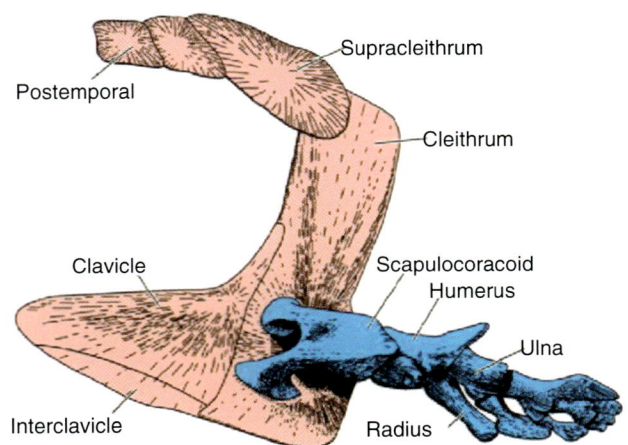

Fig. 10.64 The pectoral fins in actinopterygian fishes, seen here as *Polypterus*, is suspended from a pectoral girdle with two components. (1) 5 neural crest dermal bones (red) and mesodermal chondral bones (blue). The dermal series hugs the posterior margin of the opercular series covering the gills. Post-temporal articulates with the skull and clavicle extends to the gulars. Chondral scapulmesodermal articulates with 3 fused basal pterygiophore. These elements are discrete in chon-

drichthyans. Multiple radial pterygiophores extend outward but are united by distal arc of bone. Lepidotrichia (lepido—too many to count) complete the fin. The pectoral girdle in sarocopterygian *Eustheopteron* retains the dermal bone series but shows the etrapod form of chondral bones: stylopodium, zeugopodium, autopodium. [Courtesy of William E. Bemis]

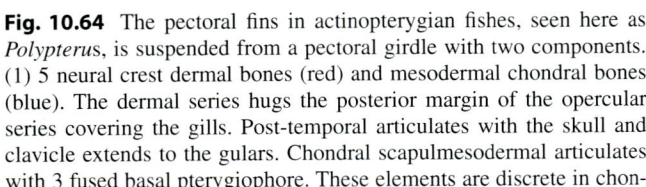

Fig. 10.65 *Eogyrinus Attheney* represents the basal tetrapod stance for terrestrial life. Pectoral girdle overlaps initial 3 cervical vertebrae. [Reprinted from Wikimedia. Retrieved from: https://commons.wikime-

dia.org/wiki/File:Eogyrinus_Attheney._Wellcome_M0006666.jpg. With permission from Creative Commons License 4.0: https://creativecommons.org/licenses/by/4.0/deed.en]

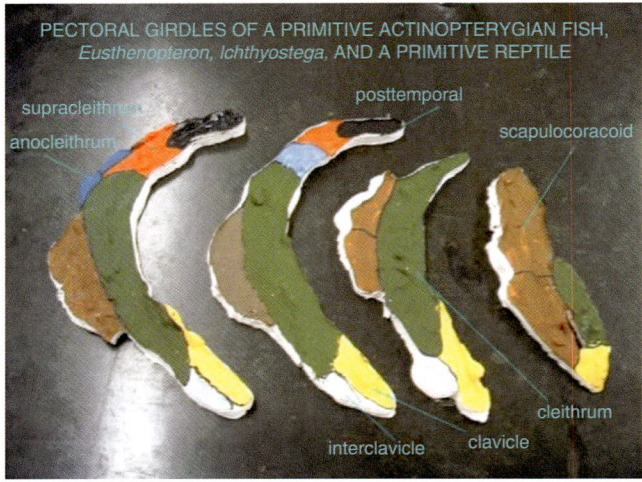

Fig. 10.66 Reduction of dermal bones. [Reprinted from: http://bio.sunyorange.edu/updated2/comparative_anatomy/anat_3/a_shoulder.htm. With permission from Dr. Walter Jahn]

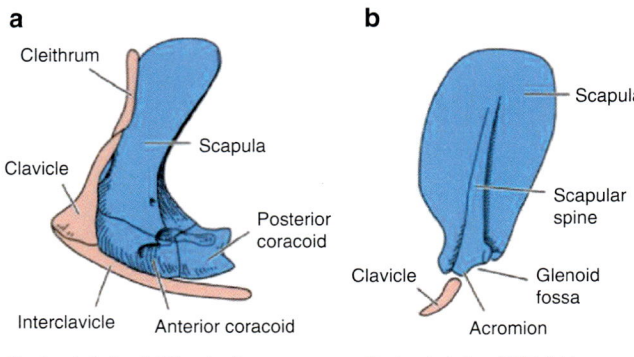

Pectoral girdle of †Dimetrodon   Pectoral girdle of Didelphis

Fig. 10.68 Comparison of pectoral girdle in amniotes from †*Dimetrodon* to primitive mammal, *Didelphusi*, the opossum. Cleithrum is relocated to dorsal aspect of scapula as the scapular spine, terminating in acromion and glenoid fossas. Anterior coracoid becomes manubrium; it maintains its relationship to calvicle. Posterior coracoid becomes the true coracoid process. [Courtesy of William E. Bemis]

Fig. 10.67 Tetrapod limb design show shift of the limb axis to the ulna. Stylopodium (purple) = purple. Zeugopodium (orange) = ulna/radius. Autopodum (yellow) = hand/foot [Courtesy of William E. Bemis]

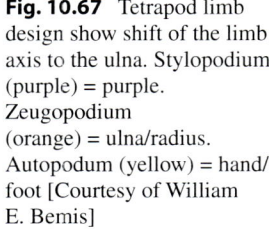

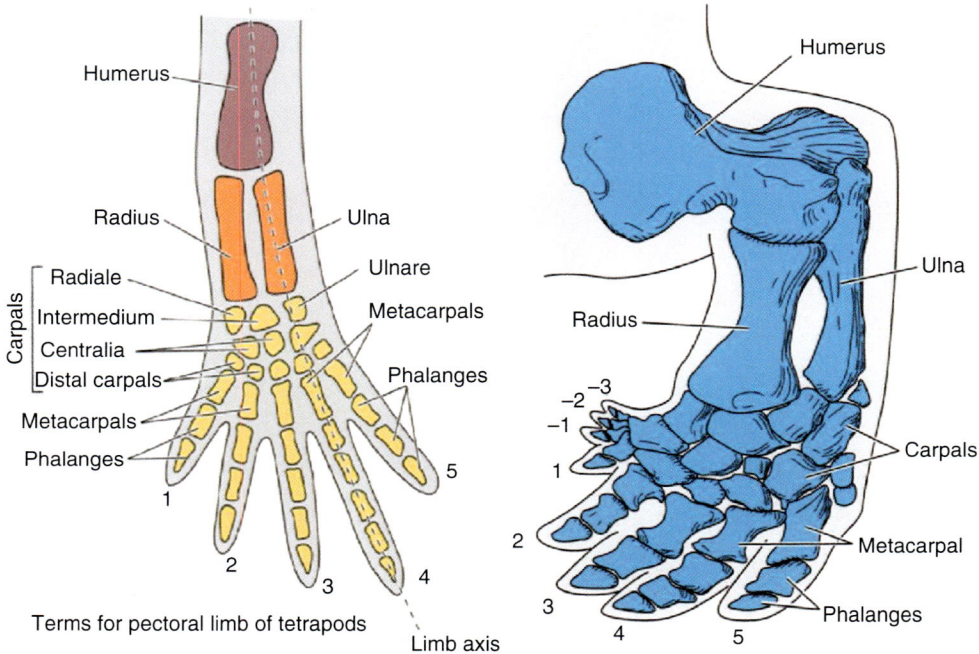

Terms for pectoral limb of tetrapods

skull and the gular bones and follow the posterior borders of the opercular series.

Primitive fishes carry a variety of projecting spines, knobs, and lobes. Fins are different: they are thin and membranous; they also are internally supported by fin rays. Like scales, fin rays develop at the epidermal/dermal interface, but subsequently sink into the dermis, where they are known as *dermal fin rays*. These take various names. *Ceratotrichia* are keratinized rods in elasmobranchs. Bony fishes have fin rays and *lepidotrichia*; these are often ossified. The proximal fin near the body is reinforced by *pterygiophores*. The proximal fin contains *basals* while slender *radials* extend from the basals into the substance of the fin. Fishes have two types of fins. Paired fins are pectoral and pelvic; these articulate with

girdles inside the body wall; the rest are singular. The upper limb arises from the pectoral fins.

Paired fins were an early innovation in piscine evolution, bringing added mobility and stability to fishes swimming within a watery environment in constant motion. They first appeared in the ventrolateral fin folds of early agnathans. Ostracoderms had only pectoral fins but later ancient fishes (placoderms, acanthodians, and the chondrichthyans) had both pectoral and pelvic girdles. Rhipidistians, such as *Eusthenopteron*, made use of fleshy fins that acted as pivots.

The tetrapod limb, formally termed the *chiridium*, is, by definition, digit-bearing. It always consists of three regions. The *stylopodium* is closest to the body; it is always singular (humerus and femur are analogous). The middle limb, *zeu-*

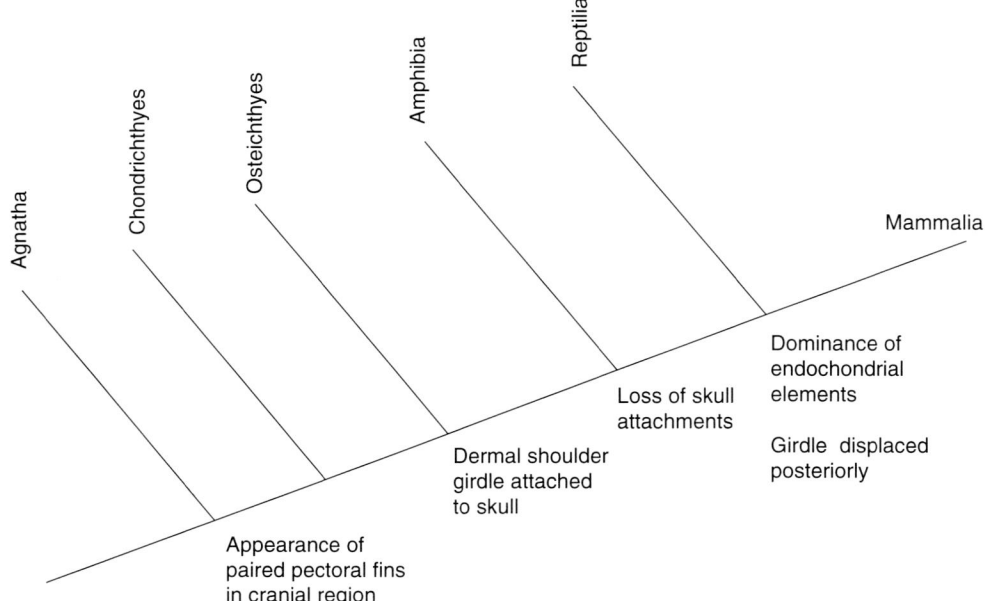

Fig. 10.69 Evolution of pectoral girdle. Tetrapods replace dermal bones of fishes with LPM. Early tetrapods lose posttemporal and gain interclavicle. Note: cucullaris in fishes inserts into anocleithrum. Loss of this bone at petoral dissociation forces cucullaris to relocate back to scapula. Note: the function of clavicles and coracoids is to serve as a brace against the sternum. Coracoid develops from embryonic coracoid plate. Anterior ossification centers give rise to *procoracoids*; posterior ossification centers give rise to *coracoids*. Clavicle absent in most non-avian reptiles; present in most mammals. [Reprinted from McGonnel IM. The evolution of the pectoral girdle *J Anat* 2001; 199:189–194. With permission from John Wiley & Sons]

gopodium, always has two elements: ulna and radius (tibia and fibula) in the forelimb. The distal upper limb, termed *autopodium*, contain the carpus and manus (tarsus and pes).

The pectoral fin was originally attached to the skull via a series of dermal bones: postempoal, supracleithrum, post-cleithrum (anocleithrum), cleithrum, clavicle, and (later) inteclavicle. The evolution of the pectoral girdle in advanced teleosts and tetrapods involved the *progressive loss of dermal elements*. In the transition to tetrapods, the endochondral element of fishes, scapulocoracoid, became dominant with two separate ossification centers: scapula and procoracoid.

Early amniotes added a brand-new endochondral element, the coracoid (=posterior coracoid). Modern tetrapods retain both "coracoids." Procoracoid becomes the dominant ventral shoulder bone in amphibians, reptiles, and birds, whereas therian mammals (that includes us) retain only the coracoid. Anterior coracoid is the likely source of the upper manubrium, where it fuses with lateral plate mesoderm to form a single structure. The cleithrum persists as the spine of the scapula.

Origin of Paired Fins: The Fin-Fold Theory and Homeotic Genes

In the mid-nineteenth century, E.M. Balfour and J.K. Thacher put forth a remarkably prescient theory regarding limb position that turned out to have a molecular rationale. Fishes were considered to have a continuous set of ventrolateral fin folds. These were reinforced by two kinds of endoskeletal

struts called *pterygiophores*. These had two components: *basal* bones were proximal and extended inward; distal to the basal were *radial* bones. Over the course of time, the basal elements fused across the midline to form the shoulder and pelvic girdles. Dermal bones, originally part of the skin of armored fishes, became "add-ons" to the pectoral girdle (but not to the pelvis) (Fig. 10.64a).

Several lines of evidence support the fin-fold theory; these range from indirect paleontologic findings to molecular mapping. Prehistoric agnathic fishes such as *Myllokunmingia* and *Haikouichthyes* had fin folds. Presumably, pectoral and pelvic fins arose at the same time. Shark embryos have continuous lateral wall ectodermal ridges.

Fins arise at specific neuromeric levels. Jarvik made two important contributions. First, he hypothesized that the mesenchymal core of the fin folds was capable of forming endoskeletal basals and radials. These, in turn, would serve as attachment sites for muscles from nearby segmental myotomes, providing control of the fin. Overlying scales became supportive dermal fin rays, an event documented in the embryogenesis of many living fishes. Dermal bone consolidation offered convenient sites of attachment along the anterior surface of the bones. At the same time, the dermal girdle would form the posterior wall of the pharynx, could protect the heart, and would provide attachment sites for pharyngeal arch muscles.

Jarvik emphasized the importance of the pectoral girdle at the transition point between axial musculature and branchial (pharyngeal) arch slits (Kardong, pp. 246–247). If we assign one mesodermal element (somitomeres or somite) per arch, then the primitive fish formula is 3 somitomeres +12 somites. The formula for sharks is 3 + 4; that for mammals is 3 + 2. Chondrichthyan fins arise at the fourth somite. Recall that the maximum number of branchial arches documented in the fossil record is 15. Thus, the primitive fish fin can arise as far back as the 12th somite. *The pectoral fin zone extends from S4 to S12.* In mammals, the S12 level marks the transition between the seventh cervical vertebra and the first thoracic. By extension, *the mammalian pectoral zone extends from the OCJ down to T1.*

Homeotic genes provide the underlying basis for the fin-fold hypothesis. Expression of *Engrailed-1* defines the ventral compartment of the embryonic body. Within the ventral zone, pre-vertebrates such as *amphioxus* express a single *T-box* gene, *AmphiTbx4/5*. This is the ancestral gene of the fin fold. In sharks, further genetic takes place, with the anterior fin defined by *Tbx5* and the posterior limb by *Tbx4*. The tetrapod limb makes use of an entirely new gene, *Sonic hedgehog*, to accomplish three revolutionary changes: (1) *Shh* promotes outgrowth of the limb from the body wall; (2) it liberates the limb from being in parallel with the body, and; (3) it sets up a proximal-to-distal limb axis.

The Pectoral Girdle in Fishes

Ostracoderms Although agnathic fishes dating back to the Cambrian had ventrolateral fin folds, paired pectoral fins made a rudimentary appearance in these armored fishes. They were definitely not buoyant. Like acanthodians, chondrichthyans, and placoderms, the ostracoderms lacked a swim bladder. These bottom-dwellers made use of their folds to undulate over the ocean floor.

Placoderms appeared in the Silurian and radiated widely, being active swimmers with strong jaws. They had a fused endoskeletal pelvis and a pectoral girdle consisting of several dermal bones attached proximally to the *postemporal bone*. These served as armor plate and reinforced a singular endoskeletal bone, the scapulocoracoid.

Chondrichthyans used their fins as stabilizers. In more derived sharks the paired basal components extended along the midline of the body and fused into a U-shaped *scapulocoracoid bar*. More caudally they formed a similar puboischiac bar as well. Chondrichthyans do not form membranous bones. From the get-go, they lost all dermal contributions to the shoulder girdle. In their modern iteration, sharks have three large pterygiophores extending off the pectoral girdle.

Acanthodians had fins with large spines along their leading edges. The sole manifestation of a pectoral girdle was a connection between a pectoral spine and the scapulocoracoid.

Actinopterygians are modern bony fishes. They have a pectoral girdle that is mostly dermal. Because they have a swim bladder, these fishes are buoyant, so they use the pectoral fins for maneuvering. The shoulder girdle is U-shaped and draped behind the gill chambers. The dermal girdle attaches to the skull by means of the *posttemporal* bone. This is followed by a *supracleithrum* to which are attached, in some fishes, additional dermal bones. Specifically, postcleithrum (also called anocleithrum) is found in this position. The largest dermal element, *cleithrum*, articulates with endoskeletal scapulocoracoid. Finally, a mixed membranous-chondral bone, *clavicle*, extends across the midline beneath the gill chambers to fuse with its opposite member as a *symphysis*. It does so via the union of paired intervening bone field, the *interclavicles*. This is *not* the source for manubrium (it originates from endochondral procoracoid).

Dermal elements of the pectoral girdle are found in bony fishes, sarcopterygians, and early tetrapods before fading from the scene. Following Kardong, we shall refer to these collectively as the "*postcleithrum*," even though this breaks with traditional nomenclature. Comparative anatomists apply the term anocleithrum in sarcopterygians and tetrapods to the postcleithrum of actinopyerigians. This redundancy makes the literature confusing.

Sarcopterygians have dermal fins supported by muscles and internal supportive elements. These are represented today by three kinds of lungfishes and the deep-dwelling coelacanth, *Latimeria*. These have reduced shoulder elements. Fortunately, fossil rhipidistians left behind a detailed record of their appendages. These points the way toward tetrapod evolution. *Eusthenopteron* in the late Devonian had dermal fins supported by bones homologous to tetrapods. Pectoral fin articulates first with scapulocoracoid and a series of supporting dermal bones. A new element, interclavicle, is first seen here. It will be retained in some later tetrapods, but not in mammals (Fig. 10.64b).

The Pectoral Girdle in Tetrapods

Early tetrapods spent the early part of their life cycles in water. Skulls of *Europs* show the continued presence of an aquatic sensory system, the lateral line canals. Between the Carboniferous and Permian periods, tetrapod girdles and limbs increased in strength as an adaptation to weight bearing. The primitive state is shown by *Eogyrinus*, a Carboniferous tetrapod. Despite a length of nearly 6 feet, it had relatively small and cartilaginous limbs. Movement was likely similar to that of salamanders, in which the limbs merely serve as pivot points around which the body swings and sways. The limbs of *Europs* were stout and densely ossified, consistent with a primary function of ambulation (Fig. 10.65).

The pectoral girdle changed significantly between rhipidistian fishes and tetrapods.

Concomitant, or shortly after, arrival on land, *tetrapods lost the posttemporal bone*. This freed up the shoulder from the skull and eliminated the constant jarring of the skull from ambulation. The dorsal series of dermal bones, such as supracleithrum and postcleithrum, are lost. The remaining elements were ventral: endochondral cleithrum, dermal clavicle, and paired interclavicles that met in the midline in front of the neck. In early amniotes cleithrum "disappears." In its place are two distinct endochondral entities: a dorsal *scapula* and a ventral *coracoid* (Fig. 10.66).

Innovations in the Tetrapod Limb

Tetrapod limbs are characterized by a common developmental plan. Although all limbs possess an axis of symmetry, the postaxial elements predominate. The most proximal element is axial and singular: the *stylopodium* (humerus and femur). A second-order *zeugopodium* consists of two elements: the postaxial ulna and preaxial radius order elements that develop from the postaxial primordial. The subsequent development of tertiary elements is very different. All branching units of the manus are postaxial in origin. The preaxial element can produce linear third and fourth-order structures but cannot branch. This genetic pattern is very ancient. Chondrichthyans, primitive actinptyerygeians, and rhipidistian fossils all show a metapterygian stem branching into postaxial elements. Comparative analysis of Hox gene expression in vertebrate limbs shows that tetrapods deform the straight metapterygial axis of fishes. The axis becomes bent anteriorly to support a distal limb such that all the digits arose from the posterior elements (Fig. 10.67).

In primitive amniotes (reptilomorphs) a brand-new endochondral element appears, confusingly named the *new coracoid*. It is located just behind the original coracoid. We can avoid getting fouled up in terminology and also track the evolution of these components by giving them separate names. The older bone element, homologous to the original tetrapod coracoid, is the *anterior coracoid* or *procoracoid*. The newer bone element is the *posterior coracoid*, or *metacoracoid*.

The three-part scapula evolved in two directions. *Sauropsids* (birds and reptiles) maintain the three-part version. Procoracoid is the dominant ventral shoulder bone. Synapsids, preserve the dermal elements to varying degrees. Clavicle is retained across the board in all taxa. Monotremes continue to possess interclavicles but *marsupials and mammals lose the interclavicles*, thus permitting direct articulation between clavicles and sternum. In mammals, the two coracoids split up, allowing the glenoid fossa to face ventrally instead of posteriorly. This positions the upper limb beneath the thorax and is useful for enhanced mobility. Posterior coracoid remains behind and fuses to the scapula as the *coracoid process*. It is positioned ventral to clavicle. Anterior coracoid shifts medially to form the *manubrium*. Mammalian evolution found dorsal repositioning of the otherwise hypaxial pectoral girdle to be a selective advantage as it gave greater leverage for limb control. *Scapula shifted dorsally to provide more efficient insertions*. Cucullaris, a branchial arch dilator (and therefore hypaxial) split apart into sternocleidomastoid and trapezius. The former remained in register with clavicle while the latter chose the dorsal margin of scapula—and an opportunistic series of insertions along the "siderails" of the thoracic spine (for details, *vide infra*) (Figs. 10.68 and 10.69).

The Sternum: Phylogeny and Development

The sternum is classified as part of the appendicular skeleton. It develops in humans by chondrification of lateral plate mesoderm in the distal interneuromeric spaces. The presternal cartilages are distal to the ribs. Because they come from LPM they do not have parasegmentation. Therefore they readily undergo longitudinal fusion (much like internal thoracic/mammary artery) to create a *sternal band*. These subsequently fuse with one another in the midline. Fusion failures result in a cleft sternum or even *ectopia cordis*. Longitudinal failures can occur as well (Figs. 10.70, 10.71).

Stenum is lacking in fishes. It is not seen in stem fossil tetrapods like *Acanthostega*, perhaps because its origins were in cartilage, but it is present in amphibians so we assume it appeared somewhere in the tetrapod transition. Amphibian ribs do not fuse with the sternum. Amniotes achieve a fused condition.

In synapsids, interclavicle is a presumed candidate as the source of the sternal bands. Interclavicles quickly become interposed between the clavicles, forcing them to retreat laterally. Shortly after the initial syapsid state, paired manubria appear caudal to the interclavicles; they fuse together into manubrium. From its very inception manubrium has an articulation. Basal theriodonts show shortening of interclavicles such that they equal the manubrium in size. In the crown state, theriodonts show complete incorporation of interclavicles into the manubrium to become the first sternal segment. Upper part of manubrium (interclavicle) therefore articulates with clavicle and lower part of manubrium articulates with the first rib (Fig. 10.72).

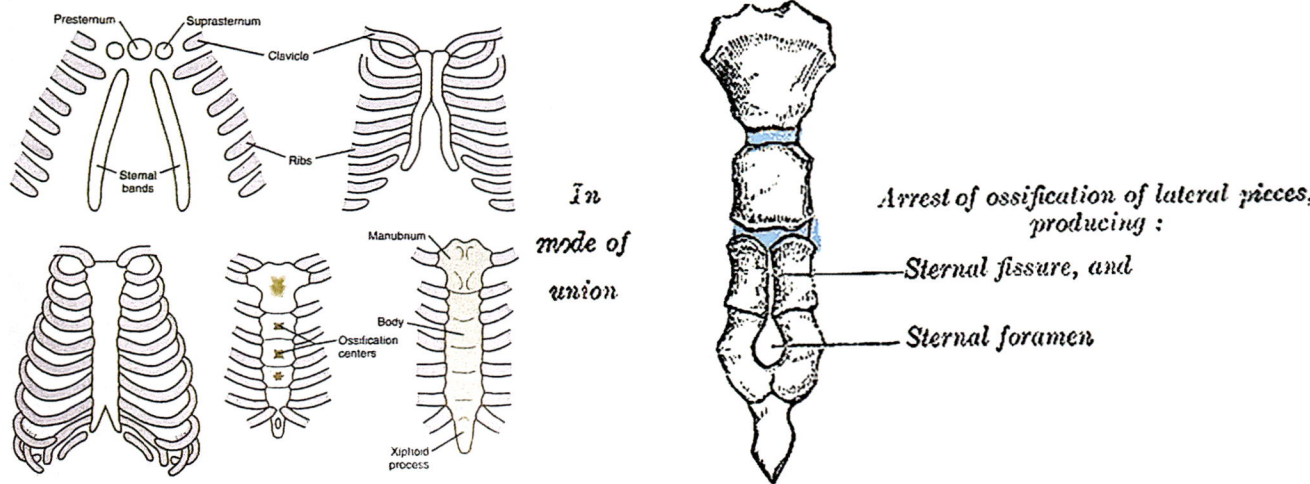

Fig. 10.70 Development of the sternum BMC. Sternum develops from ossified lateral plate mesoderm cartilages in the distal interneuromeric space. These are distal to, and separate from, the corresponding ribs. Because they are LPM (and do not come from somites) these distal components do not have parasegmentation. In the parasternal position the segments unite longitudinally. Failure of fusion of both sides is seen in sternal fissures. Left: [Reprinted from Carlson BM. Human Embryology and Developmental Biology, sixth edition. St. Louis, MO: Elsevier; 2019. With permission from Elsevier.] Right: [Reprinted from Lewis, Warren H (ed). Gray's Anatomy of the Human Body, 20th American Edition. Philadelphia, PA: Lea & Febiger, 1918]

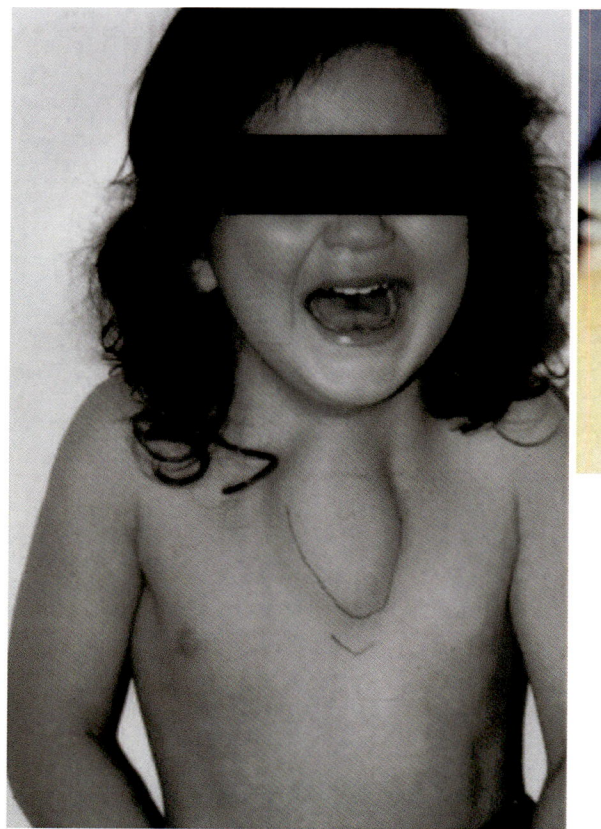

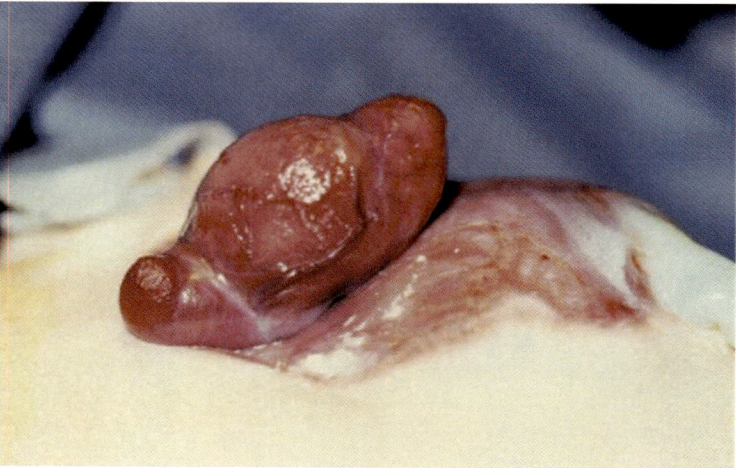

Fig. 10.71 Failure of fusion of the sternal bands can be either cranil-caudal or caudal-cranial. A superior vertical defect is seen, In severe cases this may lead to ectopia cordis. Left: [Reprinted from Acastello E, Majluf R, Barbosa LM, Peredo A. Sternal cleft: a surgical opportunity. J Ped Surg 2003; 38(2): 178–183. With permission from Elsevier.] Right: [Reprinted from Engum S. Embryology, sternal clefts, ectopia cordis, and Cantrell's pentad. Semin Ped Surg 2008; 17:154–160. With permission from Elsevier]

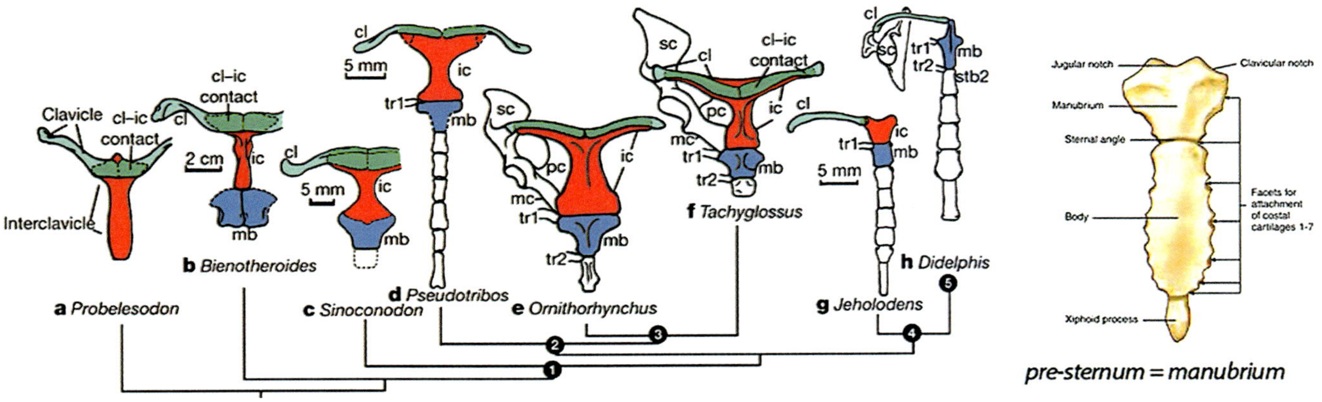

Fig. 10.72 Phylogeny of the pectoral girdle of basal mammals and their relatives. Note the presence of an interclavicle (red), clavicles (green) and a new bone, the manubrium (deep blue), which develops where the sternum develops in other tetrapods. Note zone of clavicle-interclavicle contact (dark green) in immediate pre-mammals (4) Interclavicle is interposed between manubrium and clavicle. Evolution of interclavicle and anterior sternal structure in the cynodont–mammal transition. (Placement of Pseudotribos in mammal phylogeny is shown in Supplementary Information.) **a**, Cynodont Probelesodon. **b**, Mammaliamorph Bienotheroides. **c**, Mammaliaform Sinoconodon. **d**, Pseudotribos. **e**, Monotreme Ornithorhynchus. **f**, Tachyglossus. **g**, Eutriconodont Jeholodens. **h**, Marsupial Didelphis. Apomorphies of (1) mammaliaforms: widening anterior end of interclavicle; posterior club-foot of interclavicle to the same width of manubium for extensive attachment of pectoralis muscles. (2) Yinotheria: 'T-shaped' manubrium. (3) Monotremata: gracile and elongate lateral process of inter-clavicle overlapping two-thirds the length of clavicle; enlargement of (meta)coracoid to articulate directly with the interclavicle; presence of procoracoid for strengthening the girdle. (4) Theriimorpha: shortening of interclavicle to equal manubrium; reduction of the lateral process of interclavicle; mobile claviculo–interclavicle articulation. (5) Crown Theria: incorporation of embryonic interclavicle into the manubrium, the first sternal segment in articulation of clavicle and thoracic rib 1. cl, clavicle (green); cl–ic contact zone between clavicle and interclavicle, articulation or overlap of clavicle and interclavicle (yellow); ic, interclavicle (red); mb, manubrium sternebra 1 (blue); mc, metacoracoid; pc, procoracoid; sc, scapula (grey); stb2, second sternebra; tr1 and tr2, thoracic rib costal cartilage 1 and 2, respectively. [Reprinted from Luo Z-X, Ji Q, Yuan C-X. Convergent dental adaptations in pseudo-tribosphenic and tribosphenic mammals. *Nature* 2007; 450:93–97. With permission from Springer Nature]

Neurology of the Neck

Introduction

Go on, jump in, the water is fine! People say ridiculous things like this when you are standing at the edge of a swimming pool…and you know the water is *not* warm. So why should we have to immerse ourselves in things such as the spinal cord, the cervical roots, and the cervical plexus? And what should we gain from our efforts?

Our goals are simple: we are out to discover the what, the how, and the why of the neck. Like any building, the structure of the neck depends upon a blueprint. In previous chapters, we have sought out relationships between the nervous system and peripheral tissues. We'll follow the same methodology here. We shall see that the dermal innervation patterns of the neck and its various muscles reflect a spatial blueprint within the spinal cord. When we compare the motor columns of the brain with those of the spinal cord, a remarkable and very functional system emerges. Muscle structures clustered around the vertebral axis are the most "primitive." They are supplied from the midline. As fishes became sophisticated enough to develop fins, a new system of peripheral muscles evolved to furnish control. The vertebrate body in some way anticipates the head. Thus, the ana-tomic "idea" of spatial separation of neurons seen in the spinal cord is "copied" in the design of the head muscles and their motor nuclei. We shall look at how these patterns are reiterated.

As cervical nerves leave the cord they undergo a series of simple and predictable changes that will convey them to their respective targets. There must be a way to separate out the nerve supply for axial versus peripheral structures. Muscles have to be very careful about how they find their target insertion sites, sometimes their pathways are circuitous. After all, a lot of things are going on in the neck. It's hard to find your way around! But the cervical plexus was not designed by Salvador Dali. Its goals are simple: The plexus will organize and spatially distribute *sensory nerves* to broad swatches of skin. It does so in a very logical way. The plexus will also separate out *motor nerves* for several groups of muscles: (1) muscles connecting the axial skeleton to itself, controlling movements of the head and spine, (2) the peripheral control muscle of respiration, the diaphragm, (3) muscles connecting the axial skeleton to the appendicular skeleton, that is, the shoulder girdle.

We have left out one final group of muscles, the non-striated ensheathment of the esophagus. Because it is derived from lateral plate mesoderm its motor control is from the vagus nerve. We will consider the esophagus separately.

We are left with the larger question: how and why did the neck evolve? This will be the subject of a separate later section in this chapter. It requires a multi-system approach including the development of fins and the transition to limbs, structural requirements of air-breathing, and the whole issue of where the cervical neuromeres came from in the first place. One thing is sure...the history of the neck itself is embedded in its neuromuscular design. The muscles of the neck are arranged in functional classes according to the grand evolutionary design of the region. Positioning of the head for feeding and survival is the responsibility of the axial muscles. Life on land demands an air pump. Respiratory and digestive passages must be separated. Limbs provide locomotion. So, understanding of blueprint of the cervical nerves can lead to a rock-solid appreciation for how function follows form.

A Cook's Tour of the Spinal Cord: Grey Matter

Heads versus tails...the best place to understand the vertebrate blueprint is, of course, the fish trunk. Here, developmental segments are laid out in a repetitive sequence along the spinal cord. A series of septae divide the myotomes into epaxial (dorsal) and hypaxial (ventral) components. Motor nerves follow the same terminology. Hypaxial muscles are supplied by ventral roots. Epaxial muscles receive dorsal roots.

How are these neurons laid out in the spinal cord? We all are familiar with cross sections of the cord in which grey matter is shaped like the letter "H," or like a butterfly. In the center of the H is a central canal with spinal fluid. The horizontal limb of the H straddles the canal as the dorsal and ventral commissures. At any given level, the ventral horn contains afferent somatic motor neurons whose axons will leave the cord via ventral roots. Localized collections of afferent SANS motor neurons form the lateral horn. The dorsal horn marks the termination of efferent primary sensory fibers entering the cord via the dorsal roots to create sensory columns Remember: spinal cord = neural tube = a flat neural plate rolled up on itself like a handlebar mustache or a cigar. Thus, as you "follow the curve" of spinal cord grey matter, you are merely progressing from medial to lateral. When you unroll it and flatten it out, voila! You wind up recreating the original neural plate (Figs. 10.73, 10.74, 10.75, 10.76, 10.77).

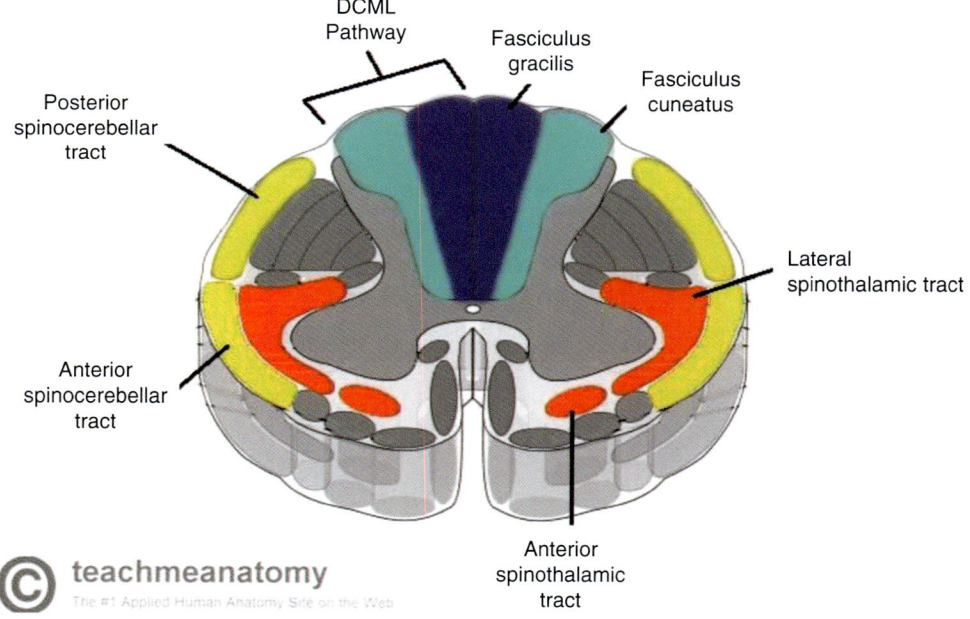

Fig. 10.73 Motor and sensory tracts of the spinal cord. Grey matter has three horns with three kinds of nuclei: ventral horn, c1 to s4 = somatic motor, (GSE) lateral horn, t1 to l3 = SANS motor, (GVE); dorsal horn (posterior columns), s5 to c1: non-nocioceptive sensory (GSA). White matter conveys 3 motor tracts (but no nuclei) from forebrain (corticospinal), midbrain (rubrospinal), and hindbrain (vestibulospinal) and and 1 nocioceptive sensory tract. Motor tracts of the spinal cord Grey matter has three horns with three kinds of nuclei: ventral horn, c1 to s4 = somatic motor, (GSE) lateral horn, t1 to l3 = SANS motor, (GVE); dorsal horn (posterior columns), s5 to c1: non-nocioceptive sensory (GSA) White matter conveys 3 motor tracts (but no nuclei) from forebrain (corticospinal), midbrain (rubrospinal), and hindbrain (vestibulospinal) and and 1 nocioceptive sensory tract. Sensory tracts of the spinal cord Touch and pressure, kinesthesia pathways: dorsal columns of the extremities (1) gracilis (blue) = leg, (2) cuneatus (green) = arm, and (2) Nocieptive pathways (red): = spinthalamic tracts (anterior and lateral) Propioceptive pathways (yellow) spinocerebellar tracts convey position information from the body to the brain: anterior (Gower's tract) and posterior (Flechsig's tract) [Reprinted from TeachMeAnatomy, courtesy of Dr. Oliver Jones].

Ventral Horn

If we look at the spinal cord as a whole, ventral horn neurons run in three vertical columns, often extending through many segments. The horns are organized into medial, central, and lateral cell columns. But remember, our focus is segmental, so the best way to appreciate functional relationships is on cross-section. Medial cell groups supply axial musculature. Lateral cell groups supply the limbs: (1) trunk-to-limb, (2) intrinsic limb, and (3) hand/foot. Don't get fooled. Some texts describe limb and foot as being dorsal to trunk. In reality, they are just "following the outer curve" of ventral horn.

Remember: the ventral horn is like a handlebar mustache. When you unroll it, you come up with the original neural plate.

Medial motor column extends the entire length of the spinal cord. The homolog of MMC in medulla is r8–r11 hypoglossal nucleus.

- *MMC medial* (MMC_M) supplies the epaxial erector spinae group. These muscles extend the head, neck, and trunk. They are innervated via dorsal primary rami.
- *MMC lateral* (MMC_L) innervates the following hypaxial groups: the prevertebral muscles of the neck, the intercostals, and anterior wall muscles of the trunk. These muscles flex the neck and trunk. They are innervated by ventral primary rami.
- *MMC cervical plexus* involves neuromeres c1–c4. It supplies the strap muscles of the neck. It provides sensation to the hypaxial neck down to clavicle.

Central motor column has two discrete parts. These are found exclusively in the cervical spinal cord and consist of neuromeres c1–c6. Its functional importance is disproportionate to its size.

- *Spinal motor column* (SMC) supplies sternocleidomastoid and trapezius. It is located at the ventral periphery of ventral horn, extending from c1 to c6. In the myology section, we shall see how this rationalizes the insertions of these two muscles. The cervical roots become confluent as the spinal root of the cranial nerve XI, the neuromeric definition of which makes it truly accessory, if not downright irrelevant. The homolog of SMC in the medulla is r6–r11 nucleus ambiguus.
- *Phrenic motor column* (PMC) supplies the diaphragm. It is interposed between SMC and MMC and runs between c3 and c5. The primary root of phrenic nerve is from C4 but it gets contributions from C3 and C5.

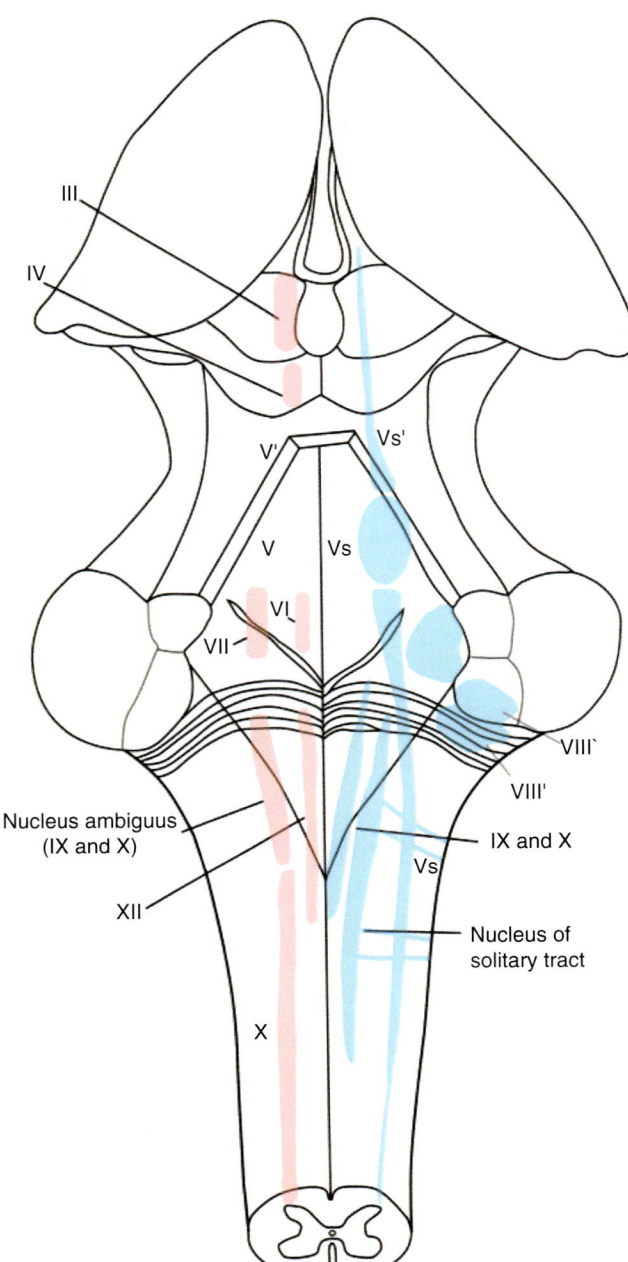

Fig. 10.74 Cranial nerve columns of the brain stem Motor = red Sensory = blue. Somatic motor column (red) supplies midline muscles unrelated to the pharyngeal arches. It is analogous to medial motor column (MMC) in the spinal cord. Dorsal vagus (blue) supplies PANS motor to glands. Nucleus ambiguus column, r6–r11 (purple) supplies pharyngeal arch muscles (palate, pharynx, larynx), continuous with central motor column (CMC) in the cervical spinal cord, but *not* analogous to lateral motor column (LMC) in the spinal cord. Somatic sensation (orange) is represented by a singe nucleus of spinal tract of trigeminal. Spinal V is somatotopic. Taste (green) is in same neuromeric zone as facial and nucleus ambiguus (facial nucleus not depicted). [Reprinted from Lewis, Warren H (ed). Gray's Anatomy of the Human Body, 20th American Edition. Philadelphia, PA: Lea & Febiger, 1918]

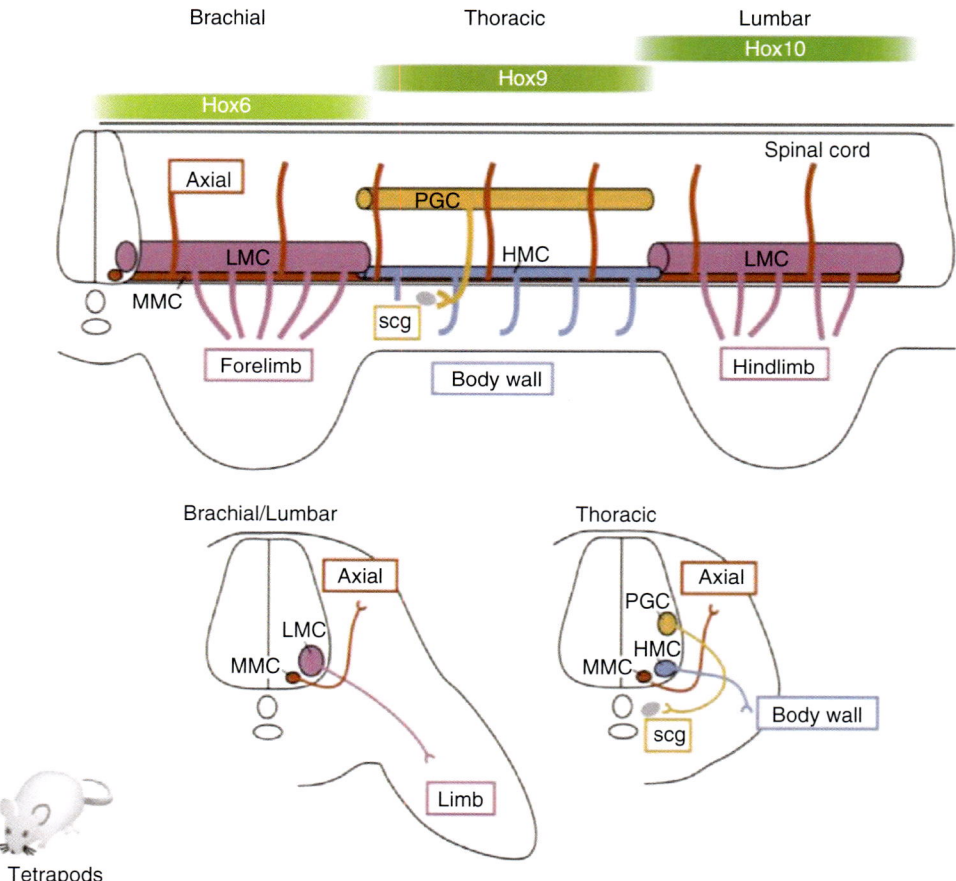

Fig. 10.75 Motor neuron columns and *Hox* proteins in the mammalian spinal cord. Motor neuron columns are generated at specific positions along the rostrocaudal axis of the spinal cord detrmined by the homeotic code. *Hox6* specifies c1–c8. *Hox9* specifies t1–t12. *Hox10* specified ls3. Lateral motor columns (LMCs) are generated at brachial and lumbar limb levels and innervate limb muscles. At thoracic levels, the preganglionic column (PGC) innervates the sympathetic ganglia (scg), whereas the hypaxial motor column (HMC) innervates the body wall muscles. Median motor columns (MMCs) are present at all levels of the spinal cord and innervate the dorsal and ventral axial musculature (note: ventral motor neurons to spine not shown here). Hox6, Hox9, and Hox10 are expressed in specific regions of the spinal cord along the rostrocaudal axis and direct the identity of motor neurons and their connectivity to peripheral targets. Hox6 controls brachial LMC identity, Hox9 controls PGC and HMC identity, and Hox10 controls lumbar LMC identity. Note [Reprinted from Murakami Y, Tanaka M. Evolution of motor neurons to vertebrate fins and limbs. *Dev* Biol 2014; 355(1):164–172. With permission from Elsevier]

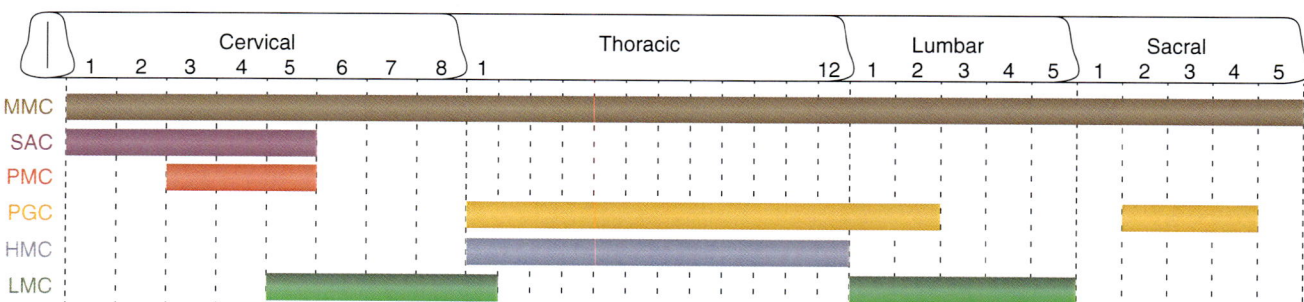

Fig. 10.76 Neuromeric organization of spinal motor columns. Schematic summarizing the segmental distribution of spinal motor columns (adapted from Dasen and Jessell, 2009). While the medial motor column (MMC, brown) is present all along the rostro-caudal axis, the spinal accessory column (SAC, purple) is restricted to the five first cervical segments (C1–C5). The phrenic motor column (PMC, red) is confined between C3 and C5. The preganglionic column (PGC, orange) extends through the thoracic segments until the second lumbar segments (L2) as well as well as between sacral segments 2 and 4 (S2–S4). The hypaxial motor column (HMC, light blue) is exclusive of the thoracic segment where as the lateral motor column (LMC, dark and light green) is located at limb levels: brachial (C5–T1) and lumbar segments (L1–L5). Note conspicuous absence of representation for the cervical plexus muscles, which belongs to MMC. Note phrenic motor nerve (unique in mammals) may have belonged originally to medial group of brachial plexus. Note cervical level is defined by Hox5. [Reprinted from Stifani N. Motor neurons and the generation of spinal motor neuron diversity. Fronters in Cellular Neuroscience 2014; 8: article 293. With permission from Creative Commons License 4.0: https://creativecommons.org/licenses/by/4.0/]

Fig. 10.77 Motor columns in the cervical spine Note the position of phrenic nerve and spinal accessory nerve intermediate between medial motor column (c1–c8) to axial muscles and lateral motor column to brachial plexus Cervical plexus strap muscles are not represented but belong to MMC, as they represent original axial muscle that pre-date the evolution of the upper extremity. LMC neurons, c5–t1, have a new set of genetic markers and occupy a new position in the cord. MMC, medial motor column; PMC phrenic motor column; SAC, spinal accessory column; LMC, lateral motor column; HMC, hypaxial motor column; PGC, preganglionic column (SANS motor) [Reprinted from Stifani N. Motor neurons and the generation of spinal motor neuron diversity. Fronters in Cellular Neuroscience 2014; 8: article 293. With permission from Creative Commons License 4.0: https://creativecommons.org/licenses/by/4.0/]

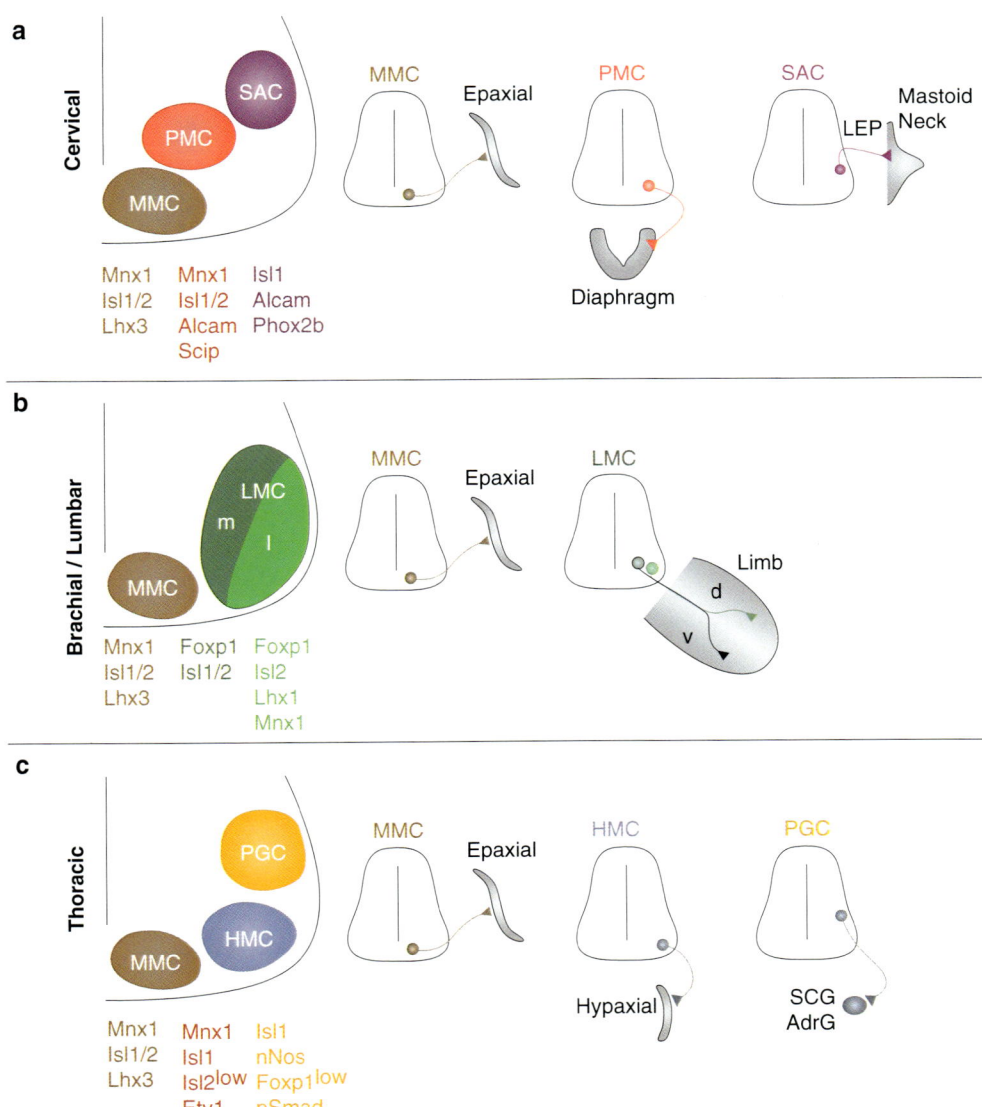

Lateral motor column is *dis*continuous. It supplies muscles that are evolutionarily new. The neurons are always ventral and they pass through a new-fangled structure…every student's nightmare…a plexus. LMC has *no* homolog in the medulla.

- *Brachial plexus* (neuromeres c5–t1): all roots are both motor and sensory to the upper extremity.
- *Lumbosacral plexus* (neuromeres t12–s4): roots l2–s3 are motor to the lower extremity; other roots supply muscles of the pelvis.
- *Coccygeal (pudendal) plexus* (neuromeres s5–cx1): strictly sensory.

Lateral Horn

Lateral horn contains the cell bodies of pre-ganglionic sympathetic autonomic nervous system (SANS) neurons. These project laterally from the "H" from the first thoracic to the second lumbar segments. This column (GVE) is lateral to LMC in the neural plate and therefore lies dorsal to it in the spinal cord. Because these neurons are motor, they must exit via the ventral root. But the SANS neurons, like escaping criminals, must return to do their time on the vertical "chain ganglia." Because the criminals have surrendered (waving a white flag) they leave the ventral root and return to the ganglia via *white rami communicantes*. The ganglia are vertically connected. In this way, SANS neurons ascend to populate three cervical ganglia, the largest one with the greatest distribution being the superior cervical ganglion. Craniofacial SANS fibers run with the branches of carotid system and therefore are in the company of the neural crest fibers that envelope these arteries. Homeotic coding of this neural crest at various junctures along the arterial system may serve as a signal for the formation of individual branches and these carry SANS fibers with them to their respective targets.

Nuclei of pre-ganglionic paraympathetic nervous system neurons (PNAS) are located in the nuclei of cranial nerves, III, VII, IX, and X and S2–S4 in the spinal cord. In the hindbrain, r6–r7 inferior salivatory nucleus and r8–r11 dorsal motor column of the vagus form a single, albeit discontinuous column which, just like SANS, is lateral to LMC in the neural plate and dorsal to it in the neural tube. In spinal cord, S2–S4 have the same location but the neurons are small and do not occasion an outward projection for the "H."

Dorsal Horn

The spinal cord conveys two types of sensory information back to the brain:

- *Proprioception* (fine touch, pressure, and position sense) are carried in grey matter.
- *Nocioception* (pain and temperature) is carried in white matter.

Grey matter contains proprioceptive nuclei and fibers in the dorsal horns. These are conveyed upward to cerebellum and higher centers for processing. Beginning caudally, the neurons add up from medial to lateral creating, in the thoracolumbosacral cord a single *nucleus gracilis*, representing the lower extremity and trunk. In the cervical cord, complex information from the upper extremity and neck necessitates an additional and physically separate *nucleus cuneatus*. The posterior columns are thus somatotopic.

A Cook's Tour of the Spinal Cord: White Matter

Three motor tracts (no nuclei) convey impulses from the tripartiate embryonic brain.

- Hindbrain is ventral (medial) and carries the *vestibulospinal* tract.
- Midbrain is intermediate and contains *rubrospinal* tract.
- Forebrain is dorsal and has the *corticospinal* tract.

These descending tracts add up from medial to lateral. Thus, primitive fibers from hindbrain occupy the midline (ventral) position while those from the cortex are the most lateral and are dorsal to the other two tracts.

White matter contains nociceptive nuclei and fibers in the ventral periphery and conveys them to the thalamus for processing. Like proprioception, these add up from medial to lateral. The anterior spinothalamic tract is medial–ventral representing the trunk and lower extremity while the posterior spinothalamic tract is lateral/dorsal and carries additional information from more rostral levels.

Spinal Cord Tracts Are Somatotopic, That Is, Neuromeric

Although the pathways conveying sensory information from the body up through the spinal cord to the brain are also very different from motor pathways, both systems make use of somatotopic organization. The peripheral nervous system is of neural crest construction (the Schwann sheath) each nerve connects a neuromeric level with target tissues produced at that level. The main center for somatic information from the head and face is *trigeminal nucleus*. From the neck on down, we see the pathways separate out. Somatic information is conveyed in the *spinothalamic tracts*. These are spatially organized by function. We'll use the neural plate model, so from medial-to-lateral, the nociceptive sequence is pressure > touch > pain > temperature.

The spinothalamic tracts are also laid out by neuromeric segments in a ventral-to-dorsal sequence: s5 > c4. The initial sacral fibers are ventral. Thus, as one proceeds upward, the fibers add up and the spinothalamic tract gets bigger. Recall that the nociceptive tract for the head and upper neck is r1–c4. The so-called trigeminal nucleus has a spinal extension down to c4. Why? Once again, we observe the developmental importance of the c4–c5 junction. It represents the primary (head-related) neck versus the secondary (trunk-related) neck.

Proprioception, exteroception, and vibration are conveyed in the *posterior columns* (dorsal funiculi). The spatial layout of these tracts is diametrically opposite that of their spinothalamic siblings. Functional modalities have a dorsal-to-ventral sequence: touch > pressure > movement > vibration > pressure. The segmental sequence is medial-to-lateral: c4 > s5. Morphologically the columns are described as fasculi. The more medial *fasciculus cuneatus* contains mid-thoracic to sacral fibers. The more lateral *fasciculus gracilis* contains mid-thoracic to sacral fibers. Thus, as one proceeds upward along the cord the posterior columns get wider.

The Unique Role of C4 as a Neuromeric "Faultline"

Pain referral to the trigeminal nucleus continues downward to C4, thereafter being referred to a separate structure for the spinothalamic tract. Is there something about the fourth cervical neuromere that separates a "head-associated neck" from a "trunk-associated neck?" Actinopterygian fishes have three occipital somites, S1–S3, each having a motor nerve to the pectoral fin. The actinopterygian head-trunk junction is S3/S4. The addition by tetrapods of a fourth occipital somite shifted the head/trunk junction backward from one neuromere to S4/S5. Liberation of the pectoral girdle from the head was accompanied by a backward transposition of the pectoral girdle to level C4. The original blueprint for brachial plexus in all tetrapods, *except mammals,* consists of four roots. These brachial plexus nerves, br1–br4, always bracket the head/trunk interface, the trunk being represented by br4. Thus, in basal amniotes (reptilomorphs) with six cervical vertebrae, the brachial plexus consists of C4–C5–C6–T1. Muscles interposed between the head and pectoral girdle continued to retain branchiomeric neural crest connective tissues and retained neurologic connections to these tissues as well. Thus, the existence spinal nucleus of V down to c4 reflects the persistence of these relationships. Somatic sensory events distal to c4 are referred to the trunk-associated spinothalamic tract.

Spinal Nerves: Anatomic Components

What could be more boring that reviewing spinal nerves? Let's spiff up this necessary section by a radical statement. There are 35 (not 31) pairs of spinal nerves: 4 occipital, 8 cervical, 12 thoracic, 5 lumbar, 5 sacral, and 1 coccygeal. [Recall that there were originally 5 coccygeal somites but these normally regress.] There is good evolutionary evidence to indicate that spinal accessory nerve is a strictly peripheral nerve that becomes incorporated into the jugular foramen during the addition of the fourth occipital somite. After all, both its target muscles, SCM and trapezius, are part of the pectoral girdle (we will consider this fascinating story later).

In any case, all spinal nerves except XI and C1 make their exit through intervertebral foramina. The first cervical nerve is designed to exit from beneath the proatlas (the vertebral product of somites 5–6). Because proatlas is absorbed into the skull, C1 exits about atlas. C2, obediently, travels below atlas. Recall from a developmental standpoint how the components of the foramina flow around the pre-existent nerves. XI is, of course, an exception. Indeed, as we shall see, it does not have a true cranial root and it is *not a cranial nerve.* The spinal roots C1–C4 of this *transitional nerve* no longer are individually inserted into the cucullaris. Instead, they link up with one

another longitudinally. The first root, C1, becomes entrapped by the incorporation of the fourth somite and winds up exiting the jugular foramen in the company of vagus. The roots of the transitional nerve keep adding up caudally and the nerve finally exits via foramen magnum.

- *Ventral (anterior) roots* are centrifugal; they contain motor neurons from anterior and lateral grey column of the cord.
- *Dorsal (posterior) roots* convey centripetal fibers carrying information back from the periphery to the brain. Each rootlet pierces the dura but remains covered by pia mater and thence by arachnoid. In the neck, dorsal roots are three times bigger than the ventral roots.
- *Spinal ganglia* are situated in the intervertebral foramina. Just behind them lies the vertebral artery.
- *Spinal nerves* refer to the zone just beyond the ganglia. Here the dorsal and ventral roots temporarily fuse.
  - *Meningeal branches* are given off at this point.
  - Note that the cervical nerves get progressively larger from C1 to C8.
- *Spinal roots* come next, that is, the dorsal and ventral spinal rami.

  - Dorsal rami travel upward to supply epaxial muscles in medial > lateral order.
  - Ventral rami immediately receive sympathetic fibers because just lateral to each ramus is a vertically oriented chain of sympathetic ganglia. The SANS fibers travel from ganglion to ventral ramus via a *grey ramus communicans.*

In summation, hypoglossal nucleus (r8–r11) in brainstem innervates hypaxial somite-derived muscles. It continues into spinal cord (c1–s4) as medial motor column all the way down the trunk innervating epaxial muscles continuously and hypaxial muscles discontinuously (not c5–t1 and not t12–l5). Nucleus ambiguus (r6–r11) supplies the hypaxial branchiomeric muscles of arches 3–5 (somitomeres 7–11) to palate, pharynx, and larynx in somatotopic fashion. It continues into spinal cord (c1–c6) as central motor column. Lateral motor column appears in the neck for the first time at levels c5–t1.

Clinical Application

Trauma to cervical spinal roots can be caused by herniation of the disc. Nerve roots live immediately next to the disc. This takes place in the root sleeve, a section of the nerve where it is still encased by dura. This condition is called lateral recess syndrome. Just outside the dura in the foramen is the dorsal root ganglion. Nerve roots can also be afflicted by neurofibromas. The symptoms of root compression are dermatome = numbness + paresthesia; myotome = pain.

Here let's confront (once again) terminology. Nerve roots always exit below their respective vertebra. The reason for this is that the caudal part of the sclerotome is made up of the loose zone. T5 is made from $T4_C + T5_R$ and the fifth thoracic nerve travels through the permissive territory of $T5_R$. However, cervical vertebrae are numbered differently due to lack of appreciation for the proatlas, Nerve roots are considered to exit *above* the corresponding cervical vertebrae. For this reason, disc prolapse between vertebrae T4 and T5 will compress spinal root T4 but the same process between C4 and C5 will compress spinal root C5.

Spinal Nerves: Functional Classification

Somatic

- *Somatic efferent* = motor control.
  - traditional terminology divides nerves to striated muscles into GSE, general somatic efferent, for garden-variety muscles and SVE, special visceral efferent (sic) for branchiomeric so-called "visceral" muscles.
- *Somatic afferent* = sensory information from skin, fascia, muscles, and joints.

Visceral

- *Visceral efferent SANS* preganglionic sympathetic neurons have their cell bodies in the lateral grey column of the spinal cord from T1 to L3. When fibers exit, they are myelinated and travel into the sympathetic trunk via white rami communicantes. Upon synapsing the ganglia, postganglionic neurons travel out to non-striated muscles or glands.
- *Visceral efferent PANS* preganglionic neurons have their cell bodies in the lateral grey column of S2 to S4.
- *Visceral afferent* neurons have their cell bodies in the spinal ganglia. Their axons run through the white rami communicantes into SANS ganglia (where they do *not* synapse) and then out to their targets in the viscera.

Meningeal

Branches to meninges are found at each vertebral level. The numbers 2–4 on each side. They all receive SANS input from grey ramus communicans. Within the spinal canal, they track along blood vessels. These mixed/SANS fibers innervate dura, blood vessels, periosteum, and ligaments. Ascending branches from C1 to C3 are distributed to dura mater of pos-

terior cranial fossa. This anatomy if important for understanding *occipital headache*.

Cervical Nerves and Plexuses

In this section, we shall discuss the general characteristics of the cervical nerves, individual cervical nerves C1, C2, C3, the cervical plexus, phrenic nerve, and the spinal accessory nerve.

Different Rami, Different Roles

Dorsal rami lead to a very monotonous existence. Beginning with C2, they all divide into medial and lateral branches. All of them supply muscles. Skin is innervated by medial branches of the second to the fifth dorsal rami. *Dorsal rami are motor for epaxial muscles of the spine supplied from the medial motor column.*

Ventral rami are more complex. Not only must they supply the needs of the body axis but they are called upon, at varying anatomic levels, to innervate the appendages. At all levels of the spinal cord, *ventral rami are larger than dorsal rami.* There are three reasons for this. First, the *volume* of hypaxial muscle controlling the axial skeleton of the neck and trunk is considerably larger. These muscles have more complex tasks. Not only do they flex and rotate the spine but they control motion (and function) of the entire thoracic cage and abdomen. Second, tetrapods an entirely *distinct set of hypaxial muscles* are assigned to the appendicular skeleton. Third, mammalin evolution converts a muscle associated with the pre-mammalian brachial plexus into an entirely novel structure, the diaphragm, which they internalize into the chest cavity. It also results in a subdivision of the cucullaris muscle into sternocleidomastoid and trapezius with extensive attachments to a new version of the pectoral girdle.

It is not surprising then that such phylogenetically new muscles should be innervated from separate zones of the spinal cord, the lateral motor column. *Thu, s ventral rami carry neurons from 3 distinct motor columns: (1) hypaxial muscles of the axial skeleton from the medial motor column, (2) the muscles of the upper extremity via the lateral motor column, and (3) the diaphragm and cucullaris derivatives via the central motor column.* This section deals with only those muscles with primary attachment to either the cervical spine or the pectoral girdle.

With exception of the first ramus, all cervical ventral rami make their debut between the guywires suspending transverse processes of the neck vertebrae, anterior intertransverse, and the posterior intertransverse muscles. Because hypaxial muscles are attached internally to the spine (and skull) and externally to the shoulder girdle, their respective motor nerves must make complex spatial decisions in order to get to their correct destinations. This results in the formation of two plexuses: the upper four cervical nerves are organized into the cervical plexus and the lower four cervical nerves are organized into the brachial plexus. All plexuses are strictly hypaxial; dorsal rami are not involved.

The contribution of the sympathetic autonomic system to the cervical nerves is uniform. They all receive SANS neurons via one or more grey rami communicantes. There are three sources for SANS: *superior cervical ganglion* > C1–C4; *middle cervical ganglion* > C5–C6; and *inferior (cervicothoracic) ganglion* > C7–C8. Thereafter the ventral rami divide; this will become the basis for understanding the cervical plexus.

Cervical Nerves C4–C8

Dorsal rami After negotiating their way around the vertebral pillars, these form medial and lateral branches directed to the epaxial muscle of the spine. *Medial branches* supply (from deep to superficial): interspinous > multifidus > semispinalis cervicis, + semispinalis capitis. *Lateral branches* (from medial to lateral) longissimus capitis > longissimus cervicis > iliocostalis cervicis. These muscles develop right away and occupy virtually all the dorsal binding sites of the cervical spine. Later on, when hypaxial muscles such as levator scapulae or the rhomboids seek a secondary attachment, they are forced to occupy more peripheral sites. Recall that scapula, a true hypaxial structure, shifts dorsally. Muscles dedicated to the pectoral girdle follow suit. Thus, although the dorsal scapular nerve is ventral, its muscles attach to the dorsal aspect of the vertebral column.

C4–C8 ventral rami all emerge reliably between scalenus anterior and scalenus medius. Ventral neurons arising from *medial motor column* continue the same pattern to supply the intervertebral muscles along the ventral aspect of the spine. Central motor column (C1–C6) produces phrenic nerve (C3–C5), a unique adaptation in mammals of a muscle formerly belonging to the brachial plexus, subcoracoideus, that migrates into the chest to form the diaphragm. It also produces and spinal accessory nerve. Those arising from the *lateral motor column* combine to form the brachial plexus (C5–T1).

Medial Motor Column: C1–C8

First Cervical Nerve

C1 is unique in that has a minimal sensory representation. Somite 5 has no dermatome but does provide PAM periosteum to ventral foramen magnum, the so-called and mis-named *"meningeal branch"* of C1. It has anterior and posterior rami and targets can be organized as follows:

1. Strap muscles (rectus cervicis)

- Geniohyoid via anterior loop, then via branch to hypoglossal. This indicates that the four-somite model of tongue muscles exists from S2 to S5.
  Thyrohyoid via anterior loop, then via branch to hypoglossal
- Sternohyoid via Ansa
- Omohyoid via Ansa

2. Muscles from atlas to skull, monosomitic

- Rectus capitis anterior
- Rectus capitis lateralis
- Rectus capitis posterior

3. Muscles from atlas/axis to skull, polysomitic

- Longus capitis
- Splenius cervicis

4. Muscle from atlas to scapula, polysomitic

- Levator scapulae

C1 dorsal ramus (suboccipital nerve) is much larger than its ventral fellow. It makes sense, the axial head extensors are bulky, whereas axial head flexors are ribbon-like. Remember that the ancient proatlas is formed by r11–c1, that is, by somites 4 and 6. C1 should rightfully lie below proatlas, but, alas, proatlas has been absorbed into the skull. Atlas is next-in-line. It is formed from c1 to c2, that is, by somites 6 and 7. Thus dorsal C1 comes out above the arch of atlas. It enters suboccipital triangle to supply five muscles: rectus capitus posterior major and minor, obliquus capitis superior and inferior, and semispinalis capitis.

C1 ventral ramus (suboccipital nerve, SON) emerges *above* the ring of the atlas. It dives forward, skirting the lateral mass of the atlas, staying medial to vertebral artery. It innervates rectus capitis lateralis and then emerges medial to it. It descends to join with the ascending branch of the second cervical ventral ramus.

Second Cervical Nerve (Greater Occipital Nerve GON)

Arises between atlas and axis. Just the opposite of C1 it has limited motor innervation, (semispinalis capitis) but extensive sensory representation to the back of the scalp, ear, and over parotid gland.

C2 dorsal ramus (greater occipital nerve, GON) is the largest of all the cervical dorsal rami…thus its name. It exits between the posterior arch of atlas and the posterior lamina of axis. This puts it directly below inferior oblique muscle, which it proceeds to innervate and then divides. *Medial branch is strongly sensory*. It travels through the attachments of semispinalis capitis and trapezius, where it is joined by dorsal C3 and by occipital artery. Next, it sends connecting branches to lesser occipital nerve. It proceeds upward to innervate scalp all the way to the vertex. On occasion, it provides sensation to the back of the ear.

Considerable developmental information is embedded in the dermatomal relationship between the greater occipital nerve versus its sidekick, the lesser occipital nerve. Both supply dermis from the second cervical dermatome. They indicate a distinction between epaxial versus hypaxial scalp skin (Fig. 10.84). Lesser occipital shares a common border with V3 while the greater occipital is co-extent with V1. Although the terms epaxial and hypaxial do not apply to the head, the boundary between these two zones replicates the territory of frontonasal neural crest.

Medial branch is motor to the most rostral of the transversospinal group: semispinalis capitis. *Lateral branch is motor to most rostral muscles of erector spinae: splenius cervicis > longissiums capitus > longissiums cervicis.*

GON neuralgia can result from (1) entrapment; (2) arthritis of the upper apophysis next to the second cervical root; and herpes indwelling in the geniculate ganglion. This is transmitted to C2 in several ways: (1) via facial nerve directly via the auricular branch of posterior auricular nerve; and (2) posterior auricular nerve makes a second connection to *lesser occipital nerve* (LON) the superficial branch of ventral C2.

C2 ventral ramus projects outward between atlas and axis. After emerging between longus capitis and levator scapulae, it divides. An ascending branch connects with the descending branch of the first cervical ventral ramus. A descending branch joins third cervical nerve.

Third Cervical Nerve

C3 dorsal ramus exits medial to posterior intertransverse muscle and divides. Medial branch ends up in lower occipital skin. Lateral branch joins with C2 dorsal. Sometimes the dorsal rami of C1 to C3 form loops known as the *occipital plexus*.

C3 ventral ramus exits between longus capitis and scalenus medius. It sends a medial branch to contribute to the phrenic nerve.

Fourth to Eighth Cervical Nerves

C4–C8 dorsal rami After negotiating their way around the vertebral pillars, these form medial and lateral branches directed to the epaxial muscles of the spine.

- Medial branches supply (from deep to superficial): interspinous > multifidus, semispinalis cervicis, semispinalis capitis.
- Lateral branches supply (from medial to lateral); longissimus capitis, longissimus cervicis > iliocostalis cervicis.

All these muscles develop right away and occupy virtually all the available binding sites along the cervical spine. Later, when hypaxial muscles such as levator scapulae or the rhomboids seek a secondary attachment, they are forced to use more peripheral binding sites. Recall that scapula in mammals shifts dorsally, bringing all of its muscles along with it.

C4–C8 ventral rami all emerge reliably between the scalenous anterior and scalenus medius. Ventral neurons arising from medial motor column take two different routes. Some continue the same axial pattern as in C1–C3 to supply intervertebral muscles along the ventral aspect of the spine. Others participate in the cervical plexus to supply the strap muscles and elevators of the scapula. Ventral neurons from central motor column also have two distinct patterns. Phrenic nerve arises from C4 via cervical plexus and dives into the chest. Spinal accessory nerve forms from separate branches of C1–C6 that do not participate with cervical plexus. Ventral neurons from the lateral motor column (C5–T1) form the brachial plexus.

The Cervical Plexus, C1–C4

Let's begin with a misleading statement. Traditional texts describe the ventral rami of C1–C4 as joining together to make up the cervical plexus. From a functional standpoint the motor components of cervical plexus are really C1–C3. All these muscles come from MMC. C4 is different. Yes, C4 is sensory to the skin of the neck. But in its original design, C4 is also motor to *subcoracoideus*, a muscle mass connecting pectoral girdle with the upper extremity. In birds, this elevates the wing. In mammals, subcoracoideus becomes the future diaphragm. Thus, from a sensory standpoint, we can talk about cervical plexus as being C1–C4 but from a motor standpoint C4 in mammals is transitioning to the brachial plexus (Fig. 10.78).

But, we get ahead of ourselves…so let's proceed with a standard description. The cervical plexus is made up of ven-

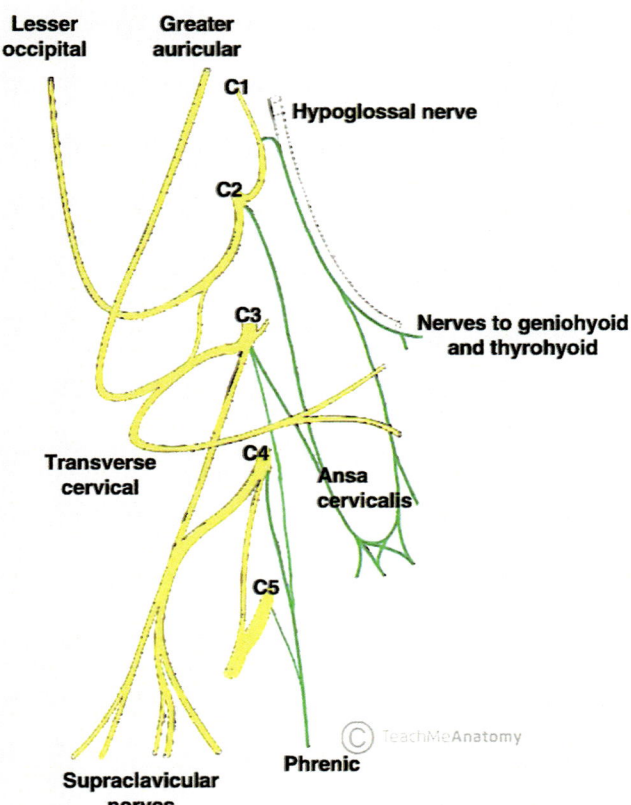

Fig. 10.78 Cervical plexus. The plexus is made up of ventral roots from spinal nerves C1–C4. These are medial motor column as they supply muscles to the axial spine. The strap muscles are derived from original rectus cervicis. C1 is strictly motor and supplies geniohyoid, the most distal strap muscle. Phrenic nerve mostly C4 reflects original tetrapod position of muscle from the neck to the pectoral girdle (see discussion on phylogeny of diaphragm). C4 is unique. It supplies the following muscle groups: (1) head-pectoral girdle (spinal accessory), (2) neck-pectoral girdle, and (3) pectoral girdle-humerus. It does *not* contribute to pectoral girdle-pharyngeal arch control. Spinal accessory nerve, cranial nerve XI innervates two hypaxial branchiomeric muscles sternocleidomastoid and trapezius. It has four–six nuclei in the central motor column, a continuation of nucleus ambiguus. Transition between cervical plexus and brachial plexus takes place at level C4. [Reprinted from TeachMeAnatomy, courtesy of Dr. Oliver Jones]

tral roots that coalesce to take the shape of an elongated ellipse. It is located at the same level as the first four cervical vertebrae; its surface markings are defined by a line running from the auricle to thyroid cartilage. The plexus is deep to sternocleidomastoid and just anterior to scalenus medius and levator scapulae.

Branches from the cervical plexus have both sensory and motor functions. Sensory branches innervate skin territories of the head, neck, and chest. Motor branches supply: (1) muscles connecting the axial skeleton to the skull, (2) muscles connecting the axial skeleton to itself, (3) muscles connecting pectoral girdle to the axial skeleton, (4) the strap muscles connecting pectoral girdle to pharyngeal arches.

The diaphragm, traditionally considered part of the cervical plexus, arises from a separate motor column and merely "goes along for the ride."

Conceptually, the cervical plexus is quite elegant. As we stated earlier, all the ventral rami divide into ascending and descending branches. The ascending branch of C1 wanders off by itself (it travels forward with hypoglossal to supply geniohyoid) but the remaining roots combine to form a series of loops. The first loop, made from C2 to C3, provides superficial branches in the form of named nerves to the head and face. These are lesser occipital, great auricular, and transverse cutaneous nerves. The second loop, made from C3 to C4, provides sensory supply to the neck and chest as the supraclavicular nerves. Thus, *superficial branches are sensory to skin*, while *deep branches supply muscle*. A single medially-directed branch, C4, off the second loop, is motor to the diaphragm. The diaphragm exists only in mammals. From an evolutionary standpoint, it represents the displacement of the subcoracoid muscle originally connecting pectoral girdle to the upper extremitiy. Thus, C4 in pre-mammals is part of brachial plexus but in mammals becomes distinct: *phrenic* nerve is really the upper-most component of the brachial plexus.

Sensory Branches: Superficial Ascending

Lesser occipital nerve (C2). It curls around the spinal accessory nerve and ascends along the posterior border of the sternocleidomastoid. Here it penetrates upward through the deep investing fascia of SCM and divides. *Auricular branch* supplies the upper posterior pinna. It connects with great auricular nerve, posterior branch. *Occipital branch* of LON supplies hypaxial posterior scalp. It interfaces with greater occipital nerve, which supplies epaxial posterior scalp.

Great auricular nerve (C2–C3) This is the largest sensory branch. It wraps around posterior border of SCM and then crosses the muscle deep to platysma. It then proceeds to parotid gland, where it divides. *Anterior branch* supplies the skin over the parotid gland. Within the substance of parotid, it connects with facial nerve. *Posterior branch* goes backward over the mastoid process. It supplies the lower posterior pinna to the occipital. It pierces the concha to supply the anterior (external) ear skin of the lobule and concha. It also connects with lesser occipital to the scalp and posterior auricular VII.

Transverse cutaneous nerves (C2–C3) These follow around the posterior border of SCM at its halfway point. *Ascending branches* innervate submandibular skin and upper neck. *Descending branches* traverse platysma to supply neck anterolateral neck skin down to sternum.

Sensory Branches: Superficial Descending

Supraclavicular nerves (C3–C4) These nerves all arise from a single trunk constructed from C3 to C4, that is, from the inferior loop. The trunk comes out from behind sterno-cleidomastoid, these tracks down beneath platysma, and then tricurcates. *Medial branches* extend to the midline and down to the second rib. They innervate sternoclavicular joint. This is important because we can neuromerically code the joint and the adjacent manubrium as potential derivatives of the c2 lateral plate mesoderm. *Intermediate branches* cross over the clavicle down to second rib. The dermatome of C3–C4 abuts against that of T2, all interme-diate dermatomes (C5–T1) being expropriated away from the trunk by upper extremity. The sensory boundary is very sharp; there is virtually no overlap. *Lateral branches* flow over trapezius and acromion process. Thus upper shoulder skin has C4 innervation.

Sensory Branches, Deep

Proprioceptive branches from C1 to C4 connect with spinal accessory nerve within the substance of sternocleidomastoid and trapezius. Recall that SAN being motor general somatic efferent is exclusively motor. Thus, these communicating branches provide much-needed coordination. They have complex central connections for head-turning, especially the startle response.

Motor Branches to Muscles of the Axial Skeleton

The cervical plexus conveys motor branches from the medial motor column to muscles connecting the axial skel-eton with itself. Four *axial muscles* are supplied by roots that reflect the spatiotemporal order of their development. All have primary insertions from the vertebra. Those that are directed upward are *rectus capitis lateralis* (C1) and *rectus capitus anterior* (C1–C2). Those directed downward are *longus capitis* (C1–C2–C3) > *longus colli* (C2–C3–C4–C5–C6). Although these muscles are hypaxial they are *not* related to the upper extremity and have their nuclei in the medial motor column. Motor nerves from C1 to C3 to these muscles are frequently depicted as part of the cervical plexus but do not fit the functional pictures. The cervical plexus is also a switch yard for muscles connecting the trunk to the pharyngeal arches. These are all medial motor column muscles that represent the ancient coracomandibu-laris in sharks.

Motor Branches to the Hypobranchial Muscles

Hypobranchial muscles come in two flavors. The *prehyoid column* originates from the occipital somites. These muscles connect the tongue with mandible and hyoid, that is, with the first, second, and third pharyngeal arches. They are inner-vated by r8–r11 hypoglossal nerve. The *posthyoid column* originates from the first three cervical somites. These mus-cles connect the appendicular skeleton (the pectoral girdle) with ventral pharyngeal arches, PA1–PA4. They are supplied by the c1–c3 cervical plexus. The motor components of cer-vical plexus are as follows.

Communicating branches pass from the C1 to C2 loop. The *hypoglossal branch* follows along XII and then splits off separately to supply thyrohyoid and geniohyoid. Together these two muscles constitute the superior strap muscles. These unite the ventral midline of pharyngeal arches 1–4. A vagal branch and sympathetic branches are described.

Ansa cervicalis is an inferiorly-directed loop from cervical plexus that supplies the inferior strap muscles, including omo-hyoid. The purpose of these muscles is to connect the midline pectoral girdle, the manubrium (ancient coracoid), and clavi-cle with the lower jaw via TH and GH. Ansa has two roots: inferior root (posterior) and superior root (anterior). *Inferior root of ansa cervicalis* comes from the medial series from a branch of C2 and a branch from C3. It links up with *superior root* just in front of common carotid. *Note the absence of C4 in motor branches.* Note: C4 motor fibers are reserved for phrenic nerve and represent the transition to brachial plexus.

Beware the masquerader! Spinal accessory nerve is com-prised of strictly segmental roots from C1 to C5 (or C6). They are not supplied by cervical plexus. Sternocleidomastoid and trapezius are developmentally unrelated to the strap muscles and have their motor nuclei in a separate functional column.

Central Motor Column: C1–C5

The rationale for a central motor column in the rostral cervical spinal cord is poorly understood. It supplies two sets of mus-cles lateral to those of the axial spine, and therefore distinct from MMC. The derivatives of cucullaris (sternocleidomas-toid and trapezius) and diaphragm are wildly different in posi-tion and function. Each has its own distinct motor column. *Spinal accessory column* (SAC) spans from c1 to c5 (some consider it to extend as caudal as c7). SAC is positioned at the lateral margin of ventral the horn. *Phrenic motor column* (PMG) belongs to neuromeres c3–5. It is intermediate in posi-tion between MMC and SAC. Yet all have a common denomi-nator, a neuromeric relationship with the pectoral girdle.

These two motor columns obviously overlap. Why is SAC lateral to PMC? This might seem trivial. After all, diaphragm is certainly internal to the scapula. The reason lies deeper (no pun intended). If we look at the organization of the LMC to the upper limb (cf Fig. 10.75) nuclei for the ventral muscles of the limb are in the medial LMC whereas those for dorsal muscles are lateral. We shall see that subcoracoideus inserts into the ventral aspect of the pectoral girdle. Trapezius is inserted on the opposite side of scapula. Thus, SMC is lateral to PMC.

We shall take now consider the innervation of the muscles of the central motor column. The nerve to the cucullaris complex, spinal accessory, comes first, as it is the more primitive of the two. Cucullaris is present in extinct placoderm fishes. Subcoracoideus does not appear in evolution until tetrapods. Its final iteration as the true diaphragm exists only in mammals.

Spinal Accessory Nerve (C1–C5)

Nucleus Ambiguus: Neuromeric Organization

Spinal accessory nerve in the spinal cord represents the physical continuation of nucleus ambiguus of the brainstem. This subject is of critical embryological and evolutionary interest because the appearance in mammals of sternocleidomastoid and trapezius, as derivatives of the ancient cucullaris muscle, marks a critical turning point in neck anatomy. Examine the neuroanatomy with care (Figs. 10.79, 10.80, 10.81, 10.82).

Nucleus ambiguus is a collection of motor neurons located in r6–r11 of the hindbrain that supply so-called "branchiomotor" muscles of the third, fourth, and fifth pharyngeal arches. The traditional name for this functional column is special visceral efferent (SVE), a term that is completely wrong on two out of three counts. There is nothing special at all about muscles that originate from somitomeres 7–11. They are striated, voluntary, with myoblasts of paraxial mesoderm origin and connective tissue stroma made from neural crest. Furthermore, it perpetuates the long-debunked proposition that, since pharyngeal arches are in continuity with the GI tract, their musculature should be "visceral" or smooth but somehow isn't. Nucleus ambiguus is lateral in the brainstem. Medial to it lies dorsal nucleus of vagus r6–r11which PANS motor to the entire presacral GI tract and represents the so-called general visceral efferent (GVE) column supplying the smooth muscle and glands of the gut. Most medial is hypoglossal nucleus r8–r11 which is generally somatic efferent to the striated tongue muscles.

The organization of NA is somatotopic. It has three neuromeric sections, but these are represented by only two cranial nerves. Rhombomeres r6–r7 are motor for the muscles of somitomere 7: stylopharyngeus, soft palate muscles, and superior constrictor and middle. These neurons can travel via one of two routes to their targets. The nuclei of r6–r7 supply two types of motor nerves. A single branch to the stylopharyngeus passes out via glossopharyngeal nerve, as it travels directly to the posterior pharynx.

All remaining muscles from Sm7 are innervated by nerves from r6 to r7 that join with those of r8–r11 to form via vagus

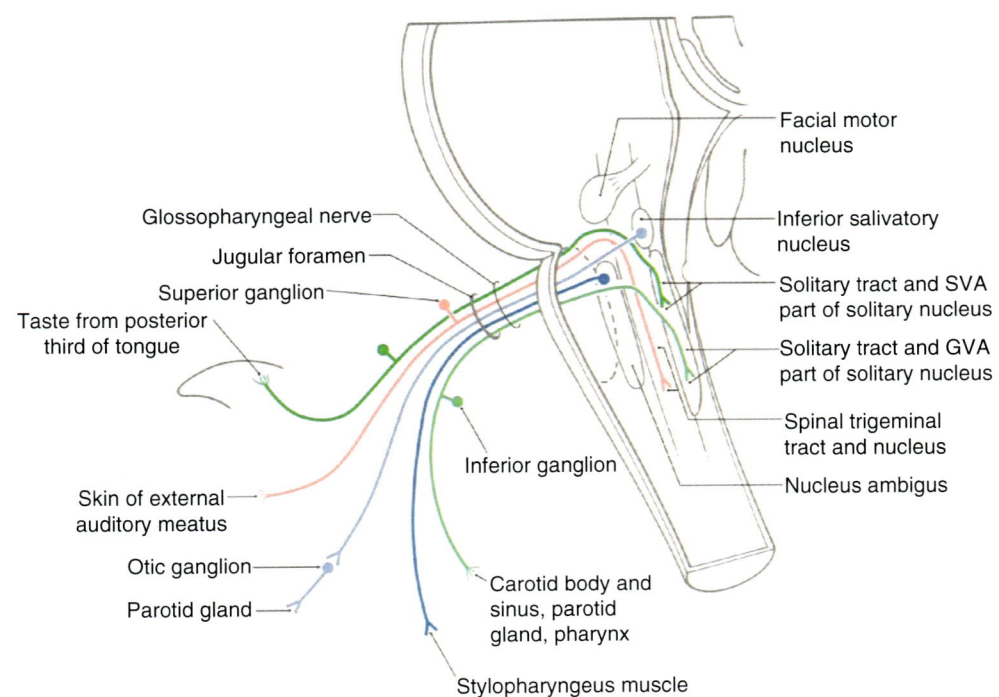

Fig. 10.79 Cranial nerve IX. Note single motor neuron from upper nucleus ambiguus directed to Sm7 stylopharyngeus. It has primary insertion into r6–r7 pharynx and secondary (retrograde) insertion into r4–r5 styloid process. Sensory distribution of IX is extensive, covering the entire third arch from the back of the fauces down to thyroid cartilage. Remainder of Sm7 muscles supplied from rostral nucleus ambiguus routing via vagus nerve. [Reprinted from Hanes DE. Fundamental Neuroscience for Basic and Clinical Applications. Philadelphia, PA: Saunders; 2006. With permission from Elsevier]

Glossopharyngeal nerve
Jugular foramen
Superior ganglion
Taste from posterior third of tongue
Skin of external auditory meatus
Otic ganglion
Parotid gland
Inferior ganglion
Carotid body and sinus, parotid gland, pharynx
Stylopharyngeus muscle
Facial motor nucleus
Inferior salivatory nucleus
Solitary tract and SVA part of solitary nucleus
Solitary tract and GVA part of solitary nucleus
Spinal trigeminal tract and nucleus
Nucleus ambigus

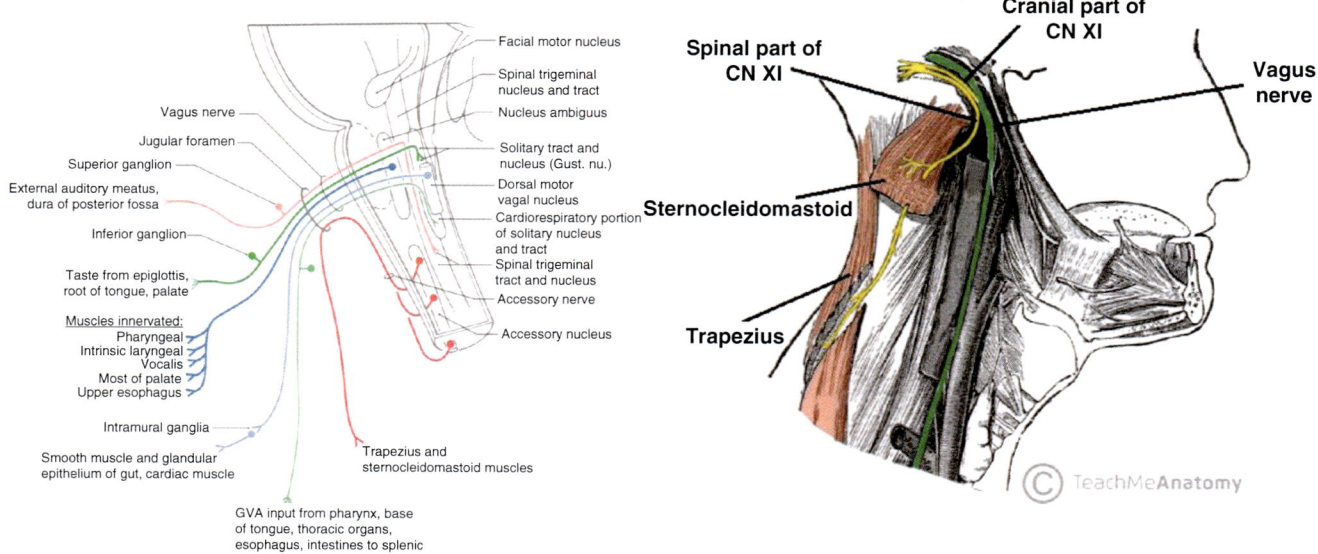

Fig. 10.80 Cranial nerves X, XI. Nucleus ambiguus is motor to the palate-pharyngo-laryngeal nerve. Vagus nerve from midsection of nucleus ambiguus supplies palatal and pharyngeal muscles, Caudal nucleus ambiguus supplies laryngeal muscles via the vagus. These rootlets have been erroneously termed the cranial root of XI. Spinal accessory nucleus does not join with vagus and does participate in pharyngeal arch innervation. Cranial nerve XI does not exist. CN XI supplies upper third of sternocleidomastoid and descends to the trapezius, indicating the phylogenetic relationships of these muscles to the first five neuromere of the spinal cord (c1–c5). So-called "cranial part" of XI given off by r10–r11 and destined for the pharynx via vaguas: palato-pahaygo-laryngeal nerve. Left: [Reprinted from Hanes DE. Fundamental Neuroscience for Basic and Clinical Applications. Philadelphia, PA: Saunders; 2006. With permission from Elsevier.] Right: [Reprinted from TeachMeAnatomy, courtesy of Dr. Oliver Jones]

nerve. Rhombomeres 8–11 are motor for middle and inferior pharyngeal constrictors via pharyngeal plexus, for the fourth arch muscles of the larynx via r8–r9 superior laryngeal nerve, and for the muscles of the fifth arch via r10–r11 inferior laryngeal nerve. We can rightly term this motor nerve of X as the palato-pharyngo-laryngeal nerve (PPL).

Herein let's debunk an anatomic anachronism. The most caudal two rhombomeres r10–r11 have been traditionally (and mistakenly) termed a separate structure, the *cranial root of spinal accessory nerve*. In reality, these two roots simply complete the PPL. Understanding the motor supply to arches 3–5 is much simpler and neuroanatomically accurate if we use the neuromeric model.

Cranial Nerve XI, Does It Exist?

Spinal accessory has traditionally been described as having two anatomically separate components. This concept has been disproven [15, 16].

Cranial part (ramus internus) has 4–5 rootlets arising in series from nucleus ambiguus, representing levels r10–r11. These are slightly displaced away from those of r6 to r9. As these rootlets approach and enter the jugular foramen, they join together just superior to the inferior (nodose) ganglion to form common vagus, that is, *palato-pharyngo-laryngeal nerve* (PPLN). The rootlets of PPLN share a combined sheath of arachnoid. PPLN in jugular foramen remains completely separate from cranial nerve IX. Glossopharyngeal nerve comes exclusively from levels r6–r7 and is sensory for nociception and taste and PANS motor to salivary glands; it has but a *single somatic motor branch to stylopharyngeus*. Within jugular foramen PPLN and glossopharyngeal nerve IX remain separate: IX has its own arachoid sheath. Spinal accessory nerve also has its own sheath of arachnoid. The periosteum of the posterior fossa is falsely labeled dura—see Chap. 12. Intracranial periosteum is continuous with extracranial periosteum so nerves IX, X (its caudal roots being the imposter XI) are all covered with "dura."

Upon exit from jugular foramen IX and PPLN immediately diverge. So too does SAN. It takes off posteriorly and laterally in the general vicinity of internal jugular vein. After crossing transverse process of C1, it is crossed by occipital

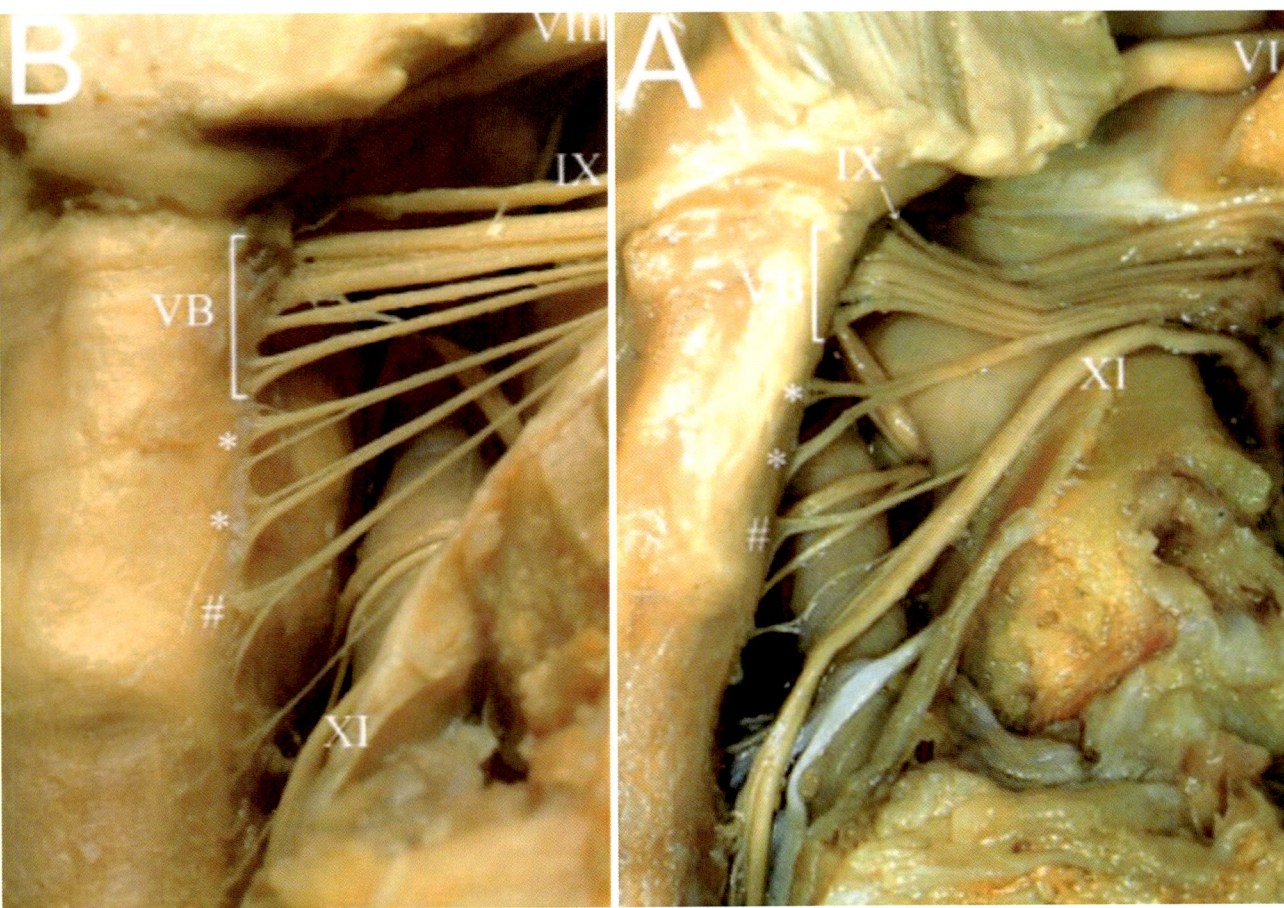

Fig. 10.81 Spinal accessory nerve in situ dissection. Left Rootlets from nucleus ambiguus indicated by * and # crossing the posterior fossa to join with vagal bundle (VB) to form common palate-pharyngo-laryngeal motor nerve. Brainstem rotated, showing olivary eminence. SAN is clearly distinct. Cranial nerve VIII, IX, and XI labeled. VB = vagal bundle. Right: inferior cerebellar peduncle sectioned, jugular foramen opened, showing SAN heading caudally toward sternoclei-domastoid and trapezius. SAN remains distinct from X without contributions from caudal medullary rootlets. Hypoglossal nerve XII (unlabeled) seen ventral to the rootlets, showing four roots converging to enter hypoglossal foramen (unlabeled). [Reprinted from Wiles CCR, Wrigley B, Greene JRT. Re-examination of the medullary rootlets of the accessory and vagus nerves. Clini Anat 2007; 20(1):19–22. With permission from John Wiley & Sons]

artery. Spinal XI follows the superior sternocleidomastoid branch of occipital artery and enters the muscle. Within SCM, spinal root has various options. It can anastomose with the *ansa of Maubrac*, that is, C2–C3 fibers from ansa cervicalis. It can also join with the *McKenzie branch*, that is, the anterior root of C1. XI usually emerges from SCM at its midpoint, called *Erb's point*. This generally occurs above (or within 2 cm) of the exit point of greater auricular nerve (C2–C3).

Spinal part (*ramus externus*) consists of 6 roots in series that coalesce to form a single peripheral nerve. Because the first root in evolution belongs to first truncal somite and because the latter is absorbed into the skull, SAN becomes entrapped by the jugular foramen. Its gross anatomy was difficult to discern in the nineteenth century and it was considered a part of what was thought to be a separate nerve from the rootlets of the posterior medulla: thus the legend of the XI was born and has persisted by default until recent neurosurgical dissections and dye studies have clarified the situation.

In sum: sternocleidomastoid and trapezius are the inheritors of cucullaris. Their myology, long misrepresented for lack of fossil evidence, has come to light revealing these muscles as likely branchiomeric but with a unique history connected with the evolution of the tetrapod neck which we shall examine in greater detail, *vide infra*.

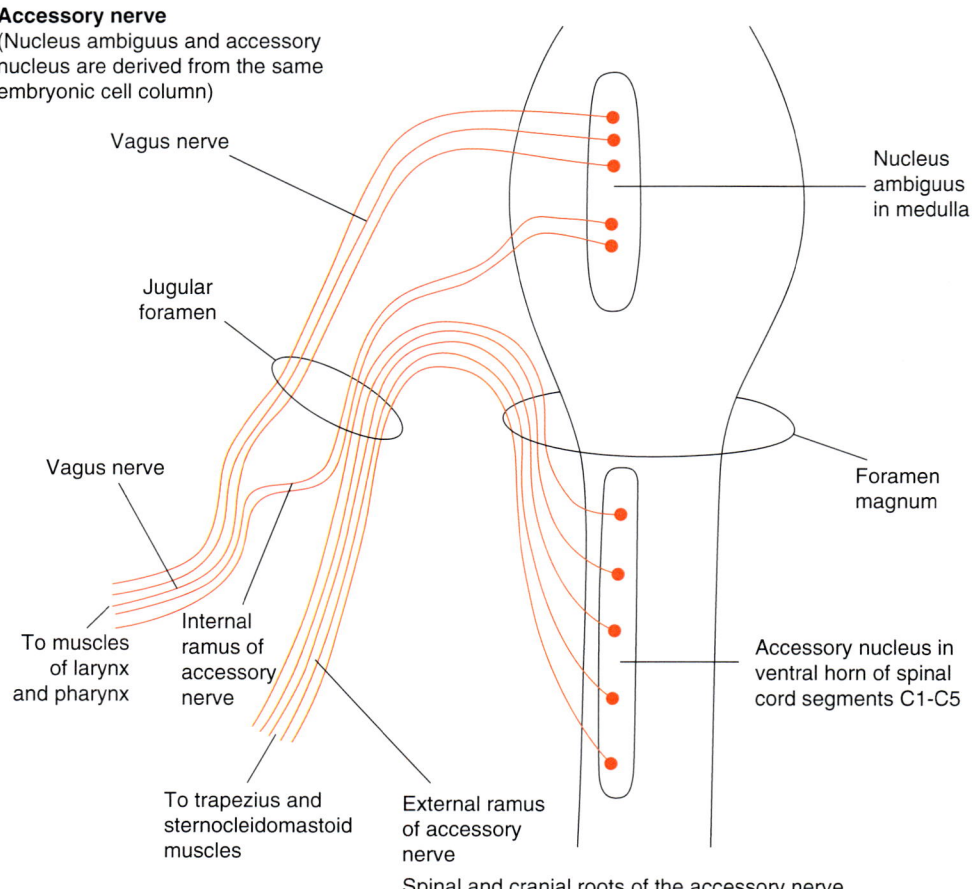

Accessory nerve
(Nucleus ambiguus and accessory nucleus are derived from the same embryonic cell column)

Vagus nerve

Jugular foramen

Vagus nerve

To muscles of larynx and pharynx

Internal ramus of accessory nerve

To trapezius and sternocleidomastoid muscles

External ramus of accessory nerve

Nucleus ambiguus in medulla

Foramen magnum

Accessory nucleus in ventral horn of spinal cord segments C1-C5

Spinal and cranial roots of the accessory nerve

Fig. 10.82 Spinal accessory nerve showing rootlets from caudal nucleus ambiguus apparently joining with spinal accessory nerve (but not sharing fascia) and eventually joining with vagus. Recall that the pleuro-peritoneal folds are associated with the pronephros sector of nephric ridge which begins at level c1 in the neck. Note that anterior scalene intervene between right phrenic n. and second part of subclavian whereas left phrenic n. descends more medially, directly over first part of subclavian. Phrenic nerve supplies diagphragm via three branches: sternal, anterolateral and posterolateral. These corresponds to congenital defects of Morgagni, eventration, and Bochdalek. Note: paired pericardiophrenic nerves penetrate to inferior surface of diaphragm; they receive pain signals from the underlying visera: right pericardiophrenic (liver), left pericardiophrenic (left lobe of liver, stomach, spleen and left adrenal). Phrenic nerve is NOT a part of cervical plexus and NOT a part of brachial plexus, as currently defined but phylogenetically it belongs to the brachial plexus. Its insertion pattern is antegrade-antegrade [Reprinted from Lewis, Warren H (ed). Gray's Anatomy of the Human Body, 20th American Edition. Philadelphia, PA: Lea & Febiger, 1918]

Phrenic Nerve (C3–C4–C5)

Diaphram is unique to mammals and, from an evolutionary perspective, is a derivative of the brachial plexus, vide infra. Its sole source of supply is phrenic nerve in which C4 is the predominant root. C4 constitute an evolutionary anatomic break-point. Recall our previous discussion of the caudal translocation of the brachial plexus that takes place in mammals. It is found along the lateral border of scalenus anterior (the most midline of the scalenii). Of course, we know phrenic nerve must follow an internal course. In order to do this, phrenic nerve must cut across planes. It penetrates the prevertebral fascia surrounding scalenus anterior and then proceeds to track across the muscle underneath its fascia.

Muscle and its fascia are coded by c3–c6 so it makes sense that c4 nerve to diaphragm follows along this plane. Its downward course takes it beneath omohyoid and between subclavian artery (behind) and vein (in front). While traveling through the neck, phrenic nerve receives SANS fibers from the cervical sympathetic ganglia. As it enters the thorax it crosses in front of internal thoracic artery (Figs. 10.83, 10.84).

Differences exist between the two phrenic nerves regarding their course within the thorax but these are not germane for our purposes. Suffice it to say that the phrenic nerves provide sensory input to the mediastinal pleura, fibrous pericardium, and the parietal (external) pericardium. Clinical correlation: The communality of C4 from the heart consti-

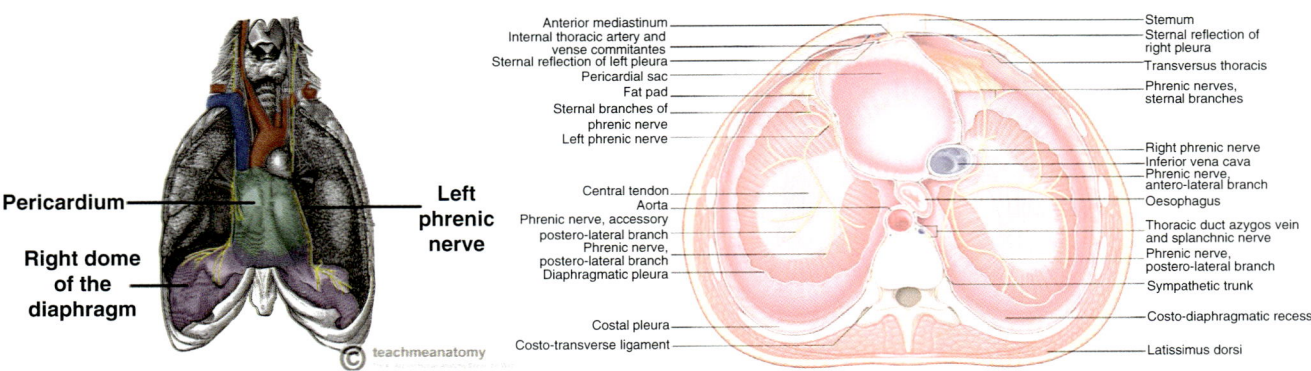

Fig. 10.83 Phrenic nerve C3–C5 Its anatomic course recapitulates entry of subscapular anlage into thorax above first rib, that is, at interface of c8–t1. Recall thath the pleuro-peritoneal folds are associated with the pronephros sector of nephric ridge which begins at level c1 in the neck. Left: [Reprinted from TeachMeAnatomy, courtesy of Dr. Oliver Jones] Right: [Reprinted from Standring S. In: Gray's Anatomy, 40th edition. Philadelphia, PA: Churchill-Livingstone; 2008. With permission from Elsevier]

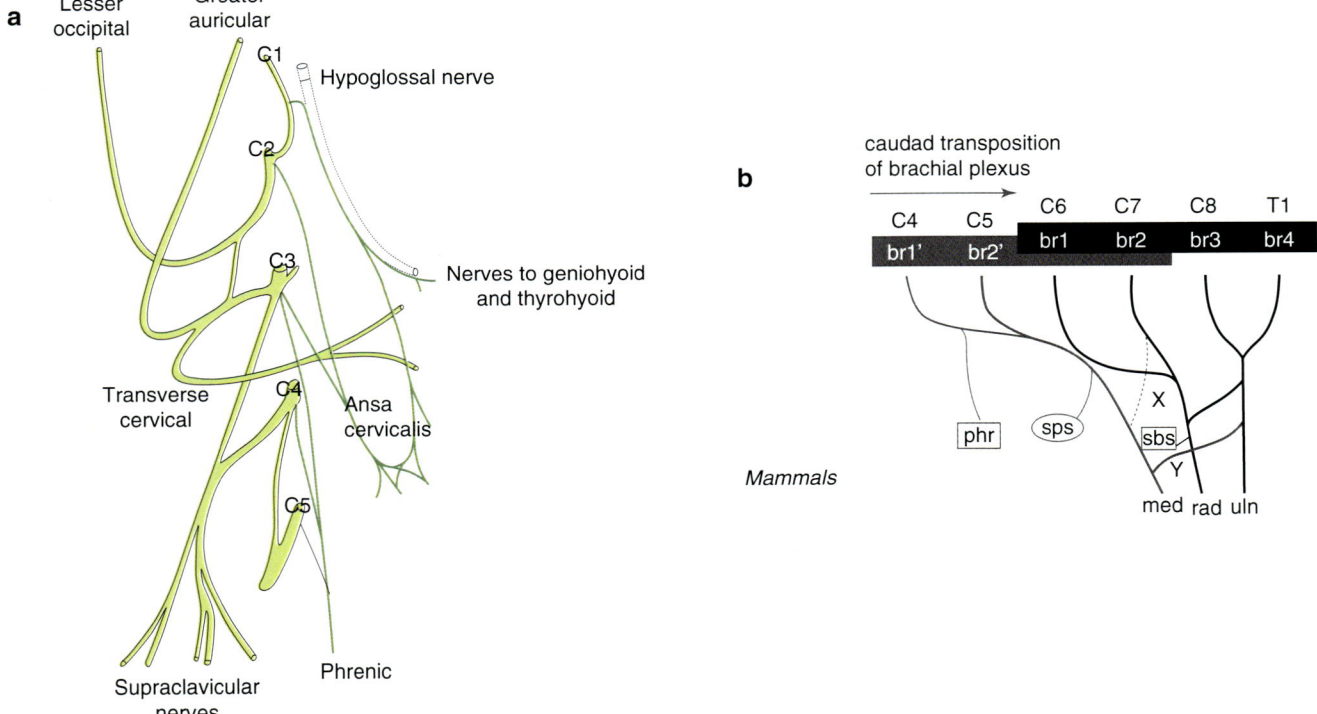

Fig. 10.84 Left: Cervical plexus with motor nerves (green) and sensory nerves (orange) demonstrates phrenic nerve arising directly from roots C3 and C5 with a contribution from C5. Phrenic nerve is NOT a part of cervical plexus and NOT a part of brachial plexus, as currently defined but phylogenetically it belongs to the brachial plexus. Its insertion pattern is antegrade-antegrade. Right: Brachial plexus. Standard description of the mammalian brachial plexus *fails to recognize the diaphragm as essentially a brachial plexus muscle*, with roots C4 and C5 innervating supracoradoideus, the precursor of diaphragm. Note the evolution of more complex motor control of the upper extremitiy results in an additional third motor branch, the median nerve. Thus median remain in the original position of radial while radial shifts backward two neuromeres to receive roots C6–C7 with a cross-branch to the union of C8–T1. Note the critical transitional nature of C4. Left: [Courtesy of Michael Carstens, MD] Right: [Reprinted from Hirasawa T, Kuratanin S. A new scenario of the evolutionary derivation of the diaphragm from shoulder muscle. *J Anat* 2013; 222:504–517. With permission from John Wiley & Sons]

tutes an important source of referred pain to the neck…a tightening of the throat.

At the central tendon of diaphragm, phrenic nerve sends branches to parietal pleura above and parietal peritoneum below. It then trifurcates. *Anterior branch* runs forward to the sternum. *Anterolateral branch* supplies the lateral leaf of the central tendon. Posterior branch supplies the crura. Below the diaphragm, branches of the phrenic nerve make clinically important connections with the celiac plexus. Sources of referring fibers are the liver, gall bladder, and the adrenals. Shoulder pain may thus reflect pathology in the thorax, hepatobiliary system, or kidneys.

Trauma to the phrenic nerve can occur in the neck at scalenus, beneath the clavicle, and from injuries to the upper brachial plexus.

Lateral Motor Column: C4–T1

Although we have foresworn venturing too deeply into the mysteries of the brachial plexus, it is very useful to dissect it apart from a functional and neuroanatomic standpoint. Although all of its muscles come from cervical somites (we must include somite 13, T1 for the sake of completeness, it is useful to consider those muscles as having a relationship with the pectoral girdle. If we do so, we can gain useful insights into the possible rationale behind the layout of brachial plexus. So be patient. We shall get into the myology again in the next section (Fig. 10.85).

Why Is a *New* Motor Column Needed in the Lowevr Neck

The neck is a mixture of the old and the new. It represents the forward continuation of purely axial trunk musculature (both epaxial and hypaxial) represented by the medial motor columns. On the other hand, the cervical region gives rise to an evolutionary innovation, the limb system. We see this transition for the first time in sarcopterygian fishes as muscles to control the fins migrate out from the body into the fin itself thus causing a bulkiness from which their name is derived. Muscles controlling the limb are, of course, ventral. But, by virtue of being in a new anatomic environment, one associated with the lateral plate mesoderm of the limb itself, the neuromuscular system makes a binary adjustment. (1) Muscles that continue to connect the pectoral girdle back to the head body at neuromeric levels c4 or more proximal than their insertion represent a reassignment of previous piscine muscles. These retain an innervation pattern based on the original motor columns, medial motor column, or as a caudal extension of nucleus ambiguus, central motor column. (2) Muscles that attach the pectoral girdle to the body wall distal to c4 or to upper extremity represent an innova-

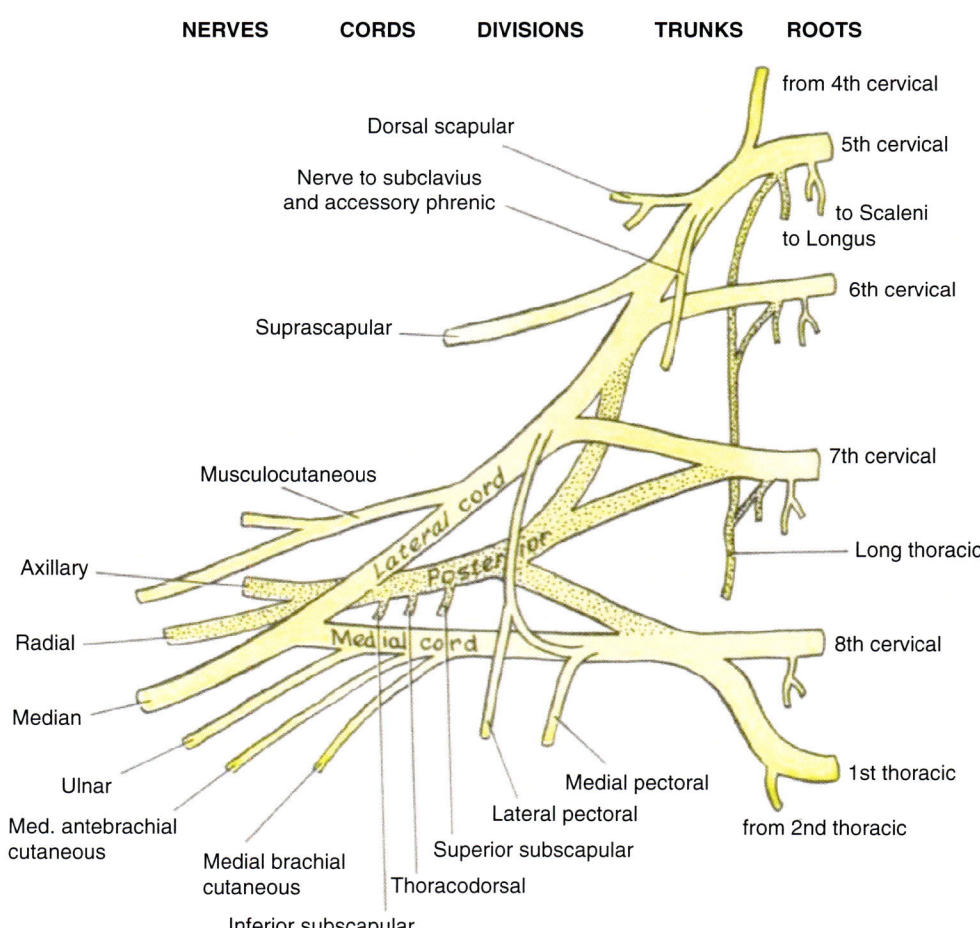

Fig. 10.85 Brachial plexus and phrenic nerve C5 sends branch to C4 phrenic nerve prior to participating with any other roots, that is, prior to entry into the plexus. All motor branches of brachial plexus supply cervical somite muscles antegrade–antegrade insertion sequence into pectoral girdle and distal neuromeres of the thorax and humerus. [Reprinted from Lewis, Warren H (ed). Gray's Anatomy of the Human Body, 20th American Edition. Philadelphia, PA: Lea & Febiger, 1918]

NERVES CORDS DIVISIONS TRUNKS ROOTS

from 4th cervical
5th cervical
Dorsal scapular
Nerve to subclavius and accessory phrenic
to Scaleni
to Longus
6th cervical
Suprascapular
7th cervical
Musculocutaneous
Lateral cord
Posterior
Long thoracic
Axillary
Radial
Medial cord
8th cervical
Median
Ulnar
1st thoracic
Med. antebrachial cutaneous
Medial pectoral
from 2nd thoracic
Lateral pectoral
Medial brachial cutaneous
Superior subscapular
Thoracodorsal
Inferior subscapular

tion. These muscles *cannot be supplied by the original primary ventral roots*; their motor nuclei become spatially distinct within the spinal cord as lateral motor column (cf Fig. 10.73).

In tetrapods, the situation becomes more complex. New genes are involved. During development, the limbs, to which these muscles attach, undergo predictable patterns of motion. In order to find their proper insertions myoblast groups must traverse complex pathways. Remember that the bones of the shoulder girdle and upper extremity possess many binding sites for muscles. These will fill up in an orderly manner, using three gradients: proximal–distal, posterior–anterior, and ulnar–radial. A new muscle seeking an available binding site will migrate in such a way as to respect the positional priority of their predecessors. This means that motor neurons will follow along previous pathways through the plexus until arriving at an appropriate exit point. Plexuses thus take on the appearance of busy switchyards. But the apparent anatomic complexity of a plexus belies the fact that it is simply a compilation of all migration pathways. *The developmental sequence of limb muscles is faithfully reproduced by its motor plexus.*

Ventral rami of C5–C8 join up with the ventral ramus of T1 to form brachial plexus. It is well described in Gray's anatomy so we shall be very selective here. Our emphasis is the neck and its connections, we will draw the line in terms of neuromuscular commentary with the pectoral girdle and its connections. This makes developmental sense. The clavicle arises from LPM of c1–c6 and scapula from c3 to c6, and it functions as a critical waystation between the trunk and the upper limb. It is a moveable platform containing muscles connecting two types of muscles: those connecting scapula with the axial skeleton and those connecting it with the humerus.

Secrets of Insertion

Pectoral girdle muscles with an *antegrade/retrograde insertion sequence* seek a secondary attachment at neuromeric sites proximal to the primary insertion. The myoblasts first gather around the pectoral anlage at the base of the limb bud, attach to their assigned sites, and then migrate backward to the midline to form secondary insertions. The secondary insertion sites follow a cranial to caudal and medial to lateral progression as the muscles seek out a(n) available binding site(s). All the most desirable parking spaces have been previously taken by the axial muscles. This pattern applies to the muscles supplied by the cervical plexus. These have secondary insertions into the skull and cervical spine.

Pectoral girdle muscles with an *antegrade/antegrade insertion sequence* have secondary attachments at neuromeric sites equal to, or distal to, the primary insertion sites. Fortunately, the upper limb follows the same pattern, with secondary binding sites on the humerus opening up in the same sequence, so that subsequent muscles insert further down on the shaft.

Functional Dissection of the Brachial Plexus to the Pectoral Girdle

Branches of the brachial plexus supply two classes of muscles, based on the insertion patterns previously described. We shall further group the targets of these nerves by insertion site. Muscles inserting into primary insertion into clavicle are indicated in red; those inserting into scapula are indicated in blue. A single muscle, deltoid, inserts into both clavicle and scapula.

(1) Antegrade/Retrograde Insertion
- Muscles that attach the clavicle (including manubrium) to the pharyngeal arches.
  - sternothyroid, sternohyoid, and omohyoid (C1–C3) = *cervical plexus.*
- Muscles that attach clavicle and scapula to the skull/vertebral column.
  - Sternocleidomastoid and trapezius (C1–C6) = *spinal accessory nerve.*
  - levator scapulae (C3–C4) = *cervical plexus* (plus C5 branch).
(2) Antegrade/Antegrade Insertions
- Muscles that attach the clavicle and scapula to the chest wall.
  - subclavius (C5–C6),
  - rhomboid major and rhomboid minor (C5),
  - serratus anterior (C5–C7),
  - pectoralis major (C5–T1) and pectoralis minor (C8–T1).
- Muscles that attach scapula to humerus, acting on the shoulder.
  - deltoid (C5–C6),
  - subscapularis (C5–C6),
  - supraspinatus (C5) and infraspinatus (C5–C6),
  - teres major (C5–C6) and teres minor (C5).
- Muscles that attach scapula to humerus acting on the arm:
  - coracobrachialis (C6–C7),
  - short head of biceps (C5–C6) and long head of biceps (C5–C6),
  - long head of triceps (C7–C8).

Motor Nerves of Brachial Plexus to Pectoral Girdle

Roots

- *Communicating to phrenic* (C5).
- *Nerve to scaleni* (C5–C8) has three parts. *Scalenus anterior* (C4–C6) is synthesized first. *Scalenus medius* (C3–C8) picks up three additional somites. Its mesenchyme is co-extensive and caudal to that of sternocleidomastoid (C1–C2). *Scalenus posterior* (C7–C8) picks up mesenchyme from the lower two roots. Its secondary insertion is into second rib because there is no further additional room. It lies deep to serratus, which forms later.
- *Nerve to longus colli* (C5–C8) also has three parts, each with a different primary insertion. *Superior oblique* LC (C2) is the smallest and is located at the atlas (a combination of $S5_C$ and $S6_R$). *Middle (vertical)* LC (C2–C4) is the largest. *Inferior oblique* LC (C5–C6) completes the sequence.
- *Dorsal scapular nerve* (C3–C5) has a very misleading name. Its primary contributor is the ventral root of C5 (with contributions from C3 and C4) and arises at the intervertebral foramen. It then penetrates scalenus medius to achieve a dorsal position. It then runs along the deep surface of *levator scapulae* which it supplies until reaching the dorsal border of scapula. From here it innervates *rhomboid major* and *rhomboid minor*. Because all three muscles are specific to the pectoral girdle, their nuclei belong to lateral motor column.
- *Long thoracic nerve* (C5–C7) This counterpart to dorsal scapular nerve supplies serratus anterior. It arises at the root level of the brachial plexus. Its nuclei are somatotopic and in line with dorsal scapular nerve. Interesting parallels exist between long thoracic and dorsal scapular nerve. Their nuclei with lateral motor column are continuous. DSN is parallel with phrenic nerve of the central motor column. All four muscles belonged at one point in evolution to a common blastema, the ancient subscapularis or subcoracoscapular. Its fragmentation created in mammals the two functional sets of muscles: those connecting pectoral girdle to the spine (levator scapulae and rhomboids) and those connecting it to the chest wall (subclavius, serratus anterior, pectoralis major and minor) plus the muscular diaphragm.

Trunks

- *Dorsal scapular nerve* (C3–C5) This nerve has a very misleading name. Its primary derivation is C5 with some contributions from C3 and C4. It arises at the interverte-

bral foramen. It then penetrates scalenus medius to achieve a dorsal position. DSN subsequently runs along the deep surface of *levator scapulae* (which it supplies) until reaching dorsal border of scapula. From here, it innervates *rhomboid major* and *rhomboid minor*. As it emanates from C3 to C5 nerve roots, dorsal scapular is sometimes not considered part picks of brachial plexus because of a mistaken model that the plexus has five roots. This mistake is corrected when we consider the phylogeny of the brachial plexus. In evolution, motor control of the upper extremity can begin as far forward as the second spinal nerve in anurans (frogs) but this is a backward step in evolution. *The basal amniote plexus begins with C4* and has four roots in the context of 5 cervical vertebrae with a formula C4–C5–C6–T1. Mammals duplicate two neuromeres C4–C5 with a formula of C4–C5–C6–C7–C8–T1 to create a brachial plexus of six roots and seven cervical vertebrae. Dorsal scapular is most certainly a full-fledged member of the plexus.
- *Nerve to subclavius* (C5) and *accessory phrenic nerve* (C5) Both these nerves are mononeuromeric. The latter can originate from subclavius n. or as a distinct nerve. Despite the difference in their names, both relate to clavicle. Recall that diaphragm is a derivative of the supracoracoideus which connects sternum to the ancient *anterior coracoid* in premammals. We distinguish this from the *posterior coracoid* which appears for the first time in the synapsid line leading to mammals. The former becomes *manubrium* and the latter the final coracoid process of the scapula. This neurology supports the origin of diaphragm as a re-directed brachial plexus muscle with C4 arising at the level of the cervical nerve and the C5 fibers traveling outward more laterally, being admixed with those supplying subclavius.
- *Suprascapular nerve* (C5–C6) is more lateral on the trunk.

Cords

- Lateral pectoral nerve (C5–C7) and medial pectoral nerve (C8–T1) arise from the lateral cords and medial cords, respectively. Pectoralis major receives both nerves while pectoralis minor is supplied by just the medial nerve.
- Upper subscapular nerve (C5–C6) is proximal, as it supplies upper subscapularis.
- Middle subscapular nerve, or thoracodorsal nerve (C6–C8) follow subscapular artery, the first branch of the third zone of axillary artery which is immediately proximal to the humerus, and descends with it for latissumus.
- Lower subscapular nerve (C5–C6) is distal, as it supplies lower subscapularis and teres major.

In Summation

Cervical plexus innervates muscles attaching pectoral girdle to the skull or cervical spine at proximal neuromeres: these are all antegrade-retrograde.

Brachial plexus innervates muscles attaching pectoral girdle to thoracic spins, thorax, or humerus at distal neuromeres: these are all antegrade–antegrade:

- only the upper trunk contributes to the pectoral girdle, and,
- no posterior division of any of the three trunks supplies the pectoral girdle.

Phyologeny of Motor Neuron to the Vertebrate Pectoral Girdle: A Preview

In the history of evolution, the fins-to-limb transition had neurologic implications. A formerly simplistic system of flexor/extensor muscles inside the body proceeding past a fixed pectoral girdle welded to the head cavity and connected to the fin undergoes a series of drastic changes. The pectoral girdle disconnects and becomes mobile, requiring motor control. New muscles extend outward into the extremity requiring new muscles supplied from an entirely different motor column. In this process, motor neurons actually migrate caudally out of the hindbrain and into the spinal cord due to a frameshift of *Hox* expression along the rostrocaudal axis. This neuromeric shift is crucial to the formation of the neck. We shall look at this process in greater detail at the conclusion of the myology section, once we have discussed the developmental anatomy of cucullaris muscle and its derivatives (sternocleidomastoid and trapezius) as well as that of the uniquely mammalian innovation, the diaphragm.

Myology of the Neck

How Cervical Muscles Originate and Insert

Cervical muscles originate from paraxial mesoderm in the myotomes of somites. The sole exception to this rule is the possibility of lateral plate mesoderm contributing to the muscles of cucllaris, sternocleidomastoid and trapezius. This difference is academic as the connective tissue stroma within these muscles arises from the neural crest of cervical neuromeres c1–c6. Using our neuroanatomic model, we can organize the muscles of the neck as follows.

- Medial motor column supplies *vertebral muscles.*
  - The anterior flexors are uniformly polyneuromeric and are innervated by ventral branches of segmental nerves C1–C6.

- The posterior extensors MMC are both mononeuromeric and polyneuromeric. They are innervated by dorsal branches of segmental nerves C1–C8.
- Cervical plexus (C1–C3) supplies *hypobranchial* muscles (cervical somites 1–3). These represent the ancient coracomandibularis which connects sternum with pharyngeal arches innervated by cervical plexus.
- Lateral motor column supplies type three types of hypaxial muscles.
  - Segmental nerves (C1–C8)—not via a plexus—supply lateral vertebral muscles arise from cervical somites 3–8, attach to ribs 1–2 and receive motor innervations from. All muscles control the upper limb girdle.
  - Cervical plexus (C1–C3) supplies hypobranchial muscles (cervical somites 1–3 and represent the ancient coracomandibularis which connects sternum with pharyngeal arches innervated by cervical plexus.
  - Brachial plexus (C4–T1) supplies the muscles of pectoral girdle and upper extremity.
- Central motor column muscles are unique.
  - Sternocleidomastoid (C1–C2), and trapezius (C2–C6) represent an evolutionary transition between the skull and pectoral girdle. They are supplied by a polyneuromeric segmental nerve, spinal accessory.
  - Diaphragm cervical C3–C5), found only in mammals, represents a pectoral girdle muscle internalized into the chest. It marks the most cranial boundary of brachial plexus.

The piscine forefin is positioned at the junction of head and trunk. Like all limbs, it is strictly hypaxial. It consists of a dermal pectoral girdle (including clavicle) and a cartilage-replacement fin (the proximal bone of which is scapula). *Over the course of evolution scapula assumes an epaxial position but, because it is inseparable from upper extremity, its neuromuscular apparatus retains the original hypaxial identity.* Leaving aside the XI-innervated sternocleidomastoid and trapezius, all other scapular muscles behave like limb muscles. They are innervated by ventral motor nerves *acting through a plexus.* Cervical plexus and brachial plexus are co-extent with C4 being the transition point.

As we proceed forward with our analysis of muscle groups, a great deal of attention will be placed on where each muscle attaches and in what order these attachments take place. You will note that the words "origin" and "insertion" are deliberately not used in these contexts. All striated muscles from level r8 and more caudal originate from the myotomes of somites. The neuromeric level(s) of origin for any muscle is identical to that of its motor nerve. Pectoralis minor is innervated by medial pectoral nerve (C8–T1). Its myoblasts therefore arise from S12–13, that is, the eighth cervical and first thoracic somites. Muscle attachments proceed along three gradients: cranial-caudal, deep-superficial, midline-lateral. Mononeuromeric muscles have priority over polyneuromeric muscles.

Let's see how this process works for the muscles of the anterior midline of the neck. Rectus capitis anterior minor (c1) has midline priority with primary insertion at mass of atlas and the medial transverse process. It gains a privileged secondary insertion site into posterior basioccipital just in front of foramen magnum. Due to its c1 predecessor, rectus capitis lateralis (c1–c2) is forced to take a primary insertion on lateral transverse process of atlas and must also seek out a more lateral secondary insertion site on jugular process of exoccipital bone. Longus capitis (c1–c3), otherwise known as rectus capitis anterior major, must take a more anterior primary insertion into basioccipital just behind basisphenoid.

Cervical Muscle Migration: Axial Muscles Versus Appendicular Muscles

Cervical muscles interconnecting the axial skeleton have a simple (somitic) embryonic origin. They can be dorsal or ventral and are always supplied by individual roots, *not by a plexus*. Their primary insertions are always local and follow the neuromeric and spatial constraints as previously described. Secondary insertions can be proximal or distal depending upon site availability. Muscles from the atlas are bidirectional, spanning upward to the skull and downward as erector spinae.

Cervical muscles associated with the pectoral girdle develop by two mechanisms described by Huang. Myoblasts move into their corresponding neuromeric zone around the lateral plate mesoderm of the clavicle and scapula. After developing their primary insertions, they follow two options: (1) The muscles translate backward to available sites on axial skeleton (skull, ligamentum nuchae, vertebral spines). (2) The muscles translate forward from pectoral girdle to humerus. The former process is called *antegrade–retrograde migration* while the latter process is referred to as *antegrade–antegrade migration*. Once again, primary insertion always takes place within the neuromeric territory of origin; secondary insertion is opportunistic.

Consider these examples. *Rhomboid minor* (C4–C5) is a somitic muscle innervated by dorsal scapular nerve. It first attaches to the c4–c5 zone of scapula just at medial border of the spine. Subsequently, it attaches to the spinous processes of the seventh cervical and first thoracic vertebrae. *Rhomboid major* (C4–C5) must attach to medial scapula distal to rhomboid minor; thus, it inserts from below the medial spine all the way down to the inferior angle of scapula. Its secondary attachments are also in sequence with the spinous processes of second–fifth thoracic vertebrae. Sternocleidomastoid myoblasts descend from S1 to S2 to posterior margin of manubrium and medial clavicle, make their attachments and return to the r6–r7 mastoid process and r8–r11 superior nuchal line.

Principles of Motor Column Analysis

With these principles in mind, let's examine the muscle groups following the same developmental schema. The epaxial muscle group is more simplistic. The suboccipital muscles and the deep back muscles are anatomically similar and form a single functional unit supplied by dorsal spinal nerves. We will then turn our attention to the hypaxial muscle group. Anterior vertebral and lateral vertebral muscles interconnect head, neck, and trunk by means of ventral spinal nerves. The more superficial infrahyoid (sic) muscles connect the pharyngeal arch system with the trunk by means of the cervical plexus. Next, we shall examine muscles of occipitocervical origin that connect upper limb to vertebral column and muscles of cervical origin that connect upper limb to thoracic wall. Finally, we conclude our survey of the cervical musculature with a discussion of muscles of having extraordinary evolutionary significance: (1) a forgotten member of the brachaial plexus, the diaphragm; and (2) the cucullaris muscle and its mammalian derivatives, sternocleidomastoid, and trapezius, including implications for the evolution of the neck.

Cervical somites differ from occipital somites in two fundamental ways. First off, they have a full complement of subunits. Recall those occipital somites have sclerotomes; their myotomes are hypaxial only, and their dermatomes are involute. Somites mature in a cranial-caudal progression. Somite 5 (C1) has an epaxial myotome, but no dermatome. Somite 6 has a fully functional dermatome supplying both hypaxial neck skin and epaxial scalp.

With the exception of esophagus, all neck muscles arise from PAM and are somitic. Esophageal muscle originates in lateral plate mesoderm. Motor columns differ by dorsal/ventral and by function: the role of pharyngeal arch muscles as "the outsider" is assumed by muscles assigned to the pectoral girdle and upper extremity.

- MMC rostral, c1–c4: *occipital cervical and upper paraspinous muscles* (S5–S8).
- MMC caudal, c5–c8: *lower paraspinous muscles* (S9–S12).
- LMC rostral, c1–c4: *cervical plexus muscles* (S5–S7) > pectoral girdle.
- LMC caudal, c5–t1: *brachial plexus muscles.* (S8–S13) > upper extremity. We should include c4/S8 in this category as well for its contribution to diaphragm, a surprising original member of the "brachial plexus breakfast club."

- CMC, c1–c6: diaphragm (S9) from phrenic motor column (PMC) and sternocleidomastoid/trapezius (S5–S9) from spinal accessory motor column (SAC).

Phylogeny of the Pectoral Girdle: A Quick Review

In the ensuing section, we shall be discussing the individual muscles of the neck from a neuromeric, neuroanatomic, and functional consideration. Because many of these muscles insert into the pectoral girdle, its evolutionary components need to be kept in mind. Although this material was presented under the section of osteology, a summary is worthwhile.

The earliest gnathostomes were armored fishes. The most basal forms, such as the osteostracans, had no jaws but did have paired pectoral fins with a shoulder girdle fused to the skull at the level of the heart. Jaws appear in the more advanced placoderms, so-called for the ring of armor surrounding the anterior trunk. These latter split into the defunct †acanthodians and the ray-finned actinopterygians, an offshoot from which became our distant ancestors, the lobe-finned sarcopterygians.

In one group of placoderms, the †arthrodires (ancestral to tetrapods), separation of the head and body is first seen using a pivot joint that permitted elevation of the head under muscle control involving the prototype of the scapula. The prototype of the pectoral girdle is desmontrated in the arthrodrthrodire †Dunkleosteus. It consisted of five dermal bones suspended from the skull by post-temporal bone and followed in succession by supracleithrum, postcleithrum, cleithrum, and the most ventral, clavicle. Immediately behind this bony suspension bridge was a mesodermal scapulocoracoid cartilage articulated between cleithrum and the forefin. Control of the arthrodire pivot joint involved two anatogonistic muscles: the dorsal levator capitis, attached to the thoracic armor and the ventrolateral cucullaris, attached to scapula. This initial genetic experiment (a homeotic switch) was abandoned in the subsequent evolution of fishes into the chondrichthyan line and osteichthyans.

In basal actinopterygians s split occurred in which scapulocoracoid split into two separate bones, scapula and coracoid. These were dorsal to clavicle. Although pretetrapod sarcoptyergians such as the osteolepiform †Eusthenopteron retained the scapulocoracoid complex, the basics of a limb are now present.

The subsequent development of the pectoral girdle in tetrapods is one in which its the dermal elements are reduced or lost and mesoderm elements become dominant. Loss of post-temporal was the fundamental step required in separation of the pectoral girdle from the head. In the early tetrapod †Europs. only cleithrum and clavicle remain (the interclavicle will be lost in mammals). A hollowing-out of scapulocoracoid creates a glenoid fossa, fit for receiving a weight-bearing extremity.

Amniotes are divided into the diapsids and synapsids; in the latter scapulocoracoid once again fragments, now into three pieces: scapula, the (original) anterior coracoid, and a new element, the posterior coracoid. Mammals position the limbs beneath the body with elbows directed posteriorly and knees anteriorly, producing a longer stride and more efficient gait. Motor needs drove change. Massive ventral muscles needed to pull the body off the ground were not needed. The shoulders being mobile versus the pelvis required expansion of dorsal muscles for stability. In some important cases, ventral muscles shifted dorsally, *without a change in their neural control*. This explains the innervation of trapezius, levator scapulae, and the rhomboids. Consequently, in therapsids the coracoids got smaller. Both are retained monotremes (the all-purpose iconoclast, the platypus) but the anterior coracoid is lost (but not really) in therians, and subsequently in their descendants, the mammals. "Loss" of anterior coracoid permitted the glenoid fossa to redirect ventrally atop the humerus. Posterior coracoid remains adjacent to glenoid fossa as the coracoid process. Cleithrum becomes repositioned over scapula as the scapular spine. Its lateral margin becomes acromion.

Interclavicle persists into synapsids as far as the monotremes but is lost in therians (mammals). Some running mammals lose the clavicle but it is retained in most as a shoulder brace. Recent evidence suggests that anterior coracoid is not really lost, but rather is reincarnated as manubrium.

Epaxial Muscles Interconnecting the Skull and Cervical Spine (MMC)

Suboccipital Muscles

These muscles connect the upper three cervical vertebrae to the skull. All four muscles in this group arise from the first cervical somite (somite 5). Depending upon their relative biologic maturity, they choose different primary and secondary sites of attachment as these become available. Taken together, they comprise an equilateral *suboccipital triangle*. Its base is transverse, in parallel with the posterior arch of the axis and is defined by *obliquus capitis inferior*. The apex of the triangle points upward toward greater occipital nerve. Its medial limb is *rectus capitis posterior major*. The medial border of its lateral limb, obliquus capitis superior, overlaps RCP major, and leads directly to greater occipital nerve. This implies that REC major is laid down first, followed by OCS. Note that three out of four of these muscles form a secondary attachment to the spinous process of the axis.

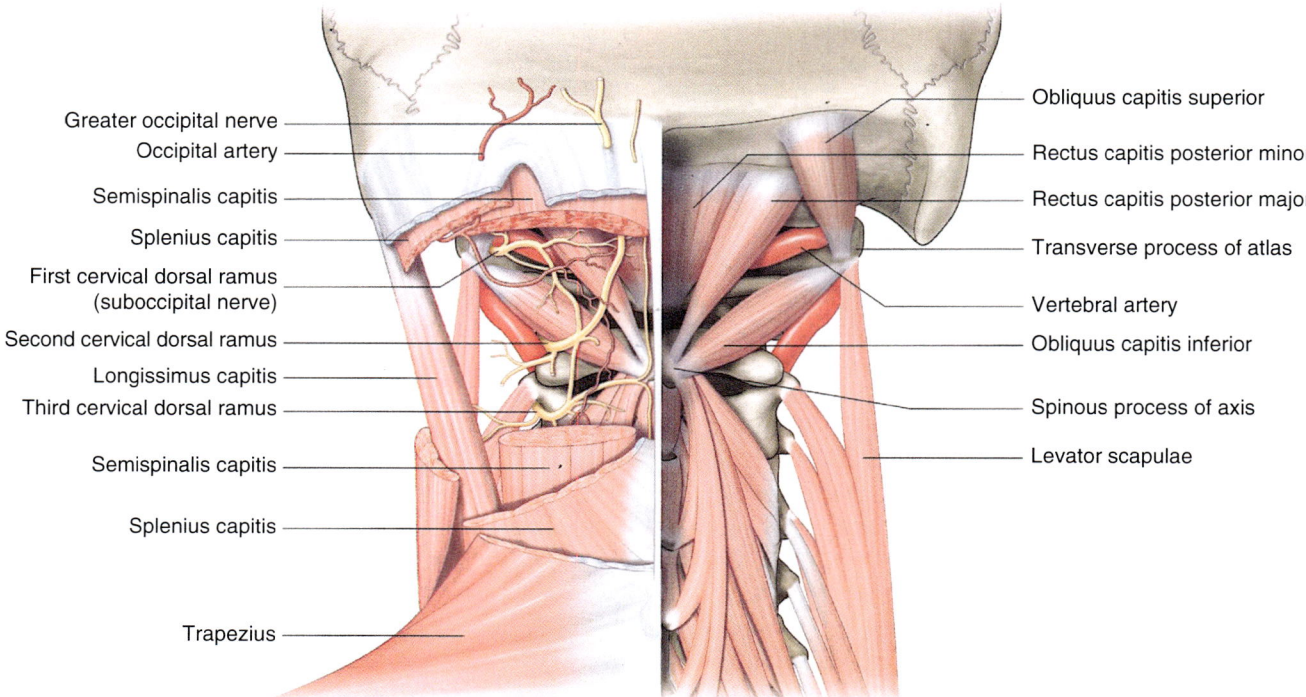

Fig. 10.86 Dorsal rami of C1–C3 C1 is strictly motor. C2 has extensive sensory distribution up the lateral aspect of the occiput while C3 gives sensory innervation to the medial zone. C3, C4, and to a lesser extent C5 supply diaphragm. [Reprinted from Standring S. In: Gray's Anatomy, 40th edition. Philadelphia, PA: Churchill-Livingstone; 2008. With permission from Elsevier]

Let's discuss them in their developmental sequence: *rectus capitus minor > rectus capitis major > obliquus capitis superior > obliquus capitis inferior* (Fig. 10.86).

The suboccipital muscles have distal attachments to occipital bone at the r11–c1 zone and higher up to r8 (superior nuchal line). Their proximal attachments are to the c1–c2 neural arch of the atlas; and to the c2–c3 zone, that is, the axis.

- *Rectus capitis posterior minor* This muscle is more midline than RCPmaj so it migrates earlier. Its primary attachment is the tubercle of posterior atlas arch; its secondary attachment is medial inferior nuchal line.
- *Rectus capitis posterior major* This muscle develops later than its minor colleague RCPmin so it assumes a more lateral course. Its primary attachment is the spinous process of axis and it secondarily attaches to lateral inferior nuchal line of chondral supraoccipital bone. It is crossed superficially by two nerves, first, the posterior division of the first cervical nerve and, superficial to that, the greater occipital nerve.
- *Obliquus capitis superior* OCS can be seen overlapping RCPmaj so we know it is laid down last. Its primary attachment is lateral mass of atlas. Its secondary attachment is occipital bone; it lies lateral to RCPmaj to fill out the lateral end of the fossa between superior and inferior nuchal lines.

- *Obliquus capitis inferior* Larger than its colleague, the primary attachment of OCI is apex of spinous process of axis. Its secondary attachment is to the dorsal part of transverse process of atlas.

Epaxial Muscles Interconnecting the Cervical Spine and Thorax (MMC)

These muscles exist in three layers. These are, from deep-to-superficial: *intrinsics, erector spinae,* and *splenii.* The extent of muscle insertions increases in the same sequence. The deep layer consists of small muscles. The splenials extend all the way down to the ribs (Fig. 10.87).

Intrinsics

Three of these four muscles are *monosomitic*; from deep to superficial and from medial to lateral, they are:

- *Multifidus* muscles have a rotatory function. They span from the *caudal surface of the laminae* outward to the *transverse process* over levels C2–C5.

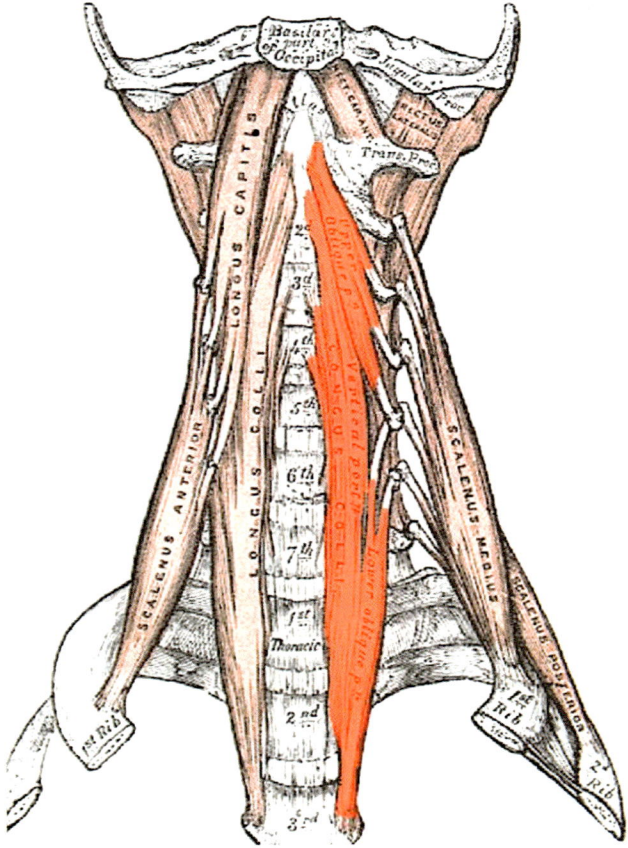

Fig. 10.87 Medial motor column supplies anterior neck muscles either directly by ventral roots or via cervical plexus. Note key role of Scalenus muscles. *Longus colli* (C1–C6) innervated via the cervical plexus, C1–C4, and as isolated branches from C5 to C6. Starts at S6–S9, sparing the dens (S5). It is in continuity with *scalenus medius*. *Longus capitis* connects r8 basioccipital to anterior neck and is in continuity with *scalenus anterior*. Both anterior and medial scalen are innervated by C4 via cervical plexus. Scaleus posterior is not supplied by C4. C4 innervates supraclavicular c4 zone. C4 from central motor column supplies diaphragm. [Reprinted from Lewis, Warren H (ed). Gray's Anatomy of the Human Body, 20th American Edition. Philadelphia, PA: Lea & Febiger, 1918]

- *Semispinalis* spans from the *spinous processes* of C2–C5 to a more lateral position on *posterior surface of transverse process* 6 segments below.
- *Interspinales* These muscles are short and stubby. They connect the *apices of spinous processes* spanning from one vertebra segmentally to the next. Thus, they exist as pairs straddled on either side of the interspinous ligament. There are three groups: cervicis, thoracis, and lumborum. Of interest to us is *interspinales cervicis*. It has six pairs. The first interspinal is between axis and C3; the last one is between C7 and T1. We don't find analogous muscles between atlas and axis, nor do they exist between

skull base and atlas. Evolutionary relationships between proatlas, atlas and axis, and atlas explain this curious absence.

- *Posterior intertransversari* These worm-like muscles connect the transverse processes of neighboring vertebrae. Although they are found along the entire length of the spine, they are best developed in the cervical region, where they exist as seven paired slips, *directed posteriorly and anteriorly*. ITC has a *medial slip* that connects to the *posterior tubercle of the next caudal transverse* process. It belongs to the *intrinsic back muscles*, so its nerve is *posterior* primary ramus. ITC also has a *lateral slip* that connects to the *anterior tubercle of the next caudal transverse process*. It is therefore *analogous to intercostal muscle*. It is therefore served by *anterior* primary ramus. Nested in between these muscles at every level is the ventral primary division of the respective cervical nerve. The biologic significance of these small muscles is as a watershed between ventral primary rami and dorsal primary rami.

Erector Spinae

The erector spinae group are *polysomitic* muscles. They are large and fleshy, extending from skull to sacrum. They are arranged in *three vertical columns* lying on either side of the vertebral column extend from skull to sacrum. These are spinalis (lateral column), longissiums (intermediate column), and iliocostalis (lateral column). Four regional groups are determined by the superior point of attachment: capitis, cervicis, thoracis, and ililumborum. Innervation is by dorsal primary division of spinal nerves.

- *Spinalis cervicis* is poorly represented. When present, it comes from *higher up on the spinous processes* of axis, C3, and C4 with secondary insertion into the midline *ligamentum nuchae* and *spinous processes* of C7 and T1(T2). On occasion, contained within spinalis cervicis is an unusual muscle, *spinalis capitis*.
- *Longissimus capitis* This muscle arises from *articular processes* of cervical vertebrae 5–8. It insets into *mastoid process* deep to splenius capitis (and to sternocleidomastoid). This implies it develops prior to splenius. Its nerve supply is c5–c8 (lower cervicals + middle).
- *Longissimus cervicis* spans from the *posterior tubercle of transverse process* from C2 to C6. The muscles slips extend into the *thoracic transverse processes* of T1–T5.
- *Iliocostalis cervicis* spans from *posterior turbercles of transverse process* of C4–C6 into the *angle of ribs* 3–6.

Splenii

- *Splenius capitis* Innervation is from posterior rami of C4–C5. Muscle originates from the first cervical somite (S5) from occipital bone immediately inferior to the lateral 1/3 of superior nuchal line and from mastoid process. It is covered by sternocleidomastoid. It swoops downward, bypassing the neck altogether. It touches down with secondary attachments to the spinous process of the seventh cervical vertebra and to thoracic vertebrae 1–3. Innervation.
- *Splenius cervicis* The primary attachment of this muscle is to the posterior tubercles of transverse processes of cervical vertebrae 2–3. The secondary attachment is to spinous processes of thoracic vertebrae 3–6.

Hypaxial Muscles Interconnecting the Skull and Cervical Spine (MMC)

The muscles under consideration are: *rectus capitis anterior*, and *rectus capitis lateralis*, *longus capitis*, and *longus colli*. These four muscles can be considered as a single functional unit. RCA and RCL stabilize the atlanto-occipital joint. RCA and LCp antagonize cervical extension and rotate the head. The most significant point is the manner in which three of four muscles "map out" the incorporation of proatlas into posterior cranium. The attachment patterns of these four muscles, based on their developmental "maturity," illustrate perfectly the rules described in the introduction. Rectus capitis (c1) has priority for attachment over rectus capitis lateralis (c1–c2). Both recti have priority over longus capitis (c1–c3) and longus colli (c2–c6).

- *Rectus capitis anterior* This muscle develops from the first cervical somite (S5). S6 may contribute. It is hypaxial, a flexor, and innervated by anterior ramus of C1 (plus, sometimes, C2). The muscle is situated directly behind longus capitis. Its primary attachment is to the posterior lip of foramen magnum, that is, r11. It passes downward and laterally to a secondary attachment to lateral mass of atlas. Together with longus capitis, RCA antagonizes the extensors of the head. As yet another example of r11–c1 mesenchymal continuity, RCA is considered a cranial continuation of the anterior intertransverse muscles of cervical vertebrae. In point of fact, *RCA unites proatlas with the true second cervical vertebra.*
- *Rectus capitis lateralis* This muscle develops from first and second cervical somites (S5–6). RCL is hypaxial, a flexor, and innervated by a branch coming off the loop between ventral C1 and C2. RCL lies lateral to RCA. Its primary attachment is to jugular foramen (lateral r10–r11). The secondary attachment is to transverse process of atlas. Its action is to laterally flex the head and, with RCA, to antagonize the extensor muscles.

- *Longus capitis* This muscle is supplied by C1–C3 therefore it originates from the first-third cervical somites (S5–7). Its primary attachment is to r10–r11 basioccipital bone lateral to pharyngeal tubercle, that is, anterior and neuromerically cranial to anterior arch of atlas. Its most cranial somite of origin is cranial to that of longus colli. Therefore, longus capitis develops and migrates prior to longus colli. Its secondary insertion is to anterior tubercles of transverse processes of cervical vertebrae 3–6. This attachment pattern makes sense for two reasons. Fourth occipital somite and first cervical somite are directly adjacent. Basioccipital binding sites mature prior to those of cervical vertebrae. The action of longus capitis is to flex the head and upper cervical vertebrae.
- *Longus colli* This muscle develops from the second-sixth cervical somites (S6–10). It is hypaxial and therefore a flexor; it is therefore innervated by the ventral rami of C2–C6.

- The muscle has three parts. (1) *Vertical portion* arises from anterior surface of bodies of cervical vertebrae 2–4 and inserts into anterolateral bodies of cervical vertebra 5–7 and thoracic vertebrae 1–3. (2) *Superior oblique portion* arises from transverse process anterior tubercles of cervical vertebrae 3–5. It ascends to the atlas, inserting into a tubercle on the anterior arch. The tubercle also bears anterior atlanto-occipital membrane ligament, representing r11–c1 proatlas. (3) *inferior oblique portion* arises from transverse process anterior tubercles of cervical vertebrae 5–6; it inserts into the anterior bodies of thoracic vertebrae 1–3.
- Development of longus colli involves a temporal-spatial gradient. The first phase occurs as mesoderm of the vertical portion flows from somites 6 to 8, "spilling" downward and centric. The second phase involves "overflow" from somites 7 to 9 in the *opposite* direction. This sequence is identified because somite 2 does not contribute and because flow is "mirror image," that is, upward and centric from the most cranial (most mature) population. The final phase involves contributions from Inferior oblique portion flows from somites 9–10 but, because midline binding sites are occupied by the previous vertical portion that has commandeered the midline, the myoblasts are forced to make use of more lateral insertion sites. The actions of this muscle are nothing to write home about. It weakly flexes, rotates, and laterally bends the neck. Taken together, longus capitis and longus colli replicate the neuromeric distinction between neck and thorax. The cranial attachment of the former duplicates the proatlas. The caudal attachments of the latter occupy the anterior vertebral column from atlas to the third thoracic vertebra.

Hypaxial Muscles Connecting Neck and Trunk (MMC)

The three scalenes have shared functions: (1) fixation of the rib cage on forced inspiration, (2) ipsilateral bending of the neck, and, due to obliquity, rotates it. All scalenes are hypaxial; they are supplied by the ventral division of the respective cervical nerve. The scalenes have a different embryologic basis and develop prior to sternocleidomastoid. Absence of the latter has been reported in isolation from the scalenes.

- *Scalenus anterior* This muscle develops from somites 8–10. It is supplied by C4–C6. Scalenus anterior lies under cover of sternocleidomastoid. Its primary insertion is as the confluence of slips from the tubercles of transverse processes of cervical vertebrae 3–5. The secondary attachment is to scalene tubercle of upper *inner* surface of first rib at the *scalene tubercle*.
- *Scalenus medius* This muscle develops from somites 7–12. It is supplied by C3–C8. The most extensive of the scalenes, its primary attachment is as confluent slips from the *posterior* tubercles of cervical vertebrae 3–8. The secondary attachment is to cranial first rib. It lies directly anterior to the sternocleidomastoid.
- *Scalenus posterior* This muscle develops from somites 10–12. It is supplied by C6–C8.

Muscles Connecting Pectoral Girdle with Pharyngeal Arches (MMC)

Posterior Hypobranchial Muscles

In contrast to the anterior hypobranchial muscle series from occipital somite and dedicated to the tongue, posterior hypobranchial muscles arise from mesenchyme from the first–third cervical somites (S5–S7). These muscles are derived from the coracomandibularis in fishes which serves as a depressor of the mandible and is supplied by occipital nerves. In tetrapods, posterior hypobranchial muscles function as stabilizers of mobile structures (hyoid and thyrohyoid) during the act of swallowing. They are innervated by cervical plexus.

The migration pattern of these muscles can be deduced from the anatomy of the *ansa cervicalis*, a component of the cervical plexus. The motor nuclei dedicated to these muscles reside in the lateral motor column of c1–c3. (Previous opinion had erroneously held these fibers to be hypoglossal… they aren't!) Infrahyoid myoblasts follow the "slime trail" laid down by pioneering tongue myoblasts from the occipital somites. Thus, they slavishly follow the hypoglossal nerve.

At the base of the tongue, they encounter hyoid, which offers them four very attractive binding sites. These are arranged along the hyoid body by order of maturity, from posterior-to-anterior: thyrohyoid. Omohyoid, sternohyoid, thyrohyoid, and geniohyoid. Thus, they insert successively into fourth, third, second, and first arches. Like jet fighters, the myoblasts (along with their attendant motor nerves) will peel off in succession as they seek attachment to the hyoid binding sites. The consequent neuroanatomy is a faithful replica of this biologic sequence.

Despite some old-fashioned and confusing terminology, ansa cervicalis is as simple as "1–2–3." (1) *Superior root of ansa* is made predominantly from one nerve, C1. (2) *Inferior root of ansa* contains two nerves, C2 and C3. Together these three nerves innervate six muscle bellies (1 + 2 + 3 = 6). The union of the two roots forms a loop lateral and ventral to common carotid. Nota bene: because C1 has been long confused with XII, the superior root is sometimes called *descendens hypoglossi* whereas inferior root has been named *descendens cervicalis*. My advice is to ignore these terms… just know where they came from.

Fibers from C1 and C2 (via a secondary loop with C1) "hitch a ride" with hypoglossal. At hyoid bone, some of them hop off, providing motor nerves to geniohyoid, thyrohyoid, and omohyoid. Other C1–C2 fibers continue downward to form the superior root of ansa. The departure of the first three motor nerves is *somatotopic*. Geniohyoid travels the farthest along XII. Next, thyrohyoid parts company with XII at the posterior border of hyglossus. Finally, omohyoid starts out with the superior root of ansa but jumps ship soon thereafter. to innervate the muscle. Fibers from C2 and C3 form the inferior root of ansa. Right where the two roots join together, two more nerves jump off to innervate first sternohyoid, then sternothyroid. Finally, at the very bottom of the loop, the nerve to the inferior belly of the omohyoid is given off. Note that C4 is sometimes attributed to this plexus but, in reality, it remains the supply for the diaphragm. Recall that C4 provided a motor supply for the ancient subcoracoideus which in the reptilomorphs, becomes internalized inside the chest and invades the lateral thoracic fold that grows inward as the anlagen of the diaphragm.

Migration sequence of the hypobranchial muscles: cranial-to-caudal.

The hypobranchial (sic) muscles originate from eight consecutive somites. Occipital somites S1–S4 give rise to the muscles of the tongue and to a single prehyoid strap muscle, geniohyoid (Chap. 9). The first four cervical somites, C1–C4, give rise to the post-hyoid strap muscles. Their function is to connect the clavicle and manubrium (the ancestral coracoid) of the pectoral girdle with the ventral aspects of the pharyngeal arches all the way forward to the mandible. As such, they act as depressors of the lower jaw.

- *Geniohyoid* The first "target" of C1, this muscle "hitches a ride" with hypoglossal nerve, thus betraying its origin from S5. For this reason, it is not considered a true part of ansa cervicalis. Geniohyoid is the forward continuation of the coracomandibularis; for this reason, it is referred to as the prehyoid strap muscle. Geniohyoid has its primary attachment on the superior surface of hyoid and greater cornu (second arch). It is deep to the V3 mylohyoid. Having a narrow belly, GH runs forward to insert on the most "mature" sector of the r3 mandible, the genial tubercle. Tightening of GH "fixes" the floor of mouth during food ingestion. It originates from a single myotome (somite 5) and is supplied by C1.

- *Thyrohyoid* The first target of ansa is also small muscle from the fifth somite. It has a primary insertion from the oblique line of thyroid cartilage (fourth arch) and spans over to the inferior surface of the hyoid greater cornu (third arch). Its action is a bit indecisive: it can either depress hyoid or elevate thyroid. True to its size, TH merits a single nerve, C1.

- *Superior belly of omohyoid* The third target of ansa is a long-distance muscle (OHs) originating from lower border of hyoid posterior to thyrohyoid and lateral to sternohyoid. It forms a central tendon which is tethered at the level of cricoid cartilage by thickening of deep investing fascia. It acts with its companion inferior belly to stabilize hyoid on the clavicle (more on this anon). As a larger three-myotome muscle, OHs are supplied by C1–C3.

- *Sternothryoid* The first target of ansa, this muscle (ST) is misnamed. Although described as "inserting" on the oblique line of thyroid cartilage, its *fibers of secondary inerstion* are confluent with the most posterior binding site on hyoid bone, deep to sternohyoid. It therefore seeks out fourth arch and third arch. ST migrates earlier than SH. Accordingly, its primary insertion is the lowest on sternum, at the manubrium and on the cartilage of first rib, spilling over to the second rib.

- *Sternohyoid* The second target of ansa, this muscle (SH) has its primary insertion into the posteromedial clavicle, that is, the c1 pole, and sternum's "backside." Its secondary insertion is into lower body of the r6 hyoid bone, that is, third arch. Swallowing activates suprahyoid muscles. These elevate hyoid and larynx. Sternohyoid provides an opposing, stabilizing force. At times, it is reinforced by an additional *cleidohyoid* muscle, forming *sternocleidohyoid*. These muscles develop in the first, second, and third cervical myotomes. They are supplied by C1–C3.

- *Inferior belly of omohyoid* The final target of ansa, second belly of the longest strap muscle (OHi) sweeps over SH and ST, passing deep to r8–r11/S1–S4 sternocleidomastoid in search of a rather elusive binding site. This is not an easy proposition. All the "parking spots" around the "sternal square" are taken. The clavicle is also off limits

because its entire superior surface is taken up by a pair of "occipital occupiers," sternocleidomastoid and trapezius. So, in desperation, omohyoid gloms onto the posterior aspect of superior border of scapula. Taken together, the action of OHsup and OHinf is not only a depressor of hyoid but also a dorsal retractor. Cervical fascia is thereby tensed.

Muscles Connecting Pectoral Girdle with Vertebral Spine: MMC, LMC, and SAC

We shall discuss these muscles in the spatiotemporal order in which they are produced. This involves analysis of the neuromeric order of muscle origin, followed by that of its primary and secondary attachments. For example, both rhomboid minor and major originate from the fifth and sixth cervical somites, S9–10. The minor is C4 and the major, being larger is C4–C5. Primary attachments for both are to medial border of scapula, a chondral bone originating from lateral plate in register with levels c4–c8. The rhomboids choose their binding sites based on the relative biologic maturity of medial scapula following a cranial-caudal gradient. Thus, minor binds more cranial than major. The area of insertion of major, being a two-myotome muscle, more extensive. Secondary insertion sites follow the same gradient as well. Thus, rhomboid minor from upper medial scapula attaches to the seventh cervical and first thoracic vertebrae while rhomboid major from lower medial scapula attaches to the second-seventh thoracic vertebrae.

These muscles then are arranged in two layers. The ventral layer consists of levator scapulae, followed by rhomboid minor and major. The dorsal layer contains trapezius (a product of cervical somites 1–5) and latissimus (a product of cervical somites 6–8). The primary attachment site of trapezius, scapular spine, is a neomorph. It was not part of the bones making up the original scapular anlage. Synthesis of this muscle group is ventral-dorsal. Hence, the medial border muscles are covered over by trapezius.

- *Levator scapulae* (MMC) This muscle develops from the third and fourth cervical somites (somites 7–8) and is innervated by C3–C4 from cervical plexus, a long lever arm. Its primary attachment is to transverse processes of atlas, axis, cervical vertebrae C3–4. The secondary attachment is to superior medial border of scapula.

- *Rhomboid major* and *minor* (LMC) Both of these muscles arise from the fifth cervical somite (somite 9) and are innervated by C5 via dorsal scapular nerve of brachial plexus. Major has its primary insertion into medial border of scapula between the root of the spine above and the inferior angle below. Its secondary attachment is from spinous processes of thoracic vertebrae 2–7 (these comprise neuromeres t1–t8). The primary insertion of rhomboid minor into scapula lies above that of major, into the triangle-shaped medial surface of spine. Its secondary attachment to the spinous process of seventh cervical and first thoracic vertebrae. The actions of the rhomboids are to adduct scapula, i.e. they stabilize it when it bears weight. Inward rotation of angle of scapula by the lower muscle helps the adduction of the arm.

- *Trapezius* (SAC) Discussed separately under muscles connecting pectoral girdle to the skull.

Muscles Connecting Pectoral Girdle with Body Wall (LMC)

Ventral Body Wall

- *Subclavius* A cylindrical muscle, subclavius arises from fifth to sixth cervical somites and is supplied by their respective segmental nerves (C5–C6), its primary attachment is the groove inferior surface of middle 1/3 clavicle. Its secondary attachment is to first rib and the first costochondral junction. It stabilizes the bone during movement.
- *Pectoralis minor* Smaller than its confrere, this muscle arises from the eighth cervical and first thoracic somite. It is innervated by *medial pectoral nerve* (C8 and T1). Its primary attachment is to medial-superior surface of coracoid process. Its secondary attachment is to the third–fifth ribs. Note that two other muscles have primary attachments to coracoid process: c5–c7 coracobrachialis and c5–c6 short head of biceps. The implication is that the coracoid process has three developmental zones, arranged from lateral to medial. Biceps (short head) binds first, coracobrachialis follows, and pectoralis minor is last.

Dorsal Body Wall

- *Serratus anterior* This thin muscle arises from fifth to seventh cervical somites; it is supplied by *long thoracic nerve of brachial plexus* (C5–C7). Its primary attachment is to the ventral surface of medial scapula. In this regard, it lies opposite the c5 rhomboids, which are attached to dorsal surface of medial scapula. The secondary attachments are to the superior surfaces of ribs 1–9. Each digitation (except the first) is intercostal, that is, it is attached to two ribs.
- *Latissimus dorsi* Not technically part of the pectoral girdle, but it has a physical relationship to the dorsal surface of scapula, being superficial to it. This large muscle is of enormous utility for reconstructive surgeons, It develops from cervical somites 6–8. It is supplied by C6–C8 thoracodorsal nerve. Its primary attachment is via 7-cm quad-

rangular tendon which passes *anterior* to teres minor and is anchored to intertubercular groove of humerus (zone c6–c8). Its secondary attachment is very extensive: to the spinous processes of T7–T12, via thoracolumbar fascia all the way down to sacral vertebrae; laterally it attaches to iliac crest. It is covered over from T7 to T12 by trapezius, as the latter muscle forms later.

Muscles Connecting Pectoral Girdle to the Upper Extremity (LMC)

Nine muscles extend from scapula to proximal half of humerus to provide control of the arm. Apart from supraspinatus of them are polyneuromeric. All derive at least part of their mesenchyme from the fifth cervical somite. Secondary attachment sites vary according to the relative biologic maturity of the muscle and of its osseous attachment site along the humerus.

- *Supraspinatus* Arising from the fifth cervical somite (S9), this muscle fills the supraspinous fossa. Secondary attachment to humerus at the most proximal fossa of greater tubercle. It is supplied by suprascapular nerve (C5), the upper division of upper trunk. Recall that the lower division contributes to lateral cord. A rotator cuff muscle stabilizes the shoulder by tractioning the humerus inward to glenoid fossa. It rotates humerus laterally. Supraspinatus also may initiate abduction, one of the reasons that pain with rotator cuff is initiated with abduction.
- *Infraspinatus* This muscle is much larger, arising from fifth and sixth cervical somites (S9–S10). It is supplied by suprascapular nerve (C5–C6). Its fibers are confluent laterally with teres minor and inferiorly with teres major. Secondary attachment to humerus is into fossa #2 (intermediate) of greater tubercle. Also rotator cuff muscle, its actions are similar but it is not involved in abduction.
- *Subscapularis* This muscle from fourth and fifth cervical somites; supplied by upper and lower subscapular nerves. Its primary insertion covers virtually the entire ventral surface of the scapular body. Secondary attachment to lesser tubercle and anterior capsule of shoulder joint. Supplied by C4–C5 upper and lower subscapular nerves from posterior cord. These are proximal to the terminal branches, axillary nerve, and radial nerves.
- *Teres minor* develops from the fifth and sixth cervical somites (S9–10) and has a broad primary insertion along upper medial (axillary) border of scapula. It is supplied by C5–C6 via the posterior cord and axillary nerve. Teres minor wraps around the border of scapula in two directions. Posteriorly, its fibers are confluent with *infraspinatus* which is innervated by C5–C6 suprascapular nerve. Anteriorly it borders with *subscapularis* which is sup-

plied by C5–C6 upper and lower subscapular nerves. These muscles plus supraspinatus, all originating from S9 to S10, constitute a functional unit known collectively as the *rotator cuff*. Its secondary insertion into the humerus consists of two tendons: upper fibers attach to fossa #3 (the most distal) of the greater tubercle. Lower fibers go distal.

- *Teres major* is larger. It develops from cervical somites 5–7 (S9–S11). Its primary insertion is to lower the medial border of scapula; it covers the angle. It inserts secondarily into the humerus at the medial lip of intertubercular sulcus. Teres major is innervated by a branch of *lower subscapular nerve*, a branch of posterior cord. It shares a common nerve supply with lower subscapularis. These two muscles plus supra/infraspinatus constitute the *rotator cuff*. They have a common root.

  - Regarding binding sites on humerus, both of the teres muscles demonstrate *spatial precedence* over latissimus. Minor binds to C4 greater tubercle, major binds to medial lip of intertubercular sulcus, whereas latissumus binds to the bottom of intertubercular sulcus (the C6–C8 zone of humerus). From here, latissiumus makes very extensive secondary attachments. The myoblasts migrate around lateral border of teres majors and spill out over the surface of underlying serratus to gain attachment to thoracolumbar fascia.

- *Coracobrachialis* This is the second muscle to have its primary insertion into coracoid process. It inserts secondarily into the medial shaft of humerus deep to and proximal to biceps, between the origin of triceps and brachialis. It is innervated by usculocutaneous nerve from upper trunk (C5, C6) and lower trunk (C7).

- *Short head of biceps brachii* This muscle is the third and most lateral of the "coracoid chorus." It joins long head of biceps to gain secondary insertion into radial tuberosity. The insertion site of short head is into tuberosity of the radius, which is more distal than that of long head of biceps. The origin is C5–C6 via musculocutaneous nerve.

- *Long head of biceps brachii* Originating from C5 to C6 (S9–S10), long head has its primary attachment to the supraglenoid tuberosity at the superior margin of glenoid labrum.

- *Pectoralis major* This muscle arises from the fifth–eighth cervical somites and the first thoracic somite; it is innervated by two nerves: *lateral pectoral nerve* (C5–8 and T1) and by *medial pectoral nerve* (C8–T1). This muscle has two heads: (1) *clavicular head* inserts into medial ½ of clavicle; (2) *sternocostal head* inserts into the junction of ribs 1–7 with sternum. The secondary insertion of pectoralis major is into the humerus as a flat tendon 5 cm in width attaching to the terminus of lateral lip of intertubercular sulcus (bicipital groove). The primary action of pectoralis major is adduction of the arm and (secondarily) as

medial rotator of humerus. Clavicular fibers flex and adduct the humerus. Sternocostal fibers extend, against resistance, the flexed humerus against the side of the chest…as in giving a backwards elbow to someone standing behind.

- *Deltoid* Arising from fifth and sixth cervical somites; deltoid is supplied by axillary nerve (C5–6). Deltoid tuberosity ½ way down shaft, lateral side. Deltoid is a "late-comer" which covers over other muscles. Furthermore, it is further out on the brachial plexus, being innervated by one of the two terminal branches of posterior cord via C5–C6 axillary nerve.

Note that the deltoid and pectoralis "layer over" the anterior scapular muscles. They accomplish this because they migrate beneath the skin, giving passage for migration of deeper muscles that achieve more direct attachment to the pectoral girdle.

Muscles Connecting the Pectoral Girdle to the Skull (SAC = CMC Lateral)

Sternocleidomastoid

Sternocleidomastoid is a hypaxial polyneuromeric somitic muscle. It develops from the myotomes of the first two to three cervical neuromeres, that is, at c1, c2, and c3. It receives motor innervation from C1 to C3 via spinal accessory nerve. The central motor column in which SAN is located is a continuation of branchiomeric nucleus ambiguus in the medulla but there is no physical connection between it and the putative intracranial division of cranial nerve XI (Fig. 10.88).

Neuromeric Coding of Clavicle and Manubrium

To understand sternocleidomastoid, we must bear in mind the neuromeric coding of its primary insertion sites, clavicle and manubrium, and of its secondary insertion sites, the mastoid and occiput. Clavicle is like a sausage-like rectangular box made of two types of mesenchyme. Its peripheral "skin" is neural crest and its core is mesoderm. The medial articulation with manubrium is c2 and the lateral articulation with scapula is c6. The dorsal and superior sides are a gradient from c2 to c6. The ventral and inferior sides are exclusively c5–c6. OK, so what about the coding for c1? Why don't we see it in clavicle? The answer lies in the mysterious manubrium.

Manubrium is considered part of sternum but appearances can be deceiving. Recall that the body of the sternum is formed by the vertical fusion of five parallel bars of lateral plate mesoderm. It forms separate articulations with manu-

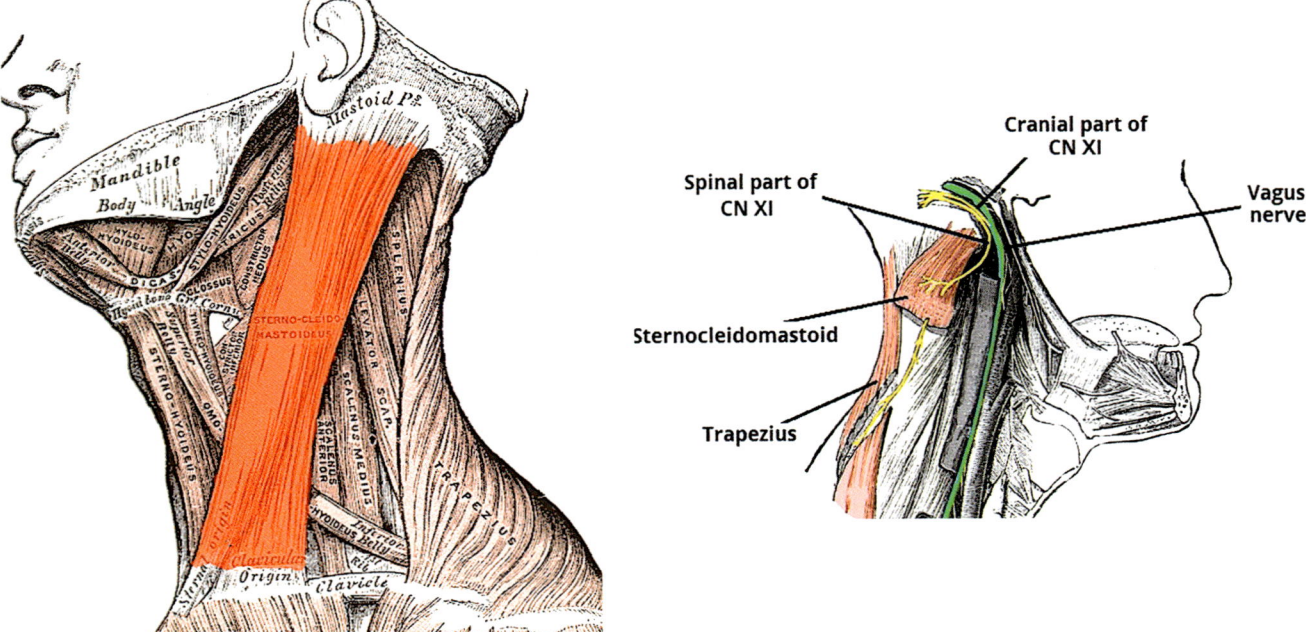

Fig. 10.88 Sternocleidomastoid. SCM has primary insertion into neuromeric sites c1–c3 along clavicle. Spinal accessory nerve CN XI develops in the lateral motor column of r8–r11 and the embryonic basal plate c1–c6. SCM is likely supplied by cranial + c1–c2 whereas trapezius receives motor contributions down to c6. Left: [Reprinted from Lewis, Warren H (ed). Gray's Anatomy of the Human Body, 20th American Edition. Philadelphia, PA: Lea & Febiger, 1918.] fig. 386. Right: [Reprinted from TeachMeAnatomy, courtesy of Dr. Oliver Jones]

brium and xiphoid. These seven segments articulate with seven ribs. Manubrium is a fused Y-shaped structure consisting of a central rectangle of 11 lateral plate mesoderm from which project two sets of shoulder-like processes. The inferior processes are located at the ¾ mark of the bone. These articulate with first rib, a combination of $C8_C$ and $T1_R$ paraxial mesoderm. They also bear on their anterior surface the upper limit of insertion for pectoralis major. The superior processes project from the upper ¼ of manubrium; they articulate with clavicle and bear the insertion of the C1 sternomastoid. It is likely that the *upper ¼ of manubrium is phylogenetically distinct*. It represents the fusion of two c1 anterior coracoid processes. For this reason, the posterior aspect of manubrium bears in its upper 1/4 the insertions of two strap muscles. Sternohyoid attaches to its upper posterior margin and to posterior clavicle. Below SH is sternothyroid which also attaches to the cartilage of first rib. Evidence for the separate nature of upper manubrium exists in running mammals that lack clavicles but present a functional sternum bearing a normal set of first ribs.

Ossification centers provide additional clues about the distinct nature of manubrium. In humans, the ossification center for manubrium is considerably larger than those of the sternal body. This perhaps indicates the coalescence of four centers. Clavicle manifests two primary centers in its shaft which ossify at weeks 5–6 of life. A separate secondary center at the sternal end of clavicle appears at ages 18–20 and unites with the rest of the bone at age 25.

Mastoid process, as previously described is derived from post-otic paraxial mesoderm corresponding to r6–r7. It develops after birth and its expansion is ascribed to traction placed on it by the neural crest fascia of sternocleidomastoid. Mastoid is confluent with parietal, interparietal and supraoccipital. The boundary between interparietal and supraoccipital is superior nuchal line. The latter bone develops from occipital somites, S1–S4 so superior nuchal line is likely encoded r8–r11. The order of insertion into mastoid is, from medial to lateral: sternomastoid, cleidomastoid, and cleido-occipitalis.

Knowing these facts, we can better understand the insertions of SCM and how they relate to those of neighboring strap and scalene muscles.

Motor Control and Insertions of Sternocleidomastoid

As previously described, in the rat model the motor nucleus of SCM within the central motor column exists in two discrete locations. *Sternomastoid* and *cleidomastoid* are supplied from the dorsomedial edge of the ventral horn between C1 and C3. Their colleague, *cleido-occipital* is supplied from the dorsolateral edge of ventral horn from C2 to C3. It shares this position with trapezius muscle C2–C6. For this reason, the secondary insertion of the three parts of SCM into the skull follows a ventral to dorsal pattern with cleido-occipital forming last and taking a higher primary insertion site. Also, SCM develops prior to its partner, trapezius, and inserts deep to it [17, 18].

Sternocleidomastoid follows an antegrade–retrograde insertion sequence. The muscle has two functional components and three primary insertions, into the ventral components of the pectoral girdle.

The sternal head is attached to the C1 manubrium whereas the clavicular head is attached to C2–C3 clavicle. Recall that manubrium represents the ancient mesodermal *anterior coracoid*. The sternal component of SCM is rounded and is attached to the upper border of the manubrium. The clavicular component is broad and flat. It is attached along the anterior surface of the medial 1/3 of clavicle. SCM may be somatotopic throughout its length. Sternomastoid and cleidomastoid are medial to occipito-mastoid with an insertion sequence of c1–c2–c3. Recall that c1–c2 manubrium is derived from the original anterior coracoid process.

As we shall see, the posterior surface of the SCM interacts with the hypobranchial strap muscles sternohyoid and sternothyroid, both of which arise from S5 to S7 and are innervated by C1–C3 via cervical plexus. These muscles develop prior to SCM are positioned posterior to it. Sternothryoid is most medial. It binds to lower posterior manubrium and opportunistically to adjacent first rib. Thus, ST lies below SH. Sternohyoid inserts into posterior manubrium above ST but below and behind sternal head of SCM. It also inserts into posterior surface of medial clavicle. This suggests that clavicular development proceeds from posterior (dorsal) to anterior (ventral).

The secondary insertion of SCM into mastoid process is perfectly logical. Recall that first cervical somite, S5, participates in the formation of the craniovertebral joint. The insertion sequence of muscles follows a strict spatiotemporal sequence. Medial motor column muscle development preceeds that of central motor column so MMC muscles get "first choice" for binding sites. [For this reason, congenital absence of SCM has been reported as a rare clinical deformity but the developmental failure of the suboccipital muscles would be catastrophic, incompatible with life.] All S5 muscles supplied by dorsal branch C1 from the medial motor column span upward from the suboccipital triangle and attach to binding sites on the chondral occipital bone complex. SCM, being thus excluded from the "occipital parking lot," has as its first available site the mastoid process of temporal bone. It inserts into the mastoid process, all the way from its superior border to the apex. In addition, cleidooccipital inserts via a thinner aponeurosis higher up on the skull, leaping over C1 epaxial muscles to find a vacant parking spot along the superior nuchal line, that is, at the border between neural crest membranous bone and PAM chondral bone. Both these insertions make neuromeric sense. Mastoid, being r6–r7, borders directly on the r8 occipital parking lot.

Superior nuchal line is also a boundary between the supraoccipital complex and the interparietal complex.

The neurovascular supply to SCM is strongly segmental and represents four distinct sources. Proximal third is supplied ECA branches representing AA2 and AA3. second arch is represented by occipital artery and third arch is represented by ascending pharyngeal artery. APA has an interesting course. It has two divisions, pharyngeal and neuromeningeal. From the latter, an *odontoid arch branch* supplies the craniovertebral junction. The latter gives off a *musculospinal branch* that connects at level C3 with ascending cervical artery from the thryocervical trunk. Middle third receives fourth arch superior thyroid artery, and third arch ECA is too far internal to lend a hand. Distal third is supplied by variable branches from thryocervical trunk: inferior thyroid, suprascapular, ascending cervical or transverse cervical. These latter represent the substitute for the defunct AA5 to the fifth arch.

Is there a hidden significance to this vascular pattern? SCM straddles from skull to pectoral girdle. Deep to its proximal 2/3 lie the structures of pharyngeal arches PA2–PA5. It makes sense that these neuromeric zones, these angiosomes include their muscle cover. Distal 1/3 of SCM covers the zone which is caudal to the fifth pharyngeal arch and includes trachea and esophagus. All these structures lie distal to the aortic arch system; accordingly, they are supplied by derivatives from second axis off the subclavians, that is, the vertebrals.

Using the Mathes/Nahai classification, sternocleidomastoid is a type V muscle. It does not have an axial vessel. Instead, it picks up arteries from surrounding structures along its course (in this case, the skin). Gracilis is another type V muscle. Because these muscles do not have a dominant pedicle they have a short arc of rotation and their usefulness for reconstructive surgery is limited [9, 19, 20].

We conclude with comments regarding congenital absence of sternocleidomastoid, an extremely rare clinical entity, with four cases reported. A review by Vajramani documents a case with MRI showing involvement of both SCM and trapezius with no other anomalies noted, including deficits higher up involving palato-pharyngo-laryngeal nerve. This supports the clinical independence of spinal accessory nerve. SCM absence has been found in isolation with a normal trapezius indicating potential autonomy of the roots. Strap muscles and scalenes are intact. Reports of congenital absence of trapezius are much more numerous. This raises the question as to why SCM is more "protected." Its synthesis could be more primordial that that of trapezius. As we shall see, mammalian division of cucullaris into these two muscles involves a concomitant process of change at the level of neuromeres c4–c5 which results in additional lengthening of the neck and expansion of the brachial plexus. Since trapezius is a c2–C6 and the only neuromere it shares in

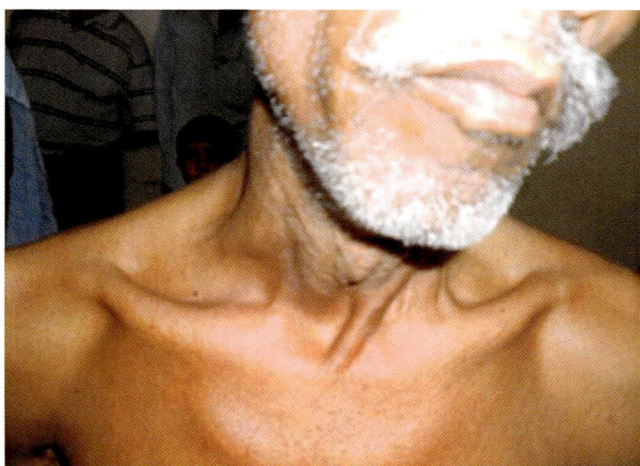

Fig. 10.89 Absent right sternocleidomastoid No other findings. Ipsilateral trapezius intact with intact spinal accessory nerve. [Reprinted from Singh HP, Kumar S, Agarwal SP. Congenital asymptomatic absence of unilateral sternocleidomastoid muscle. *BMJ Case Rep* 2014. DOI: 10.1136/bcr-2013-202,786 PMID: 24445852. With permission from BMJ Publishing Group Ltd.]

common with SCM is c2. Perhaps the pathology occurs in a posterior-to-anterior manner, with level c2 being hit last. We shall look at this topic further when we assess the spectrum of variations seen in trapezius [17, 18] (Fig. 10.89).

Trapezius

Trapezius is a hypaxial polyneuromeric muscle found only in mammalian. it develops from somites S6–S10, levels C2–C6. Recall that sternocleidomastoid arises one somite higher than trapezius, from somites S5–S6. The muscles share motor innervation is from spinal accessory nerve with serial nuclei nucleus in the cervical spinal cord from levels c1 to c6. The nucleus for trapezius is located at the posterolateral margin of anterior horn; it also supplies the cleido-occipital part of Scm (Fig. 10.90).

Trapezius and the Neuromeric Coding of Scapula

Scapula develops from lateral plate mesoderm in register with neuromeres c2–c6. Trapezius inserts along the scapular spine which is likely encoded from medial to lateral. Straddling on either side of it are the c5–c6.

Trapezius has many characteristics in common with sternocleidomastoid. (1) Both muscles are innervated from the central motor column of the cervical spinal cord, which represents the caudal continuation of nucleus ambiguus. (2) The muscles receive their blood supply, in part, from the external carotid artery and in part from subclavian. (3) Trapezius and

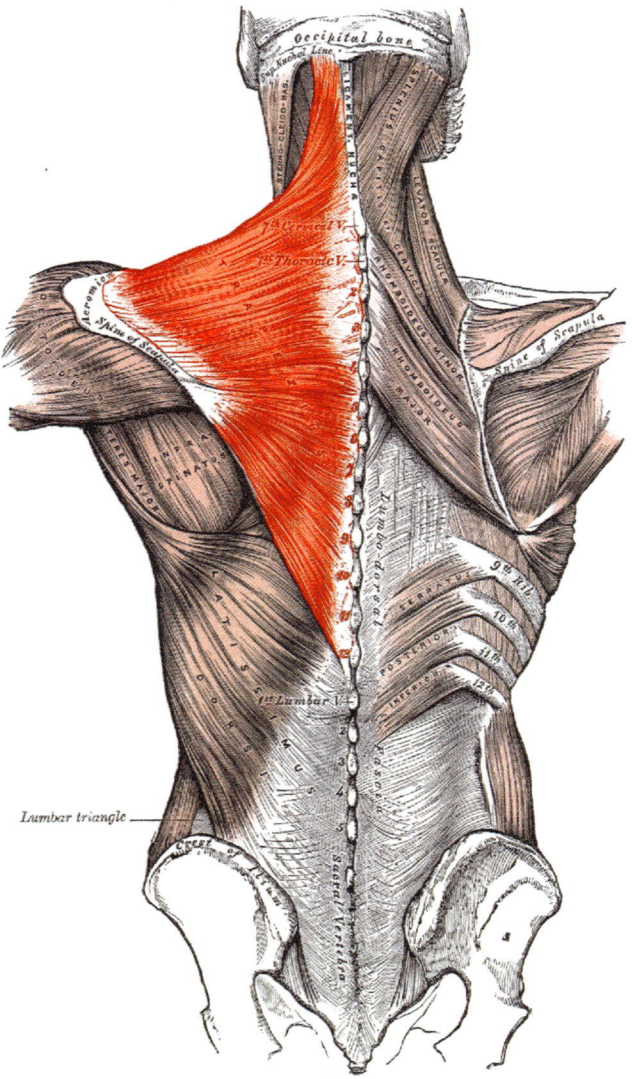

Fig. 10.90 Spino-Trapezius extends from r8 to r11, c1–c8, and T1–T12. Despite its dorsal position, trapezius is a hypaxial muscle—its motor nerves are ventral. Upper fibers (r8–r11 to c1–c7) are supplied by CNCXI and ventral branches of c2–c4; with sensory and proprioceptor innervations from c1 to c4. Middle fibers (c7–t3); lower fibers (t4–t12). The implication for motor control is that spian accessory nerve is extensive. [Reprinted from Lewis, Warren H (ed). Gray's Anatomy of the Human Body, 20th American Edition. Philadelphia, PA: Lea & Febiger, 1918]

sternocleidmastoid are descendents of the ancient cucullaris muscle, a lineage extending back 500 million years ago to the placoderms of the Devonian period. (4) Their ontology reflects the changes wrought by evolution in constructing the neck, changes that are inseparable from the transition to land.

Trapezius in mammals terminates at somite level S10… this is significant. First off, in basal amniotes, that is, the anapsids, have six cervical somites and the number of cervical vertebrae stabilizes at five. The neck-trunk interface in anapsids lies between somites S10 (sp6 or C6) and S11 (sp7 or T1). The brachial plexus in basal amniotes (and in all non-mammals) has four spinal roots (C4–C5–C6–T1). As a consequence, br1 is located at somite level S8, precisely at C4. S8 represents a fork in the road for evolution. Be it in the direction of diapsids (birds and reptiles) or synapsids (mammals), necks got longer. In contrast, brachial plexus remains invariable in diapsids. It continues to have four roots regardless of its location.

Mammals are innovators. By duplicating levels C4–C5, they create a brachial plexus of six spinal roots (C4–C5–C6–C7–C8–T1). Nonetheless, even in mammals, *C4 remains a key transition point.* It marks the first neuromere of the brachial plexus. C4 is also the site of origin for diaphragm. This puts the final mammalian somite count to 8 (and the number of cervical vertebrae to 7).

Evolution has embedded something else of deep significance at level C4. The primitive branchial arch system of armored fishes stabilizes in chondrichthyes at 7 gill arches. Given the principle of two neuromeres per arch and with the first branchial arch in register with r2–r3, the seventh branchial arch becomes located at spinal neuromere level sp3–sp4. Thus, the location of the upper extremity, at the interface between neck and trunk, becomes located in early tetrapods, at the end of the branchial arch system.

Trapezius myoblasts originate from neuromeric levels c2–c6 (S6–S10). Like SCM, trapezius is an appendicular muscle: it therefore develops later in time compared with axial counterparts from the same c1–c6 levels. Hence, trapezius must be a *master opportunist*, glomming on to whatever bindings sites have been left vacant by its predecessors.

Trapezius, like SCM, follows a *two-stage model* of antegrade and retrograde insertions. In the first stage, myoblasts and accompanying neural crest fascia from c2 to c6 migrate directly outward underneath the skin where they come into contact with their target bone, the scapula, made from c1 to c6 lateral plate mesoderm. Thus, the initial migration of myoblasts takes place with*in their neuromeres of origin.* After making primary attachments, the muscles migrate retrograde to the body axis to seek out available binding sites, sweeping from lateral-medial from r8 to r11 and downward in the midline from c1 to the S5.

Trapezius has three zones of primary attachment to upper posterior scapula from which its fibers reobound medially, fanning out to insert secondarily into the fibro-osseous structures of the vertebral midline. The extent of its attachments are surprisingly meagre. At the time of trapezius development, the bulk of dorsal scapular binding sites are already "taken" by other muscles which previously migrated to the site. These are (above the spine) levator scapulae and supraspinatus and (below the spine) rhomboids, infraspinatus, teres minor, and teres major. Poor trapezius has to settle for the "leftovers."

The primary insertions of trapezius are as follows:

(1) Superior insertion is relegated to the *lateral 1/3 of posterior clavicle.* Why such a limited zone? Recall that the anterior clavicle is completely covered over by platysma. The medial 2/3 of posterior clavicle harbors SCM and the strap muscles. So trapezius takes what remains. From there, its fibers span upward and medially to insert secondarily into the skull and the spine. Its occipital insertion is into superior nuchal line along its medial 1/3 because SCM fibers occupy the lateral 2/3. From external occipital protruberance it attaches in a vertical line downward along ligamentum nuchae from C1 to C7.

(2) Middle insertion is restricted to the *medial aspect of acromion* and *lateral scapular spine.* Despite the large surface area of the scapular bade per se, no other sites are available because it is completely covered by pre-existing muscles…but the scapular spine sticks upward and is available. From here, Tpz fibers span medially to the neural spines of C8 to T3.

(3) Inferior insertion is located along *medial scapular spine* (T4–T12). The fibers flow downward and medially; they pass over a bare area between the two fossae of supraspinatus and infraspinatus but do not attach there. They terminate medially on the neural spines of T4 to L1.

Textbook descriptions of trapezius are confusing because the muscle is described as if it had 19–20 somites of "origin." No such muscle exists in the entre body. Why should such a massive muscle be controlled by a spinal accessory nerve from only neuromeres c2–c6? Furthermore, the midline attachments of the muscle are fibroaponeurotic, not muscular. These make sense only as secondary insertions. Recall that the extent of the insertions depends strictly on (1) the mesenchymal mass of the muscle, (2) the capacity of each binding site to accept a finite amount of insertion, and (3) the total number of sites available. Recall the "toothpaste" model: a finite amount trapezius is expressed, it flows outward in all directions seeking out all available binding sites.

Trapezius, like SCM, is a type V muscle. Its vasculature is defined by its cutaneous environment. The upper 1/3 is supplied by second arch occipital artery, transverse muscle branch. The remainder of the muscle is supplied by the costocervical trunk. The middle 1/3 receives the superficial branch of transverse cervical (or the superficial cervical). The lower 1/3 is supplied by dorsal scapular artery. This reflects the relationship of the first four somites to ECA territory, whereas the terminal eight somites are beyond the distribution of pharyngeal arch circulation.

Trapezius in all non-mammalian amniotes is made from four somites. Why should trapezius expand to six neuromeric levels in mammals? Tetrapod evolution achieves changes in neck length by expropriation of truncal neuromeres into a cervical fate. Mammals are different. They lengthen their necks by duplicating levels sp4–sp5. This interposes two new cervical vertebrae and creates a six-root branchial plexus. Because the mammalian neuromeres c6–c7 are a duplication of c4–c5, two new levels of mesoderm are dedicated to making the dorsal aspect of cucullaris. Sternocleidomastoid stays the same but trapezius has more mesenchymal mass.

Development of Trapezius

Description of trapezius and SCM development are based on embryo dissections by Keibel and Mall based on crown-rump length; we have correlated them with staging as per the Carnegie Foundation.

Stages 14–15 in the 7-mm embryo trapezius develop with sternocleidomastoid as part of the common anlage of cucullaris. During stages 16–17 (11 mm) the anlage divides with trapezius being positioned dorsal. Clavicle and scapula are now distinct. At stages 18–19 (16 mm) trapezius gains attachment to scapula and extends upward toward, but is not in contact with, ligamentum nuchae. The stage 20 (20 mm) embryo demonstrates full secondary attachments of trapezius to the midline.

Trapezius is an adductor. Because of its geometry, it can be considered to have two functional components: upper fibers elevate the scapula while the lower ones draw it downward. Absence of upper trapezius has been reported as have deficiencies of the lower muscle.

Congenital Anomalies of Trapezius

Developmental defects of the trapezius muscle are also uncommon. A 2006 review found eight cases in the literature, at times accompanied by aplasias in other muscles as well. In five cases bilateral aplasia has been reported. Rarer still are deficits involving both Tpz and Scm: only four such cases have been reported [18]. Trapezius deficits occur as a spectrum and these variations provide us with a number of clues as to its development. Unilateral 50% aplasia presented as a progressive dropout of lower fibers with absent insertion below T9. This indicates that the developmental pattern of the muscle is cranial-caudal. Like paint dripping downward on a wall, when the mesenchyme runs out it simply fails to reach the most distal binding site [21]. Unilateral agenesis of the lower trapezius muscle has been reported with absent fibers from T6 to t12. In a cadaver case, a lower defect was explained by the absence of the spinal accessory nerve and blood supply. However, in the same patient, sternocleidomastoid function remained, functioning on upper roots of the SAN. This finding supports the autonomy of the two muscles with SCM originating independently from C1 to C2 roots, somites S5–S6 [22]. A clear-cut distinction seems to exist between upper cervical fibers and those of the middle and lower thirds [23]. Bergin presented a case in which ¾ of the descending (lower) trapezius appeared separate from the remainder of the muscle. The isolated part inserted into clavicle as an independent tendon [24] (Fig. 10.91).

Upper defects of trapezius also exist. In one such case, the upper fibers were replaced by a fibroaponeurosis spanning from the occiput to C4 and a normal nerve, indicating survival of neural crest connective tissue but not myoblasts.

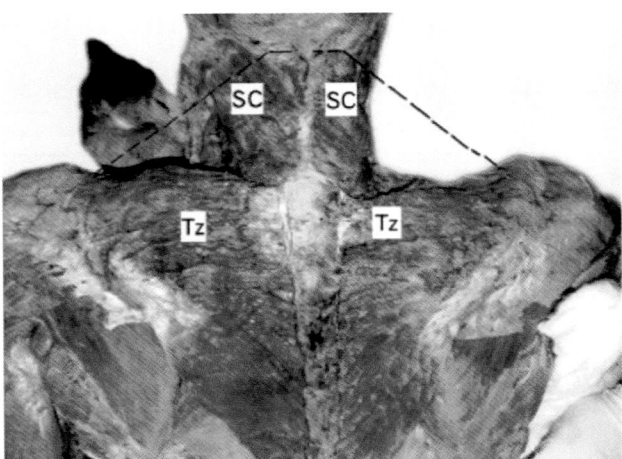

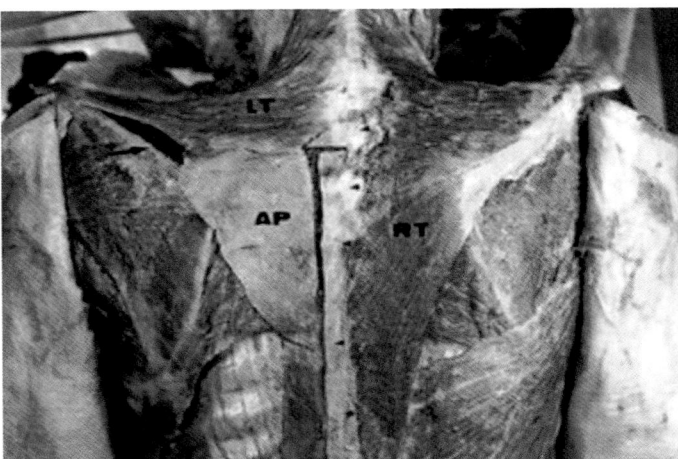

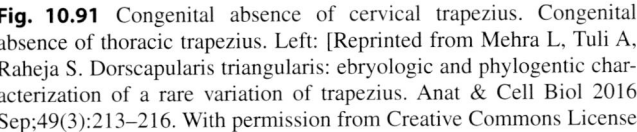

Fig. 10.91 Congenital absence of cervical trapezius. Congenital absence of thoracic trapezius. Left: [Reprinted from Mehra L, Tuli A, Raheja S. Dorscapularis triangularis: ebryologic and phylogentic characterization of a rare variation of trapezius. Anat & Cell Biol 2016 Sep;49(3):213–216. With permission from Creative Commons License 4.0: http://creativecommons.org/licenses/by-nc/4.0/.] Right: [Reprinted from Emsley JG, Davis M. Partial unilateral absence of the trapezius muscle in a human cadaver. Clin Anat 2001; 14(5): 383–386. With permission from John Wiley & Sons]

Compensatory hypertrophy of rhomboids was noted but this could reflect a diversion of mesenchyme from one muscle to the other at the level of the myotome [25]. In another case with Tpz loss down to C4 the clavicular insertion was missing [26].

Together these studies suggest that knock-outs or dysfunction may exist in either the upper somites, the lower somites, or both. An intact SAN in the absence of upper fibers implies survival of neural crest mesenchyme with persistence of all roots of SAN. Loss of the muscle "target" would be expected to lead to loss of motor neurons. When SAN and Tpz are both absent we can see that the neural crest (critical for the development of the peripheral nerves) is also absent. Finally, 17 cases of muscular anomalies concomitant with Tpz aplasia are reviewed by Yiyit, who noted a commonality with Poland syndrome [27, 28]. This suggests two basic mechanisms. (1) Common mesenchymal deficit characterizes anomalies of Tpz with Scm. (2) Common neuroangiosome or neighboring neuroangiomes along the subclavian characterize deficits of Tpz with pectoralis major (thoracoacromial) or with serratus anterior and latissimus dorsi (both thoracodorsal) [29].

Phylogeny of Sternocleidomastoid and Trapezius: The Cucullaris Muscle

The anatomy of sternocleidomastoid and trapezius in mammals is the result of two evolutionary processes: (1) the break-up of a precursor muscle, the cucullaris; and (2) the transformation of a pectoral girdle of dermal bones attached to skull to an independent claviculo-scapular assembly, a subject reviewed previously in this chapter. Understanding the anatomy of the cucullaris and its innervation through evolution is critical. Until recently, the phylogeny of cucullaris in prehistoric fishes (when it appeared and what was its function) was uncertain, as only skeletal remains were available for study. Because sharks are considered a link between ancient gnathostomes and our ancestral bony lungfishes, assumptions about cucullaris have been based on its anatomy in extent chondrichthyans. The recent discovery of placoderm fossils with preserved musculature has provided a radically different model, one offers insights into the evolution of the neck. In addition, new studies using *hox* gene analysis combined with micro CT scanning demonstrate that cucullaris anatomy and function are dynamic over time, as is its motor innervation. Analysis of these changes provide a powerful window on the evolution of the neck.

The Importance of Placoderms

During the Silurian-Devonian period, eight groups of armored fishes appeared, all of which descended from a common (unknown) basal ancestor. Their emergence marked the great divide between cyclostomes (represented today by extent hagfishes and lampreys) and jawed gnathostomes (all other vertebrates). Cyclostomes ("round mouths") have a circular oral aperture with no specialized perioral bones. Their branchial apparatus consists of 8 primitive cartilaginous branchial baskets. The "gnathostome revolution" involved the invention of branchial arches, the subsequent transformation of the first two arches into jaws, paired appendages (future limbs), a pectoral girdle, and eventually a neck region. Radical changes in muscles were required in order to control these new moving parts (Figs. 10.92, 10.93, 10.94, 10.95, 10.96).

Gnathostomes ("jawed mouth") are divided into a crown group and a stem group. All members of the crown group, be they living or extinct, have the full set of jawed vertebrate characteristics, although paired appendages are lost in snakes. All members of the stem group are extinct. The most primitive stem gnathostomes, such as ostracoderms, did not have jaws but these earliest fishes did possess pectoral fins and a pectoral girdle fused to a massive head shield. They resembled a giant vacuum cleaner, a sort of armor-plated lamprey well-suited for life on the ocean floor. Notably, ostracoderms did not have transverse abdominal muscles.

Separation of the head and pectoral girdle took place with the appearance of jaws and presumably cucullaris, to control the pectoral girdle, evolved at the same time. At the upper end of stem gnathostomes, the placoderms represent the earliest form of vertebrate anatomy with jaws. These creatures, such as *Dunkleosteus* had oral bones consisting of plates. In a more advanced iteration, *Entelognathus*, these plates became premaxilla, maxilla, and mandible. One member of this group, the †Arthrodires (Gr., arthrodes = well-jointed) had a joint between the head and the thoracic plates which allowed to the skull to be raised, increasing the gape of the jaws This vertical pivot function would require evolution of elevator and depressor muscles. When placoderms gave rise to both the chondrichthyans and the osteichthyans the original head-trunk pivot was lost and cucullaris was forced to relocate. *Transverse abdominal muscles were lost as well, not to appear again until the advent of tetrapods.*

In tetrapods, simultaneous appearance of limbs with transverse abdominal muscles is a non-trivial detail because they are involved in the control of respiration. These muscles have to travel a considerable distance to insert muscles that serve as evolutionary markers concomitant evolution of appendicular muscles and the relationship between the ancient placoderm pivot joint and the tetrapod neck. In both instances, an original muscle plan is "remembered" in evolution. This indicates the importance of studying the placoderm model because this vertebrate form is ancestral to all living tetrapods (Fig. 10.92).

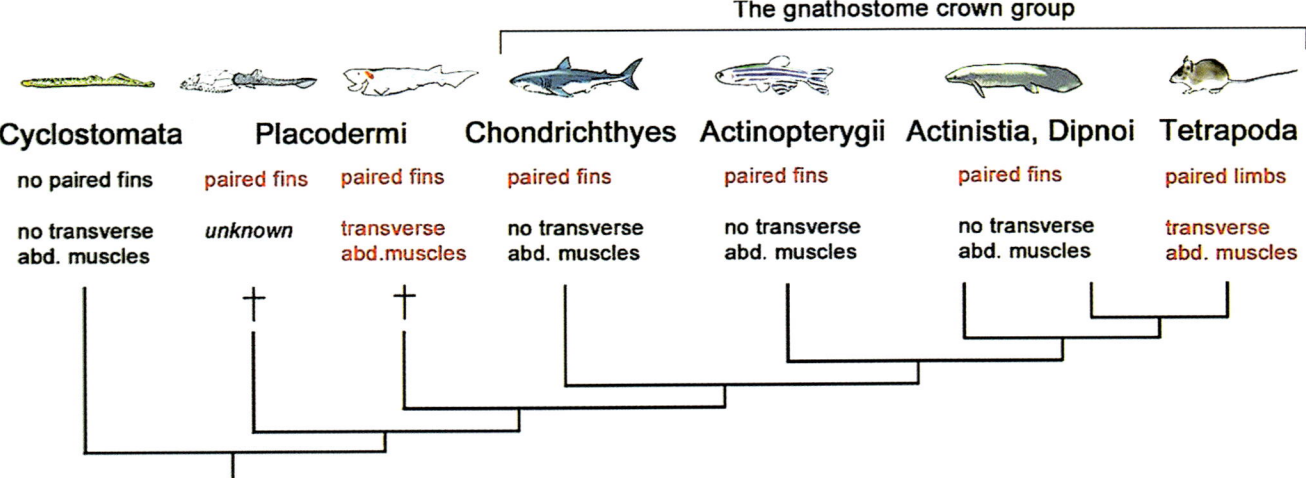

Fig. 10.92   Placoderms are in direct lineage to mammals as evidenced by four inventions: paired fins, a head–body joint, cucullaris muscle, and transverse abdominal muscles. The articulation and the transverse abdominal muscles temporarily disappear in the course of fish evolu- tion but are reincarnated in tetrapods. [Reprinted from Trinajstic K, Ahlberg, E, et al. Fossil musculature of the most primitive jawed verte- brate. *Science* 2013; 341(6142):160–164. With permission from The American Association for the Advancement of Science]

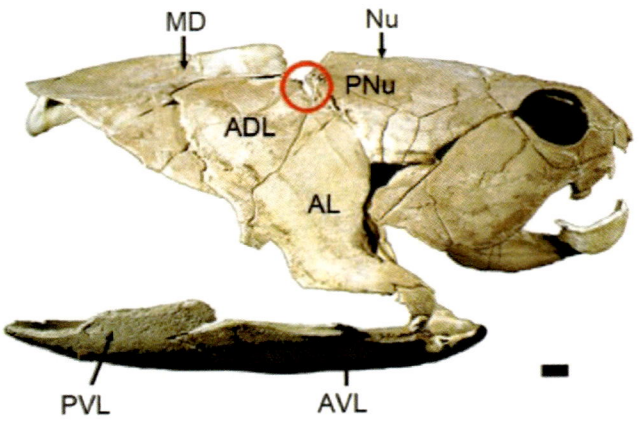

Fig. 10.93   *Eastmanosteus* skull showing the cranial-trunk joint. †*Eastmanosteus calliaspis* showing dermal neck joint (red ring). This was the original idea for the future craniovertebral joint. The placoderm skull plates have distinct terminology: ADL, anterior dorsolateral plate; AL anterior lateral plate; MD, medial dorsal plate; MV, medial ventral plate; Nu, nuchal plate; PNu, paranuchal plate, PVL, posterior ventro- lateral plate. [Reprinted from Trinajstic K, Ahlberg, E, et al. Fossil mus- culature of the most primitive jawed vertebrate. *Science* 2013; 341(6142):160–164. With permission from The American Association for the Advancement of Science]

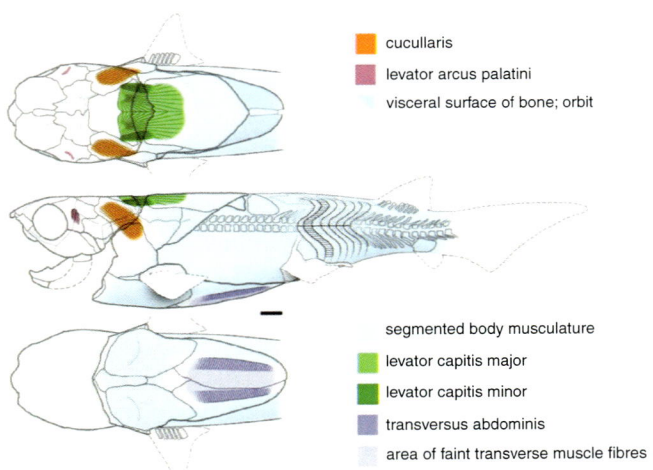

Fig. 10.94   Placoderm musculature. Cucullaris is the precursor of trapezius. Its primary insertion is into ADL, anterior dorsolateral plate. Its secondary insertion is into Pnu, paranuchal plate. [Reprinted from Trinajstic K, Ahlberg, E, et al. Fossil musculature of the most primitive jawed vertebrate. *Science* 2013; 341(6142):160–164. With permission from The American Association for the Advancement of Science]

Primordial Attachments of Cucullaris Muscle

Ostracoderms, jawless stem gnathostems basal to placo- derms had head shields with large plates and no cranio- thoracic pivot joint suggesting (1) that the shoulder girdle of placoderms results from a subdivision of the ostracoderm shield; and (2) cucullaris is not present in ostracoderms but appears for the first time in placoderms. Arthrodires have a regionalized body plan with rigid body armor and scales. The basal condition of cucullaris in placoderms is the control

of a constrained *pivot joint*, implying the need for antagonis- tic elevator and depressor muscles. Preserved neck and trunk muscles in three species of placoderms were reported by Trinajstic [30] in 2013. In these specimens, elevation of the dermal neck joint was achieved by two muscles: levator capi- tis major and levator capitis minor. These were antagonized by cucullaris which was found lateral to the joint, connecting the inner surface of the dermal girdle, *anterior dorsolateral plate* (ADL), with a hollow in the inner surface of the skull roof, *paranuchal plate* (PNu). Cucullaris in its original form

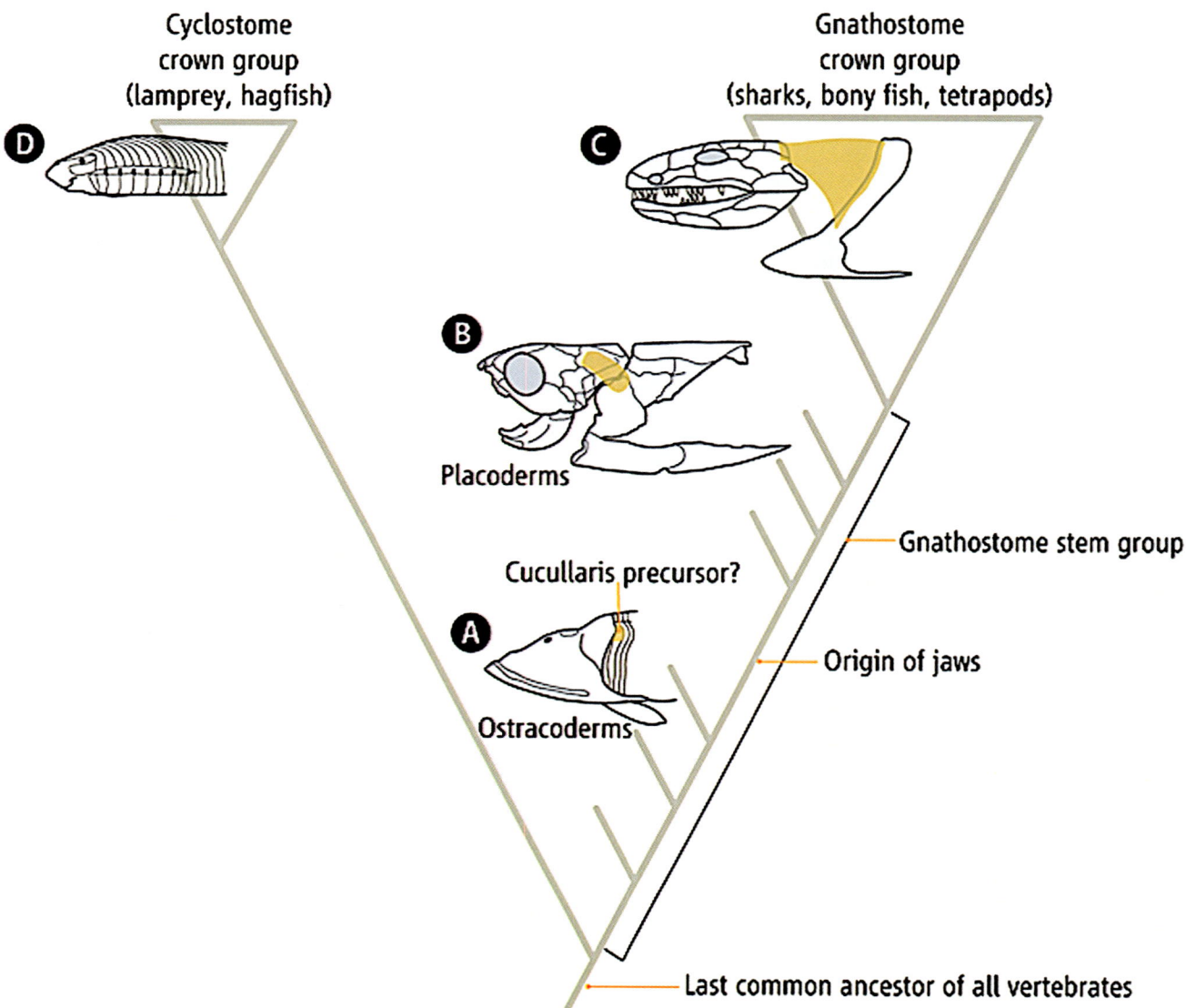

Fig. 10.95 Neck and cucullaris evolution based on Trinajstic. Cyclostomes have no jaw and no neck. Agnathans represented by *Cephalaspis*. Placoderms are considered in the stem group. (**a**) Head shield of *Cephalaspis* contains pectoral girdle precursors. Existence of cucullaris unknown. (**b**) The placoderm pectoral girdle articulates with dermal cranium. (**c**) Ancestral crown gnathostome pretetrapod has cucullaris attached to the dermal chain of bones. With tetrapods, dis-connection of the pectoral girdle from skull happens with loss of upper dermal bones. Cucullaris inserts into chondral scapulocoracoid. Cucullaris persists until mammals when it breaks up into sternocleido-mastoid and trapezius. [Reprinted from Kuratani S. A muscular per-spective on vetebrate evolution. *Science* 2013; 341(6142):139–140. With permission from The American Association for the Advancement of Science]

thus acted as a *depressor* of the skull (Figs. 10.93, 10.94, 10.95).

Although the cranial pivot joint is lost in derived fishes, cucullaris is stably conserved across gnathostomes as a link between head and neck. Cucullaris adapts to varying evolu-tionary scenarios by seeking out analogous neuromeric insertions. Neuromeric relationships are fundamental. Caecilians are limbless burrowing amphibians lacking a pec-toral girdle and gills. In these creatures, cucullaris is found to span dorsal trunk fascia forward to the otic capsule and down to the fascia of ventral trunk muscles at the same level. With the loss of branchial arches in tetrapods the proximal inser-tion of cucullaris once again returns to skull but the scapula is now disconnected from the skull and mobile. Thus, cucul-laris becomes capable of elevating it away from the ventral midline. The mammalian neck is marked by multiple exten-sors without a specific antagonist. We shall see that the sub-division of cucullaris that results in an independent sternocleidomastoid serve permits it a double function as *both flexor and rotator*.

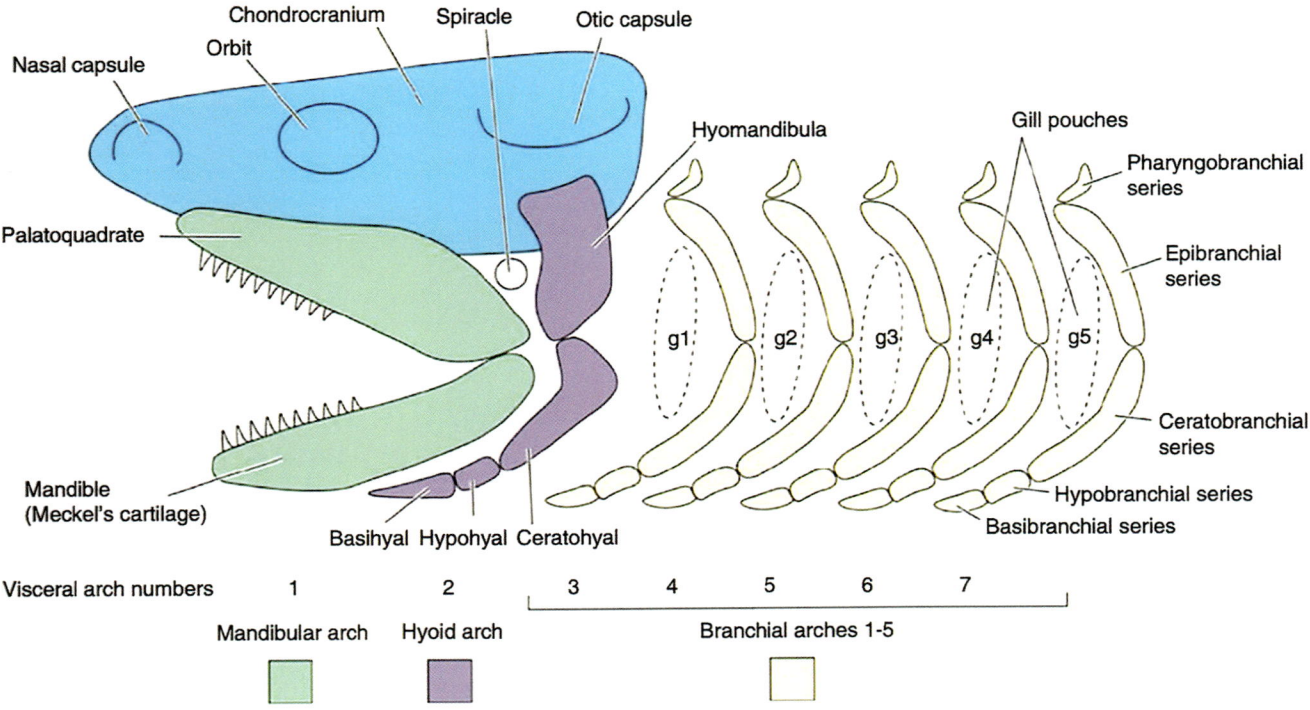

Fig. 10.96 Sharks and sarcopterygians have 7 branchial arches. Cucullaris inserts into the fourth and fifth ceratobranchial cartilages. It is in series with the levatores arcus branciales. [Courtesy of William E. Bemis]

Gross Anatomy and Developmental Origins of Cucullaris

Cucullaris is a thin flat muscle, easily disrupted by dissection and difficult to visualize in three dimensions. High-resolution micro-CT scanning offers across a spectrum of gnathostomes demonstrate remarkable uniformity of anatomy [31]. In bony fishes such as the actinopterygian bichir and the sarcopterygian lungfish, cucullaris is referred to as *protractor pectoralis*. In the lungfish *Latimeria* it does not reach the dorsal midline and being bounded by epaxial muscles. It spans from the *anocleithrum* of pectoral girdle forward and inserts into the fifth ceratobranchial cartilage (i.e., branchial arch 7). As such, it is in series with gill levators of the first 4 gill arches. Just anterior to cucullaris the levator of the fourth-gill arch is greatly enlarged, leading to speculation that this muscle should be included as part of cucullaris. This latter is a non-trivial finding. Recall from our model of two neuromeres per arch that the third-gill arch (fifth pharyngeal arch) receives its mesenchyme from neuromeres r10–r11. Therefore fourth and fifth-gill arches in fishes are located at truncal neuromeres sp1–sp4. In mammals, these correspond to levels c3–c4. As we shall see, in mammals the innervation of cucllaris mammals shifts out from hindbrain by two units, becoming completely cervical. Thus, the *territory of cucullaris is redefined levels to c1–c6*, exactly as we now see it! (Figs. 10.96, 10.97, 10.98).

Where is the source for the mesenchyme of cucullaris? Fate mapping data shows it to arise from *unsegmented* cranial mesoderm. It appears lateral to the occipital somites and extends backward as far as the fifth, sixth, or seventh arches (depending on the species). Recall that basal fishes have only three occipital somites. The fourth is added at the sarcopterygian–tetrapod transition.

This has given rise to a point of confusion. Based on work in chickens, unsegmented mesoderm fated for cucullaris has been found outside the confines of the somites; thereby some to describe cucullaris as originating within lateral plate mesoderm. This flies in the face of what we know about mesoderm for three reasons: physical, neuromeric, and neuroanatomic.

First: as we know, LPM is the source of smooth muscle, the heart, and cardiovascular system but *not* conventional striated muscle. Scanning electron microscopy demonstrates that the PAM that contributes to pharyngeal arches is continuous, like a strand of pearls, whether it is in the form of somitomeres or somites.

Second: differences in plasticity exist between branchiomeric and somitic mesoderm. Mesoderm giving rise to branchiomeric muscle, including cucullaris, when transplanted into a more posterior position cannot generate muscle whereas truncal somite muscle transplanted into the head forms pharyngeal arch muscle. This implies two types of genetic programs, one for the head and one for the trunk. The

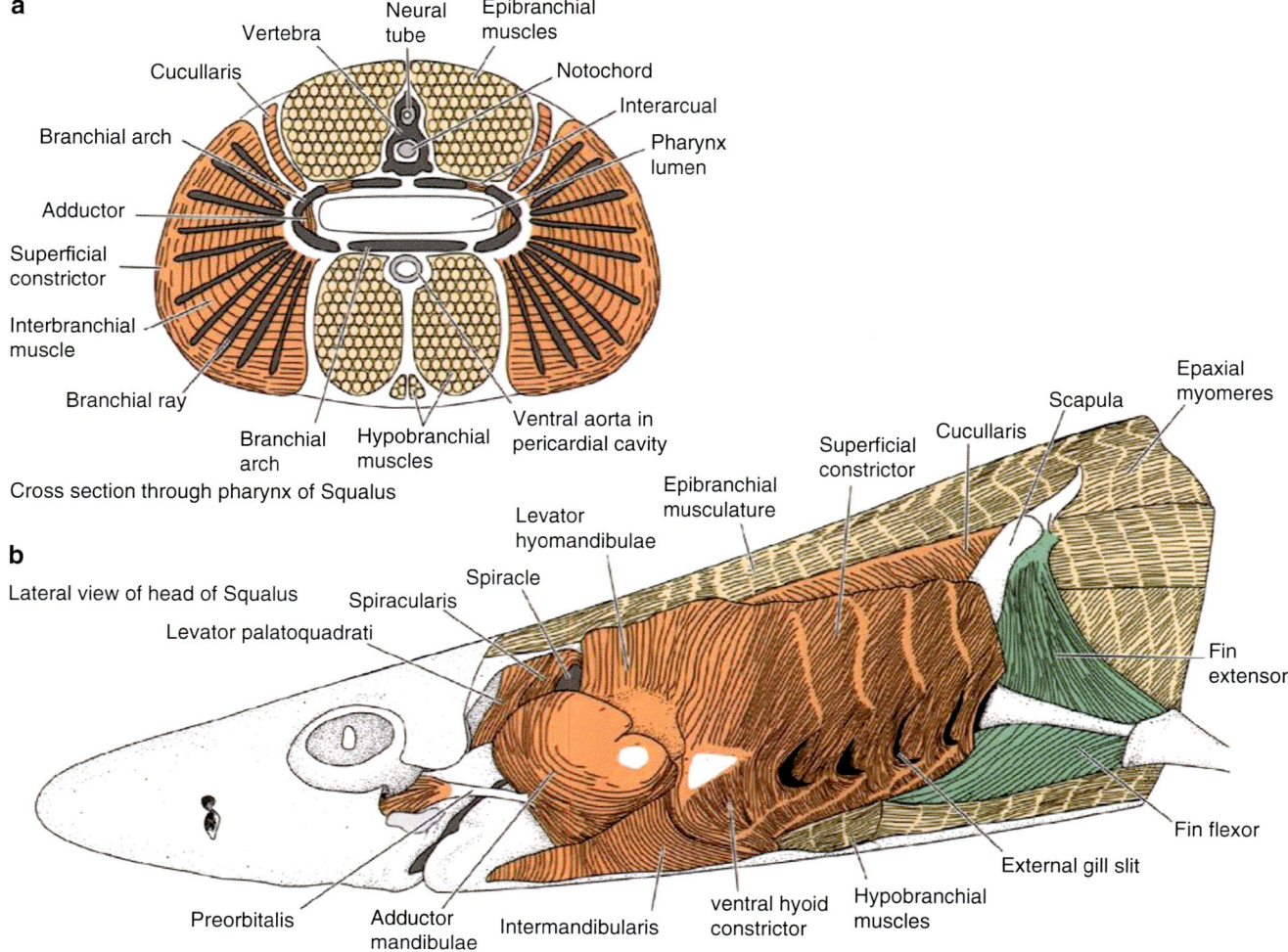

a

Cross section through pharynx of Squalus

b

Lateral view of head of Squalus

Fig. 10.97 Cucullaris is the dogfish shark, *Squalus*. [Courtesy of William E. Bemis]

connective tissues of branchiomeric muscles are always neural crest and relate back to the original 7 branchial arch system while connective tissues in somitic muscles are PAM. It has been shown that the CTs of sternocleidomastoid and trapezius are neural crests.

Finally, the motor nuclei supplying cucullaris in anamniotes reside in the nucleus ambiguus; the muscle is innervated by a branch of vagus and by spinal branches as well. In amniotes, motor nuclei for cucullaris reside within the central motor column, continuation of nucleus ambiguus into the cervical spinal cord. In both situations, neural control for cucullaris remains in the same genetic environment regardless of neuromeric level.

We can conclude (and will demonstrate) the following about cucullaris in both its ancestral and derived states:

- Cucullaris is a branchiomeric muscle throughout evolution.
- As such, the mesenchyme of cucullaris is unsegmented PAM.

- The PAM of ancestral cucullaris is lateral to occipital somites S1–S4 and is neuromeric levels r8–r11.
- The motor supply of cucullaris is located in nucleus ambiguus and/or central motor column.
- The ancestral nerve for cucullaris joins with caudal vagus nerve but shifts abruptly in birds and mammals out of medulla to spinal cord as spinal accessory nerve.
- The above change was the result of a frameshift mutation causing a posterior translocation of hox genes coding for cucullaris from rhombomeres into the spinal myelomeres.
- The mesodermal source for cucullaris moved backwards as well, becoming in register with somites S5–S10, corresponding in mammals to cervical neuromeres c1–c6.
- In mammals, the cervical spine lengthens two units by a duplication of neuromeres c4–c4. Thus, trapezius gets more mesenchyme. Instead of being a four-somite muscle, the mammalian cucullaris is a six-somite muscle.
- These changes are integral to the creation of the craniovertebral joint.

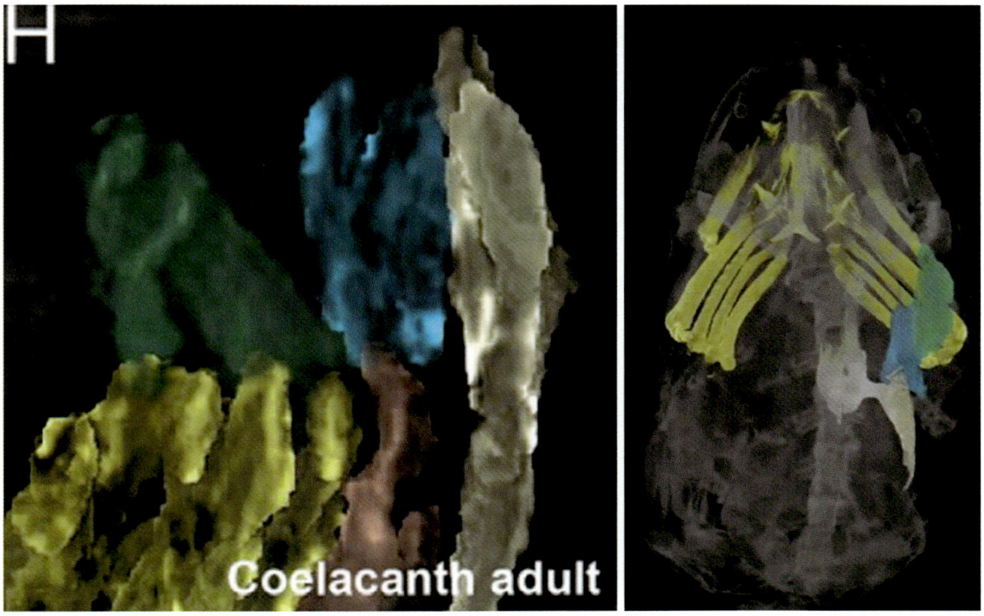

Fig. 10.98 Cranial muscle anatomy in Coelacanth. Lateral view shows: (1) pectoral girdle (white); (2) branchial arches 3–7: first–fourth ceratobranchial cartilages (yellow) and fifth ceratobranchial cartilage (pink), musculature of levatores archs branchiales (green) and cucullaris (blue). In ventral view, all ceratobranchiales are colored white. Cucullaris is clearly confluent with the gill levators at BA4–BA5. Thus, cucullaris is in register with the fifth and sixth gill arches, spinal neuromeres sp1–sp4. [Reprinted from Sefton EM, Bhullar B-A, Mohades Z, Hanken J. Evolution of the head trunk interface in tetrapod vertebrates. *eLife* 2016;5:e09972. With permission from Creative Commons License 4.0: https://creativecommons.org/licenses/by/4.0/]

Phylogeny of Spinal Accessory Nerve

The pectoral fin in stem gnathostomes was welded into the dermal bones of the braincase and was relatively immobile, serving as a stabilizer. Its mesenchyme was synthesized. The advent of a pectoral girdle distinct from the skull in placoderms was accompanied by a change in its neuromeric definition. It was now synthesized from truncal mesenchyme, both as the dermal bone chain and as mesodermal scapulocoracoid. But the muscle that evolved with it, cucullaris, remained branchiomeric; its connective tissue was neural crest from r8 to r11.

The closest we can come to the neuroananatomy of ancestral cucullaris is to study its innervation in living chondrichthyan and sarcopterygian fishes. In selachians (sharks) cucullaris is large and superficial. It is attached to the fascia of epaxial muscles connected to the skull and thence to the most posterior (7th) branchial arch and to the pectoral girdle which it serves to elevate and protract. It is innervated by a branch of vagus, *ramus accessories*, which is functional, being a special visceral efferent.

Studies in the clear nose skate, *Raja eglantera*, provide further definition of neuromeric anatomy of cucullaris in the basal state. Barry described a motor column ventrolateral to the dorsal motor nucleus of vagus (PANS) which he termed the *ventral motor nucleus of vagus*. This column extended into the spinal cord down to level Sp5 giving off motor root-lets. We now identify these structures as homologs of nucleus ambiguus and the accessory nerve. Barry also made an erroneous but prescient speculation: that the ventral nucleus of vagus represents a delamination from the dorsal nucleus… more on the signficance of this a bit later [32].

Skates are closely related to sharks but have a fusion of the proximal vertebrae, the *synarcium*, which in turn is fused to the pectoral girdle but nevertheless has a cucullaris. These fishes have three dorsal muscles (medial, intermediate, and lateral) which, together, are precursors of what will become trapezius. We can term these the *cucullaris complex*. This complex has a rostral attachment is to dorsal synarcual process. Its dorsal boundary is the midline epaxial muscle column. Laterally it is confluent with the levatores arcuum branchiales. Posteriorly it is inserted into the rostral suprascapula [33]. The lateral muscle receives an accessory branch from the caudal hindbrain that runs in the substance of the 4th (intestinal branch) of the vagus nerve. Innervation of the intermediate and medial muscles is from spinal roots (Figs. 10.99, 10.100, 10.101).

Using horseradish peroxidase (HRP) staining Sperry and Boord showed that motor neurons for cucullaris in Raja originate from caudal nucleus ambiguus, probably from levels r10–r11. None were found more rostrally. Vagus has four major roots. The accessory neurons were traced through the fourth root of X. This root is directed to the fifth arch (fifth gill arch). Accessory neurons are admixed with other motor

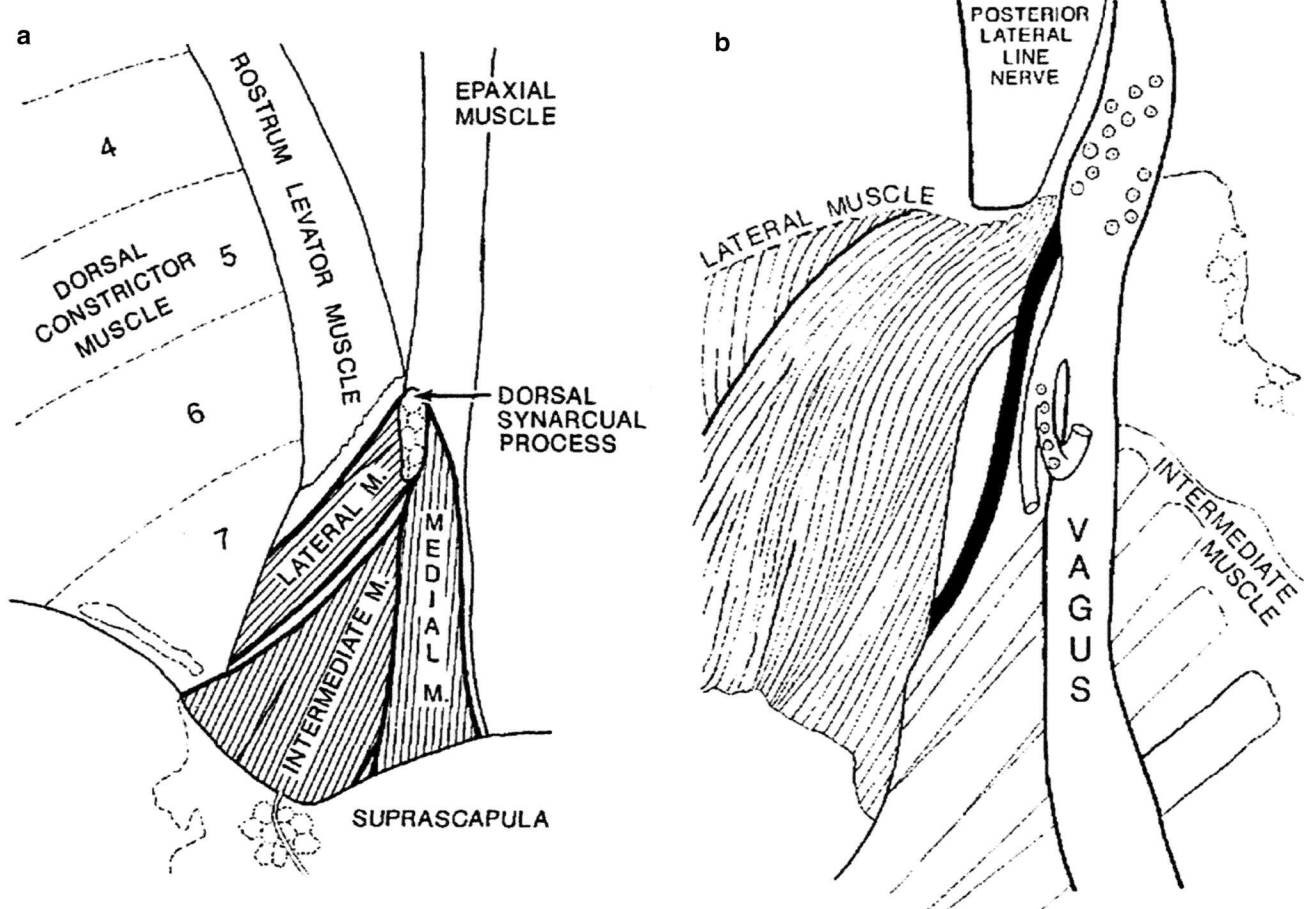

Fig. 10.99 Cucullaris complex consists of three muscles. The lateral-most one gets direct innervation from caudal ventral motor nucleus of vagus, neuromeres r10–r11. Intermediate and medial muscles are supplied from spinal nerves. Extrapolating from sharks and skates, cucullaris in the basal state of placoderms was a branchiomeric muscle spanning the head-neck joint and supplied by hindbrain. Whether neu-romeric territory extended further caudal into the spine is unknown. [Reprinted from Sperry DG, Boord RL. Central location of the moto-neurons that supply the cucullaris (trapezius) of the clearnose skate, Raja eglanteria. *Brain Res* 1991; 582:312–319. With permission from Elsevier]

neurons; these are unlabeled by HRP; in mammals, they are directed toward the muscles of the larynx. The Sperry study did not document contributions to cucullaris from sp1–sp4 and considers cucullaris the lateral muscle only. The broader definition by Marian includes all three muscles of the cucullaris complex, it makes sense that the intermediate and medial muscles represent cucullaris arising from the proximal cervical mesoderm, and innervated by the backward extension of nucleus ambiguus into levels sp1–sp4 [33–35].

Thus, the potential innervation of the cucullaris complex from mucleus ambiguus spans a total of six neuromeres across the hindbrain-spinal boundary. *This basal gnathos-tome pattern is universal in anamniota (fishes, amphibians) reptiles and persists in amniote reptiles.* This neuroanatomy changes in birds and mammals. Birds have a true accessory nerve that is physically separate from vagus but retains its hindbrain innervation. In mammals, the roots of accessory nerve are completely divorced from the medulla. Due to the *hox* gene translocation, the definition of what neuromeric levels produce cucullaris muscle shifts backwards from r10–c4 to levels c1–c6. Cucullaris now arsies strictly from cervical mesoderm. This redefines its innervation—the need for hindbrain nuclei is eliminated; accessory nerve becomes strictly spinal.

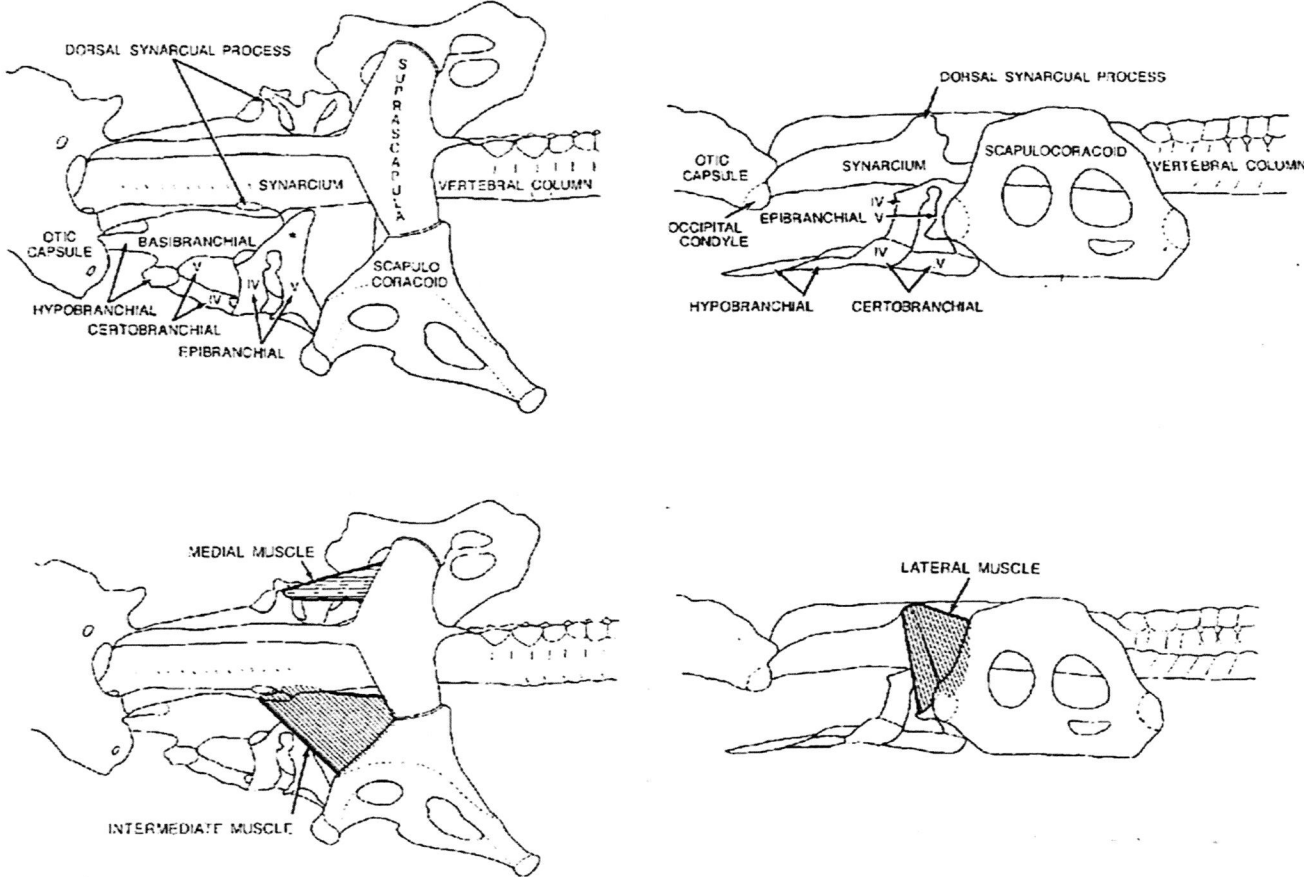

Fig. 10.100 Insertions of cucllaris into the pectoral girdle in skates. Comparative studies of primitive chondrichtyans give us a clue as to the original anatomy of cucullaris in placoderms. The *synarcium* in these creatures is a fusion is a fusion of vertebral column to the otic capsule. Scapulocoradoid is fused to synarcium, making it immobile. Lateral muscle (supplied by vagus) and intermediate muscle (supplied by ventral spinal nerves) both insert into scapulocoracoid. Medial muscle inserts solely into *suprascapula*, a transverse process fused to synarcium and unrelated to the pectoral girdle. From a functional standpoint cucullaris has two components, hindbrain and spinal. In mammals, trapezius arises strictly from cervical mesoderm, eliminating the need for hindbrain innervation. For this reason, there is a disconnect between the root of r10–r11 that join vagus 4 en route to laryngeal muscles and the spinal accessory nerve, sp1–sp6. [Reprinted from Sperry DG, Boord RL. Central location of the motoneurons that supply the cucullaris (trapezius) of the clearnose skate, Raja eglanteria. *Brain Res* 1991; 582:312–319. With permission from Elsevier]

The break-up of cucullaris into a sternocleidomastoid from c1 to c2 versus trapezius from c4 to c6 is likely related to original genetic differences between mesenchyme originally from r10 to r11 versus that from c1 and beyond. Perhaps sternocleidomastoid represents the lateral muscle mass of the cucullaris complex whereas trapezius, being spinal, is homologous to the intermediate and mediall muscles.

In summation, the cucullaris complex in mammals undergoes a homeotic shift backward two neuromeres to occupy somite levels S5–S10 (c1–c6). The muscle breaks into two units. The split between sternocleidomastoid S5–S6 (c1–c2) and trapezius S7–S10 (c3–c6) reflects the original homeotic "breakpoint" between head and neck, that is, between neuromeres r11 and sp1. It should be noted that the homeotic changes responsible for retrodisplacement of cucullaris in mammals occur much later in evolution than those required for redefinition of cervical muscles for the pectoral girdle and upper limb.

Fig. 10.101 The spinal accessory nerve is ancestral to all vertebrates. In elasomobranchs. It occupies 6 neuromeres: r10–r11, sp1–sp4. [Reprinted from Sperry DG, Boord RL. Central location of the motoneurons that supply the cucullaris (trapezius) of the clearnose skate, Raja eglanteria. *Brain Res* 1991; 582:312–319. With permission from Elsevier]

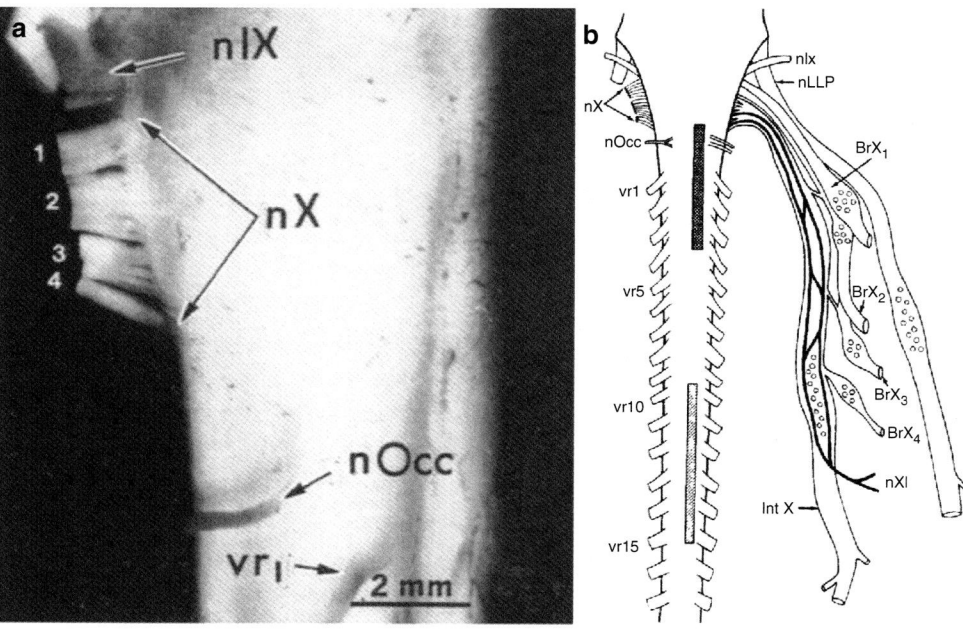

Phylogeny of the Strap Muscles and Prepectoral Muscles

In mammals, the ventral roots of C1–C4 emerge from the medial motor column to face a binary choice. They can take a direct route to innervate muscles associated with the ventral cervical spine such as rectus capitis, longus capitis, or longus colli. They can join up together to form the cervical plexus. In the latter case, they can either supply the hypobranchial strap muscles, the scalenes, or the levator scapulae. The first group of muscles is defined by secondary insertions within the axial spine (or skull). The second group of muscles has secondary insertions outside the axial spine: to pharyngeal arches, to the rib cage, or to the pectoral girdle; they are all outboard to the axial spine.

Hypobranchial muscles as so-named because of their migration below the branchial/pharyngeal arches. In fishes, these muscles originate from the myotomes of the four occipital somites (S1–S4) whereas in tetrapods they are produced from the first four cervical somites as well (S5–S8). Hypobranchial muscles in fishes are a rostral extension of hypaxial muscles supplying the pectoral girdle; they seek out insertions into the floor of the pharynx. Although not branchiomeric, they connect coracoid process with the ventral gill arches and mandible. They act as depressors of the lower jaw.

In chondrichthyans, the hypobranchial muscles are organized into two layers. Posthyoid muscles are referred to as *rectus cervicis*. These deep consist of (1) *coracoarcualis* and *coracobranchialis* that connect to the gill apparatus; and (2) *coracohyoid* that connects the pectoral girdle with the second/third arch hyoid bone. Prehyoid coracomandibularis is the most superficial layer and runs all the way forward to the lower jaw. Advanced osteichthyans show two innovations. Rectus cervicis gives rise to a rostral extension, *sternohyoideus*. This makes sense because cartilaginous fishes do not have a sternum. Coracomandibularis subdivides, giving a distinct geniohyoideus (Fig. 10.102).

Let's now consider the pectoral muscles controlling the fin. The pectoral girdle in fishes is fixed, these muscles extend outward from the scapulocoracoid, traversing a single joint to insert into the dorsal and ventral aspects of the fin. With exception of cucullaris, fishes lack the prepectoral muscles used by tetrapods to control the scapula (trapezius and levator scapulae). Recall that cucullaris in placoderms crosses the head-trunk pivot joint from paranuchal plate to anterior dorsolateral plate. Recall as well that cucullaris is subverted in crown fishes to become a branchial arch dilator. Its role reappears in tetrapods.

The origin of pectoral muscles in fishes involves four myotomes, M2–M5. M2 and M3 belong to the posterior two occipital somites and are innervated by occipital nerves Oc1 and Oc2. M4 and M5 belong to the first somites of the trunk and are supplied by spinal nerves Sp1 and Sp2. Additional myotomes contribute to the fish fin but can run as far back as sp11. The basal condition of the shoulder muscles becomes more complex in the sarcopterygians, as exemplified by Coelacanth *Latimeria*. The scapula is not mobile. Thus, the fin is controlled by a deltoid extending from cleithrum to humerus, by an adductor running from cleithrum distally, and by 11 pairs of supinator and pronator muscles (Fig. 10.103).

When pectoral girdle becomes independent the fin rays consolidate and subdivide to make stylopodium, zeugopodium, and autopodium, interconnected with additional joints.

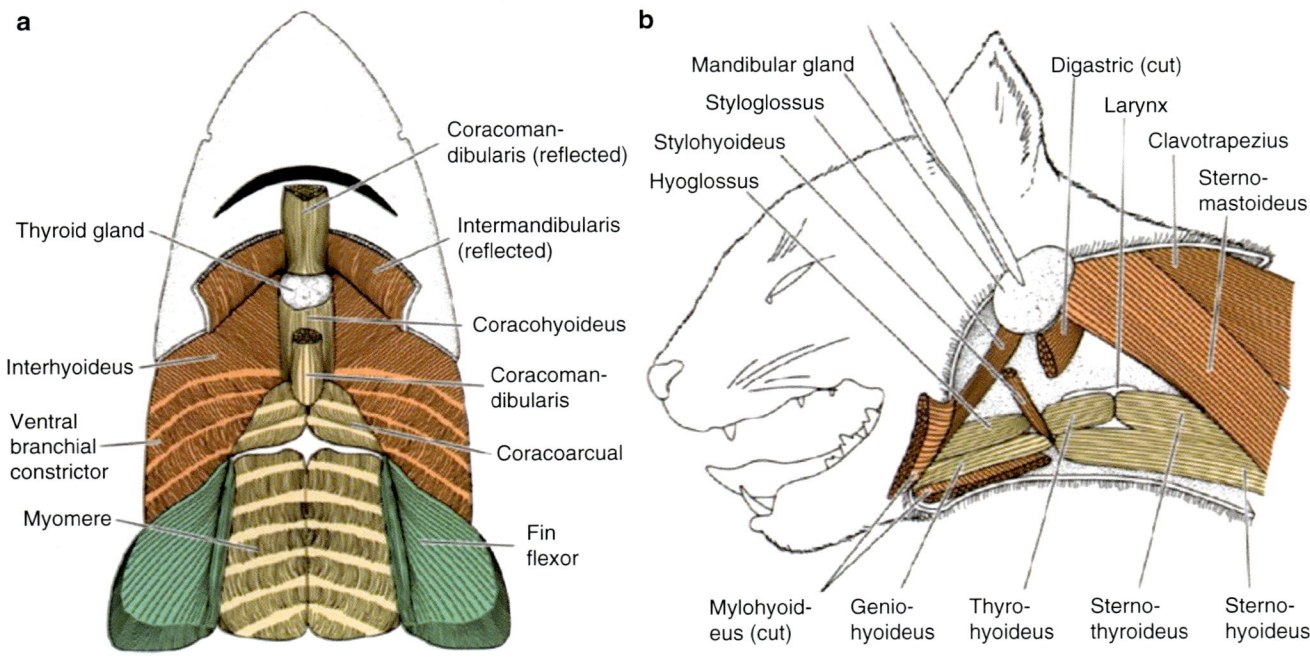

Fig. 10.102 Hypobranchial muscles in shark, *Squalus* and mammal, *Felix*. Hypobranchial muscles in fishes are in three layers. Subdivision into a tongue take place with tetrapods. [Courtesy of William E. Bemis]

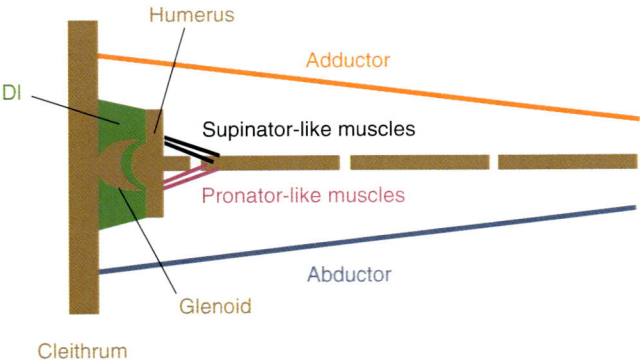

Fig. 10.103 Dorsal view of shoulder in Latimeria Cucullaris is proximal to cleithrum. Deltoid (Dl) joins cleithrum to humerus. Control of the fin consists of an antagonistic adductor and abductors and a 11 pairs of supinator and pronator muscles. These extend past one joint only, the glenohumeral joint. Cervical plexus muscle mass such as rectus cervicis appears in tetrapods. [Reprinted from Miyake T, et al. Pectoral fin muscles of the Coelacanth *Latimeria chalumnae*: functional and evolutionary implications for the fin-to-limb transition and subsequent evolution of tetrapods. *Anat Rec* 2016; 299(9):1203–1223. With permission from John Wiley & Sons]

The original muscles attached to scapula subdivide. The adductor on the ventral side of the scapula persists as serratus. On the dorsal side of scapula abductor mass splits. The lower part becomes the rhomboids. The upper part becomes the levator scapulae and assumes a new function, elevation of the scapula. Like a telescope, the autopodium expands out-

ward and the new muscles arise by subdivision. They continue to be supplied by anterior (ventral) divisions of cervical nerves.

Fishes do not have tongues. Their hypobranchial muscles are innervated by the four branches of occipital nerves. There is no hypoglossal nerve. The motor neurons for the occipital nerves arise from midline nucleus in r8–r11 that is analogous to the hypoglossal nucleus. It has been shown that primitive pectoral innervation in all fishes originated from *both the hindbrain and the spinal cord*. Furthermore, Ma et al. found the pattern conserved in sarcopterygian lungfishes in the tetrapod lineage. They hypothesized that a backward shift in Hox expression affecting both neuroepithelium and mesoderm caused muscles to relocate along the rostrocaudal axis along with their neurons. This led to a *decoupling of the motor neurons from the hindbrain*. Initially, the disconnection was osseous, as in Tiktaalik. with a neuroanatomical translocation as a secondary event [36, 37]. As we shall see, the "tetrapod shift" forced changes in the spinal cord leading to a cervical plexus from MMC and a brachial plexus from a new entity, the LMC.

Anamniote tetrapods, as represented by amphibians, undergo further subdivision of their hypobranchials into two functional groups. These recognize the relocation of the pectoral girdle and the appearance of a new structure, the tongue (although a formal hypoglossal nerve does not appear in evolution until amniota). A prehyoid group arises from *sessile* myotomes. Coracomendibularis splits into a deep *genioglos-*

sus and a superficial *geniohyoideus* that connects the second arch hyoid to first arch mandible. The posthyoid group originates from *translocated* myotomes. *Omoarcuals* continiue to connect the coracoid process of pectoral girdle with the gill arches. They are the forerunner of omohyoid. *Rectus cervicis* spans between the ventral midline of the pectoral girdle and third arch hyoid bone. Cullaris inserts into the dorsal pectoral girdle (cleithrum). *Pectoriscapularis* runs obliquely downward from the gill region. It is flanked ventrally by rectus cervicis and dorsally by cucullaris. It has two insertions: fascia dorsalis of the epaxial dorsalis trunci muscles and ventral pectoral girdle (clavicle). Pectoriscapularis is the anterior-most muscle of the pectoral girdle and inserts into clavicle, pulling the scapula upwards. As such, it is neuromerically caudal to its confrères and may be the sole muscle corresponding to level c4. Amniote evolution involves increasing complexity of the tongue supplied by a coalescence of the occipital nerves into a single hypoglossal nerve. In mammals the intrinsic *lingualis* muscles of the tongue result from a delamination genioglossus, thereby explaining their location in the center of the tongue. Geniohyoid produces *hyoglossus* and *styloglosssus*.

Between pre-mammal anapsids such as the pelycosaures †*Dimitrodon* and mammals, the two remaining dermal bones rostral to scapula undergo translocation. Dorsal cleithrum migrates to the back of scapula, taking trapezius with it. Ventral clavicle dissociates from cleithrum and moves to the extreme medial/ventral corner of scapula; causing a similar dislocation of sternocleidomastoid. Sternohyoid and sternothyroid follow the dissociation of anterior coracoid process in the development of manubrium (Figs. 10.104, 10.105).

What relationship exists between the strap muscles and the muscles of the shoulder? Shoulder muscles can be considered in three groups. *Axioscapular* muscles (trapezius, levator scapulae, rhomboids, and serratus) position the scapula, and therefore the glenohumeral joint, correctly in space.

Scapulohumeral muscles (deltoid, supra/infraspinatus, and teres major) control the position of humerus within the glenohumeral joint. *Axiohumeral* muscles (pectoralis major, subscapularis, and latissimus) position humerus independently from the scapular. Our focus here is solely on the first two axioscapular muscles.

The boundary between cervical plexus and brachial plexus in amniotes lies at neuromere c4. The levator scapulae (C3–C4) insert into medial border of scapula above (rostral to) to the spine and are supplied by cervical plexus. The rhomboids insert into medial border of scapula below (caudal to) the dorsal side (C4–C5) belong to brachial plexus. Note that scapular spine defines the breakpoint. It is the sole component of scapula derived from dermal bone. With the exception of the medial border rostral to the spine and the scapular spine itself, the entire muscle mass of scapula is supplied by c5 and below, that is, from brachial plexus. The independence of acromion is seen as a separate chondral ossification center and even as a separate *oa acromiale*. For this reason, it constitutes a breakpoint between the primary insertions of trapezius directed to the skull and deltoid, directed to humerus (Fig. 10.106).

In summation, axioscapularis muscles do not change in antero-posterior position with evolution. They control the fins in fishes and the scapula in tetrapods. For this reason, their motor nerves, although described as part of either cervical plexus or brachial plexus, in reality are just garden-variety ventral roots. Strap muscles, on the other hand, result from a homeotic shift backwards causing the cervical somites to express additional programs hypobranchial muscles "stolen" from the occipital somites. These "posteriorized" hypobranchial muscles are innervated by the medial motor column of the cervical spinal cord. The extent of the shift varies with distribution of *Pax6*. Recall that in fishes *Pax6* spans from r8 to sp5. *In the homeotic shift the definition of Pax6 is pushed back from sp1 to sp9.* In mammals, this cor-

Fig. 10.104 Phylogeny of hypobranchial muscles. Basal proximal insertion of muscles into pectoral girdle is retained. Relation to branchial arches reduced in amphibians (mudpuppy). Sternum appears in amniotes with manubrium from the anterior coracoid process. [Courtesy of William E. Bemis]

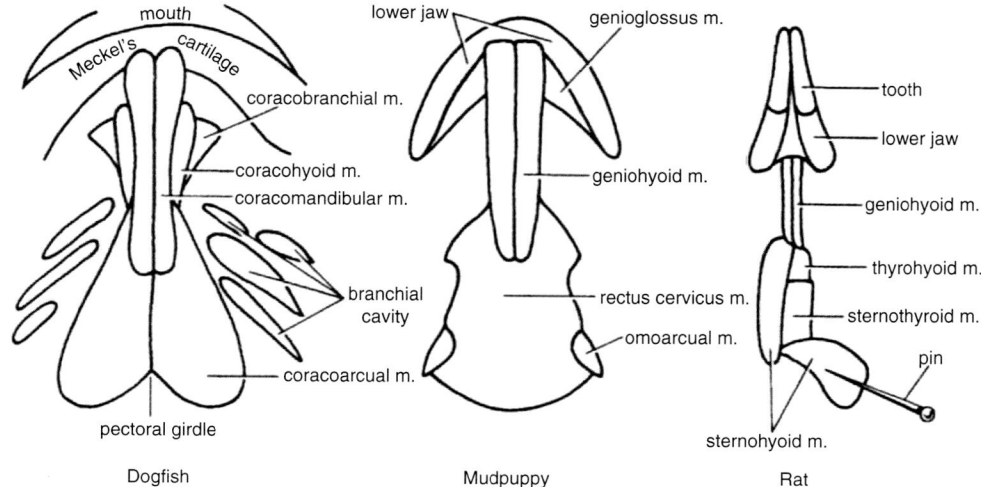

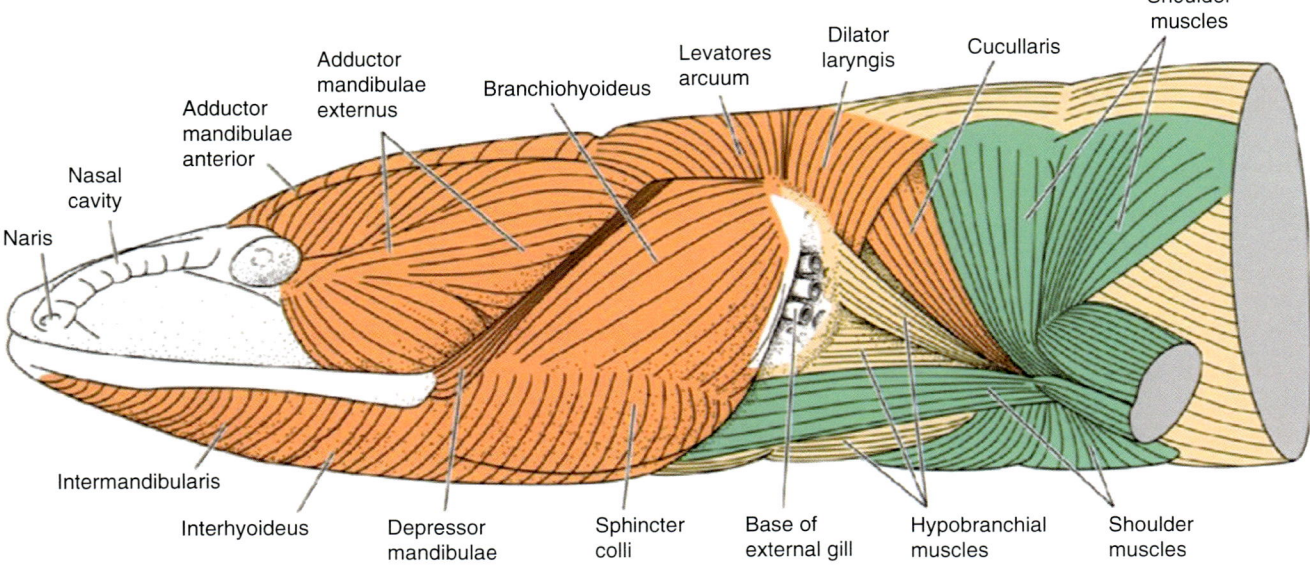

Branchiomeric and shoulder muscles of Necturus

Fig. 10.105 Dermal bones of the pectoral girdle bear insertions of hypobranchial and shoulder muscles. Into clavicle are inserted coraohyoid (green) and pectoriscapularis (tan). Cleithrum has cucullaris (the future trapezius). When cleithrum translocates to the back of the scapula it brings trapezius along with it. [Courtesy of William E. Bemis]

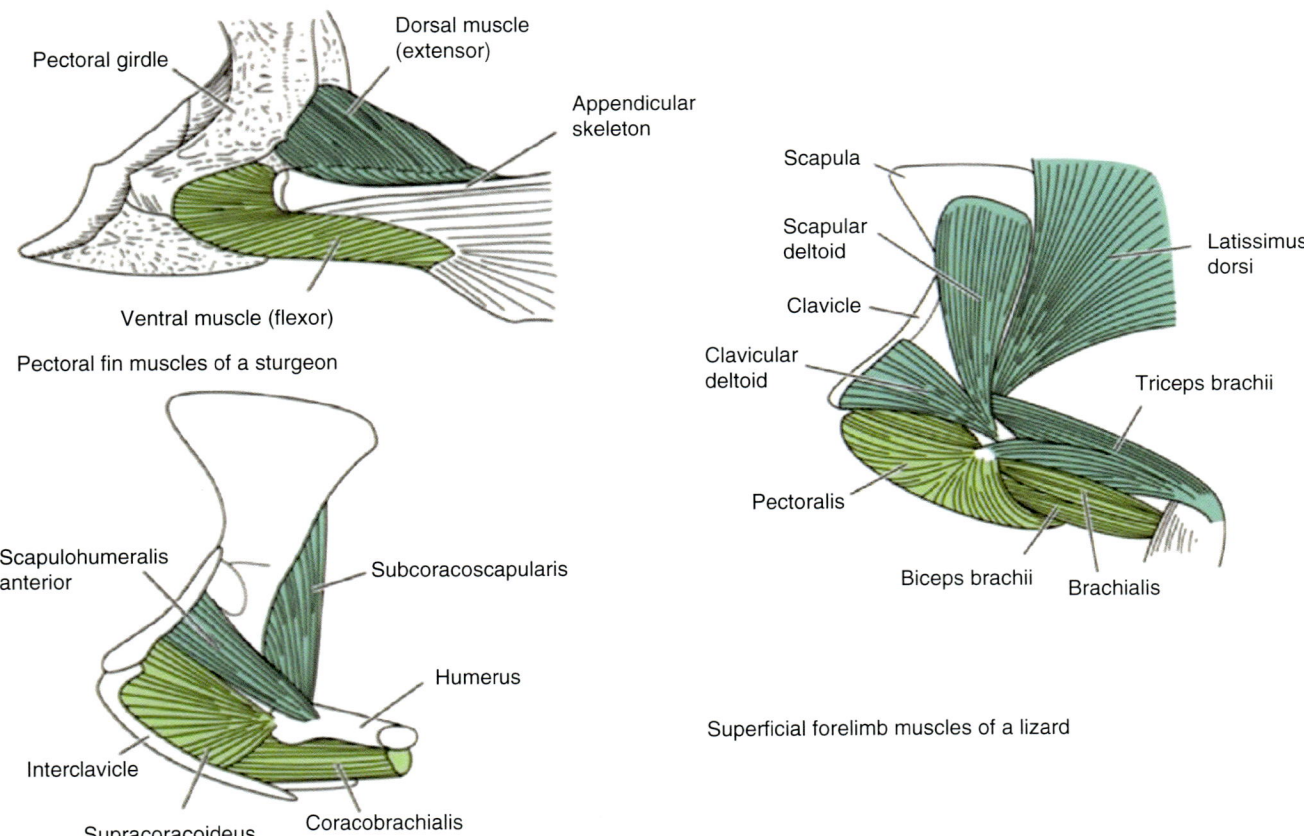

Pectoral fin muscles of a sturgeon

Deep forelimb muscles of a lizard

Superficial forelimb muscles of a lizard

Fig. 10.106 Evolution of pectoral muscles. In fishes the muscles extend past the joint to insert into the distal appendage. Tetrapods have secondary insertions into the proximal stylopodium (humerus) with a second tier of muscles extending to the zeugopodium (ulna/ventral and radius/dorsal). This activates. Dorsal muscles (dark green); ventral muscles (light green). [Courtesy of William E. Bemis]

responds to c1–t1. Thus, the prepectoral girdle musculature of the straps extends from c1 to c3, levator scapulae are supplied from c3 to c4, the diaphragm is at c4 and brachial plexus spans from c4 to t1. The appearance of the phrenic motor column, spinal accessory motor column, and lateral motor column in the cervical spinal cord reflects the evolutionary overlap of these target muscles.

Phylogeny of the Cervical Plexus

The basal form of the peripheral nervous system is seen in cyclostomes (lamprey) in which the homologs of ventral and dorsal nerve roots exist as distinct spinal nerves. Ventral nerves are segmental and somatic motor. They directly enter the myomeres. Dorsal roots are intersegmental; they contain both motor and visceral sensory neurons. They pass between myomeres and extend outward to the skin and inward toward the gut. Note: visceral motor neurons gain access to the viscera in lampreys via cranial nerves, especially the vagus [38] (Fig. 10.107).

This system changes in hagfishes forward throughout vertebrate evolution as ventral and dorsal nerve roots attached to the spinal cord at a given neuromere. Furthermore, somatic motor fibers and visceral sensory fibers (both somatic and visceral) now separate out into the ventral and dorsal roots. Evolution of motor neurons is slower. Anamniotes use both roots while amniotes concentrate all visceral motor fibers into the ventral nerve root. The territory of nerves serving the pectoral girdle and upper extremity is Hox related. In mammals, *Hox5* defines 8 cervical neuromeres and *Hox6* extends from first thoracic neuromere to the first lumbar neuromere.

In lampreys, motor neurons for the upper extremity are located in the ventral horn but without spatial orientation. Actinopterygian (ray-finned) fishes have a forefin controlled by a very simple system of dorsal extensors and ventral flexors attached to an immobile scapulocoracoid. Although they do not form motor columns they demonstrate segregation of the dorsal and ventral neurons. Their pectoral muscles have proximal attachment within the body Sarcoptyergian (fleshy finned) fishes are so-named because their muscles extend outward from the body into the "limb."

Fish fins are rather simplistic. Using a tail for propulsion the fins serve as stabilizers. These provide maneuverability and braking. In combination with the lateral line sensory system (previously discussed), the fins enable sudden, darting motions ideal for prey capture or escape. The musculature is very simple. Greater complexity is achieved by subdivision of the dorsal/ventral muscle masses.

In contrast to the simple paddle-like structure of the fin articulating from a fixed pectoral girdle, tetrapods have a mobile pectoral girdle and a three-tiered limb (stylopodium, zeugopodium, and chiropodium). The system thus advances from one joint to four joints with a consequent increase in mobility and greater complexity of muscle control. Note that proximal control of the scapula involves the subdivision of existing muscles connecting dorsal and ventral scapula to the body. Control from scapula to limb requires more elaborate set of divisions and reassignments. Innervation of this greater muscle mass is accomplished by creation of a separate lateral motor column. In mammals this extends from neuromeric levels c4 to t1.

Innervation of pectoral muscles in fish fins: relationship to the evolution of the neck.

Fishes have three somites. The first somite does not produce muscle; therefore hypobranchial muscles arise one somite backward, from S2 to S3, and are innervated by occipital nerves Oc1 and Oc2. The basal plan for muscles controlling the pectoral fin in all bony fishes involves 4 myotomes from the occipital somites S2–S3 and the first two somites of the trunk S4–S5. Let's refer to these myotomes as M1–M4. Recall that in the tetrapod transition a fourth somite is added to the braincase [37] (Fig. 10.108).

Pectoral motorneurons of all fishes arise from *both* the hindbrain and spinal cord, rather than solely from spinal cord. Mapping experiments demonstrate that this dual pattern in conserved all the way from basal forms such as *Polydon spathula* (paddlefish) through advanced teleosts, the midshipman *Porichthys notatus*. In the basal piscine pattern two occipital nerves (Oc1 and Oc2) emerging than an occipital foramen (OcF) just anterior to the craniao-vertebral junction to supply hypobranchial muscles. They also join with two spinal nerves sp1 and sp2 to supply the pectoral fin.

Lungfishes add a fourth occipital somite. They therefore have three occipital motor nerves. Oc1 and Oc 2 from somites S2 and S3 form a distinct hypobranchial nerve to the coracomandibularis-in-transition. Oc3 immediately joins Sp1–Sp3 to form the pectoral nerve. With additional complexity, fishes can have up to 11–13 spinal nerves contributing to the forefin. For our purposes, what is important is the basic tetrad (Figs. 10.109, 10.110, 10.111).

Let's introduce a definition. Prepectoral muscles are those that connect the pectoral girdle to the body. Their motor nuclei (neuromeres sp to sp4) arise from two motor columns. MMC via cervical plexus supplies the strap muscles and levator scapulae. SAC via spinal accessory nerve supplies trapezius. Postpectoral muscle are all those connecting either scapula or body wall with the upper extremity. They are sup-

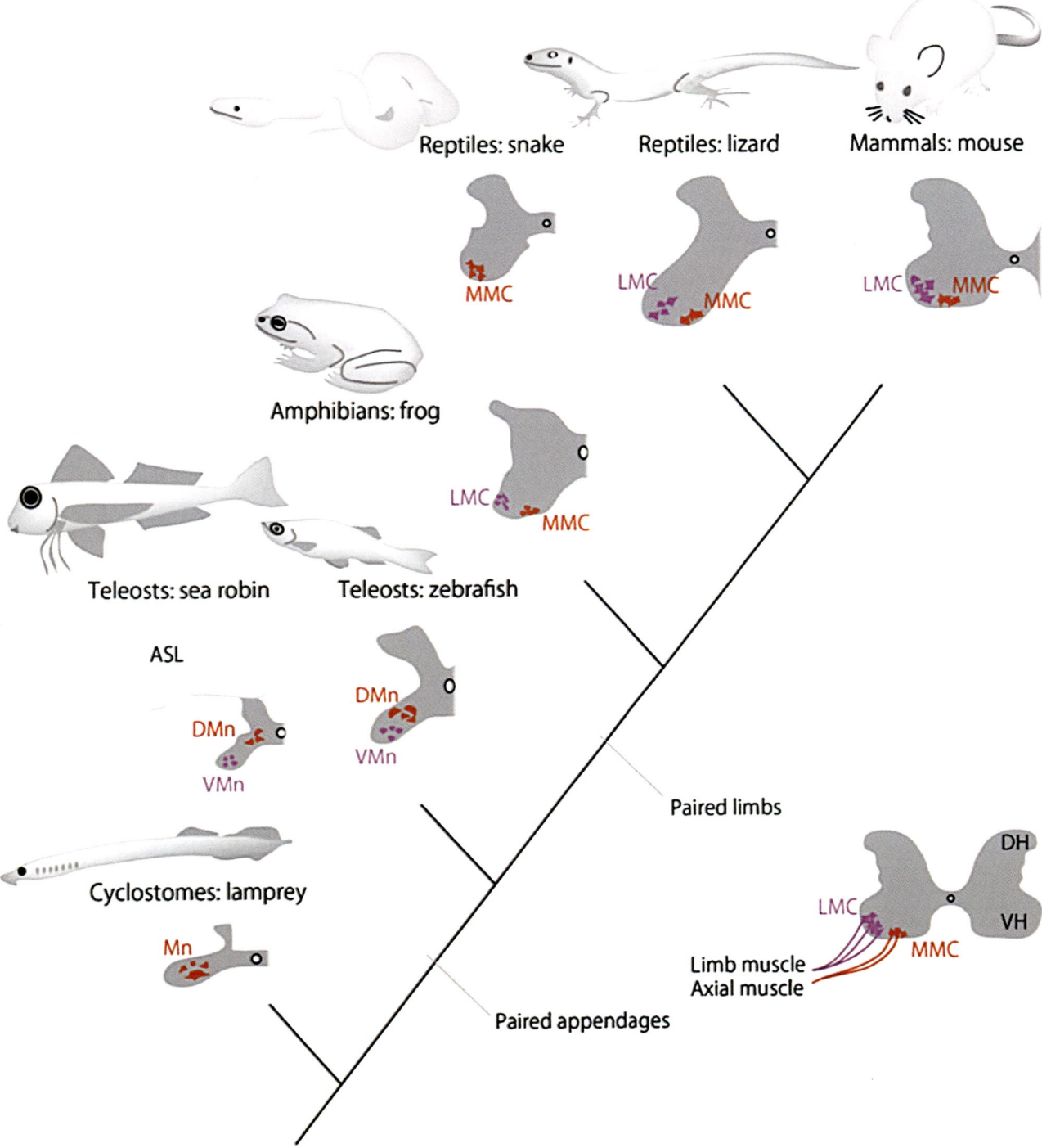

Fig. 10.107 Phylogeny of motor neurons in ventral horn (VH) of spinal cord. In lampreys, MMC-like motor neurons (Mn) are in the ventral horn but no distinction is made between dorsal and ventral. Fishes do not have motor columns in teleosts such as zebrafish and the sea robin *Prionotus carolinus*, which possess movable fin rays, are segregated into discrete pools for dorsal and ventral muscles. In the lineage of tetrapods, including amphibians, reptiles, and mammals, motor neurons located in the lateral portion of the ventral horn segregate into the LMC and innervate limbs. Motor neurons in the python snake form a single continuous MMCcolumn and lack the LMC. ASL, accessory spinal lobes; DH, dorsal horn; DMn, dorsal motor neuron; MMC, medial motor column; LMC, lateral motor column; VH, ventral horn; VMn, ventral motor neuron [Reprinted from Murakami Y, Tanaka M. Evolution of motor innervation to vertebrate fins and limbs. *Dev Biol* 2011; 355(1):164–172. With permission from Elsevier]

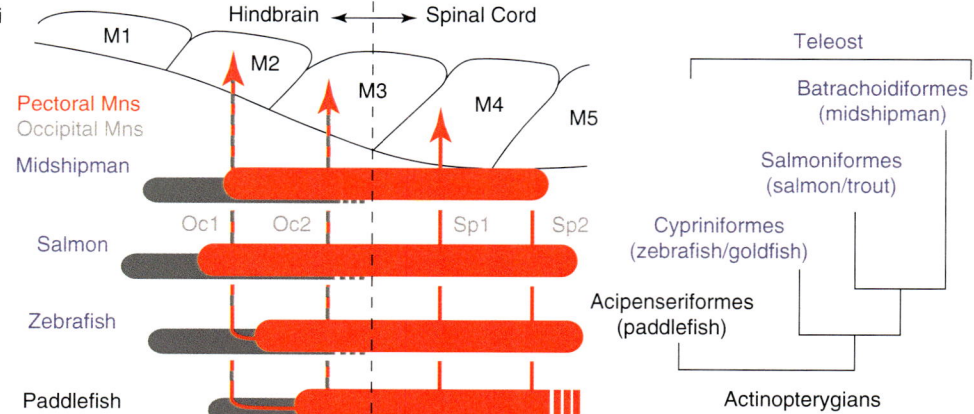

Fig. 10.108 Location of motor neurons to pectoral fin muscles in crown bony (actinopterygian) fishes. Occipital motor neurons (gray) innervate muscles of the second and third myotome producing the hypobranchal anlage (e.g., coracomandibularis). Pectoral motoneurons (red) form a single column extending from hlndbrain into at least the first two neuromeres of the spinal cord. Because the neurons are admixed in the spinal cord with standard hyaxial motor neuron to the muscles of the ventral they are functionally like MMC. [Reprinted from Ma L-H, Gilland E, Bass AH, Baker R. Ancestry of motor innervation to pectoral fin and forelimb. *Nature Communications* 2010; 1(49) doi: 10.10.38/ncomms1045. With permission from Springer Nature]

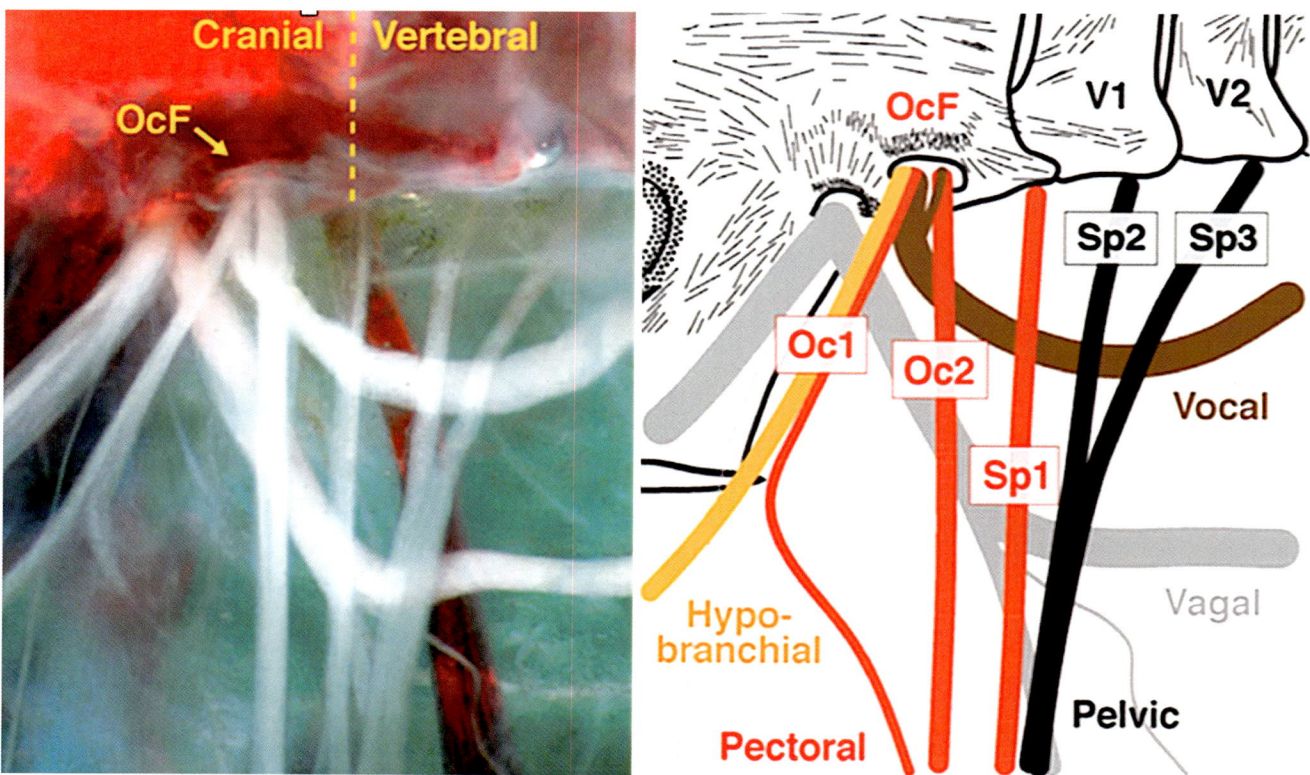

Fig. 10.109 Pectoral fin innervation in crown actinopterygian Midshipman fish. Left: hindbrain/spinal cord boundary (hatched yellow line) OcF-occipital foramen. Right: Motor control from two occipital nerves, Oc1–Oc2 and three spinal nerves, Sp1–Sp3. [Reprinted from Ma L-H, Gilland E, Bass AH, Baker R. Ancestry of motor innervation to pectoral fin and forelimb. *Nature Communications* 2010; 1(49) doi: 10.10.38/ncomms1045. With permission from Springer Nature]

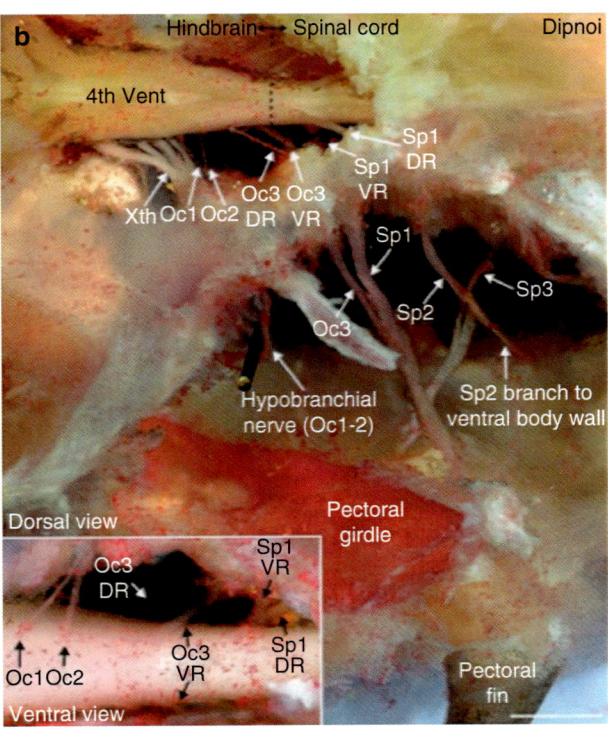

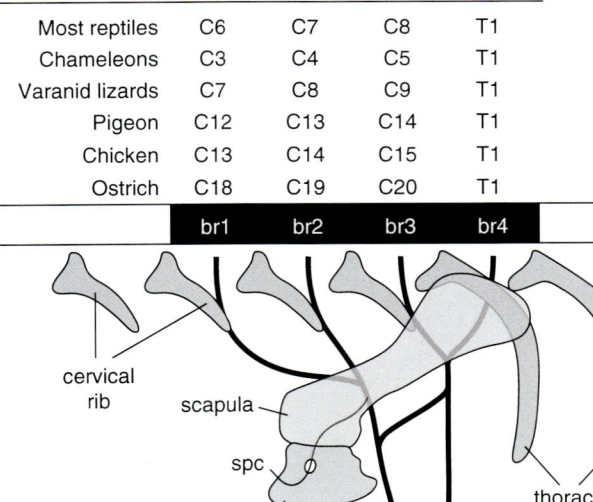

Taxon	Spinal nerve number			
Most reptiles	C6	C7	C8	T1
Chameleons	C3	C4	C5	T1
Varanid lizards	C7	C8	C9	T1
Pigeon	C12	C13	C14	T1
Chicken	C13	C14	C15	T1
Ostrich	C18	C19	C20	T1

Fig. 10.110 Pectoral innervation in Dipnoi (lungfish). We can translate the same analysis from actinoptyergian to sarcopterygian fishes. Sarcopterygians have a fourth occipital somite and consequently 3 occipital nerves. Hindbarin/spinal cord boundary is between Oc3 and Sp1. Note appearance of a distinct hypobranchial nerve Oc1–Oc2 representing somites S2–S3. Oc3 joins with Sp1–Sp3 to supply the pectoral muscles. In tetrapods, when Oc3 is displaced backwards from S4 into S5, the cervical spinal cord will have in the medial motor column four motor roots corresponding to the upper extremity. [Reprinted from Ma L-H, Gilland E, Bass AH, Baker R. Ancestry of motor innervation to pectoral fin and forelimb. *Nature Communications* 2010; 1(49) doi: 10.10.38/ncomms1045. With permission from Springer Nature]

Fig. 10.111 Diaphragm component parts. [Reprinted from Gilbert SF, Barressi MJF. Developmental Biology, 11th ed. Sunderland, MA: Sinauer; 2016. Copyright © 2016. Oxford Publishing Limited. Reproduced with permission of the Licensor through PLSclear]

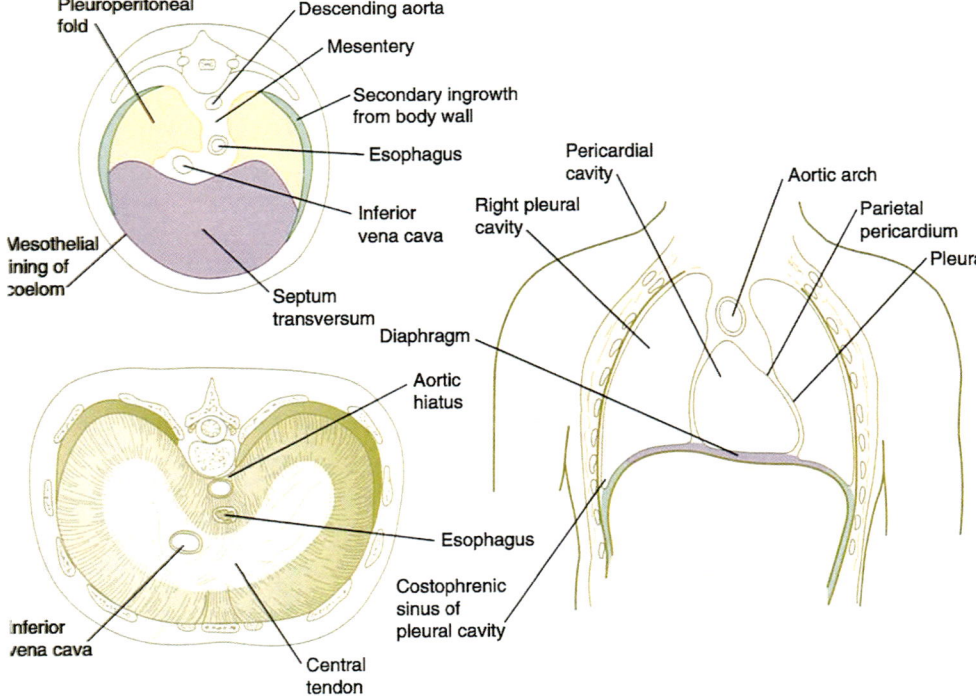

plied by a brachial plexus which, in non-mammalian amniotes, consists of four roots. The more complex mammalian brachial plexus has six roots (vide infra). The brachial plexus is physically separate from cervical plexus. It is neurologically requiring a new, autonomous motor column, the LMC [39].

Distinct motor columns do not appear in evolution until tetrapods. This caudal shift in the location of pectoral plexus and thus its motor nuclei (or vice versa) took place at the sarcopterygian–tetrapod transition, probably with *Tiktaalik*. With the creation of the neck, all four roots of the original pectoral nerve are now outside the skull. In this transition the expression of the original occipital muscles, they are split into the prehyoid lingual group which continue to be supplied from medulla and a posthyoid cervical group.

The function of occipital nerves changed as well. There were now four occipital somites and four-root hypoglossal nerve dedicated solely to the tongue. Spinal nerves sp1 and sp2 assumed a new role. Not only were they responsible for the axial musculature but they became the supply of hypobranchial muscles formerly innervated by Oc1 and Oc2. *MMC from sp1 to sp4 is thus an admixture*: the usual motor neurons for axial muscles plus retrodisplaced hypobranchial motor neurons—the future cervical plexus.

It is tempting to speculate what the consequences of this neuromeric shift were for the cervical neuromeres. Recall that in fishes neuromeres caudal to sp2 are involved in the fin field. With the "loading up" of c1–c3 with hypobranchial muscles did this push the program for a brachial plexus backward to c4–c5–c6–t1? Did this necessitate the creation of an additional lateral motor column?

Our knowledge of the cervical region in the basal tetrapods is very limited. We can surmise the situation by looking at extent amphibians. Amphibians have no atlas–axis. Frogs have only three cervical vertebrae. Their diminutive forelimbs are supplied by a rudimentary brachial plexus consisting of two spinal nerves, sp2 and sp3. Salamanders have four cervical vertebrae with a brachial plexus from root sp3 and sp4. We do know that in early tetrapods the position of the pectoral girdle was immediately behind the head. Presumably, the rectus cervis muscles are supplied by at least one spinal nerve.

Amniotes are more sophisticated. Buccal pumping was abandoned. The circulatory system was remodeled. The neck expanded to five cervical vertebrae in chameleons and subsequently to six in pre-reptiles. As we shall see, the amniote pre-reptilian brachial plexus has four roots with the final root always positioned between the last cervical vertebra and the first thoracic vertebra, *thereby indicating the location of the pectoral girdle*. When the rostral limit of the brachial plexus reaches neuromeric level c4 we see the first signs of a true three-root cervical plexus.

Pectoral Girdle Muscle Displaced into Thorax: (PMC = CMC Medial)

Diaphragm

We conclude our exploration of cervical myology with the diaphragm, the third muscle of the central motor column, and one unique to mammals. Its evolution is intimately involved with the expansion of the mammalian neck and involves the transposition of the brachial plexus [40] (Fig. 10.111).

Mammals have a high metabolic rate permitting them to function at a wide range of temperatures. This required highly efficient multi-lobulated lungs with a large surface area but with the drawback of low compliance. To ventilate the lung, mammals invented a unique muscle, the diaphragm, capable of overcoming the compliance issue. The diaphragm stems from muscles "originating" from vertebrae and the sternum at the boundary of the thorax and abdomen. Why should these muscle fibers be innervated by the C3–C5? What could explain myoblast migration into this site? To appreciate these questions, let's look at the structure of the diaphragm, blood supply and innervation, and development [41].

Structure

The diaphragm is a complex musculotendinous structure constructed from four components, each from a different tissue source. Significance.

- Septum transversum is the most significant as it makes *central tendon*. It lies ventral.
- Pleuroperitoneal folds (PPFs) fuse with two structures: dorsal mesentery of esophagus and dorsal part of septum transversum. Together, these form the *primitive diaphragm*. The membranes are dorso-lateral and constitute the portal of entry of cervical myoblasts.
- Dorsal mesentery of the esophagus (mesoesphagus) is an extension of lateral plate mesoderm hanging down into the abdominal cavity. It fuses with both septum transversum and pleuroperitoneal membranes. It forms the *midline part of diaphragm*. LPM smooth muscle surrounding the esophagus invades the dorsal mesentery to produce right and left *crura*. These are dorsal.
- Body wall gives off an internally-directed shelf of mesenchyme which forms a semi-circumferential arcade peripheral to pleuroperitoneal membranes.

Nota bene: The classical embryologic model above is under challenge. Animal models of congenital diaphragmatic hernia induced by the teratogen nitrofen demonstrate that the

PPFs constitute the major (and perhaps the sole) contributor of muscle to diaphragm [42].

Blood supply consists that two major and two minor sources.

- Phrenic arise paired or singular directly from the aorta or a common branch from the coeliac trunk. They are embryologically associated with the foregut. The phrenic arteries supply the central diaphragm. At the posterior border of central tendon they divide. Medial branch runs anteriorly to joint with pericardiophrenic and the musculophrenic from the periphery. Lateral branch travels outward to anastomose with the intercostal arteries.
- Posterior intercostal/subcostal arteries from two sources. Those for the first and second interspaces come from costocervical trunk. Those arteries supplying interspaces 3–12 arise directly from the aorta. Posterior intercostals associated with the lower 6 ribs supply the body wall component of the diaphragm.
- Pericardiophrenic arteries are small vessels that arise from the internal thoracic arteries. As such, they arise from longitudinal anastomoses of transverse segmental vessels and accompany the phrenic nerves. The neurovascular pedicles pass between the pleura and pericardium, supplying the latter.
- Musculophrenic artery is one of two terminal branches of internal thoracic, the other being superior epigastric. Both branches are anterior. Prior to its division, internal thoracic gives off anterior intercostal arteries to the upper 6 intercostal spaces. It then divides, with musculophrenic running behind the costal cartilages to supply the lower 6 intercostal spaces.

Phrenic nerves descend into the chest from C4 with minor contributions from C3 and C5. The right phrenic is forced to deviate by the presence of anterior scalene which excludes it from contact with the second segment of subclavian. Left phrenic runs a direct course and passes anteriorly to first segment of subclavian. In the thoracic cavity the phrenic nerves, accompanied by pericardiophrenic arteries, pass anterior to hilum sandwiched between fibrous pericardium and the pleura of mediastinum. Here they pick up blood supply from internal thoracic artery as its pericardiophrenic branches. Once the phrenic nerves gain access to the superior surface of diaphragm they divide into four branches. *Sternal branch* runs anteriorly. *Anterolateral branch* runs in front of the lateral lamina or leaf of central tendon. *Posterolateral branch* runs behind the lateral lamina. *Crural branch* runs posteriorly to the crura.

Neuroangiosome failure at boundary zones between neuroangiosomes correlates with diaphragmatic hernias. Hernia at the foramen of Morgagni is anterior, anterolateral defects can be large, permitting massive herniation into the chest. The Bochdalek hernia is posterolateral (Fig. 10.112).

Pain sensation is relayed in three ways. Direct transmission from the diaphragm travels via the phrenics and refers to the dorsolateral shoulder. Inflammation at the periphery of the diaphragm can be pick up by the lower six posterior

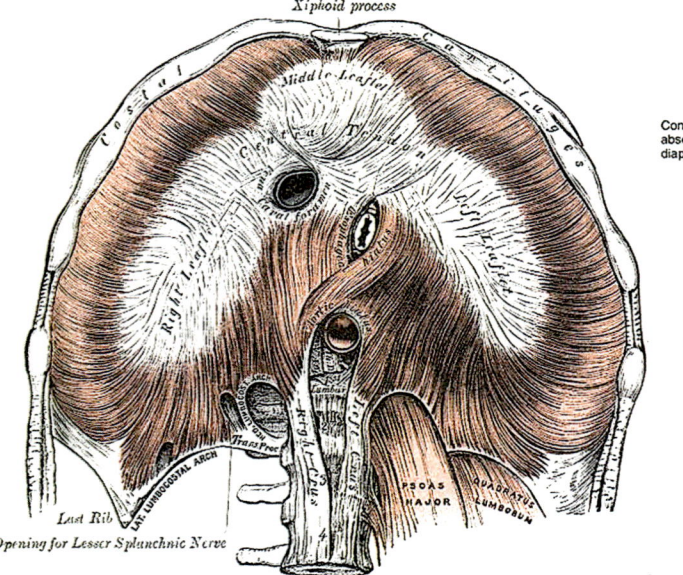
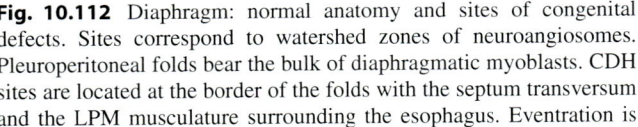
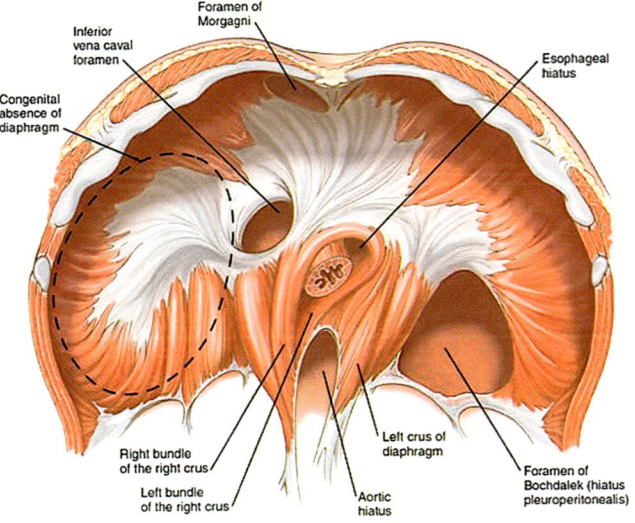

Fig. 10.112 Diaphragm: normal anatomy and sites of congenital defects. Sites correspond to watershed zones of neuroangiosomes. Pleuroperitoneal folds bear the bulk of diaphragmatic myoblasts. CDH sites are located at the border of the folds with the septum transversum and the LPM musculature surrounding the esophagus. Eventration is not a true hernia but represents upward displacement by the viscera due to hypoplasia of the scaffold. [Reprinted from Lewis, Warren H (ed). Gray's Anatomy of the Human Body, 20th American Edition. Philadelphia, PA: Lea & Febiger, 1918]

intercostals and perceived as pleuritic pain. Finally, small penetrating branches from the phrenics can detect subdiaphragmatic inflammation, on the right side from liver, and on the left side, from left lobe of liver, stomach, left adrenal, and spleen. Thus, splenic injury can refer to the left shoulder.

Congenital Defects

Diaphragmatic hernias are a well-described problem in pediatric surgery with numerous reviews available. There are three forms. *Bochdalek hernia* is most common (95%). It is posterolateral with left-side predominance (85%). This preference has been ascribed to a protective effect of the liver on the right side. Complete absence of the diaphragm has been reported [43]. The *Morgagni hernia* represents only 2% of reported cases. It is retrosternal, just behind xiphoid process. Both these defects are located along the border of septum transversum. *Eventration of the diaphragm* Is not a true hernia. It is caused by hypoplasia of the scaffold leading to an upward displacement of the diaphragm by the abdominal contents. Eventration has been reported in association with Poland syndrome [44–47] (Figs. 10.113, 10.114, 10.115, 10.116).

Development of the Diaphragm

The conceptual structure of the diaphragm consists of the creation of a platform or scaffold which subsequently is populated by myoblasts and migrates into position. Despite its position at neuromeric level t10, the diaphragm is not derived from the thorax. It is not commonly appreciated that the scaffold arises at the level of the occipital somites. During its descent, it acquires myoblasts from the myotomes of C3–C5.

Topologically the components of the diaphragm fit together l(dorsa-ventral) like the letter "T" with septum transversum as a vertical post in the midline and the pleuroperitonesal folds like the crossbar. Its three-dimensional relations with the heart, foregut, and pleuroperitoneal canals are depicted in Fig. 10.116. Lateral to the passage way between chest and lung, nephric folds run longitudinally down the embryo. These are thought to contribute to the pleuroperitoneal folds.

Cervical somite myoblasts from c3 to t1 form a common anlage. Pre-diaphragmatic myoblasts innervated from the PMC segregate out and inter the pleuroperitoneal folds and thence septum transversum, bringing their innervation from c4 to c5 as phrenic nerve.

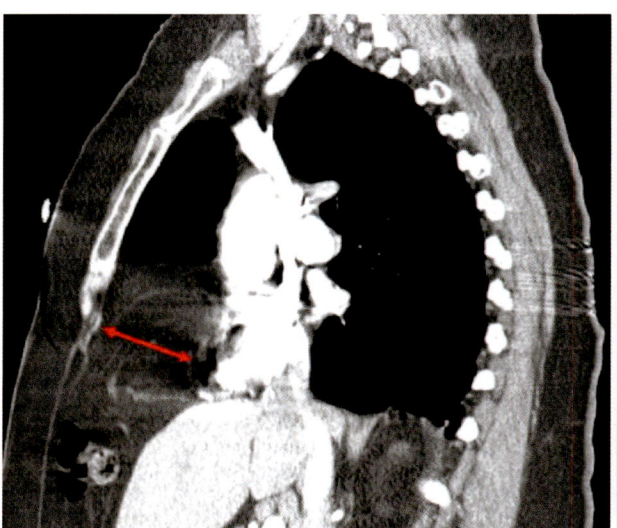
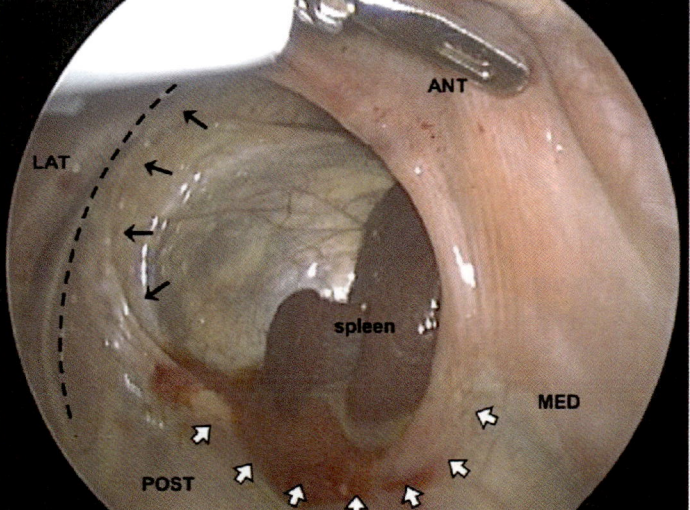

Fig. 10.113 Diaphragmatic hernia. Morgani is retrosternal; it is much less common, CDH, when large, is termed eventration of the diaphragm. It involves territory anterior to the foramen of Bochdalek. Left: [Reprinted from Wikimedia. Retrieved from: https://en.wikipedia.org/wiki/Congenital_diaphragmatic_hernia#/media/File:Morgagni_Hernia.PNG. With permission from Creative Commons License 4.0: https://creativecommons.org/licenses/by-sa/4.0/deed.en.] Right: [Reprinted from Fisher JC, Bodenstein L Computer simulation analysis of normal and abnormal development of the mammalian diaphragm. *Theor Biol Med Model*: 2006;3:9. With permission from Creative Commons License 2.0: http://creativecommons.org/licenses/by/2.0]

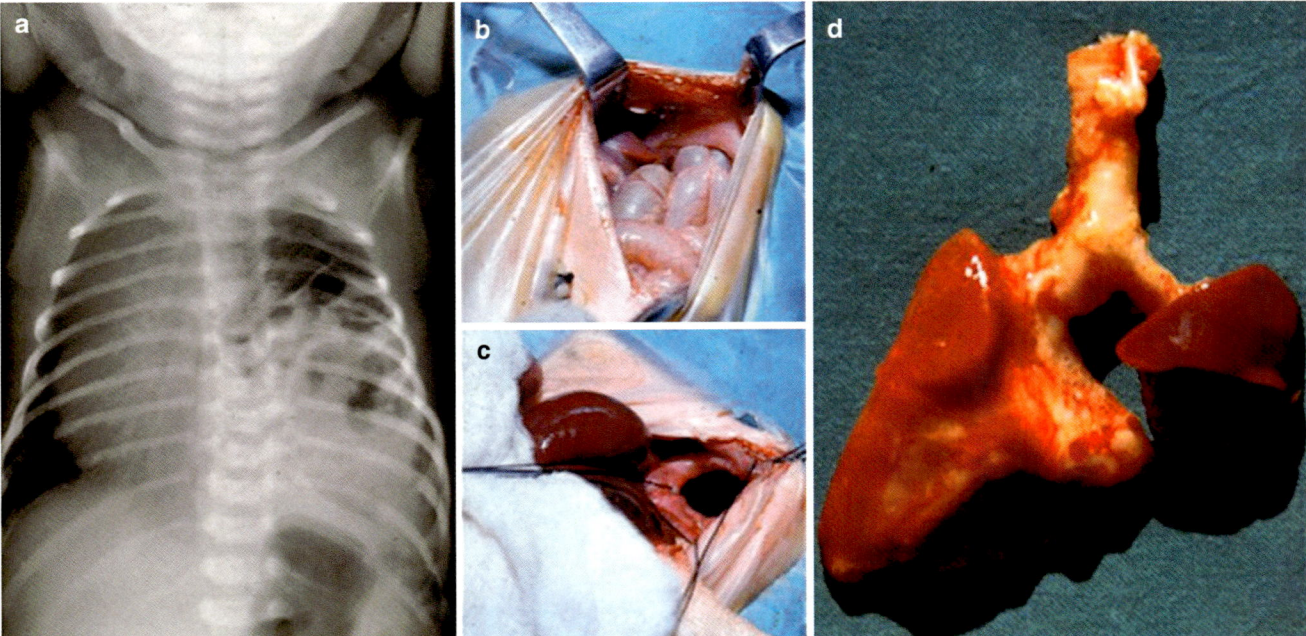

Fig. 10.114 Bochdalek hernia, *senso strictu*, refers to small postero-lateral defects on the left side. A Plain x-ray showing bowel loops in the left chest. B Loops of small bowl entering the chest through the defect. C Defect after reduction of the bowel. D Autopsy specimen showing hypoplastia of the lungs, left > right [Reprinted from Tovar JA Congenital diaphragmatic hernia. *Orphanet J Rare Dis*: 2012, 7;1. With permission from Creative Commons License 2.0: http://creativecommons.org/licenses/by/2.0]

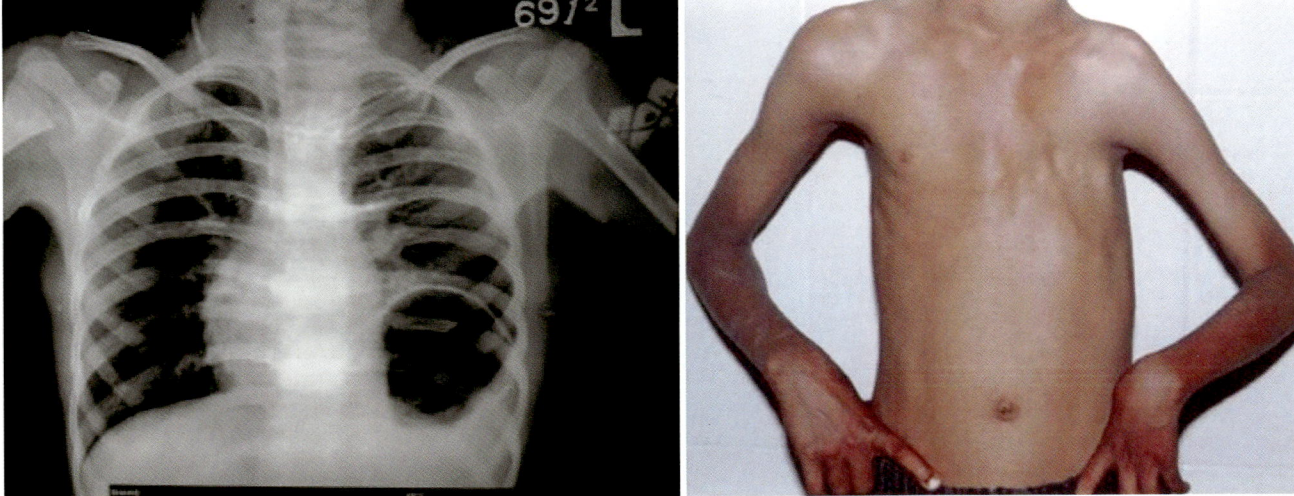

Fig. 10.115 Eventration of the diaphragm represents thinning out due to absence of myobplasts. It can be isolated or can occur as part of Poland syndrome. Left: [Reprinted from Ravisagar P, Abhinav S, Matthur RM, Anula S. Eventration of the diaphragm presenting as recurrent respiratory tract infetions—a case report. *Egyptian Journal of Chest Diseases and Tuberculosis* 2015; 64:291–293. With permission from Elsevier.] Right: [Reprinted from Kulkarni MI, Sneharoopa B, Vani HN, Nawaz S, Kannan B, Kulkarni PM. Eventration of the diaphragm and associations. *Indian J Ped* 2007; 74(2):202–205. With permission from Springer Nature]

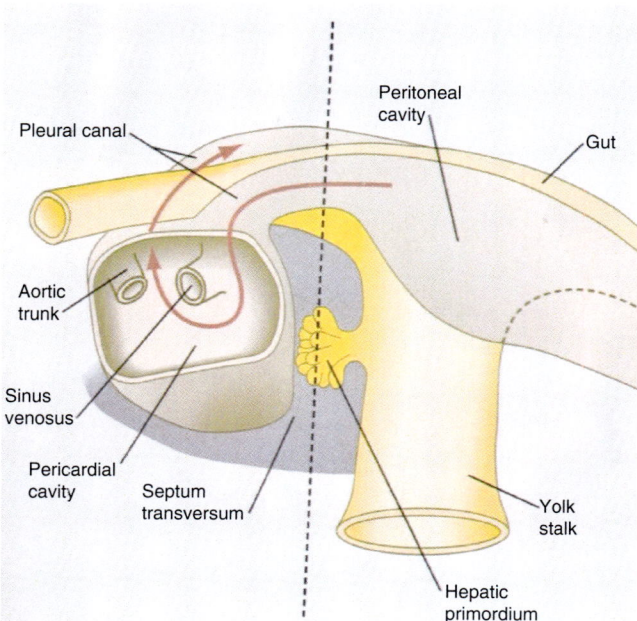

Fig. 10.116 Septum transversum (gray) is a semicircular shelf that grows from ventral body walls to ventral foregut (yellow), dividing the heart from the liver. It partially cuts apart the primary coelom, but paired pleuropericardial canals (tan) persist dosal to septum transversum. The lungs, as foregut diverticula, will invade the canals. [Reprinted from Carlson BM. Human Embryology and Developmental Biology, sixth edition. St. Louis, MO: Elsevier; 2019. With permission from Elsevier]

In mammals, the intracoelomic scaffold becomes populated *with non-cervical somite local myoblasts* throughout. This thoracic mesenchyme gives the diaphragm some additional substance but whether or not they pick up innervation from phrenic nerve is unclear. The origin of this somatic lateral plate mesenchyme is from the lateral thoracic wall. It forms a semicircular arcade wrapped completely around the pleuroperitoneal folds and partially around the septum transversum. The somatic innervation of this zone is from intercostal nerves; whether it receives phrenic nerve innervation is unclear.

Muscles that migrate great distances such as those of the tongue and extremities can be traced by means of *Pax3* expression. Extrapolations from rats using this experimental method shows give us the following developmental sequence in Carnegie stages [48, 49] (Fig. 10.117).

- Stage 8 At 22–23 days the heart has folded beneath the pharyngeal arches.
- Stage 10 Extrapolating from rat model, the muscle precursors of the diaphragm delaminate from somites in Carnegie stage 10, at which time S1–S10 are fully functional, that is, down to the sixth cervical somite (C2). Mesenchyme from the caudal part of the pericardium extends downward and spans between the ventrolateral body wall and the foregut.
- Stage 11 Motor axons enter the cervical anlage.

- Stage 13 upper limb buds appear. Septum transversum has descended from C2 to C4. It Is penetrated at its superior aspect by cervical myoblasts from C3 to C4–C5. Pleuroperitoneal folds (PPFs) project into coelom and migratory cells track along the brachial plexus but are *still within the body wall*. They have not yet entered PPFs or the limb bud.
- Stage 14 Myoblasts enter the PPFs and limb bud. Liver tissue invades the inferior aspect of septum transversum.
- Stage 15 Phrenic nerve axons enter the PPFs and innervate the myoblasts. Diaphragm begins to translocate caudally.
- Stage 16 The lung bud is growing downward surrounded by body wall and with the liver beneath. Because septum transversum is primarily a ventral structure, dorsal to it are the pleuropericardial canals, mesonephric ridges of intermediate mesoderm, and future suprarenal glands. Expanding pleural cavities will burrow into this mesenchyme dorsally. In so doing they will strip LPM_S mesenchyme away from the dorsal body wall. This will be the future source for the peripheral diaphragm.
- Stage 17 Muscle forms in the diaphragm. Phrenic nerve trifurcates. Pleuroperitoneal folds fuse to septum transversum and diaphragm closes.
- Stages 21–23 are marked by closure of the pleuroperitoneal canals under pressure from surrounding tissues (liver and adrenals). Diaphragm reaches its final position at stage 23.

Phylogeny of the Diaphragm: An Unrecognized Brachial Plexus Muscle

The mammalian diaphragm originates from a subpopulation of cells destined in pre-mammal synapsids to become supracoracoideus, connecting pectoral girdle to the humerus. The migration of this muscle is arrested. It remains within the body wall but retains its motor supply from neuromere c4. The evolution of this muscle can be understood on the basis of its motor innervation (Fig. 10.118).

Recall from our previous discussion of the cervical spine that the common ancestor of all amniotes had five cervical vertebrae and that the number of cervical vertebrae increases in each lineage. On the diapsid side, †dinosaurs had 11 to 19 (or more), birds go from 8 to 20, and crown reptiles have 8. On the synapsid line, non-therapsids had five, therapsids had six, with the final number settling at seven with cynodonts and eventually mammals (Fig. 10.119).

In all basal tetrapods, control of the upper extremity is accomplished by common muscle mass, divided into flexors and extensors and usually supplied by four spinal nerves br1–br4, the first three are always cervical and the last one thoracic. Thus, the non-mammals tetrapod formula is CX–CY–CZ–T1. Where brachial plexus begins, that is, the loca-

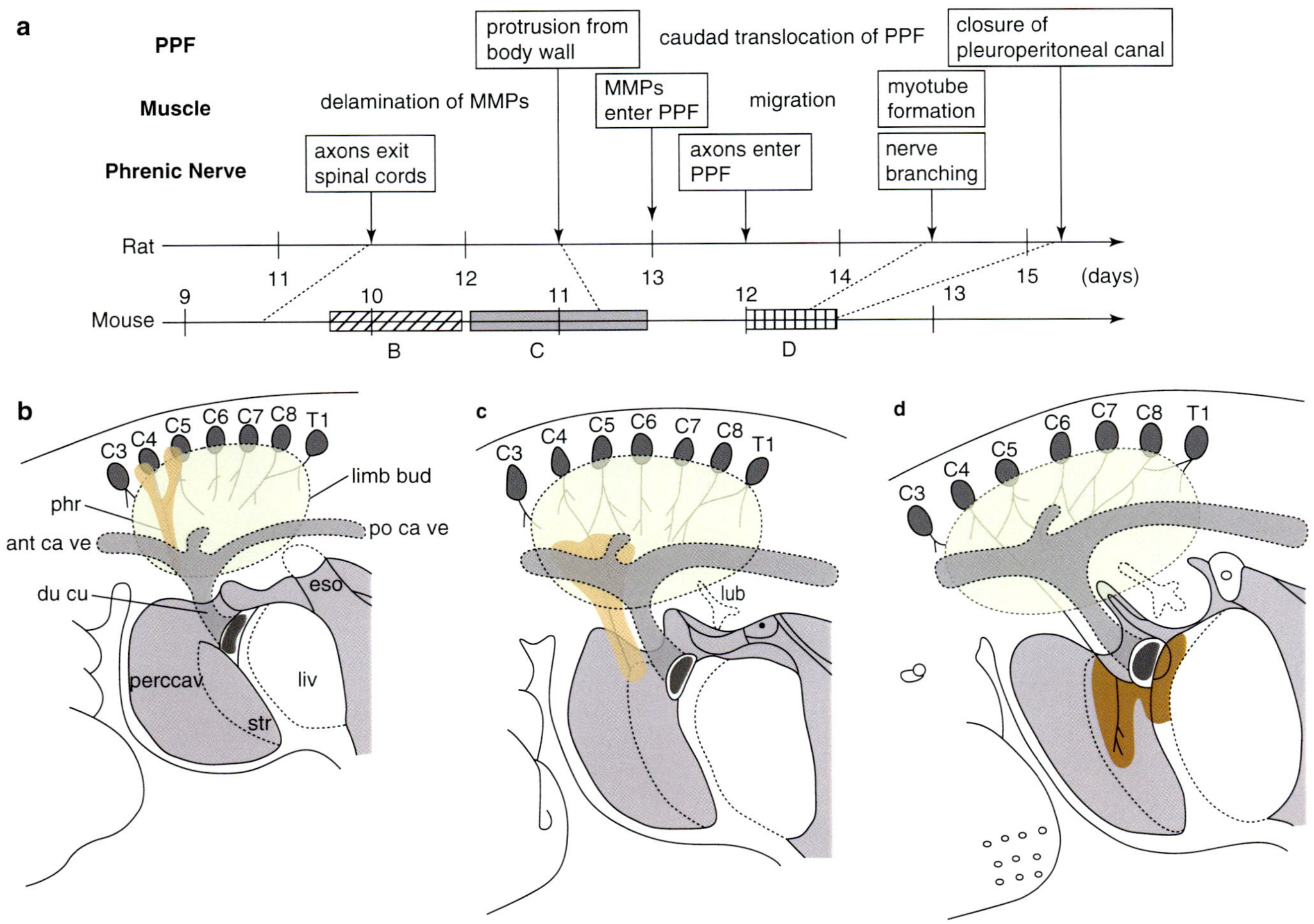

a

PPF

protrusion from body wall

caudad translocation of PPF

closure of pleuroperitoneal canal

Muscle

delamination of MMPs

MMPs enter PPF

migration

myotube formation

Phrenic Nerve

axons exit spinal cords

axons enter PPF

nerve branching

Rat

9   11   12   13   12   14   13   15   (days)

Mouse   9   10   11   12   13

B   C   D

b   C3 C4 C5 C6 C7 C8 T1

phr

ant ca ve

du cu

limb bud

po ca ve

eso

liv

perc cav

str

c   C3 C4 C5 C6 C7 C8 T1

lub

d   C3 C4 C5 C6 C7 C8 T1

Fig. 10.117 Embryonic development of the diaphragm. (**a**) Sequence of development: E11.5 = Carnegie stage 11; E12.5 = Carnegie stage 13; E13.5 = Carnegie stage 15; E14.5 = Carnegie stage 17; E15 = Carnegie stage 18. (**b**) Sagittal section of a mouse embryo at an early stage of phrenic axon elongation, (**c**), at a later stage of phrenic axon elongation, and (**d**) at an early stage of phrenic nerve branching. Dotted area indicates a cell population of migratory muscle precursors (MMPs) migrat- ing into PPF. ant ca ve, anterior cardinal vein; du cu, ductus cuvieri; eso, esophagus; liv, liver; lu b, lung bud; perc cav, pericardiac cavity; phr, phrenic nerve; po ca ve, posterior cardinal vein; s tr, transverse septum. [Reprinted from Hirasawa T, Kuratanin S. A new scenario of the evolutionary derivation of the diaphragm from shoulder muscle. *J Anat* 2013; 222:504–517. With permission from John Wiley & Sons]

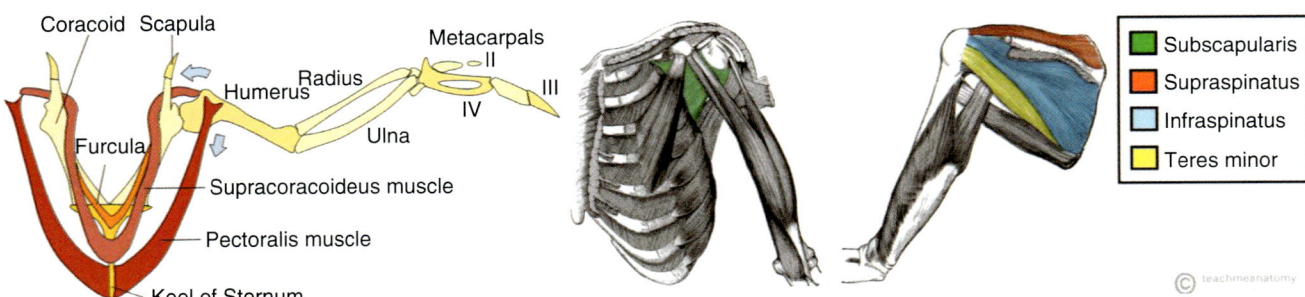

Coracoid Scapula

Metacarpals

Humerus   Radius

Furcula   Ulna

Supracoracoideus muscle

Pectoralis muscle

Keel of Sternum

- Subscapularis
- Supraspinatus
- Infraspinatus
- Teres minor

Fig. 10.118 Evolution of the pectoral girdle—leading to the diaphragm. Diaphragm originates from a subpopulation of cells that migrate toward brachial plexus yet remain within the body wall. In basal tetrapods, a common muscle mass, supplied from the first two spinal nerves br1–br2, of the brachial plexus, connects pectoral girdle to humerus. In amniotes br1 and br2 are located at C4–C5. The muscle anatomy diverges. In basal amniotes it splits into supracoracoscapularis/supracoracoideus and subscapularis. In diapsids (birds) supracoracoideus persists as the antagonist to pectoralis to elevate the wing or stabilize the upper extremity. Birds do not make a diaphragm. In synapsids (mammals) the original coracoid process is lost and is reincarnated as spine of scapula. The Pectoralis major/minor remains attached to the chest, but C4–C5 subscapularis fragments (1) part remains in situ, (2) a c5 part relocates dorsally on either side of the scapular spine as supaspinatus and infraspinatus, and (3) a C4 part attached to chest wall becomes internalized, losing its connection with pectoral girdle to become the diaphragm. Left: [Reprinted from https://commons.wikimedia.org/ wiki/File:Wing_Muscles,_color.svg. With permission from Creative Commons License 2.5: https://creativecommons.org/licenses/by- sa/2.5/deed.en.] Right: [Reprinted from TeachMeAnatomy, courtesy of Dr. Oliver Jones]

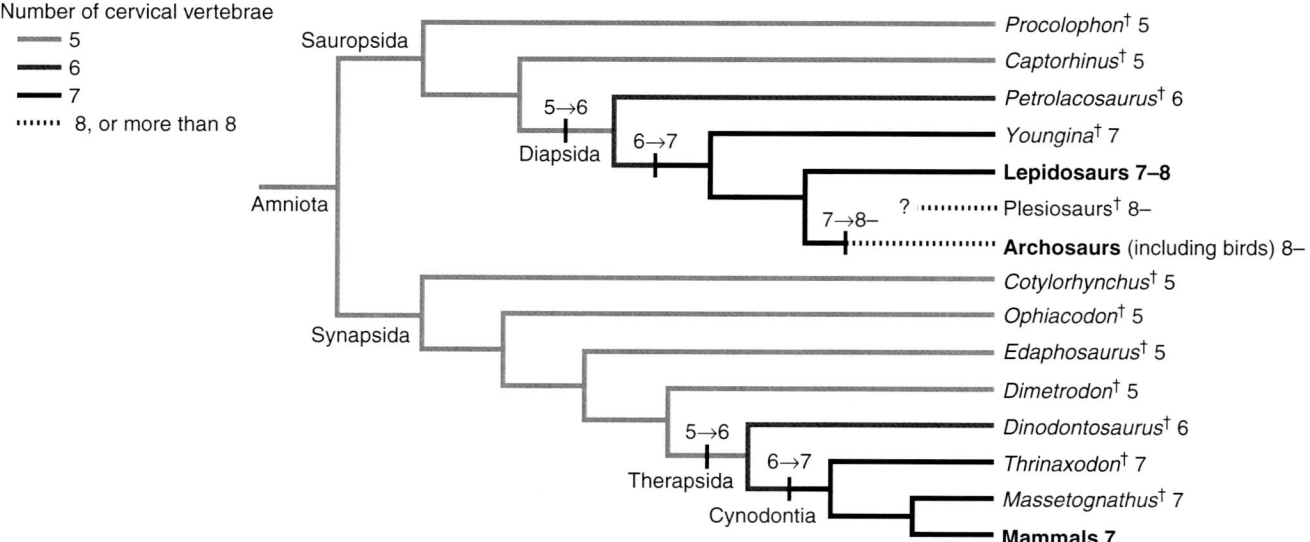

Fig. 10.119 Evolution of cervical vertebrae in amniotes. The two crown clades are diapsida and synapsida. Diapsid taxa are the lepidosaurs (modern reptiles) and archosaurs (dinosaurs and birds). Synapsids are mammals. Extant (living) taxa are in bold. [Reprinted from Hirasawa T, Kuratanin S. A new scenario of the evolutionary derivation of the diaphragm from shoulder muscle. *J Anat* 2013; 222:504–517. With permission from John Wiley & Sons]

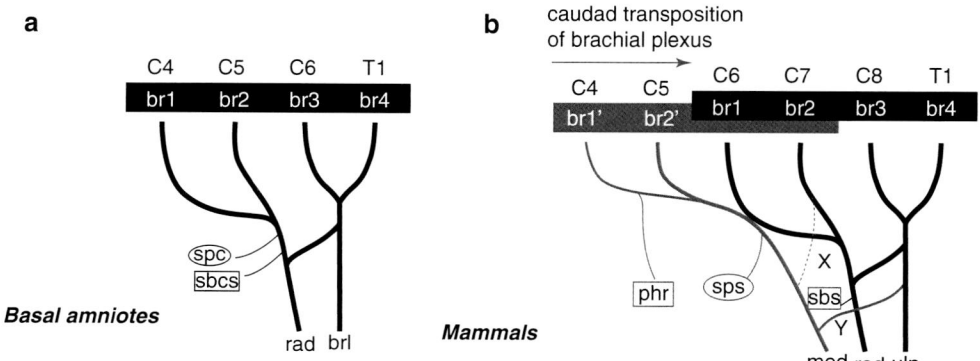

Fig. 10.120 Caudal transition of the brachial plexus. (**a**) Basic pattern of the brachial plexus of basal amniotes, based on the inference from extant reptiles and birds. (**b**) Basic pattern of the brachial plexus of mammals. br1–br4, first through fourth spinal nerves for the basic pattern of amniote brachial plexus. KEY: br1′ and br2′, remnants of the first and second, respectively, spinal nerves of the ancestral brachial plexus; brl, N. brachialis longus; C4–C8, T1 cervical spinal nerves 4–8 and thoracic spinal nerve 1; med, median nerve; phr, phrenic nerve; rad, radial nerve; sbcs, subcoraco-scapularis nerve; sbs, subscapularis nerve; spc, supracoracoideus nerve; sps, suprascapularis nerve; uln, ulnar nerve; X, communication between br2′ and br2; Y, communication between br2′ and br3–br2. Lateral nerves marked by circles; medial nerves marked by boxes. Keep in mind these homologies between amniotes and mammals: subcorascapularis = subscapularis and supracoracoidus – suprascapularis. [Reprinted from Hirasawa T, Kuratanin S. A new scenario of the evolutionary derivation of the diaphragm from shoulder muscle. *J Anat* 2013; 222:504–517. With permission from John Wiley & Sons]

tion of br1, varies with evolution. In lizards, br1 can be found at C3, in basal amniotes br1 is at C4. Advanced amniotes have longer necks. In crown reptiles, br1 is at C7 and in the ostrich, the plexus begins at C18…excellent for burying one's head in the sand!

Let's compare the situation in the chicken that that of mammals. In *Gallus* the 4-root brachial plexus extends from sp13–sp16 to the cervical–thoracic junction between vertebrae C15 and T1. Roots br1 and br2 give rise to nerves to the supracoracoid and subscapularis muscles. The same pattern is seen in the basal amniote plexus (Fig. 10.120a).

The mammalian plexus is more complex, with the *original four roots* shifted backward to the sixth–ninth spinal nerves, that is, C6–T1. On the other hand, suprascapular does not change position; it remains behind C5. This implies that mammals achieve the configuration of eight cervical neuromeres and seven cervical vertebrae by virtue of a partial duplication in the fourth and fifth spinal nerves. *Mammals are thus unique in having six spinal nerves by virtue of duplicating br1–br2.* They also have an additional nerve ord, the *median* nerve. This more sophisticated system enables enhanced motor control of the upper extremity with the formula CX–CY–CX–CY–CZ–T1. In sum, the true mammalian brachial plexus is C4–C5–C6–C7–C8–T1 (Fig. 10.120b).

The phrenic nerve can be understood as a remnant of the ancestral brachial plexus. The suprascapular nerve in mammals, by remaining a C5, *retains its original amniote identity.* In contrast, subscapular nerve moves backwards two neuromeres. Recall that subscapular or subcoracoscapular arise from br1 to br2 in the original amniote pattern at level C5. We now find them in mammals at level C7. The myoblasts of subscapularis arise from the cranial part of the muscle mass dedicated to the forelimb. In pelycosaurs, as in mammals, it attaches scapula to chest wall. In this regard, the myoblasts of diaphragm, abandoned back at levels C4–C5, pursue a similar course but penetrate the chest to enter the pleuroperitoneal folds. In the process, they lose their proximal attachment to the pectoral girdle but their innervation is retained! (Fig. 10.121).

Phylogeny of the Diaphragm

The muscle mass that will give rise to the diaphragm originates from the first two cervical somites of the plexus (C4–C5). Its spinal nerves br1–br2 connect the pectoral girdle to humerus.

In basal amniotes, the anapsids, br1 and br2 shift backward from the original tetrapod position becoming located at C4–C5. The muscle anatomy diverges into an extensor of the humerus, supracoracoscapularis/supracoracoideus, and a flexor, the pectoral mass called subscapularis. In diapsids, supracoracoideus persists as the antagonist to pectoralis to elevate the wing in birds and stabilize the upper extremity in reptiles. Note that birds do not make a diaphragm. They suffocate if the chest is compressed.

In synapsids (pre-mammals) the original coracoid process is lost and is reincarnated as spine of scapula. The muscle complex fragments. C5–T1 remains attached to the chest as pectoralis major and minor. C4–C5 relocates to the scapula, becoming worthy of its name. In mammals, the original coracoid process becomes the scapular spine. Subscapular splits. Half of the muscle remains in situ on the ventral aspect of scapula. The other half shifts dorsally to form, on either side of the spine, supaspinatus, and infraspinatus. Finally, a C4 portion migrates into the pleuroperitoneal folds to become a diaphragm, bringing its cervical nerves with it.

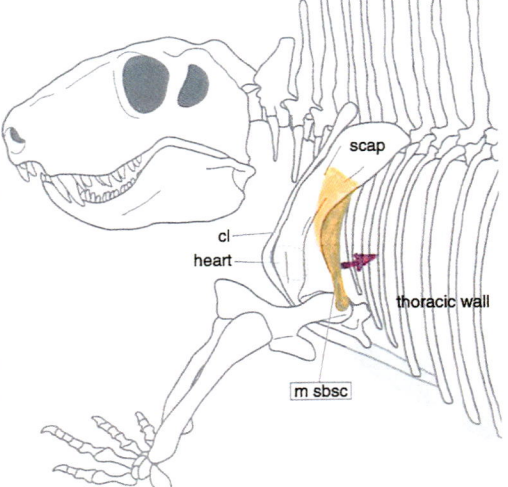

Fig. 10.121 †*Dimetrodon* and the basal diaphragm. In synapsid pelycosaur, *Dimetrodon*† show five cervical vertebra. Heart is just medial to pectoral girdle. Subscapular muscle (sbsc) becomes inserted into thoracic wall directly opposite the potential pleuroperitoneal folds. Proximalhoulder muscle derivatives enter coelomic cavity at arrow, populate the folds and bring cervical innervation with them. Left: [Reprinted from Wikimedia. Retrieved from: https://commons.wikime- dia.org/wiki/File:Dimetrodon_incisivum_01.jpg. With permission from Creative Commons License 3.0: https://creativecommons.org/ licenses/by-sa/3.0/deed.en] Right: [Reprinted from Hirasawa T, Kuratanin S. A new scenario of the evolutionary derivation of the diaphragm from shoulder muscle. *J Anat* 2013; 222:504–517. With permission from John Wiley & Sons.]

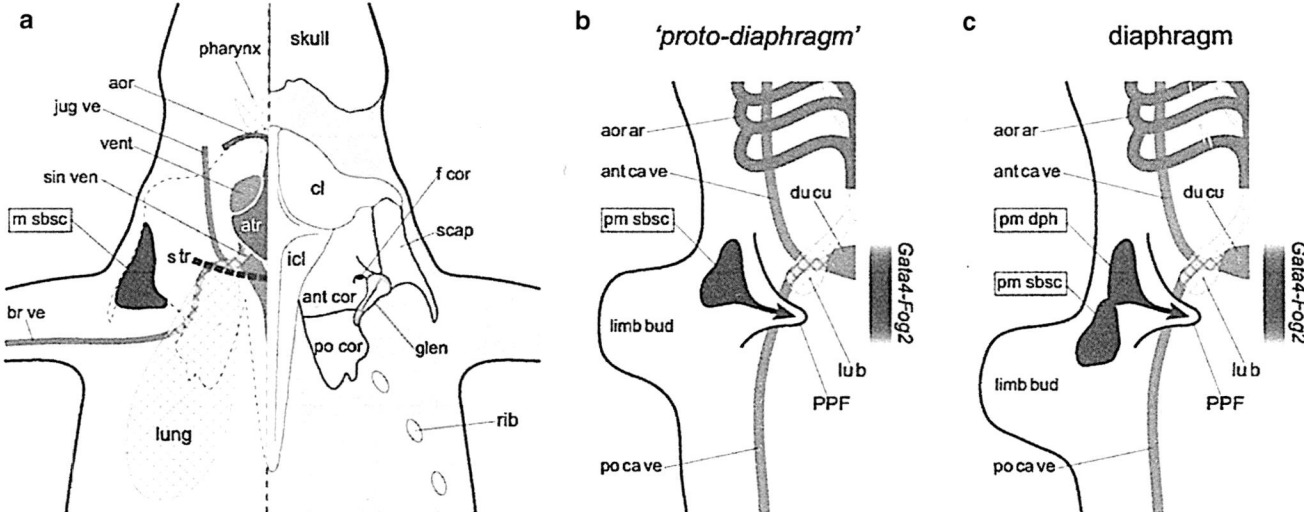

Fig. 10.122 Kuratani Hirasawa hypothesis regarding the migration of the diaphragm. Transposition of the diaphragm comes about via caudal transposition of the forelimb. Muscle mass of subscapularis will subdivide with caudal component remaining in relation to the limb and the dorsal diaphragmatic component inserting itself into the pleuroperitoneal fold. (**a**) Pelycosaur-grade synapsid pectoral region (ventral view), showing the positions of the subscapular muscle (m sbsc), the respiratory and circulatory systems (left), and the skeleton (right). The drawing of the skeleton is based on that in Romer & Price (1940). (**b**) Embryonic development of the hypothetical proto-diaphragm (ventral view). (**c**) Embryonic development of the diaphragm in extant mam-

mals (ventral view). atr, atrium; aor, aorta; aor ar, aortic arch; ant ca ve, anterior cardinal vein; ant cor, anterior coracoid; br ve, brachial vein; cl, clavicle; glen, glenoid; du cu, ductuscuvieri; f cor, coracoid foramen; icl, interclavicle; jug ve, jugular vein; lu b, lung bud; pm dph, premuscle mass for the diaphragm; pm sbsc, pre-muscle mass for the subscapular muscle; po ca ve, posterior cardinal vein; po cor, posterior coracoid; scap, scapula; sin ven, sinus venosus; s tr, transverse septum; vent, ventricle. [Reprinted from Hirasawa T, Kuratanin S. A new scenario of the evolutionary derivation of the diaphragm from shoulder muscle. *J Anat* 2013; 222:504–517. With permission from John Wiley & Sons]

Let's describe how this happens. In the basal synapsid pelycosaurs and pre-mammals subscapular muscle spans across the internal aspect of scapula (Fig. 10.121). It is located in the body wall just behind the heart, facing the coelomic cavity just opposite the pleuroperitoneal fold. This muscle mass is termed the *proto-diaphragm*. In the Hirasawa and Kuratani model, the driver for diaphragm development is the caudal shift of the forelimb bud. Seven-vertebra status was completed with the cynodonts in the Triassic, as in *Thrinaxodon†*. Sealing up the diaphragm closed down the aperture for the esophagus (a potential problem for large chunks of food). This did not present a problem as *Thrinaxodon* was a heterodont (having specialized teeth) and had a secondary hard palate indicating advanced food processing. The pre-cynodont pelycosaur myoblast population positioned beneath scapula (including primitive diaphragm) in cynodonts now formally subdivides into subscapularis and diaphragm. Thus, the HK model demonstrates the origin of diaphragm from the forelimb muscle mass assigned to the brachial plexus and supports the idea of C4 as the basic take-off point for the plexus (Fig. 10.122).

Congenital diaphragmatic hernias have received a great deal of attention in the pediatric surgery literature. The posterolateral Bochdalek form accounts for 95% of cases, the remainder divided between the hernia of Morgagni

just behind the xiphoid and eventration, a condition in which a floppy diaphragm billows upward into the chest. The *c-met* null-mutant mouse has an *amuscular* diaphragm and has been used to show that the Bochdalek defect is primarily due to mesenchymal defect in the pleuroperitoneal folds such that they do not provide a sufficient scaffold for myoblasts to populate. We can surmise that the position of the Foramen of Bochdalek indicates the perhaps the final site of pleuroperitoneal fold development, one which is last to be completed and first to demonstrate tissue insufficiency.

Phylogeny of the Phrenic Nerve and Brachial Plexus

The brachial plexus in all tetrapods (except mammals) has four spinal nerves, br1–br4. These are paired, respectively, as radial nerve (br1–br2) and long brachial nerve (br3–br4) to produce two *y*-shaped nerves. These subsequently marry up transversely to form a "double-Y" (cf Fig. 10.119).

Of critical importance is that the position of the four-nerve plexus along the neuromeric axis depends on the number of cervical vertebrae. Also, the neck trunk interface is always between br3 and br4 In chameleons, with five cervical vertebrae, the plexus begins at spinal nerve 3. Amniotes in general have six cervical vertebrae with the rostral limit of pectoral forelimb innervation being at spinal nerve 4 and the trunk beginning at sp7 Ostriches have 20 cervical vertebrae so the plexus begins at C18.

This variation in the number of cervical vertebrae is reflected in the paleontological record. The amniote bauplan begins with an anapsid skull and five vertebrae. It then bifurcates into the saurapsida leading to reptiles and birds and the synapsida leading to mammals. Saurapsids split into the dead-end eurapsids with five vertebrae and the diapsids with 6. These latter subsequently expand to seven–eight in extent reptiles and eight or more in birds. Synapsid are more sedate. The therapsid line goes from five to six. The cynodonts (immediately prior to mammals) stabilize at seven cervical vertebrae (cf Fig. 10.119).

Mammalian brachial plexus reflects the anatomic consequences of cervical expansion. The original four roots are pushed backward to levels C6–T1 maintaining their anatomy all the while. What remains behind at levels C4–C5 are two roots with exactly the same Y configuration as before. Thus, the mammalian brachial plexus consists of three Y-shaped nerves, in antero-posterior order: median, radial, and ulner. All three of the Ys are cross-linked.

Key motor nerves are affected by these anatomic shifts. In basal amniotes, the brachial plexus is at the fourth–seventh spinal nerve. Supracoroideus (spc) and subscapularis (sbcs) are supplied from the common stem of br1 and br2, that is, from the *radial nerve* (cf Fig. 10.120a).

To the amniote pattern, mammals add two additional cervical vertebrae, causing the position of the brachial plexus to shift backward to the sixth–ninth spinal nerves. This forces a *partial duplication of the original plexus* in which the original pattern of C4–C5 is repeated at levels C6–C7. Suprascapular nerve (sps) remains connected with neuromeres c5–c6; it is now innervated by a new nerve, the median. Recall the median is a mere repetition of radial and that radial and ulnar are cross-connected. Because the median repeats the ancestral condition, it too sends a cross branch backwards to connect with ulnar, bypassing the radial. It remains attached to radial nerve and is dragged backward with it to level c6–c7. Meanwhile (back at the ranch) a new nerve, phrenic, appears at the br1–br2 level; it is supplied by c4–C5 (cf Fig. 10.120b).

Implications of the Brachial Plexus Shift for Trapezius

Conflictual data exists regarding the motor supply to mammalian trapezius: does it terminate at C4 or extend down to the sixth spinal nerve? In the creation of the mammalian brachial plexus. Based on our discussion of the amniote

brachial plexus neuromuscular structures at levels c4–c5 are duplicated or altered to create an 8-nerve plexus. Subscapularis therefore moves backward from sp5 to sp7. For this reason, we propose that cucullaris simply expanded its available sources of mesenchyme and split into Scm and Tpz.

Finally, we ask the question: why are all these hypaxial muscles inserting into the skull and spine? Cucullaris in placoderms is clearly paired with an extensor antagonist. It was clearly supplied by a ventral nerve. The limbs in primitive tetrapods, later amphibians and reptile required powerful ventral muscles to control splayed-out limbs. Changes in the shoulder and hip joints allowed for mammals to positioned their limbs beneath the body. Dorsal muscles became much more important to retract the humerus. More simplistic muscle masses in reptilomorphs broke into subunits which became more sophisticated for complex and agile movements. For this reason, it is no surprise that in mammals, motor control of the dorsal scapula involves the transposition of muscles to the dorsal midline.

Myology of the Neck: Final Thoughts

We have now completed our survey of the muscles of the neck. As always, the assignment of myoblasts to a particular neuromeric level is based on the neuromeric principle of a *one-to-one correspondence between motor nerve and somite*. The origin of a muscle is determined by the root(s) of its motor nerve. Migration of myoblasts for each muscle follows a strict spatial–temporal order, an order that determines the final outcome of the system. Migration is a very physical, real-time process. Successive groups of myoblasts find a CAM "slime trail" and venture out, pushing and shoving their way along established tissue planes until they encounter an available binding site.

Important sidebar: The neuromuscular relationship is reciprocal. At each neuromeric level, a level-specific homeotic code is shared with all tissue derivatives, including somites, intermediate and lateral plate mesoderm, endoderm, neural crest, and neural tube. This resolves the issue of whether the spinal cord has individual myelomeres. Although its neuroanatomy consists of ascending and descending tracts, synaptic connections between spinal nuclei and ganglions are level-specific. The homeotic coding, seen peripherally, imposes itself on the CNS.

Muscles possess intrinsic properties; these determine four things: (1) when they migrate, (2) how they interact with their neighbors, (3) where they form a primary attachment (insertion), and (4) where they make a secondary attachment (insertion). Myoblast development is a tightly choreographed cranio-caudal dance in which the final anatomy of a muscle respects the territory of previously established muscle units.

For example, myoblasts departing from somite S_n at time t_1 are followed by those departing at t_2. Maturation of myoblasts within somite S_{n+1} occurs slightly later, therefore S_{n+1} myoblasts will always respect the pathways of their predecessors. Muscles "flow" like amoebae along spaces defined by their surroundings. Consummate opportunists, embryonic muscles always "choose" the nearest available binding site. Recognition of the binding site is the responsibility of the fascia surrounding the myoblasts. The primary attachment site is usually a bone originating from the same neuromeric level, but there are exceptions. If, after all, muscles from S_n have taken up their primary attachments, an unfilled binding site persists within the territory of neuromere, then the first available muscle from S_{n+1} will seize the opportunity and attach there. Secondary attachments develop according to the same rules. Finally, muscles insertions follow two patterns: (1) antegrade–antegrade and (2) antegrade–retrograde, as in the pectoral girdle.

Time and again we have seen how these principles are reflected in the attachment sequence of muscle units. Consider the intrinsic muscles of the larynx (somite 1) versus cricothyroid (somite 2). These are supplied by superior and inferior laryngeal nerves respectively. The intrinsics take up binding sites within the larynx while cricothryoid is forced outside, binding to r8–r9 interface between inferior border of r8 thyroid and superior border of r9 cricoid. Migration patterns of tongue and infrahyoid are very orderly. Omohyoid follows the same rules as its inferior belly seeks out a primary C2–C3 binding site on scapulae; it then doubles back to insert on hyoid. Longus capitis originates from cervical somites 1–3 whilst longus colli originates from somites 2–6; origins/insertions of longus capitis take precedence of those of longus colli. The 3-stage developmental sequence of longus colli illustrates what happens when one part of a muscle "blocks out" it neighbor. The vertical portion (S6–8) "spills" downward and centric. "Overflow" of myoblasts from somites 7–9 tracks upward in the *opposite* direction to attach at the skull base. Finally, the seemingly-absurd anatomy of the multi-layered epaxial muscles falls into a familiar (and understandable) neuromeric pattern. These muscles begin with deep-lying mono-neuromeric units (intertransverse-interspinalis) which take up available attachment sites along the midline. Additional poly-neuromeric muscles develop in progressively more superficial and lateral layers. The attachments of these muscles are determined by the principle of "first come, first serve." The origins and insertions of these later muscles utilize those sites "left over" after the synthesis of the predecessor muscles.

As one considers the organizational plan of these muscles, the most striking observation is the overall simplicity of the system. Muscles come from somitic PAM. These migrate depending upon their biologic maturity within the myotome. This is a function of their spatial location within the myotome, itself determined by homeotic genes that provide system of "coordinates" within the myotome, a sort of "street map," as it were. Axial bones come from somitic sclerotomes, whereas appendicular bones arise from lateral plate mesoderm. Each bone has a developmental sequence which is manifested as a series of collagen-II muscle binding sites laid out in spatial-temporal order. Musculoskeletal development, that is, the origin-insertion system, is a simple but elegant process in which spatiotemporal units of muscle are "matched up" with spatial–temporal system of bone. In summation, mesodermal units of muscle and bone in genetic register with one another constitute a unique and elegant system of self-assembly.

Angiology of the Neck

Introduction: Arteries of the Neck Don't Seem Segmental…But They Are

In this, our fourth and final section, we will take on the arterial system of the neck. At first glance, this anatomy seems very different from that of the head. Instead of individual arteries assigned to an organized series of pharyngeal arches what we observe is a series of long vessels extending upward from the subclavian artery in the chest. These vessels give off branches to multiple neuromeric levels. Our goal is to understand how and why the seemingly random arrangement of these arteries develops according to a Cartesian plane.

Let's begin with a simple model of neck artery development. We place a transparent graph over a segment of human embryo from neuromeres c1 to c8. Our embryo is oriented in the vertical position. The y-axis is midline of the embryo, that is, the CNS. The x-axis is remainder of the embryo away from the midline, that is, all non-CNS tissues. We start out with a simple blood supply consisting of paired longitudinal tubes (dorsal aortae) connected to a heart. From these tubes, a series of *primary horizontal segmental arteries* project medially, one for each neuromere. The resulting system looks like a ladder with the neuromeres supplied by individual rungs. Now, let's make our embryo more complex. We will add on additional tissues on either side of the neuraxis. Supporting arteries for these tissues extend outward from the rungs of our previous "ladder." These *secondary horizontal arteries* supply tissues that are genetically in register with the same neuromere. Now, let's link up our secondary segmental arteries using a series of *vertical interconnecting arteries*. The original ladder now looks like a lattice.

Let's now focus on the lowest arterial "rung" in the mammalian neck: the seventh intersegmental artery. This vessel runs between the seventh cervical vertebra and the first thoracic vertebra. It is destined to become the subclavian artery. Because it is positioned at the cervicothoracic junction, this artery will ultimately supply the upper limb. Prior to entry into the limb bud, the individual segments of the vertical interconnecting arteries are now fused into longitudinal *axial arteries* arising from the subclavian. Some ascend to supply the neck (vertebral, thyrocervical, and costocervical. Others descend to supply the sternum and chest (internal thoracic) or the muscles controlling the pectoral girdle (thoracoacromial and long thoracic). Finally, let's fuse the dorsal aortae such that they no longer supply the hindbrain and neck. The brain and peripheral tissues are now dependent on the axial arteries, each of which maintains original longitudinal segmental branches to specified tissues.

So that's our model of cervical blood supply. Now, let's look at the evolution of this system in real-time, using staged human embryo dissected and drawn by Dorcas Padget. Obviously, we are reworking the material discussed in Chaps. 7 and 8 on the vascular system. But the neck is sufficiently different in design that it makes sense to consider her work anew. Embryonic events involving the aortic arch arteries and the pharyngeal arches are rehashed…but from the perspective of tissues from c1 and below.

The embryology of the cervical arteries will be a "thrice told tale." First, we're going to review in broad brush strokes the sequence of events by which craniofacial blood supply is set up. This will include the "inside story" about how the heart and dorsal aortae are originally connected. Knowing this, the true identity of the first aortic arch artery will be revealed …are you getting curious? We will reinforce our narrative using a timetable of events relevant to the neck organized by Carnegie stages. Our second task is to develop a visual appreciation of vascular development in staged human embryos as depicted by Padget. In particular, we will focus on the embryogenesis of the vertebral artery because this model applies to all other derivatives of the subclavian. Our third iteration of this story explores the embryologic rationale behind these events. The mammalian neck is both a conduit and a container. Structures related initially to the pharyngeal arches, such as the heart and thymus, pass through it into the chest. Contained within the neck are the transition zone between pharynx and esophagus, endocrine glands (thyroid and parathyroid), the immune system (thymus), and the take-off of the respiratory system. As these structures emerge, the grid-like vascular plan of the primitive state (based strictly on neuromeres) gives way to a very different arrangement. Understanding the how and why of this process is our final goal. This requires careful study of how the great vessels develop, and in particular, the embryologic relationships that exist between them, the aortic outflow tract, and the aortic arch arteries. Background preparation for this story will involve a brief sojourn through relevant aspects of cardiac development.

In sum, the initial vascular system of the neck is simplistic and Cartesian. During the mid to late embryonic period (stages 12–23) rapid growth and differentiation force changes in the vascular system that result in an adult arrangement very different in form, but ultimately segmental in function (Figs. 10.123, 10.124, 10.125, 10.126, 10.127, 10.128, 10.129, 10.130, 10.131, 10.132).

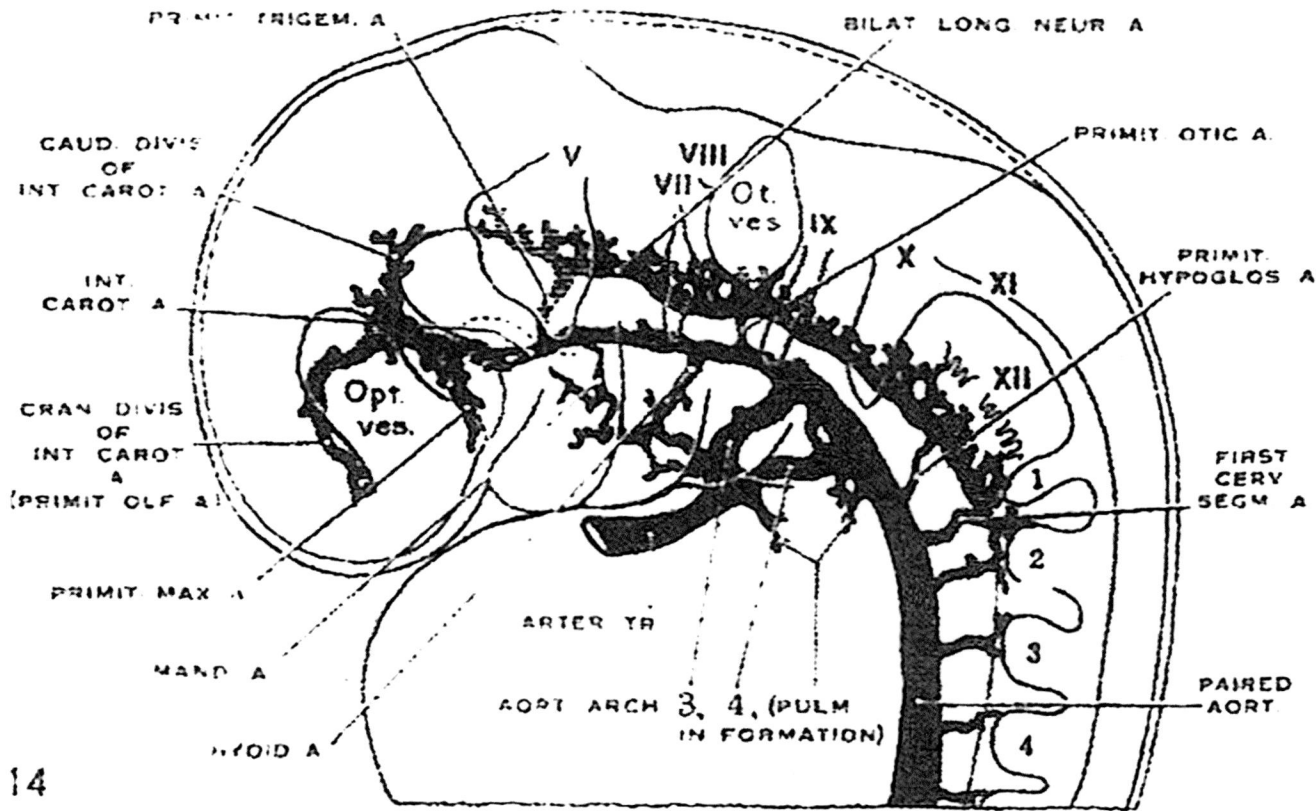

Fig. 10.123 (DHP 14). Stage 13, early Embryo 588a (4 mm) depicts baseline vascular relationships. This embryo (the first of Padget's series) has fully formed aortic arches 3 and 4. Aortic arch arteries 1 and 2 have involuted; the remnants of first and second pharyngeal arches are supplied by a tangle of vessels connected to the outflow tract. Brain circulation is as follows. The hindbrain is fed via the primitive trigeminal artery, located just above the first pharyngeal arch. This vessel marries up the dorsal aortae with the longitudinal neural artery. Note that the caudal end of the longitudinal neural artery is fed by the primitive hypoglossal and first cervical segmental arteries. The dorsal aortae *cranial to the first aortic arch artery* acquire a new name: the internal carotid arteries. These supply the midbrain and forebrain. This embryo demonstrates the earliest stages in the involution of the first two aortic arch arteries. AA1 and AA2 are relatively intact here. The tangle of vessels within the arches that will become the ventral pharyngeal artery is not yet present. Cervical somites are supplied by segmental arteries. During stage 13 the dorsal aortae fuse to form a median vessel, the descending aorta as seen in the adult. The unified dorsal aortae give rise to three groups of arteries. (1) Unpaired, ventral, visceral (splanchnic) branches supply *visceral lateral plate mesoderm*. The celiac, superior mesenteric, and inferior mesenteric arteries supply the gut. (2) Paired, lateral, visceral (splanchnic) branches supply *intermediate mesoderm*. These include the suprarenal, renal and gonadal arteries to retroperitoneal structures. (3) Dorsolateral, parietal (somatic) branches supply *paraxial mesoderm* and *parietal lateral plate mesoderm*.

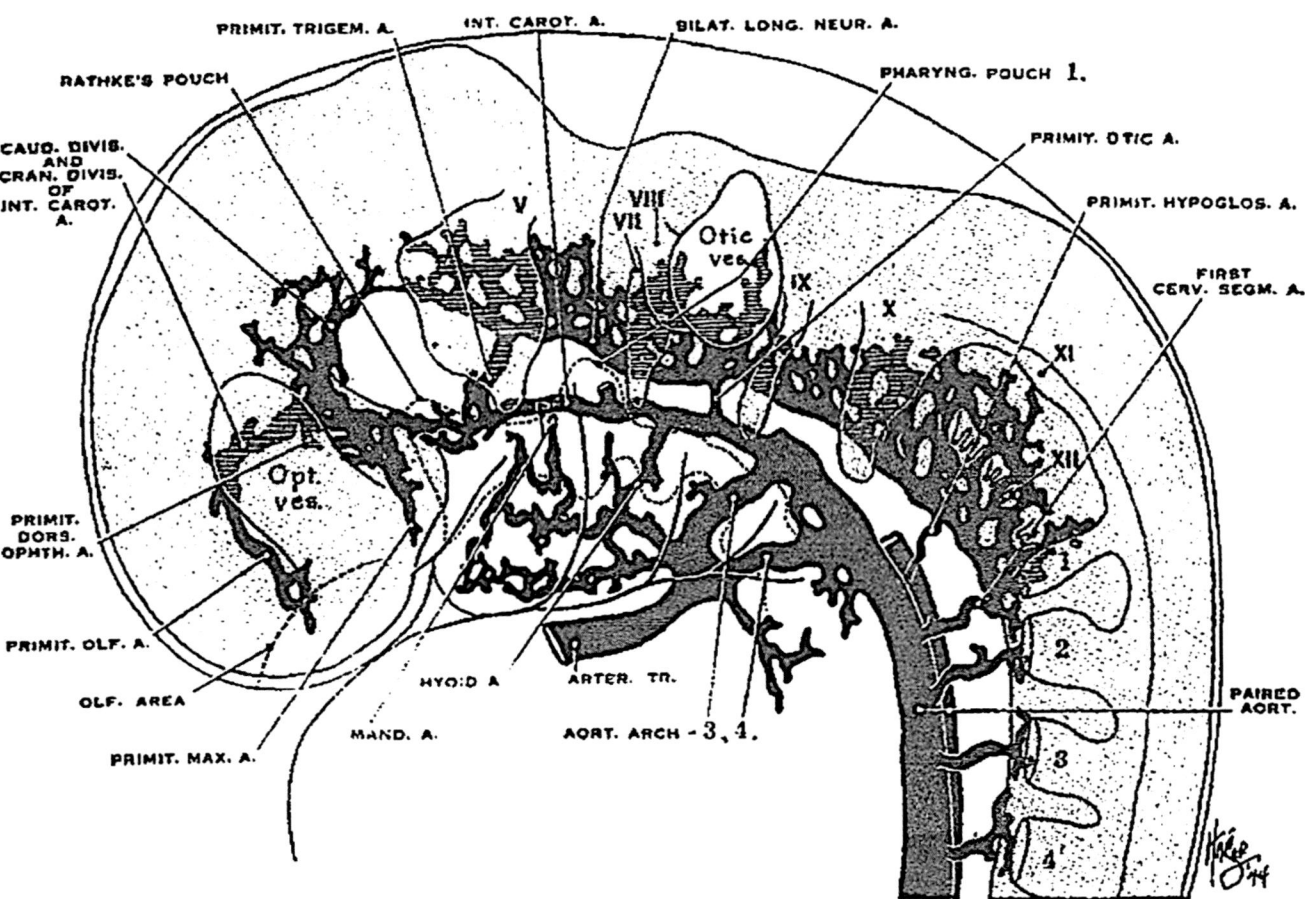

Fig. 10.124 (DHP 1a) Stage 13, late Embryo 588b (4 mm) is slightly later in development. Within pharyngeal arches 1 and 2 are vessels that will be the forerunners of the external carotid system. These connect up with the cardiac outflow tract at the junction of aortic arches 3 and 4. The former insertions of AA1 and AA2 into the dorsal aortae are marked by two remnants vessels hanging down from the dorsal aortae: the mandibular artery and the hyoid artery. The caudal end of each longitudinal neural artery is clearly supplied by the first cervical segmental artery from its respective, as-yet unfused, dorsal aorta. Because the first and second aortic arches are now interrupted, blood flow forward from the heart to the brain is entirely dependent on the third aortic arch artery. For this reason, the portion of the dorsal aortae extending forward from AA3 is now defined as the primitive internal carotid arteries. AA4 is plugged into the dorsal aorta just behind AA3. Blood flow backward into the remainder of the embryo takes place through AA4

PLATE 1

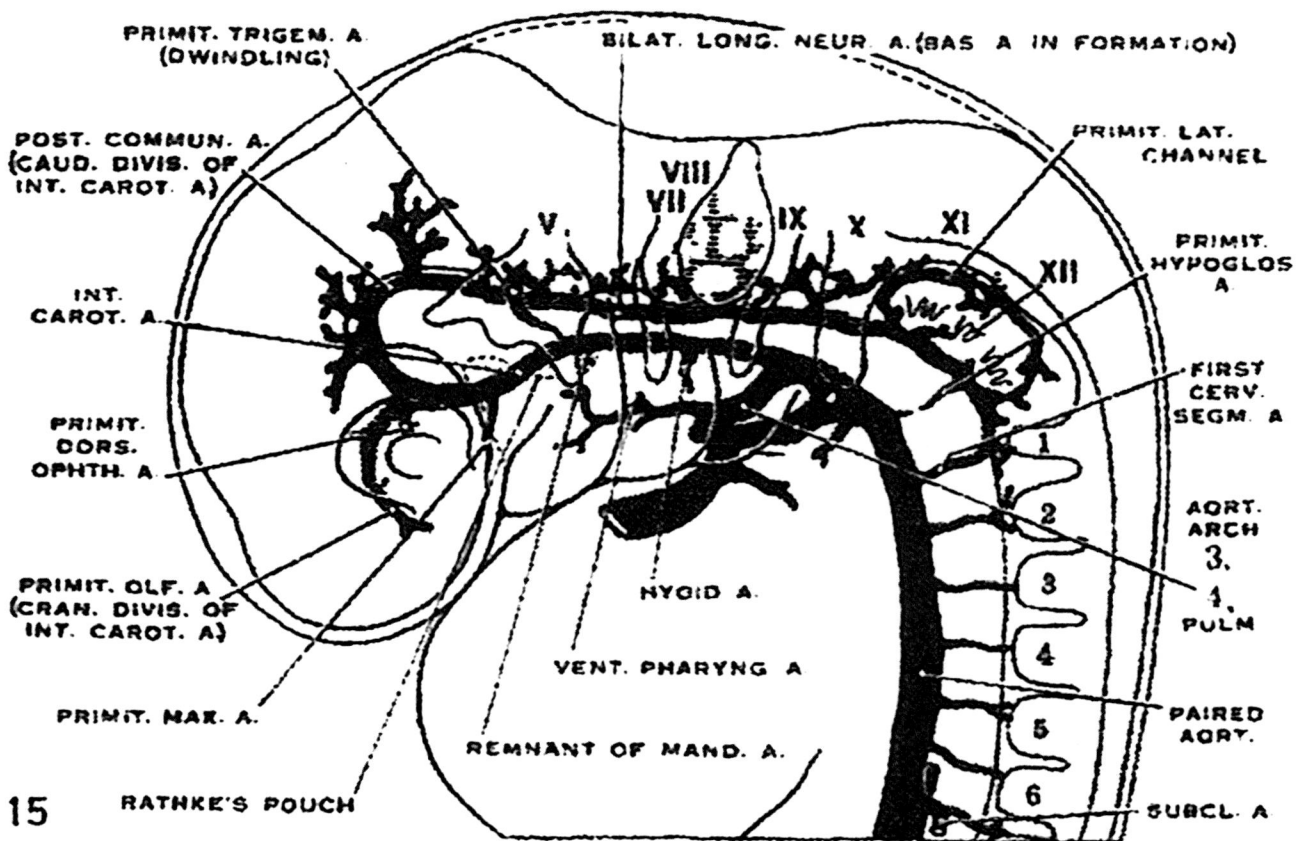

Fig. 10.125 (DHP 15) Stage 14, early Embryo 2841 (5.3 mm) demonstrates further changes in the arterial supply to pharyngeal arches. What were formerly a tangle of vessels within the arches have now consolidated into the ventral pharyngeal artery. Although the proximal take-off of the VPA is unclear, it extends as far forward as the downgrowth of V3. The *hyoid artery* (future stem of the stapedial system) is prominent.

Backward anastomosis between the caudal division of internal carotid and the longitudinal neural arteries marks the completion of brain circulation. As a consequence, smaller interconnecting arteries (primitive trigeminal and primitive hypoglossal), formerly essential for supplying the longitudinal neural artery to the hindbrain, have now involuted.

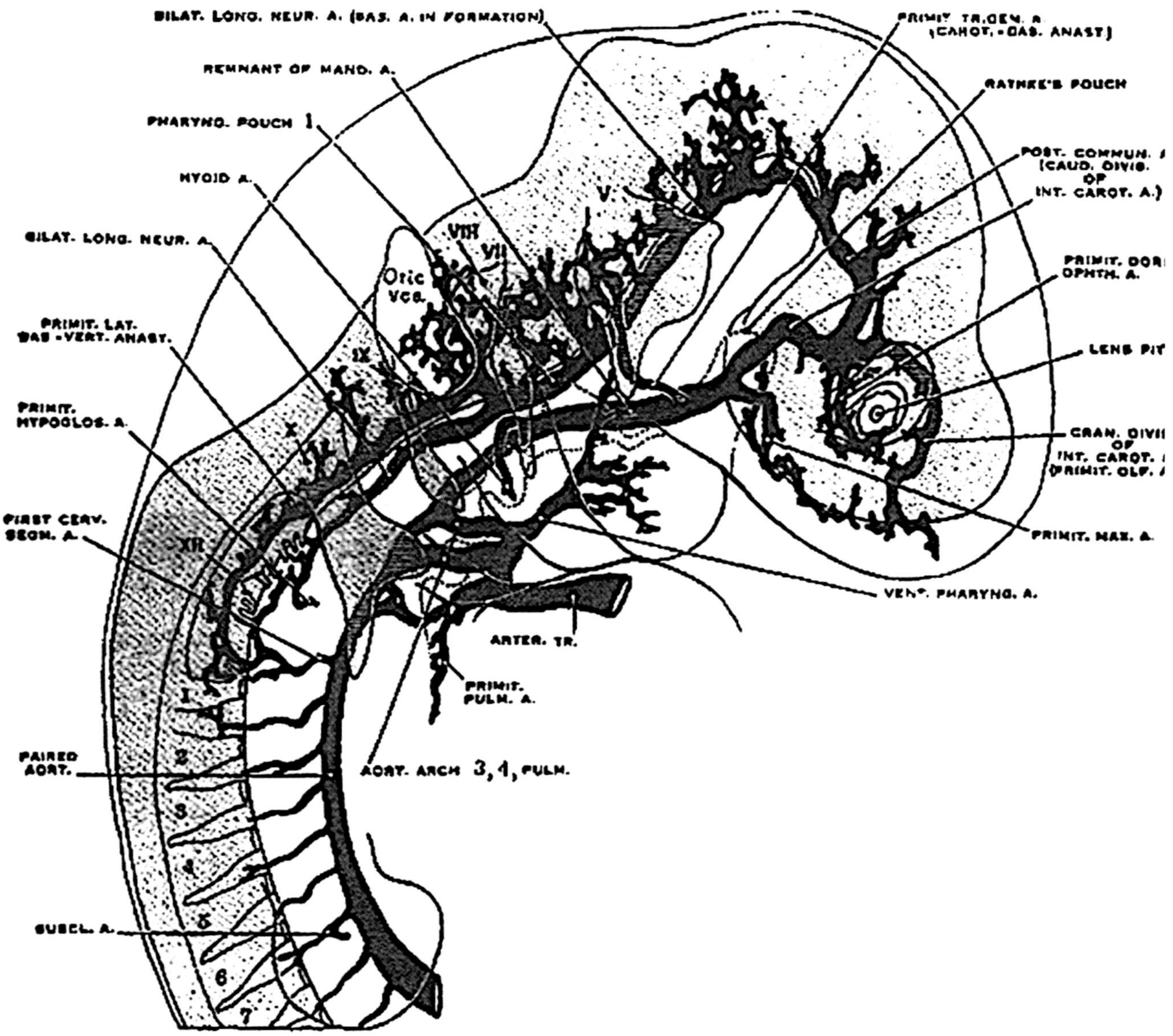

Fig. 10.126 (DHP 3a) Stage 14, late Embryo 3960 (5.5 mm) shows us more clearly how the stem of the external carotid does not take off from the third aortic arch per se. Instead, it extends forward from a segment of outflow tract *common to both* AA3 and A4. The sixth aortic arch artery, now briefly seen, has given rise to the primitive pulmonary artery. Cervical blood supply is still strictly a function of the segmental system. These arteries are clearly intersegmental. The first cervical intersegmental lies dorsal to the first cervical nerve. It supplies both the caudal termination of the longitudinal neural arteries (the anatomic boundary of the unpaired basilar arteries) and the first cervical somite. From the seventh cervical intersegmental artery, the subclavian stem of the subclavian artery is now seen. Aortic arch artery 6, often termed the pulmonary arch, arises from two sources. A *ventral stem* arises at stage 13 from the aortic sac while a *dorsal stem* arises at stage 14 from the dorsal aorta. When these unite they connect the parent structures; no intervening pharyngeal arch is present. *There is no such thing as the sixth pharyngeal arch.* At stage 14 three important structures arise from the structures encoded to r8–r9. Lung buds (endoderm) project into lateral plate mesoderm. The trachea and the esophagus differentiate with lateral plate mesoderm. A rich blood supply develops within LPM, from which arises the *pulmonary vascular plexus*. It explains why bronchial vessels are supplied separately from the pulmonary arteries per se. Once the 6th (pulmonary) arch forms, it sends out the true pulmonary artery to connect to the pulmonary vascular plexus

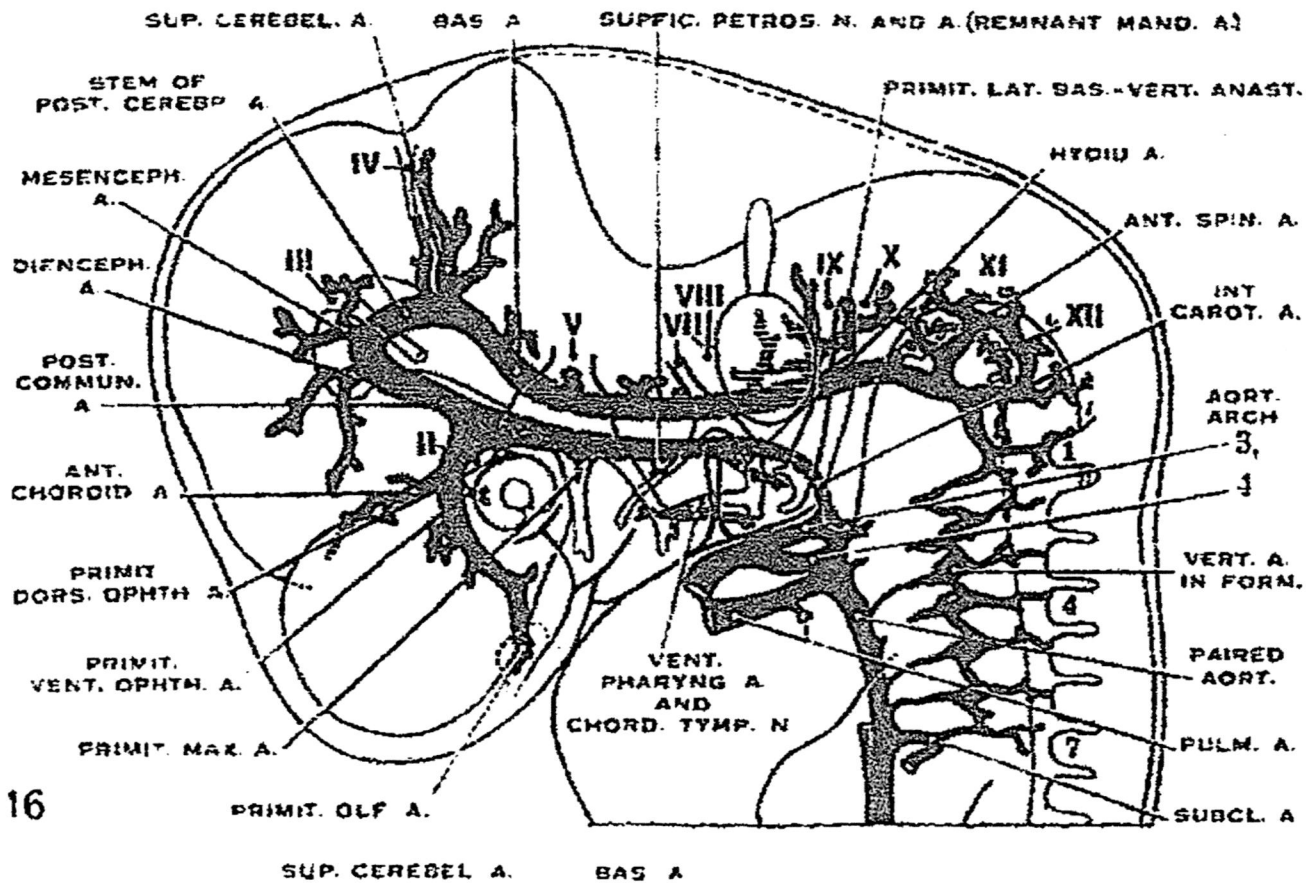

Fig. 10.127 (DHP 16). Stage 15 Embryo 163 (9 mm) demonstrates characteristic cerebral hemispheres. Transverse anastomoses have formed uniting all seven cervical segmental arteries. This is the precursor state of the vertebral artery. At their caudal termination the longitudinal neural arteries are still unpaired. Although we know that the vertebrals will eventually anastomose with the LNAs, blood flow to the hindbrain in stage 15 is still predominantly from the internal carotid. The subclavian is now prominent. It clearly relates to the paired dorsal aorta

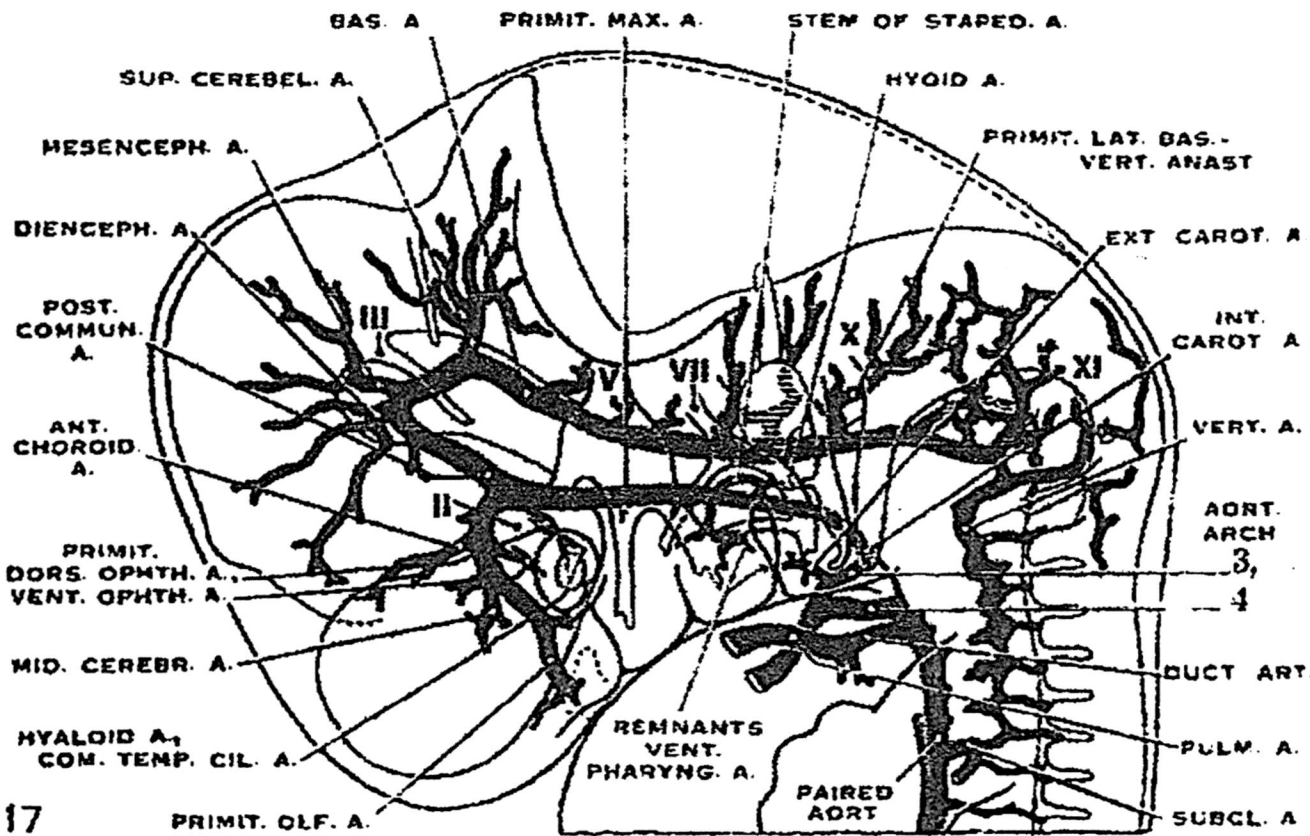

Fig. 10.128 (DHP 17) Stage 16 Embryo 1771 (12.5 mm) marks the end of the pharyngeal arch phase. *The stapedial system is developing from the hyoid artery.* Intracranial stapedial provides circulation to the meninges and the orbit and medial nasopharynx. Extracranial stapedial becomes a distal "add-on" to external carotid, the internal maxillo-mandicular artery. Its branches supply the lateral nasopharynx, oropharynx, and face. The heart has not yet descended into the chest. The future vertebral artery is irregular but clearly seen, its ventral connections to the dorsal aortae have involuted, save at the level of the seventh intersegmental artery. Here, the take-off of the subclavian artery is seen. This stage is transitional in higher vertebrates from a cerebral circulation exclusively supplied by the carotid to the later iteration in which hindbrain is supplied by the vertebrals

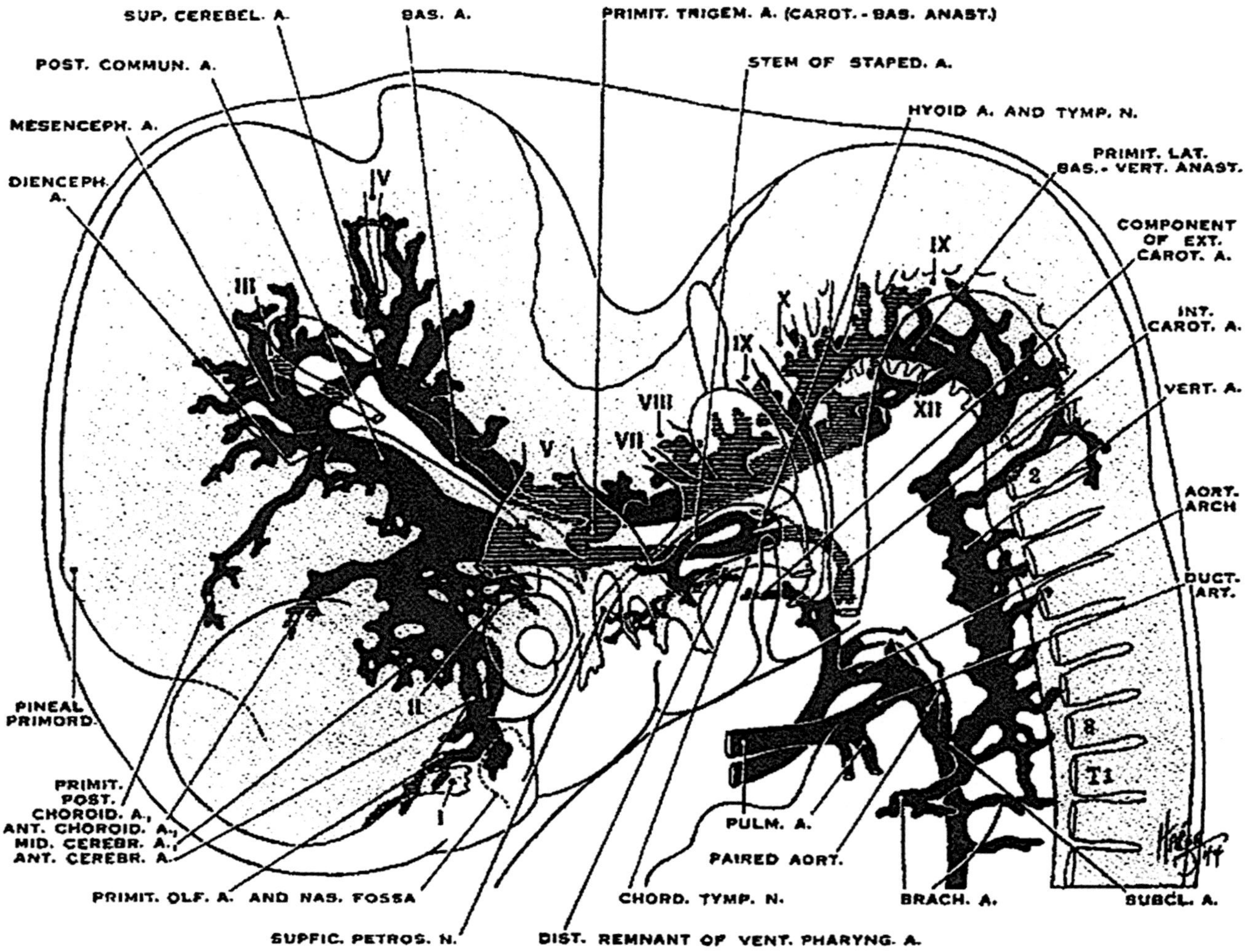

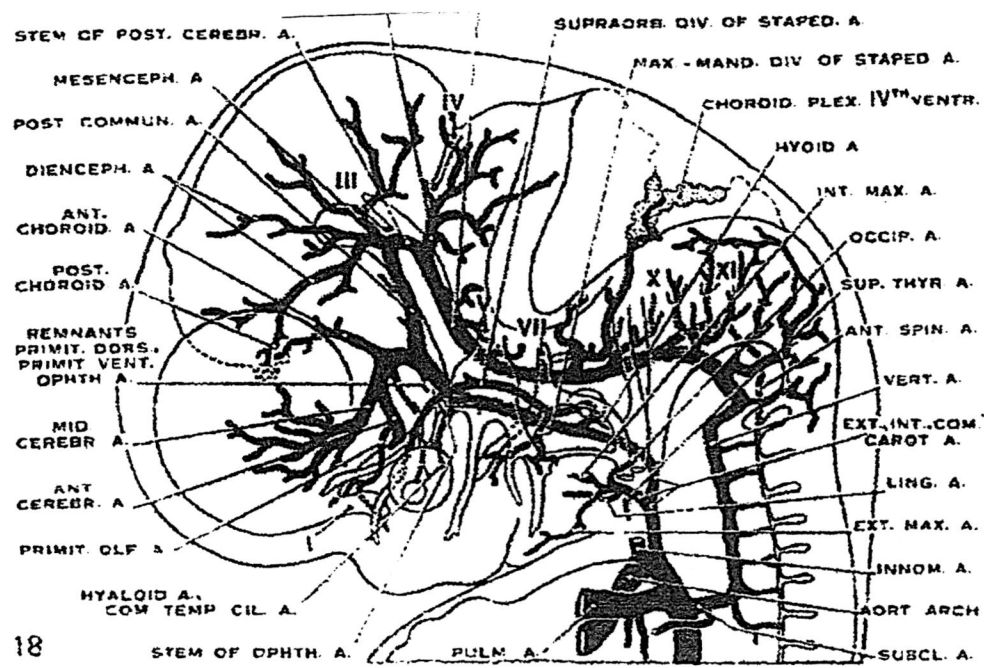

Fig. 10.129 (DHP 6) Stages 17–18 Embryo 940 (14 mm) demonstrates *annexation of the distal ventral pharyngeal* artery (external carotid) by the extracranial stapedial at the level of V3

Fig. 10.130 (DHP 18) Stages 19–20 Embryo 1390 (18 mm) shows adult form for the external carotid artery and most cerebral arteries. *The stapedial system now has orbital, maxillary, and mandibular divisions.* Subclavian now arises opposite the pulmonary arch. It is the sole source of vertebral artery. Thryocervical trunk present as well.

Fig. 10.131 (DHP 19) Stages 21–22 Embryo 632 (24 mm) has *secondary anastomoses of the stapedial artery*: StV1 orbital to the primitive ophthalmic to form definitive (hybrid) ophthalmic artery and extracranial StV2 internal maxillary to the intracranial StV2 to form middle meningeal artery. Subclavian has the full complement of branches: vertebral, thryocervical trunk, internal mammary, *costocervical trunk is not shown but is developmentally present*, and axillary (brachial) artery

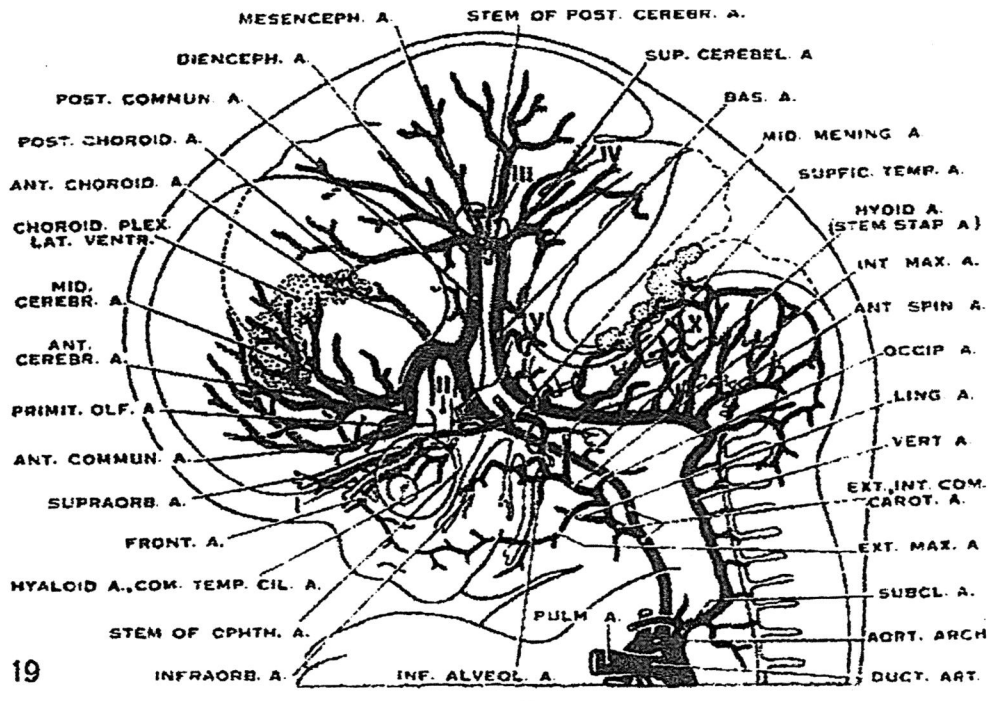

Fig. 10.132 (DHP 20) Stage 23+ Embryo 886 (43 mm) shows posterior communicating artery reversing its position from caudal-cranial 180 degrees to cranial-caudal. Stems of both inferior cerebellar arteries (anterior and posterior) are visible and dead-end in the choroid plexus of fourth ventricle. Choroid plexus of lateral ventricle supplied by anterior and posterior choroidal aa. Lacrimal branch (representing the StV1–StV2 supraorbital artery) connects with ophthalmic

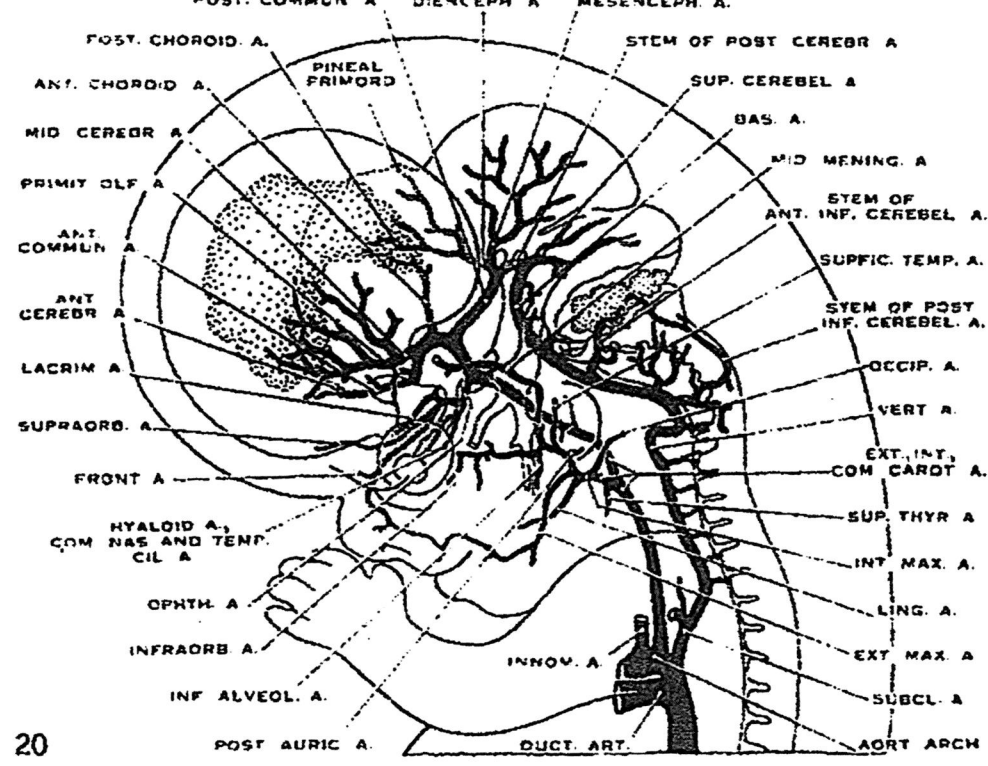

Beginning of Embryonic Circulation

In its most primitive state, the unflexed embryo has four vascular systems that develop in situ. *Vitelline vessels* arising from the yolk sac are extraembryonic. Within the embryo, the embryonic CNS (future hindbrain) is supplied by paired *primitive hindbrain channels* (PHC) while the non-CNS embryo is nourished by paired *dorsal aortae* (DA) running its entire length. The dorsal aortae develop from lateral plate mesoderm; they are situated peripherally to the paraxial mesoderm of the somitomeres and somites. At their anterior terminus, each dorsal aorta is plugged into the footplates of a horseshoe-shaped zone of *cardiogenic mesoderm.*

All other arterial systems in the embryo develop by induction and budding. To keep ourselves oriented, let's enumerate these important secondary systems in the order of their appearance.

- *Aortic arch arteries* AA1 is merely a stretched-out continuation between the ventrally-flexed heart tubes and the dorsal aortae. The six aortic arches appear between stages 9 and 15.
- *Primitive internal carotid artery* sprouting from the genu where the dorsal aortae bend downward as the first aortic arch arteries. Primitive ICA is defined as extending cranial from AA1.
- *Transverse interconnecting arteries* between the outboard dorsal aortae and the primitive hindbrain channels: *primitive trigeminal, primitive otic, primitive hypoglossal.*
- Reconfiguration and renaming of the PHCs into the *longitudinal neural arteries.*
- Breakdown of aortic arch arteries within pharyngeal arches leading to the development of the *ventral pharyngeal artery* (VPA) or *primitive external carotid.*
- Degeneration of DA between AA3 and AA4 leads to definitive *internal carotid artery.*
- Hyoid artery > *stapedial artery system* programmed by cranial nerves:
  - *intracranial stapedial* programmed by cranial nerves > meningeal arteries,
  - *extracranial stapedial* programmed by chorda tympani > internal maxilla- mandibular artery and its branches.
- Fusion of IMMA with VPA > definitive *external carotid artery.*
- Degeneration of stapedial stem > anastomoses.
  - StV1 with internal carotid > ophthalmic artery;
  - StV2 with external carotid > middle meningeal a.; other meningeal aa "originate" from ECA.

Early development of the heart takes place in the visceral layer of lateral plate mesoderm. Three phases are described. During the *plexiform phase* (stage 9) LPM$_V$ gives rise to an endothelial plexus; this is quickly surrounded by a myocar-

dial mantle. The primitive heart is non-contractile. It consists of paired heart tubes linked distally in the form of an inverted U-shaped loop. Proximally the loop is continuous with the dorsal aortae. The connection between the future aortic sac and the dorsal aortae is Y-shaped. Distally, the future sinus venosus is continuous with the vitelline vessels. In the *straight tubular heart phase* (stage 10) the endocardial plexus organizes into a single tube. It has a linear fate map, the segments of which are (from proximal to distal): aortic sac, truncus arteriosus, conus cordis, RV, LV, LA, RA, and sinus venosus. Three sets of feeding vessels plug into the sinus venosus: cardinal, omphalo-mesenteric, and umbilical vein. The *looped heart phase* (stages 10–11) takes place during flexion of the embryo. Anatomic fates are assigned to sectors of the heart. The distal loop assumes an atrial fate; it receives the future gut vessels. More proximal segments form the ventricles and the cardiac outflow tract.

During the pharyngeal arch phase, the embryonic heart lies below the pharynx. From the aortic outflow tract, a succession of six *aortic arch arteries* appears. The first five develop within the central core of the five pharyngeal arches; they connect the now-beating embryonic heart with the overlying dorsal aortae. The sixth aortic arches form the pulmonary circulation. Note that the development of the first aortic arch artery is unique. It does not "sprout" from the aortic sac. When the embryonic flexion carries the heart ventrally, the aortic sac pulls the dorsal aortae downward: a 90-degree bend is created at the fourth somitomeres, that is, at rhombomeres 2 and 3. From this site sprout: (1) the primitive internal carotid artery and (2) the primitive trigeminal artery. The *first aortic arch artery* = the vertical segment of dorsal aorta spanning from the Sm4 downward to the aortic sac.

Note that at no point in development are all aortic arch arteries present at the same time. They form and involute in strict sequence. Despite this instability, blood flow from heart to embryo is always maintained. Just when AA1 and AA2 are falling apart, AA3 and AA4 are plugged into the DAs. AA5 has a short half-life. It bequeaths its vascular territory to AA4. As the heart descends into the chest, *ventral remnants* of aortic arch arteries 1–4 reorganize into *ventral pharyngeal artery.* VPA forms the external carotid system up to the external maxilla-mandibular (facial) artery. The *dorsal remnants* of the aortic arch arteries also have important roles. For our purposes, the key player is hyoid artery of the second aortic arch. It gives rise to the stapedial arterial system, the intracranial branches of which supply the dura while the extracranial branches form the internal maxilla-mandibular artery. Thus, remodeling of aortic arches 1–4 supplies all derivatives of pharyngeal arches 1–5.

Blood supply to the remainder of the embryo is organized along very different lines. From the paired dorsal aortae segmental vessels are directed upward to supply each neuromeric zone of the embryo. The first pair of such vessels is the *primitive hypoglossal arteries*. These connect the DAs to the longitudinal neural arteries supplying r8–r11. Immediately caudal are the *segmental (cervical) arteries*. From this point backward, each somite is supplied by a segmental artery, originating from the ipsilateral dorsal aorta. Obviously, this situation is unstable. The heart will descend into the chest, the great vessels will arise and the dorsal aortae will fuse. In response to these events, the anatomic basis of blood supply to the neck will also change. We will follow this developmental sequence in staged human embryos as per Padget.

Timetable of Arterial Development by Stages

The *aortic arch artery period* lasts from stage 9 to stage 14. One arch develops per stage.

Stage 9 (first aortic arch artery): This stage is marked by the appearance of the first three occipital somites and the formation of heart tubes from lateral plate mesoderm (visceral lamina). First aortic arch artery is present at this stage.

Stage 10 (first pharyngeal arch, second aortic arch artery): A single tubular heart is present, functional embryonic circulation is established and the pharynx develops. The embryo is flexed, the heart is now ventral and the heartbeat is initiated. The second aortic arch artery is present. The heart is now connected to the remainder of the embryo via aortic arch arteries 1 and 2; and these arteries course upward through the cores of their respective pharyngeal arches to reach the dorsal aortae.

Stage 11 (second pharyngeal arch, third aortic arch artery): The oropharyngeal membrane, located at the boundary between pharyngeal arches 2 and 3, ruptures. The OPM also marks the boundary between the ectoderm of the oronasopharynx and the endoderm of the pharynx proper.

Stage 12 (third pharyngeal arch, fourth aortic arch artery) dorsal aorta begins to fuse with the process continued through stage.

Stage 13 (fourth pharyngeal arch, fifth aortic arch artery, involutes in tetrapods) ventral sprout of pulmonary artery from AA4, dorsal aorta begin to fuse distal to it.

Stage 14 (fifth pharyngeal arch, sixth aortic arch artery) dorsal sprout of pulmonary artery arises from dorsal aortae. Union of the two sprouts becomes AA6. It has nothing to do with a pharyngeal arch. Pulmonary artery appears to arise from AA6 but, in actual fact, is biologically correlated with AA4, that is, r8. Transverse branches from paired aortae supply the segment.

Stage 15 Longitudinal neural arteries fuse together from r1 to r8 to form the basilar. Interconnections form linking the segmental aortic branches. These will unite to form the vertebrals.

Stage 16 subclavian artery defined. Vertebrals now connect the subclavian to the cerebral circulation.

Stage 17 Brachiocephalic trunk and left common carotid, stapedial artery system, fusion of dorsal aortae is now complete.

Stage 18 and *Stage 19*.

Stage 20 Descent of the thymus from r8 into chest.

Stage 21 and *Stage 22*.

Stage 23 Vascularization of the thymus by internal thoracic artery.

During the aortic arch artery period the heart lies cephalad to the occipital somites, that is, cranial to r8. During stage 12, after the formation of AA4, it descends to the level of the somites r8–r11. By the eighth week of development, the heart lies at the mid-thorax.

Cervical Artery Development in Stages

The following illustrations from Padget's work demonstrate the time course of vessel development. These are based on dissections from the original Carnegie Embryo Collection. The reader is referred to Figs. 10.123, 10.124, 10.125, 10.126, 10.127, 10.128, 10.129, 10.130, 10.131, 10.132 for a stage-specific presentation of the cranial arterial development sequence.

The Carnegie collection along with contributions from many other sources has been made digital under the direction of Dr. Mark Hill at University of New South Wales http://human-embryology.org

Readers can readily access the collections: https://human-embryology.org/wiki/Carnegie_Collection

Arteries of the Neck

Arteries supplying the structures of the neck originate from two sources. Fourth and fifth pharyngeal arch structures involving the larynx are supplied from the most caudal branch of the external carotid system, the superior thyroid artery. Recall that this vessel originates from the stem of the fourth aortic arch artery; it is the right and proper supply of the fourth pharyngeal arch. However, due to the untimely death of the fifth aortic arch artery, AA4 is forced to assume the duties of supplying the fifth pharyngeal arch. Given its embryologic assignment, superior thyroid terminates within the larynx. It does not relate to the musculature of the lower pharynx, specifically to middle constrictor and below. These structures are supplied by the inferior thyroid artery.

Subclavian Artery

Subclavian artery develops from the seventh cervical inter-segmental artery. It is located at the genetic take-off of the upper extremity. The four arterial stems of the subclavian are as follows: vertebral, thyrocervical trunk, internal thoracic, and costocervical trunk. These vessels likely develop in a proximal to distal sequence.

With regard to the neck and pectoral girdle, the branches of subclavian supply: (1) structures within the neck up to the territory of fourth arch (external carotid), (2) the pectoral girdle including the rostral, medial, and rostro-lateral scapula, and (3) structures connecting the neck with the pectoral girdle (Fig. 10.133).

Vertebral Artery The first stem of the subclavian arises from the first part (medial to scalenus anterior). It avoids the transverse process of the seventh cervical vertebra but at level C6 and above it passes through foramina in the transverse processes. It has four parts: (1) From its stem to C6, (2) from C6 to C2, (3) arising medial to rectus capitus lateralis,

it curves backward and medially around the lateral mass of C1.

- *Spinal branches* enter via intervertebral foramina. They supply the spinal cord and its meninges. Paired lateral chains lie above the posterior surface of the vertebral bodies just at the attachment of the pedicles.
- *Muscular branches* are assigned to the muscles of the suboccipital triangle. Muscles arranged along the ascent of the vertebral artery are supplied from other sources.

Internal Thoracic Artery The second stem of the subclavian might seem very far afield of the neck but several of its branches have interesting developmental features. First off, internal thoracic exits from the first portion of subclavian directly opposite the thyrocervical trunk. As we discussed previously, internal thoracic is the mere continuation of a longitudinal union of a series of segmental vessels. The mesenchymal targets of these vessels include structures derived from both paraxial mesoderm and lateral plate mesoderm.

Fig. 10.133 Subclavian artery. Branches supply (1) structures within the neck up to the territory of fourth arch (external carotid), (2) the pectoral girdle including the rostral, medial and rostro-lateral scapula, and (3) structures connecting the neck with the pectoral girdle. [Reprinted from Lewis, Warren H (ed). Gray's Anatomy of the Human Body, 20th American Edition. Philadelphia, PA: Lea & Febiger, 1918]

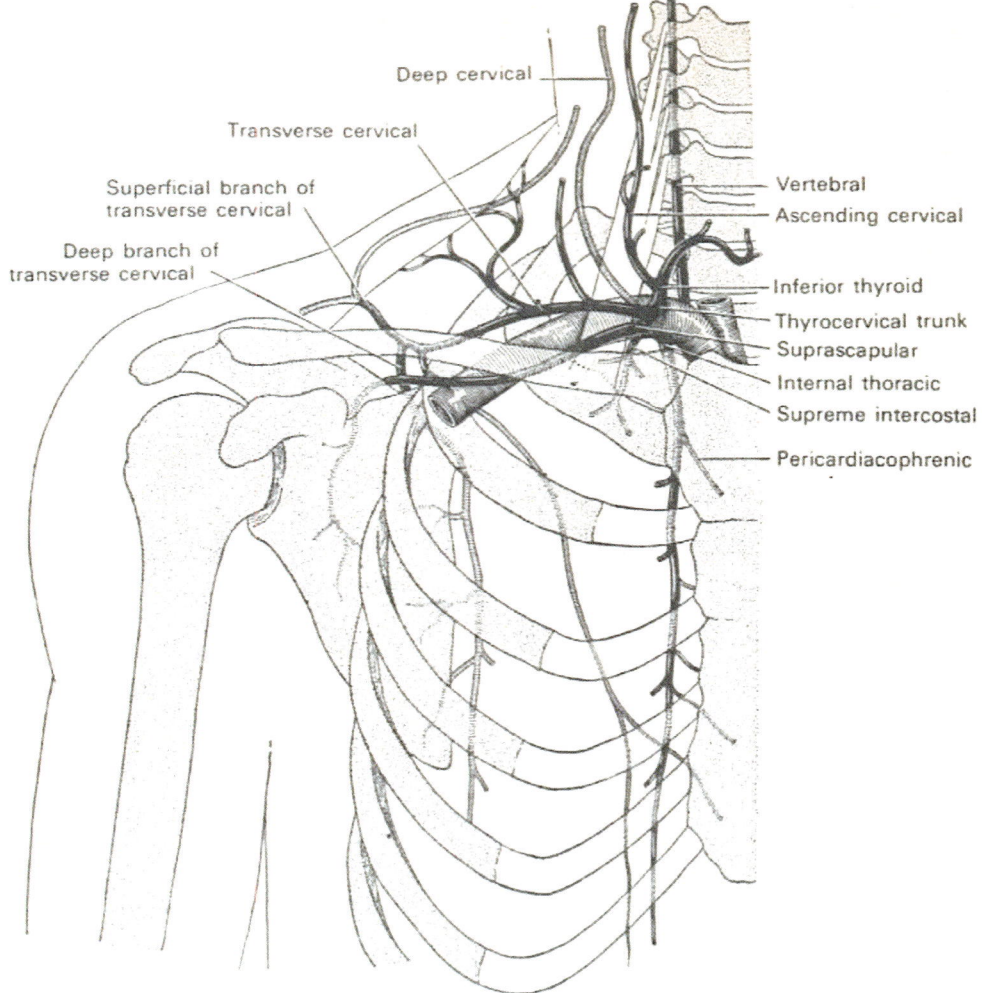

- *Pericardiophrenic artery* is the most proximal branch of internal thoracic. It accompanies (and supplies) the phrenic nerve (C4) into the chest, coursing in between pleura and pericardium. The artery is slender; one should consider it as primarily a means to support the motor nerve.
- *Mediastinal branches* are distributed anteriorly to supply the parietal (sternocostal) surface of pericardium and regional lymph nodes.
- *Thymic branches* remind us of the embryologic location of this gland. Thymus arises from the endoderm of the caudal portion of the third pharyngeal pouch, that is, between the third and the fourth arches. That places it at level r8. Primordial thymus contains two different types of tissue. The rostral part of pouch 3 produces medullary duct epithelium while caudal pouch 3 may be the site of reticular epithelium. Neural crest cells are responsible for the connective tissue framework of the thymus. They may also be responsible for inducing the pharyngeal pouch epithelium to produce and maintain the differentiation of lymphoid stem cells. At stage 20, lateral plate mesoderm penetrates the thymus, bringing in arterial supply. The thymic parenchyma becomes divided into lobules. Circulating stem cells from the bone marrow take up residence in the newly vascularized thymus. These cells produce: (1) thymocytes committed to T-cell lineage, and (2) dendritic cells. During the third trimester functionally mature thymocytes migrate from the thymus to populate peripheral lymphoid tissues permits circulating systemic dispersal of stem cells into the rest of the body. Note that thymic development cannot proceed in the absence of neural crest cells from levels corresponding to the third pouch, that is, from r7 and r8.
- *Bronchial branches* supply the lower trachea (lateral plate mesoderm) as well as neighboring bronchi.
- *Sternal branches* are six in number, the uppermost one supplies manubrium. The intermediate four branches are distributed to the four segments of the sternal body. The sixth and last branch supplies the xiphoid. [Recall that manubrium also is also supplied by suprasternal branch of suprascapular artery and by the clavicular branch of thoracoacromial artery. This explains why manubrium has six centers of ossification; three per side].
- *Anterior intercostal arteries* are also six in number. The remaining interspaces are supplied by the musculophrenic artery, the penultimate branch of internal thoracic. These arteries supply the intercostal muscles. At the midaxillary line, they anastomose with their counterpart posterior intercostal arteries from the aorta. Penetrating branches supply overlying pectoralis and breast tissue. The fourth intercostal is particularly important in that regard.
- *Perforating branches* pass through five–six interspaces at the sternal border to gain the subcutaneous plane. Here they travel laterally, accompanied by the anterior cutaneous branches of intercostal nerves. They initially run with pectoralis fascia just under the muscle. More laterally they become more superficial to supply the muscle itself and skin. Branches of the second, third, and fourth interspaces support the breast.
- *Musculophrenic artery* follows the posterior surfaces of ribs 8, 9, and 10. In like manner to the anerior intercostals it gives off branches to the seventh, eighth, and ninth interspaces. Terminal branches access the lower part of diaphragm.
- *Superior epigastric artery* is the terminal branch of internal thoracic. Its anastomosis with inferior epigastric demonstrates that the segmental neurovascular organization is continued all along the anterior abdominal wall. Internal thoracic follows the same principles as vertebral with multiple intersegmental branches linking up together longitudinally and then, due to changes in flow and the losing connection with the individual intersegmental vessels.

Thyrocervical Trunk The third stem of the subclavian, thyrocervical trunk arises from the first part of subclavian, just behind the medial edge of scalenus anterior. It provides branches to all neuromeric levels from c8 up to c1. It provides blood supply for structures lying *outboard to those supplied by the vertebral.*

Inferior Thyroid Inferior thyroid is the first artery of the thyrocervical trunk. Inferior thryoid tracks upward in front of the vertebral artery, swerves internally to run dorsal to carotid sheath, and then terminates in the thyroid gland at levels r8–r11. It has the following important branches.

- *Muscular branches* These supply infrahyoid, and three multi-segment muscles: longus colli (atlas to T3), scalenus anterior (the anterior tubercles of the third through sixth cervical vertebrae represent neuromeric levels c3–c7), and inferior constrictor (a Sm9 muscle attaching to the back of thyroid and cricoid cartilages).
- *Ascending cervical* This artery can arise in two locations: behind (dorsal) to carotid sheath from the parent artery or directly from the trunk. In any case, it skirts along the anterior aspect of the cervical vertebrae, at the level of the transverse processes. This places it between scalenus anterior and longus capitus. From this location, it sends spinal branches into the intervertebral foramina and thence into the vertebral canal. These provide segmental supply to the vertebral bodies and spinal cord *in exactly the same way* as the spinal branches given off by the vertebral artery.
- *Esophageal branches* These arteries provide segmental supply to the ventral aspect of the esophagus. They have anastomoses with segmental arteries to the dorsal esophagus arising from the aorta.

- *Tracheal branches* are distributed segmentally along the trachea. These anastomose inferiorly with bronchial arteries and superiorly with tracheal branches of superior thyroid artery (the arterial axis of the fourth and fifth pharyngeal arches). Note that bronchial arteries arise from the ventral side of thoracic aorta and from the upper intercostal arteries.
- *Inferior laryngeal artery* runs along the dorsal aspect of trachea to the dorsal larynx. In this position, it is covered by inferior constrictor. Recurrent laryngeal nerve travels along with it.

In sum: The developmental rationale of the inferior thyroid artery is to form a segmental system in parallel with the vertebral. It is responsible for structures within the core of the neck lying outboard to those supplied by the vertebral.

Suprascapular Artery The second artery of the thryocervical trunk, subscapular is the *first of three arterial systems dedicated to the pectoral girdle*. The arterial axis pursues a downward and lateral course from the thyrocervical trunk crossing in front of scalenus anterior and phrenic nerve. Sternocleidomastoid lies on top of it. It then travels underneath clavicle, lying atop subclavian artery and the brachia plexus. It can be found beneath the inferior belly of omohyoid. When it reaches the superior border of the scapula it passes backward *over* the transverse scapular ligament beneath which lies suprascapular nerve. The artery now tracks directly on the bone of the supraspinous fossa. It continues laterally around the neck of the scapula. It now passes through the great scapular notch *beneath* the inferior transverse ligament to gain access to the infraspinous fossa. Here it makes two anastomoses: (1) with transverse cervical artery via its descending scapular branch; and (2) with subscapular artery via its circumflex scapular branch.

- *Muscle branches* This artery supplies the inferior 1/3 of sternocleidomastoid, subclavius, supra- and infraspinatus. It represents the first of three angiosomes dedicated to the pectoral girdle, all of which arise from subclavian.
- *Suprasternal branch* curves over the sternal margin of clavicle to supply chest skin.
- *Acromial branch* supplies skin over the acromion, traversing trapezius in the process. Here it makes an anastomosis with the thoracoacromial artery via its own acromial branch. Recall that acromion has two ossification centers. It is reasonable to associate these with the two arterial systems. Suprascapular axis, arising more proximally from subclavian, would logically be assigned to the more medial ossification centers.
- *Articular branches* supply the joint between distal clavicle and scapula (thoracromial joint).

- *Osseous nutrient* branches supply the clavicle and scapula per se.

In sum: The developmental rationale of suprascapular artery is to supply structures related to the clavicle and rostral aspect of scapula. It supplies skin along the clavicle from manubrium as far lateral as acromion, sternocleidomastoid, omohyoid, and subclavius. It plays a supporting anastomotic role to supra and infraspinatus.

Transverse Cervical Artery The third artery of the thryocervical trunk, the course of transverse cervical runs parallels that of suprascapular artery, running above and behind it. It too crosses in front of scalenus anterior and phrenic nerve (deep to sternocleidomastoid). More laterally, it crosses in front of the trunks of brachial plexus. Just above it lies platysma, which it supplies. As it approaches the anterior border of trapezius it divides into two distinct branches, deep and superficial. It is an artery of considerable variation. Sometimes it is absent. In such cases, the regions supplied by the superficial and deep branches of TCA are served by two completely different arteries. Superficial TCA is replaced by superficial cervical artery and deep TCA is replaced by descending scapular artery. For this reason, in the descriptions below we'll have to refer to both situations.

Situation 1:

- *Superficial branch of TCA* runs deep to hypaxial trapezius and then divides. An ascending branch tracks up along trapezius to anastomose with occipital artery via its descending branch. Descending branch follows the course of spinal accessory under the trapezius.
- *Deep branch of TCA* pursues an epaxial course to supply levator scapulae and adjacent muscles. It reaches the superior angle of scapula and then runs downward along its medial (vertebral) border until reaching the inferior angle. It runs in company with dorsal scapular nerve (C5), which is motor to both levator and the rhomboids. It also supplies subscapularis and serratus posterior. Thus deep br. of TCA is the neuroangiosome of medial border of the scapula.

Situation 2:

- *Superficial cervical artery* arises directly from thyrocervical trunk and pursues a purely superficial course. There is no deep branch. It basically supplies the trapezius.
- *Descending scapular artery* is seen only when transverse cervical is absent. It arises independently from the third part of the subclavian. The structures it supplies are exactly the same as those supplied by the deep branch of TCA.

In sum: The developmental rationale of TCA is to supply the muscles of scapula from cranial to caudal and from medial to lateral: levator scapulae, rhomboids, on the dorsal side, supra and infraspinatus; on the ventral side, subscapularis and serratus posterior. These muscles connect scapula to chest wall and span from upper lateral scapula to the neck of humerus.

Costocervical trunk is the fourth stem of the subclavian. It arises from two different locations. On the left side comes off the first part of subclavian, just distal to thyrocervical trunk. On the right side, it comes from the second part of subclavian.

Supreme intercostal artery descends just in front of the neck of the first and second ribs. It gives off the first and second posterior intercostal arteries. This is compensation of an oddity of the aorta. The most superior intercostal branch of aorta supplies the third intercostal space. For developmental reasons, aorta is not capable of directly supporting the first two interspaces. Supreme intercostal to the rescue! It anastomoses with the more superior aortic branch.

Deep cervical artery is analogous to the dorsal branch of the aortic posterior intercostal artery. It ascends along the back of the neck between semispinalis capitis and semispinalis cervicis all the way up to the axis. Here are anastomoses with occipital artery (descending branch) and with branches of the vertebral.

Axillary Artery

The axillary artery represents the continuation of the subclavian into the upper extremity. It gives rise to five arterial stems: supreme thoracic, thoracoacromial, lateral thoracic, subscapular, and humeral circumflex (Fig. 10.134).

With regard to the pectoral girdle, the branches of axillary have a very limited role, supplying the following structures: (1) clavicle, (2) distal projections of the rostral scapula (acromion and the coracord process), (3) caudolateral scapula, and (4) muscles connecting pectoral girdle to the humerus and to the anterolateral chest wall.

Supreme Thoracic Artery Renowned anatomist Richard Snell at George Washington University described this structure rather pithily as "supremely unimportant."

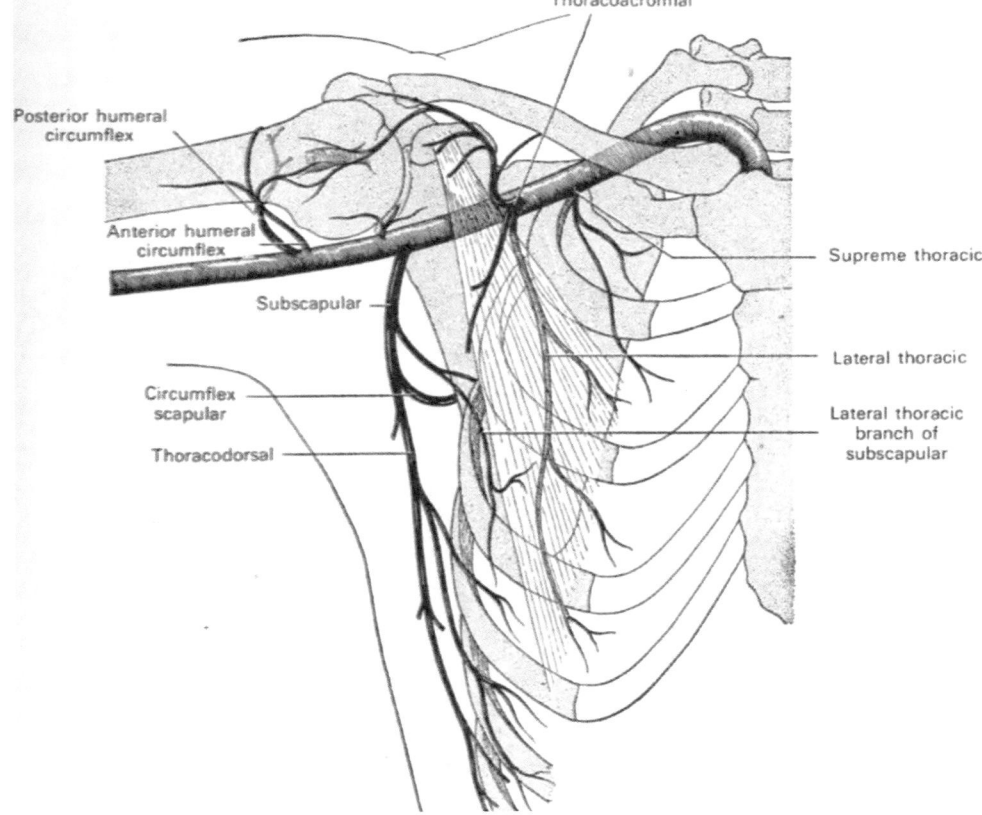

Fig. 10.134 Axillary artery. With regard to the pectoral girdle, the branches of axillary have a very limited role, supplying the following structures: (1) clavicle, (2) distal projections of the rostral scapula (acromion and the coracord process), (3) caudo-lateral scapula, and (4) muscles connecting pectoral girdle to the humerus and to the anterolateral chest wall. [Reprinted from Lewis, Warren H (ed). Gray's Anatomy of the Human Body, 20th American Edition. Philadelphia, PA: Lea & Febiger, 1918]

Thoracoacromial Artery

- *Pectoral branches* descend between pec major and pec minor supplying them and the breast.
- *Acromial branch* travels laterally under deltoid to access the acromial process where it makes a number of anastomoses to form the *acromial rete*.
- *Clavicular nutrient branch* travels upward and medially all the way to the sternoclavicular joint. It may be the arterial rationale for the secondary sternal ossification center of the clavicle.
- *Deltoid branch* supplies pectoralis major and deltoid. It ends up in the acromial rete.

Lateral thoracic artery comes off axillary artery only 30% of the time. In other cases, it comes off thoracromial or subscapular. It supplies serratus anterior and thus the anterior side of the medial (vertebral) border of scapula. For that reason, medial border has two ossification centers. The anterior one is supplied by lateral thoracic. The posterior one is supplied by the artery to the levator scapulae and the rhomboids, the deep branch of transverse cervical artery.

Subscapular Artery This is the "big boy," the most sizeable branch of the axillary artery. It arises just distal to the lateral border of subscapularis; it then quickly divides.

- *Circumflex scapular artery* is the larger of the two branches. At the lateral border of scapula is passes through the *triangular space*, bounded above by subscapularis, below by teres major, and laterally by long head of triceps. It travels along the bone of the infraspinatus fossa, supplying it. This is the likely source for one of the ossification centers of the scapular body (the other being supplied by suprascapular artery).
- *Thoracodorsal artery*.

Humeral Circumflex Arteries

- *Posterior humeral circumflex artery* arises just beyond distal border of subscapularis. It is accompanied by axillary nerve through *quadrangular space* defined above by subscapularis and teres minor, below by teres major, medially by long head of triceps, and laterally by neck of the humerus. It serves the deltoid and the shoulder joint.
- *Anterior humeral circumflex artery* arises just opposite its twin. It supplies the head of the humerus and shoulder joint.

Just like the vertebral and the thyrocervical trunk, internal thoracic was originally a series of segmental arteries coming from the dorsal aortae in the unfolded embryo. Recall that at this embryonic stage, the dorsal aortae run all the way forward and unite together into horseshoe-shape. The most distal parts of the inverted U will become paired heart tubes. The dorsal aortae give rise to three sets of arterial branches, arranged neuromerically, and assigned to specific embryonic structures. *Dorsal intersegmental arteries* give off dorsal branches supplying the spinal cord and the vertebrae. Lateral intersegmental branches from the dorsal intersegmental arteries supply the neuromeric levels of the neck and trunk. In the thorax, these are represented by the segmentals from the aorta.

Organ Systems of the Neck

Esophagus

The esophagus is a muscular tube extending from the r11 terminus of cricoid cartilage just opposite to vertebral level C6 25 cm downward to the stomach. It thus span from a c1–c8 and t1–t12, a total of 20 neuromeric units. It consists of four layers: the serosa, an external fibrous/fascial layer, the muscularis with two muscle planes, a submucosa comprised of lateral plate mesoderm admixed with sympathetic ganglia and containing blood vessels and an endodermal mucosa (Figs. 10.135, 10.136).

The lining of the esophagus is in continuity with the oropharynx. It is a non-keratinized stratified squamous epithelium which, due to its thickness, is protective in nature. It is populated by Langerhans cells which process and present antigens. Lamina propria contains lymphoid tissue and mucous glands with a strange distribution, being found only near the pharynx and at the gastroesophageal junction. This is followed by a muscularis muscosa layer (found only in esophagus and rectum) consists of longitudinal smooth muscle which is thin and whispy near the pharyx but thickens progressively. The function of this layer is to maintain the epithelial surface of the gut in motion so that material is expressed out of the crypts and that the surface lining is kept in contact with the luminal contents. The submucosa contains mucous and serous glands that communicate with the lumen. The muscular layer has a deep plane of circular muscle and a superficial layer of longitudinal muscle.

The mesenchymal composition, blood supply, and innervation of the esophagus differ from those of the midgut and hindgut. We can best appreciate its development in neuromeric terms. The entire GI tract is regionally specified through Homeotic genes. Its basic structural components of endoderm and lateral plate mesoderm, visceral layer (LPM$_V$), are neuromerically coded. Neural crest populations are segmental as well. The muscular layer of the gut has two strata of smooth muscle: the inner layer circular and the outer layer is longitudinal. The muscle layer of esophagus is unique in

Fig. 10.135 Neuromeric levels of the esophagus. Lateral plate mesenchyme, although it does not proceed through an intermediate somite, is neuromerically organized. Each segment of the esophagus is innervated by a branch from the sympathetic chain, faithfully recapitulating the individual somitic contributions to the rest of the thoracic wall. Cervical esophagus c1–c4 recapitulate the 7-arch branchial arch system. Levels BA6 and BA7 are transition zones where striated PAM muscle transitions to smooth LPM muscle. In levels c5–c8 the eosophagus becomes entirely smooth muscle. Recall that Thoracic esophagus has 10 neuromeric levels with 10 sympathetic ganglia and 10 segmental branches from the thoracic aorta. Note pathway of the left phrenic nerve coursing over the c1–c4 pericardium and its relationship to the ipsilateral vagus [Reprinted from Lewis, Warren H (ed). Gray's Anatomy of the Human Body, 20th American Edition. Philadelphia, PA: Lea & Febiger, 1918]

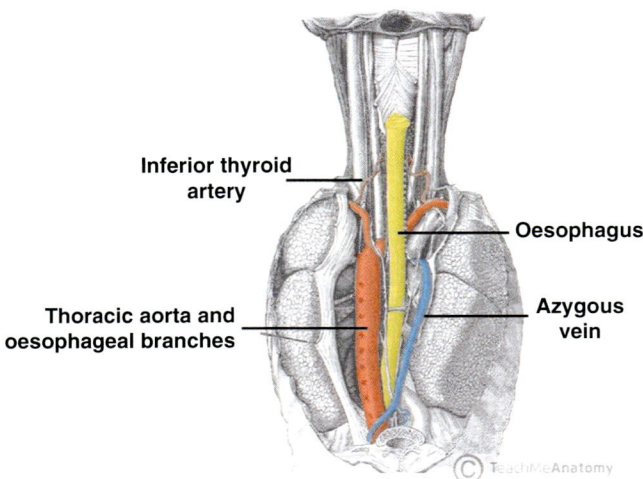

Fig. 10.136 Arterial supply to esophagus is via segmental aa. cervical = Inferior thyroid; upper thoracic = descending aorta via bronchial aa; lower thoracic; left gastric right lateral and splenic left lateral [Reprinted from TeachMeAnatomy, courtesy of Dr. Oliver Jones]

most likely represents a transformation of local LPM$_V$ from a smooth to striated fate.

Vascular supply to cervical esophagus is segmental and arises from inferior thyroid artery. As previously discussed ITA represents the unification of transverse branches originally given off by the dorsal aortae prior to the great remodeling that takes place after the pharyngeal arch period. Each neuromeric segment of esophagus is supplied segmentally from the inferior thyroid axis. Motor innervation of this muscle is supplied by nucleus ambiguus lateral motor column acting through the vagus. Nucleus ambiguus extends downward into the cervical spine at four neuromeric levels. More distal smooth muscle sector is innervated by dorsal motor nucleus of vagus.

Nerve endings for pain and stretch route through segmental spinal nerves in the cervical esophagus whereas in the thoracic esophagus they connect to celiac plexus. Thus, upper esophageal pain can be localized (vaguely) to "find the proverbial fishbone." Lower afferents convey more generalized sensation.

Note that neural crest cells migrate along the plane of the circular layer at week 6. They then penetrate internally to enter the submucosa where they form the myenteric plexus.

Neuromeric Model of the Cervical Esophagus

Paleontologic Footprint Recall that in the original agnathic vertebrates, 12 arches have been documented. This implies that 24 neuromeric levels. Recall as well that fish arches represent communications with the pharynx. They were present well beyond the head into the body. The implication is that primitive vertebrate anatomy included branchiomeric mus-

that its cervical zone has striated muscle. Moreover, the circular muscle layer is in direct continuity with inferior constrictor and is therefore intrinsically branchiomeric.

Esophagus has three sectors based on the mesenchyme of its muscular layer. Cervical esophagus is the upper third and has striated muscle from PAM. In the thorax, middle third of esophagus muscular layer is a mixture of striated and smooth muscle while the distal third is exclusively smooth muscle. Lateral plate mesoderm, visceral layer (LPM$_V$) is the source of the smooth muscle from LPM. Paraxial mesoderm provides cervical muscle. The origin of striated muscle in the thoracic esophagus is uncertain but

cles for control of the gill arches for 4 cranial and 20 postcranial segments. In humans that translates to c1–c8 + t1–t12 = 20. This represents the thoraco-abdominal boundary. Thus, the neuromeric definition of foregut versus midgut territories resides at the 20th postcranial neuromere?

The anatomy of the mammalian foregut (esophagus) is remarkably consistent with this ancient original fish model. As we have previously discussed, the pharyngeal constrictor muscles above the esophagus arise from PAM before it has organized into the somites. Thus, branchiomeric myoblasts from neuromeres r6–r11 are somitomeric. Striated muscle in the cervical esophagus arises in the same way… outside of the normal somite myotomes. PAM for striated muscle arrives to populate the cervical esophagus from somitomeres Sm9–Sm12 (neuromeric levels c1–c4) most intensely. These levels represent the ancient sixth and seventh branchial arches. Striated muscle is present to a lesser degree in the lower cervical esophagus. This can only be explained by contributions from Sm13–Sm16 (neuromeric levels c5–c8). And the only way to rationalize this is to invoke the embedded genetic capacity of c5–c8 for branchiomeric muscle, presumably a throw-back to the eighth and ninth arches.

This is not to imply that arches 6–9 exist in mammals, simply that the somitomeres Sm9–Sm16 retain the genetic capacity to send out a select population of PAM with segmental pain fibers that track superficial to LPM$_V$. Together they populate the wall of esophagus.

A neuromeric model of the esophagus begins with an endodermal tube surrounded by lateral plate mesoderm, visceral layer (LPM$_V$). The mesenchyme of muscularis mucosa and submucosa is in register with c1–c8. An external layer of striated muscle is contributed from PAM of Sm9–S12 as above. In the lower cervical esophagus LPM$_V$ smooth muscle begins to appear. Once inside the thorax no further PAM is found in the esophagus: it is surrounded by smooth muscle.

This behavior fits with observations about the unique contribution of lateral plate mesoderm to the muscular suspension of the pectoral girdle from the head: sternocleidomastoid and trapezius. These hypoaxial muscles may be possibly be derived from lateral plate mesoderm rather than PAM. As such they represent levels c1–c2 for sternocleidomastoid and c2–c6 for trapezius. Perhaps the limited availability of LPM at cervical levels for esophagus is due to "mesenchymal diversion" for the external neck.

Larynx, Trachea, and Thyroid

Key events by developmental stages have been correlated to Carnegie stages [50].

Stage 11

Ventral epithelium of the foregut thickens as respiratory primordium.

Stage 12

Diverticulum develops in the respiratory primordium. Primitive pharyngeal floor is isolated and will become epiglottis. Cephalic RD becomes the infraglottic region and gives off bronchopulmonary buds.

Stages 13–14

Bronchopulmonary buds migrate downward as the carina. Space between RD and carina becomes the trachea. Growing esophagus and trachea constitute a watershed zone vulnerable to vascular compromise leading to esophageal atresia, trachea-esophageal fistula, tracheal stenosis, or even agenesis.

Stage 15

Ventral primitive laryngopharynx is compressed bilaterally by the cartilages and muscles of the larynx.

Stage 16

Epithelial lamina obliterates primitive laryngopharynx ventral-dorsal leaving behind a narrow communication between the hypopharynx and infraglottis. A space appears, the laryngeal cecum, between the arytenoids and the epiglottis. The space deepens along the ventral aspect of the epithelial lamina. The primitive laryngopharynx has the shape of the letter "T."

Stages 17–18

Laryngeal cecum pushes downward to reach to glottis, where it stops.

Stages 19–23

Recanalization of the epithelial lamina takes place from dorsal-cephalic to ventral-caudal. The process terminates at the glottis. Failures in this process can create webs or stenosis.

Fetal Period

When recanalization of the epithelial lamina is complete, communication is established between the supraglottis and the infraglottis. Laryngeal ventricles are outgrowths of the cecum. Innnervation is established with the myenteric plexus at 13 weeks. At 16 weeks the cartilaginous vocal cords, ventricle, and saccule are formed. Fetal swallowing begins. At 6 months the epiglottis has fibrocartilage and fetal breathing is established.

Phylogeny of the Larynx and Mechanisms of Ventilation

Lungfishes have a primitive epiglottis to separate air and food. Soft tissue airway anatomy for extinct tetrapods is not available to us so we have to content ourselves with amphibians. Frogs have short nasal chambers that open through choanae anterior to palate. The amphibian glottis is suspended by lateral laryngeal cartilages that are derived from the fifth arch (no sixth arch). Behind the glottis is a small triangular laryngotracheal chamber that leads directly into the lungs. Inside the chamber are vocal cords. Inspiration is controlled by buccal pumping with expiration a combination of elastic recoil of the lungs plus the contraction of *transversus abdominis* muscle, reincarnated from placoderms (Fig. 10.137).

Neck length increases in amniotes. In reptiles, the laryngotracheal chamber subdivides into larynx, with supporting arytenoid cartilages, a cricoid, and trachea (Gr. *tracheia* = rough artery). The developmental advantage of the cartilage rings is to maintain the patency of the lengthened airway. Reptiles make use of intercostal muscles to expand and contract the chest cavity. Crocodiles have unusually paired *diaphragmatic muscle* (unrelated to mammalian diaphragm) which stretch longitudinally from liver back to the pelvic girdle and open the pleural cavity.

A secondary palate for separation air from food appears in therapsids and continues into mammals. Enlarged nasal cavities with conchae warm, humidify, and cleanse the air. The mammalian larynx, characterized by a new thyroid cartilage, houses an epiglottis and vocal folds.

It is not generally appreciated that lungs are a basic feature of all bony fishes. Air breathing is present in primitive ray-finned fishes, such as *Polypterus* and *Amia*, as well as the lungfishes. In Amia, the proportion of O_2 taken from lungs versus gills is 25% but as the water warms up this can rise to 75%. Actinopterygians use a 4-stroke buccal pumping mechanism and external oblique muscles to ventilate. Lungfishes have hearts and circulatory anatomy similar to amphibians and a primitive epiglottis to separate food and air. Ventilation is accomplished using a 2-stroke buccal pump with additional internal oblique muscles for *forced inhalation*.

The tetrapod transition took place in warm shallow waters where aqueous oxygen supply was poor so evolutionary pres-

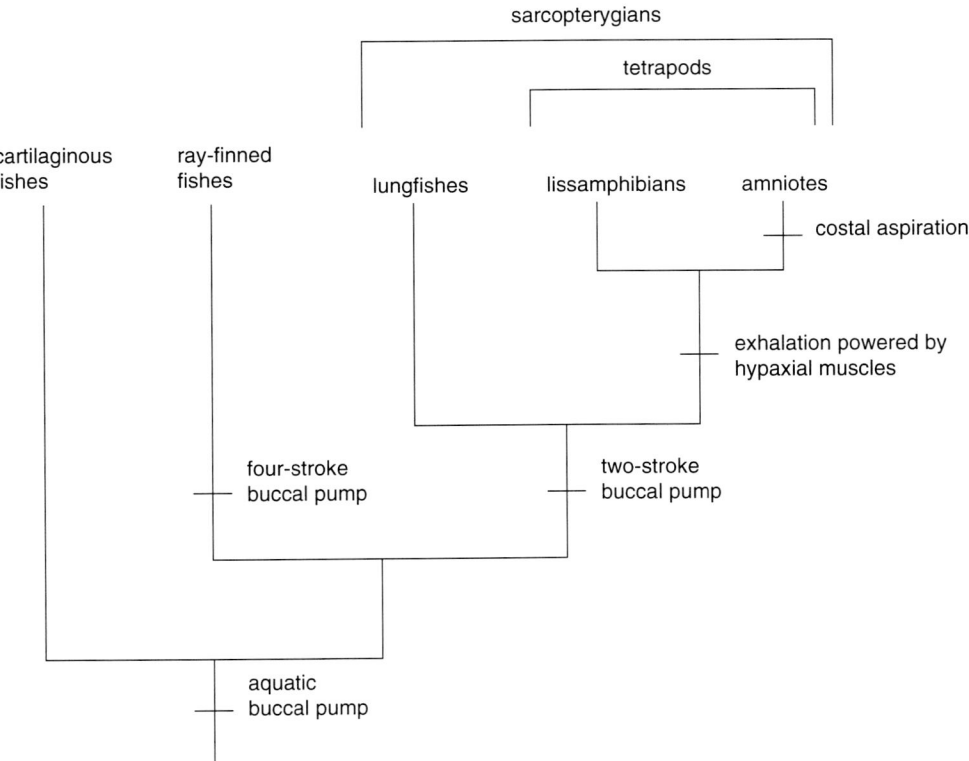

Fig. 10.137 Phylogeny of ventilation. Tetrapods invented the transverse abdominal muscles… homologous with those of placoderms. Enhancement of exhalation preceded that of inhalation. [Reprinted from Kardong K. Vertebrates: Comparative Anatomy, Function, Evolution, seventh ed. New York, NY: McGraw-Hill Education; 2015. With permission from McGraw-Hill]

sure favored the use of the lungs. Early tetrapods continued to ventilate the lungs using buccal pumping instead of ribs. This is related to the loss of opercular-gular bone series permitting the mouth to gape open more widely, and fusion between the braincase and palate. Tetrapod ventilation was made more efficient by a third muscle layer, the transverse abdominals, muscles that existed long before in placoderms but which were expressed during fish evolution. The transverse muscles assist in *forced exhalation*. Costal ventilation appeared in the amniotes and was supported by inhalation and exhalation. The final innovation is that of the mammalian diaphragm.

Phylogeny of the Lungs

Throughout vertebrate evolutions, lungs arise as endodermal outgrowths from the gut. Specifically, the transition takes place below the fifth arch, that is, as the boundary between r11 and the first truncal neuromere. In mammals, this is the r11–c1 junction.

In almost all cases, the lungs are paired, and ventral. In mammals, they connect with the pharynx via trachea (Fig. 10.138).

The primal origin of lungs is the *gas bladder* which appears with ray-finned fishes and is connected via a pneumatic duct to the gut. When functions to maintaining buoyancy, that is, the position of the fish in the water, it is referred to as the *swim bladder*. When its internal walls are infolded and heavily vascularized, it is referred to as a *respiratory bladder*, or *lung*. Swim bladders are dorsal and generally singular, with venous blood returning into the systemic circulation. Lungs are ventral and paired; their blood return and enters the heart directly. Both bladders originate from the gut and have a common form of innervation. Fish evolution reveals repeated episodes of reversal between the two forms, depending upon the respiratory versus buoyancy requirements.

Neither form of bladder is present in agnathans, placoderms, or chondrichthyans (sharks). *Lungs are actually the most basal form*. They arose in the common ancestor interposed between placoderms and the two lines of osteichthy-

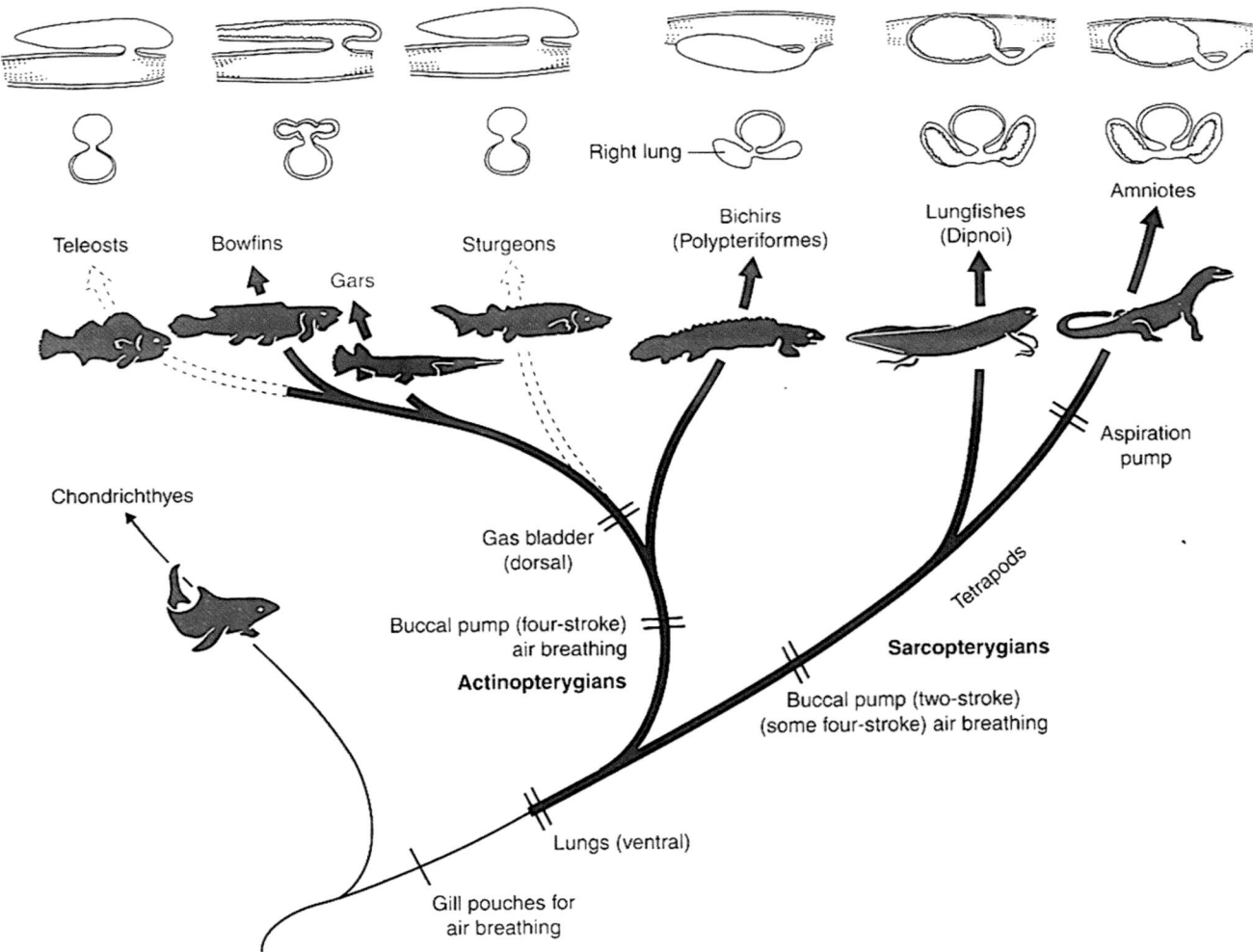

Fig. 10.138 Phylogeny of the lungs. Lungs are a feature of all bony fishes. [Reprinted from Kardong K. Vertebrates: Comparative Anatomy, Function, Evolution, seventh ed. New York, NY: McGraw-Hill Education; 2015. With permission from McGraw-Hill]

ans, actinopterygians, and sarcopterygians. Lungs begin with sarcopterygians. The primitive form, seen in the Australian lungfish is single and dorsal (one side is atrophic) but in the more derived South American and African lungfishes a dorsal–ventral hox gene places paired lungs in the final, ventral position. The swim bladder in the ray-finned fishes may have evolved separately or as a derivative of the lung.

Neuromeric Rationale of the Respiratory System

The pulmonary diverticulum from the foregut at r11–spinal1 is universal among tetrapods. Recall further that the neuro-

meric transition point for the brachial plexus and upper limb is at spinal 4 (C4 in mammals) and that cucullaris spans from sp4 forward to occiput. Recall that the pharyngeal arches share a common mapping system of *distal-less* (*dlx*) genes. Thus, neuromere c1 may be located at a transition posterior to the termination of the *dlx* system. *Bookmark this point*: interesting and instructive parallels exist between the location of the tetrapod partitioning mechanism and the respiratory system of insects (Fig. 10.139).

First off, foregut partitioning begins at the node between weeks 3 and 4 of development.

The site of the respiratory diverticulum along the ventral foregut wall at level c1 is marked by the expression of Nkx 2.1 Just opposite, the dorsal foregut wall contains the expres-

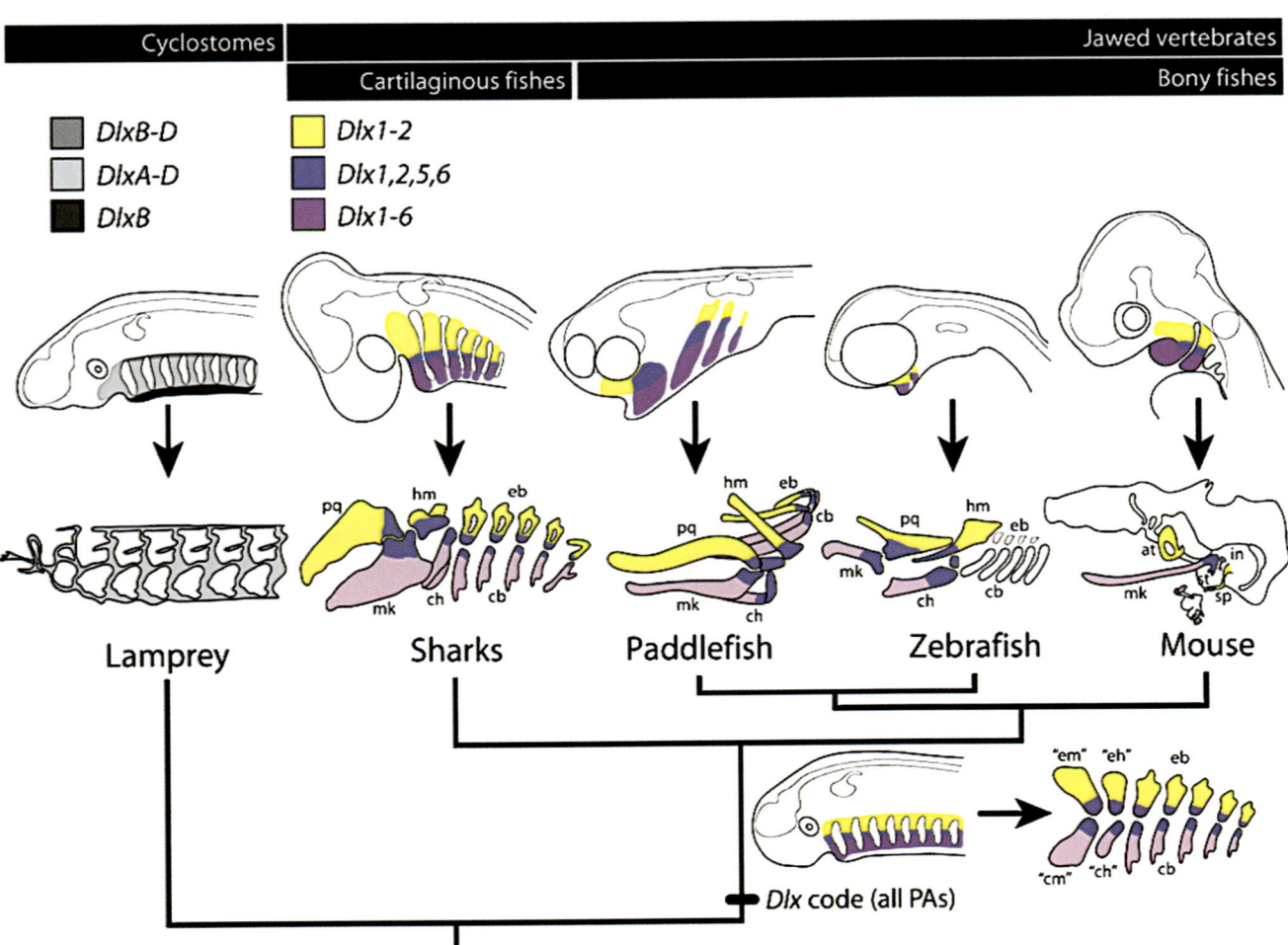

Fig. 10.139 Evolution of the *Dlx* code and serial homology of phyarngeal arch endoskeletal elements. The *Dlx* code arose along the gnathostome stem, and was primitively deployed in all pharyngeal arches. The evolutionary relationship between the *Dlx* code of gnathostomes and the nested expression of *DlxA-D* in the pharyngeal arches of lamprey remains unclear. Dorsal (*Dlx1–2*-expressing) and ventral (*Dlx1–6*-expressing) domains of the Dlx code would have primitively given rise to dorsal "epimandibular," "epihyal," and "epibranchial" elements and ventral "ceratomandibular," "ceratohyal" and "ceratobranchial" elements (in the mandibular, hyoid and gill arches, respectively), while intermediate (*Dlx1, 2, 5 and 6*-expressing) domains would have given rise to the region of articulation between these elements. The primitive role for the *Dlx* code in patterning the mandibular, hyoid and gill arch

endoskeletal segments has been conserved in elasmobranchs, and presumably in non-teleost actinopterygians (e.g., paddlefish), while posthyoid arch expression of the *Dlx* code has been modified or obscured in amniotes (e.g., mouse), and possibly in teleosts. at, ala temporalis; cb, ceratobranchials; "cb," hypothetical ceratohyal; ch, ceratohyal; "cm," hypothetical ceratomandibula; eb, epibranchials; "eh," hypothetical epihyal; "em," hypothetical epimandibula; hm, hyomandibula; in, incus; mk, Meckel's cartilage; pq, palatoquadrate; sp., styloid process; st, stapes. [Reprinted from Gillis AJ, Modrell MS, Baker CVH. Developmental evidence for serial homology of the vertebrate jaw and gill arch skeleton. *Nat Commun* 2013; 4:1436. With permission from Springer Nature]

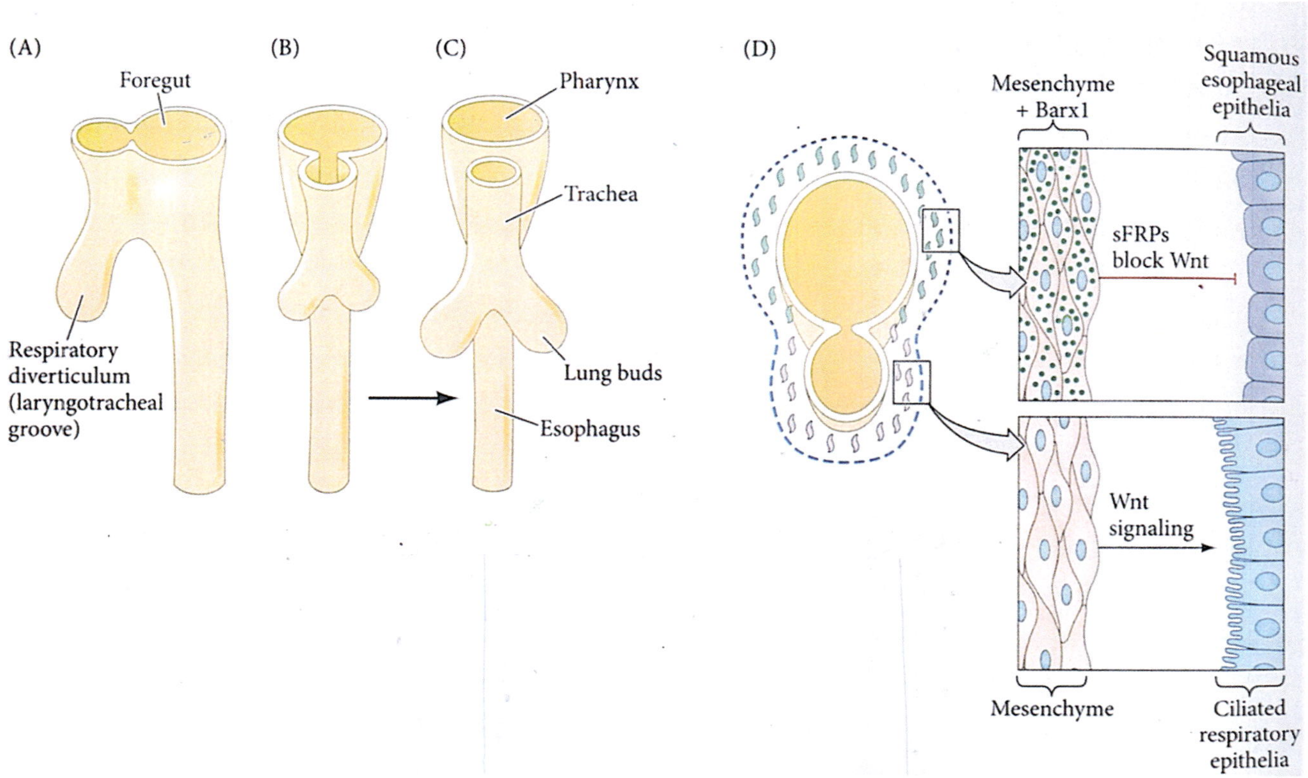

Fig. 10.140 Trachea appears immediately after lung buds at stage 13. Separation from esophagus is complete at stage 14. Tracheo-esophageal fistula can occur at this time. Descent of the trachea into the chest is complete at stage 23. D Roles of Wnt signaling and mesenchyme expressing *Barx1* causes differentiation of epithelia. Presence of Barx1

blocks Wnt and causes region to become esophageal. [Reprinted from Gilbert SF, Barressi MJF. Developmental Biology, 11th ed. Sunderland, MA: Sinauer; 2016. Copyright © 2016. Oxford Publishing Limited. Reproduced with permission of the Licensor through PLSclear]

sion of Sox-2. Wingless (wnt) signals from surrounding lateral plate mesenchyme cause Beta-catenin to accumulate in the ventral gut tube between r11 and c1. B-catenin is required for separation to occur. It acts like the Pied Piper to attract the diverticulum. If expressed ectopically, it can form extra lungs.

Meanwhile, interaction between the epithelium of the future esophagus and trachea with surrounding mesenchyme determines the type of epithelial lining. Mesenchyme surrounding the dorsal gut tube contains transcription factor Barx1 which insures the production of soluble Frizzle-related proteins (sFRPs). The sFRPs bind to Wnt so they cannot access the gut tube, thereby blocking their activity. In contradistinction, ventral mesenchyme does *not* produce sFRPs. Note how this mechanism could explain the dorsal versus ventral positioning of the gas bladder in evolution. The presence or absence of Wnt signaling directly affects the histology of the epithelium. When blocked, the lining of dorsal esophagus is squamous epithelium. When active, Wnt creates a ciliated respiratory epithelium that lines the trachea (Fig. 10.140).

In the later fourth week the posterior aspect of the respiratory diverticula produces lung buds. The lung buds in turn induce from the LPM just dorsal to the paired mesodermal ridges. These approximate one another in the midline and fuse in a dorsal to vental manner. This makes sense because

the LPM folds are posterior to the lung buds. The fusion of the folds creates a septum that separates trachea from esophagus.

Lung development is also homeotic. Combinatorial patterns of hox genes *Hoxa3* to *Hoxa5* and *Hoxb3* to *Hoxb6* are responsible for regional specification of the respiratory track such as bronchi to bronchioles and the lobes of the lung. Furthermore, coding exists in the mesoderm surrounding the tract. The more proximal zone inhibits branching and results in trachea whereas more distal mesenchyme supports branching.

A Sidebar on the Thyroid

It is significant that the NKx 2.1 zone in the c1 ventral foregut is associated with localization of the thyroid gland. Thyroid tissue likely contains r10–r11 markers from the fifth arch. Thyroid tracks toward the tongue along the hyopobranchial genetic tract, the same one used by the strap muscles running between the manubrium and the 1st/second arch interface in the tongue. Thryoid tissue probably migrates in company with the occipital myoblasts, squarely in the midline. When it returns to the neck it follows this selfsame pathway backward to manubrium and into the chest.

Lessons from *Drosophila*

Nomenclature of *Drosophila* genes: *distal-less* (*dll*), *buttonhead* (*btd*), *trachealess* (*trh*), *wingless* (*wg*), and *ventral veinless* (*vvl*).

The arthropod (insect) exoskeleton contains ectodermal tracheal placodes, numbering 10 in *Drosophila*. These become *spiracles* that penetrate the skeleton to reach a longitudinal tracheal system running the length of the body. Spiracles congregate around the three pairs of legs with the two thoracic spiracles associated with the first pair of legs. Respiratory spiracle/trachea cells and leg cells exist in a common pool and are related. Tracheal precursors develop where *trh* and *vvl* are expressed. Leg precursors appear where *Dlli* is expressed. It appears that the decision to make trachea versus legs depends on the presence of *wingless* expression. *Wg* blocks *trh* and therefore prevents trachea formation; it promotes *Dll* and causes legs to develop (Figs. 10.141, 10.142).

Primitive aquatic arthropods such as crustaceans demonstrate a similar relationship between the respiratory structures and appendages [51]. Crustacean gills develop as a

Fig. 10.141 Indian moon moth (*Actias selene*) showing respiratory spiracles transmitting O_2 through exoskeleton. These arise from placodes and lead into longitudinal internal trachea which subdivides to supply all parts of the body. It is the homolog of lungs. [Reprinted from Wikimedia. Retrieved from: https://commons.wikimedia.org/wiki/File:Actias_selene_5th_instar_spiracles_sjh.jpg. With permission from Creative Commons License 2.5: https://creativecommons.org/licenses/by-sa/2.5/deed.en]

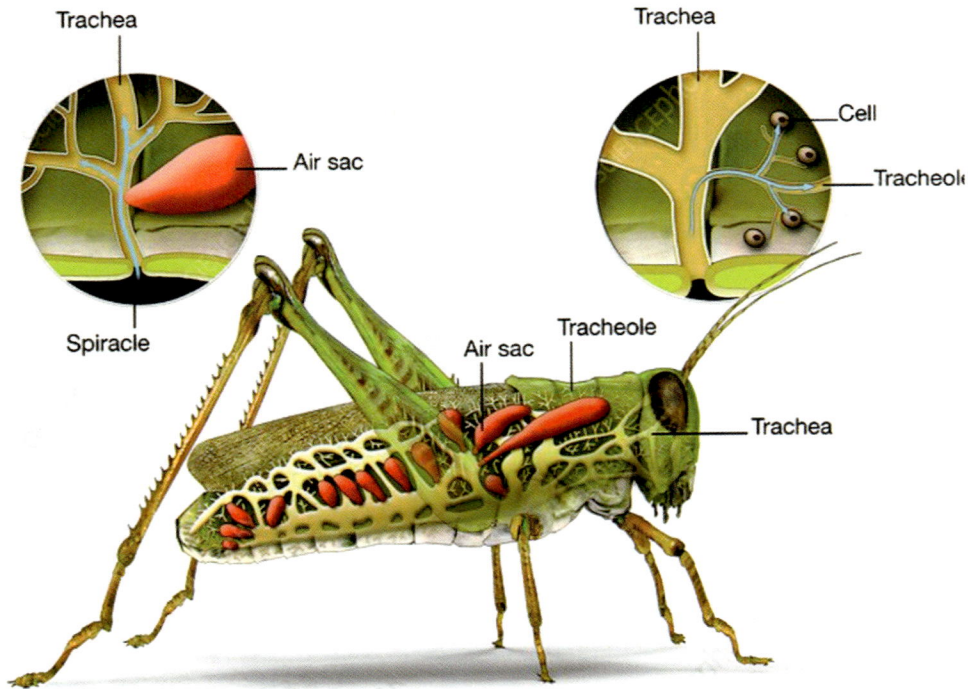

Fig 10.142 To trachea or not to trachea? Insect exoskeleton tiny valves called spiracle which admit oxygen. These are lines with hairs to keep dust out. O_2 and CO_2 are transported by tracheae; these terminate in tracheoles where gas exchange occurs. Role of wingless (wg) in repressing tracheal fate. In *Drosophila*, 10 tracheal placodes give rise to spiracles that connect with the trachea. Note two thoracic placodes are associated with first pair of legs. Tracheal cells are specified by *trachealess* (*tll*) and *ventral veinless* (*vvl*) along the body axis. Tracheal cells and leg primordia cells exist in a common pool. Legs form where *Dll* is expressed. The decision trachea versus legs depends upon the activation state of the *wingless* (wg) signaling pathway. In aquatic arthropods such as crustaceans, respiratory gills and appendages are also associated. Role of wingless (wg) in repressing tracheal fate. Genes: *distal-less* (*dll*), *buttonhead* (*btd*), *trachealess* (*tll*), *wingless* (*wg*) and *ventral veinless* (*vvl*). Wingless represents a "switch" between trachea and appendages. When not expressed respiratory system develops. Note relationship between arthropod *distal-less* (*dll*) and vertebrate *distal-less* (*dlx*) that is expressed in the pharyngeal arches. Cells of the respiratory system in Drosophila arise in close association with cells that give rise to the extremities. In aquatic arthropods such as crustaceans, respiratory gills and appendages are also associated. Genes: *distal-less* (*dll*), *buttonhead* (*btd*), *trachealess* (*tll*), *wingless* (*wg*) and *ventral veinless* (*vvl*) [Courtesy of Science Photo Library].

dorsal lobe of their appendages (epipods). Franch-Marro et al. [52] found homologs of Drosophila tracheal inducer genes expressed in crustacean epipods. This suggests that tracheal systems and gills in arthropods evolved from a common precursor.

The relationship between the forelimb and the anterior-most component of the respiratory system becomes relevant when we consider the role in humans of Wnt signals in (1) the partition of foregut into trachea and esophagus and (2) the creation of respiratory epithelium all the way down the trachea. This event takes place precisely at the neuromeric levels c–c4 (and beyond) associated with the forelimb. The primal events involved in the creation of the gas bladder likely involved similar mechanisms at the very same neuromeric level.

Coda: Thoughts on the Assembly of the Neck

In our discussion so far, we have considered the neck as a series of structural components: bones, nerves, muscles, arteries, and the endodermal-encoded tubes (the esophagus and respiratory tract). In this final section let's consider how these structures are assembled, and how this assembly changed from primitive tetrapods to the amniote condition. Some of this story is by way of review (based upon what we know from the fossil record), some of the stories is best-guess speculation…the final picture awaits future research. The neck being a truly integrative structure, our story begins with concomitant events that changed both the vertebral axis and the pectoral girdle.

Fish vertebrae are seemingly monotonous, the only difference between trunk vertebrae and those of the tail is the presence of a hemal arch in the latter as a protective device for the aorta to prevent it from kinking during swimming. Yet, beginning with transitional form *Tiktaalik*, a drastic transformation takes place in vertebral column causing specialization. From the earliest true tetrapods such as *Ichthyostega* onwards, cervical, thoracic, lumbar sacral, and caudal regions can always be identified. In the future neck, the homeotic transformation that contributes to new cervical vertebrae is an *anteriorization of identity* of truncal somites, *regardless of whether the total number of vertebrae changes or not.*

The evolution of the neck results from two fundamental processes.

- Homeotic transformation refers to the anatomic changes that results when neuromeric levels formerly responsible for the structures of the trunk are reassigned to produce new derivatives with new functions. Transformation implies just that: it is not an addition of a level but a *change in the definition* of that level from trunk to neck.

Bear in mind that the transformation of a somite with a thoracic homeotic signature into one with a cervical hox code represents not just a change in osseous structure but a change in the entire portfolio of tissues at that level.

- Homeotic duplication refers to the repetition of units within the same anatomic region resulting in new anatomic structures. It is unique to mammals where the neuroanatomic contents of levels C4–C5 are repeated to create and add the median nerve to the brachial plexus and more complex system of muscles. The result of duplication is to force the expression of anatomic structures into more posterior neuromeres.

Key Evolutionary Steps Determining the Size of the Mammalian Neck

Step 1: the first iteration of parasegmentation

In the pre-tetrapod evolution of late sarcopterygians, a fourth occipital somite appears, the result of the incorporation of the first truncal somite into the skull. Recall the first three occipital somites do not manifest parasegmentation. They are fused together as a unit. In sarcopterygians the fourth somite demonstrates a new characteristic: the rostral part of S4 centrum is loose. For the first time in evolution, it breaks apart and fuses with S1–S3. This creates the modern cranial base, S1–S2–S3–S4$_R$. The caudal part of S4 centrum is now free to interact with rostral half of S5 centrum. The recombination of S4$_C$ with S5$_R$ creates a modified spinal vertebra, the protoatlas, S4$_C$–S5$_R$. Although lungfishes do not have neck joint…we are just waiting for Tiktaalik to come along…the protoatlas creates the conditions for a joint. There is nothing special about the appearance of protoatlas. It looks like just another spinal vertebra. But its subsequent dissolution caused a revolution.

Step 2: the second iteration of parasegmentation

Lightning strike twice. Tetrapods apply the same mechanism to protoatlas. It splits apart…asymmetrically. Instead of becoming totally intracranial…in effect becoming a fifth occipital somite…its rostral centrum contributes to foramen magnum. Voila! A joint is created. This process leaves behind a fragment of the neural arch of protoatlas C0, referred to as the proatlas, intercalated above the new vertebra in contact with the skull, Atlas, now has the formula S5$_C$–S6$_R$ and a very different shape.

We have seen in the fossil record, in humans, an intact proatlas with a total of eight cervical vertebrae has been documented clinically (Fig. 10.32). Proatlas appears as part of a proatlas–atlas–axis complex followed by three additional cervical vertebrae, C3–C5. This model is seen in *Ichthyostega* and *Acanthostega*. Thus, *the common anamniote ancestor of amniotes had 5 cervical vertebrae*. The boundary between cervical and thoracic ribs is thus defined by the third spinal nerve of the brachial plexus, br3; this is spinal nerve C5. The boundary of the thorax is C5–T1. The evolution of extant amphibians such as frogs and salamanders proceeds in the opposite direction with reductions in the brachial plexus formula (vide infra).

Step 3 The amniote transition was accompanied by changes in the neck. In both the reptilian/avian (diapsid) line and the mammalian (synapsid) additional numbers of cervical vertebra were produced and a caudal shift of the thorax took place. Diapsids are more prolific. The ostrich has 20 cervical vertebrae with the first thoracic vertebra located at S25, In the mammalian line, non-therapsids have 5, therapsids have 6, and cynodonts achieve the final mammalian number of 7.

We can follow this evolutionary story by making use of two anatomic principles.

- The *neuromeric position of the pectoral fin or upper extremity* is located at the head-thorax boundary in fishes and the neck-thorax boundary in tetrapods. Anatomic differences between cervical and thoracic ribs permit the identification of the boundary in all fossil taxa.
- A *universal pattern of the brachial plexus* exists in all amniotes (mammals being an important exception) that allows one to work backwards (rostral) from the trunk to determine the identity of the neuromeres supplying the upper limb. Brachial plexus neuroatomy faithfully recapitulates the evolution of the neck.

Caudal Shift and Homeotic Duplication of the Brachial Plexus in Mammals

The brachial plexus of all non-mammal amniotes consists of four spinal nerves, three of which are of cervical origin and the last one from neuromeric level t1. These are referred to as br1–br4. Pairing of br1–br2 creates the *radial nerve* for extensors while pairing of br3–br4 creates the *brachial nerve* for flexors. Thoracic neuromere t1 is responsible for br4. Brachial and radial nerves are interconnected and supply the simple adductor and abductors muscles groups of the limb. This simple configuration goes back to the beginnings of vertebrate evolution.

Even though the mammalian brachial plexus is more complex, the same rules hold true. Mammals duplicate roots br1–br2 to create a new median nerve in a more rostral position. In the process, the rest of the plexus is pushed back to two neuromeric units, and a 6-nerve brachial plexus results. Nevertheless, the penultimate nerves c8 and t1 continue to define the neck–trunk boundary.

Let's look at how this system arose. Recall that bony fishes, osteoichthyes, evolve in two lines: actinopterygian (ray-finned) fishes and sarcopterygian (fleshy-finned) fishes.

In the basal state, both lines have only two muscles associated with the pectoral fin, one abductor and one adductor. These are innervated by a plexus of nerves, designated as either occipital (Oc) or spinal (Sp), depending upon the location of their roots in the medulla or spinal cord.

Using lipophilic dyes [37] mapped out the alignment of the fin bud, myototomes, and neuroepithelium in four species of actinopterygians, from the primitive basal paddlefish, *Polydon spatula*, to the highly derived teleost midshipman, *Porichthys notatus*. The pectoral motoneurons shared a conserved pattern with no migration during embryogenesis. *Porichthys* shows us a simple system of two occipital roots and spinal roots distributed across the hindbrain-spinal boundary. Although the number of roots could vary during development, all were derived from two distinct populations corresponding to the adductor and abductor muscles. Furthermore, these motoneurons formed a single column distinct from the posterior r10–r11 dorsal motor nucleus of vagus. Four myotomes were supplied. The anatomic parallels with tetrapods are obvious.

The pectoral fin in the sarcopterygian line is much more complex. It has recognizable bones (stylopodium, zeugopodium, and autopodium) connected with additional joints, and controlled by new muscles (arising by subdivisions for the pre-existing muscle mass). This condition exists today in the African lung fish, Dipnoi species *Protopterus dolloi*. Lungfish have *three* occipital neve roots that combine into a single pectoral nerve Oc3 which then joins spinal nerves Sp1–Sp3 to form a plexus. The presence of an additional third occipital nerve is of great significance. Although not confirmed experimentally, the *origin of the fourth occipital somite* probably takes place in late sarcopterygians just before the tetrapods.

These findings are consistent with the hypothesis that a dual hindbrain-spinal origin for pectoral motoneurons in jawed fishes was translocated backward into a spinal-only innervation of the tetrapod forelimb. The sequence Oc1–Oc2 and Sp1–Sp2 become br1–br4. We can imagine a scenario in which hox shifts cause the occiput of the fish skull to change from S3 to $S5_R$. This introduction of new programming causes a disruption in levels r10–r11. The genetic program for the peripheral hypobranchial muscles of rectus cervicis in fishes is kicked back in tetrapods to c1–c4. At the same time, the neuromeres of the tetrapod hindbrain (r8–r11) assume a more modest role in supplying the hypobranchial muscles of the tongue. The question arises, after these initial events, what happened next?

Our ability to reconstruct the life history of Devonian tetrapods is limited due to inadequate fossil evidence regarding the cervical spine and neck joint. Nonetheless, *Acanthostega* represents a model for the initial iteration of the brachial plexus, as the entire length of its skeleton has been reconstructed. It had a well-defined atlas–axis complex. The arch of atlas is small and, while there are no pleurocentra, the intercentrum is robust. No proatlas has as yet been identified. Five cervical vertebrae are present. Nothing is known about the craniovertebral junction in *Ichthyostega*. http://pondside.uchicago.edu/oba/faculty/coates/Coates_1996.pdf

Our next clues come from *Greererpeton* a crown tetrapod of the mid-Carboniferous. Here a small proatlas was present and the atlas arch embraces that of the axis. This enlarged axis supported expanded muscle insertions for greater control of the head. In the transition to the amniotes atlas pleurocentrum and axis intercentrum became firmly fused as dens to support the ring of the atlas. The neural arch of axis took the decision to marry up its pleurocentrum, a condition persisting to this day. Proatlas continues to be represented in the stem amniote Paleothyris and continues into the synapsids as far as *Dimetrodon*.

The basic pattern of four spinal nerves and the position of the br3–br4 nerves at the cervicothoracic junction is shared by virtually all living amniotes. The only difference between taxa is the total number of cervical neuromeres: the last three will always contain br1–br3. We know that amniotes start out with five cervical vertebrae with an atlas/axis complex and a brachial plexus br1 (spinal nerve 4) is located at c3. *This situation leaves three spinal nerves available for cervical plexus.* Thus: sp1 is below proatlas; sp2 is below atlas; and sp3 is below axis. *Exactly as in humans.* The sleight of hand in mammalian evolution is a duplication of br1 and br2 to produce six roots and two additional cervical vertebrae.

Frogs have rudimentary neck and a deemphasized upper limb with a two-nerve brachial plexus supplying a simple flexor/extensor. Brachial plexus consists sp2 below the axis and sp3 below first thoracic vertebra, thus conserving the piscine model with two roots straddling the cervicothoracic junction. Nota bene: The first spinal nerve *never innervates the limb*. This is probably because the forward shift of proatlas takes that neuromere out of the genetic territory of *Hox5*, the territory of upper extremity. Formerly known as the "fin fold field." With proatlas plastered up against the skull, sp1 emerges cranial to the atlas. Furthermore, the existence of sp1 above atlas demonstrates that urodeles incorporated the protoatlas, C0, with sp1 left behind. This makes sense because, in all taxa, first spinal nerve is simply not programmed for the extremity. Its purpose is to supply the axial muscles controlling the head.

- In sum: the inference of amphibians is that the presence of an atlas is evidence of the prior forward shift of proatlas.
- first spinal nerve in frogs contributes to hypoglossal.
- The amphibian brachial plexus has two nerves; it begins with second spinal nerve: (br1 = sp2 and br2 = sp3).

Pre-mammals start their brachial plexus with br1 as sp4 just below C3. With evolution, they proceed to add duplicate b1–br2 creating the formula. This forces the transfer of the original 4-root plexus from spinal nerves Sp4–Sp7 backward to Sp6–Sp9. But even though the hox code for c4–c5 changes, the neuroanatomic contents remain much the same. Thus c4–c5 combine in the same way to produce a new structure, *median nerve*. Even the original cross-bridge between radial and brachial is duplicated, with median connected to brachial (now renamed the ulnar) and radial connected to brachial.

- The formula for the mammalian brachial plexus is: br1′–br2′–br1–br2–br3–br4.
- Supracoracoideus nerve (phrenic nerve) retains its ancestral state at c4 (br1′).
- Subscapular nerve, remains attached to br1 so it shifts backward one neuromere to c4–c5.
- Median nerve br1′–br2′ at roots c4–c5 is an exact carbon copy of radial nerve br1–br2 which has been shifted backward two neuromeres at roots c6–c7.

Mesenchymal consequences of neck development: making room for mesenchyme.

In anamniote evolution, the neck remained simplistic. Salamanders have a three-spinal nerve brachial plexus with little complexity of either extremity. This is reduced to 2-spinal nerve in frogs. Snakes lose the extremities altogether. Note that they lose the lateral motor column as well. In early tetrapods, the shoulder girdle is closely tucked into the head. These animals used buccal pumping as a means to drive air into the lungs. The musculature required for this was similar to that of the gill apparatus and these muscles were inserted into the pectoral girdle. The basic structures of the esophagus and the airway (the glottis) are present in amniotes. As the neck expands, additional neuromeric units of endoderm and lateral plate, like a candy cane, simply add more stripes to the structures transmitted through the neck and into the chest. In the mammalian line lengthening of the neck was accompanied by repositioning of the extremities beneath the body expansion of the brachial plexus for more greater complex control of the forelimbs, and the invention of the diaphragm with enhanced ventilation (buccal pumping was abandoned) supporting higher oxygen demand, as in a cheetah running at top speed.

Mechanism: Is There a Common Event?

Previously, we discussed the phylogeny of the scapula and clavicle. We turn now to the evolution of their overall positioning. How does this square with the expansion of in the number of cervical vertebrae? The story of fins-to-limbs is a sequence of events: (1) the addition of the fourth occipital somite, (2a) the creation of an articulation between skull and spine, (2b) loss of bony connections between the head and pectoral girdle, (4) addition of cervical units, (3) repositioning of the pectoral girdle and brachial plexus, and (5) development of a specialized pectoral girdle. Let's look at this sequence in greater detail.

The first evidence of paired pectoral fins with attachment to the axial skeleton occurs in the agnathan armored fishes, the osteostracans. As we know, these creatures had a massive head shield of dermal bones. A single endochondral bone, the future scapulocoracoid, articulated with the posterolateral side of the head. In a later group of placoderms, the †arthrodires, two important changes are noted. First, dermal elements are incorporated into a cleithral group (supracleithrum, postcleithrum or anocleithrum, dorsal cleithrum, and ventral cleithrum) and a clavicle. The latter bone is thought to arise from a modified dermal scale. The dermal elements are much larger than scapulocoracoid. Second, the entire pectoral complex is anchored to a single cranial bone, the posttemporal.

In subsequent osteoichththyans, the dorsal and ventral cleithra unite into a single bone. The dermal bones become regularized and identifiable. Anterior to the pectoral girdle the opercular and postopercular bones, part of the 4-bone opercular complex covering the gills. Above the pectoral girdles an extrascapular series is interposed between them it and the postparietals.

Dissociation of the pectoral girdle takes place in the Devonian period with the transitional sarocopterygian, *Tiktaalik*. Because it had well-developed gills and no jaw features of tetrapods, this important creature stands between the true crown sarcoptyerygians, *Panderichthys*, and the tetrapod *Acanthostega*. This is where we shall place our attention. Gone are the intervening extrascapulars, opercular and subopercular, eliminating all bony connection between pectoral girdle and the skull. The tabulars and postparietals move into position as a replacement for the extrascapulars. These provide enhanced insertions for cervical muscles.

Many changes are noted in the skull. The postfrontal is elevated above the level of the orbital margin, forming a ridge consistent with *dorsalization of the eyes for better vision on land* (more on this concept later). The ethmosphenoid lengthens relative to the oto-occipital complex. This consistent with the development of a snout. The dermal intracranial joint is lost and the palate is less mobile. The effect of this crocodile-like head is to increase the biting surface. The middle ear of tetrapods is anticipated by a widening of fish spiracle as a potential communication into the pharynx and alterantions in the hyomandibula which is no longer bomerang-shaped but short, straight, and stout. It is the precursor for stapes in *Acanthostega*.

Bone changes also have respiratory implications. The skull is wider with a more voluminous buccal cavity. Expansion of gular plates and branchial elements support more aggressive buccal pumping of air. In obligate gill-breathers hyomandibular is involved in the forces associated with pumping of water through the gills. This function is unnecessary for terrestrial life so hyomandibula is reassigned to the ear. Decreased reliance on water pumping is associated with increased use of air breathing.

Tiktaalik has approximately 45 presacral vertebrae (Acanthostega has 30). Cervical vertebrae are not documented. The ribs are expanded and wider than those of *Panderichthys*. Plate-like flanges extend caudally. These changes presumably assisted in thoracolumbar support necessary for weight bearing.

Finally, there are immediate advantages of cranial mobility and a neck. Fishes can readily orient themselves in water to position the mouth for prey capture. On land, body position is fixed so the head position becomes independent. Increasing sophistication of control of extremities. *Tiktaalik* lived in the mud flats and was able to lift itself out of the water and snap at its target [53].

What mechanisms can be inferred from Tiktaalik? Disconnection of the pectoral girdle from the skull involves the loss of three series. The posttemporal is a dermal bone of the temporal series that directly connects the pectoral girdle with the r6–r7 opisthotic. Remaining behind in tetrapods from this series are intertemporal, supratemporal, and tabular bones. Medial and lateral *extrascapular bones* are gone as are the *opercular bones*. But the mesenchyme is likely not lost…it is merely reassigned. Experimental work tracing the homeotic signature of the posttemporal may reveal the final destiny of this mesenchyme. Could it be converted to fascia or muscle?

Addition of the fourth occipital somite and first cervical somite likely occurs as a two-step process, as suggested by the lungfish, protoatlas brought inside the braincase with the proatlas added in later. It could also have happened simultaneously. In any case, this brought additional homeotic complexity to the occiput. S4 joins with S3 to form the exoccipitals. By the time of *Greererpeton* the notochordal cotyle of the occiput disappears and the $S4_C$ moiety of the exoccipitals combines with $S5_R$ to make a single occipital condyle. Could the introduction of new paraxial mesoderm into the occiput be a contributing factor to the demise of the posttemporal?

Regarding lengthening of the neck, the selective advantages of a more cervical vertebrae giving a more mobile head for food gathering are obvious. Apart from the synapsids which fix the number at 7, diapsids are wildly diverse. The plesiosaur †*Muraenosauraus* had 76 cervical vertebrae! Snake lack them altogether. We have discussed the homeotic mechanisms previously.

The variations introduced by the reassignment of additional cervical neuromeres have little effect on the epaxial anatomy. Intersegmental muscles are simply added on. All the variations take place seen in hypaxial structures. The position of the pectoral girdle to the axial skeleton remains hypaxial, despite the apparent dorsal position of scapula. For this reason, all muscles connecting it to the head, neck, or trunk are innervated by accessory nerve r8–r11 (sternocleidomastoid, trapezius) or ventral roots of c1–c4 (the straps). Muscles from c5 to c8 (brachial plexus) that have primary insertion on scapula connect with the arm. Thus, pectoral girdle can be thought of as an island connecting two muscle groups: those of proximal cervical plexus with those of distal brachial plexus.

Dermatomal distribution is a particularly good example of hypaxial versus epaxial variation. As upper extremity is hypaxial, its entire skin envelope is hypaxial as well. Thus, ventral neck skin is mapped out into four quadrants representing C2–C4, whereas dorsal neck skin is a smooth continuum from C2 to C8 to T1–T12. Each dermatome is supplied by a segmental dorsal branch. All remaining ventral skin from C5 to T1 is expropriated into the arm. Therefore, a sharp boundary exists over the infraclavicular skin between the distribution of C4 and that of T2. The entire hypaxial region of C2–T1 is innervated by means of either the cervical or brachial plexus. Segmental ventral innervation does not recommence until level T2.

Finally, we must consider the simultaneous changes that took place within the upper limb itself. The homeotic regulation of appendicular development has been extensively documented. The homeotic code for each level consists of a combination of hox genes. These spatially distant regions may share homeotic genes common to both. What is certainly true of the tetrapod transition is that the axial patterning process controlled by homeotic genes affect patterning of multiple tissues in all three body axes during early organogenesis. Even minor disturbances can cause serious downstream effects. For this reason, the number of cervical vertebrae in mammals is highly conserved. At stake are many traits involving craniofacial structures, larynx/trachea/lungs, the kidneys, and limbs. This explains the large number of craniofacial syndromes reported.

The Evolutionary Impact of an Enhanced Visual System

Life for predators is a simple equation: you eat what you kill and you can't kill what you can't perceive. Water is a difficult medium for visual hunting. Low levels of light and the density of the medium pose extreme limitations on predation. MacIver and Schmitz applied a novel approach to understand the transition to terrestrial life: the evolution of visual acuity. Measurements of eye socket volume and the anatomic position of the spiracle in 59 tetrapodomorph taxa that span the

water-land transition yield startling results. Eyes tripled in size with vision through air improved improving a one million-fold increase in the amount of space within which objects (prey) could be perceived. The orbits, though still lateral, moved upward on the skull. At the same time changes in breathing that evolved as a compensation for the reduced oxygen content of the Devonian favored crocodilian predatory behavior by looking for prey just above the water line [54] (Fig. 10.143).

It all started with the sarcopterygians (no surprise there). The changes are seen in three stages: finned tetrapods, transitional tetrapods, and digital tetrapods. From the osteolepiform *Eusthenopteron* to the panderichthid *Panderichthys* orbital volume went up 1.42×. The transitional *Tiktaalik* had a volume increase of 1.43×. This exponential expansion peaked with Acanthostega which has an orbital volume 1.52× baseline. Enhanced socket size correlates directly with larger eyes and larger pupils (Fig. 10.144).

The depth of visual field for *Eusthenopteron* was roughly its body length, Although these evolutionary changes led to a tripling of eye size, they were not of great use underwater. However, when aerial vision became available, depth of field increased to 100 body lengths and the gain in visually monitored space under light conditions increased 1,000,000-fold.

The position of the orbits changed as well. Beginning with *Panderichthys* the eyes moved to the top of the head. The eyes of *Tiktaalik* are definitely crocodilian. Elevated bony prominences caused by the prefrontal and postfrontal bones appear at this time, the harbinger of supraorbital ridges. This proved to be a critical innovation because it permitted Tiktaalik's eyes to be out of water while at the surface. Of note, the changes seen in the orbits of Panderichthys are part of a more radical *restructuring of the face* characterized by primarily by elongated jaws, an adaptation for more successful prey capture. The palate widened and the skull became flatter, presenting a lower along the surface of the water [55, 56].

A third factor contributing to enhanced predation was respiration. Water is a difficult medium for oxygen capture compared with air. Not only is it 800× more dense, but its O_2 content is 1/30th that of air. Aqueous respiration requires a mass flux 24,000 time greater per unit of extracted oxygen. Devonian conditions were terrible due to a decline in the amount of available oxygen. Sarcopterygians responded by enlargement of the breathing passageways called spiracles. The spiracle was derived from the first branchial cleft between the first and second branchial arches in agnathic fishes. When jaws were invented and the first two arches merged together, the remnant of the cleft became the spiracle. Because it sits in front of the otic capsule (the future ear), it is also known as the otic notch. It tetrapods the spiracle is

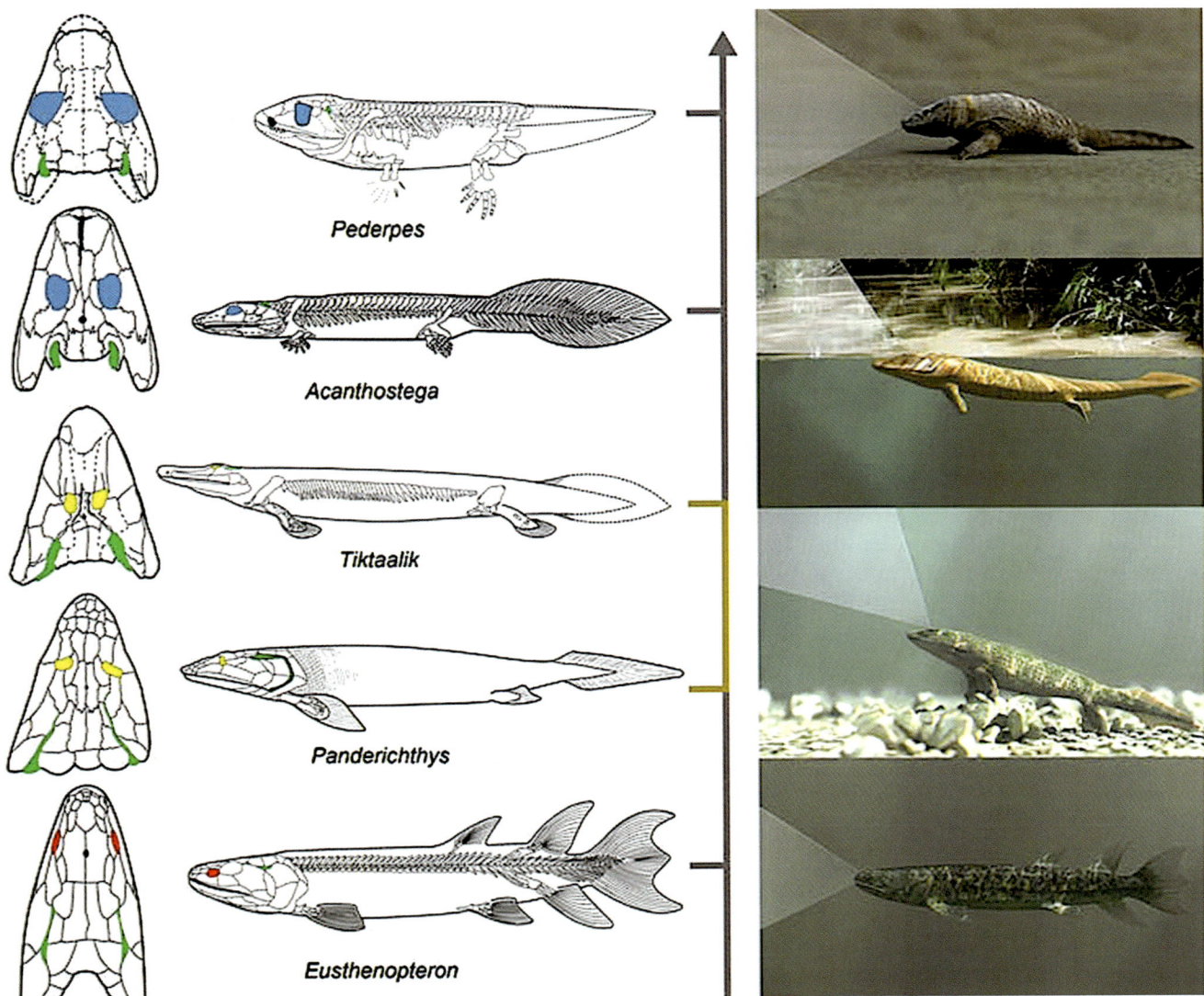

Fig. 10.143 Evolution of visual range. [Reprinted with permission from MacIver MA, Schmitz L, Mugan U, Murphy TD, Mobley CG. Massive increase in visual range preceded the origin of terrestrial vertebrates. *Proc Natl Acad Sci U S A.* 2017 Mar 21;114(12):E2375–E2384. With permission from Copyright © 2017 National Academy of Sciences]

converted into the Eustachian tube. *Repositioning of the spiracle* is seen in elpistostegalians. *Panderichthys* had a flattened skull, resembling that of a crocodile; it could lurk at the water's edge, acquire oxygen with less energy consumption, and scan the environment for prey. This provided an intermediate step between specializations for aquatic predation and innovations propitious for brief ventures onto land without a commitment to a full-time terrestrial existence.

In sum, enhanced visual acuity, a much larger depth of field, and improved respiration in the aerial environment were propitious for exploration and penetration of the terrestrial environment. Significant neuroanatomic and behav-

ioral benefits accrued. Short-range vision (single body length) permits reactions to just-in-time stimuli. Long-range permits more complex forms of decision-making using different neurocircuitry. Multiple options for pursuit of prey or evasion of predators were developed by trial-and-error behaviors. These are dependent on the hippocampus for memory and learning. Hippocampus is present in the common amniote ancestor of birds and mammals which emerged during the late Carboniferous shortly after tetrapods hit the land. We can therefore postulate that the primitive neural elements of planning were created in part for the "visual explosion" immediately prior to emergence onto land.

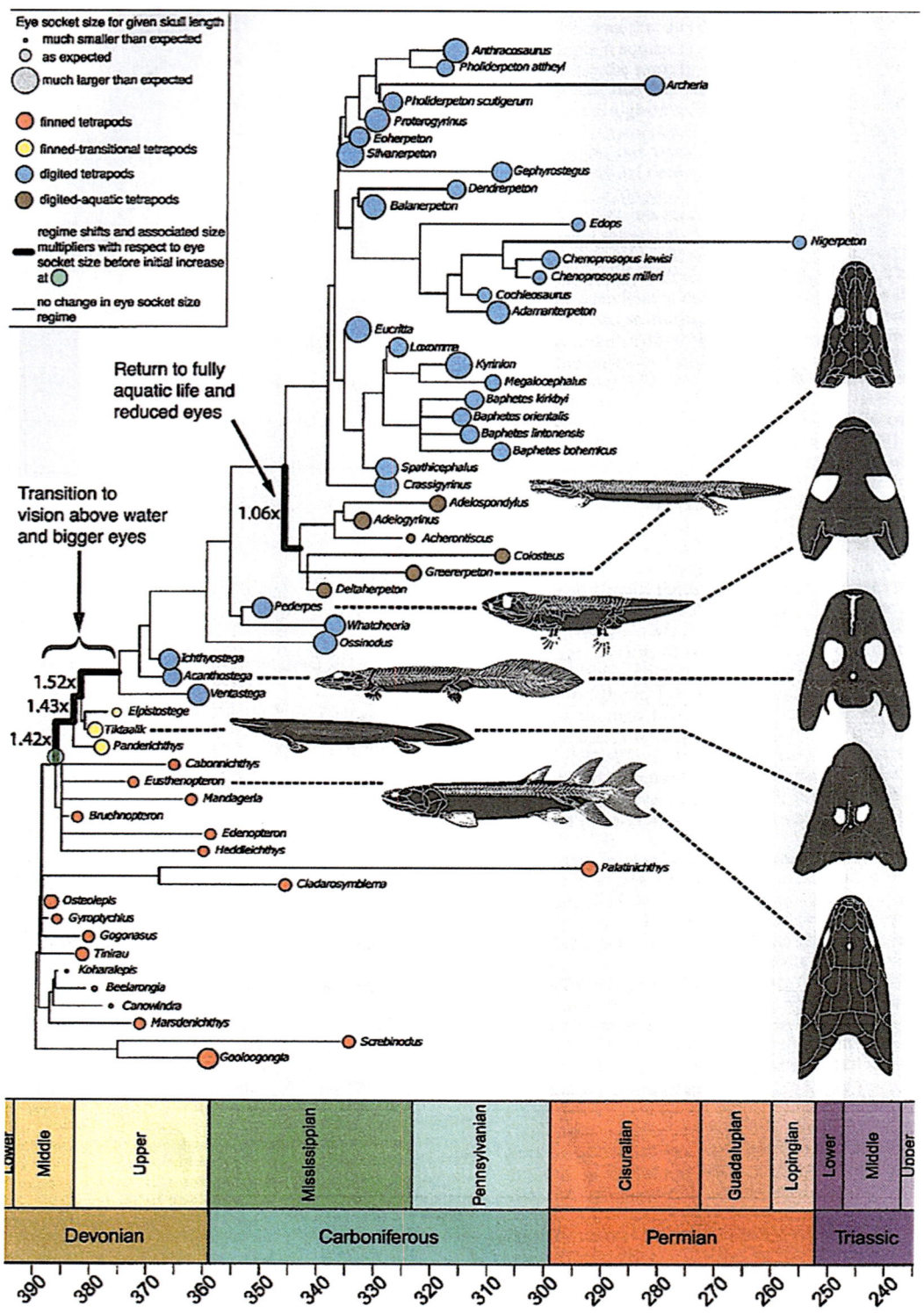

Fig. 10.144 Resizing and repositioning of the orbit and spiracle Having invaded shallow waters, with access to more, translocation of eye sockets moved to the top of the skull, providing upward vision, as shown here for *Panderichthys*. Low oxygen conditions favored, breathing through spiracles dorsalized to just behind the eyes in the elpistostegalians, as shown here for *Tiktaalik*. With continued surfacing and selection of changes to lens and cornea enabling a focused image of prey, during the 12-My transition from finned to digited tetrapods, enlarged eye sizes developed, enabling the full power of long-range vision. Simultaneously, with the selective advantages of limbs, animals like *Acanthostega* became better suited for longer forays onto land, cul-minating in more terrestrial forms, such as *Pederpes*, 30 My after *Tiktaalik*. Spiracles (green), the precursors of the Eustachian tube, were likely used for breathing at the water surface while using aerial vision. Eyes dorsalize between *Eusthenopteron* and *Tiktaalik*, and enlarge in tetrapods. S Total animal lengths are between 50 cm and 1.5 m (not drawn to scale). Age spans from 385 My for *Eusthenopteron* to 355 My for *Pederpes*. [Reprinted with permission from MacIver MA, Schmitz L, Mugan U, Murphy TD, Mobley CG. Massive increase in visual ragne preceded the origin of terrestrial vertebrates. *Proc Natl Acad Sci USA.* 2017 Mar 21;114(12):E2375–E2384. With permission from Copyright © 2017 National Academy of Sciences]

The Buena Vista Hypothesis and the Neuromeric Model

MacIver proposed that greatly enhanced vision and improved access to oxygen led to energy-efficient predation and facilitated the transition to land where new forms of information-gathering could be exploited for expansion into multiple ecologic niches [57]. In the follow-up study using orbital volumes the explosion in vision occurred in sarcopterygians while under water, an environment in which this innovation was undoubtedly of value, albeit limited, for hunting. What is striking are the changes in the rest of the face, especially involving dermal bones from r2 and r3 neural crest. Virtually all the changes seen in the orbits reside in changes in size or position of the r2 maxillary bone complex and palate which expanded the space available for a larger globe. Mirror-image changes took place in the r3 mandible which lengthened concomitantly. Because the spiracle was positioned

between the r2–r3 first arch and the r4–r5 second arch, it was passively brought forward. The sum of these innovations, plus a modest increase in aquatic vision meant that, under low oxygen conditions, surface predation brought sudden and unexpected benefits.

Restructuring of the sarcopterygian face occurred at the same time as fins became digit-bearing. It is possible that a common denominator exists between the homeotic driver of the pectoral apparatus and the altered hox code of the face. In any case, once these adaptations were present in *Panderichtys*, the stage was set for Tiktaalik to separate the pectoral girdle from the skull and convert the gas bladder to a lung-like organ. In the process, an articulation was created between the skull and pectoral girdle (Fig. 10.145). Areas for extensive cervical muscle insertion appeared along the dorsum of the skull, permitting head mobility [36, 58]. Further refinements in the craniovertebral joint took place in the digital tetrapods: the neck was born (Fig. 10.146).

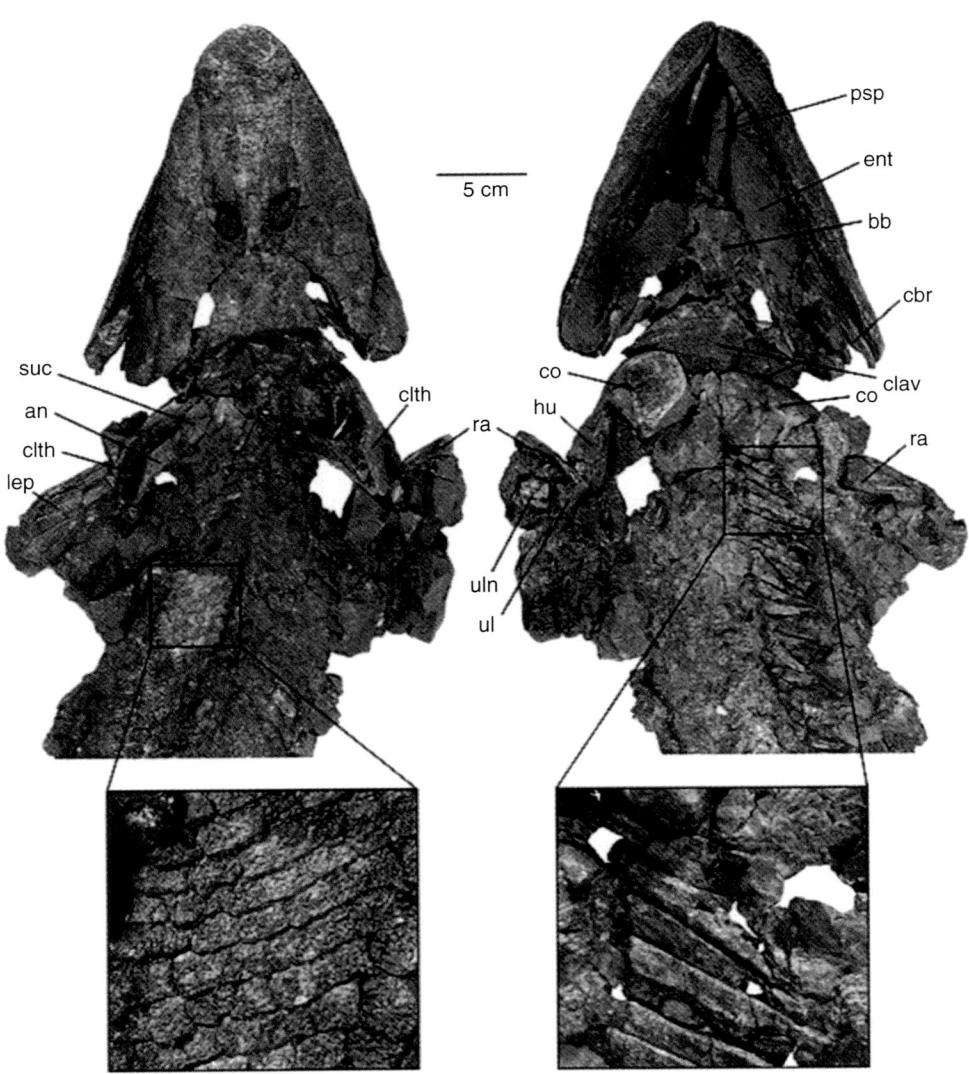

Fig. 10.145 Association of the clavicle with the first cervical vertebra. This establishes the original neuromeric coding of clavicle with c1 neural crest. Left; Dorsal view, note the presence of a cervical vertebra flanked laterally by cleithrum. The exact nature of the articulation is not well-defined. Right ventral view showing clavicle closely apposed to the skull beneath the first cervical vertebra. Abbreviations: an, anocleithrum; bb, basibranchial; co, coracoid; clav, clavicle; clth, cleithrum; cbr, ceratobranchial; ent, entopterygoid; hu, humerus; lep, lepidotrichia; mand, mandible; nar, naris; or, orbit; psp, parasphenoid; ra, radius; suc, supracleithrum; ul, ulna; uln, ulnare. Scale bar equals 5 cm [Reprinted from Daeschler EB, Shubin NH, Jenkins FA. A Devonian tetrapod-like fish and the evolution of the tetrapod body plan. *Nature* 2006; Vol 440:757–763. With permission from Springer Nature]

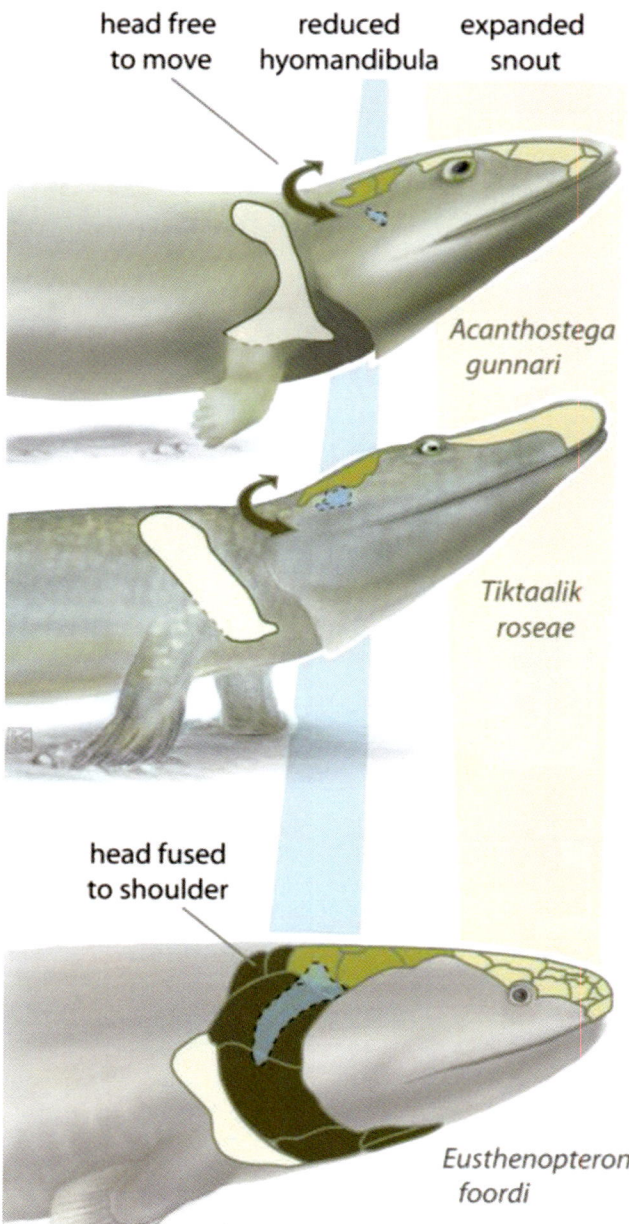

head free **reduced** **expanded**
to move **hyomandibula** **snout**

Acanthostega gunnari

Tiktaalik roseae

head fused
to shoulder

Eusthenopteron foordi

Fig. 10.146 Birth of the neck. [Reprinted from Shubin NH, Daeschler EB, Jenkins FA Jr. The pectoral fin of *Tiktaalik roseae* and the origin of the tetrapod limb. *Nature1 2006; 440:764–771*. With permission from Springer Nature]

Clinical Impact of Homeotic Transformation

A radiologic survey of autopsy material involving 1389 fetuses/infants with detectable ossification centers was carried out to assess the effects of vertebral anomalies. Homeotic pathologies were extremely common with cervical ribs present in/3 of the cases. Only 20% had normal vertebral counts (7 cervical, 12 thoracic, and 5 lumbar). The severity of malformations was correlated with the severity of vertebra aberrancy. These included craniofacial clefts and mesenchy-

mal deficiencies or excesses. Pathologies seemed to indicate local signaling problems rather than generalized defects of germ layer migration. For example, the development of cleft lip/anterior palate due to a defect of premaxilla likely has to do with downstream event involving the specific vascular axis (medial nasopalatine) rather than a problem upstream in the original neural crest population migrating from the second rhombomere [59].

Vertebral events with disturbances in A-P patterning were likely local with the majority being ipsilateral, involving genetic expressions on one side only, such as unequal ribs, and hemivertebra. These suggest strong coupling between the A–P patterning of paraxial mesoderm and the preservation of left-right symmetry. Craniofacial NC contributes strongly to the construction of multiple structures, including the heart, lungs, GI tract, and brain whereas trunk-derived neural crest has only minor, very conscripted roles. For this reason, craniofacial neural crest malformations had an extremely high degree of vertebral abnormality, more so than those involving trunk-derived neural crest. What all these points to are that early organogenesis is intensively interactive with the process of axis formation because gastrulation and segmentation lay down the neuromeric building blocks for the rest of the body.

Vertebral A–P pattern defects involving homeotic transformations are not randomly distributed over the vertebral column. Only certain regions are affected. Most changes involve frameshifts in one or at most two neuromeric levels. For example, the cervical region can vary between 6 to 8 vertebrae: anything different is lethal. The effects can be physical or biochemical. Shifts in vertebral boundaries usually do not coincide with the boundaries of outlying tissues. Compression of vascular or neural structures can result. Lumbo-sacral transitions are associated with intervertebral disc compression or degeneration and narrowing of the pelvic outlet in women affecting the birth canal. Abnormal boundaries can indicate a reduction in the mesenchyme that affect the expression of gene products critical for development either locally or peripherally. Thus, changes in a single neuromere can wreak havoc on an entire organ system.

Conclusion

In this chapter, we have examined the neck from the standpoint of both its anatomic components as well that of a multi-organ system communicating between head and trunk. In every instance, neuromeric coding was useful in understanding tissue origins and assembly. Mesenchyme from neural crest, paraxial mesoderm, lateral plate mesoderm, and epithelial tissues such as endoderm and ectoderm, all arise from specific sites of the neuromeric map with consequent restrictions placed upon the genetic expression in each neuromeric zone. We then proceeded to review the process by which the neck is constructed, taking care to respect the neuroanatomy of the brachial plexus. Although we are left with many more questions than answers the pathway forward to understanding the development of the neck will surely involve the neuromeric mapping of its parts and the tracing of the final derivatives back to their beginnings.

A somitogenesis model integrating the segmentation clock and determination front. The system of opposing gradients of FGF (fibroblast growth factor)–Wnt signaling (purple) and retinoic acid (green) signaling was proposed to position the determination front (black line) along the presomitic mesoderm (PSM). This particular level is characterized by a signaling threshold at which the cells become competent to respond to the segmentation clock signal and is conceptually similar to the wavefront of the original Cooke and Zeeman model 14. The clock signal is still poorly characterized but probably involves three signaling pathways experiencing periodic activity: FGF, Wnt, and Notch. The wave of cyclic gene expression controlled by the segmentation clock oscillator is shown in orange on the left side of the embryos. When competent cells that pass through the determination front receive the clock signal, they simultaneously activate Mesp2 (mesoderm posterior 2; shown in black), thereby defining the future segmental domain as shown on the right side of the embryos. In this model, the size of the segment (the future somite) is defined by the distance traveled by the wavefront during one oscillation of the segmentation clock. However, the role of retinoic acid in this model remains debated (see text). During the next cycle Mesp2 expression becomes restricted to the anterior compartment of S–I (grey). T, time in segmentation clock cycle unit. [Reprinted from Dequéant M-L, Pourquié O. Segmental patterning of the vertebrate embryonic axis. Nature Rev. Genet 2008; 9(5):370–382. with permission from Springer Nature]

References

1. Bénazéref B, Pourquié O. Formation and segmentation of the vertebrate body axis. Annu Rev Cell Dev Biol. 2013;29:1–26. https://doi.org/10.1146/annurev-cellbio-101011-155703. Epub 2013 Jun 26. Review PMID: 23808844.
2. Pang D, Thompson DNP. Embryology and bony malformations of the craniovertebral junction. Childs Nerv Syst. 2011;27:523–64.
3. Shishkin MA. Evolution of the cervical vertebrae in Temnospondyl amphibian and differentiation of the early tetrapods. Paleontol J. 2000;34(5):534–46.
4. O'Rahilly R, Müller F. Human embryology and teratology. 3rd ed. New York: Wiley-Liss; 2001.
5. Menezes AH, Fenoy KA. Remnants of occipital vertebrae proatlas segmentation abnormalities. Neurosurgery. 2009;64(5):945–54.
6. Spittank H, Goehmann U, Hage H, Sacher R. Persistent proatlas with additional segmentation of the cranio-vertebral junction: the Tsuang-Goehmann malformation. Radiol Case. 2016;10(10):15–23.
7. Onimaru K, Shoguchi E, Kratani S, Tanaka M. Development and evolution of the lateral plate mesoderm: compariative analysis of amphiosus, and lamprey with implircations for the acquisitions of acquired fins. Dev Biol. 2011;359:124–36.
8. Matsuoka T, Ahlberg PE, Kessaris N, Iannarelli P, Dennehey U, Richardson WD, McMahon AP, Koentges G. Neural crest origins of the neck and shoulder. Nature. 2005;436(7049):347–55.
9. Kierner AC, Zelenka AM, Reidel G, Burian M. Blood supply of SCM muscle and its clinical implications. Arch Surg. 1999;134(2):144–7.
10. Nagashima H, Sugahara F, Watanabe K, Shibata M, Chiba A, Sato N. Developmental origin of the clavicle and its implications for the evolution of the neck and the paired appendages in vertebrate. J Anat. 2016;229:536–48.
11. Ericsson R, Knight R, Johanson Z. Evolution and development of the vertebrate neck. J Anat. 2013;222:67–78.
12. Huang R, Quia Z, Patel K, Wilting J, Christ B. Contributions of single somites to the skeleton and muscles of the occipital and cervical regions in avian embryos. Anat Embryol. 2000a;202:375–83.
13. Huang R, Zhu Q, Patel K, Wilting J, Christ B. Dual origin and segmental organization of the avian scapula. Development. 2000b;127:3789–94.
14. McGonnell IM. The evolution of the pectoral girdle. J Anat. 2001;199:181–94.
15. Lachman N, Acland RD, Rosse C. Anatomical evidence for the absence of a morphologically distinct cranial root of the accessory nerve in man. Clin Anat. 2002;15:4–10.
16. Tada MN, Kuratani S. Evolutionary and developmental understanding of the spinal accessory nerve. Zool Lett. 2015;1:4. https://doi.org/10.1186/s40851-014-0006-8.
17. Singh PH, Kumar S, Agarwal SP. Congenital asymptomatic absence of unilateral sternocleidomastoid muscle. BMJ Case Rep. 2014;2014:bcr2013202786. https://doi.org/10.1136/bcr-2013-202786.
18. Vajramani A, Witham FM, Richards RH. Congenital unilateral absence of sternocleidomastoid and trapezius muscles: a case report and literature review. J Pediatr Orthop B. 2010;19(5):462–4. https://doi.org/10.1097/BPB.ObO13e32833ce404.
19. Hacein-Bey L, et al. The ascending pharyngeal artery: branches, anastomoses, and clinical significance. Am J Neuroradiol. 2002;23:1246–56.
20. Leclère FM, Vacher C, Benchaa T. Blood supply to the human sternocleidomastoid and its implications for mandible reconstruction. Laryngoscope. 2012;122(11):2402–6.
21. Emsley JG, Davis MD. Partial unilateral absence of the trapezius muscle in a human cadaver. Clin Anat. 2001;14:383–6.
22. Allouh M, Mohamed A, Mhanni A. Complete unilateral absence of trapezius muscle. McGill J Med. 2004;8:31–3.
23. Garbelotti SA Jr, de Sousa Rodrigues CF, Sgrott EA, Prates JC. Unilteral absence of thoracic part of the trapezius muscle. Surg Radiol Anat. 2001;23(2):131–3.
24. Bergin M, Elliott J, Jull G. The case of the missing lower trapezius muscle. J Orthop Sports Med. 2011;41(8):614–5.
25. Tigga SR, Goswami P, Khanna J. Congenital partial absence of trapezius with variant pattern of rectus sheath. Acta Med Iran. 2016;54(4):280–2.
26. Mehra L, Tuli A, Raheja S. Dorsoscapularis triangularis: embryological and phylogenetic characterization of a rare variation of trapezius. Anat Cell Biol. 2016;49:213–6.
27. Witbreuk MM, Lambert SM, Eastwood DM. Unilateral hypoplasia of the trapezius muscle in a 10-year old boy: a case report. J Pediatr Orthop B. 2007;16:229–32.
28. Yiyit N, Isitmangil T, Öztúrker C. The abnormalities of trapezius muscle might be a component of Poland's syndrome. Med Hypotheses. 2014;83:533–6.

29. Nooij LS, Oostra RJ. Trapezius aplasia: indications for a dual developmental origin of the trapezius muscle. Clin Anat. 2006;19:547–9. https://doi.org/10.1002/ca.20325.

30. Trinajstic K, Sanchez S, Dupret V, Tafforeau P, Ahlgren PE. Fossil musculature of the most primitive jawed vertebrates. Science. 2013;341(6142):160–4. https://doi.org/10.1126/science.1237275. Epub 2013 Jun 13. PMID: 23765280

31. Sefton EM, Bhullar B-A, Mohades Z, Hanken J. Evolution of the head trunk interface in tetrapod vertebrates. elife. 2016;5:e09972. https://doi.org/10.7554/eLife.09972.

32. Barry MA. Central connections of the IXth and Xth cranial nerves in the clearnose skate, *Raja eglantera*. Brain Res. 1987;425:159–66.

33. Boord RL, Sperry DG. Topography and nerve supply of the cucullari (trapezius) of skates. J Morphol. 1991;207:165–72.

34. Marion GE. Mandibular and pharyngeal muscles of *Acanthius* and *Raja*. Am Nat. 1905;39:891–924.

35. Sperry DG, Boord RL. Central location of the motoneurons that supply the cucullaris (trapezius) of the clearnose skate, *Raja eglanteria*. Brain Res. 1992;582:312–9.

36. Daeschler EB, Shubin NH, Jenkins FA. A Devonian tetrapod-like fish and the evolution of the tetrapod body plan. Nature. 2006;440:757–65.

37. Ma L-H, Gilland E, Bass AH, Baker RS. Ancestry of motor innervation to pectoral fin and forelimb. Nat Commun. 2010;1:49. https://doi.org/10.1038/ncomms1045.

38. Murakami Y, Tanaka M. Evolution of motor innervation to vertebrate fins and limbs. Dev Biol. 2011;355:164–72.

39. Stifani N. Motor neurons and the generation of spinal motor neuron diversity. Front Cell Neurosci. 2014;8:293. https://doi.org/10.3389/fncel.2014.00293.

40. Pickering M, Jones JFX. The diaphragm: two physiological muscles in one. J Anat. 2002;201:305–12.

41. Merrell AJ, Kardon G. Development of the diaphragm, a skeletal muscle essential for mammalian respiration. FEBS J. 2013;280(17):4026–35. https://doi.org/10.1111/febs,12274.

42. Fisher JC, Bodenstein L. Computer simulation analysis of normal and abnormal development of the mammalian diaphragm. Theor Biol Med Model. 2006;3:9. https://doi.org/10.1186/1742-4682-3-9.

43. Toran N, Eery J. Congenital bilateral absence of diaphragm. Thorax. 1891;36:157–8.

44. Clugston RD, Greer JJ. Diaphragm development and congenital diaphragmatic hernia. Semin Pediatr Surg. 2007;16:94–100.

45. Kulkarni ML, Sneharoopa B, Vani HN, Nawaz S, Kannan B, Kulkarnin PM. Eventration of the diaphragm and associations. Indian J Pediatr. 2007;74:202–5.

46. Mirza B, Bashir Z, Sheikh A. Congenital right hemidiaphragmatic agesis. Lung India. 2012;29(1):53–5.

47. Ravisagar P, Abhinav S, Mathur RM, Anua S. Eventration of diaphragm presenting a recurrent respiratory tract infections—a case report. Egypt J Chest Dis Tuberc. 2015;64:291–3.

48. Hirasawa T, Kuratanin S. A new scenario of the evolutionary derivation of the diaphragm from shoulder muscle. J Anat. 2013;222:504–17.

49. Hirasawa T, Fujimoto S, Kuratani S. Expansion of the neck reconstituted the shoulder-diaphragm in amniote evolution. Dev Growth Evol. 2016;58:143–53.

50. Kakodkar KA, Shroeder JW Jr, Holinger LD. Laryngeal development and anatomy. In: Hartnik CJ, Hansen MC, Gallaghen TQ, editors. Advances in oto-rhino-laryngology, pediatric airway surgery, vol. 73. Basel: S Karger, AG; 2012. p. 1–13.

51. Brusca RC, Brusca GJ. Invertebrates. Sunderland: Sinauer; 1990.

52. Franch-Marro X, Martin M, Averof M, Casanova J. Association of trachal placodes with leg primordia in Drosophila: implication for the origin of the insect tracheal system. Development. 2006;133:785–90.

53. Ahlberg PE, Clack JA. A firm step from water to land. Nature. 2006;440:747–9.

54. McIver MA, Schmitz L, Mugan U, Murphey TD, Mobley CD. Massive increase in visual range preceded the origin of terrestrial vertebrates. Proc Natl Acad Sci USA. 2017;114(12):E2375–84. https://doi.org/10.1073/pnas.1615563114. Epub 2017 Mar 7. www.pnas.org/cgi/doi/10.1073/pnas.1615563114.

55. Ahlberg PE, Clack JA, Luksevics E. Rapid braincase evolution between Panderichthys and the earliest tetrapods. Nature. 1996;381:61–4.

56. Schultze M, Arsenault M. The Panderichthid fish *Elpigostega*—a close relative of Tetrapods? Paléo. 1985;28(2):293–309.

57. McIver MA. Neurobiology: from morphological computation to planning. In: Robbins P, Aydede M, editors. The Cambridge handbook of situated cognition. New York: Cambridge Univ. Press; 2009. p. 480–504.

58. Shubin NH, Daeschler EB, Jenkins FA. The pectoral fin of *Tiktaalik roseaa* and the origin of the tetrapod limb. Nature. 2006;440:764–71.

59. ten Broek CMA, Bugiani M, Gallis E. Evo-devo of the human vertebral column: on homeotic transformation, pathologies, and prenatal selection. Evol Biol. 2012;39:456–71.

Developmental Anatomy of Craniofacial Skin, Fat, and Fascia

Michael H. Carstens

Introduction

Craniofacial skin is surprisingly diverse. Depending upon the tissue source, the cutaneous coverage of the head and neck has nine different developmental zones. Why should we care about the neuromeric map of skin? The primary importance of this topic is to understand the interactions between epithelium and mesenchyme so important in specifying dermal appendages, teeth, and membranous craniofacial bones and cartilages. It also lends itself to appreciating the pathology of syndromes in which mesenchymal defects are seen in association with characteristic cutaneous manifestations (e.g. the blue sclera of Osgood-Schlatter or the facial clefts seen in congenital ichthyosis). The neuromeric map of dermis (and neural crest) is fundamental to the assembly of the meninges and craniofacial blood vessels. Failure of fronto-orbital development seen in trigonocephaly involves interaction between caudal prosencephalic dermis, r1 dura and r1 neural crest. Associated problems with neural crest pericytes required for proper development of transdural blood vessels may contribute to the neuropathology (developmental delay) seen in this condition. Finally, our knowledge of anatomy is enriched when we understand the developmental diversity of human skin.

As we have previously seen, gastrulation creates a trilaminar embryo organized along its axis into neuromeric zones, beginning at r0-r1 (henceforth referred to as r1) and proceeding backwards. Endoderm and mesoderm associated with each neuromeric level share a common homeotic expression pattern. A primitive "fate map" is laid out which is roughly Cartesian in nature. Recall that the embryo at the end of gastrulation is an ovoid disc. Emanating from the midline, these mesomeric–endomeric "stripes" fan out from both sides beneath the epiblast.

Until now, nothing has been said about the post-gastrulation state of ectoderm. Recall that one is not permitted to use this term loosely. *Ectoderm is strictly those epiblast cells that remain behind after the process of gastrulation has been completed.* Thus, ectoderm, because it has not undergone gastrulation, is *not* intrinsically neuromeric. It will acquire a neuromeric definition by virtue of the mesenchyme which comes to lie beneath its surface. *Hox* genes expressed by dermis eventually appear in the epidermis. Because skin is a compound organ, with the epidermis derived from ectoderm or neural crest, and the dermis from either neural crest or mesoderm, we shall explore both layers separately, and then together.

Non-neural Ectoderm

Models of Ectodermal Organization

Gastrulation creates a trilaminar embryo. The process begins at level r0-r1 and proceeds caudally. The first wave of cells at day 14 displaces the hypoblast or primitive endoderm (tan) to become *definitive intraembryonic endoderm* (yellow). At 16 days a second wave pushes its way in between the epiblast and endoderm to create intraembryonic mesoderm, IEM (red). Residual epiblast is neural (pink), unless it receives signals from the underlying mesoderm, in which case it becomes non-neural ectoderm (blue). Hypoblast gets pushed out of the way to become *extraembryonic endoderm*, EEE; it forms the yolk sac from the outer wall of which develops *extra-embryonic mesoderm*, EEM. Blood islands in the EEM form the endothelial cells of the vascular system and the precursors of hematopoietic cells. EEM also spread dorsally over the top of the embryo, thereby becoming interposed between the extra-embryonic ectoderm, that is, the amnion, and the trophoblastic tissues. Since the vascular system develops in mesoderm, the continuity between the EEM and the IEM ensures connection with the placenta and establishes life support [1].

M. H. Carstens (✉)
Wake Forest Institute of Regenerative Medicine, Wake Forest University, Winston-Salem, NC, USA
e-mail: mcarsten@wakehealth.edu

Common Neural Plate Model: Neural Border Zone

In the model, a binary choice is made at the time of gastrulation, in which the embryonic ectoderm is subdivided into two domains: the non-neuronal ectoderm and neural ectoderm. They give rise to the epidermis and central nervous system (CNS), respectively. At the boundary between these two domains lies a third: the neural border zone (NBZ). This domain consists of two precursor fields which gives rise to two distinct cell populations. Within the NBZ, the neural crest region lies medial; lateral to it is pre-placodal region. These are established in response to signals mediated by BMPs, Wnt, and FGFs. The response of each territory to these signals is different. In the PPR, members of the Six and Eya families, neural crest-specific transcription factors, thus consolidating a placode development program. Next, the PPR becomes further subdivided into distinct placodal territory, each one with a specific menu of transcription factors that will lead to specific sensory tissue. Thus, pre-placodal region segregates into individual cranial placodes: the adenohypophyseal, olfactory, lens, trigeminal, lateral line, otic, and epibranchial placodes (from anterior to posterior). Although these are critical for craniofacial development, they are not the end of the story. Another type of placode, found throughout the skin, is essential for the development of hair, and it is these epidermal placodes that we shall now direct our attention (Fig. 11.1).

Binary Competence Model

Ectodermal development can be conceptualized by a binary competence model proposed by Schlosser et al., in which only neural ectoderm is competent to form neural crest, while only non-neural ectoderm is competent to develop into placodal fates. Furthermore, these fates depend upon the physical location of the target cells within the NNE [The reader is referred to Patthy, Pieper, and Schlosser for additional references.] (Fig. 11.2).

Prior to gastrulation, ectoderm is competent to generate all ectodermal fates, but during gastrulation the competence to form neural crest becomes restricted to dorsal ectoderm,

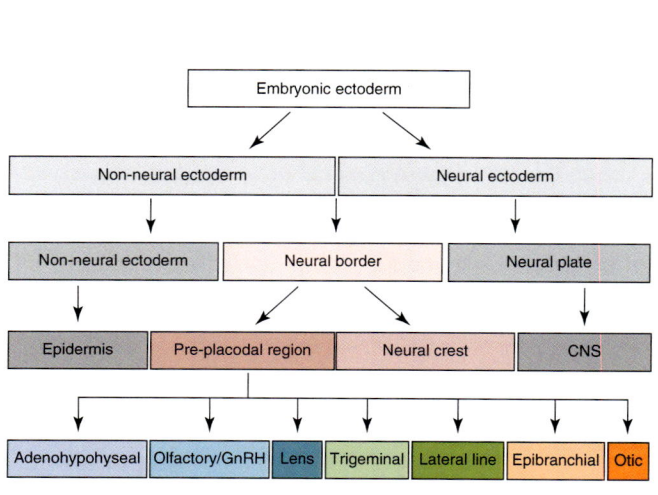

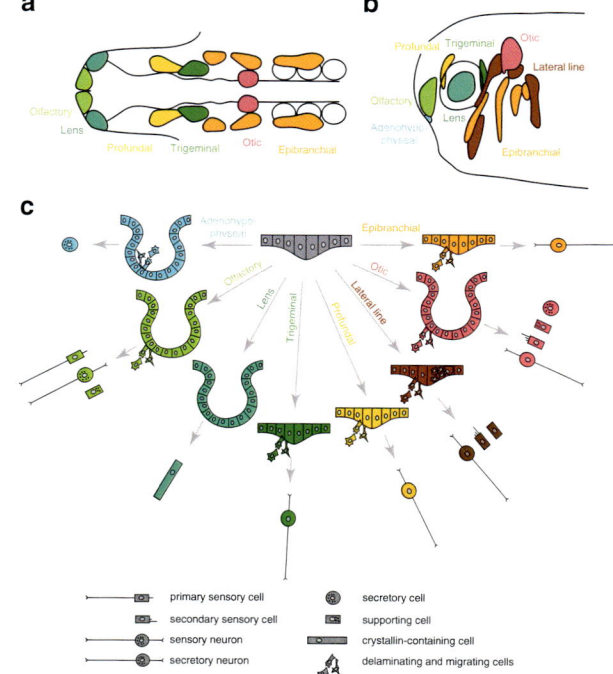

Fig. 11.1 Placodeal development: common neural plate model. Gastrulation creates a trilaminar embryo. At its conclusion the ectoderm remaining behind, the epiblast, makes fundamental choices. As to a neural versus non-neural fate. Fundamental for production of skin is the distribution of 3 components. (1) Neural crest cells from r0 to r11 produce the pericytes and pigment of craniofacial skin. Neural crest from the remaining neuromeres is primarily pigmentary. (1) Non-neural crest ectomesenchymal precursors called pre-pericytes interact with mesoderm of the somites to produce the adipose tissues of the body. The remain of the body fascia and I and placodal precursors. (3) Placodal cells are positioned outboard to neural crest. There are two models of the distribution of organization placodes: (1) the common neural plate model; and (2) the binary competence model. LEFT: Common neural plate assumes neural ectoderm subdivides between CNS and an intermediate neural border zone (IBZ) which is further divided into two strips with neural crest medial and placodes lateral. Epidermal NNE is considered distinct. RIGHT: Outboard NNE produces distinct craniofacial placodes which interacts with neural crest to form sensory ganglia of the face. Left: [Reprinted from Saint Jennet JP, Moody SA. Establishing the pre-placodal region and breaking it into placodes with distinct identities. *Dev Biol* 2014; 389(1): 13–27. With permission from Elsevier.]. Right: [Reprinted from Patthy C, Schlosser G, Shimeld SM. The evolutionary history of vertebrate placodes I: Cell type evolution. *Devel Biol* 2014; 389(1): 82–97. With permission from Elsevier]

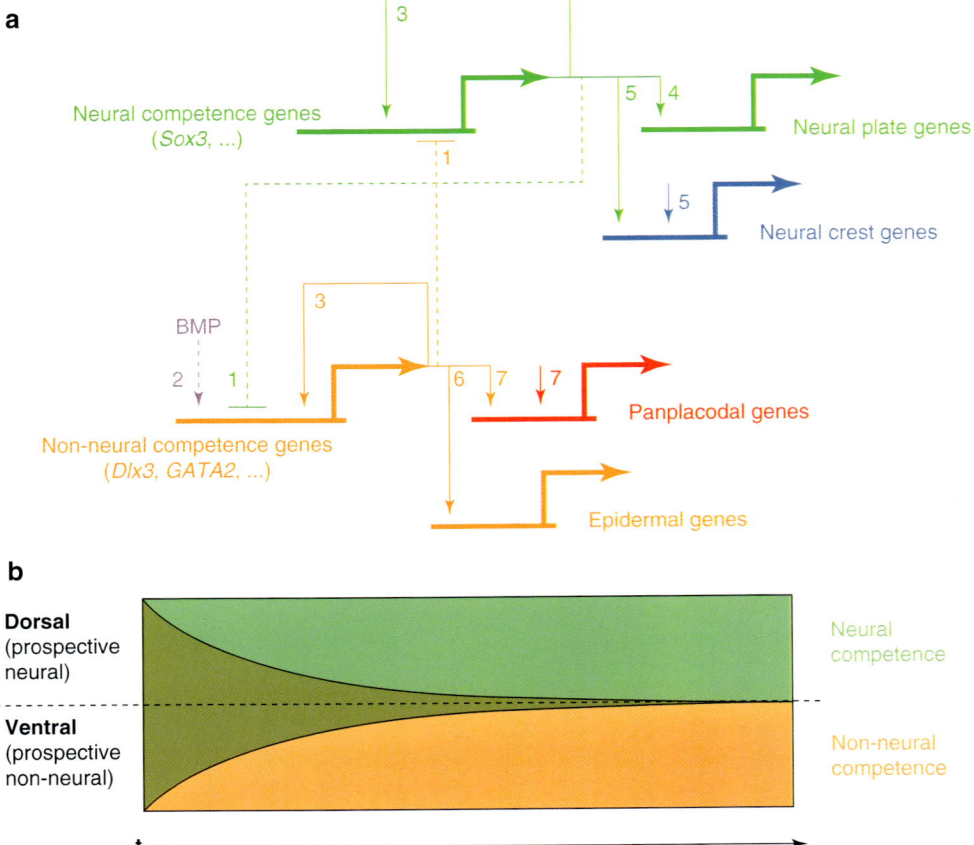

Fig. 11.2 Model for regulation of ectodermal competence. (**a**) Genes that promote non-neural competence, including *Dlx3* and *GATA2* (orange), and those that promote non-neural competence, probably including *Sox3* (green), cross-repress each others transcription (1, broken lines indicate indirect effects). Expression of non-neural competence genes is initially dependent on BMP signaling (2). In the presence of BMP, transcription of non-neural competence genes is therefore promoted over neural competence genes, whereas the reverse is true in the absence of BMP. However, persistent expression of these genes may lead to their autoactivation (3), thereby making their expression resilient to repression and BMP independent. Neural competence factors promote transcription of neural plate genes (4) or, in the presence of additional signals such as BMP, Wnt, and FGF (5), neural crest genes. Non-neural competence factors promote transcription of epidermal genes (6) or, in the presence of additional signals such as BMP inhibitors, FGFs, and Wnt inhibitors (7), panplacodal and placodal genes. (**b**) Owing to the dorsal secretion of BMP antagonists and crossrepressive interactions among competence genes, their initially overlapping expression domains will resolve into two distinct territories over time (t). [Reprinted from Pieper M, Ahrens K, Rink E, Peter A, Schlosser G. Differential distribution of competence for neural crest induction to non-neural and neural ectoderm. Development 2012; 139: 1175–1187. With permission from The Company of Biologists, Ltd.]

whereas the competence to express the panplacodal marker *Six1* becomes ventrally restricted. These changes parallel the dorsal restriction of neural competence, suggesting that, during gastrulation, two mutually exclusive competence territories are established: a <u>dorsal neural territory</u> that has *neural plate as a default state* but can be induced to form neural crest; and a <u>ventral non-neural</u> territory that has *epidermis as default state* but can be induced to form pre-placodal ectoderm. These findings confirm the predictions of the 'binary competence model'; but do <u>not</u> support models according to which a common neural plate border state is first established from which neural crest and placodes are subsequently induced [1].

Although ectoderm from early gastrulae can form all ectodermal fates, it will preferentially differentiate into neural plate or neural crest rather than panplacodal fates if exposed to signals in the neural plate border region, indicating that *neural competence overrules non-neural competence*. This suggests, that neural competence is the default state.

The Schlosser model is as follows:

- In ventral ectoderm, BMP expression promotes the expression of non-neural competence genes (e.g. *Dlx3* and *GATA2*; see below), which repress neural competence genes.

- In dorsal ectoderm, BMP inhibition relieves this repression of neural competence genes.
- Transcriptional cross-repression between neural and non-neural competence factors ensures stable boundaries.
- At the beginning of gastrulation (stage 6), the expression of neural and non-neural competence genes is still labile and BMP dependent, and the entire ectoderm, therefore, maintains neural as well as non-neural competence.
- By the end of gastrulation (stage 7), however, their expression becomes stabilized and BMP independent, e.g. due to auto-activation.
- At neural plate stages (stage 8).

    - neural competence factors promote transcription of *neural plate genes* or—in the presence of signals such as BMP, Wnt, and FGF—*neural crest genes*.
    - non-neural competence factors promote transcription of epidermal or—in the presence of signals such as BMP inhibitors, FGFs and Wnt inhibitors—*panplacodal genes*, whereas.

In agreement with this model, it has recently been shown that induction of neural crest and neural plate requires low BMP levels during gastrulation [2], whereas induction of many ventral transcription factors, pre-placodal ectoderm and epidermis requires high BMP levels. From neural plate stages onwards, the expression of many ventrally localized transcription factors becomes BMP independent, and BMP requirements for neural crest and pan-placodal induction change. BMP inhibition is now required for the induction of pre-placodal ectoderm, whereas some BMP signaling is required for neural crest induction. Because BMP signaling recedes from the neural plate border from stage 13 onwards, these later phase BMP requirements also explain why stage 12 and stage 13 NBs provide better-inducing environments for neural crest and placodes.

Epidermal Placodes: Origins

Hair is a defining characteristic of all mammals; it develops from a series of interactions between dermis and epidermis, the first step of which is the induction by the former of discrete epidermal placodes, one for each hair shaft. At the end of this chapter, we shall consider aspects of hair development and evolution but, for now, our focus is on the placodes themselves. Where did they come from? Preplacodal ectoderm directly abuts epidermal ectoderm. Neural crest cells and placodal cells both arise from the NBZ and likely share common characteristics. One possibility is that cells from the PPR have a previously unappreciated migratory ability and therefore distribute themselves into the epidermal ectoderm. Another (more likely) possibility is that placodal potential is intrinsic to all cells within the epidermal ectoderm and that

these make their appearance when appropriately stimulated from signals by mesenchymal cells in the dermis. Although hair distribution is generally uniform, unusual patterns are seen, characterized by the absence or excess of hair follicles. Such phenomena can be explained by either model but they can also occur as the result of localized deficits or excess of signals from the embryonic pre-dermal mesenchyme. In any case, the physical juxtaposition of epidermal ectoderm with the lateral region of the NBZ creates a situation in which transcription factors can have differential effects on primitive ectodermal cells depending on their physical location in the embryo.

The existence of cells with placodal potential in the epidermis thus has two different explanations, depending on one's model of ectodermal organization.

The common neural plate model presents a sorting-out in which neural crest genes with neural crest cells sort out with CNS genes and CNS cells. The so-called "third zone" or NBZ does not admix placodal genes with epidermal genes. The binary choice of NBZ cells to become neural crest versus placode cells depends on the physical location of the cells (with placodes being further lateral). This deficit means that either migration must occur, or the original model of segregation from the NBZ is incomplete (Fig. 11.1) (Saint Jennet, 2014).

The binary competence model also presents a binary choice in which neural crest genes sort out with neural plate cells and pan-placodal genes sort out with epidermal genes. The appearance of epidermal placodes during development can occur in two potential scenarios (Fig. 11.2) [3].

- Skin is a mixture of placodal (P) cells admixed with epidermal (E) cells. Thus, MSCs produce signals that selectively trigger the P cells and not the E cells.

- Placodal cells and epidermal cells are one and the same.
- The "decision" by a skin cell to adopt a P fate and become an epidermal placode is a matter of its immediate proximity to a local diffusion gradient of signals produced by MSCs.

Phylogeny of Placodes

The evolution of placode faithfully reflects increasing degrees of genetic complexity. Innovations in ectodermal patterning came about from new mechanisms and the origin of new cell types (Fig. 11.3). Important nodes are numbered and the key depicts which characters can be traced to each node. Character origination events whose placement on the tree is controversial are indicated by a question mark. AP: anteroposterior; DV: dorsoventral; TF: transcription factor [4].

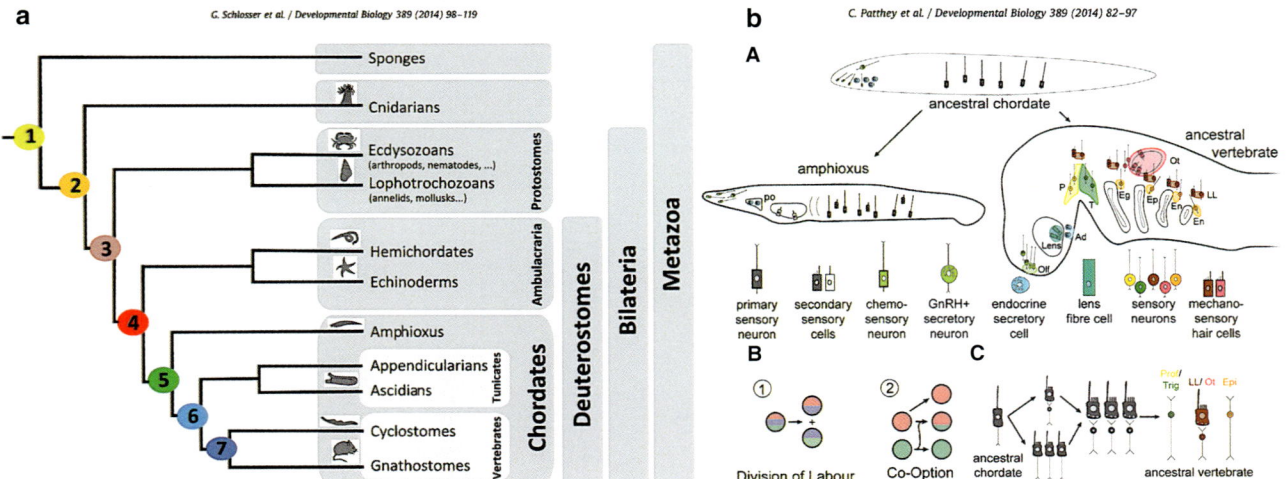

Fig. 11.3 Phylogeny of platcodes. Left (**a**): Evolutionary steps in placodal patterning and origin of new cell types. The important nodes are numbers: (1). Metazona common ancestor (yellow) (2). Cindarian-Bilaterian common ancestor (orange) (3). Bilaterian common ancestor (pink) (4). Deuterostome common ancestor (red) (5). Chordate common ancestor (green) (6). Olfatores (tunicate+vertebrate) common ancestor (blue) (7). Vertebrate common ancestor (purple). Right (**b**): Evolution of placode-derived cell types. (A) Placode-related cell types are believed to be present in the chordate ancestor, and the cell types derived from them. (B) Novel cell types appear in two possible ways. In the process of *division of labor*, a cell type becomes two sister cell types by splitting part of the transcriptomes between the two daughter cell types. In the process of *gene network co-option*, a new cell type originates by merging the transcriptome of two existing cell types. (C) Scenarios for the evolution of somatosensory (SSN: trigeminal/profundal, otic/lateral line) and viscerosensory (VSN: epi-branchial) neurons from a putative ciliated primary sensory neuron. (1) The primary sensory cell splits into a secondary sensory cell and an afferent neuron by segregation of functions. Resulting neurons and sensory cells become three distinct sister cells corresponding to the trigeminal/profundal, otic/lateral le, and epibranchial-derived neurons. (2), Three sister cell types first evolve as primary sensory neurons. Then the primary sensory cells split into secondary sensory cells and afferent neurons. Finally, in both scenarios, the secondary sensory cells of the trigeminal/profundal and epibranchial, but not otic/lateral line systems are lost. Left: [Reprinted from Schlosser G, Patthey C, Shimeld SM. The evolutionary history of vertebrate placodes II. Evolution of ectodermal patterning. *Devel Biol* 2014; 389(1): 98–119. With permission from Elsevier.]. Right: [Reprinted from Patthey C, Schosser G, The evolutionary history of vertebrate placodes I: Cell type evolution. *Devel Biol* 2014; 389(1): 82–97. With permission from Elsevier]

Summary of Placodal Evolution

1. Metazoan common ancestor.
   - Primitive genes involved in placode formation: Eya, Six1/2, PaxB.
2. Cnidarian-bilaterian common ancestor.
   - Most of the genes involved in placode development are present: Six3/6, Six 4/5, Fox1, PitX.
   - Primary sensory cells (mechano-, chemo-, photo-receptors) with differentiation dependent on: POUIVm PaxB, Six1/2, Eya.
   - Neurosecretory cells present for the first time.
   - Body axis determination depends on Wnt, BMP.
   - Some regionalized expressions in the ectoderm of TFs (eg. Hox, Six3/6).

3. Bilaterian common ancestor.
   - Dorsal-ventral ectodermal patterning depending on TFs.
     - ventral: low BMP (SoxB),
     - dorsal: high BMP (Msx, Dlx).
   - CNS is centralized on the ventral side.
   - AP ectodermal patterning: Wnt-dependent TFs.
     - anterior: Six3/6, fezf, Emx, Otx,
     - posterior: Irx, Gbx, **Hox.**
   - Left-right patterning dependent on Pitx.

4. Deuterostome common ancestor.
   - Pharyngeal pouches: Six1/2, Pax1/9, Eya.
5. Chordate common ancestor.
   - Dorsoventral axis now inverts.
   - Segregation of ectoderm.
     - dorsal neural ectoderm,
     - ventral: non-neural ectoderm,
     - neural ectoderm now contains Pax6, Pax2/5/8.
   - Specialized border territory.
   - Regionalized expression of Pax6 and Pax2/5/8.
   - New rostral neurosecretory area: Six3/6, Pax1, Pit1, Lhx, Eye (future).

6. Olfatores (Tunicates+Vertebrates) common ancestory.
   • Hair cell-like prototype secondary mechanosensory cells with cilium and microvillar collar.
   • New competence factors are recruited for non-neural ectoderm: GATA1/2/3.
   • New definition of roles for **FGF** and **BMP** in neural induction.
   • Non-neural border territory that is thickened: Six1/2, Eya, Msx, **Dlx.**
   • AP ectodermal programming now depends on Wnt, FGF, and RA.
   • New posterior ectodermal territories expressing: Fox, Pax2/5/8, Six4/5, Eya, Islet, Six1/2 and **Eya** (anterior proto-placodal domain).

7. Vertebrate common ancestor.
   • New or highly developed cell types that are placode-specific:
     – neurosecretory cells of the adenohypophysis,
     – lens cells,
     – olfactory and vomeronasal receptors neurons,
     – somatosensory and viscerosensory neurons,
     – hair cells.
   • Recruitment of new competence factors for non-neural ectoderm: Fox.
   • **Neural crest cells** appear by recruitment of TFs from ectoderm (AP2) and mesoderm (FoxD SoxE, Twist).
   • Cranial placodes zones as focused areas of proliferating progenitor cells giving rise to cells and neurons.
   • **Cranial placodes proper**: adenohypophyseal, olfactory, lens, profundal (V1)/trigeminal (V2-V3), otic, lateral line, epibranchial.
   • New roles for Six1/2 and Eye to control proliferation within placodes.
   • Recruitment of many AP-restricted TFs for new roles in placode specification.

In Summation

The above discussion should make it clear that the human skin develops from several embryonic sources. The epidermis of the face is derived from non-neural ectoderm. Its adnexal elements develop either from a selective stimulus of epidermal cells by neural crest in the dermis or as a distinct population committed to a placodal fate. Facial dermis originated from neural crest, not mesoderm. Adnexal structures are distributed according to the neural crest coding of the trigeminal system, largely eschewing the V1 distribution. Non-craniofacial epidermis is formed from non-neural ectoderm outlying the axial zones that give rise to neural crests and placodes. It likewise contains cells with placodal competence. The dermis of this skin is mesodermal, arising from somites. Subcutaneous adipose tissue arises in all sites from mesenchymal precursors cells known as pre-pericytes. With these definitions and concepts in hand, we will now consider the components of the skin, subcutaneous tissues, and fascia.

Skin: Epidermis and Appendages

Epidermis arises from two fundamentally different sources of tissue: brain and non-brain. Neural crest from the anterior forebrain (telencephalon) makes the frontonasal epidermis of the columella, nose, and forehead. All remaining epidermis of the face and body comes from the nonneural ectoderm in register with neuromeres from the hindbrain and backward.

Sources of Epidermis: Neural Crest Versus Non-neural *Ectoderm*

Neural Crest Epidermis

Recall that the forebrain, or prosencephalon, is constructed from developmental units called prosomeres. In the 2013 iteration of the Puelles model the anterior forebrain becomes telecephalon (the precursor of cortex), the eye fields, and the hypothalamus. These were formerly classified as prosomeres p6-p4 but the terminology has been dropped. Above anterior prosencephalon, *rostral neural folds* do not contain neural crest. Instead, they have specialized non-neural ectoderm (NNE), some of which form placodes and remainder frontonasal epidermis. The posterior prosencephalon develops from prosomeres p3 to p1 which contain prethalamus, thalamus, and post thalamus. Above the posterior prosencephalon, *caudal prosencephalic neural folds* contain neural crest (PNC) which produces the underlying frontonasal dermis. We shall describe how this lamination process happens shortly [Figs. 11.4, 11.5 and 11.6; neuromeric map showing prosencephalic non-neural ectoderm (epidermis) and prosencephalic neural crest (dermis)].

Because there the brain has no somatic sensory nerves forward from r1, all sensation for the frontonasal skin must come from V1. *The anatomic distribution of neural crest epidermis is defined by that of V1.* And because forebrain neural crest is incapable of forming blood vessels, the entire blood supply of this unique skin will come from arteries programmed by V1 sensory nerves. Indeed, the StV1 arterial system represents an evolutionary "add-on" to the primitive ophthalmic artery.

Fig. 11.4 Neuromeric model consists of 6 prosomeres (p1-p6), 2 mesomeres (m1-m2), and 12 rhombomeres (r0-r11). Underlying notochord from r1 backward imposes homeotic identity for each segment. From rhombomere 2 forward the alternative no-hox homeotic genes define the neuromeres. Gastrulation exposes migrating cells to homeotic imprinting from the notochord. Mesoderm produced by this process will provide a definition for the overlying ectoderm of its specific zone. [Courtesy of Michael Carstens, MD]

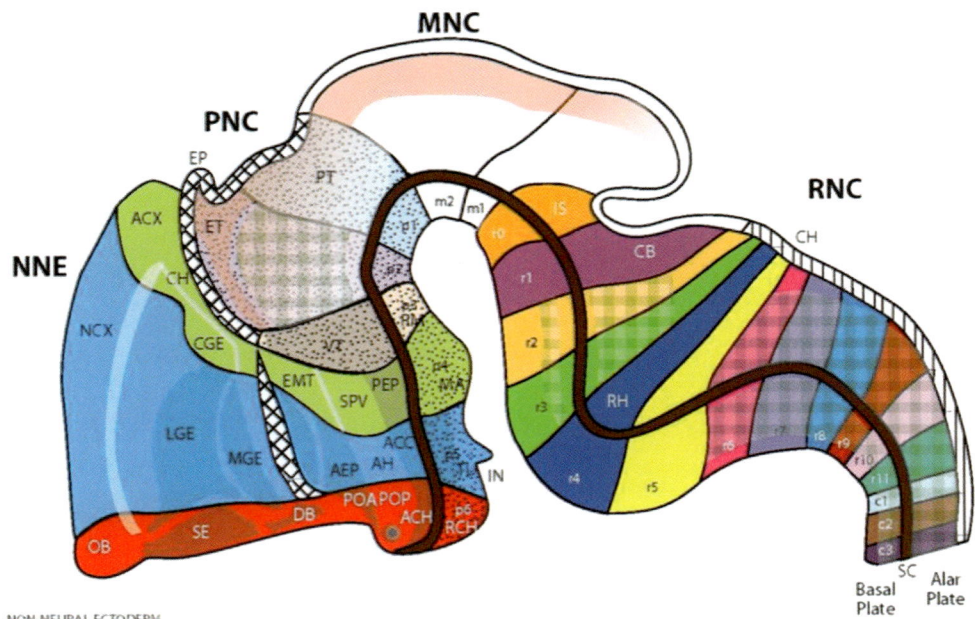

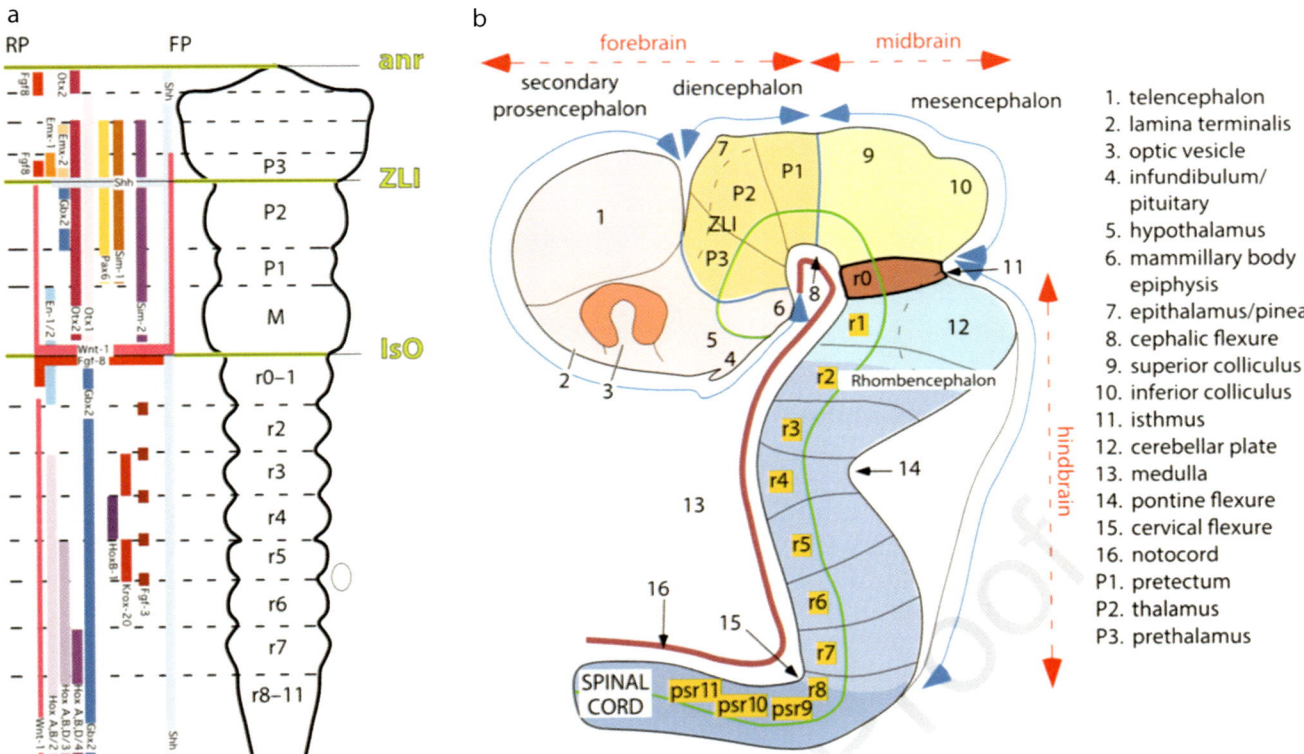

Fig 11.5 Midbrain and forebrain maturation sequence. In the original Puelles model the maturation sequence of the midbrain and forebrain proceeds forward from r0 Developmental ES stages in the mouse correlate with Carnegie stages in humans.

- ES 8.5 = Stage 8 The neural plate induces an initially flat midbrain and forebrain. Note large size of the midbrain. Floor plate terminates at rostral hindbrain. Prechordal floorplate produces Sonic hedgehog (Shh). As stage 8 progresses, the forebrain and midbrain rise out of the plane and project forward. Isthmus not present.

- ES 10.5 = stage 12 Mesomeres and prosomeres are in place. Primary (mesencephalic) flexure postions head and face in front of rostral hindbrain, r1-r3. Shh localizes to rostral forebrain, p4-p6. Isthmus is now present.

- EX 12.5 = stage 16 Secondary (cervical flexure).

[Reprinted from Puelles L, Rubenstein JLR. Expression patterns of homeobox and other putative regulatory genes in the embryonic mouse forebrain suggest a neuromeric organization. *Trends Neurosci.* 1993;16(11): 472–479. With permission from Elsevier.]

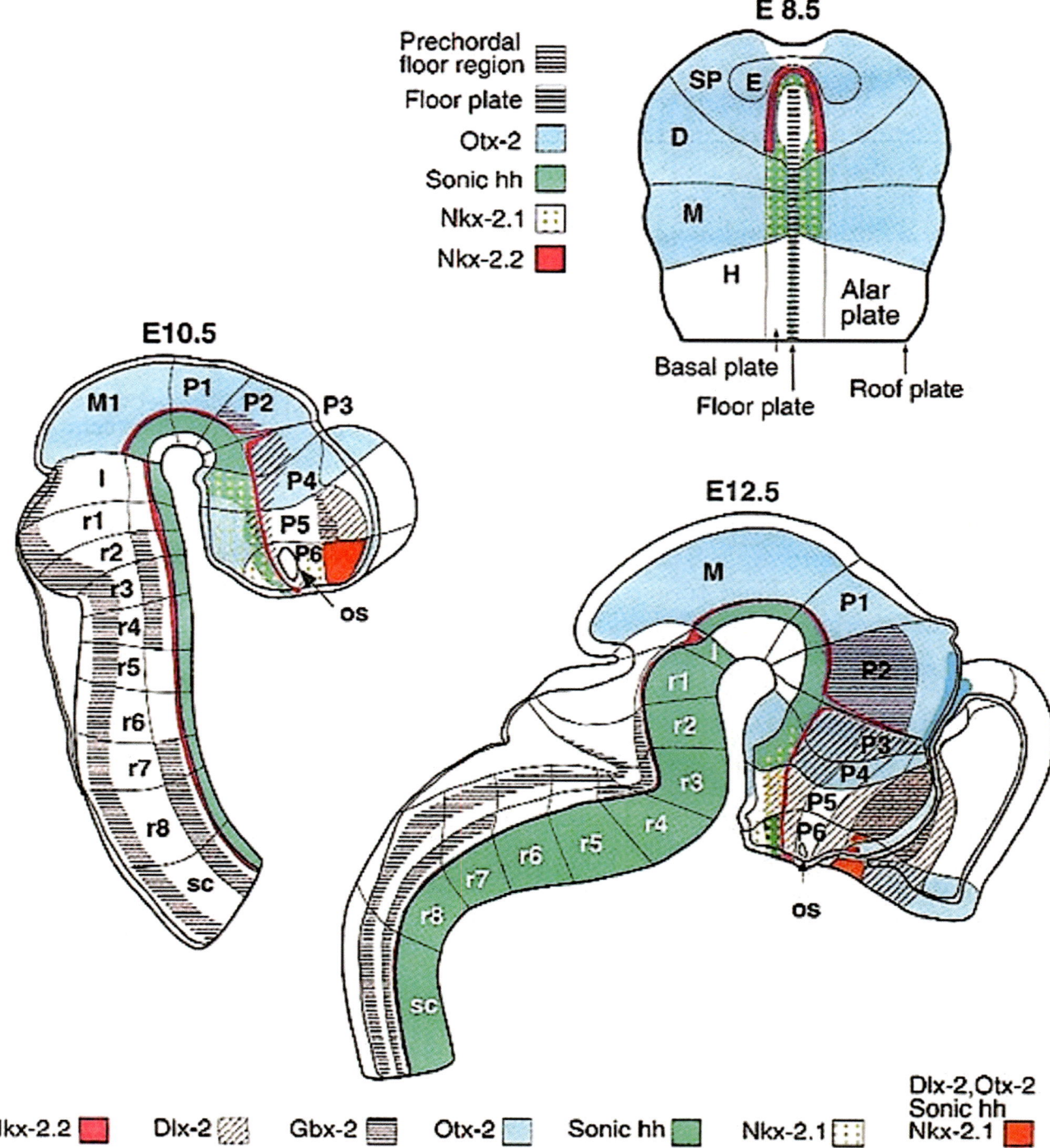

Fig. 11.6 Embryonic brain regionalization explains the order of skin development. Gene expression map supports the embryonic developmental sequence of mesoderm and neural crest migration. Note production by isthmus of *Wnt*-1 and *Otx-2* for midbrain induction and *Fgf8* for cerebellum induction. Note that *Wnt-1* continues forward until the p3 junction with secondary prosencephalon (the old p3-p4 junction). [Reprinted from Martinez S, Puelles E, Echevarria D. Ontogeny of the vertebrate nervous system. In: Galizia CG, Llego P-M (eds). Neurosciences. Berlin, Germany: Springer; 2013. with permission from Springer Nature]

Ectodermal Epidermis

Non-neural ectoderm (NNE) produces epidermis covering the entire body except the frontonasal zone. NNE acquires its neuromeric identity indirectly as a consequence of gastrulation. These cell populations are sessile and do not enter the primitive streak thus, it has not come into contact with the notochord. Initially, it does not express homeotic genes. Later, under the influence of underlying mesoderm, positional *Hox* genes appear. Thus, *the neuromeric coding of the epidermis is dependent upon that of the dermis*. It is organized as *ectomeres* in register with the notochord from r1 backward. Note: The isthmus, r0, that lies in front of r1 and separates midbrain from hindbrain, produces prechordal mesendoderm and is not relevant for mapping of epidermis. The contribution of "pure" r1 non-neural ectoderm is probably limited to skin of the upper eyelid and the epithelium of the conjunctiva.

At the same time, important growth factors produced by epidermis are vital for the development of the dermis. Later in development, both skin and gut mucosa have important programming functions for their respective mesenchymal structures. Because dermis exists in equilibrium with epidermis, let's explain why the skin-forming ectoderm represents "the layer left behind," what is left over of the epiblast after gastrulation is complete.

What Is the Origin of Ectoderm that Produces Epidermis?

During gastrulation, cells of the original primitive epiblast migrate through the primitive streak in two successive waves. The first population displaces the hypoblast laterally, creating two new structures. Cells remaining in the axial midline are known as the *mesendoderm*; these become the notochord, which in turn plays a crucial role in the organization of the overlying neural plate. Cells lateral to the midline produce the *definitive endoderm*. Following this, a second successive migration of epiblast cells exploits the potential space between epiblast and endoderm to create mesoderm. These newly-ingressed cells manufacture glycosoaminoglycans retain water. The resultant edema separates the epiblast layer from endoderm, thus creating the space for mesoderm. Once gastrulation is complete, the remaining epiblast layer takes on a new name: ectoderm.

The default expression of ectoderm, unless inhibited, is neural. The next iteration of ectoderm is a gene-driven partition of the ectoderm into two distinct zones: *neural ectoderm*, which gives rise to the neural plate and future CNS and *non-neural ectoderm* (NNE), which gives rise to the skin. The expression of neural fate is inhibited by two signals produced by mesoderm, BMP-4, and BMP-7. In the embryonic midline, Sonic hedgehog (*Shh*) produced by notochord antagonizes the inhibitory influence of BMP-4. This results in the induction of the floor plate. As one proceeds laterally from the midline, the *Shh* signal dies out; and the unopposed activity of BMP-4 and BMP-7 produced by the mesoderm results in NNE. Neural crest cells arise from the neural fold, an interface zone between NE and NNE.

Development of Epidermis

At one month's gestation the NNE is a thin layer of flat cells called *periderm*. Although the organization of underlying mesenchyme into a two-layered dermis does not take place until 12 weeks, a basement membrane layer that will give rise to postnatal epidermis and the more superficial periderm. The periderm reaches its apogee at ten weeks. Its main function is the exchange of water, electrolytes, and glucose with amniotic fluid. By sixteen weeks a basal germinative layer and an intermediate layer are identified. Periderm sloughs, producing squames in the amniotic fluid. These lead to the cellular debris seen covering the fetus at birth, the *vernix caseosa*.

As the germinative layer proliferates, successive layers of intermediate cells become interposed between it and the periderm. Initially, biosynthetic activity (as evidenced by glycogen granules) is present in all the layers but, over time, a reduction of activity occurs toward the surface leading to keratohyaline granules at 20 weeks (Fig. 11.7).

Differentiation of epidermal layers follows a spatio-temporal template. The cranial-caudal gradient reflects the spatial organization of the ectoderm after gastrulation is complete. The ectomeres overlying the pharyngeal arches develop before the cervical ectomeres. Thoracic epidermis matures before that of its lumbar counterpart. Furthermore, epidermal maturity follows the outward (ventral) development of neuro-angiosomes from the midline. Mid-axillary epidermis at any given neuromeric level matures before that of the ventral midline. Does this mean that epidermis is organized neuromerically? In a strict sense, it is not. The ectoderm remaining behind after the completion of gastrulation consists of broad geographic regions; these undoubtedly have differing genetic compositions. The epidermis overlying the future chest is distinct from that of the flanks. Furthermore, epidermal growth factors such as Fgf-8 exert a trophic effect on the underlying mesenchyme. When these are deficient, the skin itself is atrophic. This type of pathology is well illustrated by *cutis aplasia*.

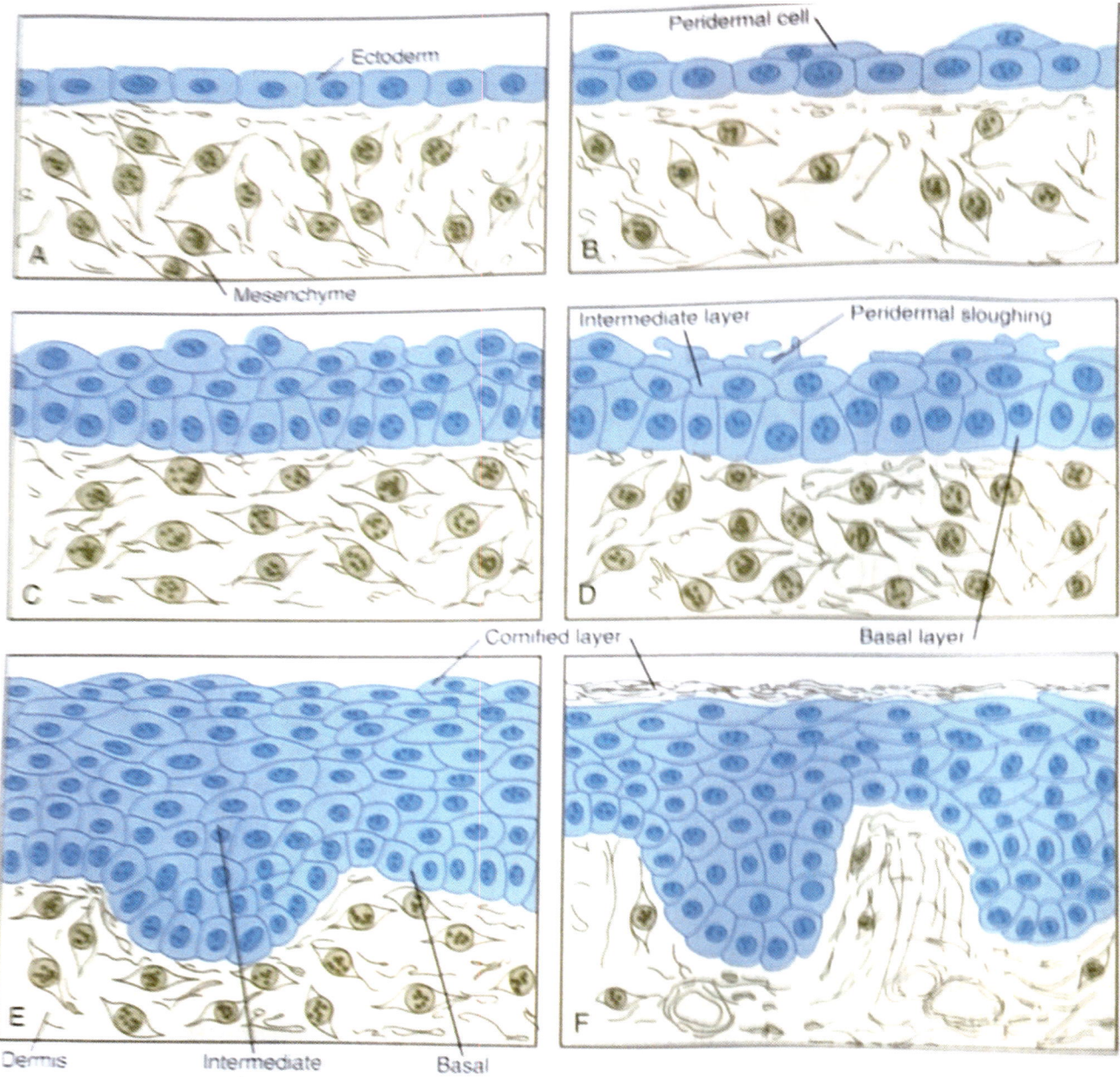

Fig. 11.7 Histogenesis of human skin. (**a**) At one month non-neural ectoderm is single layer. (**b**) At two months flattened periderm cells appear on the surface of the NNE. (**c**) the NNE thickens. (**d**) The skin had three layers, a germinitive, basal layer, an intermediate layer resulting from dividing stem cells in the basal layer, and periderm cells which slough into the amniotic fluid. (**e**) Dermis is formally identified and the individual layers of epidermis are established. (**f**) Rete pegs are present. [Reprinted from Carlson BM. *Human Embryology and Developmental Biology*, sixth edition. St. Louis, MO: Elsevier; 2019. With permission from Elsevier]

Components of Epidermis

The following specialized cells are present in epidermis. *Merkel cells* are modified keratinocytes found in the glabrous palms and soles. They appear at 8–12 weeks. Later on, they associate with dermal axon-Schwann cell complexes. *Langerhans cells* are derived from bone marrow. Development of hematopoietic population is very primitive,

and blood supply to skin via dermis is established by six weeks these cells appear in the epidermis as early as 6 weeks. Their function is immune surveillance. *Melanocytes* first appear in the cranial region at 8 weeks. The ratio of melanocytes to keratinocytes is controlled by the keratinocytes.

Pilosebacous units appear at 9 weeks. They also follow a cranial-caudal developmental gradient. The *pre-germ* refers to localized collections of cells in the basal layer of epidermis. The *hair-germ* projects downward into mesenchyme.

Cells from the latter clump around the hair germ as the *dermal papilla*. Further downward growth of the hair germ engulfs the dermal papilla like a hand grasping a ball. This is the *bulbous peg*. Three attachments develop along the shaft: an upper apocrine gland (sometimes), a middle sebaceous gland, and a lower arrector pili muscle. These arise from mesenchyme under the influence of signals from differing regions of the shaft. Eccrine sweat glands arise when epidermal down-growths into dermis induce a mesenchymal *secretory coil*. Consider the remaining components of the skin and subcutaneous tissues.

Skin: Dermis

Dermis is essential for epidermis. It provides physical support. The rete pegs are like Velcro® keeping the epidermis anchored. Ectoderm is incapable of making blood vessels. The subdermal plexus provides nutritional support. Innervation comes to the skin surface via the dermis, which also shelters. In reciprocal fashion, epidermis is essential for dermal development. It "instructs" underlying mesenchyme to become dermis.

Contact between epidermis and mesenchyme causes the latter to secrete a protein-rich matrix within which individual cell types develop. The matrix itself undergoes transformation with differential expression of glycosoaminoglycans and collagens. At 3 months, major changes occur. Collagen becomes organized, epidermal appendages appear, and the neurovascular structures develop. Superficial dermal blood vessels result from the transformation of angiogeneic MSCs. Nucleated RBCs appear at 6 weeks and the subpapillary plexus forms 2 weeks later. At 9–10 weeks a deeper *horizontal plexus* forms. Pericytes develop from neural crest MSCs. Capillary leak results in the pooling of extracellular proteinaceous fluid. MSCs organize around these pools forming encircling, interconnected channels. These will become lymphatics.

A great deal is known about epithelial-mesenchymal interactions. In the primordium of the face and limbs, chondrogenesis from MSCs within the dermis is prevented by the ectoderm. Dermal development also is epidermis dependent. Regional properties of dermis determine what appendages will be present or absent. At the same time, epidermis determines the morphology of the appendages and their cranial-caudal distribution.

Neuromeric Identify of Skin: Dermatomal vs. Non-dermatomal

How do we assign skin a neuromeric identity? Is it embedded in the epidermis from the start or is it acquired from the underlying mesenchyme? Although epidermis has a regional specification, *epidermis expresses homeobox gene produces that originate in response to signals from dermis.* This makes intuitive sense because epidermis arises from those epiblast cells left behind after gastrulation has taken place. This "straggler" population is never been in physical contact with the primitive streak; thus, it has not been exposed to notochordal signals.

Epiblast "adventurer" cells that participate in gastrulation plunge down through the primitive node or streak, at which point they acquire the homeotic "bar code" of their neuromeric level of exit. Intraembryonic germ layers of endoderm and mesoderm will bear the same bar code. Thus, *the source material of dermis (mesoderm) imparts its neuromeric identity to the overlying epidermis.* Below the craniocervical junction, an orderly pattern of cutaneous innervation is seen along the dorsal (epaxial) skin, with the continuous representation of every neuromeric level, from c2 to s5.

This apparently tidy system has two anatomic problems. First, the innervation of craniofacial skin and mucosa has a *discontinuous* neuromeric pattern. Along the posterior scalp, ear, and lower jaw dermal zones originating from r3 hindbrain neural crest abut against dermis derived from cervical c2 to c3 paraxial mesoderm (PAM). Second, epaxial skin of the entire body has a continuous sensory innervation pattern representing all neuromeres from C2 to S5. In contrast, the hypaxial skin of the <u>trunk</u> has a *discontinuous* sensory innervation pattern, with neuromeres c5-t2 and l2 to s3 confined to the extremities. These apparent discrepancies can be readily understood using developmental principles (Figs. 11.8 and 11.9; neuromeric map of the skin).

Gastrulation is a very regular process. Initially, for every neuromere there is a corresponding swatch of mesoderm. Similarly, neural crest cells develop from every neuromeric level. Why then, is the sensory pattern of the head and neck discontinuous? Let's begin by looking at classical teaching about somites. All somites are made up of paraxial mesoderm; they typically consist of three elements: (1) the *sclerotome* forms the axial skeleton, that is, the vertebra and rib; (2) the *myotome* forms the striated muscles of the body, with the exception of those developing from the first seven somitomeres; (3) and *dermatome* forms the neck, trunk, and extremities.

The Mystery of the Fifth Somite

This orderly model applies only to somites from neuromeric level c2 backward. What makes the first five somites so special is that they are more primitive…they cannot produce the full complement of mesodermal structures. As one moves caudally from the first somite (S1), a progressively greater degree of specialization takes place. The occipital somites (S1-S4) contain the sclerotomes of the posterior cranial base

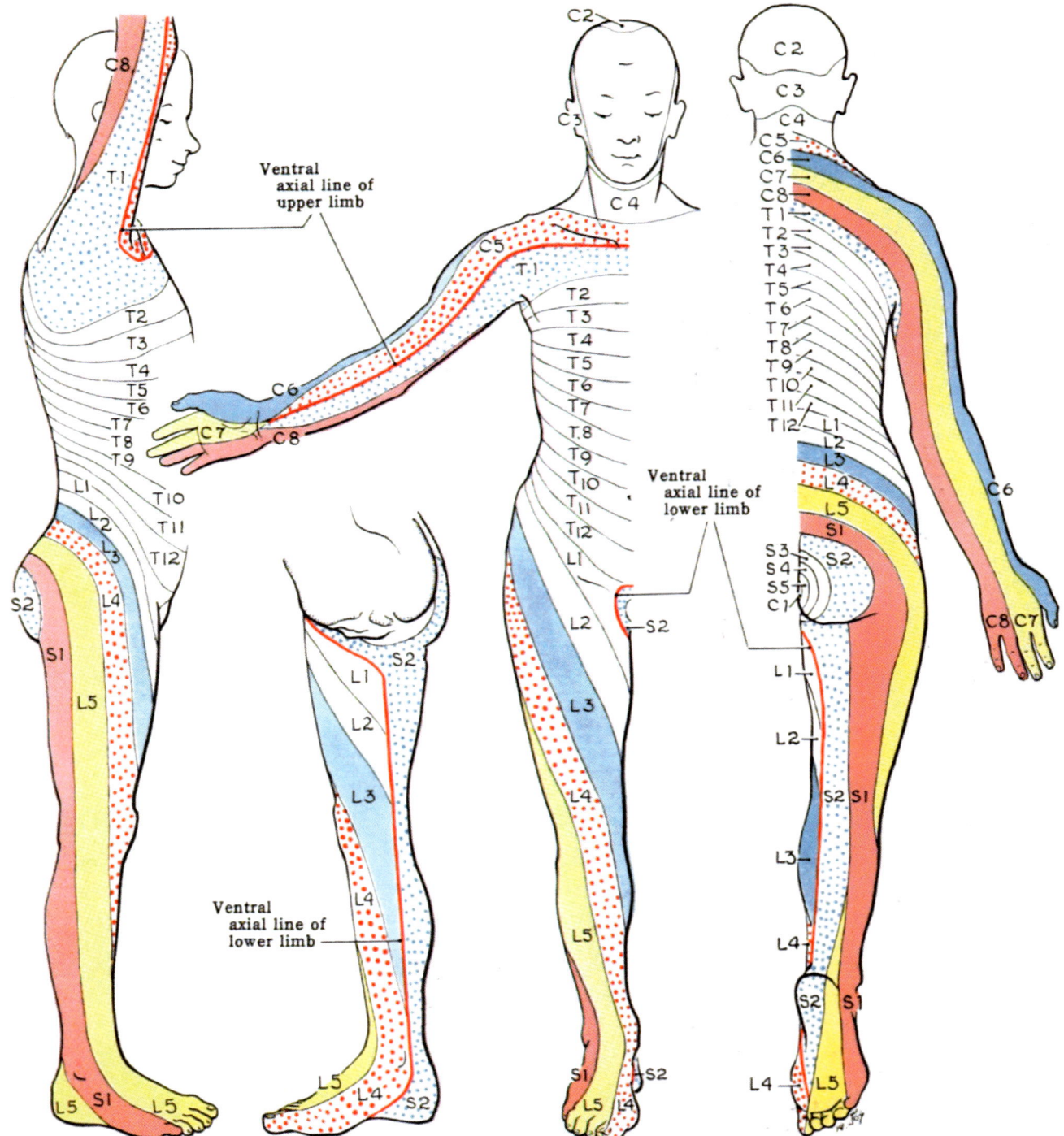

Fig. 11.8 Dermatomal map. Note the absence of representation from c1. Hypaxial (ventral) cervical dermatomes are "stolen" at level c4 with c5-t1 reassigned to the upper extremity. Epaxial (dorsal) dermatomes remain fully represented. The limbs are strictly hypaxial structures. [Reprinted from Wikimedia. Retrieved from: https://commons.wikimedia.org/wiki/File:Grant_1962_663.png]

and myotomes that give strictly ventral derivative; these create the hypaxial muscles of the tongue and the cucullaris complex. The occipital somites do not posess dermatomes.

This situation changes with the cervical somites. Somite 5 (neuromere c1) has a myotome with both ventral *and* dorsal derivatives—it is responsible for epaxial muscles controlling extension of the craniovertebral joint. These are innervated by a dorsal branch of C1. Somite 5 does have a dermatome, but it has no specific representation on the skin of the neck (although it is depicted as contributing to the cervical plexus;

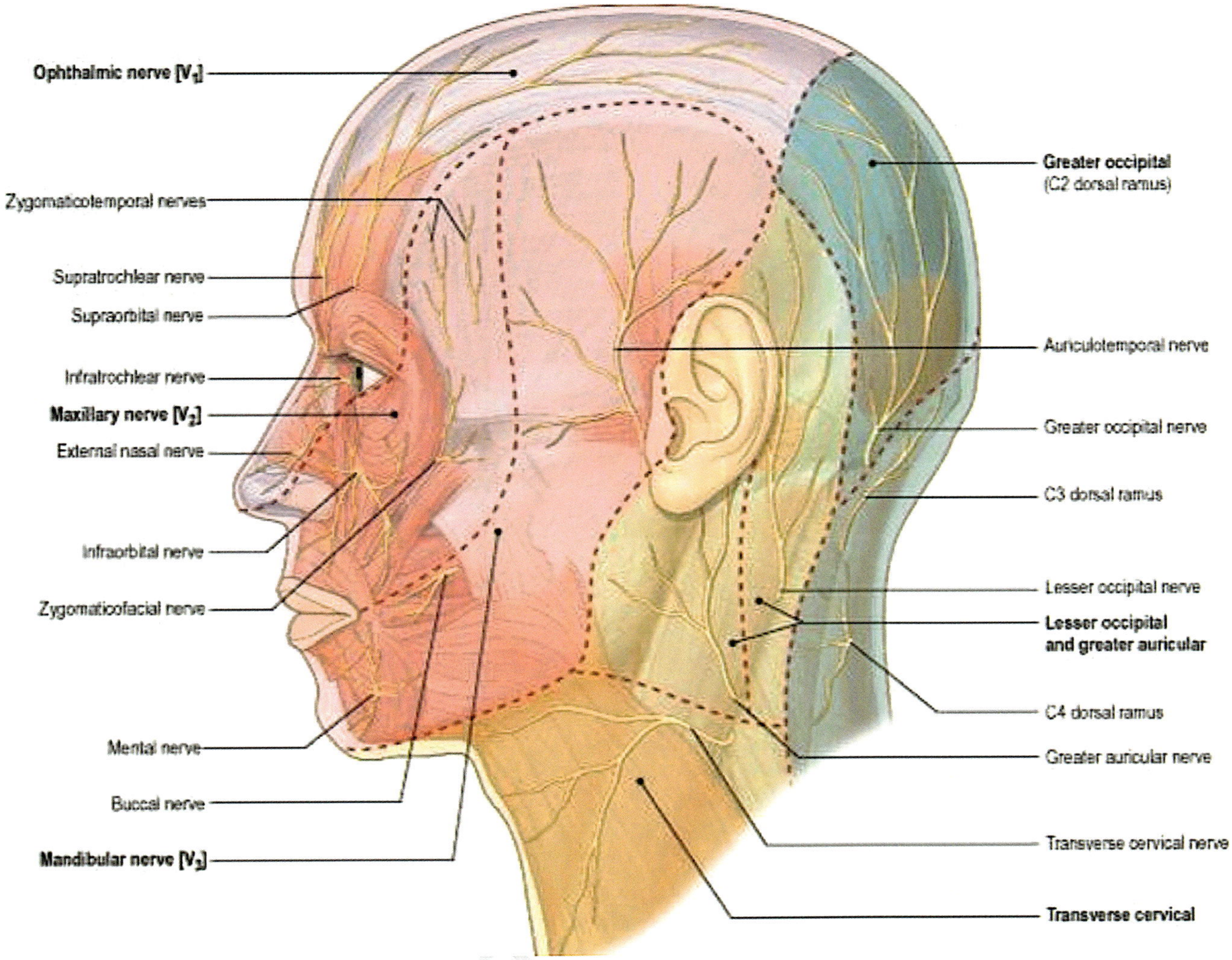

Fig. 11.9 Cutaneous innervation of the head and neck. Each half of the neck four hypaxial zones supplied through the cervical plexus: transverse cervical (C2-C3) supraclavicular (C3-C4), greater auricular (C2-C3) to the ear, and lesser occipital (C2) to the scalp. There are two epaxial zones supplied by individual roots: greater occipital (C2-C3) to the interparietal scalp and dorsal roots C2-C8 to the midline of the neck. Note the sharp divide between the hypaxial LON and the epaxial GON. [Modified from Drake R, Vogel AW, Mitchell AWM. Head and Neck. In: Gray's Anatomy for Students, third edition. Philadelphia, PA: Churchill-Livingstone. 2015: 924–1052. With permission from Elsevier]

more on this later). The true contribution of S 5 is to the dura sheath connecting atlas with foramen magnum. Somite 6 (neuromere c2) contains a complete dermatome serving both epaxial skin of the posterior scalp and hypaxial skin of the neck, as mapped out by distribution of the cervical plexus). It also produces local dura (Fig. 11.10 occipital somites).

One might ask, why are the "primitive somites" five in number? After all, mammals have four occipital somites. Why is our fifth somite anomalous? The answer has to do with the evolutionary relationship between birds and mammals. Avian embryos have, not four, but five occipital somites. Recall that occipital somites produce neither epaxial muscle nor dermis. Birds and mammals diverged from the reptilian line at about the same time, from a common ancestral form,

as yet unidentified, which possessed five occipital somites. But in the transition to the mammalian line, certain aspects of the ancestral pattern continued to persist at the level of S5. Our fifth somite behaves as if it were partly occipital and partly cervical—hence the failure to produce identifiable dermis. Mammalian neuroembryology thus recapitulates the original neuromeric arrangement of its precursor.

Dermatomes: How Zones of Hypaxial Dermis Are Stolen from the Trunk

All students of anatomy are familiar with the dermatomal system. Epaxial dermis is rather boring, as all levels from

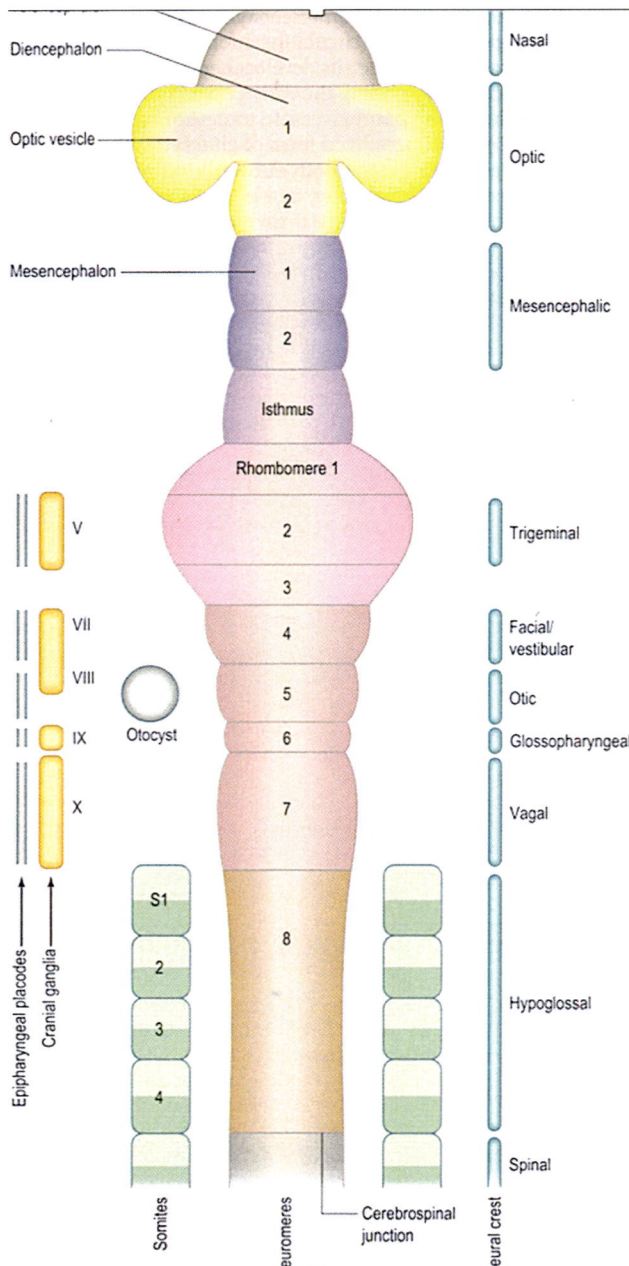

Fig. 11.10 Isthmus rhombencephali. Neuromeric map of O'Rahilly and Müller shows isthmus interposed between hindbrain and midbrain. It represents a secondary induction of r1. Isthmus neural folds do *not* produce neural crest. Basal floorplate of isthmus produces nucleus of *trochlear nerve* (CN IV). Midbrain has two neuromeres, m1 and m2. Basal plate of m1-m2 produces cerebral peduncles while *oculomotor nuclei* reside primarily in m2. Alar plate of isthmus and midbrain contain *mesencephalic nucleus of V* receiving fibers from V1-V3 for ocular and masticatory proprioception. Note: tactile fibers from V1-V3 are located at levels r1-r3. Nocioceptive fibers from V1-V3 refer to the *spinal nucleus of V*. [Reprinted from Lewis, Warren H (ed). *Gray's Anatomy of the Human Body*, 20th American Edition. Philadelphia, PA: Lea & Febiger, 1918]

neck to coccyx are innervated by dorsal nerves organized *segmentally*. Hypaxial dermis of the trunk is more intriguing. It has a central zone (t3-l3) that is quite standard, that is innervated by ventral nerves organized *segmentally*. But the hypaxial dermis of the neck (c2–4), upper extremity (c5-t2), and lower extremity (l4-s3) are supplied by ventral nerves organized into *plexuses* (cervical, brachial, and lumbosacral, respectively). Why should this take place?

The best way to understand these differences is the evolutionary model of the extremities—this can be stretched to include the neck. The limb buds are hypaxial structures located at fixed locations alomgn neuraxis of all tetrapods. Thrusting outward like a fist through a flexible membrane, the extremities drag all their associated lateral plate mesoderm away from the body wall. The neurovascular supply to the limbs is excluded as well. The upper extremity represents an "envelope" of hypaxial skin and mesenchyme stolen from the neck-upper trunk interface. The lower extremity repeats the crime by stealing hypaxial tissue from the interface between caudal trunk and the sacrum.

Within the limb buds lateral plate mesoderm makes the appendicular skeleton of the arms and legs. These bones must be populated by paraxial mesoderm muscles arising from somites *from the same neuromeric levels*. Everybody needs to know where to go. In the case of the upper limb, LPM innervated by neuromeres c5-t2 is simply no longer represented in the body wall. Paraxial mesoderm supplied by c5-t2 makes up somites S9-S15. The ventral myotomes of these 8 somites send forth hypaxial muscles that are assigned to their counterpart LPM bone structures. Thus, hypaxial PAM muscles migrate out into the limb buds, accompanied by their respective motor nerves.

During development the limb buds rotate, flex, pronate. As a consequence, the migration paths of the various myoblasts cross one another to assume a position on either the flexor or extensor surface of the limb bones. In so doing, the pathways of the peripheral nerves merge and diverge, thus creating the brachial plexus. In both upper and lower extremities, *the spatial-temporal trajectories of each all muscles in sequence are faithfully recapitulated in the anatomy of the plexus* that supplies that extremity.

The Old Neck Versus the New Neck: Neck 1.0 vs. Neck 1.1

The evolution of the neck is a fascinating topic that we shall consider in a separate chapter. For our purposes here, the rationale of the cervical plexus should be considered from an evolutionary standpoint. Fishes do not have necks. The piscine head is anchored to the body via a series of opercular and extrascapular bones. With the transition to land, the head of prototype tetrapods became independent of the trunk. This process involved a redefinition of the first four post-occipital neuromeres. The definition of the trunk moved caudally and its place was a flexible segment, the ancestral neck. Neck 1.0. was short, having four vertebrae. The muscles connecting the piscine head and trunk became the cucullaris muscle, which is the precursor for our sternocleidomastoid and trapezius. True to evolution, these modern muscles fulfill the same function as cucullaris; they span between the head and trunk, bypassing the neck altogether.

The original craniocervical joint was a peg-and-socket arrangement with the dens of the ancient first vertebra, the *proatlas* thrust into the skull base. This provided for rotation of the primitive tetrapod head but no flexion. The proatlas had two condyles articulating with the second vertebrae. With the incorporation of proatlas into the occipital bone, these two condyles now could articulate with atlas, making head flexion possible…and more effective capture of prey. Thus, as evolution proceeded, the neck developed from a primitive version (old neck) involving neuromeres c1-c4 to an advanced version (new neck) involving the re-definition of four additional neuromeres c5-t2. This "evolutionary add-on", resulted in a trunk that was pushed backward 8 neuromeric levels, leaving in its place Neck 1.1 with 8 spinal nerves, but only 7 cervical vertebrae…due to the cranial redefinition of the proatlas. As we shall see, the skin of the "new" neck is divided into two distinct neurologic zones.

Epaxial neck skin, from skull base to the trunk is organized into seven regular dermatomes (neuromeres c2-c8), each with its individual sensory nerve. Recall that Somite 5 (neuromere c1) does not produce epaxial dermis. But only four somites are responsible for hypaxial skin. In ventrolateral skin somite 5 has minimal representation. Somites 6–8 (neuromeres c2-c4) are responsible for all the dermis from the border of the mandible to downward to the thoracic inlet and backward over the sternocleidomastoid, terminating at the anterior border of trapezius. This skin is organized into four neurologic zones by the individual branches of the cervical plexus.

In the "new neck" why are somites S9-S13 (neuromeres c5-t1) not represented? *Where did all that dermis go?* The

dermis corresponding to c5-t1 is simply not present in the neck—it is expropriated by the upper extremity. How did this happen? The answer lies in the embryologic location of the upper extremity: the head-trunk junction. The anterior pectoral fin of fishes is the analog of the upper extremity. The fin is located directly behind the skull; it is constructed from the hypaxial mesoderm of somites 5–8. In the transition to land, primitive tetrapods pushed the trunk backward four neuromeric levels. In these species, somites 5–8 became re-defined to produce "neck 1.0." The primitive tetrapod limb was constructed from hypaxial mesoderm of somites 9–13 supplied by a new structure, the brachial plexus.

The neck can be considered as a mechanism to position the head in space for improved food intake. Modern birds and mammals (and in their reptilian ancestors) achieved longer, more flexible necks by re-defining five additional neuromeric levels. In "neck 1.1" the definition of the trunk was pushed still further backward to t1-t2 but *the definition of the limb remained unchanged.* Thus, in mammals, all epaxial mesoderm from somites 9–13 is dedicated to the new neck. All hypaxial mesoderm from somites 9–13 is "solen" by the limb bud. For this reason, although the dorsal skin of the neck receives mesoderm from seven cervical somite levels (S6–13) whereas the only ventral skin available for the new neck is that produced from the first four somite levels (S5-S8). The ventral skin envelope of neck 1.0 is therefore stretched downward to cover the additional five neuromeric levels. We do not encounter any representation of hypaxial dermis on the trunk until level T2 (somite 14). The dermatomes of the lower extremity have the same relationship with the trunk (Table 11.1).

Facial Dermis Is Non-dermatomal

The best way to understand the organization of craniofacial and neck dermis is to consider its two tissue sources. <u>Neural crest</u>, which produces virtually all craniofacial dermis, originates from three distinct levels. *Forebrain neural crest* makes the frontonasal dermis. *Midbrain neural crest* produces the submucosa of the upper nasal vault and the nasopharynx. *Hindbrain neural crest* produces the dermis of the eyelids, face, anterolateral scalp, and submucosa of the lower nasal

Table 11.1 Neurologic derivatives of the somites: epaxial vs. hypaxial

	S5 c1	S6 c2	S7 c3	S8 c4	S9 c5	S10 c6	S11 c7	S12 c8	S13 t1	S14 t1
Epaxial Segmental	X	X	X	X	X	X	X	X	X	X
Hypaxial Segmental										X
Hypaxial Plexus	X	X	X	X	X	X	X	X	X	

Table 11.2 Developmental potential of somites

Somite	Neuromere	Sclerotome	hypaxial myotome	epaxial myotome	hypaxial dermatome	epaxial dermatome
S1	r8	+	+	–	–	
S2	r9	+	+	–	–	
S3	r10	+	+	–	–	
S4	r11	+	+	–	–	
S5	c1	+	+	+	+	–
S6	c2	+	+	+	+	+

vault and oropharynx. Somitic mesoderm is responsible for all the remaining dermis, that is, that of the neck and the posterior scalp. The epaxial dermatomes of cervical somites 2 and 3 produce the dermis of the posterior scalp and part of upper auricle. The hypaxial dermatomes of cervical plexus, c1-c4 produce the dermis of the neck and contribute to the lower auricle and earlobe as well. We shall examine each of the forms of dermis in sequence (Table 11.2).

Sources of Dermis: Prosencephalic Neural Crest

Components of Frontonasal Skin (Figs. 11.9 and 11.10)

Frontonasal Skin Comes from Two Sources: Telencephalon and Diencephalon

The neural folds of the prosencephalon (forebrain) have been mapped by LeDouarin et al. demonstrating two major zones. The rostral prosencephalic folds correspond to the *telencephalon, eye fields, and hypothalamus*. They are in register (in the old Puelles model) with prosomeres p6-p4. Even though these terms are no longer in use, it is useful to divide the rostral folds into three distinct zones. Zone p6 has adenohypophyseal placode, nasal placode, and internal nasal (vestibular) epithelium. Zone p5 has optic placode and external nasal epithelium. Zone p4 has calvarial ectoderm which will act to program the frontal bone complex.

The caudal prosencephalic folds correspond to the *diencephalon* which contains *prethalamic, thalamic, and postthalamic nuclei*. They have three neural crest populations in register with prosomeres p3, p2, and p1. PNC flows forward beneath p6-p4 in a strict spatial-temporal sequence. Like water filling a glass PNC populates the frontonasal target areas in a distal-to-proximal fashion. Three vertical homeotic "stripes" of the dermis are created, probably from p1 medial to p3 lateral. Each one imposes a distinct genetic identity upon the overlying epidermis. *The neuromeric model of frontonasal skin is a "grid" of three developmental zones.* The supratrochlear, supraorbital, and lacrimal fields seen in the skin of the nose and forehead reflect this grid. We postu-

late that these fields represent p1, p2, and p3 dermis, respectively, but it could also be the reverse, as mapping has not been done. PNC skin, coming from forebrain per se, is devoid of the blood vessels and nerves required for its survival. Neither does it have the capacity to form bone or cartilage. The mesenchyme required for these roles must come from an external source.

Neurovascular Support for Frontonasal Skin Comes from r1 Neural Crest

The forebrain has three specific limitations. Prosencephalon cannot provide its own external coverage, is insensate, and has no means to create blood supply. The solution for all three problems is an external source of mesenchyme with neurovascular properties: mesencephalic (midbrain) neural crest. Although mesencephalon is composed of m1-m2, and r0-r1) we shall refer to mesencephalic neural crest as r1 or MNC interchangeably. They are distributed as follows: (1) r0-r1 produces forebrain dura, sphenethmoid complex, anterior cranial base, and frontal bone complex; (2) m1-m2 is assigned to cover the eye; and (3) a separate r1 population supports the upper eyelid and conjunctiva.

Nature never produces a structure without immediately providing it with blood supply for survival. Neural crest development and migration from the neural folds of the CNS follows the temporal sequence of the CNS: (1) rostral hindbrain and midbrain (2) hindbrain, (3) caudal forebrain, and (4) rostral forebrain. This makes perfect sense. Gastrulation starts at r1 and proceeds backward. Midbrain is almost simultaneously induced by r1. Forebrain induction is caudal to cranial so neural crest from the diencephalic folds (p1-p3) migrates prior to that from telencephalon (p5-p6). Thus, prior to the development of forebrain dermis from above downward, hindbrain mesenchyme has migrated from below-upward and midbrain mesenchyme is flowing in a bilaminar pattern—a deep layer (the future dura) ascending over the forebrain and a superficial layer passing through the orbit and upward as a second lamina over the deep layer. Bifucation of the StV1 neurovascular pedicles into superficial and deep components reflects this migration pattern. Frontal bone is synthesized between the two lamina of MNC. It is thus bilaminar and contains a sinus.

Once the forebrain and midbrain neural crest layers are in place all individual neuroangiosomes of stapedial system (annexed by ophthalmic) become assigned to specific developmental fields according to the organization of prosencephalic skin. The stapedial trunk attached to primitive ophthalmic has two main branches. The nasociliary axis is medial and supplies dermis from p4 and p3. The supraorbital axis is lateral and supplies dermis of p2 and p1. Let's map out these developmental fields and see how the system works.

MNC has an additional vital function: *it is the source material for all the bones and cartilages of the frontonasal vault.* In so doing, it is programmed by PNC.

Neuromeric Map of Frontonasal Skin

The Nose (Figs. 11.10 and 11.11)

Nasal skin consists of p6 vestibular epidermis and p5 nasal epidermis supported by p1 dermis. This results in two distinct developmental fields. When the face flexes, this frontonasal skin-mucosa is rolled downward and backward 180 degrees beneath the developing brain, terminating at Rathke's pouch. As a result, when the nasal chambers develop, p6 mucosa becomes tucked inside to form the nasal lining, while p5 skin remains outside as the nasal dorsum, sidewalls, philtrum, and columella.

The developmental zones of external nasal skin, and internal lining have exactly the same design. They have medial and lateral zones supplied by separate branches of the nasociliary axis. Externally, nasal dorsum is supplied by anterior ethmoid artery while the external lateral nasal skin is supplied by infratrochlear artery. Internally, septal mucosa receives septal branches of anterior and posterior ethmoid arteries while the upper lateral nasal mucosa covering superior turbinate is irrigated directly from the same arteries. From a clinical standpoint, Tessier cleft zones 13–12 in the nose are perfused by the branches from the most medial StV1 arterial, nasociliary.

The Forehead

Forehead skin consists of p5 epidermis supported by three zone of dermis (p3-p1). This creates three clinically relevant developmental fields, each with its respective arterial axis. These are medial (p1), supratrochlear, middle (p2) supraorbital, and lateral (p3), lacrimal. These fields correspond to Tessier cleft zones 13–12, 11–10, and 9.

Several lines of clinical evidence support the neuromeric model of frontonasal skin. Tessier zones 13–9 run from medial to lateral. Nasal pathology is limited to zones 13 and 12, whereas forehead pathology can be found in any of the

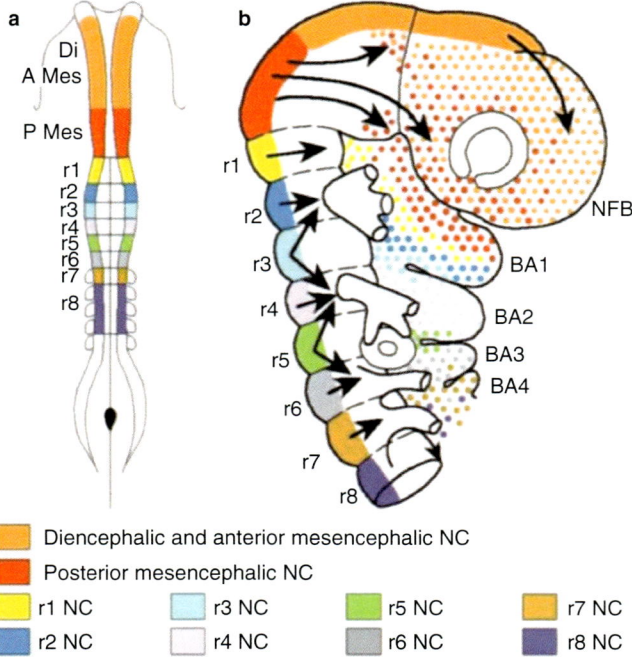

Diencephalic and anterior mesencephalic NC
Posterior mesencephalic NC

r1 NC	r3 NC	r5 NC	r7 NC
r2 NC	r4 NC	r6 NC	r8 NC

Fig. 11.11 Neural crest migration begins at stages 10–11. Diencephalic NC (orange) from neural folds in register with p1-p3 makes frontonasal dermis. Midbrain NC (red) admixed with r1 (yellow) populates the entire frontonasal zone, especially the orbit, and forms anterior cranial base. Note that r1 does *not* participate in first arch derivatives; these are derived from r2-r3. The pattern of two rhombomeres per pharyngeal arch is maintained. Neural crest cells from certain rhombomeres are blocked and need to accompany a neighboring population to gain access to the arch. Note that r8 has four subdivisions. These have been recently demonstrated as separate homeotic entities corresponding to r8-r11 and assigned as pairs to the fourth and fifth pharyngeal arches. NC from r8-r11 also interacts with the four occipital somites (S1-S4). [Reprinted from Creuzet S, Couly G, Le Douarin NM. Patterning the neural crest derivatives during development of the vertebrate head: insights from avian studies. *J. Anat.* 2005; 207:447–459. John Wiley & Sons]

three zones. Developmental field excess/deficiency states follow a medial-lateral gradient with midline zones being more frequently affected. These pathologies are based on defects of individual branches of the V1 stapedial system.

The spectrum of holoprosencephaly is based on distinct branches of the StV1 system to the nose and orbit and branches of the anterior cerebral axis in the brain. Thus, the severity of HPE follows a gradient: mild hypotelorism, heminose, arhinia, cyclopia with presence of bilateral temporal globes in a single orbit, and complete absence of globe/orbit. The take-home message is that pathologies of the stapedial system tend to involve individual branches of StV1 with midline being more frequent, whereas HPE involves a progressive medial-lateral involvment of branches of prICA. More on this subject can be found in Chap. 6.

Sources of Dermis: Mesencephalic Neural Crest

The Mesencephalon and Isthmus

The midbrain (mesencephalon) is induced by the most forward zone of the hindbrain. Recall that the gastrulation takes place through the primitive streak. The cranial limit of the streak is also the cranial tip of the notochord, located forward at r0-r1r1. The functional hindbrain is associated with the pharyngeal arches runs backward from r2-r11. Genes produced at level r1 such as *Otx-2* are responsible for induction of the midbrain.

The midbrain consists of two mesomeres, m1 and m2. These contain the two nuclei of the oculomotor nerve complex. Somitomere 1 (inferior rectus, inferior oblique, and medial rectus) is supplied by m1. Somitomere 2 (superior rectus, levator palpebrae superioris) is supplied by m2. Just behind the midbrain proper, and separating is from the hindbrain, lies the isthmus, r0. This gives rise to anterior cerebellar peduncle and the decussation of trochlear nerve.

Immediately caudal to r0 lies the first rhombomere. Although r1 is anatomically part of hindbrain it is functionally closely related to midbrain. It gives rise to the inferior cerebellar peduncle and contains the nucleus of trochlear nerve. Sm3 (superior oblique) is supplied by r1. Furthermore, running all the way forward from r1 to the anterior limit of midbrain, sensory nucleus of trigeminal nerve (called the mesencephalic nucleus of trigeminal) receives all general somatic afferents (pain, temperature, etc.) from tissues innervated by V1. Sensory nucleus of V caudal to the mesencephalic zone extends all the way down the brain stem and receives information in a somatotopic manner from the rest of the head.

The Isthmus and Hindbrain Clarified

Hindbrain technically begins with rhombomere 1. The nucleus of trochlear nerve resides in r1; it produces the posterior peduncle of cerebellum. Although the structure of hindbrain begins with r1, the functional hindbrain consists of two subunits. *Metaencephalon*, known as pons, is made up of rhombomeres 2–7. *Myelencephalon*, known as medulla, is composed of rhombomeres r8-r11.

Cutaneous Representation of Midbrain Neural Crest

What about r1 and MNC? First off, we must re-emphasize that gastrulation takes place exclusively via the primitive node/streak, the cranial limit of which is r1. By logic, there must be a zone of epidermis that is "encoded" by r1 neural crest dermis? For reasons detailed below, a strict definition of r1 "skin" is quite limited, *superior conjunctiva* being the most likely candidate.

Mesenchymal Representations of m1, m2, and r1

Little has been described mapping neural crest derivatives of r0. For this reason, we shall arbitrarily refer to neural crest from r0 and r1 as one and the same (r1 for short).

Midbrain neural crest migrates almost immediately, beginning at stages 9–10. First to migrate is r1 which flows bidirectionally along the base of the brain, forward to the anterior cranial fossa, and backward to tentorium After that m1-m2 localize around the developing eye. Because all three neuromeres contain V1, their migration pathways are defined by the individual branches of that nerve.

Accompanying these branches of V1 are the individual arteries of the StV1 stapedial complex, the stem of which is annexed by the primitive ophthalmic axis just behind the greater orbital fissure. Further clues as to the migration pathways of these three streams of MNC can be seen in the spatial positioning of the somitomeres with which they become associated. Sm1 is inferomedial, Sm2 and Sm3 is superolateral. Thus the *m1 and m2 define the insertion sites of the globe*, leaving only the lateral zone of the uncovered. This of course will receive lateral rectus originating from Sm5 and supplied by cranial nerve VI, abducens.

What we can induce from this is that MNC from m1 and m2 is buried inside the orbit where it produces non-ocular structures. Recall that the nuclei of oculomotor nerve reside in m1 and m2. Neural crest from these neuromeres is the logical source of fascia for 6 of the 7 extraocular muscles. They may well provide sclera for the globe. Thus the eye has an *internal lining of neural crest in the retina arising from the optic placode and an external* lining of neural crest as sclera originating from MNC Finally, *m1 and m2 do not produce dermis.*

When we observe the physical position of r1 neural crest we can see that it is ideally positioned to provide the *substantia propria* of the upper conjunctiva and the subcutaneous tissue of the upper eyelid. This does not include the dermis of upper eyelid which is likely to come from r1.

On the other hand, the distribution of r1 neural crest (as revealed by the innervation of V1) is very extensive. All dura innervated by V1 is an r1 neural crest derivative. first rhombomere neural crest provides the subcutaneous tissue for the mucosa of the nasal cavity. As such r1 is the mesenchymal source for the frontal, nasal, lacrimal, ethmoid, and presphenoid bones. All mesenchymal structures of the orbit are r1 derivatives. These include the sclera, connective tissue structures such as Whitnall's ligament and the septum orbitale, and intraorbital fat, and the substantia propria of the upper conjunctiva. Neural crest from the first rhombomere contributes to both upper, providing the tarsal plate and fat. Under the right circumstances, mesenchymal stem cells can convert to muscle. This is the likely source for the intrinsic muscles of the eye controlling papillary size.

In sum, the destiny of r1 neural crest is to provide support for the frontal lobe, the surrounding structures of the nasal cavity, the orbit, and the nasolacrimal system. Note that none of these structures bears any biologic relationship to the first pharyngeal arch. This is important because of number of studies have mapped the physical distribution of r1 neural crest to the superior (dorsal) aspect of the maxillary prominence (Fig. 11.11). This has given rise to the concept that the first pharyngeal arch is constructed from three neuromeric levels: r1, r2, and r3. The ultimate format of PA1 excludes V1 innervation. A possible resolution of this contradiction could be the existence of a putative *premandibular arch*, as postulated by Kuratani. For the sake of simplicity, because r1 neural crest is so heavily distributed to the orbit, *we will use MNC and r1 interchangeably.*

The importance of MNC for three pathologies of the cranial base and face cannot be overemphasized. Mesenchyme from r1 flows forward in the midline from the presphenoid to create a block of mesenchyme that lies deep to the nasal placodes, separating them and surrounding them. In normal development, the nasal placodes sink into r1 mesenchyme and then medialize. The approximation of the nasal chambers toward the midline requires apoptosis of r1 mesenchyme that separates them. *Failure of apoptosis of internasal mesenchyme results in hypertelorism*, the surgical treatment of which requires resection of ethmoid mesenchyme and orbital approximation, the facial bipartition operation. A second pathology involves the temporary filling of the nasal chambers with mesenchyme. This takes place immediately after the invagination of the nasal placodes is complete. Once again, apoptosis is required. *Failure of intranasal apoptosis results in choanal atresia.* A third pathology involves inadequate supply of r1 mesenchyme to the midline. Neural crest from r1 is responsible for dividing prechordal plate mesoderm and driving each half laterally into the future orbits. *Reduction in amount of available r1 neural crest results in hypotelorism*, a quantitative reduction of interorbital distance. This is the external manifestation of *holoprosencephaly.* Inappropriate production of gene products from the prechordal plate and/or r0-r1 affects the formation of midline neural structures such as corpus callosum, a near-constant finding in holoprosencephaly.

Sources of Dermis: Rhombencephalic Neural Crest

The hindbrain or rhombencephalon subdivides during development into two functional units. *Metaencephalon* or pons develops from rhombomeres 2–7. It supplies all derivatives of pharyngeal arches 1–3. Its floor plate contains the motor nuclei of V, VI, VII, IX, and X. Somatic sensory afferents nuclei refer to the trigeminal nucleus. RNC$_R$ is responsible for all facial dermis, the overwhelming source of which comes from r2 and r3, that is, from first pharyngeal arch. *Myelencephalon* or medulla develops from rhombomeres 8–11. Its floor plate contains the motor nuclei of X, XI, and XII. RNC$_C$ plays no role whatsoever in the formation of craniofacial skin.

Let's consider facial skin: neuromere by neuromere. The contributions of r2-r3 are well known. V2 and V3 dermis of the first pharyngeal arch is hypaxial. Note: with the face downward the skin of lateral forehead and scalp supplied by V2 and V3 is still considered hypaxial (Fig. 11.12). A small part of posterior ear canal is innervated by the facial nerve (r4-r5). Because the taste nucleus of chorda tympani resides in r4, second arch neural crest dermis likely originates from r5. The dermis of the external posterior pinna is r6-r7, that is, third arch. Recall that IX and X cohabit together in r6-r7. Thus, sensory innervation via auricular branch of vagus, X with a possible contribution from IX. The salivatory control for parotid may reside in r6, hence third arch neural crest dermis likely originates from r7.

Next, let's consider the anatomy of dermis supplied by V2 and V3. Head and neck skin has two fundamental components, both of which are supplied in similar ways. The interface between forehead skin and scalp is the boundary between hypaxial r1-r3 and epaxial C2-C3. The interface between facial skin (non-forehead) and neck skin is the boundary between hypaxial r1-r3 versus hypaxial C1-C4 (the cervical plexus).

Ear skin has an interesting composition. The pinna is constructed from six cartilaginous hillocks (Fig. 11.13). The anterior three come from the first arch; they are covered by r3 skin supplied by V3 auriculotemporal nerve. The posterior three develop from the second arch. These are *not* covered by second arch skin because *there is none available.* Instead, the posterior external auricle is supplied by cervical nerves. Its upper zone is supplied by epaxial lesser occipital nerve (C2) which is *not* a part of the cervical plexus. The lower zone is supplied by hypaxial greater auricular nerve (C2, C3) from cervical plexus. Why the need for two distinct cervical nerves? The answer has to do, once again, with epaxial vs. hypaxial structures. Of the three posterior hillocks, the upper one is epaxial and the lower two are hypaxial. The territories of lesser occipital n. and greater auricular n. reflect this fact. Finally, the skin of posterior auricle comes from somites S6-S7, that is, C2-C3. As the pinna first developed, it was flat against the mastoid, it had no deep surface. With growth, as the pinna began to project away from the head, a sulcus developed. The same two neuroangiosomes extend posteriorly, with lesser occipital supplying the epaxial half and greater auricular the hypaxial half (Fig. 11.14).

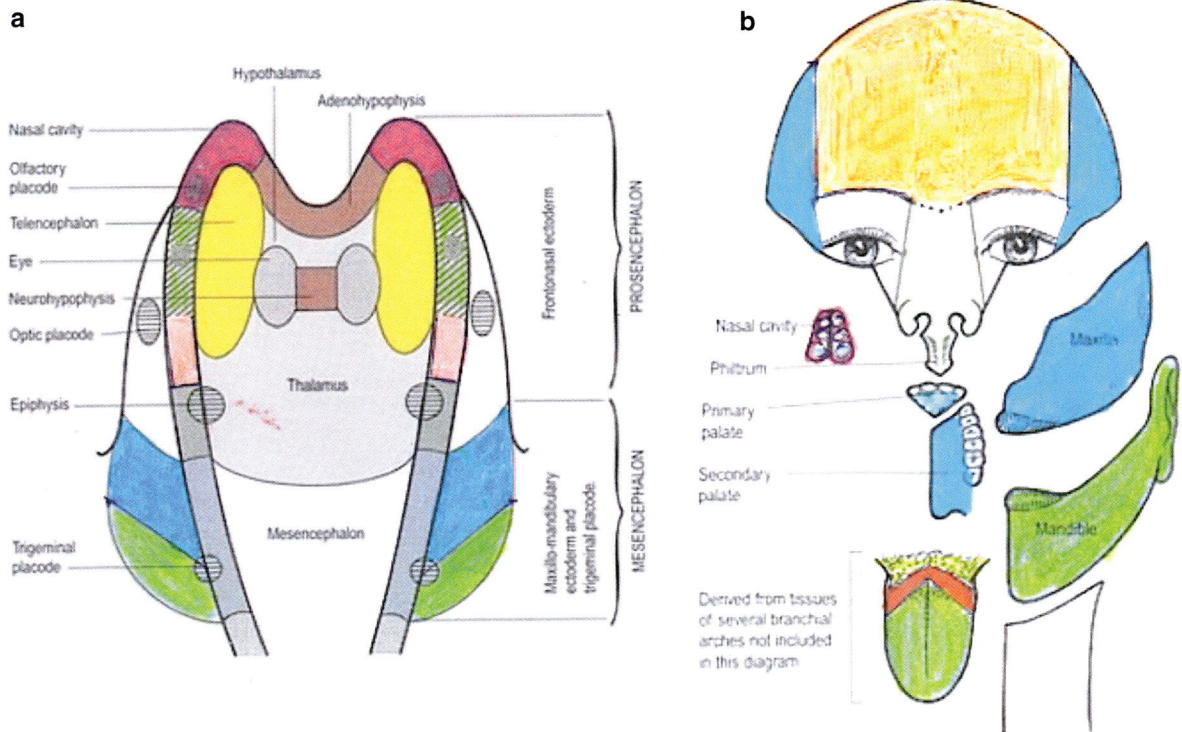

Fig. 11.12 Formation of ectomeres from forebrain and midbrain. Neural folds are color coded with p6 (red), p5 (green), and p4 pink. Frontonasal ectoderm (white) sits astride the anterior neural folds. Posterior neural folds p3-p1 (gray) provide dermis. Derivatives are seen on the right with p6 vestibular tissues (red) seen within the nasal cavity. Note that p1-p3 dermis is not sensate. Neurovascular supply of frontonasal skin depends upon V1 and V1 stapedial arteries. Source: Gray. [Reprinted from Williams PL. *Gray's Anatomy*, 38th edition. Philadelphia, PA: Churchill-Livingstone; 1997. with permission from Elsevier.]

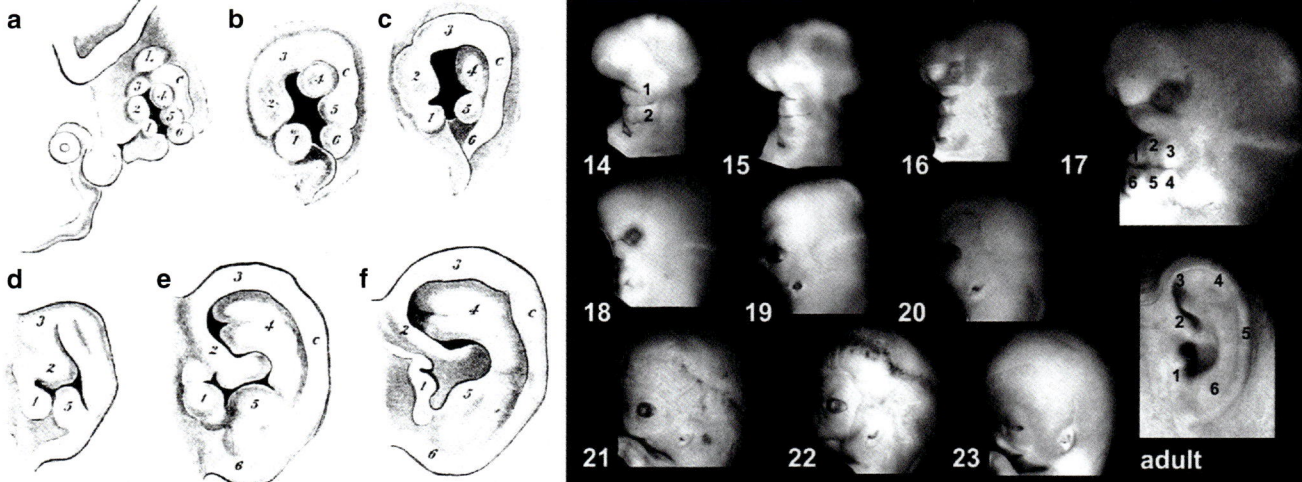

Fig. 11.13 Development of the external ear LEFT: Six stages in the development of the external ear. 1, 2, 3, elevations on the mandibular arch; 4, 5, 6, elevations on the hyoid arch. 1, tragus; 2, 3, helix; 4, 5, antihelix; 6, antitragus. c, hyoid helix or auricular fold. A, 11 mm.; B, 13.6 mm.; C, 15 mm.; D, beginning of third month; E, fetus of 85 mm.; F, fetus at term. an ingrowth takes place from the ventral portion of the groove, to form a funnel- shaped canal. The lumen of this tube is temporarily closed during the fourth and fifth months, but later re-opens. During the third month a plate of cells at the extremity of the primary auditory meatus grows in and reaches the lower wall of the tympanic cavity. During the seventh month a space is formed by the splitting of this plate, and the secondary portion of the meatus is thus developed. The tympanic membrane is formed by a thinning out of the tissue in the region where the wall of the external auditory meatus abuts upon the wall of the tympanic cavity. RIGHT: Auricular development by Carnegie stage. fifth week hillocks appear between first and second arches. sixth week six hillocks are present. seventh week development proceeds caudo-cranially (from neck upward). 12th week hillocks fuse. 20th week remodeling complete. Auriculotemporal nerve V3 covers hillocks 1–3, greater auricular (C2-C3) covers hillocks 5–6, and lesser occipital (C3) covers hillock 4. Left: [Reprinted from Charles W. Prentice Laboratory Manual and Textbook of Embryology. Philadelphia, PA: WB Saunders, 1922.]. Right: [Reprinted from The Kyoto Collection, Kyoto University Graduate School of Medicine. Courtesy of Prof. Kohei Shiota and Shigehito Yamada]

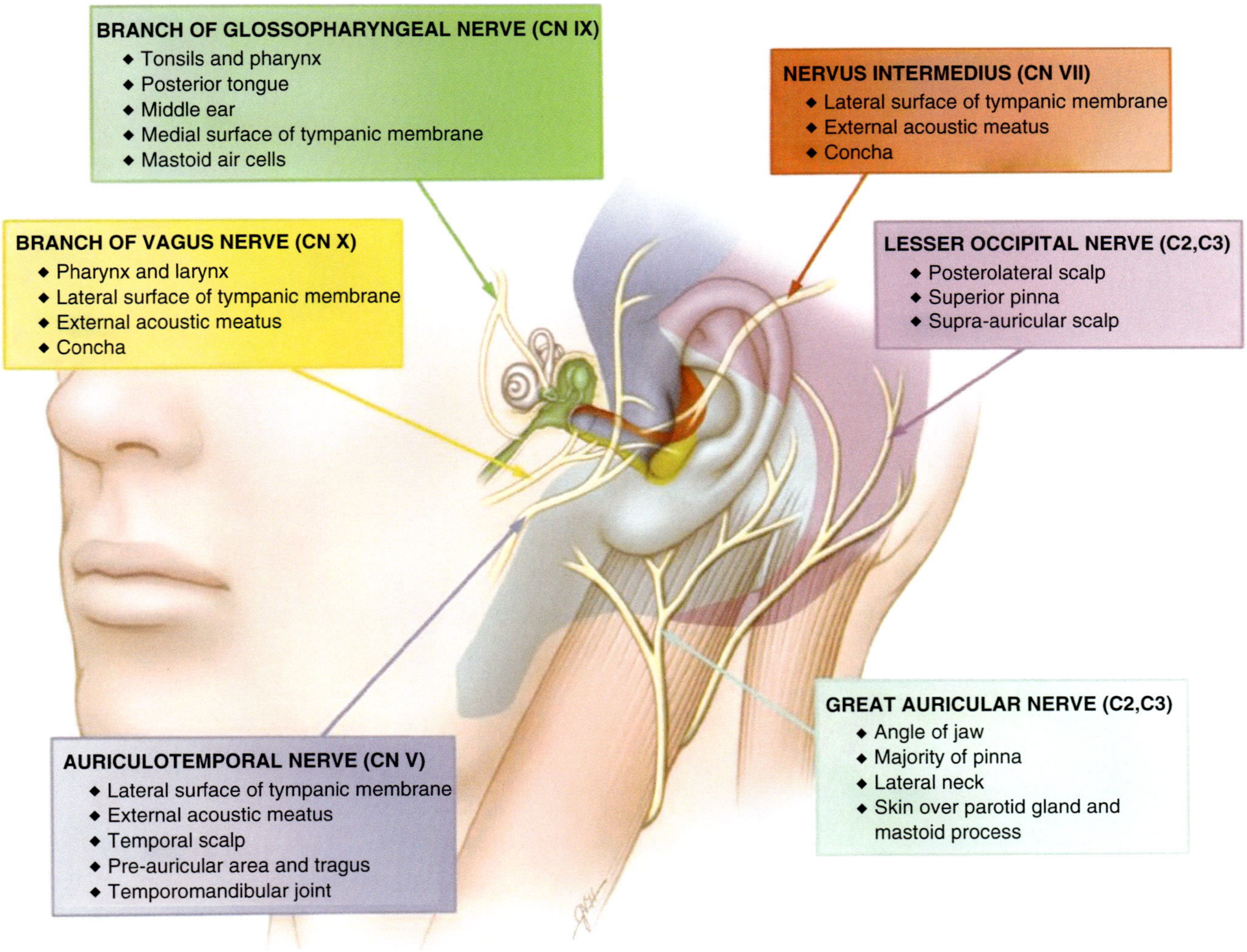

BRANCH OF GLOSSOPHARYNGEAL NERVE (CN IX)
- Tonsils and pharynx
- Posterior tongue
- Middle ear
- Medial surface of tympanic membrane
- Mastoid air cells

NERVUS INTERMEDIUS (CN VII)
- Lateral surface of tympanic membrane
- External acoustic meatus
- Concha

BRANCH OF VAGUS NERVE (CN X)
- Pharynx and larynx
- Lateral surface of tympanic membrane
- External acoustic meatus
- Concha

LESSER OCCIPITAL NERVE (C2,C3)
- Posterolateral scalp
- Superior pinna
- Supra-auricular scalp

AURICULOTEMPORAL NERVE (CN V)
- Lateral surface of tympanic membrane
- External acoustic meatus
- Temporal scalp
- Pre-auricular area and tragus
- Temporomandibular joint

GREAT AURICULAR NERVE (C2,C3)
- Angle of jaw
- Majority of pinna
- Lateral neck
- Skin over parotid gland and mastoid process

Fig. 11.14 Sensory innervation of the ear and surrounding structures. Sensory supply to the ear comes from **8** neuromeres (r2-r7 and c2-c3). Depiction of the sensory nerves shows the innervation of the ear and surrounding anatomy. Each box with its corresponding color illustrates each nerve's distribution. Sensory distributions may overlap. Note the anterosuperior aspect of the Eustachian tube is auriculotemporal (first arch). [Reprinted from DeLange JM, Garza I, Robertson CE. Clinical Reasoning: A 50-year-old woman with deep stabbing ear pain. *Neurology* 2014; 83(16): e152-e157. With permission from Wolters Kluwer Health, Inc.]

What becomes of the remaining arches? That of the second arch is the most complex. It essentially submerges into the first arch. In the process, its ectodermal epithelium disappears. The same process takes place intraorally; second arch mucosa vanishes. First arch skin now abuts cervical skin. But second arch "gets even" as its fascia, myoblasts, and motor branches of VII insinuate themselves forward into PA1, splitting around the parotid into deep and superficial components. Thus buccinator is "assigned" to oral mucosa while the muscles of facial expression remain in a superficial location. The primary attachment (origin) of the superficial facial muscles is from the superficial musculoaponeuotic system (the SMAS). The secondary attachment (insertion) is to the nearest available neighboring bone. second arch myoblasts migrate frontally beneath the skin overlying alisphenoid (zone 9) and continue beneath the frontonasal skin overlying lateral frontal bone (zones 10 and 11). Backward migration takes place to the ear and beneath the mastoid skin to produce occipitalis. The latter muscle is located over membranous (r7) supraoccipital, inserting at the membranous-chondral boundary, that is, the highest nuchal line. In sum, PA2 is completed melded into PA1, overlies PA3, and travels deep to cervical and mastoid skin populating facial muscles available zones over membranous bones of the calvarium and clavicle. fourth arch and fifth arch develop successively in stages 13 and 14. They form successive concentric rings *internal to* third arch.

This melding of pharyngeal arch mesenchyme is illustrated by the sensory innervation of the ear. The external components of the ear arise as six hillocks, three from PA1 and 3 from PA2 (Figs. 11.13, 11.14 and 11.15). Deep to them both lies PA3. Thus cranial nerves arising from rhombo-

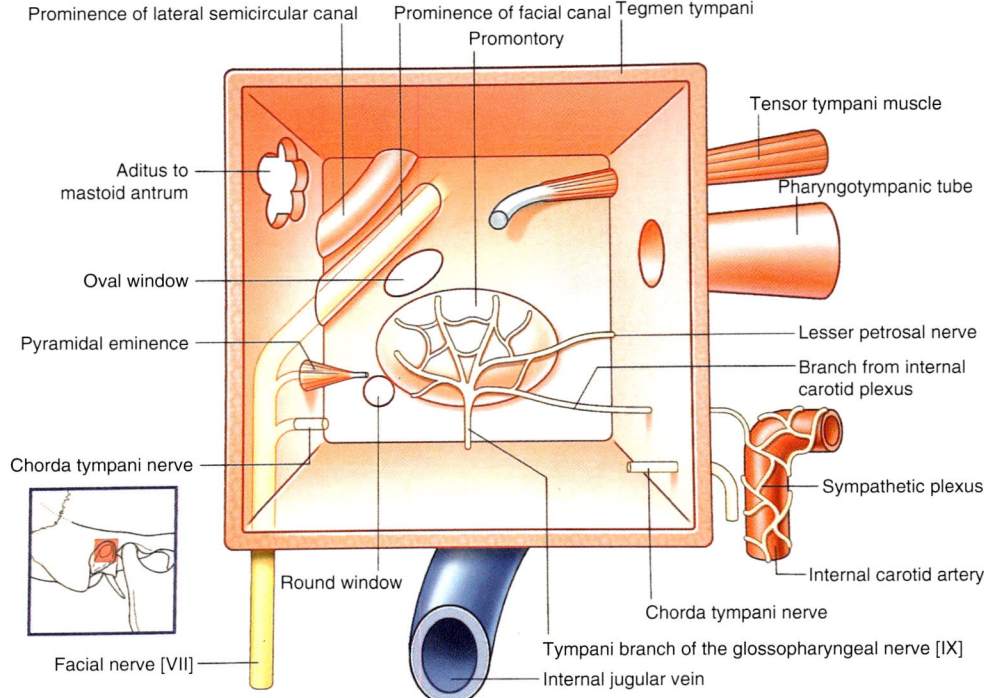

Fig. 11.15 Innervation of the tympanic cavity. **Tympanic branch of IX** sensory to the lining of middle ear, antrum, and auditory tube. Its preganglionic PANS fibers are secretomotor to parotid gland. **Caroticotympanic nerves** (sup and inferior Vasosomotor—originate from SANS plexus around the internal carotid artery. **Facial nerve:** Runs in bony canal along the medial and posterior walls of tympanic cavity: three branches: *Chorda tympani nerve:* taste anterior 2/3 tongue (except vallate papillae and secretomotor to submandibular and sublingual salivary glands. *Greater petrosal nerve*: secretomotor to lacrimal, nasal and palatal mucous glands. *Nerve to stapedius muscle* dampens down sound waves. **Mandibular nerve:** V3 tot ensor tympani muscle. Note: chorda tympani, branch of facial nerve, enters across the tympanic membrane being located lateral to the long process of the incus and medial to the handle of the malleus. It enters the tympanic cavity via the posterior canaliculus in the posterior wall and leaves via the anterior canaliculus medial to the pterotympanic fissure. [Reprinted from Drake R, Vogel AW, Mitchell AWM. *Head and Neck*. In: Gray's Anatomy for Students, third edition. Philadelphia, PA: Churchill-Livingstone. 2015: 924–1052. With permission from Elsevier]

meres r3-r7 are represented by the innervation of the aurical and tympanic cavity, Finally, there is no further cutaneous representation of neuromeres r8-r11 and c1 the skin of the posterior pinna and occipital scalp is supplied by neuromeres c2 and c3.

Adipose Tissue

Research interest in adipose tissue has undergone a metamorphosis in recent years with the original emphasis on metabolism and thermal regulation being supplemented by an intense focus on the biology of its cellular components as a source of reconstructive and regenerative cells. Despite this, concepts of its embryonic origins remain obscure. This section will outline a structural and functional model of adipose tissue as follows:

- White and brown fat: comparison and contrast.
- Pathologies of white fat.

- Evolution of adipose tissue.
- Pericytes: precursor cells for adipose tissue.
- Mesenchymal stromal and adipose-derived stromal vascular fraction (SVF).

White Fat vs. Brown Fat: Energy and Endocrinology vs. Thermoregulation

Histologic Considerations (Figs. 11.16, 11.17, 11.18, 11.19, 11.20, 11.21, 11.22, 11.23, 11.24, 11.25, 11.26, 11.27, 11.28, 11.29)

White fat is the best-known form of adipose tissue. Ubiquitous throughout the body it comprises, on average, 20% of body mass in men and 25% in women. As insulation, it is distributed in two layers superficial and deep, being separated by the superficial fascia, otherwise known by various regional terms: Internally it forms the apron of the omentum and in intrinsic to tissues such as the thymus. Adipose tissue

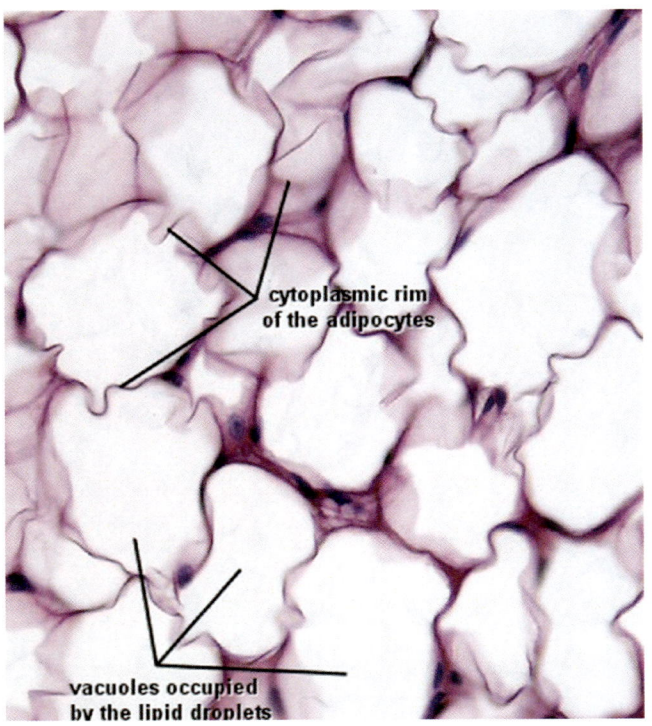

Fig. 11.16 White adipose tissue. White fat (H&E stain) has a peripheral nucleus, low number of mitochondria, and is designed for energy storage. The single large vacuole accommodates lipid and can expand its volume. Under control by SANS via NE. [Courtesy of Michael Carstens, MD]

forms deposits external to solid organs such as heart and kidney but it is notably not an intrinsic element of their parenchyma. Of note is its association with vascular bundles throughout the body as part of a trifecta in which fat accompanies and encompasses a nerve and its accompanying artery. This vascular "leash" is of use for reconstructive surgeons tracing out the blood supply for composite tissue flaps as these pass along tissue planes and from deep to superficially through intermuscular septae.

The cells of white fat are made up of a single lipid droplet which displaces the nucleus to the periphery. Important receptors in the membrane are for norepinephrine, insulin, glucocorticoids, and sex hormones. White fat is under control by the sympathetic autonomic system. White adipocytes produce two hormones: *leptin,* which, acting at the level of the hypothalamus, acts an as appetite suppressant, and *asprosin,* which stimulates the liver to release glucose (Fig. 11.16).

White adipose tissue is both dynamic (diet related) and trophic (perhaps supported by the SANS?). Newborns, especially premature, have little to no fat stores and a very sensitive to ambient temperature changes. The knock-out mouse model for *lipodystrophy* is characterized by diabetes and hepatomegaly. Loss of white fat stores, as seen in *anorexia nervosa,* involves multiple systems including the brain (memory decline and depression), skin (hair loss), muscle weakness, skeleton (bone loss), kidney (stone), intestine

Fig. 11.17 White fat is dynamic. Perry-Romberg syndrome represents an involutional state of neural crest mesenchyme affecting dermis, fat, and membranous bone within a defined neuroanatomic distribution. It occurs in childhood and "burns out" in early adulthood. [Reprinted from Raposo-Amaral CE, Denadai R, Nunes Camargo D, et al. Parry-Romberg Syndrome: Severity of deformity does not correlate with quality of life. 2013; 37: 792–801. With permission from Springer Nature]

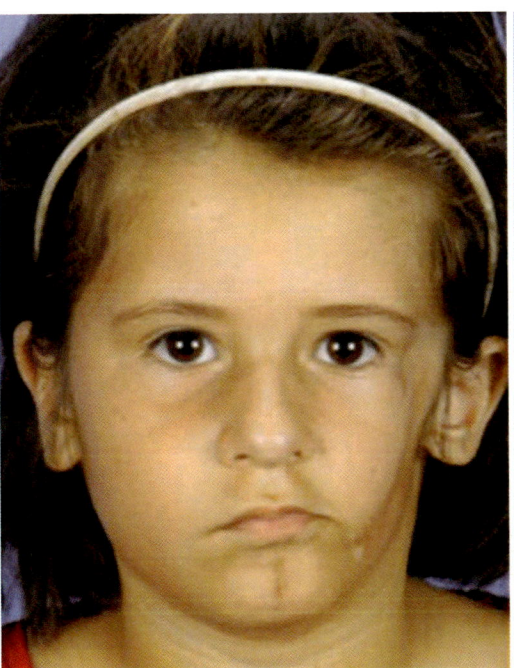
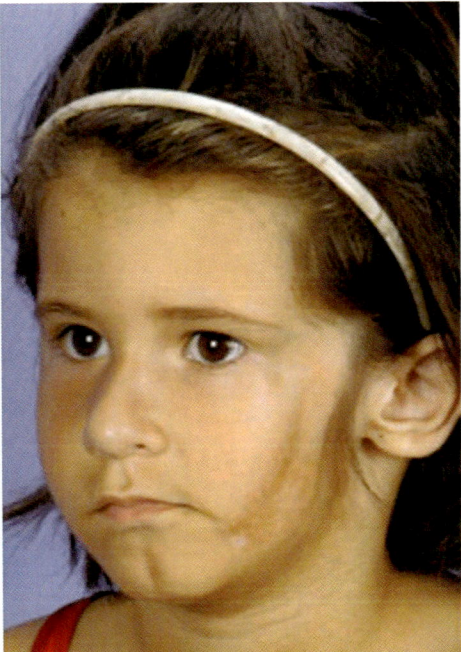

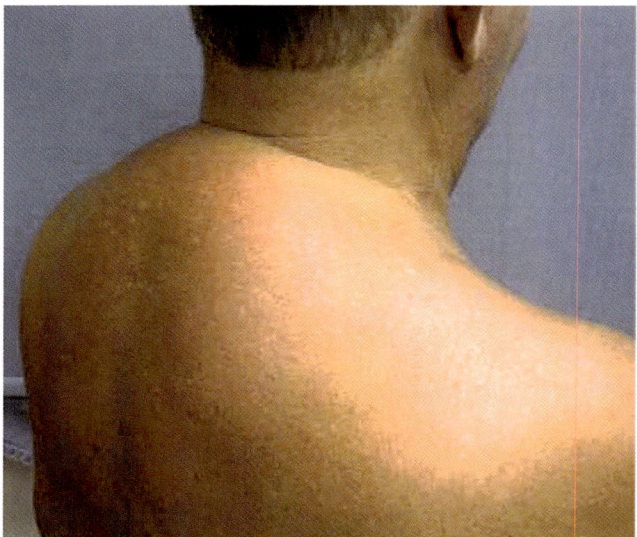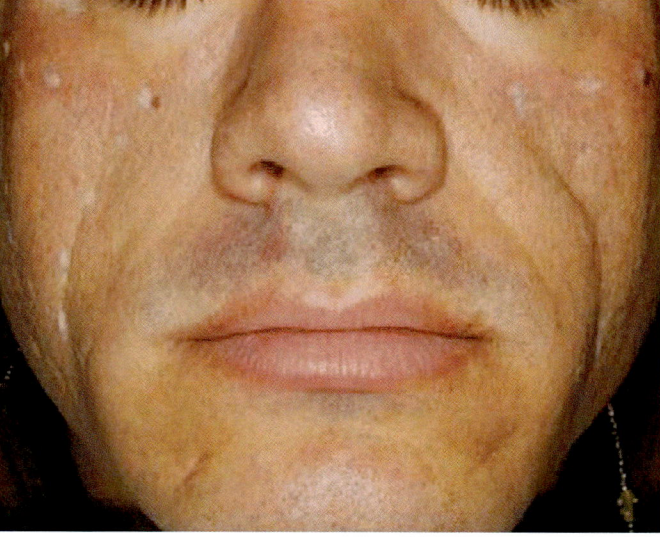

Fig. 11.18 HIV-associated liposdystrophies. LEFT: Buffalo hump" associated with protease inhibitor treatment. RIGHT: Facial fat wasting is a tell-tale finding in treated HIV patients. Left: [Reprinted from Sharma D, Bitterly TJ. Buffalo hump in HIV patients: surgical reconstruction with liposuction. Plast Reconstr Surg 2009; 62(7): 946–949.

With permission from Springer Nature.]. Right: [Reprinted from Rauso R, Gherardini G, Greco M, et al. Is buffalo hump fat the perfect filler for facial wasting rehabilitation? Reflections on three cases. Eur J Plast Surg 2012;35(7):553–556. With permission from Springer Nature]

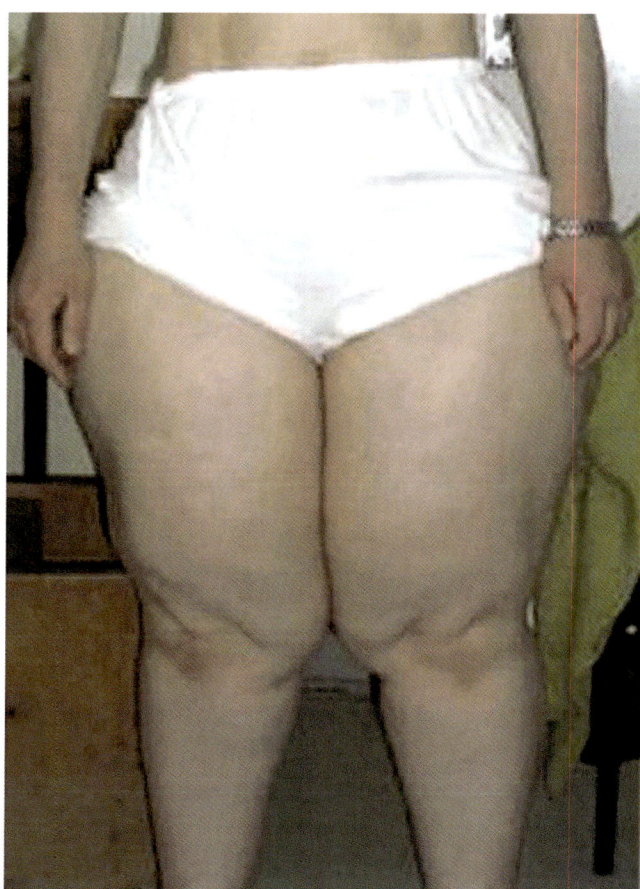

Fig. 11.19 Acquired partial lipodystrophy and lipedema. A 37-year-old woman with acquired partial lipodystrophy. Note the loss of SAT from the upper body to the waist but obesity of the hips and legs. [Reprinted from Herbst K. Rare adipose diseases (RADs) masquerading as obesity. Acta Pharmacologica Sinica 2012; 33: 155–172. With permission from Springer Nature]

(constipation), and growth retardation. This condition, found in up to 1% of adolescent girls, if uncorrected can be lethal. Parry–Romberg disease, a progressive form of volume loss involving soft tissues and bone of the face has a common denominator derivation from neural crest populations associated with branches of the trigeminal nerve (Fig. 11.17).

Adipose tissue storage can be symmetrical or asymmetrical. Obesity (BMI ≥ 30) has two common forms of symmetrical distribution. Pear-shaped obesity is subcutaneous and centers around the midriff and thighs. It is a low-risk condition regarding diabetes and metabolic syndrome. Apple-shaped obesity has increased visceral fat and presents a high risk for diabetes and diseases associated with metabolic syndrome. Lipodystrophy, on the other hand is asymmetrical, favoring localized deposits of fat in the buttocks and/or thighs that are out of proportion to the overall body fat stores [5]. HIV-related treatment may be a characteristic cervical "buffalo hump" or sites of local atrophy in the face (Figs. 11.18 and 11.19). Lipomas arising from white fat can develop in many locations (Fig. 11.20). Although most lipomas are solitary they can be multiple, as in Madelung's disease (Fig. 11.21).

Brown fat cells contain multiple lipid droplets arrayed around a central nucleus. Multiple mitochondria in the cytoplasm—which relate to the thermogenic function of this tissue—are responsible for its color. Capillary content is higher as well, perhaps reflecting the greater metabolic rate. Deposits of brown adipose tissue are found in highly vascularized sites such are the aorta, kidneys, paravertebral arteries and supraclavicular (thyrocervical, costovertebral, and thoracoacromial trunks). These arterial "switchyards" supply multiple distinct developmental fields, many of which

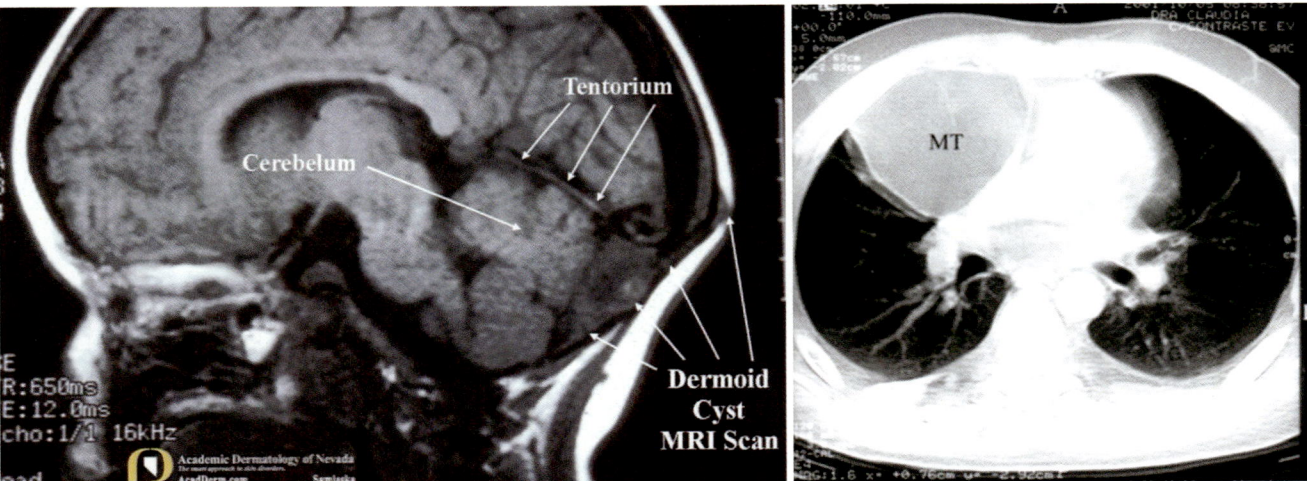

Fig. 11.20 Solitary lipomas: LEFT: Lipoma are typically found in the face and neck but they occur in many sites. MRI scan demonstrate dermoid Cyst pressing against the cerebellum and communicating through the occipital bone to the subcutaneous tissues of the scalp. RIGHT: intrathoracic thymolipoma. Left: [Courtesy of Curt Samlaska, MD, FACP, FAAD. Academic Dermatology of Nevada]. [Reprinted from Ramos Filho J; Melo RF; de Macedo M; Fiorelli LA; Costa A, Isolatto RB. Chest pain due to right atrial compression caused by a thymolipoma. Arq Bras Cardiol 2004; 82(5):484–486. With permission from Creative Commons License 4.0: https://creativecommons.org/licenses/by-nc/4.0/deed.en]

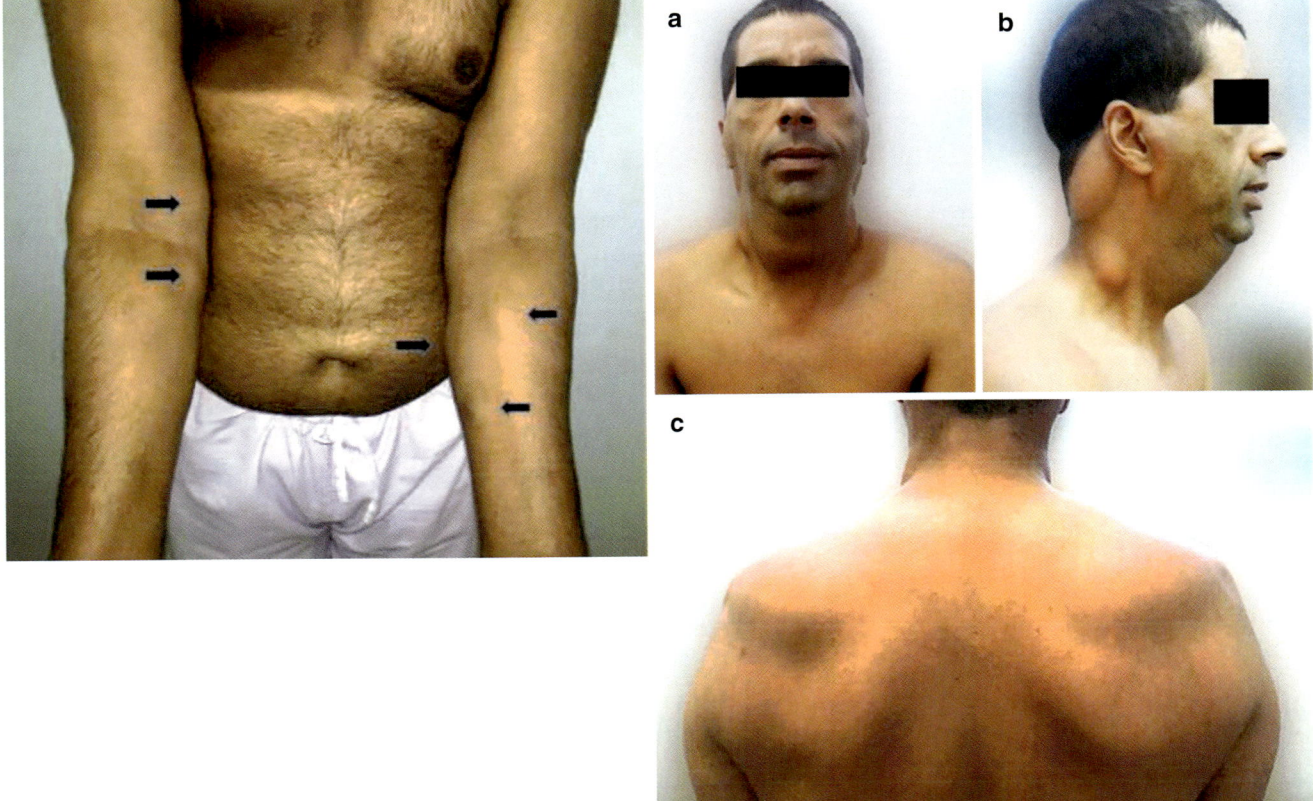

Fig. 11.21 Multiple lipomas. LEFT: [Reprinted from Reddy N, Malipatil B, Kumar S. A rare case of familial multiple subcutaneous lipomatosis with novel *PALB2* mutation and increased predilection for cancers. Hematol Oncol Stem Cell Ther 2016; 9(4):154–156. With permission from Elsevier.]. RIGHT: [Reprinted from Maximiano LF, Gaspar MT, Nakahira ES. Madelung disease (multiple symmetric lipomatosis). Autopsy Case Rep [Internet]. 2018;8(3):e2018030. With permission from Creative Commons Attribution Non-Commercial License (CC-BY-NC)]

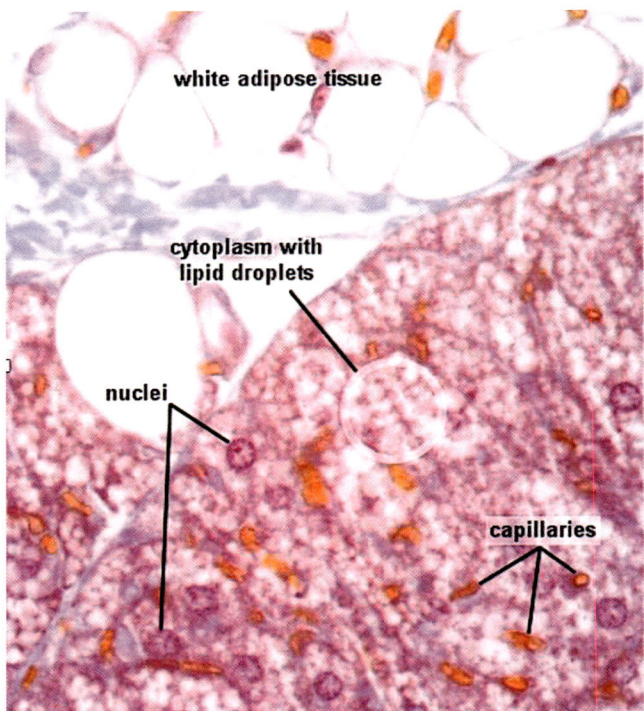

Fig. 11.22 Brown fat. BAT (trichrome stain) related to paraxial meso-derm and if PRDM16- whereas somitic muscle is PRDM16+. BAT is designed for energy production with abundant mitochondria and multiple small lipid droplets of uniform size; these are non-expansile. Control is by SANS via NE. [Courtesy of Michael Carstens, MD]

are useful as reconstructive tissue sources. The segmental location of brown fat deposits along the para-aortic branches off the aorta indicate a somitic origin with likely distinct homeotic codes for each collection of fat (Figs. 11.20 and 11.21).

Beige fat cells constitute a separate type of brown fat that develops within white adipose tissue itself as an adrenergic induction of existing pre-adipocytes interspersed within the white fat. The color difference relates to the greater relation of fat droplets to mitochondria, hence a less red color. The existence of beige fat is of developmental significance (vide infra).

Like white fat, the function of brown fat is under the control of the SANS. Mitochondria produce ATP via the enzyme *ATP synthase* which makes use of a flow of protons along a gradient of internal membranes (chemiosmosis). Once the protons are used they are pumped out of the mitochondrion by the electron transport chain (ETC). An alternative mitochondrial enzyme, *thermogenin* (otherwise known as uncoupling protein 1) provides a return route for protons previously expelled by the ETS. This couples oxidative phosphorylation so the energy of the proton motive force (PMF) is channeled into heat rather into ATP. Mammals have a unique surface/volume ratio which translates this energy release into shivering. The existence of a superficial fatty insulation layer ensures that the heat will be conserved internally.

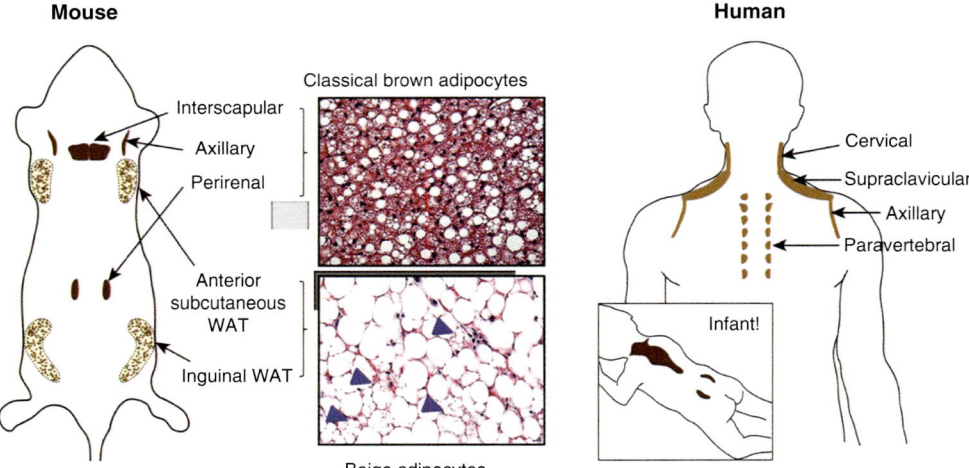

Fig. 11.23 Beige fat: "good" fat. Anatomical Locations of Thermogenic Fat in Mice and Humans. Classical brown adipocytes reside in dedicated brown adipose tissue (BAT) depots, including inter-scapular, axillary, and perirenal BAT depots in mice and infants. Beige adipocytes sporadically reside in subcutaneous white adipose tissue (WAT) depots, such as the inguinal and anterior subcutaneous WAT in mice (arrowheads indicate the multilocular beige adipocytes). In adult humans, BAT is present in multiple locations, including cervical, supra-clavicular, axillary, paravertebral, and abdominal subcutaneous regions. UCP1-positive adipocytes from the supraclavicular region show a molecular signature resembling that of mouse beige adipocytes, whereas the deep neck regions contain thermogenic fat that resembles classical brown adipocytes in mice. [Reprinted from Ikeda K, Maretich P, Kajimura S. Common and distinct features of brown and beige adipocytes. Trends Endo Metab 2018; 29(3): 191–200. With permission from Elsevier]

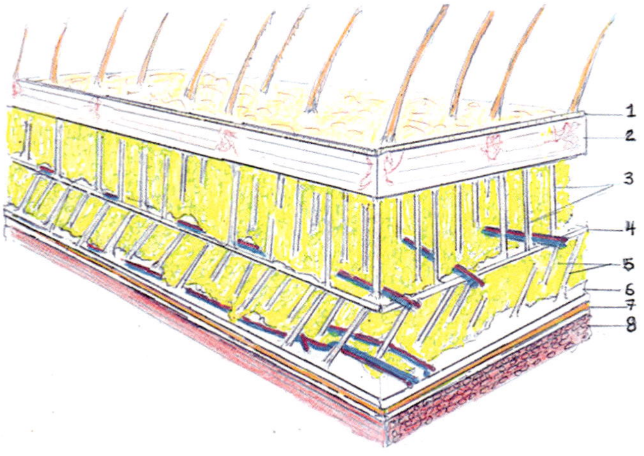

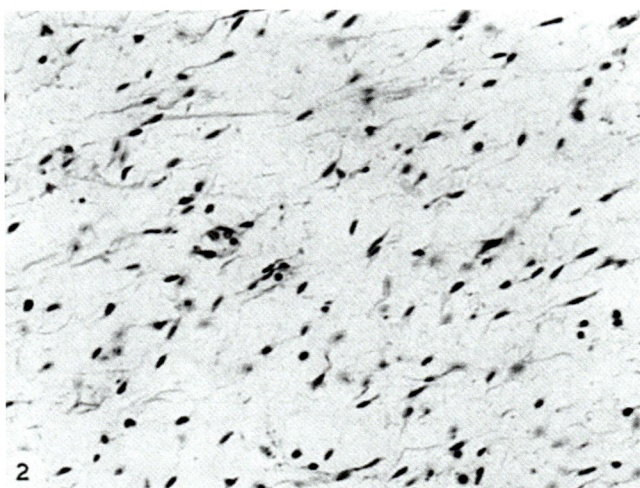

Fig. 11.24 Gross anatomy of subcutaneous fat. Note two layers of fascia. Although the nomenclature may vary between craniofacial skin and non-craniofacial skin the concepts are the same. **Superficial fascia** (SF) divides subdermal fat from deep cutaneous fat. It may be the network associated with panniculus carnosus. Retinacula cutis fibers are vertical and extend from SF to the dermis, thus creating mini-compartments. These allow for flexibility and gliding of the skin over muscle (trauma protection). Accumulations of fat within these compartments causes unsightly "cellulite." Loss of fibers allows for greater skin laxity and may be the cause of wrinkles. Key: (1) epidermis, (2) dermis, (3) superficial retinacular fibers, vertical, and superficial fat, (5) superficial or subcutaneous fascia, (6) retinacular fibers, oblique, and deep fat, (7) deep fascia, (8) hyaluronic acid layer, (9) epimysium, (10) muscle. Note: Within this adipose tissue complex (ATC) the fat cells make up 85% of the volume but, in terms of cell numbers, only 12–15% of the total count. The remaining cells are stromal and are found clustered around blood vessels and connective tissue fibers.

Deep investing fascia DIF invests voluntary muscle. DIF connects with SIF using *oblique fibers* but these are discontinuous with those of the overlying layer. The function of the superficial adipofascial plane is movement. The function of the deep adipofascial plane is energy storage and padding. Distinct vascular planes reveal different souces of embryogenesis.

Perhaps the fat layers are different embryologically?

- Superficial fat: pre-pericytes non-somitic population admixed with neural crest.
- Deep fat; prepericytes that enter the somites—perhaps later in development.

[Courtesy Michael Carstens, MD]

Gross Anatomy

In non-facial skin, white adipose tissue exists in two distinct layers separated by an intervening layer of subcutaneous fascia, likely of neural crest derivation. Although it is referred to as an investing fascia in some mammals it is related to the *panniculus carnosus*. The upper-fat layer is thinner and is interspersed with thin vertically-oriented septae known as the superficial retinaculua cutis. These fibers connect the superficial fascia with the overlying dermis, thus permitting the skin to glide. The superficial fascia is of neural crest derivation. Neurovascular structures run along its upper surface. Adipose hypertrophy in this zone creates a puckering effect

Fig. 11.25 White fat development, stage 1. 14 weeks Undifferentiated mesenchyme, no obvious vascular invasion. [Reprinted from Poissonet CM, Burdi AR, Bookstein FL Growth and development of human adipose tissue during early gestation. Early human devel 1983; 8:1–11. With permission from Elsevier]

of the overlying skin known in the popular parlance as "cellulite." (Fig. 11.24).

The lower layer is thicker; and is the site of subcutaneous fat storage. It rests on a deep (investing) fascia associated with the underlying muscle layer. In non-facial skin this fascia arises from paraxial mesoderm. Fibers of the deep reticulum cutis fibers, course obliquely upward to the superficial fascia. This layer provides insulation.

As previously stated, the neuromeric identity of skin is determined by its dermis. Outside the head and neck, skin is organized into dermatomes depending on the somitic origin of its dermis. But subcutaneous fat is not somitic. There is not dedicated "adipotome" compartment in somites. Instead, mammalian fat development follows the invasion of interposition of neural crest cells associated with neurovascular structures. As we shall now see, the true origin of white adipose tissue is vascular.

Developmental Considerations (Figs. 11.25, 11.26, 11.27, 11.28 and 11.29)

The appearance of white adipose tissue follows a developmental sequence that appears to be neuromeric: head and neck (r0-r11 and c1-c8), thorax (t1-t12), abdominal (l1-s5) and extremities (c4-t1 and l4-s3) but is not somatic. Histologic development takes place in 5 stages, each strongly correlated with blood vessel formation (Poissonet 1983, 1984)

- Stage 1 (14 weeks) undifferentiated mesenchyme: non-condensed amorphous ground substance and stellate cells and ground substance.

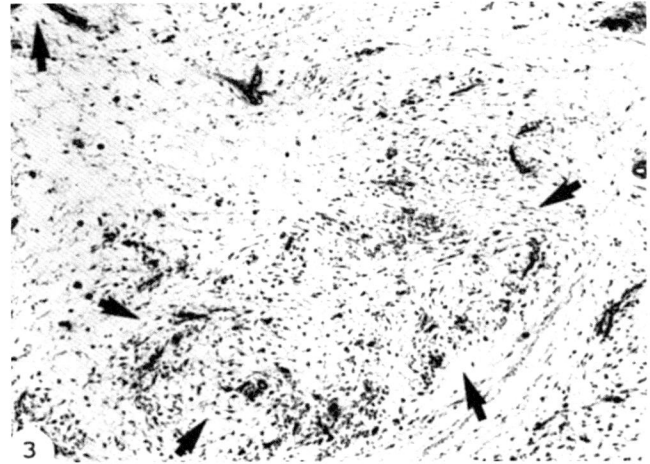

Fig. 11.26 White fat development, stage 2. 14.5 weeks Mesenchymal condensation is taking place around blood vessels. No lobules are present. [Reprinted from [26, 27]. With permission from Elsevier]

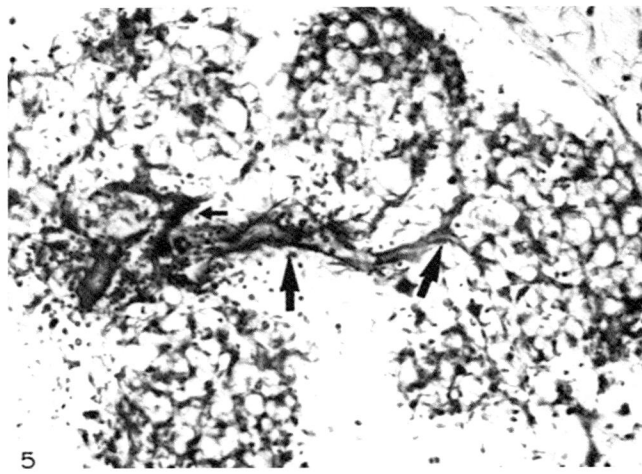

Fig. 11.28 White fat development, stage 4. 22 weeks Primitive fat lobules. Intracellular fat droplets for storage. [Reprinted from [26, 27]. With permission from Elsevier]

Fig. 11.27 White fate development, stage 3. 16 weeks Mesenchymal lobules. No intracellular fat droplets. [Reprinted from [26, 27]. With permission from Elsevier]

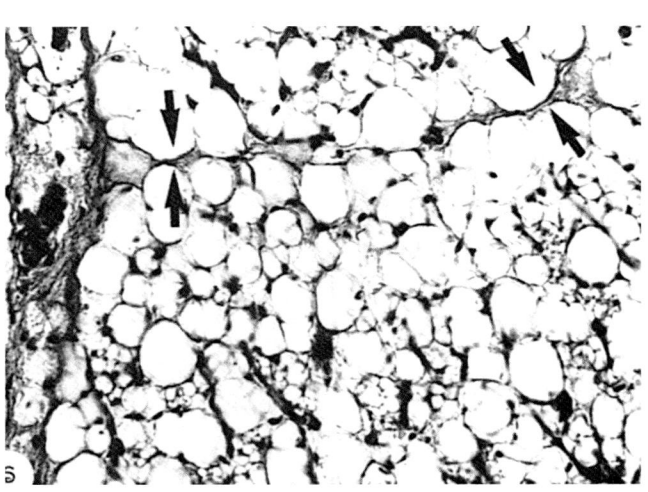

Fig. 11.29 White fat development, stage 5. Adult configuration. 28 weeks shows fat lobules clearly separated by connective tissue septae. [Reprinted from [26, 27]. With permission from Elsevier]

- Stage 2 (14.5 weeks) angiogenesis: the proliferation of primitive vessels around which fat cells will organize around primitive vessels; no lobules are present.
- Stage 3 (16 weeks) mesenchymal lobules: stellate pre-adipocytes do not contain lipid droplets.
- Stage 4 (22 weeks) primitive fat lobules: each lobule has a vascular stalk that branches off the main vessel. After 23 week number of lobules stabilizes.

- Stage 5 (28 weeks) fat lobules are well separated by stromal septae of connective tissue. In weeks 23–29, lobules increase in size and stabilize.

White fat originates from a poorly characterized precursor cell, the *pre-pericyte* which differentiates either into mature pericytes (vide infra) or into adipose tissue. Furthermore, mesenchymal stromal cells derived either from pre-pericytes or pericytes themselves have, by definition, the capacity to further differentiate into adipocytes. Pre-pericytes are likely ectomesenchymal in nature given that their pericyte offspring are responsible for SANS stimulation. Moreover, MSCs can be driven in vitro, into primitive neural tissue, indicating that pre-pericytes are NOT strictly mesodermal; more likely they share a common precursor with neural crest cells.

It is widely (and incorrectly) stated that brown fat comes from mesoderm. Pericytes and pre-pericytes are present in all somites as they are associated with the vascular axis of each neuromere. Apparently, pluripotential pre-pericytes that are physically located in a developing somite are committed by the presence of *myogenic factor 5* (myf5) into a binary choice depending on the subsequent expression of PRDM16, a zinc finger transcription factor that controls the fate of the cell to become muscle or adipose tissue. PRDM16+ cells become brown fat while PRDM16- cells are committed to myogenesis (Fig. 11.22).

Differentiation of beige fat provides clues about the embryologic origins of the precursor state. The fact that pericyte-derived pre-adipocytes can be driven to beige fat indicates that the number of mitochondria in the white fat cell line is adrenergic dependent as is the physical state of the fat droplet. In beige fat, the solitary droplet is broken down into smaller compartments to create the final histology (Fig. 11.23).

In sum, ectomesenchymal cells with multi-differentiation potential and related to neural crest, flow outward into mesenchymal environments both somitic and non-somitic. In either scenario, some of these will form pericytes and partici-pate in the vascularization of the tissue. In the case of non-somitic environment adipocytes with a metabolic fate will develop. Within somites, the pre-pericytes are exposed to a sequential influence of myf5. Those that express PRDM16 will commit to brown fat.

Evolution and Lineage of Adipose Tissue (Figs. 11.30 and 11.31)

The evolution of adipose tissue is characterized by phylogenetic changes in distribution and fat storage. Worms, such as *C. elegans*, store fat in the intestine, whereas insects such as *Drosophila* have a defined deposition site, the "fat body". Chondrichtyan fishes (sharks) are devoid of adipose tissue but store fat in the liver. Osteoichthyans such as the carp has intra-abdominal deposits of white adipose tissue (WAT) that become apparent, coinciding with the presence of leptin in boney fishes, such as carp (*Cyprinus carpio*). This pattern continues in their evolutionary derivatives, amphibians, and fast-moving reptiles. A branch point appears prior to the divergence of dinosaurs, as precursors of the avian line and the modern reptiles leading to mammals in which WAT is present both internally and subcutaneously. This fat pattern is exemplified by the chicken (*Gallus gallus domesticus*), mouse (*Mus musculus*), and human (*Homo sapiens)*. Thermoregulation also appears in the course of evolution but independently of the appearance of brown adipose tissue (BAT). Estimated to have emerged 150 million years ago, BAT is only present in higher mammals, although UCP-1 expression appears independently of BAT in bony fish.

Species	*Caenorhabditis elegans*	*Drosophila melanogaster*	*Carcharodon carcharias*	*Cyprinus carpio*	*Xenopus laevis*	*Gallus gallus domesticus*	*Mus musculus*	*Homo sapiens*
Fat storage	Stored in intestinal cells	Stored in the "fat body"	Stored in liver	Stored in WAT	Intra-abdominal WAT (no subcutaneous WAT)	Subcutaneous and internal WAT	Subcutaneous and internal WAT	Subcutaneous and internal WAT
Leptin	No	No	No	Yes	Yes	Yes	Yes	Yes
BAT	No	No	No	No	No	No	Present throughout life	Present at birth; reduced in adults
UCP	UCP-like protein (ucp-4)	No	?	UCP-1 in liver	UCP-4 in oocytes	Avian UCP in muscle	UCP-1 in BAT	UCP-1 in BAT
Thermo-regulation	Ectotherm	Ectotherm	Ectotherm	Ectotherm	Ectotherm	Endotherm Shivering and nonshivering thermogenesis	Endotherm Shivering and nonshivering thermogenesis	Endotherm Shivering and nonshivering thermogenesis

Fig. 11.30 Evolutionary phylogeny of adipose tissue. Note the sequential phylogenetic emergence of white fat (osteichthes), leptin (tetrapods), and brown fat (mammals). [Reprinted from Gesta S, Tseng Y-H, Kahn R. Developmental origin of fat: tracking obesity to its source. Cell 2007; 131(2): 242–256. With permission from Elsevier]

Fig. 11.31 From ectomesenchymoblast/ pre-pericyte to adipose lineages. Although some stages are still not clearly defined, this differentiation pathway presumably involves differentiation of the pre-pericyte to a common preadipocyte or adipoblast, which has the capacity to differentiate into either white or brown preadipocytes. The proteins and genes that represent potential molecular markers in cells within the adipocyte lineage are marked with a "+" or "−"sign, indicating the relative levels of expression of these markers. The table summarizes the relative ex-pression levels of various developmental and patterning genes in different fat depots in both humans and mice. The relative levels of expression have been graded from absent (−) to high (+++). [Reprinted from Gesta S, Tseng Y-H, Kahn R. Developmental origin of fat: tracking obesity to its source. Cell 2007; 131(2): 242–256. With permission from Elsevier]

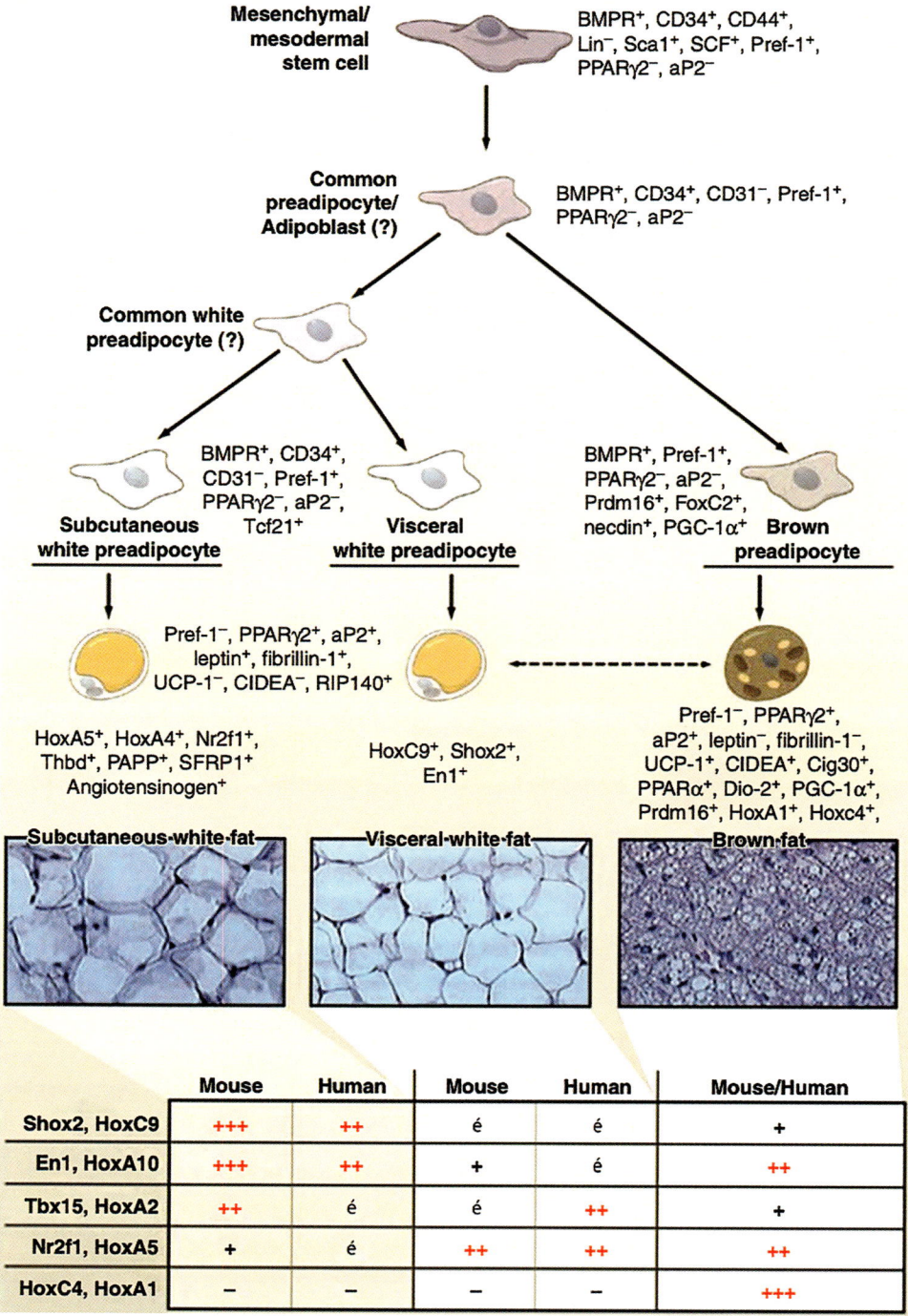

	Mouse	Human	Mouse	Human	Mouse/Human
Shox2, HoxC9	+++	++	é	é	+
En1, HoxA10	+++	++	+	é	++
Tbx15, HoxA2	++	é	é	++	+
Nr2f1, HoxA5	+	é	++	++	++
HoxC4, HoxA1	–	–	–	–	+++

Endotherm refers to an animal that produces its own heat from within versus ectotherm, an animal that does not. These terms are interchangeable with homeotherm, that is, an animal that can maintain a specific body temperature or are warm-blooded versus poikilotherm, that is, an animal that has a body temperature that varies with the ambient temperature or is cold-blooded.

The embryologic origins of adipose tissue are poorly understood. There certainly exists a mesenchymal precursor stem cell that gives rise to preadipocytes but also gives rise to pericytes. It is often assumed that this precursor arises from mesoderm but that model has developmental flaws. We will develop the argument (vide infra) that this primitive precursor cell can be better termed the **pre-pericyte** and that it is derived from a lineage of non-neural ectodermal cells, including neural crest, which is ectomesencymal. This cell line emerges from the very inception of gastrulation (stage 6), long before the intraembryonic mesodermal tissues have been defined. The derivatives of this ectomesenchymal cell line therefore contain carry genomic information which gives them the potential to differentiate, not only into mesodermal lines such as fat and cartilage but also to cells that retain a connection with the sympathetic autonomic nervous system.

For the moment, let's consider what we know about the differentiation process of fat. When triggered by appropriate developmental cues, the common pre-pericytes become committed to become pre-adipocytes (white versus brown) versus pericytes. A node exists between preadipocytes destined for white fat versus brown fat. The former carry CD34+ markers whereas the latter are Prdm16+. Note that myoblasts develop along the same line as brown pre-adipocytes but they are Prdm-. In their mature state, white and brown adipose tissues express distinct signatures Hox genes. Since brown fat deposits are associated with somites, future studies are needed to ascertain if there are distinct homeotic signatures of fat in register with specific neuromeric levels.

Pericytes

Definition (Fig. 11.32)

Pericytes (Rouget cells) are multi-functional mural cells of the microcirculation that wrap around the endothelial cells that comprise the "tubing" of all capillaries and venules throughout the body [6]. Pericytes are embedded in the basement membrane, where they communicate with endothelial cells of the body's smallest blood vessels by means of both direct physical contact and paracrine signaling to control *porosity* and *permeability* [7]. In the brain, they sustain the *blood–brain barrier*. Pericytes *regulate capillary blood flow* and dispose of cellular debris by *phagocytosis*. Pericytes *stabilize maturation of endothelial cells* which they monitor by means of direct communication between the cell membrane as well as through paracrine signaling. A deficiency of pericytes in the central nervous system can cause the blood–brain barrier to break down, leading to the leakage of high molecular weight protein down (Winkler, 2011).

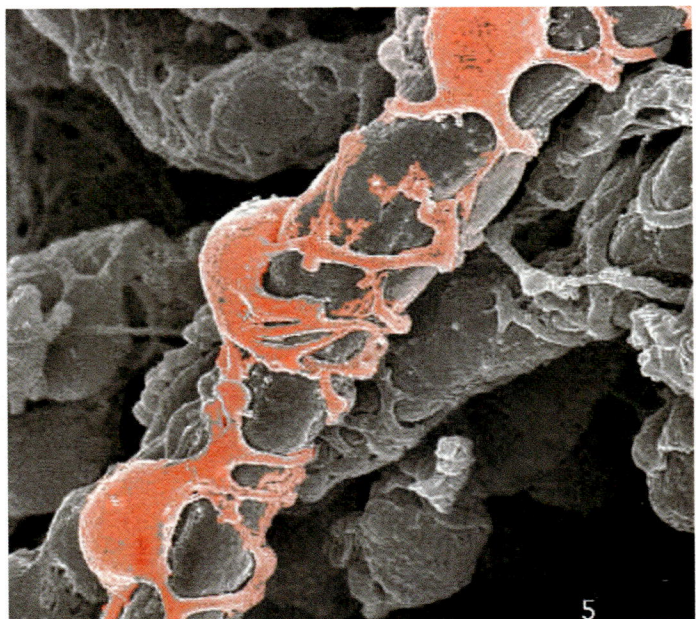 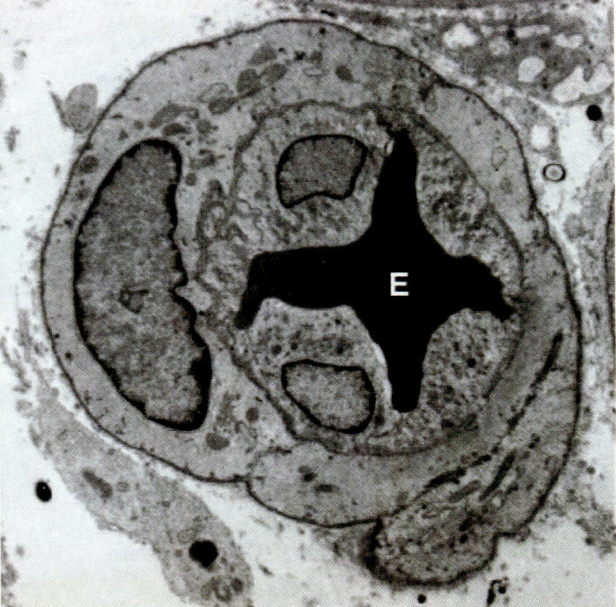

Fig. 11.32 Pericyte relationship to endothelial cells in a capillary. Pericyte wraps around the circumference of the capillary. Erythrocyte (E) surrounded by two endothelial cells with a single pericyte (nucleus left) encircling the vessel. Left: [Reprinted from Alexander RW, Harrell DB. Autologous fat grafting: use of closed syringe microcannula system for enhanced autologous structural grafting. Clinical, Cosmetic and Investigational Dermatology; 2013(6) 91–102. With permission from Dove Medical Press.] Right: [Reprinted from Wikimedia. Retrieved from: https://en.wikipedia.org/wiki/Pericyte#/media/File:Microvessel.jpg. With permission from Creative Commons License 3.0: https://creativecommons.org/licenses/by/3.0/deed.en]

Structure

Pericytes wrap around the endothelial cells that line the inside of the capillary. These two types of cells can be easily distinguished: from one another based on the presence of the prominent round nucleus of the pericytes have a prominent round nucleus whereas endothelial cells have nuclei that are flat and elongated. Pericytes have finger-like extensions that wrap around the capillary wall; these extensions are contractile, allowing for regulation of capillary blood flow. Pericytes and endothelial cells share a common basement membrane where a variety of intercellular connections are made. They communicate by means of various types of integrin molecules. Although they are dispersed along the vessel walls pericytes can form direct connections with each other and with neighboring cells by forming peg and socket arrangements in which parts of the cells interlock. At these interlocking sites, gap junctions can be formed, which allow the pericytes and neighboring cells to exchange ions and low molecular weight compounds. Intercellular connections are mediated by N-cadherin, fibronectin, connexin, and various integrins [8].

Embryology of the Pericyte (Fig. 11.33)

In the developing embryo, the vascular system appears at stage 5 (of 23) well before all others. The reason is simple: ongoing growth requires mechanisms for nutritional supply and waste removal in order for the organism to survive. Although the endothelial cells of the vascular conduits are mesodermal in nature the pericyte may originate from an ectomesenchymal precursor cell related to neural crest. Pericytes are contractile and communicate. Their adipose progeny, as white and brown adipocytes, carry out neuroendocrine functions directly related to the sympathetic nervous system [8].

Vascular Niche (Fig. 11.34)

Pericytes have a variety of mural cell types. These form a continuum along the vasculature. Smooth muscle cells form concentric rings on arterioles. Hybrid smooth muscle-pericyte cells reside on precapillary arterioles and interlock

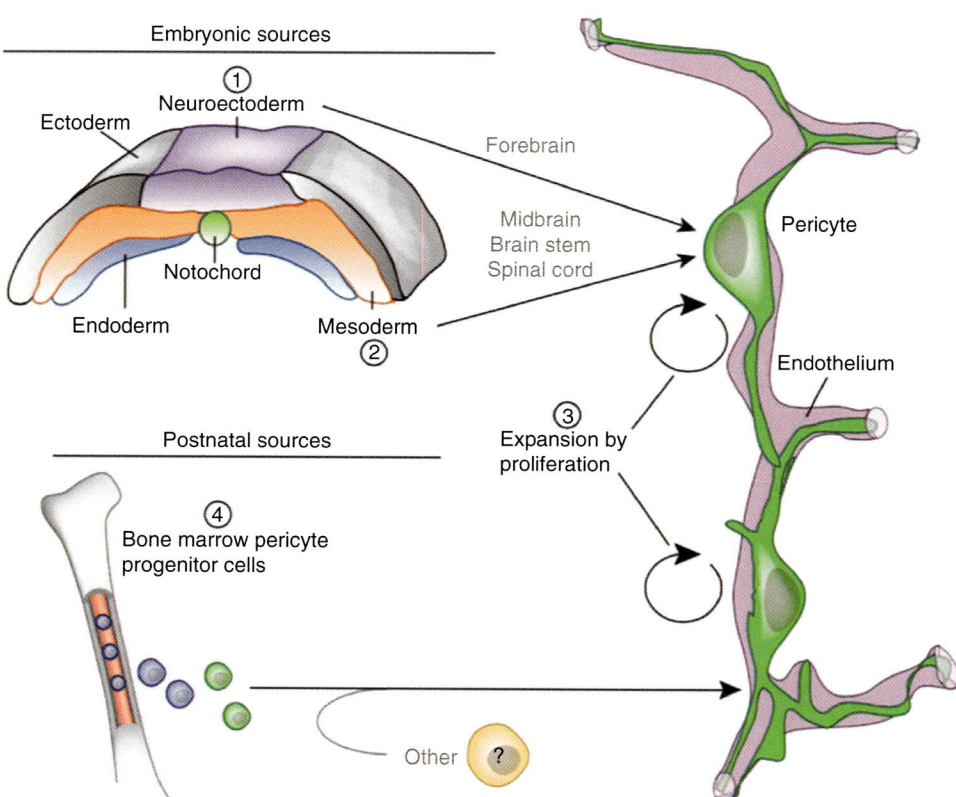

Fig. 11.33 Origin of pericytes in the CNS. The embryonic sources of pericytes include (1) *neuroectoderm-derived neural crest* cells, which give rise to pericytes of the forebrain, (2) *mesoderm-derived mesenchymal stem cells* -probably originating from pre-pericytes within the mesoderm—which give rise to pericytes in the midbrain, brain stem and spinal cord, and (3) expansion by proliferation from the newly-established *pericyte pools*. Postnatal sources of pericytes include (3) expansion by proliferation from the existing pericyte pools and (4) mesoderm-derived circulating mesenchymal stem cells (bone marrow pericyte progenitor cells) and presently undetermined 'other' sources. Note: *Neuroectoderm-derived pre-pericytes likely give rise to MSCs of (2) and (4)*. [Reprinted from [28]. With permission from Springer Nature]

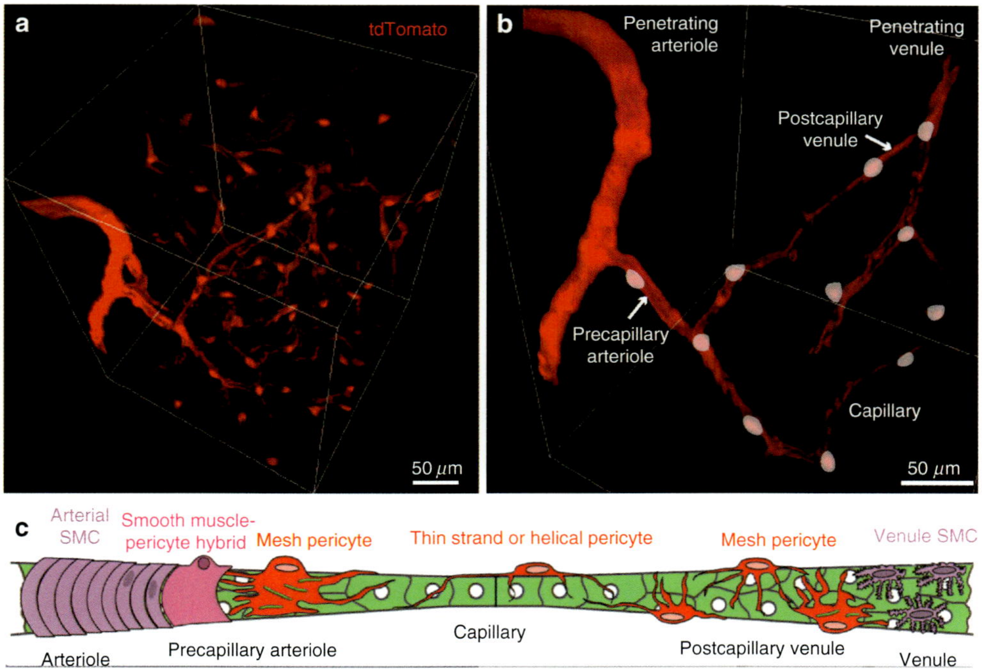

Fig. 11.34 Pericytes assume different forms depending on their location in the vascular system. Summary of mural (pericyte) cell organization in the cerebrovasculature: (**a**) A tissue volume from a PDGFRβ-tdTomato mouse imaged after optical clearing. (**b**) Distribution of pericytes and pericyte-smooth muscle hybrids in an arteriole-capillary-venule loop. The cell bodies are labeled in white. (**c**) Schematic showing the continuum of mural cell types along the cerebral vasculature. Smooth muscle cells form concentric rings on arterioles. Hybrid smooth muscle-pericyte cells reside on precapillary arterioles and interlock with mesh pericytes at the arteriole–capillary interface, which occurs as penetrating arterioles ramify into the capil- lary bed. Pericytes in capillary beds typically exhibit long processes that traverse the microvasculature in single strands or pairs that twist in a helical fashion. Mesh pericytes become more prevalent again as capil- laries turn into postcapillary venules. Stellate-shaped smooth muscle cells cover the walls of parenchymal venules. [Reprinted from Hartman DA, Underly RG, Grant RI, et al. Pericyte structure and distribution in the cerebral cortex revealed by high-resolution imaging of transgenic mice.Neurophotonics 2015; 2(4): 041402. With permission from Creative Commons License 4.0: https://creativecommons.org/licenses/by/4.0/]

with mesh pericytes at the arteriole–capillary interface, which occurs as penetrating arterioles ramify into the capil- lary bed. Pericytes in capillary beds typically exhibit long processes that traverse the microvasculature in single strands or pairs that twist in a helical fashion. Mesh pericytes become more prevalent again as capillaries turn into postcapillary venules. Stellate-shaped smooth muscle cells cover the walls of parenchymal venules.

Function

Differentiation (Fig. 11.35)

Pericytes in the skeletal striated muscle are of two distinct populations, distinguished by their ability to express Nestin, their differentiation potential, and their responses to injury by glycerol and barium chloride (BaCl2).

Type-1

- Nestin negative (PDGFRβ+CD146 + Nes-).
- Differentiates from fat cells.

- Proliferates to glycerol and to BcCl2 but differentiates only with glycerol.

Type-2

- Nestin positive (PDGFRβ+CD146 + Nes+).
- Differentiates from muscle cells.
- Proliferates to glycerol and to BcCl2 and differentiates with bpth.

Angiogenesis and Anti-Apopotosis

Pericytes are also associated with endothelial cell differenti- ation and multiplication, stimulate angiogenesis, promote survival of apoptotic signals, and travel. One subtype of peri- cytes, known as microvascular pericytes, develops around the walls of capillaries and helps to serve this function. Microvascular pericytes may not be contractile cells, as they lack alpha-actin isoforms, structures that are common amongst other contractile cells. These cells communicate with endothelial cells via gap junctions, and in turn, cause endothelial cells to proliferate or be selectively inhibited. If

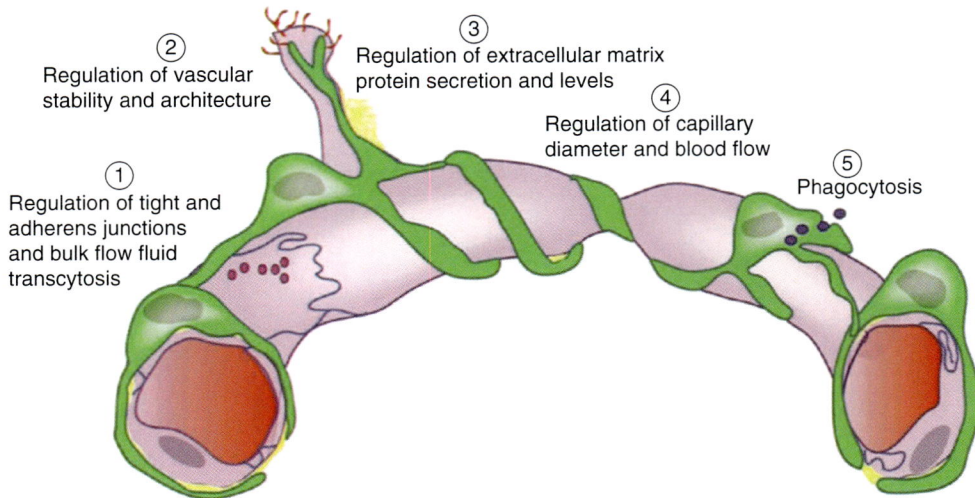

Fig. 11.35 Pericyte functions. Pericytes are multi-functional members of the neurovascular unit. Pericytes (1) control BBB integrity by regulating the orientation and abundance of endothelial tight and adherens junction proteins, as well as the rate of bulk flow fluid transcytosis (transendothelial transport of fluid-filled vesicles); (2) regulate the stability and architecture of newly formed cerebral microvessels; (3) contribute to secretion and regulate the levels of extracellular matrix proteins forming the basement membrane; (4) regulate capillary diameter and blood flow; and (5) provide clearance and phagocytotic functions in brain. [Reprinted from [28]. With permission from Springer Nature]

this process did not occur, hyperplasia and abnormal vascular morphogenesis could result. These types of pericytes can also phagocytose exogenous proteins. This suggests that the cell type might have been derived from microglia, an association that stems from their ectomesenchymal origin and thus relates them to neural crest.

Lineage relationships to other cell types have been proposed, including smooth muscle cells, neural cells, NG2 glia, muscle fibers, adipocytes, as well as fibroblasts and other mesenchymal stem cells. However, whether these cells differentiate into each other is an outstanding question in the field. Although aging affects pericytes' regenerative capacity, they remain functional throughout life. Such versatility is useful, as they actively remodel blood vessels throughout the body and can thereby blend homogeneously with the local tissue microenvironment.

Aside from creating and remodeling blood vessels, pericytes have been found to protect endothelial cells from death via apoptosis or cytotoxic elements. It has been shown *in vivo* that pericytes release a hormone known as *pericytic aminopeptidase N/pAPN* that may help to promote angiogenesis. When this hormone was mixed with cerebral endothelial cells as well as astrocytes, the pericytes grouped into structures that resembled capillaries. Furthermore, in the absence of pericytes, the endothelial cells would undergo apoptosis. Thus, proper angiogenesis depends on (1) the presence of pericytes to ensure the proper endothelial cell function, and (2) the presence of astrocytes, as these cells ensure that both other cell types remain in contact with each other. Pericytes contribute to the survival of endothelial cells,

as they secrete the protein Bcl-w, an instrumental protein in the pathway that enforces VEGF-A expression and discourages apoptosis, by inhibiting the activation of apoptosis-inducing enzymes. Two biochemical mechanisms utilized by VEGF to accomplish this would be the phosphorylation of extracellular regulatory kinase 1 (ERK-1, also known as MAPK3), which sustains cell survival over time, and the inhibition of stress-activated protein kinase/c-jun-NH2 kinase, which also promotes apoptosis.

Blood Flow

Increasing evidence suggests that pericytes can regulate blood flow at the capillary level. Retinal pericytes constrict capillaries when their membrane potential is altered to cause calcium influx. In the brain, it has been reported that neuronal activity increases local blood flow by inducing pericytes to dilate capillaries. Thus, different mechanisms regulate the constriction of capillaries by pericytes and of arterioles by smooth muscle cells. Pericytes are also important in maintaining circulation. In a study involving adult pericyte-deficient mice, cerebral blood flow was diminished with concurrent vascular regression due to loss of both endothelia and pericytes. Significantly greater hypoxia was reported in the hippocampus of pericyte-deficient mice as well as inflammation, learning, and memory.

Selective Permeability: Blood–Brain Barrier

Pericytes play a crucial role in the formation and functionality of the blood–brain barrier. This barrier is composed of endothelial cells and ensures the protection and functionality of the brain and central nervous system. It has been found that pericytes are crucial to the postnatal formation of this barrier. Pericytes are responsible for tight junction formation and vesicle trafficking amongst endothelial cells. Furthermore, they allow the formation of the blood–brain barrier by inhibiting the effects of CNS immune cells (which can damage the formation of the barrier) and by reducing the expression of molecules that increase vascular permeability [9].

Aside from blood–brain barrier formation, pericytes also play an active role in its functionality by controlling the flow within blood vessels and between blood vessels and the brain. In animal models with lower pericyte coverage, trafficking of molecules across endothelial cells occurs at a higher frequency, allowing proteins into the brain that would normally be excluded. Loss or dysfunction of pericytes is also theorized to contribute to neurodegenerative diseases such as Alzheimer's, Parkinson's, and ALS through the breakdown of the blood-brain barrier [7].

Pathologies of Pericytes (Figs. 11.36 and 11.37)

Hemangiopericytoma is a rare vascular neoplasm, or abnormal growth, that may either be benign or malignant. In its malignant form, metastasis to the lungs, liver, brain, and extremities may occur. It most commonly manifests itself in the femur and proximal tibia as a bone sarcoma and is usually found in older individuals, though cases have been found in children. Hemangiopericytoma is caused by the excessive layering of sheets of pericytes around improperly formed blood vessels. Diagnosis of this tumor is difficult because of the inability to distinguish pericytes from other types of cells using light microscopy. Treatment may involve surgical removal and radiation therapy, depending on the level of bone penetration and stage in the tumor's development.

Diabetic retinopathy is commonly associated with a loss of pericytes. Pericytes are essential in diabetic patients as they protect the endothelial cells of retinal capillaries. With the loss of pericytes, microaneurysms form in the capillaries.

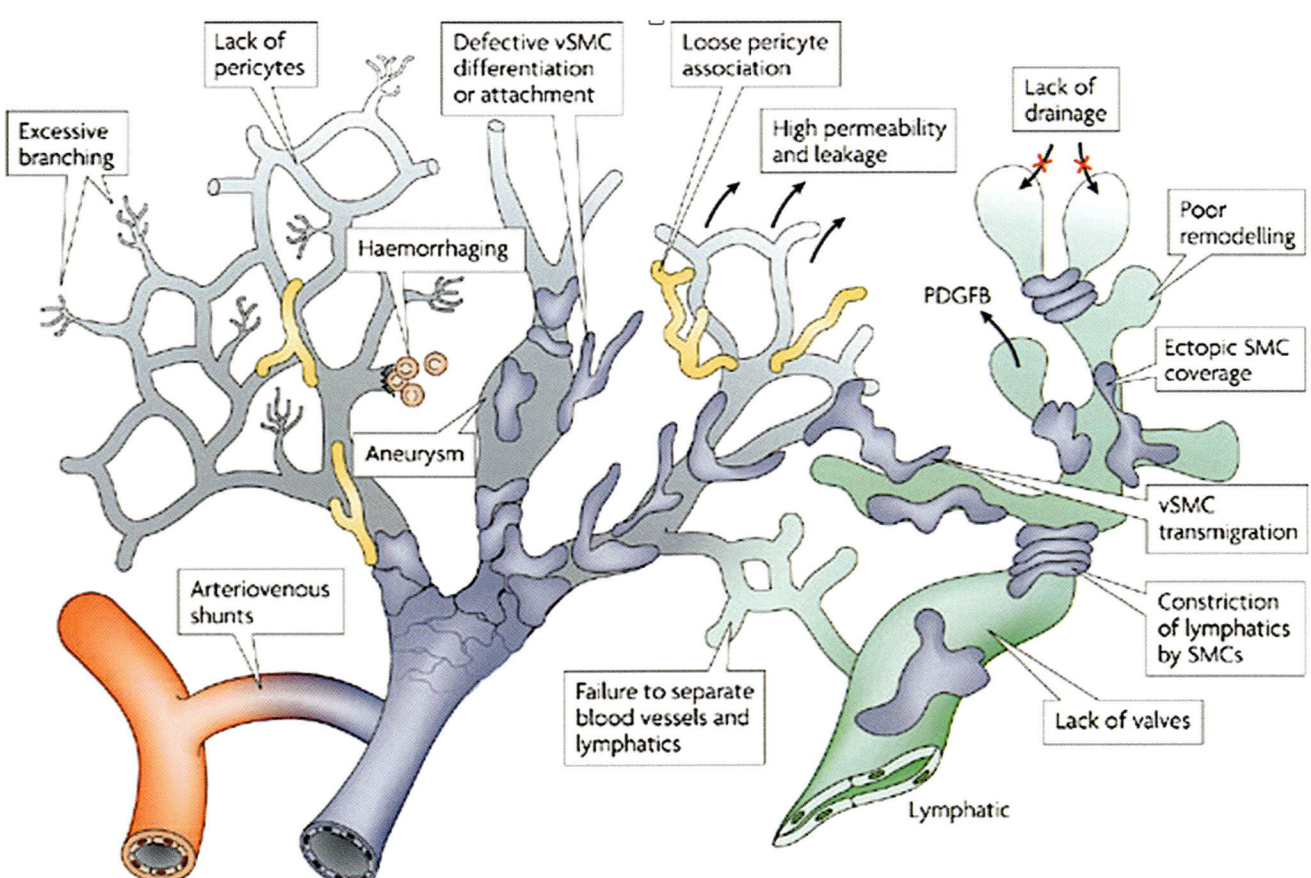

Nature Reviews | Molecular Cell Biology

Fig. 11.36 Pericyte pathologies. Defective numbers of pericytes make the vessel walls unstable: resulting in aneurysm or hemorrhage Smooth muscle may fail to attach properly to the vessel wall. Abortive attempts a new vessel formation or arteriovenous fistulae can occur. Shunting between the venous and lymphatic systems has been documented.

[Reprinted from Armulik A, Genové G, Betscholtz C. Pericytes: Developmental, physiological and pathological perspectives, problems and promises. Developmental Cell 2011; 21(2): 193–215. With permission from Elsevier]

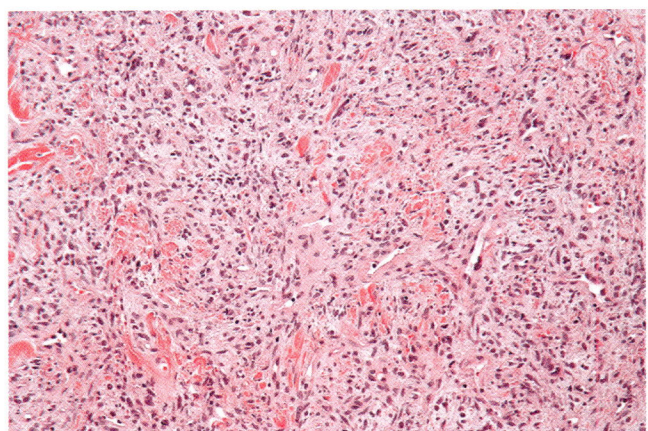

Fig. 11.37 Hemangiopericytoma. Histologic Features: Cellular solitary fibrous tumors (hemangiopericytoma pattern) are composed of plump round or oval cells lacking atypia. The cells are arranged haphazardly or in short, intersecting fascicles around a network of sinusoidal vessels with staghorn configuration. The cells have moderate amount of eosinophilic cytoplasm with indistinct borders. Mitotic activity is not increased. [Reprinted from Wikimedia. Retrieved from: https://commons.wikimedia.org/wiki/File:Solitary_fibrous_tumour_intermed_mag.jpg. With permission from Creative Commons License 3.0: https://creativecommons.org/licenses/by-sa/3.0/deed.en]

The retinal response with (1) increased vascular permeability and macular edema or with (2) neovascularity vessels that penetrate into the vitreous membrane. Either of these processes leads to a reduction or loss of vision. Pericyte loss in diabetic patients has been attributed to the stimulation of the polyol pathway by elevated glucose levels. The intracellular accumulation of sorbital, fructose, and advanced glycation products (AGEs) is toxic as it leads to osmotic imbalance, which results in cellular damage.

Interactions between Endothelial Cells and Pericytes

There are several pathways of communication between endothelial cells and pericytes. Pericyte differentiation is dependent on transforming growth factor (TGF) signaling from endothelial cells. Perictye recruitment depends on platelet-derived growth factor (PDGF) from the endothelial cells as it directs the pericytes to migrate toward and populated developing blood vessels. Blocks in this pathway lead to pericyte deficiency and instability of the blood vessel walls. Endothelial signaling with sphingosine-1-phosphate (S1P) promotes N-cadherin tracking signals in endothelial membranes; these strengthen endothelial contacts with pericytes. In turn, endothelial cells are themselves stabilized by pericyte production of angiopoietin 1 and Tie-2 signaling.

Communication defects between endothelial cells and pericytes result in vascular instability. Inhibiting the PDGF pathway leads to pericyte deficiency. This causes endothelial hyperplasia, abnormal junctions, and diabetic retinopathy. A lack of pericytes also causes an upregulation of vascular endothelial growth factor (VEGF), leading to vascular leakage and microhemorrhage. Angiopoietin 2 is antagonistic to Tie-2, thereby destabilizing the endothelial cells, resulting. Less endothelial cell and pericyte interaction. This occasionally leads to a proliferative tumor state, hemangiopericytoma. Reduction in pericytes by angiopoietin 2 reduces levels of pericytes, leading to diabetic retinopathy.

Neurodegeneration and Scarring

Studies have found that pericyte loss in the adult and aging brain leads to the disruption of proper cerebral perfusion and maintenance of the blood–brain barrier, which causes neurodegeneration and neuroinflammation. Pericyte apoptosis in the aging brain may result from failed communication between growth factors and pericyte receptors.

Immunohistochemical studies of human tissue from Alzheimer's disease and amyotrophic lateral sclerosis show pericyte loss and breakdown of the blood-brain barrier. Pericyte-deficient mouse models (which lack genes encoding steps in the PDGFB:PDGFRB signaling cascade) and have an Alzheimer's-causing mutation have exacerbated Alzheimer's-like pathology compared to mice with normal pericyte coverage and an Alzheimer's-causing mutation.

Pericytes and Mesenchymal Stromal Cells

Bone marrow contains two lines of multipotent mesenchymal cells: The first to be described was the hematopoietic line, all with CD34+ markers and with differentiation into the various types of blood cells. In the 1970s Friedenstein reported that bone marrow stroma (non-osseous structural elements such as collage fibers and blood vessels) harbored a distinct mesenchymal line of CD34- cells which were shown in vitro to differentiate into multiple mesodermal lineages such as chondrocytes, osteoblasts, myocytes, and adipocytes. This multilineage behavior gave rise to the term, *mesenchymal stem cells* (MSCs) which, however popular, is not accurate, as this differentiation does not take place in vivo. Instead, MSCs are best considered as mesenchymal stromal cells, mesenchymal signaling cells, or medicinal stromal cells [10]. In 2001 Zuk et al. described the existence of MSCs in adipose tissue. These cells have the same cell markers as those in bone marrow but exist in far greater numbers; being up to 500 times more frequent. CD marker analysis reveals that the pericyte is the likely precursor of both types of MSCs. Direct involvement of pre-pericytes in adipogenesis explains the high number of MSCs found in the stromal vascular fraction.

Comparison of cell membrane markers demonstrates a very close match between pericytes and MSCs such that pericytes can be considered the legitimate precursor cell line. Alternatively, these cell lines may represent related derivatives of a common precursor, putatively, the pre-pericyte.

[Caplan] Under conditions of stress, such as infection and inflammation, pericytes detach from their vascular niche and undergo transformation into MSCs. Furthermore, the two cell lines share an overlapping menu of documented clinical effects. From a clinical standpoint, they are literally "an injury drugstore." [11].

Adipose Tissue-Derived Reconstructive Materials

While the clinical applications of adipose-derived stem/stromal cell products are evolving rapidly, the terminology regarding fat and its reconstructive products is quite confusing; therefore, a clarification is necessary based on histology: (1) adipose tissue complex, (2) Coleman-type whole fat graft, (3) microfat whole fat graft, (4) tissue stroma graft, and (4) cellular stromal graft.

Adipose Tissue Complex (ATC) (Fig. 11.38)

In histological terms, a tissue consists of a functional component, the parenchyma, and a supporting component, the stroma. *Parenchyma* refers to an aggregate of cells specifically organized to accomplish a function. *Stroma*, a word of Greek origin ("bed"), refers to cells or structures that are not specific, but rather perform a structural or nutritional support function (connective tissues, vessels, and lymphatics) or communicate between the parenchyma and another anatomical site (nerves and ducts).

The most primitive component in the embryo is the vascular system. It is the first system to be formed; of the 23 stages in the embryonic period, the vessels are only identifiable at stage 5. And it is to be assumed because the mere survival of the embryo depends on achieving a link that guarantees nutrition. Primitive vessels have only two components: a tube made up of endothelial cells, and a network of

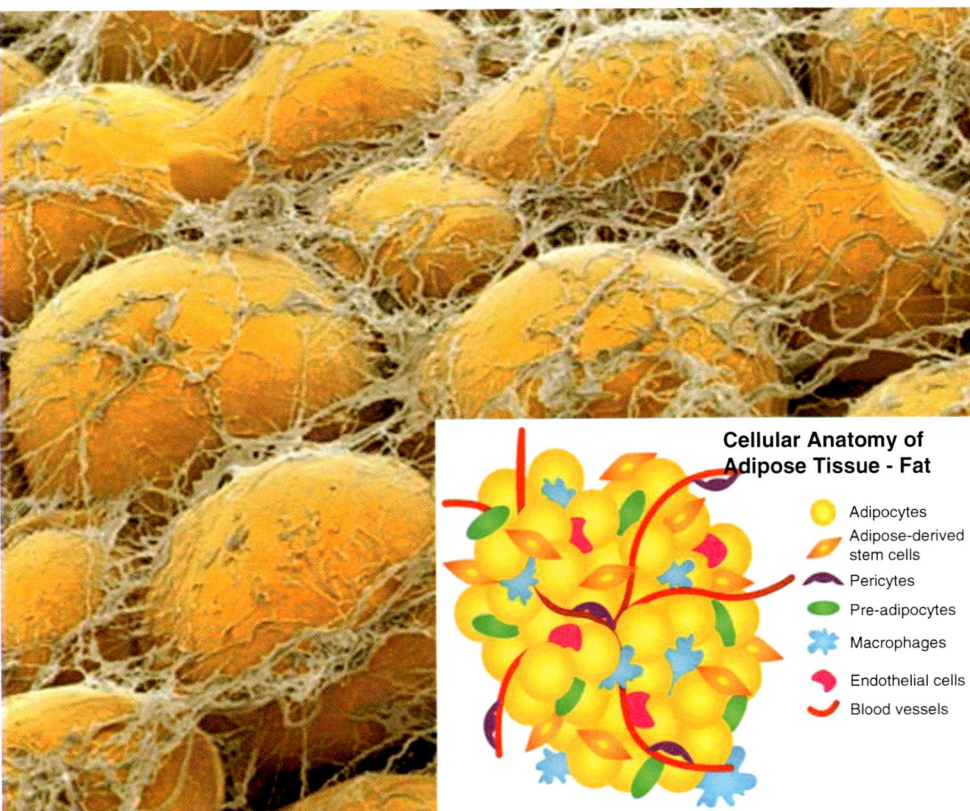

Fig. 11.38 Adipose tissue. Parenchyma consists of the functional cells of the adipose organ (adipocytes) Stroma refers to the supporting cells for the adipose organ: blood vessels, nerves, lymphatics, and structural proteins. Composition of the adipose tissue complex is 85% adipocytes by volume only 12–15% adipocytes by total cell count.
- Adipocytes (yellow): terminally differentiated, endocrine-active
- Pre-adipocytes (green): progenitor cells, near terminal differentiated form, adherent to adipocytes
- Pericytes (purple): physically related to endothelial cells of the microvascular system; important in angiogenesis.
- Endothelial cells (pink).
- Mesenchymal stem cells (orange): derived from pericytes these constitute approximately 45% of nucleated cell counts in SVF.
- Macrophages (blue) in extracellular matrix: intrinsic versus extrinsic (blood).
- Extracellular matrix: fibroblastic elements.
[Reprinted from Alexander RW, Harrell DB. Autologous fat grafting: use of closed syringe microcannula system for enhanced autologous structural grafting. Clinical, Cosmetic and Investigational Dermatology; 2013(6) 91–102. With permission from Dove Medical Press.]

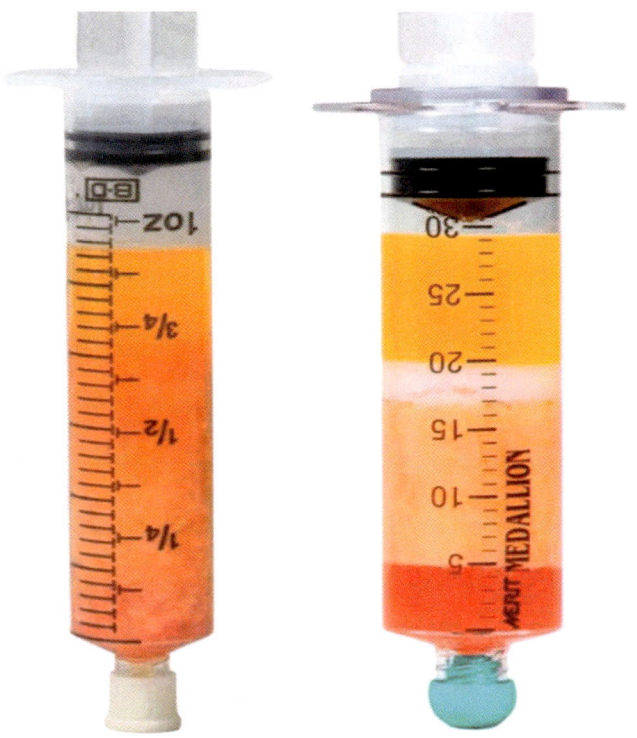

Fig. 11.39 Density Gradient Separation. Left: Standard decantation; Right: Post-centrifugation (1000 g-force for 4 minutes) Note in the latter process the clear separation of the lipid supernatant layer and the aqueous infranatant layer. [Reprinted from Alexander RW, Harrell DB. Autologous fat grafting: use of closed syringe microcannula system for enhanced autologous structural grafting. Clinical, Cosmetic and Investigational Dermatology; 2013(6) 91–102. With permission from Dove Medical Press]

regulatory cells, the pericytes. These form a network outside the actual vessel and perform functions of controlling its porosity and diameter. Therefore, they have a future connection with the sympathetic nervous system. Hence, they have an embryonic relationship with their cousins, the neural crest cells.

Being a component of every vessel in the body, pericytes are found in all human tissues, where they represent the mere origin precursors of mesenchymal cells (MSCs), the so-called "stem cells." Since the pericyte is also the precursor cell for white fat, the concentration of MSCs in fat is the richest in the body.

The adipose tissue complex (CTA) has, as its parenchymal component, adipocytes that constitute 90% of the volume plus 10% of its cellular content. These adipocyte aggregates are ≥400 microns in diameter and are supported by a multicellular stroma.

ATC fat graft (Coleman): atraumatic technique and graft placement near microvessels (Fig. 11.39).

For more than 100 years the use of whole-fat grafts has had a bad reputation. These tissue grafts had high levels of reabsorption, given the vulnerability of their parenchymal

component (adipocytes) to ischemia. In 1987 Syndey Coleman demonstrated better results using low-pressure atraumatic harvesting with smaller cannulae and graft placement in mini-rows in the circumference of the subcutaneous vascular network with greater nutritional support and greater graft retention. Thus, fat transplantation changed towards a less traumatic process with reproducible results.

The concept of separation by centrifugation, also introduced by Coleman, solved another fundamental problem: the variability in the real volume of the grafts. The aesthetic plastic surgery literature, a repository of extensive evidence regarding fat grafting, reports great variability in the rate of "resorption" observed in structural fat grafting procedures. Most of these reports do not account for the fact that these grafts *themselves* contain an appreciable amount of fluid (up to 30–40% of their total volume) due to the injection of serum with epinephrine prior to liposuction. These carrier fluid volumes have been described as a residual volume "load", and when included in the injection constitutes a diluting factor that affects the actual cellular content of the graft.

In addition, Coleman documented the trophic effects that these grafts exert on receptor sites so. These observations revealed the importance of stromal elements in adipose tissue.

- Purpose: volumetric reconstruction.
- Tissue: parenchyma + stroma.
- Adipocyte content 80%.
- Diameter: 1–3 mm.
- Processing: washing and centrifugation @ 1000 rpm for 4 minutes.
- Placement: Mini-grafts placed in multiple parallel rows.

ATC microfat graft: size reduction for better graft survival (Fig. 11.40).

The purpose of microfat is to increase graft retention by harvesting to the smallest possible diameter to allow adipocyte survival.

- Purpose: volumetric reconstruction.
- Tissue: parenchyma + stroma.
- Adipocyte content 80%.
- Diameter: 700 nm to 1.0 mm via harvest using small diameter cannulae.
- Processing and placement: similar to Coleman.

AD-tSVF graft: adipose-derived tissue SVF (Figs. 11.41, 11.42 and 11.43).

Mechanical dissociation: This technique eliminates adipocytes using a traumatic disruption, leaving as a product an emulsion in which there is a matrix of fibers, vessels, cells

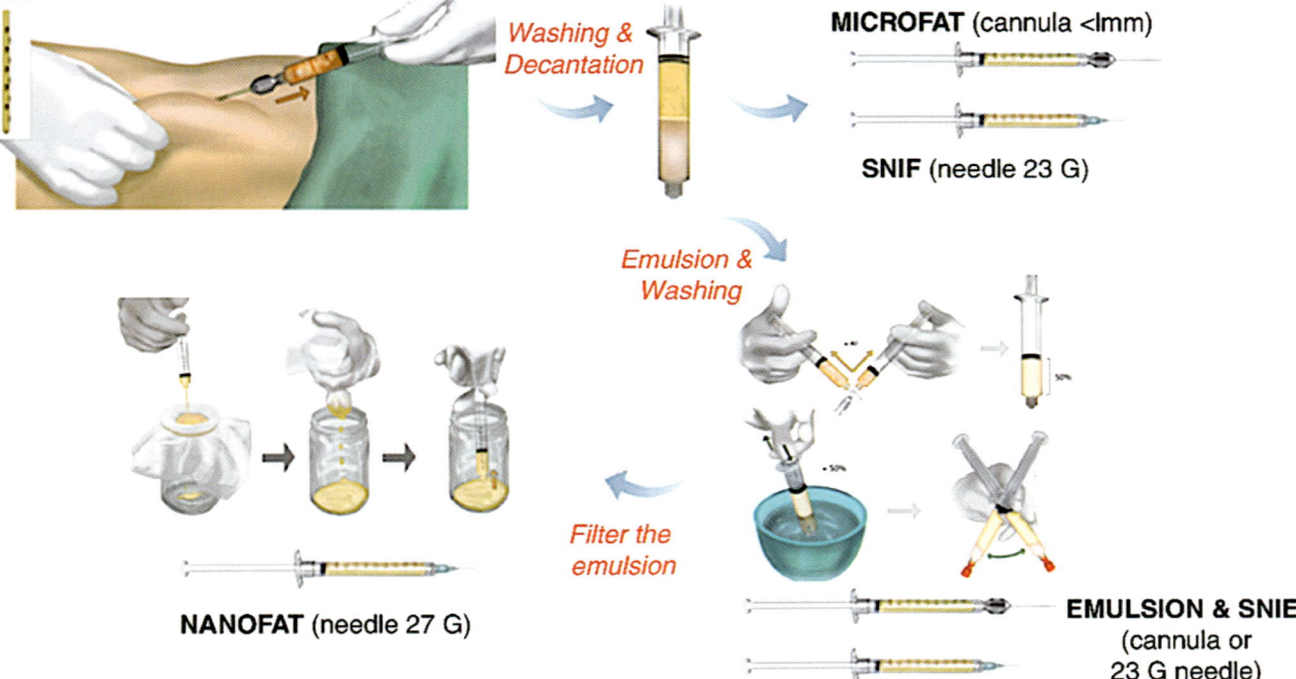

Fig. 11.40 Microfat. Microfat is produced from a mechanical separation method that maintains the viability of the adipocytes using: (1) saline wash, (2) mesh filter, (3) sedimentation by gravity vs centrifuge, (4) reduction to 1 mm Adipocytes are 400 nm so they survive. [Reprinted from Serra-Mestre J, Serra-Ramon JM. Variants of Fat Grafting: from Structural Fat Grafing to Microfat, Sharp-Needle Intradermal Fat (SNIF), Nanofat. In: Pinto H, Fontdevila J. (ed). Regenerative Medicine Procedures for Aesthetic Physicians. Springer, Cham; 2018: 81–85. With permission from Springer Nature]

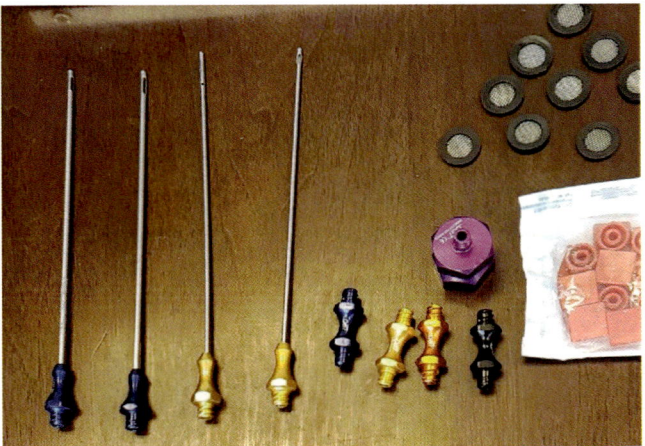

Fig. 11.41 Mechanical reduction system to produce Nanofat (AD-tSVF). 1. Tissue harvest with cannulae ≤3.0 mm 2. tissue is washed for pH control and reduction of inflammatory cells 3. centrifugation of capped (red) syringes @ 200 gm 4. serial passage of dry fat through reducing Leur-Leur connectors (3.0 mm, 2.5 mm, 2.0 mm,1.5 mm) 5. Final pass through a nanofat filteration chamber (purple) containing 400 nm mesh. [Courtesy of Michael Carstens, MD]

connected with those elements, and loose cells. Although the crude number of stem cells released in this emulsion is less than their count in a pure solution, they nevertheless maintain similar biological actions.

- Purpose: trophic effects.
- Tissue: emulsion: predominantly stroma.
- Adipocyte content 15% adipocytes.
- Diameter: 400 nm (some adipocytes survive).
- Processing: mechanical disruption of adipocytes.

The cellular composition of AD-tSVF is:

- Cell proportions do not have good characterization.

AD-cSVF graft: adipose-derived cellular SVF (Figs. 11.44, 11.45, 11.46 and 11.47).

Enzymatic dissociation: This technique removes adipocytes using enzymatic digestion by collagenases, leaving a cellular liquor as a product. AD-cSVR is a pure liquid without fibrotic structures, such as collagen fibers. Therefore, SVF of enzymatic origin is pure enough that it allows its use intravascular, but at the same time, the viable mononuclear cells in that solution lack native structures (such as elastic fibers, collagen networks, tubes) with which they previously maintained a physiological relationship:

- Purpose: trophic effects.
- Tissue: purely cellular concentrate derived from stroma.
- Adipocyte content 0%.

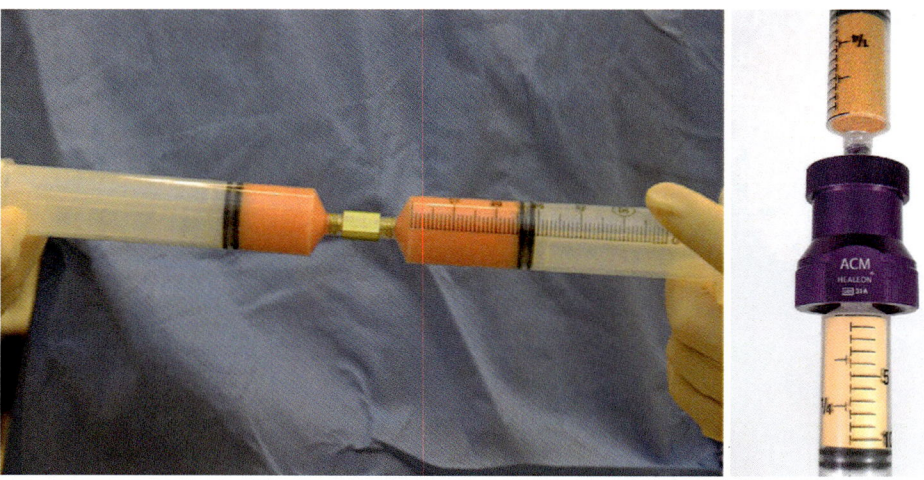

Fig. 11.42 Mechanical dissociation. Consecutive reductions (4) by Leur-Leur connector: from 3.0 mm to 1.5 mm. Each connector reduction requires 30 passes. Single-pass reduction via mesh filter: 0.4 mm. Note color change from the delivery port (left) to the recipient port (right). [Reprinted from Alexander RW, Harrell DB. Autologous fat grafting: use of closed syringe microcannula system for enhanced autologous structural grafting. Clinical, Cosmetic and Investigational Dermatology; 2013(6) 91–102. With permission from Dove Medical Press]

Microfat (emulsion, dense / color orange)

Nanofat (emulsion, creamy / color white)

Microfat 700-1000 nm non-homogeneous

Nanoofat 400 nm es homogeneous

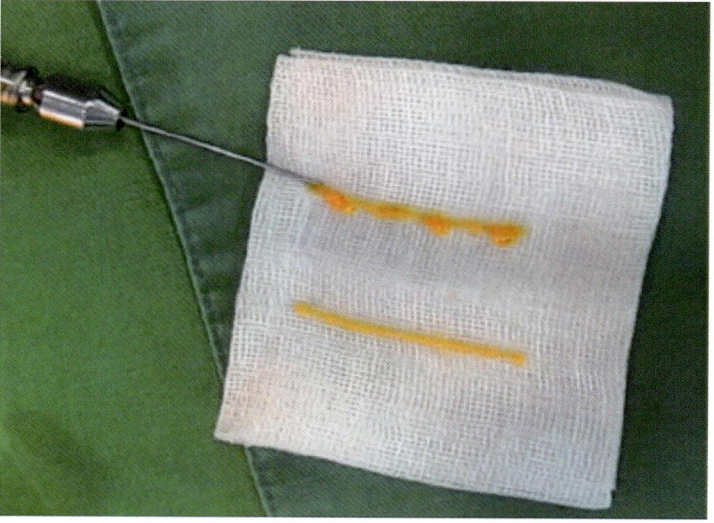

Fig. 11.43 Microfat vs Nanofat: 700 nm vs 400 nm. LEFT: microfat is a dense orange emulsion whereas nano fat is a creamy, white imulsion. RIGHT: When applied to gauze, the lumpy nature of 700 nm diameter microfat is compared with the homogeneous appearance of 400 nm nanofat. Left: [Reprinted from Rihani J. Microfat and nanofat: when and where these treatments work. Fac Plast Surg Clin North America 2019; 27(3): 321–330.with permission from Elsevier.]. Right: [Reprinted from Dong Seok Oh, Dae Hwa Kim, Tai Suk Roh, In Sik Yun, Young Seok Kim Correction of Dark Coloration of the Lower Eyelid Skin with Nanofat Grafting. Arch Aesthetic Plast Surg 2014;20(2):92–96. With permission from Creative Commons License 3.0: https://creativecommons.org/licenses/by-nc/3.0/]

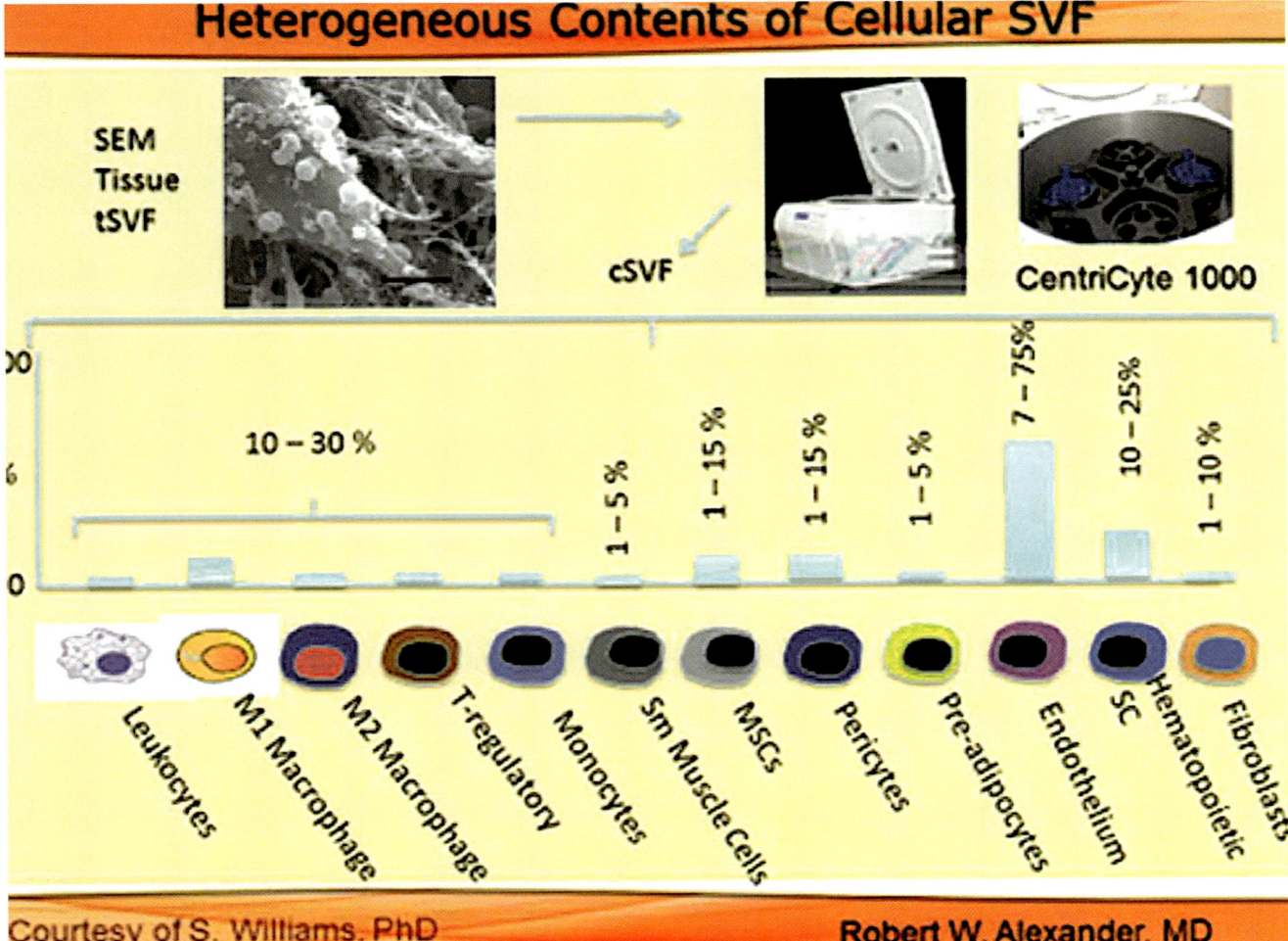

Fig. 11.44 Cell profile of adipose-derived SVF after digestion (adipocytes removed). [Reprinted from Alexander RW, Harrell DB. Autologous fat grafting: use of closed syringe microcannula system for enhanced autologous structural grafting. Clinical, Cosmetic and Investigational Dermatology; 2013(6) 91–102. With permission from Dove Medical Press]

- Diameter: less than ≤400 nm (no adipocytes).
- Processing: enzymatic digestion of adipocytes.

  The cellular characterization of AD-cSVF is:

- 10–30% leukocytes, M1 and M2 macrophages, regulatory T cells, monocytes

- 1–5% myocytes (smooth type)
- 1–15% MSCs signaling mesenchymal cells/stem cells
- 1–15% pericytes
- 1–5% preadipocytes
- 7–70% endothelial cells
- 10–25% hematopoietic stem cells
- 1–10% fibroblasts.

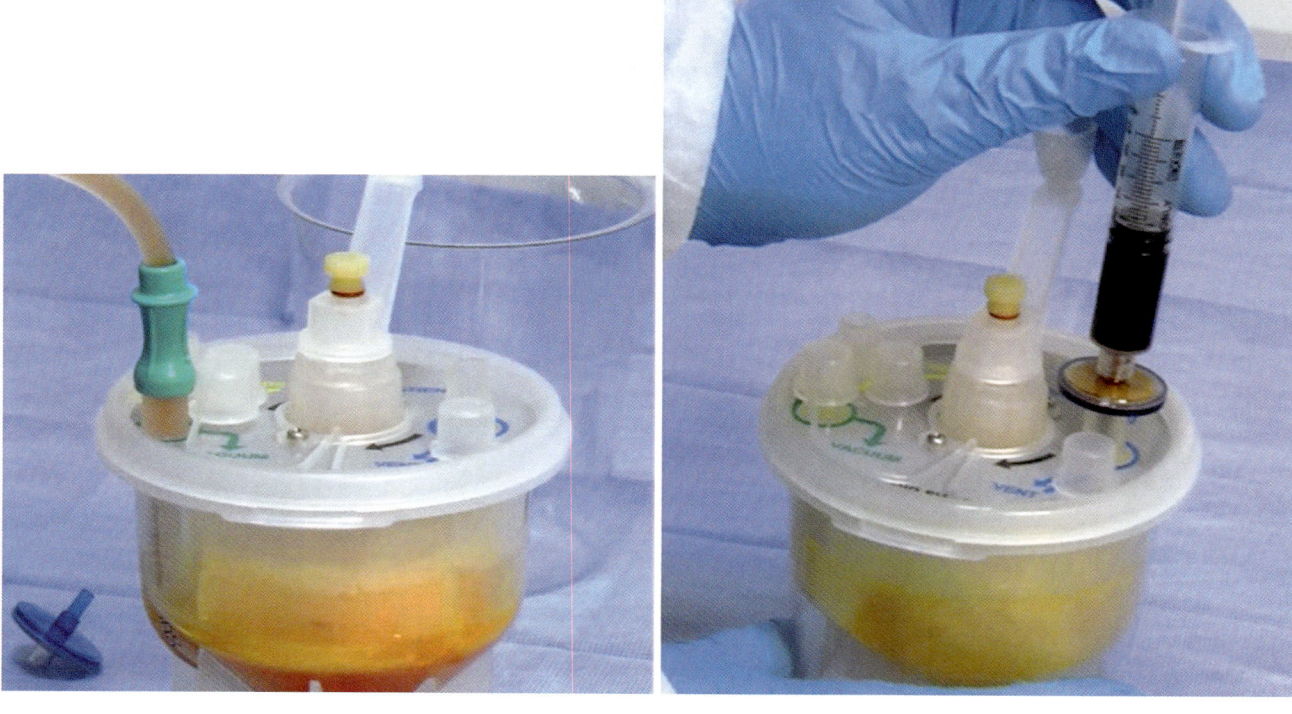

Fig. 11.45 Enzymatic reduction system to produce AD-cSVF. 1. harvest 2. wash to normalize pH and decrease inflammatory (CD 45+) cells from 60% > 10% 3. collagenase (blue) digestion 4. centrifugation 5. extraction of pellet. [Courtesy of Michael Carstens, MD]

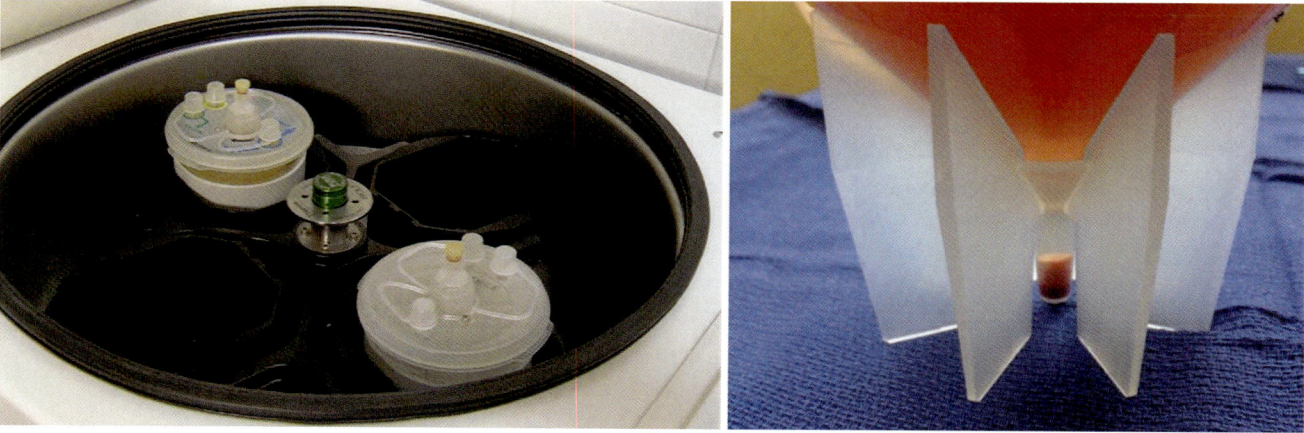

Fig. 11.46 Lipoaspirate post-digestion is a slurry of stromal tissue suspended in a slurry of ruptured adipocytes. Centrifugation at 600 g drives the stroma against the 400 nm mesh with SVF cells forced down- ward into a collection chamber to create a cellular concentrate or pellet. [Courtesy of Michael Carstens, MD]

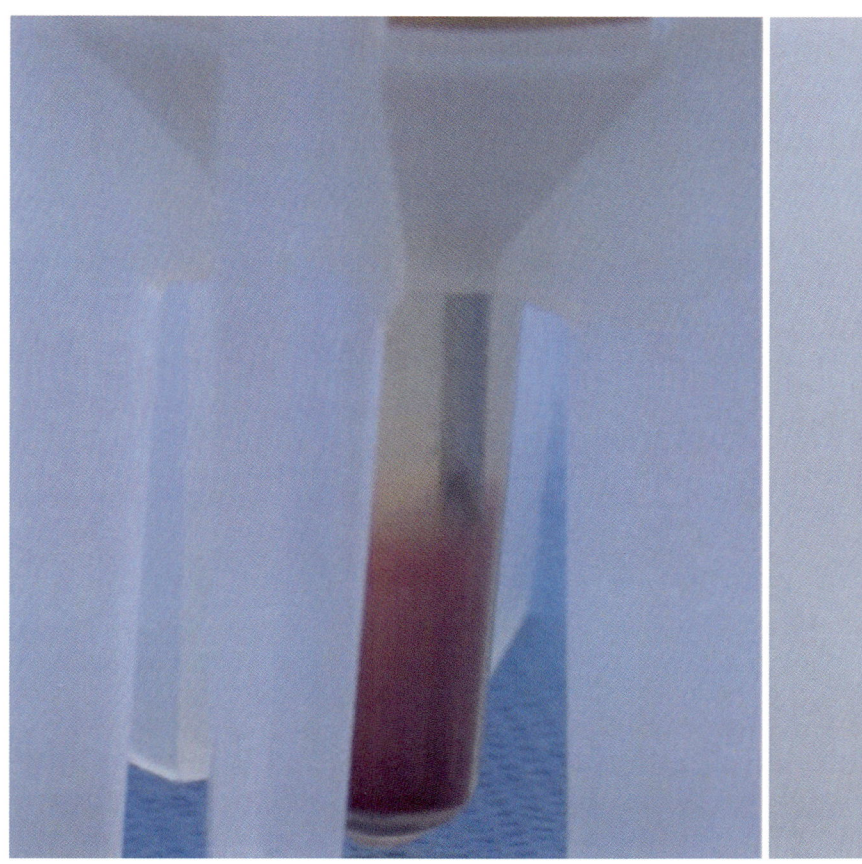
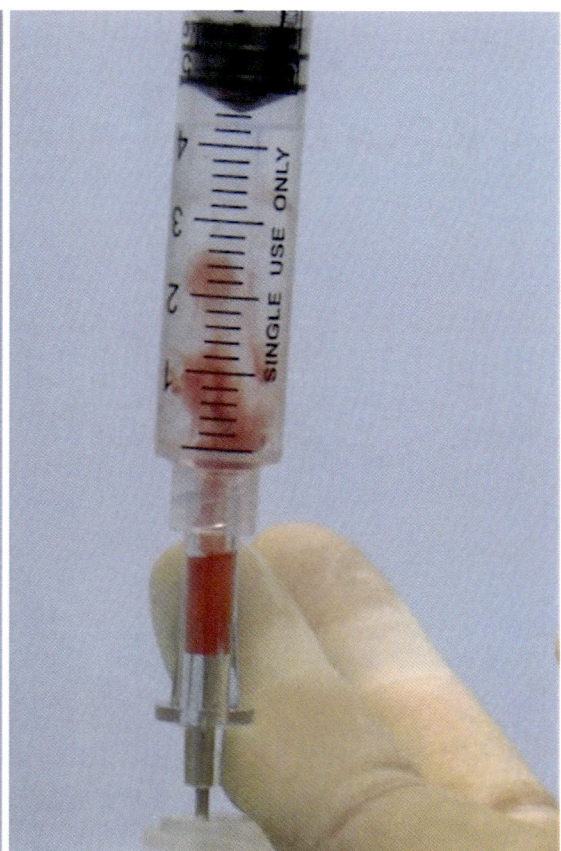

Fig. 11.47 Resuspension of the pellet. Pellet containing SVF stromal drawn upward and suspended in solution ready for administration into tissues (directly) or the intravascular space (via blood filter). [Courtesy of Michael Carstens, MD]

Stromal Vascular Fraction: AD-cSVF as Therapy

Concepts and Components

Adipose tissue, when digested with collagenase, yields an oily supernatant fluid of dissolved adipocytes and a precipitate of stromal tissue, the fibers, and blood vessels that constitute the structural framework for the adipocyte population. Clustered around the vascular component of the stromal are cells representing, adipose-derived mesenchymal signaling cells (stem cells) or ASCs, their embryologic precursors, pericytes, endothelial cell precursors capable of forming blood vessel walls, embryonic pre-adipocytes, macrophages, and other cell types. When the stromal fluid is spun down, the cellular components are separated from the fibers to form the stromal vascular fraction or SVF [12].

As polycellular mixture, SVF has multiple properties which represent the sum of its component parts. The individual characteristics of these cells are well-summarized by Nguyen and Gao. Adipose tissue contains intrinsic cells derived from an original embryologic anlage and extrinsic cells that originate from a secondary (hematogenous) source. Intrinsic cells are categorized on the basis of molecular structures of their surface membranes known as cluster of differentiation (CD) markers; these can be positive or negative. The components of SVF are as follows:

Pericytes (CD34+. CD31-, CD146+)

- 5% of SVF
- Original component of primitive vascular system; ensheath all blood vessels.
- In the embryonic state pericytes are the precursor cells of adipocytes, fibroblasts, and MSC stem cell precursors.
- Interact with endothelial cell precursors to produce new blood vessels.
- In vivo Pericytes under stress can differentiate into mesenchymal stromal cells.
- In vitro (2 weeks of cell culture) pericytes can differentiate into MSCs.

Mesenchymal stromal cells (CD34+, CD31-, CD146+)

- 30% of SVF
- MSCs have multiple broad biologic functions.

- In vivo mesenchymal stromal cells are **not** capable of differentiation.
- In vitro (2 weeks of cell culture) mesenchymal stromal cells differentiate into **true** mesenchymal stem cells (capable of multiple pathways of differentiation).

Endothelial cell precursors (CD34+ weak, CD31+)

- 5–10%
- Original component of primitive vascular system; forms lining of all blood vessels.
- Interact with pericytes to produce new blood vessels.

Fibroblasts (CD34-)

- 45–50%
- Derived from pericytes but lose the CD34+ marker.
- Fibroblasts of SVF have a physiologic role: produce normal collagen.
- Type III (organized) > type 1 (disorganized).

SVF: Mechanisms of Action

Autologous stromal vascular fraction constitutes the first known form of comprehensive management for the body's response to trauma, with actions impacting pathologies in both the acute and chronic states. We can summarize these as follows.

Inflammation SVF cells block or attenuate pro-inflammatory cytokines (acute products of injury) such as interleukins (Il-6, Il-12, and Il-17) and tumor necrosis factor. They also produce anti-inflammatory cytokines, such as Il-4 and Il-10. Finally, they antagonize pathologic growth factors in a yin-yang fashion. For example, the highly inflammatory transforming growth factor beta (TGF-b) is opposed by hepatic growth factor (HGF), and transforming growth factor alpha (TGF-a).

Immune system modulation SVF affects both cell lines of the immune response. The B cell (humoral) arm is modulated via dendritic cells the "present antigens for processing. The T cell (cellular) arm is influenced via (1) the Induction of regulatory T cells (Tregs), and (2) a phenotype shift of helper T cells from the Th1 (inflammatory) state to Th2 (anti-inflammatory) state, and (3) paracrine cytokines indirectly influence immune cells.

Anti-bacterial/viral action This property is poorly understood. SVF cells have been observed to survive and function under hostile conditions, such as cellulitis. They exercise an adjunctive role in microbial control. Direct clearance by MSCs eliminates microbes using Il-37, beta-defensin. MSCs perform indirect clearance by increasing the phagocytic (cell-eating) properties of existing macrophages. Viral control by MSCs involves: (1) surpression of viral replication and shedding, (2) production of anti-viral products: indoleamine 2,3-dioxygenase (IDO), and (3) promotion of Tregs to attack virus.

Anti-apoptosis Prevention of cell death is another property of MSCs. They stabilize cell membranes and decrease levels of attack molecules. They produce bio-active factors such as insulin growth factor (IGF) which gives protection for mitochondria (the energy-producing organelles of cells) and, by improving blood supply (VEGF), provide better metabolic support.

Tissue regeneration MSCs interact with and support regenerative cell populations that are intrinsic to every tissue. Under conditions of injury MSCS (1) reduce or prevent ongoing injury to this cell population, (2) stimulate mitosis of local cell populations, thereby replenishing tissues that are nonfunctional or lost due to injury. The multitude of interactions observed with MSCs in the presence of tissue trauma can be categorized as: immunomodulation and trophic effects [13].

Anti-fibrosis The common endpoint of pathologic wound healing is the production of scar tissue, the mechanical strength of which for any given never exceeds 70% of its baseline. Adipose tissue fibroblasts are a normal component of the stroma.

Non-adipose tissue fibroblasts are recruited (1) from local sources such as muscle or epithelium via epithelial-mesenchymal transformation o (2) from blood: monocyte > macrophages > fibroblasts > myofibroblasts. Under the condition of inflammation: platelets recruit large numbers of monocytes to the site of injury. Accumulation of myofibroblasts leads to abnormal collagen and contraction [14–17].

Matrix metalloproteinases (MMPs) control collagen production/degradation. Collagen has two primary forms: collagen I (disorganized) and collagen III (organized). In the normal state, III > I; physiologic fibers predominate. With inflammation, I > III; this indicates pathologic fibers (scar). MMPs are controlled by two main growth factors. TGF-B1 from platelets (first responders) is inflammatory whereas HGF from SVF is anti-inflammatory. In sum, the anti-fibrotic activity of MSCs takes two forms: (1) control of the number of myofibroblasts by inhibiting their proliferation and promoting their elimination by favoring apoptosis; and (2) prevention and/or reversal of pathologic fibrosis. [Pardo].

Pro-angiogenesis Vascular endothelial growth factor (VEGF) produced by MSCS and endothelial precursor cells (EPCs) has several effects. VEGF induces the sprouting of new blood vessels in which the EPCs form the substrate for new vessel walls. This represents a feedback response of MSCs to states of ischemia.

Wound Healing in the Presence of MSCs: Returning Tissues to the Fetal State

One of the most critical applications for therapy with MSCs, be they from adipose tissue, bone marrow or umbilical cord or placenta is to understand that the transition from the fetal state to the natal state involves a paradigm shift in the way tissues respond to injury. It is a transition from regeneration to repair. Regeneration involves the reconstitution of normal tissue, like replaces like, with all cellular components in place and functional. Repair inevitably involves so form of permanent tissue loss and its replacement with scar: like is not replaced by like but with a material that provides structural support but lacks function.

Why does this transition occur in the first place? As a first pass, let us consider the primary function of MSCs (1) homeostatic and therefore (2) regenerative. They act much as conductors for a complex orchestra of biological responders, directing the response to injury and/or inflammation in such a way as to restore structure and function. To this end all cell populations must be replaced, and the order of tissue microanatomy returned to its original state. An extreme example of this is regeneration of complex structures in lower orders of vertebrates such as the salamander which can replace a lost tail. Let us further postulate that, **in order for regeneration to take place within a given tissue environment, the concentration of healing cells [MSCs]**, that is, **the number of cells per cm^3 of tissue exceeds a critical threshold**.

In sum: the difference between regeneration and repair is a binary choice between healing cells and inflammatory cells:

Inflammatory cells low/healing cells high = before 24 weeks: scar-less healing.

--.

Inflammatory cells high/healing cells low = after 24 weeks: healing with scar.

Adult (inflammatory) state versus the fetal (regenerative) state.

	Adult wound healing	Fetal wound healing
Collagen content	Type 1 collagen blocks migration	Type III collagen stim migration
Hyaluronic acid	Low hyaluronic acid blocks cellular movement	High hyaluronic acid: Favors hydration and cell movement
Metalloproteinases to tissue inhibitors	Low MMP/TIMP favors the accumulation of scar	High MMP/TIMP favors fibrosis turnover
Inflammatory cells	High population	Low population
	Proinflammatory cytokines Il-6 and Il-8	Anti-inflammatory Il-10 decreases Il-6 and Il-8
Transforming growth factor	High levels b-TGF1 and b-TG-F3	Low bTGF1–2, high bTGF3
Gene expression	Delayed upregulation of genes for growth and differentiation	Upregulates genes cell driving growth & differentiation
Progenitor cells	Insufficient # to prevent scar	Sufficient # for scar-less healing

What Causes the Change in the Wound Healing Paradigm?

The answer to this may well be quite simple and is embedded in the basic difference between the embryonic period and the fetal period. As we have seen throughout this book, the 23 stages of embryogenesis involve a choreographic sequence in which tissues are induced, differentiate, and assembled into organ systems. In this scenario, the first system to arise at stage 5 is the vascular system…for upon this, the survival of the embryo depends. The primitive blood vessels of the embryo represent a marriage between two cell sources. (1) Endothelial cells arise at stage 5 from the extraembryonic mesoderm within the blood islands of yolk sac; and (2) Prepericytes arise at stage 6 from ectomesenchyme of the neural crest (or nearby cells). During gastrulation at stages 6 and 7, these two cell lines unite during vasculogenesis, thereby ensuring a source of oxygen and nutrition for the organism. Note: a semantic difference exists regarding the ectomesenchymal source for pericytes of craniofacial vessels versus those of non-craniofacial vessels. The former is classified as a formal derivative of neural crest, the latter are derivatives of cells related to the neural crest but without formal characterization.

At the start of the fetal period blood vessels have been distributed throughout the organism and an explosive period of mesenchymal growth occurs. In this process, the number of non-MSC cells expands exponentially in comparison with the number of MSCs cells. Thus [MSC], that is, #MSCs/cm^3 fetal tissue continues to drop until a critical threshold is reached at about the sixth month of life. After this threshold is passed, the biological potential of the resident MSC population within a tissue is no longer sufficient to maintain the healing potential of the early fetal period.

The Clinical Effect of MSC Transplantation: Recapitulation of the Early Fetal State

When MSCs are harvested and concentrated, either in the form of enzymatically derived SV cells delivered at the point-of-care or in the form of cultured cells produced by in vitro expansion, what is accomplished is the creation of a critical mass of regenerative cells. When transplanted into a recipient site either prior to, or consequent to, an inflammatory event, the effect of this increased [MSC] is to return the target tissue to the fetal state, one in which a sufficient number of MSCs is present such that their cytokines and growth factors are able to change the clinical behavior of the surrounding tissues.

Clinical Applications of SVF: Restoration of Homeostasis (Adult > Fetal)

Given a knowledge of the cellular components of SVF, and of their mechanisms of action, therapy with SVF cells/MSC cells can be applied to a number of clinical conditions.

Acute inflammatory states Cell therapy can counter the acute reaction of tissues to trauma, surgery, and perhaps infection. Not only are tissues under stress addressed but "innocent bystander" injury can be averted. Examples would include safer wound healing after surgery, improved local blood supply, less inflammation, and better scars. By reducing the inflammatory response faster rehabilitation can occur post-orthopedic surgery.

Chronic inflammatory states These take various forms depending on the tissues involved. *Musculoskeletal conditions* include the arthritides, both degenerative (osteoarthritis) and immunologic (rheumatoid arthritis) *Chronic kidney disease* can involve tubulointerstitial inflammation (Mesoamerican nephropathy) and glomerulonephropathies, both metabolic (diabetes) and immunologic (lupus). *Skin conditions* with inflammation managed with topic therapy such as eczema and psoriasis or immune complex attack (hair loss) may respond.

Fibrotic states *Pathologic fibrosing conditions* can include, liver fibrosis, scleroderma, and post-pneumonic scarring such as post-COVID. *Pathologic scars*: cell treatment can permit revision without recurrence for surgical scars at high-risk sites such as knees, sternum, and c-section and for keloids (almost 100% recurrence). *Periarticular soft tissue mobilization* after complex fracture surgery, or disuse (stroke). *Modulation of burn scar tissue* for better elasticity [17–19].

Ischemic states SVF cells can improve damaged microcirculation in diabetic patients for lower extremity ulcers, peripheral neuropathy, and diabetic kidney disease [20]. Angina pectoris in non-surgical candidates could benefit via neoangiogenesis Macrocirculation as in arteriosclerosis by benefiting by external application of SVF around the vessel walls affected by plaque formation. Given the extensive damage exacted by COVID-19 on pulmonary microcirculation [21, 22].

In Summation

- The origins of adipose tissue can be traced to the cell lines emerging at the earliest stages of embryogenesis during the formation of extraembryonic tissues (stage 5) and gastrulation (stages 6–7).

- Of all embryonic systems, the vascular system is the most primitive consisting of a mesodermal component, endothelial cells, and an ectomesenchymal component, the pre-pericyte.
- Pre-pericytes are a cell line synonymous with, or closely related to, neural crest.
- The ectomesenchymal derivation of pericytes explains the broad range of clinical properties they possess, both for differentiation and for the production of many bioactive molecules.
- Pre-pericytes are the precursors of pre-adipocytes and white fat.
- The biologic properties of mesenchymal stromal cells reflect their origins from the pre-pericyte/pericyte line.
- Adipose MSCs and bone marrow MSCS originate from a common source but display the difference in cell markers and relative clinical properties.
- The niche of pericytes and their MSC progeny is vascular.
- MSCs constitute a cell line dedicated to homeostasis.
- The biology of tissue response to injury prior to 6 months of intrauterine life s regenerative and reflects the high number of MSCs/cc^3 of parenchyma relative to the population of inflammatory cell populations.
- With growth the high [MSC] declines.
- The biology of tissue response to injury after 6 months of intrauterine life is reparative reflecting lesser number of MSCs/cc^3 of parenchyma with respect to inflammatory cell populations.

Reconstructive procedures that concentrate large numbers of regenerative cells either as SVF or MSCs in any form and then implant them in a recipient tissue result in a recapitulation by that tissue of its fetal state, thereby changing its healing potential from one of inflammation and fibrosis to one of regeneration.

Fascia and Blood Supply of the Head

All craniofacial fascia develops from neural crest. Dura of the forebrain and midbrain (including r1) develops from a combination of neural crest and paraxial mesoderm; the dura covering the hindbrain and spinal cord is exclusively mesodermal. This section will examine the embryologic rationale of the fascial layers and their functional significance. We shall see that the organization of craniofacial arteries is also functional and follows the same planes.

Let's make a quick summary to get oriented (Table 11.3).

Table 11.3 Fascial planes of the head and neck

Plane	Fascia	Artery	Function
CNS, deep	Pia/arachnoid	Head plexus	Blood-brain barrier
CNS, superficial FNO, palatoquadrate	Dura mucoperiosteum	Stapedial-intracranial stapedial-extracranial	Protect brain support cranial base
ECA-deep plane	Deep investing	Ext carotid-deep	Arches, oral mucosa
ECA-superficial plane	Superficial investing	Ext carotid-superficial	Facial mm, skin, scalp

Blood Supply Runs in Four Planes

The tissues that will comprise CNS deep layer of leptomeninges are in place as soon as the neural tube begins to round up. These consist of neural crest from the *local neural folds* and head mesoderm. Although the pia and arachnoid are not yet visible at stage 9, formation of the head plexus is immediate. Eventually, this will connect to the internal carotid. The spatial organization of the external carotid system is very rational, reflecting its function to supply the derivatives of pharyngeal arches 1–4 (Fig. 11.48).

At stages 12 events take place which are critical for the formation of the fascia covering the brain, the eye, and the structures that support the neurocranium (neural crest bones of the skull). By that time, the third pharyngeal arch has formed, and dramatic changes have taken place in the structure of the first and second arches. Their arterial axes have disintegrated, in each case leaving an arterial stump behind, hanging off from the dorsal aorta. The remnant of AA1 is the *primitive mandibular artery*. The remnant of AA2 is the *hyoid artery*, a critical intermediate structure that morphs into the *stem of the stapedial system*. Stapedial is the very first branch of the internal carotid before it enters the skull. Note formation of the stapedial takes place slightly before the formation of the external carotid and in a plane deep to it.

In the meantime, first and second pharyngeal arches undergo a merger into a single large entity. The bulk of PA2 is internalized within the epithelial envelope of PA1. In so doing, PA2 *loses all of its epithelial representation* save a small contribution to the external auditory meatus. The original arterial supply of the two arches, which came from the first and second aortic arch arteries, falls apart, leaving behind a tangle of comingled blood vessels. These are saved from ischemic death by a timely connecting branch sent forth from the third aortic arch artery, which comes into existence at this exact same time.

The spatial location of AA3 with the third pharyngeal arch is analogous to the arterial supply of the first two arches: runs right up the center. Unlike the first two aortic arches, AA3 does not break down—it remains connected with the dorsal aorta to form the common carotid and the external carotid system.

External carotid comes into existence at stage 12 when AA3 sends out a stem that projects forward into the 1st/second arch amalgam, where it encounters the plexus and makes an anastomosis. This intermediate vessel formed is known as the *ventral pharyngeal artery*. As it produces multiple branches VPA morphs into the external carotid system. ECA will divide into two functional units in two distinct planes.

- ***Deep plane*** branches of ECA supply all the *pharyngeal arch muscles* (except the facial series) and the *oral (non-maxillary) mucosa*. These muscles lie within the deep investing fascia, the arteries being external to the DIF (Figs. 11.49 and 11.50).
- ***Superficial plane*** branches of the ECA will supply the second arch *muscles of facial expression* and *skin and scalp*. These muscles lie are within the superficial investing fascia, or SMAS; the arteries being external to the SIF (Fig. 11.48).

External carotid artery terminates with two branches, one for each plane. Superficial temporal artery runs in the SIF/SMAS. It courses upward to supply the galea and skin of the scalp. Maxillary artery runs in the DIF to supply muscles of mastication but it is quickly joined by the external division of stapedial. The result is a composite structure, part ECA and part stapedial. It is rightly named the *maxillo-mandibular artery* (MMA) and has three parts. In the first part, stapedial contributes four arteries that enter the skull. The second part has a single stapedial branch supplying the mandible. ECA terminates here. The third part of MMA is exclusively stapedial. It supplies the maxilla and part of the oral-nasal cavity.

It will be seen that the stapedial branches of the first part of MMA do not follow a fascial plane, but run along the pterygoid plate and into the skull. In the third part of MMA, V2 stapedial plunges into the pterygopalatine fossa from whence its various branches connect with those of the V1 stapedial system that have exited the skull. Note the stapedial system occupies the ***pericranial—FNO plane*** to supply the dura. For this reason, terminal branches of stapedial that access the face through foramina proceed along two planes. Deep branches course along the surface of the bone and superficial branches connecting with SMAS and skin. And in the frontonasal skin envelope, superficial branches develop. These take over for ECA, running superficial to the plane of the SMAS to supply the muscles and make anastomoses with ECA vessels such as the angular. The nose is a perfect exam-

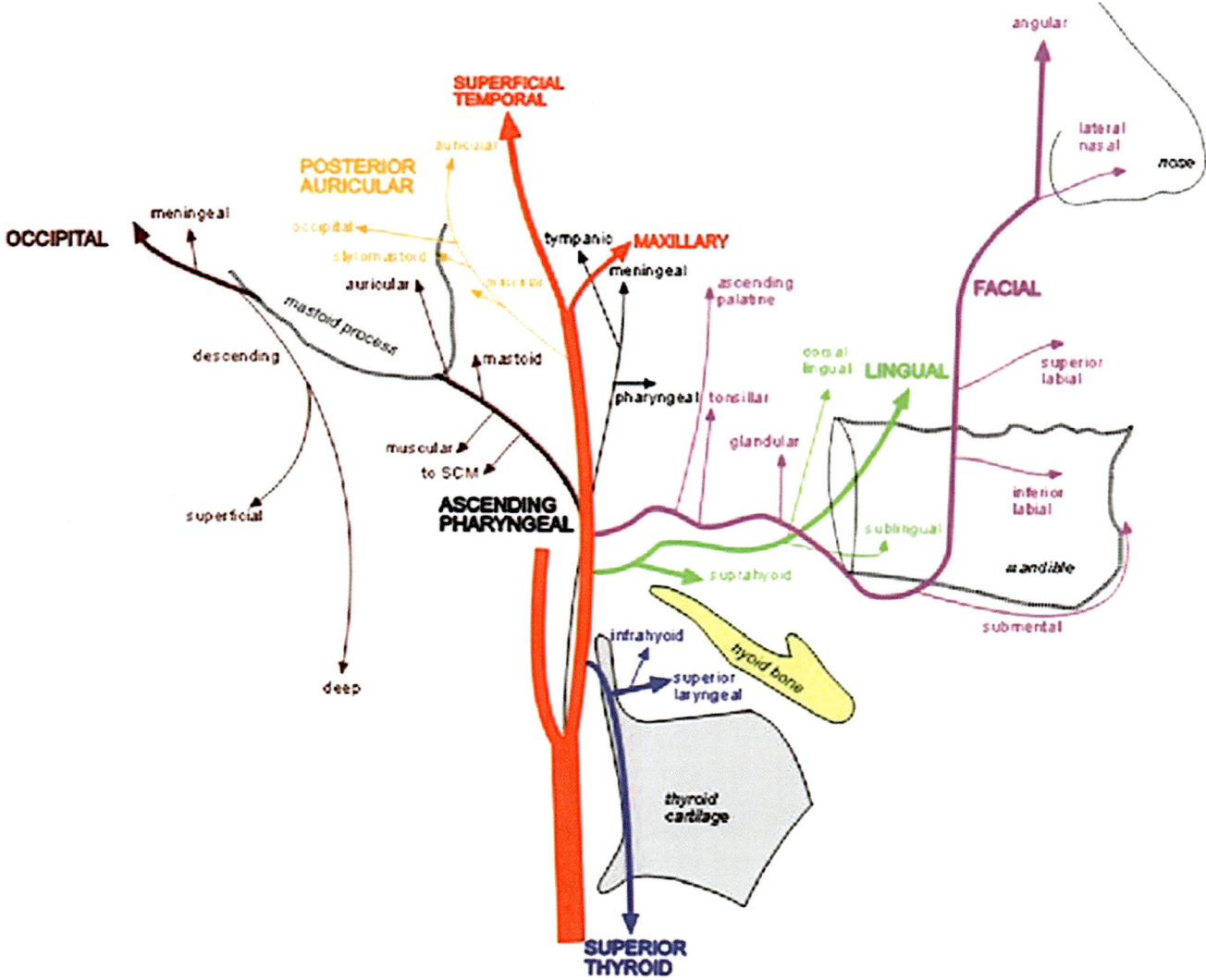

Fig. 11.48 Blood supply to *non*-FNO skin: external carotid system. SThA, superior thyroid; LA, lingual; FA, facial; APA, ascending palatine; OA, occipital; PAA, posterior auricular; STA, superficial temporal; TFA, transverse facial artery; MMA, middle meningeal; AMA, accessory meningeal; IDA, inferior dental; MDTA, middle deep temporal; Max, maxillary; ADTA, anterior deep temporal; DPA deep palatine; PSDA, posterior superior dental; IOA, inferior orbital; SPA, sphenopalatine; 1, artery of superior orbital fissure; 2, artery of foramen rotunda; 3, artery of pterygoid canal; 4, pharyngeal (pterygovaginal). [Reprinted Kiyosue H. External Carotid Artery. In: Kiyosue H. (eds) External Carotid Artery. Singapore: Springer Nature; 2020: 1–5. With permission from Springer Nature]

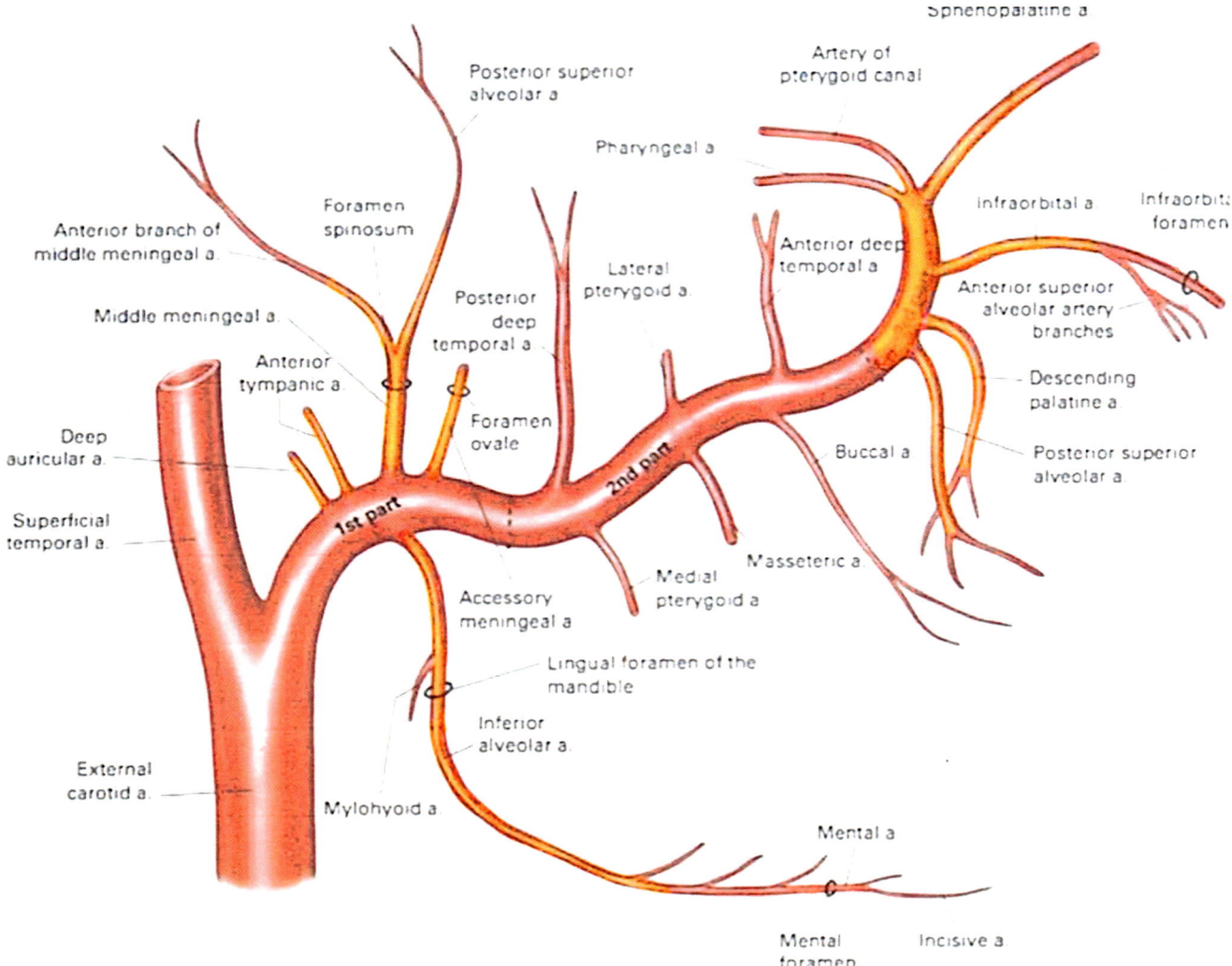

Fig. 11.49 ECA skin: IMMA system. The internal maxillomandibular system has three functional zones. IMMA is direct branch of ECA. Original components (red) <u>predate</u> the stapedial system (orange). Zone 1: stapedial branches to tympanic cavity and meninges (except: mylohyoid artery off the inferior alveolar). Zone 2: exclusively for Sm4 muscles of mastication, including Sm6 buccinator. Zone 3 exclusively for neural crest bones of midface. [Reprinted from Lewis, Warren H (ed). Gray's Anatomy of the Human Body, 20th American Edition. Philadelphia, PA: Lea & Febiger, 1918]

Fig. 11.50 ECA skin: transverse facial artery distribution. TFA is one of two terminal branches of ECA distal to internal maxillomandibular (the other being superficial temporal). It supplies infrazygomatic and lateral orbital SMAS. TFA emerges out of parotid gland and runs above the duct and below the zygomatic arch. It supplies the gland and masseter as well as SMAS. [Reprinted from Lewis, Warren H (ed). Gray's Anatomy of the Human Body, 20th American Edition. Philadelphia, PA: Lea & Febiger, 1918]

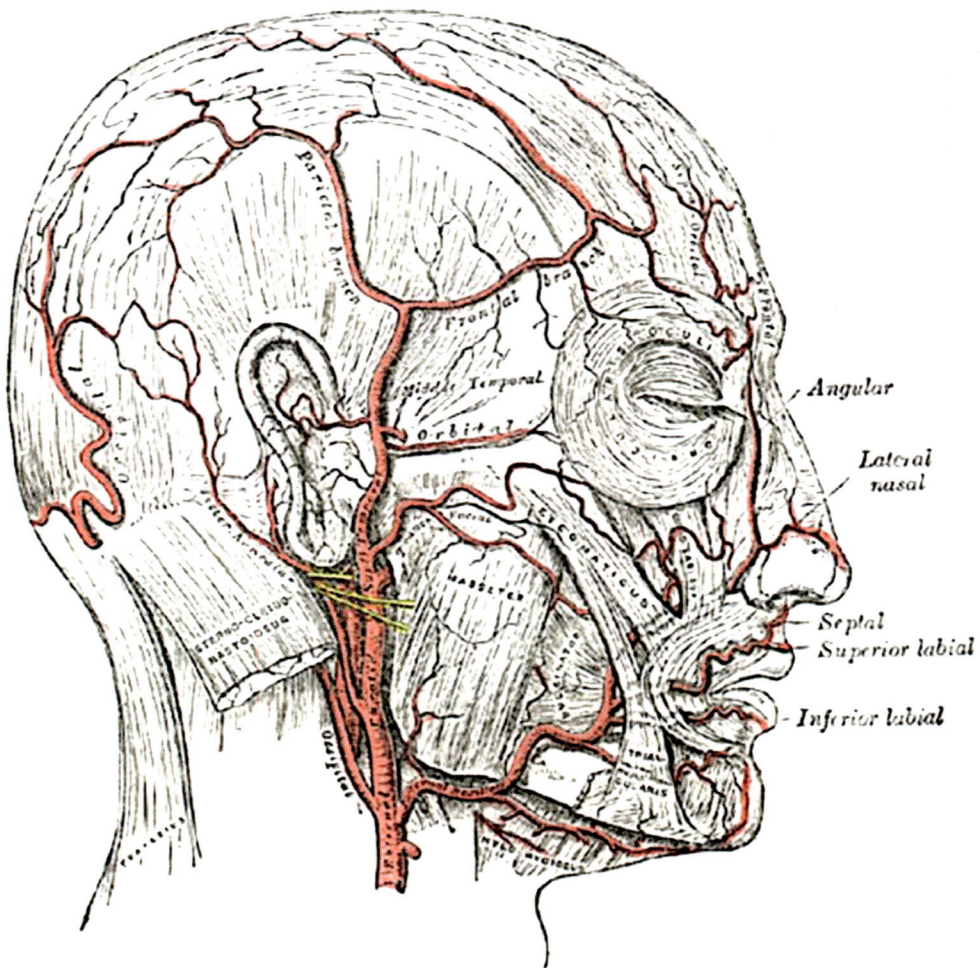

ple of this anatomy. StV1 external nasal runs deep to SMAS to access the columella and also superficial the SMAS of nasalis to supply the skin.

Thus, by the end of stage 12 two completely separate craniofacial vascular systems have been initiated. The *stapedial system* supplies all the V1-innervated fronto-nasal-orbital tissues, the zygomaticomaxillary complex, and all dura. The *external carotid system* supplies all structures of all the pharyngeal arches (skin, mucosa, muscles, fascia, and bones) and the epaxial structures of the scalp innervated by V2-V3. It also supplies all muscles originating from the second arch including auricular and occipitalis—but *none of the muscles of the forehead* (frontalis, procerus, and corrugator) (Fig. 11.51).

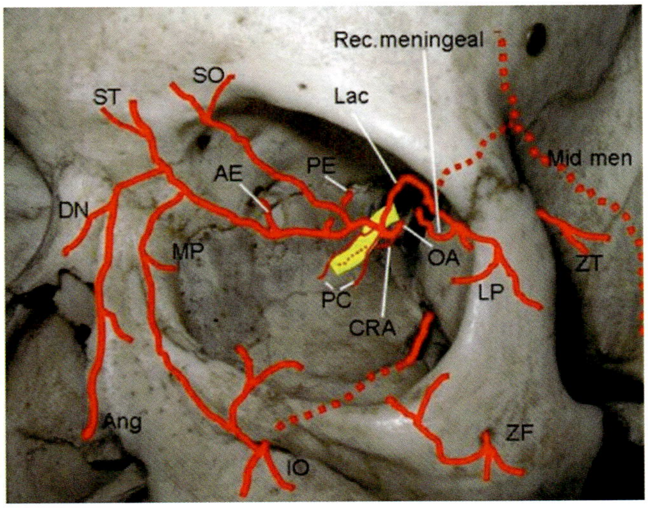

Fig. 11.51 Stapedial supply to skin: ophthalmic distribution. StV1 branches connected secondarily to the ophthalmic axis supply all V1-innervated skin. Note the role of the lacrimal artery axis in zone 9 and the anastomosis with transverse cervical branch from the external carotid. The muscular branches, short posterior ciliary, long posterior ciliary, and anterior ciliary arteries are not depicted. The yellow structure is the optic nerve. Key: OA, ophthalmic artery; CRA, central retinal artery; PC, posterior ciliary artery; Lac, lacrimal artery; LP, lateral palpebral artery; IO, infraorbital artery; ZT, zygomaticotemporal artery; ZF, zygomaticofacial artery; Mid men, middle meningeal artery; Rec meningeal, recurrent meningeal artery; Ang, angular artery; MP, medial palpebral artery; DN, dorsal nasal artery; ST, supratrochlear artery; SO, supraorbital artery; AE, anterior ethmoidal artery; PE, posterior ethmoidal artery. [Reprinted from Rene C. Update on orbital anatomy. *Eye* 2006; 20:1119–1129. with permission from Springer Nature]

Why Do we Need the Stapedial System in the First Place?

There are developmental reasons for why external carotid does not invade the forehead territory of V1 stapedial. Facial muscles migrate according to SMAS neural crest fascia "pathways" that follow a fixed spatiotemporal sequence: deep-to-superficial, lateral-to-medial, and ventral-to-dorsal. For this reason, frontalis will be one of the final muscles to migrate into position. By the time it does so, StV1 supply to FNO tissues is already fully established. The muscles are already supplied. There is no need for ECA to proceed further.

Why should these systems be so different? The distinct nature of V1-supplied FNO tissues makes sense. But have we not always considered the zygomaticomaxillary complex and mandible to be part of the first arch? Why should dura from r1-r3 neural crest be supplied by one system, whereas dermis and fascia (also from hindbrain neural crest) require a completely different source of blood supply? The answers are both (1) spatiotemporal, due to differential migrations of neural crest populations and (2) evolutionary, due to two critical inventions that revolutionized the first two branchial arches—jaws and facial muscles.

(1). Structural considerations

The structural design of the stapedial and external carotid systems reflects the anatomy of the tissues they serve. Stapedial arteries supply structures derived from the *earliest* forward migrations of MNC and RNC$_R$ (r1–3). These are *deep plane* migrations, skimming along the surface of the brain, which are dedicated to the orbit and to the creation of FNO mesenchyme from which the upper third of the face is constructed. Thus, stapedial supplies the dural protection of the brain as well as the bone fields that support the anterior cranial base and coverage of the brain itself. For this reason, in the cranium, stapedial occupies plane deep to that of external carotid. ECA arteries represent a more superficial migration, supporting tissues of the middle and lower thirds of the face as well as the pharynx. Since the brain is already enclosed with dura, the skin and subcutaneous tissues supplied by the ECA will provide the external coverage epithelial coverage of the brain. They are also essential for the synthesis of the membranous calvarium.

External carotid also matures in a temporal sequence. Pharyngeal arch muscles mature in cranial-caudal sequence. In the PA1-PA2 complex, the muscles of mastication are in place prior to those of facial expression. Thus, DIF structures precede those of the SIF. They occupy a deeper plane and are supplied by the deep branches of external carotid artery.

(2). Evolutionary Changes in the First and Second Arches.

The Invention of Jaws

Changes in oral anatomy facilitating increased efficiency of food acquisition have been favored by evolution. Transformation of the first two arches into jaws were accompanied changes in the number of respiratory organs and their arterial anatomy. Primitive *agnathic* (jawless) fishes, as represented by the lamprey, taken in food through a simple stoma. The fishes have **7** branchial arches, each of which is equipped for extracting air from water and each has its own aortic arch artery. The first *gnathic* (jawed) fishes, *chondrichthyes*, had an exclusively cartilaginous skeleton, as in sharks. In these species, the first two branchial arches stopped being gills. Their cartilaginous cores were converted into tooth-bearing jaws capable of capturing prey. The remaining five arches continued to have gill structures. Thus, sharks have **five** branchial arches. In modern bony fishes, *osteichthyces*, the last two branchial arches have morphed into primitive lungs called swim bladders. They have **3** branchial arches. Tetrapods retained jaws but, with the advent of lungs, the branchial arches (gills), no longer needed for oxygenation, were converted to pharyngeal arches. In so doing, the need for aortic arch arteries to supply the gills was also eliminated. Modern tetrapods, now with **three** pharyngeal arches, thus settled upon a revised vascular system.

Jaw-based predation demanded an innovation in the arterial supply for branchial arches 1 and 2. The result was a conversion of second aortic arch breakdown products into a new iteration: the stapedial system. Both upper and lower jaws had articulations with the skull. The relationship of the zygoma to maxilla—such that both are supplied by stapedial—stems from their derivation from a common palatoquate cartilage hinged with the skull. The mandible is attached indirectly to the skull via a second arch hyoid bone, itself suspended from the skull. All remaining structures of the first arch, *retain their supply from the external carotid* system.

The Invention of Facial Mimetic Muscles

All the muscles of the first arch plus the hyoid series of second arch transitioned from gills to mastication without change in their blood supply or their neurology. In tetrapods, the pharyngeal arches underwent radical forms of reassignment as well. Muscles formerly involved in movements of the gills became elevators and depressors of the jaws, including the hyoid. With the exception of posterior digastric these originate from first arch because they act upon first arch.

In mammals, the evolution of facial skin and hair was accompanied by the creation, from the second arch hyoid mass, of an entirely new layer of muscles and fascia accompanied by the reiteration of the facial nerve to produce the SMAS. This layer, consisting of neural crest from r4 to r5, encircles the entire face and skull and extends to the clavicles. The SMAS system functions to control movements of the skin, permit suckling, and raise the hair on one head. Note that the buccinator is part of the oral sphincter; it is not a muscle of facial expression. For this reason, the plane of the SMAS does not subdivide, it remains superficial at all times to the muscles of mastication, and, perforce to the deep plane of the stapedial system.

The anatomy of external carotid reflects these evolutionary changes. It consists of a *deep plane*, supplying all non-SMAS structures and oral mucosa and a *superficial plane* supplying the facial musculature and skin. FNO mesenchyme is the sole exception. Although SMAS is present over the nose and forehead, the muscle masses flow around, and are supplied by, V1 stapedial neurovascular pedicles; ECA is excluded.

External carotid and its branches can be categorized by their neurologic targets.

Superficial Investing Fascia (SIF): Muscles of Facial Expression and the Scalp

In 1976, a superficial musculoaponeurotic system (SMAS) was originally described in the parotid and cheek regions of the face, dividing superficial and deep adipose tissue into superficial and deep layers and constituting a surgically useful structure for enhancing results of face lift [Mitz]. Because SMAS has a regional connotation, this later is referred to as the superficial investing fascia (SIF). SIF is a second arch neural crest structure that provides primary insertion for the Sm6 muscles of facial expression. Over parotid gland, it exists as a separate layer. SIF and its muscles are supplied by the four branches of ECA derived from the original second aortic arch artery: facial, superficial temporal, posterior auricular, and occipital. Work by Carstens and Tolhurst confirmed between SIF with the galeal and subgaleal fascia over the entire skull as an enveloping epicranius layer. [23, 24] should be noted that SIF sends out retaining ligaments to the skin of the face. These divide facial fat compartments associated with specific neuroangiosomes (Figs. 11.52, 11.53 and 11.54).

Note SIF targets muscles derived from the second arch and supplied by the superficial plane of the facial nerve (deep plane of VII goes to buccinator, posterior digastric, and stapedius). The individual branches of this system are patterned upon the motor branches of the facial nerve. Although SIF is dedicated to Sm6 muscles it can be dissected even from muscle-free areas. It is a layer over the entire skull (Table 11.4).

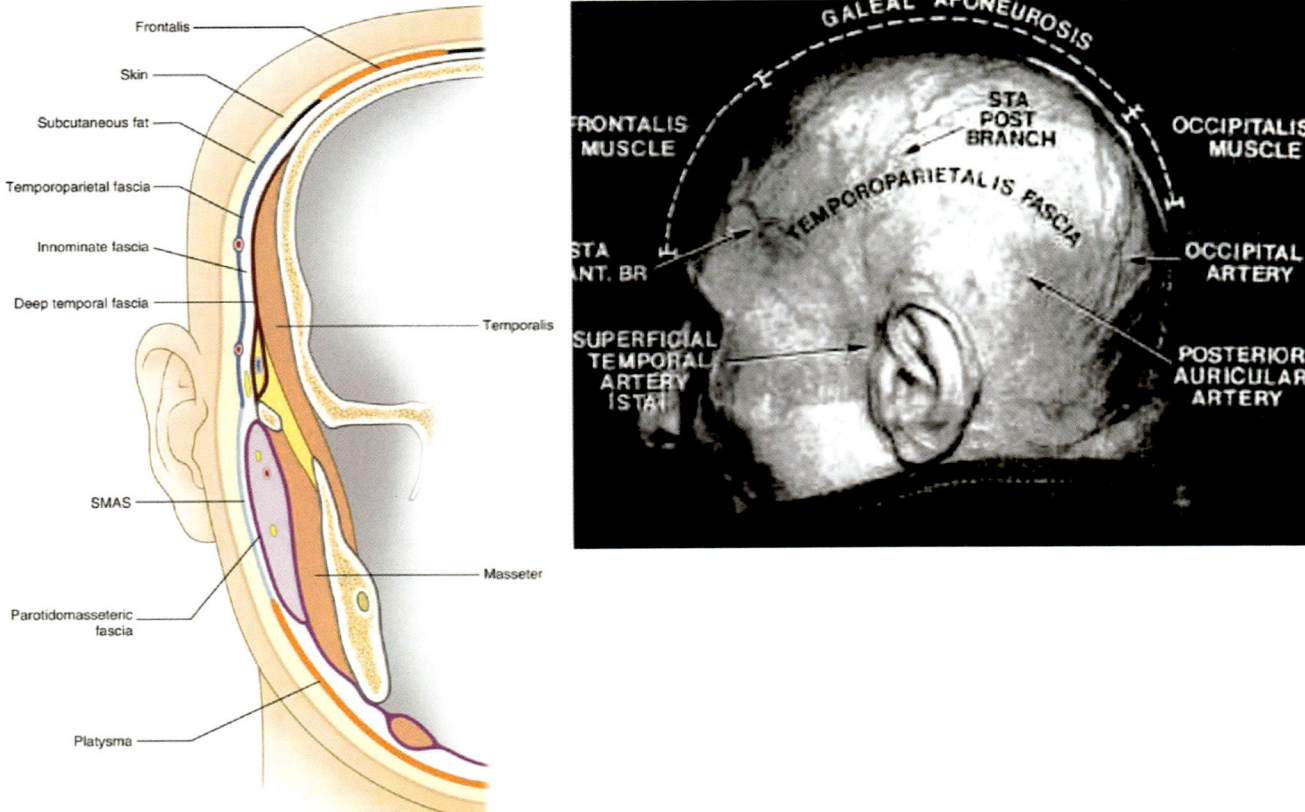

Fig. 11.52 Superficial investing fascia (SIF). LEFT: Superficial investing fascia of the face (SIF) is known as submusculoaponeurotic system (SMAS). Beneath the SMAS is an *areolar sub-SMAS* separating it from the parotidomasseteric fascia below zygomatic arch and from the temporalis muscle fascia above zygomatic arch. SMAS over the cranium is truly a global layer, known in nomina anatomica as the **epicranius;** in common parlance this layer is referred to as the <u>galea aponeurotica</u>. The areolar tissue beneath it is the <u>subgaleal fascia</u> (SGF) or innominate fascia. RIGHT: Superficial musculoaponeurotic system is covered with an external layer of fascia which is continuous with the galea, that is, the epicranius. Left: [Reprinted from Kim B, Oh S, Jung W. Anatomy for Absorbable Thread Lifting. In: The Art and Science of Thread Lifting. Singapore: Springer; 2019: 13–29. With permission from Springer Nature.] Right: [Reprinted from Tolhurst DE, Carstens MC, Greco RJ, Hurwitz DJ. Surgical anatomy of the scalp. Plast Reconstr Surg 1991; 87(4):603–612. With permission from Wolters Kluwer Health, Inc.]

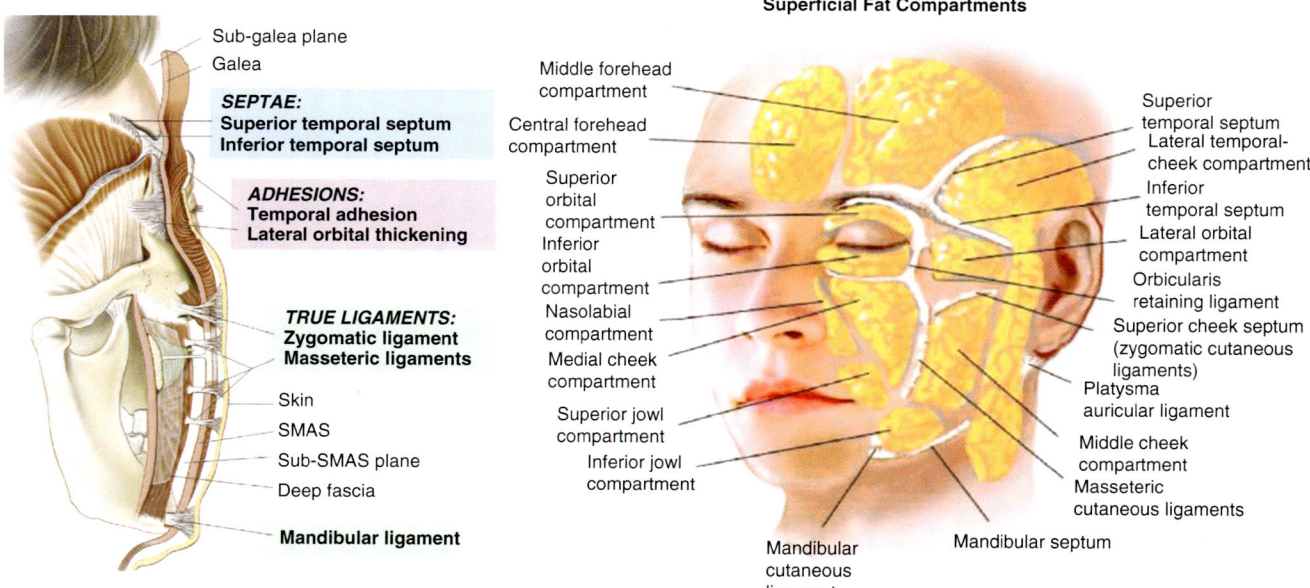

Fig. 11.53 Subgaleal fascia (SGF). Anterior and lateral dissections of the scalp demonstrating the subgaleal fascia and its relationship to other layers of fascia. Arteries were injected with orange Microfil. LEFT: SGF is shows sharply dissected from the underlying temporalis fascia RIGHT: In the view over the supraorbital rim, the artery is shown trifur-cating to supply the overlying muscle, the SGF, and the periosteum. [Reprinted from Tolhurst DE, Carstens MC, Greco RJ, Hurwitz DJ. Surgical anatomy of the scalp. Plast Reconstr Surg 1991; 87(4):603–612. With permission from Wolters Kluwer Health, Inc.]

Fig. 11.54 Retaining ligaments of the face. LEFT: In the face, DIF *ligaments* pass through the SIF to the skin connect first arch fascia/bone to first arch skin; *adhesions* and *septae* connect DIF to SIF. Surgical management of these ligaments is an important concept in facelift. RIGHT: Note that septae divide facial fat into distinct compartments with specific neurovascular definition…another example of relationship between adipose tissue and vascular development. Left: [Reprinted from Moss CJ, Mendelson BC, Taylor GI. Surgical anatomy of the ligamentous attachments in the temple and periorbital regions. Plast Reconstr Surg 2000;105: 1475. With permission from Wolters Kluwer Health, Inc.]. Right: [Reprinted from Alghoul M, Codner MA. Retaining ligaments of the face: review of anatomy and clinical applications. *AesthSurg J* 2013; 33(6): 769–782. With permission from Oxford University Press]

Table 11.4 Arterial supply of the skin

Anterior SMAS	
Superficial temporal	r4 temporal branch: V2-V3 scalp
Transverse facial	r4 zygomatic branch: V2 middle 1/3 face
Facial, superficial	r4 buccal branch: V2 cheek
	r5 marginal mandibular branch: V3 jaw
	r5 cervical branch: C1-C2 upper neck

Posterior SMAS	
Posterior auricular	r4-r5 ear
Occipital	Mastoid, occipitalis, skin over membranous occipital bone

Anterior	
Maxillary, non-stapedial	r3 cranial (PA1): muscles of mastication
Facial, deep	r3 caudal (PA1): Mucosa, masseter, ant digastric
	r4-r5 (PA2): Buccinator, stylohyoid, soft palate
Lingual	r3 (PA1): Mucosa ant 2/3 tongue
	r8-r11 (S1-S4) tongue mm. Occipital somites
Occipital, muscular	r4-r5 (PA2): Post digastric
Ascending pharyngeal	r6-r7(PA3): Mucosa post 1/3 tongue, soft palate & upper pharynx, superior/middle constrictors
Superior thyroid	r8-r11 (PA4-PA5): Mucosa larynx, hypopharynx Laryngeal mm., inferior constrictor

Posterior/epaxial	
Occipital, muscular	r8-r11 (S1–S4): sternocleidomastoid, trapezius skin over chondral occipital bone, C2-C4 neck skin

Deep Investing Fascia: Muscles of Mastication, Oral Mucosa

DIF targets all remaining muscles and mucosa of all pharyngeal arches (Sm4, Sm6-Sm11). DIF is readily apparent over the muscles of mastication but it extends backward and downward to cover the pharyngeal constrictors as well. The relationships between SIF and DIF are particularly evident at the temporal arch. As the facial nerve ascends over the zygomatic arch, it is vulnerable to injury. Subperiosteal plane exposures for trauma and deep-plane rhytidectomy take advantage of entry through the superficial temporal fat pad to encounter the dorsal rim of the arch, and from there, proceed in the subperiosteal plane, with the facial nerve preserved from injury (Figs. 11.55, 11.56 and 11.57) (Table 11.4).

In summation, we have discussed the developmental rationale of the major territories of craniofacial skin can be understood on the basis of their innervation. Fronto-naso-orbital skin and mucosa innervated by V1 are supplied by stapedial arteries associated with the nerve. Facial and scalp skin supplied by all remaining cranial nerves is supplied by external carotid, SMAS.

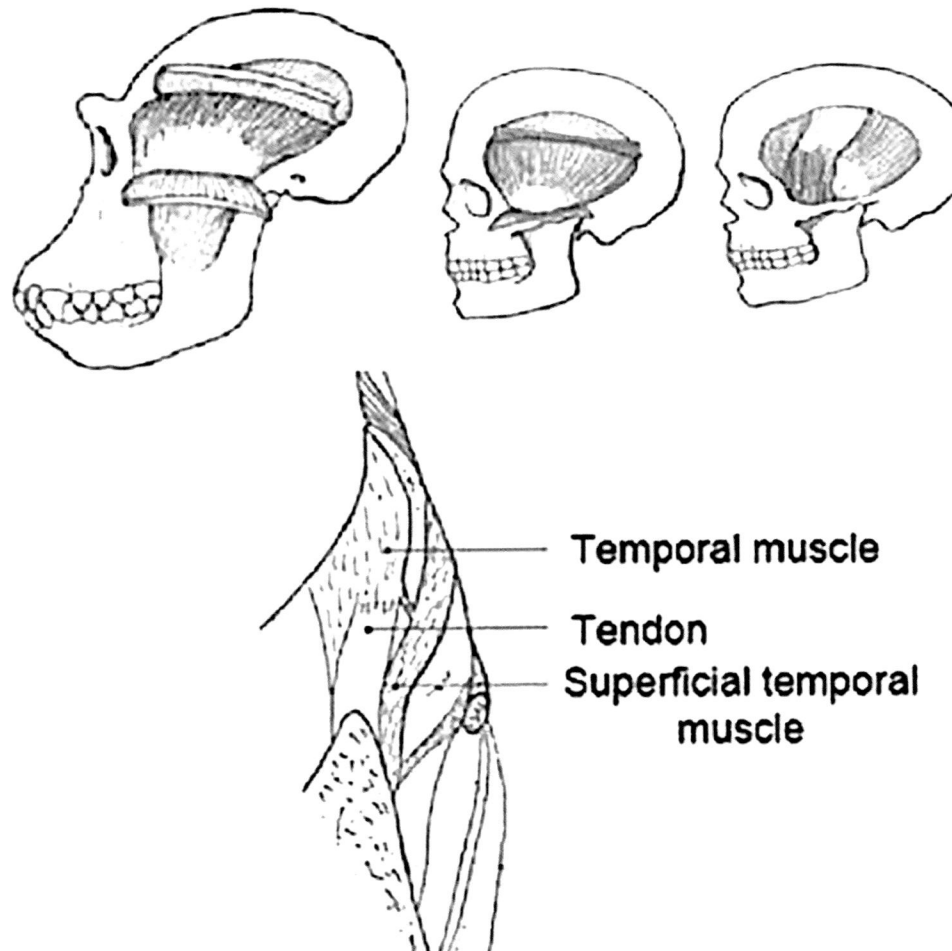

Fig. 11.55 Superficial temporalis muscle in primates. The DIF covers all the muscles of mastication. As such, temporalis muscle is covered by a cranial extension of the DIF of the face and neck. Primates have a two-layer masseter. In humans STM may be the homolog of the intermediated fat pad. Note fascia arrangement with an oblique connection between STM and DTM. [Reprinted from Guerrissi JO, Cotroneo GG. Developmental Anomalies of Temporal Muscle. *Surgery Curr Res* 2014;4: 199. doi: 10.4172/2161-1076.1000199. With permission from Creative Commons Attribution License]

Temporal muscle

Tendon

Superficial temporal muscle

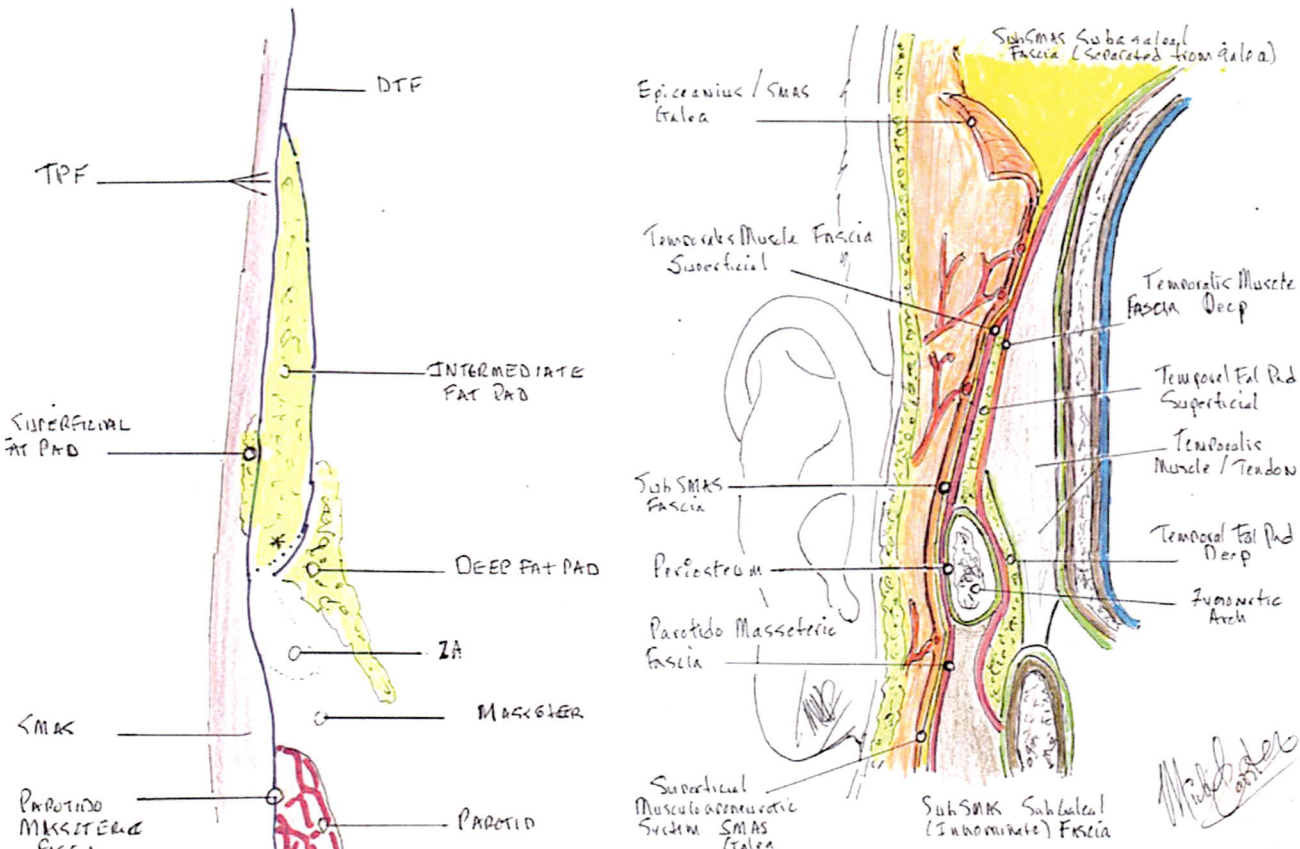

Fig. 11.56 Overview of Temporal fasciae. LEFT: <u>Preferred names of the anatomic structures</u>. The asterisk marks where the deep layer of the deep temporal fascia (D-DTF) abuts the posterosuperior surface of the zygomatic arch (ZA) on both sides (external and internal). The superficial layers and the deep layers of the deep temporal fascia (DTF) fuse at the anterosuperior surface of the Zyggomatic Arch. DFP indicates deep fat pad; IFP, intermediate fat pad; MM, masseter muscle; PG, parotid gland; PMF, parotidomasseterica fascia; S, skin; S-DTF, superficial layer of the DTF; SFP, superficial fat pad; SMAS, superficial musculo-aponeurotic system; TM, temporal muscle; TPF, temporoparietal fascia. RIGHT: <u>Management of the relationship of facial nerve to</u> <u>zygomatic arch.</u> Protection of the facial nerve at the arch is a critical point for surgeons for facial dissection. Continuity of SMAS over the zygomatic arch. SMAS has two layers: (1) an organized fascial layer (the SMAS proper) which becomes the fibrous substance of the galea and abuts directly beneath the skin of the scalp; and (2) an areolar layer (the subgaleial fascia) which, above the temporal line, received penetrating branches from the superficial temporal arterial system. <u>Although these two layers are fused at the level of the zygomatic arch, and developmentally distinctive from the periosteum, they are biologically distinct.</u> [Courtesy Michael Carstens, MD]

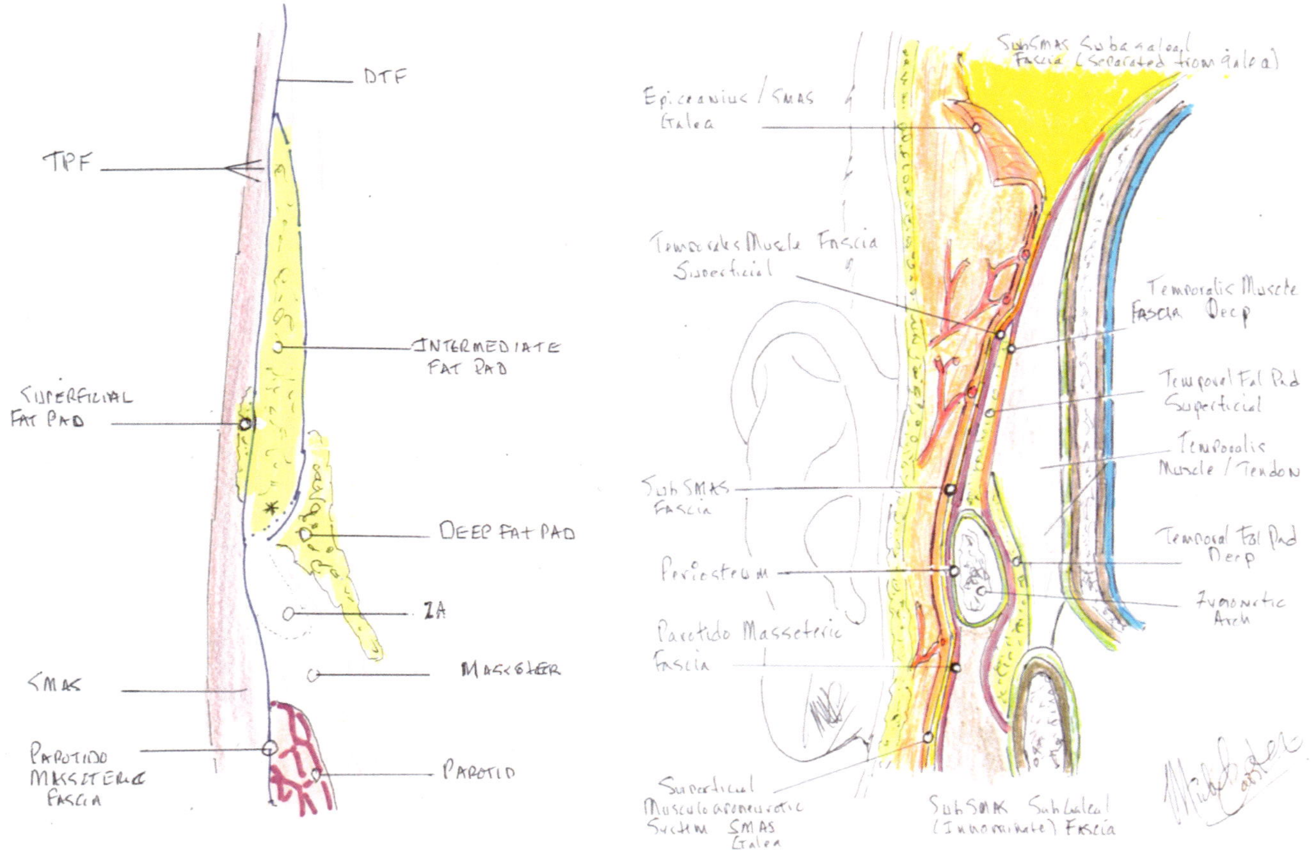

Fig. 11.57 Pathway of facial nerve: from below DIF to above SIF. The frontal branch of the facial nerve is shown exiting the superior border of parotid gland. It continues coursing superiorly before traversing from a sub-SMAS plane to a supra-SMAS plane both above the zygomatic arch and posterior to Pitanguy's line (blue line). Arrows indicate the specific region where the frontal nerve transitions from the sub-SMAS plane to an intra-SMAS plane relative to Pitanguy's line (left arrow) and the zygomatic arch (right arrow). Measurements for these arrows are in the same plane. The left arrow indicates a mean distance of 12.2 ± 4.77 mm posterior to Pitanguy's line. The right arrow indicates a mean distance of 9.6 ± 5.08 mm superior to the zygomatic arch. [Reprinted from Pankratz,J, Baer J, Mayer C, Rana V, Stephens R, Segars L, Surek CC Depth Transitions of the Frontal Branch of the Facial Nerve: Implications in SMAS rhytidectomy J Plast ReconstrAesth Surg Open. https://doi.org/10.1016/j.jpra.2019.11.00. with permission from Elsevier]

Fascia and Blood Supply of the Neck

Blood Supply to Neck Skin

The skin of the neck is supplied by two different kinds of systems, reflecting the origins of its dermis (Figs. 11.58 and 11.59).

Epaxial skin consists of seven dermatomes (c2–c8), each supplied by paired dorsal and lateral arteries directed from the aorta.

Hypaxial skin consists of four dermatomes (C1-C4) of the cervical plexus. It consists of four zones.

POSTERIOR SUPERIOR (C2-C3): descending cervical branch from occipital artery. Occipital is the distal extension of external carotid covering posterior derivatives of occipital somites, specifically sternocleidomastoid and trapezius. Both of these muscles are hypaxial in design but, with evolutionary migration of the scapula from ventral to dorsal, trapezius has assumed an apparently epaxial position. Nonetheless, its motor innervation, from hypaxial cranial nerve XI and its blood supply from hypaxial external carotid, betray its true evolutionary baseline state.

POSTERIOR INFERIOR (C3-C4): costocervical trunk

ANTERIOR SUPERIOR (C1-C2): facial artery to the level of the hyoid bone.

ANTERIOR INFERIOR (C3-C4): thyrocervical trunk.

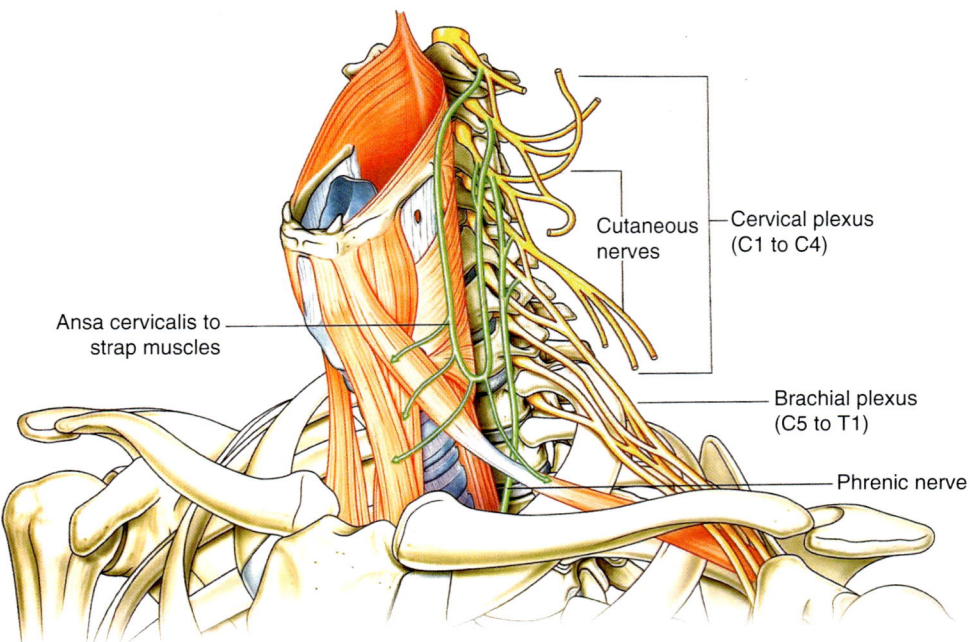

Fig. 11.58 Neural zones of the neck: plan of the cervical plexus. Cervical plexus is composed of nerves originating from the first 4 spinal segments. Its branches supply hypaxial structures originally associated with the forefin. These are in three functional groups.
(1) Motor nerves (green) supply the strap muscles associated with the ancient coracomandibularis.
(2) Sensory branches (yellow) innervate 4 zones of hypaxial skin.
• Anterior superior zone: facial artery, submandibular branch
• Posterior superior zone: occipital artery, stenocleidomastoid branch
• Anterior inferior zone: thyrocervical trunk, inferior thyroid branch
• Posterior inferior zone: thyrocervical trunk, transverse cervical branch.
(3) Mixed motor and sensory: phrenic nerve is motor to the muscular diaphragm and sensory for associated fasciae.
[Reprinted from Drake R, Vogel AW, Mitchell AWM. Gray's Anatomy for Students, third edition. Philadelphia, PA: Churchill-Livingstone. 2015. With permission from Elsevier]

Fig. 11.59 Erb's point: nerve blocks. Named for German neurologist Wilhelm Heinrich Erb, reknowned for his research on muscular dystrophies. Erb's point is located at midpoint of the posterior border of sternocleidomastoids where 4 sensory nerves of the cervical plexus emerge: greater auricular, lesser occipital, transverse cervical and supraclavicular. The point is located superficial to the junction of C5-C6 nerve roots. Spinal accessory nerve is located 1 cm above Erb's point. Local anesthesia here (red circle) can block cutaneous sensation for virtually the entire ipsilateral neck. Left: [Modified from Wikimedia. Retrieved from: https://commons.wikimedia.org/wiki/File:Wilhelm_Heinrich_Erb_(HeidICON_53028)_(cropped).jpg. with permission from Creative Commons License 4.0: https://creativecommons.org/licenses/by-sa/4.0/deed.en.]. Right: [Reprinted from Lewis, Warren H (ed). Gray's Anatomy of the Human Body, 20th American Edition. Philadelphia, PA: Lea & Febiger, 1918]

Table 11.5 Embryologic zones of the skin

Zone	Tissue source of epidermis	Tissue source of dermis	Neuromere of origin	Innervation
Fronto-nasal	NNE rostral p6-p4	PNC caudal p4-p1		Segmental V1
Eyelid, upper conjunctiva	Hypaxial Ectoderm	Hypaxial Neural crest	r1	Segmental V1
Eyelid, lower	Hypaxial Ectoderm	Hypaxial Neural crest	r2	Segmental V2
Face	Hypaxial ectoderm	Hypaxial neural crest	r2-r3	Segmental V2-V3
Scalp, lateral	Epaxial ectoderm	Epaxial Neural crest	r2-r3	Segmental V2-V3
Scalp, postauricular	Hypaxial ectoderm	Hypaxial Mesoderm	c2-c3	Plexus C2-C3
Scalp, posterior	Epaxial ectoderm	Epaxial Mesoderm	c2-c4	Segmental C2
Neck, anterior	Hypaxial ectoderm	Hypaxial Mesoderm	c1–c4	Plexus C1-C4
Neck, posterior	Epaxial ectoderm	Epaxial Mesoderm	c2–c8	Segmental C2-C8

Zones of Craniofacial Skin

Craniofacial skin zones are classified by the source of epidermis and dermis. Bear in mind that all craniofacial skin is supported by superficial investing fascia/SMAS and vascularized by branches of the superficial division of the external carotid artery (Table 11.5).

Classification of Head and Neck Skin and an Evolutionary Aside

Type 1 epidermis—neural crest//dermis—neural crest.

The unique source of frontonasal skin, these tissue derivations explain various pathologic states such as frontonasal dysplasia.

Type 2 epidermis—ectoderm//dermis—neural crest.

Lateral scalp dermis arises from small sector of epaxial V2 in the postorbital zone and large sector of epaxial V3 extensively distributed to the vertex where it has a direct interface with ascending branches of C2 and C3.

Type 3 epidermis—ectoderm//dermis—mesoderm.

For neck skin and posterior scalp there is little to add other than that the *dorsal and ventral zones have differences in innervation*. Epaxial skin is served by *segmental* sensory nerves. Hypaxial skin is served by sensory nerves that intermingle in a cervical *plexus*.

The take-home message of this section is not to add pointless trivia but to answer an important anatomic question.

Why is the neuromeric representation of craniofacial skin discontinuous? Why should we find dermis originating from the rostral hindbrain, r3 abutting directly up against dermis from the second cervical somite? And why is first cervical somite not represented?

Probably the simplest explanation is that both neural crest dermis is only produced at r1-r3. Paraxial mesoderm is incapable of producing dermis until Sm13, C2. By default, post-occipital skin will come from the first somite of the trunk. The piscine head-trunk interface is immobile. The process of resegmentation, in which the cranial half of somite merges with the caudal half of the preceding somite, did not extend forward into the occipital somites. Evolution of tetrapods involved a redefinition of the first three body somites into the primitive proatlas, which, as mentioned before, remained unincorporated into the cranial base. Incorporation of the first cervical somite into the occipital bone in reptiles permitted head flexion but eliminated its capacity to form dermis.

Several zones of mammalian skin demonstrate this evolutionary legacy. In hypaxial anterior neck, the V3/C2 interface takes place along the margin of the mandible. The ear has an anterior hypaxial zone with V3 abutting the C2-C3 greater auricular nerve from cervical plexus. It also has an epaxial posterior zone supplied by the segmental C2 lesser occipital nerve. In posterior scalp, V3 apposes that supplied by segmental C2 greater occipital nerve all the way to the vertex. The spatial positioning of C2 skin over the occipito-parietal skull occurs because of tissue expansion due to growth of the braincase.

Fasciae of the Neck

Superficial Investing Fascia (SIF) (Figs. 11.60, 11.61, 11.62 and 11.63)

SIF, as previously described, is a product of the second arch and is derived from neural crest from rhombomeres r4-r5. It covers platysma which sweeps down over the clavicles to neuromeric level c4. Platysma and the SIF sweep backward over sternocleidomastoid where with the termination of the muscle SIF blends into but is not continuous with the superficial fascia of the neck. Platysma has an anomalous relationship with the remainder of the facial muscles. It is the only one not supplied by the second arch facial arterial system, being instead supplied from the first arch. It is explicable because the blastema of the platysma, being the most caudal and posterior develops prior to the other facial muscles and thus picks up its arterial axis of the inferior alveolar. The inferior supply reflects its extension beyond its "home base."

- SIF in relationship to pharyngeal arches: submental artery from inferior alveolar, a derivative of StV3. This is the instance of a first arch artery supplying a second arch muscle.
- SIF in relationship to neck: penetrating branches of suprascapular artery, a branch of thyrocervical trunk.

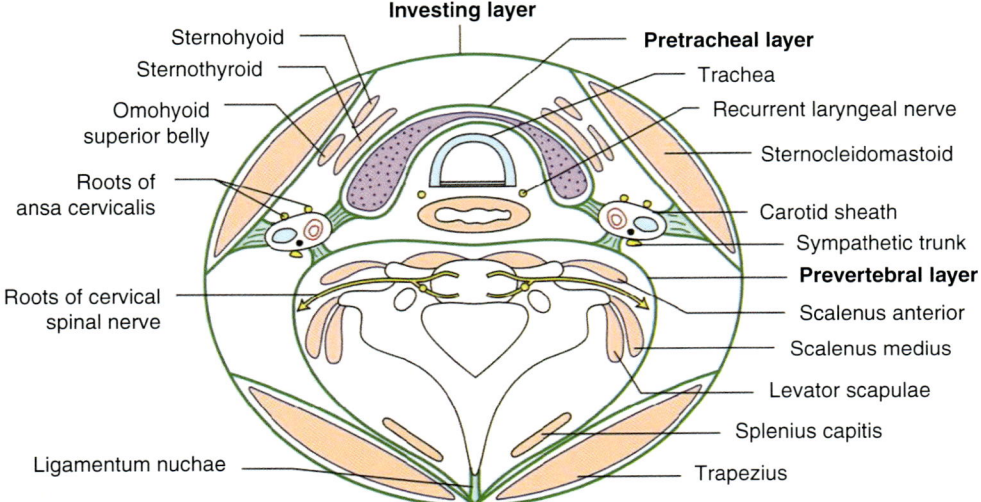

Fig. 11.60 Cervical fascia (axial at vertebra C6) SCF and SIF not labeled. Subcutaneous fascia (SCF):
- Neural crest.
- Panniculus carnosus for skin contraction.
- Arrector pili muscles—connect to hair shafts—relation.
- Platysma sometimes considered a remnant of panniculus carnosus.
Superficial investing fascia (SIF)
- Neural crest.
- Mammals only.
- Originates from r4-r5.
- Branchiomeric second arch muscles, superficial layer (muscles of facial expression).
Deep investing layer (DIF):
- Neural crest in anterior neck and <u>PAM everywhere else,</u>
- All non-mimetic pharyngeal arch muscles.
- Splits to enclose submandibular gland: superficial lamina to mandible border, deep lamina mylohyoid line.
- Splits to enclose parotid gland.

- Splits to enclose sternocleidomastoid/trapezius.
Pretracheal fascia (PTF):
- Neural crest.
- Encloses straps.
- Lateral extension to carotid sheath: connects laterally to DIF and posteriorly to the PVF.
Buccopharyngeal fascia (BPF):
- Neural crest.
- Branchiomeric constrictors of oropharynx.
- Continuous with buccinator fascia (DIF).
- Muscle function: swallowing.
- BFP (neural crest) continuous with esophageal fascia (LPM).
Pre vertebral fascia (PVF):
- PAM.
- Somitic muscles with primary insertions into cervical spine.
- Muscle function: spin-spine, spine, scapular (c4) and first rib.
- PVF continuous with to c4.
[Courtesy of Instant Anatomy (www.instantanatomy.net)]

Fig. 11.61 Cervical Fascia (sagittal).
Subdermal fascia (SDF) not shown:
• Arrector pili muscles in the scalp.
Superficial investing fascia (SIF): not shown.
Deep investing fascia (DIF):
• Superior: to superior nuchal line, mastoid, mandible.
• Inferior: jugular notch, upper clavicle (note split over manubrium and clavicle—indicating embryologic relationship).
• Anterior: symphysis, hyoid, jugular notch.
• Posterior: cervical stenocleidomastoid/trapezius.
Pretracheal fascia (PTF) lies deep in DIF:
• Saggital extent (mandible, hyoid, thyroid, thyroid gland, pericardium.
• Reproduces migration of tongue, thyroid, and thymus.
Buccopharyngeal fascia (BPF):
• Functionally related to oropharyngeal mucosa.
• Covers constrictors r2-r11.
• Changes at esophagus to LPM.
Prevertebral fascia (PVF):
• Extends downward from the skull base far as T3/T4 the upper level of pericardium.
      Danger zone: spread of oral infection into the chest.
• Extends outward as axillary sheath.
[Courtesy of Instant Anatomy (www.instantanatomy.net)]

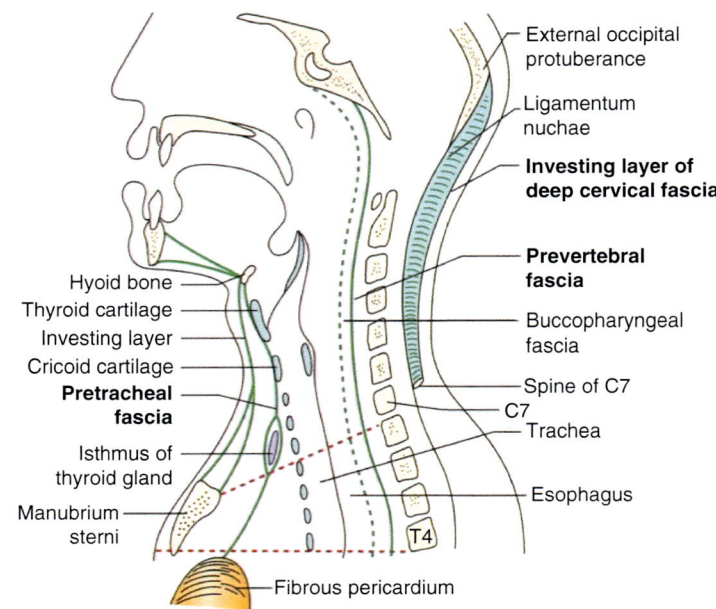

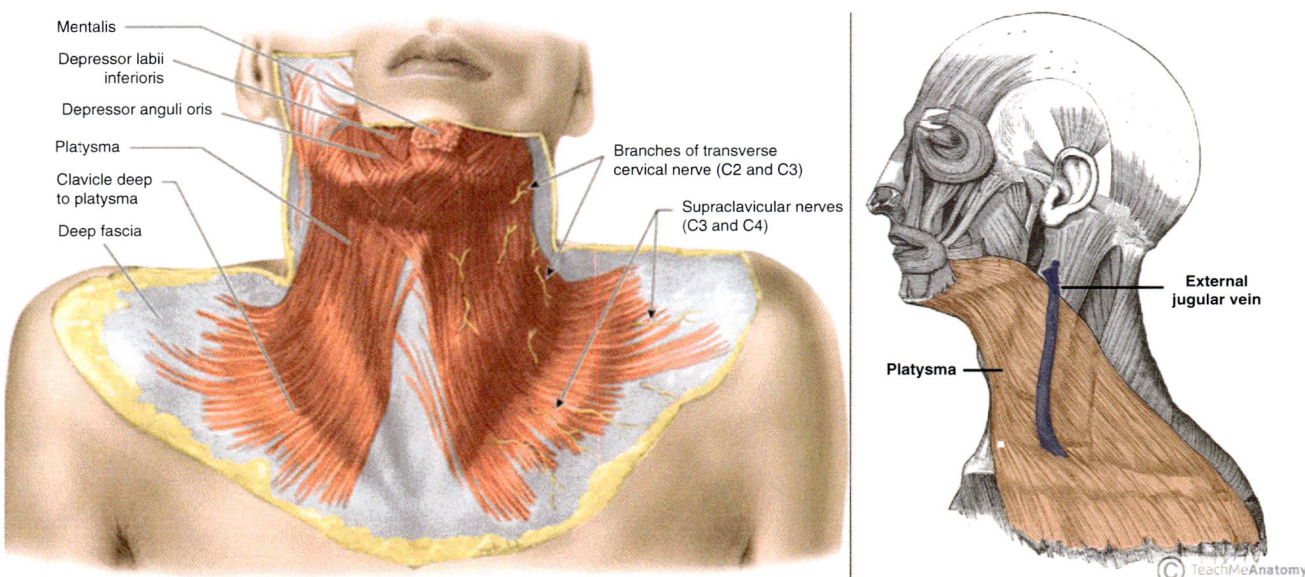

Fig. 11.62 Superficial investing fascia: branchiomeric second arch. Platsyma is a hypaxial branchiomeric muscle from Sm4-Sm5 with motor supply from the cervical branch of cranial nerve VII. It passes downward from angle of mandible beneath dermis supplied by C2-C4. Platysma exists within superficial investing fascia (SIF) or submusculo-aponeurotic fascia (SMAS), a neural crest structure from r4-r5. It advances over branchiomeric territory of deep investing fascia involving straps and sternocleidomastoid because these muscles develop neu-romeres c1-c4, which were *originally branchiomeric*. Sensory nerves C2-C3 (transverse cervical) and C3-C4 (supraclavicular) supplying their respective quadrants of skin pre-date myogenesis so platysma mesenchyme flows around them. Left: [Reprinted from JC Boileau Grant. Atlas of Human Anatomy Philadelphia: Williams & Wilkins; 1943.]. Right: [Reprinted from TeachMeAnatomy, courtesy of Dr. Oliver Jones]

Fig. 11.63 Blood supply to neck skin. Platysma is perforated by branches from facial artery (FA), superior thyroid artery (STA), occipital artery (occ), and transverse cervical artery (TCA). These penetrate platysma to supply the subdermal-dermal plexus of the skin. Musculo-cutaneous perforators (small arrows) reach the skin from sternocleidomastoid, the straps, and trapezius (small arrows). [Reprinted from Rabson JA, Hurwitz DJ, Futrell JW. The cutaneous blood supply of the neck: relevance to incision planning and surgical reconstruction. Br J Plast Surg 1985; 38:208–219. With permission from Elsevier]

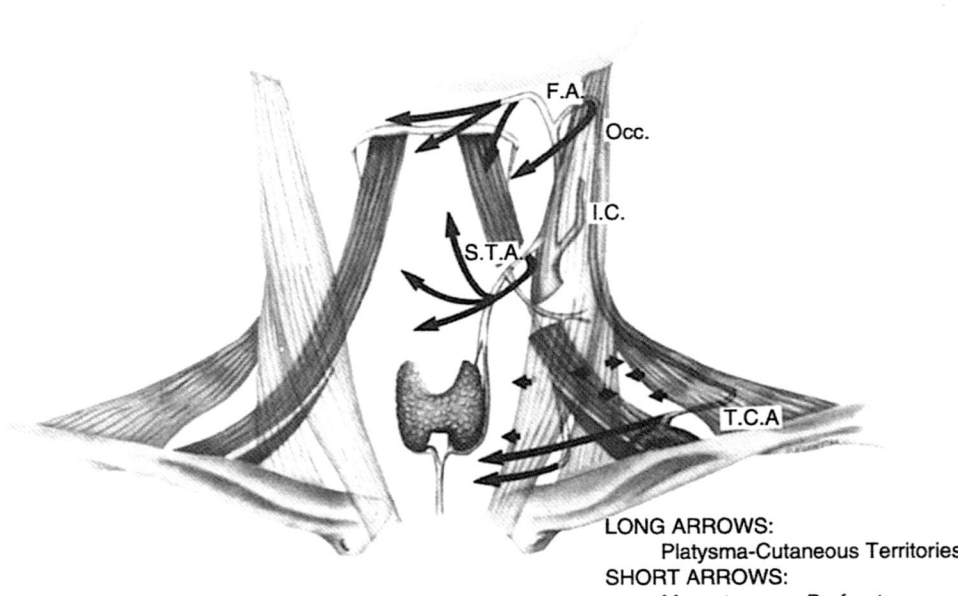

LONG ARROWS:
Platysma-Cutaneous Territories
SHORT ARROWS:
Myocutaneous Perforators

Deep Investing Fascia (DIF) (Figs. 11.64, 11.65 and 11.66)

DIF follows the same rules. This fascia remains dedicated to pharyngeal arch muscle derivatives of the ancient branchial arch system, sternocleidomastoid, and trapezius. DIF splits to enclose them and therefore encircles the neck. Recall that both these muscles are branchiomeric and descend from the ancient cucullaris muscle complex. As such they have neuro-meric elements extending from the r10 to r11 segment of the nucleus ambiguus to c4 for SCM and c8 for trapezius. Their blood supply and that of their overlying fasciae are likewise regional.

Sternocleidomastoid connects the r6-r7 mastoid process with the c1 portion of the clavicle. Its mesenchyme origi-nates from seven neuromeric levels: somitomeres 6–7, occipital somites 1–4 and the first cervical somite. Its blood supply reflects this neuromeric trifecta: its upper third is sup-plied by occipital branch of ECA representing third arch (r6-r7), the middle third receives superior thyroid from fourth arch (r7-r8), and the lower third is irrigated from the supra-scapular branch of the thyrocervical trunk (the second branch of the subclavian. The arteries reflect three distinct blocks of mesenchyme that produce this muscle.

As previously discussed, clavicular mesenchyme origi-nates from the first four spinal somites, sp1-sp4. its physical

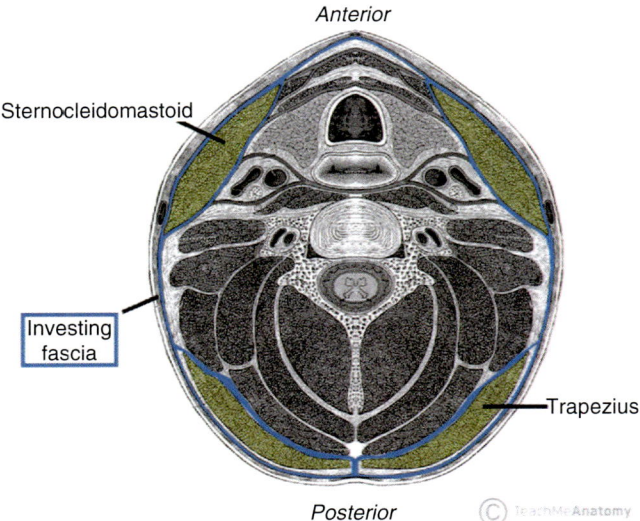

Fig. 11.64 Deep investing fascia (DIF): branchiomeric muscles. Sternocleidomastoid and trapezius originate from the ancient cucullaris, a branchiomeric derivative of arches B4-BA7 in register with neuromeres r8-r11 and Sp1-Sp4 with mesoderm derived from the first 8 somites (4 occipital and 4 spinal). [Reprinted from TeachMeAnatomy, courtesy of Dr. Oliver Jones]

disconnection from the skull reflects the changes wrought by the evolution of the neck itself. These initial spinal somites undergo a name change from sp1-sp4 to c1-c4 as clavicle migrates downward relationship to the skull migrating c1-c4/sp1-sp4) to the base of the skull, specifically the mastoid process that relates neuromerically to r6-r7. Thus, sternocleidomastoid spans from r6 to c1, that is from Sm6-.

Trapezius has an extensive territory extending from the posterior skull base at level c1/sp1 down to the scapula as low as its tip, level c7/sp7. The muscle is irrigated from the transverse cervical branch of the thyrocervical trunk. It can be elevated upward as a flap based on this arterial axis. TCA has an <u>ascending branch</u> supplying the branchiomeric (cucullaris) mesenchyme from levels sp1-sp4 (BA6-BA7) and a <u>descending branch</u> to mesenchyme from sp5-sp8 (BA8-BA9). Recall that the prototypical branchial arch system as in placoderms contains 9 arches.

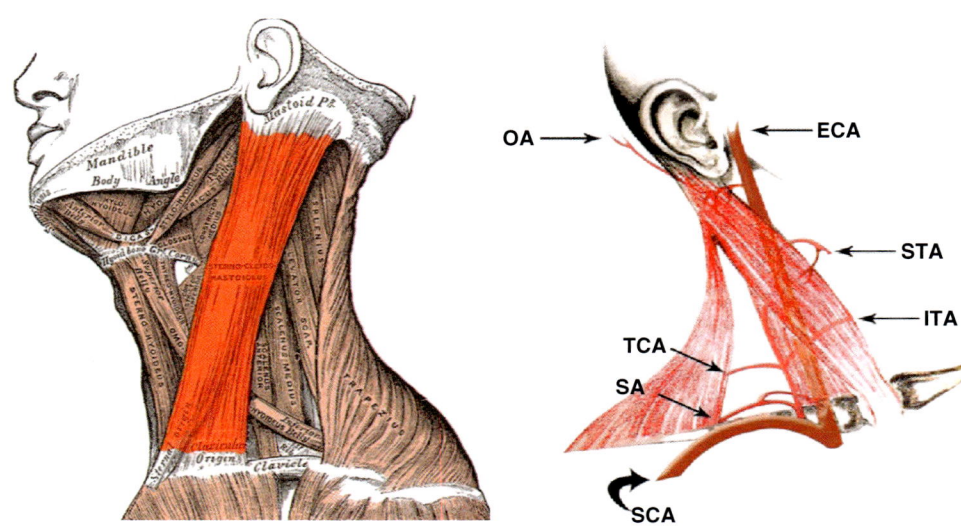

Fig. 11.65 DIF: Sternocleidomastoid. Branchiomeric muscle from multiple levels with three levels of blood supply. Left: [Reprinted from Lewis, Warren H (ed). Gray's Anatomy of the Human Body, 20th American Edition. Philadelphia, PA: Lea & Febiger, 1918.]. Right: [Reprinted from Khazaeni K, Rajati M, Shahabi A, Mashadi L. Use of sternocleidomastoid myocutaneous flap based on the sternocleidomastoid branch of the inferior thyroid artery to reconstruct extensive cheek defects. Aesth Plast Surg 2013; 37(6): 1167–1170. With permission from Springer Nature]

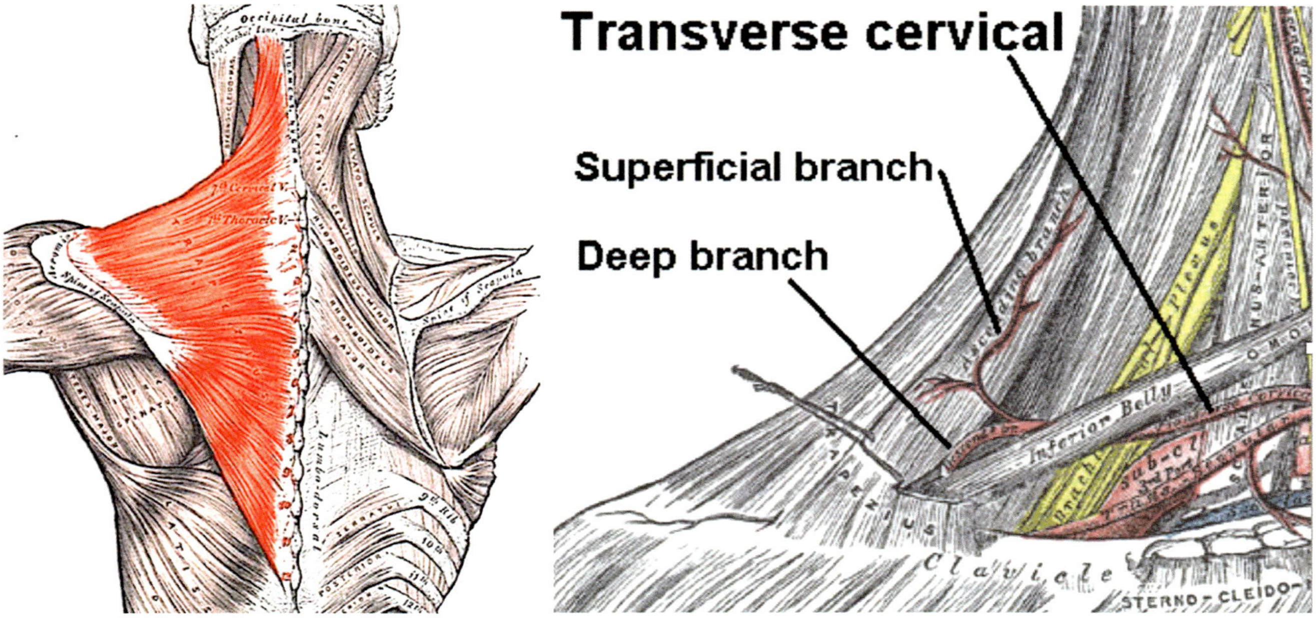

Fig. 11.66 DIF: Trapezius. <u>Transverse cervical artery</u> (TCA) is one of the four banches of thyrocervical trunk. It travels laterally in the posterior triangle, in front of phrenic nerve, the scalene muscles, and brachial plexus. It slide beneath omohyoid (inferior belly) and reaches anterior border of levator scapulae. There it divides into superficial and deep branches. <u>Superficial branch</u> (1) ascends to supply *upper 1/3 of trapezius* where it anastomoses with occipital; and (2) descends to supply *middle 1/3 and lower 1/3 of trapezius.* <u>Deep branch</u> supplies levator scapulae and the rhomboids; it contributes to trapezius. Phylogenetically, superficial branch is dedicated to a branchiomeric muscle (cucullaris) at levels sp1-sp4 and deep branch is dedicated to non-branchiomeric mus-cles connected to levels sp5-sp8 and the rhomboids; it contributes to trapezius. Phylogenetically, superficial branch is dedicated to a bran-chiomeric muscle (cucullaris) at levels sp1-sp4 and deep branch is dedi-cated to non-branchiomeric muscles connected to levels sp5-sp8. Left: [Reprinted from Lewis, Warren H (ed). Gray's Anatomy of the Human Body, 20th American Edition. Philadelphia, PA: Lea & Febiger, 1918.]. Right: [Modified from Lewis, Warren H (ed). Gray's Anatomy of the Human Body, 20th American Edition. Philadelphia, PA: Lea & Febiger, 1918]

Prevertebral Fascia (Figs. 11.67, 11.68, 11.69 and 11.70)

Prevertebral fascia (PVF) encircles the spine and contains muscles on both the epaxial and hypaxial aspects of the vertebral column.

- PVF horizontal: fascia of posterior triangle, non-branchiomeric muscles (scalenes, levator scapulae, splenius capitis). PVF lies ventral to roots of brachial plexus. As the trunks of the brachial plexus move into the upper extremity, they carry with them an extension of PVF, the axillary sheath. Note that the subclavian artery and vein are outside the sheath.
- PVF vertical: skull base in front of longus colli down to anterior ligament of t1. Neuromeric relationship with cervical neuromeres.

The epaxial (extensor) group is segmentally supplied from the vertebral system and from the <u>ascending cervical branch</u> of the costocervical trunk, the third dorsal derivative of subclavian. Recall that longitudinal nature of the ascending cervical artery is (like vertebral) really the sum of original transverse segmental branches from the dorsal aortae. There are two functional groups of hypaxial muscles. A medial vertebral group includes those muscles, such as longus capitus that are flexors; these relate segmentally to the anterior vertebrae. A lateral vertebral group consists of the three scaleni which connect the cervical spine to the first rib and levator scapulae which attach to the c4 medial superior angle. The blood supply to these two groups of muscles comes from a single axis, the thyrocervical trunk along with its overlying fascia is spatial and neuromeric. Inferior thyroid while going up to the r11-c1 junction below the cricoid ascending cervical gives off a separate ascending cervical branch. Together these two vessels supply the midline muscles.

Fig. 11.67 Prevertebral fascia (PVF): axial. PVF covers axial somitic muscles (S5-S13) from the skull base to the pericardium. [Reprinted from TeachMeAnatomy, courtesy of Dr. Oliver Jones]

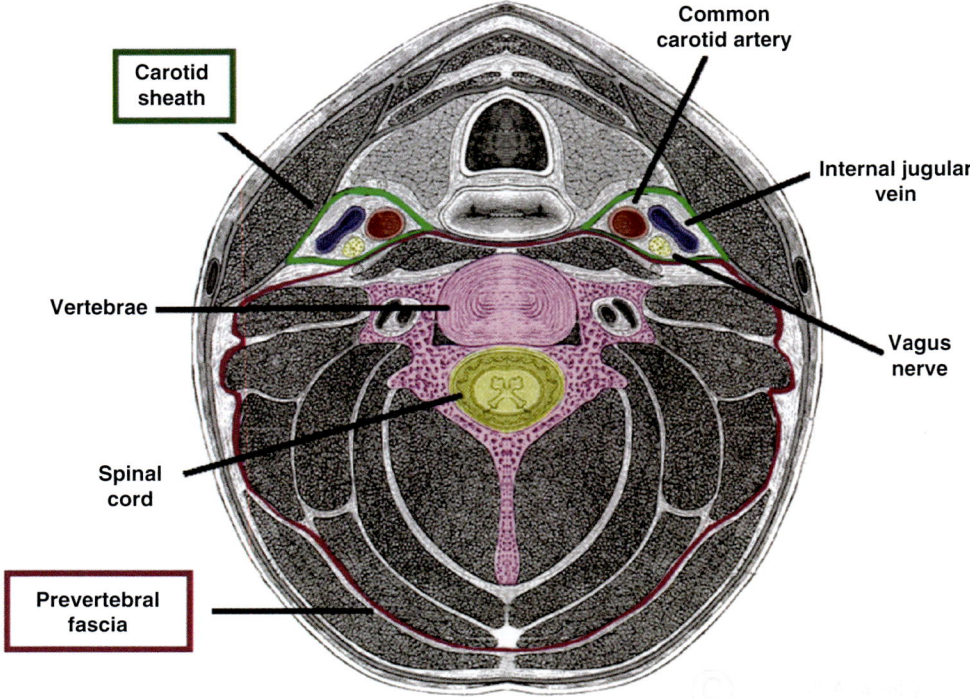

Fig. 11.68 PVF: coronal. <u>PVF horizontal</u>: fascia of posterior triangle, non-branchiomeric muscles (scalenes, levator scapulae, splenius capitis).
Lies ventral to roots of brachial plexus.
• As trunks move into upper extremity, they carry an extension of PVF, the axillary sheath.
• Subclavian artery and vein are outside the sheath.
<u>PVF vertical</u>: skull base in front of longus colli down to anterior ligament of t1
• Neuromeric relationship with cervical neuromeres.
[Courtesy of Instant Anatomy (www.instantanatomy.net). Retrieved from: https://www.instantanatomy.net/headneck/areas/fasprevertebral.html.]

PREVERTEBRAL PART OF DEEP FASCIA OF NECK

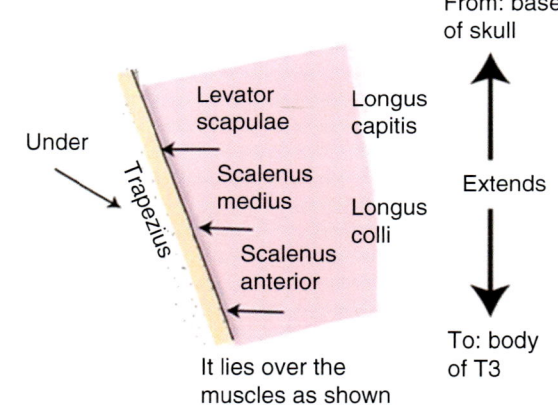

It lies over the muscles as shown

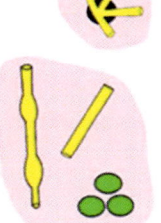

Pierced by: • Great auricular nerve
• Lesser occipital nerve
• Transverse cervical nerve
• Supraclavicular nerves
• Inferior root of ansa cervicalis

Lying on it: • Sympathetic chain
• Lymph nodes
• Spinal root of accessory nerve

Deep to it: • Cervical plexus
• Trunks of brachial plexus
• 3rd part of subclavian artery
• Phrenic nerve

It blends with the anterior longitudinal ligament. Its lower border laterally is the lower border of scalenus anterior

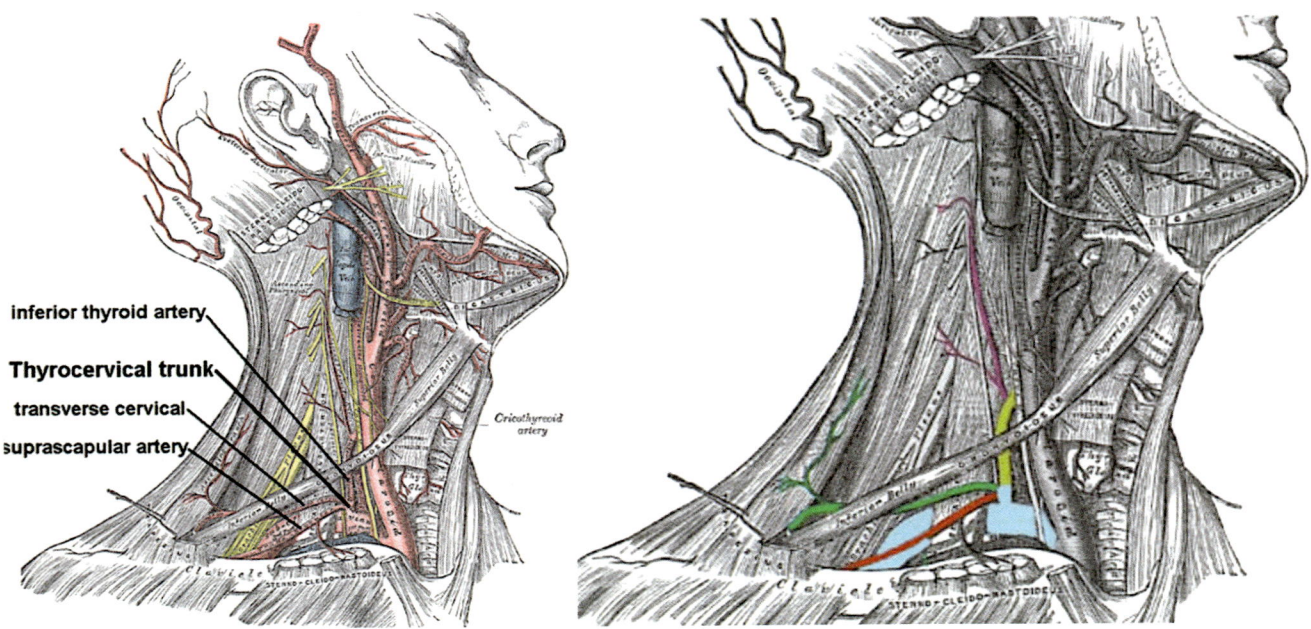

Fig. 11.69 Thyrocervical trunk: muscular relations. Thyrocervical trunk (TCT) is the predominant source of blood supply to the neck. TCT (aqua) arises from second part of subclavian (aqua). It gives off *inferior thyroid* (yellow), *ascending cervical* (magenta), *transverse cervical* (green), and *suprascapular* (red). *Dorsal scapular artery* not shown—it arises as a branch from transverse cervical or just lateral to thyrocervical trunk. Suprascapular also can arise from transverse cervical. [Reprinted from Lewis, Warren H (ed). Gray's Anatomy of the Human Body, 20th American Edition. Philadelphia, PA: Lea & Febiger, 1918.]

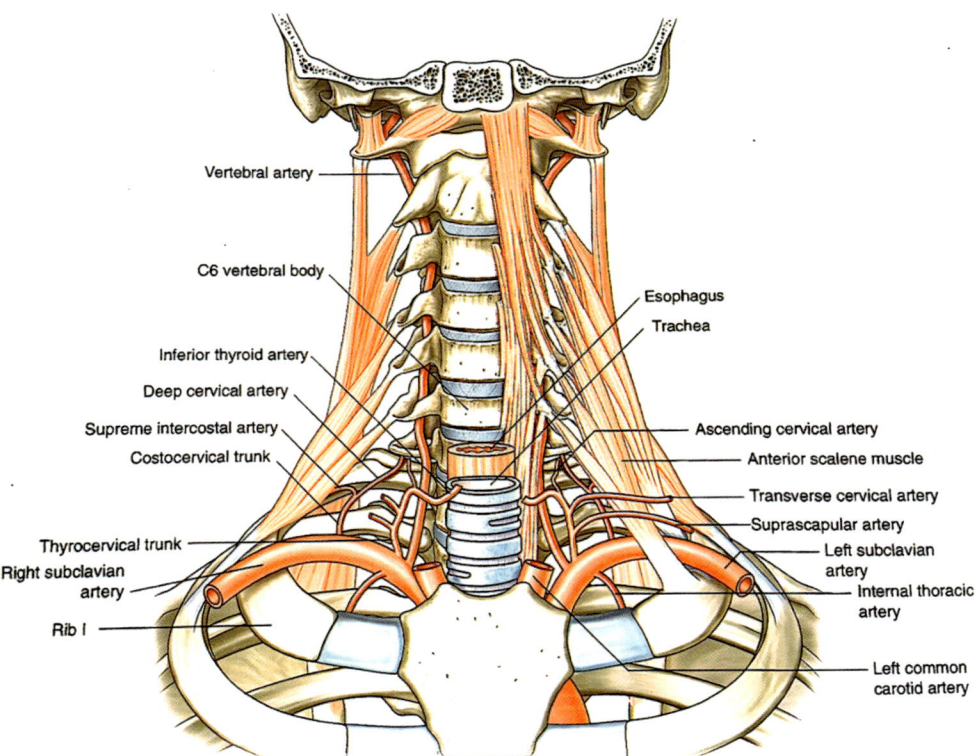

Fig. 11.70 Thyrocervical trunk: skeletal relations. TCT has 3 main branches and supplies the hypaxial muscles enclosed by all 3 cervical fasciae. (1) Inferior thyroid a. supplies the pretracheal fascia and the overlying strap muscles of ancient coracohyoideus enclose within the deep investing fascia. It gives off a deeper ascending cervical br. to the prevertebral fascia. (2) Transverse cervical branch a. supplies the deep investing fascia of the anterolateral neck. (3) Suprascapular a. supplies deep investing fascia and muscles of the scapula. Although scapula is located dorsally the muscles which control it from both the neck and trunk are all hypaxial representing its evolutionary origin as a ventral bone of the limb girdle. All branches of TCT three provide perforating branches to the overlying platysma within the superficial investing fascia. [Reprinted from Drake R, Vogel AW, Mitchell AWM. Head and Neck. In: Gray's Anatomy for Students, third edition. Philadelphia, PA: Churchill-Livingstone. 2015: 924–1052. With permission from Elsevier]

Buccopharyngeal Fascia (BPF) (Fig. 11.71)

BPF is a neural crest structure that covers the branchiomeric muscles of the pharynx (r4-r5) and the three pharyngeal constrictors (r6-r11). It extends from the skull base downward to the esophagus. This fascia is supplied by arterial branches representing each of the arches.

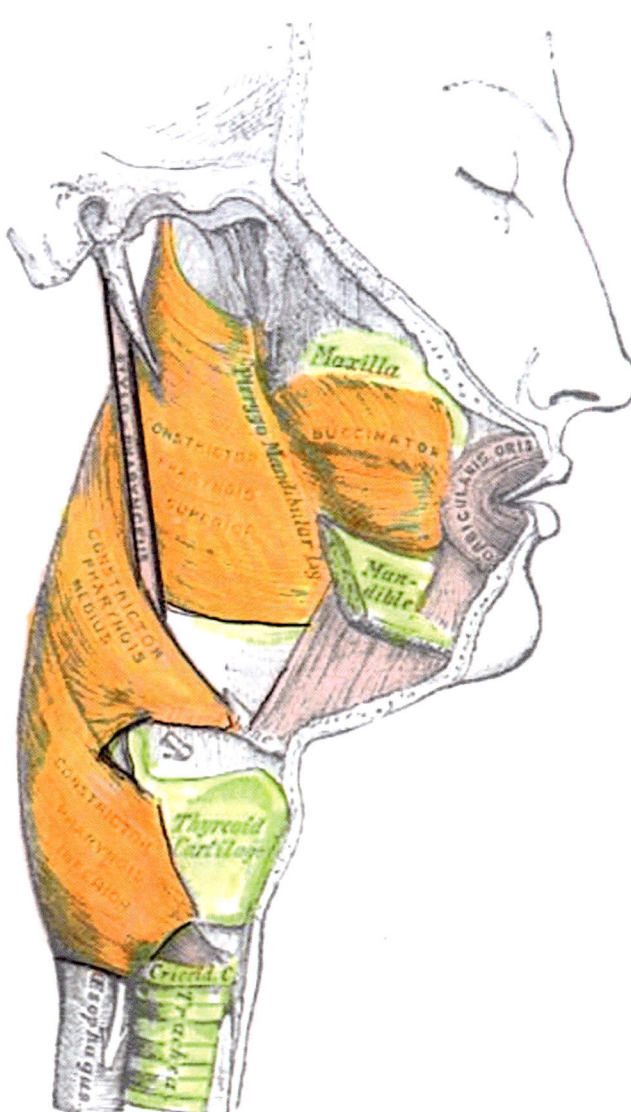

Fig. 11.71 Buccopharyngeal fascia (BPF). Neural crest structure extends from C1 down to esophagus. Covers branchiomeric all muscles lining the pharynx: buccinator (r4-r5) and pharyngeal constrictors (r6-r11). Joined in the midline with prevertebral fascia. [Reprinted from Lewis, Warren H (ed). Gray's Anatomy of the Human Body, 20th American Edition. Philadelphia, PA: Lea & Febiger, 1918]

Pretracheal Fascia (PTF) (Figs. 11.72, 11.73, 11.74, 11.75, 11.76, 11.77, 11.78, 11.79, 11.80, 11.81, 11.82, 11.83, 11.84, 11.85, 11.86, 11.87, 11.88, 11.89, 11.90, 11.91 and 11.92)

PTF is a thin layer that begins at the anterior margin of hyoid bone and extends downward to ensheath the thyroid, the infrahyoid strap muscles, and, ultimately the strap muscles. PTF recapitulates the fascia associated with coracohyoideus. Recall from our discussion of the neck that the chondrichthyan clavicle is connected to the mandible using two successive muscles coracohyoideus and geniohyoideus. Together these trace the ventral margins of branchial arches 1–7. This tract provides a guidance system for the descent of the thyroid and thymus. Furthermore, since it terminates at neuromeric level c4 (the termination of BA7), that is, at pericardium it represents the pathway by which thyroid can stay undescended in the tongue or thymus can find itself in the mediastinum.

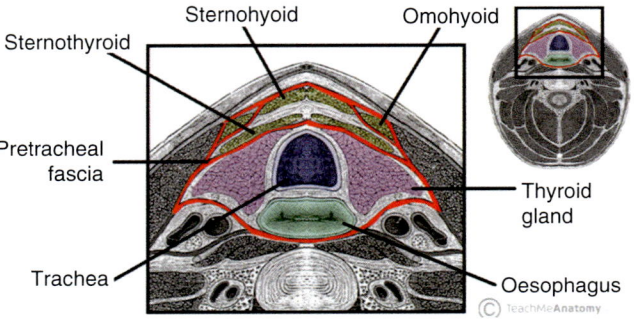

Fig. 11.72 Pretracheal fascia (PTF): infrahyoid strap muscles (C1-C3).Infrahyoid muscles (C1-C4) innervated from the cervical plexus. 4/5 muscles related phylogenetically from coracomandibularis. Innervation by the cervical plexus to hypaxial branchiomeric muscle belies ancient position of the forefin. Perhaps a remnant of more anteriorly positioned forefin? Refer to Fig. 11.57. [Courtesy of Instant Anatomy (www.instantanatomy.net). Retrieved from: https://www.instantanatomy.net/headneck/areas/faspretracheal.html]

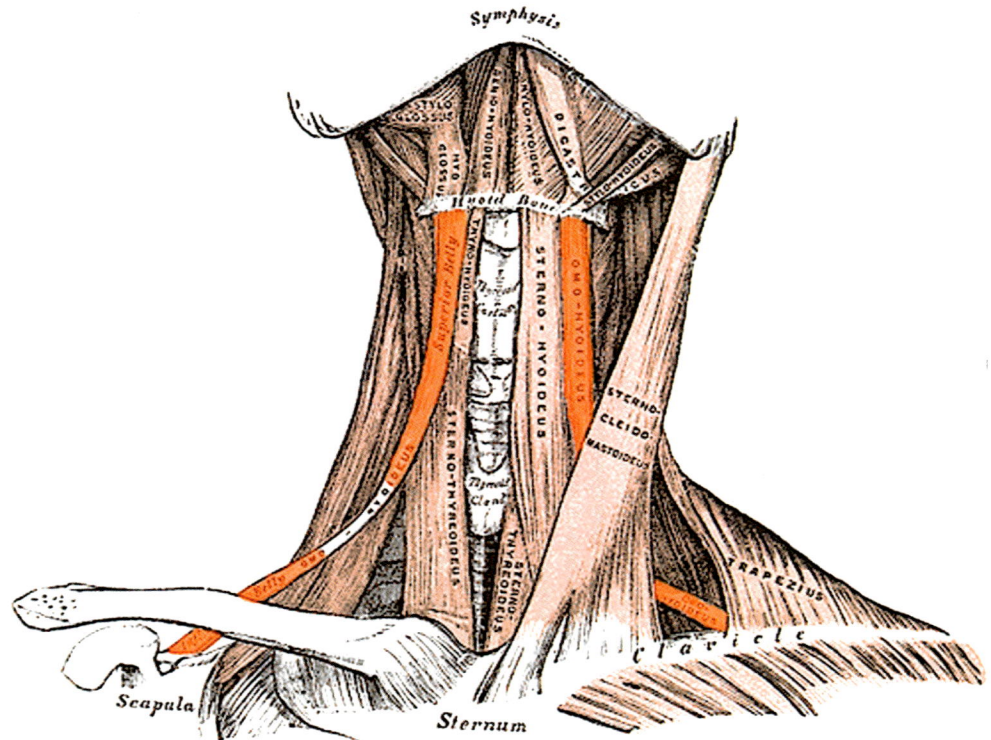

Fig. 11.73 Strap muscles and pretracheal fascia. The strap muscles are hypaxial and branchiomeric, supplying muscles from C1-C3 (S5-S7). These muscles connect the ventral margins of branchial arch structures all to way from mandible to clavicle, that is, from r2 to c2-c3. Geniohyoid and thyrohyoid recapitulate the ancient geniohyoideus which extends from BA1 to BA3. Sternohyoid and sternothyroid recapitulate ancient coracohyoideus which connects BA3 to BA6, the c1 sector of clavicle. Omohyoid connects BA3 to BA7, the c2-c3 sector of clavicle. Functionally, the muscles must be innervated by the cervical plexus because they interconnect the first five BAs belonging to the head with the last two BAs ultimately dedicated to the proximal fin/limb (pectoral girdle). [Reprinted from Lewis, Warren H (ed). Gray's Anatomy of the Human Body, 20th American Edition. Philadelphia, PA: Lea & Febiger, 1918]

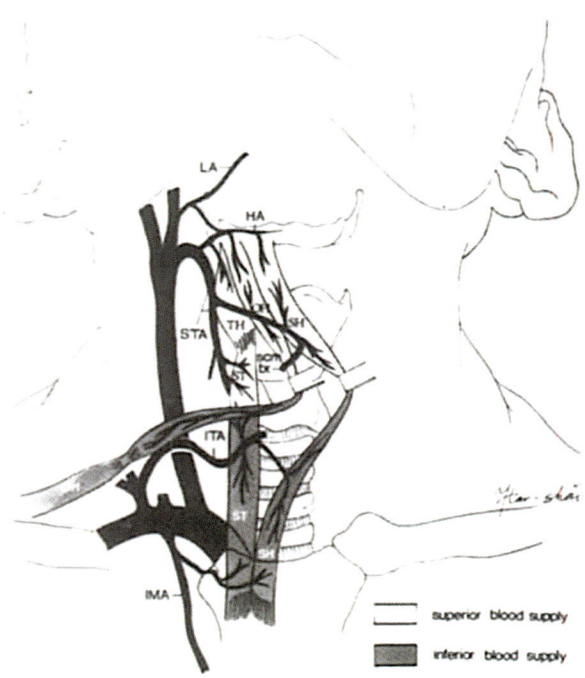

Fig. 11.74 Blood supply to PTF. Diagram of the superior and inferior blood supply to the infrahyoid muscles. *LA* lingual artery, *HA* hyoid artery, *STA* superior thyroid artery, *ITA* inferior thyroid artery, *IMA* internal mammary artery, *OH* omohyoid muscle, *SH* sternohyoid muscle, *ST* sternothyroid muscle, *TH* thyrohyoid muscle, *scmbr* ster- nocleidomastoid branch. [Reprinted from Eliachar I, Marcovich A, HarShai Y, Lindbaum E. Arterial blood supply to the infrahyoid muscles: an anatomical study. *Head & Neck Surg* 1984; 7:8–14. With permission form John Wiley & Sons, Inc.]

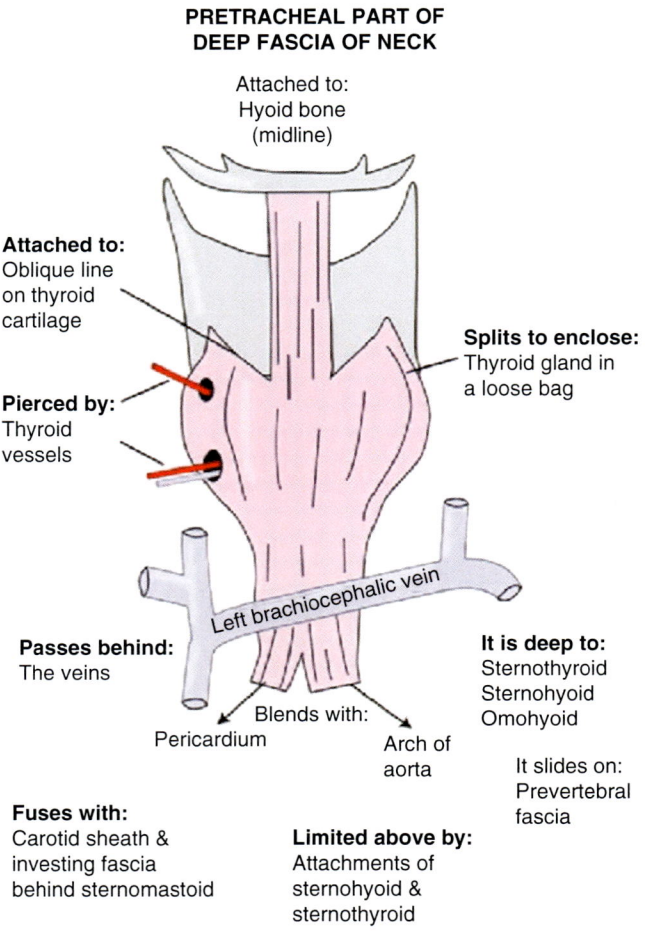

Fig. 11.75 Pretracheal fascia. Coronal: PTF is attached on its under-surface to thyroid cartilage oblique line indicating its developmental relationship with the larynx. Because it splits to enclose thyroid, PTF constitutes a potential space in which the gland descend. Thryoid neuromuscular pedicles constitute primary structures of the fourth arch anlage, around which the PTF must condense. Were this the reverse the pedicles would traverse around the fascia. PTF is narrowed by the homologs of coracomandicularis muscles, being limited by the infrahyoid strap muscles and by the paired bellies of the suprahyoid geniohyoideus. [Courtesy of Instant Anatomy (www.instantanatomy.net)]

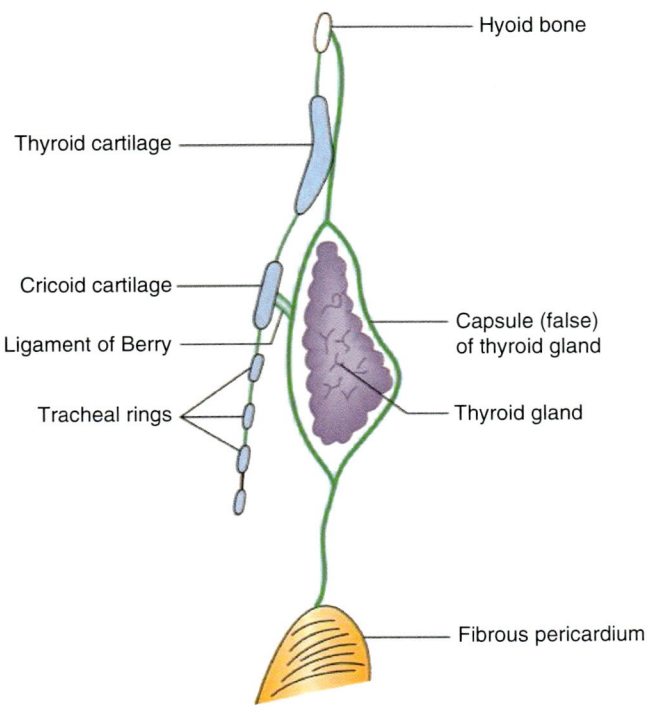

Fig. 11.76 Pretracheal fascia.

Axial: Unites with all DIF enclosing the sternocleidomastoid and the anterior wall of the carotid sheath.

Sagittal: Connected to the hyoid bone, enters thorax, fuses with the apex of the fibrous pericardium.

• PTF connects hyoid bone with thyroid and cricoid cartilages.
• Conduit for migration of tongue myoblasts, thyroid, and thymus.
• Gliding surface between trachea and esophagus during swallowing.
• Layer posterior to thyroid is thin and permits goitre to compress esophagus (dysphagia).
[Courtesy of Instant Anatomy (www.instantanatomy.net)]

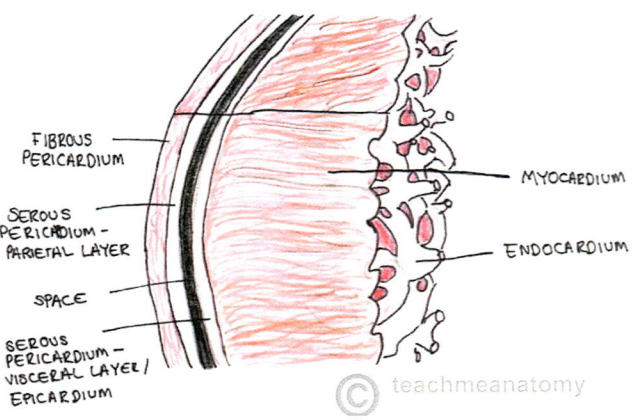

Fig. 11.77 Pericardium and endocardium. Pericardium: fibrous (cervical LPM$_S$) + serous (prechordal LPM$_S$). Epicardium: serous and fibro-adipose (prechordal LPM$_V$). [Reprinted from TeachMeAnatomy courtesy of Dr. Oliver Jones]

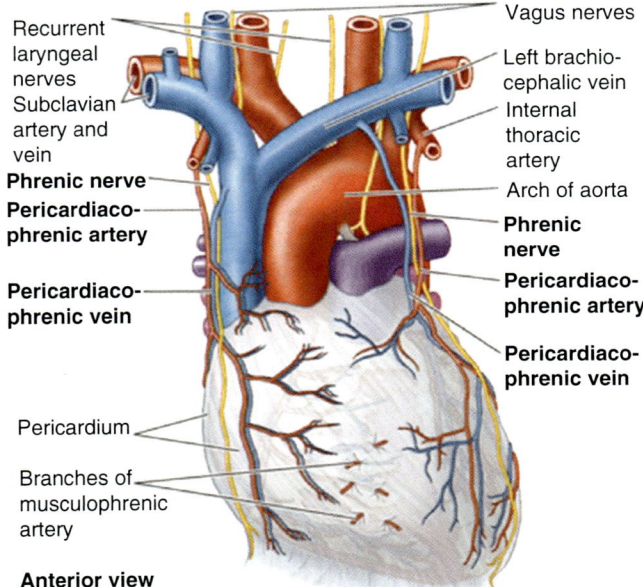

Anterior view

Fig. 11.78 Blood supply to pericardium. Cervical LPM$_S$ fibrous layer of pericardium: *pericardiophrenic arteries* accompany phrenic nerves and are distributed to the Prechordal LPM$_S$ parietal serous layer of pericardium: *Iinternal thoracic arteries* note *musculophrenic* branches connect with anterior intercostal arteries 7–9 and supply lower pericardium. LPM$_V$ epicardium (visceral serous layer + fat): *coronary arteries* supply prechordal cardiogenic mesoderm. [Reprinted from Drake R, Vogel AW, Mitchell AWM. Gray's Anatomy for Students, third edition. Philadelphia, PA: Churchill-Livingstone. 2015. With permission from Elsevier]

Fig. 11.79 Folding of the embryo 1; heart fields and septum transversum. Prechordal plate mesoderm extends forward from the primitive node to the primary heart field. Thus, when first aortic arches develop in LPM they are in intimate contact with PCM. [Reprinted from Standring S. The Back. In: Gray's Anatomy, 40th edition. Philadelphia, PA: Churchill-Livingstone; 2008: 763–774. With permission from Elsevier]

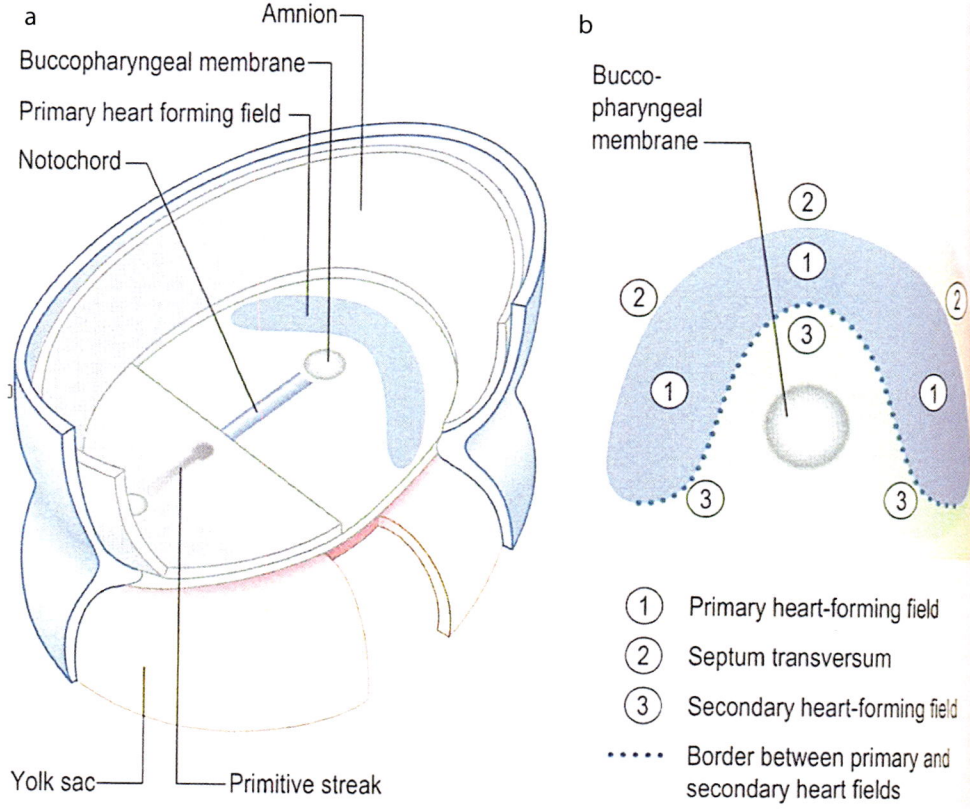

a

Amnion

Buccopharyngeal membrane

Primary heart forming field

Notochord

Yolk sac — Primitive streak

b

Buccopharyngeal membrane

① Primary heart-forming field

② Septum transversum

③ Secondary heart-forming field

•••• Border between primary and secondary heart fields

Fig. 11.80 Folding of the embryo 2: origins of dermis. Note that lateral plate mesoderm forms all ventral dermis from T2 to L1. Note that the upper extremities, although hypaxial, are PAM. The PAM/LPM interface runs down the sides of the trunk as the mid-axillary line. Key: somatic LPM, blue; visceral LPM, grey. [Reprinted from Standring S. The Back. In: Gray's Anatomy, 40th edition. Philadelphia, PA: Churchill-Livingstone; 2008: 763–774. With permission from Elsevier]

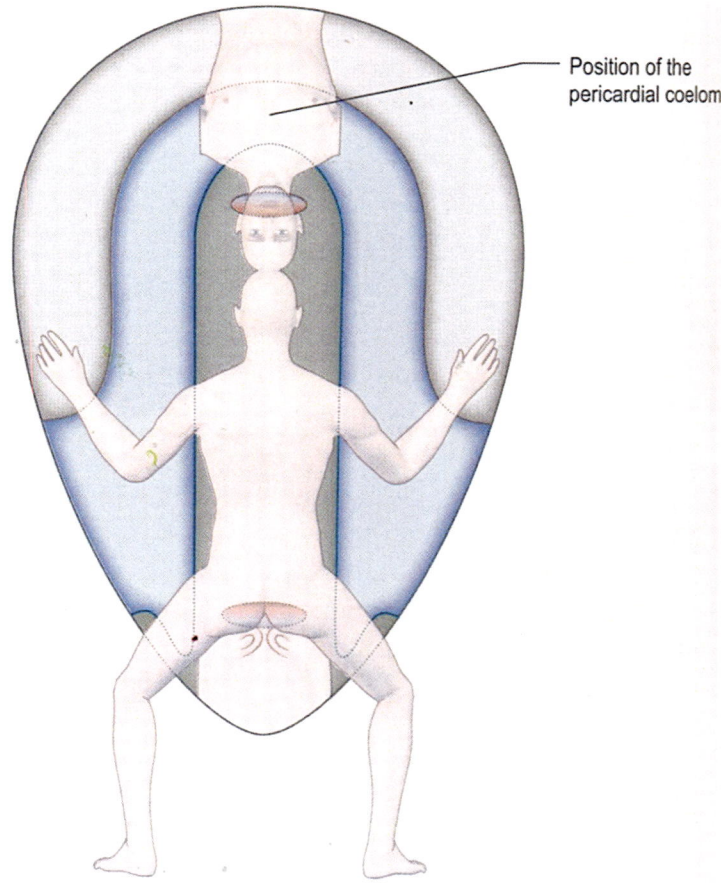

Position of the pericardial coelom

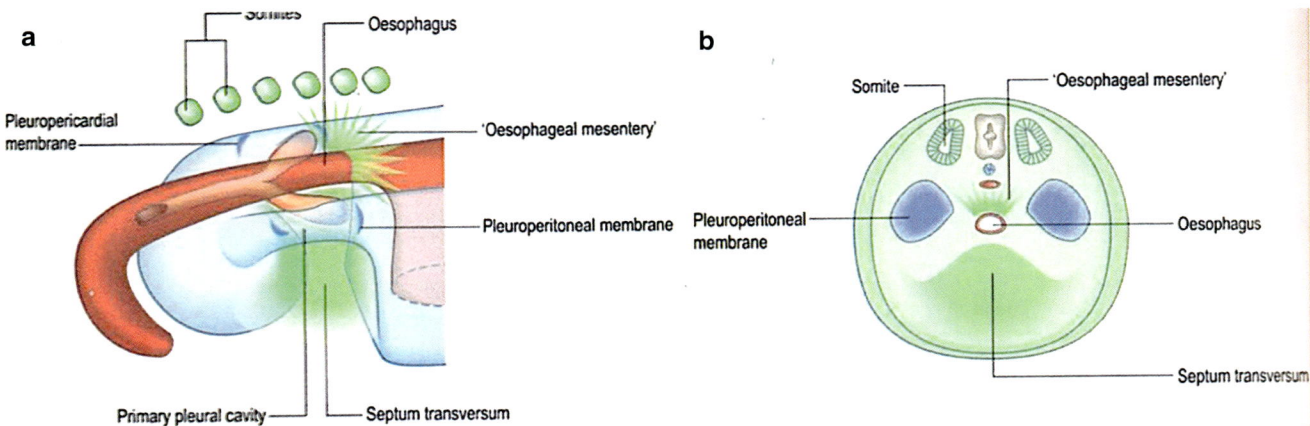

Fig. 11.81 Folding of the embryo 3. PCM comes to lie directly dosal to the first aortic arches and is ideally positioned to contribute to the cranial base at the r1 level. Pleuroperitoneal membranes define a cavity but are an important source of tissue in their own right. They contribute to the diaphragm and provide an means by which, in mammals, the C4 supracoracoideus muscle invades the chest wall to form the bulk of muscle in the diaphragm. [Reprinted from Standring S. The Back. In: Gray's Anatomy, 40th edition. Philadelphia, PA: Churchill-Livingstone; 2008: 763–774. With permission from Elsevier]

Fig. 11.82 Thyroid migration. Note association of thymus with third pharyngeal pouch and with parathyroid III. [Reprinted from Carlson BM. Human Embryology and Developmental Biology, sixth edition. St. Louis, MO: Elsevier; 2019. With permission from Elsevier]

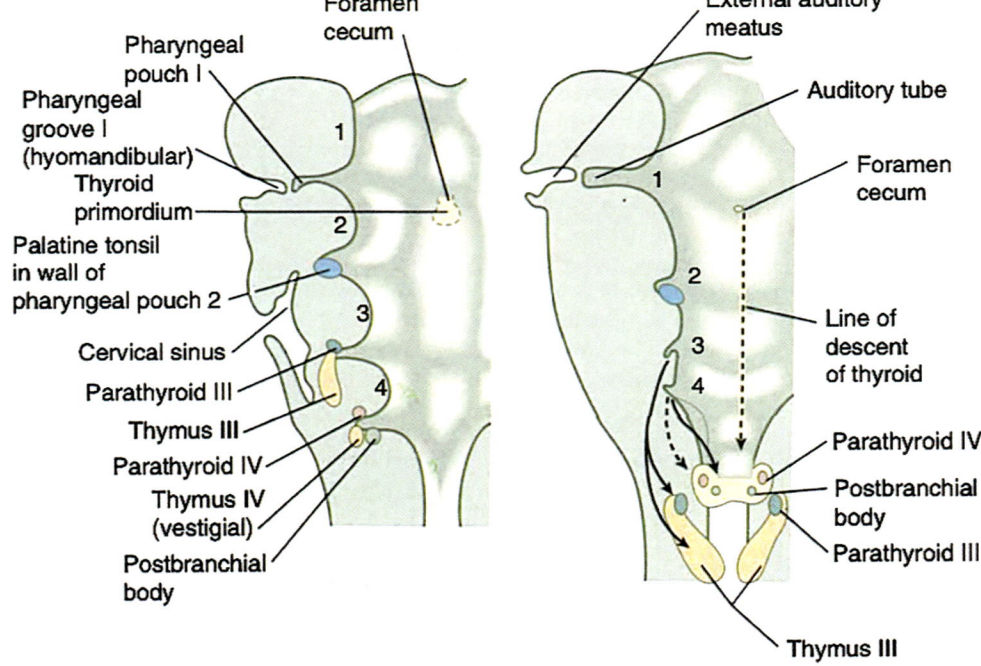

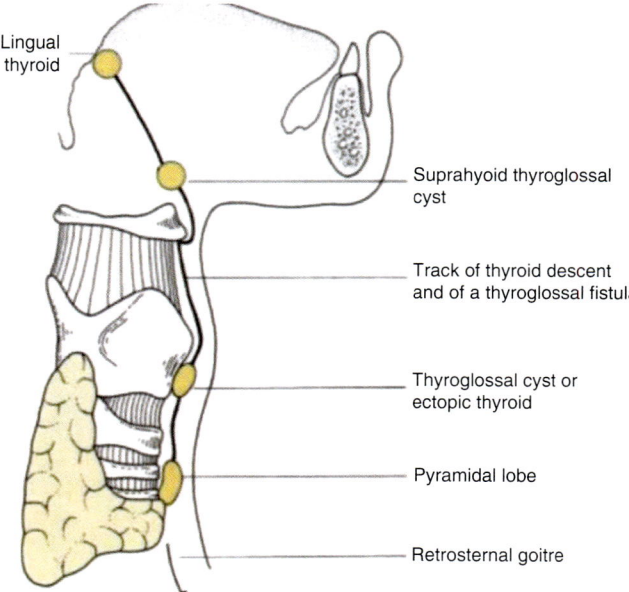

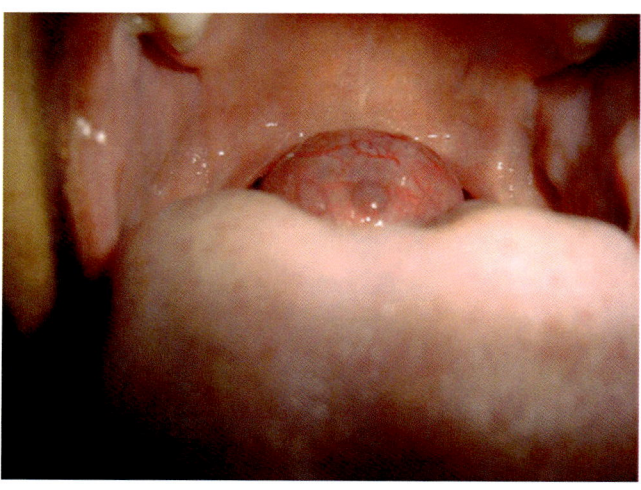

Fig. 11.84 Lingual thyroid. [Reprinted from Amr B, Monib S. Lingual thyroid: A case report. *International Journal of Surgery Case Reports* 2013; 2:313–315. With permission from Elsevier]

Fig. 11.83 Ectopic thyroid tissue. Thyroid tissue has been found within the trachea, substernal and with intra-abdominal sites reported as well. Thryoid follows the tract of PTF from the site of origin in the tongue through the hyoid bone, remaining immediately anterior to the PTF and reaching as far as level c4. [Reprinted from Fiaschetti V, Claroni G, Scarano AL, Schillaci O, Floris R. Diagnostic evaluation of a case of lingual thyroid ectopia. Radiology Case Reports 2016; 11:165–170. With permission from Elsevier]

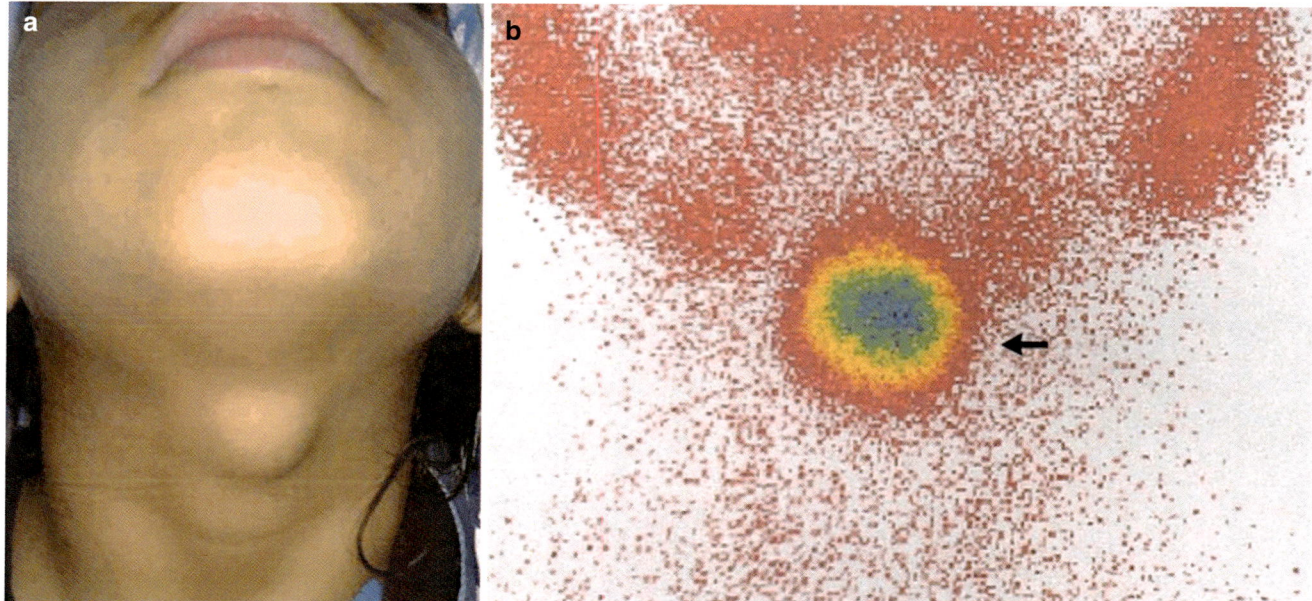

Fig. 11.85 Thyroid gland in the thyroglossal tract. A 12-year-old girl presented with a lump situated high in her neck at the midline (Panel A). She had previously been found to have hypothyroidism and had been treated with levothyroxine. There were no signs of infection. An ultrasonogram showed a solid midline mass. A technetium-99 thyroid scan showed uptake in the region of the mass but no uptake in the area of the thyroid gland (Panel B, arrow). A fine-needle aspiration biopsy did not reveal any evidence of a malignant condition. The diagnosis was an ectopic thyroid gland in a thyroglossal tract. [Reprinted from Aalaa M, Mohajeri-Tehrani MR. Ectopic Thyroid Gland. New Eng J Med 2012; 366:943. With permission from Massachusetts Medical Society]

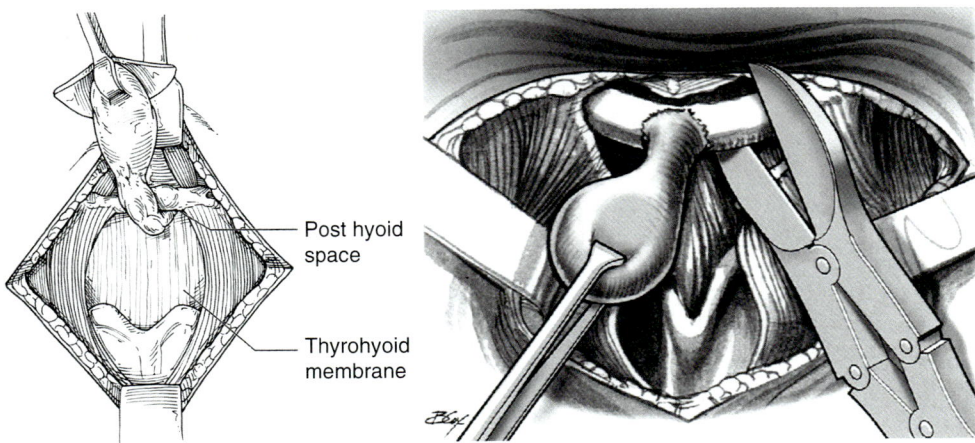

Fig. 11.86 Surgical management of thyroglossal duct cyst. LEFT: Descent of the thyroid track through the anlage of the developing hyoid bone. The cyst and duct are being elevated with a cuff of surrounding tissue off the thyrohyoid membrane. The posthyoid space is entered with blunt dissection to ensure the tract is not truncated with hyoid resection. RIGHT: Thyroglossal duct management requires at times surgical section of the hyoid bone at the anterior midline: the Sistrunk procedure. Note how fibers of the sternohyoid and mylohoid have been detached from the hyoid bone to allow for exposure of the bone. The cuts of the hyoid have been made medial to the attachment of the digastric tendon and with sufficient length so as not to cut the duct. Left: [Reprinted from Newton SS. Thyrogrloassal duct cyst. Op Tech Otlaryngol 2017; 28: 173–178. With permission from Elsevier.]. Right: [Reprinted from Goldstein H, Khan A, Pereira KD Thryoglossal duct excision—The Sistrunk Procedure Op Techn Otoloaryngol 2009; 20 (4): 256–259. With permission from Elsevier]

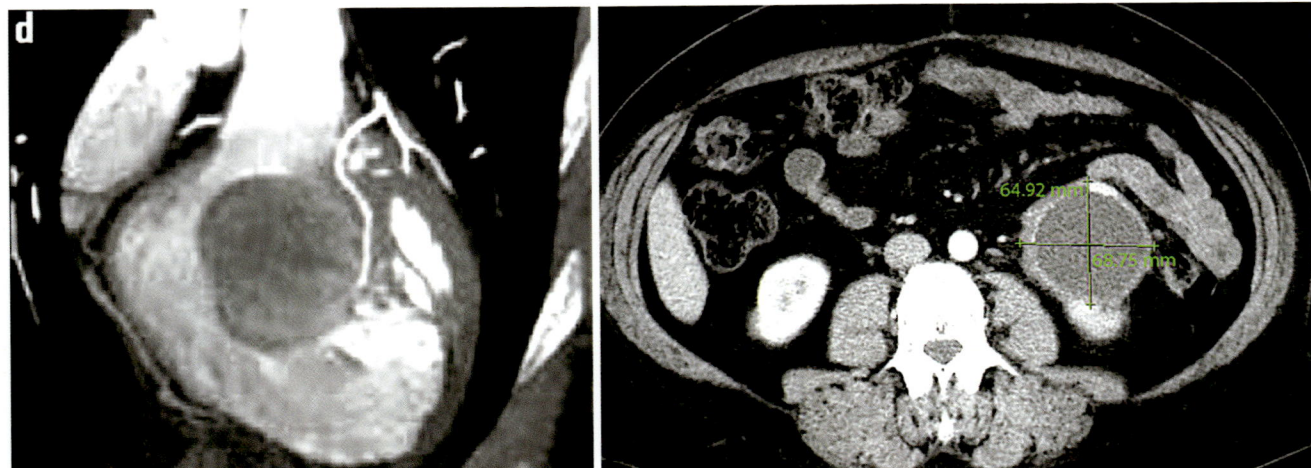

Fig. 11.87 Ectopic thyroid in viscera: right ventricle and left kidney. LEFT This case shows the neuromeric pathway of fascia connecting the anterior margins of the pharyngeal arches and the pericardial sac. It is associated with the sensory innervation of the pericardium from neuromeres c3-c5. RIGHT First-in-man report of ectopic thyroid found in the left kidney. The first iteration of renal blastema, the pronephros, develops in the intermediate mesoderm at the level of the cervical neuromeres. The close association of kidney anomalies with congenital heart disease arises from this common embryonic connection. In the same way thyroid tissue tracking along the pre-tracheal fascial plane may become incorporated into the pronephric anlage. Left: [Reprinted from Sun Y, Wang J, Cao D. Uncommon right ventricular mass: Ectopic thyroid. Anatol J Cardiol. 2018 May; 19(5): E8–E9. With permission from AJC Publications]. Right: [Reprinted from Mata Mera C, Kreutzer N, Lorenzen J, Truss M. Der setene Nierentumor: Renales ektopes Schilddrússgewebe. Urologe 2018; 57: 944–946. With permission from Springer Nature]

Fig. 11.88 Thymus development (mouse model). Stage 10 (E9.5) positioning thymic eipithlial cells arise in third pouch. Stage 13 (E11) (CS 13) initiation. Stages 14–16 (E11.5-E12.5) outgrowth and patterning. Stages 15–18 (E 12-E13.5) separation. Stage 15 to birth (12-birth) differentiation. [Reprinted from Blackburn CC, Manley NR. Developing a new paradigm for thymus organogenesis. Nature Reviews Immunology 2004; 4:278–289. With permission from Springer Nature]

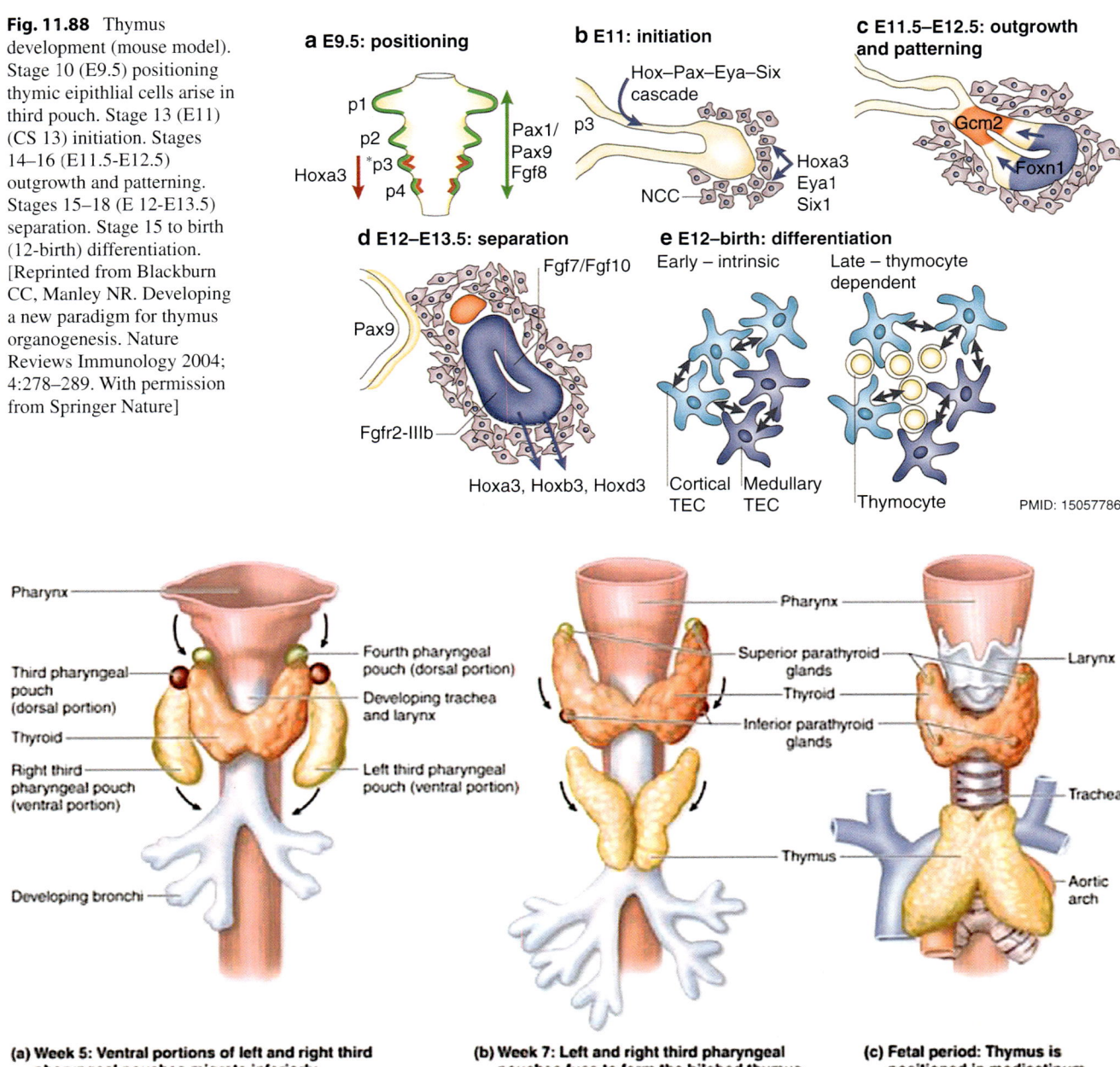

(a) Week 5: Ventral portions of left and right third pharyngeal pouches migrate inferiorly

(b) Week 7: Left and right third pharyngeal pouches fuse to form the bilobed thymus

(c) Fetal period: Thymus is positioned in mediastinum

Fig. 11.89 Descent of the thymus. Stage (6 weeks). Stage (7 weeks). [Reprinted from O'Laughlin V, McKinley M. Anatomy & Physiology: An Integrative Approach. New York, NY: McGraw-Hill Education; 2017. With permission from McGraw-Hill LLC]

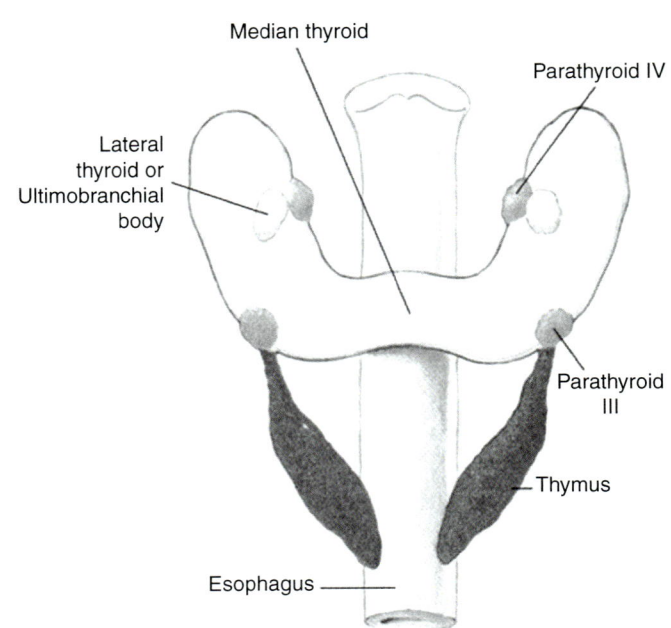

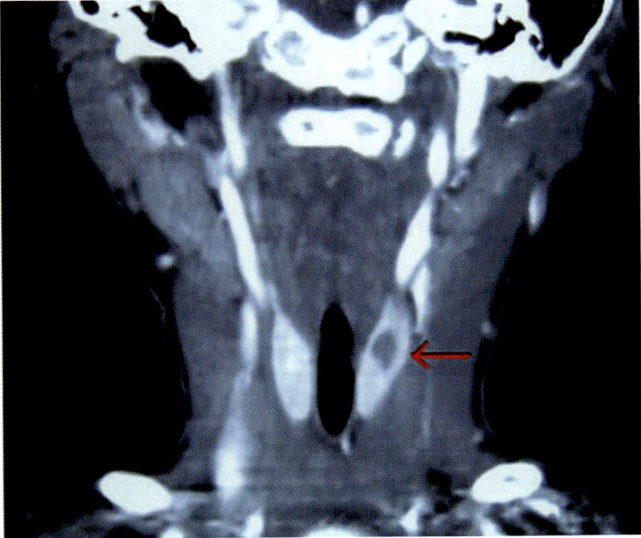

Fig. 11.90 Parathyroid displacement during thymus descent Diagrammatic representation of routes of descent of thymus, parathyroid glands, and " last branchial body " during fetal life. Parathyroid glands originating more cranially in third branchial pouch migrate in close association with thymus to reach final position more caudally with respect to parathyroid glands originating in fourth branchial pouch. Final location of last branchial body is intimately embedded in thyroid gland, to constitute parafollicular C-cells producing calcitonin. Thyroid gland originates at " blind foramen " along midline and migrates down to first tracheal ring.

- Third pouch contains thymus and parathyroid 3.
- Fourth pouch contains parathyroid 4.
- Thymus moves past thyroid dragging PT3 with it, distal to PT4.
- Thymus can occasionally get stuck in thyroid gland.
[Reprinted from Scharpf J, Kyriazidis N, Karmani D, Randolph G. Anatomy and embryology of the parythyroid gland. Op Tech Otolarygol Head Neck Surg 2016; 27(3): 117–121. With permission from Elsevier.]

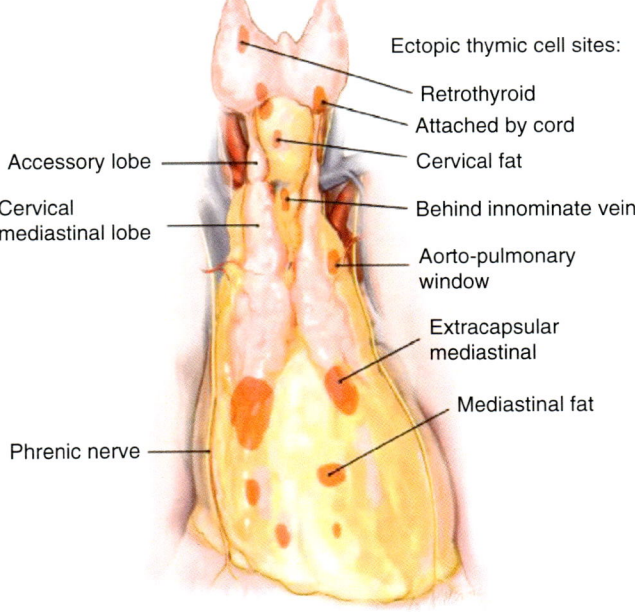

Fig. 11.92 Mediastinal thymus. Thymic remnants can reach the pericardium. Note the physical position of descent does not permit intracardiac migration of the thymus. [Reprinted from Argote-Green LM, Jaklitsch MT, Surgarbaker DJ. Thorascopic approach to thymectomy with advice on patients with myasthenia gravis. In: Sugarbaker DJ, Bueno R, Krasna MJ, Mentzer SJ, Zellos L (eds). *Adult Chest Surgery*. McGraw Hill; 2015: 1260–1265. With permission from McGraw-Hill Education]

Fig. 11.91 Thymic tissue in left lobe of thyroid. [Reprinted from Huang Y, Sheng S, Xio X. Ectopic intrathyroidal thymus in children: Two case reports and review of the literature. *J Ped Surg Case Reports* 2013; 1(11): 386–390. With permission from Creative Commons License 3.0: https://creativecommons.org/licenses/by-nc-sa/3.0/]

Evolution of Skin and Appendages

Fishes

The epidermis of living fishes is not keratinized; it is alive with a pattern of microridges that retain mucous. This mucous cuticle provides lubrication for swimming, is a barrier layer for bacteria, and can include defensive chemicals. Three unicellular glands found in fish epidermis contribute to the mucous cuticle. Club cells can produce chemicals, that once diffused into the environment constitute alarm signals. Granular cells and goblet cells also contribute to the cuticle. The latter cell is phylogenetically more recent, as it is not found in lamprey skin.

Fish dermis is very elastic allowing for bending of the body during swimming and storing energy during unbending. It produces dermal bone which in turn gives rise to scales. Fish scales have structural characteristics similar to teeth: the outer layer is made of hard acellular enamel produced by the epidermis whilst the inner layer, produced by the dermis, is dentin.

Primitive fishes The ostracoderms and placoderms lived within an exoskeleton of dermal armor. In the cranial region, these dermal bones amalgamated into a head shield; those in the caudal regions remained distinct. Ostracoderm scales, *tubercles*, resemble teeth spread over dermal bone with a deep lamellar layer and a superficial reticular vascular layer, reminiscent of dental pulp chambers. In transitional forms, hagfish, and lamprey, dermal bone generation is lost. The skin becomes smooth without scales. Pigment cells make their appearance in the dermis. Finally, the hypodermis has adipose tissue (Fig. 11.93a).

Chondrichthyes Cartilaginous fishes, not surprisingly, do not make dermal bone. The dermis has large numbers of collagen and elastic fibers for swimming and it produces *placoid scales*, which project upward through the epidermis. Sharks use these scales to reduce drag while swimming (Fig. 11.93b).

Osteichthyes Bony fishes, ultimately ancestral to us, have two layers of dermis, the upper layer being loose and the lower layer being dense. Here dermal scales are produced; although they do not actually fully penetrate through the epidermis they make the skin hard (Fig. 11.94).

a

Spaces for blood vessels, mucous glands, lateral line and electroreceptors

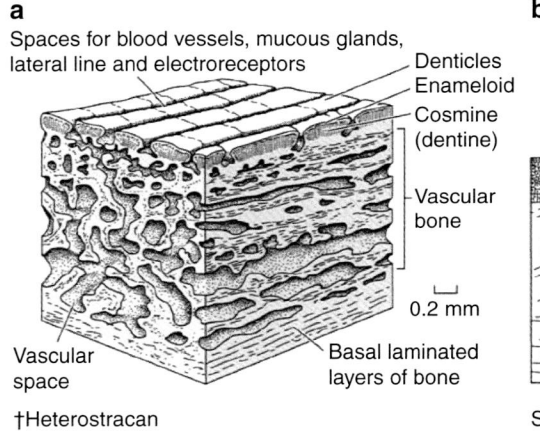

Denticles
Enameloid
Cosmine (dentine)
Vascular bone

0.2 mm

Vascular space

Basal laminated layers of bone

†Heterostracan

b

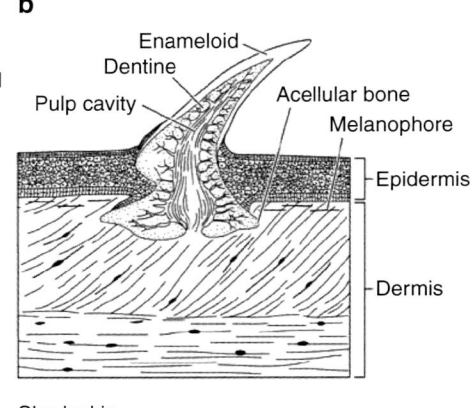

Enameloid
Dentine
Pulp cavity
Acellular bone
Melanophore
Epidermis
Dermis

Shark skin

Fig. 11.93 Evolution of fish skin 1. (**a**) Cosmoid plate of jawless fish(†herterostraca) showing cosmoid plate Cosmine is a form of dentin with well-organized dental tubules. Enameloid (sic) is a hardened form of cosmine but is dermal in origin. (**b**) Chondrichthyan fish (shark) showing adaptation of Dermal denticles become become placoid scales become spiny lacodi scalesmade of dentin covered by an unknown hard substance. [Reprinted from Kardong K. Vertebrates: Comparative Anatomy, Function, Evolution, seventh ed. New York, NY: McGraw-Hill; 2015. With permission from McGraw-Hill Education]

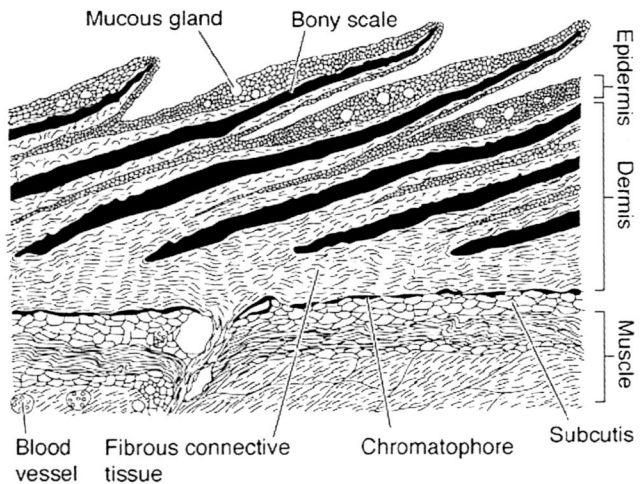

Mucous gland   Bony scale

Blood vessel   Fibrous connective tissue   Chromatophore   Subcutis

D. Teleost skin

Fig. 11.94 Evolution of fish skin 2. (**a**) Early bony fish had rhomboidal adenticulate cosmoid scales These are our ancestors. (**b**) Derived teleost fishes have imbricating cycloid scales with mucus glands and growth rings. [Reprinted from Kardong K. Vertebrates: Comparative Anatomy, Function, Evolution, seventh ed. New York, NY: McGraw-Hill; 2015. With permission from McGraw-Hill Education]

Tetrapods

Differences between Fish Skin and Tetrapod Skin

Rise of keratinization (present in fishes but to a minor degree).

Addition of lipids to the surface layer.
Increased complexity of epidermal layers.
Evolution of the pharyngeal arches: repositioning of the second arch

- Redistribution of second arch neural crest into the subcutaneous tissues.
- subcutaneous white fat (birds and mammals),
- sympathetic innervation,
- Dermal glands become multicellular and reach the surface via ducts.

Innervation of the dermis.
Subcutaneous adipose layer.
Evolution of the SMAS.
Skin appendages: feathers and hair.

Amphibians

Amphibian skin represents a radical departure from the integument of fishes. It has a multi-layer corneum adapted to terrestrial life to prevent desiccation. Amphibians shed their skins en mass under endocrine control from the pituitary and thyroid. Amphibians consist of three orders. Anura or frogs (Gk. *an* "without" + *oura* "tail") have smooth skins. Caudata, the salamanders (Lt. +*cauda* "tail"), have skins with bumps. Gymmnophiona (Gk *gymnos* "naked" + *ophis* "serpant"), known more commonly as caecilians, are limbless, wormlike creatures with skin bearing transverse folds and scales embedded in the furrows, a retention of the ancestral condition in which tetrapods had scales like the fishes from which they evolved (Fig. 11.95).

Amphibian skin is adapted to life underwater. It is delicate and permeable to water. Gas exchange can take place through the skin (cutaneous respiration). Allowing for long-term submergence, e.g. hibernation. Frogs and salamanders have two cutaneous glands, both located in the dermis and connected to the surface using ducts. Mucous glands keep the skin moist and prevent dehydration. Poison (granular) glands or parotoids are larger and store up their contents for release as needed; These foul-tasting products serve for protection.

Amphibian skin displays colors produced by three layers of pigment cells called chromatophores arranged in three layers in the dermis. Lipophores (yellow) are superficial, guanophores (blue-green) are intermediate and melanophores (brown-black) are deepest. Note that chromophores are sometimes found in epidermis. Brightly-colored amphibians are often poisonous. Unlike fishes, whose skin color is under nervous system control and can change quickly amphibian skin color is under pituitary control and therefore changes slowly. Unlike bony fish, there is no direct control of the pigment cells by the nervous system, and this results in the color change taking place more slowly than happens in fish. Amphibians do not possess a subcutaneous adipose layer (Figs. 11.96 and 11.97).

Reptiles

Reptile integument is similar to that of mammals in that it has two layers, epidermis and dermis, but is decidedly different, due to the presence of scales. Like amphibians, periodic shedding of the skin, ecdysis, or moulting serve to allow for growth and get rid of parasites.

The skin of reptiles demonstrates a greater commitment to life on land. The epidermis is characterized by a complete covering of keratin. The keratin is composed of many layers of very thin, flat cells. Cells closer to the surface become more compacted as they respond to pressure from beneath by new keratin cells being formed in the stratum germinativum. Reptilian skin has three keratin layers: (1) The *stratum corneum* is heavily keratinized and referred to as the *Oberhätchen layer*; (2) The *intermediate zone* contains beta-keratin and is composed of stratum germinativum cells in differing stages of development; and (3) the deepest layer, *stratum germinativum*, consists of cuboidal cells containing alpha-keratin.

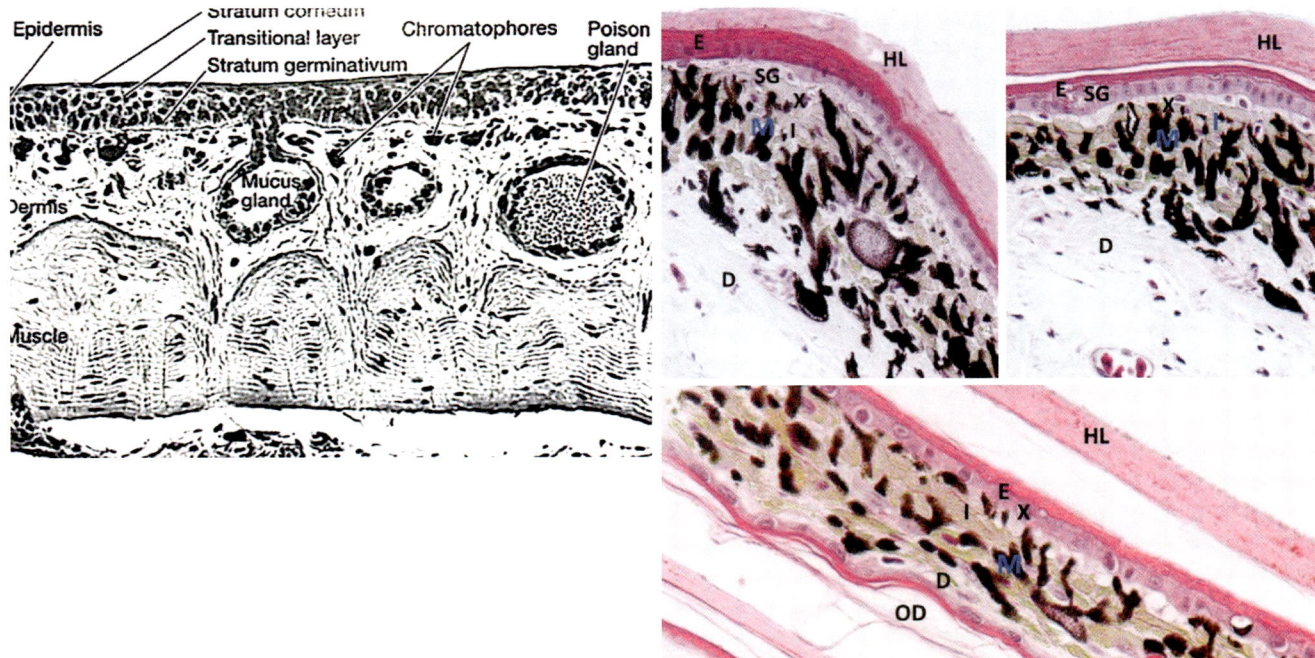

Fig. 11.95 Histology of amphibian skin. Left: Mucous cell in dermis emptying on the epidermal surface. Right: These pigment cells were observed in vertical combination, with an uppermost layer of xantho-phores, an intermediate layer of iridophores and a basal layer of mela-nophores. The ultrastructure of the melanophore is characterized by oval nucleus and numerous pigment granules, the melanosomes of different stages that remain scattered in the cytoplasm. The chromato-phores of this species contain significant information of anatomical similarity with lower as well as higher vertebrates. Left: [Reprinted from Kardong K. Vertebrates: Comparative Anatomy, Function, Evolution, seventh ed. New York, NY: McGraw-Hill; 2015. With per-mission from McGraw-Hill Education.]. Right: [Reprinted from Paray BA, Al-Sadoon MK. Ultrastructure of the dermal chromatophores in the Fringe-toed lizard, *Acanthodactylus orientalis* Zoologia 34: e11923 | DOI: 10.3897/zoologia.34.e11923. With permission from Creative Commons License 4.0: https://creativecommons.org/licenses/by/4.0/]

Fig. 11.96 Reptilian skin is notable for several types of scales, a wide variety of colors and the phenomenon of shedding. [Reprinted from Freepik.com. image: https://www.freepik.com/free-photo/closeup-shot-green-iguana_9991089.htm#page=1&query=reptile%20skin&position=10. Nature photo created by wirestock—www.freepik. com]

The epidermis of reptiles lacks supporting elements from the dermis. Dermis in reptiles consists of connective tissue. In some, there may be small bones called *osteoderms*. These are what form the distinctive specialized scales on savannah monitors and crocodilians. Glands are dermal and localized to areas such as the underside of the hindlimb. Their products serve for protection or reproduction. Reptiles do not possess a subcutaneous adipose layer. Of note: reptilian skin heals much more slowly than mammalian skin, often taking about 6 weeks for the defect to be fully restored.

Birds

Recall once again the branch point just before dinosaurs at which the lines that would eventually produce birds and mammals diverged. This is the juncture at which several analogous innovations take place; although avian and mam-malian skin differ, these commonalities are what count. Avian epidermis is thin and pliable. It has 4 well-defined lay-ers: statum basale, stratum intermedium, transitional layer, and the stratum corneum. In mammals, the two middle layers become stratum spinosum and stratum granulosum. The epidermis is lipogenic. The degree of keratinization is less than in mammal. This facilitates rapid cooling while still maintaining facultative waterproofing…i.e. the preening of feathers to distribute oil. This relatively "leaky" epidermis is because of the high metabolic rate, increased heat production

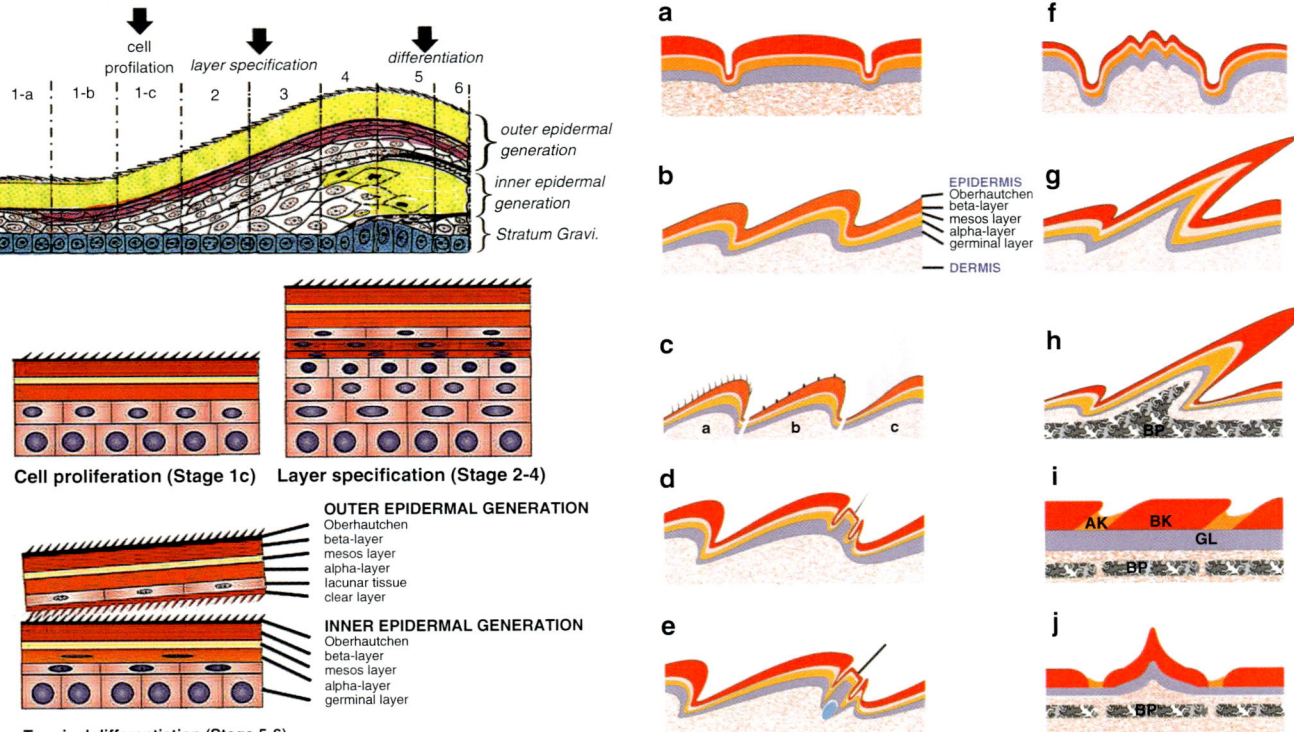

Fig. 11.97 RIGHT: Shedding of reptilian skin involves a reduplication of the layers and a separation plane of lacunar tissue plus a clear layer. LEFT: Scales (resting phase) are shown in multiple layers with names labeled in panel B. (A) Non-overlapping tuberculate type scales. (B) Overlapping scales commonly seen in squamates. (C) Variations of microstructures from the *Oberhäutchen layer* illustrating short spines in a, b and long setaes in c (such as those in the adhesive pad lamellae in geckos, Fig. 4B). (D) Pits on the scales of anole, gecko and iguana (mainly epidermal sensory organs; Fig. 4E, F). (E) Tactile sensory organ on the hinge side of a scale in Agama. Some follicle-like structures have clustered dermal cells associated to their base; Fig. 4G). (F) Scales with ridges are seen on the back of skink or the neck of anole. (G) Frills, or very elongated scales, are seen on the back of iguana

(Fig. 3B). (H) The horn on the head of chameleon contains a bony element core (osteoderm). (I) Scales on the limb of crocodilians show only minor overlapping. (J) Keeled scales with a central, elevated corneous ridge are seen on the dorsal body of crocodilians and some armored agamid lizards (e.g. Australian spiny desert lizard or molok). Legends: a, fine 'hair' on scales of anoles; b, Micro-ornamentation on scales of snakes; c, Toe pad of anole or gecko;*, dermal cells clustered at the base of sensory organs in Agama; AK, beta-keratin; BK, beta-keratin; BP, bone element. [Reprinted from Chang C Wu P, Baker R E. Reptile scale paradigm: Evo-devo, pattern formation and regeneration. Int J Develo Biol 2009; 53(5–6): 813–826. With permission from The International Journal of Developmental Biology]

during flight, and insulation by feathers. There are neither sweat glands nor sebaceous glands.

Birds have subdermal white fat. The overlying dermis layer is thicker than the epidermis and more complex than in amphibians or reptiles, having sensory nerves, blood vessels, and smooth muscle. The extension of neural control implies a proliferation of neural crest cells, required for sympathetic control. This is accompanied by the differentiation of neural crest-related pre-pericytes into smooth muscle cells. The crowning characteristic of bird skin is the presence of feathers with a highly vascularized dermal core and control via dermal muscle. Although birds do not have a SMAS layer for second arch controlled muscles, these findings indicate the dermal and subdermal anatomy of birds may be due to the subcutaneous spread of embryologic elements (from the second arch?) superficial to the underlying muscle layer (Fig. 11.98).

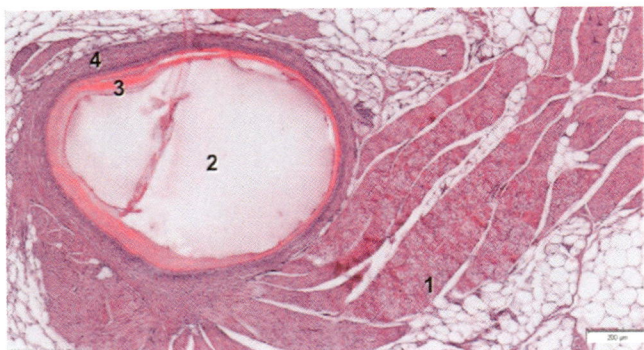

Fig. 11.98 Feathers are produced in feather follicles. The hollow basal port of the feather shaft that is implanted within the follicle is called the calamus. 1. Muscuculus pennarum 2. air space 3. Calamus wall 4. feather follicle wall. [Reprinted from Histology of Birds. Retrieved from http://www.histology-of-birds.com/galleries.php?id=32&v=. with permission from Ghent University]

Mammals (Figs. 11.99, 11.100 and 11.101)

Integument of mammals the multiple innovations that appear to recapitulate what happened in the avian line with greater overall complexity. Mammalian epidermis has five distinct layers with new cell types including Langerhans cells for antigen processing and neural crest-derived Merkel

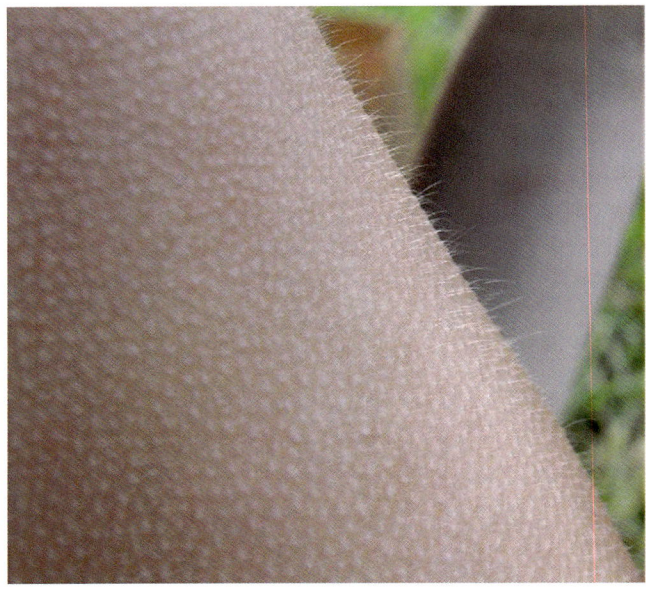

Fig. 11.99 Cutis anserine, "goosebumps". Pilomotor response indicates diffuse location of *epidermal placodes*. [Reprinted from Wikimedia. Retrieved from https://commons.wikimedia.org/wiki/File:2003-09-17_Goose_bumps.jpg. With permission from Creative Commons License 3.0: https://creativecommons.org/licenses/by-sa/3.0/deed.en]

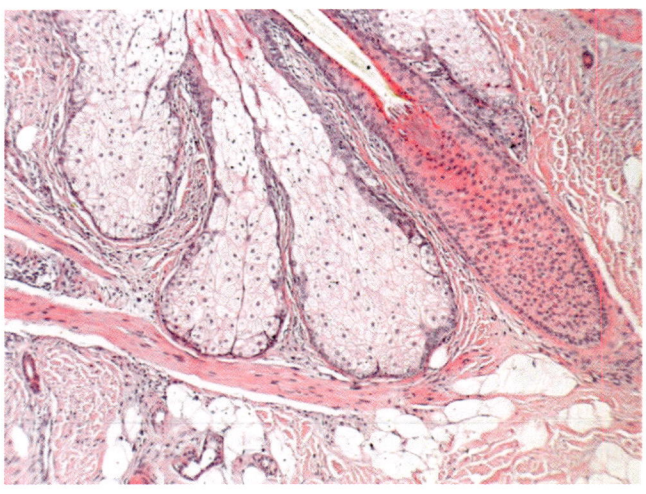

Fig. 11.100 arrector pili. Arrector pili muscle can be seen coming in from the left and inserting into the hair shaft. Its relationship to the pilosebacous unit explains how contraction of the muscle can cause sebum to be expressed. Arrector pili is under SANS control. [Reprinted from Wikimedia. Retrieved from: https://commons.wikimedia.org/wiki/File:Base_of_Pilosebaceous_Unit_10x.JPG]

display

Fig. 11.101 Chimpanzees display with arrector pili. [de Waal, Frans. Chimpanzee Politics: Power and Sex among Apes. pp. 130. © 2007 Frans de Waal. Reprinted with permission of Johns Hopkins University Press]

cells that function as mechanoreceptors. Chromatophores also arise from neural crest and populate the deeper strata of the epidermis producing pigment that is passed directly up to the surface or into the hair shafts. Dermis in mammals has two distinct layers: the papillary dermis which projects into the overlying epidermis and the reticular dermis. Blood vessels within the mammalian dermis form a subepidermal rete but do not extend into the epidermis itself. Mammals have a trifecta of glands for temperature control, waterproofing, lubrication, and lactation. Finally, dermal bone formation is retained in mammals and is used to synthesize the membranous bones of the skull, the scapula, and part of the clavicle.

The evolution of hair is a defining characteristic of mammals. The actual impetus for this innovation is unknown. One hypothesis for hair is for temperature control via insulation, as the presence of turbinates in synapsids during the late Permian is considered as evidence for endothermy. Another model considers whether small "protohairs" emerged between scales as tactile sensory devices. Therapsids, as direct precursors to the mammalian line, show evidence of pitting in the facial zones of the skull potentially associated with vibrissae.

Innovations in Skin Evolution: Strategies for Replacing the Epidermis

Among amniotes, the epidermis interacts with the harsh terrestrial environment, which can produce much wear and tear. A single-layered epithelium cannot accommodate this treatment. How to maintain an intact integument via epidermal homeostasis of stem cells, transient amplifying cells, and differentiated cells was an evolutionary challenge. The solution was a multi-layered stratified epithelium. But this layer would inevitably have to be replaced. To this end, early reptiles have developed two different strategies.

- Episodic shedding as seen in snakes and lizards (also seen in crustaceans, among invertebrates).
- Continuous renewal as seen in mammalian epidermis. This model has also been adopted by most chelonians and crocodilians.

Episodic shedding: How to replace the epidermis.

Snakes and lizards achieve epidermal replacement using an episodic strategy. This allows the formation of multiple different layers which offer different and more complete protection. Thus, the basal layer gives rise to different supra-basal layers at different times. The cost is that at a particular point, the old epidermis has to be shed, leaving the animals vulnerable. In snakes, this usually occurs simultaneously across the whole body, so the molted skin can be in one sheet. In lizards, this can occur in patches.

In the perfect resting phase, the reptilian epidermis is typically composed of four layers of keratinocytes, which have fully differentiated. Cells are dead but the mechanically strong keratin scaffolds remain. In order, from outside in, these layers are known as the *Oberhäutchen* layer, β-layer, mesos layer, and α-layer. At the base is one layer of live cells, that is, the basal layer or (germinal layer). In the lizard epidermis, these different layers are generated and specified in the temporal order. A keratinocyte will become one layer or the other. This contrasts with the mammalian epidermis where a keratinocyte must go through successive stages before being sloughed.

Birds and mammals also use this episodic strategy to generate complex cell types in feathers and hairs in cycles. Since stem cells and actively proliferating cells reside in the follicle, which is open to the external surface only via a small orifice, this does not lead to the problem of vulnerability during shedding. In the inter-follicular epidermis, the continuous replacement strategy is still used. Thus, by combining continuous and episodic strategies and by positioning each appendage with proper spacing, the integument is well protected. Furthermore, different skin regions can display appendages with different characteristics, lengths, and colors, depending on need.

Continuous Renewal

In mammals, the basal epidermal layer generates cells that make up the spinous, granular, and corneal layers. These different layers represent different stages of keratinocyte life, and a single keratinocyte develops through all these stages in.

In mammals, the basal epidermal layer generates cells that make up the spinous, granular, and corneal layers. These different layers represent different stages of keratinocyte life, and a single keratinocyte develops through all these stages in succession. The basal layer, therefore, generates new cells, which then differentiate and become supra-basal cells (basal layer → different supra-basal layers sequentially). As cells become more differentiated, they are displaced more externally to serve as the first line of defense. Eventually, they are worn and sloughed off. This same strategy is seen in alligator scales.

The Invention of Hair

The tongue is considered as one of the integuments. By comparing the topology and keratin types of scales and hairs with those of filiform papillary taste buds, gradual changes in the expression of keratins in the inner surface of the scale epidermis might have led to a transformation of the scale structure from a hemicylindrical form to the more completely cylindrical structure found in hairs. This progression could have been the origin of hair follicle invagination.

Another view focuses on the dermal side [25]. The variation in areas of dermal-epidermal interactions in the skin during evolution in different amniotes has been hypothesized to have led to the origins of scales, hairs, and feathers. According to this hypothesis, based on extensive comparative observations in reptiles, birds, and mammals, reptilian scales show extended papillae beneath the outer scale surface where the hard, β-keratin layer is formed. Therefore, no dermal condensation is generally seen during scale morphogenesis in reptiles. Based on the supposed derivation of feathers from scales, it is hypothesized that the progressive reduction of the outer surface of reptilian scales has restricted the morphogenetically active dermis previously associated with these areas, into smaller "niches", forming cell condensations. These have become dermal papillae. Hairs are hypothesized to have evolved when morphogenetically active mesenchyme invaginated near hinge regions of scales of reptilian ancestors (synapsids), a development that led to the formation of the dermal papilla.

Mammalian Skin Appendages: Hair Development Provides Evidence for the Embryonic Ectomesenchymal Precursor Cells in the Evolution of Mammalian Skin

Consider the phenomenon of "gooseflesh," the sudden elevation of body hair and scalp in response to sudden stress. The erection of the hair shafts is accompanied by a myriad of small swellings, one beneath each shaft (Fig. 11.99). This condition reveals the action of small fibers of smooth muscle, the *arrector pilae*, which span from the hair bulb to the undersurface of the overlying dermis. The muscle fibers are under SANS control (Fig. 11.100). They differentiate from ectomesenchymal precursor cells that will give rise to pericytes and white fat. Recall from our previous discussion that these cells arise from the same region of the neural fold as the neural crest. It is likely that the two cells types are cousins, sharing common surface markers but having different developmental fates. Hair erection is used to great effect by animals, such as chimpanzees, to display aggression (Fig. 11.101).

Piloerection reveals the presence of another structure: each hair shaft emerges from an individual epidermal prominence; these "bumps" arise fossils of the ancient ectodermal placodes that are to the hair. The superficial fascia in some mammals is associated with a striated muscle layer, the panniculus carnosus. In humans, this is considered homologous to platysma and the dartos muscle of the scrotum. The muscle-bearing nature of SIF is limited to SMAS (Fig. 11.102). In horses, this layer is useful for shaking off those pesky flies.

Placodes are actually fundamental for hair formation. These non-neural ectodermal cells are distributed throughout the hair-bearing epidermis. In the development of craniofacial skin and sensory organs, craniofacial placodes occupy specific locations along the neural fold and constitute populations distinct from those of neural crest (Fig. 11.103). But what can be said about the existence of placodes in skin? One model can postulate that non-neural ectoderm has two distinct populations which are admixed: garden-variety epidermal cells and placodal cells. A second model is one in which the epidermis contains a single type of non-neural ectodermal cell. Signals from mesenchymal cells in the evel-

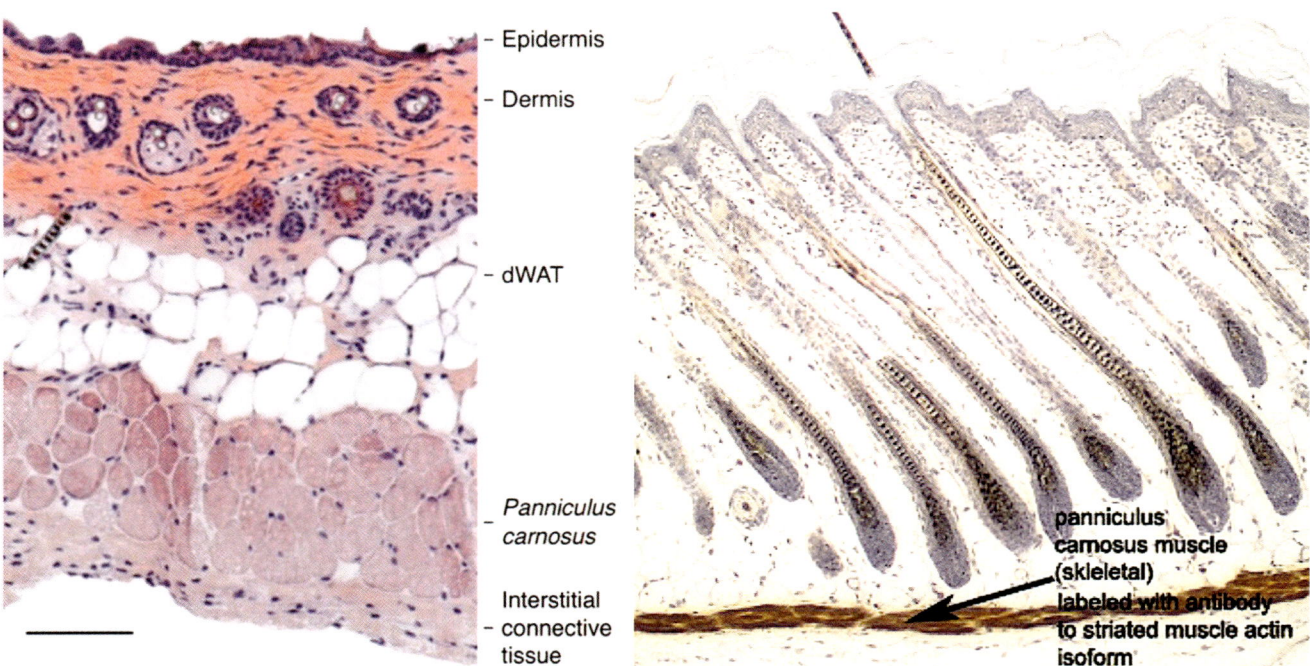

Fig. 11.102 Panniculus carnosus. Immunohistochemical labeling of newborn mouse skin for the striated muscle actin isoform only labels the panniculus carnosus muscle (PC) below the fat layer. PC is the likely precursor of superficial fascia in humans. It is not the source of arrector pili muscles. Left: [Reprinted from Izeta A, McCullagh KJA, López de Munain A, et al. The panniculus carnosus muscle: an evolutionary enigma at the intersection of distinct research fields. J Anat 2018; 233: 275–288. With permission from John Wiley & Sons.] Right: [Reprinted from Pathbase: http://eulep.pdn.cam.ac.uk/~skinbase/Hair_cycle/PANNICULUS_CARNOSUS_ANNOTAT.jpg with permission from Creative Commons License 3.0: https://creativecommons.org/licenses/by-nc-sa/3.0/]

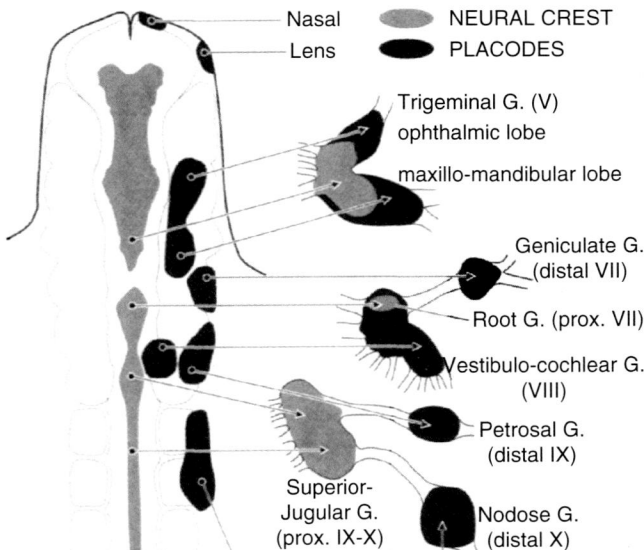

Fig. 11.103 Sensory ganglia develop from neural crest and placodes. Note that ectoderm within which placodes develop is *lateral* to that of the neural crest. V1 neurons arise from the ancient profundus placode which is in register with r1. In fishes, ancient tetrapods, amphibians and some reptiles, profundus is spatially distinct. It is exclusively sensory; but its function varies phylogenetically. In snakes V1 provides infrared thermoreception for locating prey by body temperature. In mammals the profundus placode merges with the trigeminal placode as the ophthalmic lobe of trigeminal ganglion. Thus, V1 skin is embryologically distinct from all other facial skin. It is devoid of facial hair. Boundary zones between V1 skin and neighboring territories such as V2 skin and V1 nasal mucosa are demarcated by hair-bearing structures, such as eyebrows and nasal hairs. [Reprinted from D'amico-Martel A, Noden D. Contributions of placodal and neural crest to avian cranial peripheral ganglia. Am J Anat 1983; 166(4): 445–468. With permission from John Wiley & Sons]

oping dermis seem more plausible. These cells are randomly distributed in the subepidermal plane and they produce induction signals that convert non-neural ectodermal cells from an epidermal to a placodal fate. These chemical signals from a multitude of mesenchymal sources compete with one another creating an evenly spaced pattern of placodes. This model seems well-supported by the sequence of events by which hair develops which takes place in four steps (Fig. 11.104).

- **Specification** Ectomesenchymal cells in the subepidermal tissues differentiate into fibroblasts.
- **First dermal induction** Subepidermal fibroblasts produce Wnt-11 and fibroblast growth factor (FGF); these, along with local BMPs induce a localized thickening of the epidermis, the placode. The location of these placodes may reflect the physical location of the ectomesenchymal-inducing cells or is simply a reflection of competing chemical concentration gradients.
- **Epidermal induction** Signals produced by the placode, such as Shh, cause subplacodal dermal cells to clump together into a dermal papilla.
- **Second dermal induction** Signals directed upward induce a downward proliferation of epidermal placodal cells into the papilla itself.

It should be noted that this process is not unique to hair formation. Non-neurogenic placodes are also involved in the formation of teeth, mammary glands, the lens, and feathers. Local matrix metalloproteinases produced in the dermis determine the ultimate fate of the

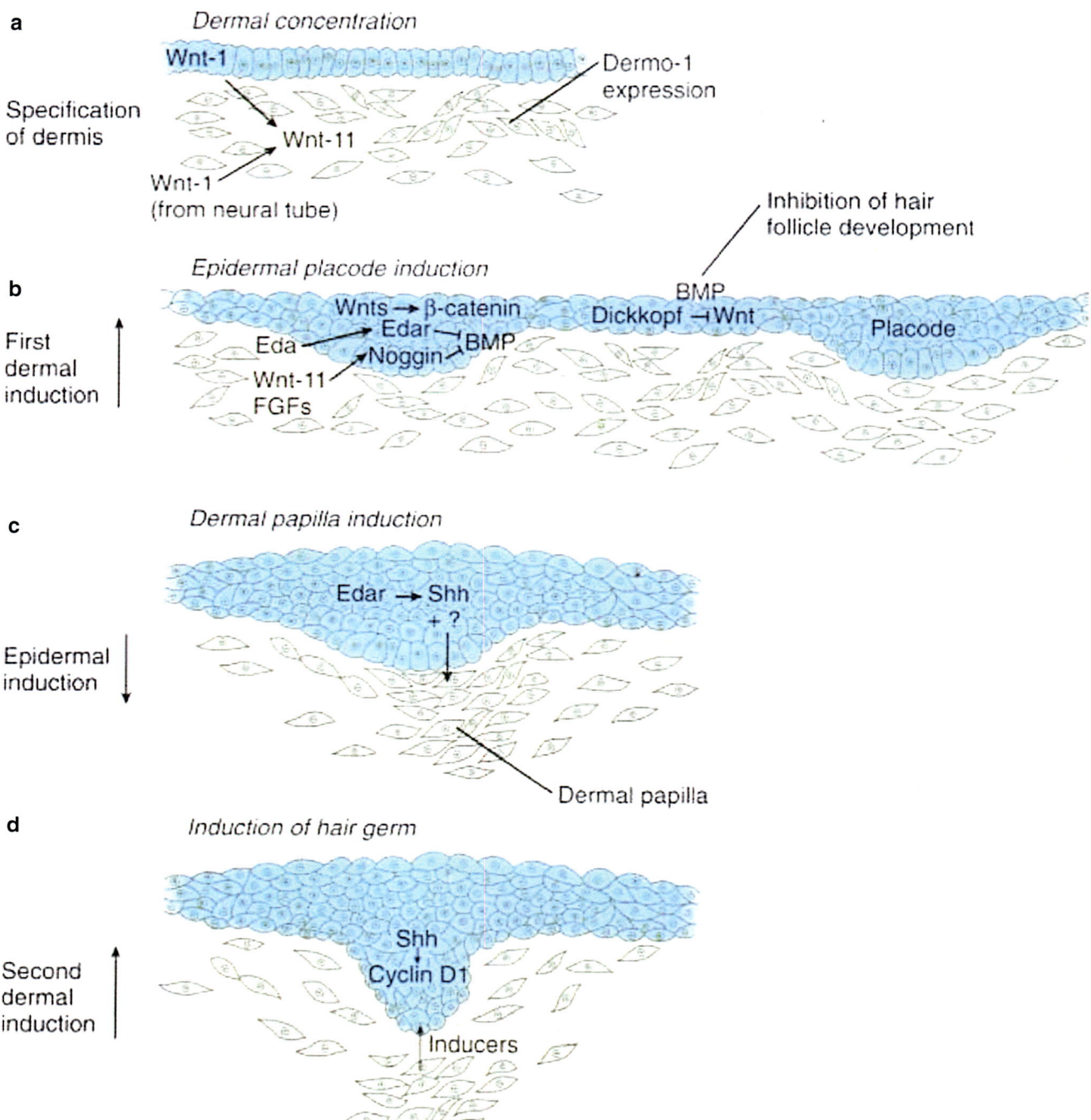

Fig. 11.104 Formation of hair follicle. Epidermal placodes the ecto-dermal source of hair shafts. The process involves 4 steps: (1) specification of subcunaeous tissue into dermis by Wnt1 inducing Wnt11; (2) primary dermal induction of placodes by dermal MSCs; (3) epidermal response via Shh to induce a papillary condensation in the dermis; and (4) production by the dermal papilla of factors inducing epidermal down-growth in the papilla in a "bell and cap" configuration. Key: BMP, bone morphgenetic protein, FGF, fibroblast growth factor; Wnt, *Wingless* genes; Eda, e*ctodysplasis*; Edar, *ectodysplasin* receptor Cyclin D, controls cell cycle—initated at G1 and drives the G1/S transition. [Reprinted from Carlson BM. Human Embryology and Developmental Biology, sixth edition. St. Louis, MO: Elsevier; 2019. With permission from Elsevier]

placodal structure. It should be noted that mesenchymal stromal cells derived from pericytes, including adipose-derived stromal vascular fraction cells produce a variety of MMPs (Fig. 11.105). In the case of hair development.

Interaction between the hair sheath with local MSCs results in the formation of an individual arrector muscle which extends upward to find an attachment site in the dermis (Fig. 11.106).

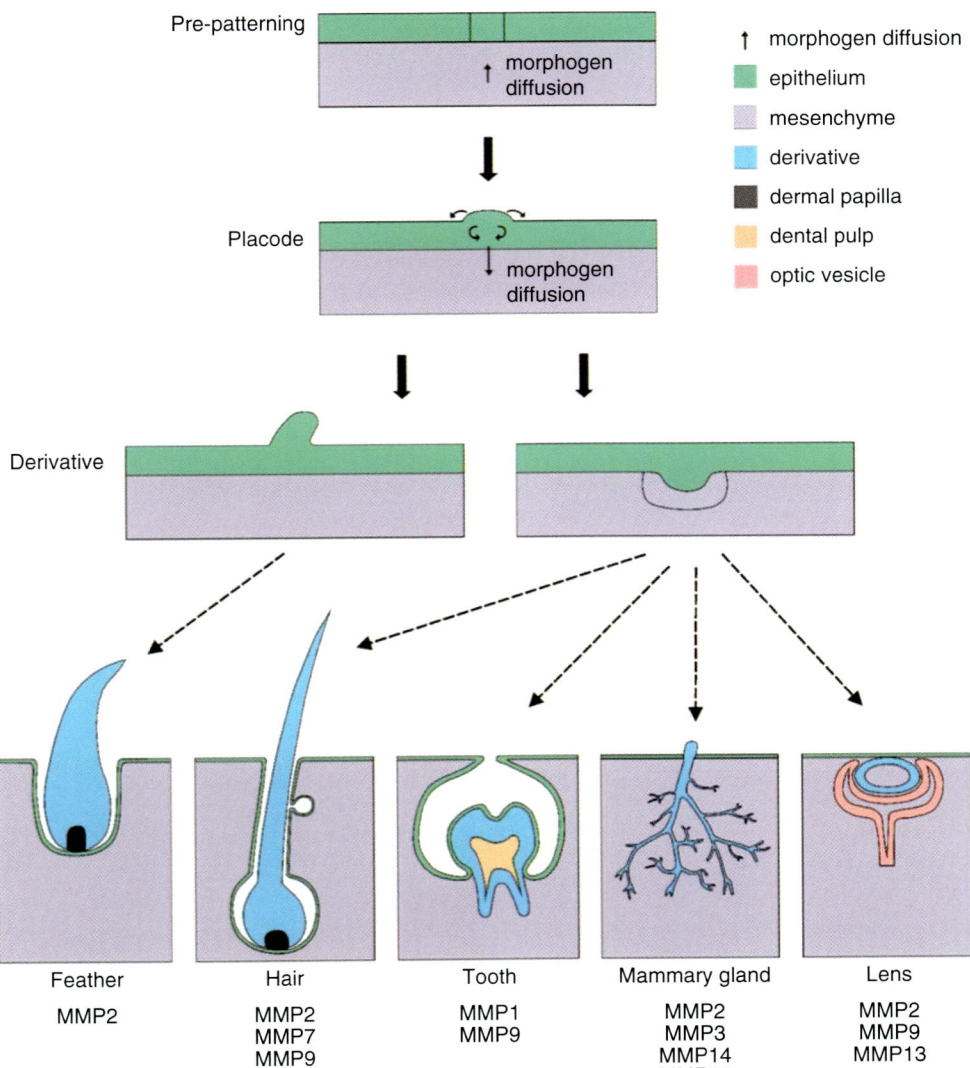

Fig. 11.105 Placode-derived structures begin development with pre-patterning of the epithelium, which involves morphogen diffusion from the mesenchyme to the epithelium. Next, a thickening of the epithelium forms a placode structure. Following the formation of a placode, morphogen signaling from the epithelium down to the mesenchyme leads to the formation of a derivative structure. Some MMPs are involved in some of these processes. [Reprinted from Drake PM, Franz-Odendaal TA. A potential role for MMPs during the formation of non-neurogenic placodes. J Developmental Biol 2018; 6: 20–33. With permission from Creative Commons License 4.0: https://creativecommons.org/licenses/by/4.0/]

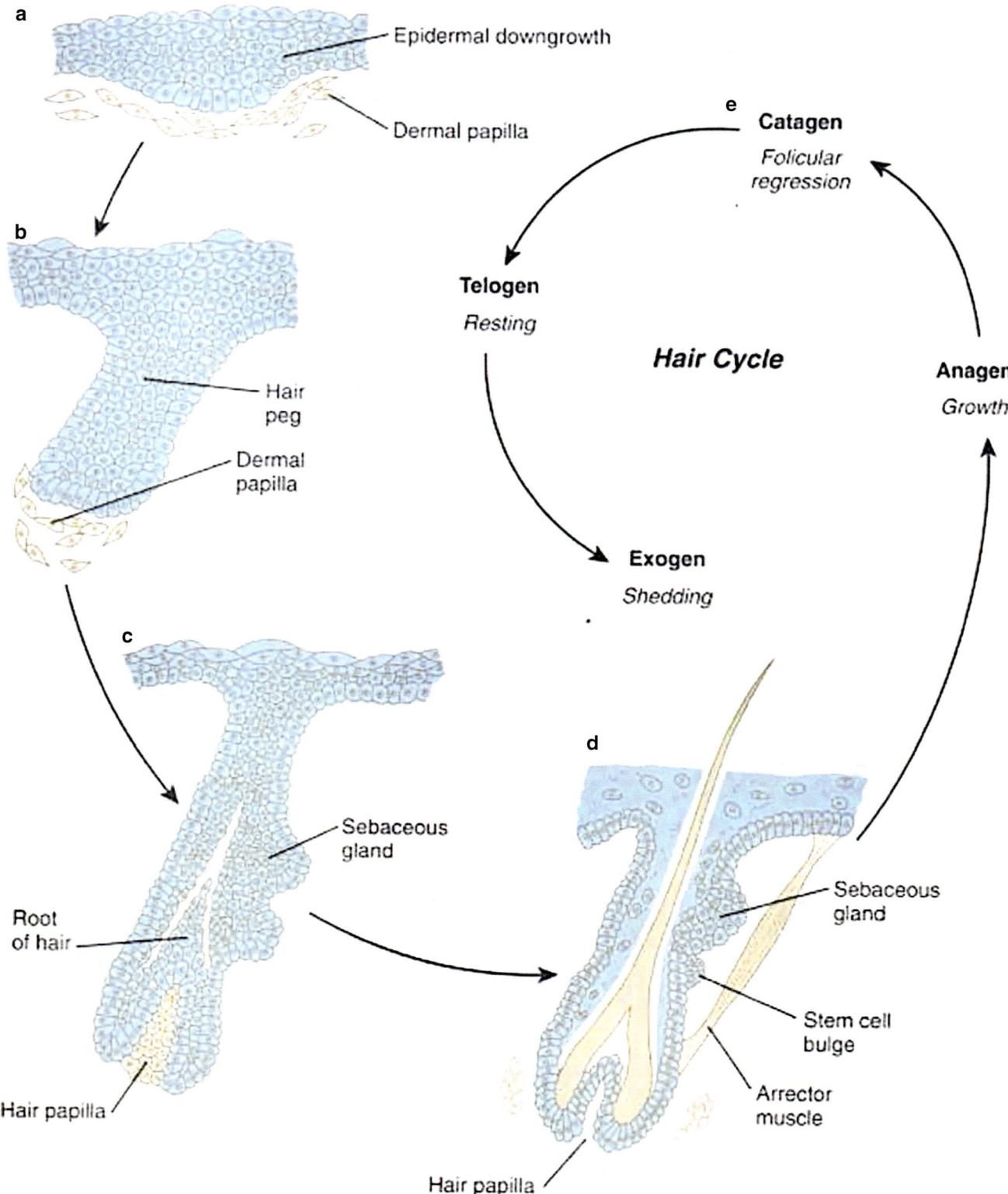

Fig. 11.106 Hair follicle differentiation. (**a**) 12 weeks; (**b**) hair primordium; (**c**) 16 weeks, hair peg; (**d**) 18 weeks, bulbous follicle; maturity; (**e**) hair cycle: anagen, catagen, telogen, exogen. Matrix metalloproteinases produced by dermal MSCs derived from prepericytes help induce the final hair structure. The smooth muscle arrector pili are of MSCs differentiation. Hair development is inseparable from the surrounding adipose tissue. [Reprinted from Carlson BM. Human Embryology and Developmental Biology, sixth edition. St. Louis, MO: Elsevier; 2019. With permission from Elsevier]

Coda: Evolutionary Aspects of Skin Appendages

Different appendages likely evolved from a primitive type of patterned skin. Odontodes (Od) are derived from both epidermis (E) and dermis (D) and are found in living chondrichthyans. Teleost scales (DSC) are mostly dermally derived and are commonly likely related to odontodes. Epidermal scales (ESCs), protofeathers (pF), feathers (F), and hairs (HFG) are proposed to be epidermally derived from a *common placodal stage*. The integument of sauropsids differs from that of synapsids by the presence of β-proteins and by the distribution of glands. Sebaceous glands are a critical element of the hair follicle: their proteolytic enzymes allow for the hair shaft to erupt. Differences between plantar and body integument are controlled by an ancient interaction between Shh and BMPs dating back to the emergence onto land (Figs. 11.107, 11.108, 11.109 and 11.110).

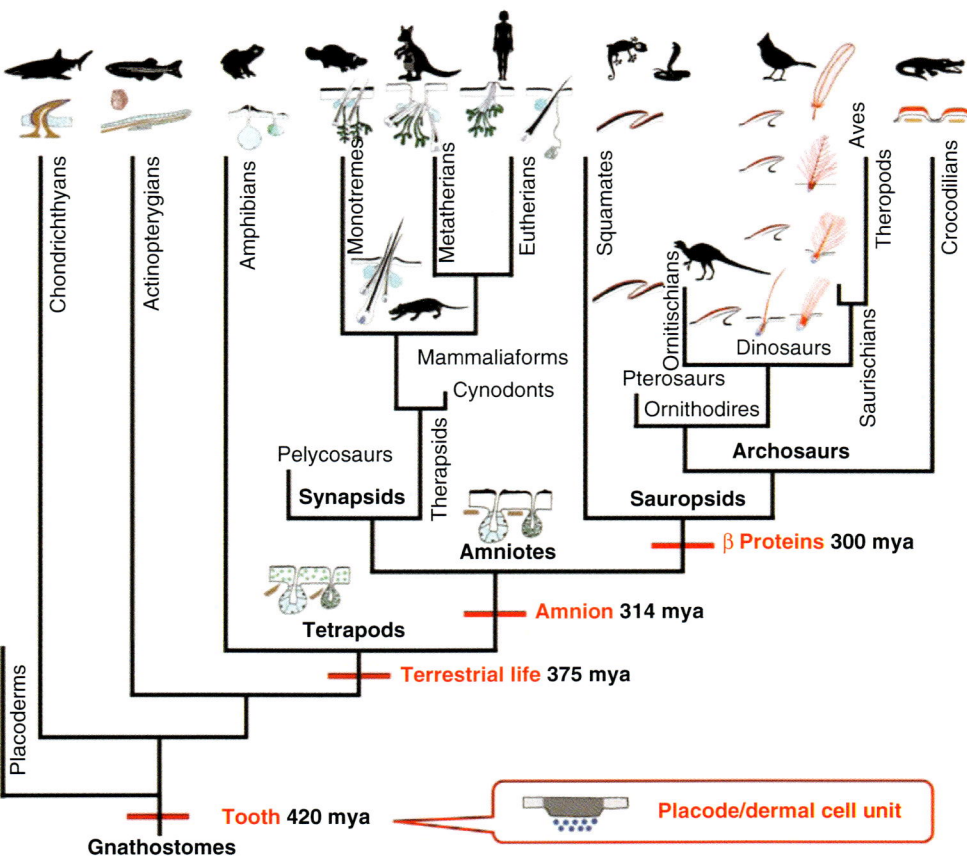

Fig. 11.107 Phylogeny of skin. Cladogram of the major groups of vertebrates. The unit composed of a placode and its associated dermal cells is ancestral and has been conserved during evolution. It represents the initial step for every type of cutaneous appendage in living vertebrates. Note that fossilized hair, primitive feathers, and epidermal scales were found dating back to the Mesozoic era (252–66 mya). A number of clades have been omitted for simplicity, and the timescale is not accurate. Red bars indicate evolutionary novelty. Note that beta-proteins (formerly beta-keratins) are an evolutionary novelty of sauropsids, while alpha-keratins appeared with the first vertebrates. [Reprinted from Dhouailly D, Godefroit P, Martin T, et al. Getting to the root of scales, feathers and hair: As deep as odontodes? Exp Derm 2017; 28(4): 503–508. With permission from John Wiley & Sons]

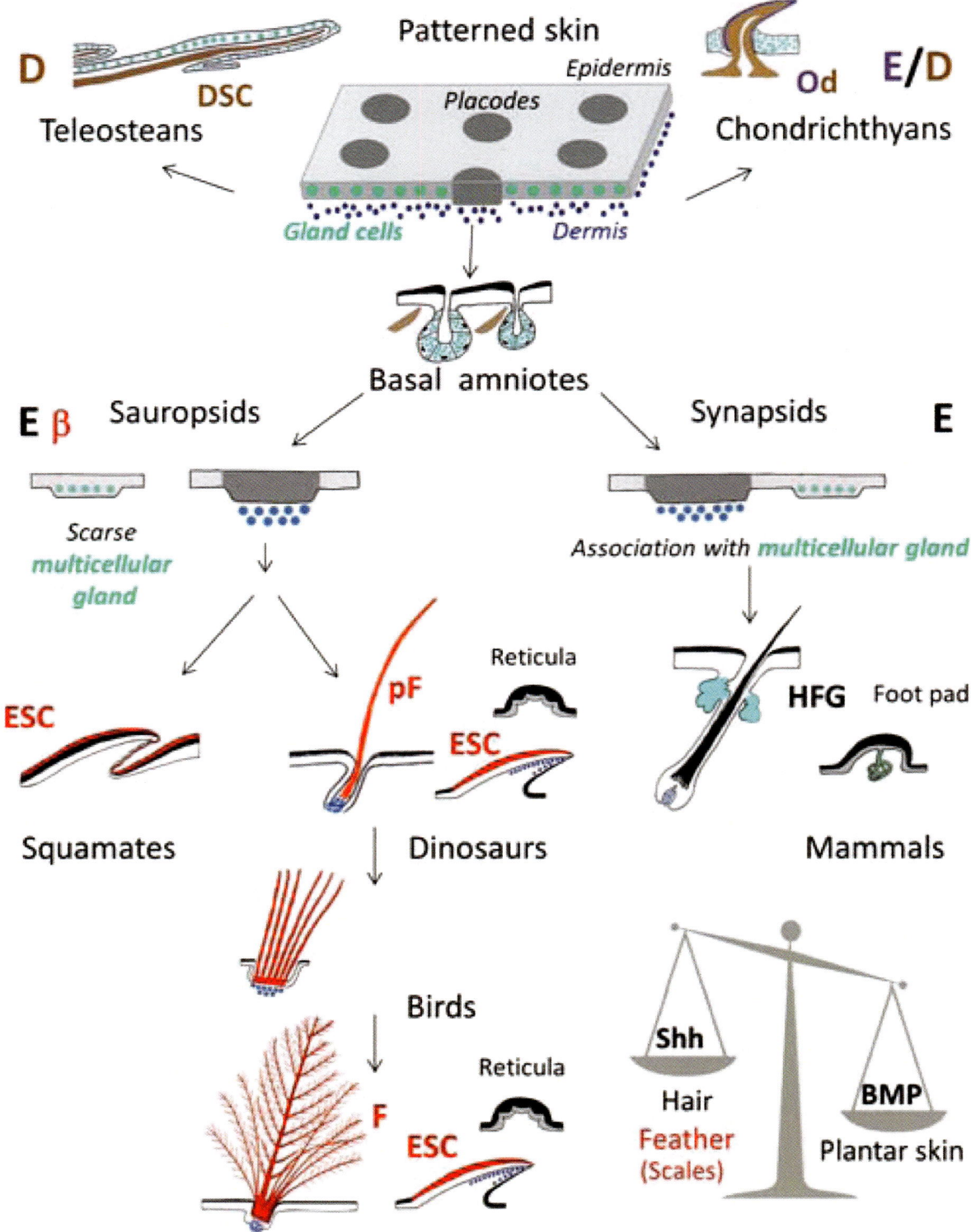

Fig. 11.108 Evolutionary relationships between integument appendages. The different appendages might have evolved from a primitive patterned skin. Odontodes (Od) are derived from both epidermis (E) and dermis (D). This type of appendage is still present in living chondrichthyans. Teleost scales (DSC) are mostly dermally derived and are commonly believed to be related to odontodes. Epidermal scales (ESCs), protofeathers (pF), feathers (F), and hairs (HFG) are proposed to be epidermally derived from a common placodal stage. The integument of sauropsids differs from that of synapsids by the presence of β-proteins (red) and by the distribution of glands (green). Sebaceous glands are an integral part of hair follicle as their proteolytic enzymes are required for the hair shaft to emerge (see text and supporting information Data S9). A balance between Shh and BMP signaling, probably established at the onset of terrestrial life, controls the difference between plantar and body integument. [Reprinted from Dhouailly D, Godefroit P, Martin T, et al. Getting to the root of scales, feathers and hair: As deep as odontodes? Exp Derm 2017; 28(4): 503–508. With permission from John Wiley & Sons]

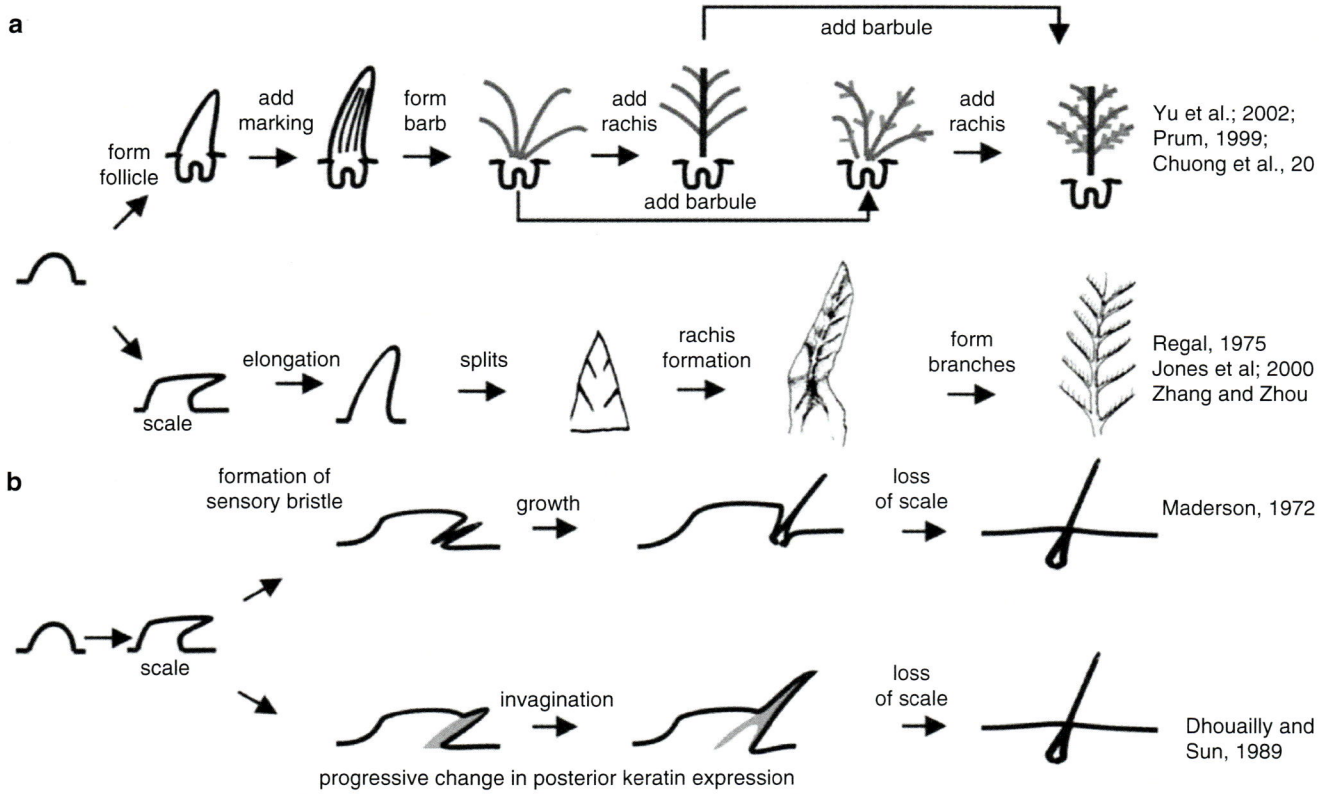

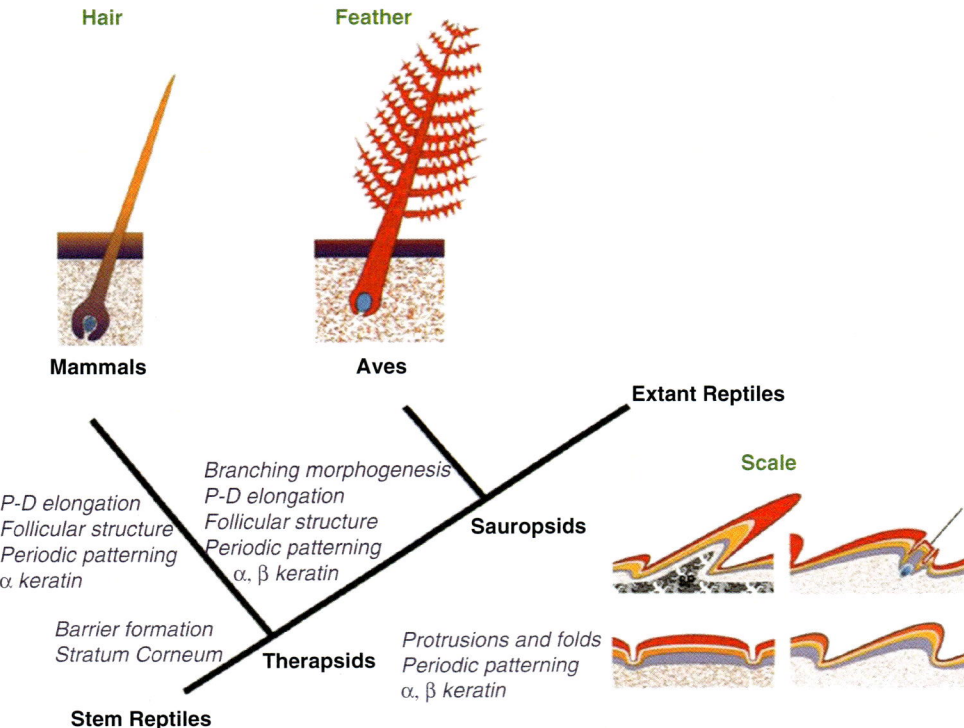

Fig. 11.109 Possible models for feather and hair evolution. (**a**) Evolution of feathers. Experiments show that the barb—rachis model is correct. (**b**) Possible models for the evolution of hairs. [Reproduced with the permission of UPV/EHU Press from Wu P, Hou L, Plikus M, et al. Evo-Devo of amniote integuments and appendages. Int. J. Dev. Biol 2004;48:249–270]

Fig. 11.110 Potential relationship among amniote skin appendages and key evolutionary novel events. The reptilian integument shows some basal characteristics in comparison to mammalian and avian integuments. Mammals evolved about 225 million years ago from Therapsid-type reptiles, while birds evolved about 175 million years ago from archosaurian Saurospids-type reptiles. Key events that led to evolutionary novelty are shown in blue italic characters. [Reproduced with the permission of UPV/EHU Press from Chang C, Wu P, Baker RE, Maini PK, Alibardi L, Chuong C-M. Reptile scale paradigm: Evo-devel, pattern formation, and regeneration. In J Devel Biol 2009; 53(5–6): 813–826]

Speculation: Hair Distribution and Baldness

From time immemorial hair loss in males has been recognized to follow fixed anatomic patterns, as can be seen by models such as the modified Norwood scale, (Fig. 11.111) Note the following types: (1) isolated V2 expanding in the anterior temple, (2), isolated V1 extending posteriorly, (3) isolated V3, (4) V1-V3 extending back to C2-C3 territory. It is productive to view these from the standpoint of neural crest distribution patterns according to scalp innervation (Fig. 11.112). The ectomesenchymal cells responsible for placode induction are themselves likely to follow the same pattern as neural crest cells. Furthermore, as the follicles are maintained by interactions with the dermal environment changes in trophic signals in localized populations can explain patterns of hair loss.

The sensory neuroananatomy of the scalp faithfully reflects developmental distinctions between hypaxial and epaxial mesenchyme. The ear is a good case in point as it demonstrates these boundaries quite clearly (Figs. 11.12, 11.56). Posterior pinna is mainly hypaxial but its upper 1/3 contains cartilage #4 of the second arch.

A useful way to think about scalp is to visualize the person in the prone position. We note the breaking point of the retroauricular hairline beginning with epaxial C3 lesser occipital nerve. The anterior temple is V2 with a ventral part being hairless and the dorsal section hair-bearing. These observations do not fit with definitions of the first arch as its derivatives are considered strictly hypaxial. But V2 and V3 in the scalp are clearly supplying non-arch r2-r3 struc-

tures. Furthermore, V1 bears no relationship to first arch. Once again, we are reminded that facial and scalp skin develops from elements that do not participate in pharyngeal arch development. The more superficial contents of the

Fig. 11.111 Male pattern baldness. All 4 patterns shown here represent degrees of failure of r2 and/or r3 neural crest cells to advance to the apex of the scalp. Skin with r1 neural crest alone cannot manufacture hair. This property may pertain to c2-c3 skin as well. [Reprinted from Koo S-H, Chung H-S, Yoo E-S, Park S-H. A new classification of male pattern baldness and a clinical study of the anterior hairline. Aesth Plast Surg 2000; 24(10): 46–51. With permission from Springer Nature]

Fig. 11.112 Innervation of facial skin and scalp. The presence or absence of hair, the defining characteristic of forehead skin versus scalp may have to do with intermingling of r1 neural crest precursors in the V1 zone with those from r2 and r3. This may mean that central pattern baldness is a problem affecting the extensions of r2/r3 into r1. If the population becomes exclusively r1 it can no longer support scalp hair and central pattern baldness appears. This places the blame for other patterns of baldness upon those specific populations. [Reprinted from Lewis, Warren H (ed). Gray's Anatomy of the Human Body, 20th American Edition. Philadelphia, PA: Lea & Febiger, 1918]

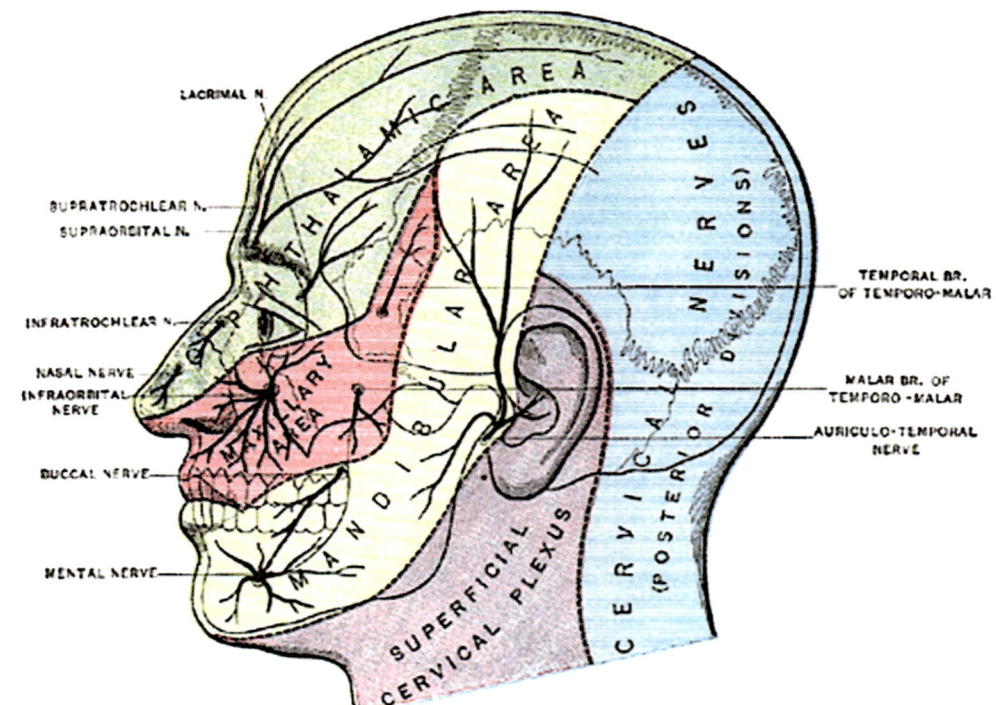

arches (muscle and fascia) merely fill in the space below a skin envelope.

We can get some further clues to this biology by looking at the position of eyebrows and nasal hairs. Eyebrows are positioned over the prefrontal and postfrontal bone fields. Recall that formed the upper tetrapod orbit but became absorbed into the frontal bone complex. Eyebrows have a distinct neuromeric zone with distinct behaviors. Hypothyroidism produces hair loss localized to the lateral third of the eyebrow, the zone perfused by StV1 lacrimal artery (Fig. 11.113). At either extreme of the eyebrows, that is, at the sutures of the ancient pre- and postfrontal bone fields, dermoids can occur, indicating a pinching off of periosteum filled with mesenchyme at those sites. Furthermore, the skin of the upper eyelids is distinct from that of the forehead. The former is strictly r1 whereas the latter is frontonasal skin. In certain individuals, the medial margins of the brow come together in the midline over the nose. This can be ethnic or seen in syndromes such as Waardenburg and Cornelia de Lange. Hair growth seems disorganized over the glabella (Fig. 11.114).

 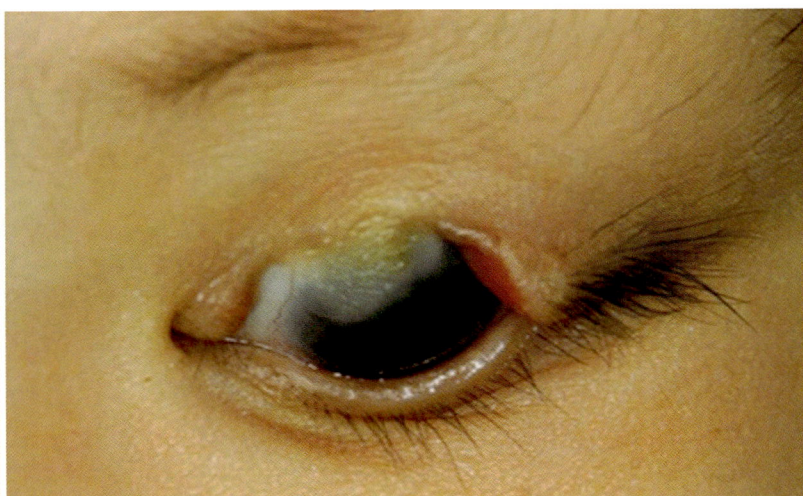

Fig. 11.113 Eyebrow pathologies reveal neuromeric boundary zones. LEFT: Loss of thinning out of eyebrows can be from system disease. The lateral margin is selectively affected in hypothyroidism, atopic dermatitis, or leprosy. Selective hair loss with hypothyroidism affects the #9 zone based on the StV1 lacrimal neurovascular pedicle. This finding, described by the pioneer Belgian endocrinologist Ludovic Christian Hertoghe, bears his name and has been putatively associated with Queen Anne of Denmark has been (uncertainly) reputed to have suffered from hypothyroidism. RIGHT: Loss of eyebrow hair may be congenital as well and can bey associated with dysplaia of eyelashes in the same zone. It is seen here in the medial margin of the #10 cleft zone. The mid zone of the eyelid is also cleft and a cutaneous pterygium is present. The #10 cleft zone centers around the StV1 supraorbital neurovascular pedicle. It affects the center of the eyebrow and upper eyelid. The lower lid is supplied by the distal branches of StV2 anterior superior alveolar artery, that is, the infraorbital n/v pedicle. Note in the upper eyelid the accumulation of soft tissue, the abnormally long and coarse eyelashes, and the atrophic soft tissue leading upward from the eyelid to the eyebrow. Left: [Reprinted from Wikimedia. Retrieved from: https://en.wikipedia.org/wiki/Anne_of_Denmark#/media/File:Anne_of_Denmark_1605.jpg]. Right: [Reprinted from Shao C, Lu W, Li J, Yao Q, Fan X, Fu Y. Manifestations and grading of ocular involvement with Tessier number 10 clefts. Eye 2017; 31(8): 1140–1145. With permission from Springer Nature]

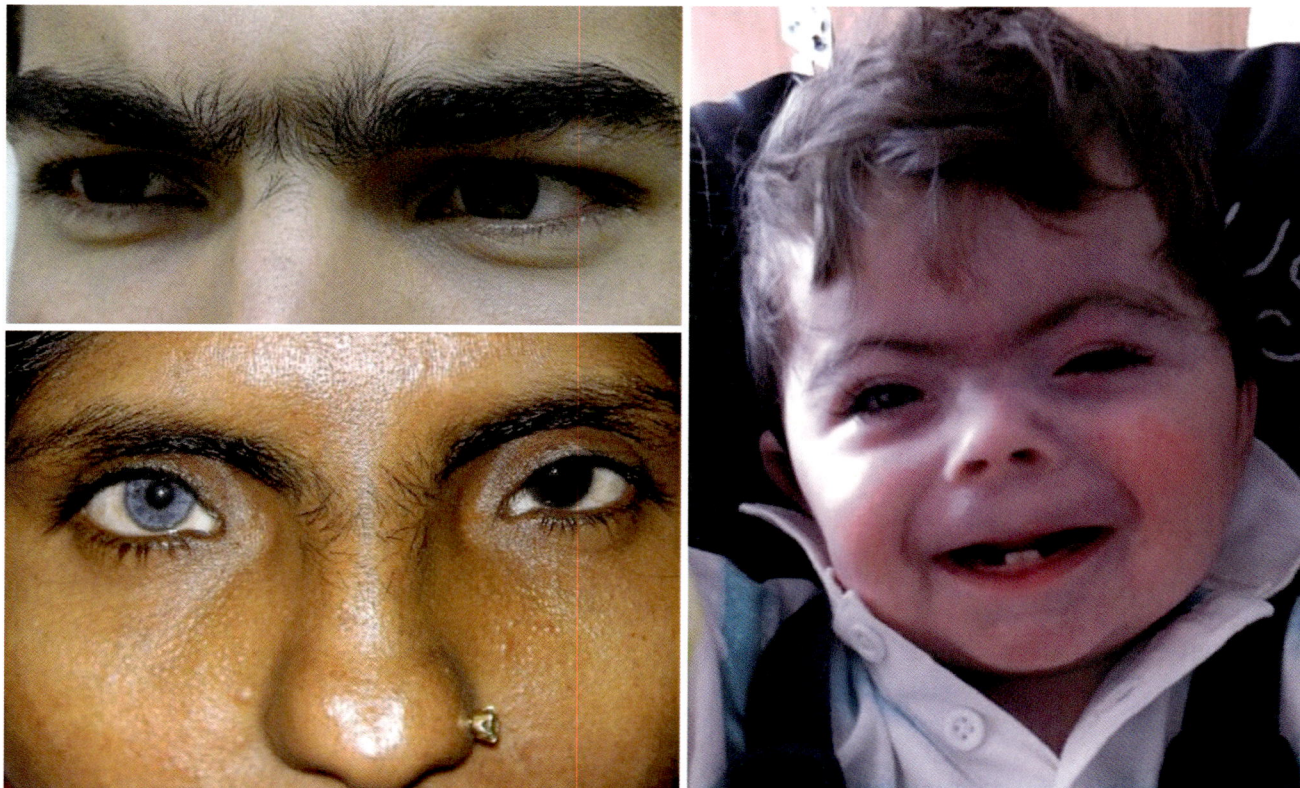

Fig. 11.114 Waardenberg syndrome. *Synophrys* (unibrow) is highly regarded in Iran. This condition is also seen in Waardenburg and Cornelia de Lange syndromes. Note the disorganized directionality of the hairs. Eyebrows demarcate two different embryologic zones of skin, forehead and upper eyelid. Top Left: [Reprinted from Wikimedia. Retrieved from: https://en.wikipedia.org/wiki/Unibrow#/media/File:Unibrow_Close_Up.jpg. With permission from Creative Commons License 3.0: https://creativecommons.org/licenses/by-sa/3.0/deed.en.].

Bottom Left: [Reprinted from Ghosh SK, Bandyopadhyay D, Ghosh A, Biswas SK, Mandal RK. Waardenburg Syndrome: a report of three cases. Indian J Dermatol,Venerol Leprol 2010; 76(5):550–552. With permission from Wolters Kluwer Medknow Publications.]. Right: [Reprinted from Wikimedia. Retrieved from: https://en.wikipedia.org/wiki/Cornelia_de_Lange_Syndrome#/media/File:Eli_CDLS.JPG. With permission from Creative Commons License 3.0: https://creativecommons.org/licenses/by-sa/3.0/deed.en]

Nasal hairs occur precisely at the external nasal valve between the nostril introitus and the caudal margin of the lower lateral cartilage. This marks the junction between nasal skin of the p5 zone and vestibular skin of the p6 zone. Due to V1 innervation, plucking these hairs induces tears (Fig. 11.115). Ear hairs can be vellus within the external auditory canal or coarse outside of the canal. Hypertrichosis is concentrated at the hypaxial c2-c3 lobule and involves the external surface and helical rim, spacing the internal skin of the auricle. This can reach extreme dimensions, referred to as *hypertrichosis lanuguiosa acquisita* and is found more frequently in men from India.

Final Thoughts

In this chapter we have explored the soft tissue coverage of the head and neck based on the neuromeric origins of its epidermis, dermis, and adipose tissue, This has necessitated some detail regarding details of innervation and blood supply. Care has been given to describe the fascial layers clearly, to define their neuromeric basis, and to rationalize the organization of the arterial systems internal and external carotid, stapedial, and subclavian which supply these regions. Attention has been given to the organization and biology of adipose tissue. Finally, phylogenetic and evolutionary aspects of mammalian skin are outlined.

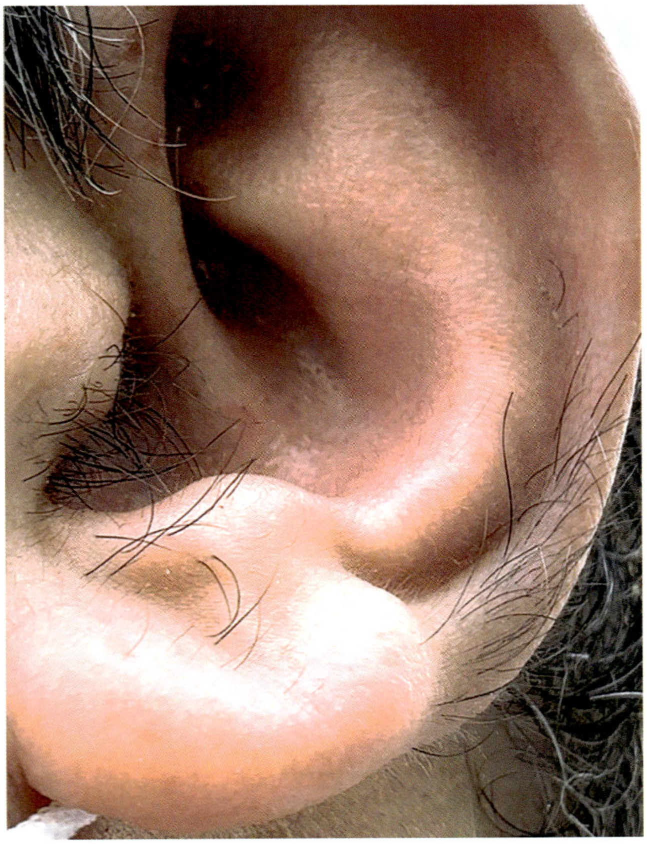

Fig. 11.115 Eyebrows and nasal hairs as boundary markers. Left: Reprinted from Wikimedia. Retrieved from: https://commons.wikimedia.org/wiki/File:Nasal_hair.jpg. With permission from Creative Commons License 3.0: https://creativecommons.org/licenses/by-sa/3.0/deed.en.]. Right: Reprinted from Wikimedia. Retrieved from: https://en.wikipedia.org/wiki/File:Long_ear_hair_on_man.png. With permission from Creative Commons License 1.0: https://creativecommons.org/publicdomain/zero/1.0/deed.en]

References

1. Jennet JP, Moody SA. Establishing the pre-placodal region and breaking it into placodes with distinct identities. Dev Biol 2014;389(1):13–27.

2. Patthey C, Schlosser G, Shimeld SM. The evolutionary history of vertebrate placodes I: cell type evolution. Dev Biol. 2014;389(1):82–97.

3. Pieper M, Ahrens K, Rink E, Peter A, Schlosser G. Differential distribution of competence for neural crest induction to non-neural and neural ectoderm. Development. 2012;139:1175–87. https://doi.org/10.1242/dev.074468.

4. Schlosser G, Patthey C, Shimeld SM. The evolutionary history of vertebrate placodes II. Evolution of ectodermal patterning Dev Biol. 2014;389(1):98–119.

5. Herbst K. Rare adipose diseases (RADs) masquerading as obesity. Acta Pharmacol Sin. 2012;33:155–72. https://doi.org/10.1038/aps.2011.153.

6. Attwell D, Mishra A, Hall CN, O'Ferrell FM, Dalkarra T. What is a pericyte? J Cereb Blood Flow Metab. 2016;36(2):451–5.

7. Armulik A, Genové G, Betsholtz C. Pericytes: developmental, physiological, and pathological perspectives, problems, and promises. Dev Cell. 2011;21(2):193–216.

8. Dore-Duffy P, Morphology CK, properties of pericytes. The blood-brain and other neural barriers. Methods Mol Biol. 2011;686:49–68. https://doi.org/10.1007/978-1-60761-938-3_2.

9. Daneman R, Zhou L, Kebede AA, Barnes BA. Pericytes are required for blood-brain barrier during embryogenesis. Nature. 2010;468(723):562–6. https://doi.org/10.1038//nature09513.

10. Caplan AI. Mesenchymal stem cells: time to change the name! Stem Cells Transl Med. 2017;6(6):1445–51. https://doi.org/10.1002/sctm.17-0051.

11. Caplan AI, Correa D. The MSC: an injury drugstore. Cell Stem Cell. 2011;9(1):11–5.

12. Brown S, Katz AJ. Adipose-derived stem cells. In: Atala A (ed). Textbook of regenerative medicine Elsevier, 3rd ed. 2019.

13. Caplan AI, Dennis C. Mesenchymal cells as trophic mediators. J Cell Biochem. 2006;98:1076–84.

14. Cahill EF, Kennelly H, Carty F, Mahon BP, English K. Hepatocyte growth factor is required for mesenchymal stromal cell protection against bleomycin-induced pulmonary fibrosis. Stem Cells Transl Med. 2016;5:1307–8.

15. Cristani B, Marchand-Adam S, Quesnel C, et al. Hepatocyte growth factor and the lung. Proc Am Thorac Soc. 2012;9(3):158–63. https://doi.org/10.1513/pats.201-202-018AW.

16. Ejaz A, Epperly MW, Hou W, et al. Adipose-derived stem cell therapy ameliorates ionizing irradiation fibrosis via hepatocyte growth factor-mediated transforming growth factor B down-regulation and recruitment of bone marrow cells. Stem Cells. 2019;37:791–802.

17. Kapur SK, Dos-Anjos Vilaboa S, Llull R, Katz AJ. Adipose tissue and stem progenitor cells. Clin Plast Surg. 2015;42(2):155–67.

18. Carstens MH, Correa D, Llull R, Gomez A, Turner E, Valladares LS. Subcutaneous reconstruction of hand dorsum and fingers for late sequelae of burn scars using adipose-derived stromal vascular fraction (SVF). CellR4 2015; 3(5): e1675-e1684.

19. Carstens MH, Perz M, Briceño H, Valladares S, Correa D. Treatment of late sequelae of burn scar fibrosis with adipose-derived stromal vascular fraction (SVF) cells: a case series. CellR4. 2017b;5(3):e204–20.

20. Zakhari JS, Zabonick J, Gettler B, Williams SK. Vasculogenic and angiogenic potential of adipose stromal vascular fraction cell populations in vitro. In Vitro Cell Dev Biol Anim. 2018;54:32–40. https://doi.org/10.1007/s11626-017-0213-7.

21. Leng Z, Zhu R, Hou W, et al. Transplantation of ACE2- mesenchymal stem cells improves the outcome of patients with COVID-19 pneumonia. Aging Dis. 2020;11(2):216. https://doi.org/10.14336/AD.2020.0228.

22. Rogers CJ, Harman RJ, Bunnell BA, et al. Rationale for clinical use of adipose-derived mesenchymal stem cells for COVID-19 patients. Int J Mol Sci. 2020;19:2532–45.

23. Carstens MH, Greco RJ, Hurwitz DJ, Tolhurst DE. Clinical applications of the subgaleal fascia. Plast Reconstr Surg. 1991;87(4):615–26.

24. Tolhurst DE, Carstens MC, Greco RJ, Hurwitz DJ. Surgical anatomy of the scalp. Plast Reconstr Surg. 1991;87(4):603–12.

25. Alibardi L. Perspectives on hair evolution based on some comparative studies on vertebrate cornification. J Exp Zool B Mol Dev Evol. 2012;318(5):325–43. https://doi.org/10.1002/jez.b.22447.

26. Poissonet CM, Burdi AR, Bookstein FL. Growth and development of human adipose tissue during gestation. Early Hum Dev. 1983;8:1–11.

27. Poissonet CM, Burdi AR, Garn SM. Chronology of adipose tissue and its appearance in the human fetus. Early Hum Dev. 1984;10:1–11.

28. Winkler E, Bull RD, Zlokovic BV. Central nervous systems pericytes in health and disease. Nat Neurosci. 2011;14(11):1398–405.

Suggested Readings

Ailhaud G. Development of white adipose tissue and adipocytes differentiation. In: Klaus S, editor. Adipose tissues. Georgetown: Landes Bioscience; 2001. p. 27–55.

Alexander RW, Harrell DB. Autologous fat grafting: use of closed syringe microcannula system for enhanced autologous structural grafting. Clin Cosmet Investig Dermatol. 2013;6:91–102.

Alexander RW. Understanding mechanical emulsification (Nanofat) versus enzymatic isolation of tissue stromal vascular fraction (tSVF) cells from adipose tissue: potential uses in biocellular regeneration medicine. J Prolother. 2016;8:e947–60.

Borrelli MR, Shen A, Lee GK, et al. Radiation-induced skin fibrosis: pathogenesis, current treatment options, and emerging therapies. Ann Plast Surg. 2019;83:S59–64.

Broughton M, Fyfe GM. The superficial musculoaponeurotic system of the face: a model explored. Anat Res Int. 2013;794682:5. https://doi.org/10.1155/2013/794682.

Caplan AI. Are MSCs pericytes? Cell Stem Cell. 2008;3(3):229–30. https://doi.org/10.1016/j.stem.2008.08.008.

Carstens MH, Gomez A, Correa D, et al. Non-reconstructable peripheral vascular disease of the lower extremity in ten patients treated with adipose-derived stromal vascular fraction cells. Stem Cell Res. 2017a;18:14–21.

Chang C, Wu P, Baker RE, Maini PK, Alibardi L, Chuong CM. Reptile scale paradigm: evo-devo, pattern formation and regeneration. Int J Dev Biol. 2009a;53(5–6):813–26.

Chang C, Wu P, Baker RE, Maini PK, Alibardi L, Chuong C-M. Reptile scale paradigm: evo-devel, pattern formation, and regeneration. In J Dev Biol. 2009b;53(5–6):813–26. https://doi.org/10.1387/ijdb.072556cc.

Dhouailly D, Godefroit P, Martin T, Nonchev S, Carguel F, Oftedal O. Getting to the root of scales, feather and hair: as deep as odontodes? Exp Dermatol. 2019;28:503–8. https://doi.org/10.1111/exd.13391.

Drake PM, Franz-Odendaal T. A potential role for MMPs during the formation of non-neurogenic placodes. J Dev Biol. 2018;6:20. https://doi.org/10.33090/jdb6030020.

Friedenstein AJ, Deriglasova UF, Kulagina NN, Panasuk AF, Rudakowa SF, Luriá EA, Ruadkow IA. Precursors for fibroblasts in different populations of hematopoietic cells as detected by the in vitro colony assay method. Exp Hematol. 1974;2(2):83–92.

Gesta S, Tseng Y-H, Kahn CR. Developmental origins of fat: tracking obesity to its source. Cell. 2007;131:242–56. https://doi.org/10.1016/j.cell.2007.10.004.

Giannandrea M, Parks WC. Diverse functions of matrix metalloproteinases during fibrosis. Dis Model Mech. 2014;7:193–203.

Guarrera M, Cardo P, Arrigo P, Rebora A. Reliability of Hamilton-Norwood classification. Int J Trichol. 2009;1(2):120–2.

Guo J, Nguyen A, Banyard DA, et al. Stromal vascular fraction: a regenerative reality? Part 2: mechanisms of regenerative action. J Plast Reconstr Aesthet Surg. 2016;69:180–8.

Hartman DA, Underly RG, Grant RI, et al. Pericyte structure and distribution in the cerebral cortex revealed by high-resolution imaging of transgenic mice. Neurophotonics. 2015;2(4):041402. https://doi.org/10.1117/1.NPh.2.4.041402.

Isaka Y. Targeting TGF-B signaling in kidney fibrosis. Int J Mol Sci. 2018;19:2532–45.

Kalev-Altman R, Monsonegro-Oman E, Sela-Donnenfeld D. The role of matrix metalloproteinases MMP2 and MMP9 in embryonic neural crest cells and their derivatives. In: Chakraborti S, Dhalla NS, editors. Proteases in physiology and pathology. Springer; 2017. p. 27–48. https://doi.org/10.1007/978-981-10-2513-6_2.

Khazaeni K, Rajati M, Shahabi A, Mashadi L. Use of sternocleidomastoid myocutaneous flap based on the sternocleidomastoid branch of the inferior thyroid artery to reconstruct extensive cheek defects. Aesthet Plast Surg. 2013;37(6):1167–70. https://doi.org/10.1007/s00266-013-0216-z.

Ligia L, et al. Multiple doses of adipose tissue-derived mesenchymal stromal cells induce immunosuppression in experimental asthma. Stem Cells Transl Med. 2020;9:250–60. https://doi.org/10.1002/sctm.19-0120.

Mitz V, Peyronie M. The superficial musculo-aponeurotic system (SMAS) in the parotid and cheek area. Plast Reconstr Surg. 1976;58(1):80–8.

Nguyen A, Guo J, Banyard DA, et al. Stromal vascular fraction: a regenerative reality? Part 1: current concepts and review of the literature. J Plast Reconstr Aesthet Surg. 2016;69:170–9.

Mok K-W, Saxena N, Heitman N, et al. Dermal concentrate niche specification occurs prior to formation and is placode progenitor dependent. Dev Cell. 2019;48:32–48. https://doi.org/10.1016/devcel.2018.11.034.

Pispa J, Thesleff I. Mechanisms of ectodermal organogenesis. Dev Biol. 2003;262:195–205. https://doi.org/10.1016/S0012-1606(03)00325-7.

Pardo A, Cabrera S, Maldonado M, Selman M. Role of matrix metalloproteinases in the pathogenesis of pulmonary fibrosis. Respir Res. 2016;17:23–33.

Stout AP, Murray MR. Hemangiopericytoma: a vascular tumor featuring Zimmermann's pericytes. Ann Surg. 1942;116(1):26–33. https://doi.org/10.1097/00000658-194207000-00004.

Tonnard P, Verpaele A, Peeters G, Hamdi M, Cornelissen M, Declercq H. Nanofat grafting: basic research and clinical applications. Plast Reconstr Surg. 2013;132(4):1017–26. https://doi.org/10.1097/PRS.0b013e31829fe1b0.

Wu P, Hou L, Plikus M, Hughes M, Scehnet J, Suksaweang S, Wideltz RW, Chuong C-M. Evo-devo of amniote integuments and appendages. Int J Dev Biol 2004;48:249–270. [PubMed: 15272390].

Xu S, Liu C, Ji H-L. Concise review: therapeutic potential of the mesenchymal stem cell derived secretome and extracellular vesicles for radiation-induced lung injury: progress and hypotheses. Stem Cells Transl Med. 2019;8:344–54.

Yagi LH, Watanuki LM, Issac C, et al. Human fetal wound healing: a review of molecular and cellular aspects. Eur J Plast Surg. 2016;39:239–46.

Zuk PA, Zhu M, Mizuno H, Huang J, Futrell JW, Katz AJ, Benhaim P, Lorenz HP, Hedrick MH. Multilineage cells from human adipose tissue: implications for cell-based therapies. Tissue Eng. 2001;7:211–28.

The Meninges

12

Michael H. Carstens

The story of the meninges is simple but complex. The basic components that make these layers are present from the moment the neural tube closes. But the different neuromeric levels from which these tissues arise impart create in the dura genetically distinct zones that are clinically relevant for the production of cranial bone. At the same time, it is impossible to discuss the dura without including the cutaneous coverage of the calvarium. Each layer has a developmentally distinct blood supply and sensory innervation. These differences have significance for the understanding of craniosynostosis and other anomalies of skeletal development.

We shall begin our discussion by first *defining the layers* of meninges as they exist in the fully formed organism. Next, a *timeline of embryonic stages* is presented, representing major events in the formation of tissue sources (neural crest and craniofacial mesoderm) and key surrounding structures (pharyngeal arches and CNS landmarks). Description of the *components of the meninges* emphasizes the dura but with a coda regarding the pia-arachnoid. A *narrative* of meningeal development follows. Finally, the *spatial relationships of the scalp, fascia, and dura* in the context of neurovascular planes.

Note that cranial meningeal development differs substantially from that of the spinal cord. The latter system is more simplistic and we shall refer to it as needed. For further information regarding the spinal subject of the spinal meninges, the reader is referred to a recent review of the subject.

Before beginning, we should bear caveat in mind when thinking about meningeal layers: (1) meninges consist of different cell populations but are initially a single stratum, (2) meningeal layers form, not as a consequence of the different cell populations, but as the result of differing distances from signals arising from either the underlying brain or the overlying soft tissues.

M. H. Carstens (✉)
Wake Forest Institute of Regenerative Medicine, Wake Forest University, Winston-Salem, NC, USA
e-mail: mcarsten@wakehealth.edu

Layers of Meningeal Strata

The strata of the meninges are divided <u>classically</u> into three developmental layers: (1) an external thick *pachymeninx* which comprises the substance of the dura mater, (2) a delicate internal *leptomeninx*, comprising the arachnoid barrier layer and the fibers making up the subarachnoid space; and (3) the *pia mater* (although in the literature pia is frequently lumped in with leptomeninx) [Figs. 12.1, 12.2, 12.3 and 12.4].

From a <u>developmental</u> standpoint, this old terminology can be simplified as follows. The original mesenchyme surrounding the neural tube is known as the *meninx primitiva*. Under signals from the underlying brain and overlying epithelium the meninx primitive develops in two layers: (1) <u>endomeninx</u>, which differentiates into the leptomeninges, pia and arachnoid; and (2) <u>ectomeninx</u>, a three-layer dura. Because dura is the primary source for calvarial bone, an old term for ectomeninx, *pachymeninx*, has been used to describe the so-called skeletogenous layer.

The mesenchymal composition of the membranes of the brain, skull, and spinal cord varies along the neuraxis. Leptomeninges are a simple matter. They cover the entire CNS and are an admixture of local paraxial mesoderm (bood vessel endothelial cells) and local neural crest (blood vessel pericytes and stroma). Dura is another matter. Supratentorial dura, including tentorium cerebelli, is again a mix with neural crest predominant but is strictly limited to forebrain. Midbrain, cerebellum, and hindbrain have no dura. Despite its name, tentorium cerebelli has no anatomic relationship with the leptomeninges of the cerebellum. Posterior fossa is lined by periosteum derived from paraxial mesoderm. It has no dura. At the orifice of the foramen magnum, dura reappears, descending to cover the leptomeninges of the spinal cord. Spinal dura does not form bone. The periosteal lining of the vertebrae is PAM originating from the somites.

Big picture idea: Periosteum represents the stem cell layer surrounding bone laid down by whatever mesenchymal source produces the bone in the first place.

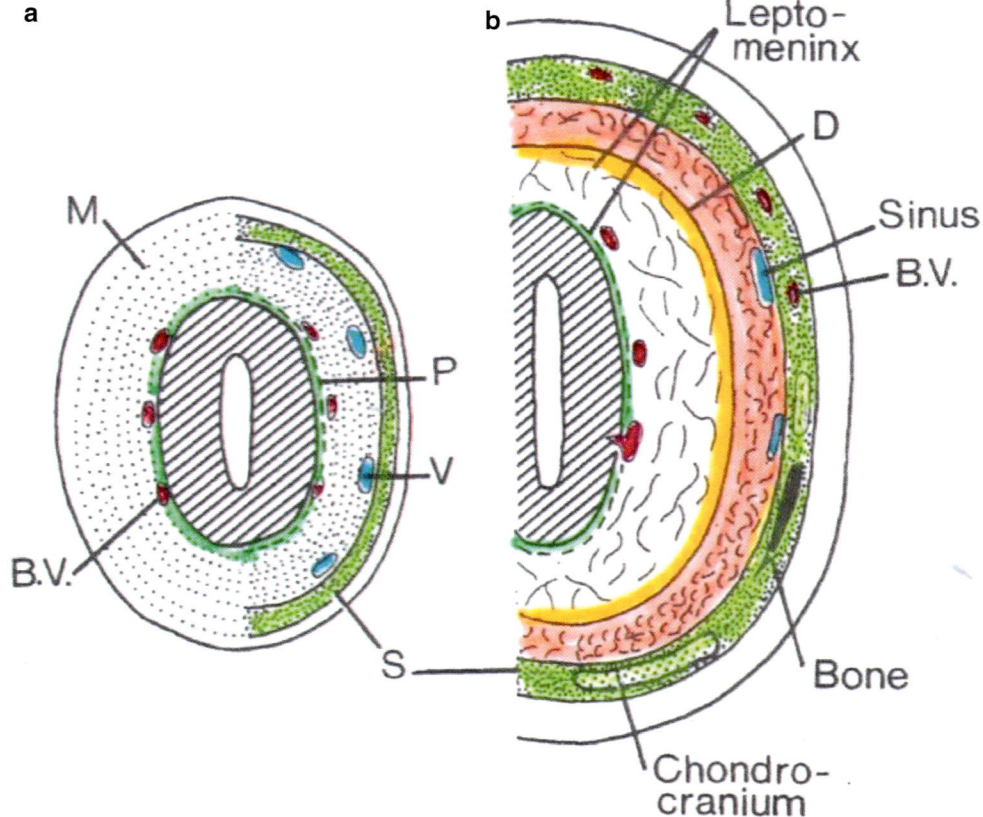

Fig. 12.1 Layers of the meninges. **a**. Mesenchyme (M) of paraxial mesoderm and (in the forebrain) neural crest surrounds the neural tube. It differentiates into layers depending upon the distance from the underlying neuroepithelium. Paraxial mesoderm is present from the very beginning. It accompanies the neural tube as soon as it rolls up during stages 8-9. It is admixed with neural crest in stages 9-10 and surrounds the neural tube. Blood vessels develop when the endothelial cells form tubes and are ensheathed by pericytes. Around the periphery, a layer of vascularized pia develops. Endothelial cells also migrate through the neuropil, seeking out the ventricular layer. They leave behind them a trail of signals attracting pericytes to follow them "down into the hole," thereby creating penetrating blood vessels between the surface and the periphery. A dense ectomeninx, skeletogenous layer (S), forms at the periphery. Deep to the ectomeninx, venous sinuses develop. **b**. Dural limiting layer D appears between the pia and the ectomeninx. Mesenchyme between pia and D becomes arachnoid. Bone or cartilage forms in the skeletogenous layer. [Reprinted from O'Rahilly R, Müller F. The meninges in human development. *J Neuropath Exp Neurol* 1986; 45(5):588–608. With permission from Oxford University Press.]

Dura mater consists of three layers. Let's describe these structures, one-by-one, moving from external to internal: periosteum (endosteum), dura propria, and dural limiting membrane. These are referred to, collectively, as *dura*.

As one reflects the skull away from the periosteum one perceives the first layer: the ***internal periosteum***, a delicate, nearly transparent endosteum. Lying subjacent to it are the meningeal arteries, lymphatics, and sensory nerves. The presence of the internal periosteal layer explains why surgical separation of the cranial bone from dura is so simple. It is virtually bloodless….but not quite. At surgery, small punctate bleeding points are seen over the surface of the periosteal layer marking the interruption of delicate vessels that extend outwards from meningeal vessels through the periosteum and into bone.

As we shall see, the meninges constitute a protective niche for stem cells. The internal periosteum represents a stem cell layer "donated" to calvarial bone from the dura. On the other hand, the external periosteum is also "donated" to the bone, this time by the neural crest fascia surrounding muscles, either from Sm4 (mastication) or from Sm6 (muscles of animation). The two stem cell layers (intracranial and extracranial) are continuous with one another via foramina and sutures. In sum, membranous calvarial bones are bilateral structures synthesized by stem cells from biologically different sources.

Meningeal arteries and veins are readily seen coursing between the thin internal periosteum and the underlying dura. These consist of endothelial tubes from PAM surrounded by pericytes of neural crest origin (*vide infra*). Although meningeal vessels send small branches outward to the endosteum and more larger branches inward into the dura mater propia, they are not fused with either layer. Epidural veins drain into dural vessels.

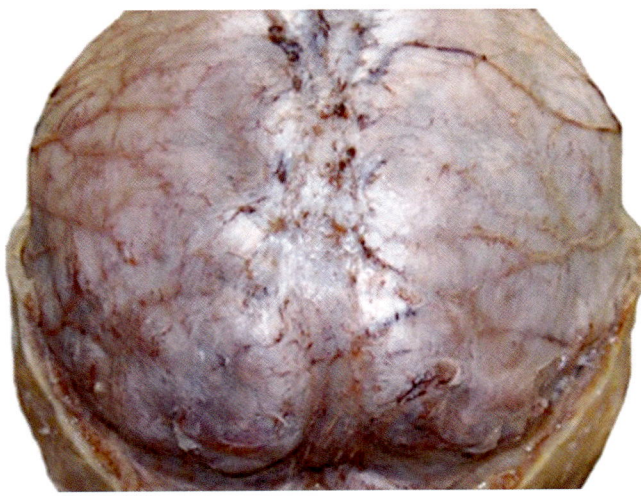

Fig. 12.2 Outer two layers of dura showing thin translucent internal periosteum covering over meningeal arteries running on the surface of dura propia. [Reprinted from Adeeb N, Mortazavi MM, Tubbs RS, Cohen-Gadol AA. The cranial dura mater: a review of its history, embryology, and anatomy. *Childs Nerv Syst* 2012; 28:827–837. With permission from Springer Nature.]

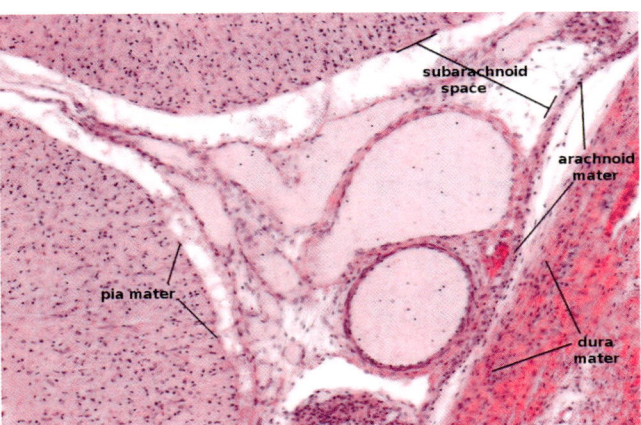

Fig. 12.4 Histology of meninges: *Dura mater propia* (intense dark pink) has distinct *dural limiting membrane* (faint pink) that is separated by a *subdural space* from *arachnoid limiting membrane* (dark pink) which is several cell layers thick. On left, vessel in the subarachnoid space penetrates brain substance. Arachnoid has no basal layer and does not follow the vessel. Pia mater ensheaths the vessel, defining the subarachnoid space into the brain and permitting circulation of the CSF. Pia is impermeable to fluid. In subarachnoid hemorrhage, red cells never have contact with the brain. [© 2020 The Regents of the University of Michigan. For information, questions or permission request please contact: Michael Hortsch, Ph.D. University of Michigan Medical School. Email: hortsch@umich.edu]

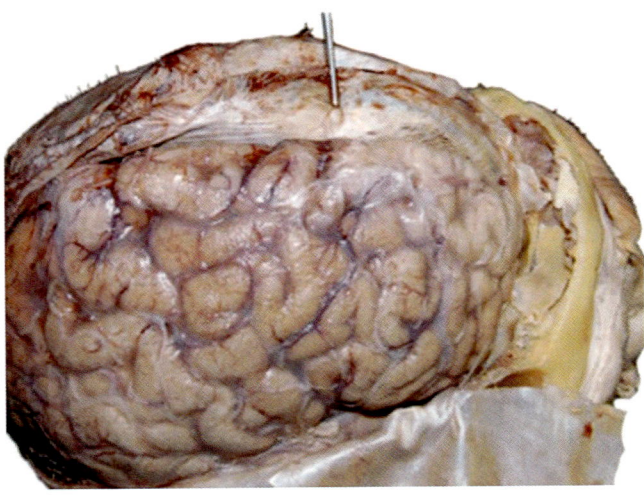

Fig. 12.3 Dura propia has dural limiting membrane fused to its undersurface. *Internal cerebral veins are* seen through the filmy *arachnoid limiting membrane*. Bridging veins between these vessels cross the *dural limiting membrane* to enter the subdural space. Here they can be traumatized to cause a subdural hematoma. [Reprinted from Adeeb N, Mortazavi MM, Tubbs RS, Cohen-Gadol AA. The cranial dura mater: a review of its history, embryology, and anatomy. *Childs Nerv Syst* 2012; 28:827–837. With permission from Springer Nature.]

The second layer is ***dura propia***. It is tough and thick. It is in continuity at foramen magnum with spinal cord dura. At sites of entry or exit of nerves and arteries, dura is continuous with epineuria, exemplified by optic nerve and adventitia. It contains veins subjacent to dura mater, and the subdural veins are encountered.

The mesenchyme of the cerebral dura is composed of neural crest and paraxial mesoderm from neuromeres (r1-r3). Dura of the falx is more simplistic, being r1 at both ends of a central r2-r3 zone. Tentorium cerebelli is an exclusively r1 structure. Beneath the tentorium, dura as a covering for the remainder of the brain ceases to exist. Midbrain, cerebellum, and hindbrain are enveloped by meninges (pia and arachnoid), they have no dura mater. The lining of the cranial fossa below the tentorium is internal periosteum, not dura, and is composed of paraxial mesoderm. Finally, beginning at the level of foramen magnum, *spinal cord dura* is exclusively paraxial mesoderm beginning with neuromeric level c2 and continuing down the length of the neuraxis.

The third and final layer of dura we encounter is the ***dural limiting membrane***. It represents the inner layer of ectomeninx. This structure is important for three reasons. First, its appearance defines the differentiation of secondary meninx from primary matrix. Second, it is readily identifiable on light microscopy and it serves as a marker for the spatiotemporal development of dura. Third, its anatomic relationship with underlying arachnoid provides an explanation for surgical bleeding.

Dural limiting membrane is the external lamina of a structure approximately two to eight cell layers thick, termed the *subdural mesothelium* or (erroneously) *arachnoid membrane*. It is composed of the external dural limiting membrane and an internal *arachnoid barrier layer*. Under normal conditions these layers are collapsed together: <u>there is no "subdural space."</u>

The leptomeninx begins with the *arachnoid barrier layer*, the outer boundary of endomeninx. Just deep into the interface layer one encounters a network of arachnoid cells interlaced with collagen fibers. The mesenchyme of these layers is likely to originate from neural crest admixed with PAM which furnishes the endothelial cells for the blood vessels that run within it. Arachnoid blends into the pia below it. There is no defining internal arachnoid layer.

Finally, the leptomeninx ends with a single-layer of *pia* abutting the surface of the brain. Pia mater is a homogenous neural crest derivative throughout the CNS. undoubtedly originating from the internal/deep migration of neural crest that immediately covers the neural tube. Pia is a source of pericytes that envelop the endothelial tubes formed in arachnoid. These blood vessels plunge into the neuropil. Pial cells are connected with desmosomes; they constitute a barrier between the subarachnoid space and the brain. In subarachnoid hemorrhage, red cells never make contact with neural tissue. Pia invests blood vessels entering into the substance of the brain. Once within the brain parenchymal, fenestrations in the pia coating permits circulation of nutrients.

Embryogenesis of Meningeal Circulation

Misunderstanding persists in so-called subdural bleeding. The dura-arachnoid boundary is easily disrupted, a fact is known to all neurosurgeons and neuropathologists; this creates a false plane. Although the dural limiting membrane and the arachnoid barrier layer can be distinguished histologically, throughout development these layers remain attached. Unlike the pleura and peritoneal cavities, a *"subdural space" does not exist*. This multicellular layer consists of stem fibroblasts (but no network of collagen fibers) and intracellular spaces. The term "dural border cell layer" or dural limiting membrane was defined by Nabeshima in 1975 and later incorporated by O'Rahilly and Müller [Fig. 12.5].

Early in development, the primitive meninx surrounding the neural tube contains a head plexus which is temporarily supplied by segmental branches from the dorsal aortae while the embryo manufactures a definitive cerebral circulation. During stages 9–10, the head plexus sends penetrating vessels into the brain. At stage 11, these connect with the nascent vessels from within the brain, a situation that makes the primitive meninx independent of external supply. That's a good thing, considering all the changes in the transient nature of the aortic arch system. At stage 17, just as meninx primitive changes into a two-layer structure, and while endomeninx remains supplied from below, a separate system of meningeal arteries from the stapedial stem arrives at the scene "in the nick of time" to supply the ectomeninx from above.

Venous drainage also differentiates into separate systems. The original system lies within the leptomeninges while newly formed meningeal veins are superficial to dura. Initially, the pial and dural layers are connected by a primitive plexus system. As development proceeds, definitive cerebral veins form in the sulci and the plexus becomes "streamlined" by apoptosis. The cerebral veins of each hemisphere coalesce into 12–18 *bridging veins* that traverse the pia-dural border. Once in the substance of the dura, they run for variable distances and then enter the superior sagittal sinus [Figs. 12.6, 12.7, 12.8 and 12.9].

It should be emphasized that the dural sinuses do not mature until several months after birth and the early fetal venous circulation is quite plastic, a useful adaptation to the spatial demands of dramatic brain growth. The upshot of this biology is the creation of a blood-brain barrier in which, through a paring process, pio-dural anastomoses become dramatically reduced in number. These vessels are quite sturdy and are not good candidates for SDH without significant trauma.

In its final iteration, the ectomeninx is supplied by meningeal arteries and veins. Meningeal arteries supply the calvarium, penetrating arteries into the substance of the dura, and arteriovenous shunts. The penetrating arteries extend all the way through dura propia to reach within μm of the dural limiting layer where they form a dense vascular plexus. As depicted by Mack these vessels are the primary suspects for non-traumatic "subdural" bleeding [Fig. 12.10].

Only the Cerebrum Has Dura

All other parts of the brain are covered with pia and arachnoid, but not dura. Posterior fossa is lined with periosteum. Supratentorial dura and tentorium are neural crest derivatives. Infratentorial periosteum is paraxial mesoderm [Figs. 12.11, 12.12, 12.13 and 12.14].

Phylogeny of the Meninges

Meninges reflect the biologic sophistication of the underlying brain. Fishes have a single membrane known as the *primitive meninx*. In non-mammalian tetrapods this structure becomes bilaminar with an outer dura mater and a thick inner *secondary meninx*. In mammals, the secondary meninx further divides into the arachnoid and the pia. Recall the metabolic demands of the highly-convoluted cerebrum. Pial vessels are worked into the depth of these cortical folds. In 1909 work, Giuseppe Sterzi demonstrated that dura behaves differently in spinal cord versus in the skull. In the calvarium, dura is fused with the internal periosteum whereas, in the spinal cord, the two layers remain separate [Fig 12.15].

Fig. 12.5 Dural limiting membrane/dural border cell layer. **a**. DLM is composed of loosely adherent cells, no extracellular collagen, and enlarged extracellular spaces. **b**. Subdural hemorrhage represents accumulates in the dural border cell layer. [Reprinted from Mack J, Squier W, Eastman JT. Anatomy and development of the meninges: implications for subdural collections and CSF circulation. *Pediatr Radiol* 2009; 39:200–210. With permission from Springer Nature.]

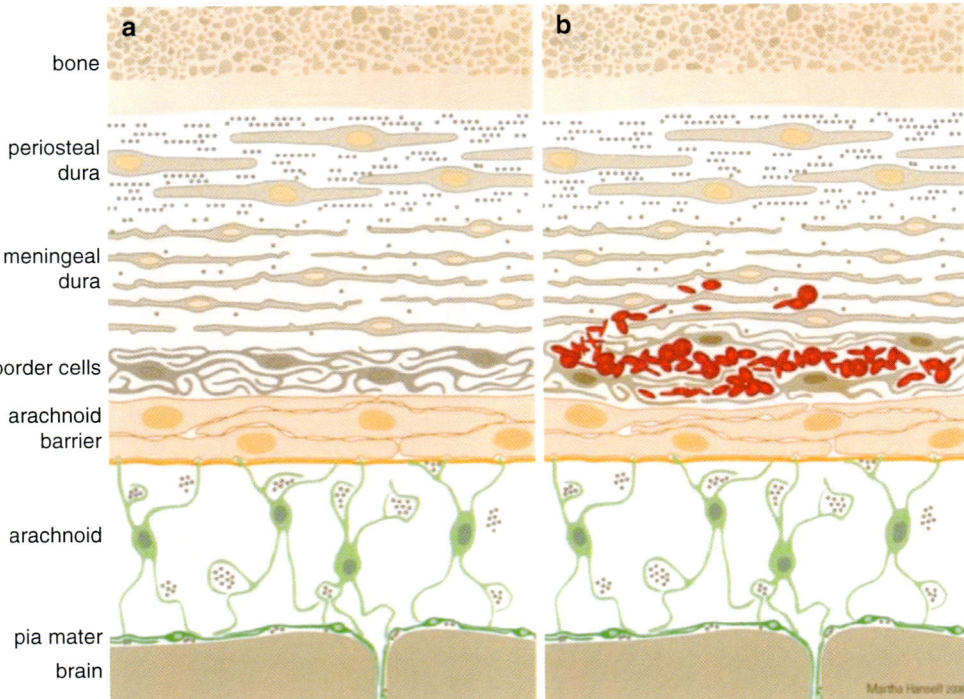

Fig. 12.6 Cerebral veins run within the leptomeninges. From these, 12–18 bridging veins connect through the arachnoid-dura interface, run for variable distances within the dura, and eventually drain into the superior sagittal sinus. [Reprinted from Wikipedia. Retrieved from: https://en.wikipedia.org/wiki/Superior_cerebral_veins#/media/File:Slide6Neo.JPG. With permission from Creative Commons License 3.0: https://creativecommons.org/licenses/by-sa/3.0/deed.en.]

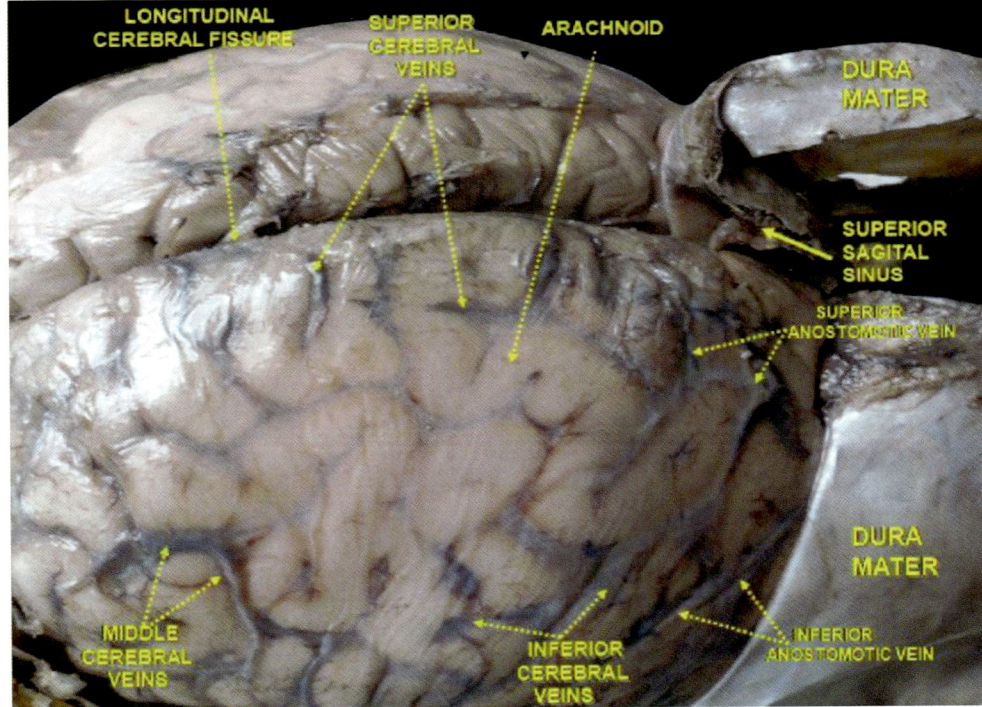

Fig. 12.7 Cerebral arteries and veins run in the arachnoid layer. Due to the pial barrier, subarachnoid hemorrhage does not penetrate the substance of the brain. Arachnoid granulations and dural bridging veins penetrate through the dura to communicate with the sinus. Histologic layers of dura are not shown here. [Reprinted from Cakmakci H. Essentials of trauma: head and spine. Pediatric Radiology 2009;39 Suppl 3:391–405. With permission from Springer Nature.]

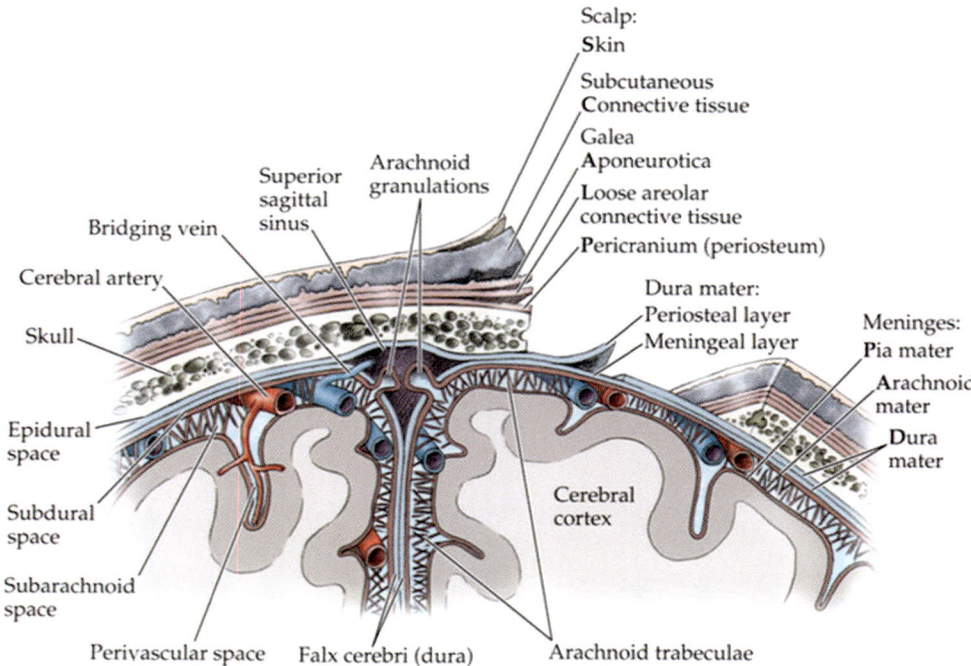

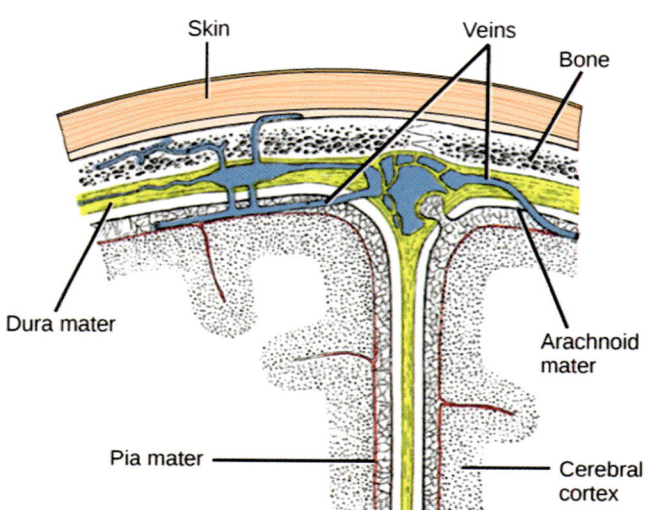

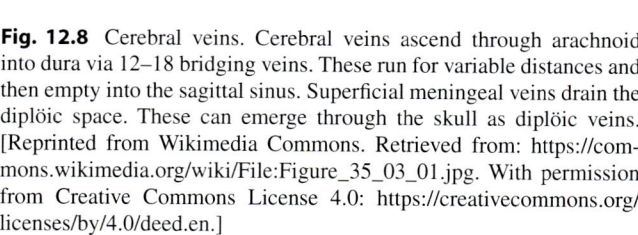

Fig. 12.8 Cerebral veins. Cerebral veins ascend through arachnoid into dura via 12–18 bridging veins. These run for variable distances and then empty into the sagittal sinus. Superficial meningeal veins drain the diplöic space. These can emerge through the skull as diplöic veins. [Reprinted from Wikimedia Commons. Retrieved from: https://commons.wikimedia.org/wiki/File:Figure_35_03_01.jpg. With permission from Creative Commons License 4.0: https://creativecommons.org/licenses/by/4.0/deed.en.]

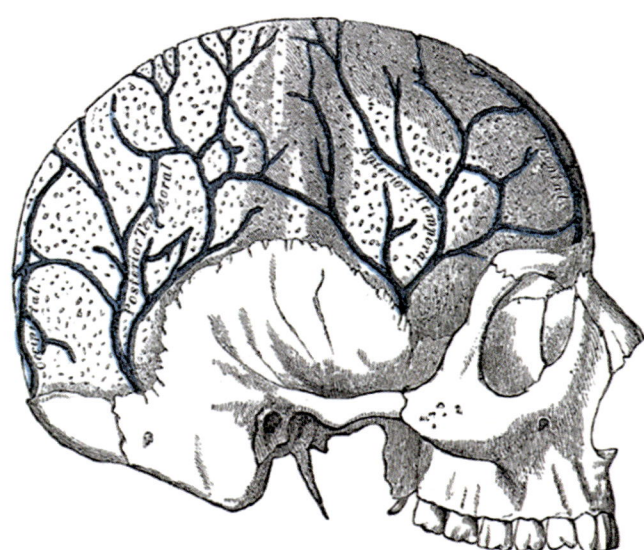

Fig. 12.9 Diplöic veins. Diplöic veins drain the neural crest membranous bones of the calvarium. They notably avoid squamosal, as it is paraxial mesoderm. Diplöic veins drain into the sagittal sinus and connect with meningeal veins running on the surface of dura, thereby establishing the relationship of calvarial bone, particularly the inner table, to dural stem cells. Emissary veins at scattered locations connect the intracalvarial diplöic veins through the outer table to the periosteum. [Reprinted from Lewis, Warren H (ed). Gray's Anatomy of the Human Body, 20th American Edition. Philadelphia, PA: Lea & Febiger, 1918.]

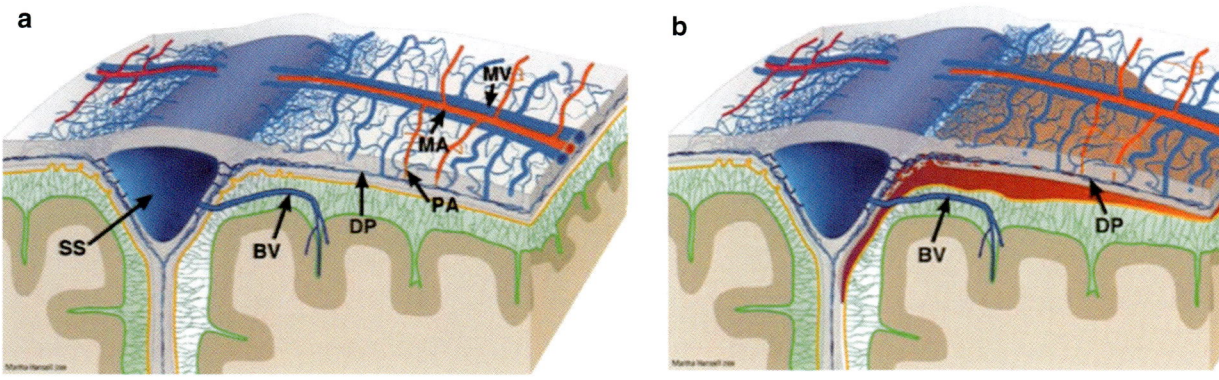

Fig. 12.10 a. Meningeal arteries and veins supply the ectomeninx, coursing between periosteum and dura propia (gray). Arachnoid proper is defined by the arachnoid limiting membrane (yellow). With the dural compartment is a dural plexus (DP) fed from above by penetrating branches from meningeal arteries. These run in the subdural compartment until reaching the sagittal sinus to which they connect. The DP adjacent to the sagittal sinus is distinct from the bridging veins (BV) which run in the arachnoid layer. **b**. Bleeding occurs within the subdural compartment causing a downward displacement of the arachnoid barrier membrane (yellow). Blood never enters the subarachnoid space. [Reprinted from Mack J, Squier W, Eastman JT. Anatomy and development of the meninges: implications for subdural collections and CSF circulation. *Pediatr Radiol* 2009; 39:200–210. With permission from Springer Nature.]

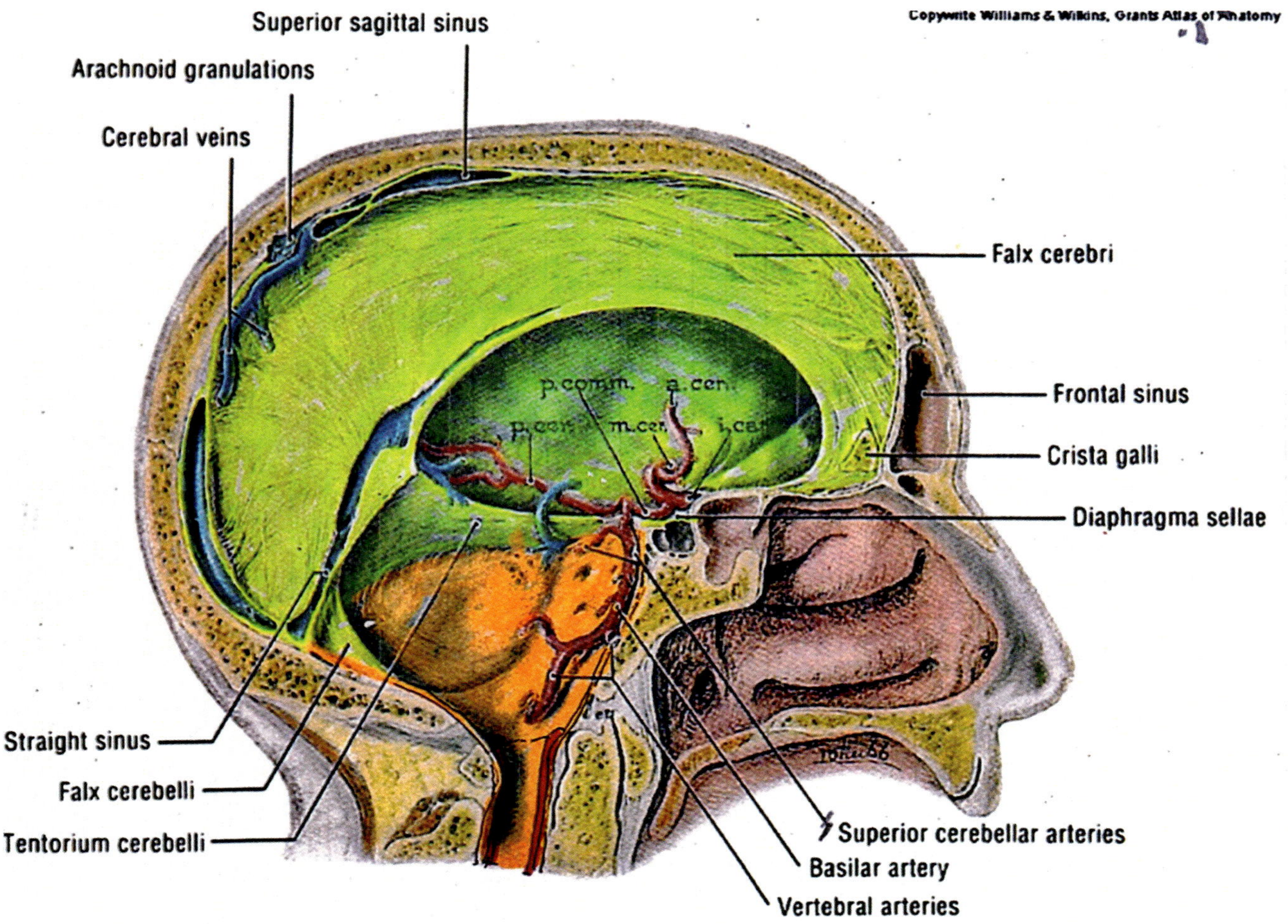

Fig. 12.11 Mesenchymal anatomy of intracranial membranes. Dura (green) is neural crest supplied by trigeminal nerves. Periosteum (orange) is mesoderm; it is not true dura. It continues all the way down the spinal canal, Mesodermal dura begins again at the foramen magnum and is a separate layer (pink). It does not participate in osteosynthesis. [Reprinted from Boileau JC. Grant Atlas of Human Anatomy Philadelphia: Williams & Wilkins 1943.]

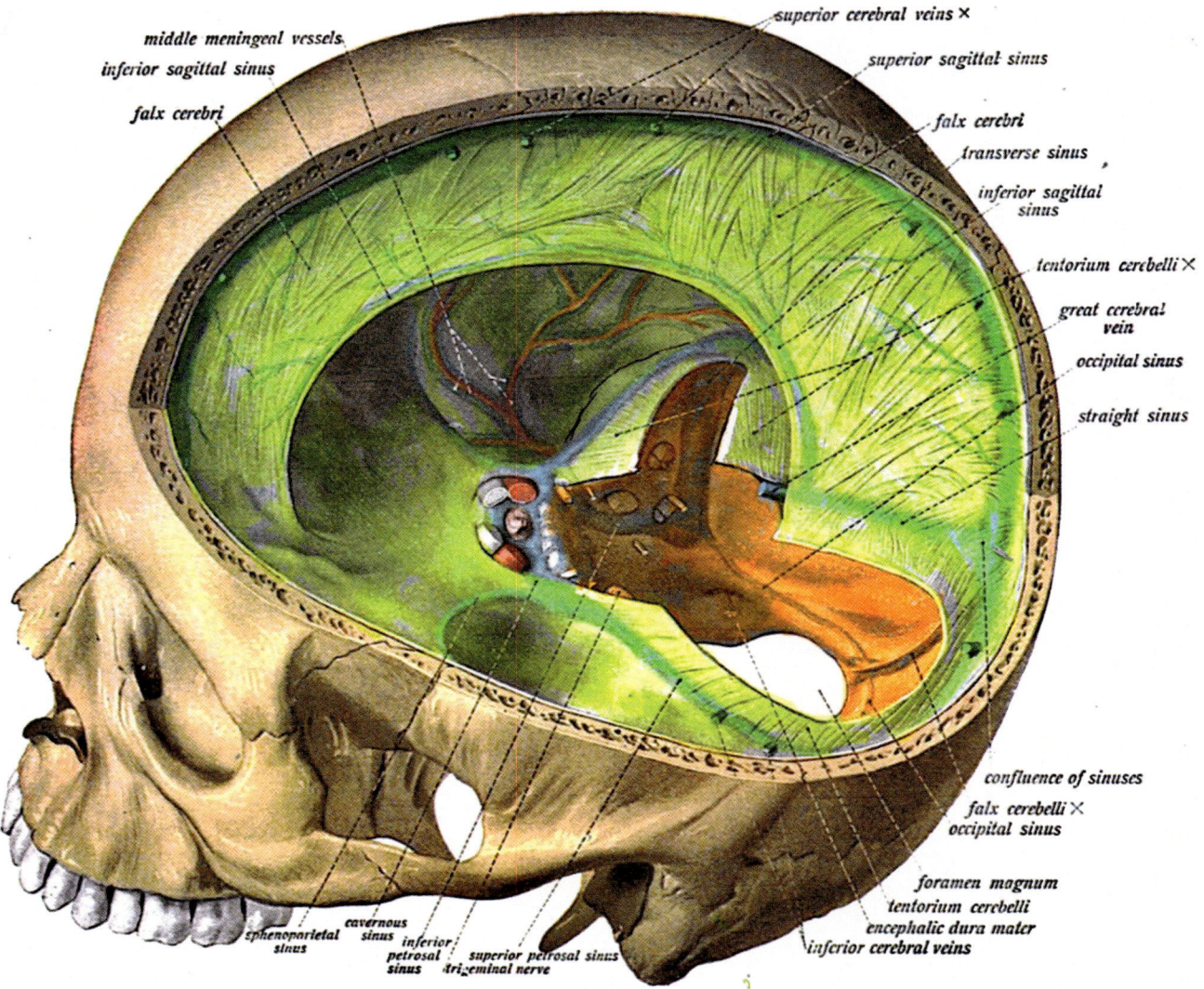

Fig. 12.12 Dura and tentorium (yellow) are classified as supratentorial. These are neural crest tissues. At sites of confluence with one another or with periosteum, venous sinuses are formed. Infratentorial periosteum (orange) is mesodermal and not bilaminar. It cannot contain venous sinuses. Posterior fossa is defined as being medial to the sinuses. [Reprinted from Sobotta J. Anatomic des Menschen. Munich, Germany: Verlag JF Lehman, 1904.]

This important observation has a developmental explanation. The membranous bones of the supratentorial calvarium require a periosteum as a source of stem cells. The blood supply to this layer comes from meningeal arteries. Therefore, the layers are fused. In the infratentorial calvarium no dura exists. The blood supply to these chondral bones runs in the substance of the lining periosteum, (sic) dura, and is *not* connected with CNS circulation. Dura reappears below the foramen magnum but is no longer required for vertebral synthesis – the layers remain separate.

Historical aspects of the meninges are summarized by O'Rahilly and Müller [1]. One can consult Sensenig's original account available online through the University of New South Wales website, http://embryology.med.unsw.edu.au.

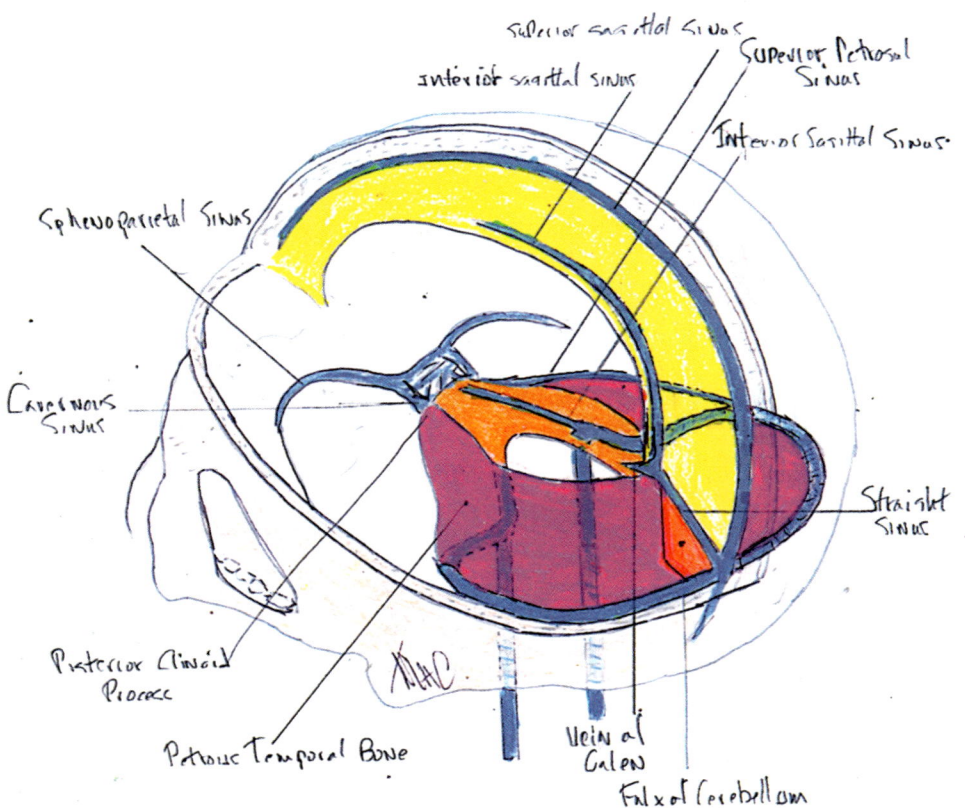

Fig. 12.13 Anatomy of the falx cerebri, tentorium cerebelli and falx cerebelli. These structures are defined by the venous sinuses that are enclosed by the bilaminar dura. *Falx cerebi* (yellow) runs from crista galli and internal frontal crest all the way back to internal occipital protruberance and the tentorium. *Tentorium cerebelliI* (pink) has fixed margins from posterior clinoid process to petrous apex laterally and thence to the transverse sinus. Its free margin allows passage of midbrain. *Falx* *cerebelli* (magenta) is a small reflection extending downward from midline of tentorium and along the inferior occipital crest. Chiari II malformations can occur in its absence. Posterior fossa periosteum (orange) extends medially from petrous apex, anteriorly along clivus, and posterolaterally below transverse sinus. [Courtesy of Michael Carstens, MD]

Fig. 12.14 Occipital dura (green) is r3 neural crest that covers posterior cerebral lobes. Posterior fossa periosteum (orange) is paraxial mesoderm and is juxtaposed to but does not cover the cerebellum. Interface between these tissues creates sinuses. Superior sagittal sinus results from the two laminae of r1 falx splitting apart and roofed over by r3 periosteum. Note the reappearance of spinal dura (pink) below foramen magnum. It has been opened to show arachnoid. [Reprinted from JC Boileau Grant Atlas of Human Anatomy Philadelphia: Williams & Wilkins 1943.]

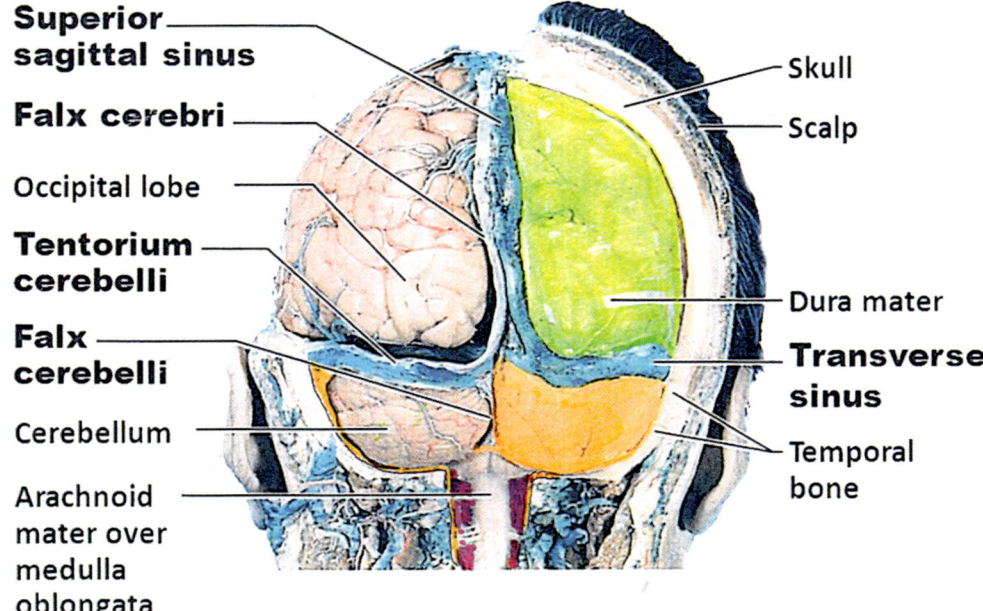

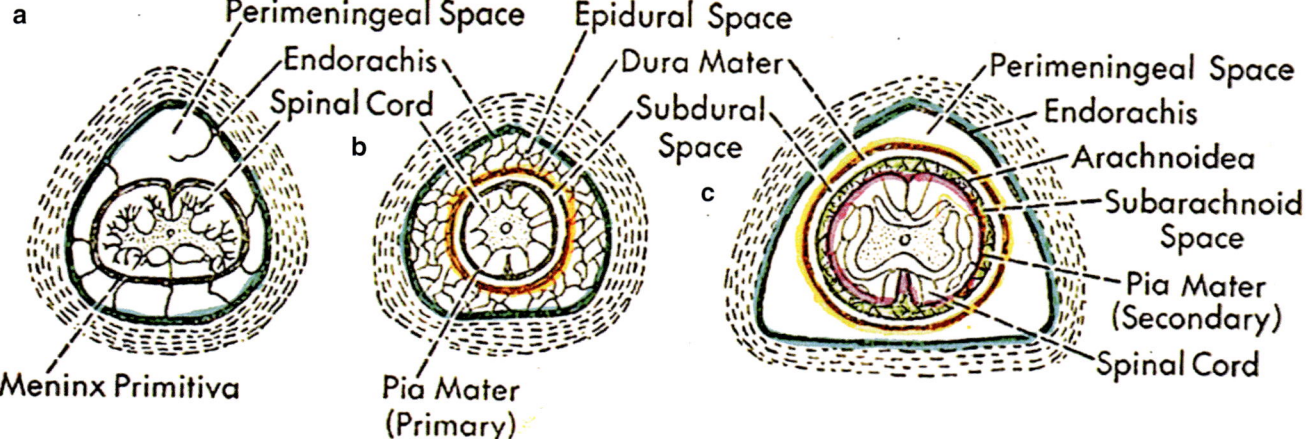

Fig. 12.15 Phylogeny of meninges. Fishes (**a**) have a single-layer primitive meninx. In non-mammal tetrapods (**b**) the primary meninx differentiates into internal lamina, the primary pia mater, and an external lamina, the dura mater. No arachnoid is present. In mammals (**c**) the primary pia mater again differentiates into a secondary pia mater and arachnoid. The dura assumes a trilaminary histology, with a dural limiting membrane defining it from arachnoid. [Reprinted from Frazer JE. Anatomy of the Human Skeleton. Philadelphia: P Blaikston,1904.]

Formation of Meninges by Developmental Stages

The purpose of the section is to review relevant events related to the production and transportation of the mesenchyme that surrounds the brain. The timeline at which various sectors of the meninges appear is defined. This material appears telegraphic and dense. Do not despair…it becomes quite useful when referring back to it from the text.

Stage 6 (13–14 days) Gastrulation starts

Primitive node and primitive streak extend from organizer forward to future level r0. At stage 6b, the dorsal aspect of PN produces prechordal mesoderm which flows forward from epiblast and hypoblast. This is the "herald event" of gastrulation in stage 7.

Stage 7 (15–17 days) CNS appears, gastrulation completed

PN at r0 produces prechordal notochord, then retreats backward, leaving additional notochord behind it. Gastrulation procedes from r0 caudally. Notochord splits PCM as far forward as future prosomere hp2. *Otx-2* from r0 induces midbrain. Notochord (assisted by PCM) induces forebrain.

Stage 8 (18–19 days) *nervous system, segmentation, and supply*

General: This stage is short, lasting 48 h. The first indications of the future brain appear. Prechordal plate mesoderm (PCM) is situated directly in front of the future notochord. From the primitive streak, cells are populating the paraxial zone. By the end of stage 8, a total of 18 somitomeres have been synthesized, the first 7 of which are cranial.

CNS: The expansion of the paraxial mesoderm causes the elevation of neural folds. PAM remains tightly connected to the folds so that, in subsequent stages, closure of the neural tube will drag mesoderm and non-neural ectoderm up and over the CNS to achieve complete coverage. Around the spinal cord, all components of dura (being 100% PAM) are thus present. At each neuromeric level, neural crest makes pia/arachnoid and PAM makes dura. Around the cerebrum, dural development requires an additional component, neural crest, which will appear at stage 9.

Stage 9 (20–21 days, 1–3 somites) *major divisions of the brain*

General: Embryo starts out as a disc. Mesencephalic flexure due to brain growth folds the future heart beneath the face and stretches anterior segment of dorsal aorta into the 1st aortic arch. As soon as the 19th somitomere is added on, the Sm8 transforms into somite 1.

CNS. Three major divisions of the brain are present but no vesicles are seen. The rolling up of the neural tube brings the brain in contact with the outlying mesoderm—this is the source of the endothelial cells for the *primitive head plexus*. Endothelial cells penetrate the embryonic neural tube and migrate through the neuropil on their way to the ventricles. As they do so, they lay down a pathway of molecular cues that will be subsequently followed by pericytes to complete nascent capillary tubes.

Mesoderm: Expansion of the forebrain in the midline displaces PCM laterally into the head mesoderm. The forebrain is coated with PAM from r1 to r3.

Neural crest: First indication of neural crest is seen in the midbrain (perhaps as early as stage 8). Neural folds of the spinal cord are beginning to develop. As they roll up like a burrito, they carry along with them associated paraxial mesoderm. The space between brain and PAM constitutes a pathway by which NC cells migrate downward.

Meninges: Physical contact between the walls of the neural tube and PAM ensures that primitive vascular channels, the head plexus, are immediately available to provide nutrition for the growing brain. The internal layer of neural crest—the future pia—surrounds the channels and provides a barrier between direct contact between mesoderm and neural tissue: the blood-brain barrier. Mesoderm: Three of four occipital somites present.

Stage 10 (22–23 days, 4–12 somites) *subdivision of forebrain, optic primordium*

General: 1st pharyngeal arch visible. 2nd aortic arch artery develops from unification of buds from dorsal aorta and from ventral cardiac outflow tract—all remaining aortic arches develop in the same way. The significance of the 1st arch for meninges is that it has already been populated by hypaxial invasion of neural crest from r2 to r3. This means that the primary epaxial migration of r1-r3 over the cerebrum has already been completed.

CNS: Diencephalon and secondary prosencephalon are now visible. Neural crest migration over the forebrain begins: Rostral hindbrain (r1-r3) NC flows around the sides of cerebrum and into the anterior cranial base. Midbrain (m1-m2) NC is directed to the eye, the future orbit, over the sidewalls of the forebrain. Neural tube fusion from the occipito-cervical junction backward rolls in the surrounding skin and buries the neural tube. This process cannot occur over the brain because no gastrulation takes place anterior to the tip of the notochord. For this reason, a *new source of epidermis* arises from the anterior forebrain neural folds.

Mesoderm: Prechordal plate mesenchyme is admixed with somitomeric paraxial mesoderm and neural crest. The significance of this for extraocular muscle development is debated (*vide infra*). Epaxial halves of somitomeric PAM remain in contact with CNS while hypaxial halves send out myoblasts for the future muscles of mastication into 1st arch…and into subsequent arches. *Epaxial PAM is responsible for creating the endothelial channel vessels supplying the primitive head plexus.*

Neural crest Migration takes place in two directions. *Segmental migration* involves NC movement within the same neromeric territory to (1) fill in pharyngeal arches and (2) interact with epipharyngeal placodes to create composite ganglia along the neural tube.

Meninges: Interaction between neural crest and primitive head plexus creates conditions necessary for the development of the *pia mater*. Pia mater is designed to form the blood-brain barrier; it will envelope all vessels penetrating into the brain parenchyma.

Stage 11 (24–25 days, 13–20 somites) *rostral neuropore closes, skin coverage complete*

General: 2nd pharyngeal arch visible. 3rd aortic arch artery develops.

CNS: Characterized by closure of the prosencephaic neural folds and the rostral neuropore. This process is *unidirectional*: upward (backward) from the optic chiasm (it was formerly thought to be bidirectional). The final location of the rostral neuropore will be the nasofrontal junction. Presence of definitive notochord established at the tip of the basisphenoid bone.

Mesoderm: Brain becomes enclosed within the mesoderm from superficial to neural crest. Blood vessels form *in situ* adjacent to the brain wall from the adjacent head mesoderm. *Penetrating blood vessels* develop when PAM endothelial cells migrate inward toward the ventricular layer. They lay down signals for pericytes which follow them into the neuropil and ensheath the endothelial tubes. Penetrating blood vessels are surrounded by pia.

Neural crest: Migration of neural crest over the forebrain takes place. Optic vesicle emerges from alar plate of prosomere hp2, to be surrounded by *optic neural crest* from m1 and m2. MNC interacts with neuroepithelium of the optic vesicle to form the sclera. Acting alone, MNC is the source of the fascia that will surround the extraocular muscles beginning at stage 12.

Meninges: First indication of meninges: *pia mater* represent local neural crest appearing at level of caudal hindbrain, that is, from rhombomeres r8–r11.

Stage 12 (26–27 days, 21–29 somites) *caudal neuropore closes*

General: 3rd pharyngeal arch visible, 4th aortic arch artery develops, 1st and 2nd arches fuse. Trigeminal and facial ganglia develop but are not yet subdivided. Neural crest from r4 to r5 has invaded the 2nd arch and becomes widely dispersed. From r4 backward, neural crest is available for leptomeninges but does *not* form dura. *Stapedial artery* forms as a remnant of the stump of the 2nd aortic arch artery, the axis of the 2nd pharyngeal arch.

CNS: longitudinal neural arteries replace the head plexus. They are constructed from PAM from levels r1-r11 like "beads on a string."

Mesoderm: Paraxial mesoderm dedicated to the orbit condensing caudal to optic vesicle. Origin of extraocular muscles was originally thought to come from PCM but has now been mapped to somitomeres 1–3, and 5. More on PCM is below.

Neural crest: Cranial neural crest migration is now complete. Crest cells *per se* cannot be identified. Forebrain is covered with neural crest from rhombomeres r1-r3. Neural crest is now found in cranial nerve ganglia, admixed with head mesoderm.

Meninges: In the CNS, pia mater from local neural crest has further matured. *Pia mater is now visible at level of the midbrain.* Whereas MNC localizes to the orbit, neural crest from r1 to r3 now advances over the entire cerebrum. although it is not yet identifiable as a separate layer, it will form a *second layer* of mesenchyme, the future pachymeninx. Pachymeninx is not a pure neural crest layer, it is admixed with mesoderm. Thus, it will have the capacity to form two separate layers of blood vessels: arteries will lie external to dura propria; veins will be located deep to it. Dura is not yet defined; the dural limiting membrane does not appear until stage 17.

Stage 13 (28–31 days, 30+ somites) *neural tube now closed, cerebellum develops from r1*

General: 4th pharyngeal arch is visible, 5th pharyngeal arch forms internal to the 4th arch, 5th aortic arch artery fails to develop. Fusion of the 2nd and 3rd arches takes place. Persistence of the 3rd aortic arch artery will form the common carotid, external carotid, and the proximal (extracranial) portion of the internal carotid artery. Scalp begins to form (first hairs of body are cranial).

CNS: Cellular sheath of notochord appears at occiput and extends to myelencephalon (medulla). It will extend caudal-to-cranial. Cranial nerve outgrowth begins. The stapedial system will be programmed by V and VII.

Mesoderm: lateral rectus appears in caudal zone.

Neural crest: internal migration surrounding the neural tube is now complete; it is segmental, each neuromere has its own distinct population. These cells populate individual peripheral nerves and autonomic nerves.

Meninges: Vessels appear between single-layer ependyma of the 4th ventricle and surface non-neural ectoderm.

Appearance of scalp hair indicates interaction between overlying ectoderm and underlying mesenchyme. Epicenter of hair formation, the *occipital whorl*, is at the boundary between hindbrain ectoderm and cervical somite ectoderm. This process indicates the induction of specialization of the outer layer of pachymeninx by its contact with overlying non-meningeal tissue.

Stage 14 (32 days) *cerebral hemispheres*

General: 6th aortic arch artery develops for pulmonary circulation. This marks the end of the pharyngeal arch period.

CNS: Cellular sheath of notochord reaches metencephalon (pons).

Mesoderm: superior oblique, then superior rectus appear in caudal-dorsal zone.

Meninges: Proliferation of vessels, especially at midbrain. Vessels, surrounded by "protective" neural tissue, penetrate brain wall, always accompanied by pia. *Pia mater matures at level of telencephalon.* Cerebral hemispheres surrounded by mesenchyme (dorsal > ventral), indicating the directionality of pia mater.

Stage 15 (33–36 days) *longitudinal zoning of the diencephalon*

General: Amalgamated 1st and 2nd arch complex. Ventral fusion at the mandibular processes indicates midline closure of the floor of mouth and anterior neck.

CNS: Three divisions of trigeminal are present. These will innervate cerebral dura. Before exit, the skull forms StV1 in orbit and forms the StV3/StV3 "add-on" to the terminus of the external carotid, that is, the internal maxillomandibular artery. The cellular sheath of notochord advances forward to its terminus at the mesencephalic flexure (r0). This sheath constitutes the axis of the skull base. Note: this is a misconception of the FMROR model. Notochord actually extends forward to the terminus of hypothalamus (hp2); it induces the entire floorplate of the forebrain (cf Ch 5). Ossification subsequently proceeds caudal-cranial. Between day 28 and day 35 the primitive skull base, *desmocranium*, develops from basal condensation of neural crest (r1) and paraxial mesoderm (r1-r11) around the notochordal sheath.

Meninges: *Primary meninx formally recognized* – loose mesenchyme is seen around most of the brain—the floor is more developed than the roof. This reflects its vascularization because arterial supply matures basal-alar, that is, ventral-dorsal. The resulting from in situ formation of channels in contact with the brain that mature into vascular structures.

Stage 16 (8–11 mm, 37–40 days) *evagination of the neurohypophysis*

CNS: All cranial nerves are identifiable. Mesencephalic tract extends from trigeminal ganglion to isthmus. This mean StV1 has reached the orbit.

Mesoderm: medial rectus and then inferior oblique appear in rostral-ventral zone.

Meninges: Cellular sheath of the notochord is clearly involved with medial leptomeninges of tentorium cerebelli. Tentorium develops from midline-laterally).

Stage 17 (41–43 days) *olfactory bulbs, amygdaloid nuclei*

CNS: Interhemispheric fissure is filled with mesenchyme. This represents neural crest and mesoderm pulled inward by the tissue expansion of the cerebral hemispheres around a fixed central point. This will be the future *falx cerebri*. *Ossification* appears in sclerotomes 1–4; this process progresses cranially. The *stem of stapedial involutes* and extensive peripheral anastomoses take place. Extracranial stapedial via chorda tympani (StVII) joins ECA maxillary. IMMA promptly sends up StV3 recurrent branch into cranium to rendez-vous with intracranial stapedial system.

Meninges: *Dural limiting membrane* found first in cranial base areas…exactly where the mesenchymal condensations of the future chondrocranium are forming. "Holes" in the limiting membrane *pori durales* represent foramina for CN 3, 4, 5, and 12. Mesenchyme above the 4th ventricle is now loose (it has proliferated). *Skeletogenous layer* of future membranous calvarium now visible. Dural lining layer lateral to diencephalon is the extension of rostrolateral tentorium. Note how extensively V1 is represented. This explains the "backward" pathway of *recurrent branch* of V1.

Stage 18 (44–47 days) *inferior cerebellar peduncles, corpus striatum, dentate nucleus*

General: Mesenchyme of the nasal septum indicates the flow of MNC into the midline. Production of premaxilla Crista galli is not seen.

Meninges: Cavitation of the primary meninx creates a meshwork between the neural wall and the skeletogenous layer: this is the future arachnoid. Floor mesenchyme under 4th ventricle thicker than the roof (relates to increase in pia)

Stage 19 (48–50 days) *choroid plexus 4th ventricle, medial accessory olive*

CNS: Choroid plexus made up of leptomeninges.

Meninges: Changes in density between the leptomeninges, the primary meninx internal to the dural limiting layer is loose, and the pachymeninx is now dense. Dural sinus present. Medial tentorium was observed positioned between basilar artery and diencephalon (on coronal section).

Stage 20 (51 days) *choroid plexus lateral ventricles, optic, and habenular commissures, interpeducular and septal nuclei*

General: Presphenoid, sella turcica, basisphenoid, and orbitosphenoid (all neural crest derivatives) are seen.

Meninges: Ventral-most hemispheric fissure contains the future falx to which is being added mesenchyme from the skeletogenous layer. The skeletogenous layer now contains very distinct parietal plate. Cartilage is distinctly seen in both basisphenoid and orbitosphenoid. Above the 4th ventricle the mesenchyme is dense but no dural limiting membrane is visible as yet above 4th ventricle.

Stage 21 (52–53 days) *cortical plate appears in the cerebral hemispheres*

General: Parietal plate and alisphenoid (neural crest). Chondrocranium shows: orbitosphenoid, basisphenoid, otic capsule.

Meninges: The primary meninx is more dense with a larger mesh pattern next to the cerebral wall and a smaller mesh beneath the skeletongenous layer. A dural limiting membrane is noted along the skull base, and at base of cerebral hemispheres but not over the sides. This fits the *caudal-cranial developmental pattern* of the vasculature. Roof of the 4th ventricle has large mesh except above the choroid plexus, where it is smaller. Medial tentorium reaches all the way to mammillary body, while lateral tentorium is distinct, both rostrally and caudally.

Stage 22 (52–55 days) *internal capsule and the olfactory bulbs*

Meninges: Dural limiting membrane now covers the roof of the hindbrain and reaches midbrain, but it does not completely cover the roof. The cerebello-medullary cistern is now defined by a limiting layer. The transverse and sigmoid sinuses are now present with the dura.

Stage 23 (56 days+) *embryonic period ends*

General: Chondrocranium replaces desmocranium, forming the primitive skull base in cartilage.

Meninges: Dural limiting membrane is now complete over hindbrain and midbrain. In forebrain it is present over the lateral fossae only. Subarachnoid tissue internal to the dural limiting layer now readily distinguished. Falx cerebri is well developed from skeletogenous layer at crista galli. Dural limiting layer also developing rostrally but tissue *between* hemispheres is still leptomeningeal (neural crest) only. Extent of dural limiting membrane correlates with chondrocranium and membranous skull. The transverse and sigmoid sinuses occupy the potential space between the skeletogenous and the pachymeninx. In the area of the otic capsule and posteror fossa, dural and skeletal components are difficult to differentiate. Perhaps this is because the petrous temporal bone is chondral, originating from r4 to r7 PAM. Most adult cisternae are now present. The medial projection of tentorium cerebelli, extending from sella turcica to mammillary body is thinning out. It creates a separation between 2 different subarachnoid zones: (a) telencephalon vs. diencephalon; and (b) cerebellum and rhombencephalon. The rostrolateral tentorium joins the otic capsule.

Migration Patterns of Neural Crest

Neural crest development begins with that of the neural folds at stage 8 and persists in the brain until the end of the pharyngeal arch period, stage 14–15. Neural crest development in the spinal cord continues for much longer. In mammals, neural crest migration starts very early, just as soon as the neural folds round upward, at Carnegie stages 9, prior to closure of the neural tube. It occurs in two forms

- **Local migration** is perineural. The NC cells follow the surface of the neural tube as it rolls upward providing an immediate covering for CNS, much like wax dripping from a candle. This neural crest admixes with paraxial mesoderm to form the primitive perineural plexus. It is the source of endomeninx: in mammals, pia mater, and arachnoid.
- **Peripheral migration** follows different routes depending on its source.

Spinal Cord Neural Crest

Neural crest originating from the neural folds above the spinal cord begins at the cranial-cervical junction at stage 9. It continues neuromere-by-neuromere in a cranial-to-caudal sequence. The arrival of spinal NC at the target site depends upon its route. Migration within the same neuromeric level means that cells travel a similar distance and arrive at approximately the same time. Polyneuromeric migrations take longer.

- *Superficial lateral migration* Neural crest cells flow outward just beneath the surface of non-neural ectoderm. This populates the entire skin with melanocytes.
- *Deep lateral migration* of neural crest flows in and around the somites. Once within the somites, neural crest provides internal organization; this is the source of neural crest fascia surrounding craniofacial muscles. Peripheral to the somites neural crest suffuses the out-lying lateral plate mesoderm.
- *Ventral migration* forms the sympathetic system and moves into the visceral/splanchnic lateral plate mesoderm provides the ganglia of the gut [Fig. 12.16].

Cranial Neural Crest

Cranial neural crest follows different pathways, which we have been previously described in terms of stages. Migrations are underway very early, from stage 9. The actual arrival of the cells at their destinations depends upon (1) the timing of their departure and (2) the relative length of the pathway. Since the physical appearance of derivatives such as bone or dura depends upon development of vascular support, deductions around neural crest populations were imprecise. The sequence was finally determined using the quail-chick chimera system and gene markers to determine trajectories and fates [Figs. 12.17 and 12.18].

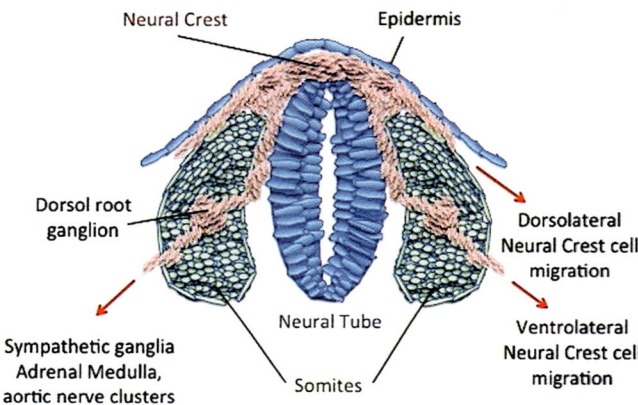

Fig. 12.16 Neural crest migration, spinal cord model. NC migrates immediately upon closure of each neuromeric segment of the neural tube. This model from the spinal cord fails to appreciate the triphasic arrival of mesenchymal populations over the cerebral wall. (1) Neural tube closure pulls up the adjacent PAM, much like pulling a hood over one's head. Since the forebrain arises from in front of level r0, it drags PAM from r0/r1, r2, and r3 over its walls to make blood vessels. This is initiated at stage 8 (2) Rhombomeric neural crest from r0 to r3 migrates forward and upward to be admixed with PAM only dermis, thus providing pericytes for the blood vessels. This happens at stage 9, that is, concomitantly or just after the arrival of PAM. This provides the dermis for all scalp skin except that in the distribution of nerve V1. This mesenchyme will be modified according to its distance from CNS signaling, from pia, to arachnoid, to dura/bone/periosteum, to dermis. [Reprinted from Ruggeri P, Farina AR, Cappabianca L, Di Ianni N, Ragone M, Merolle S, Gulino A, and Mackay AR. Neurotrophin and Neurotrophin Receptor Involvement in Human Neuroblastoma. In: Hiroyuki Shimada, (ed). Neuroblastoma" ISBN 978-953-51-1128-3, Published: May 29, 2013. With permission from Creative Commons License 3.0: http://creativecommons.org/licenses/by/3.0.]

Fig. 12.17 Stage 9 shows neuromeric origins of paraxial mesoderm and neural crest to form meninges. PAM (white) is produced by gastrulation from levels r0 backward. When the elevation of the neural folds from the plane of the embryo takes place in stage 8, hindbrain PAM from r0 to r3 forms two populations. The first population, being adherent to the neural folds, is immediately dragged over the forebrain where it makes the primitive head plexus. The second population remains *in situ*, becoming reorganized as somitomeres 1-3 which produce all extraocular muscles except the lateral rectus. Neural crest (green and blue) migration to forebrain takes place at stages 9-10 and provides meninges. RNC precedes migration from MNC into the orbit. RNC is organized by the trigeminal sensory nerves, V1-V3. [Courtesy of Michael Carstens, MD]

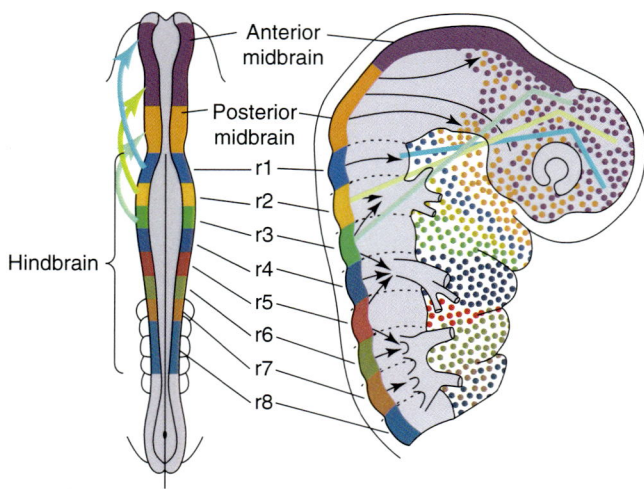

Fig. 12.18 Neural crest migration routes: r1 (blue) is distributed over basal forebrain, frontal lobe, and frontonasal mesenchyme; r2 (yellow) supplies frontolateral periorbital zone; r3 (green) supplies parietal/interparietal zone. Each the pharyngeal arch system, each arch is supplied by neural crest from two rhombomeres (even-odd). Order of migration: (1) rostral hindbrain r1-r3, followed by midbain m1-m2 and rostral hindbrain r4-r7 ; (2) caudal hindbrain r8-r11; and (3) forebrain p1-p3. [Reprinted from Gilbert SF, Barressi MJF. Developmental Biology, 11th ed. Sunderland, MA: Sinauer; 2016. Copyright © 2016. Oxford Publishing Limited. Reproduced with permission of the Licensor through PLSclear.]

Cranial Neural Crest Migration, Local

- Hindbrain

  Migration occurs just as in spinal cord. Hindbrain coverage is complete by stage 10 since the 4th occipital somites are present and pia is identified. Development of pia appears to progress seen to "develop" from caudal-to-cranial manner but this is deceptive, as it reflects the maturation pattern of the blood supply to the hindbrain.

- Midbrain and forebrain

  At stage 8 genes produced from r0 induce midbrain; additional genes from r0 and midbrain induce the forebrain. Gastrulation is complete with prechordal populations of mesoderm in place. As the neural folds of the midbrain and forebrain rise up and project forward they *drag along with them an ensheathment* of two forms of mesenchyme essential for their survival: (1) neural crest populations from r0 to r3 provide the endomeninx and pericytes; and (2) local PAM from r1 to r3 contains endothelial cells. These tissue in combination provide barrier around the brain and the components for a primitive head plexus which will sustain the CNS until a definitive blood supply can be established. Primary coverage is complete by stage 10, marked by the subdivisions of prosencephalon: diencephalon and secondary prosencephalon.

Cranial Neural Crest Migration, Peripheral

- Hindbrain (r0-r3)
  - *Hypaxial migration* stages 9–10

Neural crest from r2 to r5 enters the 1st and 2nd arches
  - *Epaxial migration* stage 11–12

Neural crest from r0 to r3 follows the walls of the forebrain providing additional mesenchyme, the ectomeninx. Dura is first seen ventrally seen at the midbrain level at stage 12. These crest populations will be innervated by V1-V3 and will supply all dura of the cerebrum as well as the dorsal aspect of V1-supplied tentorium cerebelli. Note that this coverage takes place in a plane external *to pia*. The initial migration from rostral hindbrain is probably complete by stage 12. But continued NC migration pursues a more superficial course to form two functional planes: r1-r3 produces scalp dermis and r4-r5 creates an intervening later of fascia (the future SMAS).

- Midbrain (m1-m2)
  - *Anterior migration* stages 11–12

Neural crest from m1 to m2 flows forward to populate the orbits

- Hindbrain (r4-r7)
  - *Hypaxial migration* stages 11–13

Neural crest from r4 to r11 enters the 3rd, 4th, and 5th pharyngeal arches
  - *Epaxial migration* stage 13+

Neural crest from r4 to r5 migrates massively over the entire head to form the superficial investing fascia (SIF) or superficial musculoaponeurotic fascia (SMAS) to accommodate the muscles of facial expression. Follows the previously-established pathway to cerebellum and coats the *ventral* aspect of tentorium cerebelli.

- Forebrain neural crest (p1-p3)
  - *Anterior migration* stage 14+

Forward flow beneath neural folds of hp1 and hp2 establishes (1) dermis for high-priority full-thickness skin coverage for V1-innervated calvarial and frontonasal zones; and (2) interaction with adenohypophyseal, nasal, and optic placodes

  Cerebrum has immediate epithelial coverage from NNE of the anterior forebrain folds; dermal populations can be assembled more leisurely

Forebrain Dermis Interacts with Previously Deposited MNC Structures

In any case, prior to formation of the frontal bone, we see r1 dura supporting p1-p3 dermis from diencephalon supporting hp1-hp2 epidermis from telecephalon. This topology means that MNC lies deep to the adenohypophyseal and nasal placodes, forming an initially solid block of nasoethmoid mesenchyme As the facial epithelia of the nasal processes invaginate backward into the MNC, they form cavities. The lining of the nasal cavities is non-neural epithelium from the anterior forebrain folds but the subvestibular or submucosal tissue is MNC. In response to this NNC-PNC template, the nasal capsule is converted into cartilages and membranous bones. Death of MNC in the midline ratchets the two nasal cavities toward the midline until their medial walls merge. The remaining MNC sandwiched in-between becomes the perpendicular plate of the ethmoid and the septum. Recall that in minimal cases of nasoethmoid apoptosis failure (hypertelorism) both septum and perpendicular plate can present as bifid or thickened.

It should be noted that the timing of visible appearance of certain zones of meninges as seen in human embryos is dependent on vascularization. The actual biologic migrations take place much earlier. The pathway for vascular development of the spinal cord is therefore *upward* from the ventral midline of the spinal cord. Both brain and scalp vessels appear ventrally and progressively ascend toward the vertex. Levels of scalp vasculature actually help to define stages 20–23.

Blood Supply of the Meninges

Blood Supply of Pia and Arachnoid

As discussed previously the neural tube is immediately surrounded by a primitive meninx, an admixture of "first responder" neural crest and PAM. These cells immediately produce vessels. Pia mater develops between the vessel layer and the cerebral wall. At stage 11, the vessels penetrate the brain substance seeking the ventricular layer. The vessel walls are coated with pia. In forebrain and midbrain, these vessels anastomose with the developing internal carotid system. In the hindbrain, they connect with developing longitudinal neural arteries which eventually recombine with cervical dorsal aortae to make the misnamed basilar-vertebral system. Recall that the vertebrals and former LNAs connect with each other just inside foramen magnum [Fig 12.19].

Arachnoid mater does not differentiate until stage 17, simultaneous with the beginnings of dural differentiation. The layers are kept quasi-separate by arachnoid limiting membrane and dural limiting membrane. Arachnoid is avascular, being supplied from below by the rich plexus of the pia. CSF circulate within arachnoid. Arachnoid villi penetrate into the plane between dural limiting membrane and the undersurface of dura propria. Within this subdural venous plane arachnoid granulations encounter venous sinuses thus permitting CSF from the subarachnoid space to drain into venous circulation.

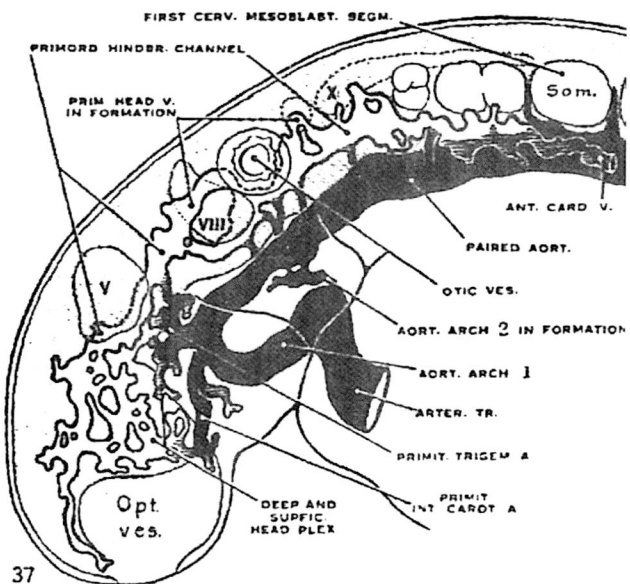

Fig. 12.19 Primitive head plexus at stage 10 The head plexus forms around the brain as soon as stage 8. By stage 9, it is connected with dorsal aortae by a series of segmental arteries, the first one, primitive trigeminal appearing at stage 9. At stage 10 the 2nd aortic arch artery is forming and multiple segmental arteries can be seen posterior to PrTg. The otic zone r4-r7 has three-four segmentals; occipital somites r8-r11 have four segmental arteries and 1st cervical segmental is identified. Beginnings of the internal carotid are seen. At stage 11 as the cerebral circulation forms, the plexus gives rise to penetrating centripetal vessels that connect with the centrifugal branches of the ICA system. Primitive head plexus remains at the surface of the brain to supply pia. In the hindbrain the plexus morphs into the longitudinal neural arteries. [Reprinted from Padget DH. The development of the cranial arteries in the human embryo. *Contribution to embryology.* Carnegie Institution 1948; 32: 205–261. With permission from Johns Hopkins School of Medicine.]

The circulations to pia mater and dura appear to be separate but actually they are connected at six different levels [Fig 12.20].

Internal Carotid, Anterior Division

1. Primitive olfactory artery has two branches. Medial olfactory is the leading edge of anterior division ICA and is the forerunner of anterior cerebral. Lateral olfactory artery gives rise to lateral striate, anterior choroidal, and is the forerunner of middle cerebral artery.
   - The *olfactory branches* represent the original medial branch of the primitive olfactory artery. These supply the pia of basal frontal lobe.
2. Anterior cerebral artery is the second branch of anterior division, although contemporary texts list it as primary.
   - Pericallosal branches extend backward to the posterior cerebral circulation. They supply the pia of the frontal lobe.

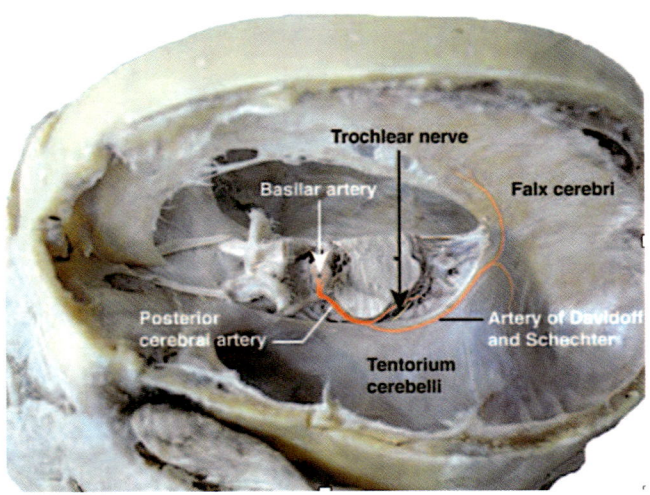

Fig. 12.20 Pia mater shares circulation with cerebral arteries. It is also supplied by direct pio-dural artery anastomoses. These are primitive olfactory (olfactory branches), anterior cerebral (pericallosal branches and anterior falcine artery), posterior cerebral artery (artery of Davidoff and Schechter), superior cerebellar artery (medial dural tentorial branch), anterior inferior cerebellar artery (subarcuate artery), posterior inferior cerebellar artery (posterior meningeal artery). The artery of Davidoff and Schechter is a normal variant, seen here supplying tentorium and falx cerebri. [Reprinted from Griessenauer CJ, Loukas M, Scott JA, Tubbs RS, Cohen-Gadol AA. The artery of Davidoff and Schechter: an anatomical case study with neurosurgical case correlates. *Br J Neurosurg* 2013; 27(6):815–181. With permission from Taylor & Francis.]

- Anterior falcine artery supplies medial pia corresponding to anterior falx, that is, frontal and parietal lobes.

Internal Carotid, Posterior Division

1. Posterior cerebral artery
   - *Artery of Davidoff and Schechter* supplies pia in region of falx and tentorium cerebelli, that is, posterior lobe

Longitudinal Neural Artery

1. Superior cerebellar artery
   - medial tentorial branch
2. Anterior inferior cerebellar artery
   - *subarcuate artery* supplies area of petromastoid
3. Posterior cerebellar artery
   - *posterior meningeal artery*

Blood Supply of the Dura and Periosteum

This subject is of great importance. Blood supply patterns are like paleontological clues giving information about the development of the meninges. We shall see that the dura and periosteum of the convexity over the cerebrum are relatively straightforward but the anatomy of dura covering the basal forebrain and the periosteum of the posterior fossa is more complex. We shall begin with general comments regarding terminology. We then proceed to look at specific arteries of importance. Finally, the developmental significance will be assessed [Fig 12.21, 12.22].

A note on the figures. For a general orientation to the meningeal arteries and a caveat regarding circulation to the anterior cranial fossa reader is advised to review figure 21 by Wolf-Heidigger. Superb work by Rhoton's group classifies the dural supplies by region and source [Figs. 12.23, 12.24, 12.25, 12.26 and 12.27]. The color scheme is simple but misleading: ICA (green), ECA (blue), and (sic) vertebrobasilar (red). *Stapedial system is not recognized.* Thus, areas in green ascribed to ICA should rightfully be only assigned to the territory of cavernous ICA. The remainder of V1 territory should really be another color. Similarly, with ECA. Here StV2 and StV3 stapedial arteries supply dura, whereas true ECA vessels (ascending pharyngeal and occipital) supply periosteum. Finally, the vertebrobasilar system is an embryologic misnomer which is explained below. The anterior and posterior meningeal arteries supposedly from VBA are really branches of the C1 segmental artery interposed between the occiput and the axis. Relationships in the posterior fossa are further amplified by Figs. 12.28 and 12.29.

General Comments

Dura mater covers the cerebrum and separates the cerebellum from posterior lobes. It does not envelop cerebellum. Midbrain and hindbrain do not have dura; they are enclosed within leptomeninges only. Dura mater starts again at foramen magnum and envelops the entire spinal cord.

Supratentorial dura develops from two tissues: (1) neural crest arising from anterior hindbrain (r0-r1, r2, and r3); and paraxial mesoderm from the same levels. Migration from levels m2 to m1 occur later and is reserved for the globe. PAM furnishes the endothelial cells required to make blood vessels for the cerebrum and cerebellum while neural crest produces the pericytes. We shall see that the neurovascular map of supratentorial dura has functional significance: it defines distinct developmental fields of the overlying calvarial bone and enables us to map out the origin of these bone fields.

Although supratentorial dura and periosteum are products of r1-r3 neural crest these tissues are NOT the "true" pharyngeal derivatives. Recall that in the gnathostome revolution, the 1st arch is so radically modified that a new vascular supply, the stapedial system, must be invented. Therefore, as we shall see, its primary blood supply does NOT arise from the original external carotid system. Instead, the predominant vascular source for the dura is the stapedial artery system.

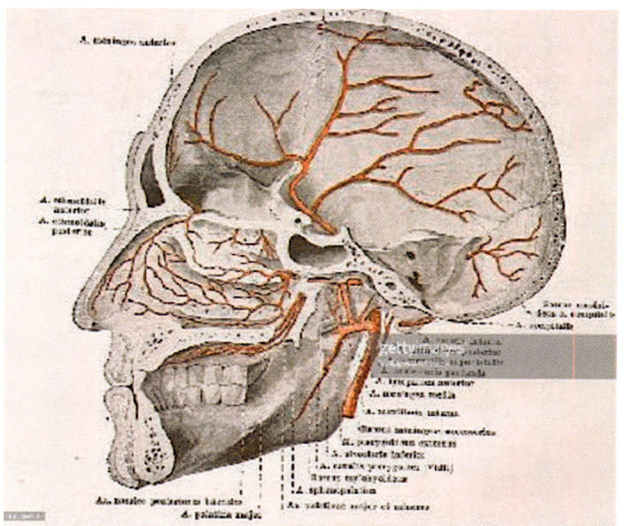

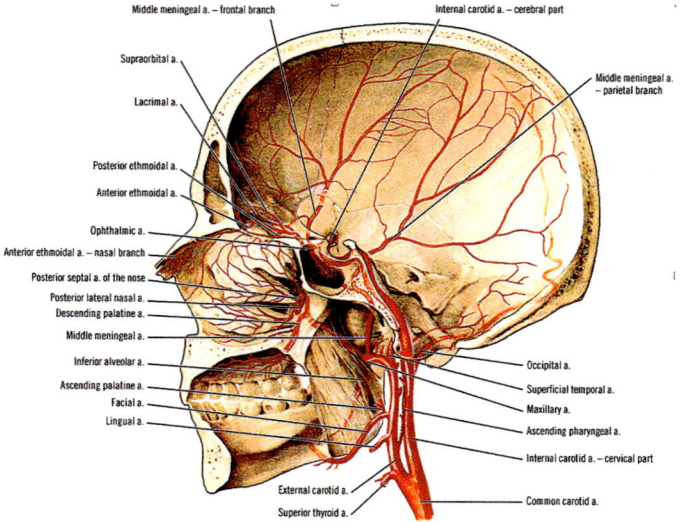

Fig. 12.21 Blood supply to the dura. Exceptional artwork depicts the anterior meningeal artery coming from StV2 but minimizes the contribution of StV prior to the orbit and anastomosis with ophthalmic. The ethmoids coming through the cribriform plate to the base of frontal lobe are shown. This is one of the few dissections showing the "recurrent meningeal artery fully developed. Penetrating branches from occipital are shown but not ascending pharyngeal. [Reprinted from Spalteholz W. Hand Atlas of Human Anatomy, Volume 2. Philadelphia, PA: JB Lippincott, 1928]

Infratentorial "dura" is a *misnomer*. Cerebellum, midbrain, and hindbrain have pia and arachnoid but not dura. Therefore, what has been historically referred to as dura in the posterior fossa is, in reality, *periosteum* with no biologic relationship to the neural structures it encloses. It is manufactured exclusively from paraxial mesoderm. From levels r4 backward, PAM acquires the ability to make pericytes so neural crest is no longer needed. Periosteum of the posterior fossa is derived from r4 to r7 mesoderm over the petromastoid complex and from r8 to r11 over the occipital complex. Levels r4-r7 relate biologically to 2nd and 3rd aortic arches therefore external carotid branch arteries occipital and ascending pharyngeal provide blood supply for this layer.

Extracranial dura mater encloses the spinal cord. It begins at foramen magnum. The mesenchymal source for spinal dura is paraxial mesoderm. Each somite beginning with C1 provides PAM to its respective neuromere. Blood supply to each zone of spinal dura is provided by paired epaxial branch from the aorta.

The Internal Carotid System, Cavernous Segment

Here we are going to concentrate on the cavernous segment of ICA and ignore the ophthalmic. In reality, the stapedial system (intracranial division), derived from the dorsal remnant of 2nd aortic arch artery, is the genetic source for meningeal supply along branches of StV1. The anastomotic relationship between stapedial has been previously detailed [Cf Fig. 12.21].

The cavernous segment of internal carotid supplies dura from r1 mesenchyme at the base of the brain, particularly that of sella turcica and tentorium cerebelli, is supplied by a seemingly confusing number of arteries arising from the cavernous internal carotid. The vessel area is organized by directionality. Medial vessels perfuse (what else?) the hypophysis and clivus. Lateral vessels supply lateral tentorium. Posterior vessels supply medial tentorium and dorsal meningeal artery. For our purposes, we shall concentrate on those vessels perfusing tentorium and clivus. An extensive review is given by Rhoton's groups [Figs. 12.22, 12.23, 12.24, 12.25, 12.26 and 12.27].

Big picture idea: The dura covering the basisphenoid and circle of Willis is r1 neural crest as is tentorium cerebelli sweeping backward from the posterior clinoid processes. These zones are continuous with the periosteum covering basioccipital which is of r8-r11 occipital somite derivation. Arteries serving these structures are programmed by V1 but are *not* part of the stapedial system.

Arteries of the Tentorium

These r1 tissues are supplied by two arteries of the *meningohypophyseal trunk*, the largest intracavernous branch of ICA. *Medial tentorial artery* (Bernasconi"s artery) travels along the free edge of tentorium all the way back to the straight sinus. *Lateral tentorial artery* travels along petrous ridge

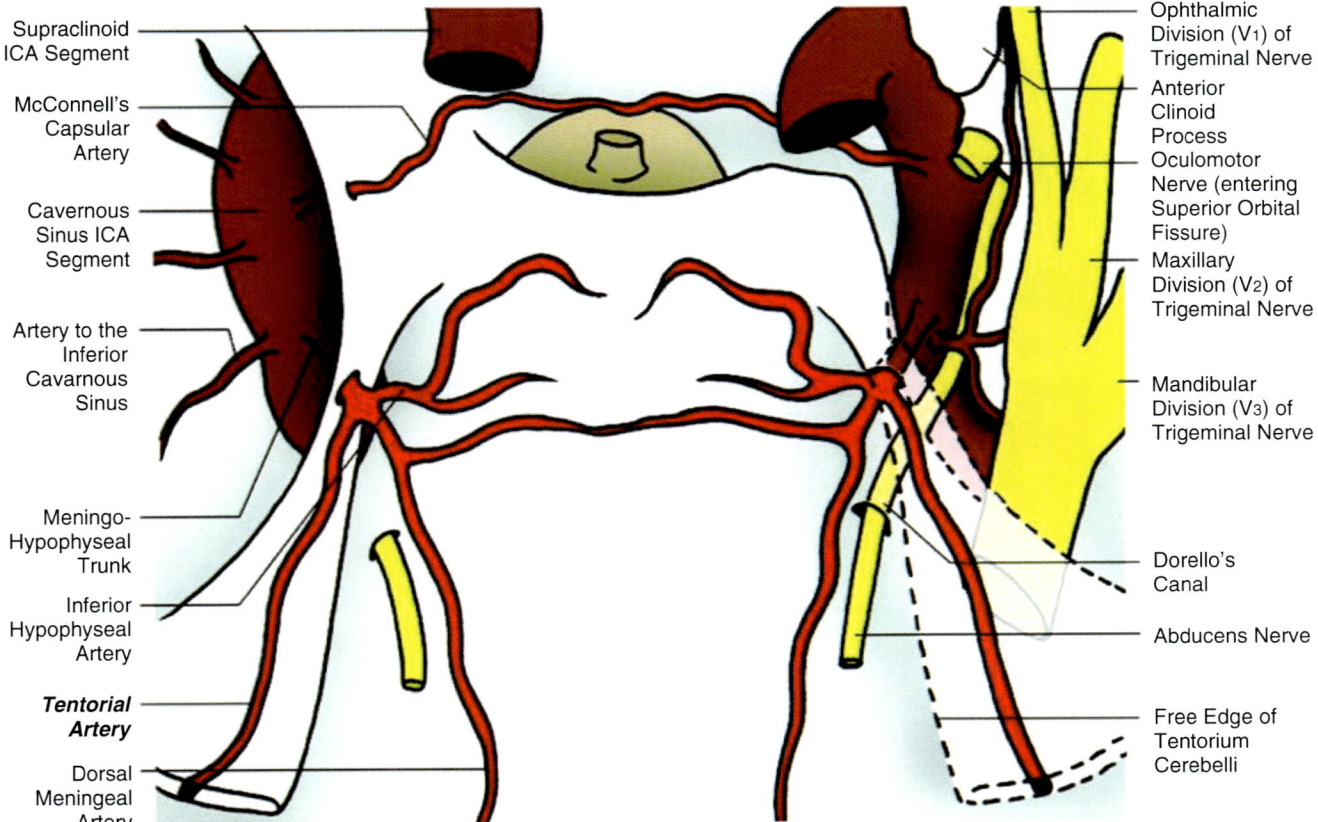

Fig. 12.22 Posterior/inferior hypophyseal artery is the embryologic remnant of the primitive maxillary artery, the very first branch of dorsal aortae. It originally was directed to the eye and served as the initial supply to the optic vesicle. Subsequently, it relocates. Here one observes anterior hypophyseal arteries directly from carotid sinus. IHA supplies the r1 tentorium and the clivus. This Arterial supply from the cavernous ICA is of vital embryologic importance because it supports the r1 mesenchyme of tentorium. [Reprinted from Banerjee AD, Ezer H and Nanda A. The Artery of Bernasconi and Cassinari: A morphometric study for superselective catheterization. *Am J Neuroradiol* October 2011; 32(9):1751–1755. With permission from American Society of Neuroradiology.]

peripheral to its medial colleague. In the posterolateral skull, it anastomoses with StV3 petrosquamous branch. [Fig 12.22]

Arteries of the Clivus

Clivus has a matched set of arteries, both with eminently forgettable names.

Lateral clival artery, AKA *dorsal meningeal artery*, It is a remnant of the primitive trigeminal artery. It also arises from meningohypophyseal trunk, which supplies dorsum sellae and divides. The medial branch goes to clivus. A lateral branch follows superior petrosal sinus along petrous ridge. Thus it remains outside the territory of tentorium. Posteriorly, it makes anastomotic loop with lateral tentorial artery.

Medial clival artery, AKA *inferior hypophyseal artery*, is a remnant of primitive maxillary artery. Yes, it supplies posterior lobe of pituitary but it then proceeds to run down the clivus for good measure.

Note that the clival arteries *descend* down the slope of basioccipital whereas, just lateral to it, flanking branches

from r6 to r7 *ascend* to terminate at the petrous apex. This vascular watershed zone is an example of boundaries amound neuromeric bone fields. Basioccipital r8-r11 is flanked by r6-r7 petrous.

Embryonic precursors

Prior to stage 9, ICA does not exist. Instead, the embryonic CNS is supplied by a primitive plexus and the body is supplied by dorsal aortae. At stage 9, with mesencephalic flexion driven by forebrain growth, the embryonic heart is tucked ventrally at level r0-r1. The connecting segments between dorsal aortae and the heart are stretched downwards and backward like parentheses to form the 1st aortic arches. From the site of flexure, forward growth produces primitive internal carotid which faces three main challenges: (1) establish a connection between dorsal aortae and the neural plexus, longitudinal hindbrain channels; (2) provide an immediate and temporary blood supply for the developing eye; and (3) continue forward to supply the brain.

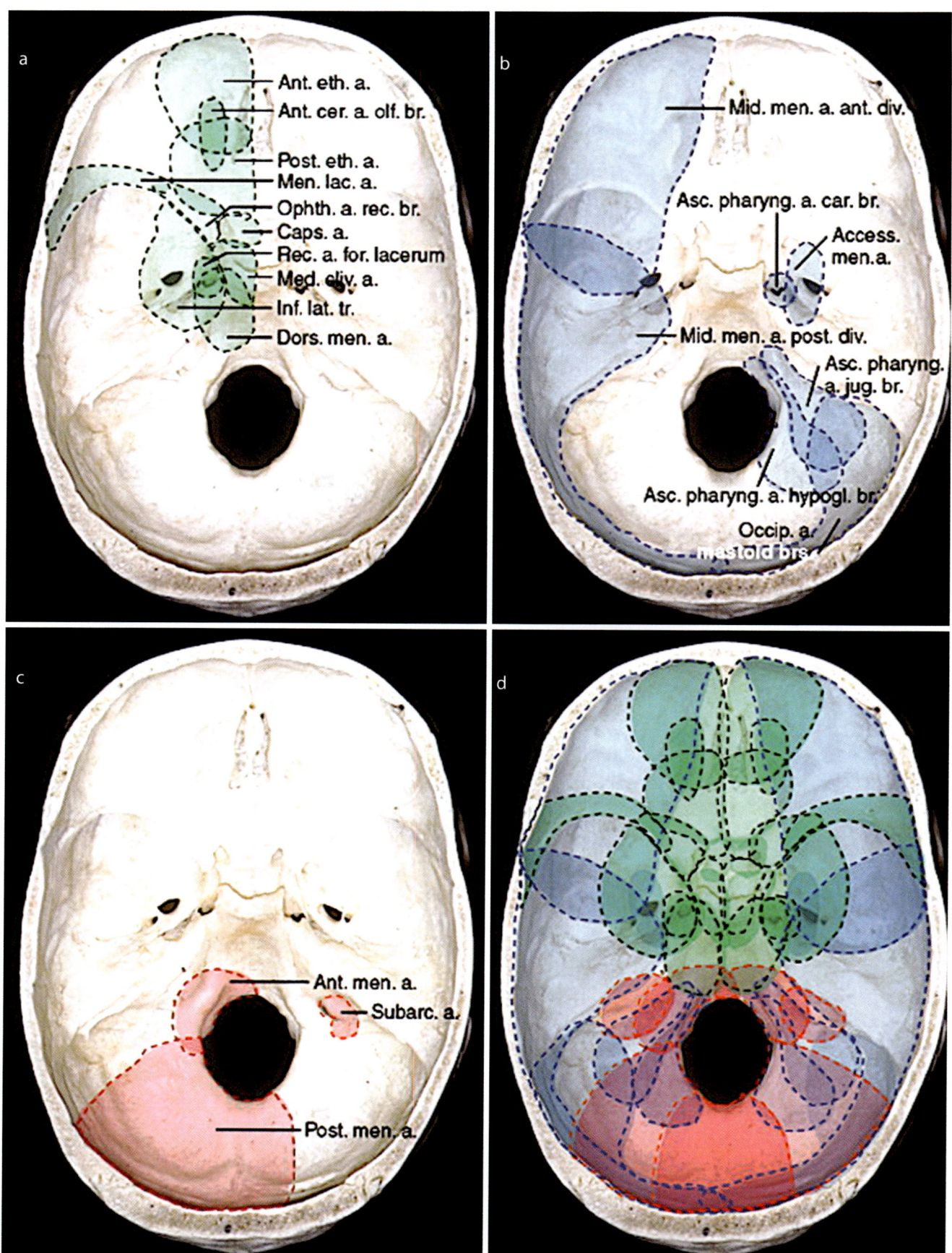

(continued)

Solution 1 is the *primitive trigeminal artery* which the heart to body supplies (dorsal aortae) and hindbrain (primitive plexus). Solution 2 is the primitive maxillary artery which is directed toward the optic vesicle as it erupts from the basal forebrain. Solution 3 is the forward axis of internal carotid. Our attention here is to primitive maxillary which is in genetic register with r0, the very tip of the notochord. As described in Chap 7 and this chapter when alternative vessels supply the eye, primitive maxillary artery is no longer needed. It returns to its original niche at Rathke's pouch, becoming inferior hypophyseal artery.

At this location, IHA is positioned at the interface between the r1 neural crest basisphenoid and the r8 segment of basioccipital. It therefore performs two distinct functions. It follows recurrent branch of StV1 to supply the posteriorly directed r1 mesenchyme of tentorium via *lateral and medial tentorial arteries*. It sends a mesial branch down the clivus to support r8-r11 occipital somitic basioccipital.

Why is this so important? First off, r1 neural crest migration along the cranial base is bidirectional. It tracks forward to form dura and periosteum of the anterior cranial fossa. It tracks backward between posterior cerebral lobe and cerebellum. These two parts of CNS have distinct blood supplies. ICA perfuses forebrain and midbrain while hindbrain *including cerebellum* is supported by the longitudinal neural system (future basilar). So which system is responsible for tentorium? IHA, being relocated at the exact interface between the systems at circle of Willis and cavernous sinus, wins the prize.

The second reason why IHA is important is that the occipital somites lose their designated segmental supply during vascular reorganization of the hindbrain. IHA takes over for basioccipital, while exoccipital and supraoccipital are rescued by occipital, ascending pharyngeal, and 1st cervical.

The Stapedial System: A Derivative the External Carotid System

This subject is covered in the previous vascular chapters, with special contributions by MK Diamond and an extensive review by Rhoton; the reader is referred to these sources for further details. For our purposes, we shall consider the organization of the individual branches and their areas of supply from a selective embryological perspective.

Big picture idea: Trigeminal nerves V1-V3 are the exclusive supply for the entire supratentorial dura, a neural crest product of r1-r3.

Big picture idea: sensory nerves of the trigeminal system accompany and program the individual arteries of the stapedial system. The neuroangiosomes that result delineate the migratory pathways of neural crest to the face and dura.

Trigeminal Anatomy, Reviewed

During stage 14 the cranial nerves develop rapidly. They (especially trigeminal and facial) determine the trajectories of the stapedial system. The innervation pattern is all-important for programming the branches of the stapedial system. *Of particular importance is the anatomic pathway for trigeminal supply to the dura*. After leaving the trigeminal ganglion V1 and V2 travel forward in the lateral wall of the cavernous sinus while V3 descends directly out of the skull. Note: *the most proximal sensory branch of each part of trigeminal is dedicated to the innervation of the dura*.

The following facts about trigeminal are relevant to understand how these nerves conduct the blood supply to the orbit:

(1) V1, upon exiting the cavernous sinus and just prior to entering the orbit, gives rise to the *anterior meningeal*

Fig. 12.23 Blood supply of the individual meningeal arteries: internal carotid (green), external carotid system (blue), and vertebrobasilar (red). **a**, Internal carotid system, here includes two forms of 1 programming: (1) V1 stapedial as ophthalmic artery branches; (2) V2 non-stapedial from the cavernous segment. The dura covering the medial part of the anterior fossa floor is supplied by StV1 and olfactory branches of the anterior cerebral artery. The cavernous internal carotid sector, through its inferolateral trunk and dorsal meningeal artery, supplies the parasellar dura part of the anterior wall of the posterior fossa and the sellar dura via paired capsular, inferior hypophyseal, medial clival, and dorsal meningeal arteries. **b**, External carotid (V2 and V3 stapedial as middle meningeal artery) supplies the dura covering the lateral skull base. The territories of StV1 anterior and posterior branches extend toward the supra- and infratentorial convexity dura and medially over the r1 falx and tentorium. The accessory meningeal and the ascending pharyngeal artery branches contribute supply periosteum between the internal carotid and middle meningeal territories on the middle and posterior fossae. The jugular and hypoglossal branches of the *ascending pharyngeal* arteries supply the inferior portion of the posterior surface of the petrous bone, lateral cerebellar dura, the midclivus, and anterolateral foramen magnum. The mastoid branch of the *occipital artery* constitutes the main supply to the lateral part of the cerebellar fossae. **c** Vertebrobasilar system. The anterior and posterior meningeal branches of the vertebral artery (1st cervical segmental) supply the foramen magnum periosteum. The posterior meningeal artery provides the major supply to the paramedial and medial portions of the dura covering the cerebellar convexity. The subarcuate artery, a branch of the anterior inferior cerebellar artery, supplies the periosteum of the posterior surface of the petrous bone and adjacent part of the internal acoustic meatus, as well as the bone in the region of the superior semicircular canal. **d** Overview. A., artery; Access., accessory; Ant., anterior; Asc., ascending; Br., branch; Brs., branches; Caps., capsular; Car., carotid; Cer., cerebral; Cliv., clival; Div., division; Dors., dorsal; Eth., ethmoidal; For., foramen; Hypogl., hypoglossal; Inf., inferior; Jug., jugular; Lac., lacrimal; Lat., lateral; Med., medial; Men., meningeal; Mid., middle; Occip., occipital; Olf., olfactory; Ophth., ophthalmic; Pharyng., pharyngeal; Pet., petrosal; Post., posterior; Rec., recurrent; Subarc., subarcuate; Tr., trunk. [Reprinted from Martins C, Yasuda A, Campero A, Ulm AJ, Tanriover N, Rhoton A. Microsurgical anatomy of the dural arteries. *Neurosurgery* 2005; 56 (ONS Suppl 2):211–251. With permission from Oxford University Press.]

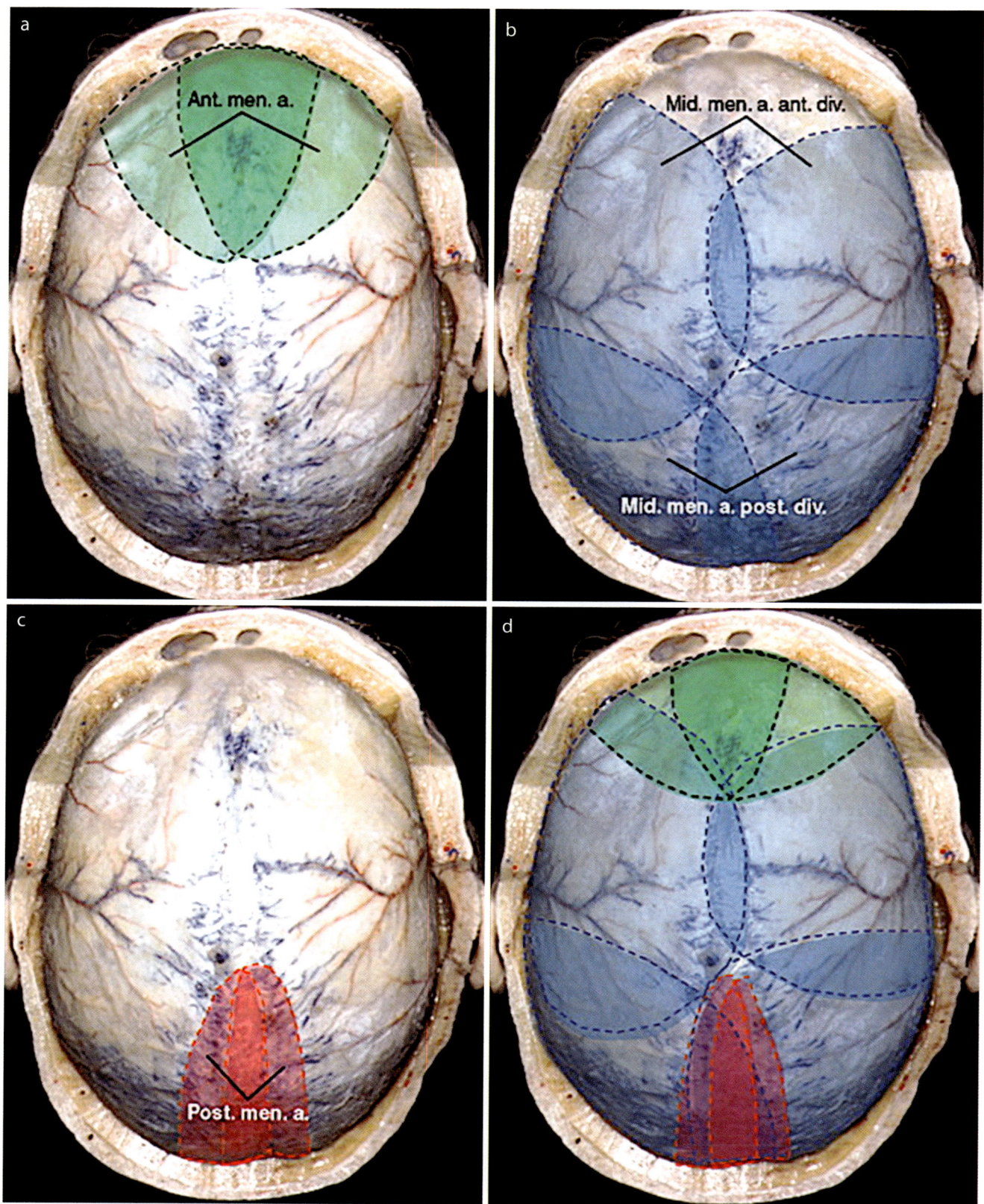

(continued)

nerve supplying the dura of the anterior cranial fossa. **Just behind the greater wing of sphenoid V1 bifurcated to produce (1) a lateral meningeal branch and (2) a medial orbital branch. The meningeal branch proceeds directly forward into the orbit lateral to superior orbital fissue. Development of alisphenoid around the nerve creates a meningo-orbital foramen.. One in the orbit, this branch is directed upward through what will become an un-named the foramen in orbital plate. The orbital branch turns medial and then** then enters the superior orbital fissure. Once in the orbit, V1 again supplies the dura via a separate *dural branch of anterior ethmoid nerve* that re-enters the cranial cavity via the cribriform fossa. A third *recurrent meningeal branch* of V1 supplies the tentorium. As we shall see, this fact is crucial for understanding development.

(2) V2, before exiting the skull via foramen rotundum gives off the ***middle meningeal nerve*** to the postorbital dura of anterolateral frontal lobe behind alisphenoid. Recall that V1 and V2 are running close to each other through the lateral wall of the cavernous sinus. Upon exiting, they part company behind the backwall of gthe orbit. V2 continues downward and lateral into the floor and lateral wall of the temporal fossa while V1 plows straight ahead to the orbits, it then bifurcates, as described above. Retrograde horseradish peroxidase label studies show tracer from the lacrimal gland back to both V1 and V2 parts of the trigeminal ganglion [2]. The V2 component reaches the gland via extratracranial zygomatic nerve from the pterygopalatine fossa. In sum: intracranial V2 supplies the lateral wall of temporal fossa including, conducting StV2 arterial supply to the dura and the inner aspect of the calvaria plate of alisphenoid. ***At no time does intracranial V2 enter the orbit.***

(3) V3, after exiting the skull via foramen ovale, V3 is distributed to lower jaw and oral cavity, but prior to doing so, the nerve is required to send its first branch to the dura. To accomplish this, V3 of the ***recurrent meningeal nerve*** supplying the temporo-parietal-occipital lobes. The nerve accompanies the StV3 middle meningeal artery upward through foramen spinosum to reach its target.

Historical note: Friedrich Arnold [3] described a recurrent branch of V1, *nervus tentorii*, having an extensive distribution to tentorium cerebelli, parieto-occipital dura of the basal posterior lobe, posterior 1/3 of falx, and superior saggital sinus [3]. Arising in r1, the recurrent meningeal nerve of Arnold is in register with the anlage of the cerebellum which arises from that neuromere. It is given off from the superior surface of V1 inside lateral wall of cavernous sinus and tracks backward in parallel with cranial nerve IV. Note that the midsection of the remainder of falx is supplied by V1 while the anterior 1/3 is innervated by anterior ethmoid nerve, arising more distally in the orbit and then tracking backward through the cribriform plate. The distribution of Arnold's nerve explains which pain from this region can be referred forward to the eye and forehead.

Stapedial Stem Divides Within the Tympanic Cavity

The stem of stapedial is the remnant of the primitive *hyoid artery* to the 2nd pharyngeal arch. When the second aortic arch to PA2 disintegrates at stage 12 it leaves behind a dorsal remnant, the hyoid artery, dangling from the dorsal aorta. With the growth of the embryo, the hyoid is repositioned backward toward the otic capsule where it eventually is identified as the sole extracranial branch of ICA.

At stage 14, the definitive circulation of the internal carotid is established. The hyoid artery remains extracranial but it gives off the stapedial stem which immediately tracks upward into the tympanic cavity where stapes form around it (henceforth the name). Directly beyond stapes, the artery bifurcates. The upper division runs directly forward to the trigeminal ganglion where it picks up V1 and V2 just as the nerves exit the cavernous sinus. The lower division follows chorda tympani out from the tympanic cavity into the face via the pterotympanic fissure. It then picks up V3 sensory nerve just below foramen ovale. All subsequent branches of the stapedial system follow the trajectories of V1, V2, or V3.

Fig. 12.24 Superior view of the convexity. Blood supply of the individual meningeal arteries: internal carotid (green), external carotid system (blue), and vertebrobasilar (red). **a** Internal carotid system. The anterior ethmoidal artery (StV1) has also been called the anterior meningeal artery when its territory extends to the dura of the frontal convexity. It gives origin to the anterior falcine artery, also called the artery of the falx cerebri, which supplies the anterior portion of the falx cerebri and adjacent dura covering the frontal pole. **b** External carotid system. The convexity dura is supplied predominantly by branches of the middle meningeal arteries (StV2 and StV3), which supply the dura of frontal, temporal, and parietal convexity and the adjacent walls of the transverse and sigmoid sinus. **c** Vertebrobasilar system. The C1 posterior meningeal artery may reach the dura of the posterior convexity in the area above the torcula. **d** Overview. The dura over the frontal convexity is supplied by the anterior meningeal branch of the anterior ethmoidal artery and branches of the anterior division of the middle meningeal artery that also reaches the dura in the anterior parietal region. The parieto-occipital and petrosquamosal branches of the posterior division of the middle meningeal artery supply the dura over the posterior convexity. A., artery; Access., accessory; A., artery; Ant., anterior; Div., division; Men., meningeal; Mid., middle; Post., posterior. [Reprinted from Martins C, Yasuda A, Campero A, Ulm AJ, Tanriover N, Rhoton A. Microsurgical anatomy of the dural arteries. *Neurosurgery* 2005; 56 (ONS Suppl 2):211–251. With permission from Oxford University Press.]

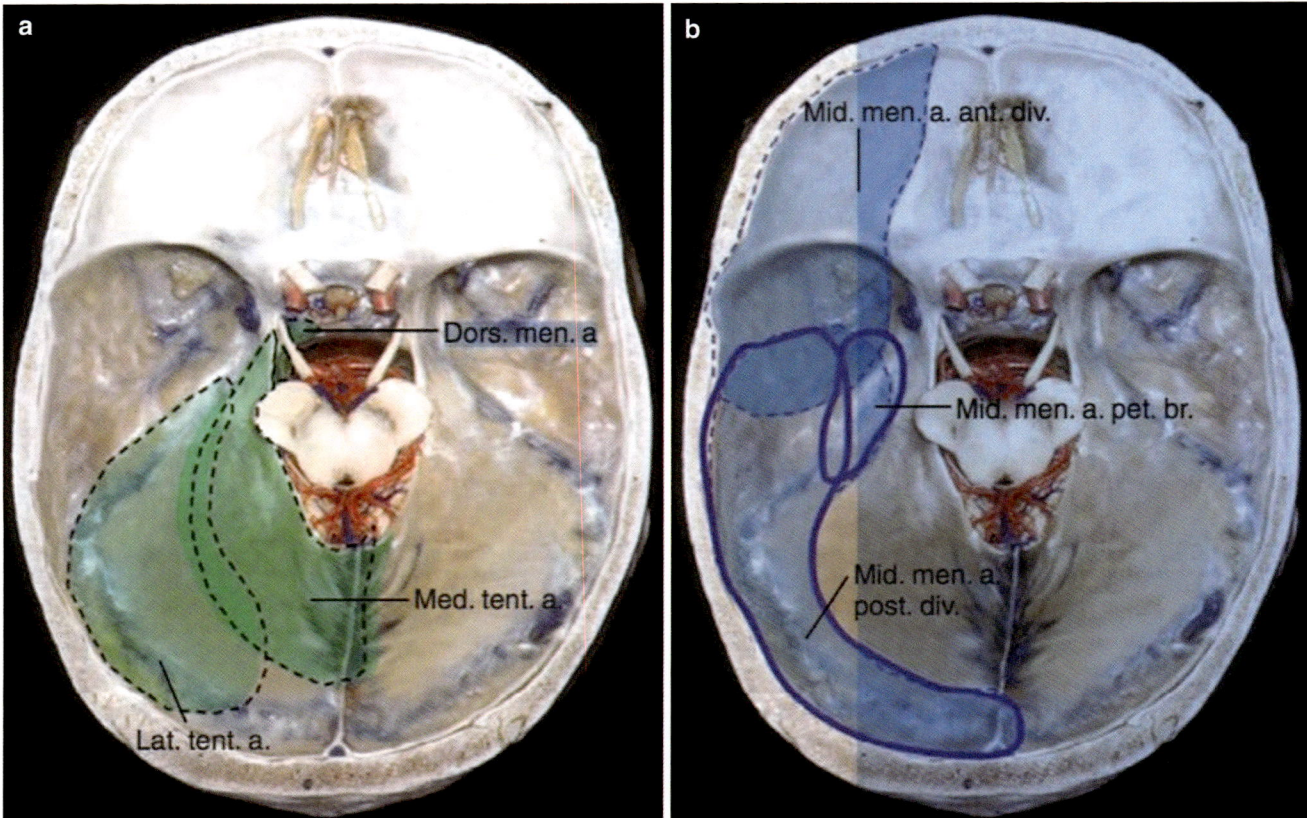

Fig. 12.25 Superior view of the tentorium. Tentorial branches from the cavernous segment of ICA (green), external carotid (blue), and the vertebrobasilar system (red). A, Cavernous ICA (StV1). From medial to lateral, the dorsal meningeal, the medial and lateral tentorial arteries supply the tentorium at its petrosal attachment. B, External carotid system. The branches of the posterior division (StV3) of the middle meningeal artery contribute to the supply of the anterolateral tentorium and extend superiorly to supply the falcotentorial junction and falx. The posterior StV3 branch of the middle meningeal artery gives rise to the *petrosquamosal branch* at the junction of the skull base and convexity and supplies the insertion of the tentorium along the petrous ridge and groove for the transverse sinus; the dura of the torcula; and the junction of the sigmoid, transverse and superior petrosal sinuses. A., artery; Ant., anterior; Br., branch; Div., division; Dors., dorsal; Lat., lateral; Med., medial; Men., meningeal; Mid., middle; P.C.A., posterior cerebral artery; Post., posterior; Tent., tentorial. [Reprinted from Martins C, Yasuda A, Campero A, Ulm AJ, Tanriover N, Rhoton A. Microsurgical anatomy of the dural arteries. *Neurosurgery* 2005; 56 (ONS Suppl 2):211–251. With permission from Oxford University Press.]

Upper division stapedial: forward to the orbit and upward to the V1 and V2 dura.

The stem artery follows greater petrosal nerve forward to the trigeminal ganglion. Here, it picks up cranial nerves V1 and V2 as they run forward in the lateral wall of cavernous sinus. Immediately upon leaving the sinus both V1 and V2 give off *dural branches* prior to exiting the skull. Upper-division stapedial artery bifurcates to provide vascular support for all branches of V1 and V2. These arteries supply their respective areas of dura in the anterior and posterior cranial fossae. Orbital branch of V1 passes through superior orbital fissure. Meningeal branch of V1 gains access lateral to the fissure.

StV1 becomes identifiable at stage 18, enters the orbit during stage 19, and forms an anastomosis with the primitive ophthalmic at stage 20. The "annexation" of stapedial takes place due to the involution of the stapedial stem. Thus, the final product, "ophthalmic artery," is a hybrid system. Primitive ophthalmic, being derived from ICA, supplies CNS tissue exclusively, that is, the optic nerve and globe. All remaining tissues of the orbit innervated by V1 are supplied by branches from the derivatives of the original stapedial system.

V1 stapedial artery follows a similar time course by stage 19 enters the orbit as *two distinct vessels*. StV1 meningeal branch follows the nerve into lateral orbit via a separate *meningo-orbital foramen* in the greater wing of sphenoid. Sometimes this foramen merely a lateral extension of the superior orbital fissure.. StV1 orbital branch goes through the fissure heading directly for the stem of ophthalmic artery. When lacrimal artery forms, both arteries connect to it but the flow changes that occur when orbital branch connects with ophthalmic make it dominant. The lateral STV1 usually involutes but sometimes it persists as the recurrent menin-

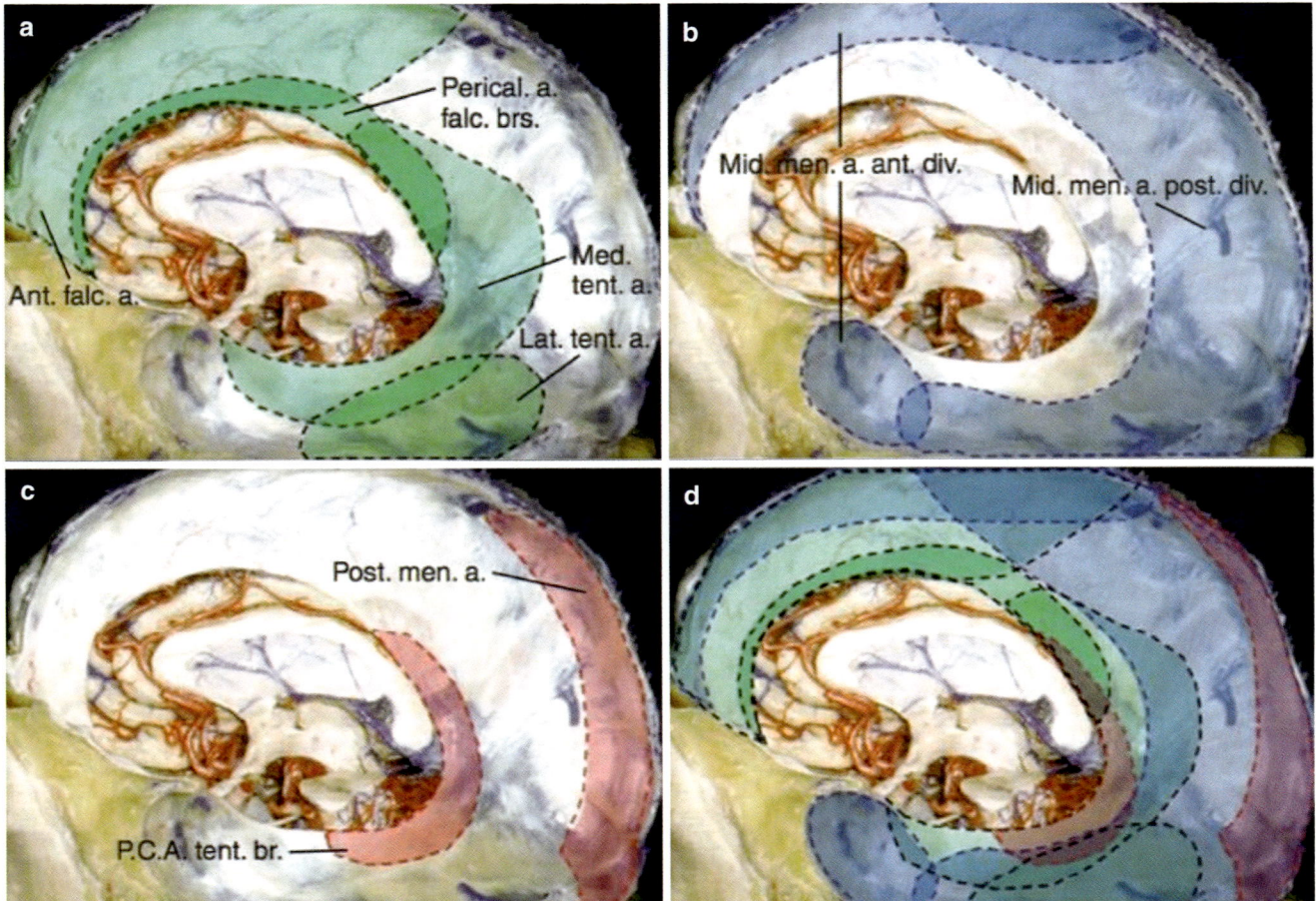

Fig. 12.26 Lateral view showing the supply of the tentorium and falx: dural branches from the internal carotid arterial system (green), the external carotid system (blue), and the vertebrobasilar system (red). **a** Internal carotid system. The *anterior falcine artery*, the distal continuation of the StV1 anterior ethmoidal artery, enters the falx at the cribriform plate and supplies the anterior portion of the falx cerebri and adjacent dura covering the frontal pole. The free border of the falx and the walls of the inferior sagittal sinus receive branches from the anterior cerebral *pericallosal arteries* anteriorly and the posterior cerebral *medial tentorial artery* posteriorly. **b** External carotid system. The anterior and posterior divisions of the middle meningeal artery (StV2 and StV3) supply the walls of the superior sagittal sinus and give rise to descending branches that are the *main supply to the falx and the falcotentorial junction*. **c** Vertebrobasilar system. The posterior meningeal arteries (C1 segmental) reach the falcotentorial junction and posterior third of the falx cerebri. **d** Overview. A., artery; Ant., anterior; Br., branch; Brs., branches; Div., division; Falc., falcine; Lat., lateral; Med., medial; Men., meningeal; Mid., middle; P.C.A., posterior cerebral artery; Perical., pericallosal; Post., posterior; Tent., tentorial. [Reprinted from Martins C, Yasuda A, Campero A, Ulm AJ, Tanriover N, Rhoton A. Microsurgical anatomy of the dural arteries. *Neurosurgery* 2005; 56 (ONS Suppl 2):211–251. With permission from Oxford University Press.]

geal artery (RMA). The dual anastomosis to lacrimal explains when, in some cases, no connection with ophthalmic transpires. In this case, the entire system of non-ocular orbital arteries is supplied by the StV1, connecting backward to the middle meningeal system In any case, two dual branches of StV1 explains why the clefts can occur in the lateral-most zone of the orbit (Tessier zone 9).

At stage 20 upper division stapedial becomes a target for a 3rd artery arising programmed by V3. This anastomosis takes place proximal to the ganglion, that is, before StV1 and StV2 have bifurcated. Extracranial stapedial (see below) passes up from below via the foramen spinosum as the mid-

dle meningeal nerve and artery. When this anastomosis takes place, the previous connection with stapedial stem involutes. StV2 now becomes the *anterior branch of the middle meningeal artery*. StV3 subsequently defines the *posterior branch of middle meningeal artery*.

Nota bene: RMA immediately connects with lacrimal. The connection generally involutes but sometimes the entire orbital stapedial system can be filled from lateral StV1 via a patent RMA. In the orbital floor zygomatic nerve and artery track laterally, giving off zygomaticofacial nerve and artery to the jugal bone field. Zgyomatic nerve continues up to lacrimal gland where, as zygomatico-temporal nerve,it joins

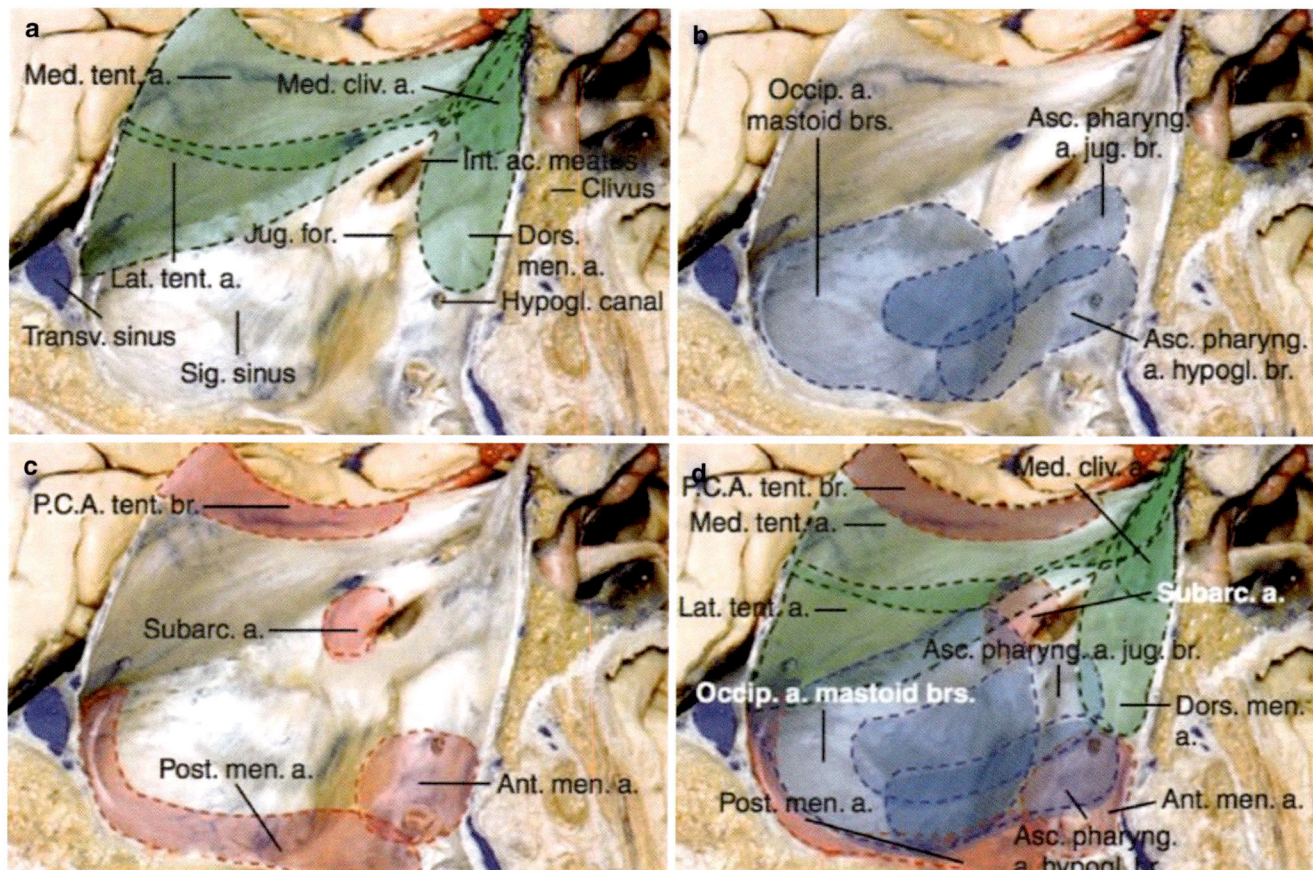

Fig. 12.27 Posterior fossa and tentorial dura. The view is directed from medially into the left half of a posterior fossa in which the cerebellum was removed. The clivus in on the right and the transverse sinus on the left. Blood supply of the individual meningeal arteries: internal carotid (green), external carotid system (blue), and vertebrobasilar (red). **a** Internal carotid system, StV1 cavernous segment. The *medial tentorial artery* supplies the medial third of the tentorium and the *dorsal meningeal* and the *lateral tentorial artery* contribute to the arcade that supply the attachment of the tentorium to the petrous ridge. The medial clival and dorsal meningeal arteries supply the dorsum sellae and upper clivus. The medial edge of the tentorium is also supplied through a branch of the posterior cerebral artery (incorrectly classified as vertebrobasilar). PCA, tentorial branch should be colored green. **b** External carotid system. The hypoglossal and jugular branches of the r6-r7 ascending pharyngeal artery supply the periosteum of petrous zone and S3-S4 supraorbital. Branches of the r4-r5 occipital artery supply the periosteum of the lateral part of the cerebellar fossa and the internal surface of the mastoid bone. The mastoid branch of the occipital artery constitutes the main supply of the lateral part of the cerebellar fossae and has a role on the supply of the lateral tentorial attachment. **c** Vertebrobasilar system. The *subarcuate artery*, a branch of the anterior

inferior cerebellar artery, supplies the posterior surface of the petrous bone above the internal acoustic meatus and surrounding the subarcuate fossa. The anterior and posterior meningeal branches of C1 segmental artery supply the foramen magnum periosteum and dura descending below foramen magnum. The posterior meningeal artery supplies the medial and intermediate portions of the cerebellar fossae periosteum. The vertebrobasilar system is given credit for suppling the medial edge of the tentorium on the false assumption that through posterior cerebral artery is not an ICA derivative (which it is). This PCA zone should be colored green. **d** Overview. Branches derived from all three arterial systems supply the dura covering the posterior surface of the petrous bone and clivus. A., artery; Ac., acoustic; Asc., ascending; Ant., anterior; Br., branch; Brs., branches; Cliv., clival; Dors., dorsal; For., foramen; Hypogl., hypoglossal; Int., internal; Jug., jugular; Lat., lateral; Med., medial; Men., meningeal; Occip., occipital; P.C.A., posterior cerebral artery; Pharyng., pharyngeal; Post., posterior; Sig., sigmoid; Subarc., subarcuate; Tent., tentorial; Transv., transverse. [Reprinted from Martins C, Yasuda A, Campero A, Ulm AJ, Tanriover N, Rhoton A. Microsurgical anatomy of the dural arteries. *Neurosurgery* 2005; 56 (ONS Suppl 2):211–251. With permission from Oxford University Press.]

lacrimal nerve to provide PANS supply to the gland. From there, ZTN swings posteriorly to supply to exit the orbit and supply the postorbital field of the zygomatic complex. In doing so, it takes along with it an artery *having its stem from lacrimal*. Thus, zygomaticotemporal artery originates from infraorbital but can also receive flow from lacrimal artery. It is possible that ZTA represents a remnant of the embryonic RCA. Despite their differing sources of blood supply, both ZFA and ZTA are StV2 neuroangiosomes.

Lower division stapedial: forward to the jaws and upward to the V3 dura

As soon as chorda tympani departs from the tympanic cavity, it makes a beeline for the sensory root of V3, where it seeks out lingual nerve by which to convey itself to the tongue. Lower division stapedial tracks along with the nerve. At the same time, the poorly-named maxillary branch from external carotid tracks forward toward V3. At stages 18–19, anastomosis takes place between lower division stapedial and ECA creates the hybrid *maxillomandibular artery* (MMA).

These anastomoses cause increase distal flow with dramatic anatomic consequences. (1) MMA now becomes a composite structure. (2) Proximal extracranial stapedial shrinks down but persists as anterior tympanic. (3) Intracranial StV2 unites with the "newcomer" intracranial StV3 to form the definitive intracranial dural artery system. (4) StV1 attaches to primitive ophthalmic and the proximal segment extending back to trigeminal ganglion involutes.

MMA has two functions and three distinct zones. Branches associated with the stapedial system supply those structures reassigned from the 1st arch: jaws, the suspensory bones of the maxilla, and the dura. Branches associated with the external carotid supply all original structures of the 1st arch, such as the muscles of mastication, fat, and glands. The zones of MMA are as follows. The *proximal (mandibular) zone* gives off stapedial derivatives that re-enter the skull to supply the tympanic cavity and the dura of the middle and posterior cranial fossa. It also sends StV3 inferior alveolar to supply the mandible. The sole ECA branch of the proximal zone supplies mylohyoid. The branches of the *middle (infratemporal) zone* are distributed exclusively to muscles of mastication. *Distal (pterygopalatine) zone* gives off StV2 branches in the pterygopalatine fossa that are subsequently distributed to the maxilla and its suspensory bones.

Big picture idea: anastomoses to the stapedial system are responsible for its demise

Reunification and disappearance of the stapedial system (Please see Figures from Chaps. 6 to 7)

Toward the end of the embryonic period, the anatomy of the stapedial system is drastically altered, making it virtually unrecognizable. Three anastomoses are responsible for these changes.

(1) At stage 20, within the orbit, StV1 is annexed by ophthalmic. Its proximal segment dies back to the bifurcation with StV2.
(2) Also at stages 18–19, inferior division stapedial joins external carotid just lateral to V3. This anastomosis creates the hybrid *maxillomandibular artery* (MMA).
(3) From MMA, middle meningeal artery follows sensory V3 middle meningeal nerve upward to supply the dura. At stage 20 middle meningeal annexes StV2. It does not annex StV1 proximal to the orbit because this segment has already undergone involution. Thus, intracranial middle meningeal becomes a hybrid system: its anterior branch arises from StV2 and its posterior branch arises from StV3.

The common denominator of these anastomoses is the exposure of the distal vessel tor higher flow; the proximal segment therefore involutes. StV1 artery dies backward from the superior orbital fissure to the bifurcation of the superior division. **StV1 meningeal artery, the RMA**, dies backward from meningo-orbital foramen to its intersection with StV3. Note that RMA in the orbit leaves behind the source of zygomatico-temporal artery to zone 8. Superior division of stapedial is eliminated completely back to stapes. Inferior division of stapedial distal to stapes persists as the *anterior tympanic artery*. Proximal to stapes the stapedial stem persists as the *caroticotympanic artery*.

Fate of the Intracranial Stapedial System

V1 stapedial first gives off a dura branch to the anterior cranial fossa. Then, having traversed the superior orbital fissure it divides into three branches, each of which supplies structures both within the orbit and outside of the orbit.

- *Supratrochlear* (frontal) and its branches supply zones 13-12 composed of the ethmoid complex, the nasal envelope, and the medial forehead between the eyebrows.
- *Supraorbital* supplies zones 11-10 consisting of the orbital roof, the upper eyelid and lateral forehead defined by the eyebrows.
- *Lacrimal* supplies zone 9 which includes alisphenoid, the lacrimal gland and then lateral corner of the upper eyelid.

V2 stapedial supplies the dura of antero-lateral temporal fosa and the lateral orbit.

Fate of the Extracranial Stapedial System

In the pterygopalatine fossa maxillomandibular artery gives off branches to the maxillary complex and palate, all of which are V2 nerves. The arteries supplying the two fields of zygoma, jugal and postorbital (see below) represent Tessier zones 7 and 8. Greater sphenoid wing originally arose as the *epipterygoid bone*, part of the ancient *palatoquadrate carti-lage* from which maxilla evolved. The original function of epipterygoid was to serve as intermediate support between palate and maxilla. As such, the blood supply for AS logi-cally relates to the structures hence its supply from the extra-cranial *zygomatic artery*. Additional support for alisphenoid comes from temporalis muscle via *anterior deep temporal branch* of the external carotid component of maxillomandibular artery. In sum, Tessier zone 9 is complex, supplied by the confluence of three distinct systems.

To complete our picture, V3 stapedial supplies two functionally-related zones: the mandible and the anterior tympanic cavity, including malleus and incus. Inferior alveolar supplies three dental units and the ramus whereas anterior tympanic is directed back to the temporal complex. This reflects the complex paleohistory of the mandibular bone complex in which its proximal components such as quadrate were brought backwards into the skull as structures of hearing.

Stapedial Branches to the Dura

Ophthalmic (V1) Branches of ophthalmic all represent the V1 stapedial added on to the primitive ophthalmic from internal carotid.

Anterior ethmoid : post-frontal dura, anterior falx
Posterior ethmoidal: medial 1/3 anterior fossa
Recurrent deep: cavernous sinus
Recurrent, superficial: lesser sphenoid wing, anteromedial middle fossa
Lacrimal: superior orbital fissure, sphenoid wings

Middle meningeal (V1, V2, and V3) The proximal 1/3 of maxilla-mandibular artery, gives rise to middle meningeal artery and accessory meningeal artery. These supply, collectively, nearly the entirety of dura over the cerebral convexities, the cerebellum and a significant sector of basal dura. They are accompanied by branches of V2-V3.

Middle meningeal artery

Petrosal branch (V1): trigeminal ganglion, geniculate ganglion
Anterior branch (V2: goes all the way forward to the back wall of the orbit
Posterior branch (V3): posterolateal middle fossa

Accessory meningeal artery

External: eustachian tube, external auditory meatus
Internal: medial middle fossa, CN III, IV, V, VI, VII

Intracranial Internal Carotid Provides Minor Branches to the Dura

Anterior cerebral artery provides minor branches to the medial 1/3 floor of anterior fossa. Posterior cerebral, in like manner supplies posterior falx and adjacent tentorium. It is of interest that neural crest cells from specific neuromeres have been mapped to the arterial walls of the internal carotid system.

External Carotid System

The branches of ECA give rise two dural arteries. The occipital a. is a derivative of 2nd aortic arch artery. Ascending pharyngeal is a derivative of 3rd aortic arch artery. Both must access the skull via foramina and they supply the posterior cranial fossa [Figs. 12.28, 12.29, 12.30 and 12.31].

Confusion arises regarding maxillary artery, as its middle meningeal branch is commonly considered to supply the dura. In reality, the stapedial system, derived from the dorsal remnant of 2nd aortic arch artery, is the genetic source for this supply. Its anastomotic relationship with IMMA have been previously detailed.

The *infratentorial periosteum (misconstrued as dura)* covers the bones of the posterior fossa below the transverse sinus. These correspond to the petromastoid complex and the occipital bone complex (supraoccipital, exoccipital, and basioccipital), both discussed at length previously. These mesodermal bones arise from r4 to r11 and ossify in carti-lage. The zones are dependent on the external carotid system.

Aortic arch arteries: precursors of the external carotid system

Human embryos possess five pharyngeal arches (PAs) with which craniofacial structures are assembled. Running through the core of each arch is a vascular axis, *aortic arch artery* (AA), that spans from the cardiac outflow tract tucked below the pharynx upward to paired dorsal aortae. The first four pharyngeal arches are important. PA5 is diminutive, producing the arytenoids, cricoid, and associated muscles. For this reason, the 5th aortic arch artery involutes almost immediately, making PA5 dependent upon the blood supply of PA4 for its survival. *There is no 6th pharyngeal arch.* Instead, the fate of the 6th and final aortic arch artery is to form pulmonary circulation.

Note: Textbook illustrations of the aortic arches are mis-leading. We can easily understand how they work by means of a little *embryonic origami*. Fact: the 1st aortic arch artery does *not* "sprout forth' from the cardiac outflow tract. At stage 8, the embryo is still a flat trilaminar disc. The embry-

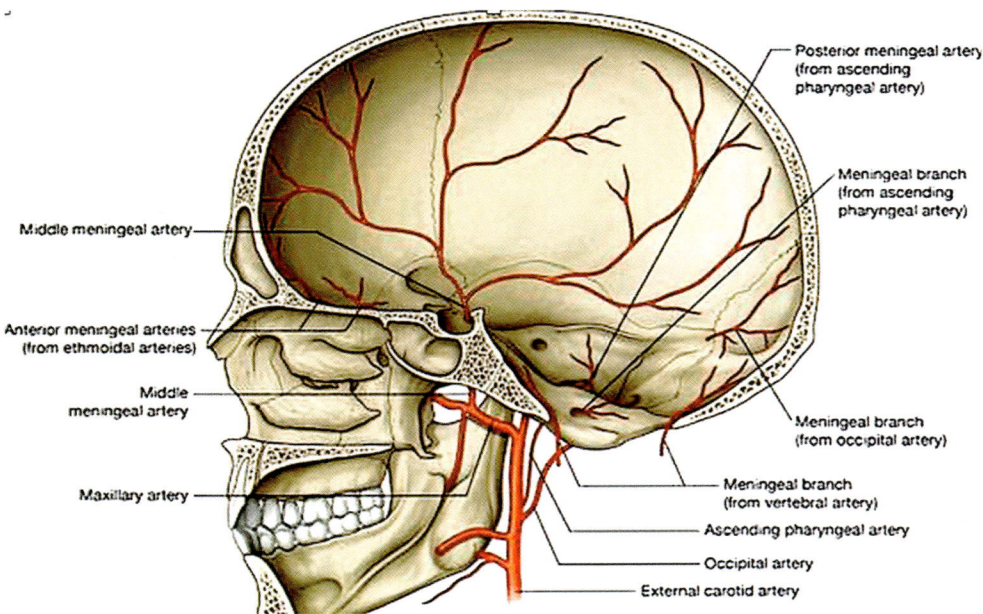

Fig. 12.28 Posterior fossa periosteum is supplied by two arterial systems. The occipital artery r4-r5 is developmentally related to 2nd aortic arch artery. It enters the skull via jugular foramen and condylar foramen. For this reason, it supplies S1–S2 that form the upper zone of supraocccipital bone. The ascending pharyngeal artery r6–r7 is developmentally related to 3rd aortic arch artery. Its neuromeningeal division enters the skull at three sites: foramen magnum to supply the clivus, hypoglossal foramen to supply foramen magnum, and jugular foramen to supply S3–S4 that form lower zone of supraoccipital. [Reprinted from Drake R, Vogel AW, Mitchell AWM. Gray's Anatomy for Students, 3rd edition. Philadelphia, PA: Churchill-Livingstone. 2015. With permission from Elsevier.]

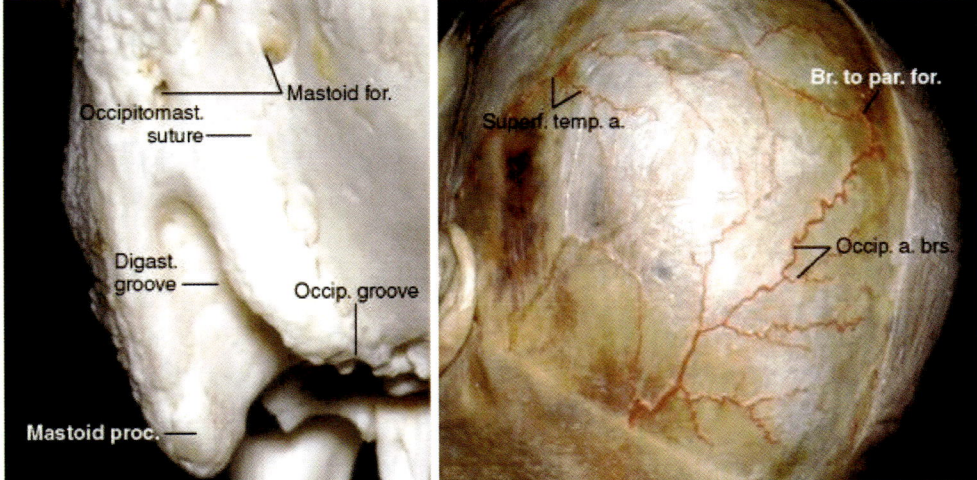

Fig. 12.29 Left: 2nd segment of occipital artery sends meningeal branch through the *mastoid foramen*, a probable boundary between r4-r5 and r6-r7 fields of mastoid. Right: 3rd segment ascends over occipitalis and perforates *parietal foramen* to supply dura of interpari-etal /parietal boundary. [Reprinted from Martins C, Yasuda A, Campero A, Ulm AJ, Tanriover N, Rhoton A. Microsurgical anatomy of the dural arteries. *Neurosurgery* 2005; 56 (ONS Suppl 2):211–251. With permission from Oxford University Press.]

Fig. 12.30 Ascending pharyngeal system r6-r7 has two trunks: pharyngeal and neuromeningeal. From the latter, clival branch ascends of basioccipital, and it is directed to the S1-S2 zone. Neuromeningeal enters the skull via hypoglossal foramen and jugular foramen. [Reprinted from Hacein-Bey L, Daniels DL, Ulmer JL, Mark LP, Smith MM, Strootmann JM, Brown D, Meyer GA, WackLyn PA. The ascending pharyngeal artery: branches, anastomoses, and clinical significance. *Am J Neuroradiol* 2002; 23:1246–1256. With permission from American Society of Neuroradiology.]

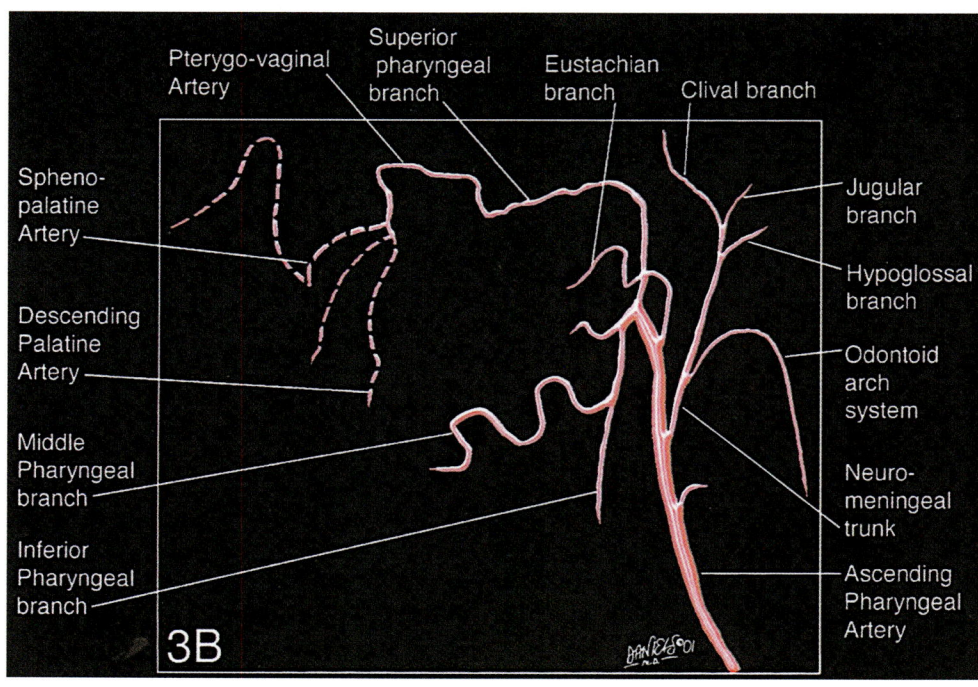

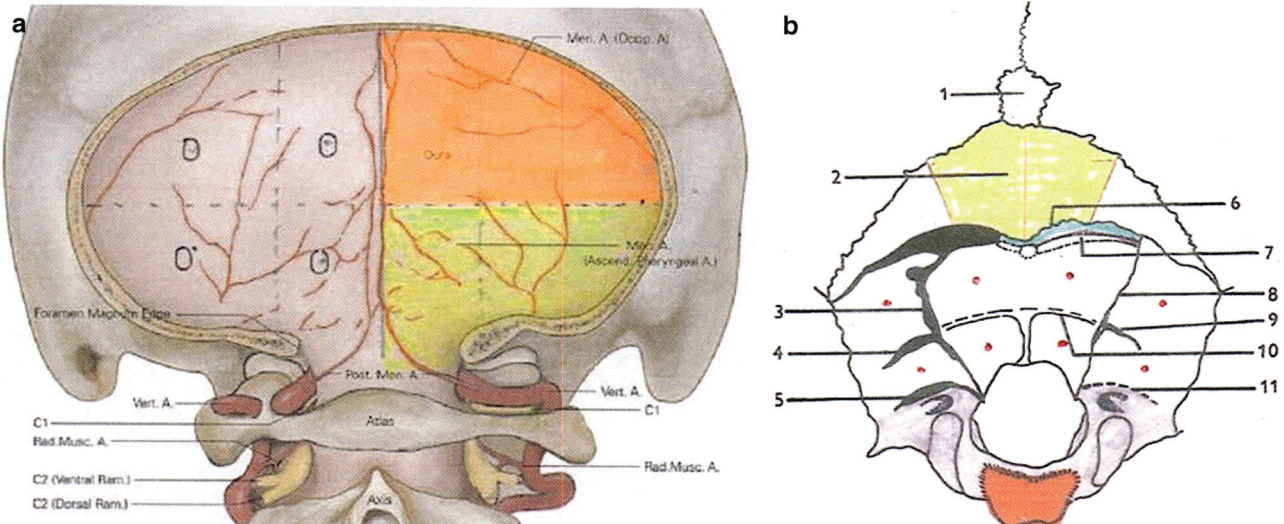

Fig. 12.31 Posterior meningeal arteries from 1st cervical intersegmental track up the midline to the torcula (not seen). Four quadrants were created which refer to the bone fields of Srivastava. A: [Reprinted from Zhao J, Feng X, Ren J, Manjila S, Bambakidis NC. Dural arteriovenous fistulas at the craniocervical junction: a systematic review. *Journal of NeuroInterventional Surgery* 2016;8:648–653. With permission from BMJ Publishing Group Ltd.] B: [Reprinted from Srivastava HC. Development of ossification centres in the squamous portion of the occipital bone in man. *J Anat* 1977; 124(3):643–647. With permission from John Wiley & Sons.]

onic heart field is located in front of the future brain and is connected to the two dorsal aortae that run passively backward along the length of the embryo. At stage 9 explosive growth of the forebrain forces the embryonic head to flex downward almost 150°. The hinge point, located at the midbrain, is called the *mesencephalic flexure*. The heart is now positioned under the neuraxis; specifically, directly below posterior pharynx.

Because the dorsal aortae are attached to the ventricles of the heart, in the process of folding they become stretched downward and backward, forming bilateral arcs connecting the heart below with the embryo above. They have now termed the *1st aortic arch arteries*. The remaining five aortic arch arteries develop one per stage. Each arch represents the union of two distinct stems. From above, the dorsal aortae send out ventrally-directed stems while, from below, the outflow tract produces paired dorsally-directed stems. The two sets of stems unite to produce the aortic arch artery. Failure of this anastomotic process explains: (1) the failure of AA5; and (2) the breakdown and reorganization of the system.

The six aortic arches appear in craniocaudal sequence in stages 9–14. The pharyngeal arch becomes visible one stage later. After their formation in stages 10–11, the 1st and 2nd pharyngeal arches merge together in stages 12–13, thus creating the tissue masses that will complete the construction of the face. In the process, their aortic arch arteries disintegrate, leaving behind arterial remnants dangling from the dorsal aortae. *Hyoid artery*, the dorsal remnant of AA2, will become the source of the stapedial system.

By the end of the pharyngeal arch period (stage 14), AA3 is transformed into the common carotid; it gives off the external carotid artery and then continues onward as the extracranial internal carotid. AA4 produces paired subclavians and, on the left side, the aortic arch. The two dorsal aortae into which AA4 inserts have merged into a single descending aorta. As a result of embryonic growth and spatial rearrangement of the pharyngeal arches the hyoid artery becomes the *first branch of the extracranial internal carotid artery* just below the otic capsule. Hyoid artery will morph into the stem of stapedial system.

The final configuration of the external carotid branches is as follows:

AA1 – 1st arch: maxillary, lingual
AA2 – 2nd arch: facial, superficial temporal, posterior auricular, occipital
AA3 – 3rd arch: ascending pharyngeal
AA4 – 4th arch: superior thyroid

Occipital Artery

This derivative of AA2 supplies structures in register with r4-r5 and supports the territories of r8-r9. It follows motor branches to facial muscles and DIF, sends penetrating branches down from SIF to underlying dura of posterior fossa. In some instances, once the bone forms, *emissary arteries* are an embryologic hold-hover of this process [cf Figs. 12.28 and 12.29].

Occipital artery has three segments: ascending cervical (described), horizontal occipital, and vertical occipital. Branches supplying posterior fossa originate from the 2nd and 3rd segments. The stem is given off from the posterior aspect of ECA at the mandibular angle. The first segment courses upward, being crossed by cranial nerve XII, running deep to digastric until it reaches mastoid process. Here it can travel either deep or superficial to longissimus capitis muscle.

Second segment *mastoid branch* enters via mastoid foramen at superior nuchal line and appears intracranially below transverse sinus, the embryologic boundary between the supraoccipital complex and interparietal complex. It takes three directions (1) It travels posteromedial to supply the lateral superior quadrant of supraoccipital. (2) It descends inferior to jugular foramen and anastomosis with ascending pharyngeal. (3) It ascends to supply the periosteum over temporal bone at the cerebello-pontine angle.

The third segment supplies the soft tissues covering the posterior cranium. It enters the skull at parietal foramen 3–5 cm cephalad to the lambda; *parietal branch* supplies dura lf the interparietal bone fields.

Ascending Pharyngeal Artery

This derivative of AA3 supplies structures in genetic register with r6-r7 and supports the territories of r10-r11. Its anterior pharyngeal division supplies soft palate, pharynx, and the tonsillar fossa. A *carotid branch* from this division supplies the periosteum of the carotid canal.

The posterior neuromeningeal division of APA must gain access to the skull via foramina. *Hypoglossal branch* follows cranial nerve XII as it exits the skull. Via the hypoglossal (anterior condylar) canal it is distributed in ring-like fashion to foramen magnum and clivus, following the mesenchyme of S4. The *jugular branch* follows cranial nerves IX, X, and XI as they exit through jugular foramen whereupon it divides. Medial branch supplies inferior petrosal sinus and hugs the territory of medial clival artery from cavernous carotid (inferior hypophyseal). It also supplies the petrous prootic field [cf Fig. 12.30].

Dorsal Aortae Segmental Arteries/the Vertebral System

1st Cervical Segmental Artery:

Two poorly named arteries are of great interest to us: anterior meningeal artery and posterior meningeal artery of extracranial vertebral are hopelessly obscure but important to understand the posterior cranial fossa. Here we shall be a bit more exacting for the sake of clarity. Both of these vessels arise from the first cervical segmental arteries, the final cross-link between the longitudinal neural arteries and the dorsal aortae prior to their demise and reconstruction [Fig. 12.31].

1. *Anterior meningeal branch of 1st cervical* supplies odontoid process, dura of the atlanto-occipital space, and anterolateral border of foramen magnum and occipital condyles. It anastomoses with ascending pharyngeal outside hypoglossal foramen.

2. *Posterior meningeal branch of 1st cervical* supplies dura of posterior antlanto-occipital space, falx cerebelli, paramedical cerebellar fossa, dura up to the transverse sinus. It ascends up the falx cerebelli, making successive anastomoses with ascending pharyngeal and occipital, thus defining the supraoccipital bone fields into four quadrants. Above the torcula, it terminates by anastomosing with two branches of StV3: petrosquamosal and parieto-occipital.

The first cervical arteries differ from those of C2 and C3 because they arise from the transverse bend of the vertebrals between atlas and the skull base. [The vertebrals subsequently ascend into the skull meet up of the longitudinal neural arteries.] In reality, 1st cervical is the original artery supplying derivatives of S5. It was designed to run *beneath* the proatlas (the original 1st cervical vertebra) but when proatlas was incorporated into the skull these arteries found themselves shipwrecked *above* the atlas. What do they supply and why?

All the derivatives of S5 are associated with the skull base. Proatlas forms the tip of the dens, a ligament from dens to basioccipital, and the posterior rim of foramen magnum and contributes (along with exoccipitals) to the occipital condyles. As S5 muscles are more numerous posteriorly, the posterior branches of c1 arteries are considerably larger than the anterior ones.

Caveat Embryologic confusion arises when these arteries are ascribed to "level C2-C3" as in the neurosurgical literature. The neuromeric levels of the occipito-cervical junction are described in detail in Chap. 10.

Longitudinal Neural/Basilar System

Cerebellum, as an r1 derivative, cannot be supplied by the internal carotid system. Recall that, immediately after giving off the posterior cerebral arteries, paired posterior communicating arteries anastomose LNAs. These come together in the form of the letter "Y" to fuse and form basilar. The midline extent of basilar is from r1 to r7, at which point the original LNAs persist although they are re-named vertebrals. The actual vascular boundary between the two systems is at level r11-c1 where the 1st cervical segmentals constitute the last connection between the old dorsal aortae and the LNAs prior to revision into the vertebral system.

Anterior Inferior Cerebellar Artery

This artery gives rise to subarcuate artery, which can also arise from the labyrinthine artery. Subarcuate anastomoses with r4-r5 stylomastid artery and r4-r5 mastoid branch of occipital artery. This helps define the petrous bone anterior to internal auditory canal as pro-otic.

Analysis of Arterial Supply

Infratentorial "dura" (sic) periosteum is synthesized from PAM originating from rhombomeres r4-r11. Its blood supply comes from meningeal branches of the two branches of external carotid plus the meningeal arteries from c1. Occipital artery represents r4-r5 and supplies epaxial paraxial meso-

derm. It logical to expect that this artery will cover the r4-r5 prootic bone. Its branches extend posteriorly to the dura covering supraccipital bone. Ascending pharyngeal represents r6-r7 and defines epaxial paraxial mesoderm dedicated to the basioccipital, exoccipital, and the relatively small intracranial zone of opisthotic.

Why is there no apparent artery dedicated to mesenchyme from rhombomeres r8 to r11, that is, the occipital somites S1-S4? Why jump from the external carotid system all the way to the vertebrals? The answer lies in the embryologic formation of the vertebral system. Recall from Chaps. 6 and 7 that the occipital somites are originally supplied by segmental branches of paired dorsal aortae, a pattern continued all the way down the neuraxis. At stages 15–16, the system undergoes a radical transformation. Subclavians develop and give off vertical vessels that run up the neck interconnecting all the transverse segmentals. The great vessels of the heart morph into a single descending aorta and the paired dorsal aortae disappear. The vertical vessels become the vertebrals, At stage 17 the process is complete. At the level of r11-c1, the first cervical branch connects the vertebral with the old longitudinal neural arteries. And by stage 18 the external carotid system is fully developed with identifiable ascending occipital and pharyngeal arteries.

Faced with this "vascular drought" occipital somites get their blood supply in various and sundry ways. Tongue muscles move into pharyngeal arch territory and are supplied from external carotid. Sternocleidmastoid and trapezius are polysomitic muscles supplied by branches arising off the subclavian. But what of the bone derivatives of S1-S4?

These somites accept a "charity donation" from the external carotid system in a highly specific manner. Recall that, in neuromeric terms, occipital artery (r4-r5) is positioned anteriorly to ascending pharyngeal artery (r6-r7). S1 and S1 therefore "adopt" the more rostral of these two vascular axes, the occipital artery, whereas S3 and S4 depend upon the more caudal axis of ascending pharyngeal artery. Recall as well that each of the four occipital somites contributes to the occipital bone complex. Huang's mapping shows that S1-S4 are laid down as concentric rings, with S1 being most peripheral and S4 at foramen magnum. For this reason, we find occipital supplying the periphery of supraoccipital and ascending pharyngeal surround the lower half of the bone.

Paleontologic note: When we examine the blood supply of the dura underlying the supraoccipital bone (see Fig. 12.31 with the bone removed) the striking pattern divides the dura into four (perhaps eight) quadrants. These correspond exactly to the subdivisions of the supraoccipital bone, each one of which bears a specific extensor muscle. Note: The zone between the superior transverse suture (superior nuchal line) and the mendosal suture (highest nuchal line) is chondral bone bearing the insertion of the only two muscles with primary insertion at the occiput: trapezius and sternocleidomastoid.

Innervation of the Meninges/Periosteum

Nota bene Pia-arachnoid layers do not have somatic sensory innervation. These tissues likely behave in register with the neuromeric segment of the brain which they supply.

Supratentorial Innervation

The supratentorial dura is a very simple affair. Its mesenchyme (with one exception) is exclusively neural crest from rhombomere r1-r3. The lower zone of parietal dura includes paraxial mesoderm from r2 and r3. Recall that these form the poorly-appreciated "parietal cartilage" which appears in the lower zone of the temporoparietal region early in the formation of the cranial base. They could well be confined to squamosal bone. Blood supply is exclusive via the stapedial system [Figs. 12.32, 12.33, 12.34, 12.35, 12.36 and 12.37].

The cerebellar tentorium is worthy of comment as it represents the boundary zone between two different mesenchymes, two different circulations, two different innervations, and two different kinds of bone formation. Tentorium is like a trampoline that supports occipital lobes above the cerebellar hemispheres. Its anterior border is an open oval beginning at the posterior clinoid process which admits the cerebral peduncles to communicate from hindbrain to forebrain. It extends peripherally and backward along the petrous ridge, enclosing superior petrosal sinus which it follows backward to transverse sinus which it also encloses.

Sinuses form at the boundary of adjacent fields. Tentorium (r1) provides the upper/external lamina and infratentorial dura (r4-r7) gives the lower/internal lamina. The two layers track along either side of the sinus and then fuse together, forming a triangle enclosing the sinus.

Tentorium houses within itself the *straight sinus*. This implies that tentorium has *two layers*. Why? Although the entire structure of cerebellar tentorium is r1 neural crest, its two laminae result from different programming signals from posterior lobes of cerebrum above and from cerebellum below.

Anatomic implication. The cerebellum is completely wrapped in meninx arising from r1. Initially, it sticks up straight in the air like a mushroom, its sole innervation being V1. However, with brain development, it will fold backward until it comes to rest against the otic capsule and the developing braincase. As it does so, it pulls its meningeal coverings along with it. Fusion of the layers occurs as levels r4-r7 resulting in a horseshoe-shaped configuration, the tentorium cerebelli.

Infratentorial Innervation

The infratentorial periosteum (IP) differs in terms of anatomic distribution, mesenchymal origin, innervation, blood supply, and the bone structures that are associated with it. As stated, it lines the posterior fossa all the way from the posterior clinoid processes to the petrous ridge, that is, superior petrosal sinus, and sweeps backward to the transverse sinus. On the other side of this line, infratentorial periosteum fuses with the r1 tentorium cerebelli. This dura is constructed from paraxial mesoderm supplied by r4-r11. It covers bones derived from r4 to r7 petromastoid and r8-r11 occipital bone complex. Why in the world should this periosteum receive innervation from cervical nerves? First, let's look first at the what of this anatomy and second, why it makes developmental sense.

Infratentorial periosteum is supplied by sensory fibers from C2 to C3 (mostly C2) plus SANS neurons from the superior cervical ganglion (SCG). The nerves have three territories of distribution in the posterior fossa, depending upon their mode of access to the skull, via the foramen magnum, the hypoglossal canal, or the jugular foramen. Nerves entering foramen magnum un-accompanied are anterior rami which track along the post-central artery and ascend on a direct course up the basioccipital almost to the posterior clinoid processes. Nerves entering the skull via hypoglossal canal or jugular foramen are also anterior rami but their course is indirect. They first pass through the upper pole of superior cervical ganglion where they pick up sympathetic nerve fibers and proceed to track along cranial nerves XII and X until gaining entry to the skull; they then follow arteries.

Zone 1 nerves ascend the clivus from foramen magnum to the posterior clinoids. In so doing, they are accompanied (and programmed) by clival branch of ascending pharyngeal. The nerves defined the basioccipital bone fields.

Zone 2 nerves accompanying hypoglossal nerve fan out from the foramen. They innervate exoccipital and foramen magnum, which is framed by the exoccipital bone fields.

Zones 3 nerves accompanying vagus nerve via jugular part company with X at the foramen. They follow sigmoid sinus backward and upward and to transverse sinus and then track medially to innervate the supraoccipital bone fields.

Why This Innervation Pattern Makes Sense
When we analyze the somatic sensory distribution of r4-r7 we find hypaxial branches extensively represented in the oropharynx. Expaxial representation is limited to the tympanic cavity. Levels r8-r11 have hypaxial representation in the larynx and hypopharynx but no epaxial branches. Neuroanatomic studies confirm the complete absence of sensory fibers to dura from levels r4 to r11. These nuclei simply do not exist in the brainstem. Tissue cannot exist without sensory repre-

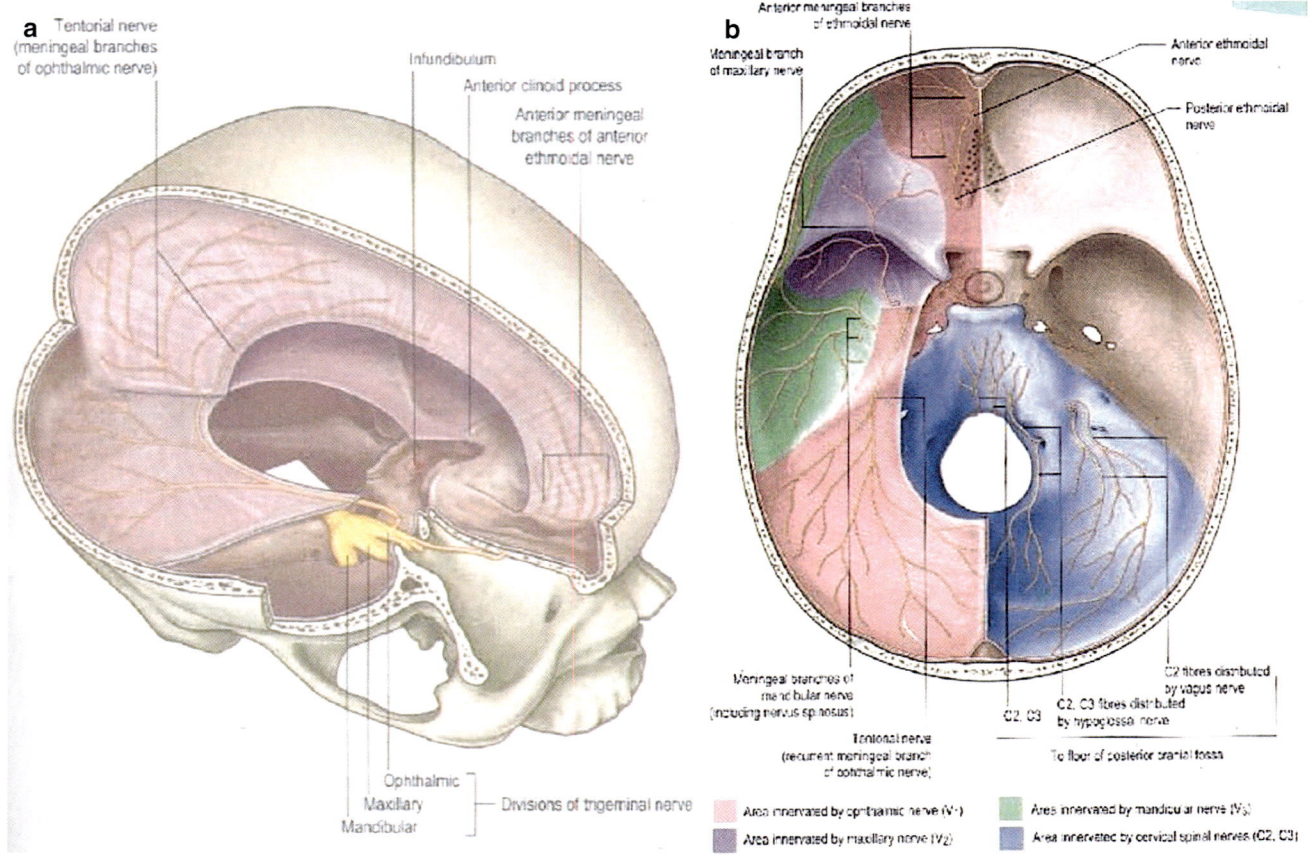

Fig. 12.32 Dura nerves. <u>Left</u>: Innervation of the dura V1 provides three branches to the falx: (1) anterior ethmoid n. to anterior third, (2) a direct branch of V1 to the middle third, and (3) recurrent branch to the posterior third and to the tentorium. <u>Right</u>: Innervation of the dura showing extensive distribution of V1 to both the anterior and middle cranial fossae. Neuromeres r4-r7 have no sensory representation so V1 and V2 must take over. Posterior fossa has no dura, being lined by *peri-* *osteum only* (blue). It is supplied by C2-C3 as a substitute for absent representation in r8-r11. Note C2-C3 access the posterior fossa by traveling with vagus (X) to the r8-r9 zones (lower supraoccipital) and with hypoglossal (XII) to zone r10-r11 (upper occipital). [Reprinted from Standring S. Gray's Anatomy, 40th edition. Philadelphia, PA: Churchill-Livingstone; 2008. With permission from Elsevier.]

sentation—the first three cervical levels are forced to "take up the slack."

Does C1 Participate or Not?

C1 is considered unique among spinal nerves because it is said to be purely motor. Both cervical skin and scalp are devoid of C1 sensory fibers. Some neuroanatomists continue to represent C1 over the clival surface of basioccipital. Does it have any role? If so, where?

The central idea is the in the transition between cranium and neck, somites get progressively larger "portfolios" in terms of what they produce. Occipital somites

S1-S4 can only make bone and hypaxial (but not epaxial) muscle. S5 is the first to make expaxial muscle but no skin. However, a fascial sleeve unites foramen magnum with axis. C1 dura could conceivable exist within this envelope, and; it could be limited to anterior to foramen magnum but not posteriorly. S6 is a full-fledged "blue ribbon" producer of both dermis and periosteum. C2 fibers are both hypaxial in the neck and epaxial in the scalp. They appear over the clivus and in the back of the skull.

It is thought that basioccipital is supplied directly by C1 (some) and by C2. C1 does *not* connect with superior cervi-

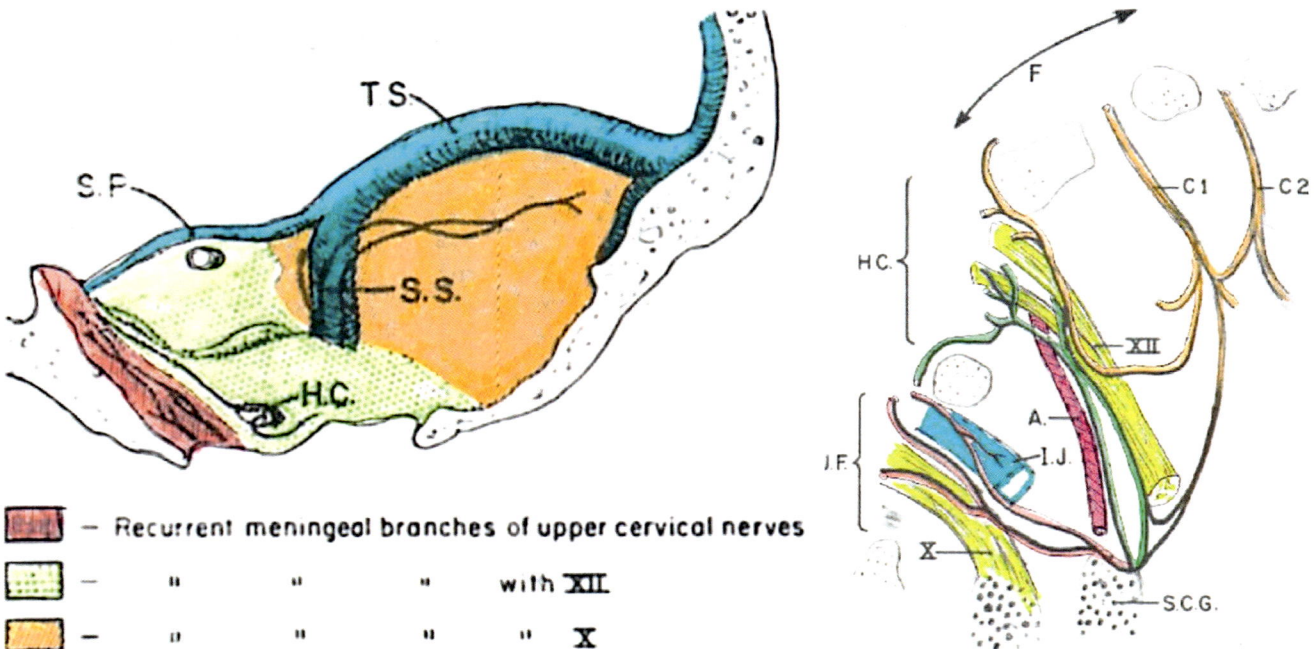

Fig. 12.33 Distribution of sensory nerves to posterior fossa. (1) Recurrent branches direct (pink) supply basioccipital and all zones S1-S4. (2) Recurrent via XII (yellow) enters via hypoglossal foramen, supplies opisthotic, exoccipital, and S3-S4 of supraoccipital. (3) Recurrent via X enters via jugular formen to supply opisthotic and S1-S2 of supraoccipital. [Reprinted from Kimmel DL. Innervation of the spinal dura mater and dura mater of the posterior cranial fossa. *Neurology* 1961; 11(9):800–809. With permission from Wolters Kluwer Health, Inc.]

Fig. 12.34 Cervical fibers arrive in three formats. (1) Isolated neurons ?c1 and c2 (not shown) climb up the clivus to supply basioccipital. (2) C2 (yellow) via superior cervical ganglion follows the more posterior CN XII via hypoglossal foramen into skull. It distributes to opisthotic, exoccipital, and S3-S4. (3) C2 (orange) via superior cervical ganglion follows the more anterior CN X through the jugular foramen. It distributes to prootic and S1-S2. [Reprinted from Kimmel DL. Innervation of the spinal dura mater and dura mater of the posterior cranial fossa. *Neurology* 1961; 11(9):800-809. With permission from Wolters Kluwer Health, Inc.]

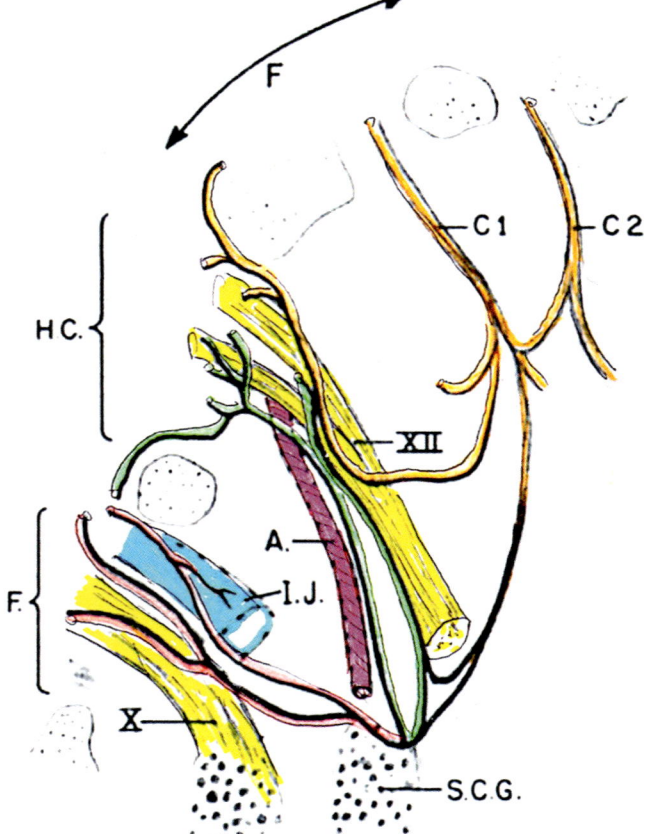

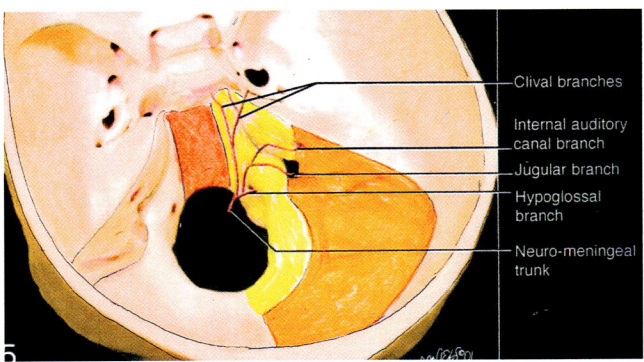

Fig. 12.35 Neural territories per Kimmel marked: (1) direct branches C1-C2 to basioccipital (pink). (2) Indirect branches of C2 distributed to prootic, medial exoccipital and lower zones (S4-S3) of supraoccipital are supplied by ascending pharyngeal representing r6-r7 (yellow). (3) Indirect branches of C2-C3 distributed to opisthotic, lateral exoccipital, and upper zones (S2-S1) are supplied by occipital artery representing r4-r5 (orange). Note the congruency of lamination between innervation and blood supply. [Reprinted from Hacein-Bey L, Daniels DL, Ulmer JL, Mark LP, Smith MM, Strootmann JM, Brown D, Meyer GA, WackLyn PA. The ascending pharyngeal artery: branches, anastomoses, and clinical significance. *Am J Neuroradiol* 2002; 23:1246–1256. With permission from the American Society of Neuroradiology.]

cal ganglion. C2 enters via jugular foramen with distribution to r4-r5 and upper zone of supraoccipital. When it enters hypoglossal foramen, it is distributed to r6-r7 and lower supraoccipital bone. This pattern makes perfect sense for the exoccipital and supraoccipital bones. The Huang model would predict that the upper zones (S1-S2) would be irrigated by branches of r4-r5 occipital artery while the lower zone (S3-4) would by supplied by r5-r6 branches of ascending pharyngeal.

An Astounding Misconception

Occipital and ascending pharyngeal supply structures of the 2nd and 3rd arches, respectively. This does NOT imply dura is as pharyngeal arch derivative. It simply means that our model of external carotid as being strictly relating to pharyngeal arches is *wrong*. We really should be talking about these arteries as in register with neuromeres. In this manner, we can explain the presence of structures such as squamous temporal bone and parietal cartilage that are clearly supplied by ECA and are unrelated to the arches.

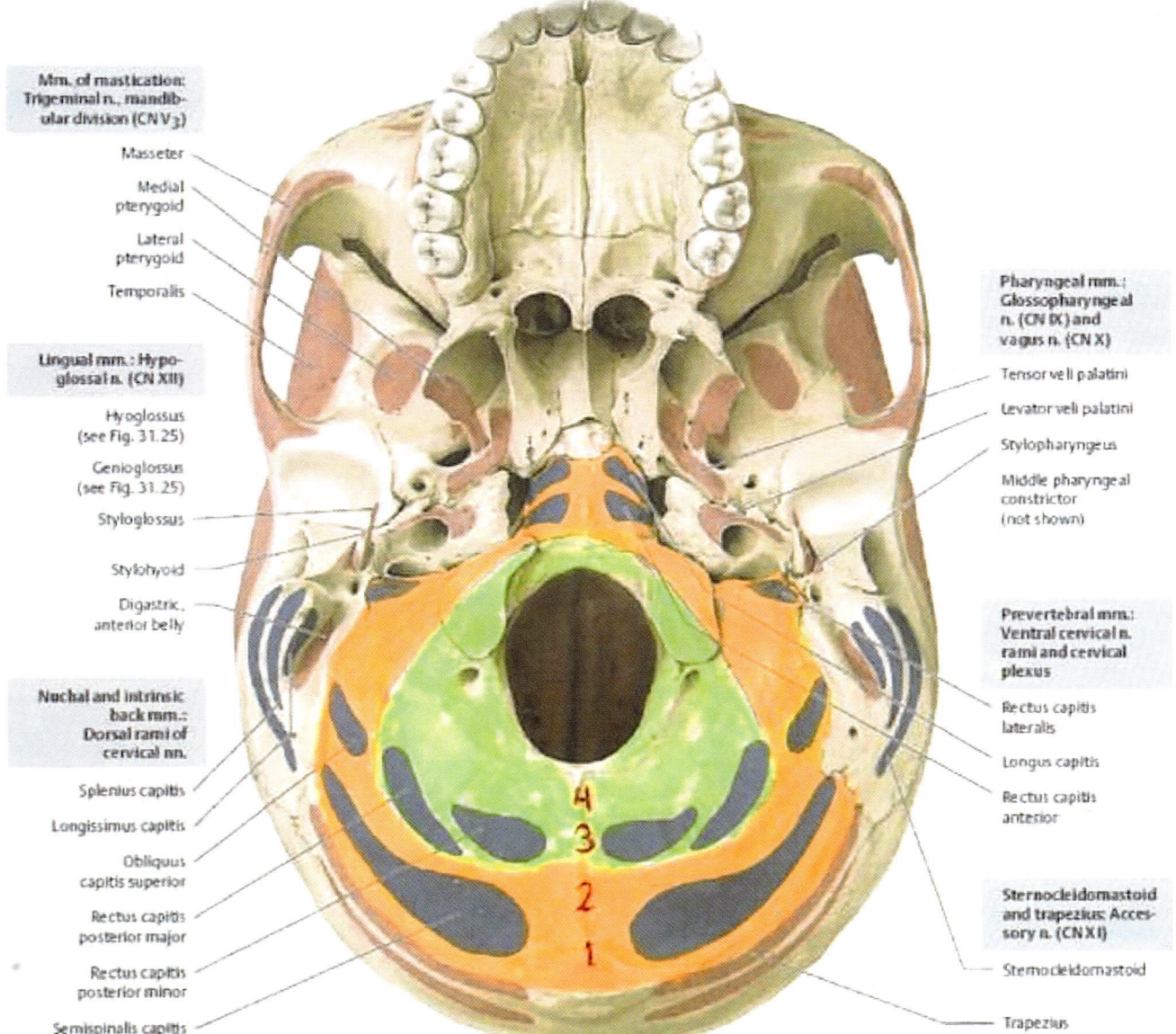

Mm. of mastication: Trigeminal n., mandibular division (CN V₃)

Masseter

Medial pterygoid

Lateral pterygoid

Temporalis

Lingual mm.: Hypoglossal n. (CN XII)

Hyoglossus (see Fig. 31.25)

Genioglossus (see Fig. 31.25)

Styloglossus

Stylohyoid

Digastric, anterior belly

Nuchal and intrinsic back mm.: Dorsal rami of cervical nn.

Splenius capitis

Longissimus capitis

Obliquus capitis superior

Rectus capitis posterior major

Rectus capitis posterior minor

Semispinalis capitis

Pharyngeal mm.: Glossopharyngeal n. (CN IX) and vagus n. (CN X)

Tensor veli palatini

Levator veli palatini

Stylopharyngeus

Middle pharyngeal constrictor (not shown)

Prevertebral mm.: Ventral cervical n. rami and cervical plexus

Rectus capitis lateralis

Longus capitis

Rectus capitis anterior

Sternocleidomastoid and trapezius: Accessory n. (CN XI)

Sternocleidomastoid

Trapezius

Fig. 12.36 Occipital somites are laminated, one upon the other, like Russian dolls. S1 is the largest and S4 is the smallest. Their bone products are in concentric rings. Somites S1 and S2 (orange) are lateral and form the superior half of supraoccipital. Somites SS3 and S4 are medial and produce lower supraoccipital. Hypoglossal foramen is between S3 and S4. Muscle insertion sites are also zonal. [Reprinted from Lewis, Warren H (ed). Gray's Anatomy of the Human Body, 20th American Edition. Philadelphia, PA: Lea & Febiger, 1918.]

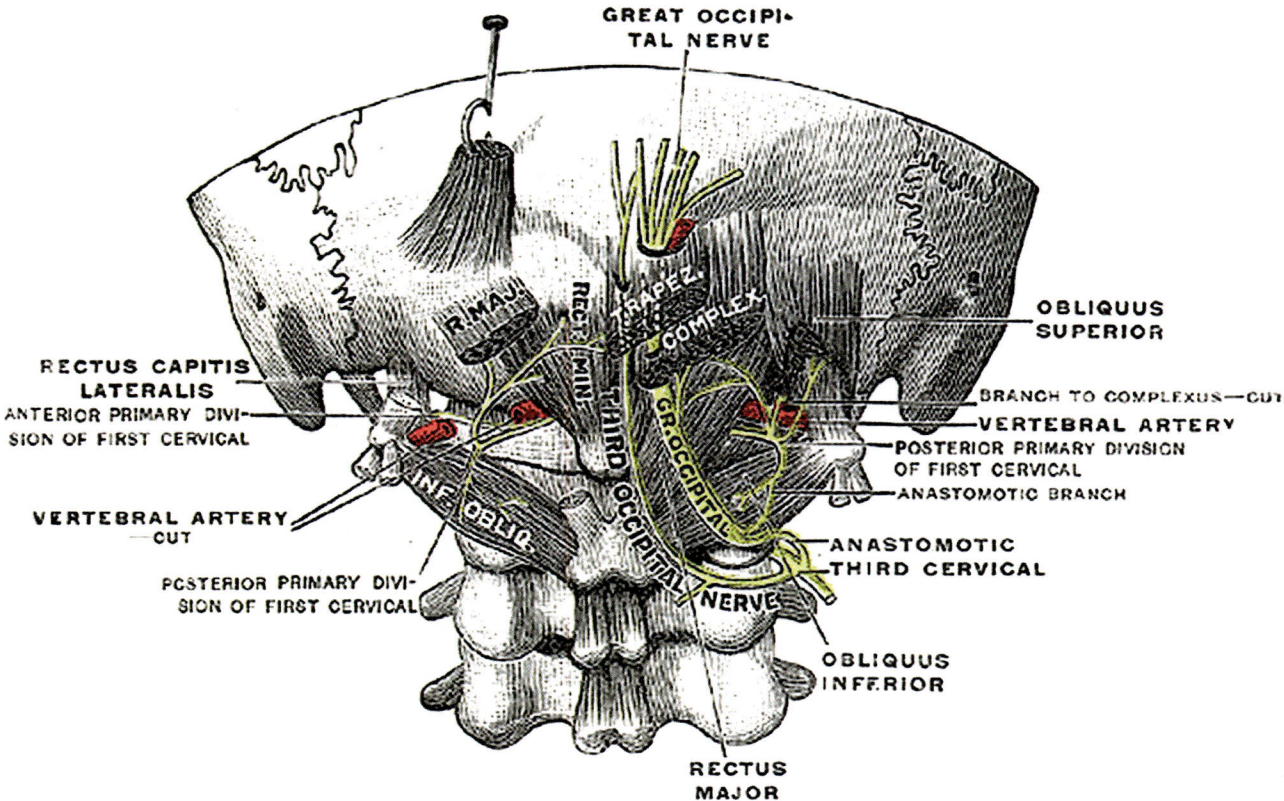

Fig. 12.37 Posterior branches of cervical nerves C1-C3. C1 suboccipital nerve is strictly motor. The anterior branch is smaller, having limited muscle targets. C2 greater occipital and C3 lesser occipital have small anterior branches contribute to cervical plexus and clival periosteum. These do not make contact with superior cervical ganglion. They enter foramen magnum directly to supply basioccipital periosteum. Posterior branches of C2 and C3 send sensory neurons to pick up SANS fibers from superior cervical ganglion. These are then are distributed to exoccipital and supraoccipital periosteum. C2 posterior division is also motor whereas that of C3 is sensory only. Sensory distribution for the membranous calvarium is C2 lateral and C3 medial. [Reprinted from Lewis, Warren H (ed). Gray's Anatomy of the Human Body, 20th American Edition. Philadelphia, PA: Lea & Febiger, 1918.]

Mesenchyme of the Meninges

Pia Arachnoid Mesenchyme

These tissues maintain vascular relationships with underlying brain on a neuromeric level. This probably determines the penetration patterns of the cortical vessels which are subsequently organized around the "feeder branches" of the cerebral circulation. Neuromeric mapping is on the basis of the underlying brain tissue.

Dural Mesenchyme

This is a relatively simple affair being provided by the PAM from levels r0-r3 and by neural crest from midbrain and r0-r3. Neuromeric mapping is on the basis of the trigeminal system.

Mesenchyme of the Posterior Cranial Fossa: Periosteum

In the absence of accurate rhombomeric mapping of these bone fields, we can make a best-guess assumption that the posterior fossa is an "equal opportunity employer" that makes use of paraxial mesoderm from all eight levels (r4-r11). In the case of temporal bone complex we know that r4-r7 are represented. Likewise, mapping by Huang in the avian model demonstrates the development of the occipital complex as a ring-like distribution of mesoderm from each occipital somite. The mammalian situation should be no different.

We have been able to account for the blood supply as the external carotid.

The dura of the occipital lobe is neural crest is r3 neural crest, innervated by V3 and supplied by StV3 meningeal arteries. Note that the occipital foramen admits 2nd arch occipital artery into the skull below tentorium. The bones above it are membranous in formation. Interparietal has two zones, the central postparietals are neural crest while the lateral tabulars are paraxial mesoderm likely from level r3 as well. What determines the "chondrality" of a bone whether it forms in membrane or not depends on the intrinsic source and the dura with which it interacts.

The infratentorial periosteum come from two sources depending on the bone fields. PAM from r4 to r7 covers petromastoid. The periosteum lining of the occipital complex is made from r8 to r11 PAM and covers basioccipital, exoccipital and supraoccipital. The territories of distribution for the occipital complex can be ascertained from the model by Huang. Mesenchyme from r8 to r9 produces occipital somites 1-2. It is distributed in an arc bordering posterior lateral exoccipital and the upper zone of supraoccipital. is likely derived from r8 to r9 while the lower zone originates r10-r11. The upper zone is supplied by branches from the occipital artery correlating with the 2nd arch while the lower zone is supplied by branches of the ascending pharyngeal correlating with 3rd arch. It is to be emphasized that these bone fields. The designated sensory supply is hyaxial C1-C2 to basioccipital and epaxial C2 to exoccipital and supraoccipital.

As previously stated, the internal periosteum covering the ones of posterior fossa represents the stem cell layer "donated" to the bone by the underlying somitomeres and somites. In the case of the temporal bone complex, its external periosteum contains stem cells from overlying mesodermal structures, such as muscle insertions, and neural crest fascia. The internal periosteum of the occipital bone complex is derived from the four occipital somites while the external periosteum bears the stem cells of the multiple muscles insertion into the posterior cranial base [Cf Figs. 12.36 and 12.37].

The fusion plane between neural crest and PAM mesenchymes results in triangular spaces that create the dural sinuses. These smooth fascial tubes have no lining endothelium and consequently no valves. *Continuity of periosteum between neural crest and mesoderm at the level of the sinuses means that the sinuses can be safely reflected away from the bone.*

The combination of arterial supply and innervation beneath supraoccipital bone divides it into four (and perhaps eight) distinct zones, each one receiving a C1 muscle. The muscles are inserted into an exterior periosteum that is continuous between the chondral supraoccipital complex and the membranous interparietal complex. The posterior fossa represents coverage over both chondral and membranous bone. It contains the suboccipital fossa of muscles from C1 to C2. It continues upward bringing C1-2 PAM into contact with r3 neural crest. There is a zone overlying the membranous occipital where dura is covered externally by C2 dermis

and DIF supplied by occipital artery. But above the highest nuchal DIF continues as the periosteum. Thus, the origin of posterior fossa dura is a mixture of neural crest as described plus PAM contributions from Sm7 to C2-C3.

In sum, the development of the posterior fossa has these key anatomic points.

(1) No dura exists below tentorium cerebelli.
(2) Venous sinuses form the superior boundaries of posterior fossa.
(3) The sinuses are boundary zones between neural crest and PAM.
(4) Posterior fossa is lined by internal periosteum: PAM from r4 to r11.
(5) Petromastoid and its periosteum are products of r4-r7, somitomeres 4-7.
(6) Occipital complex and its periosteum are products of r8-r11, somites S1-S4.
(7) The external periosteum of the occiput is a product of cervical somites C1-C3.

Themes of Meningeal Development

Stages 7–10: Mesenchyme Surrounds the Neural Tube

During early development, the component parts necessary for synthesizing an immediate blood supply to the CNS parenchyma are laid down. These are (1) mesodermal endothelial cells to form the tubular structures of blood vessels, (2) neural crest-derived pericytes to surround the tubules, and (3) neural crest to provide a primitive covering layer that coats the vessels as they enter the brain parenchyma to maintain a blood-brain barrier. Gastrulation (completed in stages 7–8) organizes mesoderm into the paraxial, intermediate, and lateral plate sectors. At stage 8 the neurectoderm forms the primordium of the brain. Immediately lateral, at the junction between the neuroectoderm and paraxial mesoderm, neural crest cells develop. Prechordal mesoderm is present just cranial to the notochord—at future level r0, it is positioned directly opposite from where the 1st pharyngeal arches will appear. By stage 9, the neural crest cells begin to migrate. Stage 10 involves further proliferation of prechordal mesoderm. Migration of MNC into the midline splits prechordal mesoderm: MNC forms the nasoethmoid mesenchyme and PCM moves out to the orbits.

Neurulation at stage 8 involves the rolling-up of the neural plate into the neural tube. At all levels from r1 backward, this pulls adjacent paraxial mesoderm upward into contact with the neural tube, thus creating an immediate source for blood vessels over the hindbrain. Hindbrain and spinal cord PAM is self-sufficient to make both endothelial tubes and pericytes—neural crest cells *are not required*. Forward from r1, paraxial mesoderm coverage of forebrain/midbrain is accomplished by a process of *tissue expansion*. As the forebrain arises from the plane of the embryo. PAM from r1 to r3 is mechanically pulled forward to cover it.

By stage 9, neural crest cells migrate over the forebrain and midbrain where they are critical for the development of the head plexus. At these levels, NCCs *are required* to produce pericytes necessary for support the blood vessels. The stage 9 brain is now completely surrounded by a primitive head plexus. Nutrients come from surrounding fluid, as the connection with the heart is not yet established.

At stage 10, in the presence of neural crest, these primitive vessels acquire pericytes and penetrate into the neuropil. The primitive head plexus is now directly supplied from the heart via segmental arteries that interconnect with dorsal aortae. Conditions are now present for development of pia mater which becomes visible at the next stage.

Stages 11–13: Peri-CNS Mesenchyme Organizes: Pia Mater

Nota bene: *As soon as a cell layer appears in between the wall of the brain and the vascular layer, this layer is identified as **pia mater**.*

By stage 11, pia mater first presents over the caudal medulla. It develops from local neural crest and contains primitive blood vessels. *Subsequent pial development proceeds cranially*, arriving at the mesencephalon by stage 12. Complete pial coverage of the telencephalon is achieved by stages 13–14.

By stage 15, the entire primary meninx is in place but the leptomeninges have not differentiated. During stages 17 and 18 *leptomeningeal development* takes place. The mesenchyme just above the brain is fluid-filled, *presumably from blood vessels of the original head mesoderm* spanning between the cerebrum and the overlying peripheral mesenchyme. This zone develops into a meshwork, first above the fourth ventricle and then spreading throughout the hindbrain/midbrain and later over the forebrain.

In summation: pia formation takes place in two ways. Hindbrain and cerebellar pia arises from the *internal migration* of neural crest which flows downward over the neural tube to achieve complete coverage. The coverage is immediate (even before closure of the neural tube) and local: neural crest participates from every neuromeric level of the neural tube, from r1 backward. Pia maturation into identifiable layers follows exactly the caudal-rostral progression of CNS vascular development. Vascular supply of the pia makes use of peritubular somitomeric mesoderm to form a succession of structures. Primitive head plexus matures into longitudinal neural arteries. The LNAs connect with the primitive internal carotid to supply each neuromere, one-by-one. They eventually connect with the vertebral arteries. The arterial supply to the brain is deep to the dura, remaining isolated from the external carotid system.

Forebrain pia is limited to the cerebral coverage because midbrain and diencephalon are enclosed within the hemispheres. PNC has no role in forebrain meninges due to its limited developmental potential—it only makes frontonasal dermis. Hindbrain neural crest (r1-r3) must be imported.

Stage 13 marks the beginning of the notochordal sheath and the consolidation of the posterior fossa cranial base.

Stages 14–17: Primary Meninx in a New Location: Tentorium Cerebelli

At stage 14 the cellular sheath of the notochord has reached the r1 and the midbrain. The forward progress of this tissue reflects the posterior to anterior development of the paraxial cranial base. By stage 15 the cellular sheath extends all the way forward.

Cerebellum arises from r0 to r1 (collectively called r1). Primary meninx (PAM and neural crest) from r1 undergoes three forms of expansion. (1) One population surrounds cerebellum; it will morph into secondary meninx with pia and arachnoid. (2) A second population extends outward as a T-shaped sheet interposed between cerebellum and forebrain and also extending backwards between the two lobes of cerebellum. The transverse limb becomes tentorium and the sagittal limb becomes falx cerebelli. (3) A third population migrates forward over both the medial and lateral surfaces of cerebrum. It will develop to frontal lobe dura and falx cerebrii.

This structure, the *protentorium*, is first observed (as expected) at the level of the mesencephalic flexure. It is medial, just behind diencephalon and forward from the termination of basilar artery, that is, exactly at rhombomere r1. This medial tissue is a placeholder. It is not yet dura. Later in development (stage 19) with development of the cerebellum, protentorium undergoes apoptosis and is replaced by lateral components.

Nota Bene At stage 16 cranial nerve outgrowth takes place. These will provide the programming for the subsequent arterials. In particular V and VII organize the stapedial system in plane immediately superficial to future dura propia.

By stage 17 lateral parts of the tentorium grow medially from the otic capsule and tentorium appears.

Stages 17–18 Secondary Meninx: Leptomeninges and Dura

Exactly how the secondary meninx develops varies depending upon its location along the neuraxis.

Mesenchyme closest to the brain forms a fluid-filled meshwork, the vessels of which are connected with veins located at the periphery. Vascularization takes place and new cellular elements appear. These will form scaffold of arachnoid. At the deep margin arachnoid is separated from pia by the blood vessel network but its definition at the perphipery depends upon events in the overlying dura.

Two external condensations take place in the dura. The more internal one becomes the identifiable *dural limiting membrane* and will define it vìs-a-vìs the arachnoid. External to that, a thicker condensation becomes the osteogenic *skeletongenous layer*. It is external to the veins. Although this layer interacts with overlying soft tissues to form the membranous calvarium, the cell source for the bone is predominantly from dura. Having given up its mesenchyme to the skull, the skeletogenous layer retreats, differentiates. Left behind immediately beneath the bone is the internal periosteum; the arteries lie just beneath. Subjacent is the dura propia and finally the dural limiting membrane.

Dura mater begins to develop at stage 17. The dural limiting membrane develops in the same caudal-to-cranial and ventral-dorsal sequence as the pia. During stage 18 the many branches of the stapedial system become dependent of the exernal carotid system for supply. This process causes the dura to mature and thicken.

Stage 19–23 Dural Venous Sinuses Appear

At stage 19 tentorium extends in the midline; the falx cerebrii is present at stage 20. The presence of parietal plate, otic capsule, obitosphenoid and alisphenoid indicates formation of periosteum, likely at the same time as the full development of tenetorium. Dural limiting membrane continues to advance forward, covering the hindbrain by stage 22 and the midbrain by stage 23. Note that forebrain will expand backward into the space such that this dura becomes basal beneath the posterior cerebral lobes. Sinuses are defined as soon as the dural limiting membrane appears in the region because of the interface with posterior fossa periosteum.

At stage 23 falx has now reached its most forward extent over crista galli. Dural lining layer, now complete basally, extends upward over the lateral walls of cerebrum.

Skin Coverage of the Brain

How does skin coverage over the nervous system take place? In the case of the spinal cord, the process is simple. The neural tube rolls up on itself like a cigar. As the two edges of non-neural ectoderm (future epidermis) approximate each other they drag along with them subjacent paraxial mesoderm as well. As these two layers fuse, the neural tube sinks beneath the skin. Local neural crest from each neuromere immediately migrates: (1) outward beneath the skin, (2) adjacent to the somites to form the sympathetic chain, (3) ventrally to populate the lateral plate mesoderm of future viscera, and (4) downward to cover the neural tube circumferentially.

Craniofacial skin is utterly different. Its epidermis comes from discontinuous sources. Frontonasal epidermis is a product of prosencephalic non-neural ectoderm from the anterior neural folds. The remaining epidermis is true ectoderm from r1 to r3, including anterior tympanic membrane. The upper eyelid epidermis and the conjunctiva are the only sites in the embryo where r1 ectoderm is expressed. Note V1 fibers supply the cornea via lacrimal gland as well as from the superior orbital fissure. This gives us a potential "roadmap" for r1 ectoderm to reach the upper eyelid. Innervation of the upper eyelid by The first 12 somitomeres (head mesoderm + occipital somites + 1st cervical somite) are incapable of producing dermis. Thus dermal tissues must come from remote sources. *Underlying all non-somitic skin is the single unifying structure of the face and skull: the 2nd arch superficial investing fascia SIF (also as the superficial musculo-aponeurotic system or SMAS).* It conveys the vessels of the external carotid system to support the overlying skin. It also interacts with the underlying dura to produce the membranous bones of the calvarium.

Frontonasal Skin and Scalp (FNO)

Recall from our previous discussion that the frontonasal skin is radically different from the other five skin types. Here, epidermis and dermis both come from non-traditional sources. This *non-neural ectoderm* (NNE) of the anterior prosencephalic folds. This ectoderm is *not* produced via gastrulation. The populations "assigned" to NNE correspond to the p6-p4 zones of the old Puelles prosomeric system. There are no neural crest cells in the anterior folds but they do contain placodes. The dermal supply for frontonasal skin is neural crest from zones p3-p1. The dermal precursor ppopulations migrate forward beneath the NNE. Together this composite tissue covers the upper face like a baseball cap pulled forward over the eyes.

Note that PNC frontonasal fields have no intrinsic neurovascular supply. The PNC dermis is layered upon the underlying r1 mesenchyme. In this way it acquires StV1 stapedial arterial vessels and V1 nerves. Later in development 2nd arch mesenchyme interposes itself between the two pre-existent layers. It brings with it intrinsic blood supply from the external carotid (transverse branch of superficial temporal) but only into the facial muscles. For this reason, both supraorbital and supratrochlear neurovascular pedicles split around the SMAS fascia. The deep branches go to periosteum and the superficial branches supply the skin. Thus, even though portions of frontalis are irrigated by ECA, the vertical pedicles run all the way upward to the vertex.

Of clinical importance is the susceptibility of frontonasal skin to underlying pathologies of the V1 stapedial neuroangiosomes. Deficits can occur in eyelids, eyebrows, and dermis. The skin is at risk for neurodegenerative states such as Perry Romberg disease and neurocutanous disorders.

Non-Frontonasal Skin

This skin covers all of the face and scalp innervated by V1, V2, and V3. Its components originate from rostral rhombomeres r1-r3. Midbrain neural crest does not participate nor is dermis produced anywhere else in the hindbrain. Gastrulation occurs at level r1. It does not supply frontonasal skin with ectoderm of neural crest. So what happens to the r1 tissues? They produce a very small component of ectoderm which is likely assigned to upper eyelid and the conjunctiva. Dermal and subconjunctival tissues from r1 are likewise extremely thin. Despite these limited efforts r1 is definitely not a "second-class citizen." Its contributions are vast, being the required mesenchyme of dura, bone, and fascia for the anterior cranial base and orbit.

Migration of PNC brings it into contact with r1 dura and the V1 stapedial neurovascular axes. At the same time, SIF invades the frontonasal zone, passing over the orbit (previously populated by midbrain neural crest) into the forehead and down over the nose. Frontal bone forms at the interface with r1 dura and SIF.

Comments About Scalp Hair

Skin coverage is timed with neural tube closure—between stage 11 when the rostral neuropore closes and closure of the caudal neuropore at stage 12. At stage 13 the first hair in the entire body appears at the parieto-occipito junction, the boundary between dermis from epaxial r3 neural crest and dermis from the epaxial dermatomes of c2 and c3. From this spot, the *occipital whorl*, hair growth radiates outward. Scalp formation demonstrates the epithelial-mesenchymal interaction required to induce dermis. One can hypothesize that the distinctions between the hair-bearing scalp and beard-forming facial skin relate to genetic differences between epaxial versus hypaxial populations of neural crest. Occipital hair could represent that neural crest from c1 to c2 does not have any hypaxial.

Bilaminar Programming of Membranous Calvarial Bone

Membranous calvarial bone is bilaminar with an intervening marrow space. This bilaminar model has to do with two sources of programming. Obviously, in the most primitive state, mseenchymel tissues of the meninges are apposed with overlying neural crest dermis from either PNC or r2-r3. Later in development neural crest fascia and muscle from the r4 to r5 differentiate and are *interposed between the dura and dermis* – this is superficial investing fascia, SIF, or galea.

So which layer is responsible for programming the external table? Is it r2-r3 neural crest dermis, r4-r5 neural crest fascia, or both? Neuromeric mapping has not been done to date but we can draw some conclusions from the embryology. The vascular supply of the scalp, primarily from superficial temporal system, appears very late in embryogenesis, ascending from in front of the ear at stage 20 and reaching the vertex at stage 23. Dissections of the epicranius by this author [4] demonstrated perforating blood vessels passing downward through the subgaleal fascia and into the external periosteum. Temporo-parietal bone flaps can survive on superficial temporal pedicle designs. Substantial galeal-subgaleal flaps can be harvested from beneath the scalp with necrosis of the overlying skin. For these reasons, it is reasonable to assign a histologic role to r4-r5 neural crest over the entire extend of the calvarium. Superimposition of these zones is clinically useful with parietal bone to explain different forms of synostosis and with the interparietal bone complex to give histologic basis to its various fields.

Fascial Planes of the Brain and Calvarium

Our final topic is of great surgical relevance as it explores planes of separation and the differential synthesis of craniofacial bones. We are going to deal with three separate systems.

In our previous discussion of the blood supply to the skin we defined the fascial and vascular planes involving the structures of the face and scalp. These are:

1. Pia/arachnoid: primitive head plexus/internal carotid
2. Dura/fronto-orbito-nasal mesenchyme/jaws: the stapedial system

3. Deep investing fascia (pharyngeal arch mm.): deep branches of ECA
4. Superficial investing fascia/galea (facial mm.): superficial branches of ECA.

The reader will note that in the face and in the non-temporal cranium only three of these planes are relevant. The temporal region is unique, as it has all four fascial layers and therefore offers the most complexity because here SIF and DIF are both present.

At this juncture, the reader may well ask, "Since the sensory nerves and arteries of the dura and scalp appear to arise from common sources of mesenchyme, by what mechanism do they develop in different planes? And why should the neurovascular supply of these two structures vary in orientation? Those of dura mater run obliquely forwards and backwards whereas the superficial temporal system is essentially vertical. Our approach to the temporal fossa follows the Rule of Fours. *The coverage of the brain consists of four distinct layers, each with its own dedicated blood supply. The four vascular systems develop in four different spatial planes and develop at four different points in time.*

Layer 1: *Endomeninx*

Pia and arachnoid arise in situ as neural crest/PAM derivatives. It is directly exposed to signal from the underlying brain. It is vascularized by the vessels of the *primitive head plexus* at stages 8–9. These vessels develop from paraxial mesoderm immediately adjacent to the neural tube and paraxial mesoderm. They ultimately connect with the internal carotid circulation. By stage 14 pial development reaches the telencephalon and the vascularization process is largely complete.

Layer 2: *Ectomeninx*

Dura is a mixture of neural crest and PAM. It is further away from the brain and receives different differentiation cues, hence its three layers. It is vascularized by the *stapedial artery system*, beginning at stage 17. Stapedial development continues through stage 23.

Layer 3: *Deep Investing Fascia*

DIF encloses muscles of mastication arising from Sm4 and innervated by V3. It forms a muscle-enclosing mesenchymal sandwich from the midline of the cranial base upward to the squamosal-parietal boundary. It is vascularized by ECA arteries from the 2nd segment of internal maxillomandibular artery corresponding to 1st arch structures. As such, it provides external coverage for the middle cranial fossa contain-

ing r4-r5 petrous complex, the synthesis of these which is chondral and unrelated to DIF. DIF development is completed by stage 17.

Layer 4: *Superficial Investing Fascia*

The non-temporal calvarium is straightforward. It is composed of r4-r5 neural crest. Depending upon the anatomic zone, galea/SIF/SMAS may or may note enclose 2nd arch muscles. The muscle-free zone between frontalis and occipitalis is a good example of a strictly fascia-only zone. This layer is vascularized by the superficial temporal branch of ECA corresponding to 2nd arch structures. These arteries run along its external surface of dura directly below the dermis. Although the vessels are tightly related to the dermis, a thin fascia separates the layers, permitting dissection in an avascular plane, save where they send out superficial branches. Below the galea is a loose layer, the subgaleal fascia, the anatomy of which has been previously described by this author. Of note, the SGF is not avascular but has a delicate pattern of vessels representing a deep division of the overlying arterial arcade. Multiple small vessels pass from the SGF into the external periosteum. SIF development takes place between stages 20 and 23.

In the temporal region, SIF splits to enclose the periosteum of zygomatic arch and then proceeds cranially to form galea. The *triangular fat pad* between the two laminae of SIF is surgically useful because it can be entered, permitting subperiosteal elevation of the external lamina fascia away from the bone, thus preserving the facial nerve branch at its crossing point. Deep investing fascia encloses the temporalis muscle which passes beneath the zygomatic arch to insert into the temporal fossa, its superior termination being the superior temporal line. The temporalis muscle is supplied by the anterior and posterior deep temporal arteries from the 2nd part of the composite maxillo-mandibular artery. This marks the cranial limit of the deep division of external carotid system. Forward (rostral) from this point there are no further derivatives of the original first pharyngeal arch. The 3rd part of MMA is exclusively stapedial; it supplies the zygomatico-maxillary complex and the fronto-naso-orbital mesenchyme

Physiologic Role of the Meninges

Development and Vasculature

The meninges are, in essence, a vascular organ responsible for the proper development and ongoing nutrition of the underlying CNS. As such they constitute a stem cell niche. Formation of the meninx primitive is essential for survival. Formation of the meninges is virtually simultaneous with that of the neural tube at stage 7. Rapid assembly of a superficial vascular plexus is accomplished by stage 8 with vascular connections with the heart beginning at stages 9–10 and with the underlying intracerebral internal carotid system at stage 11. Every penetrating vessel is accompanied by a pial sheath which constitutes a microenvironment.

Although the territory of dura is limited, leptomeninges cover the entire brain, with pia-arachnoid entending into all sulci and fissures. Meninges form non-neural structures within the brain. Choroid plexus, the roof of the 3rd ventricle, lateral ventricle, and the 4th ventricle are all pial structures. CSF manufactured by choroid at about stage 17 exactly when arachnoid begins to differentiate as does the dural limiting membrane. Intercalation of arachnoid villi into the subdural space sets up access of CSF to the venous circulation and permits fluid exchange with the brain. Every penetrating blood vessel is surrounded by pia and the arachnoid space provides for lymphatic drainage.

Forebrain meninges are required for actual development of the underlying cortex. Penetrating blood vessels receive mesoderm required for construction of their endothelial walls, while neural crest provides pericytes and connective tissue. Ablation of the posterior diencephalic and the mesencephalic neural folds causes apoptosis of the entire forebrain [5]. Meninges secrete trophic factors such as retinoic acid, stem cell proliferation factors, and are themselves responsive to mitogens, such as BMP-2. Meningeal SDF-1 guides the migration of Cajal-Retzius cells. The pial basement membrane constitutes an insertion site for the endfeet of processes sent upward by neural progenitor cell residing in ventricular zone.

Cerebral dura provides the osteogenic precursors necessary for synthesis of membranous calvarial bone. When BMP-2 added to demineralized bone matrix or tricalcium/trimagnesium phosphate is placed on dura, membranous bone produced.

The meninges, specifically the dura mater provide stem cell components that respond to injury and are required to reestablish the blood-brain barrier. Pericytes (the putative source of mesenchymal stem cells) migrate into the perivascular space and demonstrate stem cell activity. Boundary cap cells, derived from neural crest can differentiate into different neural tissues depending upon where they are located. Thus, the meninges constitute an injury-responsive stem cell niche [6].

Clinico-Anatomic Correlation: Headache, the Perigrinations of V1 Neural Crest, and the Development of Tentorium and Falx

The referral patterns of supratentorial headache from various locations in the dura demonstrates a disproportionate representation of V1 with pain perceived in the eyes, nose and forehead. The entire extent of tentorium as well as both the anterior and posterior sectors of falx, refer via V1. These tissues are located at a great distance from r1. Why should recurrent V1 pursue a *bidirectional* course? What does this tell us about the migration patterns of neural crest from r1? [Figs. 12.38, 12.39, 12.40 and 12.41].

Anterior migration of r1 NC along the basal forebrain brings it into the midline to form the sphenethmoid complex. It then sweeps upwards over the dorsum of prosencephalon, covering the future frontal lobe and then narrowing dramatically to pursue a posterior course. The narrowing represents the invasion of r2-r3 neural up the sides of the brain. With division of the forebrain, r1 NC plunges downward until it dead-ends against the midline communication between the hemispheres, corpus callosum. At the same time, recurrent V1 indicates the posterior basal migration pathway of r1 NC. The initial binding of tentorium is lateral, following along the superior petrosal sinus. It is as if r1 were following a pathway backward from r4 to r11 until it reaches posterior midline at the future torcula. Here the two columns of r1 NC join together and proceed cephalically, in retrograde fashion to form posterior falx. Thus, almost the entire falx is r1 save in the parietal-interparietal regions. There, I suspect it is admixed with r2-r3 but the latter populations are dominant and have their own nerve supply from V2 to V3.

What about the blood supply of the falx and tentorium? In Chap. 6, the development of the anterior cerebral system was reviewed. From its A5 segment becomes the *pericallosal artery* supplies the entirely of corpus callosum, extending beyond the geno to anastomose with posterior cerebral. The A4 segment of ACA, as the *callosomarginal* artery supplies the entire medial cortex of frontal and parietal lobes backward to the genu. In point of fact, ACA is the exclusive supply of the falx. Callosomarginal *paracentral branches* arch over the vertex to supply a strip of marginal cortex on either side of sagittal sinus. V2 and V3 ascend to this zone and then halt. This maps out headache and also maps out the developmental zones of the parietal bone we discussed previously. Recall that all craniofacial arteries of the ICA and ECA have neural crest cells in their walls and, from Etchever's work, these crest cells relate to specific locations along the neuraxis. We can reasonably surmise that the all the NC necessary for the synthesis of ACA originates from r1. [Fig. 12.42, cf Fig. 12.22].

Tentorium presents a similar model. Here the basal spread of r1 neural crest mesenchyme requires vascularity from the same neuromeric level. Thus, the medial and lateral tentorial arteries arise from immediately proximal to the ACA itself, that is, from the cavernous sector of internal carotid artery. This is also in keeping with the direction of maturation of the cranial base. Basisphenoid forms prior to presphenoid. Its blood supply originates more proximal than that of its distal partner and is therefore more readily available. Neural crest

Fig. 12.38 Referred pain pathways of the dura. Note the disproportionate representation of vV1 dura referring pain the eye and forehead. This reflects the contribution from r1 to the dural system. [Courtesy of Michael Carstens, MD]

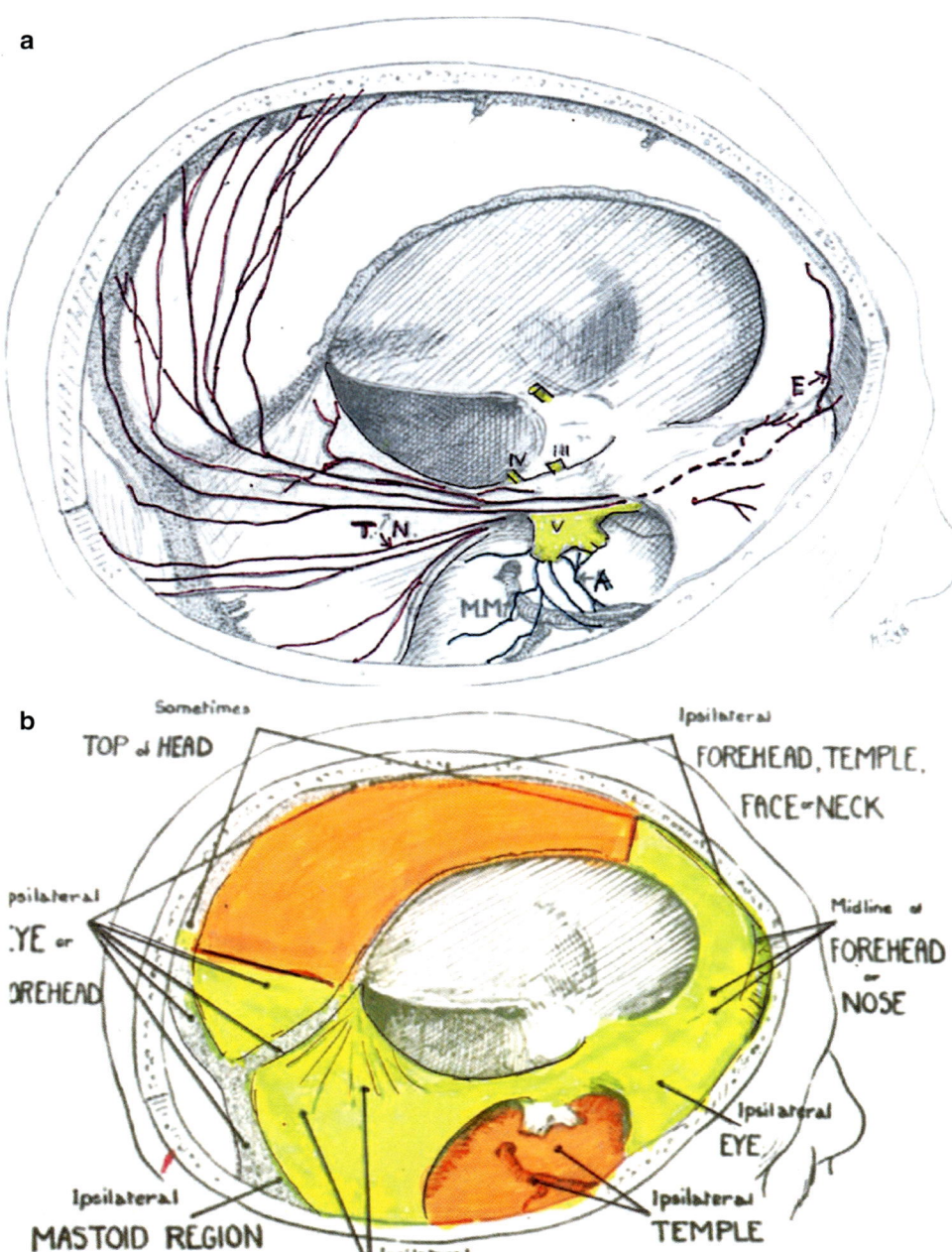

Fig. 12.39 Migration pathways of r1 neural crest. It along with r2-r3 defines the medial boundary between dura of the neural crest cranium and paraxial mesoderm periosteum of the posterior fossa. [Reprinted from Puelles L, Harrison M, Paxinos G, Watson C. Developmental ontology for the mammalian brain based on the prosomere model. *Trends in Neurosciences.* 2013;36(10):570–578. With permission from Elsevier.]

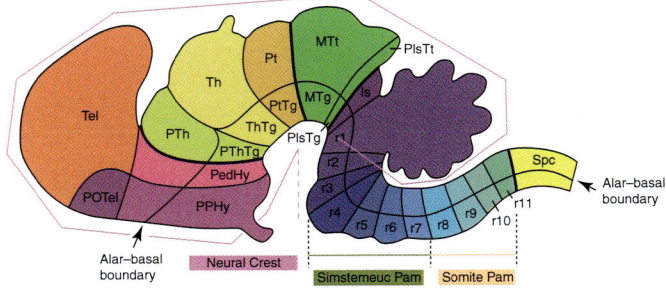

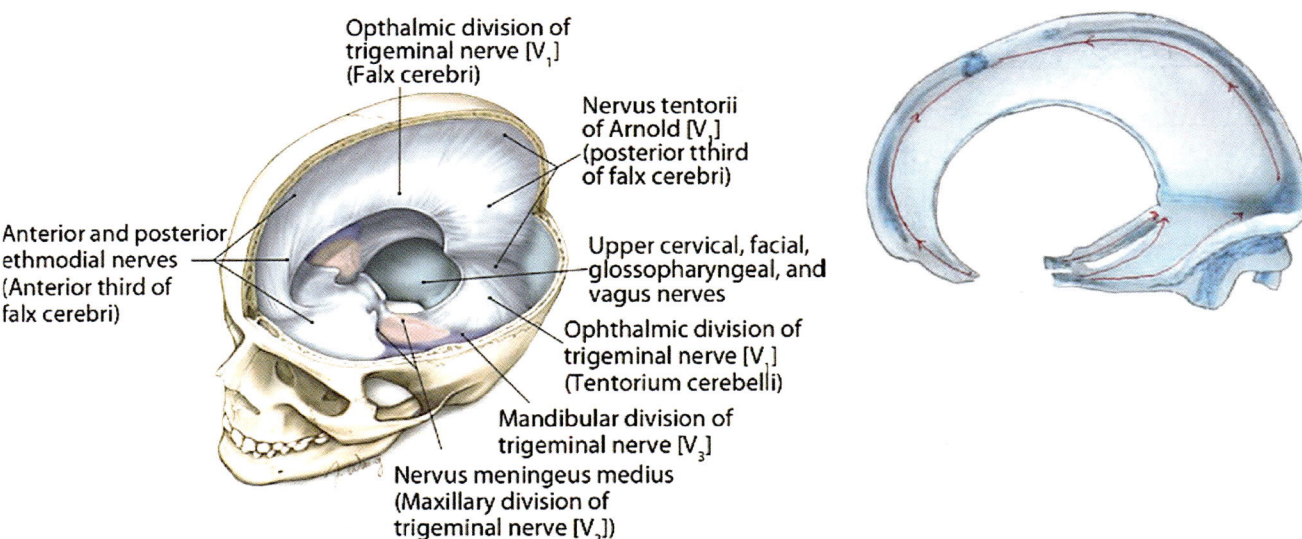

Fig. 12.40 Development of the falx and tentorium. The falx is not a structure per se but rather the fusion of opposing layers of dura, Venous sinuses develop in the folds. It can be detached from corpus callosum. Forward attachment: r1 ethmoid plate/crista gall Lateral attachments: r4 petrous apex backward to r11 internal occipital protuberance. Medial border: non-adherence to cerebellum. Starts out as a flat sheet with a partition for cerebellum. Attaches at torcula. Free border of falx cerebri follows the corpus callosum. Problems with the division of the prosencephalon affect the corpus callosum. Note flow of r1 neural crest along the base of the brain both anteriorly and posteriorly. Left: [Reprinted from Kemp WJ, Tubbs RS, Cohen-Gadol AA. Innervation of the cranial cura mater: Case correlates and Review of the Literature World Neurosurgery 2012; 78(5): 505–510. With permission from Elsevier.] Right: [Courtesy of Michael Carstens, MD]

flow from r1 backwards takes place before that along anterior cranial base. For this reason tentorial development preceeds that of falx. The tapering of saggital sinus as it runs forward indicated that falcial structures are posterior dominant.

What we can conclude from all this seemingly picayune detail is that r1 NC is a very large population that spreads along the base of the brain both forward and backward. It does seem to follow patterns of bone development, especially in the middle and posterior fossae. Its course over petromastoid recognizes a biologic boundary with r2-r3 NC which may have to do with signaling in the medial and lateral parts of the r4-r7 bone complex. It may also simply be a case of competing populations. In any case, from it knife-

edge insertion, V1 tentorium expands medially until it encounters midbrain and ceases to grow, creating the so-called *free edge* of tentorium.

What sort of mechanism could explain this? The combination of trigeminal innervated dura and tentorium versus c2-c3 periosteum defines the boundary between r1-r3 neural crest derived bone and r4-r11 paraxial mesoderm bone. The extent of migration from r1 simply reflects a large amount of mesenchyme that migrates along the base of the brain unimpeded. It recognizes and attaches to the boundary until reaching the posterior terminus. With growth of the brain, the previously flat tentorium is pulled upward like a sail, being attached via the falx along the undersurface of the expanding calvarium.

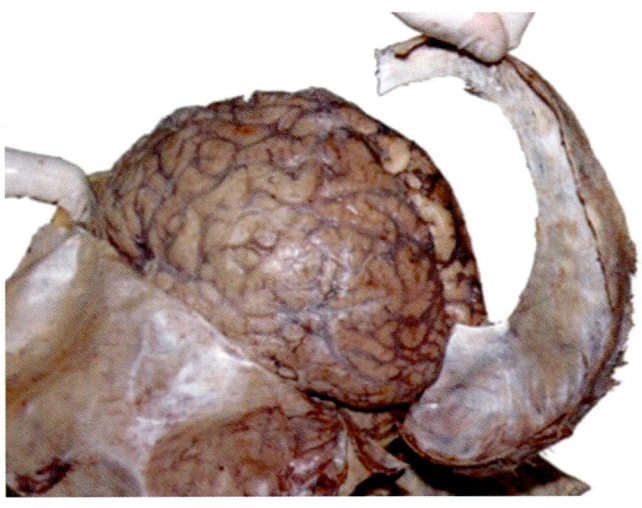

Fig. 12.41 Falx being reflected backward. At its junction with tentorium, the straight sinus is seen. Tentorium has been reflected away from falx and from its interpostion between posterior cerebral lobe and cerebellum, a small triangle of which is visible here. [Reprinted from Adeeb N, Mortazavi MM, Tubbs RS, Cohen-Gadol AA. The cranial dura mater: a review of its history, embryology, and anatomy. *Childs Nerv Syst* 2012; 28(6):827–837. With permission from Springer Nature.]

Clinico-Anatomic Correlation: Is Tentorium Bilaminar? Is the Occipital Lobe the Posterior Pole of the Brain?

Tentorium serves a dual function. It acts as a dural layer for the occipital lobe and separates cerebellum from cerebrum. It also contains the straight sinus. The only topological way the sinus can form is for tentorium to have two neural crest layers. The cranial layer, likely r3 belongs to the subjacent occipital lobe; the caudal layer is, of course, r1.

The original posterior pole of primitive telencephalon before 9 weeks is translocated inferiorly to become temporal lobe. Prior to folding, temporal is located at the posterior pole. Forward from it, the dorsal wall contains occipital and ventral wall is made up of insula. Temporal rotates downward and forward, coming into line below the insula, thus creating the Sylvian fissure. Occipital moves backwards to fill in the space. Optic radiations are connected straight back through the dorsal wall. When occipital is posteriorized its drags its visual cortex and connections backwards as well. Thus, although the initial connections between geniculate ganglia and visual cortex may have be

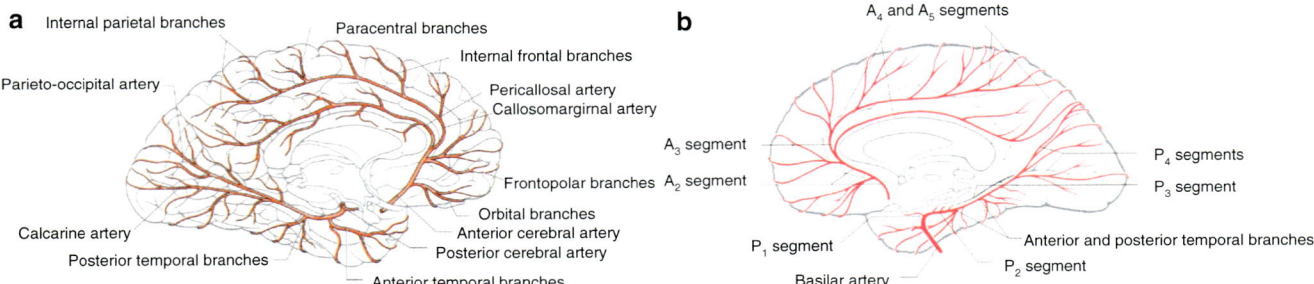

Fig. 12.42 (**a,b**) Blood supply to the falx Anterior cerebral artery and its branches are constructed from r1 neural crest. A4 callosal marginal supplies the upper half of falx and extends "over the top" to supply cerebral cortex on either side of the sagittal sinus. A5 pericallosal artery supplies the lower half of falx and the midline communicating system of corpus callosum. [Reprinted from Haines D, et al. Fundamental Neuroscience for Basic and Clinical Applications, 3rd ed. Elsevier; 2006. With permission from Elsevier.]

quite direct, the eventual pathways, running deep to parietal cortex, make sense.

We can trace the relocation of temporal and occipital lobes be comparing the initial geometry of the vascular system. Posterior cerebral is the final branch of the ICA before posterior communicating. For reasons discussed in Chap. 6 PCA is often mis-interpreted to be a branch of vertebral but this is not embryologically correct. In the initial iteration PCA has a common stem which supplies occipital lobe first and then terminates with temporal lobe. As the latter rotates downward its blood supply from PCA turns about 120 degrees downward and anteriorly. This is readily observed in the adult state [Fig. 12.43]. The common stem is seen by the arterial axes to both lobes displaced directly opposite one another.

Final Thoughts

In this chapter, we have discussed in detail the formation of meninges, with emphasis on different neuromeric zones based upon the origin and composition of its mesenchymal components. We created a "dural map" based upon sensory supply to help explain the migration patterns of neural crest. We have seen that somitomeres and somites are positioned

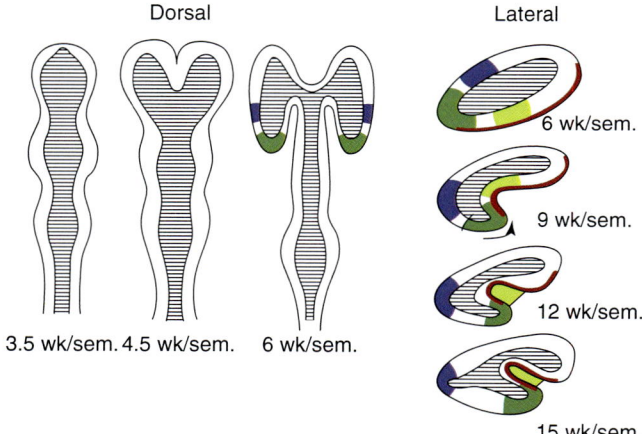

Fig. 12.43 Relocation of the occipital lobe. This drawing of the telencephalic flexure illustrates its development from a ventral bending of the telencephalic hemisphere to form the operculum and finally the Sylvian fissure. The posterior pole of the primitive telencephalon before 9 weeks becomes the temporal (green), not the occipital, lobe. The occipital lobe derives from the dorsal wall of the primitive telencephalon (blue); the insula derives from the ventral wall (yellow). Both lips of the Sylvian fissure are from the ventral margin of the primitive telencephalon, so defective ventralizing genes in the vertical axis would be expressed in the cortex forming both the frontal and temporal lips. wk/sem. 1/4 gestational weeks/semaines (French) or semanas (Spanish). (Illustration by Laura Rodríguez-Flores, graphic designer). [Reprinted from Sarnat HB, Flores-Sarnat L. Telencephalic Flexure and Malformations of the Lateral Cerebral (Sylvian) Fissure. Pediatric Neurology 2016; 63:23–38. With permission from Elsevier.]

below the brain to provide regional contributions of paraxial mesoderm sidewalls and posterior fossa. We have seen how the developmental sequence of the blood supply is spatially preserved in fascial layers of the face and skull. It is hoped that a better knowledge of how dura is synthesized will lead to insights about the mechanisms of skull osteogenesis and the pathologic states that can arise when zonal deficiency states of dura and/or SIF lead to synostosis.

SUMMARY OF DEVELOPMENTAL EVENTS (as per O'Rahilly and Müller)   Stage 07 Gastrulation

Stage 08 Future brain, surrounded by PAM

Stage 09 Cranial neural folds, 1st occipital somite

Stage 10 Prechordal mesoderm organizing, neural crest migration

Stage 11 Brain vascularization, pia in hindbrain

Stage 12 Closure of neural tube, pia lateral midbrain

Extraocular muscles organizing in Sm1, Sm2, Sm3, and Sm5

Stage 13 Lateral rectus appears

Stage 14 Cell sheath of notochord – paraxial mesoderm of cranial base

Medial tentorium condenses

Blood vessels connect in the brain

Stage 15 Entire brain covered with primary meninx

Stage 16 Cranial nerves emerge, these will program meningeal vessels

Stage 17 Pia complete as part of entire brain

Skeletogenous layer (future dura) appears

Dura limiting membrane appears (first definition of the dura)

Stage 18 Cavitation in primary meninx initiates arachnoid

Roof of 4th ventricle

Stage 19 Transverse, sigmoid sinuses indicate completion of dural layers

Choroid plexus fourth ventricle produce CSF for arachnoid

Stage 20 Cartilage in skeletogenous layer

Choroid plexus lateral ventricle

Falx cerebri – rostral-caudal development

Annexation ophthalmic

Stage 21 Completion of ophthalmic ocular and StV1 periocular system

Stage 22 Cerebello-medullary cistern

Stage 23 Bone: frontal/interparietal

Choroid plexus 3rd ventricle

References

1. O'Rahilly R, Müller F. The meninges in human development. J Neuropathol Exp Neurol. 1986;45(5):588–608.
2. Baljet B, VanderWerf F. Connections between the lacrimal gland and sensory trigeminal neurons: a WGA/HRP study in the cynomolgous monkey. J Anat. 2005;206:257–63.

3. Arnold F. Über den Ohrknoten: eine anatomisch-physiologische Abhandlung. Universität Heidberg; 1828. https://digi.ub.uni-heidelberg.de/diglit/arnold1828
4. Carstens MH, Greco RJ, Hurwitz DJ, Tolhurst D. Clinical applications of the subgaleal fascia. Plast Reconstr Surg. 1991;87(4):615–26.
5. Etchevers HC, Couly G, Le Douarin NM. Morphogenesis of the branchial vascular sector. Trends Cardiovasc Med. 2002;12(7):299–304.
6. Decimo I, Fumagalli G, Berton V, Krampera M, Bifari F. Meninges: from protective membrane to stem cell niche. Am J Stem Cells. 2012;1(2):92–105.

Suggested Readings

Adeeb N, Mortazavi MM, Tubbs RS, Cohen-Gadol AA. The cranial dura mater: a review of its history, embryology, and anatomy. Childs Nerv Syst. 2012;28:827–37.

Agur AMR, Dailey AF, editors. Grant's atlas of human anatomy. 6th ed. Philadelphia: Lippincott, Williams & Wilkins; 2012.

Diamond MK. Homologies of the stapedial artery in humans, with a reconstruction of the primitive stapedial artery configuration in Euprimates. Am J Phys Anthropol. 1991;84:433–62.

Blunt MJ. Blood supply of the facial nerve. J Anat. 1954;88:520–6.

Boghal P, Makalanda HLD, Brouwer PA, Gontu V, Rodesch G, Mercier P, Söderman M. Normal pio-dural arterial connections. Interv Neuroradiol. 2015;21(6):750–8.

Caplan AI, Correa D. The MSC: an injury drugstore. Cell Stem Cell. 2011;9(1):11–5.

David DJ, Moore MH, Cooter RD. Tessier clefts revisited with a third dimension. Cleft Palate J. 1989;26(3):168–84.

Ewings E, Carstens MH. Neuroembryology and functional anatomy of craniofacial clefts. Indian J Plast Surg. 2009;42(Suppl):S19–34.

Gans C, Northcutt G. Neural crest and the origins of vertebrates: a new head. Science. 1983;220(4594):268–73.

Gilbert S, Barressi MJF. Developmental biology. 11th ed. Sunderland: Sinauer; 2016.

Griessenauer CJ, Loukas M, Scott JA, Tubbs RS, Cohen-Gadol AA. The artery of Davidoff and Schechter: an anatomical case study with neurosurgical case correlates. Br J Neurosurg. 2013;27(6):815–181.

Huang R, Zhi Q, Ordahl CP, Christ B. The fate of the first avian somite. Anat Embryol. 1997;195:435–49.

Huang R, Christ B. Origin of the epaxial and hypaxial myotome in avian embryos. Anat Embryol. 2000;202:369–74.

Huang R, Zhi Q, Patel K, Wilting J, Christ B. Contributions of single somites to the skeleton and muscles of the occipital and cervical regions in avian embryos. Anat Embryol. 2000;202:375–83.

Hwang K, Wu XJ, Kim H, Kim DJ. Sensory innervation of the upper eyelid. J Craniofac Surg. 2018;29(2):514–7.

Kemp WJ, Tubbs RS, Cohen-Gadol AA. The innervation of the cranial dura mater: neurosurgical case correlates and a review of the literature. World Neurosurg. 2012;78(5):505–10.

Kerber CW, Newton TH. The macro and microvasculature of the dura mater. Neuroradiology. 1973;6:175–9.

Kimmell DL. Innervation of spinal dura mater and dura mater of the posterior cranial fossa. Neurology. 1961;11(9):800–9.

Mack J, Squier W, Eastmans JT. Anatomy and development of the meninges: implications for subdural collections and CSF circulation. Pediatr Radiol. 2009;39:200–10.

Marin-Padilla M. Early vascularization of the embryonic cerebral cortex: Golgi and electronic microscopic studies. J Comp Neurol. 1985;241:237–49.

Martins C, Yasuda A, Campero A, Ulm AJ, Tanriover N, Rhoton A. Microsurgical anatomy of the dural arteries. Neurosurgery. 2005;56(ONS Suppl 2):211–51.

Nabeshima S, Reese TS, Landis DM. Junctions in the meninges and marginal glia. J Comp Neurol. 1975;164:127–69.

Padget DH. The development of the cranial arteries in the human embryo. Contrib Embryol. 1948;32:205–62.

Parvi A. Dural AV fistula supplied by artery of Davidoff and Schechter. Radiol Case Rep. 2010;5(2):375–57.

Paralikar SJ, Paralikar JH. High altitude medicine. Indian J Occup Environ Med. 2010;14(1):6–12.

Penfield W, McNaughton F. Dural headache and innervation of the dura mater. Arch Neurol Psychiatr. 1940;44:43–75.

Périz-Celda, M, Martinez-Soriano F, Rhoton AL. Rhoton's Atlas of Head, Neck, and Brain: 2D and 3D Images. Thieme, 2017.

Puelles L, Harrison M, Paxinos G, Watson C. Developmental ontology for the mammalian brain based on the prosomeric model. Trends Neurosci. 2013;36(10):570–8.

Rhoton AL. Rhoton: Cranial Anatomy and Surgical Approaches: Neurosurgery. Oxford University Press. 2019.

Roland J, Bernard C, Bracard S, Czorny A, Floquet J, Race JM. Microvascularization of the intracranial dura mater. Surg Radiol Anat. 1987;9:43–9.

Sarnat HB, Flore-Sarnat L. Telencephalic flexure and malformations of the lateral cerebral (Sylvian) fissure. Pediatr Neurol. 2016;63:23–38.

Seker A, Martins C, Rhoton AL Jr. Meningeal anatomy. In: Pamir MN, Black PM, Falbusch R, editors. Meningiomas. Elsevier; 2010.

Sensenig EC. The early development of the meninges of the spinal cord in human embryos. Contrib Embryol Carnegie Inst Wash Publ. 1951;611 https://embryology.med.unsw.edu.au/embryology/index.php/Paper

Shukla V, Hayman LA, Ly C, Filler G, Taber KH. Adult cranial dura I: intrinsic vessels. J Comput Assist Tomogr. 2002;26:1069–74.

Sigenthaler JA, Pleasure SJ. We've got you "covered": how the meninges control brain development. Curr Opin Genet Dev. 2011;21(3):249–55.

Srivastava HC. Development of ossification centres in the squamous portion of the occipital bone in man. J Anat. 1977;124(3):643–7.

Standring S, editor. Gray's anatomy. 40th ed. London: Churchill Livingstone; 2008.

Streeter GL. The developmental alteration in the vascular system of the brain of the human embryo. Contrib Embryol. 1918:245–58.

Tessier P. Anatomical classifications of facial, cranio-facial and latero-facial clefts. J Maxillofac Surg. 1976;4:69–92.

Tessier P. Plastic surgery of the orbit and eyelids (trans. SA Wolfe). Masson editorial. Philadelphia: Mosby; 1981.

Tolhurst D, Carstens MH, Greco RJ, Hurwitz DJ. The surgical anatomy of the scalp. Plast Reconstr Surg. 1991;87(4):603–12.

Vignaud J, Hasso AN, Lasjaunias P, Clay C. Orbital vascular anatomy and embryology. Radiology. 1974;111:617–26.